YINGHAN HANGKONG HANGTIAN KEJI DACIDIAN

英汉航空航天科技大词典

主　编　田建国
副主编　王均松
主　审　毛军锋
编　者　（按姓氏拼音排序）
　　　　卜雅婷　常　秦　陈　洁
　　　　崔维霞　高葆华　李　丹
　　　　潘冀春　屈江丽　苏小青
　　　　孙　毅　田建国　王　娟
　　　　王均松　王　倩　王晓丹
　　　　魏晏龙　吴　燕　肖　勇
　　　　于　莉　张　鹏　张霄军

西北工业大学出版社

西安

图书在版编目(CIP)数据

英汉航空航天科技大词典/田建国主编. —西安：
西北工业大学出版社，2018.1
ISBN 978-7-5612-5231-4

Ⅰ. ①英…　Ⅱ. ①田…　Ⅲ. ①航空学—词典—英、汉
②航天学—词典—英、汉　Ⅳ. ①V2-61 ②V4-61

中国版本图书馆 CIP 数据核字(2017)第 213184 号

策划编辑：何格夫
责任编辑：何格夫

出版发行：	西北工业大学出版社
通信地址：	西安市友谊西路 127 号　邮编：710072
电　　话：	(029)88493844　88491757
网　　址：	www.nwpup.com
印　刷　者：	陕西金德佳印务有限公司
开　　本：	889 mm×1 194 mm　　1/16
印　　张：	55.25
字　　数：	2 058 千字
版　　次：	2018 年 1 月第 1 版　2018 年 1 月第 1 次印刷
定　　价：	498.00 元

前　言

本词典全面收录了航空航天领域内近十年来出现的重要词汇，收词均衡、释义量大、缩写词丰富，集实用性、专业性和权威性于一体，既可以作为航空航天专业领域技术人员的常用工具书，也可以为普通读者阅读和翻译英文版本的航空航天工程技术书刊提供参考和帮助。

本词典是基于"英汉航空航天双语平行语料库"编纂而成的，同时参考了《英汉航空词典》(《英汉航空词典》编写组编，商务印书馆，1983年)、《航空科学技术名词》(航空科学技术名词审定委员会编，科学出版社，2004年)和《航天科学技术名词》(航天科学技术名词审定委员会编，科学出版社，2005年)中的部分词汇。语料库的优势在于提供大量丰富真实的语料，通过术语提取和词频统计等方式对语料库中的词汇进行筛选，极大地缩短词典编纂周期，提高词典编纂效率。本词典共收集缩写词38 000条，总词目超过150 000条。词典内容涉及航空制造、航空电子、飞行、航天材料、动力、计算机、自动化等7个领域，涵盖航空器、飞行原理、推进技术与动力装置、飞行控制与显示系统、飞行与飞行试验、航空电子、航天器轨道动作与返回技术、试验与测控、电子与信息技术、制导与控制、航天材料与推进剂、发射与地面保障等12个学科专业。同时，考虑到阅读翻译航空航天工程技术书刊的实际需要，还收入一定数量的近代科学技术方面的词语以及工程技术书刊中主要的外来语。在选词和释义上，力求简明扼要，对于容易理解的词组、复合词和派生词，一般都未收入。

在词典的编写工作中，一些兄弟单位和科研院所给予了热情的帮助和支持。此外，陈鑫、巩海曦、于璐、王余余、贾燕、王竞秀、武娅婷、郭琪、惠戎、潘奕菲、康军娥、王培培、贺银、杨转霞、史馨妍等研究生做了大量资料收集和整理工作，在此一并表示衷心的感谢！

本词典为陕西省社科基金"基于专业双语平行语料库的词典编纂研究"(编号13K111)、西北工业大学人文社科振兴基金"航空航天科技英语语料库的建设及应用研究"(编号RW201202)、"专业双语平行语料库的建设与词典编纂研究"(编号3102014RW0015)的阶段性成果，并得到西北工业大学专著出版基金资助，特此致谢！

由于水平有限，虽经反复审定，但词典中难免存在不足之处，衷心希望广大读者在使用时提出宝贵意见。

《英汉航空航天科技大词典》编写组
2017年6月于西北工业大学

使用说明

一、词目

全部词目一律按英语字母顺序编排,列为词目的有单词、复合词和缩写词。

1. 单词

单词有多个义项时,按出现频次排序。

2. 复合词

收录的复合词有带连字符、不带连字符、连写等多种形式。例如,复合词 take off 至少有 take off 和 take-off 两种形式,在查询此类复合词的词义时,需要对词目的不同形式予以充分考虑。

3. 缩写词

缩写词的形式多种多样,同样的字母组合可能会以不同的形式呈现,意义也不尽相同,在查找时,需要注意区分和辨别。

二、符号

1. 圆括号( )

圆括号用于以下几种情况:

(1)在释义中,圆括号内是注解或同时是词义的一部分,例如,Abbe Comparator 阿贝比长仪(简称比长仪)。

(2)圆括号内是可以省略的词语,例如,fire-resistant 防火的,耐焰的(能受火焰燃烧 5 分钟以上而不燃的)。

(3)在译名后,圆括号中的"英国""美国"等字表示国名。

2. 黑方括号【 】

黑方括号用于表示词义的学科范围,例如,【天】表示"天文学",【计】表示"计算机科学技术",【物】表示"物理学",【矿】表示"矿山工程技术",【化】表示"化学",【医】表示"医学",【数】表示"数学",【气】表示"气象学",【生】表示"生物学",【力】表示"力学",【地】表示"地质学",【测】表示"测绘学",等等。

3. 尖括号⟨ ⟩

尖括号用在汉语译名之前,用来表示译名的性质,例如,⟨口语⟩⟨俚语⟩等;或表示外来语,例如,⟨法语⟩⟨西班牙语⟩等。

三、缩写词排序

首字母相同的词目排序遵循以下原则:

(1)单字母缩写,例如,D;

(2)单字母缩写+数字,例如,D 100 figures;

(3)单字母缩写+单词,例如,D criterion;

(4) 单字母缩写＋缩写词，例如，D PNL；
(5) 单字母缩写＋"&"＋缩写词，例如，D&B；
(6) 单字母缩写＋"."＋缩写词，例如，D.C trembler-operated ignition unit；
(7) 单字母缩写＋"/"＋缩写词，例如，D/A；
(8) 双字母缩写，遵循单字母缩写顺序排序，例如，DA；
(9) 多字母缩写，遵循单字母缩写顺序排序，例如，DAA；
(10) 特殊情况以实际排序为准。

目 录

A	1
B	96
C	123
D	213
E	280
F	344
G	381
H	407
I	429
J	454
K	459
L	462
M	491
N	562
O	593
P	607
Q	665

R 669

S 678

T 757

U 817

V 832

W 852

X 864

Y 866

Z 867

附录 870

 附录1 公制和英美制度量衡对照表 870
 附录2 希腊字母应用对照表 871
 附录3 美国飞机命名体系 872
 附录4 美国导弹和火箭命名体系 874

A

A 1. Annual 年度,年报,年检 2. Antenna 天线 3. Area 面积,区域,范围 4. Attendant Seat 乘务员座椅 5. Availability 可达性,可用性,有效性

A&CO Assembly and Checkout 装配与检测

A&D Accounting and Disbursing 记账和支付

A&E 1. Airframe and Engine Qualified Engineer 机体和发动机合格工程师 2. Architecture and Engineering 体系结构和工程 3. Armament and Electronics 军械与电子设备

A&I Assembly and Installation 装配和安装

A&L Approach and Landing 进近和着陆

A&M Awaiting Maintenance 等待维修

A&S Alternates and Substitutes 互换件和替换件

A/A 1. Air-to-Air 空-空 2. Angle of Attack 迎角 3. Air Abort 中断飞行,空中故障

A/ASK Antiskid 防滑

A/B Auto brake 自动刹车

A/C 1. Air Conditioning 空调 2. Aircraft 飞机 3. Assembly and Checkout 装配和检测

A/CI Approval/Contractual Implementation 批准/合同实施

A/D 1. Analog to Digital Converter 数模转换器 2. Awaiting Delivery 等待交付

A/DET Aircraft Division Technical Requirements File 飞机分部技术要求文件

A/G Air/Ground 空地

A/GS Air/Ground Sensor 空地传感器

A/L Autoland 自动着陆

A/P 1. Airplane 飞机 2. Automatic Pilot 自动驾驶仪 3. Autopilot 自动驾驶仪

A/R As Required 按需要

A/S 1. Airspeed 空速 2. Airspeed Switch 空速开关

A/T 1. Air/Temperature 大气温度 2. Auto Throttle 自动油门杆

AA 1. Absolute Altitude 绝对高度 2. Accelerated Assemblies 速成装配 3. Accelerometer Assembly 加速度计组件 4. Advertise and Award 宣传和决断 5. Air Carrier 航空器 6. Airplane Avionics 飞机电子设备 7. Airport Analysis 机场分析 8. Airworthiness Authorities 适航当局 9. Alert Availability 警告有效性 10. Alternate Aerodrome 备用机场 11. Altitude Acquire 高度警告 12. Angular Accelerometer 角加速度计 13. Airport Approach 机场进近 14. Autopilot Annunciator 自动驾驶信号器

AA/AL Automatic Approach/Auto Land 自动进近/自动着陆

AAA 1. Anti-Aircraft Artillery 高射炮 2. Air Atlantis (Portugal) 大西洋航空公司(葡萄牙) 3. Airport Advisory Area 机场咨询区 4. All American Aviation 全美航空联盟

AAAA Antique Aeroplane Association of Australia 澳大利亚古老飞机协会

AAAF Association Aeronautique et Astronautique de France 法国航空航天协会

AAAM 1. Advanced Air-to-Air Missile 先进空空导弹 2. American Association of Aircraft Manufacturers 美国飞机制造商协会

AAAR Association for the Advancement of Aeronautical Research 航空研究促进会

AAAS 1. American Association for the Advancement of Science 美国科学促进会 2. Alternative Audio Alert Selector 备用音频警报选择器

AAB 1. Army Air Base 陆军航空基地 2. Aircraft Accident Board 飞机事故调查委员会 3. Air Force Air Base 空军基地 4. Auxiliary Air Base 备用航空基地

AABM Air-to-Air Battle Management 空战管理

AABNCP Advanced Airborne Command Post 高级航空指挥基地

AABSHILT Aircraft Anti-collision Beacon System High Intensity Light 飞机防撞标志系统高亮度灯

AAC 1. Air Approach Control 空中进近控制 2. Automatic Approach Control 自动进近控制 3. Airline Administrative Control 航空公司管理部门控制 4. Acoustical Absorption Coefficient 吸音系数 5. Aeronautical Administration Communications 航空管理通信 6. Automatic Amplitude Control 自动幅度控制 7. Automatic Approach Control 自动进近控制 8. Aerial Ambulance Company 航空救护公司 9. Aeronautic Advisory Council 航空咨询委员会

AACA Alaska Air Carriers Association Inc. 阿拉斯加航空运输公司协会

AACC 1. Airport Associations Coordinating Council 机场各协会协调委员会 2. Area Approach Control Center 区域进近管制中心 3. Alternate Avionics Computer Control 备用航空电子计算机控制 4. Automatic Approach Control Coupler 自动进近控

制耦合器

AACI Airport Association Council International 国际机场联合委员会

AACOMS Army Area Communications System 军区通信系统

AACS 1. Airways and Air Communication Service 航路及航空通信服务 2. Active Attitude Control System 主动姿态控制系统 3. Attitude and Antenna Control System 姿态和天线控制系统 4. Attitude & Articulation Control System 飞行姿态与铰接控制系统 5. Auxiliary Attitude Control System 辅助姿态控制系统

AACSM Airways and Air Communication Service Manual 航路和空中通信服务手册

AACSR Airways and Air Communication Service Regulation 航路和空中通信服务规章

AACU 1. Area Approach Control Unit 区域进近管制部门 2. Antiskid/Autobrake Control Unit 防滞／自动刹车控制组件

AACV Assault Air-Cushion Vehicle 突袭气垫船

AAD 1. Active Acoustive Device 有源声装置 2. Address Adder 地址加法器 3. Advanced Ammunition Depot 前方弹药库 4. Aero Acoustic Detection System 航空声测系统 5. Assigned Altitude Deviation 给定高度偏差

AADC 1. Advanced Avionics Digital Computer 航空电子设备先进数字计算机 2. Anti-Aircraft Defence Commander 防空指挥官

AADCP Army Air Defense Command Post 陆军防空指挥所

AADHS Advanced Avionics Data Handling System 先进的航空电子设备数据处理系统

AADRP Aircraft Accident Data Reporting Panel 航空器失事资料报告专家组

AADS 1. Advanced Air Defense System 先进的防空系统 2. Army Air Defense System 陆军防空系统（美国） 3. Automatic Aircraft Diagnostic System 飞机自动诊断系统

AADSF Advanced Automated Directional Solidification Furnace 先进的自动定向固化炉

AAE 1. Aerospace Auxiliary Equipment 航空航天辅助设备 2. American Association of Engineers 美国工程师协会 3. Army Aviation Element 陆军航空兵分队

AAE&E 1. Aeronautical and Aircraft Experimental Establishment 航空与飞机实验研究院（英国） 2. Aeroplane and Armament Experimental Establishment 飞机与军械实验研究院（英国）

AAEC Attitude Axis Emergency Control 姿态轴应急控制

AAED Airborne Active Expendable Decoy 机载有源投掷一次式诱饵

AAELSS Active-Arm External-Load Stabilization System 活性武器外运稳定系统

AAERR Aerophysics and Aerospace Engineering Research Report 航空物理学和航空航天工程研究报告

AAES 1. Advanced Aircraft Electrical System 先进飞机电气系统 2. American Association of Engineering Societies 美国工程师协会

AAEWR Advanced Aircraft Early Warning Radar 先进飞机预警雷达

AAF 1. African Aviation Federation 非洲航空联合会 2. Allied Air Forces 盟国空军 3. Association Astronautique Francaise 法国宇航学会 4. Auxiliary Air Force 配属航空兵（英国）

AAFCE Allied Air Forces, Central Europe 中欧盟国空军（北约）

AAFCS Advanced Automatic Flight Control System 先进自动飞行控制系统

AAFDS Aircraft Assembly Flow Design System 飞机装配流程设计系统

AAFE 1. Aero Assist Flight Experiment 空气制动方案飞行实验（美国） 2. Advanced Applications Flight Experiment 先进应用飞行试验

AAFEA Australian Airline Flight Engineers Association 澳大利亚航空公司工程师协会

AAFIF Automated Air Facility Information File 航空器材信息库（美国）

AAFIS Advanced Avionic Fault Isolation System 先进航空电子故障隔离系统

AAFNE Allied Air Force, Northern Europe 北欧盟国空军（北约）

AAFRA Association of African Airlines 非洲航空公司协会

AAFSE Allied Air Force, Southern Europe 南欧盟国空军（北约）

AAFSS Advanced Aerial Fire Support System 先进空中火力支援系统

AAFWE Allied Air Force, Western Europe 西欧盟国空军（北约）

AAG 1. AAR Landing Gear (USA) AAR 起落架（美国） 2. Association of American Geographers 美国地理学家协会

AAGR Anti-Aircraft Gunnery Range 防空射击靶场

AAGW Air-to-Air Guided Weapon(s) 空对空制导武器

AAHA Awaiting Action of Higher Authority 等待上级决定

AAHS American Aviation Historical Society 美国航空

史学会

AAI 1. Air-to-Air Identification 空对空识别 2. Angle of Approach Indicator 进近角度指示器 3. Aircraft Accident Investigation 飞机事故调查 4. Airline Avionics Institute 航线航空电子研究所

AAIB 1. Air Accident Investigation Board 航空失事调查委员会 2. Air Accidents Investigation Branch（英国运输部）飞行事故调查处

AAIM Aircraft Autonomous Integrity Monitor 航空器自由完好性监控

AAIP Advance Avionics Integration Program 先进航空电子综合大纲

AAL 1. Above Aerodrome Level 飞机库水平面以上 2. Above Airport Level 高于机场平面 3. Aircraft Approach Light 飞机进近灯,飞机着陆灯

AALAAW Advanced Air Launched Anti-Armour Weapon 先进空射反装甲武器

AALC Amplified Automatic Level Control 放大自动电平控制

AALS Advanced Approach & Landing System 高级进近和着陆系统

AAM 1. Air-to-Air Missile 空对空导弹 2. Automatic Approach Mode 自动驾驶进近着陆模式 3. Autopilot Actuator Monitor 自动驾驶作动器监视器

AAMA Association des Amis du Musee der Air 空军博物馆之友协会（法国）

AAMP Advance Architecture Micro-Processor 先进特性微处理器

AAMRL Armstrong Aero-space Medical Research Laboratory 阿姆斯特朗航空航天医学研究实验室（美国空军）

AANCP Advanced Airborne National Command Post 国家先进空中指挥所（美国空军用此缩写词,美国防部则用 AABNCP）

AAO 1. Advanced Assembly Order 先行装配指令 2. Advanced Assembly Outline 先行装配大纲,先行装配外形图 3. Airborne Qualified of Operation 空中作战区 4. Analog Attitude Output 模拟式姿态输出 5. Anti-Air Output 防空数据输出

AAOI Advanced Assembly Outline Index 先行装配大纲索引

AAO-SR Advance Assembly Outline Ship's Record 先行装配大纲架次记录

AAO-SRS Advance Assembly Outline-Ship's Record's System 先行装配大纲架次记录系统

AAP 1. Additional Attendant Panel 附加乘员员面板 2. Aerodynamics Advisory Panel 空气动力学咨询组 3. Aging Aircraft Program 老龄飞机大纲 4. Aircraft Assembly Plant 飞机装配厂 5. Aircraft Available Productivity 飞机可用生产率

AAPD Auto Astro Position Device 自动天文定位器

AAPP Airborne Auxiliary Powerplant 机载辅助动力装置

AAPS Automated Astronomic Positioning System 自动天文定位系统

AAPU Airborne Auxiliary Power Unit 机载辅助动力装置

AAR 1. Aircraft Accident Report 飞机事故报告 2. After Action Report 事后处理报告 3. Air-to-Air Refueling 空中加油

AARA Air-to-Air Refuelling Area 空中加油区

AARB Advanced Air Refueling Boom 先行的空中加油套管,高级空中加油伸缩套管

AARL Aeronautical and Astronautical Research Laboratory 航空航天研究实验室（美国俄亥俄州立大学）

AARPLS Advanced Airborne Radio Position Location System 先进机载无线电定位系统

AARPS Air-Augmented Rocket-Propulsion System 空气加强性火箭推动系统

AARRS Air Force Aerospace Rescue and Recovery Service 空军航空空间救援与回收局

AARS 1. Attitude and Azimuth Reference System 姿态方位基准系统 2. Automatic Address Recognition Subsystem 自动寻址识别子系统 3. Automatic Altitude Report/Reporting System 高度自动报告系统

AAS 1. Aerodrome Advisory Service 机场咨询服务 2. Advanced Antenna System 先进天线系统 3. Altitude Alert System 高度警报系统 4. Automatic Addressing System 自动寻址系统

AASC Aerospace Application Studies Committee 航空航天（技术）应用研究委员会（北约）

AASIR Advanced Atmospheric Sounder and Imaging Radiometer 先进大气探测器与成像辐射仪

AASMS Advanced Air-to-Surface Missile Seeker 先进空地（舰）导弹导引头

AASP Advanced Acoustic Signal Processor 先进声音信号处理器

AASS 1. Advanced Acoustic Search Sensors 先进声测传感器 2. Automatic Abort Sensing System 自动中断检测系统

AASTS Army Air Service Tactic School（美国）陆军航空勤务队战术学校

AASU Aviation Army (Armies) of the Soviet Union 苏联空军集团军

AAT 1. Accelerated Aging Test 加速老化试验 2. Advanced Avionics Technology 先进的航空电子技术 3. Airworthiness Approval Tag 适航批准标签

AATC　1. American Air-Traffic Controllers Council 美国空中交通管制员联合会 2. Automatic Air Traffic Control (System) 自动空中交通管制(系统)

AATF　1. Airport and Airway Trust Fund 航空港与航路信托基金 2. Airworthiness Assurance Task Force 适航保证工作组

AATH　Automatic Approach to Hovering 自动进入悬停

AATMS　Advanced Air Traffic Management Systems 先进空中交通管理系统(美国)

AATS　1. Advanced Automatic Test System 先进自动测试系统 2. Advanced Automation Training System 先进的自动化培训系统 3. Alternate/Alternative Aircraft Takeoff System 备用飞机起飞系统 4. Aviation and Air Traffic Services 航空与空中交通管制系统,航空电子与空中交通服务

AATSR　Advanced Along Track Scanning Radiometer 先进沿航向扫描辐射计

AATT　Advanced Aviation Transportation Technology 先进空中交通管制系统

AATU　Association of Air Transport Unions 空中运输协会联合会

AAU　1. Audio Accessory Unit 音频附件盒 2. Absolute Alignment Update 绝对校准更新

AAUT　Alternate Attitude Update Techniques 备用姿态更新技术

AAV　Advanced Aerospace Vehicle 先进航空航天飞行器

AAVCRS　Airborne Automatic Voice Communications System 机载自动音频通信系统

AAVCS　1. Airborne Automatic Voice Communication System 航空自动通话系统,机载自动语音通信系统 2. Automatic Aircraft Vectroring Control System 飞机航向自动控制系统

AAVS　1. Automatic Aircraft Vectoring System 自动航空器引导系统 2. Aerospace Audio-Visual Service 航空航天声像业务处(美国)

AAW　Air-to-Air Warfare 空对空作战

AAWG　Airworthiness Assessment Work Group 适航性评估组

AAWS　1. Automatic Aviation Weather Service 自动空气象服务 2. Anti-Air Warfare System 防空作战系统 3. Airborne Alert Weapon System 机载警戒武器系统

AB　1. After Burner 加力燃烧器 2. Air Borne 机载的 3. Airspeed Bugs 空速游标 4. Auto-Beacon 自动信标 5. Avionics Bay 电子设备舱

ABAC　Association of British Aero Clubs and Centres 英国航空俱乐部/中心协会

abacus　算盘

abaft　1. 处于飞机尾部的 2. 在……的后部

abampere　电磁安培

abandon　弃机,离机

abandonment　抛弃,弃机

abatement　1. 减少,降低,抑制,减弱 2. 废除

ABB　Abbreviation 缩写

Abbas　阿巴斯(伊拉克地地近程弹道导弹)

Abbe comparator　阿贝比长仪(简称比长仪)

Abbey Hill　艾比山

ABBR　Abbreviated 缩写

abbreviated dial system　缩位拨号系统

ABC　1. Abbreviations and Codes 缩写与简语 2. Abc World Airways Guide 世界航线指南

ABC power unit　灯丝阳极栅极组合电源,ABC 电源组

ABCB　Air Blast Circuit Breaker 空气自动断路器

ABCC　Automatic Brightness Contrast Control 自动亮度对比控制

ABCCC　Airborne Command, Control & Communications 空中指挥控制与通信

ABCF　As-Built Configuration File 制造构型文档

ABCL　As-Built Configuration List 施工构型表,制造构型目录

abcoulomb　电磁库仑,绝对库仑

ABCR　As-Built Configuration Record 制造构型记录

ABD　1. Aboard 机上,船上,车上 2. Airbus Directive and Procedure 空客指令和程序 3. Abbreviated Dialing 缩位拨号 4. Answer-Back Device 应答设备 5. Advanced Base Depot 先进基地机库 6. Advanced Base Dock 先进基地维修厂 7. Air Bus Directive 空中客车指令

ABDL　Automatic Binary Data Link 自动二进制数据链接

abdomen　(飞机的)腹部,机腹

abdominal breathing　腹式呼吸

abdominal distension　(环境压力降低引起的)腹胀

ABDR　Aircraft Battle Damage Repair 飞机战伤修理

ABE　ARINC429 Bus Emulator ARINC429 总线仿真器

abeam　(飞机的)正侧方

Abel　"阿贝尔"作战(1944 年 8 月 27 日至 9 月 6 日,法军空降作战,以扰乱德军从法国撤退)

Abel richness class　阿贝尔富度

aberdeen cutlet　1. 未放(或收起)起落架的着陆,〈口语〉迫降 2.〈口语〉严重飞行事故

Aberporth　阿伯波思

aberration　1. 像差 2. 光行差 3. 偏差

aberration constant　光行差常数

aberration day numbers　光行差日数

aberration ellipse　光行差椭圆

aberration of light 光行差,光线像差
aberration shift 光行差位移
aberration vignetting 像差渐晕
ABES Air-Breathing Engine System 吸空气发动机系统,空气喷气发动机系统
ABF 1. Aircraft Battle Force 飞机战斗力量 2. Air Burst Fuse 空中爆炸引信
abfarad 电磁法拉,电磁亨利
Abigail "使女"计划
Abigail Rachel "雷切尔使女"作战
ABILA Airborne Instrument Landing Approach 机载仪表着陆进场
ABIS Apollo Bioenvironmental Information System 阿波罗飞船生物环境信息系统
ABL 1. Air Belgium 比利时航空公司 2. Allocated Baseline 分配基线 3. Automatic Brightness Limiter 自动亮度限制器
ablatant 烧蚀材料,烧蚀层
ablate 烧蚀,融化
ablation 烧蚀
ablation behavior 烧蚀行为
ablation chamber 烧蚀室,消融室
ablation cooling 烧蚀冷却
ablation coupling calculation 烧蚀耦合计算
ablation effect 烧蚀效应
ablation efficiency 烧蚀效率
ablation heat 烧蚀热
ablation jet 消融射流
ablation lag 烧蚀速度变慢
ablation material (防热结构中的)烧蚀材料
ablation model 烧蚀模型
ablation period 消融期
ablation plume 烧蚀羽烟
ablation pressure 消融压力
ablation process 烧蚀过程
ablating rate 烧蚀率
ablation rate transducer 烧蚀速率传感器
ablation test 烧蚀试验
ablation test of low temperature ablator 低温烧蚀材料试验
ablation testing in turbulence pipe 湍流管烧蚀试验
ablation thermal efficiency 烧蚀热效率
ablation thickness 烧蚀厚度
ablation threshold 烧蚀阈值
ablation time 消融时间,烧蚀时间
ablative 1. 烧蚀材料 2. 烧蚀的
ablative cooling 烧蚀冷却
ablative heat prevention 烧蚀防热
ablative material 烧蚀材料

ablative shock 烧蚀激波
ablative surface 烧蚀面
ablative thermal protection 烧蚀防热
ablator 烧蚀体,烧蚀屏,烧蚀层,烧蚀材料
ablator rod 烧蚀杆
ABM 1. Advanced Bill of Materials 先行物料表 2. Asynchronous Balanced Mode 异步平衡模式
ABMA Army Ballistic Missile Agency 陆军弹道导弹局(美国)
abmho 电磁姆欧
ABMIS Airborne Ballistic Missile Intercept System 机载弹道导弹拦截系统
ABN 1. Air Battle Net 空中作战通信网 2. Airborne 空中的,飞机上的
abnormal 异常的,反常的
abnormal condition 异常情况
abnormal environment 异常环境,非正常环境
abnormal fire conditions 非表定射击条件
abnormal galaxy 不规则星系
abnormal glow discharge 异常辉光放电
abnormal refraction 反常折射
abnormal spin 不正常螺旋
abnormal variable star 不规则变星
abnormality 异常,不正常
aboard 在机上,在飞行器上,在弹上,在舰上
A-boat 核潜艇
abohm 电磁欧姆
A-bomb 原子(炸)弹
abort 中止飞行,中断发射,中断试验
abort drill 中断飞行练习
abortive 1. 中断飞行的飞机 2. 未完成任务的,不成功的(出动)
above normal 超常,正常以上
above sea level 海拔(高度)
above-wing nozzle (在机翼上表面,供重力加油方法向油箱内加注燃油的)翼上加油口
ABR 1. Address Buffer Register 地址缓冲寄存器 2. Area Border Router 区域边界路由器 3. Automatic Bit Rate selection 自动比特率选择
abracadabra 高级特技
abradable coating 耐磨涂层
abradable material 耐磨材料
abradable seal 耐磨密封层
abrade 1.(用喷砂或喷丸方法)清理 2. 磨削,磨损
abrader 1. 磨光机,砂轮机 2. 磨蚀(或磨损)试验机
Abradum 抛光粉(一种氧化铝粉的商品名)
abrasion 磨损,擦伤,研磨
abrasion layer 耐磨层
abrasion performance 耐磨耗性能

abrasive 磨料
abrasive action （空间微粒子撞击航天器表面引起的）沙蚀作用
abrator 1.喷丸（或喷砂）清理 2.喷丸清理机
ABRD Advanced Base Repair Depot 先进基地修理库
abreast 1.并列，并肩 2.无高度差横队的（指编队）
abridged nautical almanac 简明航海历
ABRSV Abrasive 研磨剂，磨料的
ABRSV RES Abrasive-Resistant 抗腐蚀剂
abrupt acceleration 突然加速
abrupt asymmetric wing stall 突然的非对称机翼失速
abrupt change 突变
abrupt fault 突发性故障
abrupt junction 突变结
abrupt loss 突然损失
abrupt response 突然反应
abrupt stall 突然失速
abrupt transition 突然转变，急剧转变
abrupt wind 急风
ABS 1.Absolute 绝对的，完全的 2.Air Break Switch 空气断开开关 3.Auto Braking System 自动刹车系统
abscissa 脉横线，横坐标
ABSD 1.Advanced Base Section Dock 先进基地分段检修厂 2.Advanced Base Supply Depot 先进基地供应库
absence of cavity 无腔
absence of convection 没有对流
absence of injection 没有喷射，没有注入
absentee ratio 缺损率（轨道拦截站总数中，不能有效作战的拦截站所占的比例）
absolute 1.绝对的 2.绝对温度
absolute acceleration 绝对加速度
absolute aerodynamic ceiling 绝对升限，理论升限
absolute air mass 绝对大气质量
absolute alcohol 纯乙醇，无水酒精，纯酒精
absolute alignment 绝对对齐
absolute altimeter 真高表，绝对高度表，真实高度表
absolute altitude 绝对高度
absolute and relative alignment 绝对和相对定位
absolute and relative navigation 绝对和相对导航
absolute angle of attack 绝对迎角，零升力迎角
absolute black body 绝对黑体
absolute brightness 绝对亮度
absolute calibration 绝对标定，绝对校准
absolute catalog(ue) 绝对星表
absolute ceiling 绝对升限，绝对幂高，绝对云底高度
absolute density 绝对密度
absolute distance 绝对距离，绝对测距
absolute dud 哑核弹
absolute error 绝对误差

absolute fix 绝对位置
absolute flow angle 绝对流动角
absolute flying height 绝对航高
absolute frame 绝对坐标系
absolute frame of reference 绝对参考框架
absolute GPS 绝对全球定位系统
absolute humidity 绝对湿度
absolute inclinometer 绝对测斜仪
absolute instrument 一级基准仪器，绝对准器
absolute lethal concentration 绝对致死浓度
absolute luminosity 绝对光度
absolute magnitude 绝对星等
absolute magnitude effect 绝对星等效应
absolute measurement 绝对测量
absolute minimum 绝对极小值
absolute navigation 绝对导航
absolute obstacle 绝对障碍
absolute optical shaft encoder 绝对光学轴编码器
absolute orbit 绝对轨道
absolute parallax 绝对视差
absolute performance 绝对性能
absolute perturbation 绝对摄动
absolute photo visual magnitude 绝对仿视星等
absolute photoelectric magnitude 绝对光电星等
absolute photographic magnitude 绝对照相星等
absolute photometry 绝对光度测量，绝对测光
absolute position 绝对位置，游离位置
absolute potential 绝对电位
absolute pressure 绝对压力
absolute proper motion 绝对自行
absolute radio magnitude 绝对射电星等
absolute radiometric calibration 绝对辐射定标
absolute red magnitude 绝对红星等
absolute refractive index 绝对折射率，绝对折射指数
absolute solar flux 绝对太阳通量
absolute spectral response 绝对光谱响应
absolute spectral sensitivity 绝对光谱灵敏度
absolute spectrophotometric gradient 绝对分光光度陡度，绝对分光光度梯度
absolute stability 绝对稳定（性）
absolute standard 绝对标准
absolute star catalogue 绝对星表
absolute state 绝对状态
absolute system 1.绝对（单位）制 2.绝对系
absolute temperature 绝对温度
absolute temperature scale 绝对温标
absolute threshold 绝对阈
absolute time 绝对时间，绝对时空
absolute uncertainty 绝对不定性，绝对误差

absolute vacuum gauge　绝对真空计
absolute value　绝对值
absolute variability　绝对变率
absolute velocity　绝对速度
absolute viscosity　绝对黏度
absolute vorticity　绝对涡度
absolute zero　绝对零度
absorb　1.吸收 2.减震
absorb limit　吸收限
absorbability　1.吸收性 2.减震性,减震能力
absorbance　吸光率,吸光度,吸收系数
absorbancy　吸收率,吸收能力
absorbant　1.吸收剂,吸收体 2.吸收性的
absorbed dose　吸收剂量
absorbed energy　吸收能量
absorbed heat flux method　吸收热流法
absorbed light　吸收光
absorbency　吸收率,吸收能力
absorbent　1.吸收剂,吸收体 2.吸收性的 3.吸收性(防雷达)涂层
absorbent coverage　吸收性覆盖层
absorbent structure　吸波结构
absorber　1.吸收器 2.减震器,阻尼器 3.吸收剂,吸收体
absorber cavity　吸收腔
absorber length　吸收体长度
absorbing cloud　吸收云
absorptance　吸收率,吸收比
absorptiometer　1.(液体)吸气计 2.吸收光度计
absorption　1.吸收 2.表面吸附
absorption atelectasis　吸收性肺不张(肺膨胀不全)
absorption band　1.吸收带 2.吸收(光)谱带 3.吸收频带
absorption coefficient　吸收系数
absorption cross section　1.吸收截面 2.吸收截面积,雷达目标吸收系数
absorption dose　吸收剂量
absorption feature　吸收特征
absorption length　吸收长度
absorption line　吸收(谱)线
absorption of the photons　光子吸收
absorption peak　吸收峰值(最大值)
absorption process　吸收法
absorption pump　吸附泵
absorption spectra　吸收光谱,吸收波谱
absorption spectroscopy　吸收光谱学
absorption spectrum　吸收光谱
absorption-emission pyrometer　吸收-发射高温计
absorption-emission ratio　吸收-发射比

absorptive index　吸收指数,吸收率
absorptive power　吸收功率,吸收能力
absorptivity　吸收性,吸收率
absorptivity-emissivity ratio　吸收-发射比
abstract code　抽象码
abstract machine　抽象机
abstract system　抽象系统
abstraction　1.提取,提炼,分离 2.抽象(化)
ABT　1. About 大约的,关于 2. Air Blast Transformer 风冷式变压器 3. Airborne Tracking 空中跟踪
ABTA　Association of British Travel Agents 英国旅行代理商协会
ABTS　Airborne Beacon Test Set 机载信标测试装置
ABTSS　Airborne Transponder Subsystem 机载应答器子系统
abundance　丰度
abundance of elements　元素丰度
abuse test　非正常使用试验
abutment　支座,支撑点,接合点
abutment screw　止动螺钉
abutment sleeve　定位套筒
abutting beam　托梁
abutting collar　耳盘,凸环,凸出圆肩
ABV　1. Above 在……以上 2. Absolute Value 绝对值
abvolt　电磁伏特
AC　1. Action Code 措施代码,行动代码 2. Air Cleaner 空气清洁器 3. Aircraft Certificate 飞机适航证 4. Aircraft Commander 机长 5. Aircraft (Airplane) Configuration 飞机构型 6. Aircraft (Airplane) Contract 飞机合同 7. Alternating Current 交流(电) 8. Altitude Capability 高度能力 9. Antenna Coupler 天线耦合器 10. Above Clouds 云上 11. Absolute Ceiling 绝对升限,绝对云 12. Automatic Checkout 自动检测 13. Automatic Control 自动控制
acceleration　1.加速,加速运动 2.加速度
accelerator　1.加速器 2.加速剂 3.加速电极
AC cycle　交流周期
AC plate resistance　交流板极电阻
AC&QT　Acceptance, Conforming & Qualification Test 接收、确认和鉴定试验
AC&S　Attitude Control and Stabilization 姿态控制和稳定
AC/DC　Alternating Current/Direct Current 交直流通用,交直流两用
ACA　1. Altitude Control Assembly 高度控制组件 2. Annunciatior Control Assembly 信号器控制组件 3. Assembly Coordination Advice 装配协调方案,装配协调通知 4. Attitude Controller Assembly 姿态控制器组件

ACA　1. Automatic Circuit Analyzer 自动电路分析器　2. Air Crew Association 空勤人员协会（英国）　3. Attitude Control Assembly 姿态控制装置　4. Australian Council for Aeronautics 澳大利亚航空委员会

ACAC　Air-Cooled Air Cooler 气冷式空气冷却器

ACACS　Air Cycle Air Conditioning System 空气循环空调系统

ACADS　Alarm Communications And Display System 报警通信和显示系统

ACAM　Augmented Content-Addressed Memory 可扩充的内容定址存储器

ACAMPS　Automatic Communications And Message Processing System 自动通信和信息处理系统

ACAN　Amicale des Centres Aeronautiques Nationaux 全国航空中心联谊会（法国）

ACAP　1. Advanced Composite Airframe Program 先进复合材料机体计划　2. Aircraft (Airplane) Characteristics for Airport Planning 用于机场计划的飞机特性手册　3. Arab Civil Aviation Organization 阿拉伯民航组织

ACAPS　Advanced Configuration Automated Planning System 先进构型自动计划系统

ACARS　Aircraft Communications Addressing and Reporting System 飞机通信寻址和报告系统

ACAS　1. Airborne Collision Avoidance System 机载防撞系统　2. Autovon Centralized Alarm System 自动电话网中央报警系统

ACAV　Automatic Circuit Analyzer and Verifier 自动电路分析器和检验器

ACAW　Aircraft Control And Warning 飞机控制和警告

ACB　Air Circuit Breaker 空气断路器

ACC　1. Accessory 附件　2. Accumulator 蓄电池，储存器　3. Active Clearance Control 主动空间控制　4. Air Control Center 空管中心　5. Aircraft (Airplane) Configuration Change 飞机构型更改　6. Approach Control Center 进近控制中心　7. Area Control Center 区域管制中心　8. Area Control Code 区域控制码　9. Assembly Control Code 装配控制码　10. Automatic Chrominance Control 自动彩色控制　11. Automatic Course Control 自动航路控制　12. Area Control Center 区域控制中心

ACCA　Accounting Center of China Aviation 中国航空结算中心

ACCB　Aircraft Change Control Board 飞机更改控制中心

ACCC　Alternate Command and Control Center 备用指挥和控制中心，备用指挥和管制中心

ACCD　Aircraft Compatibility Control Drawing 飞机互换性控制图

ACCEL　1. Accelerate 加速，增进　2. Accelerater 加速度计

accel　加速

accel electrode　加速电极

accel-eration　加速作电

accel-eration voltage　加速作电压

accel grid　加速电网

accelerant　速燃剂（引起和扩大火灾的物质，通常指易燃液体）

accelerate　加快，加速，增大转速

accelerate compensation　加速度补偿

accelerate-stop　加速滑跑到停止试验

accelerated ageing　加速老化

accelerated ageing test　加速老化试验

accelerated atelectasis　加速度性肺萎陷

accelerated exposure test　加速暴露试验

accelerated factor　加速因子

accelerated flight　加速飞行

accelerated history　加速经历

accelerated life test　加速寿命试验

accelerated method　加速法

accelerated mission testing　加速任务试车，加速任务试验

accelerated random search　加速随机搜索

accelerated spin　加速螺旋（进入加速飞行时产生的螺旋）

accelerated testing　加速试验

accelerated tolerance　加速度耐力

accelerating electrode　（离子发动机的）加速电极

accelerating flame　加速火焰

accelerating pump　加速泵（活塞式发动机汽化器的）

accelerating well　（汽化器的）加速储油室

acceleration　1. 加速度　2. 加速性

acceleration atelectasis　加速度性肺不张（肺膨胀不全）

acceleration autopilot　加速自动驾驶仪

acceleration capability　加速性能

acceleration chamber　加速仓，加速室

acceleration channel　加速通道

acceleration command　加速命令

acceleration compensation　加速度补偿

acceleration component　加速度分量

acceleration constant　加速常数

acceleration constraint　加速限制

acceleration demand　加速需求

acceleration effect　加速度效应

acceleration environment　加速（度）环境，加速（度）条件

acceleration error　加速度误差

acceleration expression　加速度表达式

acceleration factor　载荷因数

acceleration feedback　加速度反馈
acceleration grid　加速电网
acceleration grid current　加速栅极电流
acceleration grid potential　加速电网潜力
acceleration guidance　加速指导
acceleration-insensitive drift rate　与加速度无关的漂移率(指陀螺漂移)
acceleration level　加速级,加速度水平
acceleration limit　加速极限
acceleration load　加速度载荷
acceleration magnitude　加速度大小
acceleration manoeuvre　加速机动
acceleration measurement　加速度测量
acceleration mechanism　加速机制
acceleration of constraint　有害加速度,抑制加速度
acceleration of following　牵连加速度
acceleration of gravity　重力加速度
acceleration of point　加速点
acceleration physiology　超重(加速度)生理学
acceleration process　加速过程
acceleration profile　加速度分布图
acceleration propulsion　加速推进
acceleration protection　加速度防护
acceleration ratio　加速度比
acceleration region　加速区域
acceleration response　加速度反应
acceleration scalar　加速度标量
acceleration section　加速段
acceleration segment　加速段
acceleration sensor　加速度传感器
acceleration simulation　加速度仿真
acceleration simulator　加速度仿真器
acceleration square-sensitive drift rate　与加速度二次方项有关的漂移率
acceleration stage　加速阶段
acceleration stress　加速度应激
acceleration stress physiology　加速度应激状态生理学
acceleration syncope　加速度晕厥
acceleration template　加速模板
acceleration test　加速试验
acceleration time of engine during start transient　发动机起动加速性
acceleration tolerance　加速度耐力
acceleration tolerance examination　加速度耐力检查
acceleration tolerance training　加速度耐力训练
acceleration transducer　加速度传感器
acceleration uncertainty　加速的不确定性
acceleration variation　加速度变化
acceleration vector　加速度向量

acceleration voltage　加速电压
acceleration zone　加速区
acceleration-invariant algorithm　加速度不变算法
acceleration-sensitive drift rate　与加速度有关的漂移率
accelerator　1.加速器 2.(航母用)弹射加速器 3.促进剂
accelerator aperture　加速器孔径
accelerator architecture　加速器结构
accelerator electrode　加速电极
accelerator grid　加速电网
accelerator grid current　加速栅极电流
accelerator module　加速器模块,加速单元
accelerator vehicle　加速器车辆
accelerator voltage　加速电压
accelerator winding　加速度绕组,二阶导数绕组
accelerograph　加速度自记仪,加速器,油门
accelerometer　1.加速度计 2.载荷因数表 3.过载表
accelerometer attachment　加速度表附件
accelerometer bias　加速度计偏差
accelerometer bias error　加速度表零位误差
accelerometer coordinate system　加速度计坐标系
accelerometer datum　加速度计数据
accelerometer experiment　加速度试验
accelerometer fault　加速度计故障
accelerometer measurement　加速度计测量
accelerometer output　加速度计的输出
accelerometer output signal　加速度计输出信号
accelerometer response　加速度计灵敏度,加速度计响应
accelerometer sensitivity　加速度计灵敏度
accelerometer signal　加速度计信号
accelerometric reentry switch　加速度表再入(大气层)开关
accendibility　着火性,可燃性
accentuation　增强
accentuation contrast　增强反差
accentuator　1.加重器 2.频率校正电路 3.声频强化器
accept　1.验收,接受 2.同意,承认
acceptability　可接受性,可验收性,验收合格性
acceptable alternative product　可替代物品(北约)
acceptable failure rate　容许故障率,合格故障率
acceptable mean life　验收平均寿命
acceptable quality level　可接受质量标准,验收质量水平
acceptable reliability level　可接受可靠性标准
acceptance　接受,接纳,受理
acceptance bogey　(验收时)抽验产品的工作时数
acceptance flight test　验收试飞
acceptance inspection package　验收检查软件包
acceptance region　接受域
acceptance requirements　验收要求
acceptance sampling plan　验收抽样计划

acceptance specifications 验收技术条件
acceptance test （订货方在场时进行的）交收试验，验收试验
acceptance test procedure 验收试验程序
acceptance trials 验收试验
accepting state 接收态
acceptor 1.接受器，带通电路 2.（半导体的）受主
acceptor atom 受主原子
acceptor of data 数据接收器
ACCESS Automated Control and Checking for Electrical System Support 电气系统自动控制和检测
access 1.进入 2.入口，舱口 3.调整孔，检查口 4.通路 5.存取，取数
access compatibility 存取兼容性
access control field 存取控制字段
access-control mechanism 访问控制机构
access door 1.检查口，工作窗 2.检查口盖，工作窗口盖
access hole 通道孔
access light 进场着陆（跑道入口）指示灯
access mask register 存取屏蔽寄存器
access panel 检查口盖
access protocol 接入规程，存取协议
access time 1.到达时间 2.到达计算器程序任何部分的时间 3.地图显示时间 4.存取时间
access window 存取窗口
accessibility 1.可达性，可接近性，开敞性 2.（飞行的）起落适应能力
accessibility difficulty 可达性困难
accessible 1.可达到的，容易接近的 2.易受影响的
accessorial services 辅助（性）勤务
accessory（accessary） 1.附件，附属品，辅助装置 2.附属的，辅助的，次要的
accessory drive 附件传动
accessory drive gear 附件传动齿轮
accessory gearbox 附件传动齿轮箱，附件传动机匣
accessory of wind tunnel 风洞辅件
accessory power supply 辅助能源
accessory support 辅助支架
accident 事故，失事，偶然事件
accident analysis 事故分析
accident board 飞行事故调查委员会
accident causation 事故原因判定
accident cause factor 事故因素
accident（data）recorder 飞行事故记录器
accident data recorder ejection 事故情况记录器的弹射
accident datum 事故数据
accident error 偶然误差
accident investigation 事故调查
accident launch 意外发射

accident location beacon 失事定位信标
accident pathology 事故病理学
accident prevention 事故预防
accident prone 易致事故的，有事故倾向的
accident rate 事故率
accident report 事故报告
accidental accuracy 随机精度
accidental coincidence 偶然符合
accidental connection 偶然关联
accidental count 偶然计数
accidental damage 意外损伤，偶然损伤
accidental decompression 突然减压，意外减压
accidental explosion 意外爆炸
accidental exposure 事故照射，偶然曝光，偶然曝射
accidental failure period 偶然失效期
accidental launch protection system （导弹核武器）意外发射预防系统
accidental maintenance 1.随遇维修，偶然性维修 2.事故损伤检修
accident-free 无事故的
accident-free flying hours 安全飞行时数
accident-protected recorder 飞行事故记录器
ACCIS Automated Command and Control Information System 自动指挥控制信息系统（北约）
acclimatization 习服
ACCLRM Accelerometer 加速表
accolade 因公死亡人员家属证书（美军）
ACCOM Accommdate 适应，供应
accommodate 1.适应 2.容纳，可供使用 3.调节
accommodation 1.（飞机）舱内乘员数，容纳量 2.客舱服务设施 3.调节，适应
accommodation coefficient 调节系数，适应系数
accommodation factor 调节系数，适应系数
accommodation ladder （乘客）登机梯
accommodation space 生活舱室
accompanying cargo/supplies 随军物资
accompanying flight 伴飞
accomplish 积累，存储，完成
accomplishment 积累，存储，完成
accord 一致
ACCOS Arms Control and Conflict Observation Satellite 军备控制和冲突观察卫星（联合国）
account 账目，账户
account manager 账户经理
accounting 会计，会计学
ACCP Area Control Code Position 区域控制码位置
accrete 吸收，增长，堆积
accreting black hole model 吸积黑洞模型
accretion 1.积聚物，增生物，积冰 2.积聚，粘连 3.吸积

accretion by black hole　黑洞吸积
accretion by neutron star　中子星吸积
accretion disk　吸积盘
accretion rate　吸积率
accretion theory　吸积理论
accretion time　吸积时间
accrue　增长,应累算,应计
ACCS　1. Air Command and Control System 空中指挥和控制系统　2. Aerospace Command and Control System 航空航天指挥与控制系统　3. Aircraft Communication System 飞机通信系统　4. Automatic Checkout and Control System 自动检测和控制系统　5. Automatic Communications and Control System 自动通信和控制系统
ACCU　Audio Central Control Unit 中央音响控制装置
ACCUM　Accumulate 累积,累计
accumulate　积累,积聚
accumulated duration(of engine)　(发动机)累计工作时间
accumulated filling throughout　积累加注量
accumulation　1.积累,储存　2.聚集物
accumulation distribution unit　累加分配器
accumulation of object　积累客体
accumulator　1.蓄电池　2.(液压系统的)蓄压器　3.(燃料系统的)储油器　4.储能电路　5.(计算机的)累加器,存储器
accumulator buffer register　蓄电池缓冲寄存器
accuracy　精确性,准确性
accuracy allocation　精度分配
accuracy calibration flight　精度校飞
accuracy class　准确度等级
accuracy distribution　精度分配
accuracy evaluation　精度鉴定
accuracy jump　定点跳伞
accuracy landing　精确着陆,定点着陆
accuracy of azimuth aiming　方位瞄准精度
accuracy of calibration model　标模试验准确度
accuracy of fire(firing)　射击准确度,射击精度
accuracy of image motion compensation　像移补偿精度
accuracy of inertial position　惯性导航系统定位准确度
accuracy of measurement　测量准确度
accuracy of practice　实际射击精度,实际射击准确度
accuracy of step angle　步距精度
accuracy of the shoot　实际射击准确度
accurate　1.准确的,正确的　2.已校准的
accurate adjustment　精调,细调
accurate aiming　精确瞄准,准确瞄准
accurate ephemeris　精密星历
accurate ephemeris model　精密星历模型

accurate estimate　准确估计
accurate measurement　精密量度
accurate missile　精确制导导弹
accurate model　精确模型
accurate navigation　精确导航
accurate orbit　精确轨道
accurate parameter　准确参数
accurate parameter estimate　准确参数估计
accurate prediction　准确预测
accurate result　精确结果
accurate solar system model　精确太阳系模型
accurate thermodynamic model　精确热力学模型
accurate time　准确时间
accurate trajectory　精确弹道
accustomatization　(对特定环境的)习惯化,逐步适应性
accustomization　适应
accu-tune　精确调谐
accu-tuned coaxial magnetron　精调同轴磁控管
ACD　1. Airways Clearance Delivery 航路放行传递　2. Automatic Chart Display 自动航图显示　3. Active Control Devices 主动控制装置　4. Aircraft Certification Division 航空器合格审定室　5. Aircraft (Airplane) Collector Drawings 飞机选项图　6. Allocated Configuration Documentation 分配构型文件　7. Antenna Control Display 天线控制显示　8. Automatic Call Distributor 自动呼叫分配器
ACDB　Aircraft (Airplane) Configuration Database 飞机构型数据库
ACDM　Association of Configuration and Data Management 构型和数据管理协会
ACDP　Antenna Control and Display Panel 天线控制和显示板
ACDPS　Automatic Cartographic, Drafting and Photogrammetric System 自动制图、绘图和摄影测量系统
ACDS　1. Advanced Command Data System 高级指令数据系统　2. Automatic Comprehensive Display System 自动综合显示系统
ACDTR　Airborne Central Data Tape Recorder 机载中央数据磁带记录器
ACDUTRA　Active duty training 预备人员的现役训练
ACE　1. Acceptance Checkout Equipment 接收测试设备,验收设备　2. Actuator Control Electronics 驱动器控制电子装置　3. Advanced Certification Equipment 先行合格审定设备　4. Aircraft Condition Evaluation 飞机状态评审　5. Air Conditioning Equipment 空调设备　6. Attitude Control Electronics 姿态控制电子装置　7. Automated Cost Estimate 自动化成本预算　8. Automatic Checkout Equipment 自动检查设备

9. Automatic Computer Evaluation 自动计算机评价

ACEBP Air-Condition Engine Bleed Pipe 空调发动机引气管

ACES 1. Airline Cost Estimation System 航空公司成本估算系统 2. Acceptance Checkout and Evaluation System 接收测试和评价系统 3. Adaptation Controlled Environment System 适应环境控制系统 4. Advanced Concept of Ejection Seat 先进的弹射座椅方案 5. Automatic Checkout and Evaluation System 自动检测和评价系统 6. Asia Cellular Satellite System 亚洲网状卫星系统 7. Automatically Controlled Electrical System 自动控制电气系统

ACESNA Agence Centrafricaine pour la Securite de Navigation Aerienne 〈法语〉中非领航安全公司

acetal plastic 缩醛塑料

acetal resin 缩醛树脂

acetaldehyde 乙醛

acetate 醋酸盐，醋酸酯（常用于涂料）

acetic acid 乙酸，醋酸

acetone 丙酮（一种溶剂，用于稀释油漆涂料，俗称香蕉水）

acetone droplet 丙酮滴

acetone mole fraction 丙酮的摩尔分数

acetylene 乙炔

acetylene ratio 乙炔比

acetylene-air stoichiometric acetylene-air 化学计量的乙炔-空气

acetylene-oxygen 乙炔氧气

Aceva/air combat evaluation 空战评定

ACF 1. Advanced Communications Function 高级通信功能 2. Access Control Field 访问控制字段 3. Area Control Facility 区域管制（控制）设施，地区管制设施

ACFS 1. Aircraft Carrier Flagship 航空母舰旗舰 2. Axial-Centrifugal Fan Shaft 轴向离心式风扇轴

ACFT Aircraft 飞机，飞行器

ACG 1. Address Coding Guide 地址编码指南 2. Aircraft Center of Gravity 飞机重心

ACGF Aluminium-Coated Glass-Fibre 镀铝玻璃纤维（干扰丝）

ACH 1. Advanced Chain Home 先进导航系统 2. Aircraft Hanger 飞机托架

Achernar 【天】水委一，阿却尔纳星

achievable limit 可达到的极限

achievable payload 有效载荷

achievable performance 可达的性能

achievable rate 可达信息率

achievable region 可达区域

achieve 完成，达到，实现

achieved availability 实（际）达（到的）可用性

achievement quotient 能力商数，成绩商数

Achilles 阿基里斯（小行星588号）

achondrite 无球粒陨石

achromate 1.消色差透镜，消色差物镜 2.色盲患者

achromatic 消色（差）的，无色的

achromatic body 消色体

achromatic condenser 消色差聚光镜

achromatic film 盲色片

achromatic lens 消色差透镜

achromatic sensation 无色差感

achromatic threshold 无色阈值

ACI 1. Actual Cost Incurred 实际承担成本，真实花费 2. Aircraft Condition Inspection 飞机状态检查 3. Air-Controlled Interception 空中引导截击 4. Air Council Instruction 空军委员会指令 5. Allocated Configuration Identification 分配构型标识 6. Automatic Card Identification 自动卡片标识 7. Automation Center International 国际自动化中心

ACIA Asynchronous Communications Interface Adapter 异步通信接口适配器

acicular nebula 针状星云

acicular structure （金相中的）针状组织

ACID 1. Aircraft Identification 飞机标识 2. Automatic Classification and Interpretation of Data 数据自动分类和说明

acid 酸，酸味物

acid battery 酸性电池

acid engine 酸式发动机

acid extraction 酸提取

acid pickling 酸渍，酸洗，除锈

acid storage battery 酸性蓄电池

acid value 酸值

acidosis 酸中毒

ACIL Automatic Controlled Instrumentation Landing 自动控制仪表着陆

ACIP 1. Aerodynamic Coefficient Identification Package 气动数据识别组件 2. Aviation Career Incentive Pay 航空专业鼓励津贴

ACIPS Airfoil and Cowl Ice Protection System 机翼与整流罩防冰系统

ACIRU Attitude Control Inertial Reference Unit 姿态控制惯性基准组件

ACIS 1. Advanced Cabin Interphone System 先进客舱内话系统 2. Aircraft Crew Interphone System 飞机机组内部通话系统，飞机机组内话系统 3. Avionics Central Information System 航空电子中央信息系统 4. Avionics, Control and Information System 航空电子、控制和信息系统

ACJ Advisory Circular-Joint 联合咨询通报（欧洲）

ACK Acknowledgement 收到,确认,承认

ackack 高射炮,高射机枪,高射炮火,高射炮弹炸点

ack-acker 高射炮兵人员

ackemma 1. 在午前(英国通信兵用语) 2. 飞机工人

Ackeret rule 阿克雷特法则

Ackeret theory 阿克雷特理论

ackety-ack 高射炮兵,高射兵器,高射炮

acknowledge character 确认字符(计算机传输控制字符)

acknowledgement 信件收悉

ACL 1. Altimeter Check Location 高度表校准点 2. Aircraft Circular Letter 飞机通报 3. Aircraft (Airplane) Configuration Library 飞机构型库 4. Allowable Container Load 容器许可负载 5. Anti-Collision Light 防撞灯 6. Allowable Cabin Load 允许客载 7. Avionics Cooling Loop 航空电子冷却环路

ACLD Aircooled 空气冷却的

ACLG Air-Cushion Landing Gear 气垫式起落架

ACLICS Airborne Communications Location Identification & Collection System 机载通信定位识别与采集系统

aclinal 水平的,无倾角的

aclinic 无倾角的

aclinic line 零磁倾角线,磁赤道

ACLO Agena Class Lunar Orbiter 阿金纳级月球轨道器

ACLS 1. Air Cushion Landing System 气垫着陆系统 2. Automatic Control and Landing System 自动控制和着陆系统

ACLT Actual Calculated Landing Time 实际计算着陆时间

ACM 1. Advanced Composite Material 先进复合材料 2. Air Cycle Machine 涡轮冷却器,空气循环机 3. Association for Computing Machinery 计算机机器协会 4. Automatic Continuous Monitor 自动连续监控器 5. Automatic Configuration Management Rules 自动化构形管理,自动化配置管理

ACME Acme Screw Thread (英制)梯形螺纹

ACMI 1. Aircraft, Crew, Maintenance and Insurance 飞机、机组、维修和保险 2. Air Combat Maneuvering Instrumentation 空战机动飞行仪表

ACMP 1. Alternating Current Motor Pump 交流电动泵,交流马达泵 2. Airframe Condition Monitoring Procedure (飞机)机体状态监控程序

ACMS 1. Advanced Configuration Management System 先进构型管理系统 2. Aircraft Condition Monitor System 飞机状态监控系统

ACN 1. Advanced Change Notice 先行更改通知 2. Aircraft Classification Number 飞机(载荷)等级数 3. All Concerned Notified 已通知有关 4. Aircraft Classification Number 飞机分类编号 5. Aircraft Number 机号 6. Automatic Celestial Navigation 自动天体导航

ACO 1. Acceptance Checkout 接收检验,验收试验 2. Aerosat Coordination Office 航空卫星协调办公室 3. Administrative Contracting Office 执行合同办公室 4. Aircraft Certification Office 航空器审定办公室

ACOC 1. Air Cooled Oil Cooler 气冷滑油冷却器 2. Area Communications Operations Center 区域通信操作中心 3. A Coefficient A系数,自发发射系数

ACOG Aircraft On Ground 停在地面的飞机,停飞待用飞机

ACOM Automatic Coding Machine 自动编码机

ACORE Automatic Check-Out and Recording Equipment 自动检测和记录设备

acorn 整流体

acorn valve 橡实管

ACOS Automatic Coding System 自动编码系统

acoubuoy 探声器,伞投监声器,声监听仪

acoumeter 声级计

acoustic 声的,听觉的,音响的;声学

acoustic absorption 声吸收

acoustic admittance 声导纳

acoustic amplitude 声波振幅,声幅值

acoustic analysis 声学分析

acoustic array 声阵

acoustic attenuation 声衰减

acoustic blanket 消声毯,吸声毯

acoustic boundary 声学边界

acoustic boundary condition 声学边界条件

acoustic boundary layer 声学边界层

acoustic capacitance 声容

acoustic cavity 声腔

acoustic characteristic impedance 声特征阻抗

acoustic cloud 声波反射云

acoustic combustion 声燃烧

acoustic combustion instability 声不稳定燃烧

acoustic communication 声通信

acoustic conductance 声导

acoustic conductivity 声导率

acoustic countermeasure 声对抗

acoustic coupling 声耦合

acoustic damper 声阻尼器

acoustic damper part 声阻尼器部分

acoustic damping device 声阻尼装置,声学吸收器

acoustic data 声数据

acoustic datum 声学数据

acoustic decay 声衰减

acoustic delay line　声延迟线
acoustic delay line storage　声延迟线存储器
acoustic detecting device　声学探测设备
acoustic detection　声探测
acoustic detector　声探测器,声检波器
acoustic dispersion　声频散
acoustic dissipation　声耗散
acoustic disturbance　声干扰,声音干扰
acoustic Doppler sounder　声学多普勒探测器
acoustic driver　声学驱动器
acoustic driving　声学驱动
acoustic echo　回声
acoustic echo sounder　回声探测器
acoustic emission　声发射
acoustic emission inspection　声发射检测
acoustic emission testing　声发射检测
acoustic enclosure　音箱
acoustic energy　声波能量
acoustic energy conservation principle　声波能量守恒原理
acoustic environment　声环境
acoustic excitation　声激励
acoustic fatigue　声疲劳
acoustic fatigue test　声疲劳试验
acoustic feedback　声反馈
acoustic field　声场
acoustic filter　声滤波器,滤声器
acoustic fluctuation　声(波)起伏
acoustic flutter　声波颤振
acoustic foam　声学泡沫
acoustic fog　声雾
acoustic forecast of shuttle launch　航天飞机发射声振预测
acoustic frequency　声波频率,音频
acoustic fuze　声引信
acoustic generator　发声器,声发生器
acoustic homing guidance　声自导引,声寻的制导
acoustic homing head　声导引头,声寻的头
acoustic homing system　声寻的系统,音响寻的系统
acoustic homing torpedo　声寻的鱼雷,音响寻的鱼雷
acoustic image　声像
acoustic imaging　声成像
acoustic impedance　声阻抗
acoustic impedance audiometry　声阻抗测听术
acoustic inertance　声惯量,声阻
acoustic input device　声音输入装置
acoustic instability　声学不稳定性
acoustic insulation material　隔音材料
acoustic intensity　声强(度)

acoustic interaction　声相互作用
acoustic jammer　声干扰机,音响干扰机
acoustic level　声学层次,声级
acoustic liner　声衬
acoustic lining　声处理衬套,吸声衬垫
acoustic load　声负载
acoustic localization plot　声定位标图
acoustic loss　声损耗
acoustic magnetic mine　音响磁性水雷
acoustic mass　声质量,声扭
acoustic material　声学材料,音响材料
acoustic measurement　声学测量
acoustic method　声波法,声学法
acoustic mixer　混声器
acoustic mobility　声导纳
acoustic mode　声学模式
acoustic model　声学模型
acoustic motion　声学运动
acoustic navigation　声导航
acoustic noise　噪声
acoustic optical receiver　声光接收机,布拉格元接收机
acoustic oscillation　声振荡,声波
acoustic oscillatory combustion　声振燃烧,声学不稳定燃烧
acoustic output unit　声音输出装置
acoustic paint　声学涂料
acoustic period　声期
acoustic perturbation　声波扰动
acoustic perturbation equation　声摄动方程
acoustic potential energy　声势能
acoustic power　声功率
acoustic prediction　声学预测
acoustic pressure　声压
acoustic problem　声学问题
acoustic property　声学性质,声特性
acoustic proximity fuze　感声近炸引信
acoustic radar　声雷达
acoustic radiation pressure　声辐射压
acoustic ratio　声学比,声强比
acoustic ray　声线
acoustic reactance　声抗
acoustic refraction　声折射
acoustic resistance　声阻
acoustic resonance　声共振,音响共鸣
acoustic resonant frequency　声共振频率
acoustic resonator　声波谐振器,共鸣器,声共振器
acoustic response　声波响应,声学响应
acoustic scattering　声散射
acoustic sensor　声敏元件,声传感器,音响传感器

acoustic signal　声频信号
acoustic simulation　声仿真
acoustic sounder　声探测器
acoustic source　声源
acoustic spectrum　声谱
acoustic speed　声波速度
acoustic splitter　消声隔板
acoustic standing　驻波
acoustic streaming　声冲流
acoustic surface wave amplifier　声表面波放大器
acoustic surface wave filter　声表面波过滤器
acoustic surface wave oscillator　声表面波振荡器
acoustic surface wave phase shifter　声表面波移相器
acoustic system　声系统
acoustic test　声学测试
acoustic theory　声学理论
acoustic torpedo　音响鱼雷,声学鱼雷
acoustic torque　声学扭矩
acoustic tracking　声跟踪
acoustic tracking system　音响跟踪系统
acoustic transmission　声传输
acoustic treatment　声学处理
acoustic tube　声管,传声筒
acoustic velocity　声速
acoustic velocity field　声波速度场
acoustic velocity perturbation　声波速度扰动
acoustic vibration　声振(动)
acoustic vibration test　声振试验
acoustic volume　声量
acoustic warfare　音响战
acoustic wave　声波
acoustic waveguide　声波导
acoustic wavelength　声波的波长
acoustic wind tunnel　声学试验风洞
acoustic(al) altimeter　声学测高仪
acoustical energy　声能量
acoustical generator　发声器,声辐射器
acoustical hologram　声频全息图
acoustical holography　声频全息照相术
acoustical profile　声剖面图,声廊线
acoustical radiometer　声辐射计
acoustic-microwave radar　声微波雷达
acoustic-pressure　声压
acoustics　1.声学 2.音响装置 3.音质
acoustic-velocity　声波速度
acoustimeter　声级计
acoustoelectric effect　声电学效应
acoustoelectric material　声电材料
acoustometer　声强计

acoustooptic ceramics　声光陶瓷
acoustooptic crystal　声光晶体
acoustooptic device　声光器件
acoustooptic effect　声光效应
acoustooptic material　声光材料
acoustooptic modulation　声光调制
acoustooptic technique　声光技术
ACP　1. Airlift Command Post 空运指挥所 2. Audio Control Panel 音频控制面板 3. Audio Control-selector Panel 音频选择板 4. Acceptance message 接受电报
ACPL　Aircraft Certification Policy Letter 航空器合格审定政策函
ACPM　ATM Central Process Module ATM 中央处理模块
ACPS　Air-Conditioning and Pneumatic System 空调和气源系统
Acquacade　(Aquacade)水技会演卫星(美国电子侦察卫星)
acquire　1.获得,取得 2.探测,发现 3.截获(目标) 4.达到
acquire sound　截获声响
acquisition　1.获取,采集 2.截获,捕获 3.探测,搜索,发现 4.(目标)显示,定位
acquisition cost　收购成本
acquisition cost model　收购成本模型
acquisition coverage　探测范围,截获范围
acquisition field of view　(导引头)捕获视场,捕获场
acquisition mode　捕获模式
acquisition of orbit　轨道探测,轨道搜索
acquisition of phase-locked loop　锁相环路捕获
acquisition probability　捕获概率
acquisition radar　捕获雷达,目标探测雷达,搜索雷达
acquisition range　捕获范围,捕获距离
acquisition round　(供飞行员用雷达瞄准具练习射击的)教练弹
acquisition sensitivity　捕获灵敏度
acquisition sequence　习得顺序
acquisition signal　捕获信号
acquisition station　目标探测站,捕获
acquisition time　(锁相环路的)捕获时间
acquisition unit　1.采集单元 2.捕获装置
ACR　1. Across 跨越,横过 2. Actrual Cost Report 实际成本报告 3. Aerodrome Control Radar 机场管制雷达 4. Access Control Register 访问控制寄存器 5. Airfield Control Radar 机场控制雷达 6. Antenna Coupler Receiver 天线耦合器接收机 7. Antenna Coupling Regulator 天线耦合调节器 8. Approach Control Radar 进近控制雷达 9. Automatic Compression Regulator 自动压气调节器 10. Avionics Communication Router

航空电子设备通信路由器

ACRC Additional Cycle Redundancy Check 附加的循环冗余码校验

ACRE 1. Automatic Call Recording Equipment 自动呼叫记录设备 2. Automatic Checkout and Readiness Equipment 自动检测和备用设备 3. Airline Composite Repair Evaluation 航空公司复合材料修理评估

acreage 英亩数,面积

ACRFLT Across Flats 对边

acrobatic 1. 特技飞行的 2. 飞机上做体育表演的

acrobatics 特技飞行

acrolein 丙烯醛

acrophobia 恐高症

across-track error 偏离航线距离

acrotorque 最大扭力

Acrux 【天】南十字二,南十字座 a 星

ACRV Armored Command and Reconnaissance Vehicle 装甲指挥与侦察车

acrylic 丙稀酸的,聚丙稀的

acrylic acid 丙稀酸

acrylic nozzle 丙烯酸喷嘴

acrylic tube 丙烯酸软管,压克力管

acrylonitrile 丙烯腈

ACS 1. Accumulator Switch 蓄电池开关 2. Acitive Control System 主动控制系统 3. Air Conditioning System 空调系统 4. Aircraft Control and Surveillance 飞机控制和监视 5. Alternating Current Synchronous 交流同步 6. Assembly Control System 装配控制系统 7. Attitude Command System 姿态指令系统 8. Attitude Control System 姿态控制系统 9. Audio Control System 音频控制系统 10. Auto Chart System 自动制图系统 11. Automated Communications Set 自动化通信装置 12. Automated Control System 自动化控制系统 13. Automatic Checkout System 自动检测系统 14. Alternating Current Supply 交流供电 15. Access Control System 接入控制系统

ACSA Allied Communications Security Agency 盟国通信安全局(北约)

ACSB Amplitude Companded Sideband 振幅压扩边带(技术)

ACSD Automatic Color Scanned Device 自动彩色扫描装置

ACSEP Aircraft (Airplane) Certification System Evaluation Program 飞机合格审定系统评审项目

ACSG Aeronautical Communications Sub-Group 航空通信子集团

ACSM Advanced Conventional Standoff Missile 防空区外发射的先进常规导弹

ACSN Advance Change Study Notice 先行更改研究通知

ACSS Automatic Contract Specification System 自动化合同规范系统

ACST Acoustic 声学的,传声的

ACSW Advanced Crew-Serviced Weapon 高级机组人员协同操作武器

ACT 1. Active 主动地,有效的 2. Active Control Technology 主动控制技术 3. Activity 活动,作用,功效 4. Advanced Computer Technology 先进计算机技术 5. Aircraft (Airplane) Configuration Tracking 飞机构型跟踪 6. Automatic Checkout Technology 自动检测技术

act 1. 动作,作用 2. 法令,规章,说明书

ACTE Actuate 致动

acteristics 特征

acterization 特性

ACTF Aeronautical Charting Task Force 航空制图工作队

ACTG Actuating 致动

acting 1. 动作的,作用的 2. 代理的

actinic balance 分光测热计,热辐射计

actinic ray 光化射线

actinide 锕系元素

actinism 感光度,光灵敏度

actinogram 日射曲线图,光能曲线图

actinograph 日射计,辅射仪

actinometer 1. 光度计,曝光表,太阳辐射计,日射表 2. 光化线强度计

actinometry 1. 日射测定学,日射测定法 2. 光化线强度测定

action 1. 作用,动作,行动,战斗 2. 作用力,作用量 3. (自动武器的)主要活动机构(如枪栓、扳机、进弹退弹机构等)

action integral 作用(量)积分

action launch 战斗起飞(舰载飞机从甲板待命状态中奉命起飞)

action network 行动网

action potential 动作电位

action principle 作用量原理

action range 作用范围

action rate 行动率

action rate zero 行动率为零

action reference 行动参考

action spring 击针(弹)簧,撞针簧,打火簧

action surface of fuze 引信作用面

action time 作用时间

action variable 作用变数

action with director control (高炮)用指挥仪射击

action with sight control (高炮)用瞄准具射击

action-oriented simulation 面向动作的仿真
activate 1.激发,开动 2.组建(部队从筹备到开始执行职能)
activated charcoal 活性炭
activated diffusion healing 活化扩散愈合
activated mine 触发地雷
activated molecular sieve 活化分子筛
activating device 触发装置
activation 1.激发,激活,活化 2.接通,开动 3.组建
activation adsorption 活化吸附,吸附作用活化
activation energy 激活能,活化能
activation energy of diffusion 扩散激活能
activation indicator (电池)激活指示器
activation of homing 接通自导系统,启动寻的系统
activation probe 活化探针
activation signal 激活信号
activator 激活剂,活化剂
active 1.主动的,积极的 2.放射性的 3.有效的,实际的 4.现役的 5.有源的
active accommodation 主动适应性
active acoustic torpedo 主动式声响鱼雷
active aerodynamic braking 主动气动减速,主动气动(反推力)制动
active air defence 积极防空
active aircraft inventory 现役(作战)飞机总数
active airfoil 活跃翼
active airfoil source control 活跃的翼源控制
active air-to-air homing guidance system 空对空主动寻的制导系统
active aluminum 活性铝
active aluminum content 活性铝含量
active antenna 有源天线,发射天线
active area 活性区,有效面积
active area of a solar cell 单体电池的有效光照面积
active array 有源阵
active attitude control 主动姿态控制
active attitude stabilization 主动姿态稳定
active balance 动态平衡
active balance return loss 有源均衡反射波损耗
active blade 活跃叶片,活跃轮片
active blade-tip 活跃叶片尖端
active cavity radiometer 主动空腔辐射计
active chromosphere 活动色球
active chromosphere star 活动色球星
active clearance 主动间隙,活跃间隙
active clearance control 主动间隙控制
active component 有源元件
active constraint 主动约束
active control 主动控制

active control stage 主动控制阶段
active control system 主动控制系统
active control technology 主动控制技术
active control time 主动控制时间
active controls 主动操纵装置
active cooling 主动式冷却
active cooling loop system 主动冷却回路系统
active cooling system 主动式冷却系统
active core 活跃核心,积极核
active corona 1.活动日冕 2.活动星冕
active corrective network 有源校正网络
active countermeasure 积极干扰,有源干扰,主动式干扰
active day 受扰日
active defect 运行缺陷,实有缺陷
active depression failure (结构的)主动抑制断裂,自行破坏
active detection 有源探测
active display 主动显示
active element 有源元件
active experiment 主动实验
active failure 1.硬故障 2.自行故障(因系统本身的问题使元件失效)
active feedback 积极反馈
active filter 有源滤波器
active flap 活跃襟翼
active flow 主动流动
active flow control 主动流动控制
active flutter 主动颤振
active flutter depression 主动颤振抑制
active fuze 主动式引信
active galactic nucleus 活动星系核
active galaxy 活动星系
active gravitational mass 主动引力质量
active guidance 主动制导
active guidance system 主动制导系统
active heat-seeking guidance 主动热辐射寻的制导
active homer 1.主动自导引导弹,自动寻的导弹 2.主动导引头,主动寻的装置
active homing 主动寻的,主动式自动引导
active homing guidance 主动寻的制导,主动式自动制导
active homing system 主动寻的系统
active inert missile 主动惯性制导导弹
active infrared line-of-sight system 主动红外瞄准线系统
active infrared tracking system 主动红外跟踪系统
active isolation 主动隔振,积极隔震
active jammer 积极干扰机
active jamming 积极干扰,有源干扰
active landing gear 主动起落架

active longitudes　活动经度
active low clock　低态有效时钟
active low inverter　低态有效反相器
active magnetic suspension　有源磁悬浮
active magnetospheric particle tracer explorer　磁层粒子主动示踪探险者卫星
active maintenance cost　实际维修费用
active maintenance downtime　实际维修停用时间,现行维修停用时间
active material　1.放射性材料(物质) 2.活性材料(物质)
active matrix　有源矩阵
active medium　激活媒质,激射工质
active metal　活泼(活性)金属
active microwave　主动微波,有源微波
active microwave remote sensing　1.主动微波遥感 2.有源微波遥感
active microwave system　主动微波系统,有源微波系统
active missile　发射后免控导弹,自导引导弹
active monitoring system　主动监测系统
active munition　主动弹药
active network　有源网络
active night vision device　主动式夜视装置
active noise　主动型噪声
active noise control　主动噪声控制
active noise control system　主动型噪声控制系统
active nutation control　主动章动控制
active nutation damping　主动章动阻尼
active obstacle　积极障碍物,主动障碍物
active operation　主动式活动
active optical fiber　激活光纤
active particle　活性粒子
active phase　活跃期
active photospheric region　光球活动区
active proximity fuze　主动式近炸引信,主动式引信
active radical　活性根,活性自由基
active relay station　有源中继站
active remote sensing　主动遥感,有源遥感
active repair time　修理实施时间
active research　活跃的研究,研究热点
active runway　现用跑道
active sail　主动航行
active sail control　主动航行控制
active sail control system　主动航行控制系统
active satellite　有源卫星,主动卫星
active satellite defense　主动卫星防御
active scattering aerosol spectrometer　有源散射气溶胶粒谱仪
active seam　活性焊缝

active search　主动搜索
active sensor　主动遥感器,有源遥感器
active set　积极集,作用集
active set method　1.活动集法 2.主动设定法
active shock absorber　主动减震器
active sonar platform　主动式声呐平台
active spare　热备份(备份部分处于工作状态)
active specie　活性物种
active stage　活动期,进行期
active steering　主动引导
active sun　活跃的太阳
active sunspot prominence　活动黑子日珥
active suppression　主动抑制,有效抑制
active system　主动系统,有源系统
active technique　主动技术,有源技术
active thermal control　主动式热控制
active time　有效时间
active tracking　有源跟踪
active twist　活跃扭
active twist blade　活跃扭叶片
active twist rotor　活跃扭转子
active vibration　主动振动
active vibration absorber　主动振动吸收器
active vibration control of space structure　空间结构主动振动控制
active volt-ampere　有功伏安
active weapon　编制武器
active-control system　主动控制系统
actively-cooled　有效冷却的
active-passive fuze　主被动引信
activity　1.活动,放射性,作用,活性 2.组织,机构
activity description　操作说明,工序说明
activity sphere　作用范围,作用球
ACTL　Actual 实际的
ACTR　Actuator 激励器,执行机构;致动器,作动筒
ACTS　1. Advanced Communications and Technology Satellite 先进通信技术卫星 2. Air Cargo Terminal System 航空货运站系统 3. Advanced Communication Technologies and Services 先进的通信技术与业务 4. Automated Change Tracking System 自动更改跟踪系统
actual aircraft　实际飞机
actual blade　实际叶片
actual climb　实际攀登
actual collision event　实际碰撞事件
actual control volume　实际控制量
actual datum　实际数据
actual distance　实际距离
actual disturbance　实际干扰

actual donor 实际供体
actual donor cell 实际供体细胞
actual engine 实际发动机
actual escape 实际逃逸
actual flight 实际飞行
actual flight usage 实际飞行的使用
actual flow 实际流量
actual flying test 实际飞行试验
actual gauge 实际测量
actual ground 实际地面
actual ground zero 实际爆心地面投影点
actual leader 实际领袖
actual loss 实际损失
actual loss of separation 分离的实际损失
actual navigation performance 实际导航性能
actual operational error 实际操作错误
actual parameter 实在参数,实际参数
actual position 真实位置
actual sector 实际部门
actual size installation model 全尺寸安装模型
actual surface 实际表面
actual system 实际系统
actual terrain 实际地形
actual time 实际时间
actual traffic 实际交通
actual traffic count 实际交通计数
actual trajectory 实际轨迹
actual transition 实际过渡
actual use 实际应用
actual value 实际价值,实际指标
actual wing 实际机翼
actual-use test condition 实际使用试验条件
actuarial analysis 保险统计分析
actuarial program 保险统计程序
actuate 开动,致动
actuated fuel 驱动燃料
actuating 执行,驱动
actuating arm regulator 力臂调节器
actuating code 执行码
actuating cylinder 作动筒,动作筒
actuating force 驱动力
actuating mechanism 执行机构
actuating unit 执行机构,致动器
actuation 1.传动 2.开动,致动,驱动 3.传动装置,致动装置作动器,致动器 4.执行机构
actuation capability of smart structure 智能结构的致动能力
actuation channel 致动通道
actuation frequency 驱动频率

actuation law of fuze 引信启动规律
actuation loss 驱动损耗
actuation point 启动点
actuation power 驱动电源
actuation scheme 驱动方案
actuation signal 启动信号
actuation system 驱动系统
actuation threshold 启动阈
actuation value 动作值
actuator 驱动器
actuator authority 执行机关
actuator bandwidth 致动器带宽
actuator beam 致动器梁
actuator chip 驱动器芯片
actuator command 致动器命令
actuator delay 致动器延迟
actuator disk 作动盘,激盘
actuator driver system 作动器驱动系统
actuator duct 执行器管
actuator dynamics 1.执行机构动力学 2.舵动态
actuator effect 驱动器作用
actuator extension 致动器扩展
actuator failure 执行器故障
actuator flexibility 致动器灵活性
actuator force 驱动力
actuator hysteresis 致动器磁滞
actuator model 执行机构模型
actuator pair 致动器电偶
actuator position 致动器位置
actuator power 致动器功率
actuator rate 致动器速率
actuator rate limit 致动器速率限制
actuator response 致动器响应
actuator response time 致动器响应时间
actuator saturation 致动器饱和
actuator system 致动器系统
actuator voltage 致动器电压
actuator-disk 促动盘
actuator-disk model 激盘模型
actuator-duct logic 执行器管逻辑
ACU 1. Air Conditioning Unit 空调装置 2. Airport Control Unit 机场控制单元 3. Annunciator Control Unit 信号器控制装置 4. Antenna Control Unit 天线控制装置 5. Antenna Coupler Unit 天线耦合器件组 6. Apron Control Unit 停机坪控制装置 7. Automatic Calling Unit 自动呼叫系统 8. Autopilot Control Unit 自动驾驶仪控制装置 9. Auxiliary Control Unit 辅助控制(操纵)装置 10. Avionics Cooling Unit 航空电子设备冷却装置 11. Avionic Control Unit 航空电子控制

组件,航空电子设备控制装置
acuity 1. 尖锐 2.（视力等）敏锐 3. 锐度,分辨能力
acutance 锐度
acute 1. 尖的,尖锐的 2. 大后掠角的 3. 敏锐的,急剧的
acute altitude hypoxia 急性高空缺氧
acute anoxia 急性缺氧症
acute dose 急性剂量
acute exposure 急性照射
acute hypoxia examination 急性缺氧检查
acute irradiation 急性辐射
ACV Air-Cushion Vehicle 气垫飞行器
ACW 1. Aircraft Control and Warning 飞机控制和警告 2. Alternating Continuous Wave 交变连续波
ACWS Aircraft Control and Warning System 飞机控制和警告系统
acyclic 1. 非周期性的 2. 单极的 3. 无环的
acyclic diagraph 非循环双图形,非循环有向图
AD 1. Access Door 检修门 2. Accidental Damage 偶然损伤 3. Administration Domain 管理范围 4. Advanced Design 先行设计 5. Advanced Development 先行开发 6. Aerodrome 机场 7. Air Data 大气数据 8. Air Density 大气密度 9. Air Defence 防空 10. Airworthiness Directive 适航指令 11. Alarm and Display 告警并显示 12. Allowable Deficiency 容许缺陷 13. Anode 阳极 14. Automatic Detection 自动探测
Ada 阿达语言（美国国防部开发的一种三军通用的高级计算机语言）
ADA 1. Air Defence Area 防空区 2. Airborne Data Automation 机载（机上）数据自动化 3. Advisory Area 咨询区 4. Automatic Data Acquisition 自动数据采集
ADAC 1. Automatic Data Acquisition Center 自动数据采集中心 2. Advise Acceptance 告知接受
ADACC Automatic Data Acquisition and Computer Complex 自动数据采集和计算机综合
ADAIRS Air Data and Inertial Reference System 航空数据与惯性参照系统
adaline 线性适应元
ADAM Air Deflection And Modulation 空中偏转和调节
Adams-Bashforth method 亚当斯-巴什福思法
Adams-Bashforth numerical integration 亚当斯-巴什福思数值积分
Adams-Moulton method 亚当斯-莫尔顿法
Adams-Moulton numerical integration 亚当斯-莫尔顿数值积分
Adams-Russell phenomenon 亚当斯-罗素现象（恒星大气）
ADAP 1. Aerodynamic Data Analysis Program 气动数据分析计划 2. Airport Development Aid Program 资助机场开发计划,机场发展援助计划
ADAPS 1. Automatic Data Acquisition and Processing System 数据自动获取和处理系统 2. Automatic Display And Ploting Systems 自动显示和标绘系统
adapt 适应,改写
ADAPT 1. Analogy Digital Automatic Program Tester 模数自动程序试验器 2. Automated Data Analysis and Presentation Techniques 自动化数据分析与显示技术 3. ATS Data Acquisition Processing and Transmission 空管服务数据录取处理和传递
adaptability 适应（合,用）性,可用性,灵活性,适应能力
adaptation 1. 适应,适合,自调正 2. 改进,改型
adaptation gain 自适应增益
adaptation kit 1. 接合装置,适配装置 2. 成套维护设备
adaptation law 适应律
adaptation layer 适应层
adaptation luminance 适应亮度
adaptation mechanism 适应机理,适应机制
adaptation parameter 适应参数
adaptation rate 适配率,自适应速率
adapted 适合（设计状态）的
adapter 1. 连接器,连接件,多级火箭级间（连接）段,星-箭连接件,航天器轨道对接接合器 2. 适配器 3. 接头,管接头 4. 架（插）座 5. 拾音器,拾波器
adapter module 过渡舱,适配舱
adapter ring （带多层密封件切槽的）接合环,（导弹上连接战斗部用）连接环
adapter section 过渡段,适配段
adapter stage 转换级
adapter-booster 传爆管
adaptive 自适应的,自调的
adaptive antenna 自适应天线
adaptive approach 自适应方法
adaptive array 自适应阵
adaptive array antenna 自适应天线阵
adaptive attitude 自适应姿态
adaptive augmentation 自适应增强
adaptive autopilot 自适应式自动驾驶仪,自调式自动驾驶仪
adaptive backstepping 自适应反推,自适应反步法
adaptive bias 自适应偏置,适应性偏差
adaptive bus 自适应总线,适配总线
adaptive communication 自适应通信
adaptive compensator 自适应补偿器
adaptive computer 自适应计算机
adaptive control 自适应控制
adaptive control law 自适应控制法
adaptive control system 自适应控制（操纵）系统
adaptive controller 适应控制器

adaptive critic   适应评价控制
adaptive delay   自适应延时
adaptive detection   自适应检测
adaptive differential PCM   自适应差分脉码调制
adaptive ejection seat   自适应弹射座椅
adaptive electronics   自适应电子设备
adaptive element   自适应单元
adaptive equalization   自适应均衡
adaptive equalizer   自适应均衡器
adaptive estimation   自适应估计
adaptive estimation system   自适应估计系统
adaptive expert system   自适应专家系统
adaptive filter   自适应滤波器
adaptive flight   自适应飞行
adaptive flight control system   自适应飞行控制系统
adaptive flow   自适应流量
adaptive flow control   自适应流量控制
adaptive flutter   自适应颤振
adaptive fuze   自适应引信
adaptive grid technique   自适应网格技术
adaptive guidance   自适应制导
adaptive input   自适应输入
adaptive law   自适应律
adaptive linear element   线性适应元
adaptive logic   自适应逻辑
adaptive maintenance   适应性维护
adaptive measurement   自适应测量
adaptive model   自适应模型
adaptive noise cancelling   自适应噪声消除
adaptive observer   自适应观测仪
adaptive optical camouflage   自适应光学伪装
adaptive optical filter   自适应滤光镜
adaptive optics   1.自补偿光学技术 2.自补偿光学设备
adaptive optimal control system   自适应最优控制系统
adaptive optimization   自适应最佳化,自适应优化
adaptive output   自适应输出
adaptive prediction   适应预报
adaptive radar   自适应雷达
adaptive recon   自适应侦察
adaptive remote control   自适应遥控
adaptive routing   自适应路由选择
adaptive sampling   适应性抽样
adaptive scheme   自适应机制
adaptive sensor   自适应传感器
adaptive signal processing   自适应信号处理
adaptive simulation   自适应模拟
adaptive smoothing   自适应平滑(技术),自适应修匀(技术)
adaptive state   自适应状态

adaptive state estimation   自适应状态估计
adaptive system   自适应系统
adaptive telemetry   自适应遥测
adaptive telescope   自适应望远镜
adaptive thresholding   自适应调阈
adaptive training   适应性训练
adaptive wall   1.自适应壁 2.自修正壁
adaptive wall wind tunnel   无洞壁干扰风洞,自适应壁风洞
adaptive weight   自适应权值
adaptive weight algorithm   自适应权值算法
adaptive wing   自适应机翼
adaptive-detector   自适应检测器
adaptometer   适应计
adaptor connector   转接连接器
ADAPTS   Analogue-Digital-Analogue Process and Test System 模拟-数字-模拟处理与测试系统
ADAR   1.Advanced Data Acquisition Routine 高级数据采集程序 2.Advanced Defense Array Radar 高级防御相控阵雷达 3.Air Defense Area Operations 防空区作战
ADAS   1.Airborne Data Acquisition System 机载数据获取系统 2.Airborne Dynamic Alignment System 机载动态校准系统 3.Automatic Data Acquisition System 自动数据采集系统
ADATS   Australian Air Defense Air Traffic System 澳大利亚防空及空中管制系统
ADAU   1.Auxiliary Data Acquisition Unit 辅助数据获取组件 2.Air-Data Acquisition Unit 大气数据采集组件
ADB   Bureau of Administration and Services 行政与服务局,行政服务局
ADC   1.Aerospace Defense Command 航空航天防御司令部 2.Air Data Computer 大气数据计算机 3.Air Defense Command 防空军,防空司令部(美国空军) 4.Aircraft Delivery Commitments 飞机交付承诺 5.Analogy-to-Digital Computer 模数计算机 6.Anology-to-Digital Converter 模数转换器 7.As-Delivered Configuration 交付构型 8.Automated Discrepancy Control 自动报废控制 9.Automatic Data Collection 自动数据汇集
ADCA   Australian Department of Civil Aviation 澳大利亚民航部
ADCAD   Airway Data Collection And Dissemination 航路数据汇集和传播
ADCAP   Advanced Capability 先进性能
ADCC   1.Air Defence Control Center 防空管制中心 2.Asynchronous Data Communication Channel 异步数据通信信道

ADCCP Advanced Data Communication Control Procedure 先行数据通信控制程序
ADCN Advanced Drawing Change Notice 先行图纸更改通知
ADCO As Design Changes Occur 因设计更改而出现
ADCOM Aerospace Defense Command 航空航天防御司令部(美国空军)
ADCS 1. Air Data Computer System 飞行数据计算机系统，大气数据计算机系统 2. Automatic Document Control System 自动文档控制系统
ADCSP Advanced Defense Communications Satellite Project 高级国防通信卫星项目
ADCSPC Air Data Computer Static Pressure Compensator 大气数据静压补偿器
ADCU Alternate Detection and Control Unit 备用探测和控制装置
add 加，增添
ADD 1. Addendum 附录 2. Automatic Document Distribution 自动文档分配
add time 加法时间，加算时间
ADDAR Automactic Digital Data Acquisition System 自动数字数据采集系统
ADDAS Automatic Digital Data Assembly System 自动数字数据汇编系统
ADDC Air Defence Direction Center 防空指挥中心
ADDD Air-Defence Data Dictionary 防空数据词典(英国)
added amount of oxygen 氧添加量
added mass 附加质量
adder circuit 加法器电路
adding interferometer 相加干涉仪
addition 1.附加，补充 2.加法 3.附加物
addition constant 加常数
addition of ethane 添加乙烷
addition of ethanol 添加乙醇
addition of mass 添加质量
addition of pitch 添加沥青
addition of radical 添加自由基
addition of red dye 添加红色染料
addition of steam 添加蒸汽
addition of sway 添加倾斜
additional ammunition 增发的弹药
additional damage 附带的破坏
additional failure 附加故障
additional pole 附加极点
additional potential 附加位
additional weight 附加重量，配重
additional zero 附加零点
additiva method 添加剂法

additive 1.添加剂，附加物 2.附加的，辅助的 3.加法的
additive drag 附加阻力
additive image enhancement 加色影像增强
additive mixture 加色混合
additive noise 相加噪声
additive of solid propellant 固体推进剂添加剂
additive perturbation 加性扰动
additive result 添加剂效果
additive uncertainty 添加剂不确定性
add-on 附加
ADDR 1. Adder 加法器，加法电路，混频器，求和装置 2. Aeroklub der Deutschen Demokratischen Republik 〈德语〉德意志民主共和国航空俱乐部
address 地址
address access time 地址存取时间
address marker 地址标识符
address matching 地址匹配
address recognition unit 地址识别装置
address register 地址寄存器
address switch register 地址切换寄存器
address synchronizing track 地址同步
addressing operation 寻址操作
ADDS 1. Advanced Data Display System 先进数据显示系统 2. Automated Digital Data System 自动化数字数据系统 3. Automatic Data Distribution System 自动数据分配系统
ade 自动设计工程
ADE 1. Address Enable 地址启动 2. Aeronautical Development Establishment 航空发展研究所(印度) 3. Automatic Design Engineering 自动化设计工程 4. Automatic Drafting Equipment 自动制图设备
ADED Advanced Engineering Division 先行工程部
ADEMS 1. Advanced Diagnostic Engine Monitoring System 预诊发动机监控系统 2. Airborne Display Electrical Management System 机载显示电子管理系统
ADEO Advance Development Engineering Order 先行开发工程指令
ADEOS Advanced Earth Observing Satellite 先进地球观测卫星
adequacy 1.适合，恰当 2.充分，足够
adequate performance 足够性能，足够绩效
ADES 1. Airport Engineering Data Sheet 机场工程数据表 2. Automatic Digital Encoding System 自动数字编码系统
ADESS Automatic Data Editing and Switching System 自动数据编辑与转接系统
ADEU Automatic Data Entry Unit 自动数据输入组件
ADEWS Air Defense Electronic Warfare System 防空

电子战系统

Adex （ADEX）Air Defense Exercise 防空演习

ADF 1. Automation Direction Finder 自动定向仪（无线电罗盘），自动测向仪 2. Aerodynamic Data File 气动数据文件

ADFC Air Defense Filter Center 防空情报鉴定中心

ADG 1. Air Driven Generator 空气驱动发电机 2. Atmospheric Dynamic Payload Group 大气动力有效载荷组

ADGB 1. Air Defence of Great Britain 大不列颠防空 2. Accessory Drive Gear Box 附件机匣

ADGES Air Defence Ground Environment System 地面防空系统

ADH 1. Adhesive 粘着的，胶粘的；黏合剂 2. Asynchronous Digital Hierarchy 异步数字系列

adherence 附着，黏附

adhesion 1. 黏附性，附着，粘接 2. 附着力，黏附力

adhesive 1. 附着的，黏附的 2. 黏合剂，胶粘剂，附着剂

adhesive for honeycomb sandwich structure 蜂窝夹层结构胶粘剂

adhesive joint 胶粘接合，附着接头

adhesive joint assembly 1. 胶接装配 2. 套装

adhesive layer 胶层

adhesive spalling （胶合板的）层间剥离

adhesive strength 胶粘强度

adhesive target 胶粘剂目标

adhesive test 胶接试验

adhesive-bonding 粘接

ADHS 1. Analog Data Handling System 模拟数据处理系统 2. Automatic Data Handling System 自动数据处理系统

ADI 1. Attitude Direction Indicator 姿态指引仪 2. Attitude Display Indicator 姿态显示指示器 3. Attitude Director Indicator or Attitude Display Indicator （飞行）姿态与方位指示器或（飞行）姿态显示器

adia thermal 绝热的

adiabat 绝热线

adiabatic 绝热的

adiabatic atmosphere 绝热大气

adiabatic case 绝热情况

adiabatic chamber 绝热室

adiabatic compressibility 绝热压缩系数

adiabatic condition 绝热条件，绝热状态

adiabatic effectiveness 绝热效率，冷却效率

adiabatic effectiveness distribution 绝热效率分布

adiabatic efficiency 绝热效率

adiabatic equivalent temperature 绝热当量温度

adiabatic expansion 绝热膨胀

adiabatic flame 绝热火焰

adiabatic flame temperature 绝热火焰温度

adiabatic flow 绝热流

adiabatic flow in pipe 绝热管流

adiabatic gas flow 绝热气流

adiabatic invariant 绝热不变量

adiabatic lapse rate 绝热直减率

adiabatic layer 绝热层

adiabatic model 绝热模型

adiabatic process 绝热过程

adiabatic section 绝热段

adiabatic shear 绝热剪切

adiabatic system 绝热系统

adiabatic temperature 绝热温度

adiabatic wall 绝热壁

adiabatic wall temperature 绝热壁温度

adiabatics 绝热曲线

adiministrative department （航母）行政部门

ADIOS Automatic Digital Input-Output System 自动数字输入-输出系统

adipic acid 己二酸

ADIRS Air Data Inertial Reference System 大气数据惯性基准系统

ADIRU Air Data Inertial Reference Unit 大气数据惯性基准装置

ADIS 1. Automatic Data Interchange System 自动数据交换系统 2. Airborne Digital Instrument System 机载数字仪表系统

ADIT Analog-Digital Integrating Translator 模数综合转换器，模数综合转换装置

ADIZ Air Defence Identification Zone 防空识别区

ADJ 1. Adjust（able）可调整的，调整 2. Adjustment 调节，调整，校正

adjacent 接近的，相邻的

adjacent aircraft 相邻飞机

adjacent blade 相邻叶片

adjacent cell 相邻区

adjacent channel interference 邻信道干扰

adjacent channel selectivity 邻信道选择性

adjacent coil 毗连线圈，连接线圈

adjacent element 相邻元素

adjacent field 相邻字段

adjacent galaxies 邻近星系

adjacent load tape 相邻载带

adjacent node 相邻节点

adjacent passage 相邻通道

adjacent sector 相邻扇区

adjacent streamtubes 相邻流管

adjacent tube 相邻的管

adjective 附属的,不独立的
adjoint 相结合的,伴随的
adjoint equation 伴随方程
adjoint operator 伴随算子
adjoint orbit 伴随轨道
adjoint source 伴随信源
adjoint variable 伴随变量
adjoint variable method 伴随变量方法
adjoint vector 伴随向量
adjust 调整,校准,校正
adjust a gun 修正枪炮的射向
adjust fire 射击修正
adjust for lead 修正提前量(前置量)
adjustable 可调整的,可调节的,可校准的
adjustability 可调(节)性,调整性
adjustable aperture 可调光圈
adjustable-thrust 推力可调的
adjustable-cistern barometer 动槽式气压表
adjustable delay 可调延时
adjustable design 可调设计,可调整式
adjustable diffuser 几何可变扩压器,可调进气道
adjustable eyepiece 可调目镜
adjustable horizontal stabilizer 可调水平安定面
adjustable inductor 可变电感器
adjustable nozzle 可调喷管
adjustable parameter 可调参量,参数
adjustable propeller (地面上)可调桨距的螺桨
adjustable spanner 活动板手,可调扳手
adjustable speed electric drive 调速电气传动
adjustable tailplane 可调尾翼
adjustable wrench 活动板手,可调板手
adjusted range correction 射程修正
adjusted rms 调整均方根
adjusted rms error 调整均方根误差
adjusted value 平差值
adjuster (adjustor) 调节器,调整装置,校准器
adjusting ring 调整环
adjusting screw 调整螺丝,调整螺钉
adjustment 1.校准,调整 2.平差
adjustment chart 射击修正量表
adjustment correction 射击诸元修正
adjustment of astro-geodetic network 天文大地网平差
adjustment of correlated observation 相关观测平差
adjustment of observations 观测平差
adjustment of sight 瞄准具校准,表尺校准
adjustment on the target 对目标的射击修正
adjustment parameter 平差参数
adjustment residual 平差残差
adjustment temperature system on the ground 地面调温系统
adjustment zone 调整带
adjutant 1.副官,人事行政参谋(美国) 2.参谋长(英国)
ADL 1. Air Data Loader 大气数据输入器 2. Aeronautical Data-Link 航空数据链 3. Airborne Data Loader 机载数据装载器 4. Aircraft Data Link 飞机数据链 5. Application Data List 应用数据表 6. Automatic Datal Link 自动数据链 7. Avionics Development Laboratory 航空电子系统开发实验室
ADLP 1. Airborne Data Link Protocol 机载数据链协议书 2. Aircraft Data Link Processor 飞机数据链处理器
ADLS Aeronautical Data Link Services 航空数据链服务
ADM 1. Air Data Module 大气数据模块 2. Automated Data Management 自动数据管理
ADMA Automatic Drafting Machine 自动绘图机
ADMC Actuator Drive and Monitor Computer 致动器驱动与监测计算机
administration 1.管理 2.行政,局,署 3.后方勤务
administrative helicopter 公务直升机
administrative loading 行政装载
administrative time 修理准备时间
administrator 1.行政官 2.管理员
admiral 海军上将
Admiralty War Office (英国的)海军部
admissible command 容许的命令
admissible error 容许误差
admissible function 容许函数,容许功能
admissible solution 允许解,容许解法
admission 1.进气,进入 2.容许,承认 3.公差,容差
admittance 导纳
admittance chart 导纳圆图
admittance function 导纳函数
admix 掺合,混合
admixer 混合器
admixture 混合,混合物,掺合剂,附加剂,外加剂,杂质
admixture instability 混合不稳定性
ADMRL Automatic Data Material Requisition List 自动数据资料需求单
ADMS 1. Airline Data Management System 航线数据管理系统 2. Advanced Data Management System 先进数据管理系统 3. Automatic Digital Message Switching Center 自动数字信息交换中心
ADMSC Automatic Digital Message Switching Center 自动数字信息交换中心
ADNC Air Defence Notification Centre 防空通报中心
ADNET Administration Data Network 管理数据网络
Adnic 铜镍合金,海军镍

ADO 1. Advanced Development Objective 先行开发目标 2. Attribute Driven Option 特性推动的选项 3. Audio Decade Oscillator 音频十进制振荡器 4. Air Defense Officer 防空军官 5. Automotive Diesel Oil 内燃机柴油

ADOCS Automated Deep Operations Coordination System 自动深度运行协调系统

Adonis 【天】阿多尼斯(小行星 2101 号)

adopt 采纳,采用

adopted latitude 纬度采用值

adopted longitude 经度采用值

adoption 采纳,采用,正式通过

Adour "阿杜尔"(法国 TH.D.1215 自动多功能导弹跟踪雷达,英国罗·罗公司制造的双轴军用涡轮风扇发动机)

ADP 1. Acceptance Data Package 验收数据包,接收数据包 2. Acoustic Data Package 音响数据包 3. Acoustic Data Processor 声数据处理器 4. Advanced Development Plan 先进的发展计划 5. Advanced Ducted Prop 先进函道螺旋桨 6. Air Data Probe 大气数据传感器 7. Air Data Processor 大气数据处理器 8. Airborne Data Processor 机载数据处理器 9. Air-Driven Pump 风动泵,气动泵 10. Airport Development Program 机场开发项目 11. Automatic Data Processing 自动数据处理 12. Automatic Destruct Program (导弹在飞行中)自毁程序 13. Automatic Diagnostic Program 自动诊断程序

ADPA American Defense Prepared-ness Association 美国国防军备协会

ADPC Automatic Data Processing Center 自动数据处理中心

ADPCM Adaptive Differential Pulse Code Modulation 自适应差分脉冲编码调制

ADPE 1. Automatic Data Processing Equipment 自动数据处理设备 2. Auxiliary Data Processing Equipment 辅助数据处理设备

adpedance 导抗

ADPES Automatic Data Processing Equipment System 自动数据处理设备系统

ADPLL All Digital Phase-Locked Loop 全数字式锁相环

ADPP Assembly/Detail Process Plan 装配/详细工艺大纲

ADPS 1. Automatic Data Processing System 自动数据处理系统 2. Automated Data Processing Software 自动数据处理软件 3. Auxiliary Data Processing System 辅助数据处理系统

ADPT Adapter 适配器,转换器,转接器

adptive control 自适应控制

ADR 1. Advisory Route 咨询航路 2. Air Data Reference 大气数据基准 3. Aircraft Destination Record 飞机目的地记录 4. Accident Data Recorder 事故数据记录器

ADRAC Automatic Digital Recording And Control 自动数字化记录和控制

ADRAS Airplane Data Recovery and Analysis System 飞机数据恢复和分析系统

Adrasteia 木卫十五

ADRDE Air Defence Research & Development Establishment 防空研究和发展局(英国)

adrenocorticotropin 促肾上腺皮质激素

ADREP Accident and Incident Data Reporting 失事和事故数据报告

ADRN Advanced Drawing Release Notice 先行图样发放单

ADRS Airfield Damage Repair Squadron 机场抢修中队(英国空军的)

ADRT Analog Data Recorder Transcriber 模拟数据记录转录器

ADS 1. Air Data Sensor 大气数据传感器 2. Air Data System 大气数据系统 3. Attitude Display System 姿态显示系统 4. Automatic Data System 自动数据系统 5. Automatic Dependent Surveillance 自动相关监视 6. Autopilot Disengage Switch 自动驾驶仪断开开关 7. Aerodynamic Deceleration System 空气动力减速系统

ADSAF Automatic Data System for the Army in the Field 野战部队的自动数据系统

ADSARM Advanced Defense Suppression Antiradiation Missile 高级防御压制反辐射导弹

ADSCOM Advanced Shipboard Communications 高级舰上通信

ADSEL 1. Address Selective 选址 2. Automatic Dependent Surveillance 自动相关监视

ADSF Automatic Dependent Surveillance Function 自动相关监视功能

ADSID Air-Delivered Seismic Intrusion Detector 空中投送地震入侵探测仪

ADSL Asymmetric Digital Subscriber Line 非对称数字环路

ADSM Aircraft (Airplane) Drafting Standards Manual 飞机制图标准手册

adsorb 吸附

adsorbate 吸附质,吸附质物

adsorbent 吸附剂,吸附物质

adsorption 吸附

adsorptivity 吸附性

ADSP (ICAO) Automatic Dependent Surveillance

Panel(国际民航组织)自动相关监视专家组

ADSS 1. Aircraft Damage Sensing System 飞机损伤传感系统 2. All-Dielectric Self-Supporting optic fiber cable 全介质自承式光缆 3. Automatic Dependent Surveillance System 自动相关监视系统

ADSW Anti Diesel Submarine Warfare 反柴油动力潜艇作战

ADT 1. Airborne Digital Timer 机载数字计时器 2. Automated Data Transmission 自动数据传输 3. Approved Departure Time 批准的离场时间 4. Average Daily Traffic 平均日交通量

ADTAM Air-Delivered Target-Activated Munitions 空气传送的目标激活弹药

ADTC Armament Development Test Center 装备研制试验中心(美国空军)

ADTN Administrative Data Transmission Network 管理数据传输网

ADTS 1. Automatic Data Test System 自动数据测试系统 2. Automatic Digital Time System 自动数字计时系统 3. Avionics Depot Test Station 航空电子系统测试基地

ADTU Auxiliary Data Translator Unit 辅助数据转换装置

ADU 1. Air Drive Unit 空气驱动组件 2. Actuator Drive Unit 作动筒驱动组件 3. Alignment Display Unit 校准显示组件

ADV Advance 先行的/先进的

Advance 阿范斯合金

advance 1. 推进,促进,前移 2. 提前,提前量 3. (螺旋桨)进距 4. 推油门 5. 提前点火

advance angle 提前角,前置角

advance control 先行(超前)控制

advance escort 先遣护航兵力

advance of periastron 近星点进动

advance of the perihelion 近日点进动

advance ratio 进速比

advance/diameter ratio 1. (飞机螺旋桨的)前进比,相对进距 2. (直升机旋翼的)工作状态特性系数

advanced 先进的(指体现最新技术的)

advanced acoustic signal processor 先进声信号处理机

advanced aerodynamic-technology 先进气动技术,先进空气动力学技术

advanced airfield (base) 前进机场(基地)

advanced atmospheric sounding and imaging radiometer 先进大气探测和成像辐射计

advanced ballistic 高级弹道

advanced ballistic computer 高级弹道计算机

advanced ballistic reentry vehicle 高级弹道再入飞行器(美国)

advanced biomedical capsule 改进型生物医学容器

advanced cloud wind 高级云风

advanced cloud wind system 高级云风系统

advanced control 先行控制,预前控制

advanced conventional ordnance 先进的常规武器

advanced cryptographic system 高级密码系统

advanced data acquisition routine 高级数据采集程序

advanced defence communications system 高级国防通信系统

advanced development 试制,高级研制,预先研制

advanced digital avionics system 先进数字式航空电子系统

advanced earth resources observation system 高级地球资源观测系统

advanced earth satellite weapon system 先进的地球卫星武器系统

advanced echelon 先行梯队

advanced engine 先进的发动机

advanced extravehicular protective system 先进舱外活动保护系统

advanced flexible insulation material 高级柔性绝热材料

advanced flexible thermal protection material 高级柔性防热材料

advanced flow-control procedure 先进流量控制程序

advanced flying training 高级飞行训练

advanced geometry 高等几何学,高等几何体

advanced guidance 高级制导

advanced hybrid computing system 先进混合计算(机)系统

advanced impact location system 先进的弹着定位系统

advanced inertial instrumentation 先进惯性仪表

advanced inertial reference sphere 高级惯性参考(基准)球,浮球式平台

advanced instruction system 高级指令系统

advanced instrument approach systems 先进仪表进近系统

advanced integrated data system 高级综合数据系统

advanced launch system 先进运载系统(美研制中的低成本运载火箭)

advanced location strike system 先进定位攻击系统

advanced lunar transportation system 先进登月运输系统

advanced meteorological satellite 高级气象卫星(美国)

advanced microwave moisture sensor 高级微波水汽传感器

advanced microwave sounding unit 高级微波探测装置

advanced mobile phone service 先进移动电话业务

advanced moisture and temperature sounder 高级温湿探测器

advanced orbital test satellite 先进轨道试验卫星
advanced pilot/vehicle interface 先进人机界面
advanced power 高级电源
advanced product 高级产品
advanced product planning operation 先进的产品规划操作
advanced project 先进（研究）计划
advanced propulsion 先进推进装置
advanced proximity scoring technique 先进的近似命中记录技术
advanced radar traffic system 高级雷达空中交通管制系统
advanced range instrumentation ship 先进靶场测量船，先进遥测船
advanced reconfigurable computer system 先进可重配置计算机系统
advanced reconnaissance satellite 先进侦查卫星
advanced rescue system 先进营救系统
advanced seal 先进的密封
advanced stall 深度失速
advanced standoff weapons 先进的防空区外发射的（空对地）武器
advanced star 高等星级
advanced tactical LADAR seeker 先进战术激光雷达导引头
advanced teleprocessing system 先进遥处理系统
advanced terminal controller 先进终端控制器
advanced terminal management system 先进终端管理系统
advanced trainer 高级教练机
advanced very high resolution radiometer 先进超高分辨率辐射计（美国）
advanced vidicon camera system 高级光导摄像管摄像机系统
advanced weapon-delivery radar 先进的武器投射雷达
advanced-core military engine 军用发动机先进核心机
advancing blade 前行桨叶
advancing blade concept 前行桨叶设计思想
advantage 1.优势,优点,利益 2.(操纵机构的)传动比
advection 平流
advection calculation 平流计算
advection fog 平流雾
advective cooling 平流冷却
advective time scale 平流时间尺度
advent 出现,到来
adventive 外来的
adversarial 敌对的
adversarial target 敌对的目标
adversary 对手

adversary aircraft 对手飞机,仿敌飞机
adversary threat 对手的威胁
adverse 1.有害的,不利的 2.逆的,相反的,反向的
adverse effect 反作用,反效应
adverse environment 恶劣环境
adverse pressure 不利压力,反压力
adverse pressure gradient 逆压梯度
adverse weather 不利天气
adverse weather aerial delivery system 恶劣天气空投系统
adverse weather condition 不利天气条件
adverse wind 逆风
advisory 1.顾问的,咨询的 2.(数据)报表,(气象)报告,(情报)通报 3.(复数)建议
advisory control 咨询控制
advisory light 机上咨询灯
advisory route 咨询航线
advisory service 咨询服务处
ADVUL Air Defense Vulnerability 防空易损性
ADW Air Defence Warning 防空警报
ADWAR Advanced Missile Warhead 新型导弹弹头
ADWCP Automated Digital Weather Communication Program 自动气象数据通信计划
ADWS Automatic Digital Weather Switch 自动数字天气转报
ADX Automatic Data Exchange 自动数据交换
AE 1. Aero-Electronics 航空电子学 2. Aviation Electronicsman 航空电子技术人员 3. Airport Elevation 机场标高
AEA 1. Aircraft Electronics Association 飞机电子学会（美国）2. Aeronautical Electronics Association 航空电子协会 3. Association of European Astronauts 欧洲航天员协会 4. Atomic Energy Authority 原子能管理局（英国）
AEB 1. Aft Equipment Bay 飞机尾部设备舱 2. Airline Engineering Bulletin 航空公司工程通告 3. Avionics Equipment Bay 航空电子设备舱
AEC 1. Airport Engineer 机场工程师 2. Assembly Effectiveness Control 装配有效性控制 3. Assembly Effectivity Control 装配有效性控制 4. Automatic Electronic Corporation 自动电气公司（美国）5. Automatic Equipment Corporation 自动设备公司（美国）
AECB Atomic Energy Control Board 原子能管理局（加拿大）
AECC Aeromedical Evacuation Control Center 航空医救撤离控制中心
AECMA Association Europeenne des Constructeurs de Materiel Aerospatial 〈法文〉欧洲航空航天器材制造商

协会

AECS Advanced Environment Control System 先进环境控制系统

AECU Audio Electronic Control Unit 音频电子控制装置

AED 1. Alphanumeric Entry Device 字母数字输入设备 2. Automatic External Defibrillator 自动外部电震发生器 3. Automated Engineering Design 自动化工程设计

AEDC Arnold Engineering Development Center 阿诺德工程发展中心

AEDPS Automated Engineering Document Preparation System 自动工程文献处理系统

AEDS 1. Airport Engineering Data Sheet 机场工程数据表 2. Airbus Enterprise Directory Service 空客企业目录服务

AEEC Airlines Electronic Engineering Committee 航空公司电子工程委员会(美国),航空电子技术委员会

AEEL Aeronautical Electronic and Electrical Laboratory 航空电子和电气实验室(美国)

AEF Airborne Equipment Failure 机载设备故障

AEG Aircraft Estimation Group 飞机评估委员会

AEHP Atmospheric Electricity Hazards Protection 天电防护

AEI 1. Areial Exposure Index 航空曝光指数 2. Air Express International 国际航空快递

AEL 1. Air Europe Ltd. 欧洲航空公司 2. Aircraft Engine Laboratory 飞机发动机实验室 3. Airplane & Engine License 飞机和发动机(维修)执照

Aelita 艾里塔卫星

AEM Airplane Energy Management 飞机能源管理

Aeneas 【天】埃涅阿斯(小行星1172号)

AEO Advance Engineering Order 先行工程指令

AEOC All Engine Operating Ceiling 全发升限

aeolotropic material 各向异性材料

aeon 1. 十亿年,京年 2. 极长时间

AEP 1. Autopilot Engage Panel 自动驾驶仪连接板 2. Aerodrome Emergency Planing 机场应急计划

AEPL Authorized Equivalent Parts List 等效代用零件表

AEPS Airlines Employment Placement Service 航空公司雇员安置服务

AER 1. Approach End of Runway 跑道进近端 2. Aeronautical Engineering Report 航空工程报告 3. Aerospace Engineering Report 航空航天工程报告 4. Air Expansion Ratio 进气膨胀比

AERA Automated En-Route ATC System 自动化航路空中交通管制系统

aerate 曝气,充气,吹风

aerated-liquid jet 加气液体射流

aeration 1. 充气,吹风,通风 2. 起沫

AERCOM Aeronautical Communication 航空通信

AERE Aeronautical Engineer 航空工程师

aeremia 空气栓塞症

aerial 1. 空气的,空中的,航空的 2. 空气构成的 3. 空中运送的 4. 天线(英国用法,美国称 antenna)

aerial（photographic）mosaic 航空(相片)镶嵌图

aerial aircraft carrier 飞机母艇

aerial ammunition 航空弹药

aerial antitank fire 空中反坦克火力

aerial archaeology 航空考古学

aerial array 天线阵

aerial bomb parachute 航弹伞

aerial burst bomb 空爆炸弹

aerial camera 航空(照)相机,航空摄影机

aerial camera gun 航空照相枪

aerial cannon 航空机关炮

aerial crane 起重直升机,空中吊车

aerial delivery system 空中投送系统

aerial depth charge 空投深水炸弹

aerial detection 1. 空气检测 2. 空中检测

aerial direct fire 对空中目标直接瞄准射击

aerial engagement 空战

aerial exploration 高空探测

aerial exposure index 航空曝光指数

aerial field artillery 航空野战炮

aerial film 航空胶片

aerial film speed 航空胶片感光度

aerial flare 航空照明弹

aerial fog 气雾

aerial fueling 空中加油

aerial gun 航空机关炮,航空机枪

aerial gunnery range 空中射击靶场

aerial hijack 空中劫持

aerial incendiary bomb 航空燃烧弹

aerial infrared imagery 航空红外成像

aerial line map 航线图

aerial machine gunnery 1. 空中射击学 2. 空中射击,空中机枪扫射

aerial map 航空图,航摄图,航测图

aerial mapping photography 航测摄影

aerial mine 1. 航空水雷 2. 空投水雷 3. 空投大型薄壳炸弹

aerial mine-laying 航空布雷,空中布雷

aerial mining 空中布雷,飞机布雷

aerial photograph 航空相片

aerial photograph interpretation 航空相片判读,航空相片解译

aerial photographic gap 航摄漏洞

aerial photographic reconnaissance　航空照相侦察
aerial photographic recording system　空中摄影记录系统
aerial photographs composite　航摄(相片)合成片
aerial photographs overlapping　航片重叠
aerial photography　空中摄影,航空摄影
aerial reconnaissance　航空侦察
aerial refueling　空中加油
aerial refueling system　空中加油系统
aerial remote sensing technique　航空遥感技术
aerial retrieval　空中回收
aerial sickness　空晕病
aerial survey　空中测绘
aerial tanker　空中加油机
aerial target warhead　攻击空中目标的弹头
aerial torpedo　航空鱼雷,空投鱼雷
aerial truck　轻型货机
aerial vehicle　飞行器
aerial warfare weapon　1.航空武器 2.机载武器
aerial window　天线窗
aerial work　空中作业
aerial-launched homing torpedo　航空自寻的鱼雷
aeriator　飞行员,航员
aerie　1.机场 2.飞行学校
aeriform　气态的
AERIS　Aeronautical En-route Information Service 航空航路情报服务
aeriscope　超光电摄像管
aero　1.空气,飞机,航空,空中 2.空气的,飞机的 3.航空事业,航空的
AERO　1. Aero 航空的 2. Air Education and Research Organisation 航空教育与研究机构 3. Azimuth, Elevation and Range Overtake 按方位角、仰角和距离捕捉
Aero Bee　(Aerobee)空蜂火箭
aero camera　航空摄影机
aero engine　航空发动机
aero transport　1.空运 2.运输机
aero vane　1.风向风速仪 2.(炸弹引信的)旋翼
aero-acoustic design　气动声学设计
aeroacoustics　气动声学,航空声学
aeroalastic tailoring　气动弹性剪裁
aeroarthrosis　航空性关节痛
aero-assist　航空协助
aero-assisted orbit　航空协助轨道
aero-dynamics　航空动力学
aero-otitis（aerotitis）　航空中耳炎,气压性中耳炎
aeroasthenia　飞行虚弱症,高空神经官能症
aeroastrodynamics　航空航天动力学
aeroastromedicine　航空航天医学

aeroatelectasis　航空性肺不张
aeroballistic reentry vehicle　航空弹道再入(大气层)飞行器
aeroballistics　航空弹道学,空气动力弹道学
aerobat　1.特技飞行 2.特技飞行员
aerobatic　特技飞行的
aerobatic battery　特技飞行用蓄电池
aerobatic delivery　(低空)特技飞行投弹
aerobatic flight　特技飞行
aerobatic oil system　特技滑油系统
aerobatics　特技飞行,特技飞行术,航空表演
aerobation　特技飞行
aerobic　需氧的,需气的
aerobic capacity　需氧量,有氧代谢能
aerobic metabolism　需氧(新陈)代谢
aerobic power　摄氧能力
aerobic respiration　需氧呼吸
aerobics　摄氧训练
aerobiology　大气生物学,高空生物学
aerobioscope　大气微生物组成分析器
aerobiotic　需氧的,需气的
aerobism　需氧性
aeroboat　飞船,船身式水上飞机
aerobomb　航空炸弹
aerobrake　空气动力制动装置,高空减速板
aerobraking　空气动力制动,高空制动,高空减速
aerobraking mission　空气动力制动使命
aerobraking phase　减速阶段
aerocannon　航空机关炮
aerocapture　高空截获,高空俘获
aerocapture maneuver　大气俘获机动
aerocapture mission　大气俘获使命
aerocapture trajectory　大气俘获轨迹
aerocar　飞行汽车
aerocarburetor　航空汽化器
aerocartograph　航空测图仪,立体测图仪
aerocele　气肿
aerochir　手术医疗飞机
aerochronometer　航空精密测时计
aeroclimatic　高空气候的
aeroclimatology　高空气候学
aeroclub　航空俱乐部
aero-control　航空控制,航空防治
aerocraft　同 aircraft,航空器,飞机
aerocruise　大气层巡航
aerocurve　曲翼面
aerocutter　电爆炸切断器
aerocycle　飞行自行车
aerocyst　气囊

aerodamping	气动阻尼
aeroderivative gas	燃气
aeroderivative gas turbine	燃气轮机
aerodermectasia	皮下气肿
aerodone	滑翔器
aerodonetics	滑翔学,滑翔术
aerodontalgic	航空性牙痛
aerodontia	1.航空牙科学 2.航空牙科
aerodontics	航空牙科学
aerodontology	航空牙科学
aerodreadnaught	巨型飞机
aerodrome	机场
aerodrome capacity	机场容量
aerodrome capacity saturation	机场饱和
aerodrome control tower	机场管制塔台
aerodrome elevation	机场标高
aerodrome forecast	机场天气预报
aerodrome meteorological minima	机场最低气象条件
aerodrome operating facility	机场运行设施
aerodrome operating minimum	机场运行最低标准
aerodrome traffic	机场交通
aerodrome traffic pattern	机场起落航线
aerodrome traffic zone	机场空中交通区
aerodrome warning	机场警告
aerodromometer	气流速度表
aeroduct	1.冲压式喷气发动机 2.离子冲压喷气发动机
aeroduster	飞机喷粉器,航空喷撒器
aerodynamic	空气动力的
aerodynamic acceleration	气动加速
aerodynamic analysis	空气动力分析
aerodynamic and structural model	气动和结构模型
aerodynamic angle	气动角
aerodynamic axis	空气动力轴
aerodynamic balance	1.空气动力平衡,气动补偿 2.(风洞内测量空气动力的)气动天平
aerodynamic balance measuremants	空气动力天平测量
aerodynamic behavior	空气动力特性
aerodynamic blade	空气动力叶片
aerodynamic braking	气动力制动
aerodynamic center	1.气动力中心 2.气动力焦点
aerodynamic channel	空气动力通道
aerodynamic characteristic	空气动力特性
aerodynamic characteristics	1.空气动力特性 2.气动特性
aerodynamic characteristics of launch vehicle	运载火箭气动特性
aerodynamic characteristics of nose	弹头气动特性
aerodynamic characteristics of vehicle	飞行器气动特性
aerodynamic characteristics of winged missile	有翼导弹气动特性
aerodynamic chord	空气动力弦
aerodynamic cleanness	空气动力完善性,空气动力净形
aerodynamic cleanup	空气动力外形净化措施
aerodynamic coef	空气动力系数
aerodynamic coefficient	空气动力系数
aerodynamic compensation	气动补偿
aerodynamic configuration	气动外形,气动构型,空气动力布局
aerodynamic configuration layout	气动力布局
aerodynamic configuration of vehicle	飞行器的气动构型
aerodynamic control	空气动力控制
aerodynamic control application	空气动力控制应用
aerodynamic control surface	气动操纵面
aerodynamic coupling	气动耦合,气动力交感
aerodynamic damping	气动阻尼
aerodynamic database	气动数据库
aerodynamic datum	气动数据
aerodynamic decelerator	气动力减速器
aerodynamic defect	气动缺陷
aerodynamic derivative	气动导数
aerodynamic design	气动设计
aerodynamic disintegration	气动解体
aerodynamic disturbance	气动力扰动
aerodynamic drag	气动阻力
aerodynamic drag coefficient	气动阻力系数
aerodynamic effect	气动效应
aerodynamic efficiency	气动效率
aerodynamic element	气动元件
aerodynamic equation	气动方程
aerodynamic facility	气动设备
aerodynamic factor	气动系数
aerodynamic feel	空气动力载荷感觉,气动感力
aerodynamic flight test	气动力飞行试验
aerodynamic flow	空气流量,气动流
aerodynamic focus	1.气动力中心 2.气动力焦点
aerodynamic force	空气动力
aerodynamic fragmentation	气动破碎
aerodynamic gas lubricated bearing	气体动压轴承
aerodynamic heat flux during ascent	上升时的气动热通量,爬高时的气动热通量
aerodynamic heat transfer	气动传热
aerodynamic heating	气动加热
aerodynamic hinge moment feedback	气动铰链力矩反馈
aerodynamic influence	气动力影响,空气动力效应
aerodynamic instability	空气动力不稳定性
aerodynamic interaction	气动力干扰
aerodynamic interference	气动力干扰
aerodynamic lag	气动滞后

aerodynamic lead-pursuit attack course 空气动力提前跟踪攻击曲线
aerodynamic lift 气动升力
aerodynamic load 气动载荷
aerodynamic loading 气动加载
aerodynamic loss 气动损失
aerodynamic matrix 气动力矩阵
aerodynamic mean chord 平均空气动力弦
aerodynamic mesh 气动网格
aerodynamic missile 飞航式导弹,有翼导弹
aerodynamic model 气动模型
aerodynamic model identification 气动模型识别
aerodynamic model mismatch 气动模型不匹配
aerodynamic model parameter 气动模型参数
aerodynamic model structure 气动模型结构
aerodynamic modeling 气动模型,气动建模
aerodynamic module 气动模块
aerodynamic moment 气动力矩
aerodynamic nonlinearities 空气动力学的非线性
aerodynamic nozzle 气动可调喷管
aerodynamic operator 气动操作
aerodynamic optimization 气动优化
aerodynamic parameter 气动参数
aerodynamic performance 空气动力性能
aerodynamic phenomenon 空气动力现象
aerodynamic power 气动功率
aerodynamic pressure 空气动力压力
aerodynamic pressure center of reentry vehicle 再入飞行器压心
aerodynamic quality 空气动力性能,空气动力特性,空气动力升阻比
aerodynamic ramp 气动斜坡
aerodynamic reentry environment 再入(大气层)空气动力环境
aerodynamic reference 气动参考
aerodynamic response 气动响应
aerodynamic resultant 气动合力
aerodynamic roll 气动辊
aerodynamic roll angle 气动转角
aerodynamic shape 气动力外形
aerodynamic shape optimization 气动外形优化
aerodynamic side 气动侧
aerodynamic side load 气动侧负荷
aerodynamic simulation 气动模拟
aerodynamic slot 气动翼缝
aerodynamic stability 气动稳定性
aerodynamic stabilization 气动稳定
aerodynamic stalling 气动失速
aerodynamic state 气动状态
aerodynamic stealth technique 气动隐形技术
aerodynamic stiffness 气动刚度
aerodynamic stiffness matrix 气动刚度矩阵
aerodynamic surface 空气动力面
aerodynamic technology 空气动力技术
aerodynamic term 空气动力学项
aerodynamic test 气动力试验
aerodynamic theory 空气动力学理论
aerodynamic torque 气动力矩
aerodynamic trail 气动(冷凝)尾迹
aerodynamic trajectory 气动力弹道
aerodynamic transfer 气动传输
aerodynamic twist 气动扭转,动扭转
aerodynamic uncertainty 空气动力学的不确定性
aerodynamic velocity 对空气的相对速度
aerodynamic work 气动工作
aerodynamical panel parachute 1.气动幅伞 2.活动幅伞
aerodynamically-advanced 空气动力改进的
aerodynamically-balanced 气动平衡的,空气动补偿的
aerodynamically-clean 空气动力净形的,具有良好空气动力外形的
aerodynamically-configured 空气动力外形良好的
aerodynamically-controlled 空气动力操纵的
aerodynamically-improved 空气动力改进的
aerodynamically-operated 空气动力操纵的
aerodynamically-shaped 流线型的
aerodynamicist 空气动力学家
aerodynamics 气动力学,空气动力学
aerodynamics analysis 空气动力学分析
aerodynamics code 空气动力学代码
aerodynamics model 空气动力学模型
aerodynamics of composite bodies 组合物体空气动力学
aerodynamics of compressible fluid 可压缩流体空气动力学
aerodynamics simulation 气动仿真
aerodyne 重(于空气的)航空器
aeroelastic analysis 气动弹性分析
aeroelastic behavior 气动弹性行为
aeroelastic characteristic 气动弹性特性
aeroelastic computation 气动弹性计算
aeroelastic control 气动弹性控制
aeroelastic coupling 气动弹性耦合
aeroelastic deformation 气动弹性变形
aeroelastic effect 气动弹性效应
aeroelastic eigenvalue 气动弹性特征值
aeroelastic energy 气动弹性能量
aeroelastic instability 气动弹性不稳定性
aeroelastic instability type 气动弹性不稳定型
aeroelastic mode 气动弹性模式

aeroelastic mode number 气动弹性模数
aeroelastic model 气动弹性模型
aeroelastic operator 气动弹性算子
aeroelastic optimization 气动弹性优化
aeroelastic performance 气动弹性性能
aeroelastic problem 气动弹性问题
aeroelastic research 气动弹性研究
aeroelastic response 气动弹性响应
aeroelastic sensitivity 气动弹性灵敏度
aeroelastic stability 气动弹性稳定性
aeroelastic system 气动弹性系统
aeroelatic tests 气动弹性试验
aeroelastically tailored composite 经气动弹性剪裁的复合材料
aeroelastician 气动弹性力学家
aeroelasticity 气动弹性,气动弹性力学
aeroelastics 气动弹性力学
aeroembolism 高空气体栓塞症,气栓
aeroembolism hazard 空气栓塞危险
aeroemphysema 航空性肺气肿
aeroengine 飞机引擎,航空发动机
aero-engine 航空发动机
aeroenterectasia 肠气胀
aeroflight mode 大气层飞行模式
aerofoil 翼型
aerofoil boat 气翼艇
aerofoil section 翼切面,翼剖面
aeroform method 爆炸成形法
aerogastria 胃积气
aerogastrocolia 高空性胃肠胀气
aerogastrocoly 高空性胃肠胀气
aerogation 空中领航
aerogator 空中领航员
aerogel 气凝胶
aerogenesis 产气
aerogenic 产气的
aerogenous 产气的
aerogram 1.高空图解 2.无线电报
aerograph 高空气象计
aerography 1.气象学(海军用语) 2.大气状况图表
aerogravity-assist maneuver 航空重力协助演习
aero-grid 航空网
aerogun 航(空机关)炮,航空机枪
aerogun-sight 航空射击瞄准具
aeroheating 空气动力加热
aerohydroplane 水上飞机
aeroil leg (起落架的)油气减震支柱
aero-isolinic wing 等迎角机翼
aerojet 喷气式飞机,空气喷射

aeroleveling 空中水准测量
aerolite 石陨星,石陨石
aerolithology 陨星学,陨石学
aerolitics 陨星学,陨石学
aerological ascent 高空探测,自由大气探测
aerological balloon 大气气球
aerological diagram 高空图
aerological instrument 高空仪器
aerological sounding 高空探测
aerological theodolite 测风经纬仪
aerology 航空气象学,大气学
aerolongation 等压面领航
aeromaneuvering 气动机动飞行
aeromania 航空狂热
aeromap 航空地图
aeromarine 1.水上飞机 2.水上飞行的
aeromechanic 航空机械员
aeromechanic(al) 航空力学的
aeromechanical stability 航空力学稳定性
aeromechanics 航空力学
AEROMED Aero-medical 航空医学的
aeromedical 航空医学的
aeromedical care 航空医学护理
aeromedical evaluation 航空医学评定
aeromedical flight 航空医学飞行
aeromedical indoctrination 航空医学训练
aeromedical monitoring 航空医学监测
aeromedicine 航空医学
aerometal 航空金属
aerometeograph (aerometeorograph) 高空气象计
aerometer 1.气量计 2.气体比重(密度)计
aeronaut 气球驾驶员,轻航空器驾驶员
aeronautic sextant 航空六分仪
aeronautical acoustics 航空声学
aeronautical application 航空应用
aeronautical beacon 航行灯标
aeronautical chart 航图,航空地图
aeronautical combustor 航空燃烧室
aeronautical earth station 飞机上的卫星通信站或导航卫星站
aeronautical engineering 航空工程
aeronautical engineering psychology 航空工程心理学
aeronautical fixed service 航空固定通信服务
aeronautical fixed telecommunications network 航空专用电传通信网
aeronautical information overprint 航空信息复印
aeronautical information service 航行情报服务
aeronautical laboratory 航空实验室
aeronautical light, beacon 导航灯标,灯光信标台

aeronautical maritime engineering satellite 航空海事工程卫星
aeronautical material 航空材料
aeronautical meteorology 航空气象学
aeronautical mile 空英里
aeronautical mobile-satellite service 卫星航空移动通信业务
aeronautical radio aids 航空无线电导航设备
aeronautical radionavigation satellite service 卫星航空无线电导航业务
aeronautical satellite 航空(导航)卫星
aeronautical satellite communications center 航空卫星通信中心
aeronautical service 航空服务
aeronautical topographic chart/map 航空地形图
aeronauticalmaterial 航空材料
aeronautics 航空学
Aeronavale 法国海军航空兵
aeronavigation 空中领航,领航学
aeronavigator 空中领航员
aeronef 重航空器,飞机
aeroneurosis 飞行员神经官能症
aeronomosphere 高层大气
aeronomy 1. 天体大气学 2.高层大气物理(学)
aeronomy satellite 高层大气物理学卫星
aero-otitis externa 航空外耳炎
aero-otitis media 航空中耳炎
aeropathy 航空病
aeropause 大气层航空边界,气圈顶,大气上界
aerophagia 吞气,吞气症
aerophare (导航)无线电信标
aerophoto grammetry 航空照相测量学
aerophotograph 航空照片
aerophotographical mosaic 航空相片镶嵌图
aerophotography 航空照相学,航空摄影
aerophototopography 航空地形照相测量学
aeroplane 飞机
aeroplane flare 航空照明弹
aeroplanist 飞行家
aeroplethysmograph 呼吸气量计,呼吸气量描记器
aeropolitics 航空政策
aeropolygonometry 空中导线测量
aeroport 航空站,航空港,机场
aeroprojector 航测投影器
aeroprojector multiplex 多倍投影测图仪
aeropropulsion 航空推进
aeropropulsion combustion 航空推进燃烧
aeropropulsive characteristic 大气推进特性
aeropulse 脉冲式喷气发动机

aeroquay 机站台,飞机码头
aeroramp 航空斜坡
aeroramp injector 航空斜坡喷射器
aeroresonator 脉动式冲压喷气发动机
aero-route surveillance radar 航路监视雷达
Aeros 1. Aero Satellite 大气卫星(德国) 2. Aeros Data Corporation 艾罗斯数据公司(美国),艾罗斯卫星
aeroseal 空气密封,气密
aeroservoelastic model 气动伺服弹性模型
aeroservoelasticity 气动伺服弹性
aeroshed (飞)机库
aeroshell 气动热防护层,气动外壳
aeroshow 航空展览(会)
aerosiderite 铁陨星,铁陨石
aerosiderolite 铁石陨星,铁石陨石
aerosimplex 简单投影测图仪
aerosinusitis 航空性鼻窦炎,气性鼻窦炎
aerosled 滑行橇
aerosol 1. 浮质 2. 烟雾剂 3. 气溶胶
aerosol bomb 浮质弹(一种喷洒杀虫剂、含菌液体等物质的容器)
aerosol electricity 气溶胶电,大气微粒电(学)
aerosol emission 气溶胶排放
aerosol flow 气雾剂流量
aerosol jet 气溶胶喷射
aerosol precursor 气溶胶前体
aerosol size 气溶胶大小
aerosol zone 气溶胶区
aerosoloscope 空气(中)微粒测定器
aerosolsonde 气溶胶探空仪
aerosonator 脉冲式空气发动机
aerosp 同 aerospace 1.航空航天 2.宇宙空间 3.大气,空气圈
aerospace 1. 大气层与太空,航空航天空间 2.航空航天的,航空、太空的
aerospace accident 航空航天事故
aerospace application 航空航天应用
aerospace biochemistry 航空航天生物化学
aerospace business environment simulator 航空航天工作环境模拟器
aerospace cardiology 航空航天心脏病学
aerospace communication 航空航天通信
aerospace community 航空航天社区
aerospace computer 航空航天计算机
aerospace craft 航空航天飞行器
Aerospace Defense Command 航空航天防御司令部(美国)
aerospace dynamics 航空航天动力学
aerospace engineer 航空航天工程师

aerospace engineering 航空航天工程	**aerostructure** 飞机结构学
aerospace environment 航空航天环境	**aerosurface** 飞机表面
aerospace ergonomics 航空航天人机工程学	**aerosurveying** 航空测量学
aerospace escape vehicle 航空航天逃逸飞行器	**aerotechnics** 航空技术
aerospace escape system 航空航天逃逸系统	**aerothermal environment** 气动热环境
aerospace forces 航空航天力量	**aerothermal heating** 气动热供暖
Aerospace Guidance and Meteorology Center 航空航天制导和气象中心(美国空军)	**aerothermal load** 气动热载荷
	aerothermal performance 气动热性能
Aerospace Industries Association 航空航天工业协会(美国)	**aerothermo chemistry** 空气热化学
	aerothermodynamic border 气动热力边界
aerospace industry 航空航天工业	**aerothermodynamic duct** 冲压式喷气发动之旧称(现用 ramjet)
aerospace instrument 航空航天仪表	
aerospace intelligence data system 航空航天情报数据系统	**aerothermodynamic heating** 气动热力学加热
	aerothermodynamics 气动热力学
aerospace maritime operations 空天海上一体战(21世纪的海上作战样式)	**aerothermoelastic effect** 气动热效应
	aerothermoelasticity 气动热弹性力学
aerospace material technology 航空航天材料工艺	**aerothermo-elasticity** 气动热弹性
aerospace medicine 航空航天医学	**aerothermoplasticity** 气热可塑性
aerospace medicine laboratory 航空航天医学实验室	**aerotire** 航空轮胎
aerospace pathology 航空航天病理学	**aero-tow** (滑翔机的)空中牵引
aero-space photogrammetry 航空航天摄影	**aerotriangulation** 空中三角测量
aerospace physiology 航空航天生理学	**aerotriangulation by independent model** 独立模型法空中三角测量
aerospace plane 航空航天飞机,空天飞机	
aerospace program-oriented language 航空航天程序专用语言	**aerotropic** 向气的,嗜气的
	aerotropism 趋气性,向气性
aerospace psychology 航空航天心理学	**aeroturbine** 航空祸轮
Aerospace Rescue and Recovery Center 航空航天救援和回收中心	**aerovehicle wing** 航空飞行器翼
	aeroweapon 空中武器
aerospace retrieval 航空航天回收	**Aeroweb** 艾罗腹板
aerospace structure 航空航天结构	**Aerozine** 混肼
aerospace system 航空航天系统	**aerozine-50** 混肼-50
aerospace system engineering 航空航天系统工程	**aery** 1.航空事业 2.空气的,航空的
Aerospace Traffic Control Center 航空航天交通管制中心	**aerylonitrile-butadiene rubber** 丁腈橡胶
	AES 1. Aerospace Electronics System 航空航天电子系统 2. Aerodrome Emergency Service 机场应急服务 3. Aerospace Electrical Society 航空航天电气学会(美国) 4. Aircraft Earth Station 飞机地面站 5. American Electronic Society 美国电子学会
aerospace vehicle 航空航天(飞行)器	
aerospace wire 航空航天导线	
aerospacecraft 航空航天(飞行)器	
aerospatial 航空航天的,航空、太空的	
aerosphere 气圈,气界,地球周围的大气层	
aerospike (气动)减阻杆,减阻器	**Aesop** "伊索"
aerospike engine 气动发动机	**AESRS** Army Equipment Status Reporting System 部队装备状态报告系统
aerospike nozzle 气尖喷嘴	
aerospike nozzle design 气尖喷嘴设计	**AESS** 1. Aircraft Environment Surveillance System 飞机环境监控系统 2. Aerospace and Electronics Systems Society 航空航天电子系统学会
aerospike rocket 气动火箭	
aerostat 轻航空器,浮空器	
aerostat meteorograph 气球气象仪	**Aestar** Ae星,A型发射线星 aestival(estival) time 夏令时
aerostatics 空气静力学	
aerostation 气球驾驶术,浮空学	**AESU** Aircraft Environment Surveillance Unit 飞机环境监控组件
aero-stretcher 航空担架	
	AETE 1. Aerospace Engineering Test Establishment 航

空航天工程试验研究所 2. Aerospace Engineering Test Establishment (Canada) 航空宇宙工程测试机构（加拿大）

AETMS Airborne Electronic Terrain Map System 机载电子地形地图系统

AETS Automatic Engine Trim System 发动机自动调整系统

AEU Auxiliary Equipment Unit 辅助设备装置

AEV Avionics Exhaust Valve 航空电子排气阀

AEVC Avionics Equipment Ventilation Computer 电子设备通风计算机

AEW 1. Airborne Early Warning 空中预警系统，机载预警系统 2. Air Electronic Warfare 航空电子战

AEWA Airborne Early Warning Aircraft 空中预警飞机

AEWIS Army Electronic Warfare Information System (USA)部队电子战争信息系统(USA)

AEWSPS Aircraft Electronic Warfare Self-Protection System 飞行器电子战争自我保护系统

AEWTF Aircrew Electronic Warfare Training Facility 空勤人员电子战争培训设备

AF 1. Air Force 空军 2. Air France 法国航空公司 3. Airway Facilities 航路设施 4. All Freighter 全货机 5. Artificail Feel 模拟感觉 6. Audio Frequency 声频 7. Automatic Following 自动跟踪

AFA 1. Aircraft Finance Association 飞机信贷协会（美国）2. Audio Frequency Amplier 音频放大器

AFAA Air Force Audit Agency 空军审计局（美国）

AFAC Airborne Forward Air Controller 机载前方空中管制员

AFALC Air Force Acquisition Logistic Center 空军采购后勤中心

AFAMRL Air Force Aero Medical Research Lab 空军航空医学研究实验室

AFAP Australian Federation of Air Pilots 澳大利亚飞行员联合会

AFATL Air Force Armament Laboratory 空军军械实验室（美国）

AFB 1. Air Force Base 空军基地 2. Air Freight Bill 航空货运单 3. Antifriction Bearing 耐磨轴承 4. Airframe Bulletin 飞机体通告

AFBMD 1. Air Force Ballistic Missile Division 空军弹道导弹局（美国）2. Antifriction Bearing Manufacturing Association 减摩轴承制造商协会

AFC 1. Active Flight Control 主动飞行控制 2. Area Forecast Center 区域预报中心 3. ATC Frequency Change Service 空中交通管制频率交换服务 4. Automatic Flight Control 自动飞行控制 5. Automatic Following Control 自动跟踪控制 6. Automatic Frequency Compensation 自动频率补偿 7. Automatic Frequency Control 自动频率控制

AFCAS 1. Advanced Flight Control Actuation System 先进飞行操纵作动系统 2. Automatic Flight Control Augmentation System 自动飞行控制增益系统

AFCE Automatic Flight Control Equipment 自动飞行控制设备，自动飞行控制装置，自动驾驶仪

AFCEA Armed Forces Communications and Electronics Association 武装部队通信和电子协会（美国）

AFCMD Air Force Contrat Management Division 空军合同管理处（美国）

AFCOMS Air Force Commissary Services 空军军需供应勤务

AFCP Advanced Flow-Control Procedure 先进流量控制程序

AFCRL Air Force Cambridge Research Laboratories 空军剑桥研究实验室

AFCS 1. Adaptive Flight Control System 自适应飞行控制系统 2. Automatic Fire Control System 自动灭火系统 3. Automatic Flight Control System 自动飞行控制系统 4. Axial Flow Compressor 轴流压缩机

AFD 1. Adaptive Flight Display 自适应飞行显示器 2. Advanced Flight Deck 先进驾驶舱 3. Autopilot Flight Director 自动驾驶仪飞行指引仪

AFDC Autopilot Flight Director Computer 自动驾驶仪飞行指引计算机

AFDD Audit Findings and Differences Database 审计结果和差异数据库

AFDS 1. Advanced Flight Deck Simulator 先进驾驶舱模拟器 2. Automatic Flight Data System 自动飞行数据系统 3. Autopilot Flight Director System 自动飞行指引系统

AFE 1. Above Field Elevation 高于机场标高 2. Authority For Expenditure 开支授权，费用批准

AFEPS ACARS Front End Processing System 飞机通信寻址和报告系统前端处理系统

AFESA Air Force Engineering and Services Agency 空军工程与勤务局（美国）

AFETR Air Force Eastern Test Range 空军东部试验场（美国）

AFF Affect 影响

AFFDL Air Force Flight Dynamics Laboratory 空军飞行动力学实验室（美国）

affect 影响，作用

AFFF Aqueous Film-Forming Foam 液体薄膜泡沫剂，水膜泡沫

affine connection 仿射连接

affine equation 仿射方程

affine function 仿射函数

affine plate constant 仿射底片常数

affine plotting 变换光束测图
affine rectification 仿射纠正
affine transformation 仿射变换
affinity 密切关系，吸引力
affinity-group charter 合伙包机
affirmative 是(无线电通话中，用来代替 yes)
affix 粘，贴，附加上，签署
afflight 1. 近月飞行 2. 月球背面的接近轨线
affordability 负担能力
AFFTC Air Force Flight Test Center 空军飞行试验中心(美国)
AFG 1. Analog Function Generator 模拟函数发生器 2. Antenna Field Gain 天线场强增益
AFGE American Federation of Government Employees 美国政府雇员联合会
AFGL Air Force Geophysics Laboratory 空军地球物理实验室(美国)
AFGS Automatic Flight Guidance System 自动飞行引导系统
AFGU Aerial Free Gunnery Unit 航空自由射击单元
AFGWC Air Force Global Weather Center 空军全球气候中心
AFI 1. African Region 非洲地区(国际民航组织) 2. Assistant Flying Instructor 飞行助教 3. Authority and Format Identifier 权限和格式标识符 4. Authority Format Identifier 权限格式标识器
AFICCS Air Force Interim Command and Control System 空军临时指挥与控制系统
AFIPS American Federation of Information Processing Societies 美国信息处理学会联合会
AFIS 1. Airborne Flight Information System 机载飞机信息系统 2. Automatic Flight Information Service 自动飞行信息服务
AFISC Air Force Inspection and Safty Center 空军检查与安全中心(美国)
AFL Action Flow 任务流
AFLC Air Force Logistics Command 空军后勤部(美国)
AFLCON Air Force Logistic Command Operations Network 空军后勤部指挥通信网(美国)
AFM 1. Aircraft (Airplane) Flight Manual 飞机飞行手册 2. Air Force Medal 空军奖章(英国) 3. Air Force Museum 空军博物馆 4. Airport Firemen 机场消防员
AFMA Anti-Fuel-Misting Additive 燃油防雾化添加剂
AFMC Advanced Flight Management Computer 先进飞行管理计算机
AFMDC Air Force Missile Development Center 空军导弹发展中心(美国)
AFMEA Air Force Management Engineering Agency 空军管理工程局(美国)

AFML Air Force Materials Lab 空军材料研究所(美国)
AFMPC Air Force Military Personnel Center 空军人事中心(美国)
AFMRL Air Force Medical Research Laboratory 空军医疗研究实验室
AFMTC Air Force Missile Test Center 空军导弹试验中心(美国)
AFN ATS Facilities Notification 空中交通服务设施通告
AFO Airport Fire Officer 机场消防官员
AFOAR Air Force Office of Aerospace Research 空军航空航天研究处(美国)
afocal 非聚的，远焦的，焦外的
afocal system 焦外系统
AFOLTS Automatic Fire/Overheat Logic Test System 自动火警/过热逻辑探测系统
AFOSI Air Force Office of Special Investigations 空军特别调查办公室(美国)
AFOSR Air Force Office of Scientific Research 空军科研处(美国)
AFP 1. Alternate Flight Plan 备份飞行计划 2. ATC Flight Plan 空中交通管制飞行计划
AFPM Association Frangaise des Pilotes de Montagne 法国山地飞行员协会
AFPR Air Force Plant Representative 空军驻厂代表
AFPRO Air Force Plant Representatives Office 空军驻厂代表办公室(美国)
AFQ Association Franpise des Qualiticiens 法国质量检验员联合会
AFQCR Air Force Quality Control Representative 空军(产品)质量控制代表
AFR 1. Abbreviated Flight Plan Routing 简式飞行计划航线 2. Acceptable Failure Rate 容许故障率 3. Actual Fuel Remaining 实际燃油剩余 4. Airframe 机体，骨架
AFRAA Association of African Airlines 非洲航空公司协会(国际组织)
A-frame A 形架
A-frame hook A 形构架拦阻钩
AFRCC Air Force Rescue Coordination Center 空军救援协调中心(美国)
AFRCSAT African Regional Communications Satellite 非洲卫星(非洲区域通信卫星)
AFRCU Air/Fuel Ratio Control Unit 空气-燃油比控制装置
AFRES (Afres) Air Force Reserve 空军后备队(美国)
Afrispace 非洲空间公司
Afristar 非洲星(国际无线电广播卫星)
AFRM airframe 1. (飞机)机体 2. (航空器)结构

AFROTC　Air Force Reserve Officers 空军后备军官

AFRPL　Air Force Rocket Propulsion Laboratory 空军火箭推进实验室(美国)

AFS　1. Advanced Flying School 高级飞行学校,高级航校　2. Aeronautical Fixed Service 航空固定服务　3. Airforce Standard 空军标准　4. Autoflight System 自动飞行系统

AFSC　1. Air Force Systems Command 空军系统司令部　2. Auxiliary Fuel System Controller 辅助燃油系统控制器

AFSCF　Air Force Systems Command Form 空军系统司令部构成

AFSK　Audio Frequency Shift Keying 音频键控偏移

AFSN　Air Forcr Stock Number 空军(库存)货号

AFSS　Automated Flight Service Station 自动化飞行服务站

AFSWC　Air Force Special Weapons Center 空军特种武器研究中心(美国)

AFT　1. After 逆航向,在后部,在……以后,后面　2. Air Freight Terminal 航空货运集散站

aft　在后部的,机尾

AFTAC　Air Force Technical Applications Center 空军技术应用中心(美国)

aft body　尾体

aft centerbody　机尾中心体

aft chamber　尾室

aft chord　尾弦

aft closure　后(堵)盖,底遮板

aft cover　尾盖

aft end　尾端

aft face　尾部表面

aft fan　尾部通风机

aft flap　后襟翼

aft flight deck　(航母上)后飞行甲板

aft isolator　尾部绝缘体

aft limit of CG　(飞机)重心后限

aft loading　尾部装载,尾部载量

aft movement　尾部运动

aft portion　尾段

aft ramp　尾部斜坡

aft ramp angle　尾部斜坡角度

aft segment　后段

aft side　尾侧

aft skirt of case　壳体后裙

aft suction surface　尾部吸力面

aft tank　尾箱

aft wall angle　尾壁角度

aft wing　后翼

AFTA　Avionics Fault Tree Analyzer 航空电子设备故障树状分析器

aftbody　尾体

AFTCLR　After Colder 后冷却器

AFTE　Authority For Tooling Expenditure 工装费授权

after flight inspection　飞行后检查

after image　残留影像,余像

after nystagmus　后遗眼球震颤

after propeller　1.(共轴螺旋桨的)后桨　2.(直升机的)尾桨　3.(在发动机后方的)推进式螺旋桨

after pulse　余脉冲

after wash　后洗流

afterbody　1.后机身,弹体尾部　2.(人造卫星的)伴体　3.再入体尾部防护舱

afterbody angle　后船体角

afterbody length　后机身长度

afterbody pressure　后机身压力

afterbody region　后机身区

afterburner　1.加力燃烧室　2.复燃室　3.补燃室

afterburner limit of center of gravity　重心后限

afterburner liner　隔热防振屏

afterburner system　加力燃烧室系统

afterburning　复燃加力,加力

afterburning cessation　补燃停止

afterburning shutdown　补燃关机

afterburning turbojet　复燃加力涡轮喷气(发动机)

afterchine　(船身式水上飞机后船身的)后舷

aftercombustion　复燃,补燃,加力燃烧

aftercooler　后冷却器

aftercooling　后冷却

aftereffect error　后效误差

afterflaming　1.残余燃烧　2.(反应堆)暂时冷却

after-flight inspection　飞行后检查

afterframe　1.后部框架　2.机身后部

afterglow　余辉(1.低气压放电管中气体在放电停止仍继续发出的辉光　2.日落后西方高空中因阳光漫射出现的微弱辉光　3.电磁输入功率中止后等离子体的瞬时衰变)

afterheat　1.衰变热,余热　2.秋老虎(指天气)

aftermarket　售后服务市场

aftersection　后段,尾部

aftershock　余震

after-treatment　后处理

aft-hinged canopy　后折式座舱盖

AFTI　Advanced Fighter Technology Integration 先进战斗机技术综合

aft-loaded wing　超临界机翼

aft-mixing chamber　超混合室

AFTN　Aeronautical Fixed Telecommunication 航空固定电信网

AFTP Aircraft Flight Test Plan 飞机试飞大纲

aft-positioned 尾部配置的

AFTR Aircraft Flight Test Requirements 飞机试飞要求

AFTS 1. Adaptive Flight Training System 自适应飞行训练系统 2. Advanced Flying Training School 高级飞行训练学校 3. Air Fuel Test Switch 空气燃料实验开关 4. Airborne Flight Test System 机载飞行试验系统

aft-swept 后冲刷

aft-wall 后壁

AFU 1. Actual Fuel Used 实际耗油 2. Artificial Feel Unit 模拟感觉组件

AFVA Air Force Visual Aid 空军直观教具

AFWAL Air Force Wright Aeronautical Laboratories 空军莱特航空实验室（美国）

AFWL Air Force Weapons Laboratory 空军武器实验室（美国）

AFWR Atlantic Fleet Weapons Range 大西洋舰队武器试验靶场

AG 1. Accessory Gearbox 附件机匣 2. Agency General 总代理 3. Air to Ground 空对地 4. Angle Gearbox 伞齿轮箱 5. Application Gateway 应用网关 6. Aviation Gasoline 航空汽油

AGACS Automatic Ground-to-Air Communication System 自动地空通信系统

Agamemnon 【天】阿伽门农（小行星911号）

AGARD （Agard）Advisory Group for Aerospace Research and Development 航空航天研究与发展顾问团（北约组织）

Agave "龙舌兰"（法国 1/J 波段直升机雷达）

AGB Accessory Gear Box 附件齿轮箱，附件传动机匣

AGC 1. Automatic Gain Control 自动增益控制 2. Automatic Gauge Controller 自动测量调整装置 3. Aerodrome Ground Movement Chart 机场地面活动图 4. Air Ground Chart 航空地形图 5. Aerographer's Mate Chief 航空气象师

AGCA Automatic Ground-Controlled Approach 地面控制自动进近

AGCS 1. Advanced Guidance and Control Systems 先进制导和控制系统 2. Air Ground Communication System 空地通信系统

AGCU Auxiliary Generator Control Unit 辅助发电机控制器

AGD 1. Air Generator Drive 风力发电机驱动 2. Axial Gear Drive 轴向齿轮驱动 3. Axial-Gear Differential 轴向差动齿轮

AGE 1. Aerospace Ground Equipment 航空地面设备 2. Aircraft (Airplane) Ground Equipment 飞机地面设备

age 1. 老化，时效 2. 使用期限，寿命

age at failure 故障工龄

age exploration 工龄探索

age factor 年龄因素

age of data 数据年龄

age of universe 宇宙年龄

A-gear arrester gear 着陆拦阻装置

age-hardening 时效硬化

age-hardening alloy 时效硬化合金

ageing （美国用 aging）1. 时效 2. 老化，陈化

ageing table 老化台

ageing test 老化试验

age-inhibiting addition 防老化添加剂

agement 老化

Agena 阿金纳

agency 1. 媒介，介质，因素 2. 作用，手段 3. 机构，代办处，局

Agency environmental test satellite 高级研究计划局环境测试卫星（美国）

agent 1. 剂，试剂 2. 介质，媒介 3. 代理人 4. 原因，因素

agent-based approach 基于代理的方法

agent-based simulation 基于代理的仿真

agent-based software 基于代理的软件

ageostrophic acceleration 非地转加速度

AGFCS Advanced Gun Fire Control System 先进机炮火控系统

agglomerate 集块岩

agglomerate size 结块的大小

agglomeration 凝聚，黏合，烧结

agglomeration multigrid 多栅的结块

agglomeration 结块，集结

agglomeration procedure 集结程序

agglutinate 烧结，黏合，附着

aggravated test 加剧试验

aggregate 1. 星集 2. 集（合）3. 凝聚

aggregate fuel 总燃料

aggregate interference 集总干扰

aggregate model 集总模型

aggregate size 星集大小

aggregated (bulk) arrivals 成批交货，成批交付的货物

aggregation 1. 集结，凝聚 2. 团块

aggregation function 聚集函数

aggregation interval 集结间隔

aggregation matrix 集结矩阵

aggressive combustion 侵蚀性燃烧

aggressive maneuver 攻击演习

aggressive mutation 积极变异

AGIL Airborne General Illumination Light 机载通用照明灯

agile 敏捷的,灵便的
Agile Eye 灵眼
agile vehicle 敏捷载机
agility (飞机的)敏捷性
agility metric 敏捷性度量
agility vector 敏捷性矢量
Agiltrac 艾吉尔特拉克
aging 1. 时效(处理) 2. 老化
aging alloy 时效合金
aging crack 时效裂纹
aging star 老化星
aging test 老化试验
AGIS Air Ground Intermediate System 空地中继系统
AGL 1. Above Ground Level 离地高度,地平面以上 2. Approach Guidance Lights 进近引导灯光
AGLT Aircraft Gun Laying Turret 飞机炮火自动瞄准器
AGM 1. Air to Ground Missile 空对地导弹 2. Air/Ground Message 空/地电报 3. Aircraft Ground Mishap 飞机地面事故
AGMC Aerospace Guidance and Meteorology Center 航空航天制导和气象中心(美国空军)
Agni 火神
AGP 1. Aircraft Grounded For Lack of Parts 因缺乏零件停飞的飞机 2. Alternate Gauging Processor 备用测量处理器,备用测量仪
AGR Air/Ground Router 空/地路由器
agravic 无重力的
agravic illusion 失重错觉
AGREE Advisory Group on Reliability of Electronic Equipment 电子设备可靠性顾问团
agreement 1. 协定,协议 2. 一致,同意
agricopter 农业直升机
agricultural 1. 农业的 2. 粗劣的
agricultural aircraft 农业飞机
agricultural airplane 农业飞机
agricultural aviation 农用航空
Agrion "阿格里昂"(法国机载雷达)
AGRISTARS Agriculture and Resources Inventory Surveys Through Aerospace Remote Sensing 航空和航天遥感技术测量农业与资源储藏量
AGRT Automatic Ground Reciever Terminal 自动地面接收机终端
AGS 1. Abort Guidance System 应急制导系统 2. Air/Ground System 空/地系统 3. Airport Ground Service 机场地面服务 4. Automatic Gain Stabilization 自动增益稳定
AGSS ACARS Ground System Standard 飞机通信寻址和报告系统地面系统标准
AGSTR Aircraft Ground Service Tie Relay 飞机地面维护连接继电器
AGT Agent 剂,代理人
AGTELIS Automated Ground Transportable Emitter Location and Identification System 自动地面可运输发射器定位和标识系统
AGTP Aircraft Ground Test Plan 飞机地面测试大纲
AGTS Air/Ground Test Station 空地测试站
AGVS Automated Guided Vehicle System 自动导航飞行器系统
AGW 1. Actual Gross Weight 实际总重 2. Allowable Gross Weight 容许起飞总重
AGWAC Australian Guided Weapons and Analogue Computer 澳大利亚制导武器和模拟计算机
AGZ Actual Ground Zero 实际爆心地面投影点
Ag-zn (Ag/Zn)银锌蓄电池
AH 1. Alert Height 警戒高度 2. Altitude Hold (ALT HOLD) 高度保持 3. Ampere/Hour 安培/小时 4. Antihunt 阻尼,定稳器 5. Artificial Horizon 航空地平仪,人工地平,地平线
AHC Attitude Heading Computer 姿态航向计算机
AHCS Automatic Hovering Control System 自动悬停控制系统
AHIP Army Helicopter Improvement Program 陆军直升机改进计划(美国)
AHM 1. Altitude Hold Mode 高度保持模式 2. Ampere-Hour Meter 安培小时计,电量计
AHQ Air Headquarters 航空兵指挥所(英国)
AHRS 1. Attitude and Heading Reference System 姿态和航向基准系统 2. Automatic Heading Reference System 自动航向基准系统
AHRU Attitude and Heading Reference Unit 姿态航向基准装置
AHS 1. Attitude Heading System 姿态航向系统 2. American Helicopter Society 美国直升机协会 3. Automated Handwritten System 自动手写系统
AHSA Aviation Historical Society of Australia 澳大利亚航空史学会
AHTR Auto Horizontal-tail Retrimming After Landing 着陆后平尾自动再配平
AI 1. Accident Indemnity 事故保险 2. Accident Injuries 失事损伤,偶然损伤 3. Air Installation 机上装置,航空设备 4. Airspeed Index 空速指标 5. Airspeed Indicator 空速指示器 6. Alternative Interogator 交替问答器 7. Altitude Intervention 高度干预 8. Anti-Icing 防冰 9. Application Integration 应用综合 10. Artificial Intelligence 人工智能
AIA 1. Aerospace Industries Association 宇航工业协会 2. Aircraft Industries Association 飞机工业协会(美

国) 3. Airbus Industries Asia 空客工业公司亚洲部

AIAA 1. Aerospace Industries Association of America 美国航空航天工业协会 2. Aerospace Industry Analysts Association 航空航天工业分析员协会（美国）3. Aircraft Industries Association of America 美国航空器工业协会 4. Area of Intense Air Activity 稠密空气层活动区 5. American Institute of Aeronautics and Astronautics 美国航空航天学会

AIAAM Advanced Intercept Air to Air Missile 先进空空截击导弹

AIAC Air Industries Association of Canada 加拿大航空工业协会

AIAEA All-India Aircraft Engineers Association 全印飞机工程师协会

AIANZ The Aviation Industry Association of New Zealand Inc. 新西兰航空工业协会

AIB 1. Accident Investigation Branch 事故调查分局 2. Airbus Industrie 空中客车工业公司（法国）3. Aeronautical Information Bureau 航空资料局

AIC 1. Aerodynamic Influence Coefficient 气动效应系数 2. Aeronautical Information Circular 航空信息通报，航行情报通报 3. Analog Interface Card 模拟接口卡

AICE American Institute of Chemical Engineers 美国化学工程师学会

AICKV Anti Ice Check Valve 防冰检查阀

AICS 1. Advanced Interior Communication System 先进内部通信系统 2. Aero-Inertial Control System 机载惯性控制系统 3. Air Intake Control System 进气口控制系统 4. Aircraft Identification and Control System 飞机识别和控制系统

AICUZ Air Installations Compatible Use Zone 空军设施适用区

AID Altered Item Drawing 变更项图

aid 帮助，助手，辅助设备

Aida "埃达"（法国机载测距雷达）

AIDAPS Automatic Inspection Diagnostic and Prognostic System 自动检测诊断与预报系统

AIDAS Advanced Instrumentation and Data Analysis System 高级仪表测量与数据分析系统

AIDATS Army In-flight Data Transmission System 部队飞行数据传输系统

AIDC 1. Aero Industry Development Center 航空工业发展中心 2. ATC Interfacility Data Communications 空中交通管制设施间数据通信

aided 半自动的

aided tracking 半自动跟踪，半自动跟踪系统

AIDES Automated Image Data Extraction System 自动图像数据提取系统

AIDS 1. Acoustic Intelligence Data System 声学情报数据系统 2. Aircraft (Airplane) Integrated Data System 飞机综合数据系统

aid to navigation 导航设备（系统、工具）

AIEE American Institute of Electrical Engineer 美国电气工程师协会

AIEM Airlines International Electronics Meeting 国际航空电子会议

AIFC Airbus Industries Finance Services 空中客车工业公司金融服务

AIFCS Airborne Interception Fire-Control System 机载拦截火力控制系统

AIFF Audio Interchange File Format 音频交换文件格式

AIFS 1. Advanced Indirect Fire System 先进间接射击系统 2. Anti Ice Flow Sensor 防冰流量传感器

AIFV Armored Infantry Fighting Vehicle 装甲步兵战车

AIG 1. Accident Investigation 事故调查 2. Address Indicating/Indicator Group 地址指示组 3. Aircraft Integrated Data System 机载综合数据系统

AIL Aileron 副翼

AILAS 1. Airborne Instrument Landing Approach System 机载仪表着陆进近系统 2. Automatic Instrument Landing Approach System 自动仪表着落进近系统

ailavator 副翼升降舵

aileron 1. 副翼 2. 侧向操纵系统的效能

aileron command 副翼控制

aileron deflection 副翼偏转

aileron device 副翼装置

aileron drag 副翼阻力

aileron effectiveness 副翼效能

aileron motion 副翼运动

aileron oscillation 副翼摆动

aileron reversal 副翼反效

aileron roll 副翼操纵横滚

aileron system 副翼系统

aileron travel 副翼偏转角，副翼偏度

aileron trim 1. 滚转配平，副翼配平 2. 滚转配平机构，副翼调整片

ailing aircraft 〈口语〉有故障的飞机

ailit 碳硅电阻材料

AILS 1. Advanced Instrument Landing System 先进仪表着陆系统 2. Automatic Instrument Landing System 自动仪表着陆系统

AILSA Aerospace Industrial Life Sciences Association 航空航天工业生命科学协会

aim 1. 瞄准，导航，引导，寻的 2. 目标，宗旨，目的

aim bias 瞄准偏差
aim dot 瞄准点
AIM parachute 1. 自动充气调节伞 2. AIM 伞
aim point 目标点
aimable cluster 可瞄集束炸弹
aimable fragment warhead 定向战斗部
aimable(-type) cluster adapter 可瞄集束炸弹架,可瞄子母弹箱
AIMD Aircraft Intermediate Maintenance Department 飞机中级维修部门
AIME 1. American Institute of Mechanical Engineers 美国机械工程师协会 2. Autonomous Integrity Monitoring Extrapolation 自主整体监控外推法
aimed fire 瞄准射击,直接瞄准射击
aimed fire opening range 直接瞄准射击开火距离
aimed guidance 瞄准具制导
aimed launcher 定向发射装置
aimed missile (非制导)火箭弹
aimed warhead (杀伤弹片定向散飞的)定向爆炸弹头
aimer 1. 引导员 2. 射击员 3. 瞄准具
AIMES Avionics Integrated Maintenance Expert System 综合航空保养专门系统
aiming 1. 瞄准,对准 2. 制导,引导
aiming angle error 瞄准角误差
aiming azimuth 瞄准方位角
aiming data 瞄准诸元
aiming device 1. 瞄准具,瞄准装置 2. 瞄准检查器,瞄准检查镜
aiming error 瞄准误差
aiming for lead 前置瞄准
aiming leeway 1. 提前量,前置量 2. 前置瞄准
aiming line 瞄准线
aiming mark 瞄准点,瞄准标志
aiming parameter 瞄准参数
aiming pip (雷达瞄准具荧光屏上的)目标光点,目标标志
aiming pipper 瞄准光点
aiming point 瞄准点,标定点
aiming post 瞄准标杆
aiming reticle 瞄准光环
aiming sight 1. 光学射击瞄准具 2. 瞄准光环
aiming signal control instrument 瞄准信号控制仪
aiming-off 1. 提前量,前置量 2. 前置瞄准
aim-off 提前角,前置角
aimpoint "瞄准点"
AIMS 1. Airplane Information Management System 飞机信息管理系统 2. Airbus Inventory Management System 空客资产管理系统 3. Automated Integrated Manufacturing System 自动综合制造系统 4. Area Inertial Navigation System 区域惯导系统
AIN Aircraft (Airplane) Identification Number 飞机标识号
AINS 1. Aircraft Information Network System 飞机信息网络系统 2. Advanced Inertial Navigation System 先进惯导系统 3. Airborne Inertial Navigation System 机载惯导系统
AINSC Aeronautical Industry Service Communications 航空工业服务通信
AIO Analog Input and Output 模拟输入输出
AIOA Aviation Insurance Offices Association 航空保险公司协会(英国)
AIP 1. Aeronautical Information Publication 航空资料汇编,航空信息出版物 2. Aeronautical Information Publication 航空信息资料汇编 3. Aeronautical Information Report 航空情报报告 4. Aircraft (Airplane) Identification Plate 飞机标识板
AIPC Aircraft (Airplane) Illustrated Parts Catalog 飞机图解零件目录
AIPS 1. Airbus Process Specification 空客工艺规范 2. Aircraft (Airplane) Ice Protection System 飞机防冰系统 3. Anti Ice Pressure Sensor 防冰压力传感器
AIPT Aircraft (Airplane) Integration Product Team 飞机集成产品组
air 1. 空气,大气,气流,气团 2. 空中,天空 3. 空军,航空兵,空中力量(air force, air arm, air power 的简写)
air (airborne) armament 1. 航空军事装备 2. 航空军械
air abort 中断飞行(起飞后中断飞行)
air accident 飞行事故,空中失事
air addition 空气添加
air admission 吸气,进气
air against airplane 飞机迎面气流
air alert 1. 空中待战,空中警戒 2. 〈口语〉空袭警报
air alignment (惯导系统的)空中对准
air angle 风角
air annual 航空年鉴
air annulus 环空
air antisubmarine 航空反潜
air approaches 进场净空区,进近空域
air around airplane 绕机气流
air assets 空中飞机
air atomizer 空气雾化器
air avenue of approach (空降兵运输机的)进近航路
air ball 球形压缩空气瓶,球形空气蓄压器
air base 1. 航空基地 2. (航空用照相测量中的)空中基线 3. 空中基线的长度
air base area 航空基地地域
air bearing 空气轴承,气浮轴承
air bearing gyroscope 空气轴承陀螺仪,气浮陀螺仪

air blast atomizer 空气雾化喷嘴
air blast wave 空气冲击波
air bleed 1.放气 2.引气
air bottle 压缩空气瓶,冷气瓶
air brake 气动力制动装置,减速板
air breather 空气呼吸器
air breathing booster 吸气式助推器
air breathing engine 空气(助燃)发动机
air bubble method 空气泡法
air buffer 空气缓冲
air cannon impact tester 航空机炮弹着测定器
air cap （流星的）空气冠
air capability 空袭兵器能力
air capture 空气捕捉
air caralry 空中骑兵
air carrier 1.航空公司,空运机构 2.运输机
air cartography 航空测绘（地图）
air case 空气箱
air catering 航空膳食
Air China 中国国际航空公司
air cleanliness 空气洁净度
air cloud 空中云层
air collection ramjet 集气式冲压喷气发动机
air column 气柱
air combat 空战
Air Combat Command 空中作战司令部
air combat game 空战游戏
air combat simulation 空中作战模拟
air combustion 空气燃烧
air combustion case 空气燃烧箱
air command automation system 空军指挥自动化系统
air compressor 空气压缩机
air condition 空气调节
air conditioning 空气调节
air conditioning cleaned tractor 空调净化牵引车
air conditioning hose 空调软管
air conditioning pipe system 空调管路系统
air conditioning purge 空调净化
air consignment note 空运提货单,空运发货单
air consumption 空气消耗
air control officer 飞机引导军官,空军前方控制军官
air control team （战术）空军引导组（引导飞机对地面部队进行近距离支援的专门组织）
air controller 空中交通管制员
air cooled 空气冷却的,气冷的
air cooled engine 气冷式发动机
air cooler 空气冷却器
air cooling 空气冷却
air corridor 空中走廊

air crew 1.空勤组 2.机组
air cushion 空气缓冲
air cycle 空气循环
air cycle cooling system 空气循环冷却系统
air damper 空气阻尼器,气压减震器
air data computer 大气数据计算机
air data system 大气数据系统
air datum 空气数据,空气基准
air decoy missile 空中诱惑导弹
air defence controller 防空引导军官,地域防空指挥官
air defence identification zone 防空识别区
air defence operations area 防空作战区
air defence post 防空哨
air defence region 防空作战分区
air defence sector 防空(作战)扇区,防空地段
Air Defence Ship 防空舰（船）
air defense 防空
air defense fire control set (system) 防空火力控制设备(或系统)
air defense fire coordination system 防空火力协调系统
air defense fire direction system crewman 防空火力射击指挥系统操纵手
air defense missile control and coordination system 防空导弹控制及协调系统
air defense missile control system 防空导弹控制系统
air defense radar 防空雷达
air defense radar system 防空雷达系统
air defense robot （自动控制）防空导弹
air defense system 防空体系
air defense weapon control case 防空武器控制法
air defense weapons control status 防空武器控制状态
air delivery 空气输送
air density 空气密度,大气密度
air density ratio 大气密度比
air department （航母）航空部门
air deposition 空气沉积
air depot 航空维修与供应基地（美国空军）
air detonation 空中爆炸
air dispatcher 空投员
air distance 1.空中段（起飞或着陆航迹的）2.空中飞行距离
air distribution board 空气配气台
air distribution system 空气分配系统
air division 1.空军师 2.分队 3.区队
air drag 空气阻力,大气阻力
air drop 空投
air ejector 空气喷射器
air embolism 空气栓塞症
air engineering department （航母上的）航空工程部门

air engineering officer　航空工程军官
air equivalence ratio　空气当量比
air evacuation　空中后送
air exchange　（闭路风洞中的）换气，空气交换
air executive　领航引导员，飞行副部队长
air faucet　1.空气旋塞 2.（接管）气嘴
air feedstock　原料气
air film accelerometer　气膜加速度计
air filter　空气过滤器
air fired　空中发射的，空中起动的
air fitter　航空发动机装配工（英国）
air flame　空气火焰
air flame operations　火焰空袭
air flame weapon(s)　空投火焰武器
air flare　空中照明弹
air flow　空气流量
air flow angle measurement　空气流量角度测量
air flowfield　空气流动场
air foam rubber　泡沫橡胶，气沫橡胶
air force　空军
Air Force atmospheric model　空军大气模式，空军标准大气（美国）
air force station　航空兵驻地
air fraction　空气分数
air gap　空气间隙（磁极间的）
air gunner　空中射击员
air handling　（导弹）导向目标，瞄准目标
air hardening　空气硬化，室温时效硬化
air heat exchanger　空气热交换器
air heat　空气热
air heated boom　空气加温全静压管（空速管）
air heater　空气加热器
air heave　气胀变形
air hostess　空中小姐，机上女服务员
air hunger　缺氧
air ignition　空气点火
air impingement starting system　气动冲击起动系统
air index table　气动分度工作台
air inflatable bomb　充气炸弹
air inflow　空气流入
air information section　空情报知组
air injection　空气喷射
air injector　空气喷射器
air inlet　进气道，进气口
air inspection　飞行检查
air intake　进气道，进气装置，进气管
air intake test　进气道试验
air intelligence processing　空中情报处理
air intercept missile　空中截击导弹

air intercept rocket　空中截击火箭
air interception　1.空中截获（雷达或目视搜索到敌机）2.空中截击
air interdiction　空中阻滞（常指切断敌之后方供应线，而不是打击地面军队。区别于战场空中阻滞 BAI）
air jet　空气射流
air launch　空（中发）射
air launched air rocket recovery　空（中发）射（的空中）火箭的回收
air launched anti-ballistic missile　空（中发）射（的）反弹道导弹
air launched cruise missile　空中发射的巡航导弹，空射巡航导弹
air launched ground mine　空中发射的沉底水雷
air launched guided missile　空中发射的导弹，空射导弹
air launched low-volume ramjet missile　空射低（燃料）容量冲压（喷气）发动机导弹
air launched torpedo attack　空（中发）射（的）鱼雷攻击
air launching　空中发射
Air League　航空爱好者协会（英国）
air leak　空气泄漏
air lever　进气操纵杆
air liaison officer　空军联络官
air light　空气光（悬浮物散射光）
air liquefaction　液化空气
air load　风荷载，气荷载
air lock (air-lock)　1.气塞 2.（航天器）气闸舱，气密过渡舱 3.（仪表的）压差室
air locked　不透气的，密封的，密闭的
air log　航空计程器
air loss　（密闭舱内的）空气损耗
air louver　空气调节孔，放气孔
air mail　航空邮件
air mass　大气质量，气团
air mass flow　空气质量流量
air mass flow rate　空气质量流量率
Air Materiel Command　空军装备部（美国）
air meter　气流表，微风表，风速表
air mile　1.空英里（长度等于海里）2.无风飞行英里
air mileage unit　空英里计数器
air mine　空投水雷，航空水雷
air ministry　空军部，航空部（英国由国防部代替）
air mixture　空气混合物
air mixture jet　空气混合物喷射
Air Mobility Command　空中机动司令部
air mobility　空运
air momentum　空气动力
air monitoring　大气监测
air movement　空中运送

air movement table 空中运送计划表
air munition 航空弹药
Air National Guard 空军国民警卫队(美国)
air navigation 空中领航
air navigation aids 1.(航空)导航设备 2.助航方法
Air Navigation Chart 领航作业图(美国)
air navigation facility 导航设施
air navigation protractor 领航(用)量角器
air node 空气节点
air nozzle 空气喷嘴
air observation spotting 空中观察与校射
air officer（Air Officer） 空军军官(特指英国空军中的高级军官)
air ordnance 航空军械
air parity 空中均势
Air Pass 空中通道
air passage 空气通道,空气道
air path 大气路径
air penetration 空气渗透
air phone 空地通话机
air pill 〈口语〉小型航空炸弹
air plan (航母)飞行计划
air plenum 空气室
air plot 1.空中领航作图 2.标图
air pocket 气潭
air pollution 大气污染,空气污染
air position 飞机位置
air position indicator 飞机位置指示器
air power gun 气动注油枪
air pressure 空气压力
air pressure drop 空气压降
air pressure test 气压试验
air propeller 空气螺旋桨
Air Pump（Antlia） 抽气筒,唧筒座
air purification unit 空气净化装置
air quality 大气质量,大气品质
air raid 空袭
air range 空中范围
Air Rank 空军高级军官,空军将官(英国)
air rate 含气率,混气率
air ratio 空气比率
air recycle system 空气再循环系统
air regenerator 空气再生器
air rescue 空中救援,空中营救
air regenerative system 空气再生系统
air route 空中航线,航路
air route surveillance 航路监视,航线监视
air route traffic control center 航路交通管制中心
air sampling 空气取样,空中取样

air sampling rig 空气取样装置
air scenarios 空情
air scoop 1.屏斗形进气口 2.冲压进气的俗称
air screw 空气螺旋桨
air seal 气封,防气圈
air section 1.气段,空气区域 2.航空部
air service 1.空军,空中勤务 2.航空运输
air shock 空气冲击
air show 1.航空表演 2.航空展
air sickness 航空病,空晕病,晕机
air side 1.供风端 2.风侧,空侧
air slaking 风化,潮解
air slewed missile 空中旋转式导弹,空中机动式导弹
air snap 气动卡规
air snatch 1.空中抓取 2.空中营救
air sounding 1.大气探测 2.大气探测记录
air source truck 气源车
air space 航空、空间的,航空、太空的,航空航天的
air speed 气流速度,空速
air speed indicator 空速表
air speed-Mach indicator 空速马赫数表
air spotting duty 空中校射任务
air stagnation 空气停滞
air stagnation temperature 停滞空气温度
air start 1.空气起动 2.(发动机)空中起动
air start button 1.空中起动按钮 2.(压缩)空气起动按钮
air station 海军航空兵基地
air stream 气流
air superiority 空中优势
air superiority fighter 制空歼击机,空中优势战斗机
air supply 供气
air supply system 气源系统
air supremacy 最高级空中优势,空中霸权
air surface zone 航空兵与(水面)舰艇配合(反潜)行动区域
air surveillance radar 对空监视雷达
air swirl 空气涡流
air system 空气系统
air target environmental warhead 攻击空中目标的沾染性弹头
air target warhead 攻击空中目标的弹头
air tax 1.出租飞机 2.(直升机)空中"滑行"
air temperature 空气温度
air throttle 节气门
air thrust measurement 空中推力测量
air tight locker 密封箱
air tight test 气密性试验
air tight valve 密闭阀

air tightness   气密性
air time   空中时间
air torpedo   1.航空鱼雷 2.滑翔炸弹
air total temperature   空气总温
air traf   空气流量
air traffic   空中交通
air traffic clearance   空中交通管制飞行许可(证)
air traffic control   空中交通管制
air traffic control center   空中交通管制中心
air traffic control radar   空中交通管制雷达
air traffic control radar beacon   空中交通管制雷达信标
air traffic control satellite   空中交通管制卫星
air traffic control service   空中交通管制勤务
air traffic controller   空中交通管制人员
air traffic coordinator   航管协调员
air traffic flow   空中交通流量
air traffic flow control   空中交通流量控制
air traffic management   空中交通管理
air traffic management system   空中交通管理系统
air traffic service   空中交通服务
air train   1.空中列车 2.气垫列车
air transit   航空转运
air transport   空运(业)风力运输,气浪
air transport operation   空中运输运营
air transportable unit   可空运的部队
air transportation   空运,航空运输
air transportation system   空运系统
air travel   航空旅行
air treatment system   空气处理系统
air tree   航空树状网
air trooping   (非战术性的)空中运兵
air trunk   1.通风干路 2.通气管路
air tube   空气管
air tube connector   气管连接器
air turnback   返回点
air umbrella   空中保护伞
air vane   空气舵,空气动力舵
air vehicle   航空器,飞行器
air vehicle operator   遥控飞机控制员
air velocity   空气流速
air ventilated clothing   通风服
air ventilated jacket   通风夹克衫
air ventilating garment   通风服
air ventilation and purification system   通风净化系统
air violation   空中入侵,侵犯领空
air ward system   空中警卫系统
air weapons control system   防空武器控制系统
Air Weather Service   空军气象局(美国)
air weight   空气的重量
air wheel   航空机轮
air work   空中作业
air zero   原子弹空中爆炸中心
air zero locator   (核弹)空中爆炸中心定位器
air zone   空域
air(-core) transformer   空(气)芯变压器
air/fuel ratio   空气-燃油比
air/land warfare   空地一体战
air/oil strut   油-气缓冲支柱
AIRAC   Aeronautical Information Regulation and Control 航空信息规则与控制
air-accident inquiry   飞行事故调查
air-actuated   气动的
air-actuated mortar   气动炮
airad airmen advisory   飞行员咨询
air-assist   空气辅助
air-assist atomizer   空气辅助雾化器
air-atomized   空气雾化的
air-augmented rocket   空气补燃火箭
airbag   安全气囊
airbag system   安全气囊系统
air-based   空基的,在母机上的,机载
air-bearing   空气轴承
air-bearing carriage   空气轴承托架
air-bearing facility   空气轴承设备
air-bearing system   空气轴承系统
air-bearing table   空气轴承架,气浮轴承平台
air-bearing test   空气轴承测试
air-bearing testbed   空气轴承测试台
airbill   空运货单
airblast   1.空气冲击流 2.空中爆炸 3.空气喷净
airblast air   雾化空气
airblast atomization   气动雾化
airblast atomizer   气动雾化器,气动喷嘴
air-blast nozzle   气动雾化喷嘴
airblast pressure   空气压力
airblast pressure drop   空气压力降
air-blast switch   气爆开关
air-blast transformer   风冷式变压器
airblast-atomized spray jet   气流雾化喷嘴
airbleed   1.放出气,引出气 2.放气口
airblower   鼓风机,增压器
airborne   1.空中的(指在大气层中飞行,不用于航天飞行) 2.机载的 3.空降的
airborne adjunct sensor   机载附加传感器,机载附加探测器
airborne alert   空中警戒
airborne anti-armour weapon   机载反装甲武器
airborne anti-ballistic missile intercept system   机载反弹

道弹截击系统
airborne anti-radiation missile 1.航空反辐射导弹 2.机载反辐射导弹
airborne anti-satellite missile 1.航空反星导弹 2.机载反星导弹
airborne antisubmarine 机载反潜
airborne antisubmarine detecting system 机载反潜探测系统
airborne antisubmarine weapon system 机载反潜武器系统
airborne anti-tank missile 1.航空反坦克导弹 2.机载反坦克导弹
airborne application 机载应用
airborne area 空降地域
airborne assault weapon 空降突击武器
airborne audible mountain-warning bepeer 机载防撞山音响警告器
airborne bombing evaluation 空中轰炸效果判定
airborne cannon 航空机炮
airborne CCN spectrometer 机载云凝结核粒谱仪
airborne cloud collector 机载云(含水量)收集器
airborne collision avoidance equipment 机载防撞设备
airborne command post 空中指挥所
airborne computer 机载计算机
airborne craft 航空运载器,航空器,飞机
airborne decoy 1.航空诱惑弹 2.机载诱惑弹
airborne early warning 1.空中预警,机载早期警报(系统) 2.空中预警机
airborne early warning radar 机载预警雷达
airborne equipment 空运装备
airborne fire control system 航空火力控制系统
airborne fire control radar 机载火控雷达
airborne force 空降部队
airborne gunlaying turret 有自动瞄准具的炮塔
airborne headquarters 空中指挥所
airborne infrared guidance system 机载红外制导系统
airborne integrated data system 机载综合数据系统
airborne intercept 机载截听
airborne interception 空中截击
airborne interception equipment 机载截击设备
airborne interception fire control system 机载截击火力控制系统
airborne interception radar and pilot attack sight system 机载截击雷达和驾驶员攻击瞄准系统
airborne interceptor 机载拦截弹
airborne interceptor missile 机载截击导弹
airborne interceptor rocket 机载截击火箭
airborne laser range finder 机载激光测距器
airborne launch control center 机载发射控制中心
airborne launch control system 机载发射控制系统
airborne lead 机载导管
airborne liaison mission 空中联络组
airborne machine gun 机载机枪
airborne mine-laying 航空布雷
airborne missile control system 机载导弹控制系统
airborne MTD radar 机载动目标检测雷达
airborne MTI radar 机载动目标指示雷达
airborne night sight 机载夜间瞄准具
airborne noise 空气噪声
airborne oil surveillance system 机载油污染监测系统
airborne operation 空运行动
airborne optical adjunct 机载光学附件
airborne optical gunsight 航空光学射击瞄准具
airborne oxygen system 航空供氧系统
airborne particle 悬浮尘粒
airborne particulate 空中悬浮微粒,浮粒
airborne platform 机载平台
airborne radar 机载雷达
airborne radar antenna 机载雷达天线
airborne radiation thermometer 机载辐射温度仪
airborne radio and flight check panel operator 机上无线电设备与检测仪表技师
airborne radio relay 空中无线电中继
airborne range and orbit determination 机载轨道测定
airborne rapid-scan spectrometer 机载快速扫描光谱仪
airborne reconnaissance radar 机载侦察雷达
airborne relay station 机载中继站
airborne rocket 1.航空火箭弹 2.机载火箭弹
airborne sensor system 机载探测器系统
airborne support equipment 机载保障设备
airborne target handover system 机载目标交接系统
airborne terrain analysis 空中地形分析,航空地形分析
airborne vehicle 飞行器/航空器
airborne vibration monitoring 机载振动监视(控)
airborne video recording system 1.航空视频记录系统 2.机载视频记录系统
airborne warning and control system 机载警戒与控制系统
airborne wave-guide slotted array 机载波导式隙天线阵
airborne weapon 1.航空武器 2.机载武器
airborne weapon system 1.航空武器系统 2.机载武器系统
airborne weapons release set 机载武器投放设备
airborne weapons release system 机载武器投放系统
airborne weather radar 机载气象雷达
airborne-delay control 机载延迟控制
airborne-qualified 有飞行资格的,有空降资格的
airbow 空气弓

airbrake （飞机的）减速板
airbrakes in 减速板收起
air-braking tachometer （空气）摩擦式转速表
airbreather 1.空气喷气发动机 2.装有空气喷气发动机的飞机
airbreathing 空气助燃的,吸气的;吸气式
airbreathing aircraft 空气喷气式飞机
airbreathing booster 吸气式增压器
airbreathing combustion 吸气式燃烧
airbreathing combustion system 吸气式燃烧系统
airbreathing combustor 吸气式燃烧室
airbreathing engine 吸气式发动机,空气喷气发动机
airbreathing launch 吸气式发射
airbreathing missile 吸气式导弹
airbreathing motor 吸气式发动机
airbreathing orbiter 吸气式探测器
airbreathing propulsion 吸气式推进
airbreathing stand-off bomb （装）空气喷气（发动机的）远射空地导弹
airbreathing vehicle 吸气式车辆,吸气式飞机
airbreathing weapon （装）空气喷气（发动机的）导弹
airbridge （旅客机的）登机（天）桥
airburst 空爆,空炸
airburst fuze 空爆引信
airburst ranging 空炸射击修正
airburst shell 带时间引信的炮弹,空炸炮弹
AIRC 1.Airlines Industrial Relations Conference 航空公司工业联系会议 2.Airworthiness Committee 适航委员会
AIRCN Air Communication Network 空中通信网
AIRCOM Air Communication 航空通信
aircom system 空军通信系统
air-combat engagement 空中交战
air-conditioning 空气调节,通风,空调
air-conditioning engine bleed pipe （座舱）空调用发动机引气管
air-conditioning heat exchanger 空调系统热交换器
air-conditioning system 空气调节系统
air-cooled machine gun 气冷式机枪
air-cooled weapon 气冷式武器
air-cooling system 气冷系统
air-core pump 气心泵
aircraft 1.飞机 2.航空器,大气层飞行器
aircraft acceleration 飞机加速度
aircraft accident 飞机事故
aircraft accident and incident 飞行事故和事件
aircraft accident psychology 飞行事故心理学
aircraft accident recorder 1.飞行事故记录器 2.黑匣子
aircraft alerting system 航空器告警系统

aircraft allowance （部队）飞机编制数
aircraft antenna 机上天线
aircraft antisubmarine attack 飞机反潜攻击
aircraft application 飞机上的应用
aircraft approach 飞机进场
aircraft armament 航空军械
aircraft assignment 飞机排班
aircraft attitude 飞机姿态
aircraft attitude perception 飞行状态知觉
aircraft attrition rate 飞机正常损耗率,飞机折旧率
aircraft automatic weapon 航空自动武器
aircraft availability 飞机（的）可利用度,飞机的可用率
aircraft avionics 航空电子设备
aircraft behavior 飞机行为
aircraft cable 航空钢索,航空电缆
aircraft carrier 航空母舰
aircraft categories 飞机种类
aircraft center of gravity 飞机重心
aircraft central conditioning system 飞机中央空调系统
aircraft certificate 飞机证件
aircraft cg aircraft center of gravity 飞机重心
aircraft characteristic 飞机特征
aircraft classification number 飞机等级数
aircraft collision 飞机相撞及接近事故
aircraft company 飞机公司
aircraft component 飞机部件
aircraft configuration 1.飞机构型 2.（设计中的）飞机总体布局
aircraft control 飞机控制
aircraft control display console 飞机控制显示操纵台,飞机显控台
aircraft control room 飞机控制室
aircraft conversion 飞机换型,飞机换装
aircraft cost 飞机成本
aircraft count 飞机计数
aircraft datum 飞机数据,航空器水平基准线
aircraft density 飞机密度
aircraft design 飞机设计
aircraft design problem 飞机设计问题
aircraft design process 飞机设计过程
aircraft designer 飞机设计师
aircraft development 飞机设计研制
aircraft direction 飞机的方向
aircraft direction of motion 飞机的运动方向
aircraft dispatcher 航行调度员
aircraft-dispensed weapon 飞机散布的杀伤剂
aircraft dope 1.飞机涂布油 2.航空燃料添加剂
aircraft drag 飞机阻力
aircraft dropwindsonde system 机载下投式测风探空仪

系统
aircraft dynamics 飞机动力学
aircraft electrical system 航空电气系统
aircraft emission 飞机排放物
aircraft engine 飞机发动机
aircraft engine business 飞机发动机业务
aircraft engine division 飞机引擎部门
aircraft engine market 飞机引擎市场
aircraft equipment 飞机设备
aircraft exhaust 飞机尾气
aircraft exposure 飞机使用经历
aircraft exterior lighting 机外照明
aircraft fabric 1.航空织物 2.飞机蒙布
aircraft fear 【医】飞机恐惧
aircraft fire control computer 机载火力控制计算机
aircraft firing capability 航空器(飞机)火力
aircraft firm 航空公司
aircraft flight 飞机飞行
aircraft fuel 航空燃油,航空燃料
aircraft fuel system 飞机燃油系统
aircraft fuel tank 飞机燃油箱
aircraft fuselage 飞机机身
aircraft gas 飞机燃气
aircraft gas turbine 飞机燃气涡轮
aircraft generator 航空发电机
aircraft geometry 飞机几何
aircraft gun laying 航空机枪瞄准,航空机炮瞄准
aircraft handler 航空机械员
aircraft handling 1.飞机操纵,飞机驾驶 2.飞机操纵特点
aircraft handling crew and chock men (航母)飞机管理人员与放轮挡人员(着蓝色服装)
aircraft handling officer (航母)飞机管理军官(着黄色服装)
aircraft hydraulic pipeline 飞机液压管道
aircraft hydraulic system 飞机液压系统
aircraft hydraulic-pipeline vibration 飞机液压管道振动
aircraft icing 航空器结冰
aircraft in service 在用飞机(数),实有飞机(数)
aircraft incident 1.飞机事故征候 2.飞机飞行事件
aircraft industry 飞机制造工业
aircraft insignia 机徽,飞机识别标志
aircraft integrity 航空器完好度
aircraft interior lighting 机内照明
aircraft intermediate maintenance department (航母)飞机中级维修部门
aircraft internal time division command/response multiplex data bus 飞机内部指令型/响应型时分多路传输数据总线

aircraft interrogator 机载询问机
aircraft landing 飞机着陆
aircraft landing training ship 飞机着舰辅助训练舰(美国海军)
aircraft lateral 飞机侧向
aircraft layout 飞机布局
aircraft life cycle cost 航空器全寿命费用
aircraft load 飞机载荷
aircraft loan guaranteed 购买飞机担保贷款
aircraft log(book) 飞机履历本
aircraft machine gun sight 航空机枪瞄准具
aircraft magnetic field 航空器磁场
aircraft management system 飞机管理系统
aircraft maneuver 飞机机动飞行
aircraft manufacturer 飞机制造商
aircraft manufacturing 飞机制造
aircraft market 飞机市场
aircraft mass 飞机质量
aircraft maximum 飞机最大极限
aircraft meteorological station 飞机气象站
aircraft missile navigation system 机载导弹导航系统
aircraft mission equipment 飞机任务设备
aircraft model 飞机模型
aircraft motion 飞机运动
aircraft motion compensation 载机运动补偿
aircraft movement 飞机运动
aircraft movement area 飞行区
aircraft movement area mark 飞行区标志
aircraft movement area reference code 飞行区参考密码
aircraft noise 航空器噪声
aircraft noise certification 航空器噪声审定
aircraft nuclear hardness 飞机核硬度,飞机核防护能力
aircraft on ground 停飞待用飞机
aircraft operating cycle 飞机使用循环,起落
aircraft operation 航空器运行,飞机操作
aircraft operator 航空器操作员
aircraft orbit 飞行器轨道
aircraft oxygen cylinder 机上氧气瓶
aircraft pair 飞机偶对
aircraft panel 飞机壁板
aircraft parameter 飞机参数
aircraft performance 飞机性能
aircraft performance penalty 航空器性能代偿损失
aircraft pitch 飞机的俯仰
aircraft pitch attitude 飞机的俯仰姿态
aircraft pitch control task 飞机的俯仰控制任务
aircraft platform 飞机平台
aircraft pod 飞机吊舱,航空吊舱
aircraft position 飞机位置,航空器位置

aircraft powerplant 航空器动力装置
aircraft price 飞机价格
aircraft problem 飞机的问题
aircraft product 飞机产品
aircraft production reference system 航空器工艺基准系统
aircraft program 飞机项目计划
aircraft program valuation model 飞机项目价值评估模型
aircraft project 飞机项目
aircraft projectile 航空炮弹
aircraft propulsion 飞行器推进系统
aircraft radar 飞机雷达
aircraft response 飞机响应
aircraft revolver-type gun 航空转膛炮
aircraft rocket launcher 机载火箭发射器,航空火箭发射器
aircraft rocket projectile 航空火箭弹,机载火箭弹
aircraft safety 飞机安全性
aircraft senior air engineering officer 主任航空机械师,机械长
aircraft serviceability 航空器出勤率
aircraft shape 飞机形状
aircraft shape function 飞机形状函数
aircraft simulation 飞机仿真
aircraft simulation test 飞机仿真试验
aircraft size 飞机大小
aircraft skin 飞行器外壳,飞机蒙皮
aircraft smoke and illumination signal 航空发烟、发光标志弹
aircraft specification 飞机技术规范
aircraft speed 飞机速度
aircraft stability 飞机的稳定性
aircraft stand marking 航空器停机位标志
aircraft state 飞机状态
aircraft stores management system 飞机外挂物管理系统
aircraft strapdown 飞机捷联
aircraft stream 飞机流
aircraft structural integrity 航空器结构完整性
aircraft structure 飞机结构
aircraft subsystem 飞机子系统
aircraft system 飞机系统
aircraft system controller 飞机系统控制机构
aircraft take-off 飞机起飞
aircraft take-off weight 飞机起飞重量
aircraft technology 飞机技术
aircraft to satellite data relay 飞机-卫星数据中继
aircraft torpedo 1.航空鱼雷 2.滑翔炸弹
aircraft track 飞机回波
aircraft track angle 飞机的航迹角
aircraft trail 1.飞机尾迹 2.飞机拉烟
aircraft trajectory 飞机轨迹
aircraft troop commander 机上部队指挥官
aircraft turbine 飞机涡轮
aircraft turn-round 飞机转身
aircraft type 飞机类型,飞机型号
aircraft usage 飞机的使用
aircraft use 飞机的使用
aircraft utilization 飞机利用率
aircraft vectoring 对飞机的引导
aircraft velocity 飞机的速度
aircraft vulnerable areas 1.飞机易损部位,飞机结构薄弱部位 2.飞机易受攻击部位
aircraft warning and ground control radar 对空警戒及地面引导雷达,地面对空警戒引导雷达
aircraft washing area 飞机洗涤场
aircraft weapon 1.航空武器,飞机武器 2.有翼导弹
aircraft weapon control system monitoring set 飞机武器控制系统监控器
aircraft weapons release set 机载武器投放装置
aircraft wear-out rate 飞机耗损率
aircraft weight 飞机重量
aircraft weighting equipment 飞机称重设备
aircraft wing 机翼
aircraft wiring 1.飞机的布线 2.飞机的导线网
aircraft-engine integration 飞机-发动机一体化
aircraft-borne 机载的,机上的
aircraft-carried normal earth-fixed system 航空器牵连铅垂地面坐标系
aircraft-controlled 机上控制的
aircraft-kilometres 飞机公里
aircraft-level testing 飞机水平测试
aircraft-like(-shaped) 飞机形的
Aircraftman,Aircraftwoman 空军列兵(英国)
aircraft-missile inference 机弹干扰
aircraft-servicing equipment 飞机维护设备
aircraftsman 1.飞机制造者,飞机制造商 2.空军士兵
aircraft-store compatibility 航空器悬挂物相容性
aircraft-towed 飞机拖航的
aircraft-type 航空器型别,机种
aircraft-type vehicle 1.飞机式飞行器 2.飞航式导弹,带翼导弹
aircrew 空勤组,机组
aircrew equipment assembly 空勤组个人装具标准配套
aircrew error 机组人员的错误
aircrew fatigue 飞行人员疲劳
aircrew helmet 飞行头盔
aircrew life support equipment 飞行人员(空勤组)生命

保障设备
air-cushion landing great 气垫式起落架
air-cushion vehicle 气垫飞行器
air-damped accelerometer 空气阻尼式加速度表
air-data 大气数据
air-delivered scatterable mine 空投散射地雷
air-derived navigation data 空中导出的导航数据
air-division 航空部
air-division valve 空气隔离阀
air-driven 气动的,风动的
air-driven horizon 气动陀螺地平仪
airdrome 机场
airdrome alert 机场待战,机场值班
airdrop 空投(包括人和物,通常用降落伞)
airdrop platform 空投平台(飞机上为大件空投而设计的平台)
airdrop system 空投系统
air-droppable expendable ocean sensor 空投抛弃式海洋遥感器
air-dry 风干的
air-drying varnish 风干漆
airedale 〈口语〉舰载飞机飞行员(美国)
airejector 1.空气喷射器 2.气动弹射器
Airep 飞行员
air-ethylene mixture 空气中乙烯的混合物
airfaring people 航空从业人员,航空界人士
airfast 不透气的
air-feed 空气喷气的(指发动机)
airfield 1.陆地机场 2.机场
airfield circuit 机场起落航线
airfield runway survey 机场跑道测量
airfield surface movement indicator 机场地面交通情况显示器
airfield surveillance radar 机场监视雷达
airfitness of passengers 旅客的适航性
airflow 1.气流,空气流 2.空气流量
airflow condition 空气流动条件
airflow fixture 空气流量测试夹具,空气流量试验器
airflow rate 空气流量,气流速率
airflow requirement 空气流量要求
airflow velocity 气流速度
airfoam 空气泡沫材料
airfoil 翼型
airfoil cascade 翼型叶栅
airfoil case 翼型案例
airfoil cavity 翼型腔
airfoil chord 翼弦
airfoil contour 翼型外形
airfoil datum 机翼数据,翼型数据

airfoil deformation 机翼变形
airfoil design 翼型设计
airfoil flow 翼型绕流
airfoil geometry 翼型几何
airfoil lift 机翼升力,翼型升力
airfoil mean line 翼型中弧线
airfoil model 机翼模型
airfoil parachute 翼伞
airfoil passage 翼型的通道
airfoil performance 翼型的性能
airfoil profile 1.翼型 2.翼剖面
airfoil rib 翼肋
airfoil row 翼型行
airfoil section 翼剖面
airfoil shape 翼型
airfoil shaped 成翼型形的,成机翼形的
airfoil source 翼型来源
airfoil surface 翼型表面
airfoil theory 翼型理论
airfoil thickness 翼型厚度
airfoil trailing-edge 翼型后缘
airfoil wake 翼型尾
airfoil wall 机翼壁
airforce 空气动力
airforwarder 航空运输行,航空货运代理商
air-frame 空气框架
airframe 弹体,箭体,机体,机身
airframe axis 机身轴线
airframe company 飞机公司
airframe concept 机身概念
airframe damage 飞机机体的损伤
airframe designer 机身设计师
airframe industry 机体制造工业
airframe maintenance 机体维修
airframe manufacturer 机身制造商
airframe material 机体材料
airframe missile 弹体导弹
airframe noise 机体噪声
airframe opening seal 飞机开口处密封胶
airframe structure 机体结构
airframe vibration 机身振动
airframe weight 机身重量
airframe-attributable 机体事故
air-free 无空气的
airfreighter 货机
airglow 1.气辉 2.夜天光
airglow photometer 气辉光度计
air-ground communication 空-地通信
Airguard 空中卫士

airhead 1.空降场 2.常规起降基地 3.空中补给及后送基地 4.最近的机场
airhog 〈口语〉飞行爱好者
airhoist 空中绞车
airiator 〈口语〉飞行员,航空员
airiness 通风
airing 1.通风 2.空气干燥 3.起泡沫 4.无线电广播
air-injection case 空气喷射箱
air-inlet controller 进气道控制器,进气道控制装置
air-intake 进气
air-intake unit 进气装置
air-interception aircraft 截击机,拦截飞机
air-interception radar 机载截击雷达
airland 机降
airlandable 可机降的
air-launched 空中发射的
air-launched air-recoverable rocket 空中发射空中回收火箭,空射空收火箭
air-launched asat 空射反卫星导弹,机载反卫星导弹
air-launched cruise missile 空射巡射导弹
air-launched mine 1.航空水雷 2.空投水雷
air-launched miniature rocket 空中发射的小型火箭
air-launched missile 空中发射导弹
air-launched rocket 空中发射火箭
air-launched surface attack missile 空射攻击地面目标的导弹
air-launched torpedo 1.航空鱼雷 2.空投鱼雷
airless 无风的,无空气的
airlift 1.空运 2.空运工具,空运的货物 3.气压起重
airline 1.有执照的航空公司 2.大圆圈航线 3.气压输送管道
Airline Deregulation Act 航空公司放松管制法
airline engine 航空发动机
airline industry 航空业
airline marketing 航线销售
airline pilot 航空公司飞行员
airline preference 航空公司优先选择
airline profit 航空公司的利润
airline reservation system 飞机航班订票系统
airline service 航空公司的服务
airline station 航空站
airline turn 航空公司周转
airliner 班机,航线客机
airliner engine 飞机发动机
airliner market 客机市场
airliner washing tunnel 旅客机随道式清洗设施
air-liquid interface 气液界面
airload 1.空气动力载荷 2.空运装载
airlock (module) 1.气闸舱 2.气密过渡舱

airmaids 〈口语〉救生艇乘员(英国)
airman 1.空军列兵(美国) 2.航空员 3.飞机工
airman protective suit 飞行防护服
airman suit 飞行服
airmanship 飞机驾驶术,飞机驾驶技巧
air-mass-type diagram 气团类型图
airmechanics 航空力学
airmedical officer 航空军医主任
Airmen Advisory 飞行情况通报
air-minded 热心航空的
airmobile (air-mobile) 空中机动的
air-mobile band 空中机动部队通信频带
air-mobile operation 空中机动作战
air-mounted 机上的
airmover 风扇
airmunition (air munition) 航空弹药
airnat 〈口语〉航空家
air-openable 空中可打开的
air-operated 气动的,风动的
AIROPNET Air Operational Network 空中操作网络
airpark 1.飞机停放场 2.私人飞机降落场
airpatch 〈口语〉机场
air-path axis system 气流坐标系
airphibian 飞行汽车
airphibious 空降的,空运的
airplane 飞机
airplane certification 飞机认证
airplane configuration 飞机结构
airplane dart 航空火箭
airplane design 飞机设计
airplane efficiency factor 飞机有效使用率
airplane in aerobatic category 特技类飞机
airplane in commuter category 通勤类飞机
airplane in normal category 正常类飞机
airplane in transportation category 运输类飞机
airplane in utility category 实用类飞机
airplane interaction 飞机的相互作用
airplane model 飞机模型
airplane state 飞机状态
airplane thrust weight ratio 飞机推重比
airplane-missile interference 机弹干扰
airpoise 空气称重器,空气重量计
airport 1.航空港 2.机场
airport advisory area 机场咨询区
airport advisory service 机场咨询服务
airport capacity 机场容量
airport code 机场代码
airport commission 机场管理委员会
airport compatibility 机场兼容性

airport information desk 机场问讯处
airport medicine 航空港医学
airport of entry 有海关及出入境管理机构的国际机场
airport of origin 始发航空站
airport surface 机场地面
airport surface detection radar 机场地面活动监视雷达
airport surveillance 机场监视
airport surveillance radar 机场监视雷达
airport survey 飞机场测量
airport traffic area 机场空中交通管制区
airport traffic control service 机场空中交通管制勤务
airport traffic control tower 机场指挥塔台
airportable 可空运的
air-pressure fuze 气压激发引信
airproof 不透气的,密封的,气密的
air-refuelable 能空中加油的
air-relief cock 放气开关
air-route 航线
air-route structure 航路结构
air-route surveillance radar 航路监视雷达
air-run landing 空中滑跑着陆
air-run take-off 空中滑跑起飞(垂直/短距起落飞机利用地面效应水平加速,然后垂直起飞)
AIRS 1. Accident Information Retrieval 失事信息检索 2. Advanced Inertial Reference System 先进惯性参考系统 3. Airport Information Retrival System 机场信息检索系统 4. Atmospheric Infrared Sounder 大气红外探测器(美国)
airscape 空中鸟瞰图
airscrew 螺旋桨(英国常用 airscrew,美国用 propeller)
airscrew deicing alternator 螺旋桨除冰发电机
airscrew-turbine engine 涡轮螺旋桨发动机
air-sea 空海两用的
air-sea (air/sea) rescue 海上航空救援,海上空中救援
air-sea survival equipment 海上飞行救生设备
air-search radar 对空搜索雷达
airsecond 时间标度
airshed (飞)机库
airship 飞艇
airship design 飞艇设计
airship equation 飞艇方程
airship-borne 飞艇载的
airside 对空面,机场控制区
airspace 1.空域 2.空运容量 3.空间,空隙
airspace capacity 空域容量
airspace class 机场飞行区等级
airspace complexity 空域复杂性
airspace denial 空域封锁
airspace limitation 空域限制

airspace management 空域管理
airspace restriction 空中禁区
airspace system 空域系统
airspace user 空域用户
airspeed 空速
airspeed command 空速控制
air-speed equivalent 空速当量
airspeed indicator 空速表
airspeed limit altitude 空速受限制的高度
airspeed tape 空速表带状度盘,空速刻度带
airspeed transducer 空速传感器
airspeed trend display 空速变化趋势显示
air-spring eject 气垫弹射,气垫抛投
airstairs 登机梯
airstart boundary 空中起动边界
airsteps 登机梯,登机踏板
airster 〈口语〉班机
airstop 降落场,着陆场
airstream 气流,空气射流
airstrike 空袭,空中突击
airstrip 简易机场
air-supplied 空中补给的
air-supported missile 飞航式导弹,有翼导弹
air-suspension gyroscope 气浮陀螺仪
airswinging 空中测罗差
air-tactical data system 空中战术数据系统
air-taxi operator 出租飞机公司
airtel 机场饭店,航空旅社
air-temperature indicator 大气温度表
air-tested 试飞过的,空中试验过的
airtight 气密的,密封的
airtight strength test 气密强度试验
air-to-air bomb 空(对)空炸弹,反飞机炸弹
air-to-air dart 空对空射击用的镞形靶
air-to-air guided missile 空(对)空导弹
air-to-air missile 空空导弹
air-to-air stores 外挂空(对)空武器
air-to-air tracer-line snap-shoot symbology 空(对)空示迹线快射瞄准符号
air-to-ground missile 空地导弹
air-to-ground store 外挂空对地武器
air-tool 气动工具
air-to-surface 空对面的,空中飞机对地(或水)面的
air-to-surface guided missile 空(对)地(舰)导弹
air-to-surface rocket-boosted bomb 空(对)地(舰)火箭助推炸弹
air-to-underwater guided missile 空(对)潜导弹,空中对水下目标的导弹
air-traffic 空中交通

air-traffic density 空中交通密度
air-traffic flow 空中交通流量
air-traffic management 空中交通管理
air-tubeoil cooler 管式滑油散热器
air-vehicle 飞行器
airwake 舰尾流
airway 航路
airway beacon 航路灯标
airway traffic control 航路交通管制
airways clearance 航路飞行许可
airways equipped 有航路行设备的
airways flight 沿航路飞行
airwing 1.机翼 2.空军联队,(航母)航空联队条令(概则)
air-wing doctrine 风翼的学说
airwoman 女飞行员
airworthiness 适航性
airworthiness approval tag 适航批准标签
airworthiness authority 适航当局
airworthiness certificate 适航证
airworthiness directive 适航性指令
airworthiness regulation 1.适航规章 2.适航标准
airworthiness regulation 适航条例
airworthiness requirement 适航要求
airworthy 适航的
Airy diffraction pattern 爱里衍射图
Airy disk 爱里斑
Airy function 爱里函数
Airy transit circle 爱里子午环
AIS 1. Academic Instructor School 航校教员学校 2. Aeronautical Information Service 航空信息服务,航行情报服务 3. Airborne Instumentation Subsystem 机载仪表分系统 4. Audio Integrated System 音频综合系统 5. Automated Information System 自动信息系统
AISF Avionics Integration/Integrated Support Facilities 航空电子综合保障设备
AISI American Iron and Steel Institute 美国钢铁协会
aisle (客舱中的)人行通道
aisle height (飞机座舱内)座位中间过道高度
aisle stand 中央操纵台
aisle width 过道的宽度
AIV 1. Accumulator Isolation Valve 储压器隔离活门 2. Annulus Inverting Valve 环形转换活门
AIWS Air Interdiction Weapons System 空军遮断武器系统
AIZ Association of Inventors 发明者协会(苏联)
AJ 1. Anti-Jam 抗干扰 2. Assembly Jig 型架,装配夹具
ajax unit (操纵)感力装置
AJC Airline Job Cards 航空公司工作卡

Ajisai 紫阳卫星
AJPO Ada Joint Program Office 语言联合计划处
Akebono 黎明卫星
AKES Aga Khan Educational Services 阿迦汗教育服务
Akita Rocket Range 秋田火箭发射场
Aktivny-IK 主动波实验卫星-国际宇宙号
AL 1. Accuracy Landing 着陆准确性 2. Airline 航空公司,航线 3. Alternate 备份,备降机场 4. Approach and Landing 进近和着陆 5. Approach Light 着陆灯 6. Autoland 自动着陆
ALA 1. Aircraft Landing Area 航空器着陆区 2. Authorised Landing Area 允许着陆区
Aladin 阿拉丁
ALAE Association of Licensed Aircraft Engineers 持证空中机械协会(英国、澳大利亚)
ALAIRS Advanced Low Altitude Infrared Reconnaissance Sensor 高级低空红外侦察传感器
Alamogordo bomb 阿拉莫高都炸弹
alane 铝烷
alane particle 铝烷粒子
alarm 警报,警报器
alarm flag 警告旗,信号旗
alarm limit 报警限
alarm line 报警线
alarm rate 报警率
ALARR Air-Launched Air Recoverable Rocket 空中发射的空中回收火箭
ALAS Approach-and-Landing Accidents 进近与着陆事故
ALB 1. Approach Light Beacon 进近灯光灯标 2. Automatic Loop Back 自动回环
albatross 信天翁,沉重负担
albedo 反照,反照率
albedo earth sensor 反照地球传感器
albedo horizon sensor 反照地平线传感器
albedo of planet 行星反照率
albedo of the earth 地球反照率
albedo of underlying surface 下垫面反照率
albedometer 反照率计
Albers projection 阿伯斯投影
Albert 阿尔伯特
albronze 铝青铜
albumin 清蛋白,白蛋白
ALC 1. Airlift Control 空运管制 2. Alclad 包铝的 3. Automatic Level Control 自动电平控制 4. Automatic Landing Control 自动着陆控制 5. Automatic Load Control 自动负载控制 6. Automatic Loading Circuit 自动加载控制线路
ALCAC Air Lines Communications Administrative

Council 航空公司通信管理委员会(美国)
Alcaid 摇光
Alcantara Launch Center 阿尔坎塔拉发射中心(巴西)
ALCC Airlift Control Center 空运控制中心
ALCE Airlift Control Element 空运控制分队
Alclad 阿尔克拉德纯铝覆面的硬铝合金,阿尔克莱德包铝
ALCM Air Launched Cruise Missile 空中发射巡航导弹
Alco metal 阿尔科金属
Alcoa 耐蚀铝合金
alcogas 汽油酒精混合燃料
alcohol 酒精,乙醇
alcoholometry 酒精测定
Alcor 辅(北斗六的一颗五等伴星)
alcosol 醇溶胶
ALCS 1. Active Lift Control System 主动升力控制系统 2. Airborne Launch Control System 机载发射控制系统
Alcumite 阿尔库麦特铜铝合金,金色氧化膜铝合金,铜铝铁镍耐蚀合金,铝青铜
ALD 1. Arbitrary Landing Distance 任意着陆距离 2. Airworthiness Liaison Division 适航联络处
ALDCS Active Lift Distribution Control Subsystem 升力分布主动控制子系统
Aldebaran 【天】毕宿五(金牛座中的一等星)
Aldecor 阿尔迪科高强度低合金钢
aldehyde 醛
Aldis 阿尔迪斯手动信号灯(专利名称)
ALDP Automatic Language Data Processing 自动语言数据处理
Aldray 阿德雷合金
aldural 阿杜拉铝
Aldurbra 阿杜不拉合金
ALE 1. Automatic Laser Encoder 自动激光编码器 2. Automatic Link Establishment 自动链接建立,自动连接功能
ALEA Airborne Law Enforcement Association 航空法实施协会(美国)
aleatory uncertainty 偶然的不确定性
ALEMS Apollo Lunar Excursion Module Sensor 阿波罗飞船登月舱传感器
alert 1.警戒状态,战斗值班 2.空袭警报 3.空中交通管制采取行动的状态 4.制造商或发证单位对硬件发生难以接受的事故征候采取的反应
Alert 45 45分钟战备状态
Alert 5 5分钟战备状态
alert area 警戒区
alert height 警戒高
alert lead 预警提前

alert lead time 预警提前期
alert reliability 待命可靠性
alert time 待命时间
alert type 报警类型
alerting center 警戒中心
alerting service 报警服务
alert-level standard 警惕级标准
alertness 战斗准备(状态)
ALERTS Airborne Laser Equipment Real Time Surveillance 机载激光设备实时监视
alerts transmission 警报传递
ALF Auxiliary Landing Field 辅助着陆机场
Alfa 1."阿尔法"单边带无线电台 2.RUR-4A型火箭助推的深水炸弹,旧称A武器
alfer 铁铝合金
ALFS Airborne Low Frequency Sonar 机载低频声纳
Alfven Mach number 阿尔文马赫数
Alfven wave 阿尔文波
ALG 1. Aircraft Landing Gear 飞机起落架 2. Aeroleasing Group 航空租赁集团 3. Autonomous Landing Guidance 独立着陆引导
algae 海藻,水藻
algal corrosion 微生物腐蚀,藻类腐蚀
algebra 代数学
algebraic code 代数码
algebraic constraint 代数约束
algebraic control 代数控制
algebraic control method 代数控制方法
algebraic equation 代数方程
algebraic expression 代数表达式,代式
algebraic part 代数部分
algebraic relation 代数关系
algebraic stress 代数应力
algebraic stress model 代数应力模型
algebraic trim procedure 代数微调程序
Algol binary 大陵(型)双星
Algol star 大陵(型)星
Algol system 大陵(型)食双星
Algol variable 大陵(型)变星
algorism 算法
algorithm 算法
algorithm accuracy 算法精度
algorithm coef 算法的系数
algorithm design 算法设计
algorithm performance 算法的性能
ALH Advanced Light Helicopter 高级轻型直升机
alhydrogel 铝胶
ALI 1. Advance Line Information 预先航线信息 2. Airworthiness Limitation Item 适航性限制条款,适

航性限制项目 3. Asynchronous Line Interface 异步线路接口

alias　别名
aliasing　混叠,走样
aliasing distortion　折叠失真
aliasing error　折叠误差(采样率过低引起的)
alidade　视准仪,照准仪,测高仪,游标盘
alift　翱翔
alight　(水上飞机)着水,水面降落
alighting channel　(水上飞机场的)降落水道
align　1. 使……成一直线,列成一列 2. 瞄准目标 3. 对准,校准 4. 定位,定中心
aligned position　协调位置
alignment　1. 对准,校直 2. 调整,调准 3. 排列 4. 同心,同心度
alignment calibration　校准
alignment error　对准误差
alignment fixture　校准定位台,校准定位型架
alignment function　定位功能,对准功能
alignment model　对齐模型
alignment optimization for strapdown system　捷联系统最佳化对准
alignment pin　定位销
alignment pose　校准位姿
alignment precision　对准精度
alignment prism　瞄准棱镜
alignment star　校准星
alignment survey　定线测量
alignment tape　调整带
alignment time　校准时间
alignment transfer　对准传递
alignment zero instrument　对零表
Alioth　【天】玉衡(星),北斗五
aliquot part charge　等分装药
aliquot propelling charge　等分推进装药
ALIRT　Advanced Large-area Infrared Transducer 高级广域红外线变频器
Alithalite　阿里萨合金
alitizing　1. 通有电流的,加有电压的 2. 活的,活跃的
Alkaid　摇光,【天】大熊座
alkal(a)emia　碱血病
alkali　碱,碱性
alkaline　(强)碱的
alkaline battery　碱性电池
alkaline cell　碱性电池
alkaline filling　碱性装填剂
alkaline primary cell　碱性原电池
alkaline secondary battery　碱性蓄电池
alkaline zinc-air battery　碱性锌-空气电池
alkaline zinc-manganese dioxide cell　碱性锌-锰氧化物电池
alkalinity　碱度,碱性
alkalosis　碱中毒
alkaluria　碱尿症
alkane　链烷,烷烃
alkane product　烷烃产品
alkene　烯烃
alkoxyl　烷氧基
alkyd　醇酸树脂
alkyd coating　醇酸涂层
alkyd varnish　醇酸清漆
alkyl　烷基,烃基
alkyl radical　烷基自由基
alkylation　烷基化
all arms air defense weapon　多兵种防空武器
all around　多用途的,多方面的
all aspect missile　全向攻击导弹
all burnt　燃尽点
all cargo carrier　货运航空公司
all coherent moving target indicator　全相参动目标指示
all flexible plate nozzle with many hinge points　多支点全柔壁喷管
all jet　1. 全喷气发动机的 2. 全喷气式飞机的
all moving fin　全动垂尾
all moving tailplane　全动平尾
all ordnance on target　全部弹药击中目标
all peripheral jet system　全周缘喷气系统
all-phase three-axis attitude stabilization　全相三轴姿态稳定
all-points addressable　所有点可寻址的,通用的,万能的
all purpose hand-held weapon　通用轻武器
all rocket　全火箭武器的,纯火箭武器的
all round looking radar　环视雷达
all services　全空运业务
all sky photography　全天空摄影
all thruster attitude control system　全推力器姿态控制系统
all up round　1. 整装弹,整装武器 2. 准备发射
all wave receiver　全波段接收机
all way guidance　全程制导
all weather flight　1. 全天候飞行 2. 四种气象飞行
ALLA　Allied Long Lines Agency 盟国远程通信局(北约)
Allan variance　阿仑方差
all-angle launcher　全向发射装置
all-aspect attack　全向攻击
all-attitude　全姿态的
all-attitude flight control indicator　全姿态飞行控制

指示器
all-attitude inertial platform 全姿态惯性平台
all-cargo 全货运的
all-chemical injection 全化学注入
all-chemical injection system 全化学注入系统
all-clear 空袭警报解除
all-course guidance 全程制导
ALLD Airborne Laser Locator Designator 机载激光定位指示器
Allen head nut 六角螺帽
Allen wrench 弯头六角套筒扳手
all-engine failure 所有的引擎故障
Allen's metal 亚伦合金,铅青铜
alleviant 衰减装置,阻尼器
alleviate 减轻,缓和
alleviation 1.减轻,缓解,缓载 2.衰减,阻尼
alleviation lag 减载滞后
alleviation technique 阻尼(衰减)技术
alleviator 衰减装置,阻尼器
all-hypersonic 纯高超声速的
Al-Li/CFC fin 铝锂-碳纤维垂尾
alliance 联盟
Alligator "鳄鱼"(一种大型进攻性机载噪声干扰机吊舱)
alligator clip 导线夹,鱼嘴(形)夹
alligator effect 鳄皮现象
alligator-hide crack 龟裂
all-in-one-piece 组成一体的,整体式的
all-laminar 所有的层流
all-laminar design 所有层设计,整体设计
all-liquid zone 全液体区
all-movable control 全动式控制,全动操纵
allocate 分配,分派
allocated value 分配值
allocation 分配,配置,分派
allocation of resource 资源配置
allocation problem 配置问题
allocation rule 分配规则
allocator 分配器,分配算符
allotment of rounds 每门炮的弹药基数
allotrope 同素异形体
allotropism 同素异形(现象)
allottor 分配器
all-out 竭尽全力的,无保留的
allowable attack zone 可攻击区
allowable bypass 容许绕过
allowable concentration 容许浓度
allowable deficiency 容许缺陷
allowable error 容许误差

allowable gross weight 最大容重,最大飞行重量
allowable velocity 允许速度
allowance 1.公差,容差 2.备份燃油量 3.加工余量,留量 4.津贴费,补助金
allowance for cutoff 关机余量
allowance for elevation 标高修正量,场面气压修正量
allowance for wind (弹道)风修正量
allowance in direction 方向修正量
allowance of ammunition 弹药基数
alloy 1.合金 2.(金的)成色,纯度 3.熔合,合铸
alloy diode 合金二极管
alloy junction 合金结
alloy plating 合金电镀
alloy powder 合金粉末
alloy transistor 合金晶体管
alloyage 合金炼制,合金工艺
all-pass network 全通网络
all-propulsive maneuver 全推进策略
all-rocket mode 全火箭模式
all-round 1.多用途的,多方面的 2.圆形的(指视界、射界等)
all-round arm 圆周射击武器,全向射击武器
all-round field of fire (高炮)圆周射界,环形射界
all-rounder 多用途飞行器
all-shot 全部命中
all-sky camera 全天空照相机
allsolid 全固体火箭发动机的,纯固体火箭发动机的
all-solid-state 全固态的
all-solid-state radio altimeter 全固态无线电高度表
all-speed aileron 全速副翼
all-subsonic 1.纯亚声速的 2.按亚声速设计的
all-supersonic 1.纯超声速的 2.按超声速设计的
all-the-way guidance 全程制导
all-the-way homing range (导弹)全程自导引距离
all-the-way tracking range 全程跟踪距离
all-transistored 全晶体管的
all-transonic 纯跨声速的
allotrope 同素异形体(现象)
allumen 锌铝合金
all-up round 总轮
all-up weight 全重,总重量
allways 全向的,各方向都起作用的
allways fuze 起爆引信
all-weather 全天候的
all-weather helicopter 全天候直升机
all-weather homing head 全天候导引头
all-weather nav/attack aid 全天候导航/攻击设备
all-weather reentry 全天候再入
all-wing aircraft 全翼飞机,飞翼

all-wood 全木(质)的

ally 盟友,同盟国

allyl 烯丙基

allylene 丙炔

ALM 1. Aircraft Landing Minima 飞机着陆最低天气标准 2. Alarm 报警器,警报

almanac 天文年历,历书,年鉴

al-manazil 马纳吉尔(月站,阿拉伯二十七宿)

Almasil alloy 阿尔马赛合金

almasilium 阿尔马赛合金

Almaz 钻石卫星

almen gauge 喷丸强度测量仪

Alminal 阿尔米纳尔铝硅系耐蚀合金

almit 铝钎料

almost periodic function 殆周期函数

ALMS Air-Lift Management System 空运管理系统

almucantar 1. 地平纬圈,等高圈 2. 高度方位仪

ALN 1. Alignment 对准状态 2. Ammunition Lot Number 弹药批号 3. Available Line Number 现有架次号

Alneon 阿尔尼翁合金

alnico 铝镍钴磁铁,阿尔尼科合金

ALNZ The Air League of New Zealand Inc. 新西兰公司航空联合会

Alochrome (涂漆前的)表面清理

Alodine 1. 涂漆前的表面清理 2. 阿洛丁(表面清理剂)

aloft 在空中,在地球大气中

anodise 阳极电镀,作阳极化处理

anodizing 阳极电镀,作阳极化处理

along and across 航迹和偏航距离显示

along-track component 沿迹分量

along-track 沿轨道

along-track bias 沿迹偏差

along-track direction 沿迹方向

along-track drift 沿迹漂移

along-track error 沿迹误差

along-track impulse 沿轨迹冲击

along-track motion 沿轨道运动

along-track runoff 沿轨道径流

along-track separation 沿轨道分离

ALOTS Airborne Lightweight Optical Tracking System 空降轻量光学跟踪系统

Aloud "大吼"(美国陆军装在UH-1H直升机上的扬声器系统,用于心理战)

Alouette 百灵鸟卫星(加拿大电离层研究卫星)

ALP 1. Air Liaison Party 对空联络组 2. Alternative Launch Point 交替发射点 3. Ambulance Loading Post 救护飞机装载机场

ALPA Airline Pilots Association 航线驾驶员协会

alpax 铝硅合金

alpha 1. 按字母顺序的(希腊字母) 2. 机翼迎角 3. (晶体管)共基极(短路)电流放大系数

alpha bombardment α粒子轰击

alpha decay α衰变,α蜕变

alpha detector α检测器

alpha floor 迎角限制系统

alpha hinge 1. (直升机的)摆振铰 2. 交叉弹簧铰

alpha irradiation α辐照

alpha particle spectroscopy α粒子谱学

alpha ray α射线

alpha spectra α粒子谱

alpha spectrometer α谱仪

alpha tape 〈口语〉迎角指示器带状度盘,迎角刻度带

alpha(nu)meric 字母数字(的)

alphabet 字母(表),电码

alphaduct 非金属软管

alphanumeric display 字母数字显示器

alphanumeric keyboard 字母数字键盘

Alphard 【天】星宿一,长蛇座α星

alphatron α管,α粒子电离压力计

Alphecca 【天】贯宿四,北冕座α星

Alpheratz 【天】壁宿二,仙女座α星

Alphonsus 阿尔芬斯环形山

ALPO Association of Lunar and Planetary Observers 月球和行星观察者协会

ALPS Aircraft Lease Portfolio Securitisation 飞机租赁文件保密

ALR Alternate Load Reduction 交流负载降低

ALS 1. Aircraft Landing System 飞机着陆系统 2. Approach Light System 进近灯光系统 3. Automatic Landing System 自动着陆系统

ALSAM Air-Launched Surface Attack Missile 空中发射地(水)面攻击导弹

ALSCU Auxiliary Level Sensing Control Unit 辅助水平感觉控制组件

Alsos "阿尔索斯"行动

ALT 1. Alteration or Alternate 改变或备用的,交替的,轮流的;交替,替代,替换 2. Altimeter 高度表 3. Altitude 高度

ALTA Association of Local Transport Airlines 地方航空运输公司协会(美国)

Altair 【天】天鹰座α星,牛郎星

altazimuth 地平经纬仪,高度方位仪

altazimuth mounting 地平装置

altazimuth telescope 地平式望远镜

ALTDS Army Laser Target Designator System 陆军激光目标指示器系统

alter 改变,更改

alterability 可变性,可改性
alteration 1.更改,改变 2.修改,改装 3.变形
alternate 1.交替,交变,交错,轮流 2.替代的,备用的 3.代用机件,替补机件 4.备用机场
alternate aerodrome 备降机场
alternate code 交替码
alternate command post 预备指挥所
alternate configuration 多用途构型
alternate direct-drive 替代直接驱动
alternate display 交替显示
alternate forecast 备降(机)场天气预报
alternate hub airport 备用民航机场
alternate humidity test 交变潮热试验
alternate inducer 交替诱导体
alternate landing site 副着陆场
alternate route 迂回路由,备用路由
alternate routing 迂回路由选择,迂回中继
alternate runway 备用跑道
alternate static air valve 备用静压阀门
alternate stress 交变应力
alternating current electricity 交流电
alternating light 彩色闪光灯
alternating load 交变载荷
alternating no-return-to zero 交替不归零制
alternation 交变,交错,交替,轮流
alternation theorem 交错定理
alternative 1.备选方案 2.可供选择的,二者取一的
alternative approach 备用方法
alternative configuration 备用配置
alternative equipment 备选设备
alternative fuel 替代燃料
alternative landing field 备降场
alternative route 迂回路由,备用路由,可选用的路由,比较(路)线
alternative routing 寻找迂回路由
Alternative Use Committee (美国军工企业)转产委员会
alternator 交流发电机
alternator voltage 交流发电机电压
alternobaric vertigo 变压性眩晕
alterroute 转路
altichamber 高空模拟室,气压实验室
alti-electrograph 高空电位计
altigraph 高度记录器
altimeter 高度表,测高仪
altimeter-calibration standard atmosphere 高度计-校准标准大气压
altimeter fatigue 高度表疲劳
altimeter lag 高度表(时间)滞后
altimeter setting 高度表拨正

altimetric point 高程点
altimetric valve 座舱高度调节阀门
altimetry 1.测高法,测高学 2.测高
altitude 1.高度 2.地平纬度 3.高程,(天体)高低角
altitude (hypobaric) chamber 低压舱
altitude acclimatization 高原适应
altitude adaptation 高空适应
altitude alkalosis 高空碱中毒
altitude alkaluria 高空碱尿
altitude amendment 高空修正,海拔修正
altitude anoxia 高空缺氧(症)
altitude axis 高度轴,水平轴
altitude azimuth 高度方位角
altitude azimuth table 高度方位表
altitude band 高度范围
altitude bin 高度仓,海拔仓
altitude boundary 高度边界
altitude breathing 高空呼吸
altitude button 高度按钮
altitude chamber 高空模拟室,低压舱
altitude chamber drill 低压舱锻炼
altitude change 高度变化
altitude characteristics 高度特性
altitude circle 地平纬圈,等高圈
altitude clearance 高度许可
altitude clearance miscommunication 高度许可误解
altitude control unit 高空调节装置,高空调节器
altitude correlation 高度相关
altitude datum 高度基准
altitude decompression sickness 1.高空减压病 2.气体栓塞症
altitude direction 高度方向
altitude disease 高空病
altitude dysbarism 高空减压障碍,高空减压病
altitude encoding air data computer 高度编码大气数据计算机
altitude error 高度误差
altitude flight 高空飞行
altitude gain 高度增益
altitude guidance 高空制导
altitude history 高度历史
altitude hold 高度保持
altitude hole 高度空白区
altitude hypoxia 高空缺氧(症)
altitude ignition 高空点火
altitude indoctrination test 高空训练试验
altitude intercept 高空截击,高空拦截
altitude layer 高度层
altitude line 高度线

altitude loss　高度损失
altitude margin　高原边缘
altitude mixture control　高空(油气)混合比调节器
altitude mode　高度模式
altitude myopia　高空近视,空虚近视
altitude prediction　高度预测
altitude profile　垂直面飞行剖面图
altitude projection　高程投影
altitude range　高度范围
altitude rate　高度速率
altitude readout window　(仪表的)高度读出口
altitude recorder　高度记录器
altitude reservation　预留高度层
altitude rocket engine　高空火箭发动机
altitude root-mean-square　高度均方根
altitude sickness　高空病
altitude signal　高度信号
altitude simulated test　高空模拟试验
altitude simulation　高空模拟
altitude slot　高度缝
altitude status　高度现状
altitude suit　高空服
altitude switch　高空电开关
altitude tape　高度表带状度盘
altitude test　高空试验
altitude test cell　高空实验室,高空试车台
altitude test chamber　高空实验室,高空试验舱
altitude tetany　高空手足抽搐
altitude tolerance　高空耐受性
altitude trajectory　高空弹道
altitude trajectory prediction error　高空弹道预测误差
altitude trim　变高度飞行时配平
altitude tunnel　高空(模拟)风洞
altitude unit　(高炮)目标高度同步传信仪
altitude-hole effect　高度空穴效应
altitude-simulation test　模拟高空条件的(地面)试验
ALTM　Altimeter 高度表
ALTN　1. Alternate 交替,轮流;交变的,备用的 2. Alteration 修正
ALTNTR　Alternator 交流电机,振荡器
altocumulus　高积云
altonimbus　高雨云
altostrato-cumulus　高层积云
altostratus　高层云
ALTR　Alternative 备用的,交变的
ALTS　1. Altimeter Setting 高度表调定数 2. Altitude Select 高度选择
ALU　Arithmetic and Logical Unit 运算逻辑组件,运算和逻辑装置

aludur　阿留杜尔合金
ALUM　Aluminum 铝
Aluman　阿留曼合金
alumel　阿鲁麦尔合金
Alumigrip　粘铝漆
Alumilite　阿留米阳极化铝
alumina　氧化铝,矾土,铝土
alumina ceramics　高铝陶瓷,氧化铝陶瓷材料
alumina droplet　氧化铝滴
alumina particle　氧化铝颗粒
alumina pellet　片粒状氧化铝
alumina powder　氧化铝粉
alumina shell　铝壳
aluminate　铝酸盐
alumina tube　铝管
alumina-titania　氧化铝-二氧化钛
aluminide　铝化物
aluminite　矾土石,铝氧石
aluminithermic weld　铝剂焊接,铝焊
aluminium　【化】铝(化学元素,符号 Al)
aluminium alloy　铝合金
aluminium boron composite　纤维铝复合材料
aluminium casting　铝铸件
aluminium dip brazing　铝浸镀(焊)
aluminium evaporation　蒸铝
aluminium lithium alloy　铝锂合金
aluminium matrix composite　铝基复合材料
aluminium powder　铝粉
aluminium-cell arrester　铝片避雷器
aluminium-lithium alloy　铝锂合金
aluminize　渗铝,含铝的
aluminized explosive　含铝炸药
aluminum　铝(北美写法)
aluminum agglomerate　铝凝聚
aluminum agglomeration　铝圈
aluminum alloy　铝合金
aluminum behavior　铝特性
aluminum boron composite　硼铝复合材料
aluminum combustion　铝燃烧
aluminum concentration　铝浓度
aluminum content　铝含量
aluminum copper alloy　铝铜合金
aluminum cylinder　铝筒
aluminum droplet　铝液滴
aluminum electrolytic capacitor　铝电解电容器
aluminum flyer　铝制飞行器
aluminum foil　铝箔
aluminum frame　铝框
aluminum graphite composite　石墨铝复合材料

aluminum hydride 氢化铝
aluminum hydride particle 氢化铝粒子
aluminum ion 铝离子
aluminum jet 铝喷射
aluminum laminate 铝层压板
aluminum laminate boom 铝层压板构架
aluminum lithium alloy 铝锂合金
aluminum magnesium alloy 铝镁合金
aluminum mass 铝的质量
aluminum matrix composite 铝基复合材料
aluminum oxidation 铝氧化
aluminum oxide 铝氧化物
aluminum panel 铝合金面板
aluminum particle 铝粒子
aluminum plasma 铝等离子体
aluminum plasma jet 铝等离子体射流
aluminum plate 厚铝板,铝合金板
aluminum powder 铝粉末
aluminum projectile 铝弹丸
aluminum rotor 铝合金转子
aluminum sheet 薄铝板,铝片
aluminum structure 铝结构
aluminum substrate 铝基板
aluminum substructure 铝的子结构
aluminum target 铝靶
aluminum titanium alloy 铝钛合金
aluminum vapor 铝蒸气
aluminum wire 铝导线
aluminum wool 铝棉
aluminum-air cell 铝-空气电池
aluminum-boron-fibre composite 铝硼纤维复合材料
aluminum-matrix composite 铝基复合材料
aluminum-silicate fiber 硅酸铝纤维
alveolar 齿槽的,小泡的;齿槽音
alveolar air 肺泡气
alveolar carbon dioxide tension 肺泡内二氧化碳分压,肺泡内二氧化碳张力
alveolar gas exchange 肺泡气体交换
alveolar nitrogen partial pressure 肺泡氮分压
alveolar ventilation volume 1. 肺泡通气量 2. 有效通气量
alveoli 肺泡(复数)
alveolus 肺泡(单数)
ALVRJ Air-Launched Low Volume Ramjet 空中发射的低容量冲压喷气发动机
ALWT Advanced Lightweight Torpedo 高级轻型鱼雷
ALY Alloy 合金
ALY STL Alloy Steel 合金钢
AM 1. Activity Modeling 行为模型 2. Agile Manufacturing 敏捷制造 3. Alter Message 警告信息 4. Amplitude Modulation 调幅 5. Approach Mode 进近方式
AMA 1. Aerospace Medical Association 航空航天医学协会(美国) 2. Aircraft Manufacturers Association 飞机制造商协会(美国)
AMASS Airport Movement Area Safety System 机场行动区域安全系统
AMB 1. Amber 琥珀色 2. Ambient 环境的,外界的
AMBT Ambient Temperature 环境温度
AMC 1. Air Material Command 空军装备司令部 2. Avionics Maintenance Conference 航空电子维修联合会
AMCS Automated Material Control System 自动材料控制系统
AMD Aircraft Maintenance Department 航空器维修部,飞机维修部
AMEND Amendment 修正
AMER STD American Standard 美国标准
AMF Aircraft (Airplane) Master File 飞机主文件
AMH Automated Material Handling 自动材料搬运
AMI Airline Modifiable Information 航线可变更信息
AML Aircraft (Airplane) Modules List 飞机模块表
AMLCD Active Matrix Liquid Crystal Display 有源矩阵液晶显示
AMM 1. Aircraft (Airplane) Maintenance Manual 飞机维修手册 2. Ammeter 安培表 3. Antimissile Missile 反导弹的导弹
AMMO Ammunition 军火
AMMORK Ammunition Rack 弹药架
AMO 1. Advance Material Order 先行材料订单 2. Aircraft (Airplane) Model Option 飞机选型
AMP 1. Airline Maintenance Program 航空公司维修大纲 2. Amperage 安培,电流强度 3. Analytical Maintenance Program 分解维修程序 4. Audio Management Panel 音频管理板
AMPL Amplifier 放大器
AMPTD Amplitude 幅度
AMR 1. Armed 准备 2. Automatic/Manual/Remote 自动/人工/远距离控制(操作)
AMRE Aircraft (Airplane) Maintenance and Repair Engineering 飞机维护和修理工程
AMRN Advance Modification Revision Notice 先行改型修正通知
AMS 1. Aeronautical Material Specification 航空材料规范 2. Air Management System 空气管理系统 3. Automated Master Schedule 自动总计划 4. Avionics Management Service 航空电子管理服务
AMSL Approved Material Substitution List 批准的材

料代用清单

AMSS Aeronautical Mobile Satellite Service 航空运动卫星服务

AMST Advance Military STOL Transport 先进军用短距起落运输机

AMT 1. Amount 总数，合计 2. Armature 电枢，衔铁

AMTOSS Aircraft Maintenance Task Oriented Support System 飞机维修任务处理支援系统，飞机维修任务支持系统

AMTS Aeronautical Message Transfer Service 航空信息传输服务

AMU Audio Management Unit 音频管理装置

AMUX Audio Multiplexer 音频多路调制器

AN 1. Airforce Navy 海军航空队 2. Aids to Navigation 导航设施，导航设备 3. Alphanumeric 字母数字 4. Area Navigation 区域导航

ANA Airforce Navy Aeronautical 空军-海军航空

ANAF Army Navy Air Force 陆-海-空军

ANAL Analysis 分析

ANC 1. Air Navigation Commission 航空导航委员会 2. Airforce Navy Civil 空军-海军-民用

ANCRS Automated Numerical Control Report System 自动数字控制报告系统

AND 1. Aircraft（Airplane）Nose Down 飞机低机头 2. Airforce-Navy Design 空军-海军飞机设计

ANDZ Anodize 阳极化

angle operating range 角操作范围

ANL Annul 退火

ANLOR Angle Order 方位角指令

ANLR Angular 角形的，角度的

ANLT Anchor Light 停泊灯

ANO Alphanumeric Output 字母数字输出

ANP 1. Actual Navigation Performance 实际导航性能 2. Area Navigation Performance 区域导航性能

ANS 1. Ambient Noise Sensor 环境噪声传感器 2. Answer 回答 3. Area Navigation System 区域导航系统

ANSI American National Standard Institute 美国国家标准协会，美国国家标准研究所

ANT Antenna 天线

ANTC 1. Advanced Networking Test Center 先进网络试验中心 2. Air Navigation Technical Committee 航空导航技术委员会

ANU Aircraft（Airplane）Nose Up 飞机抬机头

AO 1. Airworthiness Office 适航办公室 2. Assembly Order 装配大纲（指令），装配指令 3. Assembly Outline 装配大纲 4. Available Option 可选项

AO/OSP Assembly Order/Operational Sequence Planning 装配大纲/操作顺序计划

AOA 1. American Ordnance Association 美国武器协会 2. Angle Of Attack 迎角

AOC 1. Aeronautical Operational Control 航空运行控制 2. Aircraft Operational Control 飞机运行控制 3. Airline Operational Control 航空公司运行控制 4. Airport Obstruction Chart 机场障碍物图 5. Airport Operational Communications 机场运行通信 6. Automatic Overload Control 过载自动控制 7. Aviation Operator Certificate 航空承运人合格证

AOCC Airline Operation Control Center 航空公司营运控制中心

AOCD Available Option Collector Drawing 可选项图

AODT Available Option Drawing Tree 可选项图纸树

AOG Aircraft（Airplane）On Ground 飞机停飞，地面停放飞机

AOGS Aircraft（Airplane）On Ground Support 飞机地面支援

AOHE Air/Oil Heat Exchanger 空气/滑油热交换器

AOI 1. Angle Of Incidence 安装角 2. Available Option Index 可选项目录

AOIL Aviation Oil 航空滑油

AOM Aircraft Operating Manual 飞机使用手册

AOP Airline Operational Procedure 航空公司运营程序

AOPA Aircraft Owners and Pilots Association 飞机拥有人与驾驶员协会

AORN Assembly Outline Requirement Notification 装配大纲需求通知单

AO-SRS Assembly Outline-Ships Record System 装配大纲-架次记录系统

AOT 1. All Operator Telex 所有用户电传 2. Assembly Outline Tooling 工装装配大纲

AOTCR Assembly Outline Tracking Control Record 装配大纲跟踪控制记录

AP 1. Absolute Pressure 绝对压力 2. Access Panel 口盖，舱盖 3. Aircraft Productivity 飞机生产率 4. Airplane 飞机 5. Airport 机场 6. Airport Location 机场位置 7. Airworthiness Procedure 适航管理程序 8. All Passenger 全客 9. Ambient Pressure 环境压力（地面气压） 10. Assignment Process 任务过程 11. Atmospheric Pressure 大气压力 12. Autopilot 自动驾驶

AP STEER Autopilot Steering 自动驾驶仪操纵

AP/FD Autopilot/Flight Director 自动驾驶仪/飞行指引仪

APA 1. American Pilots' Association 美国飞机驾驶员协会 2. Autopilot Amplifier 自动驾驶仪放大器

APARS Automated Pilot Advisory System 自动驾驶咨询系统

APB Auxiliary Power Breaker 辅助电源断路器

APC 1. Advanced Planning Coordination 先行计划协调 2. Aeronautical Passenger Communication 航空乘客通信 3. Aeronautical Public Correspondence 航空公共信函 4. Automatic Phase Control 自动相位控制 5. Automatic Program Control 自动程序控制 6. Autopilot Computer 自动驾驶仪计算机

APD Approach Progress Display 进近过程显示(器)

APDML Airplane Product Data Markup Language 飞机产品数据标记语言

APE Aircraft (Airplane) Program Engineering 飞机项目工程

APEO Advance Process Engineering Order 先期工艺工程指令

APERT Aperture 孔,快门

APFO Automated Planning Fabrication Outline 自动(计算机)编制的制造大纲

API 1. Air Position Indicator 空中位置指示器 2. Aircraft Programming Implement 飞机计划执行 3. Aircraft (Airplane) Programming Implementation 飞机项目实施 4. Application Programming Interface 应用项目界面

APICS Automated Production Inventory Control System 生产与库存自动控制系统

APL 1. Airplane 飞机 2. Automated Parts List 自动零件表

APLGS Automated Parts List Generator System 自动零件表生成器系统

APLQ Applique 缝饰的,镶饰的

APMS Automated Performance Measurement System 自动性能测量系统

APO Automated Planning Order 自动计划指令

APP 1. Appearance 外表,外观 2. Autopilot Panel 自动驾驶仪板

APPL Application 应用

APPM Assembly Process Plan Master 装配工艺总纲

APPR Approach 进近

APPROC Approach On Course 在进近航道

APPROX Approximate 大约

APPV Approve 批准

APR 1. Actual Performance Reserve 实际性能余量 2. Automated Parts Release 自动化零件发放 3. Automatic Power Reserve 自动推力储备

APRCH Approach 进近,探讨

APRL ATN (Aeronautical Telecommunication Network) Profile Requirement List 航空电信分部要求清单

APRT Airport 机场,航空港

APS Automatic Pilot System 自动驾驶仪系统

APTO Automated Planning Tool Order 计算及编制的工装指令

APU Auxiliary Power Unit 辅助动力装置

APUC Auxiliary Power Unit Controller 辅助动力装置控制器

APUCKV Auxiliary Power Unit Check Valve 辅助动力装置检查阀

AQAB Air Quality Advisory Board 航空质量咨询部,航空质量咨询委员会

AQC Airman's Qualification Card 驾驶员合格证

AQD Aeronautical Quality Directorate 航空质量管理局

AQE Aeronautical Quality Engineer 航空质量工程师

AQF Avionics Qualification Facility 航空电子系统合格鉴定设施

AQL Acceptance Quality Level 可接受质量标准

AQLS Acceptance Quality Levels 可接受质量标准

AQP 1. Advanced Qualification Program 先进合格鉴定程序 2. Avionics Qualification Procedure 航空电子系统合格鉴定程序

AQS 1. Advanced Quality System 先进质量体系 2. Airworthiness Qualification Specification 适航合格规范

AQT Acceptable Quality System 可接受质量体系

AR 1. Acceptance Review 验收评审 2. Accidental Report 失事报告,事故报告 3. Accountability Record 责任记录 4. Accounts Receivable 可接受理由 5. As Required 按需

ARA American Radio Association 美国无线电协会

ARAC Aviation Rulemaking Advisory Committee 航空规则制订委员会

ARALL Aramid Aluminium Laminate 基于芳纶纤维层板

ARC 1. Automatic Range Control 自动距离控制 2. Automatic Remote Control 自动遥控

ARC/AAT Authorized Release Certificate/Airworthiness Approval Tag 批准放行证书/适航批准标签

ARCCOS Inverse Cosine 反余弦

ARCCOT Inverse Cotangent 反余切

ARCCSC Inverse Cosecant 反余割

ARCSEC Inverse Secant 反正割

ARCSIN Inverse Sine 反正弦

ARCTAN Inverse Tangent 反正切

ARCW Arc Weld 电弧

ARF Airline Risk Factor 航线风险因素

ARG Arresting Gear 制动,着陆阻拦装置

Argand diagram 阿尔干图

Argensat 阿根廷卫星

argentation 镀银

ARGMA Army Rocket and Guided Missile Agency 美

国陆军火箭导弹局

ARGO Advanced Research Geophysical Observatory 先进地球物理研究观测台

argon 氩

argon arc 氩弧

argon arc welding 氩弧焊

argon atmosphere 氩氛,氩保护气氛

argon dilution 氩气稀释

argon ion laser 氩离子激光器

argon laser 氩激光器

argon plasma 氩等离子体

argon shielded arc welding-pulsed 脉冲氩弧焊

argon tungsten pulsed arc welding 钨极脉冲氩弧焊

Argos 阿戈斯系统,国家天基导航系统(法国)

ARGOS Automatic Relay Global Observation System 全球观测自动中继系统

ARGS Anti-Radiation Guidance Sensor 反辐射(导弹)制导传感器

Argument "辩护"作战

argument 1.(复数的)幅角,幅度 2.(函数的)自变量,自变数 3.争论,辩论 4.角距

argument of latitude 纬度

argument of periapsis 近心点角

argument of pericenter 近心点幅角

argument of perigee 近地点角距,近地点幅角

argument of perihelion 近日点幅角

argument of perigee of satellite orbit 卫星轨道近地点幅角

Argus 百眼巨人卫星(美电子侦察卫星)

ARH 1. Active Radar Homing 主动雷达自导引 2. Advanced Reconnaissance Helicopter 先进的侦察直升机 3. Anti-Radar Homing 反雷达自导引

ARI 1. Airborne Radio Installation 机载无线电装置 2. Aileron/Rudder Interconnect 副翼方向舵交连

ARIA 1. Advanced Range Instrumentation Aircraft 先进的靶区测量飞机 2. Apollo Range Instrumentation Aircraft 阿波罗登月计划中的靶区测量飞机线

Ariane 阿里安火箭(欧洲空间局)

Arianespace 阿里安航天公司

aridextor 横向操纵输出机构,产生侧/横向力的操纵机构

Ariel 1.天卫一 2.羚羊卫星 3."阿里尔"激光跟踪器

Aries 白羊(星)座,白羊星探空火箭(美国)

ARIES 1. Airborne Reconnaissance Integrated Electronic System 空中侦察综合电子系统 2. Applied Resource Image Exploitation System 应用资源图像开发系统

ARINC Aeronautical Radio Inc. 航空无线电公司

ARIP 1. Air Refueling Initial Point 空中加油起始点 2. Automatic Rocket Impact Predictor 火箭弹着点自动预测装置

ARIS 1. Airborne Range Instrumentation System 机载区测量系统 2. Advanced Range Instrumentation Ship 先进遥测船 3. Atomic Reactors In Space 空间原子反应堆

arise 上升,出现,发生

ARISE Agricultural Resources Inventory Survey Experiment 农作物产量估测实验(印度)

arithmetic average(of measurement) (测量的)算术平均值

arithmetic code 算术码

arithmetic mean 算术平均值

arithmetic trap enable 允许算术俘获(位)

arithmetic trap mask 运算俘获(自陷)屏蔽

arithmetic unit 运算器,运算部件

arithmetic 算术,计算

arithmograph 运算图

Aritieren (钢铁表面的)渗铝法

ARJ 21 Advanced Regional Jet for the 21th Century 面向21世纪的先进涡扇支线飞机

ARJS Airborne Radar Jamming System 机载雷达干扰系统

ARL 1. Acceptable Reliability Level 可接受可靠性标准 2. Aeronautical Research Laboratories 航空研究实验室(澳大利亚) 3. Aerospace Research Laboratory 航空航天研究实验室(美国空军) 4. Aircraft (Airplane) Readiness Log 飞机状态日志 5. Astronautical Research Laboratory 航天研究实验室

ARM 1. Aircraft (Airplane) Recovery Manual 飞机恢复手册 2. Anti-Radar Missile 反雷达导弹 3. Anti-Radiation Missile 反辐射导弹 4. Armature 电枢,衔铁 5. Availability, Reliability and Maintainability 可用性、可靠性和可维护性

arm 1.臂,柄,杆 2.指针(仪表的) 3.桨叶(螺旋桨的) 4.(电线的)线担,支架 5.(电路的)支路 6.解除保险(准备发射) 7.兵种 8.力臂 9.飞机或导弹基准线至某一部件的水平距离

arm-and-hand-signal 手势信号

arm command 解除保险指令

arm elevator 旋臂升降机

arm population 臂族

arm restraint (弹射座椅上的)臂部限动器(皮带等)

arm time 1.成待发状态时间,解除保险时间 2.(导弹)发射准备时间

ARMCM Anti-Radiation Missile Countermeasures 反辐射导弹的措施,反辐射导弹对抗

armada 1.机群 2.舰队

armament 1.(复)军事装备 2.军械,武器 3.战斗部,弹

头 4.军队,武装力量 5.引爆系统
armament deactivation 军械保险
armament firing button 武器射击按钮
armament harmonization 军械校靶,武器校靶
armament panel 军械控制面板
armament station 武器悬挂位置,武器配置位置
armament trolley 武器运输(小)车
armature 1.电枢,转子 2.衔铁 3.(电缆的)铠装
armature control 电枢控制
armature reaction 电枢反作用力
ARMC Area Regional Maintenance Center 区域支线维修中心
Armco 1.阿姆科(含低碳软铁) 2.波纹金属板,波纹白铁管
Armco iron 工业纯铁,阿姆可铁
ARMDAS Army Military Damage Assessment System 陆军作战破坏程度评估系统
armed 待发射的,解除保险的,已装战斗部的
armed ammunition 打开保险的弹药
armed condition 待爆状态
armed fuze 待发引信
armed mine 待发(地、水)雷
armed period 引信待发时间,引信作用时间
armed-safe condition 爆炸保险机构状态
armed-safe mechanism 爆炸和保险机构,爆炸和不爆炸机构
armillary sphere 浑天仪,璇玑
arming 1.解除保险 2.成待发状态,备炸状态 3.战斗准备,(火箭)准备发射 4.装弹,装填(弹药),装引信
arming and fuzing system 解除保险和引爆系统
arming barrier (战斗部的)保险装置
arming circuit (引信)解除保险电路
arming command 待爆指令
arming delay 解除保险延迟时间,解除保险延迟距离
arming firing device 保险点火装置
arming train assembly 解除保险装置
arming unit 爆控机构
arming vane (炸弹)引信旋翼
arming vane stop (炸弹)引信旋翼制动器
arming-wire guide (炸弹引信)保险导丝
arming-wire retainer 保炸钩
arming-vane type fuze 旋翼保险式引信
ARMMS Automatic Reliability and Maintenance Management System 自动可靠性和维护管理系统
armor gas 装甲气
armo(u)r 1.装甲,铠装 2.防弹钢板,铠装套,防弹服 3.装甲兵,装甲部队
armo(u)red 装甲的,防弹的
armo(u)ry 1.军械库 2.兵工厂 3.整套武器

armor defeating ammunition 穿甲弹(药)
armor penetration 穿甲
armor piercer 穿甲弹
armor-piercing bullet 穿甲弹
armor-piercing discarding sabot 脱壳穿甲弹
armor-piercing discarding sabot shell 硬心穿甲(炮)弹
armor-piercing effect 穿甲效力
armor-piercing incendiary shell 穿甲燃烧(炮)弹
armor-piercing ratio (弹丸)穿甲率
armor-piercing ratio of fragment 破片穿甲率
armor-piercing warhead 穿甲战斗部
armor-defeating warhead 穿甲战斗部
armored flight deck 装甲飞行甲板
armored wire 铠装电缆,铠装导线
armor-penetrating ability 穿甲能力
armor-piercing ability 穿甲能力
armor-piercing ammunition 穿甲弹(药)
armor-piercing capability 穿甲能力
armor-piercing cartrige 穿甲弹
armor-piercing flame tracer bullet 曳光穿甲弹
armor-piercing high explosive shell 穿甲杀伤爆破(炮)弹
armor-piercing incendiary tracer 穿甲燃烧曳光弹
armor-piercing mine 穿甲(地)雷
armor-piercing shot 穿甲(炮)弹
armor-piercing tracer 穿甲曳光弹
armor-piercing weapon 穿甲武器
armourer 军械士
armrest (座椅的)扶手,肘架
ARMS 1. Aerial Radiological Measuring System 空中放射性测量系统 2. Automated Requirements Management System 自动化要求管理系统
arms 1.武器,军械,军备 2.兵种 3.兵力,军事
arms of precision 精密武器
arms rack 武器架
arm-safe order 解保-保险指令
ARMT Armament(武器)军械
ARMTS Advanced Radar Maintenance Training Set 先进的雷达维护训练设备
army 1.陆军 2.集团军 3.军队
Army Air Corps 陆军航空队(美国)
Army Air Forces 陆军航空兵(美国)
Army Air Service 陆军航空勤务队(美国)
Army airspace 陆军空域
Army SCS Army Satellite Communication System 陆军卫星通信系统
Arnold Engineering Development Center 美国阿诺德工程发展中心
ARO 1. Airport Reservation Office 机场订票处

2. Aspheric Reflective Optics 非球面反射光学
AROC Air Rescue Operation Center 空军救援活动中心,航空救援工作中心
AROD 1. Advanced Range and Orbit Determination 先进的航程与轨道测定系统 2. Aerodrome Runway and Obstruction Data 机场跑道和净空数据 3. Airborne Remotely Operated Device 机载遥控飞行器（一种小型的无人机）
AROG Automatic Roll-Out Guidance 自动着陆滑跑制导
AROM 1. Alterable Read Only Memory 交替只读存储器 2. Aromatic 芳香族的,芳香剂,芳香族燃油（包括汽油、煤油）
aromatic 芳香的,芳香族
aromatic compound 芳香族化合物
aromatic content 芳烃含量
around-the-clock operation 昼夜不停的作战活动
arousal 觉醒
arousal response 兴奋反应
ARP 1. Advanced Reentry Program 先进再入研究计划 2. Aerodrome Reference Point 机场参考点 3. Aero-Rifle Platoon 空中步枪排 4. Aerospace Recommended Practice 航空航天推荐标准,宇航推荐实施标准 5. Air Raid Precautions 预先空袭警报 6. Airborne Radar Platform 机载雷达平台 7. Aircraft Revenue Productivity 飞机实际生产率 8. Aluminum-Reinforced Polyimide 铝加强聚酰亚胺 9. Attack Reference Point 攻击参照点 10. Automatic Reserve Performance 自动备用性能
ARPA Advanced Research Projects Agency 远景研究计划局（美国）
ARPANET Advanced Research Projects Agency Net 远景研究计划局计算机网,阿帕网
ARPAS Automated Resource Planning and Analysis System 资源自动化计划与分析系统
ARPC Air Reserve Personnel Center 空军后备队人事中心（美国空军）
ARPS 1. Advanced Radar Processing System 先进的雷达处理系统 2. Aerospace Research Pilot School 航空航天驾驶员学校（美国空军）
ARPTT Air-Refueling Part-Task Trainer 空中加油部分工作练习器（美国空军）
ARR 1. Airborne Radio Relay 空中转播无线电中继通信,空中无线电中继 2. Air-Refueling Receiver 空中加油受油机 3. Automatic Repeat Request 自动重复请求装置 4. Anti-repeat Relay 反中继继电器 5. Arrival 到达
arrange 1.整理,分类,排列,布局 2.筹备,安排 3.准备,计划

arrangement 1.布局,安排,配置,布置,排列 2.装置,设备,结构 3.协议,议定书
arrangement of curve 曲线配置,曲线测设
array 1.阵列,阵,排列,布置 2.级数,数组
array and vector computer 阵列向量计算机
array antenna 1.阵列天线 2.天线阵
array circuit 阵列电路
array coupon 列阵试样
array design 阵列的设计
array element 阵单元
array factor 阵因子,排列系数
array grammar 阵列文法
array length 数组的长度
array output 阵列输出
array panel 阵列面板
array position 数组中的位置
array power 数组的乘方
array processor 数组处理程序,阵列处理机
array radar 阵列天线雷达
array retraction 阵列收缩
array spacing 阵列间距
array spin 阵列的旋转
array spin rate 阵列的旋转速度
array string 数组的字符串
array string voltage 阵列串电压
array surface 阵列的表面
array temperature 阵列的温度
array wing 阵列机翼
ARRC Aerospace Rescue and Recovery Center 航空航天救援回收中心
ARREC Accident Report Recommendations 事故报告介绍
ARRES Automatic Radar Reconnaissance Exploitation System 自动雷达侦察情报开发系统
arrest 1.阻止,制动,停止 2.延迟 3.捕获 4.止动器
arrest of unstart 制动急停
arrested landing 拦阻着舰（航母）
arrested-propeller system 螺旋桨限制系统,螺旋桨锁定系统
arrester 1.着陆拦阻装置 2.止动器,限动器 3.避雷器,放电器
arresting barrier 拦阻网
arresting cable 拦阻索
arresting gear crew （航母上）拦阻装置人员
arresting hook 拦阻钩
arresting mechanism 拦阻装置
arresting unit 拦阻器材能量吸收装置
arrestment 1.用着陆拦阻装置减速 2.机场着陆拦阻装置

ARRGP Aerospace Rescue and Recovery Group 航空航天救援和回收大队(美国空军)

arrhythmia 心律不齐

arrival 1.到达 2.(空中交通的)入境飞机

arrival aircraft 到达的飞机

arrival angle 到达角

arrival date 到达日期

arrival direction 抵达方向

arrival error 到达错误

arrival order 到达顺序

arrival period 到达时间

arrival process 到达过程

arrival rate 到达率

arrival route 到达路线

arrival runway 到达飞机专用跑道

arrival rush 到达高峰

arrival time 到达时间

arrival time difference 到达时间差异

arrival time uncertainty 不确定性的到达时间

arrival traffic 到达的流量

arrival velocity 到达的速度

arrival-time control 按时到达时间控制

ARROW Aircraft Routing Right Of Way 飞机航线上靠右边飞行

arrow 1.箭 2.箭头标志,指针 3.箭头队形

arrow diagram 矢量图

arrow plot 箭头图

arrow wing 箭形翼,大后掠(角)翼

arrowhead 1.箭头牌(指示降落方向) 2.楔队,箭头队形 3.箭头形纪念章(美国空降兵作战纪念章) 4.箭头

arrowhead method 运动线法

arrowheaded 箭形的,后掠的

ARRS 1. Advanced Rescue and Recovery System 先进的救援与回收系统 2. Aerospace Rescue and Recovery Service 航空航天救援和回收勤务处(美国空军)

ARRSC Asian Regional Remote Sensing Center 亚洲地区遥感中心

ARS 1. Action Request System 行动需求系统 2. Active Radar Seeker 主动式雷达导引头 3. Active Relay Station 有源中继传输站 4. Active Repeater Satellite 有源转发卫星,有源中继卫星 5. Advanced Reconnaissance Satellite 先进侦察卫星 6. Advanced Reentry System 先进再入系统 7. Advanced Rescue System 先进救援系统 8. Aerospace Research Satellite 航天研究卫星 9. Attitude Reference System 姿态参考系统 10. Auto-Relight System 自动重新点火系统 11. Automatic Reporting System 自动报知系统 12. Attack Radar Set 攻击雷达设备

ARS212 苏联212毫米空对地火箭

ARSC African Remote Sensing Council 非洲遥感委员会

arsenal 兵工厂,军械库,武器库

arsenic 砷

arsenide 砷化物,砷与金属的化合物

ARSP Aerospace Research Support Program 航空航天研究保障计划(美国)

ARSR 1. Air Route Surveillance Radar 航路监视雷达 2. Arrester 放电器,避雷器

ARSS 1. Airborne Rapid Scan Spectrometer 机载快速扫描光谱仪 2. Airborne Remote Sensing System 机载遥感系统

ARSTC Asian Remote Sensing Training Center 亚洲遥感培训中心

ART 1. Action Result Table 行动结果表 2. Airborne Radar Technician 机载雷达技师 3. Airborne Radiation Thermometer 机载辐射式温度计 4. Algebraic Reconstruction Techniques 代数再现技术 5. Auto Reserve Thrust 自动反推力 6. Automatic Reserve Thrust 自动推力储备 7. Automatic Range Tracker 自动距离跟踪器

ARTAC Advanced Reconnaissance Target Acquisition Capabilities 先进的目标侦察截获能力

Artads Army tactical data system 陆军战术数据系统

ARTCC Air Route Traffic Control Center 航路空中交通管制中心

ARTCLD Articulated 铰接的

ARTCS Advanced Radar Traffic Control System 先进的雷达交通管制系统

ARTEMIS Automated Reporting, Tracking and Evaluation Management Information System 自动报告、跟踪、鉴定管理信息系统

ArTEP Ariane Technology Experiment Platform 阿里安技术实验平台(西欧)

arterial hypoxia 动脉缺氧

ARTI Advanced Rotorcraft Technology Integration 先进旋翼机技术综合

article 物品,条款

article assembly 产品组装

articulated 接合的,铰接的,有关节的,摇臂式的

articulated aging 人工时效

articulated aurora 人造极光

articulated blade (直升机旋翼的)铰接桨叶

articulated celestial body 人造天体

articulated compressibility method 人工压缩法

articulated density method 人工密度法

articulated earth satellite 人造地球卫星

articulated electron belts 人工电子带

articulated robot 关节型机器人

articulated rod   副连杆,活节连杆
articulated rotor   铰接式旋翼
articulated structure   多关节结构
articulating seat   活节座椅
articulation   1. 铰接,关节 2. 清晰度
artifact   人工制品
artificer   1. 技术兵 2. 技工,工匠设计者,发明家
artificial aging   人工老化,人工时效
artificial air   人工气
artificial air cloud   人工气云
artificial airglow   人造气辉
artificial asteroid   人造小行星
artificial aurora   人造极光
artificial barium clouds   人造钡云
artificial belt   人造(辐射)带
artificial celestial body   人造天体
artificial cognition   人工识别
artificial comet   人造彗星
artificial dielectric   仿真电介质
artificial dissipation model   人工耗散模型
artificial disturbance   人为干扰
artificial electron aurora   人造电子极光
artificial feel   1. 人工感力,模拟感觉 2. 载荷感觉器,人工感力器
artificial feel system   1. 人感系统 2. 载荷感觉机构
artificial gas environment   人工气体环境
artificial gravity   人造重力,模拟重力
artificial horizon   1.（航空地平仪上的）地平线 2. 航空地平仪 3. 模拟地平线 4. 人工地平
artificial intelligence   人工智能
artificial language   人工语言
artificial line   仿真线
artificial magnetization method   人工磁化法
artificial network   仿真网络
artificial perception   人工识别
artificial planet   人造行星
artificial plasma clouds   人造等离子体云
artificial radiation   人工辐射
artificial radiation belt   人造辐射带
artificial respiration   人工呼吸
artificial satellite   人造卫星
artificial sodium cloud   人造钠云
artificial transition   人工转换
artificial viscosity   1. 人工黏性 2. 人工耗散
artificial wave   人工波
artillery   1. 炮兵 2. 火炮
artillery location radar   炮位侦察雷达
artillery reconnaissance and fire-directing radar   炮兵侦察校射雷达

artist   能手
Artois   "阿图瓦"（法国电扫描多目标导弹跟踪雷达）
ARTS   1. Automated Radar Terminal System 自动雷达终端系统 2. Automated Radar Tracking System 自动雷达跟踪系统 3. Automated Remote Tracking Station 自动遥控跟踪站
ARTU   Automatic Range Tracking Unit 自动距离跟踪装置
artwork   原图
ARU   1. Address Recognition Unit 地址识别装置 2. Altitude Rate Unit 高度变化率组件 3. Astrophysics Research Unit 天文物理研究所（英国）4. Attitude Reference Unit 姿态基准装置 5. Attitude Retention Unit 姿态保持装置 6. Automatic Range Unit 自动跟踪器 7. Auxiliary Readout Unit 辅助数字显示装置
ARV   Air Recreational Vehicle 周末旅游飞机
ary   氩
Aryabhata   阿耶波多卫星
A-S   Ampere-Second 安(培)-秒
AS   1. Adapter Section 过渡段,适配段 2. Aerospace Standard 航空航天标准（美国）3. Air Station 航空站 4. Air Start 空气起动 5. Air Superiority 空中优势 6. Aircraft (Airplane) Services 飞机服务 7. Airplane Support 航线支援 8. Airspeed 空速 9. Airworthiness Standard 适航标准 10. Alpha Speed 阿尔法速度 11. Alto-Stratus 高层云 12. Anti-Skid（也写作 A/S）防拖胎的,防滑的 13. Antisubmarine 反潜的 14. Artificial Satellite 人造卫星 15. Astronomy 天文学 16. Anti-Spoofing 反电子欺骗
ASAS   1. Active Scattering Aerosol Spectrometer 有源散射气溶胶粒谱议 2. All Sources Analysis System 全源分析系统
AS/RA   Assembly Sequence and Resource Allocation 装配程序和资源分配
AS/ALT   Airspeed/Altimeter 空速/高度表
AS/RS   Automated Storage/Retrieval System 自动存储/提取系统
ASA   1. Advanced Surveillance Aircraft 先进侦察飞机 2. Air Services Agreement 航空勤务协定 3. Aircraft Separation Assurance 飞机分离保障 4. Anti-Static Additive 抗静电添加剂 5. American Standards Association 美国标准协会 6. Army Security Agency 陆军保密局（美国）7. Astronomical Society of Australia 澳大利亚天文学会 8. Austrian Space Agency 奥地利航天局 9. Autoland Status Annunciator 自动着陆状态显示器
ASA speed   美国标准协会感光度
ASAC   1. Airborne Surveillance Airborne Control 空中监视,空中控制 2. Anti-Submarine Air Controller 反潜

空中控制员

ASAE American Society for Aerospace Education 美国航空航天教育学会

ASAIB Aerospace Safety Accident Investigation Board 航空航天安全事故调查委员会

ASALM Advanced Strategic Air-Launched Missile 先进的战略空射导弹

ASAP 1. Advanced Survival Avionics Program 先进高生存力航空电子设备计划（美国空军） 2. Aerospace Safety Advisory Panel 航空航天安全咨询委员会（美国航空航天局） 3. American Society of Aerospace Pilots 美国航空航天驾驶员协会 4. Ariane Structure for Auxiliary Payloads 阿里安多星附属有效载荷发射结构

ASAR As-Supported Aircraft (Airplane) Record 支持飞机记录

ASARG 1. Advanced Synthetic Aperture Radar Guidance 先进合成孔径雷达制导 2. Autonomous Synthetic Aperture Radar Guidance 自主式合成孔径雷达制导

ASARS 1. Advanced Synthetic-Aperture Radar System 先进的合成孔径雷达系统 2. Airborne Search And Rescue System 机载搜索救援系统

ASAT 1. Anti-Satellite System 反卫星系统 2. Anti-Satellite 反卫星（计划）

ASB 1. Accident Service Bulletin 紧急服务通报 2. Alert Service Bulletin 警戒勤务通报 3. Air Safety Board 空中安全委员会（英国） 4. Asbestos 石棉

asbestos 石棉

asbestos fuel cell 石棉燃料电池

ASBM Air-to-Surface Ballistic Missile 空（对）地弹道导弹

ASBR Aircraft (Airplane) Specific Build Record 飞机特定建造记录

ASBU Arab States Broadcast Union 阿拉伯国家广播联盟

ASC 1. Advanced Simulation Center 先进模拟中心 2. Aircraft System Controller 飞机系统控制员 3. Air Support Command 空军保障司令部（英国） 4. American Satellite Company 美国卫星公司 5. American Society for Cybernetics 美国控制论学会 6. Ascend 上升 7. Aviation Statistic Center 航空统计中心（加拿大） 8. Automatic Sensitivity Control 自动灵敏度控制

ASCAT Apollo Simulation Check-out and Training 阿波罗飞船模拟检测和训练

ASCB Avionics Synchronized Control Bus 航空电子同步控制总线

ASCC 1. Aeronautical Satellite Communications Center 航空卫星通信中心 2. Aeronautical Satellite Communications Criteria 航空卫星通信规范 3. Air Standards Coordinating Committee 航空标准协调委员会

ascend 1. 上升，爬高 2. 进入（轨道）

ascendant 上升的，爬高的

ascending line of node 升交点线

ascending node 升交点

ascending-node time 升交点时间

ascending-node control 升交节点控制

ascension 1. 上升，爬高 2. 进入（轨道） 3.【天】赤经

ascent 上升

ascent branch 上升支

ascent flight 上升段飞行

ascent load 提升载荷

ascent load case 提升负载的情况下

ascent problem 上升的问题

ascent subproblem 提升子问题

ascent trajectory 上升轨道

ascent vehicle 提升载具

ascertain 判断，确定，调查，弄清

ASCES Anti-Submarine Contact Evaluation System 反潜探测数据鉴定系统

ASCG Automatic Solution Crystal Growth 自动溶液晶体生长

ASCII American Standard Code for Information Interchange 阿士克码，美国信息交换标准代码

ASCM Anti-Ship Capable Missile 反舰导弹

ASCO Arab Satellite Communication Organization 阿拉伯卫星通信组织

ASCPC Air Supply and Cabin Pressure Controllers 供气和座舱压力控制器

ASCS Automatic Stabilization and Control System 自动稳定和控制系统

ASCTs Aircraft (Airplane) Specific Configuration Tables 飞机特定构型表

ASCU Armament Station Control Unit 武器（悬挂）位置控制装置

ASD 1. Advanced Submarine Detection 先进的潜艇探测设备 2. Aeronautical System Division 航空武器系统研究部（美国空军） 3. Aircraft Situation Display 飞机位置显示 4. Assembly Start Date 装配开始日期

ASDAR Aircraft-to-Satellite Data Relay 飞机-卫星数据中继

ASDC Aeronomy and Space Data Center 高层大气物理和空间数据中心

ASDE Airport Surface Detection Equipment 机场场面探测设备

ASDF Air Self-Defense Force 航空自卫队（日本）

asdic 声纳,超声波水下探测器

ASDIC Anti-Submarine Detection Investigation Committee 反潜探测设备研究委员会(英国)

ASDL 1. Aeronautical Satellite Data Link 航空导航卫星数据链 2. Airborne Self-Defense Laser 机载自卫激光

ASDP Advanced Sensor Development Program 高级传感器研制计划

ASDR Avionics Systems Demonstrator Rig 航空电子系统验证机试验台

ASDS 1. Aircraft-Sound Description System 飞机噪声描述系统 2. Avionics Software Development and Support 航空电子系统软件开发和支援

ASE 1. Advanced System Engineering 高级系统工程 2. Advanced Space Engines 先进空间发动机 3. Airborne Support Environment 机载保障环境 4. Airborne Support Equipment 机载保障设备 5. Aircraft Survivability Equipment 飞机生存力提升设备 6. Allowable Steering Error 容许操纵误差 7. Altimetry System Error 测高系统误差 8. Association of Space Explorers 空间探险者协会 9. Auto-Slat Extension 自动式前缘缝翼伸出

ASEB Aeronautic and Space Engineering Board 航空航天工程局(美国)

A-sector A扇区

ASEE American Society for Engineering Education 美国工程教育学会

ASEG All-Services Evaluation Group 三军鉴定组(美国海军航空试验中心)

aseismatic 耐震的,抗地震的

aseismic belt 无震带

aseismic slip 无震滑动

aseismic zone 无震区

aseptic 1. 无菌的,防腐的,消毒的 2. 防腐剂

ASESS Aerospace Environment Simulation System 航空航天环境模拟系统

ASET 1. Aeronautical Satellite Earth Terminal 航空卫星地球终端站 2. Aeronautical Services Earth Terminal 航空勤务地面站

ASEW Airborne and Surface Early Warning 空地(海)预警

ASF 1. Additional Secondary Phase Factor 附加二次相位因子 2. Aircraft Servicing Flight 飞机维修飞行(英国用法) 3. Auto-Start Flag 自动起始标记

ASFIR Active Swept Frequency Interferometer Radar 主动扫描频率干涉仪式雷达

ASFOS Air Service Field Officer School 陆军航空勤务队校官学校(美国)

ASG 1. Aeronautical Standard Group 航空标准局 2. Air Safety Group 空中安全组(英国)

ASGLS Advanced Space Ground Link Subsystem 先进的空间地面联络子系统

ash 灰,粉尘,尘埃,灰分

ASH 1. Advanced Scout Helicopter 先进侦察直升机 2. Advanced Support Helicopter 先进支援直升机 3. ASH雷达

ash bin 〈口语〉深水炸弹

ash cloud 烟灰云

asher 灰化器,碳化器

ASI 1. The Aeronautical Society of India 印度航空学会 2. Air Safety Investigation 航空安全调查 3. Air Staff Instructions 空军参谋部指示 4. Airspeed indicator 空速表,空速指示器 5. Atmosphere Structure Instrument 大气结构测量仪 6. Augmented Spark Igniter 增强火花点火器 7. Avionics System Integration 航空电子系统综合 8. Aviation Safety Institute 航空安全研究所(美国)

ASIAC Aerospace Structures Information and Analysis Center 航空航天结构信息和分析中心(美国空军)

a-siamorphous silicon 非晶硅

Asiana Airlines 亚细安纳航空公司(韩国)

ASIC Application Specific Integrated Circuit 特种应用集成电路

ASIP Aircraft Structural Integrity Program 飞机结构完整性大纲(美国国防部)

ASIR Airspeed Indicator Reading 空速表读数

ASIS Abort Sensing and Implementation System 飞行故障传感和处理系统

ASJ 1. Astronomical Society of Japan 日本天文学会 2. Automatic Search Jammer 自动搜索干扰机

ASK 1. Amplitude Shift Keying 移幅键控 2. Automatic Station Keeping 自动位置保持

ASL 1. Above Sea Level 海拔 2. Azimuth Steering Line 方位操纵线

ASLA Air Services Licensing Authority 航空勤务证件管理局(新西兰)

ASLAR Aircraft Surge Launch and Recovery 飞机松缆弹射起飞与回收

ASLR Air Surface Laser Ranger 空地激光测距仪

ASLV 1. Atlas Standard Launch Vehicle 宇宙神标准运载火箭(美国) 2. Augmented Satellite Launch Vehicle 增大推力的卫星运载火箭(印度)

ASM 1. Advanced Scatterable Mine 先进的空投散布雷 2. Advanced Systems Monitor (座舱)先进系统监控器 3. Airspace Management 空域管理 4. Air Superiority Mission 制空任务 5. Air-to-Surface Missile 空(对)地导弹 6. American Society for Metals 美国金属学会 7. Antenna Switching Matrix 天线开关矩阵

8. Association for System Management 系统管理协会（美国）9. Auto-throttle Servo Motor 自动油门杆伺服电机 10. Available Seat-Miles 可用座英里

ASMD Anti-Ship Missile Defence 对反舰导弹的防御（美国海军）

ASME American Society of Mechanical Engineers 美国机械工程师学会

A-SMGCS Advanced Surface Movement Guidance and Control Systems 先进地面运动导引与控制系统

ASMI Airfield Surface Movement Indicator 机场地面交通动态指示器

ASMK Advanced Serpentine Maneuverability Kit 先进的盘旋机动性成套设备

ASMP Aerospace Medical Panel 航空航天医学研究组（北约组织）

ASMS 1. Advanced Strategic Missile System 先进战略导弹系统 2. Advanced Synchronous Meteorological Satellite 先进同步气象卫星（美国）

ASMU Automatically Stabilized Maneuvering Unit 自动稳定机动装置

ASN Assigned Subject Number 给定主题号

ASNT American Society for Nondestructive Testing 美国无损检测学会

ASO 1. Acoustic Systems Operator 声系统操作员 2. Advanced Solar Observatory 先进太阳观测台（美国）3. Air Support Operation 空中支援行动 4. Aviation Safety Officer 航空安全军官 5. Aviation Supply Office 航空器材供应局（美国海军）6. Aviation Supply Officer 航空供应军官

ASP 1. Aerospace Plane 航空航天飞机，空天飞机 2. Airbase Survivability Program 空军基地生存能力计划（美国）3. Airborne Sensor Platform 空中传感器平台 4. Altitude Set Panel 高度设置版 5. Apollo Spacecraft Project 阿波罗飞船工程 6. Armament Status Panel 武器状况显示板 7. Audio Selector Panel 音频选择器板 8. Automated Small-batch Production 自动化小批量生产 9. Auxiliary Spacecraft Power 航天器辅助电源

aspect 1. 方面，方位 2. 状况，形势 3. 外观，面貌 4. 见解，观点

aspect angle 1. 目标进入角（从目标纵轴末端量到截击机观测线的角）2. 波弹交角（射弹纵轴与雷达波束的夹角）3. 搜索角，扫描角，视界角，视线角

aspect change 季相变化

aspect ratio 1.（导弹的）直径与弹长之比 2.（飞机的）展弦比 3.（帧的）纵横尺寸比

aspect-ratio canopy 纵横比冠层

aspect-ratio jet 纵横比喷射

aspect-stabilized 空间定向的

aspen 白杨木，似白杨的

aspen model 白杨模型

asphalt 枥青，柏油

aspherical lens 消球差透镜

aspherical surface 非球面

asphyxia 窒息

Aspide "阿斯派德"导弹

aspirant 进气的，抽吸的

aspiration 吸气，吸入

aspirator 吸气机，抽风机

Aspirin "阿斯匹林"行动

ASPR Armed Services Procurement Regulations 武装部队采购条例（美国）

ASPS 1. Acoustic Ship Positioning System 舰船声音定位系统 2. Adaptable Space Propulsion System 通用空间推进系统 3. Aerospace Physiologists Society 航空航天心理学家学会（美国）4. Annular Suspension and Pointing System 环状悬挂和指向系统

ASQC American Society for Quality Control 美国质量控制学会

ASR 1. Address Shift Register 地址移位寄存器 2. Address Switch Register 地址切换寄存器 3. Air Staff Requirement 空军参谋部要求（英国）4. Air Surveillance Radar 对空监视雷达 5. Airborne Surveillance Radar 机载警戒雷达 6. Airfield Surveillance Radar 机场监视雷达 7. Airport Surveillance Radar 机场监视雷达 8. Altimeter Setting Region 高度表调整范围 9. Approach Surveillance Radar 进场监视雷达 10. Automatic Send-Receiver 自动发送接收机

ASR/OPS Air Surveillance Radar/Operations Center System 空中监视雷达/作战中心系统

ASRAAM Advanced Short Range Air-to-Air Missile 先进近距空对空导弹

ASRC Alabama Space and Rocket Center 阿拉巴马航天和火箭中心（美国）

ASRDI Aerospace Safety Research and Data Institute 航空航天安全研究和数据所

as-received condition 接收状态

ASRI Aircraft（Airplane）Strength Research Institute 飞机强度研究所

ASRM Advanced Solid Rocket Motor 先进固体火箭发动机（美国）

ASRO Astronomical Roentgen Observatory 天文伦琴观测者

ASROC Anti-Submarine Rocket 反潜火箭

ASRP Aviation Safety Reporting Program 航空安全报告大纲

ASRS 1. Automatic Storage/Retrieval System 信息自动

储存与检索系统 2. Aviation Safety Reporting System 航空安全报告系统

ASRT　Air Support Radar Team 空中支援雷达组

ASS　1. Advanced Space Station 先进空间站 2. Aircraft Security System 飞机警戒系统，飞机安全系统 3. Aircraft (Airplane) Scheduling System 飞机计划系统 4. Airlock Support Subsystem 气闸舱保障分系统 5. Air Signalers School 机上报务员学校（英国）6. Anti-Satellite Satellite 反卫星卫星 7. Anti-Shelter Submunition 反掩体子弹药 8. Atmospheric Structure Satellite 大气结构研究卫星 9. Attitude-Sensing System 姿态敏感系统，姿态探测系统 10. Aviation Support Ship 航空兵保障船 11. Automated Scheduling System 自动计划系统，自动进度系统

ASSA　Aeronautical Society of South Africa 南非航空学会

assault　冲击，突击，强击

assault aircraft　空降突击飞机

Assault Breaker　"突击破坏者"

assault helicopter　空降突击直升机

assay　1. 化验，分析，检定 2. 试样，试料 3. 验检法

ASSEM　Assemble 装配

assemblage　1. 装配，安装 2. 装配件 3.（数学中的）集，族

assemble　1. 装配，安装 2. 集合 3.（程序）汇编

assemble line　装配线

assembled　装配好的

assembler　1. 装配工，装配技术人员 2. 装配器，收集器 3. 汇编程序

assembling frame iron wheel carriage　装配型架铁轮支架车

assembling frame vertical stand　装配型架垂直停放平台

assembling loss　组合损失

assembly　1. 装配，组装 2. 组（合）件，成套件 3. 汇编 4. 集会，会议 5. 集合号令

assembly accessories　装配附件

assembly building　装配楼

assembly cost　装配成本

assembly dimension　装配尺寸

assembly drawing　组装图，装配图

assembly hall　装配大厅

assembly hole　装配孔

assembly jig　装配型架

assembly language　汇编语言

assembly line　装配线，总装线

assembly process　装配过程

assembly routine　汇编程序

assembly stage　装配阶段

assembly step　组装步骤

assembly structure　装配式结构

assembly technique　装配技术

assembly test　组件实验

assembly-test building　装配-测试厂房

assembly tool　装配工具

assess　估计，评定，确定，评估

ASSESS　1. Airborne Science/Shuttle Experiment System Simulation 机载科学和航天飞机实验系统模拟 2. Airborne Science/Space Lab Experiment System Simulation 机载科学与空间实验室实验系统模拟

assessability　评定能力，评估能力

assessment　1. 考核 2. 评定，估价，评估

assessment of airworthiness state　适航性评定

assessment technique　评价技术

assessment test　评价试验

ASSET　1. Aerospace Structure Environmental Test 航空航天结构环境试验（美国空军）2. Airborne Sensor System for Evaluation and Test 机载评定和试验传感器系统

asset　资产，贵重的东西，优点，长处

assign　1. 分配，分派 2. 指定 3. 归因于

assigned range　试射距离

assignment　1. 分配，分派 2. 指定 3. 任务，课题

assignment of case　分配情况

assignment problem　分配问题，配置问题

assignment statement　赋值语句

assimilate　使……同化，比较，比拟

assimilation　同化，吸收

assist　1. 助推器，加速器 2. 助推，加速 3. 辅助

assistance　辅助，支援

assistant　1. 助理，副手，助教 2. 副的，辅助的

assistant electrical power source　辅助电源

assistant laboratory　实验室助理

assistant laboratory director　实验室助理主任

assistant professor　助理教授

assistant project　项目助理

assistant secretary　助理秘书长

assisted　1. 带助推器的，带加速器的 2. 助推的

ASSMT　Assignment 指定，分派

ASSN　Association 联合会，团体，公司

ASSOC　Associate 联合，使发生联系

associate　1. 联合，结合 2. 相关的，副的

associate designs　关联设计项目，有关设计

associate editor　副主编

associated (VOR, Tacan)　伏尔-塔康综合系统

associated equipment　配套设备

associated failure　牵连故障

associated model　伴随模型

association　1. 联合，结合 2. 协会，团体，公司 3.（化学

的)缔合 4.(天文的)星协
association probability 关联概率
associative dimensioning 相关尺寸
associative memory 联想记忆,相联存储器
associative memory model 联想记忆模型
associative perception 整体感
associative processor 相联处理机
association 联想机
ASSP Aerospace System Security Program 航空航天系统安全保障计划
ASST 1. Advanced Supersonic Transport 先进超声速运输机 2. Anti-Ship Surveillance and Targeting 对舰监视和导向目标 3. Assistant 辅助的,副的 4. Automatic System Self-Test 自动系统的自测试
ASSU Air Support Signal Unit 空中支援通信部队
assume 假定,设想
assumed 1.假定的 2.采用的
assumed coordinate system 假定坐标系
assumed mode 假设模态
assumed-transition-point 假定过渡点
assumed-transition-point criterion 假定过渡点标准
assumption 1.假定,设想 2.承担 3.采取
assumption of general equilibrium 一般均衡的假设
assurance criterion 保证标准
assure 使确信,使放心
assuring rate of depth datum 深度基准面保证率
ASSV Alternate-Source Select (or) Valve 备份源选择开关,代用源选择阀门
ASSY Assembly 装配件
AST 1. Address Synchronizing Track 地址同步跟踪 2. Advanced Supersonic Technology 先进超声速技术 3. Air Staff Target 空军参谋部指标 4. Amateur Space Telescope 业余爱好者空间望远镜 5. Anti-Satellite Technology 反卫星技术 6. Apollo System Test 阿波罗飞船系统试验 7. Atlantic Standard Time 大西洋标准时间
ASTA Airport Surface Traffic Automation 机场地面交通自动化
ASTAAM Advanced Seeker Technology for Air-to-Air Missiles 空对空导弹的先进导引头技术
astable 不稳定的,非稳态的
Astar 1.阿斯塔尔直升机 2."阿斯特"(美国战术监视雷达)
astatic 不稳的,不稳定的,无定位的
astatic gravimeter 助动重力仪
astatic magnetometer 无定向磁强计
astatine 【化】砹(化学元素,符号 At)
A-station A 发射台
ASTC Army Satellite Tracking Center 陆军卫星跟踪中心(美国)
asterisk 1.星号,星状物 2.加注星号
asterism 星座,星群
Asterix 试验卫星(法国)
astern 1.在机尾,在尾部,在后 2.向后,从后面
asteroid 小行星
asteroid belt 小行星带
asteroid centroiding 小行星质心
asteroid flyby mission 小行星绕越飞行任务
asteroid orbit 小行星轨道
asteroid rotation 小行星自转
asteroid rotation period 小行星的自转周期
asteroid sample 小行星样本
asteroid surface 小行星的表面
asteroid system 小行星系统
asteroid tour sequence 小行星运行序列
ASTEX Advanced Space Technology Experiment 先进空间技术试验
ASTF Aeropropulsion Systems Test Facility 空气推进系统试验部
asthenia 虚弱,无力,衰弱
asthenopia 眼疲劳,视力疲劳
asthenosphere 软流层
astigmation 象散
astigmatism 象散现象
ASTM American Society for Testing Materials 美国材料试验学会
ASTO Arab Satellite Telecommunication Organization 阿拉伯卫星通信组织
Astoff 液氧
ASTOR Antisubmarine Torpedo 反潜鱼雷
ASTP 1. Advanced Systems and Technology Programme 先进系统和技术计划(欧洲航天局) 2. Apollo-Soyuz Test Project 阿波罗-联盟号试验计划
Astra 阿斯特拉卫星
ASTRA 1. Application of Space Techniques Relating to Aviation 航天技术在民航领域的应用(国际民航组织) 2. Astronomical and Space Techniques for Research on the Atmosphere 天文和航天技术应用于大气的研究(美国华盛顿大学的计划) 3. Astronomical Space Telescope Research Assembly 空间天文望远镜研究装置
astral 1.观测天窗,天文观测窗 2.星际的,星的,星形的
Astral 天文卫星(美国)
astrionics 1.航天电子学 2.航天电子设备
ASTRO Aerodynamic Spacecraft Two-stage Reusable Orbiter 气动航天器两级重复使用轨道器
astro-geodetic deflection of the vertical 天文大地垂线

偏差
astro-gravimetric leveling 天文重力水准
astro trainer 航天员训练机
astro vehicle 航天器
astroasthenia 航天疲劳
astroballistics 天体弹道学
astrobiology 天体生物学
astrobionics 天体仿生学
astrobleme 陨星撞迹
astrobotany 天体植物学
Astro-c 天文卫星C(日本,现名Ginga)
astrochemistry 天体化学
astrochronology 天文年代学
astroclimate 天文气候
astroclimatology 天体气候学
astrocompass 天文罗盘
astrodome 天文领航舱,天文观测窗
astrodynamics 1.航天动力学,天体动力学,天文动力学 2.星际航行动力学
astroecology 宇宙生态学
astrogation 〈口语〉天文导航,天文领航
astrogator 〈口语〉天文领航仪,航天领航员
astro-geodetic network 天文大地网,国家大地网
astrogeodynamics 天文地球动力学
astrogeography 天体地理学
astrogeology 天体地质学
astrogeophysics 天文地球物理学
astrograph 天体照相仪,天体定位器
astrograghy 天体摄影学,天体照相学
astrohatch 天文观测窗,天文领航窗
astro-inertial 〈口语〉天文惯性系统
astrolabe 等高仪,星盘
astrology 占星术
Astromag 天文磁体装置
astromechanics 天体力学
astrometeorology 天文气象学
astrometric binary 天测双星
astrometric place 天体测量位置
astrometric position 天体测量位置
astrometry 天体测量学
astro-mine 航天(障碍)雷
astronaut 航天员,宇航员
astronaut camp 航天员营区
astronaut candidate 预备航天员
astronaut care doctor 航天员医师
astronaut communication headsets 航天员通信头戴
astronaut decision-making 航天员决策
astronaut flight handbook 航天员飞行手册
astronaut health care 航天员健康管理

astronaut health state assessment 航天员健康状况判断
astronaut medical certification 航天员医学鉴定
astronaut medical kit 航天员药箱
astronaut medical monitoring 航天员医务监督
astronaut medical monitoring and support 航天员医学监督与保障
astronaut medical monitoring center 1.航天员医学监督中心 2.航天员医监中心
astronaut motion 航天员的运动
astronaut nutrition 航天员营养
astronaut operation 航天员作业,航天员操作
astronaut productivity 航天员生产能力
astronaut psychological evaluation 航天员心理学评定
astronaut quarantine 航天员检疫
astronaut safety 航天员安全性
astronaut safety bunker 航天员安全掩体
astronauts selection 航天员选拔
astronaut selection and training center 1.航天员选拔训练中心 2.航天员选训中心
astronaut selection criteria 航天员选拔标准
astronaut system 航天员系统
astronaut training 航天员训练
astronaut training program 航天员训练项目
astronaut work capacity 航天员工作能力
astronaut work rest schedule 航天员作息制度
astronautical 航天的,星际航行的,宇航的
astronautics 航天学,宇宙航行学
astronaut-induced load 航天员荷载
astronauts digestion 航天员消化
astronaut's energy metabolism 航天员能量代谢
astronavigation 天文导航,天文领航
astronavigator 1.天文领航仪 2.天文领航员,航天领航员
astronics 1.航天电子学,天文电子学 2.航天电子设备
astronomer 天文学家
astronomical almanac 天文年历
astronomical body 天体
astronomical catalog 天文表
astronomical clock 天文钟
astronomical constant 天文常数
astronomical coordinate 天文坐标
astronomical day 天文日
astronomical ephemeris 天文年历
astronomical equator 天文赤道
astronomical geodesic 天文测地线
astronomical guidance 天文制导
astronomical latitude 天文纬度
astronomical longitude 天文经度
astronomical magnitude 天文级

astronomical map 天文图
astronomical meridian 天文子午线
astronomical navigation 天文导航
astronomical observatory 天文观测台
astronomical parallel 天文纬线
astronomical photography 天体摄影
astronomical photometry 天体测光
astronomical plate 天文底片
astronomical point 天文点
astronomical polarimetry 天体偏振测量
astronomical position 天文位置
astronomical refraction 大气折射
astronomical scintillation 天体闪烁(现象)
astronomical spectrograph 天体摄谱仪
astronomical spectroscopy 天体光谱学
astronomical telescope 天文望远镜
astronomical time 天文时
astronomical triangle 天文三角形
astronomical twilight 天文曙暮光,天文晨昏蒙影
astronomical unit 天文单位
astronomical year 天文年
astronomy 天文学
astronomy satellite 天文卫星
astrophotography 天体照相学
astrophysics 天体物理学
astroplane 航天飞机
astrorocket 航天火箭
ASTROS Advanced Star and Target Reference Optical Sensor 先进的恒星和目标基准光学敏感器
astrospectrograph 天体摄谱仪
astrospectrometer 天体分光计
astrospectroscopy 天体光谱学
astrotracker 天体定位仪,星体跟踪仪
astroweapon 航天武器
ASTS 1. Airborne Software Technical Standard 机载软件技术标准 2. Airport Surface Traffic Simulator 机场地面交通模拟器
ASTU Automatic Stabilizer Trim Unit 安定面自动配平组件
A-stuff 液氧
ASU 1. Aircraft Storage Unit 飞机储备部队(英国) 2. Altitude Sensing Unit 高度传感装置 3. Approval for Service Use 批准军用 4. Auto-spoiler Switching Unit 自动扰流板开关装置 5. Avionics Switching Unit 航空电子开关装置
ASV 1. Aerothermodynamic Structural Vehicles 空气热动力结构(试验)飞行器(美国) 2. Air-to-Surface-Vessel 空对水面舰船搜索雷达
ASVC Automatic Secure Voice Communications 自动保密电话通信
ASVEH Air-Surveillance Vehicle 空中监视飞行器
ASVT Application Systems Verification and Transfer (美国陆地卫星)应用系统验证和转让
ASW 1. Aft-Swept Wing 后掠翼 2. Air/Sea Warfare 空海作战 3. Air-to-Surface Weapon 空对地武器 4. Anti-Shelter Weapon 反掩体武器,反掩体炸弹 5. Anti-Submarine Warfare 反潜战
ASWCCCS Anti-Submarine Warfare Centers Command and Control System 反潜战中心指挥与控制系统
ASWE Admiralty Surface Weapons Establishment 海军部海面武器研究所(英国)
ASWEX Anti-Submarine Warfare Exercise 反潜战演习
ASWG 1. American Standard Wire Gage 美国标准线规 2. American Steel Wire Gage 美国标准线规
ASWICS Anti-Submarine Warfare Integrated Combat System 反潜战综合作战系统
ASYM Asymmetric 不对称的
asymmetric alternating current charge 不对称交流电充电
asymmetric bodies 不对称体
asymmetric clearance 不对称的间隙
asymmetric flight 不对称飞行
asymmetric loading 不对称装载
asymmetric mode 非对称模式
asymmetric set 非对称设置
asymmetric shock 不对称冲击
asymmetric shock train 非对称冲击钻
asymmetric side 非对称边
asymmetric trailing-edge 不对称的后缘
asymmetric transition 非对称转换
asymmetric vortex 非对称涡
asymmetric vortex interaction 非对称涡相互作用
asymmetric wing 非对称翼
asymmetric(al) 不对称的
asymmetrical nystagmus 不对称性眼震
asymmetry 不对称(性)
asymptote 渐近线
asymptotic analysis 渐近分析
asymptotic attitude 渐近姿态
asymptotic attitude tracking 渐近姿态跟踪
asymptotic availability 渐近可用度
asymptotic behavior 渐近行为
asymptotic domain 渐近领域
asymptotic equipartition property 渐近等分性质
asymptotic expansion 渐近展开
asymptotic expansion perturbation technique 渐近展开摄动技术
asymptotic latitude 渐近纬度

asymptotic limit 渐近极限
asymptotic longitude 渐近经度
asymptotic orbit 渐近轨道
asymptotic-periodic orbits 渐近周期轨道
asymptotic property 渐近性质
asymptotic solution 渐近解
asymptotic stability 渐近稳定性
asymptotic theory 渐近理论
asymptotic value 渐近值
asymptotic(al) 渐近的,渐近线的
asymptotics 渐近性
asynchronous 异步的,不协调的
asynchronous communication 异步通信
asynchronous communication interface adapter 异步通信接口适配器
asynchronous communication satellite 非同步通信卫星
asynchronous device 异步装置
asynchronous gyro-motor 异步陀螺马达
asynchronous satellite 非同步卫星
asynchronous switching 异步开关
asynchronous time-division multiplexing 异步时分多路转换
asynchronous transfer mode 异步转移态
asynchronous transmission 异步传输
asynchrony 协调障碍
ASZ Air Surface Zone 航空兵与舰艇配合反潜行动区域(北约,美国空军)
AT 1. Acceptance Test 交收试验,验收试验 2. Advanced Trainer 高级教练机 3. Agility Tuning 快速调谐 4. Air Traffic 空中交通 5. Air Transport 空运 6. Airtight 气密 7. Ambient Temperature 环境温度 8. Auto-Throttle 自动油门 9. Air Transmit 空中传输 10. Anti-Tank 反坦克的 11. Automatic Test 自动测试
AT/SC Auto-Throttle/Speed Control 自动油门/速度控制
AT/W Atomic Hydrogen Weld 氢原子焊
ATA 1. Actual Time of Arrival 实际到达时间 2. Advanced Tactical Aircraft 先进战术飞机 3. Advanced Test Accelerator 高级试验加速器 4. Air Transport Association 航空运输协会 5. Air Transport Auxiliary 空中运输辅助部队 6. Asynchronous Terminal Adapter 异步通信终端适配器 7. Automatic Threshold Adjust 自动阈值调整
ATAAC Anti-Torpedo Air-launched Acoustic Countermeasures 空射反鱼雷声对抗装置
ATAB Air Transport Allocation Board 空运调配处
ATABCS Aircraft Tactical Air Battle Command System 飞机战术空战指挥系统
ATAC 1. Advanced Technology Advisory Committee 先进技术咨询委员会 2. Air Transport Advisory Council 航空运输咨询委员会(英国) 3. Air Transport Association of Canada 加拿大空运输协会 4. Air Transportable Acoustic Communication 航空运输声波通讯 5. Applied Technology Advanced Computer 应用技术先进计算机
ATACCS Advanced Tactical Air Command and Control System 先进的战术空军指挥与控制系统
ATACMS Army Tactical Missile System 陆军战术导弹系统
ATACO Air Tactical Control Officer 空军战术控制军官
ATACS Airborne Target Acquisition Control System 机载目标截获控制系统
ATAFCS Airborne Target Acquisition Fire Control System 机载目标截获与火控系统
ATAL Automatic Test Application Language 自动测试应用语言
ATALS Army Transportation And Logistics School 陆军运输与后勤学校(美国)
ATAMS Advanced Tactical Attack Manned System (人工操纵的)先进战术攻击系统
atanchor 锚泊
ATAR 1. Air-to-Air Recovery 空中回收 2. Air Traffic Approval Regulations 空中交通批准条例 3. Anti-Tank Aircraft Rocket 航空反坦克火箭弹
ATARS 1. Advanced Tactical Air-Reconnaissance System 先进战术空中侦察系统 2. Army Tactical Airspace Regulation System 陆军战术空域控制调节系统 3. Automatic Traffic Advisory and Resolution Service 空中交通自动咨询和判定勤务
ATAS 1. Advanced Tactical Attack System 先进战术攻击系统 2. Advanced Target Acquisition Sensor 先进目标截获传感器 3. Air Traffic Advisory Service 空中交通咨询处(英国) 4. Automatic Three-Axis Stabilization 自动三轴稳定
ATAWS Advanced Tactical Air Warfare System 先进战术空中作战系统
ataxia 运动失调,共济失调
ATB 1. Advanced Technology Blade 先进技术叶片 2. Advanced Technology Bomber 先进技术轰炸机 3. Air Transport Bureau 空中运输局 4. Air Transport Board 航空运输委员会(美国) 5. Apollo Test Vox 阿波罗飞船试验装置
ATBM Anti-Tactical Ballistic Missile 反战术弹道导弹
ATC 1. Advanced Terminal Controller 先进终端控制器 2. Advanced Technology Component 先进技术部件 3. Aerospace Technical Council 航空航天技术委员会 4. Air Traffic Control 空中交通管制 5. Air Training

Command 空军训练部（美国空军）6. Air Training Corps 空军训练队（英国）7. Air Transport Command 空运司令部（美国空军）8. Air Transport Conference 空中运输会议（旅行代理商的）9. Approved Type Certificate 型号合格证，机型批准书（美国）10. Automatic Threat Countering 对威胁自动采取对策

ATC clearance Air-Traffic Control Clearance 空中交通管制许可证

ATCA 1. Advanced Tanker/Cargo Aircraft 先进的可空中加油货运飞机 2. Air Traffic Conference of America 美国空中交通会议 3. Air Traffic Control Assistant 空中交通管制助理（英国用法）4. Air Traffic Control Association 空中交通管制协会 5. Allied Tactical Communication Agency 盟国战术通信署（北约）

ATCAC Air Traffic Control Advisory Committee 空中交通管制咨询委员会（美国政府）

ATCAP Air Traffic Control Automation Panel 空中交通管制自动化小组

ATCC 1. Aerospace Traffic Control Center 航天交通管制中心 2. Air Traffic Control Center 空中交通管制中心（美国、英国）

ATCEU Air Traffic Control Evaluation Unit 空中交通管制鉴定单位（英国）

ATCF Air Traffic Control Facility 空中交通管制设备

ATCH Attach(ment) 相连

ATCM Advanced Technology Cruise Missile 先进技术巡航导弹

ATCO Air Traffic Control Officer 空中交通管制官（美国、英国）

ATCOM Air Traffic Communications 空中交通管制通信系统

ATCOS Atmospheric Composition Satellite 大气成分卫星（美国空军）

ATCP Advanced Technology Coorbiting Platform 先进技术同轨平台（日本）

ATCR Air Training Command Regulation 空军训练部条例（美国）

ATCRBS Air Traffic Control Radar Beacon System 空中交通管制雷达信标系统（即二次雷达）

ATCRU Air Traffic Control Radar Unit 空中交通管制雷达设备（英国）

ATCS 1. Active Thermal Control System 有源热控制系统 2. Active Thermal Control Subsystem 有源热控制子系统 3. Air Traffic Control Service 空中交通管制局（美国、英国）4. Airborne Tactical Command System 机载战术指挥系统

ATCSS Air Traffic Control Signaling System 空中交通管制信号系统

ATCT Air Traffic Control Tower 空中交通管制塔台，机场指挥塔台

ATD 1. Actual Time of Departure 实际离场时间 2. Airline Transmitted Disease 航空传播的疾病 3. Attitude 姿态，高度 4. Automatic Target Detection 自动目标探测 5. Aviation Technical Division 航空技术局（苏联）

ATDA 1. Agena Target Docking Adapter 阿金纳目标对接器 2. Augmented Target Docking Adapter 改进型目标对接器

ATDC African Telecommunication Development Conference 非洲通信发展会议

ATDM Asynchronous Time-Division Multiplexing 异步时分多路转换

ATDMA Advanced Time-Division Multiple Access 先进时分多址

ATDRS(S) Advanced Tracking and Data Relay Satellite (System) 先进跟踪和数据中继卫星（系统）（美国）

ATDS 1. Airborne Tactical Data System 机载战术数据系统 2. Air-Turbine Drive System 空气涡轮传动系统

ATE 1. Advanced Technology Engine 先进技术发动机 2. Airborne Test Equipment 机载测试设备 3. Along Track Error 距离偏差，纵向偏差 4. Automatic Test Equipment 自动测试设备

ATE/ICE Automatic Test Equipment for Internal Combustion Engines 内燃机自动测试设备

ATEC 1. Automated Technical Control Capability 自动化技术控制能力 2. Automatic Test Equipment Complex 全套自动测试设备，综合自动测试设备

ATEGG Advanced Turbine Engine Gas Generator 先进涡轮发动机气体发生器

ATEMS Automatic Tracking Equipment Management System 自动跟踪设备管理系统

ATET Advanced Technology Experimental Transport 先进技术实验运输机

ATEWS Advanced Tactical Electronic Warfare System 先进的战术电子战系统

ATF 1. Actual Time of Fall 实际落下时间（炸弹等）2. Advanced Tactical Fighter 先进战术战斗机 3. Air Task Force 空军特遣部队 4. Altitude Test Facility 高空试验设施 5. Astrometric Telescope Facility 天体测量望远镜设施（美国）6. Aviation Turbine Fuel 航空燃气轮机燃油

Atfero Atlantic Ferry Organization 大西洋渡运组织

ATFM Air Traffic Flow Management 空中交通流量管理

ATFMU Air Traffic Flow Management Unit 空中交通流量管理单位

ATGAR Anti-Tank Guided Air Rocket 反坦克航

导弹

ATGM Anti-Tank Guided Missile 反坦克导弹

ATGW Anti-Tank Guided Weapons 反坦克武器

ATH 1. Automatic Target Handoff 目标自动交接 2. Autonomous Terminal Homing 自主式末段自导引

Athena 阿西纳探空火箭(美国)

athodyd 冲压式喷气发动机

ATHS 1. Airborne Target Handover System 空中目标交接系统 2. Automatic Target Hand-off System 目标自动交接系统,目标自动处理系统

athwartship 垂直于飞行器纵轴的,垂直于龙骨的

ATI Aerial Tuning Inductance 天线调谐电感

ATIGS Advanced Tactical Inertial Guidance System 先进战术惯性制导系统

ATILO Air Technical Intelligence Liaison Officer 航空兵技术情报联络官

ation 方法,增强辐射

ation deployment 增幅调配

ation pressure 增幅压力

ation system 增幅系统

ation test 增幅测试

ATIS 1. Air Traffic Information System 空中交通信息系统 2. Airbus Technical Information System 空中客车技术信息系统 3. Automatic Terminal Information Service 自动终端信息服务 4. Automatic Terminal Information System 自动终端信息系统

ATITA Air Transport Industry Training Association 空中运输工业训练协会(英国)

ATITB Aviation and Travel Industry Training Board 航空与旅游工业训练委员会(新西兰)

ATJS Airborne Tactical Jamming System 机载战术干扰系统

ATK Aviation Turbine Kerosene 航空燃气轮机煤油

ATk 1. Anti-Tank 反坦克 2. Available Tonne-kilometer 可用吨公里

ATL Advanced Technology Laboratory 高技术实验室

ATLA Air Transport Licensing Authority 空中运输许可证管理局(中国香港)

Atlantic 大西洋卫星

Atlantic Missile Range 大西洋导弹靶场(美国)

Atlantis 阿特兰蒂斯号(美国航天飞机轨道器)

ATLAS 1. Advanced Target Location And Strike 先进的目标定位和攻击 2. Advanced Technology Large Aircraft System 先进技术大型航空器系统 3. Atmosphere Laboratory for Applications and Science 大气应用和科学实验室(美国)

Atlas 1. 宇宙神(美国运载火箭和洲际弹道导弹名) 2. 土卫十五 3. "阿特拉斯"(法国 TH.D2503 G 波段单脉冲跟踪雷达) 4. Abbreviated Test Language for Avionics System 航空电子系统简略测试语言 5. Advanced Tactical Low Arresting System 先进战术低空拦阻系统

atlas 地图集,图表集

Atlas-Able Star 宇宙神·艾布尔星运载火箭(美国)

Atlas-Agena 宇宙神·阿金纳运载火箭(美国)

Atlas-Burner 宇宙神·博纳运载火箭(美国)

Atlas-Centaur 宇宙神·半人马座运载火箭(美国)

ATLB Air Transport Licensing Board 空中运输许可证管理局(英国)

ATLIS 1. Airborne Tracking Laser Illumination System 机载激光跟踪照射系统 2. Automatic Tracking Laser Illumination System 自动跟踪激光照射系统

ATM 1. Air Traffic Management 空中交通管理 2. Air Transport Movement 空中运输动态 3. Air Turbine Motor 空气涡轮发动机 4. Apollo Telescope Mount 阿波罗飞船望远镜装置(美国) 5. Asynchronous Transfer Mode 异步传输模式,异步转移态 6. Atmosphere 大气,气氛,大气,大气压(力)

ATMDC Apollo Telescope Mount Digital Computer 阿波罗飞船望远镜装置数字计算机

ATMG Arms Transfer Management Group 武器移交管理组(美国国防部)

atmology 水汽学

atmometer 蒸发表

ATMOS Atmospheric Trace Molecule Spectroscopy Experiment 大气痕量分子分光实验

atmosphere 1. 大气 2. 大气层 3. 大气压

atmosphere absorption 大气吸收

atmosphere acoustics 大气声学

atmosphere contamination control 大气污染控制

atmosphere control system of manned spacecraft cabin 载人航天舱大气控制系统

atmosphere corrosion 大气腐蚀

atmosphere entry 进入大气层

atmosphere envelope 大气层(包围地球的)

atmosphere environment in spacecraft cabin 航天舱内大气环境

atmosphere interaction 大气的相互作用

atmosphere model 大气模型

atmosphere monitor satellite 大气监测卫星

atmosphere oblateness 大气层扁率

atmosphere optical thickness 大气光学厚度

atmosphere pressure carbonization process 常压碳化工艺

atmosphere reentry 再入大气层

atmosphere scale height 大气标高

atmosphere shower 大气簇射

atmosphere supply and pressure control system 供气调压

系统
atmospheric 大气的,大气层的,大气压的
atmospheric absorption 大气吸收
atmospheric air 大气
atmospheric attenuation 大气衰减
atmospheric backpressure 大气压力
atmospheric bands 大气谱带
atmospheric boundary layer 大气边界层
atmospheric boundary layer wind tunnel 大气边界层风洞
atmospheric braking 空气动力制动,气动减速
atmospheric circulation 大气环流
atmospheric coast 大气海岸
atmospheric coast stage 大气海岸阶段
atmospheric composition 大气成分
atmospheric condensation 大气凝结
atmospheric condition 大气条件
atmospheric constituents 大气成分
atmospheric correction 大气校正
atmospheric corrosion 大气腐蚀
atmospheric density 大气密度
atmospheric density model 大气密度模型
atmospheric diffraction 大气衍射
atmospheric dispersion 大气色散
atmospheric distortion compensation 大气畸变补偿
atmospheric disturbance 大气扰动
atmospheric drag 大气阻力
atmospheric drag perturbation 大气阻力摄动
atmospheric duct 大气波导
atmospheric eclipse 大气食
atmospheric effect 大气效应
atmospheric electric field 大气电场
atmospheric emission 大气发射
atmospheric entry 进入大气层
atmospheric environment 大气环境
atmospheric exit 退出大气层
atmospheric explosion 空爆
atmospheric extinction 大气消光
atmospheric filtering 大气层过滤
atmospheric flight 大气飞行
atmospheric flight path 大气飞行轨迹
atmospheric flow 大气流动
atmospheric friction 大气摩擦,大气摩阻
atmospheric haze 大气蒙雾
atmospheric heating 大气加热
atmospheric heave 大气升腾
atmospheric humidity 大气湿度
atmospheric interaction 大气的相互作用
atmospheric interface 大气界面

atmospheric interference 大气干扰,天电干扰
atmospheric loss 大气耗损
atmospheric model 大气模式
atmospheric noise 大气噪声,天电噪声
atmospheric opacity 大气不透明度
atmospheric optics 大气光学
atmospheric optional thickness 大气光学厚度
atmospheric oscillation 大气振荡
atmospheric parameter 大气参数
atmospheric parameter measurement 大气参数检测
atmospheric pass 大气途径
atmospheric path 大气轨迹
atmospheric path radiation 大气程辐射
atmospheric perturbation 大气摄动
atmospheric physics 大气物理(学)
atmospheric pollution 大气污染
atmospheric pollution monitoring 大气污染监测
atmospheric precipitation 大气降水
atmospheric pressure 大气静压
atmospheric pressure control 大气静压控制
atmospheric propagation correction 大气传播修正
atmospheric quality index 大气品位指数
atmospheric radiation 大气辐射
atmospheric reentry 返回大气层
atmospheric refraction 大气折射
atmospheric regime 大气规则
atmospheric regions 大气分层
atmospheric remote sensing 大气遥感
atmospheric rotation 大气循环
atmospheric scale height 均质大气高度
atmospheric scattering 大气散射
atmospheric scintillation 大气闪烁
atmospheric seeing 大气宁视度
atmospheric selectivity scattering 大气选择性散射
atmospheric self-purification 大气自净作用
atmospheric shell 大气层
atmospheric sounding 大气探测
atmospheric sounding projectile 高层大气探测火箭
atmospheric spectral transmittance 大气光谱透过率
atmospheric structure 大气结构
atmospheric temperature 大气温度
atmospheric temperature profile sounder system 大气温度廓线(分布)探测(卫星)系统
atmospheric tides 大气潮
atmospheric trace contamination control 大气微量污染控制
atmospheric transmission bands 大气透射波段
atmospheric transmittance 大气透射比
atmospheric transparency 大气透明度

atmospheric turbulence 大气涡流
atmospheric uncertainty 大气的不确定性
atmospheric visibility 大气能见度
atmospheric vorticity 大气涡度
atmospheric window 大气窗口
atmospherics 1.天电,大气干扰 2.自然产生的离散电磁波
ATMS 1. Advanced Terminal Management System 先进终端管理系统 2. Advanced Test Management System 先进的测试管理系统 3. Assembly Tracking Management System 装配跟踪管理系统 4. Automatic Teletype Message Switching 自动电传打字转报
ATN 1. Aeronautical Telecommunication Network 航空电信网 2. Advanced Tiros-N 先进泰罗斯 N 卫星(美国气象卫星)
ATO 1. Abort to Orbit 应急入轨 2. Airborne Tactical Officer 空勤战术军官 3. Air Task Order 空军特遣任务命令 4. Apollo Test and Operations 阿波罗飞船试验和操作 5. Assisted Take-Off 助推起飞 6. Assemble to Order 集中订货,为订单装配
ATOA Air Taxi Operators Association 出租飞机公司协会(英国)
ATOC Allied Tactical Operations Centers 盟军战术作战中心(北约)
ATOL Air Travel Organizer's/Operator's Licence 空中旅游组织/公司营业执照
ATOLL Assembly Test Oriented Launch Language 用于装配和测试的发射语
atom 原子
atom bomb 原子炸弹
atom concentration 原子浓度
atom frequency standard 原子频标
atom gun 原子炮
atom injection 原子注入
atom smasher 核粒子加速器
atomdef 原子防御,原子防护(atomic defense 的缩写)
atomedics 原子医学
atomic 1.原子的,原子能的 2.原子武器的
atomic airburst 原子弹空中爆炸
atomic ammunition 原子弹药
atomic and molecular oxygen 原子和分子氧
atomic arms industry 原子武器工业
atomic arms race 原子武器军备竞赛
atomic arsenal 原子武器制造厂
atomic battery 核电池
atomic-bearing capability 携带原子弹的能力
atomic biological chemical defense 对原子、生物、化学器的防御
atomic blast 原子武器爆炸的冲击波

atomic blast simulation 模拟原子武器爆炸
atomic bomb air zero locator 原子弹空中爆炸中心定位器
atomic bomb residuals 原子弹(爆炸)剩留物
atomic clock 原子钟
atomic component 原子部分
atomic concentration 原子浓度
atomic content 原子装药
atomic demolition munition 原子爆破装置
atomic demonstration 原子爆炸示范演习
atomic destruction 原子武器破坏
atomic destructive capability 原子武器的破坏能力
atomic discipline (使用原子武器应遵守的)原子武器纪律
atomic effects 原子武器(爆炸)效果
atomic emission 原子发射
Atomic Energy Commission 原子能委员会(美国)
atomic equipment 原子武器装备
atomic explosive 原子爆炸物
atomic explosive weapon 原子爆炸武器
atomic fallout 原子放射物沉降
atomic fireball 原子弹爆炸时的火球
atomic flash 原子弹爆炸时的闪光
atomic fraction 原子分数
atomic frequency standard 原子频标
atomic guided missile 原子导弹
atomic gun 原子炮
atomic hydrogen 氢原子
atomic injuries 原子(武器)伤害
atomic layer epitaxy 原子层外延
atomic lethal rays 致死的原子射线
atomic mass 原子质量
atomic mine 原子地雷
atomic mushroom 原子弹爆炸时产生的蘑菇(状)云(团)
atomic oxygen 原子氧
atomic oxygen concentration 原子氧浓度
atomic oxygen effect 原子氧效应
atomic oxygen fluence 原子氧流量
atomic oxygen fluence model 原子氧流量模型
atomic oxygen test 原子氧试验
atomic precession magnetometer 原子进动磁强计
atomic projectile 原子炮弹
atomic protection 原子(武器)防护
atomic ruin area 原子武器破坏区
atomic scale machining 原子级加工
atomic shell 原子炮弹
atomic shock wave 原子武器(爆炸时的)冲击波
atomic test 原子武器爆炸试验,原子试爆

atomic time 原子时
atomic time scale 原子时间标度
atomic torpedo 原子鱼雷
atomic underground burst 原子武器地下爆炸
atomic underwater burst 原子武器水下爆炸
atomic weapon yield 原子武器的产量
atomichran 原子钟
atomic-oxygen beam 氧原子束
atomic-oxygen density 氧原子密度
atomic-oxygen exposure 氧原子暴露
atomic-oxygen flux 氧原子通量
atomic-oxygen protection 氧原子保护
atomics 原子武器
atomite "阿托迈"(一种烈性炸药)
atomization 雾化,喷雾
atomization air 雾化空气
atomization gas 雾化气体
atomization phenomenon 雾化现象
atomization process 雾化过程
atomization quality 雾化质量
atomization surface 雾化表面
atomization zone 雾化区
atomization-air jet 雾化空气喷射
atomize 喷雾,分裂成原子
atomized metal powder 雾化金属粉末
atomizer 喷嘴,喷雾器
atomizer nozzle 雾化喷嘴
atomizing 1.雾化,喷雾 2.原子化 3.原子突击
atom-tipped missile 装原子弹头的导弹
ATOPS Advanced Transport Operating Systems 先进运输操作系统
ATOROCKET Assisted Takeoff Rocket 起飞助推火箭
ATOS Automated Technical Orders System 自动化技术命令系统
ATOT Angle Track On Target 目标角度跟踪
ATP 1. Acceptance Test Procedure 验收试验程序 2. Acquisition, Tracking and Pointing 捕获、跟踪和瞄准 3. Actual Track Pointer 实际轨迹指示器 4. Advanced Turboprop 先进涡轮机螺旋桨式飞机 5. Aspirated Tat Probe 吸气式总温探头 6. Assembly and Testing Plant 装配和测试车间(苏联) 7. Authority To Proceed/Auxiliary Tool Production 授权开始/辅助工装生产
ATP&C Air Traffic Passenger & Cargo 航空运输量
ATPCS Automatic Takeoff Power Control System 自动起飞功率控制系统
ATPL Airline Transport Pilot's Licence 航班运输机飞行员执照
ATR 1. Acceptance Test Report 验收试验报告 2. Advanced Tactical Radar 先进战术雷达 3. Air Transport Rating 运输机驾驶执照 4. Air-launched Trainer Rocket 空(中发)射(的)训练火箭 5. Air-Turbo Ramjet 空气涡轮冲压式喷气发动机 6. Analog Tape Recorder 模拟式磁带记录仪 7. Automatic Target Recognition 目标自动识别
ATRAN Automatic Terrain Recognition and Navigation 自动地形识别与导航
ATRC Air Transport/Traffic Regulation Center 空中运输/交通条例中心(美国)
ATREL Air Transportable Reconnaissance Exploitation Laboratory 可空运的侦察技术开发实验室(英国空军)
ATRIF Air Transportation Research International Forum 航空运输国际研讨会
atrium 1.心房 2.(现代建筑物开阔的)中庭,天井
atrophy 肌肉萎缩
ATRS Automatic Terminal Radar System 自动终端雷达系统
ATRT Assumed Temperature Reduced Thrust 由假设温度而得的减推力
ATS 1. Accelerator Test Stand 加速器试验台 2. Acquisition and Tracking System 截获与跟踪系统 3. Action Tracking System 措施跟踪系统 4. Advanced Tactical Support 先进战术支援 5. Advanced Technology Satellite 先进技术卫星 6. Advanced Teleprocessing System 先进远程处理系统 7. Air Traffic Services 空中交通勤务,空中交通服务 8. Aircraft Trouble-shooting System 飞机故障判断系统 9. Aircrew Training System 空勤组训练系统 10. Air Turbine Starter 空气涡轮启动系统
ATSC 1. Air Traffic Service Communication 空中交通服务通信 2. Air Technical Service Command (美国陆军航空兵)航空技术勤务部
ATSD Airborne Traffic Situation Display 空中交通情况显示器
ATSG Acoustic Test Signal Generator 声测试信号发生器
ATSJEA Automated Test System for Jet Engine Accessories 喷气发动机附件自动测试系统
ATSL Ada Technology Support Laboratory 语言技术支持实验室
ATSM Advanced Tactical Stand-off Missile 先进的防空区外发射的导弹
ATSMP Air Traffic Service Message Processor 空中交通服务信息处理器
ATSOCC Applications Technology Satellite Operations Control Center 应用技术卫星操作控制中心(美国)
ATSR Along-Track Scanning Radiometer 纵向(沿航

迹)扫描辐射计
ATSS 1. Acquisition Tracking Subsystem 截获跟踪分系统 2. Air Transport Security Section 航空运输安全处(英国空军) 3. Automatic Target Scoring Systems 自动瞄准实弹考核射击系统 4. Augmented Target Screening Subsystem 增强目标分选子系统
ATSU 1. Accelerator Test Stand Upgrade 改进型加速器实验台 2. Air Traffic Service Unit 空中交通服务装置,空中交通服务单元
ATT 1. Advanced Tactical Transport 先进战术运输机 2. Automatic Target Tracking 自动跟踪目标
ATT REF Attitude Reference 姿态基准
attach 1.配属,附加 2.连接,贴上
attached 1.附加的,附属的,配属的 2.连接的
attached shock wave 贴体波,附体激波
attached vortex 附着涡
attachment 1.附着,连接,固定 2.附件 3.接头
attachment fitting 连接接头
attachment line 连接线
attachment point 附着点
attachment resistance 附体阻力
attack 1.突击,攻击,袭击 2.(疾病)侵袭 3.着手,投入
attack aircraft 强击机,攻击飞机,突击飞机
attack airplane 1.强击机 2.攻击机
attack angle 攻角,迎角,冲角
attack assessment 攻击效果估计
attack carrier 攻击(型)航空母舰
attack director 射击指挥仪
attack fighter 攻击战斗机
attack from horizon level 水平攻击,水平轰炸
attack geometry 攻击几何图形
attack helicopter 武装直升机
attack impact 命中
attack plotter 射击标图器
attack support 攻击支援
attack value 攻击值
attacker 1.攻击者 2.攻击飞行器
attain 达到,获得
attainable equilibrium 可达到的平衡
attainable equilibrium set 可达到的平衡点集
attainable equilibrium state 可达到的平衡状态
attainable steady state 可达到的稳定状态
attained pose 达到位姿
attained pose drift 实际位姿漂移
ATTC 1. Automatic Take-off Thrust Control 自动起飞推力控制 2. Aviation Technical Training Center 航空技术训练中心(美国空军)
ATTCS Automatic Take-off Thrust Control System 起飞推力自动控制系统

ATTD Attitude 姿态
ATTEN Attenuator 衰减器,阻尼器
attendant 1.附属品 2.随行的,附带的
attention 1.注意力 2.(口令)立正
attention channelization 注意力过分集中
attention coning 注意力过分集中,注意范围缩小
attention sharing 注意力分配
attention span 注意范围
attention span test 注意广度试验
attention-getter (故障)告警显示灯
attenuant 稀释剂
attenuate 1.稀释,冲淡 2.衰减
attenuating 衰减
attenuation 1.衰减 2.减小 3.稀释 4.丧失杀伤力(指弹头杀伤体)
attenuation coef 衰减系数
attenuation compensation 衰减校正
attenuation constant 衰减常数
attenuator 衰减器
Attinello flap 阿提奈罗襟翼,吹气襟翼
ATTITB Air Transport and Travel Industry Training Board 航空运输和旅游工业训练局(英国)
attitude 姿态
attitude accuracy 姿态精度
attitude acquisition 姿态捕获
attitude and pointing control system 姿态与瞄准控制系统
attitude angle 姿态角
attitude angle transducer 姿态角传感器
attitude angular velocity 姿态角速度
attitude bandwidth 姿态频带宽度
attitude bar 1.(直升机的)姿态稳定杆 2.(仪表的)姿态指针,地平指针,地平指示线
attitude control 姿态控制(飞行)
attitude control accuracy 姿态控制精度
attitude control and stabilization 姿态控制与稳定
attitude control law 姿态控制规律
attitude control loop 姿(态)控(制回)路
attitude control mode 姿态控制模式
attitude control problem 姿态控制问题
attitude control quantity 姿态控制量
attitude control rocket engine 1.姿态控制火箭发动机 2.姿控火箭发动机
attitude control rocket motor 1.姿态控制火箭发动机 2.姿控火箭发动机
attitude control subsystem 1.姿态控制分系统 2.姿控分系统
attitude control system 姿态控制系统
attitude control thruster 姿控推力器

attitude control trainer	姿态控制训练器
attitude control unit	姿(态)控(制)装置
attitude controller	姿态控制器
attitude correction	姿态修正
attitude datum	姿态基准
attitude description	姿态描述
attitude determination	姿态确定
attitude determination accuracy	姿态确定精度
attitude determination algorithm	姿态确定算法
attitude determination system	姿态确定系统
attitude determination system for agile spacecraft	灵活航天器的姿态确定系统
attitude director	指引地平仪
attitude director indicator	1.姿态指引指示器 2.指引地平仪
attitude disturbance	姿态扰动
attitude drift	姿态漂移
attitude dynamics	姿态动力学
attitude dynamics of spacecraft	航天器姿态动力学
attitude error	姿态误差
attitude error metric	姿态误差度量
attitude error rectification	姿态误差校正
attitude estimate	姿态估计
attitude estimation	姿态估计
attitude failure flag	指引地平仪故障旗
attitude feedback	姿态反馈
attitude geometry	姿态几何
attitude gyro	姿态陀螺(仪)
attitude heading reference system	姿态航向基准系统
attitude hold	姿态保持
attitude indicator	指引地平仪,姿态指示器
attitude information	姿态信息,空间方位信息
attitude instability	姿态失稳
attitude jet	1.姿态控制喷气发动机 2.姿态控制推力器
attitude jitter	姿态(信号)跳动
attitude keeping	姿态保持
attitude maneuver	姿态机动
attitude matrix	姿态矩阵
attitude measurement	1.姿态测量 2.姿态调整 3.调姿
attitude measurement accuracy	姿态测量精度
attitude measurement system	姿态测量系统
attitude mode	姿态模式
attitude motion	姿态运动
attitude motor	姿态控制发动机
attitude observation	姿态观测
attitude of flight vehicle	飞行器姿态
attitude of satellite	卫星姿态
attitude parameter	姿态参数
attitude parameterizations	姿态参数
attitude perception	空间位置状态知觉
attitude prediction	姿态预测
attitude processing electronics	姿态(数据)处理电子设备
attitude quaternion	姿态四元数
attitude quickness	姿态敏捷
attitude rate	姿态角速度
attitude reacquisition	姿态再捕获
attitude reconstruction	姿态重构
attitude reference	姿态基准
attitude reference platform	姿态基准平台
attitude reference symbol	姿态基准符号
attitude reference system	姿态基准系统
attitude regulation	姿态调节
attitude representation	姿态表示
attitude sensor	姿态传感器
attitude solution	姿态解算
attitude sphere	姿态球
attitude stability	姿态稳定度
attitude stabilization	姿态稳定
attitude stabilization problem	姿态稳定问题
attitude state	姿态状态
attitude synchronization	姿态同步
attitude threshold	姿态界限,姿态灵敏阈
attitude thruster	姿态(控制)推力器
attitude time	姿态时间
attitude time history	姿态时间历史
attitude tracking	姿态跟踪
attitude transient	(飞行)姿态瞬变过程
attitude vector	姿态向量
attitude-determination	姿态确定
attitude-rate	姿态速率
ATTLA Air Transportability Test Loading Agency	空运适应性试验装载署(美国)
ATTMA 1. Advanced Transport Technology Mission Analysis 先进运输技术任务分析 2. Advanced Tactical Transport Mission Analysis 先进战术运输任务分析	
ATTRA Automatic Tracking Telemetry Receiving Antenna	自动跟踪遥测接收天线
attraction	引力,吸力
attractive electrostatic force	有吸引力的静电场力
attractive feature	有吸引力的特点
attractive force	吸引力
attractor	吸引子
attractor dimension	吸引子维数
attractor state	吸引子的状态
ATTRAS Automatic Telemetry Tracking Antenna System	自动遥测跟踪天线系统
attribute	属性,特征,标志

attribute data　品质数据,质量特征数据
attribute definition　属性定义
attribute definition stage　属性定义阶段
attribute testing　品质试验,质量特性试验
attrition　耗损,消耗
attrition coefficient　摩擦系数
attrition rate　损耗率
attrition test　磨损试验
ATTS　1. Assembly Target Time System 装配目标时间系统 2. Automatic Telemetry Tracking Station 自动遥测跟踪站
ATU　Automatic Tracking Unit 自动跟踪装置
ATUC　Air Transport Users Committee 航空运输用户委员会(英国民航局)
ATV　Ariane Transfer Vehicle 阿里安运货飞行器
ATVC　Ascent Thrust Vector Control 上升段推力向量控制
ATVS　Advanced Television Seeker 先进电视导引头
ATW　Aerospace Test Wing 航空航天试验联队
ATWS　Automatic Track-While-Scan 自动边扫描边跟踪
atypical characteristic　非典型特性
AU　1. Accounting Unit 货币计算单位(欧洲航天局) 2. Aircraft Utilization 飞机利用率 3. Astronomical Unit 天文单位 4. Arithmetical Unit 运算单位,运算组件
Aubert's illusion　奥伯氏错觉
AUD　Audio 声音的,听觉的
AUD SEL　Audio Selector 音频选择器
audibility　可闻度
audibility test　听力测定
audibility threshold　听阈
audible alarm　声音报警(信号、设备)
audible sound　可听声
audio　1. 声频的,音频的 2. 音响的,听觉的
audio and video monitoring　声音和图像监视
audio box　声谱控制盒
audio environment simulation　音响环境仿真
audio frequency　声频,音频
audio frequency amplifier　音频放大器
audio frequency magnetic field method　(天然)音频磁场法
audio frequency oscillator　音频振荡器
audio frequency transformer　音频变压器
audio generator　音频发生器
audio terminal　音频终端
audio warning　声音报警
audiofrequency range　声频带,声域
audiogram　听力图

audiometer　听力计,听度计
audiometric room　听力测试室
audiometry　测听术,听力测定法
audiometry sensation area　听觉区
audiorange　音频范围
audiphone　助听器
audit　1. 审计,(数据)检查 2. 结构疲劳查验
auditorium　礼堂,大会堂
auditory acuity　听力
auditory defect　听力丧失
auditory discrimination　听觉辨别
auditory display　听觉显示器
auditory fatigue　听觉疲劳
auditory perception　听知觉
auditory sensation area　听觉投射区
auditory signal　听觉信号
auditory stimulus　听觉刺激
auditory threshold　听阈,听觉阈
auditory warning　听觉告警
AUEW　Amalgamated Union of Engineering Workers 工程师联合会(英国)
Auger effect　俄歇效应
Auger electron spectroscopy　俄歇电子能谱(学)
auger in　〈口语〉在螺旋中坠毁
Auger shower　俄歇族射(宇宙线)
augment　扩大,增加
augmentability　可扩充性
augmentation　1. 增大,加强,加力 2. 增(加)量
augmentation concept　增强的概念
augmentation loop　增强环路
augmentation ratio　1. 加力比 2. 加力度 3. 升推比
augmented　1. 增大的,加强的 2. 有加力燃烧室的,加力的
augmented aircraft　增强飞机
augmented code　扩充码,增信码
augmented condition　增广的状态
augmented electrothermal monopropellant hydrazine system　增强电热式单元肼推进系统
augmented matrix　增广矩阵
augmented state　增广的状态
augmented state space　增广的状态空间
augmented system　增广系统
augmented test condition　增强测试条件
augmented transition network　扩展的转移网络
augmented turbofan　加力涡扇
augmented turbojet　加力涡轮喷气
augmenter tube　排气加力管
augmentor　1. 增强器,放大器 2. (发动机的)加力装置,加力室

augmentor wing　环量可控翼

AUM　1. Advanced Underwater Missile 先进的潜射导弹 2. Air-to-Underwater Missile 空对水下导弹,空潜导弹

auntie　〈口语〉反导弹导弹

AUOS　Automated Unified Orbital Station 自动化联合轨道站(苏联,科学卫星)

AUR　All-Up Round 整装弹,整装武器

aural　1. 听觉的,音响的 2. 电视伴音

aural acquisition　听觉目标捕获

aural alerting device　音响告警装置

aural barotrauma　耳气压伤

aural high-speed warning　音响超速告警器

aural null　听觉零点

aural vertigo　耳性眩晕

Aureole　神光晕卫星

auric　1. 金的,含金的 2.【化】正金的,三价金

Aurora　极光卫星

aurora　极光

aurora australis　南极光

aurora borealis　北极光

aurora polaris　极光,北极光

Aurora-7　曙光7号飞船

Aurorae　曙光卫星

auroral disturbance　极光干扰

auroral electro jet　极光带电集流

auroral kilometric radiation　极光千米波辐射

auroral oval　极光卵形环

auroral particles　极光粒子

auroral radar　极光雷达

auroral zone　极光区,极光带

aurum　【化】金(化学元素,拉丁文,符号 Au)

AUS　Airspace Utilization Section 空域利用处

AUSA　Association of US Army 美国陆军协会

AUSCS　Australia Communication Satellite 澳大利亚通信卫星

ausforming　奥氏体形变,形变热处理

austempering　贝氏体等温淬火

austenite　奥氏体

austenitic steel　奥氏体钢

australis　南极光

AUT　Advanced User Terminal 先进的用户终端

autonomous　自主的,独立(工作)的

AUTEC　Atlantic Undersea Test and Evaluation Center 大西洋海底试验和鉴定中心

AUTH　Authorize 授权

authentication　证实,文电鉴别

authenticator　文电鉴别符号

authority　1. 主管当局 2. (系统等的)效能(尤指可靠性) 3. 权(权力,权限,权威)

authorization　1. 授权,委任 2. 核准,审定,认可,公认

authorization bill　授权法案

authorized　核准的,制式的,公认的

authorized equipment　制式装备,制式设备

authorized explosive　制式炸药

authorizing officer　有批准权的军官

AUTO　Automatic 自动的

auto alarm　1. 自动警报器 2. 自动警报信号

AUTO APP　Automatic Approach 自动进近

auto CARP　automatic Computed Air Release Point 自动计算的空中投放点

auto flying　自动控制飞行,自控飞行

auto hover　自动悬停

auto start flag　自动起始标记

auto transformer　自动变压器

auto vac　真空箱

AUTOXFMR　Autotransformer 自耦变压器

auto(matic) trim　自动配平

auto(matic)-lean　自动贫油

auto(matic)-rich　自动富油

auto(matic)-weak　自动贫油

autoacceleration　自动加速

auto-approach　自动进场着陆,自动进近

autobalance tab　自动补偿片

autobrake system　自动刹车系统

autocartograph　自动测图仪

autocatalysis　自动催化

autochangeover　自动转换

autochrome　彩色胶片

autoclasis　自裂

autoclave　1. 高压容器,高压锅 2. 高压炉

autoclave moulding　热压釜成形

autocoarse pitch　自动调距

autocollimator　自动准直管,自动准视仪

autocontrol　自动控制,自动调节

auto-convergence　自会聚

autocorrelation　自相关

autocorrelation analysis　自相关分析

autocorrelation function　自相关函数

autocorrelator　自相关器

autocoupler　自动耦合器

auto-covariance　自协方差

autodecrement address　自减地址

Autodin　automatic digital network 自动数字网络(美国空军)

AUTODIN/AUTOVON　Automatic Digital/Voice Network 自动数字信息通信/话音传输网

autodoping　自掺杂

auto-draft	自动制图
autodyne	1.自差,自拍 2.自差接收机
autodyne oscillator	自差振荡器
autodyne radio fuze	自差式无线电引信
autodyne receiver	自差接收机
autoexitation	自激,自激振荡
autofeathering	自动顺桨
autoflare(out)	1.(着陆前)自动拉平 2.自动拉平系统
autofocus	自动聚焦
autofocus mechanism	自动聚焦机构
auto-follower	随动系统
autogate	自动波门
autogeneous pressurization	自生增压
autogeneous pressurization system	自生增压系统
autogiration	自(动旋)转
autogiro	旋翼飞机
autograph	自动测图仪
autographic instrument	自记仪器,自动图示仪器
auto-guider	自动导星装置
autogyro	自转旋翼机
autohomeomorphism	自异质同晶(现象)
autoignited combustion	自动点火燃烧
autoignited combustion condition	自动点火燃烧条件
autoignited combustion limit	自动点火燃烧极限
autoignition	1.自动点火 2.自燃
autoignition boundary	自动点火界限
autoignition correlation	自动点火相关性
autoignition delay	自动点火延迟
autoignition limit	自动点火限制
autoignition model	自动点火模型
autoignition pressure	自动点火压力
autoignition propellant	自燃推进剂
autoignition study	自燃研究
autoignition temperature	自燃温度
autoignition test	自燃测试
autoignition testing	自燃测试
autoincrement address	自增地址
autoindex	(计算机)自动变址,自动变址数
autoindex system	自动变址系统
autokinesis	自感运动错觉
autokinetic effect	动感
autokinetic illusion	自感运动错觉
AUTOLABS Automatic Low Altitude Bombing System 自动低空轰炸系统	
autoland	1.自动着陆 2.英国生产的自动着陆系统专用名
autoland system	自动着陆系统
auto-loader	自动装填器
autoloading	自动加载,自动装载
auto-manual mechanobalance	半自动机械天平
Automap	地图投影器
automat	自动装置,自动电门
automate	自动化,使自动操作
automated air defence system	自动化防空系统
automated cfd	自动计算流体力学
automated data system	自动数据系统
automated decision	自动决策
automated decision system	自动决策系统
automated design	自动设计
automated ground engine-test system	发动机自动化地面试车系统
automated method	自动化的方法
automated mission	自动化任务
automated optimization	自动优化
automated optimization loop	自动循环的优化
automated solution	自动化解决方案
automated testing	自动化测试
automated verification system	自动验证系统
automatic action mine	自发地雷
automatic aiming	自动瞄准
automatic backup	自动接替
automatic back-zero system	自动回零系统
automatic biological station	自动生物站
automatic blade-fold system	(直升机)桨叶自动折叠系统
automatic bridge	自动电桥
automatic brightness control	自动亮度控制
automatic cable hover	自动电缆悬停
automatic cartography	自动化制图
automatic coding system	自动编码系统
automatic command	自动指令
automatic command to line of sight	瞄准线自动控制,自动调整瞄准线
automatic control	自动控制
automatic control engineering	自动化控制工程
automatic controller	自动控制器
automatic coordinate plotter	自动坐标绘图器
automatic data collection buoy	自动数据搜集浮标
automatic data editing and switching system	数据自动编辑转接系统
automatic data interchange system	自动数据交换系统
automatic data processing	自动数据处理
automatic data reduction equipment	数据自动简化装置,数据自动压缩装置
automatic decode system	自动译码系统
automatic decrab system	自动抗偏流系统,自动抗侧风系统
automatic dependent surveillance	自动间接监视

automatic destruct program （导弹在飞行中）自（动炸）毁程序
automatic digital online instrumentation system 自动数字联机仪表系统
automatic direction finder 1.自动测向仪 2.全自动无线电罗盘
automatic direction finding 自动测向
automatic display and plotting system 自动显示与标图系统
automatic electronic all-weather landing system 自动电子全天候着陆系统
automatic error correction 自动纠错
automatic error correction system 自动纠错系统
automatic error request 自动检错重发
automatic escape sequence 自动应急离机程序
automatic exposure control device 自动曝光控制装置
automatic extension gear 1.自动放起落架装置 2.自动延伸装置
automatic extraction 自动录取
automatic failure monitor 故障自动检测
automatic feed mechanism 自动进弹机构，自动装弹机
automatic feed system 自动送进系统
automatic fidelity control 自动逼真度控制
automatic fine control 自动精调，自动精确控制
automatic fire 自动射击
automatic fire control system 自动火力控制系统
automatic fire power 连发火力
automatic flagman 自动信号员
automatic flight control computer 自动飞行控制计算机
automatic flight control system 自动飞行控制系统
automatic flight 自动飞行
automatic flight path recovery system 自动飞行轨迹改出系统
automatic flow control 自动流量调节
automatic fluorescent particle counter 荧光粒子自动计数器
automatic focusing device 自动调焦装置
automatic fuel control 自动燃料控制
automatic gain control 自动增益控制
automatic gear ratio changer 力臂调节器
automatic ground-controlled interception 地面自动引导截击
automatic guidance system （导弹）自动制导系统
automatic gun charger 自动装弹机构
automatic gun gas purging equipment 自动机炮排烟装置，自动机炮火药气体吹除装置
automatic gun layer 自动射击瞄准具
automatic hovering control 自动悬停控制
automatic hovering control system for helicopter 直升机自动悬停控制系统
automatic ignition 自动点火
automatic illusion 自体运动错觉
automatic inflation modulation parachute 自动充气调节伞
automatic internal diagnosis 自动内部诊断（计算机的）
automatic interplanetary station 自动行星际站
automatic kick-off-drift facility （接地前）自动消除偏流修正角装置
automatic landing 自动着陆
automatic landing positioning system 自动着陆定位系统
automatic lead computing sight 自动计算提前角的瞄准具
automatic level 自动安平水准仪
automatic load stabilization system 自动载荷稳定系统
automatic lock-on 自动跟踪
automatic longitudinal stability 自动纵向稳定性
automatic low altitude bombing system 低空自动轰炸系统
automatic maneuver device system 机动飞行自动增升装置
automatic maneuvering attack system 自动机动攻击系统
automatic manual station 自动-手动操作器
automatic measurement 自动测量
automatic mechanobalance 自动机械天平
automatic missile checkout system 导弹自动测试系统
automatic mixture-ratio control 混合比自动调节
automatic mode 自动模式
automatic monitoring system 自动监控系统
automatic navigator 自动导航仪
automatic numbering equipment 自动编号设备，自动发号设备
automatic operation mode 自动工作方式，自动工作状态
automatic overload control system 自动过载控制系统
automatic overshoot equipment 自动复飞装置
automatic parachute 1.自动开伞式降落伞 2.自动开伞
automatic phase control 自动相位控制
automatic phase lock 自动相位锁定
automatic picture transmission 自动图像传输
automatic pilot 自动驾驶仪
automatic pitch-coarsening （螺旋桨）自动变大距
automatic placement and routing 自动布局与布线
automatic plotter 自动标图器
automatic plotting 自动标图
automatic power control 自动功率控制
automatic predictive maintenance 自动预测性维护
automatic pressure breathing 自动加压呼吸
automatic pressure control 压力自动调节

automatic program control 自动程序控制
automatic programming 自动程序设计
automatic radar reporting and computing system 雷达情报自动处理与计算系统
automatic range tracking 距离自动跟踪
automatic reference system （导弹的）自动参考系统
automatic regulation 自动调节
automatic relay global observation system 全球监测自动中继系统
automatic repeat request system 自动重复请求系统
automatic reporting system 自动报知系统
automatic reset circuit breaker 自动复位断电器
automatic reverse pitch 自动反架
automatic roll-out guidance 自动滑出跑道导引装置
automatic sampler 自动采样器
automatic sear 自动扣机，装弹扣机
automatic search circuit 自动搜索电路
automatic search jammer 自动搜索干扰机
automatic selective feathering 自动选择顺桨
automatic sequencing 自动定序
automatic slat 自动前缘缝翼
automatic stall warning and stall inhibitor control system 自动失速警告和防失速控制系统
automatic submerged arc welding 自动埋弧焊，焊剂层下自动弧焊
automatic tab system 自动调整片系统
automatic target designation 自动目标指示，自动目标照射
automatic target detection 自动目标探测
automatic target finder 自动目标探测器
automatic target follower 自动目标跟踪器
automatic target handoff system 自动目标数据交接系统
automatic target recognition, identification and detection 自动目标识别、鉴定与探测
automatic telephone 自动电话
automatic television positioning system 自动电视定位系统
automatic terminal 自动化航站
automatic terminal information service 自动终端情报服务
automatic terrain avoidance system 自动地形回避系统
automatic terrain following system 自动地形跟随系统
automatic terrain recognition and navigation 自动地形识别和导航
automatic test 自动化测试
automatic test equipment 自动测试设备
automatic test module 自动测试模块
automatic test pattern generation system 自动测试模式生成系统
automatic threat countering 对威胁自动对抗
automatic timing control 自动定时控制
automatic toss 自动拉起轰炸，自动抛投
automatic touchdown release 接地自动解除装置
automatic track while scanning 自动边扫描边跟踪，自动扫描跟踪
automatic tracking 自动跟踪
automatic traffic scheduling equipment 空中交通自动调度设备
automatic transition control 自动过渡控制
automatic transition for helicopter 直升机自动过渡飞行
automatic trim system 自动配平系统
automatic trouble diagnosis 自动故障诊断
automatic tuning 自动调谐
automatic TV tracking system 自动电视跟踪系统
automatic video digitizer 自动视频数字转换器
automatic voice alerting device 自动话音告警装置
automatic weather station 自动气象站
automatic welding 自动焊
automatically activated battery 自动激活电池
automatically activated zinc-silver battery 自动激活锌银电池
automatically programmed tools 自动数控程序
automatic-detector 自动检测器
automaticity 自动度
automatics 自动学
automation 1.自动化 2.自动学
automatization 自动化
automaton 自动机
autometering 自动计量
automobile 汽车，自动车，自动的，汽车的
automodulation 自调制
automorphism group 自同构群
automotive 1.自动的，自行的，自动推进的 2.汽车的，机动车的
automotive wheel 汽车车轮
autonavigator 自动领航仪，自动领航系统
autonomous aerospace 自主飞行
autonomous aerospace vehicle 自主飞行器
autonomous analysis 自主分析
autonomous attitude determination 自主姿态确定
autonomous attitude sensor 自主式姿态传感器
autonomous capability 自治能力
autonomous channel operation 独立通道操作
autonomous control 自主控制
autonomous decision 自主决策
autonomous function 1.自主功能 2.自治函数
autonomous functionality 自主功能

autonomous guidance　自主式制导
autonomous guidance weapon　自主制导武器
autonomous helicopter　自动直升机
autonomous intelligent control　自主智能控制
autonomous landing guidance　自主式着陆引导
autonomous navigation　自主导航,自主式领航,自备式领航
autonomous navigation of satellite　卫星自主导航
autonomous operation　自主操作
autonomous rendezvous　自主交会
autonomous robot　自主式机器人
autonomous satellite　自主卫星
autonomous sensor　自主传感器
autonomous space　自主空间
autonomous space robot　自主空间机器人
autonomous station keeping　自主位置保持
autonomous system　自主系统,自治系统
autonomous trajectory　自治运动轨迹
autonomous vehicle　自主运载器
autonomous work　自治工作
autonomously guided standoff weapon　自主制导防空区外发射的(空对地)武器,自主制导远射武器
autonomy　自治权
autopilot　自动驾驶仪
autopilot all weather landing　全天候自动着陆
autopilot annunciator　自动驾驶仪信号牌
autopilot architecture　自动驾驶仪结构
autopilot design　自动驾驶仪设计
autopilot hardware　1.真实自动驾驶仪 2.自动驾驶仪部件
autopilot lag　自动驾驶仪滞后
autopilot mode annunciation　自动驾驶仪状态告示
autopilot system　自动驾驶系统
autopilot target　自动驾驶仪的目标
autopilot-disconnect　自动驾驶仪断开
autopilot-guidance　自动驾驶仪指导
autopitch vane　俯仰自动装置风标式传感器
autoplacement　自动布线,自动布局
autoplan　自动飞行方案显示器
autopolymer　(化学的)自聚物
auto-polymerization　自动聚合作用
autopower　功率自动调节装置
auto-power reserve　自动增推,自动应急推力
auto-power spectrum　自动功率谱
autopsy　1.飞机残骸检查 2.(事故调查中的)尸检,尸体剖检
AUTOPSY　Automatic Operating System 自动操作(运转)系统
autoradiography　放射自显影

autoregression　自回归
autoregression model　自回归模型
autorotation　1.自转 2.(直升机)旋翼自转 3.(飞机)失速旋转 4.(飞机非失速性)沿螺旋线滚转 5.(直升机飞行训练中)关车下降
autorotation landing　(旋翼)自转着陆
autorotational spin　自旋稳定
autorotative descent　自转下降
autorotative glide　自转下滑
autoroute　自动布线
AUTOSCAN　Automatic Satellite/Computer Aid to Navigation 自动卫星/计算机辅助导航
auto-separation　(弹射座椅的)人椅自动分离
AUTOSEVOCOM　Automatic Secure Voice Communications 自动保密电话通信(系统)
autoslat　自动式前缘缝翼
autoslot　自动调节翼缝
AUTOSPEC　Automatic Single Path Error Correction 自动单程纠错
autospectra　自动光谱
autospotter　着弹自报机
autostability　增稳,自动安定性,自稳定性
autostabilizer　增稳器
autostairs　(飞机上的)自动收放梯
autostarter　自动起动机
AUTOSTRAD　Automated System for Transportation Data 自动化数据传输系统
autosyn　自动同步机,自动整步器,交流同步器
autosynchronous　自动同步的
autotest program　自检程序
auto-throttle　自动油门
autothrottle control　自动油门控制
autothrottle equipment　自动油门装置
autothrottle system　自动油门系统
autotimer　自动计时器
auto-touchdown　自动接地
autotracking　自动跟踪
autotransformer　自动变压器
autotrim　自动配平
auto-tuning telemetry receiver　自动调谐式遥测接收机
autovibrations　自动振动
Autovon　Automatic Voice Network 自动话务网,自动语言传播网
autumnal equinox　秋分(点)
AUVS　Association for Unmanned Vehicle Systems 无人飞行器系统协会(美国)
AUW　1. Advanced Undersea Weapons 先进水下武器 2. All-Up Weight 全重,总重量
AUX　Auxiliary 辅助

AUX NAV Auxiliary Navigation 辅助导航
auxiliary 1.辅助的,副的,备份的 2.辅助装置
auxiliary armament 辅助军械,辅助武器
auxiliary attitude control system 辅助姿(态)控(制)系统
auxiliary booster grain 扩爆药性
auxiliary bus 辅助总线
auxiliary cavity 辅助腔
auxiliary electrical power source 辅助电源
auxiliary equation 辅助方程
auxiliary equipment 辅助设备
auxiliary fin 辅助垂尾
auxiliary flight control system 辅助飞行操纵系统
auxiliary fluid ignition 起动燃料点火
auxiliary fuel tank 副油箱,辅助燃料箱
auxiliary function word 辅助功能字(计算机的)
auxiliary fuze antenna 引信辅天线
auxiliary game 辅助游戏
auxiliary game of degree 辅助游戏的程度
auxiliary gun 辅助炮
auxiliary hover trim 辅助悬停配平
auxiliary inlet door 进气道辅助进气门
auxiliary parachute 引导伞
auxiliary point 辅助瞄准点
auxiliary power 辅助电源
auxiliary power generation 辅助发电
auxiliary power system 辅助电源系统
auxiliary power unit 辅助动力装置
auxiliary propulsion 辅助推进
auxiliary propulsion system 辅助推进系统
auxiliary reference pulse 辅助参考脉冲
auxiliary relationship 辅助关系
auxiliary rigging lines 辅助吊伞绳
auxiliary rocket 辅助火箭,助推火箭
auxiliary rod 副连杆
auxiliary state 辅助状态
auxiliary tank 副油箱,辅助油箱
auxiliary target 辅助目标,试射目标
auxiliary vehicle 副车
AV 1. Added Value 附加价值 2. Aerospace Vehicle 航空航天(飞行)器 3. Air Vehicle 航空器 4. Aircraft (Airplane) Validation 飞机验证 5. Audio-Visual 声像
AV/D Alternate Voice/Data 话音/数据变换
AVA Automatic Voice Advice 自动语音忠告
AVAD Automatic Voice Alerting Device 自动话音告警装置
AVADS Auto-track Vulcan Air-Defense System 自动跟踪火神式防空系统(美国)
AVAIL Available 适用的,可用的

availability 1.(飞机的)可用性 2.可用度 3.到货期
availability goal 可用性目标
availability ratio 可用率
available 1.可用的,有效的 2.可得到的,现有的
available air 现有的空气
available computational time 可用的计算时间
available control 可用的控制
available control authority 可控制权限
available datum 可用的数据
available energy 可用的能量
available exit 可用的出口
available experimental datum 可用的实验数据
available fuel 可用的燃料
available fuel budget 可用的燃料预算
available fuel-cooling capacity 可用的燃料冷却能力
available ground 有效地面
available ground time 有效地面时间
available information 有用信息
available intent 可用的意图
available machine time 机器可用时间
available mode 可用状态
available payload 有效载荷
available power 可用功率
available resource 可用资源
available seat-mile 可用客座英里
available space 可用空间
available test 有效测试
available ton-mile 可用吨英里
avalanche 1.雪崩 2.顶点带快滚的筋斗
avalanche breakdown 雪崩击穿
avalanche photodiode 雪崩光电二极管
avanlanche voltage 雪崩电压
AVASIS Abbreviated Visual Approach Slope Indicator System 简化的目标进场下滑角指示系统
AV-AWOS Aviation-Automated Weather Observation System 航空自动气象观测系统
AVC 1. Automatic Variable Camber 自动变弯度 2. Automatic Volume Control 自动音量控制
AVCAL Aviation Consolidated Allowance List 航空部件统一定额表
Avcat 航空用重煤油
AVCG Automatic Vapor Crystal Growth 自动升华晶体生长
AVCGU Automatic Vapor Crystal Growth Units 自动升华晶体生长装置
AVCS Advanced Vidicon Camera System 先进的光导摄像管摄影系统
AVD Atmospheric Vehicle Detection 航空器探测
AVE 1. Aircraft (Airplane) Validation Engineering 飞

机变型工程 2. Automatic Volume Expansion 自动音量放大
avenue 1.道路,通路 2.手段
average 平均,平均数
average acceleration 平均加速度
average age at failure 发生故障时平均工作时间,平均故障工龄
average allowable concentration 平均允许浓度
average base 平均基础
average bias 平均偏差
average control 平均控制
average cross-country speed 平均中间着陆的航线飞行速度
average delay 平均延迟
average deorbit 平均处理
average diameter 平均直径
average discharge 平均流量
average distance 平均距离
average droplet 平均液滴
average dwell 平均驻留
average dwell time method 平均驻留时间方法
average eccentricity 平均偏心
average effectiveness 平均效果
average equivalence 平均等价
average equivalence ratio 平均当量比
average error 平均误差
average estimation 平均估计
average exit 平均出口
average exit velocity 平均出口速度
average flame 平均火焰
average gas 平均气
average gas temperature 平均气温度
average heat 平均热
average heat transfer coef 平均传热系数
average injection 平均注射
average inlet 平均入口
average inlet air temperature 平均入口空气温度
average input 平均输入
average ion 平均离子
average ion charge 平均离子电荷
average length 平均长度
average length scale 平均长度尺度
average model 平均模型
average molecular weight 平均分子量
average number 平均数
average one-firing/one-missile hit probability 单发射弹/导弹平均命中概率
average outgoing quality 平均出线质量(水平)
average oxidizer 平均氧化剂
average performance loss 平均性能损失
average pore 平均孔径
average pore diameter 平均孔隙直径
average pore size 平均孔径大小
average port 平均口
average position 平均位置
average position error 平均位置误差
average power 平均功率
average power meter 平均功率计
average power spectrum 平均功率谱
average pressure 平均压强
average regression 平均回归
average regression rate 平均回归率
average rotor 平均转子
average seal 平均轴封
average shock 平均冲击
average shock velocity 平均冲击速度
average single kill probability 单发射弹平均杀伤概率
average size 平均粒径
average speed 平均速度
average spray 平均喷雾
average stage length 平均航段长度
average strain 平均应变
average strain energy 平均应变能量
average strength 平均强度
average surface 平均表面
average surrogate 平均替代
average temperature 平均温度
average thrust 平均推力
average time 平均时间
average trip length 平均旅程长度
average trip time 平均航次时间
average value 平均值
average value of sine wave alternating current 正弦波交流电平均值
average vector 平均向量
average velocity 平均速度
average velocity vector 平均速度矢量
average voltage 平均电压
average wake 平均尾流
average weight 平均重量
averaged orbital elements 平均轨道因素
averaging 平均的
averaging method 平均法
aversion decision 规避决策
AVG 1. American Volunteer Group 美国志愿航空队 2. average 平均
Avgard 阿夫加添加剂
Avgas 航空汽油

AVGDIA　Average Diameter 平均直径
AVHRR　Advanced Very High Resolution Radiometer 先进甚高分辨率辐射计(美国)
AVI　Avoid Verbal Instructions 避免口头通知,撤销口头通知,废止口头通知
aviate　驾驶飞机
Aviateknikum Moscow Aviation Technical School　莫斯科航空技术学校
aviation　1.航空 2.航空兵 3.飞机 4.飞行术,航空学
aviation accident　飞行事故,航空事故
aviation accident pathology　航空事故病理学
aviation accident reconstruction　航空事故推断
Aviation Association　陆军航空兵协会(美国)
aviation biodynamics　航空生物动力学
aviation cadet　航空学员
aviation climate　航空气候
aviation climate divide　航空气候分界
aviation climatology　航空气候学
aviation community　航空界
aviation disease　航空病
aviation emergency escapement　航空救生
aviation engineering psychology　航空工程心理学
aviation epidemiology　航空流行病学
aviation ergonomics　航空工效学
aviation facilities　(舰上)助飞设备
aviation fuel　航空燃料
aviation fuels crew　(航母)航空油料员
aviation gasoline　航空汽油
aviation goggles　航空眼镜,航空护目镜
aviation hazard weather　航空危险天气
aviation hydraulic fluid　航空液压油
aviation hygiene　航空卫生
aviation industry　航空工业,飞机制造业
aviation kerosene　航空煤油
aviation medical examiner　航空体检医生
aviation medicine　航空医学
aviation mental hygiene　航空心理卫生
aviation pathology　航空病理学
aviation physiological training　航空生理训练
aviation physiological training program　航空生理训练计划
aviation physiology　航空生理学
aviation satellite communication　航空卫星通信
aviation standards　飞行标准
aviation toxicology　航空毒理学
aviation training ship　飞行训练舰,航空训练舰
aviation weather　可飞行天气
aviator　飞行员
aviator commander　(航母上的)航空部门长
aviator psychology selection　飞行员心理学选拔
aviatorial　航空评论
aviator's vertigo　飞行员眩晕
aviatrix　女航空员,女飞行员
Aviatrust　航空工业托拉斯
Aviavnito　Aviation Department of Vnito 全苏工程技术学会航空部
avicade　(飞机的)机群
aviette　1.小型飞机,轻型飞机(体育运动用) 2.小型滑翔机 3.人力飞机,人力航空器
avigate　领航,导航
avigation　1.领航,导航 2.领航学
avigator　〈口语〉空中领航员
avigraph　自动领航仪,航行计算仪
Avimid　阿维米德热塑性聚合物复合材料(商品名)
avionic　1.航空电子的 2.飞行器电子设备的
avionic system　航空电子系统
avionics　航空电子设备,航空电子学,航空电子技术
avionics application　航空电子设备的应用
avionics architecture　航空电子系统结构
avionics bus　航空电子总线
avionics chassis　航空电子设备机箱
Avionics Laboratory　空军航空电子设备实验室(美国)
avionics navigation system　航空电子导航系统
avionics system　航空电子系统
avionics system simulation　航空电子系统仿真
avionics technology　航空电子技术
avionics test bed　航空电子试验机
AVIT　Aircraft (Airplane) Validation Integrated Test 飞机有效性集成测试
AVLAN　Aviation Local Area Network 航空电子局域网
AVLF　Airborne Very-Low Frequency 机载甚低频
AVLM　Anti-Vehicle Land Mine 反车辆地雷
AVM　1. Airborne Vibration Monitoring 机载振动监视 2. Airborne Vibration Monitor 机载振动监测器
AVNIR　Advanced Visible and Near-Infrared Radiometer 先进可见光和近红外辐射计
AVO　Avoid Verbal Orders 禁用口头命令
avoid　避免,消除
avoid curve　停车不安全曲线
avoidance　防止,避免,回避
avoidance equipment　防撞装置
avoidance maneuver　规避机动
avoidance of range blind zone　距离避盲
avoidance system　回避系统
Avpin　航空规格硝酸异丙酯,航空起动燃料
AvPOL　Aviation Petrol, Oil and Lubricant 航空油料
AVR　Additional Validation Requirements 附加认证要求

AVRADA Aviation Research And Development Activity 航空研究与开发院(美国)

AVS 1. Air Vehicle Specification 航空器规范 2. Aviation Supply Ship 航空补给船(美国海军)

AVSF Avionics Supply Filter 航空电子电源滤波器

AVT Automatic Video Tracker 自动视频跟踪器

Avtag 飞机、气轮机燃用低辛烷油

Avtur 航空涡轮用煤油

AVUM Aviation Unit-level Maintenance 单位级航空维修

AW 1. All Weather 全天候 2. Airways 航空公司,航路,航线 3. Air Weapon 空袭兵器,空战兵器,机载武器,航空武器

AWA 1. All Wave Antenna 全波天线 2. Aviation Writers Association 航空作家协会(美国)

AWACS Airborne Warning And Control System 机载预警和控制系统

AWADS 1. Adverse Weather Aerial Delivery System 复杂气象空投系统 2. All-Weather Aerial Delivery System 全天候空中投送系统

AWAR Area Weighted Average Resolution 面积权重平均分辨率

award 给与,授予

AWARDS Aircraft Wide-Angle Reflective Display System 飞机广角反射显示系统

aware advanced warning of active radar emissions 先进的有源雷达(电磁波)发射警告(装置)

awareness 警觉性

AWARS Airborne Weather And Reconnaissance Systems 机载气象与侦察系统

AWAS Automated Weather Advisory Station 气象自动咨询站

AWAVS Aviation Wide Angle Visual System 航空广视角系统

away sector 背阳扇区

AWC 1. Air War College 空军学院(美国) 2. Airworthiness Certification 适航证 3. Army War College 陆军军事学院(美国)

AWCCV Advanced Weapons Carriage Configured Vehicle 有先进的武器挂架的飞机

AWCLS All-Weather Carrier Landing System 全天候(航空母)舰上降落系统,全天候着舰系统

AWCS 1. Air Weapons Control System 对空武器控制系统 2. Airborne Warning and Control System 机载警戒与控制系统 3. Automatic Weapons Control System 自动武器控制系统

AWD Airworthiness Division 适航性部(英国民航局)

AWDREY Atomic Weapon Detection, Recognition and Estimation of Yield 原子武器探测、识别和当量估计

AWDS 1. All-Weather Delivery System 全天候投送系统 2. Automatic Weather Distribution System 自动天气分布系统 3. Automated Wire Development System 自动导线设计系统

AWE All-up Weight Equipped 全装总重

A-Weapon 原子武器

AWESS Automatic Weapons Effect Signature Simulator 武器毁伤特征自动模拟器

AWG American Wire Gauge 美国线规

AWI 1. Aircraft Weight Indicator 飞机重量指示器 2. Air Weapons Instructor 航空武器教官 3. All-Weather Intercept 全天候截击

AWIN Aviation Weather Information 航空气象信息

AWIPS Advanced Weather Interactive Processing System 先进气象交互处理系统

AWIS 1. Aircraft With Initial Shortage 带有先天缺陷的飞机,缺件出厂的飞机 2. All Weather Identification Sensor 全天候(目标)识别传感器

AWL Airworthiness Limitation 适航性限制

AWLS All-Weather Landing System 全天候着陆系统

AWM 1. Aircraft (Airplane) Wiring Manual 飞机布线手册 2. Audio Warning Mixer 音频警告混响器 3. Average Working Man 平均工作人数(机组人员的) 4. Awaiting Maintenance 等待维修

AWN Automatic Weather Network 自动气象站网

AWO Assembly Work Order 装配工作指令

AWOP All-Weather Operations Panel 全天候飞行工作组(国际民航组织)

AWOS Automated Weather Observation System 自动气象观测系统

AWP 1. Aircraft Without Parts 缺体飞机 2. Awaiting Parts 等待零备件

AWPA Australian Women Pilots' Association 澳大利亚女飞行员协会

AWR 1. Airborne Weather Radar 机载气象雷达 2. Airworthiness Requirement 适航要求

AWRA Augmentor-Wing Research Aircraft 环量可控翼研究机

AWRS 1. Airborne Weather Reconnaissance System 机载气象侦察系统 2. Aircraft Weapon Release System 飞机武器投射系统

AWS 1. Air Ward System 空中监护系统 2. Air Weapon System 航空武器系统 3. Air Weather Service 航空气象局 4. Altitude Warning System 高度告警系统 5. Amphibious Warfare Communications 两栖作战通信系统 6. Audible Warning System 音响告警系统 7. Automatic Wing Sweep 机翼自动后掠

aws angle-of-attack 航空武器迎角

aws condition 航空武器条件

aws program 航空武器计划
aws region 航空武器区域
AWSACS All-Weather Stand-off Attack Control System 全天候防空圈外攻击控制系统(美国海军)
AWSAS All-Weather Stand-off Attack System 全天候防空圈外攻击系统
AWSO All-Weather Surface Observations 全天候地面观测
AWT 1. Altitude Wind Tunnel 高空风洞 2. Automated Wire Test 自动线路测试
AWTMS All-Weather Topographic Mapping System 全天候地形测图系统
AWTSS All-Weather Tactical Strike System 全天候战术攻击系统
AWY airway 航路
AXFL Axial Flow 轴向流动
axe 斧
axes 1.坐标轴 2.坐标系(axis 的复数)
axial 轴向的,轴的
axial acceleration 轴向加速度
axial agility 轴向敏捷性
axial agility metric 轴向敏捷性度量
axial and azimuthal roughness 轴向和方位角的粗糙度
axial and radial injection 轴向和径向喷射
axial and radial velocity 轴向和径向速度
axial baf 轴向隔板
axial cable 中心索,中轴索
axial cavity 轴向腔
axial chamber 轴向室
axial chord 轴向弦
axial chord length 轴向弦长
axial clearance 轴向间隙
axial component 轴向分量
axial component of ion beam 离子束的轴向分量
axial compression load 轴向压缩载荷
axial compression test 轴压试验
axial compressor 轴流式压气机
axial cord 中心绳
axial deck 轴向甲板
axial direction 轴向的
axial distance 轴向距离
axial domain 轴向域
axial efficiency 轴效率
axial electric 轴向电
axial electric field 轴向电场
axial electron 轴向电子
axial engine 轴流式发动机
axial family 轴向族
axial firing 轴向发射

axial flow 轴向流
axial flow compressor 轴流压缩机
axial flow compressor design 轴流压缩机设计
axial flow engine 轴流式发动机
axial focusing 向轴集中
axial force 轴向力
axial fuel 轴向燃料
axial gap 轴向间隙
axial grid 轴网
axial growth 轴向生长
axial heat 轴向热
axial injection 轴向注入
axial injection motor 轴向注入汽车
axial injector 轴向喷射器
axial ion 轴向离子
axial jet 轴向射流
axial length 轴向长度
axial load 轴向载荷,轴向加载
axial loading 轴向载荷
axial location 轴向位置
axial maneuver 轴向机动
axial maneuver performance 轴向机动性能
axial mode 轴向模态
axial momentum 轴向动量
axial motion 轴向运动
axial piston motor 轴向柱塞马达
axial piston pump 轴向活塞泵
axial placement 轴向位置
axial position 轴向位置
axial pressure 轴向压力
axial pump 轴流泵
axial ratio 轴比
axial reflector 轴反射镜
axial roughness 轴向粗糙度
axial shape 轴的形状
axial spacing 轴向间距
axial station 轴向站
axial step 阶跃轴
axial stiffness 轴向刚度
axial stress 轴向应力
axial temperature 轴向温度
axial thrust 轴向推力
axial thrust balancing devices 轴向力平衡装置
axial thruster 轴向推进器
axial tube 轴管
axial variation 轴向变化
axial velocity 轴向速度
axial velocity component 轴向速度分量
axial velocity contour 轴速度等高线

axial velocity of drop 轴向速度的下降
axial volume 轴向体积
axial vortex 轴向旋涡
axial-flow 轴流式
axial-flow compressor 轴流压气机
axial-flow pump 轴流泵
axial-flow turbine 轴流式涡轮
axial-force 轴向力
axial-force coefficient 轴向力系数
axial-injector 轴向喷射器
axial-lead resistor 轴绕线电阻器,轴绕线电阻
axially adjustable gyro 轴向可调陀螺
axis 轴(线),坐标轴
AXIS Atmospheric X-Ray Image Spectrometer 大气 X 射线成像光谱仪
axis of a gun 枪轴线
axis of bore 镗孔轴线
axis of freedom 自由轴(陀螺),自由度轴线
axis of maximum inertia 最大惯性轴
axis of piece 轴件
axis of projectile 弹轴
axis of rotation 旋转轴,转轴
axis of rudder 方向舵轴
axis of scan 扫描轴
axis of sighting 瞄准轴,瞄准线
axis of symmetry 对称轴,对称轴线
axis of the bore 孔的轴线
axis of the eye 视轴
axis perpendicular 轴垂直
axis point 轴点
axis ratio 轴比
axis set 轴组
axis switching 轴切换
axis torque 轴扭矩
axis transformation 坐标轴变换
axis-switching 轴切换
axis-switching phenomenon 轴切换现象
axisymmetric and asymmetric clearance 轴对称和非对称的间隙
axisymmetric axis 对称轴
axisymmetric body 轴对称体
axisymmetric case 轴对称情况
axisymmetric clearance 轴对称的间隙
axisymmetric condition 轴对称条件
axisymmetric flow 轴对称流
axisymmetric geometry 轴对称几何
axisymmetric inlet 轴对称进气道
axisymmetric intake 轴对称进气
axisymmetric jet 轴对称射流

axisymmetric liquid sheet 轴对称液体表
axisymmetric load 轴对称载荷
axisymmetric mode 轴对称模式
axisymmetric model 轴对称模型
axisymmetric nozzle 轴对称喷管
axisymmetric projectile 轴对称弹
axisymmetric simulation 轴对称模拟
axisymmetrical flow 轴对称流
axisymmetry 轴对称
axle 轴,车轴
AXP Axial Pitch 轴向俯仰
Ayame 菖蒲卫星
azeotropic 共沸的,恒沸点的
azication (Azimuth Indication)方位指示
azide 叠氮化物
azido 叠氮
azido plasticizer 叠氮增塑剂
azimuth 1.方位,方位角 2.炸弹弹道方向偏差 3.地平经度
azimuth adder 方位电路加法器
azimuth aerial 方位天线
azimuth aiming 1.方位瞄准 2.方位
azimuth alignment 方位取齐
azimuth alignment electrotheodolite set 方位校准电经纬仪
azimuth alignment set 方位校准仪
azimuth alignment system 方位对准系统
azimuth ambiguity 方位模糊
azimuth and elevation tracking 方位角与射角跟踪
azimuth angle 方位角
azimuth ballistic correction 弹道方位修正
azimuth caging 方位锁定
azimuth compiler 方位编辑程序
azimuth digital display instrument 方位角数字显示仪
azimuth direction 方位方向
azimuth elevation indicator 方位仰角显示器
azimuth error 方位角误差
azimuth finder 方位仪,测向仪
azimuth firing angle 射击方位角,发射方位角
azimuth guidance element 方位引导单元
azimuth gyroscope 陀螺半罗盘,方位陀螺仪
azimuth landing computer 着陆航向计算机
azimuth laying set 方位角瞄准器
azimuth locking 方位锁定
azimuth marker 方位标识器,方位标线
azimuth mounting 地平式装置
azimuth navigation system 方向导航系统
azimuth of photograph 照片方位角
azimuth plane 方位平面

azimuth pointing accuracy　方位定向精度
azimuth pointing range　方位定向范围(在方位方向上能定向工作的角范围)
azimuth quadrant　地平象限仪
azimuth reference　方位基准
azimuth reference pole　方位标
azimuth resolution　方位分辨力
azimuth to future position　方位提前量
azimuth tracking　方位角跟踪
azimuth training　方位瞄准,方位转动
azimuth type range finder　测合式测距仪
azimuthal　方位(角)的
azimuthal deformation　方位角变形
azimuthal direction　方位角方向
azimuthal ion　方位角离子
azimuthal location　方位
azimuthal map　方位图
azimuthal mode　方位角模式
azimuthal projection　方位投影
azimuthal roughness　方位角的粗糙度
azimuthal span　方位角跨度
azimuthal velocity　方位角速度
azimuthal vorticity　方位角涡
azimuth-damped navigation mode　全阻尼导航模态
azimuth-distance positioning system　极坐标定位系统,方位距离定位系统
azimuth-elevation display　方位-仰角显示器
azimuth-offset processing　方位偏置处理
azine　吖嗪氮,连氮
azon　阿松式制导炸弹
azote　氮
azotizing　1.氮化(作用) 2.氮化的
azran (azimuth-range)　测向测距定位
AZS　Automatic Zero Set 1.自动归零 2.自动归零装置,自动零位调整装置
Aztec　"阿兹台克"作战
Azur　阿祖尔卫星
AZUSA　Azimuth, Speed and Altitude 方位、速度和高度

B

B　1. Blue 蓝色 2. Base 基极,基地

B damage　B级毁伤(飞机受到这种毁伤后不能返回基地)

B display　B型显示器,距离方位显示器

B H curve　磁化曲线

B power supply　乙电路

B&P　bid and proposal 投标和申请

BA　1. Backcourse Approach 反航道进近 2. Bandwidth Allocation 带宽分配 3. Barometric Altimeter 气压高度表 4. Basic Access 基本接入 5. Basic Airplane 基本型飞机 6. Beam Approach 波束引导进近着陆 7. Bell Aerosystems Company 贝尔航空系统公司 8. Bleed Air 引气 9. Boarding Advisory 登机通知 10. Brake Action 刹车动作 11. Buffer Amplifier 缓冲放大器

BAA　1. Bilateral Airworthiness Agreement 双边适航协议 2. British Airports Authority 英国机场管理公司

BAAR　Board for Aviation Accident Research 航空失事调查局(委员会)

Babbitt metal　巴比特合金,巴氏合金,轴承合金

babble　1. 复杂失真 2. 混串音,数据传输通道的感应信号,转播干扰

BABS　1. Beam Approach Beacon System 波束进近信标系统 2. Blind Approach Beacon System 仪表进场信标系统

baby　1. 小型物 2. 〈口语〉副油箱

baby A-bomb　小型原子弹

BAC　1. Back Course 背台航道 2. Bell Aerospace Corporation 贝尔航空航天公司 3. Boeing Airplane Company 波音飞机公司 4. British Aircraft Corporation 英国飞机公司 5. Bus Adapter Control 总线适配器控制 6. Boeing Process Specification 波音工艺规范

BACA　1. Baltic Air Charter Association 波罗的海包机公司协会 2. British Advisory Committee for Aeronautics 英国航空咨询委员会

BACE　Basic Automatic Check-out Equipment 基准自动检测设备

BACEA　British Airport Construction & Equipment Association 英国机场建设和设备协会

back　1. 反面,背后,向后 2. 叶背,桨叶凸面,上叶面,前桨面 3. 由上而下的情报传递 4. 后面的,反向的 5. 在后,向后

back annotation　逆向注解

back azimuth guidance element　反向方位引导单元

back beam　背台波束

back bearing　反象限角,掉头方向角

back bias voltage　反向偏压,负偏压

back focus　反焦点,后焦点

back lobe　后瓣

back nut　限动螺帽,支承螺母

back panel wiring　板后布线

back porch　1. (脉冲)后沿,返回沿 2. 后肩

back pressure　1. 反压力 2. 拉杆力(驾驶)背压力

back rake　后倾角

back scattering　反射离散,后向散射

back set　制动装置

back surface field　背面电场

back surface field solar cell　背电场太阳能电池

back trace technique　向后跟踪技术,回溯技术

back valve　回动阀,止回阀

back voltage　逆电压,反电压

backboard　后面板

back-diffusion　反向扩散,逆扩散

backdrift　后退偏航

backdrop　交流声

back-emf　反电动势

backface　底面,背面

backfire　1. 回火,逆燃 2. 逆火式战斗机

backfire antenna　背射天线

backfit　改装,改进

backflash　1. 回火,逆火 2. 逆燃,逆弧 3. 排气火舌

backflow　逆流,回流

backflush　逆流洗涤

backfolding　后折,折返

backgear　背轮,跨轮,慢盘齿轮

background　背景

background discrimination　背景鉴别

background emission　背景辐射

background limited photodetector　背景限制光电检测器

background noise　本底噪声,背景噪声

backhaul　1. (飞机或车辆的)回程,回运 2. 载货返航 3. 物资倒流

backing　垫板,垫片,支持,逆转

backing line　前级管路

backing pressure　前级压力

backing ring　垫环

backing up 1.倒退（滑行）2.回流
backing vacuum pump 前级真空泵
backkick 反转（指向燃机在手摇起动时的瞬时突然反转反向放电）
backlash 齿隙，啮合间隙
backlash charateristics 间隙特性
backlash phenomena 回差现象
backlight 1.背后照明灯 2.逆光背部照明
backlog 1.积压，搁置以待处理 2.积压的货物 3.积压未办的事情
backoff 1.补偿 2.略向后退
backorder 缺货通知单
backout 1.（火箭未能发射时的）逆序操作 2.返回 3.取消，放弃
backpack 背包式降落伞，背包式火箭发动机
backplane 底板，背板，基架
backplate 1.后挡板，隔板 2.（燃烧室的）回流隔板
backpropagation 反向传播
backrest 1.靠背 2.座椅靠背
backrolling 倒卷
backscatter 1.后向散射，反向散射 2.背景散射，背景反射
backscratcher 导弹电缆接头
backside 背面
backsighting （高炮标定时的）反觇
backspacing 倒退，退位，（打字机的）退格
backstep 向后突出部，后部台阶
backstitch 回缝，来回针脚
backstop 后挡板，后退定程挡块
backstrap 背部腰带
backstreaming 逆流
backswept 后掠，后掠角
backtrack 背台航迹，滑回，向后滑行
backup 1.备份的，备用设备 2.阻塞 3.支持性的
backup aid 备份辅助设备
backup aircraft 预备机，备用飞机
backup battery 备用电池，备用电源
backup interceptor control 截击导弹备用控制，截击机备用控制
backup system 后备系统，备用系统
backup weapon 补加的武器，备用武器
backwall temperature 后壁温度
backward 1.顺气流方向的，向后的，逆向的 2.倒的，相反的 3.落后的
backward difference 后向差分
backward diode 反向二极管
backward extrusion 反挤压
backward movement 反向运动
backward pointer 反向指针

backward reasoning 反向推理
backward supervision 反向监控
backward wave 反向波，返波
backward-wave oscillator 回波震荡器，反波振荡器
backward-wave tube 回波管，反射波管
backwash 后洗流，回流，倒流
backwater 逆（水）流，（水流中）回水
BACS Bleed-Air Control System 放气控制系统，引气控制系统
bacterial bomb 细菌炸弹
bacterial corrosion 细菌腐蚀
bactericidal action 杀菌作用
bactericidal agent 灭菌剂
bacteriological agent 细菌战剂
bacteriological attack 细菌攻击
BAD Boom-Avoidance Distance 防音爆距离
bad visibility 视线不良
bad weather approach 恶劣天气（复杂气象条件）进场（着陆）
BAD(A) Boom-Avoidance Distance on Arrival（沿到达航迹测得的）进场防音爆距离
BAD(D) Boom-Avoidance Distance on Departure（沿离场航迹测得的）离场防音爆距离
badge 符号，徽章
baffle 1.导流片，导流板 2.隔板，挡板 3.扰流片 4.（扬声器的）反射板 5.栅板，障板 6.施放干扰，对抗
baffle valve 挡板阀
baffling 起阻遏作用的，折流的
BAFO 1.Best and Final Offer 最后报价 2.British Air Forces of Occupation 英国占领军空军部队
bag 1.袋，包，囊，外壳 2.软油箱，贮器
bag tank 软油箱，袋形油箱（通常指地面储油的，而不是飞机上的）
baggage 1.行李 2.辎重
baggage compartment 行李舱
baggage hold 行李架，行李箱
bagged 击落的，击毁的
BAGR Bureau of Aeronautics General 民航总局
Ba-grease 钡基润滑脂
BAI 1.Barometric Altitude Indicator 气压高度指示器 2.Boeing Airborne Instrumentation 波音机载仪表装置 3.Battlefield Air Interdiction 战场空中折断 4.Bundle Assembly Index 线束装配目录
bail 1.跳伞 2.戽斗，提环
bail out 应急跳伞，跳伞
bailment aircraft 1.托管飞机（根据合同委托空军部以外的机构进行试验维修或改进的飞机）2.伞兵运输机
bailout bottle 跳伞用氧气瓶
bailout kit 降落伞（内的）救生包

bailout oxygen supply regulator 跳伞供氧调节器
bainite 贝氏体,贝茵体
baka bomb 神雷式飞弹(第二次世界大战中使用的一种自杀式飞机)
baked contact test 烘烤接触检验
Bakelite 1.酚醛胶木粉 2.酚醛塑料,电木
Baker-Nunn Camera 贝克纳恩摄影机,贝克纳恩照相机
BAL Balance 平衡,余额
BAL CL Balancing Coil 平衡线圈
balance 1.天平 2.秤 3.平衡 4.对称 5.秤重
balance area 平衡面积
balance axis system 天平轴系
balance character of non-return to zero under starting load 冲击不回零性
balance check 平衡检查,平衡校验,零点校验
balance circuit 平衡电路
balance point 平衡点
balance position 平衡位置
balance tab 平衡翼,平衡板
balance test 平衡试验
balance weight 平衡重量,平衡锤
balance zero 天平零位
balanced 1.平衡的,配平的,已补偿的 2.形成合理比例的
balanced bridge 平衡电桥
balanced detector 平衡检波器
balanced field length 等长机场长度,平衡跑道长度(跑道长度等于某型飞机所需的机场临界长度)
balanced hardening 均衡加固
balanced surface 平衡面,补偿面
balancer 平衡机,均衡器,配重,均压器,补偿器,平衡器
balances measuring 天平组测力
balancing 对称,平衡,补偿,定零装置,平衡的
balancing test 平衡试验
Balbo 大编队,大机群,巴尔博式队形
balked approach 进场失败
balking 1.突然停止 2.慢行
ball 1.球,珠,球状物 2.子弹,弹丸 3.燃烧着的信号弹
ball ammunition 实心弹,普通弹
ball bank unit 滚珠倾斜装置
ball bearing 球轴承,滚珠轴承
ball bearing retainer 滚珠轴承保持架
ball bonding 球焊
ball bullet 实心枪弹弹头
ball cartridge 实心弹
ball firing 实弹射击
ball inclinometer 球式倾斜仪
ball lightning 球状闪电
ball mount 球形炮塔,球形枪架

ball of fire (核弹)火球
ball piston motor 球塞式电动机,钢球式发动机
ball practice 实心弹射击练习
ball projectile 实心弹
ball turret 球形炮塔
balladromic course 命中目标的导弹航向
ball-and-socket bearing 球窝轴承
ballast 压舱物,压载物,镇重,镇流器,镇流电阻
ballast tube 稳流管(镇流电阻器)
ball-in-tube damper 管球阻尼器
ballistic 弹道温度,弹道的,弹道学的
ballistic camera 弹道照相机
ballistic case 弹道壳(体)
ballistic computer 弹道计算机
ballistic converter 弹道坐标换算器
ballistic coordinate measurement 弹道坐标测量
ballistic curve 弹道曲线,弹道轨迹,惯性飞行轨迹
ballistic deflection prediction 弹道偏差预测
ballistic delivery 沿弹道运(输)送(弹头)
ballistic elevation prediction 弹道射角预测
ballistic flight 1.弹道飞行 2.弹道飞行轨迹
ballistic function 弹道函数
ballistic galvanometer 冲击电流计
ballistic guided missile 弹道导弹
ballistic impact 弹道落点,弹道终点
ballistic limit 弹道限度
ballistic limit curve 弹道极限曲线
ballistic meteorology 弹道气象学
ballistic military target system 弹道军事目标系统
ballistic missile 弹道导弹
ballistic missile bombardment interceptor 弹道导弹轰炸截击器,反弹道导弹
ballistic missile interceptor 截击弹道导弹的导弹,弹道导弹拦截机
ballistic missile terminal defense 弹道导弹终端防御
ballistic parameter 弹道参数
ballistic path 1.弹道 2.无控飞行段
ballistic pendulum 弹道摆
ballistic period (导弹)被动飞行阶段
ballistic phase 弹道飞行阶段
ballistic photography 弹道摄影
ballistic prediction 弹道预测
ballistic range 1.弹道试验靶场,弹道试验设备 2.弹道距离
ballistic range table 弹道表
ballistic range test 弹道靶道试验
ballistic reentry 弹道式再入,弹道式返回
ballistic reentry vehicle 弹道式再入飞行器
ballistic rocket projectile 弹道火箭弹

ballistic stability	弹道稳定性
ballistic table	弹道表
ballistic technique	弹射技术
ballistic trajectory	弹道轨迹,惯性飞行轨迹
ballistic transfer	弹道传输
ballistic transistor	弹道晶体管
ballistic uniformity	弹道均匀性
ballistic vehicle	弹道式飞行器
ballistic wind	弹道风(对射弹有同等影响的假定的不变风力,实际上,射弹飞行时将遭遇不同的风力)
ballistician	弹道学专业人员,弹道学家
ballistics	弹道学,弹道特性
ballistite	巴里斯太火药
ballistocardiogram	【医】心冲击图
ballonet	1.软式补偿油箱 2.(气球或飞艇的)副气囊
balloon	气球,气舱,气室,气瓶
balloon astronomy	气球天文学
balloon barrage	气球拦阻网,气球阻塞网
balloon bed	气球系留台(气球在地面系留、放气、充气、检查及升起的场所)
balloon borne	气球载运的,气球上的
balloon cover	气球壳
balloon drag	拖曳气球
balloon fabric	1.气球布 2.气球蒙布
balloon flight	气球飞行
balloon reflector	气球反射器(制造假回波用)
balloon satellite	气球发射的卫星,气球式卫星
balloon tank	充气油箱(薄金属油箱,必须加压才能成形,用于大型洲际弹道导弹)
balloon theodolite	气球用经纬仪,测风经纬仪
balloon-borne detector	气球携带探测器
ballooning	1.驾驶气球,乘气球飞行 2.放气球升空 3.(着陆)平积
ball-proof	防弹的
ball-socked nozzle	球窝喷管
ballute	气球减速伞,气伞
ballwin	弹道风,即 ballistic wind
BALPA	British Air Line Pilots Association 英国航空公司驾驶员协会
balsa wood	轻木,软木
BAM	Business Activity Mapping 业务活动图
Bamberga	班伯加小行星,班伯加
band	1.带,范围,区域,频带,波段 2.带条,箍带(软式气球的加强带) 3.(磁盘或磁鼓上的)组槽
band filter	带式滤波器,带通滤波器
band gap	带隙
band of error	船位误差带
band of position	船位误差带,位置带
band of the Milky Way	银河带
band pass filter	带通滤波器
band reject filter	带阻滤波器,带除滤波器
band shifting	频段变换,频率移动,改频
band spread	频带宽度,波段展开
band suppression filter	带阻滤波器
banded corrosion	层次腐蚀,层状腐蚀
banded structure	带状结构,带状组织
bandit	〈口语〉敌机,(雷达)未识别的飞机
bandoleer	子弹带
band-pass	带通,通频带
band-pass filter	带通滤波器
bandpass limiter	带通限制器,带通限幅器
band-stop filter	带阻滤波器
bandwidth	带宽,频带宽度
bandwidth compression	频带宽压缩
bandwidth parameter	带宽参数
bang	重击,突然巨响
banger	拦阻气球,阻塞气球
bank	1.压坡度 2.倾斜,坡度,带坡度转弯 3.组,组合,排 4.存储单元
bank and turn indicator	转弯侧滑仪,倾斜与转向指示器
bank to turn technique	倾斜转弯技术
bank winding	迭绕线圈
banked	侧倾的,倾斜的,有坡度的,组合的
banking	倾斜,形成坡度,倾斜状态,堆积
banner	旗帜,旗,旗幅
banner cloud	旗状云(由山顶往下伸展的羽毛状云)
banner target	旗靶
banquet	宴会
bantam tube	小型管,小型电子管
BAP	Bank Angle Protection 倾侧角保护
BAPC	British Aircraft Preservation Council 英国飞机维护协会
bar	1.杆,棒,条,键,(指)针,(计算机屏幕上的)提示行 2.巴(压力单位)
bar stock	棒形钢材,钢筋储备
baralyme	二氧化碳吸收剂
barber chair	可调(俯仰的)座椅(航天员用)
barbette	炮塔,枪塔
barbette gun	炮塔炮
bare air base	无附属设备的基地,简易基地
bare electrode	1.裸极 2.无药焊条,裸焊条
bare engine	不带附属设备的发动机
bare weight	空重,皮重
bare wire	裸电线
bargraph display	条柱显示
barium	【化】钡(化学元素,符号 Ba)
barium sodium niobate	铌酸钡钠

barium star 钡星
barium titanate 钛酸钡
barium titanate ceramics 钛酸钡陶瓷
Barker sequence 巴克序列
BARO Barometric 气压
barn 靶(恩)(核子有效截面积的单位)
Barnard star 巴纳德星
barnstorm 游览飞行,空中技巧
barnstormer 游览飞行参加者
baroceptor 气压传感器,气压敏感元件,压力感受器
barochamber 气压室,气压舱,气压容器
baroclinic atmosphere 斜压大气
baroclinicity 斜压性
barocyclonometer 气压风暴表
barodontalgia 气压性牙痛,高空牙痛
barognosis 压力感,重力感
barogram 气压自记曲线
barograph 气压计,气压记录器
barometer 气压计,气压表
barometric altimeter 气压高度表,气压测高计
barometric altitude 气压高度
barometric fuze 气压引信
barometric pressure 大气压,气压
barometric pressure control 大气压力控制
barometric pressure gradient 气压梯度
barometric switch 气压开关,气压继电器
barometric tendency 气压倾向
barometrograph 气压描记器,气压自动记录器,膜盒气压记录器
barometry 气压测定法
baro-otitis 气压耳炎
baropathy 气压性疾病
baroport 静压孔
baroreceptor 压力感受器
baroscale 气压刻度
baroscope 气压指示器,验压器
barosinusitis 气压性鼻窦炎,高空鼻窦炎
barosphere 气压层
barostat 气压调节器,恒压器
baroswitch 气压开关
barotalgia 气压性耳痛
barothermograph 气压温度计,气压温度记录器
barothermohygrograph 气压温度湿度计,气压温度湿度记录器
barothermohygrometer 气压温度湿度表
barootitis media 气压性中耳炎,航空性中耳炎
barotrauma 气压性损伤
barotraumatic labyrinthitis 内耳气压性损伤,迷路气压性损伤

barotropic atmosphere 正压大气
barotropic model 正压模式
barotropy 正压(性),质量的正压分布
barrack 营房
barrage 1.拦阻,阻塞 2.拦阻轰炸 3.拦阻射击 4.空中巡逻 5.阻塞干扰
barrage balloon 阻塞气球,拦阻气球
barrage jammer 阻塞式干扰机,全波段干扰机
barrage jamming 全波段干扰,阻塞性干扰(同时干扰若干邻接的波道或邻近的频率)
barratron 非稳定波型磁控管
barred spiral galaxy 被阻塞螺旋形银河,棒旋星系
barrel 1.枪管,炮管,炮身 2.〈口语〉气缸,燃烧室,火箭发动机 3.圆筒,衬筒 4.任何近似圆截面的机体部分(锥形也可,如 F/A-18 飞机的机头锥称 nose barrel) 5.防空密集区
barrel antenna 桶形天线,余割平方天线(指产生余割平方波束的天线)
barrel assembly 炮管及其附件,枪管及其附件,炮身
barrel chamber 弹膛,药室
barrel distortion 桶形畸变,桶形失真
barrel engine 筒式发动机(活塞式发动机的气缸轴与发动机纵轴平行)
barrel finishing 滚筒清理,滚筒抛光
barrel group 炮管及其附件,枪管及其附件,枪管箍,枪管组
barrel length 炮管长度,身管长度
barrel lock 炮闩,闭锁机
barrel of monobloc construction 单层炮管,单层枪管
barrel plating 滚镀,筒式电镀
barrel receiver 机匣,炮枪管匣
barrel recoil 炮管后坐,枪管后坐
barrel recoiling aircraft cannon 管退式航空机关炮
barrel recoiling gun 管退式机关炮,管退式机枪
barrel reflector 炮膛反射镜,枪膛反射镜
barrel roll 横辊,筒形辊
barrel sleeve 枪管套筒
barrel travel 炮管行程
barrel wear 炮管磨损,枪管磨损
barreling 桶形度,鼓度,凸度
barretter 热线检流计,镇流电阻器
barrettes (机场上的)直列灯光
barrier 界线,屏障,障碍物
barrier height 势垒高度
barrier injection and transit time diode 势越二极管
bartack 加强缝法
barycenter 重心,转动物质的重心
barycentric coordinate 重心坐标
barycentric dynamical time 质心力学时

barycentric element 质心要素,重心元件
barycentric orbit 引力重心轨道,多体质心轨道
barye 微巴
baryta 氧化钡
barythymia 忧郁症
basal metabolism 基础代谢
basal orientation 基面定向,基向
bascule 平衡装置,等臂杠杆
base 1.底座,基础 2.基地 3.(飞机起落航线的)第四边 4.纵向轮距 5.(晶体管的)基极 6.(数学中的)基线,基数 7.(化学中的)碱
base altitude 基本高度(飞机在执行任务过程中,特别是在飞往目标、会合等过程中所保持的高度)
base area 1.底面面积,积面(积),底部断面面积 2.基地地区 3.根据地
base band frequency 基带频率
base case 基础方案,基础案例
base charge 基本装药,炮眼底部装药
base coordinate system 基座坐标系
base delay-action fuze 弹底延发引信
base detonating fuze 弹底起爆引信(装于弹底的引信,使射弹有充分时间穿透装甲及防御物后引起爆炸)
base drag 底部阻力,底阻
base flow 底流
base gun 基准炮
base heat protection 底部防热
base heating 基本热负荷,底部加热
base ignition 弹底引燃,弹底点火(从弹底点火的信号弹与烟火信号弹,引燃后逐渐喷出烟幕或化学制剂)
base line 基线,底线
base map of topography 地形底图
base measurement 基线测量
base metal 1.基底金属(合金中最主要的成分) 2.焊接金属(区别于焊接中加入熔合的金属) 3.贱金属 4.碱金属
base network 基线网
base of a projectile 弹底
base of shell 弹底
base of trajectory 炮口水平线,枪口水平线(自枪、炮中心点至弹道经路的降弧上与枪、炮口成水平面的水平线)
base oil 基础油
base ordnance workshop 基地军械修理所
base plug 弹底塞
base point 原点
base pressure 底座压力,底部压力
base quantity 基本量
base register 基址寄存器,变址寄存器
base spray 后飞破片(炮弹爆炸时向后飞散的破片),基线扫射
base station 基地站,基点
base time 基线时间
base turn 基线转弯
base voltage 基本电压(交流电机的额定相电压)
baseband 基带
baseband processor 基本频带信号处理机
baseband signal 基带信号
baseband switching 基带切换,基带转接
baseband transmission 基带传输
base-flow 基本流量
base-height ratio 基线与高度比
base-ignition type 弹底点火式,弹底引燃式
baseline 基线,底线
baseline curvature 1.底线弯曲 2.基线弯曲
baseline fitting 基线拟合
baseline restore 基线恢复
baseline shift 基线漂移
baseload 1.基极负载 2.基本负载
base-mounted 装在底座上的
baseplane 底板,基板(座),支承板
base-timing sequencing 进基定序,时基序列
BASH Bird Aircraft Strike Hazard 鸟撞航空器危险,鸟害
basic 1.基础的,基本的 2.(常用复数)基础训练 3.碱性的
basic access 基本接入
basic aircraft 基本型飞机(特定型号飞机的最简单的型式,可以通过增加不同的设备而形成不同用途的飞机)
basic airspeed 修正(基本)表速,基本空速
basic cloud formations 基本云状
basic color scheme 基本颜色系统(以颜色标志各种类型弹药的系统)
basic configuration 基本形状,基本布局
basic encoding unit 基本编码器
basic figures 基本射击诸元,基价,基数
basic gravimetric point 基本重力点
basic group 基本群
basic hit capability 基本命中能力
basic mapping system 基本地图制作系统
basic operating weight 基本工作重量
basic pilot training 基础飞行训练,基本操作训练
basic rack 1.主炸弹架 2.基准齿条
basic reliability 基本可靠性
basic scale 基本比例尺
basic T 仪表T形布局(在传统的飞机座舱仪表板上,主要飞行仪表呈T形布局)
basic thermal radiation 基本热辐射

basic trainer 初级教练机
basic training 1.基本训练 2.基本飞行训练,基本军事训练和驾驶员基本训练的简称
basic weapon 基准兵器
basic weight 基本重量(包括各种固定装备、燃料、滑油在内的飞机重量)
basic wing 基本机翼(不计与机身重叠部分)
basicity 碱度,碱性
basin 1.盆,水槽,水池 2.船坞
basing 驻扎,停泊,配置
basis 基础,底部,主要成分,基本原则或原理
basis of type certification 型号合格审定基础
basis vector 基本向量
BASW Bell-Alarm Switch 警铃开关
batch 批,组,批量
batch filter 间歇(分批)式过滤器
BAT 1.Battery 电瓶,电池 2.Bundle Assembly Tag 线束装配标签
BAT CHG Battery Charger 蓄电池充电器
BATC Battery Charging 蓄电池充电
BATS Business Air Transport Service 商业空运业务,商用空运服务
batten 舵夹板,桁条
battery 电池,蓄电池
battery capacity 电池容量,蓄电池容量
battle 战役,斗争
bay 舱,架,盒,段,凹槽,层,海湾
bay section 舱段
BB 1.Battery Bus 电瓶汇流条 2.Black Box 黑匣子
BBL Body Buttock Line 机身纵剖面线
BBM Break Before Make 先开后合
BBS Bulletin Board System 电子公告板
BBT Black-Box Testing 黑盒测试
BC 1.Baseline Configuration 基线构型 2.Basic Configuration 基本构型 3.Battery Charger 电瓶充电器 4.Binary Counter 二进位计数器 5.Block Check 分组检查 6.Bookcase 书架 7.Bottom Centricity 底部中心线 8.Bottom Chord 底部弦线 9.Broadcast 广播
BC superalloy 硼碳高温合金
BCAC Boeing Commercial Airplane Corp. 波音民用飞机公司
BCAG Boeing Commercial Airplane Group 波音民用飞机集团
BCAS Business and Commuter Aviation System 商务和转运航空系统
BCARs British Civil Airworthiness Requirements 英国民用航空适航要求
BCAS Beacon-based Collision Avoidance System 信标防撞系统

BCC 1.Baseline Change Control 基线更改控制 2.Block Change Concept 成组更改概念
BCD 1.Binary Coded Decimal 二十进制,二进制编码的十进制 2.Budget Change Document 预算修改文件 3.Behind Completion Date 没按期完工
BCF 1.Bulk Crystal Facility 大容积晶体生长设施 2.Bulk Cargo Fan 散装货舱风扇
BCI Binary Coded Information 二进制(代)码信息(数据)
BCL Basic Contour Line 基本外形线
BCN Beacon 信标,灯标,指向标
BCO Bridge Cutoff 电桥断开
BCP 1.Basic Control Program 基本控制程序 2.Breakcloud Procedure 穿云程序
BCPI Bar Code Part Identification 条形码零件标识
BCR 1.Battle Casualty Replacement 作战减员补充 2.Baggage Claim Request 要求行李认领 3.Blade Cooling Report 叶片冷却报告 4.Borocarbon Resistor 硼碳膜电阻
BCT 1.Bandwidth Coding Technique 带宽编码技术 2.Bandwidth Compression Technique 带宽压缩技术 3.Basic Color Term 基本色彩项 4.Bushing Current Transformer 套管式电流互感器
BCV 1.Belly Cargo Volume 机腹货舱容积 2.Bore Cooling Valve 镗道冷却活门 3.Bleed Control Valve 引气控制活门
BD 1.Background Data 背景数据 2.Base Detonating 弹底起爆(引信) 3.Blank Display 空白显示 4.Block Diagram (General System Layout) 方框图(系统原理图) 5.Blocker Door 阻流门 6.Board 板,委员会 7.Bomb Disposal 未爆炸弹处理 8.Bonner Durchmusterung【天】波恩(巡天)星表 9.Bus Drum 转动鼓轮
BDA 1.Backdrive Actuator 反驱动作动器 2.Battle Damage Assessment 战斗效果评定 3.Bomb Damage Assessment 轰炸破坏程度判定 4.Business Decision Analysis 商务决策分析
BDC Bottom Dead Center 下死点
BDD Basic Design and Development 基本设计和开发
BDE Buyer Designated Equipment 买方指定设备
BDHI Bearing Distance Heading Indicator 方位距离航向指示器
BDI Bearing Distance Indicator 方位距离指示器
BDL Bundle 导线束
BDP Breakdown Process 分支过程
BDS Broach Data Sheet 拉削数据卡
BDU Bright Display Unit 高亮度显示器
BDV Breakdown Voltage 击穿电压
BDW Bend Down(向下)弯曲

BDY Boundary 边界,界限
BE Breaker End 断路器终点,断路器末端
beachhead 滩头堡,登陆场
beaching 船只搁浅
beaching gear 1.(水上飞机的)登陆轮架 2.上下滩装置,海滩绞拖船只及水上飞机的工具
beacon 1.信标,无线电指向标 2.信标机,信标台 3.灯标
beacon acquisition time 信标捕获时间
beacon antenna 信标台天线
beacon delay 信标应答延迟
beacon receiver 信标接收机
beacon stealing 应答信标目标的丢失(由于出现另一雷达信号)
beacon tracking 信标跟踪
beacon tracking level 信标跟踪电平
bead 1.准星,环形瞄准具中心 2.卷边,弯边 3.焊珠,焊缝 4.(轮胎)胎圈,(机轮的)轮缘
bead sight 圆形准星
bead thermister 珠形热敏电阻
beading (钣材的)卷边,碾压波纹
beaker 浇铸,药柱的模子;烧杯
beallon 铍铜合金
beam 1.梁,柱,臂(特指承受弯曲和剪切应力的结构件) 2.侧方 3.光束,波束,电子束,射线,射束 4.(船身或机身的)最大宽度
beam angle 射束孔径角;波速张角
beam antenna 定向天线,波束天线
beam bracketing 在波速范围内飞行,波段
beam capture 波束截获(指导弹进入导波束)
beam climber 驾束制导导弹,波束制导导弹
beam collimation error 波束瞄准误差
beam command 驾束制导指令,波束制导指令
beam compass 航道罗盘,横杆圆规
beam confining electrode 束射屏
beam control 1.电子束控制,(示波器)辉度控制 2.波束控制
beam controlling system 1.波束控制系统 2.波控系统
beam current 射束电流,电子束电流
beam diameter 光束直径
beam direction 波束方向
beam director 波束定向器
beam divergence 电子束发散
beam dynamics 束流动力学
beam element 梁元素
beam focusing 电子束聚焦
beam follow-up 雷达波束跟踪(目标)
beam index tube 束引示管
beam lead 梁式引线

beam lead integrated circuit 梁式引线集成电路
beam path 光路,光束路径,声束路径
beam pattern 波束方向图,方向特性
beam power 波束功率,电子束功率
beam rider 波束制导导弹,驾束式导弹
beam riding 波束制导,驾束制导
beam saltus 波束跃度
beam satellite antenna 定向卫星天线
beam shape factor 波束形状因数
beam sharpening 波束锐化
beam sharpening ratio 波束锐化比
beam signal 无线电波束信号
beam solid angle 波束立体角
beam splitter 光束分离器,分光镜
beam spot 电子束光点
beam steering 波束控制,波束调向
beam switching 射束转换
beam theory 梁理论
beam trammel 长臂圆规
beam waveguide 波束波导
beam weapon 光束武器,射束武器,死光武器(利用定向能,如激光、粒子束、微波等的武器)
beam width 波束宽度,射束宽度,电子束宽度
beam with variable cross-section 变截面梁
beam zone 波束区
beam(ed) antenna 定向天线
beam-directed energy weapon 射线武器,定向能武器
beamlet 细光束,子束波,子束激光器
beam-splitter 分光板
beamwelding 电子束焊
beamwidth 波束宽度,射束宽度
bearer 支架,托架,承力构件
bearer service 承载业务
bearing 1.轴承,支座,支架 2.方向,方位,航向 3.象限角
bearing alignment 方位对准
bearing and power transfer assembly 轴承和功率传输组件
bearing angle 象限角,方位角
bearing compass 方位罗盘,定位罗盘,探向罗盘
bearing cursor 方位标,方位游标
bearing directing technique (电子)方位引导技术(指引导干扰器天线对准欲干扰的敌方电子设备的技术)
bearing discrimination 方位分辨力
bearing error 方位误差
bearing line 方位线(海图的)
bearing mark 方位标志
bearing measurement 方位(角)测定
bearing of station 电台方位

bearing plate　1.地速偏流计算器(飞机上用以推算远处目标的方位) 2.垫板
bearing rate　方位变化率
bearing tracking　方位角跟踪,方位跟踪
bearing unit　方位测定装置
beat　1.跳动,脉动,摆动 2.拍,差拍 3.音差器,频差器
beat frequency　拍频,差频
beat frequency oscillator　拍频振荡器
beat mass　拍频质量
beat period　拍合周期,拍频周期
beaten zone　被弹区,落弹地带,被射击地区
beating　差拍,跳动,脉动
beating in　合拍,进入同步
beavertail antenna　海狸尾形波束天线,平面波束(侧高)天线,刀形波束(侧高)天线
become　变成,成为
BED　Bonding and Earthing Diagram 搭接和接地图
bed　1.架,座,台,床,铺 2.地基,基础,路基 3.层
bedding correction　层面改正
bediasite　贝迪阿熔融石
bedplate　底板,台板,床板
bedside monitor　床旁监护仪
bedstead　1.试验台,试验装置 2.构架
bed-tested　试车台上试验过的
beef　〈口语〉力量,威力
beehive　有歼击机护送的轰炸机密集队形,蜂巢星团,集气架
beep box　〈口语〉遥控装置,操纵台,遥控台
beeper　无人驾驶飞机遥控员,导弹遥控员
BEF　1. British Expeditionary Force 英国远征军 2. Buyer Furnish Equipment 买方装备设备
BEI　Bundle Equipment Index 线束设备目录
begin　首创,开端
beginning of lifetime　寿命初期
behave　1.(机器等)运转,(飞机)运动,飞行 2.表现,行为,举止
behavior of mode　行为模式
behavioural science　行为科学
Beijing Aviation Simulator Development Center　北京航空模拟器开发中心
Beijing Institute of Satellite Information Engineering　北京卫星信息工程学院
bel　贝(尔)(对数的底是10时的量级单位,用于可与功率类比的量)
BEL CRK　Bell Crank 双臂曲柄,摇臂,直角(形)杠杆
bell　1.钟,铃,信号铃 2.钟形物,扩散管,锥形口 3.起落架舱 4.降落伞伞衣
bell crank lever　直角形杠杆,双臂曲柄杆
bell nozzle　钟形喷管

bell of the muzzle　炮口消焰罩
bellcrank　双臂曲柄,摇臂,双摇臂,直角(形)杠杆
Belleville washer　(装卸料机机头内的)贝氏垫圈
bellmouth　钟形口,钟形嘴
bellows　真空膜盒,波纹管,风箱
bellows pressure gauge　波纹管压力表
belly　腹部,机身腹部,机身腹部
belly down　超低空飞行
belly gun　机腹机关炮,机腹机枪
belly in　用机腹着陆
belly land　机腹着陆,收起落架着陆
bellypack　机腹容器
belt　1.弹带,带,皮带 2.地带,区域
belt box　弹带箱,弹箱
belt fed　弹带送弹式
belt feed　弹带送弹
belt feed lever　拨弹带杆
belt feed mechanism　带式送弹机构
belt feed pawl　输弹钩
belt feed slide　带传滑板
belt filling machine　弹带装弹机
belt of totality　全食带
belt stars　腰带星
belted ammunition　弹带子弹(装在弹带上的机枪用子弹)
belted rounds　弹带上的枪(炮)弹
belt-feed launcher　带式供弹发射装置
belt-loading machine　弹带装弹机(装填自动式武器子弹带的自动装弹机械)
belt-pouch　弹带盒
BELW　Brake Energy Limit Weight 刹车能量极限重量
bench　试车台,工作台,长凳子
bench check　工作台(上进行的)检验
bench mark　基准点,水准点
bench run　试验台试验(过程)
bench test　试验台试验(发动机整机静态试验)
benchboard　操纵台,控制板
benchmark　标准,基准,水准点
benchmark routine　1.基准例行程序 2.水准程序
bend　弯曲,弯头
bend allowance　弯曲余量
bending　弯曲,弯曲度
bending brake　弯板机,折板机
bending fatigue　弯曲疲劳
bending machine　弯管机
bending moment　弯矩,弯曲力矩
bending of light　光线弯曲
bending point　弯曲点
bending radius　弯曲半径

bending stiffness matrix 弯曲刚度矩阵
bending test 反复弯曲试验，弯曲试验
bending vibration 横向振动（弯曲振动）
bends 沉箱（潜函，高空）病
benefit 利益，好处
benefit-cost analysis 收益成本分析
bent 1.弯曲的 2.XX设备失灵（空中截击和近距离空中支援通话代语）
bent cone 弯曲头锥
bent sting 弯支杆
benzin(e) 汽油，挥发油
benzophenone 二苯甲酮，苯酮
benzyl bromide 溴甲苯（催泪性毒剂）
benzyl chloride 氯化甲苯（催泪性毒气）
BER Bit Error Rate 误码率
BETW Between 在……之间
BEV Bevel 斜角
BF 1.Back-Feed 反馈 2.Band Filter 带通滤波器 3.Beat Frequency 差频 4.Block Fuel 轮挡耗油 5.Bottom Face 底部表面
BFD Book Form Drawing 图册
BFE Buyer Furnished Equipment 买方提供设备，买方配置设备
BFL Balanced Field Length 平衡场长
BFO Beat-Frequency Oscillator 差频振荡器
BFR Buffer 缓冲器，减振器
BG 1.Back Gear 倒挡齿轮 2.Body Gear 机身起落架
BH Brinell Hardness 布氏硬度
BHD Bulkhead 隔框，舱壁
BHN Brinell Hardness Number 布氏硬度值
BHP Brake Horse Power 刹车马力
BI Black Iron 黑铁板
BIDG Building 建筑物，厂房
BIL Basic Impulse Level 基本脉冲电平
BIST Built-In Self Test 机内自检测，机内自检
BIT 1.Binary Digit 二进制 2.Built-In Test 机内测试
BITE Built-In Test Equipment 自检测设备，机内检测设备
bituminous paint 沥青漆
BIU Built-In Unit 内装单元
bivicon 双光导摄像管
BJ Bonding Jumper 搭接片
BK 1.Break-In Keying 插入键控 2.Brake 制动器，刹车
BKDN Breakdown 击穿，破坏，故障，事故，停炉
BKGD Background 背景，底色，本底
BKN Broken 破损的，裂开，裂云
BKSA British Kite Soaring Association 英国轻型滑翔机协会
BL 1.base line 基线 2.Barrel 桶 3.Bend Line 弯曲线 4.Between Layer 层间 5.Bill 账单，票据 6.Bleed 引气，放气 7.Buttock Line 纵剖线

black and white picture tube 黑白显像管
black and white transparency 黑白透明度
black and white TV 黑白电视
Black Arrow 黑箭（英国小型运载火箭）
black blasting powder 黑色炸药
black body 黑体（吸收而不反射辐射的物体）
black body emission 黑体辐射
black box 黑匣子，测谎器
black box model 黑盒模型
black box testing approach 黑箱测试法
black halo 黑晕
black hole 警卫室，黑洞
black hole binary 双黑洞
black hot 暗红热
black level 黑色电平，黑色信号电平，暗电平
black light 1.红外线，不可见光 2.红外线探照灯 3.信号灯有故障（在发送前已烧坏）
black matrix screen 黑色屏，黑底荧光屏
black radiation 黑体辐射
blackable-out 部分灯火管制
black-and-white film 黑白胶片，黑白影片
blackboard framework 黑板框架
blackbody 黑体（全部吸收辐射能的物体）
blackbody radiation 黑体辐射
blackbody temperature 黑体温度
black-box modeling 黑箱建模
blackening 发黑，烧坏，发黑处理
black-level clamping 黑电平箝位
blackout 1.黑视 2.无线电通信中断 3.灯火管制
blackout area 1.（飞行）黑视区 2.灯火管制区
blackout range 黑障区
blackout voltage 截止电压
blackspot 斑痕，黑点，盲点（显象管的）
BLACS Barometric Low Altitude Control System 气压低空控制系统（无人驾驶飞机上用的）
bladder 1.代偿服密封圈，抗荷衣的气囊 2.弹性囊，隔膜，薄膜
bladder anti-G suit 囊式抗荷服
bladder fuel tank 软油箱
bladder partial pressure suit 囊式部分加压服
bladder propellant tank 囊式推进剂贮箱
bladder tank 软油箱，胶囊油箱
blade 叶片，轮叶，刀刃
blade angle 叶片角，桨叶角
blade antenna 刀形天线
blade azimuth angle 叶片方位角，桨叶方位角
blade back 桨叶背，叶背

blade cyclic pitch 桨叶周期变距
blade damper 桨叶减震器(装于直升飞机的桨叶上,用以防止水平震动)
blade element 叶素,叶元
blade element theory 叶素理论
blade flapping 桨叶挥舞运动
blade flutter 桨叶颤振,叶片颤振
blade lagging 桨叶摆振
blade load (旋翼)桨叶载荷(直升机或旋翼机的总重量被桨叶总升力面积除)
blade natural frequency under rotation 叶片动频
blade noise 叶片噪声
blade passage 叶片流道
blade passing frequency 叶片通过频率
blade passing noise 叶片通过噪声
blade pitch 1.桨距 2.叶片距
blade profile 叶型,叶剖面
blade profiling 叶片造型
blade root 叶根(叶片根部非叶型部分,包括与轮盘连接的榫头)
blade root cut-off 桨根切除
blade section 桨叶截面,叶片剖面
blade section pitch 桨叶剖面安装角
blade shank 叶柄,桨叶胫
blade spring 片簧,弹簧片
blade suction side 叶片吸力面
blade suction surface 叶片吸力面
blade surface 叶面
blade sweep 叶片后掠角(少数叶片为前掠)
blade tip 叶尖,叶梢
blade twist 桨叶扭度
blade vibration 叶片振动
blade vortex interaction 桨涡干扰
blade width ratio (桨)叶宽比(螺旋桨或旋翼的桨叶平均弦长与整个直径之比)
bladed disk vibration 叶轮振动
bladed disk 叶盘系统
blade-disc coupling vibration 叶盘耦合振动
blade-section 叶片剖面
blade-tip 叶尖
blade-tip clearance 叶尖间隙,尖端间隙,齿顶间隙
blade-to-blade 叶片到叶片的
blading 1.对片组,整套叶片 2.叶片装置
blank 1.空白的 2.空白,空格 3.毛坯,坯件 4.空弹 5.消隐,截止,
blank angle of view 视场空白角
blank cartridge 空包弹
blank map 轮廓地图,空白地图,底图
blanker 1.熄灭装置 2.关闭器,抑制器

blanket 1.表面层,敷层,覆盖层,保护层 2.(空气动力)阴影 3.密云 4.施放烟幕 5.遮掩机声(一种干扰测音器的措施) 6.覆盖,掩盖,抑制 7.击中目标 8.(反应堆的)再生区
blanketing 1.(空气动力的)阴影,遮蔽 2.消隐,抑制
blanketing effect 覆盖效应
blanketing frequency 抑制频率
blanking 1.遮蔽,遮没 2.抑制,消隐 3.(钣金的)下料(指加工前将钣材切割成一定形状) 4.堵住,塞住 5.冲裁
blanking cap 堵盖(帽)
blanking plate 盖板,(板形的)堵盖
blanking zone 1.空白区 2.消隐区,消隐带
Blasius solution for flat plate flow 布拉休斯平板解
Blasius theorem 白拉斯勒斯定理,布拉休斯理论
blast 1.爆炸,爆炸气浪,冲击波,炸破 2.喷射气流,炮口火焰,排气火舌 3.喷砂
blast area 爆炸区,冲击波杀伤区,焦灼区
blast charge 爆破装药
blast cleaning 喷砂清理
blast cluster warhead 爆破弹头
blast distance 冲击波距离
blast fence 气流挡板
blast gage 鼓风计
blast line 爆炸线(地面上自爆炸点向外辐射的平面线)
blast pad 防焰垫,喷流防护垫
blast pressure 鼓风压力,风压瞬间燃烧压力,爆炸波压力,爆炸压力
blast resistant 抗炸能力
blast suppression device 炮口消焰器
blast tube 排气管,送风管
blast warhead 爆破弹头
blast wave 爆炸气浪,冲击波
blaster cap 雷管,火帽
blast-fragmentation warhead 爆破杀伤弹头
blast-hardened 防核的
blasting cartridge 爆破弹
blasting explosive 炸药
blasting gelatine 爆破胶,硝基胶
blasting oil 硝化甘油,爆炸油
blasting powder 爆破炸药
blasting time 爆破时间,爆炸时间
blastproof 防爆炸的,防冲击波的
BLATS Build-up Low-cost Advanced Titanium Structure 组件式低价先进钛结构(钛加上超级塑料等,用于发动机制造)
blaze away 开火,射击,连续发射
blazed angle 闪耀角
blazed grating 定向光栅,闪耀光栅

bleaching 1.清除效应(核爆炸产生的 X 射线能破坏大气中足够量的电子从而使 X 射线不受影响地穿过大气层的现象) 2.漂白,脱色

bleed 1.放泄(放气、放油、放水、放渣等) 2.引气,供气 3.抽取,吸取,萃取 4.将电子信号或电压降到零 5.(使飞机)降低高度,减速

bleed air 排出空气

bleed off relay 放继电器

bleed valve 引气活门,放气活门

bleeder 1.泄放器,放出管,放出开关 2.(电路中的)分压器,分泄电阻

bleeder resistor 泄放电阻器,分泄电阻器

bleeder screw 放气螺钉

blemish 缺陷

blend 掺和,混合

blend radius (转折处的)圆角半径,倒圆半径,转接半径

blended 混合的,掺和的

blended configuration 融合体布局

blended wing-body configuration 翼身融合布局

blending 混合,掺和,配料

BLF Breakeven Load Factor 盈亏平衡载运率

blind 1.按仪表操纵的,盲目的 2.低能见度的 3.遮光帘 4.未炸炮弹,失效炮弹

blind attack 盲目攻击(借助雷达瞄准具之攻击)

blind bomb 盲目轰炸,仪表轰炸

blind direction 盲向

blind firing 盲目射击(指用雷达瞄准具射击)

blind flight 盲目飞行,仪表飞行

blind fuze 未发引信,未起作用的引信

blind guidance 盲目制导,向目视观察不到的目标制导

blind nut 盲螺帽,盲盖,螺盖

blind phase 盲相

blind rivet 封闭铆接,盲铆钉

blind riveting 1.单面铆接 2.盲铆

blind search 盲目搜索

blind shell 未炸炮弹

blind takeoff 无能见度起飞,仪表起飞

blind toss 盲甩法(飞机在仪表飞行中利用雷达寻找目标投弹时的一种甩投方法)

blind zone 静区,无信号区

blinder 遮眼罩,蒙眼物

blind-landing beacon 仪表(盲)着陆系统信标

blindness 失明,没有能见度

blink 闪光,闪烁

blink comparator 闪视比较镜,闪视比较仪

blinker 1.闪光信号 2.闪光灯塔

blip 反射脉冲,尖头信号,光波,尖头脉冲信号

blip-scan ratio 尖头回波扫探比

BLIS Base Level Inquiry System 基地级查询系统(维修管理信息系统的一部分)

blisk 整体式涡轮转子级(轴向式涡轮转子轮盘和桨叶用整块材料制成),整体叶盘

blisk rotor configuration 叶盘转子结构

blister 1.球面观察窗(机身两侧用于观察或射击的球形窗) 2.(飞机上的)两侧射击装置 3.(雷达的)天线罩 4.气泡,砂眼

blister agent 糜烂性毒剂

blister gas 糜烂性毒气

blitz 1.闪击战 2.投炸弹 3.以最大力量进行武器训练 4.擦亮制服上的钮扣等

blitzee 〈口语〉空袭受害者

blizzard 雪暴

BLK 1.Black 黑色 2.Blank Or Block Or Black 空白(下料)或栏目或黑色 3.Block 块,片

BLKG Blanking 熄灭,(雷达)照明,模压,下料,坯料

BLKGD Blanking Die 下料模,冲裁模

BLKT Blanket 毯,垫

BLMTH Bell Mouth 承口,漏斗

BLO Blower 鼓风机

block 1.块,组,部件,部分,单元,组合 2.(飞机)轮挡 3.滑轮,滑车 4.(航线飞行的)阶段 5.(飞机飞行)高度配置层 6.(计算机程序的)字组 7.闭塞,封锁,挡轮挡

block address 字组地址

block adjustment 分区平差,区域网平差

block cipher 分组密码

block code 分组码

block construction 部件结构,大型砌块建筑

block diagram 方块图,框图

block encoding 块编码,分块码编码

block floating point 成组浮点

block in 停放(指飞机在停机坪或指定地点的停放,通常当日不再使用)

block out (自机轮下)拿掉轮挡

block shear test 单元剪切试验

block shipment 整体装运法(为对海外战区装运补给品的方法,以供给部队在一特定日数内所需要的储存量)

block template 模具样板

block time 轮挡时间(取或放轮挡时间,一次飞行总时间)

block transfer 字组转移,整块传送

blockade 封锁,空中封锁

blockage 堵塞,障碍,封锁

blockage effect 阻塞效应

blockage factor 堵塞系数,堵塞因子(指气流通道内附面层影响)

blockage percentage 阻塞度
blockage test 阻塞试验
blockbuster 巨型炸弹(能炸毁街道一片地区的任何巨型炸弹)
blocked force 阻塞力
blocker door 反推装置折流门
blockhouse 碉堡,掩体,地下室
blocking 阻塞,闭锁,截断
blocking capacitor 隔直流电容器,耦合电容器
blocking diameter 阻挡直径
blocking layer 闭锁层,阻挡层
blocking oscillator 阻塞振荡器,间歇振荡器
blocking time 阻塞时间,截止时间
blocking volume 阻挡体积
blockmark 块标志
blocknut 保险螺帽,防松螺帽
bloom 大钢坯,钢锭,大(初轧)方坯
blooming 1.起霜,(推进剂)晶析 2.(光学)加膜,敷霜 3.(电视图像)模糊现象
BLOS Beyond Line of Sight 超视距
blossom 1.开伞 2.开花,花
blow 1.吹,吹风 2.打击 3.(发动机)放气 4.(熔丝)烧断 5.(轮胎等)爆破
blow down propulsion system 落压式推进系统,吹下式推进系统(推进剂与挤压气体共用一容器的热气推进系统)
blow flap 吹气襟翼
blow off 1.(气体)喷出,(液体)排出 2.(发动机)熄火 3.(火箭)抛射,(火箭各段的)分离
blow pipe 1.〈口语〉喷气式飞机 2.吹管,喷枪
blowback 1.气体后泄(当枪炮射击时因受到压力而使气体向后逸出) 2.气体后坐,后坐,后坐力 3.(发动机)反吹(混合气在进气管内的反向运动)
blowby 漏气,渗漏
blowdown 吹风,吹除,吹净(发动机试验后的)
blowdown tunnel 吹风风洞,吹气风洞
blowdown turbine 冲击式涡轮
blowdown wind tunnel 放气式风洞
blowdown-ejection wind tunnel 吹引式风洞
blowdown-indraft wind tunnel 吹吸式风洞
blower 吹风机,鼓风机
blower pipe 吹送管,螺旋桨集气道
blowing 吹,吹除,吹着的
blowing dust 风尘,高吹尘
blowing sand 飞沙
blown primer 炸飞雷管
blownwing 有附面层吹除装置的机翼
blowoff 1.(气体的)喷出,(液体)排出 2.吹除(如吹除附面层),熄火(指发动机) 3.(火箭)抛射,(火箭各段的)分离
blowout 1.吹灭(由于气流速度过高造成的熄火),加力燃烧室熄火 2.爆破,喷出 3.吹除,(发动机起动系统的)冷吹
blowout disc 爆破隔膜,安全隔膜(火箭发动机的)
blowpipe 吹管,喷枪
blowtorch 1.〈口语〉喷气发动机,喷气式飞机 2.喷灯,钎焊灯
BLST Ballast 平衡器,镇流器
BLSS Base-Level Supply Sufficiency 基地级满足供应量
BLT Bolt 螺栓
BLTIN Built-In 内装的,机体内的,固有的
BLU 1. Band Laterale Unique (电信的)传输边带,单边带,单边频带 2. Blue 青,蓝 3. Bomb, Live Unit 真炸弹(美国)
blue 1.蓝色的 2.蓝色 3.发蓝处理,烧蓝
Blue airway 南向航路,蓝色航路(美国)
blue alert 1.空袭警报 2.台风警报
Blue Angels 蓝色天使(指一队具有精确编队特技的海军飞行员)
blue cross 蓝十字(喷嚏性毒气的别名)
blue gas 氰毒气,液化气,蓝煤气
blue jersey 〈口语〉水兵
blue key 蓝色说明(复制时,蓝色字体或图像不会印出,因而作为制图时的说明)
blue shift 蓝向移位,蓝色偏移
Blue steel 蓝钢(英国战略巡航导弹名)
blueprint 1.蓝图,设计图 2.计划大纲 3.蓝色版
blue-suiter 〈口语〉空军军人
bluff 非流线的,陡立的,钝头的,(前风挡玻璃)直立安装的
bluff body 非流线体,不良流线体,钝头体,阻流体
bluff shape bomb 平头炸弹(头部扁平、弹体短粗、适合低空高速轰炸的炸弹)
bluff windshield 陡斜的风挡
blunder 错误,故障
blunder factor 导航系统工作时引起严重错误的因素
blunderer 轻率粗心的人
blunt 1.钝头的 2.钝头物,(叶片的)钝尾缘
blunt body 钝头体,钝头物体
blunt leading edge 钝头进气边
blunt-body shape 钝体形状
blunting 磨钝,使钝
bluntness 钝度
bluntness effect 钝,钝度
bluntness ratio 钝度比(弹头头部与底部直径之比值)
blunt-nosed 钝头的
blunt-nosed body 钝头体
blur 模糊

blurring 影像模糊,标志模糊
blurring effect 模糊效应
blurring of the instruments 仪表指示模糊
blurring of vision 视力模糊
BLW Below 在……下面,在下面,向下
BLWS Bellows 风箱,膜盒,波纹管
BLWT Blow-Out 鼓风,放气,吹灭,熄灭,停炉
BM 1. Ballistic Missile 弹道导弹 2. Base Metal 基底金属 3. Bench Mark 水准,基准点,基准标记,标准检查程序 4. Block Mark 块标志 5. Bubble Memory 磁泡存储器 6. Buffet Margin 抖振裕度 7. Bus Monitor 总线监控器
BMD 1. Ballistic Missile Defense 弹道导弹防御 2. Bubble Memory Device 磁泡存储(器)部件
BMDATC Ballistic Missile Defence Advanced Technology Center 先进弹道导弹防御技术中心
BMEWS 1. Ballistic Missile Early Warning Satellite 弹道导弹预警卫星(美国) 2. Ballistic Missile Early Warning System 弹道导弹预警系统
BMF Business Management Finance 商务管理/财务
BMEP Brake Mean Effective Pressure 平均有效制动压力
BMI 1. Bismaleimide 双马来酰亚胺(一种耐高温树脂胶) 2. Banded Matrix Iteration 分段矩阵迭代 3. BITE Monitoring Interface 自检监控界面
BMI resin 双马来酰亚胺树脂
BMLC Basic Multiline Controller 基本多路控制器
BMMS Battlefield Mission Management System 战场任务管理系统
BMP Barrier Waterproof 防水隔板
BMR 1. Basal Metabolic Rate 基础代谢率 2. Bearingless Main Rotor 无铰主旋翼
BMS 1. Basic Meteorological Services 基本气象服务 2. Bid Work Sheet 报价工作单 3. Billing and Management System 计费和管理系统 4. Boeing Material Specification 波音材料规范 5. Bulletin Meteo Special 特殊气象通告 6. Business Management System 商务管理系统
BMTS Bandwidth Manager and Traffic Scheduler 带宽管理和业务调度程序
BMVP Barrier Moisture Vapor Proof 防潮气隔板
BN 1. Backbone Network 骨干网 2. Balancing Network 平衡网络 3. Ball Nut 球形螺母 4. Basic Number 基本号 5. Becklin-Neugebauer object 贝克林-诺伊格鲍尔天体
BND Band 带,区,范围
BNDG Bonding 搭接,焊接,连接,胶结,粘接,搭铁,接地,搭铁线,连接件
BNF Bomb Nose Fuze 炸弹头部引信

BNGS Bomb Navigation Guidance System 炸弹制导系统
BNN 1. Barium sodium niobate 铌酸钡钠 2. Boundary Network Node 边界网络结点
BNR Binary 二进制的
BNS 1. Baggage Not Seen Message 行李未寻获文电 2. Binary Number System 二进计数制 3. Bomb-Nav System 轰炸领航系统 4. Broadcast Non-profit Satellite 非商业性广播卫星
BNSC British National Space Centre 英国国家航天中心
BNSH Burnish 抛光,打磨光亮,精加工
board 1. 板,屏,台 2. 操作台,配电盘 3. 部,局,委员会 4. (船)甲板,舷,舷侧 5. 上飞机
board panel 仪表板
boarding 1. 上飞机,登机(指乘客) 2. 装货(往飞机上)
boarding gate 登机门
board-mounted connector 板装连接器
boat seaplane 船身式水上飞机
boat tail angle 船尾角
boatplane 船身式水上飞机
boattail 1. 船尾,船形尾部,收敛尾部 2. 锥形弹尾
boattailed projectile 锥形弹
boattailing (导弹、子弹等的)锥形弹尾
bobber 〈口语〉轰炸机
bobbin 1. 环,套管 2. 绕线管,绕圈架,点火线圈
bobbin oil 锭子油(一种用于高速旋转、轻载荷的润滑剂)
bobbing 目标回波起伏(指雷达示波器上目标反射点之起伏波动)
bobtailed aircraft 截尾飞机,无尾飞机,无尾航空器
bobweight 平衡重量,平衡重,平衡锤
bobweight feel system 配重式感力系统
BOC Basic Operational Capability 基本工作能力
Bode diagram 伯德图
Bode plots 伯德图标绘
body 1. 体,物体 2. 机身,弹体,火箭箭身 3. 短舱,吊舱 4. 天体
body axis system 机体坐标系
body burden 体内(放射性)积存量
body effect 衬底效应
body fixed coordinate system 本体坐标系,地固坐标系
body fluid regulation in space 航天体液调节
body fluid shift 体液转移
body force 体积力,质量力
body fuze 弹体引信
body group 枪机匣
body length 身长,体长
body link 弹体环节(在控制和干扰作用下,表示导弹运动参数变化规律的环节)

body motion　身体移动，机身移动
body of a fuze　引信体
body of rifle　枪机匣
body of shell　弹体
body plan　弹体平面图，机身平面图
body sensor　（航天员）身体情况传感器
body surface potential mapping　体表电位标测，体表电位图，体表电位分布图测量
body temperature　物体温度
body tide　固体潮
body vortex　体涡
body weight　物体重量
body-axis coordinate system　1.体轴坐标系 2.弹体坐标系
body-fitted coordinate system　贴体坐标系
bodying　1.稠化，变稠，加厚 2.增加球体积
body-mounted type solar array　壳体式太阳电池阵
body-shedding vortex　脱体涡
body-stabilized attitude control system　本体稳定姿态控制系统
boehmite　勃姆石（一种水软铝石）
boffin　（航空工程等）科学技术人员
bogey　1.未判明的敌机 2.小车式起落架 3.车架（用以支持火炮或火车车箱的有轮车架）4.承载装置，坦克的负重轮
bogey autopilot　标准特性的自动驾驶仪
bogging-down　起落架陷入泥地
bogie　小车式起落架（一个支柱上有多个机轮）
bogie beam　转向架梁，轮轴架梁
bogie landing gear　车架式起落架
bogus warning　1.（空袭）假警报 2.假报知
boil　煮沸，沸腾
boiler　1.锅炉 2.（在高函道比涡轮风扇发动机组中作为热气源的）核心发动机 3.报废的装备 4.可控火箭
boilerplate　1.锅炉钢板 2.飞行器的试验样品（指金属结构样机）
boiling　沸点温度，沸腾
boiling of body fluid　体液沸腾
boiling point　沸点
boiloff　蒸发，汽化
BOL　1.Beginning of Life 寿命初期 2.Bill of Lading 提货单 3.Boundary Light 跑道边界灯
bollard　（飞行甲板）带缆桩，系缆柱
bolometric magnitude　（星体）辐射量级，热星等
bolt　1.枪栓，机心 2.螺栓，螺杆
bolt carrier　枪机框
bolt catcher　螺栓制动器，螺栓紧固器
bolt group　枪机，枪栓
bolt mechanism　枪栓装置，枪机装置

bolt stud　双头螺栓，柱螺栓
bolt travel　机心行程
bolting　用螺栓固定，螺栓连接
BOM　1.Bill of Materials 材料清单，物料表，物料清单 2.Beginning of Message 报文开始，底部，基部
bomb　轰炸练习机，炸弹，炸弹，手榴弹
bomb aimer　轰炸瞄准具
Bomb Alarm System　核弹警报系统（北美重要目标区周围的、能报知核弹爆炸情况的自动报警系统）
bomb assembly spares　炸弹装配备用零件
bomb ballistics　炸弹弹道学
bomb bay　炸弹舱
bomb caliber　炸弹口径
bomb carpet　1.面积轰炸地域 2.面积轰炸时的投弹散布面
bomb carrier　1.炸弹架，挂弹架 2.轰炸机
bomb cemetery　未爆炸弹掩埋场
bomb cluster　子母炸箱，集束炸弹，弹束
bomb complete round　完整弹（包括引信、保险钢丝等）
bomb control　1.投弹器 2.炸弹引信装定 3.投弹控制
bomb conversion kit　炸弹附加尾翼（用于无稳定尾翼的炸弹，如凝固汽油弹）
bomb crater　炸弹坑
bomb damage　1.轰炸直接效果（指冲击波、光辐射等造成的损害）2.轰炸间接效果（指轰炸对政治、经济等所造成的损害）3.轰炸效果
bomb deactivation　使未爆炸弹失效
bomb detonator　炸弹引信
bomb dispenser　子母弹箱，集束弹箱
bomb disposal　未爆炸弹处理
bomb door　炸弹舱门
bomb dummy unit　练习炸弹
bomb dump　炸弹堆集所（通常是个露天场所，用以堆集随时可供使用的炸弹）
bomb filler　炸弹装药
bomb fin　炸弹尾翼
bomb fuze arming delay　炸弹引信解除保险延时装置
bomb fuze control　炸弹引信控制面板
bomb gear　轰炸装置（包括炸弹架及载弹、投弹等器械）
bomb hoist　绞弹机，挂弹机
bomb lift　1.炸弹升挂，挂弹 2.炸弹升降机，绞弹机
bomb lift truck　炸弹升挂车，顶弹车
bomb loading chart　载弹方案图
bomb navigation computer　轰炸导航计算机
bomb nose fuze　炸弹弹头引信，炸弹头部引信
bomb rack　炸弹架，挂弹架
bomb recess　炸弹舱
bomb release distance　投弹距离
bomb release point　投弹点

bomb rib　炸弹架肋（为飞机构造的一部分，专为安置炸弹架而设）
bomb safety line　轰炸安全线（为保护己方军队而划定的线，凡线外的目标均可轰炸，线内地区则不能轰炸）
bomb scoring　轰炸效果评定，投弹命中记录
bomb shackle　炸弹钩，挂弹钩
bomb sighting system　轰炸瞄准系统（包括矢量瞄准、同步瞄准及角速度瞄准三种）
bomb station　炸弹悬挂位置，挂弹位置
bomb strike camera　轰炸弹着照相机
bomb unloading　卸弹，卸下炸弹
bomb up　升挂炸弹，挂弹
bomb yoke　投弹杆（装于俯冲轰炸机机腹下的投弹装置）
bombable　可轰炸的，易轰炸的（指轰炸不会遇到什么严重困难，或在轰炸距离之内）
bomb-aiming　投弹瞄准，轰炸瞄准
bombard　1.轰炸 2.（火炮）轰击
bombardier　轰炸员，投弹手
bombardier-gunner　轰炸射击手
bombardier-navigator　轰炸领航员（机上同时能掌握轰炸与领航技术的人员）
bombardment　1.轰炸 2.轰炸航空兵 3.轰炸，照射
bombardment control unit　轰炸控制装置
bombardment satellite　轰炸卫星
bombay　炸弹舱
bomb-bay bin　炸弹舱工具箱（在炸弹舱不挂弹时，装在炸弹舱内用以贮存飞机的各种备份零件、工具等的工具箱，以备不时之需）
bomb-carrier hook　炸弹架挂钩
bomber　轰炸机，轰炸员，轰炸机飞行员
Bomber Command　轰炸机部队指挥官
bomber killer　反轰炸机兵器
bombex　训练轰炸，训练投弹
bombfall　1.（飞机）投下的炸弹 2.弹着点 3.炸弹散布面
bombfall line　炸弹爆炸线
bombing　轰炸
bombing and navigation optical sights　轰炸领航光学瞄准具
bombing angle　投弹角，轰炸角度
bombing bug　活动轰炸靶
bombing capacity　1.轰炸（机）能力 2.炸弹齐投量
bombing navigation system　轰炸导航系统
bombing navigational computer　轰炸导航计算机
bombing pattern　1.投弹散布面，定形轰炸弹着点形状 2.轰炸方案，轰炸法
bombing probabilities　投弹概率，投弹公算
bombing radar　轰炸雷达

bombing range　1.轰炸靶场 2.轰炸机活动距离，轰炸航程
bombing run　1.轰炸航路 2.在轰炸航路上飞行，进入轰炸航路
bombing sight　轰炸瞄准具
bombing simulated test　轰炸模拟试验
bombing simulator　轰炸模拟器
bombing system　轰炸系统
bombing table　轰炸瞄准表（提供各种情况的瞄准诸元）
bombing through overcast　穿云轰炸，云上轰炸
bombing-calculation table　轰炸计算表
bombing-navigational trainer　轰炸导航教练器
bombing-up　挂弹
bombing-up indicating lamp　挂弹指示灯
bomblet　〈口语〉小炸弹，小型炸弹
bombline　轰炸线，轰炸安全线
bombload　挂弹组，载弹量
bomb-pocked　弹坑累累的
bombproof　1.防炸弹的 2.防空洞，防空掩蔽部
bomb-release control　1.投弹器 2.投弹控制
bomb-release pip　投弹标志，投弹标记
bomb-shell　炸弹
bombsight　轰炸瞄准具，轰炸瞄准器
bombsight field　轰炸瞄准具视界
BOND　Banding 胶接
bond　1.搭接，接合，搭铁 2.黏合，粘接材料，溶合部分 3.搭铁线，接头 4.（化学中的）键
Bond albedo　邦德反照率（即球面反照率）
bond master　环氧树脂类黏合剂
Bond number　邦德数
bonded permanent magnet　粘接用磁体
bonded structure　胶接结构
bonderite　磷酸盐处理层（一种用于钢的薄膜防诱层）
bonderlube　挤压润滑剂
bonding　1.粘接，搭接，连接 2.搭铁，接地 3.搭铁线，连接件
bonding agent　粘接剂
bonding joint　胶接接头
bonding process　胶接工艺，胶接过程
bonding wire　1.焊线 2.接地线 3.等电位连接线
bondline temperature　结合面温度
bone mineral loss in space　航天骨矿物质脱失
bonedome　〈口语〉保护头盔（飞行员用的，内部有衬垫）
boneyard　废件露天堆放场
bonnet　帽，盖，机罩
Boolean add　逻辑加，布尔加
Boolean algebra　布尔代数
Boolean function　布尔函数
Boolean variable　布尔变量，布尔变数，逻辑变量

boom 1.桁,杆,梁 2.(直升机的)尾梁 3.(空中加油飞机的)伸缩套管 4.(超声速)音爆,爆音

boomer 轰炸机,轰炸机飞行员

booming 1.爆音 2.蜂鸣(交流声)

boom-type launcher 伸臂式发射装置

boondocks 荒漠无人区(海军用语),草地、牧场、田地等无跑道的降落场

boost 1.助推,加速(短时间使发动机增加推力或功率),加力 2.增压,升压 3.助推器,加速器,升压器

boost charge 急充电

boost phase 加速阶段,助推阶段

boost pressure 升压,吸入管压力

boost pump 辅助泵,供油泵

boost turbopump 增压涡轮泵

boost vehicle 运载火箭,加速火箭

boosted-high level clock generator 升压高电平时钟发生器

boosted kinetic energy penetrator 助推动能突防飞行器

booster 1.增压级(指风扇后,高压压气机前的压气机) 2.助推器,运载火箭,起动发动机 3.助力器 4.助爆装置 5.载波频率放大器,升压器 6.同频电视转播台

booster charge 1.助爆药 2.升压充电

booster engine 助推器发动机,加速发动机,助推发动机

booster explosive 传爆药,助爆药

booster grain 传爆药柱

booster horizontal checking ladder 助推级水平测试工作梯

booster horizontal sling 助推级水平吊具

booster iron wheel carriage 助推级铁轮支架车

booster motor 助推器,加速器

booster pump 增压泵,升压泵,辅助泵,充液泵

booster seven pipe connector working ladder 助推级七管连接器工作梯

booster stage 增压级

booster trailer 助推级公路运输车

booster turbopump 增压涡轮泵

booster turning sling 助推级翻转吊具

booster vacuum pump 增压真空泵

booster-cup charge 传爆管装药

boost-glide missile 助推滑翔导弹

boot 1.(飞机上的)橡胶除冰带 2.行李舱,行李箱 3.护罩

bootleg 未经批准的,非预先计划的(指飞行时间)

bootstrap 1.自举作用,自动持续作用 2.火箭发动机进入最大推力工作状态 3.〈口语〉(起飞时)突跳离地

bootstrap capacitor 自举电容器

BOOV Bar Out Of View 指引杆消失

bora 布拉风(亚得里亚海东岸的一种干冷东北风)

boraccia 强布拉风

boralloy 含硼合金

BORAM Block-Organized Random-Access Memory 块随机存取存储器

borane 硼烷

borax 硼砂,月石

border 1.边界,边缘 2.接近

border surveillance 边界监视

borderline 1.边界,界线 2.临界曲线

bore 1.炮膛,枪膛 2.口径,内径,孔径 3.气缸内径 4.镗孔,钻孔,锥眼 5.穿云飞行

bore clear 〈口语〉炮膛检查完毕

bore diameter 镗孔直径

bore premature 膛炸

borehole stressmeter 钻孔应力计

borer 1.镗床 2.镗孔刀具

boresafe fuze 膛内保险引信(附于炮弹上,使炮弹在膛内不能爆炸的安全引信)

borescope 1.孔探仪,内窥镜 2.校靶镜

boresight 1.校靶 2.校靶仪,校靶镜 3.视轴,瞄准线

boresight alignment 校靶调整(如调整飞机与靶图的位置)

boresight error 准向误差

boresight harmonization 冷校靶

boresight line 炮膛瞄准线

boresight missile 直接瞄准发射的导弹,对准目标发射的导弹

boresight weapon 直接瞄准发射的武器,对准目标发射的武器

boresighting 1.校 2.轴线校准 3.膛内瞄准

boresighting range 教靶靶场,教靶距离

boride 硼化物

borine 烃基硼

boring 1.镗孔,镗削加工扩孔 2.钻进,钻探

boring out 钻出,钻掉,钻通

Born approximation 玻恩近似

borocarbide 碳化硼

borohydride 硼氢化物(火箭燃料)

boron 硼(高能燃料)

boron fiber reinforced composite 硼纤维增强复合材料

boron nitride 氮化硼,一氮化硼

boron nitride coating 氮化硼涂层

boron trifluoride counter 三氟化硼计数管

boron-carbon superalloy 硼碳高温合金

boron-propellant 含硼推进剂

BOSS 1.Biological Orbiting Space Station 生物轨道空间站 2.Broad Ocean Scoring System 大面积海域(导弹或航天器)落区确定系统(美国)

boss 1.桨毂,轮毂 2.凸起部位(航天器安装结构的),凸耳,凸缘 3.进气道中心体

BOT Bottom 底,底部,
both-way communication 双向通信
bottle 1.瓶 2.〈口语〉火箭助推器,火箭加速器,火箭壳体
bottle bomb 酒瓶炸弹
bottleneck 临界截面,喉道(指流管的)
bottom 底,底部,下部,基础
bottom characteristics 底质
bottom cover 底盖(仪器舱的)
bottom dead center 下死点,下止点
bottom part 底部,下模
bottom phase 下相
bottom plate 底板,主夹板,下模托板
bottom section 井底部分,底部截面
bottom surface 底面,反面
bottom temperature 底部温度
bottom tube 底管
bottom wall 下盘
bottom-side sounder 底视探测仪,底部探测仪
bottom-up 自底向上,自下而上
bottom-up design 自底向上设计
bottom-up development 自下而上研制
bottom-up testing 自下而上测试
boulder 卵石,大圆石
bounce 1.跳起,反跳,(着陆)跳跃,跳弹 2.从上方突然攻击 3.无线电回波 4.(雷达荧光屏上)目标标志跳动,影像闪烁
bounce mode 跳弹模式
bounce table 冲击台振动台
bound 1.界限,边界,范围 2.限制,约束 3.跳起,弹回 4.朝向,开往……去的
bound algorithm 定界算法
bound electron 束缚电子
bound vortex 附体涡流,附着涡流,附着涡
bound vortex surface 附着涡面
boundary 1.界限,边界,范围 2.壁
boundary coding 界编码
boundary collocation 边界配置
boundary condition 边界条件
boundary constraint 边界约束
boundary control 边界控制
boundary correction 边界修正
boundary displacement 界壁位移
boundary effect 边缘效应,边壁效应
boundary element 边界元
boundary element method 边界元法
boundary enhance 边缘增强
boundary layer 1.边界层 2.附面层
boundary layer aerodynamics 边界层空气动力学

boundary layer bleed 边界层泄除
boundary layer control 边界层控制
boundary layer diffusion 边界层扩散
boundary layer displacement thickness 边界层位移厚度
boundary layer flow 边界层流
boundary layer integral relations 边界层积分关系式
boundary layer momentum thickness 边界层动量厚度
boundary layer separation 边界层分离
boundary layer stability 边界层稳定性
boundary layer theory 边界层理论
boundary layer thickness 边界层厚度
boundary layer transition 边界层转移
boundary line 边界线,疆界线
boundary node 边界节点
boundary point 边界点
boundary problem 边界问题
boundary register (存储地址的)边界寄存器
boundary stiffness matrix 边界刚度矩阵
boundary treatment 边界处理,边界处置
boundary value 边界值
boundary value problem 边界值问题
boundary velocity 界面速度
boundary wave 界面波
boundarylayer 边界层,界面层,边层
boundary-layer bleed 有附面层吸除装置的进气器
boundary-layer separation 边界层分离
boundary-value 边值,边界值
boundary-value problem 边值问题,边界值问题
boundedness 有界性,局限性
bounds register 界限寄存器
bourrelet (炮弹)定心带,定心部
bow 1.头部,艏 2.弓度
bow heavy 机头有下沉趋势的,机头重
bow loader 从头部装载的运输机
bow shock 弓形激波
bow shock wave 头波,弓形激波
bow turret 机首炮塔
bow wave 顶头波,弓形波
bowhandle 弧形把手,弓形把手
bowheaviness 俯冲趋势,机头重,(潜艇)艇首过载
bowl 1.浮子室 2.滚筒 3.碗
bowser 1.运油车(机场上运送燃料的机动车或拖车) 2.载有大量燃料的
box 1.箱,盒,外壳 2.轴(承)箱,轴衬,轴套 3.(飞机结构的)翼盒,翼箱 4.〈口语〉箱形队形 5.巡逻空域 6.(火箭的)地下贮存库
box beam 箱形梁,匣形梁
box column 箱形柱,箱柱
box diffusion 箱法扩散

box lens 盒式透镜
box level 圆水准器
box part 盒形件
box rib 盒形肋,匣形肋
box spanner 套筒扳手
box spar 箱桁,盒形翼梁
box tool 组合刀具
box type balance 盒式天平
boxcar 1.〈口语〉运输机(多指货机) 2.矩形波串(常用复数) 3.闷罐车,有盖货车
boxed dimension 全尺寸,装箱总尺寸,轮廓尺寸,总尺寸
boxing 总装配(指机体在型架上的装配过程)
box-shaped 盒形的
box-type bomb fin 盒形炸弹尾翼
BP 1. Band Pass 带通 2. Barometric Pressure 气压 3. Base Percussion 弹底着发(引信的),底部着发引信 4. Biological Processing 生物加工 5. BITE Processor 机内测试设备处理器 6. Black Powder 黑火药 7. Blueprint 蓝图 8. Boost Pump 增压泵 9. Bounded Plane 有界平面 10. Breakeven Point 盈亏平衡点 11. Business Process 业务流程
BPA Business Process Architecture 业务流程构造
BPAM Business Process Architecture Methodology 业务流程构造方法学
BPC 1. Barometric Pressure Control(燃气涡轮型)发动机压力控制 2. Back Pressure Control 反压控制 3. British Purchasing Commission 英国采购委员会
BPCU Bus Power Control Unit 汇流条电源控制器
BPF Band-Pass Filter 带通滤波器
BPM 1. Baggage Processed Message 行李处理文电 2. Biological Production Module 生物生产舱
BPP Bus Protection Panel 汇流条保护器
BPR 1. Bypass Ratio 函道比 2. Business Process Reengineering 业务流程重组
BPS 1. Balanced Pressure System 平衡压力系统 2. Beacon Processing System 信标处理系统 3. Bistable Phosphor Storage 双稳态磷光屏存储器 4. Boeing Part Specification 波音零件规范 5. Bulk Processing Subsystem 粗处理分系统
BPS Bits Per Second 每秒传送位数,比特/秒,字节/秒
BPSK Biphase Shift Keying 两相相移键控
BQS Basic Quality System 基本质量体系
BR Branch 分支
brace 1.张线,拉条,撑杆 2.手摇曲柄钻 3.系紧,拉紧,支撑
brace wrench 曲柄头扳手
brachydromic course 落后于目标的导弹航向
bracing 1.支撑 2.撑条系,拉条系,张线系

bracket 1.托架 2.夹叉射击,夹叉轰炸 3.穿越修正进近(在某点上空来回做试验性进近以建立正确进近,或来回穿过波束以建立正确航向)
bracketing 1.夹叉,夹叉射击,交叉,构成交叉 2.进入等强度信号区
bracketry 支架系统,托架系统
bradycardia 心动过缓
bradyesthesia 感觉迟钝
bradyhemarrhea 血流徐缓
bradypn(o)ea 呼吸徐缓
bradysphygmia 脉搏徐缓
braid 编织带
braided rope 辫绳,编织绳,编包索,辫带式钢索
brain 1.脑 2.〈口语〉计算机,计算装置 3.控制系统 4.发射控制室
brain bucket 防护头盔,安全帽,防震头盔
brain damage 脑损伤
brain function meter 脑功能仪
brain model 脑模型
brain unit 1.(电子)计算机 2.自动导引头
brainpower 1.科学工作者,科学技术干部 2.智力,智能
brake 1.刹车,制动 2.减速装置,刹车装置,制动器
brake control system 刹车控制系统
brake energy 刹车能量
brake energy limiting weight 1.刹车能量限制重量 2.轮胎速度限制重量
brake horsepower 制动马力(发动机输出的有效马力)
brake parachute 1.阻力伞 2.刹车伞 3.减速伞
brake pressure 刹车压力
brake pressure gage 刹车压力表
brake pressurize 制动比压
brake shoe 刹车弯块
brake speed 刹车速度
brake torque 1.刹车力矩 2.制动力矩
braking 减速技术,刹车的,制动装置
braking ellipse 减速椭圆轨道
braking orbit 制动轨道
braking parachute 着陆伞,减速伞,刹车伞
branch 1.支路,分路 2.部门 3.天体子午圈的半段
branch amplifier 分路放大器
branch pipe 支管
branched optical cable 分支光缆
branching network 分支网络
brand 1.商标,标记,烙印 2.品种
brander 〈口语〉端面燃烧的固体火箭发动机
Brans-Dicke cosmology 布兰斯-迪克宇宙论
brass 黄铜,黄铜制品
brassboard 1.仪器设备演示板,线路示教板 2.模型样件

brassboard radar 实验性雷达
brassing 1.黄铜铸件 2.镀(黄)铜
BRAZ Brazier 焊炉
BRAZ. HD Brazler Head 扁头螺钉/铜焊头
braze 硬焊,铜焊
brazier head rivet 半圆头铆钉
brazing 铜焊,硬焊,硬钎焊技术
brazing alloy 钎焊合金
brazing solder 铜焊料(硬焊料)
brazing spelter 黄铜焊条,黄铜钎焊料
BRC Base Recovery Course 返回的主航向,返航航向
BRCH Broach 拉削,拉刀
BRD 1.Bomb Release Distance 投弹距离 2.Braid 条带
BRDG Bridge 电桥
BRE Basic Release Engineering 基本发放工程
B-Rep Boundary Representation 边界表示法
break even point 1.无亏损点(指商用飞机装载率处于不盈不亏的值) 2.保本销售数(指飞机制造厂需出售到足以回收生产投资的飞机架数)
breakcord (降落伞的)拉断绳
breaker points 断电点(飞机磁电机中部件)
breakout altitude (进场着陆时)转入目视飞行的高度
breakthrough survey 贯通测量
breach 1.破口,裂口 2.破坏,违反
breadbasket (口)小型燃烧炸弹箱
breadboard 模型板,试验板,线路示教板
breadboard experiments 试验(电路)板实验
breadboarding 1.模拟板试验,模型试验 2.制造(设备的)模型
breadth-first search 广度优先搜索
break 1.破裂,断裂,断开 2.突然退出空战,突然改变航向,急剧压坡度 3.(飞机)失速坠落,翼下沉 4.克服(音障) 5.电气线路故障 6.中断线路传输 7.〈口语〉线状脉冲回波(雷达显示器上) 8.分离面,结合面
break frequency 拐点频率,折断频率
break point 1.断点,分割点,转效点 2.断点指令
break time 破译时间,断开时间
break valve 截止阀
break wire 断线装置(一种抗入侵装置,入侵者如碰断此金属线即被发觉)
breakability 易碎性,破碎性
breakage 1.损坏,破裂 2.断线
breakaway 1.脱离队形,退出攻击,脱离目标,突然改变航向 2.气流分离 3.断开,(飞机在空中加油后)脱开,摘钩 4.(飞机)开始移动
breakdown 1.详细分解,拆散 2.(电介质的)击穿 3.疲竭,恶化,中断 4.气流分离
breakdown drawing 分解图
breakdown point 破坏点,强度极限,击穿点

breakdown strength 击穿强度,断裂强度
breakdown voltage 击穿电压
breaker 1.断路器,自动保险电门 2.破碎器
breaker strip 缓冲帆布层
breakeven 无损失的,不亏本的
break-in keying 插入键控
breaking 断裂,断路
breaking test 断裂试验
breaking up 1.(机件的)完全分解,拆散 2.(液体的)雾化
breakline 断开线,分离线,接缝,对接线
breakoff 1.突然退出战斗,突然改变航向 2.(航天员的)孤寂感
breakoff phenomenon 断开现象,破坏现象
breakout 1.起始,突发 2.穿云,转入目视飞行 3.脱离战术(双机夹心战术中,攻击长机突然脱离队形,迫使防御僚机处于为难境地) 4.分支点
breakout force 始动力(指飞机停放时,向不同方向移动驾驶杆所需的最小的力)
breakout phase 突发相
breakpoint 断点,断裂点
breakthrough 1.突围,冲跨 2.挤过去,冲出来 3.(科学技术上的)突破
breaktime 休息时间,间歇期
breakup 断裂,损坏,分解,分裂
breakup length 断裂长度
breakup unit 自炸装置,自毁系统
breather 1.通气孔,通气管,通气装置 2.〈口语〉装空气喷气发动机的导弹
breathholding 屏气,屏住呼吸,屏住气
breathing 1.进气 2.呼吸,供气,换气 3.(发动机内的)压力脉动
breathing apparatus 氧气设备,呼吸机
breathing pressure fluctuation 呼吸压力波动
breathing valve 呼吸阀
breathing vibration 放气振动
breathless 气促
breech 1.炮尾 2.燃爆筒
breech chamber 弹膛,药室
breech-block 枪机炮尾栓炮闩
breechlock (航炮)机心锁,闭锁机
breeze 清劲风(蒲福氏风级 2～6 级)
bremsstrahlung effect 轫致辐射效应
brevet 1.飞行执照 2.名誉晋升
brevity 简洁,简短,短暂,短促
BRG 1. Bearing 方位,方位角,轴承 2. Bring 带来 3. Bridge 桥梁
bribe 贿赂,收买
BRIC Bomb Release Interval Control Unit 投弹间隔控

制器

brick 〈口语〉射弹(指子弹、炮弹、火箭、导弹等)

bridge 1.电桥 2.跨接 3.〈口语〉(指挥所的)操纵台

bridge arm (电)桥臂,电桥支路

bridge helicopter 架桥(用)直升机

bridged tap 桥接分支线

bridgewire 桥丝

bridging (导线)搭接,桥接,短路

bridging of model 模型连接

Bridgman method 布里奇曼方法

bridle 1.束带,籛带 2.(飞机的)牵引架,牵引杆 3.(飞艇的)系留装置

bridle line (飞机的)牵引架线

brief 1.下达简令(对机组人员作飞行前的最后指示) 2.简令,提要,摘要

brief discussion 浅谈,简令

brief lift 短时的(射击)中止

briefed 接受过指示的,接受过简令的(指机组人员)

briefing room 简令下达室

briefly 短暂地,简略地,暂时地

brigade 1.旅 2.工作队

bright annealing 光亮退火,非氧化退火

bright bolt 光制螺栓,精制螺栓

bright display 光亮显示,亮显示

bright flocculus 亮谱斑

bright giant 亮巨星

bright heat treatment 光亮热处理

bright limb 亮边缘

bright plating 光亮电镀

bright quenching 光亮淬火

brightening towards the limb 临边增亮

brightness 亮度,辉度

brightness contrast 亮度对比度

brightness discrimination 亮度辨别

brightness flicker 亮度闪烁

brightness noise 亮度杂波干扰

brightness range 亮度范围

brightness scale 亮度级,亮度标尺

brightness temperature 亮度温度

BRIL Brilliance 光彩,亮度

brilliance 1.亮度,辉度 2.高音重发,逼真度

Brilliant Eyes 智能眼卫星(美研制中的小型遥感卫星,由智能卵石派生而来)

Brilliant Pebbles 智能导弹战略,截击导弹

Brillouin diagram 布里渊图

brinell 压痕

Brinell hardness test 布里涅耳硬度试验

brinelled 有压痕的

bring around tight 小(起落)航线进场着陆

briquette 试块

brisance (炸药)震力,(炸药)猛度

brisk 有力的(指移动驾驶杆)

bristle 短而硬的毛

brittle fracture 脆性破坏,脆性断裂

brittleness 脆性,脆度

BRK Brake 刹车

BRKG Breaking 破坏,断开

BRKR Breaker 断路器,开关

BRKT Bracket 托架,括号

BRN Brown 褐色

broaching 1.拉刀(切割工具) 2.拉削(加工)

broad range 宽分布

broad region 大范围

broad search 大范围搜索

broadband 宽波段,宽频带

broadband access 宽带接入,宽频存取,宽带通信

broadband antenna 宽带天线

broadband noise 宽频带噪声

broad-band photometry 宽带测光

broad-band random vibration 宽带随机振动

broadbladed 宽叶片的,宽桨叶的

broadcast 广播

broadcast controlled air interception 广播管制飞行拦截,广播指挥截击法

broadcast ephemeris 广播星历(美国防部预报卫星轨道根数用)

broadcast ephemeris error 广播星历误差

broadcast receiver 广播收音机

broadcast satellite (用于接收和发射电视信号的)电视卫星,广播电视卫星

broadcast service 广播业务,如广播服务,播服务

broadcaster 广播公司,广播员,播送设备

broadcasting 广播

broadcasting satellite 广播卫星

broadcasting service 广播业务,广播服务,广播站

broadcasting subsystem 广播分系统

broadcasting-satellite service 卫星广播业务

broaden 扩大,变阔,变宽,加宽

broadside 1.宽边,宽面 2.机身侧部,舷侧 3.向机身侧部,从机身侧部

brodie 〈口语〉大角度俯冲,急剧俯冲

broken 1.断裂,破裂,炸裂 2.断裂的 3.〈口语〉碎云

broken base 轨底破裂

brolly 〈口语〉降落伞

bromacetophenone 溴苯乙酮(催泪性毒剂)

brombenzylcyanide 氯溴甲苯(催泪性毒剂)

bromide 1.溴化物 2.〈口语〉多波束无线电盲目轰炸系统(德国)

brominer 溴(化学元素,符号 B)
bromo-acetone 溴丙酮毒气
brontograph 雷雨计
brontometer 雷雨表
bronze 青铜
bronze weld 青铜焊
bronzing 镀青铜
Brooks comet 布鲁克斯彗星
Brown lunar theory 布朗月球运动理论
brown powder 褐色火药
Browning aircraft machine gun 勃朗宁航空机枪
brownout 减少灯光(一种节约措施)
BRS 1.Best-Range Speed 最佳航程速度 2.Brass 黄铜
BRSL Bomb Release Safety Lock 投弹安全销,投弹保险装置
BRST STR Bursting Strength 脆裂强度
BRT Brightness 亮度
BRTH Breathe 呼吸,吸入
BRTHR Breather 通气孔,呼吸阀
BRU 1.Bomb Rack Unit 炸弹架,炸弹架部件 2.Bomb Release Unit 投弹装置(挂点与弹药之间的接口)
brume 雾
brumous 雾沉沉的,多雾的
brush 1.电刷,刷子 2.刷掉
brush arm (电)刷臂
brush discharge 刷形放电,电晕放电
brush seal 刷式密封
brushing 刷光,清洁
brushless DC generator 无刷直流发电机
brushless DC torque motor 无刷直流力矩电机
brute force focusing 暴力式聚焦
brute-force 强力攻击,蛮力
brute-force filter 平滑滤波器
brute-force radar 强力雷达
BRW Brake Release Weight 开始起飞重量(松开刹车,开始起飞滑跑时的飞机重量)
BRZ Bronze 青铜
BRZG Brazing 钎焊
BS 1.Basic Standard 基础标准 2.Block Speed 轮挡速度 3.Bombsight 轰炸瞄准具 4.Body Station 壳体位置 5.Bug Speed 游标速度
BSA Block Stamp Assembly 无工厂系列号和单独检验记录的小组合件
B-SAT Broadcasting Satellite 广播卫星
BSB British Satellite Broadcast Ltd.英国卫星广播公司
BSC 1.Basic 基本的 2.Beam-Steering Computer 波束控制计算机 3.Binary Symmetric Channel 二进制对称信道
BSCU Brake System Control Unit 刹车系统控制装置

BSHG Bushing 衬套,轴瓦
BSI British Standards Institution 英国标准协会
BSKT Basket 蓝,筐,吊篮,吊舱
BSL Base Second Level 二级基地(维修服务)
BSO Blue Stellar Object 蓝星体
BSOL Below Safe Operating Limits 低于安全运行极限
BSP 1.Board Support Package 登机支援箱 2.Boeing Standard Program 波音标准程序
BSR Back Surface Reflection 背反射
BSS 1.British Standards Specification 英国标准规范 2.Boeing Specification Support Standard 波音规范辅助标准 3.Broadcasting Satellite Service 广播卫星业务
BSSEP BCAS System Software Engineering Process BCAS 系统软件工程工序
BST 1.British Standard Time 英国标准时间 2.British Summer Time 英国夏令时间
BSTA Body Station 机身站位
BSTLP Boeing Standard Thrust Limit Program 波音标准推力极限程序
BSTR Booster 助推器,增压器
BST RKT Boost Rocket 助推火箭
BSTS Boost Surveillance and Tracking System 助推段监视和跟踪系统(美国 SDI 中的项目)
BSU 1.Basic Sounding Unit 基本探测单元(泰罗斯业务垂直分布探测仪的) 2.Bypass Switch Unit 旁路开关装置
BT 1.Barium Titanate 钛酸钡 2.Balance Tab 平衡调整片,随动补偿片 3.Beta Testing β测试 4.Block Time 轮挡时间 5.Bombing Table 轰炸瞄准表 6.Bourdon Tube 包端管
BTB Bus Tie Breaker 汇流条断路器,汇流条连接接触器
BTC Bus Tie Connector 汇流条连接器
BTCE Bench Test and Calibration Equipment 基准测试与校准设备
BTD Domb Testing Device 炸弹试验装置
BTF 1.Biotechnology Facility 生物技术设施(美国空间站上的) 2.Bombtail Fuze 炸弹底部引信
BTH Beyond-the-Horizon 超视距
BTI Balanced Technology Initiative 平衡技术倡议(美国 1987 年"常规防御倡议"的组成部分,用于发展某些高技术)
BTIF Blind Target Identification Point 仪表轰炸目标标志点
BTMU Brake Temperature Monitor Unit 刹车温度监控装置
BTN Button 按钮
BTO Bombing through Overcast 穿云轰炸,云上轰炸
BTPD Body Temperature, Pressure, Dry 体内温度

（37℃）、环境压力和干燥气体状态

BTPS Body Temperature, Pressure, Saturated 体内温度（37℃）、环境压力和饱和水蒸气状态

BTR Bus Tie Relay 汇流条连接线继电器，总线继电器

BTS 1. Biotelemetry System 生物遥测系统（测量动物生理功能）2. Blue Tool Steel 蓝色工具钢 3. BPAM Tool Support 业务流程构造方法工具支持

BTT Bank-To-Turn 倾斜转弯（一种控制技术）

BTU British Thermal Unit 英制热量单位

BTWLD Buttweld 对接焊

BUA British United Airways 英国联合航空公司

BUAF British United Air Ferries 英国联合空运

bubble 1. 气泡 2. 曲线的驼峰 3. 前缘吸力式压力分布 4. 球形水准仪 5.〈口语〉座舱盖

bubble chamber 气泡室

bubble diameter 气泡直径

bubble flow visualization 气泡显示法

bubble helmet 球形头盔

bubble level 水准器

bubble memory 磁泡存储器

bubble point pressure 饱和压力

bubble sextant 气泡六分仪，水准式六分仪（利用水平仪建立当地水平基准的六分仪）

bubbling 沸腾，气泡分离，起泡，有气泡的

bubbly 起泡的，多泡的

buck 飞机掉进颠簸区，飞过颠簸区

bucket 1. 涡轮叶片（turbine bucket 之简称，美国用法）2. 戽斗，斗状物，U 形曲线 3.（弹射座椅的）椅盆 4.（发动机）火焰反射器，（反推装置）折流板 5.〈口语〉密集队形的轰炸机群

bucket shop 投机商号

bucketing 〈口语〉空中颠簸

bucking bar 抵消带，打钉杆

bucking winding 去磁绕组，补偿绕组，抵消绕组

buckle 1. 箍，卡箍，扣环 2.（结构）失稳，（薄壁）皱损，翘曲，（柱）压屈，纵弯曲

buckle pattern 翘曲图形

buckling （结构件的）皱损，翘曲，纵弯，失稳（结构件受压时产生侧向变形）

buckling test 屈曲试验

buddy aid 互救

buddy home 护送（被击伤的飞机）回场

buddy pack 伙伴油箱（一种可拆卸的油箱，装于飞机上供他机加油之用）

budget 预算，预算费

budget constraint 预算约束，预算限制

budget level 预算数额，预算标准

budget policy 预算政策

budget reserve 预备金

BUDS Back-Up Despin System 备用消旋系统

BUEC Back-Up Emergency Communication 备用应急通信

BUF Back-Up Facility 备用设施，备用设备

buffer 缓冲的人或物，缓冲器

buffer amplifier 缓冲放大器

buffer and recuperator 制退复进机

buffer battery 缓冲蓄电池

buffer control unit 缓冲控制器

buffer cylinder 制退机筒

buffer gas 缓冲气体

buffer overflow 缓冲溢出

buffer register 缓冲寄存器

buffer rod 缓冲杆

buffer storage 缓冲存储（器）

buffered FET logic 缓冲场效管逻辑

buffet 抖振

buffet boundary 抖振边界（飞机在匀速直线飞行时，不发生抖振现象的最大速度和高度范围）

buffet excitation parameter 抖振扰动参数

buffet load 抖振负载

buffet margin 抖振裕度（通常以不发生抖振的最大过载表示）

buffet test 抖振试验

buffeting 抖动，扰流抖振，抖振

buffeting load 抖振负载

buffing 抛光，擦光

bug 1.〈口语〉（常用复数）缺陷，错误，故障 2. 电键 3. 电子窃听器，雷达位置测定器 4. 空速表游标（专指空速表上定基准空速的游标），基准空速指标 5. 小飞行器，登月舱 6.（领航仪表上的）航向指标

bugged 装有窃听器的

buggy 1.〈口语〉飞机 2. 手推车，小车

BUIC Back-Up Interceptor Control 备用拦截控制（系统）（美国地面防空系统的组成部分）

build 1. 构件（软件产品的一个工作版本）2. 增强，建立

build up 1. 装配 2. 建立，形成 3. 逐次增加，增长 4. 集结

building block 积木式组件

building for leak detection 检漏厂房

buildup 增强，发展，形成，组合

buildup time 1.（火箭）进入（某一）状态时间 2.（信号的）建立时间，增长时间

build-up welding 堆焊

built in 1. 内装的，机内的，内置的，内部的 2. 嵌入的，固定的，固有的

built in field 内建电场

built-in command 内部命令

built-in diagnostic circuit 内建诊断电路

built-in potential　内置电势,自建电势
built-in range　机内油箱航程,内装燃料航程(不带副油箱,不在空中加油的航程)
built-in reliability　固有可靠性,内在可靠性,结构可靠性
built-in test　(系统)内部测试,自测试,内装测试
built-in test equipment　内装式测试设备(与被测试的功能设备结合为一整体的)
built-in twist　内置转折,内置扭曲
built-up section　组合式结构件,组合部分
bulb　球形零件
bulb angle　圆头角钢
bulge　1.鼓包,凸起部 2.凸出,膨胀
bulge X-ray source　核球X射线源
bulging　1.胀形 2.扩径旋压
bulk　1.容积,容量 2.大块,大量,大批 3.大的部件,笨重物
bulk acoustic wave　体声波
bulk charging　体充电
bulk composition　总成分
bulk effect　体效应
bulk effect diode　体效应二极管
bulk heat treatment　容积热处理
bulk material　1.基体材料(太阳电池基体并参与光电效应的材料) 2.大宗材料(指可以任意数量使用的材料,例如油漆涂料、黏胶、浆糊、油布等)
bulk modulus　体积弹性模量
bulk motion　牵连运动
bulk petroleum products　散装石油产品,散装油料
bulk processing　大容量处理
bulk recombination　体内复合
bulk redundancy　大量冗余
bulk shielding　整体屏蔽
bulk storage　大容量存储
bulk velocity　整体速度
bulkhead　隔板,隔框,舱壁,箱底,防水壁
bulk-injection　容积耗量表,容积流量计仪
bulkmeter　容积耗量表,容积流量计
bull gear　大齿轮,主齿轮
bull session　航空问题研讨会(在航空工程师与飞行人员之间对重大航空问题的非正式讨论会)
bullet　1.子弹,弹头 2.(飞机的)子弹形机头,机头整流锥 3.(发动机的)尾锥体,喷口整流锥
bullet drop　子弹下降(射出的子弹受地心吸力的影响所产生的正常降落)
bullet group　(轻武器)靶上子弹散布面
bullet hole　弹痕,枪眼
bullet inertia　子弹惯性
bullet jacket　弹头壳

bullet jump angle　弹丸带偏角
bullet pattern　子弹散布面
bullet splash　弹丸溅片
bullet squirt　机枪
bulletproof　防弹的
bumble bee　1.清扫音响水雷用的发声装置 2.冲压喷气发动机式导弹
bump　1.阵风,颠簸气流 2.颠簸(阵风引起的) 3.碰,撞,(金属板的)冲压 4.让座(指在飞机上将座位让给重要人物)5.(飞机在连续起飞着陆中的)着陆 6.凸起
bump test　颠簸试验
bumper　缓冲器,减震器,保护座,车挡
bumper bag　缓冲袋(飞艇底部的充气袋,用于着陆减震)
bumper stop　缓冲器行程限止器
bumpiness　颠簸,大气扰流
bumping　1.颠簸,碰撞 2.爆腾(剧烈沸腾) 3.非自愿取消登机(指乘客在上飞机前,因超员等原因,而被取消乘用权)
bumping hammer　开槽锤
bumpy　扰动的,不稳定的(指气流)
BUN　Block Unit Number 分段用户编码
buna　丁钠橡胶,布纳橡胶
bunaN　丁腈橡胶
buna-S　丁苯橡胶
bunaS3　丁苯(S3)橡胶
buncher　1.(电子)聚束器,聚束栅,堆栅 2.(速调管的)输入电极,调制腔,速度调制电极 3.无级变速器
bunching　1.(电子)聚束 2.成组,成群
bunching space　群聚空间
bunching voltage　聚束电压,聚群电压
bundle　1.束,捆包 2.伞投包裹(支援受困友军) 3.敷金属纸条(施放消极干扰用)
bundled cable　成束敷设的电缆
bungee　1.橡筋绳 2.弹簧筒 3.弹舱启门机(轰炸机上用液压开启弹舱门的机械)
bunker　1.掩体 2.燃料舱(舰船上的)
bunkerscope　掩体潜望镜(用以观测和拍摄火箭发射情况)
BUNO　Bureau number 部门编号(海军)
bunt　半外斤斗(飞机从正飞进入,向下作外斤斗前半圈)
buoy　1.浮标 2.救生圈 3.设置浮标,使浮起
buoyancy　浮性
buoyancy correction　浮力修正
buoyancy effect　浮力效应,弹性效应
buoyancy force　浮力
buoyancy frequency　浮力频率
buoyancy tank　浮力箱

buoyant　有合力,浮动的,能浮的

BUP　Bend Up(向上)弯曲

BUR　1. Backscatter Ultraviolet Radiometer 反向散射紫外辐射计 2.Back-Up Register 备用寄存器

burble　1.紊流,产生涡流,起气泡,冒泡 2.气流分离,气流分离区 3.层流(无涡流动)的破坏

burden　装载量

burette　滴定管,量管

Burgers vector　柏格斯矢量

buried layer　埋层

burn　1.燃烧,烧毁 2.〈口语〉发动机加力

burn out　烧断,烧尽

burn through　1.(燃烧室)烧穿 2.烧穿(雷达反干扰的一种措施)

burn up　烧尽,烧光

burned gas　燃烧过的气体

burner　1.燃烧室,火焰筒 2.喷嘴 3.火药柱 4.〈口语〉加力燃烧室

burner nozzle　喷灯嘴,燃烧器喷嘴

burner-vaporizer　汽化式喷嘴

burn-in　老化,老炼,烧(烙)上,预烧

burning　1.燃烧,(发动机)起火 2.烧毁,烧伤 3.(图像四周出现的)烧斑 4.(火箭发动机)工作

burning area　燃烧面积

burning final pressure　燃烧终点压强

burning rate　燃烧速度,燃烧率

burning rate coefficient　燃速系数

burning rate pressure exponent　燃速压强指数

burning surface　燃烧面

burning surface area　燃烧表面面积,燃烧面积

burning surface temperature　燃烧表面温度

burning surface to port area ratio　燃通面积比

burning surface to throat area ratio　燃喉面积化

burning test　燃烧试验,燃烧性试验

burning time　1.燃烧时间 2.(导弹等)起动时间 3.(火箭发动机)工作时间

burning-rate　燃烧速率

burning-rate calculation　燃烧速率的计算

burning-rate characteristic　燃烧速率的特点

burning-rate curve　燃烧速率曲线

burning-rate datum　燃烧速率数据

burning-rate enhancement　燃烧速率提高

burning-rate measurement　燃烧速率测量

burning-rate pressure　燃速压力

burning-rate pressure exponent　燃速压力指数

burning zone　燃烧层,燃烧带

burnoff　熔化

burnout　烧坏,燃料烧尽

burnout altitude　燃料烧尽高度,主动段终点高度

burnout altitude error　燃料烧尽高度误差,主动段终点高度误差

burn-out energy　抗烧毁能量

burnout interface　燃烧介面

burnout mass　燃料烧尽时质量,主动段终点质量

burnout velocity　燃尽飞行速度

burn-rate　燃烧速率

burn-rate compliance　燃烧速率服从性

burnt gas　燃烧废气

burnt gas temperature　燃烧气体温度

burnt product　烧成制品

burnt rivet　热处理柳钉(指经过热处理的)

burn-through　烧穿

burn-through range　烧穿距离

burn-through time　烧穿时间

burn-time average thrust　燃烧时的平均推力

burnup　烧尽,烧光(如指再入大气层的物体)

burr　1.毛刺,毛边 2.磨石

burring　去毛刺,去毛边

burst　猝发

burst　1.爆炸,突发,闪光 2.连发射击 3.(高炮)点射,炸点 4.(无线电的)脉冲串,(电视机的)色同步信号 5.(轮胎)爆破 6.短促快速操纵动作(如按按钮) 7.遇难,失事

burst charge　爆炸装药

burst command　爆炸指令,自毁指令

burst correcting code　纠突发错误码

burst deviation　炸点偏差

burst diaphragm　安全隔膜

burst error　突发差错

burst height　1.爆炸高度 2.炸高

burst in the bore　炸膛,膛内爆炸

burst interval　1.炸点与目标的间隔 2.炸点间隔

burst isochronous signal　突发等时信号

burst length　突发串长度

burst mode　成组方式,突发模态

burst point　爆炸点

burst point distribution density　炸点分布密度

burst pressure　爆破压力

burst range　爆炸距离,炸点距离

burst signal　(电视)色同步信号

burst test　爆破试验

burster blocks　防弹板,耐炸砖块

burster casing　起爆管

burster charge　爆炸装药

burster tube　起爆管,传爆管

bursting charge　爆炸装药

bursting cone　弹片锥形散布

bursting explosive　炸药

bursting layer　抗爆层
bursting point　爆炸点
bursting powder　炸药
bursting speed　破裂转速（超转试验）
bursting altitude　爆炸高度，炸高
bursting-type munition　爆炸式弹药（化学战与生物战的）
bury　埋葬，隐藏
bus　1.母线，干线，总线，汇流条 2.弹头母舱，卫星舱 3.公共汽车 4.〈口语〉重型飞机，旅客机
bus bar　1.母线 2.汇流条
bus control unit　总线控制器
bus controller　总线控制器
bus failure　汇流条断路
BUSH　Bushing 衬套，轴瓦
bus limiter　总线限制器，总线限幅器
bus power distribution unit　总线功率分配单元
bus spacecraft　总线航空器
bus subsystem　总线子系统
bus technology　总线技术
bus tie　汇流条，馈电线
bus topology　总线拓扑结构
bus upgrade　总线升级
bus voltage　母线电压
bush　1.衬套，衬筒，轴衬，轴瓦，绝缘管 2.〈口语〉火舌，火苗 3.（叶片）叶栅 4.装衬筒，装轴衬
bush aircraft　偏僻区飞机（供飞往遥远偏僻地区如北极地区使用的多用途飞机）
business aircraft　公务飞机（政府机构的管理和行政事务专用的飞机，包括直升机）
business airplane　公务机
business jet　商务喷射机，公务机，公务机型
business jet engine　商业喷气发动机
business jet main wing model　公务飞机机翼模型
buss　母线
buster　1.抛射药，起爆药 2.最大巡航速度飞行，以发动机额定工作状态飞行
bus-tie breaker　汇流条断路器
bustion　燃烧，燃烧现象
busy tone　忙音
butadiene　丁二烯
butadiene rubber　聚丁橡胶，丁二烯橡胶
butanal　正丁醛
butane　丁烷
butanol　丁醇
butene　丁烯
BUTN　Butane 丁烷，天然瓦斯
butt　1.粗端，枪托，根部，柄 2.冲撞，碰撞 3.对接 4.靶垛

butt rib　搭接肋
butt weld　对接焊，对接焊缝，电阻对焊接头
butt welding　对接焊
Butterfield　巴特菲尔德型金属反射体
butterfly bomb　蝶形炸弹
butterfly diagram　蝴蝶图
butterfly tail　蝶形尾翼，V形尾翼
butterfly valve　蝶（形）阀
Butterworth filter　巴特沃思滤波器
butting　拼接
button　按钮
button cell　扣式电池
button probe　按钮调查
button-on weapons　临时配备的武器
butyl　丁基
butylene　丁烯
buy　购买，采购
buyer　买主，采购方
BUZ　Buzzer 蜂音器，电气信号器
buzz　1.振动噪声 2.嗡鸣，喘振 3.振动，颤振 4.〈口语〉低空飞行，俯冲 5.飞行危及
buzzard　〈口语〉飞行员，航空员
buzz-bomb　〈口语〉飞机形导弹
buzzer　〈口语〉无人驾驶空袭兵器，飞弹
buzzing　超低空飞行，俯冲
buzz-saw noise　风扇激波噪声
BV　1. Bleed Valve 排气阀，放气活门，引气活门 2. Breakdown Voltage 击穿电压 3. Butterfly Valve 蝶形活门 4. Bypass Valve 旁通活门
BVCU　Bleed-Valve Control Unit 排气阀控制装置
BVI　Blade Vortex Interaction 桨叶涡旋互相作用
BVR　Beyond Visual Range 在视觉区以外
BVT　Boundary Values Testing 边值测试
BW　1. Basic Weight 基本重量 2. Bandwidth 宽带 3. Beam Width 波束宽度 4. Black and White 书面文字为凭 5. Bonded Warehouse 关栈仓库，保税仓库
BWC　Biological Weapons Convention 生物武器公约
BWG　Birmingham Wire Gage 伯明翰线规
BWPA　British Women Pilots Association 英国女飞行员协会
BWRA　British Welding Research Association 英国焊接研究协会
BWT　Backward Wave Tube 返波管
BYP　Bypass 旁路，支路
bypass accumulator　浮充蓄电池
by-pass capacitor　旁路电容器
bypass engine　双涵道发动机，涡轮风扇发动机
bypass filter　旁通滤波器
bypass flap　旁路风门，旁路放气门

bypass flow 旁流,旁通流量
bypass pipe 旁路
bypass ratio 1.涵道比 2.流量比
bypass section 旁路区域
bypass stream 旁路流
by-pass valve 旁通阀
byproduct 副产品
byte 【计】字节(二进制的)
byte multiplexing 【计】字节复接
byte stuffing 【计】字节填充法,塞字节法

C

C 1. Ceiling Panels 天花板 2. Locate By Coord. 按坐标定位

C criterion C 准则(航空器对纵向操纵的动态响应的时域判据)

C display C 显示器

C&W Control and Warning 控制与警告

C/A Corrective Action 纠正措施

C/A code Coarse/Acquisition Code 粗测/捕获码

C/D 1. Certificate Delivery 交货 2. Countdown 倒数计时,逆计数

C/E Course Error 航道误差,航迹误差

C/I 1. Carrier-to-Interference Ratio 载波干扰比 2. Certificate of Inspection 检验证书,检查证书

C/L Cut to Length or Change Letter 按长度切割或更改字

C/LRUS Components/LRU Specifications 部件/航线可更换装置字符

C/IM Carrier-to-Intermodulation 载波交调比

C/M 1. Configuration Management 构型管理 2. Control and Monitoring 控制和监测

C/MSR Commercial/Military Spares Release 民用/军用备件发放

C/N Carrier-to-Noise Ratio 载噪比,载波噪声比

C/O Cutout 切口,开口

C/P Check Point 检查点

C/R Command/Response 指令/响应

C/S 1. Client/Server 客户机/服务器 2. Cut to Size 切至尺寸

C/SCSC Cost and Schedule Control System Criterial 成本和设计控制系统原则

C/SKETCH Cut per Sketch 按草图切割

C/SSS Cut to Strip Stick Size 按条料尺寸切割

C/T Cut to Template 按样板切割

C° Centigrade 摄氏温度

CO^2RS CO^2 Research Satellite 二氧化碳研究卫星

C_2 Command and Control 指挥和控制

C_3 Command, Control and Communications 指挥,控制和通信

C_3I Command, Control, Communications, and Intelligence 指挥,控制,通信和情报

C_3P Combat Core Certification Professionals 战斗物资许可证专业公司

C_4I Command, Control, Communications, Computers and Intelligence 指挥,控制,通信,计算机和情报

CA 1. Cabin Altitude 座舱高度 2. Cabin Attendants 机上服务员,客舱乘务员 3. Calibrated Altitude 校正高度 4. Call Address 调用地址 5. Certification Authority 认证机构 6. Change Activity 更改活动 7. Commercial Aircraft 民用飞机 8. Configuration Audit 构型审查 9. Conformity Assessment 合格评定 10. Contamination Avoidance 污染防止 11. Continued Airworthiness 持续适航 12. Control Airspace 受管控空域 13. Control Area 管制区域 14. Control Axes 管制轴 15. Corrective Action 纠正措施,修复措施 16. Constituent Assembly 结构组件 17. Criticality Analysis 危害性分析 18. Cruise Altitude 巡航高度

CA PL Cadmium Plate 镀镉

CAA 1. Civil Aeronautics Administration 民用航空署 2. Civil Aviation Administration 民用航空管理局 3. Civil Aviation Authority 民用航空局 4. Conformal-Array Aerial 保形阵列天线 5. Computer-Aided Analysis 计算机辅助分析 6. Computer-Aided Assemble 计算机辅助装配

CAAC Civil Aviation Administration of China 中国民用航空管理局

CAADRP 1. Civil Aircraft Airworthiness Data Requirement Programme 民用航空器适航性资料要求纲要 2. Civil Aircraft Airworthiness Data Recording Program 民用航空器适航资料记录程序

CAAM 1. Civil Aeronautical Administration Manual 民航管理手册 2. Conventional Airfield Attack Missile 攻击普通机场的导弹

CAAS Computer-Assisted Approach Sequencing 计算机辅助进场着陆程序

Cab 1. 舱 2. 航空港塔台

CAB 1. Cabin 机舱 2. Civil Aeronautics Board 民用航空委员会

cabamate 氨基甲酸酯,甲氨基甲酸酯,氨基甲酸盐

cabamide 脲,尿素,碳酰二胺

cabamite 卡巴买特,碳酸胺,二乙二苯基脲

cabamonitrile 氨基化氰,氨腈

cabane (飞机的)顶架,翼间架

cabin 1. 座舱,机舱 2. 方舱

cabin depressurization 座舱减压

cabin dew point 座舱露点

cabin differential pressure 座舱压力差

cabin emergency dump valve 座舱应急卸压活门
cabin heat exchange 座舱热交换
cabin heat valve 座舱温度调节活门
cabin instrument system 座舱仪表系统
cabin interior equipment 座舱内部设备
cabin layout 1.舱室布置 2.客舱座椅配置 3.座舱布置方案
cabin pressure 座舱压力
cabin pressure altimeter 座舱压力高度表
cabin pressure control 座舱压力调节
cabin pressure failure 座舱增压故障
cabin pressure regulator 座舱压力调节器
cabin pressure relief valve 舱压安全阀
cabin pressure schedule 座舱压力制度
cabin pressure warning light 舱压下降警告灯
cabin pressurization 座舱增压
cabin pulley system 软式传动系统
cabin safety valve 座舱安全活门
cabin structure 舱结构
cabin temperature sensor 座舱温度传感器
cabin ventilation 座舱通风
cabin voice recorder 驾驶舱语音记录器
cabin window 舷窗
cabin-air heat exchanger 座舱空气换热器
cabinet 1.橱,柜 2.内室,小间,舱
CABL Customer-Authorized Buyer-Furnished (Equipment) List 客户批准的买方配置设备清单
cable 1.缆,索,钢索,锚链 2.电缆,多心导线,被覆线 3.发电报
cable angle transducer 电缆角形换能器
cable assembly 电缆组件,钢索组件
cable box 电缆箱,分线盒
cable channel 电缆通廊
cable connection box 电缆转接箱
cable cutter 割缆装置,割缆刀
cable diameter 光缆直径
cable effect 电缆效应
cable element 索单元
cable end 电缆头
cable end body 电缆端体
cable fill 电缆占用率,电缆利用率
cable guidance 有线制导
cable harness 电缆束
cable hover 放缆自动悬停
cable indicator 电缆指示器
cable member 索单元分析,索杆体系,电缆组件
cable routing 电缆走线
cable shaft 电缆井道
cable swinging rod 电缆摆杆

cable tip 电缆提示
cable trench 电缆沟
cable trough 电缆槽
cable vault 电缆地下室
cable walkway 电缆通道
cable zero calibration 有线校零
cable-control 控制电缆
cable-controlled missile 有线制导导弹
cable-drag drop 缆拖空投
cable-pulley element 电缆滑轮单元
cable-strike protection 撞电缆防护
cabling 敷设电缆
cabotage 1.航空运输权 2.沿岸航行
CACAC Civil Aircraft Control Advisory Committee 民用航空器管理咨询委员会
cacaerometer 空气污染检查器
CACC Cargo Air Control Center 航空货物管理中心,航空货运控制中心
cache 1.超高速缓冲存储器 2.隐藏所,隐藏的粮食或物资,贮藏物
cache memory 1.高速缓冲存储器 2.快取记忆体
cache storage 高速缓冲存储器
CACL Computer-Assisted Computer Language 计算机辅助计算机语言
CACRS Canadian Advisory Committee on Remote Sensing 加拿大遥感咨询委员会
CACS Computer-Assisted Communication System 计算机辅助通信系统
CACT Civil Air Carrier Turbojet 民航涡轮喷气客机
CAD 1.China Airworthiness Directive 中国民用航空总局适航指令 2.Computer-Aided Design 计算机辅助设计 3.Computer-Aided Detection 计算机辅助探测 4.Computer-Aided Diagnosis 计算机辅助诊断 5.Computer-Aided Draughting 计算机辅助制图 6.Cushion Augmentation Device（垂直或短距起落飞机的)增升装置,近地增升装置
CAD/CAM Computer-Aided Design/Computer-Aided Manufacturing 计算机辅助设计/计算机辅助制造
CADAC Computer-Aided Design and Construction 计算机辅助设计与构造
CADAM Computer-Aided Design and Manufacturing 计算机辅助设计和制造
CADAR Computer-Aided Design, Analysis and Reliability 计算机辅助设计、分析与可靠性
cadastral map 地籍图
cadastral survey 地籍测量,土地测量
cadastre 地籍
CADC Central Air Data Computer 中央大气数据计算机

CADD 1. Computer-Aided Drafting Design 计算机辅助草图设计 2. Computer-Aided Design and Drafting 计算机辅助设计与制图

CADE 1. Combined Allied Defense Experiment 盟国联合防御实验 2. Computer-Aided Design and Engineering 计算机辅助设计与工程

CADES（Cades） 1. Computer-Aided Design and Evaluation System 计算机辅助设计和评定系统 2. Computer-Aided Development and Evaluation System 计算机辅助开发与评价系统

cadet 1. 军校学员 2. 候补军官

CADF Commutated-Aerial Direction-Finder 变向天线测向器

CADM Clustered Airfield Defeat Munition 破坏机场集束弹

CADMAT Computer-Assisted Design, Manufacture And Test 计算机辅助设计，制造和试验

cadmium 【化】镉（化学元素，符号 Cd）

cadmium cell 镉电池

cadmium embrittlement 镉脆

cadmium sulfide solar cell 硫化镉太阳能电池

cadmium-air battery 镉-空气蓄电池

cadmium-mercuric oxide cell 镉-汞电池

cadmium-nickel storage battery 镉-镍蓄电池

cadmium-silver storage battery 银-镉蓄电池

CADP Central Annunciator Display Panel 中央警告信号显示板

CADRE Center for Aerospace Doctrine, Research, and Education 航空航天理论研究教育中心

CADS 1. Central Air Data System 中央大气数据系统 2. Central Attitude Determination System 中央姿态测定系统 3. Computer Analysis and Design System 计算机分析和设计系统

CADT Computer-Aided Design Tooling 计算机辅助工装设计

CADU Channel Access Data Unit 信道访问数据单元

CAE 1. Carrier Aircraft Equipment 舰载飞机设备 2. Component Application Engineer 部件应用工程师 3. Computer Application Engineer 计算机应用工程师 4. Computer-Aided Engineering 计算机辅助工程

CAED Computer-Aided Engineering Design 计算机辅助工程设计

CAEDM Community/Airport Economic Development Model 社区/机场经济发展模型

CAeM Commission for Aeronautical Meteorology 航空气象委员会

CAEM Cargo-Airline Evaluation Model 货运航空公司评定模型

CAEP Committee on Aviation Environmental Protection 航空环境保护委员会

caesium 【化】铯（化学元素，符号 Cs）

caesium clock 铯钟

caesium thermionic converter 铯热离子转换器

CAF 1. Confederate Air Force 联合空军 2. Configuration Approval Form 构型审批表

CAFH Cumulative Airframe Flight Hours 飞机机体累计飞行小时数

CAFI Commander Annual Facilities Inspection 指挥官年度设备检查

CAFMS Computer-Assisted Force Management System 计算机辅助部队管理系统

cafs whisker 灵敏传感器

CAFT Combined Advanced Field Team 联合前方外场工作组

CAG 1. Carrier Air Group 航母飞行大队，舰载航空大队 2. Civil Air Guard 民间空中防卫 3. Collective Address Group 集体地址群

CAGE 1. Commercial and Government Entity 商业和政府单位 2. Computerized Aerospace Ground Equipment 计算机化航空航天地面设备

cage 1. 锁定 2. 保持架 3. 笼子,盒,罩

cage antenna 笼形天线

cage bare 无罩笼

CAGEC Commercial and Government Entity Codes 商业和政府单位代码

cage cone 锥形笼

caging 锁定,制动,停止

caging device 锁定装置

caging knob 锁定旋钮

caging system 陀螺上锁系统

caging time 锁定时间

caging zero 锁定零位

CAH Cabin Attendant Handsets 座舱乘务员耳机

CAI 1. Computer-Aided Inspection 计算机辅助检查 2. Computer-Aided Instruction 计算机辅助教学 3. Conference Aeronautique International 国际航空会议 4. Constant Altitude Indicator 等高显示器 5. Corrective Action Investeigation 纠正措施调查

CAINS Carrier Aircraft Inertial Navigation System 航母舰载机惯性导航系统

CAIP Continuous Airworthiness Inspection Plan 持续适航检查大纲

CAIS Computer-Assisted Introduction System 计算机辅助引进系统

caisson disease 潜水员病,沉箱病

CAL 1. Caliber 口径,尺寸,卡规 2. Canadian Astronautics Ltd. 加拿大航空航天公司

CALC Caculate 计算

calcination 锻烧
calcium 【化】钙(化学元素,符号Ca)
calculated address 合成地址,计算地址
calculated burning rate 计算燃烧速度
calculated detonation velocity 计算爆速
calculated position 计算位置
calculated reaction zone length 计算的反应区长度
calculated result 计算结果
calculated value 计算值
calculated velocity 计算速度
calculation 计算
calculation domain 计算域
calculation of trajectories 分弧弹道计算,弹道的分弧解法
calculation scheme 计算方案
calculative techniques 计算技术
calculator 计算机,计算装置
calculus 微积分学,计算,演算
calculus of variation 变分学,变分法
cale gear 凯尔齿轮
calendar 1.历法,日历,月历 2.日程表 3.排列,分类,索引
calendering 班光,压制
calf bladder 小腿气囊
calfax 卡氏快卸锁扣
calf-garters 系腿带
caliber 1.口径 2.才干 3.水准
caliber ammunition 口径弹药
caliber-radius head 弹头蛋形部曲率半径
calibrate 1.校准,检定 2.定标,分度 3.测量
calibrated airspeed 修正空速
calibrated altitude 修正表高
calibrated club propeller 校准的短螺桨
calibrated focal length of camera 1.相位检定主距 2.镜箱焦距
calibrated leak 校准漏孔
calibrated pressure altitude 修正压力高度
calibration 1.校准,检查 2.定标,定分度,刻度 3.测量口径
calibration curve 校正曲线,校准曲线
calibration card 校准卡片,校准表格
calibration center of balance 天平校准参考中心
calibration certificate 校准证书
calibration chamber 检定箱,校准室
calibration check 标定检验,格值检定
calibration coef 校正系数
calibration curve 校准曲线
calibration equipment 校准装置,校准仪器
calibration error 标定误差

calibration factor 校准系数,校准因数
calibration filter 校准滤光片
calibration fixture 校准混合物
calibration flight 校验飞行
calibration flight program 校飞程序
calibration flight route 校飞航路
calibration frame 1.标定架 2.比例尺
calibration function 标定函数
calibration in dock 1.坞内标校 2.码头标校
calibration laboratory 校准实验室
calibration level 校准电平
calibration measurement 校准测量,率定测量
calibration method 校准法
calibration model 标模
calibration orifice 节流圈
calibration parameter 校准参数
calibration procedure 校准程序
calibration process 校准过程
calibration result 校准结果
calibration rig 校正架
calibration satellite 校准卫星
calibration slug 校准塞
calibration source 定标源
calibration star 定标星
calibration step 标定阶跃
calibration tape 校准带
calibration target 校准目标
calibration templet 标准模片
calibration test 校准试验
calibration timer 标准定时器
calibration tolerance 标准公差
calibration tower 校准塔
calibration value 标定值
calibration wedge 校准楔
calibration wind tunnel 校准风洞
calibration-model test 标模试验
calibrator 校准器,校准设备
calibre 1.口径,内径 2.测径规
CALI-DALE Centralized Alarm Interface-Dependent Alarm Equipment 集中告警接口-有关告警设备
California institute of Technology 加州理工学院
California Space Institute 加利福尼亚空间研究所
californium 【化】锎(化学元素,符号Cf)
calipher(caliper) 测径器,卡尺
call completing rate 接通率
call fire 召唤火力,应召火力
call for fire 火力召唤,火力要求
call loss 呼损
call number 1.呼号,调用编号 2.子程序号

call sequence 调用序列

call-back 1.回叫信号,回答信号 2.收回(产品回修),招回

calling 呼叫,呼号

Callipic cycle 卡利普周期

Callisto 【天】木卫四

callout notes 标注

callsign 呼号

caloric equation of state 量热状态方程

caloric nystagmus 1.不对称性眼震 2.温度性眼震

caloric state equation 热状态方程

calorie 卡(路里)

calorific 生热的,热量的

calorific value 热量,热值

calorifier 加热器

calorimeter 1.量热器,热量计,卡计 2.热敏式电表

calorimetric 量热的,测热的

calorimetric detector 热量探测器

calorimetry 量热术,量热学

calorizing 表面渗铝,铝化

calott 伞衣,帽状物

Calsphere calibration sphere 校准卫星

calvus(cal) 秃状

Calypso 【天】土卫十四

CAM 1. Camber/Computer-Aided Manufacture 弯度、曲度或计算机辅助制造 2. Computer-Aided Manufacturing 计算机辅助制造 3. Consolidated Aircraft Maintenance 专业合并的飞机维修,联合的飞机维修 4. Conventional Attack Missile 常规攻击导弹 5. Counter-Air Missile 防空导弹

cam 凸轮,凸块

cam follower 凸轮从动件

cam gear 凸轮传动齿轮,凸轮传动装置

cam insert 拨动导块

cam ring 凸轮环

CAMAC Computer-Aided Measurement and Control 计算机辅助测量和控制,卡马克系统

CAMAC test CAMAC测试

CAMAL Continuous Airborne Missile Alert 机载导弹持续警报

Camber 小船坞

camber 1.曲度,弯度 2.曲面 3.弧 4.弧高,中弧曲度,中弧线

camber-changing trailing-edge flaperon 变弯度后缘襟副翼

camber distribution 弯度分布

camber line 弧面曲线

cambered 曲面的,弧形的

cambox 凸轮箱,三角座

camcorder 摄录像机,便携式摄像机

CAMDS Chemical Agent Munitions Disposal System 化学战剂弹药处理系统,毒剂弹药处理系统

camera 1.照相机,摄影机,摄像机 2.镜箱,暗箱 3.室,小室

camera adapter 照相机转向镜

camera attitude 摄影机姿态

camera blanking 摄像机逆程消隐

camera calibration 相机校正

camera coverage 相机视场

camera detector 相机探测器

camera frame 摄像机框架

camera gimbal 照相机万向架

camera gun 照相枪

camera housing 照相机防护罩,摄影机外壳

camera image 摄像机图像

camera lucida 转绘仪,显微描绘仪

camera measurement 摄影测量

camera position 摄影机定位

camera station 空中摄影站

camera tilt 摄影机俯仰运动

camera transit 摄影经纬仪

camera tube 摄像管

Camflex valve 偏心旋转阀

CAMI Computer-Aided Manufacturing and Inspection 计算机辅助制造和检验

Camlock handle 坎洛克手柄,凸轮锁手柄

CAMNET Computer-Aided Manufacturing Network 计算机辅助制造网络

camouflage 1.伪装,隐蔽,伪装术 2.伪装物

camouflage code 伪装码

camouflage material 伪装材料

camouflage net 伪装网

camouflage paint 伪装涂料

camouflage target 伪装目标

camouflage technology 伪装技术

camouflage with colors 迷彩伪装

camouflage-detection film 探测伪装胶片

camouflet 1.地下爆炸弹 2.地下爆炸坑穴

campaign 1.战役 2.运动 3.作战,出征,从事……活动

campaign reconnaissance 战役侦察

CAMP Continuous Airworthiness Maintenance Program 持续适航维修大纲

Campbell diagram 1.坎贝尔图 2.共振图 3.叶片共振转速特性图

camplate 斜盘

camp-on 预占线,保留呼叫

CAMR Camera 照相机

CAMRAD Comprehensive Analytical Model of Rotor-

craft Aerodynamics and Dynamics 旋翼机气动力和动力学综合分析模型
CAMS 1. Combat Aviation Management System 作战飞行管理系统 2. Communication Area Master Station 通信区主控站
camshaft 凸轮轴
CAN 1. Change Advise Number 更改通知单 2. Combat Aviation Net 作战航空兵通信网
can 1.罐,壶,筒 2.单个火焰筒 3.导轨盒
Canadarm 加拿大机械臂
canal 1.通道,管道,波道 2.沟,槽 3.运河
Canal of Mars 火星运河
canal sickness 半规管病,空晕病
can-annular 环管式,管环式
canard 1.鸭式飞机,鸭式布局 2.前置翼,鸭翼 3.鸭式编队
canard aircraft 鸭式飞机
canard airplane 鸭式飞机
canard area 鸭翼面积
canard configuration 鸭式布局
canard control 鸭舵控制
canard design 鸭翼的设计
canard model 鸭翼模型
canard tip 鸭翼尖
canard tip vortex 鸭翼尖涡
canard vortex 鸭翼涡
canard-controlled missile 鸭式导弹
canard-delta 三角翼鸭式布局
CANC Cancel 取消
cancel 1.取消,废止 2.撤销 3.失效 4.删去,消除,对消
cancel character 作废字符
cancellation 1.撤销,取消,废止 2.删去,消除
cancellation error 冲销错误
cancellation ratio 抵消比,对消比
canceller 对消器,补偿设备
cancelling command 取消指令
candidate 候选人,候补者
candidate solution 候补解
candidate element 候选元素
candidate engine 候选发动机
candidate model 候选模型
candidate orbit 候选轨道
candidate procedure 候选程序
candidate set 候选集
candidate solution 候选方案
canister 1.箱,筒,金属容器 2.滤毒罐 3.霰弹,霰弹筒
canned cycle 固定循环
cannelure 子弹槽线,弹壳槽线
cannibalize (cannibalise) 1.串件 2.拼修

canning 装罐封存
cannon 1.机关炮 2.火炮,加农炮 3.空心轴
cannon barrel 炮管
cannon with liquid propellant 液体炮
cannon-ball 炮弹
cannon-proof 防炮弹的
cannular 环管式
cannular combustor 联管燃烧室
canoe radar 独木舟雷达
canonic grammar 规范文法
canonical conjugate 正则共轭
canonical conjugate variables 正则共轭变量
canonical constants 正则常数
canonical coordinates 正则坐标
canonical elements 正则要素,正则根数
canonical equation 正则方程
canonical extension 正则扩充
canonical form 正则形式
canonical matrix 正则矩阵
canonical maximum likelihood method 典型最大似然法
canonical momenta 正则动量
canonical quaternion 典型四元数
canonical state variable 规范化状态变量
canonical time unit 正则时间单位
canonical transformation 典型(正则)变换
canonical unit 标准单位
canonical variables 正则变量
canopy 1.座舱盖 2.透明整流舱盖 3.伞衣
canopy area 伞衣面积
canopy diameter 伞衣直径
canopy fabric 伞衣织物
canopy fabric porosity 伞衣织物透气量
canopy fluff 座舱盖爆破
canopy jettison ejection 抛盖弹射
canopy opening 伞衣开伞
canopy performance 伞衣性能
cant 1.斜面,倾斜 2.超高,铁道弯线的外轨加高 3.角隅,(建筑物)外角,切去棱角 4.四角木材,斜肋骨
cant advantage 斜面优势
cant amount 偏斜量
cant angle 偏斜角
cant two-way interaction 斜面双向互动
canted nozzle 斜喷管
canteen 1.军用水壶 2.小卖部,临时餐厅
cantilever 1.悬臂,悬臂梁 2.电缆吊线夹板
cantilever beams 悬臂梁
cantilever configuration 悬臂结构
cantilever design 悬臂式设计
cantilever ratio 悬臂比

cantilever wing　肱翼
cantilevered mode　悬臂式
cantonment　1. 驻扎, 宿营 2. 临时营房
Canukus　加努古斯标准
canvas　帆布, 帐篷
canyon　峡谷
CAO　1. Committee on Atmosphere and Oceans 大气与海洋委员会 2. Component Available for Ordering 订货可获得部件, 可订货部件 3. Contract Administration Office 合同管理办公室
CAODL　Components Available for Ordering Data List 可用于订购的构件数据表, 指令数据表可获得部件
CAOS　Completely Automatic Operating System 全自动操作系统
caoutchoid　类橡胶物
CAP　1. Capacity 电容, 容量 2. Capture 截获 3. Carrier Air Patrol 航母空中巡逻 4. Computer-Aided Production 计算机辅助生产 5. Computer-Aided Programming 计算机辅助程序设计 6. Corrective Action Programs 纠正措施方案
cap　1. 帽, 盖, 罩, 套 2. 缘条 3. 空中掩护, 战斗巡逻
cap of Mars　火星极冠
capability　1. 可能性 2. 能力, 性能, 本领, 效力 3. 可能输出功率, 生产率
capability characteristic　性能特点
capacitance　1. 电容 2. 容量, 容积
capacitance box　电容箱
capacitance bridge　电容电桥
capacitance fuze　电容引信
capacitance ratio　电容比
capacitance system tank units　电容式油量传感器
capacitance tolerance　电容量允差
capacitance transducer　电容变换器, 电容传感器
capacitance-coupled flip-flop　电容耦合触发器
capacitance-type fuel quantity indicating system　电容式油量指示系统
capacitance-type fuel transmitter　电容式油量传感器
capacitance-type sensor　电容式传感器
capacitive direct injection　电容直接注射
capacitive displacement transducer　电容式位移传感器
capacitive feedback　电容反馈
capacitive force transducer　电容式力传感器
capacitive pickoff　电容式传感器
capacitive probe　电容探测器
capacitive reactance　容抗
capacitive time constant　电容性时间常数
capacitive window　电容性窗孔
capacitometer　电容计, 电容测量计
capacitor　电容器

capacitor bank　电容器组
capacitor discharge　电容放电
capacitor sensor　电容传感器
capacitor-input filter　电容输入滤波器
capacitormic　电容传声器
capacitor-start induction motor　电容器起动式感应电动机
capacity　1. 容积, 容量 2. 能力, 本领, 效能 3. 电容
capacity altimeter　电容式高度表
capacity altimetry　电容测高
capacity of cell　电池容量
capacity payload　受容积限制的有效载重
capacity region　容量区域
capacity safety valve　承受量安全阀
capacity-related costs　运力有关成本
capadyne　电致伸缩继电器
CAPD　Computer-Aided Process Design 计算机辅助工艺设计
CAPE　China Aero-Polytechnology Establishment 中国航空综合技术研究所
Cape Canaveral　卡纳维拉尔角
Capella　【天】五车二
capillarity　毛细作用, 毛细现象
capillary　1. 毛细管 2. 毛细现象的, 毛细作用的 3. 表面张力的
capillary action　毛细作用
capillary action-shaping technique　毛细成形技术
capillary discharge　毛细管放电
capillary force　毛细管力
capillary injector　毛细管喷注器
capillary length　毛细管长度
capillary number　毛细管数, 毛管值, 毛细管准数
capillary plasma　毛细管等离子体
capillary potential　毛管势, 毛细管势能, 毛细管位能, 毛细管潜能
capillary pressure　毛细压力
capillary pump loop system　毛细抽吸回路系统
capillary standing wave　毛细驻波
capillary surface　毛细表面
capillary tube　毛细管
capillary wall　毛细管壁
capillary wave　表面张力波, 界面波, 张力波
capital contribution　资本摊缴, 出资额, 资本捐赠, 资本投入
capital investment　资本投资
capital satellite　高成本卫星, 高价值卫星
caplastometer　黏度计
capoc　木棉, 爪哇棉
CAPP　1. Carpet 毯, 毡 2. Computer-Aided Process

Planning 计算机辅助工艺规程编制 3. Computer-Aided Production Plan 计算机辅助生产规划 4. Computer-Aided Process Planner 计算机辅助工艺计划员 5. Customized Assembly Process Plan 定制装配工序大纲

CAPPI Constant Altitude Plan Position Indicator 等高平面位置指示器

capping strip 缘条

CAPRI Compact All-Purpose Range Instrument 袖珍通用测距仪

capron(e) 锦纶,卡普纶,聚己内酰胺纤维

CAPS 1. Computer Antenna Pointing System 计算机控制天线指向系统 2. Crew Activity Planning System 航天员活动编排系统

CAPSCR Cap Screw 有头螺钉,有帽螺钉

cap-shock 帽冲击

cap-shock pattern 帽冲击模式

capstan 1. 绞盘 2. 主动轮主导轴器 3. 紧身胶带,张紧装置

capstan anti-G suit 侧管式抗荷服

capstan partial pressure suit 侧管式部分加压服

capsulation 封装,胶囊包装

capsule 1. 膜盒 2. 密封舱,分离舱 3. 胶囊,容器 4. 炭精盒 5. 雷管,底火

capsule assembly 膜盒组

capsule cabin 弹射座舱,分离舱

capsule pressure gauge 膜盒压力表

capsule-type cabin 弹射座舱,分离舱

CAPT Captain 机长

captain 1. 机长,第一飞行员 2. 上尉,海军上校 3. 首领,指挥者

captain of the air station 航空站站长

captain of the flight deck 飞行甲板军士长

captive 1. 俘虏 2. 系住的,被系留的

captive balloon 系留气球

captive balloon sounding 系留气球探测

captive bolt 锁紧螺栓

captive carry flight test 系留载飞试验

captive firing 系留点火

captive firing test 台架点火试验,系留点火试验

captive launcher 静力试验发射装置,系留试验发射装置

captive test 1. 静态试验 2. 挂飞试验

captive trajectory method 捕获轨迹法

captive-test rocket 试车火箭

capture 1. 捕获,截获 2. 夺取,攻占 3. 战利品,缴获品

capture area 捕获面积

capture coefficient 捕获系数

capture cross section 捕获截面

capture effect 捕获效应,遏止效应,截获效应

capture hypothesis 俘获假说

capture latch 捕获止动销

capture orbit 捕获轨道

capture ratio 捕获比,捕获率

capture region 捕获区域

capture set 捕获集

capture spiral 捕获螺旋

capture time 捕获时间

capture trajectory 捕获轨迹

capture zone 捕获带

capturing 1. 归零,找准 2. 锁位,止位 3. 俘获,截获

CAR 1. Configuration Assessment Report (Form) 构型评估报告(表格) 2. Coordination Activity Record 协调活动记录 3. Corrective Action Request 修正措施要求

car 1. 车,汽车 2. 吊篮,吊舱

CARA 1. Combined-Altitude Radar Altimeter 组合高度雷达高度表 2. Computer-Aided Requirements Analysis 计算机辅助需求分析

CARAM Content Addressable Random Access Memory 按内容寻址随机存取存储器

carat 1. 开 2. 克拉

CARB Carburize 渗碳

carbazole 咔唑

carbide 碳化物

carbine 卡宾枪

carboblast 用木质纤维清理

carbomethene 乙烯酮

carbon 【化】碳(化学元素,符号 C)

carbon arc 碳弧

carbon arc lamp 碳弧灯

carbon atom 碳原子

carbon cloud decoy 碳微粒云(用来干扰导弹的热自导引系统)

carbon composition film potentiometer 合成碳膜电位器

carbon cycle 碳循环

carbon deposit 积碳

carbon detonation supernova 【天】碳爆炸超新星

carbon dioxide 二氧化碳

carbon dioxide (fire) extinguisher 二氧化碳灭火器

carbon emission 碳排放

carbon fiber 碳纤维

carbon flux 碳通量

carbon ion 碳离子

carbon microphone 炭粒传声器

carbon molecule 碳分子

carbon monosulfide 一硫化碳

carbon monoxide 一氧化碳

carbon nanotubes 碳纳米管

carbon nitrogen cycle 碳-氮循环
carbon number 碳数
carbon particle 碳粒
carbon phenolic 碳酚醛
carbon phenolic material 碳酚醛材料
carbon pile 碳柱
carbon powder 碳粉
carbon residue 残炭
carbon resistor 碳电阻
carbon sequence 碳序
carbon tetrachloride fire extinguisher 四氯化碳灭火器
carbon thermal barrier 碳热防护
carbon tracking 碳漏电痕迹
carbonaceous aerosol 含碳气溶胶
carbonaceous chondrite 碳质球粒陨石
carbonaceous material 碳质材料
carbonate 1.碳酸盐,碳酸酯 2.碳酸盐的,碳酸酯的
carbon-based ion optics system 碳基离子光学系统
carbon-carbon composite 碳碳复合材料
carbon-fiber-reinforced ceramics 碳纤维增强陶瓷
carbon-fibre reinforced composite 碳纤维增强复合材料
carbon-film resistor 碳膜电阻
carbon-free 不含碳的
carbon-graphite impregnated with inorganic salt 浸渍无机盐碳石墨
carbon-graphite impregnated with metal 浸渍金属碳石墨
carbon-graphite impregnated with phosphate 浸渍磷酸盐碳石墨
carbon-graphite impregnated with resin 浸渍树脂碳石墨
carbonic 碳的,含碳的
carbonification 碳化作用
carbonitriding 碳氮共渗,氰化
carbonizing 碳化
carbon-nitrogen-oxygen cycle 碳氮氧循环
carbontetrachloride 四氯化碳
carbonyl 碳酰基
carbonyl iron 羰基铁
carbonyl sulfide 硫化羰
carbon-zinc cell 碳锌电池
carborundum 金刚砂,碳化硅
carboxyh(a)emoglobin 碳氧血红蛋白
carboxyl terminated polybuadiene 末端羧基聚丁二烯
carboxyl terminated polybutadiene propellant 末端羧基聚丁二烯推进剂
carboxylate 羧酸盐,羧化物
carboy 酸坛,用木箱保护着的大玻璃瓶
carburation 1.渗碳 2.汽化

carburetor 汽化器
carburettor air 汽化空气
carburettor icing 汽化器结冰
carburization 渗碳
carburization zone 碳化层,渗碳层
carburizer 渗碳剂
carburizing 渗碳
carburizing flame 碳化火焰,渗碳火焰
CARC Central Air Rescue Center 中央航空救护中心
carcase 1.骨架,构架 2.壳体
carcinotron 返波管
CARD Computer-Aided Remote Driving 计算机辅助遥控驱动
CArD Comuting Architecture Design 计算机构造设计
card catalogue 卡片型星表,卡片型天体表
cardan 万向接头,万向节
cardan shaft 万向轴
cardan-mounted 装在万向节支架上的
cardboard 纸板
cardiac defibrillator 心脏除颤器
cardiac pacemaker 心脏起搏器
cardinal plane 基平面
cardinal point effect 方位基点效应
cardinal signs 黄道带主宫
cardinality 基数
CARG Cargo Aircraft 货机
CARDS 1.Calman Automated Routing and Design System 卡尔曼自动路由选择和设计系统 2.Computer-Assisted Radar Display System 计算机辅助雷达显示系统
CARE Computer-Aided Reliability Estimation 计算机辅助可靠性估计
care and maintenance 维护保养
CAREC China National Aero-Engine Corporation 中国航空发动机总公司
career gunner 轰炸机上专职机枪手
career reenlistment 重新应征士兵
CARES Cratering And Related Effects Simulation 炸坑效果模拟
caret 脱字符
caret wing 盖烈特机翼
caretaker 看管人,暂时行使职权者
cargo 货物
cargo airplane 货机
cargo bay 货舱
cargo compartment 货舱
cargo conversion 货机改型
cargo handling personnel 货物装卸人员
cargo loading ramp 装货跳板

cargo net　货网
cargo parachute　投物伞
cargo rocket　运输火箭
cargo spacecraft　运货火箭
cargo transport　1.运输机 2.货物运输船 3.在法国飞机分类中，代表战斗机类
cargofloor　货舱地板
cargo-handling equipment　货物装运设备
cargoliner　货运班机
cargoplane　货机
CARGUS　cargo upper stage 运货上面级
CARI　Civil Aeromedical Research Institute 民用航空医学研究所
Carme　【天】木卫十一
CARLS　Cam Roller 凸轮滚子
CARMS　Civil Aviation Radio Measuring Station 民航无线电测量站
CARNF　Charges for Airports and Route Navigation Facilities 机场及航线导航设备使用费
Carnot cycle　卡诺循环
Carnot engine　卡诺发动机
carousel　1.轮盘木马 2.行李旋转台
CARP　Computed Air Release Point 计算空投点，计算投弹点
carp　轰炸瞄准具的制导附件
Carpet　1."地毯" 2.地毯式干扰
carpet plot　曲线族，毯式曲线
Carpetbagger　"外来政客"行动
carpitron　卡皮管
carrel　座舱练习器，小隔间
carriage　1.载运，运输，输送 2.运费 3.起落架，车架，炮架
carriage return　回车，字盘返回
carriage wrench　套筒板手
carrier　1.航空母舰 2.运载工具，母机 3.运输机
carrier（beat）phase measurement　载波相位测量
carrier acquisition time　载波捕获时间
carrier air group　航母飞行大队
carrier aircraft　舰载航空器
carrier aviator　航母飞行员
carrier buffer　载频缓冲器
carrier extract　载波提取
carrier flight officer　航母飞行军官
carrier frequency　载频
carrier frequency shift　载波频移
carrier frequency synchronization　载频同步
carrier frequency tolerance　载波频率容许偏差
carrier gas　运载气体
carrier injection system　载波注入系统
carrier landing system　着舰系统
carrier leak　载漏
carrier multipath　载波多径
carrier multipath error　载波多径误差
carrier on off　1.载波通断 2.话音激活
carrier phase　载波相位
carrier phase measurement　载波相位测量
carrier qualification tests　舰载适合性试验
carrier recovery　载波恢复
carrier recovery loop　载波恢复环路
carrier rocket　运载火箭
carrier shell　特种炮弹，布雷炮弹
carrier subscribe　载波用户
carrier suppression　航母本身受抑制
carrier synchronization　载波同步
carrier task force　航母特混编队
carrier telegraphy　载波电报
carrier telephone　载波电话
carrier trend analysis form　航母趋向分析表
carrier vehicle　运载飞行器
carrier wave　载波
carrier-based（borne）　舰载的，以航空母舰为基地的
carrier-based squadron　航母舰载机中队
carrier-loader　运载车，装运设备
carrier-on-board delivery　向海上的航空母舰空运
carrier-phase datum　载波相位基准
carrier-phase measurement　载波相位测量
carrier-suppressed SSB　抑制载波单边带
carrier-to-noise ratio　载波噪声比，载噪比
carrier-to-tracking loop　载波跟踪环路
carrier-type shell　特种炮弹
Carrington longitude　卡林顿经度
Carrington meridian　卡林顿子午线
carry-around oxygen cylinder　携带式氧气瓶
carry delay time　进位延迟时间
carry lookahead adder　先行进位加法器
carry save adder　进位存储加法器
carry trials　挂载试验
carry-around oxygen bottle　轻便氧气瓶，手提式氧气瓶
carrying reliability　运载可靠性
carry-on　机上的，机载的，随身行李件
carry-on equipment　机上设备，机载设备
carryover　携带
carry-through　贯串的，传载的
CARS　Computer-Aided Routing System 计算机辅助路由选择系统
CARSRA　Computer-Aided Redundant System Reliability Analysis 计算机辅助冗余系统可靠性分析
CART　Combat Aircraft Repair Team 战伤飞机修理队

cart 手推车
cartel 同业联盟
Carterfone Decision 卡特风决议
cartesian control 笛卡儿坐标控制
cartesian coordinate 直角坐标
Cartesian coordinate system 笛卡儿坐标系
cartesian grid 笛卡儿网格
cartesian mesh 直角网格
Cartesian robot 直角坐标型机器人
cartesian space 直角坐标空间
cartodiagram method 分区统计图表法
cartogram method 分区统计图法,等值区域法
cartographic communication 地图传输
cartographic document 制图资料,图件
cartographic generalization 制图综合,地图概括
cartographic grid 地图格网
cartographic hierarchy 制图分级
cartographic pragmatics 地图语用
cartographic presentation 地图表示法
cartographic semantics 地图语义
cartographic semiology 地图符号学
cartographic symbolization 地图符号,地图图例
cartographic syntactics 地图语法
cartography 地图制图学,地图学
cartometry 地图量算
cartouche 弹药筒,装饰镜板
cartridge 1.子弹,定装药炮弹,弹药筒,弹壳 2.弹射弹,电燃药筒,火药点火器,火药包 3.盒式磁带,盒式磁盘
cartridge ball 实弹
cartridge belt 弹带
cartridge box 弹箱
cartridge case 弹筒,弹壳
cartridge clip 弹夹
cartridge drum 鼓形弹箱
cartridge extractor 退弹钩,抓弹钩
cartridge fuse 保险丝管,培丝管
cartridge igniter 火药点火器
cartridge link 弹链
cartridge pouch 弹盒
cartridge shell 弹壳,弹箱
cartridge starter 火药起动装置
cartridge tape 盒式录音磁带,盒式录像磁带
cartridge-fired ejection seat 弹道式弹射座椅
cartridge-powered launching pylon 燃爆筒投射式发射挂架
cartwheel 1.半滚跃升倒转 2.急水平转弯
cartwheel satellite 滚轮式卫星
CAS 1.Calibrated Airspeed 校正空速 2.Calculated Air Speed 计算空速 3.Canadian Applications Satellite 加拿大应用卫星 4.Castle 城堡 5.Cockpit Avionics System 座舱航空电子系统 6.Collision Avoidance System 防撞系统 7.Commercial Aviation Services 民用航空服务 8.Commercial Avionics System 民用航空电子系统 9.Crew Alerting System 机组警告系统
Cas Cassiopeia【天】仙后座
CASA Construcciones Aeronauticas Sociedad Anonima 飞机制造股份有限公司(西班牙)
CASC 1.China Civil Aviation Supplies Corporation 中国航空器材公司 2.Combined Acceleration and Speed Control 加速度与速度综合控制
cascade 1.叶栅 2.瀑布,自流放泄 3.级联,串联
cascade algorithm 级联算法
cascade compensation 串联补偿,串联校正
cascade control 级联控制,串级控制
cascade flow 叶栅流动
cascade geometry 叶栅几何参数
cascade impactor 级联硬着陆装置
cascade migration 级联偏移
cascade performance 叶栅性能
cascade reverser 叶栅式反推装置,反推导向叶片排
cascade shower 级联簇射
cascade solar cell 叠层太阳电池
cascade solidity 叶栅稠度
cascade strategy 串级策略
cascade synthesis 级联综合法,链接综合法
cascade system 串级系统
cascade transformer 级间变压器
cascade wind tunnel 叶栅风洞
cascaded code 级联码
cascaded electric drive 串级电气传动
cascode 栅地-阴地放大器,共射共基放大器,渥尔曼放大器
CASD Computer-Aided System Design 计算机辅助系统设计
CASE 1.Computer-Aided Software Engineering 计算机辅助软件工程 2.Computer-Aided System Engineering 计算机辅助系统工程 3.Computer-Assisted System Evaluation 计算机辅助系统评价
case acceleration 壳体加速度
case bonded grain 浇涛粘接药柱
case chute 抛壳道,抛壳套
case ejection 抛壳
case guide rail (炸弹)弹体导轨
case insulation 壳体绝热层
case telescoped ammunition 弹壳套进式炮弹
case-bonded 与壳体黏合的
case-hardening 表面硬化
casein 酪蛋白,干酪素

casemate　1. 防弹掩蔽部，炮塔 2. 暗炮台壳体回转技术
cash flow　现金流转
CASI　1. Canadian Aeronautics and Space Institute 加拿大航空航天学会 2. Commission Aeronautique Sportive International 国际航空体育委员会
casing　1. 机匣，外壳 2. 外胎 3. 包装
casing contour　机匣轮廓
casing endwall　机匣端壁
casing treatment　机匣处理
casing-pressure measurement　机匣压力测量
CASNUT　Castle Nut 槽顶螺母，蝶形螺母
Casper　卡斯帕号
CASS　1. Command Active Sonobuoy System 指令控制声纳浮标系统 2. Computerized Algorithmic Satellite Scheduler 计算机化算法卫星调度程序 3. Continous Analysis and Surveillance System 持续分析和监督系统
Cassegrain antenna　卡塞格伦天线
Cassegrain reflector　卡塞格伦反射望远镜
Cassegrain reflector antenna　卡塞格伦反射体天线
Cassegrain telescope　卡塞格林望远镜
cassette　1. 弹夹，炸弹箱 2. 胶卷盒 3. 盒，箱
cassette recorder　盒式录音机
cassette tape　盒式带
Cassini　卡西尼探测器
Cassini law　卡西尼定律
cassiopeids　【天】仙后流星群
cassiopeium　【化】镥（化学元素，符号 Lu）
CAST　1. Capillary Action-Shaping Technique 毛细成形技术 2. Chinese Academy of Space Technology 中国空间技术研究院
cast　1. 铸造，铸型，铸件 2. 投，抛，掷，撒 3. 计算
cast magnesium alloy　铸造镁合金
cast steel shell　铸钢炮弹
cast superalloy　铸造高温合金
cast titanium alloy　铸造钛合金
castability　可铸性
castellanus　堡状云
castellated nut　梅花螺帽，槽顶螺帽
caster　1. 脚轮，转向轮 2. 转向 3. 铸工
casting　1. 铸，铸件 2. 投，抛
casting alloy　铸造合金
CASTOR　Corps Airborne Stand-off Radar 军属机载远程雷达
Castor　海狸卫星
Castor（α Gem）【天】北河二（双子座 α）
castoring　转向的
CASU　Combat Aircraft Service Unit 作战飞机勤务部队

casual　1. 流动军人 2. 临时的，非正式的
casual target　临时目标
casuality　因果性，因果律
casualty　1. 伤亡事故 2. 损失，损坏，故障 3. 部队减员，伤员
casualty agent　致命性战剂，杀伤剂
casualty effect　杀伤效应，杀伤作用
casualty gas　杀伤性毒气
casualty-producing concentrations　杀伤浓度
casualty rate　伤亡率
CASWS　Close Air Support Weapons System 近距空中支援武器系统
CAT　1. Capsule Ariane Technologique 阿里安技术试验舱 2. Catalogue 目录一览表，目录，样本 3. Category 类，级 4. Clear Air Turbulence 晴空涡流 5. Cockpit Automation Technology 座舱自动化技术 6. Computer Aided Testing 计算机辅助测试 7. Computer Aided Theodolite 计算机辅助电子经纬仪 8. Customer Acceptance Testing 用户接收试验
cat　1. 无线电遥控靶机 2. 无线电测距系统的偏流台，"猫"台 3. 弹射器
catabolism　分解代谢
Catac　Commande Aerienne Tactique 战术空军
cataclysmic binary　激变双星
cataclysmic variable　激变变星
catadioptric telescope　折反射望远镜
catafighter　弹射起飞的战斗机
cataflier　〈口语〉弹射起飞飞机的飞行员
cata-front　下滑锋
Catalin　卡塔林塑料
catalogue　星表，天体表
catalogue astronomy　星表天文学
catalogue equinox　星表分点
catalogue number　星表号数，天体表号数
Catalogue of Galaxies and Clusters of Galaxies　星系和星系团表
catalogue of stars　星表
catalogue of the gamma-ray source　γ射线源表
catalogue parameter　星表参数
catalysis　催化
catalysis characteristics　催化特性
catalyst　催化剂
catalyst bed　催化剂床
catalyst burning　催化燃烧
catalyst layer　催化层
catalyst pellet　催化剂颗粒
catalyst reactivity　催化剂反应性
catalyst support　催化剂载体
catalyst unit　催化剂设备

catalytic combustion 催化燃烧
catalytic decomposition 催化分解
catalytic effect 催渗作用
catalytic efficiency 催化效率
catalytic ignition 催化点火
catalytic monopropellant hydrazine system 催化式单元肼推进系统
catalytic oxidizer 催化氧化器
catalyze 催化
catanator 上下飞操纵输出装置,操纵机构
cataplane 弹射起飞飞机
catapult 1.弹射装置 2.弹射
catapult and arresting gear officer 弹射器和制动机构军官
catapult crew 弹射器操作人员
catapult safety observer 弹射器安全观察员
catapulting force 弹射力
catastrophe 灾难,严重事故
catastrophe theory 灾变论
catastrophic code 恶性码
catastrophic failure 灾变故障
catatrophic fault 灾难性故障
CATCC Carrier Air Traffic Control Center 航母空中交通管制中心
catch line 截击线
catcher 1.捕捉器,收集器 2.前缘捕获栅 3.获能腔,捕获栅
catcher resonator 获能腔
catching net 捕捉网
catchment area survey 汇水面积测量
CATE Computer-Controlled Automatic Test Equipment 计算机控制的自动测试设备
catechin 儿茶酸
catechol 儿茶酚,苯邻二酚
catecholamine 儿茶酚胺
categorical variable 类别变项
categorise 加以类别
categorization 分类
category Ⅱ precision approach and landing operation Ⅱ类精确进近着陆运行
category voltage 类别电压
catena 1.连锁,链 2.位列
catenary 悬链线,悬垂线
catenary system 接触网系统
catenate 链接
catering 膳食供应
caterpillar 1.履带 2.履带车,坦克
CATH Cathode 阴极,负极
cathautograph 用阴极射线管的传真电报

cathedral 1.下反角 2.主试验间
catheter electrode 导管电极
cathode 阴极
cathode active coefficient 阴极有效系数
cathode coupling 阴极耦合
cathode current 阴极电流
cathode electrode 阴极电极
cathode fatigue 阴极疲劳
cathode follower 阴极输出器
cathode glow 阴极辉光
cathode heater 阴极加热器
cathode life 阴极寿命
cathode mass flow 阴极流量
cathode propellant 阴极推进剂
cathode ray 阴极射线
cathode spot 阴极斑点
cathode temperature 阴极温度
cathode tube 阴极管
cathodel uminescence 阴极射线致发光
cathode-ray spectroradiometer 阴极射线分光辐射计
cathode-ray tube 阴极射线管
cathode-ray tube adder 阴极射线管加法器
cathode-ray tube display 阴极射线管显示器
cathodochromism 阴极射线致变色
CATIA 1. Computer-Aided Three-dimensional Interaction Application System 计算机辅助三维相互影响应用系统 2. Computer-Graphics Aided Three-Dimensional Interactive Application 计算机辅助三维交互应用绘图
CATIC China National Aero Technology Import and Export Corporation 中国航空技术进出口总公司
cation 正(阳)离子
cation column 阳离子色谱柱
CATK counterattack 反击,反冲击
catoptric 反射(光)的
catoptric telescope 反射望远镜
catoptrics 反射光学
CATS 1. Centralized Automatic Testing System 集中式自动测试系统 2. Communication And Tracking System 通信和跟踪系统 3. Computer-Aided Telemetry System 计算机辅助遥测系统
CATV 1. Cable Television 有线电视 2. Common Antenna Television 公用天线电视
catwalk 1.狭窄过道 2.舷边狭窄通道 3.桥形通道
catwhisker 针电极,"触须"
CAU 1. Cold-Air Unit 空气冷却装置 2. Crypto Auxiliary Unit 加密辅助设备
Cauchy residue theorem 柯西留数定理
cauda 尾波

caudad acceleration 足向加速度,头向过载
caulking 1.凿密,填密,敛缝 2.堵头
causality 因果律
caustic corrosion 碱性腐蚀
caustic embrittlement 碱性脆化
caution area 关注区
CAV 1.Cavalry 机械化部队 2.Cavity 空腔,暗盒
cavalry 1.骑兵 2.机械化部队
caveat 防止误解的说明,终止诉讼手续的申请
caviation allowance 气蚀裕度
caviation specific angular speed 气蚀比转速
cavitated flow 气穴流,涡空流
cavitating venturi 气蚀文氏管
cavitation 气蚀,气穴,空穴,空隙现象
cavitation erosion 空穴腐蚀,气蚀
cavitation inception 空化初生
cavitation model 空化模型
cavitation number 空化数
cavitation parameter 汽蚀参数
cavitation zone 汽蚀区域
cavity 1.内腔,空腔,空穴,空洞 2.空隙区,气穴区 3.空腔谐振器
cavity charge 空心装药
cavity depth 空腔深度
cavity diameter 空腔直径
cavity dumping 腔倒空
cavity exit 腔出口
cavity flame holder 凹槽火焰稳定器
cavity floor 腔底面
cavity flow 空腔流
cavity front wall 空腔前壁
cavity geometry 腔体形状
cavity length 腔长
cavity mode 腔体模式
cavity oscillation 腔振荡
cavity plate 腔板
cavity pressure 模腔压力
cavity resonance 空腔谐振
cavity resonator 空腔谐振器,谐振腔
cavity shear layer 空腔剪切层
cavity volume 空腔谐振体积
cavity wall 空腔壁
cavity-based fuel 腔基燃料
cavity-trailing edge 腔尾缘
cavu 能见度极好(Ceiling and Visibility Unlimited 的缩写)
CAWA Civial Aviation Wireless Association 民航无线电协会
CAWS 1.Central Aural Warning System 中央音响警告系统 2.Caution and Warning System 注意和告警系统
Cayley Hamilton theorem 凯莱-哈密顿定理
cazin 易熔镉合金
CB 1.Cargo Bay 货舱 2.Center of Buoyancy 浮力中心 3.Chloro-Bromomethane CB 灭火剂,氯溴甲烷灭火液 4.Chloro-Bromo-type 氯溴化物 5.Circuit Breaker 断路器,跳开关 6.Citizen Band 民用频带 7.Command Bar 指令杆 8.Common Base 共基极,共用基座 9.Common Battery 通用电池 10.Cone Bolt 锥形螺栓 11.Configuration Baseline 构型基线 12.Control Board 控制盘,操纵台 13.Control Booth 控制库房
CBAL Counterbalance 1.配重,平衡块 2.移轴补偿 3.使……平衡,补偿,抵消
CBAS 1.Combined British Astronautical Societies 英国联合宇航学会 2.Commercial Bank Application System 商业银行应用系统
CBC 1.Carbon/BMI Composite 碳/双马来酰亚胺复合材料 2.Configured Build Coordinator 构型建立协调员
CBD Carbide 碳化物
CBE Cosmic Background Explorer 宇宙背景探测器
CBERS China-Brazil Earth Resources Satellite 中巴地球资源卫星
CBI 1.Component Burn-in 元件预先强化,元件老炼 2.Computer-Based Instruction 计算机化教学
CBLS Carrier, Bomb, Light Store 炸弹与轻型外挂物挂架
CBM ChloroBromoMethane 氯溴甲烷
CBMIS Computer-Based Management Information System 计算机化管理信息系统
CBN 1.Carbine 卡宾枪 2.Cabin 客舱,座舱 3.Cubic Boron Nitride 立方氮化硼
CBN TET Carbon Tetrachloride 四氯化碳
CBO 1.Channel Bus Out 通道总线输出 2.Combined Bomber Offensive 联合轰炸机进攻 3.Congressional Budget Office 国会预算局
CBOM Customized Bill Of Material 定制物料表
Cbomb 钴壳原子弹
CBORE Counterbore 扩孔,锪孔,镗孔
CBP 1.Circuit Breaker Panel 断路器板 2.Contact-Burst Preclusion 防触发爆炸装置
CBPS Combined Bracking/Correction Propulsion System 制动-修正联合推进系统
CBR Charger Battery Relay 电池充电继电器
CBS 1.Columbia Broadcasting Systems Inc. 哥伦比亚广播公司 2.Correlation Bombing System 相关轰炸系统
CBT 1.Combat 战斗,争论 2.Computer-Based Training 基于计算机的培训
CBTE Conventional Bomb Triple Ejector 常规炸弹三联

弹射器

CBU Cluster Bomb Unit 子母炸弹,集束炸弹

CBW Chemical and Biological (Bacteriological) Warfare 化学与生物(或细菌)战

CC 1. Cargo Carrier 货物托架,运货机,货机,货运航空公司 2. C-Check C检 3. Center to Center 中心到中心 4. Centering Cam 定中凸轮 5. Certified Configuration 合格的装机状态 6. Change Category 更改类别 7. Channel Controller 通道控制器 8. Coastal Command 岸防航空兵,海岸航空兵 9. Coaxial Cable 同轴电缆 10. Colour Code 颜色标识,色标 11. Combustion Chamber 燃烧室 12. Configuration Control 构型控制 13. Control Column 控制杆 14. Cross-Channel 通道跨接

C-C composite 碳-碳复合材料

CC & PPP Corrosion Control & Preventive Protection Plan 腐蚀控制和防护大纲

CCA 1. Carrier-Controlled Approach 航母控制进场 2. Circuit Card Assembly 线路板组件 3. Cluster Compression Algorithm 集聚压缩算法 4. Common Cause Analysis 共因故障分析 5. Customer Configuration Authorization 客户构型批准

CCAAB Cape Canaveral Auxiliary Air Force Base 卡纳维拉尔角附属空军基地

CCADSC China Civil Aviation Development Service Corporation 中国民航开发服务公司

CCAFS Cape Canaveral Air Froce Station 卡纳维拉尔角空军试验站

CCAR China Civil Aviation Regulation 中国民用航空规章

CCAT Communication, Command and Telemetry 通信、指挥和遥测

CCB 1. Change Control Board 更改控制委员会 2. Configuration Change (Control) Board 构型更改(控制)委员会 3. Configuration Control Board 构型控制委员会,总体构型控制委员会 4. Control Center Building 控制中心大楼 5. Converter Circuit Breaker 转换断路器

CCC 1. Camera Controller Combiner 照相机控制组合器 2. Cape Control Center 卡纳维拉尔角发射场控制中心 3. Central Communications Controller 中央通信控制机 4. Configuration Change Control 构型更改控制 5. Crash Crew Chart 应急处置用图

CCCS Command Control Communication System 指挥控制通信系统

CCD 1. Charge-Coupled Device 电荷耦合器件 2. Charged Coupled Device 充电耦合装置 3. Color Control Drawing 彩色控制图 4. Cursor Control Device 光标控制装置 5. Customer Configuration Definition 客户构型定义

CCD camera 电荷耦合器件摄像机

CCD delay line 电荷耦合器件延迟线

CCD image display instrument CCD图像显示仪

CCD image sequential switcher CCD图像切换仪

CCD memory 电荷耦合器件存储器

CCD star sensor 电荷耦合器件星敏感器

CCD star tracker 电荷耦合器件星跟踪仪

CCD sun sensor 电荷耦合器件太阳敏感器

CCD system Camouflage, Concealment and Deception System 伪装、隐蔽、欺骗系统

CCDP Continuously Computing Delivery Point 连续计算投放点

CCDS Center for the Commercial Development of Space (Program) 空间商业开发中心计划

CCE 1. Charge Composition Explorer 电荷成分探测器 2. Communications Control Equipment 通信控制设备

CCF 1. Customer Configuration File 客户构型档案 2. Central Control Function 中央控制功能

CCFG Compact Constant-Frequency Generator 小型恒频发电机

CCG Cargo Center of Gravity 货物重心

CCGE Cold Cathode Gauge Experiment 冷阴极真空规实验

CCH Close-Combat Helicopter 近距作战直升机

CCHK Contiunuity Check 连续性检查

CCI Circuit Condition Indicator 电路情况指示器

CCIA Computer and Communication Industry Association 计算机与通信工业协会

CCIL Continuously Computing Impact Line 连续计算弹着线

CCIP Continuously Computed Impact Point 连续计算弹着点

CCIP/IP Continuously Computed Impact Point/Initial Point (带预定轰炸航路起始点的)连续计算弹着点/起始点

CCIR Consultative Committee International Radio 国际无线电咨询委员

CCIRID Charge Coupled Infrared Imaging Device 电荷耦合红外成像器件

CCIS Common Channel Interoffice Signalling 共路局间信令

CCITT 1. Consultative Committee International Telephone and Telegraph 国际电话电报协会咨询委员会 2. Consultive Committee for International Telephony and Telegraphy 国际电话电报咨询委员会

CCL 1. Compliance Check List 符合性检查单 2. Critical Crack Length 临界裂纹长度

CCM 1. Configuration Control Module 配置控制模块,

构型控制模块 2. Conventional Cruise Missile 常规巡航导弹 3. Crew/Cargo Module 乘员/货物舱 4. Customer Service Coordination Memo 客户服务协调备忘录

CCMS Checkout, Control and Monitoring Subsystems 检测、控制和监控子系统

CCMTA Cape Canaveral Missile Test Annex 卡纳维拉尔角附属导弹试验场

CCN 1. Cloud Condensation Nuclei 云凝结核 2. Cost Charge Number 费用编号

CCO Component Change Order 零部件更改指令

CCOC Combustion-Chamber Outer Casing 燃烧室外壳,燃烧室外套

CCOP Committee for Coordination of Joint Prospecting for Mineral Resources in Asia Offshore Areas 亚洲近岸海域矿产资源联合勘察协调委员会

CCOS Computer-Controlled Operating System 计算机控制操作系统

CCP 1. Certification Compliance Plan 合格审定计划 2. Closed Cherry-Picker 封闭式升降舱 3. Communication Control Package 通信控制包 4. Configuration Change Processing 构型更改过程 5. Contract Change Proposal 合同更改建议 6. Crop Classification Performance 作物分类特性

CCPC Civil Communications Planning Conference 民用通信计划会议

CCPD Charge-Coupled Photodiode Device 电荷耦合光电二极管器件

CCPR Cruise Compressor Pressure Ratio 巡航状态压缩器压力比

CCR 1. Change Commitment Record 更改提交记录 2. Circulation-Controlled Rotor 环流控制旋翼 3. Control Contactor 控制接触器

CCRP Continuously Computing Release Point 连续计算投放点

CCRS Canada Centre for Remote Sensing 加拿大遥感中心

CCS 1. Cabin Communication System 座舱通信系统 2. Certification Control System 定检控制系统 3. Command Control Subsystem 指令控制分系统 4. Commercial Communications Satellite 商业通信卫星 5. Communications Control System 通信设备控制系统

CCSDS Consultative Committee for Space Data System 空间数据系统咨询委员会

CCSO Completed Closed Shop Order 已完工的车间指令

CCSS 1. Canada Centre for Space Science 加拿大空间科学中心 2. Control Center of the Symphonie Satellite System 交响乐卫星系统控制中心

CCST Consultative Committee on Satellite Telecommunications 卫星通信咨询委员会

CCT 1. Computer Compatible Tape 计算机兼容磁带 2. Configuration Change Team 构型更改小组 3. Constant Current Transformer 恒流变压器 4. Customer Core Team 客户核心团队

CCTL Continuously Computing Tracer Line 连续计算跟踪线

CCTV Closed Circuit Television 闭路电视

CCTW Combat Crew Training Wing 战斗人员训练联队

CCU Control and Compensation Unit 控制和补偿装置

CCV 1. Control-Configured Vehicle 随控布局飞机,随控布局飞行器 2. Current Change Vehicle 现行更改媒介 3. Customer Configuration Verification 客户构型验证

CCVM Configuration Change and Variance Management 更改构型和变型管理

CCW Circulation-Controlled Wing 环量控制机翼

c-cycle 全过程模拟

CD 1. Cable Duct 电缆槽 2. Case Drain 泵匣余油管 3. Certification Demonstration 满足适航证要求的演示 4. Chip Detector 金属碎削探测器 5. Circular Dispersion 圆形散布 6. Coefficient of Drag 阻力系数 7. Compact Disc 密纹唱盘,压缩磁盘,激光唱盘 8. Configuration Documentation 构型文件

Cd battery 镉镍电池

CD-2 Common Digitizer-2 通用数字读出器-2

CDA 1. CATIA Data Access 计算机辅助三维交互应用数据存取 2. Central Design Authority 中央设计管理局 3. Collision Detection and Avoidance 相撞探测与防撞 4. Command and Data Acquisition 指令和数据采集 5. Current Data Array 现行数据阵列

CDAI Centre de Documentation Aeronautique Internationale〈法语〉国际航空文献中心

CDAS 1. Central Data Acquisition System 中央数据采集系统 2. Command and Data Acquisition Station 指令和数据采集站

CDB 1. Cast Double-Base (Rocket Propellant) 浇铸双基火箭推进剂 2. Computer Data Base 计算机数据库

CDBP Command Data Buffer Program 指令数据缓冲器程序

CDC 1. Cost Driven Concept 成本驱动的概念 2. Course and Distance Computer 航向距离计算机

CDCF Cosmic Dust Collection Facility 宇宙尘收集设施

CDCP 1. Configuration Development 构型开发 2. Configuration Development and Change Process 构型开发和更改程序

CDCS Central Data Collection System 中央数据收集系统

CDD 1. Common Data Document 公共数据文件 2. Computer Data Definition 计算机数据定义 3. Configuration Description Document 构型定义文件 4. Credible Delicious Decoy 易使敌方上钩的诱饵

CDDD Components Detail Design Drawings 部件详细设计图

CDDT Countdown Demonstration Test（发射前）倒数计时演示试验

CDEL Constant Delivery 定量输入

CDES Computer Data Entry System 计算机数据输入系统

CDF Coded Digital FAX 编码数字传真

CDHS Communication and Data Handling System 通信和数据处理系统

CDI 1. Collector Diffusion Isolation 集电极扩散隔离 2. Compass Director Indicator 航道罗盘指示器 3. Conventional Defense Improvement 常规防御改进 4. Course Deviation Indicator 航道偏差指示器

CDIP Continuously Displayed Impact Point 连续显示弹着点

CDL 1. Capability Demonstration Laboratory 能力验证实验室 2. Configuration Declination List 构型偏离清单

CDM 1. CATIA Data Management 计算机辅助三维交互应用数据管理 2. Configuration Data Management 构型数据管理 3. Corona Diagnostic Mission 日冕探测任务

CDMA Code Division Multiple Access 码分多址

CDMLS Commutated-Doppler Microwave Landing System 转换式多普勒微波着陆系统

CDMS Command and Data Management Subsystem 指令与数据管理分系统

CDN 1. Certificate de Navigabilite 导航许可证书 2. Coordinating Message 协作通报

CDNC Computer-Aided Design Numerical Control 计算机辅助设计数字控制

CDND Customer-Driven, Newly Defined 客户驱动，新设计

CDOC Cash Direct Operation Cost 现金使用成本

CDOL Customer Data and Operations Language 用户数据和操作语言

CDOS Customer Data and Operations System 用户数据和操作系统

CDP 1. Cartographic Digitizing Plotter 制图数字化绘图机 2. Central Data Processor 中央数据处理机 3. Cockpit Data Printer 驾驶舱数据打印机 4. Cost Definition Phase 成本定义阶段 5. Countermeasures Dispenser Pod 电子对抗器材投放器吊舱

CDPC 1. Central Data Processing Center 中央数据处理中心 2. Central Data Processing Computer 中央数据处理计算机

CDPD Customer-Driven, Previously Delivered 客户驱动、提前交付

CDPIE Command Data Processor Interface Equipment 指令数据处理机接口设备

CDPL Cadmium Plate 镀镉

CDPS Central Data Processing System 中央数据处理系统

CDR 1. Commander 指挥长 2. Critical Design Review 关键设计评审

CDRB Cross-Divisional Review Board 跨部门评审委员会

CDRL Contract Data Requirements List 合同数据需求表

CDS 1. Central Data Subsystem 中央数据分系统 2. Cold-Drawn Steel 冷拉钢 3. Commitment Development Schedule 承诺进度计划，委托开发计划 4. Common Display System 公共显示系统 5. Command and Data System 指令和数据系统 6. Container Delivery System 子母弹投放系统 7. Customer Detail Specification 客户详细规范

CDSF 1. Commercially Developed Space Facility 商业应用空间设施 2. Customer Data Services Facility 用户数据服务设施

CDSS Compressed Data Storage System 压缩数据存储系统

CDTI 1. Centro para el Desarrollo Tecnologico Industrial〈西班牙语〉工业技术发展中心 2. Cockpit Display of Traffic Information 交通信息驾驶舱显示

CDTS Compressor Discharge Temperature Sensor 压缩机排气温度传感器

CDU 1. Cockpit Display Unit 座舱显示装置 2. Control Display Unit 控制显示组件，控制显示装置，控制显示器

CE 1. Calibration Equipment 校准设备 2. Circular Error 径向偏差，圆误差 3. Common Emitter 共用发射极 4. Comminications-Electronics 通信电子设备 5. Concurrent Engineering 并行工程 6. Conformity Evaluation 合格评价 7. Customer Engineering 客户工程

CEA 1. Circular Error Average 平均径向偏差，平均圆误差 2. Combined Electronics Assembly 组合式电子设备

CEAC 1. Committee on European Airspace Coordination（NATO）欧洲空域协调委员会（北约组织）2. Commisson Europeenne de Aviation Civile 欧洲民

航委员会
CEB 1. Cluster Effects Bomblet 集束炸弹散布的多种效果的小炸弹,多种效果子炸弹 2. Combined-Effects Bomblet 综合效能小炸弹 3. Curve of Equal Bearings 等方位曲线
CEC Central Equipment Center 中央设备中心
CECS Conference on European Communication Satellite 欧洲通信卫星会议
CED 1. Competitive Engineering Definition 竞争性工程定义 2. Continued Engineering Development 连续工程研制
CEI Contract End Items 合同最终项
ceiling 1. 升限 2. 高度限制,最大高度 3. 云底高,云幕高度
ceiling at takeoff 起飞时云高
ceiling balloon 云幂气球,测云气球
ceiling height 云幂高度
ceiling light 云幂灯
ceiling projector 云幂灯
ceiling zero 零云幂
ceilograph 云幂计
ceilometer 云幂仪,云高计
CELC Cabin Equipment Location Chart 座舱应急设备位置图
celescope Celestial Telescope 天文望远镜
celestial 天的,天空的,天上的
celestial（sphere）coordinate system 天球坐标系
celestial altitude 天体高度
celestial axis 【天】天轴
celestial background 天体背景
celestial body 天体
celestial chart 天图
celestial compass 天文罗盘
celestial coordinates 天球坐标
celestial ephemeris pole 天球历书极
celestial equator 天球赤道
celestial equator system of coordinates 天球赤道坐标系
celestial fix 天文定位的即时位置
celestial geodesy 天文大地测量学
celestial globe 天球仪,浑象
celestial guidance 天文制导,天文导航,星光制导
celestial horizon 天文地平,天文地平圈,天球地平
celestial inertial guidance 天文惯性制导,天文惯性导航
celestial latitude 【天】黄纬
celestial longitude 【天】黄经
celestial mechanics 天体力学
celestial medicine 天体医学
celestial meridian 天球子午圈
celestial navigation 天文导航
celestial navigation system 天文导航系统
celestial observation 天体观测,天文观测
celestial observation satellite 天体观测卫星
celestial parallel 天球纬圈
celestial photography 天体照相学
celestial planisphere 平面星图
celestial polar distance 天体极距
celestial pole 天极
celestial reference system 天体参考系
celestial sphere 1. 天球 2. 天球仪,浑象 3. 浑天说
celestial stems 天干
celestial system 天球坐标系
celestial-inertial integrated guidance 天文-惯性组合制导
celestial-inertial integrated navigation 天文-惯性组合导航
celestial-inertial navigation equipment 天文-惯性导航设备
cell 1. 隔间,小室,箱,格 2. 翼组 3. 气囊
cell array 电池阵
cell boundary 细胞边界
cell centroid 细胞质心
cell count 细胞计数
cell partition 区位法
cell Reynolds number 网格雷诺数
cell separation 细胞分离
cell size 1. 单元尺寸 2. 电池尺寸
cell size measurement 单元尺寸测量
cell structure 电池结构
cell surface 细胞表面
cell transmission model 元胞传输模型
cell wall 胞壁
cellar 1. 地窖,地下室 2. 后进先出存储区
cell-based data structure 网眼数据结构
Cellophane 赛璐玢,玻璃纸
cellular 1. 网眼的,蜂窝状的,格状的,细胞状的 2. 单元的
cellular convection 1. 胞状对流 2. 单元对流
cellular detonation 胞格爆轰
cellular hypoxia 细胞性缺氧
cellular material 蜂窝状材料
cellular plastic 微孔塑料
cellular plastics 泡沫塑料
cellular radio 无线电话网
cellular radio telephone 蜂窝式无线电话
cellular structure 网格结构
cellular system 蜂窝网
cellule 翼组
celluloid 1. 赛璐珞,假象牙,硝纤象牙 2. 电影胶片 3. 细

胞状的

cellulose 纤维素,植物纤维质
cellulose acetate 醋酸纤维素
cellulose nitrate 硝酸纤维素
cellulosic fuel 纤维素燃料
celonavigation 天文航行学
celsian ceramics 钡长石瓷
celsius temperature scale 摄氏温标
CELSS 1. Closed Ecologic Life-Support System 密闭生态生命保障系统 2. Closed Environmental Life Support System 闭式环境生命保障系统 3. Controlled Ecological Life Support System 受控生态生命保障系统
celtium 铪
CELV Complementary Expendable Launch Vehicles 补充性一次使用运载火箭
CEM 1. Channel Electron Multipliers 通道式电子倍增器 2. Computing Electromagnetics 计算电磁学 3. Controlled Excess Material 控制的多余材料
cement 1. 水泥 2. 胶泥,胶接剂,结合剂 3. 巩固,使结合
cementation 渗镀,扩散渗镀
cemented carbide 硬质合金
cementite 渗碳体
Cen Centaurus 半人马座
CENA Centre d'Etudes de la Navigation Aerienne 航空研究中心
censor 1. 审查,检查,删改 2. 保密检查
Centaur 【天】半人马座
Centaur G 【天】半人马座 G
centennial variation 百年变化,世纪变化
center 1. 中心,中央 2. 顶针 3. 定心
center beamlet 中心束波
center blade 中心叶片
center cue 中心指令
center dot 中心光点
center fire 1. 中央击发弹药,中央发火弹药 2. 中央击发火器
Center for Space Policy 空间政策中心
center fuel tank 中间燃料舱
center hole 中心孔
center jet 中心射流
center line survey 线路中线测量
center manifold 中心流形
center movement 轴心运动
center of attraction 引力中心
center of buoyancy 浮心,浮力中心
center of gravity 重心
center-of-gravity margin 重心位置范围,重心位置裕度
center of impact 平均弹着点,弹着点散布中心

center of mass 质心
center of mass determination 质心测定
center of mass motion 质心运动
center of pressure 压心
center of pressure coefficient 压心系数
center of rotation 旋转中心
center of thrust 推力线,推力中心
center of twist 扭心
center passage 中位油道
center point 中心点
center projection 中心投影
center section 中心剖面
center tank 中间舱
center tap 中间抽头,中点引线
center timing station 时统中心站
center track 中导板(六管炮的)
center vane 中心叶片
centerbody 中心体,中心锥
center-cell pull-down method 中心细胞下拉法
centered dipole 中心偶极子
centering 1. 定心,对正中心,中心调整 2. 置于中间位置
centering action 定心作用
centering adjustment 定中心,中心调整,中立位置调整
centerless grinding 无心磨削
centerline 中心线,轴线
centerline static pressure 中心线静压
center-lock 固定在中间位置
centerplane 中央翼
center-tapped winding 中间抽头绕组
centibar 厘巴
centimeter 厘米
centimeter height finder 厘米波测高计
centimeter wave 厘米波
centimeter-level accuracy 厘米级精度
centipoise 厘泊,10^{-2} 泊
centistroke 厘沱(动力黏度单位)
centralizer 定中心夹具
centrad 厘弧度
central 1. 中心的,主要的,中枢的 2. 电话总机,电话接线员
central aiming pipper 瞄准中心光点
central altitude reservation facility 预留高度中央管制站
central axis 管道中心轴线
central blockage 中央堵塞
central board 中央仪表板、中央操纵台
central body 中央体
central composite design 中心合成设计,星点设计
central cone 中央锥

central configuration	中心构形
central control	主控制,中央控制
central control mechanism	1.中央操纵机构 2.座舱操纵机构
central controller	引导站站长,中央控制台
central core	中央轴
central data bank	中央数据库
central difference	中心差分
central dipole pole	中心偶极子极
central distributed system	中央分布式系统
central eclipse	中心食
central electrode	中心电极
central failure	集中故障,主要故障
central field	中心视场
central fire control	集中火力控制
central force	有心力,中心力
central fuselage	中央机身
central gradient array method	中间梯度法
central gravity field	有心重力场
central information processing	中枢信息处理
central light loss	中心视觉消失
central maintenance panel	中央维修信息板
central meridian	中央子午线
central module	中央模块
central monitoring system	中心监护系统
central office	中心局
central part	中部
central pattern	中枢模式
central peak	中心峰
central pipper	中心光点
central portion	中央部分
central processing element	中央处理单元
central processing unit	1.中央处理机 2.中央处理部件
central processor	中央处理机
central readjustment	中枢重调
central recirculation region	中央回流区
central region	中心区
central section	中心截面
central standard time	中部标准时间
central station	中心站
central structure	中心结构
central support	中心架
central tactical system	中央战术系统
central terminal unit	中央终端设备
central timing unit	中心计时设备
central truss	中部桁架
central warning panel	中央警告板
central wire	中丝
central zone	中央区
centralite	火箭固体燃料稳定剂,中定剂
centrality	集中性
centralization	1.集中指挥 2.置于中间位置
centralize	1.定中心 2.置于中间
centralized approach	集中式方法
centralized control	集中控制
centralized design	集中化设计
centralized management system	集中管理系统
centralized monitoring system	集中监控系统
centralized network	集中网
centralized solution	集中解决方案
centralized traffic control	集中化交通管制
centre of burst	爆炸中心,平均弹着点,弹着点散布中心
centre of gravity	重心
centre of gyration	回旋中心
Centre Spatial de Toulouse	图卢兹航天中心
centre unit	中心单元
centreline gear	自行车式起落架,中心线起落架
centre-tap	中间抽头
centrifugal	离心的
centrifugal acceleration	离心加速度
centrifugal atomization	离心式雾化
centrifugal casting	离心铸造
centrifugal coating	离心包覆
centrifugal compressor	离心式压气机,离心式压缩器,径向式压气机
centrifugal drift	离心漂移
centrifugal field	离心力场
centrifugal flow compressor	离心式压气机,离心式压缩器
centrifugal force	离心力
centrifugal force field	离心力场
centrifugal growth	离心发育
centrifugal injection	离心式喷注器
centrifugal load	离心荷载
centrifugal pump	离心泵
centrifugal seperator	离心分离器
centrifugal stress	离心应力
centrifugal tachometer	离心式转速表
centrifugal twisting moment	离心扭转力矩
centrifugal water separator	离心式水分离器
centrifugation	离心作用,离心效应
centrifuge	离心机
centrifuge examination	离心机检查
centrifuge filter	离心过滤器,离心分离器
centrifuge test	离心试验
centrifuge training	离心机训练
centrifuging stress	离心应力
centring	1.定心,对正中心,中心调整 2.置于中间位置

centring under point	点下对中
centripetal	向心的
centripetal acceleration	向心加速度
centripetal force	向心力
centripetal turbine	向心涡轮,径向涡轮
Centrisep	发动机空气粒子分离器
centroid	形心,矩心,质心
centroid tracking	1.质心跟踪 2.形心跟踪
centurium	【化】镄(化学元素,符号 Ct)

CEO Chief Executive Officer 首席执行官,总经理,总社长

CEOA Central Europe Operating Agency 中欧管理局

CEP 1. Circle of Equal Probability 等概率圆 2. Circular Error Probable 圆概率误差,圆公算偏差 3. Cylindrical Electrostatic Probe 圆柱形静电探测器

cepheid parallax	造父视差
cepheid variable	造父变星
Cepheus	【天】仙王座

CEPS Command-Module Electrical Power System 指挥舱电源系统

cepstrum	倒频谱
ceralumin	铸造铝合金
ceramal	陶瓷金属
ceramic ablator	陶瓷烧蚀材料
ceramic capacitor	陶瓷电容器
ceramic composite	陶瓷复合材料
ceramic filter	陶瓷滤波器
ceramic insulation tile	陶瓷防热瓦
ceramic magnet	陶瓷磁铁
ceramic metallization	陶瓷金属化
ceramic model	陶瓷模型
ceramic monolith	陶瓷载体
ceramic package	陶瓷封装
ceramic packaging	陶瓷封装
ceramic pellet	陶瓷颗粒
ceramic screw	陶瓷制品螺旋
ceramic sensor	陶瓷传感器
ceramic thermal barrier coating	陶瓷隔热涂层
ceramic top coat	陶瓷表面涂层
ceramic transducer	陶瓷换能器
ceramic wall	陶瓷砖墙
ceramic wire	陶瓷线
ceramic-coated	有陶瓷敷层的
ceramic-matrix composite	陶瓷基复合材料
ceramic-metal composite	陶瓷-金属复合材料
ceramics	1.陶瓷学 2.陶瓷
ceramic-seal ring	陶瓷密封环
ceramoplastic	陶瓷塑料,陶瓷塑料的
cerapaper	陶瓷纸
cerebral anoxia	脑缺氧
Cerenkov counter	切伦科夫计数器
Cerenkov detector	切伦科夫探测器
Cerenkov effect	切伦科夫效应
Cerenkov radiation	切伦科夫辐射
Cerenkov telescope	切伦科夫望远镜

CERES Clouds and the Earth's Radiant Energy System 云和地球辐射能系统

ceria glass	铈玻璃
cerium	【化】铈(化学元素,符号 Ce)
cermet	金属陶瓷,含陶合金
cermet reactor	金属陶瓷反应器
Cerrobase	赛洛贝合金
Cerrobend	赛洛本合金
Cerromatrix	赛洛玛合金
certain	1.确实击落的飞机 2.确凿的,无疑的,可靠的
certainty	确实性,可靠性
certainty equivalence	确定性等价
certificate	证书,执照
certificate of airworthiness	适航证书
certificate of compliance	合格证书
certificate of conformity	合格证书
certificate of inspection	检验证书
certification	认证
certification flight test	型号合格审定试飞
certification test	取证试验
certification trial	取证鉴定试验
certified reference material	有证标准物质
certified value of reference material	标准物质标准值
certify	证明,保证

CERV 1. Crew Emergency Rescue Vehicle 乘员应急营救飞行器 2. Crew Escape and Reentry Vehicle 乘员逃生和再入飞行器

cervical	颈部
cervicogenic vertigo	颈源性眩晕

CES 1. Central Electronics System 中央电子系统 2. Coast Earth Stations 海岸地球站 3. Communication Engineering Standards 通信工程标准

cesium beam frequency standard	铯束频率标准
cesium clock	铯钟
cesium contact	铯接触
cesium dideuterium arsenate	砷酸二氘铯
cesium engine	铯引擎

CESM Commercial Engine Service Memorandum 商用发动机服务备忘录

CET 1. Central Earth Terminals 中央地面终端 2. Cumulative Elapsed Time 累积经过的时间

CETA Crew Equipment Translation Aid 乘员设备移动车

cetane 十六烷
cetane index 十六烷指数
CETEX Contamination by Extraterrestrial Exploration 外空探索造成的污染
CETF Critical Evaluation Task Force 关键鉴定特别工作组
CETI Communication with Extraterrestrial Intelligence 与外星智能联络
CETS 1. Conference on European Telecommunication Satellite 欧洲通信卫星会议 2. Contractor Engineering and Technical Service 承包商工程技术服务
CEU Checklist Entry Unit 检查单输入装置
CF 1. Card File 路线板盒 2. Cash Flow 现金流量 3. Check Flight 试飞,技术检查飞行 4. Cold-Finished 冷精轧的 5. Cross Functional 交叉功能的
CFA Covering Force Area 掩护部队配置地域
CFAO Conception et Fabrication Assistee par Ordinateur 计算机辅助构想及制造
CFAR Constant False Alarm Rate 恒虚警率
CFAS Commandement des Forces Aeriennes Strategiques〈法语〉法国战略空军司令部
CFC 1. Carbonfiber Composites 碳纤维复合材料 2. Centrifugal Flow Compressor 离心式压气机 3. Cycles to First Crack 裂纹初现循环数,首次裂纹周期
CFCs Chlorofluorocarbons 含氯氟烃,氯氟化碳,氯氟碳化物
CFD 1. Centralized Fault Display 集中故障显示 2. Chaff/Flare Dispenser 箔条/曳光弹投放器 3. Computational Fluid Dynamics 计算流体动力学
CFDIU Central Fault Display Interface Unit 中央故障显示界面装置
CFDS Central Fault Display System 中央故障显示系统
CFE 1. Central Fighter Establishment 战斗机研究中心 2. Contractor-Furnished Equipment 承包商提供的设备
CFEP Continuous Flow Electrophoresis 连续流电泳
CFES Continuous Flow Electrophoresis System 连续流电泳系统
CFF Critical Flicker Frequency 临界闪烁频率
CFFT Critical Flicker Fusion Threshold 临界闪融阈值
CFM Computational Fluid Mechanics 计算流体力学
CFMO Command Flight Medical Officer 飞行医学指挥官
CFP 1. Contractor-Furnished Property 承包商提供的财产(物质) 2. Cost plus Fixed Price 成本加固定价格
CFPG Cross-Functional Process Guide 交叉功能工艺指南
CFR 1. Code of Federal Regulations 联邦法典(美国) 2. Critical Failures Report 重大故障报告
CFRP Carbon-Fibre Reinforced Plastics 碳纤维增强塑料
CFS 1. Central Flying School 中央飞行学校 2. Container Freight Station 集装箱货运站 3. Cryogenic Fluid Storage 低温液体储存 4. Cubic Feet Per Second 每秒立方英尺
CFT Conformal Fuel Tank 保形油箱
CFTP Certification Flight Test Program 合格审定飞行试验项目
CFTR Cold-Fan Thrust Reverser 风扇冷气流反推装置
CG 1. Center of Gravity 重心 2. Crystal Growth 晶体生长
CGB Central Gear Box 中央减速器,中央齿轮箱
CGI Computer-Generated Image 计算机生成图像
CGIVS Computer-Generated Image Visual System 计算机成像目视系统
CGM load Combined Gust and Manoeuvre Load 阵风和机动飞行合成载荷
CGOD Configuration Guide Option Description 构型指南选择说明
CGP 1. Coal-Gasification Plant 煤气化厂 2. Computer Vision Graphics Processor 计算机视觉图形处理机
CGSP Crystal Growth and Solidification Processes 晶体生长和结晶过程
CGT Consolidated Ground Terminal 统一地面终端
CGV Computer-Generated Voice 计算机仿声,计算机产生的话音
CH 1. Case Harden 表面硬化 2. Channel 波道,通道,电路
chad 孔屑
chaetah "猎豹"式飞机
chaff 1. 敷金属条,箔条 2. 干扰物
chaff bomb 装敷金属条的炸弹,干扰物炸弹
chaff cloud 干扰云,箔条云
chaff corridor 箔条走廊
chaff dispenser 箔条投放器
chaff package 干扰包
chafing 擦伤,摩擦
chafing patch 防磨加强片
CHAG Compact High-performance Aerial Gun 结构紧凑的高性能航炮
chain 1. 链,链条,链系 2. 连锁,系列,体系 3. 电路,回路,通路
chain address method 链址法
chain code 链码
chain data address 链式数据地址
chain gun 链式炮
chain length 链长

chain of station　台链
chain segment　链段
chain-branching　链支化
chain-drive　链传动
chained aggregation　链式集结
chained list　链表
chain-operated　链传动的
chain-riveted　并列铆接的
chainstitch　避缝
chair　1.座椅,椅 2.席位,职位
chair chute　座椅伞
chair(o)dynamic(al)　弹射座椅动力学的
chair(o)dynamics　弹射座椅动力学
Chalet　农舍卫星
chalet　场外接待室
chalk　1.粉笔编号的 2.Chalk Load 或 Chalk Number 的简称 3.空降突击小组
chalk leader　空降突击组组长
chalk load　单位装载
chalk number　1.粉笔编号 2.单位装载
challenge　1.任务,问题 2.询问 3.挑战
challenge and response　一问一答座舱检查法
Challenger　挑战者号
CHAM　Chamfer 倒角
chamber　1.弹膛,药室 2.室,腔 3.舱
chamber ascent　低压舱上升
chamber backpressures　腔背压
chamber configuration　燃烧室外形
chamber diameter　室直径
chamber domain　室域
chamber effect　舱效应
chamber flight　低压舱模拟飞行
chamber pressure　燃烧室压力
chamber pressure decay　燃压衰减
chamber pressure decay time　燃压衰减时间
chamber volume　燃烧室容积
chamber wall　燃烧室壁
chambered　有高压和低压室的
chambered ammunition　定装药
chambering　送弹上膛
chamber-to-injectant density ratio　室内注入剂密度比
chamfer　倒角,倒棱
chamfer angle　倒棱前角,倒棱角
chamois　麂皮
champ　冠军,优胜者,战士
CHAN　Channel 槽形件
chance failure period　偶然失效期
chandelle　1.急上升转弯,战斗转弯 2.急垂直跃升,加速后垂直上升

Chandler period　钱德勒周期
Chandler wobble　钱德勒颤动
change　变化,改变,变换,转变
change of state　状态变化
change of velocity　速率变化
changeability　可换性,变易性,更改性
changeover　1.转换,变换 2.转换电门,转换开关
change-over rail mechanism for launch pad　发射台转轨装置
changer　1.变换器,变流机,换能器 2.转换电门,转换开关
channel　1.槽道,沟槽,管道,通道 2.槽材,槽钢 3.电路,通路,波道,信道
channel access data unit　信道访问数据单元
channel address word　通道地址字
channel associated signalling　随路信令
channel blockage　通道堵塞
channel capacity　波道容量,信道容量
channel centerline　航道中心线
channel coding　信道编码
channel command word　通道命令字
channel control　信道控制
channel cross section　槽钢截面
channel decoder　信道译码器
channel decoding　信道译码
channel demultiplexer　通道多路分解器
channel electron multiplier　通道式电子倍增器
channel encoder　信道编码器
channel encoding　信道编码
channel flow with variable mass flow rate　变流量管流
channel geometry　孔道几何形态
channel height　通道高度
channel intensifier　通道加强器
channel length　通道长度
channel midplane　通道中腔
channel nut　管道螺套
channel outer diameter　管道外径
channel protection system　管道保护系统
channel section　槽形截面
channel spacing　频道间隔,通道间隔
channel status word　通道状态字
Channel Stop　"海峡阻挡"
channel symbol　信道符号
channel traffic control（CTC）　通道通信控制
channel wall　通道壁面
channel width　道宽,沟道宽度
channeling　1.沟通作用,沟道效应 2.多路传输,频率复用,组成多路 3.开槽,沟流,引导
channeling effect　沟道效应

channelization　1.通信波道的选择 2.管道化,渠道化
channelized receiver　信道化接收机
channels ratio　道比
chaos　混沌现象
chaos control　混沌控制
chaotic motion　混沌运动不规则运动
chaotic regions　混纯区域,混乱地区
chaotic sky　混乱天空
chaotic solution　混沌解
chaotic system　混沌系统
CHAP　Chapter 章,篇
chaplain　1.随军牧师 2.军官
CHAR　Character 性格,人物,角色,字母符
char　1.炭 2.炭化,烧焦
char depth　炭化深度
char formation　成炭
char layer　炭层
character　1.符号,字母,字符 2.性格,特性,性质,特征,特点 3.征标
character display　字符显示
character display unit　字符显示器
character fill　字符填充
character generator　字符发生器
character processor　字符处理器
character register（CHR）　字符寄存器
character tendency　个性倾向
characteristic　1.特性,特征,性能 2.特性曲线特征线,特性参数 3.表示特性的,特征的,特有的
characteristic admittance　特性导纳
characteristic chamber length　燃烧室特征长度
characteristic curve　特性曲线
characteristic curve of photographic emulsion　感光特性曲线
characteristic diagram　特性图
characteristic distance　特征距离
characteristic equation　特征方程
characteristic exhaust velocity　特征排气速度
characteristic exponent　特征指数
characteristic frequency　特征频率
characteristic grammar　特征文法
characteristic impedance　特性阻抗
characteristic initial value　特征起始值
characteristic length　特征长度
characteristic locus　特征轨迹
characteristic map　特征映射
characteristic matrix　特征矩阵
characteristic of fluid　流体特征
characteristic parameter　特征参数
characteristic polynomial　特征多项式
characteristic radius　特征半径
characteristic reaction zone length　特征反应区长度
characteristic system　特征系统
characteristic test　特性试验
characteristic time　特性时间
characteristic time scale　特征时间尺度
characteristic velocity　特征速度
characteristic velocity efficiency　特征速度效率
characteristic wave　特征波
characteristics　特性
characteristics life　特征寿命
characteristics of measuring equipment　测量设备特性
characteristics of mode　模态特性
characteristics of propellant shape　火药形状特征量
characterization　表征,特性记叙
character-writing tube　字符显示管
charactron　字码显示管
charge　1.负载,负荷,载荷,定量装载物 2.充电,充气,装药,加注,装弹,装填,进气,装料 3.电荷,起电,带电
charge acceptance　充电接收能力
charge and discharge cycle　充放电循环
charge and discharge regime　充放电制
charge composition　填充成分
charge composition of primary cosmic ray　初级宇宙射线电荷成分
charge control　电荷控制
charge coupled device image sensor　电荷耦合器件影像传感器
charge coupled device star sensor　电荷耦合器件星传感器
charge coupled device sun sensor　电荷耦合器件太阳传感器
charge coupled imaging device　电荷耦合成像器件
charge coupled scanned IR imaging scanner　电荷耦合红外成像扫描器
charge distribution　电荷分布
charge efficiency　充电效率
charge exchange　电荷交换
charge exchange collision　电荷交换碰撞
charge feedback　电荷反馈
charge generation　电荷产生
charge injection device star sensor　电荷注入器件星传感器
charge measuring instrument　电荷测量仪表
charge neutrality　电荷中性,电中和
charge pump　电荷泵
charge radius　电荷半径
charge rate　充电率
charge ratio　满载系数,进料比

charge retention　充电保持能力
charge sensitive preamplifier　电荷灵敏前置放大器
charge spectrum　电荷谱
charge state　电荷态
charge storage diode　电荷存储二极管
charge transfer　电荷转移
charge transfer efficiency　电荷传输效率
charge tube　装料管道
charge valve　充气阀
charge voltage　带电电压
charge weight　装药量,装入量
charge weight ratio　装填比
chargeable fault　责任故障
charged drained battery　湿荷电蓄电池
charged ion　带电离子
charged particle　带电粒子
charged particle beam weapon　带电粒子束武器
charged-particle beam weapons technology　带电粒子束武器技术
charged particle detection　带电粒子探测
charged particle detector　带电粒子探测器
charged particle lunar environment experiment　带电粒子月球环境实验
charge-exchange　电荷交换
charge-exchange collision　电荷交换碰撞
charger　1.充电机,充电装置 2.加注设备,注油器 3.装弹机构
charge-to-mass ratio　荷质比
charge-transfer complex　电荷转移络合物
charge-transfer reaction　电荷转移反应
charging　1.充气,充电,加注,装弹,装填,进料 2.进气,增压 3.荷电,负载
charging point　灌充点
charging stand　灌充台
chargistor　电荷管
Charles Law　查理定理
Charlie　1.轻型预警和火控雷达 2.C字的代码
Charon　冥卫
Charpy（test）　摆锤(单梁)式冲击试验
charring ablator　炭化烧蚀材料
chart　1.图,图表,曲线图,计算图表 2.航图,海图,地图
chart boarder　海图图廓
chart comparison unit　图像重合显示装置
chart compilation　海图编制
chart correction　海图改正
chart datum　1.水深基准点 2.海图基准面
chart of marine gravity anomaly　海洋重力异常图
chart projection　海图投影,制图投影
chart scale　图表比例尺,海图比例尺

charter　1.租用,包租 2.包机,包机航班
charter aircraft　包机
charter flight　包机飞行
charting　制图,海图制图
CHAS　Chassis 底盘,底座,底架
CHASE　Coronal Helium Abundance Space Lab Experiment 空间实验室日冕氦丰度实验
chase　1.追击 2.炮身前部
chase mode　追逐模式
chase plane　歼击机
chaser　1.机动飞行器 2.螺纹梳刀,梳刀器 3.护航机,驱逐机
chaser orbit　截击轨道
chaser satellite　截击卫星
chaser spacecraft　追踪航天器
chassis　1.起落架 2.底盘,底板,机架
Chastise　"惩罚"作战
chatter　振动,震颤
Chaucer　"乔叟"作战
chauffeur　飞机驾驶员,汽车司机
CHC　Choke Coil 扼流圈,节流圈,节流盘管
CHCU　Channel Control Unit 通道控制器
CHD　1. Chord 翼旋,弦长 2. Crutching Heavy-Duty 重型吊架运输船
Chebyshev filter　切比雪夫滤波器
check　1.检查,检验,校验,核对,考核 2.检查飞行,考核飞行,检验飞行 3.阻止,妨碍,抑止
check action fuze　惯性引信
check before the loading　加注前检查
check bit　监督位
check digit　校验位
check ride　〈口语〉检查飞行
check standard　核查标准
check test　检查试验
check valve　1.单向阀 2.止回阀 3.检验阀
checker　1.检验器,试验装置,测试仪器 2.检验员
checkerboard　万格盘
checking bar　瞄准线检查板
checking bit　监督位
checking command　校验指令
checking datum mark　校核基点
checking fixture　检验夹具
checking of redundant substance　多余物检查
checking station　检查点,检查站位,参照站位
checking symbol　校验符号
checking table　1.检验表 2.校验台
checklist　检查顺序单,检查项目表
checkout　1.测试,检测,检验,检查,调整 2.系列工作,使飞行人员熟悉飞机的训练 3.飞行技术鉴定

checkout equipment 检测设备,测试设备
checkout equipment for satellite 卫星测试设备
checkout facility 检验设备,检测设备
checkout operation 检验操作
checkout procedure 测试程序
checkout-launch control set 检验与发射控制器
checkpoint 航线检查点
checksum 检验和
checkup 检查,检验
CHECMATE Compact High-Energy Capacitor Module Advanced Technology Experiment 密集高能电容器组件先进技术实验
chelate 螯合物,螯合的
CHEM Chemical 化学的
chemical agent 化学制剂,毒剂
chemical ammunition 化学弹药
chemical ammunition cargo 化学弹药物品
chemical analysis 化验
chemical arc jet rocket 化学电弧射流火箭
chemical cleaning 化学净化,化学清洗
chemical cloud experiment 化学云释放实验
chemical clouds 化学云
chemical composition 化学成分
chemical conversion 化学转化
chemical coprecipitation process 化学共沉淀工艺
chemical corrosion 化学腐蚀
chemical cylinder 毒气筒
chemical deposition 化学沉积
chemical double layer 化学双电层
chemical effect 化学效应
chemical energy 化学能
chemical equilibrium 化学平衡
chemical equivalent 化学当量
chemical erosion 化学侵蚀
chemical etching 化学腐蚀
chemical evolution 化学演化
chemical explosion 化学爆炸
chemical explosive 化学炸药
chemical fire extinguisher 化学灭火剂
chemical fuze 化学引信
chemical geothermometer 化学地球温度计
chemical heat treatment with rare earth element 稀土化学热处理
chemical ignition delay 化学点火延迟
chemical induction 化学诱导
chemical induction time 化学诱导时间
chemical initiator 化学引发剂
chemical injection 化学剂注入
chemical kinetics 化学动力学

chemical land mine 化学地雷
chemical laser 化学激光器
chemical liquid deposition 化学液相淀积
chemical measurement 化学测量
chemical mechanism 化学机理
chemical milling 化学铣削,化学抛光
chemical mine 化学(地)雷
chemical munition 化学弹药
chemical nonequilibrium 化学非平衡
chemical oxidation 化学氧化
chemical oxygen generator 化学氧发生器
chemical oxygen storage 化学氧贮存
chemical parameter 化学参数
chemical performance 化学性能
chemical plant 化工厂
chemical plating 化学镀膜
chemical polishing 化学抛光
chemical potential 化学势
chemical pressurization 化学增压
chemical pressurization system 化学增压系统
chemical process 化学过程
chemical process automation 化工自动化
chemical product 化学产品
chemical projectile 化学弹
chemical propulsion 化学推进
chemical protective clothing 防化服
chemical pumping 1.化学泵浦 2.化工泵
chemical quantity transducer 化学量传感器
chemical reaction 化学反应
chemical reaction flow 化学反应流
chemical reaction rate 化学反应速率
chemical remanent magnetization 化学剩磁
chemical resistant steel 耐蚀钢
chemical rocket engine 化学火箭发动机
chemical sensor 化学敏感元件
chemical shell 化学炮弹
chemical stage 化学阶段
chemical sterilization 化学灭菌法
chemical system 化学体系
chemical target analysis 化学袭击目标分析
chemical tempering 化学回火
chemical time 化学反应时间
chemical vapor deposition 化学气相淀积
chemical vapor infiltration 化学气相渗透
chemical warfare head 1.化学弹头 2.化学战斗部
chemical warhead 化学弹头
chemical weapon 化学武器
chemical-equilibrium 化学平衡
chemical-kinetic mechanism 化学动力学机理

chemically fueled 用化学燃料的
chemically mechanical polishing 化学机械抛光
chemically-powered 用化学燃料动力装置的
chemical-milled structure 化铣结构
chemi-ionization 化学电离
chemi-ionization current 化学电离电流
chemi-ionization sensor 化学电离传感器
chemiluminescence 化学发光,化合光
chemiluminescence image 化学发光图像
chemiluminescence intensity 化学发光强度
chemiluminescence measurement 化学发光测量
chemisorption 化学吸附
chemistry 化学,化学过程
chemistry interaction model 化学交互模型
chemistry of the solar system 太阳系化学
chem-mill 化学蚀刻
chemopause 光化层顶
chemoreceptor 化学受体,化学感受器
chemosphere 光化层
Chernoff bound 切尔诺夫界限
cherry rivet 空心拉杆铆钉,"彻里"铆钉
cherry riveting 抽芯铆接
chest bladder 胸部气囊
chest strap 胸带
chest to back acceleration 胸到背加速度,背到胸过载
chest type parachute 胸式降落伞
chest-mounted regulator 胸式调节器
chevron 1.人字形纹章,V形臂章 2.V形标志
CHG 1.Change 更改 2.Charge 充电
CHGR Charger 充电器,加载装置
CHI Computer Human Interface 人机界面
chick 战斗机通话代语
chicken bolts 机体结构紧固器
chief 1.长,主任,长官 2.主要的,首要的
chief engineer 总工程师,轮机长
chief location 首席地段
chief orbit 首席轨道
chill 1.冷却,冷冻,激冷,急冷 2.冷模,冷铁
chilldown 预冷
chilldown bleed valve 预冷泄出阀
chilldown system 预冷系统
chilled casting 冷硬铸造,冷硬铸件
chiller 冷却器,冷凝器,冷却装置
chimera 嵌合体
chimera grid 嵌套网格
chin 机头下部,位于机头下部的进气道
chin fin 机头下鳍片
China Aerodynamics Research Society 中国空气动力研究会

china-clay 1.陶土,瓷土,高岭土 2.瓷土法
CHINASAT 中国卫星
chine 1.舭,舭线,舭弯 2.机身边条
chin-mounted gun 机头下部机关炮,机头下部机枪
chip 1.片,屑,碎片 2.凹口,缺口 3.片,芯片,基片
chip component 片式元件
chip detector 1.屑末探测器 2.检屑器
chip inductor 片式电感器
chip microprocessor 单片微处理机
chip size 芯片尺寸
chip technology 小片工艺,小片技术
chip width 片宽
chipping 1.錾平 2.片屑 3.碎裂
chirp 展宽-压缩(技术),啁啾声
chirp excitation 线性调频脉冲激励
CHIS Center Hydraulic Isolation System 中央液压隔离系统
chisel window 机头斜口窗
CHK Check 检查
CHLD Chilled 激冷的,淬火的
chloramine 氯胺
chlorate 1.氯酸盐 2.氯化
chlorate candle 氯酸盐蜡烛
chlorate explosive 氯酸盐炸药
chlordiboron 氯二硼
chlorella 小球藻
chloride 氯化物
chloride remote sensing 氯化物遥感
chlorine 【化】氯(化学元素 Cl),氯气
chloroacetophenone gas 苯氯乙酮毒气
chloroacetophenone solution 苯氯乙酮溶液
chloroform 氯仿
chlorophyll detection 叶绿素探测
chloropicrin 氯化苦,三氯硝基甲烷
chloroprene rubber 氯丁橡胶
chlorosis 1.失绿,褪绿 2.萎黄病
chlorosufonated polyethylene 氯磺化聚乙烯
CHMBP Chamber 室,腔
chock 1.轮挡 2.制动垫快,楔形垫 3.导缆钩
chocking agent 窒息性毒剂,窒息剂
chock-to-chock 轮挡时间
CHOL Collins High Order Language 柯林斯高级指令语言
choice 选择
choice of polarization 极化选择
choke 1.扼止,抑制,阻流 2.超声速喷管,临界截面 3.节气门,节流门
choke coil assembly 扼流圈组件
choke filter 扼流圈滤波器

choke flange 扼流接头
choke joint 扼流关节
choke line 节流管线,阻塞线,阻流管线
choke piston 扼流式活塞
choke plunger 扼流式活塞
choked flow 堵塞流
choked inlet 堵塞进气道
choked technique 堵塞技术
choke-input filter 扼流圈输入式滤波器
choke-out 1.壅塞,堵塞 2.闭死
chokepoint 1.咽喉区 2.交通枢纽点
choker 节气门
Choker Ⅰ 窒息者Ⅰ计划
Choker Ⅱ 窒息者Ⅱ计划
chokes 气哽,减压病
choking 壅塞,堵塞,阻塞
choking gas 窒息性毒气
choking turns 扼流圈,抗流圈
CHTECH Chief Technician 总技师,主任技师
Cholesky decomposition 乔尔斯基分解
cholesteric liquid crystal 胆甾相液晶,多向多层液晶
cholesterin 胆甾醇,胆固醇
CHOMP changeover at midpoint 中点转换
choose 选择,选定
CHOP Change of Operational Control 改变作战指挥关系
chop 1.猛收油门 2.切断供油,迅速停车 3.迅速开关应答器
chop-out area 应急凿开部位
Chop Rest "伐木架"
chopper 1.旋翼机 2.断路器,断续装置 3.快门
chopper-borne 直升机上的
chord 1.弦,翼弦,弦线 2.边界构件
chord baseline configuration 翼弦的基线配置
chord length 1.弦长 2.弦宽
chord off-set method 弦线支距法
chord ratio 翼弦比,展弦比
chord vector 弦矢量
chordwise 弦向,沿翼弦方向
chordwise location 弦向位置
chordwise mode 弦向模式
chordwise position 弦向位置
chordwise strut 弦向支柱
choroplethic map 等值区域图,分区量值地图
chorus 合声
Christmas Island 圣诞岛
Christmas tree complex 圣诞树形综合停机坪
chroma 色饱和度,彩度,色品
chroma key 色键

chromacoder 彩色电视制式变换装置,制式转换器
chromascan 彩色飞点扫描摄像装置
chromate acid anodizing 铬酸阳极氧化
chromate enriched pellet 加富铬酸丸
chromate primer 铬酸盐底漆
chromatic aberration 色(像)差
chromatic adaptation 色觉适应,颜色适应
chromatic dispersion 多色色散
chromatic resolving power 辨色本领,色分辨率
chromatic sensation 感色能力
chromaticity 色度,色品
chromaticity diagram 色度图
chromatism 色(像)差
chromatogram 色谱图
chromatograph 色层谱仪
chromatographic analysis 1.色谱分析 2.色层分析
chromatography 色谱法
chromatron 聚焦栅彩色显像管,栅控彩色显像管
chrome 1.【化】铬(化学元素,符号 Cr) 2.铬矿石 3.氧化铬
chrome steels 铬钢
chromel-alumel thermocouple 镍铬镍铝热电偶
chromel terminal 镍铬合金接头
chromic acid 铬酸
chrominance 色品,色度
chromite 1.键铬 2.铬媒
chromizing 渗铬
chromoaluminizing 铬铝共渗
Chromoly Chromium and Molybdenum Alloy Steels 铬钼合金钢
chromoscope 彩色显像管,色度管
chromosome 染色体
chromosphere 色球
chromospheric evaporation 色球蒸发
chromospheric network 色球网络
chromospheric telescope 色球望远镜
chronaxy 时值
chronic acceleration 慢性加速度
chronic altitude hypoxia 慢性高空缺氧
chronograph 记时器测时仪
chronographing 用测时仪测定弹头初速
chronoisotherm 时间等温线
chronometer 精密计时仪
chronometer correction 表差
chronometer rate 表速
chronopher 电控报时器
CHT Cylinder-Head Temperature 气缸头温度
chuck 夹盘,卡盘,夹头
chuckhole 着陆压痕辙痕

chuffing 不稳定燃烧,不稳定工作状态
chug 发动机不稳定工作的声音,嚓嘎声
chugging 1.发动机不稳定工作的声音 2.嚓嘎声 3.低频不稳定燃烧
Chung Shan Institute of Science and Technology 中山科学研究院（中国台湾）
Churchill Research Range 丘吉尔研究发射场
churning 搅拌
chute 1.降落伞,着陆伞,伞 2.紧急滑梯 3.出口管道
chute rack 槽架
chute timer chutist 定时开伞器跳伞员,伞兵
CHY Commission for Hydrology 水文学委员会
CI 1. Cabin Interphone 座舱内话 2. Cast Iron 铸铁 3. Center of Impact 碰撞中心,平均弹着点 4. Chief Instructor 主任教官 5. Collateral Impact 间接碰撞 6. Collector Installations 安装件集成 7. Component Identification 部件标识 8. Compound Interest 复利 9. Configuration Indentification 构型标识 10. Configuration Item 构型项 11. Conformity Inspection 符合性检查 12. Continuous Improvement 持续改进 13. Critical Item 关键项 14. Customer Inspection 客户检查 15. Customer Introduction 客户推广 16. Customer Item 客户项目
CIA 1. Communication Interface Adaptor 通信接口适配器 2. Change Initiation Action Or Change Information Analysis 更改开始生效或更改信息分析 3. Concern，Issue or Action Item（Issues Database）关联、发放或行动项（发放数据库）4. Contractor Interface Agreement 承包商接口协议
CIB 1. Carry-Into Bit 进位输入位 2. Channel-In-Buffer 通道输入缓冲器 3. Channel-In-Bus 通道输入总线 4. Configuration Identification Baseline 构型标识基线
CIC 1. Combat Information Center 作战情报中心 2. Communications Intelligence Channel 通信智能信道 3. Crew Interface Coodinator 乘员接口协调人
CICE Computer，Information and Control Engineering 计算机、信息和控制工程
CICS/VS Customer Information Control System for Virtual Storage 虚拟存储用户信息控制系统
CID 1. Change Identifier 更改标识 2. Charge Injection Device 电荷注入器件
CID star sensor 电荷注入器件星敏感器
CIDA Channel Indirect Data Addressing 通道间接数据寻址
CIDEX Cometary Ice and Dust Experiment 彗星冰和尘埃实验
CIDIN Common ICAO Data Interchange Network 国际民航组织公共数据交换网
CIDRS Customer Information and Distribution Requiremnets System 客户信息和分配要求系统
CIDS 1. Cabin Interphone Distribution System 座舱内话分配系统 2. Critical Item Development Specification 关键项目研制规范 3. Customer Interior Development System 客户内部开发系统
CIEL Centre for International Environment Information 国际环境信息中心
CIEM Computer Integrated Engineering and Manufacturing 计算机一体化工程和制造
CIFS Computer-Interactive Flight Simulation 计算机配合飞行模拟
CIG Celestial Inertial Guidance 天文惯性制导,天文惯性导航
cigarette-burning 端面燃烧的
cigaretteroll 卷裹
cigar-shaped 雪茄烟形的
CIGTF Central Inertial Guidance Test Facility 中央惯性制导试验设施
CIL Configuration Item List 构型项清单
CIM Computer-Integrated Manufacturing 计算机综合制造
CIMS 1. Computer Integrated Manufacturing System 计算机集成制造系统 2. Contemporary Intergrated Manufacturing System 现代集成制造系统
CIMSS Cooperative Institutefor Meteorological Satellite Studies 气象卫星研究合作协会
CIN 1. Center Information Network 中心信息网 2. Computer Information Network 计算机信息网
CINCAFLANT Commander-in-Chief，United States Air Forces，Atlantic 美国大西洋空军司令
cine gun 照相枪
cine telescope 电影望远镜
cinecorder 影片磁性同步录音机
cinefluorography 立体荧光电影摄像术
cinematorgraphy 电影摄影术,电影摄影学,电影艺术
cinephotomacrography 低倍宏观电影摄影
cineradiography 射线电影摄影术
cinesync 影片磁性同步录音
cinetheodolite 电影经纬仪,摄影经纬仪,高精度光学跟踪器
cinetheodolite tracking system 电影经纬仪跟踪系统
cinnamene 苯乙稀
CIO Conventional International Origin 国际协议原点,国际习用原点
CIP 1. Change Implementation Point 更改实施点 2. Cold-Iso Pressing 冷均衡加压 3. Commission Internationalede Parachutisme 国际跳伞委员会 4. Critical Installation Point 关键安装点 5. Crop Identification Performance 作物识别特性

cipher 1.密码,暗号 2.零 3.译成密码 4.运算
cipher code 密码
cipher key 密钥
ciphony 密码电话学
CIQ Customer, Immigration and Quarantine 海关,移民和检疫
CIR 1. Cargo Integrated Review 货物综合评审 2. Change Incorporation Request 更改合并需求 3. Circle 圈,圆周,范围,领域 4. Coherent Imaging Radar 相干成像雷达 5. Configuration Identification Records 构型标识记录 6. Conformity Inspection Record(制造)符合性检查记录 7. Customer Inspection Requirement 客户检查要求
CIR(BOX) Completed Inspection Records(Box)完工检验记录(箱)
CIRC Circular 环形的,循环的,通报
circadian dysrhythmia 似昼夜节律障碍,时差问题
circadian rhythm 昼夜节律
CIRCE Computerized Instrument for Route Calculation and Edition 航线计算编排计算机系统,飞行任务计划系统
circle 1.圆,圆周,圈环 2.循环,周期,轨道
circle of confusion 1.散光圈,模糊圈 2.散射圆盘
circle of latitide 纬度圈,黄纬圈
circle of longitude 经度圈,黄经圈
circle of origin 基准圈
circle of right ascension 赤经圈
circle-hyperbolic system 圆-双曲线系统
circle-setting table 度盘表
circling 圆圈飞行
circling(circular) flight 起落航线飞行
circling approach 盘旋进近
circlip 簧环,卡簧
circuit 1.电路,线路,管路,回路,线路图 2.起落航线 3.环行,绕圈飞行
circuit alarm 电路故障信号
circuit analyzer 电路分析器
circuit board 电路板
circuit breaker 电路自动保险电门
circuit continuity tester 电路试验器
circuit extraction 电路提取
circuit height 起落航线高度
circuit model 电路模型
circuit pattern 1.起落航线 2.电路图形
circuit simulation 电路仿真
circuit switching 线路交换,电路交换
circuit topology 电路拓扑学
circuit voltage 电路电压,电压回路,导线间电压
circuitron 电路管

circuitry 1.电路,电路系统,整机电路 2.管路,管路系统 3.电路图,电路学
circuits and bumps 连续起落
circulant 1.循环液 2.循环的
circular 1.圆形的,循环的 2.供传阅的 3.通知,通告
circular accelerator 回旋加速器
circular antenna 圆形天线
circular approach 绕场进场降落
circular arc 圆弧
circular boundary 圆形边界
circular canopy 圆伞衣
circular cell 圆形室
circular convolution 循环卷积
circular cross section 圆形截面
circular cross-section missile 圆截面导弹
circular curve location 圆曲线测设
circular cylinder 圆柱体,圆筒
circular deviation 圆周罗差,等值罗差
circular disc 圆盘
circular duct 循环风道
circular error 径向误差,圆误差
circular error probable 1.圆概率误差 2.圆公算偏差
circular flight-path 环形飞行路径
circular formation 圆形编队
circular frozen orbit 圆形冻结轨道
circular hole 圆孔
circular injector 圆形喷射器
circular jet 圆流
circular membrane 圆环式膜片
circular motion 圆周运动
circular orbit 圆周轨道
circular path 圆周轨迹
circular plate 圆板
circular plunger 圆形柱塞
circular polarization 圆极化
circular polarization transducer 圆极化变换器,极化器
circular port 圆形口
circular probable error 圆概率误差,径向概率偏差
circular projection relative orbit 圆投影相对轨道
circular pull-up 圆上拉
circular reference orbit 圆参考轨道
circular restricted three-body problem 圆形限制性三体问题
circular scan 圆周扫描
circular scanning 圆周扫描
circular slide rule 圆形计算尺
circular target 圆形靶标
circular tube 圆管,圆形筒体
circular velocity 圆周速度,环绕速度

circular waveguide 圆波导
circular wind tunnel 圆截面风洞
circular winding 圆形绕法
circular-arc segment 圆弧段
circular-error probability 圆概率误差
circularization 通函询证,环化
circularize 分发传单,分发民意测验单
circularly polarized antenna 圆极化天线
circularly variable filter 环形渐变滤光片
circular-port injector 圆口引射器
circular-scanning reticle 圆周扫描盘
circulation 1.循环 2.环流,环量 3.流通量
circulation control aerodynamics 环流控制空气动力学
circulation control slot 环流控制翼缝
circulation distribution 环场分布
circulation integral 环积分,闭合积分
circulation measurement 循环测试
circulation model 环流模式
circulator 环行器,循环器
circulatory collapse 循环性虚脱
circulatory decompensation 循环代偿障碍
circulatory flow 1.环流 2.循环流动
circuma viator 环球飞行家
circumambient 周围的,围绕的
circumaviate 环球飞行,乘飞机绕地球一周
circumaviation 环球飞行
circum-earth orbit 环地轨道
circumference 圆周,周线
circumferential 周围的,圆周的,环绕的
circumferential aperture 环形孔径
circumferential direction 周向
circumferential length 弧周
circumferential location 周向位置
circumferential mode 圆周模式
circumferential motion 圆周运动,环流
circumferential position 周向位置
circumferential radius 圆周半径
circumferential stress 周向应力
circumferential temperature 周向温度
circumferential variation 周向变化
circumferential velocity 圆周速度
circumferential wave 环形波
circumferential wavelength 圆周波长
circumfluent 环绕的,环流的
circumgyration 旋转,回转,陀螺运动
circumlunar 环月的,绕月的
circumlunar mission 绕月飞行任务
circumlunar rocket 环月火箭
circumlunar space 环月空间

circumnavigation 环球航行,沿闭合航线航行,绕航
circumplanetary 围绕行星的
circumpolar 围绕天极的
circumpolar star 拱极星
circumscribe 限制,划边界线
circumsolar 围绕太阳的
circumsolar radiation 太阳周边辐射
circumsolar sky radiation 环日天空辐射
circumstances of eclipse 交蚀概况
circumstellar matter 星周物质
circumterrestrial 环球的,环绕地球的
circumvent 巧妙躲避,用计取胜
circus 1.特技飞行表演 2.诱惑性战斗机群
CIRE Circulate Or Circumference 循环,圆周
CIRF Consolidated Intermediate Repair Facility 合并的中级修理设施
CIRIS 1. Cryogenic Infrared Radiance Instrumentation for Shuttle 航天飞机低温红外辐射仪表系统 2. Consolidated Intelligence Resources Information System 统一情报信息资源系统 3. Completely Integrated Range Instrumentation System 综合靶场仪器测试系统
CIRM Celestial Infrared Measurement 天文红外测量
cirri 1.卷云,卷须 2.触毛(cirrus 的复数)
cirrocumulus 卷积云
cirrostratus 卷层云
cirrous 卷云的
cirrus（cloud） 卷云,藤蔓
cirvis 报告重要情报的通信规定
CIS 1. Cabin Interphone System 客舱内话系统 2. Commonality Information System 通用性信息系统 3. Communication Interface System 通信接口系统 4. Computer Interface System 计算机接口系统 5. Configuration Item State 构型项状态 6. Customer Integrated System 客户综合系统 7. Customer Introduction System 客户订货说明系统,客户推广/再购机系统
cislunar 地月间的,地月轨道间的
cislunar space 月轨内空间
cis-martian space 火地空间
cisplanetary space 行星内空间
CISS 1. Consolidated Information Storage System 统一信息存储系统 2. Contract Items Specifications and Schedule 合同项目说明书和进度表
cistern 槽,水槽,贮水器
cistern barometer 槽式气压表
CIT 1. California Institute of Technology 加利福尼亚理工学院 2. Compressor Inlet Temperature 压缩器进口温度 3. Cranfield Institute of Technology 克兰菲尔德

理工学院

CITARS Crop Identification Technology Assessment for Remote sensing 遥感农作物识别技术评价

citation 传令嘉奖,嘉奖令,荣誉状

CITE 1. Capsule Integrated Test Equipment 密舱综合测试设备 2. Cargo Integration Test Equipment 货物综合测试设备 3. Computer Integrated Test Equipment 计算机综合测试设备

CITF CDOS Integration and Test Facility 用户数据和操作系统组装测试设施

city gas 城市煤气

city pair 城市对

city-pair selection 城市对选择

CIU 1. Channel Interface Unit 通道接口装置 2. Central Interface Unit 中央接口装置 3. Central Interpretation Unit 航空照片判读中心组

CIV 1. Configuration Item Version 构型项版本 2. Crossbleed Isolation Valve 交叉排气隔离活门

CIVAVN Civil Aviation 民航

civil air navigation aids 民航导航设备

civil aircraft 民航机,民用航空器

civil aircraft airworthiness 民用航空器适航性

civil airplane 1. 民用飞机 2. 民机

civil aviation 民航,民用航空

civil aviation accident 民用航空事故

Civil Aviation Administration of China 中国民用航空局

civil aviation medicine 民航医学

civil community 公民共同体

civil day 民用日

civil engine 民用发动机

civil jet 客机

civil operator 民用航空营办商

Civil Reserve Air Fleet 民航储备机队

civil spacecraft 民用航天器

civil time 民用时,平均太阳时

civil transport 民用运输机

civil transport aircraft 民航运输机队

civil twilight 民用曙暮光

civil user 民用用户

civil year 民用年

civilian map 民用地图

civilianized 民用改型的,改为民用的

civvies 便衣,普通衣服

CIWS Close-In Weapon(s) System 近战武器系统

CKD Component Knock-Down 散装元部件

CKPT Cockpit 驾驶舱

CKT Circuit 电路

CKT BKR Circuit Breaker 断路器,跳开关

CL 1. Centreline 中轴线,中心线 2. Change Level 更改级别 3. Clearance 间隙 4. Closed Loop 闭环 5. Coefficient of Lift 升力系数 6. Compass Locator 罗盘定位台 7. Component List 部件清单 8. Crack Length 裂纹长度 9. Confidence Level 置信度

CLA Communications Line Adapter 通信线路适配器

CLAC Closed Loop Approach Control 闭合回路进近控制

clack valve 板式单向活门,瓣阀

cladding 1. 包覆层,镀层,涂层 2. 包覆,涂,镀

cladding mode stripper 包层模消除器

CLAES Cryogenic Limb Array Etalon Spectrometer 低温临边阵列标准光谱仪

claimed accuracy 要求的精度,要求的准确度

clamp 1. 饼,夹,夹具,卡钳 2. 压板 3. 箝位电路 4. 夹紧 5. 箝位

clamp voltage 箝位电压

clamped dielectric constant 受夹介电常数

clamper 箝位器,箝位电路

clamping 1. 固定,夹住 2. 电平符位,箝位电路

clamping diode 箝位二极管

clamp-on ammeter 电流互感器,夹合式电流表

clamshell 1. 抓岩机,蛤壳型抓斗 2. 蚌式挖斗装置

clandestine aircraft 隐密飞机

Clansman 克兰斯曼系统

Clapeyron equation 克拉佩龙方程

clarifier 干扰消除器

clarity 清晰性

Clarke orbit 克拉克轨道

clasp 勋章附饰,勋标

CLASP Computer Language for Aeronautics and Space Programming 航空航天程序设计计算机语言

CLASS Communications Link Analysis and Simulation System 通信链路分析和仿真系统

class 1. 阶级,班级,课 2. 等级,种类 3. 定等级,分类

class of combustion 类燃烧

class of contamination 污染等级

classic shock pulse 经典冲击脉冲

classical aeroplane 古典式飞机

classical cepheid 经典造父变星

classical control law 经典控制律

classical diffusion 经典扩散

classical flutter 二自由度颤振,古典颤振

classical information pattern 经典信息模式

classical integral 经典积分

classical lifting-line solution 古典升力线解决方案

classical method 古典机率法

classical nova 经典新星

classical orbit 经典轨道

classical trajectory 经典轨线

classification 1.分类,分级 2.类别,等级,保密等级 3.分类法
classification criterion 分类标准
classification image 分类影像
classification map 分类图
classification of hypersonic inlet 高超声速进气道分类
classifier 分类器
classify 分类
claustrophobia 幽闭恐怖症
claw 叉爪
clay 黏土,白土,陶土
CLB 1.Center Line Bend 中心线弯曲 2.Climb 爬升 3.Crash Locator Beacon 应急定位信标机
CLC Course Line Computer 航线计算机
CLCW Command Link Control Word 指令链路控制字
CLD 1.Chemical Liquid Deposition 化学液相淀积 2.Cloud 云 3.Cooled 冷却过的 4.Crutching Light Duty 轻型吊架运输船
clean air 洁净空气
clean airfoil 干净翼型
clean annealing 光亮退火
clean bench 洁净台
clean bomb 1."净形"炸弹 2.干净弹
clean condition 清洁状态
clean configuration 净形布局,洁净构型
clean fin 净形垂尾
clean flow 均匀流场,无畸变流场
clean hardening 光洁淬火
clean launch 成功的发射,正确的发射
clean nozzle 清洁喷嘴
clean out 1.整流的,使成流线形的 2.清除,除净
clean propellant 洁净推进剂
clean range 净形航程,无外挂物航程
clean room 洁净室,净化室
clean turret 清洁炮塔
clean vacuum 清洁真空
clean weapon 干净武器
cleaned up 净形的,流线型的
cleaner 1.滤清器,净化器 2.吸尘器
cleaning 1.清洁,清洗,擦拭 2.清理,净化 3.使气流平滑,改善绕流状况,改善空气动力外形
cleaning-drain flexible hose 清泄软管
cleaning-drain gas distribution board 清泄配气台
cleanliness 洁净度
cleanliness levels 洁净度等级
clean-loaded 无外挂物的
cleanness 净形,外形洁净,流线性
cleanup 1.净形改善 2.吸附除气,离子除气 3.抛光,抹平

clear air turbulence 晴空紊流,晴空湍流
clear deck 甲板无障碍物
clear dope 本色涂料
clear ice 晶冰,透明冰
clear path 无阻碍路线
clear sky 碧空
clear variation trend 明显变化趋势
clear(ance) zone 保险道,保险地带
clearance 1.清除,清理 2.间隙,余隙,缺口 3.允许,许可,批准,航行调度许可证,放飞 4.超越障碍,超越角 5.澄清试验
clearance change 间隙变化
clearance closure 间隙关闭
clearance constraint 安全间距
clearance control 余隙调节
clearance criterion 间隙标准
clearance flow 间隙流量
clearance gap 间隙
clearance hole 隔离孔
clearance limit survey 净空区测量
clearance management 间隙管理
clearance method 消除法
clearance of gun 退弹,排除停射故障
clearance plane 安全高度面,余隙面
clearance thickness 间隙厚度
clearance volume 余隙容积,间隙容积
clearance vortex 余隙涡流
cleared area 机场净空区
clearing 1.清洗,清理 2.退弹,排除停射故障
clearing cam path 退弹导槽
clearing modules 清舱
clearway 净空道
clear-weather approach 良好天气进场着陆
cleat 1.角撑 2.线夹,陶瓷夹板
cleavage 1.劈裂,裂开处 2.解理
cleavage facet 解理小面
cleft 极隙,极隙区
CLEM Cargo Lunar Excursion Module 载货登月舱
clepsydra 漏壶,漏刻
clestial maser 天体脉塞源,天体微波激射源
clevis U形夹
clevis joint 叉形接头
CLG 1.Ceiling 天花板 2.Component Locator Guide 机载位置指南
CLGE Cannon Launched Guidance Electronics 炮射制导电子设备
CLGP Cannon Launched Guidance Projectile 制导炮弹
CLI 1.Command Language Interpreter 命令语言解释程序 2.Control Language Interpreter 控制语言解

释程序
click 喀音,咔嗒声
cliff 悬崖,绝壁
climagram 气候图
climagraph 气候图表
climate 气候
climate environment 气候环境
climate experiment 气候实验
climate index 气候指标,气候指数
climate orbiter 气象卫星
climate test 气候试验
climatic 气候的
climatic chart 气候图
climatic cold 气候性寒冷
climatic extreme 气候极值
climatic plant formation 气候植物区系
climatic test 气候试验
climatization 换季工作
climatizer 气候实验室
climatography 气候志
climatological 气候学的
climatological table 气候表
climatology 气候学
climb 1.上升,爬高,攀登 2.上升距离 3.上升率,爬升率,上升能力
climb allowance 上升用油量
climb angle 爬升角
climb gradient 爬升梯度
climb illusion 上仰错觉,爬高错觉
climb performance 爬升性能
climb phase 爬升阶段
climb profile 爬升剖面
climb rate 爬升率
climb thrust 爬升推力
climb time 上升时间
climb trajectory 爬升轨迹
climb trajectory prediction 爬升轨迹预测
climb trajectory prediction accuracy 爬升轨迹预测精度
climbing distance 爬升距离
climbing limiting weight 爬升限制重量
climbing module arm 登舱臂
climbing shaft 攀登道
climbing time 爬升时间
climbout 1.起始上升 2.转入稳定上升状态 3.起落航线飞行中的上升
climograph 气象图
Climsat 气候研究卫星
clinical aviation medicine 临床航空医学
clinical evaluation 临床评定

clinical medicine of aviation 航空临床医学
clinical selection 临床选拔
clinker-built 鱼鳞式搭接的,瓦叠式搭接的
clinodromic 有固定提前角的
clinodromic course 有固定提前角的截击航向
clinometer 1.倾斜仪 2.测角器 3.磁倾计
clip 1.夹子,卡箍 2.截去,剪去,剪短 3.夹紧,夹住
clip-in（carriage）assembly 插夹式挂弹装置
clip-in rack 插夹式挂弹架
clipped wing 截尖翼,直角翼尖机翼
clipped-tail 有后掠尾翼的
clipper 削波器,限幅器
Clipper Bow 飞剪型船首
clipper-limiter 双向限幅器
clipping 削波,限幅,剪音
CLK Clock 时钟
CLKG Caulking 冲缝,砸边,填缝,嵌塞
CLL 1. Central Light Loss 中心视觉消失 2. Critical Load Line 临界负载线
CLM 1. Circum Lunar Mission 绕月飞行任务 2. Communications Line Multiplexer 通信线路多路转换器
CLN 1. Clean 清除,归零 2. Clinometer 倾斜针,量坡仪,测角器,象限仪
C-L-NAS Cobra Laser Night Attack System "眼镜蛇"激光夜间攻击系统
CLNC Clearance 放飞
clobber 1.从空中击毁地面目标 2.撞上障碍物
clobber probability 撞地坠毁概率
clock 时钟
clock angle 时针角
clock basher 〈口语〉航空仪表师
clock bias 时钟偏压
clock correction 时钟校正
clock cycle 时钟周期
clock delay mechanism 定时装置,钟表延时机构
clock drift estimate 时钟漂移估计
clock drift rate 时钟漂移率
clock error 时钟误差
clock frequency 时钟频率
clock generator 时钟发生器
clock mechanism 钟表机构
clock offset 钟差
clock offset error 钟偏误差
clock offset estimate 钟偏估计
clock position 钟点位置
clock pulse 时钟脉冲
clock rate 钟速
clock spring 定时弹簧

clock stability　时钟信号稳定性
clock time fuze　钟表时间引信
clock train　定时装置
clocking　计时,记时,记录
clock-type　时钟型,圆形度盘指针式
clockwise　顺时针方向
clockwise vorticity　顺时针涡度
clockwork　钟表机构
clockwork triggering clog　用钟表机构起动
clog　1.堵塞,阻碍 2.黏成一块 3.制动器
clogging　堵塞,堵住
clon　克隆,无性繁殖
clone　靠营养生殖而由母体分离繁殖的植物
CLOS　1.Closure 闭合,闭路,终结,隔板,围墙 2.Command to Line of Sight 瞄准线指令
CLOS guidance　瞄准线指令制导
close agreement　吻合
close approach　近区模式
close area effect　近区效应
close binary star　密近双星
close combat air-to-air missile　格斗空空导弹
close control　精确引导,具体引导
close controlled interception　精确引导截击
close earth　近地
close encounter　近距离接触
close formation　密集队形
close-look photographic reconnaissance satellite　详查型照相侦察卫星
close look reconnaissance satellite　详查型侦察卫星
close look satellites　详查型卫星
close loop　回路
close loop pole　闭环极点
close lunar satellite　近月卫星
close out　制造计划全部完成
close proximity　极为贴近
close range firing　近距射击
close spacing　密井网
close stick　密集连续投弹,齐投连续投弹
close support weapon　近距支援武器
close up　1.集合,进入战位 2.(电路)接通,愈合
close velocity　接近速度
close-controlled air interception　精确引导空中截击
close-coupled canard control surface　近距耦合鸭式操纵面
close-coupled valve　紧密耦合阀
closed air cycle cooling system　闭式空气循环冷却系统
closed ampoule vacuum diffusion　无管真空扩散
closed area marking　不适用地区的标志
closed bomb　闭合式炸弹

closed breathing circuit　闭合式呼吸回路
closed cabin　密闭座舱
closed cavity　封闭空间
closed chamber　密闭舱
closed circuit　闭回路
closed circuit oxygen equipment　闭路循环氧气设备,再生式氧气设备
closed circuit wind tunnel　闭路式风洞
closed curve　封闭曲线
closed cycle　闭式循环,闭环
closed cycle MHD generator　闭环磁流体动力发电机
closed ecological system　密闭生态系统
closed end　闭端
closed form　闭型,封闭解
closed leveling line　闭合水准路线
closed loading　闭式加注
closed loop　闭路,闭环
closed loop control system　闭环控制系统
closed loop dynamics　闭环动力学
closed loop field　闭合回线场
closed loop guidance　闭环制导
closed loop process control　闭环过程控制
closed loop stability　闭环稳定性
closed loop transfer function　闭环传递函数
closed loop voltage gain　闭环电压增益
closed loop zero　闭环零点
closed orbit　封闭轨道
closed plasma device　闭合等离子体装置
closed set　闭集
closed symbol　关闭符号
closed system　闭路系统,封闭系统
closed test section wind tunnel　闭口式风洞
closed thermosyphon　闭式热虹吸管
closed traverse　闭合导线
closed universe　闭宇宙
closed vessel　密闭容器
closed wake　闭涡
closed-chamber　闭室
closed-cycle　闭式循环
closed-cycle engine　闭式循环发动机
closed-form　封闭式的
closed-form expression　闭式表达式
closed-form solution　封闭解
closed-loop control　闭环控制
closed-loop control system　闭环控制系统
closed-loop control task　闭环控制任务
closed-loop controller　闭环控制器
closed-loop dynamics　闭环动力学
closed-loop frequency response　闭环频率响应

closed-loop guidance　闭环制导
closed-loop guidance system　闭环制导系统
closed-loop identification　闭环识别
closed-loop model　闭环模型
closed-loop performance　闭回路响应性能
closed-loop pointing　闭环式瞄准
closed-loop pole　闭环极点
closed-loop position control response　闭环位置控制响应
closed-loop remote control system　闭环遥控系统
closed-loop resonance　闭环谐振
closed-loop response　闭环响应
closed-loop sensitivity　闭环灵敏度
closed-loop simulation　闭环仿真
closed-loop stability　闭环稳定性
closed-loop steam cooling　闭式蒸汽冷却
closed-loop system　闭路系统,闭合回路系统
closed-loop telemetry system　闭环遥测系统
closed-loop throttle　闭环节气阀
closed-loop transfer　闭环传递
closed-loop transient　闭环系统的过渡过程
closed-loop vibration　闭环振动
close-fitting　贴紧的
close-in　1.近战的 2.接近中心的
close-in (air) combat　近距空战
close-in missile　近战导弹,近程导弹
close-in weapon　近战武器
closely-packed code　紧充码
closeout　收尾,出清存货
close-range photogrammetry　近景摄影测量
closeup　特写镜头,小传,近景照片,仔细的观察
closing　1.关闭 2.接近
closing error　闭合误差
closing error in coordinate increment　坐标增量闭合差
closing modules　封舱
closing rate　接近速度
closing respone time　吸合动作时间,关闭动作时间
closing time　1.闭合时间 2.接近时间
closing velocity　接近速度
closure　1.关闭,闭合,闭合差,闭合环节 2.收油门 3.截止,终结
closure disk　破片起爆
closure point　截止点
closure rate　关闭率
cloth　布,织物,擦布
cloth seal　布条式气封
clothing　1.衣服,服装 2.布罩,布套 3.被服供应
clothing assembly　成套服装,服装配套
cloth-skin structure　蒙布式结构
cloud absorption　云吸收

cloud amount　云总量
cloud attenuation　云层衰减
cloud band　云带
cloud base height　1.云底高度 2.云高
cloud chamber　云雾室
cloud chart　云图
cloud cluster　云团
cloud condensation nucleus counter　云凝结核计数器
cloud cover　云覆盖,云量
cloud droplet　云滴
cloud edge　云先锋
cloud element　云素
cloud family　云族
cloud field　云区
cloud finger　云指示计
cloud height　云底高
cloud height indicator　云高指示器
cloud identification　云的识别
cloud layer　云层
cloud mirror　测云镜
cloud photograph　云照片
cloud photography　云层摄影
cloud photopoarimeter　云光偏振表
cloud physics　云物理学
cloud point　浊点
cloud thickness　云厚
cloud top height　最高云层高度,云顶高度
cloud top height estimation system　云高估计系统
cloud top scanner　云顶扫描仪
cloud vertigo　云中眩晕
cloud wind extraction system　云风提取系统
cloud/collosion warning　云层-防撞警告
cloudbar　云堤
cloud-base height balloon　云底高度探测气球
cloud-base recorder　云底高度测录器
cloud-bound　被云遮蔽的
cloudburst　暴雨
cloud-capped　被云遮盖的
cloud-drop sampler　云滴取样器
cloud-form　云状
cloudiness　云量
cloudless　无云的,晴朗的
cloudlessness　无云,全晴
cloud-projector　云幕灯
cloudy　多云
cloudy sky　多云天空
cloverleaf　1.草花斤斗(一种特技飞行动作) 2.苜蓿叶
cloverleaf flight test　梅花瓣飞行试验
cloverleaf slow wave line　三叶草慢波线

CLP 1. Clamp 嵌位电路,卡箍 2. Communication Line Processor 通信线路处理机

CLPA Clapper 铃锤,拍板

CLPR Calipper 卡尺,圆规

CLR 1. Clear 清洁的,净的 2. Color 颜色 3. Cooler 冷却器

CLRG Collector Ring 集流环,集电环

CLRS Contract Labor Reporting System 合同劳务报表系统

CLS 1. Classify 分类 2. Contingency Landing Site 应急着陆场

clsd Closed 接通,闭合,关闭

CLSF Cutter Location Source File 刀具定位源文件

CLSI Cryogenic Limb Scanning Interferometer 低温临边扫描干涉仪

CLSS 1. Closed Life Support System 闭式生命保障系统 2. Combat Logistic Support Squadron 战斗后勤保障中队

CLT 1. Centreline Tracking 沿中心线飞行 2. Communication Line Terminal 通信线路终端

CLTC China Satellite Launch and Tracking Control General 中国卫星发射测控系统部

CLTG Collecting 搜集

CLTV Collective 集中的,聚合的

CLTU Command Link Transmission Unit 指令链路传输单元

club 1. 杆,棍 2. 俱乐部

club layout 对坐布局

club propeller 试车桨

clubmobile 流动俱乐部

clue 线索,迹象,征候,特征

clusec 流率单位

cluster 1. 集束炸弹 2. 星状烟火信号弹 3. 火箭发动机簇

cluster adapter 1. 集束炸弹架 2. 子母弹箱

cluster bomb unit cluster coding 子母炸弹集群编码

cluster bombing (bomb release) system 框式炸弹架投弹系统

cluster compression 聚类压缩

cluster configuration 集群配置

cluster controller 群控器,群集控制器

cluster crystal 簇形晶体

cluster element 簇要素

cluster engine 并联发动机

cluster galaxy 团星系

cluster launcher 多管式发射装置

cluster munition 集束炸弹,子母弹

cluster of galaxies 星系团

cluster recon 集群侦察

cluster star 团星

cluster warhead 1. 子母战斗部 2. 集束战斗部

cluster weapon 集束武器

cluster rocket 1. 集束火箭 2. 集束火箭发动机

clustered airfield defeat submunitions 攻击机场的子母炸弹

clustered atomic warhead 集束原子弹头

clustered-bomb-delivered biological (chemical) munitions 使用集束炸弹散布的生物（化学）弹药

clustering 聚类

clutter 1. 杂波 2. 地物干扰

clutter echo 杂乱回波

clutter lock 杂波锁定

clutter map 杂波图

clutter rejection 杂波抑制

cluttered environment 密集环境

CLV Clevis 挂钩,U形夹

CLW Climb Limit Weight 爬升限重

CM 1. Cargo Module 货舱 2. Center of Mass 质心 3. Central Module 中央模块 4. Change Management 更改管理 5. Chemical Milling 化学铣切 6. Composite Material 复合材料 7. Condition Monitoring 状态监控 8. Configuration Management 构型管理

CMA Common Mode Analysis 共模故障分析

Cma 【天】Canis Major 大犬座

CMAG Configuration Management Advisory Group 构型管理咨询组

CMAM Commercial Middeck Augmentation Module 中舱商业有效载荷增载舱

CMB Concorde Management Board 协和式飞机管理局

CMBD Combine 综合,联合

CMC 1. Central Memory Control 中央存储器控制器 2. Central Maintenance Computer 中央维护计算机 3. Ceramic-Matrix Composite 陶瓷基复合材料 4. Configuration Management Center 构型管理中心 5. Configuration Management Council 构型管理委员会 6. Constant Mach Cruise 恒定马赫数巡航

CMCA Cruise Missile Carrier Aircraft 巡航导弹载机

CMCCS Change Management Change Control System 更改管理更改控制系统

CMCF Central Maintenance Computer Function 中央维护计算机功能

CMCI Computer Material Configuration Index 计算机资料构型索引

CMCS Configuration Management Control System 构型管理控制系统

CMD 1. Command 指令,控制,操作 2. Configuration Management Department 构型管理部门

CMDS Cislunar Meteoroid Detection Satellite 月地空间

流星探测卫星

CME Counter-Mesures Electroniques 电子对抗

CMEA Council for Mutual Economic Assistance 经济互助委员会

CMF 1. Common Message Format 通用信息格式 2. Conceptual Military Framework 军事结构初步设想

CMFLR Can Follower 凸轮随动件

CMG 1. Control Moment Gyro 控制力矩陀螺 2. Cryosphere Monitoring Payload Group 低温层监测有效载荷组

CMH Counter Measures Homing 干扰寻的

CMI 1. Completed Maintenance Inspection 已完成的维修检查 2. Continuous Maximum Icing 最大连续结冰 3. Cruise Missile Interrogation 巡航导弹询问

CMIMC Communications Intelligent Matrix Control 通信智能矩阵控制

CMIS 1. Command Management Information System 指挥管理信息系统 2. Corporate Management Information System 联合管理信息系统

CML 1. Commercial 商业的,商务的 2. Commercial MPS Lab 商用空间材料加工实验室 3. Configuration Modeling Language 构型建模语言

CMM 1. Capability Maturity Model 能力成熟模型 2. Component Maintenance Manual 机载设备维修手册,组件维修手册 4. Commission on Maritime Meteorology 海洋气象学委员会 5. Condition Monitored Maintenance 状况监控维修 6. Coordinate Measuring Machine 坐标测量机,坐标测量仪

CMN 1. Control Matrix Network 控制矩阵网络 2. Cruise Mach Number 巡航马赫数

CMO 1. Configuration Management Office 构型管理办公室 2. Condition Monitoring Office 可靠性管理办公室

CMOS Carbon Molybdenum Steel 碳钼钢

CMOSIC Complementary MOS Integrated Circuit MOS 互补集成电路

CMP 1. Command Module Pilot 指挥舱驾驶员 2. Configuration Management Plan 构型管理大纲 3. Contamination Monitor Package 污染监测装置 4. Customer Maintenance Plan 客户维修方案

CMPD Compound 复合物,化合物

CMPNT Component 单位,元件

CMPR Compare 比较

CMPS Compass 罗盘

CMPSN Composition 成分,组成

CMPST Composite 复合材料

CMPT Compute 计算

CMPTG Computing 计算

CMPTR Computer 计算机

CMR 1. Command Register 指令寄存器 2. Contract Management Review 合同管理评审

CMR&O Configuration Management Requirements and Objectives 构型管理要求和目标

CMR&R Configuration Management Requirements and Responsibilities 构型管理要求和职责

CMRCS Command Module Reaction Control System 指挥舱反作用控制系统

CMRS 1. Colour Map-Reader System 彩色地图读出系统 2. Communication Moon Relay System 月球通信中继系统 3. Countermeasure Receiver System 对抗接收机系统

CMS 1. Cabin Management System 座舱管理系统 2. Central Maintenance System 中央维修系统 3. Command Module Simulator 指挥舱模拟器 4. Configuration Management Standard 构型管理标准 5. Continuous Monofilament Spun 单丝连续缠绕 6. Conversational Monitoring System 对话式监控系统

CMSTL Chrome-Molybdenum Steel 铬-钼钢

CMU Communications Management Unit 通信管理装置

CMX Change Management Exchange 更改管理交换

CN Celestial Navigation 天文导航

CN tear gas pot 苯氯乙酮催泪毒气罐

CNA Communications Network Architecture 通信网结构

CNAS 1. Change Number and ADCN Status 更改号和先行图纸更改通知状态 2. Civil Navigation Aid System 民用导航设施系统

CNATRA Chief of Naval Air Training 海军航空兵训练主任

CNC 1. Communication Network Control 通信网络控制 2. Computer Numerical Control 计算机数字控制

CNCL Concealed 隐蔽的,遮盖的

CNCP Control of Non-Conforming Product 不合格产品控制

CNCTRC Concentric 同心的,共轴的

CNCS Central Navigation and Control School 中央领航引导学校

CNCV Concave 凹形的

CND 1. Can Not Duplicate 不能重复,无法再现 2. Conduit 导管,导线管

CND/RTOK Could Not Duplicate, Retest OK 不能再现,重新试验良好

CNDB Customized Navigation Database 定制导航数据库

CNDCT Conductivity 导电性

CNDS Condensate 冷凝,凝缩

CNI Communication, Navigation, Identification 通信、

导航、识别

C-NITE Cobra-Night Version "眼镜蛇"夜视型

CNR Carrier to Noise Ratio 载波噪声比

CNS 1. Common Nacelle System 共用短舱系统 2. Communications, Navigation and Surveillance 通信导航警戒 3. Communications Network System 通信网络系统 4. Control vet System 控制网体系

CNS shell 催泪弹

CNSLD Consolidate 巩固,加固,压实

CNSR Comet-Nucleus Sample-Return Mission 彗核取样返回任务

CNTD Contained 包括的,含有的

CNTFGL Centrifugal 离心的

CNTOR Contactor 接触器

CNTR Counter 计数器,计算器

CNTRL Control 控制,操纵

CO 1. Central Office 中心局 2. Change Order 更改命令 3. Checkout 检验 4. Cleanout 清除 5. Custom Option 客户选择 6. Cutout 切断,中止,断路

CO laser 一氧化碳激光器

COA 1. Certificate Of Airworthiness 适航证 2. Change Order Authorization 更改指令批准 3. Circle of Ambiguity 模糊圈

COPL Copper Plate 铜板

coach 1. 客机二等舱 2. 汽车,客车 3. 座椅,卧榻

coagulation 凝聚,凝结

coagulation kernel 凝固内核

coal dust 煤尘,煤粉

coal particle 煤颗粒

coal-derived methanol 煤基甲醇

coalesce 合并,联合,接合

coalescence 1. 接合,结合 2. 合并,联合

coalescent bag 凝聚袋

coalescing filter 凝聚过滤器

co-alignment 组合对准

co-altitude 天顶距,余高度

coaming 开口边缘

Coanda effect 康德效应,贴壁效应

Coanda flap 孔安达襟翼

coannular inverted nozzle 同心倒置喷管

coarse aiming 粗瞄

coarse align 粗校准,粗校直

coarse alignment 粗对准

coarse ammonium 粗铵

coarse ammonium perchlorate 粗高氯酸铵

coarse droplet 粗液滴

coarse fine control 粗-精控制

coarse grid 粗栅格,粗网络

coarse grid level 粗网格层

coarse gyrocompass alignment 陀螺罗经粗对准

coarse mesh 粗网格,粗筛

coarse mesh volume 粗网格体积

coarse mode 粗模态

coarse orientation 粗定向,粗瞄

coarse oxidizer 粗氧化剂

coarse particle 粗分子

coarse pitch 1. 高距,大距 2. 粗螺距,大螺距

coarse reading 粗读数

coarse search 粗搜索

coarse system of bearing 方位粗测系统

coarse/acquisition code 粗码/捕获码

coarse/acquisition signal 粗测/捕获信号

coarsen 变粗糙

coarsening 1. 变大距 2. 粗糙化 3. 降低灵敏度

COAS Crew Optical Alignment Sight 航天员光学瞄准观察器

coast 1. 滑行,惯性飞行,滑翔 2. 惯性运转,惯性飞行 3. 海岸,沿海地区

coast arc 惯性飞行弹道弧

coast down time 1. 惰转时间 2. 惯性转动时间

coast line survey 岸线测量

coast(ing) time 1. 被动飞行时间,惯性飞行时间 2. 速度记忆时间

coastal monitoring 海岸监测

coastal ocean monitor satellite system 沿岸海域监测卫星系统

coastal refraction 海岸折射

coastal surveillance radar 海军监视雷达

coastal zone color scanner 沿岸水色扫描仪

coastdown 减退,下降

coaster 惯性飞行导弹,惯性飞行火箭

coasting 惯性飞行,惯性运转

coasting chart 沿海海图

coasting phase 滑行段

coasting time 滑行时间

coasting-flight phase 滑行段

coastling up 向上惯性飞行

coastwise survey 沿岸测量

COAT Corrected Outside Air Temperature 修正的外界大气温度

coat 涂层,包覆层

coated cathode 涂层阴极

coated optics 涂层光学

coated paper tape 敷层纸带

coated powder cathode 敷粉阴极

coated yarn 包覆线

coating 1. 涂层,镀层,包覆层,保护层 2. 涂料底漆 3. 涂,镀敷上胶

coating adhesion　涂层附着力
coating contamination　涂层污染
coating crack　涂层裂纹
coating degradation　涂层退化
coating flight experimenter　涂层飞行试验器
coating for EMI shielding　电磁屏蔽涂层
coating prescription　涂料配方
coating process　涂银,涂覆
coating repeatability　涂层重复性
coating sample　涂层样品
coating spall　涂层碎片
coating stress　涂层应力
coating weight loss　涂层重量损失
coating with pattern painting　迷彩涂层
COATS　Computer Operated Automatic Test System 计算机操作自动测试系统
COAX　Coaxial 同轴的,同心的
coax connection　同轴接头
coaxial antenna　同轴天线
coaxial cable　同轴电缆
coaxial cavity　同轴空腔
coaxial flow　共轴流动
coaxial helicopter　双旋翼共轴式直升机
coaxial injection　同轴喷射
coaxial injector　同轴式喷嘴
coaxial jet　同轴射流
coaxial line　同轴线
coaxial magnetron　同轴磁控管
coaxial plasma　同轴等离子体
coaxial relay　同轴继电器
coaxial rotor　同轴式旋翼
coaxial slit　同轴缝隙
coaxial spherical rotary joint　同轴球关节,微波万向接头
coaxial spray angle　同轴喷射角
coaxial spray characteristic　同轴喷射特性
coaxial swirl　同轴漩涡
coaxial thermocouple　同轴热电偶
coaxial transister　共轴型晶体管
coaxial-line switch　同轴转换开关
cobalt　【化】钴(化学元素,符号 Co)
cobalt based alloy　钴基合金
cobalt bomb　钴炸弹
cobalt-base superalloy　钴基高温合金
co-base vacuum-pumping　共底抽真空
cobbed　油门全开
COBCRM　Cobalt-Chrome 钴铬合金
COBE　Cosmic Background Explorer 宇宙背景探测器
Cobra　"眼睛蛇"

Cobra Dane radar　"丹麦眼镜蛇"雷达
Cobra Judy radar　"朱迪眼镜蛇"雷达
cobra probe　眼镜蛇探针
cobs　钟形失真
cobwed model　蛛网模型
CoC　Certificate of Conformity 符合性审定
COC　Change Of Contract 更改合同
cochannel interference　同波道干扰
cock　1.准备击发,准备投弹,击铁待发位置,待发状态 2.开关,旋塞
cockade　机徽
cocked　起飞前试车完毕的
cocked position　1.准备投弹状态 2.待发状态
cockerel　〈口语〉机载雷达识别系统
cocking　处于待发状态,解脱保险
cocking lever　拨动杆,拨机
cocking mechanism　击发装置
cocking-button　直立式按钮
cockpit　座舱,驾驶舱
cockpit control　座舱控制
cockpit cowling　驾驶舱整流罩
cockpit defogging　座舱玻璃除雾
cockpit display　座舱显示屏
cockpit light　座舱照明灯
cockpit management system　座舱管理系统
cockpit monitoring system　座舱监控系统
cockpit pressure regulator　座舱压力调节器
cockpit procedure trainer　座舱操作程序练习器
cockpit vertigo　座舱眩晕
cockpit vibration　座舱振动
cockpit voice recorder　座舱录音机,驾驶舱录音机
cocoa powder　一种褐色火药
Cocom　Coordinating Committee for Multilateral Export Controls 多边出口控制协调委员会
Cocomo　Constructive Cost Model 成本结构模型
coconut fatty acid　可可脂肪酸
coconut oil　椰子油
cocoon　密封包装壳,塑膜披盖,防护喷层
Cocraly　防氧化涂层
co-curing　共固化
COD　1. Carrier Onboard Delivery 用航空母舰运送 2. Change Object Description 更改目标说明 3. Classification Of Defects 缺陷分类 4. Coding 编码 5. Convert On Demand 要求转换
coda　尾波
CODAN　Carrier-Operated Device Antinoise 载频控制的噪声抑制器
code　1.电码,密码 2.码,代码,符号,代语 3.编码,译成电码

code audit　代码审计,代码检查程序
code beacon　闪光灯标,电码信标
code block　码组,代码块
code computer　编码计算机
code data base　代码数据库
code diversity　编码分集
code division multiple access　码分多址
code division multiple access telemetry　码分多址遥测
code division multiplex　码分多路复用
code division multiplexing telemetry　码分制多路遥测
code error tester　误码测试仪
code extension character　代码扩充字符
code generator　码生成器,代码生成程序
code length　码长
code multipath error　码多径误差
code rate　码率,编码率,码传输速度
code simulation　代码仿真
code translation　代码变换
code translator　译码器,代码转换器
code tree　码树,代码树
code value code vector　代码值码矢
code walk-through　代码走查
code word synchronization　码字同步
codebook　码本
CODEC（codec）　Coder-Decoder　编码译码器,编码解码器
co-declination　余赤纬,极距,赤纬余角
coded address　编码地址
coded program　编码程序
coded telecommand　编码遥控
coded transmitter　编码发射机
coded virtual channel date unit　编码虚拟信道数据单元
codelta　三角翼鸭式配置型式
codem　Coded Modulator/Demodulator　编码调制器/解调器
CODEM　Comet Dust Environment Monitor　彗星尘环境监测器
codeposition　共淀积
coder　1.编码器,记发器 2.编码员
code-sending radiosonde　电码式探空仪
codeword　1.密语 2.代码字
CODI　Custom Option Default Index　客户不选项目录
coding　编码,译成电码
coding circuit　编码电路
coding cycle　编码周期
coding layer　编码层
coding technique　编码法
coding theorem　编码定理
coding tube　编码管

CODMAC　Committee On Data Management And Computation　数据管理和计算委员会
CODT　Custom Option Drawing Tree　客户选项图纸树
COE　1.Certificate Of Eligibility　合格证书 2.Commitment Out of Engineering　工程部门委托
COEA　Cost and Operational Effectiveness Analysis　成本和实际效果分析
COEF　Coefficient　系数,折算率
coefficient　1.系数 2.协同因素,程度
coefficient matrix　系数矩阵
coefficient multiplier　系数乘法器
coefficient of coupling　耦合系数
coefficient of lift　升力系数
coefficient of secondary energy　次要功计算系数
coefficient of sectorial harmonics　扇谐系数
coefficient of squeezed riveting　压铆系数
coefficient of stray light　杂光系数
coefficient of tesseral harmonics　田谐系数
coefficient of thermal expansion　热膨胀系数
coefficient of viscosity　黏性系数
coefficient of zonal harmonics　带谐系数
coefficient unit　系数单位
coefficient vector　系数向量
coelonavigation　天文导航
coelostat　定天镜
coercibility　压凝性,可压缩性
coercive force　矫顽力
coevaporation　共蒸发
coexist　共存
coexistence curve　共存曲线
COF　Centrifugal Oil Filter　离心滑油滤
coffin　1.长箱形发射掩体 2.运送放射性物质的重屏蔽容器,铅箱
coffin-based　1.在长箱形发射掩体内储藏的 2.在地面发射阵地上展开的
cog　嵌齿,嵌齿轮
co-geoid　调整大地水准面,共大地水准面
cogging　1.榫 2.装齿轮装榫
cognitive behavior　认知行为
cognitive engineering　认知工程
cognitive mapping　认知制图
cognitive process　认知过程
cognitive psychology　认知心理学
cognitive science　认知科学
cognitive workload　认知工作负荷
cognitron　认知机
cognizance　1.识别,判别,观察 2.监督权
COGS　Continuous Orbital Guidance System　连续轨道制导系统

COGT　Center of Gravity Towing　在重心处牵引
cogwheel　1. 嵌齿轮 2. 大齿齿轮
cogwheel breathing　断续性呼吸，间断呼吸
coherence　1. 同调 2. 相干性，相关性 3. 黏附，附着
coherence emphasis　相干加强
coherence function　相干函数
coherence stack　相干叠加
coherent　1. 相干的，相参的 2. 黏着的 3. 连贯的
coherent acoustic Doppler radar　相干声波多普勒雷达
coherent boundary　共格界面
coherent carrier　相干载波
coherent derived unit of measurement　1. 一贯导出测量单位 2. 一贯单位
coherent detection　相干检测，相干探测
coherent detector　相干检波器，相干检测器
coherent echo　相干回波
coherent grain boundary　共格晶界
coherent integration　相干积累
coherent noise　相干噪声
coherent processing system　相干处理系统
coherent pulse　相关脉冲
coherent pulse Doppler navigation radar　相干脉冲多普勒导航雷达
coherent pulse radar　相干脉冲雷达
coherent pulsed-radar seeker　相干脉冲雷达导引头
coherent radar　相干雷达
coherent receiver　相干接收机
coherent receiving　相干接收
coherent scattering　相干散射
coherent structure　凝聚结构
coherent system　单调关联系统，相干系统
coherent system of unit (of measurement)　一贯单位制
coherent transmission　同调传输
coherent transponder　相参应答机
coherent vortex structure　相干涡结构
coherer　粉末检波器
cohesion　内聚度，内聚性
cohesion of film　胶片粘连
cohesive failure　内聚破坏
cohistory register　并列经历寄存器
coho　相参振荡器，相干振荡器
COHOE　Computer-Originated Holographic Optical Element 用计算机制成的全息光学元件
COI　Communications Operating Instructions 通信操作指令
coil　1. 线圈，绕组，线卷 2. 卷片筒 3. 蛇形管，盘管，螺旋管
coil center　卷钢中心
coil current　线圈电流
coil gun　线圈炮
coin-box paystation　投币式公用电话
coincide　恰好相合，一致
coincidence　符合，重合，一致
coincidence adder　重合加法器，组合加法器
coincidence circuit　重合电路，"与"电路
coincidence counter　符合计数器，重合计数器
coincidence counting　符合计数
coincidence effect　吻合效应
coincidency　重合度，一致性
coincident gate　符门
coining　压印加工，压花
COINS　1. Computer Operated Instrument System 计算机制仪表系统 2. Cooperative Intelligence Network System 合作智能网络系统
coke　1. 可口可乐瓶状的（机身）2. 将飞机做成细腰式机身底积碳，焦炭
COL　Column 汞柱，纵列，座架，表格栏，驾驶杆
col　气压谷，鞍形区
colatitude　余纬度
colatitude angle　余纬角
cold acclimatization　冷习服
cold acoustic test　冷声试验
cold black (background)　冷黑
cold boundary　冷却边界
cold bucket　冷空气斗
cold cathode　冷阴极
cold cathode sensor　冷阴极传感器
cold circuit　冷电路
cold cordite charge　冷态无烟火药装药
cold drawing　冷拔，冷削，冷拉
cold drop　冷投
cold end　冷接点，冷电位端
cold engine　冷发动机
cold exhaustion　冷衰竭
cold exposure　冷暴露
cold flow　冷流
cold flow condition　冷流状态
cold fuel　冷燃油
cold gas　冷气
cold gas pressurization system　冷气增压系统
cold gas system　冷气系统
cold gun　冷炮
cold gun correction　冷炮校正
cold gun error　冷炮误差
cold head　冷头
cold helium connector　冷氦连接器
cold helium connector support mount　冷氦连接器支架
cold ignition　冷态点火

cold ion	冷离子
cold jet testing	冷喷流试验
cold launch	冷发射
cold light	冷光
cold plasma	冷等离子体
cold plate	散热板,冷板
cold plug	冷电嘴
cold plume	冷焰
cold proof	冷验证
cold run	冷运转
cold setting	冷固化
cold shield	冷屏
cold shut	冷砂眼,冷隔,冷疤
cold side	冷侧
cold soak	冷渗,冷透
cold space compensation	冷空间补偿
cold spare	冷备份
cold spring	冷泉
cold stage	冷台
cold start	冷起动
cold start test	冷起动试验
cold startup	冷态启动
cold stop	冷站
cold strain	冷应变
cold stress	冷应激
cold structure	冷结构
cold temperature	低温区温度
cold test	1.冷凝试验 2.低温试验
cold tolerance	1.耐冷限 2.冷耐限
cold trap	冷阱
cold wall reactor	冷壁反应器
cold water immersion	冷水浸泡
cold welding	冷焊
cold window	冷窗
cold working	冷加工,冷作
cold-air unit	冷气机
cold-chamber test	低温舱试验
cold-cranking amps	冷起动电流,冷起动安培数
cold-fan thrust reverser	风扇冷气流反推装置
cold-flow	冷流,冷塑加工,冷变形
cold-gas	低温气体
cold-gas simulation	冷气模拟
cold-soak	冷浸
cold-solder joint	冷焊点,虚焊
cold-work	冷作,冷加工
coleopter	环翼机
collaboration	协调服务
collapse	1.倒塌,毁坏,压扁 2.折叠 3.虚脱
collapse load	损毁荷载
collapse test	破坏性试验
collapsible	可拆散的,可折叠的
collapsible helicopter	旋翼可折式直升机
collapsible radar reflector	折叠式雷达反射体
collapsing ratio	重叠比,倒塌率
collar	1.圈,环,箍,卡圈 2.凸缘,法兰盘 3.衣领
collate	1.核对,校对,对照 2.分类整理
collateral damage	旁及破坏
collation	1.核对,校对 2.整理
collect	搜集,收取
collect on delivery	货到付款
collection	1.收集,搜集,采集 2.集成,收集物
collection efficiency	收集效率
collection optics	会聚光学
collective	1.集体的,集合的,聚合性的 2.汇流的 3.油门变距杆,旋翼总距
collective address group	集址组
collective behaviour	集约行为
collective dose equivalent	集体剂量当量
collective motion	集体运动
collective oxygen regulator	集体用氧调节器
collective pitch	总距,油门变距
collective pitch angle	总距角
collective pitch stick	1.总距操纵杆 2.总距-油门杆
collective report	综合报告
collective stick	油门变距杆,总距操纵杆
collective-effect accelerator	集合效应加速器
collector	1.收集器集合器,采集器 2.整流子,换向器,集流环,集气管 3.集电极,集电器
collector area	集极面积
collector electrode	集电极
collector junction	集电结
collector plate	集电板
collector region	收集区
collector ring	集气环
collector voltage	集电极电压
collide	碰撞,猛撞
colliding galaxy	碰撞星系
collimate	1.瞄准,校准 2.调整,瞄准线 3.准直,平行校正
collimating head	视准头部
collimating lens	准直透镜
collimating light	平行光
collimating mark	框标,测标,照准标
collimation	1.准直 2.瞄准,校准,调整 3.平行校准
collimation axis	视准轴,照准轴
collimation error	准直误差
collimation line method	视准线法
collimator	准直仪,视准仪,平行光管

collimatorsight 视准式瞄准具,平行光管式瞄准具	collocated check 共点检核
collinear check 共线检核	collocation 安排,配置
collinear equilibrium 共线平衡	collocation architecture 搭配架构
collinear libration 共线平动	collodion 胶棉,火棉胶
collinear libration point 共线平动点	colloid 1.胶体,胶质 2.胶态
collinear point 共线点	colloid beam 胶体梁
collinearity 共线性	colloid thruster 胶体推力器
collinearity condition 共线条件	colloidal propellant 胶体推进剂
collinearity equation 共线方程	colloidal stability 胶体稳定性
collision 1.碰撞冲突 2.相遇,击中目标	colonel 1.陆军或海军陆战队上校,陆军,空军或海军陆战队上校 2.中校
collision angle 碰撞角	
collision avoidance 防撞	colophony 松香
collision avoidance constraint 防撞约束	color aberration 色差,颜色失真
collision avoidance functionality 防撞功能	color absorber 消色器
collision avoidance maneuver 防撞机动	color aerial film 彩色航空胶片
collision avoidance system 防撞系统	color bar signal 彩条信号
collision beacon 防撞灯	color body 彩色体
collision broadening 碰撞展宽	color burst 色同步脉冲
collision condition 冲突条件	color cell 色元
collision-course command guidance 相遇航向指令制导	color center 色心
collision course 碰撞航向	color center laser 色心激光器
collision detection 碰撞检测	color chart 地图色标
collision diameter 碰撞直径	color compensating filter 彩色补偿滤光镜
collision energy 碰撞能量	color composite image 彩色合成图像
collision event 碰撞事件	color constancy 色恒定性,色守恒现象
collision frequency 碰撞频率	color contour 彩色等值线图
collision hazard 碰撞危险	color density slicing 彩色密度分割
collision interception 相遇截击	color difference signal 色差信号
collision model 碰撞模型	color discrimination 颜色辨别
collision parameter 碰撞参数	color display 彩色显示
collision peak 碰撞峰值	color distortion 彩色畸变,彩色失真
collision probability calculation 碰撞概率计算	color enhancement 彩色增强
collision radius 碰撞半径	color field 色场
collision rate 碰撞率,碰撞强度	color film 彩色胶片
collision region 冲突区域	color grid 色栅
collision risk 碰撞危险	color hue 色调,色彩
collision threat 碰撞威胁	color infrared film 彩色红外胶片
collision trajectory 与目标碰撞弹道,相遇弹道	color kinescope 彩色显像管
collisional bremsstrahlung 碰撞韧致辐射	color layers 分层设色
collisional damping 碰撞阻尼	color manuscript 彩色样图
collisional deactivation 碰撞去活化	color matching 配色
collisional energy 碰撞能量	color picture tube 彩色显像管
collisional kinetic energy 碰撞动能	color proof 彩色校样
collisional process 碰撞过程	color purity allowance 色纯度容差
collision-free path 无碰撞路径	color reproduction 彩色复制
collision-free trajectory 无碰撞轨迹	color reversal film 彩色反转胶片
collisionless damping 无碰撞阻尼	color saturation 色饱和度
collisionless plasma 无碰撞等离子体	color scanner 电子分色机
collocate 设在同一位置	color schlieren sytem 彩色纹影仪

color segmentation 颜色分割
color sensitive material 彩色感光材料
color sensitivity 感色度
color separation 分色
color separation drafting 分色绘制
color separation filter 分色滤光镜
color spin scan cloud camera 彩色自旋扫描摄云机
color temperature 色温
color transcode 彩色电视制式转换
color transformation 彩色变换
color TV 彩色电视
coloration flow method 色流法
colored noise 有色噪声
colorimeter 比色计,色度计
colorimetry 色度学
coloring 着色
colorpexer 彩色编码器
color-ratio composite image 比值彩色合成影图像
colortron 荫罩式彩色显像管
Colossus "巨人"作战
colour 1.颜色,色彩,颜料 2.彩色标志,彩旗 3.色码 4.着色,染色
colour adaptation 色彩适应
colour adder 彩色叠加电路
colour additive viewer 彩色合成仪,加色法观察仪
colour display 彩色显示
colour excess 色余
colour flicker 彩色闪烁
colour index 色指数
colour infrared 彩色红外
colour temperature 色温度
colour-colour diagram 两色图
coloured wand 彩色棒
colour-magnitude diagram 颜色星等图
COLTS Contrast Optical Laser Tracking Subsystem 光学对比激光跟踪分系统
Columbia 哥伦比亚号
Columbus attached laboratory 哥伦布附属实验室
Columbus free flying laboratory 哥伦布自由飞行实验室
Columbus polar platform 哥伦布极轨平台
column controller 列控制器
column decoder 列译码器
column electrophoresis 柱电泳法
column sum 列和
columnar structure 柱状组织
colure 分至圈
COLPS Collapse 倒塌,失败,事故
COM 1.Cockpit Operating Manual 驾驶舱操作手册 2.Common 公共的 3.Component Object Model 组件对象模型 4.Computer Output Microfilm 计算机输出缩微胶片 5.Computer Output on Microfilm 计算机输出到缩微胶卷 6.Conceptual Object Model 概念目标模型
COM/MET/OPS Communications / Meteorological / Operations 通信/气象/运行
COMA Cometary Matter Analyzer 彗星物质分析器
coma 1.彗形像差 2.彗发 3.昏迷
coma aberration 彗差,彗形像差
COMAAS Committee On Meteorological Aspects of Aerospace System 航空-空间系统气象学委员会
COMASWWINGPAC Commander Antisubmarine Warfare Wing, Pacific 太平洋反潜机联队司令
COMAT Computer-Assisted Training 计算机辅助训练
COMB Combination 联合,组合
comb 1.梳 2.排管 3.螺纹梳刀 4.梳状天线
comb filter 梳齿滤波器
COMBAR Combat Aircraft Meteorological Report 战斗机气象报告
combat 1.战斗,作战,格斗 2.反对,斗争
combat aircraft 作战飞机
combat camera 照相枪
combat documentation motion camera 攻击照相枪(直接拍摄目标记录攻击效果)
COMBAT EDGE Combined Advanced Technology Enhanced Design G-Ensemble 综合先进技术高过载防护配套装置
combat engagement 作战任务
combat experience 战斗经验
combat fuel allowance 战斗油量标准
Combat Grande 大战斗防空系统
combat load 战斗装载
combat load displacement 战斗装载排水量
combat missile telemetry 战斗弹遥测
combat mission 战斗任务
combat model 战斗模式
combat operations center 作战指挥中心
combat plug 战斗冲刺
combat readiness 作战状态,战斗状态
combat ready 作好战斗准备的,处于战备状态的
combat-serviceable 作好战勤准备的
combat surveillance radar 战场监视雷达
combat training launch 战术训练发射
combat turn 1.战斗转弯 2.急上升转弯
combatant 1.战斗人员,战斗单位 2.战斗的,用于战斗的
combat-capable 有战斗力的,能战斗的
combi 客货两用运输机
combination 1.组合,并合,结合,混合,化合 2.组合体,

结合体
combination actuator 复合致动装置
combination aircraft 客货混合型飞机
combination control system 复合控制系统
combination decision 组合决策
combination fuze 复式引信
combination interference 组合干扰
combination of factor level 因素水平组合
combination of force 力的合成
combination of parameter 参数组合
combination of sensor 传感器组合
combination of thruster 组合推进器
combination of uncertain parameter 不确定参数组合
combination propulsion 组合式推进
combination rule 组合规则
combination storage 弹药综合储存
combination valve 复合阀门
combinatorial explosion 组合爆炸
combind training 合练
combine 兼有,使结合
combined addressing path 组合寻址通路
combined adjustment 联合平差
combined altimeter 组合式高度表
combined command 组合指令
combined company 合并公司
combined compressor 组合式压气机
combined cooling 复合冷却
combined countervelocity 反导弹导弹接近目标速度
combined cycle 联合循环,组合循环
combined cycle engine 组合循环发动机
combined display 组合显示器
combined effect 混合效应
combined electrolysis 联合电解
combined engine 组合发动机
combined environment reliability test 综合环境可靠性试验
combined environment test 综合环境试验
combined error 综合误差
combined fuze 复合引信
combined guidance 复合制导
combined guidance assembly 复合导引装置
combined heat treatment 复合热处理
combined injection 联合注射
combined injection scheme 联合注射方案
combined joint 联合的,协同的
combined lots 组批
combined maneuver 联合演习
combined method 组合方法
combined method solution 组合方法解决方案

combined mode 兼容模式
combined module 组合模块
combined motion 复合运动
combined navigation 组合导航
combined pressure and vacuum gauge 压力真空表
combined radiation 综合辐射
combined seal 复合密封
combined sketch 拼接草图
combined standard uncertainty 合成标准不确定度
combined stress 1.复合应力 2.复合刺激
combined stresses 复合应力
combined-cycle 联合循环
combined-cycle engine 联合循环发动机
combined-effects munition 综合效应弹药
combiner 1.组合器 2.组合玻璃 3.合成仪
combining 混合,组合
combining gearbox 联合减速器
combining glass display 反光玻璃显示,瞄准镜显示,组合玻璃显示
combining glass mirror 瞄准镜,反光镜
comburant 燃料
COMBL Combustible 易燃的
combust 燃烧
combustibility 可燃性,易燃性
combustible 1.可燃的,易燃的 2.可燃物,易燃物,燃料
combustible gas 可燃气体
combustible mixture 可燃混合气
combustibleness 可燃性
combustion 1.燃烧 2.氧化 3.骚动
combustion air 燃烧空气
combustion behavior 燃烧特性
combustion blockage 燃烧堵塞
combustion chamber 燃烧室
combustion chamber length 燃烧室长度
combustion chamber pressure 燃烧室压强
combustion chamber pressure roughness 燃烧室压力不稳定度
combustion chamber volume 燃烧室容积
combustion characteristic 燃烧特性
combustion condition 燃烧条件
combustion control 燃烧控制
combustion cycle 燃烧周期
combustion device 燃烧装置
combustion duct 燃烧排气管道
combustion dynamics 燃烧动力学
combustion effectiveness 燃烧完全系数
combustion efficiency 燃烧效率
combustion energy 燃烧能
combustion engine 燃烧发动机

combustion engineer 燃烧工程师
combustion enhancement 燃烧强化
combustion event 燃烧事件
combustion experiment 燃烧实验
combustion facility 燃烧设备
combustion field 燃烧场
combustion flame 燃烧火焰法
combustion flowfield 燃烧流场
combustion front 燃烧前沿,燃烧锋面
combustion gas 燃烧气体
combustion gas temperature 点火温度
combustion heat 燃烧热
combustion heat release 燃烧热释放
combustion instability 燃烧不稳定性
combustion intensification 燃烧过程强化
combustion intensity 燃烧强度
combustion irreversibility 燃烧不可逆性
combustion kinetics 燃烧动力学
combustion limit 燃烧极限
combustion loss 燃烧损耗
combustion mechanism 燃烧机理
combustion mode 燃烧模式
combustion model 燃烧模型
combustion noise 燃烧噪声
combustion noise peak 燃烧噪声峰值
combustion of heavy fuel oil 重质燃料油燃烧
combustion of hydrogen 氢燃烧
combustion oscillation 燃烧振荡
combustion parameter 燃烧参数
combustion performance 燃烧性能
combustion period 燃烧期间
combustion port 燃烧口
combustion pressure 燃烧压力
combustion process 燃烧过程
combustion product 燃烧产物
combustion ramjet 燃烧冲压发动机
combustion rate 1.燃烧率 2.燃烧速度
combustion reaction 燃烧反应
combustion region 燃烧区
combustion response 燃烧反应
combustion simulation criteria 燃烧模化准则
combustion source 1.火源 2.火箱
combustion stability 燃烧稳定性
combustion stability limit 稳定燃烧边界
combustion stability margin 燃烧稳定裕度
combustion system 燃烧系统
combustion tap-off cycle 抽气循环
combustion temperature 燃烧温度
combustion test 燃烧试验
combustion time 燃烧时间
combustion wave 燃烧波
combustion wind tunnel 燃烧风洞
combustion zone 燃烧区
combustion-flow model 燃烧流模型
combustion-induced pressure 燃烧诱发压力
combustor 燃烧器,燃烧室
combustor design 燃烧室设计
combustor dynamics 燃烧室动力学
combustor exit gas temperature 燃烧室出口燃气温度
combustor hot streak 燃烧室热斑
combustor inlet 燃烧室进口
combustor inlet condition 燃烧室入口条件
combustor instability 燃烧室的不稳定性
combustor interaction 燃烧室交互
combustor length 燃烧室长度
combustor outlet 燃烧室出口
combustor performance 燃烧室性能
combustor pressure 燃烧室压力
combustor pressure sensor 燃烧室压力传感器
combustor shape 燃烧室构型
combustor size 燃烧室尺寸
combustor stability 燃烧室的稳定性
combustor test 燃烧室试验
combustor volume 燃烧室体积
combustor wall 燃烧室壁
combustor wall heat transfer 燃烧室壁传热
COMCARGRU Commander Carrier Group 航母飞行大队队长
COMDAC Command, Display and Control 指挥,显示与控制
COMED Combined Map and Electronic Display 组合式地图与电子显示仪
comercial support 商业保障
COMET 1. Commercial Experiment Transporter 商用实验运输器 2. Computer Message Transmission 计算机信息传输
comet 彗星
comet probe 彗星探测器
comet rendezvous and asteroid flyby 彗星交会和小行星飞越探测器
cometail 彗星拖尾
cometary head 彗头
cometary nebula 彗状星云
cometary nucleus 彗核
cometary tail 彗尾
comfort chart 舒适度曲线图
comfort temperature 舒适温度
COMFAIR Commander Fleet Air 舰队航空部队司令

comfort zone 舒适区
coming-in speed 进入速度
coming-out 1.退出,改出(机动) 2.停止(工作),关闭,断开
COMINT Communications Intelligence 通信智能化
COML Commercial 商业的
COMLO Compass Locator 导航台
COMM 1.Common 公共的 2.Communication 通信
Comm Center Teletypewriter 通信中心电传打字机
comma rudder 逗号形方向舵
comma-free code 无逗点密码
command 1.指挥,指挥权 2.命令,口令 3.司令部,指挥部,空军一级的单位名称
command airspeed 目标空速,规定空速
command altitude readout window 预定高度读出口
command and control center 指挥控制中心
command and control station 指挥控制站
command and control system 指挥控制系统
command and data acquisition 指令与数据采集
command and data acquisition station 指令和数据采集站
command and telemetry system 指令和遥测系统
command arbitration 指令仲裁方
command area 指挥区
command aviation net 作战航空兵通信网
command board 命令板
command capacity 指令容量
command chain 指令链
command code 指令码,操作码
command console 指挥操纵台,控制台
command continual transmission 指令连发
command control and monitor system 指令控制和监测系统
command data acquisition (导弹)指令数据获取
command data buffer system (导弹)指令数据缓冲(存储)器系统
command demodulation 指令解调器
command destruct (火箭)指令自毁(一种靶场安全措施)
command destruct receiver (导弹)自毁指令信号接收机
command detonation 指令起爆
command dispatching equipment 指挥调度设备
command dispatching hierarchy 指挥调度体制
command dispatching system 指挥调度系统
command duty officer (航母)指挥(任务)值班军官
command ejection 指令弹射(由一名飞行员点火,所有飞行员的弹射椅皆自动击发)
command electronics 指令控制系统电子设备
command encoder 指令编码器

command encoding 指令编码
command error 指令误差
command execution unit 指令执行机构
command format 指令格式
command function 指令功能
command fuze 指令引信
command generator 指令产生器,指令产生程序
command guidance 指令制导
command guidance interrogator 指令制导询问器
command guided missile 指令制导导弹
command information 指令信息
command interval 指令间隔
command language 命令语言
command language mode 指令语言式
command length 指令长度
command level 1.命令水平 2.指挥层次 3.命令等级
command limiter 命令限幅器
command link control word 指令链路控制字
command link test 1.指令线路测试 2.指挥线路测试
command link transmission unit 指令链路传输单元
command list 指令表,命令列表
command memory 指令存储器
command model 命令模式
command module 指令舱,指挥舱
command module pilot 指令舱驾驶员
command monitoring 指令监测
command operation procedure 命令操作步骤
command pose 指令位姿
command processor 命令处理器
command receiver 指令接收机,指令接收器
command reference symbol 指令参考符号,指令基准符号
command remote control 指令遥控
command response 命令响应
command retransmission 指令重发
command sequencer 指令程序机构,指令程序装置
command service module 指挥服务舱
command signal 指令信号,控制信号
command storage 指令库
command system 指令系统,命令系统
command system of guidance 制导指令系统
command telephone 指挥电话
command terminal 指令终端
command time delay 指令时延
command to line of sight 瞄准线指令
command transmission 指令传输
command transmitter 指令发射机
command tunnel 操控指令通道
command wire guidance 有线制导

command word　指令字
command，control and communication ship　（船）指挥、控制与通信舰
commandant　1.指挥官,司令官 2.(军事学校)校长
commanded value　给定值,规定值
commander　1.指挥长,指令长,司令 2.海军中校
Commander Air　(航母)航空部门长(中校)
Commander task force　命令特遣部队
commanding officer　舰长(航母及其他军舰)
commando　突击队,袭击队员
commando helicopter operations support cell　登陆突击直升机作战保障分队
commando squadron　登陆袭击中队
commands　【计】命令,指令
command-system path　指令系统制导弹道
command-type guidance system　指令制导系统
commensurate orbit　通约轨道
commerce　1.商业,贸易,(飞机的)商业用途 2.投入市场的工业产品
commercial aircraft　商用飞机,营业飞机(指被雇用或为营利而飞行的)
commercial airline　商业航空公司
commercial airport　商业机场
commercial aviation　商用航空
commercial code　商法典
commercial communication　商用通信
commercial communication satellite　商业通信卫星
commercial communication service　商业通信业务
commercial electrics　服务性电气设备(如座舱照明等)
commercial engine　民用发动机
commercial fleet　民航飞机
commercial launch　商业发射
commercial mission　商业代表团
commercial pilot　商业飞行员
commercial risk　商业风险
commercial satellite　商业卫星
commercial service　商业服务
commercial space　商业空间,商用地方
commercial spacecraft　商用航天器
commercial transport　商业运输
commercial vehicle　商用运载器
commercial version　商用版
commingler　混合器,搅拌器
commissary　军粮库,给养部门
commissary line　补给线
commission　1.授衔令,所授军衔,任职令 2.委员会 3.委托,委任 4.代办(权),代理(权) 5.编入现役,开始服役
commissioned　投产,投入使用
commissioning　投入运行
commit　1.投入截击,使部队投入战斗 2.使承担义务,使作保证
commitment　1.由上级司令部分配的军事空运任务 2.承担义务,许诺
committee　委员会,全体委员
commodity loading　商品装载法(各类物资混装在一起但在卸载时又不致相互影响的装载方法)
commodore　1.海军准将 2.分遣舰队指挥官 3.海岸警卫队准将
common battery　中央电池组,共电制电池
common bulkhead　共用隔板,共用箱底,共底(串联配置的两推进剂储箱的共用箱底)
common cause fault　共因故障
common channel interoffice signalling　共路局间信令
common channel signalling　共路信令
common component　共同成分
common denominator　公分母
common depth point grid　共深度点网格
common design　通用设计
common-flow afterburner　(带)掺混(室的)加力燃烧室(风扇发动机两个函道气流先掺混再复燃)
common integrated processor　公共综合处理器
common mark　共用标志
common method　常用方法
common metric　对照两组的共同量尺
common mid-point stacking　共中心点叠加
common mode fault　共模故障
common module　公用舱
common nozzle　通用喷嘴
common part　1.公用部分 2.一般部件
common platform　公用平台
common pointed shell　普通尖弹
common practice　惯例,习惯作法
common problem　常见问题
common servicing　通用服务
common shell　普通炮弹
common signalling channel　公用信令信道
common subsystem　公用子系统
common termination system　通用终端系统
common year　平年
commonality　1.共同性 2.通用性(指部件或零件可通用于同一工厂生产不同系列的飞机或发动机)
common-depth-point stacking　共深度点叠加
common-mode rejection　共模抑制
common-mode rejection ratio　共模抑制比
common-user airlift　通用空运
communicate　传达,传送
communicating protocol　通信协议

communication　1.通信 2.通信设备,通信系统 3.交通,联络
communication and data management　通信和数据管理
communication channel　通信信道,通信电路
communication command and terminal system　通信命令和终端系统
communication control processor　通信控制处理机
communication control unit　通信控制器
communication countermeasures　通信对抗
communication coverage　通信覆盖
communication efficiency　通信效率,编码效率
communication link　通信线路,通信连接装置
communication map　交通图
communication network　通信网
communication network for space flight test　航天测控通信网
communication network management system　通信网管理系统
communication payload　通信有效载荷
communication protocol　1.通信规约 2.数据通信协议
communication satellite　通信卫星
communication satellite coverage　通信卫星覆盖范围
communication services　通信业务
communication station　通信站
communication subsystem　通信分系统
communication system　通信系统
communication transponder　通信转发器
communication,command,control and intelligence system　通信、指挥、控制与情报系统
communications and data handling　通信和数据处理
communications and system management　通信及系统管理
communications and tracking　通信与跟踪
communications and tracking subsystem　通信与跟踪子系统
communications module　通信舱
communications relay aircraft　通信中继飞机
communications satellite program　通信卫星计划
communications test lab　通信试验实验室
communications transponder　通信转发器
communications-electronics　通信电子学
communications-electronics cover and deception　通信电子伪装与诱骗
communications-electronics operation instructions　通信电子作业指令
communications-electronics standing instruction　通信电子现行指令
communicator　1.通信装置,通信设备 2.通信员
community reception　集体接收

commutated Doppler　波束转换多普勒雷达(一种微波着陆系统的)
commutation　1.交换,变换 2.换向,整流
commutator　1.交换机,互换机,交换子 2.转换器,转换开关
commutator bar　整流子铜条,整流条
commuter　长期车票使用者,持月票乘客
commuter aircraft　定期短途班机
commuter airline　支线航空公司
COMNAV　Communication and Navigation 通信与导航
COMNAV AIRSYSCOM　Commander Naval Air Systems Command 海军航空兵指令系统部部长
COMNAVAIRLANT　Commander Naval Air Forces Atlantic Fleet 大西洋舰队海军航空兵司令(美国)
COMNAVAIRPAC　Commander Naval Air Forces Pacific Fleet 太平洋舰队海军航空兵司令(美国)
COMO　Combat-oriented Maintenance Organization 作战维修组织
COMOC　Centralized Maintenance and Operation Center 集中维修运行中心
COMOPTEVFOR　Commander Operational Test and Evaluation Force 作战试验与鉴定部队司令
COMP　1.Compass 罗盘 2.Compensator 补偿器 3.Compressor 压气机 4.Computer 计算机
COMPA　Compressed Air 压缩空气
compact flare　致密耀斑
compact galaxy　致密星系
compact object　致密天体
compact radio source　射电致密源
compact set　紧集
compact source　致密射电源
compact star　致密星
compact subset　紧子集
compact wave　压实波
Compacta tyre　紧实轮胎(邓洛普公司生产的一种减小直径,加大宽度的飞机轮胎)
compacted powder　压实粉末
compactness　1.坚实,紧凑,紧密 2.紧密度,坚实度,密度,比重
compactron　小型电子管
compander　压缩扩展器,压伸器
companding　压缩扩展,压伸
companion body　随伴体,伴体
companion galaxy　伴星系
companion matrix　伴随矩阵
companion model　伴随模型
companion star　伴星
companion volume　成套书中的一卷
comparable computer-vision system　可比计算机视觉

系统
comparable performance 可比性能
comparative advantage 比较优势
comparative analysis 对比分析
comparative cartography 比较地图学
comparative cover 比较性空中照相（不同时间对同一目标所作的空中照相，用以发现该目标的变化）
comparative lifetime 相对寿命
comparative planetology 比较行星学
comparator 比较器，比较仪
comparing element 比较元件
comparison 比较，对照
comparison (measuring) instrument （测量）比较式仪器
comparison measurement 比较测量
comparison method 比较法
comparison of aeroelastic response 气弹响应比较
comparison parameter 比较参数
comparison spectrum 比较光谱
comparison star 定标星（卫星摄影底片上选取的背景恒星），比较星
comparison survey 联测比对
comparison with adjacent chart 邻图拼接比对
compartment 隔舱，隔间，室，舱
compartment marking 隔舱标线
compartmental model 房室模型
compartmentalization 隔舱化
compartmentation 间隔化，分成隔舱，隔舱化
compass 1. 罗盘 2. 圆规，两脚规 3. 界限，范围
compass acceleration error 罗盘加速度误差
compass adjuster 罗差校正器，罗差修正器
compass azimuth 罗盘方位角，罗方位角
compass base 罗差校正坪
compass bearing 罗方位
compass calibration 罗盘标定
compass compensation 罗盘补偿，罗经校正
compass course 罗盘航向，罗经航向
compass deviation 罗航偏差，罗盘偏差
compass deviation compensation 罗差补偿
compass error 罗盘误差
compass heading 罗航向
compass points 基点，方位基点（指东、西、南、北）
compass prime vertical 罗盘东西圈，罗盘卯酉圈
compass rose 罗盘记录盘，罗经盘
compass swing 罗盘旋转
compass system 罗盘系统
compass theodolite 罗盘经纬仪
compass turntable 转动式罗差修正器
compass variation 磁差，磁偏
compassing 定向

compatibility 1. 相容性（如进气道与发动机的匹配情况），兼容性，适合性，适应性，并存性 2. （油料的）可混性，（电视机的）两用性（指黑白与彩色信号可以兼收）
compatibility constraint 相容约束
compatibility mass matrix 相容质量矩阵
compatibility principle 相容原理
compatibility relation 相容关系
compatibility test 兼容性试验
compatibility test unit 兼容性试验设备
compatible discriminant constraint 兼容判别约束
compatible duplex system 兼容双工系统
compatible storage 弹药综合储存，相容存储
compatible time-sharing system 兼容分时系统
COMPATWING Commander Patrol Wing 巡逻机联队长
compensate 1. 补偿，平衡 2. 赔偿
compensated acidosis 代偿性酸中毒
compensated alkalosis 代偿性碱中毒
compensated cam 补偿凸轮
compensated cavity 辅助腔
compensated linear vector dipole 补偿线性向量偶极
compensated pitot-static tube 补偿式全静压管
compensated pyrradiometer 补偿式全辐射表
compensated scale barometer 补偿式定标气压表
compensating component failure 补偿元件失效
compensating error of compensator 补偿器补偿误差
compensating high-speed camera 补偿式高速摄影机
compensating magnet 补偿磁条
compensating network 补偿网络
compensating vest 代偿背心
compensation 1. 补偿，平衡 2. 赔偿 3. 补偿式高速摄影机
compensation element 补偿元件
compensation filter 补偿滤光器调整滤波器
compensation response 代偿反应
compensation theorem 补偿定理
compensator 1. 补偿器，补助器，伸缩器，校正器 2. 胀缩件，膨胀圈 3. 调相机 4. 炮口制退器
compensator level 补偿飘准仪；自动安平水准仪
compensatory display 补偿式显示
compensatory nystagmus 代偿性眼球震颤，代偿性眼震
compensatory suit 代偿服
compensatory system 补偿制度，保偿制
compensatory target-following 补偿目标跟踪
compensatory tracking test 补偿跟踪试验
COMPENSATORY zone 补偿域，补偿区域
compete 竞赛，竞争
competition 竞争，比赛，竞赛
competition decision 竞争决策

competitive equilibrium 竞争均衡	completeness 1.完全,完整,完全程度,完整性 2.完成,结束
competitive price 竞争价格	completeness of combustion 燃烧完全度
compilation 编辑,编纂	completion 完成,结束,完满
compilation sheet 底图	complex 复合的,成套装置
compile 编辑,编制	COMPLEX Committee on Planetary and Lunar Exploration 行星际和月球探索委员会
compiled original 编绘原图	
compiler 编译程序,编译器	complex acceleration 合加速度
compiler generator 编译程序用编制程序	complex block 复杂断块
COMPL Complete 完全的,完整的	complex chemistry 配位化学
complaint 控告,控诉	complex circuit 复联电路
complement 编制人数,编制人员,定额装备	complex command 复合指令
complementary acceleration 附加加速度	complex coordination test 复杂协调能力试验
complementary error function distribution 余误差函数分布	complex correction 混合改正
	complex dielectric constant 复介电常数
complementary MOS intergrated circuit 互补MOS集成电路	complex engineering 复杂工程
	complex engineering system 复杂工程系统
complementary shear 余剪应力	complex excitation 复合激振
complementary transistor logic 互补晶体管逻辑	complex experiment 复合试验,综合试验,析因实验
complementary wave 副波,余波,补偿波	complex flow 复杂流动
complete bipartite 完全偶图	complex frequency 复频率
complete colouring 完全着色	complex frequency response function 复频响应函数
complete combustion 完全燃烧	complex gain 复数增益
complete controllability 完全可控性	complex geometry 复几何
complete coverage 完全覆盖	complex impact 综合影响
complete deployment 完整部署	complex mode shape 复合振荡形式
complete diamagnetism 全抗磁性	complex modulus of elasticity 复弹性模量
complete engine 完整发动机	complex multidisciplinary system 复杂多学科系统
complete failure 完全失效	complex permeability 复数磁导率
complete functional test 全面性能检查	complex polarization ratio 复极化比
complete group 完全群	complex pole 复极
complete half-wing 全半翼	complex problem 复杂问题
complete inversion 全反转	complex resistivity method 复电阻率法
complete metric space 完备度量空间	complex response 复响应
complete observability 完全可观测性	complex structure 复合结构
complete overhaul 全面翻修	complex system parameter 系统复参数
complete oxidation 完全氧化	complexity 复杂性
complete penetration 完全侵入(指射弹完全穿透目标)	complexity level 复杂性水平
	complexity value 复杂度值
complete performance curve 完整性能曲线	compliance 1.柔度 2.柔顺性 3.合格
complete period 全周期	compliant border 柔性边界
complete reusable space vehicle 完全重复使用运载器	compliant coating 柔性涂料
	compliant member 柔性构件,顺从性部件
complete round assembled 已装配好的完整弹	compliant panel 柔性面板
complete round chart 完整弹分解图	compliant structure 顺应式结构
complete round subassembled 未完全装配好的整发弹	compliant support 顺应支承
	compliant wall 自适应表面
complete round weight 完整弹重量,全弹重,全装弹重	complicate 使复杂化
complete simulation 完全模拟	complication 1.复杂,混乱 2.并发症
complete six-dof model 完整六自由度模型	
complete solution 通解,完全解	
complete suppression 完全抑制	
complete vacuum 绝对真空	

component   1.机件,部件,元件 2.分力,分量 3.成分,组分,组元
component degradation   组件退化
component efficincy   部件效率
component element   成分组件
component failure   部件失效,元件失效
component fault   可分为元件故障
component force   分力
component level   组件层次
component life   机件寿命,部件寿命(由领导机关或制造厂给定的使用寿命,可以延长)
component map   元件地图
component matching   部件匹配
component mode   元器件模式
component of ion   组件离子
component of lift   升力分量
component of thrust   组件推力
component of turbulence   组件动荡
component of velocity   速度分量
component of velocity normal to range   横向速度分量
component of wind   风力分量
component performance   组件性能
component reliability   元件可靠性
component requirement   元件要求
component response   元件反应
component shape   元件形状
component side   元件面
component star   子星
component test   零件检验,组件试验,部件试验
component test building   单元测试楼
component values   元件参数,零件参数
component weight   零件重量
component-compensating failure   元件补偿失效
component-independent failure   (导弹的)部件自身故障
components   部件,组件,成分
compose   排字,组成
composite   1.复合材料 2.混合编队 3.复合电路 4.复合的,合成的
composite aircraft   1.混合结构飞机 2.复合材料结构飞机 3.组合式飞机(子、母机的合称)
composite beam   复合梁(不同材料胶合在一起制成)
composite cloud   混合云
composite control   组合控制
composite cylinder   复合气瓶
composite desirability function   复合期望函数
composite detonator   复式雷管
composite double-base   合成双基(火箭推进剂)
composite error   综合误差,混合误差
composite explosive   混合炸药

composite fairing   复合整流罩
composite fuselage   复合材料机身
composite hinge   复合铰链
composite in aerospace structure   航天结构复合材料
composite launch   组合发射(由一枚运载火箭发射多个卫星)
composite material   复合材料
composite modified double base   复合改性双基
composite modified double-base propellant   复合改性双基推进剂
composite motor   复合电机
composite nozzle   复合喷嘴
composite panel   复合面板
composite performance   综合性能
composite permanent magnet   复合永磁体
composite plating   复合电镀,多层电镀
composite power plant   组合式动力装置
composite primary structure   复合一级结构
composite profiling method   联合剖面法
composite propellant   复合推进剂
composite re-entry test vehicle   组合式再入试验飞行器
composite response   复合反应
composite rotation   复合旋转
composite rotor   组合式转子
composite skin   复合皮
composite stiffness   推力桩,组合刚度
composite structure   复合材料结构
composite system   复合系统
composite test   组合试验
composite test gear   成套试验仪表
composite transparency   多层玻璃
composite tube   复合材料管
composite valve   复合阀
composite wing   复合材料机翼
composite wing box   复合材料机翼盒
composite-dielectric capacitance transducer   复合介质电容传感器
composite-dielectric capacitor   复合介质电容器
composition   1.溶注炸药,可以饶铸或造型的炸药 2.成分,组成 3.组成物,合成物
composition B   B型熔注炸药(内含梯恩梯40%,黑索今60%)
composition B4   B4型熔注炸药(在新型的爆破筒和空心装药中用作主要装药)
composition C3   C3型熔注炸药(不溶于水,适合于水下爆破用)
composition C4   C4型熔注炸药(较C3型的爆破效力更高,更不易受水的侵蚀)
composition explosive   混合炸药

composition modulated alloy plating 复合可调整合金镀
composition tracer 曳光剂(曳光弹用)
composition variable 成分变量
compositional nonuniformity 成分不均匀性
compositional range 成分范围
compositron 高速显字管
compound 1.化合物,填加剂 2.组合,连接
compound aircraft (既有机翼又有升力旋翼的)组合式飞机
compound catalysis efficiency 复合催化效率
compound centripetal acceleration 科氏加速度,旋转加速度,复合向心加速度
compound control 复合控制
compound curvature 复合曲率,空间曲率
compound die 复合模
compound helicopter 复合直升机
compound pendulum physical pendulum复摆,物理摆
compound pendulum mode 复摆模式
compound sail 复合帆
compound semiconductor 化合物半导体
compound semiconductor detector 化合物半导体探测器
compound semiconductor solar cell 化合物半导体太阳电池
compound solar sail 复合太阳帆
compound stress 复合应力
compound target 复合目标
compound turbo diesel 复合涡轮柴油机
compound wing 复合翼
compound-wound generator 复激发电机,复励发电机
COMPR Compressor 压缩机
comprehensive atlas 综合地图集
comprehensive chart 详图
comprehensive geographic information system 综合地理信息系统
comprehensive map 综合地图
comprehensiveness 综合,全面性
compress 压缩,压紧,精简
compressed 压缩的
compressed air 压缩空气
compressed air disease 加压症,潜水员症,压缩空气病
compressed air pane 压缩空气控制板
compressed air starter 压缩空气起动机
compressed air tunnel 压缩空气风洞
compressed data storage system 压缩数据存储系统
compressed gas 压缩气体
compressed pulse radar altimeter 压缩脉冲雷达高度计
compressed solar wind 受压太阳风
compressibility 压缩性,压缩系数,压缩率
compressibility correction (量)压缩性修正,压缩系数校正
compressibility correction table 压缩性修正量表
compressibility effects 压缩性影响
compressibility error 压缩性误差
compressibility factor 压缩因子
compressibility wave 压缩波
compressible dynamic stall 可压缩动态失速
compressible flow 可压缩流
compressible flow equation 可压缩流动方程
compressible fluid 可压缩流体,可压缩性铃
compressible gas 可压缩气体
compressible law 可压缩法
compressible shear 可压缩剪切
compressible speed 可压缩速度
compressible stability theory 可压缩稳定性理论
compression 压缩,浓缩
compression augmentation 压增大
compression chamber 压缩室,增压室
compression corner 压缩角
compression deception jamming 压缩欺骗干扰
compression driver 压缩驱动器
compression factor 压缩因子,压缩系数
compression ignition 压缩点火
compression inlet 压缩进气道
compression lift 压缩性升力(利用下弯翼尖等迫使超声速气流在翼下加速,改善流场而产生的升力)
compression load 压缩载荷
compression member 压缩构件
compression molding 加压模制,加压造型
compression pressure 压缩压力
compression process 压缩过程
compression ratio 压缩比
compression ratio of wind tunnel 风洞压缩比
compression rib 抗压翼肋,加强翼肋
compression ring(s) (活塞式发动机活塞上的)压缩涨圈
compression shock 压缩激波
compression strength 抗压强度,耐压强度
compression system 压缩系统
compression technique of light pulse 光脉冲压缩技术
compression test 耐压缩试验
compression vacuum gauge 压缩式真空计
compression wave 压缩波
compressive 压缩的,有压力的
compressive bar 压杆
compressive bending bar 压弯杆
compressive force 压缩力
compressive strength 抗压强度
compressor 1.压缩机 2.压气机

compressor blade 压气机叶片
compressor blade row 压气机叶栅
compressor bleed 压气机放气,压气机引气
compressor blowoff 压缩机排污
compressor cascade 压气机叶栅
compressor casing 压缩机箱,压气机机匣
compressor characteristics 压气机特性
compressor configuration 压缩机配置
compressor degradation 压缩机退化
compressor design 压缩机设计
compressor diffuser 压气机扩压器
compressor discharge 机出风口
compressor efficiency 压气机效率
compressor efficiency improvement 压缩机效率改善
compressor element stage 压气机基元级
compressor entrance 压缩机入口
compressor exit 压缩机出口
compressor face 压缩机端面
compressor flow path 压气机流道
compressor impeller blade 压缩机叶轮叶片
compressor inlet 压缩器进口
compressor inlet plane 压缩机入口平面
compressor inlet temperature 压缩器进气温度
compressor instability 压气机工作不稳定
compressor intake 压缩机进气口
compressor loss 压缩机损耗
compressor map 压气机特性图
compressor model 压缩机型号
compressor noise 压缩机噪声
compressor outlet 压缩机出口
compressor passage 压气机流道
compressor performance 压缩机性能
compressor pressure 压气机气压
compressor pressure ratio 压气机压比,增压比
compressor rotor 压气机转子
compressor spool speed 压缩机轴速度
compressor stage 压缩级
compressor stall 压气机失速
compressor stator 压气机静子
compressor surge 压气机喘振
compressor torque 压气机扭矩
compressor unit 压缩机组
compressor vane 压气机静叶,压气机静子叶片
comprise 包含,包括
compromise 妥协,让步
compromise solution 折中解
COMPT Compartment 舱,间
COMPTEL Compton telescope 康普顿成像望远镜（美国）

Compton effect 康普顿效应
Compton electron 康普顿电子
Compton scattering 康普顿散射
Compton spectrometer 康普顿谱仪
Compton wave-length 康普顿波长
COMPTR Comparator 比测器,比长仪,比色计,比较器等
compulsory exercise 规定动作
computable general equilibrium model 可计算一般均衡模型
computation 计算
computation of flow 速流计算
computation scheme 计算方案
computation time 计算时间
computational aerodynamics 计算空气动力学
computational analysis 计算分析
computational approach 计算法
computational burden 计算负担
computational capability 计算能力
computational cell 计算单元
computational code 计算码,计算程序
computational complexity 计算复杂性
computational cost 计算费用
computational criterion 计算标准
computational delay 计算时延
computational domain 计算区域
computational efficiency 计算效率
computational effort 计算工作量
computational expense 计算成本
computational fluid dynamics 计算流体动力学
computational geometry 计算几何学
computational grid 计算网格
computational hypervelocity 计算超高速
computational load 计算负荷
computational mechanics 计算力学
computational mesh 计算网格
computational method 计算方法
computational model 计算模型
computational overhead 计算开销
computational particle 计算粒子
computational plasma 计算等离子体
computational power 计算能力
computational resource 计算资源
computational result 计算结果
computational simulation 计算机仿真
computational solution 计算解
computational speed 计算速度
computational study 计算研究
computational tool 计算工具

compute	计算,估算
computer	1.计算机,计算器,计算装置 2.计算员
computer address matrix	计算机地址矩阵
computer aided debugging	计算机辅助故障诊断
computer aided engineering design	计算机辅助工程设计
computer aided management	计算机辅助管理
computer aided measurement and control	计算机辅助测量和控制
computer aided testing	计算机辅助测试
computer aided trainer	计算机辅助训练器
computer code	电脑编码
computer graphics	计算机图形学,计算机制图法
computer hardware	计算机硬件
computer interface unit	计算机接口部件
computer memory	1.计算机内存 2.计算机存储器
computer model	计算机模型
computer network	计算机网络
computer output microfilm	计算机输出缩微胶片
computer program	计算机程序
computer resource	计算机资源
computer simulation	计算机仿真,计算机模拟
computer simulator	计算机模拟程序
computer system	计算机系统
computer systems engineering	计算机系统工程
computer technology	计算机技术
computer telemetry system	计算机遥测
computer time	计算机时间
computer unit	计算机组件
computer vision	计算机视觉
computer word	计算机字
computer-aided design	电脑辅助设计
computer-assisted approach sequencing	计算机辅助进场排序
computerized	装有计算机的,计算机化的
computerized tactical system	计算机战术系统
computer-limited	计算机限制的
computeron-a-chip	单片(计算)机
computer-oriented	研制计算机的,与研制计算机有关的
computer-programmable	计算机编程的
computing	计算,处理
computing electronics	1.电子计算装置 2.计算电子学
computing gunsight	计算式射击瞄准具(能自动修正瞄准误差)
computing optical gun sight	计算式光学射击瞄准具
computing reticle	(提前角)计算光环
computing tool	计算工具
COMSAT	1. Communication Satellite 通信卫星 2. Congressional Committee on Science and Astronautics 国会科学及航天委员会
COMSEC	Communication Security 通信保密措施,通信保安
COMSS	Coastal Ocean Monitoring Satellite System 沿岸海域监测卫星系统
Comstar	通信星(美国国内通信卫星)
COMTAC	Command Tactical (USN) 战术部队(美国海军)
CON	Continuous 连续
CONAIR	Commanding Officer, Naval Air Wing 海军航空兵联队长
CONC	Concentric 同心的,共轴的
concatenate	连结,使连锁
concatenated codes	级联码
concave	凹,凹的
concave angle	凹角
concave function	凹函数
concave impactor	凹撞击器
concave side	凹面
concave surface	凹面
concave wall	凹墙
concavity	凹度,凹性,凹处
conceal mineral deposits	隐伏矿体
concealment	1.隐蔽,伪装 2.隐蔽处,隐蔽物
conceive	构思,设想
concentrated attenuator	集中衰减器
concentrated force	集中力
concentrated load	集中载荷
concentration	1.浓度,密集度 2.集中,浓缩
concentration cell corrosion	浓差电池腐蚀
concentration difference	浓度差
concentration distribution	浓度分布
concentration field	浓度场
concentration gradient	浓度梯度
concentration level	浓度级
concentration measurement	浓度测量
concentration of color	色浓度
concentration of halon	卤化烷浓度
concentration of solute	溶质浓度
concentration profile	浓度(剖面)分布
concentration range	浓度范围
concentration ring	集索圈
concentrations	浓聚物
concentrator	1.集中器 2.浓缩器 3.集中者 4.专业研究者
concentrator cell	聚光电池
concentrator solar cell	聚光太阳电池
concentric conductor	同心导线,同心线
concentric resonator	共心谐振腔
concentric tube injector	同轴式喷注器

concentricity 同心性
concept 1.概念,方案 2.原理,定则
concept formulation phase 方案论证阶段
concept of operation 作业观念
concept of sensor-fused telerobotics 传感器融合遥控机器人概念
conceptual aircraft design 飞机概念设计
conceptual bias 概念上的偏见
conceptual bias estimate 概念上偏差估计
conceptual design 初步方案设计
conceptual design method 概念设计方法
conceptual design phase 概念设计阶段
conceptual design problem 概念设计问题
conceptual design stage 概念设计阶段
conceptual structure 概念结构
conceptualization 概念化
concern 涉及,关系到
concertina 蛇腹形铁丝网
concertina door 手风琴(风箱)式折门
concession 出租营业区,让步,承认
conclude 推断,决定,作结论
conclusion 结论,结局,推论
concordance 一致性(同批产品在主要性能方面的相似性)
concourse 候机楼广场,旅客集散厅
concurrent design 并行设计
concurrent engineering 并行工程
concussion bomb 震荡炸弹,气压激发炸弹
concussion fuze 气压激发引信
COND Condenser 冷凝器
condemnation 报废,作废,剔除
condensable gas 可凝性气体
condensable nature 可压缩性质
condensate 冷凝物,浓缩物
condensate and heat exchange 冷凝热交换器
condensate mass 冷凝物质量
condensate mass fraction 冷凝物质量分数
condensation 冷凝,凝聚,浓缩
condensation process 冷凝过程
condensation rate 凝结率
condensation ratio 浓缩比
condensation shock 冷凝波,凝结激波
condensation surface 冷凝表面
condensation trail 凝结尾迹
condense 浓缩,凝结
condensed explosive 凝聚炸药
condensed layer 凝缩层
condensed material 凝结材料
condensed phase 凝聚相

condensed set 凝聚集,凝结集
condensed-phase decomposition 凝聚相分解
condensed-phase reaction 凝相反应
condenser 1.冷凝器,凝结器,冷凝段(热管内工质散热部位) 2.电容器 3.聚光器
condenser discharge spot welding 电容储能点焊
condenser microphone 电容传声器
condensible propellant 可冷凝推进剂
con-di nozzle Convergent-Divergent Nozzle 收敛扩散喷管
condition 条件,情况
condition adjustment 条件平差
condition equation 条件方程
condition error 条件错误
condition monitoring 状态监控
condition number 条件数,性态数
condition of intersection 交线条件
conditional covariance 条件协方差
conditional entropy 条件熵
conditional maintenance 状态检修
conditional mapping 条件映象
conditional probability 条件概率
conditional probability table 条件概率表
conditional stability 条件稳定性
CTC Conditional Transfer of Control 条件控制转移
conditionally stable system 条件稳定系统
conditional-sum adder 条件和加法器
conditioner 调节器(常指空气的)
conditioning 调节,调定,调理,改善,使(数据、技术状况等)达到要求
conditioning controller 调节控制器
conditioning scheme 空调系统方案
condor 1.康达无线电导主系统 2.秃鹰式飞机
CONDTN Condition 状态,条件
conduct 1.引导,指导 2.实施,进行,处理 3.(电、热等的)传导 4.导管,套管 5.行为,表现
conductance 传导性,导电性
conductance method 电导法
conducted emission 传导发射
conducted susceptibility 传导敏感度
conducting epoxy resin 导电环氧树脂
conduction 1.传导,传电,导电 2.导电性,传导性 3.电导
conduction band 导带,传导带
conduction current 传导电流
conduction current of capacitor 电容器的传导电流
conduction electrons 传导电子
conduction element 导电元素
conduction emission measurement 传导发射测量

conduction emission safety factor measurement 传导发射安全系数测量
conduction equation 传导方程
conduction field 传导场
conduction sensitivity measurement 传导敏感度测量
conduction stud 导电螺栓
conductive body 导电体,导体
conductive coating 导电涂层
conductive foil 导电箔
conductive heat flow 传导热流
conductive heat flux 传导热通量
conductive heat transfer 传导性传热,传导性热传递
conductive pattern 导电图形
conductive polymer 导电聚合物
conductive tether 导电系绳
conductivity 1.传导性,传导率 2.导电性,电导率
conductivity logging 电导率测井
conductivity modulation transister 电导率调制晶体管
conductor 导线,导体
conductron 光电导摄像管
conduit 导管,导线管,管道
cone 1.锥,锥形物,锥形区 2.头锥,锥形头部 3.锥形光束
cone angle (锥形物体如扩压器或机身等)半锥角
cone apex 锥顶
cone assembly 圆锥内圈组件
cone frustum 圆锥台
cone half-angle 锥半张角
cone jet 锥面射流
cone model 锥内摆动模型
cone of escape 逃逸锥
cone of fire 1.弹道束 2.散布圆锥 3.锥形射界
cone of propagation 传播锥
cone of silence 锥形盲区,无声锥区
cone spinning 锥形变薄旋压
cone spray 锥形喷雾
cone surface 锥面
cone symmetry 对称锥
cone yawmeter 锥式偏航计
cone-cylinder combination 锥柱组合体
coned 被探照灯光束照中的
coned cutter 圆锥铣刀
cone-derived waveriders 锥导乘波体
cone-jet mode 锥喷流模态
cone-jet structure 锥喷流结构
CONF Conformance 符合,合格
conference dispatching 会议调度
confetti 〈口语〉反雷达的带状反射体,拓雷达的带状反射体

confidence 信心,信任,秘密
confidence factor 置信因数
confidence interval 置信区间,可靠区间
confidence level 置信度,置信界限
confidence limit 置信界限
confidencer 消噪送话器
confidential 参与机密的,极受信任的
CONFIG Configuration 构型
configuration 1.构型(特指装备的由技术文件所规定的功能和物理特性,包括硬件和软件) 2.外形(飞行器的气动力外形构造) 3.(飞行器总体)布局,配置(主要构件的空间配置) 4.组态(模态的配置) 5.可变外形(指可以由飞行员操纵而改变的外形,如收放起落架、襟翼) 6.系统配置
configuration audit 技术状态审核
configuration change 配置改变
configuration change control 技术状态更改控制
configuration decision 配置决定
configuration design 外形设计
configuration deviation list 状态偏离表,构型缺损清单,外形缺损清单
configuration factor 轮廓因数,形态因素
configuration identification 配置标识
configuration item 结构项目
configuration item signer crater tool 构型项目签署打包工具
configuration management 配置管理
configuration reference 配置参考
configuration simulation 外形结构仿真
configuration space 1.位形空间 2.构形空间 3.配置空间
configuration synthesis 构型综合
configuration variable 配置变量
configure 1.使成形,使具形体 2.进行空气动力外形布局,按规定要求进行设计
confine 限制,禁闭
confined area operation 禁区飞行
confinement 限制,幽禁(指人在航天舱内与外界隔离)
confinement ability 约束能力
confinement length 约束长度
confinement of vibration 振动隔离
confinement time 约束时间
confirm 证实,批准
confirmation 1.证实,证据 2.确认,批准
confirmed removal 证实性拆卸(为了证实分析的原因而进行的拆卸)
conflagration 快速燃烧,爆燃
conflict 冲突,抵触
conflict avoidance 冲突避让

conflict parameter 冲突参数
conflict prediction 冲突预测
conflict probability 冲突概率
conflict probe performance 冲突探测性能
conflict probe tool 冲突探测工具
conflict resolution 冲突消解法
conflict situation 冲突性情境
conflict start time 冲突开始时间
conflict window 冲突窗口
conflict-free trajectory 无冲突轨迹
confluence 1.汇合,合流,汇合处 2.空气压缩区,流线密集区 3.(气象)(锋区)入口处
confocal resonator 共焦谐振腔
conformability 整合性,一致性
conformal 1.(数学)保角的,保形的 2.(地图投影等)等角的,以真实形状代表小块地区的
conformal antenna 共形天线
conformal array 共形阵(雷达)
conformal array antenna 1.共形阵天线 2.保形天线
conformal carriage 保形挂载
conformal gears 正则齿轮
conformal mapping 保角(形)映射
conformal projection 正(保)形投影,等角投影
conformal radar 保形雷达
conformal transformation 保角(形)变换
conformal-array radar 保形阵列天线雷达
conformality 保形,保角,正形
conformance 符合性
conformance threshold 一致性阈值
conformity 合格
confuse 使混乱,混淆
confused 目标分辨不清
confusion 1.混乱,纷乱 2.混乱,混乱状态
confusion jamming 迷惑性干扰
confusion reflector 干扰用反射器
confusion region 混淆区
congelation 冻结,凝固
congest 拥挤,充塞
congestion 拥塞
congestus 浓云
conic 1.锥形的 2.锥曲线
conic ephemeris 圆锥星历表
conic intersection 圆锥交叉
conic nozzle 锥形喷孔
conic orbit 圆锥轨道
conic projection 圆锥投影
conic section 圆锥曲线
conical antenna 锥形天线
conical array 圆锥阵

conical beam 锥形射束
conical body 锥体
conical camber 圆锥形度
conical cavity 锥形模腔
conical convergence 锥形收敛
conical diaphragm 锥面薄片
conical diaphragm angle 锥形隔膜角
conical error 圆锥误差
conical flow 锥型流
conical flow method 锥形流法
conical inlet 锥形进气道
conical nose 锥形头部
conical nozzle 锥形喷管
conical plug 锥形塞
conical scan antenna 圆锥扫描天线
conical scan radar homing head 圆锥扫描雷达导引头
conical scanning 圆锥扫描
conical scanning earth sensor 圆锥扫描地球敏感器
conical scanning horizon sensor 圆锥扫描地平仪
conical scanning tracking 圆锥扫描跟踪
conical section 锥形截面,二次曲线,圆锥曲线
conical sheet 锥形液膜
conical shell 锥形壳体
conical shock 锥形激震
conical wave 锥形波
conically scanning 圆锥扫描
conical-scan tracking 锥扫跟踪
conical-scanning radar 圆锥扫描雷达
conics model 圆锥曲线模型
coning 1.作成圆锥形,卷于圆锥上 2.圆锥形,锥度,锥角
coning angle 成锥角
coning effect 锥效应
coning errors exhibited by inertial navigation systems 惯性导航系统的圆锥误差
CONJ Conjunction 连接
conjugate 1.结合,联合,连接,使成对,耦合 2.共轭的,缀合的
conjugate beam 共轭梁
conjugate heat transfer 共轭热传
conjugate pair 共轭对
conjugate variable 共轭物理量
conjunction 1.结合,连接 2.(天体的)会合 3.(计算机的)逻辑乘法,逻辑乘积
conjunction probability 结合概率
conjunction time 契合时间
conk (机器等)突然损坏,(发动机)突然停车
CONN Connector 连接器,插头,接头
conn 方向操纵,方向控制

CONNDIAG　Connection Diagram　连接图
connect　连接，衔接
connected structure　连体结构
connecting rod　连杆
connecting terminal　接线柱
connecting traverse　附合导线
connection　1.连接，结合，关联 2.连接物，连接法，连接件，结合处，联系手段（通信、交通等）3.电路
connection box　接线盒，接线箱
connection-oriented transmission　面向连接传输
connection point　连接点
connection point for orientation　定向连接点
connection quadrangle method　联系四边形法
connection survey　联系测量
connection trap　连接陷阱
connection triangle method　联系三角形法
connectionism　连接机制
connectionless transmission　无连接传输
connective stability　连结稳定性
connectivity　连结性
connector　1.连接器，接线器，连接装置 2.端子，接线柱，线夹，接线电缆，插头，插座，接头
conning tower　指挥台
conocyl grain　锥柱形药柱，锥柱形装药
CONOPS　Concept of Operations　运行概念
conquer　克服，征服，战胜
consecutive image　连续图像
consecutive radar sample　连续雷达样本
consensus　一致，舆论，合意
consensus vector　共识向量
consequence　结果，重要性，推论
conservation equation　守恒方程
conservation law　守恒律
conservation of (measurement) standard　（测量）标准的保持
conservation of angular momentum　角动量守恒，动量矩守恒
conservation of energy　能量守恒
conservation of information　信息守恒
conservation of mass　质量守恒，质量守恒定律
conservatism　保守主义
conservative design　保守设计
conservative force　保守力，守恒力
conservative form　保守形式
conservative system　守恒系统
conservativeness　守恒性
conservatory　机上透明炮塔
consider-covariance analysis　考察协方差分析
consigner　托运人，托运单位
consignment　交付批，交付，发货
consist　组成，存在，一致
consistency　1.一致性，相容性 2.坚固，结实 3.浓度，稠度
consistency check　一致性检验，连续检查
Consol navigation system　"康索尔"导航系统
console　1.控制台，操纵台，仪表板，综合控制台 2.悬臂架，托架，落地式支架 3.落地式（无线电）接收机
console display　控制台显示器
console floodlight　仪表板泛光灯，操纵台泛光灯
console monitor　控制台监视器
consolidated maintenance　合并维修
consolidating layer　加固层
consolidation　凝固，固结，巩固
consolidator　航空运输行，空运承包商
consolute　共溶（液），可混溶的
consonance　共鸣，谐振
consortium　财团，联合，合伙
CONST　1.Constant　恒定的，不变的，常数，恒量 2.Construction　结构，建筑物
constant　1.常数，常量，恒定值 2.不变的
constant acceleration　恒定加速度
constant altitude　定高，等高
constant altitude indicator　等高显示器
constant altitude plan position indicator　等高平面位置指示器
constant amplitude　不变幅度
constant area combustor　恒定区燃烧室
constant area duct　常数区域管
constant area passage　恒定区通道
constant boundary-layer　常数边界层
constant bypass ratio　常数涵道比
constant charge　恒压充电
constant coefficient　常系数
constant colour　常色（驾驶舱灯光警告系统常用的颜色）
constant cross-section　截面恒定
constant deceleration　恒减速
constant density　恒定密度
constant density model　恒定密度模型
constant disturbance　常值干扰
constant disturbance torque　恒定扰动转矩
constant entropy process　定熵过程，等熵过程
constant error　常差
constant failure rate period useful life　常数故障率时期有用寿命期
constant false-alarm rate　恒虚警率
constant field　恒定场
constant flow　恒流，定量流动
constant flow oxygen　恒流供氧

constant flow rate 恒定流量
constant frequency alternator 恒频交流发电机
constant gain 恒定增益
constant gradient 恒定梯度
constant heat 恒温加热
constant heat transfer coefficient 恒定传热系数
constant height 定高度
constant level balloon 定高气球
constant mass 恒重
constant multiplier coefficient unit 常系数放大装置,常系数乘法部件
constant of aberration 光行差常数
constant of gravitation 引力常数
constant of nutation 章动常数
constant oxidizer 恒定氧化剂
constant oxygen flow regulator 连续供氧流量调节器
constant parameter 恒定参数
constant percentage chord line 等百分线
constant pressure 恒压,定压
constant pressure feed system 恒压式供应系统
constant pressure process 定压过程,等压过程
constant radius 固定倒角半径
constant rate 恒速,固定比率,恒定流量
constant semimajor 常数半长
constant speed 恒速,等速
constant speed drive unit 恒速驱动装置
constant speed propeller 恒速螺旋桨
constant speed-frequency AC power system 恒速恒频交流电源系统
constant stagnation trajectory 恒动压滞止弹道
constant strength beam 等强度梁
constant tank pressure 恒槽压
constant temperature 恒温,定温,等温
constant temperature and moisture test 恒温恒湿试验
constant temperature process 定温过程,等温过程
constant temperature shift 恒温转变
constant thrust 恒定推力
constant time 常数时间
constant total pressure 恒定总压强
constant value 不变价值,常数值
constant value command 定值指令
constant vector 常向量
constant velocity 匀速,等速度,恒速度
constant volume 定容,恒定体积
constant volume balloon 等容气球
constant volume process 定容过程,等容过程
constant-airspeed 恒定空速
constant-amplitude loading 等幅加载
constantan 康斯坦镍铜合金,康铜

constantan thermocouple 康铜温差电偶
constant-area duct 恒定面积的导管
constant-bandwidth FM subcarrier 恒定带宽调频副载波
constant-bearing guidance 平行接近制导,等方位制导
constant-bearing navigation (导弹)平行接近法引导
constant-current charge 恒流充电
constant-current discharge 定电流放电
constant-duty cycle 恒定工作周期
constant-flow oxygen equipment 连续供氧设备
constant-flow oxygen system 连续供氧系统
constant-fraction discriminator 恒比鉴别器
constant-height balloon 定高气球
constant-luminance 恒定亮度
constant-pressure altitude 等压高度
constant-pressure chart 等压面图
constant-pressure thermometer 定压温度表
constant-rate phase 恒速阶段
constant-resistance discharge 定电阻放电
constant-speed drive 1.恒速传动 2.恒速传动装置
constant-speed unit 恒速装置,调速器
constant-temperature hot-wire anemometer 定温热线风速表
constant-value control system 恒值控制系统
constant-velocity 恒速
constant-velocity segment 常速段
constant-voltage charge 恒压充电
constant-volume combustion 恒定体积燃烧
constant-volume thermometer 定容温度表
constant-volume-explosion 恒体积爆炸
constant-wave radar guidance 定常波雷达制导,连续波雷达制导
constellation 星座
constellation design 星座设计
constellation health 卫星星座完好状况
constellation size 星座规模
constituent 1.组成部分,成分,组元 2.分量,分力 3.分支,支量
constituent day 有效日,分潮日
constitution 1.组织,结构,组合成分 2.制定,设立
constitutive law 本构定律
constitutive relationship 本构关系
CONSTR Construction 结构
constrain 强迫,强制,约束
constrained feed 强制馈电
constraint 1.约束,限制 2.强制力,压制因素
constraint boundary 限制式边界
constraint condition 约束条件
constraint element 约束单元

constraint equation 约束方程式
constraint error 约束错误
constraint estimation 约束估计
constraint force 约束力
constraint function 约束函数
constraint length 约束长度
constraint mode 约束模式
constraint satisfaction 约束满足
constraint value 约束值
constrict 压缩,束紧
constriction 1.收缩,压缩,颈缩 2.压缩物,阻塞物
constrictor 1.尾部收缩燃烧室 2.收缩器,压缩装置,收缩段 3.限流器,限流片
construct 修建,建造,构成
constructed profile 示意剖面图
construction 建造,结构
construction factor 施工因素
construction material 结构材料
constructor 1.设计者,设计师 2.构成机构
consultant 顾问,咨询者
consultation 磋商,协商,专家会议,会诊
consumable anode 自耗阳极
consumables 消耗材料,消耗品
consumables update 消耗品最新数据
consume 消耗,消费,耗尽
consumer 消费者,用户,顾客
consumption 消耗,消耗量
consumption function 消耗函数
consumption of fuel 燃料消耗量
consumption rate 消耗率
CONT 1.Contact 接触 2.Continue 继续 3.Control 控制,操纵 4.Controller 管制员,控制员 5.Content 内容,容量
CONT CKT Control Circuit 控制电路
CONTA Control Assembly 控制组装件
contact 1.接触,联系,接触面 2.(飞机升空后再)接地,一次起落(计算起落次数时) 3.能目视地面(飞行),目力观察(飞机上对地面的)
contact adhesion 触点黏结
contact angle 接触角,交会角
contact approach 目视进场着陆
contact binary 相接双星
contact breaker 触点电门
contact cam 接触块
contact circuit 目视飞行起落航线
contact condition 触点条件
contact corrosion 接触腐蚀
contact density 接触信号密度
contact depth datum 接触深度基准面
contact discontinuity 接触间断
contact element 1.触发敏感元件 2.杀伤体 3.接触元件
contact evaporation 蒸电极
contact exposure method 接触式曝光法
contact fault 接触不良缺陷
contact finger 接点指
contact flange 接触式法兰,连接法兰
contact flight 1.目视飞行 2.(空中加油时)接触状态的飞行
contact force 接触力
contact fuze 触发引信
contact height 目视高度
contact induced polarization method 接触激发极化法
contact initiation 触发起爆,着发起爆
contact ion thruster 接触型离子推力器
contact landing 目视着陆
contact lenses 隐形眼镜,接触适镜
contact line 接触线,地质分界线
contact lithograph 接触光蚀刻
contact load 触点负载
contact measurement 接触测量
contact navigation 地标领航,目视领航
contact penalty factor 接触罚因子
contact piston 接触式活塞
contact plunger 接触式柱塞
contact position 接触位置
contact printing 接触晒印
contact resistance 接触电阻
contact sensing 接触觉察
contact surface 接触面
contact surface burst 触地爆炸
contact terminal 接线柱,线夹
contact thermal conductance 接触热传导
contact thermal resistance 接触热阻
contact type fuze 触发引信
contact weather 能目视飞行的气象条件
contact weld 触点熔接
contact width 接触宽度
contact zone 接触区,接触地带
contacting altimeter 接点式高度表,高度信号器
contaction thruster 接触型离子推力器
contactor 触点,接触器,开关(通常是电磁遥控的)
contact-sensing device 1.触发引信,碰炸引信 2.接触敏感装置
CONTAD Concealed Target Detection 隐蔽目标探测
contain 1.包含,容纳 2.等于,相当于 3.牵制
container 1.容器,储存器(蓄电池)外壳 2.集装箱
container bomb unit 子母弹
container delivery 集装空投

container hold   下层集装箱货舱
container shelter   集装箱掩蔽部
container wall   挤压筒筒壁
containerized stand-off weapon   集装箱式远程武器
containerized weapon   集装容器化武器
containment   1.容积,容量 2.包容性,密封度 3.(反应堆的)防事故外壳
containment building   安全壳厂房
containment system   1.容载系统 2.安全壳系统
CONTAM   Contaminated 污染的,污损的,沾污的
contaminant   污染物,搀合物,杂质
contaminant layer   污染(物)层
contaminant levels   污染物等级
contaminant thickness   污染物厚度
contaminate   污染,沾染
contamination   污染,(放射性物质)沾染,污染物,沾染物
contamination control   污染控制
contamination levels   污染等级
contamination modeling   污染模拟
content   1.含量,容量,容积 2.内容,(复数)目录,(计算机的)存数
content of aluminum   铝量
context-free grammar   上下文无关文法
context-sensitive grammar   上下文有关文法
contingency   1.意外事故 2.偶然性,偶然
contingency analysis   意外事件分析
contingency food   应急食品
contingency mode   应急模式
contingency planning   应急计划
contingency rating   紧急功率
contingency retention item   应急备用物品,应急备用措施
continuation   继续,续集,延长
continuation algorithm   延拓算法
continuation method   连续法,延拓法
continuation trainer   补训教练机
continuation training   补充训练,进一步训练
continuity   连续,连续性
continuity constraint   连续性约束
continuity equation   连续方程
continuity light   通断(检查)灯
continuous airworthiness   持续适航性
continuous anode   连续阳极
continuous attenuator   连续减光板
continuous beam   连续梁,多支点梁
continuous body   连续体
continuous built-in test equipment   连续式内装测试设备
continuous buzz   连续嗡鸣

continuous climbing method   1.连续爬升法 2.直接爬升法
continuous command   连续指令
continuous configuration   连续重新构型
continuous control law   连续控制律
continuous current at locked-rotor   连续堵转电流
continuous data system   连续数据系统
continuous deceleration technique   连续减速技术
continuous descent approach   每个持续降落的方法
continuous discrete hybrid system mode   连续离散混合系统模型
continuous distribution   连续分布
continuous duty   连续工作制
continuous excitation   连续激励
continuous feedback   持续反馈
continuous flow oxygen equipment   连续供氧设备
continuous following fire   连续跟踪射击
continuous frequency   连续调频
continuous function   连续函数
continuous liquid phase   连续液相
continuous loading   均匀加载,连续加载
continuous medium   连续介质
continuous model   连续模型
continuous monitor   连续监测
continuous optimization   连续优化
continuous oscillation   连续振荡
continuous oxygen supply device   连续供氧装置
continuous path control   连续路径控制
continuous phase   连续相
continuous phase frequency shift keying   连续相位频移键控
continuous planning   连续性规划理论
continuous propulsion   连续推进
continuous radar illumination   连续雷达照射
continuous rod warhead   杆式弹头,环形弹头
continuous sampling plan   连续抽样方案
continuous self test   连续自检
continuous slot   连续槽
continuous source   连续源
continuous source of ignition   连续点火源
continuous spin   连续旋转
continuous state   连续态
continuous strip camera   带式连续照相航空照相机,连续狭条照相机(可以把飞机飞过某一段航线上的地物拍照成一条带形照片)
continuous system   连续系统
continuous telemetry parameter   遥测连续参数
continuous thrust   连续变推力
continuous tone   连续色调

continuous topping 连续补加
continuous tracking configuration 连续跟踪结构
continuous transition 连续跃迁
continuous wave 等幅波,等幅振荡,连续波
continuous wave Doppler fuze 连续波多普勒引信
continuous wave Doppler navigation radar 连续波多普勒导航雷达
continuous wave instrumentation radar 连续波测量雷达
continuous wave ionosonde (电离层)连续波测高仪
continuous wave laser 连续波激光器
continuous wave laser fuze 连续波激光引信
continuous wave laser radar 连续波激光雷达
continuous wave laser radar system 连续波激光雷达系统
continuous wave transmitter 连续波发射机
continuous waves radar tracking system 连续波雷达跟踪系统
continuous winding 连续绕组,连续式线圈
continuous wind tunnel 连续式风洞
continuous-flow system 连续供给系统(如供氧)
continuous-loop fire detection system 连续回路式火警探测系统
continuously computed impact line 连续计算命中线
continuously computed impact point 连续计算命中点
continuously computed release point 连续计算投放点
continuously computed target point 连续计算目标点
continuously displayed impact point 连续显示的弹着点
continuously pointed fire 连续跟踪射击
continuous-path machining 连续走刀切削,连续线切削加工
continuous-pressure breathing 恒压增压供氧
continuous-pressure breathing equipment 恒压增压式氧气设备
continuous-time model 连续时间模型
continuous-wave 连续波
continuous-wave Doppler radar 连续波多普勒雷达
continuous-wave Doppler radar homing head 连续波多普勒雷达导引头
continuous-wave magnetron 连续波磁控管
continuous-wave modulation 连续波调制
continuous-wave radar 连续波雷达
continuum 1.连续介质,连续流 2.连续流,闭联集
continuum approach 连续介质法
continuum assumption 连续介质假设
continuum regime 连续介质流动
contour 1.等高线,等值线 2.轮廓,外形
contour beam 等值梁
contour chart 等高线地图
contour enhancement 轮廓增强,边缘增强
contour fighter 〈口语〉攻击机,强击机
contour flight 等高线飞行(按地形的等高线低空飞行,高度略高于贴地飞行)
contour interval 等高距
contour line 等值线,等高线
contour map 等值线图,等高线地图
contour map migration 等值线图偏移
contour mapping 等高面测绘
contour of constant geometric accuracy 等精度曲线
contour plot 等值线图
contour prism 等高棱镜
contour rate 等放射剂量线
contour roll forming 型辊成形
contour size 外形尺寸,轮廓尺寸
contour template 型面样板
contoured nozzle 1.特型喷管 2.型面喷管
contoured seat 赋形座椅
contour-flying 地形跟踪飞行
contouring 造型,外形加工
contourograph 轮廓仪
contours of constant geometric accuracy 等精度曲线
CONTR Contract 合同
contra injector 反向喷射器,反向喷嘴
contra missile 反导弹导弹
contra prop 对转螺旋桨
contra rotating 同轴反转,反向旋转
contra vane 反向叶片,逆向导流叶片
contraband detector 违禁品探测器
contract 1.合同,契约 2.承包,承办 3.收缩了的,缩短了的
contract definition phase 合同论证阶段
contract image 缩小影像
contract maintenance 承包维修
contraction 1.收缩,收敛 2.收敛段,收缩段
contraction analysis 收缩分析
contraction ratio 收缩比,收敛比
contraction section 收缩段
contraction theory 1.冷缩说 2.收缩说
contractor 1.承包商 2.收敛段
contractor-furnished equipment 承包商提供的设备(通常包括硬件和软件以及某些知识和经验)
contradiction 矛盾,否认,反驳
contra-flow engine 回流式(折流式或反流式)发动机
contrail (飞机)凝结尾迹
contrail formation 凝结尾形成
contra-injection 逆向喷油,反向喷油
contra-rotating 对转式
contra-rotatory illusion 反旋转错觉
contrast 1.对比度,反差 2.对照物

contrast coefficient 反差系数
contrast compression 对比度压缩,反差压缩
contrast enhancement 反差增强,对比度增强
contrast range 对比度范围,反差范围
contrast ratio 对比度
contrast stretching 对比度扩展
contrast transfer function 对比度传递函数
contributing factor 起作用的因素
contribution 贡献,影响,作用
contribution factor 辅助因数
contributor 1.组成部分,部分 2.(造成影响的)原因之一,先决条件
contrifugal force 离心力
control 控制,管制
control accelerometer 控制加速表
control accuracy 操纵准确度,调节准确度
control activity 控制活动
control actuator 舵机
control aiming dot 瞄准光点
control algorithm 控制算法
control allocation 控制分配
control allocation algorithm 控制分配算法
control allocation problem 控制分配问题
control allocator 控制分配器
control analysis 控制分析
control and display console 控制和显示台
control and monitor panel 控制和监测板
control and reporting center 控制与报知中心
control and reporting post 控制报知站
control anticipation parameter 操纵期望参数
control approach 控制方式
control architecture 控制结构
control assembly 控制部件
control augmentation 控制增稳,操纵增稳
control augmentation concept 控制增稳概念
control augmentation system 控制增稳系统
control authority 控制权
control axis 控制轴
control bandwidth 控制带宽
control behavior 控制状态
control bias capability 控制偏置能力
control blade 舵,舵面
control board 控制板
control box 控制台,控制盒
control cabinet 控制柜
control cable 操纵钢索
control calculation 控制计算
control case 电控箱
control center 控制中心,调度室

control channel 控制通道
control character 控制字符
control characteristic 控制特性
control chart 管理图
control coating 控制涂层
control coefficient 控制系数
control column 操纵杆,控制杆
control column vibrator 振杆器
control command 控制指令
control concept 控制观念
control configured vehicle 随控布局飞行器,自控飞行品态飞行器
control constraint 控制受限
control cost 管理费,控制成本
control cubicle 控制柜,控制室
control cycle 控制周期
control decision 控制决策
control deficiency 控制缺陷
control derivative 操纵导数
control design 控制系统设计
control device 控制装置
control direction 控制方向
control dynamics engineer 控制动力工程师
control effectiveness 操纵有效性
control effectiveness matrix 控制效率矩阵
control effector 控制效应
control effector position 控制效应位置
control electrode 控制电极
control electronics 电子控制设备
control element 控制元件,控制环节
control engineer 控制工程师
control engineering 控制工程
control error 1.操纵错误 2.操纵系统误差,调节系统误差
control experiment 对照试验,控制实验
control expression 控制表达式
control failure 1.控制故障 2.无效控制
control feel 控制器官感觉,操纵感觉
control filter 控制过滤器
control force 1.操纵力 2.驾驶力
control force/displacement 操纵力和位移
control frame 控制帧
control framework 控制框架
control frequency 控制频率
control fuel system 控制式燃料系统
control function 1.控制功能 2.控制函数
control gain 控制增益
control grid 控制栅极
control grid plate transconductance 控制栅与板极间

互导
control impulse 控制脉冲
control induce moment 控制诱导时刻
control information 控制信息
control input 控制输入
control input relation 控制输入关系
control input signal 控制输入信号
control jet 控制射流,喷气舵
control law 控制律,控制规律
control law design 控制律设计
control law of flareout 拉平控制律
control limit 控制界限,管理范围
control limitation 控制限制
control locking device 控制航面锁定装置
control logic 控制逻辑
control logic design 控制逻辑设计
control loop 控制回路,控制环路
control material 物资管制
control measure 控制措施
control mechanism 控制机构
control methodology 控制方法
control missile 翼控式导弹
control mixing unit 控制系统混合器
control mode 控制方式
control model 控制模型
control module 控制舱,仪器舱
control moment 控制力矩
control moment gyro 控制用力矩陀螺
control momentum gyro 动量控制陀螺
control monitoring 控制监视
control mosaic 有控制镶嵌图
control objective 控制目标
control of aeroelastic response 气动弹性响应控制
control of combustion 燃烧控制
control of fire 火力控制,射击控制
control of flying 飞行活动的控制
control of orientation 定向控制
control of turbulent flow 湍流流动控制
control panel 控制屏
control parameter 控制参量
control performance 控制性能
control period 控制时期
control phase 1.控制相位 2.控制阶段
control plot 标有检查点的航行图
control point 控制值(热控系统控温的实际值)
control policy 控制策略
control port 控制端口
control position 操纵部位
control position transducer 航面位置传感器

control power 1.控制权 2.电源控制,功率控制
control procedure 控制程序,控制规程
control profile 控制参数文件
control program 控制程序
control range 控制范围
control rate 控制速率
control reconfiguration 控制重组态
control reporting center 控制报知中心
control resistance 控制器阻力
control rod 操纵杆
control rule 1.控制规则 2.检核尺
control scheme 1.控制方案 2.控制方式
control sector 管制扇区
control sequence 控制序列
control signal 控制信号
control simulation 控制仿真
control solution 校核液
control space 1.控制空间 2.操纵台
control stability 控制稳定性
control station 引导站,控制站
control step 控制步骤
control stick 驾驶杆,操纵杆
control stick fixed 握杆
control strategy 控制策略
control strip 1.控制航带 2.控制片/条
control structure 控制结构
control subsystem 控制子系统
control surface 操纵面,控制面,舵面
control surface deflection 操纵面偏度
control surface effectiveness 操纵面效度
control surface failure 操纵面故障
control surface loss 操纵面损耗
control surface rate 操纵面速率
control surface rate command 航面(偏转)速度控制信号
control surfaces bomb 装有控制翼的炸弹
control switch 控制开关
control synchro 控制式自整角机
control synthesis 控制器综合
control system 控制系统,操纵系统
control system design 控制系统设计
control system element 控制系统组成部分
control system model 控制系统模型
control system of rocket 火箭的控制系统
control system reconfiguration 控制系统重构
control system synthesis 控制系统综合
control systems engineering 控制系统工程
control table of pressure gas 气源控制台
control technology 控制技术

control temperature 控制温度
control terminal 控制器接线柱
control terminal unit 终端控制装置
control test 对照试验
control test vehicle 控制试验飞行器
control testing 1.控制实验 2.对照试验
control theory 控制理论
control time 1.调节时间 2.飞过检查点的计算时间
control time horizon 控制时程
control tolerance 控制容限
control torque 控制力矩
control torque coefficient 控制力矩系数
control track direction computer 控制跟踪方向计算机
control transfer 转换操纵
control transformer 1.同步接收机 2.控制变压器
control type 控制类型
control unit 控制装置,控制器,控制单元
control valve 控制阀
control vane 操纵舵
control variable 控制变量
control vector 控制矢量
control voltage 控制电压
control volume 1.控制体积 2.控制卷 3.控制容积
control weight 控制重量
control winding 控制绕组,信号绕组
control wire 1.操纵索 2.控制电缆
control word 控制字,指令
control zone 管制地带
control-allocation 控制分配
control-augmentation system 控制增稳系统
control-configured 随控布局的
control-force load 操纵力负荷,制力负荷
controllability 1.操纵性 2.可控性,可调性
controllability analysis 可控性分析
controllability index 可控指数
controllability matrix 可控性矩阵
controllable 可控制的,可操纵的,可调整的
controllable canonical form 可控规范型
controllable satellite 可控卫星
controllable set 可控集
controllable subset 可控子集
controllable-thrust 可调节推力的
controllable twist 可控扭转
controlled atmosphere 受控大气
controlled atmosphere heat treatment 可控气氛热处理
controlled bomb 可导炸弹,可控炸弹
controlled combustor 可控燃烧
controlled diffusion 可控扩散度
controlled diffusion airfoil 可控扩散叶型

controlled ditching 有控制的迫降
controlled ecological life support system 受控生态生命保障系统
controlled effects nuclear weapons 效果可控的核武器
CEC Controlled Element Computer 控制单元计算机
controlled element 1.控制单元 2.可控元件
controlled environment 可控环境
controlled flight 受控飞行
controlled fragmentation 受控(定型)弹片散飞
controlled maintenance 受控维修
controlled object 受控对象
controlled operation 有引导的作战活动,可控操作
controlled rolling 控制轧制
controlled slip 受控滑动
controlled spacecraft 可控航天器
controlled thermonuclear reaction 受控热核反应
controlled transit 控制渡越
controlled variable 被控变量
controlled ventilation 可调节通风
controlled vortex 可控涡
controlled vortex design 可控涡设计
controlled-devices countermeasure 对受控装置的电子干扰
controller 1.控制器,控制机构,操纵台,舵 2.调节装置,调节器 3.控制员,调度员,检验员引导员 4.地域防空指挥官
controller decompression 控制减压
controller design 控制器设计
controller family 控制器家族
controller function 控制器功能
controller gain 控制器增益
controller intervention 控制器干预
controller output 控制器输出
controller parameter 控制器参数
controller performance 控制器性能
controller structure 控制器结构
controller update 控制器升级
controller workload 控制器工作负载
controller-in-charge 值班管制员
control-line aircraft 线操纵飞机
controlling action 控制作用,调节作用,操纵动作
controlling jet 喷气舵,操纵用的喷流
controlman 操纵员,控制员,管理员
control-oriented model 面向控制模型
control-surface 操纵面
control-surface deflection 操纵面偏转
contromatrix network 控制矩阵网络
CONTWR Control Tower 管制塔台
CONU Control Unit 控制装置,操纵装置,管制单位

Conus OTH-B　"柯纳斯"后向散射超视距雷达
CONV　1. Convertor or Convertible 变流器或可转换的
　　2. Convention or Conventional 惯例,条约,常规的
convect　对流传热
convected disturbance　传热干扰
convection　对流
convection blocking effect　对流阻塞效应
convection cell　对流环,对流元
convection cooling　对流冷却
convection correlation　对流相关
convection current　运流电流
convection heat transfer　对流传热
convection pattern　对流模式
convection roll　对流滚轴
convection time　对流时间
convection velocity　对流速度
convection zone　对流区域,对流层
convective atmosphere　对流大气
convective base　对流基地
convective cell　对流环,对流元
convective cooling　对流冷却
convective cooling ventilated suit　对流冷却通风服
convective current　1. 对流气流 2. 对流电流
convective disturbance　对流扰动
convective effect　对流效应
convective flow　对流流动
convective heat　对流热
convective heat flow　对流热流
convective heat flux　对流热通量
convective heat loss　对流散热
convective heating　对流热
convective heating rate　对流热速度
convective instability　对流不稳定
convective motion　对流运动
convective overshooting　对流过冲
convective pattern　对流模式
convective perturbation　对流扰动
convective situation　对流情况
convective term　对流项
convective time　对流时间
convective transport　1. 对流运输 2. 传递透过
convective velocity　对流速度
convective wavelength　对流波长
convective weather　对流天气
convective zone　对流区域,对流层
conventional aileron　常规副翼
conventional aircraft　普通飞机
conventional algorithm　惯用算法
conventional ammunition　常规弹药

conventional approach　传统方法,常规方法
conventional arcjet thruster　传统电弧加热式发动机
conventional armament　常规武器,常规装备
conventional boundary-layer　常规边界层
conventional boundary-layer equation　常规边界层方程
conventional carrier　常规动力航空母舰
conventional configuration　正常式布局
conventional control　常规控制
conventional controller　常规控制器
conventional cruise missile　常规巡航导弹
conventional current　常规电流
conventional design　常规设计
conventional diffuser　常规扩散器
conventional emitter　常规发射器
conventional engine　常规发动机
conventional explosive　普通炸药,常规炸药
conventional fuselage　传统机身
conventional gain-scheduling　传统增益调度
conventional gravity bomb　常规自由落下炸弹
conventional gyro　传统陀螺
conventional international origin　国际协议原点
conventional jet fuel　常规喷气燃料
conventional layout　常规布局
conventional lightsail　传统光帆
conventional method　常规方法
conventional nozzle　常规喷嘴
conventional ordnance release computer　常规武器发射计算机
conventional particle　常规粒子
conventional projectile　常规投射物
conventional propellant　常规推进剂
conventional propellant loading system　常规(推进剂)加注系统
conventional radar　常规雷达
conventional ramjet　传统冲压发动机
conventional reference scale　约定参照标尺
conventional rotor　常规转子
conventional sail　传统风帆船
conventional scramjet　传统超燃冲压发动机
conventional solar cell　常规太阳电池
conventional spoiler　传统扰流板
conventional standoff weapon　防空区外发射的空对地常规武器
conventional take-off and landing　常规起降飞机
conventional true value　约定真值
conventional turbine　常规涡轮
conventional warhead　常规弹头
conventional weapon　常规武器
conventional weapon control system　常规武器控制系统

conventional wind tunnel　常规风洞
converge　1.收敛 2.会聚,集中
converge solution　融合解决方案
converge velocity　收敛速度
convergence　1.会聚,汇合 2.收敛(性) 3.非周期衰减运动
convergence angle　1.收敛角(喷管、经线等的) 2.会聚角
convergence belt　汇聚带
convergence bound　收敛界
convergence criterion　收敛性判定准则
convergence effect　1.集聚效应　2.趋同效应
convergence factor　1.投影常数 2.收敛系数
convergence parameter　收敛参数
convergence proof　收敛性分析
convergence property　收敛性
convergence radius　收敛半径
convergence rate　收敛速度
convergence study　收敛性研究
convergence time　收敛时间
convergence zone　汇聚带
convergent　1.收敛的,收缩的 2.会聚的 3.辐合的
convergent angle　1.收敛角 2.交向角
convergent contour　收敛形
convergent nozzle　收敛型喷管
convergent photography　交向摄影
convergent section　喷管的收敛部分
convergent type　收口式
convergent-divergent nozzle　收敛扩张喷管
converging　收敛的,会聚的
converging shock tube　收缩激波管
conversational mode　对话方式(人机对话)
conversion　1.转换,换位 2.过渡飞行(指垂直起落飞机或直升机由垂直飞行转换到水平飞行) 3.改型或改装训练
conversion angle　航线转换角
conversion circuit　变换电路
conversion device　转换器,转换装置
conversion efficiency　转换效率
conversion equipment　(数据)转换设备
conversion factor　换算因数,变换因数
conversion gain　变换增益
conversion kit fin　(燃烧弹的)附加尾翼装置
conversion of thermal energy　热能转换
conversion rate　1.转化率 2.转换率 3.兑换率
conversion transconductance　变频互导
conversion treatment　转化处理
convertawing　推力换向式机翼
converted aircraft　改型飞机
converted chemical energy　化学能转换

converted wave　转换波
converter　1.变换器,转化器 2.变流机,变流器,换流器 3.变频器
converter module　转换器模块
converter unit　译码器,变换器
convertible　1.可转变的,可变换的,可转化的 2.易改装的,改型的
convertible aircraft　客货互换型飞机
convertiplane　推力换向式飞机
convertiport　垂直起落飞机航空港
convex　1.凸面 2.凸的
convex closed-loop　凸度闭环
convex combination　凸组合
convex optimization approach　凸优化方法
convexity　1.凸,凸状 2.(焊接的)凸度
conveyor　1.传送,输送设备 2.搬运者,传达者
convolution　卷积
convolution code　卷积码
convolution method　回旋法
convolution theorem　卷积定理
convolve　1.使卷 2.使盘旋,使缠绕
convoy　1.护航,护送,护运 2.护队,护送队,护航运输队 3.被护送的军队
convoy protection　护航保护
cook off　1.自燃 2.自射(武器因膛内高温而自行发射)
cookie　大口径炸弹,巨型炸弹
COOL　Coolant冷却器
cool　冷却剂,冷冻剂,低沸点混合剂
cool flame　冷焰(亮度和温度都较低的火焰)
coolant　冷却液,冷却介质
coolant channel　冷却通道
coolant duct　冷却剂导管
coolant flow　冷却液流
coolant flow rate　冷却剂流量
coolant gas　气体冷却剂
coolant hole　冷却液孔
coolant injection　冷却剂注入
coolant inlet　冷却液进口
coolant jet　冷却剂喷嘴
coolant loop　冷却剂回路
coolant mass flow　冷却剂质量流
coolant outlet plenum　冷却剂出口腔
coolant outlet temperature　冷却剂出口温度
coolant recirculation pump　冷却液循环泵
coolant slot　冷却液槽
coolant stream　冷却剂
coolant temperature　冷却液温度
cooldown　冷却,降温
cooler　冷却器,冷凝器

cooling 1.冷却 2.冷却的 3.放射性减退
cooling effect 冷却效应
cooling gills 散热导流板,散热鱼鳞片
cooling jacket 冷却套,冷却通道
cooling passage 冷却通道,散热通道
cooling rate 冷却速度
cooling system 1.冷却系统 2.制冷系统
cooling turbine unit 1.涡轮冷却器 2.空气循环机 3.空气膨胀机
cooling-channel 冷却管道
cooling-hole 冷却孔
cooling-off 冷却,降温
CO-OP Customer On-line Order Processing 用户在线订货单的处理
coop 座舱,尤指驾驶舱,有时指座舱盖
cooper scale 库珀标准(评定飞机安定性和操纵品质等级的标准)
cooperation datum 合作基准
cooperation phase 合作阶段
cooperative aircraft 有应答的飞机(指空中交通管制中,带有二次监视雷达应答机的)
cooperative control 协同控制
cooperative game 合作博弈
cooperative guidance 协同制导
cooperative jamming 协同干扰
cooperative option 合作选择
cooperative path 协作路径
cooperative search 协同式搜寻
cooperative stand-off 合作僵局
cooperative tracking 合作跟踪
cooperative-control 合作控制
coorbital interceptor 同轨截击器
coorbital platform 同轨平台
coorbital satellite interceptor 同轨卫星拦截器
coorbiting area 共轨区
coorbiting pattern 共轨方式
COORD Coordinate(d)/Coordinating 坐标,协调的
coordinate 1.坐标,(复)坐标系 2.协调,调整 3.同等的,对等的,并列的 4.配位,配价
coordinate cadaster 坐标地籍
coordinate conversion 坐标变换
coordinate conversion computer 坐标转换计算机
coordinate conversion unit 坐标变换器
coordinate digitizer 坐标数字化仪
coordinate frame 1.坐标系 2.坐标框架
coordinate geometry language 坐标几何语言
coordinate grid 坐标格网
coordinate measurement machine 坐标测量仪
coordinate measuring instrument 坐标测量仪

coordinate strings 坐标串
coordinate system 坐标系
coordinate transformation 坐标转换
coordinate transformation device 坐标变换器
coordinate turn 协调转弯
coordinate valence 配位价
coordinated attack 协同攻击
coordinated loading 协调加载
coordinated releasing 协调投放
coordinated sideslips method 协调侧滑法
coordinated transpositions 交叉换位
coordinated universal time 世界时,格林尼治平时
coordinates of the pole 地极坐标
coordination 协调,调整,配位
coordination accuracy 协调精确度
coordination architecture 配位构造
coordination center 协调中心
coordination distance 协调距离
coordination drawing 协调图
coordination function 协调功能
coordination policy 协调策略
coordination route 协调路线
coordination strategy 协调策略
coordination variable 协调变量
coordinator 1.坐标方位仪,位器 2.协调员,协调军官
coordinator agent 协调代理
COP 1.Change Over Point 转换点 2.Copilot 副驾驶 3.Copper 铜
Copernicus 哥白尼卫星
copier 1.复印机 2.抄写员 3.模仿者
copilot 1.副驾驶员 2.自动驾驶仪
copilot's airspeed 副驾驶表速
coplanar 共面的
coplanar check 共面检核
coplanar circular orbit 共面圆轨道
coplanar elliptical orbit 共面椭圆轨道
coplanar encounter 共面交会
coplanar interceptor 共面截击器
coplanar orbit 共面轨道
coplanar transfer 共面转移
coplanarity equation 共面方程
copolarization 同极化
copolymer 共聚物,异分子聚合物
copolymerization 共聚作用,异分子聚合
copper 【化】铜(化学元素,符号Cu),紫铜,红铜
copper alloy 铜合金
copper calorimeter 铜热量器
Copper Canyon 铜谷
copper-constantan thermocouple 铜-康铜热电偶

copperizing	镀铜的
copper luminosity	铜的光度
copper nickel alloy	铜镍合金
copper plate	铜板
copper slug	铜嵌条
copper tube	铜管
copper-based catalyst	铜基催化剂
copper-clad	镀铜的,包铜的
coppering	1.镀铜 2.挂铜(射弹的弹带或弹壳留在炮膛内的金属残屑)
copper-oxide	杂的氧化铜
copperplate	镀铜,包铜
copter	〈口语〉直升机
COPUOS	Committee On the Peaceful Uses of Outer Space 和平利用外层空间委员会
copy	1.复制品,副本,拷贝 2.样板 3.复制,仿造,晒印,抄录 4.仿形,靠模加工
copy image	影(图)像复制、拷贝
copy milling	仿形铣削
copy(ing) camera	复照仪,复印机
copyright	版权,著作权
copyright protection	版权保护
COR	1. Circular Of Requirement 要求的通告,需要通告 2. Corporate Operations Review 合作工作评审
CORAL	Computer Online Real-time Application Language 联机实时应用计算机语言
CORBA	Common Object Request Broker Architecture 共同目标要求分解结构,通用对象请求代理结构
COR BD	Conner Bead 角缘,凸圆
cord	1.粗线,细绳,索 2.塞绳,软线 3.伞绳
cord mat	绳垫
cordite	柯达无烟药,无烟线状火药,无烟硝化甘油火药
cordless phone	无绳电话
cordon	警戒线,哨兵线
cords coherent on-receive Doppler system	相关接收多普勒系统
cordtex	爆炸导火索
cordwood	积木式微型组件
core	1.心,核心,主体 2.核心机(燃气轮发动机的燃气发生器部分),焰心 3.心线,铁心
core and first stage impact area	芯级一级落区
core axial velocity	芯轴向速度
core business	核心业务
core compressor	核心压缩机
core concept	核心概念
core cowl	核心整流罩
core deposits	芯区沉积
core diameter	纤芯直径
core drilling	扩孔
core duct	芯管
core engine	核心发动机,核心机
core exit	堆芯出口
core fiber	内芯光纤
core flow	岩心流动
core height	核(炉)心高度
core length	铁芯长度
core mantle coupling	核幔耦合
core module	核心模块
core noise	核心噪声
core of bullet	弹心
core parameter	磁心参数
core power	核心能力
core radius	纤芯半径
core region	核心区
core stage	芯级
core stream	堆芯束流
core temperature	核心温度
core vehicle	核心汽车
core-configured	原始生产型的(飞机)
cored-mould	(药柱内孔)芯模
core-exit gas temperature	堆芯出口气体温度
coreless	无芯的,空心的
core-mantle boundary	核幔边界
core-skin temperature gradient	核皮肤温度梯度
coring	热心流动
Coriolis acceleration correction	科氏加速度修正
Coriolis effect	科里奥利效应,科氏效应
Coriolis error	科里奥利误差,地转偏差
Coriolis force	科里奥利力,科氏力
Coriolis inertial sensor	1.科里奥利惯性传感器 2.科氏惯性传感器
Coriolis stimulation	科里奥利刺激
Coriolis stimulation effect	科里奥利刺激效应
cork	1.软木,软木塞 2.塞柱
corkscrew	1.螺旋,急螺旋下降 2.蛇形飞行
corncob	玉米棒子(俚语,指多气缸多列活塞式发动机)
corner	1.角,隅,棱 2.拐角,转角,隅角 3.带有角度的波导管
corner condition	隅角条件
corner control	拐角控制
corner cube display	角视立体图
corner flow	拐角流
corner fracturing	角断裂
corner frequency	拐角频率,转折频率
corner point	角点
corner recirculation zone	角回流区
corner reflector	角反射器,角反射体
corner separation	角区分离

corner turning memory 转角存储器
corner vortex 角区旋涡
Corogard 珂罗防水漆
Corona 花冠卫星(美国一种照相侦察卫星的代号)
corona 1.日冕 2.冕状极光,极光冕 3.电晕
corona counter 电晕计数器
corona discharge 电晕放电
corona discharge tube 电晕光放电管
corona test 电晕放电试验
coronagraph 日冕仪
coronal 日冕的
coronal condensation 日冕凝聚区
coronal fan 冕扇
coronal helmet 冕盔
coronal hole 冕洞
coronal loop 冕环
coronal mass emission 日冕物质抛射
coronal prominence 冕珥
coronal ray 日冕射线
coronal streamer 冕流
coronal transient 日冕瞬变
coronograph 日冕观测仪
corotating 共转
corotating vortex 共转涡
co-rotating wheels 共转机轮
corotation radius 共转半径
CORP Corporation 公司
corporate data 融合型数据
corporate feed 组合馈电
corporation 团体,协会,有限公司
corposant 电晕放电试验
CORPSK Correlative Phase Shift Keying 相关相移键控
CORR Correct 正确,纠正
CORRES Correspond 相当于,相应于
corpuscular cosmic ray 微粒宇宙(射)线
corpuscular eclipse 微粒蚀
corpuscular heating 微粒加热
corpuscular ionization 微粒电离(作用)
corpuscular radiation 微粒辐射
corpuscular stream 微粒流
correct coefficient of muzzle velocity 初速修正系数
corrected 1.已修正的,校正过的 2.折合的
corrected angle of attack 修正迎角
corrected atitude 修正高度
corrected azimuth 修正的方位角
corrected coefficient of peak pressure 峰值压力修正系数
corrected deflection 修正的方向,修正的方位
corrected elevation 修正的仰角

corrected field 校正场
corrected geomagnetic coordinate 修正地磁坐标
corrected result 已修正结果
corrected speed 折合转速,换算转速
corrected visual acuity 矫正视力
correcting network 校正网络
correction 1.修正,校正,改正,纠正 2.修正量,校正值 3.(数学的)补值
correction angle due to parallax 位差修正角
correction angle due to the force of gravity 抬高角
correction angle due to windage jump 侧偏修正角
correction data 修正数据,换算数据
correction factor 修正因子
correction factor for characteristic velocity 特征速度因子
correction factor for thrust coefficient 推力系数因子
correction factor table 修正因数表
correction for curvature of star image path 星径曲率改正
correction for deflection of the vertical 垂线偏差改正
correction for radio wave propagation of time signal 电磁波传播(时延)改正
correction for skew normals 标高差改正
correction for wind 风修正量
correction from normal section to geodesic 截面差改正
correction network 校正网络
correction of axis system error 轴系误差修正
correction of cross-roll error 滚转误差修正
correction of fire 射击修正
correction of scale difference 挡差改正
correction of sounding wave velocity 声速改正
correction of transducer baseline 换能器基线改正
correction of zero drift 零漂改正
correction of zero line 零(位)线改正
correction process 修正过程
correction table 修正量表
correction term 修正项
correction to time signal 时号改正数
corrective action 校正作用,修正作用,修正动作
corrective adjustment 精确调整
corrective control 校正控制
corrective loop 修正回路
corrective maintenance 修复性维修,校正性维修,事后维修
corrective network 校正网络
corrective torque 校正力矩,修正力矩
correctness 正确度,准确性
correctness proof 正确性证明
corrector 1.修正器,校正器 2.校正电路 3.改正镜

correlate 相关,关联,相关物,联系数
correlation 相关,关联
correlation coefficient 相关系数
correlation contour 相关轮廓
correlation detection 相关探测(探测飞机或宇宙飞行器的一种方法)
correlation detector 相关检测器
correlation diagram 相关图
correlation factor 相关因子,相关因数
correlation function 相关函数
correlation metric 相关性度量
correlation of measurement 相关测量
correlation of models 模型相关
correlation parameter 相关参数
correlation prediction 相关度预测
correlation processor 相关处理器
correlation protection 相关保护
correlation receiver 相关接收机
correlation requirement 相关要求
correlation routine 相关程序
correlation time 相关时间
correlation time constant 相关时间常数
correlation tracking and ranging 相关跟踪测距(系统)
correlative equation 关系值方程式
correlator 相关(分析)仪,相关器
correspond 1.通信 2.符合,一致
correspondence 1.通信(联系) 2.相适应,相应,一致, 3.对应
corresponding boundary condition 相应边界条件
corresponding chamber pressure 相应燃烧室压力
corresponding class 相应类
corresponding epipolar line 同名核线
corresponding equivalence ratio 相应等价比率
corresponding image points 同名像点
corresponding image rays 同名光线
corresponding segment 相应段
corresponding simulation 相应仿真
corresponding state 对应状态
corresponding value 对应值
corridor 走廊,纵向狭隘地形
corridor size 走廊大小
corridor width 走廊宽度
corroborate 1.证实 2.使坚固
corrode 腐蚀,侵蚀
corrodent 1.腐蚀剂 2.腐蚀的
corrodibility 腐蚀性,可蚀性
corrosion control 腐蚀控制
corrosion effect 腐蚀效应
corrosion fatigue 腐蚀疲劳

corrosion inhibitor 阻蚀剂
corrosion pit 腐蚀麻点,腐蚀斑点
corrosion prevention 防蚀
corrosion rate 腐蚀速率
corrosion resistance 耐蚀性
corrosion resistant alloy 耐腐蚀合金
corrosion resistant coating 防蚀涂层
corrosion resistant material 耐蚀材料
corrosion resistant steel 耐蚀钢
corrosion resistant titanium alloy 耐蚀钛合金
corrosion test 腐蚀试验
corrosion-fatigue 腐蚀疲劳
corrosion-proof 抗腐蚀的,耐腐蚀的
corrosion-proofing 耐腐蚀性
corrosion-resistant aluminium alloy 耐蚀铝合金
corrosive nature 腐蚀性能
corrosive salt deposits 有腐蚀性的盐沉淀物
corrosivity 腐蚀性
corrugate 使起皱,成波状
corrugated 1.波纹,波状的 2.弄皱的,起皱的
corrugated horn 波纹喇叭
corrugated nozzle 波纹状喷管
corrugated plate 波纹板
corrugated plate thrust chamber 波纹板式推力室
corrugated skin 波纹表皮,波形表层
corrugation 波纹成形,波纹加工
corrugator 波纹板轧机
corruption 【计】讹误(数据处理存储器的)
CORSA Cosmic Radiation Satellite 宇宙辐射卫星(日本X射线天文卫星)
corundum-mullite ceramics 刚玉莫来石陶瓷
COS 1.Conical Surface 锥形面 2.Cash On Shipment 装货时付款 3.Cosine 余弦
cosecant-squared antenna 余割平方天线
cosecant-squared beam antenna 余割平方波束天线
coset leader 陪集首
COSIMA Crystallization of Organic Substances In Microgravity Application 微重力条件下的有机物结晶
COSIMI Corona Sounding Interplanetary Mission 日冕探测行星际任务
cosine detector sun sensor 余弦检波太阳敏感器
cosine function 余弦函数
cosine power distribution 余弦配电
COSL Component Operation and Store Limit 附件使用和储存时限
Coslettizing 磷化处理
cosmic 宇宙的,宇航的,航天的
cosmic abundance 宇宙丰度
cosmic abundance of elements 元素宇宙丰度

cosmic age　宇宙年龄
cosmic background radiation　宇宙背景辐射
cosmic constant　宇宙常数
cosmic dust　宇宙尘
cosmic electrodynamics　宇宙电动力学
cosmic electron　宇宙射线电子
cosmic expansion　宇宙膨胀
cosmic gamma-ray burst　宇宙射线爆发
cosmic gas　宇宙气体
cosmic mapping　宇宙制图
cosmic mean density　宇宙平均密度
cosmic medicine　航天医学
cosmic microwave background radiation　宇宙微波背景辐射
cosmic noise　宇宙噪声
cosmic particle　宇宙粒子
cosmic plasma　宇宙等离子体
cosmic radiation　宇宙辐射
cosmic radio　宇宙射电
cosmic radio noise　宇宙射电噪声
cosmic radio wave　宇宙射电波
cosmic ray　宇宙(射)线
cosmic ray abundance　宇宙线丰度
cosmic ray acceleration　宇宙线加速(度)
cosmic ray altimetry　宇宙射线测高法
cosmic ray astronomy　宇宙线天文学
cosmic ray burst　宇宙线爆(发)
cosmic ray detector　宇宙射线探测器
cosmic ray intensity　宇宙线强度
cosmic ray jet　宇宙线集流
cosmic ray laboratory　宇宙线实验室
cosmic ray lifetime　宇宙线寿命
cosmic ray magnetic rigidity　宇宙线磁刚度
cosmic ray shower　宇宙线簇射
cosmic ray spectrometer　宇宙线谱仪
cosmic ray storm　宇宙线暴
cosmic ray subsystem　宇宙射线(探测)分系统
cosmic ray telescope　宇宙线望远镜
cosmic ship　宇宙飞船
cosmic velocity　宇宙速度
cosmic Virial theorem　宇宙维里定理
cosmic year　宇宙年(太阳处银河系自转一周的时间)
cosmical physics　宇宙物理学
cosmic-ray equator　宇宙线赤道
cosmic-ray knee　宇宙线膝
cosmic-ray-detection equipment　宇宙射线探测设备
cosmics　宇宙论
cosmism　宇宙进化论
cosmochemistry　宇宙化学,天体化学
cosmochronology　宇宙纪年学
cosmodrome　航天器发射场
cosmogony　天体演化学,宇宙起源论
cosmological principle　宇宙学原理
cosmological redshift　宇宙学红移
cosmology　宇宙学,宇宙论
cosmonaut　航天员,宇航员
cosmonautics　宇航学,星际航行学
cosmophysics　宇宙物理学
cosmotron　高能同步稳相加速器
COSPAR　Committee On Space Research(国际)空间研究委员会
cosputtering　共溅射
COSS　Co-Orbit Support System 同轨飞行支持系统
cossor　"科索尔"(英国二次监视雷达)
cossor IFF　"科索尔"敌我识别器(英国海军 MK10-800系列敌我识别系统)
cost　1.成本,费用 2.(复)经费 3.价值,花费
cost advantage　成本优势
cost analysis　成本分析
cost constraint　成本约束
cost effectiveness　成本效益,投资效果
cost engineering　成本管理
cost estimate　成本估算
cost estimating relationship　经费(成本)估算关系式
cost estimation model　费用(成本)估算模型
cost function　费用函数,成本函数
cost growth　成本增长
cost impact　成本冲击
cost increase　1.成本增加 2.成本超支率 3.费用超支率
cost interaction　成本交互
cost map　生产成本图
cost measure　成本计量
cost method　成本法
cost model　成本模型,费用模型
cost of deviation　成本偏差
cost of operation　1.营业成本 2.管理费 3.运转成本
cost overrun　成本超支
cost parameter　成本参数
cost per seat-mile　每座英里成本
cost performance　成本实效,性能价格比
cost plus fixed fee　成本加固定附加费
cost plus incentive fee　费用加奖金(根据承包商的表现而附加奖金的支付)
cost prediction　成本预测
cost reduction　成本降低
cost saving　成本节约
cost value　成本价值
cost weighting　成本加权

costal 翼肋的,结构肋的
costal breathing 肋式呼吸
COSTAR Corrective Optics Space Telescope Axial Replacement 空间望远镜轴向替换修正镜
Costas loop 考斯脱斯环路,同相正交跟踪环路
costate equation 协态方程
costate initialization 共态初始化
costate system 共态系统
costate variable 共态变量
costate vector 共态向量
cost-effectiveness ratio 成本效益比,费效比
cost-effectiveness trade-off 从费效比考虑的(设计)折中办法
cost-exchange ratio 费用交换比
COSTIND Commission Of Science, Technology and Industry for National Defense 国防科学技术工业委员会(中国)
cost-per-bird 〈口语〉导弹单价,火箭单价
cost-plus-award-fee contract 成本加奖金合同
cost-to-go function 值函数
COT Cotter Pin 偏销,开尾销
Cotar 相关跟踪测距系统,克塔尔测轨系统
co-tidal chart 等潮线图
cotter 销,栓,(杠的)接合榫,合板钉
cotter pin 1.开口销 2.锲形销
cotton underwear 棉背心(飞行员多层着装的第一层)
COT WEB Cotton Webbing 棉织物
COTH Hyperbolic Cotangent 双曲线余切
COTS Commercial Off The Shelf 货架商品
couch 躺椅
coulomb 库仑激励
Coulomb collision 库伦碰撞
Coulomb damping 库伦阻尼
Coulomb field 库伦(静电)场
Coulomb force 库伦力
Coulomb friction effect 库伦摩擦效应
coulometry 电量测定法
coulumb 库伦(电量单位)
CNCL Council 委员会
count 1.计算,读数 2.单个脉冲,单个尖峰信号
count rate 计数率
countdown 1.(准备发射导弹等以前)用倒数方式进行的时间计算 2.准备发射过程 3.(雷达)未回答脉冲率
countdown cycle 发射准备循环
countdown procedure 倒计时程序
counter 1.柜台,计算者 2.计数器,相对的
counter altimeter 计数器式高度表
counter anemometer 计数式风速表
counter arm 计数器指针

counter balance valve 平衡阀
counter-countermeasure 反对抗措施,反干扰
counter-drum-pointer altimeter 计数器数字鼓指针式高度表
counter flow range 逆流靶
counter flow range wind tunnel 逆流风洞
counter hodoscope 计数管描迹仪
counter pressure casting 1.差压铸造 2.反压铸造
counter pressure clothing 代偿服
counter rotating platform (仪表的)反转平台
counter rotation device 抗旋转装置
counter tank 计数存储器
counter telescope 计数管望远镜
counter weapon 1.对抗武器,防御武器 2.截击导弹
counteract 1.抵抗,抵制 2.抵销,阻碍,中和
counteraction 1.反作用,对抗作用,抵抗 2.阻碍,中和
counterair 1.争夺空中优势的 2.空战用的 3.空军中专用于对付敌空军的(部队) 4.防空的
counterair fighter 防空战斗机,截击机
counteraircraft 对飞机防御,对空防御
counterattack 反击,反冲击,反突击
CBAL Counterbalance 平衡,均衡,抵消,抗衡
counterbalancing weight 反配置
counterblow 反击
counterblow hammer 对击锤
counterbomber 反轰炸机的
counterbomber defenses 对轰炸机的防御兵器
counter-circulation 反环流
counterclockwise 反时针的
counter-controller 计数控制器
countereffort 反作用,对抗行动
counterespionage 反间谍
counterfighter 反战斗机的,反歼击机的
counterfighter defense 对战斗机防御
counterfighter maneuver 反战斗机机动
counterflow 反流,逆流
counterflow burner 逆流式燃烧器
counterflow diffusion 逆流式扩散
counterflow diffusion flame 对向流扩散火焰
counterflow flame 对向流火焰
counterforce 打击军事力量(使用战略空军或导弹部队摧毁或削弱敌人军事力量)
counterforce blow 针对军事力量的打击
counterforce targets 战略军事目标
counterforce weapon 打击军事目标的武器(打击敌方战略军事力量的武器)
counterfuze electronic countermeasure 对雷达引信的干扰,反引信电子对抗
counterglow 对日照

counterintelligence 反情报
counter-launching 反导弹导弹发射
countermeasure 对抗措施,干扰
countermeasures environment 电子对抗环境
countermissile 反导弹导弹
counter-mortar radar 反迫击炮雷达
counterpart 配重,配对物
counter-penetration 反突破
counter-pointer (计数器显示和刻度盘指针)双功能指示器
counter-pointer altimeter 计数器指针式高度表
counterpoise 配重,平衡力
counterpreparation 反准备,炮火反准备
counterpressure 反压力,背压力
counterpressure waistcoat 加压背心,代偿背心
counterradiation 反辐射,逆辐射
counterrecoil (机炮的)复进
counterrecoil cylinder 复进管,复进筒
counterrecoil mechanism 复进机,复进装置
counterrecoil velocimeter 复进速度测定仪
counterrotating 相对旋转,相反旋转
counterrotating fan 正反转双风扇
counter-rotating integrated shrouded propfan 同轴反转整体涵道式螺桨风扇
counter-rotating turbine 对转涡轮
counterrotation illusion 反旋转错觉
countershaft 副轴,中间轴
countersign 1.副署,连署 2.回令
countersink 锥形扩孔,钻孔
countersunk rivet 埋头铆钉
countertimer 计数器,计时器
countertorque 反力矩
countertrades 反信风
counter-type adder 计数器型加法器,累加型加法器
countervailing strategy 抵销战略
countervalue targets 经济战略目标,打击经济实力的目标
countervalue weapon 打击价值目标的武器
countervane 导流叶片,反向折流叶片
countervelocity 反导弹导弹飞行速度
counterweight 平衡重,配重
counterweight assembly 平衡重总成
counting accelerometer 计数器式加速度表,载荷因数计数器
counting circuit 计数电路,计算电路
counting ratemeter 计数率表
counting vial 计数管形瓶,计数瓶
countup 正数计时(从火箭起飞开始用"秒"口头报时,而在火箭起飞前用倒数报时)

couple 1.偶,力偶,电偶 2.力偶,力矩 3.连接,耦合,组合
coupled 耦合的,连接的
coupled approach 无线电交连进场着陆,用仪表着陆系统自动进场着陆
coupled cavity slow wave line 耦合腔慢波线
coupled cavity technique 耦合腔技术
coupled effect 耦合影响,耦合效应
coupled engines 并车发动机(将两个发动机并车后带动一个螺桨),成对发动机
coupled flight 耦合飞行(自动驾驶仪与无线电信号耦合操纵)
coupled flutter 耦合颤振
coupled longitudinal vibration 纵向耦合振动
coupled mode 耦合模态
coupler 1.连接器 2.耦合器
coupling 1.连接,结合,匹配 2.耦合,交连 3.交感(作用) 4.连接器,联轴节,联轴器
coupling agent 耦联剂,耦合剂
coupling aperture 耦合孔
coupling capacitor 耦合电容器
coupling coefficient 耦合系数
coupling control 耦合控制
coupling dynamics 耦合动力学
coupling effect 耦合效应
coupling efficiency 耦合效率
coupling factor 耦合因数
coupling hole 耦合孔
coupling impedance 耦合阻抗
coupling loop 耦合环
coupling mechanism 耦合机制
coupling of orbit and attitude 轨道和姿态耦合
coupling parameter 耦合参数
coupling probe 耦合探针
coupling term 耦合项
coupling torque 连接力矩
coupling unit 耦合部件,耦合部分,耦合器,交联装置
coupling variable 耦合变量
coupling vibration 耦合振动
coupling voltage 耦合电压
coupon 凭证,票据
courier 1.军邮班机,传令军官 2.信使,送急件的人
course 1.(CRS)航道,路线 2.无线电指标波束 3.航线角 4.预定航向
course(deviation)bar 航道偏离针
course angle 航向角,航线角
course arrow 航道指针
course correction 航向修正
course deviation 航向偏差

course interception 切入航道,切入航线
course line 航线,轨道
course made good 航向保持正确
course selector 航向选择器
course transmitter 航向传感器
course-line computer 航线计算机
court 1.法院,法庭 2.基地营区
court-martial 1.军事法庭 2.军法审判
court-martial order 军事法庭判决书
COV Cover 盖,包括
covariance 协方差
covariance analysis 协方差分析
covariance error matrix 协方差误差矩阵
covariance expression 协方差表达式
covariance function 协方差函数
covariance matrix 协方差矩阵
covariance method 协方差法
covariance model 协方差模型
covariance propagation 协方差推算
covariance shaping 协方差成形
cove 凹圆线,凹口
covector mapping 余向量映射
cover 1.(COV)覆盖,盖上,包括 2.(CVR)盖子,盖上 3.负担支付 4.弥补
cover distance 覆盖距离
cover gas pressure 覆盖气体压力
cover glass 1.防护玻璃罩 2.盖玻片
cover slide 盖板滑轨
coverage 1.覆盖范围,作用距离 2.台站分布范围 3.覆盖区,摄影区 4.可达范围,视界
coverage area 1.覆盖范围 2.有效区 3.作用范围
coverage boundary 覆盖边界
coverage chart (雷达)探测范围图
coverage cycle 覆盖周期(卫星周而复始地对地球观测一遍所需的时间),重复观测周期
coverage diagram 1.(雷达)探测范围图,覆盖范围图 2.制图资料分布地区示意图
coverage factor 包含因子,覆盖率
coverage gap 覆盖缺口
coverage index 照相侦察范围索引图
coverage of survey area 测区范围
coverage rate 覆盖率
coverage region 覆盖区
coverage reliability 覆盖区域可信度
coverage window 覆盖窗口
coverage zone 覆盖区
coveralls 工作服(连衫裤)
COVERED Covered 覆盖
coverglass 盖片(加在单体太阳电池上表面的透光玻璃,具有保护电池的作用)
coverplastic 塑料套,塑料
cover-pulse jamming 覆盖脉冲干扰
covert 隐藏的,暗藏的
covert strike radar 隐蔽进攻雷达
covert survivable in-weather reconnaissance strike 隐身全天候侦察及对地攻击飞机
covery capsule 生物医学回收舱
covey 〈口语〉飞机编队,飞机队形
covolume 协体积
cowl 1.整流罩,罩,盖上整流罩 2.整流罩,包皮
cowl angle 整流罩角
cowl closure 整流罩关闭
cowl door 整流罩舱门
cowl drag 整流罩阻力
cowl exit 整流罩出口
cowl flaps 整流罩阻力板,整流罩鱼鳞片
cowl hole 整流罩孔
cowl length 整流罩长度
cowl lip 整流罩前缘
cowl shock 整流罩冲击
cowl surface 整流罩表面
cowl wall 整流罩壁
cowling 整流罩,(发动机)包皮
cowl-side 整流罩侧
CP 1.Capacity Planner 能力计划员 2.Capacity Planning 能力计划 3.Certification Plan 合格审定大纲 4.Clock Pulse 时钟脉冲,同步信号脉冲 5.Cockpit Printer 驾驶舱打印机 6.Command Pulse 指令脉冲 7.Computer Programming 计算机编程 8.Condition Process 条件过程 9.Constant Potential 恒电位 10.Constant Pressure 恒压 11.Contracts Procedure 合同程序 12.Control Panel 控制板 13.Corrosion Prevention 防腐蚀 14.Critical Part 关键件 15.Critical Point 临界点
CPA 1.Cathay Pacific Airways 国泰航空公司 2.Closed Packed Array 密集矩阵 3.Closest Point of Approach 最近进近点 4.Computing Process Architecture 计算次序结构 5.Continuous Patrol Aircraft 持续巡逻飞机 6.Control Pannel Assembly 控制板组件 7.Critical Path Analysis 关键路径分析
CPACS Coded-Pulse Anticlutter System 编码脉冲反杂波系统
CPC 1.Cabin Pressure Controller 座舱压力控制器 2.Cargo Processing Contractor 货物处理承包商 3.Call Processing Control 呼叫处理控制 4.Computer Program Components 计算机程序元件 5.Collaborative Product Commerce 产品协同商务 6.Controller Pilot Communication 驾驶员通信控制器

7. Cursor Position Control 光标位置控制
CPCI Computer Program Configuration Item 计算机程序构型项,计算机项目构型项
CPCM Corrosion Prevention and Control Manual 腐蚀预防与控制手册
CPCP Corrosion Prevention and Control Plan 腐蚀预防与控制大纲
CPCS Cabin Pressure Control System 座舱压力控制系统
CPD 1. Charge Priming Device 电荷引发器件 2. Circuit Protection Device 线路保护装置 3. Collaborative Product Development 协同产品开发 4. Concurrent Product Definition 并行产品定义 5. Continuing Product Development 持续的产品开发
CPDLC Controller Pilot Data Link Communications 驾驶员数据链通信控制器,管制员和飞行员数据链通信
CPDM Collaborative Product Development Management 协同产品开发管理
CPE 1. Central Processing Element 中央处理单元 2. Chief Project Engineer 项目主任工程师 3. Circular Probable Error 圆概率误差 4. Contractor Performance Evaluation 合同方性能评估 5. Customer Premises Equipment 用户端设备
CPF 1. Central Programme Function 中央编程功能,中央计划功能 2. Change Process Flow 更改流程
CPFF Cost Plus Fixed Fee 成本加固定附加费,成本加固定费用
CPFSK Continuous Phase Frequency Shif Keying 连续相位频移键控
CPG 1. Call Progress 呼叫进展 2. Clock Pulse Generator 时钟脉冲发生器 3. Co-Pilot Gunner 副驾驶炮手
CPGS Cassette-Preparation Ground Stations 任务资料磁带地面准备站
CPI 1. Center Pressure Index 中心压力指数 2. Change Package Identification 更改包装标记 3. Characters Per Inch 字符数/英寸 4. Cost Performance Index 成本效率指数 5. Common Part Indicator 公共部分指示 6. Crash Position Indicator 失事位置指示器
CPIF Cost Plus Incentive Fee 成本加鼓励费
CPILS Correlation Protected Instrument Landing System 抗干扰仪表着陆系统
CPIN Computer Program Identification Number 计算机程序标识号
CPL 1. Commercial Pilots License 商用飞行员执照 2. Computer Program Library 计算机程序库 3. Couple 耦合 4. Component Parts List 零部件清单
CPL/H Commercial Pilot's Licence/Helicopter 商用直升机驾驶员许可证
CPL/IR Commercial Pilot's Licence/Instrument Rating 有仪表飞行等级的商用飞机驾驶员许可证
CPL/SEL Commercial Pilot's Licence/Single-Engine Limitation 单发动机商用飞机驾驶员许可证
CPLD Coupled 耦合的,连接的
CPLG Coupling 联接的,管接头
CPLR Coupler 耦合器
CPM 1. Call Processing Model 呼叫处理模式 2. Call Progress Message 呼叫进行消息 3. Capacity Passenger-Miles 容量客英里 4. Circles Per Minute 转数/分 5. Consumable Product Manual 消耗性产品手册,消耗品手册 6. Core Processor Module 核心处理模块 7. Cycle Per Minute 周/分,每分周期数
CPN 1. Collins Part Number 柯林斯件号 2. Corporation 公司,法人 3. Closed Private Network 专用闭环网络 4. Critical Path Network 关键路线网络 5. Customer Premises Network 用户驻地网
CPP 1. Columbus Preparatory Programme 哥伦布空间站筹备计划(西欧) 2. Cost Per Passenger 每张机票价
CPPD Capped 加盖的
CPR 1. Coherent-Pulse Radar 相干脉冲雷达 2. Compressor Pressure Ratio 压缩机增压比 3. Chirp-to-Power Ratio 啁啾与功率比 4. Crack-Propagation Rate 裂纹扩展速度 5. Cost-Performance Ratio 价格性能比
CPRA Compressed Pulse Radar Altimeter 压缩脉冲雷达高度计
CPRS Compress 压缩
CPRSN Compression 压缩
CPS 1. Cabin Pressure Sensor 座舱压力传感器 2. Cabinet Pressurization System 客舱增压系统 3. Call Privacy Service 呼叫保密业务 4. Cargo Portal Service 网上货运跟踪 5. Cathode Potential Stabilized tube 阴极电位稳定管 6. Cycle Per Second 周/秒
CPSE Crew and Passenger Support Equipment 机组和旅客保障设备
CPSL Capsule 包套,(密封)舱
CPSR Compressor 压缩机
CPT 1. Cellular Paging Telecommunications 蜂窝寻呼电信业务 2. Check Point 检查点 3. Cockpit Procedure Trainer 飞机座舱程序训练机 4. Cockpit Trainer 驾驶舱训练器 5. Compartment 舱,室 6. Control Power Transformer 电源控制变压器
CPU Central Processing Unit 中央处理组件,中央处理器
CPUAX CPU-Arithmetic Extended 扩充运算中央处理机
CPUNCH Counterpunch 冲孔埋头;反击,回击
CQI Continuous Quality Improvement 持续质量改进

CR 1.Calling Rate 呼叫率 2. Carriage Return (Teletype) 字盘返回,回车 3.Change Record 更改记录 4.Change Report 更改报告 5.Change Request 更变要求,更改请求,更改需求 6.Cold Rolled 冷轧 7.Company Route 公司航路 8.Configuration Rule 配置规则 9.Configuration Report 构型报告 10.Corrosion Resistance 腐蚀防护 11.Credit 信用,赊销 12.Crew 机组人员

CR VAN Chrome Vanadium 铬钒

CRA 1.Composite Research Aircraft 复合材料研究飞机 2.Conflict Resolution Advisory 冲突解决咨询 3.Crew Rest Area 机组休息区 4.Customerized Record Announcement 客户规定的记录通知

crab 1.航向法偏流修正 2.起重吊车 3.巨蟹座,蟹座

crab angle 偏航角,漂移角,偏流角

crab nebula 巨蟹座星云,蟹状星云

crack 1.裂纹,裂缝,开裂 2.(化学的)裂化 3.(起动前)稍稍打开油门

crack absorption 裂纹吸收

crack arrest 止裂

crack arrest hole 止裂孔

crack arrest temperature 止裂温度

crack arrestor 阻裂材,止裂器

crack extension 裂纹扩展

crack growth 裂纹扩展

crack growth rate 裂纹扩展速率

crack growth resistance 裂纹扩展阻力

crack initiation 裂纹萌生

crack initiation life 裂纹形成寿命

crack length 裂纹长度

crack line 裂缝线

crack propagation 裂纹扩展

crack propagation life 裂纹扩展寿命

crack shape factor 裂纹形状因子

crack tip 裂纹尖端

cracking 开裂,龟裂

crack-stopper 止裂构件

crack-up 〈口语〉失事毁机,飞机摔毁

cradle 1.支架,火箭助推器束 2.支架,托架

cradle base 1.支承基架 2.摇篮基座,调整头

craft 飞行器,船舶

CRAMD Cosmic Ray Anti-Matter Detector 宇宙射线反物质探测器

cramps 痛性痉挛,抽筋

crane 1.吊车 2.起重机,用起重机吊

crane helicopter 起重直升机

Cranfield Institute of Technology 克兰菲尔德理工学院(英国)

crank 曲柄,摇把

crankcase 曲轴箱,曲轴机匣

cranked wing 倒海鸥式机翼

crankpin 曲柄销

cranking 1.起动,摇转 2.用曲柄启动或转动

crankshaft 曲轴

crankweb 曲臂,曲柄臂

crash 1.坠毁,失事 2.碰撞,粉碎

crash accelerometer 应急加速度计,应急加速表

crash and salvage crew (航母)失事飞机抢救人员(着红装)

crash arch 安全拱顶(飞机失事时保护飞行员头部的结构)

crash barrier 跑道拦阻装置

crash development 〈口语〉紧急研制,限期研制

crash equipment 应急救生设备,抢救设备

crash exit 应急出口

crash gate 抢救门

crash helmet 防碰撞头盔

crash injury 飞机坠毁损害,坠机损伤

crash locator beacon 失事飞机定位信标

crash pan 防震货盘

crash protection 1.当机防护 2.防坠毁装置

crash recorder 事故记录器

crash safety 坠毁安全性

crash sensor unit 事故情况传感器

crashboat 水上救生艇

crashproof 防摔坏的,不易摔坏的

crashproof tank 防摔油箱

crashworthiness 坠毁承受性

crashworthiness requirement 耐撞性要求

crashworthy 有抗坠毁能力的

crashworthy seat 1.耐坠毁座椅 2.抗坠毁座椅

crate 板条箱,〈口语〉旧飞机

crater 1.焊口,弹坑 2.环形坑,火山口

crater charge 漏斗孔装药,辐射装药(一种破坏表面设施的爆炸装药)

crater test 炸坑试验

cratering bomb 炸坑炸弹

cratering charge 漏斗孔装药,辐射装药

cratering effect 炸坑效果,炸坑效力

craterlet 小火山口,小火口

crawfish 小龙虾投弹器(用这种投弹器,领队的轰炸机可以同时投下编队中飞机的全部炸弹)

crawler 1.履带车 2.履带式发射装置

crawler transporter 履带运输车,履带输送器

crawlerway 1.履带车通道 2.(发射场专为运输火箭或航天器而建造的)慢速道

craze 银纹,发裂

CRB Change Review Board 更改评审委员会

CRC 1. Cyclic Redundancy Code 循环冗余代码 2. Cyclic Redundancy Checking 循环冗余检查

CRCMF Circumference 圆周,四周,周围,周边

CRD 1. Call Rerouting Distribution 重选呼叫路由分布 2. Card 卡片 3. Clock Recovery Device 时钟恢复设备 4. Collision Resolution Device 碰撞检测设备 5. Control Rod Drive 操纵杆传动

CRDA Corporative Research and Development Agreement 合作研究与开发协议

CRDM Control Rod Drive Mechanism 操纵杆传动机构

CRDS Customer Requirement and Distribution System 客户要求与分配系统

CRE 1. Cell Reference Event 信元参考事件 2. Component Responsible Engineering 部件可靠工程 3. Corrosion-Resistant 不锈的,耐蚀的,抗腐蚀

crease 1. 折缝,皱痕 2. 弄皱

credibility 可靠性,可信性

credibility assessment 可信度评估

credible accident 设想事故,可信事故

creep 蠕变

creep compliance 蠕变顺度,蠕变柔量

creep deformation 蠕变变形,蠕变应变

creep fatigue 蠕变疲劳

creep off the target 炸点逐渐远离目标

creep slip 蠕滑

creep strength 蠕变强度

creeping 蠕动,蠕升

creeping adjustment (高炮)逐次接近法射击校正(使炸点逐次接近目标)

creeping method of adjustment 由远到近试射法,炸点逐次接近目标法试射

creeping of grease 润滑脂蠕升

creepmeter 蠕变仪

creep-resistant titanium alloy 1. 抗蠕变钛合金 2. 热强钛合金

creep-rupture test 蠕变破裂试验

crepe ring 薄饼环,C环(土星光环)

CRES Corrosion Resistance Steel 抗锈蚀钢,耐蚀钢,不锈钢

cresceleration 按指数律改变的速度级数,幂次加速度

crescent wing 镰形机翼,新月形机翼

CRESS Center for Research in Experimental Space Science York 空间科学实验研究中心(加拿大大学)

CREST 1. Comprehensive Radar Effects Simulator Trainer 综合雷达效应模拟训练器 2. Crew Escape Technology 机组人员逃生技术(研制弹射座椅用)

crest 1. 顶,牙顶,齿顶 2. 波峰,最大值 3. 饰章

crest factor 峰值振幅因数,峰值因数

crest value 峰值,巅值,最大值

crest voltage 峰(值电)压

crevice corrosion 1. 裂隙腐蚀 2. 隙间腐蚀

Crevv Training Complex 航天员训练场(西欧)

crew 1. 驾驶术 2. 空勤组,机组 3. 空勤组,空勤人员

crew chief 飞机地勤组组长,机工长

crew duty time 机组人员值班时间

crew escape 机组人员逃生

crew member 乘务员

crew ratio (一个单位中)机组人数与飞机数之比

crew size 1. 班组规模 2. 职工数 3. 人员数量

crew station 成员组太空站

crew subsystem 乘员分系统

crew support equipment 乘员生命保障设备,(空勤组)乘员装具

crew trainer 机组教练机

crew training 乘员组训练

crewbunk 空勤人员铺位(飞行器上的)

crewchief (飞机)地勤组长,飞机机械师,空勤机械员

CREWE Coordinated Radio Electrons and Wave Experiment 射频电子和波协同实验

crewing 1. 机组成员,空勤组成员 2. 炮手,操纵手,号手 3. 机组的使用

crewmate 同机组人员,机组同事

crewmember 1. 空勤组成员 2. 炮手,操纵手

CRF 1. Channel Repetition Frequency 信道重复频率 2. Composite Rear Fuselage 复合材料后机身 3. Compressor Rear Frame 压气机后框架 4. Cosmic Ray Flux 宇宙射线通量

CRG Computer Release Group 计算机发放组

CRIB Computerized Resources Information Bank 计算机化的资源信息库

crimp contact 压接接触

crimpage 波纹,皱褶

crimp-on terminals 卷曲终端(采用挤压成形的导线终端)

CRIMS Comet Retarding Ion Mass Spectrometer 彗星减速离子质谱仪

cripple 〈口语〉战斗损坏的飞机,飞机在战斗中的损坏

crippling failure 局部失稳断裂

CRISD Computer Resources Integrated Support Document 计算机资源综合支持文件

crisis stability 危机平衡

CRISP Computer Resources Integrated Support Plan 计算机资源综合支援计划

crisp logic 明晰的逻辑

crispening 勾边,勾边电路

CRIT Critical 临界的,极限的

criterion 准则,标准,尺度,判别式

criterion of strength 强度准则

criterion value 价值准则
critic network 评价网络
critical 临界的,危急的,关键性的
critical altitude 临界高度
critical amount 临界量
critical angle 临界角
critical angle of attack 临界迎角
critical area 临界面积
critical argument 临界幅角
critical Bond number 临界邦德数
critical case 严重情况,危险情况(特指飞机的动力装置和操纵系统的综合情况达到危险程度)
critical charge 临界负荷
critical collision frequency 临界碰撞频率
critical component 关键部件
critical condition 临界条件
critical crack 临界裂纹
critical crack depth 临界裂纹深度
critical current 临界电流
critical damping 临界阻尼
critical damping ratio 临界阻尼比
critical decision 临界判定
critical defect 致命缺陷,严重缺陷
critical density 临界密度
critical diameter 临界直径
critical distance 临界距离
critical disturbance torque 临界干扰力矩
critical dynamic pressure parameter 临界动态压力参数
critical endspeed 临界端速
critical energy 临界能量
critical engine 临界发动机
critical equipotential surface 临界等位面
critical error 临界误差
critical fault 危险性故障
critical flicker fusion threshold 临界闪光融合阈值
critical flight 临界飞行
critical flow 临界流
critical flow coefficient 临界流量系数
critical flutter 颤振临界速率
critical flux 临界流量
critical fuel flux 临界燃料流量
critical grid voltage 临界栅(极)电压
critical heat flux 临界热通量
critical hypoxia 临界性缺氧
critical ignition 临界点火
critical inclination 临界倾角
critical initiation 临界启动
critical initiation condition 临界启动条件
critical initiation energy 临界起爆能量
critical level of atmospheric escape 大气逃逸临界高度
critical load 临界负载
critical load line 临界负载线
critical Mach number 临界马赫数
critical magnetic bond number 临界磁邦德数
critical mass 临界质量
critical mode 临界模式
critical net positive suction head 临界净正抽吸压头
critical number 临界值
critical opening speed 临界开伞速度
critical parachute open altitude 临界开伞高度
critical parachute open dynamic pressure 临界开伞动压
critical parameter 临界参数
critical parts 关键件,关键部位
critical phase 临界相
critical point 临界点
critical power 临界功率
critical pressure 临界压力
critical radius 临界半径
critical reactor 临界反应堆
critical region 临界区域,拒绝域
critical rotor speed 转子临界转速
critical speed 临界转速,临界速度
critical speed map 临界转速图
critical stability 临界稳定性
critical stable state 临界稳定状态
critical strain 临界应变
critical stress 临界应力
critical surface 1.关键表面(有洁净度要求的航天器上设备的表面) 2.临界面
critical temperature 临界温度
critical thermal load test 临界热载荷试验
critical threshold 临界阈限
critical time of exposure 安全暴露时间
critical transmission 临界传输
critical value 临界值
critical velocity 临界速度
critical velocity of atmospheric escape 大气逃逸临界速度
critical void fraction 临界空隙度
critical wavelength 临界波长
critical width 临界宽度
criticality 1.临界性,临界 2.致命度,危险度(对故障后果严重程度的量度) 3.关键性(一种产品或某种功能的必要性的量度)
criticality analysis 临界性分析,致命度分析,危险度分析
criticality category 危害度类别
criticality factor 临界因数

critical-length crack 临界长度裂纹
critical-mass 临界物质
CRM 1. Call Recording Monitor 呼叫记录监视器 2. Chemical Release Module 化学物质释放舱 3. Chrome 铬 4. Cockpit Resource Management 驾驶舱资源管理 5. Collision Risk Modeling 碰撞危机模型 6. Credit Risk Mitigation 信用风险缓释技术 7. Crew Resource Management 机组资源管理 8. Customer Relations Management 客户关系管理
CRMOLY Chrome Molybdenum 铬钼合金,铬钼钢
crop identification performance 作物识别特性
crop information systems 作物信息系统
cropped-fan engine 截尖风扇发动机
cropped surface 截尖翼面
cross 1. 十字,十字队形 2. 叉形,叉形标志 3. 横向的,相反的,互相矛盾的
cross acceleration 交叉加速度
cross ahead interception 前交叉截击
cross beam velocity 射束横向移动速度
cross behind interception 后交叉截击
cross contamination 交叉污染
cross coupling 交叉耦合,交感(作用)
cross deck 相互使用飞行甲板
cross derivative 交叉导数
cross feed 1. 交叉输油,交输 2. 交叉供电,交叉馈电 3. 横进刀
cross fin 十字形稳定翼,十字形安定面
cross flow 横向流
cross fracture of cartridge 弹壳横断
cross hair 1. 交叉瞄准线 2. 十字线 3. 十字型光标
cross modulation 交叉调制,交调
cross passage 横通道
cross planform 交叉平面图
cross point 交叉点
cross polarization 交叉极化,横向极化
cross product 1. 交叉乘积 2. 向量积
cross range 横向距离
cross section 横截面
cross spectrum 互谱
cross stream 横断流
cross stream direction 交叉流方向
cross talk 串扰,串音
cross track distance 偏航距离
cross track scanning 横向扫描,垂直于航向地面轨迹的扫描
cross utilization training 交叉利用训练,兼工训练
cross velocity 横向分速度
cross wind 侧风,逆风,横风
cross wire 十字线,十字标线

crossbar 1.(跑道的)十字灯,灯光地平线 2.(漆在跑道上的)标志横线
crossbar switch 纵横交换机
crossbar transformer 交叉变换器,横杆式变换器
cross-check (飞行中的)反复核对,综合判断(根据仪表指示综合判断飞行情况)
cross-color 串色
cross-correlated phase-shift keying 互相关相移键控
cross-correlation 互相关
cross-correlation analysis 互相关(函数)分析
cross-correlation detection 互相关检测
cross-country flying 越野飞行,中途着陆的长途飞行
cross-country navigation 越野航行
cross-country speed 越野速度
cross-coupling 交叉耦合
cross-coupling error 交叉耦合误差
crosscut vibrographing 横向振动测量
cross-deck pendant wire 甲板拦阻索
cross-derivative 交叉导数
crossed controls 交叉控制
crossed field plasma accelerator 交叉场等离子体加速器
crossed polarization jamming 交叉极化干扰
crossed-field device 正交场器件
crossed-field amplifier 正交场放大管
crossed-field magneto-hydrodynamical accelerator 交叉场磁流体动力加速器
crossed-spring balance 交叉弹簧平衡(风洞中的)态
cross-entropy 互熵,交叉熵
crosser 侧行目标
cross-feed valve 交输阀
crossfire 1. 串报 2. 串扰电流 3. 交叉火力,交叉射击
cross-flow 横向流动
crossflow flame 横流式火焰
cross-flow velocity 横流速度
crossflow vortex 横向气流涡
cross-hair 横标线,十字线
crosshairs 1. 十字准线 2. 瞄准器
crosshatching 画截面线
cross-ignition tube 联焰管,传焰管
crossing 1. 十字路口 2. 杂交 3. 横渡 4. 横道
crossing time 穿越时间
crossing-runway approach 交叉跑道进场着陆
crossline 联络测线
crosslink range 交联范围
cross-linked double base propellant 交联双基推进剂
cross-linking agent 交联剂,交链剂
cross-linking density 交联密度
cross-linking density increase 交联密度增加
cross-loading 交叉乘载

cross-manifold 燃油交输导管
cross-monitoring 交叉监控
crossover 1.交叉,交叠,跨越 2.上方交叉飞过 3.目标通过航路捷径
crossover exhaust 跨越排气
crossover frequency 转换频率,交越频率
crossover limit 交叉限制
crossover point 1.交叉点 2.跨越点
crossover position 交叉位置
crossover probability 交叉概率
crossover speed 转换速度
crossover struts 径向气体斜通道
crossover temperature 横跨温度
crossover voltage 平衡电压,交点电压
cross-passage 区间隧道
cross-plane 横面
cross-plot 综合图表
crosspointer 交叉指针式指示器,十字指针式指示器
cross-pointing 1.(火箭)直角方向改变 2.(航海)双针定位,交点定位
cross-polarization discrimination 交叉极化鉴别
cross-power spectrum 互功率谱
crossrange 侧偏距离,偏航距离,横向航程
cross-range 1.横向航程 2.横向距离
crossrange control 横偏控制,横向偏差修正
crossrange resolution 横向距离分辨
cross-reference 1.互相参照 2.交叉引用
cross-referencing 互相关联,互相参照,交叉引用
cross-refering 十字交叉线
crossrelate 对照,相互关联
cross-roll error 滚转误差
cross-section 横剖面,横截面
cross-section airframe 截面机身
cross-section baseline 截面基线
cross-section missile 截面导弹
cross-section profile 横断面图
cross-section survey 横断面测量
cross-sectional area 横断面积
cross-sectional shape 横截面形状
cross-sectional size 横截面尺寸
cross-servicing 交叉勤务,互助勤务
cross-software test 交叉软件测试
cross-spectral density 互谱密度
cross-spectrum 互功率谱
cross-spectrum phase-angle 互谱相位角
cross-stream velocity 交叉流速度
cross-switch 交叉转接,交叉转换
crosstalk 1.串话,串音 2.(雷达的)串扰 3.交扰,相互影响(通路的)

cross-told data 交叉转送数据
cross-track 航迹侧向
cross-track component 横向分量
cross-track deviation 水平航迹偏离
cross-track direction 航迹方向
cross-track error 横侧偏差,偏离航线距离
cross-track maneuver 交叉轨道机动
cross-track motion 交叉轨道运动
cross-track position error 偏离航线距离测定误差
cross-track resolution 横向分辨率
crosstrail 1.(投弹)横偏长 2.方向偏差
crosstrail angle (投弹)横偏移角
crosstrail error (炸弹的)横偏长误差
cross-turn 交叉转弯
crossunder 下方交叉飞过
cross-validation 交叉验证
cross-validation error 交叉验证误差
crosswind 横风,侧风
crosswind angle 侧风角
cross-wind control 侧风修正
crosswind landing gear 侧风起落架(侧风或侧滑着陆时,主轮可以转向的起落架)
crosswind leg (起落航线)第二边,侧风边
crosswind runway 侧风跑道(作为辅助跑道)
crosswind vortex 侧风涡
cross-wire 十字线
crotchet 1.小钩,钩状物 2.叉架,叉柱
crouch position 蜷缩姿态
crow fly distance 直线距离
crowbar 1.电弧保护开关 2.撬棒
crown 1.凸起部,顶部,圆顶 2.轮周,齿冠,叶冠 3.(降落伞)伞衣上部
CRP 1. Capacity Requirements Planning 能力要求计划 2. Crimp 卷边,曲折
CRPA 1. Controlled Reception Pattern Antenna 手控图形接收天线 2. Controlled Radiation Pattern Antenna 定向辐射天线
CRPL Chrommium Plate 镀铬
CRR Component Retest Request 零部件复试请求
CRS Cold-Rolled Steel 冷轧钢
CRSN Corrosion 腐蚀
CRSV Corrosive 腐蚀性的,生锈的
CRSVR Crossover 跨越,(立体)交叉,截面,跨接结构
CRT 1. Cathode Ray Tube 阴极射线管 2. Correction 更正,改正
CRTG 1. Cartridge 子弹,弹药筒 2. 弹射弹 3. 滤芯
CRU Card Reader Unit 读卡组件
CRUD Create, Read, Update, Delete 创建、读取、更新、删除

crucible	坩锅
cruciform	1.十字形的 2.十字形配置
cruciform canard	1.十字翼鸭式布局 2.鸭式十字翼火箭
cruciform grain	十字形药柱,十字形装药
cruciform tail	十字形尾
cruciform winged rocket	十字翼火箭
cruise	巡航
cruise condition	巡航条件
cruise configuration	巡航布局,巡航构形
cruise drag	巡航阻力
cruise efficiency	巡航效率
cruise engine	巡航发动机
cruise flight	巡航飞行
cruise fuel	巡航燃料
cruise missile	巡航导弹
cruise mode	巡航工作方式
cruise path	1.巡航航迹 2.(远程导弹的)主动段弹道
cruise phase	巡航阶段
cruise range	巡航距离
cruise relief autopilot	巡航自动驾驶仪
cruise speed	巡航速度
cruise stage	巡航阶段
cruise thrust	巡航推力(巡航状态下发动机的推力)
cruise vehicle	巡航飞行器
cruise-climb	巡航上升
cruiser	1.远程(带翼)导弹 2.巡航机,巡洋舰,大型快船,游艇
cruiser-destroyer group	巡洋舰驱逐舰大队
cruising attitude	巡航姿态
cruising phase	巡航段
cruising regime	巡航状态
cruising speed	巡航速度
crumbling	1.碎件 2.粉碎
crump	〈口语〉爆裂弹
CRUP	Cosmic Ray Upset Program 宇宙射线扰动观测计划
crush	压碎
crush force efficiency	压碎力效率
crushing strength	挤压强度
crushing stress	挤压应力
crust	地壳,外壳
crustal deformation	地壳形变
crustal structure	地壳构造
crustal transfer function	地壳传递函数
crutches	止动叉杆
Crux	南十字座
CRWS	Change Request Work Statement 更改需求工作说明
crycon	晶件变频器
CRYG	Carrying 装载的,承载的
cryochemistry	低温化学
cryocondensation	低温冷凝
cryocooler	1.低温冷却剂 2.制冷机
cryocooler system	制冷机系统
cryocooling	低温冷却
cryoelectronics	低温电子学
CRYOG	Cryogenic(100℃以下)低温物理学,低温实验法
cryogen phase	致冷剂相
cryogen thrust	致冷剂推力
cryogenerator	低温发生器,深冷致冷器
cryogenic	低温的,深冷的
cryogenic absorption	低温吸附
cryogenic accelerometer	低温加速度表,超导加速度计
cryogenic ball value	低温球阀
cryogenic boil-off	低温蒸发
cryogenic cooling unit	低温冷却装置
cryogenic electronics	低温电子学
cryogenic engine	低温发动机(指用液氢等燃料的发动机)
cryogenic fluid	1.低温溶液 2.低温流体 3.冷却剂
cryogenic foam	低温泡沫
cryogenic fuel	低温燃料
cryogenic heat pipe	深冷热管
cryogenic liquid	低温冷却液
cryogenic loading system	低温加注系统
cryogenic material	低温材料
cryogenic measurement	低温测量
cryogenic processing	低温处理
cryogenic propellant	低温推进剂
cryogenic propellant rocket engine	低温推进剂火箭发动机
cryogenic propellant tank	低温推进剂储箱
cryogenic rocket engine	低温火箭发动机
cryogenic stage	低温级,低温燃料级
cryogenic storage	低温蓄冷
cryogenic storage tank	低温储罐
cryogenic superconducting gyroscope	低温超导陀螺
cryogenic system	低温系统
cryogenic tank	低温储罐
cryogenic tank structure	低温储罐结构
cryogenic technology	低温技术
cryogenic temperature	1.制冷温度 2.低温度
cryogenic temperature sensor	低温温度传感器
cryogenic transfer	低温传输
cryogenic treatment	深冷处理
cryogenic wind tunnel	低温风洞

cryogenics 低温学,低温实验法
cryopanel 深冷板
cryoprotective agent 防冷剂
cryopump 1.低温泵 2.深冷法供气,深冷抽气
cryopumping 低温抽气
cryoquenching 液态气体中淬火,冷冻液中淬火
cryorefrigerator 低温制冷机,低温冷冻器
cryosar 雪崩复合低位开关,低温雪崩开关
cryosistor 低温晶体管
cryostat 低温恒温器
cryosublimation trap 低温升华阱
cryosurface 低温面
cryotron 低温管,冷子管
cryovalve 低温液活门
cryptanalysis 密码分析
crypto 密码
cryptocenter 密码中心
cryptochannel 密码信道
cryptoequipment 密码设备
cryptogram 密文,密码电文
cryptographic system 密码体制
cryptography 密码术
cryptography code 密码
cryptography key 密钥
cryptoguard 密码保护
cryptology 密码术
cryptomaterial 密码器材
cryptomorphic 隐形的
cryptopart 密码段
cryptosecurity 密码保密措施
cryptosystem 密码体系
cryptotechnique 密码技术
cryptotext 密码电文
crystal 晶体
crystal defect 晶体缺陷
crystal conduction detector 晶体电导型探测器
crystal display 液晶显示器
crystal fiber 晶体光纤
crystal filter 晶体滤波器
crystal growth 晶体生长
crystal habit 1.晶体习性 2.晶体惯态
crystal mixer 晶体混频器
crystal oscillator 晶体振荡器
crystal pickup 晶体感头,晶体传感器,晶体拾音器
crystal size 晶粒大小,晶体粒度
crystal structure 晶体结构,结晶组织
crystal transducer 1.晶体变换器,晶体换能器 2.压电晶体传感器
crystal triode 晶体三极管换能器,压电晶体传感器

crystallinity 结晶度
crystallite size 1.雏晶大小 2.晶粒尺寸
crystallization 结晶化
crystallization process 结晶过程
crystallization remanent magnetization 结晶剩磁
crystal-melt interface 固液凝固界面
CRZ Cruise 巡航
CS 1.Carbon Steel 碳钢 2.Clock Signal 时钟脉冲 3.Coaxial Shaft 同心轴 4.Common Service 公共服务,通用服务 5.Compliance Statement 符合性报告 6.Compressor Stall 压气机失速 7.Compresso Surge 压气机喘振 8.Computer Software 计算机软件 9.Configuration Specification 构型规范 10.Conformity Statement 符合性报告 11.Conformity Surveillance 合格监督 12.Control Surface 操纵面 13.Coordination Sheet 协调单 14.Customer Satisfaction 客户需求 15.Customer Service 客户服务 16.Customer Support 客户支持
CSA 1.Computing System Architecture 计算系统结构 2.Configuration Status Accounting 构型状态纪实
CSAR 1.Component Similarity and Analysis Report 部件相似性和分析报告 2.Configuration Status Accounting Report 构型状态纪实报告
CSC 1.Cargo System Controller 载货系统控制器 2.Computer Software Component 计算机软件部件 3.Computing Services Center 计算服务中心
CSCD Component Specification Control Drawings 部件规范控制图
CSCI 1.Computer Software Configuration Identification 计算机软件构型标识 2.Computer Software Configuration Item 计算机软件构型项 3.Control of Sensitive Company Information 公司敏感信息控制
CSCP Cabin System Control Pannel 座舱系统控制板
CSD 1.Component Specification Drawing 部件规范图 2.Computer Software Documentation 计算机软件文档 3.Constant Speed Drive 恒速转动装置 4.Constant Speed Drive or Control Surface Display 匀速驾驶或操纵面显示 5.Customer Services Division 客户服务部
CSDB Commercial Standard Digital Bus 商用(飞机)标准数字总线
CSDC Computer Signal Data Converter 计算机信号数据转换器
CSDM Computer System Diagnositc Manual 计算机系统诊断手册
CSDP Cold Soak Development Plan 冷浸开发大纲
CSDR Common Strategic Doppler Radar 普通战略多普勒雷达
CSDRN Central Satellite Data Relay Network 中央卫星数据中继网(联)

CSDS　Cargo Smoke Detector System 货舱烟雾探测系统

CSEU　Control Systems Electronics Unit 操纵系统电子装置

CSFDR　Crash Survival Flight Data Recorder 抗坠毁飞行数据记录器

CSG　Constructive Solid Geometry 体素构造法,构造实施几何

CSH　1. Called Subscriber Hold 被叫用户保持 2. Combat Support Helicopter 战斗保障直升机

CSHAFT　Crankshaft 曲轴

CSII　Centre for Study of Industrial Innovation 工业革新研究中心(英国)

CSINK　Countersink 锪窝,埋头孔

CSIR　Council for Scientific and Industrial Research 科学和工业研究委员会(南非)

CSIRO　Commonwealth Scientific and Industrial Research Organization 联邦科学和工业研究组织(澳大利亚)

CSIS　Center for Strategic and International Studies 战略和国际研究中心(美国)

CSKH　Countersunk Head 埋头

CSL　Console 控制台,操纵台,托架

CSM　1. Cabin System Module 座舱系统组件 2. Cabin Service/Management System 客舱服务/管理系统 3. Computer Software Manual 计算机软件手册 4. Computational Structural Mechanics 计算结构力学

CSMA/CA　Carrier Sense for Multiple Access/Collision Aviodance 用于多路存取/防止冲突的载波检测(一种计算机网络或数字通信网络的访问控制方式)

CSMM　Crash Survival Memory Modules 耐坠撞存储模块

CSMU　Cabin System Management Unit 座舱系统管理装置

CSN　Cycles Since Number 新件使用循环数

CSO　Cycles Since Overhaul 翻修以来的循环周期

CSOC　Consolidated Space Operations Center 联合空间操作中心(美国)

CSOS　Customer Specific Option Selection 客户特定选型

CSP　1. Certification Support Plan 合格审定支持大纲 2. Computer Security Program 计算机防护项目

CSPA　Canadian Sport Parachuting Association 加拿大跳伞运动协会

CSS　1. Configuration Specification System 构型规范系统 2. Customer Satisfaction Surveys 用户满意度调查

CSSB-AM　Companded Single Side Band AM 压扩单边带

CSSS　Combat Service Support System 战斗勤务保障系统

CSTC　Consolidated Space Test Center 统一空间试验中心(美国)

CsTe　Caesium Telluride 碲化铯

CSTG　Casting 铸件

CSTL　Castellated 造成城堡形,槽顶的

CSTR　Canister(金属)罐,筒,箱

CSTS　Cross-Software Test System 交叉软件测试系统

CSU　1. Command and Signalling Unit 指令和信号装置 2. Computer Software Unit 计算机软件单元

CSUNK　Countersunk 钻孔装埋,打埋头孔

CT/CR　Chord of Tip/Chord of Root 梢根比

CTA　1. Companion Trainer Aircraft 陪练飞机 2. Controlled Time of Arrival 控制的到达时间

CTAF　Common Traffic Advisory Frequency 公共交通咨询频度

CTAI　Cowl Thermal Anti-Icing 整流罩热防冰

CTAM　Climb to and Maintain 爬升后保持高度

CTC　Cabin Temperature Controller 座舱温度控制器

CTD　1. Coated 镀层 2. Charge Transfer Device 电荷转移器件

CTE　1. Cable Termination Equipment 电缆终端设备 2. Channel Translating Equipment 信道转换设备 3. Customer Terminal Equipment 用户终端设备

CTG　Cuttering 切削,切割

CTK　1. Capacity Tonne Kilometre 吨千米容量 2. Center Tank 中央油箱

CTL　1. Central 中心的,中央的 2. Control 控制

CTLST　Catalyst 催化剂

CTMO　Centralized Air Traffic Flow Managemnt Organization 空中交通流量集中管理组织

CTN　Carton 纸板箱,卡片纸,靶心白点

CTO　1. Conventional Take-Off 常规起飞 2. Chief Technology Officer 首席技术主管 3. City Ticket Office 城区票务办事处

CTOL　1. Contract Tools On Line (System)在线(系统)契约工具 2. Conventional Take Off and Landing 常规起飞和着陆

CTP　1. Connection Terminal Point 连接终端点 2. Critical Technical Parameter 关键技术参数

CTR　1. Center 中心 2. Cutter 刀具

CTRD　Configuration Test Requirements Document 构型试验要求文件

CTRG　Centering 定中心

CTRL　Control 控制

CTS　1. Cabin Temperature System 座舱温度系统 2. Clear To Send 清除发送 3. Component Technical Specifications 部件技术规范

CTSHFT　Countershaft 中间轴,副轴

CTSO　China Technical Standard Order 中国技术标准规定
CTSS　Compatible Time-Sharing System 兼容分时系统
CTU　Cabin Telecommunication Unit 座舱电信系统
CTV　Curved Trend Vector 曲向矢量
CTVS　Cockpit TV Sensor 座舱电视传感器
CTWT　Counterweight 平衡,配重
CTZ　Control Zone 管制地带
CU　1. Control Unit 控制组件,管制单位 2. Cubic 立方的
CUFT　Cubic Feet or Cubic Foot 立方英尺
cub　闭舱式轻型飞机
cubage　货舱容积
cube　1. 立方 2. 立方体,立方形
cube tree algorithm　多维数据集树算法
cube tree model　多维数据集树模型
cube tree trajectory　多维数据集树轨迹
cubic fit　三次拟合
cubic hexapod　立方六足体
cubic model　立方模型
cubic stamping　体积冲压
cubic stiffness nonlinearity　三元刚度非线性
cubic surface　立方面
cubicle　小室,隔间
cubiform　立方形的
cue　1. 感示(感觉的模拟) 2. 提示,暗示(指令) 3. 插入信号(电视的)
cuff　桨叶柄整流袖套
CUIS　Common User Interface Service 公共用户界面服务器
culmination　1. 顶点,极点 2.(天文的)中天
CUM　Cumulative 累加的
cumene　枯烯,异丙基苯
cumulative bomb　聚能穿甲炸弹
cumulative damage rule　累积损伤法则
cumulative delay　累积延时
cumulative deviation　累积离差
cumulative distribution　累积分布
cumulative distribution function　累积分布函数
cumulative equivalence　累积等效
cumulative error　累积误差
cumulative failure probability　累积失效概率
cumulative flight time　累计飞行时间
cumulative frequency　累积频数
cumulative frequency histogram　累积频数图
cumulative missile kill probability　导弹总毁伤概率
cumulative probability　累积概率
cumulative sum chart　累积和图
cumulative time　累积时间

cumuliform　积云状
cumulogenitus　积云性
cumulonimbus　积雨云(符号)
cumulus　积云(符号)
cup　1. 杯,帽,盖,罩 2. 前室,喷注室
cup anemometer　风杯风速表
cupola　1. 炮塔 2. 伞衣
CUR　1. Component Unscheduled Removals 部件非计划拆换 2. Current 电流,现时的
curative　1. 固化剂(燃料的) 2. 治疗物,药品 3. 医疗的,治病的
cure　1. 固化 2. 硫化 3. 治愈
cure condition　固化条件
curie　居里(放射性强度的旧计量单位,符号 Ci)
Curie point　居里点
Curie temperature　居里温度
curing　1. 固化 2.(橡胶的)硫化
curing agent　固化剂
curing catalyst　固化催化剂
curium　【化】锔(化学元素,符号 Cm)
curium isotope　锔同位素
curling die　卷边(压)模
CURR　Component Unscheduled Removals Rate 部件非计划拆换率
current　1. 流,气流,水流 2. 电流 3. 流行的,通用的,现时的
current address　现用地址
current amplifier　电流放大器
current division ratio　电流分配系数
current electricity　电流电(以电子流为特征的电)
current electrode　供电电极
current feedback　电流反馈
current-limiting resistor　限流电阻(器),保险电阻(器)
current source inverter　电流(源)型逆变器
current temperature coefficient　电流温度系数
current temperature coefficient of a solarcell　太阳电池电流温度系数
current transformer　交流器,电流互感器
current user　现有用户,当前用户
current value　现行值
current yield　电流量
current-balance accelerometer　电流平衡式加速度表
current-free electric double layer　无电流电双层
current-free field　无电流场,势场
current-generation　电流产生
current-limited mode　电流限制模式
current-mode　电流模式
current-switching mode logic　电流开关型逻辑
current-voltage characteristic　电流电压特性

cursive writing　笔画式书写(显示器显示字符方式之一)
cursor　1. 指针,指示器,游标 2. 光标(显示器),刻度线
CURT　Curtain 幕,屏
curtailed inspection　截尾检查
curtain　1. 帘,窗帘,屏 2. 幕,烟幕,屏幕
curtain shutter　帘幕快门,焦面快门
curvature　1. 曲率 2. 弯曲,弯曲部分
curvature agility metric　曲率敏捷性度量
curvature code　曲率代码
curvature correction　弹道弯曲修正角,弹道弯曲修正量
curvature discontinuity　曲率不连续
curvature drift　曲率漂移
curvature effect　曲率效应
curvature maneuver　曲率机动
curvature maneuver performance　曲率机动性能
curvature of field　场曲率,地面曲率,象场弯曲
curve　曲线,弯曲
curve approach　1. 曲线拟合 2. 曲线分析法
curve channel　曲线通道
curve diagram of surface movement　地表移动曲线图
curve factor　曲线因子
curve family　曲线族
curve fitting　曲线拟合
curve gain regulation　曲调
curve of growth　生长曲线
curve of trajectory　弹道曲线
curve slope　曲线斜率
curved-finger gloves　可弯指手套
curvilinear target approach　曲线接近目标法
CUS　Customer 用户
CUSH　Cushion 垫子,缓冲器
cushion　1. 垫(子),软垫,减震垫,减震器 2. 减震,缓冲
cushion layer　缓冲层
cushioncraft　气垫运载器(泛指气垫的车、船和飞行器)
cushioning action　缓冲作用,减震作用
cusp　1. 尖顶,尖端,尖点 2. 小海角
cusp electrode　尖端电极
cusp region　极尖区
custody　保管,监管
customer　1. 〈口语〉受油机(空中加油的) 2. 用户,买方,顾客 3. 消耗器,耗电器
customize　(根据用户的要求)定做
customized wavelet　自定义小波
cusum chart　累积和图
cut　1. 减少,缩短 2. 切割,割纹
cut plane　剖面
cut set　割集
cutaway　1. 露空部分 2. 剖面,切去

cutback distance　斜切距离
cut-back nozzle　斜切喷管
cut-bombing　遮断轰炸
cut-in　1. 接通,开动 2. 插入,切入,割入
cutlass　弯刀形编队
cutoff　1. 断开,切断,截止 2. 停车,关机 3. 切半径(拦截) 4. 结束工作
cutoff accuracy　关机精度
cutoff angle　熄火点弹道倾角,主动段终点弹道倾角
cutoff attenuator　截止式衰减器
cutoff diameter　截止直径
cutoff energy　截止能量
cutoff equation　关机方程
cutoff filter　截止滤光片
cutoff frequency　截止频率
cutoff impulse　后效冲量
cutoff ports　关机口
cutoff rigidity　截止刚度
cutoff valve　断流活门,停车开关
cutoff velocity　截止速度
cutoff voltage　截止电压,断电电压
cutoff wave number　截止波数
cutoff waveguide　截止波导
cutoff wavelength　截止波长
cutout　1. 切开,切断,断绝,终止 2. 关闭(发动机),停车 3. 断流器,保险装置 4. 切口,仪表上的小窗口
cutout switch　断路器,断流器,切断开关
cutset code　割集码
cutter　1. 刀具,铣刀 2. 切割机 3. 割纹头(机械录音头) 4. 小汽艇,快艇,小型沿海武装艇(美国)
cutting　切削,切断
CV　1. Casecade Vane 叶栅叶片 2. Charging Valve 充气活门,添加活门 3. Check Valve 单向活门,检查活门 4. Computer Version 计算机图像 5. Crossfeed Valve 交输活门,交输阀 6. Customer Variables 客户变型
CV/DFDR　Cockpit Voice and Digital Flight Data Recorder 驾驶舱语音和数字式飞行数据记录器
CVD　Chemical Vapour Deposition 化学气相沉积
CVE　Compliance Verification Engineer 符合性验证工程师
CVFR　Controlled Visual Flight Rules 可控视觉飞行规则
CVR　Cockpit Voice Recorder 驾驶舱录音机,飞机座舱录音机
CVRCP　Cockpit Voice Recorder Control Panel 驾驶舱录音机控制板
CVRSN　Conversion 转换,改变,转换
CVS　1. Correlation Velocity Sensing 相关速度敏感 2. Customer Variable Selections 客户可变选装

CVSD Continuously Variable Slope Deltamodulation 连续可变斜率增量调制
CVW Carrier Air Wing（航空母舰）航空联队的代号（英国）
CVWR Carrier Air Wing Reserve 预备役航空母舰航空联队的代号（英国）
CVX Convex 凸的,凸形的,凸面,凸圆
CW 1. Carrier Wave 载波 2. Clockwise 顺时针 3. Continuous Wave 等幅波
CW defence radar 连续波警戒雷达
CW Doppler fuze 连续波多普勒引信
CW instrumentation radar 连续波测量雷达
CW laser 连续波激光器
CW laser fuze 连续波激光引信
CW laser radar 连续波激光雷达
C-washer 开口垫圈,C 形塑圈
CWC 1. Customer Web Center 用户网络中心 2. Comparator Warning Computer 比较警告计算机
CWCS Common Weapons Control System 共用武器控制系统
CWE Caution and Warning Equipment 注意和警告设备
CWFS Crash Worthy Fuel System 碰撞安全的燃料系统
CWG 1. Charges Working Group 货运计费工作组（国际空运协会的）2. Commercial Working Group 商用工作组（美国空间站商用项目评选工作组）
CWI CATIA-WIRS Interface CATIA-WIRS 界面
CWP 1. Central Warning Panel 中央告警面板 2. Compact When Packed 收拢时体积很小的（天线）
CWS 1. Caution and Warning System 提醒和警告系统 2. Central Warning System 中央警告系统 3. Container Weapon System 集装武器系统,箱式武器系统,子母弹武器系统 4. Control Wheel Steering 驾驶盘操纵
CWU Caution and Warning Unit 注意和告警器
CWV Crest Working Voltage 峰值工作电压
CWW Cruciform-Wing Weapon 十字翼型武器
CXR Coaxial Rotor 1. 同轴双旋翼（直升机的布局）2. 同轴转子
CY 1. Calendar Year 历年 2. Cycle 周期,循环
cyanate 氰酸酯
cyaniding 氰化,加氰硬化
cyano 含氰基的
cyanoacetylene 丙炔腈
cyanoethylene 丙烯腈
cyanogen 氰
cyanogen radical 氰基
cyanometer 天空蓝度测定仪,天蓝计
cyanometry 天空蓝度测定法

cyanosis 发绀,青紫
cybernetic machine 控制论机器
cybernetics 控制论
cycle 1. 周期,周,一转 2. 循环,一个操作过程 3. 天体运转轨道
cycle ambiguity 波未定值
cycle amplitude 循环振幅
cycle analysis 周期分析
cycle calculation 周期计算
cycle diagram 循环图
cycle efficiency （热力）循环效率
cycle engine 二冲程发动机
cycle frequency 循环频率
cycle gas turbine 闭循环燃气轮机
cycle graph 循环图
cycle life 循环寿命
cycle of gun charging 装弹循环
cycle of oscillation 周期振荡
cycle performance 循环性能
cycle period 循环周期
cycle skipping slip 跳周,滑步
cycle slip detection 周跳探测
cycle temperature 循环温度
cycle time 循环时间
cycler search 循环搜索
cycler trajectory 循环轨迹
cycles-to-failure 无故障使用次数
cyclic（pitch）control 1.（直升机的）周期变距操纵 2.（直升机的）驾驶杆
cyclic channel （旋翼的）变距通道
cyclic code 循环码
cyclic code subframe synchronization 循环码副帧同步
cyclic control 周期距控制
cyclic hysteresis error 循环迟滞误差
cyclic loading 1. 循环荷载 2. 周期荷载 3. 交变载荷
cyclic magnetization 循环磁化（强度）
cyclic mode 循环模式
cyclic nitramine 环杂硝胺
cyclic permutation 循环排列
cyclic pitch 周期螺距
cyclic product code 循环乘积码
cyclic rate 发射率,发射速度
cyclic redundancy check 循环冗余校验
cyclic redundancy code 循环冗余码
cyclic remote control 循环遥控
cyclic symmetry 循环对称
cyclic testing 循环试验
cyclic torque 周期性力矩
cyclic twist 循环扭

cyclic variability　循环变动
cyclical　1.循环的 2.黑子周的 3.太阳周的
cyclical redundance correction　循环冗余校正
cyclic-pitch stick　1.周期变距操纵杆 2.驾驶杆
cycling　1.循环,周期变化,摆动 2.循环测试 3.周期工作的
cyclocrane　翼桨式旋转飞艇
cyclogenesis　气旋生成
cyclogiro　横轴旋翼机
cyclogirocycloidal gear　摆线齿轮
cyclogyro　旋翼机
cyclohexane　环己烷
cycloidal gear system　摆线齿轮系
cycloidal pump　摆线泵
cyclolysis　气旋消失
cyclometer　测圆弧器,跳字转数表
cyclone　气旋,旋风
cyclonite　旋风炸药,三次甲基三硝基胺
cyclopentadienyl　环戊二烯基
cyclopentane　环戊烷
cyclophon　旋调管
cyclosrophic　旋衡的
cyclostasis　节律稳定
cyclotron　回旋加速器
cyclotron damping　回旋阻尼
cyclotron frequency　回旋频率
cyclotron radiation　回旋辐射
cyclotron resonance　回旋共振
cyclotron-resonance heating　回旋共振加热
cyclotron wave　回旋波
Cygnus　1.【天】天鹅座 2.天鹅属(鸟类)
CYL　Cylinder 圆柱,油缸
cylinder　1.气缸 2.(液压或气缸)作动筒 3.圆筒,圆柱形气瓶
cylinder block　气缸柱,气缸座,气缸体
cylinder cavity　缸腔
cylinder head　气缸头,缸头
cylinder head thermometer　气缸头温度表
cylinder liner　气缸套,气缸筒
cylinder panel　气缸面板
cylinder rotation　油缸转动
cylinder study　气缸研究
cylinder wall　气缸壁
cylinder with locking device　带锁紧装置的液压缸
cylindrical aperture　圆柱孔
cylindrical array　柱面阵
cylindrical cadmium-nickel battery　圆柱型镉镍蓄电池
cylindrical charge　柱状装药
cylindrical combustor　圆柱形燃烧室
cylindrical container　圆柱形容器
cylindrical detonation wave　圆柱形爆轰波
cylindrical duct　圆柱管
cylindrical geometry　圆柱几何体
cylindrical helix　柱面螺旋线
cylindrical hole　圆柱形孔眼
cylindrical jet　圆柱射流
cylindrical motor　圆柱形电机
cylindrical pipe　圆柱管
cylindrical projection　圆柱投影
cylindrical reflector　圆柱形反射器
cylindrical reflector antenna　柱形抛物面反射体天线
cylindrical robot　圆柱坐标型机器人
cylindrical section　圆柱段
cylindrical shell　柱状壳体
cylindrical slip ring　圆柱导电环
cylindrical tube　柱形管
cylindrical vessel　圆筒形容器
cylindrical wave　柱面波
Cymbeline　辛勃林(一种探测追击炮位置的轻型便携式雷达,也用来调整炮火)
Cyrano　"西拉诺"(法国机载单脉冲火控雷达)
CYSA　Cape York Space Agency 约克角航天局(澳大利亚)
cytohyperoxia　细胞内氧气过多,细胞内氧气过多症
cytohypoxia　细胞缺氧症
CZ　Control Zone 管制范围
CZCS　Coastal Zone Color Scanner 沿海水色扫描仪
Czochralski method　切克劳斯基法(一种全新的晶体生长方法)

D

D 1. Date 日期 2. Day 白天,昼间 3. Deleted 删除 4. Departure 起飞,离港 5. Desk 书桌 6. Diameter 直径 7. Diode 二极管 8. Distance 距离 9. Drag 阻力

D 100 figures 名义速度为100英尺/秒时的阻力

D criterion D准则

D layer 电离层D层(高度60~80千米)

D PNL Distribution Panel 配电盘

D&B Docking and Berthing 对接和停泊

D&C Delay and Cancel 延误和取消

D&D 1. Design and Development 设计和研制 2. Diesel and Dye 柴油与颜料 3. Distress and Diversion 遇险与飞向备降场(空中交通管制用语)

D&PD Design and Drafting Directives 设计和制图指南

D&F Determination and Findings 测定与结论

D&O Description and Operation 说明和操作

D&PD Definition and Preliminary Design 技术条件确定和初步设计

D&V Demonstration and Validation 演示和证实

D.C trembler-operated ignition unit 直流断续器操纵的点火器

D/A 1. digital/analog 数字/模拟 2. Digital To Analog Converter 数模转换器

D/M Demodulate-Modulate 解调/调制

D/W Depth-to-Width ratio 深宽比

DA 1. Data Analysis 数据分析 2. Database Audit 数据库审核 3. Decision Altitude 决断高度 4. Deck Altitude 甲板待战 5. Delayed-Action 延时爆炸的(炸弹) 6. Delta Airlines 德尔塔航空公司,三角航空公司 7. Design Assurance 设计保证 8. Diplomatic Authorization 外交当局 9. Direct Action 瞬发 10. Director of Aircraft 飞机指挥员 11. Documentation Audit 文档审核 12. Double Attack 双向攻击 13. Drift Angle 偏流角 14. Dual-Alloy 双合金的

DA/H Decision Altitude/Height 决断高度/高

DAA 1. Data Access Arrangement 数据存储装置 2. Digital Analog Adapter 数字模拟转接器 3. Directorate of Air Armament 空中武器装备处

DAACM Direct Airfield-Attack Cluster Munition 直接攻击机场集束弹药,直接攻击机场子母弹

DAAT Digital Angle-of-Attack Transmitter 数字式迎角发送器

DAB 1. Defense Acquisition Board 国防采购局(美国) 2. Delay(ed)-Action Bomb 延时炸弹 3. Digita Audio Broadcasting 数字式声频广播

Dabber welding 达伯焊接

DABNLS Discrete Address Beacon Navigation and Landing System 离散地址信标导航和着陆系统

DABNS Discrete Address Beacon and Navigation System 离散地址信标和导航系统

DABS Discrete Address Beacon System 离散地址信标系统,离散型信标系统

Dabsef Dabs Experimental Facility 达布斯研究所

DAC 1. Data Acquisition Camera 数据采集摄影机 2. Data Acquisition Computer 数据采集计算机 3. Data Acquisition Center 数据获取中心 4. Douglas Aircraft Company 道格拉斯飞机公司 5. Digital to Analog Converter 数字-模拟变换器

DACS 1. Data Acquisition and Computer System 数据采集和计算机系统 2. Data Acquisition and Control System 数据采集和控制系统 3. Directorate of Aerospace Combat Systems 航空航天作战系统管理局(加拿大)

DACT 1. Dissimilar Air Combat Training 异型空战训练 2. Direct Acting 直接行动

DAD 1. Deep Air Defence 纵深防空 2. Dual Authority Dataset 双权数据集

DADC Digital Air Data Computer 数字式航空(大气)数据计算机

DADS Digital Air Data System 数字式大气数据系统

Daedalians 军用飞机驾驶员互助会成员(美国)

Daedalus 1. 代达罗斯星际飞船 2. 代达罗斯空间平台(欧洲)

DAF 1. Drawing Action to Follow 随后要进行的绘图工作 2. Delay(ed)-Action Fuze 延期引信

DAFICS Digital Automatic Flight Inlet Control System 数字式自动飞行入口控制系统

DAG Deutsche Astronautische Gesellschaft 德国航天学会

dag 石墨粉

daily cost of operation 日运转成本

daily demand 每日需求量

daily diurnal variation 日变化,周日变化

daily extremes 日极值

daily mean sea level 日平均海面

daily means 逐日平均,日均值

daily motion 日运动

daily rate 日速

daily system operability test 系统战备状况每日测试

DAIS 1. Data Acquisition and Interpretation System 数据采集和判读系统 2. Digital Avionics Information System 数字式航空电子设备信息系统 3. Direct Access Intelligence System 直接存取智能系统

daisy chain 1. 多机集体仪表进场 2. 串联空投包

daisy-clipping 〈口语〉掠地飞行

daisy-cutter 〈口语〉杀伤炸弹

DAL Digital Accelerometer Loops 数字式加速度计回路

Dalmatian effect 达尔马蒂亚效应

Dalton's Law 道尔顿定律

daltonism 色盲

DAM 1. Data Addressed Memory 数据定址存储器 2. DECT Authentication Module DECT 认证模块 3. Dollars per Aircraft-Mile 飞机每飞行一英里的美元费用数

dam 1. 水坝,堤,堰,障碍物,屏障 2. 断,拦阻,抑制,封闭

DAMA Demand-Assigned Multiple Access 按需分配多址

damage 损伤,破坏,击毁

damage agent 1. 战斗装药 2. 杀伤体

damage area 受灾面积

damage assessment 损坏程度评定,毁坏情况估计,破坏效果评估

damage control (舰艇)损(害)管(制)

damage controller 损失控制,损害管制,损害防控,损坏管理制

damage cycle 毁损期

damage detection 灾害探测

damage effect 毁伤效应

damage function 损害函数

damage gettering technology 损伤吸除工艺

damage induced defect 损伤引生缺陷

damage limitation 限制损害

damage location 破损位置

damage mark 破损标记

damage metric 损伤指标

damage parameter 破坏参数

damage potential 破坏潜力

damage probability 毁伤概率

damage radius 破坏半径

damage rate 损伤率

damage ratio 损坏率

damage risk contour 损伤危险等值线

damage threshold 损伤阈值,损坏阈

damage tolerance 容损设计(要求系统在损坏容限值内仍能工作的设计原则)

damage tolerance design 损伤容限设计

damage tolerant 损伤容限的,损伤可容的

damage volume 杀伤区域,杀伤范围

damage volume of missile 导弹的威力范围

damage-based approach 基于损害方法

damaged aircraft 受损飞机

damaged area 烧毁面积,损毁面积

damaged blade 受损叶片

damaged panel 损坏面板

damaged plant 受损设备

damage-equivalent fluence 损伤等效通量

DAME 1. Data Acquisition and Measuring Equipment 数据采集和测量设备 2. Data Acquisition and Monitoring Equipment 数据采集和监视设备 3. Designated Aviation Medical Examiner 指定航空体检医生,指定航空医学体检医生,指定航空医学检验员 4. Division Airspace Management Element 划分空域管理小组

damp 1. 阻尼,缓冲,减幅,衰减 2. 湿气,潮湿 3. 弄湿 4. 潮湿的

damped 阻尼的,衰减的

damped dynamical system 动态阻尼系统

damped mounting rack 减震安装架

damped natural frequency 有阻尼的自然频率,有阻尼的固有频率,自然衰减率

damped oscillation 阻尼振荡

dampen 1. 阻尼,减幅,衰减 2. 弄湿

dampener 缓冲器,阻尼器,抑制器

damper 1. 阻尼器,减震器,缓冲器 2. 减震配重

damper model 阻尼器模型

damper tube 阻尼管

damping 阻尼,衰减

damping action 衰减作用,阻尼作用,缓冲作用

damping adjustment 阻尼调整,阻尼调节

damping capacitance 阻尼容量

damping control 阻尼调整,阻尼控制

damping derivative 阻尼导数

damping due to rotation 旋转阻尼

damping factor 减幅因数,阻尼因(系)数,衰减系数

damping fluid 阻尼液

damping grease 阻尼脂

damping material 阻尼材料

damping matrix 阻尼矩阵

damping radiation 阻尼辐射

damping ratio 阻尼比

damping ratio of slosh 晃动阻尼比

damping sleeve 阻尼筒

damping structure 阻尼结构

damping term 阻尼项
dampness 潮湿,湿度
damp-proof 防潮的,抗湿的
dampy 潮湿的潮的,抗湿的
DAMS Drum Auxiliary Memory Subunit 磁鼓辅助存储器装置
daN Load (DecaNewton) 负载(十牛顿)
DANAC Decca Area Navigation Airborne Computer 台卡区域导航机载计算机
danger 危险,危险物
danger area 危险区(该地区在一定时间内有危及飞行的活动)
danger beacon 危险灯标
danger zone 危险区,危险地带
danger zone boundary 危险区域边界
dangerous fragment 危险的碎片
dangle 拖曳角
DAO Defence Attache Office 武官处
DAP 1. Data Acquisition and Processing 数据采集和处理 2. Data Access Point 数据接入点 3. Digital Array Processor 数字式阵列处理机 4. Distortion of Aligned Phases 已调整相位的失真 5. Digital Avionics Processor 数字式航空电子处理器
DAPE Developed Armament Probable Error 武器概率误差
DAPR Digital Automatic Pattern Recognition 数字式自动模式识别
daptation 适应(作用)
DAR Design Assurance Review 设计保证审查
DARA Deutsche Agentur for Raumfahrt 德国航天局
daraf 拉法(法拉的倒数)
DARC Digital Access Radar Channel 雷达数字式存取通道
DARD Directorate of Armament Research and Development 军械研究发展局(英国)
daredevil 〈口语〉莽撞的飞行员
dark 黑的,暗的,隐藏的
dark adaptation 对黑暗的适应
dark burn 暗伤
dark characteristic curve 暗特性曲线
dark cloud 乌云,黑云,暗云
dark cockpit 〈口语〉指一切正常,即座舱内所有的警告灯都不亮
dark companion 暗伴星
dark core 黑核
dark core length 黑核长度
dark current 暗电流
dark discharge 暗放电
dark field 暗视野,暗视场

dark lane 暗带
dark limb 暗临边
dark matter 暗物质
dark nebula 暗星云
dark noise 暗噪声
dark of the moon 1. 新月暗期 2. 无月光期
dark radiation 暗辐射
dark satellite 秘密卫星,哑卫星,失去作用的卫星
dark smoke 黑烟
dark space 暗区
dark star 暗星
dark zone 暗区,预备区
dark zone reaction 暗区反应
dark-core length 暗区长度
dark-eclipsing variable 暗食变星
darken 使变暗,使模糊
darkened conditions 全黑(暗)条件下
darkening 1. 变音,变黑 2. 灯火管制
darkfire 暗中发射,隐密发射
darkness visual acuity 暗视力
darkroom 暗室
dark-smoke zone 黑烟区
dark-trace 暗迹
Darlington pair 达林顿对
Darlington power transistor 达林顿功率管
DARPA Defence Advanced Research Projects Agency 国防远景研究计划局(美国国防部)
DARS 1. Data Acquisition and Recording System 数据采集和记录系统 2. Digital Adaptive Recording System 数字式自适应记录系统 3. Digital Attitude-Reference System 数字姿态基准系统
Darsonval meter 达松瓦尔电流计
DART 1. Directional Automatic Realignment of Trajectory 轨迹方向自动重新对准 2. Dual-Axis Rate Transducer 双轴速率传感器
dart 1. 标枪 2. 〈口语〉航空火箭 3. 箭形
DART stabilization system 1. 达特(稳定)系统 2. 弹道方向自动调准系统
Darter 海鲫(南非研制的空空导弹)
DARTS 1. Derivative Aircraft (Airplane) Requirements & Tabulantion System 变形飞机要求和制表系统 2. Design Approach for Real-Time Systems 实时系统设计法
Darwin ellipsoid 达尔文椭球体
DAS 1. Data Acquisition System 数据采集系统 2. Data Analysis System 数据分析系统 3. Defensive Avionics System 防御型航空电子系统 4. Design Assurance System 设计保证系统 5. Digital Avionics Systems 数字电子设备系统 6. Digital Assembly Sequnce 数字化

装配过程设计

DASA Defense Atomic Support Agency 国防原子能支援局（美国）

DASC Direct Air Support Center 直接空中支援中心

DASD Data Access Storage Device 数据存取装置

DASH Drone Anti-Submarine Helicopter 无人驾驶反潜直升机

dash 1.猛冲，冲撞 2.冲刺 3.飞溅

dash board 仪表板

Dash One 〈口语〉飞行员驾驶守则，飞机飞行使用手册

DASH weapons system 无人驾驶反潜直升机武器系统

dashpot 缓冲器，阻尼器，缓冲筒

DASS 1. Defence Aids Support System 防御手段保障系统 2. Demand Assignment Signaling and Switching Unit 按需分配信令交换单元

DAT Datum 数据，资料，基准点

data （datum 的复数）1.数据，诸元，资料，信息，特性 2.已知数，论据 3.基准面

data acquisition 数据采集，数据获取

data acquisition and interpretation system 数据采集和解释系统

data acquisition bus 数据采集总线

data acquisition logging system 数据采集记录系统

data acquisition system 数据采集系统

data acquisition tracking and telecom-mand station 数据采集、跟踪和遥控指令站

data adaptive monitor 数据自适应监测器

data age 取数据时间

data analysis 数据分析

data analyzer 数据分析仪

data and control interface unit 数据和控制接口装置

data association 数据关联

data bank 库集（计算机库的集合）

data base 数据库

data broadcasting 数据广播

data buffer unit 数据缓冲装置，数据缓冲器

data buoy 数据浮标

data bus 数据总线

data bus control unit 数据总线控制装置，数据总线控制器

data bus protocol 数据总线规约

data capsule 数据舱

data capture 数据捕取，数据采集

data cell unit 磁卡片机

data channel 数据通道

data circuit 数据电路

data circuit terminating equipment 数字电路终接设备

data collection 数据收集，数据采集

data collection and location system 数据收集和定位系统

data collection and platform location system 数据收集和观测平台定位系统

data collection satellite 数据收集卫星

data collection subsystem 数据收集分系统

data command 数据指令

data communication 数据通信

data communication input buffer 数据通信输入缓冲器

data communication network 数据通信网

data compression 数据压缩

data conversion 数据转换

data conversion routine 数据转换程序

data converter 数据转换器

data correction 数据修正

data correction and transfer system 数据修正和传输系统

data correlation 数据相关

data direct programming system 数据直接编程系统

data display system 数据显示系统

data driven 数据驱动

data element 数据元

data encryption 数据加密

data error analyzer 数据误差分析仪

data exchange center 数据交换中心

data exchange control unit 数据交换控制装置

data extrapolation 【数】数据外插（法），【数】数据外推（法）

data file 数据文件

data filtering 数据滤波

data flow computer 数据流计算机

data flow diagram 数据流图

data fusion 数据汇合

data generator 数据发生器

data ground station 数据地面站

data handling subsystem 1.数据管理分系统 2.数管分系统

data handling unit 数据处理器

data identifier 数据标识符

data in 输入数据，输入诸元

data independence 数据独立性

data injection 数据注入

data insertion converter 数据插入转换器

data integrity 数据完整率，数据净差错率

data interface 数据接口

data interpretation 数据解释

data link 数据传输线，数据链，数据链路

data link equipment 数据输送设备

data link escape character 数据链转义字符

data link guidance 数传引导

data link layer 数据链路层
data link layer data logger 数据记录器
data loading 数据注入
data logging 数据记录
data management 数据管理
data memory 数据存储器
data modem 数据调制解调器
data multiplexer 数据复用器
data object 数据体
data out 输出数据，输出诸元
data output multiplexer 数据输出多路传输装置，数据输出复接器
data packet acknowledgement 数据包应答
data packet size histogram 数据包容量直方图
data packet switching network 数据分组交换网
data parameter 数据参数
data path parallelism 数据通路并行性
data plane 数据平面，数据层
data point 数据点
data preprocessing 数据预处理
data processing 数据处理
data processing antenna 数据处理天线
data processing center of remote sensing 遥感数据处理中心
data processing security 数据处理保密
data processing system 数据处理系统
data processor 数据处理机
data product 资料产品
data rate 数据率
data read out 数据读出
data reasonableness test 数据合理性检验
data receiver 数据接收机
data record 数据记录，资料记录
data recording 数据记录
data reduction 数据简化，数据压缩
data reduction error 数据简化误差
data relay satellite 数据中继卫星
data relay station 数据中继站
data retention period 数据保存周期
data retrieval 数据检索
data sampling 数据取样
data service 数据服务，资料服务
data set 数据集，数据传输设备
data smoothing 数据平滑
data snooping 数据探测法
data source 数据源，数据发送器
data stable platform 数据稳定平台
data storage 数据存储，数据存储器
data storage unit 数据存储器

data stream 数据流
data stream computer 数据流计算机
data structure 数据结构
data study 数据研究
data synchronization 数据同步
data system 数据系统
data tablet 数据输入板
data tape 数据（记录）带
data terminal equipment 数据终端设备
data timing transfer 数据定时传输
data tracks 数据磁道
data transcription equipment 数据转存设备
data transducer 数据变换器
data transfer equipment 数据传送设备
data transfer unit 数据传送装置
data transformation unit 数据转换装置
data translation 数据转换
data transmission equipment 数据传输设备
data transmission interface 数据传输接口
data transmission rate 数据传输速率
data transmission subsystem 1.数据传输分系统 2.数传分系统
data transmission systems 数据传输系统
data transmitter 数据传感器，数据发送器
data type default rule 数据类型缺省规则
data type designator 数据类型标志符
data type rule 数据类型规则
data unit 数据单位
data validity 数据有效性
data way 数据路
data window 数据窗口
data word 数据字，作数据的字
data-acquisition 数据采集，信号采集
data-acquisition equipment 数据收集设备
data-analysis time 数据分析时间
data-association problem 数据采集问题
database 数据库，资料库
DATAC Digital Autonomous Terminal Access Communications 数字自主终端访问通信
data-collecting device 数据收集器
data-gathering satellite 数据收集卫星
datagram 数据（电）报
data-input unit 数据输入器
datalink 数据链，数据链路，数据传输线路
data-link 数据传输器
data-logger （工作）数据记录器
data-logging computer 数据记录计算机
data-logging equipment 数据记录设备
data-management unit 数据管理单元

datamation 数据自动处理
data-only satellite earth station 只传输数据的卫星地球站
data-output unit 数据输出器
dataphone 数据电话机
dataplotter 数据标绘器
data-processing equipment 数据处理设备
data-reduction 数据简化,资料缩减
data-reduction equipment 数据简化设备
Datasat 数据卫星
dataset 资料组
datasets 数据集,数据集集合,数据集对象(dataset 的复数形式)
data-takeoff equipment 数据读出设备
DATAX Data Transmission 数据传输,信息转输
date 日期,年代,历史时期
date line 日界线
dated item 1.限时使用件 2.不耐用产品
datin data inserter 数据插入程序,数据输入器
DATIS Digital Automatic Terminal Information System 数字式自动终端信息系统
DATS Despun Antenna Test Satellite 消旋天线试验卫星(美国空军)
DATSA Depot Automatic Test System for Avionics 修理工厂用的航空电子设备自动测试系统
DATT Defense Attache 国防武官
DATTS 1.Data Acquisition,Telecom mand and Tracking Station 数据采集、遥控和跟踪站(德国) 2.Data Acquisition, Telecommunication and Tracking Station 数据采集、通信和跟踪站
datum (data 的单数)基准线,基准点,基准面读数起点,给定值
datum acquisition 数据采集
datum acquisition system 数据采集系统
datum aircraft 基准航空器
datum airspeed 基准空速
datum analysis 数据分析
datum association 数据联系
datum bus 数据总线
datum case 基准实例
datum collection 数据收集
datum comparison 基准比较
datum design 基准设计
datum diffuser 基准扩散器
datum dropout 原始数据丢失
datum error 原始数据误差,基值误差
datum flow 基准流
datum format 数据格式
datum fusion 数据融合

datum length 基准长度
datum level 基准面
datum link 数据链接
datum model 模型基准,模型基准面
datum of elevation 高程基准
datum point 基准点
datum quality 数据品质
datum rate 基准利率
datum record 数据记录
datum recording 数据记录
datum reduction 数据减少
datum service 数据服务
datum set 基准设置
datum simulation 数据模拟
datum space 基准空间
datum static correction 基准面静校
datum study 基准研究
datum transmission 数据传送
datum value 基准值
datum window 数据窗口
datum word 基准词
DAU 1.Data Acquisition Unit 数据采集装置 2.Directly Administered Units 直属单位(英国空军)
daughter ion 子离子
daughter payload 子负荷
daughter satellite 子卫星
daughter spacecraft 子飞船,子航天器
daughter vehicle 子运载工具
Dauphin 海豚(法国直升机名)
DAVC Delayed Automatic Volume Control 延迟自动音量控制
Davis barrier 戴维斯拦阻装置
Davis wing 戴维斯翼
dawn side 晨侧
dawn-dusk electric field 晨昏电场
dawn-to-dusk 从拂晓到黄昏
day 1.日 2.白天,白昼
day and night distress signal 昼夜遇险信号弹
day assumption 昼间假设
day condition 昼间状况
day fighter 昼间战斗机
day flight 昼间飞行
day hemisphere 向阳半球
day of operation 昼间操作
day of self-pressurization 自我加压时间
day of year 积日
Day Ranger "白日流浪者"行动
day shift 日班
day to day bias 逐日偏移

day to day stability 逐日稳定性
day tracer 曳光弹
day with fog 雾日
day/night 昼夜练习器
daybreak 黎明,拂晓
Day-Glo 昼晖颜料
dayglow 日辉
daylight 1.昼光 2.昼间 3.清早
daylight IFR flight 昼间复杂气象飞行,昼间仪表飞行
daylight saving time 夏令时
daylight VFR flight 昼间简单气象飞行,昼间目视飞行
day-night effect 昼夜效应
day-off 休息日
dayplane 昼航飞机
dayside （行星）的光面
day-to-day drift 逐日漂移
dazer 致眩器
dazzle （太阳等）眩眼,耀眼
dazzle camouflage schemes 迷彩伪装色
DB 1.Data Base 数据库 2.Data Bus 数据总线 3.Decibel 分贝（声强单位） 4.Depth Bomb 深水炸弹 5.Distribution Box 配电箱,分线盒 6.Double 两倍 7.Drag Brace 阻力撑杆 8.Dry Battery 干电池
db of reduction 还原数据库
DB propellant 双基推进剂
db reduction 数据库还原
DB/DC Data Base/Data Communication 数据库/数据通信
dba 1.database administrator 数据库管理人员 2.decibel adjusted 调整分贝 3.decibels absolute 绝对分贝
DBA Dynamic Bandwidth Allocation 动态带宽分配
DBC 1.Data Bus Coupler 数据总线耦合器 2.Data Bus Check 数据总线检查 3.Denied Boarding Compensation 让座补偿金
DBF Demodulator Band Filter 解调器带通滤波器
DBFM Defensive Basic Flight Maneuvers 防御性基本飞行动作
DBIA Data Bus Isolation Amplifier 数据总线隔离放大器
DBL 1.Development Baseline 设计基线,研发基线 2.Diameter Bolt Circle 圆形螺杆直径 3.Double 双,加倍
DBLR Doubler 倍频器
dbm 毫瓦分贝
DBM 1.Data Base Management 数据库管理 2.Dynamic Balance Mechanisms 动力平衡装置
DBM/C Data Bus Monitor/Controller 数据总线监控器
DBMS 1.Data Base Management System 数据库管理系统 2.Doppler Bombing Navigation System 多普勒轰炸导航系统
D-bomb 深水（航空）炸弹
DBS 1.Database Storage 数据库存储 2.Direct Broadcast Satellite 直接广播卫星（简称:直播卫星）
DBSC Direct Broadcast Satellite Corp. 直播卫星公司（美国）
DBSS Direct Broadcast Satellite Service 直播卫星广播业务
DBT 1.Diffusion-Bonded Titanium 扩散结合钛 2.Design-Build Team 设计建造组 3.Double Base Transistor 双基极晶体管
DBTF Duct-Burning Turbofan 外涵道燃烧加力涡扇发动机
DBTT Ductile-to-Brittle Transition Temperature 韧脆性转变温度
DBU Database Unit 数据库装置
DBUP Defence Build-Up Plan (Programme) 国防发展计划（大纲）（日本）
DBV Diagonally Braked Vehicle（跑道）摩擦力试验车
DBW Decibel-Watts 分贝-瓦,瓦分贝
dbw model 瓦分贝模型
DC 1.Data Collection 数据采集 2.Departure Control 离场控制 3.Depth Charge 深水炸弹 4.Design Change 设计更改 5.Design Control 设计控制 6.Design Cost 设计成本 7.Development Configuration 研制构型 8.Development Cost 开发成本 9.Direct Cycle 直接循环 10.Direct Current 直流电（流） 11.Double Contact 双接点 11.Dispersion Constant 色散常数 12.Drag Control 阻力控制
dc arc 直流电弧
dc arc discharge 直流电弧放电法
dc discharge 直流放电
dc electric drive 直流电气传动
dc electric field 直流电场
dc electrode 直流电极
dc generator-motor set drive 直流发电机-电动机组传动
dc power 直流电源
dc power failure 直流电源故障
dc shift 直流漂移
DCA 1.Data Correction Amplifier 数据校正放大器 2.Decimal Classification of Astronaucs 航天文献十进制分类法 3.Deck Carrier Assembly 上承货架装置（美国空间站上的） 4.Defense Communications Agency 国防通信局（美国） 5.Defense Contre Avions 防空,对空防御（法国） 6.Defensive Counter-Air 防御性反航空兵（作战） 7.Department of Civil Aviation 民航部（巴西） 8.Directorate of Civil Aviation 民航局 9.Director of Civil Aviation 民航局长 10.Document Change

Analysis 文件修改分析 11. Document Content Architecture 文件内容结构(IBM 公司) 12. Double Circular Arc 双圆弧(叶型) 13. Dual Capable Aircraft 双重能力飞机

DCA EUR Defense Communication Agency Europe 欧洲国防通信局

DCAA Defense Contract Audit Agency 国防合同审计局(美国)

DCAC Define and Control Airplane Configuration 飞机构型定义和控制

D-carts decoy cartridges 诱馆弹，假目标弹

DCAS 1. Defense Contract Administration Service 国防合同管理处(美国国防部) 2. Deputy Chief of Air Staff 空军副参谋长(英国) 3. Digital Core Avionics System 数字式核心航空电子设备系统 4. Digitally Controlled Audio System 数控音频系统

DCC 1. Data Collection Center 数据收集中心 2. Data Collection Coding 数据收集编码 3. Dataset Configuration Control 数据集构型控制 4. Development Configuration Control 开发构型控制 5. Digital Computer Complex 数字计算机综合体 6. Direct Computer Control 直接计算机控制 7. Discrimination and Control Computer 鉴别与控制计算机 8. Drone Control Center 无人机控制中心

DCCS Distributed Capability Computing System (计算)能力分布式计算系统

DCD 1. Data Collector and Diagnoster 数据搜集器和诊断器 2. Dimensional Control Drawing 尺寸控制图 3. Double Channel Duplex 复合双通道

DCDI Digital Course Deviation Indicator 数字航道偏差指示器

DCDR Decoder 译码器

DCDS Distributed Computing Design System 分布式计算设计系统

DCE 1. Data Communications Equipment 数据通信设备 2. Display Central Equipment 中央显示设备 3. Distributed Computing Environment 分配的计算设施

DCFS Digitally Controlled Frequency Service 数字控制频率业务

DCH Destination Change 航线终点变更

DCHEM Dry Chemical 化学干燥

DCICB Distributed Computing Implementation Co-ordination Board 分配的计算工具协调委员会

DCIS Distribution Construction Information System 分布结构信息系统

DCIU Digital Control and Interface Unit 数字控制接口部件

DCL 1. Defence Contractors List 国防承包商清单 2. Direct Communications Link 直接通信线(华盛顿—莫斯科1978年启用的首脑卫星通信热线) 3. Drawing Change List 图纸更改单

DCLR Decelerate 减速

DCLS Data Collection and Location System 数据收集与定位系统

DCME Digital Circuit Multiplication Equipment 数字话路复用设备

DCMF Data Conmmunication Management Function 数据通信管理功能

DCMS 1. Data Communication Management System 数据通信管理系统 2. Distributed Cost Management System 分配的成本管理系统

dcmsnd decommissioned 退役的

DCMU Digital Colored Map Unit 彩色数字地图显示设备

DCN 1. Design Change Notice 设计更改通知 2. Distributed Computer Network 分布式计算机网络 3. Documentation Change Notification 文件更改通知 4. Drawing Change Notice 图纸更改通知 5. Drawing Change Notice Or Design Change Notice 图纸更改通知或设计更改通知

DCNR Dataset Classification, Notes and Revisions 数据集分类、标注和修订

DCO 1. Delivery Configured Orbiter 运货轨道器(美国) 2. Detailed Checkout 详细检查

Dcode D码(在飞行设计中表示有测距仪)

Dcontroller 微分控制器

DCP 1. Data Collection Platforms 数据收集平台 2. Data and Command Processor 数据和指令处理器 3. Design Change Proposal 设计更改建议 4. Display Control Panel 显示控制板 5. Distributed Communications Processor 分布式通信处理机

DCPLS Data Collection and Platform Location System 数据收集平台定位系统

DCPRS Data Collection Platform Radio Set 数据收集平台无线电台

DCPT Data Collection Platform Transponder 数据收集平台应答机

DCPU Display Control Power Unit 显示控制功率单位

DCR 1. Data Conversion Receivers 数据转换接收器 2. Destruct Command Receiver 自毁指令接收机 3. Differential Correlation Radiometer 微分相关辐射计 4. Digital(ly) Coded Radar 数字编码雷达 4. Design Compliance Reports 设计符合性报告 5. Drawing Change Request 图纸更改申请

DCS 1. Data Collection System 数据收集系统 2. Data Command System 数据命令系统 3. Data Communication System 数据通信系统 4. Defense

Communication System 国防通信系统 5. Despin Control System 消旋控制系统 6. Digital Communication System 数字通信系统 7. Distributed Computer System 分布式计算机系统 8. Double Channel Simplex 单工双通道

DCS/R&D Deputy Chief of Staff for Research and Development 主管科研的空军副参谋长

DCSP Defense Communications Satellite Program 国防通信卫星计划(美国)

DCSS Defense Communication Satellite System 国防通信卫星系统(美国)

DCT 1. Digital Communications Terminal 数字式通信终端 2. Direct Cosine Transform 直接余弦变换

DCTE Data Circuit Terminating Equipment 数字电路终端设备

DCTL Direct Coupled Transistor Logic 直接耦合晶体管逻辑

DCTRB Distributed Computing Technical Readiness Board 分配的计算技术准备状态委员会

DCU 1. Data Concentrator Unit 数据集中装置 2. Data Control Unit 数据控制装置 3. Dedicated Control Unit 专用控制装置 4. Digital (Engine) Control Unit 数字(发动机)控制装置 5. Digital Correlation Unit 数字相关单元

DCV 1. Demonstrated Crosswind Velocity 指示侧风速度 2. Directional Control Valve 单向控制阀

DCWS Debris Collision Warning Sensor 空间碎片防撞报警传感器(美国)

DD 1. Data Delivery 数据发送 2. Data Demand 数据要求 3. Design Document 设计文档 4. Digital Display 数字显示 5. Direct Drive 直接传动 6. Dispatch Deviation 缺件放行形态变动 7. Dynamic Data 动态数据

DD&T Design, Development & Test 设计、研制与试验

DDA 1. Design Deviation Authority 设计修改权,设计偏差批准权 2. Design Deviation Authorization 设计偏差负责部门 3. Distance Data Adapter 远距数据适配器

DDAPS Digital Data Acquisition and Processing System 数字数据采集和处理系统

DDAS Digital Data Acquisition System 数字数据采集系统

DDBS Distributed Data Base System 分布式数据库系统

DDC 1. Decision, Design and the Computer 判定、设计和计算机 2. Digital Display Console 数字显示操纵台,数字显控台 3. Direct Digital Control 直接数字控制

DDCS Digital Data Calibration System 数字数据标定系统

DDCU DC to DC Convertor Unit 直流-直流变换器装置(美国空间站上的)

DDD 1. Deadline Delivery Date 交货截止日期 2. Dual Disk Drive 双盘驱动

DDDRSS Department of Defense Data Relay Satellite System 国防部数据中继卫星系统(美国)

DDE 1. Digital Data Exchange 数字化数据交换 2. Direct Data Entry 直接资料输入

DDI 1. Digital Display Indicator 数字式显示器 2. Direct Digital Interface 直接数字接口

DDL 1. Data Description Language 数据描述语言 2. Data Dialog (System) 数据对话系统 3. Down Data-Link 下行数据链

DDM 1. Design, Drafting and Manufacturing 设计、绘图与制造 2. Design-Driven Manufacturing 设计直接传控制造(技术) 3. Design Decision Memo 设计决定备忘录 4. Design Definiton Memo 设计定义备忘录 5. Difference in Depth of Modulation 调制深度差 6. Drop Dynamics Module 液滴动力学实验舱(西欧Spacelab中的)

DDMS Drawing Data Management System 图纸数据管理系统

DDP 1. Declaration of Design and Performance 设计与性能公告 2. Design Data Package 设计数据包 3. Demage Detection Period 损伤探测周期 4. Detail Design Phase 详细设计阶段

DDPE Digital Data Processing Equipment 数字数据处理设备

DDPG Dispatch Deviation Procedure Guide 放飞偏离程序指南,派遣偏差指南

DDPS 1. Digital Data Processing System 数字数据处理系统 2. Discrimination Data Processing System 鉴别数据处理系统

DDR 1. Data Descriptive Record 数据描述记录 2. Detail Design Review 详细设计评审 3. Diluter Demand Regulator 稀释氧气调节器 4. Draft Document Review 文件初稿评审 5. Drawing Data Record 文件数据记录 6. Dual Differential Radiometer 二元微分辐射计

DDR&E 1. Defense Development Research & Engineering 国防研究与工程署 2. Director of Defense Research & Engineering 国防研究与工程处处长(美国国防部)

DDRM Data, Document and Records Managemsnt 数据、文件和记录管理

DDRS Digital Data Recording System 数字数据记录系统

DDS 1. Data Display System 数据显示系统 2. Direct Digital Synthesizer 直接数字式合成器 3. Digital Data

Service 数字数据业务 4. Display and Debriefing Subsystem 显示与分析子系统 5. Drawing Data Sheet 图纸数据单

DDT 1. Direct Digital Targeting 直接数字式瞄准 2. Downlink Data Transfer 下链数据传输 3. Dynamic Debugging Technique 动态调试技术

DDT&E 1. Design, Development, Test and Engineering 设计、研制、试验和工程 2. Design, Development, Test and Evaluation 设计、研制、试验和鉴定

DDTS Digital Data Transmission System 数字数据传输系统

DDVR Displayed Data Video Recorder 显示数据视频记录仪

DE 1. Data Entry 数据输入 2. Decision Element 判定元件 3. Design Engineering 设计工程 4. Directed Energy 定向能 5. Dynamics Explorers 动力学探险者卫星（美空间物理探测卫星）

de effective drag 除有效阻力，无效阻力

de excitation 去激活，去活化作用

de Laval nozzle 拉瓦尔喷管，收扩喷管

DEA Delegated Engineering Authority 指定工程负责部门

deaccentuator 去加重电路，减加重电路，校平器

deactivation 1.(化学)减活化(作用)，钝化(作用) 2. 消除(放射性)沾染 3. 遣散，撤销 4.(枪炮、爆炸物等的)保险，卡死，失灵

dead 1. 静的，停滞的 2. 失效的，断电的，熄火的(指发动机) 3. 死的，无放射性的

dead band 不灵敏区，死区

Dead Duck "死鸭"(苏联海军敌我识别系统的北约代名)

dead engine 死发动机

dead fluid zone 气流死区

dead point 死点，静点

dead reckoning 1. 推测领航，航位推测法 2.(计算机的)推算定位

dead rise 1. 底部升高量 2. 船底倾斜度

dead room 静室，消声室

dead space 1. 盲区 2. 静区 3.(高炮的)死角 4. 空舱位

dead spot 静点，死角点，不灵敏区

dead stick 〈口语〉1. 对停车飞机的驾驶 2. 停车着陆

dead time 1. 空载时间，静寂时间，空闲时间 2. 延迟时间，滞后时间，补给品拨给时间 3. 射击前瞄准时间，(高炮)发射延迟时间 4. 死时间，失效时间

dead time correction 死时间校正

dead volume 死体积

dead vortex 死涡

dead weight tester 破码式(压力)测试仪

dead zone 1. 盲区 2. 禁射区 3.(飞机)超声探伤无效区 4. 死区

dead zone of infrared homing head 红外导引头盲区，红外导引头不灵敏

deadband 死区，死谱带，静带

dead-band control 死区控制

deadband region 死区区域

dead-beat 1. 速示的，不摆的，不振荡的，非周期的 2. 速示(仪表指针)，不摆，无差拍

deadbeat condition 非周期状态

deadbeat observer 有限拍观测器

deaden 堵塞，减震，减弱

deadeye 穿孔滑车

deadhead 1. 送修飞行 2. 搭机飞行 3. 空载返航

deadline 1. 最后限期 2. 确定不适用

deadlock (计算机)死锁，无效锁定

deadly embrace 死锁，锁死

deadman 锚桩，吊货杆牵索

dead-reckoning 推测航行法

dead-reckoning navigation 推测领航

dead-reckoning plot 推测领航的图上作业

dead-rise angle 底部升高角，侧缘角

dead-stick landing 停车着陆

deadzone 提供输出为零的区域，死区

deaeration 排除空气，排气

deaerator 油气分离器，除气器，脱氧器

deafen 消音，隔音

deafener 消音器

dealer 商人，经纪人

Deal-Grove model 迪尔格罗夫模型

dealkalization 脱碱(作用)

death date (人造卫星)坠落日期

death organ 〈口语〉机枪

DEB Detector Electronic Box 探测器电子设备盒

debark 1. 下飞机 2.(从飞机上)卸货

debond 脱黏，脱胶

debond angle 脱黏角

debonding 脱黏

deboost 1. 制动，阻尼，减压，减速 2. 降低增压 3. 限制电压 4. 回力，减小助力

debooster 1. 减速器，制动器，减压器 2. 限压器 3. 回力机构

debriefing 飞行后汇报，飞行后分析

debris 碎片，碎屑，空间残骸

debris cloud 碎片云

debris distribution model (太空)碎片分布模式

debris environment 碎片环境

debris flux model (太空)碎片通量模式

debris fragment 残骸碎片

debris growing rate 碎片增长率

debris growth 碎片增多
debris hazard （太空）碎片危害
debris impact 碎片碰撞
debris object 碎片物体
debris particle 碎片颗粒
debris pattern 碎片形状
debris population 碎片群
debris size 碎片大小
debug 1.调理,调整 2.排错,排除故障,调试(计算机程序) 3.寻找并排除窃听器
debugging 调试
debugging software package 调试软件包
debunching 1.散束,电子离散 2.散焦
deburring 去毛刺
Debye length 德拜长度
Debye shielding distance 德拜屏蔽距离
Debye temperature 德拜温度
DEC 1. Decimal 小数,十进的 2. detached experiment carrier 离机实验台
deca 倍乘字,用作前缀,代表10倍,同
deca metric 波长相当于10米的,高频(无线电)波的
decade 十年,十年期,十
decade resistance box 十进电阻箱
decading counting tube 十进制计数管,十进管
decal 识别标记
DECAL Decalcomania 移面印花法
decalage 翼角差,机翼安装角差
decalcification 脱钙
decalescence 钢材吸热现象
decalescence point （钢条)吸热点
decalin 萘烷,十氢化萘
decan 黄道十度分度
decane 癸烷,十炭矫质
decane conversion 癸烷转换
decane-air 癸烷空气
decane-air mixture 癸烷空气混合物
decans 旬星
decant 放泄的
decarbonization 脱碳(作用),失碳
decarbonizer 除碳剂,脱碳剂
decarbonylation 脱碳作用
decarburizing 脱碳,除碳(钢的)
decay 1.衰减 2.衰变,蜕变(放射性物质的) 3.余辉 4.电荷减少
decay coef 衰减系数
decay constant 衰变常数
decay factor 衰减系数
decay product 衰变产物
decay rate 1.衰减率 2.衰变率

decay time 1.衰减时间 2.衰变时间
decayed object (轨道)衰减物体
Decca flight log "台卡"导航设备飞行记录
Decca lane "台卡"导航系统航道
Decca positioning system 台卡定位系统
decelerate 减速,降低速度
decelerating electrode 减速电极
deceleration 减速度
deceleration limit 1.负过载限度 2.减速极限
deceleration magnitude 减速量级
deceleration module 减速模块
deceleration parachute 减速伞
deceleration parameter 减速因子
deceleration-time profile 减速时间曲线图
decelerator 减速装置,减速器
decelerator grid 减速器网格
decelerometer 减速计
deceleron 减速副翼
decelostat (机轮的)防拖胎装置,刹车压力自动调节装置
December solstice 1.冬至 2.冬至点
decentrality 分散性
decentralization 分散化
decentralized approach 分散处理方法
decentralized control 分散控制,分散指挥
decentralized control law 分散控制律
decentralized controller 分散控制器
decentralized estimation 分散估计
decentralized estimator 分散估计器
decentralized filter 分散滤波器
decentralized robust control 分散鲁棒控制
decentralized stabilization 分散镇定
decentralized state 分散状态
decentralized state estimator 分散状态估计量
decentralized stochastic control 分散随机控制
deception 欺骗,迷惑,伪装,遮盖
deception jammer 欺骗式干扰机
deception jamming 欺骗干扰
deception meaconing 欺骗模拟干扰
de-chirping 去线性调频(脉冲)
decibel (dB)分贝(声强单位)
decibel above reference noise 超过基准噪声的分贝数
decidability 可判定性,可判定能力
decimal 1.小数,十进小数 2.十进位的
decimal adder 十进制加法器
decimal addition 十进制加法
decimeter wave 米波
decipherable map 易读地图
deciphering 解释,解密

decision 判决,决策
decision altitude 决断高度
decision analysis 决策分析
decision criterion 决策基准,决策准则
decision cycle 决策周期
decision function 决策函数
decision height 决断高度
decision height abuse 决断高度违例
decision instant 决策瞬时
decision level 判定电平,决策阶层,决策位准
decision logic 判定逻辑
decision maker 决策者
decision maker preference 决策者的偏好
decision making 决策
decision model 决策模型
decision point 决策点,决定点
decision process 判定过程
decision program 决策程序
decision space 决策空间
decision speed 起飞决断速度
decision support 决策支持
decision support system 决策支持系统,判定支援系统
decision support tool 决策支持工具
decision table 判定表
decision theory 决策(理)论
decision threshold 判定阈
decision time 抉择时刻,决定时间
decision tree 决策树
decision variable 决策变量,决策变数
decision-directed adaptive estimation 决策引导自适应估计
decision-directed adaptive scheme 决策引导自适应策略
decision-maker 决策者
decision-making 决策
decision-making process 决策程序
decisive action 果断的动作
deck 1.甲板,地板,层舱(大型飞机的) 2.驾驶舱(运输机的) 3.最低飞行高度 4.平台,场地 5.【气】(云)盖 6.(多翼机的)一层机翼 7.(卡片)组,叠
deck alert state 甲板战备状态
deck bearing 甲板方位角
deck coordinate system 甲板坐标系
deck department (航母)甲板部门
deck edge 甲板边缘
deck elevation 甲板高低角
deck handling 甲板作业
deck landing trial 甲板降落试验
deck letter 甲板符号
deck limitation 甲板的有限性

deck park 甲板停机坪,甲板停放区
deck plate 台面板,甲板板
deck run (航母上,非弹射起飞的)甲板起飞滑跑距离
deck-based 舰基的
deck-edge elevator 甲板缘升降机
Decker "装饰者"
decking 机身顶面
deck-landing 在甲板上降落
deck-launched 甲板弹射起飞的
DECL Declaration 申报
declaration 投入兵力规模
declarative language 申请语言
declared 申报资料
declared alternate 申报备降机场
declared destination 申报目的地
declared force 特遣部队
declared life 宣称寿命
declared thrust 公告推力
declared weight 申报重量
declination 1.偏差,倾斜,偏向 2.偏角,磁偏角 3.赤纬
declination axis 赤纬轴
declination circle 1.赤纬圈 2.赤纬度盘
declination compass 1.赤纬计 2.磁偏仪
declination parallel 赤纬圈,等纬圈
declinator 测斜仪,磁偏计,赤纬计
declinometer 磁偏计,赤纬计,测斜仪
declutch 分离,断开,分开离合器
DECM 1. Deception Electronic Counter-Measures 欺骗电子对抗措施 2. Defense Electronic Counter-Measures 防御电子对抗措施
decode 译码,解码
decoder 译码器,译码员
decoder pair 译码器对
decoding 译码
decohesion 检波器恢复常态,减聚力
Decometer 台卡计
decommutation equipment 译码设备,解密设备
decommutator 分路器
DECOMN Decompression 减压,分解
decompensation 1.代偿失调 2.失代偿
decompose 分解
decomposition 1.分解 2.蜕变,衰变
decomposition-aggregation approach 分解-集结法
decomposition behavior 分解行为,分解特性
decomposition capacity 分解能力
decomposition delay 分解延迟
decomposition distance 分解的距离
decomposition event 分解事件
decomposition flame 分解火焰

decomposition gas　分解气体
decomposition instability　分解的不稳定性
decomposition method　分解法
decomposition of olefin　烯烃分解
decomposition of silanes　硅烷分解
decomposition pathway　分解途径
decomposition performance　分解性能
decomposition process　分解过程
decomposition product　分解产物
decomposition rate　分解率
decomposition reaction　分解反应
decomposition temperature　分解温度
decomposition time　分解时间
decomposition-based approach　基于分解的方法
decomposition-coordination approach　分解-协调法
decomposition-product　分解产物
decompound　1.分解 2.再混合
decompression　减压,释压
decompression accident　座舱失压事故,减压事故
decompression chamber test　低压舱试验
decompression experiment　减压实验
decompression exposure　减压暴露
decompression hazard　减压危险
decompression sickness　减压病
decompression stress　减压应激
decompression susceptibility　减压易感性
decompressor　减压器
deconditioning　失调
decongestion　疏散,分散,减小分布密度
decontamination　去污,消除沾染
DECONTN　Decontamination 去污
deconvoluted temperature　去卷积温度
deconvolution　解卷积,去卷积
deconvolution technique　反卷积技术
decoration　1.装饰,装潢 2.装饰品 3.勋章,饰带
decorrelation　去相关,解相关
decouple　减弱震波,去耦
decoupled pylon　去耦挂架
decoupled system　解耦系统
decoupled zero　解耦零点
decoupler　1.断开器,分离器 2.去耦电路
decoupling　解耦,去耦
decoupling epoch　退耦期
decoupling filter　去耦滤波器
decoupling method　解耦算法
decoy　诱饵,假目标
decoy bird　诱惑导弹,假导弹
decoy return　假目标回波信号
decoy warhead　假弹头,诱惑弹头

DECR　Decrease 减少,降低
decrab　做修正转弯以对准跑道,消除偏流修正角
decrab initiation　启动修正转弯以对准跑道
decrab maneuver　修正转弯以对准跑道演习
decrab rate　消除偏流修正角比率
decrab window　消除偏流修正角窗口
decrabbing facility　(接地前)消除偏流修正角装置
decrabbing time　消除偏流修正角时间
decrease　1.减少,减小 2.减小量
decreasing failure rate　逐减故障率,渐降故障率
decrement　1.减少,缩减 2.减量,缩减量
decrement gradient　衰减梯度
decremeter　减量计,减幅计,衰减计
decrescent　下弦(月)
decrypt　解密
Dectra　"台克特拉"导航系统,"台卡"跟踪和测距导航系统
DECU　1. Development and Educational Communication Unit 发展和教育通信部门 2. Digital Engine Control Unit 数字式发动机控制单元,数字式发动机控制装置
DED　Dedendum 齿根,齿高
DEDA　Data Entry & Display Assembly 数字输入和显示设备
DEDD　Diesel Electric Direct Drive 柴油机电力直接传动
dedicate　专用
dedicated　专用的(只在某一个指定的用途上可用的)
dedicated air store　专用航空备件
dedicated case　专用的情况
dedicated computer　专用计算机
dedicated flight　专用飞行
dedicated runway　仪表着陆跑道
dedicated satellite communication network　专用卫星通信网
dedicated space lab　专用空间实验室(欧洲,专供单项科学实验用的)
dedicated spare　专用热备盘
deduce　推论,推断,演绎
deductive modeling method　演绎建模法
deductive-inductive hybrid modeling method　演绎与归纳混合建模法
dedud　清除未爆炸弹(炮弹)
DEE　Direct Engineering Estimate 直接工程估算
DEEC　Digital Electronic Engine Control 发动机数字电子控制,发动机数控调节
de-emphasis　1.去加重,减加重 2.信号还原,削弱
de-emphasis network　去加重网络
de-energize　断电,断开,关断,去激励
deep cavity　深腔

deep cavity projectile 深腔弹头
deep creep attack 深水突击
deep cycling 1.过(电)压通电试验 2.深循环
deep discharge 深放电
deep energy level 深能级
deep level center 深能级中心
deep level transient spectroscopy 深能级暂态谱(学)
deep operation 纵深作战
deep penetration 1.深远突防 2.深度穿透
deep penetration bomb 深穿甲炸弹
deep research vehicle 深空研究飞行器
deep shell 深海贝壳
deep space 深空(内行星以外的空间)
deep space communication 深空通信
deep space instrumentation facility 深空探测设备,深空(飞行器)跟踪与测量设备
deep space maneuver 深空操作
deep space measurement 深空测量
deep space network 深空跟踪网(美国)
deep space probe 深空探测器
deep space station 深空站
deep space surveillance 深空监视
deep space telemetry 深空遥测
deep stall 深度失速
deep-cooled turbojet 深冷却涡轮喷气飞机
deep-draft ship 大型舰只,深吃水船
deep-drawn 深拉的,模压的
deepening 加深,深化
deep-space (太阳系以外的)远太空,深空,远空,宇宙空间,外层空间
deep-space detecting and tracking system 深空探测与跟踪系统
deep-space mission 深太空任务
deep-space science mission 深空科学任务
deepspace tracking system 深空(间)跟踪系统
deep-space vacuum 深空真空
deep-stall 严重失速,深沉失速
deep-stall region 严重失速地区
deep-UV lithography 远紫外光刻术
DEF 1.Defective 有缺陷的 2.Definition 定义,说明,规定
DEFA Direction des Etudes et Fabrications Apartment 武器研制管理局(法国)
defatting 1.脱脂作用 2.脱脂的
default 约定,缺省(值),隐含值
default security level 系统设定安全级
default selection 缺省选择
default trajectory 默认的轨迹
Defcon (DEFCON)defence contracting 国防产品合同

(英国国防部用语)
DEFCS Digital Electronic Flight Control System 数字电子飞行控制系统
DEFDARS Digital Expandable Flight Data Acquisition and Recording System 数字式可扩展飞行数据采集和记录系统
defeat 1.战胜,击败 2.失败,挫折
defect 1.缺陷,故障 2.缺乏,不足 3.挠曲,倾斜,偏斜,偏转
defect density 缺陷密度
defect skip 缺陷跳越
defective 1.不合格品,有缺陷件 2.有缺陷的
defective item 不合格品
defectoscopy 探伤(法)
defence 防御(护,备,务,卫),保卫(战),保护(层),辩护,答辩,(复数)防御工事(设施)
defence radar 防御雷达
defence suppression 压制(敌)防空火力
defend 保卫,防守
defender 防卫者,守卫者
defender guidance 防卫的指导
defender guidance law 防卫指导律
defender missile 防御导弹
defense 1.防御,防护,保卫 2.防务,防御物,(复数)防御工事
defense budget 国防预算
Defense Mapping Agency 国防地图绘制局(美国)
defense ministry 国防部
defense sector 防空分区
defense suppression 防御压制
defense suppression rocket 压制防空武器的火箭弹
defense surveillance system 防空监视系统
defense-dominant strategy 防御为主战略
defense-related payload 军用有效载荷
defensible region 可防御区域
defensive 1.防御态势 2.适于防御的,作防御用的
defensive aid 防御援助
defensive aid subsystem 防御援助子系统
defensive combat spread 防御性战斗横队
defensive electronics 防御电子设备
defensive guided missile 防御性导弹
defensive gunnery (飞机的)防御性射击
defensive missile 防卫性导弹,截击导弹
defensive power 防御强化
defensive radar 防空警戒雷达
defensive satellite 防御卫星
defensive spiral 防御性盘旋下降
defensive split 防御性半滚倒转
defensive turn 防御性盘旋

deferability 可缓性
deferent 传送物,导管
deferred maintenance 推迟维修
deferrization 除铁
deficiency 1.失效(一种故障状态) 2.缺陷(状态指飞机有故障或怀疑有故障、缺件等不完好的状态) 3.不足,亏数
deficit 缺乏,亏损,赤字
defilade 1.遮蔽 2.遮蔽物,障碍物
define 定义,使明确,规定
defining 最典型的,起决定性作用的
definite matrix positive definite matrix 正定矩阵
definition 1.(显示器上)图像清晰度,分辨率 2.(投标者对产品的)评细说明 3.论证 4.定义
definitive ephemeris 测定星历
definitive method of measurement 定义测量方法
definitive orbit 测定轨道
definitive orbit determination system 最终轨道测定系统
DEFL 1.Deflect 偏转,偏移 2.Deflection 偏离,偏差
deflagrate 使……爆燃,使……突然燃烧
deflagration 爆燃,突燃
deflagration speed 爆燃速率
deflate 放气
deflation vent 放气口,放气孔
deflect 使转向,使偏斜,使弯曲
deflect body flap 偏转主副翼
deflect spoiler 偏转扰流板
deflected slipstream 偏转喷流
deflected-thrust 推力方向可改变的,推力换向式
deflecting electrode 偏转电极,致偏电极
deflecting magnet 致偏磁体
deflecting magnetic field 致偏磁场
deflecting yoke 1.偏转线圈组 2.偏转系统,偏转装置
deflection 1.偏斜,偏转,偏移 2.偏度,偏差,(高炮的)提前修正量,(轰炸的)横偏 3.弯曲,挠度,垂度,变形
deflection angle 转角,转折角
deflection angle of vane 叶片偏转角
deflection assembly 方向瞄准装置
deflection bar (高炮瞄准具)方向提前量装定刻线
deflection coefficient 偏转系数
deflection cone 导流锥
deflection crash switch 猛撞触发开关
deflection distortion 偏转畸变
deflection error 方向误差,横向误差
deflection factor 偏转因数
deflection gyro assembly 方向瞄准陀螺装置
deflection input 偏转输入
deflection limiter 挠度限制器
deflection observation 挠度观测

deflection of the vertical 垂线偏差
deflection offset 炸弹落点偏差
deflection prediction (射击)总修正角水平分量
deflection ratio 下沉率,偏转率
deflection winding 偏转线圈
deflection yoke 1.偏转系统,偏转装置 2.偏转线圈组
deflectometer 挠度计,弯度计
deflector 1.折流板,导流片 2.偏转器,偏导装置
deflector nozzle 换向器喷管
deflectron 静电致偏(电子)管
defocus mode 散焦模式
defocused star image 散焦星图像
defocusing 散焦
defogging 除雾,散雾,驱雾
defogging system 除雾系统
defoliant 脱叶剂,落叶剂
deform 使变形,使成畸形
deformability 变形性
deformable membrane 变形薄膜
deformable mirror 可变形反射镜
deformation 变形
deformation error 变形误差
deformation mode 变形模式
deformation model 变形模型
deformation observation 变形观测
deformation of film 胶片变形
deformation of vertically aligned phase 垂直排列相畸变(模式)
deformation process 变形加工
deformation test 变形试验
defroster 除霜器,除霜装置
defrosting system 除霜系统
defruiting 异步回波滤除
deftecting 检测,检定
deftection 侦查,探测
deftection angle 探测角度
defueling 放(燃)油,抽油,(从油箱)排出燃料
defueling and jettison system 放油系统
defuze 卸除引信,使……不能起爆
DEG 1.Degeneration 老化作用,衰减 2.Dressed Engine Gearbox 发动机减速器总成
deg degree 度,程度
deg angle of attack 度攻角
DEG C Degree Centigrade 摄氏度
deg circumferential position 度圆周位置
DEG F Degree Fahrenheit 华氏度
deg inclination 度倾斜
deg injector 度注射器
deg interaction 度交互

deg jet　度飞机
DEG N　Degree North 北纬度数
deg of sweepback　后掠角度
deg orientation　度方向
deg peak-to-peak　度峰间值
deg plane　平面度
deg profile　度剖面
deg range　度范围
deg sideslip angle　度侧滑角
deg stagger　度交错
deg sweep　度扫
degarbling　消除窜扰（雷达的）
degassing　除气
degauss　消磁，去磁，退磁
degausser　退磁器，去磁器
degaussing　去磁，退磁
degaussing computer　消磁计算机
degeneracy　1.衰退，变异 2.【物】简并性，简并度
degenerate　使退化，恶化
degenerate gas　简并气体
degenerate mode　简并模（式）
degenerate orbit　退化轨道（圆轨道或赤道轨道）
degenerate semiconductor　简并半导体
degeneration　1.退化，衰退，变异 2.负回授 3.【物】简并化
degenerative feedback　负反馈
degradation　1.退化 2.降解 3.降级，递降，衰减 4.破坏，摧毁
degradation failure　恶化故障（使机件性能逐渐或局部恶化的）
degradation from shadowing power　阴影功率下降
degradation limit　退化极限
degradation testing　老化试验
degrade　使……降级，使……降解
degraded mode　降级工作状态
degraded performance　性能衰退，（长期使用后）衰退的性能
degraded product　次品
degraded recovery　降级恢复
degraded surface　降级道面
degrading　（性能）衰退
degreaser　1.除油器，启油封装置 2.脱脂剂
degreasing　除油，去油，洗去油脂，启油封
degree　1.度，度数，程度 2.级，等级 3.【数】次，次数
degree correction　度调整
degree field model correction　度场模型调整
degree of accuracy　精确度
degree of advancement　提升水平
degree of atomization　雾化度

degree of charge　1.充电深度 2.充电程度
degree of conformance　一致性程度
degree of consistency　均匀度
degree of continuity　连续性程度
degree of controllability　操纵度
degree of coupling　耦合度
degree of cure　固化度
degree of damping　阻尼度
degree of decay　衰减度
degree of dispersion　分散度
degree of dissociation　离解度
degree of dustiness　含尘度
degree of freedom　自由度
degree of fuel-air　燃气度
degree of homogeneity　均匀度，均质度
degree of latitude　纬度
degree of longitude　经度
degree of mixedness　混合度
degree of nonneutrality　非中性程度
degree of obscuration　食分
degree of observability　可观测度
degree of out-of-roundness　不圆度，失圆度
degree of overdrive　超速程度
degree of premixing　预混合度
degree of reaction　反力度（叶轮机设计参数）
degree of redundancy　冗余度，余度，重复度，静不定度
degree of regulation　调节精度
degree of saturation　饱和度
degree of swirl　旋涡度
degree of vacuum　真空度
degree of voltage rectification　电压整流度
degree unity　统一度
degree variance of gravity anomaly　重力异常阶方差
degree zero　零度
degree-of-freedom　自由度
degrouping　射束上靶不准
DEGT　Deviation of Exhaust Gas Temperature 发动机排气温度偏差
dehorn　〈口语〉卸除引信，使……失效
dehumidifier　脱湿器，干燥器
dehydrated food　脱水食物
dehydration　失水，脱水
dehydrogenation　除氢（作用），脱氢作用
dehydrogenation step　脱氢步骤
DEI　Deliverable End Item 可交付的最终项
deice　除冰，防止结冰，装以除冰装置
deicer　除冰器，除冰设备
deicing　除冰
deicing system　除冰系统

deicing trunk 除冰系统管路
Deimos 【天】火卫二
deinhibiting 去掉保护层,启封
deion circuit breaker 消电离断路器
deionization 消电离
deionization time 消电离时间
deionized water 去离子水
dejam 消除无线电干扰,压制无线电干扰
dekatron scaling unit 十进(制计数)管计数部件
del 倒三角形
DEL 1. Delay 延误(指文电,同 DLA,均为国际民航组织代码) 2. Delete 删除,取消 3. delete character 删除字符 4. Delegation 代表团 5. Delineation 描绘,叙述,轮廓,草图,图解
delamination 1. 脱层 2. 分层
Delaunay element 德朗奈(轨道)根数,德朗奈要素
Delaunay's lunar theory 德朗奈月球理论
delay 延时,滞后,延迟
delay angle 滞后角,延迟角
delay circuit 延时电路
delay collar (引信的)延期药座
delay constant 延时常数
delay control 延时控制,延迟控制
delay cost 延迟成本
delay distortion 延时畸变
delay element (引信的)延期药
delay feature 延时装置
delay firing pin (炸弹引信的)延时撞针
delay graze fuze 短延时着发引信
delay impact fuze 延时着发引信
delay increase 延迟增加
delay launching 滞后发射
delay line 延迟线
delay line storage 延迟线存储器
delay lock technique 延迟锁定技术
delay percussion fuze 延期着发引信
delay plunger 延时引信惯性击针
delay primer detonator 延期起爆雷管
delay rate 延误率
delay ratio 慢波比
delay reduction 延迟减少
delay time 延迟时间,滞后时间
delay timer 延时装置
delay unit 延迟装置,延时器
delay-Doppler mapping 延迟多普勒测绘
delayed action 延迟动作,迟滞反应
delayed aftereffect 延迟后效应
delayed arming 1. 延迟打开保险 2. 成迟缓待发状态
delayed clearance 延迟清除

delayed command 延时指令
delayed drop 延迟降落
delayed exploding 延时引爆
delayed feedback 延迟反馈
delayed flap approach 迟放襟翼进场着陆
delayed mechanism 延时机构
delayed neutron 缓发中子
delayed relaying communication satellite 延迟中继通信卫星
delayed repeater satellite 延迟中继卫星
delayed start 延迟起动
delayed telemetry 延时遥测
delayed-action bomb 延期炸弹
delayed-action fuze 延发引信
delayed-action impact fuze 延时触发引信
delayer 1. (火箭、导弹)缓燃剂 2. 延迟器,延时器
delay-superquick fuze 延时与瞬发两用引信
delay-time telemetry 延时遥测
delete 1. (硬件的)淘汰 2. (技术资料的)废除,废止,删除
delete image 删除影(图)像
deleterious 有害的,有毒的
deleterious effect 有毒影响
deletion 删除,缺失,删除部分
deliberate attack 预有准备的攻击,周密计划的进攻
delimit 定界
delimiter 定界符
delineate 描绘,描写,画……的轮廓
delineation 描绘,略图,草图
delinker 1. 除链机 2. 拆链器
delinking 拆弹链
deliquescence 潮解,融解
deliver 1. 投放(炸弹等) 2. 交付(飞机、设备等) 3. 运送 4. 提供
delivery 1. 交付 2. 输送供给 3. 投放(炸弹等)
delivery error 投射误差
delivery flight test 出厂试飞
delivery system error (武器系统)投射误差
delivery test 1. 交付试车 2. 提交试车
delivery vehicle 运载器
deloxing 放出液氧
DELPHI Discriminating Electrons with Laser Photon Ionization 激光光子电离电子识别(美国一项识别来袭目标和诱饵的研究计划)
Delrac "台尔锐克"导航系统
Delta 德尔塔运载火箭(美国)
delta 1. 三角形 2. 〈口语〉三角翼,三角翼飞机 3. 希腊字母 δ(大写 Δ)的读音
delta aircraft 三角翼飞机

delta function δ函数,脉冲函数
delta hinge 直升机旋翼叶片的挥舞铰,水平铰,水平关节
delta matching transformer 三角形匹配(阻抗)变换器
delta modulation 增量调制
delta pseudorange 伪距增量
Delta Star 德尔塔星(美国SDI计划中的导弹搜索跟踪卫星)
delta tab 三角翼调整片,三角翼补翼,三角翼阻力板
delta v V型三角翼
delta velocity 三角翼速率
delta wing 三角翼
delta wing model 三角翼模型
delta winged 三角(机)翼的
delta-correction 三角翼修正
delta-v V型三角翼
delta-v dispersion V型三角翼分散
delta-v magnitude V型三角翼大小
delta-VEGA delta-V Earth gravity assist 地球重力提供速度增量
deltoid 三角形的
deluge 1.喷水,喷洒 2.泛滥,满溢 3.洪水,暴雨
delusion 幻觉,错觉
DEM Dynamic Enterprise Modeler 动态企业造型者
demagnetization 去磁,退磁
demagnetization curve 退磁曲线
demagnetize 退磁,去磁
demagnetizer 退磁器,去磁机
demand 1.需要,要求,要求知道,查问 2.指令,控制信号 3.所要求的
demand assignment 事物,需要量,按需分配
demand assignment multiple access 按需分配多址
demand mask 肺式供氧面罩,呼吸式面罩
demand oxygen 肺式供氧
demand oxygen equipment 断续供氧设备,肺式自动供氧设备
demand oxygen regulator 肺式氧气调节器
demand oxygen supply 需求式供氧(设备)
demand scenario 需求方案
demand staging 请求分段
demand system 供氧系统(只用于吸气)
demanded-trip length 需要的行程长度
demate 分开,拆开
dematron 代玛管(分布放射磁控放大管)
Dember effect 丹倍效应
deme 同类群
demerit 缺点,缺陷
demerit index 品质指数,性能指标
demeshing 脱开,分离

Demeter 得墨忒耳(掌农业、结婚、丰饶的女神)
demineralization 脱矿物质
demineralized water 无矿物水
demise event 死亡事件
demist 除雾
demister 除雾装置,除雾器
demisting equipment (玻璃)除雾设备
demix 分层,分开,反混合
DEMOD Demodulator 解调器,反调制器,反调幅器
demodulation 解调,反调制
demodulator 1.解调器,反调制器 2.检波器
demoisturization 除湿,除潮
demolition 1.破坏,拆毁 2.爆破
demonstrate 1.演示 2.验证
demonstrate aircraft 验证(飞)机,演示(飞)机
demonstration 演示
demonstration/validation 演示/验证
demonstration ammunition 表演(示范)用弹药
demonstration and shakedown operation 演练性操作试验
demonstration engine 验证发动机
demonstration flight 演示飞行
demonstration mission 演示任务
demonstrator 1.示范飞行的飞机,表演飞机 2.示范表演的样品
demonstrator engine 表演的发动机,展出的发动机
demountable 拆卸,卸下
demounting 拆卸,分解
DEMS Dynamic Environment Measurement System 动力环境测量系统
demultiplex 多路解编
demultiplication 倍减
denaturant 变性剂
dendrite grain 枝状晶体
dendrite tip 枝晶尖端
dendritic crystal 枝晶,枝晶体,枝状晶体,枝状冰晶
dendritic growth 枝状生长
dendritic solidification 枝晶凝固
DEngRD Directorate of Engine Research and Development 发动机研究与发展局(英国)
denial 制止,限制,遏制
denier 否认者,极微量
denim 1.斜纹粗棉布,帆布,苦布 2.(复数)蓝斜纹布工作服
denitrification 脱氮(作用)
denitrogenation 排氮,脱氮(作用)
denominator 分母
denote 表示,指示
Denpa 电波卫星(日本科学卫星)

DENS Density 密度,浓度
dense gas 稠密气体
dense plasma 稠密等离子体,稠密电浆
dense tungsten 密集的钨
dense-output 密集型输出
dense-spray 密集型喷射
densification 1. 密实(化),压实 2. 稠化,增浓
densified wood 高压层板
densitometer 1. 光密度计,显像密度计 2. (燃料)比重计
densitometry 显像测密法
density 1. 密度,浓度 2. 比重,重度 3. 场强
density airspeed 密度空速
density altimetry 密度测高(法)
density altitude 密度高度
density aneroid 密度膜盒
density beam 密度电波
density change 密度改变
density contour 等密度线
density datum 密度基准值
density difference 密度差
density dispersion 密度分散
density distribution 密度分布
density drop 密集雾滴
density enhancement 密度强化
density error 空气密度误差
density evolution 密度演化
density fluctuation 密度起伏,密度波动
density function 密度函数
density gauges (大气)密度计
density gradient 密度梯度,密度陡度,密度涨落
density measurement 密度测量
density model 密度模型
density of loading 装填密度
density peak 最大密度,峰值密度
density percent 密度百分比
density perturbation 密度扰动
density profile (大气)密度剖面
density ratio 密度比
density result 密度结论
density scale height 密度标高
density slicer 密度分割仪
density slicing 密度分割
density specific impulse 密度比冲
density structure 密度结构
density uncertainty 密度的不确定性
density value 密度值
density variation 密度差异
density versus altitude profile 密度(高度)剖面
density wave 密度波

density-tapered array antenna 密度递减阵天线
dent 1. 凹痕,压痕 2. 凹进,使凹入 3. 消弱(影响)
dent depth 凹陷深度
dented 有凹痕,有压痕的
DEO Document Engineering Order 文件工程指令
deodorant 除臭剂,去臭剂
deorbit 轨道脱离
deorbit device 轨道脱离引导设备
deorbit package 脱轨程序包
deorbit phase 离轨段
deorbit time 脱轨时间
deorbiting 脱离轨道
deorbiting maneuver 离轨机动
deoscillator 阻尼器,减振器
deoxidation 脱氧,去氧,还原
deoxygenation 排氧,脱氧
deoxygenation integral index 脱氧积分指数
DEP 1. Departure 出发,离场 2. Department of Employment and Productivity 就业与生产率司(英国) 3. Departure Point 出发点,起航点 4. Deposit 存储,沉淀 5. Design Eye Position 设计(飞行员)眼睛位置
DEP CON Departure Control 离场管制
depart 1. 飞出,离开,出发 2. 偏离(预定方向) 3. 翼尖下坠,失去安定性和操纵性
department 1. 部,部门,科,局 2. 系,学部 3. 车间
departure 1. 出航,起飞,离场 2. 始发机场,航线起点 3. 翼尖失速,翼下沉 4. 航线偏移量,偏航距离 5. 扰动自动增大 6. 失控状态(如自动抬头、失速、螺旋) 7. (频率的)漂移 8. (电子的)飞出
departure aircraft 出航飞机
departure airport 离港时间,出发机场
departure alternate 备降场
departure angle 1. 掷角 2. (火箭)主动段末端弹道倾斜角 3. (电波、电子射线的)出射角
departure control 离场(空中交通)管制
departure count 离场计算,出航计算
departure date 启程日期,出航日期
departure delay 出航延迟
departure epoch 出航时期
departure instructions 起飞指示,离港指示
departure lane 飞离航道
departure pattern 离场起落航线
departure plan (班机的)离场计划(由航空公司拟定)
departure planet 出航星球
departure point 航线起点,出发点(领航检查点之一)
departure procedures 离场程序
departure profile 离场航线剖面
departure rate 离开率
departure resistance 离场阻力

departure route　离场航线
departure runway　离场使用跑道
departure strip　（空中交通管制的）离场记录条（记录呼号、预计起飞时间及航线）
departure susceptibility　离场磁化系数
departure time　离场时间
departure track　离场航迹
departure traffic　离场交通量
departure trajectory　转迹漂移
departure velocity　离场速度
depend　1.依赖,依靠 2.信赖,信任 3.从属,隶属
dependability　可信性,可信度
dependence　1.依靠,依赖,依赖性,相关性 2.信任,信赖 3.从属,隶属
dependence method　依数法
dependency　属国,从属,从属物
dependent　1.军人家属 2.相关的,依靠的,从属的 3.下垂的,悬吊的
dependent approach　相关的方法
dependent behavior　依赖行为
dependent equatorial coordinate system　时角赤道坐标系,第一赤道坐标系
dependent failure　牵连故障,从属故障（故障统计分析用语）
dependent fault　从属故障
dependent measure　因变量
dependent variable　因变量,应变数
deperm　消磁,退磁
depict　描述,描画
depiction　描写,绘图
depigram　露压图
deplane　离机（飞机到达终点,离开飞机）
deplete　耗尽,用尽,使衰竭
depleted shutdown　（推进剂）耗尽关机
depletion　消耗,用尽;耗减,倒空
depletion approximation　耗尽近似
depletion area　阻挡区,耗尽区
depletion layer　耗尽层,阻挡层
depletion mode field effect transistor　耗尽型场效管
deploy　1.部署,展开 2.开伞
deployability　可部署性
deployable antenna　可展开天线（航天器的）
deployable hinge　可部署的铰链
deployable missile control fin　可部署的导弹控制飞翅
deployable missile　可部署的导弹
deployable radiators　展开式辐射器
deployable solar array　可展开式太阳能电池阵
deployable structure　展开式结构
deployable support structure　可部署的支撑结构

deployable truss　展开式桁架
deployables　散播性
deployer　部署者
deployer friction　部署人员的摩擦
deployment　1.部署,调度,配置 2.（部队前沿的）扩展,前伸 3.展开（从集结到形成战斗队形的转变过程）4.（降落伞）开伞 5.（弹头）投放,释放 6.（航天器）打开太阳电池板
deployment bag　开伞袋,拉直袋
deployment characteristic　部署的特点
deployment drive　部署驱动器
deployment duration　部署时间
deployment dynamics　展开动力学
deployment force　部署力量
deployment force margin　部署力量界限
deployment force profile　部署力量配置文件
deployment load　部署装载量
deployment location　部署位置
deployment mechanism　展开机构
deployment module　释放舱（释放弹头用）
deployment motion　部署运动
deployment phase　1.（弹头）释放（阶）段 2.使用期
deployment problem　飞机调度问题
deployment process　展开过程
deployment rate　部署率
deployment reliability　部署可靠性
deployment repeatability　部署可重复性
deployment scheme　展开计划
deployment sequence　展开顺序
deployment signature　部署签名
deployment strategy　部署策略
deployment structure　部署结构
deployment test　部署测试
deployment testing　部署测试
deployment time　展开时间
deployment transient　部署瞬态
deployment tube　部署管
depolarization　退极化,去极化,消偏振
depolymerization　（化学）解聚（合作用）
deposit　1.存款,保证金,沉淀物 2.使沉积,存放
deposit thickness　沉积层厚度
deposition　沉积物,矿床
deposition measurement　沉积测量
deposition of seed material　种子材料的沉积
deposition rate　淀积率
deposition surface　沉积面
depositional feature　沉积特征
depositional remanence　沉积剩磁
depositional remanent magnetization　沉积剩磁

depot 1. 维修与供应基地,大修厂（美国军队）2. 仓库 3. 兵站,补给站
depot level maintenance 维修基地级维修,后方维修,工厂维修
DEPR Depression 降低,减压
depreciation 1. 减小,降低 2. 折旧
depressant 1. 抑制剂 2. 抗凝剂
depressed pole 俯极
depressed trajectory 低飞弹道,压低高度的弹道
depression 1. 减小,降低,抑制 2. 减压,抽空 3. 凹部,（襟翼等的）偏度,下垂 4.【气】低气压,低压区 5. 偏侧角,【天】俯角
depression angle 1. 俯角 2. 下偏角,倾角 3. 垂度
depression effect 减压效应
depression syndrome 抑郁综合症
depressor 抑制器,抑制剂
depressurization 降压,释压,（座舱）解除密封
depressurization test 降压测试
depressurize 减压
deprivation 剥夺,消失（除）,脱离
DEPT Department 部门,部,系,科,处
depth 1. 纵深,深度洲 2. 厚度,高度 3. 能见度极限
depth bomb 深水炸弹
depth calibration 深度校准
depth calibration curve 深度校准曲线
depth charge 深水炸弹
depth contours 等深线
depth datum 深度基准面
depth discrimination 深度辨别
depth distribution 深度分布
depth migration 深度偏移
depth of charge 1. 充电深度 2. 充电程度
depth of compensation 补偿深度
depth of discharge 放电深度
depth of field 景深
depth of focus 焦深
depth of penetration 突破纵深
depth perception 深度知觉,深度感
depth pro 沿深度掺杂分布图,深度剖面
depth probe 测深
depth ratio 深度比
depth-first search 深度优先搜索
depuration 纯化,净化,提纯,精炼
deputy 1. 代理（人）,代理,副（司令、队长、主席、主任等）2. 代表
deputy orbit 副轨道
deputy position 副职
deputy project 代理项目
deputy project manager 项目副经理
deputy satellite 伴随卫星
deputy vehicle 代理运载工具
DER Designated Engineering Representative 工程委任代表
derandomizer 解随机化器
derate 降低功率,降低推力
derated engine 降功率使用的发动机
derating 降额
derating curve 降负荷曲线
derating factor 降额因数
derby 竞赛
dereefer （降落伞的）解除收口装置
deregulation 1. 减少对……的干预 2.（民用航空管理部门对航线、设备及航空运输公司的）撤销,注销
dereverberation 解混响
derivation 1. 推导,求导,导出 2. 分支,分路 3. 偏转,偏差 4. 衍生
derivation tree 导出树,派生树
derivative 1. 导数,微商 2. 方案,（飞机等的）改型 3.【化】衍生物
derivative action 微商作用（按被调参数变化速率进行调节）
derivative engine 导出发动机,派生发动机,变型发动机
derivative estimation 导数估计
derivative feedback 导数反馈,微商反馈,微分反馈
derivative of pitching damping moment coefficient 俯仰阻尼力矩系数导数
derivatives 派生物
derive 1. 推导,求导,导出 2. 分支,分路 3. 偏转,偏差 4. 衍生
derived envelope 导出包络
derived map 派生地图
derived quantity 导出量
derived type filter 导型滤波器
derived unit（of measurement） 导出（测量）单位
derivometer 测偏仪
derma teen 漆布
derogative matrix 减次矩阵
derrick 人字起重机
Derry "德里"作战
DES 1. Data Encryption Standard 数据加密标准 2. Design Environmental Simulator 设计环境模拟器（美国空军）3. Descent 下降 4. Destination 目标,目的地 5. Douglas Equipment Specification 道格拉斯设备规范
desalting 脱盐（作用）
desalting kit 海水淡化器
desaturation 去饱和,减饱和
DESC Defense Electronic Support Center 国防电子支

援中心(美国)
descend 1.下降 2.伞降
descending arc 弹道降弧
descending branch of trajectory 弹道下降段
descending data 1.引导数据 2.引导量
descending line of node 降交点线
descending node 降交点
descending parameter calculation 引导参数计算
descent 1.下降,降落 2.斜坡,坡道
descent algorithm 下降算法
descent allowance 下降用油量
descent calculation 下降计算
descent engine 下降发动机
descent fuel 下降燃油
descent guidance 弹道下降段制导,降落制导系统
descent instrument penetration 仪表穿云下降
descent performance 下降性能
descent phase 下降阶段,下降段
descent problem 下降的问题
descent rate 下降率,飞机低头和掉高度
descent stage battery 下降段用电池组
descent trajectory 下降弹道
descent vehicle 下降的运载工具
descrambler 解扰器
describe 1.描画,作图 2.做盘旋动作
describing function 描述函数
description 1.描述,描写 2.类型 3.说明书
description of relative motion 相对运动的描述
descriptive astronomy 描述天文学,通俗天文学
descriptive equation 描述性方程
descriptive text 图说,地图的文字说明
descriptor 描述符号
descriptor string 串描述项
descriptors 1.描述符,叙述元 2.叙述语(descriptor 的复数形式)
desensitization 灵敏度降低
desensitize 减少感光性,脱敏
desiccant 防潮剂,干燥剂
desiccator 干燥器,防潮器
design 1.设计,计算 2.计划,企图 3.图样,图纸
design activity 设计行为
design adequacy 设计良好度
design alternative 设计备选方案,设计替代,设计选择
design analysis 设计分析
design analysis method 设计分析方法
design analyzer 设计分析器
design angle of attack 设计攻角
design approach 设计方法,设计方式
design approval data 设计审定数据

design approval test 设计认可试验
design aspect 设计方面,设计方向
design bureau 设计局
design burst 设计破裂
design burst pressure 设计爆破压力
design candidate 设计备选
design challenge 设计竞赛,设计挑战,设计挑战赛
design change 设计更改
design chart 设计图,设计图表
design choice 设计选型
design code 设计规范,设计准则
design compensation 设计补偿
design concept 设计理念,设计构思,设计原理
design condition 设计条件
design configuration 设计构型
design consideration 设计根据,设计依据
design constraint 设计约束
design convergence 设计融合
design cost 设计成本
design criterion 设计准则
design cycle 设计流程
design data package 设计数据包
design datum 设计基准
design decision 设计方案
design diving speed 设计俯冲速度
design driver 1.驱动器 2.驱动设计 3.驱动程序设计
design effort 设计工作
design engineer 设计工程师,设计师
design engineering 工程设计
design environment 设计环境
design error 设计误差
design example 设计实例
design eyes 设计眼位,基准眼位
design factor 设计因数
design feasibility test 设计可行性试验
design file 设计文件
design flexibility 设计适应性
design flow 设计流程,设计量
design flow rate 设计流量
design framework 设计框架
design frequency 设计频率,设计重现期
design fuze 制式引信
design goal 设计目标
design gross weight 设计总重
design guideline 设计方针,设计准则
design implementation 设计执行,设计实现
design input 设计输入
design instance 设计实例
design intent 设计意图

design issue 设计上的问题
design iteration 设计迭代
design landing weight 设计着陆重量
design layout 设计图表,设计布局
design leader 1.设计组长 2.设计主持国
design level 设计水位,设计水平
design life 设计寿命,计算寿命,设计使用期
design lifetime 设计寿命
design lifetime increase 设计寿命增加
design lifetime requirement 设计寿命的要求
design limit load 设计极限载荷
design load 设计载荷(即最大可能载荷)
design load factor 设计载荷因数
design load test 设计载荷试验
design loading 设计荷载,设计负荷
design margin 设计余量,设计裕度
design maximum weight 最大设计重量
design method 设计方法,设计法
design methodology 设计方法学
design mission 设计使命
design model 设计模型
design modi 设计修改
design objective 设计目标
design of environment 环境设计
design of experiment 实验设计,实验计划
design of satellite 卫星设计
design of space 空间设计
design office 设计局,设计室
design optimization 设计优化
design optimization algorithm 设计最优化算法
design option 设计方案
design output 设计输出
design parameter 设计参数,设计规范
design phase 设计阶段
design plant 设计车间
design point 设计点(作为设计过程依据的各种有关变量的具体组合)
design position 设计职位
design practice 设计原则
design prediction 设计评估
design prescription 设计指示
design pressure 设计压力,设置压力
design problem 设计问题
design procedure 设计程序,计算方法
design process 设计过程,设计程序,设计加工
design profile 设计剖面
design proof cycle 设计检查周期
design range 设计范围
design rationale 设计原则

design requirement 设计要求
design result 设计成果
design review 设计评审
design risk 设计风险
design scheme 设计方案
design search 设计搜索
design section 设计部门
design segment 设计部分
design service 设计服务
design service life 设计使用寿命,设计服役寿命
design space 设计空间
design specification 设计规范
design spectrum 设计谱
design speed 设计速度,构造速度,计算行车速度
design stage 设计阶段
design strategy 设计策略
design study 设计研究
design system 设计系统,设计体系
design target 设计目标
design team 设计团队,设计小组
design technique 设计技术,设计方法
design time 设计时间
design to budget 根据预算进行设计,按经费来设计的
design to cost (按)成本设计
design to life-cycle cost 按全寿命费用设计
design tool 设计工具
design trade 设计行业
design tradeoff 设计折中,设计权衡
design trend 设计趋势
design uncertainty 设计不确定性
design value 设计值,结构参数,计算值
design variable 设计变数,设计变量
design variable group 设计变量组
design vector 设计矢量
design verification 1.设计验证 2.设计验证品
design walk-through 设计初审
design weight 设计重量
design wheel load 机轮设计载荷
design wing area 设计翼面积
design/off-design points 设计点-非设计点
designate 1.指定,指派 2.标出 3.把……定名为
designated rendezvous area 指定会合区
designated target 指定的目标
designating optical tracker 目标指示光学跟踪器,激光照射仪(美国)
designating system 引导系统
designation 1.指定,指明 2.标记,名称,番号 3.选派,任命
designation device 标示装置

designation marking (激光或其他照射器使用的)指示标记
designation radar 引导雷达
designation range 引导范围
designator 1.符号、代号,(计算机)标志符 2.指示器
designed storage life 设计储存期
designer 设计者,设计师
design-induced pilot error 设计缺陷造成的飞行员(操纵)错误
design-point 设计点
design-type test 设计定型试验
desirability 可取性
desirability function 渴求函数;期望函数
desirable feature 需要的功能
desirable set 理想的设置
desire 期望,要求
desired effects 期望(预定,最佳)效果
desired ground zero (核武器)预期爆心(地面)投影点
desired heading 应飞航向,预定航向
desired track 所需航迹,预定航迹
desired-axis system 预定的轴分析
deSitter static model 德西特静态宇宙模型(荷兰天文学家德西特提出的)
deSitter universe 德西特宇宙
desk checking 桌面检查
desktop 桌面,台式机
desktop analysis 桌面分析
desktop computer 台式电脑
desmolysis (化学)解链作用,碳链分解作用
Deso (DESO) Defence Export Services Organization 国防出口机构组织(英国)
desorb 释出被吸收之物,使解除吸附
desorption 解吸(被材料吸附的气体的释放现象),脱吸附作用
despatch 1.跳伞(或空投)调度 2.放飞调度
despatch deficiency 放飞允许缺陷
despatch delay 放飞延误(通常指班期飞行的5或15分钟以上的延误)
despin 停止旋转,反自转
despin control system 消旋控制系统
despin system 消旋系统
despinner 消旋体
despun antenna 消旋天线
despun orbits 消旋期轨道
dessyn synchro 同步(商标名)
DEST Destination 目的地
destabilization 减稳(作用),不稳定,不安定
destabilize 使动摇
destabilizer 减稳装置,减稳器
destabilizing action 减稳作用
destabilizing effect 失稳作用,失稳效应
destination 目的地,航线终点,目标
destination airport 目的地机场,到达机场
destination planet 目的地的星球
DESTN Destination 目的地,终点
DESTR Destruction 破坏
destroy 摧毁,击毁,破坏
destroyed 被击毁的
destroyer 1.破坏者 2.驱逐舰
destroyer flight 驱逐舰飞行小队
destroying satellite 截击卫星
destruct (火箭、导弹失灵后的)自毁
destruct command (导弹)自毁指令
destruct command receiver 自毁指令接收机
destruct initiator 自毁(系统)起爆器
destruct line 1.自毁线 2.禁止通行线 3.炸毁边界线
destruct package 1.战斗部 2.战斗装药
destruct system 自毁系统
destruction 破坏,歼灭,摧毁,毁灭
destruction behavior 破坏行为
destruction line 炸毁线
destruction system (导弹等的)自毁系统
destructive charge 爆破装药
destructive corrosion 破坏性腐蚀
destructive fault 破坏性故障
destructive payload 弹头,战斗部
destructive test 破坏性试验
destructiveness 1.摧毁能力,破坏程度 2.破坏能力
destructor 自毁器,自炸器
desulfate 脱硫
desulfurizadon 脱硫(作用)
desultory bombing 不瞄准投弹(轰炸)
desynchroneity 非同步性
desynchronization 失调,去同步
DET 1. Design Evaluation Testing 设计评定试验 2. Detail 零件,细节,详细,详(分,零件,分件,细部)图 3. Detailed ET 详细检查 4. Detachment 分遣队 5. Detector 探测器 6. Direct Energy Transfer 电能直接传输(系统)
detach 1.拆分,分开 2.派遣,分遣
detachable 可拆开的,可卸的
detachable cabin 分离座舱,分离式救生舱
detached binary 不接双星
detached shock wave 脱体激波,脱体波,离体激波
detached system 不接(双星)系统
detached-eddy 分离涡
detached-eddy simulation 分离涡模拟
detachment 1.分离,脱离,拆开 2.独立小分队,分遣队

detachment point 分离点,拆卸点
detail 1.零件,元件 2.详细,细目 3.(摄影的)清晰度,影像的细节 4.小分遣队 5.派遣,特派,详述,画详图 6.详细的,明细的
detail design 细节设计
detail design phase 施工图设计阶段
detail fatigue rating 细节疲劳额定强度
detail part 细节部分
detail point 碎部点
detail specification 详细规范,细节规范(美国的一种正式规范性文件)
detail strip (将枪、炮)完全分解
detail survey 碎部测量
detail time 逐一处理时间,细目处理时间
detailed aerodynamic analysis 详细的空气动力学分析
detailed analysis 细部分析,详细分析
detailed calculation 详细计算
detailed chemistry 详细机理
detailed comparison 详细的对比
detailed definition 详细的定义
detailed description 详细描述,详细说明
detailed design 详细设计,具体设计
detailed discussion 细节描述
detailed flow 细化流程
detailed information 详细信息,详细情报
detailed kinetic mechanism 详细的动力学机制
detailed kinetic model 详细的动力学模型
detailed measurement 详细测量
detailed mechanism 详细的机理
detailed model 详细模型
detailed report (photographic interpretation)详情判读报告(相片判读)
detailed schedule 明细进度表
detailed structure 细部结构
detailed study 重点研究,详细研究
detailed trajectory 详细的轨迹
detect 1.发现 2.探测,检测 3.检波
detect and control system 检测与控制系统
detectability 1.探测能力,检验能力 2.可检性(机件易于接受检测的特性)
detectable crack 可察觉裂纹
detectaphone 窃听电话机,侦听器
detection 1.检波 2.探测,检测 3.发现,发觉
detection algorithm 检测算法
detection coverage 探测范围
detection criterion 检测准则
detection delay 发现延迟
detection limit 探测范围,检测极限
detection method 检测方法

detection of intake 检测的摄入量
detection performance 检测性能
detection probability 发现概率
detection range 检测范围,探测范围,探测距离
detection rate 破案率
detection region 检测区域
detection system 检测系统,探测系统
detection test 检测试验
detection threshold 探测限界,检测阈
detection time 检测时间,探测时间
detection unit 1.探测部件,敏感元件 2.导引头
detective cell 探测元件
detective field of optical fuze 光学引信探测角
detective field of view angle 探测视场角
detective quantum efficiency 探测器量子效率
detectivity 探测灵敏度,探测率
detector 1.检测器,探测器,探伤器 2.检波器 3.敏感元件,传感器
detector array 阵列探测器
detector conformity 探测器一致性
detector generation 检测器生成
detector set 检测装置
detector-grade silicon 探测器级硅
detent 1.棘爪,掣子,定位销,卡销 2.制动器,擒纵装置 3.操纵杆的固定位置
deter 制止,威慑
detergent 1.清洁剂,消毒剂 2.除垢剂
detergent cleaning 洗漆剂清洗
deteriorate 恶化,变坏
deterioration 1.恶化,劣化,变质,性能降低 2.磨损,损伤,损耗
determinant 1.行列式 2.决定因素
determination 确定,测定,轨道测定,测轨
determination of orbit 轨道测定,轨道确定
determinative star 距星
determine 决定,判决
determining function 决定函数,母函数
deterministic algorithm 确定性算法
deterministic analysis 确定性分析
deterministic automaton 确定性自动机
deterministic behavior 确定性特性
deterministic control system 确定性控制系统
deterministic input 确定性输入
deterministic optimization 确定性最优化
deterministic service 保证服务
deterministic service time 确定的服务时间
deterministic solution 定解
deterrent 1.阻碍 2.威慑力量 3.缓燃剂
detin 脱锡,去锡

detonability 可爆炸性
detonability limit 爆轰界限
detonable and inert gas 易爆的和惰性气体
detonable gas 易爆的气体
detonable mixture 易爆的混合物
detonate 爆炸,起爆
detonating agent（composition）powder 起爆剂(药)
detonating cap 雷管
detonating charge explosive 起爆药
detonating cord 导爆索
detonating rate 起爆速度
detonation 1.起爆,爆炸 2.(发动机)爆震,爆燃
detonation burning 爆震燃烧
detonation cell 爆轰细胞
detonation cell size 爆炸单元格大小
detonation cell width 爆轰细胞宽度
detonation chamber 爆震室
detonation characteristic 爆轰特性
detonation combustion 爆轰燃烧
detonation combustor 爆震室
detonation cycle 爆轰波周期
detonation engine 爆震发动机
detonation failure 熄爆
detonation front 1.爆轰波前锋 2.爆炸面 3.爆震波前
detonation hazard 爆轰危害
detonation initiation 爆轰起爆
detonation initiation energy 爆震起爆能量
detonation instability 爆轰不稳定性
detonation kernel 爆炸的内核
detonation limit 爆轰极限
detonation mode 传爆方式,起爆方式
detonation parameter 爆震参数
detonation peninsula 爆轰半岛
detonation phenomenon 爆震现象
detonation point 爆点,爆震点
detonation pressure 爆震压力,爆轰压力,爆轰波压
detonation process 爆轰过程
detonation product 爆炸产物
detonation product gas 爆炸产品气体
detonation propagation 爆炸传播
detonation property 起爆性,爆轰性
detonation reaction 爆炸反应,起爆反应
detonation reinitiation 爆炸再引发
detonation run-up 爆炸之前
detonation run-up distance 爆炸之前的距离
detonation sensitivity 爆轰感度
detonation shock 爆轰冲击
detonation simulation 爆炸模拟
detonation solution 爆炸的解决方案

detonation spraying 爆轰枪喷涂
detonation stability 爆轰稳定性
detonation structure 爆震波结构
detonation transition 爆轰转变
detonation transition limit 爆轰转变极限
detonation transmission 爆炸传递
detonation tube 爆轰管
detonation tube exit 爆震管口
detonation tube impulse 爆轰管脉冲
detonation velocity 爆炸速度,起爆速度
detonation velocity measurement 爆速测量仪
detonation wave 爆震波
detonation wave ramjet 爆震波冲压喷气
detonation-degenerated shock 爆震冲击
detonation-driven mhd 爆震磁流体动力
detonation-initiation 爆轰波的形成过程,起爆过程
detonation-tube 爆震管
detonation-tube nozzle 爆震管管口
detonator 起爆管,雷管,起爆剂
detonator holder 传爆药管,起爆药管
detonator pellet 引信火帽座
detonator-safe 膛内保险,起爆保险
detonator-safety device 起爆保险机构
detonator-safety fuze 隔离雷管型引信
detotalizing counter 总剩余量计数器(如燃料等)
detoxify 消毒,消除沾染
detrimental effect 有害作用
detrimental torque 有害力矩
detrital magnetic particle 碎屑磁颗粒
detrital remanence 碎屑剩磁
detrital remanent magnetization 碎屑剩磁
detritus 碎屑(常指磨损或碎落掉下的)
detuner 1.(机场)消声器,排气减音器 2.(曲轴的)动力减振摆,解谐器
detuning 失谐
DEU Display Electronics Unit 显示电子部件
deuterate 1.氘水合物,重水合物 2.氘化 3.使……重氢化
deuterium 氘,重氢
deuteron 氘核,重氢核
deuteroxide 重水
DEV 1. Development 发展,研制,扩大显影,冲洗 2. Deviation 偏差,偏离
DEV LG Developed Length 展开长度
devaporation 止汽化(作用),蒸汽冷凝(作用)
devastation 破坏,践踏
develop 1.发展,研制 2.展开,扩展,开发 3.提高,改进,改型 4.导出,提出 5.叙述,推理 6.(数学公式的)推导,(函数的)展开 7.显影,冲洗(底片或照片),显像

8.整理汇集(情报资料等的)
developed armament probable error 武器概率误差
developed trajectory 实际弹道
developer 显示剂,(摄影)显影剂,(化学)显色剂
developing system 发展中系统
development 1.研制,发展(尤指飞机从首飞到批生产的过程) 2.(用计算、计算机图表或作图法)确定非平坦部件的大小、形状等有关特性 3.(降落伞)开伞 4.显影
development activity 开发工作,开发活动
development aircraft (军用飞机的)原型机
development algorithm 开发算法
development assist test 研制辅助试验
development batch (飞机等产品的)试制批
development contract 研制合同
development cost 研制费用(成本),发展费用(成本)
development date 开发期
development engine 研制的发动机,实验发动机
development examination 加密探测
development flight test 调整试飞
development funding 开发基金
development group 开发组,研发组
development method 开发方法,研制方法
development methodology 开发方法学
development of spacecraft 宇宙飞船的开发
development process 发展进程,显色法,开发流程,开发程序
development program 研制计划,发展计划
development stage 研制阶段,开发阶段
development test 研制试验(初样阶段的)
development test satellite 研制试验卫星
development time 研制时间,程序调试时间
Developmental Sciences 科学开发公司(美国科尔·西格勒公司的航天电子设备部门)
deviant scenario 非正常情况
deviate 背离,偏离
deviated pursuit guidance 以固定提前角沿追击曲线制导
deviation 1.(统计中的数字)偏差,(理论与实际的)差值 2.偏移,(频率)漂移 3.(飞行轨迹)偏离 4.罗差
deviation alarm 偏差报警器
deviation angle 1.落后角 2.偏角
deviation card (磁罗盘的)罗差表
deviation compensator 罗差补偿器,罗差校正器
deviation distortion 频移失真
deviation indicator 偏离指示器
deviation level 偏离程度
deviation of light 光偏差
deviation of mass centre 质心位置偏差

deviation of projectile 射弹偏差
deviation of the vertical 垂线偏差
deviation point 偏差点
deviation probability 偏差概率
deviation ratio 频偏比
deviation table 罗差表
deviative absorption 偏移吸收
deviator 致偏装置,变向装置,折流板
deviatoric strain 偏差应变
device 1.装置,设备,机构,器件,仪器 2.方法,手段,措施 3.计划,设计变向装置,折流板
device control character 设备控制字符
device height 设备高度
device model 器件模型
device nuclear 核装置
devil 【气】小尘暴,(印度和南非的)尘旋风
deviometer 偏航指示器
devitrification 1.透明消失 2.反玻璃化
DEVN Deviation 偏离,偏差
DEW 1. Delivery Empty Weight 交机空重,交付空机重量 2. Directed-Energy Weapon 定向能武器,射线武器(可产生大功率定向射束,以其能量毁伤目标的武器) 3. Distant Early Warning 远程预警(系统),远程早期警戒
dew cell 湿敏元件
DEW line Distant Early-Warning line 远程预警线
dew point test 露点试验
Dewar flask 杜瓦瓶
de-waxing 脱错,排错
dew-cap 露罩
Dewline chain 远程预警线
DEWS Distance Early Warning System 远期预警系统
dexilator 左右(横向)操纵输出装置
dexterity 1.灵巧,敏捷,机敏 2.提高防御效果和远程攻击力
dexterous manipulator 提高防御效果和远程攻击力调制器
dextran 右旋糖酐,葡萄聚糖
dextrorotatory 向右旋转的,顺时针转动的
deyaw 消除偏航
Deyoung-Axford model 德扬-埃克斯福德(星系)模型
df 1. detonating fuze 起爆引信 2. direct fire 直接瞄准射击
DF 1. Digital Filter 数字式滤波器 2. Direction Finding/Finder 定向/测向器 3. Direction Finding 定向 4. Drawing Folders 图形夹
DF/GA Day Fighter/Ground Attack 昼间战斗机/对地攻击机
DFA 1. Delayed-Flap Approach 迟放襟翼进场着陆

2. Design For Assembly 面向装配的设计 3. Deutsche Flug-Ambulanz gemeinnutzige GmbH 德国航空救护福利公司 4. Direction Finding Antenna 定向天线

DFAD Digital-Feature Analysis Data 数字特征分析数据

DFBW Digital Fly-By-Wire 数字电传操纵系统

DFC 1. Digital Fuel Control 数字式燃油调节 2. Direct Force Control 直接力控制 3. Distinguished Flying Cross 空战有功飞行十字励章 4. Dual Feed Channel 双向输送通道

DFCL Directorate of Flight-Crew Licensing 飞行人员执照管理处

DFCS Digital Flight Control System 数字式飞行控制系统

DFD 1. Data Flow Diagram 数据流程图 2. Digital Frequency Discriminator 数字式监频器 3. Digital Flight Display 数字飞行显示(器)

DFDAF Digital Flight Data Acquisition Function 数字式飞行数据采集功能

DFDAU Digital Flight Data Acquisition Unit 数字式飞行数据采集装置(组件)

DFDR Digital Flight Data Recorder 数字式飞行数据记录仪

DFDRS Digital Flight Data Recorder System 数字式飞行数据记录系统

DFDS Data Flow Description 数据流程说明,资料流程说明

DFDU Digital Flight Data Unit 数字式飞行数据装置

DFE Douglas Furnish Equipment 道格拉斯提供的设备

DFF Display Failure Flag 故障显示牌

DFGC 1. Digital Flight Guidance Control 数字飞行引导控制 2. Digital Flight Guidance Computer 数字式飞行引导计算机

DFGS Digital Flight Guidance System 数字式飞行制导系统

DFH-communications satellite 东方红 2 号通信卫星(中国)

DFI 1. Development Flight Instrumentation 研制性飞行仪器(美国航天飞机上的) 2. Digital Fax Interface 数字传真接口

DFIC Duty-Free Import Certificate 免税品进口许可证

DFIDU Dual Flight Interactive Display Unit 双套飞行交互显示装置

DFIU Dual Flight Instrument Unit 双套飞行仪表装置

DFL Deflating 放气,减压

DFLD Definitely Loaded 肯定已装上飞行

DFLS Day Fighter Leader's School 昼间战斗机机长学校(英国空军)

DFM 1. Design For Manufacturing 面向制造的设计 2. Digital Frequency Measurement 数字频率测量,数字测频 3. Direct Flight Mode 直接飞行方式 4. Direct-Force Mode 直接力工作方式 5. Distinguished Flying Medal 杰出飞行奖章 6. Douglas Furnish Material 道格拉斯提供的材料

DFMACH Drafting Machine 绘图机,制图机

DFMS Digital Fuel-Management System 数字式燃油管理系统

DFP 1. Design For Production 面向生产的设计 2. Detailed Flight Planning 详细飞行计划 3. Dorsal Fin 背鳍

DFQI Digital Fuel-Quantity Indicator 数字式油量表

DFR 1. Departure Flow Regulation 离场流量控制 2. Defrost 解冻 3. Digital Flight Recorder 数字式飞行记录 4. Dynamic Flap Restraint 动力襟翼限动器

DFRC Dryden Flight Research Center 德赖登飞行研究中心(美国航空航天局)

DFRR Detailed Functional Requirements Review 性能要求详细方案的评审

DFS 1. Detail Finish Specification 零件表面处理规范 2. Deutsche Fernmelde Satelliten 德国通信卫星 3. Depth First Search 深度优先搜索 4. Digital Frequency Synthesis 数字频率合成 5. Digital Frequency Select 数字式频选 6. Discrete Fourier Series 离散傅里叶级数 7. Distributed File System 分布式文件系统

DFSM 1. Distribution Fast Simulation and Modeling 配电快速仿真与模拟 2. Dispersion Flattened Single Mode 色散平坦单模

DFT 1. Discrete Fourier Transform 离散傅里叶变换 2. Distance From Threshold 距跑道头的距离 3. Douglas Furnish Tooling 道格拉斯提供的工装

DFTI Distance-from-Touchdown Indicator 接地后滑跑距离指示器

DFTR Deflector 1. 偏转器,转向器,偏导装置 2. 导流板,折流板,导流片

DFTSMN Draftsman 绘图员

DFU Detail Follow-Up 零件分工路线

DFV Deutsche Flugdiensberater Vereinigung 德国航空公司顾问协会

DFVLR Deutsche Forschungs und Versuchsanstalt for Luft und Raumfahrt 德国航空航天试验研究院(1988 年 11 月起改名为 DLR)

DFW Disk File Write 磁盘文件写出

DFWD Discrete Flight Warning Display 离散式飞行告警显示器

DFWES Direct-Fire Weapons Effects Simulation(不用指挥仪的)武器直接瞄准射击效果模拟

DG 1. De-gassing 除气,排气 2. Degaussing 去磁,消磁

3. Dangerous Goods 危险品 4. Directional gyro 陀螺半罗盘,航向陀螺仪,方位陀螺

DGA 1. Delegation Generale pour l'Armement〈法语〉武器装备代表处(法国国防部) 2. Dispersed Ground Alert 分散区警报

DGAC 1. Direction Generale Aviation Civile 民航总局(法国) 2. Directorate-General of Air Communications 航空通信总局(印度尼西亚)

DGCA Director General of Civil Aviation 民航局长,民航总局局长

DGI Directional Gyro Instrument (indicator) 航向陀螺仪,陀螺半罗盘

DGLR Deutsche Gesellschaft for Luft-und Raumfahrt 航空航天协会(德国)

DGM Distance-Gone Meter (已经过距离)测距仪(多普勒)

DGNSS Diffenrential Global Navigation Satellite System 差分全球导航卫星系统

DGP Dangerous Goods Panel 危险货物标志牌

DGPS Differential Global Positioning System 差分式全球定位系统

DGR Degreaser 脱脂,除油

DGRR Deutsche Gesellschaft for Raketentechnik und Raumfahrt〈德语〉德国火箭技术和航天协会

DGS Disc-Generated Signal 磁盘产生的信号

DGSI Drift and Ground Speed Indicator 偏流地速指示器

DGTL Digital 数字的

DGU 1. Directional Gyro Unit 航向陀螺装置(组件) 2. display generator unit 显示(字符)发生器

D-gun Detonation Gun 爆轰枪

DGVS Doppler Ground Velocity System 多普勒地速系统(仪)

DGZ Desired Ground Zero (核武器)期望爆心(地面)投影点

DH 1. Data Handling 数据处理 2. Decision Height 决断高度 3. Design Handbook 设计手册

DHC Data Handling Center 数据处理中心

DHD Drop Hammer Die 落锤模

DHDG Desired Heading 应飞航向,所需航向

DHF Data Handling Function 数据处理功能

DHMY Dehumidify 去潮湿

DHO Daily Hours of Operation 每日运行小时数

DHS Data Handling System 数据处理系统

DHUD Diffraction-optics HUD 衍射光学平视显示器

DI 1. Daily Inspection 每日检查,日常检查 2. Data Input 数据输入 3. Dataset Identification 数据集标识 4. De-Icing 除冰 5. Defence Inteligence 国防情报局(英国) 6. Demand Indicator 用量指示表 7. Discrepancy Item 偏差项目 8. Density Indicator 密度指示器

DI controller 显示控制器

DI/DO Data Input/Data Output 数据输入/数据输出

DIA 1. Defense Intelligence Agency 国防情报局(美国) 2. Density Alarm Indicator 密度报警指示器 3. Diameter 直径 4. Documentation Internationale des Accidents 国际事故文献 5. Document Inter-change Architecture 文件交换结构(美国 IBM 公司) 6. Dual Interface Adapter 双接口适配器

diabatic 非绝热的

diabatic process 传热过程

DIAC Data Interpretation Analysis Center 数据判读分析中心

diac 二端交流开关元件

Diademe 王冠卫星(法国测地卫星)

DIAG Diagonal 对角的

diagnosis 1. (对故障的)诊断(尤指通过仪器等手段进行的) 2. 调查分析,识别

diagnostic 诊断的,特征的,侦错

diagnostic analysis 诊断分析

diagnostic approach 诊断方法

diagnostic channel 诊断通道

diagnostic facility 诊断设施

diagnostic method 诊断法

diagnostic model 诊断模型

diagnostic payload 诊断的有效载荷

diagnostic prediction 诊断预报

diagnostic result 诊断结果

diagnostic routine 诊断程序

diagnostic routing equipment (故障)诊断程序设备

diagnostic sensor 诊断传感器

diagnostic system 诊断系统

diagnostic technique 诊断技术

diagnostic test 诊断试验

diagnostic tool 诊断工具

diagnostics (故障)诊断学

diagonal 1. 对角线 2. 对角撑杆,斜杆 3. 对角线的,对顶的,斜的

diagonal element 对角元素

diagonal eyepiece 对角目镜,弯管目镜

diagonal input 对角线的输入

diagonal matrix 对角矩阵

diagonal member 对角构件

diagonal rib 交叉肋

diagonal-flow compressor 斜流式压气机,混流式压气机

diagonally dominant matrix 对角主导矩阵

DIAGR Diagram 图,图表

diagram 图,图表,图解,曲线图

diagrammatic(al) 图示的,图解的

diagraph 1.（机械）仿型仪,绘图仪 2.分度划线仪,分度尺
Dial 戴尔卫星（由法国运载火箭发射的德国科学卫星）
dial 1.度盘,刻度盘 2.拨号（码,字）盘 3.转盘 4.数字盘
dial face 刻度盘,刻度盘面
dial gauge 英制千分表,度盘式指示器,刻度盘
dial your weight 〈口语〉便携式重量计算机
Dialcom 拨号通信（一种电子邮政业务）
dialing 拨号
Dialloy 戴洛伊硬质合金
dialman （高炮）对正手,测合手
dialoque 〈口语〉无线电通话,无线电通信
DIALS Digital Integrated Automatic Landing System 数字式综合自动着陆系统
diamagnet 抗磁体,反磁体
diamagnetic 抗磁的,反磁的
diamagnetic cavity 抗磁性腔
diamagnetic force 抗磁力
diamagnetic moment 抗磁矩
diamagnetism 抗磁性,反磁性
Diamant 钻石（法国运载火箭）
diameter 1.直径 2.光学镜头的放大倍数
diameter chamber 直径室
diameter grinding 直径研磨
diameter of bore 枪炮口径
diameter ratio 直径比,内外径比,直径螺距比
diameter tube 直径管
diametral pitch 节距（齿轮的齿数与节径之比）
diametric path 正相反的路径
diamide 肼,联氨;二酰氨
diamidogen 1.肼,联氨 2.二胺
diamine 1.二（元）胺 2.肼,联氨 3.双胺染料
diaminoethane 二氨基乙烷
diamond 1.金钢石 2.菱形,菱形段,菱形区 3.菱形队形
diamond(shaped) dot/pip 菱形光点
diamond antenna 菱形天线
diamond cutter 金刚石刀具
diamond cutting 金刚石刀削
diamond fillet 钻石角
diamond injector 钻石注射器
diamond landing gear 四点式起落架
diamond region of test section 试验段菱形区
diamond wave 菱形圈
diamond-port 菱形港口
diamonds 菱形激波
diamond-shaped fillet 钻石形的角
diamond-shaped injector 菱形注射器,钻石形注射器

diamond-shaped orifice 菱形孔
Diana 【天】月神星（小行星78号）
dianegative 透明负片
Diapason 音域卫星（法测地卫星）
DIAPH Diaphram 隔片,隔板
diaphragm 1.隔膜,隔板,膜片 2.光阑,光圈
diaphragm pressure gauge 膜片压力表
diaphragm propellant tank 隔膜式（推进剂）贮箱
diaphragm rupture 膜片破裂
diaphragm type accelerometer 膜片型加速度计
diaphragm valve 隔膜阀
diaphragm-ring filter 膜环滤波器
diary 日志,每日天象（年历）
DIAS 1. Digital Integrated Avionics System 集成数字航空电子系统 2. Dublin Institute for Advanced Studies 都柏林高级研究所（爱尔兰空间科学研究所）
diastrophism 地壳变动
diathermy （高频）电热法,透热（疗）法
diation 蓝色辐射
diazo copying 重氮复印
DIB De-icer Boot 除冰带（套）
DIBA Digital Internal Ballistic Analyzer 膛内弹道数字分析器
dibber 侵彻混凝土跑道以后爆炸的穿破混凝土武器
dibber bomb 混凝土穿破炸弹
dibit 双比特
DICASS Directional Command Activated Sonobuoy System 主动式指令定向声纳浮标系统
dice 切割（成的）硅片（芯片）
dichotomizing search 二分法检索
dichotomy 1.【天】（上,下）弦 2.二（两）分（法）,均分
dichroic 二向色的
dichroic mirror 分色镜,二向色反射镜
dichroic subreflector 二色性副反射器
dichromate 重铬酸盐
Dicke radiometer 狄克辐射计
Dicke receiver 1.零平衡接收机 2.迪克接收机
Dickens "魔鬼"作战
dicovery-class mission 探索任务
dictate 命令,口述,使听写
diction 用语,措词
DID 1. Data Insertion Device 数据插入装置 2. Data Item Description 数据项说明 3. Defense-In-Depth 纵深防御 4. Dust Impact Detection 尘埃碰撞检测
DIDACS 1. Digital Data Acquisition And Control System 数字数据采集和控制系统 2. Digital Data Communications System 数字数据通信系统
DIDSY Dust Impact Detection System 尘埃碰撞检测系统

didymium 钕(及)镨(即钕镨两元素混合物)
die 1.钢型,硬模 2.板牙 3.冲模,锻模 4.管芯
die casting 1.压力铸造 2.压铸
die forging 模锻
die pressing 模压加工
die quench-forming 模内淬火成形
dieaway 消失,衰减,降低,减弱
DIEL Dielectric 非传导性的,不导电的,绝缘的
dielectric 电介质,绝缘体,非导体
dielectric absorption (电)介质吸收,介质损耗
dielectric antenna 介质天线
dielectric barrier 介质阻挡
dielectric breakdown 介质击穿
dielectric constant 电容率,介电常数
dielectric fatigue 电介质疲劳
dielectric heating 介质加热
dielectric isolation 介质隔离
dielectric loss 介电损耗
dielectric loss angle 介质损耗角
dielectric loss factor 介质损耗因数
dielectric material 介电材料
dielectric monitering 介电监控
dielectric phase angle 介质相角,介电损耗角
dielectric power factor 介质损耗因数
dielectric properties 介电性能
dielectric strength 介电强度
dielectric tape (电)介质带
dielectric waveguide 介质波导
diene 二烯烃
diepoxide 双环氧化合物
diergol 双组分火箭燃料,二元火箭燃料
diergolic 非自燃的(指燃料)
diesel 柴油机,狄塞尔(发动)机,压燃(狄塞尔)发动机
diesel engine 柴油机
diesel fuel 柴油,柴油机燃料
diesel particulate filter 柴油机微粒过滤器
diesel ramjet 自动点燃冲压发动机,自燃冲压发动机
dieseling 压缩点火
diester 二酯(合成润滑材料)
diet 饮食
diethy diphenylurea 二乙基二苯基脲
diethy lamine 二乙胺
diethyl 二乙基的
diethylenetriamine 二亚乙基三胺,二乙撑三胺
diethylphthalate 酞酸二乙酯
dif diiodofluorescein 二碘荧光素
DIF 1.Data Interchange Format 数据互换格式 2.Data Interface Facility 数据接口设备 3.Domsat Interface Facility 国内通信卫星接口设备

DIFAR 1.Direction-Finding And Ranging 定向与测距 2.Directional Frequency And Ranging 定向频率与范围
DIFF Difference 差值
DIFF PRESS Differential Pressure 压差
diffeomorphism 微分同胚映射
differ 使……相异,使……不同,相异
difference 1.差别,区别 2.(数学词)差,差数,差分
difference approximation 差分近似法,差分法,差分近似
difference beam 差波束
difference equation 差分方程,差分方程式
difference equation model 差分方程模型
difference formula 差分公式
difference method 差分法,差动法,差别法
difference scheme 差分格式
difference slope 差斜率
difference threshold 差异阈
difference vector 差矢量
difference-set code 差集码
differencing 差分,【数】差分化
different 不同的,个别的
different source calibration 异源校准
different term repeatability 不同期重复性
differentiable 可微分的
differential 1.微分,差 2.微分的,差动的,(有)差别的
differential aberration 光行差较差
differential acceleration 差别促进
differential action 1.差动作用 2.微分作用
differential ailerons 差动副翼(一般向上最大偏度大于向下的)
differential amplifier 差分放大器
differential astrometry 较差天体测量
differential atmospheric absorption 较差大气吸收
differential biphase mark code 差分双相传号码
differential biphase space code 差分双相空号码
differential capacitor 差动电容器
differential carrier 差速器壳,差速器座架
differential carrier phase 差分载波相位
differential catalogue 较差星表
differential charging 不等量充电
differential circuit 微分电路
differential coil 差动线圈
differential constraint 微分约束,微分拘束
differential control crank arm 差动操纵摇臂
differential controls 1.差动控制,差动操纵 2.微分控制
differential correction (导航系统的)差分修正,微分改正
differential cross 差速器星轮十字轴

differential cross section   差分截面
differential deflection   微分偏转
differential directional flux   1.微分单向通量动控制,差动操纵 2.微分控制
differential discriminator   微分甄别器
differential Doppler effect   微分多普勒效应
differential drag   微分阻力
differential energy flux   微分能通量
differential energy spectrum   微分能谱
differential equation   微分方程
differential equation model   微分方程模型
differential equation of motion   运动微分方程
differential evolution   差分进化,差分演化,微分进化
differential fare   （航空公司因飞行时间长或飞机旧式而给予的）差价
differential flexure   较差弯沉
differential gain   微分增益
differential galactic rotation   银河系较差自转
differential game   微分对策
differential gear   差动齿轮
differential gear adder   差动齿轮加法器
differential global positioning system   差分全球定位系统
differential GPS   1.差分全球定位系统 2.差分GPS
differential gravity   微分重力
differential gravity effect   微分引力效应
differential heating   差温加热
differential interferometry   差动干涉测量法
differential ionosphere   微分电离层
differential lift   微分抬升
differential Loran   差分式罗兰系统
differential method of measurement   微差测量方法
differential method of photogrammetric mapping   分工法测图,微分法测图
differential mobility   微分迁移率
differential number spectrum   微分（粒子）数谱
differential nutation   较差章动
differential observation   较差观测
differential Omega navigation system   差分式欧米茄导航系统
differential Omega system   差分式欧米茄系统
differential operator   微分算子,微分算符
differential orbit improvement   微分轨道改进,差动轨道改进
differential path delay   微分路径延迟
differential phase   微分相位
differential phase shift keying   差分相移键控
differential photometry   较差测光
differential piston   差动活塞,异径活塞,级差活塞
differential precession   较差岁差

differential pressure conditioner   压差变换器
differential pressure feedback   差压反馈
differential pressure liquid level indicator   压差液面计
differential-pressure range of aircraft cabin pressurization   飞机座舱增压压差范围
differential pressure transducer   差压传感器,差压变送器
differential rate   区别运价率
differential rectification   微分纠正
differential refraction   较差（大气）折射
differential rotation   较差自转
differential satellite tracking   差分卫星跟踪
differential scanning calorimeter   示差扫描量热仪
differential spoilers   差动扰流片
differential stabilator   微分全动平尾
differential stabilator channel   全动平尾差通道
differential star catalogue   较差星表
differential tailplane   差动平尾
differential thermal analysis   热差分析（热控技术）
differential transformer   差动变压
differential transformer displacement transducer   差动变压器式位移传感器
differential vacuum gauge   压差式真空计
differential winding   1.差动线圈,差动绕组 2.差动绕法
differential-drive tachometer   差动驱动式转速表
differentially-compounded electric motor   差动复励电动机
differential-mode   差模
differential-voltage reverse-current relay   差压反流继电器,差压反流割断器
differentiate   1.区别,区分 2.使（如某种生物）分化 3.求……的微分
differentiated   分化型,已分化的,可区分的
differentiating   微分的
differentiating unit   微分部分
differentiation   微分法,微分
differentiation approach   分化的方法
differentiation element   微分环节
differentiation positioning   差分法定位
differentiation process   分异过程
differentiator   1.微分器 2.差分器
difficulty index   难度指数
difficulty index value   难度指数值
diffluence   1.【气】分流 2.流线稀疏区 3.峰区出口式自适应记录系统
diffract   衍射,使……分散,碾碎
diffraction   1.折射,偏转（射线）2.衍射,绕射
diffraction contrast image   衍衬象
diffraction distance   衍射距离

diffraction grating 衍射光栅
diffraction loss 衍射损耗
diffraction optics HUD 衍射式平视仪,带衍射光学装置的平视仪
diffraction theory 衍射理论
diffractive beam 衍射光束
diffractive HUD 衍射式平视显示器
diffractometer 衍射学
DIFFUS Diffusing 扩散,弥散,漫射,散布,传播
diffuse 1.扩散,渗出 2.漫射 3.传播,散布
diffuse nebula 弥漫星云
diffuse radiation 漫辐射,扩散辐射
diffuse reflection 漫反射
diffuse reflection spectra 漫反射光谱
diffuse reflector 漫反射体,漫反射器
diffuse sky radiation 天空漫辐射
diffuse skylight 漫射天光
diffuse sound 漫射声
diffuse surface 漫反射面
diffuse terrain 漫射型地域
diffused-alloy transistor 扩散合金型晶体管,载流子漂移型晶体管
diffuser 1.扩压器,扩散器 2.(发动机的)进气道 3.漫射体
diffuser area 扩散器面
diffuser area ratio 扩压器(出口/进口)面积比
diffuser design 扩压器设计
diffuser efficiency 扩压器效率,扩散器效率
diffuser exit 扩压器喷管
diffuser of area 扩散器面积
diffuser of area ratio 扩散器的面积比
diffuser vanes (压气机)扩压器叶片,扩散器叶片
diffusibility 扩散能力,扩散系数
diffusible 可扩散的,可扩压的
diffusion 1.扩散 2.漫射 3.气流滞止
diffusion annealing 扩散退火
diffusion bonding 1.扩散焊 2.扩散连接
diffusion boundary layer 扩散边界层
diffusion brazing 扩散钎焊
diffusion capacitance 扩散电容
diffusion chamber (免疫)扩散盒,(高能)扩散室
diffusion coefficient 扩散系数
diffusion combustion 扩散燃烧
diffusion combustion condition 传播燃烧条件
diffusion control 扩散控制
diffusion effect 扩散效应,扩散效果
diffusion factor 扩散因子(压气机设计参数)
diffusion flame 扩散火焰
diffusion length 扩散长度,扩散距离

diffusion loss 扩散损失
diffusion model 扩散模型
diffusion of fuel 燃料扩散
diffusion of gas 气体扩散
diffusion of impurities 杂质扩散
diffusion of radical 原子团扩散
diffusion potential 扩散势
diffusion process 扩散过程
diffusion pump 扩散泵(一种真空泵)
diffusion rate 扩散率,扩散速度
diffusion technique 扩散工艺
diffusion technology 扩散工艺
diffusion transistor (载流子)扩散型晶体管
diffusion tube 扩散管
diffusion velocity 扩散速度
diffusion welding 1.扩散焊 2.扩散连接
diffusional transport 扩散运输
diffusion-controlled combustion 扩散控制燃烧
diffusion-controlled growth 扩散控制生长
diffusion-controlled reaction 扩散控制反应
diffusion-limited ablation 扩散限制消融
diffusion-tube 扩散管
diffusive 漫射的,扩散的
diffusive convection 扩散对流
diffusive equilibrium 漫射平衡,扩散平衡
diffusive instability 扩散不稳定性
diffusive transport 扩散性运输
diffusivity 扩散性,扩散系数,扩散率
diffusor 1.扩压器,扩散器 2.(发动机的)进气道 3.漫射体
difluoride 二氟化物
difluoroamine 二氟胺
difluorodiazine 二氟二嗪
difluoromethane 二氟甲烷,甲叉二氟
DIFMR Digital Instantaneous Frequency Measurement Receiver 数字式瞬时测频接收机
DIG 1.Directional Gyro 航向陀螺 2.Display/Indicator Group 显示器/指示器组
dig 挖,掘
digest 摘要,文摘,汇编
digibus 数字式多路数据总线
DIGICOM(S) Digital Communication System 数字通信系统
digicon 二极管阵列
digisonde 数字式测高仪
digit 1.食分(太阳或月球直径的十二分之一) 2.数字,位数,位
digit path 数字通路
digit winding 位绕组

Digitac 1. Digital airborne computer 机载数字计算机 2. Digital tactical aircraft control 数字式战术飞机控制（美国空军）
digital 数字的
digital adder 数字加法器
digital air data computer 数字式航空数据计算机
digital air data system 数字式大气数据系统
digital air data-computer system 数字式航空数据计算机系统
digital aircraft 采用数字式电子设备的飞机
digital architecture 数字建筑
digital attitude sensor 数字式姿态敏感器
digital automatic pattern recognition 数字式自动模式识别
digital automatic tracking and ranging 数字自动跟踪和测距
digital autopilot 数字自动驾驶仪
digital avionics information system 数字式航空电子信息系统
digital avionics system 数字式航空电子系统
digital broadcasting 数字广播，数位广播
digital camera 数字摄像机，数字照相机
digital capture loop 数字归零回路
digital cartography 数字地图制图
digital chirp 数字线性调频（脉冲）
digital circuit 数字电路
digital circuit tester 数字电路测试器
digital command 数字指令
digital command communication system 数字指令通信系统
digital communication 数字传输，数字通信
digital communication system 数字通信系统
digital compensation 数字补偿
digital component 数字分量
digital computation 数值计算，数字计算
digital computer 数字电脑
digital computing 电子计算
digital control valve 数字控制阀
digital controller 数字控制器，数控装置
digital correlation unit 数字相关单元
digital data acquisition system 数字式数据采集系统
digital data management and diagnostic system 数字式数据管理和诊断系统
digital data processor 数字式资料处理机
digital data service 数字式数据业务
digital data servo 数字式数据伺服系统
digital data transmission network 数字式数据传输网
digital datum 数字式数据
digital demodulation technique 数字解调技术

digital demultiplexer 数字分接器
digital demultiplexing 数字分接
digital designation 数字引导
digital display 数字显示
digital domain measurement 数域测量
digital eletronic engine control 数字电子式发动机控制
digital elevation model 数字高程模型
digital engine 数字引擎，数位引擎
digital engine control unit 数字式发动机控制单元
digital equalizer 数字式均衡器
digital equipment 数字（计算）设备，数字式仪器
digital error 数字错误
digital fax interface 数字传真接口
digital file 数字化文件
digital filter 数字滤波器
digital filter bank 数字滤波器组
digital fire control computer 数字火力控制计算机
digital flight control system 数字式飞行控制系统
digital flight path control system 数字式飞行轨迹控制系统
digital guidance system 数字制导系统
digital holography 数码激光全息，数字全息术
digital image 数字影像，数字图像
digital image processing system 数字图像处理系统
digital instrument 数字仪器
digital integrated circuit 数字集成电路
digital integrating accelerometer 数字积分加速度计
digital integration 数字积分，数字积分法，数位积分
digital interface 数字接口，数字界面
digital internal ballistic analyzer 内弹道数字分析器
digital least square filtering 数字最小二乘方滤波
digital level 数字水准仪
digital map system 数字式地图系统
digital measuring instrument 数字式测量仪器
digital micropropulsion 数字微推进
digital microwave communication system 数字微波通信系统
digital mosaic 数字镶嵌
digital multimeter 数字式多用表
digital multiplex equipment 数字复用设备
digital multiplexing 1.数字复接 2.数字复用设备
digital multiplier unit 数字乘法器
digital multispectral scanner 数字式多谱段扫描仪
digital network 数字网
digital number 数字数值，数字值
digital oscilloscope 数字示波器
digital phase-locked loop 数字锁相环路
digital plotter 数控绘图机
digital processing 数字处理，数字加工，数位处理

digital processor   数字处理器
digital projector   数字投影仪
digital radio   数字无线电,数字广播
digital radio system   数字无线系统
digital range tracker   数字距离跟踪器
digital rate gyro   数字速率陀螺
digital recorder   数字录音机
digital resolving   数字分解
digital satellite   数字卫星传播,数字卫星流传
digital scan convertor   数字扫描变换器
digital sensor   数字(式)传感器
digital service   数字化服务
digital signal   数字信号
digital simulation   数字仿真
digital simulation facility   数字模拟设备
digital solar aspect sensor   数字式太阳方位遥感器
digital speech interpolation   数字话音内插
digital subscriber filter   数字用户线滤波器
digital sun sensor   数字式太阳敏感器
digital surface model   数字表面模型
digital symbol generator   数字式字符发生器
digital symbology generator   数字符号发生器
digital system   数字系统
digital technology   数字技术
digital telemetering system   数字遥测系统
digital telemetry   数字遥测
digital television system   数字电视系统
digital terrain model   数字地形模型,数值地型
digital thermal infrared scanner   数字式热红外扫描仪
digital time dissemination   数字时间发播
digital to analog converter   数模变换器
digital to analogy conversion   数模转换
digital tracing table   数控绘图桌
digital tracker   数字跟踪装置
digital transducer   数字传感器,数字转换器
digital transmission   数字信息传输
digital variable   数字变量
digital video integrator and processor   数字视频积分处理器
digital voltmeter   数字电压表
digital VTR   数字录像机
digital weather chart recorder   数字式天气图记录器,数字式天气图传真收片机
digital wire   数位线
digital-data transmitter   数字数据传感器,数字数据发送器
digitalization   数字化
digitalization of flow picture   流动图像数字化
digital-to-analog transducer   数字模拟变换器

digitalyzer   模拟数字转换器,模数转换器
digitatron   图像位置读出器
digitization   1.量化 2.数字化
digitize   使数字化
digitized cloud map   数字云图
digitized image   数字化影(图)像
digitized video   数字影(图)像
digitizer   数字化仪(数字计算机一种输入装置)
diglycol   二甘醇,一缩二乙二醇
dignitary   (尤指政界、宗教界的)职位高的人,达官贵人,要人,要员,显要的人物,显贵
digram   双字母组
digraph   有向图
digress   离题,走向岔道
DIGS   1.Delta Inertial Guidance System 德尔他火箭惯性制导系统(美国),三角惯导系统 2.Digital Image Generation Simulator 数字式图像产生模拟器
dihedral   反角
dihedral angle   上反角
dihedral effect   上反角效应
dihedral vane   倾斜叶片
dihydronaphtalene   二氢(化)萘
dike   1.沟,渠 2.堤,围墙,护墙
DIL   1.Digital Integrated Logic 数字集成逻辑 2.Dilution 稀释 3.Doppler-Inertial-LORAN 多普勒-惯性-罗兰组合导航系统
Dilag   Differential Laser Gyro 差动激光陀螺
dilatancy   膨胀
dilatancy hardening   膨胀硬化
dilatancy-diffusion model   膨胀-扩散模式
dilatation   膨胀,扩展
dilatation effect   时间变慢效应,钟慢效应(引力理论)
dilatational wave   膨胀波
dilate   膨胀,扩展
dilatometer   膨胀仪
dilatometry   膨胀测量法,膨胀测量术
DILS   Doppler Instrument Landing System 多普勒仪表着陆系统
diluent   稀释剂,冲淡剂
diluent amount   稀释剂数量
diluent mixture   稀释剂混合物
dilute   1.冲淡,稀释 2.冲淡的,稀释的
diluted mixture   稀释混合物
diluter   冲淡器,稀释器
diluter demand regulator   稀释肺式调节器
diluter-demand oxygen system   混合氧断续供氧系统
dilution   1.冲淡,稀释 2.淡度,稀释度
dilution air   稀释空气
dilution factor   稀化因子

dilution hole 掺混孔
dilution of precision 误差放大因子，精度因子
dilution ratio 稀释率，稀释比例
dilution zone 掺混区，掺合区，降温区
dilvar 迪尔瓦镍铁合金
DIM 1. Dimension 尺寸 2. Dimmer 亮度调节器
DIMACE Digital Monitor and Control Equipment 数字监控系统
dimazine 偏二甲肼
dimension 1. 尺寸，尺度 2.【数】维 3.【物】量纲，因次 4. 范围，方向 5.（复数）面积，容积
dimension of a quantity 量纲
dimension of parameter vector 参数矢量的维数
dimension of state vector 状态矢量的维数
dimension of system 系统维数
dimension transducer 尺度传感器
dimensional 尺寸的，因次的，量纲的，空间的，维的
dimensional analysis 量纲分析
dimensional calculation 维度计算
dimensional coefficient 尺寸系数
dimensional issue 尺寸问题
dimensional resonance 尺寸共振
dimensional similarity 量纲相似
dimensional stability 尺寸稳定性
dimensional tolerance 尺寸公差
dimensionality 维度，幅员，广延
dimensionality of data 数据的维数
dimensionally-accurate 尺寸精确的
dimensionless 无因次的，无量纲的
dimensionless force 无量纲力
dimensionless group 无因次群
dimensionless impulse 无量纲脉冲
dimensionless loss 无因次损失
dimensionless parameter 无量纲参数，无维参数
dimensionless receiver 无量纲的接收机
dimensionless time 无因次时间
dimensionless value 无量纲值，无因次值
dimensionless variable 无因次变量，无量纲变量
dimensionless velocity 无量纲速度
dimer 二聚物，二量体
dimethyl hydrazine 二甲肼
dimethyl beryllium 二甲基铍
dimethyl lamine 二甲胺
dimethyl ether 二甲醚
diminish 1. 减少 2. 缩小
diminution 减少，下降（如压力）
DIMM Dual-part Integrated Memory Monitor 双组件综合存储监控器
dimmer 调光器，亮度调节器

dimming rheostat 调光变阻器
Dimond C C类钻石奖
dimorphic 双晶（的），（同质）二形的
dimorphism 双晶现象
dimorphous 同 dimorphic 二态的
dimout 1. 昏暗 2. 半灯火管制
dimple 微坑
dimple depth 窝深度，窝的深度
dimpled tyre 窝纹轮胎
dimpler 压窝器（铆接用）
dimpling 1. 波纹板（金属）2. 开凹槽（用非切削方法）3. 造窝，打窝
DIMS Distributed Intelligence Microcomputer System 分布式智能微计算机系统
DIN Deutsche Industrie Normen〈德语〉德国工业标准
DINAS （Dinas）Digital Inertial Nav/Attack System 数字式惯性导航/攻击系统
ding 刺激（产生行动）
dinghy 橡皮救生船，充气救生船，机载救生艇，救生筏（飞机落入水中时供救生用）
dinitramide 二硝托胺
dinitrate 二硝酸盐
dinitrobenzene 二硝基苯
dinitrocellulose 二硝基纤维素
dinitroglycol 二硝基乙二醇，乙二醇二硝基酸
dinitromethane 二硝基甲烷
dinitrophenol 二硝基酚
dinitrotoluene 二硝基甲苯
dinitrotoluol 二硝基甲苯
dinking 靠模打孔
dinol 重氮硝基酚
DINS 1. Digital Inertial Navigation System 数字惯性导航系统 2. Dormant Inertial Navigation System 休眠型惯性导航系统
DIO Diode 二极管
DIODE Digital Input/Output Display Equipment 数字输入/输出显示设备
diode 二极管
diode bypass network of battery 蓄电池组二极管旁路网络
diode discharge 二极管放电
diode gun 二极管电子枪
diode lamp 二极管灯，发光二极管
diode laser 二极管激光器
diode pumping 二极管泵浦，二极管抽运
diode sensor 二级传感器
diode thyristor 双向两端可控硅，二极管晶闸管
diode triode 二极三极复合管
diode-laser 二极管激光器

diode-pentode 二极五极管
diode-transistor logic 二极管-晶体管逻辑
DIOI Digital Input/Output Interface 数字输入/输出接口
DIOL Drawing Index On Line (System) 在线图纸目录（系统）
Diomedes 【天】狄奥墨德斯(小行星1437号)
Dione 【天】土卫四
Dione B 【天】土卫十二
dioptric light 屈光,反射平行光
dioxide 二氧化物
DIP 1. Digital Image Processing 数字图像处理 2. Digital Inboard Profile 数字化内部轮廓 3. Dual In-line Package 双列直插式组件 4. Diplomacy 外交
dip 1. 倾角,磁倾角,天地线倾角 2. 浸,蘸,把(布或海绵等)放入(液体中)又迅速取出 3. 下沉,下降,急剧掉高度 4. 腐蚀剂,酸洗,(化学法)镀层液 5. 摇晃机翼(以作为信号) 6. (火箭)将舵面放入燃气流 7. 摆动机翼(表示向观众致敬) 8. (抢风调向时)降升(四角帆斜桁) 9. (飞机为获得上升动力而)作短暂骤降
dip angle 1. 磁倾角 2. 弹着角,落角
dip brazing 浸渍钎焊
dip circle 磁倾圈
dip equator 倾角赤道
dip move-out 倾斜时差校正
dip of horizon 地平俯角
dip orientation 倾向定向
dip pole 磁倾极,磁极
dip position 吊放位置
dip soldering 浸渍钎焊
DIPEC Defense Industrial Plant Equipment Center 国防工业工厂设备中心(美国)
dipentene 二聚戊烯
diphase 双相(的),二相(的)
Diphda 【天】土司空,鲸鱼座星
dipheny guanidine 二苯胍,促进剂,密苯胺
diphenyl 联(二)苯,二苯基
diphenylamine 二苯胺
diphenylurea 二苯脲
diplex transmission 两信伴传,双工传输
diplexer 双工器,天线分离滤波器,天线共用器
dipole 偶极子
dipole antenna 偶极子天线,对称振子天线
dipole electrode array 偶极排列
dipole electrode sounding 偶极测深
dipole magnet 由磁偶极
dipole meridian 偶极子午圈,偶极子午线
dipole model 偶极子模型
dipole moment 偶极矩

dipole vector 偶极矢量
dipole-dipole array 偶极排列
dipole-dipole array method 偶极排列法
dipper 燃料调整器
dipping 1. 浸渍 2. (电镀前)腐蚀金属
dipping needle 磁倾针
dipping procedure 吊放程序
dipping sonar operations 吊放式声纳作业
dipping sonobuoy 吊放式声纳浮标(直升机将声纳浮标悬于海上选定的位置上,并浸入水中,而不是投放到水中)
DIPS 1. Dipole Inches Per Second (投掷箔条的)每秒偶极子英寸数 2. Dynamic Isotope Power System 动力同位素发电系统(美国)
DIR 1. Deficiency Investigation Report 缺陷调查报告 2. Direct 直接,直流,正向 3. Direction 方向,方位,指示
DIR FLT Directional Filter 定向滤波器
direct 1. 指导,引导 2. 操纵,指挥 3. 指向 4. 直接的
direct access 随机存取,直接存取
direct adaptive control scheme 直接自适应控制方案
direct addition 直接相加
direct address 直接地址
direct addressing 直接寻址
direct allocation 直接分配
direct allocation problem 直接分配问题
direct and indirect combustion 直接和间接燃烧
direct approach 1. 直接法 2. 直接进场 3. 直接接近法
direct ascent 直接起飞,直接进入轨道
direct attack 直接进攻
direct bomb hit 炸弹直接命中
direct bombing 直接瞄准轰炸(以目标为瞄准点)
direct broadcast 卫星直播
direct broadcasting satellite 直接广播卫星
direct collocation 直接配点
direct combustion noise source 直接燃烧噪声源
direct command wire guidance 1. 直接指令有线制导 2. 目视跟踪有线指令制导
direct comparison 直接比较
direct conductivity 直接电导率
direct coupled amplifier 直接耦合放大器
direct coupling 1. 直接耦合 2. 直接连接
direct current 直流
direct current amplifier 直流放大器
direct current arc weld 直流电弧焊
direct current cleaning 直流(场)清洗
direct current relay 直流继电器
direct current sputtering 直流溅射
direct detonation 直接起爆

direct differentiation method　直接微分法
direct differentiation　直接分化
direct digital interface　直接数字接口
direct display storage tube　直接显示存储管
direct drive　直接驱动
direct energy transfer system　直接能量传输系统
direct entry　直接款目,直接进入
direct evaluation　直接鉴定
direct feedback　直接反馈,刚性反馈,硬反馈
direct file organization　直接文件组织
direct flame　直接火焰
direct force control　直接力控制
direct gap semiconductor　直接带隙半导体
direct hit　直接命中
direct impingement　直接冲击,直接碰撞
direct initiation　直接起爆
direct initiation of detonation　直接引发爆炸
direct initiation of spherical detonation　球形爆炸的直接启动
direct injection　直接喷射,直接注入,直接射出法
direct instruction　直接指令
direct interpretation mark　直接判读标志
direct lift control　直接升力控制
direct line explosive train　直列式传爆系列
direct linear transformation　直接线性变换
direct maintenance　直接维修
direct measurement　直接测量,直接测定
direct memory access　存储器随机存取
direct metal mask　直接金属掩模
direct method　直接法,绝对摄动法
direct method of measurement　直接测量法
direct motion　顺行
direct multiple shooting method　直接多重打靶方法
direct numerical control　直接数控(群控)
direct numerical procedure　直接数值计算程序
direct numerical simulation　直接数字仿真
direct operating cost　直接使用费用
direct optimization　直接最优法
direct optimization method　直接优化法
direct orbit　顺行轨道
direct photograph　直接照相法
direct photography　直接摄影,直接取景
direct picture　(遥测数据范围内拍摄的)卫星照片
direct point fire　直接瞄准射击
direct pointing　直接瞄准
direct procedure　直接程序
direct reception　直接领受
direct recombination　直接复合
direct recording　直接记录

direct reentry　直接再入,直接返回
direct reentry return mode　直接进入法返回
direct satellite probing　卫星直接探测
direct search　直接检索
direct search algorithm　直接搜索算法
direct sequence　直接序列
direct side force control　直接侧力控制
direct sight　1.直接瞄准 2.直接瞄准具
direct signal　直接信号
direct simulation　直接模拟
direct solar radiation　太阳直接辐射
direct stationary　顺留
direct steering strategy　直接指导策略
direct thrust　直接推力
direct thrust control　直接推力控制
direct trajectory　直接轨道
direct transcription　直接转录
direct transcription method　直接转录法
direct transfer　直接移植法
direct viewing storage tube　直观存储管
direct wave　直达波,直射波
direct-action fuze　瞬发引信
direct-allocation　直接分配
direct-ascent satellite interceptor　直接上升式卫星截击器
direct-connect　直连
direct-connect experiment　直连实验
direct-cranking starter　手摇起动机
direct-detection receiver　直接检波式接收机
direct-drive option　直接传动的选择
directed graph　定向图,有向图
directed inspection　深入检查
directed jamming　引导式干扰
directed slipstream　定向滑流
directed takeoff　1.使用指令仪表起飞 2.风扇滑流吹过整个机翼以实现短距起落
directed-energy weapon　射线武器,定向能武器
directed-flow thrust reverser　喷流换向反推装置,推力换向装置
direct-entry　直接进入
direct-entry trajectory　直接进入轨道
direct-fire missile　直接瞄准发射的导弹
direct-hit missile　碰炸式导弹
directing radar　引导雷达
direction　1.方向 2.指挥,引导,指导 3.(常用复数)指出,用法,说明(书)
direction angle　方向角,方位角
direction cosine　方向余弦
direction cosine matrix　方向余弦矩阵

direction error 方向误差,方向偏差
direction finder 1.测向器,定向器 2.无线电罗盘
direction finding 测向,定向
direction finding antenna 测向天线
direction gain 指向增益
direction gradient filter 定向梯度滤波器
direction index 指向系数
direction matrix 方向阵
direction of arrival 到达方向
direction of flow 流向,流体流向
direction of motion 运动方向
direction of rotation 旋转方向
direction of rotor rotation 转子旋转方向
direction post 引导哨,引导站
direction probable error 方向概率误差
direction selection 方向选择
direction sight 方向瞄准
direction sighting by multiple-angle method 倍角法定向瞄准
direction sighting by navigational method 航行法定向瞄准
direction sighting by vector method 矢量法定向瞄准
directional 1.定向的,指向的 2.航向的 3.方向的
directional aids 定向设备
directional antenna 定向天线
directional antenna of fuze 引信定向天线
directional automatic realignment of trajectory system 1.达特(稳定)系统 2.弹道方向自动调准系统
directional buoy 定向浮标
directional control 航向操纵
directional control valve 方向控制阀
directional coupler 定向耦合器
directional crystallization blade 1.定向结晶叶片 2.定向凝固叶片
directional dynamics 方向动力学
directional filter 分向滤波器
directional fragment warhead 定向战斗部
directional gyro 航向陀螺仪,陀螺半罗盘
directional gyro unit 航向稳定器
directional gyroscope 陀螺半罗盘,航向陀螺仪
directional initiation 定向引爆
directional microphone 定向传声器
directional motion 方向移动
directional property 定向性,方向性
directional sighting 1.定向瞄准 2.方向瞄准
directional solidification 定向凝固
directional sonobuoy 定向声纳浮标
directional stability 方向稳定性
directional trim 方向配平
directional warhead 定向作用弹头
directional warhead fuze 定向弹头引信
directionally solidified eutectic superalloy 定向凝固共晶高温合金
directionally solidified superalloy 定向凝固高温合金
directionally-stable 航向稳定的
direction-annexed traverse 方向附合导线
direction-determining board 正方案
direction-finding bearing 无线电方位,电台相对方位角
directive 1.命令,指令 2.定向的,有方向性的
directive antenna 定向天线
directive effect 方向效应
directive gain 方向性增益
directivity 指向性,方向性
directivity angle 辐射仰角
directivity function 方向性函数
directivity of antenna 天线方向性
directly-heated cathode 直热式阴极
director 1.指挥仪,指引仪,控制仪 2.导向器,引向器 3.操纵台长,指挥员 4.引导站,指挥站 5.无源定向偶极子 6.射击指挥装置,军舰上的射击指挥塔
director amplifier 指引仪放大器
director command 主管指令
director control 主管控制
director element 引向器单元
director horizon 指引地平仪
director output error 指挥仪输出误差
director radar 引导雷达
director sight 指挥仪式瞄准具
director trailer 指挥仪车
director wire 指引仪的指示线
directorate 董事会,理事会,指挥部
director-type computer 射击指挥仪型计算机
director-type fire control 射击指挥仪型火力控制
directory 手册,指南,索引簿
directory routing 地址路由选择
direct-reading accelerometer 直读式加速度计(仅用于飞行试验)
directrix 准线
direct-to-home 直接到户
direct-view 直观
direct-vision pane 直接观察窗
direct-vision tube 直视显像管
dirigible 1.飞艇,气艇 2.可操纵的
dirigible bomb 可导炸弹
DIRP Defense Industrial Reserve Plant 国防工业后备工厂(美国)
DIRRECT Digital Information Repository for Release, Exchange, Control, and Tracking 发放、交换、控制和

跟踪数字信息的储存
dirt 污物,灰尘
dirt trap 吸尘罩,防尘罩
dirthole 1.砂眼,砂孔 2.夹渣
dirty 非净形的
dirty bomb 肮脏弹
DIS 1. Defense Investigative Service 国防调查研究部(美国国防部) 2. Distance 距离 3. Distributed Intelligence System 分布式智能系统 4. Douglas Inspection Standard 道格拉斯检验标准 5. Draft Interim Standard 初稿临时标准
disability 失能,能力丧失
disable 使失去能力,使无资格
disabled aircraft 残缺飞机
disabled bird 有故障的飞行器,有故障的导弹
disaccommodation factor 减落因数
disadvantage 缺点,不利条件,损失
disaggregation 解裂
disaggregation algorithm 解集的算法
disagreement 失调,失谐
DISAP Disapprove 不批准,不同意
disappear 消失,失踪,不复存在
disappearance point 消失点(流星)
disarm 卸除引信,使……不能起爆
disarmament 裁减军备
disassemble 争解,拆卸
DISASSM Disassemble 拆卸,分解
disassociation 1.离解(作用) 2.分离,游离
disaster 事故,失事,灾难
disaster relief 赈灾,灾难援助
disaster victim identification 事故死难者识别
disastrous accident 致命事故,机毁人亡事故
disastrous disorientation 致命性定向障碍
disband 解散,遣散,除名,退役
disbarism 减压症
DISC Disconnect 拆开,拆卸,断开
disc 1.轮盘,盘,圆板 2.磁盘 3.唱片 4.激光唱盘
disc burst speed 轮盘破裂转速
disc loading (直升机旋翼的)桨盘载荷
disc unit 磁盘储存器
discard 1.报废,丢掉 2.废器,废弃物 3.(燃料燃尽后火箭级的)脱落
discard task 报废,报废工作
disc-drum rotor 盘鼓式转子
DISCH Discharge 放电,放油,卸载,排放
discharge 1.卸载,排出,排放,卸货,卸下,放出 2.排出口,排出管 3.流出,流(出)量,排出量 4.(枪炮)发射 5.退弹 6.解雇 7.放电
discharge arrester 放电避雷器
discharge cathode 辉光放电阴极
discharge cathode assembly 放电阴极组装
discharge cathode ignition time 阴极放电点火时间
discharge cell 气体放电元件
discharge chamber 排气室,放电室
discharge chamber wall 放电室壁
discharge channel 泄水渠,排水道,放出沟
discharge characteristic 放电特性,流量特性
discharge characteristic curve 放电特性曲线
discharge coef 排放系数
discharge coefficient 流量系数,放电系数
discharge current 放电电流
discharge development 放电过程
discharge duration 放电持续过程
discharge efficiency 放电效率
discharge energy 放电能量
discharge energy level 放电能量水平
discharge indicator disk (高压灭火瓶)耗量指示盘
discharge lamp 放电灯
discharge loss 放电损失
discharge loss change 放电损失变化
discharge loss increase 放电损失增加
discharge model 放电模型
discharge oscillation 流量振荡
discharge performance 放电性能
discharge plasma 放电等离子体
discharge power 放电功率
discharge pressure 排气压力,排放压力,出口压力
discharge propagation 放电传播
discharge pulse 放电脉冲
discharge rate 放电率
discharge region 放电区
discharge temperature 排气温度
discharge tube 1.排气管,排出管 2.辉放电管 3.闸流管
discharge valve 安全溢出阀
discharge voltage 放电电压,放电电位
dischargeable weight 投弃重量
discharged drained battery 湿放电蓄电池
discharger 1.发射装置,发射架 2.排气装置,排出装置 3.放电器,避雷器 4.卸载器
Disclaim "放弃"行动(1942年2月5日,英军第133空中联络组在萨拉热窝附近空降,以便与南斯拉夫抵抗运动部队联系)
discolor 使变色,使褪色
discoloration 变色,褪色
discoloured 变色的,褪色的
DISCOM Digital Selective Communications 数字选择通信
discomfort 不适(感)

discomposition 原子位移
DISCON Defence Integrated Secure Communication Network 国防综合保密通信网（澳大利亚）
discone antenna 盘锥天线
disconnect 1.拆开，断开，断接 2.装卸接头，断接器，接线盒
disconnection 拆开，断开，脱开，断线
disconnector 断路器，切断开关，装卸接头
discontinue 撤销（某一番号，机构等），停止使用
discontinuity 1.不连续（性、点），间断（性、点）2.突变性，突跃，骤变
discontinuous command 断续指令
discontinuous diagonal 不连续的对角线
discontinuous fibre 不连续纤维
discoordination 失协调
Discos Disturbance compensation system（卫星上的）扰动补偿系统
discount rate 贴现率，折扣率
discounting 削价销售
discourage 阻止，使气馁
discoverer 发现者
discovery 1.发现，发觉 2.（当事人必须）透露（事实真相或有关文件内容）3."发现者号"航天飞机（美国航空航天飞行学）
discovery efficiency 发现效率
DISCR Discrepancy 不符合，偏差
discrepancy 偏差，误差，偏离，缺陷，失调
discrepancy between channels 路际差异
discrepancy between twice collimation error 二倍照准部互差，二倍照准差
discrepancy function 数据之差距函数
discrete 离散的，不连续的
discrete analysis 离散分析
discrete approximation 离散逼近
discrete attitude 离散姿态
discrete branch 离散分枝
discrete command 离散指令
discrete component 分立元件
discrete control 离散控制
discrete control variable 离散控制变量
discrete design 离散设计
discrete distribution 离散分布
discrete event 离散事件
discrete event dynamic system 离散事件动态系统
discrete Fourier series 离散傅里叶级数
discrete Fourier transform 离散傅里叶变换
discrete frequency 离散频率
discrete function 离散函数
discrete geometric combination 离散几何组合

discrete gust 离散突风
discrete Hilbert transform 离散希尔伯特变换
discrete hole 离散孔
discrete Kalman filter 离散卡尔曼滤波器
discrete optimal control law 离散最优控制律
discrete optimization 离散优化
discrete passage 离散通道
discrete path 离散路径
discrete phase 不连续相，分散相
discrete phase locked loop 离散锁相环
discrete point 离散点
discrete radio source 分立射电源
discrete roughness 离散的粗糙度
discrete sail 离散帆
discrete search 离散搜索
discrete search strategy 离散搜索策略
discrete solution 离散解
discrete spectrum 分立光谱
discrete state 离散状态
discrete state space 离散状态空间
discrete system 离散系统，集块参数系统
discrete time 离散时间
discrete tone 离散音频
discrete transmission 非连续发射
discrete value 离散值
discrete variable 离散变量
discrete wave 不连续波
discrete wave number 离散波数
discrete wave number method 离散波数法
discrete-address beacon system 离散地址信标系统
discrete-event simulation model 不连续事件的仿真模型
discrete-frequency noise 离散频率噪声
discreteness 1.不连续性 2.对比
discrete-phase 离散相
discrete-time signal 离散时域信号
discrete-time system 离散时间系统
discretionary wiring method 选择布线法，选择连接法
discretization 离散化
discretization architecture 离散化的体系结构
discretization error 离散误差，离散化误差
discretization method 离散化，离散方法
discretization scheme 离散图表
discretization technique 离散化技术
discretization timescale 离散化的时间尺度
discretize 使离散，离散化
discretized equation of motion 离散运动方程
discretized expansion 离散扩张
discriminability 判别力，分辨力
discriminant 判别式

discriminant function 判别函数
discriminating comparator 鉴别比较装置
discrimination 1.鉴别,区分 2.分辨力,鉴别力
discrimination（threshold） 鉴别力(阈)
discrimination between the earth and the moon light 地球月球信号鉴别
discrimination between the earth and the sun light 地球太阳信号鉴别
discrimination reaction time test 鉴别反应时间试验
discriminator 1.鉴频器,鉴相器 2.假信号抑制器
discriminator voltage 鉴频器电压,鉴相器电压
discus （变几何形翼的)弧形滑槽
disdrometer 雨滴谱仪
disease 疾病
diseased crop 病害农作物
disembarkation （旅客)下机
disengage button 断开按钮
disengage the autopilot 关断自动驾驶仪
disengagement 脱离,解脱,断开,分离
disequilibrium 失去平衡
disexplosion 解爆
disguise 伪装
dish 抛物面反射器
dish antenna 抛物面天线
dish out 〈口语〉1.飞机向压坡度方向大角度侧滑下降或在盘旋(转弯)中侧滑下降 2.由于不正确操纵造成的飞行轨迹下降和偏向一边
dished grid 碟形网格
dishing 表面凹陷,凹弯
disilicide 二硅化物
disiloxane 二硅氧烷,二甲硅醚
disintegrate 使分解,使碎裂,使崩溃,使衰变
disintegrated warhead 集束弹头,多弹头
disintegrating links 散弹链
disintegration 1.(核)衰变,蚁变 2.分裂,碎裂,分解,离解
disintegration structural process 解体结构工艺,离解
disk 磁盘,磁碟片,圆盘,盘状物
disk area 圆盘面积,盘面积,视盘面积
disk design 盘面设计
disk galaxy 盘星系
disk height 碟片高度
disk laser 盘形激光器
disk loading 桨盘载荷
disk margin 磁盘边缘
disk model 偏心盘模型,偏心盘模型
disk of galaxy 星系盘
disk population 盘族
disk shock 磁盘冲击

disk shock location 磁盘冲击位置
disk shock structure 磁盘结构冲击
disk size 磁盘容量
disk solution 光盘解决方案
disk vibration 圆盘振动
disk-blade interface 磁盘叶片接口
disk-gap-band parachute 盘缝带伞
disk-type magazine 鼓轮式弹仓(弹盒)
dislocation 1.转换位置,错位 2.位错(晶体晶格发生畸变的现象)
dislocation density 位错密度
dislocation free crystal 无位错晶体
dislocation loop 位错环
DISM Dismantal 分解,拆卸
dismantle 1.分解,拆解 2.拆除,撤除 3.粉碎
dismount 拆卸,卸下
dismountable 可拆卸的
disorbit 逸出轨道,离开轨道
disorder 无序
disorientation 定向障碍,迷航
disorientation accident 定向障碍事故,失定向事故
DISP Dispenser 分配器
disp display 显示器
disparity 不同,不一致,不等
dispart （瞄准线与炮膛轴线)不平行
dispatch 1.跳伞(或空投)调度 2.放飞调度
dispatch deviation 放飞允许偏差
dispatch office 调度室
dispatch reliability 正点放飞率
dispatcher 1.调度员 2.空投组组长(也可写作 despatcher)
dispatcher telephone 调度电话
dispatching loudspeaker set 扬声调度单机
dispense 分配,分发
dispenser 1.投放器 2.子母弹箱,集束弹箱 3.分配器 4.加注器
dispenser bomb 子母炸弹
dispenser cathode 储备式阴极
dispensing 加油(特指通过加油栓给飞机加添燃料)
dispensing sequence 投放程序
dispersal areas 疏散区(特指机场上供战时飞机和保障设备疏散停放的区域)
disperse 分散,弥散,疏散,散开
disperse phase 弥散相
disperse trajectory 分散的轨迹
dispersing agent 分散剂
dispersion 1.分散,散布,色散,弥散,频散 2.(高炮)射弹散布,(化学弹)弥散,(火箭)圆形散布,(电磁辐射)散射 3.漂移,偏移,偏差,(炸弹离瞄准点的)平均偏

差量
dispersion analysis 方差分析
dispersion area of landing point 着陆点散布范围
dispersion azimuth angle 飞散方位角
dispersion behavior 分散行为
dispersion characteristics 色散特性
dispersion diagram 散布图
dispersion error 散布误差
dispersion hardening 弥散硬化
dispersion mechanism 分散机理
dispersion of fire 射弹散布
dispersion of fuze actuation angle 引信启动角散布
dispersion orbit 弥散轨道
dispersion pattern （射弹）散布面
dispersion pressure 压力弥散
dispersion rate 分散率
dispersion rectangle 弹着散布矩形图表
dispersion relation 色散关系,分散关系,频散关系
dispersion ring 弥散环
dispersion strengthening 弥散强化
dispersion strengthening composites 弥散强化复合材料
dispersion warhead 散射弹头,散射战斗部
dispersion wave 频散波
dispersive medium 弥散介质
dispersive wave 频散波
DISPL Displacement 位移,偏移
displace 1.移动,转移 2.置换,取代
displaced phase center antenna 相移中心天线
displaced vacancy 位移空位
displacement 1.位移,变位,置换 2.（目标与截击机航迹间的）航迹间隔 3.（船身或浮筒的）排水量,（气球或飞艇的）排（空气）量（以体积或重量计）4.（气缸）排量,工作容积 5.（操纵面）偏度,（驾驶杆）行程
displacement aircraft rocket launcher 移动式机载火箭发射装置,收放式机载火箭发射装置
displacement amplitude 位移幅度
displacement autopilot 比例式自动驾驶仪,有差式自动驾驶仪
displacement bridge 测位移电桥
displacement capacity 排量
displacement current 位移电流
displacement feedback 位移反馈,位置反馈
displacement follow-up unit 位移随动部件,位置反馈部件
displacement gyro assembly 三自由度陀螺仪组
displacement manometer 差动压力计,位移压力计
displacement meter 位移计
displacement observation 位移观测
displacement of image 像点位移
displacement of spectral line 谱线位移
displacement response spectrum 位移反应谱
displacement sensor 位移传感器
displacement thickness 排移厚度,位移厚度
displacement transducer 位移传感器,位移变换器
displacement vector 位移矢量
displacement vibration amplitude transducer 位移振幅传感器
displacer 置换剂,取代剂,平衡浮子,排代剂
display 1.显示,指示 2.显示器,指示器 3.示范飞行 4.表演
display access 显示取数
display and control module 显示和控制舱
display augmentation 显示增强
display augmentation concept 显示增强的概念
display background 1.显示背景（色彩）2.显示后台
display bank 显示器组,指示器组
display board 1.空情显示板 2.仪表板,信号盘 3.信号显示板
display configuration 显示配置
display console 显示器操纵台
display controller 显示（器）控制台
display device 显示器件
display elements 显示要素
display field 显示范围,数据显示区（平视仪）
display foreground 1.显示前景（色彩）2.显示前台（显示图像和要素的集合）
display format 显示格式,显示形式
display generator 显示发生器
display instrument 显示仪器
display menu 显示选择单
display method 显示方法
display monitor 显示监视
display panel 显示板
display refresh rate 显示刷新率
display screen 显示屏
display surface 显示面
display symbology 显示标记
display system 显示系统
display technology 显示技术
display unit 显示部件,显示器,显示装置
display with extracter 录取显示器
displaying (measuring) instrument 显示式（测量）仪器
disposable lift 可用升力（总升力小于气球重量）
disposable load 可卸载荷（载重）,消耗性载重
disposal 1.配置,布置,安排 2.处理,处置 3.消除,排除
disposal orbit 配置轨道
disposed satellite 处理卫星
disposed vehicle 处理车辆

disposition　1.配置,布置,安排,部署　2.倾向
disposition of air defence weapons　防空兵器配置
disposition of electronic equipment　电子设备配置
dispositive　(事件,行为等)决定性的
disproportion　不均衡,不相称,不成比例
disproportionate pyrolysis　不成比例的热解
disqualification　不合格,报废,废品
DI-SR　Discrepant Items-Ship's Record 偏差项目架次记录
disrepair　1.失修　2.错误修理
disrupter-type spoiler　分裂型扰流板
disruption　1.溃消(电流) 2.崩裂(磁力线),分裂,爆炸　3.击穿
disruptive burning　破坏性的燃烧
disruptive explosive　破坏炸药,猛性炸药
disruptive innovation　破坏性创新
dissection　解剖,分析
dissector　析象管,解剖者,解剖学家,解剖器
dissemination　通报,传播
dissemination efficiency　散布效率
dissemination of the value of quantity　量值传递
dissent circuit　否决电路
dissimilar air-combat training　模拟敌机空战训练
dissimilar redundancy　非相似余度
dissipate　浪费,使……消散
dissipation　1.耗散,浪费,消散,损耗　2.下降,消失
dissipation factor　耗散因数
dissipation function　耗散函数
dissipation length　消散长度
dissipation power　耗散功率,功耗
dissipation rate　耗散速率
dissipation term　耗散项
dissipative forces　耗散力
dissipative structure　耗散结构(理论)
dissipative system　耗散系统
dissociate　游离,使分离,分裂
dissociation　离解,脱节
dissociation energy　离解能
dissociation fraction　分离部分
dissociation temperature　解离温度
dissociative excitation　离解激发
dissociative recombination　离解性复合
dissociative sensations　分离感觉
dissociety　离解度
dissolubility　溶解度,溶解性
dissolution　溶解,分解
dissolution and wetting bond　溶解与润湿结合
dissolvant　溶剂,溶媒
dissolve　溶,溶解
dissolvent　溶剂,溶媒
dissymetrical transducer　不对称换能器
DIST　Distance 距离,航程
distance　1.(两点之间的)距离　2.航线距离　3.(时间的)间隔　4.远方
distance bar　牵引杆
distance between rotor centers　旋翼中心间距
distance classifier　距离分类器
distance curve　时间曲线,距离曲线
distance datum　距离数据
distance decision function　距离判决函数
distance finder　测距仪
distance indicator　示距天体
distance marker　1.(跑道两侧所明的)长度数字标记(通常以千英尺为单位) 2.(雷达荧光屏上的)距离标记(以同心圆表示,又称 range marker) 3.距离标识器
distance measuring equipment　(DME)测距仪,测距设备
distance measuring gate　测距门
distance modulus　距离模数
distance of landing run　着陆滑跑距离
distance of take-off run　起飞滑跑距离
distance performance　距离性能
distance profile　距离轮廓
distance readout window　(仪表的)远距读出口
distance scale　距离尺度
distance thermometer　遥测温度表
distance to go　未飞距离,剩余距离
distance to liftoff　起飞滑跑距离
distance wadding　非爆炸装填料(装在弹壳内的惰性物质)
distance-measuring gate　测距门
distance-measuring satellite　测距卫星
distant early warning radar　远程预警雷达
distant geomagnetic field　远地磁场
distant-reading tachometer　远读转速表
distillate　馏出物,馏出液
distillate fraction　蒸馏物部分
distillation　蒸馏
distillation curve　蒸馏曲线
distillation range　馏程
distillator　蒸馏器
distilled fuel　馏分燃料油
distilled water　蒸馏水
distiller　1.蒸馏器　2.蒸馏者,制酒者
distinct　1.独特的,性质不同的　2.清楚的,明显的
distinct region　显著区域
distinct torque　不同的转矩
distinction　1.差别,区别　2.特征,特性
distinguish　1.区分,辨别　2.使杰出,使表现突出

distort 1.使失真,使变形 2.施放干扰 3.歪区,曲解
distorted sector 变形部分
distortion 1.(流场)畸变 2.失真,变形,扭曲 3.(燃气涡轮)流速或温度变化
distortion analyzer 失真分析仪
distortion coefficient 失真系数,畸变系数
distortion energy 畸变能
distortion factor of a wave 电波失真因数
distortion generator 失真发生器
distortion generator speed 畸变发生器速度
distortion index 畸变指数
distortion isograms 等变形线
distortion level 失真水平
distortion map 畸变地图
distortion measure 失真测度
distortion of projection 投影变形
distortion parameter 变形参数
distortion pattern 畸变图谱
distortion rate function 失真信息率函数
distortion screen 扭曲的屏幕
distortion tolerance 畸变容限
distortionless 无畸变(的)
DISTR 1. Distortion 变形 2. Distribute or Distribution 分配,分派,分布
distraction 注意力分散,分心,精神涣散
DISTRAM Digital Space Trajectory Measurement System 空间轨道数字测量系统
distress 危急,遇险,事故
distress beacon 失事信标
distribute 分配,分布,分类
distributed Bragg reflection type laser 分布布拉格反射型激光器
distributed capacitance 分布电容
distributed charge 分离式装药
distributed computer 分布式计算机
distributed computer control system 分布式计算机控制系统
distributed computer system 分布式计算机系统
distributed computer telemetry 分布式计算机遥测
distributed data-processing 分布式数据处理
distributed emission crossed-field amplifier 分布发射式正交场放大管
distributed feedback laser 分布式反馈激光器
distributed force 分布力
distributed function computer 功能分布式计算机
distributed interaction klystron 分布作用速调管
distributed joint 分布式关节
distributed mass-balance 分布质量平衡
distributed parameter 分布参数
distributed parameter system 分布参数系统(参数用空间和时间的函数描述的系统)
distributed pole motor 分布极电动机
distributed processing 分布式处理
distributed target 分布目标
distributed telemetry system 分布式遥测系统
distributed warhead 定向爆破弹头
distributed-feedback semiconductor laser 分布反馈半导体激光器
distributer 分配器,配电盘
distributing (distribution)/distributor box 配电盒,分线盒
distributing board 配电盘
distribution 1.分布,分配 2.配电
distribution constant 分布恒量
distribution for population inversion 粒子数反转分布
distribution function 分布函数,分配函数
distribution manifold 分配歧管
distribution of debris mass (太空)碎片质量分布
distribution of droplet 微滴分布,粒径分布
distribution of electric potential 电势分布
distribution of flame 火焰分布
distribution of flow 水流的分配
distribution of particle 颗粒分布
distribution of pressure 压力分布
distribution of response 响应分布
distribution of temperature 温度分布,气温分布
distribution system 分配制度,配电系统
distributor 分配器,配电盘
distributor plate 配电盘,配电板
district 区,管区
disturb 扰动,妨碍,使不安
disturbance 1.扰动(对飞机飞行来说,常指迎角的意外变化) 2.(飞行)失控(如自动上仰、失速等) 3.(风的)突变(常指气旋、低气压) 4.变位,摄动 5.故障,失调 6.损伤,破坏,断裂 7.(电、磁、声等)干扰
disturbance (disturbing) torque 扰动力矩
disturbance acceleration 扰动加速度
disturbance attenuation 干扰抑制
disturbance cell 干扰细胞
disturbance crossover 干扰交叉
disturbance crossover frequency 干扰交叉频率
disturbance decoupling 扰动解耦
disturbance effect 干扰效应
disturbance energy 微扰能
disturbance environment 干扰环境
disturbance equation 扰动方程
disturbance field 干扰场
disturbance force 扰动力

disturbance frequency 扰动频率
disturbance input 扰动输入
disturbance load 负载扰动，负荷扰动
disturbance mode 干扰模式
disturbance model 扰动模型
disturbance open loop 干扰开环
disturbance parameter 扰动参数
disturbance period 扰动时期
disturbance quantity 扰动量
disturbance rejection 干扰抑制
disturbance signal 干扰信号
disturbance source 干扰源
disturbance task 干扰任务
disturbance torque 干扰力矩
disturbance torque by the fuel slosh 燃料晃动干扰力矩
disturbance vector 干扰向量
disturbance wave 扰动波
disturbance zone of sun 太阳干扰区
disturbance-corrupted datum 破坏干扰基准
disturbance-feedback 干扰-反馈
disturbance-feedback strategy 干扰反馈策略
disturbance-rejection task 抑制干扰任务
disturbances 干扰，失调（disturbance 的复数形式）
disturbed body 受摄体
disturbed coordinates 受摄坐标
disturbed daily variation 扰日日变化
disturbed day （地磁）扰日
disturbed motion 受摄运动
disturbed motion of two-bodies 受扰两体运动，两体扰动
disturbed orbit 受摄轨道
disturbed reticle 活动光环
disturbed sun 扰动太阳
disturbed trajectory 干扰弹道，扰动弹道
disturbing acceleration 干扰加速度
disturbing aircraft field 飞行器造成的扰动（磁）场
disturbing body 摄动体
disturbing effect 摄动效应
disturbing force 摄动力
disturbing function 摄动函数
disturbing gravity 扰动重力
disturbing mass 扰动质量
disturbing potential 扰动位
disulfide 二硫化物
DIT Dynamic Integrated Test 动力学综合试验
DITACS Digital Tactical System Radar 数字战术系统雷达（美国机载雷达 APS-128 的改进型）
DITC Department of Industry, Trade and Commerce 工商贸易部（加拿大）
ditch groove 沟槽

ditching 水上迫降，水上紧急降落
ditching brief 水上迫降简令
ditching characteristic 水上迫降特性
ditching device （遥控飞机的）降落手段
ditching drill （机上人员的）水上迫降应急程序，水上迫降练习
ditching exit 水上迫降应急出口
DITEC Digital Television Encoding 数字电视编码
dither 1. 高频振动 2. 高频脉冲信号
dither bearing 抖动轴承，抖动支承
dither signal 高频脉动信号
dither tuned magnetron 抖动调谐磁控管
DITS 1. DAIS Integrated Test System 数字式航空电子设备信息系统的综合测试系统 2. Data Information Transfer Set 数据信息传输装置 3. Data Information Transfer System 数据信息传输系统
dits digital information transport standard 数字信息传输标准
ditty bag 水手包，杂物袋
diurnal 周日的
diurnal aberration 周日光行差
diurnal apparent motion 周日视运动
diurnal bulge 日隆
diurnal change 周日变化，日变化
diurnal circle 周日（平行）圈
diurnal clock rate 周日钟速
diurnal density 日密度
diurnal inequality （月球运动的）日差
diurnal libration 周日天平动
diurnal modulation 周日调制
diurnal motion 周日运动
diurnal nutation 周日章动
diurnal parallax 周日视差
diurnal parallel circle 周日平行圈
diurnal range 周日变幅
diurnal rhythm 日周期节律，昼夜节律
DIV 1. Detailed Inspection Visual 详细目视检查 2. Diverter 分流器，换向器
DIVADS Division Air-Defense System 师防空系统（美国）
dive 1. 俯冲 2. 潜水，潜航（潜艇的）3. 突然从视野中消失
dive bombing 俯冲轰炸
dive brakes 俯冲减速板
dive guidance 潜水指导
dive illusion 俯冲错觉，潜水错觉
dive phase 潜水阶段
dive spiral 急盘旋下降
dive-bombing sight 俯冲轰炸瞄准具

diver 1.俯冲飞机 2.潜水员 3.潜水艇
dive-recovery flap 改出俯冲襟翼
diverge 分歧,偏离,分叉,离题
divergence 1.变形扩大 2.变形发散 3.纵向运动发散,纵向不安定性
divergence angle 发散角,扩张角
divergence belt 发散带
divergence control (滤波)发散控制
divergence lateral 侧向运动发散,侧向不安定性
divergence length 散度长度
divergence speed 发散速度
divergence wave 发散波
divergent combustor 相异燃烧室
divergent duct 扩散管
divergent nozzle 渐扩喷嘴,扩散喷管
divergent part 发散部分
divergent port 不同的港口
divergent section 扩散段
divergent type 扩散类型
diverse array 分集阵列
diversify 1.换场着陆 2.(飞行任务的)改变 3.换向,转向,改变航线 4.变更航班
diversion 改航
diversion airfield 备降机场
diversion trough 导流槽
diversionary missile 诱馆导弹,牵制性导弹
diversity 1.分集 2.多样,参差 3.不同,相异 4.疏散(性)
diversity combiner 分集合成器
diversity factor (照强)差异因数
diversity gate 【计】"异门"
diversity module 多样性模块
diversity radar 分集式雷达
diversity receiver 分集接收机
diversity receiving 分集接收
diversity reception 分集接收
diversity stack 反比叠加,花样叠加
diversity technique 分集技术
diversity telemetry 分集遥测
divert 转移,使……转向
divert engine 转向发动机
divert propulsion 转向推进
diverted-thrust 推力(向下)转向的(指升力巡航发动机)
diverter 1.偏向器,推力转向器 2.分流电阻 3.避雷针
diverter height 分流器的高度
diverter wedge 分流器楔
dive-strafer 俯冲扫射飞机
dive-toss bombing 1.改出俯冲轰炸 2.俯冲拉起轰炸

DIVIC Digital Variable Increment Computer 可变增量数字计算机
divide 1.划分,除,分开 2.使产生分歧
divide-and-conquer 分而治之
divided attention 注意力分配
divided shielding 多层防核辐射屏蔽
divider 1.分配器 2.分压器,分频器 3.减速器,减压器 4.除法器 5.座舱隔板 6.(复数)双脚规,分线规,圆规 7.除法逻辑电路
diving medicine 潜水医学
divisibility 可分性
division 1.划分,分开,分割 2.分配,分派,分布 3.分度,刻度,标度 4.部分,部门,部,处,科 5.隔板,隔拦,挡板,阻挡层 6.师,海军航船分队 7.除法
division for fighter planes 歼击机机场守卫师
division of airspace 空域划分
division of panzer-Grendiere paratroops 坦克步兵伞降师
dizziness 头晕
DJE Deception Jamming Equipment 欺骗干扰设备
DJOEO Development Job Outline Engineering Order 研制任务大纲工程指令
DK docked 已对接的(航天器之间)
DKC Design Knowledge Capture 设计资料归档
DL 1. Deciliter 分升 2. Data Link 数据传输线路 3. Data List 数据表 4. Datum Level 基准面 5. Delay Line 延迟线 6. Developed Length 展开长度 7. Dome Light 舱顶灯
DLA Defense Logistics Agency 国防后勤局(美国)
DLB Data-Link Buffer 数据链路缓冲器
DLC 1. Data Link Control 数据链路控制 2. Digital Loop Carrier 数字环路载波 3. Direct Lift Control 直接升力控制 4. Downlink Communication 声呐浮标下行通信线路 5. Dynamic Load Control 动态负载控制
DLCI Data Link Control Identifier 数据链控制标识
DLE Data Link Entity 数据链机构
DLF Design Load Factor 设计载荷因数
DLG Divided Landing Gear 分置式起落架
DLGF Data Load Gateway Function 数据加载途径功能
DLI Deck-Launched Intercept 甲板起飞截击
DLINDG Dial Indicating 刻度指示
DLIR Downward Looking Infrared 下视红外(装置)
DLL 1. Data Link Library 数据链库 2. Design Limit Load 设计最大使用载荷
DLM Declarative Language Machine 申请语言计算机
DLME 1. Data Link and Message Engineering 数据链与信息工程 2. Direct Lift and Manoeuvre Enhancement 直接升力和机动性增强
DLMS Data Link Management System 数据链管理

系统

DLMU Data Link Management Unit 数据链管理装置

DLOD Duct Leak and Overheat Detection 管路泄露与过热探测

DLP 1. Data Link Processor 数据链处理器 2. Deck Landing Practice 甲板降落训练 3. Distribution List Process 清单发放过程

DLS 1. Data Link Subsystem 数据链路子系统 2. Data Load System 数据加载系统 3. DME Landing System 测距仪着陆系统 4. Dynamics Limb Sounder（卫星上的）动力学临边探测器

DLSC Defense Logistics Service Center 国防后勤服务中心（美国）

DLTS Deep Level Transient Spectroscopy 深能级暂态谱（学）

DLU 1. Data Logger Unit（工作）数据记录装置 2. Down Load Unit 下载装置

DLUE Delivery Load Uninstalled Equipment 非安装设备的移交装载

DLVY Delivery 交付，交货

DLW Design Landing Weight 设计着陆重量

DLX Deluxe 上等的，高级的

DLYCK Daily Check 每日检查，日检

DM 1. Data Management 数据管理 2. Data Modeling 数据建模 3. Delta Modulation 增量调制 4. Dependent Modules 相关模块 5. Disconnected Mode 断开状态 6. Docking Module 对接舱 7. Drafting Manul Or Design Manual 制图手册或设计手册 8. 2D Drawing 二维图

DMICS Design Methods for Integrated Control System 综合控制系统的设计方法

DMIS Data Management Information System 数据管理信息系统

DMA 1. Defense Mapping Agency 国防测绘局（美国） 2. Despin Mechanical Assembly 消旋机械装置 3. Direct Memory Access 直接存储器存取

DMC 1. Data Management Center 数据管理中心 2. Direct Maintenance Cost 直接维修费用

DMD 1. Deployment Manning Documents 部署人员配备表 2. Digital Message Device 数字信息装置 3. Diamond 金刚石

DMDB Digital Map Data Base 数字地图数据库

DME 1. Distance Measurement Equipment 测距器 2. Distance Measuring Equipment 测距仪

DME fix 测距装置定位点

DME seperation 测距仪间隔

DME transponder 测距应答器

DMEA Defect Mode and Effect Analysis 故障模式和影响分析

DMED Digital-Message Entry Device 数字信息输入装置

DMES Deployable Mobility Execution System 疏开机动执行系统

DMF Digital Matched Filter 数字匹配滤波器

DMGZ Demagnetize 去磁，退磁

DMI Department of Manufacturing Industry 制造工业部（澳大利亚）

DMIR Designated Manufacturing Inspection Representative 生产检验委任代表

DML Demolition 拆毁，毁坏

DMM 1. Data Memory Module 数据存储模块 2. Digital Multimeter 数字式多用表

DMN Data Multiplexing Network 数据多路网

DMNRLZR Demineralizer 脱矿质器，软化器

DMO 1. Data Management and Operations 数据管理和操作 2. Data Management Office（或 Officer）数据管理处（或人员） 3. Dependent Meteorological Office 直属气象处 4. Development Manufacturing Organization 发展制造组织

DMOR Digest of Mandatory Occurrence Reports 偶发事件报告摘要

DMOS Diffusive Mixing of Organic Solutions（在美国航天飞机上进行的）有机溶液扩散混合实验

DMP 1. Data Management Plan 数据管理大纲 2. Display Management Panel 显示管理面板

DMPI Desired Mean Point of Impact 期望平均弹着点

DMPP Display and Multipurpose Processor 显示和多功能处理器

DMPR Damper 阻尼器，缓冲器

DMR 1. Delayed Multipath Replica 延迟多通道复制光栅 2. Diagram Manual Report 图册报告 3. Differential Microwave Radiometer 差动微波辐射计

DMS 1. Data Management System 数据管理系统 2. Data Monitoring System 数据监控系统 3. Digital Measuring System 数字测量系统 4. Distributed Management System 分布式管理系统 4. Douglas Material Specification 道格拉斯材料规范

DMSP Defense Meteorological Satellite Programme 国防气象卫星计划（美国）

DMSS 1. Data Multiplex Subsystem 多路数据传输分系统 2. Display Mode Selection Switch 显示状态选择开关

DMTI Digitized Moving Target Indicator 数字式动目标显示器

DMTZR Demagnetizer 去磁器，退磁装置

DMU 1. Data-Management Unit 数据管理装置，数据管理元 2. Digital Mockups 数字样机 3. Distance-Measurement Unit 测距仪

DMX Data Multiplex 数据多路传输
DMZ Demilitarized Zone 非军事区
DN 1. Debit Note 借记通知单 2. Down 向下，降低
DNA 1. Defense Nuclear Agency 核武器局（美国）2. Does Not Apply 不能适用
DNC 1. Direct (Digital) Numerical Control 直接（或数字）数控 2. Distributed Numerical Control 分布式数字控制
DNCMS Data Network Configuration Management System 数据网构型管理系统
DNCS 1. Day/Night Camera System 昼夜摄像系统 2. Distributed Network Control System 分布式网络控制系统
DNI Director of National Intelligence 国家情报局长
DNIF Duty Not Involving Flying 不担任飞行的值勤
D-nose D形前缘
DNS 1. Direct Numerical Simulation 直接数字模拟 2. Distributed Network System 分布式网络系统 3. Domain Name System 域名系统
DNSS 1. Defense Navigation Satellite System 国防导航卫星系统（美国）2. Doppler Navigation Satellite System 多普勒导航卫星系统
DNTKFX Down Track Fix 航线前方定位点
DNW Deutsch-Niederlandischer Wind-kanal 德国-荷兰风洞
DO 1. Data Object 数据对象 2. Design Organization 设计组织 3. Diversion Order or Design Order 转换指令或设计指令 4. Drop-Out 取下（面罩）
DOA 1. Design Organization Approval 设计组织批准 2. Designated of Authority 委任代表权
DOB Dispersed Operating Base 分散作战基地
Dobson spectrophotometer 多布森分光光度计
DOC 1. Delayed-Opening Chaff 延时撒开的干扰丝 2. Department Of Communications 通信部（加拿大）3. Department Of Commerce 商务部（美国）4. Designed Operational Capability 设计作战能力 5. Direct Operating Cost 直接使用费用（成本）6. Document 文件 7. Documentation 文件，文档
dock 1. 机库，船坞 2. 飞机检修场，检修架 3.（航天器）对接
docking 1. 对接（两个航天器的接合）2. 入坞，进机库 3. 进厂检修
docking axis 对接轴
docking corridor 对接走廊
docking mechanism 对接机构
docking module 对接舱（航天飞行器）
docking port 对接口（两航天器的对接面）
docking tunnel 对接通道
doctrine 1.（指导军队行动的）基本原则，概则，准则 2. 主义，学说
document 文件，文档
document recording 实况记录
documentation 1. 文件（集）2. 文件编制，文件管理 3. 使用记录文件（飞机的）
docuterm 【计】检索字
DOD 1. Department Of Defense 国防部（美国）2. Depth Of Discharge 放电深度（蓄电池的）
dodecane 十二烷
dodging 规避机动
DODISS The Department of Defense Index of Specifications and Standards 国防部规范与标准目录
dodo 〈口语〉未放单飞的学员
DODS Definitive Orbit Determination System 最终轨道测定系统
DOE 1. Department Of Energy 能源部（美国）2. Department Of the Environment 环境部（英国）
DOF 1. Degree Of Freedom 自由度 2. Direction Of Flight 飞行方向
dog 1. 钩，弹钩，抓钩，夹头，挡块 2.〈口语〉指一种型别飞机中质量或飞行品质最差的一架 3. 机场导航指向标 4.〈俚语〉弹道导弹
dog course 纯跟踪航线，纯跟踪曲线
dog days 伏日
Dog star 犬星，天狼星
dog tag 〈口语〉别标记
dogbone 骨形连杆
dogfight 空中格斗，混战
dogfight missile 格斗导弹
dogleg 1. 折线（飞行航线）2. 航迹转向（用于修正航天器轨道倾角），折线形变弹道（轨道）
dogleg course 曲折航线
doglight bombardment 〈口语〉夜间轰炸
dogtooth （机翼）前缘据齿
DOH 1. DOHA 多哈（卡塔尔）2. Department Of Health 卫生署
DOI 1. Descent Orbit Insertion 进入下降轨道 2. Docking Orbit Insertion 进入对接轨道 3. Department Of Industry 工业部（现已改为贸易和工业部）（英国）
Doirseye 多尔氏眼
DOIU Digital Output Interface Unit 数字量输出接口组件
DOL Dispersed Operating Location 分散的操作站
DOLARS Doppler Location and Ranging System 多普勒定位与测距系统
doldrums 赤道无风带
DOLE Detection of Laser Emitters 探测激光发射体，探测激光发射源

dolly 1.(铆接用的)顶铁,抵座 2.轮式吊车 3.行李挂车(机场上运送行李的拖挂车的一节) 4.活动摄影车 5.航空(机载)数据传输设备
dolly roll 转动台架车
Dolphin "海豚"(英国海军目标探测雷达,用于"海上哨兵"武器系统)
DOM 1. Digital Orthophoto Map 数字正射影像图 2. Disk On Module 模组 3. Document Object Model 文档对象模型
domain 1.区,域,领土 2.(影响的)范围,(学术的)领域 3.磁畴
domain decomposition 域分解,区域分解
domain knowledge 领域知识
domain length 域名长度
domain of attraction 吸引范围
domain of interest 感兴趣的领域
domain size 区域尺度
domain transformation 域变换
domain transformation technique 确认技术加工
domain wall resonance 畴壁共振
dome 1.圆顶室,圆顶建筑物 2.整流罩,航行舱 3.半球形隔框
dome closeout 圆顶收尾
dome error 圆顶错误
dome error slope 圆顶误差斜率
dome less solar refractor 无圆顶太阳折射望远镜
dome material 音膜材料
dome phased array antenna 圆顶相控阵天线
dome type helmet 圆顶头盔
domestic brief 飞行前简令飞机部分(有关飞机分配、停放地点及呼号)
domestic communication 国内交通
domestic paraffin 国产石蜡
domestic reserves 国内燃油储备(供国内定期航班使用)
domestic satellite 国内(通信)卫星
domestic service 1.国内保养(航空公司飞机的) 2.国内航线飞行
dominance 优势,支配,控制,统治
dominant body 主导人物
dominant fault 支配性故障
dominant frequency 优势频率,突出频率
dominant heat-transfer mechanism 主导传热机制
dominant low mode 主导低模式
dominant mechanism 主要机理,显性机制
dominant mode 主模(式)
dominant motion 主运动
dominant noise 占主导地位的噪声
dominant parameter 主要参数
dominant pole 主导极点

dominant reflector 占主导地位的反射器
dominant regime 支配性体制
dominant role 主要角色
dominant tone 主色调
dominant wave 主波,优势波
dominant wave number 占主导地位的波数
dominate 控制,支配,占优势,在……中占主要地位
domination 支配,统治,控制
DOMSATCOM Domestic Satellite Communication 国内(通信)卫星通信
Donati comet 多纳提彗星
donor 施主
donor atom 施主原子
donor cell 供体细胞
donor cell search algorithm 供体细胞的搜索算法
donor section 捐赠部分
door 1.门,舱门,口盖 2.入口,舱口,通道 3.(雷达显示器上的)炮弹爆炸信号
door bundle 空投装载
door configuration 门配置
door hinge rotor 折页铰接式旋翼
DOORS Dynamic Object Oriented Requirements System 动态目标导向需求系统
DOP 1. Design Opereating Procedures 设计操作程序 2. Dilution Of Precision 精度降低因子,精度冲淡系数 3. Division Operating Plan 部门操作大纲
DOPAA Description Of Proposed Actions and Alternatives 对计划行动及其备分计划的描述
dopant 掺杂剂
dope 1.航空涂料,腻子(燃油的)添加剂,防爆剂 3.(合金冶炼)加入合金元素 4.上涂料,涂油漆 5.掺入,掺杂 6.黏稠物,胶状物 7.〈口语〉非官方最新消息
doped oxide diffusion 掺杂氧化物扩散
doped polycrystalline silicon diffusion 掺杂多晶硅扩散
doping 1.掺杂质(半导体中) 2.加添加剂(燃烧中) 3.加元素(合金中) 4.(给飞机布质蒙皮)上涂料 5.(向活塞式发动机气缸)注油
Dopper radio fuze 多普勒无线电引信
Doppler 1.多普勒(奥地利物理学家) 2.多普勒雷达,多普勒测速仪
Doppler antenna 多普勒雷达天线
Doppler beam sharpening 多普勒波束锐化
Doppler broadening 多普勒致宽
Doppler contour 多普勒轮廓
Doppler core 多普勒(谱线)核心
Doppler count 多普勒计数
Doppler drift 用多普勒雷求得的偏流角
Doppler effect 多普勒效应
Doppler error 多普勒误差

Doppler fluctuation error 多普勒波动误差
Doppler frequency shift 多普勒频率漂移
Doppler hover 多普勒悬停
Doppler navigation 多普勒导航
Dopper navigation system 多普勒导航系统
Doppler phase shift 多普勒相移
Doppler point positioning 多普勒单点定位
Doppler positioning by the short arc method 多普勒短弧法定位
Doppler radar 多普勒雷达
Doppler radio fuze 多普勒无线电引信
Doppler radio proximity fuze 多普勒无线电近炸引信
Doppler radiosonde system 多普勒无线电探空仪系统
Doppler range rate measurement 多普勒测速
Doppler ranging 多普勒测距
Doppler scanning technique 多普勒扫描技术
Doppler shift 多普勒频移
Doppler shifts of carrier signals 载频信号的多普勒漂移
Doppler tracking 多普勒跟踪
Doppler tracking system 多普勒跟踪系统
Doppler translocation 多普勒联测定位
Doppler velocity and position 多普勒测速和定位
Doppler VOR 多普勒伏尔
Doppler-Fizeau effect 多普勒-裴索效应
Doppler-navigated 有多普勒导航系统的
dopplometer 多普勒频率测量仪
dormant inertial navigation system 休眠型惯性导航系统
dormette 飞机躺椅
dorsal 机身背部的,机背的
dorsal fin 背鳍
dorsal spine (飞机机身的)背脊骨架
dorsal turret 机身上部炮塔,上炮塔
DOS 1. Department Of Space 航天部(印度) 2. Department Of Supply 供应部(澳大利亚) 3. Disk Operating System 磁盘操作系统
dosage 剂量
DOSC Direct Oil-Spray Cooled 直接喷滑油冷却
dose 剂量
dose equivalent 剂量当量
dose equivalent commitment 剂量当量负担
dose equivalent rate 剂量当量率
dose rate (辐射)剂量率,测量单位
dose rate contour line (辐射)剂量率等值线
dose rate effect 剂量率效应
dose rate meter 剂量率计
dose rate response 剂量率响应
dose-equated radiation phantom 等效辐射假人
dosemeter 剂量计

dose-response model 剂量反应模型
dosimeter 1.剂量计 2.紫外线(测量)计
dossier 卷宗
DOT 1. Day Of Training 训练日 2. Department Of Transport 运输部(加拿大等国的) 3. Department Of Transportation 运输部(美国)
dot 1.点,圆点 2.小数点,相乘符号 3.打上点
dot grid 网点板
dot mapping software 点描法制图软件
dot method 点值法
dot of light 光点
DoTI Department of Trade and Industry 贸易和工业部
dot-matrix plotter 点阵绘图仪
dot-projection photogrammetry 点投射摄影测量
dot-projection technique 点投射技术
dot-projection videogrammetry 点投射视觉测量
dots 网点
dotted line (尤指文件上指示签名处的)虚线,点划线
double 1.双,对,复 2.加倍,倍增 3.双重的
double acting electromagnetic valve 双向电磁阀(活门)
double astrograph 双筒天体照相仪
double attack 双向攻击
double axis rate gyro 1.双轴速率陀螺 2.双轴速率陀螺仪
double base 双基
double bubble 双环截面
double circular arc 双圆弧(叶型)
double curvature 双曲率
double degree of freedom gyro 双自由度陀螺
double delta 双三角翼
double designation 两家指定航空公司(政府指定两家航空公司为有国旗标志的公司飞同一国际航班)
double diffusion 双扩散
double drift 双偏流求风法,曲折飞行求风法
double electric layer capacitor 双电层电容器
double elliptical system 双重椭圆星系
double equatorial 双筒赤道仪
double fail-correcting actuator 双余度故障自动修正致动器
double feed 双进弹
double feedback 双反馈
double galaxy 双重星系
double gimbal wheel 双框架飞轮(用内外框架支承的飞轮)
double gimbaled flywheel 双框架飞轮
double heterojunction laser 双异质结激光器
double integrating accelerometer 二重积分式加速度表
double integrating angular accelerometer 双重积分角加速度计

double integrator 二重积分器	doubled hoop 双跳电路
double launch 双星发射的	double-decked 双层甲板的,双层舱的
double lift harness 双人吊起的背带(绞车作业时使用)	double-decker 1.〈口语〉双翼机 2.双层舱飞机,双层甲板船
double loop coordination strategy 双环协调策略	double-delta-winged 双三角翼的
double magneto 双磁电机	double-density recording 双倍密度记录
double manned 双人驾驶的(飞机)	doubled-headed warship 上层建筑前后均装有舰空导弹的军舰
double nuclear resonance magnetometer 双重核共振磁力仪	double-difference 双差
double o-ring 双密封圈	double-difference carrier 双差分载波
double peak 双峰雨型	double-difference carrier-phase 双差分载波相位
double pendulum 双摆	double-differenced ambiguity 双差模糊
double perimeter defense system 双层环形防空系统	double-differenced integer 双差整数
double perimeter radar system 双层环形警戒雷达系统	double-diffusive convection 双扩散对流
double plate 带孔底板,双板,垫板	double-diode 双二极管
double plate model 双板模型	double-duty 两用的
double precision 双倍精度	double-edged 双刃的
double probe 双探针	double-end beacon 双光束灯标
double pulse ranging 双脉冲测距	double-ended system 双端系统
double radio source 双射电源	double-engine 双发动机
double return wind tunnel 双回路风洞	double-entry compressor 双面进气压气机(离心式)
double roll-out solar array 双卷式太阳能电池阵	double-faced 双面的
double root blade 双根叶片(两端均固定的涡轮叶片)	double-flow engine 双路发动机,双函道发动机
double rotor pneumatic gyro 双转子气动陀螺	double-focusing mass spectrometer 双聚焦质谱仪
double row 双列,双行	double-fuselage 双体机身
double row of hole 双排孔	double-gimbaled momentum wheel 双框架动量轮
double sampling 双重取样,二次取样	double-governor 双重调节器的
Double Sandwich 双机夹心战术	double-hinged rudder 双铰链方向舵
double separate busses 双独立母线	double-humped 双峰的
double sided board 双面板	double-integrating gyro 双重积分陀螺
double slot 双槽式翼	double-integrating range unit 重积分距离部分
double star 双星,重星	double-integrator 双积分器
double system 1.双星 2.双重星系 3.双重天体系统	double-layer 双层的
double taper 双梯形度(翼)	double-line spectroscopic binary 双谱分光双星
double transit 二次中天	double-needle-width turn (转弯侧滑仪指针)两针宽转弯
double transition 双沿传输	double-nozzled 双喷口的
double triode 双三极管	double-peak 双峰
double wedge aerofoil profile 菱形翼型	double-piston 双活塞的
double-acting 双向作用的	double-pole single-throw switch 双极单投开关,双刀单掷开关
double-arm spectrometer 双臂谱仪	double-protection honeycomb 双重防护蜂窝结构
double-barrel 双管炮,双管枪	doubler 1.加强板 2.倍增器,倍压器,倍频器
double-barrelled gun 双管枪(炮)	double-row aircraft 双排货机(货物双排放置)
double-base matrix 双基矩阵	double-seater 双座飞机的
double-base propellant 双基推进剂	double-shear 双剪切的
double-beam beacon 双光束灯标	double-shock detonation 双激波爆炸
double-charged ion 双带电离子	double-sideband modulation 双边带调制
double-clutch 双离合器	double-sideband signal 双边带信号
double-cone 双锥区(天线)	
double-convex 双凸形的	
double-curvature channel 双曲通道	

double-sideband transmission 双边带传输
double-slab 双板
double-slit 双窄缝
double-slit injection 双开缝式注入
double-slotted flap 双开缝式襟翼
double-stall phenomenon 双停滞现象
doublet 1.偶极子 2.偶极天线 3.双重线 4.复制品
doublet antenna 对称振子天线
doublet distribution 偶流分布,偶极分布
doublet flow 偶极子流
doublet lattice 偶极子网格
doublet lattice method 偶极子网格法
double-throw circuit breaker 双掷断路器
doublet-lattice code 双重晶格代码
doublet-lattice method 双重晶格方法
doublet-lattice model 双重晶格模型
double-to-single ion 双单离子
double-travel feed 双行程进弹
double-wall 双重壁,双层墙
double-wedge aerofoil 双楔形翼剖面,菱形翼剖面
double-wedge charge 双楔形装药
doubling 加倍,并线,折回,防护板
doubly averaged orbital element 二次平均轨道要素
doubly-charged ion 双电荷离子
doughnut 1.回旋加速器室 2.〈俚语〉低压(小)轮胎
DOV Discreet Operational Vehicle 谨慎使用的飞行器
DOVAP Doppler Velocity And Position 多普勒测速和定位,多瓦普(系统)
DOVAP ELSSE 多瓦普轨道偏差指示器,多瓦普电子天空监视设备
DOVE Digital Orbiting Voice Encoder 轨道数字声频编码卫星
dovetail root 燕尾形榫头
DOW Dry Operating Weight 净操作重量
dowel 1.销,定位销 2.定缝销钉,暗钉,木钉
Dowgard （道氏化学公司生产的)航空乙二醇
down 1.向下,低落 2.关闭 3.(将飞机等)击落 4.〈口语〉停飞,禁止飞行 5.(因缺损或故障而)无法使用的
down and locked 起落架放下并锁好
down converter 下变频器
down elevator travel 升降舵向下偏(角)度
down link 下行链路
down range 下靶场(位于靶场弹着区),弹道下段射程
down sweep 降频扫描
down time 1.不能工作时间 2.停机时间
down-aileron 下偏副翼
downburst 突发下降气流
down-conversion 下变频
downcutting 下切侵蚀

downdraft 下曳气流,下沉气流,气潭
downdraft carburetor 下吸式汽化器
downdraft wind 下坡风
downed aircraft 迫降飞机(尤指因战伤而迫降的飞机)
down-elevator 下偏的升降铊
downflow 下冲,下洗流
downforce 下压力
downgrade 1.降低保密等级 2.(文职人员的)降级
downgraded resolution 降低分辨力
downlead (天线的)引下线
downleg （弹道)下降段
downleg trajectory （弹道)下降段轨道
downlink 下行线,向下链路,向地传输
download 1.下向载荷(如负迎角时的翼载荷) 2.(作战任务完成后或中断任务后)卸除载荷
downlock 下位锁
Down-look radar 俯视雷达,下视雷达
downpour 倾盆大雨
downrange 下靶场,弹道下段射程
downrange error （弹着)距离误差
downrange position 试验位置
downrange velocity error 试验速度误差
downrange velocity estimate 试验速度估计
downslide 向下侧滑
downstage 降级(信号、振动、液体流量、压力等的)
downstaire 〈口语〉在低层大气中,在大气低层
downstream 1.下游 2.顺流的,顺流地
downstream angle 下游角
downstream blade row 下游叶片排
downstream blade 下游叶片
downstream blockage 下游淤塞河道
downstream boundary 下游闭塞界限
downstream cavity 下游腔
downstream cavity trench 下游腔沟
downstream combustion 下游燃烧
downstream combustion ramjet mode 下游燃烧冲压喷气模式
downstream combustor 下游燃烧室
downstream condition 下游的条件,下游状况
downstream diffuser 下游扩散器
downstream direction 流方向
downstream distance 下游距离
downstream domain 下游领域
downstream ejector placement 下游喷射位置
downstream electrode 下游的电极
downstream end 下游端
downstream face 下游面
downstream flame tube 下游火焰管
downstream flap 下游皮瓣

downstream flap deflection 下游襟翼偏转
downstream flow 顺流
downstream fuel 下游的燃料
downstream fuel injection 下游燃油喷射
downstream geometry 下游几何
downstream influence 下游影响
downstream jet 下游飞机
downstream layer 下游层
downstream location 下游方位
downstream neutral density 下游中性密度
downstream placement 下游位置
downstream portion 下游部分
downstream potential 下游的潜力
downstream pressure 下游压力,阀后压力
downstream propagation 下游传播
downstream region 下游控制区
downstream rotor 下游转子
downstream rotor blade 下游转子叶片
downstream row 下游排
downstream section 下流段,下游段,下游区段
downstream shock 下游冲击
downstream span 下游范围
downstream spray 下流喷射
downstream stator 下游定子
downstream stator blade row 下游静叶片排
downstream straight duct 下游直接管
downstream straight section 下游直管段
downstream surface 下游面
downstream throat 下游的喉咙
downstream torch 下游的火炬
downstream vane 下游叶片
downstream vane row 下游叶片排
downstream wall 下游墙
downstream-electrode current 下游电极电流
downstroke 下行程,下行冲程
downsun 背阳方向
downtail convection 尾向对流
downtime 1.停用时间,不运转时间,检修停工期 2.(弹药的)装车时间,(供应品)补给时间 3.非战斗状态时间(导弹或地面设备的)
downtrack 下行路轨
downward acceleration 自由落体加速度
downward annual budget pressure 年度预算下行压力
downward budget 向下的预算
downward dihedral (翼面的)下反角
downward direction 下坡方向
downward ejection 向下弹射
downward ejection seat 向下弹射座椅
downward identification light 下标志灯(即白色标志灯)

downward looking Doppler radar 俯视多普勒雷达
downward looking radar 俯视雷达
downward maneuver 向下回旋余地
downward pressure 下向压力,向下压力
downward travel 下偏量,向下偏(角)度
downward-fired (发动机)在轨道下降段点火的
downwash 1.气流下洗,下洗流 2.(升力螺旋桨或旋翼的)滑流
downwash correction 下洗修正
downwash in wing wake 翼后下洗,机翼尾流中的下洗
downwash induced velocity 下洗诱导速度
downwash study 向下运动的研究
downwind 1.顺风,顺流,顺风飞行 2.(起落航线的)第三边 3.反着陆航向(飞行)
downwing side (转弯时)下垂机翼一侧飞行
DOZ Dozen 一打
DP 1.Data Package 数据包 2.Data Processing 数据处理 3.Differential Protection 差动保护(装置) 4.Direction des Poudres 火药局(法国) 5.Display Processor (计算机的)显示处理器 6.differential pressure 压差,座舱压力差 7.Dimeric Polymer 二聚物
DPA Diphenyl Amine 二苯胺
DPC Data Processing Center 数据处理中心
DPCC 1.Data Processing Control Center 数据处理控制中心 2.Data Processing Control Console 数据处理控制台
DPCL Dedicated Payload Communications Link 专用有效载荷通信链路天线
DPCM Differential PCM 差分脉码调制
DPFG Data Processing Functional Group 数据处理功能组
DPI 1.Designed Point of Impact 计算命中点 2.Desired Point of Impact 预期命中点 3.Direct Petrol Injection 直接注入汽油
DPLL Digital Phase-Lock Loop 数字式锁相环
DPM 1.Digital Pulse Modulation 数字脉冲调制 2.Digital Processing Module 数字处理模块
DPMA Data Processing Management Association 数据处理管理协会(美国)
DPP Dual Processor Program 双处理器程序
DPS 1.Data Processing System 数据处理系统 2.Descent Propulsion System 下降推进系统(行星登陆舱或登月舱) 3.Differential Phase Shift 差分相移 4.Dynamic Pressure Sensor 动压敏感元件,动压传感器
DPSK Differential Phase Shift Keying 差分相移键控
DPU Data Processing Unit 数据处理器
DQ Design Qualification 设计资格
DQAB Defence Quality Assurance Board 国防装备质量

管理委员会(英国)
DQE Detective Quantum Efficiency 探测器量子效率
DQPSK Differential Quadrature Phase Shift Keying 差分四相相移键控
DR 1. Data Requirement 数据要求 2. Dead Reckoning 推测领航 3. Dispense Rate(箔条或浮质的)投放率(以磅/秒计)
DR Drift 偏流角
DRA 1. Defence Research Agency 国防研究局(英国) 2. Dual Rail Adapter 双轨接头(能使一个挂架挂两枚空空导弹)
dracone 拖囊,拖行软罐
draconitic month 交点月
draconitic revolution 1. 交点转动 2. 交点周
draft 1. 通风,风力 2. 气流(尤指垂直方向的气流) 3. 吃水深度(美国) 4. 草图,草稿,草案 5. 斜度 6. 牵引,拖 7. 征兵(美国)
draftsman 1. 制图员 2. 起草人
drag 1. 阻力,阻碍 2. 拉,拖,拖长 3. 制动,减速 4. 低空通过
drag acceleration 阻力加速度
drag accelerometer 纵向负加速度表
drag analysis 拖曳分析
drag area 阻力区
drag at incidence 有迎角时的阻力
drag brake 阻力刹车,棘轮闸
drag bucket 反铲,牵斗
drag buildup 阻力增加
drag buildup technique 拖动积累技术
drag characteristic 阻力特性
drag chute 阻力伞
drag coef 阻力系数,曳引系数
drag coefficient 阻力系数
drag coefficient ratio 阻力系数比
drag component 阻力分量,拖曳元件
drag compression 阻力压缩
drag count 反向阻力
drag datum 基本阻力
drag device 增阻装置
drag divergence 阻力突增
drag due to lift 升致阻力
drag dynamics 拖动力
drag effect 阻力效应
drag efficiency 拖动效率
drag enhancement 阻力增强
drag error 拖动错误
drag force 阻力
drag hinge (直升机旋翼叶片的)摆振铰,阻力铰,垂直关节

drag increase 阻力增加
drag index 阻力指数
drag link 阻力拉杆,阻力张线
drag loss 阻力损失
drag map 可以拖拉地图
drag member 承阻构件
drag parachute 1. 阻力伞 2. 刹车伞 3. 减速伞
drag penalty 阻力增大,阻力带来的损失
drag plate 下型箱板
drag prediction 阻力预测
drag profile 剖面阻力,断面曳力
drag rate 阻力速率
drag ratio 升阻比
drag reduction 减阻
drag reference 拖动参考
drag result 阻力结果
drag rib 承阻肋,压力肋
drag rise 阻力突增
drag rope 阻力绳
drag sail (用帆布制成的)海锚,浮锚,流锚
drag value 阻力值
drag variation 阻力变化
drag weight ratio 阻力重量比
drag wires (机翼或气球上的)阻力张线
drag-cup tachometer 感应式转速表
drag-divergence 拖拽分叉
drag-enhancement 拖拽提升
drag-free control 畅行控制
drag-free controller 畅行控制器
drag-free loop 畅行循环
drag-free mission 畅行的使命
drag-free operation 畅行操作
drag-free satellite 无阻力卫星
dragging 拖曳(利用图形输入装置引导亮点在显示屏上拖动)
dragless antenna 无阻力天线
drag-rise 阻力增长
dragulator 阻力调节器
drain 1. 放泄 2. 放泄管,放泄口 3. (场效应管)漏极
drain conductance 漏极电导
drain flight valve 排水飞行阀
drain pipeline 泄漏油管
drain trap 1. 漏油盒(仪器的) 2. 沉液槽 3. 放泄弯管
drain wire 加蔽线,排扰线
drainage 1. 排水,放出 2. 排水设备
drain-cock 放水开关,放水旋塞,水龙头
drainer 放泄器
dramatic effect 戏剧效果
Draper catalogue 德雷伯星表

drastic change 激烈变化,重大变化
drastic maneuver 剧烈的动作
draw 1.拉,拖 2.划,画,绘制,描写 3.拔出,抽出,汲出 4.(船只)吃水
drawback 缺点,不利条件,退税
drawing 1.图纸,图样,图画 2.绘图 3.拉制,拉拔,拔丝,压延 4.回火,退火
drawing-board-to-operational status cycle 工程研制周期
drawing grid 绘图网格
DRB Defense Resources Board 国防资源委员会(美国)
DRC Defence Review Committee 防务审查委员会(北约)
DRCC Data Referencing and Conditioning Centre 数据核对和加工中心(欧洲)
DRD 1.Data Requirements Description 数据要求描述 2.Data Requirements Document 数据要求文件 3.Dry-Runway Distance 干跑道滑跑距离
DRDB Dual-Redundant Data Bus 双余度数据总线
DREA Defence Research Establishment Atlantic 大西洋国防研究所(加拿大)
dreadnought 无畏级战舰
dress 1.打磨,修整,整形 2.服装
dressed 1.装挂齐全的(指飞机上所有设备和系统管路都完整) 2.配备齐全的(指发动机的外部附件、管路和控制装置都完整)
dressed engine 齐装发动机
DRETS Direct Re-entry Telemetry System 直接再入遥测系统
DREWS Direct Readout Equatorial Weather Satellite 直接读出数据赤道气象卫星
DRF Dualrole Fighter 双重任务歼击机
DRFM Digital RF Memory 数字式无线电频率存储器
DRI 动态响应指数
dribble 1.滴流 2.点滴
dribble volume 运球卷
drift 1.偏流,偏流角,测偏流 2.偏移,偏移速度 3.漂移(如仪表指针的误差,陀螺自旋转离其固定基准的角位移或半导体载流子在电场中的运动) 4.频移 5.偏航 6.冲头,打孔器
drift angle 1.偏流角 2.偏离角
drift axis (航向陀螺的)框架垂直轴,漂移轴
drift bar 偏流针
drift chamber 漂移室
drift correction 偏流修正
drift correction angle 偏流修正角
drift down 逐步下降
drift error 1.漂移误差 2.偏流差
drift indicator 偏移指示器
drift klystron 漂移速调管

drift magnet 偏移磁铁
drift mobility 漂移迁移率
drift of pose accuracy 位姿精度漂移
drift of ship position 船位漂移
drift orbit 漂移轨道
drift rate 漂移率
drift region 漂移区
drift shell 漂移壳
drift space 漂移空间
drift transistor 漂移晶体管
drift tube 漂移管,漂流管
drift value 漂移量
driftage 偏流,偏移,漂移距离
drift-compensated 修正偏流的
drift-correction table 偏流修正量表
driftmeter 偏流器
drill 1.训练,练习 2.编队飞行训练 3.守则(飞行特定阶段所采取的正确程序) 4.钻头
drill bit 钻头,【矿】钎头,强钻头
drill card 程序卡(飞行特定阶段正确程序的检查单)
drill cartridge 练习弹,教练弹
drill round 教练弹,教练导弹,教练炮弹
drill shell 练习炮弹
drill string 钻柱
drilling 钻孔
drilling mechanism 钻进机理
Dring D形吊索(气球或其他轻航空器上用的,形状如字母D) 2.D形拉环,(降落伞上的)开伞拉环
drip band 滴水带
drip strip 液滴
drip-roof test 淋雨试验,淋水试验
dripshield 集液槽,集液盘
drip-stick fuel gage 滴流杆式油量表
DRIR Direct Readout Infrared 直读式红外(辐射仪)
DRIRU Dry Rotor Inertial Reference Unit 干式转子惯性基准部件
drive 1.传动,驱动 2.传动装置 3.驾驶(车辆) 4.推进 5.激励
drive circuit 驱动电路
drive gas 驱动气体
drive mechanism 驱动机构,传动装置
drive plate 拨盘,传动盘,驱动圆盘
drive point 驱动点
drive pressure 驱动压力
drive program 驱动程序
drive shaft 驱动轴,主动轴
drive surface 行进面
drive system 驱动系统,传动系统
drive torque 驱动力矩,驱动转矩,传动转矩

drive wheel 主动轮
drive winding 驱动绕组
drive wire 驱动线
driven 从动的,传动的
driven element 激励单元
driven head 传动机头,套管头,打头
driver 1.传动装置,传动器,驱动器,驱动轮 2.激励器,末级前置放大器 3.汽车驾驶员,司机,飞机驾驶员 4.(暂冲式风洞的)驱动气体
driver gas 启动程序气
driver gas mixture 驱动气体混合物
driver pressure 驱动压力
driver pressure ratio 驱动压力比
driver section 驾驶部
driver-gas contamination 启动程序气污染
drive-shaft 传动轴,主动轴
driving band 1.弹带(炮弹上的软金属带) 2.导环,准环 3.传动带
driving circuit 驱动电路,激励电路
driving clock 转仪钟
driving element 驱动元件,激励单元
driving mechanism 驱动机构
driving spring (航炮)复进簧
Drivmatic riveter 自动钻孔铆钉机,全自动高速铆钉机(一种专利产品)
drizzle 毛毛雨,细雨
DRL Data Requirements List 数据要求表
DRLMS Digital Radar Land Mass Simulation 数字雷达陆地模拟
DRM Ducted Rocket Motor 冲压火箭发动机,涵道火箭发动机
DRO 1. Daily Routine Orders 日常命令 2. Drone Recovery Officer 无人机回收军官
drogue 1.减速伞,稳定伞 2.(空中加油的)锥形套,喇叭罩 3.锥形拖靶
drogue chute 阻力伞
drogue parachute 1.稳定减速伞 2.稳定伞
drogue parachute deployment 制动伞开伞
drogue position 浮标的位置
drogue recovery 制动(降落)伞回收
drogue retarded weapon (用)降落伞减速(的)武器
drogue-probe system 探头锥套式空中加油设备
drome 〈口语〉飞机场
drone 1.无人驾驶飞机 2.无人驾驶航空器 3.靶机
drone-assisted torpedo 无人驾驶飞机助放鱼雷
droop 1.下垂,(前缘)下弯,下偏 2.前缘襟翼 3.舱门下垂量(机身下舱门或检查口门因重力下垂,对总阻力起重大影响)
droop balk 前缘襟翼连锁装置

droop nose 下垂机头
droop snoot 1.下垂机头 2.机头伸长下垂的飞机
droop stop (旋翼的)下弯止动器,下弯限制器
drooping ailerons 下偏副翼,襟副翼
drooping leading edge 下偏前缘
droops (机翼前缘的)下偏段
droops shoot 1.水平轰炸机 2.水平轰炸
drop 1.下降,使飞机下坠接地 2.空投,伞降 3.降低(速度、压力等) 4.下垂量,下落距离 5.〈口语〉放起落架 6.(空中校射员指示减小火炮射程的)修正量
drop accuracy 精度下降
drop aircraft launcher 机载投放式发射装置
drop altitude 1.空投高度 2.空投真高
drop ammunition 航空炸弹
drop back 后缩,后退
drop bar 接地棒,短路棒
drop collector 水滴集电器(测空中电位梯度用)
drop diameter 滴径
drop forging 冲锻,落锤锻造
drop generator 滴式发动机
drop height 空投真高,空投高度
drop in 冒码,混入(信息)
drop interval 1.空投时间间隔 2.(炸弹的)连投间隔
drop message 空投信件
drop package 摔箱
drop regime 放弃政权
drop shock test machine 落下式冲击试验机
drop short 近弹
drop size 水滴大小,粒子尺寸
drop size distribution 水滴大小分布,分布,粒径分布
drop speed 坠落速度,点滴速度
drop tank 副油箱
drop test 1.(起落架的)落震试验 2.(飞机的)自由下落试验 3.(机件结构的)抛掷试验
drop the cookies 〈口语〉投弹
drop tower 落塔(塔式失重试验设施)
drop tower test 落塔试验
drop tube 落管(管式失重试验设施)
drop tube test 落管试验
drop wire 引下线
drop zone 伞降区,空投区
drop-launching 飞机发射,机上发射
droplet 滴
droplet breakup 液滴破碎
droplet burning 液滴燃烧
droplet coalescence 液滴聚结
droplet collection efficiency 水滴收集效率
droplet combustion 液滴燃烧
droplet density 雾滴密度

droplet density wave	液滴密度波
droplet diameter	雾滴直径
droplet drag	液滴拖
droplet evaporation	液滴蒸发
droplet flame	液滴的火焰
droplet formation	点滴形成
droplet generator	液滴发生器
droplet heating	液滴加热
droplet impact	熔滴冲击力
droplet impingement parameter	水滴撞击参数
droplet interior	液滴内部
droplet lifetime	液滴寿命
droplet mean size	液滴尺寸
droplet number	液滴数量
droplet number density	液滴数密度
droplet sedimentation	释义微滴沉降
droplet shadowed zone	水滴遮蔽区
droplet size	油雾粒度
droplet solution	液滴的解决方案
droplet stream	液滴流
droplet surface	液滴表面
droplet temperature	液滴温度
droplet time	液滴时间
droplet trajectory	水滴轨迹
droplet transport	液滴运输
droplet vaporization	液滴蒸发
droplet velocity	熔滴颗粒速度
droplet volume	液滴体积
dropmaster	1.空投调度员,空投组组长(负责准备空投物质,检验、装载等) 2.空投联络员(负责在飞行员和跳伞组长之间传递信息)
drop-off	投放,脱离,降落,减少
dropoff rate	衰减速率,衰减率
drop-out	漏码,丢失(信息)
dropout error	掉落误差,失漏误差(磁带录音时的)
dropout voltage	(继电器)回动电压,失步电压
droppable	可抛放的,可空投的
droppable pyrotechnic flare system	投掷焰弹系统
dropping	1.抛下,投下,滴下 2.空投,空降,伞降 3.点滴,落下物
dropping angle	(轰炸)瞄准角,投弹角
dropping corrosion test	点滴腐蚀试验
dropping gear	空投装置,投弹装置
dropping plug	脱落插头
drop-pod	可抛式短舱
drop-shaped	滴状的
drop-size distribution	滴谱
drop-size spectrum	滴谱
dropsonde	下投式探空仪
dropsonde datum	下投式探空仪数据
dropsounding	用空投探空仪探空
drop-test	降落伞试验,从飞机上空投试验,(起落架)落震试验
droptower	冲击塔
dropwindsonde	下投式测风探空仪
dropwise	点滴的
DROS	Direct Readout Satellite 直接读出数据卫星
drosometer	露量表
DRP	Design Review Panel 设计审查组
DRS	1. Data Receiving Station 数据接收站 2. Data Recording System 数据记录系统 3. Data Relay Satellite 数据中继卫星(欧洲) 4. Dead-Reckoning Subsystem 推测航行分系统 5. Direct Receive Sets 直接接收装置
DRSS	Data Relay Satellite System 数据中继卫星系统
DRU	1. Data Recovery Units 数据复原装置 2. Data-Retrieval Unit 数据恢复单元 3. Direct Reporting Unit 直接报告单元 4. Dynamic Reference Unit 动态基准组件
drum	1.鼓轮(压缩器转子的) 2.鼓,鼓状物,磁鼓 3.油桶,筒,滚筒 4.绕线架,钢索轮 5.桶(油量单位)
drum altimeter	数字鼓式高度表
drum assembly	(转膛炮的)转膛部件
drum body	(转膛炮的)转膛体
drum cradle body	转膛拱架体
drum magazine	鼓形弹仓
drum-and-disc construction	鼓盘式构造
drum-container feeding	鼓形弹箱供弹
drum-pointer altimeter	数字鼓指针式高度表
drum-shaped	鼓形的
DRVS	1. Doppler Radar (Radial) Velocity Sensor 多普勒雷达(辐射)速度传感器 2. Doppler Radar Velocity System 多普勒雷达测速系统 3. Doppler Radial Velocity System 多普勒径向速度测量系统
dry	1.干的(指未使用加力),净的(如净重) 2.(发动机)不加力 3.(机翼内)无燃油的,不能挂副油箱的(指悬挂接头)
dry adiabatic lapse rate	干绝热(温度)直减率
dry battery	干电池
dry bulb	干球温度表
dry cell battery	干电池(碳锌电池)
dry charged battery	干电荷蓄电池
dry chemical (fire) extinguisher	干粉化学灭火器
dry contact	干结合(空中加油机与受油机已接通加油管线但还未放油)
dry corrosion	干腐蚀,气体腐蚀
dry discharged battery	干放电电池
dry engine	未加燃料、滑油和液压油等的发动机

dry etching 干法腐蚀
dry film lubricant 干膜润滑剂,固体润滑膜
dry filter 干过滤器
dry flashover voltage 干闪络电压,干飞弧电压
dry fog 干雾(灰尘和烟雾形成的霾)
dry friction damping 干摩擦阻尼
dry friction test 干摩擦试验
dry fuel 干燃料,固体燃料
dry growth 干成长,干增长
dry gyro 1. 干式转子(用机械轴承的) 2. 干式陀螺
dry gyroscope 干陀螺,烧性陀螺
dry hub (直升机旋翼的)干桨毂,无润滑剂桨毂
dry hydrogen bomb (无重水或其他气化合物的)干氢弹
dry immersion 干浸法
dry lease 干租赁
dry leasing（lease） 1. 干租,无机组包机 2. 分期付款购机
dry mass 干质量
dry mass fraction 干燥的质量分数
dry plate 干板
dry powder extinguisher 干粉(粉末)灭火器
dry power 干功率
dry run 1. 干运转 2. 演习,预演 3. 练习轰炸进入
dry run bombing 模拟投弹练习
dry shooting 空弹射击,模拟射击
dry soot 微粒中的干碳烟
dry squeeze (系统或装置的)模拟运转,模拟工作
dry stores 干给养(食物及不含酒精的饮料)
dry sump 干机匣,干收油池,干油池
dry thrust 不加力推力,干推力
dry tuned gyroscope 干式调谐陀螺,烧性调谐陀螺
dry tuned rotor gyro 干式调谐转子陀螺
dry weight 干重
dry winding 干法缠绕
dry-bulb temperature 干球温度
dry-bulb thermometer 干球温度计
dry-charged battery 干充电池(组)
Dryden Flight Research Center 德赖顿飞行研究中心(美国)
dryer 干燥器
dry-extruding 干挤压
dryheat sterilization 干热灭菌法
dry-ice 干冰
drying 烘干,干燥
dryness 1. 干燥 2. 干燥度
dry-oxygen oxidation 干氧氧化
dry-sealed vacuum pump 干封真空泵
dry-star 金星,晨星,启明星
Drzewiecki theory 杰维茨基原理(计算螺旋桨叶片用)

DS 1. Data Sheet 数据单,数据表 2. Documented Sample 有文件的样品 3. Document Signed 文件已签字 4. Dynamic Simulator 动态模拟器
DSA 1. Data System Architecture 数据系统体系结构 2. Defence Supply Agency 国防供应局(美国) 3. Distributed System Architecture 分布式系统结构
DSAC 1. Defence Scientific Advisory Council 国防科学咨询委员会(英国) 2. Digital Simulated Analog Computer 数字仿真模拟计算机
DSAR Data Sampling Automatic Receiver 数据取样自动接收机
DSARC Defense Systems Acquisition Review Council 国防系统采购评审委员会(美国防部)
DSAS Digital Solar Aspect Sensor 数字式太阳方位传感器
DSASO Deputy Senior Air Staff Officer 空军参谋机构副科长(英国空军)
DSB 1. Defense Science Board 国防科学委员会(美国) 2. Double Sideband 双边带
DSC 1. Defect-Survival Capability 抗缺陷的生存能力 2. Defense and Space Center 国防和航天中心 3. Differential Scanning Calorimeter 差动扫描量热计 4. Digital Scan Convertor 数字扫描变换器 5. Digital Selective Calling 数字式选择呼叫 6. Direct Satellite Communication 卫星直接通信 7. Directional Solidification Crystals 定向结晶晶体 8. Distinguished Service Cross 杰出服役十字勋章(英国的) 9. Dynamic Standby Computer 动态备用计算机
DSCG Directional Solidification Crystal Growth 定向结晶晶体生长
DSCP Defense Satellite Communications Program 国防卫星通信计划(美国)
DSCS Defense Satellite Communications System 国防卫星通信系统(美国)
DSD Defence Support Division 防务保障司(北约)
DSE Directionally Solidified Eutectialloys 定向固化晶合金
DSEO Defense Systems Evaluation Office 国防武器系统鉴定局(美国)
DSES Defense Systems Evaluation Squadron 国防武器系统鉴定中队(北美防空司令部)
DSF Domestic Supply Flight 国内供应航班(英国空军)
DSFC Direct Side Force Control 直接侧力控制
DSI 1. Data Stream Interface 数据流接口 2. Digital Speech Interpolation 数字话音内插
DSID Data Set Identification 数据集标识
DSIF Deep Space Instrumentation Facility 深空探测设备,深空(飞行器)跟踪与测量设备(设在地面上)
DSIPS Digital Satellite Image Processing System 数字

卫星图像处理系统
DSIR Department of Scientific and Industrial Research 科学与工业研究部（新西兰、英国）
DSISCO Dual Spectral Irradiance and Solar Constant Orbiter 双光谱辐射和太阳常量轨道器（西欧太阳天文卫星）
D-skin （机翼等）前缘蒙皮
DSL 1. Depressed Sightline 下俯瞄准线 2. Digital Simulation Language 数字模拟语言
DSLC Dynamic-Scattering Liquid Crystal 动态散射液晶
DSM Dynamic Scattering Mode 动态散射模式
DSMAC 1. Digital Scene-Matching Area Correlation 数字式景物匹配区域相关（技术）2. Digital-Scene Matching Area Correlator 数字式景物匹配区域相关器
DSMC Direct Seat-Mile Cost 座英里直接运营成本
DSN Deep Space Network 深空跟踪网（美国行星际探测器跟踪网）
DSNS Division of Space Nuclear Systems 空间核系统部（美国原子能委员会）
DSO 1. Defence Sales Organization 国防销售组织（英国）2. Defense System Operator 国防武器系统销售商（美国）3. Distinguished Service Order 军功章，优异服役勋章（英国）
DSP 1. Defense Support Program 国防支援计划（美国早期预警卫星计划）2. Digital Signal Processing 数字信号处理
DSPN Deep Space Planetary Network 深空行星际跟踪网
DSR 1. Digital Shunt Regulator 数字式并联（分路）调节器 2. Director of Scientific Research 科研处长 3. Dynamic Space Reproducer 动态空间再现装置
DSS 1. Data Storage System 数据存储系统 2. Data Subsystem 数据子系统 3. Data Switching System 数据切换系统 4. Decision Support System 决策支持系统 5. Dedicated Support Structure 专用支撑结构 6. Deep Space Station 深空探测站 7. Department of Space Science 空间科学部（欧洲航天研究和技术中心的）8. Diagnostic Simulation System 诊断仿真系统
DSSD Data Structure and Systems Development 数据结构和系统发展
DSSP Deep Submergence System Program 深潜系统程序
DST 1. Daylight Saving Time 夏令时 2. Dry Specific Thrust 干比推力
DSTO Department of Defence Science & Technology Organization 国防科学技术组织部（澳大利亚）
DSTU Digital Signal Transfer Unit 数字式信号传递装置

DSU 1. Direct Support Unit 直接保障单位，直接支援部队（美国空军）2. Dynamic Sensor Unit 动态传感装置
DSV Deep Space Vehicle 深空飞行器
DSVT Digital Secure Voice Terminal 数字保密话音终端
DT Day Television 昼间电视
DTA 1. Differential Thermal Analysis 热差分析 2. Dynamic Test Article 动态试验件
DTAD Demonstrated Technology Availability Date 技术验证成功日期
DTAT Direction Technique des Armements Terrestres 陆军武器技术局（法国，属武器装备部）
DTC 1. Decision Threshold Computer 判（决）断阈计算机 2. Design To Cost（按）成本设计
DTCS Drone Tracking and Control System 无人机跟踪和控制系统
DTD 1. Damage Tolerant Design 损伤容限设计 2. Data Transfer Device 数据传送设备 3. Digital Terrain Data 数字地形数据 4. Directorate of Technical Development 技术开发局（英国）
DTE 1. Data Terminal Equipment 数据终端设备 2. Development Test and Evaluation 研制试验和鉴定
DTED Digital Terrain Elevation Data 数字化地形高程数据
DTG 1. Distance To Go 待飞距离（常见于导航系统控制显示器面板）2. Dynamically Tuned Gyro 动力调谐陀螺
DTI 1. Department of Trade and Industry 贸易工业部（英国）2. Digital Timebase Interval 数字时基间隔
DTLCC Design To Life Cycle Cost（按）全寿命（寿命周期）费用设计
DTM Digital Terrain Model 数字地形模型，数值地型
DTMC Direct Ton-Mile Cost 直接吨英里费用，吨英里直接成本
DTMDS Digital Terrain Management Display System 数字式地形处理显示系统
DTMF Dual-Tone Multifrequency 双音多频
DTN Data Terminal Network 数据终端网络
DTO Detailed Test Objective 详细试验目标，具体试验目标
DTP Design To Price（按）价格设计，最低价格设计
DTR Damage Tolerance Rating 损伤容限品级
DTS 1. Data Transmission System 数据传输系统 2. Defensive Technologies Study 防御技术研究 3. Development Test Satellite 研制试验卫星 4. Dynamic Test Station 动力试验台
DTSC Digital Transmission Standards Committee 数字传输标准委员会（英国）

DTST　Defensive Technology Study Team 防御技术研究组(美国)
DTU　Data-Transfer Unit 数据传送装置
DTUPC　Design To Unit Production Cost (按)单位生产费用设计
DTVM　Differential Thermocouple Voltmeter 差动热电偶电压表
DTVOR　Doppler Terminal VOR 多普勒航站伏尔台(小功率)
DTWA　Dual Trailing-Wire Antenna 双拖曳线天线
DU　Depleted Uranium 贫(化)铀
dual　1. 双的,双重的,双套的 2.(飞机的)双套操纵装置(可供正、副驾驶,或飞行教员、学员各自独立操纵的) 3. 在双座飞机上进行飞行训练
dual accelerometer　双联加速度表
dual action starter　双重作用起动机
dual actuators　双重致动器,双重助力器
dual algorithm　对偶算法
dual bell　双钟
dual bell nozzle　双钟喷嘴
dual chamber　双燃烧室
dual code　对偶码
dual coding　双份编码
dual color homing head　双色导引头
dual compressor　双转子压气机,双轴压气机
dual control　双重控制,复式控制,双操纵
dual cusp　双尖点
dual data collection　双重数据收集
dual designation　双指定
dual differential radiometer　二元微分辐射计,双微分辐射计
dual diffused MOS integrated circuit　双扩散型 MOS 集成电路
dual expander engine　双膨胀发动机
dual fail-off actuator　双余度故障自动断开致动器
dual firing capabilities　双重射击能力(既能单发发射,又能连发发射)
dual flattened-cone charge　双扁锥形装药
dual frequency　双频
dual frequency phase locked receiver　双频锁相接收机
dual frequency range rate instrumentation　双频测速仪
dual frequency receiver　双频接收机
dual gate field effect transistor　双栅场效应管
dual ignition　双重点火,双电嘴点火(由两个独立电源供电)
dual indicator　双(参数)指示器(用在双发动机仪表)
dual injection　复式喷射
dual injection system　复式喷射系统
Dual Keel　"双龙骨"空间站(美国空间站初期设计方案之一)
dual lane slide　双道滑梯(两队旅客分别从两个滑梯同时紧急离机)
dual magneto　双(高压)输出磁电机
dual mode　双重模式,两种方法
dual mode EPU　双模式应急动力装置
dual mode propulsion system　双模式推进系统
dual mode rocket engine　双模式火箭发动机
dual mode traveling wave tube　双模行波管
dual modulation　双重调制
dual modulation telemetering system　双重调制遥测系统
dual network　对偶网络
dual node　对偶结点
dual node configuration　对偶结点配置
dual persistence　双余辉
dual port access　双通道存取
dual principle　对偶原理
dual propellant　双推进剂
dual pseudolite system　双伪卫星系统
dual rotor　双转子
dual sounding　双探空仪探空
dual spin attitude control system　双自旋姿态控制系统
dual spin spacecraft　双自旋航天器
dual spin stabilization　双自旋稳定
dual spin turn　双自旋转向
dual spinner　双自旋器
dual thermometer　双标温度表
dual thrust　两级推力
dual thrust rocket motor　两级推力火箭发动机
dual transverse injection system　双横向喷射系统
dual-axis rate transducer　双轴速率传感器
dual-band　双频段
dual-beam oscilloscope　双线示波器
dual-bell　双钟形
dual-bell nozzle　双钟形喷嘴
dual-burn　双层燃烧
dual-capable weapon　两用武器,双重能力武器
dual-cell rubidium magnetometer　双元铷蒸汽磁强计
dual-code　双代码
dual-code strategy　双代码策略
dual-control　双重控制,双重操纵
dual-cycle　双循环
dual-destination mission　双宿任务
dual-element transducer　双元件传感器
dual-frequency measurement　双频测量
dual-grid reflector　双栅反射器
dual-in-line package　双列直插式封装
dual-inline type　双排线型
duality　二重性,二元性,对偶性

dualized 双重的,双套的
dual-missile system (同时制导两枚导弹的)双弹制导系统
dualmode 双重工作状态的,两种工作方式的
dual-mode combustion 双模式燃烧
dual-mode combustor 双模式燃烧器
dual-mode engine 双模式引擎
dual-mode operation 双模式操作
dual-mode propellant 双模推进剂
dual-mode ramjet 双模式冲压式喷气发动机
dual-mode scramjet 双模式超声速燃烧冲压喷气发动机
dual-mode seeker (红外及雷达两种制导系统并用的)复式导引头
dual-mode system 双模式系统
dual-polarized antenna 双极化天线
dual-port 双端口
dual-port torch 双端口火炬
dual-positioning satellite 双星定位卫星
dual-purpose 双用途的,双效的,两用的
dual-purpose aerial rocket (攻击空中和地面目标的)两用航空火箭弹
dual-rail adapter 双轨接头
dual-rate lunar position camera 双速月球定位照相仪
dual-redundant inertial measurement unit 双余度惯性测量组件
dual-rotation 同轴反转,同心反转
dual-rotor 双转子
dual-rotor front 双转子前端
dual-rotor speed 双转子速度
dual-rotor system 双转子系统
dual-satellite launch 双星发射(一枚火箭发射两颗卫星)
dual-speed magnetic transducer 双速磁传感器
dual-spin 双自旋
dual-spin satellite 双自旋卫星
dual-thrust motor 双推力发动机
dual-tone multifrequency 双音多频
dual-trajectory algorithm 双轨道算法
dual-tube case 双管箱
dub 配音
Dubhe 【天】天枢,北斗一
DUC Dense Upper Cloud 高空密云
duck 1.突然低头,突然潜水,闪避 2."前方有目标"(空中截击通话代语) 3.〈口语〉水上救援飞机 4.水陆两用飞机,水陆两用车
duck under 降到下滑线下面,在规定高度层下面飞行
duct 1.导管,涵道,管道,槽 2.波导 3.(沿导管)输送或增压
duct acoustics 管道声学

duct area 管道面积
duct axis 管轴
duct branch 支管,分支通风管
duct burner 1.外涵加力燃烧室 2.管道加力燃烧室
duct entrance 电缆沟入口
duct exit 管出口
duct height 波导高度
duct noise 管道噪声
duct part 管道部分
duct propagation (大气)波导传播
duct radius 管半径
duct section 管道部分
duct system 管道系统,风管网路
duct trench 电缆沟,管道沟
duct volume 通道体积
duct wall 管壁
duct-burner 涵道燃烧室
duct-burning turbofan 外涵道燃烧加力涡轮风扇发动机,外涵加力涡轮风扇发动机
ducted cooling 管道式冷却
ducted fan 导管风扇
ducted propeller 导管螺旋桨
ducted radiator 安装于涵道内的散热器(可获得推力)
ducted rocket 冲压火箭发动机,涵道火箭发动机
ducted rocket combustor 管火箭燃烧室
ducted tail rotor 1.涵道尾桨 2.涵道风扇式尾桨
ducted-fan 涵道风扇
ducted-fan engine 涵道风扇发动机,涡扇发动机
ducted-propulsor 涵道推进器(变距风扇在涵道内旋转产生推力,低速飞机上适用效率高,噪声低)
ductile 1.可锻的,可塑的,韧性的 2.易拉长的,可展伸的,易变形的
ductile-to-brittle transition temperature 韧脆性转变温度
ductility 1.延(展)性,韧性 2.可塑性,可锻性
ducting 1.管道,管系,导管系统 2.用管道输送,用管道连接 3.大气波导现象
ducting system 管道系统
duction 转向
dud 〈口语〉未爆炸弹,未炸炮弹,哑弹
dud ammunition 未爆弹药
dud shell 未爆炮弹
duel 战斗
dugout 地下掩蔽部,防空洞,地下室
dull 1.钝的 2.暗淡的 3.弄钝,钝化,减弱
dull switch 暗淡开关(座舱内的按钮开关因从不使用而色泽暗淡)
dumb 迟钝(武器)(指非制导的、不灵活的弹药)
dumb bogie (空战中)不重要的敌机

dumb terminal 不灵活的终端设备
dumb weapon 迟钝武器（指非制导武器）
dumbbell galaxy 哑铃状星系
dumbbell inertia model 哑铃式惯性模型
dumbbell model 哑铃模型，铃模型
dum-craft 千斤顶，起重机
dumdum(bullet) 达姆弹，柔弹头
dumet 代用白金（一种铁镍合金）
dummy 1.假的，教练用的 2.模仿物，假弹，假目标，假人，人体模型 3.哑元，哑变量，伪变量
dummy antenna 仿真天线，假天线
dummy cartridge(round) 假炮弹，教练弹
dummy cell 虚设单元，空单元
dummy deck 假甲板（在陆上标出的航母模拟甲板）
dummy ejection 人体模型弹射
dummy fuze 假引信，练习用引信
dummy guided missile 假导弹，教练导弹
dummy hot run 冷开车，假热开车
dummy instruction 伪指令
dummy launch 假发射（使用惰性装药）
dummy load 仿真载荷，假载荷
dummy parameter 虚参数
dummy projectile(shell) 假弹，练习弹
dummy propeller 试车（螺旋）桨
dummy round 假炮弹，教练弹
dummy run 1.试验性着陆 2.模拟进入（俯冲而不投弹）
dump 1.堆集所(场) 2.（计算机的）清除，转储 3.倾卸，(飞机)应急放油 4.〈口语〉投弹，空降物资 5.(航天器上)排污（排废水及垃圾等）6.切断电源
dump combustor 突张式燃烧器
dump cooling 排放冷却
dump diffuser 突扩扩压器
dump door 1.倾卸门 2.快卸舱门
dump gap 转储差距
dump plane 空降物资飞机，应急放油飞机
dump plenum 转储充气
dump valve （应急）放油阀，排泄活门
dumper 卸除器，卸减器，卸货车
dumping 1.应急放油 2.卸除，卸减
dump-plane injection 应急放油飞机注入
dunk （把系绳或通信电缆）沉入水中
dunnage 垫木，枕木
dunnite 邓氏炸药，D 型炸药（一种含有苦味酸铵的炸药）
duotone 双色的（迷彩配色有两种主要颜色）
duodiode 双二极管
duoplasmatron 双等离子管，双等离子流发生器
duopoly 两家卖主垄断市场（的局面），双头垄断市场
duotriode 双三极管

duplex 双工的，二重的，双联的
duplex autolanding 双重操纵系统自动进场着陆
duplex burner 双路喷嘴
duplex computer 双计算机
duplex computer system 双工计算机系统
duplex management program 双工管理程序
duplex operation 双工操作，双工运行
duplex pistol 双击发引信
duplex system 双工系统，双重系统
duplex tube 孪生管
duplex wind tunnel 双试验段风洞
duplexer 双工器，天线（收发）转换开关，天线共用器
duplicated artificial feel 复式人工感力器，复式载荷感觉器
duplicated autoflare(out) 双套自动拉平系统
duplicated crank 复合摇臂
duplicate-monitored 双（套）监控的
duplication 1.复制，复制品 2.重复 3.加倍
duplicator 1.倍增器，加倍器 2.复制装置，复印机，靠模装置
duplicatus 复（云）（符号 du）
durability 1.耐久性，持久性（常指结构抗恶化的能力）2.耐航力，续航力 3.耐用年限，疲劳寿命
durability design 耐久性设计
durable 1.耐久的，持久的 2.可长期使用的，可长时间工作的
duralumin 1.硬铝合金 2.杜拉铝
Duramold 杜拉胶合木
duration 1.持续时间，使用期限 2.续航时间
duration model 耐航（比赛）模型飞机
duration of annular phase 环食时间
duration of eclipse 掩食时间，食长
duration of totality 全食时间
Durestos 杜里斯托斯复合材料
during operation check 运行中检查
duriron 硅铁，铁硅合金
durite hose 夹布胶管
durometer 硬度计
dusk 黄昏，薄暮，幽暗，昏暗
dusk orbit 黄昏轨道
duskside magnetosphere 昏侧磁层
dust 1.灰尘，尘埃 2.粉末，屑 3.喷粉，除尘
dust cloud 尘埃云，(天)尘云，尘雾
dust counter 尘埃计数器
dust cover 防尘罩
dust detonation 粉尘爆轰
dust exhaust 出尘道
dust explosion 尘爆，尘末爆炸
dust impact detection system 尘埃撞击探测系统

dust lane	尘埃带
dust nebula	尘埃星云
dust particle	尘粒,微尘
dust storm	尘暴
dust-air	含尘空气
dust-air flame	含尘空气火焰
dustbin	〈口语〉下炮塔
dust-devil	尘卷
duster	1.喷粉器,(农用)撒粉飞机 2.除尘器
dust-proof	防尘的
dust-proof device	防尘装置
dust-proof test	防尘试验
dustry	中断程序工业
dustsonde	尘埃探空仪
duststorm	尘暴
duststorm roof	尘暴防尘的
dutch roll	横滚
Dutch roll mode	荷兰滚模态
DUTE	Digital Universal Test Equipment 数字式通用检测设备
duty	1.负载,能率,(在给定状态的)功 2.工作状态,工作循环 3.职务,勤务,任务,值勤 4.(复)职责
duty controller	值班引导军官,防空分区指挥所值班军官(英国)
duty crew	特殊任务机组,值班空勤组
duty cycle	占空因数,负载循环,工作循环
duty factor	占空因数,占空系数,占空比
duty officer	值班军官
duty ratio	占空比(雷达脉冲功率的),负载比
duty swimmer	值勤蛙人
DUV	Data Under Voice 话音数据,音频数据
DV	1. Direct-Vision 直接观察(窗) 2. Distinguished Visitor 贵宾(美国用法)
DVA	Design Verification Article 设计验证项目,设计论证书
DVCS	Data-Voice Communications System 数据-话音通信系统
DVE	Design Verification Engine 设计验证发动机
DVFR	Defense Visual Flight Rules 防空识别区目视飞行规则(美国)
DVI	Direct Voice Input 直接语音输入
DVIP	Digital Video Integrator and Processor 数字视频积分处理器
DVM	Digital Voltmeter 数字电压表
DVO	Direct-View Optics 直观式光学装置
DVP	Digital Video Processor 数字式视频处理机
DVST	Direct Vision Storage Tube 直视存储管
DW	1. Drop Weight 下落重量,落锤重量 2. Double Wedge 双楔形(翼型),菱形(翼型)
dwarf cepheid	矮造父变星
dwarf galaxy	矮星系
dwarf nova	矮新星
dwarf sonobuoy	袖珍声纳浮标
dwarf star	矮星
dwell	1.小停顿(机器运转中有规则的) 2.保压(保持最大的压塑力) 3.上死点,下死点(活塞发动机冲程的) 4.(航空自动武器闭锁机构工作时的)闭锁时间
dwell mark	停顿斑
dwell time	驻留时间
dwell time of detector	探测器驻留时间
DWG	Digital Word Generator 数字式字产生器
DWI	Directional Wireless Installation 定向无线电装置(用于探矿)
DWS	1. Dispenser Weapon System 子母弹箱武器系统 2. Doppler Wind Sensor 多普勒风敏感器
DWSS	Disaster Warning Satellite System 灾难报警卫星系统
DY	dysprosium 镝
DYA	Dynamic Assignment 动态分配
dyad	1.双 2.【生】二分体 3.一对 4.【化】二价元素 5.二的,双的
dyadic grid	二元网格
dye	1.染色 2.染料,颜料,着色剂
dye cell	染料池
dye laser	染料激光器
dye marker	着色标志
dye penetrant flaw testing	着色渗透检测
dye Q-switching	染料 Q 开关
dye sensitizer	染料增感剂
dye tracers	染色示踪
dyeline proof	彩色线划校样
dye-sensitization	染料增感
DYN	dyne 达因(力的单位)
Dyna Soar dynamic soaring	动力滑翔飞行器(美国)
dynaforming	爆炸成形
dynamic accuracy	动态精度,动态准确度
dynamic accuracy evaluation	动态精度估计
dynamic accuracy test system	动态精度测试系统
dynamic airspace	动态空域
dynamic amplifier	动态放大器
dynamic amplitude	动态振幅
dynamic analysis	动态分析,动力特性分析
dynamic analysis of structure	结构动力分析,结构动强度分析
dynamic and erosive burning	动态和侵蚀燃烧
dynamic anisoinerda error	动态不等性误差
dynamic aquaplaning	动力滑水
dynamic assignment	动态分配

dynamic assignment multiple access 动态分配多路存取	dynamic environmental test 动力学环境试验
dynamic automatic monitoring 动态自动监测	dynamic equation 动态方程式,动力方程
dynamic balance 1.动态平衡 2.飞行中力或力矩的平衡	dynamic erosion 动态侵蚀
	dynamic error 动态误差
dynamic balance mechanism 动平衡机构	dynamic event 动态事件
dynamic balance test 动平衡试验	dynamic exactness 动态吻合性
dynamic balancing machine 动平衡机	dynamic extinction 动态消光
dynamic balancing test 动平衡试验	dynamic factor 动力因数
dynamic balloon 动力气球	dynamic factor of the earth 地球动力因子
dynamic bandwidth allocation 动态带宽分配	dynamic feedback matrix 动态反馈矩阵
dynamic behavior 动态行为,动态特性,能动行为	dynamic film suspension accelerometer 动压膜支承加速度表
dynamic braking 能耗制动	
dynamic burning 动态燃烧	dynamic filter 动态滤波器
dynamic bus control 动态总线控制	dynamic flexibility 动态柔性,动态烧性
dynamic calibration of balance 天平动校(准)	dynamic flight simulator 动态飞行模拟器
dynamic ceiling 动力升限,动升限	dynamic force 动力
dynamic characteristic 动态特征曲线	dynamic formation 动态建成
dynamic characteristics of flight vehicle 飞行器动态特性	dynamic fracture toughness 动态断裂韧性
	dynamic function survey meter 动态功能检查仪
dynamic checkout unit 动态测试设备	dynamic game 动态对策
dynamic compaction 冲击压塑	dynamic geodesy 动力大地测量学
dynamic component (直升机的)动力转动部件	dynamic guidance error 动态制导误差
dynamic computation 动态计算法	dynamic halo model 动力晕模型(宇宙线的模型)
dynamic condition 动态,动态条件	dynamic head 动压头
dynamic constraint 动态约束,与动态约束条件	dynamic head displacement measuring system 头部动态位移测量系统
dynamic control 动态控制	
dynamic convergence voltage 动态会聚电压	dynamic height 1.动高度 2.动力高度
dynamic coupling 动态耦合,动力耦合	dynamic holography 动态全息摄影
dynamic damper 动力减震器	dynamic image analysis 动态影(图)像分析
dynamic damping 动力减震	dynamic imbalance 动不平衡(度)
dynamic debugging technique 动态调试技术	dynamic inflow 动态入流
dynamic decoupling 动态解耦	dynamic inflow formalism 动态流入形式主义
dynamic deformation 动态变形	dynamic inflow model 动态入流模型
dynamic degree 动态度	dynamic input-output model 动态投入产出模型
dynamic density 动态密度	dynamic instability 动力不稳定,动态不稳定性
dynamic derivative 动导数	dynamic inversion 动态逆
dynamic derivative balance 动导数天平	dynamic inversion control law 动态逆控制律
dynamic derivative testing 动导数试验	dynamic inversion technique 动态反演技术
dynamic design 动力学设计	dynamic killing zone of warhead 战斗部动态杀伤区
dynamic detonation 动态爆炸	dynamic lag error 动态滞后误差
dynamic deviation 动态偏差	dynamic libration 动力学天平动
dynamic directional stability 动方向稳定性	dynamic load 动态载荷
dynamic docking test facility 动态对接测试设备	dynamic load characteristics 动载荷特性
dynamic drift 动态漂移	dynamic load factor 动载因子
dynamic effect 动态效应	dynamic loading 动力载荷
dynamic effect of liquid core of the earth 地球液核动力效应	dynamic maneuver 动态操作
	dynamic map 动态地图
dynamic ellipticity of the earth 地球动力扁率	dynamic measurement 动态测量
dynamic environment 动力环境,动态环境	dynamic measurement of opaque membranes 动态测量不

透明膜
dynamic meteorology 动力气象学
dynamic meter 动力计
dynamic mode 动态模式
dynamic model 动力模型，动态模型(飞机的一种模型)
dynamic modeling 动态模型，动态建模
dynamic modulus of elasticity 动态弹性模量
dynamic moment 动态力矩
dynamic moment constraint 动态力矩约束
dynamic motion 动态移动
dynamic multiplexing 动态多路复用
dynamic noise filter 动态噪声滤波器
dynamic optimization 动态优化，动态最佳化
dynamic orbit solution 动力学轨道解
dynamic parallax 动力学视差
dynamic parameter 动态参数
dynamic part 动态部分
dynamic particle filter 动力固体粒子分离器
dynamic performance 动态特性
dynamic phase 动态相位
dynamic photographic resolution 动态摄影分辨率
dynamic positioning 动力定位，动态定位
dynamic pressure 1.动压 2.速压
dynamic pressure feedback servo valve 动压反馈伺服阀
dynamic pressure measurement 动压测量
dynamic pressure ratio 动压比
dynamic pressure transducer 动压传感器
dynamic problem 动力问题，动态问题
dynamic programming 动态规划
dynamic propagator 动态传播算子
dynamic property 动力特性，动力特性，动力特性
dynamic pump 动力式泵
dynamic pump gain 动态泵获得
dynamic random access memory 动态随机存取存储器
dynamic range 1.动态范围，动态量程(仪表输出量程与阈值之比) 2.动力射程
dynamic reconfiguration 可随动态重新部署(配置)的(防御系统)量程与阈值之比
dynamic replanning 动态重新规划
dynamic response 动态响应，动力学响应
dynamic response index 动态响应指数
dynamic response test stand 动态响应试验台
dynamic roughness 动力糙度
dynamic sampling 动态抽样
dynamic satellite geodesy 卫星动力测地
dynamic scale 动力相似尺度，动力学比例尺度
dynamic scaling 实现动态缩放
dynamic scattering mode 动态散射模式
dynamic seal 动密封，动密封件

dynamic segment 动态段，动态分段技术
dynamic sensitivity 动态灵敏度(光电管的)
dynamic sensor 动态传感器
dynamic separation 动态分离
dynamic shape 动态形态
dynamic shift register 动态移位寄存器
dynamic similarity 动力学相似性，动态相似性
dynamic simulation 动态仿真
dynamic soaring 动力翱翔(利用空气流动的动能来滑翔)
dynamic spectrum 动态谱，谱时曲线
dynamic speed 动力学速度
dynamic stability 动稳定性，动态稳定性
dynamic stability model 动(态)稳定性模型
dynamic stall 动态失速
dynamic stall event 动态失速的事件
dynamic stall vortex 动态失速涡
dynamic standby computer 动态备用计算机
dynamic stiffness 动态刚度，动安定性，动态稳定性
dynamic storage lit 动态存储点(电子数据处理的)
dynamic strength 动(态)强度
dynamic stress 动(态)应力
dynamic structural analysis 动态结构分析，结构动强度分析
dynamic structure 动态结构
dynamic suspension 动力悬浮
dynamic system 动力系统
dynamic temperature 动力温度
dynamic test 动态试验(在负载下进行的试验)
dynamic test panel 动态测试板
dynamic test vehicle 动力试验飞行器，动态试验飞行器
dynamic tester 动态测试仪
dynamic testing 动态测试
dynamic theory 动力学理论，动态说
dynamic thrust 动力，飞行推力
dynamic time scale 力学时标
dynamic trim 动力配平
dynamic tuned gyroscope 动力调谐陀螺
dynamic tuning 动力调谐
dynamic uncertain environment 动态不确定环境
dynamic vertical 动态垂线，表观垂线
dynamic vibration absorber 动力吸振器
dynamic viscosity 动(力)黏度，动力黏滞率
dynamic visual acuity 动态视力
dynamic wake 动态尾迹
dynamic wake model 动态尾流模型
dynamic wave 动力波
dynamic yield 动态屈服
dynamic(al) 1.动力的 2.动力学的自 3.动态的

dynamical constraint 动态约束	**dynamics of atmospheric flight** 大气层飞行动力学
dynamical cosmology 动力宇宙学	**dynamics of fluid** 流体动力学
dynamical equation 动力(学)方程	**dynamics of gas** 气体动力学
dynamical equilibrium 动力(学)平衡	**dynamics of machinery** 机械动力学
dynamical equinox 力学分点	**dynamics of multiple rigid body spacecraft** 多刚体航天器动力学
dynamical flattening 力学扁率	
dynamical friction 动摩擦	**dynamics of orbits** 轨道动力学
dynamical mean sun 力学平太阳	**dynamics problem** 动力问题,动态问题
dynamical mechanical magnification 动态机械放大倍数	**dynamics program** 动态程序,动程序
dynamical method of satellite geodesy 卫星大地测量动力法	**dynamics simulation** 动力学仿真
	dynamics test 动力学试验
dynamical model 动态模型(一种飞机模型,它的线性尺寸、重量、转动惯量等与原型成比例地相似)	**dynamite** 达纳炸药
	dynamite bomb 达纳炸弹
dynamical modeling 动态建模	**dynamo** 发电机
dynamical parallax 力学视差	**dynamo region** 发电机区
dynamical reference system 动力学参考系	**dynamograph** 功率自记器,自记测力计
dynamical system 动力系统	**dynamometer** 测力计,功率计
dynamical time 力学时	**dynamometer system** 测功器仪表系统
dynamically similar model 动力(学)相似模型	**dynamometer test** 测功器试验
dynamically tuned flexure gyro 动力调谐式挠性陀螺	**dynamometric** 测力的,计力的
dynamically tuned gyro 动力调谐陀螺,挠性陀螺	**dynamometry** 测力法
dynamically tuned gyro for a strapdown 捷联式动力调谐陀螺	**dynamotor** 直流升压机,电动发电机
	dynatron (打拿)负阻管
dynamically unstable 动不安定的	**dynatron effect** 负阻效应
dynamically-slender 动力学细长体的,质量沿纵轴集中配置的	**dynatune gyro** 动力调谐陀螺
	dyne 达因(力的单位)
dynamicist 动力学家	**dynistor** 逆电压控制的半导体开关
dynamicizer 动态转换器	**dynode** 倍(增)器电极,中间极,打拿极
dynamic-response factor 动(态)响应系数	**dynode system** 倍增系统
dynamics 1.动力学,力学 2.动力学特性 3.动力,动态 4.动力转动部件(直升机的)	**dysbarism** 气压病
	dysbolism 代谢障碍
dynamics analysis 动力学分析	**dysfunction** 功能不良
dynamics board 动态板	**dysgraphia** 书写困难
dynamics controller 动力学控制器	**dyskinesia** 运动困难
dynamics modeling 动态模型,动态建模	**dyspnea** 呼吸困难
dynamics of a flexible body 柔性体动力学,挠性体动力学	**dysprosium** 【化】镝(化学元素,符号 Dy)
	dysrhythmia 节律障碍
dynamics of a particle 质点动力学	**dystaxia** 共济失调
dynamics of a rigid body 刚体动力学	**dystrophy** 营养障碍

E

E 1. East 东 2. Electronic/Electronics 在美国军用飞机命名代码中,代表电子预警机 3. Electronic Counter Coutermeasure 在英国军用飞机型号中,代表"电子对抗训练"飞机 4. Electronic Reconnaissance 在美国军事任务符号中代表电子侦察 5. Emitter (Electron Device) 发射机(电子装置),发射极输出器

E frame 1. E 型结构 2. Electronic Frame 电子相框

E galaxy Elliptical Galaxy 椭圆星系

E layer E 层(地球电离层中层)

E&C Electronics and Control 电子设备与控制

E&E 1. Electrical and Electronic Compartment 电气电子设备舱 2. Electrical and Environmental 电气和环境 3. Emergency and Evacuation 紧急疏散,应急疏散 4. Escape and Evasion 脱离与规避

E&S Executive and Support (空中交通管制勤务)行政与保障

E&SP Equipment and Spare 设备和备件

E&ST Employment and Suitability Test Program 使用性和适用性试验计划(美国空军)

E/E 1. Edge Enhancement 边缘增强 2. Electrical/Electronic 电气/电子 3. Emergency Equipment 应急设备 4. Electrical Engineering 电气工程 5. Energy Efficiency 能量系数,能效,能源效应 6. Engagement Effectiveness (攻防)对战作战效能 7. Engineering Evaluation 工程(质量)评价 8. Envelope Expansion 飞行包线扩充

EA 1. Each 每一个 2. EATON (CPA and PWM Dimming Control) 依通公司(控制板组件) 3. Electronic Asseembly 电子装置,电子装配,电子集成 4. Electronic Attack 电子攻击 5. Electromagnetic Anomaly 电磁异常 6. Elevation Angle 升运角,倾斜角,仰角 7. Enemy Aircraft 敌机,敌方飞机 8. Engine-Attributable 发动机方面的(原因) 9. Engagement Authorization 交战许可 10. Engineering Authorization 工程授权 11. Engineering Authority 工程部门,工程机务部门 12. Environmental Assessment 环境评价

EAA 1. Edge Anti-aliasing 边缘抗锯齿失真 2. Engineer in Aeronautics and Astronautics 航空航天工程师 3. Experimental Aircraft Association 实验飞机协会(美国) 4. Export Airworthiness Approval 出口适航批准书 5. European Aviation Agency 欧洲航天局

EAAS East Antrim Astronomical Society 东安特里姆天文学会

EAC 1. Earth Central Angel 地心角 2. Electric Control Amplifier 电子控制放大器 3. Electrical Control Assembly 电子控制组件 4. Endpoint Admission Control 端点准入控制 5. Enhanced Attitude Control 增强的姿态控制 6. Enter Control Area 进入控制区 7. Equipment Condition Analysis 设备状态分析 8. Estimated Approach Control 估计进近控制 9. Expected Approach Clearance 预期进场许可 10. Expected Approach Clearance Time 预计许可进场时间 11. Experimental Apparatus Container 实验设备箱 12. Export Airworthiness Certificate 出口适航证 13. European Astronauts Centre 欧洲航天员中心 14. European Airlift Center 欧洲空运中心 15. European Civil Aviation 欧洲民航组织 16. European Combat Aircraft 欧洲战斗机

EACC 1. Error Adaptive Control Computer 误差自适应控制计算机 2. European Airlift Coordination Cell 欧洲空运协调室

EACS 1. Electronic Automatic Chart System 自动电子地图显示系统 2. Emergency Air Cleaning System 应急空气净化系统 3. Expeditionary Air Control Squadron 远征航空管制中队(美国空军)

EADB 1. Elevator-Angle Deviation Bar 升降舵角偏移杆 2. Enhanced Area Distribution Box 增强型区域分配盒 3. European AIS Data Base 欧洲航空信息系统数据库

EADI 1. Electronic Attitude Director (Display) Indicator 电子姿态指引仪,电子指引地平仪,电子垂直位置指示器 2. Electronic Airborne Data Indicator 机载电子数据指示器 3. Electronic Attitude Director Indicator 电子姿态指引仪,电子姿态指引指示仪 4. Electronic Attitude Direction Indicator 电子式姿态指引仪

EADS Engineering Administrative Data System 工程管理数据系统

EAEM European Airlines Electronics Meeting 欧洲航空公司电子学会议

EAF 1. Engineering Analysis Facility 工程分析设施 2. Expand Aluminum Foil 膨胀铝箔

EAFB Edwards Air Force Base 爱德华兹空军基地(美国)

EAFR Enhanced Airborne Flight Recorder 增强型机载飞行记录器

EAGE Electrical Aerospace Ground Equipment 航空航天地面电子设备
Eagle 1. 美国 Apollo-11 号飞船登月舱 2. 瑞典火控雷达
Eagle fire control radar "鹰"式火控雷达
EAH Emergency Artificial Horizon 应急地平仪
EAI 1. Engine Anti-Ice 发动机防冰 2. Enterprise Application Integration 企业应用集成
EAL 1. Electromagnetic Amplifying Lens 电磁放大透镜 2. Engine Assignment Letter 发动机调拨书
EALS 1. Ejector Augmented Lift System 引射增升装置 2. Emergency Airfield Lighting System 机场应急照明系统
EAM 1. Emergency Action Message 紧急措施信息 2. Engineering Administration Manual 工程管理手册 3. Eurocontrol Airspace Model 欧洲空管空域模型
EAMDS European Airlines Medical Directors Society 欧洲航空公司医务经理协会
EAMR 1. Engineering Advance Material Releases 工程先行资料发放 2. Engineering Advance Material Request 工程先期物料需求
EANS 1. Empire Air Navigation School 帝国空中领航学校(英国) 2. Emergency Action Notification System 紧急行动通知系统(美国白宫电传打字电报机网)
EAP 1. Each Additional Part 每个附加零件 2. Effective Air Path 有效航线,有效航道 3. Enhanced Authentication Protocol 可扩展的身份验证协议 4. Emergency and Accident Procedure 紧急与事故程序 5. Engine Alert Processore 发动机告警处理器 6. Equivalent Air Pressure 等效空气压力,等效气压 7. Experimental Aircraft Program 实验机计划(亦用作此实验机之代名,EAP 是 EFA 的验证机)(英国)
EAR 1. Electronic Audio Recognition 电子声音识别 2. Electronically Agile Radar 电子捷变雷达 3. Engineering Action Request 工程措施要求 4. Engineering Analysis Report 工程分析报告 5. Espace Aerien Reglemente 空间管理(法国) 6. Expanded Area Ratio 伸张面积比
ear pressure test 耳压试验
ear protector 护耳器
EARB European Airlines Research Bureau 欧洲航空研究局
EARC 1. Electric Armaments Research Center 电子军事装备研究中心 2. Extraordinary Administration Radio Conference(国际电信联盟的)特别无线电行政会议
EARL European Advanced Rocket Launcher 欧洲先进火箭运载器
Early Bird 晨鸟卫星(第一个国际通信卫星)

early bird 早到的导弹(先于目标到达计算相遇点的导弹)
early burst 早炸
early design 早期设计,最初设计
early failure period 早期故障期,早期失效期
early flight interception 初始段拦截
early study 早期研究
early system Early Warning System 早期预警系统
early transition metals 前过渡金属
early turn 提前转弯(未到航路转弯点提前转弯,以修正航迹,避免越出航路边界)
early universe 早期宇宙
early warning aircraft 预警机
early warning airplane 预警机
early warning radar 预警雷达
early warning satellite 预警卫星
early warning system 预警系统
early-bird encounter 提前交会,导弹先于目标通过计算相遇点
early-type cluster 早型星系团
early-type galaxy 早型星系
early-type star 早型星
early-type spiral galaxy 早型旋涡星系
early-warning radar intelligence 预警雷达情报
early-warning radar screen 预警雷达警戒网
EAROM Electrically Alterable Read Only Memory 电控可变只读存储器
EARS 1. Electronics and Amateur Radio Society 电子和业余无线电爱好者学会(英国) 2. Electromagnetic Aircraft Recovery System 电磁飞机拦阻系统 3. Emergency Airborne Reaction System 机载应急反应系统 4. Enviroment Anolog Recording System 环境模拟记录系统 5. Expeditionary Air Refueling Squadron 远征空中加油中队
EMALS Electromagnetic Aircraft Launch System 电磁飞机弹射系统
EARSL European Association of Remote Sensing Laboratories 欧洲遥感实验室协会
earth 地球,地面,接地,地线
earth acceleration 地球加速度,地球重力加速度
earth acquisition 地球捕获
earth albedo 地球反照率,地球反射率
earth albedo factor 地球反照因数
earth angle 地球角
earth approach asteroid 近地小行星
earth attraction 地球引力
earth axis 地球自转轴
earth-based support equipment 地面保障设备
earth centre phase shift 地中相移

earth coverage　（卫星）地面覆盖范围
earth coverage antenna　全球覆盖天线
earth current　大地电流,接地电流
earth curvature correction　地球曲率改正
earth curvature rectification　地球曲率校正
earth dynamic ellipticity　地球力学椭率
earth dynamics　地球(动)力学
earth elasticity　地球弹性
earth electrostatic field　地球静电场
earth ellipsoid　地球椭球体
earth entry module(also re-entry module)　返地舱,返回舱
earth equatorial plane　地球赤道(平)面
earth escape circle　地球逃逸轨道
earth escape velocity　第二宇宙速度,逃逸地球速度
earth exploration satellite service　地球卫星探测服务
earth fixed station　固定地球站,地面固定电台
earth flattening　地球扁率
earth funnel　地磁极漏斗
earth geodesic satellite　测地卫星,大地测量卫星
earth gravitational potential　地球重力势,地球引力势
earth infrared radiation factor　地球反照角系数
earth intercept zone　地球交截区域
earth light　大球反射/照光,地球反照,地照
earth limb background　地球外缘背景
earth limb radiance　地球临边辐射率
earth line　接地线
earth lunar shuttle　地月往返飞船
earth magnetic field　地球磁场
earth magnetosphere　地球磁层
earth mean orbital speed　地球平均轨道速度
earth model　地球模型
earth momentum of inertia　地球惯性矩
earth oblateness　地球扁率
earth observation satellite　对地观测卫星
earth orbit　地球轨道
earth orbiter　地球轨道飞行器,地球轨道卫星
earth out　地出
earth pendulum　地球摆(即舒拉摆)
earth penetrator　（海湾战争中美国使用的)钻地弹
earth pole　地极
earth potential field　地引力势场
earth radiation budget　地球辐射收支,地球辐射平衡
earth radiation budget satellite　地球辐射收支卫星
earth radiation budget scanning radiometer　地球辐射收支扫描辐射仪
earth radius　地球半径
earth rate　地球自转速率
earth rate unit　地球角速率单位,地球角速度单位

earth reference ellipsoids　地球参考椭球
earth resistance meter　地电阻表,接地电阻计,接地电阻表
earth resource　地球资源
earth resource survey　地球资源调查
earth resources satellite　地球资源卫星
earth return　接地回路
earth rotation　地球自转
earth rotation parameter　地球自转参数
earth rotation rate　地球自转速率
earth satellite　地球卫星
earth satellite network　地球卫星网
earth science　地学
earth segment　地球段
earth sensor　地球传感器
earth shadow　地影
earth shield　接地屏蔽
earth simulator　地球仿真器,地球模拟器(由日本宇宙开发事业团、日本原子能研究所以及海洋科学技术中心共同开发的矢量型超级计算机)
earth spheroid　地球扁球体,地球椭球体
earth station　地球站,地面站
earth stationary orbital keeping　地球同步轨道保持
earth system　地球系统
earth tide　陆潮,固体潮
earth tilt　地倾斜
earth viewing (equipment) module　地球观察(设备)舱
earth year　地球年
earth/moon liberation centre　地球/月球天平动中心
earth-atmosphere system radiation budget　地-气系统辐射收支,地-气系统辐射平衡
earth-fixed axis system　地面固定坐标系
earth-fixed coordinate system　地固坐标系
earth-fixed frame　地固坐标系,地固坐标框架
earth-flattening approximation　地球变平近似
earth-flattening transformation　地球变平换算
earth-in　地入
earthing wire　接地线
earth-ionosphere wave guide　地-电波导,地球-电离层波导
Earth-Mars trajectory　地球-火星飞行轨道
earth-orbit rendezvous　绕地轨道交会
earth-orbit transfer　绕地轨道转移
earth-orbital photography　地球轨道摄影
earth-orbiting satellite　地球轨道卫星
earth-pointing　指向地球,对地的
earthquake bomb　巨型穿地炸弹,地震炸弹
earth-rate correction　地球自转速率(度)修正
earth-resources remote sensing　地球资源遥感

earthrise 地出
earth's gravity 地球引力，地球重力
earth's pear shape 梨形地球
earth's polar axis 地球极轴
earth's rotation vector 地球自转(速度)矢量
earthspeed 合成地速
earth-sun rotation angle 地球-太阳两面角
earth-synchronous orbit 地球同步轨道
earth-to-satellite link 地球-卫星通信链路
Earth-Venus trajectory 地球-金星飞行轨道
Earthwatch 地球(环境)监察(一种拟议中的全球环境监察网站)
EARTS Enroute Automated Radar Tracking System 航路自动雷达跟踪系统
EAS 1. Emergency Alert System 紧急报警系统(美国联邦通信委员会(FCC)，联邦应急管理局(FEMA)和国家气象服务(NWS)三个部门建立的当国家处于紧急状况时迅速通告全民的通信方式)(美国) 2. Engineering Administration Services 工程管理服务 3. Equivalent Air Speed 等效空速 4. Essential Air Service 基本航空服务(美国实施的一项旨在为小社区和偏远地区提供全面而又方便的航空服务计划) 8. Extensive Air Shower 空气(广延)簇射 5. Espace Aerien Superieur 上层空域(法国) 6. European Astronomical Society 欧洲天文学会 7. Executive Air Service 公务飞行服务
EASCON Electronics and Aerospace System Conference 电子仪器和航空航天系统会议
EASE 1. Equal-Area Scalable Earth (Grids) 等面积可扩展地球网格(NSIDC—National Snow and Ice Data Center，即美国国家冰雪数据中心使用的一种主要用于卫星数据表达的方式) 2. Evolutionary Acquisition for Space Efficiency 太空效率渐进式采购(美国空军计划施行一个新的采购战略) 3. Electronic Airbetne System Evaluation 机载电子系统评定器
EASM Explicit Algebraic Stress Model (力学)显式代数应力模型
EASM calculation (力学)显式代数应力模型计算
EASM formulation (力学)显式代数应力模型表达式
EASM solution (力学)显式代数应力模型解
east European time 东欧时
east longitude 东经
east longitude date 东经日
EASTCON Electronics and Aerospace Systems Technical Convention 电子仪器和航空航天系统技术会议(美国)
easterly variation 东磁差
eastern (standard) time 东部标准时间(美国)
eastern elongation 东距角(行星、卫星),(恒星视运动)东大距
eastern greatest elongation 东大距
eastern quadrature 东方照
eastern standard time zone 东部标准时区(美国)
Eastern Test Range 东部试验靶场(原太平洋靶区，美国导弹试验场，从卡纳维拉尔角越过大西洋进入印度洋和南极洲)
Eastport Study Group 东港研究组(美国 SDI 即 Strategic Defense Initiative Organization 中研究作战管理计算机的研究小组)
EASTT Experimental Army Satellite Tactical Terminals 陆军实验卫星战术终端(美国)
east-west station keeping 东西位置保持
EASY 1. Engine Analyzer System 发动机分析仪系统 2. Engineering of Automation Systems 自动化系统工程 3. Enhanced Avionics System 增强的航空电子系统(由法国达索公司和霍尼韦尔公司联合倡导的驾驶舱概念) 4. Exception Analysis System 异常分析系统
EAT 1. Eletronic Angle Tracking 电子角度跟踪 2. Estimated Approach Time 预期进场时间 3. Expected Approach Time 预期进场时间
EAU Extended Arithmetic Unit 延伸算术单元，扩充运算器
EAV Effective Angular Velocity 有效自动交换
EAX Electronic Automatic Exchange 电子自动交换
EB 1. Electron Beam 电子束 2. Electronics Bay 电子设备舱 3. Electronics Box 电子设备盒 4. Employee Burnout 雇员短缺 5. Engineering Bulletin 工程通报 6. Essential Bus 主汇流条(为保证飞机安全的重要设备供电) 7. Expendable Booster 可丢弃助推器 8. Equipment Bay 设备舱 9. External Bremsstrahlung 外韧致辐射 10. Extraterrestrial Bases 地外基地，空间基地
EBA 1. Electronics Bay Assembly 电子舱集成 2. Emergency Breathing Apparatus 应急呼吸器 3. Engine Bleed Air 发动机放出(或引出)的空气
EBA/H Engine Bleed Air and Hydrazine 发动机引气和肼
EBAA 1. European Business Aviation Association 欧洲商业航空协会 2. European Business Aircraft Association 欧洲商用飞机协会 3. Experimental Balloon and Airship Association 实验气球及飞船协会
EBB Extra Best Best 最高质量
EBC 1. Electronic Business Communications 电子商业通信 2. Embedded Battle Command 嵌入式作战指挥 3. Enroute Battle Command 航路战斗指挥
EBCDIC 1. Extended Binary Coded Decimal Interchange Code 扩充二-十进制交换码 2. Extended Binary Coded Decimal Information Code 广义二进制

编码的十进制信息码

EBF 1. Eigen-Beam Forming 特征波束成形 2. Electronic Beam Forming 电子波束成形 3. Externally Blown Flap 外吹式襟翼

EBFD Electronic Brake Force Distribution 电子制动力分配系统

EBIC 1. Electron-Beam-Induced Conductivity 电子束感应电导(率) 2. Electron Beam Induced Current 电子束感应电流,电子束诱生/发电流 3. Electron-Bombardment-Induced Conductivity 电子轰击感应电导率

EBL 1. Eextragalactic Background Light 河外背景光 2. Electronic Bearing Line 电子方位线 3. Exterior Ballistics Laboratory 外弹道实验室

Eblis "埃布利斯"激光探测器(法国研制,来配合BGL激光制导炸弹的投放)

EBM 1. Electron-beam Machining 电子束加工 2. Electron Beam Method 电子束法 3. Electron Beam Melted 被电子束熔化的 4. Electron-beam Microanalysis 电子束微量分析(显微分析) 5. Electronic Battle Management 电子战管理 6. Engineering Business Management 工程商务管理

EBMA Engine Booster Maintenance Area 发动机助推器保养场

EBMD Electron Beam Mode Discharge 电子束放电

EBMF Electron Beam Microfabricator 电子束微加工机

EBOM Engineering Bill of Materials 工程物料表

ebonite 硬质胶,硬橡皮,胶木,硬质橡胶

EBP 1. Earth Bonding Point 接地连接点 2. Effects-Based Planning 基于效果的(战役)规划 3. Electric Bilge Pump 舱底电力泵 4. Electron Beam Pumping 电子束抽运 5. Electron-Beam Perforated 电子束穿孔 6. Exhaust Back Pressure 排气反压力,排气背压

EBPA Electron Beam Parametric Amplifier 电子束参数放大器

EB-PVD Electron Beam-Physical Vapor Deposition 电子束物理气相沉积

EBR 1. Electron Beam Recording 电子束录像 2. Electron Beam Recorder 电子束记录器 3. Electron Beam Regulator 电子束调节器 4. Engineering Business Report 工程工作报告

EBRM Electronic Bearing and Range Marker 电子方位和距离信标

EBSC 1. Enhanced Base Station Controller 增强型基站控制器 2. European Bird Strike Committee 欧洲防鸟撞击委员会

Ebsicon 硅加强目标移像(北约的标准用语),增强硅靶摄像管

EBSV Engine-Bleed Shutoff Valve 发动机放气断开阀

EBT 1. Effective Brightness Temperature 有效亮度温度 2. Electron Beam Technology 电子束工艺,电子束技术 3. Engine Block Test 发动机气缸排气试验 4. Excess Baggage Ticket 超额行李付费票联

EBU 1. Engine Build-up 发动机装配,发动机配套装配 2. European Broadcasting Union 欧洲广播联盟

ebullism 起泡(液体火箭推进剂因外界压力降低而引起的),体液沸腾,气肿(人体因处于高空等导致外界压力降低而引起的)

EBW 1. Effective Band Width 有效带宽 2. Electron-Beam Welding 电子束焊 3. Entropy-Based Warfare 基于信息熵的战争 4. Exploding Bridge Wire 电桥式传爆线

EC 1. Eddy Current 涡流,涡电流 2. Electrolysis Cell 电解电池 3. Electronic Combat 电子战 4. Electrostatic Collector 静电收集器 5. Electronic Commerce 电子商务 6. Engine Control 发动机控制 7. Engine Change 换发 8. Engineering Change 工程更改 9. Equipment Center 设备中心(舱) 10. Escadrille de Chasse 歼击(机)中队(法国) 11. Error Correction 误差修正 12. Etch Characteristics 刻蚀特性 13. Echo Cancellation 回声消除 14. Economic Commission (ICAO Assembly) 国际民航组织经济委员会 15. Edge Connector 边缘连接器 16. Elasticity Coefficient 弹性系数 17. Electron Capture 电子捕获 18. Electron Coupling 电子耦合 19. Experimental Certificate 试验许可证 20. Extra Control 额外控制,附加控制

EC/LS 1. Environmental Control and Life Support System 环境控制与生命保障系统 2. Extracorporeal Life Support System 体外生命维持系统

ECA 1. Earth Central Angle 地心角 2. Electronics Control Amplifier 电子控制放大器 3. Enter Control Area 进入控制区 4. Equipment Condition Analysis 设备状态分析 5. European Combat Aircraft 欧洲战斗机

ECAC 1. Electromagnetic Compatibility Analysis Center 电磁兼容性分析中心 2. European Civil Aviation Conference 欧洲民航会议 3. European Civil Aviation Commission 欧洲民航委员会

ECAE European Committee for Aviation Electronics 欧洲航空电子委员会

ECAM 1. Electronic Caution Alert Module 电子告警模块 2. Electronic Centralized Aircraft Monitoring 飞机综合电子监控,飞机电子中央监控,飞机电子集中监控 3. Electronically Computer Aided Manufacturing 电子计算机辅助制造

ECAPS Emergency Capability System 应急能力系统

ECASS Electronically Controlled Automatic Switching System 电子控制系统自动转换系统

ECAT　Equipment Category 设备分类

ECAV　Electronic Combat Air Vehicles 电子战飞行器

ECB　1. Electrically Controlled Birefringence Mode 电控双折射模式 2. Electronic Control Box 电子控制盒 3. Electrically Controlled Birefringence (LCD driving mode) 电控双折射 4. Economic Cruising Boost 经济巡航助推 5. Electronic Components Board 电子元件板 6. Eddy Current Brake 涡流制动器 7. Etched Circuit Board 刻蚀电路板

ECC　1. Eccentric 偏心，离心 2. Engine Control Center 发动机控制中心 3. Engineering Change Classification 工程更改分类 4. Environmental Compatibility Certificate 环境兼容性合格审定 5. Error Correction Code 误差校正码，差错改正码

eccentric angle　偏心角，离心角

eccentric anomaly　偏近点角

eccentric dipole　偏心偶极(子)

eccentric latitude　偏心纬度

eccentric orbit　偏心轨道

eccentric reference orbit　偏心参考轨道，偏心基准轨道

eccentricity　偏心率，偏心距，偏心

eccentricity effect　偏心效应

eccentricity error　偏心误差

eccentricity function　偏心率函数（如 orbital eccentricity function 轨道偏心率函数）

eccentricity growth　偏心率增加（如 orbital eccentricity growth 轨道偏心率增加）

eccentricity of ellipsoid　椭球偏心率

eccentricity of satellite orbit　卫星轨道偏心率

eccentricity state detection　偏心状态检测，离心态检测

eccentricity vector　偏心(率)矢量

ECCM　1. Electronic Counter-Countermeasures 电子反对抗，抗干扰，电子反干扰 2. Electronic Counter Counter Measures 电子反干扰设备，反电子干扰措施

ECCM improvement factor　Electronic Counter Counter Measures Improvement Factor 反电子干扰改善因子

Eccosorb　艾科吸波材料（现有烧结铁淦氧雷达吸波材料中主要的一族）

ECD　1. Eddy-Current Distribution 涡流分布 2. Electrochromic Display 电致变色显示 3. Electron Capture Detector 电子捕获检测器 4. Electrochemical Deburring 电解去毛刺 5. Electron-Chemical Drilling 电化学钻孔 6. Energy Conversion Device 能量转换设备 7. Equivalent Circuit Diagram 等价电路图，等效电路图 8. Equivalent Circulating Density 当量循环密度 9. Estimated Completion Date 预计完成日期 10. Excusable Contract Delay 允许的合同延误，可原谅的合同延误(可不罚款的)

ECDES　Electronic Combat Digital Evaluation System 电子战斗数字评定系统

ECDL　Engineering Change Differences Lists 工程更改偏差清单

ECE　1. Economic Commission of Europe 欧洲经济委员会(联合国) 2. Electron Cyclotron Emission 电子回旋加速器发射 3. Environmental Control Equipment 环境控制设备 4. External Combustion Engine 外燃发动机

ECEF　Earth-Centered Earth-Fixed 地心固连(坐标系)

ECF　1. Effective Cutoff Frequency 有效切断频率 2. Eletro-Chemical Forming 电化学成形 3. Elevation Correction Factor 高程校正因数 4. Enhanced Connective Facility(系统网络分析的)加强结合装置 5. Envelope Check Fixture 包络检查夹具 6. Externally Caused Failure 外因造成的故障，外因故障

ECFR　Engine Condition and Fault Reporting 发动机状态和故障报告

ECG　1. Electroepitaxial Crystal Growth 电外延晶体生长 2. Electrochemically Assisted Grinding 电解研磨（或 Electro Chemical Grinding 电解磨削，电化学磨削）3. Electronic Combat Group 电子战大队（美国空军）

ECH　Echelon 梯队，阶梯

Echeloning　梯次配置

Echelonment　梯形配置

echo　回波，回声

echo box　回波箱，回波谐振器，回波谐振盒

echo box actuator　回波箱激励器

echo box performance monitor　雷达空腔监控器

echo canceller　回声消除器，回波消除器，回波抵消器

echo lock　回波锁定

echo suppression　回波抑制，回声抑制

echo suppression circuit　回波抑制电路，回声抑制电路

echo suppressor　回波抑制器

echo talker　回波干扰(信号)，回送干扰

ECIF　1. Electronic Components Industry Federation 电子元器件工业协会（英国）2. Empirical Cumulative Intensity Function 经验累积强度函数

ECIM　1. Electronics Computer-Integrated Manufacturing 电子设备的计算机综合加工制造 2. Exterior Communications Infrastructure Modernization 外部通信基础设施现代化

ECIS　1. Earth-Centered Inertial System 地心惯性系统 2. Error Correction and Information System 纠错与信息系统

ECJC　Engineering Change Justification Code 工程更改论证代码

ECL　1. Eddy Current Loss 涡流损耗 2. Electronic Check List 电子检查单 3. Electronic Components Laboratory 电子元件实验室 4. Emitter Coupled Logic

发射极耦合逻辑 5. Equipment Check List 设备检查单 6. Equipment Component List 设备零件表,设备元件明细表 7. Exchange Control Logic 交换控制逻辑 8. Exercise Caution while Landing 着陆时注意 9. External Cavity Laser 外腔激光器

eclipse 食,交食,(雷达显示器)遮蔽,发暗,日食或月食
eclipse cycle 食周
eclipse effect (日)食效应
eclipse of satellite 卫星食
eclipse of the moon 月食
eclipse of the sun 日食
eclipse year 食年
eclipsing binary 食双星,光度双星,交食双星,重叠双星
eclipsing binary star 交食双星
eclipsing binary system 食双星系统
eclipsing variable star 食变星
eclipsing X-ray star X射线食变星
ecliptic 1. 黄道 2. 黄道的 3. 月(日)食的
ecliptic armillary sphere 黄道经纬仪
ecliptic coordinates 黄道坐标
ecliptic coordinate system 黄道坐标系
ecliptic diagram 黄道图
ecliptic limits 食限
ecliptic latitude 黄纬
ecliptic longitude 黄经
ecliptic obliquity 黄赤交角,黄道斜度
ecliptic of date 瞬时黄道
ecliptic orbit 黄道轨道
ecliptic plane 黄道面
ecliptic pole 黄极
ecliptic reference frame 黄道坐标系
ecliptic system of coordinate 黄道坐标系
ecltic time 地影时间
ECLSS Environmental Control and Life-Support Subsystems 环境控制与生命保障子系统
ECM 1. Electric Coding Machine 电动编码机 2. Electron Cyclotron Maser 电子回旋微波激射器 3. Electronic Counter-Measures 电子对抗,电子干扰,电子干扰措施 4. Electronic Countermeasures Mission 电子对抗任务 5. Electronic Configuration Management 电子构型管理 6. Electronic Control Module 电子控制模块 7. Electrochemical Machining 电化学加工,电解加工 8. Emergency Conservation Measures 应急保存措施 9. Engine Condition Monitoring 发动机状态监视(监控) 10. Engine Control Module 发动机(电子)控制组件 11. Engineering Change Management 工程更改管理 12. Engineering Change Memo 工程更改备忘录 13. Engineering Coordination Memo 工程协调备忘录,工程协调章程 14. Equipment Change Management 设备变更管理 15. Error Correction Mode 纠错模式
ECM aircraft 电子对抗飞机
ECM pod 电子战吊舱
ECMA 1. European Computer Manufacturing Association 欧洲计算机制造协会 2. Extended Constant Modulus Algorithm 扩展恒模算法,扩展常数模算法,扩展常模算法
ECME 1. Electronic Checkout Maintenance Equipment 电子检测维修设备 2. Electronic Countermeasures Equipment 电子对抗设备 3. Electronic Countermeasures Environment 电子对抗环境
ECMEA European Conference of Meteorological Expert for Aeronautics 欧洲航空气象专家会议
ECMJ Electronic Counter Measures Jammer 电子对抗干扰机
ECMO 1. ECM Officer 电子对抗军官 2. Electronic Countermeasure Observer 电子对抗观察员
ECMR Engineering Configuration Management Requirements 工程构型管理要求
ECMS 1. ECM Simulator 电子对抗模拟器 2. Electrical Contactor Management System 电路接触器管理系统 3. Electronic Component Management System 电子部件管理系统 4. Engine Configuration Management System 发动机构造管理系统
ECN 1. Electronic Communication Network 电子通信网 2. Emergency Communications Network 应急通信网络 3. Engineering Change Note 工程更改通知单,工程变更说明 4. Explicit Congestion Notification 显性拥塞通知
ECO 1. Electron-Coupled Oscillator 电子耦合振荡器 2. Emergency Changeover Order 紧急转换命令 3. Engine Checkout System 发动机检测系统 4. Engine Cut Off 发动机停车 5. Engineering Change Order 工程技术更改令
ecological equipment 生态学设备
ecological sign 生态标志
ecological sign stability approach 生态标志稳定性方法
ecological sign-stable matrix 生态标志稳定矩阵
ECOM 1. Electronics Command 电子装备司令部 2. Electronic Control Unit Communications Module 电子控制单元的通信模块
ECOMS 1. Early Capability Orbital Manned Station 初期试验载人轨道站 2. Jeppesen Explanation Of Common Minimum Specification 杰普逊共用最低标准的解释
ECON Economy 经济
economet 镍铬铁合金,埃康诺梅特耐热耐蚀铁镍铬铸造合金

economic cruising rating 经济巡航状态
economic life 经济年限,经济寿命,符合经济效益的使用年限
economic speed 经济航速
economic(al)airspeed 经济空速,经济航速
economizer (燃料系统的)节油器,节能器,节热器,废气预热器
economizer valve 节油器阀门
economy 经济,节省,节约措施,经济(性),组织,系统
Ecosat ecological satellite Ecosat 环境生态监测卫星(意大利阿莱尼亚喷气推进(Alenia Spazo)公司提出研制一颗搭载有大型 X 波段合成孔径雷达和光学敏感器的平台型卫星)
ECP 1. ECAM Control Panel 飞机综合电子监控控制板 2. Effective Candlepower 有效烛光 3. EICAS Control Panel 发动机指示和机组告警系统控制板 4. Electronic Circuit Protector 电子电路保护装置 5. Electrical Contact Plate 电接触板 6. Engine Control Program 发动机控制大纲 7. Engineering Change Priorities 工程更改优先次序 8. Engineering Change Proposal 工程修改建议 9. Engineering Cost Proposal 工程成本建议书 10. Error-Correction Parsing 纠错剖析,误差校正剖析
ECPLC Engineering Computer Program Library Control 工程计算机程序库控制
ECPNL 1. Effective Continuous Perceived Noise Level 有效连续感觉噪声级 2. Equivalent Continuous Perceived Noise Level 等值连续探测噪声级
ECPT Engineering Change Proposal Types 工程更改建议类型
ECR 1. Afterburner Exhaust Collector Regulator 加力燃烧室排气收集器调节器 2. Electrical Contact Resistance 电接触阻力 3. Electron Cyclotron Resonance 电子回旋共振(推进技术) 4. Electrochemical Reduction 电解还原 5. Electronic Combat and Reconnaissance 电子对抗与电子侦察 6. Embedded Computer Resources 嵌入式计算(机)资源 7. Engine Condition Review 发动机状况复查 8. Engineering Change Request 工程更改需求 9. Excessive Closure Rate (to terrain) 飞机起飞后以过大的速度接近地面 10. Explict Cell Rate 显性信元率
ECRH 1. Eddy Current Rotating Headstock 涡流旋转主轴箱 2. Electron Cyclotron Resonance Heating 电子回旋共振加热
ECS 1. Electronic Cooling System 电子(设备)冷却系统 2. Electronic Chronometer System 电子计时系统 3. Electronic Countermeasures System 电子对抗系统 4. Electronic Countermeasure Subsystem 电子干扰子系统 5. Energy Conversion Subsystem 能量转化分系统 6. European Communications Satellite 欧洲通信卫星 7. Engine Control System 发动机控制系统 8. Engine/Environmental Control System 发动机/环境控制系统 9. Engineering Compiler System 工程编译系统 10. Engineering Configuration Statements 工程构型说明 11. Environmental Control System 环境控制系统
ECSC European Communication Satellite Committee 欧洲通信卫星委员会
ECSO European Communication Satellite Organization 欧洲通信卫星组织
ECSS European Communication Satellite System 欧洲通信卫星系统
ECSVR Engine-Caused Shop Visit Rate 发动机故障返修率
ECT 1. Echo Cancellation Technique 回波消除技术 2. Engine Cutoff Timer 发动机停车计时器
ECTS Electric Circuit Test Set 电路测试装置
ECU 1. EICAS Control Unit 发动机指示和机组告警系统控制装置 2. Electronic Control Unit 电子控制装置 3. Engine Control Unit 发动机控制装置 4. External Compensation Unit 外部补偿装置
ED 1. Edge Distance 边距 2. EICAS Display 发动机指示和机组告警系统显示器 3. Electrostatic Discharger 静电放电器 4. Engineering Dataset 工程数据集 5. Engineering Design 工程设计 6. Engineering Document 工程文件 7. Engineering Division 工程部门 8. Environmental Damage 环境损伤
EDA 1. Electronic Design Automation 电子设计自动化 2. Equipment Disposition Authorization 设备布置图批准
EDC 1. Engineering Data Control 工程数据控制,工程资料控制 2. Error Detection and Correction 差错探测与纠正
EDD Engineering Dataset Derivatives 工程数据集衍生
EDDS Electroic Document Distribution Service 电子文件分配服务
EDES Equipment Definition Evolution Sheet 设备定义进展单
EDGW Edgewise 延边,在边上,从旁边
EDI 1. Electronic Data Interchange 电子数据互换 2. Engine Data Interface 发动机数据界面
EDIF Engine Data Interface Function 发动机数据界面功能
EDIU Engine Data Interface Unit 发动机数据界面装置
EDLG Economic Design Life Goal 经济设计寿命目标
EDM 1. Engineering Data Management 工程数据管理 2. Ethical Decision Making 做出合乎职业道德的决定

EDMS　Electronic Data Management System 电子数据管理系统

EDO　Engineering Deviation Order 工程偏离指令

EDOC　Effective Date Of Change 更改有效日期

EDP　1. Electronic Data Processing 电子数据处理 2. Engine Display Panel 发动机显示板 3. Engine Driven Pump 发动机驱动泵 4. Engineering Development Pallet 工程发展台

EDPC　Electronic Data Processing Center 电子数据处理中心

EDPR　Engineering Design Process Requirements 工程设计工艺要求

ECU　1. Electrical Control Unit 电气控制组件 2. Electronic Control Unit 电子控制组件 3. Electronic Conversion Unit 电子转换装置 4. Elevator Control Unit 升降舵控制组件 5. Engine Change Unit 发动机更换装置 6. Engine Control Unit 发动机控制组件 7. Environmental Control Unit 环境控制装置 8. Exercise Control Unit 演习指挥组，演练控制分队 9. External Compensation Unit 外部补偿单元

ECW　1. Electronic Combat Wing 电子作战连队 2. Elliptical Cylindrical Waveguide 椭圆形圆柱波导

ECWL　Effective Combat Wing Loading 有效作战机翼负载

ED　1. Edge Device 边缘设备 2. Edge Distance 边缘距离 3. Electro-Dynamics 电动力学 4. Emergency Distress Signal 紧急遇险信号 5. Emergency Distance 紧急距离 6. End of Data 数据结束 7. End of Descent 下降结束 8. Ending Delimeter 结束定界符 9. Engineering Development 工程研制（设计阶段的一部分） 10. Envelop Drawing 包络图 11. Error Detection 错误检测，误差检测 12. Expanded Display 扩充显示器 13. Expedited Data 加速数据 14. Explosive Device 爆炸装置

EDASS　Environmental Data Acquisition Subsystem 环境数据采集子系统

EDAU　Engine Data-Acquisition Unit 发动机数据采集装置

EDB　1. Engineering Database/Engineering Data Bank 工程数据库 2. Equipment Data Base 设备数据库

EDBMS　Engineering Data Base Management System 工程数据库管理系统

EDC　1. Emergency Descent Configuration 应急降落形态 2. Engineering Data Control 工程数据控制 3. Engineering Document Control 工程文件管理 4. Error Detection Circuit 差错检测电路 5. Error Detection Code 差错检测码 6. European Defense Community 欧洲防御共同体

EDCAS　1. Equipment Designers' Cost Analysis System 装备设计成本分析系统 2. Equivalent Designers' Cost Analysis System 等效设计成本分析系统

EDCC　1. Electronic Data Communications Centre 电子数据通信中心 2. Environmental Detection Control Center 环境监测控制中心 3. European Dependable Computing Conference 欧洲可信计算会议

EDCT　Estimated Departure Clearance Time 预计离港起飞放行时间

EDD　1. Electronic Data Display 电子数据显示 2. Envelop Delay Distortion 包线延迟失真

Eddington limit　埃丁顿极限

EDDS　1. Early Docking Demonstration System 早期对接演练系统 2. Electronic Data Display System 电子数据显示系统

eddy　涡动，涡旋，涡流，涡度，涡团

eddy current　涡流，涡电流

eddy (current) losses　涡流损耗，涡电流损耗

eddy current testing　涡流检测，涡流探伤

eddy current thickness meter　（电）涡流测厚仪，电涡流厚度计

eddy current torquer　涡流式力矩器，润流修正机，涡流转矩装置

eddy diffusivity　涡流扩散度，紊流扩散率，涡流扩散度，涡流扩散率，涡团扩散率

eddy dissipation　涡流耗散，涡流消散

eddy impaction　涡流碰撞

eddy Mach wave radiation　涡流马赫波辐射

eddy simulation　涡流模拟

eddy viscosity　涡流黏度，湍流黏滞性

eddy-current detector　（电）涡流探伤仪

eddy-sonic inspection　涡流超声检测

EDG　1. Electrical Density Gauge 电子密度计 2. Electrical-Discharge Grinding 放电研磨，放电抛光 3. Electronically Drive Gyro 电动陀螺仪 4. EOS (Earth Observing System) Data Gateway (NASA satellite database) EOS（地球观测系统）数据网关（NASA卫星数据库）

edge　缘，边缘，边界，棱，端

edge alignment　（螺旋桨的）边缘对位距离（指桨叶的某一剖面中心线到前缘的距离），桨叶缘配准，边对齐

edge aperture　边缘孔径

edge board contacts　板边插头

edge channel　边（缘）通道

edge controller　翼控制器

edge defined film-fed growth　限边馈膜生长

edge detection　边缘检测

edge device　边界设备，边缘设备

edge diffraction　边缘绕射，边缘衍射

edge dislocation　刃位错，棱位错，刃型位错，（晶体）边

缘位错,刃差排

edge displacement 边缘位移,边缘移位
edge effect 边沿效应,边缘效应
edge enhancement 边缘增强,轮廓增强
edge etching 腐蚀边
edge flare (图像)边缘闪光,(能源科学)卷边对接
edge flame (发动机)边缘燃烧
edge flap 襟翼(如 leading edge flap 前缘襟翼, trailing edge flap 后缘襟翼)
edge fog 图像边缘模糊
edge gain 边缘增益
edge graph 边图
edge influence coefficient 边缘影响系数
edge length 棱长,边缘长度,边长
edge limiter 边缘限制器
edge node 边缘节点
edge of rotor (直升机的)旋翼边缘
edge of stator (发动机)定子边缘
edge of the format 图廓
edge plasma 边缘等离子体
edge protection 边缘加固叶缘的保持
edge rate 边缘速率,缘速率,边际率
edge rounding 边缘钝化,边缘圆化
edge sharpness 边缘锐度,轮廓清晰度
edge thickness 边缘厚度
edge tone 边棱音
edge tracker 边缘跟踪器,缘追踪器
edge tracking 边缘跟踪,边界跟踪,轮廓提取
edge-on object 侧向天体
edge-socket connector 边缘插座连接器
edge-type elevator (位于航母舷侧)甲板缘升降机
edging 磨边,使……锋(锐)利,嵌入,逐渐移进
EDI 1. Edge-Directed Interpolation 边缘定向插值 2. Electron Diffraction Instrument 电子衍射仪 3. Electron Drift Instrument 电子漂移仪器 4. Electronic Data Interchange 电子数据交换,电子资料交换 5. Electronic Data Interface 电子数据接口 6. Electronic Document Imaging 电子文档影像 7. European Defence Initiative 欧洲防御倡议
EDIP European Defence Improvement Programme 欧洲防务提升计划(北约)
EDIPS EROS (Extremely Reliable Operating System) Digital Image Processing System 高性能安全操作系统数字成像处理系统
editor 编辑,编辑程序,编辑器
EDL 1. Electric-Discharge Laser 光电放电激光器 2. Electronic Defense Laboratory 电子防务实验室 3. Engage/Disengage Logic 接通断开逻辑电路 4. Equipment Deficiency Logs 设备缺陷记录 5. Ethernet Data Link 以太网数据链 6. Executive Data Link 执行数据链
EDM 1. Electrical-Discharge Machining 电火花加工 2. Electromagnetic Distance Measuring 电磁测距 3. Electronic Data Management 电子数据管理 4. Engine Data Multiplexing 发动机数据多路传输 5. Engineering Development Mission 工程研制任务
EDMA Engineering Drawing and Model Application 工程图和模型应用库
EDMI Electromagnetic-Wave Distance Measuring Instrument 电磁波测距仪
EDMS 1. Engineering Data Management System 工程数据管理系统 2. Engineering Documentation Management Services 工程文档管理服务 3. Experimental Data Management System 实验数据管理系统
EDO 1. Enhanced Data Output 增强数据输出 2. Extended Duration Orbiter 延长运营时间轨道器
EDP 1. Electric Diffusing Process 电渗处理 2. Electronic Data Processing 电子数据处理 3. Embedded Data Processor 嵌入式数据处理机 4. Engine-Driven Pump 发动机驱动的泵,发动机主液压泵 5. Experimental Data-Processor 实验数据处理机
EDPS Electronic Data Processing System 电子数据处理系统
EDR 1. Electronic Decoy Rocket 电子诱惑火箭 2. Engine Data Retrieval 发动机数据检索 3. Engineering Data Release 工程数据发放 4. Engineering Design Rules 工程数据规则 5. Engineering Drawing Release 工程图发放 6. Environmental Data Record 环境数据记录 7. Environmental Damage Rating 环境损伤等级 8. Estimated Date of Return 估计返回日期 9. Equipment Defect Report 设备缺陷报告 10. Expected Departure Release 预期允许分离,预期放行 11. Experiment Data Record 实验数据记录
EDRS 1. Engineering Data Retrieval System 工程数据检索系统 2. European Data Relay Satellite 欧洲数据中继卫星
EDRTS Experimental Data Relay and Tracking Satellite 实验型数据中继和跟踪卫星
EDS 1. Electrical Distribution Subsystem 配电分系统 2. Emergency Detection System 应急故障探测系统 3. Enamel Double Silk 双丝漆包线 4. Engine Data Submittal 发动机数据提交 5. Engine Diagnostic System 发动机故障诊断系统 6. Environmental Data Service (美国国家海洋和大气局的)环境数据资料服务处 7. European Distribution System (美国空军的)欧洲分配系统 8. Explosives Detection System 爆炸物

检测/探查系统

EDSD Electrostatic Display Sensitive Devices 静电放电敏感元件

EDT Eastern Daylight Time 东部夏令时（美国）

EDU Electronic Display Unit 电子显示装置

EDUR Engineering Drawing Usage Record 工程图使用记录

EDW Earth Departure Window 脱离地球窗口

EDWA Erbium Doped Waveguide Amplifier 掺铒波导放大器

EE 1. Earth Entry 进入地球大气层 2. Electrical/Electronic 电气/电子 3. Electrical and Electronic 电气和电子 4. Electrode Effect 电极效应 5. Electrodynamic Explorer 电动力学探测卫星 6. Electronic Equipment 电子设备 7. Electronics Equipment（Bay）电子设备（舱）8. Errors Excepted 允许误差 9. Expected Errors 预期误差 10. External Environment 外部环境

EEA Electronic Engineering Association 电子工程协会（英国）

EEC 1. Earth Entry Capsule 进入地球大气层的舱 2. Electric Echo Canceller 电回波消除器 3. Electronic Equipment Compartment 电子设备舱 4. Emergency Electrical Configuration 应急供电构型 5. Engine Electronic（Electronic Engine）Control 发动机电子控制，电子式发动机控制 6. Extendable Exit Cone 可延伸（喷管）出口锥

EECS Electrical/Electronics Cooling System 电气/电子冷却系统

EED 1. Electrical-Electronic Drawings 电气电子图 2. Electro-Explosive Device 电起爆装置 3. Electronic Explosive Device 电子爆炸装置 4. Emergency Escape Device 应急逃逸装置 5. Empirical Eigenvalue Distribution 经验特征值分布 6. Energy Efficient Design 能效设计，节能设计 7. Equipment Engineering Department 设备工程部

EEDR Engineering Electronic Data Retention 工程电子数据保存

EEE 1. Earth and Environmental Engineering 地球与环境工程 2. Electrical and Electronics Engineering 电气与电子工程 3. Electrical Electronic and Electromechanical 电气电子和机电 4. Electromagnetic Environmental Effects 电磁环境效应 5. Energy-Efficient Engine 节能发动机 6. Energy Efficient Ethernet 高效节能以太网

EEE parts 电气和电子机电零件

EEG Enterprise Encryption Gateway 企业级加密网关

EEI 1. Electrical Engineering Instruction 电气工程指南 2. Engineering Estimating Initiative 工程估算倡议 3. Essential Elements of Information 情报要素，信息要素

EEL 1. Emergency Exposure Limit 应急暴露极限 2. Electron Energy Loss 电子能量损失，电子能耗 3. External Elastic Lamina 外弹性膜

EEMAC 1. Electrical & Electronic Manufacturers Association of Canada 加拿大电气和电子制造商协会 2. Electrical Equipment Manufacturers Association of Canada 加拿大电气设备制造商协会

EEP 1. Earth Equatorial Plane 地球赤道（平）面 2. Elliptical Error Probable 椭圆误差概率 3. ETOPS（Extended-Range Operations）Entry Point 延程运行进入点 4. Experimental Electronics Package 实验电子设备盒

EEPGS Enhanced Electrical Power Generation System 增强型发电系统

EEPPI Electrical/Electronic Preferred Process Implementation 电气/电子优先工序执行

EEPROM Electrically Erasable Programmable Read Only Memory 电擦除可编程序只读存储器

EEROM Electrically Erasable Read Only Memory 电可擦只读存储器

EES 1. End-to-End Signaling 端到端信令 2. Ejection Escape Suit 弹射救生衣 3. Electro-Explosive Subsystem 电爆分系统 4. Electromagnetic Emission and Susceptibility 电磁发射和敏感度 5. Emergency Ejection Suits 应急弹射服 6. Emergency Evacuation System 应急撤离系统，应急疏散系统 7. Emergency Exhaust System 应急排放系统，紧急排气设备

EESS 1. Earth Exploration Satellite Service 地球勘探卫星服务 2. Earth Environmental Sensor Systems 地球环境传感器系统网络 3. Environmental Effect on Space System 环境对空间系统的影响

Eestra "埃斯特拉"（瑞典 I/J 波段搜索和火控雷达）

EET 1. Estimated Elapsed Time 预计经过时间，预计已飞时间，估计的航程时间 2. Estimated Enroute Time 预计途中时间

EEV 1. Earth Entry Vehicles 进入地球（大气层）飞行器 2. Electronic Expansion Valve 电子膨胀阀 3. Emergency Escape Vehicle（NASA）紧急逃生飞行器

EEW Equipped Empty Weight 带设备空重，飞行器自重

EF 1. Electrical Fuse 电熔丝 2. Energy Factor 能量因数（因子）3. Exposed Facility 暴露设施（日本 JEM 空间站舱的组成部分）

EFA 1. Engineering Failure Analysis 工程故障分析，工程失效分析 2. Engineering Flight Activity 工程飞行活动 3. European Fighter Aircraft 欧洲战斗飞机（含直

升机)

EFAS 1. Electronic Flash Approach System 电子闪光进近系统 2. Enroute Flight Advisory Service 飞行途中咨询业务

EFATO Engine Failure After Take-Off 起飞后发动机停车

EFC 1. Effective Full Charge 有效全充电 2. Elevator-Feel Computer 升降航操纵感力计算机,升降舵感觉计算机 3. Equivalent Full Charge 全装药当量 4. European Federation of Corrossion 欧洲腐蚀联合会

EFCS Electrical Flight Control System 电飞行控制系统,电传式飞行操纵系统

EFD Electronic Flight Display 电子飞行显示器

EFDAS Electronic Flight Data System 电子飞行数据系统

EFDC 1. Early-Failure Detection Centre 早期故障探测中心 2. Environmental Fluid Dynamics Code 环境流体动力学模型

EFDR Expanded Flight Data Recorder 扩展的飞行数据记录器

EFDS Engine Fault Display System 发动机故障显示系统

EFDU Engine Fault Display Unit 发动机故障显示装置

EFEO European Flight Engineers Organization 欧洲飞行工程师组织

EFF Efficiency 效率,有效系数

effect 效应,效果,影响,作用,结果,实行,实施,完成,产生

effect of acoustic oscillation 声波效应

effect of aerodynamic lift 气动升力效应

effect of chamber diameter 发动机燃烧室直径(的影响),发动机燃烧室直径效应

effect of compressibility 压缩(性)效应

effect of end-wall motion 端壁运动效应

effect of erosive burning 侵蚀燃烧效应

effect of evolution 演化效应,演进效应

effect of forcing-function 强制函数效应,外力函数效应

effect of freestream turbulence 自由流湍流效应,未扰动流湍流效应

effect of injection modeling 注射成型法(效应)

effect of injection pressure 喷注压力效应,喷射压力效应

effect of large deflection 大变形效应,大扰度效应

effect of liquid viscosity 液(体)黏(性)效应

effect of nuclear electromagnetic pulse 核电磁脉冲效应

effect of oscillation amplitude 振(荡)振幅效应

effect of outboard camber 外(侧)弧效应,外曲面效应

effect of preferential diffusion 优选(先)扩散效应

effect of projectile acceleration 抛射体加速效应

effect of surface roughness 表面粗糙度效应,表面光洁度效应

effect of undersampling 欠(抽)采样效应,采样过疏效应

effective ablation heat 有效烧蚀热

effective acceleration of gravity 有效重力加速度

effective activation energy 有效活化能

effective address 有效地址

effective ampere 有效安培(数)

effective angle of attack 有效迎角,有效攻(击)角(度)

effective aperture 有效孔径

effective aspect ratio 有效展弦比

effective casualty radius 有效杀伤半径

effective charge 有效装药,有效电荷

effective contraction ratio 有效收缩比,有效锥度比

effective current 有效电流

effective displacement(capacity) (航天流体系统的)有效排量

effective dose equivalent 有效剂量当量(放射性的)

effective efficiency 有效效率,有效功能

effective emittance of multilayer insulation 多层隔热层有效发射率

effective equivalence ratio 有效当量比,有效等值比

effective exhaust velocity 有效排气速度

effective fragment 有效破片

effective gravitational coefficient 有效重力系数

effective gun bore line 有效炮孔线

effective heat of ablation 有效烧蚀热

effective helix angle 有效螺旋线角

effective horsepower 有效功率,有效马力

effective isotropic radiated power 等效全向辐射功率

effective kill radius 有效杀伤半径

effective launcher line 有效发射线

effective magnetic induction 有效磁感应

effective mass 有效质量

effective measurement noise 有效测量噪声

effective navigation gain 有效导航增益

effective navigation ratio 有效导航率(比)

effective peak velocity 有效峰值速度

effective perceived noise level 有效可感噪声水平

effective permeability 有效磁导率

effective physiological back angle 有效生理背角

effective pitch 有效桨矩

effective pitch ratio (螺旋桨的)有效相对桨距(即桨距与直径之比)

effective profile drag 有效翼型阻力,有效型阻

effective propeller thrust 螺旋桨有效拉力(推力)

effective proximity 有效杀伤半径

effective radiated power 有效辐射功率

effective range 有效射程
effective reflection heat 有效反射热
effective span 有效翼展(左右翼尖弦之间的水平距离)
effective specific impulse 有效比冲(比冲与有效装药量之比)
effective temperature 实感温度,有效温度
effective terrestrial radiation 有效地面辐射
effective thermal conductivity 有效热传导
effective thrust 有效推力,净推力
effective torque 有效扭矩
effective true airspeed (等压面领航的)有效真空速,假定真空速
effective value 有效值
effective velocity ratio 有效相对速度
effective warhead radius 弹头有效毁伤半径
effective wavelength 有效波长
effective width 有效宽度
effectiveness 效能(特指一种装备满足给定的定量特性和使用要求的能力,通常是设计达到的可用性、可信性和性能的函数),效用,效益,效果
effectiveness distribution 效能分布
effectiveness of boundary-layer 附面层效能(有效性),边界层效能(有效性)
effectiveness of thermal protection structure 结构防热设计效果
effectiveness of white-light 白光(效能/有效性)
effectiveness theory 效益理论
effectivity 技术适应性,有效度,效率
effector 效应器,导弹操纵装置,火箭操纵装置,飞行器操纵装置
efficiency 效率,效能,功效,效果,效力,实力,有效性
efficiency factor 效率系数,效率因子
efficiency of charge weight 装药利用系数
efficiency of solar array 太阳电池阵效率
efficiency of specific impulse 比冲力效率,比冲量效率
efficient polygon overlay algorithm 高效率多边形叠置算法
EFFO Efficiency Overall 总效率
EFFTAS Effective True Airspeed 有效真空速
effusive flow 分子泻流(一种气体流态)
EFG 1. Earth Fixed Geocentric 地固地心(的) 2. Electric Field Gradient 电场梯度
EFH 1. Earth Far Horizon 远地平线 2. Engine Flight Hours 发动机空中工作小时数 3. Equivalent Flight Hours 当量飞行小时
EFI Electronic Flight Instrument 电子飞行仪表
EFIC Electronic Flight Instrument Controller 电子飞行仪表控制器
EFIP Electronic Flight Instrument Processor 电子飞行仪表处理器
EFIS 1. Electronic Flight Instrument System 电子飞行仪表系统 2. Electronic Flight Information System 电子飞行信息系统 3. Electronic Fuel Injection Simulator 电子燃油喷射模拟器
EFL 1. Emitter Follower Logic 射极跟随器逻辑 2. Entry Flight Level 进入大气层飞行高度 3. Error Frequency Limit 错误频率极限
EFM 1. Electromagnetic Frequency Management 电磁频谱管理 2. Electrostatic Force Microscopy 静电力显微镜 3. Enhanced Fighter Manoeuvrability 歼击机强化机动性(如使用推力矢量控制及滚转耦合机身瞄准工作方式),高机动性战斗机(指X-31A战斗机)
EFP Enroute Fuel Planning 航路燃油计划
EFPD Effective Full Power Days 有效满功率天数
EFPH Effective Full Power Hours 有效满功率小时
EFR 1. Early Failure Rate 早期故障率 2. Engine Firing Rate 发动机爆燃速度 3. Electronic Failure Report 电子故障报告 4. Extended Flight Rules 扩展飞行规则
EFSSS Engine Failure Sensing and Shutdown System 发动机故障检测与停车系统
EFT 1. Earliest Finish Time 最早完成时间 2. Early Field Trial 早期野外试验,早期现场试验 3. Electronic Funds Transfer 电子转款 4. Elementary Flying Training 初级飞行训练 5. Engine Fuel Tank 发动机燃料箱 6. Engineering Feasibility Test 工程可行性试验 7. Engineering Flight Test 工程飞行试验 8. Estimated Flight Time 预计飞行时间 9. External Fuel Tank 外部燃料箱,外部油箱
EFTS 1. Elementary Flying Training School 初级飞行训练学校(英国) 2. Enhanced Flight Termination System 增强飞行终止系统
EGA 1. Earth Gravity Assist 地球重力借力飞行,地球重力助推 2. Exhaust Gas Analyser 排气(成分)分析器
EGADSB Electronic Ground Automatic Destruct Sequencer Button 地面自毁自动电子程序装置按钮
Egeria 【天】芙女星(小行星13号)
egg 〈口语〉航空炸弹,深水炸弹,水雷,地雷,手榴弹
eggbeater "打蛋机"(对带有两个交错旋翼的直升机的俗称)
E-glass E型玻璃(专门用作纤维增强的标准高强度玻璃,主要用于先进工程结构的复合材料,不用于商用石墨增强材料)
EGO 1. Eccentric Orbiting Geophysical Observatory 偏心轨道地球物理观测台 2. Equator Geophysical Observatory 赤道地球物理观测台 3. European Gravitational Observatory 欧洲引力波天文台
egoless programming 非自我程序设计
EGP 1. Exhaust Gas Pressure 排气压力 2. Ex-

perimental Geodetic Payload 实验测地卫星（日本）

EGPWS　1. Enhanced Ground Proximity Warning System 增强近地警告系统　2. Enhanced Ground Proximity Warning System 增强型近地警告系统

egress　出口，出路，外出，终切，外出，出舱，【天】终切，凌日

egress handle　出口座椅弹射把手，应急离机把手

egress maneuver　出舱行动

egress trainer　出舱练习器，离舱训练器

EGRET　Energetic Gamma Ray Experiment Telescope 高能伽玛射线实验望远镜（美国）

Egrett　埃格雷特（美国和北约某些国家合作研制中的电子战飞机）

EGRS　1. Electronic Geodetic Ranging System 电子测地测距系统　2. Experimental Geodesic Research Satellite 实验测地研究卫星

EGS　1. Electronic Guidance System 电子导航系统　2. Elementary Gliding School 初级滑翔学校　3. Experimental Geodetic Satellite 实验测地卫星（日本）　4. European Geophysical Society 欧洲地球物理学会

EGSE　1. Electrical Ground Support Equipment 电气地面保障设备　2. Electronic Ground Support Equipment 电子地面保障设备

EGT　1. Engine Gas Temperature 发动机燃气温度　2. Eye-Gaze Tracking 视线跟踪　3. Estimated Ground Time 预计地面停留时间　4. Exhaust-Gas Temperature 排气温度

EGTP　External Ground Test Program 外部地面试验大纲

egy　埃及（EGY）

Egyptian Goose　"埃及鹅"（美国机载 I/J 波段脉冲多普勒雷达）

Eh　总能量（如规定飞行高度及速度下的总能量）

EH　1. Eastern Hemisphere 东半球　2. Electric Hydraulic 电动-液压（的）　3. Extrusion Honing 挤压珩磨

E-H tuner　E-H 调配器

EHA　1. Electro Hydrostatic Actuator 电动液压传动装置，电液作动筒　2. European Helicopter Association 欧洲直升机协会

EHD　Electro Hydro Dynamics 电动-液压-动力（的）

EHD pump　电液动力泵

ehd thruster　电液动力推进器，电液动力发动机

EHDD　1. Electronic Head-Down Display 电子下视显示仪　2. Enhanced Head-Down Display 增强型下视显示仪

EHF　1. Extremely High Frequency 极高频　2. Extra High Frequency 特高频

EHL　Environmental Health Laboratory 环境安全实验室（美国空军）

EHM　1. Engine Health Monitor(ing) 发动机状态监控器，发动机完好性监控，发动机健康监视（控）　2. Engine Heavy Maintenance 发动机大修　3. Equipment Hydraulic Module 设备液压模块　4. Extreme High Modulus 特高模量

EHOC　European Helicopter Operators' Committee 欧洲直升机承运人委员会

EHP　1. Electrtical Horsepower 电功率　2. Equivalent Horsepower 当量马力

EHPL　Electro-Hydraulic Primary Lock 电子液压主锁

EHR　Engine History Recorder 发动机历程记录仪

EHSD　Electronic Horizontal Situation Display 电子式水平位置显示器，电子式水平情况显示器

EHSI　1. Electronic Horizontal Situation Indicator 电子水平位置指示器，电子水平姿态指引仪　2. Elevator Horizontal Situation Indicator 升降舵水平位置指示器

EHT　1. Electrothermal Hydrazine Thruster 电热肼推力器　2. Extra High Tension 超高压

EHTV　European Hypersonic Transport Vehicle 欧洲超声速运输飞行器

EHV　1. Electro-Hydraulic Valve 电液阀门　2. Extreme High Vacuum 极高真空　3. Extremely High Voltage 极高电压

EI　1. Earth Interface 地球（大气）界面　2. Engineering Index 工程索引　3. Electrical Integration 电子集成　4. Electronic Interface 电子接口　5. Engine Instruments 发动机仪表　6. Entry Interface 入口接口　7. Enviornmental Interface 环境界面　8. External Inspection 外部检查

EIA　1. Electronic Industries Alliance 电子工业联盟（美国）　2. Electronic Industries Association 电子工业协会（美国）　3. Environmental Impact Assessment 环境影响评估　4. Extended Interaction Amplifier 分布作用放大器

EICAS　1. Engine Indication and Crew Alerting System 发动机（参数）指示与机组告警系统　2. Engine Instrument and Crew Alerting System 发动机仪表与机组告警系统

EICD　Electrical Interface Control Document 电气界面控制文件

EICMS　Engine In-flight Condition Monitoring System 发动机空中状态监控系统

Eicon connector　埃尔康连接器（用于大电流输出的蓄电瓶与飞机电网系统之间的电联结）

EICS　1. Engine Integrated Control System 发动机综合控制系统　2. Equipment Identification Coded System 设备识别编码系统

EID 1. Electrically Initiated Device 电起爆装置,电子启动装置 2. Electron Impact Desorption 电子碰撞脱附 3. Electron-Induced Desorption 电子诱导解吸

EIDI Electro-Impulse-De-Icing 电脉冲除冰

EIF 1. ECCM Improvement Factor 反电子干扰改善因子 2. Equivalent Initial Flaw 当量初始缺陷

EIFS Equivalent Initial Flaw Size 当量初始缺陷尺寸

eigen analysis 特征（本征）分析（如 Standard Eigenanalysis 标准本征分析）

eigen axis maneuver 拟欧拉轴姿态机动

eigen axis slew 拟欧拉轴旋转

eigen direction 特征方向

eigen frequency 特征频率,本征频率,特有频率

eigen function 本征函数

eigen mode 本征模

eigen mode analysis 本征模分析

eigen pair 本征对,特征对

eigen problem 本征问题

eigen structure 本征结构

eigen structure assignment 本征结构配置,特征结构配置

eigen system 特征系统,本征系统

eigen system assignment 特征系统赋值,本征系统赋值

eigen system realization 特征系统实现,本征系统实现

eigen value 本征值,特征值（如 Aeroelastic Eigenvalue 气动弹性本征值）

eigen value distribution 本征值分布,特征值分布

eigen value extraction 本征值提取,特征值提取

eigen value locus 本征值位点（轨迹）

eigen value perturbation 本征值扰动（摄动）

eigen value perturbation model 本征值扰动模型

eigen value problem 本征值问题

eigen vector 本征矢（向）量,特征向量

eigen vector matrix 本征向量矩阵,特征向量矩阵

eigen vector sensitivity 本征向量灵敏度,特征向量灵敏度

eigen-vector 特征向量,本征向量

eight point roll 八点间歇横滚（每滚转45°作短暂停顿）

eight pylon 八杆机动飞行（飞机绕八根标杆飞行,每次翼尖似乎是绕一根标杆旋转,用于飞行竞赛）

eighth-order 第八阶

eikonal equation 程函方程（也有称短时距方程或几何光学方程）

EIM Engineering Information Management 工程信息管理

EIMS 1. End Item Maintenance Sheet 最终项目维修分析单 2. Energetic Ion Mass Spectrometer 高能离子质谱仪

Einstein effect 爱因斯坦效应

Einstein shift 爱因斯坦位移,谱线引力红移

Einstein tensor 爱因斯坦张量

Einstein universe 爱因斯坦宇宙

Einstein X-ray Observatory 爱因斯坦 X 射线天文台

Einstein's model 爱因斯坦模型

einzel lens 单透镜

EIO Extended Interaction Oscillator 分布作用振荡器（管）

EIP 1. Enterprise Information Portal 企业信息门户 2. Equipment Interface Panel 设备接口面板

EIPC Engine Illustrated Parts Catalog 发动机图解零件目录

EIPM Engine Interface Power Management 发动机接口电源管理

EIPMU Engine Interface Power Management Unit 发动机接口电源管理组件

EIRP Equivalent Isotropically (Effective Isotropic) Radiate Power 等效全向辐射功率,有效各向同性辐射功率（决定卫星信号强度的主要因素）

EIS 1. Ejection Initiation Subsystem 弹射起动分系统 2. Electronic Instrument System 电子仪表系统 3. End Item Specification 最终项目技术规格 4. Engine Indication System 发动机指示系统 5. Entry Into Service 投入使用,列装（列入装备） 6. Environmental Impact Statement 环境影响（指污染）报告 7. Executive Information System 执行信息系统

EISCAT European Incoherent Scatter Scientific Association 欧洲非相干散射学学会

EISF 1. Elastic Incoherent Structure Factor 弹性非相干结构因子 2. Engine Initial Spares Factor 发动机原始零备件因素

EISW Equivalent Isolated Single-Wheel Load 当量孤立单轮载荷

EIU EFIS/EICAS Interface Unit EFIS/EICAS 界面装置

EJ 1. Electronic Jamming 电子干扰 2. Expendable Jammer 投掷式干扰机

eject 喷射,弹射,抛壳,抽壳,退弹,吹洗炮膛

eject escape 弹射逃逸

eject momentum 喷射动量

eject motor 弹射发动机,抛置发动机

ejectability 可弹射性,弹射能力

ejectable cabin 弹射舱

ejectable cockpit 弹射座舱,分离座舱

ejectable radiosonde 弹射式无线电探空仪,弹射雷达

ejectable recorder unit 弹射式记录器

ejected case chute 抛壳道,抛壳套,退壳槽

ejected link chute 抛链道,抛链套

ejected nosecap 弹射式端头帽,弹射式端头罩

ejecting recovery 弹射回收
ejection 喷射,引射,排射,弹射,抛出,抛壳,抽筒,退弹,炮膛吹洗
ejection acceleration 弹射加速度
ejection angle （座椅）弹射角
ejection bailout 弹射跳伞
ejection capsule 弹射舱,弹射回收记录仪器盒
ejection chute 弹射（降落）伞
ejection condition 弹射条件
ejection dummy 弹射假人
ejection escape 弹射救生,弹射逃逸
ejection escape device 弹射救生装置
ejection fatality 弹射死亡事故
ejection height 弹射高度,喷射高度
ejection injuries 弹射损伤
ejection launcher （导弹）弹射发射装置
ejection mechanism 弹射机构
ejection port（slot） （枪、炮的）排壳槽,抛壳口
ejection power plant 弹射动力装置
ejection ramjet 冲压式火箭发动机
ejection seat 弹射座椅
ejection seat trainer 座椅弹射训练器
ejection sequence 弹射程序
ejection sequence control unit 弹射程序控制装置
ejection survival 弹射救生
ejection test in flight 飞行中弹射试验
ejection test tower 弹射试验塔（台）
ejection test vehicle 弹射试验机
ejection training 弹射训练
ejection trajectory 弹射轨迹
ejection vector 弹射速度矢量
ejection velocity 弹射速度
ejection with canopy 带盖弹射
ejector 引射器,喷射器,引射泵,顶推器,退弹器,抽筒子,退壳簧,弹射器
ejector augmented lift 引射增升
ejector augmenter 引射增装置
ejector bomb rack 弹射式挂弹架
ejector configuration 喷射器结构,引射器结构
ejector diameter 喷射器直径,引射器直径
ejector diffuser 喷射扩散
ejector duct 引射涵道
ejector entrance 喷射（泵）入口,引射（器）入口
ejector exhaust 引射排气
ejector exit 喷射（泵）出口,引射（器）出口
ejector geometry 引射器结构
ejector inlet 引射喷管入（进）口
ejector length 引射喷管长度
ejector lift 引射升力（喷气发动机高能热气流经喷嘴排管向下引射而获得动力升力）
ejector mode 引射模态
ejector nozzle 引射喷管
ejector performance 喷射器性能
ejector piston 弹射杆
ejector rack 弹射式挂弹架
ejector rack adapter 弹射式挂弹架过渡梁
ejector radius 弹射半径（如 Optimum Ejector Radius 最大弹射半径）
ejector ramjet 冲压火箭复合式发动机
ejector rocket 引射火箭
ejector section 弹射面
ejector system 喷射系统,弹射系统
ejector thrust 喷射推力,喷射推进
ejector vacuum pump 喷射真空泵
ejector wall 引射器壁
ejector-jet 引射喷气
ejector-jet mode 引射喷气模式
EJOTF Earth-Jupiter Orbiter Transfer Flight 地球-木星轨道器转移飞行
EJS 1. Ejection Seat 弹射座椅 2. Enhanced JTIDS 改进的联合战术信息分配系统
EJU Equipment Job Update 设备工作更新
Ekman layer 埃克曼（过滤）层,螺旋层（附面层和自由大气之间的）
ekranoplane 地效飞机
EKW Electrical Kilowatts 千瓦
EL 1. Ejector Lift 引射动力装置产生的升力（垂直推力）2. Elastic Limit 弹性极限 3. Electroluminescent 场致发光的,电致发光的 4. Elevation 上升,举起,提高 7. Engineering Liaision 工程联络 5. Emitter Locator 发射极定位器 6. Empty-Load 空载 8. Error Level 误差高度 9. Equipment List 设备清单 10. Extra-terrestrial Life 地外生命
EL/FCG Electronic Logbook and Fault Correction Guide 电子记录本和故障纠正指南
EL/VT Ejector Lift/Vectored Thrust 引射升力,矢量推力
ELAC Elevator Aileron Computer 升降舵副翼计算机
elapsed-time counter 累计时间计数器
ELAR Electrical Load Analysis Report 电气负载分析报告
Elara 【天】木卫七
ELAS Elastic 有弹性的,有伸缩的,灵活的
elastance 倒电容（电容的倒数）
elastic 弹性的,橡皮绳
elastic axis 弹性轴（悬臂式机翼的展向直线,沿此线的载荷对机翼只产生弯曲而无扭转）,减震轴
elastic axis location 弹性轴位置

elastic centre 弹性中心,刚心
elastic collision 弹性碰撞
elastic compliance constant 弹性顺服常数
elastic coupling 弹性联轴节,弹性联轴器
elastic damping 弹性阻尼
elastic deformation 弹性变形,弹性形变
elastic element 弹性元件
elastic energy 弹性能
elastic energy degradation 弹性能(量)降级
elastic energy density 弹性能量密度
elastic fatigue 弹性疲劳
elastic force 弹力
elastic fuselage 弹性机身
elastic instability 弹性不稳定性(度)
elastic limit 弹性极限
elastic line 线弹性
elastic line model 线弹性模型
elastic mechanical system 弹性机械系统
elastic membrane 弹性膜(片),弹力膜(片)
elastic memory 弹性记忆,形状记忆
elastic mode 弹性模态
elastic model 弹性模型
elastic modulus 弹性模量,弹性模数,杨氏模量,弹性组件,弹性计量单位
elastic precursor wave 弹性前驱波
elastic rebound 弹性回跳,弹性回跃
elastic recovery (力)弹性复元,弹性恢复,弹性回复
elastic restraint coefficient 弹性约束系数
elastic scattering 弹性散射
elastic scattering collision 弹性散射碰撞
elastic scattering cross section 弹性散射截面
elastic scattering resonance 弹性散射共振
elastic scattering submatrix 弹性散射子矩阵
elastic shell 弹性壳(体)
elastic stability 弹性稳定(性)
elastic stiffness constant 弹性刚度常数,弹性劲度常数
elastic stop 弹性限制销,弹性制动爪
elastic stop nut 弹性防松螺母,弹性锁紧螺帽
elastic structure 弹性结构
elastic twist 弹性扭曲
elastic twist distribution 弹性扭曲分布(分配)
elastic vibration 弹性振动
elastic vibration mode 弹性振动模态
elasticity 弹性,顺应性,伸缩性,弹性力学
elasticity factor 弹性因数
elasticizer 增韧剂,增弹剂,致弹剂
elastic-plastic fracture 弹塑性断裂
elastic-plastic fracture mechanics 弹塑性断裂力学
elastivity 倒电容系数,倒介电常数

elastodynamics 弹性动力学
elastohydrodynamic condition 弹性流体状态
elastohydrodynamics 弹性流体动力学
elastomer 弹性体,类橡胶件
elastomeric bearing 弹性体轴承,弹性轴承
elastomeric component 弹性组件
elastomeric damper Baseline Elastomeric Damper 弹性减摆器
elastoplastic deformation 弹(性)塑性变形
elastoplasticity 弹塑性
elastoslide element 弹性滑动单元/元件
elastostatics 弹性静力学
el-az Elevation/Azimuth 俯仰/方位
ELB 1. Elbow 压(形弯)管,弯管接头 2. Electronic Log Book 电子记录本 3. Emergency Locator Beacon 应急定位信标
ELBA Emergency Location Beacon Aircraft 紧急定位信标飞机
elbow 肘形管,肘拐,弯头
ELC 1. Emergency Launch Capability 应急发射能力 2. Engine Life Computer 发动机寿命计算机
ELCA Earth Landing Control Area 返回地球着陆控制区
ELCCR Electrical Liaison Change Commitment Record 电气联络更改提交记录
ELCTLT Electrolyte 电解液,电解质
ELCTRN Electronic 电子
ELCU 1. Electrical Control Unit 电气控制盘,电气控制组 2. Emergency Lighting Control Unit 应急照明控制单元
ELD Electroluminescent Display 电致发光显示,电致发光显示器
ELDO European Launcher Development Organization 欧洲运载器发展组织
eldorado "埃尔多拉多"(法国 I/J 波段低空跟踪雷达)
electret 驻极体(一种永久极化的电介质)
electret microphone 驻极体传声器
electric 电的,电气的
electric(al) fuse box 保险丝盒
electric (wave) filter 滤波器
electric actuator 电动执行机构,电力致动装置,电动式舵机
electric aircraft 全电飞机
electric and magnetic field 电磁场
electric arc welding 电弧焊
electric battery 电池,电池组
electric beacon (电)灯标
electric blasting cap 电起爆雷管,电雷管
electric blasting valve 电爆阀

electric bomb fuze 电发炸弹引信
electric bomb release 电动投弹器
electric bonding 电连接,电接地
electric boresight 光轴与电轴的平行校准
electric breeze 离子风
electric cable 电缆
electric capacity altimeter 电容式高度表
electric cartridge 电发火炮弹
electric cell fusion 电细胞融合
electric charge 电荷
electric circuit 电路
electric conductance method 电导法
electric conductive oxidation 导电氧化
electric connector 电连接器,电插头,电插座
electric contact material 电极材料,电接触材料
electric current 电流
electric detonator 电雷管
electric dipole 电偶极子
electric discharge 放电
electric discharge anemometry 放电风速测定法
electric discharger 放电器
electric drive 电气传动
electric drive control gear 电气传动控制设备
electric earth-orbit-raising （电力驱动）地球轨道变轨
electric effect 电效应
electric eye 电眼
electric field 电场
electric field strength 电场强度
electric firing head 电点火头
electric firing lock 电点火装置
electric ground power system 地面电源系统
electric gun 电动枪
electric heater 电热器
electric hydraulic converter 电液转换器
electric hydraulic pump 电动液压泵
electric igniter 电点火头,电点火器
electric ignition 电点火
electric ignition bomb fuze 电点火炸弹引信
electric installation rehearsal 电气合练
electric insulation anodizing 绝缘阳极（氧）化
electric interlock 电气联锁
electric machine 电机
electric match 电点火头,电点火器
electric motor 电动机,电滚子
electric pneumatic converter 电气转换器
electric potential 电位,电势
electric potential gradient 电位梯度
electric power 电功率,功率
electric power consumption 电力消耗（量）,耗功

electric power distribution 配电
electric power generation 发电
electric power input 输入（电）功率
electric power line 电力线
electric power mode 电力（气）模式
electric power plant 发电站
electric power-generating equipment 发电设备
electric pressure 电压
electric primer 电发火器,电起爆管,电底火（Lead-Free Electric Primer 无铅电发火器,无铅电起爆管,无铅电底火）
electric propeller 电动（变距）螺旋梁
electric propulsion 电推进（利用静电或电磁力使推进剂加速产生推力的推进方式）
electric propulsion system 电推进系统
electric resistance belt heater 电阻带式加热器
electric resistance tube heater 电阻管式加热器
electric sear 电动扣机,电烧灼
electric servo actuator 电动伺服舵机
electric signal gun 电动信号枪
electric squib 电发火管,电爆管,电气导火管,电点火（器）,电爆装置
electric stall warner 电动失速告警器
electric starter 电起动机,电起动器
electric starter button 电动起动按钮,电启动机按钮
electric steel 电炉钢
electric tachometer 电动转速表,电力转速计,电测速计,电动转速计,电流速计
electric tachometer generator 电动测速机（测量转轴速度）
electric tachometer indicator 电气式回转速度计指示器
electric test equipment 电测试设备
electric test model 电性试验模型
electric thruster 电推力器
electric transducer 电换能器,电能变换器,电转换器,换流器
electric trigger 电动扳机
electric tug 电动牵引车（机场特种车辆）
electric vacuum gyro 电真空陀螺
electric vector 电场矢量,电矢量
electric vehicle lead acid battery 电动车辆铅酸蓄电池
electric vibrator 电动振捣器,电动振动器,电振动器,电振动子
electric welding 电焊
electric wiring 布线,配线,导线网
electric(al) angle 电角度
electric(al) data transmission 电动数据传输
electric(al) spot welding （电）点焊
electric(al) technician 电气技师

electrical actuation	电驱动,电传动
electrical analog(ue) equipment	电模拟设备
electrical anemometer	电传风速表
electrical boresight	电轴
electrical breakdown	电击穿,电网失电
electrical cage switch	（瞄准具）稳环按钮,电锁定按钮
electrical capacitance levelmeter	电容液位计
electrical circuit	电路
electrical command	电信号传输指令,电指令
electrical conductance levelmeter	电导液位计
electrical conductivity	电导率,导电率,导电性
electrical conductivity of wall	壁电导率
electrical connection	电线接头,电（路）连接
electrical corona	电晕
electrical despun antenna	电消旋天线
electrical discharge	电火花
electrical discharge machining	电火花加工
electrical energy	电能
electrical engagement length	电啮合长度
electrical engineering	电气工程
electrical equipment (electrics) bay	电气设备舱
electrical error of null position	零位误差
electrical failure	电气故障
electrical feedback	电反馈
electrical firing device	电点火装置
electrical firing head	电点火头
electrical flight path line	电航迹线
electrical four-pole swing	电动四柱秋千
electrical function test	电路功能试验
electrical fuze	电引信
electrical ground	搭铁,接地线
electrical igniter	电点火头,电点火器
electrical ignition	电点火
electrical ignition bomb fuse	电点火炸弹引信
electrical input	电输入
electrical interface	电接口,电气接口装置
electrical interference	电干扰
electrical isolation	电隔离,电绝缘
electrical micro-machine	微电机
electrical mobility	电迁移率
electrical mode	电气模态
electrical model test	电性模型测试
electrical parameter	电量参数
electrical performance	电（气）特性,电（气）性能
electrical power	电力
electrical power characteristics	电源特性
electrical power generating system	电源系统
electrical power source	电源,发电装置
electrical power supply system	供电系统
electrical power transmission/distribution system	输配电系统
electrical property	电特性,电性能
electrical property test of satellite	卫星电性能测试
electrical property test of satellite subsystem	卫星分系统电性能测试
electrical propulsion vehicle	电推进飞行器
electrical resistance temperature bulb	电阻测温包
electrical resistivity	电阻率
electrical resolver	电分解器
electrical rocket	电火箭
electrical rocket engine	电火箭发动机
electrical short	电短路
electrical spring	电弹簧
electrical stimulation	电刺激
electrical supply box	供电接线盒
electrical suspended gyro	静电陀螺仪
electrical system	电气系统
electrical thruster	电力推进（力）器
electrical transmitting thermometer	遥测电传温度表,温度传感器
electrical trim unit	电动配平机构
electrical work	电气工作,电工作业,（复数）发电站,电气企业
electrical zero	电气零位（指传感器输出零电量时的角位置）
electrical, electronic and electromechanical part	电气、电子和机电零件（EEE 零件）
electrical-bay door	电气（设备）舱门
electrically conductive film	导电薄层
electrically fired explosive charge	电燃爆炸装药
electrically heated boom	电加温全静压管（空速管）
electrically heated wire	电加热丝（点火器）
electrically power subsystem	电源分系统,电源子系统
electrically primed ammunition	电点火炮弹
electrically programmable read only memory	电可编程只读存储器
electrically propelled spacecraft	电推进航天器
electrically scanning microwave radiometer	电扫描微波辐射仪
electrically signalled flying controls	电信号飞行操纵装置,电传飞行操纵系统
electrically tunable filter	电调滤波器
electrically tunable oscillator	电调振荡器
electrically-actuated	电力致动的,电动的
electrically-deiced	电加温除冰的
electrically-erasable programmable read only memory	电可擦除可编程只读存储器
electrically-fired	电点火的

electrically-ignited 电点火的
electrically-powered 装电火箭发动机的,电驱动的
electrically-similar vehicle 飞行器电动模型,电模拟飞行器
electro explosive actuated valve 电爆阀
electro-optical countermeasures 光电对抗
electro-optical distance measuring instrument 光电测距仪
electro valve 电动阀,电磁阀(活门)
electro vectorcardiography 心电向量描记器
electroacoustic 电声(学)的
electroacoustic transducer 电声换能器
electroactive polymer 电活性聚合物
electro-anaesthesia 电麻醉
electro-bath 电镀浴
electrocardiography 心电描记器,心电图学
electrocardiophonograph 心音电描记器
electrocatalyst 电催化剂
electrochemical analysis 电化学分析
electrochemical and chemical transition thermal control coatings 电化学和化学转换型热控涂层
electrochemical battery 电化学电池
electrochemical cell 电化学电池
electrochemical cleaning 电化学清洗
electrochemical coating 电化学涂层
electrochemical corrosion 电化学腐蚀
electrochemical enamelizing 瓷质阳极化
electrochemical equivalent 电化当量
electrochemical machining 电化学加工,电解加工
electrochemical oxidation 电化学氧化
electrochemical power source 电化学电源
electrochemical sensor 电化学传感器
electrochemical sonde 电化学探空仪
electrochemical transducer 电化学传感器,电化学换能器
electrochemical treatment 电化学处理
electrochemical valve 电化阀,电解阀
electrochemichromism 电化致变色
electro-chemistry 电化学
electrochromatography 电色层分析法
electrochromic display 电致变色显示
electrochromism 电致变色
electrochronograph 电动精密记时计
electrocision 电切术
electrocladding 电键包覆,电镀保护层
electrocoagulation 电凝法
electrocoating 电(泳)涂(层),电化涂敷
electrocochleography 耳蜗电描记术
electrocorticography 脑皮层电描记术
electrocystography 膀胱电描记术
electrode 1.(电,焊)极 2.焊条(电焊用的)
electrode admittance 电极导纳
electrode array 电极排列
electrode bias 电极偏压,偏压电极
electrode case outer electrode case 外电极(壳体)
electrode current 电极电流
electrode erosion 电极烧蚀
electrode gap 电极间隙,电极间距
electrode geometry 电极几何形状
electrode loss 电极损耗
electrode pattern (机电)电极图形
electrode potential 电极电位,电极电势
electrode power supply 电极供电
electrode reaction 电极反应
electrode shape 电极形状,电极成形
electrode (contact) surface 电极(表)面(Electrode Contact Surface 电极接触面)
electrode temperature 电极温度
electrode voltage 电极电压
electrodeless plasma accelerator 无电极等离子体加速器
electrodeposition 电解沉积,电极淀积
electrodermography 皮肤电阻描记术
electro-discharge machining 放电加工
electro-display 电子显示
electrodynamic 电动的,电动力的,电动力学的
electrodynamic damping 电动阻尼
electrodynamic earphone 电动耳机
electrodynamic effect 电动力学效应
electrodynamic exciter 电动主控振荡器
electrodynamic force (电磁)电动力,电动势
electrodynamic loudspeaker 电动扬声器
electrodynamic microphone 电动传声器
electrodynamic tether 电动力缆绳
electrodynamic tether system 电动力缆绳系统
electrodynamics 电动力学
electrodynamics vibration generator 电动振动台
electrodynamometer 电功率计,力测电流计,双流作用计,电动测力计
electroencephalic mapping 脑电分布图测量
electroencephalograph 脑电描记器
electroencephalograph in space medicine 航天医学脑电描记法
electroencephalography 脑电描记术
electroendosmosis 电(内)渗
electro-erosion 电浸蚀,电蚀
electro-erosive machining 电蚀加工
electro-explosive device 电爆装置
electro-explosive subsystem 电爆分系统

electro-explosive valve 电爆阀
electrofax 电传真,电子摄影,电子照相
electroform 电铸
electroforming 电冶,电铸,电成形
electrogasdynamic 电气体动力学的
electrogasdynamics 电气体动力学
electrogastrography 胃电描记术
electrogilding 电镀
electrograph 电记录器,电刻器,传真电报,电图
electrographic 1.电记录器的 2.传真电报的
electrohydraulic 电动液压的,液压电动的,电液的
electrohydraulic actuator 电动液压作动器,电液执行机构,电动液压致动装置
electrohydraulic control 电动液压控制
electrohydraulic exciter 电动液压激发器
electrohydraulic forming 电动液压成形
electrohydraulic servovalve 电动液压伺服阀,电液伺服阀
electrohydraulic transducer 电液式换能器
electrohydraulic unit 电动液压附件
electrohydraulic valve 电动液压阀,电动液压活门
electrohydrodynamic generator 电流体动力发电机
electrohydrodynamics 电流体动力学
electrojet 电喷流,电急流(电离层中流动的强电流束),电子流,电流
electrokinetic transducer 电动传感器
electrokinetics 电动学,电动力学
electroless plating 化学镀(膜),非电解镀层,无电镀
electrolithotrity 电碎石法
electroluminescence 电致发光,场致发光
electroluminescent device 电致发光器件
electroluminescent display 电致发光显示
electroluminescent film 电致发光层,场致发光层
electrolyser 电解装置
electrolysis 电解(作用)
electrolysis unit 电解装置
electrolyte 电解液,电解质
electrolyte creepage 爬碱
electrolyte solution 电解质溶液
electrolytic bath 电解液槽
electrolytic capacitor 电解电容器
electrolytic cell 电解电池
electrolytic corrosion 电解腐蚀
electrolytic derusting 电解除锈
electrolytic forming 电解型腔加工
electrolytic integrating accelerometer 电解集成加速度计,带电解积分仪的摆式加速度计
electrolytic machining 电解加工
electrolyze 电解,用电蚀法除去

electrolyzer 电解剂,电解器,电解槽
electromagnet 电磁铁,电磁体
electromagnet configuration 电磁铁形状,电磁体形状
electromagnetic accelerator 电磁加速器,电磁导轨炮
electromagnetic analog(ue) 电磁模拟装置
electromagnetic compatibility 电磁兼容性,电磁相容性
electromagnetic compatibility measurement 电磁兼容性测量
electromagnetic compatibility test 电磁兼容性试验
electromagnetic control 电磁控制
electromagnetic coupling 电磁耦合
electromagnetic cyclotron wave 电磁回旋波(垂直于磁场传播方向的电磁波)
electromagnetic delay line 电磁延迟线
electromagnetic effect 电磁效应
electromagnetic environment 电磁环境
electromagnetic environment experiment 电磁环境实验
electromagnetic environment observation satellite 电磁环境观测卫星
electromagnetic exciter 电磁激发器
electromagnetic explosive device 电磁爆炸装置
electromagnetic field 电磁场
electromagnetic flow transducer 电磁流量传感器
electromagnetic force 电磁力
electromagnetic forming 电磁成形
electromagnetic fuze 电磁引信
electromagnetic hydrodynamics 电磁流体动力学
electromagnetic induction method 电磁感应法
electromagnetic interaction 电磁相互作用
electromagnetic interference 电磁干扰
electromagnetic interference emission 电磁干扰发射
electromagnetic interference measurement 电磁干扰测量
electromagnetic intrusion 电磁入侵,电磁干扰
electromagnetic jammer satellite 电磁干扰卫星
electromagnetic jamming 电磁干扰,电磁干扰技术
electromagnetic launcher 电磁发射器,电磁炮
electromagnetic measurement 电磁测量
electromagnetic noise 电磁噪声
electromagnetic pouring 电磁浇注
electromagnetic probe 电磁探测器
electromagnetic probing 电磁探测
electromagnetic propulsion 电磁推进
electromagnetic pulse 电磁脉冲
electromagnetic pulse shielding 电磁脉冲屏蔽
electromagnetic pump 电磁泵
electromagnetic radiation 电磁辐射
electromagnetic radiation detection 电磁辐射探测
electromagnetic railgun 电磁导轨炮
electromagnetic relay 电磁继电器

electromagnetic release unit 电磁释放机构,电磁锁
electromagnetic riveting 电磁铆接
electromagnetic rocket 电磁火箭
electromagnetic scattering 电磁波散射
electromagnetic sensitivity measurement 电磁敏感度测量
electromagnetic sensor 电磁传感器
electromagnetic shaker 电磁振动台
electromagnetic shielding 电磁屏蔽
electromagnetic simulated test 电磁模拟试验
electromagnetic spectrum 电磁波谱
electromagnetic suspension balance 电磁悬挂天平
electromagnetic suspension system 电磁悬浮系统
electromagnetic technique 电磁技术
electromagnetic transparency 透电磁波
electromagnetic transparent 透电磁波的,不反射电磁波的
electromagnetic unit 电磁单位
electromagnetic unit balance 电磁元件天平
electromagnetic valve 电磁阀(活门)
electromagnetic vibrator 电磁振动器
electromagnetic vulnerability 电磁波受扰性,易受电磁波干扰性
electromagnetic wave 电磁波
electromagnetic wave transparent material 透(电磁)波材料,不反射电磁波的材料
electro-magnetic distance measurement 电磁波测距
electro-magnetic distance measuring instrument 电磁波测距仪
electromagnetic weapon 电磁武器
electromagnetically operated valve 电磁(驱动)阀
electromassage 电按摩
electromechanical 电动机械的
electromechanical actuation 电动机械传动
electromechanical actuator 电动机械致动装置
electromechanical coupling factor 机电耦合系数
electromechanical servo system 机电伺服系统
electromechanical tachometer 机电式转速表
electromechanical transducer 1.机电传感器 2.机电换能器,机电变换器
electromedication 电透药法
electrometallurgy 电冶金,电冶金学
electrometeor 带电流星
electrometer 静电计
electrometer tube 电表管,测量用电子管
electromigration 电迁移
electromotive force 电动势
electromyelography 脊髓电描记术
electromyogram 肌动电流图

electromyograph 肌电描记器
electromyography 肌电描记术
Electron 电子号卫星(苏联测磁层的卫星系列)
electron 1.电子 2.轻质镁基合金
electron accelerator 电子加速器
electron acceptor 电子受主
electron affinity 电子亲和势
electron albedo 电子反照,电子反照率
electron attachment 电子附着,电子附件
electron aurora 电子极光
electron back bombardment 电子回轰
electron backflow 电子回流
electron backstreaming 电子返流
electron beam 电子束,电子注
electron beam accelerator 电子束加速器
electron beam energy 电子束能量
electron beam evaporation 电子束蒸发
electron beam exposure equipment 电子束曝光机
electron beam exposure system 电子束曝光系统
electron beam flow visualization 电子束流动显示
electron beam fluorescence technique 电子束荧光技术
electron beam injection 电子束注入
electron beam lithography 1.电子束光刻 2.电子束金属版印刷品
electron beam machining 电子束加工
electron beam parametric amplifier 电子束参量放大器
electron beam perforation 电子束打孔
electron beam pumped semiconductor laser 电子束抽运半导体激光器
electron beam pumping 电子束抽运
electron beam recording 电子束录像,电子束记录
electron beam resist 电子束光刻胶
electron beam semiconductor 电子束半导体(器件)
electron beam slicing 电子束切割
electron beam welder 电子束焊机
electron beam welding 电子束焊(接)
electron belt 电子带
electron block 电子块
electron bomb 镁壳燃烧炸弹
electron bombardment 电子轰击
electron bombardment thruster 电子轰击型推力器
electron bunching 电子聚束,电子群聚
electron capture 电子俘获
electron charge 电子电荷
electron chemistry 电子化学
electron collision 电子对撞
electron compartment 电子设备舱
electron concentration 电子(密)浓度
electron confinement 电子约束

electron content	电子含量,电子浓度
electron current	电子电流
electron cyclotron	电子回旋
electron cyclotron drift instability	电子回旋漂移不稳定性
electron cyclotron harmonic instability	电子回旋谐波不稳定性
electron cyclotron maser	电子回旋微波激射器
electron cyclotron resonance heating	电子回旋共振加热
electron cyclotron wave	电子回旋波
electron decay rate	电子衰变率
electron density	电子密度
electron density distribution	电子密度分布
electron device	电子仪器,电子装置
electron diffraction	电子衍射
electron discharge	电子放电
electron dose	电子剂量
electron double resonance spectroscopy	电子双共振谱(学)
electron drift	电子漂移
electron dynamics	电(子)动力学
electron echo experiment	电子回声实验
electron emission	电子发射
electron emitter	电子发射装置(器,体)
electron energy	电子能(量)
electron events	电子事件
electron excitation function	电子激发函数
electron eye	(机器人的)电眼
electron flare	电子耀斑
electron flux	电子通量
electron flux density	电子通量密度
electron gas	电子气
electron gun	电子枪
electron image tube	电子变像管,移像管
electron impact	电子碰撞
electron impact desorption	电子碰撞脱附
electron lens	电子透镜
electron loss spectroscopy	电子损失谱(学)
electron microprobe	电子探针
electron microscope	电子显微镜
electron microscopy	电子显微术
electron mobility	电子迁移率
electron momentum	电子动量
electron motion	电子运动
electron nuclear double resonance spectroscopy	电子核双共振谱(学)
electron number	电子数
electron number density	电子数密度
electron optics	电子光学
electron physics	电子物理学
electron plasma	电(子)等离子体
electron precipitation	电子沉降
electron radiation	电子辐射
electron reconnaissance satellite	电子侦察卫星
electron saturation	电子饱和
electron scattering	电子散射
electron source	电子源
electron spectroscopy for chemical analysis	化学分析电子谱(学)
electron spin resonance spectroscopy	电子自旋共振谱(学)
electron telescope	电子望远镜
electron temperature	电子温度
electron temperature probe	电子探温头,电子式温度传感器
electron thermal current	电热电流
electron thermal velocity	热电子速度
electron trajectory	电子轨迹
electron transfer	电子转移
electron transfer coefficient	电子转移系数
electron transit time	电子渡越时间,电子飞越时间
electron transport	电子输运
electron trap	电子陷阱
electron tube	电子管,真空管
electron unit	电子电荷单位
electron volt	电子伏特
electronally scan technology	电扫描技术
electron-beam	电子束,电子注
electron-beam tube	电子射束管,电子束管,电子注管
electron-beam valve	电子束管,电子注管
electron-beam window	电子束窗
electron-bombardment	电子轰击
electron-bombardment thruster	电子轰击型推力器(一种离子推力器)
electronegativity value	电负值
electroneurography	神经电描记术
electronic	电子的
electronic analogy	电子模拟
electronic art	电子艺术
electronic attachment	电子附着
electronic attitude director	电子姿态指引仪,电子指引地平仪
electronic batching scale	电子配料秤
electronic belt conveyor scale	电子皮带秤
electronic bombsight countermeasure	对雷达轰炸瞄准具的干扰
electronic camouflage	电子伪装
electronic ceramics	电子陶瓷

electronic chart	电子海图
electronic circuit board	电子线路板
electronic combat	电子战斗
electronic commutator	电子交换子,电子转换器
electronic component	电子元件
electronic contact fuze	电子触发引信
electronic co-pilot	电子副驾驶
electronic correlation	电子相关
electronic counter counter-measures	电子反对抗,抗干扰,电子反干扰
electronic counter countermeasures	反电子干扰措施
electronic counter measures	电子对抗措施,电子干扰
electronic countermeasures equipment	电子对抗设备
electronic countermeasures satellite	电子对抗卫星
electronic countermeasures vulnerability	易受电子干扰性
electronic data processing equipment	电子数据处理设备
electronic deception	电子欺骗
electronic defence evaluation	电子防御评定
electronic despin of antenna	天线电子消旋
electronic detonator	电子起爆管
electronic device	电子设备
electronic efficiency	电子效率
electronic engine	电控引擎
electronic engineering	电子工程学
electronic engraver	电子刻图机
electronic equipment cooling air exhaust port	电子设备冷却气排气口
electronic excitation	电子激发
electronic exploder	电子起爆装置,电子爆炸装置
electronic failure	电子设备故障
electronic fire-control equipment	电子火控设备
electronic fix	用电子设备定位
electronic flight imitator	电子飞行模拟器
electronic flight instrument system	电子飞行仪表系统
electronic flight-control unit	电子飞控部件
electronic fuzing device	电子引信起爆装置
electronic guidance system	电子制导系统
electronic hopper scale	电子料斗秤
electronic horizontal situation indicator	电子水平状态指示器
electronic imaging system	电子成像系统
electronic industry	电子工业
electronic integrated display system	电子综合显示系统
electronic intelligence	电子侦察,电子情报
electronic jamming	电子干扰
electronic library system	电子资料库系统
electronic line of sight	电子瞄准线
electronic line scanner	电子行扫描器
electronic location-finder	电子定位器
electronic mail	电子邮政,电子邮件
electronic measurement	电子(学)测量
electronic measuring instrument	电子测量仪器
electronic mechanically steering	机电调向
electronic message system	电子报文系统
electronic microscope	电子显微镜
electronic missile acquisition	导弹电子截获
electronic packaging	电子封装
electronic part	电子元件
electronic plane-table	电子平板仪
electronic planimeter	电子求积仪
electronic printer	电子打印机
electronic printing equipment	电子印刷设备
electronic probe	电子探针,电子探测器
electronic processing equipment	电子(数据)处理设备
electronic data processing center	电子数据处理中心
electronic range finder	雷达测距仪
electronic reconnaissance	电子侦察
electronic reconnaissance satellite	电子侦察卫星
electronic scanner	电子扫描仪
electronic scanning	电子扫描
electronic scanning antenna	电扫描天线
electronic search	电子搜索
electronic shell	电子炮弹
electronic shutter	1.电子开闭器 2.电子快门
electronic skyscreen equipment	电子空网设备
electronic state	电子态
electronic stopping power	电子阻止本领
electronic storage	电子存储器
electronic support	电子支援
electronic support measure	电子支援措施
electronic surge arrester	电涌放电器,防过载电子放电器
electronic surveillance	电子监视,电子侦察
electronic surveillance equipment	电子监视设备
electronic switching system	电子交换系统
electronic tachometer	全站型光电速测经纬仪,全站型电子速测仪
electronic tail fuze	弹尾电子引信
electronic telephone circuit	电子电话电路
electronic teleprinter	电子电传机
electronic temperature probe	电子温度探测仪
electronic test	电子设备试验
electronic time fuze	电子定时(时间)引信(利用电子装置计时的引信),电子时间引信
electronic timer	电子式计时器,电子式计时装置
electronic tracking	电子跟踪
electronic transducer	电子传感器

electronic tuning unit	电子调谐器
electronic typewriter	电子打字机
electronic voltmeter	电子电压表，电子管电压表，场效应管(式)电压表
electronic warfare	电子战
electronic warfare airplane	电子战飞机
electronic warning station	警戒雷达站
electronically phased array sectorscan	电子相控阵扇形扫描
electronically scanned optical tracker	电子扫描光学跟踪装置
electronically scanned radar	电扫雷达
electronically scanning microwave radiometer	电扫描微波辐射计
electronically steerable antenna	电子可控天线
electronically steerable array radar	电控天线阵雷达
electronically-clean	电子设备净形的(飞机上电子设备和天线等突出在飞机外的部分很少，阻力很小的)
electronically-controlled	电子操纵的，电子控制的
electronics	电子学，电子设备
electronics bay	电子设备舱
electronics compartment	电子设备舱
electronics manufacturer	电子设备生产商
electronic-scanned pressure sensor	电子式压力扫描传感器
electron-induced desorption	电子诱导解吸
electron-negative	负电的，阴电性的
electron-neutral collision	电子中性粒子碰撞
electronograph	电子显像机，电子显像
electronographic camera	电子照相机
electronography	电子显像术
electron-optical guidance	光电制导
electron-ray tube	电子射线管
electron-recording tube	电子记录管
electronystagmography	眼震电描记术
electrooculography	眼电描记术
electrooptic ceramics	电光陶瓷
electro-optic converter	电场光学转换器
electro-optic coupler	光电耦合器
electro-optic coupling	光电耦合
electrooptic crystal	电光晶体
electrooptic crystal light valve	电光晶体光阀
electro-optic deflector	电光偏转器
electrooptic effect	光电效应
electro-optic modulation	电光调制
electro-optic Q-switching	电光Q开关
electrooptic sensor	光电传感器
electrooptic transfer characteristic	电光转换特性
electro-optical combined radar	光电复合雷达
electro-optical detector	电光探测器
electro-optical glide bomb	光电飘滑炸弹
electro-optical guidance	光电制导
electrooptical photography	电子光学摄影
electro-optical pickoff	光电式传感器
electro-optical reconnaissance system	光电侦察系统
electro-optical sight	电子光学瞄准具
electro-optical target point	光电显示目标点
electro-optical tracking system	光电追踪系统
electro-optics	1.电(场)光学 2.电(场)光(学)仪器
electro-optics weapons system	电(子)光(学)武器系统
electropathy	电疗法
electrophonocardiograph	电心音图仪
electrophoresis	电泳现象
electrophoresis coating	电泳涂层，电泳涂层工艺
electrophoretic display	电泳显示
electrophoretic migration	电泳迁移
electrophoretic mobility	电泳迁移率
electrophoretic separation	电泳分离
electrophotometry	光电光度学，电测光度学
electrophysiology	电生理学
electropneumatic	电动气压的，电动气动的
electro-pneumatic transducer	电气动式换能器
electro-pneumatic valve	电动气动阀，电动气动控制阀
electropneumography	电子呼吸描记术
electropolar	电极性，电极化的
electropolating	电镀
electropolishing	电抛光
electroprobe	电笔，电测器
electropuncture	电针术
electropyrometer	电阻高温计，电测高温计
electroretinography	视网膜电描记术
electroshock	电休克
electroslag welding	电渣焊
electrosonde	电位探空仪
electrosphygmomanometer	电子血压计
electrospirometer	电子肺活量计
electrospray	电喷射，电喷雾
electrospray beam	电喷雾(粒子)束
electrospray chamber	电喷雾室
electrospray droplet	电喷雾液滴，电喷射液滴
electrospray emitter	电喷雾装置，电喷雾喷头
electrospray operation	电喷操作
electrospray particle	电喷雾颗粒物
electrospray performance	电喷射性能
electrospray propellant	电喷推进的
electrospray propulsion	电喷推进
electrospray source	电喷雾源
electrospray thruster	电喷推进器

electrospraying 电喷射,电喷雾
electrostatic 静电的
electrostatic aerosol sampler 静电式气溶胶取样器
electrostatic analyzer 静电分析器
electrostatic capacity 静电容
electrostatic charge 静电电荷,静电荷
electrostatic control 静电控制
electrostatic deflector 静电致偏器
electrostatic discharge 静电放电
electrostatic discharge damage 静电放电损伤
electrostatic discharge interference measurement 静电放电干扰测量
electrostatic discharge sensitivity measurement 静电放电敏感度测量
electrostatic double probe 静电双探针
electrostatic energy 静电能(量)
electrostatic field 静电场
electrostatic force 静电力
electrostatic gyro 静电陀螺
electrostatic gyro strapdown navigation system 静电陀螺捷联式惯性导航系统
electrostatic lens 静电透镜
electrostatic memory 静电存储器
electrostatic mic 电容传声器
electrostatic orbit 静电(场)轨道
electrostatic plotter 静电绘图机
electrostatic powder spraying 粉末静电喷涂
electrostatic pressure 静电压(力)
electrostatic printer 静电打印机,静电印刷机
electrostatic probe 静电探针
electrostatic protection 静电保护
electrostatic proximity fuze 静电近炸引信
electrostatic rocket 静电火箭,离子火箭
electrostatic rocket engine 静电火箭发动机
electrostatic shield 静电屏蔽
electrostatic spraying 静电喷涂,静电喷漆
electrostatic storage 静电存储器
electrostatic storage tube 静电存储管
electrostatic storage unit 静电存储器
electrostatic support accelerometer 静电加速度计
electrostatic suspension 静电悬浮
electrostatic tractor 静电牵引(车)
electrostatically focused klystron 静电聚焦速调管
electrostatically suspended accelerometer 静电支承加速度计
electrostatically suspended gyroscope 静电悬浮陀螺
electrostatics 静电学
electrostethophone 电扩音听诊器
electrostimulation 电刺激

electrostriction 电致伸缩(反压电效应)
electrostrictive ceramics 电致伸缩陶瓷
electrostrictive grating 电致伸缩光栅
electrothermal 电热的
electrothermal hydrazine engine 电热肼发动机
electrothermal monopropellant hydrazine system 电热式单元肼推进系统(由电加热器加热分解成高温高压气体经喷管排出产生推力的装置)
electrothermal thruster 电热推力器
electrothermic 电热(的)
electrothermometer 电子体温计
electrotonograph 电子眼压计
electrotyping 电铸术
electrotyping process 电铸成形
ELEM 1. Element 元件,单元,元素 2. Elementary 初步的,基本的,本质的,单质的
element 元素,要素,根数,元件,构件
element analysis 元分析(如 Finite Element Analysis 有限元分析)
element balance 元素平衡
element differences 元差分(如 Finite Element Difference 有限元差分)
element error 元误差(如 Finite Element Error 有限元误差)
element formulation 元公式,元方程式(如 Finite Element Formulation 有限元公式,有限元方程式)
element mass 单元质量
element method 元(方)法
element model 元模型(如 Finite Element Model 有限元模型)
element modeling 元建模(如 Finite Element Modeling 有限元建模)
element of rectification 纠正元素
element of structure 结构元件
element of type k K型元件
element set 元素集(合)
element stiffness matrix 单元刚度矩阵
element target 元素靶
elemental aluminum 元素铝
elemental aluminum mass fraction 铝元素质量分数
elemental filter 基本滤波器
elemental force 单元力,单元内力
elemental force increments 单元内力增量
elemental maneuver 基本(战术)动作,基本操作动作
elemental reaction 基元反应
elemental semiconductor 元素半导体
elemental transport 元素迁移
elementary guided bomb 弹道末段制导炸弹
elementary particle 基本粒子

elementary pilot training 驾驶员初级训练
elementary step 1.基本步骤 2.初步(始)阶段
elementary step function 初等阶梯函数
elementary storm 元磁暴
elementary surveying 普通测量学
elementary trainer 初级教练机
element-based reduced-order 基于有限元的降阶,有限元降阶
elements of a fix 定位要素
elements of absolute orientation 绝对定向元素
elements of centring 归心元素
elements of eclipse 交食要素
elements of exterior orientation 外方位元素,相片外方位元素
elements of orbit 轨道要素,轨道根数
elements of relative orientation 相对定向元素
elephant 指一种标准的绘图纸尺寸,即28英寸×32英寸,〈口语〉加固板
elephant ear 1.加固板(火箭或导弹壁上开口处的) 2.机身进气道
elerudder 升降方向舵
ELEV Elevate or Elevator 升降或升降舵
elevate 加大仰角,增大射角,抬高,举起,升高
elevated background radiation 增强背景辐射
elevated load 高负荷
elevated pole 上天极
elevated pressure 增压,高压
elevated pressure range 增压范围
elevated temperature 高温
elevated temperature structure test 高温结构试验
elevated trajectory 高弹道
elevated-line-source biological munitions 空中线状生物弹药
elevated-temperature test 高温试验,高温测试
elevated-temperature zone 高温区(域)
elevating ladder 升降工作梯
elevating rail (发射装置的)升降导轨
elevation error 仰角误差,高低角误差
elevation 标高,海拔高,仰角,高低角,上升,升高,立视图,正视图
elevation adder 仰角电路加法器
elevation angle 1.仰角,高角 2.高低角 3.射角,目标角 4.高度角 5.天体高度
elevation assembly 高低瞄准装置
elevation carriage (炮塔)高低回转架
elevation correction 高低角修正量,射角修正量
elevation deviation 上升漂移
elevation error 仰角误差,高低角误差
elevation firing angle 发射仰角

elevation guidance element 仰角引导单元
elevation gyro assembly 高低瞄准陀螺装置
elevation look 正视图,垂直投影
elevation mask angle 遮掩高度角,上仰遮蔽角
elevation mirror pickoff (瞄准具)反射镜俯仰传感器
elevation of sight 视线高程
elevation point 高程点
elevation point by independent intersection 独立交会高程点
elevation prediction (射击时)总修正角垂直分量
elevation rudder 升降舵(旧用名,现均用elevator)
elevation training 高低瞄准,(发射装置的)高低转动
elevation unit 1.仰角单元 2.仰角引导单元
elevator 升降舵
elevator angle 升降舵安装角
elevator angle per gram 每克升降舵偏角
elevator artificial feel unit 升降舵载荷感觉器
elevator command 升降舵(上升或下降)指令
elevator control fixed 升降舵固持
elevator control free 升降舵松浮
elevator deflection 升降舵偏转
elevator failure 升降舵故障
elevator fault 升降舵故障,升降舵缺陷
elevator feel 升降舵操纵感力
elevator illusion 升降机错觉
elevator lock-in-place 升降机卡死(故障)
elevator operator (航母)升降机操作员
elevator saturation control 升降机饱和控制
elevator tab 升降舵调整片
elevator travel 升降舵偏度(偏转角)
elevator trim 1.纵向配平,升降舵配平 2.纵向配平机构,升降舵调整片
elevator trimmer 升降航配平机构,升降舵调整片,升降舵修正片
elevon 升降副翼(既可操纵升降又可操纵滚转,常用于三角翼)
elevon deflection 升降副翼偏转
ELEX Electronics 电子学,电子设备
ELF 1.Electrostatic Levitator Facility 静电悬浮器设施 2.Elevator Load Field 升降舵载荷场 3.Extremely Low Frequency 甚低频,极低频
ELFA Elevator Load Feel Actuator 升降舵载荷探测执行机构
Elfin (容纳仪表、电子设备等的)航空运输无线电标准组件
ELGB Emergency Loan Guarantee Board 紧急贷款担保委员会
ELGO European Launcher for Geostationary Orbit 欧洲地球静止轨道运载器

ELGRA European Low Gravity Research Association 欧洲低重力研究协会

ELI 1. Emitter Location and Identification 发射体定位与识别 2. Equipment List Index 设备清单目录 3. Equipment List Identification 设备清单识别

eliminate stray light 消杂光

ELINT Electronic Intelligence 电子侦察，电子情报

ELINTS Electronic Intelligence Satellite 电子侦察卫星

elinvar 埃林瓦合金（一种铁、镍、铬合金）

ELJ External Load Jettison 外部载荷投放，外挂物投放

ellipse 椭圆，椭圆轨道

ellipse correction 椭圆修正

ellipse of zero velocity 零速度椭圆

ellipsoid 椭球面，椭球

ellipsoid appoximation 椭圆近似法

ellipsoid energy 椭球体能量

ellipsoid of inertia 惯量椭球

ellipsoid of revolution 旋转椭球，旋转椭圆面，旋转椭圆体

ellipsoid surface 椭球面

ellipsoidal binary 椭球双星

ellipsoidal configuration 椭球结构

ellipsoidal coordinates 椭球坐标

ellipsoidal distribution of velocities 速度椭球分布

ellipsoidal geodesy 球面大地测量学

ellipsoidal height 大地高，椭球体高程

ellipsoidal variable 椭球变星

ellipsoid-of-revolution family 旋转椭球体（族），旋转椭球族

ellipsometer 椭圆率测量卫星系统，椭率计

ellipsometry method 椭圆偏振法

elliptic aberration 椭圆光行差

elliptic blast wave 椭圆冲击波

elliptic chief orbit 主椭圆轨道

elliptic coordinates 椭圆坐标

elliptic frozen orbit 椭圆冻结轨道

elliptic integral 椭圆积分

elliptic jet 椭圆射流

elliptic modulus 椭圆模量

elliptic orbit 椭圆轨道

elliptic point 椭圆点

elliptic reentrance orbit 椭圆再入轨道

elliptic restricted problem 椭圆型限制性问题

elliptic slot 椭圆形裂隙

elliptic velocity 第一宇宙速度

elliptic waveguide 椭圆波导

elliptic wind tunnel 椭圆形截面风洞

elliptic wing 椭圆形机翼

elliptical orbit 椭圆轨道

elliptical combustor 椭圆形燃烧室

elliptical cone 椭圆锥体

elliptical equatorial orbit 椭圆赤道轨道

elliptical error 椭圆误差

elliptical galaxy 椭圆星系

elliptical ground 椭圆面（的）

elliptical loading （机翼的）椭圆形载荷分布

elliptical motion 椭圆运动

elliptical orbit trajectory 椭圆轨道轨迹

elliptical polarization 椭圆极化，椭圆偏振

elliptical reference orbit 椭圆参考轨道

elliptical reflector 椭圆反射镜，椭圆反光镜

elliptical space 椭圆空间

ellipticity 椭率

ellipticity correction 椭率改正

elliptic-slot jet 椭圆形（缝/槽）射流

ELM 1. Experimental Logistic Module 实验后勤舱（日本空间站舱 JEM 的组成部分）2. Extended Lunar Module 扩展的登月舱

ELMS 1. Earth Limb Measurement Satellite 地球临边测量卫星 2. Earth Limb Measurement System 地球临边测量系统 3. Electrical Load Management System 电负载管理系统 4. Engineering Lunar Model Surface 工程月球模型表面

ELNG Elongate 拉长，伸张度，延伸率

ELOISE European Large Orbiting Insrumentation for Solar Experiments 欧洲太阳实验大型轨道仪器

elongate 拉长，延长

elongation 拉长，延长，延长部分，延伸率，（行星的）距角，（拱极星的）大距

elongation angle 距角

elongation of circumpolar star 拱极星大距

ELP 1. Electro Luminescent Panel 场致发光板 2. Elliptical 椭圆形的

ELPH Elliptical Head 椭圆头

ELR Environmental Lapse Rate 环境（温度）递减率

ELRAD Earth-Limb Radiance Experiment 地球临边辐射实验

ELRAC Electronic Reconnaissance Accessory 电子侦察辅助设备

ELS 1. Earth Landing System 返回地球着陆系统 2. Electron Loss Spectroscopy 电子损失谱（学）3. Electronic Library System 电子图书馆系统 4. Emergency Landing Site 应急着陆场 5. Emergency Landing Strip 应急着陆跑道 6. Emitter Location System 发射体定位系统

Elsa Electronic Lobe-Switching Antenna 电子波瓣转换天线

ELSS 1. Emergency Life-Support System 应急生命保

障系统 2. Emplaced Lunar Scientific Station 月球表面科学空间站 3. Extravehicular Life Support System 舱外生命保障系统

ELSSE Electronic Sky Screen Equipment 电子天空监视设备（用于指示火箭偏离预定弹道的情况）

ELT 1. Emergency Locator Transmitter 应急定位器发射机 2. Emergency Locator Transponder 应急定位器应答机

ELTM Eliminate 消除

ELV Expendable Launch Vehicle 一次使用的运载火箭

EM 1. Element Management 元件管理 2. Electrical Model 电性星，电试验（模型）星 3. Electro Magnetic 电磁的 4. Electro-Magnetic Energy 电磁能 5. Emergency Maintenance 紧急维护 6. Engine Manual 发动机手册 7. Enhanced Map 增强型地图 8. Engineering Model 工程模型 9. Enterprise Modeler 企业造型者 10. Expanded Metal 多孔金属网

EM characteristics measurement of wave-absorption materials 吸波材料电磁特性测量

EM detonation and firing safety factor measurement 电磁引爆引燃安全系数测量

EM measurement 电磁测量

EM pump 电磁泵

EM sensitivity measurement 电磁敏感度测量

EM shielded room 电磁屏蔽室

EM signature 电磁特征

EMA 1. Electron Micro Probe Analysis 电子探针微量分析 2. Electronic Missile Acquisition 导弹电子截获（用无线电信号相位比较法测定导弹角坐标） 3. External Mounting Assembly （直升机的）外装组件

EMAC Energy Management Analog Computer 能量管理模拟计算机

EMAD Engine Maintenance, Assembly and Disassembly 发动机维修、装配与拆卸

E-mail Electronic Mail 电子邮件

EMAS Electromechanical Actuation System 电动机械传动系统

emanate 发出，散发，放射，起源

emanation survey 射气测量

EMB 1. Emboss 凸台 2. Extendable MAD Boom 伸缩式磁异常探测仪连杆

embarkation 登机（船），装载

embed 把……嵌入

embedded 嵌入式的

embedded computer system 嵌入式计算机系统

embedded Newtonian flow theory 内伏牛顿流理论

embedded software 嵌入式软件

embedded thermocouple 埋入式热电偶

embedding 灌封

embedding substance 埋封材料

embolismic year 闰年

embossing 模压加工

EMBRIT Embrittlement 氢脆

embrittlement 脆化，脆变，氢脆

embrittlement point 脆变点，脆变温度，脆变温度

EMC 1. Electromagnetic Compatibility 电磁兼容性，电磁相容性 2. Entertainment Multiplexer Controller 娱乐多路调制器控制器 3. External Magnetic Coupler 外磁耦合器

EMC measurement 电磁兼容（性）测量

EMC test of system 系统电磁兼容性试验

EMCCP EMC Control Plan 电磁兼容性控制大纲

EMCD Equipment Maintenance Cost Dispatch 设备维修成本分派

EMCDB Elastomer-Modified Cast Double-Base Propellant 弹性体改进浇注双基推进剂

EMCFA Electromagnetic Compatibility Frequency Analysis 电磁兼容性频率分析

EMCM Engineering Material Control Manual 工程物料控制手册

EMCON 1. Electronic Measure Control 电子测量控制 2. Emergency Condition 应急状态 3. Emission Control 电子辐射控制，发信控制

EMCS 1. Electromagnetic Compatibility Society 电磁兼容学会 2. Electromagnetic Compatibility Study 电磁兼容研究 3. Electronic Measuring Control System 电子测量控制系统 4. Energy Monitoring and Control System 能量监视与控制系统

EMD 1. Emergency Distance 应急距离 2. Energy Management Display 能量管理显示器，能量管理显示 3. Engine Management Display 发动机管理显示器 4. Engine Model Derivative 发动机改型方案 5. Engineering Manufacturing Development 工程制造阶段，工程制造研制 6. Equilibrium Mode Distribution 稳态模分布

EMDa Emergency Distance Available 可利用的中断起飞滑跑距离

EMDM Enhanced Multiplex-Demultiplex Unit 强化多路传输信号分离装置

EMDP Engine Model Derivative Program 发动机改型计划

EMDR Emergency Distance Required 必要应急距离

EME 1. Electromagnetic Energy 电磁能 2. Electromagnetic Effects 电磁效应 3. Electromagnetic Emission 电磁辐射 4. Electromagnetic Environment 电磁环境 5. Electrical & Mechanical Engineers 电气和机械工程师 6. Environmental Measurements Experiment 环境测量实验

EMED Engine Manufacturing Engineering Definition 发动机制造工程定义
EMER Emergency 紧急情况，应急
emergence 紧急情况，应急
emergencies brief 紧急情况简令
emergency 紧急情况，应急
emergency action 应急措施
emergency air 应急压缩空气
emergency airdrome equipment 机场抢救设备
emergency and rescue equipment 应急、救援设备
emergency armed release 应急爆炸投弹
emergency auxiliary release 应急投弹
emergency battery 应急电池(组)，应急蓄电池
emergency brake 紧急刹车
emergency brake system 应急刹车系统，紧急刹车系统
emergency call 紧急呼叫，呼救信号
emergency cartridge 应急传爆管
emergency case 急救药箱
emergency ceiling 应急升限
emergency combat capability 应急作战能力
emergency communication net 紧急通信网，应急通信网
emergency communication terminal 应急通信终端
emergency cutoff 紧急关机
emergency descent 应急下降，紧急下降
emergency detection system 应急探测系统
emergency drill 应急训练
emergency egress chute 安全出舱滑道
emergency egress team 应急出舱抢救队
emergency ejection 应急弹射
emergency electrical escape system 应急离机系统
emergency electrical power source 应急电源
emergency electrical power supply 应急供电
emergency elliptic hatch 应急舱门
emergency elliptic oxygen mask assembly 应急氧气面罩装置
emergency escape 应急离机，应急逃逸
emergency escape equipment 应急离机设备
emergency escape system 应急逃逸系统
emergency evacuated equipment 应急撤离设备
emergency evacuation 应急撤离，紧急撤离，紧急疏散
emergency evacuation plan 紧急撤离计划，紧急疏散计划
emergency evacuation procedure 应急撤离程序
emergency evacuation route 紧急疏散路线
emergency evacuation slide 应急撤离滑梯
emergency evacuation slide of aft area 后区应急撤离滑梯
emergency exit 应急出口
emergency exposure limit 应急暴露限值
emergency flight termination 应急飞行终止
emergency flotation gear 应急浮囊
emergency handling 应急处理
emergency landing 应急着陆，紧急着陆
emergency landing survival accident 紧急着陆人员幸存事故
emergency landing zone 应急着陆区
emergency launch 紧急起飞，紧急发射
emergency life supporting system 应急生命保障系统
emergency lighting battery 应急照明电池(组)
emergency locator/transmitter 应急定位/发射机
emergency low visibility approach 低能见度应急进近(降落)
emergency oxygen apparatus 应急氧气设备
emergency oxygen equipment 应急供氧设备
emergency oxygen pack 应急氧气装置(包)
emergency oxygen supply 应急供氧(设备)
emergency oxygen supply system 应急供氧系统
emergency oxygen tank 应急氧源
emergency oxygen rate valve 紧急供氧阀(活门)
emergency oxygen supply valve 应急供氧阀
emergency parachute 应急伞，备份伞
emergency power unit 应急动力装置，备用动力装置
emergency processing program 应急处理程序
emergency rating 应急功率额定值
emergency ration 应急口粮，应急食物配给
emergency recompression 应急复压(加压)
emergency recovery sequence 应急回收程序
emergency reentry 应急返回
emergency release handle 紧急投弹手柄
emergency repressurization system 紧急复压系统
emergency rescue station 应急救生站
emergency return 应急返回
emergency scramble 紧急起飞
emergency shutoff valve 应急切断阀，应急关闭阀
emergency stop 异常停止，紧急停止，紧急停运，事故停机
emergency sun reacquisition 紧急太阳捕获，应急太阳捕获
emergency survival equipment 应急救生设备
emergency switch 应急开关
emergency training 应急训练
emergency turbogenerator 应急涡轮发电机
emergency wheel lock (舰载机的)紧急刹车
emersio 缓始
emersion 复现(掩星的)
EMF 1. Electromotive Force 电动势 2. EM Levitator Facility 电磁能悬浮器设施 3. Emergent Manufacturing Facility 紧急制造设施

EMG 1. Electromagnetic Gun 电磁炮 2. Electromyograph 肌动电流图 3. Electro-Magnetic Gyro 电磁陀螺 4. Emergency 紧急情况,应急 5. Environmental Management Guidelines 环境管理准则
EMGE Electronic Maintenance Ground Equipment 电子维修地面设备
EMI 1. Electromagnetic Impulses 电磁脉冲 2. Electromagnetic Interference 电磁干扰
EMI diagnosing technology 电磁干扰诊断技术
EMI measurement 电磁干扰测量
EMI/EMP EMI/EMP Resistant Data Bus 抗电磁干扰/电磁波传播的数据总线
EMIH 1. Eletromagnetic Impulse Hardening 抗电磁脉冲加固 2. Electromagnetic Interference Hardening 抗电磁干扰加固
EMINT 1. Electromagnetic Intelligence 电磁信息,电磁情报 2. Exercise Message Intercept System 演习信息截获系统
EMIS Electromagnetic Intelligence System 电磁情报系统
EMIO Egyptian Military Industrialization Organization 埃及军事工业化组织
EMISM Electromagnetic Interference Safety Margin 电磁干扰安全系数,电磁干扰安全裕度
emission 辐射,发射
emission angle (导弹弹头杀伤体或弹片的)散飞角
emission capability 发射本领,排放性能
emission characteristic 发射特性
emission component 发射部件,辐射元件
emission computerized tomography 发射型计算机断层成像
emission control 辐射控制,电磁辐射管制
emission current 放射电流
emission datum 排放数据
emission image 辐射图像,发射图像
emission index 排放指数
emission intensity 发射强度,排放强度,辐射强度
emission law 发射定律
emission level 发射水平,排放水平,辐射水平
emission line 发射谱线,辐射谱线
emission nebula 发射星云
emission plume 喷射羽流
emission policy 发射策略
emission ratio 发射率
emission security 发射加密
emission signal 发射信号
emission spectra 发射(光)谱
emission spectroscopy 发射光谱学
emission spectrum analysis 发射光谱分析

emission-line galaxy 发射线星系
emission-line star 发射线星
emissions control 电磁辐射管制
emissive electrode 电子发射的,发射电极
emissive membrane ion thruster 放射性膜离子推进器,放射性膜离子发动机
emissive probe 发射探头,发射探针
emissive surface 发射面,辐射面
emissivity 发射率,比辐射率,发射率
emissivity control ability of louver 百叶窗发射率变化率
emit 发出,发射
EMIT 1. Electromagnetic Intelligence 电磁情报,电磁信息 2. Engineering Management Information Technique 工程管理信息技术
emittance 发射率,辐射率
emittance head 发射头
emitter 发射体,辐射体,辐射源,(晶体管的)发射极
emitter array 发射阵列(如 Field Emmiter Array 场发射阵列),射极阵列
emitter coupled logic 发射极耦合逻辑
emitter dipping effect 发射极陷落效应
emitter follower 射极输出器,射极跟随器
emitter geometry 发射体几何结构
emitter junction 发射结
emitter radius 发射半径
emitter region 发射区
emitter side 发射区侧
emitter size 发射极面积,发射极尺寸
emitter spacing 发射极间距,滴头间距
emitter temperature 辐射器温度,发射极温度
emitter tip 发射体尖(端)
emitter tip radius 发射体尖(端)半径
emitter voltage 射极电压
emitter-coupled trigger 射极耦合触发器
EML Equatorial Magnetosphere Laboratory 赤道磁层实验室(美国)
EMMA Engineering Mock-up and Manufacturing Assembly 全尺寸实体模型和制造件
EMOS 1. Earth's Mean Orbital Speed 地球平均轨道速度 2. European Meteorological Satellite 欧洲气象卫星
EMP 1. Electromagnetic Pulse 电磁脉冲 2. Electric Motor Pump 电动机泵 3. Empennage 尾翼
EMP resistance 电磁脉冲阻抗
EMPAR "恩帕"多功能相控阵雷达(英国、意大利、法国合作研制,用于对付空舰导弹)
EMPASS Electromagnetic Performance of Air and Ship System 飞机与舰船系统的电磁性能
EMPC Enhanced Model Predictive Control 增强模型预测控制

empennage   尾翼
empennage interaction   尾翼相互作用
emphasis network   加重网络
emphasis transmission   提升传输
emphysema   气肿
EMPIRE   Early Manned Planetary-Interplanetary Round-trip Expedition 早期行星和行星际载人往返探险
empirical   经验的，实验的
empirical analysis   实证分析
empirical approach   实证方法，实验方法
empirical autocorrelation   经验自相关
empirical average   经验平均值
empirical code   经验符，经验代码
empirical coefficient   经验系数
empirical correlation   经验关联，经验关系式，经验关联式
empirical datum   实验数据
empirical density model   （大气）密度的经验模式
empirical distribution   经验分布
empirical equation   经验公式，经验方程，经验方程式
empirical formula   经验公式
empirical information   经验信息，先验信息，实证资料
empirical input   经验输入
empirical model   经验模型，实验模型
empiricism   经验主义，实证主义，实证论
employment   1.（使，雇）用 2.服（业，职）务，工作职业
emptiness problem   空虚性问题，空集问题
empty field vision   无方位视野
empty load   空载
empty magazine   空弹仓，空弹匣
empty medium   空白媒体（不含数据而只作为基准格式的数据媒体）
empty round   空弹壳
empty string   空串
empty the gun   退弹
empty tunnel   空风洞
empty visual field   无参考目标的视野，空虚视野
empty weight   飞机空重
empty wind tunnel operation   空风洞运行
empty-field myopia   空旷近视
empty-space myopia   空旷近视
empty-weight center of gravity   飞机净重重心
empty-weight center of gravity range   飞机净重重心范围
EMR   1. Electromagnetic Radiation 电磁辐射 2. Electromechanical Relay 机电继电器 3. Electromagnetic Resonance 电磁共振 4. Electromagnetic Response 电磁响应 5. Electromagnetic Reconnaissance 电磁侦察 6. Electromagnetic Riveting 电磁铆接 7. Emergency Management and Response 应急管理与响应 8. Equipment Maintenance Record 设备维修记录 9. European Midcourse Radar 欧洲中段雷达 10. European Model Rocketry 欧洲模型火箭
EMRLD   艾姆尔德激光器
EMRU   1. Electro-Magnetic Release Unit 电磁释放装置，电磁释放机构，电磁锁 2. Electromechanical Release Unit 电动释放装置
EMS   1. Electromagnetic Surveillance 电磁监视 2. Electronic Mail System 电子邮件系统 3. Electronic Management System 电子管理系统 4. Electronic Message Services 电子报文业务 4. Electronic Message System 电子报文系统 6. Emergency Medical Service 急诊勤务（直升机的），紧急医疗服务 7. Engine Management System 发动机管理系统 8. Engine Monitoring System 发动机监视（控）系统，发动机监测系统 9. Entry Monitor System 进入（大气层）监测系统 10. Environmental Monitoring Satellite 环境监测卫星
EMSG   European Maintenance System Guide 欧洲维修制度指南
EMSI   1. Electronic Manufacturing Service Industry 电子制造服务业 2. European Manned Space Infrastructure 欧洲载人空间基础设施
EMSP   1. Embedded Machine Signal Processing 嵌入式机器信号处理 2. Enhanced Modular Signal Processor 增强型模块化信号处理机
EMSS   1. Electronic Message Service System 电子信息服务系统 2. Emergency Manual Switching System 应急人工交换系统 3. Emergency Mission Support System 应急任务支持系统 4. Experiment Mobile Satellite System 实验移动通信卫星系统 5. Experimental Manned Space Station 实验性载人空间站 6. European Mobile Satellite Services 欧洲移动卫星业务 7. European Mobile Satellite System 欧洲移动卫星系统
EMT   1. Equivalent Megatons 当量百万吨 2. Equivalent Megatonnage 等效百万吨当量 3. Elapsed Maintenance Time 维修历经时间 4. Elapsed Mission Time 任务历经时间 5. Electromagnetic Transducer 电磁探头，电磁变换器 6. Eye-Movement Tracking 眼动追踪 7. Electromagnetic Tomography 电磁层析成像 8. Electromagnetic Transient 电磁暂态 9. Engineering Model Thruster 工程模型发动机 10. European Mean Time 欧洲标准时间 11. Engine Materials Technology 发动机材料技术 12. Energy Management Technology 能源管理技术
EMT model   Electromagnetic Topology Model 电磁拓扑

模型
EMTR 1. Electronic Modular Test Rig 电子模块化测试平台 2. Emitter 发射源，发射机
EMU 1. Electromagnetic Unit 电磁单位 2. Engine Maintenance Unit 发动机维护装置 3. Engine Monitoring Unit 发动机监测装置 4. Extravehicular Maneuvering Unit 舱外航天服 5. Extravehicular Mobility Unit 舱外灵便式航天服
EMUDS Extravehicular Maneuvering Unit Decontamination System 舱外机动飞行装置减污染系统
emulate 仿真
emulation 仿真
emulation job 仿真作业
emulator 仿真器，仿真程序
emulator trap code 仿真陷阱码
emulator trap processor 仿真陷波处理机，仿真陷阱处理程序
emulsion （药）乳剂，（物化）乳状液，感光乳剂，乳胶
emulsion interference filter 乳胶干涉滤光片
emulsion plate 乳胶版
emulsion speed 乳剂感光度
EMUX 1. Electrical Multiplexing 电多路传输 2. Electronic Multiple Bus 电子多路数据总线 3. Electronic Multiplexing System 电子多路传输系统
EMWR Eddy Mach Wave Radiation 涡流马赫波辐射
EN 1. Engineer 工兵，机械师，工程师 2. Envelope 包络线，包迹
en route 在飞行途中
en route aircraft 在航线上的飞机
ENA 1. Engineering Next Assembly 下一工程装配件 2. Equipment Network Access 设备网络接入 3. Exhaust Nozzle Area 排气喷口面积
enable register 允许(启用)寄存器
enablers 促成事情成功的人或因素，促成因子
enabling pulse 启动脉冲
enabling signal 启动信号
ENAC Ecole Nationale de l'Aviation Civile 国立民航学校(法国)
ENAM Enamel 搪瓷，瓷漆，珐琅
enamel paint 搪瓷，瓷漆，珐琅，涂瓷漆，上珐琅
ENAP Energetic Neutral Atom Precipitation Experiment 高能中性原子淀析实验
ENB Effective Noise Bandwidth 有效噪声带宽
en-block construction 单元体结构，积木式构造
enbrittle 脆化
ENCAP Encapsulation 密封，封装，用胶囊包(起来)
encapsulate 压缩，将……装入胶囊，将……封进内部，封装
encapsulated ejection seat 密闭式囊形弹射座椅
encapsulating layer 气密限制层
encapsulation 囊封(太阳电池的包装工艺)，封装
encapsulation protocol data unit 包装规约数据单元
encapsulation service 包装业务
encapsulation test 密封性试验
encase 把……装入箱中，包起
encastre 固定梁(两端固定而非铰接的结构梁)，端部固定
ENCC Emergency Network Control Center 应急网络控制中心
ENCD Encode 编码，译码
ENCDR Encoder 编码器，编码号
Enceladus 【天】土卫二
enceladus system 土卫二系统
encipher 译成密码，编码
encircle 包围，环绕
encirclement (能源)圈闭，包围，环绕
Encke comet 【天】恩克彗星
Encke division 恩克环缝
Encke's method 恩克法
ENCL Enclosure 包围，围绕，封入，附件，封入物
enclave defense 飞地式防御，包围式防御
enclose 围绕，装入，放入封套，封闭
enclosed accumulator 封闭式蓄电池，密封式蓄电池
enclosed ejection 封闭式弹射
enclosed helmet 密封头盔
enclosed launcher 内装式发射装置
enclosure 盖，罩，外壳，包入物，封入物，附件，封入，围绕
encode 编码
encode address 编码地址
encoded sun sensor 编码式太阳敏感器(其输出是光学码盘的"编码"信号的数字式太阳敏感器)
encoder 编码器
encoder ambiguity 编码器的模糊性
encoding 编码
encoding altimeter 编码式高度表
encoding law 编码律
encoding the response 应答编码
encompass 围绕，环绕，围住，包围，包封，封入，包裹
encounter 相遇，交会
encounter angle 交会角，遭遇角
encounter b-plane 交会B平面
encounter condition 交会条件
encounter cylinder 相遇圆柱体
encounter day 交会日
encounter frame 接触框架
encounter mode 交会方式，会合方式
encounter phase of trajectory 轨道遭遇段

encounter plane 相遇飞机
encounter probability 交会概率,(飞机空中)相遇概率
encounter sequence 交会程序
encouraging agreement 相当一致(的)
encrypt 译成密码,加密,编码
encrypted message 密文
encryption 加密
encryption key 加密钥
encryption protocol 加密协议
ENCSD Encased 包装起来的,在箱内的,嵌入的
END 1. End of Data 数据结束,数据结尾 2. European Nuclear Disarmament(旨在反对美国 SDI 计划的)欧洲核裁军运动
end 端,末端,终结,最后,目的,目标,结果,限度
end area 端面积
end bend blading 端弯叶片,过弯叶片
end bending 端弯叶片,过弯
end burning 端面燃烧,香烟式燃烧
end cap 端盖
end cell 尾电池,末端电池
end cell switch 末端电池转换开关
end clearway 端净空
end cross section 端横截面
end effect 端部效应(指叶片端部)
end effector 末端执行器
end fixture 末端固定
end frame 端框
end game 最终的较量,决战拦截器(天基动能杀伤武器斯需要的高速拦截器)
end instrument 传感器,敏感元件,(计算机的)终端装置
end item 终端件,终端项目,成品
end mass (末)端质量
end of eclipse 层拦截器,食终
End of Heat 处暑
end of lifetime 寿命末期
end of medium character 媒体结束字符
end of totality 全食终
end of transmission block character 块传输结束字符(表示数据分块传输结束的计算机字符)
end of transmission character 传输结束字符
end organ 灵敏元件,传感器
end overlap 航向重叠
end packet 结束(数据)包
end packing 末端密封,端部密封
end part 尾部,(船舶锚泊系统)末端链节
end plate latch 后挡板锁键
end play 轴向间隙(指非正常的轴的纵向活动),端隙
end point 终(端)点

end product 成品,最后产品,燃烧产物
end speed 末速(尤指飞机脱离弹射器时与航母的相对速度)
end state 结束节点,最终状态,结束状态
end system 终端系统
end thrust 轴向推力,轴端推力
end time 终止时间,结束时间
End Tray "末盘"(苏联地面雷达的北约代名)
end user 最终用户,直接用户
end value 最终值,结果值
end view 端视图,侧视图
end voltage 终端电压
end wall (涡轮机或汽轮机的)端壁
end wall boundary 端壁边界
end wall boundary layer 端壁边界层
end wall boundary-layer 端壁边界层
end wall boundary-layer fluid 端壁边界层流体
end wall effect 端壁效应
end wall flow 端壁流动
end wall heat transfer 端壁热传导
end wall ignition 端壁点火
end wall ignition time 端壁点火时(间)
end wall loss 端壁损失
end wall pressure 端壁压力
end wall pressure distribution 端壁压力分布
end wall region 端壁区
end-around carry 循环进位
end-around shift 循环移位
end-burning 端面燃烧
end-burning grain 端面燃烧药柱
end-cell 末端电池
endeavor 尽力,努力,尝试,力图
Endeavour 1. 奋进号(美国 Apollo-15 号飞船指挥舱) 2. 奋进号(美国航天飞机轨道器)
end-effector 末端执行器,末端操作器,端部执行器
end-effector coupling device 末端执行器耦合装置
end-equipment 末端设备
end-face clearance 端面间隙
endfire 轴向辐射的,端射的(天线阵)
end-fire array antenna 端射阵天线
endgame (导弹的)末端博弈(导弹的末制导段的最后阶段),终端末制导,终端对策,终端制导,末端控制
end-game phase 终端末制导段,末端制导段
endgame trajectory 终端末制导弹道轨迹,末端制导弹道轨迹
end-initiated warhead 端部起爆弹头
endnote 1. 参考文献目录管理软件(美国科学信息所研制开发) 2. 尾注
endoatmosphere 稠密大气层

endoatmospheric 大气层内的,稠密大气层的
endoatmospheric engagement 稠密大气层交战
endoatmospheric intercept 稠密大气层拦截,稠密大气层截击
end-of-life （产品或设备已到使用寿命,不能再继续使用）报废
endogenous steam 内生蒸汽
endogenous variable 内生变量
end-on object 端向天体
Endo-NNK Endoatmospheric Nonnuclear Kill 大气层内非核杀伤
endoscope 内窥镜,孔探仪
endotherm 吸热
endothermic chemical reaction 吸热化学反应
endothermic fuel system 吸热型燃料(油)系统
endothermic fuel 吸热燃料
endothermic heat 吸入热
endothermic melting 吸热熔化
endothermic reaction 吸热反应
endplate 端板,端翅（靠近尾翼端部的小型辅助翼片）
endplate igniter 端板点火器
endpoint monitoring 终点监测
end-pumping 端抽运
end-state 结束节点,最终状态,结束状态
end-to-end 端到端的
end-to-end data system 终端站间数据系统
end-to-end simulation 全链路仿真,一体化模拟,端到端仿真
end-to-end test 完整检查,全程序检查
endurance 续航时间,续航力,最大续航时间（不包括空中加油）,持久性,耐久性
endurance limit （结构的）耐久极限,疲劳极限
endurance on station 最大巡逻时间
endurance test 长期试车,持久试车,耐久性试验
end-user 直接用户,终端用户
end-user device 终端用户设备
end-view 端视图,侧视图
end-view image 端视图成像,侧视图像
end-view structure 端视图结构,侧视图结构
ENE East North East 东北东
ENEC 1. European Norms Electrical Certification 欧洲标准电器认证 2. Extendable Nozzle Exit Cone 可延伸喷管出口锥
en-echelon structure 雁列(行)构造
enemy aircraft 敌方飞机
enemy defence 敌方防御
enemy radar 敌方雷达
energetic binder 高能黏合剂,含氧黏合剂
energetic compound 含能化合物,高能化合物

energetic interaction 能量(的)相互作用
energetic ion 载能离子,高能离子
energetic material 含能材料,高能材料
energetic nanocomposite 高能纳米复合材料
energetic particle event 高能粒子事件
energetic performance 能量性能,能量特性
energetic plasticizer 含能增塑剂
energetics 能量学,力能学
Energia 能源号
Energia/Buran 能源号/暴风雪号
Energiya 能源号/暴风雪号
energize 使活跃,给与……能量,给与……电压
energy 能,能量,能力,活力
energy absorber 吸能机构
energy absorption 能量吸收
energy absorption behavior 能量吸收特性,吸能特性
energy absorption test 能量吸收试验,减震试验
energy accommodation 能量调节
energy accommodation coefficient 能量调节系数,能量适应系数
energy addition 能量增加
energy analysis 能量分析,能值分析,能耗分析
energy analysis technique 能量分析方法,能值分析方法,能耗分析方法
energy analyzer 能量分析器
energy availability 可用能量
energy availability analysis 可用能量分析
energy availability loss 可用能量损失,可用能量损耗
energy availability rate 能量可用率
energy balance 能量平衡
energy bypass 能量旁路
energy bypass ratio 能量旁路比
energy calibration 能量刻度
energy change 能变,能量变化
energy channel 能量通道,能量频带
energy compensation 能量补偿
energy component 能部件（如 Absorbing Energy Component 吸能部件,Negative Energy Component 负能级分量）
energy conservation 节能
energy consumption 能耗
energy content 能含量,内能,含能量
energy conversion 能量转换
energy conversion efficiency 能量转换效率
energy cost 能量消耗,能源成本
energy coupling 能量耦合
energy density 能量密度（单位体积的声能）
energy deposition 能量沉积
energy difference 能差,能量差,能级差

energy dispersal signal 能量扩散信号
energy dissipation 能量耗散
energy distribution 能量分布,能量分配
energy efficiency 能效
energy electron 能电子
energy ellipsoid 能量椭球
energy equation 能量方程
energy exchange 能量交换
energy expenditure 能量消耗
energy expression 能量表达式
energy flux 能通量
energy function 能量函数(的)
energy gain 能量增益,能量收益
energy height 能量高度
energy identity 能量恒等式
energy increment 能量增加,能量增量
energy input 能量输入
energy integral 能量积分
energy level 能量级,能级
energy loss 能量损失,耗能
energy loss factor 能量损失因子,耗能因子
energy loss thickness of boundary layer 边界层能量损失厚度
energy management 能量管理
energy management system 能量管理系统
energy manoeuvrability 能量机动力
energy metabolism 能量代谢
energy of deformation 变形能
energy of electromagnetic field 电磁场能量
energy output 能量输出
energy profile (物理化学)能线图,能量分布(图)
energy propagation 能量传播
energy propulsion 能量推进(如 Beam-Energy Propulsion 光束能推进,Solarenergy Propulsion 太阳能推进)
energy range 能区,能量范围
energy rate 能量比
energy ratio 能量比
energy ratio of blowdown tunnel 吃气式风洞能量比
energy ratio of continuous tunnel 连续式风洞能量比
energy recovery 能量回收,能量恢复
energy recovery system 能量恢复系统
energy release 能量释放
energy release rate 能量释放率
energy representation 能量表现,能量表象
energy requirement 能量需求,能量需要
energy reserve 能量储备,能源储备
energy sink 能量库,能量槽,能量耗散,消耗能源
energy source 能源
energy spectrum 能谱

energy spectrum analysis 能谱分析
energy spectrum density function 能谱密度函数
energy spectrum function 能谱函数
energy spectrum width 能谱宽度
energy stability 能量稳定(度)
energy stability theory 能量稳定性理论
energy state 能量状态
energy storage 能量储备,蓄能
energy storage capacitor 储能电容器
energy surface 能面的
energy technology 能源技术
energy theory 能量理论,能量原理
energy threshold 能量阈值,阈值能量
energy transfer 能量变换,能量转换,能量传递
energy transfer path 能量变换路径,能量转换路径,能量传递路径
energy transformation 能量变换
energy utilization 能量利用
energy value 能量值,热值
energy-conversion coefficient 能量转换系数
energy-efficient engine 节能发动机
energy-matching condition 能量匹配条件
energy-transfer reaction 能量转换反应
ENF 1. Expected Number of Failures 预期失效值,预期故障次数 2. External Noise Factor 外部噪声指数 3. External Noise Field 外部噪声场
enforce 加强,实施,执行
ENG 1. Energize 供能,激励 2. Engage 衔接,接通,啮合,接合 3. Engine 发动机
ENGA Engaging 啮合,连接
engage (下令)攻击目标(用于空中截击),与拦阻索或拦阻网接合(连住)
engage button 接通按钮
engage the autopilot 接通自动驾驶仪
engagement 交战
engagement altitude 接通高度(如接通自动驾驶仪,投弹按钮等),投入战斗高度,开火高度
engagement control 交战控制
engagement simulation 攻防对抗仿真
engagement state 交战状态
engagement time 射击持续时间,开火时间
engaging force 啮合力,插拔力
engaging speed 飞机与拦阻网接合时的相对速度
engaging switch bar 电门连接条(用以同时接通自动驾驶仪各舵机)
engin 〈法语〉导弹
engine 发动机
engine/airframe matching 发动机-机体匹配
engine acceleration 发动机加速性

engine altimeter　测量发动机增压的高度计
engine altitude characteristic　发动机高度特性
engine altitude simulated test facility　发动机高空模拟试车台
engine bleed air system　发动机引气系统
engine calibration test　发动机校准试验
engine car　发动机短舱,发动机吊舱
engine chilldown system　发动机预冷系统
engine compartment　发动机舱
engine compartment checking ladder　发动机舱测式工作梯
engine cutoff　发动机关机
engine deceleration　发动机减速性
engine display　发动机显示器,发动机参数显示器
engine dry mass　发动机干质
engine fire extinguisher　发动机灭火器
engine flight test bed　发动机飞行试验台,空中试车台
engine ignition timer　发动机点火定时机构
engine indication and crew alerting system　发动机仪表和机组告警系统
engine life computer　发动机寿命计算机
engine line characteristic　发动机管路特性
engine maximum negative thrust　发动机最大负推力
engine mixture ratio　发动机混合比
engine mount　发动机架
engine mounting　发动机架,发动机安装节
engine negative thrust　发动机负推力
engine nozzle　发动机喷管
engine operability　发动机适用性
engine operating duration　发动机工作时间
engine performance　发运机性能
engine performance parameter　发动机性能参数
engine performance test　发动机性能试验
engine pressure ratio　发动机压比,发动机压力比,发动机增压比
engine rating　发动机状态,功率状态
engine reliability　发动机可靠性
engine reliability test　发动机可靠性试验
engine reverse thrust　发动机反推力
engine specific impluse　发动机比冲
engine speed　发动机转速
engine speed sensing unit　发动机转速传感器
engine stability margin　发动机稳定性裕度,发动机喘振裕度
engine starting　发动机起动
engine starting contactor　发动机起动接触器
engine starting control　发动机起动控制
engine starting fuel　发动机起动燃料
engine starting gear　发动机起动装置

engine starting handle　发动机起动手柄
engine starting mechanism　发动机起动机构
engine starting system　发动机起动系统,引擎起动系统
engine structural integrity program　发动机结构完整性大纲,发动机结构完整性计划(发动机结构设计应遵循的大纲)
engine test　发动机试验
engine test bed　发动机试车台
engine test in simulated altitude condition　发动机高空仿真试验
engine test stand　发动机试车台
engine throttle characteristic　发动机节流特性
engine thrust　发动机推力
engine thrust frame　发动机(推力)架
engine thrust-mass ratio　发动机推质比
engine torque indicator　发动机扭矩指示器
engine vibration monitoring system　发动机振动监视系统
engine wet mass　发动机湿质
engine-airframe integration　发动机与机身一体化
engine-attributable　发动机故障引起的,发动机故障造成的
engine-driven alternator　发动机驱动的交流发电机
engineer on-board　飞行工程师
engineering　研制部门,(硬件的)设计和研制,工程研制
engineering ceramics　结构陶瓷,工程陶瓷
engineering cost　工程成本
engineering cybernetics　工程控制论
engineering department　轮机部门
engineering fixes　工程安装,技术安装
engineering management　工程管理
engineering mock-up　全尺寸实体模型
engineering model　工程模型,初样星(航天器研制阶段的初样星模型,用来评定航天器重量、尺寸和电气性能的协调性)
engineering module　工程舱
engineering prototype　工程样机
engineering psychology　工程心理学
engineering survey　工程测量
engineering test panel　工程试验板(盘,台)
engineering test satellite manipulator　技术实验卫星机械手
engineering test vehicle　技术试验飞行器,工程试验飞行器
engineering time　工程研制工时
engine-gauge unit　发动机三用表
engine-off in flight　空中停车
engine-plus　(Fuel Weight)发动机加燃料重量
engine-speed method　转速法

english system of unit 英制单位
ENGR Engineer 工程师
ENGRG Engineering 工程,设计,技术
ENGRV Engrave 雕刻
ENH 1. Enhancement 增强,加强 2. Earth Near Horizon 近地平线
enhance 提高,增加,加强
enhanced accuracy 增强精度
enhanced data access features 增强的数据访问特性
enhanced data encryption schemes 增强数据加密机制
enhanced data rate 增强型数据速率
enhanced datum 增强数据
Enhanced Flight Screener 强化飞行筛选机
enhanced gas turbine 增强型燃气轮机
enhanced heat loss 增强热损耗,大热损耗
enhanced image 增强图象
enhanced infrared satellite cloud picture 增强显示红外卫星云图
enhanced radiation (太阳的)增强辐射
enhanced reactivity 增强反应活性
enhancement 增强,加强,提高,改善
enhancement factor 增强因子(的),强化系数
enhancement mode field effect transistor 增强型场效应管
enhancement technique 增强技术
enhancement-depletion mode logic 增强-耗尽型逻辑
ENJJPT Euro-NATO Joint Jet Pilot Training 欧洲与北约联合的喷气式飞机飞行员训练
enlarge 扩大,增大
enlarged gap 差距扩大,裂(缝)隙扩大
enlargement factor 放大因数,放大率
ENLG Enlarge 扩大
ENLGD Enlarged 扩大的
ENN Expand Nonstop Network 无休止扩展网络
ENNK Endo-atmospheric Nonnuclear Kill 大气层内非核杀伤
ENOC Engineering Network Operations Center 工程网络运行中心
enough energy 足够能量
enough time 足够时间
enplanements 旅客登记数
enquiry character 询问字符
ENRGZ Energize 激磁,激励
enrichment 富油,浓缩,富集(Oxygen Enrichment 富氧)
enroute 在航线上,飞行中,沿航线,航线飞行
en-route area 航线区域
enroute base 途中基地
enroute climb 沿航线爬高,航线飞行上升

enroute height 航线高度
enroute maintenance 航线维修,短停维修
enroute navigation 沿航线领航,沿航线航行
enroute navigational facility 航线导航设备
en-route reliability 航行可靠率
enroute report 航线天气报告
enroute support team 途中保障队
enroute time 途中时间
enroute traffic control service 途中空中交通管制勤务
enroute weather 航线天气
ENS Euler/Navier-Stokes 欧拉-纳维尔斯托克斯(方程)
ensemble average 总体均值,总平均值
ensemble domain 集合域
ensemble-averaged velocity 总体平均速度
ENSIM Environment Simulation 环境模拟
ENSIP Engine Structural Integrity Program 发动机结构完整性计划(美国空军)
ensurance period 保险期
ensure 担保,保证,使安全,确保
ENT 1. Enter 装订 2. Entrance 入口 3. Entry 输入 4. Equivalent Noise Temperature 等效噪声温度
entanglement 纠缠,缠结(如 Quantum Entanglement 量子纠缠)
entanglement density 缠结密度
Entel Empresa Nacional de Telecomunicaciones 国营电信公司(阿根廷、智利等国的)
entering-aircraft 进场飞机
entering-aircraft bearing 进场飞机航向
entering-aircraft position 进场飞机方位
entering-aircraft position angle 进场飞机方位角
Enterprise 创业号
ENTG Euro/NATO Training Group 欧洲与北约联合训练团
enthalpy 焓,热焓(热力学单位)
enthalpy change 焓变(化)
enthalpy deficit 失焓,热焓损失
enthalpy deficit factor 失焓因子
enthalpy drop 焓降
enthalpy extraction 焓取出
enthalpy extraction ratio 焓取出比
enthalpy loss 焓损失
enthalpy method 焓法,热焓方法
enthalpy of formation 生成焓,形成焓
enthalpy ratio 焓比
enthalpy-entropy diagram 焓-熵图
entire assembly 整体装配,总装
entire blade 完整叶片
entire domain 全频域
entire earth 整个地球

entire engine 整台发动机
entire field of view 全视域,全视野
entire flight 整个飞行(过程)
entire frequency range 整个频率范围
entire propellant 全部推进剂
entire pulse 完整脉冲,整个脉冲
entire range 整个范围
entire satellite 整颗卫星
entire search region 整个搜索区域
entire set (数)全集
entire time interval (天)时间间隔
entire wake 整体唤醒
entity 实体,实际存在物,本质
entity set model 实体集模型
entomopter 振翼机,仿昆虫飞行器
entrain 带走,夹带
entrainment 夹带,确定周期,吸入,带走,雾沫,以雾沫状夹带走
entrainment rate 流量率,夹带率
entrainment ratio 喷射系数
entrance 进流段,进入,入口,进风口
entrance cone 风洞加速段(收敛段),风洞进口锥
entrance height 入口高度
entrance hole 注入口
entrance liner 入(进)口
entrance perturbation 入(进)口扰动
entrance pupil 入射光瞳
entrance separation bubble 入口分离泡
entrance spillage 入口溢出,进口溢出
entrance velocity 进口速度,入口速度
entrant 进入者,进入物
entropy 熵
entropy assumption 熵假设
entropy coding 熵编码
entropy contour 熵轮廓
entropy disturbance 熵扰乱
entropy gain 熵增益
entropy generation 熵产(生)
entropy gradient 熵梯度
entropy increase 熵增
entropy layer 熵层
entropy layer swallowing 熵吞
entropy maximum 熵极大
entropy method 熵(法)
entropy mode 熵模型,熵模式
entropy power 熵功率
entropy production 熵产(生)
entropy representation 熵表象,熵表示,熵表征
entropy rise 熵增长

entropy wave 熵波
entry (再)进入(如航天器从外层空间进入行星大气),入口,入口段,登记,记入,条目,项目
entry condition 进入条件
entry constraint 进入约束(条件)
entry corridor 进入通道,进入走廊
entry flight (再)进入飞行
entry gate 进入点(从此点进入航站机动飞行区)
entry guidance (再)进入制导
entry guidance system (再)进入制导系统
entry heating (飞行器)进入大气层时升温
entry interface 进入边界
entry interval 进入间隔
entry mass 进入质量
entry mission 进入任务
entry phase 进入阶段
entry point 进入点
entry problem 进入问题
entry sequence 进入程序
entry speed 进入速度
entry state 进入状态
entry system 输入系统,登入系统
entry time 进入时间,入口时间
entry time distribution 进入时间分配
entry trajectory 进入轨道
entry vehicle (再)进入航天器,(再)进入飞行器,再入航天器,再入飞行器
entry velocity 进入速度
ENU East-North-Up 东-北-上(一种坐标系)
ENU coordinate system 东-北-上坐标系
enumerate 列举,枚举,数
enumeration 计数,列举,细目,详表,点查
ENV Envelope 封皮,外套,包迹线
envelope 外壳,封皮,包层,(软式飞艇的)气囊,(艇的)蒙皮,(武器的)有效使用范围,包络,包层
envelope constraint 包络约束(条件)
envelope detection 包络检波,包络检测
envelope diameter 飞艇外壳直径
envelope growth selection 增幅选择
envelope growth selection circuit 增幅速率选择电路
envelope limit 包线限制
envelope method 包络法
envelope of feasible future position 未来可能方位范围
envelope of relative wind 相对风范围
envelope power 峰包功率
envelope shock 包络击波
envelope velocity 包络速度,群速度
enveloping balloon 包络气球
enveloping body 包络体

enveloping curve 包络曲线
enveloping ring 包络环
enveloping surface 包络面
ENVIR Environment 环境
environment 环境
environment design margin 环境设计余量
environment dispersion 环境扩散
environment factor 环境因数
environment interaction 环境互作(与环境的相互作用)
environment management 环境管理
environment model 环境模型
environment monitoring 环境监控
environment prediction 环境预示
environment simulation 环境模拟
environment simulator 环境模拟器
environment state 环境状态
environment stress 环境应激
environment test equipment 环境试验设备
environmental (control) system 环境控制系统
environmental appraisal 环境评价,环境鉴定
environmental capacity 环境容量
environmental chamber 环境实验室,环境模拟室
environmental characteristics 环境特性
environmental compatibility 环境适应性,环境兼容性
environmental conditions 环境条件
environmental control 环境控制
environmental control and life support subsystem 环境控制和生命保障分系统
environmental control system 环境控制系统
environmental corrosion (周围介质)接触腐蚀,环境腐蚀
environmental disaster control 环境灾害监测
environmental disturbance 环境扰乱
environmental dosemeter 环境剂量计
environmental effect 环境效应,环境影响
environmental engineering 环境工程
environmental engineering of flight vehicle 飞行器环境工程
environmental exposure 环境照射量
environmental factor 环境因子
environmental flight 有空中交通管制的飞行
environmental impact 环境效应,环境影响
environmental issue 环境问题
environmental laboratory 环境实验室
environmental lapse rate 环境温度递减率(测量高度增加时温度降低的速率)
environmental limit 环境影响限度
environmental medicine 环境医学
environmental mock-up 环境模装舱(用来帮助设计环境控制系统的模装座舱)
environmental noise 环境噪声
environmental observation mission 环境观测飞行任务
environmental perturbation 环境扰动
environmental pollution control 环境污染控制
environmental pressure 环境压力
environmental pressure decrease 环境压力减轻
environmental psychology 环境心理学
environmental regulation 环境管理,环境规制,环境管制
environmental satellite 环境卫星
environmental simulation 环境仿真
environmental state 环境态
environmental stress 环境应力
environmental stress screening 环境应力筛选
environmental stress screening test 环境应力筛选试验
environmental surveillance 环境监测
environmental survey satellite 环境勘测卫星
environmental system 环境系统
environmental temperature 环境温度
environmental test 环境试验,环境测试
environmental test program 环境试验计划
environmental testing 环境测试
environmental torque 环境力矩
environmental variables 环境变量
envision 想像,预见,展望
enzyme electrode 酶电极
enzyme sensor 酶敏感元件,酶传感器
EO 1. Earth Observation 对地球观测 2. Earth Orbit 地球轨道 3. Electrooptic 电光的 4. Electro-Optical 电子光学的 5. Engineering Opeartions 工程运作 6. Engineering Order 工程指令 7. Escape Orbit 逃逸轨道 8. Explosive Ordnance 弹药,爆炸物
EO guidance Electro-Optical guidance 电子光学制导
EO-AAO Engineering Order-Advance Assembly Order 工程指令-先行装配指令
EOAP Electronic Overhead Annunciator Panel 顶部电子告示板,电子过热指示面板
EOAR European Office of Aerospace Research 欧洲航空航天研究处(美国空军)
EOARD European Office of Aerospace Research and Development 欧洲航空航天研究和开发处(美国空军)
EOB 1. Electronic Order of Battle 电子作战命令 2. End Of Block (符)块结束
EOC 1. Earth Orbital Capsule 环地球轨道舱 2. Eletro-Optic Converter 电/光转换器
EOCM Electro-Optical Countermeasure 电子光学对抗,电子光学干扰
EOD 1. Every Other Day 隔天 2. Explosive Ordnance

Disposal 未爆弹药处理,爆炸物处理
EoD　End of Descent 下降结束
EOE　1. Earth Orbit Ejection 地球轨道入轨 2. Elasto-Optical Effect 弹性光学效应
EOFC　Electro-Optical Fire Control 电子光学火控
EOG　Electro-Oculogram 眼电图
EOGB　Electrooptically Guided Bomb 电子光学制导炸弹
EOGO　Eccentric Orbiting Geophysical Observatory 偏心轨道地球物理观测台
EOI　1. Electro-Optical Instrumentation 电子光学仪表 2. Engineering Operating Instructions 工程运作说明书 3. Exchangeability Of Items 项目互换性
EOIS　Electro-Optical Imaging System 光电成像系统
EOL　1. Effective Operational Length 有效工作时间 2. End Of Life 寿命末期 3. Engine-Off Landing 发动机关机的着陆
EOL power　End-Of-Life power 寿命末期功率
Eole　风神卫星(法国气象卫星)
EOM　Earth Observation Mission 地球观测任务
EOMS　Earth Orbital Military Satellite 地球轨道军用卫星
Eonnex　绒耐克丝(一种航空织物的商品名)
EOP　1. End Of Part 零件端头 2. Engine Oil Pressure 发动机滑油压力 3. Engine Operating Point 发动机工作点 4. Engineering Operating Procedure 工程运作程序 5. Experiment of Opportunity Payload 搭载有效载荷试验
EOPP　Earth Observation Preparatory Programme(欧洲航天局)地球观测筹备计划
EOR　1. Earth Orbital Rendezvous 地球轨道交会 2. End Of Record 记录结束,录音结束 3. Evidence Of Release 发证证据 4. Extend/Off React 放下(或伸出)/关断/收起(或缩入)(例如起落架收放开关位置)
EORBS　Earth Orbiting Recoverable Biological Satellite 地球轨道可回收生物卫星
EORSAT　Elint Ocean Reconnaissance Satellite 电子情报型海洋侦察卫星,苏联海洋监视卫星
EOS　1. Earth Observation Satellite 地球观测卫星 2. Earth Observation System 地球观测系统(美国、西欧、日本联合的对地观测卫星) 3. Earth to Orbit Shuttle 地球与轨道间航天飞机 4. Electrical Over Stress 电超载 5. Electro-Optical Scanner 电-光学扫描仪 6. Electro-Optical System 光电系统,电子光学系统 7. Electrophoresis Operations in Space 空间电泳作业 8. Embedded Operating System 嵌入式操作系统 9. Emergency Operations System 应急操作系统 10. Emergency Oxygen System 应急氧气系统 11. End Of Signaling 信令结束 12. Equation Of State 状态方程 13. External Observation System 外部观测系统
Eos　【天】曙神星(小行星221号)
EOSAT　Earth Observation Satellite Corp 地球观测卫星公司(美国)
EOSP　Electro-Optical Signal Processor 电子光学信号处理器
EOSS　Earth Orbiting Space Station 地球轨道空间站
EOT　End Of Transmission Character 传输结束符
EOTAD　Electro-Optical Targeting And Designation 电光瞄准和识别
EOTDS　Electro-Optical Target Detection System 电子光学目标探测系统
EOTV　Electric Orbit Transfer Vehicle 电推进轨道转移飞行器
EOVL　Engine-Out Vertical Landing(发动机)停车垂直降落
EOW　Electro-Optical Warfare 电子光学战
EP　1. Electric Primer 电火帽 2. Electrical Power 电力,电源 3. Engineering Procedure 工程程序 4. Explosion-Proof 防爆 5. External Power 外电源
EP system　Electrical Power System 电力系统,电源系统
EPA　1. Engineering Parts Approval 工程零件批准 2. Environmental Protection Agency 环境保护局(美国) 3. Epoxy Polyamide 环氧聚酰胺
epact　【天】闰余,岁首月龄
epagomenal day　闰日
EPC　1. Engineering Process Council 工程工艺委员会 2. External Power Contactor(Connector) 外接电源插座,机场电源插座,外电源连接器
EPCC　Engine Performance Control Committee 发动机性能控制委员会
EPCG　Engine Performance Control Group 发动机性能控制组
EPCM　Engineering Process Control Manual 工程工艺控制手册
EPCO　Emergency Power Cutoff 紧急电源断开
EPCR　External Power Control Relay 外电源控制继电器
EPCS　Engine Propulsion Control System 发动机推进控制系统
EPCU　Electric Power Control Unit 电源控制装置
EPD　Electrophoretic Display 电泳显示
EPDC　Eletrical Power Distribution and Control 电能(力)分配和控制
EPDD　Engineering Product Definition Data 工程产品定义数据
EPDWO　Engineering and Product Development Work

Order 工程和产品研制指令
EPE 1. Energetic Particle Explorer 高能粒子探险者（卫星） 2. Estimated Position Error 估计位置误差
EPG 1. Electronic Proving Ground 电子验证场 2. European Participating Governments (Groups) 欧洲参加国政府集团
EPGS Electrical Power Generation System 发电系统
ephemeris 【天】星历
ephemeris accuracy 星历精度
ephemeris accuracy goal 星历精度目标
ephemeris calculation 星历计算
ephemeris capture 星历获取，星历捕获
ephemeris cycler 星历循环
ephemeris day 星历表日，历书日
ephemeris determination technique 星历确定技术
ephemeris error 星历误差
ephemeris estimation 星历预测
ephemeris hour angle 天文时角
ephemeris longitude 天文历经度，历书经度
ephemeris meridian 天文子午线，历书子午线
ephemeris model 星历模型
ephemeris problem 星历问题
ephemeris second 历书秒
ephemeris sidereal time 天文恒星时，历书恒星时
ephemeris solution 星历解算
ephemeris time 历书时，天文历时
ephemeris track 星历跟踪，星历追踪
ephemeris trajectory 星历轨道
ephemeris transit 历书中天
EPI 1. Engine Performance Indicator 发动机性能指示器 2. Electronic Position Indicator 电子位置指示器 3. Elevator (Surface) Position Indicator 升降舵位置指示器 4. Engineering Program Integration 工程项目集成 5. Expand Position Indicator 延伸位置指示器 6. Expected Point of Impact 预期弹着点，预期命中点
EPIA European Photovoltaic Industry Association 欧洲光电设备工业协会
EPIC 1. Electronic Preassembly Integration in CATIA (System) CATIA 系统中的电子预装配集成 2. Emergency Procedures Information Centre 应急措施信息中心（英国） 3. Engineering and Production Information Control 工程和生产信息管理 4. Epitaxial Passivated Integrated Circuit 外延钝化集成电路
epicycle 【天】本轮
epicycle element 星历根数
Epimetheus 【天】土卫十一
epipolar constraint 极线约束，核线约束
epipolar correlation 核线相关
epipolar line 核线
epipolar plane 核面
epipolar ray 核线
epipole 核点
EPIRBS Emergency Position-Indicating Radio Beacon 应急定位无线电信标
epistemic uncertainty 认知不确定性
epitaxial growth 外延生长
epitaxial isolation 外延隔离
epitaxial process 外延生长过程
epitaxial transister 外延型晶体管
epitaxy 外延
epitaxy defect 外延缺陷
epitaxy stacking fault 外延堆垛层错
epitaxy technique 外延工艺
E-PKG Electronics Package 电子设备盒
EPL 1. Electronic Packing List 电子打包系列 2. Engine Power Lever 油门杆
EPLD Electrically Programmable Logic Device 电气可编程逻辑装置
EPMaRV Earth-Penatrating Manoeuvring Re-Entry Vehicle 进入稠密大气层机动弹头（再入飞行器）（美国空军）
EPMS Engine Performance Monitoring System 发动机性能监控系统
EPN European Participating Nations 欧洲参加国
EPNDB 1. Effective Perceived Noise in Decibels 有（效）感知噪声分贝 2. Equivalent Perceived Noise Decibel 等效感知噪声分贝
EPNL Effective Perceived Noise Level 有效感觉噪声水平
EPO Earth Parking Orbit 地球停泊轨道，地球驻留轨道
epoch 【天】历元，纪元，世，时期，初相，（卫星）过地点时刻
epoch state vector 历元状态向量
epoxy 1. 环氧的 2. 环氧树脂
epoxy composite 环氧树脂复合材料
epoxy fabric 环氧树脂纤维（材料）
epoxy matrix 环氧树脂基体
epoxy panel 环氧树脂（面）板
epoxy resin 环氧树脂
epoxy transistor 塑封晶体管
epoxy-silicone paint 环氧有机硅漆（涂料型热控涂层）
EPP 1. Emergency Power Package 应急动力包，应急电源包 2. Engineering Program Plan 工程项目大纲
Eppler 埃普勒翼剖面簇
EPR 1. Engine Pressure Ratio 发动机压力比 2. Engineering Part Release 工程零件发放 3. External Power Receptacle 外接电源插座

EPRF　Environmental Prediction Research Facility 环境预测研究中心（美国海军）
EPRL　Engine Pressure Ratio Limit 发动机压力比极限
EPROM　1. Electrically Programmable Read Only Memory 电可编程序只读存储器 2. Erasable Programmable Read-Only Memory 可擦（可重写）可编程只读存储器
EPS　1. Electric Power Storage 蓄电池 2. Electrical Power System 电源系统 3. Emergency Power System 应急电源系统 4. Energetic Particle Satellite 高能粒子探测卫星
EPSCS　Enhanced Private Switched Communications Service 增强型专用交换通信业务
EPSOC　Earth Physics Satellite Observation Campaign 地球物理学卫星观测运动
EPSU　European Public Service Union 欧洲公共业务联盟
EPT　External Pipe Thread 外管螺纹
EPTA　Expanded Programme of Technical Assistance 技术援助发展计划署（联合国）
EPU　1. Emergency Power Unit 应急动力装置 2. Estimated Position Uncertainty 估计的位置不确定性
EPWR　Emergency Power 应急电源
EQ　Equation 公式，方程式
EQA　Equipment Quality Analysis 设备质量分析
EQC　1. Engineering Quick Change 工程快速更改 2. Equipment Code 设备代码
EQL　1. Equalizer 补偿器，平衡器 2. Equally 等同的，相同的
EQL SP　Equally Spaced 等距器的
EQPT　Equipment 设备
EQR　1. Equivalence Ratio 当量比 2. External Quality Review 外部质量评估
equal and opposite value　绝对值相等而符号相反的值
equal area　等面积（法）
equal deflections　等偏移（原则）
equal gain combiner　等增益合成器
equal latitude　等纬度
equal mass　等质量
equal nodal rate condition　等波节率条件
equal number　等量
equal precision measurement　等精度测量
equal ripple approximation　等波纹逼近
equal value gray scale　等值灰度尺
equal weight　等重，等权
equal-altitude method of multi-star　多星等高法
equal-amplitude surface　等幅面
equality constraint　等式约束（条件）
equalization　均衡
equalized maintenance program　均分维修大纲，均衡维修大纲
equalizer　均衡器，平衡器，均压器，均值器，补偿器，补偿电路，均压线
equalizer circuit　均衡电路
equalizing lattice　均压网
equalizing pulses　平衡脉冲
equally tilted photography　等倾摄影
equal-probable error ellipse　等概率误差椭圆
equate　使相等，相当于，等同
equation　相等，平衡，公式，等式，方程（式）
equation clock　时差钟
equation formulation　方程式提出
equation in conservation form　守恒型方程
equation in nonconservation form　非守恒型方程
equation of aeroelastic stability　气（动）弹（性）稳定性方程
equation of center　（椭圆运动）中心差
equation of continuity　连续方程
equation of equinoxes　二分差，赤经章动
equation of light　光行时差
equation of motion　运动方程
equation of motion n-body problem　n体问题运动方程
equation of radiative transfer　辐射转移方程
equation of relative motion　相对运动方程
equation of satellite motion　卫星运动方程
equation of state　状态方程，物态方程
equation of state of gas　气体状态方程
equation of time　时差（真太阳时与平太阳时的时刻之差）
equation system　方程组，方程系统，方程体系
Equator　赤道卫星（监测地球周围环流的卫星）
equator　赤道圈，地球赤道，中纬线
equator of date　瞬时赤道
equator of epoch　历元赤道
equator reference frame　赤道坐标系
equatorial　1. 赤道仪 2. 赤道（附近）的
equatorial anomaly　赤道导常
equatorial armillary sphere　赤道经纬仪
equatorial bulge　赤道隆起（部分）
equatorial coordinate system　赤道坐标
equatorial crossing time　穿越赤道时间
equatorial day　恒星日
equatorial electrojet　赤道电集流
equatorial elements　赤道诸元
equatorial enhancement　赤道增强
equatorial horizontal parallax　赤道地平视差
equatorial mounting　赤道装置

equatorial orbit 赤道轨道
equatorial orbital elements 天球(赤道)轨道要素,天体轨道参数
equatorial parallax 赤道视差
equatorial plane 赤道面
equatorial point 二分点
equatorial points of earth-moon system 地-月系对称点
equatorial radius 赤道半径
equatorial rectangular coordinates 直角坐标
equatorial region 赤道带
equatorial ring current 赤道环流
equatorial satellite 赤道卫星
equatorial system of coordinates 赤道坐标系
EQUIL Equilibrium 平衡,均衡
equilibrium 平衡,均势,平静
equiaccuracy chart 等精度(曲线)图
equiangular positioning grid 等角定位网格
equidistant hole 等距孔
equidistant projection 等距投影
equilateral 等边的,等面的
equilateral points of earth-moon system 地-月系对称点
equilateral triangle 等边三角形
equilateral triangle point 等边三角形点
equilibrate 平衡,补偿,均衡
equilibration 平衡,平衡化,平衡作用
equilibrium 平衡,均衡(复数 equilibria)
equilibrium airspeed 平衡空速,空速稳定值
equilibrium calculation 平衡计算
equilibrium carrier 平衡载流子
equilibrium charge 平衡电荷,均衡充电
equilibrium chemistry 化学平衡
equilibrium combustion 均衡燃烧
equilibrium composition 平衡成分
equilibrium computation 配平计算
equilibrium condensation 平衡冷凝,平衡凝结
equilibrium condition 平衡条件
equilibrium configuration 初始平衡构型
equilibrium detonation 均衡爆震
equilibrium distance 平衡间距
equilibrium end state 平衡终结状态,均衡末态
equilibrium enthalpy addition 热焓均匀增加(大)
equilibrium equation 平衡方程
equilibrium expansion 均匀扩张
equilibrium flow 平衡流动
equilibrium growth 均衡增长
equilibrium height 平衡高度
equilibrium hypothesis 平衡假设,平衡假定
equilibrium ionization 平衡电离,均衡电离
equilibrium line 均衡线,平衡曲线
equilibrium location 均衡点,平衡位置
equilibrium mhd 平衡磁流体
equilibrium model 平衡模式
equilibrium mole fraction 摩尔分数平衡
equilibrium performance 平衡特性
equilibrium point 平衡点
equilibrium position 平衡位置,均衡位置
equilibrium pressure 平衡压力
equilibrium procedure 平衡过程
equilibrium region 平衡区域
equilibrium separation factor 平衡分离因数
equilibrium separation process 平衡分离过程
equilibrium set 平衡集
equilibrium shape 平面平衡形态
equilibrium solution 平衡解
equilibrium state 平衡态
equilibrium temperature 平衡温度
equilibrium test 平衡试验
equilibrium theory 均衡论,平衡理论
equilibrium tide 平衡潮
equilibrium value 平衡值,均衡值
equilibrium vapour 平衡蒸汽,平衡蒸汽压
equilibrium operating 平衡运行,均衡运行
equilibrium operating temperature 平衡工作温度
equilibrium-gas effect 平(均)衡气体效应
equilong circle arc grid 等距圆弧网格
equinoctial 1.昼夜平分的 2.昼夜平分线,天赤道
equinoctial colure 二分圈
equinoctial day 昼夜平分日
equinoctial orbital element 赤道轨道根数
equinoctial points 二分点
equinoctial system of coordinates 分至坐标系
equinoctial year 分至年
equinox 二分点
equinox correction 春分点改正(星表)
equinox inversion manoeuvre 二分点倒飞机动
equinox mean 二分点平均值
equinoxes 二分点
EQUIP Equipment 设备
equiperiod orbit 等周期轨道
equi-period transfer orbit 等周期转移轨道
equiphase surface 等相面
equiphase zone 等相位区
equipment 1.设备,装备 2.(某一特定空运航线上使用的)飞机机型(或飞机等级)
equipment bay 仪器舱,设备舱
equipment interchange 装备交换使用
equipment operational ready 装备完成使用准备,完成使用准备的装备,良好装备

equipment section　仪器舱,设备舱
equipment shelf　仪器支架
equipment status parameter　设备状态参数
equipment to be flown　要做飞行试验的设备
equipotential　等势的,等位的
equipotential surface　等位面
equipotential surface of gravity　重力等位面
equipped empty weight　带设备空重
equiscalar　等标量的
equispaced　等间隔的,均匀分布的
equity　股权
equity funds　股权基金
equity standards　公平标准
EQUIV　Equivalent 等效的,当量的,当量
equivalence　当量,等效,等价
equivalence criterion　等效原则,等价准则
equivalence function　等效函数,等价函数
equivalence partitioning　等价类划分
equivalence ratio　恰当的和实际的油气比的比值,当量比
equivalence ratio oscillation　当量比振动
equivalence ratio perturbation　当量比扰动
equivalence ratio range　当量比范围(区间)
equivalence techniques　等效技术
equivalent　1.当量,等效,等值,克当量 2.相当的,相等的,等效的,等价的,等值的
equivalent airspeed　当星空速,等效空速
equivalent antenna　等效天线
equivalent available seat-miles　当量可用座英里
equivalent beam　等效梁,等值梁
equivalent brake horsepower　当量轴马力
equivalent cabin altitude　座舱高度(根据座舱压力折算出高度)
equivalent circuit of a solar cell　太阳电池等效电路
equivalent coil　等效线圈
equivalent con　(计)等价构型,等价构图,等价配置,(计)等效构型,(物)等价组态,等效组态
equivalent current system　等效电流系
equivalent delay　等效时延
equivalent device　等效器,等效装置,模拟器
equivalent diameter　等效直径,当量直径
equivalent divergent angle　当量扩张角
equivalent drag area　等效阻力面积
equivalent duct　等效管
equivalent equator　等价赤道
equivalent flat-plate area　等效(当量)平板面积,等效阻力面积
equivalent focal length　等值焦距
equivalent gap　等效隙缝

equivalent horsepower　当量马力,当量功率
equivalent isotropic radiated power　等效全向辐射功率
equivalent jet diameter　等效喷口直径
equivalent level of safety　等效安全水平
equivalent mass　等效质量
equivalent megatonnage　等效百万吨当量
equivalent monoplane　等效单翼机
equivalent monoplane aspect ratio　(双翼机的)等效单翼机展弦比
equivalent network　等效网络
equivalent noise temperature　等效噪声温度
equivalent parasite drag　等效寄生阻力
equivalent parasite drag area　等效寄生阻力面积
equivalent pendulum　当量摆
equivalent planar fin configuration　等效平面翼结构
equivalent planar flame speed　等效平面火焰传播速度
equivalent potential temperature　相当位温(与绝热等效温度相对应的位温)
equivalent power　当量功率
equivalent projection　等积投影
equivalent radiant temperature　等效辐射温度
equivalent ratio　恰当的和实际的油气比的比值,当量比
equivalent shaft horsepower　当量轴马力
equivalent single-wheel load　当量单轮载荷
equivalent skin-friction　等效表面摩擦
equivalent skin-friction coefficient　等效表面摩擦系数
equivalent solar hour　当量太阳小时
equivalent source theorem　等效电源定理
equivalent sources　等效源
equivalent state　等价态
equivalent system　等效系统
equivalent temperature　等效温度
equivalent thermal conductivity of multi-layer insulation　多层隔热层当量导热系数
equivalent time delay　等效时延
equivalent transducer　等效传感器
equivalent unit　当量单位,等效部件
equivalent viscous damping　等效黏性阻尼
equivalent-altitude method　等量高度法
equivalent-barotropic atmosphere　相当正压大气
ER　1. Echo Ranging 回波测距,回声测距法,回声定位 2. Engine Reliability 发动机可靠性 3. Engineering Release 工程发放 4. Engineering Reports 工程报告 5. Event Review 事故评审 6. Extend 延程型 7. Extended-Range 增程的,扩大范围的,延程的 8. Extraterrestrial Resources 地外资源 9. E/R Extend/Retract (起落架等的)收放
ER range　回波测距范围,回声定位范围
ER/DL　Extended-Range/Data Link 加大距离数据线

路,距离增大的数据链路

ERA 1. Elastic Recoil Analysis 反弹分析(用于分析气含量) 2. European Regional Airlines Association 欧洲地区航空公司协会 3. Exobiology and Radiation Assembly 外空生物学和辐射装置

era 时代,年代,纪元,阶段

era of colloid thruster research 胶体推进器研究的时代(胶体推进器是电推进的一种,属于电推进中的静电推进)

eradiation 地球辐射

ERAM 1. Extended Range Active Missile 增程主动制导导弹 2. Extended-Range Anti-Armour Munition 增程穿甲弹药 3. Extended-Range Anti-Tank Mine 大范围反坦克(地)雷

ERAP Earth-Resources Aircraft Program 地球资源勘测飞机计划(美国)

erase 擦除,清洗,抹除(清除存储器中的信息)

erasing head 擦除头

erasure 擦除,消磁,删去

erasure channel 删除信道

ERAU Embry-Riddle Aeronautical University 恩布里里德尔航空大学(美国)

ERB 1. Earth Radiation Budget 地球辐射收支 2. Enhanced Radiation Bomb 增强型辐射弹(指的是中子弹) 3. Engineering Review Board 工程评审委员会 4. Executive Review Board 执行检查委员会

ERBE Earth Radiation Budget Experiment 地球辐射收支实验

ERBI Earth Radiation Budget Instrument 地球辐射收支仪

erbium 【化】铒(化学元素,符号 Er)

erbium laser 铒激光器

ERBM 1. Electronic Range/Bearing Marker 电子距离/方位标志 2. Extended-Range Ballistic Missile 延长射程的弹道导弹,增程弹道导弹

ERBOS Earth Radiation Budget Observation Satellite 地球辐射收支观测卫星

ERBS 1. Earth Radiation Balance Satellite 地球辐射平衡卫星 2. Earth Radiation Budget Satellite 地球辐射收支测量卫星(美国) 3. Earth Radiation Budget Sensing System 地球辐射收支遥感系统

ERC Electronics Research Center 电子设备研究中心(美国航空航天局)

ERCG Erecting 竖立,垂直安装

ERCR Erector 安装工人,拖车的升降架

ERCS 1. Emergency Rocket Communications System 应急火箭通信系统(美国空军) 2. ECM-Resistant Communication System 抗电子对抗通信系统

ERD 1. Elastic Recoil Detection 弹性反冲探测 2. Emergency Return Device 应急返回装置 3. Engine Research and Development 引擎研究和开发(部门)

ERDA Energy Research and Development Administration 能源研究和开发局(美国)

ERDC Earth Resources Data Center 地球资源数据中心

ERDE Explosive Research and Development Establishment 爆破器材研究发展中心(英国)

ERDRI Earth Resources Development Research Institute 地球资源开发研究所

ERE Effective Resolution Element 有效分辨(单)元(量)

ERECT Erection 竖立,建立

erect 1. 竖起,竖直,升起,安装,装配 2. 直立的,竖直的

erectable truss (空间站的)装配式桁架

erecting 1. 竖起,竖直,升起,安装,装配 2. 直立的,竖直的

erecting hall 起竖厅

erection (陀螺的)修正,起竖

erection torque (陀螺仪的)修正转矩

erection turn error 修正机构转弯误差

erector 1. 竖起架,升降架 2. (陀螺)修正器 3. 安装工,装配工

erector transporter 垂直运输器

ered Environmental Residue Effects Database 环境残留影响数据库

EREP 1. Earth Resource Experiment Package 地球资源实验组件 2. Earth Resources Experiment Program 地球资源实验计划(美国)

ERF Ephemeris Reference Frame 星历参考系

ERFA Conference on Economics of Route Air Navigation Facilities and Airports 航线领航设施与机场经济性会议

ergodic process 遍历过程

ergodic source 遍历信源

ergometer exerciser 测功计锻炼器,功量计运动器

ergonomical laboratory 工效学实验室

ergonomics 工效学,人机工程学

ergosphere 能层

ERGP Extended-Range Guided Projectile 增程制导射弹(导弹),增程制导炮弹

ERHAC Enroute High-Altitude Chart 航线高空图

ERI Extravehicular Reference Information 舱外活动参照信息

ericsson "埃里克森"(瑞典机载 J 波段海上监视雷达)

ERINT 1. Extended Range Intercept Technology 加大射程拦截技术 2. Extended Range Interceptor 远程截击导弹

ERIPS Earth Resources Interactive Processing System 地球资源人机对话处理系统

ERIS Exoatmospheric Reentry Vehicle Interceptor System 大气层外再入弹头拦截器系统(美国)
ERJ External-Combustion Ramjet 外燃冲压发动机
erk 〈口语〉地勤人员,航空机械员(英国)
ERL 1. Electronics Research Laboratory 电子设备研究实验室(澳大利亚) 2. Environmental Research Laboratories 环境研究实验室(美国国家海洋大气局)
ERL/RFC Environmental Research Lab/Research Facilities Center 环境研究实验室-研究设施中心
ERLAC Enroute Low-Altitude Chart 航线低空图
ERM 1. Experimental Rocket Motor 试验火箭发动机 2. Earth Re-entry Module 地球(大气层)再入舱
ERMS Equipment Resource Management System 设备源管理系统
ERN 1. Emergency Response Network 应急响应网络 2. Engineering Release Notice 工程发布通知
ERO Extended Relationship Object 扩展关系对象
erode 腐蚀,侵蚀,磨蚀
erodibility 腐蚀性,浸蚀性,侵蚀性
EROS 1. Earth Resources Observation Satellite 地球资源观测卫星(美国) 2. Earth Resources Observation System 地球资源观测系统
erosion 侵蚀,浸蚀,磨蚀,腐蚀
erosion contribution 侵蚀影响
erosion level 侵蚀剥削面,侵蚀水平面,磨损度
erosion mechanism 侵蚀机制(理)
erosion of bore 炮膛磨蚀
erosion profile 侵蚀剖面
erosion rate 侵蚀率,侵蚀速度
erosion rate distribution 侵蚀率分布
erosion rate of polyethylene 聚乙烯磨损率
erosion resistance 抗腐蚀性
erosion resistance rate 抗腐蚀率
erosion resistant material 抗侵蚀材料
erosion test 磨损试验,侵蚀试验,冲刷试验
erosion yield 侵蚀率
erosive burning 侵蚀燃烧
erosive burning effect 侵蚀燃烧效应
erosive burning model 侵蚀燃烧模型
erosive effect 侵蚀效应
erosive pressure peak 侵蚀压强峰
erosivity 侵蚀系数,侵蚀度
ERP 1. Effective Radiated Power 有效辐射功率 2. Enterprise Resource Planning 企业资源计划(规划) 3. Eye Reference Point 眼位基准点 4. Eye Reference Position 目视参考位置
ERR 1. Engine Removal Report 发动机拆换报告 2. Engineering Release Record 工程发放记录 3. Engineering Reliability Review 工程可靠性评审

erratic error 不规则误差
erratic firing 射速不稳定,不规则发火
erratum 误符
ERRB 1. Enhanced Radiation, Reduced Blast 加强辐射,减少冲击波 2. Enhanced Radiation, Reduced Blast Weapon 增强辐射、减弱冲击波的武器
ERRC Error Correction 误差修正
ERRDF Earth Resources Research Data Facility 地球资源研究数据设施(美国航空航天局)
erroneous trajectory 错误轨道
error 错误,误差,偏差,失调
error analysis 误差分析
error angle 误差角,错位角,失调角
error assessment 误差评估(如 Independent Error Assessment 独立误差评估)
error attenuation 误差衰减
error band 误差带,定值范围,误差范围
error bar 误差棒
error bound 误差界,误差范围,误差界限
error box 误差框,偏位框
error budget 误差分配
error case 错误情况
error characteristic 误差特性(征)
error characteristic statistics 误差特性统计
error checking 误差校验
error checking code 检错码
error coefficient 误差系数
error command 误指令
error command probability 误指令概率
error compensation 误差补偿
error condition 误差条件
error convergence 误差收敛
error correcting code 纠错码,错误校正码
error correction 纠错(如 Forward Error Correction 前向纠错)
error correction code 纠错码
error corridor 误差通道
error covariance 误差协方差
error covariance matrix 误差协方差矩阵
error density 误差密度
error detecting 检错,错误检测
error detecting code 检错码
error detection 检错,错误检测
error display 误差显示器
error distribution 误差分布,误差分配
error due to background 背景造成的误差
error due to curvature of earth 地球弯曲差
error dynamics 误差动力学
error ellipse 误差椭圆

error equation 误差方程
error field 误差场
error figure 误差值
error fitting 误差拟合
error free channel 无错误信道
error function 误差函数
error in address 地址错误
error in label 标号错误
error in operation 操作错误
error interrupt request vector 错误中断请求向量
error limit 误差限度
error loop 错误循环
error magnitude 误差值,误差幅度
error mechanism 误差机制
error metric 误差测度,误差度量
error minimization 误差最小化
error model 误差模型
error model identification 误差模型辨识
error multiplication 误码增殖,差错增殖
error of direction 方向误差
error of focusing 调焦误差
error of indication 示值误差
error of measurement 测量误差
error of observation 观测误差
error of pivot 轴颈误差
error of scale 刻度误差,比例尺误差
error of scintillation 闪烁误差
error of test data 试验数据误差
error on the safe side 偏于安全的误差,偏于增大安全系数的误差
error out of plane 偏离(轨道)平面的误差
error pattern 错误型
error probability 误差(概)率,差错率,误码率
error propagation 误差传播,错误传播
error rate 出错率,误码率
error recovery 错误恢复,错误校正,差错校正
error response 误差响应
error seeding 错误撒播
error separation 误差分离
error signal 误差信号,错误信号,出错信号
error source 误差来源,错误来源
error spread 差错扩散,误码扩散
error term 误差项(的)
error test 误差检验
error theory 误差(理)论
error threshold 误差阈(值)
error tolerance 容错,容差
error transfer function 误差传递函数
error vector computer 误差向量计算机

error voltage 误差(信号)电压
error-circular radius 误差圆半径
error-correcting parsing 纠错剖析
error-estimation 误差估计
error-locator 错误定位子
ERRSAC Eastern Regional Remote Sensing Applications Center 东部地区遥感应用中心(美国航空航天局)
ERS 1. Earth Resources Satellite 地球资源卫星(日本) 2. Earth Resources Survey 地球资源勘测 3. Environmental Research Satellite 环境研究卫星(美国小型科学卫星) 4. Erased 抹去,涂去 5. European Remote Sensing Satellite 欧洲遥感卫星 6. External Reflection Spectroscopy 外反射谱(学)
ERSATS Earth Resources Survey Satellites 地球资源勘测卫星
ERSOS Earth Resources Survey Operational Satellite 地球资源勘测应用卫星
ERSP Earth Resources Survey Program 地球资源勘测计划(美国)
ERSPRC Earth Resources Survey Program Review Committee 地球资源勘测计划评审委员会(美国航天局)
ERST Error State 异常状态,误差状态
ERTS 1. Earth Resources Technology Satellites 地球资源技术卫星 2. Earth Resources Test Satellite 地球资源试验卫星
ERU 1. Earth Rate Unit 地球转速单位,地球角速度单位 2. Ejector Release Unit for External Stores 外挂物弹射投放装置
eruptive galaxy 爆发星系
eruptive prominence 爆发日珥
eruptive variable 爆发变星
ERV Earth Return Vehicles 地球返回飞行器(把火星样品送回地球的飞行器)
ERVIS Exoatmospheric Re-entry Vehicle Interception (Interceptor) System 大气层外再入飞行器拦截系统
ERW 1. Enhanced Radiation Warhead 中子弹弹头 2. Enhanced Radiation Weapon 增强辐射武器(即中子弹)
ERWE Enhanced Radar Warning Equipment 改进型雷达警戒设备
erythemal spectrum dosimeter 红斑谱辐射剂量仪
erythemal dosimeter 红斑辐射剂量仪
ES 1. Econ Speed 经济速度 2. Electrical Schematic 电气原理图 3. Electrostatic 静电 4. Electronic Scanner 电子扫描仪 5. Eletronically Scanned 电子扫描的 6. Engineering Schedule 工程计划 7. Engineering Specification 工程规范 8. Engineering Standard 工程

标准 9. Engineering Suppliers 工程供应商 10. Escape Slide 紧急离机滑道(梯) 11. Eutectic Solder 易熔焊料 12. Evident Safety 显性安全性 13. Expert Systems 专家系统

ES/RO Engineering Stop/Resume Order 工程停止/复工指令

ESA 1. Earth Station-Arabia 阿拉伯地球站 2. Enhanced Signal Average 增强信号平均(值) 3. European Space Agency 欧洲航天局

ESACS Enhanced Saftey Assessment of Complex Systems 改进的复杂系统安全性评价系统

ESA-IRS European Space Agency Information Retrieval Service 欧洲航天局信息检索服务处

ESANET European Space Agency Network 欧洲航天局网络

ESARS Earth Surveillance And Rendezvous Simulator 地球监视和交会模拟器

ESAS 1. Electronic Situation Awareness System 电子位置认知系统 2. Electronically Steerable Antenna System 电子控制天线系统 3. Expert System for Avionics Simulation 航空电子设备模拟专家系统

ESB 1. Earth Station-Brazil 巴西地球站 2. Elevating Sliding Bridge 升降滑动式登机桥

ESC 1. Earth Station-Congo 刚果地球站 2. Excess Survey Card 余量调查卡 3. European Space Conference 欧洲空间会议 4. Evanescent Space Charge 短暂空间带电

ESCA Electron Spectroscopy for Chemical Analysis 化学分析电子谱(学)

esca pe trajectory 逃逸轨道,应急脱离轨道

escalation (战争的)升级,(由计算错误、通货膨胀、技术进步、技术规格改变等原因造成的)成本上升

escapability 逃生能力,逃逸能力

escape 离机,紧急脱离,脱险,逃逸,脱离引力范围,泄漏,漏出,流出,排出,(计算机的)代码转换,编码转换,换码

escape capsule 逃逸舱,救生舱

escape character 转义字符

escape chute 1.应急离机滑梯 2.应急离机口 3.(应急)救生伞

escape data injection 逃逸参数注入

escape depth 逸出深度

escape envelope 1.救生性能包线 2.安全弹射包线

escape equipment 应急离机设备,航空救生设备

escape exit 应急离机出口

escape factor 逃脱/逸因子,逸出因子

escape guidance strategy 应急离机指导策略

escape hatch 应急离机舱口(盖)

escape maneuver 逃逸机动(飞行),脱离机动

escape orbit 逃逸轨道

escape parachute 救生伞

escape pathlength of cosmic rays 宇宙线逃逸路径长度

escape rocket 逃逸火箭,救生火箭

escape rocket motor 逃逸火箭发动机

escape slide 应急离机滑梯

escape spiraltrajectory 螺旋逃逸轨道

escape spoiler 紧急离机挡风板

escape stage 逃逸(阶)段

escape strategy 逃逸策略,逃生策略,躲避策略(飞机面对导弹攻击)

escape system 逃逸系统,逃生系统

escape time 逃逸时间,逃生时间

escape tower 逃逸塔,救生塔

escape tower assembly and test building 逃逸塔装配测试厂房

escape trajectory 逃逸轨道

escape vehicle 逃逸飞行器

escape velocity 逃逸速度,第二宇宙速度

escape warning 逃逸告警

escaper 泄放器

ESCES Experimental Satellite Communications Earth Station 实验卫星通信地球站(印度)

ESCO Earth Station-Colombia 哥伦比亚地球站

escort 护航,护送,护航部队,护航飞机,护送卫兵,武装仪仗队

escrow (包租飞机的)银行保证金协议

ESCRI Eliminate, Simplify, Combine, Rearrange and Increase 取消、简化、合并、重排和新增

ESCS 1. Emergency Satcom System 应急卫星通信系统 2. Electronic Spacecraft Simulator 航天器电子模拟器

ESD 1. Electronic Software Distribution 电子软件发布 2. Electronic Systems Division 电子系统部(美国空军系统司令部) 3. Electrostatic Discharge 静电放电 4. Emergency Shut-Down (核反应堆)紧急停堆,紧急停车,紧急停机 5. Energy Spectral Density 能谱密度 6. Energy Storage Device 能量存储装置

ESDAC European Space Data Analysis Centre 欧洲空间数据分析中心

ESDP Evolutionary System for Data Processing 数据处理调优系统

ESDRN Eastern Satellite Data Relay Network 东部卫星数据中继网(苏联)

ESDS 1. Elemental Standard Data System 基本标准数据系统 2. Engineering Standard Distribution System 工程标准分发系统

ESD-S Electrostatic Discharge-Sensitive 静电放电敏感(电子设备)

ESE 1. East Southeast 东东南 2. Extra Vehicular

Support Equipment 舱外保障设备
ESEC Earth Station-Ecuador 厄瓜多尔地球站
ESEP 1. Earth Science Education Program 地球科学教育计划 2. Electrical System Enterprise Procedure 电气系统企业程序
ESF Epitaxy Stacking Fault 外延堆栈层错
ESFC Equivalent Specific Fuel Consumption 当量燃料比耗
ESG Electrostatically Suspended Gyro 静电悬浮陀螺
ESGM Electrostatically Suspended Gyro Monitor 静电悬浮陀螺监控器
ESH Electric Surface Heating 表面电加热
ESHP Equivalent Shaft Horsepower 当量轴马力,等效轴输出功率
ESI 1. Engineering Staff Instruction 工程人员须知 2. Environmental Sensitivity Index 环境敏感性指标图
ESIC Environmental Science Information Center 环境科学信息中心(美国国家海洋大气局)
ESID Engine and System Indication Display 发动机和系统指示显示器
ESIP Engine Structural Integrity Program 发动机结构完整性大纲(发动机结构设计应遵循的大纲)
ESIS Engine and System Indication System 发动机和系统指示系统
ESIT Electrical System Integrated Test 电系统综合测试
ESJ Equivalent Single Jet 等效单发喷气式飞机
ESL 1. Earth-Science Laboratories 地球科学实验室(美国国家海洋大气局) 2. Experimental Space Laboratory 实验性空间实验室
ESLAB European Space Laboratory 欧洲空间实验室
ESLE Electronic Survivor-Location Equipment 遇险人员定位电子设备
ESLO European Satellite Launching Organization 欧洲卫星发射组织
ESLR Electronically Scanned Laser Radar 电子扫描激光雷达
ESM 1. Electronic Support Measure 电子支援措施,电子信号截获设备 2. Electronic Surveillance Measures 电子监视措施 3. Electronic Surveillance Monitoring 电子监视监控 4. Equipment Support Module 设备保障舱 5. Escort Mission 护航任务
ESMC Eastern Space and Missile Center 东部航天与导弹中心(美国空军)
ESMO Electronic Support Measures Operator 电子支援措施操作员
ESMR Electrically Scanning Microwave Radiometer 电扫描微波辐射计
ESMRO Experiments for Satellite and Materials Recovery from Orbit 从轨道回收卫星和材料的实验
ESN Engine Serial Number 发动机序列号
ESO Engineering Standard Order 工程标准法规(美国联邦航空局)
ESOC 1. Earth Station Operations Committee 地球站操作管理委员会 2. European Space Operations Centre 欧洲航天操作中心
ESP 1. Electronic Standard Procedure 电子标准程序 2. Equipment Specifications 设备规范 3. External Starting Power 外接起动电源
Esprit 1. European Strategic Programme for Research into Information Technology 欧洲信息技术研究的战略计划 2. Eye-Slaved Projected Raster Inset 眼控投影光栅插入件
ESR 1. Earth Science Research 地球科学研究 2. Electro-Slag Refined 电渣精炼,电渣重熔 3. Electro-Slag Remelt 电渣重熔 4. Equivalent Service Rounds 当量发射弹数
ESRANGE European Space Range 欧洲航天发射场
ESRIN European Space Research Institute 欧洲空间研究所
ESRO European Space Research Organization 欧洲空间研究组织
ESRRD E-scope Radar Repeater Display E 型显示器雷达重复显示
ESRS Electron Spin Resonance Spectroscopy 电子自旋共振谱
ESS 1. Electronic Security System 电子保密系统,电子安全系统 2. Electronic Switching System 电子开关系统 3. Emergency Exit System 应急出口系统(美航天飞机固定服务架上的缆车救生系统) 4. Energy Storage Subsystem 能量储存分系统 5. Engineering Statusing System 工程状态控制系统 6. Environmental Stress Screening 环境应力筛选 7. Equipment Support Structure 设备支撑结构 8. Essential 重要的 9. European Space Station 欧洲空间站 10. Experiment Support System (航天器的)实验保障系统
ESSA 1. Environmental Science Services Administration 环境科学服务局(美国) 2. Environmental Survey Satellite 艾萨卫星,环境勘测卫星
ESSD Electrostatic Sensitive Devices 电敏感装置
ESSDE European Space Software Development Environment 欧洲航天软件开发环境
essential air defence 要地防空
essential boundary condition 本质边界条件
essential bus 重要电气设备汇流条
essential feature 重要特征,本质特征
essentially singular point 本性奇点
ESSL Emergency Speed Select Lever 应急速度选定杆

ESSM　1. Evolved Sea Sparrow Missile 改进型海麻雀导弹　2. Emergency Ship Salvage Material System 船舶应急救捞器材系统

ESSP　1. Environmental Stress Screening Procedures 环境应力筛选程序　2. European Satellite Services Provider 欧洲卫星服务运营商　3. European Space Station Programme 欧洲空间站计划（即 Columbus 空间站计划）

ESSPEC　Engineering Standards Specification 工程标准规范

ESSS（ES3）　External Stores Support System 外挂载荷支承系统（直升机上的挂架）

Esswacs　Electronic Solid-State Wide-angle Camera System 电子固态广角相机系统

EST　1. Eastern Sandard Time 美国东部标准时　2. Elevation，Slope，Temperature 海拔、坡度、温度　3. En-route Support Team 中途保障队（美国空军）　4. Equipment Status Telemetry 设备（装备）状态遥测计

ESTA　1. Earth Sciences Technologies Association 地球科学技术协会（美国）　2. Electronically Scanned Tacan Antenna 电扫描塔康天线　3. Estimated 预计的，评估的

ESTC　European Space Tribology Centre 欧洲空间磨损学中心

ESTEC　European Space Research and Technology Centre 欧洲空间研究和技术中心

esterification　酯化作用

estimation　估计，估算，评价

estimate　估计，估算，评价，评论，估量，估价

estimate error　估计（值）误差

estimate of downrange velocity　下降速度（导弹从发射中心和沿着试验航向的逐步下降的速度）估计（值）

estimated attitude　估计姿态

estimated departure time error　预计离场时间误差

estimated disturbance　预计扰动，预计干扰

estimated error　预计误差

estimated parameter　预计参数，估计参数

estimated position　预计位置，推算（出的飞行器）位置，预定位置

estimated signal　估计（的）信号

estimated state　估计（的）状态，预计状态，预定状态

estimated storage life　估算储存期

estimated target　估计目标

estimated time　预计时间，估计时间

estimated track　推算航迹线

estimated uncertainty　估计不确定度

estimated value　估计值

estimating method by engineering item　工程估算法

estimation　估计，估量，概算

estimation accuracy　估算精（确）度

estimation algorithm　估计算法，估值算法

estimation delay　预计延时，估计延时

estimation error　估计误差，估算误差

estimation lag　预计滞后，估计滞后

estimation method　估算方法

estimation model　预测估算模型

estimation model system　预测估算模型（体）系（统）

estimation of loss　损失估算

estimation problem　估计问题

estimation process　估计（估算）过程

estimation system　估测系统（如 Adaptive Estimation System 自适应估测系统）

estimation technique　估测方法，估算方法，估计方法

estimator　估计器，估计值，估计量，估计式，估计方法

estimator performance　估计（器或方法的）性能

estimator-guidance-law　估算法制导律

ESTP　Electronic Satellite Tracking Program 电子卫星跟踪计划

ESTRACK　European Space Satellite Tracking and Telemetry Network 欧洲卫星跟踪和遥测网

ESU　Environmental Sensor Unit 环境敏感装置

ESV　Earth Satellite Vehicle 地球卫星运载器

ESWR　1. Engineering Scheduled Work Release 工程计划工作发放　2. Engineering Scheduled Work Report 工程计划工作报告

ESWS　Earth Satellite Weapon System 地球卫星武器系统

ET　1. Elapsed Time 经过的时间　2. Eletrical Turbulence 电涡流检验

ETA　1. Effective Turn Angle 有效转折角　2. Ejector Thrust Augmentor 引射器加力装置　3. Estimate 预计的　4. Estimated Time of Acquisition 预计截获时间　5. Estimated Time of Arrival 预计到达时间

eta patch　艾搭补片（气球气囊上的扇形布片）

ETAC　Environmental Technical Applications Center 环境技术应用中心（美国空军）

ETACCS　European Theatre Air Command and Control Study 欧洲战区空军指挥和控制的研究

ETADS　Enhanced Transportation Automated Data System 扩大的运输自动化数据系统

Etalon　1. 频谱分析干涉仪　2. 基准星（苏联被动式激光反射测地卫星）

ETB　End of Transmission Block 传输块结束

ETC　Engineering Tool Coordination 工程工装协调

etch　蚀刻

etch cutting　腐蚀切割

etch disk　蚀刻磁盘

etchant 浸蚀剂,蚀刻剂
etched circuit board 印刷电路板
etched-disk filter 蚀刻圆盘过滤器
etching 刻蚀,蚀刻
etching pattern 侵蚀形态,侵蚀图,蚀刻图像
etching rate 侵蚀速度,蚀刻速度
ETCO External Tank Corporation 外贮箱公司(美国)
ETD 1. Estimated Time of Departure 预计出航时间,预计离港时间 2. Estimated Time of Departure 预计起飞时间
ETDRS Experimental Track and Data Relay Satellite 实验型跟踪和数据中继卫星(日本)
ETE 1. Environmental Test and Evaluation 环境试验和评定 2. Estimated Time En-route 预计航线(航路)飞行时间
ETEC Expendable Turbine Engine Concept 一次性涡轮发动机设计思想
ETER 1. Estimated Time En Route 预计途中时间 2. Estimated Time of Engagement Time 预计接战时间
ETH Extra Terrestrial Hypothesis 宇宙假说,地外假说
ethane-methane-air mixtures 乙烷-甲烷-空气混合气体/物
ethanol 乙醇
ethanol mass 乙醇质量
ethanol mass fraction 乙醇质量分数
ethyl 乙(烷)基,四乙铅,防爆(抗震)剂
ethylene fuel 乙烯燃料
ethylene fueling 乙烯燃料加注
ethylene ignition 乙烯点火
ethylene jet 乙烯燃料发动机
ethylene mole fraction 乙烯摩尔分数
ethylene-air mixtures 乙烯-空气混合气体/物
ETI Elapsed Time Indicator 经过时间指示器
ETM 1. Effectively/Tabulation Management 有效性配套管理 2. Elapsed Time Measurement 经过时间测量 3. Enhanced Thematic Mapper 改进型主题测绘仪
ETMS Enhanced Traffic Management System 改进的交通管理系统
ETNAS Electro-Level Theodolite Naval Alignment System 海军电经纬仪校准系统
ETOA Estimated Time Of Arrival 预计到达时间
ETOP Engineering Technical Operating Procedure 工程技术操作程序
ETOPS 1. Extended Range Operation with Two-Engine Airplanes 双发延程运营 2. Extended-Range Twin (Engine) Operations 延长航程的双发动机运营
ETOT Estimated Time Over Target 预计飞越目标时间
ETP 1. Environmental Test Plan 环境试验大纲 2. Environmental Test Procedures 环境试验程序 3. Estimated Time of Penetration 预计突防时间
ETPS Empire Test Pilots School 帝国试飞员学校(英国)
ETR 1. Eastern Test Range 东部(导弹)试验靶场(美国) 2. Emergency Test Request 紧急试验申请
ETS 1. Electronic Test Set 电子测试设备 2. Engineering Test Satellite 工程试验卫星 3. Engineering Trade Studies 工程经营研究 4. ERINT Target System 加大射程拦截技术目标系统 5. Experimental Test Site 实验导弹试验场
ETSTA External Tank Static Test Article 外贮箱静试验件
ETT Electrothermal Thrusters 热电推力器
ETW European Transonic Windtunnel 欧洲跨声速风洞
E-type device E型器件
EUAFS Enhanced Upper Air Forecast System 增强高空天气预报系统
EUCLID 欧几里得计划(即欧洲长期防务合作倡议,俗称军事尤里卡计划,是北约欧洲国家的国防高技术研究与发展计划)
euclidean space 欧几里得矢量空间
EuIG Europium Iron Garnet 铕铁石榴石
Euler equation 欧拉方程
Euler number 欧拉数
Euler parameter 欧拉参数
Euler rate 欧拉速率
Euler theorem on rotation 欧拉转动定理
Euler vector 欧拉矢量
Euler viewpoint 欧拉观点,欧拉法
euler-axis 欧拉轴
Eulerian angle 欧拉角
Eulerian coordinate 欧拉坐标
Eulerian equation of motion 欧拉运动方程
Eulerian period 欧拉周期(极移)
Eulerian variable 欧拉变量,欧拉值
Euler-Lagrange equation 欧拉-拉格朗日方程
Euler-Lambert equation 欧拉朗伯方程
EULMS Engine-Usage Life Monitoring System 发动机使用寿命监控系统
Eumetsat European Meteorological Satellite Organization 欧洲气象卫星组织
Eumilsat Europe's Military Communications Satellite 欧洲军事通信卫星
Eumilsatcom European Military Satellite Communications 欧洲军事卫星通信
EUMS Enginge-Usage Monitoring System 发动机使用监控系统
EUPS External Uninterruptible Power Supply 外部不

间断供电源

Euram European Research on Advanced Materials 欧洲对先进材料的研究

Eureca 1. European Retrievable Carrier 尤里卡平台,欧洲可回收平台 2. European Research Coordination Agency 欧洲研究协调机构(其研究计划即称为尤里卡计划)

Eureka 1. 尤里卡信标,地面应答信标 2. 尤里卡高电阻铜镍合金

EURICAS European Research Institute for Civil Aviation Safety 欧洲民航安全研究所

EUROCONTROL European Organization for the Safety of Air Navigation 欧洲空中航行安全组织("欧管")

Euro Hermes 欧洲赫尔墨斯(航天飞机)工业集团

Euro-Apache 欧洲阿帕奇

Eurocard 欧洲电路(标准单面印刷电路板)

Eurofar European Future Advanced Rotorcraft 欧洲未来先进旋翼机(欧洲五国合作研制的偏转旋翼飞机)

Eurofighter 欧洲战斗机

Euroflag European Future Large Aircraft Group 欧洲未来大飞机集团(英、法、德、意、西五国合作研制运输机)

Eurogroup 北约欧洲集团(北约十三国国防部长非正式集团)

Eurojet 欧洲喷气发动机

Euromep European Mission Equipment Package 欧洲任务设备组件(Euro-copter 研制的直升机夜视仪等设备)

Euromissile 欧洲导弹公司

Europa 【天】木卫二

European airframe 欧洲(制造的)飞机

European airline 欧洲航线

European airliner 欧洲航空公司

europium 【化】铕(化学元素,符号 E)

EUROSPACE European Industrial Space Study Group 欧洲工业空间研究集团

eutectic 共晶的

eutectic alloy 共晶合金

eutectic point 共晶点,低共熔点

EUTELSAT European Telecommunications Satellite Organization 欧洲通信卫星组织

EUV 1. Extreme Ultra-Violet 极远紫外 2. Extreme Ultraviolet Radiation 极远紫外辐射

EUVE Extreme Ultraviolet Explorer 极远紫外探测器

EV 1. Earned Value 赢得价值 2. Electron-Volt 电子伏 3. Engineering Verification 工程验证 4. Equivalent Velocity 当量空速 5. Extravehicular 舱外的

EVA 1. Extravehicular Activity 舱外活动(航天员出舱工作) 2. Electronic Velocity Analyzer 电子速度分析器

EVA/IVA Extra-/Intra-Vehicular Activities 舱外和舱内活动

EVAC 1. EVAC(Water/Waste)艾瓦克公司(水/废水) 2. Evacuate 抽空,排空,消除

evacuate 排气,排空,疏散

evacuated bellows 真空膜盒

evacuating 排气,排空,疏散

evacuation 排气,排空,疏散

evacuation capacity 1. 疏散能力 2. 排气能力,排空能力

evacuation hospital 后送医院,转运医院

evacuation performance 排气性能,排空性能

evacuation process 1. 疏散过程 2. 排气过程

evacuation quality 疏散质量,排气质量

evacuation time 1. 安全疏散时间,撤离时间 2. 排气时间,排空时间

evacuee 撤离者,被疏散者

EVADE Escape Vehicle Analysis and Definition for EMSI(欧洲载人空间基础设施的)乘员逃生飞行器分析和初步设计

evade 逃避,躲避,避开,规避

evader 逃避者,规避者

evading 规避机动

EVAL Earth Viewing Applications Laboratory 地球观测应用实验室

evaluate 评价,估价,计算,求……的值

evaluation (对情报的)评估,鉴定

evaluation accuracy 评定精度,评价准确度,鉴定精度

evaluation criterion 评价标准,评估标准,鉴定标准

evaluation environment 求值环境

evaluation flight test 鉴定试飞

evaluation model 评价模型,评估模型,鉴定模型

evaluation system 评价系统

evaluation task 评价任务

evaluation technique 评价技术

evaluator 求值器,评估器

evanescent mode 衰逝模(式)

evanescent wave 消散波

evanescent waveguide 衰逝波导

Evans method 埃文斯方法

EVAP Evaporate 蒸发,汽化

evaporate (使某物)蒸发掉,挥发,消散

evaporating deposition 蒸发沉积

evaporation boat 蒸发舟(真空蒸发工艺中,用以放蒸发源的器皿)

evaporation of the antireflective coating 蒸减反射膜(用真空蒸发法,沉积减反射膜的工艺)

evaporation process 蒸发过程

evaporation rate 蒸发速度,蒸发速率,蒸发率,汽化率

evaporative cooling 蒸发（式）冷却
evaporative cooling ventilated suit 蒸发冷却通风服
evaporative heat loss 蒸发性散热
evaporative ice 蒸发冰
evaporative loss 蒸发（热量）损失，蒸发（热量）损耗
evaporator 蒸发器，蒸发段
evaporator assembly 蒸发器总成
evaporator fluid 蒸发（器）流体
evaporator temperature 蒸发器温度
EVAS 1. Emergency Vision Assurance System 应急视觉保证系统 2. Enhanced Vortex Advisory System 增强型涡旋咨询系统 3. Extravehicular Activity System 舱外活动系统
evasion 逃避，规避，躲避
evasion game 规避对策
evasion problem 规避问题，躲避问题
evasion strategy 回避策略，规避策略
evasion zone 规避区
evasive maneuver 躲避机动
evasive satellite 逃避卫星，机动（飞行）卫星
EVC Error Vector Computer 误差向量计算机
EVCS Extravehicular Communications System 舱外通信系统
EVD Elementary Vortex Distribution 涡流基本分布
evection 出差，月动差
E-vector 矢量，电场矢量
even function 偶函数
even-harmonic component 偶数谐波分量
evenness 平滑，平滑度，均匀性，一致性
event 事件，事件，冲程，间隙
event chain 事件链
event horizon 事件水平线
event parameter 事件参数
event telemetry parameter 遥测指令参数
event time 事件时间
event tree analysis 事件树分析
event trigger 事件触发，事件触发器，事件触发机制
event upset 事件扰乱因素
event-oriented inspection 对事件的检查
event-oriented simulation 面向事件的仿真
Evergreen Airways 长荣航空公司（中国台湾）
Everling number 埃弗林数
Evett's Field 埃维特机场（澳大利亚武器研究院所属飞机场）
EVG Electrostatically Supported Vacuum Gyro 静电支承真空陀螺
evidence theory 证据理论，证据推理
evidence-theory-based state 基于证据理论的状态
EVM 1. Earth Viewing Module 对地观测舱 2. Engine-Vibration Monitor 发动机振动监控器 3. Extended Virtual Machine 扩充的虚拟机
EVO Extravehicular Operation 舱外作业
evoked potential 诱发电势，诱发电位
evoked response 诱发反应
evolution equation 发展方程，演化方程
evolution of ramjet 冲压式发动机的发展/演进
evolution process 演化过程，演变过程，发展过程
evolutionary algorithm 进化算法，演化算法，改良算法
evolutionary operation 改良操作，寻优运算
evolutionary step 进化步骤，发展步骤
evolutionary system 发展中系统
evolutionary track 演化程
EVR Electronic Video Recording 电子视频信号记录
EVS 1. Effectivity Verification Sheet 有效期核实单 2. Electro-Optical Viewing System 光电观察系统 3. Enhanced Vision System 增强视觉系统
EVSD Electronic Vertical Situation Display 电子式垂直位置显示器，电子式垂直情况显示器
EVSS Extravehicular Space Suit 舱外活动航天服
EVT 1. Educational and Vocational Training 对复员军人的教育和假期训练 2. Extravehicular Transfer 舱外转运
EVVA Extravehicular Visor Assembly 舱外活动面盔装置
EW 1. Early Warning 预警，早期报警 2. Electronic Warfare 电子战 3. Equivalent Weapons 等效武器
EW pod Electronic Warfare pod 电子战吊舱
EW&C Early Warning and Control 预警和控制
EW/GCI Early Warning and Ground Controlled Intercept 预警及地面控制截击
EWA Engineering Work Authorization (Form) 工程工作审批（表格）
EWACS Electronic Wide-Angle Camera System 电子广角照相机系统
EWAISF Electronic Warfare Avionics Integrated Support Facility 电子战航空电子设备综合保障设施
EWAMS Early Warning and Monitoring System 预警监视系统
EWAS Electronic Warfare Analysis System 电子战分析系统
EWAU Electronic Warfare Avionics Unit 电子战航空电子设备（美国）
EWD Electrical Wiring Diagrams 电气线路图
EWEDS Electronic Warfare Evaluation Display System 电子战判定显示系统
EWEP Electronic Warfare Evaluation Program 电子战鉴定计划（美国空军）
EWO 1. Electronic Warfare Officer 电子战官

2. Emergency War Order 紧急作战命令 3. Engieering Work Order 工程工作指令
EWPA European Women Pilots' Association 欧洲女飞行员协会
EWR Early Warning Radar 预警雷达
EWS Engineering Work Statement 工程工作说明
EWSM 1. Early-Warning Support Measures 预警保障措施 2. Electronic Warfare Support Measures 电子战支援措施 3. Electronic-Warfare Surveillance Measures 电子战监视措施
EWT Effects of Weapons Targets 武器对目标的杀伤破坏作用
exact aiming 精瞄
exact analytical result 正确的分析结论
exact collision detection 精确碰撞检测
exact differential 正合微分,完整微分,恰当微分
exact distribution 精确分布,恰当分布
exact integer 精确整数值
exact knowledge 精确知识
exact model 精确模型,正合模型
exact numerical solution 数值解,数值计算
exact orbit 精确轨道
exact search algorithm 精确搜索算法
exact solution 精确解,准确解,严格解
exactness 正确,精确
exact-point symbology 激点显示
EXAFS Extended X-ray Absorption Fine Structure 广延 X 射线吸收精细结构
EXAM 1. Examination 考试,审查 2. Experimental Aerospace Multi-Processor 实验航空航天多处理机
examine 检测,检查,检验
example application 实例应用,应用实例
example motion 运动实例
example of operational missile 作战导弹实例
example of ramjet 冲压发动机实例
example optimization 示例优化
example problem 算例,工程实例
EXC Excitation 刺激,激磁,励磁
Excap 1. Expanded Capability 已提高的能力 2. 膨胀能力
exceed 超过,超越
exceedence 超限记录器
excellent performance 出色的性能
exception analysis system 异常分析系统
exception handling 异常处理
excess air 余气
excess air coefficient 余气系数
excess air ratio 余气比
excess carbon 富碳(的)
excess carrier 过剩载流子
excess electron 多余电子
excess flow 过流量,过剩流量
excess noise ratio 超噪(声)比
excess oxidizer coefficient 余氧系数
excess power 剩余功率(可用功率和需用功率之差)
excess speed 过高速度,超出(规定或允许的)速度
excess temperature 过高的温度,超出(规定的)温度
excess velocity 过高速度,超出(规定或允许的)速度,速度过高
excess velocity vector 过高速度向量,超出(规定或允许的)速度向量
excessive bank 多余的坡度,超过预定的坡度
excessive feedback 过度回授
excessive pressure 余压
excessive rate 过度,过高
excessive travel (船面的)过量偏度
exchange collision 交换碰撞
exchange momentum 交换动量
exchange rates 换装影响因素
exchange ratio 交换(比)率
exchangeability 互换性
excimer 激励态二聚物,准分子
excimer laser 准分子激光器
excimer lithography 准分子激光光刻
excitation 激励,刺激(作用于人身时)
excitation (exciting) voltage 激励电压,激发电压
excitation beam 激发光
excitation collision 激发碰撞
excitation force 激振力
excitation frequency 激磁频率
excitation function 激励函数
excitation input 激励输入
excitation laser 激发激光
excitation magnitude 励磁电源幅值,激励振级
excitation mode 励磁方式,激励方式,激磁方式,激发方式
excitation point 激励脉点,激发点,激振点
excitation pulse 激励脉冲,激发脉冲
excitation signal (电)激励信号
excitation slot 激发槽/孔
excitation temperature 激发温度
excitation velocity 激发速度
excited species 受激态物质
excited state 激发态,受激态
excited-state specie 激发态种类,受激态种类
exciter 激励器
exciter armature 励磁机电枢
exciter location 激励器位置

exciter set 励磁机组,激励器组
exciter tube 激励管,主振管
exciting star 激发星
exciting time 激发时间
exciting voltage 励磁电压,激发电压
excitron 单阳极汞整流器,汞气整流管,激励管
EXCL Exclusive 除外的,排除的
exclusion of modes 模态截断
exclusion region 1.排除范围,排外范围 2.无线通信冲突域
exclusivity 排他性,独占性
EXCON Executive Control 执行控制
excrescence drag 机身凸出部位产生的阻力
EXCTR Exciter 激励器
excursion 偏移,位移的峰值,幅度
excursion fare 差额票价
excursion level 1.偏差度(在下滑道上,中心线电压/信号的最大垂直变化量或角变化量) 2.最小安全角
Exdrone expendable drone 一次性(使用的)无人机
EXEC 1.Execute 执行 2.Executive 执行的
execute 执行,实施,运行
execute control cycle 执行控制循环
execution angle 执行角(由参考脉冲到执行脉冲的相位滞后角)
execution command 执行指令
execution status 运行状况
execution system 执行系统
execution time 执行时间,完成时间
executive agent 执行代理(人)
executive aircraft 公务机
executive airplane 公务机
executive communication exchange 执行通信交换(机)
executive control system 执行控制系统
executive council 执行委员会
executive jet 公务机
executive officer 副舰长,执行官,执事、行政主任
executive procedure 执行程序,实施程序,行政程序
executive routine 执行程序,检验程序
EXER Exercise 练习,实习
exercise option 实现选择(将选择变为坚决的命令)
exercise test 运动试验
exercise tolerance 运动负荷
exercise tolerance test 运动负荷试验
exercise warhead 练习弹头
exergy 热能
exergy destruction 热能损失,热能损耗
exergy loss 热能损失,热能损耗
exergy transfer 热量传递
exert 发挥,运用,施加,努力,行使

ex-factory review 出厂评审
exfoliation corrosion 剥落腐蚀
EXH Exhibit 展览,展出,附件
exhaust 排气,排放,放出
exhaust aftertreatment 排气后处理
exhaust branch 排气支管
exhaust capability 排气能力,排气量
exhaust collector ring 环形排气收集器,环形排气总管
exhaust component 1.废气成分 2.排气部件
exhaust cone 排气锥(排气管内涡轮后的锥体),喷管内锥,尾喷管
exhaust configuration 排气结构
exhaust diffuser 排气扩压器
exhaust duct （后风扇发动机的)排气管
exhaust duct cover 喷管堵盖
exhaust emission 废气排放,尾气排放
exhaust energy 排气能量
exhaust flame-damper 排气管消焰器
exhaust flow 排气流
exhaust flow rate 排气流速(率)
exhaust gas 排气,废气,尾气
exhaust gas analyser 燃气分析仪,废气分析仪(主要分析燃烧后气体成分以确定燃烧效率、排污情况等)
exhaust gas composition 废气成分
exhaust gas pressure gage 1.排气压力表 2.功率损耗表
exhaust gas temperature indicating system 排气温度指示系统
exhaust gas thermometer 排气温度表/计
exhaust heat 废热,余热
exhaust impulse 排气冲量
exhaust jet 尾喷流,燃气喷流
exhaust matrix 排气矩阵
exhaust nozzle （排气)喷管,尾喷管
exhaust nozzle exit 尾喷口
exhaust nozzle flaps 喷管鱼鳞板
exhaust pipe 排气道,排气管
exhaust plane 排气面(nozzle exhaust plane 喷管出口面)
exhaust plug 排气口塞
exhaust plume 排气羽流,羽焰
exhaust pollution 排放(气体)污染
exhaust port 排气口,排出孔
exhaust process 排气过程
exhaust product 排出的燃烧产物
exhaust region 排气区域,排气侧
exhaust reheater 排气再热器
exhaust sample 排气样本
exhaust speed 排气速度
exhaust stack 排气管

exhaust stator blade　排气导流叶片
exhaust stream　排气气流
exhaust stub　排气短管
exhaust system　排气系统
exhaust temperature　排气温度
exhaust turbocharger　排气涡轮增压器
exhaust valve　排气门，排气阀
exhaust vane　（火箭、导弹）燃气舵
exhaust velocity　排气速度
exhaust-duct　排气道
exhausted cutoff　耗尽关机
exhaust-gas thermometer　排气温度表
exhaustion　衰竭，耗竭，耗尽
exhaustive search　穷举搜索，穷尽搜索
exhaust-refill　排气回灌
exhibit　展览，显示，提出
exiguous triangle method　微三角形法
exist　存在，生存，生活，继续存在
existing aircraft　现有飞机
existing practice　现行做法
existing product line　现有产品线，现有产品系列
exit air　排出气体
exit angle　喷束锥角，出口角
exit area　出口面积
exit boundary　出口边界
exit condition　（叶栅）出口条件
exit cone　1.尾喷管（发动机的）2.扩散段（风洞的）
exit criteria　放行准则
exit cross section　出口横截面
exit diameter　出口直径
exit end　出口端
exit energy　排出能量
exit expansion wave　出口膨胀波
exit flow angle　出口气流角
exit freestream　自由气流
exit hole　出口孔
exit jet　（出口）（喷射）射流
exit mirror　出光镜面
exit mirror ratio　出光反射系数
exit nozzle　喷嘴，喷口
exit nozzle throat　喷管临界截面
exit of combustor　燃烧室出口
exit of stator　静子出口，定子出口
exit phase　退出阶段
exit plane　出口（平）面
exit pressure　出口压（力）
exit pupil　出射光瞳
exit region　出口区，（气象）急流出口区
exit ring　出口圈

exit section　出口截面，出口段
exit side　出口侧
exit temperature　出口温度
exit temperature distribution　出口温度分布
exit time　空投物品离舱时间
exit tube　排气管
exit velocity　1.空投物品离舱速度 2.（流体如气流水流或湍流等）出口速度，流出速度
ex-nova　爆后新星
EXO　European X-ray Observatory 欧洲 X 射线观测卫星
exoatmosphere　外大气层，外逸层
exoatmospheric　大气层外的
exoatmospheric engagement　大气层外交战
exoatmospheric intercept　大气层外拦截
exoatmospheric interceptor　大气层外拦截器，大气层外拦截弹
exobase　逸散层底
exobiology　地外生物学，外（层）空（间）生物学
exogenous disturbance　外部干扰，外界干扰
exogenous input　外源输入，外部输入，外因输入
exogenous noise　外部噪声，外源噪声
exogenous variable　外生变量
Exo-NNK　Exoatmospheric Nonnuclear Kill 大气层外非核杀伤
exopause　逸散顶层，外逸层顶
EXOSAT　ESA X-Ray Observatory Satellite 欧洲航天局射线观测卫星
Exosat　European X-ray Observatory Satellite 欧洲 X 射线天文卫星（欧洲航天局）
exosphere　外大气层，外逸层，外大气圈
exosphere temperature　外大气层温度
exospheric satellite　外大气圈卫星
exotherm　放热
exothermal center　放热中心（点）
exothermic chemical reaction　放热化学反应
exothermic decomposition　放热分解
exothermic event　放热事件
exothermic fuel　放热燃料
exothermic hump　放热峰（值）
exothermic peak　放热峰（值）
exothermic process　放热过程
exothermic reaction　放热反应
exothermicity　放热性
EXP　1. Expand 延伸 2. Exposed 暴露的 3. Executive Platinum 白金商务飞行服务（美国航空公司为经常出差的常客提供的一种会员服务）
EXP JT　Expansion Joint 伸缩接缝，伸缩接头，涨缩接合

EXPAN　Expansion 扩张
expand　扩展，发展，张开，展开，膨胀
expandable space structure　可展开空间结构，延伸空间结构，可展开的航天器结构
expanded foam　膨胀泡沫材料，膨胀发泡，膨化泡沫
expanded uncertainty　扩展不确定度
expander　扩展器，扩展电路，膨胀器
expander cycle　膨胀循环
expanding arm　膨胀臂
expanding balloon　可变容积气球
expanding brake　胀闸，扩张制动器
expanding bullet　裂开弹
expanding reamer　胀开式铰刀
expanding shock tube　扩张激波管
expanding universe　膨胀宇宙
expanding universe model　膨胀宇宙模型
expansion　膨胀，扩展
expansion angle　膨胀角
expansion corner　膨胀角
expansion efficiency　膨胀效率
expansion fan　膨胀（波）扇（Prandtl-Meyer Expansion Fan 普朗特梅耶尔膨胀波扇）
expansion head　膨胀头
expansion in powers of height　高度的幂级数展开式
expansion joint　膨胀接头
expansion of air　气体膨胀
expansion part　膨胀部分，扩展部分
expansion plane　膨胀面
expansion process　膨胀过程
expansion ramp　膨胀斜面（Single Expansion Ramp Nozzle, 单膨胀斜面喷管）
expansion ratio　膨胀比（喷管）
expansion shock　膨胀激波
expansion slot　【计】扩充插槽，扩展插槽
expansion system　膨胀系（统）
expansion wave　膨胀波（超声速气流特有的重要现象，超声速流在加速时要产生膨胀波）
expansive phase　膨胀相
expect　期望，预料，要求
expectation of low fuel price　（航空企业管理）期望较低的燃料价格
expectation value　期望值
expected characteristics　期望特性
expected trajectory　（卫星的）期望轨道
expected value　期望值，预期值
EXPED　Expedite 加快，加紧，促进
expedite　加快进展，迅速完成
EXPEN　Expendable 可消费的，消耗的
expend　花费，耗尽，用光
expendable　可消耗的，消耗性的，可丢弃的
expendable cluster aircraft rocket/launcher　一次使用的机载多管火箭发射器，机翼火箭发射器
expendable construction　分段抛弃结构
expendable engine　短寿命发动机，一次使用的发动机
expendable equipment　消耗性装置，一次使用装置，不可回收装置
expendable jammer　投掷式干扰机
expendable kick stage　一次性启动（阶段）
expendable launch vehicle　一次（使用）运载器，一次（使用）运载火箭
expendable recoverable capsule　一次性返回器
expendable rocket　一次性火箭
expendable weight　耗重
expendable(s)　1.消耗件，一次使用件 2.投放物（指可以在飞行中投掉的，包括外挂物、燃油、弹药等）
expended case　废弹壳，射击样的药筒
expended link compartment　弹链收集箱
expense　费用，消耗
experience　经验，体验，经历
experience correction method　经验修正法
experience rating　经验费率法，经验定额
experienced test　有经验的测试（人员）
experiment　实验，试验，做实验，进行试验，尝试
experiment correction method　实验修正法
experiment ensure plan　试验保障方案
experiment environment　实验环境
experiment representative　典型实验
experiment setup　实验设置
experimental　实验的，根据实验的，试验性的
experimental aerodynamics　实验空气动力学
experimental aircraft　实验飞机，研究机（供基本研究或供研制开发用的），验证机
experimental analysis　实验分析，试验分析
experimental and numerical method　实验及数值（模拟）方法
experimental and numerical result　实验及数值（模拟）结果
experimental and numerical study　实验研究及数值模拟
experimental animal decompression　试验动物减压
experimental apparatus　实验仪器，实验装置
experimental approach　实验研究法
experimental astronomy　实验天文学
experimental burning rate　实验中的燃烧速度
experimental campaign　实验活动
experimental casestudy　实证性个案研究
experimental characterization　实验表征，实验特征
experimental condition　实验条件，试验条件
experimental configuration　实验装置，实验配置，实验

构型
experimental data 实验数据
experimental database 实验数据库
experimental datum 实验数据,实验数据
experimental detail 实验细节
experimental error 实验误差,试验误差
experimental evaluation 实验评价,实验评估
experimental evidence 实验证据
experimental facility 实验装置
experimental flap optimization 试验(性)襟翼优化
experimental flight 试飞
experimental heat transfer 实验传热(学)
experimental information 实验数据信息,实验资料
experimental investigation 实验(试验)研究
experimental mean pitch 平均实验桨距(在无力情况下旋转一周前进的距离),实际平均螺距
experimental measurement 实验测量
experimental model 经验模型,实验模型,实验型号
experimental observation 实验观测,实验观察
experimental parameter 实验参数,经验参数
experimental procedure 实验程序,实验步骤
experimental propulsion system 实验性推进系统
experimental research 实验研究
experimental response 实验响应时间,试验响应
experimental result 实验结果
experimental run 试验运行,实验运转,实验性生产
experimental setup 实验装置,实验设置
experimental shocktube 实验激波管
experimental specimen 实验样本
experimental standard deviation 实验标准偏差
experimental standard deviation of weighted arithmetic average 加权算术平均值的实验标准偏差
experimental study 实验(性)研究
experimental support 实验(方面的)依据
experimental system 试验系统,实验系统,实验体系
experimental technique 实验技术,实验方法
experimental test 试验性检验,实验测试
experimental test case 实验检验实例
experimental time 实验时间
experimental uncertainty 实验不确定性,实验不确定度
experimental uncertainty band 实验不确定度区间
experimental validation 试验验证,实验验证,实验校验
experimental value 试验值,实测值(在实验中测到的值)
experimental velocity 实验速度,实测速度
experimental verification 实验验证,试验验证
experimental work 实验研究,实验工作
experimental results 实验结果
experimenter 实验者

experiment-repetition theorem 重复试验定理
expert control system 专家控制系统
expert data mining system 专家数据挖掘系统
expert opinion 专家意见
expert system 专家系统
expiration term of service 最大使用期限,使用期满
expiratory reserve volume 补呼气量
expiratory resistance 呼气阻力
expired satellite 报废卫星,失效卫星
explicit equation mechanization 显式方程编排
explicit function 显函数
explicit guidance 显式制导
explicit identification 显式标志
explicit path programming 显路径编程
explicit product definition 明确的产品定义
explicit reservation 显示预约
explicit scheme 显格式,显式积分方法
explicit solution 显(性)解
explode 爆(突)发,发怒,激增,迅速扩大,(使)爆炸
exploder 起爆装置,雷管,引信,爆炸装置,爆炸物
exploding 引爆
exploding bridgewire 爆炸桥线(测量电桥中的)
exploding bridge-wire initiator 桥式爆炸发火器
exploding fuse 导爆索
exploding pipe 导爆管
exploding primer 爆炸底火(火帽)
exploding reflector 爆炸反射面
exploding wire 爆炸导线
exploit 开采,开拓,利用(……为自己谋利),剥削
exploration 探索,探测,勘探,研究
exploration architecture 探险规划
exploration geophysics 勘探地球物理(学)
exploration mission 探险任务,探索任务
exploration program 探险计划,探索计划,研究计划
exploration seismology 勘探地震学
exploration staging post 探险补给站,探险中转站
exploration system 物探系统,探测体系
exploratory development 勘探开发,应用研究,探索性研究,探究性研究
explore 勘查,探测,勘探,探查,探索,研究
Explorer 探险者卫星
explosion 爆炸,炸裂,爆发
explosion analysis 爆炸分析
explosion center 爆心
explosion chamber 灭弧箱,(脉动式空气喷气发动机的)燃烧室
explosion condition 爆炸条件
explosion energy 爆能,爆炸能
explosion event 爆炸事件

explosion experiment	爆炸实验
explosion（explosive）forming	爆炸成形
explosion heat	爆热,爆炸热
explosion jack	火药机构
explosion length	爆炸时长
explosion limit	爆点,爆炸极限
explosion line	爆炸线,爆破线
explosion on orbit	在轨爆炸
explosion pressure	爆(炸)压
explosion seismology	爆炸地震学
explosion shock	爆炸冲击
explosion source	爆炸源
explosion turbine	爆燃式燃气轮机
explosion valve	爆破阀
explosion volume	爆容
explosion-proof equipment	防爆炸设备
explosion-proof motor	防爆型电动机
explosion-proof switch	防爆开关
explosive	爆炸物,炸药,（复）爆破器材,爆炸(性)的,爆发(性)的
explosive actuator	火药起动器
explosive aerodynamics	爆炸气体动力学
explosive assembly	爆炸装置,(导弹的)战斗部
explosive bolt	爆炸螺栓
explosive bomb	爆炸性炸弹（炸药为主要装药,区别于化学炸弹、燃烧炸弹）
explosive bullet	爆破弹
explosive cartridge	电爆管,爆破管
explosive charge	1.爆炸装药 2.发射药,抛射药 3.射孔弹
explosive cladding	爆炸包覆
explosive D	D型炸药
explosive decompression	爆发性减压,突然失压,迅速减压
explosive device	引信,引爆装置
explosive element	爆炸装药
explosive filler	爆炸装药
explosive filling	爆炸装药,战斗装药
explosive forming	爆炸成形
explosive galaxy	爆发星系
explosive gas gun	气退式机关炮,气退式机枪
explosive handling wharf	火工品装卸码头（潜地导弹专用的）
explosive initiation	起爆
explosive mass	炸药(爆炸物)质量
explosive mass fraction	炸药(爆炸物)质量分数
explosive mixture	1.爆炸混合物,混合炸药 2.可爆燃混合气体 3.爆炸性混合物
explosive molecule	炸药分子
explosive nut	爆炸螺母
explosive ordnance	爆炸物
explosive ordnance disposal	爆炸物处理人员
explosive plasma	爆致等离子
explosive plasticizer	炸药增塑剂
explosive powder	(爆炸)火药
explosive power	爆炸力
explosive pumped laser	爆炸抽运激光器
explosive rivet	爆炸铆钉
explosive shackle	爆炸锁扣
explosive shower	爆发簇射
explosive surface	爆炸(破)面
explosive train	传爆序列
explosive valve	电爆阀门
explosive welding	爆炸焊(接)
explosive yield	爆炸威力,梯恩梯当量
explosive-actuated valve	爆破作动阀
explosively driven shock tube	爆炸驱动激波管
explosive-pellet warhead	装有爆破杀伤圆珠的弹头
explosive-type generator	爆燃式发电机
exponent	指数,说明者,样品,说明物
exponent value	指数值
exponential	指数的,指数
exponential atmosphere	指数大气
exponential atmospheric density model	大气密度指数模型
exponential convergence	指数收敛
exponential growth	指数增长
exponential line	指数线
exponential model	指数模型
exponential time	指数时间
exponential window	指数窗
export license	出口许可证
export sale	外销
expose	揭露,揭发,使暴露,使遭受,使曝光
exposed sample	经处置的样品（经过环境试验的试验样品）
exposed wing	外露翼
exposed zone	暴露区
expost calculation	事后计算
expost licensing	事后许可
expost tests	事后测试
exposure	暴露,曝光
exposure condition	曝光条件
exposure duration	曝露时间,曝光时间,暴露时间
exposure interval	曝光间隔
exposure latitude	曝光宽容度,曝光时限
exposure level	噪声感知等级,接触程度
exposure limit	暴露极限

exposure meter 照射量计
exposure ratemeter 照射量率计
exposure station 摄站,曝光点
exposure suit 暴露服,防护服
exposure test 暴露试验
express package 特快货运,特快空运,特快邮件
expression 解析式,分析式,表达式
expurgated code 删信码
exsiccant 干燥剂,脱水剂
extend 伸长,加长,扩展
extendable nozzle 可延伸(出口锥)喷管
extended atmosphere 厚大气
extended boundary-layer equation 扩展边界层公式
extended code 扩展码
extended duration photographic reconnaissance satellite 延长寿型照相侦察卫星
extended envelope 延伸包层
extended function code 扩充功能(操作)码
extended interaction amplifier 分布作用放大器
extended interaction klystron 分布作用速调管
extended interaction oscillator 分布作用振荡器
extended Kalman filter 广义卡尔曼滤波器,扩展式卡尔曼滤波器
extended least square 广义最小二乘法
extended life 延长使用寿命
extended life cycle 扩展的生命期
extended life superplasticizer 长期有效的增塑剂
extended linearization control method 增强线性化控制方法
extended main store 扩充主存储器
extended memory area 扩充存储区
extended mission 增加的任务,额外任务
extended network services 扩充网络服务
extended over water operation 远岸水面作战
extended perturbation problem 增强扰动问题,增强摄动问题
extended phase space 扩充相空间
extended radio source 射电展源
extended range 增程,加大射程
extended relational memory 扩充相关存储器
extended target 广延目标,展开目标,扩展目标
extended target effect 扩展目标效应
extended testing 延长测试
extended transfer matrix 扩阶传递矩阵,扩展传递矩阵
extended X-ray absorption fine structure 广延 X 射线吸收精细结构
extended X-ray source X 射线展源
extended-root blade 伸根叶片(叶片叶身和榫头间有延伸段)

extendible ballistic measuring system 1.外弹道测量系统 2.外测系统
extendible nozzle 可延伸(出口锥)喷管,延伸喷管
extensibility 延伸率,伸长性,可扩展性,可延长性,展开性
extensible tether 可扩展系绳
extension 伸长,加长,扩展
extension contract 增补合同,延伸合同,延长期限的合同
extension flap 后伸式襟翼
extension length 加长长度,伸展长度
extension nozzle 延伸式喷嘴,可延伸喷管
extension of flight envelope in flight test 飞行包线扩展试飞
extension part 扩张段,延伸部分
extension pipe 延伸管,尾管,延长管
extension tube 延伸管,外接管
extension-bending 屈伸弯曲
extension-twist 弯扭耦合
extension-twist coupling 弯扭耦合作用
extensive coherent shower 广延相干簇射
extensive variable 广延变数
extensometer 伸长计(能测量微小的变形),伸缩仪,应变仪,变形测定器
extent 范围,程度
extent of laminar flow 层流流动范围
exterior 1.外部,表面的,外在的 2.外部,表面,外型,外貌
exterior (external) ballistics 外弹道学,膛外弹道学
exterior check 外观检查
exterior injector 外喷嘴,外引射器
exterior injector hole 外喷嘴口,外引射器口
exterior orientation 外部定向,外方位
exterior orientation element 外方位元素
exterior trajectory measurement 1.外弹道测量 2.外测
exterior trajectory measurement system 1.外弹道测量系统 2.外测系统
external aileron 外副翼
external air 外界空气
external airfoil 外翼
external armament 外挂式武器
external augmentor 远方增升装置
external ballistic telemetry 外弹道遥测
external base 外基区
external blast warhead 外爆(破)弹头
external bond 外粘接,外结合,外联结
external bond coefficient 外粘接系数,外结合系数
external cathode 外阴极(的)
external characteristics 1.外特性 2.负荷特性 3.转

速特性
external clock 外部时钟
external coil 外(部)线圈
external compression 外压式
external compression inlet 外压式进气道
external compression shock system 1.外压式激波系统 2.外压式冲击系统
external cooling 外冷却
external cooling circuit 外部冷却回路
external crack 表面裂纹
external device 外部设备
external disturbance 外扰
external ejector 外引射器
external energizer 外接电源电动机(向发动机惯性起动机提供原动力的小电动机)
external error 外在作用误差,外界影响误差
external event 外部事件
external event horizon (黑洞)外视界(线)
external excitation 1.外激励,外部激励 2.外扰
external expansion nozzle 外部膨胀喷管
external field 外场
external field protection 外场保护
external flow 【力】表面流,外流
external flow field 外流场
external force 外力
external fuel tank 外部燃料箱,(飞船)外部燃料罐
external fueling 外(卡)加油
external heat sink 外部散热器,外置散热片
external input 外部输入,外部输入信号
external input stream 外部输入信号流
external insulation 1.外绝缘,外部绝缘 2.外保温
external interface 外部接口
external lens heating unit 外(部)透镜加温机构,外(部)透镜加温装置
external load 机外挂载,外载荷
external loss 1.外损失 2.外部损耗
external margin 外图廓
external mixing type atomizer 外混式雾化喷嘴
external mode 外模式,外部模式
external modulation 外部调制,外调制
external moment 外力矩,外弯矩
external noise 外部噪声
external nozzle 外喷嘴,外喷管
external or internal disturbance 外/内部扰动,外/内部干扰
external perturbation 外部扰动
external photoelectric effect 外光电效应
external power 外接电源
external pressure test 外压试验

external pressure tester 外压试验器
external quality factor 表观品质因数,外品质因数,外品质因子
external quantum efficiency 外量子效率
external ramp 外(部)斜
external reflection spectroscopy 外反射谱(学)
external region 外部区域
external shock 外部冲击
external shock system 外冲击系统
external standard 外标准,外标物
external steering coil 外导向线圈
external storage 外存贮器,外部存储器,外存储器
external store 外挂物
external stores configuration 外挂物布局,外挂物配置
external stores testing 外挂物试验
external strain gage balance 外式应变天平
external stream 外界气流
external structure 表面结构,外部结构
external supercharger (装在发动机之外)外增压器,外装增压器
external support 外支承,外支架
external surface 外表面
external temperature 外部温度
external test facility 外部测试设备
external torque 外力矩,外转器
external torque vector 外力矩向/矢量,外转矩向/矢量
external trim coil 外补偿线圈,外修正线圈
external velocity 外(部)速度
external-combustion ramjet 外燃冲压发动机
externally blown flap 外吹式襟翼
externally carried bomb 外挂炸弹
externally coherent moving target indicator 外相干动目标显示器,外相干动目标指示
externally-powered machine gun 外动力(工作的)机枪
externally-powered weapon 外动力武器
externally-programed test equipment 外部编程控制的试验设备,外部编程控制的测试试设备
external-mixing 外混
extinct elements 灭绝元素
extinction 熄灭,消灭,消弧,熄火,消声,消光,衰减
extinction coefficient 消光系数
extinction index 消光系数,消光率
extinction model 消光模式,衰减模式
extinction strain rate 熄火拉伸率
extinguish 灭火,灭绝,消除,偿清
extinguishant 灭火剂
extinguisher 灭火器,灭火瓶;消除器
extinguishing agent 灭火剂
extra address 附加地址

extra contour 助曲线,辅助等高线
extra fuel 额外燃料,备用燃料
extra fueltank 备用油箱
extra high modulus 特高模量
extra motion 额外运动,附加运动
extra motion cue 额外运动线索
extra oxygen 备用氧气,额外的氧气
extra oxygen unit 备用氧气装置,额外的氧气单元
extra planetary space 行星外空间
extra pressure 外加压力,附加压力
extra schedule flight 加班飞行
extra section 1.加班(飞行)(航空公司在正常班期以外的客货运飞行,同 extra section flight) 2.附加部分,外加部分
extra time 额外时间
extract 萃取,提取,提取物,提取液
extract type mass fluxmeter 推导式质量计
extraction 1.退弹,抽壳,抽筒 2.抽出,拔出,排出 3.(雷达情报)录取, 4.提取,析取 5.萃取剂,提取剂
extraction electrode 引出电极
extraction force (炮弹的)抽壳力
extraction grid 引出栅
extraction mark 录取标志
extraction of model parameters 模型参数提取
extraction parachute 牵引伞,分离降落伞
Extraction Parachute Ejector System 降落伞弹出系统
extraction precision 录取精度;提取精度
extraction zone 拖曳空投区
extractive sampling 抽取采样
extractor 1.退弹钩,抓弹钩,退弹簧,退壳器 2.提取器,抽出器 3.分离器,分离装置 4.录取设备
extractor cam 抽壳凸部
extractor electrode 【医】萃取电极
extractor hook 抓弹钩钩部,抽筒子钩
extractor lip 抓弹钩
extractor ring 分离器环
extractor spring 抓弹钩簧
extractor striker (转膛炮的)退壳撞击器
extractor vacuum gauge 分离型(电离)真空计
extractor voltage 萃取电压
Extradop Extended Range Dovap 增程多瓦勒,增程多普勒测速和定位系统
extrados 拱顶面,机体上表面,拱背(拱的外缘线)
extragalactic astronomy 星系天文学,河外天文学
extragalactic cosmic ray 河外宇宙射线
extragalactic extended radio sources 河外射电源
extragalactic radio radiation 河外射电
extra-meridian observation 非中天观测,近子午线观测,近中天观测

extraneous vibration 附加振动
extranuclear electron 核外电子,原子中外层电子
extraordinary inspection 临时检查
extraordinary ray 非常射线(电波在磁离子介质中分裂成的两个特征波中的一个)
extraordinary ray traces 非常光描迹(频高图上非寻常波回波的频高曲线)
extraordinary wave 非常波,非寻常波,异常波
extrapolate 外推,推断
extrapolated position 外推位置
extrapolation 外推(法),插值(法)
extrapolation type satellite guidance system 外推法卫星制导系统
extrasensory perception 超感官知觉,超感知觉
extraterrestrial 地(球)外的,行星际的,外星人(的)
extraterrestrial biology 地外生物学,外星生物学
extraterrestrial civilization 地外文明,外星文明
extraterrestrial disposal 地(球)外(放射性废物)处置(法),外层空间处置(法)
extraterrestrial environment 行星际环境
extraterrestrial inteligence 地外智慧,外星智慧
extraterrestrial life 地外生命,外星生命
extraterrestrial matter 地球外物质
extraterrestrial noise 地球外噪声
extraterrestrial radiation 地外辐射,行星际辐射
extraterrestrial resource 地外资源,外星资源
extraterrestrial topography 天体形貌学
extravehicular activity 舱外活动,出舱活动
extravehicular activity robot 舱外活动机器人
extravehicular gloves 航天器舱外工作手套
extravehicular mobility unit 舱外活动设备
extravehicular space suit 舱外航天服
extra-vehicular suit 舱外活动(防护)服
extrema (extremum 的复数形式)极值
extremal 1.极值曲线 2.极值的 3.致极曲线 4.致极函数
extreme 极限,极值
extreme dimension 极限尺寸,极限规格
extreme firing range 最大射程,极限射程
extreme low-altitude flight 超低空飞行
extreme operating conditions 极限工况,极端使用条件,极端操作条件,最恶劣使用条件
extreme trim 过度配平,调整片偏转过度
extreme ultraviolet 极(远紫)外,超紫外(线)
extreme ultraviolet glancing incidence telescope 极(远)紫外水平入射望远镜
extreme ultraviolet laser 极紫外激光器
extreme ultraviolet photometer 极(远)紫外光度计
extreme ultraviolet radiation detector 极远紫外辐射探

测器
extreme value 极值
extremity （复数）肢体,极端,极度,尽头,绝境,非常手段
extremum 极值
extremum control 极值控制
extremum control system 极值控制系统
extremum principle 极值原理
extremum-seeking algorithm 极值搜索算法
extrinsic calibration （仪器仪表）外部标定（的）
extrinsic engine 外（部）引擎
extrinsic feedback 外反馈
extrinsic parameter 外参数,外部参数,非固有参数,非本证参数
extrude 挤压,挤出
extruder 挤出机,挤压机,压出机,挤塑机
extrusion 1.挤压（加工）,挤出,挤出成型 2.挤压,冲压,挤压型材
exudation 1.渗出,渗漏,(推进剂)汗析 2.渗出,渗出物
EXW 1. Expeditionary Warfare 远征作战 2. Explosive Welding 爆炸焊
eye 1.孔眼 2.吊环,接头耳环 3.眼,圈,眼环结,绳环,锚眼
eye base 眼基线（眼距）,眼基距
eye closure 眼图闭合
eye control 目视控制,目视操纵
eye observation 目测
eye pattern （脉冲编码调制的）眼孔图样,眼图（接收数字信号用同步示波器显示的连续重叠波形）
eye protection 1.眼防护 2.护目用具
eye relief 1.目视暂留（观测者能清晰看到整个视场的像时眼球与夜视目镜的间距）2.镜目距
eyeball 1.眼球 2.〈口语〉用目视搜索,盯住目标 3.通气小球（客舱中乘客头项上方用来调节新鲜空气的）
eyeball/shooter 目视搜索/攻击机动
eyeballs down 1.〈口语〉把飞行员压向椅盆的过载 2.向上加速
eyeballs in 1.〈口语〉把飞行员压向椅背的过载 2.向前加速
eyeballs left 1.〈口语〉把飞行员压向座椅左侧的过载 2.向右加速
eyeballs out 1.〈口语〉使飞行员离开椅背的过载 2.向前减速
eyeballs right 1.〈口语〉把飞行员压向座椅右侧的过载 2.向左加速
eyeballs up 1.〈口语〉使飞行员离开椅盆的过载 2.向下加速
eye-bar 眼杆
eyebrow lights 眉灯（仪表刻度盘的小照明灯）
eyebrow panel 上仪表板,上视仪表盘（板）
eyelids 1.可调节喷口的半圆形调节片 2.眼睑
eyenut 吊环螺帽,接耳螺帽,吊环螺母
eyepiece 1.目镜 2.眼罩
eyepiece with micrometer 测微目镜
eyeshield 防护眼罩,护目罩

F

F　1. Fahrenheit 华氏温度 2. Farad 法拉 3. Fire 射击，发射，火力 4. Floor 地板 5. Force 力，力量 6. Frictional Force 摩擦力 7. Fuel 燃油

F&E　Facilities and Equipment 装置和设备

F&F　Fire and Forget 发射后免控

F/A　Fix/Attack 定位/攻击

F/G　Fiber Glass 玻璃纤维

F/O　First Officer 副驾驶，大副

F/P　Flat Pattern 展开形状，平面展开图

F/S　1. Fast/Slow 快慢 2. Full Size 全尺寸

F°　Fahrenheit 华氏温度

FA　1. Fabrication Assembly 制造组合件 2. Fan Air 风扇气流 3. Final Approach 最后进近 4. Flame Arrestor 火焰阻止器 5. Free-Air 开口式风洞 6. Frontal Aviation 前线航空兵（苏联）7. Function Area 功能区 8. Flaperon Angle 襟副翼偏转角

FAA　1. Federal Aviation Administration 联邦航空局（美国）2. Fleet Air Arm 舰队航空兵（英国）

FAA-CAB　Federal Aviation Agency, Civil Aeronauticl Board 联邦航空局民航委员会

FAAD　Front Area Air Defense 前沿地区防空

FAA-DMI　Federal Aviation Administration Designated Manufacturing Inspector 联邦航空局委任制造检查员

FAAM　Future Air-to-Air Missile 未来空空导弹

FAAMMEL　FAA Master Minimum Equipment List 联邦航空局主最低设备清单

FAAR　Forward Area Alerting Radar 前方地域警戒雷达

FAATC　FAA Techincal Center 联邦航空局技术中心

FAATDC　FAA Techincal Development Center 联邦航空局技术发展中心

FAB　1. Fabricate or Fabrication 制造 2. Functional Auxiliary Block 功能辅助舱（苏联"量子号"空间实验舱的组成部分）

FABL　Fire Alarm Bell 火警铃

fabric　1. 布,织物 2.（飞机、飞艇）蒙布

fabric count　织物密度

fabric for parachute canopy　降落伞伞衣织物

fabrication　1. 制造,装配生产 2.（焊件的）组装

FABTEM　Fabrication Target Evaluation Master 制造目标评估标准

FAC　1. Factor 因素,因数 2. Federal Airports Corporation 联邦航空港公司（澳大利亚）3. Final Approach Course 最后进近航道 4. Flight Augmentation Computer 飞行增稳计算机 5. Forward Air Controller 前方空中控制员,前方飞机引导员 6. Frequency Analog Computer 频率模拟计算机

FACC　Ford Aerospace and Communication Corporation 福特航空空间与通信公司

FACCL　First Article Conformity Check List 首件合格检验清单

face　1. 面,表面,正面,端面,饰面,液面 2. 表盘 3. 荧光屏

face alignment　叶面校准距

face curtain　面帘

face guard　防护面罩

face seal　端面密封,端面密封件

face sheet　面板

faceblind firing　面帘打火

face-on object　正向天体

faceplate　1. 面板 2. 平面卡盘,花盘 3. 安装座 4. 阴极射线管萤幕,荧光屏

face-protection blind　防护面罩,防护布帘

FACETS　Future Anti-Air Concepts Experimental Technology Seeker 未来防空思想实验技术导引头

face-up bonding　正焊法,正面结合

FACI　First Article Configuration Inspection 首件构型检验

FACIL　Facility 设备

facility performance category　地面导航设施性能等级,仪表降落系统性能

FACO　Final Assembly and Check-Out 总装与检验

FACP　Forward Air Control Post 空军前方控制站

FACRI　Flight Automatic Control Research Institute 飞行自动控制研究所（中国）

FACS　Fine Attitude Control System 精确姿态控制系统

FACT　Facility of Automation, Control and Test 自动化、控制和测试设备

fact base　事实库

factor of safety　安全因数

factored field lengths　机场安全使用长度

factory loaded　在厂内装药的,交付前装药的

factory test　1. 工厂试车 2. 验收试车

FACTS　FLIR-Augmented Cobra Tow Sight 前视红外装置增强式"眼镜蛇"陶式导弹瞄准具

FAD　1. Fast-Action Device 速动装置 2. Fighter

Aerodynamics Development 战斗机空气动力学研究 3. Fleet Air Defence 舰队防空（美国）

FADA 〈西班牙语〉Federacion Argentina De Aeroclubes 阿根廷航空俱乐部联合会

FADD Fatigue And Damage Data 疲劳和损伤数据

fade chart 盲图，衰落区图

fade indicator 衰落指示器

fade zone 消失区，静区，盲区

FADEC 1. Full-Authority Digital Engine Contorol 全权数字式发动机控制 2. Full-Authority Digital Electronic Control 全权数字式电子控制

fading margin 衰落裕量，衰减速边际

fading memory 淡化记忆

FADS Flexible Air-Data System 接口可变型大气数据系统

FAE Fuel/Air Explosive 油气炸药，油气爆炸物

FAEJD Full Automatic Electronic Judging Device 全自动电子判定器

FAER First Article Engineering Review 首件工程评审

FAF Final Approach Fix 第五边定位点，最后进近定位点

FAFT Fore/Aft Fuselage Tankage 箭体前/后贮箱

FAGC Fast Automatic Gain Control 快速自动增益控制

FAI First Article Inspection 首件检验

fail hardover 失控故障（自动飞行控制系统一种故障，其现象为舵面急剧偏转到极端位置后保持不动）

fail link 脆弱环节

fail operational/fail-operation 故障后保持工作的，带故障运行的

fail safe 故障安全

fail safe structure 破损安全结构

fail safety 故障自动防护，破损安全，可靠性

fail tree analysis 失效树分析

fail-diagnosis 故障诊断

fail-functional 故障后保持功能的

failing load 破损载荷，破坏载荷

fail-operation 故障工作

fail-operative automatic flight control system 故障运行式自动飞行控制系统

fail-passive 故障降级，故障被动防护的

failsafe 故障自动防护的

failsafe automatic flight control system 故障自动防护式自动飞行控制系统

fail-soft 故障自动缓和的，故障有限影响的

fail-soften 故障弱化

fail-steady 故障状态稳定（的）

failure 1. 故障，失效 2. 破损，断裂 3. （发动机的）停车

failure accommodation 容错

failure analysis 故障分析，失效分析

failure condition 1. 断裂条件 2. 故障情况

failure criteria 失效判据，故障判据，故障准则

failure damping system 故障阻尼系统

failure detection 故障探测

failure detector 故障探测器

failure diagnosis 故障诊断，失效诊断

failure effect 失效影响

failure flag 故障信号旗

failure handling program 故障处置预案

failure identification 故障标识

failure isolation 故障分离，故障隔离

failure load test 破坏载荷试验

failure location 故障定位

failure mechanism 失效机理

failure mode 失效形式，故障形式，故障模式

failure mode and effect analysis 失效模式与影响分析

failure monitoring system 故障监视系统

failure operation 故障运行

failure rate 失效率，故障率，失败率

failure reconfiguration 故障重构

failure report 失效报告

failure simulation 故障模拟

failure substitution 故障取代

failure telltale 故障信号装置，故障显示装置

failure to cross-check 综合判断失误，注意力分配不当

failure-free operation 无故障运行，正常运行

faint blue object 暗蓝天体

faint companion 暗伴星

faint star 暗星

FAIP 1. First Article Inspection Plans 首件检验大纲 2. First-Assignment Instructor Pilot 第一（指令）飞行教官

FAIR 1. Failure Analysis Information Retrieval 故障分析信息检索 2. Fairing 整流片

fair drawing 清绘

fair over 加整流罩（套）

fairing 1. 整流罩，整流片，整流包皮 2. 曲线或数据中取平均值

fairing assembling frame 整流罩装配型架

fairing casing 整流罩

fairing install room 整流罩扣罩间

fairing jettison motor 抛整流罩发动机

fairing rail transporter 整流罩铁路运输车

fairing separation test 整流罩分离试验

fairing wire 飞艇蒙皮的整形张线

fairing working ladder 星罩工作梯

FAJ 1. First (Final) Assembly Jig 初（总）装夹具 2. Floor-Mounted Assembly Jig 地面安装装配夹具

faker 假想敌机（防空演习用）

FAL Final Assembly Line 总装线
Falac Forward-Area Liaison and Control 前方地带联络和控制
Falconer "鹰猎者"（英国精密进场着陆雷达）
fall back 1.回落原地（弹道火箭垂直发射失败后的） 2.放射性回降物（核爆炸升空物降回地球的）
fall of temperature 温度下降，降温
fall of the sea 落潮
fall time 下降时间
fallaway section 分离段
fallback area 回落区，导弹临发时前的人员撤离区
fallback programme 备用方案，第二方案（前一个方案失败之后采用的）
fall-down test 跌落试验
falling angle 落角
falling out of synchronism 失步
falling-sphere method 落球法
falloff 急剧下降，脱离编队
fallout 1.放射性沉降物 2.沉降 3.技术副产品
fallout contours 等辐射强度线
fallout pattern 放射沉降物分布图
fallout prediction 放射性沉降物预测
fallout safe height 放射性沉降物安全高度
fallout wind plot 沉降风矢量图
false add 无进位加
false alarm probability 虚警概率
false alarm rate 虚警率
false alarm time 虚警时间
false brinelling 虚假硬度，摩擦腐蚀压痕
false call 误呼叫
false cirrus 伪卷云
false code 非法代码
false color composite 假彩色合成
false color film 彩色红外片，假彩色片
false color photography 假彩色摄影
false command 虚指令
false command probability 虚指令概率
false cone of silence 假静锥区，假静区
false conic error 假圆锥误差
false contouring 假轮廓
false feel 错误操纵感
false glidepath 假下滑道
false heat 假热
false lift 假升力
false nosing 前缘假肋
false ogive 炮弹或炸弹的减阻帽，风帽
false output 伪输出，假输出
false ribs 假肋（机翼前缘主肋之间用以支撑布质蒙皮外形用的肋）
false spar 副梁，纵墙（与机身不连接的第二梁，用于安装活动翼面）
false start 假起动（燃气涡轮的起动循环）
false synchronization 虚同步
false target 假目标
false zero 虚零点
false-alarm 虚警
false-color composite 假彩色合成
false-color rendition 假彩色还原，假彩色再现
false-colour film 假彩色胶片
false-target generator 假目标产生器
FALW Family of Air Launched Weapons 空中发射武器族
FAM Flight Attendant Manual 乘务员手册
FAMG Field Artillery Missile Group 野战炮兵导弹群（美国）
familiarization 熟悉训练
family of characteristics 特性曲线族
family of curves 曲线族
FAMIS Full Aircraft Management Inertial System 全飞机管理/惯性系统
FAMO Facilities Asset Management and Operations 设施器材管理和操作
FAMOS Fleet Air Meteorological Observation Satellite 舰队航空气象观测卫星
fan engine 风扇发动机
fan exit case 风扇排气机厘，扇出口罩
fan inlet 通风进气口
fan jet 涡轮风扇发动机
fan lift 风扇升力（垂直或短距起落飞机上用专门的风扇产生的升力）
fan mapping 扇形航测制图
fan marker 扇形指点标
fan nacelle 风扇整流罩
fan pressure ratio 风扇压比
fan straightener 梳直风扇后气流的整流叶片，扇形导流栅
fan stream burning 风扇气流复燃
fan system 风扇系统
fan-failure clutch 风扇故障离合器
fan-in 扇入
fanned-beam antenna 扇形波束天线
fanning beam 扇形波束
Fano coding 费诺编码
Fano decoding algorithm 费诺译码算法
fan-out 扇出
fantail 1.尾扇（一种新设计，用来代替直升机的尾桨） 2.（航母）舰突（甲板拦阻索由此伸向甲板）
Fantrainer 涡扇教练机（德国制造的教练机）

FAP 1. Failure Analysis Program 故障分析程序 2. Fleet Average Performance 机群平均性能 3. Forward Attendant Panel 前乘务员面板

FAPA Future Aviation Professionals of America 美国未来航空专业人员

FAR 1. Fatigue Report 疲劳分析报告 2. Federal Aviation Regulations 联邦航空条例

Far Eastern Air Transport Co. 远东航空公司(中国台湾)

far field 远场

far field boom 远场声爆

far infrared scanner 远红外扫描装置

far side of the moon 月球背面

far ultraviolet photometer 远紫外光度计

far ultra-violet radiation 远紫外辐射

far ultraviolet spectrograph 远紫外摄谱仪

FARA Formula Air Racing Association 方程式飞机竞赛协会

Faraday cylinder 法拉第筒

Faraday effect 法拉第效应

faradmeter 法拉计

fare dilution 机票收入减少

fare structure (航空公司的)收费一览表

far-end crosstalk 远端串话

far-field 远源场

far-field body wave 远场体波

far-field noise 远场噪声

far-field region 远场区

far-field surface wave 远场面波

far-infrared laser 远红外激光器

far-infrared region 远红外区

farm 天线场

farm-gate operations 援外作战训练

Farnborough indicator 范堡罗记录仪,能连续记录活塞式发动机压力循环的第一代记录器

Farnham roll 法汉滚压机(滚轧机),弯板机

FARS First Aircraft (Airplane) Reporting System 首架飞机报告系统

far-ultraviolet camera 远紫外相机

far-ultraviolet space telescope 远紫外空间望远镜

FAS 1. Final Assembly Scheduling 总装计划 2. Flexible Assembly System 柔性装配系统 3. Flight-Attendant Station 飞行乘务员舱位 4. Forward Acquisition Sensor 前视探测传感器 5. Frequency-Agile Subsystem 频率捷变分系统

FASA Friendly Aircraft Simulating Aggressors 模拟敌机

FASCO Foreign Airlines Service Corporation 外航服务公司

FASE Fast Auroral Snapshot Explorer 极光速摄探测器(美国)

FASS Forward-and-Aft Scanner System 前后扫描系统

Fast 1. Fan and Supersonic Turbine 风扇超声速涡轮 2. Flight Advisory Service Testing 飞行咨询服务测试 3. Flying Ambulance Surgical Trauma 航空救护外伤

FAST 1. Facility for Automatic Sorting and Testing 自动分类与测试设备 2. Fast Acquisition Search and Track 快速截获搜索和跟踪 3. Fence Against Satellite Threats 防卫星雷达情报网 4. Forward Area Support Team 前方地域保障队

fast angular variable 快变角变量

fast arming fuze 快速解除保险引信

fast burn booster 速燃助推器(导弹反拦截的一种措施,在主动段加速飞行以防止拦截)

fast changing parameter 速变参数

fast charge 快速充电

fast erection (陀螺)快速修正

fast Fourier transform 快速傅里叶变换

fast freight line 快运货物直达运输线

fast frequency-shift keying 快速频移键控

fast ion conduction 快离子导电

fast mode 快变模态

fast mover 喷气式作战飞机

fast nova 快新星

fast reaction time 快速反应时间

fast response time 快速响应时间

fast state 快变状态

fast synchronizing button 快速协调按钮

fast thermocouple 小惯性热电偶

fast time control 快时间控制

fast variation telemetry parameter 遥测速变参数

fast varying parameter 速变参数

fast-action device 速动装置

fastener 1. 紧固件 2. 锁扣,钩扣 3. 接线柱,线夹

fast-response transducer 快响应传感器,小惯性传感器

faster-than-real-time simulation 超实时仿真

fast-screen tube 短余辉显像管

fast-sequencing architecture 快速时序结构

fast-sinking depth charge 速沉深水炸弹

fast-slow separation 快慢分离

FAT 1. Factory Acceptance Test 工厂验收试验 2. Final Assembly Test 总装试验 3. Flechett Anti-Tank 小型杀伤反坦克导弹

fat vector 宽向量

fatal accident 致命事故

fatal crash 机毁人亡

fatal dose 致死剂量

fatal injury 致死性外伤

fatality　死亡性事故
FATDL　Frequency And Time-division Data Link 频分与时分数据链
FATE　1. Factory Acceptance Test Equipment 工厂验收试验设备 2. Fuze Arming Test Experiment 引信解除保险试验 3. Fuzing, Arming, Test and Evaluation 引信解除保险和起爆试验鉴定
FATG　Fixed Air-To-Ground 空对地固定目标
fathom　英寻（1英寻＝6英尺＝1.83米，水深单位）
fathometer　回声测深计
fatigue accommodation　疲劳积累
fatigue assessment　疲劳评定
fatigue crack　疲劳裂纹
fatigue crack growth rate　疲劳裂纹扩展速率
fatigue curve　疲劳曲线
fatigue factor accident　疲劳性事故
fatigue failure　疲劳失效
fatigue index　疲劳指数
fatigue life　1.疲劳寿命 2.安全寿命
fatigue limit　疲劳极限
fatigue load　疲劳载荷
fatigue load spectrum　疲劳载荷谱
fatigue notch factor　切口疲劳因数
fatigue performance　疲劳性
fatigue quotient　疲劳商数，疲劳系数
fatigue reduction　疲劳减缓（主动控制系统在阵风作用下的一种减轻结构疲劳的功能）
fatigue strength　疲劳强度
fatigue striation　疲劳条带
fatigue test　疲劳试验
fatigue test at high temperature　高温疲劳试验
fatigue test at low temperature　低温疲劳试验
fatigue test under increasing load　加载疲劳试验
fault case　故障案例
fault cause　故障原因
fault countermeasure　1.故障预案 2.故障对策
fault data　故障数据
fault detect and warning　故障检测与报警
fault detect rate　故障检测率
fault detection　故障检测
fault diagnosis　故障诊断
fault direction　故障搜索方向
fault effects　故障影响
fault evidence　故障迹象
fault handling　错误处理
fault insertion　缺陷插入
fault isolation　故障隔离，故障定位
fault isolation and diagnosis　故障隔离和诊断
fault isolation monitoring　故障定位监视（控），故障隔离监视（控）
fault isolation rate　故障隔离率
fault localization　故障定位
fault mode　故障模式
fault orientation　错误定向（空间）
fault protection　故障保护，防止故障
fault rate　故障率
fault signal　故障信号
fault signature　故障表征
fault tolerant　容错的，故障容限的
fault tolerant computing　容错计算
fault tree　故障树
fault tree analysis　故障树分析
fault-free　无故障的，无缺陷的
fault-free system　无故障系统
fault-tolerance　容错
fault-tolerance capability　容错能力，故障容限能力
fault-tolerant communications processor　容错通信处理机
fault-tolerant computer　容错计算机
fault-tolerant processor　容错处理机
fault-tolerant software　容错软件
fault-tolerant system　故障容限系统
faulty burst　故障突发
FAV　1. Fan Air Valve 风扇气阀 2. First Article Verification 首件验证
favorable pressure gradient　顺压梯度
favorable wind　顺风，惠风
FAW　Fighter All Weather 全天候歼击机（战斗机）
fax chart　传真图
fax transmitter　传真发送机，传真发射机
faying surface　搭接面（蒙皮与飞机外露边缘的搭接部分），结合面，接触表面
FB　1. Fighter-Bomber 歼击轰炸机（英国） 2. Flat Bar 扁条，板片 3. Fuel Burn 燃油消耗 4. Flying Bomb〈口语〉飞弹，带翼火箭，滑翔炸弹 5. Fragmentation Bomb 杀伤炸弹
FBL　1. Fly-By-Light 光传操纵 2. Functional Baseline 功能基线
FBM　Fleet Ballistic Missile 舰载弹道导弹
FBMS　1. Fleet Ballistic Missile Submarine 舰队弹道导弹潜艇 2. Fleet Ballistic Missile System 舰队弹道导弹系统
FBO　1. File Bus Out 文件输出总线 2. Fixed-Base Operator 固定机场为基地的航空公司 3. Flights Between Over-hauls（两次）翻修间飞行次数
FBR　Fiber 纤维
FBS　1. Financial Budget System 财务预算系统 2. Fixed-Base Simulator 固定模拟器，固定基地模拟器

3. Forward-Based System 设在前方基地的系统
FBT Feedback Technology 反馈技术
FBV Force Balance Valve 力平衡活门
FBW 1. Fly-By-Wire 电传操纵遥控飞行,遥控自动驾驶仪 2. Fly By Wire 电传操纵
FC 1. Failure Causes 故障原因 2. Failure Condition 失效状态 3. Fan Cowl 风扇整流罩 4. Feel Computer 感觉计算机 5. Filiform Corrosion 线状腐蚀 6. Fire Control 火灾控制 7. First Class 一等,上等,第一流的,最好的,头等舱 8. Fire Control 火力控制,发射控制,射击指挥 9. Fixed Cost 固定成本 10. Flight Configuration Change 飞行构型更改 11. Flight Control 飞行管制,飞行操纵 12. Flight Controls 飞行操纵面 13. Flight Cycle 飞行起落,飞行循环 14. Foot Candles 英尺烛光 15. Fuel Cell 燃油箱,燃油舱(或室),燃料电池 16. Fuel Conservation 节约燃油 17. Functional Check 功能检查 18. Functional Characteristics 功能特性 19. Functional Collectors 功能集成件
FCA 1. Functional Configuration Audit 功能构型审核,功能配置检查(软件) 2. Future Cycle Accumulation 未来周期贮存
FCB 1. Flight Configuration Change Board 飞行构型更改委员会 2. Frequency Co-ordinating Body 频率协调机构 3. Functional Configuration Baseline 功能构型基线
FCC 1. Federal Communications Commission 联邦通信委员会(美国) 2. Fire Control Computer 点火控制计算机 3. Flat Conductor Cable 扁导体带状电缆 4. Flight Control Centre 飞行控制中心 5. Flight Coordination Center 飞行协调中心
FCcost Fligh-Crew Cost 飞行人员费用,空勤组费用
FCCS Flight Control Computer System 飞行控制计算机系统
FCCSET Federal Coordinating Council for Science, Engineering, and Technology 联邦科学、工程、技术协调委员会(美国)
FCD 1. Frequency Compression Demodulator 频率压缩解调器 2. Functional Configuration Documentation 功能构型文件
FCDC Flight Control Digital Computer 飞行控制数字计算机
FCE 1. Fire-Control Equipment 火(力)控(制)设备,射击指挥器材 2. Flight Control Electronics 飞行控制电子设备
FCEP Flight Control Electronics Package 飞行控制电子装置
FCES Flight-Control Electronics System 飞行控制电子设备系统

FCF Functional Check Flight 功能(检查)试飞
FCG 1. Facing 面饰,车端面 2. Fatigue Crack Growth 疲劳裂纹扩大
FCI 1. False Color Image 假彩色图像 2. Flight Command Indicator 飞行指令指示器 3. Fuel-Consumed Indicator 耗油量指示器 4. Functional Configuration Identification 功能构型标识
FCL Flight-Crew Licensing 颁发飞行人员执照
FCLP Field Carrier-Landing Practice 陆上着舰练习
FCM Fiber Composite Matter 纤维合成材料
FCNP Fire Control Navigation Panel 火力控制导航控制板
FCO Formal Change Order(合同的)正式更改通知书
FCOC Fuel Cooled Oil Cooler 用燃油冷却的滑油冷却器
FCOM Flight Crew Operating Manual 飞行机组操作手册,飞行员操作规程
FCP 1. Flight Control Panel 飞行控制板 2. Flight Control Processor 飞行控制处理器 3. Fluid and Chemical Processing 流体和化学处理 4. Fuel Cell Powerplant 燃料电池动力装置 5. Fuel Consumption Projections 燃料消耗量预测 6. Function Control Package 操作控制包
FCPF Foreign Commodity Production Forecasting 国外商品产量预报
FCR Flight Control Room 飞行控制室
FCRLS Flight Control Ready Light System 飞行控制准备灯系统
FCS 1. Facsimile Communications System 传真通信系统 2. Fire Control Simulator 火(力)控(制)模拟器 3. Fire-Control System 火灾控制系统,火(力)控(制)系统,射击指挥系统 4. Flight Control System 飞行控制系统,飞行操纵系统 5. Fuel Control System 燃油调节系统 6. Fast Circuit Switching 快速电路交换 7. Fiber Channel Standard 光纤信道标准 8. Frame Check Sequence 帧校验序列
FCSS Fire-Control Sight System 火控瞄准系统
FCST Federal Council for Science & Technology 联邦科学技术委员会(美国)
FCSU Fire Control Switching Unit 火(力)控(制)转换装置
FCT 1. First Configuration Test 初级构型试验 2. Flight Control Team 飞行控制小组 3. Flight Crew Training 飞行机组培训
FCTR Fan/Core Thrust Ratio 风扇与核心发动机推力比
FCU 1. Fighter Control Unit 战斗机控制组 2. Fire Control Unit 火(力)控(制)装置 3. Fuel Control Unit 燃油控制装置,燃料控制装置

FCV Flow Control Valve 气流控制阀
FD 1. Face of Drawing 图面 2. Fatigue Damage 疲劳损伤 3. Flap Down 襟翼放下 4. Flight Direct 飞行指引 5. Flight Director 飞行指引仪 6. Flight Dynamics 飞行力学 7. Flight Deck 驾驶舱,飞行甲板 8. Frequency Domain 频率范围 9. Frequency Duplex 频道双工制 10. Fuel Drain 放油 11. Functional Design 功能设计 12. Fixed Displacement 固定位移 13. Free Delivery 免费交运
FDAF Flight Data Acquistion Function 飞行数据采集功能
FDAI Flight Director Attitude Indicator 飞行指引仪姿态指示器
FDAMS Flight Data Acquistion and Management System 飞行数据采集和管理系统
FDANA Frequency-Domain Automatic Network Analyzer 频域自动网络分析器
FDAS Flight Data Acquisition System 飞行数据采集系统(飞行记录仪的)
FDAU Flight Data Acquisition Unit 飞行数据采集设备
FDBK Feedback 反馈,回授
FDC 1. Flight Director Computer 飞行指挥计算机 2. Frequency-to-Digital Converter 频率-数字转换器
FDCS Flight Deck Control System 驾驶舱控制系统
FDDI Fiber Distributed Data Interface 光纤分布式数据接口
FDE 1. Flight Deck Effects 驾驶舱影响 2. Fire Detection and Extinguishing 火警探测与灭火/消防 3. Flight Dynamics Engineer 飞行动力学工程师
FDEO Flight Development Engineering Order 飞行研究工程指令
FDEP Flight Data Entry and Print-Out 飞行数据输入和打印输出
FDF 1. Fachverband Deutscher Flugdatenbearbeiter 德国飞行数据处理人员同业公会 2. Flight Data File 飞行数据文件 3. Flight Dynamics Facility 飞行动力学设施
FDH Flight Deck Handset 驾驶舱耳机
FDI Failure Detection and Isolation 故障检测和隔离
FDIO Flight Data Input/Output 飞行数据输入/输出
FDIR Fault Detection, Isolation, and Reconfiguration 故障检测、隔离和重新配置
FDL 1. Fast Deployment Logistic (Ship) 快速供应船 2. Flight Dynamics Laboratory 飞行动力学实验室 3. Full-Dracon Line (在显示器上)实线标志
FDM Frequency Division Multiplex 频分多路传输
FDMA Frequency Division Multiple Access 频分多址
FDMDA Full Diameter Motorized Door Assembly 全直径电动门装置

FDME Frequency-Division Multiplex Equipment 频分多路传输设备
FDO 1. Firing Order 发令订单 2. Flight-Deck Officer 飞行甲板军官 3. Forecast Demand Ordering 预测需求订单
FDP Funded Delivery Period 长交付期
FDPS Flight Data Processing System 飞行数据处理系统
FDR 1. Fatigue Damage Range 疲劳伤等级 2. Fixed Depression Reticle 固定压低角光环 3. Flat-Deck Runway 平甲板跑道 4. Flight-Data Recorder 飞行数据记录仪
FDRS Flight Data Recorder System 飞行数据记录系统
FDS 1. Fence Disturbance Sensor 翼刀扰流探测器 2. Flight Data Subsystem 飞行数据分系统 3. Flight Director Pop-Up 飞行指引仪自动显示 4. Flight Display System 飞行显示系统 5. Function Distributed System 功能分布系统
FDSO Full Dispersed Site Operations 全部疏散配置的作战
FDSS Fine Digital Sun Sensor 精密数字式太阳敏感器
FDSU Flight Data Storage Unit 飞行数据存储器
FDTE Force Development Test and Experimentation 兵力展开试验
FDU Flux Detector Unit 磁感应传感器,磁通量探测器装置
FE 1. Far Encounter 远交会阶段 2. Ferro Electric 铁电的,铁电体 3. Field Elevation 机场标高 4. Flight Engineer 飞行工程师
FEAF Far East Air Force 远东空军
FEAR Failure Effect Analysis Report 故障后果分析报告
feasibility report 可行性报告
feasibility study 可行性论证,可行性研究
feasible region 可行域
feasible trajectory 可行轨
feather head 〈俚语〉莽撞的飞行员
feathering angle 顺桨桨叶角,顺桨桨距
feathering hinge 1.变距铰 2.轴向铰
feathering pitch 顺桨桨距
feathering pump 顺桨泵
feathers 〈口语〉机翼"羽毛"(指机翼上各式各样的活动翼片,包括副翼、襟翼、扰流片等)
feature codes 特征码
feature codes menu 特征码选择单
feature coding 特征编码
feature detection 特征检测
feature extraction 特征抽取,特征提取
feature modeling 特征造型

feature of emission spectrum　发射波谱特征
feature-line overlap　地物线重叠
FEB　1. Free Electron Bolometer 自由电子辐射热测量计 2. Functional Electronic Block 电子功能块
FEBA　Forward Edge of Battle Area 作战地域前沿
FEC　1. Failure Effect Category 故障影响类型 2. Forward Error Correction 前向纠错
fect　电场效应
FED　Federal 联邦
Federal Aviation Regulations　联邦航空条例（规范）
FED-STD　Federal Standards 联邦标准
feed　1. 供给,送料 2. 馈电,供电 3. 进刀 4. 进弹
feed assist mechanism　拨弹机,辅助进弹机
feed belt　弹带
feed block　进弹机
feed chute　输弹道,输弹套
feed lever　拨弹板,拨弹杆
feed line　馈线
feed rotor　装弹轮,喂入轮
feed source　馈源
feed sprocket　拨弹轮
feed stock　原料
feed system　1. 输送系统 2. 馈电系统
feed track　输弹道
feedback adjustment　反馈调整
feedback admittance　回输导纳,反馈导纳
feedback check　反馈校验
feedback coefficient　反馈系数
feedback compensation　反馈补偿
feedback control　反馈控制
feedback controlled heat pipe　反馈控制热管
feedback decoding　反馈解码,反馈译码
feedback edge set　反馈边集合
feedback effect　反馈效应
feedback element　反馈环节
feedback factor　回输因数,反馈因数
feedback gain　反馈增益
feedback loop　反馈回路
feedback potentiometer　反馈电位计
feedback ratio　回力比
feedback solution　反馈解
feedback system　反馈系统
feedback type rate gyro　反馈式速率陀螺
feedback voltage　回输电压,反馈电压
feedback winding　回输线圈,反馈绕组
feeder　1. 进弹机,装弹机,托弹盘 2. 馈电线,馈电板 3. 供给装置 4. 支线航空公司（经营通往国内和国际干线的支线航班）
feeder liner　支线客机

feeder link　馈送链路
feeder route　支线航路,过渡航路
feeder sprocket shaft　（航炮）拨弹轮轴
feederliner　支线航班飞机
feedforward　前馈
feedforward compensation　前馈补偿
feedforward control　前馈控制
feedforward path　前馈通路
feedline　供油管路
feedstock　进料
feedthrough capacitor　馈通式电容器
feedthrough type power meter　穿心式功率计
feedway　1. 进弹口,进弹窗 2. 供给道输送道
feel jack　载荷感觉器,操纵感力器
feel of the stick　杆力感觉
feel pressure computer unit　（助力操纵）感力系统计算装置
feel system　人工感力系统,载荷感觉装置
feel trim actuator　（操纵）感力卸除机构,卸载机构,调整片效应机构
feel unit　载荷感觉器,操纵感力器,载荷机构
feeler　1. 灵敏元件,传感器 2. 千分垫,测隙规 3. 探针,探头,探测器
feeling threshold　感觉阈
feel-simulation unit　感力模拟机构,载荷感觉器
FEET　Flight Emergency Egress Test 飞行应急出舱试验
FEF　Flight Evaluation Folder 飞行评定卷宗
FEFA　Future European Fighter Aircraft 欧洲未来战斗飞机（方案）
FEFET　Ferro Electric Field Effect Transistor 铁电场效管
FEFI　Flight Engineers Fault Isolation 空勤工程师故障隔离（技术或手册）
FEIA　Flight Engineers' International Association 随机工程师国际协会（美国）
FEID　Flight Equipment Interface Device 飞行设备接口装置
fence　1. 电子篱笆 2. 雷达预警线（网）3. 翼刀 4. 导流片,隔流片
fenestron　窗隔式尾桨,窗式涵道尾桨,涵道尾桨（直升机尾桨由细长桨叶组成,在短涵道中旋转）
FEO　Federal Energy Office 联邦能源局（现改用 FEA）（美国）
FEP　1. Far End Package 远端设备（美国系绳卫星试验中,栓在航天飞机的系绳上施放到远处的设备）2. Front-End Processor 前端处理器
FERD　Facility and Equipment Requirements Document 设施和设备要求文件

Fermi acceleration 费米加速
Fermi level 费米能级
Fermi mechanism 费米机制
ferret receiver 侦察接收机
ferret satellite 电子侦察卫星
ferristor 铁磁电抗器
ferrite antenna 铁氧体天线
ferrite core memory 铁氧体磁芯存储器
ferrite memory core 铁氧体记忆磁芯
ferrite modulator 铁氧体调制器
ferrite permanent magnet 铁氧体永磁体
ferrite phase modulator 铁氧体调相器
ferrite phase shifter 铁氧体移相器
ferroelectric crystals 铁电晶体
ferroelectric display 铁电显示
ferroelectric hysteresis loop 铁电电滞回线
ferroelectric semiconducting glaze 铁电半导体釉
ferrofluid particle gyro 铁磁流体粒子陀螺
ferrograph 铁磁示波器
ferrography 铁谱分析,铁谱
ferromagnetic display 铁磁显示
ferromagnetic resonance 铁磁共振
ferromagnetic resonance linewidth 铁磁共振线宽
ferromanganese 锰铁,铁锰合金
ferry 转场,渡运,空运
ferry range 空载转场航程
ferry tank 转场副油箱
fertile material 增殖性物质,可转变成裂变物质的材料
FES Fluid Experiment System 流体实验系统(空间实验室中的)
FESC Forward Electrical/Electronic Service Center 前方电气/电子服务中心
FESS Flight Experiment Shielding Satellite 辐射屏蔽飞行试验卫星
FET Field-Effect Transistor 场效应晶体管
FETT First Engine Type Test 第一次发动机型号试验
FEW 1. Fighter Escort Wing 护航战斗机(歼击机)联队 2. Fleet Empty Weight 机队空重
FEWG Flight Experiment Working Group 飞行实验工作组
FEWP Federation of European Women Pilots 欧洲女飞行员联合会
FEWS Follow-on Early Warning System 后继早期预警系统
FF 1. Fatigue Failure 疲劳断裂 2. Feeder Fault 馈电线故障 3. Feeder Fix 支线定位点 4. Ferry Flight 渡运飞行 5. Flip-Flop 触发器 6. Folding Fin 折叠式尾翼 7. Forward Firing 向前发射的 8. Free-Flyer 自由飞行器(航天器) 9. Functional Failure 功能性故障 10. Fuel Flow 燃油流量,燃油耗量
FFAR 1. Folding-Fin Aircraft Rocket 拆叠尾翼式航空火箭弹 2. Forward Firing Aerial Rocket 向前发射的航空火箭弹 3. Free-Flight Aircraft Rocket 无控航空火箭弹,自由飞行航空火箭弹
FFB Free-Fall Bomb 自由落下炸弹
FFC 1. Fan-Failure Clutch(涡扇发动机)风扇故障离合器 2. Fuze Factor Correction 引信因数修正量
FFCC Forward-Facing Crew Cockpit 机组面向前方的驾驶舱
FFCS Free-Fall Control System 自由下落控制系统
FFF Free Float Facility 自由漂浮设施
FFH Fleet Flight Hours 机队飞行时数
FFHA Final Function Hazard Analysis 最后功能危险分析
FFM 1. Future Flight Management 未来飞行管理 2. Fuel Flow Meter 燃油流量表
FFN Far-Field Noise 远场噪声
FFO 1. Fixed-Frequency Oscillator (固)定频(率)振荡器 2. Furnace Fuel Oil 炉用燃料油
FFPB Free-Fall Practice Bomb 自由下落练习(炸)弹
FFR 1. Flight Feasibility Review 飞行可行性评审 2. Free Flight Rocket 非制导火箭 3. Fuel-Flow Regulator 燃油流量调节器 4. Full Fight Regime 全航程工作方式
FFRAT Full-Flight-Regime Auto Throttle 全航程工作自动节流门
FFRDC Federally Funded, Research and Development Center 联邦投资研究开发中心(美国)
FFRR First Flight Readiness Review 首次待飞状态评审,首飞准备状态评审
FFS Full Flight Simulator 全动飞行模拟器
FFSK Fast Frequency-Shift Keying 快速频移键控
FFT 1. Fast Fourier Transform 快速傅里叶变换 2. Free Fall Test 自由落体试验
FFTP Fast Fourier Transform Processing 快速傅里叶变换处理(改进雷达探测概率的处理方法)
FFWS Free-Flyer Work Station 自由飞行器工作站
FG Filament Ground 灯线接地
FG&CS Flight Guidance & Control System 飞行导航及控制系统
FGA Fighter, Ground Attack 对地攻击歼击机
FGC Flight Guidance Computer 飞行导引计算机
FGCP Flight Guidance Control Panel 飞行导引控制板
FGMDSS Future Global Maritime Distress and Safety System 未来全球海难救援系统
FGS Fine Guidance Sensors 精密制导传感器
FGV Field-Gradient Voltage 场梯度电压
FH 1. Flight Hour 飞行小时 2. Frequency-Hopping 跳

频 3. Fuel Heater 燃油加热器

FHA 1. Fleet Hour Agreement 机队工作时间协议 2. Function Hazard Analysis 功能危害分析 3. Functional Hazard Assessment 功能危害评估

FHP Fractional Horsepower 相对马力,小功率,分马力

FHU Force Helicopter Unit 特遣部队直升机部队(派遣独立作战部队的直升机部分)

FI 1. Fatigue Index 疲劳指数 2. Fault Isolation 故障隔离,差错隔离 3. Fluid Injection 液体喷射 4. Functionl Identification 功能标识 5. Functional Interface 功能界面

FIAT First Installed Article Test 安装首件试验

fiber cladding 光纤包层

fiber diameter 纤维直径

fiber gyroscope 光纤陀螺

fiber hybrid composite 混杂纤维复合材料

fiber laser 纤维激光器

fiber metal 纤维金属

fiber optic cladding 光纤包覆

fiber optic communication 光纤通信

fiber optic gyro 光纤陀螺仪

fiber optic multiplex optical transmission system 光纤多路光传输系统

fiber optic sensor 光纤传感器

fiber optic technology 光纤技术

fiber optic transmission system 光纤传输系统

fiber orientation 纤维取向,纤维定向

fiber radio burst 纤维射电爆发

fiber reinforced composite 纤维增强复合材料

fiber reinforced metal matrix composite 纤维增强金属基复合材料

fiber thickness 纤维粗度

fiber weight of unit area 单位面积纤维质量

fiber wetness 纤维浸润性

fiber-optic guidance 光纤制导

fiber-optic gyroscope 光纤陀螺

fiber-optic photo transfer 光纤图片传送

fiber-reinforced ceramic-matrix composite 纤维增强陶瓷基复合材料

fiber-reinforced plastic 纤维增强塑料

fibrescope 光纤孔探仪,纤维内窥镜

FIC 1. Film Integrated Circuit 薄膜集成电路 2. Finance Committee 财政委员会(国际民航组织) 3. Frequency Interference Control 频率干扰控制

FICS Fault Isolation Checkout System 故障隔离检测系统

fictitious year 假年(贝塞尔年)

FIDS 1. Fault Identification and Detection System 故障鉴别与探测系统 2. Flight Information Display Set 飞行情报显示器(空中交通管制雷达的)

fiducial mark 相机框标,框标在照片上的标记

FIELD First Integrated Experiment for Lunar Development 首次月球开发综合实验

field 1. 机场,战场,作战训练或演习区域,现场,领域 2. 场(电场、磁场、力场等) 3. 客运服务的 4. 在前方基地以简易设施进行工作的

field alignment error 1. 视场校准误差 2. 电场调整误差

field application 1. 野外使用 2. 现场应用

field book 外场工作记录本

field coils 磁场线圈

field corrector 像场校正镜

field curvature 场曲

field data code 军用数据码(美国)

field effect 场效应

field effect transistor 场效应(晶体)管

field elevation 机场海拔高度,机场标高

field emission 场致放射

field emission microscopy 场致发射显微(镜)学

field excitation 场激励

field frame 励磁线圈架,场线圈架

field galaxy 场星系

field gradient 场梯度

field guidance 场制导(利用场特性使导弹到场内某点的制导,场可以是重力场,如无线电场)

field induced junction 场感应结

field ion mass spectroscopy 场致离子质谱(学)

field joint connector 现场连接器,外场连接器

field length 场长

field line 场线

field maintenance test equipment 野战维修测试设备

field mapping 野外制图

field modification 现场改造

field of divergence 辐散场,散度场

field of flow 流场

field of force 力场

field of gravity 重力场

field of regard 视界角覆盖面

field of search 搜索区,探测区

field of view of optical fuze 光学引信视场角

field operation 野战使用

field oxide 场区氧化层

field pattern 1. 辐射图 2. 田区规划

field performance 起飞着陆性能

field service 野战勤务

field sketch 现场草图

field solver 场测定仪,场解算器

field star 场星

field stop 场阑,场栏,视场光阑

field strength　1. 场强　2. 信号强度（雷达）
field strength meter　场强测量仪
field strip　1. 外场拆卸　2. 不完全分解（特指枪炮擦洗时，只对主要部件的分解）
field support equipment　野外辅助设备
field test　机场内试验，野外试验
field training detachment　教导队，训练分遣队（美国空军的，主要是对维修人员在使用现场进行新装备的技术训练）
field usage factor　外场使用因数
field variable　场变量
fieldbus　现场总线
field-effect varistor　场效应变阻器
Fieldguard/Skyguard　战场卫士/空中卫士（瑞士固态搜索雷达）
field-handling frame　抬运把手架（连接在飞艇身上的框架，供众多人员在地面抬动之用）
fieldistor field joint　场控晶体管现场连接，场控晶体管现场连接器
field-programmable software　现场（外场）编制程序软件
FIES　Factor of Initial Engine Spares 发动机初始备件因数
FIF　Fluorescent Inspection Fluid 荧光检验液
FIG　1. Fibre Interferometer Gyroscope 光纤干涉仪陀螺　2. Fighter Interceptor Group 截击机大队　3. Figure(s) 图，数据
fighter　战斗机，歼击机
fighter affiliation　战斗机（歼击机）参与
fighter control course　战斗机控制课
fighter cover　战斗机（歼击机）掩护
fighter escort　战斗机护航
fighter sweep　歼击机扫荡
fighter-bomber　歼击轰炸机
fighter-direction aircraft　歼击（战斗）机引导机
fighting aircraft　战斗航空器，战斗飞机
fighting top　顶部炮塔
figure memorization test　图形记忆试验
figure of merit　1. 品质因数　2.（仪表）灵敏度
figure-ground discrimination　图形背景辨别
figure-reader electronic device　电子图形阅读器
FIH　Flight Information Handbook 飞行情报手册
FIL　1. Filament 灯丝，阴极　2. Filter 滤波器，过滤器　3. Fountain-Induced Lift（垂直起飞的）热气柱诱导升力
filament motor assembly　丝带式转子组件
filamentary metal　纤维增强金属
filamentary nebula　纤维状星云
filamentary transister　线状晶体管
FILE　Feature Identification and Location Element 特征识别和定位系统（美国航空航天局）
filed flight plan　备案飞行计划
FILH　Fillister Heed 有槽圆头（螺钉）
fill factor　1. 填充因数　2. 曲线因子
filled and charged battery　注液充电蓄电池
filler pipe　加注管
filler pulse　填充脉冲
filling pressure　加注压强
film badge　胶片式射线计量器
film cooling　薄膜冷却
film inductor　薄膜电感器
film integrated circuit　膜集成电路（薄膜与厚膜集成电路的总称）
film plating　镀膜
film reader　胶片判读仪
film recording　胶片记录
film resistance　膜电阻
film scanner　胶片扫描器
film second surface mirrors　薄膜型二次表面镜（航天器热控涂层）
film speed　胶片感光速度
film-cooled combustion chamber　薄膜冷却燃烧室
filmistor　薄膜电阻
film-plating machine　镀膜机
film-recovery photographic reconnaissance satellite　胶片回收型照相侦察卫星
filter　1. 过滤器　2. 滤波器，滤声器　3. 滤光镜，滤色镜　4. 过滤　5. 鉴定（情报）6. 防空情报筛选
filter assembly　过滤装置
filter bank　过滤机组
filter capacitor　滤波电容器
filter divergence　滤波发散
filter factor　滤波（滤光）因数
filter pulse　填充脉冲
filtered plot　核对过的目标航迹图
filtergram　太阳单色像
filtering capacitor　滤波电容器
filter-press　压滤
FIM　1. Fault Isolation Manual 故障隔离手册　2. Fault Isolation Monitoring 故障定位监视（监控），故障隔离监视（监控）
fimbaled nozzle　摆动喷管
FIMS　Fault Isolation Monitoring System 故障隔离监控系统
fin blade　（火箭、导弹）尾翅，（炸弹）尾翼翼片
fin box　（炸弹的）尾翼撑杆
fin flash　尾标色带
fin line　鳍线
fin servo　（导弹）舵机

fin sleeve （炸弹）尾翼套筒	**fine orbit correction** 精密轨道修正
fin stabilized rocket 尾翼稳定火箭	**fine pitch** 小距,低距
fin stabilizer （火箭或炸弹的）稳定尾翼,安定器	**fine pitch stop** 低距止动点（螺旋桨的）
final approach 1.（起落航线）第五边 2.在第五边上进入着陆 3.最后逼近	**fine powder** 细粉
	fine reading 细读数
final approach altitude （起落航线）第五边高度	**fine structure** 精细结构
final approach fix 第五边进入点	**fine system of bearing** 方位精测系统
final assembly 最后装配,总装	**fine tracking** 精确跟踪
final assembly drawing 总装图	**fine trimming** 精确配平
final attack solution 最后攻击方案	**fine-follow circuit** 精确跟踪电路
final breakout 出云进入（起落航线）第五边	**fineness ratio** 1.长细比,长径比（流线体长度与最大直径之比）2.展弦比
final controller 着陆雷达控制员	**finger** 1.指,爪,销 2.指针 3.航站大楼上下机走廊桥
final equilibrium 最终平衡	**finger dexterity test** 手指灵活性试验
final heat treatment 最终热处理	**finger-bar controller** 指式驾驶杆
final inspection 最终检查	**finger-tight** 用手拧紧的,松散装配的
final lock up 最终锁定（太阳能电池帆板的）	**fingertip formation** 四机指尖队形
final map 成图	**finish** 1.（飞机表面的）漆层,合层 2.抛光,光制,光洁度 3.（对飞机的）表面察看
final mass 最终质量（火箭发动机的）	
final minification 精缩	**finishing after firing** 烧后加工
final orbit 终轨	**finite amplitude** 有限振幅
final peak sawtooth shock pulse 后峰锯齿冲击脉冲	**finite automaton** 有限自动机
final pressure 终压	**finite beam** 有限射束
final replacement 终期置换	**finite burn** 有限点火
final run 进入第五边,进入攻击	**finite difference migration** 有限差分偏移
final squint angle 最终瞄准偏斜角	**finite difference scheme** 有限差分格式
final stage vehicle 末级飞行器	**finite difference theory** 有限差分理论
final state 终态	**finite element** 有限元
final steady-state 最终稳定状态	**finite element analysis** 有限元分析
final terminal 1.终端站 2.终端接头	**finite element formulation** 有限元公式表示法
final trajectory 最后弹道	**finite element mesh** 有限单元网格
final trim （火箭和导弹为达到所需关机速度进行的）最终调整	**finite element method** 有限元法
	finite element model 有限元模型
final value 终值	**finite element solution** 有限单元解法
final vectoring error 末段导引误差	**finite element technique** 有限元（素）法
final voltage 终点电压	**finite firing** 有限点火
finance lease 分期付款（购买飞机）,融资租赁	**finite impulse response** 有限脉冲响应
finder 1.探测器,定向器,测距器 2.瞄准装置 3.取景器,寻像器 4.寻星镜	**finite impulse response model** 有限脉冲响应模型,有限脉冲特性模型
finding chart 证认图	**finite moving source** 有限移动源
fine adjustment 精调,细调	**finite nonempty set** 有限非空集
fine aiming 精瞄	**finite number** 有限数
fine alignment 精对准	**finite order model** 有限阶模型
fine blanking 精密冲裁	**finite rate** 有限率
fine data channel 精确数据通道（弹道测量系统的）	**finite rotation** 有限转动
fine drawing (fair drawing) 精绘	**finite set** 有限集
fine fit 精配合	**finite span wing** 有限翼展机翼
fine grid 密网格,精密栅板	**finite state acceptor** 有限状态接收器
fine leak test 精检漏试验	**finite state generator** 有限状态生成器
fine mesh 细孔,细筛孔,细网目	

finite state machine	有限状态机
finite transducer	有限变换器
finite volume solution	有限容积溶液
finite volume method	有限体积法
finite wing	有限翼展机翼,有翼尖的机翼
finite-amplitude convection	有限幅度对流
finite-amplitude disturbance	有限振幅扰动
finite-difference	有限差分
finite-difference approximation	有限差分近似
finite-difference model	有限差分模式
finite-displacement stick	定量位移驾驶杆
finite-length launcher	有限长度的发射装置,导轨式发射装置
finiteness correction	有限性校正
finiteness factor	有限性因子
finiteness transform	有限性变换
finned air rocket	翼式火箭
finned bomb	有尾翼炸弹
finned projectile	尾翼式导弹,尾翼式炮弹
finocyl grain	翼柱形药柱(推进剂)
fin-stabilized projectile	尾翼稳定式投射物,尾翼稳定式射弹(如炸弹、火箭弹等)
FIR	1. Finite Impulse Response 有限脉冲响应 2. Flight Information Region 飞行情报区 3. Full Indicator Reading 千分表全读数,全指示读数
FIRAMS	Flight Incident Recorder And Monitor System 飞行事故征候记录仪和监控系统
fire annihilator	灭火器
fire area	射界
fire at full automatic	连发射击,全自动射击
fire attack	使用燃烧弹的空袭
fire balloon	热气球
fire bomb	火焰炸弹
fire code	费尔码(纠正突发错误的一种码)
fire control computer	点火控制计算机
fire control material	火(力)控(制)设备,射击控制设备
fire control radar	火控雷达
fire control sight system	火(力)控(制)瞄准系统
fire control simulator unit	火控模拟装置
fire control system coordinator	点火控制系统协调程序
fire deluge system	遥控灭火系统(导弹和航天发射台区域用的)
fire detection system	火焰检测系统
fire detector	火警探测器
fire direction computer	火控计算机
fire direction system	发射指挥系统,射击指挥系统
fire dispersion	火力散布,炮弹散布
fire distribution in unlocked position	锁膛不到位击发
fire efficiency	射击效率
fire helicopter	消防直升机
fire in clusters	(火箭弹)齐射
fire in ripples	点射
fire in salvoes	齐射
fire in turn	依次射击
fire interrupter	射击限制器
fire load	火荷载,可燃材料荷载
fire point	着火温度,燃点
fire pulse	1.(火箭的)发射冲信号 2.触发脉冲起动脉冲
fire ratardancy	阻燃剂
fire rate	射速,发射速度
fire stop mechanism	停射装置
fire support weapon	火力支援武器
fire suppression system	火警系统,火灾抑制系统
fire unit	发射分队(地对空导弹的)
fire watcher	〈口语〉火警信号器,火警探测系统
Fire Wheel	"火轮"(苏联炮瞄雷达的北约代名)
fire zone	火警区
fire-and-forget guidance system	(载机)发射导弹后免控(即刻脱离)的制导系统
fire-and-forget missile	发射后免控导弹,具有自寻的能力的导弹
fireball	1.火球(核爆炸的) 2.火流星
fireblocker	防火布,挡火帘
fire-control coupler	1.雷达瞄准具与自动驾驶仪交连部分 2.火控系统计算装置
fire-control system	点火控制系统,射击控制系统,发射控制系统
fired out	空空导弹全部发射完毕的战斗机
fireguard	消防员(特指发动机起动时站在飞机一侧,持灭火器的人员)
fire-in	(校靶后进行的)试射
fire-out	发射
firepower	火力
fireproof	防火
fire-protection equipment	防火设备
fire-resistant	防火的,耐焰的(能受火焰燃烧5分钟以上而不燃的)
fire-retardant paint	防火涂料
firewall	防火壁(墙)
firewire	火警(探测)线
firewire control unit	火警线控制装置
FIREX	Free-flight Imaging Radar Experiment 自由飞行成像雷达实验
firing altitude	1.(高射炮)发射瞬间目标高度 2.(火箭、导弹)发射高度 3.(发动机空中)起动高度
firing and guidance unit	发射与制导装置
firing angle	射角

firing attitude 点火姿态（变轨发动机点火所要求的姿态）

firing button 射击（发射）按钮

firing chamber 1. 弹膛，药室 2. 燃烧室

firing console （火箭发动机的）点火操纵台

firing cycle 1.（枪炮的）射击循环，发射间隔，射速 2. 起动周期

firing data 1. 射击诸元，发射数据 2. 火箭发动机的崩车数据

firing data computer 射击诸元计算机

firing equipment 发射设备，点火设备

firing gear 击发装置

firing order 点火顺序（活塞式发动机的）

firing pin extension （枪炮的）撞针尾杆

firing pit 1. 发射试验井（火箭发动机的）2. 发射人员掩体

firing potential 发射势

firing readiness 射击准备状态，发射准备（状态）

firing sequence 点火顺序

firing set 点火装置，点火器

firing signal light 发射信号灯

firing signal transfer path （弹头）传爆电路

firing solenoid 1. 发射螺线管 2. 电打火（发射）机

firing system tester 点火系统试验器，发射系统试验器

firing test 点火试验（火箭发动机的静态点火试验）

firing time 1. 点火时间 2. 射击时间，发射时间 3. 火箭发动机工作时间

firing tone 允许发射音响信号

firing trigger 1. 扳机，射击按钮 2. 弹射手柄

firing tube 发射导管，发射管

firing unit 点火装置

firing voltage 点火电压

firing zone boundary 射击范围边界，发射范围边界

firmware 固件（软件与硬件结合为一体）

firmware circuitry 固件电路

firmware package 固件程序包

first aid 急救

first aid apparatus 急救设备

first aid in accident 事故营救

first aid in flight 空中急救

first aid service 急救服务

first approximation 第一近似

first carrier 第一载波

first configuration test 首次构型试验

first cosmetic velocity 第一宇宙速度

first flight spacecraft 首飞航天器

first hop 首次（地面）滑行试验

first Lagrangian point 【天】第一拉格朗日点

first law of thermodynamics 【物】热力学第一定律

first level address 一级地址，直接地址

first level of packaging 一级组装

first line life 服役寿命

first line servicing 一级维修保障勤务

first motion 起始运动，起飞

first pilot 机长，正驾驶（员）

first round hit probability 首发命中率

first-angle projection 第一象限投影，立面图

first-burst accuracy 首发精确度（航炮射击的）

first-choice route 首取路由

first-ended first-out 先结束先输出

first-in first-out buffering 先进先出缓冲法

first-in last-out 先进后出

first-look, first-kill opportunity "先敌发现"机会（尤指机载雷达设备能在敌人发现我机之前先发现敌人的能力）

first-time-around echo 第一周期扫描回波

FIS 1. Federal Inspection Service 联邦检查局 2. Fighter Interceptor Squadron 截击机中队 3. Flight Information Service 飞行信息（情报）服务 4. Flight Information System 飞行信息系统 5. Flight Instrument System 飞行仪表系统 6. Flight Instructors School 飞行教员学校 7. Floating (Point) Instruction Set 浮点指令集

FISA 〈法语〉Federation Internationaledes Societes Aerophilatelique 国际航空集邮协会联合会

FIS-B Flight Information Services-Broadcast 飞行信息服务广播

fishbone antenna 鱼骨天线

Fisher information matrix 费希尔信息矩阵

FIT 1. Fault Isolation Test 故障隔离测试 2. Filament 丝，灯丝 3. First Indication of Trouble 首次故障指示 4. Flexible Interface Technique 灵活接口技术

FITCK Flight Check 飞行检查

fitness figure 1. 舒适度图 2. 适合度因数

fitness function 适应度函数

FITR Filter 过滤，滤波

fitter 1. 装配工，钳工 2. 发动机装配工

five-component balance 五分力天平

five-minute oscillation 五分钟振荡

five-minute rating 5 分钟额定（放电）值

five-minute signal （导弹发射前的）五分钟准备信号

fix 1. 定位 2. 即时位置，已知位置，定位点 3. 固定，安装 4. 调整，修理 5. 确定，决定

fix rate 定位率（单位时间内从系统可得到的独立位置坐标或数据点）

fixation 固定，定影

fixation axis 固定轴

fixed address 固定地址

fixed address offset　固定地址偏移量
fixed ammunition　固定式弹药,定装式弹药(弹头与弹壳固定结合为一体的弹药)
fixed angle-optical bombing　用光学瞄准
fixed angle-radar bombing　用雷达瞄准
fixed annulus gear　固定内齿圈
fixed assignment　固定赋值
fixed base　静底座
fixed base training simulator　固定基(训练)仿真器
fixed bed　固定试验台
fixed block　固定块
fixed block architecture device　固定块结构设备
fixed capacitor　固定电容器
fixed cloud-top altitude　固定云顶高度
fixed cloud-top temperature　固定云顶温度
fixed connector　固定连接器
fixed contact terminal　固定触点接线柱
fixed displacement motor　定量马达
fixed distance marker　固定距离指点标(常指距跑道头300米处的指点标)
fixed drift rate　常值漂移率
fixed earth station　固定地球站
fixed echo interference　固定回波干扰
fixed electron　束缚电子,固定电子
fixed error　固定误差,系统误差
fixed feed chute　硬式输弹道
fixed flow restriction　固定节流(流通面或长度不可调的节流)
fixed frequency sounder　固频探空仪
fixed gravity drop　弹道降低固定修正量
fixed gun　固定机枪,固定机炮(固定于飞机上的机枪或机炮)
fixed gun mode　(用于近距快射的)固定机枪(炮)工作状态
fixed gun sight　固定机枪瞄准具
fixed gunnery　(战斗机)固定机枪射击术
fixed head star tracker　固定探头式星跟踪器(利用自身光电探测器的电子扫描实现星跟踪)
fixed hook　固定索钩
fixed inductor　固定电感器
fixed landing gear　固定式起落架(不能收放的)
fixed light　光度不变的灯光(设备)
fixed loading round　定装式炮弹
fixed machine gun　固定机枪(炮)
fixed mean pole　固定平极
fixed mode　固定模
fixed munition　固定式弹药
fixed phase drift　固定相移
fixed pipper　固定环中心
fixed pitch propeller　定距螺桨
fixed point　定点,不动点
fixed point computer　定点计算机
fixed preassignment　固定预分配
fixed resistor　固定电阻器
fixed reticle sight　固定环瞄准具
fixed round　定装炮弹
fixed satellite　对地静止卫星
fixed satellite service　卫星固定业务
fixed sequence manipulator　固定顺序机械手
fixed service tower　固定式勤务塔
fixed set point control　定值控制
fixed sight　固定式瞄准具
fixed sight depression angle　瞄准具的固定俯角
fixed slat　固定式前开缝翼
fixed station　固定电台(航空通信的)
fixed target　固定目标
fixed telemetry station　固定遥测站
fixed trajectory weapon　固定弹道式武器
fixed trimmer　固定调整片,修正片
fixed wall nozzle　固壁喷管
fixed weight　固定重量(不带燃油、滑油等消耗物、可投放物和业载时的重量)
fixed yoke　固定偏转线圈组
fixed-area nozzle　定面积喷管,面积不可调喷管,不可调喷管
fixed-base cockpit　静底座舱
fixed-base operator　固定基地航空公司(以通用航空飞机场为固定基地,为私人飞机业主提供各种服务,包括销售、修理、日常维护、加装设备及飞行训练等)
fixed-base simulator　固定基底模拟器
fixed-control computer　固定控制计算机
fixed-displacement pump　固定流量泵,定量泵
fixed-gain damping　固定增益阻尼
fixed-geometry　定几何形
fixed-gimbal　固定框架
fixed-length packet　(固)定长(度)包
fixed-point accuracy　定点精度
fixed-point arithmetic　定点运算
fixed-point calculation　定点运(计)算
fixed-stick stability　握杆安定性
fixed-tolerance-band computer　固定容错范围计算机
fixed-wing autopilot　定翼机自动驾驶仪
fixed-wing flight control　固定翼飞机飞行控制
fixer network　定位台站网
fixing aids　定位设备
fixing error　定位误差,测定(目标)位置误差
fixture　夹具,(小型)型架,定位器
FJS　Fuel Jettison System 应急放油系统

FL 1. Fan Lift 风扇升力 2. Flag 旗 3. Flight Level 飞行高度（按百英尺计，如 FL96＝9 600 英尺），高度层 4. Flight Line 起机线，起飞线 5. Fluorescent Light 荧光 6. Flush 冲洗，洗涤

FL/W Flush Welding 闪光焊

FLA Future Large Aircraft 未来大型运输机（欧洲 Euroflag 集团研制中的货机）

Flade Fan Blade 风扇叶片

FLAG Floor Level Above Ground 飞机地板离地面高度

flag 1. 标识，标志位 2. 故障警告牌

flag alarm 1. 警旗 2. 警告牌

flag event 标志事件

flag flip-flop 标志触发器

flag register 标志寄存器

flag signal 旗语，旗帆信号

flag stop 计划外特定停运（美国空军执行空运时，临时停止空运）

flag window （仪表上）信号旗窗口

flagman 1. 司旗员 2. 标杆员，花杆员

flagplane 领队机，指挥机

flagpole antenna 1. 刀状天线 2. 桅杆式天线

flailing 扑打，甩动，气流抽打

flailing injury 气流吹袭伤，扑打伤

flair point 识别点，明显地物点

flak 高射炮火，高射炮

flak blob 高射炮弹爆烟

flak burst 高射炮弹炸点

flake 薄片，剥落

flake propellant 片状火药

flak-suppression 压制高炮的火力

FLAM Fault Location And Monitoring 故障定位与监视

FLAME Fast, Lightweight, Agile Missile 轻型快速敏捷导弹

flame acceleration 火焰加速

flame area 火焰面积

flame arrester 1. 灭火器 2. 隔焰屏，消焰筒

flame attenuation 火焰衰减

flame axis 焰轴

flame blowoff 熄火

flame brazing 火焰钎焊

flame bucket 排焰偏转器

flame chute 火焰导出（流）槽

flame coating 火焰喷镀

flame extinction 火焰熄灭，火焰淬熄

flame float 烟火浮标（在水面燃烧的烟火照明标志）

flame front 1. 火焰前峰 2. 火焰前沿 3. 火焰面

flame hardening 火焰硬化（一种表面硬化加工方法）

flame height 火焰高度

flame holder 火焰稳定器

flame instability 火焰失稳

flame kernel 火焰中心

flame length 火焰长度

flame path 火道

flame propagation 火焰传播

flame radiation 火焰辐射

flame resistance 1. 阻燃性 2. 防燃烧性

flame resistant 阻燃

flame shape 火焰形状

flame spray coating 火焰喷涂

flame spread 火焰传播

flame stability 火焰稳定性

flame stabilizer 火焰稳定器

flame stretch 火焰扩张

flame structure 火焰结构

flame tip 焰舌

flame trap 挡焰板

flame tube 火焰筒

flame velocity 火焰传播速度

flame zone 燃烧区，火焰带

flame-arrester vent plug 防火气塞

flame-front propagation 火焰锋面传播

flameholder 火焰稳定器

flameholding 火焰稳定

flamelet model 火焰面模型

flameout 熄火，燃烧终止

flame-resistant material 阻燃材料

flame-spray 火焰喷涂

flamestat 感温器，温度自动控制电门

flammability 可燃性，易燃性

flammability limit 可燃极限

flammability point 燃点，发火点

flammable 易燃

flange 1. 法兰盘 2. 凸缘

flange connection 法兰连接

flanging machine 折边机，外缘翻边机

flap 1. 鱼鳞板，鱼鳞片 2. 襟翼

flap angle 襟翼偏转角

flap blowing 襟翼区附面层吹除

flap fan 襟翼扇片（实验用）

flap position transmitter 襟翼位置传感器

flap valve 翻板式活门，瓣状活门

flaperon 襟副翼

flaplet 1. 广义地指任何小的翼片 2. 有圆弧形前缘，上下面是平的窄襟翼（柯恩达环量控制机翼上用）

flapping （直升机旋翼桨叶的）挥舞

flapping angle （旋翼的）挥舞角

flapping hinge （旋翼的）挥舞铰，水平铰，水平关节

flapping plane 挥舞平面（垂直于旋翼每个水平铰轴的平面）

FLAPE Force-Level Automated Planning Experiment （盟国中欧空军的）自动制订兵力水平计划的实验

FLARE Fault Locating And Reporting Equipment 故障定位与报告设备

flare 1.耀斑，闪耀，火苗 2.扩张，（导管的）喇叭口，锥形管，裙部 3.地面或水面固定照明光源（过去用燃烧煤油或其他燃料的方法，现已废弃），照明弹，曳弹，火炬 4.拉平（着陆前最后的上仰，以便把下降率减小到接近零）5.外侧、外飘（水上飞机浮筒或机身底部从中心线向外倾斜的侧面）6.（直升机）瞬时增距 7.明亮的雷达回波

flare adapter 照明弹架

flare antenna 拉平段天线（微波着陆系统中用于在飞机拉平段引导飞机的天线）

flare computer 拉平计算机

flare demand 自动着陆编码信号指令拉平

flare dud 发光的无效弹

flare guidance （着陆系统对）着陆拉平段引导

flare guidance system 拉平引导系统

flare nimbus 耀斑晕

flare out 1.（飞机着陆前）拉平 2.（直升机着陆前）减小下降速度

flare particle 耀斑粒子

flare star 耀星

flare-aircraft 照明（飞）机

flared body 裙体

flared fitting 扩口式接头

flared landing 照明着陆

flareout 拉平

flare-out elevation guidance element 微波着陆系统的拉平仰角引导单元

flare-path dinghy 助降小艇（位于水上飞机拉平降落航线上的水面小艇）

flare-related effects 耀斑效应

flareup 转入爬高

flash 1.（飞机的）尾徽（漆在垂直尾翼上）2.帽徽

flash blindness 闪光盲

flash burn 闪光烧伤，光辐射烧伤

flash flood warning 暴洪警报

flash flood watch 暴洪监视

flash lamp 闪光信号灯

flash light 闪光灯

flash point 闪点

flash reducer 消焰罩，消焰器

flash resistant 抗闪燃

flash spectrum 闪光谱

flash suppressor 消焰器，火焰抑制器（用来减少或抑制可见光辐射的炮口装置）

flash test 1.瞬间高压试验 2.油料的，闪点试验

flash vaporization 闪蒸

flash welding 闪光焊，电弧焊

flashback 回火，逆燃

flashback voltage 反闪电压，逆弧电压

flashbomb 闪光（炸）弹

flasher 〈口语〉角反射器（用于雷达轰炸训练），闪光灯

flasher unit 闪光标灯

flashess cordite （柯达）无烟火药

flashing beacon 闪光灯标

flashing code beacon 闪光密码灯标

flashing off （漆膜的）晾干

flashing the field 对励磁线圈充磁，对场线圈充磁

flashless charges 无烟火药

flashless nonhygroscopic powder 无焰防潮（不吸湿）火药

flashless powder 无烟火药

flash-light 1.闪光灯（灯塔等上的）2.摄影闪光灯 3.手电筒

flashover 1.闪络（绝缘子表面放电）2.跳火，飞弧 3.闪燃

flash-over discharge 飞弧放电

flashover voltage 闪络电压

flash-tube 闪光灯泡

flat absorber 全吸收表面

flat Bellani cup radiometer 贝拉尼板罩式辐射仪

flat cable 扁平电缆，带状电缆

flat cathode-ray tube 扁平阴极射线管

flat copy facsimile scanner 平面拷贝传真扫描器

flat diameter 降落伞的衣铺平直径

flat die forging 平面模锻

flat flame 无光焰

flat flange 平板法兰

flat four engine 卧式四缸发动机

flat gain regulation 平调，平增益调整

flat grid 平格栅

flat head rivet 平头铆钉，埋头铆钉

flat pack 扁平组件，扁平封装

flat packaging 扁平封装

flat panel electronic instrument 扁平型电子仪表，平板型电子仪表

flat parachute 平面伞

flat plate 平板

flat plate aerofoil 平板翼型

flat plate mode 平板模式

flat plate module 平板式组件

flat plate radiometer 平板辐射计

flat riser 平放式垂直起落飞机

flat six 平直对列六缸活塞式发动机
flat spectrum 平谱
flat spin recovery 平旋恢复
flat tip 平头电极
flat trajectory 平直弹道
flat transmission 等幅传输
flat tuning 粗调
flat wall 平直模壁
flat zone 平滑区（航道扇区或仪表着陆系统下滑道扇区指示的特征曲线斜率为零的区域）
flat-band voltage 平带电压
flatbed aircraft 巨型集装箱运输机
flat-bed scanner 平板(台)扫描仪
flat-bed trailer 平板拖车
flat-faced tube 扁平显像管
flatness 平直度
flat-plate boundary layer 平板边界层
flat-spectrum source 平谱源
flatten mechanism 展平机构
flattening 1.扁率 2.整平
flattening factor for the earth 地球扁率
flat-top antenna 平顶天线
flattop decline 平顶降落
flavor sensor 味敏传感器
FL-BE Filter-Band Eliminator 带阻滤波器
FL-BP Filter-Band Pass 带通滤波器
FLC Full-Load Current 满载电流
FLCH Flight Level Change 飞行高度改变
FLDNG Field Management Shop 机场维修厂（车间）
FLDT Floodlight 泛光照明
FLE Flap Leading Edge 襟翼前缘
flechette 小型杀伤导弹，箭形弹（一种细长的有硬金属弹心的子弹）
flechette warhead 箭形弹头（火箭弹上的）
FLED Forward Line of Enemy Defence 敌防御前线
FLEEP Flying Lunar-Excursion Experimental Platform 月球旅行试验平台
fleet 1.机队，机群 2.舰队 3.滑过，飞过
Fleet Air Arm 海军航空兵
fleet ballistic missile 舰队弹道导弹
fleet equipment 机队共用设备
fleet leader 领先使用的飞机
fleet noise level 机群噪声水平
Fleet Satcom 舰队通信卫星
fleld-reversa 场（致）反向
FLEM Fly-by Landing Excursion Mode 飞越地标着陆偏离形式
Flettner 弗莱特纳补翼（飞机上的调整片、补偿片和操纵片的统称）

FLEX 1.Flexible 柔性的/可变的 2.Flexure 弯曲，曲率
flex 1.挠曲 2.花线，皮线
flex joint seal 柔性连接密封
flex lead 软导线
flexbeam 柔性梁
flex-free-rotor gyro 柔性自由转子陀螺
flexibility matrix 柔性矩阵
flexible base material 软性基材
flexible bearing 柔性轴承
flexible bearing nozzle 柔性喷管
flexible blade （旋翼的）柔性桨叶（尤指后缘带有补偿片的）
flexible body 柔性体
flexible ceramic magnet wire 软陶瓷磁线
flexible chute 软式输弹道
flexible component 柔性构件
flexible conductor 软导线
flexible control 柔性控制，软操纵
flexible control computer 弹性控制计算机
flexible die deep drawing 弹性凹模深压延
flexible die forming 软模成形
flexible dynamics 【物】柔性动力学
flexible felt 柔性毡
flexible flight deck 活动飞行甲板
flexible gun 活动机枪(炮)，旋转机枪(炮)
flexible hose 软管
flexible hose assembly 软管组件
flexible hose for loading 加注软管
flexible joint nozzle 柔性喷管（通过柔性接头实现摆动的喷管）
flexible machine gun 活动式机枪
flexible manufacturing system 柔性制造系统（自动化制造系统的）
flexible metallic conduit 金属软管
flexible mode 柔性模式
flexible multibody dynamics 【物】柔性多体动力学
flexible pavement 柔性道面
flexible plate 柔性底板
flexible plate nozzle 扰性板喷管，柔壁喷管
flexible polyurethane foams 软质聚氨脂泡沫塑料
flexible printed board 柔性印制板
flexible response 灵活反应（战略）
flexible rotor 柔性转子
flexible shaft 弹性轴，软轴
flexible solar cell array 柔性太阳电池阵
flexible space 弹性空间
flexible space structure 挠性空间结构
flexible space suit 软式航天服
flexible structure 柔性结构

flexible support　1.弹性支承 2.柔性支承
flexible take off　弹性起飞
flexible tank　软油箱
flexible trailing edge　柔性后缘
flexible turret　活动炮塔
flexible waveguide　软波导
flexible wing　可折叠翼
flexiplast　柔性塑料，挠性塑料
flexlead hysteresis effect　软导线滞后效应
flexlead torque　软导线力矩
flexseal　柔性密封件
flexural axis　弯轴（如机翼的刚性轴，加载后只产生弯曲，而不产生扭转）
flexural mode　挠曲型
flexural rigidity　弯曲刚性
flexural strength　抗弯强度，挠曲强度
flexural wash-out　挠性负扭转（后掠翼受力向上挠曲时，从翼根至翼尖迎角逐步减小）
flexure accelerometer　挠性加速度表
flexure deformation　挠曲变形
flexure gyro gyroscope　挠性陀螺
flexure spring suspended gyro　挠性弹簧支承陀螺
flexure suspension　挠性支承（用刚度很小的弹性元件做的支承）
FLFT　Full-Load Frame Time 满载帧时间
FLG　Flange 法兰盘
FLH　Flat Head 扁头
FLHLS　Flashless 无闪光的
flick　照见瞬间
flicker discrimination　闪光辨别
flicker vertigo　闪烁眩晕
flickering　闪变
FLIGA　Flight Incident or Ground Accident 飞行事故或地面事故
flight　1.航班 2.小队编队
flight abortion criteria　飞行中止标准
flight accident　飞行事故
flight advisory　飞行咨询
flight altimetry　飞行高度测量（法）
flight altitude　飞行高度
flight analyzer　飞行分析器
flight angle　腾起角
flight article　飞行样机
flight assist　飞行救助
flight at the controls　自动驾驶飞行
flight attendant　机上服务员
flight attitude　飞行姿态
flight augmentation computer　飞行增稳计算机
flight axis　飞行轴线，自动驾驶仪控制通道

flight azimuth　飞行方位角
flight block　摄影分区
flight boundary control system　飞行边界控制系统
flight by flight spectrum　飞续飞谱
flight calibration work　飞行校正作业（飞机与地面导航设备的）
flight characteristic angle　飞行特征角
Flight Chart　（百万分之一比例尺）航图（美国）
flight commander　飞行指挥官，飞行小队长
flight computer　飞行计算机
flight condition　飞行条件
flight configuration　飞行形态
flight control box　飞行控制盒
flight control headquarter　1.飞行控制指挥部 2.飞控指挥部
flight control instrument　飞行控制系统电子设备
flight control law　飞行控制律
flight control room　飞行控制室（载人航天计划的飞行任务指挥部）
flight control software　飞行控制软件
flight control system　1.飞行控制系统 2.飞控系统
flight control tester　飞行控制仪表检验器
flight control unit　飞行控制装置
flight controller　飞行管制员
flight corridor　飞行通道，飞行走廊
flight coupon　飞行奖券
flight crew　飞行人员，机组成员
flight cycle　飞行循环（机体、推进系统及其他系统的工作程序，这些程序加在一起完成一次飞行）
flight data acquisition system　飞行数据采集系统
flight data recorder　1.飞行参数记录器 2.飞行记录器
flight deck　1.飞行甲板 2.驾驶舱
flight deck lights　飞行甲板灯光
flight deck net　飞行甲板安全网
flight deck officer　飞行甲板军官
flight deck rail　飞行甲板滑轨
flight deck television camera　飞行甲板电视摄像机
flight despatcher　航行调度员
flight director　机长
flight director approach　用飞行指引仪进场（着陆）
flight director signal adapter　飞行指引仪信号换接器
flight director system　飞行指引系统
flight director warning flag　飞行指引仪警告旗
flight dispatch　飞行签派
flight display　飞行显示器
flight distance　飞程
flight dossier　飞航气象要览
flight duration　续航时间，飞行时间
flight duty period　飞行值勤期，空勤组当班时间（即

Crew Duty Time)
flight dynamic code　航班动态代码
flight dynamic model　飞行动力学模型
flight dynamic pressure　飞行动态压力
flight dynamics　飞行动力学
flight dynamics model　飞行动力学模型
flight engineer　空勤工程师,空勤机械员
flight envelope　飞行包线
flight envelope monitoring　飞行包线监控
flight environment　飞行环境
flight equipment　机上设备,机载设备
flight experiment　飞行实验
flight facilities equipment　飞行保障设备
flight fatigue　飞行疲劳
flight flutter kit　飞行颤振发生器,飞行颤振试验装置（飞机在飞行中进行颤振试验用）
flight flutter test　飞行颤振试验
flight flying prototype　可飞行的原型机(样机)
flight following　飞行跟踪
flight garment　飞行服
flight gyroscope　飞行陀螺仪
flight hardware　飞行器
flight hazard　飞行危险
flight height　航高,飞行高度
flight helmet　飞行头盔
flight idle　空中慢车(飞行中发动机最低转速)
flight indicator board　飞行出入港显示板
flight information　飞行情报
flight instruction program　飞行指令程序
flight instrument　飞行仪表
flight integrated data system　飞行综合数据系统
flight integrity　飞行整体性(指两架作战飞机为互相支援进行机动飞行时的密切关系)
flight level　飞行高度(按海平面标准气压计的飞行高度,常用百英尺计,如 235 指高度 23 500 英尺)
flight line　飞行路线
flight load　飞行载荷
flight management system　飞行管理系统(飞机上的计算机化的自动飞行控制系统)
flight maneuver　飞行动作,飞行操纵
flight manual　飞行教范,飞行教令
flight measurement　飞行测量
flight mechanics　飞行力学
flight medicine　飞行医学
flight mission　飞行任务
flight mission analysis　飞行任务分析
flight mission profile　飞行任务剖面
flight mission training　飞行任务训练
flight mode annunciator　飞行状态信号牌,飞行状态指示器
flight model　飞行计算模型
flight navigator　飞行领航员
flight number　（民航飞机的)航班号码
flight nurse　航空护士,飞行护士
flight operation　航空器运行,航务运行
flight operation team　飞行操作团队
flight orientation　飞行定向
flight panel　飞行仪表板
flight parameter　飞行参数
flight path　飞行路线,航线
flight path angle　飞行轨迹倾角,弹道倾角
flight path measurement　飞行轨迹测量
flight path recovery　航迹恢复
flight performance　飞行性能
flight permission criteria　放飞标准
flight phase　飞行阶段
flight plan　飞行计划
flight plan of aerial photography　航摄计划
flight platform　飞行站台
flight point　飞行点
flight procedure　飞行程序
flight profile　飞行(轨迹)剖面图
flight progress strip　飞行进程记录条
flight range　航区
flight range rehearsal　航区合练
flight rating test　飞行人员等级测试(按执照要求测试其能力)
flight readiness　飞行准备
flight recorder　飞行记录器
flight recorder unit　机载记录器
flight reference stabilization system　飞行基准稳定系统
flight refuelling　空中加油
Flight Refuelling Ltd.　空中加油有限公司
flight regime　飞行状态
flight region　航区
flight research program　飞行研究项目
flight risk index　飞行风险指数
flight route　飞行路线
flight rule　飞行规则
flight safety　飞行安全
flight segment　航段
flight sequence　飞行程序,飞行顺序
flight service station　飞行服务站(美国联邦航空局提供飞行救助服务的机构)
flight servicing　飞行维修保养(区别于日常维修保养)
flight simulation　1.模拟飞行 2.模飞
flight simulation bed　飞行模拟转台
flight simulation model　飞行仿真模型

flight simulator 飞行模拟器
flight sister 航空女护士
flight software 飞行软件
flight space 飞行空间
flight speed 飞行速度
flight speed control 飞行速度控制,飞行仿真
flight stability 飞行稳定性
flight status 飞行态势(表明飞机是否需要空中交通管制中心给予特殊的服务)
flight stewardess 空中女乘务员
flight strip 1.简易机场(特指在公路附近的私人机场、农用机场等) 2.着陆场
flight surgeon 航空医生,航医(负责飞行人员在地面的保健工作,不是外科医生)
flight syncope 【医】空中晕厥
flight system 飞行系统
flight table 飞行时刻表
flight team 飞行小组
flight technical error 驾驶技术误差
flight telerobotic servicer 飞行遥控机器人服务器
flight termination system 飞行终止系统
flight test 1.飞行试验 2.试飞 3.飞行员等级考试
flight test of aircraft static electricity 航空器静电试飞
flight test procedure 飞行试验程序,试飞程序
flight test program 飞行试验大纲
flight test support system 飞行试验支援系统
flight test vehicle 飞行试验运载器(特殊飞机、导弹或其他运载器,专供飞行试验用)
flight time 飞行时间(从飞机用自身动力开始移动起,至结束飞行、着陆停止不动为止)
flight time series 飞行时串
flight training 飞行训练
flight training psychology 飞行训练心理学
flight trajectory 飞行弹道
flight trial evaluation 飞行结果分析
flight tunnel 飞行走廊
flight vehicle 飞行器
flight velocity 飞行速度
flight vibration measurement 飞行振动测量
flight visibility 飞行能见度,空中能见度(从驾驶舱内看前方的能见度,以轻型飞机的一般飞行高度估算)
flight watch 空中守听(用122兆周保持与飞行服务站接触,以便进行航线飞行咨询服务)
flight weather condition 飞行气象条件
flight weather forecast bulletin 飞行天气预报表格
flight weight 起飞重量
flight-control 飞行控制
flight-control electronics 飞行控制电子学,飞行控制系统电子设备

flight-controlling mode 自动驾驶仪工作方式,飞行自动控制方式
flight-instrument amplifier 飞行仪表放大器
flight-instrument board 飞行(驾驶)仪表板
flight-level pressure altitude 修正压力高度
flightpath 飞行轨迹,弹道,航线
flight-path axis system 航迹坐标系
flight-path azimuth angle 航迹方位角
flight-path computer 航线计算机
flight-path predictor 航迹预测
flightpath recorder 飞行轨迹记录仪
flightpath sight 飞行轨迹瞄准(在平视显示仪上瞄准飞机将要通过的远方点)
flightpath vector 飞行轨迹矢量(在上指示的未来飞行轨迹)
flight-related mishap 与飞行有关的事故
flight-safety channel 飞行安全走廊
flight-safety control system (飞行)安全控制系统
flight-safety judgment criterion (飞行)安全判断准则
flight-safety region 飞行安全区
flight-safety self-destruction determine mode (飞行)安全自毁判定模式
flight-safety self-destruction system (飞行)安全自毁系统
flight-simulation equipment 飞行模拟装置
flight-tested equipment 经过飞行试验的设备
flightworthy 适航的,适于飞行的,做好飞行准备的
FLING Fueling 加油
FLIP 1. Flight Information Publication 飞行资料汇编 2. Floated Lightweight Inertial Platform 轻型液浮惯性平台
flip-flop 双稳态多谐振荡器,触发器
flip-flop circuit 双稳态触发电路
flip-flop counter 触发计数器
flipper 1.〈口语〉升降舵 2.(轮胎的)钢丝圈包布
flipper door 〈口语〉起落架舱门
FLIT Fighter Lead-In Training 战斗机过渡训练(从航校毕业到战斗部队服役之间的训练)
FLL 1. Flight Line Level 航线高度层(英国及北约用语) 2. Flap Load Limiter 襟翼载荷限制器
FL-LP Filter-Lowpass 低通滤波器
FLLW Field Length Limit Weight 场长极限重量
FLM 1. Flight Line Maintenance 飞行在线维护 2. Flight Line Mechanic 外场机械员(美国)
FLMB Flammable 易燃的
FLMPRF Flameproof 防火的
FLMT Flushmount 平接安装
FLMTS 1. Flight Line Maintenance Test Set 航线维修试验台 2. Future Land Mobile Telephone Service 未来

陆地移动电话业务
FLN Fuel Line 燃料管道
Flo Trak 浮轮板（一种附加在轮胎上类似履带的装置，供飞机在软场地上着陆用，专利技术名）
float 1.浮筒（水上飞机起落装置）2.（接地前）平飘，（着陆）平飘距离 3.（操纵面在气流中）飘浮 4.浮子，浮标
float alignment 浮子定轴，浮法线
float angular momentum 浮子角动量
float assembly 浮子组件
float displacement 浮筒排水量（水上飞机浮筒全部浸入水中的排水量）
float flare 海上标志弹
float flowmeter 浮子流量计
float gear 浮筒式起落架
float gudgeon pin 游动活塞销
float levelmeter 浮子液位计
float light 飘浮照明灯，伞降照明弹
float mounting connector 浮动安装连接器
float oscillation 浮子振荡
float type gyroscope 悬浮陀螺仪
float valve 浮子活门
float volume 浮力容积比
float zone 浮区
floatation 1.浮动，漂浮 2.浮游选矿，浮选
floatation bag 浮囊
floatation balloon 漂移气球
floatation fluid 浮液
floatation gear 漂浮装置
floatation suspension 液浮支承
floated accelerometer 悬浮加速度表
floated gyro 液浮陀螺仪
floated gyroscope 悬浮陀螺
floated inertial platform 液浮惯性平台
floated pendulous accelerometer 液浮摆式加速度表
floated rate gyro 液浮速率陀螺
floater 〈口语〉载荷轻的飞机
floating accumulator 浮点累加器
floating add 浮点加
floating address 浮动地址，可变地址
floating base 悬空基极
floating battery 浮充蓄电池
floating charge 浮接充电，浮充
floating earth 浮地
floating lines 浮动线
floating mark 1.立标（测距仪或测高仪的）2.浮标
floating nozzle 浮动喷管
floating point accumulator 浮点累加器
floating point arithmetic package 浮点运算程序包

floating point hardware 浮点硬件
floating point instruction set 浮点指令系统
floating point unit 浮点部件
floating potential 浮动电势，漂游电位
floating radiosonde 漂浮探空仪
floating ratio 悬浮比（液浮速率陀螺的）
floating reticle 活动光环（在瞄准具视场内活动的图像）
floating seal 浮动环密封
floating voltage 浮动电压
floating zenith telescope 浮动天顶仪
floating-point 浮点
floating-point binary 浮点二进制
floating-point calculation 浮点运算
floating-point package 浮点程序包
floating-point status vector 浮点状态向量
floating-point truncation 浮点截断，浮点截尾
floating-zone 浮区
floating-zone grown silicon 悬浮区熔硅
floating-zone method 悬浮区熔法
floatplane float seaplane 浮筒式水上飞机（在水上由独立的浮筒支撑的飞机）
float-type carburetor 浮子式汽化器
FLOC Fault Locator 故障定位器
FLOLS Fresnel-Lens Optical Landing System（航母上）菲涅耳透镜光学着陆系统
flood flow 大流量，洪流
flood gun 泛射式电子枪
flood light 泛光灯
flood valve 溢流活门
floor vents 地板通气道（将舱内空气通往增压或非增压的机身下层的通道）
FLOT Flotation 漂浮
flotation jacket 充气救生衣
flow angle 气流角
flow area 流动面积
flow asymmetry 流动非对称性
flow augmenter 1.引射器，喷射器 2.进气导流器
flow behavior 流动特性，流动情况
flow bleed valve 泄流阀（火箭发动机用）
flow blockage 流动阻塞
flow calculation 流量计算
flow channel 流道，流动水槽
flow characteristic 流动特性，量特性
flow chart 流程图，生产过程图解，作业图
flow coefficient 流量系数
flow combining valve 集流阀（流量控制）
flow compressor 轴流式压缩机
flow computation 流量计算
flow condition 流动状态，流动条件

flow conductance   流导
flow conductance method   流导法
flow control   1.流量控制 2.(数据)流控制 3.交通量控制
flow control valve   流量控制阀
flow controller   流量控制器
flow core   湍流核
flow diagram   流程图,程序框图
flow direction probe   流向探头
flow direction uniformity   气流方向均匀性
flow disrupter   扰流板(铰接的可收放小板用来使飞机故意失速)
flow distortion   流场畸变,气流畸变,流动变形
flow distribution   气流分布
flow disturbance   流动扰动,流扰动
flow divider and combiner valve   分流集流阀
flow domain   流域
flow duct   流管
flow duration   流量历时
flow dynamics   流体动力学
flow effect   流动效应
flow equation   流体运动方程,流动方程
flow expansion   流动扩散,气流膨胀
flow factor   流量因数
flow feature   流动特征
flow field   流场
flow field calibration   流场校测
flow field distortion   流场畸变
flow field quality   流场品质
flow field similarity   流场相似
flow field survey   流场测量
flow injection   流动注射
flow instability   流量失稳
flow loss   流动损失
flow measurement   流量测量
flow mechanism   渗流机理,流动机理
flow meter   流量计
flow method   流机制,流水作业法
flow model   流程模型,流动模型
flow motion   液化流动效果
flow network   流动网络
flow network solver   流网络解算器
flow passage   通流部分,流路,流道
flow path   气流通道,流路,流道
flow pattern   流谱
flow penetration depth   气流穿透深度
flow perturbation   气流扰动
flow process diagram   流程图,程序框图
flow profile   流量剖面

flow property   流动性,流动性能
flow pulsation quality   气流动态品质
flow reactor   【化】连续反应器,【化】流动反应器
flow reattachment   气临附
flow recirculation   回流
flow regime   (流体)流动方式(如连续流、层流等)
flow region   流动区域
flow regulator   流量调节器
flow resistance   流阻
flow response   流量回应
flow reversal   回流,逆流
flow rule   流动法则,流动规则
flow sensor   流量传感器
flow separation   气流分离,气流离体
flow simulation   流体分析,流动分析,流量模拟
flow speed   流速
flow splitter   分流器
flow stability   流动稳定性
flow stagnation   停滞,流动滞止
flow state   流态
flow structure   流动结构
flow switch   流量开关
flow system   流动系统,流水作业方式
flow temperature   流动温度
flow test   流通试验,流量试验
flow theory   流动理论
flow totalizer   流量加法器
flow transition   明满过渡流
flow transmitter   流量变送器
flow tube   流量管,流量测量管
flow turning   强力旋压
flow variable   流量变数
flow variation   径流量变化
flow vector   气流矢量
flow velocity   流速,流动速度
flow visualization   流动显形,流场显示
flow visualization by luminescence   辉光放电流动显示
flow(-rate) transducer   流量传感器
flowback   回流(特指在机翼结冰时,水从机翼前缘的回流)
flowchart   1.流程图,程序框图 2.流量图,流动图,流线谱
flowchart package   流程图包
flow-measuring apparatus   流量表
flowmeter   流量计
flowmeter transmitter   流量表传感器
flow-rate equipment   流率测量设备
flow-sensing unit   流量计
FLP   Fleet Lead Plan 领先机队大纲

FLPS Flight Loads Preparation System 飞行载荷准备系统

FLR 1. Floor 地板 2. Forward Looking Radar 前视雷达

FLRG Flaring 闪光,照明,爆发

FLRS Forward Looking Radar Set 前视雷达装置

FLRT Flow Rate 流动的速度

FLS 1. Financial Labor System 工时财务系统 2. Future Launch System 未来(航天)发射系统

FLT 1. Fault Location Technology 故障定位技术 2. Flashlight 闪光 3. Flight 飞行

FLT CONT Flight Control 飞行控制

FLT DIR Flight Director 飞行指引仪

FLT ENG Flight Engineer 随机工程师,飞行机械师

FLT INPH Flight Interphone 飞行内话

FLT INST Flight Instrument 飞行仪表

FLTA Forward Looking Terrain Avoidance 前视地形回避

FLTG Floating 浮动,漂浮

FLTR Filter 过滤器

FLTS Fault-Locating Tests 故障定位测试

FLU Fault Locating Unit 故障定位设备

fluctuation intensity 波动强度

flue gas 废气,烟气,管道气

fluid conductance 流导

fluid density 流体密度

fluid diode 射流二极管

fluid dynamic property 流体动力特性

fluid dynamics 流体动力学

fluid film bearing 液膜轴承

fluid four 四机流动队形

fluid impedance 流抗

fluid induced vibration 流体诱导振动,流体诱发振动

fluid injection 流体引射

fluid jet 流体射流

fluid loop 液体回路

fluid mechanics 流体力学

fluid movement 液体运动

fluid particle 流体质点

fluid phase 流体相位,液相

fluid physics module 流体物理舱

fluid resistance 流阻

fluid rotor gyroscope 流体转子陀螺仪

fluid shift 体液移位

fluid statics 流体静力学

fluid temperature 介质温度

fluid volume compensator 浮液体积补偿器

fluidic accelerometer 流体加速表,液压系统过载传感器

fluidic amplifier 流控放大器,流体放大器

fluidic autopilot 1. 液压式自动驾驶仪 2. 气动式自动驾驶仪 3. 射流式自动驾驶仪

fluidic control 射流控制

fluidic display 射流显示器

fluidic flight control system 射流式飞行控制系统

fluidic gyroscope 射流陀螺(仪)

fluidic oscillator 流体振荡器

fluidic rate sensor 射流速率传感器

fluidic sensor 射流传感器

fluidic technique 射流技术

fluidics 射流技术,流控技术

fluidized bed 射流试验台,流化床

fluid-solid coupling 流固耦合

FLUOR Fluorescent 荧光的,日光(灯)

fluorescence induced by laser 激光诱导荧光

fluorescence linewidth 荧光线宽

fluorescence microtuft method 荧光微丝法

fluorescence signal 荧光信号

fluorescence spectrophotometer 荧光分光光度计

fluorescent character-display tube 荧光数码管

fluorescent crack detection 荧光探伤

fluorescent dye 荧光染料

fluorescent map 荧光地图

fluorescent penetrant flaw detection 荧光渗透探伤

FLUT Flutter 颤振,高频抖动

flutter analysis 颤振分析

flutter boundary 颤振边界

flutter characteristic 颤振特性

flutter condition 颤振状态

flutter control 颤振控制

flutter damper 颤振阻尼器

flutter exciter 颤振激励器

flutter frequency 颤动频率

flutter instability 颤振失稳

flutter margin 颤振余量

flutter mechanism 颤振机理

flutter mode 颤振模态

flutter mode suppression 颤振模态抑制

flutter model 颤振模型(其颤振品质与全尺寸飞机相似)

flutter speed 颤振速度(发生颤振的最低当量空速)

flutter wind tunnel 颤振风洞

flux 1. 助熔剂,焊剂,焊药 2. (磁、电等的)流量,通量

flux conserver 通量保藏

flux density 流量密度

flux divergency 通量辐散

flux growth method 助熔剂法

flux of space debris 空间碎片通量

flux pulling technique 助熔剂提拉法

flux range 流量范围
flux ratio 通量比
flux unit 流束单位,流量单位
fluxgate 感应式磁传感器
fluxgate compass 感应式磁罗盘
fluxgate magnetometer 饱和式磁强计,磁通门磁强计
fluxmeter 磁通计
fluxplate 热流板,热通量板
FLV Finite Logical View 有限逻辑图
FLW 1. Flat Washer 扁平垫片 2. Forward Looking Windshear Radar 前视风切变雷达
FLWP Follow-Up 跟踪装置,随动系统,伺服系统
fly around 绕飞
fly ball speed governor 离心式调速器,飞球调速器
fly before buy 先飞后买
fly bomb 飞弹
fly by light (FBL) 光传操纵
fly by wire (FBW) 电传操纵
fly fritz 〈口语〉飞弹,飞航式导弹
fly space 飞行空间
flyaway 空运物资
fly-away cost 现货飞走价格
fly-away disconnects 旋转断接器
flyaway ejection seat 飞行弹射座椅
flyaway kit （部队自带）空运成套备件
flyback （阴极射线管中）回扫线,光的回程
flyback booster 可飞回助推器
fly-back period 返回周期
flyback time 回扫时间,返回时间
fly-back vehicle 返回式航天器
flyby 绕天体飞越
fly-by mission 绕越飞行任务
fly-by-light 光传操纵
fly-by-light control system 光传飞行控制系统
fly-by-light flight control system 光传飞行控制系统
fly-by-wire control 电传飞行控制
Flycatcher 京燕(荷兰造的雷达系统)
Flycatcher radar 京燕雷达
fly-home capability 返航能力
fly-in 1.飞行集会 2.(作战中)进入目标区
flying bazooka 〈口语〉飞机火箭发射筒
flying boat 飞船,船身式水上飞机(一般为平底)
flying control 1.飞行控制部位 2.飞行控制室
flying coveralls 飞行服
flying decompensation syndrome 飞行代偿失调综合症
flying diameter 飘降伞衣直径(指降落伞下降时张开的最大直径)
flying grading 飞行等级的划分
flying height 航高,飞行高度

flying illusion 飞行错觉
flying machine 飞行机,有动力的重航空器
flying operation 飞行活动
flying personnel 飞行人员
flying platform 飞行平台
flying position (飞机的)水平姿态
flying programme 飞行计划
flying qualification 飞行合格,飞行资格
flying quality simulator 1.飞行品质仿真器 2.工程模拟器
flying reliability 飞行可靠度
flying rigging 飞行悬索
flying shears 飞刀剪床(工件高速移动,刀具转动)
flying skill 飞行技能
flying skill error 飞行技巧错误,飞行反应动作错误
flying speed 飞行速度(广义上指飞机保持平飞的最小速度)
flying spot pick-up device 飞点摄像机
flying spot scanner 飞点扫描仪
flying stovepipe 〈口语〉冲压式喷气发动机
flying stress 飞行应力
flying syllabus 飞行大纲
flying tail 1.全动平尾,可操纵安定面 2.(靠)平尾飞行(以整个平尾作为主操纵面的飞行简化)
flying test bed 空中试验飞行器,空中实验室,飞行试验台
flying torpedo 航空鱼雷,滑翔炸弹
flying unit 飞行部队
flying weight 飞行重量(发动机的)
flying wing 飞翼,翼形飞机
flying wires 机翼张线,升力张线
flying wreck rate 飞行事故毁机率
flying-crane helicopter 起重直升机,空中吊车
flyingdart 〈口语〉自由飞行火箭,惯性飞行火箭
flying-qualities 飞行品质
flying-tab control 随动补偿片操纵
flyover 飞越,通场(仪表或其他装置上,用以显示发生故障的小牌,通常为红色,有故障时才显现)
fly-over noise 飞越噪声(指飞机接近机场、进场与离场航道上的噪声)
flyover test for fuze 引信绕飞试验
flyup 高于下滑道飞行
flyway 航路
flywheel 飞轮
flywheel bearing unit 飞轮轴承组件
FM 1. Failure Mode 故障模式 2. Fan Marker 扇形指点标 3. Flight Management 飞行管理 4. Flight Mechanic 随机机械师 5. Flight Manual 飞行手册 6. Frequency Modulation 调频

FM radio fuze 调频无线电引信
FM ranging fuze 调频测距引信
FM recording 调频记录
FM sideband fuze 调频边带引信
FMA Flight Mode Annuciator 飞行状态显示器
FMAD Fluid Management and Distribution 流体管理和分配
FMC 1. Flexible Manufacturing Cell 柔性制造单元 2. Flight Management Computer 飞行管理计算机
FMCF Flight Management Computer Function 飞行管理计算机功能
FMCS Flight Management Computer System 飞行管理计算机系统
FMD 1. Financial Management Division 财政管理处 2. Flight Management Display 飞行管理显示器
FMDS Flight Management Data System 飞行管理数据系统
FMEA Failure Modes and Effects Analysis 故障模式和影响分析
FMECA Failure Modes, Effects and Criticality Analysis 故障模式、影响和危害性(致命度)分析
FMF Flight Management Function 飞行管理功能
FMFB Frequency Modulation Feedback 调频反馈
FMGC Flight Management Guidance Computer 飞行管理引导计算机
FMGS Flight Management Guidance System 飞行管理引导系统
FMI 1. Flexible Modular Interface 可变模块化接口 2. Functional Management Inspection 功能管理检查
FMICW Frequency Modulated Intermittent Continuous Wave 调频间断连续波
FMO Flight Management Officer 飞行管理军官
FMOF First Manned Orbital Flight 首次载人轨道飞行
FMP Flight Mode Panel 飞行状态版
FMPT First Materials Processing Test 首次材料加工试验
FMRT Final Meteorological Radiation Tape 最终气象辐射带
FMS 1. Federation of Materials Societies 材料学会联合会 2. Field Maintenances Squadron 野战维修中队(美国) 3. File Management System 文件管理系统 4. Flexible Manufacturing System 柔性制造系统 5. Flight Management System 飞行管理系统 6. Fluid Management System 流体管理系统 7. Foreign Military Sales 对外军品销售(美国国防部) 8. Fuel Management System 燃油控制系统
FMTAG Foreign Military Training Affairs Group 外国军事训练事务组
FMTG Fluid Management Technology Group 流体管理技术组
FMU Fuel Metering Unit 燃油流量表,燃油调节装置,燃油测量装置
FMW Fixed Momentum Wheel 固定式动量飞轮
FNCP Flight Navigation Control Panel 航行管制小组
FNL Fleet Noise Level 机队平均噪声级
FNSG Flight Navigation Symbol Generator 飞行导航信号生成器
FO 1. Fabrication Order 制造指令 2. Fabrication Outline 制造大纲 3. Fail Operational 带故障工作
foam carpet 泡沫毯,泡沫铺地
foam core 泡沫芯层
foam density 泡沫密度
foaming space 泡沫空间
foam-laying equipment 泡沫剂喷洒设备
FOBS Fractional Orbit Bombardment System 部分轨道轰炸系统
FOC 1. Flight Operation Control 飞行运行控制 2. Focal 焦点的 3. Fuel/Oil Cooler 燃油/滑油冷却器
FOD 1. Flight Operation Domain 飞行操作范围 2. Foreign Object Damage 外来物损伤
FODB Fiber Optic Data Bus 纤维光学数据总线
FODCS Fibre Optics Digital Control System 光纤数字控制系统
FODT Fibre Optics Data Transmission 光纤数据传输
FOEC Flight Operation Estimation Committee 飞行运行评估委员会
FOG Fiber Optical Gyro 光纤陀螺
fog scale 雾级标度
fog signals 雾警设备,雾号
fog visiometer 雾天能见度仪
fog-gauge 雾量计
FOG-M Fiber Optic Guided Missile 光纤制导导弹
FOI Follow On Intercepter 后继截击机
FOIA Freedom Of Information Act 信息自由法(美国)
FOL Follow 跟随,随动,遵守
folding fin 1. 折叠式尾翼 2.(飞机的)折叠式垂直尾翼
folding fin aircraft rocket 折叠翼航空火箭弹
folding tail 可折叠的机身尾段
folding wing 折叠翼
fold-out type solar array 折叠式太阳电池阵
foldover effect 折叠效应
following error 跟踪误差,随动误差
follow-on developmental test 后继研制试验
follow-on production 后继生产
follow-up device 1.跟踪装置,随动装置 2.(导弹)导引头 3.目标位标器
follow-up potentiometer 跟踪电位计
follow-up report 补充报告

follow-up study 追踪调查	forcer 力发生器(产生作用力的装置)
follow-up unit 随动装置,随动部件	force-sensing controller 感力控制器,操纵力传感器
FONAC Flag Officer Naval Air Command 海军航空兵司令	forcing function 1.强制函数,外力函数 2.强加功能(给设备施加某种机械的或环境的输入,以影响其设计、使用寿命和工作能力)
footprint 1.足迹 2.(航天器)着陆地带	
foot-thumper 脚感失速警告装置,振脚器	forcing term 扰动项,加力项
foot-to-head acceleration 由足至头的加速度	Fordsat 福特卫星(美国福特公司的通信卫星)
FOP 1.Flight Operations Plan 飞行操作程序(航天器的) 2.Forward Operating Pad 直升机前方起降场	fore-and-aft accelerometer 纵向加速度表
	fore-and-aft level 前后水准仪,纵向测斜仪
FOPT 1.Fiber Optic Photo Transfer 纤维光电变换 2.Fixed Operating Point Flight 固定作业点飞行	fore-and-aft motion signal 前后运动信号
	fore-and-aft overlap 航向重叠
FOQA Flight Operations Quality Assurance 飞行运行质量保证	fore-and-aft trim 纵向配平
	forebody 1.前机身 2.前体(浮筒的滑行底部或机身断阶的上游部分)
FORACS Fleet Operational Readiness and Calibration System 舰队作战准备及检查系统	
	forecast bulletin 天气公报
forbidden line 禁线	forecast ensemble 预报集成
forbidden transition 禁戒跃迁	foreflap 襟翼前段(双开缝或三开缝式襟翼的最前部分)
force balance accelerometer 力平衡加速度计	
force balance pressure transducer 力平衡式压力传感器	foreign flag carrier 有外国国旗标志的航空公司,经营国际航线的外国航空公司
force combat air patrol 特遣队空中战斗巡逻(战斗机为保护特遣部队免遭空袭而进行的巡逻)	
	foreign military sales 对外军事销售
force control unit 杆力调节机构,加载机构	foreign-object damage 外物打伤,外物损伤
force defect 弹性不良力亏损	forest aerial photography 森林航空摄影
force density 力密度	forest aerial reconnaissance 森林航空勘查
force direction 作用力方向	forest aerial survey 森林航空调查
force measurement 力值测量	forest border 林缘
force model 力学模型	forest fire surveillance 森林火灾监视
force ratio 力比	forest interpretation templet (template) 森林判读模片
force rendezvous 兵力会合点	forest photo plot area templet 森林相片样地面积模板
force restored accelerometer 力复原式加速度计	forest photo stereogram 森林相片立体样片
force sensor 力传感器,感力器	forest site interpretation 森林立地条件判读
force structure 兵力结构	FORG Forging 锻造,锻件
force transducer (测)力传感器	forging aluminium alloy 锻铝合金
force value 强制值	fork mounting 叉式装置
force/power feedback 力/功率反传	fork ring 叉形环
force-balance transducer 力平衡式传感器	forked rod 叉形(连)杆
force-balance type pressure transducer 力平衡式压力传感器	form block 成型模
	FORMAT Fortran Matrix Abstraction Technique Fortran 矩阵分离技术
forced air cooling 强迫空气冷却	
forced convection 强迫对流	formation centre 飞行情报中心
forced disengagement 强制脱落机构	formation flight 编队飞行
forced landing 迫降,强迫着落	formation lights 编队灯(便于夜间编队飞行)
forced oscillation 强迫振荡,强迫振动	formation model 生长模型
forced return 强制返回	formation reconfiguration 队形重构
forced transition 受迫跃迁	formation-flying 编队飞行
force-free relative motion 无外力相对运动,自由相对运动	formation-keeping light 航迹灯
	form-drag 型阻力
force-multiplication factors 力量倍增因素(指可使武器命中精度、作战能力等大大增强的军事卫星)	formula cost 公式费用(按照美国空运协会统一规定的公式计算出的运输机直接使用费用)

formula translation 公式翻译,公式转换
FORS 1. Fabrication Operations Requirements System 制造工作需求系统 2. Fibre Optics Rotation Sensor 光纤旋转传感器
forsterite ceramics 镁橄榄石瓷
forward aeromedical evacuation 前方航空医疗后送(从战场上空运伤员至战斗地域最近的一个医疗站)
forward area support team 前方地域保障队
forward bulkhead 前舱壁
forward closure 前封头,前堵盖
forward deployment 前沿部署
Forward Deployment Strategy 前沿部署战略
forward difference 前向差分
forward direction (电子)正向
forward error correction 前向纠错
forward gun 前枪,前炮
forward integration 向前整合,向前合并
forward intersection 前方交会
forward kernel 正向核,正变换核(对一个函数进行变换时所用的变换核)
forward lap 纵向重叠,前后重叠
forward launched aerodynamic missiles 前射有翼导弹,前射飞航式导弹
forward limit of center of gravity 重心前限
forward link 前向链路(从地面经中继卫星向航天器传送信号的无线电链路)
forward lock 前位锁
forward looking infrared fire control system 前视红外火力控制系统
forward loss 正向损耗
forward movement 前向运动
forward operating base 前方作战基地(没有充分保障施的前方飞机场)
forward overlap 前向重叠,航向重叠
forward skirt of case 壳体前裙
forward supervision 前向监控
forward supply point 前进供应站
forward sweep 前掠式(翼)
forward thrust 正推力,向前推力
forward tilt (直升机旋翼的)倾角
forward tilt wing (slewedwing) 前偏转翼(指斜翼机或剪式翼的向前偏转的半翼)
forward tilting angle of rotor shaft 旋翼轴前倾角
forward transition 垂直飞行向水平飞行过渡,正过渡
forward twin pointer 向前孪生指针
forward type cycle 前向式循环
forward velocity 前行速度
forward wave 前向波
forward-based weapons 前方基地(核)武器,海外基地(核)武器
forward-looking infrared system 前视红外系统
forward-swept wing 前掠翼
FOS 1. Faint Object Spectrograph 暗天体光谱仪 2. Fibre Optics Sensor 纤维光学传感器 3. Forecast Option Selection 预选项选择
FOSP Fabrication Outline Special Purpose 特殊目的制造大纲
FOSS Fibre Optic Sensor Systems 光纤传感系统,光纤敏感系统
FOT FAL Operational Testing 总装线运行试验
FOT&E Follow-On Test and Evaluation 后继试验与鉴定
Foucault pendulum 傅科摆
four bank eight 四点转弯8字飞行(8字特技圆圈的外侧部分,不是圆形,而是两个45°转弯加一段短距直线飞行)
four poster 四喷管喷气式发动机
four-ball test 四球试验
four-stroke engine 四冲程式发动机
fourth line servicing 四线维修,四线保障勤务(英国空军飞机维修体制中最高一级,包括翻修、改装等)
four-vector 四维向量
four-wave mixing 四波混频
FOV Field of View 视场,视野
FOW Family of Weapon 武器系列,武器族
Fowler flap 富勒(式)襟翼
FP 1. Fail Passive 故障消极防护 2. Fatigue Performance 疲劳特性 3. Flight Plan 飞行计划 4. First Pilot 机长 5. Flight Profile 飞行剖面 6. Flight Progress 飞行进展情况,飞行进度 7. Food Preparation (Area)(载人航天器上的)配餐区 8. Freezing Point 冰冻点
FP area Food-Preparation Area 配餐区
FP/DF Fluid Physics/Dynamics Facility (美国空间站上的)流体物理学和动力学设施
FPA 1. Failure Probability Analysis 故障概率分析 2. Fire Protection Association 消防协会(英国) 3. Flight Path Accelerometer 飞行轨迹加速度计 4. Flight Path Angle 飞行轨迹倾角,航迹倾角 5. Flying Physicians' Association 航空医生协会(美国) 6. Focal Plane Arrays 焦平面阵 7. Forward Pitch Amplifier 前俯仰放大器
FPAC Flight Path Acceleration 航线加速度
FPAS 1. Flight Profile Advisory System 飞行剖面咨询系统 2. Focal Plane Array Seeker 焦面阵天线导引头
FPB 1. Floating-Point Buffer 浮点缓冲器 2. Fast Patrol Boat 巡逻快艇
FPC Flight Path Control (Command) 飞行轨迹控制

（指令）

FPCC Flight Propulsion Control Coupling 飞行动力控制连接器

FPCD Functional Product Collector Drawings 功能产品选项图

FPD Flight Planning Document 制订飞行计划的文件

FPG Functional Process Guide 功能工艺指南

FPI 1. Fabry Perot Interferometer 法布里-珀罗干涉仪 2. First Product Inspection 首件产品检查 3. Fixed Price Incentive 固定价格加奖励 4. Fluorescent Penetrant Inspection 荧光透检查,荧光探伤

FPL 1. Flight Plan Message 飞行计划通知 2. Fluctuating Pressure Level(s) 气压波动的高度（层） 3. Friction Pressure Loss 摩擦压力损耗 4. Full Performance Level 全性能级（联邦航空局规定的空中交通管制员等级） 5. Full Power Level 全推力水平,（火箭的）全功率级

FPM 1. Feet Per Minute 英尺/分钟（最好使用 ft/min） 2. Flight Path Miles 飞行轨迹的英里数 3. Fluid Physics Module （空间实验室的）流体物理舱

FPN Fix Pattern Noise 定型噪声

FPOV Fuel Preburner Oxidizer Valve 燃料预燃室氧化剂活门

FPP Fixed Pitch Propeller 定距螺旋桨

FPPM Fuel Pipe Repair Manual 燃油导管修理手册

FPPS Flight Plan Processing System 飞行计划处理系统

FPR 1. Fan Pressure Ratio 风扇压比 2. Flat Plate Radiometer 平板辐射计 3. Floating Point Register （计算机）浮点寄存器

FPRM Flight Phase Related Mode 飞行阶段有关方式

FPS 1. Feet Per Second 每秒英尺 2. Fine Pitch Stop （螺旋桨）小距限动钉 3. Fire Protection System 防火系统 4. Fixed Program Send 固定节目发送

FPSS Fine Pointing Sun Sensors 精定位太阳敏感器

FPT Female Pipe Thread 内管螺纹

FQI 1. Fuel Quantity Indicator 燃油量指示器 2. Fuel Quantity Indication 燃油量指示

FQIS 1. Fuel Quantity Indicating System 燃油量指示系统 2. Fuel Quantity Indicator Switch 燃油量指示器开关

FQPU Fuel Quantity Processor Unit 燃油量处理装置,燃油量处理机

FQR Formal Qualification Review 正式鉴定评审,正式合格评审

FQS Fuel Quantity System 燃油量系统

FQT Formal Qualification Testing 正式合格试验,正式（质量）鉴定试验

FQTI Fuel Quantity Totalizer Indicator 总燃油量指示器

FR 1. Failure Report 故障报告 2. Failure Rate 故障率 3. Flight Recorder 飞行记录仪 4. Flight Refuelling 空中加油 5. Front 前面,正面

FRA Flap Retraction Altitude 收襟翼高度

FRAC Fractional 分数的,小数的

FRACAS Failure Reporting, Analysis and Corrective Action System 故障报告、分析和纠正措施系统

frag fragment 破片,弹片

frag (**bomb**) cluster 集束杀伤弹,杀伤弹束,杀伤弹箱

fragility fairing 可碎性整流罩

fragment beam angle 破片散飞角

fragment cloud 碎片云

fragment warhead （破片）杀伤弹头

fragmentary order 简令,要令

fragmentation ratio 破碎率

fragmentation unit （战斗部的）杀伤部分

fragmentation warhead （破片）杀伤弹头

fragmenting warhead rocket （破片）杀伤弹头火箭弹

FRAM Fleet Rehabilitation And Modernization 机队整顿与现代化

frame alignment recovery time 帧定位恢复时间

frame camera 画幅式相机,分幅式相机

frame check sequence 帧校验序列

frame format 帧格式

frame-dependent control 帧相关控制

frameset 框式支架

framing format 帧格式,帧内信息编排

framing pulse 帧脉冲

framing rate 帧变化率

FRC Flight Research Center 飞行研究中心

FRCI 1. Fibre Reinforced Composite Insulation 纤维增强复合材料绝热层 2. Fibrous Refractory Composite Insulation 耐高温纤维复合材料绝热层 3. Flexible Reusable Carbonin Insulation 柔性可重复使用碳绝热层

FRCMC Fiber Reinforced Ceramic Matrix Composite 纤维增强陶瓷基复合材料

FRCS Forward Raction Control System （美国航天飞机的）前反作用控制系统

FRD Functional Requirements Document 功能要求文件

free azimuth inertial navigation system 自由方位式惯性导航系统

free azimuth platform 自由方位平台（方位轴稳定在惯性坐标系中的水平平台）

free canopy 不可抛座舱盖,自由舱盖,非弹射座舱盖

free connector 自由端连接器

free fall 1. 自由下落（无制导,无推力,无减速装置）

2.沿开普勒轨道下落(重力与惯性力平衡) 3.开伞前降落(伞降中降落到预定高度由跳伞员自行开伞或自动开伞) 4.标准重力加速度(其值为9.8米/秒)

free fall landing gear 自由放下(式)起落架(可以依靠本身重量或风力放下)

free flight 自由飞行(无引导,自动驾驶仪除外)

free flight dart 〈口语〉自由飞行火箭,惯性飞行火箭

free flight model 自由飞行模型

free flight phase measurement 自由(飞行)段测量

free flight wind tunnel 自由飞风洞

free fluid surface 自由液面

free flying robot 自由飞行机器人

free gun 活动式机枪

free gunnery 活动式机枪射击术

free gyroscope 自由陀螺仪(不受外力作用的两自由度陀螺仪)

free jet 喷射气流,喷管出口射流,自由射流

free machine gun 活动式机枪

free mass 非脂肪重量

free molecular flow 自由分子流

free oscillation 自由振荡

free progressive wave 自由行波

free radical 自由基,游离基

free rocket 无控火箭,自由飞行火箭

free rotor gyroscope 自由转子陀螺仪

free shaft 自由轴

free shear layer 自由剪切层

free sphere 自由球,(航天器指向控制系统的)反作用球

free stream 自由流

free streamline 自由流流线,势流(无旋流)流线

free turbine 自由涡轮

free valence 游离价,自由价

free vibration 自由振动

free volume 自由体积

free vortex 自由涡

free vortex surface 自由涡面

free-air anomaly 自由空气异常值

free-air correction 自由空间校正

free-air overpressure (爆炸冲击波引起的)自由空气超压

free-air reduction 自由空气衰减

free-air temperature 外部气温,大气气温

free-air tunnel 自由大气风洞(模型放在飞机或滑轨车上,使模型运动,进行气动试验)

free-balloon net 自由气球吊篮承载网(用来把吊篮载荷分配到球囊上部表面去)

FREEBD Freeboard 超(出水)高,余幅,净空

freedom to mix 自由调配权(在符合协议限制的条件下自由调配核武器库组成成分的权利)

freedoms 自由通航协定(特指两个国家之间经谈判允许相互在对方领空自由航行的协定,包括过境权、停留权等)

free-expansion 自由膨胀

free-fall 自由投放

free-fall altimeter 跳伞高度表

free-fall descent 自由坠落

free-fall model 自由落体(试验)模型

free-fall testing 自由落体试验

free-fall(ing) bomb 自由下落炸弹

free-firing armament 活动射击武器

free-flight projectile range 自由飞行导弹试验靶场,弹道导弹试验靶场

free-flight trajectory 自由飞行弹道,弹道被动段

free-flow 自由流动,自重流动

freeflyer 自由飞行器

free-flying robot 自由飞行机器人

free-flying system 自由飞行系统

free-free state 自由-自由状态

free-free transition 自由-自由跃迁

free-gyro 自由陀螺仪

freejet 自由喷流

freelance "游猎"(美国国防部规定的代语,意为已使用自己控制的空中截击)

free-lance fighter 游猎(歼击)机

freely-released weapon 自由投放武器

free-molecule temperature 自由分子流温度

free-return trajectory (失效航天器的)自由返回轨道

free-running blocking oscillator 自激间歇振荡器

free-standing propellant (与壳体不黏合的固体火箭)独立药柱

free-stream value 自由流值,未扰动流值

freestream velocity 来流速度

free-surface 自由表面

free-vortex 自由涡流,自由涡,自由旋涡

free-vortex flow 自由旋流,自由涡流

free-wake 自由尾迹

freewheeling 自转(旋翼系统与发动机断开时旋翼的自动转动)

FREQ Frequency 频率

FREQ CHG Frequency Changer 变频器

FREQ CONV Frequency Converter 变频器

FREQM Frequency Meter 测频仪

frequency division multiplexing 频分多路传输

frequency division multiplexing telemetry 频分制多路遥测

frequency division telemetry 频分遥测

frequency domain 频域

frequency domain model reduction method 频域模型降阶法
frequency drift 频率漂移
frequency error 频率误差
frequency histogram 频数直方图
frequency hopping 跳频
frequency hopping spread spectrum 跳频扩频
frequency identification unit 波长表
frequency interval 频程
frequency inversion 频带倒置
frequency jitter 频率抖动
frequency measurement 测频,频率测定
frequency memory 频率记忆
frequency meter 频率计
frequency modulated radar 调频雷达
frequency modulated transmitter 调频发射机
frequency modulation radio fuze 调频无线电引信
frequency modulation ranging fuze 调频测距引信
frequency modulation recording 调频记录
frequency modulation sideband fuze 调频边带引信
frequency swing 频率摆动
frequency synchronization 频率同步
frequency synthesis technique 频率合成技术
frequency synthesizer 频率合成器
frequency test 频数检验
frequency tolerance 频率容差
frequency tone 频率音调
frequency tracker 频率跟踪系统
frequency translation 频率变换
frequency up-conversion 频率上转换
frequency value 频值
frequency-agile 频率捷变
frequency-agile magnetron 捷变频磁控管
frequency-agile radar 频率捷变雷达
frequency-modulated fuze 调频引信
frequency-modulating laser 调频激光器
frequency-scan array 频率扫描天线阵
frequency-scan radar 频扫雷达
frequency-sharing 频分割
frequency-wavenumber filtering 频率-波数滤波
frequency-wavenumber migration 频率-波数偏移
fresh charge 新鲜充量
Fresnel contour 菲涅耳等值线
Fresnel lens （航空母舰上）菲涅耳透镜
Fresnel number 菲涅耳数
FRF Flight Readiness Firing （火箭）飞行前点火试验，（发动机）飞行前检查起动
FRFI Fuel-Related Fare Increase 燃料涨价引起的机票涨价

FRG Federal Republic of Germany 联邦德国
FRH Flap-Retraction Height 收襟翼高度
FRI First Record Issue 首次记录发放
FRICT Friction 摩擦
friction coefficient 摩擦系数
friction drag 摩擦阻力
friction factor 摩擦因数
friction force 摩擦力
friction fuze 摩擦引信
friction horsepower 摩擦马力(指示马力减去制动马力)
friction influence 摩擦影响
friction lock 摩擦制动器,摩擦闸
friction loss 摩擦损失
friction performance 摩擦性能
friction power 摩擦功率
friction sensitivity 摩擦感度
friction tachometer 摩擦式转速表
friction wake 摩擦拌流
friction welding 摩擦焊
friction(al) error 摩擦误差
frictional force 摩擦力
frictional loss 摩擦损耗,摩擦损失
frictional power 摩擦功率
frictional torque 摩擦力矩
frictionless simulator 无摩擦模拟台
Friedmann cosmological model 弗里德曼宇宙模型
friendly force 友军
FRIG Floated Rate-Integrated Gyro 悬浮速率积分陀螺
frigate flight 护卫舰飞行
frigid zone 寒带
FRM/FIM Fault Reporting/Isolation Manual 故障报告/隔离手册
FRMR Former 早先的,前任的,样板
front course sector （着陆航向信标台的）前航道扇区
front ignition （固体火箭发动机）前端点火
front line squadron 前线中队
front mask 车头盖
front side 进料侧,正视图
front station 前置站
front telemetry station 前置遥测站
front track guide （转膛炮进弹机的）前输弹导
front unlocking cam （航炮）前开锁块
front view 1.前视图,正视图 2.对景图
frontal approach （从）前半球接敌,（从）前半球进入攻击
frontal area 迎面面积
frontal attack 迎面攻击,前向攻击(截击机与目标的航向差大于 135°)
frontal weather 锋面天气(锋面可能产生云、雨、温度突

变等现象)
front-end cost 前期成本
front-line airfield 前线机场
frontogenesis 锋生(作用),锋生过程
frontolysis 锋消(作用)
FROST Food Reserves On Space Trip 航天食物储备
frozen flow 冻结流动
frozen orbit 冻结轨道(地球资源卫星常用的一种轨道)
frozen-in field 冻结(流)场,零电阻介质中的磁流体场
FRP 1. Federal Radio-navigation Plan 联邦无线电导航计划 2. Fibre Reinforced Plastics 纤维加强塑料 3. Flap Reference Plane 襟翼基准平面 4. Flight Refuelling Probe 空中受油探管
FRPA Fixed Reception Pattern Antenna 固定式方位接收天线
FRR 1. Failure and Rejection Report 故障和拒收报告 2. Flight Readiness Review 待飞状态评审,飞行准备状态评审
FRR/SR Failure and Rejection Report/Ship's Record 故障拒收/架次记录
FRS 1. Failure Report System 故障报告系统 2. Fan Rotation Speed 风扇转速 3. Ferret Reconnaissance Satellite 无线电侦察卫星 4. Fighter, Reconnaissance, Strike 战斗机,侦察机,对地攻击机 5. Fleet Reliability Summary 机群可靠性总结 6. Functional Requirements Summary 功能要(需)求总览
FRSI Flexible Reusable Silica Insulation 柔性可重复使用二氧化硅绝热层
FRTV Forward Repair and Test Vehicle 前方修理测试车
FRU Field Replaceable Unit 现场可换部件
FRUSA Flexible Rolled-Up Solar Array 柔性折叠式太阳电池阵
FrW Friction Welding 摩擦熔焊
FRWK Framework 骨架,框架
FS 1. Factor of Safety 安全系数 2. Fail Safe 破损安全的,故障自动防护 3. Fail Safety 故障自动防护,破损安全(性),可靠性 4. Faying Surface 接合面 5. Field Service 机场服务 6. Fighter Squadron 战斗(歼击)机中队 7. Flap Setting 襟翼位置设定 8. Flight Simulator 飞行模拟器 9. Flight Spoiler 飞行扰流板 10. Flight Surgeon 航空医生 11. Float Switch 浮子开关 12. Frame Station 隔框站位(安装位置) 13. Front Spar 前梁 14. Full Size 全尺寸 15. Fuselage Station 机身站位,机身测量点
FSA 1. Final Squint Angle 最终瞄准偏斜角(雷达波束基准方向与最大辐射方向的细小差别) 2. Flight Safety Analyses 飞行安全性分析 3. Force Structure Aircraft 主战飞机,主力飞机 4. Fuel Saving Advisory 节油咨询
FSAA Flight Simulator for Advanced Aircraft 高级飞机飞行模拟器(美国航空航天局)
FSAGA First Sortie After Ground Alert 地面待战首次出动
FSAS 1. Flight Service Automation System 飞行服务自动系统 2. Fuel Savings Advisory System 节油咨询系统
FSAT 1. Ford Satellite 福特卫星(美国通信卫星) 2. Full Scale Aerial Target 全尺寸空中靶标
FSB 1. Fan Stream Burning 风扇涵道气流燃烧 2. Fasten Seat Belts 系好座椅安全带 3. Field Service Bulletin 外场服务通告 4. Fleet Satellite Broadcast 舰队卫星广播 5. Forward Support Battalion 前方保障营(保障旅一级空降作战的战斗勤务保障队)
FSC 1. Fault Simulation Comparator 故障模拟比较器 2. Fault Symptom Code 故障征兆码 3. Flight Standard Committee 飞行标准委员会 4. Force Sensing Controller 感力控制器,力传感器 5. Fullscale 全刻度,全比例
FSD 1. Fabrication Start Date 制造开始日期 2. Full Scale Deflection 满刻度偏转 3. Full Scale Development 全尺寸研制(阶段),全规模(全面)研制(阶段)
FSDP Final Safety Data Package 最终安全数据包
FSDPS Flight Service Data Processing System 飞行服务数据处理系统(美国联邦航空局)
FSE 1. Field Service Engineer 外场服务工程师 2. Field Support Equipment 外场支援设备 3. Flight Support Equipment 飞行保障设备
FSED Full Scale Engineering Development 全尺寸(全规模、全面)工程研制(阶段)
FSEO Flight System Engineering Order 飞行系统工程指令
FSEU 1. Flap Slat Electronics Unit 襟缝翼电子装置 2. Flap System Electronic Unit 襟翼系统电子装置
FSF Flight Safety Foundation 飞行安全基本原则
FSFT Full Scale Fatigue Test 全尺寸疲劳试验,整机疲劳试验
FSG Fluid Sphere Gyro 液浮陀螺
FSI 1. Fault Symptom Index 故障征兆(码)索引 2. Field Service Instruction 外场维护细则 3. Free Space Isolator 自由空间隔离器 4. Full Screen Interface 全屏幕接口
FSII Fuel System Icing Inhibitor 燃料系统防冻剂,燃料系统结冰抑制剂
FSK Frequency Shift Keying 频移键控
FSL 1. Flight Simulation Laboratory 飞行模拟实验室 2. Full Stop Landing 着陆至停放
FSLP First Space Lab Payload 首批空间实验室有效

载荷

FSM 1. File Server Module 文件服务器模块，文件盘模块 2. Forward Scatter Meter 前向散射计（仪）3. Frequency Shift Modulation 移频调制

FSN Federal Stock Number 联邦材料标号

FSOV Fuel Shut Off Valve 燃料关断活门，停车开关

FSP 1. Fragment Simulator Projectiles 碎片模拟弹（穿甲试验）2. Full Screen Processing 全屏幕处理

FSPUPT Flight Screening Program and Undergraduate Pilot Training 飞行甄选计划及未取得执照的飞行员训练

FSRB Flight Safety Review Board 飞行安全检查局（委员会）

FSS 1. Fixed Service Structure 固定服务结构（塔架）2. Flap Speed Schedule 襟翼速度计划 3. Flight Service Station 飞行服务站 4. Flight Service Station 飞行保障勤务站 5. Flight Standards Service 飞行标准处（美国联邦航空局）6. Flight Support Station 飞行支援站（美国航天飞机上用于卫星维修等任务的停靠和操作平台）7. Flight Support System 飞行支援系统 8. Flying Selection Squadron （美国空军）飞行选拔中队 9. Flying Spot Scanner 飞点扫描仪 10. Freedom Space Station 自由号空间站（美国）11. Frequency Selective Surface 频率可选面 12.（Satellite）Field Services Station 卫星服务站

FSSE Forward Service Support Element 前方勤务保障分队（营规模空降突击部队的战斗勤务保障组织）

FSSP Forward Scattering Spectrometer Probe 前向散射粒谱仪探头

FSSR Federal Safety Standard Regulation 联邦安全性标准条例（美国）

FST 1. Fire Safety Technology 点火安全技术 2. Flight Simulation Test 飞行模拟试验 3. Forged Steel 锻钢 4. Full Scale Tunnel 全尺寸风洞，实物风洞

FSTNR Fastener 紧固件

FSTS Future Space Transportation System 未来空间运输系统

FSV Flow Sensor Ventura 流量传感器细缩管

FSVS Flight Software Verification System 飞行软件校验系统

FSW 1. Flight Soft Ware 飞行软件 2. Forward Swept Wing 前掠翼

FSWD Full Scale Weapons Delivery 实际尺寸（的）武器投射，全尺寸（的）武器投射

FT 1. Fabrication Tool 零件制造工艺装备 2. Fast Track 高速线路 3. Fault Tolerant 故障容限的，容错的 4. Feature Technology 特征技术 5. Flight Termination 飞行终止 6. Flight Test 试飞，飞行试验 7. Flight Training 飞行训练 8. Flight Time 飞行时间 9. Full Throttle 全油门，最大油门 10. Functional Test 功能试验

FTA 1. Fatigue Test Article 疲劳试验试件 2. Fault Tree Analysis 故障分析树 3. Flow Time Analyzer 流程时间分析员

FTAE First Time Around Echo 第一周期扫描回波

FTAJ Frequency/Time Ambiguity Jamming 频率/时间模糊干扰

FTAM File Transferand Access Method 【计】文件传送与存取方法

FTB Flying Test Bed 飞行试验台

FTBO Flow Time Between Operation 停工等待时间

FTC 1. Fast Time Control 快时间控制 2. Fault Tolerant Computer 容错计算机 3. Flight Test Configuration 试飞构型 4. Fast Time Constant 短时间常数 5. Federal Trade Commission 联邦贸易委员会（美国）

FTCA Future Tactical Combat Aircraft 未来战术战斗飞机

FTCR Flight Test Certification Report 试飞合格审定报告

FTD 1. Field Training Detachment 现场训练支队（美国空军）2. Fitted 装配的 3. Flight Training Device 飞行培训设备 4. Functional Test Documents 功能试验文件

FTE 1. Factory Test Equipment 工厂测试设备 2. Flap Trailing Edge 襟翼后缘 3. Flight Technical Error 飞行技术差错，技术性飞行误差 4. Flight Test Engineer 飞行试验工程师 5. Flight Test Engineering 试飞工程

FTEP Full Time Echo Protection 全时回波保护

FTF 1. Flared Tube Fitting 闪光管装配 2. Functional Test Flight 功能试验飞行

FTG 1. False Target Generator 假目标发生器 2. Floated Type Gyro 液浮陀螺 3. Footing 基础，底座

FTH Full Throttle Horsepower 油门全开功率，（发动机）最大功率（马力）

FTHRD Female Thread 内螺纹

FTI 1. Fixed Target Imagery 固定目标成像 2. Fixed Target Indicator 固定目标指示器 3. Fixed Time Interval 固定时间间隔 4. Flight Test Inventory 试飞前清点

FTIR Flight Test Instrumentation Requirements 飞行试验仪表要求

FTIS Flight Test Instrumentation System 飞行试验仪表系统

FTIT Fan Turbine Inlet Temperature 风扇涡轮进口温度

FTK 1. Freight Tonne Kilometre(s) 货运吨千米 2. Fuel Tank 燃油箱

FtL　Foot Lambert 英尺朗伯(亮度单位,1英尺朗伯＝1流明/平方英尺)

FTLB　Flight Time Limitation Board 飞行时间限制委员会(英国)

FTLO　1. Fast Tuned Local Oscillator 快速调谐本机振荡器 2. False Target Lock On 假目标截获

FTM　1. Flexible Theater Missile 活动发射装置发射的战区导弹,活动式战区导弹 2. Flight Test Manual 飞行试验手册 3. Flight Test Mission 飞行试验任务 4. Frequency/Time Modulation 频率/时间调制

FTMR　Flashweld Test Material Request 闪光焊试验材料申请

FTMS　1. Fabrication Tracking and Management System 制造跟踪/管理系统 2. Fluid Transfer Management System 流体输送管理系统

FTO　Flexible Take Off 弹性起飞

FTOI　Flight Test Operation Instruction 飞行试验操作说明书

FTP　1. File Transfer Protocol 文件传输协议,文档传递协定书 2. Flight Test Procedure 飞行试验程序 3. Fly To Point 飞至目的地(的) 4. Fuel Transfer Pump 燃料输送泵 5. Full Throttle Position 节流阀全开位置 6. Functional Test Procedures 功能试验程序

Ftr　Fighter 战斗机,歼击机

FTS　1. Federal Telecommunications System 联邦电信系统(美国) 2. Flexible Turret System 活动炮塔系统 3. Flight Telerobot Servicer 飞行遥控机器人服务器 4. Flight Test Specification 试飞规范 5. Flying Training School 飞行训练学校 6. Fourier Transform Spectrometer 傅里叶变换分光光度计 7. Free Time System 自由时间系统

FTSA　Fault Tolerant System Architecture 容错系统结构

FTTS　Fabrication Target Time System 制造目标时间系统

FTV　1. Flight Test Vehicle 飞行试验飞行器 2. Functional Test Vehicle 功能试验飞行器

FTWS　Flight Test Work Sheet 试飞工作单

FTX　Field Training Exercise 野战训练演习

FTZ　Foreign Trade Zone 对外贸易区

FU　1. Fan Unit 通风机,风扇组件 2. Fuel Uplifted 已装上的燃油,加油量 3. Fuel Used 已用去的燃油,耗油量

fuel air explosive　燃料空气炸药,液体燃料炸药,云爆装药

fuel air ratio　燃料空气比

fuel atomizer　燃油雾化喷嘴

fuel battery　燃料电池

fuel blending facility　燃油掺合设备(用来按规定标准在燃油中掺入添加剂)

fuel bypass　燃油旁路

fuel capacity　载油量

fuel cell　燃料电池

fuel cell electrode　燃料电池电极

fuel cell performance　燃料电池的性能

fuel cell powerplant　燃料电池动力装置

fuel cell system　燃料电池系统

fuel cell water　燃料电池水

fuel concentration　燃料浓度,燃料浓缩

fuel concentration distribution　燃油浓度分布

fuel consumption　燃油消耗量,耗油量

fuel content transmitter　燃料容量传感器

fuel control unit　燃油控制单元

fuel cut off　断油装置,燃油关断装置,停止供油

fuel delivery　燃油供给

fuel detonation suppressant system　防爆系统

fuel dipper　燃料调整器

fuel dispersion　可燃物分散性

fuel distribution　燃料分布,燃料分配

fuel endurance　燃料用尽的续航时间,受油量限制的续航时间

fuel energy　燃料能源,燃料能量

fuel equivalence ratio　燃料当量比

fuel filling port support mount　燃烧剂加注口支架

fuel film　燃油膜

fuel filter　燃烧剂过滤器

fuel flow　燃油流量

fuel flow meter　1. 燃油流量表 2. 燃油流量计

fuel flow rate　燃烧剂流量

fuel flow totalizer　燃料流量加法器

fuel flow（meter）transmitter　燃料流量(表)传感器

fuel flow-rate equipment　燃料流率测量设备

fuel jettison　应急放油(在紧急状态下将飞机上的燃料尽快放出)

fuel lag　燃料供给滞后,推进剂供给滞后

fuel layer　燃料层

fuel level transmitter　燃料量传感器,油量表传感器

fuel line　燃油管,燃油输送管路,燃料管路,燃料导管

fuel load　载油量

fuel loading connector　燃烧剂加注连接器

fuel loading flexible hose　燃烧剂加注软管

fuel loading system　燃烧剂加注系统

fuel manifold　燃油总管,燃油歧管(多支管),输油圈

fuel mass at given height　规定高度的燃料质量

fuel measuring equipment　燃料测量设备

fuel mixture　燃料混合物,可燃混合气

fuel monitoring system　燃油监控系统

fuel nozzle　燃料喷嘴

fuel oil　燃油

fuel outlet　放油口
fuel overflow connector　燃烧剂溢出连接器
fuel overflow flexible hose　燃烧剂溢出软管
fuel penalty　燃料附加损耗(有限点火和脉冲点火的燃料消耗量之差)
fuel performance　燃料性质
fuel photoelectric sensor　燃烧剂光电传感器
fuel port　燃料输入孔
fuel preburner　燃料预燃室
fuel pressure switch　燃油压力开关
fuel pressure transmitter　燃料压力传感器
fuel property　燃料特性
fuel pump　燃油泵
fuel quantity gauge (indicator)　(燃)油量表
fuel quantity measurement system　(燃)油量测量系统
fuel quantity meter　燃油油量表
fuel rate　燃料(每秒)消耗量
fuel reactor　浓化燃料反应器
fuel reduction　燃料减少
fuel regression rate　燃面退移速率
fuel residence time　燃料滞留堆芯时间
fuel sample　燃油样品
fuel saving　节约燃料
fuel section　燃料段
fuel shutoff terminals　燃料关断活门接头,防火开关接头
fuel shut-off/fuel cut-off　燃料关断装置
fuel sink　剩油热容,燃油热汇(作为机上剩余能量的燃料总热量,以焦耳计)
fuel slab　燃料板
fuel slosh disturbance torque　燃料晃动干扰力矩(航天器机动时燃料晃动对星体形成的干扰力矩)
fuel sloshing　燃料晃荡
fuel spray　燃料雾化
fuel storage　燃料储藏
fuel stream　燃料流
fuel sulfur content　燃油含硫量
fuel supply system　燃料供给系统
fuel surface　燃料表面
fuel swirler　离心式燃料喷射器
fuel system　燃油系统
fuel tank　1.燃烧剂箱 2.油箱
fuel tape　燃料带
fuel test equipment　燃料试验设备
fuel-air explosive bomb　油气炸弹,燃料空气弹
fuel-air mixture analyser　油气比分析仪,混合比分析仪
FUL　Fulcrum 支点,转轴
full annealing　完全退火
full arctic clothing　全套防寒服

full authority control　全权限控制
full authority flight control　全权限飞行控制
Full authority fly-by-wire flight control system　全权限电传飞行操纵系统
full auto-bonding system　全自动焊接系统
full autolanding　完全自动着陆(接地以前全部自动操纵)
full automatic processing　全自动处理
full cabin air flow　座舱空气总流量
full hands-on stick and throttle facility　手不离油门和驾驶杆的操纵设备
full illumination　全部照射
full life test　全寿命试车
full lift　卷扬全高
full load current　满载电流
full matrix　全矩阵
full mission capable　可以执行全部任务的(指飞机的状态),全任务的
full needle width turn　(转弯侧滑仪指针)一针宽转弯(窄针表示作 360°转弯需 4 分钟,宽针表示作 360°转弯需 2 分钟)
full network synchronization　全网同步
full operational capability　全部作战能力
full order observer　全阶观测器
full performance level　全性能级(美国联邦航空局给空中交通管制员定的最高等级)
full pressure suit　全身加压服,高空密闭服
full range altimeter　大量程高度表
full range test　全程(飞行)试验
full scale wind tunnel　全尺寸风洞
full screen interface　全屏幕接口
full screen processing　全屏幕处理
full service round　全装药炮弹
full shroud　全齿高加强板齿轮
full stock　全部储备
full thrust　最大推力
full time echo protection　全时回波保护
full up air craft　全部装配好的飞机(区别于买散件后自己装配的飞机)
full-mission simulator　全任务飞行模拟器
full-mission space flight simulator　全任务航天训练仿真器
full-mission-capable rate　全任务出勤率
full-order system　完备的系统
full-potential equation　全速势方程
full-resolution picture　全精度影(图)像,高分辨率影(图)像
full-scale development　全尺寸(全面、全规模)研制
full-scale space flight　真实飞行器(航天)飞行

full-scale wind-tunnel data	全尺寸（模型）风洞试验数据
full-system	全系统的
full-thrust duration	最大推力工作时间，全推力持续时间
full-wave theory	全波理论
fully active magnetic bearing	全主动磁轴承
fully allocated costs	全部分配成本
fully articulated rotor	全铰接式旋翼
fully automatic compiling technique	全自动编译技术
fully automatic doppler hover	全自动多普勒悬停
fully automatic ejection seat	全自动弹射座椅
fully automatic network	全自动网络
fully buffered channel	全缓冲通道
fully connected primary satellite system	全边主卫星系统
fully connected single primary satellite system	全边单用主卫星系统
fully developed flow	完全发展流（黏性流体流过固体表面附面层已充分发展其厚度、流速分布等不再变化）
fully distributed costs	完全分散式成本，全分配费用
fully expanded nozzle	完全膨胀喷管
fully-active homing	全主动寻的，全自动引导
fully-charge condition	全充电态
fulminating altitude hypoxia	暴发性高空缺氧
fulminating anoxia	暴发性缺氧症
fulminating powder	起爆药
function of astronaut location	航天员定位功能
function of control	调控职能
function of engine	发动机工作性能
functional agent	功能助剂
functional area	功能区，职能范围
functional ceramics	功能陶瓷
functional chain	功能链
functional composite material	功能复合材料
functional condition code	弹药性能情况代号
functional damage	功能损坏
functional gradient material	功能梯度材料
functional group	【化】官能团，功能团（化合物中参加化学反应的活性基团）
functional kill	功能杀伤，软杀伤
functional material	功能材料
functionally distributed computer system	功能分布计算机系统
function-distributed system	功能分布系统
function-graded material	梯度功能材料
functioning delay	起爆延迟时间，起爆延迟距离（由引信起爆算起，到弹头爆炸为止）
functioning efficiency	（生物战）作用效率（散布的生物战剂实际起作用的百分比）
function-permitting maintenance	保持功能的维修
function-preventing maintenance	妨碍功能的维修
fungus proof	防菌
fungus test	霉菌试验
fungus-resistant	抗霉的，防霉的
funicular polygon	索(状)多边形
funnel	1.安全进近区，漏斗区（在仪表着陆系统下滑道上，上下各 0.5°，左右各 2°，在此区内，飞机可以着陆）2.第五边（1950 年以前用法，尤指有灯光或引导系统时）
FUPRO	Future Production 未来生产
FUR	Furnace 炉子
FURN	Furnish 提供
furnishing	客舱设备（包括座椅、过道、行李架等）
furnishing mock-up	有客舱设备的全尺寸模型
FUS	Fuselage 机身
fuse	保险丝
fuselage datum	机身水平基准线，机身纵轴
fuselage drag	机身阻力
fuselage engine	安装在机身上的发动机
fuselage fairing	机身整形片，机身整流装置
fuselage fineness ratio	机射长细比
fuselage length	机身长度
fuselage model	机身模型
fuselage reference line	机身基准线
fuselage skin	机身蒙皮
fuselage structure	机翼
fusible alloy	易熔合金
fusible pattern molding	熔模铸造
fusible plug	易熔塞
fusion	1.聚变 2.合成
fusion (-type) weapon	热核武器
fuze actuation zone	引信启动区
fuze adapter	引信连接螺套，引信套筒
fuze alarm	保险丝烧断信号，熔线报警
fuze antenna	引信天线
fuze antenna pattern	引信天线方向图
fuze beam incidence	引信方向图倾角
fuze calibration	（高炮）引信校准
fuze cap	引信帽，信管帽
fuze captive carrying test	引信挂飞试验
fuze cavity	引信口(室)，信管口(室)
fuze combinations	复式引信，组合引信
fuze counter jamming	引信抗干扰性
fuze dead zone	引信盲区
fuze delay	延期引信
fuze delay time	引信延迟时间
fuze desensitizing cap	引信钝感帽
fuze drop test	引信跌落试验
fuze dynamic simulation	引信动态仿真

fuze electromagnetic full scale dynamic simulation 引信电磁全尺寸动态仿真
fuze electromagnetic scaled dynamic simulation 引信电磁缩比动态仿真
fuze factor 引信修正系数
fuze firing circuit 引信执行电路
fuze function probability 引信功能概率,引信起爆概率
fuze function range 引信作用距离
fuze hydro-acoustic simulation 引信水声仿真
fuze instantaneity 引信瞬发度
fuze matching error 引信匹配误差
fuze mathematical simulation 引信数学仿真
fuze physical simulation 引信物理仿真
fuze power supply 引信电源
fuze prediction time 引信提前时间
fuze preset 引信预置(预先设置引信的工作方式)
fuze quasi-dynamic simulation 引信准动态仿真
fuze radar range equation 引信雷达方程
fuze range 引信作用距离
fuze reaction zone 引信反应区
fuze routine test 引信例行试验
fuze running time 引信作用时间
fuze safety test 引信安全性试验
fuze sensitivity 引信灵敏度
fuze sensitivity test 引信灵敏度测试
fuze setting 引信装定
fuze shell 已装上引信的炮弹
fuze shock test 引信冲击试验
fuze simulation 引信仿真
fuze static simulation 引信静态仿真
fuze strike test 引信敲击试验
fuze target discrimination 引信对目标的分辨力
fuze time-delay characteristic test 引信延时性能试验
fuze timer 引信定时器
fuze unit test 引信单元测试
fuze vibration test 引信振动试验
fuzed bomb 定时炸弹
fuze-warhead coordination 引战协调性
fuze-warhead matching 1.引信与战斗部配合 2.引战配合
fuzing accuracy 引信起爆精度
fuzing control system 引爆控制系统
fuzing system 引爆系统
FV 1.Final Value 终值 2.Float Valve 浮子活门 3.Front View 正视图
FVD Fluorescent Vacuum Display 荧光真空显示器
FW 1.Failure Warning 失效警告 2.Fire Wall 防火墙
FWC 1.Factory Work Code 工厂工作代码 2.Filament-Wound Case 丝绕发动机壳体 3.Filament Wound Composites 纤维缠绕复合材料 4.Filament Wound Cylinder 纤维缠绕气缸 5.Flight Warning Computer 飞行告警计算机
FWD Forward 前面的,向前
FWG Following 下列的,接着的
FWIC Fighter Weapons Instructor Course 歼击(战斗)机武器教官课程
FWO 1.Fabrication Work Order 制造工作指令 2.Flight Work Order 飞行工作指令
FWOC Forward Wing Operation Centre 前线联队作战中心(英国空军)
FWO-SR Flight Work Order-Ships Record 飞行工作指令-架次记录
FWR Full Wave Rectifier 全波整流器
FWS 1.Failure Warning System 故障警告系统 2.Fighter Weapons School 歼击(战斗)机武器学校(美国空军) 3.Filter Wedge Spectrometer 楔形滤光片光谱仪 4.Flight Warning System 飞行警告系统
FXD Fixed 安装的,固定的
FXP Fixed Point 固定点

G

G 1. Acceleration Due To Gravity 重力加速度（海平面为 9.806 65 米/秒）2. Galley 厨房 3. Gate 整流栅，控制极 4. Glider 滑翔机 5. Unit of Gravitational Acceleration 重力加速度组件 6. Guard 保护，挡板，保护罩

G Acceleration Gravity Acceleration 重力加速度

G.H.A. Greenwich Hour Angle 格林尼治时角

G/A Ground-to-Air 地对空

G/BMI Graphite Bismaleimide Composite 石墨双马来酰亚胺复合材料

G/E Ground Engineer 地面工程师，地面机械师

G/S 1. Glide Slope 下滑道 2. Ground Slope 下滑坡度 3. Groundspeed 地速

G/VLLD Ground/Vehicular Laser Locator Designator 地面和飞行器激光定位（照射）器

GA 1. General Aviation 通用航空 2. Gauge 仪表，量规 3. Generic Aircraft 新增等级飞机 4. Glide Angle 下滑角 5. Go Around 复飞 6. General Aircraft Corporation 通用航空器公司，通用飞机公司

GAA General Aviation Association 通用航空协会（美国）

GAACC General Aviation Airworthiness Consultative Committee 通用航空适航性顾问委员会（英国）

GaAlAs Gallium Aluminium Arsenide 镓铝砷（三元化合物半导体）

GaAs Galnum Arsenide 砷化镓

GAC 1. General Automatic Control 全自动操纵（飞行操纵系统的）2. Global Area Coverage 全球覆盖 3. Go-Around Computer 复飞计算机 4. Gust Alleviation Control 阵风衰减控制

GACC 1. Ground-Attack Control Capability 对地攻击控制能力 2. Ground Attack Control Center 地面攻击控制中心 3. Guidance Alignment and Checkout Console 制导校准和测试控制台

GACW Gust Above Constant Wind 大于定常风的阵风

GAD General Assembly Drawing 总装图

GADL Ground-to-Air Data Link 地空数据链

GADO General Aviation District Office 通用航空区办公室（美国联邦航空局）

GAEO General Assembly Engineering Order 通用装配工程指令

gage 1. 压力计，真空计 2. 标准度量，计量器，量规 3. 精确计量

gain 1. 放大 2. 增益 3. 获得

gain adjustment 1. 增益控制 2. 增益调节，增益调整

gain coefficient of war head 战斗部增益系数

gain control 增益控制

gain factor 1. 增益系数 2. 增益因数，增益（因子），放大系数

gain margin 1. 增益裕量 2. 增益边际 3. 增益容限

gain matrix 增益矩阵

gain pattern 增益方向图

gain reduction 增益衰减

gain saturation 增益饱和

gain scheduling 1. 调参 2. 增益调度 3. 程序调参

gain set 增音机

gain slope 增益斜率

gain vector 增益，向量

gain-phase 增益相位

gain-phase diagram 增益相位图

gain-phase plane 增益相位平面

gain-schedule 1. 增益调度 2. 变增益 3. 增益策略 4. 增益匹配

gain-scheduled controller 1. 增益调度控制器 2. 增益调度调节器

gain-scheduling 增益调度

gain-scheduling approach 增益调度方法

gait analysis system 步态分析系统

Gal 1. Gallon 加仑 2. GEMINI AIRLINES LTD. 双子星航空公司

gal 1. Gallon 加仑 2. 伽

galactic anticenter 反银心方向

galactic bulge 银河系核球

galactic center 银河中心，银心

galactic concentration 银河系密度，银聚度

galactic coordinate system 银道坐标系

galactic corona 银河晕

galactic cosmic ray 银河系宇宙射线

galactic disk 银河圆盘，银河盘

galactic dynamics 星系动力学

galactic equator 银河赤道，银道

galactic halo 银河晕圈

galactic kinematics 星系运动学

galactic latitude 银纬，银河纬度

galactic longitude 银经

galactic nebula 银河星云，银河内星云

galactic noise 银河噪声
galactic nova 银河新星
galactic nucleus 银河核心,星系核
galactic plane 银河面,银道面
galactic pole 银河极,银极
galactic radio spur 银河射电支
galactic space 星系空间
galactic structure 星系结构,银河系结构
galactic supernova 银河超新星
Galactic System 银河系
galacticrotation 银河旋转,银河系自转
Galaxy 银河系
galaxy 星系
galaxy clustering 星系成团
galaxy count 星系计数
galaxy evolution 星系演化
galaxy formation 银河系形成
gale warning 大风警报,强风警报
Galilean satellites 伽里略卫星(木星的四大卫星)
Galilean telescope 伽利略望远镜
Galileo 伽利略
galling 金属磨损,黏结,使烦恼的,难堪的
gallium 镓铝砷
gallium arsenide 砷化镓
gallium arsenide solar cell 砷化镓太阳电池
gallium 【化】镓(化学元素,符号 Ga)
gallon GL 加仑(容量单位)
gallop 1.迅速运输 2.不正常运转,发动机不正常运转,运转不稳定,飞奔
GALV Galvanize(d)镀锌
galvanic cell 1.原电池 2.伽伐尼电池
galvanic corrosion 原电池腐蚀
galvanic skin response 皮肤电反应
galvanizing 镀锌
galvanolysis 电解
galvanometer 1.又称"灵敏电流计" 2.检疗 3.电流计,检流计
GAM 1. General Aeronautical Material 通用航空材料 2. Graphic Access Method 图形存取法 3. Ground-to-Air Missile 地对空导弹 4. Guided Aircraft Missile 机载导弹,航空导弹
GAMA General Aviation Manufacturer's Association 通用航空制造商协会(美国)
game 博弈
game model 博弈模型
game theory 博弈论
game tree 对策树
gaming simulation 博弈模拟,对策模拟
Gamma 伽玛

gamma 1.灰度系数 2.微克,伽马
gamma camera 伽马照相机
gamma distribution 伽马分布
gamma spectroscopy 伽马能谱学
gamma spectroscopy system 伽马能谱学系统
gamma-ray 伽马射线的
gamma-ray laser 伽马射线激光器
gamma-ray pulsar 伽马射线脉冲星
GAMP Global Atmospheric Measurement Program 全球大气测量计划
GAMTA General Aviation Manufacturers and Traders Association 通用航空制造商和销售商协会
gang 组,套
gang drill 排钻床
gang plank 1.跳板 2.登机梯 3.软梯
gantry 1.门式起重机,门字架 2.龙门架(起重机),塔架,雷达天线
gantry robot 龙门型机器人(亦作 Gantry Type Robot)
GANTT General Agreement on National Trade and Tariffs 关税及贸易总协定
ganymede 【天】木卫三
GAO 1. Gasper Air Outlet 乘客用空气出口 2. General Accounting Office 总审计局(美国国会) 3. Government Accounting Office 政府会计署(美国联邦)
gap 间隙,空隙
gap coding 1.间隙编码 2.中断编码
gap filler 裂缝填充物
gap filler radar 填隙雷达
gap region 间隙区域
gap size 间隙大小
gap spacing 间隙间距
gap width 1.缝隙宽度 2.空隙宽度
GAPA Ground-to-Air Pilotless Aircraft 地面起飞无人驾驶飞机
GAPAN Guild of Airpilots and Air Navigators 飞机驾驶员与领航员协会(英国)
gaping 豁开,张口
gapless connection 无缝连接
garbage 1.干扰,杂音,噪声 2.垃圾
garbling 窜扰
garboard strake (水上飞机)机身列板,龙骨翼板
GARD General Address Reading Device 通用地址读出装置
GARP Global Atmospheric Research Programme 环球大气研究计划
GARS Gyrocompassing Attitude Reference System 陀螺罗盘姿态基准系统
GARTEUR Group for Aeronautical Research and

Technology in Europe 欧洲航空研究技术组
GAS 1. Gasoline 汽油 2. Ground Analysis Station 地面分析站
gas 透气性,气体
gas bag 气袋,气囊
gas ballast vacuum pump 气镇真空泵
gas ballast valve 气镇阀
gas bearing 气体轴承
gas blast 气吹
gas bomb 毒气炸弹,气窜,气体旁路;气体钢瓶
gas bottle depot 气瓶库
gas bottle set 气瓶组
gas box 煤气柜
gas bubble 1. 气泡 2. 水下爆炸气泡
gas cap 气顶,气帽
gas carburizing 气体渗碳
gas chamber 1. 煤气室 2. 毒气室
gas chromatograph 1. 气相色谱 2. 气相色谱仪
gas chromatography 1. 气相色谱法 2. 气相色谱分析
gas combustion 气体燃烧
gas composition 1. 气体成分 2. 气体组成
gas compressibility 1. 气体压缩性 2. 气体压缩
gas compression 气压缩
gas compressor truck 气体压缩机车
gas constant 气体常数
gas cooler 1. 气体冷却器 2. 燃气降温器
gas cooling 1. 气冷 2. 气体冷却
gas cushion pressure 气枕压强
gas cylinder 1. 瓦斯筒 2. 气体前冲筒,燃气前冲筒
gas deflation assembly 排气组件
gas density 1. 气体密度 2. 气体浓度
gas detonation 1. 气体爆轰 2. 瓦斯传爆
gas diffusion 1. 气体扩散 2. 天然气扩散
gas dilution 气体稀释
gas discharge 气体放电
gas discharge tube 气体放电管
gas discharging radiation counter tube 气体放电辐射计数管
gas distribution board of booster 1. 助推器配气台 2. 氦气瓶库配气台
gas distribution room 配气间
gas distributor 气体分配器
gas dynamic laser 气动激光器
gas dynamics 气体动力学
gas escaping 漏气
gas exchange 气体交换
gas exchanger 气体交换器(指生物系统)
gas expansion 1. 气体膨胀驱动 2. 溶解气驱
gas explosion 气体爆炸

gas film 气膜
gas film cooling 气膜冷却
gas filter 1. 煤气过滤器输出管 2. 气体过滤器
gas flow 1. 燃气流,气流 2. 燃气流量,气体流量
gas flow detector 气流探测器
gas generating 1. 生气 2. 瓦斯发电
gas generation 1. 生气 2. 气体发生 3. 天然气发电 4. 天然气生成
gas generation rate 1. 产气率 2. 产气速率
gas generator 1. 燃气发生器 2. 燃气涡轮核心发动机
gas generator cycle 燃气发生器循环
gas heating 气体加热,煤气供暖
gas heating value 煤气热值
gas hood 通气罩(气艇上的通气口)
gas horsepower 1. 燃气马力 2. 燃气功率
gas hydraulic assembly 气动液压组件
gas hydrodynamic bearing 气浮流体动力轴承
gas hydrogen combustion pool 氢气燃烧池
gas hydrogen exhaust tower 氢气排放塔
gas ingestion 燃气回吞
gas injection 1. 注气 2. 气体喷射 3. 注天然气
gas inlet degree 进气度
gas interface 气体界面
gas ionization 气体电离
gas ionization chamber 气体电离室
gas ionization potential 气体电离电位
gas jet 煤气喷嘴,喷气
gas jet attitude control 喷气姿态控制
gas jet orientation 用燃气舵定向
gas laser 气体激光器
gas layer 气体层
gas leak inspection 1. 气密性检查 2. 气检
gas leak inspection of tank 贮箱气检
gas leak inspection of valve 活门气检
gas leak quantity 漏气量
gas leak rate 漏气率
gas main (飞艇上通往各气囊的)主充气管
gas mass 气体质量
gas measurement 1. 天然气计量 2. 气体检测 3. 燃气计量
gas measuring apparatus 气体流量表
gas metal arc welding 气体保护金属极电弧焊
gas misalignment 燃气喷流偏心度
gas molecule 气体分子
gas motor 气体发动机,高压燃气马达
gas munitions 化学弹药,毒气弹药
gas nozzle 1. 气体喷嘴 2. 排气喷管 3. 煤气喷嘴
gas oxygen exhaust tower 氧气排放塔
gas path 气路

gas phase 气相
gas plating 气相淀积
gas powered aircraft cannon 气推式航空机关炮
gas pressure 1.气体压力 2.气压 3.瓦斯压力
gas pressurization 1.气体加压 2.气体挤压(用于输送火箭推进剂)
gas product 气体产物
gas property 气体性质
gas pulse 1.气脉冲 2.燃气脉冲 3.气体脉冲
gas reaction 1.气相反应 2.气体反应 3.气反应
gas recirculation 1.气体再循环 2.烟气再循环 3.废气循环
gas region 气区
gas release 1.气体释放 2.瓦斯泄出 3.气喷
gas replacement 气体置换
gas rudder 燃气舵
gas sample 气体取样,气体样品
gas sampling 1.气体取样 2.气体采样 3.采气样
gas scrubbing 气体净化,气体洗漆,涤气过程
gas seal 气封
gas sensor 1.气(体)敏感器 2.气敏元件
gas side 1.燃气侧 2.烟侧
gas simulation 1.气体模拟 2.气体仿真 3.气模拟
gas source diffusion 气态源扩散
gas speed 1.气速 2.电场风速 3.气流速度
gas spring 1.气弹簧 2.气体弹簧 3.氮气弹簧
gas start system 气体起动系统
gas starter 1.压缩空气起动机 2.燃气起动机
gas state 1.气体状态 2.气态 3.气体的状态
gas stay time 燃气停留时间
gas storage 气体储存
gas stream 气流
gas supply assembly 供气组件
gas supply regulator 供气调节器
gas supply station 气源站
gas supply system 供气系统
gas suspension 气浮支承
gas system 1.煤气系统 2.气体系统 3.气系统
gas tank 气瓶,气体贮箱
gas temperature 气体温度
gas tester 气体测试仪
gas tetrode 充气四极管
gas tightness 不透气性
gas transfer system 气体传输系统
gas triode 气三极管,闸流管
gas trunk 排气涵道
gas tube 1.充气管,离子管 2.导气管
gas tungsten arc welding 钨极惰性气体保护焊
gas turbine 1.燃气轮机 2.燃气涡轮 3.燃气涡轮发动机
gas turbine combustor 燃气轮机燃烧室
gas turbine engine 燃气涡轮发动机
gas turbine installation 燃气轮机装置
gas turbo alternator 燃气涡轮交流发电机组
gas vane(jetvane) (火箭的)燃气舵
gas velocity 气流速度
gas vent 气体出口,排气管道
gas volume (气艇在海平面的)气体体积
gas welding 气焊
gas yield 1.产气率 2.气体产率
gas-bag net 气囊定位索网
gas-bag wiring 气囊承力线网
gas-bearing accelerometer 气浮加速度计
gas-bearing support 气浮轴承支架
gas-buffered heat pipe 气体缓冲热管
GAS/W Gas Weld 气焊
GASC 1.General Aviation Safety Committee 通用航空安全委员会 2.Ground Air Support Command 空中对地支援司令部(美国陆军航空兵)
gaschromatography 气体色谱法
gas-cooled reactor 气冷(反应)堆
gas-damped accelerometer 气体阻尼加速度计
gas-discharge 气体放电
gas-dust complex 气尘复合体
gasdynamics 气体动力学
gaseous acidosis 气性酸中毒
gaseous alkalosis 气性碱中毒
gaseous combustion 1.气体燃烧 2.气相燃烧
gaseous component 气相成分
gaseous detonation 气体爆炸
gaseous electronics 气体电子学
gaseous environment 气体环境
gaseous fuels 气体燃料,气态燃料
gaseous helium system 气氦系统
gaseous hydrogen 气态氢系统
gaseous jet 气体射流
gaseous methane 甲烷气体
gaseous mixture 1.气体混合物 2.混合气体
gaseous nebula 气体星云
gaseous nitrogen warm-up system 气氮加热系统
gaseous oxygen 1.气态氧 2.气氧 3.气相氧
gaseous oxygen bottle 氧气瓶
gaseous pollutant 气体污染物
gaseous pollution 气体污染
gaseous product 气体产物
gaseous propellant 1.气态推进剂 2.推进剂的气态组分
gaseous reaction 1.气相反应 2.气化反应 3.气态反应
gaseous rocket 气体推进剂火箭

gaseous state 气态

gaseous tube 离子管,充气管

GASES Gravity Anchored Space Experiment Satellite 重力稳定空间实验卫星

gas-filled rectifier tube 充气整流管

gas-filled surge arrester 充气电涌放电器

gas-filled tube 充气管

gas-filled voltage regulator tube 充气稳压管

gas-fired heater 燃气加热器

gas-free environment 无火药气体的环境

gas-generator 1.煤气发生炉 2.气体发生器 3.燃气发生器

gasification 1.气化 2.煤气化

gasifier 气化器,热气流发生器

Gasil General-Aviation Safety Information Leaflet 通用航空安全资料小册子(英国民航局)

gasket 密封垫,垫片

GASL General Applied Science Laboratories 通用应用科学实验室(美国航空航天局)

gasoline 汽油

gasoline bomb (凝固)汽油(炸)弹

gasoline proof grease 耐汽油润滑脂(不能被汽油溶解或变质的)

gas-operated 气体传动的,气动的,气退式(武器)

gas-operated weapon 气退式武器

GASP 1. Grand Accelerated Space Platform 大型加速空间平台 2. Graph Algorithm Software Package 图示算法软件包

gas-phase 1.气相 2.气相法 3.气相扩散

gas-phase reaction 1.气相反应 2.气相化学反应

gas-phase reaction mechanism 气相反应机理

gas-phasemass transfer coefficient 气相质量转移系数

gas-pressures to rage vessel 高压气瓶

GASR Guided Air-to-Surface Rocket 制导空对(地、水)面导弹

GASS 1. Geophysical Airborne Survey System 航空地球物理勘测系统 2. Gravitational Acceleration Simulation Suit 重力加速度模拟服

gas-solid interface 气体-固体界面

gas-to-particle 气粒转化

gas-turbine numerology 燃气涡轮发动机(内的)站位编号

GAT 1. General Air Traffic 通用空中交通 2. General Aviation Terminal 通用航空终点站 3. German Airborne Tow 德国机载"陶"式导弹 4. Greenwich Apparent Time 格林尼治视(太阳)时 5. Go-Around Thrust 复飞推力

gat 狭窄航道,狭航道

GATCO Guild of Air Traffic Control Officers 空中交通管制员协会(英国)

GATE Get-Away Tether Experiment 搭载容器施放系绳卫星实验

gate 1.(航站的)乘客登机门 2.(民航飞行的)始发点 3.(油门等操纵机构的)限动器 4.空中走廊,狭长通道 5.选通,开启 6.进场门 7.(空中截击代语)"在短时期内用最大可能的速度飞行"

gate array 门列阵

gate array method 门阵列法

gate guardian 门卫飞机

gate position (航班的)编号登机门

gate propagation delay 门传输延迟

gate time 门时刻

gate tube 门电子管,选通管

gate valve 插板阀,闸式阀

gate-arrival 1.门阵列 2.门阵 3.与门阵列

gated 同步选通脉冲的

gated fulladder 选通全加器

gated integrator 门控积分器

gated-beam tube 射束开关管,选通电子束管(一种锐截止和阳极电流极大的五极管)

gateway 网络连接器,门径,入口

gateway computer 网间连接计算机

gateway controller 网间连接控制器

gateway protocol 网间连接协议

gateway station 入口站(两个不同通信网的连接点)

gather 收集,搜集

gathering 1.导入,引入 2.聚集,收集 3.增长

gating 1.(电磁场的)选通,开启 2.(激光)在规定时间间隔内控制其发射 3.油门限动

Gatlinggun 加特林机枪,加特林机关炮

GATM Global Air Traffic Management 全球空中交通管理

Gator "盖托"流线型小炸弹

GATT Gate-Assisted Turn off Thyristor 门辅助切断的闸流晶体管

GAU 1. GPS Antenna Unit 全球定位系统天线装置 2. General Acquisition Unit 通用探测部件 3. Gun Aircraft Unit 航空机枪,航空机炮

gauge 1.仪表,传感器 2.量规,样板,标准规格,标准(比例)尺 3.(金属板的)厚度,(金属丝的)直径,(枪炮的)口径 4.(两侧的)轮缘,(铁道的)轨距

gauge board 仪表板

gauge factor 仪器灵敏度系数

gauge of nuclear capability 核能力衡量尺度(核武器数与目标数之比)

gauge pressure 表压(指系统压力超过大气压数量的读数),计示压力

gauged fuel (油量表)指示燃料量

gauging 用规检验,测量,测定
Gauss grid convergence 高斯平面子午线收敛角
gauss meter 高斯计磁强计
Gauss method 高斯法
Gauss mid-latitude formula 高斯中纬度公式
Gauss plane coordinate 高斯平面坐标
Gauss projection 高斯投影
Gauss theorem 高斯定理
Gaussian beam 高斯光束
Gaussian distribution 正态分布,高斯分布
Gaussian gravitational constant 高斯引力(重力)常数
Gaussian method of orbit determination 高斯轨道测定法
gaussian noise 高斯噪声
gaussian process 高斯过程
Gaussian random noise 高斯随机噪声
Gaussian test of the mean 平均值的高斯试验
Gaussian test of the variance 方差的高斯试验
Gaussian whitenoise 高斯白噪声
Gauss-Kruger projection 高斯-克吕格投影,高斯投影
Gauss-Newton method 高斯-牛顿法
GAVC Ground Air Visual Code 地空视频通信编码(美国联邦航空局)
GAVRS Gyro Compassing Altitude and Velocity Reference System 陀螺罗盘姿态和速度基准系统
GB 1. Generator Breaker 发电机断路器 2. Glide Bomb 滑翔炸弹
GBD Gamma-Ray Burst Detector 伽马射线爆发探测器
GBDM(S) Ground-Based Data Management (System) 陆基数据管理(系统)
GBFEL Ground-Based Free Electron Laser 地基自由电子激光器(美国)
GBHE Ground-Based Hyper Velocity Gun Experiment 地基超高速炮实验(美国)
GBHRG Ground-Based Hyper Velocity Railgun 地基超高速磁轨炮
GBI Ground-Based Interceptor 地基拦截器
GBL 1. Government Bill of Lading 政府提货单,政府运货单 2. Ground-Based Laboratory 地基实验室 3. Ground-Based Laser 地基激光器
GBMD Global Ballistic-Missile Defence 全球弹道导弹防御
GBMI Ground-Based Midcourse Interceptor 地基中段拦截器
GBOM General Bill Of Material 通用物料表
GBR Ground-Based Radar 地基雷达(美国)
GBS General Business System 通用商业系统
GBU Guided Bomb Unit 制导炸弹
GC 1. Galvanic Corrosion 电解腐蚀 2. Gigacycle 千兆周 3. Guidance Computer 制导计算机 4. Gyro-Compassing 陀螺定向
GCA Ground-Controlled Approach 地面控制进场着陆(系统),地面指挥进近系统
g-capsule 抗荷分离舱
GCAS Ground Collision Avoidance System 地面防撞系统
GCB 1. Generator Circuit-Breaker 发电机断路器 2. Gun Control Box (直升机的)航炮控制箱
GCbrg Greatcircle Bearing 大圆圈方位
GCC 1. Goggles-Compatible Cockpit 夜视镜兼容座舱 2. Graduated Combat Capability 分级战斗能力 3. Ground Cluster Controller (ACARS) 地面综合控制员(飞机通信寻址和报告系统) 4. Ground Control Center 地面控制中心 5. Guidance & Control Computer 制导和控制计算机 6. Gulf Cooperation Council 海湾合作委员会
GCD GAS Control Decoder 有效载荷搭载容器控制译码器
GCE 1. Ground Communications Equipment 地面通信设备 2. Ground Control Equipment 地面控制设备
GCF Ground-Conditioning Fan 地面空调风扇
GCI Ground-Controlled Interception 地面引导截击
GCIvector 地面引导截击系统给定方向
GCMS Gas Chromatograph-Mass Spectrometer 气体色谱质谱仪
GCMTS Great-Capacity Mobile Tele-communications System 大容量移动通信系统
g-counter 过载计数器
GCP 1. Geostationary Communications Platform 静止轨道通信平台 2. Global Capacity Planning 全球能力计划
GCR 1. Galactic Cosmic Ray 银河宇宙射线 2. Generator Control Relay 发电机控制继电器 3. Great Circle Route 大圆航线 4. Ground Control Radar 地面控制雷达
GCRSS Geological Committee on Remote Sensing from Space 空间遥感地质委员会
GCS 1. Ground Cluster Suppression 地面综合抑制 2. Ground Control Station(s) 地面控制站
GCSS Global Communication Satellite System 全球通信卫星系统
GCT 1. Government Competitive Test (为确定选型由政府主持的)竞争性试验 2. Greenwich Civil Time 格林尼治民用时
GCU 1. General Control Unit 通用控制单元 2. Generator Control and Protection Unit 发电机控制与保护组件
GD 1. Gear Door 起落架舱门 2. Gear Down 起落架放

下 3. General Document 普通文档 4. Guard 警戒,保护

GDAS 1. Global Data Assimilation System 全球数据同化系统 2. Ground Data Acquisition System 地面数据采集系统

GDBMS Generalized Data Base Management System 综合数据库管理系统

GDB General Duties Branch 一般勤务分部(英国空军管空勤组)

GDBS Global Data Base System 全局数据库系统

GDC 1. Geophysical Data Center 地球物理数据中心(美国) 2. Gyro Display Coupler 陀螺显示耦合器

GDCP GOES Data Collection Platform 地球静止环境业务卫星数据收集平台

GDCS Ground Distributed Control System 地面分配控制系统

GDDS GOES Data Distribution System 地球静止环境业务卫星数据分布系统

GDE 1. Gas-Discharge Element 气体放电元件 2. Guided 指引的,导航的

GDHS Ground Data Handling System 地面数据处理系统

GDIL General Data Interchange Language 通用数据交换语言

GDL 1. Gas Dynamics Laboratory 气体动力学实验室 2. Gas Dynamic Laser 气动激光器 3. Gear Down Latch 起落架下位锁

GDLP Ground Data Link Processor 地面数据链处理器

GDM Generalized Development Model (全球定位系统接收机的)通用发展型

GDOP 1. Geodesic Degradation Of Performance 短程性能下降 2. Geometric Dilution Of Precision 几何精度因子,几何精度冲淡系数,几何精度淡化

GDPS Global Data Processing System 全球观测数据处理系统(世界气象组织)

GDS 1. Graphic Data System 图形数据系统 2. Ground Data Systems 地面数据系统

GDT Graphic Display Terminal 图形显示终端

GE 1. General Electric 通用电气 2. Graphite Epoxy 石墨环氧树脂 3. Ground Equipment 地面设备

GEADGE German Air Defence Ground Environment "盖其"防空系统(德国地面防空系统)

GEAE General Electricic Aircraft Engines 通用电气公司发动机分公司

GEANS Gimballed Electrostatic Aircraft Navigation System 平衡环式静电(悬浮)飞机导航系统

gear 起落架(Landing Gear 的口语)

gear transmission 齿轮传动

gear up warning 未放起落架警告(信号或装置)

gear walk 起落架走步现象

gearbox 齿轮箱,传动机匣

gearbox fairing 减速器整流罩

geared engine 带减速器的发动机

geared fan 齿轮传动风扇

geared supercharger 齿轮传动增压器,增速增压器

geared tab 随动补偿片(飞机舵面上的)

gearing ratio 传动比

geartrain 齿轮组

gear-type pump 齿轮泵

GEC 1. General Electric Company 通用电气公司 2. Graphite Epoxy Composite 石墨环氧树脂复合材料

GECS 1. Gemini Environmental Control System 双子星飞船环境控制系统(美国) 2. Ground Environmental Control System 地面环境控制系统

GeCu Germanium/Copper 锗/铜(红外探测器)

GeeG 导航系统,"奇异"双曲线导航系统(第二次世界大战前期英国的一种中距离双曲线导航系统)

Gee-H G-H 导航系统(一种雷达导航系统,是 G 导航系统的改进)

GEEIA Ground Electronics Engineering Installation Agency 地面电子设备工程安装局(美国)

Ge-gallium arsenide solar cell 锗砷化镓太阳电池

GEH Graphite-Epoxy Honeycomb 石墨环氧蜂窝结构

Geiger counter 盖格计数器

Geiger-Muller region 盖格弥勒区

GEL 1. General Emulation Language 通用仿真语言 2. Graphite-Epoxy Laminate 石墨环氧(树脂)层压板

gel 1. 凝胶 2. 凝胶体

gel formulation 1. 凝胶制剂 2. 凝胶外用制剂 3. 凝胶剂处方

gel propellant 1. 凝胶推进剂 2. 膏状推进剂 3. 膏体推进剂

gel structure 1. 凝胶结构 2. 胶状结构 3. 冻胶结构

gelatin dynamite 胶质炸药

gelatination 胶凝作用

geliograph 1. 日照计 2. 日光仪 3. 日光信号机

GELIS Ground-Emitter Location and Identification System 地面发射机定位和识别系统

gellant 胶凝剂

gelled propellant 胶体推进剂

gelled propellant rocket motor 胶体(推进剂)火箭发动机

GEM 1. Generic Electronic Module 通用电子设备舱 2. Global Exponential Model (大气密度的)全球指数模型 3. Graphite Epoxy Motor 石墨环氧树脂(固体火箭)发动机 4. Ground-Effect Machine 地效(飞)机,地面效应飞行器,地面效应运载工具 5. Guidance Evaluation Missile (鉴定制导系统用的)制导试验(导)弹

Gemini 双子星飞船(美国载人飞船)
GEMS 1. General Energy Management System 总能量管理系统 2. Geostationary European Meteorological Satellite 欧洲地球同步气象卫星(欧洲空间研究组织) 3. Global Environment Monitoring System 全球环境监测系统 4. Grouped Engine Monitoring System 组合发动机监控系统
Gen 1. Generator 发电机 2. General Information 总述,概况
gene modification 1.基因修饰 2.基因改造
gene value 基因值
general(GEN) 总的,通用的,一般的
general address 总地址
general airsituation 总空情
general airtraffic 通用空中交通
general approach 一般方略
general arrangement 总体布置
general assembly 总装配
general assembly drawing 总装图
general astronomy 普通天文学 通用航空
general aviation(G) 通用航空
general aviation accident 通用航空事故
general breathing oxygen system 一般供氧系统
general buckling 总体失稳,一般压曲
general calling 全呼
general cargo 普通货物(指未使用集装箱或货盘的零散货物),一般货物
general circulation 大气环流
general class 大类
general console for statellite 卫星综合控制台
general corrosion 总体腐蚀,全面腐蚀
general design requirement 总体设计要求
general energy management system 总能量管理系统
general equilibrium mode 一般均衡模型
general equilibrium theory 一般均衡理论
general flying procedures 一般飞行程序
general flying training 一般飞行训练
general inference (对气象形势和未来预报的)一般推断
general information test 一般信息测试
general inspection 总检查
general load 总载荷
general modular redundancy 通用模块冗余(技术)
general notice 通告集
general perturbation 普通摄动,普遍摄动
general precession 总岁差
general precession constant 总岁差常数
general precession in longitude 黄经总岁差
general problem solver 通用问题解算机
general purpose 通用

general purpose discrete simulator 通用离散模拟器
general purpose frigate 通用护卫舰
general purpose multiplex system 通用多工传输系统
general selection 基础选拔
general shape 1.总体形态 2.一般形状 3.总体外形
general structure system 总体结构系统
general system 统摄系统
general test facility 通用测试设备
general time 1.通用时间 2.广义时间 3.泛时
general trend 一般趋势,一般走向
general upkeep 一般维修保养
general user system 普通用户系统
general-arrangement drawing 总体配置图(通常为三面图)
generality 1.通用性 2.共性 3.一般性
generalization 普遍,推广
generalize 一般化,概括
generalized active force 广义主动力
generalized algebraic translator 通用代数翻译程序
generalized analysis 广义分析
generalized coordinate 广义坐标
generalized displacement 广义位移
generalized equation 广义方程
generalized excitation 广义扰动力
generalized force 广义力
generalized inertia force 广义惯性力
generalized internal force 广义内力
generalized least squares estimation 广义最小二乘估计
generalized main sequence 广义主序
generalized mass 广义质量
generalized modelling 广义建模
generalized potential equation 全位势方程
generalized ray 广义射线
generalized ray theory 广义射线理论
generalized similitude parameter 广义相似参数
generalized speed 广义速度
generalized transmission function 广义传递函数
general-purpose aircraft 多用途(军用)飞机
general-purpose bomb 普通炸弹,杀伤爆破炸弹
general-purpose buffer 通用缓冲器
general-purpose computer 通用计算机
general-purpose oscilloscope 通用示波器
general-purpose warhead 通用(爆破)弹头
general-purpose workstation 通用工作站
general-utility helicopter 多用途通用直升机
generate 产生,形成,生成,生育
generated address 合成地址
generating 展成法
generating 1.生成 2.产生 3.发电

generating function 生成函数,母函数
generation 发生,振荡
generation rate 1.产生速率 2.产生率 3.产气速率
generator 1.发生器,产生器,振荡器 2.直流发电机 3.发送器,传感器 4.生成程序 5.【数】母线,生成元
generator circuit breaker 发电机电路断路器
generator line contactor 发电机线路接触器
generator trailer 电源拖车
generator voltmeter 发电机电压表
generic fault 类属性故障
generic model 1.一般模型 2.基本模式 3.通用模型
generic set 生成集,类集
generic system software 类属系统软件
generic unit 类属单元
genetic algorithm 1.遗传算法 2.遗传算法的 3.基因算法
genetic code 遗传密码
GENL General 一般,通用的,总的
gentle turn 小坡度盘旋(坡度25°以下)
GEO 1.Geostationary Earth Orbit 地球同步轨道,地球通信卫星 2.Geosynchronous Earth Observatory 同步地球观测台(美国) 3.Geosynchronous Earth Orbit 地球同步轨道 4.Geosynchronous Equatorial Orbit 地球同步(赤道)轨道
geoastrophysics 地球天体物理学
geobiont 地面植物群落
geobotanical cartography 地植物学制图
geobotanical chart 地植物学图,地植物图
geoceiver 大地接收机
geocentric 地心的,以地球为中心的
geocentric angular separation 地心角(间)距
geocentric apparent motion 地向视动
geocentric celestialsphere 天球
geocentric colatitude 地心余纬
geocentric coordinate 地心坐标
geocentric coordinate system 地心坐标系
geocentric diameter 地心直径
geocentric ephemeris 地心历表
geocentric gravitational constant 地心引力常数
geocentric inertial coordinates 地心惯性坐标系
geocentric latitude 地心纬度
geocentric longitude 地心经度
geocentric orbit 地心轨道
geocentric parallax 地心视差
geocentric rectangular coordinate 地心直角坐标
geocentric solar ecliptic system 地心太阳黄道系统
geocentric solar magnetospheric system 地心太阳磁层系统
geocentric system 地心体系
geocentric system of coordinate 地心坐标系
geocentric vertical 地心垂线,几何垂线
geocentric zenith 地心天顶
geochemical environment 地球化学环境
geocorona 地冕,地华
geodesic 1.大地测量学的 2.最短线的 3.短程线,测地线,大地线
geodesic configuration 测地方案
geodesic coordinate 1.短程线坐标 2.大地坐标 3.测地坐标
geodesic equation 测地方程
geodesic equations 测地线方程组
geodesic line 1.(最)短程线,测地线,大地线 2.大圆航线
geodesic satellite 大地测量卫星,测地卫星
geodesy 大地测量学
geodetic astronomy 测地天文学
geodetic azimuth 大地方位角
geodetic baseline 大地测量基线
geodetic constant 大地常数
geodetic construction 最短线网状结构
geodetic coordinate 大地坐标
geodetic coordinate system 大地坐标系
geodetic datum 1.大地基准 2.大地原点,大地基准点
geodetic equator 1.大地赤道 2.测地赤道
geodetic height 大地高
geodetic instrument 大地测量仪
geodetic latitude 大地纬度,测地纬度
geodetic line 大地测量基线,测地基线
geodetic longitude 大地经度,测地经度
geodetic measurement 大地测量
geodetic meridian 1.大地子午线 2.测地子午线
geodetic norms 大地测量法式
geodetic origin 大地原点
geodetic position 大地位置,测地位置
geodetic precession 大地岁差,测地岁差
geodetic refraction 地平大气折射
geodetic satellite 测地卫星,大地测量卫星
geodetic subsystem 测地分系统
geodetic survey 大地测量
geodetic survey error 大地测量误差
geodetic surveying 大地测量术
geodetic triangle 大地定位三角形
geodetic zenith 大地天顶
geodimeter 光速测距仪,光电测距仪
GEODSS Ground-Based Electro-Optical Deep Space Surveillance 地基电子光学深空监视网(美国)
geodynamics 地球动力学
geoelectric 地电的

geoelectric cross section	地电断面
geofluid	地热流体
geographic (map)-vertical	地理垂线（大地水准面法线的方向线）
geographic coordinate	地理坐标
geographic coordinate system	地理坐标系
geographic distribution	1.地理分布 2.地域分布 3.区位分布
geographic display system	地理显示系统
geographic equator	地理赤道
geographic graticule	地理坐标网
geographic landscape	地理景观
geographic pole	地极
geographical medicine	地理医学
geographical pole	地（理）极
geographical position	地理位置
geographical viewing distance	地理视距
geoheat	地热
geoid	大地水准面，地球体
geoidal height	大地水准面高（度）
geoidal horizon	大地水准面地平圈，海面水平面
geoidal surface	大地水准面
geoidal undulation	大地水准面起伏
geolocation	1.地理定位 2.地球定位
geologic orbital photography	地质轨道摄影
geological interpretation of photograph	相片地质判读，相片地质解译
geological photomap	影像地质图
geology	地质学
geology of Mars	火星地质
geology of Mercury	水星地质
geology of Moon	月球地质
geology of moon	月质学（即月球地质学）
geology of satellite	卫星地质
geology of Venus	金星地质
geomagnetic	地磁的
geomagnetic activity index	地磁活动指数
geomagnetic ally quiet time	（地）磁宁静期
geomagnetic cavity	地磁腔
geomagnetic chronology	地磁年代学
geomagnetic coordinates	地磁坐标
geomagnetic dipole	地磁偶极子
geomagnetic disturbance	地磁扰动，磁扰
geomagnetic equator	地磁赤道
geomagnetic excursion	地磁漂移
geomagnetic field	地磁场
geomagnetic index	磁情指数，地磁指数
geomagnetic physics	地磁物理学
geomagnetic polarity reversal	地磁极性反向
geomagnetic poles	地磁极
geomagnetic storm	地磁暴
geomagnetic tail	磁尾
geomagnetic tide	地磁潮
geomagnetic torque	地磁力矩
geomagnetism	1.地磁（性） 2.地磁学
geometric calculation	1.几何计算 2.几何学计算法
geometric code	几何码
geometric correction	几何校正
geometric cross section	几何截面
geometric displacement	几何排量
geometric factor	几何因子
geometric geodesy	几何大地测量学
geometric imperfection	几何缺陷
geometric inertial navigation system	几何式惯性导航系统
geometric interpretation	几何解释
geometric measurement	几何量测量
geometric orientation	几何定向
geometric path	几何路径
geometric pitch	（螺旋桨的）几何桨距
geometric rectification	几何校正
geometric rectification error	几何校正误差
geometric rectification of imagery	图像几何纠正
geometric registration of imagery	图像几何配准
geometric spreading	几何扩散
geometric standard deviation	1.几何标准偏差 2.几何标准差
geometric system mechanization	几何式系统机械编排
geometric transition absorber	几何结构跃迁的雷达吸波材料
geometric twist	（机翼的）几何扭转（翼弦与基准面的夹角沿翼展变化）
geometric variable	几何变量
geometric(al)	几何学的
geometrical axis	几何轴
geometrical method of satellite geodesy	卫星大地测量几何法
geometrical parameter	螺旋桨几何参数
geometrical similarity	几何相似
geometry	几何形状，几何图，几何学
geomorphological map	地貌图
geomorphology	地形学
geophone	地震检波器
geophone array	组合检波
geophones	1.地震传感器 2.地声器（测量声波在地壳内传播的灵敏声探测器）
geophysical anomaly	地球物理异常
geophysical exploration	地球物理勘探

geophysical fluid dynamics laboratory 地球物理流体动力学实验室

geophysical prospecting 地球物理勘探

geophysical research satellite 地球物理研究卫星（美国空军）

geophysical satellite 地球物理卫星

geophysical year 地球物理年

geophysics 地球物理学

geopotential 地球重力势，地球位（势），大地位，位势高度（用于气象目的时，大多数情况下，位势高度与几何高度相等，但位势米＝0.98米）

geopotential altitude （地球）重力势高度，地球位势高度

geopotential function （地球）重力势函数

geopotential height （地球）重力势高度

geopotential meter 地（球重力）势测量计

geopotential model 1.重力场模型 2.地球位模型 3.地球重力场模型

geopotential number 地球位数

geopotential surface 地球重力等势面，等地势面

geoprobe 地球物理探测火箭

georeceiver 大地接收机

Georef 全球地理参数系统

GEOS 1.Geodetic Earth Orbiting Satellite 地球轨道测地卫星 2.Geodetic Satellites 测地卫星，大地测量卫星（美国，又名 Geodetic Explorers） 3.Geodynamics Experimental Ocean Satellite 地球动力学实验海洋卫星（美国）4.Geostationary Earth Orbiting Satellite 通信地球轨道卫星 5.Geosynchronous Earth Observation System 地球同步观测系统

GEOSAR Geosynchronous Earth Orbit Synthetic Aperture Radar 地球同步轨道合成孔径雷达

GEOSAT Geological Satellite 地质卫星

geoscientist 地球科学家

geospace 地球空间

geosphere 1.陆界，陆圈 2.地圈（地球固态及液态部分的总称，即包括岩界 Lithosphere 与水界 Hydrosphere）

geostationary meteorological satellite 1.地球静止气象卫星 2.地球同步气象卫星

geostationary orbit 地球静止轨道，对地静止轨道

geostationary orbit satellite 地球静止轨道卫星

geostationary satellite 地球静止轨道卫星，对地静止卫星

geostationary transfer orbit 地球静止（同步）转移轨道

geostrophic divider 地转风风速分析器

geostrophic drag 地转曳力

geostrophic momentum 地转动量

geostrophic potential vorticity 地转位涡

geostrophic wind 地球自转风

geostrophic windlevel 地转风高度

geosynchronous earth observation system 地球同步对地观测系统

geosynchronous meteorological satellite 地球同步气象卫星

geosynchronous orbit 地球同步轨道

geosynchronous orbit satellite 地球同步轨道卫星

geosynchronous satellite 地球同步卫星

geosynchronous synthetic aperture radar 地球同步合成孔径雷达

geosynchronous transfer orbit 地球同步转移轨道

Geotail 地磁尾卫星（美国、日本合作的科学探测卫星）

geotherm 等地温面

geothermal activity 地热活动

geothermal energy 地热能

geothermal field 地热田

geothermal fluid 地热流体

geothermal gradient 地温梯度

geothermally-anomalous area 地热异常区

geothermics 地热学

geothermometer 地球温度计

geotropism 向地性

germanium 【化】锗（化学元素，符号 Ge）

germanium transister 锗晶体管

germbomb 细菌炸弹

germweapon 细菌武器

gerotor pump 盖劳特泵，摆线泵，转子泵，游星齿轮泵

GERTS 1.General Remote Terminal System 通用远程终端系统 2.General Remote Transmission Supervisor 通用远程传输管理程序

GES Ground Earth Station 地球地面站

GET Ground Elapsed Time 地面经历时间（从起飞开始记录）

get-away 离水（水上飞机离开水面，相应于飞机离地）

get-away speed 离水速度

GETI Ground Elapsed Time of Ignition 地面点火经历时间

getter 1.吸气剂 2.吸气器 3.吸气，消气

getter ion pump 吸气剂离子泵

getter pump 吸气剂泵

get-you-down（pressure）suit 保证下降到安全高度的加压服

GEV Ground Effect Vehicle 地效飞行器

GEW Ground Effect Wing 地效机翼

G-excess illusion 超重错觉

geyser 热水锅炉，热水器

GF 1.Generator Field 发电机励磁 2.General Function 全局功能 3.Germanium Content Fiber 含锗光纤 4.Ground Fault 地面故障

GFAC Ground Forward Air Controller 前方地面空中管制员

GFAE 1. Government Furnished Aerospace Equipment 政府提供的航天设备 2. Government Furnished Aircraft Equipment 政府装备的机载设备

GFCI Ground Fault Circuit Interrupter 地面故障电路中断器

GFD Government Furnished Data 政府提供的数据

GFE Government Furnished Equipment 政府提供的设备,政府配置设备

Gfield 重力场

GFM Government Furnished Material 政府提供的器材

gforce 1. 重力 2. 惯性力(常以重力加速度的倍数表示)

GFP 1. General Function Platform 总功能平面 2. Global Function Plane 全局功能平面 3. Government Furnished Parts 政府提供的零件 4. Government-Furnished Property 政府提供的资产 5. Ground Fine Pitch 地面小距(螺旋桨桨距)

GFT 1. General Flight Test 一般飞行试验 2. Generalized Fast Transform 广义快速变换 3. Ground Functional Test 地面功能测试

GFY 1. Glider Flying 滑翔机飞行 2. Glider Flying/Glider Flight 滑翔机飞行 3. Government Fiscal Year 政府财政年度

GG 1. Gas Generator 燃气发生器,气体发生器 2. Graphics Generator 图形产生器 3. Gravity Gradient 重力梯度 4. Ground Guidance 地面制导

GGC Ground Guidance Computer 地面制导计算机

GGS 1. GPS(Global Positioning Sysgtem) Ground Station 全球定位系统地面站 2. Ground Guidance System 地面制导系统

GGTFM Ground-Ground Traffic Flow Management 地-地交通流量管理

GH 1. Grid Heading 网格坐标航向 2. Ground Handling 地面操作 3. Gyro Horizon 陀螺地平仪

GHA Greenwich Hour Angle 格林尼治时角

GHOST 1. 重影 2. 双重图像

ghost reflection 虚反射

GHP Gross Horse Power 总马力

GI Group Identifier 组别标识器

Gi 1. Guilder 盾(货币单位) 2. Gill 吉耳(液量单位)

giant 巨星

giant branch 巨星支

giant elliptical galaxy 巨椭圆星系

giant galaxy 巨星系

giant pulse technique 巨脉冲技术

giant pulse laser 巨脉冲激光器

giant scale display 巨屏幕显示

giant star 巨星

GIB GNSS Integrity Broadcase 全球导航卫星系统完整性广播

GIC GNSS Integrity Channel 全球导航卫星系统完整性频道

GICO GEO Index of Cloud Opacity 同步卫星云阻光指数

GIE Ground Instrumentation Equipment 地面仪表设备

GIF Graphic Interchange Format 图形内部转换格式

giga(10g) 吉(倍乘词,用于前缀,指 10 亿倍或千兆倍,符号为 10^9),京

giga cycle computer 吉周计算机

GIGS Gemini Inertial Guidance System 双子星飞船惯性制导系统

GIL Green Indicating Lamp 绿色指示灯

Gilbert 吉伯(磁通势单位,约等于 0.8 安匝)

Gilberts per centimeter 吉伯/厘米(磁场强度单位,同奥斯特)

gills 鱼鳞板,导风板(通常指活塞式发动机后方的)

GIM 1. General Introduction Manual 通用简介手册 2. Generalized Information Management 综合信息管理,广义信息管理

GIMADS Generic Integrated Maintenance Diagnostics 通用综合维修诊断

gimbal 1. 万向支架,常平架,万向接头,平衡环,陀螺支架,陀螺框架 2. 装在万向支架上,在万向支架上转动

gimbal angle 1. 框架角 2. 摆动角

gimbal assembly 万向架组件,常平架组件

gimbal axis 万向支架轴,常平架轴

gimbal bearing 1. 常平座,万向支座 2. 常平座轴承,万向支架轴承

gimbal bellow 摇摆波纹管,摇摆软管

gimbal caging 框架锁定(使框架固定于零位或预定位置)

gimbal deflection 框架偏转,常平架偏转

gimbal frame 万向支架

gimbal freedom 常平架自由度

gimbal gain 常平架传动比

gimbal lock 1. 框架自锁 2. 常平架自锁

gimbal mount 常平座

gimbal mounted hub (旋翼的)半刚接式桨毂,万向接头式桨毂

gimbal mounting 常平座

gimbal nozzle 万向喷管,摆动喷管

gimbal pickup 万向架位置传感器

gimbal servo offset 框架伺服系统偏差

gimbal stabilizing loop 框架稳定回路

gimbal stress analysis 框架应力分析

gimbal torquer 常平架转矩产生器,常平架修正电动机(陀螺仪的)

gimbaled flywheel 框架飞轮
gimbaled hub 万向接头式桨毂
gimbaled inertial navigation system 平台式惯性导航系统
gimbaled rocket engine 摇摆火箭发动机
gimbaled star tracker 框架式星跟踪器
gimballed chamber 万向架支承的推力室，变向燃烧室
gimballed engine 万向架支承的发动机
gimballed motor 万向架支承的(固体)火箭发动机，万向悬挂式(固体)火箭发动机
gimballed startracker 框架式星跟踪器
gimballed wheel 框架飞轮(用框架支承的飞轮)
gimballess inertial system 捷联式惯性系统，无常平架惯性系统
gimballing 万向支架连接，万向支架转动
gimballock 1.(陀螺)框架锁定 2.常平架锁，万向支架锁
G-induced loss of consciousness 过载引起的意识丧失
gion 引力场量子，引力子
GIS Generalized Information System 通用情报信息系统
GL 1.Gallon 加仑(容量单位) 2.Glass 玻璃 3.Ground Level 地平面 4.Ground Location 地面定位
GLA Gust Load Alleviator 阵风负载缓和装置
gland 1.(轴孔的)封严套，封严压盖 2.(飞艇的)缆绳滑套
gland nut 套管螺帽
glare 闪光，眩光
glareshield (驾驶舱的)遮光屏，防眩板
glass aircraft 玻璃钢飞机
glass bulb 玻壳
glass ceramic coating 玻璃陶瓷涂层
glass cloth 玻璃布
glass cloth laminate 玻璃布层压制品
glass cockpit 全显示驾驶舱，〈口语〉玻璃座舱(指舱内用显示屏代替传统的大量仪表)
glass envelope 玻璃泡，玻璃容器(封套)
glass epoxy 玻璃环氧
glass fiber 玻璃纤维(直径一般为0.025毫米)
glass fiber reinforced plastic 玻璃纤维增强塑料
glass laser 玻璃激光器
glass microball reflector 玻璃微珠反射体
glass microsphere 玻璃微球
glass packaging 玻璃封装
glass plate 玻璃板
glass semiconductor 玻璃半导体
glass slide 1.载玻片 2.玻片 3.玻璃载片
glass yarn 玻璃线
glass-ceramic 微晶玻璃，玻璃陶瓷

glass-insulated wire 玻璃绝缘导线
glassivation 玻璃纯化
glass-plastic warhead 玻璃塑料壳战斗部
glass-reinforced 玻璃纤维增强的
glasswoo 玻璃棉，玻璃绒，玻璃丝
Glauert factor 格劳渥(升力增量)因数
G-layer G层，自由电子层(电离层中的)
glaze 1.镶以玻璃 2.抛光，上釉 3.坚冰，薄冰层
GLB Glassblock 玻璃块，镜片
GLC Generator Line Contactor 发电机馈线接触器，发电机供电接头
GLD Gold 金子
GLE Government-Loaned Equipment 政府贷给的设备
glide 1.下滑，滑降，滑翔 2.下滑道
glide angle 下滑角
glide beacon 下滑信标
glide illusion 下滑错觉
glide landing 下滑降落，无拉平着陆
glide mode 下滑工作状态
glide path 1.下滑道 2.〈口语〉下滑信标台
glide path bend 下滑道弯曲，下滑轨道弯曲
glide path indicator 下滑道指示器(仪表着陆系统的指示仪)
glide path landing system 下滑航迹着陆系统
glide path localizer 下滑道航向信标台
glide path sector 下滑道扇形区
glide ratio 滑翔比，下滑比(下滑水平距离与损失高度之比)
glide reentry vehicle 滑翔再入飞行器
glide slope 1.下滑道(着陆时的) 2.下滑信标台 3.下滑波束(仪表着陆系统的)
glide torpedo 滑翔鱼雷
glider 1.滑翔机 2.滑翔器 3.再入(大气层)滑翔航天器
glider bomb 滑翔炸弹
glider rocket bomb 火箭助推滑翔炸弹
glider train 滑翔列车(连成一串的滑翔机，可用于运输)
glider tug 滑翔机牵引机(拖曳机)
gliding parachute 滑翔伞
gliding ratio 下滑比，滑翔比
gliding turn 下滑转弯，盘旋下降
g-limiting accelerometer 重力限制加速度计
glimlamp 微光灯，暗灯
glint 1.闪光，反光(常指飞机金属表面反射阳光) 2.回波起伏(目标反射面的迅速变化而引起)
glint error 闪烁误差
glint noise 闪烁噪声
glitch 1.电压波动对敏感装置造成的影响 2.小技术问题，小故障 3.自转突变

GLNS GPS (Global Positioning System) Landing and Navigation System 全球定位系统着陆和导航系统
GLNU GPS (Global Positioning System) Landing and Navigation Unit 全球定位系统着陆和导航装置
g-load 过载,载荷因数
global address 全局地址
global analysis 整体分析
global and diffuse solar radiation integrator 总辐射和太阳漫射辐射积分器
global beam 全球波束
global beam antenna 覆盖波束天线
global circulation 全球环流
global communication system 全球通信系统
global coordinator 全局协调者
global data assimilation system 全球数据同化系统
global database 全局数据库,综合数据库
global decision maker 全局决策者
global display address 全局显示地址
global expression 全局表达
global irradiance 1.总辐照度 2.太阳辐照度
global lock management 全局闭锁管理
global mass matrix 整体质量阵
global mode 总体模态
global navigation 1.气温 2.全球导航
global navigation chart 全球导航图
global navigation satellite system 全球导航卫星系统
global network 全局网络
global operational seasurface temperature computation 全球海面温度计算(系统)
global optimal solution 1.全局最优解 2.全局最小解 3.全局的最优解
global optimization 1.全局优化 2.全局最优化 3.全局最优
global optimum 全局最优值
global orbiting navigation satellite system 全球轨道卫星导航系统
global performance 1.全域性能 2.全面性能
global positioning 全球定位
global positioning satellite 全球定位卫星
global positioning system 1.全球定位系统 2.定时测距导航系统
global prediction system 全球预报系统
global problem 1.全球问题 2.全球性问题 3.全局问题
global program control 全局程序控制
global radiation 总辐射
global rawinsonde 全球无线电探空测风仪
global reaction 宏观反应
global register optimization 全局寄存器优化
global satellite system 全球卫星系统
global satellite tracking and control facility 全球卫星跟踪与控制设备
global search 全局搜索
global simulation 1.全局仿真 2.全球模拟 3.全局数值模拟
global solution 1.整体解 2.全局解
global stiffness matrix 整体刚度阵
global surface 球弧
global temperature 全球性温度
global tracking network 全球跟踪网
global truncation error 1.整体截断误差 2.整体截断误差限
global variable 1.全局变量 2.全程变量 3.全局变量的
global weather experiment 全球天气试验
global weather reconnaissance 全球天气侦察
global wind system 全球风系
globalization 全球化,全球性
globar 碳硅棒,碳化硅,电热棒
GLONASS Global Navigation Satellite System 全球轨道卫星导航系统
glove vane 扇翼
glovepylon 戴套挂架
GLPG Glowplug 热线点火塞
GLS GPS (Global Positioning System) Landing System 全球定位系统着陆系统
GLT Glide Light 导航灯
GLU GPS (Global Positioning System) Landing Unit 全球定位系统着陆装置
glue 胶水,黏合剂
gluing 黏合
glycerol 甘油,丙三醇
glycidyl azide polymer propellant 缩水甘油叠氮聚合物推进剂
GM 1.Geometric Mean 几何平均数,等比中项 2.Green Maintenance 绿色维修 3.Greenwich Meridian 格林尼治子午线 4.Guidance Material 指导材料
GMBL Gimbal 万向接头
GMC Ground Movement Control 地面运动控制
GMR Ground Mapping Radar 地面标记雷达
GMRES Generalized Minimal Residual Algorithm 广义最小残差算法
GMT Greenwich Mean Time 格林尼治标准时间
GNAS General National Airspace System 国家总体空域系统,国家综合空域系统
GNC 1.General Navigation Computer 通用导航计算机 2.Global Navigation Chart 全球导航图 3.Graphic Numerical Control 图形数字控制
GND Ground 地,接地
GNE Gross Navigation Error 导航总误差

GNR　Global Navigation Receiver 全球导航接收机
GNS　Global Navigation System 全球导航系统
GNSS　Global Navigation Satellite System 全球导航卫星系统
GO　Generic Option 生成选项
goal 目标值
goal coordination method 目标协调法
goal directed programming 目标指向编程
goal location 目标定位
goal state 1.目标状态 2.状态性目标
go-around computer 复飞计算机
go-around control system 复飞控制系统
go-around mode 复飞工作方式
GOE　Ground Operation Equipment 地面操作设备
Goethert rule 格特尔特法则
GOFR　General Operating and Flight Rules 通用操作和飞行规则
goggle　Goggles 防护眼镜
gondola 吊舱,吊篮
GOP　Ground Operations Planning 地面操作大纲
GORD　General Operations Requirement Document 通用技术要求文档
gore 伞衣幅,降落伞三角布
GOS　Grade Of Service 服务等级
Gottingen-typetunnel 哥廷根型风洞(工作段为开式的有回流风洞)
GOV　Governor 调整器,控制器
GOVT　Government 政府,支配
GOX　Gaseous Oxygen 气态氧
GP　1.Gage Pressure 压表 2.General Purpose 通用用途 3.Glide Path 下滑道,下滑轨迹 4.Gross Performance 总性能 5.Grounding Point (Terminal) 接地点(接头)
Gpa　Giga pascal 十亿(吉)帕斯卡,吉帕,旧称千兆帕
GPAC　General Purpose Analog Computer 通用模拟计算机
GPACK　General Utility Package 通用实用程序包
GPADIRS　Global Positioning, Air Data, Inertial Reference System 全球定位、大气数据、惯性基准系统
GPCU　Ground Power Control Unit 地面电源控制组件
GPD　Gallons Per Day 加仑/天
GPH　Gallons Per Hous 加仑/小时
GPI　1.Ground Positon Indicator 地面位置指示器 2.Gear Positon Indicator 起落架位置指示器
GPM　Gallon Per Minute 每分钟加仑数
GPP　General Purpose Processor 通用处理器
GPS　1.Gallon Per Second 加仑/秒 2.Geophysical Processor System 地球物理处理机系统 3.Global Positioning System 全球定位系统 4.Ground Playback Station 地面重放站

GPS navigation　GPS 导航
GPS orbit determination 1.GPS 轨道确定 2.GPS 定轨
GPS position　GPS 定位
GPS receiver　GPS 接收机
GPS tracking　GPS 跟踪
GPS-inertial integrated guidance 1.全球定位系统-惯性组合制导 2.GPS-惯性组合制导
GPS-inertial integrated navigation 1.全球定位系统-惯性组合导航 2.GPS-惯性组合导航
GPSSU　Global Positioning System Sensor Unit 全球定位系统传感器装置
GPU　Ground Power Unit 地面电源装置
GPWC　Ground Proximity Warning Computer 近地警告计算机
GPWS　Ground Proximity Warning System 近地警告系统
GR　1.Generator Relay 发电机继电器 2.Gradient 梯度 3.Ground Router 地面路由器
GR WT　Gross Weight 毛重
GRA　Gray 灰色,灰色的
grabline (飞艇或水上飞机的)拖缆,攀绳
graceful degradation 柔性降级,故障弱化,缓慢降级
gradation 灰度
grade location 坡度测定
graded fiber (标准)等级纤维布
graded index 渐变折射率
graded index fiber 渐变光纤
gradient 1.梯度(飞行路线梯度是高度与距离之比,以百分数表示) 2.坡度 3.(曲线的)斜率 4.变化率 5.斜面,斜坡
gradient algorithm 梯度算法
gradient control 梯度控制
gradient heating furnace 梯度加热炉
gradient information 1.梯度信息 2.梯度图像
gradient of position line 位置线梯度
gradient search 梯度搜索
gradient template 梯度模板
gradient wind 梯度风
gradient wind speed 梯度风速
grading certificate 定级证书
grading curve 品级曲线(确定螺旋桨性能用的)
gradiometer 重力仪,磁力仪,坡度仪,梯度测定仪
gradual failure 1.渐变性故障 2.耗损性故障
gradual fault 渐变故障
gradual onset rate 慢增长率
graduate 分度,刻度
graduation 分度
Grafil 格拉斐(碳纤维原料的注册名称)
grain 1.药柱,装药 2.晶粒

grain boundary 晶界
grain boundary crack 晶界裂纹
grain burning rate 1.药柱燃烧速率 2.药柱燃速
grain configuration 药柱形状
grain core 药柱型芯
grain design （发动机）药柱设计
grain orientation 1.药柱方向性 2.晶粒取向
grain port area 药柱通气面积
grain powder 粒状火药
grain web thickness 药柱肉厚（药柱在燃烧过程中烧掉的燃层厚度）
gram 克
Granat 石榴石卫星（天文卫星，苏联、法国、保加利亚、丹麦合作）
granularity 颗粒度（巨大的电子数据处理系统量度其结构之用，视处理机的数量而定）
granulation 1.米粒组织（一种日面结构）2.（火药）颗粒度
graph 1.图表 2.图，曲线 3.曲线图
graph algorithm software package 图算法软件包
graph grammar 图文法
graph theoretic code 图论码
graph theory 图论
graph typer 图文打字机
graphic access method 图形存取法，图像存取法
graphic application subroutine package 图形应用子程序包
graphic data structure 图形数据结构
graphic design system 图形设计系统
graphic determination 绘图测定
graphic display 图形显示
graphic display resolution 图像显示分辨率
graphic display unit 图形显示装置
graphic language 图像语言
graphic library 图形库
graphic mapping control point 图解图根
graphic method 图解法
graphic output unit 图形输出设备
graphic package 图像程序包
graphic part programming 零件图形程序编制
graphic processor 图形处理机
graphic representation 图示
graphic scale 图解比例尺
graphic transform package 图形变换程序包
graphic transmission 图像传输
graphic variable 图形变量
graphical automatically program med tool 自动程序控制绘图机
graphical display system 图形显示系统

graphical identification 图像识别
graphical manipulating package 图解处理程序包
graphical output system 图形输出系统
graphical rectification 图解纠正
graphical solution 图解
graphical symbol （显示器）图形符号
graphics 图示，图解法
graphics compatibility system 图形兼容系统
graphics imulation 图解模拟
graphics software system 图形软件系统
graphics terminal system 图形终端系统
graphite 石墨
graphite composite 石墨复合材料
graphite electrode 石墨电极
graphite epoxy 石墨环氧，石墨环氧树脂
graphite fiber 石墨纤维
graphite grease （含）石墨润滑脂
graphite resistance heater 石墨电阻加热器
graphite susceptor 石墨接受器
graphite vane 石墨（燃气）舵
graphite yarn 石墨线
graphite-epoxy composite 石墨环氧复合材料
graphite-seal ring 石墨密封环
graphitization 石墨化
grappling 锚定
grate 1.格栅 2.（固体发动机）挡药板，后支承件
graticule mesh 网格
graticule projecting screen 瞄准具视准仪
grating 光栅
gravimeter 1.比重计 2.重差计，重力仪
gravimeter drift 重力仪零飘
gravimeter drift correction 重力仪零漂改正
gravimetric baseline 重力基线
gravimetric deflection of the vertical 重力垂线偏差
gravimetric specific energy 重量比能量
gravimetric specific power 重量比功率
gravimetry 1.重测量学 2.重力测量
graving 擦地而过
gravipause 重力分界（两个天体的重力大小相等而方向相反的空间分界）
gravireceptor 重力感受器
gravireceptors 人体重力感受器（人体上所有姿态、重力和加速度传感器的总称）
gravisphere 引力范围
gravitational 1.万有引力的 2.重力的，受重力作用的 3.重力效应
gravitational acceleration 引力加速度
gravitational acceleration simulation suit 重力加速度模拟服

gravitational constant	引力常数
gravitational cue	重力信号
gravitational deflection	引力偏折,引力弯曲
gravitational drop	1.弹道降低量 2.弹道降低修正量
gravitational effect	1.引力效应 2.重力作用
gravitational field	1.引力场 2.重力场 3.万有引力场
gravitational force	1.引力 2.重力
gravitational gradient torque	重力梯度力矩
gravitational instability	引力不稳定性
gravitational lens	引力透镜
gravitational mass	引力质量
gravitational perturbation	重力摄动
gravitational physiology on spaceflight	航天重力生理学
gravitational potential	引力势,引力位,重力势,重力位
gravitational potential energy	引力势能
gravitational radiation	引力辐射
gravitational redshift	引力红移
gravitational torque	重力转矩
gravitational vertical	重力作用线,重力垂线
gravitational wave	引力波
gravito-inertial force	重力
gravitometer	重差计(测比重用)
graviton	重(力)子,引力子
gravity	1.重力,(月球或行星的)引力 2.重量
gravity aided control	引力辅助控制
gravity anomaly	重力异常
gravity anomaly due to magnetic body	磁源重力异常
gravity anomaly error	重力异常误差
gravity center	重心
gravity coefficient	重力系数
gravity compensation	1.重力补偿 2.重力补偿的
gravity constant error	重力常数误差
gravity correction	重力订正,重力校正
gravity effect	重力效应,重力影响
gravity fall gear	重力放下式起落架
gravity feed	重力输送
gravity gradient acquisition	重力梯度捕获(重力梯度稳定卫星的姿态捕获)
gravity gradient attitude control system	重力梯度姿态控制系统(利用重力梯度力矩控制姿态)
gravity gradient boom	重力梯度杆
gravity gradient stabilization	重力梯度稳定
gravity gradient torque	重力梯度力矩
gravity loss	重力损失
gravity model	1.引力模型 2.重力模型 3.重力场模型
gravity refueling	1.重力加油 2.开式加油
gravity vector	1.重力矢量 2.重力向量 3.引力向量
gravity wave	重力波
gravity-turn	1.引力转弯 2.重力转弯
gravo-inertial force environment	重力-惯性力环境
Gravsat (Earth) Gravity Survey Satellite	地球重力测量卫星(美国)
gray area	灰区(常指多数导航系统不采用的频率段)
gray body	灰体(对任何波长的辐射的吸收率保持不变并恒小于1的物体)
Gray code	格雷码
gray level	灰度级
gray line	灰线
gray scale	灰阶,灰色级谱,灰度级
grayout	灰视(超重下头部缺氧引起视力模糊)2.〈口语〉局部灯火管制
graze burst	着发,瞬发
graze pellet	(引信的)惯性撞针座
graze percussion mechanism	着发引信惯性装置
graze sensitive	着发灵敏的,瞬发灵敏的(指射弹即使斜向击中目标仍能触发引信)
graze sensitive fuze	瞬发灵敏引信
grazing angle	擦地角,掠射角,入射余角
grazing impact machanism	擦地炸机构
grazing incidence	掠入射
grazing incidence X-ray telescope	掠射X射线望远镜
GRBX Gearbox	齿轮箱
GRC	1. Gearcase 齿轮箱外壳 2. Glass Reinforced Composite 强化玻璃复合材料 3. Ground Radio Communications 地面无线电通信
GRD	1. General Requirements Document 通用要求文件 2. Grind 研磨 3. Ground 地面 4. Ground Resoluted Distance 地面分辨距离
GRE	1. Ground Radar Equipment 地面雷达设备 2. Gamma Ray Explorer 伽马射线探测器
great circle	大圆圈航线的,大圆圈
great circle arc	大圆弧
great circle course angle	大圆航线角
great circle route	大圆航线
great circle sailing chart	大圆航线图
great half axis	长半轴
great inequality	中心差(月球运动的)
great orbit axis	轨道长轴
great radiation belt	大辐射带(地球的)
Great Red Spot	大红斑(木星的)
great-circle track	大圆圈航迹
greatest eastern elongation	东大距(行星的)
greatest elongation	大距(行星的)
greatest north latitude	最大北黄纬(行星的)
greatest south latitude	最大南黄纬(行星的)
greatest western elongation	西大距(行星的)
green aircraft	基本可用飞机,堪用飞机
Green airway	东向航路,绿色航路(美国)

Green Archer "绿衣射手"(英国迫击炮定位雷达)
green arrow 绿灯箭头(允许着陆信号)
green cross 绿十字毒气,窒息剂
green density 湿密度(烧结前的紧密粉末的密度)
Green Flag 绿旗演习(战术空军作战演习,特别强调电子战)
green line 轰炸线,敌我分界线
green run 首次试车,首次运转(指新的或翻修后的发动机或其他设备机件)
green zone 绿区(绿色航路与交叉航路的交叉处,此时应注意查对高度间隔)
greenfield site 待建机场场地
Greenie 地面航空技术勤务组
Greenwich apparent civil time 格林尼治视民用时
Greenwich apparent noon 格林尼治视正午
Greenwich apparent sidereal time 格林尼治视恒星时
Greenwich apparent time 格林尼治视(太阳)时
Greenwich civil time 格林尼治民用时
Greenwich hour angle 格林尼治时角
Greenwich lunar time 太阴世界时,格林尼治太阴时
Greenwich mean astronomical time 格林威治平天文时
Greenwich mean sidereal time 格林尼治平恒星时
Greenwich mean time 格林尼治平时,世界时
Greenwich meridian 格林尼治子午线,格林尼治子午圈
Greenwich sidereal date 格林尼治恒星日期
Greenwich sidereal day number 格林尼治恒星日数,儒略恒星日数
Greenwich sidereal time 格林尼治恒星时
Greenwich time 格林尼治时
Greenwich zone 格林尼治时区,零时区
Gregorian calendar 格里历
Gregorian year 格(雷果)里年(365日5时49分12秒)
grenade 榴弹(火箭测风用)
grenade bomb-bomblet dispenser 子母弹箱
grenade method 榴弹测风法
grens 透镜棱栅
grey body radiation 灰体辐射
grey hole 灰洞
grey level 灰度
grey wedge 减光板,灰楔
GRG 1. Gearing 齿轮传动,啮合 2. Ground-Roll Guidance 着陆滑跑引导
grid 1.舰上固定格栅 2.坐标方格,方格网 3.网(如通信网) 4.(蓄电池的)隔板 5.栅极
grid aperture 门极孔径
grid approach 网格搜索法
grid assembly 格架组件
grid attachment 贴栅(硫化镉太阳电池生产中将金属网状电极贴在成结的电池上的工序)

grid bearing 坐标方位角,网格方位角
grid bias voltage 栅偏压
grid cell 网格单元
grid control tube 栅极控制管
grid convergence 网格收敛角(网格北与真北之间的夹角),平面子午线收敛角
grid coordinate system 格网坐标系
grid cut-off voltage 截止栅压
grid declination 真网格差(真北与网格北之间的夹角)
grid distance 格距
grid fin 1.格栅翼 2.栅格尾翼
grid generation 网格生成(计算流体用)
grid generation technique 网格生成技术
grid glow tube 栅控辉光放电管
grid heading 网格航向(从网格北顺时针量至飞机纵轴方向的角度)
grid leak detector 栅漏探测器
grid length 格距
grid line 1.栅格线 2.网格线
grid magnetic angle 网格磁偏角,方格磁角
grid material 板栅材料
grid mesh 1.栅极网孔 2.栅网 3.网格
grid method 网格法
grid network 1. grid 网络 2.栅极网络 3.网格网络
grid number 网格数
grid parameter 网格参数
grid pool tube 带有栅极的汞弧管,栅控汞弧整流器
grid reference 网格坐标,网格基准
grid resolution 网格分辨率
grid search 1.网格搜索 2.栅格搜索 3.网格搜索法
Grid shield "方格护板"(苏联海军雷达的北约代名)
Grid Sphere 网格球卫星(美国空军大气阻力测量卫星)
grid structure 网格结构
grid survey 1.网格法地形测量 2.网格法测量 3.网测量
grid system 1.网格系统 2.网格体系 3.坐标方格系统
grid ticks 方格记号
grid transformer 栅极变压器
grid transverse Mercator 横轴墨卡托格网
grid type 1.栅极型 2.网格型 3.栅型
grid universal transverse Mercator 通用横轴墨卡托格网
grid voltage 栅(极电)压
grid zone 栅格区
gridding 网格化;绘格线
grid-generation 网格生成
gridlock 网格同步
grid-plate transconductance 栅极与板极互导
grid-point method 网点板法
Griffith wing 格里非斯翼(一种上表面带有附面层吸除

缝的亚声速机翼）
grill 格栅，网格
grinding 磨削
grip nut 紧扣螺帽，夹紧螺帽
grip range 紧固厚度范围（用铆钉或其他紧固件将材料夹固在一起的厚度大小）
gripper 手形爪
grit blast 吹砂，喷砂清理
GRK Gear Rack 齿条
GRL 1. Goodrich Hella (Lighting) 古德里奇公司（照明）2. Grille 栅，栅格
GRLP Ground Lamp 地面指示灯
GRM Ground-Roll Monitor 着陆滑跑监控器
GRN Green 绿的，绿色
GRO Gamma Ray Observatory 伽马射线观测台（美国科学卫星）
GROM Grommet 护孔圈
gromagnetic trapped radiation 地磁场捕获辐射
grommet 孔圈，锁环
groove 沟，槽
groove type spray suppressor 喷溅抑制槽
grooved runway 开槽跑道
groover 开槽机，挖沟机（有轮子，用于在跑道上开槽沟）
Groshawk 苍鹰（美国空军45高级教练机之俗名）
gross adjustment 粗调
gross area 总面积（机翼总面积指其投影面积，包括机翼通过机身、吊舱等部分的面积在内）
gross bombing error 1. 轰炸过失误差 2. 轰炸超差
gross dry weight 总干重
gross error 粗差
gross error detection 粗差检测
gross flight path 总飞行轨迹（爬高阶段的飞行剖面）
gross flight performance 单机实测飞行性能
gross flight performance profile （飞机）飞行剖面图
gross height 总高（总飞行轨迹图上任何一点的高度）
gross leak test 粗检漏试验
gross lift 总升力（气球充气后在标准情况下及容许湿度情况下的浮力）
gross thrust 总推力，毛推力（未计入进气动量阻力）
gross weight 总重（飞机的总重过去都指最大允许飞行重量，现在都指最大起飞重量）
gross wing 内插翼
ground airconditioning unit （机场）地面空调设备
ground airvehicle 短距近地飞行器
ground alert 地面待战
ground angle 停放角（飞机在地面停放时，纵轴与水平面的夹角）
ground antenna 地面天线

ground attack 1. 对地攻击 2. 地面部队攻击 3. 陆攻
ground avoidance radar 地面回避雷达
ground clearance 离地距离，距地间隔（飞机停放时，某一部分与地面之间的距离，尤指直升机停放时，旋翼翼尖与地面的距离）
ground clutter 地面干扰，地物反射波
ground contact 1. 与地面的目视联系 2. 看地面的能见度 3. 飞机接地
ground control 1. 按地面指令操纵，地面引导，地面制导 2. 地面引导设备，地面制导设备 3. 地面交通管制
ground control mode 地面控制模式
ground control point 地面控制点
ground controlled approach system 地面指挥进近系统
ground cooling equipment 地面冷却设备
ground crew 地勤（机）组
ground data display equipment 地面数据显示设备
ground data handling system 地面数据处理系统
ground distributor 地面配电器
ground dynamic ejection test 地面动态弹射试验
ground echo 地面回波，地物回声
ground effect 地面效应
ground effect test 地面效应试验
ground effect testing 地面效应试验
ground effect vehicle 地效飞行器
ground electrical power source 地面电源
ground environment 1. 地面环境（指地面设备周围的环境）2. （由地面站造成的）电子环境（用于防空目的）3. 防空系统
ground environment equipment 地面环境模拟设备
ground equipment 1. 地面装备（航空武器系统中不升空的装备，如空导弹发射架）2. 地面保障设备（保证飞行所需的地面设备）
ground equipment bus 地面设备总线
ground fine pitch 地面小距（无反推装置的螺旋桨工作状态，以增大阻力）
ground fine pitchs top 地面小距限动锁（桨毂上的锁定机构，起落架压缩或受其他信号控制时可解脱锁钩）
ground fire 地面防空火力（多数人仅把此词用于小口径高炮及高射机枪）
ground flight trainer 地面飞行训练器
ground fog 地面雾
ground follow-up 地面查证
ground guidance 地面引导，地面制导，用地面设备导航，按地标驾驶飞机
ground half-coupling 地面半交联
ground handling equipment 地面装运设备（用于大部件，如机身、机翼等在地面移动的设备）
ground hold 地面等待（空中交通管制用语，指飞机等待批准后才开车）

ground impact   着陆冲击,撞地
ground influence mine   空投沉底非触发水雷
ground infrastructure   地面基础设施
ground instruction   地面讲授
ground interception   地面中断
ground liaison   地面部队对空联络官
ground load   地面载荷
ground loiter   地面待机
ground loop   (飞机)打地转
ground maintenance equipment   地面维护设备
ground mapping   地形绘制,地形显示
ground mapping radar   地形测绘雷达
ground movement control   地面运输管制机构(管理空军部队地面运输的军事单位)
ground moving target indication   地面活动目标显示(利用地面活动目标的多普勒频移从杂波背景中将其区分出来)
ground nadir   地面天底点(照相曝光时在照相机透视中心正下方的地面点)
ground navigation aids   地面导航设备
ground observer   地面观察员
ground operating equipment   地面操作设备
ground operation order   地面操作指令
ground operations aerospace language   航空航天地面操作语言
ground plane   1.地平面,地球平面 2.接地平面
ground plate   1.履带板 2.地板
ground plot   按航迹求推算位置,地面推算标绘图
ground position   地面位置(飞行器地面投影位置)
ground position system   地面定位系统
ground power supply for satellite   卫星地面电源
ground pressure breathing training   地面加压呼吸训练
ground proximity warning system   近地告警系统
ground radiation   地面辐射
ground range   地面距离,水平距离
ground range display   地面距离显示
ground readiness   地面备战状态
ground readout equipment   地面数据读出设备
ground receiving radius   地面接收半径
ground reference navigation   地标领航
ground reflection   地物反射波
ground relay   地面中继
ground resolution   影像分辨力,像元地面分辨力,地面分辨率
ground resolution cell   地面分辨单元
ground resolved distance   地面分辨距离
ground resonance   (直升机的)地面共振
ground resonance test   地面共振试验
ground return   地面反射,地面杂乱回波

ground roll   着陆滑跑距离
ground rule   1.程序规则 2.场地使用规则
ground run   1.地面运转,地面试车 2.起飞滑跑距离(从松开刹车到飞机离地)
ground safety control   地面安全控制
ground school syllabus   地面学校教学大纲
ground segment   地面段
ground sheet   地面气帘(垂直起飞时在地效范围内悬停,飞机下面地面上有"气流组成的辐射状的帘子")
ground signals   地面信号
ground simulation test   地面仿真试验
ground slope in aerial photo measurement   航空摄影测量中的地面坡度
ground spare satellite   地面备份卫星
ground specific impulse   地面比冲
ground speed   地速,对地速度
ground speed-drift angle indicator   地速偏流表
ground spoiler   地面扰流器
ground stabilised plot   地面坐标图
ground start   地面起动式发射(运载火箭的)
ground station electronic equipment   地面站电子设备
ground station of remote sensing   遥感地面接收站
ground support   1.(空中)对地支援 2.地面保障设备
ground support equipment   1.地面辅助设备 2.地面支持设备
ground swinging   地面测罗差
ground syncope   地面晕厥
ground system   地面系统
ground system for earth resource satellite   地球资源卫星地面系统
ground target   1.地面目标 2.大地目标 3.地靶
ground telemetering equipment   地面遥测设备
ground temperature   地温
ground terminal   地线接线柱,接地端
ground test   (航空装备或系统的)地面试验
ground test coupling   地面试验交联(被试验设备与测试设备之间的连通)
ground thrust   地面推力
ground tilt measurement   地倾斜观测
ground timing system   地面定时系统
ground track   地面轨迹,(航天器的)地面航迹
ground track angle   地面航迹角
ground truth   地面真值,地面实况
ground upset   地面倾翻(指轻型飞机或其他车辆、设备被气流吹翻)
ground vector control   (对歼击机的)地面引导
ground vehicle   地面车辆
ground vibration   地面振动
ground vibration test   地面振动试验

ground visibility 地面能见度(由专职观察员报告的或跑道能见距离测定仪测定的地面主要范围能见度)
ground wash (大型飞机发动机或翼尖在地面产生的)地洗流
ground wave 地表波
ground wind load 地面风载荷
ground wind test 地面风载荷试验
ground wind vortex sensing system 地面风涡遥感系统
ground wreck rate 地面事故毁机率
ground zero (原子弹)爆心(地面)投影点
ground-adjustable propeller 地面可调(桨距)螺旋桨
ground-air-ground load cycle 地-空-地载荷循环
ground-based 地基,地基的
ground-based astronomy 地面天文学
ground-based laser 地基激光器
ground-based surveillance and tracking system 地基监视跟踪系统
ground-based training device 地面训练器
ground-contour matching 地形匹配(制导)
ground-control equipment 地面控制设备
ground-controlled approach 地面控制进场着陆(用雷达监视飞机进场,用无线电指挥其着陆)
ground-controlled approach system 地面指挥引进系统
ground-controlled intercept 地面控制拦截
ground-derived navigation data 地面导出的导航数据(由陆地或海上测量获得的导航数据)
grounded 1.停飞的,禁止飞行的 2.接地的
ground-effect 地面效应
ground-elapsed time 地面经历时间
ground-electrical powersource 地面电源(航天器地面测试用的外接电源)
grounding 1.停飞,禁止飞行,飞行淘汰 2.接地,接地装置,地线
grounding and lapping resistance measurement 接地和搭接电阻测量
grounding electrode 接地电极
grounding lattice 接地网
grounding resistance 接地电阻
grounding transformer 接地变压器
ground-line-source biological munitions 地面线源生物战剂弹药
ground-mapping radar 地形测绘雷达
ground-performance aircraft 地面性能试验飞机
ground-power breaker 地面电源断路器
ground-power unit 地面电源设备(模拟卫星星上电源供电的直流稳压电源设备)
ground-proximity extraction system 近地拖投系统(低空空投装在货盘上的物资时,地面的减震装置挂住货盘下的吊钩,把货盘拖下)

ground-ranging equipment 地面测距器
ground-referenced navigation data 地面基准导航数据
groundspeed 地速
ground-speed computer 地速计算机
ground-speed mode (飞行控制系统的)恒定地速(飞行)方式
ground-support simulation computer 对地支援模拟计算机
ground-support vehicle 地面保障车辆
ground-test 地面试验
ground-testing 地面试验
ground-to-air bomber destroyer 地对空反轰炸机导弹
ground-to-air datalink 地对空数据链路
ground-to-space 地对空间,地对天
groundtrack 1.星下点轨迹 2.地面轨迹 3.地面径迹
group 1.大队 2.飞机主要部件(如机翼) 3.飞机主要部件研制组
group action 1.群作用 2.集团诉讼 3.群体诉讼
group address 组地址
group codes 群码
group control 群控
group dataset 组数据集
group decision 群决策
group delay 群时延
group delay distortion 群时延失真
group flashing light 闪光灯组(地面一组闪光灯,定时闪出摩尔斯码)
group frequency 群频率,频率组合
group repetition interval 组重复间隔
grout 水泥涂层,薄浆
growing 日益增长的
growl 隆隆声(飞行员耳机中听到的,表示导弹的红外导引头已截获目标)
growler 〈口语〉短路测试仪
growth behavior 生长行为
growth factor 生长因子
growth hillock 生长丘
growth model 生长模式
growth orientation 生长取向
growth percentage 生长率
growth process 1.增长过程 2.生长进程
growth rate 1.生长率 2.生长速度 3.增长率
growth temperature 生长温度(液相外延生长新单晶的温度)
growth tube 生长管道
growth velocity 生长速度
growth rate 生长率
GRS Grease 润滑脂,油脂
grunt manoeuvre 〈口语〉高过载机动

GRV Groove 槽,沟,凹线
GRVD Grooved 刻槽的
GRWT Gross Weight 总的重量
Gryphon 格里风通信系统(导弹核潜艇与海岸的通信系统)
GS 1. Glide Slope 下滑,下滑道,下滑坡度,下滑台 2. Ground Speed 地速 3. Ground Spoiler 地面扰流板 4. Guarded Switch 带保护罩的开关
GS/GPIA Glide Slope/Glide Path Intercept Altituede 下滑道/下滑航径切入高度
GSA 1. General Services Administration 综合服务管理局 2. Ground Service Agreement 地面服务协议 3. Ground Speed Ahead 前方地速
GSAPR Ground Service Auxiliary Power Relay 地面维护辅助电源继电器
GSARS Global Search And Rescue System 全球搜寻与救援系统
GSC 1. Ground Station Control 地面站控制 2. Ground Station Controller (ACARS) 地面站控制员(飞机通信寻址和报告系统) 3. Ground Switching Center 地面交换中心
GSD 1. Global Shape Deformation 整体外形修改 2. Ground Station Data 地面台数据
GSE Ground Support Equipment 地面保证设备,地面支援设备,地面辅助设备
Gsensing 加速度测量,载荷因数测量
GSEPR Ground Service External Power Relay 地面维护外电源继电器
GSI 1. Glide Slope Intercept 下滑道截获(切入) 2. Government Source Inspection 政府资源调查
GSIM Generic System Implementation Methology 生成系统执行分类法
GSIU Ground Standard Interface Unit 地面标准接口设备
GSKT Gasket 垫圈,密封垫
GSM 1. General Structure Manual 通用结构手册 2. Generative Shape Modeling 通用外形造型 3. Global Systems for Mobile Communications 移动通信全球系统(全球通)
GSMS Ground Station Management System 地面站管理系统
GSO Ground Systems Operations 地面系统操作
GSP Glare Shield Panel 遮光板,遮光罩
GSPU GPS Sensor Processor Unit 全球定位系统敏感处理部件
GSQA Government Source Quality Assurance 政府提供质量保证(美国)
GSR 1. Global Shared Resource 全球共享资源 2. Ground Service Relay 地面服务继电器 3. Ground Speed of Return 反航时地速 4. Ground Surveillance Radar 地面监视雷达
GSS 1. Geo Stationary Satellite 对地静止卫星 2. Global Surveillance System 全球对空观察系统 3. Gyro Stabilizing System 陀螺稳定系统
GSSF Government Satellite Services Facility 政府卫星服务设施
GSTDN Ground Satellite Tracking and Data Network 地面卫星跟踪和数据网
Gt 雷达天线增益(分贝)
GT 1. Gas Tight 气密的 2. Gear Train 齿轮箱 3. Grid Track 网格航迹 4. Group Technology 成组技术 5. Ground Transit 地面过境 6. Ground Transmission 地面传输
GTA General Terms Agreement 通用项协议
GTACS Ground-Target Attack Control System 地面目标攻击控制系统
GTC 1. Ground Terminal Computer (数据链)地面终端计算机 2. Gyro Time Constant 陀螺时间常数
GTDS Goddard Trajectory Determination System 哥达德定轨系统
GTF Ground Test Facility 地面试验设施
GTO Geostationary Transfer Orbit 地球静止转移轨道 2. Geosynchronous Transfer Orbit 地球同步转移轨道
gtolerance 过载容限,容许过载
GTOS Ground Terminal Operations Support 地面终端操作保障
GTOSS Generalized Tethered Object System Simulation 广义绳系物体的系统模拟
GTOW Gross Takeoff Weight 起飞总重,起飞总质量
GTP Graphic Transform Package 图形变换程序包
GTPE Gun Time Per Engagement 机枪(机炮)在一次攻击中射击持续时间
GTR 1. Galley Transfer Relay 厨房转换继电器 2. General Technical Requirements 通用技术要求 3. Ground Test Report 地面试验报告 4. Gulf Test Range 海湾试验场
GTRB Gas Turbine 燃气涡轮
GTS 1. Gas-Turbine Starter 燃气涡轮起动机 2. General Technical Specifiation 通用技术规范 3. Geostationary Technology Satellite 地球静止轨道技术卫星 4. Glider-Training School 滑翔机训练学校 5. Global Telecommunication System 全球通信系统
GTT Ground Test Time 地面试验时间
GTW Gross Takeoff Weight 起飞总重
GTY 1. Ground-Test Vehicle 地面试验飞行器(指直升机) 2. Guidance Test Vehicle 制导试验飞行器
g-type tester 过载试验台
GU Gear Up 起落架收起

GUAR Guarantee 保证

guarantee (GTEE),保证,担保

guaranteed rating 保证功率,保证推力(生产厂家保证提供的最低功率或最小推力)

guaranteed service life 保证使用寿命,保用期

guard band 防护频带,防扰频带

guard frequency 防护频率(防相互干扰的),应急甚高频波道(频率)

guard gates 防护波门

guard ring 保护环

guard ship 1.护航武装直升机 2.(航母上的)值班救援直升机

guard vacuum 真空保护层

guarded hot plate conductmeter 护热式平板热导仪

guarded switch 保护电门

guess field 推测场

guest-host effect 宾主效应

GUI Graphic User Interface 图解/用户界面

guidance 1.制导,引导,导航 2.指引,指导,导向 3.控制,操纵 4.制导系统,导航系统

guidance accuracy 制导精度

guidance and control system 制导和控制系统

guidance and navigation computer 制导与导航计算机

guidance axis 导向轴

guidance channel 导引槽

guidance checkout 制导系统测试

guidance command 制导指令

guidance computer 制导计算机

guidance counter measure 对制导系统的干扰,制导对抗

guidance cutoff 制导关机

guidance device 导引装置

guidance electronics 制导电子设备

guidance engine cut off 制导发动机停火

guidance equation 制导方程

guidance equation error (制导)方程误差

guidance equipment 1.导航设备 2.制导设备

guidance error 制导误差

guidance failure 制导(系统)故障

guidance field 引导区域,引导场,制导场

guidance fuze 制导引信

guidance law 制导律,导引规律

guidance loop 制导系统回路

guidance model 制导模型

guidance noise 1.制导起伏误差 2.制导系统噪声(干扰)

guidance nose 〈口语〉制导头部,导引头

guidance operation 制导操作

guidance parameter 制导参数,制导参量

guidance platform assembly 制导平台装置

guidance position tracking 制导位置跟踪

guidance positioning assembly 制导定位装置

guidance radar 制导雷达

guidance scheme 1.制导方案 2.导引方案 3.指导方案

guidance sensor 制导传感器

guidance sight 制导瞄准具

guidance signal processor 制导信号处理机

guidance slot 导向槽

guidance strategy 1.指导策略 2.诱导策略 3.制导策略

guidance system 制导系统

guidance system technology 制导系统技术

guidance system test 制导系统测试

guidance technique 制导技术,制导方法

guidance test 制导系统试验,制导设备测试

guidance test vehicle 制导系统试验飞行器

guidance time 制导持续时间

guidance time-base 制导时基

guidance tolerance 制导容许误差

guidance unit 制导装置

guidance van 车载活动制导台,制导车

guide 1.制导,引导,控制 2.导向装置,导轨,导套 3.指南,手册 4.波导,波导管

guide face 导流面

guide finger (六管炮的)导向爪

guide flare (炸弹的)制导曳光管

guide rope 拖引索(从气球上扔下,当刹车用,即阻力索)

guide vane 导流叶片,导流片

guide wave length 波导波长

guide aileron 导引副翼,辅助副翼

guided air-to-underwater missile 空对潜导弹

guided antitank missile 反坦克导弹

guided bomb 1.制导炸弹 2.灵巧炸弹

guided cluster weapon 集束制导武器

guided flight 制导飞行

guided flight vehicle 制导飞行器

guided folding fin aircraft rocket 机载折叠翼可控火箭弹(导弹)

guided missile 导弹

guided missile explosive harness 导弹战斗部(自毁)导火线

guided missile service unit 导弹维护部队

guided missile system 导弹系统(包括导弹及其发射与制导设备)

guided missile unit 导弹部队

guided modular weapon 模式制导武器

guided space vehicle 制导航天器

guided stand-off weapon 远射空(对)地导弹,在防空区外发射的空(对)地导弹

guided trajectory 制导飞行弹道,制导轨迹
guided trial (在制导系统工作条件下的导弹)制导试验
guided wave 导波
guided weapon 制导武器
guide-surface canopy 可操纵的降落伞(伞衣从背包中放出后,可操纵运动方向)
guide-surface parachute 导向面伞
guide-wire (有线制导导弹的)制导线
guiding 制导,引导,控制,导向,波导
guiding center 1.引导中心 2.导向中心 3.导心
guiding centre plasma 引导中心等离子体
guiding device 导星装置
guiding fin 1.安定面,尾翼 2.(散热)导流片
guiding star 引导星
guiding telescope 导星镜(主望远镜镜筒上附加的)
Guidonia 圭多尼亚(意大利航天研究中心,1944年前属国防部)
Guier plane 吉尔平面(卫星在"近站点"时,卫星相对于测站方向及其变率所组成的平面)
Guier theorem 吉尔定理(用卫星近站点参数表示一次卫星通过信息的定理)
guillotine factor 截断因子
guillotine firing unit 切割器点火设备
gull wing 海鸥式机翼
gull-wing door 弯翼式舱门(向两面打开后,有如海鸥张翼,呈倒W形)
gum 树胶,胶,胶质物(常指汽油因氧化而生成的残余物)
Gummel-Poon model 根摩尔-普恩模型
gun 1.炮,火炮 2.枪,步枪,手枪 3.〈口语〉筒形发射导轨 4.(活塞式发动机的)油门
gun bore line 枪膛轴线,炮膛轴线
gun breech firing cap 燃爆筒打火帽
gun clear lockin relay 机关炮退弹自保(持)继电器
gun clearing 1.退弹 2.排除停射故障
gun cross 机炮瞄准十字线
gun feeder 进弹机
gun fire coverage 射击范围
gun fire dispersion 射弹散布
gun jump 定起角,发射差角(射击时炮膛轴线与炮弹离开炮口时的弹迹之间的夹角)
gun launching 炮式发射
gun metal 1.炮铜,锡粹青铜 2.古铜色
gun muzzle 炮口,枪口
gun operation alcontrol (高炮)射击指挥
gun pack 炮组
gun post (轰炸机上的)炮塔,射击装置
gun powder 黑色火药,有烟火药
gun probe 炮射探测系统

gun purge system 枪炮排烟系统,枪炮火药气体吹除系统
gun recoil 炮管后坐,炮身后坐
gun ring 机枪旋转架,回转炮塔
gun synchronizing gear 机枪射击协调装置
gun tube 1.火炮身管 2.身管
gunner 射击员,炮手
gunnery 1.射击学,射击技术 2.射击 3.枪炮操作
gunpod 航炮吊舱
guns battery-hold back switch 机心释放装弹和后退扣住开关
gunship 武装直升机(特指一种细长双座机身的,带有防御和进攻武器的直升机)
gunsight 射击瞄准具
gunsight lamp 瞄准具灯(对太阳方向目标射击用)
gunsight line 瞄准线
gunstrike camera 攻击照相枪(直接拍摄目标,记录射击效果)
guntype weapon 枪(炮)式武器,炮式核武器(以最大速度同时发射两个或两个以上亚临界可裂变物质而激发的核武器)
gununit 炮组(炮、炮架、弹箱等整体装置)
gunweld 半自动电弧焊接机,焊枪,焊钳
Guppy 〈口语〉指机身肥大的飞机,通常用来装运战略导弹或其他大型装备
GUSS Gusset 角撑板
gusset 角撑板,结点板(结构转角处或连接处的小块加强构件)
gust 1.风速突增,阵风 2.(突然遇到的)上升或下降气流区
gust alleviation 阵风缓和动态系统
gust alleviation factor 阵风缓和因子(英国民航适航证要求估计此因子为0.61,即阵风对结构载荷的影响按真实的锐边阵风61%来计算)
gust alleviation system 阵风缓和(减缓)系统
gust curve 阵风曲线(阵风相对于周围空气的风速对未扰动空气至阵风风速峰值位置的水平距离的曲线图)
gust envelope 阵风包线(此包线图的纵轴为结构安全系数,横轴为空速,用于飞机设计)
gust load 阵风载荷
gust load alleviation 阵风载荷减缓
gust response 阵风响应
gust sonde 阵风探空仪
gust speed 阵风速率
gust velocity 1.阵风速度 2.阵风风速 3.脉动风速
gust wind factor 阵风因子
gust wind tunnel 阵风风洞
gustiness factor 阵风风速增量因子(最大阵风风速与无阵风时风速之差,以平均风速的百分数计)

gustload alleviation 阵风减载,阵风缓和(主动控制系统在阵风作用下的一种乘座品质控制和减轻结构疲劳的功用),阵风载荷减缓

gustload factor 阵风载荷因数

gustloading 阵风载荷(阵风作用使结构承受的载荷)

gustlocks (操纵系统的)锁定装置(飞机地面停放时用)

gutta-percha 马来橡胶,古塔波橡胶

gutter 稳定槽(火焰稳定器的)

guy 拉线,牵(支,稳,拉,张)索

GV 1. Gate Valve 闸板活门 2. Groove 凹槽 3. Ground Valve 地面槽

GVE Graphics Vector Engine 图解矢量发动机

GVI General Visual Inspection 一般目视检查,一般外观检查

GVPF Geared Variable-Pitch Fan 齿轮传动变距风扇

GVS Ground Velocity Subsystem (对)地速(度)分系统

GVSC Generic VHSIC Spaceborne Computer 通用甚高速集成电路星载计算机

GVT Ground Vibration Test 地面振动试验

GVTA Ground Vibration Test Article 地面振动试验件

GVW Gross Vehicle Weight 飞行器总重量

GW 1. Gross Weight 全重,毛重 2. Ground Wave 地(面)波 3. Guided Weapon 制导武器,导弹

GWE Global Weather Experiment 全球天气试验

GWEN Ground Wave Emergency Network 地波应急网络

GWJ Garnet Waterjet 加奈特水射流(装置用于高速切割金属的)

GWS 1. Graphical Weather Services 图解气象服务 2. Guided Weapon System 制导武器系统

GWT Gross Weight 总重,毛重,(飞机的)起飞重量,满载重量

GWVSS Ground Wind Vortex Sensing System 地面风涡遥感系统

GYP Gypsum 石膏

GYRO Gyroscope 陀螺仪

gyro accelerometer 陀螺加速度表

gyro amplifier 回旋放大管

gyro angling gain 陀螺定向角增益

gyro azimuth 陀螺方位角

gyro backup 陀螺仪修正

gyro characteristic time 陀螺特征时间

gyro clinometer 陀螺倾斜仪

gyro cluster 陀螺组件

gyro compass 1.陀螺罗盘 2.陀螺半罗盘 3.感应式陀螺磁罗盘

gyro compass alignment 陀螺罗经对准

gyro compassing 陀螺定向,陀螺罗盘对准

gyro compassing time 陀螺罗盘工作准备时间,陀螺罗盘起动时间

gyro compensation 陀螺补偿

gyro compliance 陀螺顺从性

gyro control amplifier 陀螺控制放大器

gyro coordinate system 陀螺坐标系

gyro copter 自转旋翼机

gyro correlation time 陀螺相关时间

gyro cradle 陀螺支架

gyro drift rate 陀螺漂移率

gyro dyne aircraft 螺旋桨拉进式垂直起飞飞机

gyro effect 陀螺效应

gyro electric medium 旋电媒质

gyro envelope 陀螺外壳

gyro erected optical navigation 陀螺垂直光学导航

gyro erection 陀螺的修正

gyro error 陀螺仪误差

gyro failure 陀螺故障

gyro float 陀螺浮子

gyro fluid 1.陀螺浮油 2.陀螺液

gyro foot 陀螺安装脚

gyro frame 陀螺框架

gyro frequency 旋转频率(指电子绕某指示电离层中磁场方向的线旋转的频率)

gyro gimbal 陀螺框架

gyro graph 记转器,陀螺漂移记录器

gyro gunsight 陀螺射击瞄准具

gyro horizon 1.陀螺地平仪,回转水平仪 2.人工地平

gyro housing 陀螺外壳

gyro input axis 陀螺仪输入轴

gyro klystron 回旋速调管

gyro laser 激光陀螺

gyro level 陀螺水平仪,陀螺倾斜仪

gyro link assembly 陀螺连杆装置

gyro log 陀螺仪记录器(计算并记录陀螺漂移)

gyro loop Nichols chart 陀螺回路尼氏图

gyro magnetic 回转磁的,旋磁的(有关回转带电粒子的磁性)

gyro magnetic compass 陀螺磁罗盘,回转磁罗盘

gyro magnetic device 旋磁器件

gyro magnetic effect 旋磁效应

gyro magnetic filter 旋磁滤波器

gyro magnetic limiter 旋磁限幅器

gyro magnetic medium 旋磁媒质

gyro magnetic oscillator 旋磁振荡器

gyro magnetic ratio 旋磁比

gyro magnetic shield 陀螺磁屏蔽

gyro meridian 陀螺仪子午线

gyro motor 陀螺马达,陀螺电动机

gyro mounting interface 陀螺安装面

gyro null　陀螺零位
gyro oscillator　回旋振荡管
gyro out approach　陀螺半罗盘故障条件下进场着陆
gyro output axis　陀螺仪输出轴
gyro peniotron　回旋潘尼管
gyro pickoff　陀螺角度敏感器,陀螺传感器
gyro pilot　陀螺驾驶仪,自动驾驶仪
gyro platform　陀螺平台
gyro ramp drift　陀螺斜坡漂移
gyro rebalance loop　陀螺再平衡回路
gyro reference　陀螺基准,陀螺传感器
gyro reflector　（瞄准具的）陀螺（反射）镜
gyro resonance　陀螺谐振（陀螺章动）
gyro rotor　陀螺转子
gyro runup time　陀螺起动时间,陀螺加速转动时间
gyro scope assembly　陀螺组件
gyro servo test　1.陀螺伺服试验 2.转台反馈试验
gyro spin axis　陀螺自转轴
gyro spinner　陀螺旋转器
gyro test　陀螺试验
gyro theodolite　陀螺经纬仪
gyro time constant　陀螺时间常数
gyro torque　1.陀螺力矩 2.陀螺转矩传感器
gyro torque gasymmetry　陀螺力矩不对称性
gyro torque rebalance test　1.陀螺力矩反馈试验 2.速率反馈试验
gyro torque-feedback testing　陀螺力矩反馈试验
gyro torquing rate　陀螺力矩修正速率
gyro transfer table system　陀螺传递台系统
gyro tron　振动陀螺仪,陀螺振子
gyro tumbling test　陀螺翻滚试验
gyro turn table　陀螺转台
gyro TWA　回旋行波放大管
gyro unbalance　陀螺失衡
gyro unit　陀螺环节,陀螺组件
gyro vector axis　陀螺转动矢量轴
gyro vertical　垂直陀螺仪,陀螺垂线
gyro vibrated drift　陀螺振动漂移
gyro viscous damping coefficient　陀螺黏性阻尼系数
gyro wheel　陀螺仪转子
gyro wobbling drift　陀螺摇摆漂移
gyro yoke　陀螺磁轭

gyrocompass　1.陀螺罗盘 2.陀螺罗经
gyrocompass alignment　陀螺罗盘对准
gyro-compassing alignment　陀螺罗盘对准
gyrodyne　螺旋桨拉进式直升机
gyro-magnetron　回旋磁控管
gyroradius　回转半径
gyro-rotor　陀螺转子
gyroscope　陀螺,陀螺仪,回转仪
gyroscope axis　陀螺自转轴
gyroscope bearing　陀螺轴承
gyroscope stabilized platform　陀螺稳定平台
gyroscopic action　陀螺效应,回转作用
gyroscopic apparatus　陀螺装置
gyroscopic body　陀螺体
gyroscopic drift　陀螺漂移
gyroscopic effect　陀螺效应
gyroscopic float　陀螺浮子
gyroscopic gun sight　陀螺射击瞄准具
gyroscopic inertia　回转惯性
gyroscopic moment　陀螺力矩
gyroscopic precession torque　陀螺进动力矩
gyroscopic sensor　陀螺传感器
gyroscopic sight　陀螺瞄准具
gyroscopic torque　陀螺转矩
gyroscopic wander　陀螺漂移
gyroservo-turntable testing　陀螺伺服转台试验
gyroset coordinate system　陀螺坐标系
gyro-sextant　陀螺六分仪
gyrostabilization unit　陀螺稳定装置
gyrostabilized　陀螺稳定的（利用陀螺定轴性稳定的）
gyrostabilized mirror　陀螺稳定镜
gyrostabilizer　陀螺稳定器
gyrostat　1.陀螺体（刚体上带固定轴的高速旋转轴对称转子所构成的物体）2.陀螺稳定体（休斯公司的双自旋稳定方案）3.陀螺仪回转仪 4.卫星自转技术（休斯公司研究的,长度很大的人造卫星绕一短轴自转的技术）
gyrostatic orientation survey　陀螺定向测量
gyrosyn　感应式陀螺磁罗盘
GZ　Ground Zero 地面零点
Gz-induced loss of consciousness　Gz 引起的意识丧失

H

H 1. Hour 小时 2. Height 高度 3. High Potential Side of Output(Input) 输出(入)线路高电位端

H/W Hardware 硬件

HA 1. Hazard Analysis 危害分析 2. High Altitude 高空,高度 3. Hour Angle 时角

HAA 1. Height Above Aerodrome 高出机场高度 2. Height Above Airport 机场上空高度 3. High Angle of Attack 大迎角

HAARS Heading and All-Attitude Reference System 航向和全姿态基准系统

HAB 1. Hazards Analysis Board 险情分析委员会 2. High Altitude Burst 高空爆发,高空爆炸

habit face 惯态面

habitability in space 太空可居住性

habitation module 居住舱

habituation 习惯,熟习

HAC 1. Heading Alignment Circle 航向校准圆 2. Heavy Antitank Convoy 重型(飞机)反坦克护航 3. High Altitude Calibration 高空校准 4. High Altitude Chart 高空图

hacienda 〈口语〉航空航天研究局(美国空军)

hack 1. 役用飞机 2. 使能完成 3. 劈,砍,劈砍工具,劈痕 4. 格架

hackly fracture (金属断裂的)锯齿状断

HAD 1. Hardware Architecture Document 硬性特性文件 2. Hybrid Analog/Digital 模/数混合

hadron era 强子期

HADS 1. Helicopter Air-Data System 直升机空气数据系统 2. High-Accuracy Digital Sensor 高精度数字敏感器

HADTS High-Accuracy Data Transmission System 高精度数据传输系统

haematogenic hypoxia 血源性缺氧

hafnia 二氧化铪

Hafnium 铪

Hagelbarger code 哈格伯尔格码(纠正突发错误的一种码)

HAH Hot And High 高温高原

HAI Helicopter Association International 国际直升机协会

hair absorber 发垫吸波材料

hair line 1. 标线 2. 细缝 3. 细微的区别

hair line crack 发丝裂纹,发裂

HAL 1. Handle 手柄 2. Helicopter Attacksquadron Light 轻型攻击直升机中队 3. Holding and Approach-to-Land 等待与进近着陆

halation 光晕

HALEX Halogen Lamp Experiment 卤灯实验

half adder 半加(法)器

half duplex 半双工

half duplex link 半双工链路

half duplex transmission 半双工传输

half echo suppressor 半回波抑制器

half flexible plate nozzle with many hinge point 多支点半柔壁喷管

half flexible plate nozzle with single hinge point 单支点半柔壁喷管

half floated rate gyro 半液浮速率陀螺仪

half life 半衰期

half model 半模型

half model balance 半模型天平

half model test 半模型试验

half residence time 半留时间(武器碎片在大气中沉降量达到原始值一半所需的时间)

half roll and half loop 1. 半滚倒转 2. 下滑倒转

half selection noise 半选噪声

half thickness 强度减半厚度(使辐射强度减半的吸收介质厚度)

half time emitter 半时发送器

half value thickness 半值厚度(指吸收体)

half view 半视图(只绘出对称物体的一半)

half width 半宽

half-angle 半角的

half-concentric resonator 半共心谐振腔

half-coneangle 半锥角

half-confocal resonator 半共焦谐振腔

half-duplex channel 半双工信道

half-fairing iron wheel carriage 整流罩半罩铁轮支架车

half-fairing trailer 整流罩半罩运输车

half-fairing turning sling 星罩半罩翻转吊具

half-frequency 半频

half-interval contour 间曲线,半距等高线

half-life 半衰期,半排出期

half-loaded 半装弹的

half-model test 半模实验

half-period 半周期,半周,半衰期

half-power points rings　半功率点环(指雷达天线的辐射功率)
half-power width　半功率点宽度
half-scale model　半尺寸模型
half-section network　半节网络
half-select pulse　半选脉冲
half-shadow　半影,半荫
half-sine shock pulse　半正弦冲击脉冲
half-spherical resonance gyro　半球谐振陀螺
half-thickness　半厚度(把电离层电子密度最大值所在高度以下的电子密度分布看成抛物线分布而得的等效厚度)
half-time value　半衰期值
halftone　半色调
half-twist　半扭转的
half-wave dipole　半波振子
half-wave length　半波长
half-wave rectification　半波整流
half-wave transformer　半波(整流)变压器
half-wing　半机翼
halide leak detector　卤素检漏仪
Hall conductivity　霍尔电导率
Hall current accelerator　霍尔电流加速器
Hall device　霍尔器件
Hall displacement transducer　霍尔位移传感器
Hall mobility　霍尔迁移率
Hall-effect　霍尔效应
hallucination　幻觉
HALO　High-Altitude Long Operation 高度远程运行
halo cell　晕细胞
halo effect　晕圈效应(超声速飞机营运后能使该公司在同一航线上营运的亚声速航班的头等舱客人增加)
halo orbit　晕轨道
halo population　晕族
halogen counter　卤素计数管
halyard　吊索,升降索
Hamilton principle　哈密顿原理
Hamilton sequation　哈密顿方程
Hamiltonian　1.哈密顿函数 2.哈密顿算符
Hamiltonian function　哈密顿函数
hammer　锤击
hammer test　锤击试验
hammerhead　跃升下坠倒转
hammerhead fairing　锤头形整流罩
Hamming bound　汉明界
Hamming code　汉明码(一种纠错码)
Hamming distance　汉明距离(描述检错纠错码性能的一种量度)
Hamming weight　汉明权(重)

HAMOTS　High Accuracy Multiple Object Tracking System 高精度多目标跟踪系统
HAMS　Hot Air Management System 暖气管理系统
HAND　Handling 搬运
hand anemometer　手提风速表,轻便风速表
hand bumping　手冲(加工),手工钣金,人工锤击
hand controller　手操纵器(特指自动或半自动系统如平显、多工作方式雷达等人机接口用的手柄、滚球等)
hand cranking　手摇起动(活塞式发动机或小型燃气涡轮机)
hand flying　手操纵飞行
hand forging　手工锻造,打铁
hand forming　手工成形
hand held　自备电源的话筒
hand held console　手持控制台
hand inertia starter　手摇惯性起动机
hand lay-up　手糊成形
hand linker-delinker　手工弹链装拆器
hand operated mechanobalance　手动机械天平
hand staff　1.联轴齿轮 2.(仪表的)指针轴
hand starter　手摇起动机
hand tracking　手控跟踪
hand writing test　书写试验
handbook problem　飞行员手册问题
hand-colored map　手工着色地图
hand-crafted　1.手工制的 2.特制的
hand-drawn original　精绘原图
handed　左右对称的(指飞机左面有的机件,右面也有相同的机件)
handed propellers　左右对转的(两副)螺旋桨
handheld　手动的,人工的
hand-held thermal imager　手提式热成像器
hand-held radio　手提式电台
handhold　握住,把柄,线索,交给
handiness　(操纵的)应手性,灵便性
handing hoisting device working ladder　装卸吊具工作梯
handle　柄;处理,办理;手柄,把手
handling　1.操纵,驾驶,控制,调节 2.对某一飞机操纵面响应的主观印象,操纵特点,操纵性 3.处理,装卸,操作 4.维护,保管,管理
handling error　操纵(处置)错误
handling line　1.(地面人员的)气球操纵索 2.(水上飞机重心上方的)吊索
handling pilot　实际操纵飞机的飞行员
handling quality　(飞机的)驾驶品质
handling squadron　基本飞行性能试飞中队
handoff　控制转交(地面雷达操作员将目标交给另一个雷达的操作员)
handover　指挥交接(把对飞机的指挥从一个指挥机构

移交给另一个指挥机构)

handshaking 数据交换,符号交换过程(同步传输系统的)

hands-off 1.松杆飞行 2.用自动驾驶仪飞行 3.(地面人员)松开气球吊篮

handsoff flight 飞机飞行

hands-on 1.握(驾驶)杆的 2.(电子数据处理系统)人机接口的

hands-on throttle and stick 握杆控制

hands-on control 飞行员自己操纵

hand-starter magneto 手摇起动磁电机(用于活塞式发动机)

hang 挂

hang glider 悬挂式滑翔机

hangar 机库

hangar flying 〈口语〉指飞行人员在地面或机库谈论飞行情况

hangar Queen 〈俚语〉机窝飞机(指那些经常出故障,长期在机库停放维修的飞机)

hangar refuelling point 机库加油口

hangarage 1.机库容量 2.在机库停放 3.机库群

hangarette 小机库

hanger fitting 起吊接头

hangfire device 击发延迟测试仪

hang-up 1.转速悬挂(指喷气发动机推油门时转速不增加) 2.(外挂或内挂物)投放不下去 3.中止,挂起,意外停机

hang-up prevention 意外停机预防

hangwire 炸弹保险丝(用于投弹时自动启开引信保险)

Hansen coefficient 汉森系数

Hansen theory 汉森理论

hant (仪表着陆系统或微波着陆系统)天线离地面的高度

Hanzi code 汉字代码

Hanzified multilink system 汉字化多链路系统

HAO High Altitude Observatory 高空观测台

HAPI Helicopter Approach Path Indicator 直升机进近航道指示器

Hapshelmet-angle position sensor (头盔瞄准具的)头盔角位置传感器

haptic feedback 1.触觉反馈 2.接触反馈 3.力觉反馈

haptic interface 触觉感知接口

HARA High-Altitude Radar Altimeter 高空雷达高度计

harass 1.空中骚扰 2.接近骚扰

harbor radar system 港口雷达系统

harbor surveillance radar 港口监视雷达

Hard 哈德(瑞典三坐标搜索雷达)

hard alloy 硬质合金

hard aluminium alloy 1.硬铝合金 2.杜拉铝

hard anodizing 硬质阳极(氧)化

hard body 硬体(指飞行中的导弹和红外敏感器探测到的飞行目标)

hard constraint 硬约束

hard copy 硬拷贝,硬性复制件

hard deck 硬甲板(即指地面,用于低空空战)

hard flutter 硬颤振,突发颤振

hard hose for loading 加注硬管

hard image 高反差影像

hard iron 硬铁

hard kill 硬杀伤

hard lander 硬着陆航天器,硬登陆舱

hard landing 1.硬着陆,粗猛着陆 2.硬登陆舱落在月球或行星表面

hard limited integrator 硬限量积分器

hard limiter 硬限幅器

hard limiting transponder 硬限幅转发器

hard memory failure 存储器硬件故障

hard point 1.硬点 2.硬质点

hard pressure suit (外层为金属材料制成的)硬质加压服

hard radiation 硬辐射,贯穿辐射

hard recovery 硬回收

hard rubber 硬橡胶

hard science 硬科学,自然科学

hard sifter injector 多孔材料喷注器

hard space 硬太空(充满宇宙线、射线和紫外线等硬辐射线的空间环境)

hard space suit 硬式航天服

hard start 硬启动

hard stop 1.硬停机 2.急停

hard target 硬目标,加固的目标

hard temper 冷作软化,冷作韧化

hard time maintenance 定期维修

hard tube 高真空管,硬性(真空)管

hard tube switch modulator 硬性管调制器

hard turn 急转弯

hard vacuum 高真空

hard water 硬水

hard weapon 硬武器

hard wing 硬前缘翼,前缘无缝翼

hard wired controller 硬连线控制器

hard wiredinter connection 硬连线互连

hardcore fault 核心硬件故障

harden 使变硬,使坚强,使冷酷,变硬,变冷酷,涨价

harden antenna window material 加固天线窗材料

hardenability 加固能力,加固程度

hardened 1.有核防护设施的,防原子的 2.地下的,坚

固防护的 3. 硬化的, 淬火的
hardened target 加固目标, 设防目标
hardening 1. 硬化, 淬火, 凝固 2. 硬化剂 3.（抗核、激光、粒子云、辐射等的）加固
harder spectrum 较硬能谱
hard-iron magnetism 硬铁磁性
hardness 1. 硬度（金属材料的）2. 核防护能力, 防核坚固度
hardness allocation 加固分配
hardness design 加固设计
hardness test 硬度试验
hardness testing fixture 硬度试验夹具
hardness value 硬度值
hardness verification test 加固（程度）验证试验
hardover 1.（被调参数）急增, 急偏,（舵面）急剧偏转 2. 失控, 突然偏离预定轨道 3.（调节系统）信号过强
hardover failure 1. 失控故障（指飞行器运动参数急剧偏离预定值）2. 信号过强故障（指调节系统）
hardover runaway 急偏失控
hardover signal 过强信号
hardpoint 承力点, 承力接头
hardpoint target 硬点目标
hard-sphere 1. 硬球 2. 刚球
hard-sphere mode 硬球模型
hardtime 定时的, 按严格时限的
hardwall hose 硬壁软管
hardware 1.（计算机等的）硬件 2. 硬设备 3. 金属制品
hardware address 机器地址
hardware address control 机器地址控制
hardware arbiter 硬件仲裁器
hardware associative memory 硬件相联存储器
hardware breakpoint 硬件断点
hardware component 硬件部件, 硬件成分
hardware context 硬件关联
hardware debug 硬件调试
hardware deficiency 硬件缺陷
hardware development 硬件开发
hardware dump 硬件转储
hardware enhancement 硬件增强
hardware failure 硬件故障
hardware implemented fault tolerance 硬件实现的容错（技术）
hardware in loop fuze simulation 引信半实物仿真
hardware interrupt system 硬件中断系统
hardware logic diagram 硬件逻辑图
hardware logic simulation 硬件逻辑模拟
hardware monitor 硬件监视器
hardware rategyro 硬件速率陀螺
hardware real-time monitor 硬件实时监控器

hardware reliability 硬件可靠性
hardware stack 硬件栈
hardware vector to raster 向量光栅变换硬件
hardware virtualizer 硬件虚拟器
hardware-in-the-loop simulation 半实物仿真
Hares and Hounds 双机夹攻战术
HARM Highspeed Anti-Radiation Missile 高速反辐射导弹
harmonic 1. 谐波, 谐波的 2. 正弦（曲线）的 3.（复数）调和函数
harmonic analyser 简谐分析器, 谐波分析仪
harmonic analysis 谐波分析
harmonic balance 1. 谐波平衡 2. 谐波平衡法 3. 谐波抑制
harmonic component 1. 谐波分量 2. 谐波 3. 谐波成分
harmonic content 1. 谐波含量 2. 谐波 3. 谐波成分
harmonic conversion transducer 谐波变换换能器
harmonic distortion 谐波失真
harmonic excitation 谐波励磁
harmonic filter 谐波滤波器
harmonic frequency 1. 谐振频率 2. 谐波频率 3. 振动频率
harmonic motion 谐运动
harmonic oscillator 1. 谐振子 2. 谐波振荡器
harmonic oscillator model 1. 简谐振子模型 2. 谐振子模型 3. 振子模型
harmonic polynomial 1. 调和多项式 2. 多项式调和
harmonic SAR 合成孔径谐波雷达
harmonic solution 调和解
harmonic suppression 谐波抑制
harmonic suppression filter 谐波抑制滤波器
harmonics 1. 谐波 2. 谐波振荡
harmonious deviation 和谐偏差
harmonious strategy 和谐策略
harmonious variable 和谐变量
harmonization 校靶, 谐和
harmonization tool 安装校准工具
harness 1.（降落伞）背带, 编织带 2.（座椅）安全带 3. 导线束, 电缆 4.（集装箱或货盘）固定索, 系留索具
HARP 1. Helicopter Advanced Rotor Program 先进的直升机旋翼计划 2. High Altitude Research Project 高空研究计划
harpoon 鱼叉（舰载直升机上的着舰装置）
harpoon and grid decksecuring system （舰载直升机）鱼叉及格栅甲板固定系统
harpoon penetrator 插入固定叉（苏联"火卫一"着陆探测器的固定装置）
harpoon system 倒钩系留系统, 锚式系留系统
Hartmann generator 哈特曼噪声发生器

HAS 1. Hardened Aircraft Shelter 飞机核防护掩体,加固的飞机掩体 2. Hover Augmentation System 悬停增稳系统
Hasell check 六项检查,"六查"(飞行员进行螺旋等大难度机动动作之前必须完成六个方面的检查:高度、机体、安全、发动机、飞机位置、向外观察)
hash 脉冲干扰,地物回波
hash file 散裂文件
hash file system 散列文件系统
hash line 散列线
hash transformation 散列向量
HASP High-Altitude Sampling Plane 高空采样飞机
HASTE Helicopter Ambulance Service To Emergencies 直升机紧急救护(美国)
Hastelloy 哈斯特合金
HAT Height Above Touchdown 离接地点高度
hat 帽盖,端盖
hatch 舱口
hatchback 客货运输机
hatchway 舱口
Hatol horizontal-attitude takeoff and landing 水平姿态起飞和着陆
HATS Helicopter Automatic Targeting System 直升机自动瞄准系统
HAWC Homing And Warning Computer 归航和告警计算机
Hawfcar helicopter adverse-weather fire control acquisition radar 直升机恶劣气候火控探测雷达
HAZ Hazardous 危险的,有害的
hazard alert 危险通告
hazard analysis 危险分析
hazard assessment 1.危险性评估 2.灾害评估
hazard avoidance 危险规避
hazard beacon 危险灯标
hazard control 危险控制
hazard level 危险等级
hazard severity 危险程度,后果严重性
hazard free flip-flop 无险触发器
hazardous duty life jacket 危险工作救生服
hazardous weather 灾害性天气
hazardous weather advisory 危险天气报告
hazardous weather warning 危险天气警报
haze 烟雾
haziness (浑)浊度(常用于汽油等油料的检查)
HB 1. High Blower 增压泵 2. Hand Book 手册
HBM Held By Manufacturing 加工停顿
HBPR High Bypass Ratio 高涵道比,大涵道比
HC 1. Hairline Crack 细缝裂纹 2. Heating Cabinet 加热箱 3. Helicopter Council 直升机委员会

HCA 1. Helicopter Club of America 美国直升机俱乐部 2. Hot Compressed Air 热压缩空气
HCCS High Capacity Communication System 大容量通信系统
HCE Human Cause Error 人为误差,人为差错
HCF 1. Height Correction Factor 高度修正因数 2. High Cycle Fatigue 高循环疲劳,高周疲劳 3. Hermetically Coated Fiber 密封涂覆光纤
HCI Human Computer Interface 人机(计算机)界面
HCL Horizontal Center Line 水平中心线
HCR Human Cognitive Reliability 人的认知可靠性
HCS 1. Header Check Sequence 信头检查序列 2. Hierarchical Control System 分级控制系统 3. High-Carbon Steel 高碳钢 4. Helicopter Computer System 直升机计算机系统 5. Host Computer System 主计算机系统
HCSHT High-Carbon Steel Heat Treated 热处理的高碳钢
HD Height Difference 高度差
HDBK Handbook 手册
HDCR Hard Chromium 硬铬
HDD Head Down Display 俯视显示器
HDG Heading 顶镦,方向,标题,航向
HDG HOLD Heading Hold 航向保持
HDG SEL Heading Select 航向选择
HDISK Hard Disk 硬盘
HDL 1. Handle 搬运,处理 2. Hybrid Data Link 混合数据链
HDLC High-level Data Link Control 高端数据链控制
HDLG Handling 处理,搬运
HDLMS Hybrid Data Link Management System 混合数据链管理系统
HDLS Headless 无头
HDN Harden 硬化
HDNS Hardness 硬度
HDOP Horizontal Dilution Of Precision 水平精度淡化
HDP Hardware Development Plan 硬件开发大纲
HDRM Headroom 舱内顶高,顶部空间
HDW Hardware 硬件
HDWC Hardware Cloth 硬件外罩
HDWD Hardwood 硬木
HE 1. Heat Exchanger 热交换器 2. High Explosive 高爆性炸药
head 1.(HD),题目,头,顶 2.落差,水头 3.水头,压头
head coefficient 压头系数
head down display 1.下视显示器 2.下视仪
head restraint 护头装置
head up tilt 头高位倾斜

head-end　前端系统
header　1.标题 2.集管,联管箱
head-free casting　无冒口铸造
heading　1.航线(向),(飞行)方向,方位 2.镦头(锻),顶锻,镦粗 3.标题,项(标,题)目
heading and attitude reference system　航向姿态基准系统
heading hold　航向保持
heading of station　1.电台航向 2.无线电航向
heading reference　航向基准
headquarters　司令部,指挥部,总部
head-up display　1.平视显示器 2.平视仪
headwind　逆风,逆风滑行
heat　热,热量;热,热学;吸热,经受长时间热处理
heat absorption　1.热吸收 2.吸热 3.吸热量
heat addition　供热
heat capacity　热容量
heat capacity ratio　热容比
heat conduction　热传导
heat conduction coefficient　热导率
heat exchanger　1.换热器 2.热交换器
heat exchanger of cold helium　冷氦热交换器
heat feedback　1.热反馈 2.反馈热量
heat flux　1.热通量 2.热流 3.热流密度
heat flux calculation of equivalent cones　等价锥热流计算
heat flux per unit time　热流密度
heat generation　1.热源 2.发热,热生成量
heat generator　1.热发生器 2.热风炉 3.高频加热器
heat injection　注热
heat input　热收入,供热
heat insulation layer　绝热层
heat intensity　热强度
heat iso-hydrostatic diffusion welding　热等静压扩散焊
heat isostatic pressing　热等静压
heat knife　1.热刀 2.热损失
heat leak　1.热漏泄 2.热渗透
heat load　1.热负荷 2.热载荷 3.热负荷的
heat load of anti-icing　防冰表面热载荷
heat loss　1.热阻 2.热损失
heat of combustion　燃烧热
heat of formation　1.生成热 2.形成热 3.合金形成热
heat of fusion　1.熔化热 2.熔解热 3.熔融热
heat of reaction　反应热
heat of vaporization　蒸发热
heat pipe　热管
heat protection　防热
heat pump　热(力)泵,蒸汽泵
heat rate　耗热率,燃料燃烧效率
heat ratio　热比
heat receiver　吸热体,受热器
heat recirculation　热循环
heat regenerative exchanger　再生式热交换器
heat rejection　热排出,热损失
heat release　放热
heat release rate　放热率
heat resistance　1.热阻 2.热损失
heat resistant steel　耐热钢
heat shield　防热层
heat short　热短路
heat sink　1.冷源 2.热沉
heat source　热源
heat storage　蓄热
heat storage capacity　蓄热能力,蓄热特性
heat tolerance　1.耐热限 2.热耐限
heat transfer　换热,热传导
heat transfer coefficient　热传递系数
heat transfer enhancement　强化传热
heat transfer mechanism　热传输机理
heat transfer rate　导热速度,比热流
heat treatment　热处理
heat treatment in fluidized bed　流态床热处理
heated gas-heat exchanger pressurization system　热交换器加热气体增压系统
heater　加热器
heater power　加热功率,灯丝功率
heat-exchanger　热交换器
heat-flux　热流
heating　暖气装置(设备);加热,采暖
heating characteristic　采暖特性
heating element　发热元件
heating load　供热量
heating loss　加热减量
heating method　加热法
heating motor　暖风电机
heating process　加热过程
heating rate　加热速率,单位热流
heating system　供暖系统,暖气系统,加热装置
heating time　加热时间
heating value　热值,发热量
heatload　热负荷,热载荷
heat-loss　热损失
heat-pipe　热管
heat-reflecting layer　热反射层
heat-rejection　排热,热损耗,散热
heat-release　放热
heat-sensitive recorder　热敏记录仪
heatsink　散热器

heat-transfer	传热
heat-transfer calculation	传热计算
heat-transfer coefficient	传热系数,换热系数,车体的传热系数
heat-transfer enhancement	强化传热
heave	起伏,上下飞
heaving measurement system	升沉测量系统
heavy alloy	重合金
heavy dropping	重型物件空投
heavy ferret	重型搜索者卫星
heavy flight clothing	冬季飞行服
heavy fuel	重质燃料
heavy fuel oil	重燃油
heavy gas	重气体
heavy landing	重着陆,粗暴着陆
heavy left(right)	重左(右)梯队
heavy lift	重量货物,重件
heavy maintenance	大型维修
heavy oil	重油
heavy particle	重粒子
heavy permeable clothing	厚透气服
heavy route	重路由
heavy torpedo	重型鱼雷
heavy traffic	繁忙运输,繁重交通
Heavy Wagon	"重货车"航路
heavy-case bomb	厚壳炸弹
heavy-lift booster	重型助推器,重型运载火箭
heavy-route circuit	重路由电路
hedgehog	刺猬弹(一种反潜深水炸弹)
hedge-hopping flight	超低空飞行
heeling	1.倾斜,倾斜飞行 2.(飞机滑行中转弯造成的)侧翻转
heeling error	(磁罗盘在飞行中的)倾斜误差
height	1.高度,海拔,垂直距离 2.高处,高地,顶点 3.高程
height above airport	机场上空高度
height above touch down	接地区上空的高度
height adjustment	高度修正,高度调整
height and range finder	测高测距仪
height anomaly	高程异常
height datum	高程基准面,水准零点
height direction	高度方向
height engage button	高度稳定器接通按钮
height equivalent	高度当量,等效高度
height error	高度误差
height lock	高度锁定
height ring	高度环(机载雷达荧光屏上由飞机正下方地面的回波形成的亮环)
height system	高程系统
height traverse	高程导线
height vertigo	高空眩晕
height zoning	按高度作层次配置
height/velocity curve	高度/速度曲线
height-above-obstacle altimeter	真高度表,真实高度表
height-finding radar	测高雷达
height-keeping accuracy	保持高度的准确度
height-keeping error	高度保持误差
height-of-burst probable error	炸高概率误差
HEL	1. Helicopter 直升机 2. High Energy Laser 高能激光器 3. Header Extension Length 信头扩展长度
Heli Coil	螺旋钢丝圈
Heliarc	氦弧焊
heliborne	1.直升机空运(的) 2.直升机机降
helical antenna	螺旋天线
helical compressor	螺旋压气机
helical gear	螺旋齿轮
helical mode	螺旋模式
helical potentiometer	螺旋电位器
helical scan recorder	旋转头磁记录器
helical trajectory	螺旋形轨迹
helical vortex	螺旋涡
helical winding	螺旋绕法
helical wrap	螺线圈
helicity	螺旋性
helicogyro	直升机
helicon plasma	螺旋波等离子体
helicopter	直升机
helicopter actuator	直升机作动器
helicopter controlled approach	直升机引导进场降落
helicopter controller	直升机控制员
helicopter cruiser	直升机巡洋舰
helicopter deck-landing devices	直升机着舰装置
helicopter ditching	直升机水上迫降
helicopter engine	直升机发动机
helicopter fire control laser system	直升机火力控制激光系统
helicopter floatation gear	直升机着水装置
helicopter forbidden region	直升机回避区
helicopter ground resonance	直升机地面共振
helicopter intercom	直升机舰内通话装置
helicopter landing gear	直升机起落装置
helicopter lane	直升机安全走廊
helicopter laser range finder	直升机激光测距器
helicopter LSE	(航母)直升机着舰信号兵
helicopter model	直升机模型
helicopter plane captain	(航母)直升机器材检查员
helicopter power loading	直升机功率载荷
helicopter power utilization coefficient	1.直升机功率传

递系数 2.直升机功率利用系数
helicopter rotor 直升机旋翼
helicopter securing standards 直升机系留标准
helicopter service ceiling 1.直升机前飞升限 2.直升机动升限
helicopter vibration 直升机振动
helicopter-mounted laser weapons 直升机机载激光武器
helidrone 无人驾驶直升机
helidrop （直升机）悬停空投
helio centric 日心的
heliocentric constant 日心常数,太阳中心常数
heliocentric coordinate 日心坐标
heliocentric distance 日心距离
heliocentric gravitational constant 日心引力常数
heliocentric inertial coordinates 日心惯性坐标系
heliocentric longitude 日心经度
heliocentric orbit 日心轨道
heliocentric orbit rendezvous 日心轨道会合
heliocentric parallax 日心视差
heliocentric system 日心体系
heliocentric system of coordinate 日心坐标系
heliocentric transfer 日心轨道转移
heliocentric velocity 日心速度
heliograph 1.日光仪,日照计 2.太阳摄影机 3.太阳光度计 4.反光镜（救生设备）
heliographic chart 日面图
heliographic coordinate 日面坐标
heliographic latitude 日面纬度
heliographic longitude 日面经度
heliographic system of coordinates 日面坐标系
heliopause 太阳风层顶,日球层顶
Helios 1.太阳神卫星（德国空间探测器）2.赫利俄斯卫星（法国侦察卫星）
helios 回照器
heliosphere 阳风层,日球,日球层
heliospheric current sheet 日球层电流片
heliostat 定日镜
heliosynchronous orbit 太阳同步轨道
helipad 直升机起降场,直升机起降台
heliport 直升机航站,直升机航空港
helistat 旋升气球,旋翼浮空器
helistop 民用直升机起降场
helium 【化】氦（化学元素,符号 He）
helium blow-off 氦气吹除
helium bottle truck 氦气瓶车
helium bubble method 氦气泡法
helium burning 氦燃烧
helium compressor truck 氦压缩机车
helium concentration 氦浓度

helium cryopanel 氦深冷板
helium flash 氦闪
helium gas 氦气
helium gas distribution board 氦气配气台
helium injection 氦引入,喷射氦,通氦
helium leak test 氦检漏试验
helium mass spectrometer leak detector 氦质谱检漏仪
helium mass spectrum leak detection 氦质谱检漏
helium mass-spectrometer detecting system 氦质谱仪检漏系统
helium neon laser 氦氖激光器
helium purge 氦吹除
helium replacement 氦气置换
helium seal 氦封
helium shielded arc welding 氦弧焊
helium star 氦星
helium tank 氦气瓶
helium turbine 氦气轮机
helium wind tunnel 氦气风洞
helium-cadmium laser 氦镉激光器
helium-neon 氦氖气
helium-rich core 富氦核
helium-rich star 富氦星
helium-strong star 强氦星
helium-weak star 弱氦星
helix antenna 螺旋天线
helix slow wave line 螺旋慢波线
helix-coupled vane circuit 螺旋线耦合叶片线路
helmet 1.飞行帽,头盔,盔式面罩,防护帽 2.机罩,罩
helmet display 头盔显示器
helmet enclosure 头罩,头盔外壳
helmet sight 头盔瞄准器(具)
helmet visor 密封头盔观察窗
helmet-mounted display 头盔显示仪(器)
helmet-mounted oxygen regulator 头盔氧气调节器
helmet-mounted sight 头盔瞄准具
Helmholtz resonator 海姆霍兹谐振器
helmsman 舵手,操舵机构
helper 助手,辅助机构
helper spring 辅助弹簧副钢板
hemisphere 半球
hemisphere engine 半球形燃烧室发动机
hemispheric fine mesh 半球细网格
hemispheric prediction 半球预报
hemispherical absorptance 半球吸收比
hemispherical emittance 半球发射率
hemispherical reflectance 半球反射比
hemispherical resonance gyroscope 半球谐振陀螺
hemispherical resonant gyro 半球谐振陀螺仪

hemispherical shell 半球形壳体
hemispherical transmittance 半球透射比
hemodynamic response 血液动力反应
HEO 1. High Earth Orbit 高地球轨道 2. High Elliptical Orbit 高椭圆率轨道
heptane 庚烷
heptyl 庚基
hermaphroditic connector 无极性连接器
hermaphroditic contact 无极性接触件
hermetic integrating gyro-scope 密封式积分陀螺仪
hermetic package 密封包装
hermetic zipper 气密拉链
hermetically sealed battery 密封电池(组)
hermetically sealed cell 全密封电池
Hermite polynomial 赫密特多项式
Hermitian matrix 赫密特矩阵
HERO Hazards of Electromagnetic Radiation to Ordnance 电磁辐射对武器的危险
herringbone gear 人字齿轮,双螺旋齿轮
hertz 赫,周秒
hesitation roll 间歇横滚
Hessian matrix 黑森矩阵(二阶偏导数矩阵)
Hessian Matting 黑森道面(一种预制沥青道面板,用于修筑机场临时跑道)
heterocycli ccompound 杂链聚合物
heterodyne 1. 外差(法),外差作用 2. 外差振荡器
heterodyne conversion transducer 外差式变频器
heterodyne detection 外差探测
heterodyne method 外差法
heterodyne oscillator 外差振荡器
heterodyne receiver 外差接收机
heteroepitaxy 异质外延
heterogeneity 异质性,不均匀性
heterogeneous combustion 复相燃烧
heterogeneous flow 非均质流
heterogeneous group 非同质组异类组
heterogeneous light 杂色光,多色光
heterogeneous mixture 不均匀混合物
heterogeneous nucleation 非匀相成核
heterogeneous oxidation 多相氧化(作用)
heterogeneous propellant 异质推进剂
heterogeneous reaction 非均相反应
heterogeneous surface 非均匀表面
heterogeneous system 多相体系,非均匀体系
heterojunction 异质结
heterojunction cell 异质结电池
heterojunction solar cell 异质结太阳电池
heterojunction transistor 异质结晶体管
heterophoria 隐斜视
heterosphere 非均质层
heterostructure 异质结构
heuristic argument 渐近角
heuristic model 启发模式
heuristic routing 试探性路由选择
heuristically-programme dalgorithmic computer 启发式程序算法计算机
HEX Hexagon 六角
HEX HD Hexagonal Head 六角头
hexaboride 六硼化物
hexad 六价原子
hexadecane 十六烷
hexagon 六角形,六边形
hexahedra 六面体的
hexane 己烷
hexogen(RDX) 黑索金(一种烈性炸药)
HEXSOC Hexagon Socket 六角插座
HEXSOCH Hexagon Socket Head 六角插座端头
HF 1. High Frequency 高频 2. Holding Fixture 固定夹具 3. Hydraulic Fuse 液压保险
HFDL High Frequency Data Link 高频数据链
HFDM High Frequency Data Modem 高频数据调制解调器
HFDR High Frequency Data Radio 高频数据无线电
HFE Held For Engineering 等待工程处理
HFEC High Frequency Electro-turbulence Check 高频涡流检查
HFM Held For Material 停工待料
HFNPDU High Frequency Network Protocol Data Unit 高频网络协议数据装置
HFO High Frequency Oscillator 高频振荡器
HFR Height Finding Radar 测高雷达
HFS 1. High Fidelity Simulator 高逼真度模拟器 2. High Frequency System 高频系统
HFT Held For Tooling 停工等待工装
HGA High Gain Antenna 高增益天线
HGALV Hot-Galvanize 热电镀,热镀锌
HGC Head-up Guidance Computer 平显导引计算机
HGR Hanger 搬运吊架
HGS Head-up Guidance System 平显导引系统
HGT Height 高度
HGU Horizon Gyro Unit 陀螺地平仪
HHLD Heading Hold 航向保持
HHT Hand-Held Terminal 手持终端
HI 1. Handling Instruction 搬运保持 2. Honer Indicator 悬停状态显示器
HI region 中性氢区
hibernate 休眠,使(尽可能多的分系统)处于关闭状态
hibernating spacecraft 休眠航天器

HIC　Head Injury Criterion　头部伤害判据
hidden bit　隐（藏）位
hidden buffer　隐式缓冲器
hidden computer　隐式计算机
hidden data　隐式数据
hidden fault　隐患性故障
hidden function　隐蔽功能
hidden layer　隐蔽层
hidden mass　隐质量
hidden oscillation　隐蔽振荡
hide　掩蔽所、伪装网等
hierarchical abstract computer　分级抽象计算机
hierarchical architecture　分层体系结构
hierarchical chart　层次结构图
hierarchical control　递阶控制
hierarchical cosmology　等级式宇宙论
hierarchical decomposition　层次结构分解
hierarchical design method　层次设计法,分级设计法
hierarchical method　层序法,非层序法
hierarchical network　分级网络
hierarchical network architecture　层次网络结构
hierarchical optimization　分级优化
hierarchical organization　等级结构
hierarchical parametere estimation　递阶参数估计
hierarchical planning　递阶规划
hierarchical storage manager　分级存储管理程序
hierarchical structure　递阶结构
hierarchy　1.层次结构,层次,分级,等级 2.体系（制）,谱系
HIF　High Fidelity 高保真度的,高度灵敏的,易感的
high　高压区,反气旋,高气压
high accuracy　高准确度,高精确度
high accuracy measurement corridor　高精度测量带
high altitude　高空,高度
high altitude antiradiation missile　高空反辐射导弹,哈姆导弹
high altitude compensating suit　高空代偿服
high altitude nuclear explosions　高空核爆炸
high altitude pressure suit　高空加压服
high altitude protective assembly　高空防护装备
high amplitude　高振幅
high and low temperature cycling treatment　高低温循环处理
high angle of attack　大攻角,大迎角
high angle of attack aerodynamics　大迎角空气动力学
high antishearing riveting　高抗剪铆接
high aspect ratio　大长宽比
high bypass　高涵道
high capacity bomb　高爆炸弹,大威力炸弹,重型爆破炸弹（英国用法）
high chamber pressure　高室压
high combustion efficiency　高燃烧效率
high compressor　高压压气机
high concentration　高浓度
high control　较高的控制,高控制
high correlation　高度相关
high cost　高成本
high current　高安培电流,高强度电流;强流
high data　高数据,高速数据
high density electronbeamoptics　强流电子光学
high density fragment warhead　重金属破片弹头
high density fuel　高密度燃油,大比重燃油
high density seating　密排座席
high differential cabin　高压差制座舱
high drag　大阻力
high dynamic pressure　高动压,动高压技术
high dynamics　高动态
high eccentric orbit　高偏心率轨道,大椭圆轨道
high efficiency　高效率
high electronmobility transistor　高电子迁移率场效管
high end　高端
high energy　高能量,高能率
high energy astrophysics　高能天体物理
high energy beam machining　高能束加工
high energy firing unit　高能射击装置
high energy fuel　高能燃油,高能燃料
high energy ignition unit　高能点火装置
high energy laser weapon　高能激光武器
high energy particle spectrometer　高能粒子谱仪
high energy rate forming　高能成形
high energy solar proton　高能太阳质子
high enthalpy　高热焓
high enthalpy wind tunnel　高焓风洞
high explosion　高空爆炸
high explosive　高能炸药,烈性炸药
high explosive ammunition　高爆弹药,杀伤弹药
high explosive bomb　1.烈性（炸药）炸弹 2.高爆炸弹
high explosive gas shell　高爆毒气弹
high explosive incendiary cartridge　爆破燃烧弹
high fineness　高细度
high flow　高流动性,洪流
high flowzone　高流量区,高消耗区
high frequency　高频率,频繁出现
high frequency amplifier　高频放大器
high frequency transformer　高频变压器
High Frontier Program　天疆计划,高边疆计划（美国1982年提出的建立全球弹道导弹防御系统的方案）
high fuel　高燃料

high G telemetry 高 G 遥测
high gain 高增益
high gain antenna 高增益天线
high gravity 超重力,高重力环境
high gravity environment 高重力环境
high g trim 大过载配平
high harmonic control 高次谐波控制
high heat 高热量
high heat load 高热负载
high heating load 高热负载
high heating rate 高加热率
high heliocentric orbit 高日心轨道
high intensity flare 高亮度曳光弹
high ion 高电离离子
high latitude 高纬度
high level 高电平,高级,高能级,高水位
high level data link control 高级数据链路控制
high level language 高级语言
high level navigation 高空航行
high level altimetry 高度测量(法)
high life device 增升装置
high lift 高升力
high light tracking 最亮点跟踪
high load 高负荷,高负载
high loading 高负荷,高加载
high loss 高损耗
high Mach buffet 高马赫数抖振
high Mach flight 高马赫数飞行(指高亚声速飞行)
high Mach trimmer 高马赫数配平机构
high maneuverability 高机动性
high mode 高次模,高模态
high natural frequency accelero graph 高固有频率加速度记录器
high nitrogen 高氮,高氮肥
high pass filter 高通滤波器
high performance (HP),高性能
high performance communications adapter 高性能通信适配器
high performance external gun 高性能机身外炮,(变距螺旋桨)高距冲大距
high pitch 1.大距 2.高距
high plateau voltage 高阶电压
high point 高点
High Pole "高杆"(苏联海军敌我识别应答器的北约代名)
high polymer 高聚物,高分子化合物
high power 高倍,高功率,大功率的,强功率
high precision air pressure generator 1.高精度气压发生器 2.大气静压模拟器

high pressure 高压,高气压,高度紧张
high pressure admission 高压进气
high pressure chemical vapor deposition 高压化学气相沉积
high pressure compressor 高压压气机
high pressure decompression sickness 高压减压病
high pressure gas bottle 高压气瓶
high pressure gas storage 高压气态贮存
high pressure hose 高压软管
high pressure impregnation 高压浸渍
high pressure oxidation 高压氧化
high pressure oxygen system 高压氧气系统
high pressure saltwater system 高压海水输送系统
high pressure test ammunition 高压试验弹药
high pressure turbine 高压涡轮机
high pressure vessel 高压气瓶
high pressure water jet cutting 高压水切割
high rate 高速,高效率
high ratio engine 高涵道比发动机
high reliability 高可靠性
high removable worktable 高可移动工作台
high resolution 高分辨率
high resolution Bragg spectrometer 高分辨率布雷格光谱仪
high resolution facsimile 高分辨率传真
high resolution global measurement of atmospheric ozone 高分辨率全球大气臭氧测量仪
high resolution infrared radiometer 高分辨率红外辐射计
high resolution plate 高分辨率板
high Reynolds number wind tunnel 高雷诺数风洞
high route 高空航路
high rudder 1.上舵 2.上偏(方向)舵 3.方向舵上段
high satellite 高轨卫星
high sensitivity 高灵敏度
high shear 高剪切
high shear rate 高切速
high side 高压侧,高边坡
high slew rate 高旋转速率
high specific impulse 高比冲,高比推力
high speed 高速,高速的,高速度
high speed alloy 高速合金
high speed anti-radar missile 高速反雷达导弹
high speed anti-radiation missile 高速反辐射导弹
high speed bus adapter 高速总线适配器
high speed camera 高速摄影机
high speed electrodeposition 高速电镀
high speed gun 高速炮
high speed photograph application in wind tunnel 风洞

高速摄影技术
high speed photography house 高速摄影间
high speed warning 高速告警
high speed wind tunnel 高速风洞
high spool 高压转子
high stiffness 高刚度
high strength aluminium alloy 1.高强铝合金 2.超硬铝合金
high strength steel 高强钢
high stress 高应力
high subsonic 高亚声速的
high subsonic missile 高亚声速导弹
high subsonic speed 高亚声速
high sulfur 高硫,高含硫
high super charger gear 高压增压器齿轮装置
high surface area 高比表面积,高表面积
high swirl 高涡流
high temperature alloy 高温合金
high temperature capability 高温性能
high temperature chamber 高温舱
high temperature coating 高温涂层
high temperature oxidation-resistant coating 高温抗氧化涂层
high temperature oxidation-resistant coating for refractory 难熔金属高温抗氧化涂层
high temperature protection coating 高温防护涂层
high temperature tempering 高温回火
high temperature test 高温试验
high tension electrode 高压电极
high tension transformer 高压变压器
high time 长服役期的
high trajectory test 高弹道(飞行)试验
high trial 1.高原机场试验,高原条件试验 2.高空试验
high turbine 高压涡轮机
high utility 较高的实用
high vacuum 高(度)真空
high velocity anti-tank shell 高速反坦克炮弹
high velocity drop 高速空投(指不用降落伞的空投,下降速度在10米/秒至自由落体收尾速度之间)
high voltage silicon stack 高压硅堆
high yield weapon 高能量武器
high/low-voltage double insulated bus 高、低压双隔离母线
high-acceleration interceptor 高加速度截击器,高加速度截击导弹
high-accuracy 高精度
high-altitude aero drome 高原机场
high-altitude altimeter 高空高度表
high-altitude balloon 高空气球

high-altitude bombing 高空轰炸
high-altitude breathing apparatus 高空呼吸设备,高空氧气设备
high-altitude burst (核武器)高空(3万米以上)爆炸
high-altitude chamber 高空(模拟)室
high-altitude electromagnetic pulse 高空电磁脉冲
high-altitude nuclear test 高空核试验
high-altitude operation platform 高空作业平台
high-altitude satellite 高轨卫星
high-amplitude 强振幅
high-apogee 高远地点
high-atmosphere wind 高空风
high-bandwidth 高带域
high-brightness 高亮度
high-burst ranging 空炸射击修正
high-contrast 高反差,硬调
high-contrast target 高反差目标,高对比度目标
high-cycle fatigue 高周疲劳
high-density focal-plane array 高密度焦面阵(有数千个二维红外元集成的电子光学传感器)
high-density tunnel 高密度风洞
high-drag 高阻力
high-drag bomb 高阻炸弹
high-efficiency 高效
high-energy 高能
high-energy beam machining 高能束加工
high-energy density beam welding 高能束焊接
high-energy fuel 高能燃料
high-energy additives 高能添加剂
high-energy ignitor 高能点火器
high-energy-rate forming 高速高能成形
higher mode 高阶振型
higher moment 高阶矩
high-explosive armor-piercing shell 高爆穿甲弹
high-explosive payload 爆破装药
high-fidelity 高保真度,高置信度
high-fidelity engineering 高保真工程
high-fidelity propagation 高保真传播
high-fidelity simulation 高拟真情境模拟
high-frequency 高频率
high-frequency combustion 高频燃烧
high-frequency combustion instability 高频燃烧不稳定性
high-frequency combustionin stability 高频振荡燃烧,高频不稳定燃烧
high-frequency discharge 高频放电
high-frequency noise 高频噪声
high-frequency pressure 高频压力
high-frequency resistance welding 高频电阻焊(接),高

频接触焊
high-frequency starter 高频起动器
high-frequency starter circuit 高频起动器电路
high-frequency state 高频状态
high-gain antenna （High-gain Antenna）高增益天线
high-gain control 高增益控制
high-intensity explosion 高强度的爆炸
high-latitude ionosphere 高纬度地区电离层
high-level accelerometer 大量程加速度表
high-level data link controller 高级数据链路控制器
highlift 增升的（系统，装置或构型，使飞机升力比净形或巡航构型时要大）
high-lift 高升力
high-lift airfoil 高升力机翼
high-lift device 增升装置
high-lift system 高升力系统
highlight 最精彩的部分，最重要的事情，加亮区
high-loss 高损耗
high-low temperature test 高低温试验
high-luminosity star 高光度恒星
highly elongated orbit 大扁度轨道
highly parallel computer 高度并行计算机
highly-elliptical orbit 大扁率椭圆轨道
highly-instrumented 装大量测量仪表的
high-mach-number 大马赫数
high-modulus yarn 高模数线
high-momentum 高动量
high-mutube 高放大系数管，高管
high-oblique 倾斜照相（侦察）
high-octane 高辛烷值（表示燃料品级）
high-order 高阶，高端
high-order dynamics 高阶动力学
high-order loop 高阶环路
high-order mode 高阶模
high-order propagator 高阶传播算子
high-performance aircraft 高性能飞机
high-performance avionics 高性能航空电子设备
high-power 大功率的
high-power laser 大功率激光
high-power radio-frequency weapon 大功率射频武器
high-power satellite 高功率卫星
high-pressure air 高压空气
high-pressure ammunition 高压弹药
high-pressure compressor 高压空压机
high-pressure cryogenic storage vessel 高压低温贮存器
high-pressure ratio 高压比
high-pressure region 高压区域
high-pressure stator 高压定子
high-pressure turbine 高压汽轮机

high-recovery thermo couple 高恢复系数热电偶
high-reliability 高可靠性
high-resistance voltmeter 高阻电压表
high-resolution 高分辨率
high-resolution radar 高分辨力雷达
high-risk conjunction 高风险的结合
high-speed bus 高速总线，高速汇流排
high-speed camera 高速照相机
high-speed controller 高速控制器
high-speed core 高速磁心
high-speed data acquisition 高速数据采集
high-speed datum 高速数据
high-speed digital camera image 高速数码相机图像
high-speed duct 高速管
high-speed film 高速感光胶片
high-speed flight 高速飞行
high-speed flow 高速流
high-speed image 高速图像
high-speed imaging 高速成像
high-speed imaging resolution 高速图像分辨率
high-speed massmemory 高速大容量内存储器
high-speed missile 高速导弹
high-speed photometry 高速测光
high-speed pressure 高速压力
high-speed schlieren 高速纹影
high-speed stall 高速失速
high-speed stream 高速流
high-speed switched digital service 高速切换数字业务
high-speed target 高速目标
high-speed testing technique 高速测试技术
high-stress combat environment 高强度作战环境
high-sulfur 高硫
high-tailed aircraft 高平尾飞机
high-temperature ignition 高温点火
high-temperature soaking strength 高温浸润强度
high-temperature composite 高温复合材料
high-temperature wire 耐热导线，（耐）高温导线
high-temperature zone 高温区域
high-tension wire 高压线
high-tension wiring 高压导线网
high-test peroxide 高级过氧化物
high-vacuum tube 高真空电子管
high-velocity armor-piercingshell 高速穿甲弹
high-velocity star 高速星
high-voltage 高电压
high-voltage mode 高压模式
high-voltage pulse 高压脉冲
high-voltage switch 高电压开关
high-volume particulate sampler 大容量微粒取样器

high-volume production	大量生产
highway	数据总线（电子数据处理系统中的数据电路）
highwing	上单翼（机翼在机身顶部）
highyo-yo	高摇摇特技
HII region	电离氢区
HIHUM	High Humidity 高湿度
HIMP	High Impact 高冲击力的
HIN	High Intensity 高强度
hinge	铰链，折页；关键，转折点；枢要，中枢
hinge line	枢纽线
hinge moment	铰链力矩
hinge moment balance	铰链力矩天平
hinge moment coefficient	铰链力矩系数
hinge moment derivative	铰链力矩导数
hinge moment testing	铰链力矩试验
hinged bearing	悬挂式轴承
hinged wing	铰接翼
hingeless rotor	1. 无铰式旋翼 2. 刚接式旋翼
HINI	High Intensity 高强度，高密度
HIPOT	High Potential 高电位
HIPOTT	High Potential Test 高电位试验
HIRF	1. High Intense Radiated Fields 高强辐射场 2. High Intensity Radio Frequency 高强射频
HIRFQTR	HIRF Qualification Test Report 高强辐射场鉴定试验报告
histohyperoxia	组织内氧气过多
histohypoxia	组织缺氧症
histotoxicanoxia	组织中毒性缺氧
hit probability	命中概率
hit rate	命中率
hitband	降落带图（降落在月球或行星表面的航天器起始速度对起始轨迹角曲线图）
Hitch Hiker	搭载星（美国电子侦察卫星，搭载在"大鸟"侦察卫星上发射）
hit dispersion	落点散布，弹着点散布
hit the silk	〈口语〉弃机跳伞
hit tile	直接命中式导弹
hit-to-kill	杀伤
hit-tolerance	有抗弹能力的
HJ	Hose Jacket (Insult) 软管护套
HKA	HongKong Airways Ltd. 香港航空公司
HKP	Hookup 挂钩，悬挂，试验线路，电路耦合
HL	1. Highline 高压线路 2. Height Lift 增升装置
HLCL	Helical 螺线，螺旋的
HLCPS	Helical Compression 螺旋压缩
HLCPTR	Helicopter 直升机
HLCS	High Lift Control System 高升力控制系统
HLD	Hold 保持
HLDG	Holding 保持，把持
HLDN	Hold Down 保持，使保持向下
HLDR	Holder 夹具，炳，把，架座，保持架，持有人
HLEXT	Helical Extension 螺旋延伸
HLL	High Level Language 高级语言
hmax	最大波高
HMDS	Hexamethyl Disilazane 六甲基二硅胺烷
HMI	Human Machine Interface 人机界面
HMNC	Harmonic 谐波的，调和的
HMOS	High-density Metal Oxide Semiconductor 高密度金属氧化物半导体
HMR	Hammer 锤子
HMS	Health-Monitoring System 良好状况监视系统
HMU	Height Monitoring Unit 高度监控装置
HND CONT	Hand Control 手动控制
HNDRL	Hand Rail 扶栏，栏杆
HNDST	Hand Set 听筒，送受话器
HNG	Hanging 悬挂
HNS	Hexadecimal Numbering System 十六进制编号系统
HNYCMB	Honeycomb 蜂窝（结构）
HO	1. Handoff 松开 2. Hard-Over 急偏
hoar frostre mover	除霜液
hoarfrost	白霜，树冰
HOB	Height Of Burst 爆高，爆炸高度
HOC	1. High Oil Consumption 高滑油消耗量 2. Highest Outgoing Channel 最高输出信道
hogging	1. 锻件加工 2. 拱起，翘曲（一种应力状态，如飞艇的变形，两端下沉，中间拱起，机身头尾下垂等）
hoghorn antenna	弯状喇叭天线
Hohmann orbit	霍曼轨道（共面圆轨道之间的最小能量转移轨道）
Hohmann transfer	霍曼转移
Hohmann transfer ellipse	霍曼转移椭圆（轨道）
hoisting	吊装
hoisting tool bogie	吊具小车
hold down test	台架试验，静态点火试验
hold mode	保持模式（防止太阳电池阵转动的工作模式）
HOLD P	1. Holding Pattern 等待航线 2. Holding Procedure 等待程序
hold torque	保持力矩
hold-down mechanism	牵制-释放机构
HOLDF	Holding Fix 等待定位点
holding area	等待区（有等待点的区域）
holding bay	等待坪，待机坪
holding course	空中等待航线
holding fix	等待位置，等待点
holding fue	等待油量
holding off	（飞机着陆前）保持平飘

holding pattern 等待航线
holding point 等待点
holding procedure 等待程序
holding room 候机楼大厅
holding side 等待航向上有等待航线的一侧
holding time 保持时间,占用时间
holding vacuum pump 维持真空泵
holdover time 保持时间
holdup 滞留量
hole and slot resonator 孔槽形谐振腔
hole interface 孔接口
hole number 孔数
hole pair 电子空穴对
hole periphery 周边孔
hole tone 音程音
hole-burning effect 灼孔效应
holemobility 空穴迁移率
holetrap 空穴陷讲
holing through survey 贯通测量
hollow particle 空心粒子
hollow plate 空心板
hologram information light 全息信息灯
hologram memory system 全息存储系统
hologram photography 全息摄影
hologram technique 全息技术
hologrammetry 全息摄影测量
hologrampage 全息图面
holograph 全息照相
holographic antenna 全息天线
holographic display 全息显示
holographic filter 全息滤波器
holographic headup display 全息平视显示器
holographic information storage 全息、信息存储
holographic mask technology 全息掩模技术
holographic memory 全息照相存储器
holographic microscopy 全息显微术
holographic radar 全息雷达
holographic technology 全息技术
holographical display 全息显示
holography 全息学,全息摄影(术),全息照相
holoscope 全息照相机
HOM Homing 归航,归位
home 1.寻的(自动向辐射源飞行),自导引 2.归航(人工操纵使飞机对准导航点或其他辐射源飞行)
homebuilt 业余设计者制作的(常指有动力的飞机)
homeotropic alignment 垂面排列
homer 1.导航台,引导站 2.归航信标,(电台)附加归航装置 3.导引头,寻的导弹
hometaxial-base transister 轴向均匀基极晶体管

homing 1.导航,归航,向电台飞行 2.寻的,自导引(制导飞行器自动飞向特定的发射源)
homing action 1.归航 2.寻的,自导引,自导引系统的工作 3.回复原位,还原动作
homing activation 自导引系统接通,转入寻的状态
homing active guidance 主动寻的制导
homing aids 归航台,导航台
homing airborne guidance 弹上自导引装置,弹上寻的制导装置
homing beacon 归航台,归航信标
homing device 寻的装置,自导引装置(拦截武器上的)
homing equipment 自导引装置,寻的装置,归航设备
homing guidance 1.寻的制导 2.自动导引
homing guidance package 寻的制导装置,自导引装置
homing head 1.导引装置 2.导引头
homing head blind zone 1.导引头盲区 2.(导引头)非灵敏区
homing head range 导引头作用距离
homing head resolution 导引头分辨率
homing missile ECCM 寻的导弹反干扰
homing optical guidance 光学寻的制导
homing overlay experiment 大气层外自动寻的实验(美国)
homing radar 寻的雷达,自导引雷达
homing rendezvous 归航会合
homing semi-active guidance 半主动寻的制导
homing sensor 寻的探测器,寻的器,寻的头
homing station 归航台
homing torpedo 自导鱼雷,自导雷
homodyne 1.零差,零拍 2.同步检波,零差检波
homodyne detection 零差检波
homoentropic flow 匀熵流
homoepitaxy 同质外延
homogeneity 同质性,均匀性
homogeneous atmosphere 均质大气
homogeneous flow 均匀流
homogeneous fluid 均质流体
homogeneous function 齐次函数
homogeneous medium 均匀介质
homogeneous mixture 均匀混合物
homogeneous model 均相模型
homogeneous nucleation 匀相成核
homogeneous propellant 均质推进剂,胶质推进剂
homogeneous shear 均匀切变
homogenize 均化,匀化
homogenizing annealing 均匀化退火
homogenous solution 齐次解
homographic filtering 同态滤波
homologous flare 相似耀斑

homologous image points 同名像点
homojunction laser 同质结激光器
homojunction solar cell 同质结太阳电池
homologue 同调(指变量取相应的、但不一定是相等的值)
homomorph 同态像
homomorphic filter 同态滤波器
homomorphic model 同态模型
homomorphic systems 同态系统
homopause 均质层顶
homosphere 均质层,均匀层
honeycomb 1.蜂巢,蜂巢状之物 2.蜂窝结构,蜂窝材料 3.(风洞内的)格栅(消除紊流用)3.蜂房线圈
honeycomb structure manufacturing process 蜂窝结构工艺
honeycomb core 蜂窝芯材,蜂巢状芯轴,蜂窝状中心
honeycomb launcher 蜂窝式发射装置,24管发射装置(美国)
honeycomb panel 铝蜂窝板
honeycomb plate 波导通风板
honeycomb sandwich 蜂窝夹层,蜂窝夹层结构
honeycomb sandwich construction 蜂窝夹层结构
honeycomb structure 蜂窝结构
hood 1.罩,座舱盖,整流罩 2.暗舱罩
hooded flight 暗舱飞行
hooded penetration 暗舱穿云,模拟仪表穿云
hooded windshield 带暗舱罩的风挡
hook 1.钩 2.着陆拦阻钩 3.钩住,挂住 4.钩取
hook and cablecluster 集束(炸)弹架
hooked riveting with lock rivet 环槽铆钉铆接
hookman 拦阻钩操作员
hoop direction 环向
hoop strain 圆周应变
hoop stress 环向应力,箍应力,环形电压
hoop stress distribution 环向应力分布
hoop winding 环向缠绕
hop transmitting 跳频发射(通信反干扰技术)
hopper 1.漏斗 2.滑油箱油井,热油隔间,小型插座(滑油箱上供起动滑油冲淡用的)3.料斗,粉箱(农业飞机上装药粉或种子的容器)
hopper vehicle 料斗车
Hops helmet optical position sensor 头盔光学位置传感器
HORAD Horizontal Radar Display 雷达水平显示器
HORI Horizontal 水平的
Horizon 地平线卫星(以色列技术试验卫星)
horizon 1.天地线,水平线,地平线,地平圈 2.水平仪,地平仪 3.(水平)视距 4.层顶
horizon bar 1.摇摆指示器 2.(仪表上的)人工地平线标志
horizon camera 地平线摄影机
horizon circle 地平经仪
horizon control approach 地平线控制方法
horizon control scheme 地平线控制方案
horizon crossing 地平穿越
horizon crossing indicator 地平穿越式地球敏感器
horizon flattening 层位拉平
horizon flight director 水平飞行指挥仪
horizon gyro unit 陀螺地平仪
horizon photograph 地平线相片
horizon sensor 水平传感器,水平探测器
horizon step size 地平线步长
horizon system of coordinate 地平坐标系
horizon trace 像地平线,合线,真水平线
horizon tracker 1.水平跟踪器 2.地平仪传感器
horizon tracking 地平跟踪(地平仪视场轴对地平的自动跟踪)
horizon transmission 直视距离传输,视距传输
horizontal axis 水平轴,横轴
horizontal checking ladder 水平测试工作梯
horizontal control network 平面控制网,水平控制网
horizontal controlpoint 平面控制点
horizontal coordinate 地平坐标
horizontal coordinate system 地平坐标系
horizontal decomposition 横向分解
horizontal deflecting electrode 水平偏转电极,水平致偏电极
horizontal deflection yoke 水平偏传系统
horizontal dimension 水平尺寸
horizontal error 水平误差
horizontal flow chart 水平流程图
horizontal flyback 水平(面)回描
horizontal gradient of gravity 重力水平梯度
horizontal gyro 水平陀螺仪
horizontal landing 水平着陆
horizontal launch 水平发射
horizontal layer 水平岩层
horizontal line 水平线,横线
horizontal linear polarization 水平线极化(一种电波极化取向)
horizontal loop method 水平回线法
horizontal nystagmus 水平性眼震
horizontal orientation 水平定向
horizontal output transformer 行扫描输出变压器
horizontal parallax 1.水平视差 2.地平视差
horizontal plane 水平面,地平,(潜水艇)水平舵
horizontal polarization 水平极化
horizontal positioning 水平位置,横焊位置,平仰卧位

horizontal reference　水平基准
horizontal refraction error　水平折光差,旁折光差
horizontal route　水平线
horizontal scanning　水平扫描,行扫描
horizontal separation　水平错距,水平错开,水平离距,水平分距
horizontal separation ratio　水平分离比
horizontal situation display　1.水平状态显示器 2.导航显示器 3.电子航道罗盘
horizontal situation indicator　水平位置指示器
horizontal stabilizer　水平安定面
horizontal sweep voltage　行扫描电压
horizontal tail　1.水平尾翼 2.平尾
horizontal tailplane　水平尾面
horizontal test　水平测试
horizontal transit circle　水平子午环
horizontal trim　水平配平
horizontal turn　水平转弯
horizontal universal winder　卧式万能绕线机
horizontal velocity　水平速度
horizontal wind　水平风
horizontal-damped navigation mode　水平阻尼导航模态
horn　1.(舵面的)操纵摇臂 2.微波天线耦合波导管 3.声发射管(截面变化以控制声阻抗及方向性) 4.(舵面)突角补偿 5.角,角状物,突出部
horn aerial　喇叭天线
horn antenna　喇叭天线
horn arrester　角形避雷器
horn balance　(飞机舵面的)突角补偿,突角配重
horn radiator　喇叭辐射体
Horn Spoon　"角匙"(苏联舰载导航雷达的北约代名)
horn-paraboloid antenna　抛物面天线
horse race acquisition studies　赛马式方案竞争
horsepower　马力
horsepower loading　马力载荷
horseshoe　马蹄形机窝
horse-shoe vortex　马蹄涡
hose reel unit　软管卷盘装置
host computer　主机
host processor　主处理机
hostile　(尤指在远距显示器上看到的)敌性目标
hostile condition　敌对环境,敌对状态
hostile environment　敌对环境
hostile fire indicator　(直升机用的)敌方火力位置指示器(通过探测弹丸的激波来指示火力位置)
hostile track　敌性航迹(根据规定标准可确定为空中威胁的敌机、弹道导弹及轨道飞行器等)
HOT　Hot Oil Temperature 热滑油温度
hot agglomeration　热聚集

hot air　热风
hot airstream　热的气流
hot and high　高温高海拔(指机场海拔高且场面气温高,起飞距离加长情况)
hot area　辐射区
hot boundary　热边界
hot carrier diode　热载流子二极管
hot cathode ionization gauge　热阴极电离真空计
hot combustion product　热的燃烧产物
hot core　热核
hot dimpling　热压凹坑,热压波纹,热打孔(将工件预先加热,以防加工时开裂)
hot die forging　热模锻
hot electron　热电子
hot diping　热浸镀
hot end　(燃气轮发动机的)热端,热部件
hot electron transistor　热电子晶体管
hot exhaust　热排气
hot exhaust duct　热排风管道
hot film　热膜
hot film anemometer　热膜风速仪
hot fire　烈火
hot gas　热气
hot gas region　热气体区域
hot gas system　热气系统
hot ionization gauge　热电离计
hot launch　热发射
hot line sight　"热线"瞄准具,快速射击瞄准具
hot mission　对周围有危险的任务
hot isostatic pressing　热等静压
hot nozzle　热嘴
hot outer surface　热表面
hot plasma　热等离子体
hot plasma cloud　热等离子体云
hot potato routing　快速路由选择
hot pressing　热压
hot probe method　热探针法(测量半导体材料导电类型的方法)
hot product　热点产品
hot refuelling　热加油(飞机着陆后,发动机不停车的状态下给飞机加燃油,保证快速地再次出动)
hot rock　(口语)有经验的新飞行员
hot rocket　热火箭(指燃料正在燃烧着的火箭)
hot round　热弹(指发动机正在工作的火箭或导弹)
hot shot tunnel　热射风洞
hot spare　热备份
hot spot　热班,热点,过热部位
hot spraying　1.热喷涂(工艺) 2.热风烘干法
hot standby　备用卫星

hot start　过热起动
hot start life test　热起动寿命试验
hot streak　火舌加力起动
hot structure　热结构
hot test　热试验
hot vehicle　〈口语〉导弹，火箭
hot wall　热壁
hot weather construction　防热结构（对航天器返回地面时的气动热进行热防护的结构）
hot winchback　热进库（飞机在发动机工作时进入核防护飞机掩体）
hot wire　电启动，热线，火线
hot wire anemometer　热线风速仪
hot-air balloon escape system　热气球救生系统
hot-air windshield　热气加温的风挡
HOTAS　Hands On Throttle And Stick Controls 手握油门和侧杆操纵
hot-cycle　热压缩传动的（指喷气式旋翼）
hot day　标准温度日
hot-electron triode　热电子三极管
hot-extrusion　热挤压
hot-fire condition　点火条件
hot-fire test　点火测试
hot-fire testing　热火测试
hot-gas　热气
hot-gas temperature　高温气体温度
hot-gas ejection system　热气弹射系统
hot-gas system　燃气系统
hot-gas valve　燃气活门（用来调节燃气压力以达到推力矢量控制）
hot-pressed silicon nitride　热压氮化硅（天线窗材料）
hot-shot　高能
hot-shot ignition　热射流点火，火舌点火
hot-shot wind tunnel　热射式风洞，高温脉冲式风洞
hot-stream nozzle　热涵道喷管（指双涵道发动机的内涵道喷管），内涵喷管
hotwall reactor　热壁反应器
hot-water rocket　蒸汽火箭
hotwell　热油隔间，(滑油箱的)热油井
hot-wire　热丝钨极惰性气体保护焊
hot-wire measurement　热线式测量
hot-wire probe　热线探针
hot-wire technique　热线测试技术
hot-wire transducer　热丝传感器（利用热丝阻抗的变化来探测和测定声波）
hotwork　加热工作，热作
hour-angle　【天】时角
hour circle　【天】时圈，子午线
hour-glass　沙漏、水漏

hour-glass mode　沙漏模式
hour-glass viscosity　沙漏黏度
hour rate　时率（以充放电时间表示的充放电速率）
house aircraft　内部科研飞机，研究飞机
house resolution　室内分辨力
hover　1. 悬停飞行 2. 零空速飞行 3. 在地球同步轨道上 4.（飞机）在目标上盘旋 5.（滑翔机）翱翔 6. 乘坐气垫飞行器
hover condition　气垫悬浮状态
hover flight　高降投掷
hover height　悬停高度
hover test　悬停试验
hovercoupler　悬停连接器（直升机在海上全自动悬停的装置，以便夜间进行吊放式声呐作业）
hovercraft　气垫（飞行）器，地效飞行器
hovering　悬停
hovering ceiling　1. 悬停升限 2. 直升机静升限
hovering efficiency　悬停效率
hovering flight　悬停飞行
hovering indicator　悬停指示器
hovering point　悬停点（尤指垂直/短距起落飞机不着陆装卸载荷的地点）
hovering rig　悬停试验台（用空间构架式结构装上喷气式升力发动机，组成自由飞试验台，用来研制低空悬停飞行的操纵系统）
hoverprint　直升机垂直起落标志点
hovership　气垫船
Howe truss　霍氏构架（一种框架结构形式，有水平和垂直构件和从下方构件的中点开始的对角斜撑）
howgozit　飞行计划图
HP　1. High Pressure 高压 2. Holding Pattern 等待航线 3. Horse Power 马力
HPA　1. Hectopascals 百帕 2. High Power Amplifier 高功率放大器
HPC　High Pressure Compressor 高压压缩机
HPF　1. Horizontal Position Finder 水平位置探测仪 2. Hazardous Processing Facility 危害处理设施
HPFE　Horizontal Position Fix Error 水平位置坐标误差
HPGC　High Pressure Ground Connector 高压地面连接器
HPHR　Horsepower-Hour 马力--小时
HPI　Hydraulic Pressure Indicator 液压指示器
HPL　High Power Laser 大功率激光器
HPLB　High-Power Laser Blinding 大功率激光致盲
HPM　High Performance Motor 高性能发动机
HPOT　High-Pressure Oxidizer Turbopump 高压氧化剂涡轮泵
HPOX　High Pressure Oxygen 高压氧

HPQY High Purity Quartzyam 高纯石英丝
HPRFW High-Power Radio Frequency Weapons 大功率射频武器
HPRL Human Performance Research Laboratory 人类行为特性研究实验室（美国国家航空航天局）
HPRP High-Power Reporting Point（雷达）大功率报知点
HPS Horizon and Pitch Scale 地平与俯仰刻度
HPSN Hot-Pressed Siliconnitride 热压氮化硅
HPSOV High-Pressure Shutoff Valve 高压切断阀
HPSRM High-Performance Solid Rocket Motor 高性能固体火箭发动机
HPT 1. High Pressure Test 高压试验 2. High Pressure Turbine 高压涡轮
HPV High Pressure Valve 高压阀
HQA Hardware Quality Assurance 硬件质量保证
HR 1. Hose Rack 软管托架 2. Hour 小时
HRD 1. Hardware Requirements Document 硬件要求文件 2. High Rate Discharge 高速放电
HR-FAX High Resolution Facsimile 高分辨率传真
HRL Horizontal Reference Line 水平基准线
HRM Human Resource Management 人力资源管理
HRP 1. Human Resource Planning 人力资源规划 2. Hypothalamic Regulatory Peptides 下丘脑调节性多肽
HRR High Resolution Radar 高分辨率雷达
HRS 1. High Resolution Spectrometer 高分辨率光谱仪 2. Hot-Rolled Steel 热轧钢
HRZN Horizon 水平线
HS 1. Heat Shield 隔热罩,隔热屏 2. Hidden Safety 隐性安全性 3. High Speed 高速 4. Horizontal Stabilizer 水平安定面
HSACE Horizontal Stabilizer Actuator Control Electronics 水平安定面驱动器控制电子装置
HSDB High-Speed Data Bus 高速数据总线
HSDL High-Speed Data Link 高速数据链路
HSDP High Speed Data Processor 高速数据处理机
HSE Hamilton Sundstrand（EPS)汉胜公司（电源系统）
HSG Housing 外壳,外罩
HSH Hamilton Sundstrand（HL)汉胜公司（增升装置）
HSI Horizontal Situation Indicator 水平状态指示器,航道罗盘
HSIF Hardware/Software Integration Facility 硬件/软件合成设施
HSIT Hardware and Software Integration Test 硬件和软件综合试验
HSL Heading Select 航向选择
HSM 1. Heading Select Mode 航向选择模式 2. Hierarchical Storage Management 分级存储管理

HSR High Speed Reader 高速阅读器
HSRP Hot Standby Routing Protocol 热备路由协定书
HSS High-Speed Steel 高速钢
HST 1. High Speed Taxiway/Exit/Turnoff 高速滑行道/出口/滑出道 2. Hoist 升起,提高,升降机,卷扬机
HSTA Horizontal Stabilizer Trim Actuator 水平安定面配平驱动器
HSTH Hose Thread 软管螺纹
HT 1. Hard Time 定时 2. Heat Treat(ment)热处理
H-T High Tension 高张力
HT RES Heat Resisting 耐热的
HTCI High-Tensile Cast Iron 高强度铸铁
HTD Heated 加热的
HTG Heating 加热
HTPB propellant 端羟基聚丁二烯推进剂
hub 轮毂
human centrifuge 1.载人离心机 2.人体离心机
human control 人为管控
human interaction 人际互动
human reliability 人员可靠性
human visual system 人类视觉系统
humidity control 湿度控制
humidity test 潮湿试验
humidity-heat test 湿热试验
hump 驼峰,驼背,圆形隆起物
hump effect 驼峰效应
hybrid access 混合接入
hybrid approach 混合算法
hybrid assembly language 混合汇编语言
hybrid atomization 混合雾化
hybrid atomizer 混合动力喷雾器
hybrid bearing 动静压轴承
hybrid coating 复合镀层
hybrid coil 混合线圈
hybrid comparison 混合动力的比较
hybrid composite 混杂复合材料
hybrid compressor 混合动力压缩机
hybrid computer 混合计算机,混合电脑
hybrid computer simulation 混合计算机模拟,混合计算机仿真
hybrid coordinates 混合坐标系
hybrid data acquisition system 并合数据采集系统
hybrid detonation 混合爆炸
hybrid diffuser 混合扩散器
hybrid engine 组合发动机
hybrid error control 混合差错控制
hybrid frequency 混频
hybrid fuel 混合动力
hybrid hologram 混合全息图

hybrid injector 混合喷射器	hydraulic flight control 液压飞行控制
hybrid junction 混合接头	hydraulic forming 液压成形
hybrid liquid-solid rocket motor 混合固液火箭发动机	hydraulic hammer 液压锤,水压锤,水力冲击
hybrid multi-shock shield 混合多次冲击防护屏	hydraulic jump 水跃
hybrid navigation 混合导航	hydraulic lock 液(压)锁
hybrid navigation system 混合导航系统	hydraulic missile tester 液压导弹测试仪
hybrid network 混合网络	hydraulic motor 液压马达,液压发动机
hybrid optimal control problem 混合最优控制问题	hydraulic oil 液压油
hybrid optimization 混合最优化	hydraulic pipeline 液压管路
hybrid oxidizer 固液混合氧化剂	hydraulic power unit 液压动力组,液压泵站,液压动力装置,液动压力机构
hybrid power source 混合电源	hydraulic pressure transmitter 液压传感器
hybrid problem 并合问题	hydraulic redundancy control 液压余度控制
hybrid procedure 镶嵌治疗	hydraulic seal 液压密封
hybrid propellant 固液混合推进剂	hydraulic step motor 液压步进马达
hybrid propellant rocket engine 1.混合(推进剂)火箭发动机 2.固液火箭发动机	hydraulic system 液压系统
hybrid propulsion 混合燃料推进	hydraulic tank 液压油箱
hybrid pseudorandom code and side tone ranging 1.伪码侧音混合测距 2.码音混合测距	hydraulic transmission 液压传动
hybrid reaction 混合反应	hydraulic vibration generator 液压式振动台
hybrid reaction model 混合反应模型	hydraulic winch 液压卷扬机
hybrid relay 混合继电器	hydrazine 肼,联氨,酰肼
hybrid rocket 混合式火箭	hydrazine arcjet 肼电弧加热发动机
hybrid rocket engine 混合式火箭引擎	hydrazine engine 肼发动机
hybrid satellite system 综合卫星系统	hydrazine exposure 肼曝光
hybrid simulation 【计】混合模拟	hydrazine propellant 联氨推进剂
hybrid simulation system 并合仿真系统	hydrobooster 液压加力器
hybrid solar array 混成太阳电池板	hydrocarbon 碳氢化合物
hybrid structure 混杂结构	hydrocarbon combustion 碳氢化合物燃烧
hybrid wave 混合型波	hydrocarbon emission indices 碳氢化合物排放指标
hybridization frequency 杂化频率	hydrocarbon fuel 烃类燃烧
hydrate 水合物,氢氧化物	hydro-drawing 液压拉延
hydraulic accessory integration 液压附件集成	hydrodynamic equation 流体动力学方程
hydraulic actuating unit 液压致动机构	hydrodynamic frequency 水动力的频率
hydraulic actuator 液压执行器,液动装置	hydrodynamic gas bearing 水压气体轴承
hydraulic amplifier 液压放大器,液压助力器	hydrodynamic gas bearing gyro 动压气浮陀螺仪
hydraulic analysis 水力分析	hydrodynamic gas bearing motor 动压陀螺电机
hydraulic autopilot 液压自动驾驶仪	hydrodynamic impact 水动力冲击
hydraulic balance 液压平衡	hydrodynamic instability 流体动力学不稳定性
hydraulic booster 液压助力器	hydrodynamic layer 水动力层
hydraulic brake system 液压刹车系统	hydrodynamic losses 水动力损失
hydraulic capacitor 蓄压器	hydrodynamic sliding bearing 水动力滑动轴承
hydraulic circuit 液压回路	hydrodynamic strain 水动力应变
hydraulic control 液压控制,水力操纵,液压式控制	hydrodynamic thickness 水动力厚度
hydraulic diameter 水力直径,水压直径	hydrofoil boat 水翼船
hydraulic divider 液压分配器	hydrogen 【化】氢(化学元素,符号 H)
hydraulic drawing 液压拉延	hydrogen arcjet 氢电弧
hydraulic feedback 液压反馈	hydrogen barbotage 氢鼓泡
hydraulic filter 液压油滤	hydrogen bomb 氢弹
	hydrogen bubble method 氢气泡法

hydrogen combustion　氢燃烧
hydrogen combustion heated facility　氢燃烧加热设备
hydrogen coolant　氢冷却剂
hydrogen datum　氢基准
hydrogen detection system　氢泄漏检测系统
hydrogen economy　氢经济
hydrogen flow　氢气流
hydrogen fuel　氢燃料
hydrogen gas　氢气
hydrogen isocyanide　异氰化氢
hydrogen leakage　漏氢
hydrogen mixture　氢混合物
hydrogen mole　氢摩尔
hydrogen mole fraction　氢的摩尔分数
hydrogen particle　氢粒子
hydrogen pellet　氢颗粒
hydrogen side　氢侧
hydrogen sulfide　硫化氢
hydrogen test　氢含量试验
hydrogen vent auto-disconnect coupler support mount　氢排气自动脱落连接器支架
hydrogen-deficient star　缺氢星
hydrogen-nickl battery　氢镍蓄电池
hydrogen-oxygen　氢氧
hydrogen-oxygen fuel cell　氢氧燃料电池
hydrogen-oxygen mixture　氢氧混合物
hydrogen-oxygen vent auto-disconnect coupler　氢氧排气自动脱落连接器
hydrogeologic map　水文地质图
hydroglider　水上滑翔机
hydrogyro　流体浮悬陀螺仪
hydrojet engine　水力喷射引擎
hydromagnetic theory　磁流理论
hydromagnetics　磁流体动力学
hydrometer　液体比重计,浮秤
hydrophobic coating　疏水涂层
hydrophobicity　疏水性
hydrophone　水诊器,检漏器,水中听音器
hydroplane　水上滑艇,水上飞机,水平舵
hydroplaning　滑水现象,水面滑行,水漂;湿路滑胎
hydropower　水力发出的电力,水力发电
hydropress　液压机,水压机
hydrospace　海洋世界
hydrostatic bearing gyro　静压轴承陀螺
hydrostatic bomb fuze　静水炸弹引信
hydrostatic drive　静液压传动
hydrostatic equation　流体静力学方程
hydrostatic extrusion　静液挤压
hydrostatic fuze　水压引信
hydrostatic gas bearing gyro　静压气体轴承陀螺
hydrostatic gas bearing PIGA　静压气浮陀螺加速度计
hydrostatic liquid bearing gyro　静压液压陀螺仪
hydrostatic liquid bearing PIGA　静压液浮陀螺加速度计
hydrostatic pressure　液体静压力
hydrostatic test　水压试验,静水压试验,流体静压试验
hydrothermal convection system　水热对流系统
hydrothermal effect　湿热效应
hydro-thermal method　水热法
hydrothermal resources　水热资源
hydrothermal system　水热体系
hydroxide　氢氧化物,羟化物
hydroxy terminated polybutadiene propellant　端羟基聚丁二烯推进剂
hydroxyl　羟基,氢氧基
hydryzing　氢流热处理
hyetograph　雨量分布图,雨量计
hygristor　湿敏电阻
hygrograph　自动湿度计
hyper　超过
hyper ballistics　高超声速弹道学
hyper focal distance　超焦距
hyperbaric　施压力以供氧气的,高比重的
hyperbaric airlock　高压气闸
hyperbaric chamber　高压氧舱,(航)(船)高压舱
hyperbaric oxygen chamber　高压氧舱
hyperbarism　超压的,高压(处理)的,(尤指在医学手术或实验时)使用高气压的
hyperbolic lens　双曲透镜
hyperbolic navaids　双曲线助航设施
hyperbolic navigation chart　双曲线导航图
hyperbolic navigation system　双曲线导航系统
hyperbolic orbit　双曲线轨道
hyperbolic positioning　双曲线定位
hyperbolic positioning grid　双曲线格网
hyperbolic reentry　双曲线再入
hyperbolic speed　双曲线轨道速率
hyperbolic trajectory　双曲线轨道
hyperbolic umbilic catastrophe　双曲脐点剧变
hyperbolic velocity　双曲线轨道速度
hyperboloid　双曲面,双曲线体
hyperboloid surface　双曲面
hypercapnia　血碳酸过多症
hypercube　超立方体
hypercycle theory　超循环理论
hyperellipsoid　超椭圆体
hyperextension fracture　过度伸展型骨折
hyperflexion fracture　过度弯曲型骨折

hyperfocal distance 超焦距
hypergeometric function 超比函数,超几何函数
hypergolic fuel 双组分火箭燃料
hypergolic propellant 自燃推进剂
hypergolic propellant rocket engine 自燃推进剂火箭发动机
hypergranulation 超米粒组织
hypergranule 超米粒
hypersonic 极超声速的,特超声速的
hypersonic aerodynamics 特超声速空气动力学
hypersonic airbreathing 高超声速吸气式
hypersonic airbreathing engine 超声速喷气引擎
hypersonic aircraft 高超声速飞机
hypersonic boundary layer 高超声速边界层
hypersonic cruiser 高超声速巡洋舰
hypersonic facility 高超声速设施
hypersonic flight 高超声速飞行
hypersonic flow 高超声速流
hypersonic flow solver 高超声速流解算器
hypersonic inlet model 高超声速进气道模型
hypersonic inlet 高超声速进气口
hypersonic nozzle 高超声速喷管
hypersonic reentry 高超声速再入
hypersonic shock layer 高超声速激波层
hypersonic similarity law 高超声速相似律
hypersonic static margin 高超声速静态余量
hypersonic vehicle 高超声速飞行器
hypersonic wake 高超声速尾迹
hypersonic wind tunnel 高超声速风洞
hypersonic wind-tunnel facility 高超声速风洞设施
hyperspace 超空间,多维空间
hyperstability 超稳定性
hyperstereoscopy 超立体观察,超体视术
hypervelocity aircraft rocket 极高速航空火箭
hypervelocity armor piercing rocket 超高速盔甲穿火箭
hypervelocity collision 超高速碰撞
hypervelocity flow 超高速流
hypervelocity impact 超高速撞击
hypervelocity impact test 超高速撞击试验
hypervelocity launcher 高速发射装置
hypervelocity oxygen fuel spray painting 极高速氧燃料喷漆
hypervelocity projectile 超高速射弹(指炮弹、导弹、火箭等)
hypervelocity wind tunnel 超高速风洞
hypobaric and temperature test chamber 低压温度试验舱
hypobaric chamber 低压舱
hypobaric hypoxia 低压缺氧
hypobaric hypoxia examination 低压缺氧检查
hypobaric susceptibility test 低压敏感性检查
hypobaric thermal chamber 低压温度舱
hypobarism 低比重
hypodynamic state 低动力状态
hypoglycemic syncope 低血糖晕厥
hypogravic 减重的
hypogravics 减重
hypometabolism 基础代谢率减退
hypotaxia 控制力减弱,协调减退
hypotension 低血压,血压过低
hypotenuse 直角三角形的斜边
hypothermia 降低体温,低体温症
hypothesis 假设
hypothesis of isostasy 地壳均衡假说
hypothesis space 假设空间
hypothesize 假设,假定
hypothetical reference circuit 假设基准电路
hypothetical scenario 假设性情境分析
hypo-wakefulness 低醒觉状态
hypoxia alarm 缺氧警告
hypoxia time 缺氧时间
hypoxia tolerance 缺氧耐力
hypoxic convulsion 低氧惊厥
hypoxic hypoxia 低氧性缺氧
hypoxic syncope 缺氧晕厥
hypoxidosis 氧过少
hypsography 测高学,标高图,地势图
hypsometric layer 分层设色法
hypsometric map 地势图
hypsometric tints 分层设色
hysteresis behavior 磁滞行为
hysteresis behavior of mode transition 磁滞行为模式的转变
hysteresis cycle 滞后回线,磁滞循环,滞后周期,磁滞回线
hysteresis effect 滞后效应
hysteresis error 滞后误差,回程误差,磁滞误差
hysteresis gyro motor 磁滞陀螺电机
hysteresis loop 滞后回线,磁滞回线
hysteresis loss 磁滞损失
hysteresis model 滞后模型
hysteresis nonlinearity 磁滞非线性
hysteresis phenomenon 滞后现象
hysteresis phenomenon of mode transition 滞后现象的模式转变
hysteresis synchronous gyro motor 陀螺磁滞同步电动机
hysteresis synchronous motor 磁滞同步电动机
Hytex 高超声速技术试验机
Hz 频率

I

IA 1. Inactive Aircraft 非现役飞机，退役飞机 2. Inertial Altitude 惯性高度 3. Inertial Autopilot 惯性自动驾驶仪 4. Information Access 信息存取 5. Information Adaption 信息适配 6. Input Axis 输入轴 7. Inventory Accounting 库存统计 8. Inspection Authorization 检查核准权（美国联邦航空局用语）9. Installation Approval 装机批准书 10. Instruction Address 指令地址 11. Intelligent Agent 智能代理 12. Intermediate Altitude 中间高度 13. Internal Authentication 内部认证 14. Interface Adapter 接口适配器 15. International Airport 国际机场 16. Interstage Assembly 级间组件

IAAA 1. International Airforwarder & Agents Association 国际航空运输行和代理商协会 2. International Airfreight Agents Association 国际航空货运代理商协会

IAAC 1. International Agricultural Aviation Centre 国际农业航空中心 2. International Association of Aircargo Consolidators 国际航空运输行协会

IAAE Institution of Automotive & Aeronautical Engineers 汽车与航空工程师学会（美国）

IAAEES International Association for Advancement of Earth & Environmental Sciences 国际地球与环境科学促进会

IAAEM International Association of Aircraft Equipment Manufacturers 国际飞机设备制造商协会

IAAFA Inter-American Air Forces Academy 泛美空军学院（美国）

IAAI 1. Indonesian Aeronautical & Astronautical Institute 印度尼西亚航空航天研究会 2. International Airport Authority of India 印度国际机场管理局

IAASM International Academy of Aviation & Space Medicine 国际航空航天医学院

IABA International Aircraft Brokers Association 国际飞机经营人协会

IABCS Integrated Aircraft Brake Control System 飞机刹车综合控制系统

IAC 1. Immediate Action Command 立即作用指令，快速行动指令 2. Indian Airlines Corporation 印度航空公司 3. Initial Approach Course 初始进近航线 4. Iranian Airways Co. 伊朗航空公司 5. Irish Air Corps 爱尔兰空军 6. Integration Assembly & Checkout 集成装配和检验 7. Intelligence Analysis Center 情报分析中心（美国）8. International Aerological Commission 国际高空气象学委员会 9. International Air Convention 国际航空公约 10. International Astronautical Congress 国际宇航会议 11. Interstate Aviation Committee 洲际间航空委员会 12. Instructor Aircraft Commander 飞机指挥员教官 13. Instrument Approach Chart 仪表进近图

IACA 1. International Air Carrier Association 国际航空运输公司协会 2. International Air Cargo Association 国际航空货运协会

IACARS International Aircrft Communication Addressing & Reporting System 国际通用飞机通信寻址和报告系统

IACES International Air Cushion Engineering Society 国际气垫工程学会

IACG Inter-Agency Consultative Group 航天局际顾问团

IACS 1. Inertial Attitude Control System 惯性姿态控制系统 2. Integrated Armament Control System 综合军械控制系统 3. Integrated Avionics Control System 综合航空电子控制系统 4. Intermediate Altitude Communication Satellite 中等高度通信卫星 5. International Annealed Copper Standard 国际退火铜标准

IADB Inter-American Defense Board 泛美防务委员会

IADS Integrated Air-Defence System 联合防空系统

IAE 1. Infrared Astronomy Explorer 红外天文探测卫星 2. Institute for Advancement in Engineering 工程改进研究所（美国）3. International Aero Engines 国际航空发动机公司 4. Institution of Aeronautical Engineers 航空工程师协会

IAEA 1. Indian Air Engineer Association 印度航空工程师协会 2. International Atomic Energy Agency 国际原子能机构

IAEDS Integrated Advanced Electronic Display System 先进综合电子显示系统

IAES Institute of Aeronautical Sciences 航空科学学会（英国）

IAF 1. Image Analysis Facility 图像分析设备 2. Imperial Air Force 皇家空军 3. Indian Air Force 印度空军 4. Indonesian Air Force 印度尼西亚空军 5. Intermediate/Initial Approach Fix 中间/起始进近定位点 6. International Astronautical Federation 国际

航天联合会 7. Iraqui Air Force 伊拉克空军

IAFA International Airfreight Forwardersr Association 国际航空货运商协会

IAFU Improved Assault Fire Unit 改进的突击火力单元

IAG International Association of Geodesy 国际大地测量学协会

IAGC Instantaneous Automatic Gain Control 即时自动增益控制

IAGS Inter-American Geodetic Survey 泛美大地测量

IAI 1. Information Acquisition & Interpretation 信息获取和判读 2. Intake Anti-Ice 进气道防冰 3. Israel Aircraft Industries 以色列飞机公司

IAIN International Association of Institutes of Navigation 国际导航学会联合会

IAIP Integrated Aeronautical Information Packages 综合航行情报包

IAL 1. International Aeradio Ltd. 国际航空无线电有限公司 2. International Airtraffic League 国际空中交通联合会 3. Instrument Approach & Landing 仪表进近和着陆

IAM Indicated Airspeed Mode 指示空速模式

IAOPA International Council of Aircraft Owner and Pilot Associations 飞机拥有者和驾驶员协会国际理事会

IAP 1. Immediate After Passing 飞越后立即 2. Improved Accuracy Program 改进精确度计划 3. Initial Aiming Point 起始瞄准点 4. Initial Approach Procedure 起始进近程序 5. Inlet Absolute Pressure 进口绝对压力 6. Instrument Approach Procedure 仪表进近程序 7. Integrated Actuator Package 集成作动器组件 8. Integrated Aeronautical Program 综合航计划 9. International Aero Press 国际航空出版社 11. Internet Access Point 网际访问点 10. International Airport 国际机场 12. Internet Access Provider 因特网接入服务供应商

IAPA International Aviation Photographers' Association 国际航空摄影师协会

IAPC International Airport Planning Consortium 国际航空港规划财团

IAPCH Initial Approach 起始进近

IAPO International Association of Physical Oceanography 国际物理海洋学协会

IAPS 1. Integrated Actuator Package System 组合制动器组件系统 2. Integrated Aviation Processing System 综合航空电子处理系统 3. Ion Auxiliary Propulsion System 离子辅助推进系统

IAPSO International Association for the Physical Sciences of the Ocean 国际海洋物理科学协会

IAQG International Aerospace Quality Group 国际航空航天质量组

IAS 1. Impact Attenuation System 冲击缓减系统 2. Independent Airlift Support 独立空运保障 3. Indicated Air Speed 指示空速 4. Institute for Atmospheric Sciences 航空科学院,大气科学研究院 5. Instrument Approach System 仪表进近系统 6. Integrated Access Server 综合接入服务器 7. Integrated Acoustic Structure 集成声学结构 8. Interactive Application Server 交互应用服务器 9. International Aircraft Standards 国际飞机标准 10. International Application Satellite 国际应用卫星 11. Interplanetary Automated Shuttle 行星间自动穿梭飞行器

IASA International Air Safety Association 国际航空安全协会

IASC 1. Integrated Air System Control 综合空气系统控制器 2. Integrated Air System Controller 综合空气系统控制器 3. Industrial Aviation Service Corporation 中国民航工业航空服务公司

IASEL International Automatic Space Ecological Laboratory 国际自动空间生态学实验室

IASY International Active Sun Year 国际活动太阳年

IAT 1. Indicated Air Temperature 表温,指示气温 2. Inspection Apply Template 检验用样板 3. Intake Air Temperature 进气道进口空气温度

IATA 1. International Air Transport Association 国际航空运输协会 2. International Airline Telecommunications Association 国际航空系统电信协会

IATS Intermediate Automatic Test System 中间自动测试系统

IATSC International Aeronautical Telecommunications Switching Center 国际航空电信转换中心

IAU 1. Interface Adaptor Unit 接口适配器 2. International Accounting Unit 国际核算机构 3. International Astronomical Union 国际天文联合会

IAVC Instantaneous Automatic Volume Control 瞬时自动音量控制

IAVW International Airways Volcano Watch 国际航路火山灰观察

IAWG Industrial Avionics Working Group 工业航空电子设备工作组

IB 1. Identification Beacon 识别信标 2. In Bond 归航的 3. Inboard 内侧的,机内的 4. Inbound 入境的,进港的 5. Iron Bird 铁鸟

IBA 1. Igniter Booster Assembly 点火器助推器组件 2. Inbound Boom Avoidance 避免进港音爆

IBAA 1. International Business Aircraft Association 国际公务飞机协会(欧洲) 2. Italian Business Aviation

Association 意大利公务航空协会

IBAC International Business Aviation Council 国际公务航空理事会,国际商业航空委员会

IBC Information Bearer Channel 信息承载信道

IBCN 1. Integrated Broadband Communication Network 综合宽带通信网 2. International Broadband Communication Network 国际宽带通信网络

IBD Ion Beam Deposition 离子束淀积

IBDN Integrated Building Distribution Network 楼宇综合布线网络

IBEN Incendiary Bomb with Explosive Nose 带爆炸头的燃烧炸弹

IBIS 1. Image Based Information System 图像信息系统 2. International Bird-hazard Information System 国际民航组织撞鸟资料系统

IBM International Business Machine 国际商用机器公司

IBOM Integrated Bill Of Material 集成物料表

IBRD Inflated Ballute Retarding Device 充气伞减速装置

IBRL Initial Bomb Release Line 起始投弹线

IBS 1. Intelsat Business Service 国际通信卫星商用业务 2. Ionospheric Beacon Satellite 电离层信标卫星

IBSS Infrared Background Signature Survey 红外背景特征探测器(美国)

IBU 1. Independent Back-up Unit 独立备用组件 2. Integrated Ballast Unit 集成式镇流器组件

IBW Ion Beam Weapon 离子束武器

IC 1. Inductance-Capacitance 感应电容 2. Integrated Circuit 集成电路 3. Intercabinet 机箱之间 4. Interface Control 界面控制 5. Isobaric Condition 等压状况

ICAM 1. Integrated Computer-Aided Manufacturing 综合计算机辅助制造 2. International Civil Aircraft Marking 国际民用航空器标记

ICAN International Commission on Air Navigation 国际空中导航委员会

ICAO International Civil Aviation Organization 国际民用航空组织

ICAOTAM International Civil Aviation Organization Technical Assistance Mission 国际民航组织技术援助任务

ICAS International Council of the Aeronautical Sciences 国际航空科学理事会

ICASS International Confidential Aviation Safety System 国际航空安全保密系统

ICB International Competitive Bidding 国际竞争性投标

ICBD Ionized-Cluster Beam Deposition 离子团束淀积

ICBE Ionized-Cluster Beam Epitaxy 离子团束外延

ICBM Inter-Continental Ballistic Missile 洲际弹道导弹

ICC 1. Inner Combusion Chamber 内燃烧室 2. International Chamber of Commerce 国际商会 3. International Computing Center 国际计算中心 4. International Conference on Communications 国际通信会议 5. Instrument Control Center 仪表控制中心 6. International Control Commission 国际控制委员会

ICCAIA International Coordinating Council of Aerospace Industries Association 航空航天工业协会国际协调理事会

ICCC International Conference on Computer Communication 国际计算机通信会议

ICCD Intensified Charge Coupled Device 增强型电荷耦合器件

ICCP Institute for the Certification of Computer Professionals 计算机专业证书学会

ICD 1. Interface Control Document 接口控制文件 2. Interface Control Drawing 界面控制图 3. Integrated Control/Display 综合控制/显示器 4. Installation Control Drawing 安装控制图

ICDS 1. Integrated Control & Display System 综合控制显示系统 2. Integrated Configuration & Data Support 集成构型和数据支持 3. Inventory Control Distribution System 库存管理分发系统

ICE 1. Instrument Checkout Equipment 仪表检验设备 2. Internal Combustion Engine 内燃机

ICEM Intergovernmental Committee for European Migrations 政府间欧洲移民委员会

ICES International Committee for Earth Sciences 国际地球科学委员会

ICFD Interdisciplinary Computational Fluid Dynamics 多学科计算流体动力学

ICI 1. Initial Capability Inspection 起始能力检查 2. Intelligent Communications Interface 智能通信接口 3. International Commission for Illumination 国际照明委员会 4. Inter-Carrier Interference 载波间干扰

ICLECS Integrated Closed-Loop Environmental Control System 综合闭环式环境调节系统

ICLS Instrument Carrier-Landing System 航空母舰仪表降落系统

ICM 1. Image Compression Manager 图像压缩管理器 2. Improved Conventional Munition 改良常规弹药 3. Incoming Call Management 来话呼叫管理 4. Information Coordination Memo 信息协调备忘录 5. Intercontinental Missile 洲际导弹 6. Intelligent Composite Material 智能复合材料 7. Interface Control Model 界面控制模型 8. Institution of Configuration Management 构型管理委员会 9. Interface Coordination Memo 界面协调备忘录

ICMP Internet Control Message Protocol 因特网控制

信息协议

ICMUA International Commission on the Meteorology of the Upper Atmosphere 国际高层大气气象学委员会

ICN Indicator Coupling Network 指示器耦合网络

ICNI Integrated Communication-Navigation-Identification 通信、导航和识别综合系统

ICNIA Integrated Communications, Navigation & Identification Avionics 通信、导航和识别综合航空电子系统

ICNS Integrated Com/Nav System 通信导航综合系统

ICO Interim Circle Orbit 中高度圆轨道

ICPA Indian Commercial Pilots Association 印度商业飞机驾驶员协会

ICR 1. Initial Cell Rate 初始信元率 2. Integrated Communication Resource 集成通信资源 3. Inductance-Capacitance Resistance 感应电容电阻

ICRC Interiors Configuration Review Committee 内部构型评审委员会

ICRH Ion Cyclotron Resonance Heating 离子回旋共振加热

ICRPG Interagency Chemical Rocket Propulsion Group 跨部门化学火箭推进研究组

ICS 1. Improved Composite Structure 改进的复合材料结构 2. Improved/Integrated Communications System 改进的／综合的通信系统 3. Intercommunicatin System 机内通话系统 4. Interim Contractor Support 过渡性承包商保障 5. Internal Countermeasures System 内部对抗措施系统 6. Inverse Conical Scanings 逆圆锥形扫描 7. International Commuter System 国际短程飞机系统 8. Inventory Control System 库存管理系统

ICSC 1. Interim Communications Satellite Committee 临时通信卫星委员会 2. International Communications Satellite Corporation 国际通信卫星公司

ICSMA Integrated Communications System Management Agency 综合通信系统管理机构

ICSS Integrated Communications Switching System 综合通信转换系统

ICST Institute for Computer Science & Technology 计算机科学和技术学会(美国)

ICSU International Council of Space Unions 国际空间联盟理事会

ICT 1. Interface Control Tooling 接口控制仪 2. Information & Communication Technology 信息和通信技术 3. Implementation/Coordination Team 实施/协调组

ICU 1. Instrument Comparator Unit 仪表比较器装置 2. Interface Control Unit 交连控制组件 3. Isolation Control Unit 隔离控制组件

ICWAR Improved Continuous Wave Acquisition Radar 改进的连续波探测雷达

ICWG Interface Control Working Group 界面控制工作组

ICWRO Inter-Component Work Release Order 分厂间工作发放指令

ID 1. Installation Drawings 安装图 2. Interface Drawings 界面图 3. Internal Diameter 内径

IDA 1. Integrated Digital Avionics 数字式综合航空电子系统 2. Interactive Design Applications 交互式设计应用 3. Input Data Assempler 输入数据组合器 4. Interchange of Data between Administrations 机构间的数据交换 5. Inter-Divisional Agreement 部门间协议

IDAS 1. Integrated Defensive Aids System 综合防御辅助系统 2. Integrated Design Automation System 综合自动设计系统 3. Integrated Digital Avionics System 综合数字航空电子设备系统

IDC 1. Image Dissector Camera 析像管摄像机 2. Imperial Defence College 帝国国防学院 3. International Data Company 国际数据公司 4. Internet Data Center 因特网数据中心

IDCSP Initial Defense Communication Satellite Program 初级国防通信卫星计划

IDD Interface Design Document 界面设计文件

IDDS Image Data Digitizer System 图像信息数字化系统

ideal filter 理想滤波器
ideal fluid 理想流体
ideal frequency domain filter 理想频域滤波器
ideal gas 理想气体
ideal gas constant 理想气体常数
ideal gas law 理想气体定律
ideal time domain filter 理想时域滤波器
identifiability 可辨识性,可识别性
identification friend or foe 敌我识别
identification friend or foe equipment 敌友识别装置,敌我识别器

IDEP Interagency Data Exchange Program 部门间资料交换程序

IDG Integrated Drive Generator 综合驱动发电机

IDHS Intelligence Data Handling System 侦察数据处理系统

idling 空转,零负载转

IDM Integrated Data Management 集成数据管理

IDMS 1. Information & Data Management System 信息和数据管理系统 2. Integrated Database Management System 综合数据库管理系统

IDN 1. Integrated Digital Network 综合数字网络 2. Intelligent Data Network 智能数据网络 3. Interactive Data Network 交互式数据网络 4. International Directory Network 国际目录网络 5. Isolated Data Network 独立数据网络

IDO International Development Organization 国际开发组织

IDP Integrated Data Processing 综合数据处理,积累数据处理,累计数据处理

IDPM Institute of Data Processing Management 数据处理管理学会

IDPS Image Data Processing Station 图像数据处理站

IDR 1. Instrument Departure Route 仪表离场路线 2. Interim Design Review 中间设计评审,临时设计评审 3. Intermediate Data Rate 中等数据速率

IDSCS Initial Defence Satellite Communications System 初级国防卫星通信系统

IDT 1. Image Dissecting Tube 图像分析象管 2. Integrated Digital Terminal 综合数字终端 3. Intelligent Data Terminal 智能数据终端 4. Interactive Data Terminal 交互数据终端

IE 1. Igniton Exciter 点火激励器 2. Industrial Engineering 工业工程,工艺工程 3. Initial Equipment 初始装备 4. Institution of Electronics 电子学会 5. Instrument Error 仪表误差 6. Ionosphere Explorers 电离层探险者卫星

IEA 1. International Environmental Agreements (IEAs) 国际环境协定 2. International Ergonomics Association 国际人机工程学协会,国际工效学协会

IEC 1. Integrated Environmental Control 综合环境控制 2. Integrated Ethernet Chip 集成以太网电路芯片 3. Inter-Exchange Carrier 局间载波 4. International Electric Committee 国际电气委员会 5. International Electrotechnical Commission 国际电子技术委员会

IECM Induced Environmental Contamination Monitor 诱导环境污染监测器

IECMS In-flight Engine-Condition Monitoring System 飞行中发动机状态监控系统

IED Improvised Explosive Device 简易爆炸装置

IEE Institution of Electrical Engineers 电气工程师协会

IEEE 1. Institute of Electrical & Electronic Engineers 电气、电子工程师协会 2. Institute of Electrical & Electronic Engineering 电气和电子工程学院

IEOF Ignore End Of File 忽略文件末尾

IEPG 1. Independent European Programme Group 独立欧洲项目组 2. Internet Engineering & Planning Group 因特网工程和规划组

IEPR Integrated Engine Pressure Ration 集成发动机压力比

IER 1. Inherent Equipment Reliability 设备固有可靠性 2. Institute of Engineering Research 工程研究所 3. Institutes for Environmental Research 环境研究协会

IERE Institution of Electronic & Radio Engineers 电子和无线电工程师协会

IES 1. Industrial Evaluation Sheet 产业评价表 2. Institute of Environmental Societies 环境科学研究所 3. ISDN Earth Station 综合业务数字网络地球站 4. Ion Engine System 离子发动机系统

IESD Instrumentation & Electronic Systems Division 仪表和电子系统局

IET Initial Entry Training 起始进入训练

IETC International Explosive Technical Commission 国际炸药技术委员会

IETF 1. Initial Engine Test Firing 发动机初期点火试验 2. Internet Engineering Task Force 因特网工程任务组

IEU Interface Electronics Unit 接口电子器件

IF Intermediate Frequency 中频

IFA 1. Integrated Fault Analyzer 综合故障分析器 2. Intermediate Frequency Amplifier 中频放大器 3. International Federation of Airworthiness 国际适航性联合会

IFAA International Flight Attendants Association 国际机上服务人员协会

IFALPA International Federation of Airline Pilots Association 航线驾驶员协会国际联盟

IFAPA 1. International Federation of Airline Pilots Association 航空公司驾驶员协会国际联合会 2. International Foundation of Airline Passengers Association 民航旅客协会国际基金会

IFAST Integrated Facility for Avionics System Test 航空电子设备系统试验综合设施

IFATCA International Federation of Air Trafic Controllers Associations 空中交通管制员协会国际联合会

IFATE International Federation of Airworthiness Technology & Engineering 国际适航性技术与工程联合会

IFB Integrated Forebody 总装的前机身

IFBP Inflight Broadcast Procedure 飞行广播程序

IFC 1. In-Flight Collision 空中相撞 2. In-Flight Control 飞行控制 3. Inflight Calibration 飞行校准 4. Inflight Computer 飞行用计算机 5. Instantaneous Frequency Correction 瞬时频率修正 6. Instrument Flight Center 仪表飞行中心

IFE 1. In-Flight Emergency 飞行中的紧急情况 2. In-

Flight Entertainment 飞行中的文娱活动,空中娱乐
IFF 1. Identification Friend or Foe 敌我识别 2. Institute of Freight Forwarders 航空货运公司协会 3. International Flying Farmers 国际飞行承包商
IFFAA International Federation of Forwarding Agent Association 航空货运代理商协会国际联合会
IFM 1. In-Flight Monitor 机载监控器 2. Instantaneous Frequency Measurement 瞬时测频 3. Instrument Flight Manual 仪表飞行手册 4. Integrated Flow Management 综合流量管理 5. Interface Memorandum 接口备忘录
IFMA In-Flight Mission Abort 飞行中中断任务,中途停止执行任务
IFMR Instantaneous Frequency Measurement Receiver 瞬时测频接收机
IFNC Integrated Flight/Navigation Control 综合飞行及导航控制
IFO 1. International Field Office 国际地区办公室 2. Intermediate Fuel Oil 中级燃油(用于海洋产业指定燃料等级)
IFP 1. Initial Flight Path 起始飞行路线,航迹起始段,弹道起始段 2. Intenmediate Frequency Preamplifier 中频前置放大器
IFPM In-Flight Performance Monitor 飞行中性能监控
IFR 1. Increasing Failure Rate 故障率递增型 2. In-Flight Refuelling 空中加油 3. In-Flight Repair 空中修理,飞行中修理 4. Instrment Flight Recovery 仪表飞行改出 5. Instrument Flight Rules 仪表飞行规则
IFRB 1. International Frequency Registration Board 国际频率登记委员会 2. International Frequency Registration Bureau 国际频率注册局
IFSA 1. In-Flight Service Association 空中勤务协会 2. Inflight Food Service Association 航空食品服务协会
IFSAU Integrated Flight System Accessory Unit 综合飞行系统附加组件
IFSD In-Flight Shutdown 空中停车
IFSDR In-flight Shut Down Rate 空中停车率
IFSS 1. Instrumentation & Flight Safety System 仪表测量和飞行安全系统 2. International Flight Service Station 国际飞行服务站
IFT Integrated Functional Test 综合功能试验
IFTA In-Flight Thrust Augmentation 飞行中增大推力
IFTO International Federation of Tour Operators 旅行社国际联合会
IFTU Intensive Flying Trails Unit 加强飞行轨迹组件
IFU 1. Interface Unit 接口器件 2. Interworking Functional Unit 互通功能单元
IG 1. Igniter 点火器 2. Imperial Gallons 英加仑 3. Inaugural Guests 被邀请参加首航典礼的客人 4. Interactive Graphics 交互式图形 5. Inertial Guidance 惯性制导,惯性导航 6. International Gateway 国际网关
IGA 1. Integrated Gate Array 集成网关矩阵 2. International General Aviation 国际通用航空
IGB Inlet Gear Box 进口齿轮箱
IGBP International Geosphere-Biosphere Programme 国际陆圈-生物圈计划
IGE 1. In Ground Effect 受地面效应作用,有地效 2. Instrumentation Ground Equipment 地面仪表设备 3. International General Electric Corporation 国际通用电器公司
IGES 1. Initial Graphics Exchanges Standard 原始图形交换标准 2. International Graphics Exchange Standard 国际图像交换标准
ignitability 可燃性,易燃性
ignite 点火,着点
igniter 点火器,点火剂
igniter plug 点火电嘴,火花塞
ignition advance angle 点火提前角
ignition altitude 点火高度
ignition current 引燃电流
ignition delay 点火延迟,点火滞后
ignition harness 点火线,导火线
ignition lag 点火滞后,点火延后
ignition limit 点火边界
ignition system 点火系统
ignition test 点火试验
ignition timer 点火计时器
ignition voltage 着火电压
ignitor firing time 引燃时间
IGS Inertial Guidance System 惯性导航系统
IGV Inlet Guide Vanes 进口导向叶片,进气导流叶片
IH 1. Initial Heading 最初航向 2. Inhibition Height 禁止高度
IHAS Integrated Helicopter Avionics System 直升机综合航空电子系统
IHC Integrated Hand Control 一体化手操纵
IHP Indicated Horsepower 指示马力
II 1. Interchangeable Item 可互换项目 2. Interal Inspection 内部检查
IIA 1. Initial Installation Approval 初次安装批准 2. Interactive Instructional Authoring 交互式教学写作 3. Internet Image Appliance 网络影像家电
IIAL Institute of International Air Law 国际航空法学会
IIRS Instrument Inertial Reference Set 仪表惯性基准装置

IIS 1. Integrated Information System 综合信息系统 2. Integrated Instrument System 综合仪表系统 3. Internet Information Service 因特网信息服务

IISA Integrated Inertial Sensor Assembly 综合惯性敏感器组件

IISP Interim Inter-switch Signaling Protocol 临时的交换机间的信令协议

IISS Integrated Information Support System 综合信息支援系统

IIU Input Interface Unit 输入接口组件

IKPT Initial Key Personnel Training 关键人员初级训练

IL 1. Index List 目录索引，索引清单 2. Insertion Loss 插入损耗

ILA 1. International Law Association 国际法律协会 2. Instrument Low Approach 仪表低空进近 3. International Language for Aviation 国际航空用语 4. Instrument Landing Approach 仪表着陆进场

ILAS Instrument Landing Approach System 仪表着陆进场系统

ILAF Identical Location of Accelerometer & Force 加速度计指示与力的对应

ILC 1. Intelligent Line Card 智能线路卡 2. Inner Loop Control 内环路控制

ILD 1. Injection Laser Diode 注入式激光二极管 2. Insertion Loss Deviation 插入损耗偏差

illuminance 照度，照明度

illuminate 照亮，照明

illumination 照明，照度，照射

illuminator 照射雷达，施照体，照视器，照明器

illuminometer 照度计

ILM 1. Independent Landing Monitor 自主式着陆监控器，独立式着陆监控设备 2. Intermediate Level Maintenance 中级维护

ILS 1. Instrument Landing System 仪表着陆系统 2. Integrated Logistics System 综合后勤系统 3. Integrated Logistic Support 综合后勤保障 4. Instrumentation Level Simulator 仪表等级模拟器

ILSS Intigrated Life Support System 综合生命支持系统

ILSWG Integrated Logistics Support Working Group 综合后勤保障工作组

ILT Installation Lead Time 安装提前期

IM 1. Improvement Maintenance 改进性维修 2. Independent Modules 独立模块 3. Information Management 信息管理 4. Inner Marker 近距指点标 5. Instrument Measurement 仪表测量 6. Interline Message 航空同业文电 7. Intermediate Maintenance 中间级维修，二级维修 8. Intermediate Missile 中程导弹 9. Inventory Management 库存管理

IMA Integrated Modular Avionics 集成模块航空电子系统

IMAD Integrated Multisensor Airborne Display 机载多传感器综合显示器

image averaging 图像取均值

image channel 图像通道

image contrast 图像对比度

image converter 图像光电变换器变像管

image converter tube 光电图像变换管，变像管

image degradation 图像退化

image display 图像显示

image element 像素

image encoding 图像编码

image enhancement 图像增强

image exposure 图像曝光

image formation 图像形成

image frequency interference 镜像频率干扰

image intensifier 像增强器，变像管，像亮化器

image magnification 图像放大

image matching 图像匹配

image matching guidance 影像匹配制导

image motion 影像移动

image motion compensation 图像运动补偿

image parameter 影像参数

image processing 图像处理

image quality 图像品质，图像质量

image quality indicator 图像质量指示器，像质指示器灵敏度

image recognition 图像识别

image reconstruction 影像重建

image recovery mixer 镜频回收混频器

image rejection ratio 镜像抑制比

image resolution 图像清晰度，图像分解力

image restoration 图像恢复

image rotation 图像旋转

image segmentation 图像分割

image sharpening 图像锐化

image smoothing 图像平滑

image synthesis 图像合成

image theory 镜像原理

image transform 图像变换

imager 成像器

imaging plane 镜像平面

imaging radar 成像雷达

IMC 1. Image Motion Compensation 图像运动补偿 2. Image Movement Compensator 影像位移补偿器 3. Instrument Meteorological Conditions 仪表飞行气象条件 4. International Meteorological Center 国际气

象中心 5. Indirect Maintenance Cost 间接维修成本 6. Information Management Committee 信息管理委员会

IMD 1. Indian Meteorological Department 印度气象局 2. Intermodulation Distortion 交叉调制失真,互调失真

IMDB Integrated Maintenance Data Base 综合维修数据库

IMDR Item Master Definition Request 项目主定义需求

IMEP Indicated Mean Effective Pressure 指示平均有效压力

IMF 1. Integrated Maintenance Facility 综合维修设施 2. International Monetary Fund 国际货币基金组织 3. Item Master File 项目主文件

IMFS Item Master File System 项目主文件系统

IMI 1. Imbedded Message Identifier 嵌入的信息标志符 2. Improved Manned Interceptor 改进型有人驾驶截击机 3. Installation & Maintenance Instructions 安装及维修说明 4. Instrument Memory Indicator 仪表存储指示器 5. Intermediate Maintenance Instruction 中级维修指示,中级维修细则 6. Israel Military Industries 以色列军用工业公司

IMIP Industrial Modernization & Improvement Program 工业现代化和改进计划

IMIS 1. Integrated Maintenance Information System 综合维修信息系统 2. Integrated Management Information System 综合管理信息系统

IML Inside Mold Line 内模线

IMLI Import Licence 进口执照

immersion lens 浸没透镜

immersion objective lens 浸没物镜

immersion of infrared detector 红外探测器的浸没

immersion suit 抗浸服,暴露服

immersion thermocouple 浸没式热电偶

imminent to danger 危险接近

immittance 导抗

immittance bridge 导抗电桥

immune 免除的,不受影响的

immune sensor 免疫敏感器

immunity 免除,豁免;免除性,抗扰性,不敏感性,免疫性

immunity to interference 抗扰性

IMO 1. International Maritime Organization 国际海事组织 2. International Meteoroloigcal Organization 国际气象组织 3. International Money Order 国际汇票 4. Inventory Management Organization 库存管理机构

IMP 1. Impact 碰撞 2. Impedance 阻抗 3. Individual Mission Plan 单独飞行任务计划 4. Integrated Mechanism Problem 综合机构问题 5. Integrated Memory Processor 综合存储处理机 6. Integration Management Plan 综合管理计划 7. Interface Message Processor 接口信息处理器

impact load 撞击负荷

impact mechanism 触发机制

impact parameter 冲晃数

impact parameter approximation 冲晃数近似

impact point 落点

impact prediction 撞击预测

impact pressure 冲还力,撞击压力

impact sensitivity 冲击感度,撞击感度

impact test 冲击试验

impact toughness 冲坏,冲击韧性

impactor 硬着陆航天器

impatt diode 碰撞雪崩渡越时间二极管

impatt oscillator 碰撞雪崩渡越时间二极管振荡器

IMPD Interactive Multi-Purpose Display 交互式多功能显示器

impedance bridge 阻抗电桥

impedance cardiography 阻抗心动描记法

impedance chart 阻抗圆图

impedance matching transformer 阻抗匹配变压器

impedance meter 阻抗计

impedance plethysmography 阻抗容积描记法

impeller 叶轮

imperfection 缺陷,不完美

impetus 推动力,动力

impinge 撞击,碰撞

impingement area 撞击,冲击

impingement cooling 冲击冷却

implant 离子注入

implanted electrode 植入式电极

implosion 内向爆炸

impregnant 浸渍剂

impregnate 使充满,使饱和,浸渍

impregnating 浸渍

impregnation 浸渍,浸透

impressed voltage 外加电压

impulse 冲量,冲击,脉冲

impulse function 脉冲函数

impulse invariance 冲激不变法

impulse maneuver 利用推力冲量机动

impulse noise 脉冲噪声

impulse radar 无载波雷达,冲击雷达

impulse turbine 脉动透平,冲击式透平机(汽轮机)

impurity band 杂质带,杂质能带

impurity concentration 杂质浓度

impurity concentration transition region 杂质浓度过渡区

impurity energy level 杂质能级
IMR 1. Indicated Mach Reading 指示马赫数 2. Interlock Monitoring Relay 互锁监控继电器
IMRL 1. Individual Material Readiness List 个人器材准备单 2. Interlink Manufacturing Reference List 互连制造基准清单
IMS 1. Inertial Measuring System 惯性测量系统 2. Information Management System 信息管理系统 3. Institutional Management System 体制管理系统 4. Integrated Mission System 综合任务系统 5. Integrated Multiplex System 综合多路系统 6. Interactive Multimedia Service 交互式多媒体服务 7. Integrity Monitoring System 完好性监视系统 8. Integrated Master Schedule 综合总进度
IMT 1. Intelligent Multimode Terminal 智能多模式终端 2. Intermediate Maintenance Trainer 中级维护训练器
IMTA Intensive Military Training Area 集中军事训练区,密集的军事训练区
IMTS Improved Mobile Telephone Service 改进型移动式电话业务
IMU Inertial Measurement Unit 惯性测量装置
INA 1. Information Network Architecture 信息网体系结构 2. Initial Approach 起始进近 3. Integral Network Arrangement 整体网络布局 4. Integrated Network Architecture 综合网络体系结构
inaccurate missile 制导准确性低的导弹
INAS 1. Inertial Navigation & Attacks System 惯性导航与攻击系统 2. Integrated Navigation Attack System 综合导航攻击系统
INB Inbound 归航的,进站的,入境的
INBD Inboard 机上的,内测的
INCA Incorporated Nonconformance & Corrective Action 综合不符合和纠正措施
INCAND 白炽的,炽热的
incandescent display 白炽显示
incapacitation 损坏,失效,丧失能力,无资格
INCAS Integrated Navigation & Collision Avoidance System 综合导航与防撞系统
incendiary 燃烧剂,燃烧弹
inching 点动
incidence 安装角,倾角,入射角,入射,落下的方式
incidence angle 入射角
incipient failure period 早期失效期
INCLN Inclined 倾斜的
inclination 倾斜角,磁倾角,倾斜
inclination of orbit 轨道倾角,轨道交角
inclined orbit 倾斜轨道
inclinometer 倾斜仪,磁倾计

INCLR Intercooler 中间冷却器
INCLS Inclosure 罩,壳
INCND Incendiary 放火的,煽动性的,燃烧弹
incoherent detection 非相干检测
incoherent grain boundary 非共格晶界
incompatibility 不兼容性
incompatible element 不相容的,不协调的;不相容元素,不协调元素
incompressibility 不可压缩性
incompressible boundary layer 不可压缩边界层
incompressible flow 不可压缩流(体)
incompressible fluid 不可压缩流体
incremental control system 增量控制系统
incremental encoder 增量编码器
IND SENS Indicator Sensitivity 指示器灵敏度
index error 指数误差,指示误差
indexed address 变址选址
INDIC Indicator 指示器
indicated airspeed 指示空速,表速
indicated altitude 指示高度,仪表指示高度
indicated horsepower 指示马力
indicated pressure altitude 指示压力高度
indicated temperature 指示温度
indicated value 指示值
indicator diagram 指示器图,指示图
indirect gap semiconductor 间接带隙半导体
indirect recombination 间接复合
indirect wave 间接波
indirectly-heated cathode 间热式阴极,旁热式阴极
individual baseline 独立基线
INDTR Indicator-Transmitter 指示器传感器
induced drag 诱导阻力,导出阻力
induced drag coefficient 诱导阻力系数
induced environment 诱发环境
induced velocity 诱导速度
induced voltage 感应电压
induced voltage test 感生电压试验
inductance 1.电感,感应 2.进气
induction field 感应场
induction hardening 感应加热淬火
induction heating 感应加热,感应加热器
induction manifold 吸气歧管,进油歧管
induction period 诱导期,感应时间
induction phase 感应移相器
induction stroke 进气冲程
induction system 进气系统,吸气系统
induction valve 进气阀
inductive coupling 电感耦合
inductive coupling factor 感应耦合因数

inductive feedback 电感反馈
inductive reactance 感抗,感应电阻
inductor 电感器,感应器,感应物,感应体
inductosyn 感应式传感器
inductosyn characteristics measurement 感应同步器特性测试
industrial electronics 工业电子学
industrial interference 工业干扰
industrial robot 工业机器人
INE 1. Intelligent Network Element 智能网元素 2. Inertial Navigation Equipment 惯性导航设备
INEA International Electronics Association 国际电子学协会
inert gas atomized powder 惰性气体雾化粉末
inert particulate 惰性微粒子
inert region 惰性区域
inert tracer 惰性追踪器
inert tracer gas 惰性追踪器气体
inertia 惯性,惰性,惯量
inertia accelerometer 惯性加速度计
inertia coupled 惯性交感动作
inertia starter 惯性起动器,惯性起动机
inertia switch 惯性开关
inertial 惯性的,惰性的
inertial coordinate system 惯性坐标系
inertial coupling 惯性耦合
inertial cross-coupling control system 惯性交叉耦合控制系统
inertial flight 惯性飞行
inertial flight phase 惯性飞行段
inertial force 惯性力
inertial frame 惯性系
inertial guidance 惯性制导,惯性导航
inertial mass 惯性质量
inertial measurement unit 惯性测量单元
inertial navigation 惯性导航
inertial navigation system 惯性导航系统
inertial navigation system mechanization 惯性导航系统机械编排
inertial navigation unit 惯性导航组件
inertial platform 惯性平台
inertial platform aiming 惯性平台瞄准
inertial reference frame 惯性基准坐标系,惯性参考坐标系
inertial reference unit 惯性基准部件
inertial sensor 惯性敏感器
inertial sensor assembly 惯性传感器装置
inertial sensor display unit 惯性感应器显示组件
inertial sensor system 惯性传感系统

inertial space 惯性空间,惯性作用区
inertial system 惯性坐标系
inertial system display unit 惯性系统显示组件
inertial-DME Integrated Navigation System 惯性-测距器综合导航系统
inertialess scanning 无惯性扫描
INEWS Integrated Electronic Warfare System 综合电子战系统
INF 1. Inferior 低劣的 2. Infinite 无限的,无穷的 3. Information 情报,资料,信息 4. Interface 交连,交接面
infancy 初期,幼年时期
infant mortality 初期失败率,早期损坏率
inferior conjunction 下合
inferior planet 内行星
infighting missile 近战导弹
infiltration 渗入,渗透
infinite impulse response 无限冲激响应
infinite span wing 无限翼展机翼
inflammability 可燃性,易燃性
inflatable 可膨胀的,可充气的
inflatable structure 充气结构
inflate 使充气,使膨胀
inflation sleeve 充气套管
inflection 弯曲,屈曲
inflight alignment 空中对准
inflight fuel jettison 空中应急放油
inflight ignition 空中点火
inflight interruption rate 空中中断率
inflight maintenance message 飞行中维护信息
inflight maneuver 机动飞行,特技飞行
inflight modification 飞行中改变几何形状
inflight monitor 机载监控仪
inflight monitoring 飞行监测,飞行中监控,机上监控
inflight shutdown rate 空中停车率
inflight start 空中起动
inflight starting EGT 空中启动状态燃气温度
influence line 影响线
influence quantity 影响量
INFORM Information for Optimum Resource Management 最佳资源管理信息
information capacity 信息容量
information code 信息码
information content 红外发射机
information display 信息显示
information display rate 信息显示速率
information extraction 信息提取
information flow 信息流
information processing 信息处理

information rate	信息率,信息传输速率
information science	信息科学
information sink	信宿
information source	信源,信息源
information symbol	信息码元
information system	信息系统
information theory	信息理论,信息论
information unit	信息单位
INFORSAT Information transfer Satellite	信息传输卫星
infrare camera	红外相机
infrared	红外线的,红外技术
infrared antijamming	红外干扰
infrared astronomy	红外天文学
infrared bonding	红外键合
infrared countermeasure technique	红外对抗技术
infrared detector	红外探测器
infrared focal plane array	红外焦平面阵列
infrared fuze	红外引信
infrared galaxy	红外星系,红外银河
infrared gas analyzer	红外气体分析器
infrared guidance	红外制导
infrared heater	红外加热器
infrared imaging	红外线成像
infrared inhibition	红外抑制
infrared interference method	红外干涉法
infrared jamming	红外线干扰
infrared line scanner	红外行扫描仪
infrared microscope	红外显微镜
infrared night vision system	红外夜视系统
infrared optical material	红外光学材料
infrared optical system	红外光学系统
infrared radar	红外雷达
infrared radiation	红外辐射
infrared scanner	红外扫描装置
infrared search & track device	红外搜索跟踪器
infrared source	红外线源
infrared star	红外星,红外线星
infrared stealth material	红外隐形材料
infrared track system	红外跟踪系统
infrared TV homing head	红外电视导引头
infrared window material	红外窗口材料
infrasound	次声,亚声
infrastructure	基础设施,基本设施
ingestion	吸入压力,摄入反压力
ingot	锭,晶锭,金属锭,铸锭
inherent availability	固有可用性
inherent delay	固有延迟角
inherent error	固有误差
inherent filtration	固有滤过
inherent weakness failure	本质失效
inhibit circuit	禁止电路
inhibit gate	禁止门
inhibiting	抑制
inhibiting action	抑制效应
inhibitor	抑制剂
inhomogeneous broadening	非均匀展宽
INI Intelligence Network Interface	智能网络接口
iniature switchboard	小型配电盘
initial alignment	初始校准
initial approach fix	起始进近定位点
initial breakdown interface	设计分离面
initial charge	初充电
initial erection	起始竖立
initial error	初始偏差
initial estimate	初始估(计)值
initial heading	初始航向,起始方向
initial inverse voltage	起始逆电压,起始反电压
initial mass	初始质量,火箭发射前质量
initial nuclear radiation	早期核辐射
initial phase	初相,初相位
initial point of bombing run	轰炸航路起点
initial pressure	初压,初始压力
initial state	初态
initial temperature	初始温度
initial value	初值
initial velocity	初速度
initial velocity of fragment	弹片初速
initial voltage	起始电压
initialization	起始,初始化
initiated self test	引发自检
initiating mechanism	引燃装置
initiation	1.开始,创始 2.起燃
initiation surface of fuze	引信起爆面
injectant	喷入物
injection	注射,注入
injection carburet	喷射式汽化器
injection carburetor	喷射汽化器,喷射式汽化器
injection electroluminescence	注入电致发光
injection error	入轨误差
injection locking technique	注入锁定技术
injection moulding	注射成形,注塑,热压铸
injection point	入轨点
injection pressure	喷射压力,喷注压力
injection pump	引射泵
injection pumping	注入式泵浦
injection station	注入站
injection system	灌注系统

injection velocity 燃油喷出速度
injector 注射器
inlet air temperature indicator 进气温度表，进口空气温度指示器
inlet air 进气道空气
inlet buzz 进气道嗡鸣，进气道喘振
inlet distortion 进气畸变，进气变形
inlet duct 进气管
inlet dynamic response 进气道动态响应
inlet external drag 进气道外阻
inlet guide vane 入口导向叶片，进口导叶
inlet guide vane actuator 进气道导向叶片作动器，进气导流叶片定位器
inlet guide vane (IGV) positioner 进气导流叶片定位器
inlet lip 进气道唇口
inlet lip radius 进气道唇缘半径
inlet pressure 进口压力，入口压力
inlet pressure loss coefficient 进口压力损失系数
inlet pressure recovery 进气道压力恢复
inlet stability margin 进气道稳定裕度
inlet swirl flow distortion 进气旋流畸变
inlet temperature 进口温度，入口温度
inlet throat 进气道喉道，进气道唇口
inlet total pressure recovery 进气道总压恢复
inlet total pressure recovery coefficient 进气道总压恢复系数
inlet total temperature distortion 进气总温畸变
inlet valve 进气阀
inlet-engine compatibility 进气道-发动机相容性
inlet-engine compatibility flight test 进气道-发动机相容性飞行试验
inline 直列式的
INMARSAT International Maritime Satellite Organization 国际海事卫星组织
inner corona 内电晕
inner electron 内层电子
inner lead bonding 内引线焊接
inner marker 近距指点标
inner planet 内行星
inoperative 不起作用的
in-orbit experiment 在轨道飞行中进行的实验
inorganic resist 无机光刻胶
INRTL Inertial 惯性的
input 输入，输入量
input axis 输入轴
input axis misalignment 输入轴安装误差
input impedance 输入阻抗
input/output control element 输入／输出控制元件
input/output controller 输入／输出控制器

input range 输入信号变化范围
input rate 输入速率，进气速率
input resistance 输入电阻
input terminal 输入端子
input variable 输入变量，输入变数
input winding 输入绕组
INS 1. Immigration & Naturalization Service (US) 美国移民局 2. Inches 英寸 3. Inertial Navigation System 惯性导航系统 4. Information Network System 信息网络系统 5. Instrument 仪表 6. Integrated Navigation System 综合导航系统 7. Insurance 保险 8. International Navigation System 国际导航系统
INSAT International Satellite 国际卫星
INSARST International Search & Rescue Satellite 国际搜寻救援卫星
INSCS Integrated Navigation Steering & Control System 综合航行操纵和控制系统
insensitivity 不敏感度
INSEP Inseparable 不可分割的
INSIT Initial Situation 起始情况，最初情况
in-slot signalling 时隙内信令
INSOL Insoluble 不可溶的
insolation 日光浴，日射
inspectability 可检查性
inspection & repair as necessary 按需检修，定期维修，必要的检修
inspection by attributes 计数检查
inspection by variables 计量检查
inspection frequency 检查频度
inspiratory resistance 吸气阻力
INST RLY Instantaneous Relay 瞬动继电器
instability 不稳定，不稳定性
installation 安装
installation error of the armament 武器安装误差
installation error of the tracking device 跟踪设备的安装误差
installation ring 安装环
installed thrust 安装推力
instantaneous acceleration 瞬时加速度
instantaneous automatic gain control 即时自动增益控制
instantaneous axis of rotation 瞬时旋转轴
instantaneous conditions 瞬时条件
instantaneous failure rate 瞬时失效率
instantaneous field of view 瞬时视场，导引头瞬时视场
instantaneous frequency measurement receiver 瞬时测频接收机
instantaneous overload 瞬时过载
instantaneous value 瞬时值
INSTL 1. Install 装配，安装 2. Installation 安装

INSTM　Instrument　仪表
INSTPN　Instrument Panel　仪表板
instruction address　指令地址
instruction for continuous　持续适航文件
instruction register　指令寄存器
instruction unit　指令部件
instruction word　指令语
instructor station　教员台
instrument　仪表,仪器
instrument approach chart　仪表进场图,盲目降落临场导航图
instrument approach & landing chart　仪表进近与着陆图
instrument approach procedure　仪表进场程序,盲目降落临场程序
instrument approach runway　仪表进场跑道,盲目降落临场跑道
instrument compartment　仪器舱
instrument error　仪表误差,仪器误差
instrument error of airborne fire control system　航空火力控制系统工具误差
instrument flight　仪表飞行
instrument flight center　仪表飞行中心
instrument flight rating　仪表飞行执照
instrument flight rules　仪表飞行规则
instrument flight trainer　仪表飞行训练器
instrument landing system　1.自动着落系统,仪表着陆系统 2.仪表着陆方式 3.盲目设备
instrument landing system critical area　仪表着陆系统关键区
instrument meteorological conditions　仪表天气情况,仪表气象条件
instrument panel　仪器板,仪表板
instrument rating　仪表飞行执照(授予批准进行仪表飞行的飞行员)
instrument rating examiner　仪表飞行等级考核员
instrument rating renewal　仪表额定更新
instrument runway　仪表滑行跑道,盲目飞行起落跑道
instrumentalist　仪表专业人员
instrumentation　检测仪表,仪表设备,测试设备
instrumentation & illumination subsystem　仪表与照明分系统
instrumentation radar　测量雷达
insulated gate field effect transistor　绝缘栅场效管
insulation resistance　绝缘电阻
insulation resistance meter　绝缘电阻表
insulator　绝缘体,绝缘子,隔电子
INSTR XFER　Instrument Transfer　仪表转换
INT　1.Intake 入口,进气道 2.Integral 完整的,整体的;积分,套数 3.Integrating 整体的,综合的 4.Interim 暂时的 5.Intermittent 间歇的 6.Internal 内部 7.Internatioal 国际的 8.Interphone 内话,对讲机,内话机 9.Interrogator 询问机 10.Interrupt 中断 11.Intensity 强度 12.Intersection 交叉点
INT CON　Internal Connection　内部连接
intact　完整的,无损伤的,原封不动的
intake　进气道,进气管,进气装置
intake pressure　进气压力
intake spike position　进气整流锥位置指示器
intake stroke　进气冲程
intake valve　进气阀
INTCHG　Interchangeable　互换的
INTCP　Intercept　切入,截获,截取,相交
INTEG　Integrateing　完整的,积分的
integral action　积分酌
integral action coefficient　积分酌系数
integral action control　积分动控制,积分酌控制
integral action controller　积分型控制器,积分控制器
integral fuel tank　整体油箱
integral panel　整体壁板
integral performance criterion　积分质准则
integral structure　整体结构
integral structure manufacturing process　整体结构工艺
integral tank　整体油箱,整体箱,整体型油箱
integrand　被积函数
integraph　积分仪
integrate　整合,使成整体
integrated actuator package　自主式伺服舵机
integrated avionics system　综合航空电子系统
integrated circuit　集成电路,积分电路
integrated communication navigation & identification　通信、导航和识别综合系统
integrated data processing　综合数据处理,集中数据处理
integrated digital network　综合数字网络,综合数字网
integrated diode solar cell　集成二极管太阳电池
integrated drive generator　整体驱动发电机
integrated electronics　集成电子学
integrated EW system　综合化电子战系统
integrated flight/propulsion control　飞行/推力综合控制
integrated hydraulic actuator　液压复合舵机
integrated inductor　集成电感器
integrated navigation　组合导航
integrated navigation computer　组合导航计算机
integrated optical circuit　光集成电路
integrated optics　集成光学
integrated optics technique　集成光路技术
integrated optics technology　集成光路工艺学

integrated optoelectronics　集成光电子学
integrated product team　集成产品小组
integrated satellite system　综合卫星系统
integrated sensor　集成传感器,集成敏感器
integrated service digital network　综合业务数字网
integrated system　综合系统,成套系统,集成系统
integrated system test　综合系统测试
integrated tactical navigation system　综合战术导航系统
integrated test article　综合测试工具
integrated test provision for oxygen equipment　供氧设备综合试验器
integrated test specification　综合测试规格
integrated test vehicle　综合测试飞行器
integrated trajectory system　综合式轨道测量系统,综合弹道系统
integrated wideband communication system　综合宽带通信系统
integrating accelerometer　综合加速测量仪,积分加速度表,积分仪
integrating circuit　积分电路,积分网路
integrating factor　积分因子,积分因数
integrating gyroscope　积分陀螺,积分陀螺仪
integrating gyroscopic accelerometer　积分式陀螺加速度表
integrating instrument　积算仪,积分仪
integrating sphere　积分球,累计球
integrating unit　积分装置
integration test　综合实验室
integration time　积分时间
integrator　积分器,积分装置,积分仪,积分电路
integrity　集成度,完整性
intellectual technology　智能工艺学,智能技术
intelligence gathering equipment　情报搜集设备
intelligence quotient　智商,智力指数
intelligent automation　智能自动化
intelligent cable　智能电缆
intelligent computer　智能计算机
intelligent control　智能控制
intelligent control system　智能控制系统
intelligent controller　智能控制器
intelligent digitizer　智能数字化仪
intelligent instrument　智能仪器
intelligent material　智能材料
intelligent network　智能网
intelligent peripheral　智能外设
intelligent peripheral controller　智能外围设备控制器
intelligent robot　智能机器人
intelligent sensor　智能敏感器,智能传感器
intelligent terminal　智能终端

INTELSAT　International Telecommunication Satellite Consortium 国际通信卫星组织
intelsat　国际通信卫星
intensity modulation　光强调制,辉度调制;亮度灯
intensity scale　烈度表,强度标,强度量表
interacting galaxy　互扰星系
interaction　1.相互作用,相互影响 2.干扰
interaction parameter　相互作用参数,相互酌参数
interaction prediction approach　交互作用预测法,关联预测法
interactive computer　交互式计算机
interactive computer graphics system　对话式计算机图像仪系统
interactive computer-aided design　交互式计算机辅助设计
interactive computer-aided technology　交互计算机辅助技术
interactive graphics　交互式图形
interactive layout system　交互式布图系统
interactive mode　交互模式,对话方式
interactive processing　人机交互处理
interactive programming　交互式程序设计
interactive restoration　人机对话复原
interactive simulation　交互仿真对话式仿真
interactive system　交互系统
interblock gap　数据块间隙
intercardinal plane　基间平面
intercept　截击,拦截,截取,侦听,截断
intercept attack　拦截攻击
intercept attack course　拦射攻击路线
intercept point　截击点,拦截点
intercept probability　截获概率
interceptibility　截击率,截击能力
interception　截击,拦截,侦听,窃听,截断,横穿,切入
interception radar　拦截雷达
interception time　拦截时间
interceptor　1.截击器,截击机,截击导弹 2.窃听器,窃听台 3.阻流片,扰流片 4.拦截雷达
interceptor missile　拦截导弹
interceptor satellite　截击卫星
interchange　互换,交替
interchangeability　互换性,交换性
interchannel crosstalk　信道间串音
inter-clutter visibility　杂波间可见度
INTERCO　International Code of Signals 国际信号编码
intercomparison　互相对比,相互比较
interconnected system　互联系统
interconnecting device　转接设备
interconnection　相互连接

interconnection network　互连网络,互连图
interconnector　1.传焰管 2.联络线路,内部连接线
intercooler　中间散热器,中间冷却器
intercrystalline corrosion　晶间腐蚀
intercrystalline fracture　晶间断裂,沿晶界断裂
intercrystalline rupture　晶间断裂,粒间断裂
intercycle　中间循环的,中间周期的
interdiffusion　相互扩散
interdigitated　相互成一定角度的,按相对角度布局的
interdiscipline　跨学科,交叉学科
interface　交界面,接口,相互作用
interface control document　接口控制文件,互连控制文件
interface driver　接口驱动器
interface frame　对接框
interface reaction　界面反应
interface reaction-rate constant　界面反应率常数
interface specification　接口规范
interface trapped charge　界面陷阱电荷
interface unit　接口部件,接口装置
interference drag　干扰阻力,干涉阻力
interference factor　干扰因子,干扰系数,干扰因素
fnterference filter　干扰滤波器
fnterference fit　干涉配合
interference measuring set　干扰测量仪
interference squealing　干扰哨声
interfering torque　干扰力矩
interferogram　干涉图
interferogram technique　1.干涉图法 2.干涉图技术
interferometer　干涉仪,干扰计
interferometer antenna　干涉仪天线
interferometer tracking system　干涉仪跟踪系统
interferometric binary　干涉双星
interferometric spectroscopy　干涉测量光谱学
interferometric test　干涉仪试验
interframe　帧间
interframe predictive coding　帧间预测编码
intergalactic cloud　银河系际云
intergalactic matter　星系际物质
intergalactic space　星系际空间
intergranular attack specimen　晶间腐蚀试样
intergranular corrosion　晶间腐蚀
intergranular crack　晶间裂纹,晶粒间裂缝
interior ballistics　内弹道学
interior orientation　内部定向,内方位元素
interior planet　1.内行星 2.行星内部
interlace flicker　行间闪烁
interlaced code　交织码
interlacing　1.隔行扫描 2.交错

interlaminar shear strength　层间剪切强度
interlayer　1.界层,中间层 2.中间层材料
interleave　交叠,交错
interleaved pulse train　交叠脉冲列
interline　航空公司之间的
interlock　1.联锁,连接,联动,闭塞 2.联锁器,联锁装置,保险设备 3.相互关系
interlocking point　互换点
interlocks　联动装置
interloper　截击机
intermediary　1.仲裁者,调解者 2.媒介
intermediate approach area　中间进近区
intermediate approach segment　中间进场阶段
intermediate assignment　中间赋值
intermediate band photometry　中带测光
intermediate contour　基本等高线
intermediate flow　中间流
intermediate frequency amplifier　中频放大器
intermediate frequency preamplifier　前置中频放大器
intermediate frequency transformer　中频变压器
intermediate orbit　中间轨道,转移轨道,过渡轨道
intermediate pressure　中间压力
intermediate range　中程,中间区段,中间范围中程
intermediate range ballistic missile　中程弹道导弹
intermediate rating　中间状态
intermediate reduction gearbox　中间减速器
intermetallic compound　金属互化物,金属间化合物
intermittency　间歇现象
intermittency factor　间歇因子
intermittent discharge　间歇放电
intermittent duty　断续工作制
intermittent fault　间歇故障
intermittent movement　间歇运动
intermittent wind tunnel　暂冲式风洞,脉动风洞
intermix　混合物
intermodal　联运的,综合运输的,联合的
intermodulation　互调,相互灯
intermodulation distortion　互调失真,互撑真
intermolecular force　分子间力,分子力
internal aerodynamics　内流空气动力学,管流空气动力学
internal air system　内部空气系统
internal balance　内部平衡
internal calibration　1.内部校准法 2.内部刻度
internal calibrator　内部校准仪,内部校准器
internal combustion engine　内燃机,内燃发动机
internal compression inlet　内压式进气道
internal cooling　内冷,内部冷却
internal crack　内部裂纹

internal disturbance 内扰,内部干扰
internal efficiency 内效率
internal field 内场,内极式同步发电机,旋转磁场式同步发电机
internal gas detector 内气体探测器
internal gun 内部枪
internal heat source 内热源
internal pressure test 内压试验,耐压试验
internal standardization 内标准化
internal stress 内应力,内部应力
internal volume 内体积
internals 内部构件,堆内构件,内部组织
international atomic time 国际原子时,国际原子时间
international cloud atlas 国际云图
international ellipsoid 国际椭球体,国际参考椭球体
international ellipsoid of reference 国际参考椭球体
international geomagnetic reference field 国际标准地磁场
international geophysical year 国际地球物理年
international nautical mile 国际海里
international organization for standardization 国际标准化组织
international polar motion service 国际极移服务
international projective chart 国际投影图
international reference atmosphere 国际标准大气压,国际参考大气压
international standard atmosphere 国际标准大气
international system of units 国际单位制
international waters 国际海域,国际水域公海
interocular distance 瞳孔间距离
interoffice trunk 局间干线
interoperability 互用性,相互工作能力,互通性
interpenetrating polymer network 互穿聚合物网络
interphase boundary 相间边界,异相边界
interphase transformer 相间变换器,相间变压器
interphone 机内通话器,内部互通电话,内部通信装置,内部电话机
interplane strut 翼间支柱
interplanetary 行星间的,星际间的,行星际的
interplanetary & interstellar navigation 星际航行
interplanetary magnetic field 行星际磁场,星际间磁场
interplanetary matter 行星际物质
interplanetary medium 行星际介质
interplanetary meteor 行星际流星
interplanetary navigation 星际导航
interplanetary propulsion 星际推进
interplanetary rocket 行星际火箭
interplanetary scintillation 行星际闪烁
interplanetary shock 行星际激波,星际碰撞
interplanetary space 星际空间,太空,行星际航天器,星际间宇宙飞船
interplanetary spacecraft 行星际航天器,星际间宇宙飞船
interplanetary station 行星际站
interplanetary transfer orbit 行星际转移轨道,星际间转移轨道
interpolate 插入,窜改
interpolation 插入法,内插法
interpolation polynomial 插值多项式
interpretability 解释能力
interpreter 口译者
interpretoscope 判读仪,解译镜
interprocess communication 进程间通信
interrecord gap 记录间隙,记录间隔
interrelationship 相互关系
interrogation 询问,查询
interrogation mode 询问模式,询问方式,询问工作状态
interrogation pulse 询问脉冲
interrogator 询问机,询问器
interrupt flip-flop 中断触发器
interrupt mask 中断屏蔽,隐中断字中断屏蔽
interrupted projection 分瓣投影
interrupter 断路器,中断器,引信保险装置
interruptibility 可中断性
interruption 断续,中断
intersatellite communication 卫星间通信
intersection 交叉,横断,相交
interservice 军种间的
interstage 级间的
interstage section 级间段(连接多级火箭各级的部段)
interstellar cloud 星际云
interstellar communication 星际通信
interstellar dust 星际尘埃
interstellar extinction 星际消光
interstellar gas 星际气体
interstellar matter 星际物质
interstellar medium 星际介质
interstellar reddening 星际红化
interstellar space 恒星际空间,宇宙空间
interstellar travel 星际旅行
interstice 间隙
interstitial diffusion 间隙扩散
intersymbol interference 符号间干扰,码间干扰
interval time-out 区间超时
intervalometer 时间间隔计,定时器
intervening galaxy 居间星系
intervertebral disk 椎间板
INTFC Interface 交界面,接口

INTIPS Integrated Information Processing System 综合信息处理系统
INTK Intake 吸入,进气
INTL Internal 内部的
INTLK Interlock 互锁
INTMED Intermediate 中介的
INTMT Intermittent 断断续续的
INTPH Interphone 机内电话
INTR Interior 内部
intra-clutter visibility 杂波内可见度
intraconnection 内部连接
intrados 拱腹线,拱腹
intransit 在旅途中的
intrascope 内窥镜
INTRF Interference 干涉
intrinsic efficiency 内禀效率
intrinsic electroluminescence 本征电致发光
intrinsic error 固有误差
intrinsic quality factor 固有品质因数
intrinsic redshift 内禀红移
intrinsic semiconductor 本征半导体
intrinsic variable 内因变星,内蕴变量,固有变量
INTRO Introduction 介绍,导言
intruder 入侵者,入侵飞机
INTS 1. Integrated National Telecommunications System 国家综合通信系统 2. International Switch 国际交换 3. Inter-Network Time Slot 网络内部时隙
INTSE Intelligent System Environment 智能系统环境
INTST Intensity 强度,亮度,密度,明暗度
INTVL Interval 间隔
INU Inertial Navigation Unit 惯性导航仪,惯性导航部件
INV Inverter 变换器
INVAL Invalid 无效
INVAR 不变式,不变量;恒定的
invar 因瓦合金,殷钢,不胀钢
invar tape 因瓦基线尺,殷钢基线尺
invariable plane 不变平面
invariance 不变性,不变式
invariance principle 不变性原理
inventory control system 存货控制系统
inventory management system 库存管理系统,库存管理制度
inventory theory 库存论,存储论,编目理论
inverse channel 逆信道
inverse coaxial magnetron 反同轴磁控管
inverse code 逆代码
inverse compton effect 逆康普顿效应
inverse discrete cosine transform 离散反余弦变换

inverse discrete fourier transform 反演离散傅里叶变换
inverse dispersion 反波散逆频散
inverse duplex circuit 逆双工电路
inverse feedback 负反馈
inverse feedback filter 倒反馈滤波器
inverse filter 逆滤波器
inverse gain jamming 逆增益干扰
inverse Laplace transformation 拉普拉斯逆变换
inverse nyquist diagram 逆乃奎斯特图,反幅相曲线
inverse peak voltage 反峰电压
inverse residue code 反剩余码
inverse square aw 平方反比律
inverse synthetic aperture radar 逆式合成孔径雷达
inverse system 逆向系统反向系
inverse tacan 逆式"塔康"导航系统
inverse voltage 反向电压,负极性压
inverse Z-transform 反 Z 变换
inversion 逆温,倒转,反向
inversion point 逆转点,转化点,反演点
inversion theorem 反演定理
inversor 反演器,控制器
invert 倒置,翻转,使颠倒,使回翻
inverted engine 倒缸发动机,倒立式发动机
inverted file 倒排文件,倒排文档
inverted flight 倒飞,逆飞行
inverted flight fuel tank 倒飞油箱
inverted gull wing 逆鸥型机翼,倒海鸥形翼
inverted receiver 倒置接收机
inverter 1.变换器,换流器,变流机 2.倒相器,反相器 3.转换开关,变换电路 4.反演器
inverter buffer 反相器缓冲器,倒相缓冲器
investment casting 熔模铸造,失蜡铸造
invigilator 监视器
inviscid 无黏性的,非黏性的
inviscid flow 无黏性流
invisible matter 隐物质,不可见物质
invisible range 不可视区
involute gear 渐开线齿轮,模数铣刀
involute gear pump 渐开线齿轮泵
invulnerability 耐损性,耐攻击性,不易破坏性
inward flux 输入通量,进入通量
inward-turning 内向旋转
IOAT Indicated Outside Air Temperature 指示外界气温
IOAU Input/Output Access Unit 输入/输出存取装置
IOC 1. Indirect Operating Cost 间接使用成本,间接运行成本,间接运费用 2. Initial Operational Capability 初始作战能力,初始工作能力,初始操作能力 3. In-Orbit Checkout 在轨测试 4. Input/Output Channel 输

入/输出通道 5. Input/Output Console 输入/输出控制台 6. Input/Output Controller 输入/输出控制器 7. Input/Output Control Module 输入/输出控制模块 8. Input/Output Converter 输入/输出转换器 9. INTELSAT Operations Center 国际通信卫星操作中心 10. Intergovernmental Oceanographic Commission 政府间海洋法委员会 11. International Order of Characters 飞行员级别国际标准 12. Inter-Orbit Communication Experiment 轨道间通信试验 13. International Olympic Committee 国际奥林匹克委员会

IOCS Input/Output Control System 输入/输出控制系统

IOIC Integrated Operational Intelligence Center 综合作战情报中心

IOIS 1. Input-Output Interface System 输入/输出接口系统 2. Integrated Operation Intelligence System 综合作战情报系统

IOL Inboard Open/Loop 内开/环路

IOLA Input/Output Link Adapter 输入/输出链路适配器

IOLC Input/Output Link Control 输入输出链路控制

IOMS Input/Output Mangement System 输入/输出管理系统

ION Institute Of Navigation 美国导航学会

ion altimeter 电离层高度表
ion beam coating 离子束镀,离子束涂敷
ion beam epitaxy 离子束外延
ion beam etching 离子束腐蚀
ion beam evaporation 离子束蒸发
ion beam ithography 离子束光刻
ion beam machining 离子束加工
ion beam milling 离子铣
ion beam polishing 离子束抛光
ion beam pro 离子束探针
ion beam sputtering 离子束溅射
ion bombardment 离子轰击
ion burn 离子斑
ion engine 离子发动机
ion gas aser 离子气体激光器
ion gauge 离子压力计,电离真空计
ion gun 离子枪
ion impingement 离子撞击
ion implantation 离子注入
ion implantation apparatus 离子注入机
ion implanter 离子注入机
ion microanalysis 离子微分析
ion microprobe 离子探针谱
ion migration 离子迁移,离子徙动

ion neutralization spectroscopy 离子中和谱
ion number density 离子数密度
ion oscillation 离子振荡
ion plating 离子镀,离子镀敷,离子电镀法
ion rocket 离子火箭
ion scattering spectroscopy 离子散射谱学,离子散射分光法
ion sensitive FET 离子敏场效管
ion sensor 离子敏感器
ion source 离子源
ion thruster 离子推力器
ion trap 离子捕集器
ionic nitriding 离子氮化
ionic tube 离子管
ionicity 离子性
ionization 电离作用,离子化
ionization arc-over 电离闪络
ionization chamber 电离室,电离箱
ionization energy 电离能,电离能量
ionization gauge 电离真空计
ionization ratio 电离比
ionization smoke detector 离子感烟探测器
ionization temperature 电离温度
ionization time 电离时间
ionize 电离化,使电离,离子化
ionizer 电离器
ionizing radiation 电离辐射,致电离辐射,电离性辐射
ionogram 电离图,频高图
ionopause 电离层顶
ionosonde 电离层探测仪,电离层探测火箭
ionosphere 电离层,电离圈
ionosphere storm 电离层干扰
ionospheric backscatter radar 电离层后向散射雷达,后向散射超视距雷达
ionospheric blackout 电离层信号消失
ionospheric critical frequency 电离层临界频率
ionospheric disturbance 电离层干扰,电离层扰动
ionospheric eclipse 电离层食
ionospheric refraction correction 电离层折射校正
ionospheric storm 电离层暴,电离层暴风雨,电离层骚扰

IOP Input/Output Processor 输入/输出处理机

IOPDS Integrated-Optic Position/Displacement Sensor 集成光学位置/位移传感器

IOPU Input/Output Interface System 输入/输出接口系统

IOR 1. Immediate Operational Requirement 即刻运营要求 2. Indian Ocean Region 印度洋地区

IOS 1. Input/Output System 输入/输出系统

2. Instructor Operating System 教员操纵台 3. Integrated Observation System 综合观测系统 4. Integrated Operational System 综合作战系统 5. Intelligent Office System 智能办公室系统 6. Interactive Operating System 交互式操作系统 7. Interceptor Operator Simulator 截击机驾驶员模拟器 8. International Organization for Standardization 国际标准化组织 9. Internet Operating System 因特网操作系统 10. Internetwork Operating System 网间操作系统

IOSA IATA (International Air Transport Association) Operational Safety Audit 国际航空运输协会运行安全审计

IOT 1. Input-Output Termination 输入/输出终止 2. Intra Office Trunk 局内中继

IOT&E Initial Operational Test & Evaluation 初始运行试验和评价

IP 1. Identification Point 投弹识别点 2. Identification Position 识别位置 3. Identification Pulse 识别脉冲 4. Image Processor 图像处理机 5. Impact Point 弹着点,命中点,碰撞点 6. Index Plan 定位大纲,定位计划 7. Information Protection 信息保护 8. Initial Point 起始点 9. Installation Plan 安装大纲,安装计划 10. Instructor Pilot 飞行教员,驾驶教员 11. Intellectual Property 知识财产 12. Interface Processor 接口处理机 13. Intermediate Pressure 中压 14. International Priority 国际优先快递服务 15. Internet Protocol 国际协议 16. Internetwork Protocol 网络间协议 17. Iron Pipe 铁导管 18. Issue Paper 问题纪要

IPA 1. Independent Parallel Approach 独立平行接近 2. Inspection Program Approval 检验程序批准书 3. Image Processing Algorithm 图像处理算法

IPACS Integrated Power Attitude Control System 综合动力姿态控制系统

IPARS International Programmed Air-lines Reservation System 国际定期航班订票系统

IPAS Integrated Pressure Air System 综合压缩空气系统

IPBD Illustrated Parts Breakdown 图解零件分解,图解零件明细表

IPC 1. Illustrated Part Catalog 图解零件目录 2. Industrial Process Control 工业过程控制 3. Information Processing Code 信息处理代码 4. Integrated Peripheral Channel 集成外围通道 5. Integrated Processing Cabinet 综合处理机箱 6. Intelligent Peripheral Controller 智能外设控制器 7. Intermediate Pressure Compressor 中间压力压缩机 8. Intermittent Positive Control 不连续无反向力操纵 9. Inter-Process Communications 进程间通信 10. Intermediate Pressure Compressor 中间压力压缩机

IPCC 1. Information Processing in Command & Control 指令与控制信息处理 2. Intergovernmental Panel on Climate Change 国际气候变化研究组

IPCS 1. Input Port Controller Submodule 输入端口控制器的子模块 2. Integrated Propulsion Control System 综合推进控制系统

IPCV Intermediate Pressure Check Valve 中间压力单向阀

IPD 1. Illustrated Provisioning Document 图解备件供应文本 2. Initial Performance Data 起始性能数据 3. Initial Provisioning Data 初始备件供应数据 4. Integrated Product Definition 集成产品开发 5. Issue Priority Designator 发布优先权指定器

IPE 1. Improved Performance Engine 改进性能的发动机 2. Industrial Plant Equipment 工业工厂设备 3. International Petroleum Exchange 国际石油交换 4. In-band Parameter Exchange 带内参数交换

IPF 1. Image Processing Facility 图像处理中心 2. Integrated Process Flow 集成工艺流程

IPG 1. Interactive Program Guide 交互式节目指南 2. Inter-Packet Gap 分组信息间隙

IPI 1. Initial Protocol Identifier 初始协议标识符,初始协议标识器 2. Intelligent Peripheral Interface 智能外围接口

IPJT Intrplant Job Ticket 厂内工作票

IPL 1. Illustrated Parts List 图解零件清单 2. Initial Program Load 初始程序装入 3. Interchangeable Parts List 互换件清单

IPM 1. Inspection Procedure Manual 检查程序手册 2. Inter-Personal Messeging 人际传信

IPME Inter-Personal Messaging Environment 人际传信环境

IPMS 1. Inter-Personal Messaging System 人际传信系统 2. Inter-Personal Messaging Service 人际传信业务

IPN 1. Instant Private Network 瞬时专用网络 2. Inter-Personal Notification 人际通知

IPO Interplant Planning Order 厂内工艺计划指令

IPPD Integrated Product & Process Development 综合产品与过程开发

IPR Internet Protocol Router 因特网协议路由器

IPS 1. Image Processing System 图像处理系统 2. Incorrect Phase Sequence 逆相序 3. Information Presentation 信息表达系统 4. Instrument Pointing System 仪器定向系统 5. Integrated Program Summary 综合程序摘要 6. International Pipe Standard 国际导管标准

IPT 1. Inadvertent Power Transfer 意外电源转换 2. Information Processing Technique 信息处理技 3. Information Providing Terminal 信息提供终端 4. Integrated Product Team 集成产品小组,产品协同设计组,综合产品组 5. Iron Pipe Thread 铁导管螺纹 6. Intermediate Pressure Turbine 中压涡轮 7. Internal Pipe Thread 内管螺纹

IPWO Interplant Work Order 厂内工作指令

IR 1. Ice on Runway 跑道上有冰 2. Implementation Rule 执行法 3. Incident Report 事故征候报告 4. Inertial Reference 惯性基准 5. Information Retrieval 信息检索 6. Inside Radius 内半径 7. Inspection Report 检查结果报告 8. Instrument Rating 仪表飞行等级 9. Instrument Runway 仪表跑道 10. Insulation Resistance 绝缘电阻 11. Integrated Range 综合航程 12. International Reserves 国际航线备用油量 13. Island Reserves 岛屿备用油量

IR Infrared 红外线

IRA 1. Inertial Reference Assembly 惯性参考组件 2. Intermediate Range Aircraft 中程飞机

IRACQ Infrared Acquisition Radar 红外探测雷达

IRAD Independent Research & Development 独立研究的发展

IRAN Inspection & Repair as Necessary 按需检修

IRASA Internation Radio Air Safety Association 国际无线电航空安全协会

IRBM Intermediate Range Ballistic Missile 中程导弹

IRC 1. Inbound Radar Control 进场雷达管制 2. Information Retrieval Center 信息检索中心 3. Initial Rate of Climb 初始爬升率 4. International Radiation Commission 国际辐射委员会 5. International Red Cross 国际红十字会 6. International Rescue Committee 国际救援委员会 7. International Route Charge 国际航路费 8. Internet Relay Chat 因特网中继交谈

IR-CAF Internal Release Configuration Approval Form 内部发放构型审批表

IRCC International Radio Consultative Committee 国际无线电咨询委员会

IRCS 1. International Research Communications System 国际研究通信系统 2. Intrusion-Resistant Communications System 抗干扰通信系统

IRD 1. Infrared Detector 红外探测器 2. Integrated Receiver/Decodrer 集成接收/解码器 3. Interface Requirements Document 接口要求文件

IrDA Infra-red Data Association 红外数据协会

IRDC Interface Remote Data Concentrator 接口远程数据采集器

irdome 红外穹门,红外导流罩

IRDS Information Resources Dictionary System 信息资源词典系统

IRE 1. Institution of Radio Engineers 无线电工程师协会 2. Instrument Rating Examiner 仪表飞行等级检查员 3. Ireland 爱尔兰 4. Irregular 不规则的,不定期的 5. Internal Roll Extrusion 内滚挤压

IRI 1. International Reference Ionosphere 国际参考电离层 2. International Research Institute for Climate Prediction 国际气候预测研究所 3. Infra Red Image 红外图像

iridium 铱

iridium anomaly 铱反常现象

IRIG Inter-Range Instrumentation Group 靶场间仪表组

iris 光圈,可变光阑

iris photometer 光瞳光度计

IRL Infrared Lens 红外光透镜

IRLAP 1. Infrared Link Access Protocol 红外链接存取协议 2. IrDA Link Access Protocol IrDA 链路接入协议

IRLS Infra-Red Line Scan 红外行扫描

IRM 1. Information Resources Management 信息源管理 2. Integrated Reference Model 综合参考模型 3. Intelligent Robotics Manufacturing 人工智能机器人制造

IRMP 1. Infrared Measurement Program 红外测量计划 2. Inertial Reference Made Panel 惯性基准组件面板 3. Inertial Reference Mode Panel 惯性基准状态面板

IRMS Information/Informational Retrieval & Management System 信息检索及管理系统

IRN 1. Information Resource Network 信息资源网络 2. Intermediate Routing Node 中间路由选择节点

iron accumulator 碱铁蓄电池

iron aluminide 铁铝化合物

iron ball 铁球

iron bird 铁鸟

iron soldering 烙铁钎焊

iron-base superalloy 铁基高温合金

ironing 变薄压延,减径挤压加工

ironless 无铁心的,无铁的;无电枢,空心电枢

IRP 1. Integrated Refuel Panel 综合加油面板 2. Intelligent Resource Planning 智能资源计划 3. Internal Reference Point 内部参考点 4. International Routing Plan 国际路由规划 5. Irregularity Report 非正常情况报告

IRR 1. Improved Rearming Rate 改进的装弹率 2. Irrevocable 不可改变的 3. Instrument Rating Renewal 仪表额定更新 4. Internal Rate of Return 内

部收益率法

irradiance 辐照度,辐射通量密度
irradiate 辐照,辐射
irradiation sickness 放射性病,射线病
irradiation test 辐照试验
irrecoverable error 不可恢复错误
irreducible polynomial 不可约多项式
irreflexive relation 非自反关系
irregular galaxy 不规则星系,不规则星云
irregular variable 不规则变量,不规则变光星
irregularity 不规则性,不均匀度;异常
irreversible boosted mechanical control 不可逆助力机构操纵
irreversible control 不可逆操纵,不可逆操纵系统,不可逆操纵杆
irreversible element 不可逆元件
irreversible process 不可逆过程
irreversible transformation 不可逆变换
irritability 应激性,过敏性
irrotational flow 无旋涡流,无旋流
irrotational wave 无旋波
IRS 1. Inertial Reference System 惯性基准系统 2. Inspection Report System 检查报告系统 3. Interface Requirements Specification 接口要求规格
IRST Infrared Search & Track 红外线搜索与跟踪
IRT Instrument Rating Test 仪表额定性能测试
IRTF Internet Research Task Force 因特网研究任务工作组
IRTS 1. Initial Radar Training Simulator 初级雷达训练模拟器 2. Infra-Red Target Seeker 红外目标导引头 3. Infra-Red Target Sounder 红外目标探测仪 4. Infra-Red Temperature Sounder 红外温度探测仪
IRTU Intelligent Remote Terminal Unit 远程智能终端装置
IRU Inertial Reference Unit 惯性基准装置
IRUE Inertial Reference Unit Electronics 惯性基准电子元件
IRVR Instrumented Runway Visual Range 仪表的跑道能见度
IS 1. Ignition System 点火系统 2. Increment Scheduling 增长速度 3. Industry Standard 工业标准 4. Information System 信息系统 5. Integrated Scheduling 集成进度计划 6. Interface Standard 接口标准 7. International Standard 国际标准
ISA 1. Industry Standard Architecture 工业标准体系结构 2. Inertial Sensor Assembly 惯性感应组件 3. Information System Architecture 信息系统结构 4. Instruction Set Architecture 指令集体系结构 5. Instrument Society of America 美国仪表学会 6. International Standard Atmosphere 国际标准大气 7. International Standardization Association 国际标准化协会
ISADS Integrated Strapdown Air-Data System 综合捷联式大气数据系统
ISAE International Society of Air-breathing Engineers 国际空气喷气式发动机工程师协会
isallobar 等变压线
ISAR 1. Information Storage & Retrieval 信息存储和检索 2. Institute Space Atmosphere Research 空间大气研究所
ISASI International Society of Air Safety Investigators 国际航空安全调查员协会
ISA-SL International Standard Atmosphere at Sea Level 海平面国际标准大气
ISAT Integrated Shipping & Assembly Tool 集成运输和装配工具
ISAW International Society of Aviation Writers 国际航空作家协会
ISB 1. Independent Sideband 独立边带 2. Inspection Service Bulletin 检查服务通报 3. Intelligent Signaling Bus 智能信令总线 4. Interface Schduling Block 接口调度块 5. Interchangeability Survey Board 互换性考察委员会 6. Inter-System Bus 相互系统总线
ISC 1. Industry Steering Committee 工业指导委员会 2. Integrated Semiconductor Circuit 集成半导体电路 3. Integrated System Controller 综合系统控制器 4. Instrument System Corporation 仪表系统公司 5. Inter Stellar Communications 星际通信 6. International Switching Centre 国际转接中心 7. Internet Software Consortium 因特网软件联盟
ISCAN International Sanitary Convention for Air Navigation 国际空中航行环境卫生公约
ISCS 1. Integrated Sensor Control System 综合传感器控制系统 2. International Satellite Communications System 国际卫星通信系统
ISD 1. Information System Development 信息系统开发 2. Information System Division 信息系统部门 3. Initial Search Depth 初始搜索深度 4. Issued 发行的,填开的,发布的
ISDAS In Service Data Acquisiton System 使用数据采集系统
ISDN Integrated Services Digital Network 综合业务数字网,整合服务数码网络
ISDOS Information System Design & Optimization System 信息系统设计和优化系统
ISDU Inertial System Display Unit 惯性系统显示部件
ISE 1. Interconnected Stabilizer/Elevator 安定面/升降舵铰接 2. Integrated Switch Element 综合交换单元

3. Intelligent Synthesis Environment 智能综合环境
ISEC Internet Service & Electronic Commerce 因特网服务和电子商务
isentrope 等熵线,等位温线
isentropic 等熵;熵的;等熵线,等熵条件,等熵关系式,等熵温度
isentropic chart 等熵图
isentropic compression 等熵压缩
isentropic compression inlet 等熵压缩进气,绝热压缩
isentropic flow 等熵流,等熵线
ISFD Integrated Standby Flight Display 综合备用飞行显示,集成备用飞行显示器
ISG 1. Inflatable Survival Gear 充气救生器材 2. Instrument Symbol Generator 仪表符号发生器
ISGN Insignia 国徽,证章,标志,识别符号(拉丁语)
ISI 1. Institute for Scientific Information 科学情报学会 2. International Sales Indicator 国际销售代号
ISL 1. Inactive-Status List 停用状态单 2. Initial Spare Parts List 原始备件表 3. Inter-Satellite Links 卫星间通信链路 4. Interface Signal List 界面信号清单
ISLN Isolation 隔离,隔绝,孤立,绝缘
ISM 1. Igniter Safety Mechanism 点火器安全机构 2. Illumination Sensor Module 照明传感器组件 3. Industrial Security Manual 工业安全手册 4. Inertial System Mount 惯性系统架 5. Interactive Storage Media 交互式存储媒体 6. Intelligent Synchronous Multiplexer 智能同步复用器 7. Interface Subscriber Module 用户接口模块 8. Internet Server Manager 因特网服务器管理器
ISMLS Interim-Standard Microwave Landing System 临时标准微波着陆系统
ISO 1. International Science Organization 国际科学组织 2. International Standardization Organization 国际标准化组织 3. Isolation 隔离
ISO AMP Isolation Amplifier 隔离放大器
isobar 等压线,等压线,异序素
isobaric chart 等压面图
isochoric 等容的,等体积的
isochoric change 等容变化
isochoric process 等容过程,定容过程
isochrone 等时线
isochronous transmission 等时传输,同步发送
isocyanic acid 异氰酸
isodops 等值多普勒频移
isoelectronic center 等电子中心
isoentrope 等熵线
isoentropic flow 等熵流动
isoga 等重力线,等磁力线
isogam 等磁力线

isogeotherm 等地温面
isogon 等磁偏线,等角多边形,正多边形,同风向线
isogonal 等角的,等偏角的;等偏角线
isogonic 等偏角的
isogonic chart 等磁偏线图,等线图,等磁差图,等偏图
isogram 等值线图,等频率线
isogrid 地磁等变线,地磁等偏线图
isogriv 等网格差线,等座标磁偏角图,等重力线图
isohypse 等高线
isokinetic 等动能线
isokinetic sampling 等动能取样法
isolated amplifier 隔离放大器
isolation 隔离,隔绝,隔离度
isolation information system 分割信息系统
isolation transformer 隔离变压器
isolation valve 隔离活门
isolator 隔离器,缘体,隔离者,隔离物
isoline 等值线
isoline map 等值线图
isomagnetic chart 等磁图,等磁力线图
isomagnetic map 等磁线图
isomer 同分异构体,同分异构物
isomers 等雨率线,等比值线
isometric 等体积的,等容的
isometric mapping 等距映象
isometric space 等距空间
isomorphic image 同构图形
isonif 等雪量线
isooctane 异辛烷
isopach 等厚线
isophane 等物候线,同时开花线
isophotometry 等光度测量
isopiestics 等压线
isoplanar isolation 等平面隔离
isoplanar process 等平面工艺
isopleth 等值线,等浓度线
isoporic 等磁变线
isoporic line 等磁变线,地磁等年变线
isopotential 等势线
ISOS Isosceles 等边的,等腰的
isostasy 地壳均衡
isostasy hypothesis 地壳均衡假说
isostatic adjustment 均衡调整
isostatic anomaly 均衡异常
isostatic correction 均衡改正,均衡校正
isostatic pressing 等静压,等静压压制
isostere 等比容线
isosurface 等面,等值面
isotherm 等温线

isothermal 同温的,等温的
isothermal annealing 等温退火
isothermal atmosphere 等温大气
isothermal forging 等温模锻
isothermal hardening 等温淬火
isothermal process 等温过程
isothermal quenching 等温淬火
isothermal remanent magnetization 等温剩余磁化
isothiocyanic acid 异硫氰酸
isotone mapping 保序映象
isotope 同位素
isotope abundance 同位素丰度
isotope power source 同位素电源
isotropic 各向同性的
isotropic antenna 无方向性天线
isotropic etching 各向同性腐蚀
isotropic material 各向同性材料,蛤同性材料
isotropic medium 各向同性媒质
isotropic radiator 各向同性辐射器
isotropic turbulence 各向同性紊流度,各向同性湍流
isotropy 各向同性,等方向性
ISP 1. Integrated Support Plan 综合保障计划 2. Integrated Switching Panel 集成转换板 3. Inertially Stabilized Platform 惯性稳定平台 4. Intermediate Service Part 中间业务部分 5. Internet Service Provider 国际互联网接入服务供应商 6. International Signaling Point 国际信令点 7. International Standardized Profile 国际标准化规格 8. Interoperable Systems Project 可互操作系统计划
ISPA Informal South Pacific ATS Coordination Group 非正式南太平洋空中交通服务协调小组
ISPI Integrated System Processes & Infrastructure 综合系统工艺和基础结构
ISPRS International Society for Photo-grametfy & Remote Sensing 国际摄影测绘和遥感学会
ISR 1. Initial Submission Rate 初始提供速率 2. Integrated Surveillance Resource 集成监视资源 3. Interim System Review 临时系统评审 4. International Simple Resell 国际简单转卖 5. Interrupt Service Routine 中断服务程序
ISRC International Search & Rescue Convention 国际搜索与救援公约
ISRO In-Service Reportable Occurrence 运营中值得报告的事件
ISS 1. Inertial Sensor System 惯性敏感元件系统,惯性传感器系统 2. Information Systems Services 信息系统服务 3. Information & Support Service 信息和支持服务 4. Infrared Sensor System 红外传感器系统 5. Integrated Surveillance System 综合监视系统 6. Instrument Subsystem 仪表子系统 7. Intelligent Support System 智能支持系统 8. International Space Station 国际空间站
ISSN International Standard Serial Number 国际标准序号
ISSP Instrument Source Select Panel 仪表源选择板
ISSPU Integrated Surveillance System Processor Unit 集成监视系统处理器单元
ISSR Independent Secondary Surveillance Radar 独立的二次监视雷达
IST 1. International Standard Thread 国际标准螺纹 2. Interstellar Travel 星际航行
ISTAR Image Storage, Translation & Reproduction 图像存储、解译和再现系统
ISTIP 信息系统技术综合委员会
ISTP 1. 瞬间地面处理压力 2. Integrated Systtem Test Procedure 完整系统试验程序
ISU 1. Idle Signal Unit 空闲信号单元 2. Inertial Sensor Unit 惯性敏感组件 3. Initial Signal Unit 初始信号装置 4. Integrated Sight Unit 综合瞄准装置 5. Integrated Surveillance Unit 综合监视组件 6. Interface Surveillance Unit 接口监视装置 7. Isochronous Slot Utilization 等时隙利用
ISV 1. Independent Software Vendor 独立软件销售商 2. Isolation Valve 隔离活门
ISVR Inter Smart Video Recorder 灵巧型视频录像机
ISWL Isolated Single Wheel Load 单轮当量载荷
IT 1. Information Technology 信息技术 2. Insulating Transformer 隔离变压器 3. Integrated Tank 整体油箱 4. Integration Test 综合试验 5. Integration Testing 集成测试 6. In Transit 在运输中 7. Items 项目
ITA 1. Indicated True Altitude 绝对表高 2. Institute of Air Transport 航空运输学会 3. Instituto Tecnologico de Acronautica〈法语〉航天技术研究所 4. International Telegraph Alphabet 国际电报字母表 5. International Touring Alliance 国际旅游联盟
ITAR International Traffic in Arms Regulations 国际武器贸易条例(美国)
ITAS Indicated True Airspeed 指示真空速
ITCAN Inspect Test & Correct As Necessary 按需要检验,试验和改正
ITCS Integrated Target Command System 综合目标指挥系统
ITD 1. Initial Temperature Difference 起始温差 2. Interaural Time Difference 声源到达听者两耳的时间差
ITDM Intelligent Time-Division Multiplexer 智能时分多路复用
ITDP International Technology Demonstrator Pro-

gramme 国际技术示范项目
ITE　1. Information Technology Equipment 信息技术设备　2. Inner Thermal Enclosure 内热控隔层　3. International Telephone Exchange 国际电话交换台
ITEC　International Turbine Engine Corporation 国际涡轮发动机公司
ITED　Illustrated Tool & Equipment Drawing 图解工具和设备图纸
ITEM　Illustrated Tool & Equipment Manual 图解工具和设备手册
ITMPC　Integrated Test Management, Planning & Control 综合试验管理,计划和控制
iterated integral　迭代积分
iteration　重复,迭代
iteration solution　迭代解法
iterative addition　迭代加法
iterative array　迭代阵列
iterative filter　链形滤波器,累接滤波器
iterative impedance　重复阻抗,累接阻抗
ITF　1. Information Transport Function 信息传送功能　2. International Transport Workers Federation 国际运输工人联合会
ITGS　Integrated Track Guidance System 综合跟踪制导系统
itinerary　旅程,路线;旅行的,旅程的
ITIP　International Technical Integration Panel 国际技术综合委员会
ITL　1. Integration Test Lab 综合实验室　2. Intent To Launch 准备发射
ITM　ISDN Trunk Module 综合业务数字网中继模块
ITMC　International Transmission Maintenance Center 国际传输维护中心
ITO　1. Information Technology Operations 信息技术运用　2. Independent Test Organization 独立试验组织　3. Instrument Take-Off 仪表起飞　4. International Trade Organization 国际贸易组织
ITP　1. Initial Trail Phase 初始阶段　2. Integration Test Plan 综合试验大纲
ITPC　International Television Program Center 国际电视节目中心
ITPS　International Test Pilots School 国际试飞员学校
ITR　1. Incremental Tape Recorder 增量磁带记录器　2. Initial Trouble Report 最初故障报告　3. Instantaneous Transmission Rate 瞬时传输速率　4. Internet Talk Radio 因特网无线对话
ITS　1. Independent Television Service 独立电视服务　2. Inertial Timing Switch 惯性定时开关　3. Information Transfer Satellite 信息传输卫星　4. Information Transfer System 信息转换系统　5. Information Transmission System 信息传输系统　6. Insertion Test Signal 插入测试信号　7. Instrumentation & Telemetry System 仪表和遥测系统　8. Integrated Test Specification 综合测试规格　9. Integrated Trajectory System 综合弹道测量系统
ITSO　International Telecommunications Satellite Organization 国际电信卫星组织
ITT　1. Interstage Turbine Temperature 涡轮级间温度　2. InterToll Trunk 长途电话中继线　3. International Telephone & Telegraph Corp. 国际电话电报公司　4. Inter-Turbine Temperature 涡轮内燃气温度
ITTCC　International Telegraph & Telephone Consultative Committee 国际电报电话顾问委员会
ITU　International Telecommunications Union 国际电信联盟
ITV　1. Idling Throttle Valve 慢车油门开关,慢车节流活门　2. Integrated Test Vehicle 综合测试飞行器　3. Interactive Television 交互式电视　4. Internal Transfer Vehicle 场内搬运车
ITVF　Integration, Test & Verification Facility 组装、测试和检验设施
ITWS　Integrated Terminal Weather System 一体化终端气象服务系统
ITX　Inclusive-Tour Excursion 全包旅游团
IU　1. Input Unit 输入器,输入装置　2. Interface Unit 接口单元
IUAI　International Union of Aviation Insurers 国际航空保险联盟
IUC　Initial User Capability 初始用户能力
IUGG　International Union of Geodesy & Geophysics 国际测地和地球物理协会
IUMS　International Union of Marine Science 国际海洋科学联合会
IUOTO　International Union Of Tourist Organizations 国际旅游组织联合会
IURAP　International Users Resource Allocation Panel 国际用户资源分配委员会
IUT　Instructor Under Training 见习教官
IV　1. Initial Velocity 初速　2. Interactive Video 交互式视频　3. Interface Vector 接口向量　4. Invoice Value 发票价值金额　5. Isolation Valve 隔离活门
IVA　1. Initial Video Address 初始视频地址　2. Input Video Amplifier 输入视频放大器
IVAD　Integrate Voice & Data 综合话音和数据
IVALA　Integrated Visual Approach and Landing Aids 目视进近着陆综合导航设备
IVMS　Integrated Voice-Messaging System 综合语音信息系统
IVR　1. Instrumented Visual Range 仪表可见距离

2. Integrated Voice Response 综合语音响应
3. Interactive Voice Response 交互式语音应答
IVRS Interim Voice Response System 过渡性话音响应系统
IVSI Instantaneous Vertical Speed Indicator 瞬时升降速度表
IVV Instantaneous Vertical Velocity 瞬时垂直速度
IW 1. Information War 信息战 2. Interceptor Warning 截击机警告
IWC 1. Indoor Wireless Channel 室内无线信道 2. Interferometric all-optical Wavelength Converter 干涉全光波长变换器 3. Interferometric Wavelength Converter 干涉波长变换器
IWCS Integrated Wideband Communication System 综合宽带通信系统

J

J 1. Jamming 干扰 2. Jet 喷气机 3. Joule 焦耳 4. Journal 日记,杂志 5. Junction 连接,接头

JA 1. Japan 日本 2. Joint Account 共同账户,联合账 3. Java Applications Java 应用程序 4. Judge Advocate 军法官

JAA 1. Japan Aeronautic Association 日本航空协会 2. Japan Asia Airways CO. LTD. 日本亚洲航空公司 3. Joint Airworthiness Authorities 联合适航当局 4. Joint Aviation Administration(Europe) 联合航空局(欧洲) 5. Joint Aviation Authorities 联合航空当局

JAAA Japan Ag-Aviation Association 日本农业航空协会

JAC 1. Joint Aircraft Committee 联合飞机制造委员会 2. Joint Airworthiness Code 联合适航性法规 3. Joint Airworthiness Committee 联合适航委员会 4. Junta de Aeronautica Civil (智利)民航委员会 5. Job Assignment Card 工作分配卡

jack 1. 插孔,插座 2. 千斤顶,起重器 3. 插孔,塞孔

jack box 配电箱

jacket 1. 外套,上衣 2. 外壳,套箱 3. 套,罩,短外衣

jacketed bullet 包壳弹

jackpost 轴柱

jackscrew 止动螺杆,螺旋千斤顶

jackstay 支索,撑杆

JACMAS Joint Approach Control Meteorological Advisory Service 联合进近管制气象咨询勤务

Jacobi matrix method 雅科比矩阵法

Jacobian elliptic function 雅科比椭圆函数

Jacobian matrix 雅科比矩阵

Jacobi's integral 雅科比积分

Jacobi's integral of earth-moon trajectory 地月轨道雅科比积分

Jacobi's integral of restricted three-body problem 有限三体问题雅科比积分

JADC Japan Aircraft Development Company 日本飞机开发公司

JAFE Joint Advanced Fighter Engine 联合先进战斗机发动机(美国)

JAIC Japanese Aircraft Industry Council 日本飞机工业理事会

jalopy 破旧飞机

jam 1. 堵塞,干扰 3. 干扰,抑制 4. 拥挤,阻塞

jam nut 保险螺帽,防松螺帽

Jamac Joint Aeronautical Materials Activity 联合航空器材局(美国空军)

jammer 1. 干扰发射机 2. 干扰机 3. 人为(电气,接收)干扰 4. 簧丝芯撑

jamming 1. 干扰,人为干扰,干扰杂音(噪声) 2. 堵塞,卡住

jamming equation 干扰方程

jamming signal 干扰信号

jamming transmitter 干扰发射机,干扰台

jamming vulnerability 弱抗干扰性,低抗扰性,抗扰性不良

jam-proof 抗干扰的

jam-proof guided missile 抗干扰导弹

JAN January 一月

JANAF Joint Army-Navy-Air Force 陆海空三军联合

Jansky noise 宇宙噪声

Janus 【天】土卫十,土卫土

Janus configuration 两面神配置

Janus system 两面神系统

JAR Joint Aviation Requirements, European 欧洲适航条例

JAS 1. Japan Air System CO. LTD. 日本航空公司 2. Japan Air System 日本航空系统 3. Job Analysis System 作业分析系统 4. Joint Airmiss Section 空中相撞联合工作组 5. Journal of Aeronautical Science 航空科学学报

JASDF Japan Air Self-Defense Force 日本航空自卫队

JASSM Joint Air-to-Surface Standoff Missile 空对地远程导弹

JASU Jet Aircraft Starting Unit 喷气机起动装置

JATCRU Joint Air Traffic Control Radar Unit 联合空中交通管制雷达设备

JATE Joint Air Transport Establishment 联合航空运输公司

JATO Jet-Assisted Take-Off 喷气助推起飞,起飞助推器

JATS Joint Air Transportation Service 联合空运勤务

JAWG Joint Airmiss Working Group 空中相撞联合工作组

JAWS 1. Jamming and Warning System 干扰和预警系统 2. Jet Advance Warning System 喷气飞机预警系统 3. Joint Airport Weather Studies 联合机场气象研究 4. Joint All Weather Seeker 共用全天候导引头

JB 1. Jack Box 配电箱 2. Jet Barrier 喷气机阻栏栅 3. Junction Box 接线盒

JBD James Brake Decelerometer 詹姆斯刹车减速表

JBI James Brake Index 詹姆斯刹车指数

JC 1. Jettison Control 放油控制 2. Jitter Compensation 抖动补偿 3. Joint Compound 黏合剂,密封胶

JCA Java Component Architecture 组件结构

JCAB Japan Civil Aviation Bureau 日本民航局

JCAP 1. Joint Conventional Ammunition Program 联合常规弹药计划 2. Joint Committee on Aviation Pathology 航空病理学联合委员会

JCL Job Control Language 作业控制语言

JCN Job Control Number 作业控制编号

JCS Joint Chiefs of Staff 参谋长联席会议

JCSAT Japan Communication Satellite 日本通信卫星

JCT Junction 联结

JCU James Cook University 詹姆士库克大学

JDA 1. Japanese Defense Agency 日本防卫署 2. Japanese Domestic Airlines 日本国内航空公司

JDAM Joint Direct Attack Munition 联合直接攻击弹药,联合制导攻击武器

JDL Job Drawing List 作业图形表

JDP Joint Development Project 联合开发计划

jeep 吉普车,小型航空母舰

jeep trailer 吉普车的拖车

JEFM Jet Engine Field Maintenance 喷气发动机外场维修

JEM Japanese Experimental Module 日本试验舱

JEMTOSS Jet Engine Maintenance Task Oriented Support System 喷气发动机维修工作支持系统

JEOS 1. Janus Earth Observation Satellite 贾纳斯地球观测卫星,法国地球资源卫星 2. Japanese Earth Observation Satellite 日本地球观测卫星

Jeppesen 杰普逊公司(波音主要子公司之一)

Jeppesen Chart 杰普逊航图

Jeppesen Inflight Database 杰普逊飞行数据库

Jeppesen Integrated Tool Kit 杰普逊综合工具箱

jerk 1. 急牵,急撞 2. 冲击 3. 急促利舵,急推驾驶杆

jet 1. 喷射,喷注 2. 喷气式发动机

jet advisory service 喷气式飞机咨询服务

jet age 喷气机时代,喷气时代

jet aircraft 喷气式飞机

jet airliner 喷气客机,喷气式班机

jet attitude control system 飞机飞行姿态控制系统

jet axis 射流轴(线)

jet blast deflector 射焰偏转器

jet boundary 喷流边界

jet bow shock 喷流弓形激波

jet breakup 射流断裂

jet centerline 射流中心线

jet control 喷流调节,喷射流的控制

jet core 射流轴心

jet correlation 飞机相关

jet damping 喷射阻尼,喷流阻尼,射流阻尼

jet damping moment 喷流阻尼力矩

jet deflection 喷流偏斜,喷流造斜

jet deflection system 喷流偏转系统

jet effect 喷流影响,喷流效应

jet engine 喷气式发动机,空气喷气发动机

jet engine modulation effect 喷气发动机(叶片)调制效应

jet exhaust 排气喷口,喷气口

jet exit 喷嘴出口,射流出口

jet expansion 射流胀大

jet facility 喷气机设备

jet field 射流场,喷射流场

jet fighter 喷气式战斗机

jet flame 喷气火焰

jet flap 喷气襟翼

jet flow 喷气流,急流

jet fluid 喷射液流

jet force 喷射力

jet front 喷油柱前锋

jet fuel 1. 喷气机燃油,喷气发动机用燃料 2. 喷气燃料

jet galaxy 喷射星系

jet hole 喷嘴口,喷孔

jet impingement 射流冲击,喷气冲击,射流撞击

jet injection 1. 射流注射 2. 射流喷射

jet injector 喷射注水器,喷射器

jet lag 飞行时差反应

Jet misalignment 发动机喷流偏心度

jet momentum coef 射流动量系数

jet noise 喷流噪声

jet nozzle 喷嘴,尾喷管,喷射管

jet penetration 射流穿透深度

jet periphery 喷气边缘

jet pipe 射流管,喷口管,喷射管

jet potential core 喷气式飞机的潜在的核心

jet pressure 喷射压力

jet propulsion 喷气推进,喷射推进

Jet Propulsion Laboratory 喷气推进实验室

jet pulsation frequency 射流脉动频率

jet pulse 喷气脉冲

jet pulse condition 喷气脉冲条件

jet pump 喷射泵,引射泵

jet reaction control system 喷气反作用控制系统,喷气驱动控制

jet route 喷射机航路,喷气航路

jet shear layer 射流剪切层
jet sheet 喷流面
jet shoes 喷气鞋
jet size 喷管直径
jet speed 喷流速度,射流扩展速度
Jet Squalus 喷气鲨鱼
jet stir reactor 射流搅拌反应器
jet stream 射流,急流
jet structure 喷流结构
jet tab 塔盘的喷射舌片
jet testing 喷流试验
jet trainer 喷气教练机
jet vane 喷气舵,喷气导流控制片
jet velocity 喷出速度,喷射速度
jet-atmosphere interaction 飞机大气的相互作用
jetavator 喷流偏转器,燃气舵,喷流偏转舵
jet-driven wind tunnel 喷气发动机驱动风洞
JETDS Joint Electronics Type Designation System 联合电子型号标识系统
jetevator 喷气流偏转器转动式喷管,喷气流,排气流
jet-fuel starter 喷气发动机燃油起动机,主发动机起动机
Jethete 杰塞特耐热钢板
jet-impingement fluidic oscillator 射流冲击射流振荡器
jet-liftoff 火箭发射的起飞,启动
jetlike protrusion 喷流状突起
jetocopter 喷气式直升机
jetpen-predicted fuel 预测燃料
jetprop 1. 涡轮螺旋桨动力装置 2. 装涡轮螺旋桨发动机的飞机
jetstream 1. 喷气流,射流 2. 对流层顶急流 3. 急流射流 4. 西风急流
jet-stream 喷射涡流,喷射气流,高速气流
jettison 抛,投弃,抛射
jettison control module 投弃控制微型组件
jettison testing 投放试验
jet-to-freestream momentum 喷射到自由流的动量
jetty 停机桥,旅客登机桥
jetway 登机道,喷气式飞机跑道,喷气式飞机乘客上下走廊
jet-with-cavity 射流腔
JEWC Joint Electronic Warfare Center 联合电子战中心
jewel bearing 宝石轴承
Jezebel 杰泽贝尔空投声呐浮标
JFC Jet Flow Control 喷气(发动机燃油)流量控制
JFS Jet Fuel Starter 喷气(燃油)起动机
JFTO Joint Flight Test Organization 联合飞行试验组织

JG Jitter Gain 抖动增益
JGSDF Japan Ground Self-Defence Force 日本陆上自卫队
JHSU Joint Helicopter Support Unit 联合直升机保障部队
JIAWG Joint Integrated Avionics Working Group 电子设备工作组
JIC 1. Jet Induced Circulation 引射环流 2. Joint Industrial Company 联合工业公司
jig 1. 装配架,型架 2. 夹具,钻模
Jikiken 磁源,日本科学卫星
jink 闪避,进行反高炮机动
JINTACCS Joint Interoperability of Tactical Command Control Systems 战术指挥与控制联合互通系统
JIO Joint Intelligence Organization 联合情报组织(澳大利亚)
JIP Joint Interface Program 联合接口计划
JIS Japanese Industrial Standards 日本工业标准
JISS Japanese Ionospheric Sounding Satellite 日本电离层探测卫星
JIT Just In Time 准时
jitter 1. 抖动,颤动,振动,晃动 2. (电子)跳动,(图像)跳动
JJPTP Joint Jet Pilot Training Programme 喷气式飞机驾驶联合训练计划
JK Jack 千斤顶
JMCC Johnson Mission Control Center 约翰逊飞行任务控制中心
JMEM Joint Munitions Effectiveness Manual 共用弹药效能手册(美国三军)
JMRC Joint Mobile Relay Center 联合机动转报中心,联合机动中继站
JMSAC Joint Meteorological Satellite Advisory Committee 气象卫星联合咨询委员会(美国)
JMSDF Japan Maritime Self-Defence Force 日本海上自卫队
JMSNS Justfication for Major System New Start 主系统重新起动可行
JMTSS Joint Multi-channel Trunking and Switching System 联合多路中继和交换系统
JOAC Junior Officers Advisory Council 基层军官顾问委员会
job virtual memory 作业虚拟存储器
job-flow control 作业流控制
job-lot control 批量控制
JOC Joint Operation Centre 联合作战中心
jockey weight 活动砝码
Johnson noise 约翰逊噪声
joined-wing strutless biplane 连接翼无支柱双翼机

joined-wing aircraft 连接翼飞机
joined-wing concept 连接翼的概念
joined-wing design 连接翼的设计
joined-wing model 连接翼模型
joined-wing test article 连接翼测试条
joint 关节,接合处
joint airborne training 联合空中训练,联合空降训练
joint coordinate system 关节坐标系
joint distribution 联合分布
joint effect 联合效应
joint length 接头长度
joint model 组合模型,联合模型
joint rate 联合运费,联运运价率,共同运费
joint strike 联合罢工
joint strike fighter 联合攻击战斗机,联合打击战斗机,雷神战机,联合攻击机
joint structure 接榫结构,连接结构,节理构造,缝结构
joint support 关节支持,保护关节,联合支护,连接支撑
joint surveillance system 联合监视系统(美国)
joint tactical information distribution system 联合战术信息分发系统
joint time 综合时,综合时号改正数
joint torque 关节力矩
joint velocity 关节速度
joint wing 连翼飞机
Joker 燃油警告值(预先选定的燃油量,作为发出警告的时机)
JONA Joint Office of Noise Abatement 消减噪声联合办公室(美国运输部和国家航空航天局)
JOP 1. Joint Operating Procedures 联合操作程序 2. Junior Officer Pilot 尉官飞行员
Josephson Tunneling Logic 约瑟夫森隧道逻辑
Joule 焦耳
joule heating 电阻加热
Joule-Thomson effect 焦耳-汤姆森效应
journal 1. 日态,协议 2. 轴颈 3. 杂志,日记账
Journal of Aircraft 航空学报(中国)
journey 1. 旅行,旅程日期 2. 历程,过程
journey time 行程时间
JOVE Jupiter Orbiting Vehicle for Exploration 木星轨道探测飞行器
JOVIAL Jules Own Version of International Algorithmic Language 国际算法语言的朱尔斯文本
Jovian planet 木星型行星,类木行星
joystick 驾驶杆,操纵杆
JP 1. Jamming Pulse 干扰脉冲 2. Jet Propulsion 喷气发动机
JPL Jet Propulsion Laboratory 喷气发动机实验室
JPO Joint Program Office 联合计划处

JRE Java Runtime Environment Java 运行环境
JRC Joint Research Centre 联合研究中心
JRIA Japan Radioisotope Association 日本放射性同位元素协会
JRMET Joint Reliability and Maintainability Evaluation Team 可靠性与维修性联合签定组
JSAE Japan Society of Aeronautical Engineering 日本航空工程学会
JSA 1. Japanese Sommeliers Association 日本斟酒服务员组织 2. Jet Standard Atmosphere 喷射标准大气 3. Joint Security Area 联合保密区
JSAC Joint Strategy and Action Committee 联合战略与行动委员会
JSC Johnson Spaceflight Center 约翰逊航天飞行中心
JSG Jump Strut(Landing) Gear 活动支柱起落架
JSIPS Joint Surveillance Imagery Processing System 联合监视图像处理系统
JSLC Jiuquan Satellite Launch Center 酒泉卫星发射中心
JSME Japan Society of Mechanical Engineers 日本机械工程师学会
JSOP Joint Strategic-Objective Plan 联合战略目标规划
JSOW Joint Stand-Off Weapon 联合远距攻击武器
JSTARS Joint Surveillance Target Acquisition and Reconnaissance System 联合侦察目标定位与监视系统
JSTPS Joint Strategic Target Planning Staff 战略目标规划机构
JTDE Joint Technology Demonstrator Engine 联合技术演示发动机
JTF 1. Jitter Transfer Function 抖动传递函数 2. Joint Task Force 联合特遣部队
JTFA Joint Time-Frequency Analysis 联合时间-频率分析
JTIDS 1. Joint Tactical Information Distribution System 联合战术信息分布系统 2. Joint Tactical Interoperable Data System 联合战术可调数据系统
JTRU Joint Tropical Research Unit 联合热带研究组
JTST Jet Stream 喷流,急流
Judy 我已发现目标,并着手截击(空中截击通话代语)
jug 〈口语〉气缸,副油箱
Julian calendar 儒略历
Julian century 儒略世纪
Julian date 儒略日期
Julian day 儒略日
Julian day calendar 儒略日历
Julian day number 儒略日数
Julian ephemeris date 儒略历书日期
Julian epoch 儒略纪元

Julian era 儒略纪元
Julian period 儒略周期
Julian year 儒略年
jumbo 大型喷气式飞机
jumbo-group 巨群
jump 1. 跳伞,从飞机上投伞兵 2. 跳伞,跳跃
jump address register 转移地址寄存器
jump condition 转移条件,跳跃条件
jump correction 跳偏修正角
jump jet 垂直起落的喷气式飞机
jump master 跳伞组长,跳伞长
jump phenomenon 跃变现象,跳跃现象
jump seat 安全门座位
jump strut gear 活动支柱起落架
jump take off 1. 垂直起飞 2. 起始跃升
jump trace 跳转踪迹
jump transfer 跳转转移
jumper 1. 跨接线,跨接 2. 跳伞员
jumper training 跳伞训练
jumper wire 跨接线
jumpseat 弹射座椅电子侦察卫星
junction 联结,源极衬底结
junction battery 结型电池
junction box 1. 接线箱 2. 电缆接线箱
junction capacitance 阻挡层电容
junction circulator 结环行器
junction depth 结深
junction field effect transistor 结型场效晶体管
junction point of traverses 导线节点
junction resistance 结电阻
junction transistor 面结型晶体管
junction transposition 接线换位
juncture 连接,接合点,连接接头
June Solstice 夏至
junior aircrew officer 初级飞行军官
junior technician 初级技师
Juno 【天】婚神星,小行星3号
Jupiter 【天】木星
Jupiter cycle 岁星纪年
JURA Joint-Use Restricted Area 公共限制区,联合使用限制区
jury strut 保险撑杆,辅助撑杆,副支柱
JUSMAG Joint United States Military Advisory Group 美国联合军事顾问小组
just noticeable difference 最小可觉差
justifiable 正当的
justifiable digit time slot 可调数字时隙
justification 码速调整
justification ratio 码速调整比
justifying digit 调整数字
JUT Jet Utility Transport 通用喷气运输机
juvenile gas 初生气
juvenile water 初生水,原生水
JV 1. Joint Venture 合资企业 2. Journal Voucher 转账传票,日记账凭据
JVM Java Virtual Machine Java 虚拟机
JVC Joint Vane Control 喷气舵操纵
JW 1. Jacket Water 水套 2. Jamming War 电子干扰战

K

K 1. Kilohm 千欧 5. Karat 开（黄金成色），克拉（宝石重量单位） 2. Kerosene 煤油 3. Kilo 公斤，公里 4. Knots 海里每小时

KA 1. Keyed Address 键入地址 2. Knowledge Acquisition 知识获取

KAC 1. Kaman Aircraft Corporation 卡曼飞机公司 2. Kuwait Airways Corporation 科威特航空公司

KADS Knowledge Acquisition Data System 知识获取数据系统

KAI Kharkov Aviation Institute 哈尔科夫航空研究所（俄罗斯）

Kalman cycle 卡尔曼循环，卡尔曼周期

Kalman filtering 卡尔曼过滤器

Kalman filtering technique 卡尔曼过滤技术

Kalman period 卡尔曼周期

KARI Korea Aerospace Research Institute 韩国航空航天研究所

Karman vortex street 卡门涡道

Karman-Tsien formula 卡门-钱学森公式

Karp line 卡普线

KAS 1. Killed on Active Service 因公牺牲，殉职 2. Knowledge Acquisition System 知识获取系统

KAT Key to Address Transformation 键-地址变换

katabatic wind 山风，下降风

KB 1. Kilobyte 千字节 2. Knowledge Base 知识库

kbps Kilobits Per Second 千位每秒

KBSA Knowledge Based Software Assistant 人工智能软件辅助

KBU Keyboard Unit 键盘装置

KC Key Characteristics 关键特性

KCAB Korean Civil Aviation Bureau 韩国民航局

KCAL Kilocalorie 千卡（热量单位）

KCAS Knots Calibrated Air Speed 校准空速（节）

KDC 1. Key Distribution Center 密钥分配中心 2. Knock Down Component 分解部件

KE 1. Kinetic Energy 动能 2. Knowledge Engineering 知识工程

KEAS Knots Equivalent Air Speed 等效空速（海里每小时）

keel 龙骨，下部纵梁

keel area 龙骨面

keelson 内龙骨

keep alive 保活

keeper 永磁衔铁，锁紧螺母

keeper current 保持电流

keeper electrode 定位电极

keeper voltage 保持器电压

keg 小桶

Kell factor 凯尔系数

kelmet 油膜轴承，油膜轴承合金

Kelvin temperature 绝对温度，开氏温度

Kelvin theorem 开尔文定理

KEP 1. Key Emitter Parameters 发射体主要参数 2. Kinetic Energy Penetrator 动能穿透器

Keplerian orbit 开普勒轨道

Kepler's euaqtion 开普勒方程

Kepler's law 开普勒定律

kermet 铅铜轴承，铅铜轴承合金

kernel 核心

kerosene 煤油

kerosene engine 煤油发动机

kerosene jet 煤油喷气

Kerr effect 克尔效应

ketone 酮

KEV Thousand Electron Volts 千电子伏

Kevlar 芳纶，凯夫拉（聚对苯二甲酰对苯二胺纤维，亦称纤维B）

Kevlar fiber 凯夫拉纤维

key 1. 键，开关 2. 钥匙 3. 关键的

key board 键盘

key click filter 电键喀喀声消除器

key filter 1. 键路火花消除器 2. 键噪滤波器

key hierarchy 密钥分级结构

key parameter 关键参数

key person 关键人物

key register 钥匙登记簿

key switch 按键开关

key technology 关键技术

keyboard 键盘

keyboard switch 键盘开关

keyboard tape punch 纸带键盘凿孔机

keyed address 键入地址

keyed option 键控选择

keyhole 锁眼，钥匙孔

keyholing 弹头翻滚

keying 1. 键控 2. 按键，自动开关

keysets　1.转接板 2.配电板
keyshelf　键座,电键盘
keystone distortion　梯形畸变
KF　Krueger Flaps 克鲁格襟翼
KFS　Kaspersky File System 卡巴斯基文件系统
KG　Kilogram 千克
KG PS　Kilogram Per Second 千克每秒
khz　Kilohertz 千赫兹
KIAS　Knots Indicated Air Speed 指示空速(节)
kick　1.突跳 2.跑,反冲
kickback　1.反转 2.回扣,佣金
kicker　脉冲式火箭发动机,空投员
kill　杀伤,摧毁,击落
kill probability　毁伤概率
kill rate　杀伤率,击毁率
kill ratio　击毁比率,杀伤比率
kilo pound　千磅
kilometer　千米,公里
kilometer scale　千米尺
kiloton　千吨
kilovolt　千伏
kilowatt　千瓦
kinematic coupling　运动耦合
kinematic equation　动态方程
kinematic model　动态模型
kinematic similarity　运动学相似性
kinematic velocity　运动速度
kinematic viscosity　运动黏度
kinematics　运动学,运动学弹道
kinescope　电子显像管
kinetheodolite　电影经纬仪,摄影经纬仪
kinetic constant　动力学常数
kinetic energy　1.动能 2.动能系数 3.毁伤动能
kinetic energy weapon　动能武器
kinetic filter　动态滤波器
kinetic model　动力学模型
kinetic modeling　动力学模型
kinetic noise　动力噪声
kinetic parameter　动力参数
kinetic potential　动势
Kinetic theory of gas　气体动力学理论,气体分子运动论
kinetic vacuum pump　动量传输泵
kinetics　动力学
kingpost　单支柱桁架
kink　扭结,扭折
kiosk　配电亭,公用电话间,报摊
kip　千磅
Kirchhoff　基希霍夫定律
Kirksite　卡克塞特锌合金

Kirkwood gaps　柯克伍德空隙
KIS　Kick-In Step 开始步骤,第一步
kit　箱,成套工具
kite　风筝,(轻型)飞机
kite balloon　(系留在地面或车上的)系留气球
Kitty Hawk　小鹰号航空母舰
kj　kilojoule 千焦耳
KKV　Kinetic Kill Vehicle 动能杀伤飞行器
KL　Kiloliter 千升
KM　1.Kilometer 公里,千米 2.Knowledge Management 知识管理 3.Knowledge Map 知识映射图
KN　Kilo-Newton 千牛(牛顿)
knee　1.膝,膝状物 2.升降台
knee point　曲线弯曲点,拐点
knee sensitivity　拐点灵敏度
kneeboard　矮墙
kneeling　飞机下蹲
knife edge　刀口,刃状物
knob　按钮,旋钮,调节钮
knock　1.爆震 2.撞击
knock rating　抗爆值,抗爆品级
knockdown　价格的压低,船的破损
knot　1.海里每小时,节 2.结,连接
knotting　打结,结绳
knowledge　了解,知识
knowledge acquisition　知识收集,知识的获得
knowledge aided protocol automation　知识辅助协议自动化
knowledge assimilation　知识同化
knowledge base　知识库,知识汇流台
knowledge discovery　知识获取,知识发掘
knowledge dissemination　知识传播
knowledge engineering　知识工程
knowledge industry　知识工业
knowledge inference　知识推理
knowledge management　知识管理
knowledge repository　知识仓库
knowledge representation　知识表达
knowledge transfer　知识传送
knowledge-based approach　基于知识的方法
knuckle pin　转向销,节销盖
knuckled　铰键结合的,单悬臂铰接的
Knudsen number　克努森数
KP　1.Kick Pipe 反冲管 2.Kick Plate 反冲板 3.Kilopound 千磅,千克力
KPA　1.Key Process Area 关键过程域 2.Key Pulse Adapters 键控脉冲适配器 3.Kilopascal 千帕斯卡
KPH　Kilometers Per Hour 千米每小时
KPL　Kilometers Per Liter 公里每公升

KPPS Kilo Pulses Per Second 每秒千脉冲数
KPS Knowledge Processing System 知识处理系统
Kriging 克里格法
Kristall "晶体号"技术舱
Krueger flag 克鲁格襟翼
kryptoclimate 室内小气候
kryptoclimatology 室内小气候学
krypton 【化】氪（化学元素，符号 Kr）
KS 1. Knowledge Set 知识集合 2. Knowledge System 知识系统
KSC Kennedy Space Center 肯尼迪航天中心
KSI Kips per Squre Inch 每平方英寸千磅数
KST Keyseat 键槽，销槽
KT 1. Kiloton 千吨 2. Knowledge Tracing 知识跟踪
KTF Kaluga Turbine Factory 卡卢加涡轮机厂
KTS Key Telephone System 按键电话系统
kuroshio 黑潮，日本暖流
kurtosis 峰度，峭度，峰态
kutta flow 库达流
kutta joukowsky hypothesis 库达儒科夫斯基假说
KV Kilo Volts 千伏
KVA Kilovolt Ampere 千伏安
KVAR 1. Kilovars 千伏安 2. Kilovolt Amperes Reactive 无功千伏安
kw kilowatt 千瓦
KWE Kilowatts of Electrical Energy 千瓦电力
kwh kilowatt hour 千瓦时
k-words kilo-words 千字
kwt kilowatts thermal 千瓦热量
KWY Keyway 键槽
KYBD Keyboard 键盘
kymograph 转筒记录器，波形自记器
Kyokko 极光卫星（日本）
kytoon 小型风筝式系留气球

L

L 1. Late 晚 2. Latitude 纬度 3. Lavatory 盥洗室 4. Left 左 5. Light 灯 6. Line 线 7. Litre 升 8. Locator 定位器,探测 9. Long 长的
L/M List of Materials 材料清单
L/N Line Number 生产线号
LA 1. Lateral Axis 横轴 2. Lead Angle 导角 3. Lightning Arrester 避雷器 4. Linear Accelerometer 线性加速器 5. Letter of Advice 送货通知 6. Longitudinal Axis 纵轴
LAAD Low Altitude Air Defence 低空防空
LAADS Low Altitude Aircraft Detection System 低空飞机探测系统
LAANC Local Authorities Aircraft Noise Council 飞机噪声委员会区域机构
LAAT Logistics Assessment and Assistance Team 后勤评估与援助小组
LAB 1. Label 标签 2. Laboratory 实验室
labelled graph 标定图
labelled image 有标号图像
labelled storage 有标号存储器
labelled tree 标记树
labor load 劳动负荷
laboratory analysis 试验分析
laboratory module 试验舱
labour cost 人工成本,工价
LABS 1. Low Altitude Bombing System 低空轰炸系统 2. Low Altitude Bombing Sight 低空投弹瞄准器
LABS maneuver 利用低空轰炸系统轰炸的机动
labyrinth 篦齿,迷宫
labyrinth seal 迷宫汽封,曲径汽封
labyrinthine function 迷路功能
labyrinthine motion stimulus 迷路运动刺激
labyrinthine sensitivity test 迷路敏感性试验
LAC 1. Local Area Controller 本地区管制员 2. Local Area Controller 区域控制器 3. Lockheed Aircraft Corporation 洛克希德飞机公司
lace 全列穿孔操作
laceration 划破,撕裂
lachrymator 催泪剂
LACIE Large Area Crop Inventory Experiment 大面积作物估产实验
lacing 栅格桁架
lacing tape 编织带

LACO Laser Communications 激光通信
lacrimatory 催泪的
lacrimatory gas 催泪性毒气
Lacrosse 1. 长曲棍球卫星 2. 军事测距系统军事测距系统
lactoprene 聚酯橡胶,乳胶
Lacus 月面湖
LACW Leading Aircraftswoman 上等空军地勤女技师
LAD Ladder 梯子,阶梯
LADA Laser Radar 激光雷达
LADAR Laser Detection And Ranging 激光探测与测距
ladder approximation 梯形近似法
ladder diagram 梯形图顺序控制图
ladder network 梯形网络
LADGPS Local Area Differential GPS 局域差分全球定位系统
LADS 1. Local Area Distributed System 局部分布式系统 2. Low Altitude Detection System 低空探测系统
LAE 1. Licensed Aircraft Engineer 持有执照的飞机工程师 2. Low-altitude Extraction 低空拖投
LAFTS Laser and FLIR Test Set 激光与前视红外试验设备
lag Lagging 滞后
lag case 滞后情况
lag compensation 滞后补偿
lag correction 延迟性修正量
lag damper 滞后阻尼器
lag dynamics 滞动力学
lag equation 滞后方程
lag hinge 摆振铰,垂直铰,垂直关节
lag mode 滞后模式
lag model 滞后模型
lag network 滞后网络
lag of wash 洗流时差
lag window 滞后窗
lag-lead compensation 滞后超前补偿
lagrange bracket 拉格朗日括号
lagrange duality 拉格朗日对偶性
lagrange viewpoint 拉格朗日观点
lagrangian analysis 拉格朗日分析
lagrangian bracket 拉格朗日支架
lagrangian formulation 拉格朗日公式
lagrangian marker 拉格朗日标记

lagrangian model	拉格朗日模型
lagrangian points	拉格朗日点
lagrangian tracking	拉格朗日跟踪
LAH	Level Alarm-High 高位警报
LAHS	Low Altitude and High Speed 低空和高速
LAHS mission	低空大速度飞行任务
LAHS handling qualities	低空高速操纵品质
LAHSL	Land and Hold Short Lights 着陆和等待灯
LAL	Level Alarm-Low 低位警报
LALD	Low Angle Low Drag 小角度低阻力
LAM	1. Laminate 分层，夹层，层压的 2. Landing Attitude Modification 着陆姿态修正 3. Lateral Autopilot Module 横向自动驾驶组件
lamb dip	兰姆凹陷
lamb noise silencing circuit	兰姆消噪电路
lambda	希腊字母λ，希腊字母第十一字
lambert	朗伯，兰伯特投影
Lambert Law	朗伯定律
lamella	薄片，薄层
lamellar domain	层状畴
lamellar structure	片状组织，层状结构
lamilloy	层板
lamina	薄板，层次
laminar boundary layer	层流附面层，片状边界，层流边界层
laminar boundary layer separation	层流边界层分离
laminar combustion	层流燃烧
laminar diffusion	层流扩散
laminar electron beam	层流电子束
laminar flame	层流焰
laminar flow aerofoil profile	层流翼型
laminar flow wing	层流机翼
laminar flow control	层流控制
laminar gun	层流电子枪
laminar region	层流区
laminar separation	层流分离
laminar separation bubble	层流分离气泡
laminar sublayer	亚层流，层流底层
laminar viscous flow	黏性层流
laminar wake	层流尾流
laminate	层制品
laminate propellant	叠层推进剂
laminated	分层的
laminated blade	层板叶片
laminated glass	层合玻璃
laminating molding	层压成形
LAMMR	Large Antenna Multifrequency Microwave Radiometer 大型天线多频微波辐射计
lamp	灯
lamp house	光源
lampholder	灯座
lamplighter	照明飞机
LAMPS	Light Airborne Multipurpose System 轻型机载多用途系统
LAMS	Light-Aircraft Maintenance Schedule 轻型飞机维修时间表
LAN	1. Inland 内地 2. Local Area Network 局域网，本地网
LANAC	Laminar Air Navigation and Anti-Collision System 拉纳克航行方法，无线电空中导航及防撞系统，拉纳克系统
land arm	地杆，陆军
land capability map	土地规划图
land clutter	地面杂波
land fall	地崩
land mobile-satellite service	卫星陆地移动业务
land resource map	土地资源图
land roll	镇压器
land satellite	陆地卫星
land use mapping	土地利用制图
land use monitoring	土地利用监测
landability	着陆可能性
landau damping	朗道阻尼
lander	正在着陆的飞机
landfill	填埋，填埋区
landform map	地形图
landing	着陆，降落，着陆方法
landing aids	着陆辅助设备，着陆导航设备
landing angle	着陆角，降落角
landing approach	着陆进场
landing approach sink rate	进场着陆时的下降率，第五边的下沉速度
landing area	着陆场，飞机场，降落区
landing area floodlight	着陆探照灯
landing bounce	着陆跳跃
landing categories	着陆标准
landing compass	着陆罗盘
landing control center	着陆控制中心
landing distance	着陆距离
landing distance available	可用着陆距离
landing energy	着陆功量
landing footprint	着陆地带，可能着陆轨道区域
landing forecast	着陆预报，着陆场天气预报
landing gear	起落架，着陆装置
landing gear drop test	起落架落震试验
landing gear operating mechanism	起落架收放机构
landing gear system	起落装置系统
landing gear with two stage shock absorber	双腔起落架

landing gross weight 落地全重
landing ground 着陆场
landing guidance system 着陆制导系统
landing impact 着陆撞击
landing impact attenuation rocket 着陆缓冲火箭
landing impact tolerance 着陆冲击耐力
landing instrument 着陆仪表
landing light 着陆灯
landing limiting weight 着陆限制重量
landing mat 拆装式着陆道面板
landing operation 着陆操作
landing pad 降落场
landing party 先头登陆队
landing point 登陆点
landing point accuracy 着陆点精度
landing point designator 着陆点指示器
landing precision 着陆精度
landing radar 着陆雷达
landing runway 着陆跑道,着陆地带
landing signal officer 着陆信号官
landing simulation 模拟着陆
landing site 着陆场
landing site latitude 着陆场高度
landing speed 着陆速度,着水速度
landing standard 着陆标准
landing subsystem 着陆分系统
landing surface 乘坐面
landing system 着陆系统
landing weight 着陆重量
landing wires 着陆张线
landing zone 降落地带,着陆区
landing zone of emergency rescue 应急救生着陆区
LANDIS Low Approach Navigation Director System 低空进近导航指引仪系统
LANE Low Altitude Navigational Equipment 低空导航设备
landmark 地标,检查点
landmark beacon 陆标标志灯,地点灯标
landmark discrimination 地标识别
landmark discrimination test 地标识别试验
landmass 陆地
landplane 陆上飞机
Landsat 地面通信卫星
landscape map 景观地图
landside 对陆面
lane 巷道,低空导航设备,等相位区,航线
lane identification 巷道识别
lane width 巷宽
Langley 兰(符号 ly,太阳辐射单位,等于卡/厘米^2)

Langley type 兰利型
langmuir probe 兰米尔等离子测量仪
language interface 语言接口
LANS Land Navigation System 陆地导航系统
lanthanide 镧化物,镧族元素
lanthanum 【化】镧(化学元素,符号 La)
lanthanum hexaboride 六硼化镧
lanyard 牵索
lap 适用出版物清单,搭接
lapping 缠绕,重迭
large amount of process noise 大量过程噪声
large amplitude 大振幅
large angle maneuver control 大角度机动控制
large aperture 大孔径
large array 大数组,大队列
large bandwidth 大带宽
large capacity storage 大容量存储器
large computer 大型计算机
large deflection 大挠度
large deformation 大变形
large diameter 大直径的
large displacement 大位移,大排水量
large eddy simulation 大涡模拟
large error 大误差
large flexible spacecraft attitude control 大型挠性航天器的姿态控制
large flexible spacecraft vibration control 大型挠性航天器的振动控制
large flow 大流量
large flow rate filling 大流量加注
large format camera 大像幅摄像机
large grain 粗粒的
large number hypothesis 大数假说
large potential 大电位,大电势
large rotation 大循环,大旋转
large scale 大规模的
large scale array 大规模阵列
large scale chart 大比例尺海图
large scale compound intergration 大规模集成电路
large scale structure of the universe 宇宙大尺度结构
large scale system cybernetics 大系统控制论
large scale topographical map 大比例尺地形图
large screen projector 大屏幕投影仪
large space structure 大空间结构
large transport 大型运输机
large-amplitude-roll maneuver 大幅度滚转机动动作
large-amplitude-slosh liquid dynamics 大幅液体晃动力学
large-particle method 大粒子方法

large-scale experiment　大规模实验
large-scale integration　大规模集成化
large-signal analysis　大信号分析
Larmor precession　拉莫尔旋进
Larmor rotation　拉莫尔旋动
LARS　1. Local Area Radio System 局域无线电系统 2. Low-Altitude Radar System 低空雷达系统
LAS　1. Landing Area Security 着陆区安全措施 2. Low-Alloy Steel 低合金钢 3. Lower Airspace 低空空域
LASA　Large-Aperture Seismic Array 大孔径地震检波器组合
lasant　激射物
LASC　Lead Angle Steering Command 前置角驾驶指令
lase　放射激光
lasecon　激光转换器
laser　激光
laser ablation　激光烧蚀
laser absorption　激光吸收
laser accelerometer　激光加速计
laser acquisition device　激光截获装置
laser aiming instrument　激光瞄准仪
laser alignment　激光准直
laser altimeter　激光测高计
laser anemometer　激光风速计
laser annealing　激光退火
laser antenna　激光天线
laser automatic tracking　激光自动跟踪系统
laser beacon　激光信标
laser beam　激光束
laser beam cutting　激光切割
laser beam guidance　激光束导向
laser beam image reproducer　激光束图像重现器
laser beam machining　激光束加工
laser beam perforation　激光穿孔
laser beam welding　激光束焊接
laser bombing system　激光轰炸系统
laser bonding　激光焊接
laser boresight　激光校靶
laser cavity　激光腔
laser ceilometer　激光测云仪
laser channel marker　激光航道标
laser chemical vapor deposition　激光化学气相沉积
laser collimator　激光准直仪
laser communication and tracking　激光通信与跟踪
laser computer　激光计算机
laser conductometer　激光热导仪
laser corner reflector　激光角反射体
laser countermeasure　激光对抗
laser cutting　激光切割
laser cutting machine　激光切割机
laser damage　激光损伤
laser data transmission　激光数据传输
laser dazzle sight　激光伪装瞄准
laser deposition　激光淀积
laser designator　激光指示器
laser detection and ranging　激光探测与定位
laser device　激光装置
laser diode　激光二极管
laser diodes matrix　激光二极管矩阵
laser displacement sensor　激光位移传感器
laser distance measuring system　激光测距系统
laser Doppler anemometer　激光多普勒风速计
laser Doppler radar　激光多普勒雷达
laser Doppler velocimeter　激光多普勒测速仪
laser drilling　激光钻孔
laser dye　激光染料
laser dynamic balancing　激光动平衡
laser echo ratio　激光回波率
laser energy　激光能
laser evaporation　激光蒸发
laser fracturing　激光破碎
laser fusion　激光核聚变
laser fuze　激光引信
laser grooving　激光刻槽
laser guidance　激光制导
laser gyro　激光陀螺仪
laser gyroscope　激光陀螺
laser hardening　激光硬化
laser heat treatment　激光热处理
laser heating　激光加热
laser holographic interferometer　激光全息干涉仪
laser holography　激光全息照相
laser homing and warning system　激光寻的与警戒系统
laser ignition　激光点火
laser illuminator　激光照射器
laser imaging　激光成像
laser inertial reference system　激光惯性参照系统
laser information display system　激光信息显示系统
laser interferometer　激光干涉仪
laser interferometer gravitational wave observatory　激光干涉仪引力波观测台
laser intersatellite communication　卫星间激光通信
laser isotope separation　激光分离同位素
laser level　激光水准仪
laser light detector　激光光探测器
laser linewidth　激光线宽
laser locator　激光定位器
laser medium　激光媒质

laser melting coating	激光熔覆
laser memory	激光存储器
laser microprobe mass analyser	激光探针质量分析仪
laser missile tracking system	激光导弹跟踪系统
laser optical modulator	激光光学调制器
laser oscillation condition	激光振荡条件
laser oscillator	激光振荡器
laser physical vapor deposition	激光物理气相沉积
laser plane	激光平面
laser plumbing	激光投点
laser plummet apparatus	激光铅垂仪
laser power	激光功率
laser processing	激光加工
laser propulsion	激光推进
laser protection	激光防护
laser pulse	激光脉冲
laser pulse energy	激光脉冲能
laser pulse repetition frequency	激光脉冲重复频率
laser pumping	激光激励,激光激励系统
laser radar	激光雷达
laser radiation	激光辐射
laser range finder	激光测距器
laser ranger	激光测距仪
laser ranging	激光测距
laser ranging device	激光测距仪
laser ranging reflector	激光测距反射器
laser receiver system	激光接收系统
laser receiving set	激光接收机
laser recognition	激光识别,用激光装置识别
laser recrystallization	激光再结晶
laser reflector	激光反射器
laser satellite	激光卫星
laser satellite tracking installation	激光卫星跟踪装置
laser screen	激光屏幕
laser screen generator	激光光网探测器
laser seeker	激光寻的器
laser semi-active missile	激光半主动寻的导弹
laser sensor	激光传感器
laser sounder	激光测深仪
laser source	激光源
laser speckle	激光散斑
laser speckle velocimeter	激光斑点速度计
laser speckle interferometer	激光散斑干涉仪
laser spectroscopy	激光光谱
laser storage	激光存储器
laser surface modification	激光表层改性
laser swinger	激光扫平仪
laser system	激光系统
laser target designation	激光目标指示
laser target designator	激光目标指示器
laser target recognition system	激光目标识别系统
laser target scouting system	激光目标搜索系统
laser theodolite	激光经纬仪
laser thermal weapon	激光武器
laser threshold	激光阈值
laser topographic position finder	激光地形仪
laser tracker	激光跟踪器
laser tracking	激光跟踪
laser tracking measurement system	激光跟踪测量系统
laser transformation hardening	激光淬火
laser transmission	激光传输
laser trimming	激光微调
laser trimming technique	激光微导术
laser velocimeter	激光速度仪
laser vibrometer	激光振动计
laser wavelength	激光波长
laser weapon	激光武器
laser-beam	激光束
laser-Doppler anemometer	激光多普勒风速计
laser-guided bomb	激光制导炸弹
laser-produced plasma	激光引发等离子体
laser-screen method of flow visualization	激光屏显示
LAT	1. Landing Approach Trainer 着陆进近练习器 2. Lateral 横向 3. Lattitude 纬度
latax	侧加速度
latch	销存器,闭锁,闩
latch register	锁存寄存器
latching relay	自保持继电器
latching valve	自锁阀
latch-up	闩锁效应
late burst	迟炸
late potential	迟电位
late-model	新型的
latency	等待时间
latent energy	潜能
latent heat	潜热
latent heat of condensation	凝结潜热
latent injury	潜在性损伤
latent instability	潜在不稳定
latent tetany	潜伏性手足抽搐
lateral	横向的,侧面的
lateral acceleration	侧向加速度
lateral acceleration autopilot	过载自动驾驶仪
lateral axis	横轴,俯仰轴
lateral boundary	边线
lateral component	侧向分力
lateral control	横向操纵
lateral control actuator	横向控制作动器

lateral control departure parameter 横向操纵偏离参数
lateral control qualities 横向操纵品质
lateral control surface 横向操纵面
lateral deviation 横向偏差
lateral dimension 横向尺寸
lateral direction 横向
lateral dynamics 横向动力学
lateral error of traverse 导线横向误差
lateral excursion 侧移动
lateral extent 横向伸展
lateral force 横向力,侧力
lateral force coefficient 侧力系数
lateral homing depth charge 水平自动寻的深水炸弹
lateral jet 喷流横向
lateral load 横向负荷
lateral migration 侧向迁移
lateral mode 横模
lateral motion 横向运动
lateral oscillation 横向振荡
lateral overlap 横向重叠
lateral parasitic transistor 横向寄生晶体管
lateral separation 横向分离
lateral spread 横向散布
lateral stability 横向稳定性
lateral steering 横向导引
lateral stiffness 侧向刚性
lateral tell 横向信息传送器
lateral tilt 横向倾斜
lateral touchdown 侧向着地,外侧触地
lateral trajectory divergence rocket 侧向轨道发散火箭
lateral transport 侧向运输
lateral velocity 横向速度
lateral vibration 横振动
lateral-directional motion 横侧运动
late-type galaxy 晚型星系
late-type star 晚型星
lathe 车床
latin hypercube 拉丁超立方
latin square 拉丁方
latitude 纬度
latitude effect 纬度效应
latitude of reference 基准纬度
latitude rate 纬度率
latitude variation 纬度变化
latitude-longitude grid 经纬度网格
LATL Lateral 横向的
lattice 格子
lattice algorithm 格点算法
lattice constant 晶格常数
lattice curvature 点阵曲率
lattice defect 晶格缺陷
lattice filter 格型滤波器
lattice fin 栅格翼
lattice match 晶格匹配
lattice mismatch 晶格失配
lattice orientation 晶向
lattice plane 晶面
lattice structure 晶格结构
latus rectum 正焦弦
launch 发射,投放(鱼雷等),(水上飞机)下水
launch abort guide simulation 发射中断制导模拟
launch angle 发射角度
launch area 发射区
launch azimuth 发射方位角
launch command and control center 发射指挥控制中心
launch commit criteria document 发射实施标准文件
launch complex 发射设施
launch complex equipment 发射综合场地设备
launch control and monitoring system 发射监控系统
launch control unit 发射控制装置
launch cost 发射费用
launch data 发射诸元
launch day 发射日
launch drill 发射演练
launch enable system 发射防误系统
launch environment 发射环境
launch escape 发射逃逸
launch escape subsystem 发射逃逸分系统
launch escape tower 发射逃逸塔
launch experiment outline 发射试验大纲
launch guidance 发射段制导
launch leading team 发射领导组
launch load 发射载荷
launch mode 发射方式
launch monitor equipment 发射监控设备
launch month 发射月
launch operating team 发射操作队
launch operations control center 发射操作控制中心
launch opportunity 发射机会
launch order 发射顺序
launch pad 发射台
launch pad rail 发射台导轨
launch pad support 发射台支架
launch pad transportation 发射台折倒臂
launch plan network chart 发射计划网络图
launch point 发射点
launch preparation time 发射准备时间
launch range 发射距离

launch rate 发射速度
launch readiness review 发射准备状态评审
launch release mechanism 发射释放机构
launch reserve scheme 发射预案
launch sequence control 发射顺序控制
launch site 发射场
launch site for manned space flight 载人航天发射场
launch site television monitor system 发射场电视监视系统
launch system 发射系统
launch tower 发射塔
launch tube 发射管
launch umbilical tower 发射操纵缆塔
launch vehicle 运载火箭
launch vehicle assembly and test building 运载火箭装配测试厂房
launch vehicle coordinate 箭体坐标
launch vehicle environmental test 运载火箭环境试验
launch vehicle guidance simulation 运载火箭制导模拟，运载火箭导航模拟
launch vehicle horizontal sling 箭体水平吊具
launch vehicle iron wheel carriage 箭体铁轮支架车
launch vehicle rail transporter 箭体铁路运输车
launch vehicle railway platform truck 箭体铁路运输车
launch vehicle telemetry 运载火箭遥测
launch vehicle to launch satellites 卫星发射运载工具
launch vehicle trailer 箭体公路运输车
launch vehicle transfer 运载火箭转运
launch vehicle turning sling 箭体翻转吊具
launch vehicle verticality 火箭垂直度
launch window 发射时限
launch window beginning 发射窗口前沿
launch window ending 发射窗口后沿
launch working team 发射工作队
launcher 发射装置
launcher barrel 发射筒
launching and control center 发射控制中心
launching coordinate system 发射坐标系
launching direction 发射方向
launching level ground 发射场坪
launching pad leveling mechanism 发射台调平机构
launching phase 发射阶段
launching phase guidance 起飞段制导，发射段制导
launching position indicator 发射位置指示器
launching reliability 发射可靠度
launching service support 发射勤务保障
launching service tower 发射勤务塔
launching shock 发射冲击
launching site 发射场地

launching success rate 发射成功率
launching test 发射试验
launching time 发射时间
launching workplace 发射工位
LAV 1. Lavatory 盥洗室 2. Least Absolute Value 最小绝对值 3. Light Armored Vehicle 轻型装甲飞行器
Laval nozzle 拉瓦尔喷管
law of area 面积定律
law of cosine 余弦定律
law of mass action 质量作用定律
law of motion 运动定律
law of superposition 叠覆律，层序律
law of thermodynamics 热力学原理
lawrencium 铹
LAWS Light Aircraft Warning System 轻型飞机警报系统
Lawson criterion 劳森判据
lay 加工痕迹方向，绞
lay shaft 中间轴，副轴
laydown 沉淀作用
laydown bombing 沉积轰炸，极低空轰炸
layer 层
layer thickness 层厚度
layer type 层型
layer-built dry cell 叠层干电池
layered structure 层次结构
layered system 层状体系
layout 草案，平面图
layout design automation system 自动布图设计系统
layout rule check 布图规则检查
layover 中途短暂的停留
layup 叠层
LBA 1. Logical Bus Application 逻辑汇流条应用 2. Luftfahrt Bundes Amt 德国民航管理机构
LBI Low-Band Interrogator 低波段询问机
LBL Label 标签
LBM Locator Back Marker 背台示位信标
LBNP Lower Body Negative Pressure 下体负压
LBO Landing Burn Off 着陆耗油
LBR 1. Local-Base Rescue 本场救援 2. Low Bit Rate 低位速率 3. Lumber 木材
LBRV Low Bit Rate Voice 低比特率语音
LBS 1. LAN Bridge Server 局域网桥服务器 2. Laser Bombing System 激光轰炸系统 3. Lateral Beam Sensor 横向波束传感器
LBTY Liberty 自由
LC 1. Landing Chart 着陆图 2. Labour Cost 人工费用 3. Leased Channel 租用信道 4. Letter of Contract 合同函件 5. Life Cycles 生命周期 6. Low Carbon 低碳

7. Lower Comparment 下货舱 8. Load Control 载荷控制

LCB Little Change Board 小更改委员会

LCC 1. Landing Control Center 着陆控制中心 2. Launch Control Center 发射控制中心 3. Leased Circuit Connection 租用电路连接 4. Life Cycle Cost 寿命周期费用 5. Line Concentration Controller 集线控制器 6. Link Controller Connector 链路控制连接器

LCCA 1. Lateral Central Control Actuator 横向中央控制作动器 2. Life Cycle Cost Analysis 寿命周期成本分析

LCC-CTR Liaison Change Commitment Center 联络更改提交中心

LCCR Liaison Change Commitment Record 联络更改提交记录

LCCS Large Capacity Core Storage 大容量磁心存储器

LCD 1. Liquid Crystal Detector 液晶检测器 2. Liquid Crystal Diode 液晶二极管 3. Liquid Crystal Display 液晶显示器

LCE Link Control Entity 链路控制实体

LCF 1. Large Cargo Freighter 大型货物运输 2. Launch Control Facility 发射控制设施 3. Local Control Facility 局域控制设施 4. Local Currency Fare 当地货币票价 5. Low Cyclic Fatigue 低循环疲劳损伤

LCFS Last Come First Served 后来先服务

Lch 1. Latch 锁 2. Logical Channel 逻辑信道

LCHG Latching 锁住

LCHR Launcher 发射器

LCIA Local Check-In Assistant 当地值机助理系统

LCIP Load Compressor Inlet Pressure 负载压缩机进气口压力

LCIT Load Compressor Inlet Temperature 负载压缩机进气口温度

LCL 1. Line Check List 航线检修清单 2. Local 当地 3. Lower Control Limit 下控制界限

LCLV Low Cost Launch Vehicle 低成本发射飞行器

LCM 1. Large Capacity Memory 大容量存储器 2. Life Cycle Manager 生命周期经理 3. Line Control Memory 线路控制存储器 4. Logic Control Module 逻辑控制模块,逻辑控制组件

LCMR Liaison Change Management Record 联络更改管理记录

LCN 1. Load Classification Number 载荷等级编号 2. Local Communications Network 局域通信网络

LCOS Line Class Of Service 线路业务等级

LCOSS Lead Computing Optical Sight System 前置角计算光学射击瞄准系统

LCP 1. Laboratory Control Procedure 实验室控制程序 2. Lighting Control Panel 导光控制板 3. Lighting & HIRF Company Policy 公司闪电和高强辐射场政策 4. Link Control Protocol 链路控制协议 5. Local Control Panel 就地控制柜

LCPOF Large Core Plastic Optical Fiber 大芯径塑料光纤

LCPT Light Cargo Passenger Turboprop 轻型客货两用涡轮螺桨飞机

LCS Longitudinal Control System 纵向操纵系统

LCSTB Low Cost Simulation Testbed 低成本模拟试验台

LCSW Latch Checking Switch 自锁检查开关

LCT Locate 定置,固定,控制

LCU Light Control Unit 灯光控制装置

LCV 1. Load Control Valve 负载控制阀 2. Logic Control Variable 逻辑控制变量

LCXT Large Cosmic X-ray Telescope 大型宇宙X射线望远镜

LCZR Localizer 着陆航向信标台

LD 1. Leading 领导的,导向的 2. Logic Diagram 逻辑电路示意图

LDA 1. Landing Distance Available 可用着陆距离 2. Loading Advice 装载通知

LDAP Light Weight Directory Access Protocal 轻量级目录访问协议

LDAR Liaison-Design Action Request 联络设计动作需求

LDASGN Loading Assignment 装载分配

LDB Limited Data Blocks 有限数据程序块

LDBLC Low Drag Boundary Layer Control 减阻附面层控制

LDBRK Load Breakdown 装载分配

LDC 1. Local Door Controller 区域门控制器 2. Long Distance Call 长途电话 3. Low Density Control 低密度交通管制

LDDC London Docklands Development Corporation 伦敦多克兰开发公司

LDDS Low Density Data System 低密度数据系统

LDG 1. Landing 着陆 2. Loading 加载

LDG GR Landing Gear 起落架

LDGPS Local Area Differential Global Positioning Satellite 局域差分全球定位卫星

LDM Load Message 载量电报

LDMCR Lower Deck Mobile Crew Rest 下舱灵活机组休息室

LDMI Laser Distance Measuring Instrument 激光测距仪

LDMN Leadman 领导者

LDMX Local Digital Message Exchange 本地数字报文交换

LDNS Laser Doppler Navigation System 激光多普勒导航系统

LDO 1. Lease-Development-Operate 租赁, 开发, 经营 2. Limited-Duty Officer 有限勤务军官

LDOC Long Distance Operational Control 远距运行控制

LDP 1. Label Distribution Protocol 标记分配协议 2. Language Data Processing 语言数据处理 3. Laser Designator Pod 激光照射器吊舱

LDPA Landing Performance Application 着陆性能应用

LDR 1. Ladder 梯子 2. Landing Distance Required 所需着陆距离 3. Loader 装货器, 装换手,【计】输入程序 4. Low Data Rate 低数据传输速率

LDS 1. Local Digital Switch 本地数字交换机 2. Local Distribution System 本地分配系统 3. Load Sharing 负荷分担

LDSL Low Bit-Rate Digital Subscriber Line 低比特率数字用户线

LDT 1. Linear Differential Transformer 线性差动变压器 2. Long Distance Telephone 长途电话

LE Leading Edge 前缘

LEA 1. Launcher Electronics Assembly 发射器电子设备 2. Long-Endurance Aircraft 续航时间长的飞机

lead 引线, 螺纹导程, 螺距, 电线头, 导管

lead accumulator 铅蓄电池

lead acid battery 铅酸蓄电池

lead aircraft 长机

lead angle 升角, 导程角

lead azide 氮化铅

lead bonding 引线焊接

lead code 前导码

lead collision attack distance 拦截攻击距离

lead collision course 提前角碰撞路线

lead compensation 超前补偿

lead computing optical sight 计算提前角的光学瞄准

lead down 减少提前量, 减少前置量

lead explosive 导爆药柱

lead firing 前置射击, 有提前角的射击

lead frame bond 引线框式键合

lead network 超前网络

lead pursuit approach 沿追踪曲线接近

lead storage battery 铅蓄电池

lead time 提前时间

lead zirconate titanate 锆钛酸铅

lead zirconate titanate ceramics 锆钛酸铅陶瓷

lead-acid storage battery 铅酸蓄电池

lead-angle guided missile 按前置法引导的导弹

lead-bias control 超前偏置控制

leaded fuel 含铅燃油

leader cable 主电缆, 引线电缆

lead-in 天线引入线

leading beacon 导标

leading edge 前缘, 前沿, 上升边

leading edge droop 前缘下垂

leading edge flap 前缘襟翼

leading edge notch 前缘缺口

leading edge radius 前缘半径

leading edge sawtooth 前缘锯齿

leading edge slat 前缘缝翼条

leading edge suction 前缘吸力

leading fire 前置射击

leading launching 提前发射, 前置发射

leading petty officer 上士

leading sunspot 前导黑子

leading-edge flap 前缘襟翼

leading-edge slat 前缘缝翼

leading-edge vortex 前缘涡流

lead-lag hinge 超前-滞后铰接

lead-lag motion 超前滞后运动

LEAF Large Effective-Area Fiber 大有效面积光纤

leak 渗漏, 泄漏

leak detection 泄漏探测

leak detector 泄漏检录器

leak localization 泄漏定位

leak measurement 泄漏测量

leak rate 泄漏率

leak test 检漏试验

leak tightness 密封性, 密闭度

leakage 泄漏, 分散

leakage detector 检漏器

leakage flow 泄流

leakage inductance 漏感

leakage magnetic flux 漏磁通

leakage noise 直漏干扰

leakage power 漏过功率

leakage rate 漏网率

leakage test 密封试验

leakage vortex 泄漏涡

leakage-check test 检漏试验

leaking mode 波能漏失

leaky coaxial cable 漏泄同轴电缆

leaky mode 泄漏

lean 贫油的

lean body weight 无脂体重

lean combustion 稀薄燃烧

lean mixture 贫油混气

leans 倾斜

leap day 闰日

leap month 闰月
leap second 闰秒
leap year 闰年
leapfrog 跳步法
learning control 学习控制
learning machine 学习机
learning matrix 学习矩阵
learning method 学习方法
learning process 学习过程
learning rate 学习率
learning system 学习系统
lease 出租,租赁
lease back 将财产产权出售
leasecraft 租赁飞机
leased network service 租用网络业务
leased private channel 租用专用信道
leased-line network 租用专线通信网
least cost input 最小成本投入
least square error approximation 最小二乘差逼近
least square method 最小二乘法
least squares correlation 最小二乘相关
least squares criterion 最小二乘准则
least squares estimation 最小二乘估计
least-square 最小二乘方
LED 1. Leading-Edge Devices 前缘增升装置 2. Light-Emitting Diode 发光二极管
LEED Low Energy Electron Diffraction 低能电子衍射
leeward side 背风面
LEF Leading Edge Flap 前缘襟翼
LEFM Linear Elastic Fracture Mechanics 线性弹性断裂力学
LEFT Left Turn 左转
left aileron 左副翼
left characteristic 左特性
left luggage 寄存行李
left seat 左侧座椅
left wing 左翼,左边锋
left-hand corner 左转
left-hand rotation 左旋
left-hand rotation propeller 左旋螺桨
leg belt 腿带
leg negative pressure 下肢负压
leg restraint 腿部固定位置
legal metrology 法制计量学
legal time 法定时
legal unit of measurement 法定计量单位
legend 图例
Legendre function 勒让德函数
Legendre polynomials 勒让德多项式

LEM Lunar Excursion Module 登月舱
LEMAC Leading Edge of Mean Aerodynamic Chord 平均空气动力弦前缘,前缘平均气动弦
lemma 辅助定理
length of command code 指令长度
length of real aperture 真实孔径长度
length of synthetic aperture 合成孔径长度
length overall 全长,总长
length scale 长度规
length unit 长度单位
length-to-diameter 长度直径比,长径比
lens 透镜
lens antenna 透镜天线
lens coating 透镜镀膜,透镜涂层
lens distortion 透镜畸变
lens shutter 透镜光闸,镜头快门
lenticular 双凸透镜状的
lenticular galaxy 扁豆状银河
LEO Low Earth Orbit 近地轨道
LER Leading Edge Radius 前缘半径
LES 1. Land Earth Station 陆地地球站 2. LAN Emulation Server 局域网仿真服务器 3. Leading Edge Slat 前缘缝翼 4. Leading Edge Station 前缘站位
LESS 在较小的程度上
lessee 承租人
Lesser Fullness 小满
lessor 出租人
let-down vessel 放压容器
lethal agent 致死剂
lethal area 杀伤面积
lethal concentration 致死浓度
lethal dose 半致命剂量
lethal gas 致命性毒气
lethal index 致命指数
lethal zone 杀伤区
lethality 杀伤力
letter of intent 意向书
LEU Leading Edge Up 前缘收上
LEV Lever 杠杆,连杆
LEV DET Level Detector 油量传感器,电平检测器
level 水准仪,水平仪
level bombing 水平轰炸
level controller 位面控制器
level curve 阶层曲线
level detector 水平测试仪
level flight 水平飞行
level flight ceiling 平飞升限
level flight indicator 水平飞行指示器
Level Flight Segment 平飞航段

level flight speed	平飞速度
level indicator	液位指示器
level meter	电平表
level of abstraction	抽象程度
level of automation	自动化程度
level of free convection	自由对流高度
level set	水平集
level straight acceleration flight method	水平直线加速飞行法
level straight deceleration flight method	水平直线减速飞行法
level surface	水准面
level switch	水平开关
level transmitter	能级传送级
level tube bubble	水准管起泡
level turn	水平转弯
level-based system	分层系统
leveling error	调平误差
leveling line	水准路线
leveling of model	模型置平
leveling origin	水准原点
leveling staff	水准尺
levelling	水平测量
level-off	水平飞行位置
level-out	拉平高度
level-setting resistor	电平调整电阻器
level-shifting diode	电平漂移二极管
lever	杆,杠杆
lever arm	杠杆臂
leverage	杠杆率
leveraged lease	融资租赁
levered suspension	杠杆式悬置
levered suspension landing gear	摇臂式起落架
levitate	漂浮,悬浮
LEVL	Leading-Edge Vortex Lift 前缘涡升力
Lewis Number	刘易斯数
LEX	Leading-Edge Extension 前缘突出部,前缘锯齿
Lexan	勒克森聚碳酸酯纤维
LF	1. Laser Finder 激光探测器 2. Left Front 左前 3. Line Feed 换行 4. Load Factor 装载系数 5. Low Frequency 低频
LFA	1. Landing Fuel Allowance 着陆燃油许可量 2. Latching Fault Annunciation 未上锁警告,扣锁失灵信号 3. Low-Flying Area 低空飞行区 4. Loss of Frame Alignment 帧失步
LFAS	Low-Frequency Active Sonar 低频主动声呐
LFC	Laminar Flow Control 层流控制
LFF	Load Factor for Flight 飞行载荷因数,飞行载运比率
LFICS	Landing Force Integrated Communications System 登陆部队完整通信系统
LFL	Landing Field Length 着陆场长
LFLM	Lowest Field Level of Maintenance 外场维修最低级别
LFO	Low Frequency Oscillator 低频振荡器
LFR	Low Frequency Range 低频范围
LFRJ	Liquid Fuelled Ramjet 液体燃料冲压喷气发动机
LFRR	Low Frequency Radio Range 低频无线电指标
LFT	Lifting 升力
LFSMS	Logistic Force Structure Management System 后勤保障部队结构管理系统
LFV	Lunar Exploration Flying Vehicle 月球探测飞行器
LG	1. Landing Gear 起落架 2. Laser Gyro 激光陀螺 3. Length 长度 4. Logic Gate 逻辑门
LGA	Low Gain Antenna 低增益天线
LGB	Laser Guided Bomb 激光制导炸弹
LGC	1. Landing Gear Cable 起落架钢索 2. Logic 逻辑
LGCIS	Landing Gear Control Indication System 起落架控制指示系统
LGCIU	Landing Gear Control and Interface Unit 起落架控制和接口组件
LGD	Landing Gear Doors 起落架系统
LGE	1. Large 大的 2. League 团,同盟
LGEC	Lunar Geological Exploration Camera 月球地质勘探摄影机
LGERS	Landing Gear Extension and Retraction System 起落架收放系统
LGM	Logistics Module 后勤舱
LGMS	Landing Gear Management System 起落架管理系统
LGP	Local Galley Panels 本地厨房面板
LGPI	Landing Gear Position Indicator 起落架位置指示器
LGS	1. Landing Gear System 起落架系统 2. Landing Guidance System 着陆引导系统
LGSC	Linear Glid-Slope Capture 保持直线下滑道
LGSP	Landing Gear Sampling Plan 起落架抽样方案
LGT	1. Light 轻的 2. Lighted 有照明的 3. Landing Gear Tread 起落架轮距
LGW	Landing Gross Weight 降落全重
LGWSS	Landing Gear Well Surveillance System 起落架舱监视系统
LH	1. Labour Hour 工时 2. Left Hand 左手方向 3. Left Hydraulic 左液压系统 4. Light Helicoper 轻型直升机
LHA	Landing Helicopter Assault 直升机机降攻击
LHCP	Left-Hand Circular Polarization 左旋圆极化

LHT　Long-Haul Transoceanic　越洋长途通信
LHTE　Light Helicopter Turbine Engine　轻型直升机涡轮发动机
LHV　Lower Heating Value　低热值
LI　1. Letter of Intent 意向书 2. Limit Indicator 极限指示器 3. Link Interface 链路接口 4. Load Index 载量指数 5. Low Intensity 低强度
LIA　Liaison　联络
liability　负债,债务
liability convention　责任公约
liaison　联络
library graphics　图形库
library program　程序库
libration　天平动
libration damping　天平动阻尼
libration in latitude　纬度摆动
libration in longitude　经秤动
libration point　天平动点
librational motion　天平动
librium　利眠宁
LIC　1. Licence 执照,许可证,特许 2. Line Interface Card 线路接口插件板 3. Low-Intensity Conflict 低强度冲突 4. Lowest Incoming Channel 最低入局信道
license agreement　许可证协议
LID　Link Interface Device　链路接口设备
Lidar　Light Detection And Ranging　光探测与测距
Lie series　【天】李级数
lieutenant commander　少校
LIFE　Linear Integrated Flight Equipment　线性综合飞行设备
life　寿命
life cycle　生活周期
life cycle cost　全寿命期费用
life cycle cost analysis　寿命周期费用分析
life extension　寿命延长
life jacket　救生衣
life profile　寿命剖面
life saver　救生圈
life saving apparatus　救生器具
life saving equipment　救生设备
life saving waistcoat　救生背心
life scatter factor　寿命分散系数
life science　生命科学
life support　生命保障
life support system　生命支持装置
life support system manager　生命保障系统处长
life support subsystem　生命支持子系统
life test　寿命试验
life test information　寿命试验信息

life unit　寿命单位
life vest　救生衣,救生背心,救生防护衣
life vest stowage　救生背心存放处
lifeboat　救生艇
life-cycle cost　寿命周期成本
life-expectancy analysis　寿命预计分析
life-limit element　到寿件
liferaft　救生船
liferaft light marker　充气救生船灯光标志
life-support equipment　生命保障设备
lifetime　寿命
lifetime extension　延寿
LIFO　Last In First Out　后进先出
lift　升降机,电梯
lift augmentation　升力增大,增升
lift characteristic　升力特性
lift coefficient　升力系数
lift curve　升力曲线倾斜
lift damper　减升板
lift distribution　提升分布
lift dumper　减升板
lift engine　升力发动机
lift enhancement　增升
lift fan　升力风扇
lift force　升力
lift interference　升力干扰
lift line theory　升力线理论
lift motor　电梯用电动机
lift off　起飞
lift off speed　离地速度
lift surface theory　升力面理论
lift vector　升力矢量
lift wire　升力张线
lift-drag ratio　升阻比
lift-drag ratio of reentry vehicle　再入飞行器升阻比
lifting body　升力体
lifting cage　吊篮
lifting cage sling　吊篮吊具
lifting car　吊篮
lifting effect　举升效果
lifting line　升力线
lifting re-entry　利用升力再入
lifting rotor　升桨
lifting surface　升力面,浮升面
lifting system　升力产生装置
lifting table　平行升降台
lifting-body　升力体再入
lifting-line theory　升力线理论
lifting-surface theory　升力面理论

lift-off claming strip 起飞压板
lift-off contact 起飞触点
lift-off drift 起飞漂移
lift-off mass 起飞质量
lift-off support plate 起飞托盘
lift-off technology 剥离技术
lift-off zero 起飞零点
lift-to-drag 升力阻力比,升阻比
ligament 系带
light adaptation 光适应
light airborne ASW vehicle 轻型机载反潜火箭
light aircraft 轻型飞机
light alloy 轻合金
light annealing 光亮退火
light antitank weapons 轻型反坦克武器
light assault weapon 轻型攻击武器
light atomic ordnance 轻型原子武器
light beam 光束
light curve 发光曲线
light deflection 光偏转
light emission 光发射
light equation 光变方程
light gas 轻瓦斯气体
light gas gun 轻气炮
light guide 光导航
light gun 光枪
light homer 光寻的头
light homing guidance 光寻的制导
light homing head 光自动寻的头
light house lamp 灯塔灯
light intensity 光强度
light modulation 光调制
light modulator 调光器
light navigation 灯标导航
light pen 光笔
light permeable clothing 轻透气服
light pipe 光管
light pressure 光压
light product 轻产品
light radiation 光辐射
light range 灯光照距
light ray 光线
light scattering 光散射
light self-trapping 光自陷
light sensitivity 光敏度
light sheet 薄钢皮
light signal 光信号
light snow 小雪
light source 光源

light time 光时
light transmissivity 透光性
light turbulence 弱湍流
light valve 光阀
light warning radar 轻型警戒雷达
light weight 车辆自重,空车自重
light year 光年
light-attack tactical navigation and bombing system 轻型攻击战术导航与轰炸系统
lightbulb 电灯泡
light-case bomb 薄壳炸弹
lightcraft 光船
light-dark cycle 明暗周期
light-day 光日
light-emitting diode(LED) 发光二极管
lightening hole 减重孔
lighting condition 照明条件
lightly doped drain technology 轻掺杂漏极技术
lightmeter 照度计
lightning 闪电
lightning strike 雷击,闪电式罢工
lightning test 雷击试验
lightoff 点火
lightplane 轻型飞机
Lightsat 轻量卫星
light-scattering 光散射
light-spot 光点
light-wave guidance 光波制导
lightweight all weather missile 轻型全天候导弹
like-impinging injector element 自击式喷嘴
likelihood 可能,可能性
likelihood estimation 似然估计
likelihood function 似然函数
likelihood method 相似法
likelihood ratio 似然比
LIL Light Intensity Low 低强度灯
LIM 1. Laser Intensity Monitor 激光强度监测器 2. Light Intensity Medium 中强度灯 3. Limit 极限,极限尺寸 4. Limousine Service 客车服务 5. Locator Inner Marker 近示位信标和指点标
LIM SPD Limited Speed 限制速度
LIM SW Limit Switch 极限开关,终点开关
LIMAC Large Integrated Monolithic Array Computer 大规模集成单片阵列计算机
limacon 蚶线
limb 分度弧
limb brightening 临边增亮
limb darkening 临边昏暗
limb restraint 四肢约束装置

limb scanning method 临边扫描法
limb-scanning pressure modulated radiometer 临边扫描压力调制辐射仪
limit alarm 极限报警器
limit angle of ricochet 跳弹极限角
limit cycle 极限环
limit error 极限误差
limit load 极限载荷
limit of fire 射击安全角
limit of proportionality 比例极限,屈服点
limit state 极限状态
limit switch 限制开关
limit value 极限值
limit velocity 极限速度
limit-cycle oscillation 极限环振动
limited life item 有限寿命产品
limited torque motor 有限力矩电动机
limiter 限幅器
limiting 限制
limiting accuracy machining 极限精度加工
limiting amplifier 限幅放大器
limiting condition 极限条件
limiting dynamic pressure 限制动压
limiting exposure 极限曝光时间
limiting line of submerged approach 潜艇接近限度线
limiting load 极限负载
limiting Mach number 限制马赫数
limiting magnitude 极限星等
limiting quality 极限质量
limiting resolution 极限分辨率
limiting speed 极限速度
limiting velocity 极限速度
LIN 1. Linear 线性的 2. Liquid Nitrogen 液态氮
LIN FT Linear Feet 直线英尺
Linac 直线加速器
linchpin 销
line 行,行列
line broadening 谱线变宽
line concentrator 用户集线网,线路集中器
line core 线核
line defect 线缺陷
line displacement 线位移
line feed 换行
line filter balance 线路滤波器平衡
line fitting 线路配件
line identification 谱线证认
line inspection 线路检查
line inversion 线反演
line loop 线路回路

line loss 线路损耗
line maintenance 航线维修
line microphone 线列传声器
line of aim 瞄准线
line of apsides 拱点线
line of arrival 弹着线
line of bomb release 投弹线
line of constant Doppler shift 等多普勒频率线
line of declination 赤纬圈
line of fall 落地线
line of impact 击中线
line of nodes 交点线
line of position 位置线,定位线
line of right ascension 赤经圈
line of separation 分离线
line of sight 瞄准线,视线
line of sight propagation 视距传播
line of sight rate 视线角速度
line of tension 应力线
line output transformer 行输出变压器
line paire per millimeter 每毫米线对数
line pairs 线对
line pressure 线压,管线压力
line profile 谱线轮廓
line reconstruction 行重构
line relaxation method 逐行松驰法
line replaceable unit 航线可换组件
line scanning 行扫描
line search 线搜索
line segment 线段
line service 航线服务
line shape 线形
line smoothing 线光顺
line source 线源
line speed 线路速度
line splitting 谱线分裂
line spread function 线扩展函数
line squall 线飑
line strength 谱线强度
line stretcher 延伸线
line symbol 线状符号
line transformer 线间变压器
line vortex 线状漩涡
line width 线宽
lineage structure 嵌晶结构脉理构造
linear ablative rate 线烧蚀率
linear acceleration 线加速度
linear accelerometer 线性加速度计
linear actuator 线性执行机构

linear analysis   线性分析
linear approximation   线性近似,线性逼近
linear array   线性阵
linear array detector   线阵探测器
linear array scan   直线阵列扫描
linear bearing   直线轴承
linear block code   线性分组码
linear burning rate   线性燃烧速率
linear cascade   平面叶栅
linear code   线性码
linear combination   线性组合
linear constraint   线性约束
linear control   线性控制
linear control system   线性控制系统
linear control theory   线性控制理论
linear controller   线性控制器
linear crosstalk   线性串音
linear deceleration   直线减速度
linear dependence   线性相关
linear detection   线性检波
linear difference equation   线性差分方程
linear displacement   线位移
linear displacement gauge   线性位移测量计
linear elastic fracture mechanics   线弹性断裂力学
linear element   线性元件
linear equalizer   线性均衡器
linear equation   线性方程
linear error   线性误差
linear filter   线性滤波器
linear filtering   线性滤波
linear FM   线性调频
linear FM pulse compression technique   线性调频脉冲压缩技术
linear formation   一字队形
linear function   线性函数,一次函数
linear gate   线性门
linear graded junction   线性缓变结
linear growth rate   线性生长速度
linear independence   线性无关
linear inequality   一次不等式
linear integrated circuit   线性集成电路
linear interpolation   线性内插
linear intersection   边交会法
linear load   直线力,单位长度荷载
linear mapping   线性映象
linear matrix   线性矩阵
linear model   线性模式,线性模型
linear modulation   线性调制
linear momentum   线性动量

linear motion   直线运动
linear motion electric drive   直线电气传动
linear motion valve   直行程阀
linear network   线性网络
linear parameter-varying   线性变参数
linear perturbation   线性扰动
linear phase filter   线性相位滤波器
linear plant   线性设备
linear polarization   线极化
linear polynomial   线性多项式
linear prediction   线性预测
linear predictive coding   线性预测编码
linear program   线性规划,线性程序
linear quadratic Gaussian problem   高斯线性二次型问题
linear quadratic regulator problem   线性二次型调节器问题
linear rate   线速度
linear region   线性区
linear regression   线性回归
linear relationship   线性关系
linear response   线性响应
linear revolver   线性旋转变压器
linear scanning   线性扫描
linear shaped charge   线形装药
linear sliding potentiometer   直滑电位器
linear smoothing   线性平滑
linear solution   线性解
linear state-space   线性状态矢量空间
linear sweep rate   线性扫描率
linear system   线性系统
linear theory   线性理论
linear time-invariant   线性时不变
linear time-varying   线性时变
linear triangulation chain   【测】线形锁
linear triangulation network   【测】线形网
linear velocity transducer   线性速度传感器
linear vibration   线性振动
linear viscous damping   线性黏性阻尼
linear-angular intersection   边角交会法
linearity   线性,直线性
linearization   线性化
linearization control   线性化控制
linearization family   线性化族
linearization technique   线性化方法
linearize   直线化,线性化
linearized theory   线性化理论
linearizer   线性化电路
linearizing circuit   线性化电路
lineman   巡线工

lineman climbers 线路工脚口
line-of-sight 视线
line-of-sight angle 视线角
line-of-sight rate 瞄准线移动角速度
line-of-sight transmission 视距离信号传输
line-oriented flight training 航线熟悉飞行训练
liner 衬垫
lineshape 谱线形状
linewidth 线宽,谱线宽度
linguistic variable 语言变量
lining 衬里
link 通信信道,连接
link belt 链带
link budget 链路设计
link ejector 除链机
link exchange 链路交换
link guide 输链导
link layer 连接层
link loss 链路损耗
link management 链路管理
link quality 链路质量
link trainer 环状训练舱
linkage 连接,交连
linkage computer 联动计算机
linked ammunition 装链接的弹药
linked ejection 联动弹射
linked list 连接表
linker 装链器
linker-delinker 装链器
linking 装弹链
linking up 操纵杆加速操作
linoleum 漆布,亚麻油毡
LINS Laser Inertial Navigation System 激光惯性导航系统
lip 刀刃
lip seal 唇式密封
LIPA List of Interchangeable Parts and Assemblies 互换零件清单
Lippmann holography 李普曼全息术
LIPS Laser Image Processing Scanner 激光图像处理扫描器
LIQ Liquid 液体,液态
liquefaction 液化
liquefy 液化
liquid air 液态空气
liquid atomization 液体雾化
liquid bearing nozzle 液浮喷管
liquid chromatograph 液色谱法
liquid column 液柱

liquid control valve 液体控制阀
liquid cooled engine 液冷发动机
liquid cooled suit 液冷服
liquid cooling 液冷却
liquid cooling garment 液冷服
liquid core 液相穴
liquid crystal 液晶
liquid crystal display 液晶显示器
liquid crystal shutter 液晶开关
liquid drop 液滴
liquid dynamics 液体动力学
liquid encapsulation technique 液封技术
liquid entrainment 液体雾沫
liquid explosive 液体炸药
liquid filament 细水流量
liquid film 液膜
liquid film cooling 膜冷却
liquid film thickness 液膜厚度
liquid floated gyroscope 液浮陀螺
liquid floated pendulous accelerometer 液浮摆式加速度计
liquid flow rate 液体流量
liquid fluorine 液氟
liquid flux 液体溶剂
liquid fraction 含液率
liquid fuel 液体燃料
liquid helium 液态氦
liquid hydrocarbon 液烃
liquid hydrogen 液氢
liquid hydrogen fill-drain auto-disconnect coupler 液氢加泄自动脱落连接器
liquid hydrogen fill-drain auto-disconnect coupler support mount 液氢加泄自动脱落连接器支架
liquid hydrogen fueled aircraft 液氢动力飞机
liquid hydrogen loading connector fitting 液氢加注连接器接头
liquid hydrogen loading controller 液氢加注控制机
liquid hydrogen loading liquid line system 液氢加注液路系统
liquid hydrogen loading measuring and control system 液氢加注测控系统
liquid hydrogen loading microcomputer station 液氢加注微机站
liquid hydrogen loading monitoring system 液氢加注监测系统
liquid hydrogen loading system 液氢加注系统
liquid hydrogen loading valve checking ladder 液氢加注活门测试工作梯
liquid hydrogen pump 液氢泵

liquid hydrogen railway loading vehicle 液氢铁路加注运输车
liquid hydrogen tank 液氢箱
liquid inertia vibration eliminator 液体惯性减振器
liquid injection thrust vector control 液体喷射推力向量控制
liquid interface 液体界面
liquid jet 液体喷射
liquid laser 有机染料激光器
liquid level control 液面控制
liquid level manometer 液位压力计
liquid lithium 液体锂
liquid loading 井底积液
liquid meniscus 液体弯月面
liquid metal 液态金属
liquid metal forging 横压铸造
liquid metal fuel cell 液态金属燃料电池
liquid monopropellant 液态单元推进剂
liquid nitrogen 液氮
liquid nitrogen distribution system 液氮用气系统
liquid nitrogen loading and topping vehicle 液氮加注补加车
liquid nitrogen loading liquid line system 液氮加注液路系统
liquid nitrogen loading measuring and control system 液氮加注测控系统
liquid nitrogen loading system 液氮加注系统
liquid nitrogen test 液氮试验
liquid oxygen 液氧
liquid oxygen apparatus 液氧装置
liquid oxygen container 液氧瓶
liquid oxygen converter 液氧转换器
liquid oxygen fill-drain auto-disconnect connector 液氧加泄自动脱落连接器
liquid oxygen fill-drain auto-disconnect connector support mount 液氧加泄自动脱落连接器支架
liquid oxygen fill-drain gas distribution board 液氧加泄配气台
liquid oxygen ground gas distribution system 液氧地面用气系统
liquid oxygen jet 液氧喷射
liquid oxygen loading and topping truck 液氧加注补加车
liquid oxygen loading connector fitting 液氧加注连接器接头
liquid oxygen loading controller 液氧加注控制机
liquid oxygen loading liquid line system 液氧加注液路系统
liquid oxygen loading measuring and control system 液氧加注测控系统
liquid oxygen loading system 液氧加注系统
liquid oxygen loading test-control desk 液氧加注控制台
liquid oxygen nitrogen distribution board 液氧氮配气台
liquid oxygen pump 液氧泵
liquid oxygen storage tank 液氧固定贮罐
liquid oxygen supercooler 液氧过冷器
liquid oxygen tank 液氧箱
liquid oxygen/kerosene rocket engine 液氧煤油火箭发动机
liquid oxygen/liquid hydrogen rocket engine 液氧液氢火箭发动机,氢氧火箭发动机
liquid phase 液相
liquid phase epitaxial growth 液相外延生长
liquid phase epitaxy 液相外延
liquid post 液柱
liquid potassium 液钾
liquid pressure 液压
liquid pressure drop 液体压力降
liquid propellant 液体推进剂
liquid propellant rocket engine 液体推进剂火箭发动机
liquid property 流体性质
liquid ratio 流动比率
liquid residue 液状残渣
liquid ring vacuum pump 液环真空泵
liquid rocket 燃料推进火箭
liquid rocket ablative motor 有烧蚀防热系统的液体火箭发动机
liquid rocket propellant 液体火箭推进剂
liquid sample 液体试样
liquid sensor 液体传感器
liquid sheet 液膜,液层
liquid shell 液体壳层
liquid slosh test 液体晃动试验
liquid sloshing 液体晃动
liquid sloshing dynamics 液体晃动力学
liquid sloshing load 液体晃动载荷
liquid slug 液塞
liquid source diffusion 液态源扩散
liquid spray 液雾
liquid spring 液体弹簧
liquid start system 液体起动系统
liquid state 液态
liquid stream 液流
liquid surface 液体界面
liquid turbine 液涡轮
liquid viscosity 液体黏性
liquid volume 液态体积
liquid water content 含水量

liquid zone 液区
liquid-cooled helmet 液冷头盔
liquid-cooled suit 液冷服
liquid-hydrogen 液氢
liquidity 流动性
liquid-solid interface 液-固相界面
liquidus 液相线
liquid-vapor interface 液-蒸汽界面
LIR Lion Air 雄狮航空公司
LIRL Low-Intensity Runway Light 低亮度跑道灯
LIRTS Large Infrared Telescope on Spacelab 空间实验室大型红外望远镜
LIS 1. Localizer Inertial Smoothing 定向器惯性校平 2. Lisbon 里斯本 3. Logically-Independent IP Subnet 逻辑上独立的 IP 子网 4. Loran Inertial System 罗兰惯性导航系统
lissajous figures 李沙育图形（示波图）
list 明细表，清单
list code 列表码
list of radio beacon 无线电指向标表
list processing 表处理
listener 听众
LIT 1. Lead-in Training 入门培训 2. Lira 里拉 3. Litre 升 4. Little 小的
LITAS Low Intensity Two Colour Approach System 低强度双色进近坡度系统
lithergol 液固混合推进剂
lithium 【化】锂（化学元素，符号 Li）
lithium battery 锂电池
lithium carbonate 碳酸锂
lithium cell 锂电池
lithium germanium oxide 锗酸锂氧化物
lithium iodate 碘酸锂晶体
lithium niobate 铌酸锂
lithium storage battery 锂蓄电池
lithium tantalate 钽酸锂晶体
lithium-doped solar cell 掺锂太阳电池
lithium-drift detector 锂漂移检测器
lithium-metal sulfide cell 锂金属硫化物电池
lithium-sulfur dioxide cell 锂二氧化硫电池
lithium-thionyl chloride cell 锂亚硫酰氯电池
LITHO Lithograph 平板面
lithography 光刻
lithometeor 大气尘粒
lithosphere 岩圈
LITVC Liquid Injection Thrust Vector Control 液体喷射推力向量控制
LIVE 有效的
live bomb 真炸弹

live cartridge 实弹
live launch 实弹发射
live recording system 实况记录系统
live system 实用系统
live test 真人试验，活体试验
live time 活时间
Lixiscope 利克斯仪
LJ 1. Lap Joint 搭接 2. Life Jacket 救生衣
LK Link 链环，开关，边杆，连接
LK WASH Lock Washer 防松垫圈
LKD Locked 被锁定的
LKD C Locked Closed 锁闭
LKD O Locked Open 锁开
LKG Locking 锁定的
LKGE Linkage 连接
LKNT Locknut 防松螺母
LKR Locker 锁，橱，柜，箱，机架，锁头，锁扣装置
LKT Lookout 注意，监视
LKWR Lockwire 安全锁线，锁紧丝，保险丝
LL 1. Latitude/Longitude 纬度/经度 2. Left-Lateral 左侧 3. Leverage Lease 杠杆租赁 4. Life Limit 寿命 5. Live Load 工作负载 6. Logo Light 标志灯 7. Lower Limit 下限 8. Low Level 低度高层
LLA 1. Latitude-Longitude-Altitude 纬度-经度-高度坐标系 2. Logical Layered Architecture 逻辑分层结构
LLAD Low-Level Air Defence 低空防空
LLC 1. Lift-Lift/Cruise 升力-升力/巡航 2. Logic Link Control 逻辑链路控制 3. Low Layer Compatibility 低层兼容
LLCC Lowest Life Cycle Costing 最低全寿命周期费用
LLD Light Landing Device 灯光降落装置
LLDC Low Level Direct Current 低电平直流电
LLDF Low Level Discomfort Factor 低空不适因素
LLF 1. Low Layer Function 低层功能 2. Low Level Flight 低空飞行
LLGB Launch-and-Leave Guided Bomb 投下后免控制导炸弹
LLLGB Low-Level Laser Guided Bomb 低空激光制导炸弹
LLLTV Low-Light Level Television 微光电视
LLM 1. L-band Land Mobile 频段陆地移动通信 2. Local Loopback Management 本地环回管理 3. Lower Layer Module 低层模块
LLMS Liquid Level Measurement System 液位测量系统
LLNL Lawtrence Livermore National Laboratory 劳伦斯·利弗莫尔国家研究所
Lloyd's underwriter 伦敦劳埃德保险公司保险商成员
LLP 1. Life Limited Parts 寿命限制件，有时限的零部

件 2. Lightning Location and Protection 雷电定位和避雷 3. Logical Lightwave Path 逻辑光波路径 4. Low Level Protocol 低级协议

LLR Leased Loaded Routing 最小负荷选路

LLRF Lunar Laser Range-Finder 月球激光测距仪

LLSS Low Level Sounding System 低空声测系统

LLTV Low-Light Television 微光电视

LLV Lower Limit of Video 视频下限,影像下限

LLWAS Low Level Wind Shear Alert System 低空风切变警告系统

LLWS Low level Windshear 低高度风切变

LLZ Localizer 航向信标台

LM 1. Last-Minute（Cargo）最后装上的货物 2. Load Monitor 负载监控器 3. List of Material 器材清单 4. Lockheed Martin 洛克德国-马丁公司 5. Logic Module 逻辑模块 6. Long Module 长舱 7. Lunar Module 登月舱

LMA 1. Lean Manufacturing Analysis 精益制造分析 2. Loss of Multiframe Alignment 多帧失步

LMAC Leicestershire Microlight Aircraft Club 雷斯特微型飞机俱乐部（英国）

LMDS 1. Local Microwave Distribution System 本地微波分配系统 2. Local Multipoint Distribution System 本地多点分布式系统

LMDE Lunar Module Descent Engines 登月舱下降发动机

LME 1. Layer Management Entity 层管理实体 2. Line Monitoring Equipment 线路监视设备 3. Link Management Equipment 链路管理设备

LMES 1. Land Mobile Earth Station 陆地移动地球站 2. Loss of Main Electrical Supply 主电源供给损失

LMF 1. Large Mode Fiber 大模光纤 2. Location Management Function 定位管理功能 3. Low Medium Frequency 低中频

LMG Left Main Gear 左主起落架

LMI 1. Line Move Item 生产线移动项 2. Logical Management Inferface 逻辑管理界面

LMM Locator Middle Marker 中指点标,航向中指点标,中示位信标和指点标

LMP 1. Lamp 灯 2. Left Middle Plug 左中部插头 3. Line Maintenance Part 航线维修件

LMPBLK Lampblack 黑灯,黑烟

LMS 1. Land-Mobile Satellite 陆地移动卫星 2. Landing Monitor System 着陆监控系统 3. Leakage Measurement System 渗漏测量系统 4. Least Mean Square 最小均方 5. Local Message Switch 本地信息交换机 6. Lunar Mass Spectrometer 月球质谱仪 7. Lunar Module Simulator 登月舱模拟器

LMSS 1. Land Mobile Satellite System 陆上移动卫星系统 2. Lunar Mapping and Survey System 月球测图和测量系统

LMT 1. Limit 限制,极限 2. Local Maintenance Terminal 本地维护终端 3. Local Mean Time 当地平时,当地时间

LN Logarithm 自然对数

L-N Line to Neutral 相电压

LNA 1. Label Not Available 没有标签 2. Local Network Architecture 局部网络结构 3. Logical Network Address 逻辑网络地址 4. Low-Noise Amplifier 低噪声放大器

LNG 1. Lengthening 延长 2. Long 长,大起落航线 3. Length 长度

LNP Local Network Protocol 本地网络协议

LNR Linear 线性的,直线的

LNS Land Navigation System 地面导航系统

LNYD Lanyard 小索,拉火绳

LO 1. Left Outboard 左外侧 2. Local Office 当地办事处 3. Local Oscillator 局部振荡器 4. Lockout 锁定 5. Low 低

LOA 1. Leave Of Absence 休假 2. Length Over All 全长 3. Letter of Offer and Acceptance 交货验收单 4. Level-Off Altitude 改平高度 5. Loss Of Alignment 帧失步

LOAC Low Accuracy 低精度

load 负载,负荷

load analysis 负载分析

load balancing 负载平衡

load case 载重方案

load cell 测力传感器

load characteristic 负载特性

load characteristic curve 负载特性曲线

load classification number 负载分类号

load condition 载荷情况

load control 载荷控制

load cycle 负荷循环

load design 载荷设计

load diagram 负载图,负载曲线

load distribution 载荷分布

load factor 负载系数

load history 载荷历程

load image 装入映象

load impedance 载阻抗

load leveling 负载均衡

load limit 负荷极限,负荷限度

load locus 负载轨迹

load manifest 货物清单

load module 输入模块

load path 加载路径

load profile of static strength test　静强度试验载荷图
load reconstruction　负荷重构
load resistance　负荷电阻,负载电阻
load ring　集索图
load sensor　负载检测器
load sheet　货物单
load splitting　负载分割
load spreader　负载分散器
load supporting system　承力系统
load tape　装带
load tide　负荷潮
load torque　负载力矩
load transfer　负荷转移
load transfer ability　传荷能力
load up ageing　加荷时效
load voltage　负载电压
load waterline　载重吃水线
load-carrying ability　承载能力
load-carrying capability　承载力
loaded cylinder　承力筒
loaded quality factor　有载品质因数
loaded-rocket mass　加注后总质量
loader　装载机,载货设备,升降车
loading　加负载,装载
loading bridge　桥式起重机,桥式装载机
loading chart　装料图表
loading coil　加感线圈
loading condition　载荷状况
loading control room　加注试验控制室
loading density　装填密度
loading diagram　加载图式
loading plan　装载计划
loading ramp　装卸货场
loading rehearsal　加注合练
loading signal board　加注信号台
loading signal box　加注信号箱
loading signal unite test　加注信号联试
loading spectrum　载荷谱
loading test-control desk　加注试验控制台
loading tray　装料盘输弹槽
loading up-down the temperature　加注升降温
loadmaster　装卸长
load-out　装载隧道
LOAL　Lock-On After Launch 发射后锁定
lobby　大厅
lobe　波瓣
lobe switching　波瓣转换
lobed mixer　波瓣掺混器
lobed nozzle　波瓣喷管
lobing　波瓣的形成
lobing radar　波瓣转换式雷达
lobing rate　锥形扫探频率
LOC　1. Local 当地的 2. Local Cargo 当地货物 3. Localizer 航向信标台,定位器 4. Locate Or Location 定位或位置 5. Location 位置 6. Locator 示位信标
local anomaly　局部异常
local anoxia　局部缺氧症
local apparent time　地方视时
local area network　局域网
local asymptotic stability　局部渐近稳定性
local automation　局部自动化
local buckling　局部弯曲
local civil time　当地民用时间
local cluster of galaxies　本星系团
local commanding　本地发令
local computer network　本地计算机网络
local controller　本地控制器
local convergence　局部收敛
local data base　局部数据库
local decision maker　局部决策者
local distribution system　局部配电系统
local elastic instability　局部弹性不稳定性
local equalize　局部补偿
local error　局部误差
local estimation　区域性评价,局部估计
local extrema　局部极值
local failure　局部破坏
local fare　管内票价
local feedback　局部反馈
local flight　本地飞行
local gradient　局部梯度
local group of galaxies　局部银河组
local heat treatment　局部热处理
local horizon　视地平
local hour angle　当地时角
local injection　局部喷射
local interaction　局部相互作用
local level　当地水平面
local material　地方材料
local maxima　局部最大值
local maximum　局部极大
local mean time　本地平均时
local measurement　局部量测
local meridian　地方子午线
local microcode compaction technique　局部微码紧致技术
local minimum　局部最小值
local mode　局部模态,本地方式

local noon 当地正午
local optimal solution 局部最优解
local optimization technique 局部优化技术
local optimum 局部最优
local oscillation 局部振荡
local oscillator 本机振荡器
local pressure 局部压力
local resonance 局部共振
local search 区域搜索法
local sidereal time 地方恒星时,当地恒星时
local solar time 当地太阳时
local solution 局部解
local stability 局部稳定
local standard of rest 当地静止标准
local standard time 当地标准时
local strain method 局部应变法
local strain rate 局部应变率
local telephone 市内电话
local telephone network 市内电话网
local thermal equilibrium 局部热平衡
local thermodynamic equilibrium 局部热动力学平衡
local time 地方时
local true time 地方真时
local vacuum electron beam welding 局部真空电子束焊
local velocity 局部速度
local vertical 当地垂线,局部垂线
local-area 局部地区网
localize 本地化
localized arc 局部电弧
localizer 航向信标台
localizer beacon 着陆航向信标台
localizer unit 定位单元
Locap Low Combat Air Patrol 低空战斗巡逻
locate 定位
locating mechanism 定位机构
location 定位技术
location diagram 位置简图
location hole 定位孔
location map 定位图
location notch 定位槽
location survey 定线测量
locator 甲波方向探测器定位器
LOCE Loss-Of-Coolant Experiment 冷却剂丧失实验,失水实验
locep 局部外延
LOCGS Localizer Glide Slope 定位器下滑道
lock 锁定装置,锁定,闭锁,固定,同步
lock image 封锁图像
lock management 加锁管理

lock nut 防松螺母
lock nut washer 扣定螺母垫圈
lock register 锁定寄存器
lock time 击发间隙
lock up relay 闩锁继电器
lock valve 锁闭活门
lock washer 锁紧垫圈
Lockalloy 洛卡洛伊合金
locked 锁闭的
locked-rotor characteristic 堵转特性
locked-rotor exciting power 堵转励磁功率
locked-rotor torque 堵转转矩
lock-in 锁定
lock-in frequency 锁定频率
lock-in rate 闭锁速率
locking block 定位块
locking latch 插接锁定
locking mechanism 闭锁机构
locking relay 锁定继电器
locking wire 保险丝
lock-on 捕捉,自动追踪
lockplate 锁片
lockup 锁住,闭
locomotive 机车,火车头
locomotive robot 移动式机器人
locus 位置,轨迹
LOD 1. Learning On Demand 教学点播 2. Letter Of Definition 定义函,定义书
LOE 1. Lane Of Entry 进场走廊 2. Level Of Effort 努力程度
LOEP List Of Effective Pages 有效页清单
LOES Lower-Order Equivalent System 低阶等效系统
LOF 1. Lift-Off 升空,离地 2. Line Of Flight 飞行路线 3. Local Oscillator Frequency 本振频率 4. Loss Of Frame 帧丢失 5. Low On Fuel 燃油不足 6. Lowest Observed Frequency 最低观测频
LOFAR Low Frequency Acquisition & Ranging 低频搜索与测距
LOFT Line Operation Flight Training 航线运行飞行训练
loft 放样间
loft and bombing system 甩投轰炸系统
loft bombing 上仰轰炸
lofting 放样,模线绘制
LOG 1. Logarithm 对数 2. Logarithmic 对数的 3. Logic 逻辑
log 协议记录表
log magnitude-frequency characteristics 对数幅频特性
log magnitude-phase diagram 对数幅相图

log periodic antenna 对数周期天线
log phase-frequency characteristics 对数相频特性
logarithm 对数
logarithm periodic antenna 对数周期天线
logarithmic amplifier 对数放大器
logarithmic decrement 对数衰减,对数减量
logarithmic frequency sweep rate 对数频率扫描率
logarithmic ratemeter 对数率表
logarithmic scale 对数刻度
logbook 飞行日志,记录簿,履历簿
logic 逻辑,逻辑电路
logic analyzer 逻辑分析仪
logic bomb 逻辑炸弹
logic card 逻辑插件
logic chart 逻辑流程图
logic command 逻辑指令
logic computer 逻辑计算机
logic control 逻辑控制
logic controller 逻辑控制器
logic decoder 逻辑译码器
logic diagram engineering 工程逻辑图
logic line group 逻辑线路组
logic link layer 逻辑链路层
logic probe 逻辑探针
logic schematics 逻辑框图
logic signature analyzer 逻辑特征分析仪
logic simulation 逻辑模拟
logic state analyzer 逻辑状态分析仪
logic static hazard 逻辑静态冒险
logic swing 逻辑摆幅
logic timing and sequencing 逻辑定时和排序
logic tree 逻辑树
logical address 逻辑地址
logical channel 逻辑通道
logical circuit 逻辑电路
logical computer 逻辑计算机
logical data base 逻辑数据库
logical data path 逻辑数据通路
logical element 逻辑元件
logical flow chart 逻辑流程图
logical overlap 逻辑重叠
logical terminal pool 逻辑终端组
logistic delay 后勤延误
logistic delay time 保障资源延误时间
logistic support 后勤保障
logistic support analysis 后勤保障分析
logistic support area 勤务区
logistics module 后勤舱
logitron 磁性逻辑元件

log-likelihood 对数似然函数
Logmars Logistics Markings and Reading Symbols 后勤标号与阅读符号
lognormal distribution 对数正态分布
Logo Logographic 标志,航徽
log-periodic antenna 对数周期天线
LOH 1. Light-Observation Helicopter 轻型观察直升机 2. Line OverHead 线路开销
lohilo 低高-低飞行剖面图
LOI 1. Letter Of Intent 意向书 2. Level Of Involvement 牵连程度 3. Low Order Interface 低阶接口
loiter 空中巡逻
LOL Local 当地的,面部的
LOM 1. Level of Maintenance 维修水平 2. Light-Optic Microscope 光电显微镜 3. Locator Outer Marker 远指点标,外示位信标和指点标
LON 1. Longitude 经度 2. London 伦敦
long 1. Longitude 经度 2. Longitudinal 经度的,纵向的
long axis 长轴
long base-line interferometer 长基线干涉仪
long distance arming 远程待爆
long distance arming time 远程待爆时间
long distance measurement 远距离测量
long distance transmission 长途输电
long distance transmission line 长途输电线
long duration 长期荷载
long duration exposure facility 长期照射设备
long haul 长远距离的
long haul network 广域网
long lead time 长交货时间项目
long line 长线
long line effect 长线效应
Long March 长征号运载火箭
long period cepheid 长周期造父变星
long period perturbation 长周期摄动
long period variable 长周期变量
long periodic perturbation 长周期摄动
long playing record 密纹唱片
long range navigation 远程导航,劳兰导航
Long Range Navigation Equipment 远程导航设备
long range objectives 远景目标
long segment type arc heater 长分段式电弧加热器
long shot tunnel 长射式风洞
long shot wind tunnel 长射风洞
long term drift 长期漂移
long term memory 长期储存器
long time horizon coordination 长时程协调
long wave 长波

long-dated 远期的
long-duration mission 长续航时间飞行任务
long-duration motor 长燃时固体火箭发动机
long-endurance vehicle 长期运行飞行器
longeron 大梁,机身桁梁,长桁
longeron structure 桁梁式结构
longevity 寿命
longitude 经度
longitude circle 经圈,子午圈,黄经圈
longitude of ascending node 上升点经度
longitude of descending node 降交点经度
longitude of node 交点黄经
longitude of periastron 近星点经度
longitude of perigee 近地点黄经
longitude of perihelion 近日点经度
longitudinal 纵向的,经度的
longitudinal acceleration 纵向加速度
longitudinal acoustic mode 纵向声学模式
longitudinal antenna fuze 纵向天线引信
longitudinal axis 纵轴
longitudinal bulkhead 纵舱壁
longitudinal conductance 纵向电导
longitudinal control 纵向操纵
longitudinal control qualities 纵向操纵品质
longitudinal crack of cartridge 弹壳纵裂
longitudinal direction 轴向,纵向流变
longitudinal displacement 纵向位移
longitudinal dynamics 纵向动力学
longitudinal error of traverse 导线纵向误差
longitudinal force coefficient 纵向力系数
longitudinal maneuver 纵向机动,纵向运动,平面内机动,距离机动
longitudinal mode 纵向运动形式
longitudinal motion 纵向运动
longitudinal oscillation 纵向振动
longitudinal overlap 纵向重叠率
longitudinal plane 纵向平面,断面
longitudinal profile 纵剖面
longitudinal propagation 纵向传播
longitudinal redundancy check 纵向冗余校验
longitudinal separation 纵向间距
longitudinal strengthening 纵向强化
longitudinal stress 纵向应力
longitudinal tilt 航向倾斜
longitudinal turbulence 纵向涡流
longitudinal vibration 纵向振动
longitudinal vortex 纵向涡
longitudinal wave 纵波
longitudinal winding 纵向缠绕

LONGN Longeron 长桁
long-range 远程的
long-range acquisition radar 远程探测雷达
long-range air search radar 远程对空搜索雷达
long-range aircraft 远程飞机
long-range navigation 长距离导航
long-term 长期的,长周期的
long-term aging 长期老化
long-term prediction 长期预测
long-term stability 长期稳定性
long-term storage 长期库容
long-wave communication 长波通信
long-wire antenna 长线天线
look angle 观察角,视角
look direction 观察方向
look-ahead strategy 先行策略
looks aback upon 回顾
loom 翼梁腹板,织机
loop 环路,回线,半波
loop antenna 环状天线
loop computer network 环形计算机网络
loop control 环路控制
loop direction finder 环状天线测向器
loop filter 环路滤波器
loop gain 无循环定向图
loop gain characteristic 环路增益特性
loop inversion 循环反演
loop of retrogression 逆行圈,逆行环状飞行
loop prominence 环状日珥
loop resistance value 回路阻值
loop time 回路时间
loop-around test line 环路测试线
loop-free directed graph 无循环定向图
loop-gain matrix 环路增益矩阵
loop-mile 环线英里
loose powder 松粉,疏松粉末
loose tooling forging 胎模锻
LOP 1. Line Of Position 位置线,位置域 2. Local Operating Procedure 本机操作程序 3. Log Of Pages 记录本 4. Low Oil Pressure 低滑油压力 5. Lubricating Oil Pump 滑油泵
LOPA Layout Of Passenger Accommodations 旅客设施布置
LOPS Layout Of Payloads System 商载配置系统
LOR 1. Low Frequency Omnidirection Radio Range 低频全向无线电信标 2. Lunar Orbit Rendezvous 月球轨道会合
LORA Level Of Repair Analysis 修理水平分析
LORAC 1. Long Range Accuracy 远距精度 2. 罗拉克导

航系统,劳拉克
LORAN　1. Long Range Navigation 远距导航 2. Long Range Navigation(system) 罗兰远距导航系统
Loran chain　罗兰电路
Loran communication　罗兰通信
Loran retransmission　罗兰转发
LORAN-C　罗兰-C
Loran-C alarm　罗兰-C 告警
Loran-C antenna　罗兰-C 天线
Loran-C envelope of pulse　罗兰-C 脉冲包络
Loran-C ground station　罗兰-C 地面地台
Loran-C system　罗兰-C 系统
Loran-C timing　罗兰-C 授时
LORET　Loran retransmission 罗兰转发
LORL　Large Orbital Research Laboratory 大型轨道研究实验室
Loroc　Long-Range Optical Camera 远距离光学摄像机
LOROP　Long-Range Oblique Photography 远距倾斜摄影
LOS　1. Line Of Sight 视线 2. Local Operating System 本机操作系统 3. Loss Of Signal 信号丢失
Loschmidt number　洛喜米特常数
losing lock　失锁
LOSS　Landing Observer Signal System 着陆观察员信号系统
loss angle　损耗角
loss coefficient　损失系数
loss event　损失事件
loss factor　功耗因素
loss function　损耗函数
loss measurement　损耗测定
loss mechanism　损耗机理
loss model　损耗模型
loss modulus　损耗模量
loss of cabin pressure　座舱减压
loss of central vision　中心视觉丧失
loss of consciousness　意识丧失
loss of control　失控
loss of coordination　协调障碍
loss of energy　能量丧失
loss of equilibrium　失平衡
loss of evaporative heat　蒸发散热
loss of grasping power　握力丧失
loss of hearing　听力丧失
loss of noise induced hearing　噪声性听力丧失
loss of orientation　失定向,定向障碍
loss of peripheral light　周边光亮丧失
loss of peripheral vision　周边视觉丧失
loss of personnel capability　人员能力损失

loss of situational awareness　情景警觉丧失
loss peak　损耗峰
loss rate　损失率
loss ratio　赔付率
loss reduction　损失减少
loss system　损耗系统
lossless network　无损网络
lost balloon　测风气球
lost wax casting process　失蜡铸造法
lost-wax molding　熔模法造型
lot　组,批量
lot cargo　大宗货物
lot inspection　批检
lot size　批量
lot tolerance percent defective　批容许不合格率
LOTAWS　Laser Obstacle Terrain Avoidance Warning System 激光障碍地形回避警告系统
loudspeaker　扬声器
lounge　高级客舱,休息室
louver　百叶窗,放热孔
low absorption　低吸收
low altitude　低空
low altitude bombing　低空轰炸
low altitude bombing system　低空轰炸系统
low altitude flight　低空飞行
low altitude satellite　低空卫星
low altitude surveillance radar　低空搜索雷达
low amplitude　低振幅
low angle　低角度
low angle loft bombing　低角向高处投弹
low atmospheric pressure test　低气压试验
low blower　低压增压器
low body negative pressure test　下体负压试验
low cloud　低云,低层云
low density ablator　低密度烧蚀材料
low density electron beam optics　弱流电子光学
low density wind tunnel　低密度风洞
low distortion　轻度失真
low drag bomb　低阻炸弹
low earth orbit　近地轨道
low earth orbit environment　低地轨道环境
low emission　弱发射
low energy　低能源的
low energy electron diffraction　低能电子衍射
low energy ion scattering　低能离子散射
low expansion superalloy　低膨胀高温合金
low explosive　低级炸药
low flow rate　小流量
low flow rate filling　小流量加注

low frequency 低频率的
low frequency amplifier 低频放大器
low gate 低限阀
low level flight 低空飞行
low light level television camera 微光电视摄像机
low load balance 微量天平
low loss 低损耗
low molecular weight 低分子质量
low noise amplifier 低噪声放大器
low oblique photograph 低斜角摄影
low orbit 低轨道
low orbiting satellite 低轨道卫星
low order 低值位
low oxygen 低氧
low pass filter 低通滤波器
low pitch 低距
low power 低功率
low power range 小功率范围
low pressure chamber 低压舱
low pressure compressor 低压压缩机
low pressure hose 低压软管
low pressure plasma deposition 低压等离子体淀积
low pressure test 低压试验
low resistance high efficiency solar cell 低阻高效太阳电池
low resolution facsimile 低分辨率传真
low season 淡季
low shear rate 低切速率
low side tone 低侧音
low smoke propellant 微烟推进剂
low speed flow 低速流
low speed fuel air explosive 低速油气弹
low speed wind tunnel 低速风洞
low strain 弱应变
low temperature chamber 低温室
low temperature heat pipe 低温热管
low temperature lubricant 低温润滑剂
low temperature sealing 低温密封
low temperature test 低温试验
low thrust 小推力
low trajectory test 低弹道试验
low vacuum 低真空
low velocity drop 低速空投
low voltage 低压
low voltage power supply 低压电源
low wing 下单翼
low wing airplane 低翼机
low-alkali ceramics 低碱瓷
low-altitude 低空的
low-altitude bomb 低空炸弹
low-altitude fighter direction radar 低空战斗机引导雷达
low-altitude satellite 低高度卫星
low-altitude short-range missile 低空近程导弹
low-angle loft bombing 小角度低空甩头炸弹
low-cycle fatigue 低周疲劳
low-density 低密度
low-energy 低能量的
low-energy fuel 低能燃油
lower airspace 低空空域
lower airspace radar advisory 低空雷达咨询服务
lower atmosphere 低层大气
lower body negative pressure test 下体负压试验
lower culmination 下中天
lower edge 下边缘
lower heating value 低热值
lower level problem 下级问题
lower mantle 下覆盖层
lower sideband 下边带
lower stage 下面级
lower transit 下中天
lower-level threshold 下阈
lower-order 低次的
lowest launching condition 最低发射条件
low-fidelity 低保真度的
low-frequency 低频的
low-frequency combustion instability 低频燃烧不稳定性
low-frequency linear vibration table 低频线振动台
low-latitude 低纬度的
low-level flight 低空飞行
low-level protocol 低层次结构规程
low-level transmission 低电平传输
low-light level 弱照度条件下的
low-light level television 微光电视
low-loss 低损耗
low-order 低位的,低阶的
low-order finite element scheme 低阶有限元法
low-pass filter 低通滤波器
low-pressure chemical vapor deposition 低压化学气相沉积
low-resistance 低阻,低电阻
low-tech 低技术,低技术的
low-temperature and low-atmospheric pressure test 低温低气压实验
low-temperature property 低温抗裂性
low-thrust 小推力飞行任务
low-velocity 低速
low-voltage ion reactive plating 低压反应离子镀

low-voltage mode 低压电模式
low-wing 下单翼
low-wing monoplane 下单翼机
low-Z material 低原子序数材料
LOX Liquid Oxygen 液氧
lox apparatus 液氧装置
loxing 加注液氧
loxodrome 等角线
loxygen Liquid Oxygen 液氧
LOZ Liquid Ozone 液态臭氧
LP 1. Lean Production 精益生产 2. Light Plate 照明板
LPA 1. Linear Power Amplifier 线性功率放大器 2. Load Planning Assistant 装卸计划助理系统 3. Load Power Analysis 负载功率分析 4. Log Periodic Antenna 对数周期天线 5. Logical Pack Area 逻辑包装区 6. Low Pin Actuator 锁定销作动器
LPB Landing Planning and Balance 载重平衡
LPBA Lawyer Pilot's Bar Association 律师飞行员协会
LPC 1. Linear Predictive Coding 线性预测编码 2. Low Pressure Compressor 低压压缩机 3. Low Pressure Core 低压轴 4. Luftfahrt Presse-Club 航向报刊俱乐部
LPCVD Low Pressure Chemical Vapor Deposition 低压化学气相淀积
LPD 1. List of Program Definition 程序定义清单 2. Lockheed Private Data 洛克希德专用数据 3. Low-Performan Drone 低性触遥控无人驾驶飞机 4. Low Probability of Dectection 低概率检测
LPDTL Low Power Diode Transistor Logic 小功率二极管晶体管逻辑
LPG Lapping 搭接,重叠,研磨
LPH Landing Platform for Helicopter 直升机降落平台
LPI Lateral Path Integrator 横向通道积分器
LPICT Loop In Coming Trunk 环路来话中继
LPIR Low Probability of Intercept Radar 低截获概率雷达
LPM 1. Landing Path Monitor 着陆轨迹监视仪 2. Linearly Polarized Mode 线性偏振模 3. Liter Per Minute 升/分
LPN 1. Local Packet-switched Network 本地分组交换网 2. Low Pass Network 低通网络
LPO 1. Landings Per Overhaul 每次翻修的着陆数 2. Low Power Output 低功率输出
LPOM Low Order Path Overhead Monitoring 低阶通道开销监视
LPOX Low Pressure Oxygen Service 低压氧气服务
LPRE Liquid Propellant Rocket Engine 液体推进剂火箭发动机
LPRSVR Life Preserver 保存寿命,存储寿命
LPS Laser Power Supply 激光能源

LPT 1. Low Pressure Test 低压试验 2. Low Pressure Turbine 低压涡轮
LPTV Low Profile Transfer Vehicle 低翼型运输机
LPW Lumens Per Watt 流明/瓦
LQA Link Quality Analysis 传输线路质量分析
LQG Linear Quadratic Gaussian 线性二次高斯
LQTR Lightning Qualification Test Report 闪电鉴定试验报告
LR 1. Last Received 最后收到的 2. Left Rear 左后 3. Link Request 链路请求 4. Loads Report 负载报告 5. Location Register 位置寄存器 6. Long Radius 大活动半径 7. Long Range 远程
LRA 1. Line Replaceable Assembly 航线可更换组件 2. Local Regulatory Authority 当地管理当局 3. Low-Range Radio Altimeter 无线电低高度表
LRAACA Long Range Air Antisubmarine Capability Aircraft 远程空中反潜飞机
LRBM Long Range Ballistic Missile 远程弹道导弹
LRBS Laser Ranging Bombing System 激光测距轰炸系统
LRC 1. Langley Research Center 兰利研究中心(美国国家航空航天局) 2. Long Range Cruise 远程巡航,远程巡航速度 3. Longitudinal Redundancy Check 纵向冗余度校验
LRCA Location Register Coverage Area 位置寄存器有效区域
LRCO Limited Remote Communication Outlet 输出端有限遥控通信
LRD 1. Labled Radar Display 示踪雷达显示器 2. Laser Ranger Designator 激光照射器 3. Lightning and Radiation Detector 闪电与辐射探测器 4. Lightning and Radio Emission Detector 闪电与无线电发射探测器 5. Lockheed Requirements Document 洛克希德要求文件 6. Long Range Data 远程数据
LRE Low Rate Encoding 低速率编码
LRF Laser Range Finder 激光测距仪
LRFAX Low Resolution Facsimile 低分辨率传真
LRI 1. Line Replaceable Item 航线可更换件 2. Link Remote Inhabit 链路远端禁止 3. Liquid Resin Infusion 液态树脂溶渗
LRIR Low Resolution Infrared Radiometer 低分辨力红外线探测仪
LRL Lunar Receiving Laboratory 月球信息接收实验室
LRM Line Replaceable Module 航线可更换模块
LRMA Long Range Marine Aircraft 远程海上飞机
LRMP Long-Range Maritime Patrol 远程海上巡逻
LRN Long Range Navigation 远程导航
LRPA Long-Range Patrol Aircraft 远程巡逻机
LRR Long Range Radar 远距雷达

LRRA Low Range Radio Altimeter 无线电低高度表
LRS 1. Load Range Schedule 长期计划,远景规划 2. Load Relief System 卸荷系统
LRSS Long Range Survey System 远程测量系统
LRU Line Replaceable Unit 航线可换组件
LRV Liquid Relief Valve 液体释放阀
LS 1. Landing System 着陆系统 2. Last Sending 最后发出的 3. Left Side 左侧 4. Less 少于 5. Light Switch 照明开关 6. Limit Switch 极限电门 7. Line Stretcher 拉线器 8. Loudspeaker 扬声器
LSA 1. Label Switched Application 标记交换应用 2. Limited-Space-Charge Accumulation 限制空间电荷积累 3. Link State Advertisement 链路状态广播 4. Logistics Support Analysis 后勤保障分析 5. Low Speed Aileron 低速副翼
LSAO Line Station Assembly Order 生产线装配指令
LSAP 1. Link Service Access Point 链路业务接入点 2. Loadable Software Airplane Part 可装载飞机软部件
LSAPL Loadable Software Airplane Part Librariay 可装载飞机软部件库
LSAR Logistic Support Analysis Records 后勤保障分析记录本
LSB 1. Least Significant Bit 最低有效位 2. Lower Side Band 下边带,低边带
LSB communication 下边带通信
LSC 1. Least Significant Character 最低有效字符 2. Line Service Center 线路服务中心 3. Link Set Control 链路组控制 4. Logistic Support Costs 后勤保障费用
LSCP Large Scale Computer Project 大型计算机工程
LSCS Large Scale Computer System 大型计算机系统
LSD 1. Large Screen Display 大屏幕显示 2. Least Significant Digit 最低有效数字 3. Low Speed Data 低速数据
LSDS Large Screen Display System 大屏幕显示系统
LSE 1. Layer Service Element 层服务单元 2. Line Signaling Equipment 线路信令设备 3. Link Switch Equipment 链路交换设备 4. Local System Environment 本地系统环境
LSECS Life Support and Environmental Control System 生命保障和环境控制系统
LSFFAR Low Speed Folding Fin Aircraft Rocket 低速折叠翼飞机火箭
LSHG Lashing 绳套,鞭打
LSI 1. Large Scale Integration 大规模集成 2. Large Scale Integrated/Integration/Integrator 大刻度积分的/积分/积分器
LSIC Large Scale Integrated Circuit 大规模集成电路
LSIT Large Scale Integration Technology 大规模集成技术
LSJ Lifesaving Jacket 救生衣
LSK Line Select Keys 行选键
LSL 1. Laser Stereo Lithography 激光立体制版,激光造型 2. Link Support Layer 链路支持层 3. Low Speed Logic 低速逻辑
LSLN Low Speed Local Network 低速局部网络
LSMIS Logistics Support Management Information System 后勤支援管理系统
LSML Low Sodium, No Salt Added Meal 低钠无盐餐
LSO Landing Signal Officer 降落信号员
LSP 1. Label Switched Path 标签交换路径 2. Layer Service Primitives 层服务原语 3. Limit Select Panel 极限选择板 4. Line Select Panel 线选择板 5. Line Symchronizing Pluse 行同步脉冲 6. Logistics Support Plan 后勤支援活动计划 7. Low Speed 低速
LSR Load Shedding Relays 负载减少继电器
LSS 1. Lightning Sensor System 闪电传感系统 2. Logistic Support System 后勤支援系统 3. Local Synchronization Subsystem 本地同步子系统
LSSM Luna Surface Scientific Module 月面科学舱
LST Local Standard Time 当地标准时间
LT 1. Layout Template 模线样板 2. Light 灯 3. Local Time 当地时 4. Litre 升 5. Left 左 6. Left Turn 左转弯 7. Less Than 少于 8. Level Trigger 电平触发器
LTA 1. Large Transport Airplane 大型运输机 2. Left Train Arm 左配平准备 3. Lighter Than Air 轻于空气 4. Line Terminal Adapter 线路终端适配器
LTBT Limited Test Ban Treaty 有限禁止核试验条约
LTC 1. Life Time Control 寿命时限控制 2. Local Telephone Circuit 本地电话电路 3. Lowest Two-way Channel 最少双向信道
LTCBJ Long-Term Credit Bank of Japan 日本长期信贷银行
LTD Limited 有限的,极限的,受到限制的
LTDP Long Term Defence Programs 长期防务规划
LTDR Laser Target Designator Receiver 激光目标照射器接收机
LTDS Launch Trajectory Data System 发射弹道数据系统
LTE 1. Late 迟的,晚的 2. Launch To Eject 发射弹射 3. Lightware Terminal Equipment 光端机 4. Line Terminating/Termination Equipment 线路终端设备 5. Line Test Equipment 线路测试设备 6. Local Thermal Equilibrium 局部热平衡 7. Local Thermodynamic Equilibrium 局部热动力平衡
LTER Long-Term Ecological Research 长期生态研究
LTG 1. Lightening 照亮,亮度 2. Lightening or Lighting 减轻的或点燃的

LTGH Lightening Hole 减轻孔
LtHo Lighthouse 灯塔,灯台
LTIT Low Turbine Inlet Tmperature 低压涡轮入口温度
LTM 1. Livestock Transportation Manual 家畜运输手册 2. Local Traffic Management 地方交通管制
LTMA London Terminal Control Area 伦敦航站管制区
LTMR Laser Target Marker/Ranger 激光目标识别器/测距器
LTN 1. Line Terminating Network 线路终端网络 2. Litton Company 利登公司 3. Long Ton 长吨
LTP 1. Left Top Plug 左上电阻 2. Link Terminating Point 链路终接点
LTS 1. Lights 灯光 2. Load Transfer Signal 负荷传递信号 3. Load and Trim Sheet 载重和平衡表 4. Local Telephone System 本地电话系统
LTV 1. Light Vessel 灯船 2. Loan To Value 贷款与价值比率
LU Logical Unit 逻辑单元
LUAR Laboratory Unit Acceptance Review 实验室单位授权审核
lubber 标志
lubber line 航向标线
lube 润滑油
lubricant 润滑剂
lubricant compatibility 润滑剂可混性
lubricating material 润滑材料
lubricating oil 润滑油
lubrication 润滑,润滑法
lubrication system 润滑系统
lubrication system capacity 润滑系统容量
LUCC Land Use and Cover Change 土地利用/土地覆盖变化
lucimeter 亮度计
lucite 人造荧光树脂,有机玻璃
LUF Lowest Usable Frequency 最低可用频率
lufbery 卢氏圆形队
Lufthansa 汉莎航空公司
lug 连接盘,接线片
luggage compartment 行李舱
lumen 流明
lumenmeter 流明计
luminaire 光源
luminance 亮度,照度
luminary 发光体
luminescence 发光,荧光
luminophore 发光体
luminosity 发光度

luminosity class 光度级
luminosity evolution 光度演化
luminosity function 光度函数
luminous 发光的
luminous diode 发光二极管
luminous efficiency 发光效率
luminous flame 发光焰,光焰
luminous flux 光通量
luminous flux density 光通量密度
luminous giant 亮巨星
luminous intensity 发光度
luminous intensity distribution curve 发光强度分布曲线
luminous nebula 亮星云
luminous zone 发光带,发光区
lump 块,团
lumped mass 集中质量
lumped mass system 集中质量系统
lumped parameter control system 集总参数控制系统
lumped parameter model 集中参数模型
lumped parameter network 集总参数网络
lumped-parameter 集总参数
Luna 月神,月亮
Luna, Venus and Mars program 月球、金星和火星探测计划
lunar calendar 阴历,农历
lunar capture 月球捕获
lunar crater 月面环形山
lunar eclipse 月食,月蚀
lunar escape ambulance pack 月球逃逸救护包
lunar exploration engineering system 月球探测工程系统
lunar exploration mission 探月飞行任务
lunar exploration module 月球考察舱
lunar exploration satellite 月球探测卫星
lunar geodesy 月面测量学
lunar gravity 月球重力
lunar gravity simulator 月面重力模拟器
lunar landing module 月球着陆舱
lunar landing trajectory 登月轨道
lunar laser ranging 激光测月
lunar libration 月球天平动
lunar module 登月舱
lunar occultation 月球星蚀
lunar orbit 月球轨道
lunar orbit reentry 从月球轨道再进入
lunar orbit rendezvous 绕月轨道会合
lunar orbital rendezvous mode 月球轨道会合方式
Lunar Orbiter 登月轨道飞行器,月球轨道
lunar phase 月相
lunar pole 月极

lunar probe	月球探测器
lunar rover	绕月飞行器
lunar satellite	月球卫星
lunar seismogram	月震图
lunar seismology	月震学
lunar spacecraft	月球宇宙飞船,月球飞船,月球探测器
lunar surface	月面
lunar topology	月志学
lunarnaut	登月航天员
Luneberg lens	楞勃透镜
lung collapse	肺萎陷
lung damage threshold	肺损伤阈
lung distension	肺胀
lung irritant	肺部刺激物
lunicentric	以月球为中心的
Lunik	月球探测器
lunisolar calendar	阴阳历
lunisolar gravitational perturbation	日月引力摄动
lunisolar perturbation	日月摄动
lunisolar precession	日月岁差
lunokhod	月球车,月行车
LUT	Local User Terminal 当地用户终端
lutetium	【化】镥（化学元素,符号 Lu）
lux	勒克斯（照度单位）
luxmeter	勒克计,照度计
LV	1. Laser Videodisc 激光视盘 2. Leave 离开,起飞 3. Level 水平,层 4. Light and Variable 微风而风向不定,轻而多变的 5. Low Voltage 低电压
LVA	1. Large Vertical Aperture 大垂直孔径 2. Logic Virtual Address 逻辑虚拟地址
LVC	Label Virtual Circuit 标记虚拟电路
LVDA	Launch Vehicle Data Adapter 运载火箭数据转接器
LVDC	Low Voltage Direct Current 低压直流
LVDT	Linear Variable Differential Transformer 线性可变差动变压器
LVIS	Low Voltage Ignition System 低电压点火系统
LW	1. Landing Weight 着陆重量 2. Last Word 最后一字 3. Left Wing 左大翼 4. Light Weight 轻型的 5. Long Wave 长波
LWC	Light Wave Communication 光波通信
LWD	1. Left Wind Down 左翼向下 2. Lowered 起落架襟翼等已放下
LWF	Light Weight Fighter 轻型战斗机
LWR	Lower 下部的,较低
LWS	Left Wing Station 左翼站位
LWSA	Lavatory Water Supply Assy 洗手间供水组件
LWST	Lowest 最低的
LWSD	Laser Weapon System Demonstrator 激光武器系统示教器
LWTR	Licence Without Type Rating 无机型等级执照,基础执照
Lyapunov's direct method	莱阿波诺夫直接法
lycopodium	石松子,石松属
lyman bands	赖曼谱线带
lynchpin	车轴销
Lynx	天猫座,天猫星座
Lysithea	【天】木卫十
LZ	Landing Zone 着陆地带
LZCU	Liquid Zone Control Unit 液体区域控制装置
LZCC	Landing Zone Control Center 着陆区控制中心
LZT	Local Zone Time 区时

M

M 1. Mach number 马赫数 2. Maxwell 麦克斯韦,磁通量单位 3. Meteorological 气象的 4. Missile 导弹 5. Moment 力矩,弯曲力矩

M(m) Meter 米,公尺

M band M 波段

m probe 测量探针,丈量探针

m probe location 测量探针的位置

M wing M 形机翼

M&O 1. Maintenance and Overhaul 维修与翻修 2. Maintenance and Operation 维修与使用

M&R Maintenance and Repair 维护和修理

M&S 1. Maintenance and Supply 维护和供给 2. Marred and Scarred 有擦伤和划痕的

M&U Make and Use 制造和使用

M/A 1. Mach/airspeed 马赫数/空速 2. Maintenance Analysis 维护分析

M/ASI Mach/Airspeed Indicator 马赫数/空速指示器

M/C 1.〈口语〉Machine 飞机 2. Machined 机械加工过的

M/D Miscellaneous Data 其他数据

M/S Meters per Second 米/秒

M/T Maintainenability Testability 维修性/可测试性

MAA 1. Manchester Astronautical Association 曼彻斯特宇航协会 2. Manufacturers 制造商,制造者 3. Max. Authorized Altitude 最大规定高度,最大批准仪表飞行规则高度 4. Minimum Approach Altitude 最低进场高度

MAAG Military Assistance Advisory Group 军事援助顾问团(美国)

MAAH Minimum Asymmetric Approach Height 不对称推力进场最低高度

MAB 1. Marine Air Base 海军陆战队航空兵基地 2. Marine Amphibious Brigade 海军陆战队两栖旅 3. Missile Assembly Building 导弹总装厂房,导弹装配间

MABES Magnet Bearing Flywheel Experimental System 磁轴承飞轮试验系统(日本军事通信技术试验卫星)

MAC 1. Maintenance Action Classification 维修行动分类 2. Maintenance Allocation Chart 维修项目分配图 3. Maintenance Analysis Center 维修分析中心 4. Media Access Control 媒体存取控制 5. Medium Access Controller 媒体接入控制器

macaviator 自动驾驶仪

MACC 1. Military Air Control Center 军用空中管制中心 2. Multiple Applications Control Center 多种应用控制中心

MACCS Marine Air Command and Control System 海军陆战队航空兵指挥与控制系统

Mace Command and Control System 马斯指挥控制系统

maceration 浸软,浸解,浸渍

Mach 马赫

Mach angle 马赫角

Mach buster 超声速飞机

Mach cone 马赫锥

Mach disc 马赫盘,圆盘形激波

Mach effect 1. 马赫效应 2. 马赫数影响,压缩性影响

Mach front 马赫扰动面,马赫阵面

Mach hold 马赫数保持稳定

Mach intersection 马赫相交

Mach limited ceiling 马赫数限制的升限

Mach line 1. 弱激波,扰动线 2. 马赫线,马赫波

Mach lock 保持马赫数稳定

Mach meter 马赫数表

Mach number 马赫数

Mach number control 马赫数控制

Mach number trim system 马赫数配平系统

Mach stem 马赫波前

Mach trim 马赫数配平

Mach trim coupler 马赫数配平耦合器

Mach trim system 马赫数配平系统

Mach trimmer 马赫数调整片

Mach tuck 马赫下俯(近声速飞行时)速度安定性勺形特性

Mach wave 马赫波,弱扰动特征线

machinability (材料的)易切削性,可加工性

machinable 可以机械加工的

machine 1.〈口语〉飞机 2. 机器,机械装置 3. 电腐蚀机,电火花机床

machine addressing 【计】机器编址,机器寻址,机器地址

machine alarm 机器报警信号

machine cognition 机器识别

machine fuze setting 引信机械装定

machine intelligence 机器智能

machine interface 机器接口

machine language 计算机语言,计算机机器语言,机器

语言

machine learning 机器学习
machine tool 机床，工具机，工作母机，机械工具
machinegun ammunition 机枪弹药
machinegun barbette 机枪炮塔
machinegun blister 机枪旋回座，机枪塔
machining 机械加工，切削加工
machining centre （机床）加工中心
Mach-limited 马赫数限制的边界
Mach-Zehnder interferometer 马赫曾德尔干涉仪
MACIMS Military Airlift Command Integrated Management System 军事空运司令部综合管理系统（美国）
Macker type arc heater 叠片式电弧加热器
mackerel sky 鱼鳞天
MACMIS Maintenance And Construction Management Information System 维护和工程管理信息系统
macro code 宏代码
macro economic model 宏观经济模型，常量经济模式
macro engineering 1.宏工程，宏观工程学 2.大工程学
macro expansion algorithm 宏扩展算法
macro level 宏指令级，总体设计级
macro scale 大尺度，宏观尺度，大型化，宏观规模
macroaddress bus 宏地址总线
macroblock 宏功能块，宏模块
macroclimate 大气候
macroetch 宏观腐蚀
macroinstruction 宏指令
macrometeorologist 大气象学家
macrometeorology 大尺度气象学
macromodel 宏观模型
macromodular computer 宏模组件计算机
macromolecule 大分子
macroparticles 大粒子
macrophotogrammetry 超近摄影测量
macroprocessor 宏处理程序，宏加工程序，宏处理器
macroscale circulation 大尺度环流
macroscopic 1.宏观的，大范围的 2.肉眼可见的 3.低倍放大的
macroscopic defect 宏观缺陷，明显缺陷
macrosegregation 宏观偏析，严重偏析
macroshrinkage 宏观收缩，宏观缩孔
macrostructure 宏观结构，低倍组织
MACS 1. Marine Air Control Squadron 海军陆战队航空兵控制中队 2. Medium Altitude Communication Satellite 中高度通信卫星 3. Modular Attitudecontrol System 模块式姿态控制系统 4. Monitoring And Control Station 监控站 5. Multiple Applications Control System 多用途控制系统

MACSAT Multiple Access Communications Satellite 多址通信卫星
MACSS Medium Altitude Communication Satellite System 中高度通信卫星系统
MACSV Multi-purpose Airmobile Combat Support Vehicle 多用途空中机动战斗支援飞机
MAD 1. Magnetic Airborne Detector 机载（反潜）磁性探测器 2. Magnetic Anomaly Detection 磁异常探测器 3. Maintenance Access Door 维修工作口 4. Manufacturing Assembly Drawing 制造装配图 5. Marine Aviation Detachment 海军陆战队航空支队 6. Mass Air Delivery 密集空投 7. Minimum Approach Distance 最短进近距离 8. Mutual Assured Destruction 相互确保摧毁（战略）
MADAR Malfunction Analysis Detection And Recording 故障分析与记录
MADGE Microwave Aircraft Digital Guidance Equipment 机载微波数字制导装置
MADL Module Application Data List 模块应用数据表
MADS 1. Meteorological Airborne Data System 航空气象数据系统 2. Multiple Access Data System 多路存取数据系统 3. Multiple Access Digital System 多路存取数字系统
MAE 1. Mean Absolute Error 平均绝对误差 2. Mean Area of Effectiveness 平均有效面积 3. Medical Air Evacuation 医疗空运后送
MAF Marine Amphibious Force 海军陆战队两栖部队
MAFCO Magnetic Field Code 磁场代码
MAFFS Modular Airborne Firefighting System 模块式机载灭火系统
MAFIS Mobile Automated Field Instrumentation System 机动式外场自动仪表测量系统
MAFT Major Airframe Fatigue Test 机体主要部件疲劳试验
MAG 1. Marine Air Group 海军陆战队航空兵大队 2. Military Assistance Group 军事援助团 3. Mobile Arresting Gear 机动式拦阻装置
mag 1. Magnetic 磁铁，磁(性)的 2. Magneto 磁电机
mag drop 磁电机掉转速
magazine feed 1.弹箱供弹 2.仓库送料 3.自动传输带
magazine-fed rocket discharger 带燃料储箱的火箭启动装置
Magcom 1.磁差对比地形跟踪导航系统 2.美康地磁导航术
Magellan 麦哲伦探测器
Magellanic Clouds 麦哲伦云，麦哲伦星云
MAGGS Modular Advanced Graphics Generation System 组合式先进制图系统
magic number 幻数

MAGIIC　Mobile Army Ground Imagery Interpretation Center 机动式陆军地面图像判读中心

MAGIS　1. Marine and Atmospheric Geographical Information System 开发型海洋大气地理信息系统　2. Megawatt Air to Ground Illumination System 兆瓦级空地照明系统

magjet　美格喷气式发动机

maglev　Magnetic Levitation 磁(力)悬浮,磁浮

Magnaflux　1. 荧光磁力探伤器 2. 磁粉探伤,磁粉探伤机,磁通量 3. 寻求金属表面和内表面缺陷的磁性方法

magnalium　镁铝合金

Magnamite　1. 马格纳复合材料 2. 石墨碳纤维

magnescale　磁尺

magnesia-lanthana-titania system ceramics　镁镧钛系陶瓷

magnesium　【化】镁(化学元素,符号 Mg)

magnesium air battery　镁空气电池

magnesium alloy　镁合金

magnesium aluminum alloy　镁铝合金

magnesium anode　镁阳极

magnesium battery　镁电池

magnesium cell　镁电池

magnesium ferrite　镁铁氧体,铁酸镁

magnesium incendiary bomb　镁燃烧弹

magnesium lithium alloy　镁锂合金

magnesium reserve battery　镁储备电池

magnesium silver chloride battery　镁氯化银电池

magnesiun fluoride　氟化镁

magnesyn　磁自动同步机,磁同位器,永磁转子自同步机

magnet　磁铁,磁石,磁体

magnet assembly　磁系统

magnet ring　磁环

magnet surface　磁面

magnet yoke　1. 磁轭,定子机座 2. 磁偏转系统

magnetic actuator　磁悬浮激励器

magnetic aftereffect　磁后效,剩磁效应

magnetic aging　磁老化,磁心老化,磁心陈化

magnetic alloy　磁性合金

magnetic amplifier　磁放大器,磁性放大器,磁增强器

magnetic annular shock tube　环状磁激波管,磁性环形激波管

magnetic anomaly　磁异常,地磁异常

magnetic anomaly area　磁力异常区

magnetic anomaly detector　磁异常探测器,磁探仪,磁导探测仪

magnetic balance　磁秤,磁悬挂天平

magnetic bearing　1. 磁轴承 2. 磁方位角,磁象限角 3. 磁方位,磁向位

magnetic bearing fly wheel　磁轴承飞轮

magnetic bearing of station　电台磁方位

magnetic bearing reaction momentum wheel　磁轴承反作用动量轮

magnetic bomb　磁性炸弹

magnetic bond number　磁邦德数

magnetic brake　电磁制动器

magnetic braking　磁力制动

magnetic bubble　磁泡

magnetic bubble memory　磁泡存储器

magnetic card　磁卡

magnetic character figure　地磁特性图

magnetic charging method　磁充电法

magnetic chart　地磁图,磁性图

magnetic chip detect　磁性金属屑探测,磁碎屑探测

magnetic circuit　磁路

magnetic cleanliness　磁洁净

magnetic coating　磁性涂层

magnetic coil　电磁线圈

magnetic colatitude　磁余纬

magnetic compass　磁罗盘,磁针罗盘

magnetic confinement　磁约束

magnetic control　磁铁控制

magnetic cooling　磁致冷,磁性冷却,退磁法冷却

magnetic coordinate system　磁坐标系统

magnetic core　磁芯,铁芯

magnetic crotchet　地磁扰动,磁扰,磁差,磁鼻

magnetic curvature　磁曲率

magnetic damping　磁阻尼

magnetic decision element　磁判定元件,磁性决定组件

magnetic declination　磁偏角,磁偏移

magnetic deflection mass spectrometer　磁偏转质谱仪

magnetic detector　磁性检波器,磁性探测器

magnetic diffusivity　磁扩散率

magnetic dip angle　磁倾角

magnetic dipole　磁偶极子

magnetic dipole field　偶极磁场,磁偶极场

magnetic dipole moment　磁偶极矩

magnetic disturbance torque　磁干扰转矩,磁干扰力矩

magnetic dumping　磁卸载

magnetic environment　磁场环境

magnetic field　磁场

magnetic field configuration　磁场形态,磁场位形

magnetic field datum　磁场数据

magnetic field effect　磁场效应

magnetic field line　磁力线

magnetic field model　磁场模型

magnetic field strength　磁场强度

magnetic field topography　磁场地形

magnetic flux　磁通量

magnetic flux density　磁感应强度, 磁通密度
magnetic flux tube　磁性涨潮管, 磁通管
magnetic force　磁力
magnetic fuze　磁引信
magnetic heading　磁航向
magnetic heat treatment　磁场热处理
magnetic impulse machining　磁脉冲加工
magnetic induction　磁感应强度
magnetic influence exploder　磁性雷管
magnetic iron　磁铁
magnetic isoanomalous line　等磁异常线
magnetic isoclinic line　等磁倾线
magnetic latching relay　磁保持继电器, 磁性闩锁继电器
magnetic layer　磁层
magnetic layer type　磁层类型
magnetic lens　磁透镜
magnetic level　磁平
magnetic local time　磁地方时
magnetic loop system　环状电磁天线系统, 磁回路系统
magnetic memory　磁存储器
magnetic meridian　磁子午线
magnetic mirror　磁镜
magnetic mirror effect　磁镜效应
magnetic mirror point　磁镜点
magnetic modulator　磁调制器
magnetic moment　磁（力）矩
magnetic multiaperture element　多孔磁元件
magnetic navigation　磁导航
magnetic north　磁北
magnetic north pole　磁北极
magnetic nozzle　磁喷管
magnetic nozzle region　磁喷嘴区域
magnetic orange pipe　磁铁管扫雷器, 磁性橙色管
magnetic overprinting　磁叠印
magnetic particle clutch　磁粉离合器
magnetic particle inspection　磁粉探伤, 磁粉检验, 磁力线探伤
magnetic permeability　磁导率, 导磁率, 导磁系数
magnetic perturbation　磁扰动, 磁扰
magnetic plug　磁塞（吸除金属屑用）
magnetic pole　磁极, 磁荷
magnetic powder　磁粉
magnetic probe　磁探针, 探磁圈
magnetic property　磁性
magnetic prospecting　磁法勘探, 磁性探矿
magnetic quiet zone　地磁低缓带, 磁静带
magnetic radio　确定无线电方位角的基准电台
magnetic RAM　Magnetic Random Access Memory 磁荷随机存储器

magnetic reconnection　磁场重联
magnetic recording　磁性录制, 磁性记录
magnetic recording medium　磁记录媒质
magnetic refrigerator　磁性致冷器
magnetic resonance　磁共振
magnetic resonance gyroscope　核磁共振陀螺
magnetic resonance imaging　磁共振成像
magnetic rigidity　磁刚度, 磁刚性
magnetic ring　磁环, 磁性钢领
magnetic sail　磁化帆
magnetic screen　磁屏蔽
magnetic sector　扇形磁场
magnetic semiconductor　磁性半导体
magnetic sensor　磁性传感器
magnetic separation　磁选, 磁性分离, 磁力分离
magnetic shell　磁壳
magnetic shielded gun　磁屏蔽电子枪
magnetic shielding system　磁屏蔽系统
magnetic sounder　磁测深仪
magnetic spectrograph　磁谱仪
magnetic stabilization　磁稳定
magnetic star　磁性星体
magnetic steel　磁性钢
magnetic storage　磁存储器
magnetic storm　磁暴
magnetic storm satellite　磁暴（探测）卫星
magnetic stratigraphy　磁性地层学, 古地磁地层学
magnetic substorm　磁亚暴
magnetic substorm simulation facility　磁亚暴仿真设备
magnetic susceptibility　磁化率
magnetic susceptibility meter　磁化率计, 磁化率测定器
magnetic suspension　磁悬浮
magnetic suspension accelerometer　磁悬浮加速度表, 磁悬浮加速度计
magnetic suspension balance　磁悬天平
magnetic suspension gyroscope　磁悬浮陀螺仪
magnetic suspension technique　磁悬浮技术
magnetic system　磁系
magnetic tachometer　电磁测速装置, 电磁转速表
magnetic test　磁检验, 磁性测量
magnetic test facility　磁性测量设备
magnetic torquer　磁力矩器
magnetic track　磁航迹角
magnetic tube　磁控制管
magnetic variation　1. 磁差 2. 磁偏角
magnetic vector　磁场矢量
magnetic vector potential line　磁矢位线
magnetic viscosity　磁黏滞性, 磁黏滞度
magnetic washing　磁清洗

magnetic wiggler	磁摆动器
magnetic zenith	磁天顶
magnetically disturbed day	磁扰日
magnetically filter arc deposition	磁过滤电弧沉积
magnetically quiet day	磁静日
magnetics	1.磁学 2.磁性元件,磁性材料 3.磁力学
magnetism	磁性,磁力,磁学
magnetism theodolite	地磁经纬仪
magnetizability	磁化能力,可磁化性,磁化强度
magnetization	1.磁化 2.磁化强度 3.磁化作用
magnetize	磁化,起磁,使磁化
magnetizer	1.磁化器 2.感磁物,传磁物
magneto	磁电机(供点火用),永磁发电机
magneto ionic	磁离子的
magnetoacoustic coupling	磁声耦合
magnetoaerodynamics	磁空气动力学
magnetocaloric effect	磁致热效应
magnetocardiography	心磁描记术,心磁图仪
magnetoconductivity	磁导率,导磁率,磁致电导率,导磁性
magnetocrystalline anisotropy	磁晶各向异性
magnetodrop	磁石式交换机吊牌
magnetodynamic	磁(性)动力的
magnetoelastic	磁致弹性的,磁弹性
magnetoelastic force transducer	磁弹性式力传感器
magnetoelectric relay	磁电式继电器
magnetoelectric tachometric transducer	磁电式转速传感器
magnetoelastic weighing cell	磁弹性式称重传感器,磁致弹性称重传感器
magnetoencephalography	脑磁图描记术
magnetofluid dynamics	磁流体动力学
magnetogasdynamics	磁气体动力学
magnetogram(m)	磁强记录图,地磁自记图,磁图
magnetograph	磁强(自动)记录仪,地磁记录仪,磁力记录计
magnetohydrodynamic acceleration	磁流体动力加速度
magnetohydrodynamic generation	磁流体动力发电
magnetohydrodynamic propulsion	磁流体动力推进
magnetohydrodynamic wind tunnel	磁流体动力风洞
magnetohydrodynamics	磁流体动力学
magnetoionic theory	磁离子理论
magnetometer	磁力计,磁强计,地磁仪
magnetometer bias	磁强计偏差
magnetometer calibration	磁力仪校准
magnetometer datum	磁强计数据
magnetometer error	磁强计误差
magnetometer measurement	磁强计测量
magnetometer model	磁强计模型
magnetometer-based orbit	基于磁强计的轨道
magnetometry	磁力测定术,测磁法,测磁强术,测磁学
magnetomotive force	磁动势,磁通势
magnetomyography	肌磁描记术,肌磁波描记术
magnetooculography	眼磁描记术
magnetooptic display	光磁显示
magnetooptic storage	磁光存储
magnetooptical effect	磁光效应
magnetooptical memory material	磁光存储材料
magnetooptical modulator	磁光调制器
magnetooptics	磁光学
magnetopause	磁层顶
magnetopause current	磁层顶电流
magnetoplasma	磁性等离子体
magnetoplasma thruster	磁性等离子体推进器
magnetoplasmadynamic propulsion	磁等离子体动力推进装置
magnetoplasmadynamics	磁(性)等离子体动力学
magnetoplasmadynamics engine	磁等离子体动力发动机
magnetoplumbite type ferrite	磁铅石型铁氧体
magnetoresistance effect	磁阻效应
magnetoresistor	磁动电阻器
magnetorheological	磁流变
magnetorheological brake	磁流变制动器
magnetoscope	验磁器,验阻器
magnetosensor	磁敏元件
magnetosheath	磁鞘,防磁套,磁套
magnetosonic wave	磁声波
magnetosphere	磁层,磁气圈,磁大气层
magnetospheric bow shock	磁层弓形激波
magnetospheric cavity	磁层内腔
magnetospheric cavity compression	磁层内腔压缩
magnetospheric cleft	磁层裂缝
magnetospheric storm	磁层暴
magnetospheric substorm	磁层亚暴
magnetospherics	磁层物理学
magnetostatic problem	静磁问题
magnetostatic pump	静磁泵
magnetostatic surface wave	静磁表面波
magnetostatic wave	静磁波
magnetostatics	静磁学
magnetostratigraphy	磁性地层学,磁层学
magnetostriction	1.磁致伸缩 2.磁致形变
magnetostriction transducer	磁致伸缩式换能器
magnetostrictive effect	磁致伸缩效应,磁效伸缩效应
magnetotail	磁尾
magnetotelluric method	磁大地电流法
magnetoturbulence	磁性素流,磁性湍流
magnetrol	磁放大器

magnetron 磁控管
magnetron sputtering 磁控溅射,磁控管溅射
magnification 放大
magnification factor 倍率系数,放大因数,放大因子
magnify 放大,扩大
magnistor 磁变管,(电)磁开关
magnitude 1.幅度 2.大小,量级 3.星等,星级 4.震级,光度
magnitude equivalence 量值等效
magnitude frequency characteristics 幅频特性
magnitude margin 幅值裕度
magnitude of burst 爆炸强度
magnitude of eclipse 日蚀大小,月蚀大小
magnitude of threshold 阈值星等,级的阈值
magnitude phase characteristics 幅相特性
magnitude scale 震级标度,大小标尺,量值,量表
magnitude scale factor 幅值比例尺,幅度比例尺,幅值缩尺
Magnum 1.大酒瓶 2.大酒瓶卫星
Magnus balance 马格努斯天平
Magnus effect 马格努斯效应
Magnus force 马格努斯力
Magnus moment 马格努斯力矩
magslip 1.旋转变压器 2.无触点式自动同步机,无触点自整角机
Mag-Thor Magnesium Thorium 镁钍合金
MAGVAR Magnetic Variation 磁差
MAID Manufacturing Assembly and Installation Data 制造装配和安装数据
MAHRSI Middle Atmospheric High Resolution Spectrograph Investigation 中层大气高分辨率摄谱仪研究
mail priority 邮件优先权
Mailstar 邮政星
MAIN 1.Maintenacen 维护,维修 2.Material Automated Information Network 材料自动信息网络
main acceleration 主加速度
main advantage 主要优点
main anode 主阳极
main anode temperature 主阳极温度
main anode voltage 主阳极电压
main beam 主声束,主光束
main beam efficiency 主波束效率
main bearing 主轴承
main belt 干线皮带运输机,主传送带,主小行星带
main blade 主旋翼
main body 主体,主要部分,主船体,机身
main burner 主燃烧器,主喷嘴
main cabin door 主舱门
main category 大类

main chamber 主舱,主燃烧室
main check comparison 主检比对
main coil 主线圈
main combustor 主燃烧室
main component 主要成分,主要部件
main computer 主计算机
main connecting rod 脂杆,主连杆
main crack 主裂纹
main data package 主数据包,主数据分组
main detonation chamber 主爆震室
main diagonal 主对角线,衷角线
main difference 主要区别
main discharge 主放电,主排出管
main droplet 水解液滴
main effect 主效应
main element 主要元素,主振子
main engine 主机,主发动机
main field 1.永久机场 2.主磁场
main float 主浮筒
main flow 主流(量)
main fluid 主要的流体
main fuel 主燃料
main fuel control 主燃料调节器
main fuze antenna 主要引信天线
main gear 主传动齿轮,主起落架
main gear tire 主起落架轮胎
main gearbox 主减速器
main goal 主要目标
main idea 主要意思,主旨
main injector 主喷嘴
main jet 主喷嘴,工作喷嘴
main landing 1.主攻部队的登陆 2.基站,主层站
main (landing) gear 主起落架
main line 1.主线,干线 2.数据总线 3.主管路,总管 4.(飞艇的)系留索
main lobe 主波瓣,主叶,主瓣
main mixer 主机
main motor 主动机,主电动机
main nozzle 主喷嘴
main objective 主目标
main oxygen equipment 主供氧装备
main parachute 主伞
main parameter 主要参数
main part 主件,主要部分
main peak 主峰
main phase 主相,主要阶段
main plasma 主要的等离子体
main plenum 主要充气
main plenum flow rate 主要充气流量

main propellant valve 主推进剂阀
main propulsion 主推进装置
main pump 主泵
main reaction 主要反应,知反应,主反应
main recirculation 主循环
main recirculation region 主要循环区域
main recovery 主要回收
main recovery parachute 主要回收降落伞
main reflector 主反射器,主反射体
main result 主要成绩,主要结果
main rotor （直升机的）主旋翼,主转翼
main rotor blade 主旋叶
main rotor gearbox 主旋翼减速器（直升机用）
main rotor head 主旋翼桨毂
main runway 主跑道
main satellite 主星
main section 干线段
main sequence 主序列,主星序
main sequence star 主序星,主星序
main shock 主震,本震
main spacecraft 大航天器或宇宙飞船
main spacecraft body 大航天器或宇宙飞船的主体
main spar 主梁,大梁（尤指机翼的）主桁架,主翼梁
main spring 1.主弹黄 2.撞针簧
main stage 主要阶段
main step 主断阶
main storage unit 主存储器
main stream 干流,主干流
main transverse 主隔框
main tube 主显像管
main undercarriage 主起落架
main unit 主要部件
main vehicle 主车
main vortex 柱后旋涡
main wing 主翼,主机翼
main-beam 主波束
main-beam clutter 主波束杂波
main-beam ion 主波束离子
main-belt asteroid 主带小行星
main-belt asteroid sample return 主带小行星样本返回
main-element 主成分
main-fuel 主燃油
mainplane 主翼,机翼
mainstage 1.主级 2.主发动机 3.主发动机推进 4.主推进阶段
mainstream （风扇发动机的）内涵道气流
mainstream combustion 主流燃烧
mainstream flow 主流（流量）
mainstream flow path 主流的流动路径

mainstream turbulence 主流湍流
MAINT Maintenance 维护,维修
MAINT MON Maintenance Monitor 维护监控器
maintain 维修,维护,保养
maintainability 可维修性,可维修度
maintainability allocation 维修性分配
maintainability analysis 维修性分析
maintainability assurance 维修性保证
maintainability assurance program 维修性保证大纲
maintainability index 可维护性指数,可维护指数
maintainability management 维修性管理
maintainability model 维修性模型
maintainability prediction 维修性预计
maintainability program plan 维修性工作计划
maintaining period 保管期
maintenance accessibility 维修可达性,维护方便
maintenance action 维修工作,维修活动
maintenance activity 维护活动
maintenance analysis referral 维修分析结果报告
maintenance and depreciation cost 维修及折旧费
maintenance break 维修分离面
maintenance burden 1.总维修费用 2.间接维修费用
maintenance complaints 维修意见
maintenanee complex 维修部门
maintenance concept 1.维修思想 2.维修方案
maintenance contact 维护联系人
maintenance cost 维修费用,维修成本
maintenance crew 1.（航母）维修人员 2.维修队,养护工队
maintenance cycle 维修周期
maintenance data panel 维修资料显示板
maintenance diagnostic logic 维护诊断逻辑
maintenance diagnostic unit 维护诊断设备
maintenance distinguish 维修鉴别性
maintenance downtime （因）维修（而）停用（的）时间
maintenance effectiveness 1.维修有效性 2.维修效能 3.维修优化
maintenance engineering 1.维修工程 2.机务工程 3.保养工程
maintenance error 维修差错
maintenance event 维修事件
maintenance factor 维修系数,维修因子
maintenance float 维修备件周转量
maintenance level 维修级别
maintenance logistics 保养勤务技术保证
maintenance man-hours 维修工时
maintenance man-hours per flying hour 每飞行小时的维修工时
maintenance manual 维修规程,维护规程,维护手册,维

修指南
maintenance operation 维修作业,维护操作
maintenance point 1.(飞机上的)维修点 2.(地面的)维修工作地,维修站
maintenance production 维修生产
maintenance program 维修大纲,维修程序
maintenance rate 维修率(规定时期内的维修次数)
maintenance ratio 维修比
maintenance recorder 检测自记器,维修记录器
maintenance release 维修放行
maintenance service 维护,修理勤务维护业务
maintenance status 维修状态
maintenance task distribution 维修工作分配比
maintenance telemetry 维修遥测
maintenance tests 维修前检测
maintenance time 维修时间
maintenance tree 维修逻辑图,维修树
maintenance unit 1.维修部队 2.维修单位 3.内窥镜维修器 4.维修装置
maintenance-free 不需维修的,无维修的
maintenance-free battery 免维护蓄电池
maintenance-free reliability 不维护可靠性,不维修可靠性
mainwheel 主轮
major accident 严重事故
major airlines 大型航空公司,骨干级航空公司
major alteration 主要变动
major axis 主轴,长轴
major axis direction 主轴方向
major body 主体
major carrier 大型航空公司,主要载体
major component 主要部件,主要元件
major contributor 主要因素
major decision 首要决定,主要决定
major difference 重大差异,主要区别
major effect 重大影响,主要效应
major encounter 主要遭遇战
major engine 主发动机
major feature 主要特点、性能或特色
major frame 主体结构,主框架
major inspection 大检查
major installation 主要地面设施
major landing site 主着陆点
major limitation 主要限制条件或因素,主要局限
major lobe 主瓣,主波瓣,吱瓣
major loop 主回路,主循环
major loop validation 大环比对,主回路验证
major operation 战役,大规模作战,主操作,主运算
major planet 大行星,类木行星

major planetary satellite 主要的行星,卫星
major product 大积
major radius of ellipsoid 椭球长半径
major repair 大修
major semi-axis of the earth 地球长半轴
major source 主要来源
major species 主要物种
major subsystem 主要子系统
major system 基本法,主系统
major vitiation 主要的损害
major-axis plane 长轴面
major-axis spinner 长轴微调器
major-axis spinner case 主轴微调框
majority agreement 多数航空公司支持的协议
majority carrier 多数载流子
majority decoding 大数判决译码
majority rule 择多原则
majority vote method 多数判决法,多数表决法
majority voting system 多数表决系统
majority-logic decoding 择多逻辑解译码
make 制造,接通,闭合,构成,制定
make ready time 生产准备时间
make-and-break 通断,开闭,接离
maladjustment failure 失调故障
maldistribution 分配紊乱,分配失调,分布破坏
maldrop 误投
male contact 阳接触件,刀头触片,插头,插塞,插塞接点
malfunction 故障,机能不良,失灵,工作不正常
MALLAR Manned Lunar Landing And Return 载人宇宙飞船月球着陆和返回
Mallory cell 马洛里电池
MALM Master Air Loadmaster 空勤装卸长(主任)
MALP Manufacturing Assembly Labor Performance 制造装配件劳动效率
MALS 1. Medium-intensity Approach Light System 中等亮度进场照明系统 2. Miniature Air Launcher System 小型空中发射系统
MALU Mode Annunciator Logic Unit 模式显示逻辑部件
MAM Milliammeter 毫安表
MAMDT Mean Active Maintenance Down Time 平均有效维修停用时间
mamma tus 乳房状层积云
MAMOS Marine Automatic Meteorological Observing Station 海洋自动气象观测站
MAMS 1. Manufacturing Applications Management System 制造应用管理系统 2. Missile Altitude Measurement System 导弹高度测量系统 3. Mobile

Air Movements Squadron 机动空运中队
MAN 1. Manual 手动的,手册 2. Maintenance Alert Network 维护警报网
man simulator 人体模型
man space 个人占用空间
manacle 手铐,手铐式缩环
manacle ring 固定环
management control system 管理控制系统
management decision 管理决策
management information system 管理信息系统
management office 管理处
management phase 管理阶段
management program 管理程序
management system 管理系统
management team 管理团队
man-airplane combination 人机组合
man-breathing simulator 人呼吸模拟装置
Manchester's code 曼彻斯特编码
man-computer system 人机系统
mandate 命令,委任
mandatory 指定性的,强制性的
mandatory clearance read-back 空中交通管制许可的强制性复诵
mandatory directive removal 指令性拆卸
mandatory replacement item 强制更换件
mandrel 1. 型芯,芯棒,芯子 2. 心轴
maneton 夹紧螺栓
maneuver 机动飞行,机动,演习,策略,调遣
maneuver constraint 机动约束
maneuver control 操纵控制,机动控制
maneuver cycle 机动周期,操作周期
maneuver design 机动设计
maneuver detector 机动探测器
maneuver direction 机动方向
maneuver dynamics 机动飞行动力学
maneuver flap 机动襟翼,空战襟翼,操纵襟翼
maneuver load 机动载荷
maneuver load alleviation 机动负载减缓
maneuver load control 机动载荷控制
maneuver margin 机动限度,机动能力储备
maneuver model 机动模型
maneuver parameter 操作参数
maneuver path 操作路径
maneuver performance 机动性能
maneuver plane 机动飞机
maneuver plane turn rate 机动飞机转弯角速度或偏转速率
maneuver point 机动点
maneuver rate 航向变化率

maneuver region 机动域
maneuver sequence 机动序列
maneuver state 机动状态
maneuver strategy 机动策略
maneuver threshold 机动阈值
maneuver time 机动时间,机动时机
maneuver tracking 机动跟踪
maneuver type 机动型
maneuver velocity 机动速度
maneuverability 机动性,机动能力
maneuverability polyhedron 可操作性多面体
maneuverable aircraft 机动飞机
maneuverable anti-radiation vehicle 机动反辐射飞行器,反辐射导弹,反雷达导弹
maneuverable missile 机动导弹
maneuverable reentry vehicle 机动再入弹头,机动再入飞行器
maneuverable warhead telemetry 机动弹头遥测
maneuver-induced wake 机动飞行尾流
maneuvering 机动技术,机动,操纵
maneuvering aircraft 机动飞机
maneuvering capability 操纵能力,机动能力
maneuvering chart 机动飞行图
maneuvering condition 机动的条件
maneuvering control 机动飞行控制
maneuvering effect 机动效果
maneuvering energy 机动能量
maneuvering flight 机动飞行
maneuvering free wake model 无机动尾流模型
maneuvering logic 操作逻辑
maneuvering phase 机动阶段
maneuvering target 机动目标
MANF Manifold 总管,支管
manganese content 锰含量
manganese oxide 氧化锰
manganin 锰铜,(锰铜镍线)锰镍铜合金
manger 鬼宿星团,链舱底泄水板
Mangler transformation 曼格勒变换
Manhattan Project 曼哈顿计划
manhole 入孔,入口,检查孔
manifest (运输机的)舱单(记载一次飞行乘载的全部人员和货物),(空运)装货清单,旅客舱单
manifold 1. 歧管,多支管,排管 2. 总管,集流环 3. 进气管
manifold connector 歧管连接器
manifold cut 流形切割
manifold of libration 歧管的天平动
manifold of point 歧管的点
manifold path 多重路径

manifold pressure　1.(活塞式发动机的)进气压力 2.支管压力 3.管道内压力
manifold pressure gage　进气压力表
manifold tube　歧管管路
manikin　人体模型,假人
man-in-loop simulation　人机在环仿真
manipulate　操作,控制,利用,应付,假造
manipulation　1.机械手 2.操作机 3.操作,控制
manipulation scheme　处理方案
manipulator　1.机械手,机械臂 2.边界层气流增速器
manlock　人用气闸
man-machine communication　人机通信,人机联系
man-machine dialog(ue)　人机对话
man-machine function allocation　人机功能分配
man-machine functional comparison　人机功能比较
man-machine integration　人机一体化
man-machine intelligence system　人机智能系统
man-machine interaction　人机交互
man-machine interface　1.人机接口 2.驾驶员运载器接口
man-machine interface design　人机接口设计
man-machine symbiosis　人机共栖,人机共存
man-machine system　人机系统
man-machine system design　人机系统设计
man-machine system engineering　人机系统工程
man-machine-environment system engineering　人机环境系统工程
man-machine-sight system　人机瞄准系统
man-made fault　人为故障
man-made fault tolerance　人为容错
man-made noise　人为噪声,工业噪声,人为干扰
manmade object　人造物体
man-made screen　人造屏幕
man-mounted oxygen regulator　佩戴式氧气调节器
mannable　可住人的
manned　载人的,有人驾驶的
manned aircraft　有人驾驶飞机
manned aviation　载人航空
manned balloon　载人气球
manned capsule　载人(航天)舱
manned flight　载人飞行
manned maneuvering unit　载人机动装置
manned orbital station　载人轨道站
manned orbiting laboratory　载人轨道实验室
manned remote work-station　载人遥控工作站
manned rocket　载人火箭
manned run　真人试验,载人运行
manned space center　载人航天中心
manned space engineering system　载人航天工程系统

manned space flight　载人航天,载人太空飞行
manned space surveillance system　载人空间监视系统,载人航天监测系统
manned space technology　载人航天技术
manned spacecraft　1.载人飞船 2.载人航天器
manned spacecraft assembly and test building　载人飞船装配测试厂房
manned spacecraft engineering　载人飞船工程
manned spacecraft launch　载人飞船发射
manned spacecraft loading room　载人飞船加注间
manned spacecraft recovery　载人航天器回收
manned spacecraft system　载人飞船系统
manned spacecraft telemetry　载人飞船遥测
manned vehicle　载人飞行器
manned vibrator　载人振动实验设备
mano voltmeter　毫微伏表
manoeuver point　飞行机动点
manoeuvrable re-entry vehicle　机动式再入飞行器,机动再入弹头,洲际弹道导弹分弹头
manoeuvre ceiling　机动升限
manoeuvre diagram　机动飞行图
manoeuvre enhancement　机动增强能力
manoeuvre load factor　机动负荷系数
manoeuvre margins　机动边界
manoeuvre-induced error　(仪表的)机动飞行误差
manoeuvring area　机场机动区,运转区
manoeuvring ballistic re-entry vehicle　机动式再入飞行器,机动再入弹头,洲际弹道导弹分弹头
manoeuvring factor　载荷因数,过载
manoeuvring speed　机动速度
manometer　压力计,压力表
manometer bank　压力表阵
manometric lock　气压仪表参数锁定
MANOP　1.Manual of Operation 作战使用手册 2.Manually Operated 手控的,人工操作的
man-out-of-space-easiest　最简便空间救生
manpack　1.便携式的,单人可携带的 2.便携式容器,背负式设备,全球定位系统背负式接收机
MANPAD　Man-Portable Air Defense 便携式防空系统,单兵携带式防空系统
manpower fluctuation　人力需求波动
man-rate　安全评定
man-rated　适于载人的,可载人的
man-rating　适宜于载人航天的,作过载人航天评定的
man-rating test　宜人试验
man-robot system　人-机器人系统
MANS　Missile And Nudet Surveillance 导弹与核爆炸探测监视
man-tended　(临时)有人照料的

man-tended free flyer　有人照料的自由飞行平台
mantle　1.机套,外皮 2.罩 3.地幔 4.等离子体幔
mantle convection　地幔对流
mantle convection cell　地换对流环
mantle heat flow　地幔热流
mantle plume　地幔热柱,地幔羽,地幔涌流
manual bomb hoist　人工挂弹机
manual bomb release　人工投弹
manual caging lever　手控锁定杆,(瞄准具的)选环杆
manual clearing of gun　人工退弹
manual command　人工指令
manual control　人工操纵
manual control behavior　人工或手动控制行为
manual data input programming　手工数据输入编程,手动数据输入编程
manual escape sequence　手动应急离机程序
manual extraction　人工录取,人工回输
manual feedback　人工控制反馈
manual hover　手操纵悬停
manual ignition　手动点火
manual intervention　人工干预
manual iteration　人工迭代
manual iteration method　人工迭代法
manual jump pattern　手操纵"点水"悬停航线
manual mode　手动模式
manual operation mode　人工操作方式,手操作方式
manual override　手动控制装置,人工控制装置
manual pip control　人工脉冲控制器
manual release lever　1.手动投放杆 2.(炸弹钩)手动开放杆 3.(弹射座椅背带系统的)人工开锁手柄
manual reversion　恢复手控
manual shutoff valve　手动断油活门
manual station　手动操作器
manual test　1.手动测试 2.人工测试
manual tracking　手动跟踪,手动追踪,瞄准,人工跟踪
manual welding　手工焊接,手焊
manually steerable directional antenna　手动可操纵的定向天线
manufactured head　铆钉头
manufacturing　制造业,工业,制造,生产
manufacturing automation protocol　制造自动化协议
manufacturing break　制造工艺分离面
manufacturing company　制造企业
manufacturing cost　生产成本,造价
manufacturing process　制造过程,制造工艺,生产过程
manufacturing technology　制造技术,制造工艺
manufacturing tolerance　制造公差
manuscript　数字控制指令表,手稿,手抄本,加工图
man-vehicle system　人-飞行器系统

many body problem　多体问题
MAO　Mature Aircraft Objective 成熟飞机目标
MAOT　Mobile Air Operations Team 流动的航空兵作战小组(英国陆军)
MAP　1.Machine Assembly Program 飞机装配程序 2.Maintenance Assessment Panel 维护评估板 3.Manifold Absolute Pressure 歧管绝对压力 4.Manufacturing Assembly Plan 制造装配大纲 5.Manufacturing Automation Protocol 制造自动化议定书 6.Mars Atmosphere Probe 火星大气探测器 7.Maximum Allowable Probabilities 最大允许概率 8.Missed Approach Point 错失进近点,复飞点 9.Mode Annunciator Pannel 状态通告板
map　地图,天体图
map appearance　地图整饰,图示特证
map border　图廓
map code　1.地图代码 2.变换码,映象码
map color atlas　地图色谱
map color standard　地图色标
map compilation　地图编制,编图
map decoration　地图整饰
map digitizer table　地图数字化桌
map display　地图显示器
map interpretation　地图判读,读图
map interpretation statistics　图像判读统计
map layout　图面配置,图面配置装备摆设,地图布局
map legibility　地图易读性
map load　地图负载量
map measure　量度规,测图器,量图规
map out　在地图上标出,绘图制订
map overlay analysis　地图叠置分析
map perception　地图感受
map plotting　填图
map projection　地图投影
map projector　地图投影仪
map reading　航图测读,地图判读,地标定位
Maple Flag　枫叶旗演习
map-matching　地图匹配
map-matching guidance　地图匹配制导
mapper　测绘仪,绘图仪,变换器
mapping　1.地图绘制,制图作业 2.(铸件的)包砂,夹层
mapping accuracy　测图精度
mapping algorithm　映象算法,演算法
mapping control　图根控制,测图控制
mapping control point　图根点,图根控制点
mapping function　映射函数,变换功能
mapping hyperplane　映射超平面
mapping method with transit　经纬仪测绘法
mapping mission　制图卫星

mapping orbit　映射轨道
mapping photograph　制图照片
mapping radar　地形测绘雷达
mapping recorded file　图历簿
mapping unit　填图单元,制图组
MAPS　1. Measurement of Air Pollution Sensor（航天飞机）传感器测量空气污染　2. Measurement of Atmospheric Pollution from Satellites 卫星测量大气污染　3. Mobile Aerial Port Squadron 机动航空站中队　4. Modular Azimuth Position System 方位位置系统模块
Mapsat　测图卫星
mapsheet　图幅
MAPt　Missed Approach Point 复飞点
MAR　1. Manufacturing Action Request 制造工作申请　2. Marine 海上的
mar　损伤,划痕,擦伤
maraging steel　马氏体时效钢,特高强度钢,高镍合金钢
Marangoni number　马兰戈尼数
Marathon　马拉松卫星系统
Marcopolo　马可波罗卫星
MARCS　1. Marine Computer System 海运电脑系统　2. Military Airlift Reaction Communication System 军事空运司令部反应通信系统
Mardis Air Defence Network　马迪斯防空网
mare　1.〈拉丁文〉海　2.（月亮、火星表面的阴暗区）海　3. 马尾云天气
Mare Crisium　危海,危难海
Mare Foecunditatis　丰富海,月面
Mare Frigoris　冷海
Mare Humorum　湿海
Mare Imbrium　雨海
Mare Nectaris　酒海
Mare Nubium　云海
Mare Serenitatis　澄海
Mare Tranquillitatis　宁静海
Mare Undarum　浪海
Mare Vaporum　汽海（亦作 Sea of Vapors）
MARECS　Maritime European Communication Satellite 欧洲海事通信卫星
Mareng tank　马伦软油箱
MARENTS　Modified Advanced Research Environmental Test Satellite 改进型环境测试研究卫星
margin　1. 余量,裕度,储备量　2. 边缘,界限,图廊　3. 差数,幅度
margin capacity　备用容量,富裕容量
margin of lift　1. 剩余升力　2. 升力限度
margin of safety　（构件载荷的）安全裕度
marginal check　1. 边缘检验　2. 拉偏检验

marginal component　1. 边缘部件,临界机件　2. 临界分量
marginal effectiveness　边际效益
marginal performance　起码性能,边缘性能（仅仅能满足适航性要求或能保证安全飞行的性能）
marginal physiological protection　临界生理防护
marginal stability　极限稳定性,临界稳定性
marginal stability curve　临界稳定曲线
marginal test　边界测试,裕量测试
marginal visibility condition　最低能见度条件
marginal weather　边缘天气,边缘气象
marginal weather approach　极限气象条件进场着陆
marginalization　边缘,忽视,排斥
marginalize　使边缘化,忽略,排斥
marine air（mass）　海洋气团,海洋空气
marine aircraft　水上飞机,水上航空器
marine application　海上应用
marine atmosphere　海洋大气
marine atmosphere corrosion　海洋大气腐蚀
marine condition　海洋状况
marine engine　船用发动机,船用引擎
marine environment　海洋环境
marine geodesy　海洋大地测量学
Marine Landflieger Abteilung　（一次世界大战时期德国的）海军陆基航空分队
marine leveling　海洋水准测量
marine life saving　海上救生,船用救生
marine magnetic survey　海洋磁力测量
marine marker　海上浮标
marine observation　海洋（天气）观测
marine radar　航海雷达,船用雷达
marine radio facsimile recorder　船用无线电传真记录器
marine science　海洋科学
marine vehicle　水上运载工具
Mariner　水手号探测器
Marisat　1. marine communications satellite 海事通信卫星（美国）　2. maritime satellite 海事卫星
maritime aircraft　海上航空器,海上飞机
maritime mobile-satellite service　卫星海事移动业务,卫星水上移动业务
maritime patrol　海上巡逻
maritime patrol aircraft　海上巡逻机
maritime radionavigation-satellite service　卫星海事无线电导航业务
maritime rescue　海上营救,海上救护
maritime satellite　海事卫星,航海卫星
maritime search　海上搜索
maritime service　海上业务,海上服务
MARK　Material Accountability and Robotic Kitting 器

材盘点与自动配套系统
mark 飞机的型,改型
mark dip 进入点水航向
Markarian galaxy 【天】马卡良星系或马卡林星系
marked target 被(激光等)照射的目标
marker 1.指点标,指向标 2.标记,标志器,指示器 3.照明弹 4.反射标准层
marker beacon 示标电台,无线电信标台,指点信标
marker beacon light 飞越信标台指示灯
marker template 标记(截面)样板
market capture 市场占有
market demand 市场需求
market development 市场发展,市场开发
market leader 市场领袖,市场上的主导公司
market mechanism 市场机制,市场调节作用
market niche 市场利基,市场定位,市场补缺基点,市场补缺者
market opportunity 市场机会
market outcome 市场产生的结果,市场回报
market penetration 市场渗入,行销渗透
market position 市场地位,市场定位
market price (股票的)市场价格争夺市场的要求
market requirement 市场需求
market segment 细分市场,市场区隔,分块市场
market share 市场占有率
market uncertainty 市场不确定性
market value (有别于账面价值的)市场价值
marketable 市场的,可销售的,有销路的
marketable range of radius 销售范围
marketing (航空公司通过宣传来)开拓市场
marketing department 市场部,销售部
marketing people 营销人员
marking 用信号板等来传递航空信息
marking of rounds 弹药标记
marking off 划线,标出刻度,用界线隔开
marking panel 信号板(地面军队用来为己方飞机提供识别信号),标志盘
marking team 信号组(空降分队,用来建立导航设施或其他电子装置),标示部队
marking up 标出,标示
Marksman 神枪手
Marmon clamp 马门夹箍(一种专门用于导管连接处的环形夹的专利名称)环形夹
MARRES Manual Radar Reconnaissance Exploitation System 人工操纵的雷达侦察系统
marrying up 匹配组装
MARS 1. Maintenance Analysis and Reports Section 维护分析与报告科 2. Management Analysis Reporting System 管理分析报告系统 3. Maritime Satellite System 海事卫星系统 4. Military Affiliated Radio System 军用附属无线电系统 5. Multimedia Audiovisual Retrieval Service 多媒体声视检索服务 6. Multiple Aperture Reluctance Switch 多孔磁阻开关 7. Manned Aerodynamic Reusable Spaceship 可重复使用气动载人飞船 8. Microwave Atmospheric Remote Sensor 微波大气遥感器 9. Mid-Air Recovery System 空中回收系统 10. Minimally Attended Radar Station 少人值守雷达站 11. Mobile Automatic Reporting Station 机动式自动报知站 12. Modular Airborne Recorder Series 模块式机载记录器系列
Mars balloon relay experiment 火星气球继电器实验
Mars geoscience orbiter 火星地质学轨道器
Mars Observer 火星观测者
marshal 1.起飞顺序调度员,离场飞机调度员 2.陆空军高级将官,典礼官,职行官 3.消防队长,消防部门
marshaller 停机坪调度员,信号员
marshalling point 地面调度点(进场或离场调度员的指挥位置),集结点
marshalling wand 停机指挥旗,信号旗
Marsokhod 火星车,火星探索飞行器
Mars-orbit-insertion 进入火星轨道
marsquake 火星地震
Marte "马特"直升机武器系统(意大利)
Martello "马特洛"
martempering 马氏体等温淬火,马氏体分级淬火
martensite 马氏体,硬化铁炭
martensite phase 马氏体相
martensite phase fraction internal state variable 马氏体相分式内部状态变量
martial dust storm 火星尘暴
martian atmosphere 火星大气
martian capture 火星捕获
martian capture sequence 火星捕获序列
martian channel 火星河床
martian orbit 火星轨道
Marv 机动重返大气层运载工具,机动再入飞行器
MAS 1. Microwave Active Spectrometer 主动式微波波谱仪 2. Middle Airspace Service 中间空域勤务 3. Military Agency for Standardization 军用标准化局 4. Military Area Services 军事区域勤务 5. Military Assistance Sales 军方协助销售
mascon Mass Concentrations 1.质量密集 2.质量瘤
maser source 微波激射源
MASF 1. Military Assistance Service Fund 军事援助勤务经费 2. Mobile Aeromedical Staging Facility 流动空运医疗中转机构
MASH Mobile Army Surgical Hospital 陆军流动外科医院

MASI Mach Airspeed Indicator 马赫空速表
mask aligner 光刻机掩模对准器
mask alignment 掩模对准,掩模校准,掩模重合
mask correlation 掩模相关
mask excess pressure 面罩内余压
mask holder 掩模架
mask hose disconnection 面罩软管断开
maskable interrupt 可屏蔽中断
masked diffusion 掩蔽扩散
masking interrupt 屏蔽中断
mask-making technology 制版工艺
mask-mounted regulator 面罩供氧调节器
Masonite 绝缘纤维板,梅索奈特纤维板
MASPS Minumum Aviation System Performance Standards 最低航空系统性能标准
MASR Microwave Atmospheric Sounding Radiometer 大气探测微波辐射计
MASS 1. Marine Air Support Squadron 海军陆战队空中支援中队 2. Minimum Aeronautical System Standards 最低航空系统标准 3. Multiple Access Sequential Selection 多路存取顺序选择法
mass ablative rate 质量烧蚀率
mass addition 质量并入,质量增加
mass air penetration 集中兵力空中突防
mass attenuation coefficient 质量衰减系数
mass attraction 质量引力
mass attraction vertical 质量引力垂线
mass averaging 质量平均
mass axis 主惯性轴 重心轴线
mass balance 质量配重
mass balance weight 重量平衡配重
mass capability 质量能力
mass capture 流量
mass capture ratio 流量系数
mass center 质量中心
mass center location 质量中心位置
mass center of reentry vehicle 再入飞行器的质心
mass change 质量变化
mass concentration 质量浓度
mass condition 质量条件
mass conservation 质量守恒,质量不变
mass convergence 质量辐合
mass damper 质量减震器
mass decrease 质量下降
mass diffusion 质量扩散
mass discharge coefficient 质量排出系数,质量流出系数
mass discrimination experiment 质量鉴别实验
mass distribution 质量分布
mass drop-off 质量下降
mass efficiency 质量效率
mass ejection 质量抛射
mass entrainment 卷吸
mass estimation 质量估计
mass exchange 质量交换
mass expulsion control 质量排出式控制
mass expulsive device 质量排出装置
mass expulsion system 质量排出系统
mass flow 质量流量
mass flow gain factor 质量流量增益系数
mass flow point 质量流量点
mass flow rate 质量流率
mass flow ratio 质量流率,流量系数
mass fraction 质量比,质量分数
mass function 质量函数,质量分布函数
mass inflow 质量入流
mass injection 质量注入
mass loading ratio 质量负载率
mass loss 质量损失
mass loss rate (烧蚀材料的)质量损失率
mass loss ratio 失重率
mass luminosity relation 质光关系,星的质量与光度关系
mass market 大众市场,大规模的市场
mass matrix 质量矩阵
mass mean diameter 质量平均直径
mass measurement 质量测量
mass model 质量模型,密集模态
mass movement control 质量运动控制
mass multiplier 质量乘子
mass number 质量数
mass of air 气团
mass parameter 质量参数
mass performance 质量性能
mass property 质量特性
mass pulse 质量脉冲
mass ratio 质量比
mass removal 质量脱除
mass sampling 质量取样,群集抽样
mass spectra 质量谱
mass spectrograph 质谱仪
mass spectrometer 质谱仪,质谱分析器
mass spectrometer incoherent scatter model 质谱计非相干散射模型
mass spectrometry 质谱法,质谱学
mass storage 海量存储器
mass taper (旋翼或螺旋桨桨叶的)质量梯度
mass transfer 传质,质量传递

mass transfer cooling 传质冷却
mass transportation 质量输运
mass unbalance gyroscope 质量不平衡陀螺
mass unbalance torque 质量不平衡力矩
mass unit 质量单位
mass utilization 质量利用
mass utilization efficiency 质量利用效率
mass value 质量值
massive body 巨大天体
massive bomb 强力炸弹,大规模破坏性炸弹
massive planet 巨大行星
massive star 大质量星
massive toxic bomb 大型毒气炸弹
massive weapon 大规模破坏性武器
massless 无质量的
massless particle 无质量粒子
mass-to-area ratio 质量面积比
mass-to-light ratio 质光比
mass-to-radius 质量对半径
mast 1.(直升机)旋翼主轴,主轴外伸杆 2.排液管,排泄管
mast bumping 旋翼主轴冲击
mast line (飞艇的)系留钢索
mast mounted sight 旋翼轴顶瞄准具
master 主导装置,原图,主机
master alloy 母合金,重铸键块
master arm selector 军械总电门
master bomb control panel 投弹主控制板
master caption panel 主信号板
master caution signal 主注意信号
master (connecting) rod 主连杆(活塞式发动机的)
master contour template 主外形样板
master control program 主控程序
master control station (全球定位系统)主控站
master control set 主控装置
master curve 主曲线,理论曲线量板,通用曲线
master curve of solid propellant 固体推进剂主曲线
master data 主数据,基本数据
master diversion airfield 主备降机场
master fail relay 主故障继电器
master grid 主网格
master group 主群
master international frequency list 国际频率总表
master mask 母版,母掩模
master navigator 领航主任,领航长,主导航器
master node 主网点
master oscillator 主控振荡器
master pilot 主控驾驶仪,总信号灯
master processor 主处理器,主处理机

master program 主程式,主规划,主控程序
master radar station 主雷达站
master rating 空勤最高技术级
master responding 主台回答(信号)
master rod 主联杆,主连杆
master router template 加工用平面样板
master service sight 标准瞄准具
master slice 母片,主截片
master slice method 母片法
master station 主台
master surface 主尺寸表面模型
master switch 总开关,主控接线机
master template 主模板,划线样板
master timing pulse 主控定时脉冲
master tooling 标准工艺装备
master-slave flip-flop 主从触发器
masthead bombing 超低空轰炸,桅顶轰炸
mastication 素炼,塑炼
MAT 1. Maintenance Access Terminal 维修接近终端 2. Message Acknowledge Time 收信时间 3. Military Aircraft Types 军用飞机型号 4. Multidrop Access Trunk 多点分出接入中继线
Matador Mobile and Tridimensional Air Defence Operations Radar 机动式三坐标防空作战雷达
match filtering 匹配滤波法
match head 匹配头
match point 匹配点
matched filter 匹配滤波器
matched load 匹配负载
matched nozzle 匹配喷嘴
matched print 立体象对,匹配相片,接边相片
matched termination 匹配终端
matcher 匹配器,四面刨制榫机
matching ability of fuze 引信协调性
matching check 匹配检查
matching criterion 匹配准则
matching package 匹配包
matching section 匹配节,匹配段
matching template 匹配模板
matching test 匹配试验
MATCON 1. Microwave Aerospace Terminal Control 航空航天微波终端控制 2. Military Air Traffic Control 军事空中交通管制
MATE 1. Modular Automatic Test Equipment 模块式自动测试设备 2. Multi-system Automatic Test Equipment 多系统自动测试设备 3. Multiterminal Access and Transfer Equipment 多终端存取与替换设备
material 航材,物料

material alternative 代用材料
material arm 物质臂
material behavior 材料特性
material compatibility with propellants 推进剂和材料相容性
material contamination 材料污染
material cost 材料成本，原料成本
material degassing 材料去气
material degradation 材料降解
material dispersion 材料色散
material failure 材料破坏
material function 材料函数
material handling system 材料吊运系统
material interface 物料接口
material mass 物质质量
material mass density 材料质量密度
material mass loss 材料质量损失
material measure 实物量具
material model 材料模型
material outgassing 材料放气
material point 质点
material processing 材料加工
material property 物料性质
material reference 材料参考，参考材料
material reference frame 物质参考系
material science 材料科学
material stiffness 材料刚度
material surface 实质面
material system 物质系统，材料系统
material temperature 材料温度
material thermal response 材料热响应
material type 材料类型，物料类型
material weight 最大材料重量，承重，最大料量
material weight loss 材料重损
material wrinkle 资材皱折
material-based actuator 基于材料的制动器
materials processing in space 空间材料加工
MATH Mathematical 数学的
mathematical analysis 数学分析
mathematical biology 数理生物学
mathematical cartography 数学制图学，理论图学
mathematical formalism 数学形式体系
mathematical formulation 数学公式
mathematical model 数学模型
mathematical optimization 数学优化
mathematical procedure 数学过程
mathematical simulation 数学模拟法
mathematical simulation of fuze warhead matching 引战配合数学仿真，数学仿真的引信弹头匹配

mathematical singularity 奇点
mathematics model method 数学模型法
mating frame 对接框
mating surface 对接面，接合面
matrix algebra 矩阵代数
matrix burning 矩阵燃烧
matrix cracking 基质裂化
matrix display 矩阵显示
matrix equation 矩阵方程
matrix extinction 基膜消光
matrix form 矩阵形式，矩阵组织
matrix function 矩阵函数
matrix inequality 矩阵不等式
matrix interaction 矩阵互动
matrix inverse 反矩阵，矩阵倒式
matrix lamina 矩阵合板
matrix lamina thickness 矩阵合板厚度
matrix material 黏结材料，胶结材料，胎体材料，复合材料
matrix memory 矩阵存储器，矩阵式存储器
matrix method 矩阵法
matrix multiplication 矩阵乘法
matrix notation 矩阵符号
matrix of alternative 替代矩阵
matrix of order 阶矩阵
matrix property 基体性能
matrix receiver 矩阵接收机
matrix representation 矩阵表示
matrix sandwich 中间矩阵
matrix sequence 矩阵次序
matrix surface 基体表面
matrix theory 矩阵理论
matrixer 矩阵变换电路
matrix-fraction 基体分数
matrix-valued function 矩阵值函数
MATS Military Air Transport Service 军事空运局（美国）
matte 冰铜，无光粗糙层，影像形板
matter dominated era 物质占优期
matter-antimatter cosmology 物质反物质宇宙论
maturity 成熟，成熟性，成熟期
MATZ Military Aerodrome Traffic Zone 军用机场交通地带
MAU Marine Amphibious Unit 海军陆战队两栖部队（美国）
MAV 1. Mars Aerocapture Vehicle 火星大气制动飞行器 2. Mars Ascent Vehicle 火星上升飞行器
MAW 1. Medium Assault Weapon (Anti-Tank) 中程攻击武器（反坦克） 2. Military Airlift Wing 军事空运联

队 3. Missile Approach Warning 导弹接近警告
MAWCS Mobile Air Weapons Control System 活动航空武器控制系统
MAWD Mars Atmospheric Water Detector 火星大气水分探测器
MAWP Missed Approach Waypoint 错失进近航路点
MAWS 1. Missile Approach Warning System 导弹来袭警告系统 2. Modular Automated Weather System 模块式自动气象观测系统
MAWTS Marine Aviation Weapons and Tactics Squadron 海军陆战队航空兵武器及战术中队
max maximum 最大
MAX CAP Maximum Capacity 最大容量
MAX FLT Maximum Flight 最大飞行
max speed 最高转速,最大速度,最高速度
Maxaret 马克萨勒刹车防滑系统
maximal allowable concentration 最大允许浓度
maximal clockwise vorticity 最大的顺时针旋转涡度
maximal controllable set 最大可控集
maximal oxygen consumption 最大耗氧量,最大氧耗量
maximal permissible dose 最大容许剂量
maximal power generation 最大发电量
maximal pressure 最高压
maximal shooting range 最大射程
maximal wind 最大风
maximal wind condition 最大风条件
maximally flat amplitude approximation 最平幅度逼近
maximally flat delay approximation 最平延时逼近
maximal-ratio combiner 最大比合成器
maximax 大中取大
maximax approach 大中取大方法
maximax result 大中取大结果
maximin criterion 极大极小判据
maximin method 极大极小法
maximin strategy 极大极小策略
maximin technique 极大极小技术
maximization 极大化,最大化
maximize 最佳化,极大化
maximum 极点,最大量,极大
maximum a posteriori estimate 最大后验估算
maximum acceleration 最大加速度
maximum actuator voltage 最大的致动器电压
maximum allowable concentration 最高容许浓度,最大容许浓度,最大允许浓度
maximum allowable pressure 最高容许压力
maximum altitude 最大高度
maximum amount 最高额
maximum amplification 最大放大
maximum amplitude 最大振幅,振幅

maximum angle of attack 最大攻角
maximum angular acceleration 最大角加速度
maximum angular rate 最大角速度
maximum augmentation 最大限度的增加
maximum augmented power 最大增强力量
maximum authorized altitude 最大核定高度
maximum axis principle 最大轴原理
maximum backpressure 最高背压
maximum bank 最大倾斜
maximum bank angle 最大倾斜角度
maximum baseline 最大基线
maximum blade 最大桨叶
maximum boom-free speed 无爆音最大速度
maximum breathing capacity 最大供氧能力
maximum burning 最大燃烧
maximum burning rate 最大燃烧速率
maximum chamber pressure 最大膛压
maximum climb 最大爬升
maximum coefficient 最大系数
maximum cold thrust 最大冷推力
maximum concentration 最大浓度
maximum concentration level 最大浓度标准或含量
maximum conjunction probability 最大的联合概率
maximum contingency 极大列联
maximum continuous 最大连续
maximum continuous rating 最大连续状态
maximum cruise 最大巡航
maximum cruise rating 最大巡航功率额定值,最大巡航工作状态
maximum cycle 最大循环
maximum cycle pressure 最大循环压力
maximum cyclic load 最大循环负荷
maximum dark time orbit 最大阴影时间轨道
maximum deflection 最大偏转
maximum deflection of Mach number 马赫数最大偏差,马赫数的最大挠度
maximum density of reentry heat flow rate 再入最大热流密度
maximum deviation 最大误差
maximum diameter 最大直径
maximum difference 最大差值
maximum diving speed 最大俯冲速度
maximum drag 最大阻力
maximum drop to arm 引信解除保险最大落下距离
maximum dry thrust 最大干推力
maximum dynamic pressure 最大动态压力
maximum dynamic pressure load 最大动压载荷
maximum dynamic response 最大动态响应
maximum efficiency 最大效率

maximum effort 全力以赴
maximum electron 最大电子
maximum elevation 最大高程，最大仰角
maximum entropy estimation 最大熵估计
maximum erosion 最大冲刷深度
maximum error 最大误差
maximum error magnitude 最大误差级
maximum except take-off 小于起飞功率的最大状态
maximum fillet 最大圆角
maximum flame speed 最大火焰速度
maximum flight 最大飞行
maximum flight range 最大飞行距离
maximum flow 最大流量
maximum force 最大力
maximum force magnitude 最大力级
maximum fuel 最大燃料
maximum fuel temperature 最高燃料温度
maximum fuel-rich 最大燃油浓度
maximum g 1. maximum gravity 最大重力 2. maximum gradability 最大爬坡度
maximum g loading 最大重力负载
maximum gimbal 最大常平架
maximum gimbal angle 最大常平架角，万向架角，万向接头角
maximum gross weight 最大总重量
maximum growth 最大增长量
maximum growth rate 最大的增长率
maximum heat 最大发热量
maximum heat flux 最大热通率
maximum heat transfer capability 最大传热能力
maximum heating 额定功率，稳定功率
maximum height 最大高度
maximum hoop 最大环，最大周
maximum hoop stress 最大周向应力
maximum image visibility 最大影像可见度
maximum impulse 最大脉冲
maximum incidence 临界入射
maximum landing 最大着陆
maximum landing weight 最大着陆重量
maximum length null sequence 最长零序列
maximum level speed 最大平飞速度
maximum lifetime 最大化生命周期
maximum lifetime orbit 最大轨道寿命
maximum lift 最大举升力，最大升力，最高升程
maximum lift coefficient 最大升力系数
maximum lift-to-drag 最大升阻
maximum lift-to-drag ratio 最大升阻比
maximum likelihood 极大似然，最大似然率
maximum likelihood classification 最大似然分类

maximum likelihood criterion 最大似然准则
maximum likelihood decoding 最大似然译码
maximum likelihood estimate 最大似然估计
maximum likelihood estimation 最大似然估计，最大相似估计法
maximum likelihood method 最大似然法
maximum load 最大负载
maximum local equivalence ratio 最大局部等价比率
maximum mass 最大质量
maximum mean camber （翼剖面）最大平均中弧高
maximum mole fraction 最大摩尔分数，最大克分子分数
maximum moment 最大力矩
maximum noise 最大噪声
maximum normal strain theory 最大正应变理论，最大法向应变理论
maximum normal stress theory 最大正应力理论，最大法向应力理论
maximum number 最大数
maximum on ground （军用运输机）起降场地总面积
maximum operating range 1. 最大工作范围 2. 最大作用距离
maximum operation limit speed 最大运行限制速度
maximum orbit 最大轨道
maximum oscillation frequency 最高振荡频率
maximum output power of servo 伺服系统的最大输出功率
maximum output torque 最大输出扭矩
maximum overpressure 最大过压
maximum overshoot 最大超调量
maximum passenger capacity 最大旅客容量，飞机（能容纳的）最多座位数
maximum payload 最大容许载重，最大商载
maximum peak 最大峰值
maximum penetration 最大穿透深度
maximum percent overshoot 最大百分超调量
maximum performance 最大性能
maximum performance take-off 最大性能起飞
maximum permissible concentration 最大容许浓度
maximum permissible error 最大允许误差
maximum permitted life （飞机的）最大允许寿命
maximum permitted mileage 最大允许里程
maximum position 最大位置
maximum position error 最大的位置误差
maximum power 最大功率
maximum power altitude 允许使用最大功率的高度
maximum power case 最大功率情况
maximum power consumption 整机最大功耗，最大功耗
Maximum power point 最大功率点

maximum power setting	最大功率设置
maximum power transfer	最大功率输送
maximum power transfer theorem	最功率传输定理
maximum predicted environment	最高预示环境
maximum pressure	最大压力
maximum principal stress	最大主应力
maximum principle	极大值原理
maximum profit programming	最大利润规划
maximum propagation	最大传输
maximum propellant utilization efficiency	最大推进剂利用效率
maximum pulsation	最大脉动
maximum ramp weight	最大停机坪重量，最大滑行重量，最大机坪重量
maximum range	最大射程，最大探测距离
maximum rate	最高率，最大定额
maximum rating	最大额定值，最大功率，最高级别
maximum reaction rate	最高反应速率
maximum reaction torque	最大反作用力矩
maximum refrigerating capacity current	最大致冷量电流
maximum response	最大响应频率
maximum revolution	最大转速，最大转数，最大传数
maximum roll	最大横倾
maximum safe current	最大安全电流，最大不发火电流
maximum shock response spectrum	最大冲击响应谱
maximum sled	最大雪橇
maximum space	最大空间
maximum specific impulse	最大比冲量
maximum speed	最高速度，最大转速
maximum squeeze	最大收缩
maximum squeeze force	最大收缩力
maximum stagnation pressure	最大滞止压力
maximum steady state burn	最大恒稳态燃烧
maximum steady state burn time	最大恒稳态燃烧时间
maximum stress	最大应力
maximum structural payload	最大结构商载
maximum structural temperature	最高结构温度
maximum structure	最大结构
maximum subcooling	最大过冷度
maximum sunlit orbit	最大日照轨道
maximum surface	最大界面
maximum surface temperature	最高表面温度，最高地表温度
maximum take-off	最大起飞
maximum take-off mass	最大起飞质量
maximum take-off weight	最大起飞重量
maximum tariff rulemaking	最高收费的立法
maximum taxi weight	最大滑行重量
maximum temperature	最高温度
maximum tether	最大范围
maximum tether line load	最大范围线负载
maximum thickness	最大厚度
maximum throttle	最大油门
maximum thrust	最大推力
maximum thrust-to-weight	最大推力重量
maximum tolerance concentration	最大耐受浓度
maximum torque	最大扭矩
maximum total	合计
maximum total weight authorized	法定最大总重量
maximum tracking angular speed	最大跟踪角速度
maximum trim capability	最大纵倾能力
maximum trim incidence	最大纵倾入射
maximum turbine	最大涡轮，最大涡轮机
maximum turn	最大转向
maximum turn rate	最大转速率
maximum uncertainty	最大的不确定性
maximum uncertainty direction	最大的不确定性方向
maximum undistorted output	最大无失真输出
maximum upward acceleration	向上最大加速度
maximum usable frequency	最高可用频率
maximum utility	最大效用
maximum value	最大值
maximum vectoring altitude	雷达引导最大高度
maximum velocity	最大速度，最高速度
maximum vertical speed	最大垂直速度
maximum wall	最大壁
maximum wall temperature	最高壁温
maximum weak-mixture	最大贫油混合气功率
maximum wheel	起重机总重
maximum width	最大宽度
max-sum	最大和
max-sum method	最大和法
Mayak	灯塔卫星
MB	1. Magnetic Bearing 磁方位 2. Marker Beacon（无线电）指点标 3. Mass Breakdown 质量分解 4. Mercury Bromide 溴化汞（激光）
Mb	Megabit 兆位
mb	Millibars 毫巴
MBA	1. Marker Beacon Antenna 信标台天线 2. Minimum Basic Altitude 最低基本高度
MBAR	1. Main Belt Asteroid Rendezvous 主带小行星交会任务（美国）2. Multiple Beam Acquisition Radar 多波束探测雷达
mbar	Millibar 毫巴
MBB	Make Before Break 中断前完成，先接后断
MBC	1. Master Brightness Control 主亮度控制 2. Master Bus Controller 主总线控制器 3. Meteor

Burst Communication 流星脉冲群通信 4. Multicasting Balancing Circuit 多播平衡电路

MBCS Manoeuvre-Boost Control System 快速机队操纵系统

MBD Motor Belt Drive 电动机传送带驱动

MBE 1. Material Bin Evaluation 器材库评估 2. Maximum Brake Energy 最大刹车能量 3. Multiple Bit Error 多位误差,重码错误

MBL Mobile 易流动的,运动物体,汽车

MBM Magnetic Bubble Memory 磁泡存储器

MBO Management By Objectives 目标管理

MBOM Manufacturing Bill Of Material 制造物料表

MBRV Manoeuvrable Ballistic Reentry Vehicle 机动弹道再入飞行器,弹道导弹的分弹头

MBS Marker Beacon System 指点标系统

MBTS Meteorological Balloon Tracking System 气象(气球)追踪系统

MBU Manufacturing Business Unit 制造作业单元

MCA 1. Medium Combat Aircraft 中型战斗机 2. Minimum Crossing Altitude 最低穿越高度,穿越航线最低高度 3. Ministry of Civil Aviation 民用航空部

MCAS 1. Marine Corps Air Station(US)海军陆战队航空站(美国) 2. Midland Countries Aviation Society 内陆国家航空学会

MCB Metal Corner Bead 金属边角卷边

MCBF Mean Cycles Between Failures 平均故障间隔周期

MCC 1. Main Combustion Chamber 主燃烧室 2. Maintenace Control Center 维修控制中心 3. Maintenace Control Computer 维修控制计算机 4. Maintenace Coverage Code 维修范围代码 5. Manual Control Center 人工控制中心,手控中心 6. Material Class and Code 材料等级和代码 7. Mercury Control Center 水星飞船控制中心(美国) 8. Meteorological Communications Centre 气象通信中心 9. Miscellaneous Contract Commitment 其他合同提交

MCCS Multifunction Command and Control System 多功能指挥控制系统

MCD 1. Machine Control Data 机器控制数据 2. Magnetic Chip Detector 金属屑探测器,磁塞 3. Management Collector Drawings 管理选项图 4. Marine Craft Detachment(RAF)水上飞机分遣队(英国空军) 5. Master Control Drawing 主控制图 6. Mission and Communications Display 任务与通信显示器 7. Module Configuration Data 模块构型数据

MCDP Maintenance Control and Display Panel 维修控制显示面板

MCDS Management Communications and Data System 管理通信和数据系统

MCDU Multifunction Controller Display Unit 多功能控制显示器

MCE 1. Modular Control Element 模块式控制元件,积木式控制元件 2. Modular Control Equipment 模块式控制设备,积木式控制设备 3. Motor Control Electronic 电子马达控制

MCF 1. Master Control Facility 主控站 2. Modular Combustion Facility 模块式燃烧装置

McGee tube 麦克吉管

McGill fence 麦吉尔雷达预警线

MCGS Microwave Command Guidance System 微波指挥引导系统

MCID Manufacturing Change In Design 制造中更改设计

MCIP Module Critical Installation Point 模块关键安装点

MCIS Material Control Identifier System 材料控制标识系统

MCK Mission and Communications Keyboard 任务与通信键盘

McKinnon Wood bubble 麦伍氏压力分布图

MCL 1. Major Component List 主要装机设备清单 2. Master Call (Caution) Light 主呼叫(提醒)灯 3. Master Configuration List 主构型清单,主要设备配置表 4. Maximum Climb 最大爬升 5. Minimum Crossing Level 最低穿越高度层

MCLOS Manual Command to Line Of Sight(导弹)手控瞄准线制导

MCM 1. Management Coordination Memo 管理协调备忘录 2. Manipulation and Control Mechanization 操作和控制机械化 3. Manufacturing Control Memo Or Machine Control Medium 制造控制备忘录或机床控制工号 4. Mars Cruise Module 火星巡航舱 5. Master Controlled Medium 机床数控介质 6. Mission Communications Manager 飞行任务通信负责人

MCM/AFM Machine Control Media for Automated Fastening Machines 自动紧固机床的机床控制工具

MCN 1. Manufactureing Control Number 制造控制号 2. Material Change Notice 材料更改通知

MCO 1. Mapping and Communication Orbiter 测绘和通信轨道器 2. Mars Climate Orbiter 火星气象卫星

MCOAM Material Control Order Additional Material 补充材料控制指令

MCP 1. Magnetic Compass Pilot 磁罗经导航 2. Maintenance Control Panel 维修控制板 3. Master Control Program 主控程序 4. Missile Control Panel 导弹控制板 5. Mode Control Panel 工作方式控制板,工作状态控制板 6. Multi-Channel Plate(图像增强器)多路板

MCPF Modular Containerless Processing Facility (美国空间站上的)模块式无容器加工装置

MCR 1. Master Change Record 主更改记录 2. Maximum Cruise Rating 最大巡航额定功率 3. Multispectral Cloud Radiometer 多谱段测云辖射计

MCS 1. Magnetic Control System 磁控制系统 2. Maintenance Control System 维修操纵系统,维护控制系统 3. Management Control System 管理控制系统 4. Maneuver Control System 机动控制系统 5. Maritime Communications Subsystem 海事通信分系统 6. Master Control Station 主控制站 7. Miniature Control System 微型控制系统 8. Minimum Control Speed 最小控制速度 9. Minimum Cutting Set 最小割集 10. Missile Control System 导弹控制系统 11. Mission Control Segment 飞行任务控制部分

MCSR Mission Completion Success Rate 成功完成任务的概率

MCSS Military Communication Satellite System 军事通信卫星系统

MCT 1. Master Contour Template 标准外形样板 2. Maximum Climb Thrust 最大爬升推力 3. Maximum Continuous Thrust 最大连续推力 4. Mercury Cadmium Telluride 碲镉汞 5. Mobile Communications Terminal 机动型通信终端

MCTA 1. Militarily Critical Technology Agreement 军用关键性技术协议

MCU 1. Management Control Unit 管理控制设备 2. Marine Craft Unit 水上飞机小队(英国空军) 3. Modular Concept Unit 模块概念装置 4. Motor Control Unit 机电控制装置(用于自动油门杆) 5. Multifunction Concept Unit 多功能概念装置

MCUR Mean Cycles between Unscheduled Removals 平均非预定拆卸间隔循环数

MCW 1. Modulated Carrier Wave 已调载波 2. Modulated Continuous Wave 已调连续波

MD 1. Magnetic Detector 磁航向传感器 2. Magnetic Deviation 磁差 3. Magnetidc Disk 磁盘 4. Main Deck 主舱 5. Manufacturing Data 制造数据 6. Mean Deviation 平均偏差 7. Motor Direct 直流电动机

MD simulation Molecular Dynamics simulation 分子动力学模拟

MDA 1. Master Diversion Airfield 主备降机场 2. Materials Dispersion Apparatus 材料弥散装置 3. Mechanically Despun Antenna 机械消旋天线 4. Milestone Decision Authority 里程碑决策机构 5. Minimum Decision Altitude 最低决断高度 6. Minimum Descent Altitude 最低下降高度,仪表下降最低高度

MDAIS McDonnell Douglas Aerospace Information Services 麦道宇航信息服务公司

MDAP 1. Mars Data Analysis Program 火星数据分析计划 2. Mutual Defense Assistance Program 共同防御援助计划

MDAS Miniature Data Acquisition System 小型数据采集系统

MDB propellant Modified Double Base propellant 改性双基推进剂

MDBMS Multimedia Database Management System 多维数据库管理系统

MDC 1. Main Deck Cargo 主甲板载货 2. Main Display Console 主显示器控制台 3. Maintenance Data Center 维修数据中心(英国空军) 4. Maintenance Data Collection 维护数据收集 5. Maintenance Data Computer 维修数据计算机 6. Maintenance Diagnostic Computer 维修诊断计算机 7. McDonnel Douglas Corporation 麦道公司 8. Missile Development Center 导弹发展中心(美国空军) 9. Motor Driven Compressor 电机驱动压缩机 10. Multiple Drone Control 多架无人机控制

MDCRS Meteorological Data Collection and Reporting Service 气象数据收集和报告系统

MDCS Maintenance Data Collection System 维护数据收集系统

MDD 1. Mate Demate Device 轨道器装卸设施 2. Master Dimension Definition 主尺寸定义 3. Mission Description Document 任务描述文件

MDE Magnetic Decision Element 磁判定元件

MDF 1. Manipulator Development Facility 机械臂研制设施 2. Mission Degradation Factor 任务降级因素

MDGT Midget 小型物,微型物

MDH Minimum Descent Height 最低下降高度

MDHS Meteorological Data Handling System 气象数据处理系统

MDI 1. Magnetic Direction Indication 磁航向指示器 2. Manual Data Input 人工数据输入 3. Master Dimension Identifier 主尺寸标志

MDL 1. Maintenance Diagnostic Logic 维护诊断逻辑 2. Middle 中间的 3. Minimum Descent Level 最低下降高度层 4. Multipurpose Data Link 多用途数据链

MDMD Meteosat Data Management Department 欧洲气象卫星数据管理部

MDN Median 中央的,中线的,(数)中线

MDOF Multiple Degrees Of Freedom 多自由度

MDR 1. Magnetic-field Dependent Resistor 磁场电阻 2. Mandatory Defect Reporting 规定的故障缺陷报告 3. Mission Definition Review 任务定义评审 4. Multi-channel Data Recorder 多通道数据记录器 5. Multichannel Data Register 多通道数据寄存器

MDRC　Material Development and Readiness Command 作战物资器材研制及战备部(美国)

MDRL　Mandrel 心轴,芯棒

MDS　1. Malfunction Detection System 故障检测系统 2. Master Dimensin Surface 理论外形曲面和基准数据模型,主尺寸表面 3. Materiels De Servitude 辅助设备装置 4. Measurement and Debriefing System 测量与敌情汇报系统 5. Meteoroid Detection Satellite 流星体探测卫星 6. Minimum Detectable Signal 最小可探测信号 7. Minimum Discernible Signal 最小可辨别信号 8. Minimum Discriminable Signal 最小可识别信号

MDSF　Manipulator Dynamics Simulator Facility 机械臂动力学模拟设施(加拿大)

MDSS　Mass Digital Storage System 大容量数字存储系统

MDT　Maintenance Display Terminal 维修显示终端

MDU　1. Maintenance Diagnostic Unit 维护诊断设备 2. Microwave Distribution Unit 微波分配器

MDW　1. Mars Departure Window 飞离火星窗口 2. Mass Destruction Weapon 大规模破坏性武器

MDWP　Mutual Defense Weapon Program 共同防御武器计划

ME　1. Maintenance Engineering 维修工程 2. Manufacturing Engineering 制造工程 3. Micrometeoroid Explorer 微流星体探险者卫星(美国科学卫星) 4. Miter End 榫接端,斜接端 5. Multi-engine 多发动机 7. Munitions Effectiveness 弹药效能

MEA　1. Maintenance Engineering Analysis 维修工程分析 2. Materials Experiment Assembly 材料实验装置 3. Minimum Enroute Altitude 最低航路高度

measured quantity　测定量,被测量

measurement　测量,度量,尺寸,量度制

meacon　1. 模拟干扰 2. 虚造干扰设备,干扰信号发出设备

MEADS　Maintenance Engineering Analysis Data System 维修工程分析数据系统

mean　平均值

mean aerodynamic chord　1. 平均空气动力弦 2. 平均气动弦长

mean albedo　平均反照率

mean amplitude　平均振幅

mean and standard deviation　平均标准差

mean and typical deviation　平均典型偏差

mean angle　平均角

mean anomaly　平近点角

mean apparent viscosity　平均表观黏度

mean axis　中轴,平均轴,中间轴

mean axis system　中轴系统,平均轴系统,中间轴系统

mean blade width ratio　平均叶宽比

mean burning　平均燃烧

mean center of the Moon　平均月心

mean chamber pressure　平均燃烧室压力

mean chord　平均翼弦

mean collapse load　平均损毁荷载,平均极限载荷,平均破坏负荷,平均临界纵向荷载

mean density　平均密度

mean diameter　平均直径,中径,二次平均直径

mean difference　平均差

mean discharge　平均流量,平均放电,平均排放

mean down time　平均停歇时间,平均故障时间,平均停机时间

mean drift rate　平均漂移率

mean drop size　液滴平均尺寸

mean earth ellipsoid　平均地球椭球,均匀地球椭球,平均地球椭球体

mean ecliptic　平均黄道,平黄道

mean element space　平均元间距

mean equator　平赤道,平均赤道

mean equinox　平春分点,平均分点

mean equivalence　平均对等

mean error　平均误差

mean error magnitude　平均误差幅度

mean exhaust　平均排气

mean exhaust velocity　平均排气速度

mean flame　平均火焰

mean fleet performance　机队平均性能

mean flow　平均流量

mean flow velocity　平均流速

mean flowfield　平均流场

mean force　平均力

mean force value　均力值

mean free path　平均自由程

mean frequency　平均频率,中频

mean fuel pressure　平均燃油压力

mean gas temperature　气体平均温度

mean geometric chord　平均几何弦

mean geometric pitch　平均几何螺距

mean heat　平均热力

mean incidence　平均冲角

mean interdiurnal variability　日际变化平均值,平均日际变率

mean jump value　平均跃值

mean life　平均寿命,平均使用年限

mean life criteria　平均寿命判据

mean line　中线

mean location　一般位置

mean logistic delay　平均后勤延误时间

mean longitude　平黄经

mean longitude at epoch 历元平黄经
mean loss 平均损耗
mean man-hours to repair 修理平均工时
mean mass 平均质量
mean mission duration time 平均任务持续时间
mean motion 平均运动
mean node 平均节点
mean noise 平均噪声
mean noon 平正午,平均正午,平午
mean number 平均数
mean obliquity 平均黄赤交角
mean observatory 平均天文台
mean orbit 平均轨道
mean orbit element error 平均轨道要素偏差
mean orbital element 平均轨道根数,平均轨道要素
mean orbital heat flux 平均外热流
mean parallax 平均视差
mean parameter 平均参数
mean particle 平均颗粒
mean performance 平均性能
mean period 平均周期
mean point of impact 平均弹着点
mean point of impact error 平均弹着点误差
mean pole 平极,平均极
mean pole of the epoch 历元平极
mean position 平均位置,平位置
mean pressure 平均压力,平均有效压力
mean profile 平均轮廓
mean radius of curvature 平均曲率半径
mean reaction 平均反应
mean reaction rate 平均反应速率
mean reattachment 平均再附着,平均再接触
mean reattachment point 平均再附着点,平均再接触点
mean response 平均响应
mean sea level 平均海平面
mean semimajor axis 平均半长轴
mean separation 平均分隔距离
mean sidereal time 平恒星时
mean side-view 平均侧视
mean side-view structure angle 平均侧视结构角
mean size 平均直径,平均大小,平均尺寸
mean solar day 平太阳日
mean solar hour 平太阳时数
mean solar time 平太阳时
mean solar year 平太阳年
mean speed 平均速度
mean square error 均方差,中误差
mean square error of a point 点位中误差
mean square error of angle observation 测角中误差

mean square error of azimuth 方位角中误差
mean square error of coordinate 坐标中误差
mean square error of height 高程中误差
mean square error of side length 边长中误差
mean square value 均方值
mean structure 均值结构
mean temperature 平均温度
mean thickness 平均厚度
mean time 平均时间
mean time between downing event 平均不工作事件间隔时间,停机事件间的平均时间
mean time between failures 平均故障间隔时间
mean time between maintenance 平均维护间隔时间,平均能工作时间
mean time between overhaul 平均大修间隔时间,大修平均间隔时间
mean time between repairs 平均修理间隔时间
mean time to detection 平均故障检测时间
mean time to failure 平均故障发生时间
mean time to repair 平均修复时间
mean time to restore 平均故障修复时间
mean traffic 平均流量
mean transinformation content 平均信息传送量
mean transmissivity 平均透过率
mean transmitter level 平均发射机电平
mean up-duration 平均持续在用时间
mean value 平均值
mean velocity 平均速度
mean velocity profile 平均速度剖面
mean wind 平均风速
mean-burning propellant 平台推进剂
meander 曲流
meandering coefficient of traverse 导线曲折系数
mean-field 平均场
meaningful result 有意义的结果
meanline 等分线
meanline analysis 等分线分析
mean-mass 平均质量
mean-mass diameter 质量中间直径
mean-mass particle 平均质量颗粒
mean-mass particle diameter 颗粒平均直径
means of compliance 符合性方法
means of navigation 导航方式
mean-time clock 平时钟
mean-time signal 平凡式时号
mean-to-peak pressure 平均到峰值压力
MEAR Manufacturing Engineering Action Required 制造工程分析
MEAS Mechanical Engineering Aircraft Squadron 飞机

机械工程中队（英国空军）
Measat Malaysian East Asia Satellite 马来西亚东亚卫星
measurable magnitude 可测度数量
measurable quantity 可测量
measurand 被测变量，被测性能，被测情况
measure 测量，措施，程度，尺寸
measure of control 控制措施
measure of control activity 控制活动或行为的测量
measure of distance 衡量距离
measure of dynamic distortion 动态畸变的测量
measure of effectiveness 有效性测量
measure of merit 效果测试
measure of uniformity 一致性测量
measured performance 实测性能
measured thrust 实测推力
measured thrust coefficient 实测推力系数
measured trajectory 实测弹道，实测轨迹
measurement ambiguity 测量模糊或歧义
measurement analysis program 测量分析程序
measurement and control system of wind tunnel 风洞测量控制系统
measurement angle 测量角度
measurement area 测量区
measurement battery 测量电池
measurement bias 计量偏误
measurement covariance 计量协方差
measurement datum 测量基准
measurement delay 量测滞后
measurement density 密度测定
measurement during mooring 停泊测量
measurement during sailing 航行测量
measurement effectiveness 效能测试
measurement effectiveness time 测量时间有效性
measurement element 测量元件
measurement epoch 纪元或时代测量
measurement equation 测量方程
measurement error 测量误差，量度误差
measurement history 计量历史
measurement in a closed series 组合测量
measurement information 量测资讯，计量资料
measurement information processing 测量信息加工
measurement information receive 测量信息接收
measurement location 测量定位
measurement matrix 测量矩阵
measurement model 度量模型
measurement module 量测模组
measurement noise 测量噪声
measurement object 度量对象
measurement of membrane 膜测量

measurement of transition 相变测量
measurement partials matrix 测量偏导数矩阵
measurement plane 测量平面
measurement point 测量点
measurement population 测量总体
measurement port 测量端口
measurement precision 测量精度
measurement procedure 测量程序
measurement process 测量过程
measurement process control 测量过程控制
measurement range 测量范围
measurement residual 测量残差
measurement result 测量结果
measurement sample 测量样本
measurement signal 测量信号
measurement spot 点测量
measurement standard 计量标准
measurement station 测点
measurement system 测量系统
measurement technique 测量技术
measurement time 测量时间
measurement uncertainty 测量不确定性
measurement unit 计量单位，测量单位
measurement update 测量校正
measurement vector 测量向量
measurement volume 测量卷，测量体积
measuring 测量，衡量
measuring chain 测量链
measuring element 测量元件
measuring equipment 测量设备
measuring instrument 测量仪器
measuring object 测量对象
measuring range 测量范围
measuring segment 测量段或环节
measuring system 测量系统
MEB 1. Main Electronics Box 主电子设备盒 2. Moisture Extracting Blade 去湿叶片
MEBUL Multiple Engine Build-Up List 多发动机配套装配表
MEC 1. Main Engine Control 主发动机控制 2. Main Equipment Center 主设备中心 3. Materials Experiment Carrier 材料实验台 4. Manual Emergency Control 人工应急操纵 5. Metrology Engineering Center 气象工程中心
Mecaplex 麦卡有机玻璃
MECH Mechanism 机械，机构
mechanic refrigerator 机械致冷器
mechanical alloy 机械合金
mechanical and thermal property 机械性能和热性能

mechanical behavior   机械特性,力学特征,机械行为
mechanical blockage   机械堵塞
mechanical bond   机械结合
mechanical caged reticle   机械锁定光环
mechanical commutator   换向器,机械转换器
mechanical de-icing   机械除冰
mechanical denuding   机构剥蚀
mechanical design   机械设计,构造设计,机械设定
mechanical despin of antenna   天线机械消旋
mechanical efficiency   机械效率,力学效率
mechanical energy   机械能
mechanical engineering   机械工程,机械工程学
mechanical environment   机械环境,力学微环境
mechanical equilibrium   力学平衡机械平衡
mechanical equivalent of heat   热功当量
mechanical erosion   刻蚀,机械浸蚀,机械腐蚀
mechanical fastener   机械固定件
mechanical fatigue   机械疲劳
mechanical fault diagnosis   机械故障诊断
mechanical fiducial mark   机械框标
mechanical fuze   机械引信
mechanical hander   机械把装置
mechanical hinge   机械铰链
mechanical impedance   力阻抗,机械阻抗
mechanical interface   机械接口,结构接口
mechanical interface coordinate system   机械接口坐标系
mechanical joining technique   机械连接工艺
mechanical load   机械负荷,机械载荷,机械负载
mechanical lung   1.机械肺 2.假肺
mechanical measurement   机械法测量,力学测量
mechanical model of slosh   晃动力学模型
mechanical noise   机械噪声
mechanical polishing   机械抛光
mechanical pressure scanner   机械式压力扫描仪
mechanical projection   机械投影
mechanical property   机械性能,力学性质
mechanical pulsator   机械搅拌器或振动器,机械脉动机
mechanical quality factor   机械品质因数
mechanical reactance   力抗
mechanical remanence   机械剩磁
mechanical resistance   力阻
mechanical scanning   机械扫掠,机械工作台,机械扫描
mechanical sector scan   机械扇形扫描
mechanical selection of aircrew   空勤人员医学选拔
mechanical shock test   机械冲击试验
mechanical shutter   机械快门
mechanical simulation test   力学环境模拟试验
mechanical stop   1.机械限动器 2.(转膛炮的)阻弹板 3.机械停车

mechanical stress   机械应力,力学应力
mechanical system   机械系统,力学系统,力学体系
mechanical test   力学试验,机械试验
mechanical type balance   机械式天平
mechanical vibration   机械振动
mechanically activated battery   机械激活电池
mechanically alloyed dispersion strengthened material   机械合金化弥散强化材料
mechanically despun antenna   机械消旋天线
mechanics of rarefied gas   稀薄气体力学
mechanism model   机理模型
mechanism of propagation   传播机理
mechanism of transition   转变机理
mechanism rehearsal   机械合练
mechanistic understanding   机械的理解
mechanization   机械化,机动化
mechanobalance dynamic stability   机械天平动稳定性
mechanobalance restoring moment   机械天平恢复力矩
mechanobalance static stability   机械天平静稳定度
mechatronics   机电一体化,机械电子学
MED   1. Manufacturing Engineering Documentation 制造工程文件 2. Manufacturing Enhanced Dataset 制造拓延数据集 3. Maximum Exposure Dose 最大照射剂量 4. Microelectronic Device 微电子装置 5. Momentum Exchange Devices 动量交换装置
MEDA   1. Maintenance Error Decision Assistance 维修错误决策辅助 2. Military Emergency Diversion Airfield 应急军用备降场
Medevac   Medical Evacuation 医疗后送,伤病员空运后送
medevac helicopter   医疗后送直升机,救护直升机
media-access control   介质访问控制,媒体存取控制,媒体访问控制,媒体接入控制
median   中位数,中动脉,中值,中线
median filter   中值滤波器
median lethal concentration   半数致死浓度
median lethal dose   半致死剂量
median radius   中数半径
median selection   中值选择
median volume   中值体积
medical air evacuation   空中医疗后送
medical assessment   医疗审定
medical care   医疗保健,医疗护理
medical certificate   诊断书,健康状况证明
medical cyclotron   医用回旋加速器
medical disability   医学失能
medical electron linear accelerator   医用电子直线加速器
medical electronics   医学电子学
medical evaluation   医学鉴定

medical evaluation of medicine 内科鉴定
medical evaluation of surgery 外科鉴定
medical examination 体检
medical examiner 体检医生
medical grounding 医学停飞
medical information form 旅客健康情况登记表
medical monitoring 医学监测
medical monitoring of astronauts 宇航员医务监督
medical selection 医学选拔
medical standards 医学标准
medical treatment facilities 医疗设施
medium 媒质,介质
medium air traffic bub 中型空中交通枢纽
medium altitude 中空
medium bird 中等鸟
medium bomber 中型轰炸机
medium caliber 中等口径
medium caliber ammunition 中等口径弹药
medium career commission 中期服役
medium cloud 中层云
medium field-of-view radiometer 中等视场辐射仪
medium fire 中等速度射击
medium frequency 中频
medium gun 中口径炮
medium haul 中程运输
medium launch vehicle 中型运载火箭
medium level map reading 中空地图判读
medium machine gun 中型机枪
medium resolution infrared radiometer 中分辨率红外辐射计
medium scale integrated circuit 中规模集成电路
medium support helicopter 中型支援直升机
medium term planning 中期规则
medium turn 中等坡度转弯
medium wide-bodied freighter 中程宽机身货机
medium-accuracy 中等精度
medium-altitude 中高度
medium-angle loft bombing 中角度上仰轰炸
medium-capacity bomb 普通炸弹,中等威力炸弹(英国)
medium-case bomb 中厚壳炸弹
medium-lift booster 中型助推器,中型运载火箭
medium-range 中程的
medium-range ballistic missile 中(近)程弹道导弹
medium-range cover 中距离掩护
medium-range positioning system 中程定位系统
medium-range transport 中程运输机
medium-scale data utilization station 中型(卫星)资料利用站
medium-scale map 中比例尺地图

medium-wave communication 中波通信
MEECN Minimum Essential Emergency Communications Network 最低限度基本应急通信网
MEED Medium Energy Electron Diffraction 中能电子衍射
MEF 1. Minimum Essential Facilities 最低限度基本设施 2. Maintenance Entity Function 维护实体功能
MEFC Manual Emergency Fuel Control 应急人工控制燃油
mega 兆,百万
megacycle 兆周,百万周
megadeath 一百万人死亡
megahertz 兆赫
megaline 兆力线
megaton 1. 兆吨 2. 百万吨级
megaton weapon 百万吨(梯恩梯当量)级(的核)武器
megawatt 兆瓦,百万瓦
megger 高阻表,兆欧表,摇表,迈格表
Meggitt decoder 梅吉特译码器
MEGO Megohm 兆欧
MEGV Million Volts 百万伏
MEGW Megawatt 兆瓦
MEI 1. Main Engine Ignition 主发动机点火 2. Maintenance Engineering Inspection 维修工程检查 3. Maintenance Event Information 维护事件信息
MEIS Modular Engine Instrument System 模块式发动机仪表系统
MEK Methyl Ethyl Ketone 甲基乙基酮溶液,丁酮
MEL 1. Minimum Equipment List 最低设备清单,最小限度设备清单 2. Multi-Engine Licence 多发动机证书
melt 熔融,熔化,熔化物
melt infiltration 熔融态浸溃法,熔体浸渗法
melt layer 熔化层
melt snowfall amount 融化的雪量
melted metal squeezing 液态模锻
melting 熔解,融化的,熔化的
melting interface 熔化分界面
melting point 熔点
melting time 熔化时间,熔断时间
melt-through 全部熔化
MEM 1. Maintenance Enginnering Manual 维护工程手册 2. Mars Excursion Module 火星登陆舱(美国) 3. Materials Experimentation Module 材料实验舱 4. Memory 记忆,内存储器
member galaxy 成员星系
membership function 从属函数,属籍函数
membrane 膜片,薄膜,隔膜
membrane aeroshell 膜减速伞

membrane aeroshell model 膜减速伞模型
membrane edge 膜边界
membrane effect 薄膜效应
membrane equation 薄膜方程
membrane filtration 膜过滤
membrane ion 薄膜离子
membrane layer 薄膜层
membrane material 膜材,膜材料,膜材质
membrane mirror 薄膜反射镜
membrane power supply 膜电源
membrane stress 膜应力,薄膜应力
membrane strip 膜条
membrane structure 膜结构,薄膜结构
membrane substrate 膜介质
membrane surface 细胞膜
membrane temperature 膜温度
membrane tension 膜片张力
membrane theory 薄膜理论
memistor 存储电阻器,电解存储器
MEMO Memorandum 备忘录
memorandum of understanding 谅解备忘录
memoristor 忆阻器
memory 1.存储,存储器 2.记忆,记忆力
memory address register 存储器地址缓冲寄存器
memory alignment 存储对准
memory bandwidth 存储带宽
memory circuit 记忆电路,存储电路
memory economy 存储器经济性
memory effect 记忆效应,存储效应
memory effect of battery 蓄电池的记忆效应
memory footprint 内存占用,内存印迹
memory guida nee 存储制导
memory logic unit 存储器逻辑单元
memory margin 存储器容限,记忆裕度
memory resource 存储资源
memory retention (电源切断时)存储信息保存期
memory sharing 存储器共享
memory tracking 记忆跟踪
memory unit 存储单元
memory window 存储窗
memorycoil spring switch 螺簧交换机,螺簧接线器
memoryless source 无记忆信源
memory-replay telemetry 记忆重发遥测
MEMPT Memory Ponit 记忆点
mend 改进,改良,修补处
menders 报废零件
meniscus 弯月面,弯月形透镜
mental capacity 智能,心智容量
mental confusion 精神错乱,意识模糊

mental fatigue 心理疲劳,精神疲劳
mental map 心理地图,意境地图,心象地图,意象图
mental model 心智模式,心智模型,心理模型
mental performance 心理作业
mental test 智力试验
menu selection mode 菜单选择式
menu-driven scenario 菜单驱动说明
MEO 1. Manned Earth Observatory 载人地球观测台 2. Mass in Earth Orbit 地球轨道上的质量 3. Medium Earth Orbit 中地球轨道,地球中轨道 4. Memo Engineering Order 备忘录工程指令
MEOL Manned Earth Orbiting Laboratory 载人地球轨道实验室
MEOP Maximum Expected Operating Pressure (固体火箭发动机)预期的最大工作压力
MEOS Multidisciplinary Earth Observation Satellite 多用途地球观察卫星(欧洲空间局)
MEOSS Monocular Electro-Optical Stereo Scanner 单筒光电立体扫描仪
MEOTBF Mean Engine Operating Time Between Failures 发动机平均故障间隔时间
MEP 1. Marine Environmental Protection 海洋环境保护 2. Mean Effective Pressure 平均有效工作压力 3. Mission Equipment Package 成套任务设备 4. Multi Engine Pilot 多发飞机驾驶员
MEPED Medium Energy Proton and Electron Detector 中能质子电子探测器
MEPS Medium Energy Particle Spectrometer 中能粒子能谱仪
MEPU Monofuel Emergency Power Unit 单元燃料应急动力装置
MEPW Minimum Electrical Pulse Width 最小电脉冲宽度
MER 1. Meridian 子午线(圈) 2. Mission Evaluation Room 飞行任务评审室 3. Multiple Ejection Rack 复式弹射挂架,多外挂物弹射挂架
MERA Molecular Electronics for Radar Applications 雷达应用的分子电子学
Mercator mosaic 麦卡托拼图
Mercator projection 麦卡托投影
mercurous chloride 氯化亚汞,甘汞
Mercury 水星飞船
mercury 汞
mercury battery 水银电池
mercury cadmium telluride detector 汞镉碲检测器,碲镉汞探测器
mercury ion 汞离子
mercury level 水银水平仪
mercury pool cathode 汞池阴极

mercury pool rectifier 汞池整流管
mercury vapor tube 汞汽整淋,汞汽整流管,水银充气整流管
mercury-arc rectifier 汞弧整流管
merge 双机会合近距攻击,使合并,使并入
merge of galaxy 星系吞并
merged indication 机群目标合批显示
merged zone 合并后的区域
merger 合并,吞并
merger location 合并位置
meridian 1. 子午线,经线,子午圈 2. 子午线的 3. 子午面
meridian altitude 过子午圈高度,中天高度
meridian circle 子午环(天文光学仪器)
meridian deflection 子午线偏差
meridian of longitude 经度
meridional parts 经线弧长
meridional plane 子午线断面
meridional ray 子午光线
MERIS Medium Resolution Imaging Spectrometer 中分辨率成像光谱仪
merit factor 优质率,质量因数,品质因素
merit factor of receiving system 接收系统品质因素
merit function 优值函数,优化方程
Merkel number 麦克尔数
MERM 1. Manufacturing Engineering Requirement Model 生产工程需求模型 2. Material Evaluation of Rocket Motor 火箭发动机材料鉴定
Merritt Island 梅里特岛
MERS 1. Magnetic Energy Recovery Switch 磁能恢复开关 2. Multi-Element Radiometer System 多元件辐射计系统
MERTO Maximum Energy Rejected Take Off 最大能量中断起飞
MES 1. Main Engine Starting 主发动机起动 2. Major Equipment Supplier 主要设备(或装备)供应商 3. Manned Escape System 载人救生系统
MESA 1. Manned Environmental System Assessment 载人环境系统评价 2. Mobile Exploration System for Apollo 阿波罗飞船移动探测系统 3. Modular Equipment Stowage Area 模块式装备储存区
mesa 台地,平顶山
mesa burning 台面燃烧
mesa transistor 台面型晶体管,台面式晶体管
Mesar Multifunction Electronically Scaned Adaptive Radar 多功能电子扫描自适应雷达
MESC Mid Electrical Service Centre 中央电器服务中心
MESFET Metal Semiconductor Field Effect Transistor 金属半导体场效应晶体管
MESG Micro Electrostatically Suspended Gyro 微电子静电悬浮陀螺仪
mesh cathode 网状阴极
mesh density 网目密度,网格密度,密目密度
mesh design 网目(格)设计
mesh dimension 网格尺寸
mesh double bumper shield 双层网格防护屏
mesh generation 网格生成
mesh grid 编织网,细网,网状栅极
mesh interface 网格界面
mesh level 网络级
mesh network 网状网络
mesh of finite element 有限元网络
mesh point 网格点,网点
mesh refinement 网格细化
mesh resolution 网格分辨率
mesh scale 网格尺度,网格距
mesh screen filter 网式过滤器
mesh size 筛孔尺寸,网格大小,目径
mesh solution 网格解决方案
mesh volume 网格体积
mesh-free method 无网格法
mesocline (大气)中间层的下层,(大气)中间层升温段
mesodecline (大气)中间层的上层,(大气)中间层降温段
mesometeorology 1. 中尺度气象学 2. (大气)中间层气象学
mesomorphic states 介晶态,液晶状态
meson 介子,重电子
meson telescope 介子望远镜
mesopause 中间层顶,中气层顶
mesopeak (大气)中间层最高温度点
mesopic vision 黄昏视觉,中间视觉
mesoscale model 中尺度模式,中间比例模型
mesosphere 中间层
message 1. 文件 2. 信息 3. (通信系统中的)起始调制波
message authenticator code 信息鉴别码
message data 信息数据
message processing 信息处理
message rate 消息速率
message sequence 信息序列,消息序列
Messenger "信使"(英国单脉冲二次监视雷达)
messenger and telephone talker (航母)传令兵和电话兵
MESSOC Model to Estimate Space Station Operations Costs 空间站运行成本估算模型
MESSR Multispectral Electronic Self Scanning Radiometer 多谱段电子自扫描辐射计
MEST Mobile Earth Station Facility 移动式地球站设施

met briefing 飞行前飞行员索取气象资料，向飞行员下达气象情况指示
met data meteorological data 气象资料，气象诸元
metabolic clock 代谢钟，生理时钟
metabolic imaging 代谢成像
metabolic simulation device 代谢仿真装置
metabolism simulator 代谢模拟装置
metacartography 形而上制图学，元地图学
metacode 元代码
meta-deficient star 贫金属星
metagalaxy 宇宙，总星系
metaknowledge 元知识
metal air battery 金属空气电池
metal brush seal 金属刷密封
metal ceramic 金属陶瓷
metal concentration 金属浓度
metal content 金属含量
metal corrosion 金属腐蚀
metal deactivator 金属钝化剂
metal emitter 金属发射器
metal fiber reinforced metal matrix composite 金属纤维增强金属基复合材料
metal film 金属薄膜，金属膜电阻
metal film potentiometer 金属膜电位器
metal foam 泡沫金属
metal foil 金属箔
metal fouling 炮管碎片，炮膛挂铜
metal fuel 金属燃料
metal glaze potentiometer 金属玻璃釉电位器
metal hose 金属软管，柔性管，金属蛇形管
metal ignition 金属点火
metal inert gas welding 金属焊条电极惰性气体掩弧焊
metal layer 金属层
metal loading 金属加载
metal matrix 金属基体，金属模版
metal matrix composite 金属基复合材料
metal matrix particulate composite 微粒增强金属基复合材料
metal melting 金属熔炼
metal oxidation 金属氧化
metal oxide 金属氧化物，金属氧化物电阻，金属绝缘膜
metal particle 金属微粒
metal powder 金属粉末
metal propellant 金属推进剂
metal rotor 金属转子
metal sample 金属试样
metal screen filter 金属网过滤器
metal skin 金属皮
metal soft magnetic material 金属软磁性材料，金属软磁材料
metal spinning 旋压加工，金属旋压
metal spraying 金属喷镀，喷金
metal strip 铁带，钢带，金属条
metal structure 金属结构
metal tape 金属磨条，金属磁带，钢卷尺
metal temperature 金属温度，管壁金属温度计算，料温
metal vapor laser 金属蒸气激光器
metal-air cell 金属空气电池
Metalastik 麦塔拉斯减振装置
metal-clad plate 覆箔板
metal-hydrogen cell 金属氢电池
metal-hydrogen nickel battery 金属氢化物镍电池
metal-isolator-semiconductor solar cell 金属绝缘体半导体太阳能电池
metalize 使金属化，使硬化，使矿化
metallic element 金属元素
metallic exchanger 金属交换器
metallic fuel 金属燃料
metallic hose 金属软管
metallic packaging 金属封装
metallic particle 金属颗粒
metallic phase 金属相
metallic plate 金属导体板
metallic platform 金属平台
metallic propellant 金属推进剂
metallic thermal protection system 金属热防护系统
metallic uranium 金属铀
metallic-line star 金属线星
metallics 1.金属基体复合材料 2.金属物质，金属粒子
metallization 金属化，敷金属，金属喷镀
metallize 用金属处理，使金属化
metallized paper capacitor 金属化纸介电容器
metallized parachute 金属化降落伞
metallizing 金属喷涂，表面合金化
metallographic analysis 金相分析
metallographic examination 金相检验
metallographic inspection 金相检查
metallurgical defect 冶金缺陷，金相缺陷
metallurgical-grade silicon 原料冶金硅，冶金级硅
metal-nitride-oxide semiconductor 金属氮化物氧化物半导体
metal-oxide 金属氧化物
metal-oxide semiconductor 金属氧化物半导体
metal-poor star 贫金属星，贫金属恒星
metalster 金属膜电阻器
metal-to-metal adhesive 金属胶，金属黏合剂
metamodel 元模型
metamorphic water 变质水，变质水分

metascope 红外线指示器,红外线显示器
metastable 亚稳的,相对稳定的
metastable intermolecular composite 亚稳态分子间复合材料
metastable ion 亚稳离子
metastable propellant 亚稳定推进剂
metastable state 亚稳状态
METEC Meteoroid Technology 流星体技术
meteo groundspeed 按气象风求出的地速
meteogogical orifice 节流嘴
meteogogical report 气象报告
meteogogical watch office 气象监视台
meteor 流星
Meteor 300 "流星300"(德国的气象雷达)
meteor bumper (航天器的)防流星屏,防流星罩
meteor burst communication 流星短脉冲通信
Meteor-Priroda 流星自然卫星
meteor radar 流星雷达
meteor shock 流星撞击
meteor shower 流星雨
meteor stream 流星群
meteor trail 流星余迹,流星尾迹
meteor trail communication system 流星余迹通信系统
meteoric dust 流星尘,陨星尘,燎埃,燎尘
meteoric trail communication 流星余迹通信
meteoric water 雨水,大气水,天落水
meteorite 陨星,陨石
meteorite crater 陨石坑,陨星坑
meteorite shower 陨石雨,陨星雨
meteoritic tail 陨星尾
meteoritics 流星学,陨星学
meteorogical collecting center 气象收集中心
meteorogical elements 气象要素
meteorogical observation 气象观测
meteorogical parameter of icing 结冰气象参数
meteorogical prevision 气象预报
meteorogram 气象(要素曲线)图
meteorograph 气象计,气象自记仪
meteoroid 流星体
meteoroid flux 流星体通量
meteoroid model 流星体模型
meteorologic chart 气象要素图
meteorologic map 气象要素图
meteorological airborne data system 航空气象数据系统
meteorological balloon 气象(探测)气球
meteorological battery 气象电池
meteorological codes 气象电码
meteorological condition 气象条件
meteorological correction 气象修正

meteorological data handling system 气象数据处理系统
meteorological datum plane 气象基准平面
meteorological impact statement 气象冲击的声明
meteorological officer 气象主任,气象官
meteorological optical range 气象光学距离,气象光学视距,气象能见度
meteorological parameters 气象参数
meteorological radar 气象雷达
meteorological reconnaissance flight 气象侦察飞行
meteorological representation error 气象代表误差
meteorological research flight 气象研究飞行
meteorological rocket 气象火箭
meteorological rocket facility 气象火箭站
meteorological rocket sonde 气象火箭探空仪
meteorological satellite 气象卫星
meteorological sensor 气象传感器
meteorological support system 气象保障系统
meteorological visibility 大气能见度,气象能见度
meteorological wind 气象风
meteorological-satellite service 卫星气象业务
meteorology 气象状态,气象学
meteorology information collecting system 气象情报收集系统
Meteosat 气象卫星
meter 1.表,计,仪,计量器 2.米(长度单位) 3.计量
meter level 米级
meter wave fuze 米波引信
metering 1.节流,计量 2.(空中交通管制)疏导到达航班,调整到达航班顺序
metering fix 疏导定位点
metering jet 1.定量限流喷嘴 2.定量射流
metering orifice 限流孔,定量孔
metering pin 调节油针,计量油针,量油杆
metering roller 量片辊
metering valve 计量阀,限压阀
methane booster 甲烷增压器
methane concentration 甲烷浓度
methane engine 甲烷发动机
methane feedstock 沼气原料
methane flame 甲烷火焰,瓦斯火焰
methane flow rate 甲烷流量
methane fuel 甲烷燃料
methane injection 注甲烷气
methane stream 甲烷水蒸气
methane-air 甲烷空气
methane-air mixture 沼气
methane-based mixture 基于甲烷混合物
methanol decomposition 甲醇裂解
methanol droplet 甲醇液滴

method by hour angle of Polaris 北极星任意时角法
method by series 方向观测法
method of analysis 分析法,解析法
method of attributes 特征法,属性检验法
method of characteristic 特征线法
method of chord deflection distance 弦线偏距法
method of deflection angle 偏角法
method of direction sighting by drift angle supplied automatically 自动提供偏流角法定向瞄准
method of equal weight substitution 等权代替法
method of finite fundamental solution 有限基本解法
method of image 镜像法,像法
method of line 直线法,线方法,线性法
method of measurement 测量方法,计量方法
method of moment 矩量法,力矩法,矩法
method of navigation 导航方法
method of plate constant 底片常数法
method of plate orientation elements 底片方位元素法
method of separation of variable 变数分离法
method of singularities 奇异点法
method of solution 方法描述,解法
method of tension wire alignment 引张线法
method of thermal node network 热节点网络法
method of time determination by star transit 恒星中天测时法
method of time determination by Zinger star pair 津格尔(星对)测时法,东西星等高测时法
method of turns 盘旋法
methodical error 方法误差
methyl 甲基,木精
methyl alcohol 甲醇
methyl bromide 甲基溴,溴化甲烷
methyl cellosolve 甲基溶纤剂
methyl cubane 甲基立方烷
methyl ester 甲酯
methyl formate 甲酸甲酯,甲酸薄荷醇酯
methyl mercaptan 甲硫醇
methylacetylene 丙炔,甲基乙炔
methylamine 甲胺
methylenimine 甲亚胺
METO 1. Maximum Engine Take Off (Power) 发动机最大起飞功率 2. Maximum Except Take Off (Power) 除起飞外最大额定功率
metric 公制的,米制的,公尺的,度量标准
metric accuracy 公制准确性
metric and photometric information 度量和光度信息
metric calibration 测量校正
metric camera 测量相机,星测摄影机
metric information 米制信息,计量信息

metric performance 性能指标
metric photography 测量摄影,测量照相
metric space 度量空间
metric threshold 指标阈值
METRL Meteorological 气象的
METROC Meteorological Rocket 气象火箭
metrological assurance 计量保证
metrological assurance system 计量保证体系
metrological confirmation 计量确认
metrological management 计量管理
metrological supervision 计量监督
metrology (精密)计量学
metrology system 量测系统
metroplex routes 市际高空仪表飞行航线
metrosonde (边界层)气象探空仪
metry 测定,测量
METS 1. Mobile Engine Test Stand 活动式发动机试车台 2. Modular Engine Test System 模块式发动机试验系统
METSAT Meteorological Satellite 气象卫星
METT Microwave Energy Transmission Test 微波能量传输试验
MEV 1. Mega Electron Volts 兆电子伏特,百万电子伏特 2. Million Electron Volts 百万电子伏特
MEW 1. Manufactured Empty Weight 制造空重 2. Megawatt Early Warning 兆瓦级预警雷达
MEXE Military Engineering Experimental Establishment 军用工程实验研究院(英国)
mexometer 测距仪
MEZ Missile Engagement Zone 导弹使用区
MF 1. Main Force 主力 2. Main Frame 主框架 3. Medium Frequency 中频 4. Medium Frequency or Millifarad 中频或毫法拉 5. Make From 由……制造
MFBF Multifunction Bomb Fuze 多功能炸弹引信
MFC 1. Main Fuel Control 主燃油控制 2. Main Fuel Controller 主燃油控制器 3. Maximum Fuel Capacity 最大燃油容量
MFCP Multifunction Control Display Panel 多功能控制显示板
MFCS 1. Manual Flight Control System 人工飞行控制系统 2. Missile Fire Control System 导弹发射控制系统
MFD 1. Magnetofluid Dynamics 磁流体动力学 2. Manufactured 制造的 3. Multi-Function Display 多功能显示仪,多功能显示器
MFDS Multifunction Display System 多功能显示系统
MFDU Multifunction Display Unit 多功能显示装置
MFE Magnetic Field Explorer 磁场探险者(美国地磁场探测卫星)

MFES　Main Fixed Earth Station 固定式主地球站

MFF　1. Master Freight File 主要运货档案 2. Mixed Fighter Force 战斗机混成部队

MFFC　Mixed Fighter Force Concept 战斗机混成部队思想

MFG　1. Manufacturing 制造，生产 2. Miniature Flex Gyro 微型柔性陀螺 3. Munitions Family Group 弹药种类分组

MFG INFO　Manufacturing Information 制造数据

MFHBF　Mean Flight Hours Between Failures 平均故障间隔飞行小时

MFIT　Mean Flight Isolation Time 平均故障间隔时间

MFK　Multifunction Keyboard 多功能键盘

MFL　Minimum Field Length 最短机场长度

MFLI　Magnetic Fluid Level Indicator 流体液面磁指示器

MFM　Maintenance Fault Memory 维修故障存储器

MFMA　Multi-Function Microwave Aperture 多功能微波孔径

MFO　Multinational Force and Observers 多国部队及观察员

MFP　1. Main Fuel Pump 主燃油泵 2. Maintenance Facility Plan 维修设备计划 3. Mean Free Path 平均自由程

MFR　1. Manipulator Foot Restraint 机械臂足固定器（美国航天飞机）2. Manufacture 制造 3. Multi-Function Radar 多功能雷达（美国的武器控制雷达）

MFS　1. Multifunction Sensor 多功能传感器 2. Multi-frame Synchronizer 复帧同步器 3. Multiple Frequency Shift 多频位移

MFSK　Multi-Frequency Shift Keying 多频移键控

MFT　1. Master Fuel Trip 主燃料跳闸 2. Mean Flight Time 平均飞行时间 3. Missile Facilities Technician 导弹设备技师 4. Mission Flight Trainer 任务飞行训练器

MFV　Main Fuel Valve 主燃油活门

MFW　Maximum Fuel Weight 最大设计燃油重量

MG　1. Machine Gun 机关炮，机枪 2. Master Gauge 主量规,主仪表,标准测量仪,校对规 3. Main Gear 主起落架 4. Motorglider 动力滑翔机 5. Motor Generator 电动发电机

Mg　1. Magnesium 镁 2. Megagramme 兆克,千千克,吨

mg　毫克(milligram)

MGA　1. Middle Gimbal Angle 中常平架角 2. Middle Gimbal Axis 中常平架轴

MGC　1. Manual Gain Control 人工增益控制 2. Maximum Guaranteed Capacity 最大保证功率 3. Mean Geometric Chord 平均几何弦 4. Missile Guidance Computer 导弹制导计算机

MGCC　Missile Guidance and Control Computer 导弹制导控制计算机

MGCO　Mars Geoscience and Climatology Orbiter 火星地质学和气候学轨道器

MGCS　1. Meteosat Ground Computer System 气象卫星地面计算机系统（西欧）2. Missile Guidance and Control System 导弹制导和控制系统 3. Mobile Ground Control Station 流动地面控制站

MGCU　Main Generator Control Unit 主发电机控制器

MGE　1. Maintenance Ground Equipment 地面维修设备 2. Missile Ground Equipment 导弹地面设备

M-generator　Main Generator 主发电机

MGGB　Modular Guided Glide Bomb 模式化制导滑翔炸弹

MGIR　Motor Glider Instrument Rating 动力滑翔机仪表飞行执照

MGN　Magneto 磁电机

MGO　1. Main Geophysical Observatory 地球物理观象总台（苏联）2. Magnesium Oxide 氧化镁 3. Marine Gas Oil 船用汽油

MGR　Manufacturing Generated Recorders 制造部门制定的记录

MGRS　Military Grid Reference System 军用网格参考系,军用网格坐标系

MGS　1. Missile Guidance Set 导弹制导装置 2. Missile Guidance System 导弹制导系统 3. Mobile Gateway Switch 移动网关交换

MGSCU　Main Gear Steering Control Unit 主起操纵控制装置

MGSE　Mechanical Ground Support Equipment 地面机械保障设备

MGT　1. Minimum Ground Time 地面最少停留时间 2. Motor Gas Temperature（固体火箭）发动机燃气温度

MGTOW　Maximum Gross Takeoff Weight 最大起飞总重量

MGTP　Main Gear Touchdown Point 主起落架着陆接地点

MGU　1. Midcourse Guidance Unit（导弹）中段制导装置 2. Motor Governor Unit 电动调节单元

MGW　Maximum Gross Weight 最大总重量

MH　1. Magnetic Heading 磁航向 2. Magnetic Head 磁头 3. Minimum Height 最低高度 4. Mobile Host 移动式主机

mh　1. Milihenry 毫亨 2. Milihour 毫小时

MHA　1. Maintenance Hazard Analysis 维修危险性分析 2. Minimum Holding Altitude 空中等待最低高度

mhd　Magnetohydrodynamics 磁流体力学

mhd accelerator　流体动力加速器

mhd action	磁流体动力作用
mhd bypass	磁流体动力旁路
mhd bypass engine	磁流体动力旁路发动机
mhd case	磁流体动力情况下
mhd channel	磁流体动力学通路
mhd device	磁流体动力设备
mhd energy	磁流体动力能源
mhd energy interaction	磁流体动力能量交互
mhd engine	磁流体动力引擎
mhd flow	磁流体动力流
mhd generator	磁流体动力发电机,磁流体发生器
mhd interaction	磁流体动力相互作用
mhd model	磁流体动力学模型
mhd power	磁流体动力
mhd power generator	磁流体动力发电机
mhd scheme	磁流体动力格式
mhd system	磁流体动力系统
mhd work	磁流体动力工作

MHF Medium High Frequency 中高频

MHI Master Heading Indicator 主航向指示器

MHIR Manufacturing Hookup and Installation Report 制造接线和安装报告

MHM Machine Hold Mode 机器停机状态

MHP Main Hydraulic Pump 主液压泵

MHR Master Heading Recorder 主航向记录仪

MHRS Magnetic Heading Reference System 磁航向基准系统,磁航向参考系统

MHS Microwave Humidity Sounder 微波湿度探测器

MHT Microwave Hologram Techniques 微波全息技术

MHV Mean Horizontal Velocity 平均水平速度

MHX Main Heat Exchanger 主热交换器

mhz Megahertz 兆赫兹

MI 1. Maintenance Index 维护索引 2. Magnetic Indicator 磁指示器 3. Maintenance Instructions 维护说明 4. Medium Intensity 中等强度 5. Model Improvement 模型改进 6. Moment of Inertial 惯性力矩 7. Mutual Inductance 互感 8. Minimum Impulse 最小冲量

MIA 1. Minimum Intercept Altitude 最低截击高度 2. Missing In Action 作战失踪 3. Multiplex Interface Adaptor 多路传输接口适配器

MIB Management Information Base 管理信息库

MIC 1. Management Information Center 管理信息中心 2. Maximum Input Current 最大输入电流 3. Message Identification Code 消息识别码 4. Micrometer 测微计 5. Microphone 话筒,耳机 6. Microwave Integrated Circuit 微波集成电路 7. Mineral Insulated Cable 无机绝缘电缆

MICAM Microameter 微安计,微安表

Micap Mission Capable 能执行(战斗)任务的

Micarta 胶合云母纸板,米卡他绝缘板

MICD Mechanical Interface Control Document 机械界面控制文件

mice 小凸块(用于插入流体截面内以调整流体截面积,尤指燃气涡轮喷气管内的)

MICE Microwave Integrated Checkout Equipment 微波综合检测设备

Michigan height 密执安高度

MICNS Modular Integrated Communication and Navigation System 模块化通信导航综合系统

MICOM Missile Command 导弹部队(美国陆军)

Micon Missile Contraves 米康地空导弹(瑞士)

MICR Microscope 显微镜

micrad Microwave Radiometer 微波辐射计

micro air	微量空气
micro air vehicle	微型飞行器,微型飞机,小型飞行器
micro actuators	微致动器,微执行器
micro adjustable valve	微调阀
micro adjuster	磁差校正器,微调器
micro aluminized propellant	微型镀铝推进器
micro ampere	微安培,微安
micro arcing	微型电弧作用
micro atmosphere	微大气
micro balance	微量天平,微量秤
micro balloons	微形空心球
micro bar	微巴
micro barograph	微气压计
micro barom	微气压图
micro barometer	微气压表
micro beam	微光束
micro channel plate detector	微通道板探测器
micro economic model	微观经济模型
micro inertial measurement unit	微惯性测量单元,微惯性测量组合
micro level	1. 软件的最终设计 2. 微级
micro omega	微型奥米加导航系统
micro programmer	微程序,微程序控制
micro switch	微动电门,微动开关
micro tacan	微型"塔康"导航系统
Micro vision	1. 本迪克斯公司研制的全天候着陆辅助系统的商品名 2. 微型版 3. 微波观察仪
micro volter	微伏表
microbial contaminant control	微生物污染控制
microbial sensor	微生物传感器
microbiological corrosion	微生物腐蚀
microburst	微型下冲气流
microburst encounter	微爆气流遭遇
microburst escape	微爆气流逃生

microburst escape maneuver 微爆气流逃生演习	**microfoams** 微泡沫
microchannel plate 微通道板	**microgeometry** 微观几何形态
microchannel plate cathode-ray tube 微通道板示波管	**micrograph** 显微照片,显微图,微动描记器
microchannels 微通道	**microgravimeter** 微重力仪
microcircuit 微电路	**microgravimetry** 微重力测量
microcircuitry 微型电路系统	**microgravity** 微重力
microclimate 微小气候,小气候	**microgravity condition** 微重力状态
microclimate in spacecraft cabin 航天器座舱小气候	**microgravity environment** 微重力环境
microcode 微码	**microgravity experiment** 微重力实验
microcode system 微码系统	**microgravity experimentation** 微重力实验
microcomputer 微计算机	**microgravity platform** 微重力平台
microcomputer controlled 微(型)计算机控制的	**microgravity processing** 微重力状态下处理
microcomputer system 微计算机系统	**microgravity sciences** 微重力科学
microcomputer-controlled autopilot 微计算机控制的自动驾驶仪	**microgravity simulation** 微重力模拟
microcontroller 微控制器	**microgravity test** 微重力试验
microcontroller unit 单片机,微控制器	**microgravity test platform** 微重力实验平台
microcopying 缩微摄影	**microgravity time** 微重力时间
microcosm 微观世界,小宇宙,作为宇宙缩影的人类,缩图	**microgroove record** 密纹唱片
	microhall thruster 微霍尔推进器
microcoulomb ozone meter 臭氧微库仑分析仪	**microimage** 微影像
microcracking 微裂纹	**microinch** 微英寸(百万分之一英寸)
microdefect 微缺陷	**microinstability** 微观不稳定性
microdensitometer 测微密度计,显微光密度计	**microinstruction register** 微指令寄存器
microdial 精密标度盘,微量刻度	**Microlab** 微型实验室
microdischarge 微放电	**microlaser ablation** 微型激光器消融
microdisplay 微显示器,微型显示器,投影	**microlayer** 微表层
microdot 缩小影印文件,小粒迷幻药,微粒照片	**microlight** 超轻型飞机
microdroplets 微滴	**micromaching** 微切削加工
microelectromechanical system 微机电系统	**micromachining** 精微机械加工
microelectronic packaging 微电子组装,微型电子封装,微电子封装	**micromanometer** 测微压力计,精测流体压力计
	micromanufacturing technology 微加工工艺
microelectronics 微电子学,微电子技术	**micromechanical accelerometer** 微型机械加速计
microelement 微型元件,微量元素	**micromechanical gyro** 微型机械陀螺仪
microemitter 微型发射器	**micrometeorite** 微陨星,微陨石,陨尘
microenvironment 人造环境,微环境	**micrometeorite simulator** 微流星仿真器
microexplosion 微爆	**micrometeorograph** 微气象记录仪,微气象计
microextraterrestrial object 地外物体	**micrometeoroid** 微流星体
microfabricated emitter 超微型发射器	**micrometeoroid hypervelocity impact** 微流星体超高速冲击
microfabrication 精密加工,微制造,微型品制造	**micrometeoroid protection garment** 微流星防护服
microfarad 微法拉	**micrometeoroids explorer** 微流星体探险者卫星
microfeed 微量进给	**micrometer** 测微计,千分尺,千分表
microfiche 缩微平片,缩微胶片	**micromirror** 微镜
microfilm 缩微胶片,缩微胶卷	**MICROMIX** Microminiature 超小型,微型
microfilm map 缩微地图	**micromodule** 微型组件,微模块
microfilmer 缩微胶卷照相机,缩微电影摄影机	**micromounting** 微组装
microfilming 缩微摄影	**micron** 微米(百万分之一米)
microflamelets 超微火苗	**micronewton** 微牛顿
microflames 微火焰	**micronewton cold-gas** 微牛顿冷气体

micronic 微米的
micronlitre 微米升
micronozzle 微喷管
micronozzle geometry 微喷管几何结构
micronsized aluminum 微米尺寸铝
micronsized particle 微米尺寸微粒
microphone boom 传声器架
microphone output 送话音频输出
microphotogrammetry 显微摄影测量
microphotography 缩微照相术
microphotometer 显微光度计
microphysical process 微物理过程
microphysics 粒子物理学,微观物理学
microplasma arc welding 微束等离子弧焊
micropower integrated circuit 微功耗集成电路
microppt micropipette 微量吸液管
micropressure sensor 微压传感器
microprobe 显微探针
microprobe sampling 微探针取样
microprocessor 微处理机
microprocessor development system 微处理机开发系统
microprocessor terminal 微处理机终端
microprogram 微程序
microprogrammed control unit 微程序控制器
microprogramming technique 微程序设计技术
micropropulsion 微推进
micropropulsion system 微推进系统
micropulsation 微脉动,(地)磁性(微)脉动
micropulsation unit 微脉冲发生器
micropulse 微脉冲
microradian 微弧度
microradiometer 微辐射计
microreactor 微反应器
microregionalization 小区划分
microrocket 微型火箭
microrocket engine 微型火箭发动机
Microsat 微型卫星
microsatellite 微卫星,小随体
microscale application 微尺度应用程序
microscale thruster 微型推进器
microscan receiver 微扫接收机
microscope 显微镜
microscope image 显微图像
microscopic analysis 显微镜分析,显微分析,微分析
microscopic casualty 微观因果性,微观伤亡
microscopic crack 微裂纹,显微裂纹
microscopium 显微镜(星)座
microscopy 显微镜检查,显微镜使用,显微镜学
microsecond 微秒(百万分之一秒)
microsecond discharge 微秒级放电
microsecond spark 微秒的火花
microsegregation 显微偏析
microseism 脉动,微弱的震动
microseismic storm 微地震扰动,脉动暴
microseismograph 微震仪
microsensors 微传感器
microshaving 微顶头
microslug 微斯勒格(质量单位)
microspacecraft 微型航天器
microsphere 微球体,中心体,微滴
microstrip 微波传输带,微带(线)
microstrip antenna 微带天线
microstrip antenna array 微带天线阵
microstrip array 微带阵
microstrip circuit 微带(线)电路
microstrip dipole 微带偶极子
microstrip line 微带线
microstructure 微观结构,显微结构
microswitch 微动开关,微型电门
microsyn 微动同步器
microsyn torquer 微动同步力矩器
microtabular display 小表格显示
microthrust 微推力
microthruster 微型喷气发动机
microthruster design 微型喷气发动机设计
microthruster operation 微型喷气发动机操作
microthruster performance 微型喷气发动机性能
microtorr 微托(百万分之一托,压强单位)
microtube 微型管
microvalve 微型阀
microvolcano 微型火山
microvolt 微伏(特)
microwatt 微瓦(特)
microwave 微波
microwave absorbing material 微波吸收材料
microwave aircraft 微波飞机
microwave atmospheric sounding radiometer 微波大气探测辐射计
microwave background 微波背景
microwave cataract 微波所致白内障,微波性白内障
microwave communication 微波通信
microwave coupling efficiency 微波耦合效率
microwave course beacon 微波导标
microwave digital guidance equipment 数字式微波引导飞机设备
microwave discharge 微波放电
microwave electronics 微波电子学
microwave electrothermal thruster 微波电热推进器

microwave energy 微波能量
microwave ferrite 微波铁氧体
microwave frequency 微波频率
microwave fuze 微波引信
microwave gas discharge duplexer 微波气体放电天线开关
microwave generation 微波生成
microwave generation subsystem 微波生成子系统
microwave hologram 微波全息图
microwave hologram radar 微波全息雷达
microwave holography 微波全息摄影
microwave hybrid integrated circuit 微波混合集成电路
microwave input 微波输入
microwave integrated circuit 微波集成电路
microwave ion 微波离子
microwave ion source 微波离子源
microwave ion thruster 微波离子推进器
microwave landing system 微波着陆系统
microwave landing system coverage 微波着陆系统覆盖
microwave limb sounder 微波临边探测仪
microwave magnetics 微波磁学
microwave monolithic integrated circuit 微波单片集成电路
microwave network 微波网络,微波通信网
microwave plasma 微波等离子体
microwave power 微波功率
microwave power supply 微波电源
microwave pulse 微波脉冲
microwave radiation 微波辐射,微波射线
microwave radiation thermal effect 微波辐射的致热效应
microwave radio relay communication 微波接力通信,微波中继通信
microwave radiometer 微波辐射计
microwave refractometer 微波折射计
microwave remote sensing 微波遥感,微波遥测
microwave remote sensor 微波遥感器,微波传感器
microwave satellite 微波卫星
microwave scatterometer 微波散射计
microwave single sideband modulator 微波单边带调制器
microwave sounding unit 微波探测装置
microwave thermography 微波热成像
microwave tube 微波管
microwave united carrier system 微波统一载波系统
microwave weapon 微波武器
microwave wind scatterometer 测风微波散射计
microwave windfield scatterometer 微波风场散射计
microzonation 小区划

MICS 1. Management Information and Control System 管理信息和控制系统 2. Material Inventory Control System 材料资源控制系统 3. Multiple Internal Communication System 多路内部通信系统
mid airway flight 中空飞行
mid and high case 中、高情况下
Mid Atlantic route 中大西洋路线
Mid Asian route 中亚航线
Mid Canada chain 中期加拿大链
midair collision 空中相撞
midair retrieval 空中回收
MIDAS 1. Management Information and Development Aids System 管理信息和开发辅助系统 2. Manufacturing Information Distribution and Acquisition System 制造信息分配与采集系统 3. Missile Intercept Data Acquisition System 导弹截击数据采集系统
midbody 中体,中间体,船体中部
midcareer officer 在职军官
midchamber 中室,中膛
midchannel 中水道,中航道
midchord 弦线中点
midcourse 1. 弹道中段 2. 轨道中段
midcourse and terminal guidance system 飞行中段与末段制导系统
midcourse control 中程控制,(弹道)中段控制
midcourse correction manoeuvre 中途修正机动,中制导
midcourse defense 中段防御
midcourse guidance 中制导,中段制导
midcourse intercept 中段截击,中段拦截
midcourse phase 中段
midcruise 半程巡航
midcruise weight 半程重量
middle airspace 中空空域
middle and rear part 中后部
middle atmosphere 中层大气
middle ear pressure 中耳压力
Middle East route 中东航线
middle infrared 中红外
middle injector 中间喷油器
middle latitude 中纬度,中间纬度
middle locator 中间定位器,中距导航台,中示位信标
middle marker 中距指点标
middle or alto clouds 中高云
middle surface 中曲面
middle temperature heat pipe 中温热管
middle tube 中筒
middle ultraviolet 中紫外,中紫外线
mid-flap 中襟翼,襟翼中段

midheight line 中高线
midhigh wing 中高(单)翼
midinfrared 中红外
midinfrared advanced chemical laser 先进中红外化学激光器
midlevel 中间层
mid-ocean ridge 洋中脊
midplane 中平面盆腔,中段平面
midpoint 中点,正中央
midpoint node 中点节点
midpoint voltage 中点电压
midpressure 中压法
midpressure extinction 中压消光
MIDR Maintenance Incident and Defect Report 维修事件和缺陷报告
midrange approximation 中列数逼近
midrange approximation method 中列数逼近方法
mid-range ballistic missile 中程弹道导弹
MIDS 1. Management Information and Decision Support 管理信息与决断保障 2. Multifunctional Information Distribution System 多功能信息分布系统
midsection 1. 中段 2. 中间截面,中间剖视
midspan 中跨,跨距中点,档距中间,开度中间
midspan plane 中跨平面
midspan shroud 叶片中间箍环
midstream 中流,河流正中
midterm 期中的,中间的,中间
MIDU 1. Missile-Ignition Delay Unit 导弹点火中继装置 2. Multi-input Interactive Display Unit 多输入干扰显示组件
midvalue logic 中值逻辑
midwing 中单翼
midwing fairing 中单翼整流罩
MIE 1. Manoeuvre-Induced Error 机动(引起的)误差 2. Micromedia Information Exchange 微媒体信息交换
Mie scattering 米氏散射
MIEC Meteorological Information Extraction Centre 气象数据分选中心(西欧)
MIF Module Integration Facility 舱段总装设施
MIFL Master International Frequency List 国际频率总表
MIG Miniature Integrating Gyro 微型积分陀螺
migration (分子内)原子迁移,离子迁移,(两个固体间的)分子迁移
migration frequency 迁移率
migration radius 迁移半径
migration rate 迁移率,迁徙率,迁移速度
migration region 迁移区域
MII Method Improvement Item 方法改进项目

mike 扩音(微音)器,千分尺,测微器
MIL 1. Mileage 里程 2. Military 军用的,军事的
mil 1. 密耳,千分之一英寸 2. 角密耳,密位
MIL specification Military specification 军事器材规格(美国三军通用),美国军用规范
MILCOMSAT Military Communication Satellite 军用通信卫星
Milcon military construction 军事建筑
mild anoxia 轻度缺氧症
mild hypoxia 轻度缺氧
mileage prorate 按里程分配收费的
MIL-HDBK Military Handbook 美国军用手册
MILIRAD Millimeter Radar 毫米波雷达
military 军事的,军人的,适于战争的
military aircraft 军用飞机
military and commercial engine 军用和商用发动机
military application 军事应用
military assistance service funded 有经费的军援勤务
military authority assumes responsibility for seperation of aircraft 由军事当局负责飞机间隔
military aviation 1. 军事航空 2. 军用航空
military avionics 军事航空电子学
military charter 军用包机
military climb corridor 军用飞机爬升走廊
military community 军事共同体,营区
military computer family 军用计算机系列
military computer-based system 军用计算机控制系统
military customer 军事客户
military department 军事部,军种部
military engine 军用发动机
military environment 军事环境
military environment microprocessor 军用环境微处理机
military equipment 军事装备
military fighter 战斗机
military grid 军用网格
military industry 军事工业
military jet 喷气式战斗机
military market 军品市场
military medicine 军事医学
military meteorological support 军事气象保障
military mission 军事任务
military occupation specialties 军事专业,军职专长
military operation 军事作战,军事行动
military platform 军用平台
military posture map 军事态势地图
military power 军用功率
military procurement 军事采购
military production 军火生产
military productivity 军事生产率

military program　军火生产计划
military psychology　军事心理学
military qualification test　军品合格试验
military reference material　军用标准物质
military role　军事作用,军事参与
military satellite　军用卫星
military service　兵役,军役
military space engineering system　军用航天工程系统
military space technology　军事航天技术
military specification　军用规范,军用规格
military system　兵制,军制
military technology　军事技术
military training routes　军事训练航路区
military transport　军事运输机(船)
military turbofan　军用涡轮风扇发动机
military use　军事用途
military use of outer space　外层空间的军事应用
military user　军事用户
Military Wing　陆军联队(英国皇家飞行队)
milk run　〈口语〉例行的飞行,战时没有危险性的飞行勤务
Milky Way　银河
mill　1.磨碎机,碾磨机,粉碎机 2.密耳(反应性单位),千分之一英寸
Miller integrating circuit　密勒积分电路
milliammeter　毫安表,毫安计
milliampere　毫安
millica　米里卡
milliearth rate unit　毫地速,千分之一地球速率单位
millifarad　毫法(拉)
milligal　毫伽(千分之一伽,重力加速度单位)
milligram　毫克
millilitre　毫升
millimeter wave　毫米波
millimeter wave communication　毫米波通信
millimeter wave fuze　毫米波引信
millimeter wave sensor　毫米波探测器,毫米波传感器
millimicron　毫微米
milling　铣削
milling fluted thrust chamber　铣槽式推力室
million accounting units　百万计算单位(欧洲)
milliradian　(雷达)毫弧度
millisecond　毫秒
millitorr　毫托,微米汞柱
MILS　Microwave Instrument Landing System 微波仪表着陆系统
MILSICC　Military Standard Item Characteristics Coding 军用标准项特性编码
MILSTD　Military Standard 军用标准(美国)

MILU　Missile Interface and Logic Unit 导弹接口与逻辑装置
Milvan　军用集装箱
Mimas　【天】土卫一
MIMD　Multiple Instruction Multiple Data 多指令多数据流
MIMEO　Multiple Input Memo Engineering Order 多输入备忘录工程指令
MIMO　Multiple-Input Multiple-Output 多输入多输出
MIMP　Manufacturing Interface Management Plan 制造界面管理计划
MIMR　Multifrequency Imaging Microwave Radiometer 多频成像微波辐射计
MIMU　Micro Inertial Measurement Unit 微惯性测量单元,微惯性测量组合(装置)
min　1. minimum 最小值,极小值 2. minute 分(量度角的分值用),分钟
mination　不可测定系数
mine　1.地雷,水雷 2.布雷
mine countermeasure　反雷措施
mineral oil　矿物油
mineral wool　矿物棉,玻璃棉
mineralogical composition　矿物组成,矿物成分
minerogenetic prognostic map　成矿预测图
MINFAP　Minimum Facilities Project 最低限度设施的计划
mini flare　小照明弹
miniature data acquisition system　小型数据采集系统
miniature demand regulator　微型肺式调节器
miniature detonating cord　微型爆破索
miniature engine　微型发动机
miniature flex gyro　小型挠性陀螺
miniature homing vehicle　小型寻的飞行器
miniature ion　微型离子
miniature ion thruster　微型离子推进器
miniature neutralizer　微型中和器(剂)
miniature positive pressure demand regulator　微型肺式加压供氧调节器
miniature rocket　微型火箭
miniature seat mounted regulator　微型座椅供氧调节器
miniature vehicle　小型寻的飞行器
miniaturization　小型化
miniaturize　使小型化,使微型化
minicomputer　小型计算机,微型电脑
minicontroller　微型驾驶杆
Minigun　"米尼冈"机枪
Minilab　小型实验室
minima　极小值(minimum 的复数),最小数
minimal and maximal wind　最小和最大风力

minimal computer load 计算机最小负荷
minimal cut 最小割截,极小截量
minimal flight path 最短飞行路线
minimal realization 最小实现
minimal set 极小集
minimal turn 最小转弯
minimal turn radius 最小转弯半径
minimal wind 最小的风
minimal wind condition 最小风力条件
minimally manned 最小载人
minimax 极小极大,极小化极大
minimax approach 极小极大方法,使对方得点减到最低以使自己得最高分的战略
minimax controller 极小极大控制器
minimax principle 极小化极大原理
minimax problem 极小极大问题
minimax solution 极小极大解
minimax system 极小极大系统
minimax technique 极小化极大技术
minimise 尽量减少,使最小化
minimization 减到最小限度,估到最低额
minimization technique 最小化技术,最小化方法
minimum 最小值,最低限度,最小化,最小量
minimum acceptable value 最低可接受值
minimum aircraft 最轻型飞机
minimum airplane 最小飞机
minimum altitude 1.最低高度 2.低空
minimum approach distance 最小接近距离
minimum arming altitude 引信解除保险最低高度
minimum augmentation rating 最小加力状态
minimum autoignition 最低自燃
minimum budget 最低预算
minimum centerline 最低中心线
minimum centerline potential 最低中线潜力
minimum charge 最低收费
minimum clearance 最小余隙,最小间隙
minimum combustor 最低燃烧室
minimum constraint 最小约束
minimum control 最小控制
minimum cost programming 最小成本规划
minimum crab angle 最小偏流修正角
minimum cross section 最小横截面
minimum crossing altitude 最低通过高度
minimum curvature 最小曲率,最小曲率法
minimum decision altitude 最低决断高度
minimum descent altitude 1.仪表下降最低高度 2.(最低)决断高度
minimum descent height 最低下降高度
minimum design 于自然力的最少设计

minimum design weight 最小设计重量
minimum detectable signal 最小可探测信号
minimum detectable signal-to-noise ratio 最小可检测信噪比
minimum detectable temperature difference 最小可探测温差
minimum diameter 最小直径
minimum discriminable signal 最小可识别信号
minimum distance 最小距离
minimum distance classification 最小距离分类
minimum distance function 最小距离函数
minimum distortion 最小失真
minimum disturbance 最小干扰
minimum drag 最小阻力
minimum drag coef 最小阻力系数
minimum dwell 最小滞留
minimum dynamic pressure boundary 最小波动压力边界
minimum eccentricity 最小离心率
minimum eccentricity point 最小偏心点
minimum electrical pulse width 最小电脉冲宽度
minimum energy 最低能量
minimum energy Hohmann transfer 最小能量霍曼变轨
minimum energy orbit 最小能量轨道,霍曼轨道
minimum energy solution 能量最省解决方案
minimum energy trajectory 最小能量弹道
minimum energy transfer 最低能量转移
minimum energy transfer orbit 最小能量转移轨道
minimum enroute altitude 最低航线高度
minimum entropy decoding 最低熵译码
minimum equipment item 最低设备项目,起码设备项目
minimum fire current 最小发火电流
minimum flying speed 最小飞行速度
minimum frequency shift keying 最小频移键控
minimum fuel 最少燃油量
minimum fuel case 最低油量案例
minimum fuel load 最低油量
minimum fuel problem 最低油量问题
minimum fuel solution 最低油量解决方案
minimum fuel spiral 最低油量盘旋
minimum fuel transfer 最低油量转移
minimum gauge 最小厚度
minimum gliding speed 最小滑翔速度
minimum ground speed system 最小地速系统
minimum holding altitude 最低等待高度
minimum horizontal separation 最低水平分距
minimum ignition energy 最小点火能,最小点燃能量,最小点火能量
minimum impulse 最小脉冲

minimum impulse bit （火箭发动机）单脉冲的最小冲量
minimum impulse bit at MEPW MEPW（Minimum Electrical Pulse Width）最小电子脉冲宽度下的单个脉冲的最小冲量
minimum impulse limit cycle 最小冲量极限环
minimum induction 最小诱导
minimum induction length 最小感应长度
minimum information method 最小信息法
minimum initial mass 最小初始质量
minimum irradiance time 最低辐照时间
minimum latency 最小等待时间,最低延时
minimum lease payments 承租（飞机）方应付的最低租金
minimum lethal concentration 最小致死浓度
minimum level speed 最小平飞速度
minimum line of detection 最低探测线
minimum line of interception 最近截击线,最迟拦截线
minimum material 最小材料
minimum military requirement 最低军用要求
minimum natural frequency 最低固有频率,最小的自然频率
minimum norm 最小范数
minimum norm inverse 最小范数逆
minimum norm solution 最小范数解
minimum normal burst altitude （防空核弹头）正常爆炸最低高度
minimum number 最小数
minimum obstruction clearance altitude 最低安全高度,最低越障高度
minimum order realization 最小阶实现
minimum path cover 最小路径覆盖
minimum periapsis 最低近拱点
minimum periapsis distance 近拱点最小距离
minimum phase shift keying 最小相移键控
minimum phase shift network 最小相移网络
minimum phase system 最小相位系统
minimum plane change 最小平面变化
minimum point 最低点
minimum positioning time 最小定位时间
minimum potential 最小电势
minimum power 最小功率
minimum pressure 最小压力
minimum principle 最小值原理
minimum range 最小作用距离,最小射程
minimum rate of benefit 最低收益率
minimum reception altitude 最低可接收信号高度
minimum resolvable angle 最小可鉴角
minimum resolvable temperature 最低溶解温度
minimum resolvable temperature difference 最小可分辨温差
minimum response 最小响应
minimum resultant correction （射击的）最小总修正角
minimum runway 最短跑道
minimum safe altitude 最低安全高度
minimum safe distance 最小安全距离
minimum safe flight altitude 最低安全飞行高度
minimum safety altitude warning 最低安全高度警告
minimum sector altitude 最低扇形区高度
minimum separation 最小间距,最小间隔
minimum separation standard 最小间隔标准
minimum set 最小集
minimum shift keying 最小频移键控
minimum size 最小尺寸
minimum smoke 微烟
minimum smoke propellant 微烟推进剂
minimum speed 最小速度
minimum spillage 最低溢出量
minimum spillage configuration 最低溢出配置
minimum state 最小状态
minimum TAS minimum True Airspeed 最小真空速
minimum temperature 最低温度
minimum time 最短时间,最少时间
minimum time-delay 最小延时
minimum tolerable single-shot kill probability 最低可以容忍单发杀伤概率
minimum turn 最小转向
minimum turn radius 最小转弯半径
minimum uncertainty 最小不确定性
minimum value 【数】极小值,最小值
minimum vectoring altitude 最低引导高度
minimum velocity 最小速度
minimum voltage 最小电压,最低电压
minimum wall 下限
minimum warning time 最短预警时间
minimum weight 最小重量
minimum wind 最小风
minimum-landing-error 最小着陆误差
minimum-landing-error problem 最小着陆误差的问题
minimum-noise 最低噪声
minimum-noise trajectory 最低噪声轨迹
minimum-order 最小订单
minimum-phase network 最小相位网络
minimum-radius 最小半径
minimum-radius level 最小水平半径
minimum-risk decision maker 最小风险决策
minimums 最低气象条件
minimum-takeoff-gross-weight 最小起飞总重量
minimum-time 最小时间

minimum-time problem 最小时间问题
minimum-time solution 最小时间的解决方案
minimum-variance estimation 最小方差估计
minimun temperature 最低温度
minimun warning time 最短警告时间
mining 采矿,布雷
mining system 采矿系统,布雷系统
mini-RPV 小型(或迷你型)RPV(Remotely Polited Vehicle 用于战斗侦察的遥控飞机)
mini-satellite 超小卫星
mini-shuttle 小型航天飞机
minisonde 小型探空仪(火箭)
Ministry of Transport and Civil Aviation 运输和民用航空部
minisub 小型潜艇
MINITAT Minimum Tactical Aircraft Turret 小型战术飞机炮塔
Minitrack 卫星跟踪装置
minitrack station 卫星跟踪站,电子跟踪站
minivehicle 微型汽车
MIN-MAX Minimum-Maxium 最小最大
min-max 最低-最高、极小-极大
min-max controller 极小极大控制器
min-max fuel 最小-最大燃料
min-max selection 最小-最大选择
min-max selection strategy 最小-最大选择策略
minor accident 轻微事故,一般事故
minor angle method 小角度法
minor axis 短轴,小惯量轴
minor axis direction 短轴方向
minor body 镜像实体
minor difference 微小的差别
minor effect 微效果,微效应
minor equipments 次要设备
minor frame 子帧
minor lobe 副瓣,旁瓣
minor loop 副回路,小回路,局部回路
minor loop validation 小环比对
minor planet 小行星
minor species 次要物种
minor tone 小全音
minor triangulation 小三角测量
minor unit 1.分队 2.次要刻度
minor-axis plane 短轴面
minor-axis side 短轴侧
minor-axis spinner 短轴旋转
minor-axis spinner case 短轴旋转的情况
minority carrier 少数载流子
minority component 次要元件

MINS 1. Management Information Network System 管理信息网络系统 2. Miniature Inertial Navigation System 微型惯导系统
minus 负号,减号,负数,负的
minute marker (角的度量单位)分的符号
minute of latitude 纬度分
minute of longitude 经度分
minute volume of respiration 每分换气量
Minuteman 民兵导弹(美国地-地洲际弹道导弹)
MINWR Minimum Weapon Radius 武器最小作用半径
MIO Manual Inputs Operator 人工输入操作员
MIP 1. Missile Impact Prediction 导弹弹着点预测 2. Missile Impact Predictor 导弹弹着(点)预测器 3. Module for Insertion of Parameters 参数插入模块
MIPAS Michelson Interferometric Passive Atmospheric Sounder 迈克尔逊无源大气探测干涉仪
MIPI Material In Process Inventory 使用中的材料库存
MIPR Military Interdepartmental Purchase Request 军事部门间采购申请
MIPS 1. Maintenance Information Planning System 维修信息、计划制度 2. a Million Instruction Per Second 每秒百万次指令 3. Missile Impact Prediction System 导弹弹着点预测系统 4. MSS Image Processing System 多谱段扫描仪图像处理系统 5. Multiband Imaging Photometer for SIRTF 多谱段成像光度计
Mira Miniature Infrared Alarm 微型红外警报器
MIRACL Mid-Infrared Advanced Chemical Laser 中红外先进化学激光器(美国)
Miranda 1. 米兰达卫星(英国技术试验卫星) 2.【天】天卫五
MIRAS Mir Infrared Atmospheric Spectrometer 和平号红外大气光谱仪
MIRR Material Inspection and Receiving Report 材料验收报告
mirror 镜,反射器
mirror cell 镜室
mirror diameter 镜子直径
mirror furnace 反光炉
mirror grinding 镜面磨削
mirror machine 磁镜装置
mirror magnet 镜子磁铁
mirror nephoscope 测云镜
mirror ratio 磁镜比,反射系数,镜化
mirror reverse 反像
mirror sight 助降反光镜
mirror support 镜像支架
mirror surface 镜面,镜像曲面
mirror symmetry 镜面对称
mirror test 镜子测试

mirror turning 镜面车削
MIRTS Modularized Infrared Transmitting Set 模块式红外发射装置
mirved warheads 多弹头分导弹头
MIS 1. Management Information System 管理信息系统 2. Manufacturing Information System 制造信息系统 3. Metal-Isolator Semiconductor 金属-绝缘体-半导体 4. Meteorological Impact Statement 气象影响通报
misalignment 1. 不同心度,不同心 2. (直线)不对准 3. 角偏差
misalignment angle 对准误差角,失调角
MISC Miscellaneous 杂项,其他
miscalibration 刻度错误,校准误差
miscellaneous charge order 运费收据
miscibility 可混性,可混物,可混合性
miscibility gap 混溶隙,溶混性间隔,不相混溶区
miscommunication 错误传达
misconception 误解,错觉,错误想法
misconvergence 失会聚
MISDAS Mechanical Impact System Design for Advanced Spacecraft 先进航天器机械着陆系统设计
MISDS Multiple Instruction Single Data Stream 多指令单数据流
mismatch 错配,不协调,失配
mismatch error 失配误差
mismatching 不匹配,不协调
MISP Man-In-Space Progrom 载人航天计划
MISR Mechanical Inoperative Summary Report 机械故障汇总报告
MISS 1. Missile Intercept Simulation System 导弹截击模拟系统 2. Model Integrated Suspension System 模型综合悬挂系统
miss burst probability 漏爆概率
miss direction 脱靶方向
miss launch opportunity probability 失机概率
miss-and-hit score recorder 射击效果记录装置,脱靶与命中记录装置
miss-distance 1. 脱靶距离 2. 脱靶量
miss-distance indicator 脱靶距离指示器
miss-distance scorer 脱靶距离指示器
missed approach 进场失败,中断进场着陆并复飞
missed approach point 进场失败复飞点
missed command 漏指令
missed command probability 漏指令概率
missed synchronization 漏同步,失步
missile 1. 导弹 2. 火箭弹 3. 发射物,投射物
missile acceleration 导弹加速度
missile acceleration limit 导弹加速度极限
missile acquisition zone 导弹截获区

missile aerodynamic configurations 导弹气动构型,导弹空气动力布局
missile aerodynamic heating 导弹(空)气动(力加热)
missile age 1. 导弹寿命 2. 导弹时代
missile aging test 导弹时效试验
missile air range 导弹飞行距离,导弹射程
missile aircraft compatibility 导弹与运载飞机的适应性
missile airframe 导弹机身
missile application 导弹应用
missile armament 1. 导弹武器 2. 导弹战斗部
missile arming panel 导弹发射前准备点火控制板
missile assembly building 导弹总装厂房,导弹总装车间
missile assembly-checkout facility 导弹总装检验设施
missile attack envelop 1. 导弹攻击区 2. 导弹允许发射区
missile attitude demand 导弹姿态控制指令
missile automatic checker 导弹自动测试装置
missile autopilot 导弹自动驾驶仪
missile avoidance 导弹规避
missile avoidance problem 导弹规避问题
missile base 导弹基地
missile blast 导弹喷射流
missile blind area 导弹盲区
missile body 导弹弹体
missile body decoupling 弹体解耦
missile body fixed frame 弹体坐标系
missile boresight symbol 导弹校准符号
missile borne computer (导)弹载计算机
missile borne equipment 弹载设备
missile borne power (导弹)弹载电源
missile borne radio guidance assembly (导弹)弹载无线电导引装置
missile brain 导弹制导系统,导弹电子自控装置
missile calibration command 导弹校准指令
missile capacity 导弹战斗性能
missile capture 导弹截获
missile carriage 导弹运输
missile carrier 导弹运载工具
missile case 导弹壳体
missile center of gravity 导弹重心
missile characteristic 导弹特点
missile cleaner 导弹清洁器
missile command coder 导弹控制编码器
missile compensation 导弹振动补偿
missile concept 导弹的概念
missile configuration 导弹构型
missile configuration design 导弹构型设计
missile construction 导弹结构
missile control 导弹控制

missile control fin	导弹操纵舵
missile control joy stick	飞机驾驶杆式的导弹控制手柄
missile control radar	导弹控制雷达
missile control receiver	导弹控制接收机
missile control tracking radar	导弹控制跟踪雷达
missile control unit	导弹控制装置
missile cost	导弹成本
missile datum	导弹基准,导弹数据
missile defense	导弹防御
missile defense radar	导弹防御雷达
missile design	导弹设计
missile destruct	导弹自毁
missile development	导弹研制
missile diameter	导弹直径
missile dynamics	导弹动力学
missile effect	导弹散飞物体的杀伤作用
missile effective rate	导弹有效率(能发射的导弹占导弹总数的百分比)
missile electrical harness tester	导弹电控测试装置
missile electrical system test	导弹电气系统测试
missile engagement	导弹使用
missile engagement zone	导弹使用区
missile engine launching	导弹发动机起动
missile envelope	1.导弹外壳 2.导弹作用范围
missile fire control computer	导弹火控计算机,导弹发射控制计算机
missile fire control system	导弹火控系统,导弹发射控制系统
missile flight	导弹射击、飞行
missile flight range	导弹射距
missile fuel	导弹燃料
missile fuel resistant coating	防导弹燃料腐蚀的保护层
missile ground equipment	导弹地面设备
missile ground interface	弹地接口
missile guidance	导弹制导
missile guidance by acoustic means	导弹(利用)声波制导
missile guidance computer	导弹制导计算机
missile guidance control system	导弹制导控制系统
missile guidance laser radar	导弹制导激光雷达
missile guidance law	导弹导引律
missile guidance set fault locator	导弹制导设备故障探测器
missile guidance system band width	导弹制导系统的带宽
missile holdback	(发射架上的)导弹牵制器,导弹锁定装置
missile impact location system	导弹落点定位系统
missile indicator	导弹发射准备情况指示器
missile integration terminal equipment	导弹综合终端设备
missile intercept	导弹拦截
missile intercept data acquisition system	导弹截击数据采集系统
missile intercept zone	导弹拦截区
missile kill	〈口语〉用导弹摧毁目标
missile killer	反导弹武器
missile lateral	导弹横侧向
missile launch	导弹发射
missile launch detection infrared system	导弹发射的红外探测系统
missile launch weight	导弹发射重量
missile launcher	导弹发射架
missile location system	导弹定位系统
missile maneuverability	导弹机动性能
missile mass	导弹质量
missile mass ratio	导弹质量比
missile mass-only model	导弹质量模型
missile model	导弹模型
missile monitor system	导弹监控系统
missile noise	导弹信号起伏,导弹噪声
missile off boresight launch	导弹离轴发射
missile optimization	导弹最优参数的确定
missile order	导弹装定指令,导弹程序指令
missile orientation survey	导弹定向测量
missile payload	1.导弹战斗部 2.导弹有效载重
missile penetration aids	导弹突防手段
missile performance envelope	导弹作用范围,导弹性能范围
missile performance measurement system	导弹性能测量系统
missile plume	导弹火舌
missile power supply system	导弹电源系统
missile production	导弹生产
missile production cost	导弹生产成本
missile propulsive plant	导弹发动机,导弹动力装置
missile protective clothing	导弹防护衣
missile range	导弹射程
missile ranging	导弹射程测定
missile reaction	1.导弹发射准备时间 2.导弹反应
missile readout system	导弹坐标读出系统,导弹数据传送系统
missile reference line	1.(从飞机上)发射导弹起始线 2.导弹基准线,导弹轴线
missile requirement	弹药需求
missile response	导弹的感扰性,导弹的响应,飞弹反应
missile retainer	导弹限动器,导弹保持器,导弹发射架
missile ring	1.导弹安装环 2.导弹(弹体)隔框
missile round	导弹整发弹

missile safety system 导弹安全装置,导弹自毁系统	**missing command** 漏指令
missile scoring system 导弹脱靶距离测量系统	**missing mass** 短缺质量,损失质量,丢失质量
missile seeker 导弹导引	**missing-man flyby** 失踪者飞越
missile seeker head 导弹导引头	**mission** 1.任务,一次任务 2.出动,出动架次 3.(导弹或火箭的)基本功能
missile shaping antenna 弹体赋形天线	
missile strategy 导弹战略	**mission achievement** 任务完成
missile subsystem 导弹分系统	**mission adaptive cockpit avionics** 任务适应座舱电子设备
missile suspension system 导弹悬挂系统	
missile system 导弹系统	**mission adaptive wing** 任务适应机翼
missile targeting equipment 导弹瞄准设备	**mission analysis** 任务分析
missile technology 导弹技术	**mission and traffic model** 任务及架次模型
missile terminal lethality 导弹终点杀伤力,导弹在与目标相遇时的杀伤力	**mission application** 任务应用
	mission assurance 任务保证
missile test and readiness equipment 导弹测试和准备状态试验设备	**mission availability rate** (飞机的)出勤率
	mission capability rate 任务可用率
missile test range 导弹试验场	**mission capable rate** 1.能执行任务的(飞机的)比率 2.任务可用率
missile thrust fitting 导弹承力接头,导弹承力固定组合件	
	mission completion success probability 任务成功概率
missile time 导弹时间	**mission computer** 任务计算机
missile time constant 导弹时间常数	**mission concept** 使命观
missile tone 导弹音响	**mission cost** 飞行任务费用
missile tracking and guidance radar 导弹跟踪与制导雷达	**mission crew** 空勤组
	mission datum 飞行任务数据
missile trajectory 导弹弹道	**mission decision** 飞行任务决策
missile trajectory measurement 导弹弹道测量	**mission degradation factor** 任务降格因素
missile trajectory measurement system 导弹弹道测量系统	**mission design** 飞行任务设计
	mission design space 飞行任务设计空间
missile transponder set 导弹应答器	**mission designer** 飞行任务设计者
missile umbilical cable 导弹发射控制电缆	**mission director** 飞行控制主任
missile velocity 导弹速度	**mission doctor** 随船医生
missile warhead cart 导弹弹头对接车	**mission document** 发射任务书
missile warning 导弹告警	**mission dose** 飞行剂量
missile weight 导弹的重量	**mission duration** (飞行)任务持续时间
missile wing 弹翼,导弹联队	**mission effectiveness** 飞行任务有效性
missile zero-in 导弹归零校正	**mission element** 飞行任务要素
missile-bearing beam 导弹制导波束	**mission engineer** 随船工程师
missile-body 弹体	**mission equipment** 任务专用设备
missile-body telemetry 弹体遥测	**mission example** 飞行任务实例
missile-booster combination 带助推器的导弹	**mission failure rate** 任务失效率
missile-borne radome 弹载天线罩	**mission function** 飞行任务职能
missile-jato combination 带助推器的导弹	**mission goal** 飞行任务目标
missil(e)ry 1.导弹技术,导弹学 2.导弹	**mission integrated transparency system** 综合任务座舱玻璃系统
missile's rearward-facing antenna 导弹尾部天线	
missile-target encounter point 导弹与目标相遇点	**mission level** 飞行任务等级
missile-target engagement range 导弹截获目标距离	**mission lifetime** 航天器的寿命
missile-target relative movement simulator 弹体目标相对运动仿真器	**mission load** 飞行任务负载
	mission maintainability 任务维修性
missile-tracking chronograph 导弹跟踪记时器	**mission management** 飞行任务的管理
missile-tracking laser radar 导弹跟踪激光雷达	**mission management team** 飞行任务的管理团队

mission manager 任务经理
mission mirror 作战反射镜,拦截反射镜
mission mix 任务混频
mission mode 任务模式,任务形式
mission objective 任务目的
mission operation 任务操作
mission opportunity 任务时机
mission option 任务选项
mission paradigm 任务范式
mission parameter 任务参数
mission performance 任务效能
mission phase 飞行阶段
mission plan 任务方案
mission planner 任务规划器
mission planning 任务规划,飞行任务计划
mission planning system 飞行任务计划系统
mission profile 飞行剖面,任务剖面图
mission program 飞行程序
mission proposal 任务提案
mission radius 任务活动半径
mission reliability 任务可靠性
mission requirement 任务要求
mission review report 任务回顾报告
mission scenario 任务方案
mission simulation 任务飞行仿真
mission software 任务软件
mission specialist 任务计划专家
mission success 任务成功
mission success criterion 任务成功标准
mission support system 任务支持系统
mission system 任务系统
mission team 任务组
mission time 任务时间
mission to planet Earth 地球使命计划(美国)
mission trainer 任务练习器
mission type 任务类型
mission utility 任务的实用程序
mission value 任务价值
mission versatility 任务变通性
mission-oriented items 按任务需要的硬件项目
miss-point 脱靶点
mis-stability 失稳
mist 轻雾,烟云,湿度
mistake 错误(常指人为操作错误或设备故障)
misting 1. 发雾 2. 雾化程度
Mistral 西北风导弹(法国)
MISTRAM Missile Trajectory Measurement System 导弹弹道测量系统
mistuned blade 失谐叶片

mistuned bladed disk 失谐叶片轮盘
mistuned forced response 失谐强制响应
mistuned mode 失谐模式
mistuned response 失谐响应
mistuned system 失谐系统
mistuning 失谐,误调,失调
mistuning model 失谐模型
mistuning pattern 失谐模式
misuse failure 误用失效,误用故障
MITI Ministry of International Trade and Industry 通商产业部(日本)
mitigate 使缓和,使减轻
mitigation 减轻,缓和,平静
mitigation approach 缓解方法
mitigation measure 缓解措施
mitigation technique 缓和技术
MITO Minimum-Interval Take-Off 最短间隔(时间)起飞
MIU Missile Interface Unit 导弹接口装置
MIV Manifold Interconnect Valve 总管连接活门
MIX Mixture 混合
mix 配制,混淆,使混和,使结交
mixed admission 混合进气
mixed airspace 混合空域
mixed amine 混胺(由三乙胺和二甲苯胺)混合成的液体火箭燃料
mixed amplifier 混合放大器
mixed autoregressive-moving-average 混合自回归移动均值
mixed charter 混合包机
mixed class 混合等级
mixed compression 混合压缩
mixed compression inlet 混压式进气道
mixed cycle 混合循环
mixed fleet 混合机队
mixed flow 混流
mixed flow compressor 混流式压气机,斜流式压气机
mixed flow exhaust 混流式排气
mixed fluid 混合液
mixed fluid composition 混合液成分
mixed fluid fraction 混合液分数
mixed gas welding 混合气体保护焊
mixed gravity anomaly 混合重力异常
mixed ice 混合冰
mixed integer 混合整数
mixed mode 混合工作方式
mixed optimization 混合优化
mixed optimization algorithm 混合优化算法
mixed optimization method 混合优化方法

mixed optimization problem　混合优化问题
mixed oscillatory pattern　混合振荡模式
mixed oxidizer　混合氧化剂
mixed path　混合路径
mixed power aircraft　混合动力(装置的)飞机
mixed propellant　混合推进剂
mixed radix　混合基数
mixed reactant　混合反应剂
mixed reactant surface　混合反应剂表面
mixed service　混合服务
mixed strategy　混合战略
mixed stream　混合流
mixed stream engine　混合流引擎
mixed stream turbofan　混排涡扇发动机
mixed traffic　混合交通
mixedness　混合,混合度
mixer　1. 混频器,合路器 2. 掺合室,混合室 3.(飞行控制信号)混合箱
mixer exit　混合器出口
mixer height　混合机的高度
mixer inlet　混合机入口
mixer power　搅拌功率
mixer section　混频器的部分
mixer throat　混合器喉口
mixer-ejector　喷射混合器
mixer-ejector exit　喷射混合器出口
mixer-ejector inlet　喷射混合器入口
mixer-ejector model　喷射混合器模型
mixer-ejector nozzle　喷射混合器喷嘴
mixing　混频,混合
mixing box　(飞机操纵系统的)差动机构,混合箱
mixing chamber　掺合室,混合室
mixing length theory　混合长理论
mixing rule　混合定律
mixing-controlled combustion　混合控制燃烧
mixing-controlling　混合控制
mixing-layer　混合层
mixing-layer thickness　混合层厚度
mixing-plane　混合平面
mixture　混合,混合物,混合剂
mixture component　混合物组分
mixture composition　混合物的组成
mixture condition　混合物情况
mixture density　混合料密度
mixture flow　混合流体
mixture fraction　混合分数
mixture ignition　混合物点火
mixture increase　混合物增加
mixture mass　混合物的质量

mixture molecular mass　混合物分子质量
mixture of formamide　甲酰胺的混合物
mixture of fuel　混合燃料
mixture of gas　混合气体
mixture ratio　混合比
mixture ratio regulator　混合比调节器
mixture ratio stabilizator　混合比稳定器
mixture reactivity　混合物反应性
mixture resolution imager　混合分辨率成像器
mixture sensitivity　混合物的敏感性
mixture strength　混合强度
mixture temperature　混合气温度
mixture-based specific impulse　基于混合物的比冲量
mix-up, caution　(指挥引导用语)敌我飞机混淆,注意
mj　1. Mastic Joint 玛帝脂接缝 2. megajoule 兆焦耳
MJB　Main Junction Box 主接线盒
MJD　1. Management Job Description 管理工作描述 2. Mass Joggle Die 大型小陷模 3. Modified Julian Day 修正儒略日
MJS　Mariner-Jupiter-Saturn 木星、土星探测飞船"水手"号的发射火箭
MK　1. Mark 标志 2. Master Key 主密钥
MKC　Multiple-Kill Capability 多目标杀伤能力,多目标击毁能力
MKR BCN　Marker Beacon 指点信标
MKUP　Make-up 修补,组织,结构,配件,制作
MKV　Multiple-Kill Vehicle 多目标杀伤器
ML　1. Maintenance Levels 维修等级 2. Missile Launcher 导弹发射装置 3. Mold Line 模型线
ml　milliliter 毫升
ml of hexane　毫升己烷
ML/V　Memory Loader/Verifier 存储装入程序/校对机
MLA　1. Manoeuvre Load Alleviation 机动载荷缓和 2. Maneuver Limited Altitude 机动限制高度 3. Multispectral Linear Array 多谱段线阵(探测器)
MLB　1. Main-Lobe Blanking 主波瓣消除 2. Multi-Layer Board 多层板
MLBM　Modern Large Ballistic Missile 现代大型弹道导弹
MLC　Maneuver Load Control (system) 机动载荷控制(系统)
MLCCM　Modular Life-Cycle Cost Model 单元体全寿命费用模型
MLD　1. Maintenance Logic Diagram 维修逻辑图 2. Mold(ed) 制模,造型
MLDG　Molding 模铸,模铸件
MLE　1. Manned Lunar Exploration 载人月球探险 2. Maximum Likelihood Estimation 最大可能性估计 3. Mesoscale Lightning Experiment 中规模闪电实验

MLG Main Landing Gear 主起落架

MLI Multilayer 多层绝热（缘）

MLL Manned Lunar Landing 载人登月

MLLP Manned Lunar Landing Program 载人登月计划

MLLR Manned Lunar Landing and Return 载人登月和返回地球

MLLV Medium-Lift Launch Vehicle 中型运载火箭

MLMS Multipurpose Lightweight Missile System 多用途轻型导弹系统

MLO Manned Lunar Orbiter 载人月球轨道器

MLP Mobile Launch Platform 活动发射平台

MLR Monodisperse Latex Reactor 等弥散乳胶反应器

MLRS Multiple Launch Rocket System 多管火箭发射系统

MLRV Manned Lunar Roving Vehicle 载人月行车

MLS 1. Main Landing System 主起落架系统 2. Metal Slitting 金属切口 3. Microwave Landing System 微波着陆系统 4. Microwave Limb Sounder 微波临边探测器 5. Mirror Landing System（航空母舰甲板）助降镜着陆系统 6. Missile Launching System 导弹发射系统 7. Multi-Level Secure 多级保密

MLS categories 微波着陆系统等级（一级，进场高度离接地点至少200英尺，跑道能见度至少1 800英尺，二、三级均属未判明，等待搜集数据与分析）

MLSP Multiple-Link Satellite Program 多路通信卫星计划

MLT 1. Mission Life Test 使用寿命试验 2. Module Lead Time 模块提前期

MLU 1. Mid-Life Update（在全）寿命中期（的）改进 2. Miscellaneous Live Unit 综合居住单元

MLV 1. Medium Launch Vehicle 中型运载火箭 2. Mobile-Launch Vehicle（导弹）活动发射车，活发射装置

MLW 1. Maximum Landing Weight 最大着陆重量 2. Mean Low Water 平均低水位

MM 1. Magnetic Media 磁介质 2. Maintenance Manual 维护手册 3. Manueuver Margin 机动性余度 4. Mass Memory 大容量存储器 5. Master Model 主模线 6. Mega-Mega 兆兆 7. Middle Marker 中距指点标

mm Millimter 毫米

MM&T Manufacturing Method and Technology 制造方法与工艺

MMA 1. Maximum Mean Accuracy 最大平均精度 2. Multimission Aircraft 多任务飞机

M-marker Middle marker 中距指点标

MMB Maintenance Manual Bulletin 维修手册通告

MMC 1. Metallic-Matrix Composite 金属基复合材料 2. Monitor Mach Computer 马赫数监控计算机

MMDT Mean Mission Duration Time 平均任务持续时间

MME Modular Mounting Enclsoure 模块安装箱

MMEA Multiple Model Estimation Algorithm 多模型预测算法，多模型判断算法

MMEL Master Minimum Equipment List 主最低设备清单，最低设备总目表

MMF 1. Magnetomotive Force 磁动势，磁通势 2. Maximum Midexpiratory Flow 最大呼气中期流速 3. Multimode Fiber 多模光纤

MMFF Multi Mode Fire and Forget 发射后不管的多工作方式的（导弹），发射后不管的多用途（导弹）

MMI 1. Man/Machine Interface 人机界面（接口） 2. Mandatory Modification and Inspection 指令性改装与检查

MMIC 1. Millimetre-wave Integrated Circuit 毫米波集成电路 2. Monolithic Microwave Integrated Circuit 单块（片）微波集成电路

MMICS Maintenance Management Information and Control System 维修管理信息与控制系统（美国空军）

MMⅡ Mariner Mark Ⅱ 水手Ⅱ型探测器（美国空间探测器）

MMIPS Multiple Mode Integrated Propulsion System 多模式综合推进系统

MML Man-tended Multipurpose Laboratory 有人照料多用途实验室

MMM 1. Mars Mission Module 火星探测任务舱 2. Monolithic Main Memory 整体主存储器

MMO 1. Materials Management Organization 材料管理组织 2. Module Master Object 模块主对象

MMOD Micromodule 微型组件

M-mode ultrasonic scanning M型超声扫描

MMP 1. Manufacturing Management Plan 制造管理大纲 2. Metallic Material Processor 金属材料加工机械 3. Multiple Micro-Processors 多微处理机（系统）

MMPF Microgravity and Materials Processing Fnacility 微重力材料加工装置

MMR 1. Machmeter Reading 马赫数表读数 2. Minimum Military Requirement 军事人员及军需品最低需要量 3. Multi-Mode Receiver 多模式接收器

MMRH/FH Mean Maintenance and Repair Hours per Flight Hour 每飞行小时平均维修小时

MMS 1. Man-Machine System 人-机系统 2. Maintenance Management System 维修管理大纲 3. Mast-Mounted Sight 桅杆式（旋翼主轴式）瞄准具 4. Missile Management System 导弹管理系统 5. Modular Multimission Satellite 模块式多用途卫星（美国）6. Modular Multispectral Scanner 组合式多谱段扫描仪 7. Momentum Management System 动量管理系统 8. Multimission Modular Spacecraft 多任务

（用途）组合式航天器 9. Multimission Spacecraft 多用途航天器（美国）

MMSE　1. Man-Machine System Engineering 人机系统工程学 2. Minimum Mean Square Error 最小均方误差 3. Multi-Mission Support Equipment 多用途保障设备

mmse estimator　minimum mean square error 最小均方误差，最小均方差估计

MMSS　1. Manned Maneuverable Space System 机动载人航天系统 2. Multi-Module Space Station 多舱空间站

MMT　1. Mean Maintenance Time 平均维修时间 2. Multiple Mirror Telescopes 多镜面望远镜（美国）

MMTS　1. Manned Military Test Station 载人军事试验空间站 2. Mixing Manifold Temperature Sensor 混合气管温度传感器

MMU　1. Man（manned）Manoeuvring Unit 载人机动装置 2. Modular Maneuvering Unit 模块式机动飞行装置

MMW　1. Mean Maximum Weight 平均最大重量 2. Millimeter Wave 毫米波

MMW weapon　毫米波制导武器

MN　1. Mach Number 马赫数 2. Magnetic North 磁北 3. Main 主要的 4. Minimum 最小 5. Minute 分，分钟 6. Model Number 模块号

mn content　manganese content 锰含量

mn oxide　manganese oxide 氧化锰

MNB　1. Menadione Nicotinamide Bisulfite 亚硫酸氢烟酰胺甲萘醌 2. Multiple Narrow Beams 多道窄波束

MND　Mission Need Documents 任务需要文件（北约）

mnemonic　1. 易于记忆的，帮助记忆的 2. 飞行前检查的助记口诀

mnemonic code　助记码

mnemonic symbol　助记符

mnemonics symbolic language　助记符计算机语言

MNET　Measuring Network 测量网络

MNFP　Multinational Fighter Program 多国歼击机（研制）方案（美国）

Mnm　Minimum 最小

MNOSFET　Metal-Nitride-Oxide Semiconductor Field Effect Transistor 金属-氮化物-氧化物-半导体场效管

MNPA　Minimum Navigation Performance Airspace 最低导航性能空域

MNPS　Minimum Navigation Performance Specification 最低导航功能规范

MNRL　Mineral 矿物质，矿物的

MNS　Mission Need Statement 任务需求说明

MNTV　Mercury Network Test Vehicle 水星飞船跟踪网试验飞行器

MO　1. Magneto Optial 磁光 2. Maintenance Operation 维护操作 3. Manual Operation 人工操作 4. Mars observer 火星观测器（美国）5. Master Object 主对象 6. Month 月

MO LEM　Mobile Lunar Excursion Module 活动式登月舱

MO&DSD　Mission Operations & Data Systems Directorate 飞行操作和数据系统管理处（美国）

MOA　1. Make On Assembly 装配件制造 2. Memorandum Of Agreement 议定书，协定备忘录 3. Method Of Averaging 平均法 4. Military Operations Area 军事行动区，作战区

MOB　Main Operational Base 主要作战基地

MOBCOMSAT　Mobile Communication Satellite 移动通信卫星

MOBIDIC　Mobile Digital Computer 移动式数字计算机

mobile aeronautical station　移动航空电台

mobile air defence center　机动型防空中心

mobile air defence missile fire direction system　移动式防空导弹射击指挥系统

mobile air movements team　流动空运组

mobile communication　移动通信

mobile communication satellite　移动通信卫星

mobile defect　可动缺陷

mobile land station　移动地面站

mobile launch　机动发射

mobile launch pad　活动发射台

mobile launcher　机动发射车

mobile lever　动态水准仪

mobile lounge　移动休息室

mobile maritime station　移动海事站

mobile missile　机动导弹

mobile piston　移动活塞

Mobile Quarantine Facility　活动检疫设施

mobile radio station　移动无线电台

mobile radio unit　移动式无线电台，流动无线电台

mobile robot　1. 移动式机器人 2. 移动式遥控装置

mobile（robot）servicing system　移动式机器人服务系统

mobile satellite　移动卫星

mobile satellite communication　移动卫星通信

mobile satellite service　移动卫星业务，移动卫星服务

mobile satellite system　移动卫星系统

mobile service　移动业务，手机服务

mobile service satellite communication　移动式服务卫星通信

mobile service tower　1. 活动式发射服务塔 2. 移动式勤务塔

mobile station　移动电台，移动式工作站

mobile system　流动系统，机载系统

mobile telemetry station　活动遥测站，车载遥测站

mobile user 移动用户
mobile workshop （航空公司的）流动技术服务车
mobile-to-satellite return channel 移动点至卫星转回信道
mobility 1.机动性 2.流动率,迁移率 3.导纳
mobility aids 助行器
mobilization production requirement 动员后对产量的要求
MOBS Multiple Orbit Bombardment System 多轨道轰炸系统
MOBSS Mobility Support Squadron 机动保障中队
MOC 1. Maintenance Operational Check 维修作业检查 2. Mars Observer Camera 火星观测器摄影机 3. Minimum Operational Characteristics 最低工作性能
MOCA Minimum Obstruction Clearance Altitude 最低越障高度,最低净空高度
MOCC 1. Meteosat Operations Control Centre 气象卫星运行控制中心（西欧）2. Mission Operations Control Center 任务行动控制中心,运行控制中心
mock firing 模拟射击,模拟发射
mockup 1.实物模型,全尺寸模型,实体模型 2.假雷达
MOCP Missile Out of Commission for Parts 导弹因缺零件不能使用
MOCR Mission Operations Control Room 任务行动控制室
MOCS Multichannel Ocean Color Sensors 多通道海洋彩色遥感器
MOD 1. Manufacturing Operation Data 制造操作数据 2. Ministry Of Defense 国防部 3. Mission Operations Directorate（飞行）任务"管理处" 4. Model 模型 5. Modification飞机改型编号,飞机改装文件号
Mod Center （飞机制造公司的）改进改型中心
modal aggregation 模态集结
modal amplitude 模态振幅
modal amplitude coherence 模态幅值相干系数
modal analysis 模态分析,振动型分析
modal balancing 1. 模态平衡 2.振型平衡
modal characteristic 模态特性
modal coupling 模式耦合
modal datum 模态数据,模态基准
modal density 模态密度
modal dispersion 模（式）色散,模间色散
modal displacement 模态位移
modal efficient mass 模态有效质量
modal error 模态误差
modal excitation 模态励磁
modal force 模态力
modal frequency 模态频率,众数频率

modal function 模态函数
modal identification 模态辨识
modal mass 模态质量
modal matrix 模态矩阵
modal noise 模式噪声
modal numbers 模态数
modal parameter 模态参数
modal power 模功率
modal response 动态反应
modal sensitivity 模态灵敏度
modal sensitivity matrix 模态灵敏度矩阵
modal shape 模态振型
modal stiffness 模态刚度
modal structure 模态结构
modal superposition technique 模态叠加技术
modal survey 模态测量
modal synthesis 模态综合
modal test 模态试验
modal testing 模态试验,模态测试
modal transformation 模态变换
modal value 模态值
modal velocity 模态速度
modal wave 模态波
modal-interaction 模态交互
modal-interaction approach 模态交互方法
modality 模态,程式
MODAP Modified Apollo 改进型阿波罗飞船（美国）
MODAS Modular Data Acquisition System 模块化数据采集系统
mode 1.模式,样式 2.模,型 3.波型,振荡模,振荡型 4.传输型,传输模
mode activity 模式活动
mode change 模式变化,模式变换
mode combustion 模式（态）燃烧
mode competition 模（式）竞争
mode control 模式控制,模态控制
mode control law 模式控制律
mode control scheme 模式控制方案
mode controller 模式控制器
mode converter 模式变换器
mode coupling 模式耦合
mode D transponder D型应答机
mode damping 模式阻尼,振型阻尼
mode degeneracy 模（式）简并,波模简并度
mode disturbance observer 模式扰动观测器
mode estimate 模式预测
mode filter 模式滤波器
mode firing 模式点火
mode frequency 波模频率

mode function 模式函数
mode hopping 跳模
mode jump 跳模
mode localization 工况定位
mode locking 锁模,波模锁定,振荡型同步
mode magnitude 模式量级
mode method 模式方法
mode number 模数
mode of collapse 崩塌模式
mode of combustion 燃烧模式
mode of failure 破坏模式
mode of free vibration 自由振动模式
mode of instability 失稳模式
mode of maximum growth 最大增长模式
mode of maximum growth rate 最大增长率模式
mode of motion 运动方式
mode of operation 操作方式,运行方式
mode of oscillation 振荡模式,振动型,振动模式
mode of rotation 旋转模式
mode of valve 阀模态
mode of vibration 振动模态
mode pair 模态对
mode participation term 模式参与期限
mode probability 模式概率
mode pulling effect 模(式)牵引效应
mode ratio 模式比
mode response 模式响应
mode S transponder S模式应答器
mode scrambler 搅模器,模态码器,扰模器
mode selection by short cavity 短腔选模
mode selection switch 模式选择开关
mode selection technique 选模技术
mode selector 状态选择器
mode separation 模(式)分隔
mode shape 振型,振荡形式
mode shape correlation coefficient 振型相关系数
mode shape density 振型密度
mode shape energy 振型能量
mode shape slope 振型斜率
mode switch 模转换器,方式开关,波模转换开关
mode switching 波型转换开关
mode transducer 模式变换器
mode transition 模态跃迁
mode update 模式更新
mode volume 模体积
model 模型,典型
model account 台账
model adjustment 模型调整
model atmosphere 模式大气

model attitude 角度
model balance sting system 风洞天平支架
model base 模型库
model basin (水上飞机的)模型试验池
model cart 模型字
model centerline 中心线模型
model chamber 模型槽
model checking 备型校验
model coef 模型系数
model coefficient 建模型系数
model combustor 模型燃烧室
model confidence 模型置信度
model constant 模型常数
model control 模型控制,模型对照组,对照组
model controller 模式控制,模态控制
model coordination method 模型协调法
model correction 模型修正
model cycler 模型周期计
model debugging 模型除错
model decomposition 模型分解
model description 模型描述
model domain measurement 模态域测量
model engine 模型引擎
model error 模型误差
model estimate 模型估计
model fidelity 模型逼真度
model fit 模型拟合
model following controller 模型跟踪控制器
model free-flight test 模型自由飞行测试
model fuselage 机身模型
model identification 模型辨识
model in wind tunnel 风洞试验模型
model inversion 模型反演
model inversion controller 模型反演控制器
model likelihood 模型可能性
model loading 模型装载
model mismatch 模型失配
model modification 模型修改
model of aircraft 模型飞机
model of detonation 爆轰模型
model order 模型的阶
model output 模式输出
model pad method 模型垫块法
model parameter 【数】模型参数
model parametrization 模型参数化
model phase colinearity 模相同线性
model prediction 模型预测法
model problem 模型问题
model qualification test 型号合格试验,型号定型试验

model reduction 模型降阶
model reduction method 模型降阶法
model reduction theory 模型降阶理论
model reference 模型参考
model reference adaptive control 模型参考自适应控制
model reference adaptive controller 模型参考自适应控制器
model reference technique 模型参考技术
model refinement 模型改进
model response 模型响应,模型反应
model result 模型结果
model rig 模型试验装置
model rocket 模型(火)箭
model rotor 模型转子
model scale 模型比例尺
model scale factor 模型比例因子
model scramjet 超燃冲压发动机模型
model sensitivity 模型灵敏度
model sequence 模型序列
model spacecraft 航天器模型
model structure 模型结构
model support 模型支架
model surface 模型表面
model tank 模型箱
model testing 模型检验
model track 轨道模型
model transformation 模型变换
model tunnel 模型风洞
model uncertainty 模型不确定性
model validation 模型确认,模型验证
model verification 模型验证
model weight 模型权重,模型重量
model-based algorithm 基于模型的方法或算法
model-based control 模型控制
model-controlled system 模型控制系统
modeler 塑造者,制造模型者
modeling 建模
modeling approach 建模方法
modeling error 建模误差
modeling methodology 模型建立的方法论
modeling of combat 实战建模
modeling process 模型建立的过程
modeling result 建模结果
modeling technique 建模技术
modeling tool 建模工具
modeling uncertainty 建模不确定性
modelings 构模
mode-locked 锁模
model-scale rotor 模型比例尺转子

MODEM Modulator Demodulator 调制器-反调制器
modem 调制解调器
modem moderator-demodulator 调制解调器
moderate 适度的,中等的
moderate amplitude 中等响度(中等幅度)
moderate angle of attack 中等攻角
moderate anoxia 中度缺氧症
moderate flare climb-out 一段平飞后爬高到稳定上升
moderate hypoxia 中度缺氧
moderate risk 中度风险
moderate traffic 中量交通
moderate turbulence 中度颠簸
moderate visibility 中常能见度
moderator 1.减速器 2.阻滞剂,慢化剂
mode-ray duality 振型射线双重线
modern aircraft 现代飞机
modern application 现代应用
modern control theory 现代控制理论
modern control 现代控制
modern fighter 现代战斗机
modern high-bypass 现代高涵道
modern radar 现代雷达
modes change-over disturbance 模转换干扰
mode-shape error 模态振型误差
mode-shape measurement 模态振型测量
mode-shape uncertainty 模态振型不确定性
mode-transition 模态切换
modex 美国海军飞机别代码
MODFET Modulation Doped Field Effect Transistor 调制掺杂场效晶体管
modifiability 可改性
modification 改变(进、善、良),变更(化、换)改进了的形式
modification center 飞机改进中心
modification kit 改装成套器材
modification process 改性加工
modified cast 变性铸造
modified close control 改良精确引导
modified conical projection 改良圆锥投影
modified cut 修饰型
modified cylindrical projection 改良圆柱投影
modified design 改变标准设计
modified double-base propellant 改性双基推进剂
modified engine 改装过的引擎
modified equation 转换后的方程
modified flight control system jump pattern 使用部分飞行控制系统的半自动"点水"悬停航线
modified inertia parameter 修正惯性系数
modified Julian date 修正儒略日,约简儒略日

modified linkup 修改连接
modified linkup model 修改连接模型
modified mission 修改后的任务
modified Newtonian equation 修正牛顿公式
modified operational missile system 改进的作战导弹系统
modified PAR 改进型 PAR (Precision Approach Radar 精密进近雷达)
modified pursuit course 前置跟踪曲线,跟踪射击曲线
modified strip theory 修正的切片理论
modified version 修改版本
modified wing 改进后的机翼
modifier 1.改性剂,改良剂,调节剂 2.变换装置 3.变址数
modify 修改,修饰,更改
MODILS Modular Instrument Landing System 模块化仪表着陆系统
MODS 1. Manned Orbital Development Station 载人轨道开发站（美国航宇局）2. Military Orbital Development System 军用轨道开发系统（美国空军）3. Missile Offense/Defense System 导弹进攻与防御系统 4. Modification Sheet 更改单
modular 单元体,模块,模块化的,组合的
modular aircraft 模块式结构飞机,单元体结构飞机
modular architecture 模块结构,积木结构
modular attitude control system 模块化姿态控制系统
modular automated weather system 模块式自动气象观测系统
modular avionics 模块化航空电子学（设备）
modular boxes （航空电子设备系统的）模块匣
modular computer 模块化计算机
modular construction 模块式结构,单元体结构
modular data system 模块化数据系统
modular data transfer system 模块式数据传送系统
modular decomposition 模块分解
modular design 模块化设计,单元体设计
modular information processing equipment 模块化信息处理装置
modular intelligent terminal 模块化智能终端
modular lounge 活动休息室（机场内的）
modular programming 模块化程序设计
modular redundancy 模数冗余度
modular tactical missile 模式战术导弹
modular weapon 模式武器
modularity 模块性
modularization 模块化,积木化,单元体化
modulate 调节,(信号)调制,调整
modulated continuous wave 已调等幅波
modulated wave 已调(电磁)波,灯波,调幅波

modulation 调制
modulation distortion 调制失真
modulation format 调制格式
modulation frequency 调制频率
modulation index 调制指数
modulation rate 调制速率
modulation sensitivity 调制灵敏度
modulation technique 调制技术
modulation transfer function 调整传递函数
modulation-demodulation 调制解调
modulation-doped field effect transistor 调制掺杂场效晶体管
modulator 调制器
modulator vacuum gauge 调制型真空计
module connection 模块连接
module efficiency 组件效率
module integration facility 舱段总装设施
module interconnection language 模块互连语言
module management language 模块管理语言
modulus 系数,模数
modulus of elasticity 弹性模量
MOE Measure Of Effectiveness 有效性测量
MOEAS Multi Objective Evolutionary Algorithms 多目标进化算法
MOF Manned Orbital Flight 载人轨道飞行
mofette 【地】碳酸喷气孔
MOG Maximum On Ground 大型场地
MOH Major Overhaul 大翻修
MOI 1. Maintenance Operating Instruction 维修作业指示（美国空军部队）2. Mars-Orbit-Insertion 进入火星轨道
moire 网纹干扰
moist air 湿空气
moisture 湿度
moisture apparatus 测湿器,水分仪
moisture indicator 湿度指示器
moisture separator 水气分离器,去湿器
moisture test 潮湿试验,耐湿试验
MOKM Module-to-Option Knowledge Map 模块至选项经验图
MOL Manned Orbiting Laboratory 载人轨道实验室
mol 摩尔,克分子
molar 克分子的,摩尔（浓度）的
molar concentration 体积摩尔浓度
molar flow rate 摩尔流率
molar fraction 克分子分数,摩尔分数
molar mass 摩尔质量
molar ratio 摩尔比率,摩尔浓度,克分子比,克分子比率

molar specific heat 摩尔比热
mold line 型线
molding 模塑,铸造,装饰用的嵌线
mole 摩尔式管道测弯仪,塑孔,防波堤,克分子
mole fraction 摩尔分数,【化】克分子分数
mole ratio 摩尔比,克分子比
molecular aerodynamics 分子空气动力学
molecular beam epitaxy 分子束外延
molecular biology 分子生物学
molecular clock 分子钟
molecular cloud 分子云
molecular contamination 分子污染
molecular datum 分子数据
molecular device 分子器件
molecular diffusion 分子扩散
molecular drag pump 牵引分子泵
molecular dynamics 分子动力学
molecular effect 分子效应
molecular effusion 分子焊流
molecular electronics 分子电子学
molecular flow 分子流(一种气体流态)
molecular flow rate 分子流率
molecular flux 分子流率
molecular gas laser 分子气体激光器,分子气体镭射
molecular hydrogen 分子氢
molecular mass 分子质量
molecular motion 分子运动
molecular nitrogen 分子态氮
molecular oscillator 分子振荡器
molecular oxygen 分子氧
molecular pump 分子泵
molecular scale 分子尺度
molecular sieve 分子筛
molecular sieve oxygen generation 分子筛制氧
molecular sieves trap 分子筛陷阱报文
molecular structure 分子结构
molecular transport 分子输运
molecular transport process 分子输运过程
molecular weight 分子量
molecular weight close 分子量接近
molecular-beam 分子束
molecularized digital computer 分子化数字计算机
molecule 分子
mole-fraction 克分子比,克分子分数
moleskin strip 头蓝内衬带
Mollier diagram 给熵图,莫里尔图,莫里尔蒸汽图
molten binder 熔融黏结剂
molten carbonate fuel cell 【电】熔融碳酸盐燃料电池
molten phase 熔融状态

molten surface 熔液表面
molten surface layer 熔液表层
molten-layer thickness 熔化层厚度
molten-salt electrolyte 熔融盐电解液
molten-salt electrolyte cell 熔融盐电池
molten-salt growth method 溶盐法
MOLY Molybdenum 钼
moly 〈口语〉二硫化钼润滑剂
molybdenum 【化】钼(化学元素,符号 Mo)
molybdenum gate technology 钼栅工艺
molybdenum grid 钼栅极
molybdenum plate 供应钼板
molybdenum-oxide 氧化钼
molybdenum-oxide system 氧化钼系统
MOM 1. Mars Orbital Module 火星轨道舱 2. Metal-Oxide-Metal 金属-氧化物-金属 3. Methanol Oxygen Mix 甲醇/氧混合液 4. Modified Operation Missile 改进的作战导弹
moment 力矩
moment arm 力臂,矩臂,力矩臂
moment coef 力矩系数
moment coefficient 力矩系数
moment compensation 力矩补偿
moment constraint 力矩约束
moment datum 力矩数据
moment distribution 力矩分配,弯矩分配
moment equation 力矩方程
moment excursion 力矩偏移
moment index 力矩计算指数
moment matching method 时矩匹配法
moment model 力矩模型
moment of area 面积矩
moment of inertia 惯性矩,转动惯量
moment of momentum 动量矩,角动量
moment of momentum of gyro 陀螺动量矩
moment reference center of balance 天平力矩参考中心
moment span 矩跨距
moment variation 力矩变化
momentum 动量
momentum balance 动量平衡
momentum bias attitude control system 偏置动量姿态控制系统
momentum bias control 偏置动量控制
momentum bias system 偏置动量系统
momentum boundary 动量边界
momentum change 动量变化
momentum coefficient 动量系数
momentum coupling 动量耦合
momentum coupling coefficient 动量耦合系数

momentum dump 动量卸载
momentum effect 动量效应
momentum equation 动量方程
momentum exchange 动量交换
momentum exchange device 动量交换装置
momentum exchange maneuver 动量交换机动
momentum fluid 动量流体
momentum flux 动量通量
momentum flux measurement 动量通量测量
momentum flux ratio 动量通量比
momentum flux sensor 动量通量传感器
momentum gain 动量增益
momentum integral equation 动量积分方程式
momentum integral method 动量积分法
momentum loss 动量损失
momentum loss thickness of boundary layer 边界层动量损失厚度
momentum magnitude 动量级,动量值
momentum method 动量法
momentum of fuel jet 燃油喷射的动量
momentum radius 动力半径
momentum ratio 动量比
momentum separation 动量分离
momentum sphere 动量球
momentum storage 动量存储
momentum storage capability 动量储存能力
momentum theory 动量原理
momentum thickness 动量损失厚度值
momentum transfer 动量交换,动量转移
momentum transfer method 动量转移法,冲量法
momentum unloading 动量卸载
momentum vector 动量矢量
momentum wheel 动量轮
momentum wheel and gimbal desaturation 动量飞轮和万向框架去饱和
momentum wheel fault 动量轮故障
MOMS 1. Map, Operation and Maintenance Station 地图、操作和维修站 2. Modular Optoelectronic Multispectral Scanner 模块式光电多谱段扫描仪
MON 1. Monitor 监听器,监视装置,监控员 2. Monitoring 监控,监视 3. Monument 纪念碑 4. Motor Octane Number 发动机辛烷值
MON PWR Monitor Power 监控器电源
monatomic gas 单原子气体
monatomic oxygen 单原子氧
monergol 单组元火箭燃料
Monel 蒙耐尔合金
MONISAR Monitoring with Synthetic Aperture Radar 合成孔径雷达监测计划(荷兰)

monitor 1.监控器,监视器,监听器,检测器 2.监视,监控
monitor amplifier 监控放大器
monitor and display system 监视显示系统
monitor event simulation system 监控事件模拟系统
monitor of ultraviolet solar energy 太阳紫外能监测仪
monitor of ultraviolet solar radiation 太阳紫外辐射监视仪
monitor program 监控程序,管理程序
monitor station (全球定位系统)监控站
monitored system 受监控系统
monitoring 监控,监测,监听,监视
monitoring and control station 监控站
monitoring criterion 监测标准
monitoring guidance 监控提示
monitoring location 监控位置
monitoring process 监视进程
monkey drop 〈口语〉直升机软梯空降
monkey wrench 活动扳手,螺旋钳
monoball mount 单球座架,单球安装点
monobloc 1.单块(工)件 2.全动式操纵面
monobloc battery 整体蓄电池组
monoblock 整体,直板,单块
monochromatic aberration 单色像差
monochromatic absorptance 单色吸收率
monochromatic emittance 单色发射率
monochromatic image 太阳单色像
monochromatic light 单色光
monochromatic solar radiation 单色太阳辐射
monochrome display 单色显示,单色显示器
monochrome image 单色图像
monochrome TV 黑白电视
monocolar vision 单眼视觉
monocomparator 单片坐标量测仪
monocoque 硬壳,硬壳式结构,硬壳式飞机
monocoque construction 硬壳式构造
monocoque structure 硬壳式结构
monocular accommodation 单眼调节
monocular camera 单目摄影机
monocular vision sensor 单眼视觉传感器
monodisperse 单分散的
monodromy 单值
monofuel 单元燃料,单组分燃料
monohud 单眼平视显示器
monolayer 单层
monolith 整块石料,庞然大物
monolith catalyst 整体式催化剂
monolith catalyst bed 整体式催化剂床
monolith support 整体支撑

monolithic array logic	单片阵列逻辑
monolithic ceramic	整体陶瓷,单片陶瓷
monolithic ceramic capacitor	单片陶瓷电容器
monolithic computer	单片计算机
monolithic integrated circuit	单片集成电路
monolithic material	整块材料,整体材料
monolithic microprocessor	单片微处理机
monolithic microwave intergrated amplifier	单片微波集成放大器
monolithic rocket	单块式火箭,整体药柱火箭
monolithic storage	单片存储器
monolithic structure	整体结构,单片结构,独石结构
monolithic thruster	单片推进器
monolobe scanner	单瓣扫描仪
monolook processing	单视处理
monomer	单(分子)体,单基物,单元结构
monomethyl hydrazine	一甲基肼(火箭推进剂)
monomode fiber	单模光纤
monophone	单声
monoplane	单翼机,单翼飞机
monoplane wing	单翼机的机翼
monopole	单极子
monopole antenna	单极子天线
monopole mass spectrometer	单极质谱仪
monopoly	垄断,垄断者,专卖权
monopropellant	单元推进剂
monopropellant chamber pressure	单组元室压力
monopropellant limit	单元推进剂限制
monopropellant propulsion system	单组元推进系统
monopropellant rocket	单组元推进剂火箭
monopropellant rocket engine	单组元推进剂火箭发动机
monopropellant thruster	单组元推进器
monopropellant thruster prototype	单组元推进器原型
monopulse	单脉冲
monopulse antenna	单脉冲天线
monopulse null depth	单脉冲零深
monopulse radar	单脉冲雷达
monopulse radar homing head	单脉冲雷达导引头
monopulse radar homing system	单脉冲雷达寻的系统
monopulse radar tracking system	单脉冲雷达跟踪系统
monopulse technology	单脉冲技术
monopulse tracking	单脉冲跟踪
monopulse tracking receiver	单脉冲跟踪接收机
monorail	伞钩单轨
monoslabs	水泥板块预制件
monostatic microwave imaging	单静态微波成像
monostatic radar	单基地雷达
monostation locating	单站定位
monotectic alloy	偏晶合金
monotone	单调,单音调
monotone mode	单调模式
monotonic Boolean function	单调布尔函数
monowing missile configuration	单翼导弹构型
monoxide	一氧化物
monsoon	1.季节风,贸易风 2.印度的季节雨
Monte Carlo method	蒙特卡洛法(用随机数对数值所求近似解的方法)
MOO Manned Orbital Observatory	载人轨道观测台
Moon Agreement	月球协议,月球条约
moon boundary	月球边界
moon crawler	月球车
moon dog	幻月,假月
moon flight trajectory	月球飞行轨道
moon probe	月球探测器
moon recognition	月球识别
moon resonance	月球共振
Moon rocket	月球火箭
moon seismograph	月震仪
moon system	月球系统
moon triangular libration point	月球三角平动点
moon triangular point	月球三角点
moon walk	月球漫步,月面行走
moonbuggy	月球车
mooncraft	月球飞船
moondown	月落
moon-earth trajectory	月球-地球轨道
moonfall	登月
moonflight	月球飞行
Moonik	月球火箭,月球卫星(苏联)
moonman	登月航天员
moonmark	月球陆标
moonport	月球火箭发射站
moonquake	月震
moonscooper	月球标本收集器
moonscope	(人造)卫星观测(望远)镜
moonset	月落
moonship	月球飞船
moonshot	1.月球探测器 2.向月球发射
moontrack	卫星跟踪
moonward(s)	往月球的
moonwatch	卫星监视站
moored	可停泊在水面的飞机
mooring band	(气球的)系留索带
mooring cable	(飞艇)地面系留索
mooring drag	气球的拖绳
mooring guy	(气球的)固定绳,系留索
mooring line	(气球的)固定绳,系留索

mooring mast 系留柱

MOOSE 1. Manned Orbital Operations Safety Equipment 载人轨道作业安全设备 2. Man-Out-Of-Space-Easiest 最简便空间救生(设备)

MOP 1. Major Outside Production 重要的外厂生产 2. Manned Orbital Platform 载人轨道平台 3. Meteosat Operational Programme Satellite 气象业务计划卫星(欧洲的气象卫星)

MOP costs Maintenance and Overhaul Personnel cost 维修与翻修的人事费用

MOPITT Measurement Of Pollution In The Troposphere 对流层污染测量

MOPP Mission-Oriented Protection Posture 与任务相适应的保护措施

MOPR Minimum Operational Performance Requirements 最低使用性能要求

MOPS Minimum Operational Performance Standard 作战性能最低标准,最低使用性能标准

MOPTAR Multiple Object Phase Tracking And Ranging 多目标相位跟踪与测距

mop-up equalization 全程均衡,扫余均衡

MOR 1. Mandatory Occurrence Report 规定的情况报告 2. Mars Orbital Rendezvous 火星轨道交会 3. Military Operational Requirement(s) 作战要求

MORAG Mandatory Occurrence Reporting Advisory Group 编写规定情况报告顾问组

Morelos 莫雷洛斯卫星(墨西哥国内通信卫星)

Morko 幽灵(芬兰造的歼击机)

MORO Moon Orbiting Observatory 环月观测台

morphing 变形

morphing-wing weight 变形机翼重量

morphogenesis 形态发生

morphological instability 形态不稳定性

morphological stability 形态稳定性

morphology 形态学

morphometric map 地貌形态示量图

Morse automatic decoder 自动莫尔斯译码器

Morse code 莫尔斯码

Morse code photographic reconnaissance satellite 莫尔斯码型照相侦察卫星

MORT 1. Missile Operational Readiness Test 导弹战备状态试验 2. Morse Taper 莫氏锥度

mortality 故障率,死亡率

mortar 1. 射伞枪,深水炸弹发射器 2. 迫击炮

mortar locating radar 迫击炮定位雷达

mortgage clause 抵押条款

MOS 1. Manned Orbital Station 载人轨道站 2. Maritime Observation Satellite 海洋观测卫星(日本) 3. Metal Oxide Semiconductor 金属氧化物半导体

MOS memory Metal Oxide Semiconductor (silicon) memory 金属氧化物半导体(硅)存储器

MOS solar cell Metal Oxide Semiconductor (silicon) solar cell 金属氧化物半导体(硅)太阳能电池

mosaic 1. (航空照片、红外图像、光导摄像管等的)镶嵌图 2. 感光嵌镶幕,嵌银光电阴极 3. 嵌花式的,镶嵌式的,马赛克的

mosaic array 镶嵌阵列

mosaic assembly 航空像片嵌拼图

mosaic index (像片)镶嵌索引图

mosaic mirror telescope 镶嵌镜面望远镜

MOSFET Metal Oxide Semiconductor Field Effect Transistor 金属氧化物半导体场效应三极管

MOSL Manned Orbital Space Laboratory 载人轨道空间实验室

MOSP Multi-Optical Stabilized Payload 多光学传感装置稳定有效载荷

Mosquito 蚊式教练机(意大利帕尔迪那维亚飞机公司生产)

MOSS 1. Manned Orbital Space Station 载人轨道空间站 2. Manned Orbital Space System 载人轨道空间系统

MOST Metal Oxide Silicon (Semiconductor) Transistor 金属氧化物硅(半导体)晶体管

most economic control theory 最经济控制理论

most economic observing theory 最经济观测理论

most remote failure 罕见故障

most significant package 最大有效程序包

MOT 1. Manned Orbital Telescope 载人轨道望远镜 2. Maximum Overhaul Times 最大翻修时间

MOTE Multinational Operational Test and Evaluation 多国作战试验与鉴定

moth eating test 虫蛀试验

mother satellite 母卫星

mother ship 1. 控制遥控飞机的母机,运载飞机,运载飞船 2. 母舰

mother spacecraft 航空母机,母航天器

motherboard 底板,母板

mothercraft 太空母船

motion 运动,移动

motion afterimage 运动后像

motion base training simulator 基于运动的训练仿真器

motion capture system 动作捕捉系统,运动捕捉系统

motion compensation 运动补偿,动态补偿

motion component 运动部件,运动成分

motion condition 运动状态

motion constraint 运动约束

motion control 运动控制,拖动控制,动作控制,移动控制

motion coupling　运动耦合
motion drive　运动驱动
motion ellipse　运动椭圆
motion equation　运动方程
motion error　运动误差
motion estimation　运动估计
motion feedback　运动反馈
motion fidelity　运动保真度
motion fidelity rating　运动保真度等级
motion filter　运动滤波
motion gain　运动增益
motion intensity　运动强度
motion level language　动作级语言
motion model　运动模型
motion of satellite　卫星运动
motion perception　运动知觉
motion perception gain　运动知觉获得
motion plan　运动规划
motion planner　运动规划器
motion planning　运动规划
motion platform　运动平台
motion sickness　运动病,晕机病,晕动病
motion sickness drug　抗运动病药
motion sickness susceptibility　运动病敏感性
motion sickness testing　运动病试验,晕机病试验
motion simulator　运动仿真器
motion solution　运动解
motion space　可动空间,动作空间
motion stability　运动稳定性
motion system　运动系统
motion system characteristic　运动系统特性
motion trajectory　运动轨迹
motion variable　运动变量
motion-based algorithm　动态的算法
motion-filter　运动滤波器
motion-filter gain　运动滤波器增益
motion-filter natural frequency　运动滤波器自然频率
motion-gap　运动间隙
motion-perception　运动知觉
motion-perception gain　运动知觉获得
motion-planning algorithm　运动规划方法
motion-planning problem　运动规划问题
motivator　1.操纵输出装置 2.改变飞行轨迹的装置(操纵面、襟翼、矢量喷管等)
MOTNE　Meteorological Operational Telecommunications Netcwork in Europe 欧洲气象业务通信网
motor　发动机,马达,汽车
motor action time　发动机工作时间
motor boating　1.(类似摩托艇发动机的)亚声频振荡 2.低频振荡
motor burning　发动机燃烧,发动机烧毁
motor burnout mass　发动机燃尽质量
motor case　电动机壳
motor chamber　发动机燃烧室
motor controller　电动机控制器
motor design　电机设计
motor firing　发动机的点火
motor flow　电机过流
motor glider　摩托滑翔机,带发动机的滑翔机
motor initial mass　发动机初始质量
motor mass　发动机质量
motor mass fraction　发动机质量比
motor noise　发动机噪声,电动机噪声
motor operation　电动机运行
motor pressure　发动机燃烧室压力
motor program memory　运动程序记忆
motor rotating test　发动机旋转试验
motor six component test　发动机六分力试验
motor specific impulse　固体火箭发动机比冲(量)
motor structure mass　发动机结构质量
motor test　发动机试验
motor torque　电机扭矩,马达转矩,电动机转矩
motor-converter　电动机发电机组,电动变流机
motorpump　电动泵
Mott-barrier diode　莫脱势垒二极管
mottle　杂斑,模纹,日芒
MOTU　Maritime Operational Training Unit 海上作战训练部队
MOTV　Manned Orbital Transfer Vehicle 载人轨道转移飞行器(美国)
MOU　Memorandum Of Understanding 谅解备忘录
mould dimension　铸型尺寸
mould growth　霉菌生长
mould test　霉菌试验
mouldy　〈口语〉航空鱼雷,空投鱼雷
mount　安装点,架,架置,安装,装配,固定
mount engine　安装发动机
mountain sickness　高山病
mounting　1.(飞机上的)安装物 2.座,架,框,固定装置 3.搭载
mounting pad　(发动机或附件传动机匣的)环形安装凸台
mounting technology　装架工艺
mouse　1.〈美国口语〉火箭 2.最低轨道运行的人造地球卫星
moustaches　前置小翼
mouth　1.孔,进口,入口 2.伞口
mouth lock　伞口闭锁装置

mouth-to-mouth resuscitation 口对口人工呼吸
MOV 1. Main Oxidizer Valve 氧化剂主阀 2. Manned Orbiting Vehicle 载人轨道飞行器 3. Metal Oxide Varistor 金属氧化物调节电阻 4. Motor Operated Valve 马达操作活门
movable image reticle 活动光环影象
movable measuring device 移测装置
movable nozzle 可动喷管
movable wall 移动墙
movables 活动操纵面
move 运动,移动
move distance 移动距离
move limit 移动限制
move off blocks 1.移开轮挡 2.开始移动时间
movement 军事空运行动
movement area 飞机地面活动区
movement boards 飞机进港出港时刻公告板
movement coordination test 运动协调试验
movement of fluid 流体运动
moveout 时差
movie-screen 电影屏幕
movie-screen antenna 电影屏幕天线
moving alignment 动基座对准
moving armature speaker 移动电枢扬声器
moving average method 移动拟合法,移动平均法
moving base flight simulator 活动基座飞行模拟器
moving cluster 移动星团
moving coil earphone 电动耳机
moving coil loudspeaker 电动扬声器
moving coil microphone 动圈传声器
moving coil pickoff 动圈式传感器
moving coil transducer 动圈式变换器
moving coil vibrator 动圈式振荡器
moving conductor mic 电动传声器
moving mass roll control 移动质量滚动控制
moving reference coordinate system 1.动基准坐标系 2.动参考坐标系
moving reticle 活动光环
moving seal 动密封
moving source method 动源法
moving target 活动靶标
moving target detection 动目标探测
moving target indication 移动目标指示
moving target indication radar 移动目标指示雷达
moving window display 快视移动窗
moving wings 活动翼
moving-base alignment 动基座对准
moving-base simulator 活动座舱飞行模拟器
moving-base-derived navigation data 动基导出的导航数据
moving-base-referenced navigation data 动基基准导航数据
moving-bug system 飞机标志移动型系统
moving-coil speaker 动圈式扬声器
moving-iron instrument 动铁式仪表
moving-target selection 动目标选择
moving-wall 移动墙
moving-wall effect 移动墙效果
MP 1. Main Processor 主处理器 2. Maintenance Period 维修周期 3. Maintenance Platform 维修平台 4. Maintenance Program 维护大纲 5. Manifold Pressure 进气压力(歧管内压力) 6. Manoeuvre Progammer(遥控飞机的)机动飞行编程器 7. Metal Point 金属点
MPA 1. Man-Powered Aircraft 人力飞机 2. Maritime Patrol Aircraft 海上巡逻机 3. Multiple Payload Adapter 多个有效载荷连接器
MPAA Multibeam (Multifunctional) Phased-Array Antenna (Aerial) 多波束(多功能)相控阵天线
MPAC Multipurpose Applications Console 多用途控制台
MPAG Man-Powered Aircraft Group 有人驾驶的机群
MPAR Modified Precision Approach Radar 改进型精确进场雷达
MPAYL Maximum Payload 最大有效载荷
MPB 1. Maximum Pressure Boost 最大增压 2. Mean Point Bursts 爆炸中点 3. Multiple Pencil Beam 多锐方向性波束
MPC 1. Military Personnel Center 军事人员中心 2. Multiple Payload Carrier 多个有效载荷运输器(美国) 3. Multi Purpose Console 多用途操纵台
Mpc Mega-parsec 兆秒差距
MPCD Multi Purpose Colour Display 多用途彩色显示器,多功能彩色显示器
MPCS Maintenence Power Control System 维修电源控制系统
MPD 1. Magneto-Plasma Dynamics 磁等离子体动力学 2. Magnetosphere Particle Detector 磁层粒子探测器 3. Maintenance Planning Document 维修计划文件 4. Manufacturing Plan Document 制造计划文件 5. Maximum Permitted Dose 最大允许辐射剂量
MPDB Main Power Distribution Box 主电能分配盒
MPDR 1. Mobile Pulse Doppler Radar 机动式脉冲多普勒雷达 2. Mono Pulse Doppler Radar 单脉冲多普勒雷达
MPDS Missile Penetrating Discarting Sabot 导弹穿透时脱落的外壳
MPDU Multiplexing Protocol Data Unit 多路复用协议

数据单元
MPE Mission to Planet Eearth 地球使命计划
MPED Multipurpose Electronic Display 多用途电子显示板
MPEL Maximum Permissible Exposure Level 最大允许暴露高度
MPESS Mission Peculiar Experiment Support Structure（美国航天飞机货舱中的）飞行任务特殊实验支撑结构
MPF Materials Process Facility 材料加工装置（美国航天飞机上的）
MPflt Mission Planning Flight 任务计划飞行（英国空军）
MPFS Microwave Position-Fixing System 微波定位系统
MPG 1. Maintenance Program Group 维修方案组 2. Miles Per Gallon 每加仑燃料的英里数 3. Moulded Propellant Grain 铸型推进剂药柱（装药）
MPH Miles Per Hour 英里/小时
MPI 1. Magnetic Particle Inspection 磁粉检验,磁粉探伤 2. Magnetic Position Indicator 磁位置指示器 3. Manufactruing Process Instruction 制造工艺说明书 4. Mean Point of Impact 平均落点,平均弹着点 5. Merdian Precision Indicator 子午线精密指示器 6. Mission and Payload Integration Office 飞行任务和有效载荷综合处（美国航空航天局）
MPL 1. Manufactruing Parts List 制造零件表 2. Master Parts List 零件总清单 3. materials processing laboratory 材料加工实验室 4. Maximum Payload 最大商载 5. Mid-Pacific landing 中太平洋降落 6. Modules Parts List 模块零件表
MPLM Mini Presssurized Logistics Module 微型增压后勤舱
MPM 1. Message Processing Module 信息处理装置 2. Multipurpose Missile 多用途导弹（美国空军）
MPMS 1. Missile Performance Measurement System 导弹性能测量系统 2. Missile Performance Monitoring System 导弹性能监控系统
MPNL Master Parts Number List 主要零件号码单
MPO Mission Planning Officer 任务计划军官（英国空军）
MPP 1. Maitainability Program Plan 维修性项目大纲 2. Maintenance Program Proposal 维修大纲建议（书）3. Master Phasing Plan 总阶段大纲 4. Massively Parallel Processor 大规模并行处理器 5. Material Processing Platform 材料加工平台 6. Maximum Power Point 最大功率点 7. Most Probable Position 最可能位置
mppa million passenger per annum 年百万乘客数

MPPL Manufacturing Part Parts List 制造件零件表
MPPS Multi Purpose Pylon System 多用途外挂（架）系统
MPR 1. Master Parts Record 零件总记录 2. Military Photo Reconnaissance 军事照相侦察
MPS 1. Main Propulsion System（Stage）主推进系统（级）2. Master Production Schedule 主生产计划 3. Master Program Sheet Or Master Production Schedule 主程序单或生产进度 4. Materials Processing in Space 太空材料加工 5. Materials Processing System 材料加工系统 6. Material and Process Specifications 材料和工艺规范 7. Megtacycle Per Second 兆周/秒
MPSI Material Process Specification Index 材料工艺规范目录
MPSR Monthly Progress Status Report 进度状况月报
MPT 1. Main Propulsion Test 主推力测试 2. Microwave Plasma Thruster 微波等离子推力器 3. Male Pipe Thread 外螺纹导管
mpt system mpt（Main Propulsion Test 主推力测试）系统
MPTO Maximum Performance Take Off 最大性能起飞
MPTV Multi-Programming Television 多节目电视广播
MPU Missile Power Unit 导弹动力装置,导弹发动机,导弹电源装置
MQAD Materials Quality Assurance Directorate 材料质量保证管理处（英国防部）
MQF Mobile Quarantine Facility 流动式隔离设施
MR 1. Management Review 管理评审 2. Manned Reentry 载人再入 3. Maritime Reconnaissance 海上侦察
mr milliroentgen 毫伦（琴）,千分之一伦琴
MRA 1. Major Replaceable Assembly 主要可换件 2. Maximum Relight Altitude 重新点火最大高度 3. Minimum Reception Altitude 最低接收高度 4. Minimum Release Altitude 最低投弹高度 5. Mission Reliability Analysis 任务可靠性分析
MRAAM Medium Range Air-to-Air Missile 中距空（对）空导弹
MRAC Model Reference Adaptive Control 模型参考自适应控制
mrad millirad 1. 毫弧度 2.【医】毫拉德（辐射剂量单位）
MRAG Metallics Research Advisory Group 金属粒子研究咨询组
MRB 1. Main Rotor Blade 旋翼奖叶 2. Maintenance Review Board 维修审查委员会 3. Material Review Board 材料评审委员会,不合格品审理委员会
MRBM Medium（mid）Range Ballistic Missile 中程弹道导弹
MRBR Maintenance Review Board Report 维修大纲

MRC 1. Machine Readable Code 可机读码 2. Manufacturing Responsibility Center 制造责任中心 3. Materail Requirement Change 材料需求更改 4. Maximum Raange Cruise 最大航程巡航速度 5. Modular Radio Cabinet 模块式无线电舱 6. Multirole Recoverable Capsule 多用途返回舱

MRCA Multi-Role Combat Aircraft 多任务作战飞机

MRCC 1. Maintenance Reliability Control Committee 维修可靠性控制委员会 2. Materials Research Consultative Committee 材料研究咨询委员会 3. Material Review Control Center 材料评审控制中心

MRCL Mercurial 水银的

MRCO Maintenance Reliability Control Office 维修可靠性控制办公室

MRCS Multiple Rate Circuit Switching 多速率电路交换

MRD 1. Main Rotor Diameter 主旋翼直径 2. Material Requirement Drawing 材料要求图 3. Mission Requirements Document 飞行任务要求文件

MRD/FT Missile Restraint Device/Field Tester 导弹固定装置/野外试验台

MRDB Mission Requirements Data Base 飞行任务要求数据库

MRDU Multichannel Receiver/Decoder Unit 多通道接收/解码装置

MRE Mean Radial Error(武器投放)平均径向误差

MRF 1. Meteorological Research Flight 气象研究飞行 2. Multi-Role Fighter 多用途战斗机

MR-FI Malfunctional Report-Functional Item 故障报告-功能项目

MRG Master Reference Gyro 主基准陀螺

MRGB Main Rotor Gearbox 旋翼减速器

MRH Manufacturing River Head 机械制造的铆钉头

MRI 1. Magnetic Resonance Inspection 磁共振检查 2. Medium-Range Interceptor 中程截击机

MRLC Moon Rocket Launching Center 月球火箭发射中心

MRLS Minimum Requirements List for Serialization 系列化最低要求清单

MRM 1. Manufacturing Resource Management 制造资源管理 2. Medium Range Missile 中程导弹

MRMU Mobile Remote Manipulating Unit 移动式遥控机械臂装置

MRN 1. Manufacturing Revision Notice 制造修订通知 2. Meteorological Rocket Network 气象火箭网

MRO Maintenance, Repair and Overhaul 维修、修理和翻修

MROSE Multiple-tasking Real-time Operating System Executive 多任务实时操作系统执行

MRP 1. Manned Reusable Payload 载人可重复使用有效载荷 2. Manufacturing Resource Planning 制造资源计划,制造资源计划工作 3. Materials Requirements Planning 物料需求计划,制定器材要求计划 4. Material Resource Planning 物料资源规划 5. Medium-angle Rocket Projectile 中等角度火箭弹道(空对地瞄准装置) 6. Military Rated Power 军用额定功率 7. Mission Requirement Package 任务要求组件

MRP RELEASE Manufacturing Resource Planning Release 制造资源计划发放

MRR 1. Manufacturing Revision Request 制造修正需求 2. Maritime Radar Reconnaissance 海上雷达侦察 3. Mechanical Reliability Report 机械可靠性报告 4. Miscellaneous Retrofit Requirement 其他改型要求 5. Modification Revision Record 改进修订记录

MRRPV Medium-Range Remotely Piloted Vehicle 中程遥控飞机

MRRV Manoeuvrable Re-entry Research Vehicle 机动再入研究飞行器

MRS 1. Maintenace Reporting System 维修报告系统,维护报告系统 2. Maintenance Requirement System 维修要求系统 3. Manned and Retrievable System 载人和回收系统 4. Manned Reconnaissance Satellite 载人侦察卫星 5. Manned Reusable Spacecraft 可重复使用载人航天器 6. Microwave Ranging System 微波测距系统 7. Mobile Remote Servicer 移动式遥控服务器 8. Multispectral Resource Sampler 多谱段资源取样仪(扫描仪)

MRSA Mandatory Radar Service Area 指定的雷达使用区

MRSE 1. Microwave Remote Sensing Equipment 微波遥感设备 2. Microwave Remote Sensing Experiment 微波遥感试验

MRSP Multifunction Radar Signal Processor 多功能雷达信号处理器

MRSR Mars Recover Sample Return 火星车取样返回(美国研究中的火星取样计划)

MRSV Maneuverable and Recoverable Space Vehicle 可机动和回收的空间飞行器

MRT 1. Maximum Reheat Thrust 最大复燃(加力)推力 2. Maximu Repair Time 最长修理时间 3. Mean Response Time 平均响应时间 4. Military Rated Thrust 军用额定推力 5. Miniature Receive Terminal 微型接收终端

MRTD Minimum Resolvable Temperature Difference 最小可分辨温差

MRTF Mean Rounds To Failure(机炮)发生故障的平均发弹数

MRTFB Major Range and Test-Facility Base 主要靶场

与试验设施基地

MRU 1. Materials Recovery Unit 物资器材回收队 2. Mobile Receiving Unit 机动数据接收系统

MRUASTAS Medium Range Unmanned Aerial Surveillance and Target Acquisition System 中程无人驾驶空中监视与目标截获系统

MRV Multiple Re-entry Vehicles 集束式多弹头，掷弹式多弹头

MRW 1. Maintenance Requirement Worksheet 维修要求项目分析表(工作单) 2. Maximum Ramp Weight 最大停机重量

MRWS Manned Remote Work Station 载人遥控工作站

MS 1. Man Systems 人员系统 2. Maintenance Schedule 维修计划 3. Manned Satellite 载人卫星 4. Margin of Safety 安全余量 5. Market Share 市场占有率 6. Material Specifications 材料规格，材料规范 7. Material Standard 材料标准 8. Meteorological Sensors 气象遥感器 9. Military Standard 军用标准 10. Micro Switch 微动电门 11. Mission Specialist 飞行任务专家 12. Mircrosoft 微软公司

ms millisecond 毫秒

ms pulse 毫秒脉冲

MS/SP Material Science/Space Processing 材料科学和空间加工

MSA 1. Maritime Safety Agency 海上保安厅(日本) 2. Minimum Safe Altitude 最低安全高度 3. Minimum Surface Area 最小表面积

MSA Division Microgravity Science and Applications Division 微重力科学和应用部(美国航空航天局)

MSAM 1. Mobile Surface-to-Air Missile 机动式地(对)空导弹 2. Medium Surface-to-Air Missile 中程地空导弹

MSAT Mobile Satellite 移动卫星

MSAW Minimum Safe Altitude Warning 最低安全高度警告

MSB Most Significant Bit 最高有效位

MSBLS Microwave Scanning Beam Landing System 微波扫描波束着陆系统

MSC 1. Maintenance Steering Committee 维修指导委员会 2. Major Sub-Contract Change 主转包合同更改 3. Manned Spacecraft Center 载人航天器中心

MSCC Major Subcontractor Change Coordination 主转包合同更改协调

MSCOP Missile System Checkout Program 导弹系统检测程序

MSCP 1. Mean Spherical Candlepower 平均周边烛光强度 2. Missile System Checkout Programmer 导弹系统检测程序装置

MSCR 1. Machine Screw 机床螺钉 2. Manufacturing Specification Control Record 制造规范控制记录

MSCR/A Major Subcontractor Change Request/Approval 主转包合同更改申请/批准

MSD 1. Mass Storage Device 大容量存储器 2. Minimum Safe Distance 最小安全距离 3. Model Special Document 机型特别文件 4. Multi-Sensor Display 多传感显示器

MSDG Multi Sensor Display Group 多传感显示器群

MSE 1. Mean Square/Squared Error 均方误差 2. Minimum Single-Engine Speed 单发飞行最小操纵速度 3. Mobile Subscriber Equipment 移动用户设备

MSEP Maintenance Standardization and Evaluation Program 维修标准化与鉴定大纲

MSER Multiple Stores Ejection Rack 弹射式多外挂物挂架，多外挂弹射架

MSEs Military Spaceflight Engineers 军事航天工程师

MSF 1. Militarily Significant Fallout 影响军事行动的放射性沉降 2. Multi-Sensor Fuse 多传感器引信

MSFC Marshall Space Flight Center 马歇尔航天中心

MSFEF Manned Space Flight Education Foundation 载人航天教育基金会(美国)

MSFL Manned Space Flight Laboratory 载人航天实验室

MSFN Manned Space Flight Network 载人航天跟踪网

MSFP Manned Space Flight Program 载人航天计划

MSFS 1. Manned Space Flight Simulator 载人航天模拟器 2. Manned Space Flight System 载人航天系统

MSFSG Manned Space Flight Support Group 载人航天保障大队(美国空军)

MSFTP Manned Space Flight Test Plan 载人航天试验计划

MSG 1. Maintenance Steering Group 维修指导组 2. Message 文电 3. Meteosat Second Generation 第二代气象卫星

MSG-3 Maintenance Steering Group-3 维修指导小组-第三特别工作组制定的分析方法

MSH Metastable Helium 亚稳氦

MSI 1. Maintenance Significant Item 重要维修项目(指飞机上必须进行维修的) 2. Medium Scale Integration 中规模集成

MSIC Medium Scale Integrated Circuit 中规模集成电路

MSIF Multi-Systems Integration Facility 多系统总装设施

MSIP Multinational Staged Improvement Program 多国分阶段改进计划

MSIS Manned Satellite Inspection System 载人卫星检验系统

MSK Minimum Shift Keying 最小频移链控

MSL 1. Main Supply Line 主要电源线，主供电线路 2. Maintenance Supply Liaison 维修供应联络站（美国空军）3. Mapping Sciences Laboratory 地图绘制科学实验室 4. Material Science Lab 材料科学实验室（美国航天飞机上的）5. Mean Sea Level 平均海平面 6. Missile 导弹

MSLA MSL（Mean Sea Level）Altitude 平均海拔高度

MSLC Maintenance Stock Level Case 维修用（长条）型材水平分格架

MSLS Maneuvering Satellite Landing System 机动卫星着陆系统

MSM 1. Matrix Stackable Module 矩阵式堆叠模块 2. Message Switching Multiplexing 信息交换复用 3. Mobile Station Modem 移动站调制解调器 4. Monitoring System Module 监视系统模块 5. Multiwavelength Simultaneous Monitoring 多波长同步监视

MSN Manufacturing Serial Number 制造系列号

MSO 1. Manager of Shop Operations 车间主任 2. Multiple Service Operator 多业务运营商

MSOCC Multi-Satellite Operation Control Center 多卫星操作控制中心

MSOG Molecular Sieve Oxygen Generation 分子筛制氧

MSOGS Molecular-Sieve Oxygen Generation System 分子筛制氧系统

MSOL Manned Scientific Orbital Laboratory 载人轨道科学实验室

MSOW Modular Standoff Weapons 制式远射（防空区外发射空对地）武器

MSP 1. Magnetic Speed Probe 磁速率探头 2. Maintenance Service Plan 维修服务计划 3. Mars surface probe 火星表面探测器 4. Material Space Platform 空间材料平台 5. Maximum Structural Payload 最大结构的有效商载

MSR 1. Manage Service Request 管理服务需求 2. Manufacturing Service Request 制造服务申请 3. Microwave Scanning Radiometer 微波扫描辐射计 4. Millimetric Surveillance Radar 毫米波监视雷达（英国）5. Missile Site Radar 导弹场雷达

MSRF Microwave Space Research Facility 微波航天研究设备

MSS 1. Magnetic Storm Satellite 磁暴探测卫星 2. Manned Space Station 载人空间站 3. Mission Sequence System 任务程序系统 4. Missile Sight Subsystem 导弹瞄准具子系统 5. Mission Specialist Station 任务专家站 6. Mission Support System 任务保障系统 7. Mobile Servicing System 活动服务系统 8. Mode Selector Switch 状态选择开关 9. Modular Space Station 模块式空间站 10. Multi-Spectral Scanner 多谱段（多光谱）扫描仪

MSSA Major Supplier Stress Analysis 重要供应商应力分析

MSSC Maritime Surface Surveillance Capability 海面监视能力

MSSCC Multicolor Spin Scan Cloudcover Camera 多色自旋扫描摄云相机

MSSCS Manned Space Station Communication System 载人空间站通信系统

MSSD Missiles and Space System Division 导弹及空间系统部门

MSSR 1. Mars Soil Sample Return 火星土壤取样返回 2. Mars Surface Sample Return 火星表面取样返回

MSSS 1. Man Seat Separation System 人椅分离系统 2. Manned Static Space Simulator 载人静态空间模拟器 3. Multi-Spectral Scanner System 多频段扫描仪系统

MST 1. Machine Steel 机床钢 2. Maritime Shore Terminals 海事卫星海岸终端 3. Missile Surveillance Technology 导弹监视技术 4. Mobile Service Tower 活动发射服务塔

MST&E Multi Service Test and Evaluation 多军种试验与鉴定

MSTC 1. Mastic 胶，膏，树脂，胶黏剂 2. Manned Spacecraft Test Center 载人航天器试验中心（美国）

MSTI Miniature Seeker（Sensor）Technology Integration 小型搜索器（遥感器）技术综合卫星

MSTS Meteorology System Test Satellite 气象系统试验卫星

MSU 1. Mass Storage Unit 大容量存储器 2. Microwave Sounding Unit 微波探测装置 3. Mode Selection Unit 工作方式（状态），模式选择装置

MSUP Master Schedule Update Program 主进度更新程序

MSW Master Switch 主电门，总开关

MT 1. Mach Trim 马赫配平 2. Magnetic Test 磁粉检验 3. Maintenance Training 维修培训 4. Manual Trip 手动断开 5. Master Tool 标准工艺装备 6. Maximum Torque 最大力矩 7. Megaton 百万吨，百万吨 TNT 当量（爆炸威力）8. Metric Ton 米制吨 9. Missed Target 未击中的目标 10. Missile Technician 导弹技术军士 11. Mobile Transporter 移动运输器 12. Mountain Time 山区时间（美国）

MTA 1. Maintenance Task Analysis 维修任务分析 2. Military Training Airspace 军事训练空域

MTACS Marine Tactical Air Control System 海军陆战队战术空中控制系统

MTAD Multi Trace Analysis Display 多轨迹分析显示

MTAR Mine/Torpedo Aviation Regiment 水雷/鱼雷航

空兵团

MTAT Mean Turn-Around Time 平均再次出动准备时间

MTBCF 1. Mean Time Between Component Failure 平均部件故障间隔时间 2. Mean Time Between Critical Failures 平均致命（严重）故障间隔时间，平均危险性故障间隔时间

MTBCM Mean Time Between Corrective Maintenance 平均恢复性维修间隔时间

MTBD Mean Time Between Demands 平均需求间隔时间

MTBDE Mean Time Between Downing Events 平均停用事件间隔时间

MTBF 1. Maximal Time Between Faults 最大无故障工作时间 2. Mean Time Before Failure 故障前平均时间 3. Mean Time Between Failures 平均故障间隔时间，平均维修间隔时间

MTBFRO Mean Time Between Failures Requiring Overhaul 需要翻修的故障平均间隔时间

MTBI Mean Time Between Incidents 平均事故征候间隔时间

MTBM Mean Time Between Maintenance 平均维修间隔时间

MTBMA 1. Mean Time Between Maintenance Actions 平均维修措施间隔时间 2. Mean Time Between Mission Aborts 平均任务中止间隔时间

MTBO 1. Mean Time Between Outages 平均停机间隔时间 2. Mean Time Between Overhauls 平均翻修寿命，平均翻修时限，平均翻修间隔时间

MTBR 1. Mean Time Between Removals（Replacement）平均拆卸（换件）间隔时间 2. Mean Time Between Repairs 平均修理间隔时间

MTBUR Mean Time Between Unscheduled Removals 计划外更换部件的平均间隔时间，平均非预定拆卸间隔时间

MTC 1. Mach Trim Compensator 马赫数配平补偿器 2. Mach Trim Computer 马赫数配平计算机 3. Magnetic Tape Container 磁带盒 4. Maintenance Task Cards 维修工卡 5. Maintenance Terminal Cabinet 维修终端箱 6. Man-Tended Capability（临时）有人照料能力 7. Mars Transport Craft 火星运输飞船 8. Mean Transinformation Content 平均传送信息量 9. Mobile Tactical Computer 移动式战术计算机

MTCA Minimum Terrain Clearance Altitude 离地最低（飞行）高度

MTCHD Matched 匹配的

MTCT Manipulator/Teleoperator Control Technology 机械臂与遥控操作器控制技术

MTD 1. Maintenance Terminal Display 维系终端显示

2. Mounted 安装的 3. Moving Target Detection 移动目标检测

MTDS Missile Trajectory Data System 导弹弹道数据系统

MTE 1. Maximum Tracking Error 最大跟踪误差 2. Megaton Equivalent 百万吨（TNT）当量 3. Mesosphere Thermosphere Explorer 中间层-热层探险者 4. Missile Targeting Equipment 导弹目标（瞄准）设备 5. Modern Technology Engine 现代技术发动机

MTEX Mission Template Expert 飞行任务模型专家

MTF 1. Maintenance Terminal Function 维护终端功能 2. Mean Time to Failure 平均故障间隔时间 3. Mechanical Time Fuze 机械定时引信 4. Modolutaion Transfer Function 调制传递函数

MTFF Man-Tended Free-Flyer 有人照料自由飞行平台（欧洲航天局）

MTFP Man Tended Flying Platform 有人照料的飞行平台

MTG Mounting 安装

MTGC Mounting Center 安装中心

MTGW Maximum Taxi Gross Weight 最大滑行总重（量）

MTI Moving Target Indicator 活动目标指示器

MTIP Maintenance Training Improvement Program 维修训练改进大纲

MTIRA Machine Tool Industry Research Association 机床工业研究协会（英国）

MTL 1. Materials Technology Laboratory 材料技术实验室 2. Mean Transmitter Level 平均发射机电平

MTLM Major Throttle-Level Movement 主油门杆行程

MTM 1. Maximum Take-off Mass 最大起飞质量 2. Million Ton-Miles 百万吨英里（短吨）3. Mission and Traffic Model 任务与飞行架次模型 4. Module Test and Maintenance 模块测试与维修

MTM/D Million Ton-Miles per Day 百万吨英里/日

MTMA 1. Military Terminal Manoeuvring Area 军事终点航站机动飞行区 2. Military Terminal Movement Area 军事终点航站飞行区

MTMC Military Traffic Management Command 军事交通管理（司令）部

MTO 1. Make To Order 为订单生产 2. Man-Tended Operation 有人照料作业 3. Master Tool Order 标准工装订货单

MTOGW Maximum Take Off Gross Weight 最大起飞重量

MTOP Maintenance Task Operating Plan 维修工作执行计划

MTOR Master Task Order Request 标准工装订货请求

MTOS　Maintenance Task Operation System 维修工作执行系统

MTOW　Maximum Take Off Weight 最大起飞重量

MTP　1. Maintenance Test Panel 维修测试（控制）板 2. Mandatory Technical Publications 指令性技术文件 3. Meteosat Transition Programme satellite 气象卫星过渡计划卫星 4. Microwave Temperature Profiler 微波温度廓线仪

MTR　1. Main and Tail Rotors 旋翼与尾桨 2. Mean Time of Remove 平均拆卸时间 3. Mean Time to Repair 修理前平均工作时间 4. Metering 计量 5. Missile Tracking Radar 导弹跟踪雷达

MTRDN　Motor Driven 电动机驱动的

MTRE　Missile Test and Readiness Equipment 导弹测试和准备状态试验设备

MTS　1. Maintenance Training Simulators 维修训练模拟器 2. Make To Stock 为库存生产 3. Meteoroid Technological Satellite 流星体技术卫星 4. Microwave Temperature Sounder 微波温度探测器 5. Mobile Test Set 流动试验装置 6. Mobile Tracking Station 移动式跟踪站 7. Mobile Training Sets 流动训练装置 8. Motion/Time Survey 运转/时间测定

MTSL　Microelectronics Technology Support Laboratory 微电子技术支持实验室

MTSS　1. Manned Test Space Station 载人试验空间站 2. Military Test Space Station 军用试验空间站

MTST　Magnetic Tape Selective Typewriter 磁带选择打字机

MTTA　Machine Tool Trade Association 机床贸易协会（英国）

MTTD　Mean Time To Detection 平均故障检测时间

MTTDA　Mean Time To Dispatch Alert 平均发出警告前时间

MTTFF　Mean Time To First Failure 平均首次故障时间

MTTFSF　Mean Time To First System Failure 平均系统首次发生故障时间

MTTM　Mean Time To Maintenance 平均维修前时间

MTTR　1. Maximum Time To Repair 修理前最长工作时间 2. Mean Time To Removal 平均拆卸时间 3. Mean Time To Repair 平均初次修理时间，平均修复时间 4. Mean Time To Restore/Restoration 修复前平均时间

MTTRF　Mean Time To Restore Functions 平均恢复功能时间

MTTRS　Mean Time To Restore Service 平均恢复使用前时间

MTTS　1. Mean Time To Service 平均维修时间 2. Mobile Target Tracking System 机动式目标跟踪系统 3. Multi Task Training System 多任务训练系统

MTTUR　Mean Time To Unscheduled Removal 平均非计划拆卸前时间

MTVC　1. Manual Thrust Vector Control 人工推力矢量控制 2. Motor Thrust Vector Control （导弹）发动机推力矢量控制

MTW　Maximum Taxi Weight 最大滑行重量

MTWA　Maximum Take-off Weight Authorized 批准的最大起飞重量

MTZ　Motorized 装有发动机的，机动的

MTZT　Multiple Time Zone Travel 多时区飞行

MU　1. Maintenance Unit 维修单位，维修部队 2. Mockup （制造）模型，制造样板

MU/MT　Multi User/Multi Task 多用户多任务

MUD　Mock-Up Drawings 样机图

mud crack　泥纹裂纹

MUF　Maximum Usable Frequency 最高使用频率

muff　1. 排气热交换器 2. 套筒，燃烧室外筒

muffler　消声器

muhipie gun mounting　1. 联装炮架 2. 联装机枪架

muldenhancement　多次增强

muldperforated powder　多孔火药

mule　1. 加油车（美国用语）2.〈口语〉液压试验设备 3. Modular Universal Laser Equipment 模块式通用激光设备

Mulite　一种烧蚀材料的商品名

mullite　多铝红柱石

mullite coating　莫来石涂层

mullite monolith　莫来石巨石

mullite powder　莫来石粉

multiaccess　多路存取

multialkali photocathode　多碱光电阴极

multiapplication computer　多用途计算机

multiattributive utility function　多属性效用函数

multiaxial stress condition　多轴应力状态

multiaxis　多轴

multiaxis shaker　多轴振动器，多轴振子

multiband　多频带

multiband color photography　多谱段彩色摄影

multibank engine　多排汽缸的（活塞式）发动机

multibarrel machine gun　多管机枪

multibarreled gun　多管炮，多管机枪

multibeam antenna　多波束天线

multibeam phased array antenna　多波束相控阵天线

multibeam radar　多波束雷达

Multibeam survivable radar　多波束耐久雷达

multiblock　多模块

multibody　多体的

multibody control　多体控制

multibody dynamics	多体动力学,利用多物体运动学
multibody freighter	多体货机
multibody model	多体模型
multibody problem	多体问题
multibody program	多体程序
multibody system	多体系统,基于多体系统
multibody system model	多体系统模型
multibody technique	多体技术
multibogey	有许多敌机的空战形势
multiburn	多次燃烧的,多次起动的
multiburst signal	多波群信号
multibus	多总线
multicapable weapons	多种能力武器
multicarrier	多载波
multicavity magnetron	多腔磁控管
multicell	多室的,多舱的
multicellular foam	多室泡沫材料,多穴泡沫材料
multichamber engine	多管发动机,多燃烧室发动机
multichamber liquid rocket engine	多推力室液体火箭发动机
multichannel	多波段的,多通话线路的
multichannel analyser	多通道分析器
multichannel carrier	多路载波
multichannel coincidence system	多通道重合系统
multichannel communication system	多信道通信系统
multichannel infrared radiometer	多通道红外辐射仪
multichannel ocean color sensor	多通道海洋彩色遥感器
multichannel perception	多通道感知
multichannel photometer	多频道光度计
multichannel pilot	多通道试验
multichannel pilot model	多通道试验模型
multichannel pyrheliometer	多频道日射强度计
multichannel receiver	多路接收机
multichannel selector	多信道选择开关
multichannel signal	多路话音信号
multichannel telemetry	多路遥测
multichannel transmitter	多路发射机
multichannel-per-carrier	群路单载波
multichip circuit	多片电路
multichrome penetration screen	多色穿透屏
multicolor spin scan cloud camera	多色自旋扫描摄云相机
multicolour photometry	多色测光
multicolour system	多波长系统
multicombiner	多面组合玻璃
multicommodity	多重物资
multicommodity network	多物网络
multicommunications service	多用途通信服务
multicomponent balance	多分量(力)天平
multicomponent film	多组分薄膜
multicomponent method	多组分方法
multicomputer system	多机计算机系统
multicoupler	多路耦合器
multicriteria	多(重)判据
multicriteria optimization	多准则优化
multicurve fits	多曲线拟合
multicycle	多循回,多周期的,多循环的
multicycle operation	多循环操作
multicyclic control	多循环控制
multicyclic control input	多循环控制输入
multidate photography	多日期摄影
multi-degree-of freedom system	多自由度系统
multidetector homing head	多元探测导引头
multidimensional datum	多维数据
multidimensional pursuit test	多元空间追踪试验
multidimensional space	多维空间
multidimensional system	多维系统
multidisciplinary	多学科的
multidisciplinary analysis	多学科分析
multidisciplinary design	多学科设计,分为多学科设计
multidisciplinary optimization	和多学科优化,多学科优化
multidisciplinary problem	多学科的问题
multidisciplinary requirement	多学科的要求
multidisciplinary system	多学科系统
multidomain grain	多畴颗粒
multidomain thermal remanence	多畴热剩磁
multielement	多元素,多元件
multielement airfoil	多元素机翼
multielement detector	多元探测器
multielement detector array	多元探测器阵列
multienvironment system	多环境系统
multiequation	多元方程
multiflow wing	多种流线型翼
multifoil wing	多翼片机翼
multiformator	多帧照相机
multiframe	复帧,多帧
multifrequency microwave radiometer	多频微波辐射计
multifrequency ring bus	多频振铃总线
multifrequency shift keying	多频移键控
multifunction	多功能,多用途,一体机
multifunction display	多功能显示器
multifunction radar	多功能雷达
multifunction spacecraft	多用途飞船
multifunction strapdown inertial sensor	多功能捷联式惯性传感器
multifunction switching/display	多功能转换/显示装置
multifunction transducer	多功能传感器

multifunctional aileron	多功能副翼
multifunctional horizontal sling	多功能水平吊具
multifunctional rotating chair	多功能转椅
multifunctional structure	多功能结构
multifunctional structure design	多功能结构设计
multigraph	多重图
multigrid	多栅的,多重网格的
multigrid cycle	多重网格循环
multihead telemetry	多弹头遥测
multihole	多孔的
multihole film	多孔膜
multihole film cooling	多孔膜冷却
multihole unit	多孔发射筒
multihop satellite link	多跳卫星链路
multiimpulse	重复点火
multiinertial sensor	多功能惯性敏感器
multiingredient model	多因素相关模型
multiinput-multioutput	多输入多输出
multijet	多喷嘴的
multijet mode	多股射流模式
multijunction	多结,多接点
multijunction gallium arsenide solar cell	多结砷化镓太阳电池
Multijunction solar cell	多结太阳电池
multilane airway	多走廊航路
multilateral	多边的,多国参加的
multilateration	多点定位
multilayer board	多层印制板
multilayer defense system	多层防御系统
multilayer dielectric passivation	多层介质钝化
multilayer hierarchical control	多层递阶控制
multilayer insulation	多层隔热
multilayer insulation material	多层隔热材料
multilayer interconnection	多层互连,多层布线
multilayer printed board	多层印制板
multilayer sheet	多层板
multilayer wiring	多层布线
multilayered (Hg,Cd,Te) infrared detector	多层汞镉碲红外探测器
multilens camera	多镜头相机
multilevel computer control system	多级计算机控制系统
multilevel decision	多级决策
multilevel hierarchical control	多级递阶控制
multilevel metallization	多层金属化
multilevel optimization	多级优化
multilevel procedure	多级程序
multilevel process	多级过程
multilevel resist	多层光刻胶
multilevel security	多级安全性
multilook	多视
multiloop	多环的,多回路的,多匝的
multiloop control	多回路控制
multiloop identify	多回路识别
multiloop network system	多环网络系统
multimanometer	多管压力计
multimedia classroom	电化教室,多媒体教室
multimegawatt	多兆瓦的
multimeter	万用表
multimicroprocessor flight control system	多微机飞行控制系统
multimirror telescope	多镜面望远镜
multimission	多任务的
multimission modular spacecraft	模块式多用途航天器
multimodal control behavior	多通道控制行为
multimodal pilot	多通道试验
multimodal pilot control behavior	多通道导频控制行为
multimodal pilot model	多通道试验模型
multimodal spectrum	多峰谱
multimode	多状态,多方式
multimode autotracking	多模自跟踪
multimode feed	多模馈源
multimode fiber	多模光纤
multimode fire control system	多用途火控系统
multimode guidance	1.多模制导 2.多工制导
multimode radar	多方式雷达,多模雷达
multimode system	多模系统
multimoment display	多矩显示
multimunition bomb	多弹药炸弹
multinested mesh (grid) model	多重嵌套网格模式
multinodal	多节点的,多途径的
multinomial distribution	多项分布,多项分配
multinotch	大幅度
multiobject spectrograph	多目标摄谱仪
multiobjective decision	多目标决策
multiobjective function	多目标函数
multiobjective optimization	多目标优化
multiobjective optimization problem	多目标优化问题
multiorbit integration algorithm	多圈积分法(积分步长为轨道周期的整数倍的数值积分法)
multiorifice laminated material	多孔层压材料
multioutlet ventilating garment	多孔式通风服
multipass	多通道的
multipass cell	多程吸收池
multipassage	多路
multipath	多路径,多通道的,多途径的
multipath delay	多路延迟
multipath effect	多路径效应

multipath environment 多径环境
multipath error 多路径误差
multipath transmission 多路径传输
multipe impulse welding 脉冲点焊
multiperforated grain 多孔形药柱
multipersistence penetration screen 多余辉穿透荧光屏
multiphase detonation 多相爆轰
multiphase flow 多相流
multiphoton absorption 多光子吸收
multiplane 1.多翼飞机 2.多套翼面的(尤指尾翼组)
multiple 多重的,多样的,许多的,倍数,并联
multiple accelerometer 复式加速度计
multiple access 多址方式,多路存取
multiple access channel 多路存取通道
multiple access communication 多址通信
multiple aim point deployment 多瞄准点部署方式
multiple air weapon control system 多种航空武器控制系统
multiple airborne target tracking system 多个空中目标跟踪系统
multiple aircraft 多种飞机
multiple algorithm 多重算法
multiple antiaircraft machine gun 多管高射机枪
multiple arcs homing 多弧段寻的
multiple array processor 多阵列处理机
multiple autonomous vehicle 复试无人驾驶汽车
multiple beam antennae 多波束天线
multiple burn 多次套晒
multiple carriage bomb rack 复式炸弹架
multiple carriage clip in assembly 多弹插夹式挂弹装置
multiple carrier waveform 多个载波波形
multiple choice reaction test 多次选择反应试验
multiple computer complex 多计算机复合系统
multiple contoured beam 多成形波束
multiple control 复式控制
multiple courses 多(无线电指向标)航向
multiple decision 多重判定
multiple detonation tube 多个爆管
multiple echo 多次回波,多重回声,多重回波
multiple ejection rack 复式挂弹架,复式发射架
multiple ejector rack 弹射式多弹挂弹架
multiple event 多重事件
multiple failure 多重故障
multiple fault 多重故障
multiple flame 多焰
multiple fragment wounds 多弹片杀伤
multiple galaxy 多重星系
multiple gravity assist 多重力辅助
multiple gun 1.联装炮 2.联装机枪

multiple image 多重图像
multiple impulse fuse 重复冲压引信
multiple independently targeted reentry vehicle 多弹头分导再入飞行器,分导式多弹头
multiple input data acquisition system 多路输入数据采集系统
multiple installation 多弹悬挂装置
multiple integral 多重积分
multiple kill capability 多种杀伤能力,多重猎杀能力
multiple launch 齐射
multiple lift 多轮空运
multiple lifting-line 多个升力线
multiple line controller 多重线路控制器
multiple load path concept 多路承力的设计思想
multiple manometer 多管压力计
multiple modulation radar fuze 复合调制雷达引信
multiple nozzle 多管喷管
multiple options (兵力部署的)多种选择
multiple pencil beam 多个锐方向性波束,多个笔形波束
multiple perturbation 多重摄动
multiple phantom target 多假目标
multiple phenomenology 多现象(观测)学
multiple project 多项目,多投射
multiple projector system 多投影系统
multiple protective shelters 多掩蔽所防护法
multiple pure tone noise 多个纯音噪声
multiple rate 倍率
multiple reentry vehicles 集束式多弹头
multiple reference 多参考帧
multiple ribbon growth 多带生长
multiple rigid body spacecraft dynamics 多刚体航天器动力学
multiple rocket launcher 多管火箭筒
multiple run 多个运行
multiple runway 多个跑道
multiple sampling 多次取样
multiple satellite 多卫星
multiple shock 多重震源
multiple shooting 多重打靶
multiple shooting approach 多重射击的方法
multiple simulation 多个模拟
multiple solution 多重解
multiple spacecraft 多个航天器
multiple stakeholder 多重利益相关者
multiple star 聚星
multiple station rack 复式炸弹架
multiple subsystem 多子系统
multiple surrogate 多个代理
multiple surrogate model 多个代理模式

multiple system networking	多系统网络连接
multiple target	多目标
multiple target instrumentation radar	多目标测量雷达
multiple target tracking	多目标跟踪
multiple terminal system	多终端系统
multiple time	多重时帧
multiple trailing-edge	多个机翼后边缘
multiple trajectory	多元弹道
multiple transfer	多路传送
multiple unit of measurement	倍数(测量)单位
multiple units steerable antenna	(方向图)可控多元天线
multiple vehicle	多个车辆
multiple-aperture	多孔径
multiple-beam klystron	多注速调管
multiple-beam radar	多波束雷达
multiple-burn trajectory	多次燃烧轨迹
multiple-burn	多次燃烧
multiple-cathode	多个阴极
multiple-hop connection	多跳连接
multiple-kill air-to-air missile	多重杀伤空-空导弹
multiple-kill vehicle	多重杀伤器
multiple-look technique	多视技术
multiple-point-source bioiogical munitions	多点源生物弹药
multiple-point-source chemical munitions	多点源化学弹药
multiple-point-source release	多点源施放
multiple-ram forging	多压头锻造
multiple-simulation	多重模拟或仿真
multiple-threat	多重威胁
multiple-user information theory	多用户信息论
multiple-warhead telemetry	多弹头遥测
multiplet	多重态,多重谱线
multiplex	多路传输,多路复用,多工
multiplex autopilot	多余度自动驾驶仪
multiplex communication	多路(复用)通信
multiplex data bus	多路传输数据总线
multiplex modulation	多路通信调制
multiplexed implementation	多路复用的实现
multiplexer	多路转接器,多路开关选择器
multiplexer access point	复路器访问点
multiplexing	多路调制,多路传输
multiplexing liquid-crystal displays	多路传输液晶显示器
multiplexing protocol data unit	复路规约数据单元
multiplexing service	多路复用业务,复路业务
multiplication	乘法,增加
multiplication constant	乘常数
multiplicative factor	乘积因子
multiplicative uncertainty	乘性不确定性
multiplier	1.倍增器,乘法器 2.扩(量)程器 3.乘数
multiplier phototube	光电倍增管
multiplier rule	乘数法则
multiplier system	倍频系统
multiply	1.增加,倍增,多样,多重,乘,并联 2.多层材料
multiplying valve	倍增阀
multipoint	多点,多点的,多位置的
multipoint constraint	多点约束
multipoint design	多点设计
multipoint ignition	多点点火
multipoint inverse	多点逆
multipoint link	多点链路
multipoint loading system	多点加载系统
multipoint restriction	多点约束
multipoint vibration excitation system	多点激振系统
multipolarization photography	多向偏振摄影
multipole fields	多极场
multi-port network	多口网络
multiprobe	多探测器宇宙飞船,多功能探针
multiprocessing	多重处理
multiprocessor	多道程序处理机,多处理机
multiprocessor array	多处理机阵列
multiprocessor series	多处理机系列
multiprogramming	多道程序设计
multiproject automated control system	多元自控系统
multipulse	多脉冲
multipurpose acquisition and control system	多用途采集与控制系统
multipurpose carrier	多用途航空母舰
multipurpose display panel	多用途显示板
multipurpose photography	多用途摄影
multipurpose satellite	多用途卫星
multipurpose weapons	两用武器,多用途武器
multiradix computer	多基数计算机
multiresolution	多分辨率
multirib construction	多肋构造,多肋结构
multirocket engine cluster	并联火箭发动机
multisatellite	多卫星
multisatilite colocation	多星共位
multiscaler	多路定标器
multiseasonal image	多季相影像
multiseasonal-multispectral remote sensing	多季节(多时相)多谱段遥感
multisegment	多节的
multisegment model	多段模型
multisensor combined displays	多传感器综合显示器,多敏感元件综合显示器

multisensor/multifunction displays　多传感器/多功能显示器
multiservice connector　多用连接器
multishaker system　多振动系统
multishell curing　分层固化
multishock shield　多次冲击防护屏
multisine　多重正弦波
multispacecraft　多用途航天飞机
multispecies　多种群
multispectral　多光谱的,多谱段的
multispectral bathymetry　多谱段测深法,多谱段海洋测深学
multispectral camera　多光谱相机
multispectral classification　多谱段分类
multispectral image　多谱段影像
multispectral image unitary transform　多谱段影(图)像面变换
multispectral imagery　多波段云图
multispectral imaging　多谱段成像
multispectral infrared radiometer　多谱段红外辐射仪
multispectral mapping system　多谱段制图系统
multispectral remore scanner　多谱段扫描仪
multispectral remore sensing　多光谱遥感
multispectral scanner　多光谱扫描仪
multispectral sensing　多光谱敏感
multispectral sensor　多光谱敏感器
multistage compressor　多级压气机
multistage decision process　多段决策过程
multistage ignition　多级点火
multistage interpretation　多阶段判读
multistage least squares estimation　多步最小二乘估计
multistage opening　1.多级开伞 2.多次(充气)开伞
multistage photography　多阶段摄影
multistage radar　多基地雷达
multistage rectification　多级纠正
multistage rocket　多级火箭
multistar equal altitude method　多星等高法
multistart rocket　多次起动火箭
multistart rocket engine　多次起动火箭发动机
multistate controller　多位控制器
multistate logic　多态逻辑
multistatic radar　多基地雷达
multistation joining tracking system　多站联用追踪测量系统
multistation Doppler radar　多站多普勒雷达
multistation incoherent scatter radar　多站非相干散射雷达
multistation system　多站系统
multistation triggering　多站触发

multistations intersection　多站交会
multistep　1.多级的,多阶的 2.多级火箭
multistep avalanche chamber　多步雪崩室
multistep controller　多位控制器
multistep maneuver　多步机动,多步操作
multistratum control　多段控制
multistratum hierarchical control　多段递阶控制
multitarget　多目标,多靶
multitarget measurment　多目标测量
multitarget processing　多目标处理能力
multitarget tracking　多目标跟踪
multitasking　多任务,多进程
multitemporal analysis　多时相分析
multitemporal remote sensing　多时相遥感
multiterminal network　多端网络
multithreading　多线处理
multitone command　多音指令
multitrace analysis display　多迹分析显示器
multitube　复极真空管,多真空管
multitube design　多管设计
multitube launcher　多管发射器
multiunit shaped-charge warhead　多单元聚能装药弹头
multiunits calibration　多元校准
multivalue decision　多值决策
multivariable case　多变量情况
multivariable control　多变量控制
multivariable control system　多变量控制系统
multivariable disk　多变量磁盘
multivariate histogram　多变量直方图
multivariate optimum interpolation　多元最优插值
multivariate sensitivity　多元灵敏感度
multivibrator　多谐振荡器
multiwavelength nephelometer　多波长浑浊度仪
multiwavelength sunphotometer　多波长太阳光度计
multiway plug　多孔插头,多脚插塞
multiwire proportional chamber　多丝正比室,多线配比室
multiwork station　多工作站
multizone　多层的,多区的
mu-meter　跑道摩擦因数测定仪
munition　军需品,军火
munitions　1.弹药 2.军械 3.军需品,军用品
mural circle　墙仪
mural quadrant　墙象限仪
Murman-Cole scheme　穆曼-科尔格式
MURP　Manned Upper-stage Reusable Payload 可重复使用载人上面级有效载荷
MUS　Minimum Use Specification 最低使用规范
MUSC　Microgravity User Support Center 微重力用户

保障中心（德国）

muscle atrophy 肌肉萎缩
muscle atrophy in space 航天肌肉萎缩
muscular fatigue 肌肉疲劳
muscular incoordination 肌肉运动失调
MUSE Monitor of Ultraviolet Solar Energy 紫外太阳能监测器
mush 半失速飞机
mushy zone 糊状区
music 电子战发射，电子干扰
Musketball 步枪弹卫星
MUSR Monitor of Ultraviolet Solar Radiation 太阳紫外辐射监视仪
MUST 1. Mobile Undersea Systems Test 机动式水下武器系统试验 2. Multimission UHF Satcom Terminal 多任务超高频卫星通信终端
mustard (gas) 芥子气
mustard T-mixture (gas) 芥子气T混合剂
mustard-lewisite mixture 芥子气与路易氏气混合剂
mutation 突变，变化
mutation coef 变异系数
mutation operator 变异算子
mute 消减（噪声）
mute switch 断接器
MUTES Multiple Threat Emitter System 多威胁发射源系统
mutual coupling factor 相互耦合系数
mutual designation 互引导
mutual gravity 相互的重力作用
mutual induction 互感（应）
mutual information 互信息
mutual interaction 相互作用
mutual interception （友机间）相互截击
mutual strategic security 相互战略安全
mutually synchronized network 互同步网
mutually visual zone 共视范围
MUX 1. Multiplex 多路传输，倍增，多路，多工，负重，通道 2. Multiplexer 多路调制器，倍增器
MUX data bus Multiplex data bus 多路传输数据总线
muzzle 枪口，炮口，筒口
muzzle angle 炮口角
muzzle bell 炮口消焰罩
muzzle brake 炮口制退器
muzzle burst 炮口爆炸
muzzle cap 炮口罩，枪口帽
muzzle closure 筒口堵盖
muzzle compensator 炮口制退器
muzzle energy 炮口动能
muzzle horsepower 炮口马力

muzzle power 炮口功率
muzzle velocity （炮弹）初速，炮口速度
muzzle velocity radar 炮口速度雷达
MV 1. Magnetic Variation 磁差 2. Mean Variation 平均偏差 3. Metering Valve 计量活门 4. Miniature Vehicle 小型飞行器，小型寻的飞行器 5. Mix Valve 混合活门 6. Multivibrator 多谐振荡器 7. Muzzle Velocity（炮弹）初速，炮口速度
MVA 1. Mega Volt Amperes 兆伏安 2. Minimum Vectoring Altitude 雷达引导最低高度 3. Multi Variate Analysis 多元分析，多变量分析 4. Mutual Visual Systems 互视系统
MVAR Magnetic Variation 磁差
MVB Motor V-Belt 电动机V型传送带
MVBL Movable 可活动的，移动式的，可拆卸的
MVC Manufacturing Variability Control 制造可变性控制
MVD 1. Map and Visual Display 地图的视觉显示 2. Migration Velocity Determination 撤动速度测定 3. Motor Voltage Drop 电动机电压降
MVDF Medium And Very High Frequency Direction Finder 中频和甚高频定向台
MVEE Military Vehicles Experimental Establishment 军用交通工具实验研究所
MVG Moving 运动的，活动的
MVGVT Mated Vertical Ground Vibration Tests 整机垂直地面振动试验
MVL 1. Mid Value Logic 中值逻辑 2. Motion Video Library 移动视频库
MVM Muzzle Velocity Measure(ment)（火炮）初速测量
MVO Module Version Object 模块版本对象
MVOD Minimum Variance Orbit Determination 最小偏差轨道测定
MVP 1. Master Verification Plan 总检验计划 2. Media Vision Pocket 媒体视觉器 3. Methl Violet Paper 甲基紫色纸
MVR Master Verification Requirements 总检验要求
MVT Malfunction Verification Test 故障验证试验
MVUE Man Pack Vehicular User Equipment 背负式车载用户设备
MW 1. Medium Wave 中波 2. Megawatts 兆瓦，百万瓦 3. Methanol/Water 甲醇/水 4. Microwave 微波 5. Milliwatts 毫瓦 6. Mine Warfare 水雷战 7. Multipurpose Weapon 多用途武器
mw beam Micro Wave Beam 微波束
mw discharge Micro Wave or Medium Wave 中波或微波放电
mw energy 兆瓦能源，兆瓦能量

mw heating 微波加热
mw power 微波功率
mw radiation 微波辐射
MWA Multiple Weapons Adaptor 多弹转接架,多弹过渡梁
MWCC Master Warning Caution Controller 主警告控制器
MWCS Multiple Weapons Carrier System 多种武器挂架系统
MWDP 1. Master Warning Display Panel 主警告显示板 2. Mutual Weapons Development Program 共同武器研制计划
MW(E) Megawatts (Electrical) (电功率)兆瓦
MWE 1. Manufactured Weight Empty 出厂空重 2. Maximum Weight Empty 最大空重
MWIR Medium Wavelength Infrared 中波长红外
MWL Millwatt Logic 毫瓦逻辑电路
MWP Maneuvering Working Platform 机动飞行工作平台
MWR Microwave Radiometer 微波辐射计
MWS 1. Master Warning System 主告警系统 2. Microwave Windfield Scatterometer 微波风场散射计 3. Mini Workstation 小型工作站
MW(T) Megawatts (Thermal) 兆瓦(功率)
MWVS Mission Weapon Visionics System 任务武器光学电子系统
MY 1. Man Year 人年 2. Multi Year 多年度的
mylar 聚酯薄膜
mylar balloon 聚酯纤维气球
mylar corner reflector 聚酯薄膜角形反射器
myoelectric control 肌电控制
myoelectric potential 肌电位
MYP 1. Multi Year Procurement 多年采购 2. Multi Year Programmes 多年计划
myriametric wave communication 超长波通信
MZFW Maximum Zero Fuel Weight 最大无燃油重量

N

N array　N 阵列
N wave　N 型波(声波,音爆的特征,压力对时间或压力对直线距离的剖面为 N 形)
N&G　Navigation and Guidance 导航和制导
NA　1. Nautical Almanac 航海历书 2. Naval Air 海军航空兵(的) 3. Naval Aviator 海军飞行人员 4. Noise Abatement 噪声抑制
NAA　1. National Aeronautic Association 国家航空协会(美国) 2. National Aviation Academy 国家航空研究院(美国) 3. Navy Air Arm 海军航空兵
NAAA　National Agricultural Aviation Association 全国农业航空协会(美国)
NAACS　National Association of Aircraft and Communication Suppliers 国家飞机和通信供应厂商协会
NAAF　Naval Auxiliary Airfield 海军辅助机场
NAAFI　Navy, Army and Air Force Institutes 陆海空三军协会(英国)
NAAP　Netherlands Agency for Aerospace Promotion 荷兰航空航天发展局
NAAS　1. National Association of Aerospace Subcontractors 航空航天分包商全国协会(美国) 2. Naval Auxiliary Air Station 海军三级空军站
NAATS　National Association of Air Traffic Specialists 空中交通专业人员全国协会(美国)
NABS　NATO Air Base Satcom 北约空军基地卫星通信
NAC　1. NASA Advisory Council 航空航天局咨询委员会(美国) 2. Naval Avionics Center 海军航空电子中心(美国) 3. Noise Advisory Council 噪声咨询委员会(英国) 4. North Atlantic Council 北大西洋理事会
NACA　1. National Advisory Committee for Aeronautics (美国)国家航空咨询委员会 2. National Air Carrier Association 全国航空运输公司协会(美国)
NACA cowling　NACA 低阻整流罩
NACA section　NACA 翼型
nacelle　发动机短舱,吊舱,导流罩
nacelle angle　吊舱角
nacelle geometry　发动机短舱几何
nacelle length　发动机短舱长度
nacelle spray ring　发动机短舱灭火环
nacelle surface　发动机短舱表面
NACF　Navy Air Combat Fighter 海军空战歼击机
NACISA　NATO Communications and Information Systems Agency 北约通信和信息系统局

nacreous cloud　珠母云
NACS　Nonlinear Automatic Control System 非线性自动控制系统
NAD　Not on Active Duty 非现役(的)
NADAC　Navigation Data Assimilation Computer 导航数据类比计算机
NADB　Netherlands Aircraft Development Board 荷兰飞机发展委员会
NADC　1. Naval Air Development Center 海军航空发展中心(美国) 2. Nuclear Affairs Defence Council 核防御委员会(北约)
Nadezhda　希望号卫星
NADIN　National Airspace/Aeronautical Data Interchange Network 国家空域数据交换网
nadir　天底
nadir angle　天底角
nadir distance　天底距
nadir photograph　垂直摄影照片
nadir point plot　天底点略图
nadir return　天底点回波
nadir solar backscatter　天底太阳后向散射计
NADL　Navy Avionics Development Laboratory 海军航空电子发展实验室(美国)
NADUC　Nimbus/ATS Data Utilization Center 雨云卫星/应用技术卫星数据利用中心
NAEC　1. National Aerospace Educational Council 国家航空航天教育委员会(美国) 2. Naval Aviation Engineering Center 海军航空工程中心
NAECON　National Aerospace Electronics Conference 全国航空航天电子学会议(美国)
NAES(U)　Naval Aviation Engineering Service (Unit) 海军航空兵工程勤务(分队)(美国)
NAEW　NATO Airborne Early Warning 北约空中预警(机)
NAF　Naval Air Facility 海军二级航空站
NAFAG　1. NATO Air Force Advisory Group 北约空军顾问团 2. NATO Air Force Armaments Group 北约空军军械组
NAFDU　Naval Air Flying Development Unit 海军飞行研究发展部队
NAFI　National Association of Flight Instructors 飞行教员全国协会(美国)
NAFP　National Aeronautical Facilities Program 国家航

空工程设施规划

NAGARD NATO Advisory Group for Aeronautical Research and Development 北约航空研究与发展顾问组

NAI Negro Airmen International 黑人飞行员国际组织（美国）

NAIC National Astronomy and Ionosphere Center 国家天文和电离层中心（美国）

NAIF Navigation Ancillary Information Facility 导航辅助信息设备

naked 1.净形的,无外挂物的 2.裸露的,无保护层的

naked-eye 肉眼,裸眼式

naked-eye observation 目视观测,肉眼观测

NAL 1. National Aerospace Laboratory 国家航空航天实验室 2. Naval Air Logistics 海军航空物流

NALNET NASA Library Network 航空航天局图书馆网（美国）

NAM Nautical Air Miles 海里航程

nameplate 名牌,铭牌,标牌

NAML Naval Aircraft Materials Laboratory 海军航空器材实验室（英国）

NAMP Naval Aviation Maintenance Program 海军飞机维修大纲

NAMRL Naval Aerospace Medical Research Laboratory 海军航空航天医疗实验室（美国）

NAMS Network Administration and Management System 网络经营及管理系统

NAMSA NATO Maintenance and Supply Agency 北约维修与供应局（卢森堡）

NAMSO Navy Maintenance Support Office 海军维修保障处

NAMT Naval Air Maintenance Trainer 海军航空维修练习器

NAMTC Naval Air Missile Test Center 海军防空导弹试验中心（美国）

nano 纳,毫微

nano watt electronics 纳瓦功率电子学

nanoaluminized propellant 纳米铝推进剂

nanoaluminum 纳米铝

nanoaluminum powder 纳米铝粉

nanoaluminum-based propellant 纳米铝基推进剂

nanocluster 纳米团簇

nanocomposite 纳米复合材料

nanocomposite material 复合纳米材料

nanocomposite powder 纳米复合粉末

nanocomposite sample 纳米复合材料样本

nanoindentation 纳米压痕

nanolaminates 纳米层压板

nanometal particle 纳米金属粒子

nanometre 纳米,毫微米

nanoparticle 纳米颗粒,毫微粒

nanoparticle additive 纳米粒子添加剂

nanoparticle suspension 纳米颗粒,毫微粒

nanophotogrammetry 电子显微摄影测量

nanorods 纳米棒

nanosatellite 纳卫星,纳米卫星

Nanosats 超小型卫星

nanoscale 纳米级

nanoscale composite 纳米复合材料

nanoscale particle 纳米粒子

nanoscale thermites 纳米铝热剂,灼热剂

nanosecond 纳秒

nanosecond discharge 纳秒放电

nanosecond initiation 纳秒启动或起爆

nanosecond pulse 纳秒脉冲

nanosized additive 纳米添加剂

nanosized particle 纳米微粒

nanostructure 纳米结构

nanotechnology 纳米技术

nanothermite 纳米铝热剂

nanothermite composite 纳米铝热剂复合材料

NAP 1. Naval Air Priorities 海军空运优先等级 2. Noise-Abatement Procedure 噪声抑制程序 3. Normal Acceleration Point（超声速运输机的）正常加速点

nap 局部地形剖面

napalm bomb 凝固汽油弹

napalm rocket 凝固汽油火箭

napalm-filled head 凝固汽油弹头

NAPC Naval Air Propulsion Center 海军航空推进中心（美国）

nape 〈口语〉用凝固汽油弹去对付

naphtha 石脑油,粗汽油

naphthenic acid 环烷酸,环酸

NAPL National Air Photo Library 国家航空摄影图片库（加拿大）

NAPO NASA Pasadena Office 国家航空航天局帕萨迪纳办事处

nap-of-the-earth flight（flying） 掠地飞行,超低空飞行

NAPP National Association of Priest Pilots 全国牧师飞行员协会（美国）

NAPR NATO Armaments Planning Review 北约军备计划评审

NAPTC Naval Air Propulsion Test Center 海军航空推进装置试验中心（美国）

NAR 1. North American Routes 北美航路 2. North Atlantic Route 北大西洋航路

NARCOM North Atlantic Relay Communication

satellite 北大西洋中继通信卫星

NARF 1. Naval Aerospace Recovery Facility 海军航空航天回收设施(美国) 2. Naval Air Reserve Force 海军航空兵后备队(美国) 3. Naval Air Rework Facility 海军飞机返修厂 4. Nuclear Aerospace Research Facility 航天核能研究设施(美国空军)

narrow angle camera 窄角相机
narrow annular gap 环形狭窄通道
narrow band amplifier 窄带放大器
narrow band communication system 窄带通信系统
narrow band filter 1.窄带滤波器 2.窄带滤光片
narrow band filter of base frequency 基准频率窄带滤波器
narrow beam radiogoniometer 无线电窄束定向器
narrow body 窄机身客机
narrow chamber 窄腔
narrow channel effect 窄沟效应
narrow corridor 狭长的走廊
narrow edge 窄边
narrow gap semiconductor 窄带隙半导体
narrow gate 窄门
narrow range 窄分布
narrow region 缩小区域,狭窄区域
narrowband 窄带
narrowband photometry 窄带测光
narrowband radiation 窄带辐射
narrowband random vibration 窄带随机振动
narrowband reduction 窄带减少
narrowband secure voice 窄带保密话音
narrowband source 窄带源
narrowbeam antenna 窄波束天线
narrow-body 窄机身
narrow-body aircraft 窄机身飞机
narrow-field scanning system 窄视场扫描系统
narrow-sector recorder 窄带记录器

NARTEL National Air Radio Telecommunication 全国空中无线电通信(网)(美国)

NAS 1. National Academy of Sciences 美国科学院 2. National Aerospace Standard 国家航空航天标准 3. National Airspace System 全国空域系统

NASA National Aeronautics and Space Administration 国家航空航天局(美国)

NASAO National Association of State Aviation Officials 全国各州航空官员协会(美国)

NASARR North American Search And Ranging Radar 北美搜索测距雷达

NASC 1. National Aeronautics and Space Council 国家航空航天委员会(美国) 2. National Aerospace Standards Committee 国家航空航天标准委员会 3. National Aircraft Standards Committee 国家飞机标准委员会 4. National Association of Spotters' Clubs 全国观测员俱乐部协会

nascent oxygen 新生态氧,初生态氧

NASCOP NASA Communications Operating Procedures 国家航空航天局通信操作程序(美国)

NASD Naval Air Supply Depot 海军航空兵供应站

NASEMM National Aeronautic & Space Engineering & Manufacturing Meeting 全国航空航天工程和制造会议

NASF Navigation and Attack Systems Flight 导航及攻击系统训练飞行

NASIS NASA Aerospace Safety Information System 国家航空航天局航空航天安全信息系统(美国)

NASM National Air and Space Museum 国家航空航天博物馆(美国)

NASN National Air Sampling Network 全国空中取样网

NASO 1. National Astronomical Space Observatory 国家空间天文观测台(美国) 2. Naval Aviation Supply Office 海军航空器材供应处(美国)

NASP 1. National Aerospace Plane 国家空天飞机 2. National Airspace System Plan 全国空域管制系统计划(美国) 3. Navy Airship Program 海军飞艇计划(美国)

NASPG North Atlantic Systems Planning Group 北大西洋武器系统计划组

NASREM NASA Standard Reference Model 美国航空航天局标准参考模型

NASS 1. Naval Anti-Submarine School 海军反潜学校 2. Navigation Satellite System 导航卫星系统

NASSA National Aerospace Service Association 全国航空航天勤务协会(美国)

NAST Nuclear Accident Support Team 核事故支援队(加拿大)

nastran code 国家航空航天局的结构分析程序代码(美国)

NASWDU Naval Air/Sea Warfare Development Unit 海军空中和海上作战研究发展部队

nat 奈特

NATA 1. National Air Taxi Association 全国出租飞机同业公会(英国) 2. National Air Transportation Association 全国航空运输协会 3. National Aviation Trades Association 全国航空同业工会(美国)

NATB Naval Air Training Base 海军航空兵训练基地

NATC 1. National Air Transportation Conferences 国家空中运输会议(美国) 2. Naval Air Test Center 海军航空试验中心(美国) 3. Naval Air Training Center 海军航空兵训练中心(美国)

NATCC National Air Transport Coordinating Com-

mittee 全国航空运输协调委员会(美国)
NATCS National Air Traffic Control Services 国家空中交通管制局
national aerospace plane 国家空天飞机
national airlines 国家级航空公司
national airspace 领空
national airspace system 国家空域管制系统
national defence metrology 国防计量学
national emergency 国家紧急状态
National Flight Data Center 全国飞行数据中心
national flight data digest 国家飞行数据文摘
National Gas Turbine Establishment 国立燃气轮机研究院（英国）
national model 国家模型
national program 国家计划
National Qualification Authority 国家质量认证机构
national range 国家试验场
national responsibility 国家责任
National Search and Rescue Plan 国家搜索救援计划
national strategic target list 国家战略目标清单
national team 国家队
national technical means 国家(核查)技术手段
National Weather Service 国家气象局(美国)
nationality mark (飞机上的)国籍标志
Nationals 国家级空运企业
nationwide service 全国性服务
native 【计】本机的
native language 【计】本机语言
native mode 【计】本机方式
NATO North Atlantic Treaty Organization 北大西洋公约组织
NATOSAT NATO Satellite 北约卫星
NATS 1. National Air Traffic Services 全国空中交通服务处(英国) 2. Naval Air Transport Services 海军航空运输处(美国)
NATSF Naval Air Technical Services Facility 海军航空兵技术勤务处
NATTC Naval Air Technical Training Center 海军航空兵技术训练中心
natural aging 1.自然时效 2.自然老化
natural boundary survey 自然边界测量
natural buffet 自然抖振
natural calamity 自然灾害
natural climate test 天然气候试验
natural convection 自然对流
natural coordinates 自然坐标
natural disaster 自然灾害
natural environment 自然环境
natural finish 自然光面,未涂漆的

natural frequency 自然频率,自振频率
natural frequency measurement 固有频率测量
natural gas 天然气
natural gas composition 天然气成分
natural gas flame 天然气火焰
natural gas fuel cell 天然气燃料电池
natural laminar flow 天然层流
natural laminar flow aerofoil profile 自然层流翼型
natural landscape 自然景观
natural language 自然语言
natural language interface 自然语言接口,自然语言前端
natural linewidth 自然线宽
natural mode 固有模态
natural mode of vibration 固有模态
natural orbit 天然轨道
natural remanence 天然剩磁
natural remanent magnetization 天然剩磁
natural satellite 天然卫星
natural scale 实物大小,全尺寸
natural shape 自然形态
natural shape balloon 自然形状的气球
natural stability 固有稳定性
natural storage life 固有储存期
natural tendency 自然趋势
natural transition 自然转变
natural vortex 自然涡流
natural wavelength 自然波长,固有波长
natural wind 自然风
natural year 自然年
naturally aspirated 自然吸入的(不增压)
nautical almanac 航海天文历
nautical astronomy 航海天文学
nautical mile 海里
nautical twilight 航海曙暮光
nav/attack aid 导航/攻击设备
Navaglobe "纳瓦格洛布"远程无线电导航系统
navaid 助航设备,航行设备,导航设备
navaid dropwindsonde 下投式导航测风探空仪
navaid wind-finding 导航测风
naval air 海军航空兵(美国)
Naval Air Facility 海军空中发射设施
naval air squadron 海军航空兵飞行中队
naval air station 海军航空站
naval aircraft 海军飞机
naval architect 造船工程师,造船技师
naval architecture 造船学,造船工程,海洋建筑工程,海事工程
naval auxiliary air station 海军辅助航空站

naval campaign 海军为主,加上地面部队、水下部队、两栖部队、航空兵等参与的夺取制海权的战役
naval flying standards flight 海军飞行标准检查小队
naval force 海军力量
naval gunfire support 舰炮火力支援
naval or marine air base 海军航空基地
naval variant 海军版
Naval Wing 海军联队
Navar 导航雷达,雷达导航和控制系统
Navarho "纳瓦尔霍"远程无线电导航系统
Navascope 导航雷达机上显示器
Navascreen 导航雷达地面屏幕
NAVAVNLOGCEN Naval Aviationlogistics Center 海军航空兵后勤中心
NAVAVNWEPSFAC Naval Aviation Weapons Facility 海军航空兵武器站
navex navigation exercise 领航练习
NAVFAC Naval facility 二级海军站
Navier-Stokes Equations 纳维尔-斯托克斯方程
navigable 适于航行的,可通航的,适于领航的
navigable airspace 通航空域,适航空域
navigate 领航,导航,航行,驾驶
navigation 1.导航,航行 2.领航学,航海术
navigation accuracy 导航精度
navigation aid 助航设备,导航设施
navigation algorithm 导航算法
navigation and weapon delivery computer system 导航与武器投射计算机系统
navigation attack system 导航攻击系统
navigation bias 导航偏差
navigation bombing 导航轰炸
navigation by space references 空间(基准)导航
navigation capability 导航能力
navigation chart 导航图
navigation computer 导航计算机
navigation coverage 导航设备有效工作区域,导航设备覆盖区
navigation datum 导航数据
Navigation department (航母)航海部门
navigation dispersion 导航离差
navigation display 领航仪
navigation dome 天文领航舱
navigation duty 领航勤务,导航勤务
navigation equation 导航方程
navigation error 导航误差
navigation filter 导航滤波
navigation frame 导航框架
navigation function 导航功能
navigation gain 导航增益

navigation iteration 导航迭代
navigation iteration interval 导航迭代时间间隔
navigation light 航行灯
navigation log 航行记录簿,航行日志
navigation message 导航电文
navigation mode 导航模式
navigation obstruction 航行障碍物
navigation of aerial photography 航摄领航
navigation parameter 导航参数
navigation performance 航行性能
navigation processing software 导航处理软件
navigation radar 导航雷达
navigation ratio 1.导航比 2.导航常数
navigation receiver 导航接收器
navigation satellite 导航卫星
navigation satellite network 导航卫星网
navigation satellite system 导航卫星系统
navigation sensor 导航传感器
navigation sensor error 导航传感器误差
navigation signal 导航信号
navigation solution 导航解算
navigation stars 导航星
navigation state 航态
navigation station location survey 导航台定位测量
navigation subsystem 导航分系统
navigation system 导航系统
navigation system of stationary satellite 静止卫星导航系统
navigation team 导航团队
navigation technique 导航技术
navigation unit 导航装置
navigational 领航的,导航的,航行的
navigational accuracy 导航精度
navigational aids 导航设备,助航方法
navigational datum 航行基准面
navigational error 导航偏差
navigational lighting aid 助航灯光
navigational planets 导航行星
navigational stars 导航星,航行星
navigational visual aid 目视助航设施
navigation-and-weapon aiming system 导航与武器瞄准系统
navigator 1.领航员 2.导航仪,导航系统
NAVPRO Naval Plant Representative Office 海军厂代表处(美国)
NAVSAT Navigation Satellite 导航卫星
NAVSTA Naval Station 海军站
Navstar 导航星
NAVWASS Navigation and Weapon Aiming Subsystem

导航及武器瞄准分系统

NAWAU National Aviation Weather Advisory Unit 国家航空天气咨询处(美国)

Nawtol Night/all-weather Take Off and landing 夜间/全天候起飞着陆

N-axis 零向量

NB 1. Narrow Band 窄(频)带 2. Narrow Beam 窄波束 3. Navigation Base 导航基地 4. Navigator Bombardier 领航轰炸员 5. Neutral Buoyancy 中性浮力

NBAA National Business Aircraft Association 全国勤务飞机协会(美国)

NBC 1. Nuclear Biological Chemical (warfare) 核生化(战) 2. Noise Balancing Control 噪声平衡控制

NBC protective suit 防核生化服

NBFM Narrow-Band Frequency Modulation 窄带调频

N-bomb Nuclear bomb 核弹

NBR Nitrile Based Rubber 腈基橡胶

NBS 1. National Bureau of Standards 国家标准局(美国) 2. Neutral Buoyancy Simulator 中性浮力模拟器

NBST National Board for Science and Technology 国家科学技术委员会(爱尔兰)

NBSV Narrow Band Secure Voice 窄带保密话音

NBT Nimbus Beacon Transmitter 雨云气象卫星信标发射机

n-butane 正丁烷

NC nitrocellulose 硝化纤维(也可写作 Nc)

NC relay Normally Closed relay 常闭继电器

NCA National Command Authorities 国家指挥当局

NCAA National Council of Aircraft Appraisers 全国飞机鉴定委员会(美国)

NCAP Night Combat Air Patrol 夜间空中战斗巡逻(也写作 night cap)

NCAR National Center for Atmospheric Research 国家大气研究中心

NCAS Nomex Core, Aluminium Skin 铝皮蜂窝夹芯材料

NCB Network Connect Block 网络连接块

NCC Nickel-Coated Carbon 锁键碳

NCCI Numerically Controlled Configuration Identification 数控配置标识

NCD Network Cryptographic Device 网络密码装置

NCDU Navigational Control and Display Unit 导航控制与显示装置

NCE Non-Cooperative Emitter 非合作式发射源

NCF 1. Network Configuration Facility 网络配置设备 2. Network Connection Failure 网络连接故障

NCI 1. Network Command Interpreter 网络命令解释器 2. Non Code Information 非编码信息

NCISA NATO Communication and Information Systems Agency 北约通信及信息系统局

NCL 1. Navigation Control Language 导航控制语言 2. Network Control Language 网络控制语言

NCMA National Contract Management Association 全国合同管理协会

NCO Numerically Controllerd Oscillator 数控振荡器

N-code 云量

NCOS National Commission On Space 国家空间委员会(美国)

NCPRSC National Cartographic Photogrammetric and Remote Sensing Center 国家地图摄影测量与遥感中心(菲律宾)

NCQR National Council for Quality and Reliability 全国质量及可靠性委员会(英国)

NCR Nonlinear Correlation Receiver 非线性相关接收机

NCS Numerical Control Society 数控学会(美国)

NCT Non-Cooperative Target 非合作式目标

NCTR Non-Cooperative Target Recognition 非合作式目标识别

ND Navigation Display 导航显示

Nd neodymium【化】钕(化学元素,符号 Nd)

ND point Nominal Deceleration point 规定减速点(超声速飞机的)

NDA National Defense Area 国家防御地域(美国)

NDB 1. Navigational Data Base 导航数据库 2. Non-Directional Beacon 全向信标,导航台

NDDS Nuclear Detonation Detection System 核爆炸检测系统

NDE Non-Destructive Evaluation 非破坏性鉴定

n-decane 正癸烷

NDER National Defense Executive Reserve 国家国防执行预备队

NDEW Nuclear Directed Energy Weapon 核定向能武器

NDHF Nimbus Data Handling Facility 雨云卫星数据处理中心(美国)

NDHS Nimbus Data Handling System 雨云卫星数据处理系统

NDI 1. Non-Destructive Inspection 无损检验(探伤) 2. Non-Development Item 非发展项目

NDIA National Defense Industrial Association 国防工业协会

NDIC National Defence Industries Council 国防工业委员会(英国)

NDM Noise Definition Manual 噪声定义手册

n-dodecane 正十二烷

NDPF NASA Data Processing Facility 美国航空航天局数据处理中心

NDRC National Defence Research Committee 国防研究委员会
NDS 1. Navigational Development Satellite 导航发展卫星 2. Nuclear Detection Satellite 核探测卫星 3. Nuclear Data Section 核数据处
NDT 1. Non-Destructive Testing 非破坏性试验 2. Non-Destruction Test 无损探伤
NDTA National Defence Transportation Association 国防运输协会(美国)
NDTM Non-Destructive Testing Manual 无损检测手册
NDV 1. NASP Derived Vehicle 国家空天飞机派生运载器(美国) 2. Nuclear Delivery Vehicle 核武器投放运载器
Ne 1. neon【化】氖(化学元素,符号 Ne) 2. number of engines 发动机数目
NEA 1. National Electronics Association 全国电子协会(美国) 2. Nuclear Energy Agency 核能局
NEAC Noise and Emission Advisory Committee 噪声及噪声传播咨询委员会
NEACP National Emergency Airborne Command Post 国家紧急空中指挥所(美国)
NEAR Near Earth Asteroid Rendezvous 近地小行星交会(探测)任务(美国)
near blowout 近距爆裂
near blowout condition 近距爆裂条件
near collision course 小提前角相遇航向
near coupled canard 近(距)耦(合)鸭翼
near encounter 近距交会
near field 1. 近(扰动源)激波区 2. 最接近超声速飞机航迹的音爆区 3. 近场
near hover 近地盘旋
near infrared 近红外
near infrared camera 近红外相机
near miss 1. 靠近弹 2. 近距脱靶,近距爆炸击中 3. 几乎相撞 4. 相撞险情
near net-shape 近净成形
near polar orbit 近极地轨道
near polar sun-synchronous orbit 近极地太阳同步轨道
near side of the moon 月球正面
near stall point 近失速点
near surface burst 接近地面爆炸,接近水面爆炸
near synchronous equatorial orbit 近同步赤道轨道
near wake 近尾迹
near-accident analysis 险情分析,事故征候分析
nearby explosion 附近爆炸
nearby galaxy 近邻星系
near-circular orbit 近圆轨道
near-design 接近设计
near-design condition 接近设计条件
near-design loading 接近设计荷载
near-disaster 近乎坠毁事故
near-earth 近地
near-earth space 近地空间
near-electrode 电极附近
near-equatorial orbit 近赤道轨道
nearer-term 近期,近期的
nearest-neighbor 最近邻
nearfield 近场
near-field effect 近区效应
near-field region 近场区
near-gaussian pulse shaping 近高斯脉冲成形
near-in reflection 近区反射
near-injector 喷油器附近
near-limit 接近极限
nearmiss accident (飞行)事故征候
near-nozzle region 近喷嘴区域
near-optimal aircraft 近似最佳飞机
near-optimality 近优性
near-parabolic orbit 近抛物线轨道
near-real-time 近实时
near-real-time capability 近实时传送信息能力
near-real-time impact calculation 准实时落点计算
near-real-time reconnaissance 近实时侦察
near-space 近空间
near-stall 近失速
near-stall condition 近失速条件
near-sun trajectory 近日轨迹
near-surface 近地表
near-surface layer 近地表层
near-surface part 近地表部分
near-term 近期,近期的
near-term mission 近期任务
near-tip 近尖端
near-wake 近尾迹
near-wake region 近尾迹区域
near-wall 近壁
near-wall region 近壁区
near-wall resolution 近壁分辨率
near-zero 接近零
neat resin 净树脂
neatlines 1. 准线,图表边线 2.(地图上的)经线和纬线
NEB 1. National Energy Board 全国能源委员会(加拿大) 2. National Enterprise Board 全国企业委员会(英国)
nebula 星云
nebular variable 星云变星
necessary condition 必要条件
necessary minimum diameter 必要的最小直径

necessity measure 必然性测度
NECI Noise Exposure Computer/Integrator 噪声发觉计算机/积分器
neck 1.短管,管颈 2.气球球囊下端管状部分
neck bladder 颈部气囊
neck of cartridge 药筒口,弹壳口
neck of case 药筒口,弹壳口
neck region 颈区
neck ring 颈圈
necking 缩颈
necking in spindown 缩径旋压
NEE Near-End Error 近端错误
NEEC Noise-excluding Ear Capsule 消噪声耳罩,隔音耳罩
NEEDS NASA End-to-End Data System 国家航空航天局端-端数据系统(美国)
needle (表盘式仪表的)指针
needle and ball 针球仪,转弯倾斜仪
needle emitter 针发射器
needle pellet 撞针头
needle split 转速表指针错开
needle tip 针尖
needle voltage 针电压
needles 仪表读数
Neel temperature 奈耳温度
NEF Noise Exposure Forecast 噪声干扰预报(指数)
NEFC NATO Electronic-warfare Fusion Cell 北约电子战合成分队
NEFD Noise Equivalent Flux Density 噪声等效通量密度
negate 否定,取消,使无效
negation gate 非门,负闸
negative (空中通话用语)不,不许可,说得不对
negative absorption 负吸收
negative acceleration 负加速度,减速
negative acknowledge character 否认字符
negative altitude 负高度,低度
negative angle 【数】负角
negative angle of attack 负冲角
negative angular position 负角位置
negative anomaly 负异常
negative area (气象图或温熵图上的)负区
negative base 负基底
negative camber 负弯角
negative charging 负带电
negative contact 1.看不见前面的飞机 2.不能在某一频率上与空中交通管制机构通话
negative curvature 负曲率
negative diagonal element 负对角元素

negative differential mobility 负微分迁移率
negative dihedral (机翼)下垂
negative divergence 负背离,顶背离,正形成负逆差
negative electrode 负极,负电极,阴极
negative electron affinity 负电子亲和势
negative electron affinity cathode 负电子亲和势阴极
negative end 负端
negative fault 负故障
negative feedback 负反馈
negative feedback amplifier 负反馈放大器
negative fuel 负燃料
negative gradient 负梯度
negative hydrogen ion 负氢离子
negative image 阴像
negative incidence 负倾角,负入射
negative jet 负压脉冲射流
negative justification 负码速调整
negative lift 负升力
negative lift enhancement 负升力增强
negative logic 负逻辑
negative panic (紧急情况下的)被动反应
negative phase of the shock wave 冲击波负压期,核爆炸后气压的负压期
negative photoresist 负性光刻胶
negative plane change 负平面变化
negative polarity 负极性
negative potential 负电位,负电势
negative pressure 负压
negative quantity 负量
negative rate 负率
negative rate of pressure 负压率
negative relative 负相关
negative resistance oscillator 负阻振荡器
negative roll 负辊
negative roll axis 负轧辊轴线
negative rolling moment 负滚转力矩
negative shear angle 负剪切角
negative sheath 负鞘层
negative slope 负坡度,复斜率,负斜率
negative stability 负稳定性,负安定性
negative stagger 负斜罩
negative stall 负过载失速
negative static stability 负静稳定性
negative sweep 前掠(翼)
negative terminal 负极,负端,负极端子
negative thrust duration 负推力持续时间
negative value 负值
negative vorticity 负涡度
negative yaw 负偏航,向左偏转(英、美制),向右偏转

(苏联制)
negative-torque signal 负扭矩信号
negator 非门,非元件,倒换器
negentropy 负熵,负平均信息量
negligible effect 影响可以忽略不计
negotiation threshould 谈判门槛
neighbor 邻居,邻近值,邻国
neighbor cell 邻区,相邻小区信息
neighborhood 邻近邻域
neighborhood method 邻元法
neighboring aircraft 相邻飞机
neighboring center 邻近的中心
neighboring orbit 相邻的轨道
neighboring rib 相邻肋
neighboring sensor 邻近的传感器
neighboring vehicle 相邻的车辆
NEL 1. National Engineering Laboratory 国家工程实验室 2. Network Element Layer 网元层
NEM 1. Noise Exposure Map 曝噪示意图 2. Not Elsewhere Mentioned 别处无提及
nematic 液晶,液晶的
nematic liquid crystal 向列相液晶
NEMO Naval Experiment Manned Observatory 海军载人实验观测台
NEMP Nuclear Electromagnetic Pulse 核电磁脉冲
NEMS 1. Nano Electro Mechanical System 纳米电子机械系统 2. Near-Earth Magnetospheric Satellite 近地磁层探测卫星 3. Nimbus-E microwave Spectrometer 雨云-E卫星微波谱仪
NEO 1. National Energy Outlook 国家能量观 2. Near-Earth Orbit 近地轨道 3. Nuclear Equipment Operator 核设备操纵员
neodymium 【化】钕(化学元素,符号 Nd)
neodymium crystal laser 钕晶体激光器
neodymium glass laser 钕玻璃激光器
neodymium pentaphosphate laser 过磷酸钕激光器
neodymium rod 钕棒
neodymium-doped glass 掺钕玻璃
Neomax 钕铁硼磁性材料
neon 【化】氖(化学元素,符号 Ne)
neon bulb 氖灯
Neoprene 氯丁橡胶
neotectonic map 新构造图
NEP 1. Noise Equivalent Power 噪声等效功率 2. Nominal Entry Point 标称进入点 3. Nuclear-Electric Propulsion 核电推进
Nepal drop 尼泊尔式空投
neper 奈培
neph chart 云层分析图

nephelometer 1. 能见度测定表 2. 光散浊度计
nephelometry 油度测定法
nepheloscope 1. 云室,成云器,测云器 2. 云滴凝结器 3. (气体迅速压缩或膨胀时)气温改变指示器
nephograph 云图拍摄机
nephohypsometer 云高计
nephometer 测云量计
nephoscope 测云器(测云的运动方向),反射式测云器,云速计
nephros function meter 肾功能仪
nephsystem 云系
Neptune 【天】海王星
neptunium 【化】镎(化学元素,符号 Ne)
ner discretization 尼珥离散化
NERAC Nuclear Energy Research Advisory Committee 美国核能研究顾问委员会
NERAM Network Reliability Assessment Model 网络可靠性评价模型
NERC 1. National Environmental Research Center 全国环境研究中心(美国环境保护局) 2. National Environmental Research Council 全国环境研究理事会(英国)3. Natural Environment Research Council 自然环境研究委员会(英国)
Nereid 【天】海卫二
NERO 1. National Energy Resources Organization 全国能源组织(美国) 2. Near Eirth Rescue and Operations 近地(轨道)救援和操作
NERVA Nuclear Engine for Rocket Vehicle Application 火箭飞行器核发动机
nerve agent 神经(性)毒剂
nerve gas 神经毒气,中毒性毒气
NES Not Elsewhere Specified 不另说明
Nesa glass 透明导电薄膜半导体,玻璃(可以防结冰)
NESC 1. National Environmental Satellite Center 国家环境卫星中心(美国国家海洋与大气管理局) 2. Naval Electronics Systems Command 海军电子系统司令部(美国)
NESCTM National Environmental Satellite Center Technical Memoranda 国家环境卫星中心技术备忘录(美国)
NESDIS National Environmental Satellite, Data and Information Service 国家环境卫星数据和信息业务处
NESN NATO English-Speaking Nations 北约英语国家
NESS National Environmental Satellite Service 国家环境卫星处
ness 海角,突端
NEST 1. Naval Experimental Satellite Terminal 海军实验卫星终端 2. Nuclear Emergency Search Team 核紧急事故搜救队(美国)

nested function 嵌套函数
nested grid model 嵌套网格模式
nested nozzle 套装喷管
nested operation 嵌套运算
net 1.网,网状物 2.气球与有效载荷之间的吊网 3.(覆盖指定地域的电子、光学等)通信网
net area 净面积,机翼净面积
net circulation 净旋量
net defense capability 净防御能力
net drag 净阻
net dry weight 净干重
net effect 合成串音,有效效应
net exchange radiometer 净辐射表
net flightpath 基本飞行轨迹
net force 净力
net heat 净热
net heat input 净输入热量
net heating flux 净热流密度(防热层的)
net height 有效飞行轨迹上某点的高度
net mass 净重,净含量,净质量
net mole 净摩尔
net passenger cost 净客运成本
net peak 净峰
net performance 净性能
net positive suction head 净正抽吸压头
net power 净功率
net propulsive force 净推力
net pyrradiometer 辐射平衡表
net radiation factor 有效辐射系数
net structure 网状组织(金相学的)
net test 网络测试
net thrust 净推力,有效推力
net torque 净力矩
net transport 净输送
net weight 净重
NETCON Network Control 网络控制
NETD Noise Equivalent Temperature Difference 噪声等效温差
netic field 稳定磁场
NETT Network Technique 网络技术
nettage decrement 净减率
netting 组网
network 1.网络,电路 2.网状系统
net-work 净功,纯功
network address 网络地址
network analog 网络模拟
network analysis 网格分析
network analyzer 网络分析仪
network architecture 网络体系结构

network communication 网络通信,网络传播
network computer interface 网络计算机接口
network control mode 网络控制方式
network control unit 网络控制部件
network controller 网络控制器
network coordination station 网络协调站
network error 网络误差
network function 网络函数
network interface 网络接口,网络界面
network layer 网络层
network level 网络级
network model 网络模型
network node 网络节点
network operating system 网络操作系统
network operator command 网络操作员命令
network operator console 网络操作员控制台
network optimization 网络优化,网络最佳化
network optimization approach 网络优化方法
network parameter 网络参数
network performance 网络性能
network planning 网络规划(法),统筹法
network segment 网络段
network server 网络服务器
network service 网络服务,网络通信
network service frame 网络服务帧
network socket 网络套接字
network structure 网络结构
network synthesis 网络综合
network system 网络系统,网状制,网络系统供电网系统
network terminal protocol 网络终端协议
network termination processor 网络端处理器
network topology 网络拓扑结构
network weight 机器重量,网络权值
network-based design 网络化设计
networking 计算机网络的设计,一种互助性的网络体系
network-optimized 网络优化
neural assembly 神经集合
neural controller 神经控制器
neural guidance 神经引导
neural mismatch theory 神经匹配不当学说
neural model 神经模型
neural net 神经网络
neural network computer 神经网络计算机
neural network model 神经网络模型
neural network parameter 神经网络参数
neural network system 神经网络系统
neural network-based design 神经网络设计

neural-network 神经网络
neural-network method 神经网络方法
neural-network-based control 基于神经网络的控制
neuro-circulatory collapse 神经循环性衰竭
neuro-cybernetics 神经控制论
neuro-vegetative response 植物神经反应
neurological dysbarism 神经性减压障碍,神经性减压病
neuromuscular dynamics 肌肉神经动力学
neuromuscular frequency 肌肉神经频率
neuromuscular system 肌肉神经系统
neuron 神经元,神经单位
NEUS Nuclear-electric Unmanned Spacecraft 不载人核电航天器
neut 中子弹
neutral 中间的,中立的,中性的,中和的
neutral abnormality 中性失常
neutral acceleration 中性加速度
neutral atmosphere 中性气氛,中性大气
neutral atom 中性原子
neutral axis 中轴
neutral buoyancy 中性浮力
neutral buoyancy facility 中性浮力设备
neutral buoyancy simulator 1.拟池 2.中性浮力模拟器
neutral buoyancy test 中性浮力试验
neutral buoyancy trough 中性浮力槽
neutral burning 中性燃烧,等面燃烧
neutral collision 中性碰撞
neutral density 中性密度,减光镜,黑白密度
neutral density distribution 中性密度分布
neutral density filter 中性密度滤光片
neutral engine 中性发动机
neutral equilibrium 中性平衡,随遇平衡
neutral filter 中性滤光片
neutral flame 中性焰(燃料与氧都不过量)
neutral flow 中性流
neutral gas 惰性气体,中性气体
neutral gas density 中性(惰性)气体密度
neutral hole 中性孔
neutral loss 中性丢失扫描,中性丢失扫描模式
neutral mass spectrometer 中性(成分)质谱仪
neutral number 中和值
neutral particle 中性粒子
neutral particle beam generator 中性粒子束发生器
neutral particle beam weapon 中性粒子束武器
neutral particle density distribution 中性粒子密度分布
neutral particle distribution 中性粒子分布
neutral particle jet 中性粒子喷射
neutral pion 中性介子
neutral plane 中和平面

neutral point 1.中性点,中立重心点 2.拉格朗日点 3.飞机焦点
neutral position 空档位置,中间位置,中性位置,空挡
neutral position of the rotating magnet 旋转磁铁的中性位置(磁电机的)
neutral pressure 中性压力,中和压力
neutral propellant 中性推进剂
neutral sheet 中性片(磁尾赤道附近磁场为零的平面)
neutral stability 中性稳定(度)性
neutral stability curve 中稳态曲线
neutral stability parabola 中稳态抛物线
neutral static stability 中性静稳定性
neutral temperature 中性温度
neutral vapor 中性蒸汽
neutral velocity 中性流速
neutral voltage 相对中性点电压
neutral wind 中性风
neutral xenon 中性氙
neutraliser 缓冲器,中性化器
neutralization 中和,中和作用,中立状态
neutralization mechanism 中和机制
neutralization process 中和法
neutralize 压制(使失去作用或无法使用)
neutralized controls 操纵机构处于中立位置
neutralized track 空中截击代语,目标无意义
neutralizer 中和剂,中和器
neutralizer assembly 中和器总成
neutralizer body 空档器阀体
neutralizer cathode 中和器阴极
neutralizer coupling 中和器耦合
neutralizer coupling voltage 中和器耦合电压
neutralizer flow 中和剂流
neutralizer flow rate 中和剂流率
neutralizer keeper voltage 中和器守护电压
neutralizer neutral 中和剂中性
neutralizer neutral loss rate 中和剂中和损失率
neutralizer operation 空档器操作
neutralizer plasma 中和器等离子
neutralizer voltage 中和器电压
neutrino 微中子
neutrino astronomy 中微子天文学
neutron 中子
neutron activation analysis 中子活化分析
neutron bomb 中子(炸)弹
neutron capture 中子俘获
neutron counter tube 中子计数管
neutron detector 中子探测器
neutron generator 中子发生器
neutron graphy 中子照相法

neutron monitor 中子监测器,中子堆
neutron radiography 中子射线照相术
neutron star 中子星
neutron transmutation doping 中子嬗变掺杂
neutron warhead 中子弹头
neutron yield 中子产额,中子产率
neutropause 中性层顶
neutrosphere 中性层
never-exceed speed 极限速度,永不超越速度
new moon 新月
new aircraft 新型飞机
newly formed star 新生星
Newton 牛,牛顿
newtonian 信仰牛顿学说的人,牛顿(学说)的
Newtonian cosmology 牛顿宇宙论
Newtonian flow 牛顿流,自由分子流
newtonian impact 牛顿撞击
newtonian impact theory 牛顿撞击理论
Newtonian mechanics 牛顿力学
newtonian plateau 牛顿高原
newtonian result 牛顿结果
Newtonian static margin 牛顿静稳定度
Newtonian theory 牛顿理论
newtonian time 牛顿时间,牛登时间,绝对时间
newtonian time-domain 牛顿时域
NEWTS Naval Electronic Warfare Training System 海军电子战训练系统(美国)
next accelerator 下一代加速器
next engine 下一代发动机
next equation 次方程
next grid 下一代网格
next higher assembly 上级装配件
next ion engine 下一代离子发动机
next ion thruster 下一代离子推进器
next launch 下一次发射
next planet 下一个星球
next section 下一个部分
next thruster 下一代推进器
NF 1. Night Fighter 夜间歼击(战斗)机 2. Nose Fuze 弹头引信,头部引信 3. Negative Feedback 负反馈 4. Noise Factor (无线电的)噪声系数
NFAC National Full-scale Aerodynamics Complex 国家全尺寸空气动力综合实验设施(美国国家航空航天局)
NFCS Nuclear Forces Communications Satellite 核部队通信卫星
NFDC National Flight Data Center 国家飞行数据中心
NFF No Fault Found 未发现故障
NFH NATO Frigate Helicopter 北约护卫舰舰载直升机
NFIP National Foreign Intelligence Program 国家国外情况计划(美国)
NFN Near Field Noise 近场噪声
NFO Naval Flight Officer 海军飞行军官(美国)
NFOV Narrow Field Of View 窄视场,小视场
NFP Net Flight Path 基本飞行轨迹
NFPA National Fire Protection Association 全国消防协会(美国)
NFR No Further Requirement 无进一步要求
NFRL Naval Facilities Research Laboratories 海军设施研究实验室(美国)
NFS 1. National Fire Service 国家消防局 2. Network File System 网络档案制度
NFSN NATO French-Speaking Nations 北约法语国家
NG 1. Narrow Gage 窄选通电路 2. Nitroglycerin(e) 硝化甘油(炸药),甘油三硝酸酯 3. Nose Gear 前起落架 4. Narrow Gauge 窄轨距
Ng 气体发生器转速(转/分)
NGB National Guard Bureau 国民警卫队局(美国陆军)
NGC Nylon Graphite Composite 尼龙石墨复合材料
NGDC National Geographic Data Center 全国地理数据中心(美国国家海洋与大气管理局)
NGE Non-Ground Effect 无地(面)效(应)
NGF Navigation Flare 航行灯
NGIMS Neutral Gas and Ion Mass Spectrometer 中性气体离子质谱仪
NGL Natural Gas Liquids 液化天然气,气体汽油,凝析油
NGPS NAVSTAR Global Positioning System 导航星全球定位系统
NGR 1. Navstar Geodetic Receiver 导航星测地接收机 2. Night Goggle Readable (红外)夜视镜
NGS 1. Naval Gunfire Support 舰炮火力支援 2. National Geographic Society 全国地理学会 3. National Geodetic Survey 国家大地测量局
NGSDC National Geophysical and Solar-Terrestrial Data Center 国家地球物理与日地数据中心(美国)
NGSP National Geodetic-Satellite Program 国家测地卫星计划(美国航空航天局)
NGT 1. NASA Ground Terminal 航空航天局地面终端(美国) 2. New Generation Trainer 新一代教练机,新一代练习器
NGTE National Gas Turbine Establishment 国立燃气轮机研究院(英国)
NGV Nozzle Guide Vane 涡轮导向叶片
NGW Nuclear Gravity Weapon 自由下落式核武器
NHC Navigator's Hand Controller 导航仪手操纵装置
n-heptane 正庚烷

n-hexane 正己烷
n-hexane spray 正己烷喷雾
NHGA National Hang Gliding Association 全国乘风滑翔协会
NHP Nominal Horse Power 额定马力
NHpowder Non-Hygroscopic powder 防潮火药
NHR Net Heat Rate 净热耗
ni 镍
NIAG NATO Industrial Advisory Group 北约工业咨询组
NIAST National Institute for Aeronautics and Systems Technology 全国航空及系统技术学会(南非)
nib 1.轴向尖头整流包皮,(两个并列尾喷管间的)尖尾整流包皮 2.(变后掠飞机机翼)前缘突齿尖头延伸件
nibble 近失速点
nibble address buffer 分时地址缓冲器
nibbler 切片机,毛坯下料机,步冲轮廓机
NIC Navigation Information Centre 航引情报中心
Nicad 镍镉蓄电池
Nicalloy 高导磁铁镍合金,镍铁合金
NICAP National Investigation Committee on Aerial Phenomena 全国空中现象调查委员会(美国)
nicarbing 碳氮共渗
Nicerol 耐西罗泡沫灭火剂
niche 小生态环境
Nichrome 镍铬合金
nichrome wire 镍铬线,镍铬合金线
nickel 【化】镍(化学元素,符号 Ni),镍币,五分镍币
nickel aluminide 镍铝化合物
nickel based alloy 镍基合金
nickel matrix cathode 海绵镍阴极
nickel screen 镍网
nickel-base superalloy 镍基高温合金
nickel-cadmium battery 镍镉蓄电池
nickel-cadmium diffused plating 镍镉扩散镀(层)
nickel-hydrogen battery 氢镍蓄电池
nickel-iron storage battery 镍铁电池
nickelplate 镀镍
ni-coated powder 镍包覆粉末
ni-coated sample 镍包覆样本
NICS NATO Integrated Communications System 北约一体化通信系统
NIFA National Intercollegiate Flying Association 全国大学飞行协会(美国)
NIFS Nuclear Influence Fuzing System 核感应引信起爆系统
night airglow 夜间气辉
night and all-weather 夜间及全天候战斗机
Night CAP Night Combat Air Patrol 夜间空中战斗巡逻
night effect 夜间效应,夜间误差
night fighter 夜间战斗机(截击机)
night flight 1.夜间飞行 2.夜航
night flight control system 夜间飞行控制系统
night flying chart 夜间飞行航图
night glow 夜天光
night hemisphere 背阳半球
night jump 夜间跳伞
night light-intensifying sight 夜间增亮瞄准具
night myopia 夜间近视
night observation gunsight system 夜间观察射击瞄准系统
night of optical tracking 夜间光学追踪
Night Owl 夜间对地攻击任务
night sight 夜间瞄准具
night sky light 夜天光
night vision 夜间视力,夜视
night vision aids 夜视设备
night vision goggles 夜视镜
night vision sight 夜视瞄准具
night vision threshold 夜视阈,暗视阈
night vision trainer 夜视训练器
night visual acuity 夜视力,夜视敏度
Nighthawk "夜鹰"(即 F-117A 隐形战斗机)
nightside 夜侧,背阳侧
night-time vision device(system) 夜视装置(系统)
Nimbus 雨云卫星(美国气象卫星)
Nimocast 镍模铸合金
Nimonic 镍蒙尼,镍基耐热抗腐蚀合金
Nimrod "猎迷"(英国拟研制的预警机)
NIMS 1. NASA Interface Monitoring System 航空航天局接口监测系统(美国) 2. Near-Infrared Mapping Spectrometer 近红外测绘光谱仪
Nimtan "尼姆坦"机载夜间瞄准具(英国)
ninety-minute rule 90 分钟规定(双发客机适航证要求,在越洋飞行时其航区内到达应急备降场的时间不准超过 90 分钟)
niobate system ceramics 铌酸盐系陶瓷
niobium alloy 铌合金
niobium based alloy 铌基合金
niobium titanium aluminide 铌钛铝化合物
NIOSH National Institute for Occupational Safety and Health 国家职业安全与职工保健研究所(美国)
NIP Normal Incidence Pyrheliometer 垂直入射绝对日射表
nip 挤压配合紧度
NIS NATO Identification System 北约识别系统
NISO Netherlands Industrial Space Organisation 荷兰

工业空间组织
Nitralloy 渗氮合金
nitramine 硝胺
nitrate compound 硝酸盐复合物
nitrate dope 硝基漆,透布油
nitrate ester 硝酸酯
nitrate ester plasticized polyether propellant 硝酸酯增塑聚醚推进剂
nitration 硝化,用硝酸处理,硝基置换
nitric acid 硝酸
nitric oxide 一氧化氮
nitric sulfide 一硫化氮
nitridation 氮化
nitride 氮化物
nitriding 渗氮
nitrile rubber 腈橡胶
nitro 硝基
nitro compound 硝基化合物
nitro group 硝基
nitrocellulose 硝化纤维素,硝化棉,纤维素硝酸酯
nitrocellulose powder 硝化纤维(素)火药,硝化棉火药
nitrocellulose propellant 硝化纤维素推进剂
nitrocompound 硝基化合物
nitroexplosive 硝化火药
nitrogen atom 氮原子
nitrogen blow-off 氮气吹除
nitrogen case hardening 渗氮
nitrogen concentration 氮浓度
nitrogen cycle 氮循环
nitrogen desaturation 1.缺氮症 2.氮脱饱和 3.(从体内)排氮
nitrogen dilution 氮稀释
nitrogen dilution level 氮气稀释程度
nitrogen dioxide 二氧化氮
nitrogen feedstock 氮原料
nitrogen fill valve 充氮阀
nitrogen fixation 固氮作用
nitrogen gas 氮气
nitrogen gas distribution board 氮气配气台
nitrogen injection 注入氮气
nitrogen jet 氮射流
nitrogen material 氮材料
nitrogen mixture 氮气混合气
nitrogen molecular laser 氮分子激光器
nitrogen mustard gases 氮化芥子气
nitrogen oxide 氧化氮,氮的氧化物
nitrogen partial pressure 氮分压
nitrogen plasma 氮等离子体
nitrogen plasma jet 氮等离子体射流

nitrogen propellant 氮推进剂
nitrogen replacement 氮气置换
nitrogen sequence 氮序
nitrogen tetroxide 四氧化二氮
nitrogen tetroxide railway tank transporter 四氧化二氮铁路运输车
nitrogen torch feedstock 氮火炬原料
nitrogen wind tunnel 氮气风洞
nitrogen-rich compound 富氮化合物
nitroglycerine 硝化甘油,丙三醇三硝酸酯
nitroglycerine explosive 硝化甘油炸药
nitroguanidine 硝基胍
nitromethane 硝基甲烷
nitrostarch 硝化淀粉
nitrous acid 亚硝酸
nitrous oxide 一氧化二氮,氧化亚氮
nitrous-oxide system 一氧化二氮气体系统
nity 保护软体
NIU Network Interface Unit 网络接口装置
NLC Noctilucent Cloud 夜光云
NLCM Non-lethal Counter Measures 非杀伤性对抗措施
NLF 1. Natural Laminar Flow 自然层流 2. Normal Load Factor 法向载荷因数
nlf airfoil 独立机翼
nlf panel 独立委员会
NLG Nose Landing Gear 前起落架
NLL No Load Lubrication 无负载润滑
NLM 1. Network Link Module 网络链路模块 2. Network Load Module 网络装入模块 3. Network Loadable Module 网络可加载模块
NLOS Non-line-of-sight 非瞄准发射的,间接瞄准的
NLOS missile 非瞄准发射型导弹
NLP 1. Network Layer Packet 网络层信息分组 2. Non-Linear Processor 非线性处理器 3. Normal Link Pulse 链路正常脉冲
NLVP National Launch Vehicle Program 国家运载火箭计划
Nm Newton-meter 牛顿米
nm beam 纳米梁
nm experiment 纳米实验
nm of average pore 纳米的平均孔隙
nm oxide 纳米氧化
nm ultraviolet 纳米紫外线
NMAB National Materials Advisory Board 国家材料咨询委员会(美国)
NMC 1. Naval Missile Center 海军导弹中心(美国) 2. Non-conforming Material Complaint 不合格材料(造成)的缺陷

NMCC National Military Command Center 全国军事指挥中心(美国)
NMCS National Military Command System 全国军事指挥系统(美国)
NMF Network Management Facility 网络管理设施
NMG Numerical Master Geometry 数字主尺寸系统,NMG 数模系统
NMI NASA Management Instruction 航空航天局管理规程(美国)
NMOS Negative(N-type) Metal Oxide Semiconductor N 型金属氧化物半导体
NMP 1. Navigation Microfilm Projector 导航缩微胶卷投影仪 2. Network Management Protocol 网络管理协议
NMR Nuclear Magnetic Resonance 核磁共振
NMS 1. Navigation Management System 导航管理系统 2. Neutral Mass Spectrometer 中性粒子质谱仪
Nms Newton-meter-second 牛顿米秒
NMT Network Management Terminal 网络管理终端
NNA Neutral and Non-Aligned 中立位置未对准
NNE 1. Noise and Number Exposure 噪声和暴露次数指数 2. North-North-East 东北北
NNI Noise and Number Index 噪声和次数指数
NNK Non-Nuclear Kill 非核杀伤,非核摧毁
NNMSB Non-Nuclear Munition Safety Board 非核弹药(常规武器)安全委会
NNRDC National Nuclear Rocket Development Center 国家核火箭发展中心
NNSS Navy Navigation Satellite System 海军导航卫星系统(美国,又称子午仪系统)
NNTS Navy Navigation Technology System 海军导航技术系统(美国)
no in-flight attention equipment 飞行中无须照管的设备
NOA 1. Non-Operational Aircraft 非作战飞机,不使用的飞机 2. Not Organizationally Assigned 未下发的飞机,储备的飞机
NOAA National Oceanographic and Atmospheric Administration 1.国家海洋和大气局(美国) 2.诺阿卫星(美国国家海洋和大气局气象卫星)
NOACT National Overseas Air Cargo Terminal 全国海外空中货运枢纽站(美国海军)
no-autoignition 没有自燃
nobelium 锘
noble gas 惰性气体,稀有气体
no-bleed case 无出血情况
no-break supply 不中断供应
NOC Network Operation Center 网络运行中心
NOCC Network Operation Control Center 网络操作控制中心
nocoplanar trajectory 不共面轨迹
no-crossover 无交叉
no-crossover case 没有交叉情况
noctilucent cloud 夜光云
nocturnal radiation 红外辐射,夜间辐射
NOD Night Observation Device 夜间观察装置
nodal analysis method 节点分析法
nodal beam 波节梁
nodal degree of freedom 节点的自由度
nodal diameter 节径
nodal diameter mode 节直径模式
nodal force 节点力
nodal geometry 节点的几何形状
nodal history 节历史
nodal increment 截距,交点经距
nodal line 节点线,波节线,【物】节线
nodal line of the orbit 轨道交点线
nodal month 交点月
nodal period 波节期,交点周期,节点周期
nodal plane 节面
nodal point 节点
nodal rate 节点率
nodal regression 节点回归
nodal value 节点值
nodalization 波节减振,节线减振
Noda Matic 诺达节振阻尼系统(一种用于直升机的旋翼振动阻尼系统的专利名称)
NODC National Oceanographic Data Center 国家海洋数据中心(美国)
nodding 点头挠动
nodding aerial 点头天线
noddy cap 〈口语〉傻瓜帽(精密导弹头部的保护罩)
node 1.波节,节点,节线,节面 2.交点 3.(网络的)终点分支交点 4.卫星轨道与原始轨道面的交点 5.(人员物资运输系统的)输送起点、处理站或终点
node angle 节点角
node bus 节点总线
node configuration 节点配置
node correction 节点校正
node line 节点线
node location 节点定位,节点位置
node module 连接舱
node point 结点
node processor 节点处理器
node representation 节点表示
node-point camera 节点式相机
nodical elongation 交点距角
nodical month 交点月

nodimensional parameter 无量纲参数,无维参数,无因次参数
no-disruptive penetration 无破碎侵彻
no-draft forging 挤压模锻
NODS Night Observation and Detection System 夜间观察与探测系统
nodular cast iron 球墨铸铁,球状石墨铸铁
NOE Nap Of Earth 掠地飞行
NOESS National Operational Environmental Satellite System 国家业务环境卫星系统(美国)
noetic science 思维科学
no-failure life 无故障寿命,无故障工作时间
no-feathering axis (旋翼自动倾斜器的)无周期变距运动轴
no-feathering plane 1.桨距不变平面 2.旋翼等效平面
no-fire area 禁射区
no-flare landing 无平飘着陆
no-flow 不流动,无流量
no-flow model 无流动模型
no-fly zone 禁飞区
no-fuel 不助燃
no-fuel condition 没有燃料条件
no-go item 不放飞机件,不放飞项目
NOGS Night Observation Gunship System 夜间观察武装运输机(直升机)
NOI Notice Of Inquiry 调查通知,询问通知
no-injection 没有注入
no-injection case 没有注入情况下
NOISE National Organization to Insure a Sound-controlled Environment 全国确保无噪声环境组织(美国)
noise 1.噪声 2.杂声 3.干扰,杂波
noise abatement 噪声抑制
noise abatement climb procedure 噪声抑制爬高程序
noise abatement procedure 噪声抑制程序
noise allowance 噪声容限
Noise and Number Index 噪扰指数
noise angle 噪声角
noise annoyance 噪声烦恼度
noise assignment 噪声分配
noise attenuation 噪声消声措施
noise background 背景噪声,声底数值
noise carpet 噪声影响(地)带
noise certificate 噪声合格证
noise certification 噪声合格鉴定
noise characteristic 噪声特性
noise contour 噪声区图
noise control 噪声控制
noise criteria 噪声标准

noise damage 噪声致害,噪声损伤
noise datum 噪声数据
noise elimination structure 噪声消除结构
noise equivalent bandwidth 噪声等效带宽
noise equivalent exposure 噪声等效曝光量
noise equivalent flux density 噪声等效通量密度
noise equivalent power 噪声等效功率
noise equivalent radiance 等效噪声辐射率
noise equivalent reflectance change 等效噪声反射变化
noise equivalent reflectance difference 噪声等效反射比差
noise equivalent temperature difference 噪声等效温差
noise exposure 噪声暴露量
noise factor 噪声系数
noise figure 噪声因数
noise figure meter 噪声系数测试仪
noise floor 最低背景噪声级
noise footprint 噪声影响区
noise frequency 噪声波频率
noise generation 噪声的产生,噪声发生
noise generator 噪声发生器
noise grid 噪声网格
noise hazard 噪声危害
noise immunity 抗扰性,抗扰度
noise impact 噪声冲击
noise index 噪声指标,噪声指数
noise intensity 噪声强度
noise intensity contour 噪声强度曲线
noise jammer 杂波干扰机
noise jamming 噪声干扰
noise level 噪声级,噪声电平
noise margin 噪声容限
noise mitigation 减低噪声
noise model 噪声模型
noise nuisance 噪声伤害
noise objectives for telephone circuit 话音噪声标准
noise power ratio 噪声功率比
noise power spectral density 噪声功率谱密度
noise prediction 噪声预估
noise process 相关的噪声过程
noise radar 噪声雷达
noise radar fuze 噪声雷达引信
noise radiation 声辐射
noise ratio 噪声比
noise reduction 噪声降低,噪声消减
noise reference points 噪声测定点
noise removal 去噪
noise shielding 噪声屏蔽
noise signal 噪声信号
noise signature 噪声特点

noise source	噪声源
noise spectrum	噪声谱
noise statistic	统计噪声
noise storm	噪暴
noise suppression	噪声抑制
noise suppression gasket	消声衬
noise suppression nozzle	消声喷管
noise suppressor	消声器
noise temperature	噪声温度
noise threshold	噪声门限
noise transmission	噪声传播
noise tube	噪声管
noise，spike and transients	噪声、尖脉冲和暂态
noise-abatement	噪声控制
noise-annoyance	噪声烦恼度
noise-annoyance deviation	噪声滋扰偏差
noise-cancelling mic	抗噪声传声器
noise-induced deafness	噪声性耳聋
noise-induced hearing loss	噪声性耳聋
noise-induced vertigo	噪声眩晕
noiseless channel	无噪信道
noiseless coding theorem	无噪编码定理
noise-metallic	金属（线路）噪声
noise-modulated fuze	噪声调制引信
noise-optimal trajectory	噪声最优轨迹
noiseproof	抗噪声的，抗干扰的，隔音的，防杂音的
noise-reduction rating	消噪声额定值
noise-sensitive location	噪声敏感的位置
noise-shield aircraft	噪声屏蔽飞机
noise-stop	抗噪声
noisy measurement	嘈杂的测量
noisy phenotype	嘈杂的表型
noisy signal	噪声信号
noisy water	大噪声水域
NOL Naval Ordnance Laboratory	海军军械实验室（美国）
no-lift	"禁止举托"（刻在飞机上，告知地面人员该处不准施加举托力）
no-lift angle	禁举托
no-lift direction	零升力方向
no-lift wire	零升力张线
NOLO No Live Operator	无人操纵机
no-load	空载
no-load test	空载试验
nologies	生物医学仪器
nology	拉丝模工艺
nom	名义上的
nom vane	叶片
nomenclature of asteroids	小行星命名
nomenclature of comets	彗星命名
nomenclature of the atmosphere	大气层次命名法
Nomex	1.诺麦克斯蜂窝材料 2.诺麦克斯材料
nominal	1.标称的，额定的 2.标称的，名义的
nominal acceleration	标称加速
nominal acceleration point	标称加速点
nominal angle of attack	标称攻角
nominal angular momentum	标称角动量
nominal ascent	标称提升
nominal burst	标准原子弹爆炸
nominal capacity	额定容量，公称容积
nominal case	标称情况
nominal climb	标称上爬
nominal closed-loop	标称闭环
nominal closed-loop system	标称闭环系统
nominal condition	标称工况
nominal configuration	标称配置
nominal control	额定控制
nominal control force	标称控制力
nominal deceleration point	标称减速点
nominal design	标称设计
nominal dimension	标称尺寸，名义尺寸
nominal edge	标称边
nominal entry	报名单
nominal flight	标称飞行
nominal flow	公称流量
nominal frequency	标称频率
nominal gas capacity	标准气体容量
nominal load	额定负载
nominal Mach number of nozzle	喷管名义马赫数
nominal margin	标称容限
nominal mean element	标称平均元素
nominal mission	标称任务
nominal model	标称模型
nominal momentum	标称角动量
nominal motion	上行额定
nominal on-orbit	标称轨道上
nominal orbit	标称轨道
nominal orbital position	标称轨道位置
nominal path	额定弹道
nominal performance	标称性能
nominal pitch	标准螺距
nominal prediction	标称预测
nominal range	标称范围
nominal reference	标称参考
nominal route	标称路线
nominal scaling	名义量表
nominal separation	标称分离
nominal shape	标称形状

nominal solution　标称解决方案
nominal spacecraft bus voltage　航天器额定母线电压
nominal speed　标称转速
nominal spin　标称自旋
nominal spin axis　标称自旋轴
nominal stagger　标称交错
nominal stress method　名义应力法
nominal tension　额定张力
nominal thrust　额定推力
nominal trajectory　标称弹道
nominal value　标称值
nominal vehicle　标称车辆
nominal voltage　标称电压
nominal vortex　标称涡
nominal weapon　标准核武器
nomograph　列线图
NOMSS　National Operational Meteorological Satellite System　国家业务气象卫星系统(美国)
non-acoustic technique　非声学技术
non-acoustic detecting device　非声学探测设备
non-active maintenance downtime　非现行维修停用时间
non-active redundancy　非工作贮备,非在用余度
non-adaptive combination　非自适应组合
non-air transportable　非航空运输
non-alignment　不结盟
non-aluminized counterpart　非镀铝对应
non-approach control tower　非方法控制塔
non-autonomous attitude sensor　非自控姿态传感器
non-autonomous sensor　非自控传感器
non-autonomous system　非自治系统
non-axisymmetric body　非轴对称体
non-based variable　无基变量
non-blocking switch　无阻塞交换
non-boiling convection　非沸腾对流
non-boiling natural convection　非自然对流沸腾
non-boresafe fuze　非保险型引信
non-boresight　偏离武器轴线的,离轴的,间接瞄准的
non-central chi-square　非中心卡方
non-central chi-square distribution　非中心卡方分配
non-chargeable fault　非责任故障
non-classical information pattern　非经典信息模式
non-closure　非合闸
non-coherent echo　非相干回波
non-coherent scattering　非相参应答机
non-coherent system　非单调关联系统
non-coherent transponder　非相干应答机
non-communications jamming　非通信干扰
non-competitive tender　非竞争性投标
non-concentric float　不同心浮子

non-condensable gas　非可凝性气体
non-conformity　不合格
non-congeneric redundancy　非同类余度
non-conservative force　非保守力
non-conservative scattering　非守恒散射
non-contact　无触头,无触点
non-contact fuze　非触发引信
non-contact measurement　非接触测量
non-contactable object　1.非接触测量技术 2.非接触式活塞
non-contiguous multiplexer　不邻接多工器
non-continuum　不连续流
non-controlled airport　非控制的机场
non-controlled airspace　非管制空域
non-convex　非凸
non-convex thrust constraint　非凸推力约束
non-cooperative aircraft　非合作飞机
non-cooperative game　非合作对策
non-co-operative scorer　独立弹着偏差指示器,非协同弹着记录器
non-coplanar beams　非共面波束
non-coplanar encounter　非共面交会
non-coplanar transfer　非共面转移
non-critical failure　非关键故障,无危险故障
non-critical state　非临界状态
non-datalinked configuration　非数据与配置
non-delay fuze　不延迟信管
non-destructive flaw detection　无损探伤
non-destructive inspection　无损检验
non-destructive test　无损探伤,无损检测
non-destructive testing　无损检验
non-deterministic model　非确定性模型
non-deviative absorption　非偏移吸收
non-differential spoiler　(机翼上的)非差动扰流片
non-dilatable balloon　非膨胀式气球
non-dimensional force　非维力,无量纲力
non-dimensional inlet　非空间的入口
non-dimensional length　无量纲长度
non-dimensional libration　非维天平动
non-dimensional neutral density　非维中性密度
non-dimensional range　非空间的范围
non-dimensional speed　无量纲飞行速度
non-dimensional time　无量纲时间
non-dimensional unit　无维性单位
non-dimensionalized　未量纲化的,未因次化的
non-directional beacon　1.无方向性信标 2.全向信标
non-directional radio beacon　导航台,全向无线电信标
non-director disturbed reticle sight system　非指挥仪式活动光环瞄准系统

non-dispersed infrared analyzer 非色散
non-dominated solution 非支配解
non-draft forging 精密锻造
non-effective sortie 无效出动,无效出动架次
non-electrical parameter 非电量参数
non-elementary part 非基本部分
non-emissive electrode 非发射电极
non-energetic propellant 非高能推进剂
non-engineering system simulation 非工程系统仿真
non-equilibrium 非平衡态
non-equilibrium carrier 非平衡载流子
non-equilibrium chemistry 非平衡态化学
non-equilibrium computation 非平衡计算
non-equilibrium effect 非平衡效应
non-equilibrium flow 非平衡流动
non-equilibrium ionization 非平衡态电离
non-equilibrium plasma 非平衡态等离子体
non-equilibrium process 非平衡过程,非准静态过程
non-equilibrium scheme 非平衡方案
non-equilibrium state 非均衡态,非平衡状况
non-equilibrium system 非均衡系统
non-expanding balloon 非膨胀式气球
non-explosive warhead 非爆破弹头,惰性装药弹头
non-failure operating time 无故障运转时间
non-fatal crash 无死亡的坠机事故
non-ferrous 非铁的,有色金属的,非铁基的(尤指青铜、黄铜)
non-flying prototype 不飞行的原型机(一切均按飞行的要求制作,但不能或不用于飞行,只做地面试验用)
non-frangible wheel 不碎涡轮盘
non-functioning interval 停机持续时间
non-galvanic corrosion 非电化腐蚀
non-gradient method 非梯度法
non-gravitational acceleration 非重力加速度
non-gravitational effect 非引力效应
nongravitational force 非重力
non-grey atmosphere 非灰大气
non-grey coating 非灰体涂层
non-grey emissivity 非灰体辐射率
non-guided missile 火箭弹
non-handling pilot 不驾驶飞机的飞行员
non-holding side 非待命边(在飞向着陆等待点的进场等待航线的左边)
non-hygroscopic 不吸湿的
non-hygroscopic powder 防潮火药
non-hypergolic propellant 非自燃推进剂
non-hypergolic propellant rocket engine 非自燃推进剂火箭发动机
non-ideal cycle 实际循环

non-ideal gas 非理想气体
non-ideal sail 非理想的帆
non-ignition 不着火
non-impact 非冲击式
non-impact algorithm 非触发算法
non-impact docking 非触发对接
non-impact fuze 非触发引信
non-impinging injector 非撞击射流喷头
non-initiation 漏炸
non-integer 非整数
non-integrated method 非整合方法
non-integrated optic flow measurement 非整合光学流量测量
non-integrated optic flow method 非整合光流法
non-intelligible cross talk 不可懂串音
non-interchangeable socket 不可互换插座,非标准多头插座
non-interruptive checking 不停机检查,运转中检查
non-intrusive measurement 非侵入测量
non-invasive technique 非损伤性技术
non-invasive test 非损伤性试验
non-isotropic warhead 定向爆破弹头
nonius 游标,游尺
non-keplerian orbit 非开普勒轨道
non-landing section (跑道的)非着陆段
non-lethal agent 非杀伤性化学战剂
non-lethal countermeasures 非杀伤性对抗措施
non-lethal mission 非杀伤性任务(指侦察、无线电通信中转、电子对抗等任务)
non-lethal weapons 非致命武器
non-level bombing 非水平轰炸
non-leveraged lease 非杠杆租赁
non-lift balloon 无升力气球
non-linear 非线性的
non-linear aerodynamics 非线性空气动力学
non-linear aeroelastic analysis 非线性气动弹性分析
non-linear aeroelastic characteristic 非线性气动弹性特性
non-linear aircraft 非线性飞机
non-linear analysis 非线性分析
non-linear approach 非线性逼近
non-linear autopilot 非线性自动驾驶仪
non-linear beam-column 非线性梁柱
non-linear boundary-value problem 非线性边值问题
non-linear characteristics 非线性特性,非线性特征
non-linear clearance 非线性间隙
non-linear control 非线性控制
non-linear control law 非线性控制律
non-linear control method 非线性控制方法

non-linear control system 非线性控制系统	**non-linear solver** 非线性求解器,非线性优化求解程序
non-linear controller 非线性控制器	**non-linear spring** 非线性弹簧
non-linear damping 非线性阻尼	**non-linear stability** 非线性稳定性
non-linear design 非线性设计	**nonl-inear state** 非线性状态
non-linear distortion 非线性失真	**nonl-inear stiffness** 非线性刚度
non-linear dynamic inversion 非线性动态逆	**non-linear sweep** 非线性扫描
non-linear dynamical system 非线性动力学系统	**non-linear system** 非线性系统
non-linear dynamics model 非线性动力学模型	**non-linear term** 非线性项
non-linear effect 非线性效应	**non-linear terminal** 非线性终端
non-linear element 非线性环节	**non-linear time series model** 非线性时间序列模型
non-linear engagement 非线性接触	**non-linear two-point** 非线性两点
non-linear equation 非线性方程	**non-linear uncertainty** 非线性不确定
non-linear estimation 非线性估计	**nonlinear variation** 非线性变参数
non-linear estimator 非线性估计量	**non-linearity** 非线性(度)
non-linear feedback 非线性反馈	**non-linearity index** 非线性指数
non-linear filter 非线性滤过器,非线性系统滤波	**non-local sensor** 非局部传感器
non-linear filtering technique 非线性滤波技术	**non-local thermodynamic equilibrium** 非局部热动(态)平衡
non-linear flight 非线性飞行	**non-local variable** 非局部变量
non-linear flutter 非线性颤振	**non-magnetic** 非磁性的
non-linear function 非线性函数	**non-maintained ensemble** 无维修系统
non-linear guidance law 非线性制导律	**non-matched data** 不匹配数据
non-linear guidance method 非线性制导方法	**non-metallized propellant** 非镀金属推进剂
non-linear integrated circuit 非线性集成电路	**non-methane** 非甲烷
non-linear lift 非线性升力	**non-metric camera** 非量测摄影机
non-linear mapping 非线性映射	**non-minimum phase** 非最小相位
non-linear model 非线性模型	**non-minimum phase system** 非最小相位系统
non-linear multibody 非线性多体系统	**non-mission-capable** 不能执行任务的
non-linear network 非线性网络	**non-mixing region** 非混合区
non-linear optical crystal 非线性光学晶体	**non-monotonic logic** 非单调逻辑
non-linear optical effect 非线性光学效应	**non-munition item** (军械设备中的)非弹药项目
non-linear optics 非线性光学	**non-neutrality** 非中性
non-linear optimal control problem 非线性最优控制问题	**non-nominal condition** 非名义上的地位
non-linear optimization 非线性最优化	**non-nuclear airmunitions** 无核武器的航空弹药
non-linear optimization problem 非线性最优化问题	**non-nuclear kill** 非核杀死
non-inear orbit 非线性轨道	**non-nuclear weapons fire control system** 无核武器的武器火控系统
non-linear path 非线性路径	
non-linear phenomenon 非线性现象	**non-objective grating** 非物端光栅
non-linear photomixing 非线性光混频	**non-operating active aircraft** 在编不在队的飞机
non-linear pressure gradient 非线性压力梯度	**non-operating soak temperature** 非工作均匀温度
non-linear problem 非线性问题	**non-operational consequence** (故障)非使用性后果
non-linear programming 非线性规划	**non-operation(al) firing** 1.非战斗发射 2.非使用起动
non-linear region 非线性区	**non-operative time** 不工作时间
non-linear regulation 非线性正则化	**non-optical imagery** 非光学图像(红外图像与雷达图像)
non-linear regulator 非线性调节器	
non-linear relative dynamics 非线性相对动力学	**non-optimum expansion** (喷管)非最佳扩张
non-linear response 非线性响应	**non-orthogonality** 非正交性
non-linear simulation 非线性仿真	**non-oscillatory and non-free-parameter dissipation difference scheme** 非振荡非自由参量耗散差分格式
non-linear simulation model 非线性仿真模型	
non-linear solution 非线性求解	

non-oxidizing 无氧化性的
non-parametric detection 非参量检测
non-parametric training 非参数训练
non-parametric uncertainty 非参数不确定性
non-participating receiver 非参与接收机
non-pendulous accelerometer 非摆式加速度计
non-perforated honeycomb adhesive 无孔蜂窝黏结剂
non-periodic disturbing torque 非周期性干扰力矩
non-persistent（chemical）agent 暂时性（化学）战剂
non-persistent effect 非长效
non-persistent gas 暂时性毒气
non-planar network 非平面网络
non-planar wing 非平面翼
non-planarity 非平面度
non-porous flat plate 无孔的平板
non-powered flight 无动力飞行
non-precision approach 非精密进近
non-precision approach procedure 非精密进近程序
non-precision instrument runway 提供非精密仪表着陆设备的跑道
non-premixed case 非预混合的情况下
non-premixed flame 非预稳定火焰
non-pressurized helmet 非密封飞行帽，非增压飞行帽
non-program aircraft 编外飞机
non-pulsating 非脉动
non-pulsating jet 非脉动喷射
non-pulsating jet flame 非脉动喷射火焰
non-pumping mode 非脉动模式
non-radar 无雷达的，不用雷达的
non-radar airspace 非雷达领空
non-radar approach 非雷达进近
non-radar approach control 非雷达进近控制
non-radar route 无雷达航线
non-radar seperation 非雷达航线间隔
non-radial pulsation 非径向脉动
non-radiative recombination 无辐射复合
non-reacting 无反应
non-reacting air 非反应空气
non-reacting calculation 非反应计算
non-reacting case 非反应情况
non-reacting flow 非反应流
non-reactive section 不反应的部分
non-reactive shock 不反应的冲击
non-reactive trajectory 不反应的轨迹
non-real-time processing 非实时处理
non-reciprocal network 非互易网络
non-reciprocal phase-shifter 非互易移相器
non-recoverable interceptor 一次使用截击兵器，不可回收的截击兵器

non-recoverable vehicle 一次使用飞行器
non-recur 一次性的
non-recurring cause 不能再现的原因
non-recurring cost 一次性成本，非经常性成本
non-recurring cycle 非递归循环
non-reflecting boundary 无反射边界条件
non-reflective cell 无反射电池
non-reflective missile 不反射（电磁波的）导弹，有防雷达涂层的导弹
non-regenerative life support system 非再生式生命保障系统
non-regulate 未经调节的，未校准的
non-regulated busbar 非稳压母线
non-relevant failure 无关联故障，非外因故障
non-relevant fault 非关联故障
non-repairable item 不可修复产品
non-reversible electric drive 不可逆电传动
non-rigid airship 软式飞艇
non-rigid spacecraft 非刚性（柔性）航天器
non-rotating 不旋转
non-routine maintenance 不定期维修，非例行维修
non-scheduled maintenance 非预定维修
non-scheduled service 非定期航班
non-segregated airspace 非隔离空域
non-self 异物
non-selfcontain servo 非自主航机
non-self-maintained discharge 非自持放电
non-sequenced chain 非顺序链
non-single value nonlinearity 非单值非线性
non-singular element 非奇异元素
non-singular perturbation 非奇异摄动
non-sinusoidal radar 非正弦波雷达
non-sked 〈口语〉无计划的，非计划安排的（即 non-scheduled 的简称，美国用）
non-skid 1. 防滑的（指机轮）2. 有胎面花纹的（指轮胎）
non-solution anticipation toss delivery 无解预告拉起投弹
non-spherical earth perturbation 地球形状摄动
non-spherical gravity 非球形引力
non-spheroidicity of the Earth 地球非球形
non-spinning 不自旋
non-spinning body 非自旋体
non-spinning missile 非旋转导弹
non-spinning reentry space vehicle 非旋转的再入航天器
non-stable star 不稳定星
non-stall 未失速
non-stall region 非失速区
non-started flow 非开始流
non-stationary airloads 非平稳飞机机翼空气动力负荷
non-stationary channel 非平稳信道

non-stationary random process	非平稳随机过程
non-stationary time series	非平稳时间序列
non-structural	非结构件
non-structural mass	非结构化质量
non-swirling	非旋转
non-swirling case	非旋转的情况下
non-swirling condition	非旋转状态
non-synchronous multiplex system	非同步多路复用制
non-synchronous vibration	非同步的振动
non-system job	非系统作业
non-systematic code	非系统码
non-tactical missile	非战术导弹
non-tax oriented lease	非节税租赁（又称 conditional sale,实际是分期付款的购买）
non-terminal	非终止
non-terminal character	非终止符
non-thermal plasma	低温等离子体
non-thermal radiation	非热辐射
non-threshold logic	非阈逻辑
non-tilted dipole	非偶极子倾斜
non-topographic photogrammetry	非地形摄影测量
non-toxic agent	非毒性药物
non-traditional machining	特种加工
non-traffic stop	非业务停留
non-trivial solution	非平凡解,非无效解
non-uniform condition	不均匀的情况
non-uniform contrast	非均匀的对比
non-uniform exit	非均匀退出
non-uniform flow	不等速流
non-uniform grid	非均匀网格
non-uniform inflow	非均匀流
non-uniform nozzle-exit	非均匀喷嘴出口
non-uniform quantization	非均匀量化
non-uniform wind	非均匀风
non-uniformity	不均匀性
non-uniqueness	非唯一性,不唯一
non-volatile	1.非易失性的,永久的（计算机断电后仍能工作的,如非易失性存储器）2.不易挥发的
non-volatile memory	非易失性存储器
non-volatile particle	非易失性的粒子
non-volatile residues	非易失性残留物
non-volatile semiconductor memory	非易失性半导体存储器
non-volcanic geothermal region	非火山地热区
non-wetting liquid	非浸润性液体
non-wire wound potentiometer	非线绕电位器
non-zero angle	非零角
non-zero angle of attack	非零攻角
non-zero probability	非零概率
nonzero-sum	非零和情况
non-zero sum game model	非零和对策模型
non-zero value	非零值
non-zero-mean error	非零平均误差
noon-midnight meridian plane	正午子夜子午面
NOPR	Notice Of Proposed Rulemaking 建议制定规则的通知
norator	任意子
Norden gear	诺登装置
Norden sight	诺登瞄准具
nordo approach	不使用无线电设备进场着陆
NOREC	1. No Record 无记录 2. Not Received 未接收的
norm	规范,基准,定额,分配之工作量
normal	1.正常的,常态的 2.标准的,额定的,标称的 3.垂直的,正交的,法向的 4.法线
normal acceleration	法向加速度
normal alkane	正烷烃
normal attack offset bomb mode	正常间接瞄准轰炸工作状态（方式）
normal attitude	正常姿态
normal axis	法向轴,垂直轴
normal bifurcation	正常的分歧
normal burning	正常燃烧
normal burst	正常爆炸
normal case photography	正直摄影
normal category	标准级（飞机型号证书中的标准级,不包括特技飞行）
normal charge	正常装药,标准装药量
normal concentration	规定浓度,标准浓度,（克）当量浓度
normal configuration	正常式布局
normal direction	法线方向,正常方向
normal dispersion	正频散
normal distribution	正态分布
normal earth-fixed axis system	铅垂地面坐标系
normal electrical power supply	正常供电
normal ellipsoid	正常椭球
normal emittance	法向发射率
normal environment	正常环境
normal equation	法向方程式,正规方程
normal flight path	正常弹道
normal force	法向力
normal force coefficient	法向力系数
normal force coefficient derivative	法向力系数导数
normal freezing	正常凝固
normal fuel	正常燃油
normal fuel injection	正常的燃油喷射
normal galaxy	正常星系
normal glide	有利下滑,最小下滑角下滑

normal glow discharge　正常辉光放电
normal gravitational potential　正常引力位
normal gravity　法重力,标准重力,正常重力值
normal gravity formula　正常重力公式
normal gravity potential　正常重力位
normal gross weight　正常总重
normal hole　标称孔径
normal horsepower　额定马力
normal impact　正碰撞
normal impact effect　正常弹着效应
normal impingement　正常碰撞
normal incidence pyrheliometer　垂直入射绝对日射表
normal injection　正常注入
normal landing　正常落地,正常着陆
normal law　正态律
normal line　法线
normal load　正常负载
normal loop　正常舶斗
normal matrix　正规矩阵,正规阵,范真值表
normal mode　1.正常型 2.正常振荡型,固有振荡模,自然振荡
normal mode initialization　正交模初值化
normal mode method　正则模方法
normal mode shape　主振型
normal moveout correction　动校正
normal operating condition　正常工作条件
normal operation　正常运行,常规操作
normal outsize cargo　特大尺寸货物
normal paraffin　正链烷
normal polarity　正向极性
normal pressure drag　压差阻力
normal projection　正轴投影
normal propeller state　通常螺旋梁状态
normal rated power　额定功率
normal rating　额定状态
normal recoil length　正常后坐距离
normal recovery sequence　正常回收程序
normal return　正常返回
normal section　法截面,法截线
normal shock　正激波
normal shock density ratio　正常冲击密度比
normal shock train　正常冲击火车
normal shock wave　正激波
normal spin　正常螺旋(从正飞位置进入螺旋,杆舵中立或反舵就能改出的螺旋)
normal spiral galaxy　正常旋涡星系
normal state simulation　正常状态仿真
normal steering　法向导引
normal strain　正应变,法向应变

normal stress　正应力
normal temperature　正常体温,标准温度
normal temperature connector　常温连接器
normal turn　标准转弯(两分钟转360°)
normal value　正常值
normal vector　法向量,法向向量,法向矢量
normal velocity　正常速度
normal voltage　额定电压
normal wash　普洗
normal-air　标准空气
normal-air cabin　增压舱
normal-force　正交力,法向力
normal-force coefficient　法线力系数
normal-gravity　正常重力
normality　常态,当量浓度,规定浓度
normalization　规格化,归一化,标准化
normalization impulse　标准冲量
normalization state　归一化状态
normalized detectivity　归一化探测率
normalized impedance　归一化阻抗
normalized power spectrum　归一化功率谱
normalized quadrature spectrum　归一化求积谱
normalized rate　归一变化率
normalized total gravity gradient　归一化重力梯度
normalizing　正火
normally closed valve　常关阀门
normally off device　正常关断器件
normally on device　正常开启器件
normally open valve　常开阀门
normal-mode　简正方式
normal-mode frequency　简正模频率
Norsk Romsenter　挪威空间中心
North American Air Defense　北美防空司令部
North American route　北美航线
north celestial pole　北天极,上天极
north magnetic pole　磁北极
north mode　指北方式(多指地图显示系统中地图北向向上的工作方式)
north pointing inertial navigation system　指北式惯性导航系统
north polar sequence　北极星序
north polar spur　【天】北银极支
north pole　北极
northerly turning error　(罗差的)北转误差
Northern Cosmodrome　北部航天发射场
northern hemisphere　北半球
north-finding instrument　寻北器
north-south station keeping　南北位置保持
NOS　1. Night Observation Sight 夜间观察视野 2. Night

Observation Surveillance 夜间观察监视
NOSC NATO Operations Support Cell 北约作战支援小分队
nose 1.(飞行器的)头部 2.(机翼的)前缘部分
nose agents（gas） 喷嚏性毒剂
nose arming circuits 头部引信解除保险电路
nose arming wire （炸弹）头部引信保险钢丝
nose art 机头彩绘,机头装饰画
nose batten 飞艇头部(外表面的环形)加强条
nose bluntness 头部率直
nose cap 机头罩,桨帽
nose cone （火箭或飞弹）前锥体,头锥
nose dive 垂直俯冲,大角度俯冲
nose down 1.推机头(从平飞进入下滑或俯冲) 2.机头下沉姿态(不一定掉高度)
nose entry （从空气动力学及审美观点评定的）飞机头部形状记载
nose fairing 机头整流罩
nose fineness 头部细度
nose gear 前起落架
nose gearbox 前传动机匣,前减速器
nose graze fuze 弹头瞬发引信
nose inlet 头部进气道
nose irritant 喷嚏性毒气
nose landing 机头着陆
nose landing gear 前起落架
nose leg 前起落架支柱
nose over （后三点飞机的）拿大顶,飞机颠覆
nose over angle 防倒立角
nose percussion fuze 弹头着发引信,弹头碰炸引信
nose plane 机头小翼
nose plug （弹头）端头塞,端头帽
nose probe （反坦克弹的）头部探针
nose radius 球头半径,刀尖半径
nose ribs （机翼的）前缘翼肋
nose roll moment 前横滚力矩
nose slots 机头开缝（在机头高速气流形成的低压区部位开孔,以便排出冷却空气等流体）
nose spray 前飞破片
nose strake 机头边条
nose tip 尖锥端部
nose tow 机头牵引,前起落架牵引装置
nose up 拉机头,抬头,机头上仰
nose wheel 前轮,机头前轮,前舱
nose wheel shimmy 前轮摆振
nose whistler 鼻哨
nose wind cap （弹头）风帽
nose with control wing 控制翼弹头
nose with small asymmetry 小不对称弹头
nose-cone 机头锥,鼻锥,锥形机头
nose-cone light balloon decoy 模拟弹头的轻型气球假目标
nose-down 俯冲
no-slip 无滑移,无滑动边界
no-slip condition 无滑移条件,无滑动边界条件,动条件
NOSP Network Operations Support Plan 网络操作保障计划
no-spillage 运行平稳
NOSS 1. National Oceanic Satellite System 国家海洋卫星系统（美国）2. Navy Ocean Surveillance Satellite 海军海洋监视卫星 3. Nimbus Operational Satellite System 雨云实用气象卫星系统（美国）
no-strut 没有支撑
Notal not to all 毋需收发
NOTAM Notice To Airmen 飞行通报
Notam code 飞行通报代码
Notar no tail rotor 没有尾桨的（直升机）
notation 1.标记,符号,标志 2.注释,注记 3.计数法,符号表示法
notch 锯齿（槽）,切口（特指在翼型头部的弦向缺口或沟槽）
notch aerial 开槽天线
notch filter 陷波滤波器
notch sensitivity 缺口敏感度
notched cone nozzle 开槽锥形喷管
notching control 下凹控制
notching test 缺口试验,带谷试验
not-exceed-limit maintenance 不超限维修,不超期维修
notice to airmen 飞行人员通告
notifiable accident 必须报告的事故（如有人受伤,第三者财产受损失等）
notional datum 概念基准
notional item 概念项
notional item datum 概念项基准
not-operating time 不使用时间,待用时间
no-transgression zone 禁入区
NOTS 1. Naval Ordnance Test Station 海军军械试验站（美国）2. Nuclear Orbit Transfer Stage 核动力轨道转移级
not-so-stock 不太先进的（设备、系统等）
Nova 新星（美国海军导航卫星,是子午仪卫星的改进型）
Nova laser 新星激光器
nova-like variable 类新星变星
novel approach 新方法
Novoview 计算机产生图像系统的目视距离
NOWB Navy Oceanographic Weather Buoy 海军海洋天气浮标

no-wind position 无风位置
NOX 1.(一般指无大气情况下供人呼吸用的)氮氧混合气 2.氮氧化物(有害的燃烧排放物)
nox emission 氮氧化物排放
noy 诺伊(一种噪声分级标准)
nozpos nozzle positions 喷口位置
nozzle 1.喷管,喷嘴 2.风洞试验段进口 3.(轴流式燃气涡轮)涵道进口
nozzle admittance 喷嘴导纳
nozzle apex 喷嘴尖
nozzle apex angle 喷嘴顶角
nozzle area 喷嘴面积,喷嘴截面面积
nozzle area contraction ratio 喷管面积收缩比
nozzle area expansion ratio 喷管面积扩张比
nozzle area ratio 喷嘴面积比
nozzle base drag 喷管底阻
nozzle blade 导向器叶片
nozzle block (风洞的)试验段进口
nozzle boundary 喷嘴边界
nozzle box 喷嘴室,喷嘴箱
nozzle bucket (涡轮)喷嘴环,导向叶片环
nozzle cavity 喷管空腔
nozzle closure 喷管盖,喷口盖
nozzle closure opening pressure 喷管堵盖打开压强
nozzle computation 喷嘴的计算
nozzle concept 喷嘴的概念
nozzle contour 喷嘴型线
nozzle contraction ratio 喷管(流通面积)收缩比
nozzle design 喷管设计
nozzle diameter 喷嘴直径
nozzle diaphragm 1.润轮导向器 2.(固体火箭)喷管爆破隔膜
nozzle divergence half angle 喷管半扩张角
nozzle edge 喷嘴斜切边
nozzle efficiency 喷管效率
nozzle end 喷嘴端部
nozzle entrance 喷管入口
nozzle entrance plane 喷管入口平面
nozzle erosion 喷嘴腐蚀
nozzle exhaust 喷管排气
nozzle exit 喷管出口
nozzle exit area 喷管出口面积
nozzle exit diameter 喷嘴出口直径
nozzle exit plane 喷管出口面
nozzle exit pressure 喷管压力
nozzle exit velocity 喷嘴出口速度
nozzle expansion 喷管膨胀
nozzle expansion ratio 喷管膨胀比
nozzle flow 喷管流

nozzle force 射嘴接触力,射嘴推力,喷管接触力
nozzle geometry 喷管几何形状
nozzle guide vane 1.导向器叶片 2.导叶 3.涡轮静叶
nozzle head 喷嘴头
nozzle impulse 冲动喷嘴
nozzle initial divergence angle 喷管初始扩张角
nozzle inlet 喷管进口
nozzle insert 喷管衬套
nozzle interaction 喷嘴的交互
nozzle jet 气嘴,喷嘴
nozzle length 喷嘴长度
nozzle liner 喷管衬垫
nozzle loss 喷嘴损失
nozzle mass 喷管质量
nozzle material 喷嘴材料
nozzle model 喷头型号
nozzle of length-to-diameter ratio 喷嘴的长径比
nozzle outlet plane 喷嘴出口平面
nozzle performance 喷嘴性能
nozzle pivot point 喷管摆心
nozzle pivot point drift 喷管摆心漂移
nozzle polar fitting 喷管极线拟合
nozzle position indicator 喷口位置表
nozzle power 喷嘴功率
nozzle pressure 喷嘴压力,喷射压力
nozzle pressure losses 喷管压力损失
nozzle pressure ratio 喷管膨胀比,喷管落压比
nozzle ramp 喷管膨胀
nozzle reservoir 喷嘴储层
nozzle reservoir pressure 喷嘴储层压力
nozzle ring 喷管环,涡轮导向器
nozzle section 喷管段
nozzle segment 喷嘴弧段,喷嘴组
nozzle shape 喷管形状
nozzle side 舷侧喷口
nozzle side load 喷嘴边荷载
nozzle size 喷嘴尺寸
nozzle slew rate 喷管摆动速率
nozzle stagnation 喷嘴停滞
nozzle startup 喷嘴的启动
nozzle structure 喷嘴结构
nozzle surface 水面式喷口
nozzle survey 喷嘴的调查
nozzle swing moment 喷嘴摆动力矩
nozzle swing rate 喷嘴摆动速度
nozzle swivelling 可旋喷嘴
nozzle temperature 喷嘴出口温度
nozzle test 喷嘴试验
nozzle throat 喷管喉道

nozzle throat area 喷管喉道截面积
nozzle throat diameter 喷嘴喉部直径
nozzle throat inlet 喷嘴喉部入口
nozzle thrust 喷口推力
nozzle thrust coefficient (火箭发动机)喷口推力系数
nozzle tip 喷嘴
nozzle total pressure 喷嘴总压强
nozzle total temperature 喷嘴总温度
nozzle vibration 加装振荡的喷嘴
nozzle vibration power 喷嘴振动力量
nozzle wall 喷嘴壁,喷管壁
nozzle weight 喷嘴的重量
nozzle wide edge 喷嘴宽边
nozzle-end 喷嘴端部
nozzle-exit condition 喷嘴出口条件
nozzleless rocket motor 无喷管火箭发动机
nozzleless solid rocket motor 无喷管固体火箭发动机
nozzle-passing 喷嘴激振
nozzle-rim 喷嘴环
nozzle-to-plate distance 喷嘴板的距离
NPA 1. National Packaging Authority 国家包装管理局（英国）2. Notice of Proposed Amendment 建议修正的通知）
NPB 1. Nadge Policy Board 奈其系统政策局 2. Neutral Particle Beam 中性粒子束 3. Nuclear Powered Bomber 核动力轰炸机
NPC 1. NASA Procurement Circular 国家航空航天局采购通报 2. Notice of Proposed Change 建议更改通知单
NPD NASA Program Director 航空航天局计划负责人（美国）
NPDU Network Protocol Data Unit 网络协议数据单元
NPE Navy Preliminary Evaluation 海军原型机性能初步鉴定
NPG Nuclear Planning Group 核计划小组（北大西洋公约组织）
NPIC National Photographic Interpretation Centre 国家照片判读中心（美国）
NPL National Physical Laboratory 国家物理实验室（英国）
NPLOs NATO Production and Logistics Organizations 北大西洋公约组织生产与后勤组织
n-point correlation function n点相关函数
NPR 1. No Power Recovery 无动力回收 2. Noise Power Ratio 噪声功率比 3. Nozzle Pressure Ratio 喷管压力比
NPRM Notice of Proposed Rule Making 建议制定规则的通知
NPS 1. Nuclear Power Source 核能源 2. Numerical Plotting System 数字绘图系统
NPSH Net Positive Suction Head 净正抽吸压头
NPT Non-Proliferation Treaty 不扩散条约
NPTR National Parachute Test Range 国家降落伞试验场
NRA 1. Non-Repair Assembly 不可修理的配件 2. Nuclear Reaction Analysis 核反应分析
NRAG Non-metallics Research Advisory Group 非金属研究咨询组
NRAO National Radio Astronomy Observatory 国家射电天文台（美国）
NRC 1. National Research Council 全国科学研究委员会（美国、加拿大）2. Noise Reduction Coefficient 减噪系数
NRCC National Research Council of Canada 加拿大国家研究委员会
NRDC National Research Development Corporation 全国研究发展公司（英国）
NRDS Nuclear Rocket Development Station 核火箭研制站（美国）
NRE Nuclear Rocket Engine 核火箭发动机
NRIS Natural Resource Information System 自然资源信息系统
NRL Naval Research Laboratory 海军研究实验室（研究所）（美国）
NRLA 1. Network Repair Level Analysis 网络式修理级别分析 2.修理级别分析网络（模型）
NRO National Reconnaissance Office 国家（卫星）侦察局（美国）
NROSS Navy Remote Ocean Sensing System 海军海洋遥感系统
NRP 1. Navigation reference point 导航基准点 2. Normal Rated Power 标准额定功率
NRSA National Remote Sensing Agency 国家遥感局（印度）
NRSC National Remote Sensing Centre 国家遥感中心（英国）
NRT Near Real Time 准实时,近实时
NS Nacelle Station 吊舱站位
ns nanosecond 毫微秒或纳秒
NSA 1. National Security Agency 国家安全局（美国）2. Node Switching Assembly 节点转接装置
NSBF National Scientific Balloon Facility 国家科学探测气球中心（美国）
NSC 1. National Security Council 国家安全委员会（美国）2. National Space Council 国家空间委员会（美国）3. Norwegian Space Centre 挪威航天中心
NSCA National Safety Council of America 美国国家安全委员会

NSCI NASCOM System Control Interfaces 航空航天局通信网系统控制接口(美国)

NSDD National Security Decision Directive 国家安全决策条令(美国)

NSE 1. Network Systems Engineer 网络系统工程师 2. North Steaming Error 向北航行误差

nsec nanosecond 毫微秒或纳秒

N-sector N 扇区(无线电指向标能听到摩尔斯电码 N 的扇区)

NSEN Network Simulations Engineer 网络模拟工程师

NSF National Science Foundation 国家科学基金会(美国)

NSG Nuclear Suppliers Group 核供应国集团

NSGr 夜间近距支援大队

NSI NASA Standard Initiator 航空航天局标准点火器(美国)

NSIA National Security Industrial Association 国家安全工业协会(美国)

NSIS NASA Software Information System 航空航天局软件信息系统(美国)

NSN National Signaling Network 国内信令网

NSO National Service Officer 现役军官

NSP 1. NASA Support Plan 航空航天局保障计划(美国) 2. Navigational Satellite Program 导航卫星计划

NSR 1. Narrow Sector Recorder 窄带记录器 2. No Scheduled Removal 不定期拆卸(件)

NSRP National Search and Rescue Plan 国家搜索与救援计划(美国)

NSS 1. National Seismic Station 国家地震站(美国) 2. Navigation Satellite System 导航卫星系统 3. Near Source Simulation 近源模拟 4. Neutral Shift Sensor 中立变换探测器

NSSC 1. NASA Standard Spacecraft Computer 航空航天局标准航天器计算机(美国) 2. Naval Ship System Command 海军舰船武器系统部(美国)

NSSCC National Space Surveillance Control Center 国家空间监视控制中心(美国)

NSSDC National Space Science Data Center 国家空间科学资料中心

NSSL Normal Steady State Limit 正常稳定状态极限

NSTI NASCOM Simulation Traffic Interfaces 航空航天局通信网模拟通信接口

NSTP National Space Technology Programme 国家空间技术规划(英国贸易和工业部)

N-strut N 形翼间张线,N 形构架

NSTSPO National Space Transportation System Program Office 国家空间运输系统计划处(美国)

NSV Negative Sequence Voltage 反相电压

NSW Nominal Specification Weight 额定要求质量

NSWC Naval Surface Weapons Center 海军地面武器中心(美国)

NSWP Non-Soviet Warsaw Pact 非苏联华沙条约的部队或国家

nt nit 尼特(表面亮度单位)

NTAS Norad Tactical Autovon System 北美防空司令部战术自动话务网系统

NTB National Test Bed 国家试验台(美国模拟星球大战计划的计算机和视频操作系统)

NTC National Training Centre 国家训练中心

NTCC Nimbus Technical Control Center 雨云气象卫星技术控制中心(美国)

NTD Neutron Transmutation Doping 中子嬗变掺杂

NTDS Naval Tactical Data System 海军战术数据系统

NTF 1. National Test Facility 国家试验设施 2. New Tactical Fighter 新战术战斗机 3. No Trouble Found 未出故障

NTFWTC NATO Tactical Fighter and Weapons Training Centre 北大西洋公约组织战术战斗机和武器训练中心

NTIA National Telecommunications and Information Administration 国家电信和信息管理局(美国)

NTIS National Technical Information Service 国家技术情报处

NTM National Technical Means 国家技术手段

NTMV National Technical Means of Verification 国家技术核查手段

NTO Normal Take Off 正常起飞

NTOS No Time On Station 1. 来不及到达指定截击位置 2. 来不及作好攻击准备

NTRL NASA Technology Readiness Level 航空航天局技术准备状态等级(美国)

NTS 1. Navigation Technology Satellite 导航技术卫星 2. Negative Torque Signal 负转矩信号 3. Nightvision Targeting System 夜视瞄准系统,夜视目标捕获系统

NTSB National Transport Safety Board 国家运输安全委员会(美国)

NTTF 1. National Tracking and Test Facility 国家跟踪与测试中心(美国航空航天局) 2. Network Test and Training Facility 网络测试与训练中心

n-tuple n 元组

NTV Network Television 网络电视

NU Navigation Unit 导航组件

nuclear airburst 空中核爆炸

nuclear aircraft 1. 核武器飞机 2. 核动力航空器,核动力飞机

nuclear auxiliary power unit system 辅助核电源系统

nuclear battery 核电池

nuclear battery concept 核电池的概念

nuclear biological and chemical protective suit 防核生化服
nuclear biological chemical and damage training 核生化和破坏训练
nuclear bomber 核轰炸机
nuclear burst detection instrument 核爆炸探测仪
nuclear burst observation post 核爆观测哨所
nuclear burst optical radiation 核爆热光辐射
nuclear burst power 核爆炸威力
nuclear cloud 核爆炸云
nuclear column 核爆炸水柱
nuclear cosmochemistry 核宇宙化学
nuclear cross section 核反应截面
nuclear defence 核防御
nuclear delivery system 核武器投射系统
nuclear depth bomb （反潜）深水核炸弹
nuclear detection satellite 核爆炸探测卫星
nuclear detonation detection and reporting system 核爆炸探测报知系统
nuclear directed-energy weapon 定向能核武器
nuclear dud （口语）未爆炸核弹
nuclear electromagnetic pulse 核电磁脉冲
nuclear electromagnetic pulse coupling 核电磁脉冲耦合
nuclear electronics 核电子学
nuclear emulsion 核感光乳剂，核乳胶
nuclear energy 核能
nuclear environment 核环境
nuclear exchange 互投核弹
nuclear explosion center 核爆中心
nuclear explosive 核装药，核炸药
nuclear fission 核裂变，原子核分裂
nuclear flash blindness 核闪光盲
nuclear force 核力
nuclear fuel 核燃料，原子核燃料
nuclear fusion 核聚变
nuclear gyro 核子陀螺
nuclear hardening 抗核加固
nuclear hardness level 核加固程度，核加固等级
nuclear hardness specification 核加固技术规范
nuclear incident 核事件
nuclear interceptor 核拦截器
nuclear magnetic resonance 核磁共振
nuclear magnetic resonance computerized tomography 核磁共振计算机断层成像
nuclear magnetic resonance detector 核磁共振探测器
nuclear magnetic resonance gyro 核磁共振陀螺
nuclear magnetic resonance spectrometer 核磁共振谱仪
nuclear nations 核国家
Nuclear Non-proliferation Treaty 核（武器）不扩散条约
nuclear ordnance items 核军械项
nuclear parity 核均势
nuclear power system 核电源系统
nuclear powered carrier 核动力航空母舰
nuclear powered satellite 核动力卫星
nuclear propulsion 核推进（利用核变能量产生推力的推进方式）
nuclear pumping 核泵浦
nuclear radiation 核辐射
nuclear radiation environment 核辐射环境
nuclear radiation levelmeter 核辐射物位计
nuclear radiation thickness meter 核辐射厚度计
nuclear reactor 核反应堆
nuclear reactor thermoelectric generator 核反应堆温差发电机
nuclear rocket 核火箭
nuclear round 核（炮）弹
nuclear safety line 核辐射安全线
nuclear stalemate 核僵局
nuclear stethoscope 核听诊器
nuclear surface burst 地面核爆炸
nuclear survivability 核生存能力（在核环境下的）
nuclear test 核试验
nuclear transmutation energy 核转变能
nuclear underground burst 地下核爆炸
nuclear vulnerability 核易损性
nuclear warhead 核弹头，核战斗部
nuclear weapon 核武器
nuclear weapon damage effect 核武器毁伤效应
nuclear weapon debris 核武器碎片
nuclear weapon degradation 核武器降级
nuclear weapon effects research 核武器效能研究
nuclear weapon employment time 1.核武器反应时间 2.投放核武器需用时间
nuclear weapon exercise 核武器演习
nuclear weapon maneuver 核武器演练
nuclear weapon package 核武器编组
nuclear weapon subpackage 核武器分组
nuclear weapon(s) accident 核武器事故
nuclear weapons acceptance inspection 核武器验收
nuclear weapons storage facility 核武器储存设施
nuclear weapons surety 核武器安全保障（措施）
nuclear yield 核威力，核当量
nuclear/heater propulsion 核能加热器推进装置
nuclear-hardened 抗核加固的
nuclear-light-bulb rocket 核光子火箭
nucleate 有核的
nucleate pool 核态池
nucleation 成核

nucleation site 成核位置
nucleation theory 成核作用理论,凝结理论
nucleogenesis 核生成
nucleon 核子
nucleonics 应用核子学,原子核工程
nucleosynthesis 核合成
nucleus 1.原子核 2.(大气水蒸气凝结及冻结的)核心微粒 3.心,核心,中心
nucleus counter 核子计数器
nuclide 核素
nude gauge 裸规
nudger (飞机上的)自动推杆器
nudging 轻推,轻轻接触
nugget 〈美国口语〉塞紧器,气密装置
NUI Network User Identification 网络用户识别
nuisance alarm 干扰警报
nuisance malfunction 扰乱性失常
null 1.(无线电罗盘接收机天线在)无声区位置,零信号区 2.直接升力控制角度调定在扰流器振荡位置(约7°) 3.(空间的)无重力点 4.零(位),空的
null adjustment 零位调整,零校准装置
null argument 空变元
null bias 零点偏差
null character 空白字符
null coil 空线圈
null depth of difference beam 差波束零(值)深(度)
null electric current 零电流
null ephemeris table 零值星历表
null frame 零帧
null hypothesis 零假设
null link 空连接
null lobes 零瓣
null method of measurement 零位测量方法
null miss 零脱靶距离
null motion 零速运动
null offset 零位偏移
null position 零位,零点位置,零位置
null space 零空间
null valtage 零位电压
null vector 零矢量
nullator 零子
nullcline 零斜率线
nullor 零任偶
nullspace 零空间
null-space injection 零空间注入
null-space measurement 零空间测量
null-space measurement update 零空间测量更新
null-steering antenna system 零控制天线系统
null-type 指零式

null-type direction measuring system 零点型测向系统
number airfoil 多翼型
number approximation 近似数
number assumption 数的假设
number capability 钻进能力数
number case 案件编号,箱号,症例番号,案件号码
number component 组分数
number contour 标高列注记,等高线海拔注记
number density 数密度
number density of molecules 分子数密度
number density value 数密度值
number dependence 数量的依赖
number distribution 数目分布
number effect 数目效应
number flight regime 数字飞行状态
number flow 流速,流数
number increase 数量的增加
number level 水平数
number of analysis 数字系统分析
number of anomaly 异常数
number of arrival 到货数量
number of bias 偏离数量
number of blade 叶片数量
number of block 拉拔道次,加热块数
number of burn 燃烧的数量
number of candidate 人数要求
number of cell 细胞数
number of channel 信道数量
number of chemical reaction 化学反应
number of chromosome 染色体数目
number of coil 叠层层数
number of collision 碰撞数量
number of conflict 冲突点数目
number of connection 连接数量
number of control point 控制点数量
number of crossover 交叉数
number of cycle 工作循环次数
number of data point 数据点数量
number of debris 碎片数量
number of decision 决策变量数量
number of decision variable 决策变量数量
number of degree 度数
number of design 设计变量数
number of design variable 设计变量数
number of device 设备编号
number of disturbance 干扰的数量
number of element 单元数
number of emitter 发射器的数量
number of evacuee 撤离数量

number of facet　小(平)面数量
number of factor　因素数量
number of failure　故障数量
number of false alarm　假警报数量
number of flap　副翼数量
number of flight　飞行架次
number of fragment　碎片数量
number of function　函数数量
number of function call　函数调用数量
number of grid　网格数量
number of grid point　网格点数量
number of hardware set　硬件设置数量
number of hidden layer　隐层数
number of inequality　不平等数
number of iteration　迭代数
number of Knudsen　努森数
number of laser pulse　激光脉冲数
number of load　载荷数
number of magnetic ring　磁环数量
number of measurement　测量数量
number of mode　模数
number of mole　摩尔数
number of multigrid cycle　多栅的周期数
number of natural mode　自然模式的数量
number of neuron　神经元数量
number of night of optical tracking　晚上的光学跟踪数
number of nodal diameter　节的直径数量
number of node　叉点数
number of orifice　孔数
number of oscillation　振荡次数
number of parameter　方法参数的总数
number of pass　流程数
number of passenger　乘客人数,额定乘客人数
number of phase　相数
number of pivot　支点数量
number of point　节点的数量
number of pole　极数
number of processor　共有人数
number of pulse　脉波数
number of repeated experiment　重复实验数
number of revolution　旋转次数,转数
number of ring　振铃次数
number of rotor　转子导体数目
number of row　行数
number of sample　样品号,加样数
number of soft constraint　软约束数量
number of spacecraft　宇宙飞船数
number of spectral line　谱线数
number of target　目标配数目

number of transponder　应答器数
number of unknown　未知数
number of user　用户数量
number of variable　变量数
number of vertical rivet　垂直铆钉数
number of way　方式
number of year　年数
number range　数值范围
number regime　编号制度
number region　数字区域
number test　数字测试
numerator coef　分子系数
numeric coding　数字编码
numeric computing　数值计算
numeric display　数字显示
numerical algorithm　数值算法
numerical analysis　数值分析
numerical and experimental result　数值和实验结果
numerical aperture　数值孔径
numerical approach　数值方法
numerical computation of turbulent flow　湍流数值计算
numerical continuation　数值延拓
numerical control machining　数控加工
numerical convergence　数值收敛
numerical correlation　数字相关
numerical cosmology　数值宇宙学
numerical differentiation　微分计算,数值微分,数值微分法
numerical diffusion　数值扩散
numerical dissipation　数值耗散
numerical error　数值误差
numerical flow　数字流
numerical grid　数值网格
numerical instability　数值不稳定性
numerical integration　数值积分,积分计算
numerical integration method　数值积分方法
numerical iteration　数字迭代
numerical lifting-line　数值升力线
numerical method　数值方法,数值计算法
numerical minimization　数值极小化
numerical model　数值模式,数值模型
numerical modeling　数值模拟
numerical noise　数值噪声
numerical optimization　数值优化
numerical orbit integration　轨道数值积分法
numerical parameter　数值参数
numerical photogrammetry　解析摄影测量
numerical prediction　数值预测
numerical procedure　计算方案

numerical propagation 数字传播
numerical rebuilding 数字改造
numerical reconstruction 数值再现
numerical result 数值结果
numerical scheme 数值化方案
numerical schlieren 数值纹影
numerical simulation 数值模拟
numerical simulator 数值模拟模型
numerical solution 数值解,近似解
numerical solution of motion equation 运动方程数值解
numerical stability 数值稳定性
numerical sweep 数字扫描
numerical technique 数值技术
numerical test matrix 数值测试矩阵
numerical tool 数控
numerical weather 数字气象
numerical-continuation 数值延拓
numeric-simulation-based method 基于数值仿真方法
numerous simulation 大量的仿真
numerous targets vicinity 无数的目标位置附近
nurse balloon 补气囊
NUSAT Northern Utah Satellite 北犹他卫星
Nusselt number 努塞尔数(符号为 Nu)
Nut near-unity probability 概率接近于 1
nut plate 托板螺帽
nut runner 1.千斤顶大螺帽 2.螺帽拧紧(机动)工具
nutating antenna 盘旋馈入天线
nutation 1.章动 2.卫星自旋轴与角动量轴不重合时产生的自旋轴绕角动量轴的摆动 3.自转天体的自转轴在空间的周期性微小变化 4.旋转体转轴的摆动
nutation damper 章动阻尼器
nutation experiment 转头运动实验(对植物进行的一种实验)
nutation frequency 章动频率
nutation in longitude 【天】黄经章动
nutation in obliquity 【天】交角章动,倾角章动
nutation in right ascension 【天】赤经章动
nutation mode 垂头式
nutation sensor 章动敏感器
nutcracker 1.(起落架的)防扭臂,扭力臂 2.机身自动定向的垂直或短距起落飞机
NUTI NASCOM User Traffic Interfaces 航空航天局通信网用户通信接口(美国)
NV thrust Nominal Vacuum Thrust 额定真空推力,标称真空推力
NVG Night Vision Goggles 夜视镜
NVG/HUD Night Vision Goggle/Head Up Display 夜视镜/平视显示器
NVIS Near Vertical Incident Skywave 近垂直入射天波

NW Nuclear Weapon 核武器
NWA Nuclear Weapon(s) Accident 核武器事故
NWC 1. Naval Weapons Center 海军武器中心(美国) 2. National War College 国家军事学院(美国)
NWDS Navigation and Weapon Delivery System 导航与武器投放系统
NWEF Naval Weapons Evaluation Facility 海军武器鉴定处(美国)
NWEO Nuclear Weapons Employment Officer 核武器使用军官
NWER Nuclear Weapon Effects Research 核武器效能研究
NWET Nuclear Weapon Effects Test 核武器效能试验
NWI Nuclear Weapons Inventory 核武库,核武器总数
NWP Naval Warfare Publication 海战出版物
NWR No Weight Restriction 无重量限制
NWS 1. National Weather Service 国家气象处 2. North Warning System 北方警戒系统 3. Nose Wheel Steering 前轮转向
NWSC National Weather Satellite Center 国家气象卫星中心(美国)
NWSS National Weather Satellite System 国家气象卫星系统
NWSSG Nuclear Weapon System Safety Group 核武器系统安全组
NX Nonexpendable Item 非消耗品
Nycote 尼龙拉克油保护层(专利名称)
nyctohemeral activity 昼夜活动
Nyl 尼龙,锦纶
Nylafil 玻璃纤维增强尼龙(商品名)
nylon 耐纶,酰胺纤维,尼龙
nylon fiber-reinforced phenolics 尼龙纤维增强酚醛塑料
nylon lashing 尼龙绳索
nylon strap 尼龙索
nylonic acrylic fiber 耐纶丙烯腈系纤维
Nyquist criterion 奈奎斯特判据
Nyquist diagram 奈奎斯特图
Nyquist frequency 奈奎斯特频率
Nyquist rate 奈奎斯特速率
nystagmograph 眼球震颤描记器
nystagmus 眼球震颤
NZ New Zealand 新西兰
NZAPA New Zealand Airline Pilots Association 新西兰民航飞行员协会
NZCA NZ College of Aviation 新西兰航空学院
NZGA NZ Gliding Association 新西兰滑翔协会
NZMAA NZ Model Aeronautical Association 新西兰航模协会

O

O 1. Orange 橙色 2. Open 打开 3. Operating 操作,运行 4. Organizational 组织的 5. Oxygen 氧气

OA 1. Office Automation 办公自动化 2. Olympic Airline 奥利匹克航空公司 3. On Account 暂付,记账 4. Operational Analysis 操作分析,运行分析 5. Optical Amplifier 光放大器 6. Overhead Approach 飞越机场进近 7. Option Attribute 选项特性

OAD 1. Orbital Aerodynamic Drag 轨道气动阻力 2. Output Analog Discrete 输出模拟离散(信号)

OADD Optically Amplified Direct Detection 光放大直接检测

OADG Open Architecture Development Group 开放体系结构开发组

OADM Optical Add-Drop Multiplexer 光分插复用器

OADMT Overall Aircraft Design Management Team 整体飞机设计管理工作组

OADS 1. Omnidirectional Air Data System 全向大气数据系统(直升机用) 2. One Atmosphere Diving System 常压潜水系统

OAE 1. Office Automation Equipment 办公室自动化设备 2. Optimized After Erosion 侵蚀后外形优化(指直升机旋翼桨叶的叶型)

OAFS Open Apron, Free Standing 自由停放空旷停机坪(指没有登机桥等设备,因而无固定停机地点)

OAFU Observers Advanced Flying Unit 先进飞行装备观察员

OAG Official Airline Guide 官方航线指南

OAI Omnidirectional Airspeed Indicator 全向空速指示器

OAM Office of Aviation Medicine 航医处(美国联邦航空局)

OAM&P Operation, Administration, Management & Provision 运行、管理、维护和供给

OAMC Operation, Administration and Maintenance Center 运行、管理和维护中心

OAMC-MF OAMC Management Function 运行、管理和维护中心管理功能

OAMC-OS OAMC Operation System 运行、管理和维护中心运行系统

OAMS 1. On-board Asynchronous Messaging Service 机载异步信息服务 2. Optical Attenuation Measuring Set 光衰耗测量仪

OANS 1. Observers Air Navigation School 观察员空中领航学校 2. On-board Airport Navigation System 机载机场导航系统

OAP 1. Office of Aircraft Production 飞行生产处 2. Oversale Auction Plan (机票)超售竞卖计划

OAPEC Organization of Arab Petroleum Exporting Countries 阿拉伯石油输出国组织

OAPP Office of Aviation Policy and Plans 航空政策与计划处(美国联邦航空局)

OAR 1. Off-Axis Rejection 转偏离抑制 2. Office of Aerospace Research 宇航研究署 3. Optically Amplified Regenerator 光放大再生器 4. Overhaul And Repair 翻修和修理

OARN Off-Airway R-Nav 航路外区域导航

OART Office of Advanced Research & Technology 先进研究和技术处(美国国家航空航天局)

OAS 1. Obstacle Assessment Surface 障碍物评价面 2. Oceanic Automation System 海洋自动化系统 3. Offensive Air Support 进攻性空中支援 4. Office Automatic System 办公自动化系统 5. Oman Aviation Services 阿曼航空服务公司 6. Onboard Authentication Service 机上(软件)验证服务 7. Operational Announcing System 现用通告系统 8. Optical Access System 光接入系统 9. Originating Access Situation 始端接入情况

OAT 1. Operating Ambient Temperature 工作环境温度 2. Operation Acceptance Test 使用验收试验 3. Operational Air Traffic 可供使用的空中交通 4. Optional Auxiliary Terminal 可选辅助终端 5. Outside Air Temperature 外界大气温度

OATP On Aircraft Test Procedure 飞机测试程序

OATS 1. Optical Amplifier Transmission System 光放大器传输系统 2. Optical Attitude Transfer System 光学姿态变换系统 3. Optimum Aerial Targeting Sensor 最佳空中目标瞄准传感器 4. Orbit and Attitude Tracking Subsystem 轨道与姿态跟踪子系统 5. Oxford Air Training School 牛津飞行训练学校

OB 1. Object 物品,反对,目的 2. Objection 反对,异议 3. On-Board 有搭机,在机上

OBA 1. Off-Bore sight Angle 目标偏离武器轴线的角度 2. Oxygen Breathing Apparatus 供氧设备

OBC 1. On-Board Checkout 机上检查 2. Optical Barrel Camera 筒式光学照相机(SR-71飞机上的特种侦察传感设备)

OBCE　On-Board Control Equipment　机载控制设备

OBCO(S)　On-Board Cargo Operations (System)　机载货物作业（系统）

OBD　1. Optical Beam Deflection　光束偏转　2. Optical Branching Device　光分路器

OBDMS　On-Board Data-Monitoring Systems　机载数据监控系统

OBE　1. Off-Board Expendables　非机载消耗品　2. Outerback End　外后端　3. Overtaken By Events　偶然出现的

OBECO　Out-Board Engine Cut Off　外侧发动机熄火

Oberon　天卫四，控制炸弹的雷达系统

OBEWS　1. On-Board Electronic Warfare Simulation　机上电子战模拟（模拟从尾后警戒雷达或其他探测器见到的外部威胁）2. On-Board EW System(s)　机载电子对抗（电子战）系统

OBI　1. Omni Bearing Indicator　全向无线电导航指示器　2. Optical Beat Interference　光差拍干扰

oblate　扁球状的，扁（圆）的

oblateness　1. 扁率　2. 扁圆形

oblateness effect　章动效应

oblique　斜的，倾斜的

obliquity of the ecliptic　黄道斜度

OBND　Out Bound　离场，背台

OBOE　Offshore Buoy Observing Equipment　近海浮标观测系统

Oboe　"欧波"雷达领航和仪表轰炸系统

OBOGS　On-Board Oxygen Generation System　机载制氧系统

OBP　On-Board Processor　机载处理机

OBS　1. Observe　观察，观测，遵守　2. Observer　观察员，观测员　3. Observation　遵守，观测，观察　4. Omni Bearing Selection　全方位选择　5. Obsolete　作废的，陈旧的　6. Omni Bearing Selector　全方位选择器　7. On-board Simulation　机载模拟器　8. Orbital Bombardment System　轨道轰炸系统　9. Outdoor Base Station　室外基站

OBSC　Obscured　不清楚的，阴暗不明的

obscuration　灯火管制，遮蔽

observability　可观测性

observables　可观测量

observation　遵守，观测，观察

observation equation　观测方程

observatory　天文台，气象台，瞭望台

observe　观察，观测，遵守

observed altitude　观测高度

observer　1. 观测器　2. 观察员

obsolescence　正在被废弃，逐渐淘汰

obstacle　障碍，妨碍，障碍物

obstacle avoidance laser radar　激光防撞系统

obstacle clearance　障碍物清除

obstacle clearance altitude　超障净空高度

obstacle clearance height　超障净空高度

obstacle free airspace　机场净空

obstacle limiting weight　越障限制重量

obstacle restrictive surface　障碍物限制面

obstruction　障碍物，阻塞，堵塞

obstruction clearance surface　障碍物净空面

obstruction light　障碍灯

obstruction marker　障碍物标志

obtain　获得，得到

obturator　封闭器，紧塞器

obturator ring　紧塞圈

OC　1. Object Code　目标代码　2. Ocean　海洋　3. Of Course　当然　4. On Condition　以……为条件　5. On Course　在航道上　6. Open Circuit　开路　7. Operating Certificate　运行合格证　8. Operations Channel　操作信道　9. Optical Circulator　光环行器　10. Order Canceled　订单取消　11. Order Card　指令卡　12. Overcharge　过载，超载　13. Over Current　超电流　14. Overcurrent　过流　15. Our Cable　我方电报　16. Overload Control　过载控制　17. Oversale Cost（机票）过售损失　18. Optical Channel　光纤通道

OCA　1. Obstacle Clearance Allowance　允许越障高度　2. Obstacle Clearance Altitude　超障净高度　3. Oceanic Control Area　海洋管制区　4. Offensive Counter-Air　进攻性反航空兵作战　5. Optical Channel Analyzer　光信道分析仪

OCAMS　On-board Check-out And Monitoring System　机上检查与监控系统

OCB　1. Oil Circuit Breaker　滑油断电器　2. Overload Circuit Breaker　超载断电器

OCC　1. Occupied　占据，占领　2. Occulting Light　明暗灯　3. Operational Control Centre　航行管制中心　4. Optical Cable Connector　光缆连接器　5. Optical Cross Connect　光交叉连接　6. Outer Communication Channel　外部通信信道　7. Outer Communication Control　外部通信管理　8. Orthogonal Convolutional Code　正交卷积码　9. Other Common Carrier　一般电信公司，普通运营商　10. Ozone Catalytic Converter　臭氧转换器

occluded front　【气】锢囚锋，封锁前线

occlusion　1. 吸气酐　2. 锢囚

occultation　【天】掩星

occulting　掩蔽杆

occupancy　1. 占有，占用　2. 载运人数（乘客与机组人员总数）3. 占有率

occupant　1. 乘员　2. 占有者，居住者

occurrence　出事，发生，出现

OCDMS On-board Checkout and Data Management System 机载检查和数据管理系统

OCE Ocean Color Experiment （美国航天飞机上的）海洋水色实验

ocean 海洋

ocean observation satellite 海洋观测卫星

ocean satellite 海洋卫星

ocean surveillance satellite 海洋侦察卫星

oceanic clearance 海洋放行许可

OCF Operational Control Facility 运行控制设备，操作控制和显示设备

OCH Obstacle Clearance Height 越障高度

OCI Outside of Clearance Indicator 净空界外指示器（微波着陆系统）

OCIP Option Critical Installation Point 选型关键安装点

OCL 1. Obstacle Clearance Limit 超障净空极限，越障限制，越障安全高度界限 2. Operation Control Language 运行控制语言 3. Operational Cable Load 光缆操作负荷 4. Operational Check List 操作检查单 5. Operational Control Level 操作管理水平 6. Operator Command Language 操作员命令语言 7. Optical Confinement Layer 光限制层 8. Optimum Cruising Level 最佳巡航高度 9. Overall Connection Loss 总连接损耗 10. Outer Compass Locator 外罗盘定位台

OCM 1. Ocean Color Monitor 海色监视仪 2. Oceanic Clearance Message 海洋放行许可信息 3. On-Chip Monitor 芯片上监测器 4. On-Condition Maintenance 视情维修 5. Ongoing Call Management 去话呼叫管理 6. Optical Counter Measures 光学干扰，光学对抗 7. Ortho Conjugate Mirror 正交共轭镜 8. Out of Control Months 不可控制月份

OCR 1. Oceanic Control Region 海上管制区 2. Oil Circuit Recloser 滑油路自动开关 3. On-Condition Replacement 视情更换（零件） 4. Operation Control Reports 运行管理报告 5. Optical Character Reader 光学字符阅读机 6. Optical Character Reading 光学字符读取，光学特性读取 7. Optical Character Recognition 光学字符识别 8. Optical Code Reader 光代码读取机 9. Order Control Record 指令控制记录 10. Overhaul Component Requirement 部件大修要求 11. Over-current Relay 过电流继电器

OCS 1. Obstacle Clearance Surface 超障净空面 2. Ocean Color Scanner 海洋水色扫描仪 3. Off Center Sweeps 偏心扫描 4. Officer Candidate School 预备军官学校（美） 5. On-board Checkout Systems 机上检查系统 6. Operating Control System 运行指挥系统 7. Operational Control Segment 操作控制段 8. Operations Computing System 操作计算系统 9. Optical Character Scanner 光字符扫描器 10. Optical Cohenrent System 相干光系统 11. Optical Contrast Seeker 光学对比导头 12. Optical Control System 光学控制系统 13. Optimum-Cost Speed 最佳经济速度 14. Orbit Control System 轨道控制系统 15. Originating Call Screening 发端去话筛选 16. Overseas Communications Service 海外通信业务处 17. Overall Customer Satisfaction 顾客总体满意度

OCT 1. October 十月 2. Octane Number 辛烷值 3. Optical Current Transducer 光流换能器

octa 1. Oceanic Control Area 海洋管制区 2. Outside Control Area 在管制区以外

octal base 八脚管座（电子设备的标准座）

octant 八分仪（能测量90度角的气泡六分仪）

octave 倍频程，八音度

octet 八位位组

OCTS Optical Cable Transmission System 光缆传输系统

OCU 1. Office Channel Unit 局内信道单元 2. Operational Control Unit 操作控制单元

ocular 接目镜

oculogravic（optogravic）illusion 眼重力错觉，重力异常

oculogyral（optogyral）illusion 眼旋转错觉

OD 1. Outside Diameter 外径 2. Outside Dimensions 外围尺寸 3. Operations Directive 操作指示，作战指令 4. Optical Demultiplexer 光解复用器 5. Optical Detector 光检测器 6. Order 指令 7. Ordnance Data 军械数据

ODA 1. Operational Data Analysis 操作数据分析 2. Overseas Development Administration 海外开发管理局（英国）

ODAPS 1. Oceanic Display & Planning System 远洋飞行显示和规划系统 2. Oceanic Display & Processing System 海洋显示与处理系统 3. Over Ocean Display and Positioning System 越海显示及定位系统

ODAR Organizational Designated Airworthiness Representive 机构指定适航代表

ODB Object Database 目标数据库

ODBC 1. Open Data Base Connection 开放数据库互连 2. Open Data Base Connectivity 开放数据库互连

ODC 1. Oceanographic Data Center 海洋资料中心（苏联） 2. Office of Defense Cooperation 防御作战协同指挥室 3. On Deck 甲板着陆 4. Operations Data Control 操作数据控制 5. Operational Discrete Command 操作离散指令 6. Operational Dispatch Center 运行签派中心 7. Optical Data Corrector 光学数据校正器 8. Other Direct Cost 其他直接成本

ODE 1. Object Database and Environment 目标数据库

与环境 2. Open Development Environment 开发环境 3. Orientation Dependent Etch 定向腐蚀剂，定向相关刻蚀

ODEB Option Data Entry Builder 选项数据登陆员

ODF 1. Optical Distribution Frame 光纤配线架 2. Origin-Destination Fares 全程票价

Odin 奥丁卫星（瑞典研制的射电天文和大气物理研究小卫星）

ODL 1. Object Definition Language 目标确定语言 2. Oceanic Data Link 海洋数据链 3. Optical Data Link 光数据链

ODLF On-board Data Load Function 机载数据装载功能

ODLI Open Data Link Interface 开放式数据链路接口

ODM 1. Object Data Manager 对象数据管理程序 2. Office of Defense Mobilization 国防动员办公室 3. One-Day Mission 一日（飞行）任务 4. Operating Data Manual 使用数据手册，运行数据手册 5. Operational Data Message 操作数据通报 6. Operational Development Model 使用发展模型

odograph 航向航程自记仪

odometer 数字仪表

ODR 1. Order 订货，订票，命令 2. Origin Dependent Routing 由发端位置选路

ODS 1. Obstacle Detection System 障碍探测系统 2. On-board Data System 机载数据系统 3. Open Data Service 开放式数据服务 4. Operational Discrete Status 操作离散状态 5. Optical Data System 光数据系统 6. Optical Display Sight 光学显示瞄准具 7. Oxide-Dispersion Strengthened 氧化扩散强度的

ODT 1. On-line Debugging Technique 在线调试技术 2. Optical Data Transmission 光数据传输 3. Optical Distance Terminal 光纤远程终端 4. Outside Diameter Tube 管路外径

ODU 1. Out-Door Unit 室外设备 2. Optical Display Unit 光纤显示单元

ODW Optimum-Drag Windmilling（旋翼的）最小阻力风转

OE 1. Open End 开口端，开路端 2. Operating Envolope 使用包线 3. Operating Experience 使用经验 4. Operational Error 操作误差 5. Operator Error 操纵员误差 6. Opportunity Evaluation 机遇评价 7. Opportunity & Evaluation 机会和评估 8. Opt-Electronic 光学电子的 9. Option Expression 选型表达 10. Output Enable 允许输出 11. Over Excitation 过激励 12. Overrun Error 超限运转错误

OEB Operational Engineering Bulletin 运行工程通报

OECD Organization for Economic Cooperation & Development 经济合作与发展组织

OEI One Engine Inoperative 一台发动机不工作

OEIC 1. One Engine Inoperating Ceiling 单发失效上限 2. Optical Electronic Integrated Circuit 光电子集成电路

OEM Original Equipment Manufacturer 原设备制造厂，原始设备制造商

OEO Optical-Electrical-Optical 光-电-光

OEP Office of Emergency Preparedness 紧急战备处

OER 1. Operational Effectiveness Rate 使用效能变化率 2. Operator Error Recording 操作错误记录

oersted 奥斯特（磁场强度单位）

OES Orbital Escape System 轨道逃逸（救生）系统

OEU Overhead Electronics Unit 顶部电子装置

OEW Operational Empty Weight 使用空机重量

OF Overfrequency 超频，过频

OFAS Overseas Flight Assistance Service 海外飞行支援服务

OFATS Overseas Foreign Aeronautical Transimitter Station 海外外国航空发射机台（站）

OFB Output Feedback 输出反馈

OFBD Optical Fiber Branching Device 光纤分路器

OFBG Optical Fiber Bragg Grating 光纤布拉格光栅

off-boresight 离轴

off-boresight angle 离轴角（目标偏离武器轴线的角度）

off-boresight launch 离轴发射

offline 脱机，离线

off-operational fault detection 1. 脱机故障检测 2. 离线检测

off-route 偏航

offset 1. 偏置，补偿，偏移 2. 残余误差

offset hinge 外伸铰链

offset parallel runways 错列平行跑道（中心线平行，但长短不齐）

offset pass 1. 试射进入（目标）2. 按辅助瞄准点进入轰炸（目标）

offset printing 胶印，胶版印刷

offset QPSK 交错四相相移键控，偏量四相相移键控

offset range 偏置距离

offset track 偏置航迹

offset voltage 失调电压

offshore procurement 海外采购（美军用语，指用国防费用在美国以外的地方采购军用品）

off-site 1. 非现场的 2. 装置外的

off-site maintenance 发射场外维护

offspring 产物，结果

off-system unit（of measurement） 制外（测量）单位

offtake （动力源的）输出通道，（热空气的）排放通道

off-the-shelf 现成的，买来不用改就用的

off-the-shelf item 现货项目，现有项目（不需修改或稍

加修改即可供用户使用）
off-the-shelf product 货架产品
off-track error 越（离）轨误差
OFO Orbiting Frog Otolith 蛙耳石卫星（美国生物卫星）
OFP 1. Operational Flight Profile 作战飞行剖面图 2. Operational Flight Program 作战飞行程序,营运飞行程序,操作飞行程序 3. Original Flight Plan 原始飞行计划
OFPP Office of Federal Procurement Policy 联邦采购政策办公室（美国）
OFR 1. Open Failure Report 公开的故障报告 2. Operational Failure Report 操纵故障报告
OFS 1. Ocean Flight Satellite 海洋侦察卫星 2. Operational Flying School 作战飞行学校（英国海军） 3. Operations Flight Software 操作飞行软件 4. Optical Fibre Sensor 光纤传感器 5. Optical Fiber System 光纤通信系统 5. Optical Fuzing System 光引信系统 6. Optical Frequency Shifter 光移频器 7. Offset 偏离
OFT 1. Office of Fair Trading 公平贸易处（英国） 2. Operational Flight Test 使用飞行试验,作战飞行试验 3. Operational Flight Trainer 作战飞行教练机 4. Operational Flight Training 作战飞行训练 5. Orbital Flight Test 轨道飞行试验
OFTS 1. Operational Flight and Tactics Simulator 作战飞行与战术模拟器 2. Optical Fiber Transmission System 光纤传输系统
OFV Outflow Valve 放泄活门,溢流活门,放气阀
OFZ Obstacle Free Zone 无障碍物区
OG 1. Observation Group 观察组（美国陆军航空中心,美国陆军航空兵） 2. On Ground 在地面
OGA 1. Out Gimbals Axis 外平衡环轴 2. Outer Gimbal Angle 外框架角 3. Outer Gimbal Axis 外部万向支架轴,陀螺外框轴
Ogasawara 小笠原群岛（日本航天跟踪站）
OGE 1. On-Gimbals Electronics 陀螺电子设备 2. Operations Ground Equipment 营运地面设备,作战用地面设备 3. Out of Ground Effect 无地效（直升机远离地面,旋翼拉力不受地面影响）
ogee S形曲线的
ogee planform wing S形前缘翼（指平面形状）
ogive 1. 尖拱形曲线（其曲率半径逐渐增大,最终成一直线） 2. 哥德式机翼（翼平面由尖拱形曲线逐步变成两边与纵轴平行的直线） 3. 由尖拱形曲线形成的旋转体,尖拱卵形体
OGL Outgoing Line 引出线,出发航线
OGO Orbiting Geophysical Observatory 轨道地球物理观测台（美国空间物理探测卫星系列,根据轨道不同,分为 EOGO 和 POGO）
OGV Outlet Guide Vane 出口导向叶片
OGW Overload Gross Weight 超载总重
ogy strategy 技术战略
OH 1. Oil Heater 油加热器 2. On Hand 现有 3. Operational Hardware 作战武器 4. Opposite Hand 另外一侧,异侧 5. Overhaul 翻修,大修 6. Overhead 越顶的,从头顶上空飞过,企业管理费 7. Overheat 过热
OHA 1. Operating Hazard Analysis 运转危险性分析 2. Overhead Access 开销接入
OHC Operating Hours Counter 使用小时记数器
OHM 1. Ohmmeter 欧姆表,电阻表 2. Overhaul Manual 翻修手册
ohmic bridge 电阻电桥
ohmic contact 欧姆接触
ohmic heating 欧姆加热
ohmmeter 欧姆表,电阻表
Ohsumi 日本第一颗用自制火箭发射的试验与研究卫星
OID 1. Object Identifier 对象标识符 2. Outline Installation Drawing 外形安装图
OIDA Original Image Data Array 原始图像数据阵列
OIIX 德黑兰（区域管制中心/飞行情报中心）
OIL Orange Indicating Light 橙色指示灯
oil 油,滑油
oil circuit breaker 滑油断电器
oil film light valve 1. 油膜光阀 2. 油膜光阀管
oil film light valve projector 油膜光阀投影器
oil filter 滤油器,加油口
oil flow technique 油流法
oil heat exchanger 滑油通风器
oil pump 油泵,送油泵
oil ring 油环
oil water separator 油水分离器
OIP Optimum Implementation Plan 最佳执行计划
OIPS Optical Image Processing System 光学图像处理系统
OIS 1. Office Information Systems 办公室信息系统 2. On-board Information System 机载信息系统 3. On-line Information Service 在线的信息服务 4. Operational Information Services 业务信息服务 5. Operational Information System 业务信息系统 6. Operator Interface Station 操作员接口站 7. Order Item Split 定单项目分解
OIT Operation Information Telex 使用信息电传
OJT 1. On Job Training 在职培训 2. Overwater Jet Navigation 喷气式飞机水上领航
okta 八分之一
OL 1. Off Load 卸载 2. Oil Level 油位 3. Operating

Lease 经营性租赁 4. Operating Life 工作寿命,运转寿命 5. Operating Location 作业地点 6. Our Letter 我方函件 7. Overall Length 全长,总长 8. Overload 超载

OLAN On-board Local Area Network 机上局域网

OLC 1. On-Line Code 联机码,在线码 2. Outer Loop Control 外回路控制

OLD On-Line Data 在线数据

OLDI On-Line Data Interchange 在线数据交换

OLDP On-Line Data Processing 联机数据处理

olefin 烯族烃,烯烃

oleo 油,油液减震的

oleo gear 油压减振起落架

OLF Outlying Field (Navy) 海外机场(海军)

OLG Open Loop Gain 开环增益

olive 球面卡套

OLP On-Line Planning 在线计划

OLS 1. Oil Level Sensor 滑油油量传感器 2. On-Line Service 在线服务 3. Optical Landing System 光学着陆系统 4. Outgoing Line Signaling 去话线路信令

OLT Option Lead Time 选型提前期

Olympus 奥林匹斯

OM 1. Object Management 对象管理 2. Operation Manual 使用手册 3. Operations Manual 飞机操作手册 4. Optical Multiplexer 光复用器 5. Option Management 选项管理 6. Option Maturity 选项成熟度 7. Operation Mode 操作模式 8. Our Message 我方函电 9. Outer Marker 外指点标,外信标台 10. Overbooking Management (机票)超售管理 11. Overhaul Manual 翻修手册

OMA 1. Object Management Architecture 目标管理特征,对象管理体系结构 2. Organizational Maintenance Activity 结构维修工作 3. Orthogonal Multiple Access 正交多址接入

OMB 1. Office of Management and Budget 经营与预算办公室(联邦航空局) 2. Operation and Maintenance Block 运行和维护功能块 3. Operations Manual Bulletin 使用手册通告

OMCM Operational and Maintenance Configuration Management 使用与维修中的装备构型管理(主要是对改装的控制)

OMD On-board Maintenance Documentation 机上维修文件

OME Operating Mass Empty 使用空重

Omega (OMEGA) Very Low Frequency Waveband Navigation 甚低频波束导航系统(欧米伽)

Omega segment synchronization 欧米伽导航系统段同步

Omega sky wave correction table 欧米伽天波修正表

Omega system 欧米伽系统

omegatron 欧米伽器回旋质谱计

OMG Object Management Group 目标(对象)管理组

OMI 1. Omnibearing Magnetic Indicator 全向磁指示器 2. Open Messaging Interface 开放式信息界面 3. Operations and Maintenance Instruction 使用和维修说明书 4. Optical Modulation Index 光调制指数

OMIN Optical Multistage Interconnected Network 光多级互联网络

OMIT Omitted 省去的

OMKM Option-on-Module Knowledge Map 选项至模型经验图

OML Outside Mold Line 外模线

OMM Overhaul Manual Manufacturer 生产厂家大修手册

OMNI Omnidirectional 全向的

omni 全,总

omnibearing 全方位的,按全向指向标定位

omnidirectional 全向的,无方向性的

omnidirectional antenna 全向天线

omnidirectional microphone 全向传声器

omnidirectional range 全向无线电信标

omnirange 全向指向标

OMP 1. Open Management Protocol 开放的管理协议 2. Operation & Maintenance Processor 运行维护处理器 3. Operator Maintenance Plan 承运人维修大纲

OMRS Operations and Maintenance Requirements Specification 操作和维护要求规范

OMRSD Operations and Maintenance Requirements Specification Document 操作和维护要求规范文件

OMS 1. Object Management System 目标管理系统 2. On-board Maintenance System 机上(机载)维修系统 3. Open Management System 开放管理系统 4. Optical Multiplexer Section Layer 光复用段层 5. Opto-electronic Multiplex Switch 光电子复用转换 6. Option (Maintenance) Management System 选项管理系统 7. Order Management System 指令管理系统 8. Outage Management System 停电管理系统

OMT 1. On-board Maintenance Terminal 机载维护终端 2. Object Modeling Technique 目标(对象)建模技术

OMTBF Observed Mean Time Between Failure 观察的故障间隔平均时间

OMV 1. Orbital Maneuvering Vehicle 轨道机动飞行器(美国曾设计的一种可执行多种在轨任务的飞行器) 2. Overhaul Manual Vendor 协作厂大修手册

on course 在航道上

ONA 1. Off-Net Access 网外接入 2. Office of Noise Abatement 降低噪声办公室(美国联邦航空局) 3. Open Network Architecture 开放式网络体系结构 4. Optical Navigation Attachment 光学领航辅助装置

5. Optical Network Analyzer 光网络分析仪
6. Overseas National Airways 国家海外航空公司
on-board 机上的,弹上的,在飞机上的
on-board maintenance 机上维护,空中维护
on-board maintenance system 机上维修系统
on-board oxygen generation 氧气系统
on-board tracking and safety control system 箭上跟踪与安全控制
once command 一次指令
oncoming missile 接近目标的导弹
on-condition maintenance 视情维修,按需维修
oncourse 在航线上,在航道上
one and a half stage rocket 一级半火箭
one-dimensional steady channel flow 一维定常管流
one-wheel landing 单轮着陆
on-line 1. 在航线上,在现场 2. 联机 3. 在线
on-line diagnostics 联机诊断
on-off ratio 占空因数,占空系数
on-orbit 轨道工程卫星
on-orbit failure 在轨故障
on-orbit repair 在轨(道上)修理
ONR Office of Naval Research 海军研究局(美国)
ONS 1. Omega Navigation System 欧米伽导航系统 2. On-line Notifying Server 在线通知服务器 3. Open Networking Supportware 开放式组网支撑件
ONSCEN Omega Navigation System Center 欧米伽导航系统中心
onset 开始,突然开始
on-site measurement 1. 现场测量 2. 原位维修
on-site repair 现场修理
ontarget 已捕获目标
on-the-deck dash range 极低空冲刺距离
ontop 云上的
ONVL Over-the-Nose Vision Line 机头上部视线
OO Object-Oriented 面向对象
OOA 1. Object Oriented Analysis 面向对象分析 2. Out Of Action 不工作,发生故障
OOBOM Object-Oriented Bill Of Material 面向对象物料表
OOD 1. Object-Oriented Design 面向对象的设计 2. Out Of Detent 脱离卡挡
OODB Object-Oriented Data Base 面向对象的数据库
OODBMS Object-Oriented Data Base Management System 面向对象数据库管理系统
OOEU Outboard Overhead Electronics Unit 外顶部电子组件
OOF 1. Out Of Flatness 不平整 2. Out-Of-Frame 帧失步
OOKM Option-to-Option Knowledge Map 选项至选项经验图
OORAM Object-Oriented Role Analysis Method 面向对象的任务分析方法
OOTB Out-of-the-box 快速实施产品
OP 1. Oil Pressure 滑油压力 2. Oil Pump 油泵 3. Opalescent 乳色的 4. Open 开放,打开 5. Operating Procedure 操作程序,操作步骤 6. Operation 工作,运行,操作,航行 7. Operational 使用的,操作的,运行的 8. Opposite 对面的,相对的,相反的 9. Outer Perimeter 外周长 10. Output 输出
OPA Operating Procedure Agreement 操作程序协议
opacity 不透明性,阻光性
opaque photocathode 不透明光阴极
opaque plasma 不透明等离子体
OPC 1. Operation Check 使用检查 2. Operation Code 操作码 3. Operational Control 航行管制 4. Operational Program Configuration 操作程序组合 5. Optical Phase Conjugation 光相位共轭 6. Organic Photo Conductor 有机光电导体,光敏电阻 7. Originating Point Code 始端(信令)点代码
OPCEN Operations Center 操作中心
OPCON Operational Control 作战控制
OPCR Operation Plan Change Request 操作大纲更改需求
OPD 1. Opened 开放的,公开的 2. Operating Procedure Directive 运行程序指令 3. Optical Path Difference 光程差
OPDR Optical Polarization Domain Reflectometry 光偏振域反射测量法
OPEC Organization of Petroleum Exporting Countries 石油输出国组织
open 1. 开路,断开 2. 开放,打开
open circuit termination 开路终端
open circuit voltage 开路电压
open circuit wind tunnel 开路式风洞
open cluster 疏散星团
open cycle 开式循环
open delta connection 开放三角形结线
open die forging 自由锻造
open ejection 战斗(导)弹
open ended coil 终端开路线圈
open ended spanner 开口扳手
open jet wind tunnel 开口式风洞
open loop transfer function 开环传递函数
open loop voltage gain 开环电压增益
open reel tape 盘(式磁)带
open source 开放源码
open source software 开源软件
open system 开口系统

open universe 开式宇宙，开宇宙
open wiring 眉线
open-circuit line 开路线
opening shock 开伞冲击，阀门开启时的冲击
opening speed 回程速度，开伞速度，开断速度
open-jet wind tunnel 开式喷流风洞，开口式风洞
open-loop control 开环控制
open-loop frequency response 开环频率响应
operability 战斗适用性，运行性
operand 操作数（计算机参与操作的数据），运算对象
operate 操作，工作，经营，飞行
operating crew monitoring 空勤组监控
operating current 工作电流
operating line 共同工作线
operating point 工作点，工作状态
operating range of infrared system 红外系统作用距离
operating range without degradation 保精度工作范围
operating ratio 运转时间比，利用比
operating slide 滑板导轨
operating slide guideway rail 滑板导轨（转膛炮的）
operating speed 有效巡航速度，运转速度
operating stand 作业区（停机坪上的一个区域，用于装卸货物及机务保障）
operating temperature 运行温度
operating temperature range 工作温度范围
operating time 运行时间，动妆间
operating voltage 工作电压
operating weight empty 使用空机重量
operation 工作，运行，操作，航行
operation flight program 氧气余压表
operation test program 操作飞行程序
operational aircraft 1.作战飞机（英国用法）2.做好使用准备的飞机
operational amplifier 运算放大器
operational availability 使用可用性
operational characteristics 使用性能参数，使用特性（说明设备功能的参数，例如电子设备的使用性能参数包括频率范围、波道、调制类型和发射特性）
operational environment 工作环境
operational flight plan 航务飞行计划
operational flight test 作战飞行试验，作战飞机试验
operational flight trainer 作战飞行训练装置，战斗机练习器
operational load 使用载荷监控系统
operational missile 1.战斗导弹 2.实弹
operational mission 作战任务
operational mode of Doppler navigation system 多普勒导航系统的工作状态
operational program configuration 运行程序构型

Operational Program Software 运行程序软件
operational readiness 1.战备完好状态，使用准备状态（装备的）2.战备完好性（部队接到作战命令时，对作战计划的响应能力，是实力、装备的可用性以及训练、供应等各方面情况的综合反映）
operational readiness inspection 战备状态检查，战备检查
operational recording 操作记录
operational reliability of fuze 引信作用可靠性
operational requirement and specification 战术技术要求
operational satellite 作战卫星，军事卫星
operational suitability 作战适应性
operations manual 使用手册
operations research 运筹学
operations specification 运营规范
operator 经营者，报务员，用户，承运人，调度员，空运公司
operator factor 观察者系数
OPF Off Peak Fare 非高峰票价
OPLE Omega Position Location Equipment 欧米伽定位设备
OPM 1. Office of Personnel Management 人员管理处 2. Operations Per Minute 每分钟操作次数，每分钟动作次数 3. Order Point Method 订货点方法
OPMET 1. Operational Meteorological Information 可用气象资料 2. Operational Meteorology 航务气象
OPN 1. Open 打开，开放 2. Open or Opening or Opened 开启 3. Operation 运行，工作，操作，航行 4. Operational Planning Notes 业务规划指出 5. Opinion 观点，意见 6. Option 选择
opposed piston engine 对置活塞式发动机
opposite effect 异性效应，相反的结果
opposite side 对边，对方，对面
opposition 反对，对立，抗击，反抗
OPR 1. Operation 工作，运行，航行 2. Operator 经营者，报务员，用户 3. Operate, Operated 操作，操作的 4. Optical Preamplifier Receiver 光预放大接收机
OPS 1. Operational Program Software 操作程序软件 2. Operations or Operates 运行，工作，操作，运转，航行，运行，航务，运用 3. Optical Smoothing 光平整
OPSEC Operational Security 作业安全
OPSP Outside Production Specification Plan 外部生产规范计划
OPSPECS Opereational Specifications 运行规范
OPT Optimum 最佳，最优
optic fiber gyroscope 光纤陀螺
optical alignment 光学对准
optical attenuator 光衰减器
optical axis 光轴，视轴

optical biasing 光偏置
optical bistable device 光学双稳态器件
optical cable connector 光缆连接器
optical character reader 光学字母阅读装置
optical coating 光学涂层,光学涂膜
optical communication 光通信
optical communication parts 光学通信部件
optical communication receiver 光通信接收机
optical countermeasures 光学对抗措施
optical coupler 光耦合器
optical damping coefficient 光学阻尼系数
optical decoy 光假目标
optical density 光密度(photographic transmission density 即照片透光密度)
optical depth 光深,光学厚度
optical detector 光辐射探测器
optical direction finding 光学测向
optical distortion 光学畸变
optical distortion of helmet faceplate 头盔观察面板光学畸变
optical double star 光学二重星,光学双星
optical fiber 光导纤维
optical fiber cable 光缆
optical fiber connector 光纤连接器
optical fiber dispersion 光纤色散
optical fiber sensor 光纤敏感器
optical fiber splice 光纤固定接头
optical fiber transducer 光纤传感器
optical filter 光滤波器
optical frequency standard 光频标
optical gyro 光学陀螺仪
optical heterodyne detection 光外差探测
optical heterodyne receiver 光外差接收机
optical homodyne detection 光零差探测
optical identification 光学识别
optical information processing 光学信息处理
optical isolator 光频隔离器
optical landing system 光学着陆系统
optical maser 光学微波激射器,激光器,激光(同 laser)
optical measurement 光学测量
optical micrometer 光测微计(测量伤痕深度的精密仪器)
optical microscope 光学显微镜
optical modulator 1.光灯器 2.光调制器
optical nutation 光学章动
optical parametric amplification 光参量放大
optical parametric oscillation 光参量振荡
optical path 光路,光迹
optical path length 光程长度

optical path difference 光程差
optical path distortion 光路畸变
optical processor 光处理机
optical projection exposure method 光学投影曝光法
optical projection lithography 光学投影蚀刻
optical projection master 光学投影原版
optical pump 光泵
optical pumping 光泵激,光抽运
optical pumping magnetometer 光泵激磁力计
optical pyrometer 光学高温计,光测高温计
optical radar 光雷达
optical radiation 光学范围辐射
optical range 直视距离,光学距离
optical resonator cavity 光空腔谐振器
optical sighting 光学瞄准
optical storage 光学存储
optical surface curvature 光学表面曲率
optical switch 光开关
optical system 光学系统
optical target seeker 光学目标位标器
optical telescope 光学望远镜
optical time domain reflectometer 光时域反射仪
optical tracking 光学追踪,光学跟踪
optical tracking satellite 光学跟踪卫星
optical tracking system 光学追踪设备,光学跟踪系统
optical transfer function 光学传递函数
optical transmitter 光发送机
optical-mechanical scanner 光机扫描仪
optics 光学
optimal control 最优控制
optimal control theory 最优控制理论
optimal design 优化设计
optimal flight 最佳飞行
optimal flight control system 最佳飞行控制系统
optimal stagger ratio 最佳参差比
optimum 最佳聚束,最佳的,最合适的,最适宜的,最优值
optimum coupling 最佳耦合
optoelectronics 光电子学
OPU Overspeed Protection Unit 过速保护装置
OQA Order Quantity Adjustment 订单数量调整
OQAF Order Quantity Adjustment Form 订单数量调整表
OQPSK Offset Quadriphase Shift Keying 交错四相相移键控
OR 1.Operational Reliability 运行可靠性 2.Operational Requirements 操作要求,使用要求 3.Optical Reflectance 光反射比 4.Orange 桔黄色 5.Order 订货,订票,命令 6.Other 其他 7.Outgoing Route 输出

路由 8. Over Run 安全道 9. Oversale Risk 机票超售风险

ORA Operational Readiness Assessment 运行准备状态评价

ORACLE 1. Operational Research and Critical Link Evaluation 运筹学与统筹法评定 2. 奥拉克尔(甲骨文公司,全称甲骨文股份有限公司,是全球最大的数据库软件公司,总部位于美国加州的红木滩)

Orads Optical Ranging and Detecting System 光学测距和探测系统

oralloy 橙色合金

Orange 桔黄色

ORB 1. Omnidirectional Radio Beacon 全向无线电信标 2. Object Request Broker 对象请求代理,对象需求分配器

orbit 轨道,轨道运行

orbit altitude 轨道高度

orbit control 轨道控制

orbit determination 轨道决定,轨道确定

orbit estimation 卫星轨道估计

orbit module 轨道舱

orbit parameter 轨道参数

orbit period 轨道周期

orbital 轨道的

orbital acquisition 轨道捕获

orbital control 轨道控制

orbital decay 轨道下降,轨道衰减

orbital eccentricity 轨道偏心

orbital electron 轨道电子

orbital element 轨道要素,轨道参数,轨道根数

orbital mechanics 轨道力学

orbital period 轨道周期,运行周期

orbital plane 轨道平面

orbital region 轨道区域

orbital rendezvous and docking 轨道交会对接

orbital rendezvous mission 轨道会合任务

orbital speed 轨道速度

orbital stability 轨道稳定性

orbiter 轨道飞行器,轨道卫星

orbitron 轨道管,轨旋管

ORBS Orbital Rendezvous Base System 轨道交会基地系统

ORC 1. Optimal Retransmission Control 最佳转发控制 2. Originating Region Code 始发地区代码 3. Outbound Radar Control 离场雷达管制

Orchidee 兰花(法国直升机载远距离雷达,用以向战场指挥员提供战术信息)

ORD 1. Order 订货,订票,命令 2. Ordinary 普遍的,一般的 3. Operational Readiness Demonstration 运行准备就绪示范 4. Operational Requirement Document 使用要求文件

order 1. 订货,订票 2. 点,订购 3. 命令 4. 队形

order of magnitude 数量级

order of reflection 反射级

order parameter 有序参量

ordinary ray 普通射线

ordinary wave 寻常波(电波在磁离子介质中分裂成的两个特征波中的一个)

ordinate 纵坐标

ordnance 军械

ordnance work rate 军械使用率(某种军械在某一战区或对某一目标的使用频率)

ORE 1. Ocean Research Equipment 海洋研究设备 2. Operational Readiness Evaluation 战备评估 3. Optical Reading Equipment 光读出设备 4. Optical Repeater Equipment 光转发设备 5. Overall Rreference Equivalent 全程参考当量

organic 有机的,结构的,有组织的,有系统的,根本的,固有的

organic chelate liquid laser 有机整合物液体激光器

organic functional material 有机功能材料

organic semiconductor 有机半导体

organization 组织,编制

organometallic 有机金属化合物(金属原子与有机根的结合)

ORI Operational Readiness Inspection 工作准备状态检查

orient 定方位,定向

orientation 定位,定向

orifice 量孔板,出口,出孔

orifice meter 孔板流量计,节流式流量计,定量油孔

origin 始发站,原点,起源,始点

origin of the solar system 太阳系起源

origin planet 星球起源

original data 原始数据

original incident wave 初始入射波

originate 发源,发生,引起,创始

originator 发射机,创始人,签发

Oring 圆形截面密封圈

Orion 猎户星座,猎户座

ORIS Online Rejection Information System 生产线拒收信息系统

ORL Overrun Area Edge Lights 安全道区域边缘灯

ornithopter 扑翼机,扑翼飞机

orographic 山形的,地形的

OROS Optical Read Only Storage 光学只读存储器

Orotron 奥罗管

ORP 1. On-request Reporting Point 申请报告点

2. Output Routing Pool 输出布线区
ORR Omnidirectional Radar Range 全向雷达信标
orr sommerfeld equation 奥尔拴菲方程
orrery 太阳系仪，天象仪
ORS 1. Option Relationship Specification 可选关系规范 2. Others 其他的 3. Overhaul and Repair System 翻修和修理系统
orsat analyser 奥尔氏气体分析仪
ORT 1. Owner Requirement Table 物主需求表 2. Operational Readiness Test 战备检查
orthicon 正析象管，正摄象管
orthoferrite 正铁氧体
orthogonal polarization 正交极化
orthogonal scanning 正交扫描
orthogonality 正交性，直交性
orthometric height 绝对高度
ORU 1. Optical Receive Unit 光接收单元 2. Optical Repeater Unit 光中继单元
OS 1. Office System 办公室系统 2. Oil Separator 滑油分离器 3. Oil Strainer 滑油筛 4. Oil Switch 油开关 5. On Schedule 按时，按预定计划 6. Operating System 操作系统 7. Optical Section 光纤段 8. Optical Sender 光发射机 9. Optical Soliton 光孤子 10. Optical Switch 光交换 11. Operations Specifications 运行规范 12. Order Sheet 订单 13. Out of Service 不能使用，停止使用 14. Outplant Supplier 外场供应商 15. Over Station 过台 16. Overspeed 超速
OSA 1. Optical Spectrum Analyzer 光谱分析仪 2. Office System Automation 办公系统自动化 3. Open System Architecture 开放系统体系结构
OSAR Operations Suitability Assessment Report 操作适用性评估报告
OSC Oscillate, Oscillator 振荡，振荡器
OSCD Operations Support Computing Division 操作保障计算处
OSCE Operation Spares Configuration Environment 操作备用构型环境
OSCF Operations Support Computing Facility 操作保障计算设备
oscillate 振动，振荡，摆动，摇摆
oscillating combustion 振荡燃烧
oscillating combustion pressure measurement 振荡燃烧压力测量
oscillation 振荡，振动，摆动
oscillation mode 振荡模（式）
oscillator 振荡器
oscillator noise 振荡器噪声
oscillator strength 振子强度
oscillatory combustion 振荡燃烧

oscillatory motion 振动，摆动
oscillogram 示波图，波形图
oscillograph 示波器
oscillometer 示波器
oscilloscope 示波器
oscilloscope tube 示波管
osculating element 接触要素，接触单元
osculating orbit 接触轨道，密切轨道
osculating plane 接触面，密切面
OSCS Out-of-Sequence Control System 超序控制系统
OSCU Oxygen System Control Unit 氧气系统控制装置
OSD 1. Office of Secretary of Defense 国防部长办公室（美国）2. Operational Sequence Diagram 操作顺序图 3. Optical Scanning Device 光扫描装置
OSDM Optical Space Division Multiplexing 光空分复用
OSE 1. Observing System Experiments 观测系统试验 2. Open Systems Environment 开放系统环境
OSEM Office of Systems Engineering and Management 系统工程与管理办公室（美国联邦航空局）
OSF 1. Open Software Foundation 开放式软件基金会 2. Operating System Function 操作系统功能 3. Operation Service Function 操作业务功能
OSFB Operation System Function Block 运行系统功能块
OSHA 1. Occupational Safety and Health Administration 职业安全和卫生局（美国）2. Occupational Safety & Health Act 职业安全与健康法
OSI 1. Open System Interconnection 开放系统互连 2. Open System Interface 开放系统界面（接口） 3. Operating System Interface 操作系统接口 4. Other Service Information 其他服务情况，其他服务资料
OSID Operations System Interface Document 操作系统接口文件
OSIE OSI Environment 开放系统互连环境
OSL 1. Oil Seal 油封 2. Optical Signal Level 光信号级
OSM 1. Onboard Storage Management 机载储存器管理 2. Oscillator Strength Modulation 振荡器强度调制 3. Outgoing Switch Module 出局交换模块
osmium 【化】锇（化学元素，符号 Os）
osmosis 渗透
OSO 1. Office of Space Operations 空间活动处，航天操作处 2. Offensive System Operator 攻击系统操作员 3. Operations Scheduling Office 操作进度处 4. Orbiting Solar Observatory 轨道太阳观测卫星
OSP 1. Open Settlement Protocol 开放结算协议 2. Optical Saturation Parameter 光饱和量 3. Optical Signal Processing 光学信号处理 4. Optical-switched Service Provider 光交换业务供应商
OSPF Open Shortest Path First 开放式最短路径优先

OSS 1. Open Simulation System 开放式模拟系统 2. Operation-Support System 运行支持系统 3. Operating System Software 操作系统软件 4. Operating System Storage 操作系统存储器 5. Operator Service System 话务员业务系统 6. Over Station Sensor 过台传感器

OSSE Observing Systems Simulation Experiment 观测系统模拟实验

OST 1. Operational Suitability Testing 作战适用性(操作适用性)试验 2. Optical Section Termination 光纤段终端 3. Optical Soliton Transmission 光孤子传输

OT 1. Object Technology 对象技术 2. Oil Temperature 滑油温度 3. Optical Tapoff 光分接 4. Optical Terminal 光终端 5. Other Time 其他时间 6. Ought 应该 7. Our Telex 我方电传 8. Over Temperature 超温 9. Overtime 加班

OT&E Operational Test and Evaluation 运行测试和评价

OTA Office of Technology Assessment 技术评审局(美国)

OTC 1. Operating Telephone Companies 运营电话公司 2. Originating Toll Center 长途始发中心 3. Originating Toll Circuit 长途始发电路 4. Originating Trunk Center 发话中心 5. Outgoing Trunk Circuit 出中继电路 6. Overseas Telecommunications Company/Corp. 海外通信公司

OTD 1. On Time Delivery 及时运送 2. Optical Time Domain 光时域 3. Origin To Destination 始发地至目的地

OTDL Object Type Definition Language 对象类型定义语言

OTDM Optical Time Division Multiplexing 光时分复用

otection 保护角

OTER Over-Temperature Emergency Rating 超温应急额定功率

OTF Optical Transfer Function 光学传递函数

OTFP Operational Traffic Flow Planning 运行的交通流量计划

OTLX Our Telex 我方电传

OTP 1. On Top 在顶部 2. One Time Password 一次性口令 3. One Time Programmable 一次性可编程

OTS 1. Off The Shelf 货架(产品) 2. On-line Terminal System 在线终端系统 3. Organized Track Structure 编组航迹结构 4. Organized Track System 编组航迹系统 5. Others 其他的

OTW Over The Wing 在机翼表面,机翼涵盖的客舱座位

OU 1. Operations Unit 作战部队 2. Organizaiton Unit 组织单元 3. Outlet Unit 出口装置 4. Overhead Unit 头顶组件

OUCD Operations and Utilization Capability Development 操作和应用能力开发

OUT 1. Outlet 出气口 2. Output 输出

out of alignment 桨叶错位的,未对准的

out of bound probability 操作测试程序

out of phase 不同相的

out of pitch 桨叶错距(螺旋桨或旋翼在同一半径处,一个桨叶的桨距与其他桨叶的桨距不同)

out of step 失步

out of track 桨叶错轨迹的,不同锥度的

outage 1. 断电,运转中断 2. 耗油量,预留容积

outboard 1. 外侧的(指远离飞机对称面的) 2. 外装的,外部的 3. 翼侧的(从中心线到翼尖之间的)

outbound 飞出去的,离开电台的

outburn (燃烧物的)烧尽,燃尽

outburst (辐射)爆发,突发,溃决,冲破,闪光

outcome 结果,成果,出口

outdiffusion 外扩散

outer fix 外侧定点,外定位

outer lead bonding 外引线焊接

outer locator 外示位信标,远台

outer loop 外侧环路

outer marker 1. 外指点标 2. 远台

outflow 流出,外流

outgas 除气,排气,放气

outgassing 脱气,出气,除气

outline 外形(线、图),轮廓,略图,剖面,周线,回路

out-of-control 失控

out-of-focus 不聚焦,焦点失调,离焦

output 输出(发动机功率输出)

output feedback 输出反馈设计

output impedance 输出阻抗

output layer 输出层

output meter 输出仪表

output range 输出量程

output resistance 输出电阻

outside air temperature 外界大气温度

outside loop 1. 侧飞翻 2. 外部环状飞行,外部循环

out-to-out 总尺度,全长

outweigh 在重量上超过,在价值上超过

OV 1. Outflow Valve 外流活门 2. Over 超过 3. Overvoltage 超压

overall arrangement 总体布置图

overall height 1. 全高 2. 停机高度

overall length 全长,外形长度

overbooking 超出飞机座位的预定

overbunching 过聚束

overcast 阴天,密云
overcoat 覆盖层
overcontrol 操纵过量,舵面偏转过量,操纵过量,过度调节
overcorrect 修正动作过量,过度修正(偏差)
overcrowd 使过度拥挤,把……塞得太满
overcurrent protection 过电粒护装置
overdamping 衰减过度,过阻尼,衰减过度
overdischarge 过放电
overdrive 超速传动,过度激励
overestimate 估计过高,评价过高
overexcitation 过激励,过激磁
overexpansion 过度膨胀
overexposure 过曝光
over-fins diameter (火箭、导弹的)尾翼翼展
overflight 飞越上空,飞越上空,从上方飞过
overflow 外溢,溢流,充满,充斥,过剩
overfly 飞越
over-frequency 超频(率)
overhang 1.突出 2.悬臂梁,伸臂梁 3.吊挂,悬吊,突出部,伸出物
overhaul 翻修,全面检修
overhaul life 大修寿命
overhead 1.顶部,飞越上空 2.在跑道上空作转弯进入着陆 3.在云上,高于云 4.(飞机)顶部构件
overhead bin 顶部行李箱
overhead stowage 头顶行李箱
overheat 1.过热,超温 2.燃烧殆尽
overhung 外挂的,悬臂的,悬垂的
overlap 1.中性区,中间带 2.重叠 3.防空重叠区的情报传递
overlay 1.镀层 2.透明片,透写图,透明图,透明图板
overload 超负荷,过负荷,过载
overload circuit breaker 超载断电器
overload factor 1.过负荷系数 2.过载系数
overload weapon 超载武器(为执行特殊任务或加强火力而额外加装的机载武器)
overnight service 过夜维护
overpass 通过,越过
overpressure 过压,超压
overpressurization 超压,过量增压
override 1.超控(超越或绕过预设指令动作的正常限度,以得到额外的响应,如超越自动驾驶仪的手控) 2.人工代用装置
overrun 1.安全道,冲出跑道,保险道 2.超限运动
overrunning clutch 1.单向离合器 2.超控离盒器
oversale 超过机座或客房数量的销售
overseas 海外的
overshoe 除冰套,除冰带

overshoot 1.过冲,酌过度 2.上冲 3.超调 4.过调量,复飞,过调,着陆目测高,着陆时越过指定地点
oversight 严密监视,失察
overspeed 超转(转速超过规定的最大转速),超速
overspeed test 1.超速试验 2.超额定速度试验
overspeed warning 超速警告
overstability 1.超稳定性 2.过度安定性
overswing 1.过冲,酌过度 2.罗盘磁针摆动,超越,过摆
overtemperature test 超温测试
over-the-horizon radar 超视距雷达,超地平线雷达
overview 视界,视野,概要
overvoltage 超高压,超压,过电压,超电压
overvoltage protection 过电压保护装置
overvoltage protector 过压保护器
overwatch 掩护
overwater 1.水上的,水面上的 2.海上轰炸任务 3.水上飞行距离,水上靶区
overweight 超重,使超重
overwing pylon (在机翼上方的)翼上挂梁
OVHT overheat 过热
OVID Obviously Visible Impact Damage 明显的可见撞击损伤
OVS Overhead Video System 顶部音响系统
OW 1.On Wing 装机状态 2.One Way 单程,单向 3.Order Wire 联络线 4.Outer Wing 机翼外部
OWE Operating Weight Empty 空机使用重量(最大起飞重量减去全部业务载荷、可用燃油、滑油及其他消耗品后的重量)
OWF Optimum Working Frequency 最佳工作频率
OWL On-Wing Life (主发动机的)不卸下连续使用寿命
OWS 1.Obstacle Warning System(直升机上的)障碍告警系统 2.Ocean Weather Station 海洋气象站 3.Office Work Station 办公室工作站
OWSF Oblique-Wing Single Fuselage 斜置翼单机身
oxidant 氧化剂
oxidation 氧化(作用)
oxide trapped charge 氧化层陷阱电荷
oxidizing flame 氧化焰(用于气焊和切割)
oximeter 血氧(测量)计
oximetry 血氧饱和度测定术
OXY Oxygen 氧气
oxyacetylene 氧乙炔的
oxygen bottle 氧气瓶,高压氧气筒
oxygen delivery capacity 1.供氧能力 2.供氧高度
oxygen demand 氧气用量调节装置,断续供氧调节器
oxygen excess 供氧能力
oxygen flow indicator 氧气示流器,光纤传感器

oxygen jet　氧气喷嘴，氧气射流
oxygen mask　氧气机罩，氧气面具，氧分压
oxygen mask adapter　氧气面罩连接器
oxygen overpressure　氧气调节器
oxygen overpressure indicator　氧气余压表
oxygen partial pressure　1.氧气压力比 2.氧分压传感器 3.氧分压
oxygen partial pressure control　氧分压控制
oxygen partial pressure depression　氧气压降低
oxygen pressure ratio　氧气余压
oxygen regulator　氧气调节器，机载制氧
oxygen supply altitude　1.供氧高度 2.氧气面罩

oxygen system　氧气系统，供氧系统，敞开式弹射
oxygen system charging adapter　氧气系统充气接头
oxygen system control unit　氧气系统控制单元
oxygen system controller　氧气系统控制器
oxygen system set　氧气系统装置
oxygenate　氧化，充氧
oxyhydrogen　氢氧气
ozone profile　臭氧分布图
ozonometer　臭氧计
ozonopause　臭氧层，臭氧层上限
ozonosphere　臭氧层
OZR　Overall Zonal Rating 区域总等级

P

P 1. Panel 板,控制板 2. Patent 专利 3. Pilot 飞行员 4. Pressure 压力 5. Probe 探头,探针,传感器

P charge 发射药弹头

P to F Permit to Fly 准许飞行

P&C Performance and Control 性能和控制

P&E 1. Propellant and Explosive 发射药和炸药 2. Pyrotechnics and Explosives 烟火和炸药

P&ES Personnel and Equipment Supply 人员和装备供应

P&L Power and Lighting 电源和照明

P&L DISTR Power and Lighting Distribution 电源和照明配电

P&O Price and Offerability 价格和让价

P&P 1. Payments and Progress 支付款项和进度 2. Plans and Programs 计划和纲要

P.T. Pacific Time 太平洋时间(美国)

P/A Payload/Altitude(Curve) 有效载荷/飞行高度曲线

P/CASS Picture Computation and Storage System 图像计算与存储系统

P/L 1. Pay Load 有效载荷,业载,净载荷,战斗部 2. Plain Language 明文,明码

P/LN Program/Line Numver 项目/架次号

P/N Part Number 零件号

P/P Pupil Pilot 飞行学员

P/PATM Passengers Per Air Transport Movement 每次空运活动的载客人数

P/PFRT Pupil-Pilot Flight Rating Test 飞行学员飞行等级考试

P/R/Y Pitch/Roll/Yaw 俯仰/横滚/偏航

P/RST Press To Reset 按压复位

P/S Pitot/Static 动/静

P1 First Pilot 正驾驶员,机长

P2 Second Pilot 副驾驶员

PA 1. Parametric Amplifier 参数放大器 2. Passenger Address 旅客广播 3. Pitch Axis 俯仰轴 4. Performance Analysis 性能分析 5. Power Amplifier 功率放大器 6. Preliminary Analysis 初步分析 7. Preparing Activity 编写单位 8. Pressure Altitude 压力高度 9. Procurement Agreement 采购协议 10. Production Airplane 投产飞机 11. Pulse Amplifier 脉冲放大器

Pa Pascal 帕(斯卡)

PA/CI Passenger Address/Cabin Interphone 旅客广播/座舱内话

PAA Primary Aircraft Authorization 初级飞机核准

PABST Primary Adhesively Bonded Structure Technology 基本的胶接结构技术

PAC Pilotless Aircraft 无人驾驶飞机

PACAF Pacific Air Forces 太平洋空军

PACCS Post-Attack Command and Control System 原子袭击后指挥控制系统

PACE 1. Performance And Cost Evaluation 性能与成本评定 2. Precision Analog Computing Equipment 精密模拟计算设备 3. Prelaunch Automatic Checkout Equipment 发射前自动检测设备

pacer 基准飞机,标准飞机

Pacific time 太平洋时间

pacing item 进度关键项目

PACIS Passenger Address and Communication Interphone System 旅客广播和通信内话系统

pack 1. 包,组件箱 2. 把……装箱,整理

pack hardening 堆集硬化

package 1. 总检,打包,部件 2. 包裹,包,组合程序,程序包 3.(国际合作研制飞机或出售飞机的)一揽子交易合同

package aircraft 一揽子交易的飞机,一揽子报价的飞机

package factor 封装因子

package gun 包壳机枪

package sealing with laser 激光密封

package tour 1. 一揽子报价的航空旅游 2. 公开招揽的航空旅游 3. 出发前收齐费用的航空旅游

package with nitrogen 充氮包装,充氮封存

packaged 组装好的,已包装的

packaged actuator 组合舵机,自主式伺服舵机

packaged bulk petroleum products 桶罐散装油料

packaged circuit 封装电路

packaged petroleum products 桶装石油制品

packaged propellant 预封装推进剂

packaged technology 成套技术

packaging 封装

packaging density 组装密度

packaging technique 组装技术

packet 1. 包,束,小包 2.(定期)游轮,班轮 3. 打成小包

packet channel 分包信道

packet filter 包滤波器

packet flow 包流

packet interleaving 分组交织
packet multiplexer 分组多路复用器
packet multiplexing technique 包多路复用技术
packet number 包编号
packet reservation system 分组预留系统
packet routing address 包路由地址
packet service 分包业务
packet size 分组长度
packet switcher 包交换器
packet switching 分组交换,分组接转
packet switching network 分组交换网
packet switching services 包交换业务,包交换服务程序
packet switching unit 包交换单元
packet telemetry 分包遥测
packet-based protocol 基于分组的协议
packing algorithm 包装算法
packing density 1. 组装密度,装填密度 2. 存储密度
packing fraction 1. 聚集率 2. 紧束分数 3. 敛集系数
packing pressure 保压压力
packplane 主货舱可更换的飞机
PACRAD Primary Absolute Cavity Radiometer 一级绝对腔体辐射仪
PACS 1. Passive Attitude Control System 被动姿态控制系统 2. Pitch Augmentation Control System 俯仰增稳控制系统 3. Pointing And Control System 定向和控制系统
Pacsat Packet Satellite 信息包卫星
PACT 1. Portable Automatic Calibration Tracker 便携式自动校准跟踪雷达 2. Precision Aircraft Control Technology 飞机精确操纵技术 3. Purchase Article Configuration Traveler 采购件构型单据
PAD 1. Packet Assembler Disassembler 分组拆装 2. Parts Availability Date 零件可用日期,零件可供日期 3. Photo Amplifier Detector 光敏放大检测器
pad 1. 发射坪,发射台 2. 降落场 3. 起降平台
pad abort 发射前中断发射
pad aligning at the central point 对中定位
pad deluge 冲水冷却
pad output 安装座输出,附件传动功率
padding 填充,统调
padding capacitor 微调电容器,垫整电容器
paddle 1. 桨叶,叶片 2. 桨形包装尺 3. 阻尼片
paddle blade propeller 宽叶螺旋桨
paddle switch 叶板电门
paddle type solar array 桨叶式太阳电池阵
paddlefoot 空军中除飞行人员以外的人员,尤指地勤人员
paddleplane 横轴旋翼机
paddlewheel satellite 带翼卫星,帆板卫星,桨轮式卫星

PADL Part and Assembly Description Language 零件和装配描述语言
PADS 1. Performance Analysis Display System 性能分析显示系统 2. Pneumatic Air Distribution System 气动空气分配系统 3. Position and Azimuth Determining System 定位和定方位系统
PAE 1. Port of Aerial Embarkation 启运航空港,装运机场 2. Preliminary Airworthiness Evaluation 初步适航性评定
PAFAM Performance And Failure Assessment Monitor 性能故障分析监视器
PAFS Primary Air Force Specialty 空军基本专业分类(美国)
PAG 1. Precision Alignment Gyrocompass 精密校准陀螺罗盘 2. Programmable Automatic Gauge 可编程序自动量规
page mode 页模式
page teleprinter 纸页式电传机
paged system 分页系统
page-nibble mode 页分段模式
PAGEOS Passive Geodetic Earth-Orbiting Satellite 地球轨道被动测地卫星(美国)
paging technique 分页技术
PAH Polycyclic Aromatic Hydrocarbon 多环芳烃
PAID Parked-Aircraft Intrusion Detector 停放飞机警卫信号器
paint 1. 图像 2. 识别图像,显示雷达回波
painting 涂装,涂漆,着色
paint-on bag 真空贴合软袋
paint-stripe loading 按色带装载
PAIR Precision-Approach Interferometer Radar 精密进场着陆干涉仪雷达
pair 1. 对,双 2. 电偶,3. 偶,对
pair of galaxies 星系对
pair of pictures 像对
pair of stations 台对
pair of stereoscopic pictures 立体像对
pair production 粒子偶的产生
PAIRC Pacific Air Command 太平洋空军(美国)
pairing 成对,并行
pairs 双机组
pairwise comparison 成对比较
Paksat Pakistan satellite 巴基斯坦卫星
PAL 1. Passive Augmentation Lens 无源放大透镜 2. Permissive Action Link 核武器启动连接装置,允许行动联系 3. Programmable Array Logic 可编程排列逻辑
palaeogeomagnetic equator 古地磁赤道
palaeogeomagnetic intensity 古地磁强度

palaeogeothermics 古地热学
palaeolatitude 古纬度
palaeolongitude 古经度
palaeomaenetic direction 古地磁方向
palaeomagnetic field 古地磁场
palaeomagnetic pole 古地磁极
palaeomagnetism 古地磁(学)
palau 钯金合金
PALC Passenger Acceptance and Load Control 旅客接纳和载荷调整
PALEA Philippine Airlines Employees Association 菲律宾航空公司雇员协会
pallet 1.货盘,集装托板,保形吊架 2.底座 3.标准尺寸底座 4.标准尺寸台架
palletisation 1.货盘,集装托板 2.货盘运输
PALM Precision Altitude and Landing Monitor 精密高度及着陆监控设备
palm 手掌,掌状物
Palmer scan 巴耳莫扫描
palnut 薄板螺帽,薄形保险螺帽
PALS 1.Precision Approach and Landing System 精确进近和着陆系统 2.Program Automated Library System 计划自动化数据库系统(美国)
palus 月沼
Palus Somnil 梦沼
PAM 1.Performance Assessment Monitor 性能评定监控器 2.Pitch Autopilot Module 自动驾驶俯仰组件 3.Pulse Amplitude Modulation 脉冲幅度调制
PAMA Professional Aviation Maintenance Association 职业航空维修协会(美国)
PAM-A Payload Assist Module-Atlas Centaur 宇宙神级有效载荷助推舱
PAMC Provisional Acceptable Means of Compliance 临时执行措施
PAM-D Payload Assist Module-Delta 德尔他级有效载荷助推舱
PAMIRASAT Passive Microwave Radiation Satellite 被动微波辐射卫星
pan 全,泛,总
Pan Amsat Pan American satellite 泛美卫星
pan film 全色胶片
pan head screw 盘头螺钉
PANAFTEL Panafrican Telecommunication 泛非通信网
PANAR Panoramic Airborne Radar 机载全景雷达
pancake 1.失速平坠着陆 2.空中截击代语:"我想着陆" 3.上表面扁平的机身
pancake ammo 请求着陆装弹
pancake fuel 请求着陆加油

pancake landing 平坠着陆
panchromatic film 全色胶片
panchromatic photography 全色摄影
pancratic lens 可调节物镜
pancreatic system 变焦系统
Pandora 【天】土卫十七
panel 1.板,壁板 2.面板,配电板,仪表板,控制板 3.翼段,机翼,翼板,翼片
panel code 目视地对空通信的标准代码
panel design 面板设计
panel display 屏面显示
panel elements 仪表板上的器件
panel flutter 壁板颤振
panel lighting 仪表板照明
panel method 1.面元法 2.板块法
panel monitor 屏面监控器
panel power 面板电源
panelboard 1.仪表板,配电板 2.绘图板 3.镶板,压制纸板
panelled 装备仪表和航空电子设备的
panel-mounted regulator 板式调节器
panex 帕内克斯碳纤维
panhandle 椅边手柄
pannier 小型火箭发射装置
panoramic barograph 全量程气压计
panoramic camera 全景照相机
panoramic display 全景显示器
panoramic distortion 全景失真
panoramic photo 全景相片
panoramic radar 全景雷达
panoramic receiver 全景接收机
panoramic spectrum analyzer 全景频谱分析仪
panoramic thermograph 全量程温度计
PANS Procedures for Air Navigation Services 空中导航中心用程序(英国),航空导航服务程序
Panther 黑豹(法国的多用途直升机,也可改为强击直升机)
panting 振动,脉动,波动,晃动
pantobase aircraft 任意场面起落飞机
pantograph 缩放仪
pants 起落架整流罩(此词主要用于美国,又称 spats)
pants duct 分叉管,Y 形进气道
PAP Precision Approach Procedure 精密进近程序
paper 材料,资料,文件
paper honeycomb 纸蜂窝,纸蜂窝结构
paper tape reperforator 纸带复凿机
paper-lined dry cell 纸板干电池
paper-pencil intelligence test 纸笔智力试验
paper-study follow-on 改进型设计方案研究

paperwork 书面工作
PAPI Precision Approach Path Indicator 精密进近航路指示器
PAPM Pulse Amplitude and Phase Modification 改变脉冲幅度和相位
PAR 1. Performance And Reliability 性能和可靠性 2. Precision Aircraft Reference 精确飞行器参考坐标 3. Precision Approach Radar 精密进近雷达 4. Preventive Aircraft Repair 飞机预防性修理
para 空降兵,伞兵
para phase amplifier 倒相放大器
para rescue team 伞降救援分队,跳伞救援小组
para sheet 多边形降落伞
para visual director 侧视指引器
paraballoon 1.充气天线 2.跳伞用氧气瓶
parabay 伞投舱,带伞空投舱
parabola 抛物线
parabola orbit 抛物线轨道
parabolic antenna 抛物面天线
parabolic approximation 抛物线近似,抛物逼近
parabolic flight 抛物线飞行
parabolic flight test aircraft 失重试验飞机
parabolic nozzle 抛物线型喷嘴
parabolic orbit 抛物线轨道
parabolic reflector 抛物面反射器
parabolic surface 抛物面
parabolic trajectory 抛物线轨道
parabolic velocity 抛物线速度
parabolize 形成抛物线
parabolized equation 抛物化方程
paraboloid 抛物线体,抛物面,抛物面反射器
paraboloidal mirror 抛物面镜
paraboloidal reflector 抛物面反射器
parabomb 1.带伞炸弹 2.带伞空投容器
parabrake 减速伞(即 parachute brake)
parachute 1.降落伞 2.用降落伞降落,跳伞
parachute bay 伞舱
parachute bomb 带伞炸弹
parachute brake 减速伞
parachute canopy 伞衣
parachute deployment 开伞
parachute deployment height 开伞高度
parachute descent 降落伞降落
parachute dummy 试伞用假人
parachute flare 降落伞照明弹,带伞照明弹
parachute fragmentation bomb 带伞杀伤炸弹
parachute gore 伞衣幅,伞衣三角布
parachute harness 伞带,降落伞背带
parachute magnetic mine 伞投磁性水雷

parachute opening shock 开伞冲击
parachute opening shock tolerance 开伞冲击耐力
parachute pack 伞包
parachute prejump training 伞前训练
parachute radiosonde 下投式无线电探空仪,降落伞无线电探空仪
parachute rigger 降落伞装配工,叠伞员
parachute rigger certificate 叠伞员资格证书
parachute system 伞系
parachute tower 1.跳伞塔 2.伞塔
parachute training 跳伞训练
parachute tray 1.伞盘 2.包伞座
parachute troops 降落伞部队,伞兵
parachute vent 伞衣顶孔
parachute-retarded bomb 带减速伞炸弹
parachutist 跳伞者,伞兵,跳伞运动员
para-circular orbit 近圆轨道
paracone 降落锥
paracone ejection seat 降落锥弹射座椅
paradigm 范例,样式
paradox 自相矛盾,悖论
paradrag drop 空投下落,减速伞阻力空投
paradrop 空投,伞降
paraffin 石蜡,链烷烃夫烷烃
paraffinic 石蜡族的
parafoil 翼伞
parafoil recovery 用翼伞回收
parafoils 可转向降落伞,翼伞
parafrag 伞投杀伤炸弹
parafrag cluster 带伞集束杀伤弹
para-hydrogen 仲氢,并转氢
parallactic angle 视差角
parallactic displacement 视差位移
parallactic ellipse 视差椭圆
parallactic inequality 均视差,视差不等量
parallactic libration 视差天平动
parallactic motion 视差运动,视差移动
parallactic refraction 视差折光差,折射视差
parallax 视差,位差
parallax angle 视差角
parallax bar 视差杆
parallax error 视差误差
parallax wedge 视差量测楔
parallel 1.平行,并联 2.平行的,并联的
parallel access multiple distribution 并行存取多址分配
parallel actuators 并联致动器
parallel aerofoil 长方形机翼
parallel approach 平行接近法
parallel arithmetic unit 并行运算器

parallel beam expand device   扩束装置
parallel buffer   并行缓冲器
parallel burning   平行燃烧
parallel circle   平行圈
parallel communication link   并行通信链路
parallel compression   并行压缩
parallel computer   并行计算机
parallel computing   并行计算,并行信息处理技术
parallel double-wedge   平行双楔翼剖面
parallel efficiency   平行效率,并行效率
parallel electric field   平行电场
parallel failure   并生故障
parallel framework   并行框架
parallel fuzing system   并联引爆系统
parallel grid   平行滤线栅
parallel implementation   并发运行,并行执行,并行计算
parallel loading   平行负载,并行加载
parallel multiple incremental computer   并行多重增量计算机
parallel of altitude   等高度线,地平纬圈
parallel of declination   等赤纬线,赤纬圈
parallel of latitude   纬度圈,等纬圈,黄纬平行圈
parallel of origin   基准黄纬圈
parallel offset route   平行航道
parallel optical computer   并行光学计算机
parallel pattern processor   并行模式处理机
parallel polarization   平行极化
parallel processing   平行处理,并行处理
parallel program   并行程序
parallel reduction   平行缩减
parallel redundancy   并联式余度
parallel resonant circuit   平行谐振电路
parallel runway   平行跑道
parallel scan   并联扫描
parallel servo   并联伺服装置
parallel target approach   平行接近目标法
parallel thread   平行线索,直螺纹
parallel transmission   并行传输
parallel version   并列版本
parallel yaw damper   并联偏航阻尼器
parallel-averted photography   等偏摄影
paralleled component   可代用元件,同类元件
parallelepiped   平行六面体的
parallel-flow   平行流,平行流式
parallel-heading square   方形航线
paralleling generators   并联发电机组
parallelism   1.平行,类似,对应 2.并行性,并行度,平行性
parallelization   平行化

parallelogram   平行四边形
parallelotope   超平行体
paramagnet   顺磁体
paramagnetic   顺磁的
paramagnetism   顺磁性,常磁性
paramatta   帕拉玛塔
parameter   参数,参量
parameter adjustment   参数平差,间接平差
parameter adjustment with constraint   附条件参数平差,附条件的间接平差
parameter bound   参数界限
parameter calibration   参数标定
parameter estimate   参数估算
parameter estimation   参数估计
parameter identification   参数辨识
parameter model   参数模型,参数模式
parameter monitoring   参数监控
parameter optimization   参数优化
parameter optimization problem   参数优化问题
parameter packet   参数包
parameter perturbation   参数扰动法
parameter potentiometer   参数电位计,参数电势计
parameter pre-processing   参数预处理
parameter requiring high response   速变参数
parameter requiring low response   缓变参数
parameter selection   参数选择
parameter set   参数集
parameter space   参数空间
parameter system   指标体系
parameter uncertainty   参数不确定性
parameter update   参数更新
parameter value   参数值
parameter variation   参数变化
parameter vector   参数向量
parameterization   参数化
parameterize   使……参数化
parameters at injection   入轨参数
parameters estimation   参量估计
parametric amplifier   参数放大器
parametric analysis   参量分析,参数分析
parametric detection   参量型检测
parametric frequency converter   参量变频器
parametric method   参数法
parametric mixer   参量混频器
parametric model   参数模式,参数模型
parametric noise   参数噪声
parametric noise source   参数噪声源头
parametric optimization   参数优化
parametric oscillator   参量振荡器

parametric representation 参数表示,参数表现
parametric solution 参数解
parametric study 参数分析
parametric take-off number 参数起飞系数
parametric test 参数检验
parametric uncertainty 参数不确定性
paraplane 翼伞,翼伞飞行器
para-retarded 带减速伞的
parasite 寄生飞机,机载飞机
parasite drag 废阻力,寄生阻力,干扰阻力
parasitic amplitude modulation 寄生调幅
parasitic capacitance 寄生电容
parasitic coupling 寄生耦合
parasitic echo 寄生回波
parasitic element 寄生元件,寄生元
parasitic emission 寄生发射
parasitic feedback 寄生反馈
parasitic frequency 寄生频率
parasitic leakage 寄生漏电
parasitic loss 寄生损失,附加损失
parasitic oscillation 寄生振荡,无源振荡
parasitic output 寄生输出
parasitic solution 寄生解
parasiting 采用寄生元件的技术
parasol monoplane 伞式单翼机
parasol radiometer 伞式辐射计
parasol wing 伞式翼
para-tellurite crystal 对位黄碲矿晶体
paravane 1. 翼伞 2. 扫雷器 3. 空中回收装置
paravulcoon 充气减速伞
parawing 翼伞,翼状降落伞,翼伞飞行器;滑翔降落软翼机
parcel 流体部分
pardop Passive Ranging Doppler 无源测距多普勒雷达
pare 削减,缩减
parent 原生放射性同位素
parent body 母星体,轨道母体
parent company 母公司,总公司
parent compound 母体化合物
parent fuel 原生燃料
parent galaxy 母星系
parent rack 主炸弹架,主挂弹架
parent ship 母舰,载舰
parent squadron 原中队
parenthesis 圆括号,圆括弧
Pareto-optimal solution 帕累托最优解
parity 1. 对称性,宇称性 2. 均势,均等 3. 奇偶(性)
parity check 奇偶校验
parity check matrix 奇偶校验矩阵,一致校验矩阵

parity generator 奇偶生成器,奇偶发生器
parity vector 奇偶矢量
park 1. 停机场 2. 停车坪
parker 停放牵引车
Parker Kalon 派克卡隆螺钉,派克螺钉
Parkerizing 磷酸盐处理,派克化
parking 停靠
parking brake 停留刹车,停车制动,停车制动器
parking catch 停放刹车转换装置
parking gate 门位
parking meter 停靠表
parking orbit 停车轨道,驻留轨道
parking stand 停机点
parking trajectory 停泊轨道
Parmod Progressive Aircraft Rework Modification 渐进性飞机拆修改装
parrot 敌我识别应答器
PARS 1. Photoacoustic Raman Spectroscopy 光声喇曼谱(学) 2. Programmed Airline Reservation System 航空公司程序订票系统
parse 解析
parsec parallax second 秒差距
PARSECS Program for Astronomical Research and Scientific Effects Concerning Space 天文研究及有关航天的科学活动计划
part 部分,局部,零件,部件,元件
part charter 定期航班包机
part count 零件计数,部分计数
part load 部分负荷,部分载荷
part number 零件号码,零件编号
partial 一部分的,局部的
partial closure 局部封闭,部分封闭
partial communication 部分交流
partial derivative 偏导数,偏微商
partial differential 偏微分
partial differential equation 偏微分方程
partial dislocation 不全位错,局部错位
partial eclipse 偏食
partial failure 局部故障,部分失败
partial fault-tolerance 部分容错
partial feedback 部分反馈
partial focused processing 部分聚焦处理
partial illumination 局部照射
partial inversion 部分反转
partial load 部分负荷,分载
partial match pattern 部分匹配模式
partial oxidation 部分氧化
partial pressure 分压,分压力
partial pressure analyser 分压分析器

partial pressure depression   分压降低
partial pressure helmet   部分加压头盔,代偿头盔
partial pressure oxygen sonsor   氧分压传感器
partial pressure vacuum gauge   分压真空计
partial priority   准优先权
partial response   部分响应,部分回应
partial reusable space vehicle   部分重复使用运载器
partial shielding   单向屏蔽
partial simulation   部分仿真
partial sum register   部分和寄存器
partial task trainer   部分任务训练器
partial thermoremanent magnetization   部分热剩磁
partial thickness crack   非穿透裂纹
partial vacuum   未尽真空,部分真空
partial-admission   部分进汽,部分进气
partial-admission turbine   部分进气涡轮
partial-equilibrium   局部均衡
partial-pressure suit   部分加压服,高空代偿服
participant   参与者,参加者
participant factor   参与因数
participate   1.参与,参加 2.分享,分担
participating receiver   跟踪接收机
participation   1.参与,参加 2.分享,分担
particle   1.粒子,微粒 2.尘粒,粉粒 3.质点
particle accelerator   粒子加速器
particle agglomerate   粉粒结块
particle beam   粒子束
particle beam weapon   粒子束武器
particle cloud   粒子云
particle collision   颗粒碰撞
particle concentration   颗粒浓度,粒子浓度
particle contamination   尘粒污染,粉粒污染
particle density   粒子密度
particle deposition   粒子沉积,颗粒沉降
particle detector   粒子探测器
particle diameter   粒径,颗粒直径,粒子直径
particle dispersion   颗粒扩散
particle distribution   颗粒分布,粒子分布
particle drift   粒子漂移
particle dynamics   质点动力学
particle emission   粒子发射
particle energy   质点能量
particle erosion   粒子浸蚀
particle fall-out   尘粒沉降
particle field   粒子场
particle flux   粒子通量
particle gyro   粒子陀螺
particle image   粒子图像
particle image velocimetry   粒子图像测速

particle impingement   粒子碰撞
particle injection   粒子注入
particle irradiation test   粒子辐照试验
particle levels   尘粒等级,粉粒等级
particle loading   粒子浓度
particle method   粒子法,质点法
particle monitor   粒子监测仪
particle motion   颗粒运动,质点运动
particle movement   粒子运动,质点运动
particle number   粒子数
particle orbit theory   单粒子轨道理论
particle phase   颗粒相
particle position   粒子位置
particle radiation   粒子辐射
particle radiation damage   粒子辐射损伤
particle radiation detector   粒子辐射探测器
particle radius   粒子半径
particle sedimentation   颗粒沉降
particle segregation   颗粒分离
particle separator   颗粒分离器
particle shape   粉粒形状,颗粒形状
particle size   粉末粒度
particle size distribution   粒度分布,粒径分布
particle static   粒子静电干扰
particle surface   颗粒表面
particle swarm   颗粒群
particle system   粒子系统,质点系统
particle temperature   粒子温度
particle trajectory   粒子轨道,颗粒轨道
particle transport   粒子输运
particle type   粒子类型
particle velocity   粒子速度,质点速度
particle weight   粒子量
particle-beam generator   粒束发生器
particle-in-cell   粒子模拟
particle-in-cell simulation   粒子模拟
particles   微粒,颗粒,极少量
particles & fields subsatellite   粒子和场的子卫星
particle-size distribution   颗粒大小分布
particular   1.特殊的,特定的 2.个别的,各个的 3.详细的,精密的
particular case   特例,特定情况
particular design   详细设计
particularity   1.特殊性,特征,特质 2.详细,细致,精确性
particulate   微粒,微粒状物质
particulate contaminants   尘粒污染物
particulate contamination   尘粒污染
particulate emission   颗粒排放,微粒排放

particulate filter 微粒过滤器
particulate matter 微粒物质,悬浮微粒
particulate radiation 粒子辐射,微粒辐射
particulate reinforced composite 颗粒增强复合材料
particulate size 颗粒大小,颗粒物粒度
parting line 分模线
parting strip 1.热刀 2.分离带
partition 1.划分,分开 2.分割,隔墙,隔离物
partitioned data set 分区数据集
part-load 部分负载,部分载荷
part-load operation 部分负荷运转
partner 1.合作伙伴 2.椎孔加固板
partnership 合股企业,合伙关系
parts catalog 零件目录
parts manufacturer approval 零部件制造人批准书
parts per million 百万分之几,百万分率
part-span shroud of blade 叶片阻尼凸台
part-task trainer 部分任务训练器
part-task trainer simulator 专项训练模拟器
party line 合用线
PAS 1. Passenger Address System 旅客广播系统 2. Phase Angle Selector 相角选择器 3. Primary Alert System 主警报系统
pascal 帕斯卡
Paschen curves 帕邢曲线
Pasiphae 【天】木卫八
PASS 1. Parked-Aircraft Security System 停放飞机保安系统 2. Passive Aircraft Surveillance System 被动式飞机监视系统 3. Passive and Active Sensor Subsystem 被动式与主动式传感器分系统,无源式与有源式传感器分系统
pass 1.通过,飞过 2.孔型,轧槽 3.乘客 4.通过
pass matrix 卫星过站参数矩阵
pass point 加密点,标志点
passage 旅行,旅程,航程,长途旅程的费用
passage flow 流量
passage lanes 穿越通道
passage of command 指挥交接
passage point 穿越点
passage shock 通道激波
passage vortex 通道涡
passageway 1.出入口 2.走廊,过道,通路
passband 通(频)带
passenger 旅客
passenger airplane 客机
passenger boarding bridge 旅客登机桥
passenger cabin 客舱
passenger coupon 飞机票收据
passenger cover 旅客人身安全保险

passenger door 客舱门
passenger embarkation 旅客登机
passenger enplanement 登机旅客总人数
passenger load 乘客载荷
passenger loading bridge 旅客登机桥
passenger mile 乘客英里
passenger profile 旅客分布图
passenger service charge 旅客离境费
passenger ton-mile (乘)客吨英里
passenger trip length 旅客行程
passenger uplifted 收费旅客人数
passenger weight 旅客重量
passenger yield 客运人均收益
passenger/cargo service 客货运服务
passengers carried 客运人数
passing 通过,许可,传送
passivating 钝化,钝化涂层
passivation 钝化
passivation layer 钝化层,保护层
passivation technology 钝化工艺
passive 1.无源的 2.被动的,消极的
passive aerodynamic braking 被动气动减速
passive air defense 消极防空
passive antenna 无源天线
passive anti-submarine tactics 被动反潜战术
passive attitude control 被动姿态控制
passive attitude stabilization 被动姿态稳定
passive cavity 无源谐振腔
passive communication satellite 无源通信卫星
passive component 无源元件
passive control 被动控制
passive cooling 被动冷却
passive countermeasure 消极干扰,无源干扰,被动式干扰
passive detection 无源探测
passive display 被动显示
passive drag 被动阻力
passive effect 消极影响,负面效应
passive electrical circuit 无源电路
passive elint system 被动式电子情报系统
passive failure 消极故障,软故障
passive fault-detection 无源故障检测
passive fuze 被动式引信
passive guidance 无源制导
passive heterodyne radiometer 被动外差辐射计
passive homer 1.被动自导引导弹,被动寻的导弹 2.被动导引头,被动寻的装置
passive homing 被动自导引,被动寻的
passive identification device 被动识别装置

passive imaging　被动成像
passive infrared night equipment　无源红外夜视设备
passive infrared rangefinder　无源红外测距器
passive infrared sighting system　无源红外瞄准系统
passive intermodulation products　无源交调产物
passive jamming　无源干扰
passive landing gear　被动式起落架
passive location　无源定位
passive magnetic suspension　无源磁悬浮
passive method　1. 被动法 2. 绝缘法
passive microwave remote sensing　1. 被动微波遥感 2. 无源微波遥感
passive microwave sensing　无源微波遥感
passive mode　无源工作状态
passive mode locking　被动锁模
passive munition　被动弹药
passive network　无源网络
passive nutation damping　被动章动阻尼
passive operation　被动式活动
passive paralleling　被动式并联双余度
passive proximity fuze　被动式近炸引信,被动式引信
passive radar　无源雷达,被动雷达
passive radar target search head　无源雷达目标搜索头
passive ranging　无源式测距,被动式测距
passive relay station　无源中继站
passive safety　被动安全性
passive satellite　无源卫星
passive scalar　被动标量
passive sensor　无源遥感器,被动遥感器
passive sensor detonating system　被动式感应起爆装置
passive sonar　无源声呐
passive sonobuoy　被动式声呐浮标
passive stability　被动稳定
passive stabilization　平衡稳定性
passive system　1. 被动系统 2. 无源系统
passive thermal control　消极热控制,被动式热控
passive tracking　无源跟踪,被动跟踪
passivity　1. 被动性,被动结构 2. 无源性,无源化
password protection　口令保护
PAST　Passenger Address Side Tone 旅客广播侧音
past experience　过去的经验
past research　过去的研究
past study　以往的研究
past work　过去的工作
paste　1. 糊,胶,膏 2. 糊化,糊料
paste foods　半固体食品
pasty propellant　膏体推进剂
pasty propellant rocket engine　膏体推进剂火箭发动机
pasty propellant rocket motor　膏体推进剂火箭发动机

PAT　1. Patent 专利 2. Pilot Applications Terminal 驾驶员应用终端 3. Production Acceptance Testing 生产验收试验
patch　一小块,片,块
patch state　补丁状态,补丁状况
patched conics technique　匹配圆锥曲线法
PATCO　Professional Air Traffic Controllers Organization 职业空中交通管制员组织
PATEC　Portable Automatic Test Equipment Calibrator 便携式自动测试设备校验器校验器
patent　1. 专利 2. 专利品 3. 知识产权
path　1. 航道,轨迹 2. 路程 3. 航径,轨线,路径
path acceleration　路径加速度
path accuracy　路径精度
path angle　航向角
path attenuation correction　路径衰减校准
path constraint　路径约束
path following　路径跟踪
path information unit　通路信息单位
path length　软通路长度,路径长度,轨迹长度
path length fuze　弹道长度引信
path line　1. 迹线 2. 路径
path loss　路径损耗
path number matrix　通路编号矩阵
path of electric current　电流的路径
path planning　航路规划
path radiance　路径辐射
path repeatability　路径重复率
path service　路径业务
path stretching　延长航迹
path velocity　路径加速度
path velocity accuracy　路径加速度精度
path velocity fluctuation　路径加速度波动量
path velocity repeatability　路径速度重复精度
pathfinder　1. 空中引导人员 2. 机载导航雷达 3. 目标引导组,空降信号员
pathfinder aircraft　领航飞机,带队飞机,探路飞机
pathfinder drop-zone control　探路飞机空投区控制站
pathfinding　引导,领航
path-following　路径跟踪
pathlength　弹道长度,路程长度
pathlines　迹线
path-planning　路径规划
path-planning strategy　路径规划策略
pathway　1. 路,道 2. 途径,路径
PATP　Product Acceptance Test Procedure 产品验收试验程序
Patrick Air Force Base　帕特里克空军基地(美国)
patrol　巡逻,巡逻队,侦察队

patrol aircraft 巡逻飞机
patrol boat 巡逻船
patrol camera 巡天照相机
PATS 1. Precision Automated Tracking Station 精确自动跟踪站 2. Primary Aircraft Training System 初级教练机系统
pattern 1. 图型,流谱,模型 2. 飞行航线图 3. 天线发射强度图 4. 方式,形式规范
pattern aircraft 样板飞机,标准样机
pattern airspeed 起落航线飞行速度
pattern analysis 模式分析
pattern bombing 定形轰炸,地毯式轰炸
pattern classification 模式分类,类样分类
pattern detection 图形检测,模式检测
pattern distortion 图形畸变
pattern distortion caused by radome 天线罩波瓣畸变
pattern factor 出口温度分布系数
pattern generator 测视图案信号发生器,图形发生器
pattern identification 模式标识
pattern imaging guidance 图像匹配制导
pattern information processing system 模式信息处理系统
pattern matching guidance 图像匹配制导
pattern of bursts 炸点群集,炸点散布
pattern of dispersion 射弹散布面
pattern of fire 射弹散布面
pattern of flow 流动类型
pattern of lethal density 密集杀伤区射弹散布面
pattern of shrapnel 弹片散布面
pattern primitive 模式基元
pattern recognition 模式识别,图形识别
pattern recognition of remote sensing 遥感模式识别
pattern recognizer 模式识别器
pattern segmentation 1. 图形分割 2. 模式分割
pattern velocity 图案速度
PATTS Programmed Auto Trim/Test System 程控配平及试验系统
PATU Panafrican Telecommunicatioin Union 泛非通信联盟
PATWAS Pilots Automatic Telephone Weather Answering Service 飞行员自动电话气象问询服务
PAWE Performance Assessment and Workload Evalution 性能评估和工作量评价
PATWING Patrol Wing 巡逻机联队
PAUC Program Acquisition Unit Cost 计划采购单位成本
paulin 防水布
pause 停顿,中止,暂停
pavement classification number 道面等级号

pavement groving 道面槽
PAW Plasma-Arc Welding 等离子弧焊接
pax 乘客,旅客
pay 1. 付,支付,缴纳 2. 付清,偿还 3. 工资,薪水,付款,报答
pay load 收费装载量,有收入的载重
Paycoms Payload Communicators 有效载荷通信员
payload 1. 有效载荷,业载 2. 有效负载,净载重量
payload assist module 有效载荷助推舱
payload bay 1. 有效载重舱 2. 有效载荷湾
payload capability 有效负载能力,有效载荷装载能力
payload capacity 业载能力,有效载重量
payload design 有效载荷设计,有效负载设计
payload fairing 有效载荷整流罩
payload flexibility 有效载荷灵活性
payload fraction 有效载重量与车重之比
payload integration 弹头装配,有效载荷装配
payload jettison 有效载荷抛投
payload mass 有效负载质量
payload mass ratio 有效载荷质量比
payload mission training 载荷任务训练
payload module 有效载荷舱
payload of opportunity 机会性搭载有效载荷,额外有效载荷
payload preparation room 有效载荷准备间
payload range 有效载荷试验场
payload ratio 有效载荷比
payload separation 有效载荷分离
payload specialist 有效载荷专家,业载工程师
payload transfer 有效载荷转运
payload weight 有效载荷重量
payload-range diagram 业载-航程曲线图
payoff 报酬,结果;发工资,结算
payoff function 支付函数
pay-per-view programming 每观看一次即收费的电视节目
PB 1. Passenger Bridge 旅客登机桥 2. Preflight Bulletin 飞行前公告 3. Push Button 按钮
PB STA Push Button Station 按钮板
PBAA Polybutadiene Acrylic Acid 聚丁二烯丙烯酸
PBAA propellant 聚丁二烯丙烯酸共聚物推进剂
PBAN Polybutadiene Acrylonitrile 聚丁二烯丙烯腈
PBAN propellant 聚丁二烯丙烯腈推进剂
PBATS Portable Battlefield Attack System 轻便式战场攻击系统
PBB Passenger Boarding Bridge 旅客登机桥
PBC Printer Board Connector 印刷电路板连接器
PBCS Post-Boost Control System 助推段后控制系统
PBD Product Baseline Definition 产品基线定义

PBE　Piggyback Experiment 搭载实验
PBI　Passenger Boarding Information 旅客登机通知
PBIT　Power-up Built In Test 电源机内测试
PBJ　Partial-Band Jammer 部分波段干扰机
PBL　1. Passenger Boarding List 旅客登机单 2. Product Baseline 产品基线
PBOM　1. Process Bill Of Material 工艺物料表 2. Production Bill Of Material 生产物料表
PBP&E　Professional Books，Papers and Equipment 专业书籍、论文及设备
PBPS　Post-Boost Propulsion System 助推段后推进系统
PBSI　Pushbutton Selector/Indicator 按钮控制选择开关/指示器
PBS　Process Breakdown Structure 工艺分解结构
PBV　Post-Boost Vehicle 助推后飞行器,弹头母舱
PBW　Particle-Beam Weapon 粒子束武器
PBX　Private Branch Exchange 专用电话交换机
PC　1. Parts Control 部件控制 2. Personal Computer 个人计算机 3. Phase Check 分段检查 4. Physical Characteristics 物理特性 5. Pressure Controller 压力调节器 6. Printed Circuit 印刷线路 7. Product Conformance 产品一致性 8. Production Certificate 生产许可证 9. Production Change 生产更改 10. Production Configuration 生产构型 11. Prototype Configuration 飞行试验机构型
pc　parsec 秒差距
PCA　1. Parts Control Area 零件控制区 2. Physical Configuration Audit 物理构型审核 3. Positive Control Area 绝对管制区 4. Power Control Actuator 动力控制作动器,动力控制作动筒 5. Practical Critical Area 实际重要区域
PCB　1. Play Control Block 播放控制块 2. Printed Circuit Board 印刷电路板
PCBW　Provisional Combat Bomb Wing 暂编战斗轰炸联队
PCC　1. Personal Code Calling 个人代码呼叫 2. Pilot Controller Communication 驾驶员管制员通信 3. Printed Circuit Card 印刷电路板 4. Production Change Control 生产更改控制 5. Production Control Center 生产控制中心
PCCS　Power Conditioning and Control System 电源调节和控制系统
PCD　1. Performance Compliance Document 性能符合文件 2. Plasma-Coupled Device 等离子体耦合器件 3. Procurement Control Drawings 采购控制图 4. Product Collector Drawings 产品选项图 8. Production Configuration Data 生产构型数据
PCDN　Process Change Design Notice 设计进程更改通知
PCE　Pre-Cooler Exchanger 预冷交换器
PCF　1. Passenger Cum Freighter 客货飞机 2. Port Core Function 端口核心功能 3. Program Control Facility 程序控制设备
PCG　Protein Crystal Growth 蛋白质晶体生长实验
PCGF　Protein Crystal Growth Facility 蛋白质晶体生长设备
PCGU　Protein Crystal Growth Unit 蛋白质晶体生长装置
PCI　1. Passenger Clearance Information 旅客放行通知 2. Peripheral Component Interconnection 外围部件互连 3. Product Configuration Identification 产品构型标识 4. Program Controlled Interruption 程序控制中断 5. Protocol Control Information 协定书控制信息
PCID　Preliminary Change In Design 设计的初步更改
PCL　Product Controlled Library 产品受控库
PCM　1. Pilot Control Module 飞行员控制组件 2. Photo Contact Master 明胶板结构模线 3. Power Conditioning Module 动力调节组件 4. Practitioner of Configuration Management 构型管理专业人员 5. Program Coordination Memo 项目协调备忘录 6. Pulse Code Modulation 脉冲编码调谐
PCMTS　Pulse Code Modulation Telemetry System 脉码调制遥测系统
PCN　1. Pacific Communications Network 太平洋通信网络 2. Pavement Classification Number 道面等级数 3. Personal Communication Network 个人通信网
PCO　1. Point of Control and Observation 控制观察点 2. Program Control Office 项目控制办公室
PCOS　Purchaseing Control Online System 采购控制联机系统,生产线采购控制
PCR　1. Planning Change Request 计划改变请求 2. Peak Cell Rate 峰值信元率 3. Preventive Cyclic Retransmission 预防性循环重发
PCRT　Projection Cathode Ray Tube 投射式阴极射线管,投影管
PCS　1. Personal Communication Satellite 个人通信卫星 2. Power Conditioning System 电力调节系统 3. Project Control System 计划管理系统
pcs　1. package 包裹 2. prices 价格 3. pieces 片
PCSA　Production Configuration Status Accounting 生产构型状态纪实
PCSV　Pilot to Controller Service 飞行员对管制员的服务
PCT　1. Partial Compatibility Test 局部兼容性试验 2. Photometric Calibration Target 光度定标测量,光度定标核准 3. Portable Control Terminal 便携式控制终端

PCU　1. Parameter Control Unit 参数控制单元 2. Passenger Control Unit 旅客控制装置 3. Power Control Unit 动力控制装置 4. Pressure Control Unit 压力控制装置 5. Priority Control Unit 优先权控制单元

PD　1. period 时期, 周期 2. Potential Difference 电位差 3. Premodule Drawing 预模件图 4. Price Discrimination 多级票价, 差别定价 5. Product Definition 产品定义 6. Product Development 产品开发 7. Protect Device 保护装置, 保护设备 8. Profile Descent 下降包线 9. Program Directives 项目指令 10. Purchased Document 采购文件

PDA　1. Personal Digital Assistant 个人数字助理 2. Preliminary Destination Airport 初步预定的目的地机场 3. Premature Descent Alert 过早下降警告

PDADS　Passenger Digital-Activated Display System 乘客数字起动显示系统

PDAR　Preferential Department and Arrival Route 优选进出港航线

PDB　Performance Data Base 性能数据库

PDC　1. Pre-Departure Clearance 离港前许可 2. Performance Data Computer 性能数据计算机 3. Power Distribution Control 配电控制

PDCA　Plan-Do-Check-Act 策划-行动-检查-改进

PDCS　Performance Data Computer System 性能数据计算机系统

PDD　1. Package Design Document 封装设计文件 2. Preliminary Design Directive 初步设计指令 3. Prerelease Design Data 预发放设计数据 4. Product Definition Data 产品定义数据

PDDD　Product Design Definition Data 产品设计定义数据

PDDI　Product Definition Data Interface 产品定义数据界面

PDDP　Preliminary Declaration of Design and Performance 设计和性能初步说明

PDE　Path Defination Error 航迹清晰度误差

PDES　1. Product Definition Exchanges Standard 产品定义交换标准 2. Pulse Doppler Elevation Scan 脉冲多普勒仰角扫描

PDF　1. Pavement Depth Factor 铺筑面厚度系数 2. Primary Display Function 主显示功能 3. Probability Density Function 概率密度函数

PD-fuzed ammunition　头部引信弹药

PDG　1. Polarization Dependent Gain 偏振相关增益 2. Procedure Design Gradient 程序设计梯度

PDI　1. Pictorial Deviation Indicator 图像偏航指示仪 2. Picture Description Instruction 图形描述指令 3. Polarization Dependent Isolator 偏振相关隔离器 4. Proviately Developed Item 私人研制的项目

PDL　1. Parallelling and Default Logic 平行和暂定的逻辑图 2. Picture Description Language 图像描述语言 3. Polarization Dependent Loss 偏振相关损耗 4. Portable Data Loader 便携式数据加载器 5. Program Design Language 程序设计语言

PDM　1. Possible Duplicate Message 可能重复的电报 2. Product Data Management 产品数据管理 3. Product Data Manager 产品数据经理 4. Product Development Management 产品开发管理 5. Proposal Definition Memo 建议定义备忘录

PDME　Precision Distance Measuring Equipment 精密距离测量设备, 精密测距器

PDMF　Programmable Digital Matched Filter 可编程数字匹配滤波器

PDMM　Pulse Doppler Map Matching 脉冲多普勒地图匹配

PDMS　1. Point Defense Missile System 要地防空导弹系统 2. Precision Distance Measuring System 精密距离测量系统 3. Product Definition Management System 产品定义管理系统

PDMT　Pulse Duration Modulation Telemetering 脉宽调制遥测

PDN　Public Data Network 公共数据网络

PDNES　Pulse Doppler Non-Elevation Scan 脉冲多普勒无仰角扫描

PDO　1. Power Door Opening 动力开门 2. Pendulum Dynamic Observer 振动体动力观测仪 3. Power Discrete Output 电源离散输出

PDOP　Position Dilution Of Precision 位置精度扩散因子, 精确位置淡化

PDP　1. Packet Data Protocol 分组数据协议 2. Policy Decision Point 决策点 3. Predetermined Point 预定点 4. Product Definition Plan 产品定义大纲 5. Program Definitin Phase 项目定义阶段

PDPS　Pack Discharge (Anti-Icing) Pressure Sensor 储器释放(防冰)压力传感器

PDQ　Photo-Data Quantizer 光学数据数字传输

PDQE　Product Development Quality Engineer 产品开发质量工程师

PDR　1. Pilots Display Recorder 飞行显示记录器 2. Pre-Determined Routes 预定航线 3. Preferred Departure Route 优选的离港航路 4. Preliminary Design Review 初步设计评审 5. Product Definition Release 产品定义发放

PDRC　Procurement Data Reference Card 数据获得基准卡

PDRCB　Preliminary Design Review Configuration Baseline 初步设计评审构型基线

PDRD Program Definition and Requirements Document 计划规定的定义和要求文件

PDRJ Pulse-Doppler Radar Jammer 脉冲多普勒雷达干扰机

PDS 1. Passive Double Star 无源双星 2. Premises Distribution System 综合布线系统 3. Primary Display System 主显示系统 4. Process Documentation System 工艺文件体系 5. Product Data Sheet 产品数据单 6. Product Development System 产品开发系统

PDSD Point Detonating Self-Destroying 弹头起爆自毁

PDSTT Pulse-Doppler Single Target Track 脉冲多普勒单个目标跟踪

PDT 1. Pacific Daylight Time 太平洋夏令时 2. Panoramic Design Technique 全景设计技术 3. Processor Diagnostic Test 处理机诊断测试

PDTS Pack Discharge Temperature Sensor 储压器释放温度传感器

PDU 1. Packet Data Unit 分组数据单元 2. Pilot's Display Unit 驾驶员显示组件，驾驶员显示器 3. Power Distribution Unit 配电装置 4. Power Drive Unit 动力驱动组件

PDUS Primary Data-User Stations 原始数据用户站

PDV 1. Path Delay Value 通路延迟值 2. Pressurizing and Drain Value 增压释压量

PDVOR Precision Doppler VOR 精密多普勒伏尔系统

PE 1. Performance Enhancement 性能增强 2. Pilot Error 驾驶员误差，驾驶员失误 3. Port of Embarkation 登机机场 4. Product Engineering 产品工程 5. Project Engineer 项目工程师 6. Purchased Equipment 采购设备

PEACESAT Pan Pacific Education and Communications Experiments by Satellite 泛太平洋卫星教育和通信实验（美国）

peacetime 和平时期，平时

peacetime utilization rate （军用运输机）和平时期利用率

peak 1. 峰值，最大值 2. 山峰，顶点，高峰 3. 生产高峰期，最高生产率

peak acceleration 峰值加速度，最大加速度

peak amplitude 高峰振幅，最高振幅，振幅峰值

peak control power at locked-rotor 峰值堵转控制功率

peak cp 尖峰浓度

peak current at locked-rotor 峰值堵转电流

peak deceleration 峰值减速

peak detection 峰值检波，峰值检测

peak diameter 最大直径

peak displacement 峰值位移

peak distribution analyzer 峰值分布分析仪

peak efficiency 最高效率，峰值效率

peak envelope power 包络线峰值功率，峰包功率

peak flow 洪峰流量，尖峰流量

peak frequency 峰值频率，高峰频率

peak heat 峰值热流，燃烧峰值

peak heat rate 峰值热率

peak heating 峰值热流，燃烧峰值

peak height 峰高

peak location 峰值位置

peak overpressure 最大超压，最大剩余压力

peak overshoot voltage 峰值上冲电压

peak performance 最佳表演，最佳性能

peak position 峰位

peak power 峰值功率，巅值功率

peak power meter 峰值功率计

peak power output 峰值输出功率

peak pressure 峰压，最大压力，最高压力

peak ratio 力矩摆动系数

peak response 最大灵敏度，峰值响应

peak searching 寻峰

peak signal 最大信号，峰值信号

peak suction 最小压力

peak temperature 峰值温度

peak to peak noise 峰-峰噪声

peak to peak value 峰-峰值

peak undershoot voltage 峰值下冲电压

peak value 峰值

peak velocity 峰值速度

peak voltage 峰值电压，尖峰电压

peak volume of inspiratory flow rate 吸气流率峰值

peaker 1. 微分电路 2. 峰化器

peaking circuit 1. 峰化电路 2. 信号校正电路

peak-notch equalizer 峰谷均衡器

peak-seeking 寻峰

peak-to-peak 峰间值，峰到峰，峰间值最大电压波动正负峰间波动，振荡总振幅

peak-to-peak amplitude 峰间幅值，峰间振幅

peak-to-peak jitter 峰-峰抖动

peak-to-peak voltage 峰-电压值

peak-to-valley 峰谷值

peak-to-valley ratio 峰-谷比

peaky 尖峰翼型

peanut fares 特低票价

pear 梨形物

pearlite 珠光体，球光体

pebble-bed heater 卵石床加热器

PEC 1. Personal Equipment Connector 随身设备连接器 2. Photo Electric Cell 光电池 3. Pressure Error Correction 压力误差修正

pecked line （图上）虚线

Peclet number 皮克里特数，皮克里特准数
PECM Passive Electronic Counter Measures 消极（或无源）电子对抗
PECO Pitch Erection Cut Off 俯仰修正切断
PECT Positron Emission Computerized Tomography 正电子发射型计算机断层成像
peculiar galaxy 特殊星系
peculiar motion 本动
peculiarity 特性，特质
PED 1. Peripheral Equipment Data 外围设备数据 2. Portable Electronic Document 可移植的电子文档 3. Physical Evidence Drawing 物理数据制图员 4. Production Engineering Document 生产工程文件
PED/D Purchased Equipment Drawings/Documents 采购设备图/文件
pedal 踏板，脚蹬板
pedal force 踏板力
pedal input 踏板输入
PEDAS Potentially Environmentally Detrimental Activities in Space 空间环境潜在放射性危害
pedestal 台基技术，轴承座，柱脚，操纵台，托架，支座，底座
PEE Packet Entry Event 分组进入事件
PEEK Poly Ethyl Ethyl Ketone 聚醚醚酮
peel 剥……皮，削……皮，表皮脱落，脱皮
peel off 1. 从平飞中滚转脱离，俯冲脱离 2.（附面层）分离
peel strength 抗剥强度，剥离强度
peel-coat 包膜的
peeling fatigue fracture 剥落型疲劳断裂
PEEP Pilot's Electronic Eye-level Presentation 飞行员平（视）显（示器）
peep type sight 规孔式瞄准具
peer review 工作中的评估，同行互查，同行评议
PEF Pylon Extension Fairing 吊架延伸整流罩
PEG polyethyleneglycol 聚乙二醇
pelagic survey 远海测量
pellant 去垢的，清除的
pellet 1. 小球 2. 小子弹
pellet ablation 弹丸消融
pellet bed 颗粒床，丸片床
pellet density 1. 球团密度 2. 粉体压实密度
pellet injection 芯块注入
pellet powder 球状火药
pellet size 颗粒大小
pellet surface 芯块表面
pelletizing 粒化，使成丸状
pels picture elements 象素，象元
PELSS Precision Emitter Location Strike System 精密发射体定位攻击系统
Peltier effect 佩尔蒂埃效应，电-热效应
PEM 1. Performance Engineers Manual 性能工程师手册 2. Privacy Enhanced Mail 增强保密的邮件 3. Program Element Monitor 程序单元监视器
pen 1. 掩体，机窝 2. 隔间
penaids 突防手段，突破辅助装置
penalize 处罚，处刑，使不利
penalty 处罚，刑罚，罚款
penalty factor 惩罚因子
penalty function 罚函数，补偿函数
penalty function method 罚函数法
penalty parameter 惩罚因子，惩罚参数
penalty term 惩罚项
pencil 铅笔
pencil-beam 笔形波束，窄波束
pencil-beam antenna 笔形波束天线，定向波束天线
pendant 悬架式操纵台
pendular robot 摆动式机器人
pendulosity 摆性
pendulous accelerometer 摆式加速度表，摆式加速度计
pendulous axis 摆轴
pendulous axis of accelerometer 加速度计摆轴
pendulous float 摆式浮子
pendulous gyroscope 悬锤陀螺仪，摆式陀螺仪
pendulous integrating gyro accelerometer 摆式积分陀螺加速度表
pendulous linear accelerometer 摆式线性加速度表
pendulous moment 摆性力矩
pendulous pulse rebalance accelerometer 摆式脉冲再平衡加速度表
pendulous vibration 悬摆阵动
pendulum 摆
pendulum damper 摆式阻尼器
pendulum experiment 摆实验
pendulum mode 钟摆式
pendulum motion 摆运动，摆动
pendulum stability 摆稳定性
pendulum test 单摆实验，吊摆实验，钟摆实验
pendulum valves 摆式阀，摆式活门
penetrant flaw detection 渗透探伤
penetrate 1. 渗透，穿透，刺透 2. 弥散于，扩散于
penetrating earth telemetry 穿地遥测
penetrating examination 渗透检验
penetrating force 侵彻力，贯穿力
penetrating particle radiation 贯穿粒子辐射
penetrating radiation 贯穿辐射
penetration 1. 穿越，穿过 2. 渗透性，洞察力 3. 突防穿透，穿透深度

penetration aids　突防设备
penetration area　突防区
penetration course　穿云着陆航向,穿云着陆航线
penetration degree　针入度,渗入度
penetration depth　1.透入深度,有效肤深 2.浸透深度
penetration distance　穿透距离,渗透距离
penetration factor　穿透系数,穿透因子
penetration function　渗透功能,透过函数
penetration height　穿透度
penetration probability　穿透概率
penetration radiation　贯穿辐射
penetration warhead　穿地弹头
penetrative convection　穿透对流,贯穿对流,透射对流
penetrativity　1.突防能力 2.穿透性
penetrator　1.突防飞机 2.穿透切割器 3.硬度计压头
penetrometer　1.硬度计,透度计 2.针入度计,针穿硬度计
penguin suit　企鹅服
peninsula　1.半岛 2.突出到水中的陆地,半岛状的地方
peniotron　潘尼管,超高频放大管
pennant　系留索
penny farthing　〈口语〉正常构型的直升机
penocfic disturbing torque　周期性干扰力矩
pentad　五,五个一组,五年间,五价物
Pentagon　五角大楼
pentane　五面棱镜
pentaprism　五角棱镜
pentavalent element　五价元素
pentene　戊烯
penthouse roof　斜坡顶
pentium　美国英特尔公司生产的微处理器,中文译名为"奔腾"
pentode　五极管
penumbra　半影,黑影周围的半阴影
penumbra cone　半影锥
penumbral eclipse　半影食
PEO　Program Executive Officer 计划执行官(美国)
Peole　佩奥利卫星
people　人,人类,民族,公民
PEP　1. Peak Envelope Power 峰值包络功率 2. Performance Engineers Programs 性能工程师方案 3. Productivity Enhancement Program 提高生产率计划
PEPE　Parallel-Element Processing Ensemble 并行部件处理复合计算机
PER　1. Packed Encoding Rules 分组编码规则 2. Packet Error Rate 分组错误率 3. Part Evaluation Record 零件评价记录 4. Premature Engine Removal 发动机提前拆卸

per　每,经,按照,每一
per unit　每单位
perambulator　测距仪
perature　脆性转化温度
perceive　察觉,感觉,理解,认知
perceived model　视模型
perceived noise level　感觉噪声级
perceived noisiness　噪度
percent　百分比,百分率,部分,百分数
percent change　百分比变化
percent difference　百分比差
percent frequency effect　百分频率效应
percent time in sunlight　受晒因子
percentage　百分比,百分率,百分数
percentage elongation　伸长率
percentage error　百分误差,误差百分率
percentage modulation　调制度
percentage variation　百分比变化
percentage-thrust indicator　推力百分比表
percentile　1.按百等分分布的数值 2.百分位,百分位数值 3.百分比下降点
perceptible band　感测谱段
perception　知觉,感知,理解
perception modeling　感知建模
perceptive-motor performance　感知运动能力
perceptivity　感受性,知觉,理解力
perceptron　视感控器,感知器
perceptual aftereffect　知觉后效应
perceptual effect　感受效果
perceptual error　知觉错误
perceptual judgement　知觉判断
perceptual process　知觉过程
perceptual time constant　知觉时间常数
perch　栖息,就位,位于,使坐落于
perchlorate　高氯酸盐
percussing device　着发机构
percussion　1.碰撞,敲打,叩击 2.震动,冲击
percussion bullet　爆破枪弹
percussion cap　刺发雷管,着发火帽
percussion cap composition　引火药
percussion charge　底火药,击发点火药
percussion composition　击发爆炸药
percussion detonator　击发雷管
percussion device　1.击发装置 2.着发引信
percussion fire　着发射击
percussion gun　驱鸟枪
percussion lock　(枪、炮等)击发装置,击发机;撞击式枪机
percussion mechanism　击发机构

percussion shell 着发炮弹
PERF Performance 性能
perfect dielectric 理想电解质
perfect dislocation 完全位错
perfect fluid 理想流体
perfect gas 理想气体,完全气体,完美气体
perfect information 完全信息,完美信息
perfect information feedback 全信息反馈
perfect knowledge 完全知识,完备知识
perfect medium 理想媒质
perfect platform 理想平台
perfect radiation 完全辐射,理想辐射
perfect-gas 理想气体,完美气体
perfect-gas law 理想气体定律
perfectly diffuse reflector 理想漫反射器,全漫反射体
perforated 有孔的,多孔的
perforated plate 多孔板,冲孔板
perforated surface 多孔表面
perforated wall 开孔壁
perforating ground telemetry 穿地遥测
perforation 穿孔,打孔,贯穿,打眼
perforation diameter 炮眼直径
perforation pattern 炮眼排列分布
perform 执行,履行,实施,完成,机器运转
performability 可运行性
performance 性能,执行,完成,行为,工作,成绩,绩效,飞机性能,工作性能,运行特性,工况,效果,运行,实行
performance advantage 性能优势
performance analysis 技术性能分析
performance and cost evaluation 性能成本评定
performance benefit 性能好处
performance bonds 履约保证书
performance bound 性能指标边界
performance calculation 性能计算,特性计算,功率计算
performance calibration flight 性能校飞
performance change 性能变化,性能差异
performance characteristic 性能特性
performance characterization 性能表征
performance chart 操作图,性能图,工作特性图,操作性能图
performance coefficient 性能系数
performance comparison 性能比较
performance constraint 性能约束,操作限制
performance/cost ratio 性能价格比
performance criterion 性能判据,性能标准
performance curve 性能曲线
performance datum 性能数据
performance degradation 性能衰退,性能降低

performance design 性能设计
performance diagram 工作特性图
performance distribution 操作能力分配
performance error 执行误差
performance evaluation 性能鉴定
performance evaluation missile 性能鉴定(导)弹
performance factor 1.性能因素 2.正确动资,正确动作率
performance function 性能函数
performance gain 性能收益,性能获益
performance groups 飞机(性能)分类
performance improvement 性能改进
performance increase 性能提升,性能提高
performance index 性能指标
performance indicator 绩效指标,业绩指标,成绩衡量指标
performance indices 性能指标
performance issue 性能问题
performance level 性能水平,性能水准,性能等级
performance limiting conditions 1.性能限制条件 2.性能极限状态
performance loss 性能损失
performance management system 性能管理系统
performance map 性能(曲线)图
performance measure 衡量工作表现的方法,性能测量,工作指标
performance measurement 工作状况的测定,绩效衡量,性能测定,性能测量
performance metric 性能指标
performance model 性能模型,应用模式
performance number 特征数,功率值,品度值
performance parameter 性能参数,效能参数,特性参数
performance penalty 性能代偿
performance prediction 性能估计,性能预测,效果预测
performance proof cycle 性能验证期
performance reduction 性能换算
performance requirement 性能要求,技术特性要求
performance result 性能结果
performance robustness 品质鲁棒性,性能鲁棒性
performance scalability 性能可伸缩性
performance simulation 性能仿真,功能仿真,性能模拟
performance specification 性能说明,性能规格
performance standard 性能标准,生产定额
performance test 性能试验,出力试验
performance tolerance 能力耐限,性能容限
performance tolerance limit 工作能力容限,性能容限
performance trend 性能趋势
performance value 绩效评价
performance variable 操作变量

performance weight　绩效权重
performance-command system　性能优化系统
performance-type glider　高性能滑翔机,翱翔机
PERGS　Portable Earth Resources Ground Station 机动式地球资源地面站
PERI　perigee 近地点
periapse　近质心点
periapsis　近星点,近拱点
periapsis altitude　近拱点高度
periapsis distance　近拱距
periapsis passage　通过近心点
periapsis radius　近拱点半径
periastron　近星点
pericenter　近星点,近心点
pericenter altitude　近心点的高度
pericenter control　近心点控制
pericynthian　近月点
perifocus　近焦点
perigee　近地点
perigee advance　近地点前移
perigee altitude　近地点高度
perigee boost　近地点助推
perigee boost motor　近地点发动机
perigee height　近地点高度
perigee injection　近地点注入
perigee kick rocket engine　近地点火箭发动机
perigee kick rocket motor　近地点火箭发动机
perigee motor　近地点发动机
perigee passage　通过近地点
perigee propulsion　近地点推进
perigee regression　近地点回归
perigee speed　近地点速度
perigee stage　近地点级
perigee-to-perigee period　近地点周期
perihelion　近日点
perijove　近木星点,近木木点
perilune　近月点
perilune altitude　近月点高度
periments　科学实验
perimeter　周,周界线,周长
perimeter acquisition radar　边境截获雷达,环形搜索雷达
perimeter track　周边滑行道
period　1.周期 2.时间间隔,一段时间,阶段
period doubling　周期倍增
period meter for reactor　反应堆周期仪
period of light variation　光变周期
period of operation　运转时期
period of restricted availability　（航母）有限使用期

period of time　一段时间,时段
period oscillation　短周期振动
periodic boundary　周期边界,周期性边界,周期型边界
periodic boundary condition　周期性边界条件
periodic breathing　周期性呼吸
periodic control　周期控制,周期性控制
periodic distribution　周期分布,周期性分布
periodic disturbance　周期干扰,周期扰动
periodic disturbing torque　周期性干扰力矩
periodic duty　周期工作制
periodic error　周期误差
periodic excitation　周期激励
periodic health assessment　定期健康评定
periodic inspection　定期检查
periodic maintenance　定期维修,定期检修
periodic monitor　定时监测
periodic motion　周期运动
periodic nonstationarity　周期非平稳性
periodic orbit　周期轨道
periodic perturbation　周期摄动
periodic physical examination　定期体检
periodic sequence　周期序列,周期时序
periodic solution　周期解
periodic structure　1.周期性结构,周期组织 2.循环结构
periodic term　周期项
periodic trajectory　周期轨迹
periodic variable　周期变量
periodic vibration　周期振动
periodicity　周期性
periodicity condition　周期性条件
period-luminosity relation　周期光度关系
periodogram　周期图
period-spectrum relation　周谱关系
peripheral asymmetric configuration　非周向对称布局
peripheral field　周边视野,外围视野
peripheral hem　伞衣前缘
peripheral interface　外部接口,外围接口
peripheral module　外围部件
peripheral processing unit　外围程序单元
peripheral vision　周边视觉,周围视觉
periphery　外围,边缘,圆周,圆柱体表面
periscope depth　潜望镜深度
peritectic structure　包晶组织,包晶结构
Permalloy　坡莫合金,透磁性铁镍合金
permanent damage　永久性损伤
permanent deformation　永久变形,残留变形
permanent disability　永久失能
permanent disqualification　永久不合格
permanent echo　固定目标的回波

permanent effect 永久效应
permanent grounding 永久停飞
permanent hearing loss 永久性听力丧失
permanent magnet 永磁体,永久磁铁
permanent magnet DC torque motor 永磁直流力矩电机
permanent magnet gyro motor 永磁陀螺电机
permanent magnet torquer 永磁式力矩器
permanent magnetic material 永磁材料
permanent magnetic synchronous motor 永磁同步马达
permanent memory 永久存储器,固定存储器
permanent mold casting 1.冷硬铸造,冷硬铸件 2.金属型铸造
permanent repair 经常修理,永久性修理
permanent solution 永久性解决方案,长久之计
permanent threshold shift 永久性阈移
permanent-data storage technology 固定数据存储技术
permanganate 高锰酸盐
PERME Propellants, Explosives and Rocket Motor Establishment 推进剂、炸药和火箭发动机研究所(英国国防部)
permea meter 磁导计
permeability 1.渗透率,渗过性,渗透性,透气性 2.导磁性,导磁率,磁导率
permeability factor 渗透因子,渗透因素
permeability measurement 磁导率测量
permeability tuning 磁导系数调谐
permeable base transistor 可渗基区晶体管
permeation 渗透
Permendur 坡明杜合金
Permenorm 一种铁镍合金,波曼诺铁镍合金
Perminvar 坡明瓦合金,波尼瓦尔铁镍钴合金
permissible error 允许误差,容许误差
permissible flight envelope 允许飞行包线
permissible interference 允许干扰
permissible load factor 可用过载
permissible stress 容许应力,许用应力
permissive action link 容许动作连系
permissive matching 容错匹配
permit 许可,允许
permit escape 允许逃逸
permitted blockage percentage 允许阻塞度
permitted transition 容许跃迁
permittivity 1.介电常数,电容率 2.介电常数分散
permly 永久性地
permutation 排列,置换
permutation group 置换群
permutation matrix 置换矩阵
peroxide 过氧化氢,过氧化物
peroxide fuel 过氧化氢燃料

perpendicular 1.成直角的,正交的,垂直的 2.垂线,垂直的位置
perpendicular polarization 垂直极化
Pershing 潘兴式导弹
persist 持续,持久,存留
persistence 余辉时间,余辉
persistence of regular reflection 持久性的常规反射
persistent (chemical) agent 持久性化学战剂
persistent gas 持久性毒气
persistent radioactivity 持久放射性,持久性放射作用
persistent vesicants 持久性糜烂(毒)剂
persistent war gas 持久性毒气
persistron 图像保持管,持久显示器
personal and instrumental equation 人仪差
personal characteristics 个人特性
personal communication 1.人员传播,个人沟通 2.个人通信,专用通信
personal computer 个人电脑,个人计算机
personal computer telemetry station 个人计算机遥测站
personal cooling system 个体冷却系统
personal effects floater 个人财物保险
personal equipment connector 飞行人员装具插头(氧气、通信、抗荷衣、电子对抗的快卸插头)
personal error 人为误差,人为差错
personal locator beacon 个人定位信标,人员示位信标
personal mobile communication 个人移动通信
personal protection 个体防护
personal protection equipment 个人防护装备
personal relationship 个人关系,人际关系
personal rescue system 个人营救系统
personal survival equipment 个人救生装备
personal thermal conditioning 个体热调节
personality change 个性变化,人格变化
personality clash 人格冲突
personality tendency 人格倾向,个性倾向
personality test 人格试验,性格试验
personnel 全体人员,全体职员,员工
personnel beacon 个人定位信标
personnel bomb 杀伤炸弹
personnel locator beacon 个人定位信标,航空救生信标
personnel monitoring 人体辐射剂量检查,个人放射性沾染监测
personnel rescue beacon 人员救援信标
perspective 1.观点 2.远景 3.透视,透视图
perspective flight-path 飞行路径透视
perspective flight-path display 飞行路径透视显示方法
perspective grid 透射格网
perspective illusion 透视错觉
perspective projection 透视投影

perspective traces 透视截面法
perspective view 透视图,鸟瞰图
Perspex 坡斯培有机玻璃
PERT 1. Programme Evaluation Review Technique 计划评审技术,统筹法,网络计划管理 2. Program Evaluation Review Technique 程序评价评审技术
pertain 属于,关于,适合
pertinax 层压纸板,胶纸板,电木,酚醛塑料
perturb 1. 扰乱,使……混乱,使……心绪不宁 2. 使摄动
perturbation 1. 干扰,杂音,噪声 2. 扰动,摄动
perturbation approach 摄动法
perturbation by earth's atmosphere 地球大气摄动
perturbation by earth's figure 地球外形摄动
perturbation by moon's figure 月球外形摄动
perturbation by solar radiation pressure 太阳辐射压力摄动,太阳风摄动
perturbation by the third body 第三体摄动
perturbation calculation 摄动计算,微扰计算
perturbation dynamics 振动动力学
perturbation effect 扰动效应
perturbation equation 微扰方程,扰动方程
perturbation guidance 摄动制导
perturbation method 1. 摄动法 2. 微扰法,微扰理论
perturbation model 摄动模型
perturbation problem 扰动问题
perturbation solution 摄动解
perturbation technique 摄动技术,摄动法
perturbation term 摄动项
perturbation theory 微扰理论,摄动理论
perturbation velocity 扰动速度
perturbation velocity potential 扰动速度势
perturbative acceleration 扰动加速度
perturbative effect 扰动效应
perturber 1. 使摄动 2. 微扰剂
perturbing force 扰动力
perturbing function 扰动函数,微扰函数
perturbing potential 摄动位,扰动位
perturbing-disturbing function 摄动-扰动函数
perveance 导电系数,导流系数
PES 1. Passenger Entertainment System 旅客娱乐系统 2. Personal Earth Station 个人地球站 3. Product Engineering Standard 产品工程标准 4. Production Engineering Specification 生产工程规范
PESO Product Engineering Services Office 产品工程服务处
PET 1. Pacific Engineering Trials 太平洋工程试验 2. Point of Equal Time 等时点
petal 折动板,鱼鳞板

petal cowling 瓣状整流罩
PETN Pentaerythritol Tetranitrate 季戊四醇四硝酸酯
petrol 汽油
petrol bomb 凝固汽油弹
petrol bowser 汽油加油车
petroleum 石油
PETT Project Engineers & Technologists for Tomorrow 研究未来的工程师和技术专家
petticoat 百褶套
PEU 1. Power Electronics Unit 电源电子装置 2. Processing Electronics Unit 加工电子设备
PF 1. Pattern Flight 按起落航线飞行 2. Pilot Flying 操纵飞机的驾驶员,驾驶员飞行 3. Pluse Frequency 脉冲频率 4. Power Factor 功率因数 5. Process Flow 工艺流程图
PFA 1. Popular Flying Association 飞行爱好者协会 2. Post Flight Analysis 飞行后分析
PFAT Preflight Assurance Test 飞行前保证试验
PFB Preliminary Flying Badge 初级飞行证章
PFC 1. Preliminary Flight Certification 初步飞行审定 2. Passenger Facility Charges 旅客附加费,旅客设施费 3. Primary Flight Computers 主飞行计算机
PFCES Primary Flight-Control Electronic System 主飞行控制电子系统
PFCS Primary Flight Control System 主飞行控制系统
PFCU Primary Flight Control Unit 主飞行控制组件
PFD 1. Pilots Flight Display 驾驶员飞行显示器 2. Primary Flight Director 主飞行指引仪 3. Primary Flight Display 主飞行显示器
PFE Path Following Error 航路跟踪误差
PFF Pre-First Flight 首航试飞
PFFT Parallel Fast-Fourier Transform 平行快速傅里叶变换
PFH Per Flight Flying Hour 每飞行小时
PFHA Preliminary Function Hazard Analysis 初步功能危险分析
PFI Post Flight Inspection 飞行后检查
PFL 1. Planned Flight Level 计划飞行高度 2. Practice Forced Landing 练习性迫降
PFLA Practice Forced Landing Area 迫降练习场
PFLD Pilot's Fault List Display 飞行员驾驶错误细目显示器
PFM 1. Performance Figure of Merit 性能品质指数 2. Pulse Frequency Modulation 脉冲频率调制
PFO Particle Fall-Out 尘粒沉降
PFPM Production Flight Procedures Manual 生产性飞行程序守则
PFQT Preliminary Flight Qualification Test 飞行资格初试

PFR 1. Part Failure Rate 部件故障率 2. Passenger Flow Rate 客流率 3. Pay From Receipt 凭收据支付 4. Permitted Flying Routes 允许飞行航路 5. Power Failure Recovery 电源故障恢复

PFRT Priliminary Flight Rationing Test 额定功率预飞试验

PERT/CPM Project Evaluation Review Technique/Critical Path Method 计划评审技术-关键路径法

PFS 1. Planned Flying and Servicing 有计划的飞行和维护 2. Pre-Feasibility Study 可行性研究

PFSV Pilot Forecaster Service (Military) 军用移动气象预报服务

PFTA Payload Flight Test Article 有效载荷飞行试验件

PFTS 1. Permanent Field Training Site 永久性野战训练场 2. Production Flight Test Schedule 生产中飞行试验计划

PG 1. Permanent Grade 正式军衔,永久军衔 2. Pressure Governor 压力调节器

PGA 1. Power Generating Assembly 发动机起动装置 2. Programmable Gain Amplifier 可编程增益放大器 3. Programmable Gate Array 可编程门阵列

PGCF Premixed Gas Combustion Facility 预混合气体燃烧设备

PGHM Payload Ground Handling Mechanism 有效载荷地面处理装置

PGM Precision Guided Munitions 精确制导武器

PGN Passenger-Generated Noise 旅客(产生的)噪声

PGOC Payload Ground Operations Contractor 有效载荷地面操作承包商

PGRM RGTR Program Register 程序寄存器

PGRV 1. Post-boost Guided Reentry Vehicle 助推结束后制导再入飞行器 2. Precision-Guided Reentry Vehicle 精确制导再入飞行器

PGS Payload Ground Support 有效载荷地面保障

PGSC 1. Payload and General Support Computer 有效载荷和通用保障计算机 2. Personnel Guide Surface Canopy 人用导向面伞衣

PGSE 1. Payload Ground Support Equipment 有效载荷地面保障设备 2. Peculiar Ground-Support Equipment 特殊型号地面保障设备

PH 1. Packet Handling 分组处理 2. Parker Hydraulics 派克公司(液压系统) 3. Per Hour 每小时 4. Phase 相位,相

pH 酸碱值,酸碱度

pH gradient 酸碱梯度

PHA Preliminary Hazard Analysis 初级危险分析

phanotron 热阴极充气二极管

phantom 1. 幻象,假象 2. 仿真,模型,体模

phantom drawing 虚线图,透视图
phantom lines 虚线,点划线,断划线
phantom member 虚设构件
phantom order 虚设定单
phantom target 假目标
phantom target trajectory 假目标轨迹
phantom track 虚假航迹
phase 1. 阶段,周期的一部分 2. 相,相位,位
phase advance 相位超前
phase alternation line system 相变线
phase ambiguity 相位模糊,相位多值性
phase angle 相位角
phase array 相控阵
phase bandwidth 相位带宽
phase boundary 相界面,相界线
phase calculation 相位计算
phase center 相位中心
phase change 相变
phase change material 相变材料
phase change material device 相变材料装置
phase check 阶段检查
phase code 相位编码
phase coherence 相位相干
phase comparator 相位比较器
phase comparison monopulse 比相单脉冲
phase comparison position determining system 比相定位系统
phase comparison radar fuze 比相雷达引信
phase conjugation 相位共轭
phase contrast 相位对比,位相衬度
phase control 相位控制
phase curve 相位曲线
phase cycle 相位周,巷
phase cycle value 相位周值,巷宽
phase datum 相位基准
phase delay 相时延,相位滞后
phase demodulator 相位解调器
phase density 相密度
phase detection 检相,相位检测
phase detector 鉴相器,相位比较器,相位检波器
phase deviation 相位偏差,相位偏离
phase diagram 相图,相位图
phase difference 相位差,相差
phase differential 相位差
phase discriminator 相位鉴别器,鉴相器
phase equalizer 相位均衡器
phase equilibria 相平衡
phase equilibrium 相平衡,相位平衡
phase error of satellite 卫星相位差

phase history	相位历程
phase induced polarization method	相位激发极化法
phase integral	相积分
phase interface	相界面
phase interference	相位干涉
phase inverter	倒相器,反相器
phase jitter	相位抖动
phase jump	相位跃变,相位突变
phase lag	相位滞后
phase linearity	相位线性度
phase lock	锁相
phase lock demodulator	锁相解调器
phase lock doppler tracking system	锁相多普勒跟踪系统
phase lock technique	锁相技术
phase locked crystal oscillator	锁相晶体振荡器
phase locus	相位轨迹
phase margin	相位余量,相位裕度
phase matching angle	相位匹配角
phase measurement	相位测量,相量测量,相角测量
phase meter	相位计
phase modulated transmitter	调相发射机
phase modulation	相位调制,调相
phase modulator	调相器
phase noise	相位噪声
phase nulling optical gyro	零相位光学陀螺
phase of flight	飞行阶段
phase of operation	飞行事故阶段
phase of the moon	月相
phase out	分阶段淘汰,逐步淘汰
phase plane	相位平面,相平面
phase plane portrait	相平面图
phase portrait	相位图,相图
phase pre-equalization	相位预均衡
phase rate	相位变率
phase reference voltage	相位基准电压
phase resonance	相位共振
phase rule	相律
phase sensitive amplifier	相敏放大器
phase sensitive demodulator	相敏解调器
phase separation	相位分离
phase shift	相移,移相,相位差
phase shift keying	相移键控
phase shift oscillator	移相振荡器
phase shifter	移相器
phase shifter package	移相器组件
phase skip	相位跳越
phase space	相空间
phase spacing	相间距离,相空间
phase stabilization	相位稳定
phase stabilizer	相位稳定器
phase step	相位阶跃
phase synchronization	相位同步
phase system	相位系统
phase trajectory	相轨迹
phase transfer function	相位传递函数
phase transformation	相位变换
phase transition	相变,相位变换
phase variation	相转变,相变异
phase velocity	相速,波速
phase viscosity	相黏度
phase volume	相体积
phase volume ratio	相体积比
phase-angle	相位角
phase-change	相变
phase-comparison monopulse	比相单脉冲
phased array	相控阵
phased array antenna	相控阵天线
phased array instrumentation radar	相控阵测量雷达
phased array jammer	相控阵干扰机
phased array radar	相控天线阵雷达
phased array space tracking radar	相控阵航天跟踪雷达
phased-array radar homing head	相控阵雷达导引头
phase-delay difference frequency	差频相位延迟
phase-frequency characteristics	相频特性
phase-frequency distortion	相位频率失真,相频畸变
phase-insensitive	对相位变化不灵敏的,相钝的
phase-lag	相位滞后
phase-lag network	相位滞后网络
phase-lead	相位超前
phase-lead network	相位超前网络
phaselock technique	1.锁相技术 2.相位锁定技术
phaselocked frequency discriminator	锁相鉴频器
phase-locked interferometer	锁相干涉仪
phase-locked local oscillator	相位锁定本振
phase-locked loop	锁相环
phase-locked oscillator	锁相振荡器
phase-locked receiver	锁相接收机
phaseout date	截止日期
phase-plane	相平面
phase-plane controller	相平面控制器
phase-scan radar	相扫雷达
phase-scanning	相位扫描
phase-shift	相位移
phase-shift constant	相移常数
phase-shift discriminator	相移鉴频器
phase-shift driver	相移激励器
phase-shift network	相移网络

phase-space　相空间
phasometer　相位计
phasor　相矢量，相量
phatanx close-in weapons system　密集阵近战武器系统
PHD　Pilot Horizontal Display 驾驶员水平显示器
phenogram　物候图
phenol　酚，石碳酸
phenolic　酚醛的，(苯)酚的
phenolic resin　酚醛树脂
phenolic/epoxy　酚醛环氧树脂(胶)
phenolics　酚醛塑料
phenological chart　物候图
phenomenon　现象
phenyl　苯基
pheromone　信息素，外激素
pheromone trail　信息素追踪
PHI　1. Position and Heading Indicator 位置及航向指示器 2. Protocol Handling Input stream 协议处理输入码流 3. Packet Handling Interface 分组处理接口
Phillips entry　菲力浦前缘形状
phlebography　静脉波描记术，静脉造影术
phlegma　冷凝液
Phobos　【天】火卫一
Phoebe　【天】土卫九
phone　1. 电话 2. 耳机，听筒 3. 头戴受话器
phonetic alphabet　字母发音单字表，音标字母
phonic warning　语音告警
phonograph　1. 留声机，电唱机 2. 电话电报，话传电报
phonograph record　唱片
phonometer　声强计，音波测定器，测声机
phos　无机磷，磷酸盐
phosgene　光气，碳酰氯
phosgene smoke screen aerial bomb　光气毒烟幕空投炸弹
phosphate　磷酸盐，皮膜化成
phosphate esters　磷酸酯
phosphate paint　磷酸盐漆
phosphating　磷化
phosphor　1. 磷，黄磷 2. 荧光物质，磷光体
phosphor bomb　磷弹，燃烧弹
phosphor bronze　磷青铜
phosphor screen　荧光屏
phosphorescence　磷光，磷光现象
phosphoric acid electrolyte fuel cell　磷酸电解质燃料电池
phosphoric acid fuel cell　磷酸燃料电池
phosphorus　磷
phosphosilicate glass　磷硅玻璃
phot　辅透，厘米烛光

photaceram　光敏微晶玻璃
photicon　光电摄像管，高灵敏度摄像管
photo alidade　摄影照准仪
photo amplifier detector　光敏放大检测器
photo angulator　摄影量角仪
photo base　像片基线
photo coordinate system　像平面坐标系
photo data quantizer　光电数据量化器
photo document　相片文书
photo duplicate　复制相片
photo enlargement　照相放大
photo goniometer　像片量角仪，光电测角仪
photo interpretation　像片判读，图像判读，照片判读
photo ionization　光致电离
photo luminescence detector　光致发光探测器
photo mosaic　像片镶嵌
photo nadir point　像底点，照片底点
photo orientation elements　像片方位元素
photo perspectograph　摄影透视仪
photo planimetric method of photogrammetric mapping　综合法测图
photo reconnaissance satellite　照相侦察卫星
photo rectification　像片纠正
photo sensitive coating　感光层
photo sensitivity　感光性
photo transformer　照相纠正仪，像片纠正仪
photo valve　光电管
photo warhead　摄影战斗部
photoabsorption　光吸收
photoacoustic Raman spectroscopy　光声喇曼谱(学)
photoacoustic spectroscopy　光声光谱(学)
photoactivated　光(能)启动的，光致动的，光驱动的
photoactive　光敏的，感光的
photoactor　光电变换器
photoamplifier　光电放大器
photocartograph　立体测图仪
photocathode　光电阴极
photocell　光电管，光电池
photochemical　光化学的，光化的
photochemical equilibrium　光化学平衡
photochemistry　光化学
photochromic　1. 光致变色的，光彩色的 2. 光敏材料
photochromic glass　光色玻璃
photochromism　光色性
photochromy　彩色摄影术，彩色照相
photoconduction　光电导
photoconductive　光电导的
photoconductive detector　光电导探测器
photoconductive effect　光电导效应

photoconductive solid-state detector　光电导型固态探测器
photoconductivity　光电导性
photoconductivity decay　光电导衰退
photoconductor　光电导体
photo-control　光控的
photocurrent　光生电流,光电流
photodetachment　光致脱离
photodetection　光探测,光检测
photodetector　光电探测器
photo-digital store　光数字存储器
photo-digitizing system　图像数字仪
photodiode　光电二极管,光控二极管
photodiode array　光电二极管阵列
photodissociation　光致离解
photodoping　光掺杂
photodrafting　照相放样
photodrone　无人驾驶照相侦察机,无人驾驶航测飞机
photoeffect　光效应,光电效应
photoelastic coating　光弹涂层
photoelasticity　光弹性学,光弹
photoelasticity test　光弹性试验
photoelectric　光电的
photoelectric astrolabe　光电等高仪
photoelectric cell　光电管,光电池
photoelectric collimating tube　光电准直管
photoelectric collimator　光电准直管
photoelectric colorimetry　光电比色分析
photoelectric conversion efficiency　光电转换效率
photoelectric device　光电器件
photoelectric effect　光电效应
photoelectric emission　光电发射
photoelectric guider　光电导星装置,光电导星镜
photoelectric material　光电材料
photoelectric photometry　光电测光
photoelectric proximity fuze　光电近炸引信
photoelectric scanning　光电扫描
photoelectric sensor　光电式传感器,光电敏感器
photoelectric smoke detector　光电感烟探测器
photoelectric system　光电系统
photoelectric tachometric transducer　光电式转速传感器
photoelectric target seeker　光电自动导引头
photoelectric telescope　光电望远镜
photoelectric theodolite　光电经纬仪
photoelectric transit instrument　光电中星仪
photoelectric width meter　光电式宽度计
photoelectrochemical cell　光电化学电池
photoelectron　光电子
photoelectron spectroscopy　光电子能谱(学)
photoelectronic imaging　光电成像
photoelectronics　光电子学
photoelement　光电池,光电管
photoemission　光电发射
photoemissive　光电发射的
photoemissive detector　光电发射探测器
photoetching　光蚀刻加工
photo-excitation　光致激发
photofet　光控场效应晶体管
photofixation　先固定
photoflash　1.照相闪光灯 2.闪光灯照片 3.照明弹
photoflash bomb　照相闪光弹,照相炸弹,摄影闪光弹
photoflash cartridge　低空照相闪光弹
photofluorogram　荧光屏图像照片
photo-generated current　光生电流
photo-generated voltage　光生电压
photogeology　摄影地质学
photogrammeter　摄影经纬仪
photogrammetric camera　摄影测量相机
photogrammetric compilation　摄影测量编图
photogrammetric control　摄影测量控制
photogrammetric coordinate system　1.摄影测量坐标系 2.像平面坐标系
photogrammetric fixing position　摄影测量定位
photogrammetric interpolation　摄影测量内插
photogrammetric map　摄影测量地图
photogrammetric mapping　摄影测图
photogrammetric measurement　航空摄影测量
photogrammetric sketch　像片略图,摄影测量草图
photogrammetric stereocamera　摄影测量立体相机
photogrammetric survey　摄影测量
photogrammetry　摄影测量学,照相制图
photograph　照片,相片
photograph meridian　相片主纵线
photograph nadir　摄像底点,相片底点
photograph perpendicular　相片垂线,摄影机主光轴
photograph pyramid　相片锥形法
photographer　(航母)照相员
photographic astrometry　照相天体测量学
photographic baseline　摄影基线
photographic coordinate system　摄影坐标系
photographic dry plate　(照相)干板
photographic emulsion　感光乳胶
photographic film　胶片
photographic flight line　照相飞行路线
photographic fog　底片雾
photographic frequency　摄影频率
photographic image　摄影影像
photographic intelligence　摄影情报,照相情报

photographic interpretation 照片判读	photomicrography 显微摄影
photographic intersection 摄影交会	photomicrometer 光电测微计
photographic layout drawing 照相放样图纸	photomisson 光电发射
photographic lofting 照相放样	photomontage 相片镶嵌
photographic magnitude 照相星等	photomultiplier 光电倍增器,光电倍增管
photographic photometry 照相测光	photomultiplier tube 光电倍增管
photographic radiant 摄影辐射点	photon 光子,光电子,光量子
photographic reading 照片阅读	photon count 光子计数
photographic reconnaissance 摄影侦察,照相侦察,摄影勘察	photon counter 光子计数器
photographic reconnaissance satellite 照相侦察卫星	photon echo 光子回波
	photon engine 光子发动机
photographic record 照片记录法,照相记录	photon integrating nephelometer 光子积分浑浊度表
photographic registration 相片配准,摄影记录	photon rocket 光子火箭,光子火箭发动机
photographic resolution 1.摄影鉴别率 2.摄影分辨率 3.相片清晰度	photo-navigator 航测领航员
	photon-counting 光子计数
photographic scale 摄影比例尺,照片比例	photonegative 负光电导的
photographic sketch 相片草图	photonegative characteristic 负光电特性
photographic sortie 飞机摄影架次,航摄飞行架次	photopair 相片对
photographic speed 摄影速度,感光度	photopic adaptation 对光适应
photographic star catalogue 照相星表	photopic vision 亮视觉,白昼视觉,可见光视觉
photographic strip 1.航空摄影胶卷 2.单连续航空照片镶嵌图 3.摄影航带	photoplan 相片平面图
	photoplane 1.航摄飞机,航测飞机 2.相片平面图
photographic sunshine recorder 摄影日照计	photoplasticity 光塑性
photographic telescope 照相望远镜	photoplotter 光学绘图机
photographic transmission density 胶片透光密度	photoplotting 照相测图
photographic vertical circle 照相垂直环	photopolarimeter 光偏振仪
photographic zenith tube 照相天顶筒	photopolymer 光聚合物,光敏聚合物,感光聚合物
photography 摄影学	photoreading 照片判读
photoheliogram 太阳全色照片	photo-recombination 光致复合
photoimpact 光冲量	photorectifier 光电二极管,光电检波器
photoionization 光致电离,光化电离	photoreducer 缩小仪
photolithography 光刻	photoresist 光刻胶
photology 光学	photoresist coating 涂胶
photoluminescent dosemeter 光致发光剂量计	photoresistor 光敏电阻
photolysis 光分解,光解,光解作用	photoresponse non-uniformity 光响应度不均匀性
photomacrography 低倍放大摄影	photosensitivity 感光性,光敏性
photomagnetic effect 光磁效应	photosensor 光敏器件,光电传感器
photomagnetic memory 光磁存储器	photosmoke method 光学烟度测定法
photomap 照相地图,照相制图	photosmoke unit 光学烟度计
photometer 光度计,曝光表	photosphere 光球,光球层
photometric analysis 光度分析	photostability 耐光性,光稳定性
photometric binary 测光双星	photostage 光(照)阶段
photometric measurement 光度测量	photostereograph 立体测图仪
photometric observation 光度观测	photosynthesis 光合作用
photometric sequence 测光序	photosynthetically active radiation 光合作用有效辐射
photometric solution 测光解	phototheodolite 照相经纬仪
photometric system 测光系统	photothyristor 光敏闸流管
photometry 光度学,测光学,测光法	phototimer 曝光计,摄影记时器
photometry aperture 测光孔径	phototransister 光电晶体管

phototriode 光电三极管
phototube 光电管,光电池
phototypesetter 照相排字机
photounit 光电元件
photovaristor 光敏电阻
photoviscoelasticity 光黏弹性
photovisual magnitude 仿视星等,光视星等
photovoltage 光生电压
photovoltaic cell 光生伏打电池,阻挡层光电池
photovoltaic concentrator module 聚光太阳电池组件
photovoltaic concentrator solar array 聚光太阳电池方阵
photovoltaic detector 光伏探测器
photovoltaic device 光伏器件
photovoltaic effect 光生伏打效应,光伏效应
phraseology 术语,用语
PHST Packaging, Handling, Storage, and Transportation 包装、搬运、储存和运输
pht phenothiazine 吩噻嗪
phugoid 1.长周期振动的 2.长周期俯仰振荡,起伏运动 3.长周期的
phugoid mode 沉浮模态
phugoid motion 起伏运动
phugoid oscillation 长周期振荡
physic 医学,药品(特指泻药)
physical and mechanical property 物理力学
physical anthropology 体质人类学
physical approximation 物理近似
physical behavior 物理行为,身体行为
physical bomb damage 直接轰炸效果
physical boundary 体力限度
physical change 物理变化
physical channel 物理信道
physical characteristic 物理特性,体型特征
physical characterization 物理表征
physical double 物理双星
physical effect 物理效应
physical electronics 物理电子学
physical examination 体检
physical exercise 体育锻炼
physical explanation 物理解释
physical factor 物理因素,身体因素
physical fitness 体能,健康
physical fitness training 体能训练
physical geodesy 物理大地测量学,大地重力学
physical heat 物理热沉
physical interpretation 物理解释
physical layer 物理层,实体层
physical level 物理级
physical libration 物理天平动

physical limit check 物理合理性检验
physical limitation 体力限度
physical map 物理图谱,自然地理图
physical measurement 1.物理测量 2.人体测量
physical mechanism 物理机制
physical model 物理模型
physical modeling 物理建模
physical motion 物理运动
physical orientation 物理定向
physical parameter 物理参数,人体生理参数
physical phenomenon 物理现象
physical power source 物理电源
physical process 物理过程,自理过程
physical processes of the solar interior 太阳内部物理过程
physical property 物理性质,实物财产
physical quantity 物理量
physical record 物理记录,实际记录
physical security 积极保安措施,物理安全性
physical sensor 物理敏感元件
physical simulation 物理模拟
physical simulation test 实物仿真试验
physical standard 体格标准
physical structure 物理结构,物质结构
physical symbol system 物理符号系统
physical system 物理系统,实体系统,因果系统
physical terminal 物理终端,实际终端
physical testing 物理试验,物理检验法
physical timer 物理定时器
physical training 体育训练
physical unit 物理单元,实际设备
physical unit block 物理设备块
physical vapor deposition 物理气相淀积
physical variable 物理变量
physical workload 体力劳动负荷
physicochemical analysis 物理化学分析
physics 物理学,物理现象
physics of failure 故障机理学,失效机理
physics of the earth 地球物理学
physics package 核装置,核战斗部
physiological acceleration 生理加速度
physiological adaptation 生理性适应
physiological age 生理年龄
physiological ceilling 生理升限
physiological clock 生理钟
physiological demand 生理需要
physiological disturbance 生理障碍,生理失调
physiological effect 生理效应
physiological effects of angular acceleration 角加速度生

理效应

physiological effects of Coriolis acceleration 科里奥利加速度生理效应

physiological equivalent altitude 生理等效高度

physiological factor 生理因素

physiological inert gas 生理惰性气体

physiological limit 生理极限

physiological monitoring 生理监视

physiological monitoring system 生理监视系统

physiological reserve 生理储备

physiological rhythms 生理节律

physiological saturation deficit 生理饱和差

physiological sensor 生理传感器

physiological stress 生理应激,生理应力

physiological stress tolerance limit 生理负荷耐限

physiological temperature regulation 生理性体温调节

physiological test 生理试验

physiological threshold 生理阈值

physiological tolerance limit 生理耐受限度,生理耐限

physiological training 生理学训练

physiological unconsciousness 生理性意识丧失

physisorption 物理吸附

phytotron 人工气候室

PI 1. Parameter Identifier 参数标识 2. Performance Inflight 空中性能 3. Peripheral Interface 外围接口 4. Physical Interface 物理界面 5. Production Illustration 生产图,生产图解 6. Products Identification 产品标识 7. Protocol Identifier 协议标识符

PI/O Processor Input/Output 处理器输入/输出

PIA 1. Peripheral Interface Adapter 辅助接口转接器 2. Performance Integrity and Avialability 性能完整性和可用性 3. Pilots International Associate 国际驾驶员协会 4. Programmable Interface Adapter 程序接口适配器 5. Proprietary Information Agreement 专有信息协议书

PIAG Propulsion Installation Advisory Group 推进装置咨询组

piano hinge 长条铰链,琴键式铰链

piano keys 跑道头黑白标志

piano wire 钢琴丝

PIAT Precision Image Annotation Tape 精制影像注记磁带

PIBS Polar Ionospheric Beacon Satellite 极轨电离层信标导航卫星

PIC 1. Person In Charge 主管人 2. Personal Identification Code 个人识别码 3. Pilot In Command 机长 4. Product Improvement Change 产品改进更改 5. Production and Inventory Control 生产与库存管理

pic simulation 粒子模拟

PICAO Provisional ICAO 临时国际民航组织

piccolo actuator 短笛式微型致动器

piccolo tube 短笛式防冰导管

pick 拾取,精选,采摘,掘

pick-and-place 拾取和放置

picket 警戒哨,巡逻飞机

picket ship 哨船,雷达哨舰

picketing 系留

pickle 航空鱼雷,投放

pickle altitude 〈口语〉投放高度,投弹高度,投鱼雷高度

pickle barrel bombing 对小面积目标轰炸,极精确轰炸

pickle button 投放按钮

pickling 酸浸,酸蚀,酸洗,封藏

pickoff 传感器,敏感元件,传感器发送装置

pickoff assembly 传感器组件

pickoff excitation voltage 传感器激励电压

pickoff sensitivity 传感器灵敏度

pickoff stator 传感器定子

pickup 传感器;发现,地面装载,取货

pickup tube 摄像管

pico 皮可(简称:皮),微微(用于前缀)

pico-farad 微微法

picosecond 皮可秒,微微秒

picric acid 苦味酸,三硝基酚

picrite 辉石,橄榄岩,苦橄岩

PICS 1. Parts Inventory and Control System 零件库存和控制系统 2. Production (Parts) Inventory and Control System 生产(零件)库存和控制系统 3. Protocol Implementation Conformance Statement 协议实现一致性陈述,协定书执行符合性说明

pictoline 等密度线

picto-line map 浮雕影像地图

pictorial information digitizer 画面信息数字化器

picture 图像,照片,图画,影片,景色,化身

picture assembly multiplexer 图像组合多路器

picture detail 图像细节,图像细部

picture distortion 图形畸变

picture element 像元,像素

picture encoding 图像编码

picture format 像幅

picture frame 像框,图幅,帧

picture interpretation 图像判读

picture manoeuvre 图案机动,造型特技

picture pair 像片对,像对

picture tube 显像管

picture-edge distance 像片边距

PID 1. Process Identifier 工艺标识 2. Process Interface Description 工艺界面说明 3. Proportion Integration

Differentiation 比例积分微分
pid control 1.比例积分微分控制 2.集散控制
pid controller 比例积分微分控制器
PIDP 1. Pilot Information Display Panels 驾驶员信息显示板 2. Programmable Indicator Data-Processor 程序控制指示器数据处理器
PIDS Prime-Item Development Specification 主要产品发展规范,主项目研制规范
piece 碎片,切片,断片,段,枝,零件
piecewise approximation method 逐段逼近法
piecewise linear approximation 逐段折线逼近法
piecewise linear function 分段线性函数
pier 码头,廊桥,候机楼登机走廊
pierce 钻孔,冲孔,穿孔,刺穿
piercing 冲孔
pieze 皮兹
piezo electric impact fuze 压电着发引信
piezoceramic material 压电陶瓷材料
piezo-coupler 压电耦合器
piezocrystal 压电晶体
piezodiode 压电二极管
piezo-effect 压电效应
piezoelectric 压电效应
piezoelectric accelerometer 压电加速度计
piezoelectric actuator 压电传动装置
piezoelectric ceramic delay line 压电陶瓷延迟线
piezoelectric ceramics 压电陶瓷
piezoelectric constant 压电常数
piezoelectric crystal 压电晶体
piezoelectric device 压电机构
piezoelectric effect 压电效应
piezoelectric flowmeter 压电流量计
piezoelectric fuze 压电引信
piezoelectric gyroscope 压电陀螺
piezoelectric microphone 压电传声器
piezoelectric patch 压电片,压电补片
piezoelectric polymer 压电高分子,压电聚合物
piezoelectric pressure 压电压力
piezoelectric pressure sensor 压电压力传感器
piezoelectric pressure transducer 压电式压力传感器
piezoelectric sensor 压电式传感器
piezoelectric transducer 压电式转换器
piezoelectric tuning fork 压电音叉
piezoelectric type balance 压电式天平
piezoelectric vibrator 压电振子
piezoelectricity 压电现象,压电效应
piezoelectric-thermal analogy 热弹性比拟
piezoid 石英片,石英晶体
piezoluminescence 压电发光现象
piezomagnetic effect 压磁效应
piezomagnetic material 压磁材料
piezometric efficiency 示压效率
piezo-oscillator 压电振荡器
piezophony 压电送受话器
piezoquartz 压电石英
piezo-remanence 压剩磁,化磁线
piezo-remanent magnetization 压剩磁化强度
piezoresistance 压敏电阻
piezoresistant effect accelerometer 压阻效应加速度计
piezoresistive accelerometer 压阻式加速度计
piezoresistor accelerometer 压阻加速度计
piezoresonator 压电谐振器
piezosensors 压电传感器
PIF 1. Pilot Information File 飞行员须知 2. Program Information File 程序信息文件
PIFET Piezoelectric Field-Effect Transistor 压电式场效应晶体管
PIG Program Image Generator 程序图像发生器
PIGA Pendulous Integrating Gyro Accelerometer 摆式积分陀螺加速度表
piggyback 1.搭载,驮载 2.复合飞机,子母飞机
piggyback payload in satellite 卫星搭载有效载荷
PIGMA Pressurized Inert Gas Metal Arc 增压惰性气体金属电弧
pigment 1.颜料,色料 2.色素
PIGS Passive Infrared Guidance System 无源红外制导系统
pigtail 1.抽头,抽出一段 2.弯头导管
PIIVT Production Implementation, Installation, Verification, and Training (Team) 生产实现、安装、验证和培训(团队)
pileup effects 堆积效应
pileup rejection 反堆积
pill 〈口语〉小型炸弹,炮弹
pillow 轴枕,垫座
pillow tank 软油罐,柔性贮液罐
pilot 1.飞行员,正驾驶,机长 2.自动驾驶仪,操纵系统 3.驾驶方法
pilot accident 驾驶员事故,飞行员事故
pilot astronaut 航天驾驶员
pilot atlas 引航图集
pilot authority 飞行员职权
pilot balloon 测风气球
pilot balloon plotting board 气球测风绘图板
pilot balloon self-recording theodolite 测风气球自记经纬仪
pilot bomb 照明炸弹
pilot burner 1.起动喷嘴 2.起动燃烧室,点火室

pilot canopy 引导伞
pilot cell 领示电池,指示电池,引示电池
pilot certificate 飞行员驾驶执照,飞行员证书
pilot control 导频控制,引导控制,领示控制
pilot diet 飞行员膳食
pilot disorientation 飞行员迷航
pilot error 飞行员差错
pilot error accident 驾驶员失误事故,飞行员失误事故
pilot error fatality 飞行员操纵错误死亡事故
pilot factor 飞行员人为因素
pilot factor accident 飞行员责任事故
pilot fatigue 飞行疲劳症
pilot fitness 飞行员的适应性,飞行员体能
pilot flame 引燃火焰,起火焰
pilot fuel 副油道燃油
pilot fuel injection 调节喷射燃油
pilot grading 飞行员等级划分
pilot hole 导向孔,定位孔,装配孔,辅助孔
pilot hydrogen 引导氢
pilot in command 机长,正驾驶员
pilot induced oscillation 驾驶员诱发振荡
pilot injection 引燃喷射,前导喷射,启喷压力
pilot interpreted system 飞行员自主判断的系统
pilot intervention 飞行员人工干预
pilot judgement 飞行员判断
pilot model 试选样品,试验模型
pilot negligence 飞行违章
pilot nozzle 主喷油孔
pilot number 引示号码
pilot operation procedure 驾驶员操作程序
pilot opinion rating 飞行员意见评估
pilot parachute 1.飞行员降落伞 2.降落伞的导伞
pilot performance 飞行员能力
pilot phase 试验阶段
pilot plane 前导翼,前置子翼
pilot pocket aid 飞行袖珍指南
pilot pressure 控制压力,导向压力
pilot production 试生产,试制
pilot psychological selection 驾驶员心理选拔
pilot pushing 违章迫使飞行员加班飞行
pilot rating 飞行员技术等级,飞行员检定,驾驶执照
pilot scale 中试规模,试验性规模
pilot seat 响导座
pilot selection 飞行员选拔
pilot signal 监控信号,导频信号
pilot stage 样品试制阶段
pilot station 引航站,飞行员位置
pilot stick force transducer 驾驶杆力传感器
pilot system 引示系统,试验系统

pilot system of afterburner 加力预燃装置
pilot technique 驾驶技术
pilot training 飞行员训练
pilot valve 分油活门,滑动活门
pilot view 飞行员视图
pilot warning indicator 驾驶员告警指示器
pilot wind tunnel 引导性风洞
pilot wings 飞行员双翼袖标(英国海军)
pilot workload 飞行员工作负荷
pilotage 地标航行,地标领航,目视飞行驾驶术
Pilotage Chart 地标领航图(美国)
pilot-aid system 驾驶员助手系统
pilot-astronaut 航天驾驶员
pilot-caused accident 飞行员责任事故
piloted 有人驾驶的,人工操纵的
pilot-flame ignition 预燃火焰点火,火舌点火
pilot-induced oscillation 驾驶员诱发振荡
piloting 1.地标领航 2.驾驶,引导
piloting navigation 地标领航,目视领航
pilot-in-the-loop 人在环
pilot-in-the-loop experiment 人在环实验
pilotless aircraft 无飞行员的飞机,无人驾驶飞机
pilot-navigator 飞行员兼领航员
pilot's associate 飞行员助手系统
pilot's associate system 飞行员辅助系统,飞行员助手系统
pilot's discretion 飞行员自行决定
pilot's failure 飞行员动作失误
pilot's false perception 飞行员错觉
pilots personality 飞行员的人格
pilots reference eye position 飞行员参考眼位
pilot's transfer function 飞行员的传递函数
PILS Payload Intergration Library System 有效载荷总装信息库系统
PIM 1. Personal Identity Module 个人标识模块 2. Personal Information Management 个人信息管理 3. Priority Interrupt Module 优先中断模块
PIN 1. Part Identification Number 零件识别号 2. Position Indicator 位置指示器 3. Program Identification Number 项目标识号 4. Program Item Number 进程项目号,项目项号
pin contacts 销钉接触,插头接触
PIN diode PIN 二极管
pin fin 扰流柱
pin joint 插销连接,枢轴铰接
PIN junction diode PIN 结二极管
PIN photodiode PIN 光电二极管
pinboard 插接板
pinch 箍缩,收缩

pinch point 夹点
pinch roller 压带轮,夹送轮(辊),压轮
pinchoff 夹断
pinchoff voltage 夹断电压
pincushion distortion 枕形畸变
PIND Particle Impact Noise Detection 微粒撞击噪声探测
PINE Passive Infrared Night Equipment 被动式红外夜视仪
pinger 弹上音响信标
pingly 〈口语〉1.反潜直升机 2.反潜直升机飞行人员
ping-pong snow 乒乓球大小的聚合体群
ping-pong test 乒乓球实验
pinhole 针孔
pinhole filter 针孔滤波器
pinhole lens 针孔透镜
pinion 小齿轮
pin-joint 别针连接
pink and green 〈口语〉美国陆军航空兵制服
pink noise 粉红噪声
pinked 将边缘剪成锯齿的,穿孔的
pinking 1.爆震,爆燃, 2.穿孔,扎眼
pinlite 微型光源
pinning 阻塞,锁住
pinpoint 航空照片,鉴别
pinpoint navigation 精确领航
pinpointing 1.精确定位,精确定点 2.定点轰炸,准确命中(目标) 3.〈口语〉瞄准
pint 品脱
pintle 销,针
pintle hole 扣钉孔
pintle injector 轴针式喷油器
PIO 1. Pilot-Induced Oscillations 飞行员诱发的振动 2. Program Integration Office 项目综合办公室
pion 介子
Pioneer 1.先锋(无人驾驶飞机) 2.先驱者探测器
pioneer 1.先驱者,先锋,倡导者,创始人 2.〈口语〉探测月球的无人太空船
pioneering technology 先导技术
PIP 1. Pack Inlet Pressure 储压器进口压力 2. Partner Interface Processes 伙伴接口流程 3. Programmable Interface Processor 可编程接口处理机
pip 1.尖头信号,尖波 2.小峰,小突起,射电描记曲线
PIPA Pulse Integrating Pendulous Accelerometer 脉冲积分摆式加速度计
pipe 导管
pipe bending 弯管
pipe diameter 管直径,管径
pipe section 管段,管道,管路部分,管节

pipe vibration 管道振动
pipeline 管路,管道
pipeline aircraft 维修周转飞机
pipeline vibration 管系振动,管道振动
piping 管网
pipper (瞄准具的)中心光点,环心,环板中心孔
PIPS Post-Injection Propulsion System 入轨后推进系统
PIR 1. Packet Insert Rate 分组插入率 2. Pilot Incident Report 驾驶员事故症候报告 3. Pressure Interstage Return 级间回油压力
piracy 非法使用飞机,劫机
Pirani gauge 皮拉尼真空计
PIRAZ Positive-Identification and Radar Advisory Zone 可靠识别与雷达咨询区
piston 活塞
piston area 活塞面积,活塞工作面积
piston displacement 活塞行程,活塞位移
piston effect 活塞效应
piston ejection ram 弹射杆
piston engine 活塞式发动机
piston model 活塞模型
piston pin 活塞销
piston ring 活塞杆,活塞环
piston skirt 活塞裙,活塞侧缘
piston speed 活塞速度,活塞速率
piston theory 活塞理论
piston vacuum pump 往复真空泵
piston-cylinder 活塞缸套,活塞圆筒
piston-ring seal (活塞的)密封涨圈
PIT 1. Pack Inlet Temperature 储压器进口温度 2. Pilot Instructor Training 飞行教员训练
pit 坑,铸锭坑,小凹陷,疤痕
pit size 地坑尺寸
pitch 节距,排距,俯仰,音调,绕横轴运动,俯仰平面,纵向运动
pitch acceleration 俯仰加速度
pitch angle 俯仰角
pitch attitude 俯仰姿态,俯仰高度
pitch attitude acquisition 俯仰姿态捕获
pitch axis 俯仰轴,横轴
pitch bucking 俯仰颠簸
pitch channel 俯仰通道
pitch circle (齿轮的)节圆
pitch cones 1.节锥 2.分圆锥,分度锥
pitch control 俯仰控制
pitch control task 啮合曲线,分度曲线
pitch curve(s) 1.节锥曲线 2.(齿轮)节线,啮合曲线 3.分度曲线

pitch deflection angle 俯仰摆角
pitch deformation angle 纵摇变形角
pitch diameter 1.(齿轮的)节径 2.(螺纹的)中径 3.(电缆的)半均直径
pitch fin order (导弹)俯仰指令
pitch gimbal 纵摇悬架
pitch gyroscope 俯仰陀螺仪
pitch heave 1.变距铰 2.轴向铰
pitch hinge (旋翼的)变距铰,轴向铰
pitch jet 俯仰操纵喷流
pitch lever 1.(直升机的)总距油门杆,油门变距杆 2.(自动驾驶仪的)俯仰通路手柄
pitch line 1.(齿轮或齿条的)节线 2.基线,中心线 3.分度线
pitch moment 变桨力矩
pitch motion 俯仰操纵机构
pitch motion cue 纵摇
pitch plane 节平面
pitch pointing 俯仰定向
pitch pointing accuracy 俯仰定向精度
pitch pointing range 俯仰定向范围
pitch program 俯仰程序
pitch program angle 俯仰程序角
pitch range 间距范围
pitch rate 俯仰速度
pitch rate control 俯仰速度控制
pitch ratio 螺距直径比,(螺旋桨)相对螺距
pitch response 俯仰反应
pitch rotation 轮流调换摊位
pitch setting (螺旋桨或旋翼的)桨距调定,桨距调整
pitch size 桨叶在某一时刻的实际桨距
pitch speed 节距速度,桨距速度
pitch stability 俯仰安定性
pitch steering order 俯仰(控制)指令
pitch tracking 基音跟踪,基音周期提取
pitch trim compensator 俯仰配平补偿器
pitch upset 失去俯仰安定性
pitch-and-roll stabilized system 俯仰倾侧稳定系统
pitch-flap coupling coefficient 挥舞变距耦合系数
pitching 纵摇
pitch(ing) illusion 俯仰错觉
pitch(ing) up illusion 上仰错觉
pitching moment 俯仰力矩
pitching-moment coefficient 俯仰力矩系数
pitchout 急转弯,急转弯点
pitchover (火箭垂直上升后的)按程序转弯
pitch-pause point 1.(火箭的)程序俯仰转动 2.(飞机在失速转弯时)从最高点作俯仰转动 3.(飞机为创高度记录)上仰急跃升

pitch-up 自动抬头,自动上仰,尾倾
pitot 空速管,全静压管,皮托管
pitot bomb 皮氏弹
pitot comb 总压梳状管
pitot distribution 梳状全静压管,全压排管
pitot head 皮托头
pitot inlet 全静压受感头,全压头
pitot pressure 皮托管压力,皮托压力
pitot probe 边界层皮托探针
pitot rake 梳状全压管
pitot static system 皮托静压系统
pitot static tube 皮托静压管
pitot traverse 冲量法测定阻力,皮托排管测量法
pitot tube 1.皮托管 2.空速管
pitot-pressure 全静压系统
pitting 1.凹坑的形成,凹痕 2.点状锈蚀,锈斑
PIU 1. Passenger Information Unit 旅客信息组件 2. Path Information Unit 路径信息单元 3. Product Interface Unit 产品内连组件
PIV 1. Peak Inverse Voltage 反峰电压,峰值反压 2. Pressure Isolating Valve 压力隔离活门
pivot 1.照片,影片,沥青,焦油 2.枢轴,中心点,旋转运动
pivot terminal character 中枢终结符
pivot-jewel clearance 宝石轴承间隙
pixel (显示器或电视机图像的)像素
pixel intensity 像素强度
pixel location 像素定位
pixel plane 像素平面
pixel registration 像元配准
pixel reset clock 像元复位时钟
pixel resolution 像元分辨率
pixel size 像素尺寸,像素大小
pixel value 像素值
PJ 1. Para Jumper(直升机)伞降救援人员 2. Plasma Jet 等离子流,等离子发动机
PJF Partially Jet-borne Flight 部分用喷气发动机飞行
PK 1. Private Key 专用密钥 2. Probability of Kill 命中率,杀伤率 3. Production Kit 生产工具 4. Pulse Keyer 脉冲键控器
PKD Public Key Distribution 公钥分配
PKI Public Key Infrastructure 公钥基础结构
PKS 1. Process Knowledge System 过程知识系统 2. Public Key System 公共密码系统
PL 1. Parameter Length 参数长度 2. Parts List 零件表 3. Permanent Line 永久线路 4. Plug 插头 5. Physical Layer 物理层 6. Probability of Loss 损失概率
PLA 1. Power Level Angle 油门杆角度 2. Post-Landing Attitude 着陆后姿态 3. Practice Low

Approach 低空进近训练 4. Programmable Logic Array 可编程逻辑排列
placard value 标牌值
PLACE Position Location and Aircraft Communication Experiment 定位和飞机通信实验
place 座（如 4 place ship 为四座飞机）
place in orbit（to） 射入轨道
placement 铺放
PLADS Parachute Low Altitude Delivery System 低空伞投系统
plage 海滨，谱斑
plain bearing 滑动轴承，平面轴承，普通轴承，滑体轴承
plain conductor 普通导线，裸导线
plain fin 光滑翅片
plain flange 对接法兰，平面法兰
plain flap 简单襟翼
plain joint 平接关节
plain nozzle 普通喷嘴
plain orifice atomizer 直接喷嘴
plain tube 光管
PLAN Payload Local Area Network 有效载荷局部区域网络
plan 1.打算，设计 2.平面图，图样
plan area 计划区域，规划区域
plan display 平面显示器
plan execution 策划执行，计划执行
plan of a zone 带状平面图
plan range 侦察平面距离
plan view 平面图，俯视图，主视图
planar 平面的，在一个平面上的
planar array 平面阵
planar array antenna 1.平面阵天线 2.二维阵天线
planar computer 平面计算机
planar defect 面缺陷
planar detonation 平面爆震
planar detonation wave 平面爆震波
planar diode 平面二极管
planar ferrite 平面型铁氧体
planar interface 平直界面，平界面
planar network 平面网络
planar osculating angle 平面掩星角
planar shock 平面冲击
planar technique 平面工艺，平面技术
planar technology 平面工艺
planar transistor 平面晶体管
planar truss 平面桁架
planar winding 平面缠绕
planar-array radar 平面阵列雷达
Planck law 普朗克定律
Planck's constant 普朗克常数
Planck's radiant body 普朗克辐射体
Planck's radiation law 普朗克辐射定律
Plandtl number 普朗特数
plane 1.飞机 2.机翼 3.平面，面
plane albedo 平面反照率
plane array 焦平面阵列
plane cascade 平面叶栅
plane change 变轨道面
plane change engine 变轨道面发动机
plane component 扁平子系
plane curve location 平面曲线定位
plane director （航母）飞机引导员
plane flying 平面飞行
plane guard 舰载警戒飞机
plane hole survey 飞机洞库测量
plane jet 平面射流
plane of focus 焦点平面
plane of polarization 极化平面，偏振（平）面
plane of symmetry 对称面
plane polarization 平面极化，平面偏振
plane profile 平面轮廓
plane subsystem 扁平次系
plane surface 平面，平表面
plane table 平板仪
plane wall 平直墙
plane wave 平面波
plane-change 变轨道面
plane-jet 平面射流
planemaker 飞机制造厂（商）
plane-mate 登机车，旅客接送车
plane-photogrammetry 交会摄影测量学
plane-polarized 平面极化的，(平)面偏振的
planer 刨床，平路机
planet 行星
planet exploration 行星探测
planet pinion 行星小齿轮，游星小齿轮
plane-table traverse 平板仪导线
planetarium 天象仪
planetarium projector 天象投影仪
planetary 行星的，有关行星的，行星传动
planetary aberration 行星光行差
planetary airglow 行星气辉
planetary atmosphere 行星大气
planetary atmospheric composition 行星大气成分
planetary boundary layer 行星的边界层
planetary circulation 行星环流，大气环流
planetary configuration 行星方位
planetary entry 行星坠入

planetary entry vehicle	进入行星大气层飞行器
planetary evolution index	行星演化指数
planetary exploration	行星探索,行星探测
planetary exploration robot	行星考察机器人
planetary frame	行星架,行星框架
planetary friction layer	行星边界层
planetary gear	行星齿轮
planetary geodesy	行星测量学
planetary geological function	行星地质作用
planetary geology	行星地质学
planetary gravity	地心引力
planetary lander	行星降落器
planetary mission	行星飞行任务
planetary nebula	行星状星云
planetary orbit	行星轨道
planetary passage	飞越行星,通过行星
planetary precession	行星岁差
planetary probe	行星探测器
planetary quarantine	行星检疫
planetary radiation budget	行星辐射收支
planetary reduction gear	行星减速齿轮,游星减速齿轮
planetary satellite	行星卫星
planetary scale	行星尺度
planetary seismology	行星震学
planetary space	行星空间
planetary system	行星系
planetesimal	星子,微星
planetocentric	以行星为中心的,行星中心的
planetocentric coordinate	行星心坐标,行星面坐标
planetography	行星表面学
planetoid	小行星
planetology	行星学
planet-wide geothermal belt	全球性地热带
planform	(从上向下看物体的)平面图,俯视图
planform area	(机翼)平面形状面积
planimeter	面积仪,测面仪,求积仪
planing bottom	滑水船底
planing step	断阶
planish	(金属板的)打平,锤光,精轧
planispheric astrolabe	星盘
plank wing	〈口语〉平直机翼
planned flight	计划飞行
planned load	计划装载
planned route	计划路线
planned target	计划内目标
planner	计划者,设计者,策划者
planning	规划,计划编制,规划计划预算系统
planning algorithm	规划算法,计划编制算法
planning horizon	规划周期,计划期距,规划时轴
planning problem	规划问题
planning software	规划软件
planning system	规划系统,规划体系
planometer	测平仪,平面规
plant	设备,装置,(电)站,厂,车间
plant design	工厂设计,装置设计
plant dynamics	植物动态
plant efficiency	设备效率,工厂效率
plant model	工厂模型
plant output	工厂产量
plant-community map	植物群落图
PLAP	Power Lever Angle Prime 油门杠杆注油角度
plaque	基板,板,平板
PLASI	Pulse Light Approach Slope Indicator 脉冲光束进近坡度指示器
plasma	等离子体,等离子区
plasma acceleration	电浆加速度
plasma accelerator	等离子体加速器
plasma actuator	等离子激励器
plasma arc	等离子弧,等离子体电弧,电浆弧
plasma arc furnace	等离子电弧炉,等离子炉
plasma arc welding	等离子(体)弧焊
plasma arc welding with adjustable polarity parameters	变极性等离子弧焊
plasma asher	等离子灰化器
plasma astrophysics	等离子体天体物理学
plasma beam	等离子束流,等离子体束
plasma bridge	胞间丝
plasma chamber	电浆腔
plasma cleaning	等离子清洗
plasma cloud	等离子体云
plasma coating	等离子喷涂
plasma column	等离子体柱
plasma density	等离子体密度
plasma diagnostic	等离子体诊断
plasma display	等离子体显示
plasma display panel	等离子体显示板
plasma dynamics	等离子体动力学
plasma emission	红外成像光谱分析
plasma energy	等离子能
plasma engine	等离子体发动机
plasma etching	等离子腐蚀,等离子蚀刻
plasma flow	等离子体流
plasma frequency	等离子体频率
plasma generation	等离子体产生
plasma generator	等离子体发生器
plasma gun	等离子体枪
plasma ignition	等离子点火,等离子体点火器
plasma immersed modification	等离子体浸没改进

plasma injection　等离子体注入
plasma injector　等离子体注入器
plasma instability　等离子体不稳定性
plasma ion-assisted deposition　等离子体离子辅助沉积
plasma jet　1.等离子体发动机 2.等离子体射流,等离子流
plasma layer　大气压等离子体层
plasma machining　等离子加工
plasma mantle　等离子体耦合器件
plasma metal spray　等离子体金属喷镀
plasma model　等离子体模型
plasma oxidation　等离子氧化
plasma pane　等离子显示板
plasma panel　等离子体仪表板
plasma parameter　等离子体参数
plasma plume　等离子体羽辉
plasma potential　等离子体势
plasma power　等离子电源
plasma pressure　等离子体压力,等离子体压强
plasma production　电浆生产,等离子体生产
plasma propulsion　等离子体推进
plasma pulse　等离子脉冲
plasma radiation　等离子体辐射
plasma rocket　等离子(体)火箭
plasma rocket engine　等离子火箭发动机
plasma screen　电浆电视,等离子屏幕
plasma sheath　等离子体包层
plasma sheet　等离子体片
plasma source　等离子体源
plasma source ion implantation　等离子体源离子注入
plasma spraying　等离子体喷涂
plasma spraying method　等离子喷涂法
plasma sputtering　等离子溅射
plasma stability　等离子稳定性
plasma system　等离子手术系统
plasma temperature　等离子体温度
plasma torch　电浆炬,等离子体焰炬,等离子体喷枪
plasma torus　等离子体环
plasma voltage　等离子体电压
plasma wave　等离子体波,电浆波
plasma wind tunnel　等离子风洞
plasma-jet　等离子体射流,等离子流
plasma-jet experiment　等离子体射流实验
plasmapause　等离子体层顶
plasmasphere　等离子体层
plasmoid　等离子粒团
plastic　塑料,可塑物质,塑料的
plastic anisotropy　塑料各向异性
plastic dynamite　可塑甘油炸药
plastic effect　(相位失真)浮雕效应
plastic explosive　可塑炸药
plastic film capacitor　塑料膜电容器
plastic flow　塑性流,塑性变形流
plastic fracture　塑性断裂
plastic instability　塑性失稳,塑性不稳定
plastic map　塑料地图,立体地图
plastic package　塑料封装
plastic relief map　塑料立体地图
plastic strain　塑性应变
plastic zone　塑性区
plastic zone size　塑性区尺寸
plasticity　塑性
plasticize　塑化,使成为可塑体
plasticizer　增塑剂,柔韧剂
plastics　塑料
plastics factor　塑性因子
plastique　可塑炸弹
PLAT　Programmed Lateral Automatic Test 程序纵向自动测试
plat　土地图,地区图,地籍图
plate　1.板,中厚板材 2.电子管的屏极,阳极,(蓄电池的)极板 3.机场资料活页纸
plate brake　薄片制动器
plate collision　板块碰撞
plate constant　底片常数
plate copying　晒版
plate element　板单元,板元
plate for ultra-microminiaturization　超微粒干版
plate impact test　平板撞击试验
plate length　平板长度
plate membrane　平板式膜
plate movement　板块运动
plate scale　底片比例尺
plate shell theory　板壳理论
plate spring accelerometer　弹簧片式加速度计
plate stiffness　加劲板
plate surface　板表面,电镀表面
plate tectonics　板块(大地)构造学
plateau　坪,平台
plateau mountain area flying　高原山区飞行
plateau pressure　平台压力,平衡氢压
plateau propellant　平台推进剂
plateau region　坪区,平坦区
plateau slope　坪斜
plated-through hole　金属化孔
plated-wire memory　镀线储存器
platelet　1.片晶,片板 2.多孔片
platelet injection　层板式喷注器

platelet injector 多层薄片式喷嘴
platform 1. 平台 2. 起降平台 3. 空投平台
platform absorber 平台减振器
platform alignment 平台校准,平台对准
platform alignment prism 平台瞄准棱镜
platform architecture 平台架构
platform balance 台式天平
platform base 平台基座
platform cluster 平台组件
platform coordinate 平台坐标
platform coordinate system 平台坐标系
platform coordinate transformation device 平台坐标变化器
platform cover 平台外罩
platform drop (货物)平台空投
platform dynamics 平台运动参数
platform electronic assembly 平台电子箱
platform erection 平台扶正
platform face (涡轮叶片)叶根平台面
platform for remote sensor 遥感平台
platform inertial guidance system 平台式惯性制导系统
platform inertial navigation system 平台惯性导航系统
platform inner gimbal 1. 平台内框架 2. 平台内常平架
platform mirror 平台反射镜
platform motion 平台运动
platform noise 无线平台杂讯
platform outer gimbal 1. 平台外框架 2. 平台外常平架
platform position computer 平台位置计算机
platform servo loop 平台稳定回路
platform slipping ring 平台导电滑环
platform stabilized loop 1. 平台伺服回路 2. 平台稳定回路
platform structure (卫星)平台结构
platform switch lock check 平台开闭锁检查
platform transfer 平台转运
platform vibration isolator 平台减震器
platform-computer guidance 平台计算机制导
platinite 铁镍合金
platinoid 假铂,假白金,赛铂金
platinotron 铂管
platinum 【化】铂(化学元素,符号Pt),白金
platinum cobalt alloy 铂钴合金
platinum resistance thermometer 铂丝电阻温度表
platinum spark plug 铂丝火花塞
PLATO Preliminary Lethality Assessment Test Object 最初杀伤力评定试验项目
platometer 求积仪,侧面仪
playback 播放,放像,重现
playboy 〈口语〉夜间战斗机

player 播放器,参加者
playing area 工作面积
PLB Passenger Loading Bridge 搭机空桥
PLC 1. Power Line Communication 电力线通信方式 2. Power Loading Control 电源负载控制 3. Programmable Logic Controller 可编程逻辑控制器
PLCP Physical Layer Convergence Protocol 物理层会聚协议
PLD 1. Precision Laser Designator 精密激光照射器 2. Rule-Length Discrimination 脉宽鉴别
PLE 1. Path Length Efficiency 路径长度效率 2. Phase Loading Entry 阶段载入入口 3. Program Listing Editor 程式列表编辑器
plenum 1. 压力通风 2. 压力通风系统,除冰空气管道 3. 管线元件
plenum chamber 1. 驻室 2. 送气室,增压室,静压箱
plenum flow 腔室中的流动
plenum volume 充气部容积
plenum-chamber door 风扇外函集气室舱门
plethysmogram 体积描记图
plethysmography 体积描记法
plex grammar 交织文法
plexiglas 有机玻璃
PLF 1. Passenger Load Factor 载客率 2. Present Leg Fault 当前航段故障
PLH 1. Payload Handling 有效载荷处理 2. Propeller Load Horsepower 螺旋桨加载马力
PLI 1. Pilot Location Indicator 飞行员位置指示器 2. Pitch Limit Indicator 俯仰极限指示器 3. Private Line Interface 专用线路接口
plication 皱纹,细褶皱
plidar polychromatic lidar 折扇式多色激光雷达
PLGR Precision Lightweight GPS Receiver 轻型精确全球定位系统接收机
PLL 1. Payload Limit 业载限制 2. Phase Lock Loop 相位锁定环路,锁相环
PLM 1. Passenger Load Message 旅客载量电报 2. Payload Level Multiplexing 净荷级复用 3. Product Lifecycle Management 产品生命周期管理
PLN Path Layer Network 通道层网络
PLOC Payload Operations Contractor 有效载荷操作承包商
plot 小块土地,标定,标图,绘制
plot extraction 图像录取,点迹录取
plot extractor 图像录取器,点迹录取设备
plottable error 展绘误差
plotter 标图器,绘迹器,描绘器,绘图仪
plotting 标图,标绘,标定
plotting board 标图板

plotting file 绘图文件
plotting instrument 绘图仪
plotting machine 绘图仪,绘图机
plotting sheets 领航作业图
plotting tablet 标图板
ploughing （水上飞机的）水上慢速滑行
plow 犁,似犁的工具,北斗七星
ployment 策略,活动,工作
PLP Pipeline Patrol 输油管巡逻
PLRO Plain-Language Readout 明文(明码)读出
PLRS Precision (or Position) Location (and) Reporting System 精确定位报告系统
PLS 1. Personnel Launch System 人员运载系统 2. Point Level Sensor 点式平面传感器
PLSS 1. Payload Support Structure 有效载荷保障结构 2. Portable Life Support System 轻便式生命保障系统
PLT Pilot 驾驶员
PLU Position Location Uncertainty 定位不详
plug 1.机身加长段 2.空中加油接头 3.(进气道或喷管的)中心体,锥体
plug base 插头底座,插座
plug board 插接板
plug braking 反接制动,反相序制动,反相制动
plug calorimeter 塞形测热计
plug door （飞机的）堵塞式舱门
plug flow 塞流
plug gauge 塞规
plug inlet 带中心锥的轴对称超声速进气道
plug nozzle 塞式喷管
plug nozzle rocket engine 塞式喷管火箭发动机
plug ring 导风环
plug tap 中丝锥,二锥
plug weld 电铆焊,塞焊
plug welding 塞焊
plug window 堵塞式应急出口窗
plug-in board 插件
plug-in discharge tube 插入式放电管
plumb line 铅垂线
plumb-bob vertical 铅垂垂线
plumbicon 氧化铅光导摄像管
plumbing 〈口语〉液体管路
plumbum 【化】铅（化学元素,符号 Pb）
plume 羽毛,羽状烟柱,烟缕,(烟囱)羽状排气,烟羽,羽毛状物
plume area 羽状风区,卷流区
plume attenuation 喷焰衰减
plume avoidance maneuver 防喷流机动
plume deflector （导弹或火箭喷焰的）导流器,导流装置
plume expansion 羽流膨胀

plume flow 羽流
plume flowfield 羽流流场
plume impingement 羽流冲击,羽流撞击
plume measurement 卷流测量
plume model 烟流模式
plume particle 羽流微粒
plume plasma 等离子体羽,等离子体云
plume structure 灰岩,羽痕构造
plume testing 羽流试验
plume-impingement environment 喷流撞击环境
plume-mode 烟流模式
plunge 1.投入,插进 2.下降,急降 3.垂直颠动
plunger 1.柱塞 2.棒式铁心
plunger pump 柱塞泵
plus count 加法计数,往上计数
Pluto 【天】冥王星
plutonic water 深成水,深成岩水
plutonium 【化】钚（放射性元素,符号 Pu）
plutonium bomb 钚弹
PLV 1. Presentation Level Video 显示级视频 2. Production Level Video 制作级视频
PLWS Planned Layup Work Statement 计划接合工作说明
ply 厚度,板层,褶
ply angle 铺设角
Plymetal 木质夹铝层板
PM 1. Part Master 零件主数据 2. Performance Monitoring 性能监控 3. Phase Modulation 相调制,调相 4. Pilot Monitoring 监控飞机的飞行员 5. Post Meridian 下午,午后 6. Process Management 工艺管理 7. Procurement Management 采购管理 8. Product Manager 产品经理 9. Production Management 生产管理 10. Production Memos 生产备忘录 11. Program Management 项目管理 12. Project Management 项目管理
PMA 1. Part Manufacturer Approval 零部件制造人批准书 2. Parts Manufacturing Approval 零件制造批准 3. Permanent Magnetic Alternator 永磁交流发电机 4. Plane Manufacturers Association 飞机制造商协会 5. Portable Maintenance Aid 便携式维修设备 6. Preliminary Maintainability Analysis 初步维修性分析
PMAD Power Management And Distribution 电能管理和分配
PMAR Preliminary Maintenance Analysis Report 初步维修分析报告
PMAS Performance Measurement Analysis System 性能测试分析系统
PMAT Portable Maitenance Access Terminal 便携式维

修存取终端

PMB Process Management Board 工艺管理委员会

PMC 1. Performance Management Computer 性能管理计算机 2. Power Management Control 动力管理控制 3. Process Material Control 工艺材料控制

PMCF Post-Maintenance-Check Flight 维修后的检查飞行

PMD 1. Program Memory Device 进程存储装置 2. Physical Medium Department 物理介质门类 3. Polarization Mode Dispersion 偏振模色散

PMED Propulsion Manufacturing Engineering Document 动力装置制造工程文件

PMFA Permanent Magnetic Focusing Assemblies 永磁聚焦装置

PMFS Pulsed Magnetic Field System 脉冲式磁场系统

PMG Permanent Magnet Generator 永磁发电机

PMH Per Man Hour 每人时

PMI 1. Partner Managed Inventory 由合作伙伴管理的库存 2. Principal Maintenance Inspector 主任维修检查员

PMIR Pressure-Modulated Infrared Radiometer 压力调制红外辐射计

PML Preferred Materials List 优选物料清单

PMM Principles of Material Management 材料管理原则

PMMA polymethylmethacrylate 聚甲基丙烯酸甲酯（有机玻璃）

PMO 1. Part Master Object 零件主对象 2. Present Mode of Operation 当前运行模式 3. Program Management Organization 程序管理机构 4. Project Management Office 产品管理办公室

PMOS P-Type Metal Oxide Semiconductor P型金属氧化物半导体

PMP 1. Performance Monitoring Program 性能监控程序 2. Point to Multipoint 点到多点 3. Primary Maintenance Process 主维修工艺 4. Project Management Plan 项目管理大纲

PMR 1. Performance Maintenance Recorder 性能维护记录器 2. Preliminary Material Review 初步材料评审 3. Private Mobile Radio 专用移动无线电 4. Public Mobile Radio 公用移动无线电

PMRF Pacific Missile Range Facility 太平洋导弹靶场

PMRT Program Management Responsibility Transfer 型号（大纲）管理责任移交

PMS 1. Performance Management System 性能管理系统 2. Polaris Missle System 北极星导弹系统 3. Probability of Mission Success （飞行）任务成功概率

PN 1. Part Number 零件号 2. Pseudo-Noise 伪噪声 3. Public Network 公共网

PN code 伪噪声码，PN 码

P-N junction PN 结

P-N junction diode PN 结二极管

PN junction isolation PN 结隔离

PNA Point of No Alternate 不能去原定备降场的点

PNC Pneumatic Nozzle Control 气压式喷管控制

PNCC Partial Network Control Center 部分网络控制中心

PNCP Peripheral Node Control Point 周边交点控制点

PNCS Performance Navigation Computer System 性能导航计算机系统

PND 1. Perceived Noise Decibels 能感应的噪声分贝 2. Primary Navigation Display 主导航显示器

PNEU Pneumatic 气动，供气

pneumatic 气压的，气动的，压缩空气传动的

pneumatic actuator 气动执行机构，气动作动器

pneumatic altimeter 气压式高度表

pneumatic analog computer 气动模拟计算机

pneumatic bearing 压缩空气轴承

pneumatic brake system 气压刹车系统

pneumatic console 气动控制台

pneumatic control 气动控制

pneumatic control system 气动控制系统

pneumatic drill motor 气动钻孔马达

pneumatic electric converter 气-电转换器

pneumatic expulsion tube 气推发射管

pneumatic feedback 气压反馈

pneumatic gun charger 气压装弹机构

pneumatic logic 气动逻辑线路

pneumatic plug nozzle 气动塞式喷管

pneumatic quick-disconnect coupling 气动快速脱落连接器

pneumatic servomechanism 气动伺服机构

pneumatic system 1. 冷气系统 2. 气压系统，气动系统

pneumatic tubing 气动导管，气压输送管，气压管，输气管

pneumatic valve 气动阀，气动活门，气动调节阀

pneumatic-hydraulic shock absorber 气-液减震器

pneumatics 气动力学

pneumodynamic 压缩空气动力的，运动中由气体力作用的

PNF Pilot Not Flying 不操纵飞机的飞行员

PNG 1. Passive Night（Vision）Goggles 被动夜视眼镜 2. Portable Network Graphics 可移植的网络图像文件格式

PNI Pictorial Navigation Indicator 图像导航指示器

PNJ Pulse(d) Noise Jamming 脉冲噪声干扰

PNL 1. Panel 板 2. Passenger Name List 旅客名单

3. Perceived Noise Level 能感应的噪声程度
PNP 1. Positive Negative Positive 正负正（晶体管） 2. Precision Navigation Processor 精确导航处理机 3. Preliminary Network Plan 最初网络计划 4. Programmed Numerical Path 程序数字控制飞行轨迹
PNP transistor PNP型晶体管
PNR 1. Passenger Name Record 旅客姓名记录 2. Point of No Return 不能返航点 3. Prior Notice Required 需事先通知
PNVS Pilot Night Vision System 飞行员夜视系统
PO 1. Post Order 邮政汇票 2. Purchase Order 订货单，订购单
POA 1. Point Of Arrival 到达机场 2. Production Organization Approval 生产机构批准书，生产组织批准 3. Purchased On Assembly 装配件采购
POB 1. Persons On Board 机上人员 2. Post Office Box 邮政信箱 3. Pressure Off Brake 松刹车压力
POC Proof Of Concept 概念验证
POCC Payload Operations Control Center 有效载荷操作控制中心（美国）
POCCNET Payload Operation Control Center Network 有效载荷操作控制中心网（美国）
pocket 旋翼桨叶后段件
pocket dosimeter 袖珍式辐射剂量计
POD 1. Point Of Departure 起飞机场 2. Port Of Debarkation 下机（卸货）机场 3. Port Of Destination 到达地
pod 1. 吊舱，短舱，分离舱，可卸货舱 2. 箱，容器
pod formation （干扰）吊舱编队
pod gun 吊舱炮
podded 装在吊舱内的
podding 采用吊舱的设计思想和技术
poded cannon 吊舱炮
PODS 1. Personnel and Organizational Disaster Services 个人和组织的灾难协助服务 2. Pneumatic Overheat Detection System 压缩空气过热探测系统
podularization 装备吊舱，装备外挂
POE 1. Port Of Embarkation 出境机场（港口） 2. Port Of Entry 入境机场（港口） 3. Probability Of Error 错误的可能
POET Portable Opto-Electronic Tracker 便携式光电跟踪器
POF 1. Phase Of Flight 飞行阶段 2. Plastic Optical Fiber 塑料光纤 3. Polymer Optical Fiber 聚合物光纤
POGO Polar Orbiting Geophysical Observatory 极轨地球物理观测台
POGO effect 波哥效应，纵向耦合振动
POGO vibration 纵向耦合振动

POH Pilot's Operating Handbook 驾驶员操作手册
Pohlhausen method 波尔豪森法
POI 1. Point Of Initiation 起始点 2. Principal Operations Inspector 主管运行检查员
Poincare sphere 庞加莱球
point 点，尖端，站
point blast 点爆炸
point contact diode 点接触型二极管
point contact solar cell 点接触太阳电池
point defect 点缺陷
point defence alert radar 点防御警戒雷达
point discharge 尖端放电
point graph 点图
point grid 测点网
point identification 点像辨认
point light 1. 点状灯光 2. 点光源
point load 集中荷载，点荷载，集中载重
point mode 多点式，点集式，点式
point of attachment 附着点
point of comparison survey 联测比对点
point of entry 1. 输入点，进线点 2. 入口
point of ignition 着火点，燃烧点
point of impact 弹着点，碰撞点
point of incidence 入射点
point of intersection 交点，交叉点，交会点
point of no return 航线临界点
point of separation 分离点
point of square control network 方格网点
point performance 1. 设计点性能 2. 优化点法
point scatter 点散射，实验点的分散
point spread function 点扩展函数
point symbol 点状符号
point target 点目标
point transfer device 转点仪，刺点仪
point vortex 点涡
point-designation grid 点标示网格
pointer 指针，指示器
pointer barometer 指针气压计
pointing accuracy 目标指示精度
pointing and attitude control system 指向和姿态控制系统
point-shaped landmark 点状地标
point-to-point communication 点对点通信
point-to-point control 端点控制系统
point-to-point control system 点对点控制系统
point-to-point data link 点对点数据链路
pointwise availability 逐点可应用度
poise 1. 泊（非公制动黏度单位，等于1达因-秒/厘米^2） 2. 砝码

Poiseuille flow 泊肃叶流(流过圆截面长导管的层流黏滞流)
poisoning 中毒
poisoning of cathode 阴极中毒
poisonous gas bomb 毒气炸弹
Poisson bracket 泊松括号
Poisson distribution 泊松分布
Poisson's equation 泊松方程
Poisson's ratio 横向变形系数,泊松比
POL 1. Petrol, Oil and Lubricant 汽油、滑油与润滑剂,油料 2. Probability Of Loss 损耗概率
polar 1. 极,极线,极线图 2. 极的,极性的 3. 极化的,北极的,地极的 4. (特指高级滑翔机的)性能曲线
polar air 极地气团
polar angle 极角
polar aurora 极光
polar cap 极冠,极地冰冠
polar capacitor 极性电容器
polar control 极坐标法控制
polar coordinate 极坐标
polar curve 极坐标曲线
polar distance 极距
polar front 地极前沿,极锋
polar moment of inertia 惯性极矩
polar motion 极运动
polar navigation 地极导航,极区航行
polar orbit 极地轨道,极轨道
polar orbit meteorological satellite 极轨气象卫星
polar orbiting satellite 极地轨道卫星,极轨卫星
polar pantograph 极坐标缩放仪
polar phase shift 极相漂移
polar plot 极坐标曲线图
polar plume 极羽(日晃中的)
polar region 极区(地球两极附近的区域,无确定界限)
polar semiconductor 极性半导体
polar stereographic mosaic 极射赤面(投影)拼图
polar stereographic projection 极射赤面投影
polar sun-synchronous orbit 极地太阳同步轨道
polar triangle 极三角形
polar wander 极移
polar wind 极风
polarimeter 偏振计,旋光计,极化计
Polaris 1. 北极星 2. 北极星导弹
polarisation 极化
polariscope 偏(振)光镜,偏光计
polarity 极性
polarizability 1. 极化性 2. 极化率
polars 高偏振星
polar-wander curve 极移曲线
polar-wander path 极移路径
pole 1. 极坐标原点,极点 2. 电极,磁极 3. 地极 4. 杆,柱
pole assignment 极点配置
pole of angular momentum 角动量极动量矩极
pole of ecliptic 黄极
pole of rotation 旋转极,自转极
pole piece 极靴,极片
poled 接入的,已接通的
pole-dipole array 单极-偶极排列
pole-on object 极向天体
pole-on star 极向恒星
poleward 极向迁移
pole-zero cancellation 零极点相消
police radar 警察用雷达
policy statement 政策声明
polishing 研磨,抛光
poll 轮询,顺序询问
pollutant 污染物
pollutant emission 污染物排放
pollution control 污染防治,污染控制
pollution monitoring 污染监测
Pollux 【天】北河三,双子座 β 星
POLO 1. Parallel Optical Link Organization 并行光链路组织 2. Polar Orbiting Lunar Observatory 月球极轨观测台
poloidal 角向磁场
poly crystalline silicon solar cell 多晶硅太阳能电池
polyatomic molecule 多原子分子
polybenzimidazole 聚苯并咪唑
polybenzimidazole fibre 聚苯并咪唑纤维(商品名为 PBI 纤维)
polybutadiene acrylic acid 聚丁二烯丙烯酸(固体火箭推进剂)
polybutadience acrylic acid copolymer propellant 聚丁二烯丙烯酸共聚物推进剂
polybutadiene acrylonitrile 聚丁二烯丙烯腈(固体火箭推进剂)
polybutadience acrylonitrile propellant 聚丁二烯丙烯腈推进剂
polybutadiene composite propellant 聚丁二烯复合推进剂
polybutadiene resin 聚丁二烯树脂
polybutylene 聚丁烯
polycaprolactone 聚已酸内酯
polycaprolactone polyol 聚己内酯多元醇
polycarbonate sheet 聚碳酸酯板
polycell method 多元胞法
polycide gate 多晶硅-硅化物栅

polyconic projection 多圆锥投影
polycrystal 多晶体，多晶
polydiacetylene 聚丁二炔，聚二乙炔
polyester 聚酯
polyester coating 聚酯漆包线
polyetheretherketone 聚醚醚酮
polyethylene 聚乙烯
polyethylene film 聚乙烯薄膜
polyethyleneglycol 聚乙二醇（一种固体推进剂）
polyfocal projection 多焦点投影
polygon 多边形，多角形物体
polygon overlay 多边形重叠
polygon structure 多边形结构
polygonal surface 多边形表面
polyimide 聚酰亚胺（耐高温结构材料）
polyimide fiber 聚酰亚胺纤维
polyimide film 聚酰亚胺薄膜
polyimide membrane 聚酰亚胺膜
polyimide plastic 聚酰亚胺塑料
polyimide resin 聚酰亚胺树脂
polymer 聚合物
polymer composite 聚合复合材料
polymer film 聚合物膜，高分子膜
polymer material 高分子材料，聚合材料
polymer property 聚合物性质
polymer resin 聚合树脂
polymer surface 聚合物表面
polymer system 聚合物体系
polymeric composite 高分子复合材料
polymeric material 聚合材料，聚合物材料
polymerization 聚合（作用）
polymeter 1. 多能湿度表（测量绝对湿度、相对湿度、露点和温度）2. 多能测量仪
polyol 多元醇，多羟基化合物
polyolefin 聚烯烃石油，聚烯油
polyphase alternating current 多相交流电
polyphase electric motor 多相电动机
polyphase structure 多相组织
polysaccharide 多糖，多聚糖
polysilicon emitter transistor 多晶硅发射极晶体管
polystation Doppler system 多站多普勒系统
polystyrene 聚苯乙烯
polysulfide 聚硫化物，多硫化合物
polysulfide propellant 聚硫推进剂
polytetrafluoroethylene 聚四氟乙烯，特氟隆
polytope 多胞形，多面体，可剖分空间
polytrope 多变曲线
polytropic atmosphere 多元大气
polyurethane 聚氨基甲酸乙酯，聚亚胺酯，聚氨酯

polyurethane foam 聚氨酯泡沫塑料，泡沫聚氨酯，聚氨酯泡沫体，聚乌拉坦泡沫胶
polyurethane propellant 聚氨酯推进剂
polyurethane-foam plastic 聚（氨基甲酸）酯泡沫塑料
polyvinyl 乙烯基聚合物的，聚乙烯基的，聚乙烯化合物
POM 1. Printer Output Microfilm 打印机输出缩微胶片 2. Production Organization Manual 生产组织手册 3. Program Objective(s) Memorandum 计划目标备忘录（美国国防部）4. Purchase Order Management 采购订单管理
POMCUS Prepositioned Overseas Materiel Configured in Unit Sets 按用户分组配置的海外物资储备
POMO Production-Oriented Maintenance Organization 以生产为主的维修组织，面向生产的维修组织（美国空军）
POMROS Power-Off Minimum Rate Of Sink 无动力最小下沉率
ponderomotive force 有质动力
ponding 雨后跑道积水
PONO Project Office Nominated Official 规划办公室指定官员
pontoon 浮筒，浮囊（可用作直升机起落架的充气浮囊）
POO Payload Of Opportunity 备分业载
pool 1. 合营，合办 2. 水池，沉淀池
pool fire 池式起火
pooling 合作经营
poor visibility 不良能见度
POP 1. Perpendicular to the Orbit Plane 轴面垂姿控方式 2. Point Of Presence 现时点 3. Polar Orbiting Platform 极轨平台 4. Probability Of Precipitation 降水概率 5. Product Optimization Programme 产品优化方案 6. Program Operation Plan 大纲实施计划 7. Purchased Outside Production 生产之外采购
pop rivets 波普空心铆钉
poppet 盘形活门，菌状活门，提动阀芯，座阀芯，锥阀芯，碟形阀芯
POPS Position and Orientation Propulsion System 推进系统的位置
population 人员，人口，总人数
population inversion 粒子数反转
pop-up 1. 弹出式 2. 跃升机动加俯冲
pop-up maneuver 跃升动作
porcelain 陶瓷，瓷料
pore 气孔，细孔
pore diameter 孔径，孔隙直径
pore size 孔径，孔隙大小，气孔尺寸
pore volume 孔体积
porosity 多孔性，透气性，透气量
porous friction asphalt 多孔毛面沥青（跑道面上第一

层,可防止跑道积水)
porous glass　多孔玻璃
porous layer　多孔岩层,多孔质层
porous material　多孔材料,多孔性材料,疏松材料
porous medium　多孔介质,疏松介质
porous medium analysis　多孔介质分析
porous metal　多孔金属
porous plate　多孔板
porous sample　孔隙扰动体
porous shell　多孔外壳
porous surface　多孔表面
porous tungsten　活性钨
porous wall　多孔壁
Porro prism　波罗棱镜,直角棱镜
port　港口
port area　码头区,港口区,喷口面积
port diameter　油口直径
portability　可移植性(软件从一个系统或环境转移到另一个的容易程度)
portable battery　便携式蓄电池
portable infantry radar　携带式步兵雷达
portable life support system　便携式生命保障系统,背包式生命保障系统
portable magnetic tape recording　佩带式磁带记录法
portable oxygen equipment　便携式供氧装备
portable traffic light　手提式交通管制灯
porthole　1.检查孔,观察孔 2.圆形眩窗口
portion　部分,一段
portray　描绘
portrayal　描绘,展示
portside　1.左舷,港口地区 2.码头边,滨水区
POS　1. Peacetime Operating Stocks 平时作战物资库 2. Permanent Orbital Station 长久性轨道站 3. Polar-Orbiting Satellite 极轨(道)卫星
POS REF　Position Reference 位置基准
POS SW　Position Switch 定位开关
pose　盘问,作……姿势
posigrade　正向加速的,顺行的
posigrade rocket　(顺行)加速火箭
POSINIT　Position Initialization 位置初始化
position　位置
position angle　位置角
position change　位置变化
position control　位置控制
position controller　位置控制器,席位控制器
position correction　位置校正
position datum　位置基准
position detection sensor　位置检测传感器
position determination　定位,位置测定

position dilution of precision　定位几何误差因子
position drift of geostationary satellite　地球静止卫星位置漂移
position error　1.位置误差,定位误差,位置错误 2.气动激波修正量 3.空气动力误差
position error coefficient　位置误差系数
position estimate　位置估计
position feedback　位置反馈
position finding　测位
position fixing　定位
position indicator　位置指示器
position light　航行灯
position line　位置线
position measurement　位置测量,定位技术,姿态测量
position of equilibrium　平衡位置
position pulse　定时脉冲,定位脉冲
position radar　目标位置测定雷达
position repetitive error　定位重复误差
position servo system　位置伺服系统
position test　钟表换位测试
position transducer　位置检测器,位置传感器
position uncertainty　位置不确定性
position update　位置更新
positional vertigo　位置性眩晕
positioner　位置控制器
position-error　位置误差
positioning　定位,引导
positioning accuracy　定位精度,定位准确度
position-sensitive detector　位置灵敏探测器
positive　积极的,正的,阳性的,确定的,肯定的,实际的,真实的,绝对的
positive acceleration　正加速度
positive anomaly　正异常
positive charge　正电荷,阳电荷
positive control　绝对空域管制
positive coupling　正耦合
positive definite　正定的
positive definite matrix　正定矩阵
positive effect　积极的效果,明显效果
positive electrode　正极
positive feedback　正反馈
positive G　正过载
positive image　正像
positive lift　正升力
positive photoresist　正性光刻胶
positive picture　正片,正像
positive pitch　正桨距
positive positioning　主动体位
positive potential　正电势,正电位,阳电势

positive pressure　正压力
positive pressure breathing　加压呼吸
positive pressure oxygen supply　加压供氧
positive pressure oxygen system　加压供氧系统
positive rate　阳性率
positive recovery　安全改出
positive rolling moment　正滚转力矩,右滚力矩
positive stability　正稳性
positive terminal　正极端子
positive transparency　正透明度
positive yaw　正偏航力矩
positive-displacement pump　容积泵正排量泵
positron　正电子,阳电子
positron emission tomography　正电子放射层扫描术
positron spectroscope　正电子谱仪
POSS　1. Passive Optical Satellite Surveillance 无源光学卫星监视 2. Power-Off Stall Speed 无动力失速速度
possess　控制,使掌握,持有,拥有,具备
possibility theory　可能性理论,确定性理论
POST　1. Passive Optical Seeker Technique（Technology）被动式光学导引头技术 2. Production-Oriented Scheduling Technique 生产调度技术
post　1. 主垂直构件,柱,杆 2. 主起落架支柱 3. 岗位,哨所,营区
post boost vehicle　助推后飞行器
post crash survivability　摔机生存性
post mission processing　事后数据处理
post stall　过失速
post-boost control system　助推段后控制系统
postbuckling　后屈,后期压曲
postcollision distribution　碰撞周分布
postcombustion　补充燃料,二次燃烧
postdecompression shock　减压后休克(高空减压病的一种症状)
post-deflection acceleration　偏转后加速
postdescent shock　下降后休克
postdetection recording　检后记录
posterior　1.(时间上)以后的,较迟的,随后的 2.(位置上)位于后部的,后面的,(尤指)背面的,尾部的,臀部的
postflight adaptation　飞行后适应
postflight check　飞行后检查
postflight medical analysis　飞行后医学分析
postforming　二次成形
posthypoxic reaction　缺氧后反应
post-injection propulsion system　入轨后推进系统
post-integration　总装后(指航天运载系统总装完毕至发射前)
postpass　卫星通过后(瞬间)的,过顶后(通常指航天器刚从头顶上空飞过)
postposition launching　后置发射
postprocessing　后加工,后部工艺
postprocessor　后处理程序,后外理机
poststack migration　叠后偏移
poststall　失速
poststall maneuver　过失速机动
post-takeoff check　起飞后的检查
posture　姿态
postweld treatment　焊后处理
post-welding heat treatment　焊后热处理
POT　1. Potential 电位,电势 2. Potentiometer 电位计 3. Proximity Operation Trainer 靠近操作训练器(航天员训练用的)
pot　1. 罐,筒,盆 2. 熔锅 3.(活塞式发动机的)汽缸
potassium bicarbonate　碳酸氢钾(干粉灭火剂)
potassium niobate　铌酸钾
potential　1. 势,位(能) 2. 电势,电位,电压 3.(大气)热力势
potential electrode　测量电极
potential energy　势能,位能,电位能
potential equation　位方程式,电位方程
potential field　势场,位场
potential flow　1. 位流(流场中涡强为零的流动) 2. 无旋流
potential of centrifugal force　离心力位
potential of the earth　地球势能
potential profile　电势分布图,位能图形
potential slope　位梯度
potential theory　势论,势能理论,位势理论
potentiometer　电位计,电位器,分压器
potentiometric analysis　电位分析,电势分析
potentiostat　稳压器,恒电势器
pothole　地面深穴
potted circuit　封闭式电路
Potter horn　玻特喇叭
potting compound　灌注胶
POTV　Personnel Orbit Transfer Vehicle 乘员轨道转移飞行器
pour point　浇灌点
pouring foaming　浇注发泡
POV　1. Peak Overshoot Voltage 峰值上冲电压 2. Proximity Operation Vehicle 靠近操作飞行器
powder　1. 火药,炸药 2. 粉,粉末
powder blend　粉合料
powder electroluminescence　粉末电致发光
powder forging　粉末锻造
powder metal　粉末金属
powder metallurgy　粉末冶金

powder metallurgy superalloy 粉末高温合金
powder metallurgy titanium alloy 粉末冶金钛合金
powdered rubber 粉末橡胶
power 1.功率,2.电源
power amplification 1.功率放大 2.电功率放大倍数(输出交流电功率与输入电功率之比)
power amplifier 1.功率放大器 2.助力器
power available 可用功率
power brake 动力操纵制动器
power bus 电源总线
power circuit 电力电路,电源电路
power coefficient 1.功率系数 2.动力系数
power condition 动力状态
power conditioner 1.电力调节器 2.动力调节器
power conditioning 电源调节,脉冲功率调节
power connector 电源连接器
power constraint 功率束缚,功率限制
power consumption 能量功耗,能量消耗,用电量
power control electronics 电源控制电子设备
power conversion 功率换算,能量变换
power conversion efficiency 功率转换效率,电光转换效率,能量转换效率
power conversion system 功率转换效率,电力变换系统,电网接入系统
power cutoff 1.断电 2.(发动机的)关机
power cycle 动力循环
power delivery 功率输出,动力输送
power demand 功率需量,电力需量
power density 功率密度(单位体积的功率,如用于电子设备或核反应堆上)
power deposition 功率沉积
power dissipation index 功耗指数
power distribution 配电,功率分布
power distribution system 配电系统
power distribution unit 配电装置,配电部件
power divider 功率分配器
power efficiency 能量利用效率,功率效率
power electronics 1.功率电子学 2.电力电子学
power equipment part 电源设备部件
power extraction 功率削减,电能提取
power factor 1.功率因数 2.力率
power failure 电源故障,电源中断,动力故障
power flight phase 动力飞行段
power flux density 功率通量密度
power gain 功率增益,放大系数
power generation 发电,发电量,发电机设备
power generation system 发电系统
power generator 电力发电机
power head 动力头

power in, power out 滑进滑出(飞机不用牵引车而用自身的动力进出停机坪或停放地点)
power in, push out 滑进推出(飞机用自身动力滑入停机坪,用牵引车顶推到滑行道)
power input 电源输入,功率输入
power jet 动力喷口,主射流,强射流,动力射流
power law 幂次法则,幂次定律
power level 功率级,(电子)功率电平,功率位准
power load 电力负荷,功率载荷,动力负载
power loading 动力负载,功率载荷
power loss 功率损耗
power loss indicator 排气压力表
power loss parameter 功率损耗参数
power meter 功率计
power mode 动力模式
power operation 动力操作,带电操作,功率操作
power output 功率输出,电源输出,输出功率
power performance 动力性,动力性能
power performance index 发动机性能指数
power plant 发电厂,动力装置,电源设备,发电设备
power plant control console 动力控制台
power plant relay cabinet 动力继电器柜
power production 产生功率,发电
power rammer 自动输弹机
power range 功率范围,功率区段
power rating 功率定额,功率额定值
power ratio 功率比,功率输出输入比
power reduction 功率降低,功率减小
power required 需用功率
power requirement 电源要求,功率需求
power reserve 后备功率,备用功率
power (return) slip ring 功率环,回路环
power saving 节电,节能,功率节省
power setting 1.动力调整 2.动力装置
power shaft 动力轴,传动轴
power shear 动力剪切机
power source 电源,能源
power spectral density 功率谱密度
power spectrum 1.密度谱 2.功率谱
power spinning 强力旋压
power splitter 功率分配器
power supply 电源,电源供应器
power supply subsystem 电源分系统
power synthesis 功率合成
power system 电网,电力系统,动力系统
power take off 1.功率提取 2.动力输出
power to weight ratio 功重比
power tracking 功率跟随,能量跟踪
power train (运载工具中的)动力传动系(可包括驱动

轴、离合器、变速器和差速器）

power transfer　能量输送,(车辆)动力分配装置

power transfer unit　(液压)动力传输装置(特指在两个分离的液压系统之间)

power transformer　电源变压器

power transmission　电力传输,能量传输

power traverse　动力回旋

power turbine　动力涡轮

power turret　动力炮塔

power unit　1.供电设备 2.动力头

power value　功率值,权值

power weight ratio　功率重量比,功重比

power-assisted flight control　电力辅助飞行控制

powered ascent　动力(推进)上升,(火箭)主动段上升

powered controls　功率控制机构,动力控制机构

powered descent　1.(飞机)带油门下降 2.航天器有动力下降

powered flight　动力飞行

powered lift　动力起落

powered missile　动力导弹,有发动机的导弹

powered phase guidance　主动段制导

powered-flight phase　主动段

power-frontal area ratio　功率与迎风面积之比

power-law　幂次法则,幂次定律

power-off　1.断电 2.关机(发动机的)

power-on　1.通电,通动力 2.(发动机)工作

power-pack　电源组,动力单元

powerplant　1.动力装置(发动机及其附属系统的总称),推动系统 2.发电站

powerstat　可调的自耦变压器

powertrain　动力系统,传动系统

Poynting vector　玻印亭矢量

PP　1. Parallel Process 平行过程 2. Parameter Patch 参数曲面片 3. Peak-to-Peak 峰间值 4. Pilot Production 试生产,零批生产 5. Power Plant 动力装置 6. Present Position 当前位置 7. Production Protection 产品防护 8. Production Permit 生产许可证 9. Production Phase 生产阶段 10. Production Plan 生产计划 11. Program Partners 项目合作方 12. Project Plan 工程计划 13. Proprietary Protection 产权保护

PP/S　Project (Program) Plan/Schedule 项目大纲/计划

PPA　1. Passengers Per Annual (航空公司的)每年乘客人数 2. Preplanned Attack 预先有计划的攻击,按计划攻击

PPB　Program, Planning, Budgeting 规划、计划与预算

PPBES　Program, Planning, Budgeting Evaluation System 规划、计划、预算评估系统

PPBS　Planning, Programming and Budgeting System 策划、计划和预算系统

PPC　1. Patrol Plane Commander 巡逻机机长 2. Permitted Parts Catalogue 许可件项目 3. Production Possibilities Curve 生产(量)可能性曲线 4. Program Planning and Control 项目计划和控制

PPCR　Production Planning Change Request 生产计划更改需求

PPD　Programming Planning Directives 程序设计指示

PPDL　Piont-to-Point Data Link 点对点数据链路

PPDS　Pilot Planetary Data System 实验行星数据系统(美国)

PPE　Passengers (per Year) per Employee (航空公司)每个职工每年服务的乘客数(即职工数与乘客数之比)

PPF　Payload Processing Facilities 有效载荷处理设施

PPFC　Process and Procedure Flow Chart 工艺和程序流程图

PPFRT　Prototype Preliminary Flight Rating Test 原型机最初飞行等级试验

PPH　Pulse per Hour 脉冲/时

PPI　1. Plan Position Indicator 平面位置显示器 2. Power Performance Index 发动机性能指数

PPL　1. Plasma Propulsion Laboratory 等离子体推进实验室(美国) 2. Polar Plasma Laboratory 极轨等离子体实验室(美国) 3. Polypropylene 聚丙烯 4. Processor-to-Processor Link 处理器-处理器链接

PPL/H　Private Pilot's Licence/Helicopter 直升机私人驾驶执照

PPM　1. Performance Programs Manual 性能程序手册 2. Pulse Position Modulation 脉冲位置调制,脉位调制

PPP　1. Peak Pulse Power 峰值脉冲功率 2. Point to Point Protocol 点到点协议

PPPL　Planetary Physical Processes Laboratory 行星物理过程实验室(美国)

PPR　1. Payload Preparation Room 有效载荷准备室 2. Plans, Progrmmes, Requirements 计划、程序、要求

PPROM　Personality Programmable Read Only Memory 终端专用可编程只读存储器

PPS　1. Passenger Processing System 乘客调整系统 2. Payload Pointing System 有效载荷定向系统 3. Picture Perception System 图像感觉系统 4. Pilot's Performance System 飞行员考绩制度 5. Pulse Per Second 每秒钟脉冲数

PPSI　Probe Powered Speed Indicator 动力传感器速度表

PPSL　Program Parts Selection List 项目零件选择单

PPT　1. Peak Power Tracking 峰值功率跟踪 2. Perishable Portable Tool 手提易损工具

PQA　Project Quality Assurance 项目质量保证

PQAR　Product Quality Action Request 产品质量措施申请

PQC　Picture Quality Control 图像质量控制

PQE　Product Quality Engineer 产品质量工程师

PQTP　Preliminary Qualification Test Procedures 初步鉴定试验程序

PR　1. Performance Requirements 性能要求 2. Pitch Rate 俯仰率 3. Photo Reconnaissance 照相侦察 4. Ply Rating（轮胎帘线层的）层级 5. Poly-sulphide Rubber 聚硫橡胶 6. Pressure Ratio 压力比,增压比 7. Problem Report 问题报告 8. Properietary Rights 专有权 9. Public Relations 公共关系 10. Purchase Request 采购申请

PRA　1. Particular Rish Analysis 特定风险分析 2. Planetary Radio Astronomy 行星际射电天文学 3. Probabilistic Risk Assessment 概率风险评估 4. Program Risk Analysis 项目风险分析

PRACA　Problem Reporting And Corrective Action 问题报告和改正措施

practicability　可实行性,实用物

practical astronomy　实用天文学

practical efficiency of solar array　太阳电池阵的实际效率

practicality　实用性,实际性,实际,实例

practice depth charge　练习深弹

practice missile　教练导弹,实习导弹

PRADS　Parachute, Retrorocket Air Drop System 降落伞,减速火箭空投系统

Prandtl number　普朗特数

prang　飞机相撞

PRAT　Production Reliability Acceptance Test 生产可靠性验收试验

Pratt truss　普拉特构架

PRAWS　Pitch, Roll Attitude Warning System 俯仰、滚转姿态告警系统

PRC　Program(me) Review Committee 计划审查委员会

PRCR　Preliminary Request for Customer Response 买方反应的初步要求

PRCS SPEC　Process Specification 工艺规范

PRDS　Processed Radar Display System 已处理过的雷达显示系统

preamble　报头

preamp　前置放大器（由 preamplifier 简化而得）

preamplifier　前置放大器

preassigned multiple access　预分配多址

precede　先于……,位于……之前,比……优先,比……重要

precedence graph　前趋图

preceding sunspot　前导黑子

precess　产生进动,（按岁差）向前运行,旋进

precession　岁差,先行,优先

precession cone　旋进锥

prechamber　预燃室

precious metal　贵金属（指金、银或铂）

precipitate　析出物,脱溶物,沉淀物

precipitation　落下,降水

precipitation attenuation　降水衰减

precise alignment　精密准直

precise injection　精确入轨

precise leveling　精密水准测量

precision　精密（度）,精确（度）

precision airdrop　精确空投

precision alloy　精密合金

precision approach　精确进场

precision approach procedure　精密进近程序,精确进场程序

precision approach radar　精密进场雷达

precision casting　精密铸造

precision drop glider　精确空投滑翔机

precision machining　精密机械加工

precision spin　精确螺旋

precision strike　精确打击

precision timing system　精确定时系统

precision VOR　精密伏尔

precision welding　精密焊接

preclude　排除,妨碍,阻止

precombustion　预燃

precombustion chamber　预燃室

precompression　预先压缩,预加压力

preconditioning　预处理

precooler　预冷器,预先冷却器

precursor　先驱者,前导,先进者

predecessor　前导

predelivery test　出厂试验

predetermine　预先确定,预先决定,预先表明

predetonation　过早点燃,先期爆炸,预爆轰

predict　预报,预言,预知

predictability　可预言性,可预报性

predicted barrage　移动拦阻射击

predicted fire　预测提前量的射击

predicted guidance law　预测导引律

predicted shooting　提前量已测定的射击

prediction　预报,预测,预告

prediction angle　总修正角

prediction error　预测误差

prediction interval　预测区间,预测数的变化范围

prediction method　预测方法,预报方法

prediction mirror 瞄准具修正角反射镜
predictive analysis 预测分析,法预期分析
predictive control 预测控制,预估控制,预防控制
predictive encoding 预测编码
predictive value 预测值
prediffusion 预扩散
predominant height 支配高度
pre-emphasis 1.预校 2.预修正 3.预加重 4.预先加强
preetching 预腐蚀
preexist 先存在,先于……而存在
prefeasibility study 初步可行性研究
preferential runway 优先使用的跑道
preferred orientation 择优取向
preferred parts list 所需零件清单
prefilter 初滤器
prefix 1.地区号码 2.电报报头 3.前缀
prefix code 前缀码
prefix method synchronization 词头法同步
preflight 1.飞行前的 2.飞行前检查 3.起飞前滑行
preheat 预热,预先加热
preheat temperature 预热温度
preheated 预先加热的
preheated air 预热空气
preheater 预热器(或炉)
preheating 预热,加温
preignition 提前点火,早燃
preinitiation 提前起爆
prelaunch (宇宙飞船等)发射之前的,准备发射之前的,下水前的
preliminary amplifier 前置放大器
preliminary design 初步设计
preliminary flight rating test 飞行前规定试验
preliminary inspection period 首次检查期
preliminary orbit 初轨,最初轨道
preloading 预先加载
premature 先期爆炸,过早爆炸
premature fuze action 引信过早起爆
premature ignition 先期点火
pre-mission zero calibration 事前校零
premix 预混合料,预拌和料
premixed burner 预混燃烧器
premixed combustion 预混燃烧
premixed combustion condition 预混燃烧条件
premixed combustor 预混燃烧室
premixed gas 预混合气
premixer 预混合器,预先混合器
preoxygenation 吸纯氧,吸氧排氮
PREP 1. Partreliability Enhancement Programme 加强部分可靠性计化 2. Pilot's Reference Eye Position 飞行员基准眼位
prepackaged rocket engine 预包装火箭发动机
prepaid ticket advice 预付票款通知
preparation 准备工作
preplanned air support 规划航空支援,预有计划的空中支援
preplanned mission request 预定任务之申请
prepositioning 预先部署,预先到位,前置
prepreg 预浸渍材料,预浸胶体
preprocessing 预处理,预加工,加工前的
preprocessor 预处理器,【计】预处理程序
preproduction 生产前,试生产,预生产,小批生产
prerotating vane 预扭导流片
prerotation 预转,预旋
prerotation gear 机轮预转的起落架(飞机放出起落架接地之前,机轮预先旋转)
prerotation vanes 预旋叶片(置于风洞中风扇的上游)
prescribe 指令,指示,规定,命令,法规
prescribed value 已知数,规定值
preselector 预选器
presence 1.存在,出席,参加,风度,仪态 2.驻外部队(指经双方协议驻在对方国家的军队),(某国在国外的)政治(或军事、经济等)势力
presentation 存在,出现,提出,呈递
presentation layer 表示层
preset angle 超越角
preset firing distance 预定射击距离
preset guidance 预置制导
presetting 预先装定
PRESS Project Review, Evaluation and Scheduling System 项目审查、鉴定与计划安排系统
press fit 压配合
press riveting 压铆
press-brake forming 闸压成形
pressing 压模,压制
pression 压力
pressurant 受压物体
pressure 1.压强 2.压力强度
pressure accumulator 蓄压器,压缩空气箱
pressure adaptation 压觉适应
pressure altitude 气压高度
pressure amplitude 压力变化幅度,压力振幅
pressure axis 压力轴
pressure balance valve 压力平衡阀
pressure bomb 火花塞试验器
pressure bottle 高压气瓶,耐压瓶
pressure boundary 压力边界,承压边界
pressure breathing 加压呼吸
pressure bulkhead 气密隔板,耐压舱壁

pressure cabin 压力坐舱
pressure cell 压力受感装置,压敏(压力)元件,膜盒室气压计
pressure center 压力中心
pressure chamber 压力室,气压实验室
pressure change 压力变化,气压变化
pressure change rate 压力变化率
pressure chart 压力曲线图
pressure coefficient 压力系数
pressure constraint 压力约束
pressure control 压力控制
pressure control subsystem 压力控制分系统
pressure cooling 加压冷却
pressure correction 压力修正,压力校正
pressure decay 压力降低,压力衰减
pressure die casting 1.压力铸造 2.压铸
pressure difference 压力差,差压,压力补偿器,气压梯度
pressure differential (指容器内外的,例如增压舱与外界大气的)压(力)差
pressure distribution 压强分布
pressure distribution test 压强分布试验
pressure face 压力面,承压面
pressure flap 压力活门
pressure fluctuation 压力脉动,压力波动
pressure force 压力,气压力
pressure fuze 气压式引信
pressure gauge 压力计
pressure gloves 增压手套,密封手套
pressure head 压力头,压强水头
pressure height 气压高度,压头
pressure helmet 密封头盔,增压头盔
pressure impulse 压力脉冲
pressure injection 压力灌注,加压回灌
pressure input 压力输入
pressure instrument 压力式仪表
pressure intensifier 1.增压器 2.压力增强器
pressure jacket 代偿背心
pressure jet 加压射流
pressure load 抗压应力,压力载荷
pressure loading 抗压应力,压力载荷
pressure loss 压力损失,压力损耗
pressure lubrication 压力润滑,加压润滑
pressure manometer 压力计,压力表
pressure measurement 压力测量
pressure microphone 声压式话筒
pressure orifice 测压孔
pressure pattern flying 气压型飞行,等压面飞行
pressure pump 压力泵,增压泵

pressure rake 测压排管
pressure ratio 压力比,增压比
pressure recovery 压力恢复
pressure reducer 减压器
pressure refueling 压力加油
pressure regulation 调压压力调节
pressure regulator 调压器,压力调节器
pressure rigid airship 压力硬式飞艇
pressure schedule of cabin 座舱压力制度
pressure seal 加压密封
pressure sensor 压力传感器
pressure spike 压力尖峰
pressure suit 增压服,加压服
pressure suit gloves 高空加压服手套,代偿手套
pressure surface 1.压力面 2.推进面
pressure switch 压力操纵开关
pressure test 测压试验,压力试验
pressure thrust 压力推力,增压推力
pressure transducer 压力变换器,压力传送器
pressure transmitter 压力传送器,压力传感器,传压器,压力传递器
pressure vessel 压力容器
pressure-feed system 压力进给系统
pressure-sensor array 气压传感器阵列
pressure-volume diagram 压力容积曲线图
pressurization 1.高压密封 2.加压,增压 3.压力输送
pressurization air source 增压气源
pressurize (使高空飞行的飞机机舱)增压,密封
pressurized 增压的,密封的
pressurized bay 增压舱
pressurized cabin 增压舱,气密座舱
pressurized feed 加压输送
pressurized module 增压舱,压力舱
pressurized tank (依靠油箱内充入的气压将油液输出的)增压油箱
pressurized-gas reaction control system 压缩气体反作用控制系统
presswork 压力加工,压制成品
prestage 前置级
pre-start check 起动前检查
prestress 预应力,预先拉伸,镇压固结力
prestrike recon 攻击前侦察
pretension 预加拉应力,预拉
pretensioned system 先张拉设备,预拉系统
pretest 事先试验,(对产品等的)预先测试
pretreatment 预备处理
PREV Previous 先前的
prevail 1.胜,胜过;优胜,获胜 2.生效,有效
prevailing wind 主风,盛行风

prevarication 多义度
preventive maintenance 预防性维修
preventive perimeter 预防性外围防御
preventive war 预防性战争（已知战争不可避免，首先发动战争以减少损失）
prewhirl 预旋
PRF 1. Parachute Refurbishment Facility 降落伞整修设施（美国） 2. Primary Reference Fuel 正标准燃料 3. Pulse-Repetition Frequency 脉冲重复频率
PRFD Pulse Repetition Frequency Distribution 脉冲重复频率分布，脉冲重复谱
PRFJ Pulse Repetition Frequency Jitter 脉冲重复频率抖动
PRFS Pulse Repetition Frequency Stagger 脉冲重复频率摆动，交错变脉冲重复频率
PRGM Program 程序
PRI 1. Primary 基本的，主要的 2. Pulse-Repetition Interval 脉冲重复间隔
primacord 爆炸导火索，导爆索
primary 主要的，初级的，基本的
primary air 主气流，一次气流，一次流，主流
primary area 一次区域，主区
primary body 主天体，初生体
primary bomb damage 直接轰炸效果
primary cell 一次电池，原电池（一次性使用电池）
primary circulation 一次循环，一级环流，初级环流
primary color unit 基色单元
primary combustion 初次燃烧，一次燃烧，预燃
primary cosmic rays 原宇宙线
primary development model 初始发展模型，原始模型
primary dump system 初级转储系统
primary electrical power source 主电源，一次电源
primary electron 一次电子，原电子
primary failure 独立失效
primary flat 主平面
primary flight control system 主飞行操纵系统
primary flight controls 主飞行控制器
primary flight display 1. 主飞行显示器 2. 垂直状态显示仪
primary glider 初级滑翔机（结实而简单的，不具备翱翔能力）
primary magnetization 原生磁化（强度）
primary maintenance process 主维修工艺
primary member 主要构件，基本构件
primary nozzle （喷嘴的）主喷油孔
primary radar 一次雷达，主要雷达
primary radiator 初级辐射器
primary reaction 初级反应，原初反应
primary remanent magnetization 原生剩余磁化

primary runway 主跑道
primary satellite 主用卫星
primary standard 标准原器，原标准器
primary standard solar cell 一级标准太阳电池
primary stress 初始应力
primary structure 主要构造，一级结构
primary target 主目标，初始目标
primary trainer 初级教练机
primary wave 初波，一次波
primary zone 主燃区
prime 1. 最初的，原始的 2. 涂底层
prime airlift 实际空运飞机数
prime commutator 主交换子
prime contractor 主要承包商，主承包单位，主合同单位
prime focus 主焦点
prime mover 1. 原动机 2. 牵引车，拖船
primer 1. 起动注油器 2. 底漆，底胶 3. 雷管，火帽，底火
primer cap 起爆雷管
primer seat 底火座
primitive 原始的，粗糙的
primitive code 本元码
primitive orbit 初（始）轨（道）
principal axis 主轴，主轴线
principal axis of inertia 惯性轴
principal component analysis 主成分分析
principal component transformation 主分量变换
principal inertia axis 主惯性轴
principal inertia coordinates 主惯性坐标系
principal moment of inertia 惯性主矩
principal strain 主形变，主应变
principal stress 主应力
principal tensile stress 主拉应力
principal vertical 主垂线（倾斜照相中，通过像主点并和真地平线垂直的线）
principal wave 主波
principle of equivalence 等效原理，当量原理
printthrough 磁带透印
prior permission 预先许可，事先批准
prior probability 先验概率
prior-to-service program 使用前大纲
priority 优先权
priority target 优先目标，首要目标
prism 棱镜，棱柱
prismatic astrolabe 棱镜
prismatic joint 移动关节，柱状节理
privacy key 保密钥匙
private aircraft 私人飞机，私有飞机
private automatic branch exchange 专用自动小交换机
private communication 私人通信

private pilot　私人飞机驾驶员
private satellite　私人卫星
privileged instruction　特许指令
PRM　1. Presidential Review Memorandum 总统审查备忘录 2. Product Relationship Model 产品关系模型
PRMD　Pilot's Repeater Map Display 飞行员回答式地图显示器
PRN　Pseudo-Random Noise 伪随机噪声
PRNSA　Pseudo-Random Noise Signal Assembly 伪随机噪声信号装置
probabilistic decoding　概率译码
probabilistic machine　随机元件计算机
probabilistic search　概率搜索法
probability　概率
probability density　概率密度
probability density function　概率密度函数
probability density of failure　故障概率密度
probability distribution　概率分布
probability distribution function　概率分布函数
probability of damage　破损概率
probability of detection　发现概率,检测概率
probability of failure　故障概率
probability of hitting　命中概率
probability of kill　杀伤概率
probability of success　成功概率
probability of survivability　存活率
probability of survival　生存概率,存活概率
probability value　概率值
probable　大概的,很可能的
probable error　概率误差
probable value　可能值,或是值
probe　动静压探头
probe adapter　探测器转接器
problem　问题
problem analysis　问题分析
problem trend analysis　问题趋势分析
PROC　Procedure 程序
procedure　程序,过程,方法,步骤,手续
procedure activation　过程活动,过程激励
procedure package　过程包
procedure turn　程序转弯
proceed　开始,继续进行,发生,行进
process　进程,加工
process acceptance　工艺验收文件
process automation　生产过程自动化
process average　过程平均
process control　过程控制
process evaluation pilot testing　过程性评价前导测试
process flowchart　进程流图

process inspection　工序检验,过程检查
process lens　分色镜头
processability　工艺性,可加工性
processing　处理,加工
processing condition　加工条件,工艺条件
processing monitoring　工艺过程监测
processing parameter　工艺参数
processing program　处理程序
processing simulation　工艺模拟
processor　处理机
processor sharing　处理机共享
processor terminal　处理机终端
procure　（努力）取得,（设法）获得,把……弄到手,得到
procurement　购买,获得,采办
procurement lead time　预定提前时间
prod　探针,刺针
produce　1. 生产,出产 2. 制造,制作
producibility　可生产性
product　产品,结果,乘积,作品
product assurance　产品保证
product certification　产品合格证
product detector　乘积检波器
product gas　析出气体
product improvement　产品改进
product inspection　成品检验
product liability　产品责任
product of inertia　惯性积
product quality　产品质量
product service　产品服务
product specification　产品规格
product support　产品支援,产品支持
production aircraft　生产型飞机,批生产的飞机（区别于原型机）
production base　生产基地
production break　工艺分离面
production capacity　生产能力,生产力
production certificate　生产许可证
production cost　生产成本,生产费用
production development　产品开发,生产发展
production engine　大批量生产发动机,正常系列发电机
production facility　生产设备,生产设施
production function　生产函数
production lot　生产批
production permit　制造许可证
production phase　生产阶段
production rule　生产式规则
production turnaround　生产周转期
productive potential　生产潜力
productivity　生产率

PROF Profile 翼型,包线
profile 轮廓,外形,剖面,侧面
profile descent 剖面下降
profile drag 理想铃中的阻力,形状阻力,翼型阻力
profile drag coefficient 形状阻力系数
profile map 剖面图
profile mean line 翼型中线
profile milling 仿形铣削
profile survey 剖面测量
profile thickness 翼型厚度,翼剖面厚度
profiler 靠模铣床
profilometer 表面粗糙度仪,轮廓测定仪
profilometry 轮廓测定法
prognostic 预兆,预言,预测,预知
prograde 同向旋转运动,顺行
program 程序,计划
program block spectrum 程序块谱
program clock 程序时钟
program command 程序指令
program control 程序控制
program data set 程序数据集
program designation 程序引导
program directive 程序指令
program flight control system 程序飞行控制系统
program flow chart 程序流程图
program generation system 程序生成系统
program graph 程序图
program guidance 程序制导
program identification code 程序识别码
program index register 程序索引寄存器
program information code 程序信息代码
program interrupt control 程序中断控制
program memory 进程存储装置
program monitoring and diagnosis 程序监测和诊断
program portability 程序可移植性
program pulse 程序脉冲
program register 程序寄存器
program test system 程序检验系统
programmable 可编程序的
programmable attenuator 程控衰减器
programmable calculator 可编程计算器
programmable instrument 程控仪器
programmable logic array 可编程逻辑阵列
programmable logic controller 可编程序逻辑控制器
programmable logic device 可编程逻辑器件
programmable network 可编程网络
programmable read only memory 可编程序只读存储器
programmable signal generator 程控信号发生器
programmable telemetry 可编程遥测
programmable telemetry system 可编程遥测系统
programmable transversal filter 可编程横向滤波器
programmer 程序员,程序规划员
program(m)ing 计划,规划,程序设计,程序编制
programming panel 程序设计板
programming transparency 程序设计透明性,程序透明性
progress 1.前进,进展,进程,行进 2.发展,进步,改进,上进,生长 3.(向更高方向)增长,扩展
progress payment 按进度付款
progress report 进度报告,进展报告
progression 1.前进,行进,发展 2.(行为、动作、事件等的)连接,连续,一系列 3.行星由西向东的运行
progressive burning 集聚烧除
progressive damage 累进破坏
progressive die 顺序冲模
progressive orbit 顺行轨道
progressive wave 前进波
prohibited area 禁飞区,飞行禁区
project 方案,计划,工程,工程项目
project engineer 项目责任工程师,主管工程师
project file 项目文件
project network analysis 计划网络分析
project review 项目审核
projected area 投影面积
projected blade area 桨叶投影面积
projected range 投影射程
projectile 射弹,抛射体
projectile depth charge 射弹式深水炸弹
projectile filling 装药
projectile line 射弹线
projectile nose 弹头
projectile velocity 射弹速度
projection 投射,投影
projection mask aligner 投影光刻机
projection tube 投影管
projection TV 投影电视
projector 投影器
PROM Programmable Read-Only Memory 可编程只读存储器
prompt nuclear radiation 瞬间的核辐射
prony brake 普郎尼制动动力计,普郎尼测功器
proof 1.证明,证据,校样,考验,验证,试验 2.(武器等的)试验强度,试验场所
proof mass 检测质量
proof of loss 损失证明
proof pressure 耐压试验压力
proofing 防护,防护剂
proof-of-concept 概念验证

propagate 传播,分布,蔓延
propagation 1.传播,繁殖,增殖 2.传播,传送
propagation constant （波的）传播常数
propagation velocity 传播速度（电磁波的传播速度为每秒30万千米）
propagator 传播者,分布函数
propane 丙烷
propane flame 丙烷火焰
propane gas 丙烷气
propargyl 炔丙基
propel 推进,驱使,激励,驱策
propellant 推进剂,火箭燃料
propellant additive 推进剂添加剂
propellant binder 推进剂黏合剂
propellant booster 燃料起飞发动机,火箭助推器
propellant burning 推进剂燃烧
propellant burning rate 1.推进剂燃烧速率 2.推进剂燃速
propellant burning time 推进剂燃烧时间
propellant combustion 推进剂燃烧
propellant composition 燃料成分
propellant consumption 推进剂消耗量
propellant feed 推进剂输送
propellant feed system 1.推进剂供应系统 2.推进剂输送系统
propellant flow rate 燃料消耗率
propellant formula 推进剂配方
propellant grain 火箭火药柱
propellant injection 喷气燃料注入
propellant mixture 推进剂混合
propellant particle 发射药颗粒
propellant spray 燃料喷注
propellant storage depot 推进剂供应
propellant supply 推进剂供应率
propellant supply rate 推进剂表面
propellant tank 推进剂箱
propellant utilization efficiency 推进剂利用效率,推进剂效率
propeller 螺旋桨,推进器
propeller advance ratio 螺旋桨进距比
propeller area 螺旋桨桨叶面积
propeller blade 螺旋桨叶片
propeller blade angle 桨叶角
propeller brake 螺旋桨刹车装置
propeller camber ratio 螺旋桨叶剖面相对曲度
propeller cavitation 螺旋桨旋转真空
propeller characteristic 螺旋桨特性
propeller disc 螺旋桨旋转面,螺旋桨桨盘（桨叶扫过的圆面积）

propeller efficiency 螺旋桨效率
propeller feathering 螺旋桨顺桨
propeller governor 推进器调速器,螺旋桨调速器
propeller hub 螺旋桨桨毂
propeller interference 螺旋桨干扰
propeller pitch 推进器倾斜,螺旋桨螺距
propeller pitch indicator 桨距表
propeller rake 螺旋桨倾角
propeller reversing 反桨
propeller slipstream 螺旋桨滑流
propeller solidity ratio 螺旋桨实度
propeller speed governor 螺桨调速器
propeller torque 螺旋桨转矩
propeller turbine 螺旋桨式透平机
propeller wash 螺旋桨洗流
propelling nozzle 推力喷管
proper motion 固有运动
property 1.所有权,所有 2.性质,性能,特性,属性
propfan 螺桨风扇
propfan engine 桨扇发动机
propjet 涡轮螺桨发动机,涡轮螺旋桨喷气发动机
proportional command 比例指令
proportional control 比例控制
proportional counter 比例计数器
proportional limit 比例极限
proportional navigation 比例导航,比例引导
proportional region 正比区,比例范围,比例区域
proportional revolver 比例式旋转变压器
proportionality 比例性
proportionality factor 比例系数,比例因子
propose 1.提议,建议,提出（行动,计划或供表决的方案等）2.打算,计划,意欲
proposition 1.提议,建议,主张,提案,计划 2.议题,论点,陈述 3.命题,（待证明的）定理;待解决的问题
propulsion 1.推进,被推进 2.推进力,推进器,推进法
propulsion bay 推进装置舱
propulsion branch 主动段
propulsion control 推进器控制
propulsion device 推进器,推进设施
propulsion efficiency 推进效率,螺旋桨效率
propulsion engine 喷气发动机,推进发动机
propulsion mode 推进方式
propulsion module 动力装置舱
propulsion subsystem 推进分系统
propulsion system 推进系统
propulsion system of rocket 火箭推进系统
propulsion technology 推进技术
propulsion test 推进实验,推进测试
propulsion thruster 推力器

propulsion unit 推进装置
propulsion wind tunnel 推进风洞
propulsive 推进的,有推进力的
propulsive deceleration 推进减速
propulsive deorbit 推进脱轨
propulsive deorbit device 推进脱轨设计
propulsive duct 推进管（即无压缩器式喷气发动机,一般是脉动式冲压或亚声冲压）
propulsive efficiency 推进效率
propulsive force 推力,推进力
propulsor 推进器,推进发动机
propylene 丙烯
propyne 丙炔
proration 按比例分配,分摊
prospecting line profile map 勘探线剖面图
prospecting line survey 勘探线测量
protected zone 防护带,被保护区域
protection check 保护检验
protection circuit 保护电路
protection mechanism 保护机构,保护机制
protection of laser hazard 激光防护
protection of solar radiation 太阳辐射防护
protection of space debris 太空碎片防护
protective atmosphere heat treatment 保护气氛热处理
protective breathing equipment 呼吸保护设备
protective clothing 防护服
protective coating 保护涂层
protective cover 防护堵盖,护盖
protective cutoff 保护关机
protective device 防护装备,防护措施
protective earthing 保护接地
protective goggles 护目镜
protective grounding 保护性接地
protective helmet 防护盔,防护帽
protective suit 防护服
protective treatment 防护处理
protocol 规约,协议,规程
protocol analyzer 协议分析仪
protocol control information 规约控制信息
protogalaxy 原始银河,原星系
proton 质子
proton exchange membrane fuel cell 质子交换膜燃料电池
proton flare 质子耀斑
proton-precession magnetometer 质子旋进磁力仪
protoplanet 原始行星,原行星
protostar 原始星,原恒星
prototype 原型机,样机,原型
prototype design 初样设计

prototype filter 原型滤波器
prototype flight test 原型机试飞
prototype model 初样模型,试验星,初样星
prototype rocket 模样火箭
prototype system 原型系统
prototype test 样机试验,初样试验
protractor 分度规,量角器
protrude 突出,伸出,吐出
protuberance 突起,突起部
prove 1.证明,证实 2.检验,试验,考验
provide 1.预备,准备 2.提供,供给 3.装备,装备起来
proving flight 验证试飞
proving ground 试验场,检验场,射击场
provision 规定,条款,设备,装置
provisioned spares 提供,准备
PROX Proximity 接近
proximity 接近,接近度
proximity exposure method 接近式曝光法
proximity focusing 近贴聚焦
proximity fuze 近炸引信
proximity switch 近感应电子开关（一种取代暴露在外用以指示机件位置的微动电门的新式器件,通常是一种向微电子组件输入信号的变电抗受感器）
PRP 1. Personnel Reliability Program 人员可靠性计划 2. Power-deployed Reserve Parachute 自动开伞的备份伞 3. Premature Removal Period 提前拆卸周期 4. Project Requirements Planning 项目要求计划 5. Proprammed Random Process 程序随机过程 6. Pulse Recurrence Period 脉冲重复周期
PRR 1. Preliminary Requirements Review 初步要求评审 2. Premature-Removal Rate 提前拆卸率 3. Production Readiness Review 生产准备状态评估 4. Production Revision Record 生产修改记录 5. Program Requirements Review 计划要求评审 6. Pulse Repetition Rate 脉冲重复率
PRRC pitch, roll rate changer assembly 俯仰、滚转率变换装置
PRRFC Planar Randomly Reinforced Fiber Composite 同平面无规则增强纤维复合材料
PRR-R Production Revision Record-Retrofit 生产修正记录-改型
PRS 1. Pattern Recognition System 模式识别系统 2. Pressure-Ratio Sensor 压力比传感器,压力比敏感元件
PRSDS Power Reactant Storage and Distribution Subsystem （美国航天飞机上的）（燃料电池）发电反应剂贮存和分配分系统
PRSG Pulse-Rebalanced Strapdown Gyro 脉冲再平衡捷联式陀螺（惯性导航系统中的陀螺,利用电子再平

衡回路来减小其动力误差）

PRSOV Pressure Regulating and Shut-Off Valve 调压与断油活门，压力调节和关断活门

PRT 1. Pattern Recognition Technique（目标）图形识别技术 2. Plastics Rubbers Textiles 塑料，橡胶，纺织物 3. Platinum Resistance Thermometer 铂丝电阻温度表 4. Pulse Recurrence Time 脉冲重复时间，脉冲周期 5. Pulse Rise Time 脉冲上升时间

prudent limit of endurance 安全续航时间持久极限，安全续航时间范围

prudent limit of patrol 实用巡查极限，安全巡逻范围

prune 消减（预算的），删除

PRV 1. Paraglider Research Vehicle 滑翔伞研究飞行器 2. Pressure Regulation Valve 调压阀

PS 1. Part Standards 零件标准 2. Performance Standards 性能标准 3. Pico Second 皮秒（10^{-12} 秒） 4. Position Shift 移位 5. Power Supply 电源 6. Pressure Switches 压力转换 7. Process Specification 工艺规范 8. Process Standard 过程标准，工艺标准 9. Procurement Specifications 采购规范 10. Product Software 产品软件 11. Product Standard 产品规范 12. Product Structure 产品结构 13. Product Support 产品支援 14. Production Schedules 产品计划 15. Program Schedule 项目计划 16. Provincial Standard 地方标准 17. Proximity Sensor 接近传感器 18. Proximity Switch 接近电门

PSA 1. Part Stress Analysis 零件应力分析 2. Payload and Servicing Accommodations 有效载荷和服务条件保障 3. Pilot's Associate 副驾驶员 4. Power Supply Assembly 电源组件 5. Pneumatic Sensor Assembly 气动敏感组件 6. Prefab Semi-permanent Airfield 预制半永久性机场 7. Preliminary Safety Analysis 初步安全性分析 8. Propellant Storage Assembly 推进剂贮存装置 9. Proximity Sensor Actuator 接近传感器动作器

PSAC 1. Plan for Software Aspects of Certification 软件状况合格审定大纲 2. Presidential Science Advisory Committee 总统科学顾问委员会（美国）

PSAI Public Safety Aviation Institute 公共安全航空研究所（美国）

PSAS 1. Pitch Stability Augmentation System 俯仰增稳系统 2. Primary Stability Augmentation System 主增稳系统

PSBLS Permanent Space Based Logistics System 长久性天基后勤系统

PSC 1. Parliamentary Space Committee 议会航天委员会（英国） 2. Platform Support Center 平台保障中心 3. Pneumatic System Control 压缩空气系统控制 4. Polar Stratospheric Clouds 极区平流层云 5. Product-Support Committee 产品保障委员会 6. Program Support Contract 项目计划保障合同（美国国家航空航天局用语） 7. Program Switching Center 程序交换中心

PSCN Program Support Communications Network 计划保障通信网

PSCRD Program Support Communications Requirements Document 计划保障通信要求文件

PSCS Public Services Communications Satellite 公共业务通信卫星

PSD 1. Phase Sensitive Demodulator 相敏解调器 2. Physiological Support Division 生理保障分队 3. Port Sharing Device 端口共享装置 4. Power Spectral Density 功率频谱密度 5. Process Specification Departure 工艺规范偏离 6. Production Salvage Drawings 生产补救图纸

PSDE Payload and Spacecraft Development and Experimentation 有效载荷和航天器研制试验计划（欧洲航天局）

PSDL Pricture Sheet Data List 图纸数据表

PSDN Packet Switched Data Network 包转换数据网

PSDP 1. Preliminary Safety Data Package 最初安全数据包 2. Programmable-Signal Data-Processor 可编程信号数据处理器

PSE 1. Passenger Service Equipment 乘客服务设备（指旅客机上的阅读灯，招唤服务员的按钮等等） 2. Passive Seismic Experiment 无源地震实验 3. Principal Structural Elements 主要（关键）结构元件

PSEP Passive Seismic Experiments Package 无源地震实验装置（美国）

PSEU Proximity Sensor Electronic Unit 进近传感器电子装置

pseudo color picture treatment 伪彩色图像处理

pseudo color slicing 伪彩色（密度）分割

pseudo fly-by-wire 准电传操纵（飞机操纵系统中至少有一个轴的至少一个点上是电传的，整个系统可转换为手操纵或超控）

pseudo noise 伪噪声码

pseudo noise code 伪噪声码

pseudo random binary sequence 伪随机二进制序列

pseudo random sequence 伪随机序列

pseudo satellite 伪卫星

pseudo-azimuthal projection 伪方位投影

pseudo-cylindrical projection 伪圆柱投影

pseudo-noise interferometer 伪噪声干涉仪

pseudo-noise sequence 伪噪声序列

pseudo-random noise 伪随机噪声

pseudo-velocity response spectrum 伪速度反应谱

pseudocode 伪码

pseudocontrol vector 伪控制向量

pseudodensity 视密度
pseudo-front-end system 伪前端系统
pseudogravity anomaly 磁源重力异常
pseudoimpulse 伪脉冲
pseudoinverse 伪逆法
pseudo-isoline map 伪等值线地图
pseudolite 伪卫星
pseudopressure 拟压力
pseudo-random 伪随机的
pseudo-random code 1.伪随机码 2.伪随机序列
pseudo-random code frequency-modulated fuze 伪随机码调频引信
pseudo-random code modulation fuze 伪随机码调制引信
pseudo-random code phase-modulated fuze 伪随机码引信
pseudo-random noise signal assembly 伪随机噪声信号组件
pseudo-random number 伪随机数
pseudo-random pulse position fuze 伪随机脉位引信
pseudorange 假设距离,伪距
pseudo-range measurement 伪距测量
pseudo-rate increment control 伪速率增量控制
pseudosection map 拟断面图
pseudoshock 伪冲激波
pseudosphere 伪球面
pseudostate 伪状态
pseudo-steady state 伪稳态,准定常状态,准稳态
PSF 1. Personnel Services Flight 人员服务飞行(英国空军) 2. Phosphosilicate Fiber 磷硅酸盐纤维(光纤) 3. Polystyrene Foam 聚苯乙泡沫
PSFP Pre-simulator Familiarization Panel 前置模拟器练习板
PSG 1. Passenger Safety Guide 乘客安全须知 2. Post-stall Gyraton 失速后旋转 3. Pulse Signal Generator 脉冲信号发生器
PSH Preselected Heading 预选航向
PSI 1. Pounds per Square Inch 磅每平方英寸 2. Pressure Sensitive Instrument 压力敏感仪表
PSIT Process and System Integration Team 工艺和系统综合团队
PSK Pulse Shift Keying 相移键控法
PSK demodulator PSK 解调器(从相移键控信号中恢复出调制信号的装置)
PSK modulator PSK 调制器(实现相移键控调制的单元)
PSM 1. Power Supply Modules 电源模块 2. Product Structure Management 产品结构管理 3. Product Structure Modeling 产品结构建模
PSN 1. Packet Switched Network(数据)包交换网络 2. Packet Switching Node 分组交换节点 3. Personal Server Network 个人服务器网络 4. Phase Shift Network 相移网络 5. Processor Sharing Node 处理器共享节点
P-S-N curves P-S-N 曲线
PSO 1. Pilot System Officer 导航军官(前后座作战飞机中的领航员或其他后座人员) 2. Protective Service Operations 保护性作战任务
psophometer 噪声计
PSP 1. Personal (or personnel) Survival Pack 个人救生包 2. Pierced Steel Planking(铺跑道用的)预制穿孔钢板 3. Product Support Programme 产品保障计划 4. Programmable Signal Processor 可编程信号处理机 5. Program Support Plan 大纲保障计划,项目保障计划
PSPL Preferred Standard Parts List 优先标准件清单
PSR 1. Phase Sensitive Rectifier 相敏整流器 2. Point of Safe Return 安全返航点 3. Primary Surveillance Radar 一次监视雷达
PSRM Pressurization Systems Regulator Manifold 增压系统调速器歧管
PSS 1. Proximity Sensor System 进近传感器系统 2. Part Standards and Specifications 零件标准和规范
PSSA 1. Pilot Stick Sensor Assembly 飞行员驾驶杆传感器组件 2. Primary System Safety Assessment 初步系统安全性评估
PSSC Public Service Satellite Consortium 公用业务卫星国际财团
PST 1. Pacific standard time 太平洋标准时间 2. Pitot Static Tube 总静压管 3. Planetary Spectroscopy Telescope 行星光谱望远镜 4. Propeller Stol Transport 短距起落螺旋桨运输机
PSTM Payload System Test Machine 有效载荷系统测试设备
PSTN Public Switching Telephone Network 公共交换电话网
PSU 1. Passenger Service Units 旅客服务设备,旅客服务组件 2. Photo Smoke Unit 光学烟度计(用于测量喷流的发烟度)3. Pilot Screening Unit 飞行员屏蔽装置 4. Power Supply Unit 电子设备的取电装置,稳压电源装置
psychrograph 干湿计
psychrometer 干湿球温度计,干湿表
psychrometric 干湿表的
PT 1. Penetration Test 渗透检验 2. Performance Testing 性能测试 3. Pitot Tube 皮托管 4. Power Turbine 动力涡轮 5. Primary Training 初级训练 6. Procedure Turn 程序转弯
PTA 1. Part-Throttle Afterburning 部分油门复燃,部

分加力 2. Pilotless Target Aircraft 无人驾驶靶机

PTAB Project Technical Advisory Board 项目技术咨询委员会

PTAG Portable Tactical Aircraft Guidance 便携式战术飞机引导设备

PTC 1. Part-Through Crack 未穿透的裂纹 2. Passive Thermal Control 被动热控制 3. Pitch Trim Compensator 仰俯配平补偿器 4. Positive Temperature Coefficient 正温度系数 5. Programming and Test Center 编程和测试中心

PTCS Passive Thermal Control System 无源热控系统

PTD Performance Test Domain 性能试验范围

PTE Production Test Equipment 生产测试设备

PTEH Per Thousand Engine Hours 每一千发动机小时

PTF 1. Phase Transfer Function 相位传递函数 2. Programmable Transversal Filter 可编程横向滤波器

PTFCE polytrifluorochlorethylene 聚三氟氯乙

PTFE polytetrafluoroethylene 聚四氟乙烯,特氟隆

PTIT Power-Turbine Inlet Temperature 动力涡轮进口温度

PTL 1. Primary Target Line 主目标线 2. Primary Technical Leaflet 技术简介

PTM 1. Performance Technical Memorandum 技术性能备忘录 2. Pulse-Time Modulation 脉冲时间调制

PTO 1. Participating Test Organization 参加试验的单位 2. Patent and Trademark Office 专利和商标局(美国) 3. Place, Time, Object 地点,时间,对象 4. Power Take-Off 动力输出轴

PTP 1. Paper Tape Punch 纸带穿孔 2. Point-To-Point 点对点 3. Present Target Position 即时目标位置

PTR 1. Paper Tape Reader 纸带读出器 2. Part-Throttle Reheat 部分加力,部分油门复燃 3. Performance Technical Report 性能技术报告 4. Power-Turbine Rotor 动力滑轮转子 5. Preliminary Test Report 初步试验报告 6. Punched Tape Reader 穿孔带读数器

PTRP Propfan Technology Readiness Programme 桨扇发动机工艺准备大纲

PTS Photogrammetric Target System 目标摄影测量系统

PTSA Prior To Sample-Approval 样品批准前

PTT 1. Part-Task Trainer 部分任务练习器(模拟器) 2. Press To Transmit 按键发送(发射) 3. Push To Talk 按压通话 4. Push To Test 按压试验

PTTER Preliminary Tactical Technical Economical Requirement 初步战术,技术,经济要求(瑞典确定新项目之前先做战术技术经济要求的草案,然后提出初步要求,接着就是生产原型机)

PTU 1. Power Transfer Unit 功率传输装置 2. Pressure Transducer Unit 压力传感器组件

PTV 1. Parachute Test Vehicles 降落伞试验飞行器 2. Propulsion Technology Validation 推进技术核准

P-type semiconductor P型半导体

PU 1. Peripheral Unit 外圈部件 2. Pick-Up 拾音器 3. Pluggable Unit 插件 4. Polyurethane 聚氨基甲酸酯 5. Position Updating 位置修正 6. Power Unit 动力装置

PU propellant 聚氨酯推进剂

public address system 公共广播系统

public aircraft 公用飞机

public airport 公用机场(公有、公营、供公众使用的机场)

public charter 公众包机(一种新的包机制度,可单程包租,不限最低人数,并可提供折扣)

public parking area 公用停机坪

publication 1. 发表,公布 2. 发行物,出版物,刊物

PUD Power Unit Deicing 动力装置除冰

pugging 捏练,捏揉

PUGS Propellant Utilization and Gaging System 推进剂输送及量度系统

pull 拔,拉

pull broaching 拉削

pulley 滑轮,皮带轮,滑车

pull-in 捕捉

pull-in range 捕捉带,捕获带

pull-off 1. 拔出,拉出,脱离 2. 扣压扳机 3. 跳伞时靠滑流帮助跳伞员脱离飞机

pull-off strength 拉脱强度

pull-out bombing (俯冲)拉起轰炸

pull-up 1. 急跃升,拉起,急剧上仰 2. 停止,刹住,拔起

pull-up bombing (平飞)拉起轰炸

pulsar 脉冲星,脉冲射电源

pulsar timing 脉冲星计时法

pulsatance 角频率

pulscope 脉冲示波器

pulse 1. 脉冲,脉动,冲动 2. 半周 3. 使产生脉动 4. 一个固体推进剂药柱(在火箭发动机中如有两个药柱,工作就会有两次冲动,two-pulse rocket 也就是双药柱火箭)

pulse address multiple access 脉冲寻址多址联接

pulse amplifier 脉冲放大器

pulse amplitude 脉冲幅度,脉冲振幅

pulse amplitude modulation 脉冲幅度调制

pulse analog modulation 脉冲模拟调制

pulse arrival 脉冲波至

pulse back edge 脉冲后沿

pulse capacitor 脉冲电容器

pulse code 脉冲编程,脉冲代码

pulse code modulation 脉冲编码调制
pulse code modulation telemetry 脉码调制遥测系统
pulse combustor 脉动燃烧器,脉冲燃烧器
pulse compression 脉冲压缩(一种雷达探测方法,压缩后可获得高的脉冲功率)
pulse compression radar 脉冲压缩雷达
pulse compression ratio 脉冲压缩比
pulse compression receiver 脉压接收机
pulse compression technology 脉冲压缩技术
pulse counter 脉冲计数器
pulse decay time 脉冲后沿持续时间,脉冲衰减时间
pulse detection 脉冲检波
pulse detonation 爆震
pulse detonation engine 脉冲爆震发动机
pulse digital modulation 脉冲数字调制
pulse discharge 脉冲放电
pulse Doppler elevation scan 脉冲多普勒仰角扫描
pulse Doppler non-elevation scan 脉冲多普勒无仰角扫描
pulse Doppler radar 脉冲多普勒雷达
pulse Doppler spectrum 脉冲多普勒频谱
pulse driver 脉冲驱动器
pulse duration 脉冲持续时间,脉冲宽度(单个脉冲超过指定值的时间,指定值一般为峰值的10%,国际民航组织规定为50%)
pulse duration rate 占空系数,脉宽周期比
pulse encode modulation 脉冲编码调制
pulse energy 脉冲能量,脉波能
pulse equivalency 脉冲当量(每个脉冲所相当的输入或输出值)
pulse fall time 脉冲下降时间
pulse firing 脉冲点火
pulse frequency 脉冲频率,脉率,脉搏率,脉冲重复频率
pulse frequency modulation control system 脉冲调频控制系统
pulse frequency modulation telemetry 脉冲调频遥测
pulse front edge 脉冲前沿
pulse generator 脉冲发生器
pulse height discriminator 脉(冲振)幅鉴别器
pulse instrumentation radar 脉冲测量雷达
pulse interval 脉冲间隔
pulse interval modulation 脉冲间隔调制
pulse jet 脉动式喷气发动机,脉动式喷气飞机
pulse jet engine 脉动式喷气发动机
pulse laser 脉冲激光,脉冲激光器
pulse laser fuze 脉冲激光引信
pulse laser radar 脉冲激光雷达
pulse length 脉波长,脉冲持续时间
pulse mode 脉冲模式,脉冲方式

pulse modulation 脉冲调制
pulse modulator 脉冲调制器
pulse motor 脉冲发动机,脉冲马达
pulse pair processor 脉冲对处理器
pulse per second 秒脉冲
pulse period 脉冲周期
pulse phase system 脉相系统
pulse plating 脉冲电镀
pulse position modulation 脉冲位置调制
pulse power 脉冲功率
pulse profile 脉冲包络
pulse radar 脉冲雷达
pulse range gate fuze 距离选通脉冲引信
pulse rebalance 脉冲再平衡
pulse rebalanced rate gyro 脉冲再平衡速率陀螺
pulse recurrence frequency 脉冲重复频率
pulse recurrence time 脉冲重复时间,脉冲周期
pulse repeater 脉冲转发器
pulse repetition 脉冲重复
pulse repetition frequency 脉冲重复频率
pulse repetition interval 脉冲间隔,脉冲重复间隔
pulse repetition rate 脉冲重复率
pulse repetition time 脉冲重复时间,脉冲周期
pulse reshaping 脉冲整形
pulse response 脉冲响应特性曲线,脉冲响应
pulse rise time 脉冲上升时间
pulse rocket motor 脉冲式火箭发动机
pulse separator 脉冲分离装置
pulse sequence 脉冲序列
pulse shape discriminator 脉冲形状鉴别器
pulse shaper 脉冲整形器
pulse shaping 脉冲成形
pulse shift keying 脉冲移位键控,脉移键控
pulse size 脉冲振幅
pulse solar simulator 脉冲式太阳仿真器
pulse solid rocket engine 脉冲固体火箭发动机
pulse sounder 脉冲探测仪
pulse spike 脉冲尖峰
pulse steering circuit 脉冲引导电路
pulse stretcher 脉冲展宽器
pulse stuffing 脉冲塞入
pulse technique 脉冲技术
pulse telemetry parameter 遥测脉冲参数
pulse test 脉冲测试
pulse time multiplex 脉冲定时多路转接
pulse time-division multiplex 脉冲分时多路转接
pulse train 脉一群,脉冲链,脉冲序列
pulse transformer 脉冲变压器
pulse transmission 脉冲传输

pulse transmitter 脉冲发射机
pulse tube 脉冲管
pulse voltage 脉冲电压
pulse waveform 脉冲波形
pulse width 脉冲宽度,脉冲持续时间
pulse width modulator 脉冲幅度调制器
pulse wind tunnel balance 脉冲风洞天平
pulse-control 脉冲操纵
pulse-counting sonde 脉冲计数式探空仪
pulsed gasdynamic laser 脉冲气动激光器
pulsed laser deposition 脉冲激光沉积
pulsed laser range-finder 脉冲激光测距器
pulsed magnetron 脉冲磁控管
pulsed plasma thruster 脉冲等离子体推力器
pulsed specific impulse 脉冲比冲
pulsed spot welding 脉冲点焊
pulsed torquing 1.脉冲施矩 2.脉冲加矩
pulsed-laser-hardened materials synthesis 耐脉冲激光合成材料(能承受激光武器的照射)
pulsed-plasma arc welding 脉冲等离子弧焊
pulse-echo 脉冲反射波,回波脉冲
pulse-frequency modulation 脉冲频率调制
pulsejet 脉冲喷射发动机,脉冲喷射飞弹
pulse-jet 脉动式空气喷气发动机
pulse-jet engine 间歇燃烧喷射引擎,脉动喷气式发动机
pulse-length modulation 脉长调制
pulse-phase modulation 脉冲相位调制
pulser 脉冲发生器
pulse-time modulation 脉冲时间调制
pulse-width modulation 脉冲宽度调制
pulse-width modulation inverter 脉冲宽度调制转换器,脉宽调制逆变器
pultrusion 拉出成形
pultrusion process 挤拉成形
pulverization 喷雾,粉碎
pump 1.脉动,脉冲 2.泵激器,抽运器,汽油加油泵
pump cavitation 泵气蚀,泵空化
pump cavitation coefficient 泵气蚀系数
pump cavity blow-off 泵腔吹除
pump design 泵系统设计
pump efficiency 泵效率
pump fluid 泵工作液
pump head 泵扬程
pump material 水泵材料,水管材料
pump parameter 泵参数
pump pressure 泵压力,电机功率
pump rate distribution 泵速率分布
pump specific angular speed 泵比转速
pump stall 泵失速

pump unit 泵组
pumpability 泵送能力,泵送性能
pumpage 抽运量,送量
pumping 1.抽运,泵浦(泵压、泵汲) 2.唧筒作用(例如活塞涨圈在运动中将滑油压入气缸燃烧室) 3.(气压指示的)急剧升降
pumping efficiency 抽运效率
pumping rate 抽运速率
pump-up time 充气时间(放气式风洞贮气罐充满气的时间)
punched tape 穿孔带
punching 冲孔
punch-through 穿通现象,击穿现象(晶体管内的)
punch-through discharge 击穿放电
PUP 1.Performance Update Programme 性能更新方案,性能现代化大纲 2.Pop-up Point 跃升点(执行近距支援任务时从低空进入目标区的飞机从此点跃升,以占领攻击的位置)
purchase 1.(副翼的)传动摇臂固定点 2.(复数)复滑车,滑轮组
purchase cable (甲板拦阻装置的)滑轮钢索
purchase price 买价,进货价格
purchasing 1.采购,购买 2.依靠机械力移动(或举起),靠机械力抓紧
pure gas 净化气体,单一气体,纯煤气
pure gravity anomaly 纯重力异常
pure helicopter 常规直升机,普通直升机
pure helium 纯氦
pure jet 纯射流
pure methane 纯甲烷标准
pure nitrogen 纯氮
pure oxygen 纯氧
pure research 纯学理研究
pure strategy 纯策略
pure substance 纯净物,纯物质,纯正物
pure system 纯系统,纯体系
pure time 纯时间
pure time delay 纯时间滞后
pure tone 纯音
pure tungsten 纯钨极
purge air 冲放空气
purge flow 清洗流
purge gas 净化气体,吹扫用的气体
purge system 清洗系统,净化系统
purging pressure 吹除压强
purging system (发动机)吹洗系统,排烟系统
purification 净化,提纯
purification technology 净化技术
purity 1.纯度 2.气球内升力气体在气囊内所占体积的

百分比
purity of carrier frequency 载频纯洁度
puron 高纯度铁
purpose 1.意图,目的,意向 2.意志,决心,决断
purser 乘务长
pursuit attack 追踪攻击
pursuit course 追逐曲线(法),追踪航线
pursuit tracking 尾随追踪
pursuit-course missile 按追踪曲线引导的导弹
push 撞击,打击
push broom 推式路帚
push fit 轻推配合
push up list 先进先出表
push/pull (轻型飞机用的)推拉式油门
pushback 飞机推迟起飞,推迟飞机的起飞时间
push-broom sensor 推扫式传感器
push-button 1.按钮 2.按钮式的 3.远距离操纵的,按钮操纵的
pushbutton indicator 按钮控制(照相)亮度指示器
push-button switch 按钮开关
pushdown automaton 下推自动机
push-down list 后进先出表
pushdown storage 下推存储器
pusher 推进式,推进器,推动杆
pusher propeller 推动式螺桨
pushover 推杆,(指由推杆器指令完成的)推机头作机动飞行
push-pull power amplifier 推挽功率放大器
push-pull rod system 推拉杆系统
pushrod 推杆
pushswitch 液动开关,气动开关
pushup storage 上推存储器
PUT 1.Pilot Under Training 飞行学员 2.Pop-Up Test (弹道导弹)发射试验,其他冷发射系统的发射试验
put 1.推,送,发射 2.掷,投 3.促使……去做,使从事 4.出发,航行
putizze 硫化氢气孔
PUV Peak Undershoot Voltage 峰值下冲电压
PV 1.Pack Valve 空调组件活门 2.Part Version 零件版本 3.Photovoltaic 光电的,光生伏打的 4.Present Value 现值 5.Prevailing Visibility 有效能见度,主导能见度 6.Priority Valve 优先活门 7.Purge Valve 除尘活门
PVA 1.Plane View Area 平视图区域(精密锻造用) 2.Polyvinyl Acetate 聚酯酸乙烯酯
PVAD Pulsed Vacuum Arc Deposition 脉冲真空电弧沉积
PVC 1.Parameter Virtual Circuit 参数虚拟电路 2.Polyvinyl Chloride 聚氯乙烯 3.Position and Velocity Computer 位置和速度计算机
PVD 1.Para Visual Director 侧视指引器 2.Para Visual Display 超视距显示
PVF Polyvinyl Fluoride 聚氟乙烯
PVO 1.Part Version Object 零件版本对象 2.Pioneer-Venus Orbiter 先驱者金星轨道器(美国)
PVR Premature Voluntary Retirement 自愿提前退役
PVS Pilot's Vision System 飞行员视力系统
PVT 1.Position,Velocity,Time 位置、速度、时间 2.Product Verification Test 产品鉴定试验
PVTOS Physical Vapour Transport of Organic Solids (美国航天飞机上的)有机颗粒蒸发输运实验
PVU 1.Portable Ventilator Unit 轻便通风装置 2.Precision Velocity Update 精确速度更新
PVV Proof Vertical Velocity (主起落架破坏试验中的)验证垂直速度
PW 1.Packed Weight 包装重量 2.Passenger Weight 乘客重量 3.Potable Water 饮用水 4.Pulse Width 脉冲宽度
PWC Primary Work Code 主工作码
PWD Pulse Width Discriminator 脉冲宽度鉴别器
PWI 1.Pilot Warning Indicator 飞行员告警指示器 2.Pilot Warning Instruments 飞行员告警仪表 3.Power Warning Indicator 电源告警指示器 4.Preliminary Warning Instruction 预警指令 5.Proximity Warning Indication 防撞警告指示
PWM Pulse-Width Modulation 脉冲宽度调制
PWMI Pulse-Width Modulation Inverter 脉宽调制转换器
PWP 1.Peak Working Pressure 峰值工作压力 2.Pylon Weight Plug 挂架重量插头
PWR Passive Warning Radar 无源警戒雷达
PWRS Programmable Weapon Release System 可编程武器投放系统
PWS 1.Performance Work Statement(航空人员)工作考绩评语 2.Plasma Wave Subsystem 等离子体波分系统 3.Proximity Warning System 防撞警告系统(用于直升机)
PWT Propulsion Wind Tunnel 发动机试验风洞
PX 1.Post Exchange 陆军消费合作社(美国) 2.Private Exchange 用户交换机,专用小交换机
PY Program(me) Year 计划年度
pylon 吊架,挂架
pylon ejection mechanism 塔弹射机构
pyramid 角锥状物,角锥形,角锥体,棱锥体
pyramidal absorber 角锥形吸波材料
pyramidal balance 塔式天平
pyramidal horn 角锥形喇叭筒
pyramid-type system 金字塔型系统

pyranometer 日辐射强度计
pyrex 耐热硬质玻璃,硼硅酸玻璃
pyrgeometer 地面辐射强度计
pyrheliometer 日射表,太阳热量计
pyro(technic) shock 火工品(爆炸)冲击,爆炸分离冲击
pyroceramic 玻璃结晶材料
pyroelectric ceramics 热释电陶瓷
pyroelectric detector 热电检测器
pyroelectric vidicon 热电摄像管
pyrofoam 泡沫焦性石墨
pyrogenic technique of oxidation 加热合成氧化技术
pyrolysis 热解,高温分解,加热分解
pyrolysis experiment 热解实验
pyrolysis gas 裂解气,热解气
pyrolysis gas mass flow rate 热解气体生成率
pyrolysis law 裂解规律
pyrolysis plane model 热解模型飞机
pyrolysis process 热解过程
pyrolysis product 裂解产物,热解产物
pyrolysis rate 热解速度
pyrolysis response 热解反应
pyrolysis temperature 热解温度
pyrolysis tube 插入管
pyrolytic carbon 热解碳
pyrolytic deposition 热解淀积
pyrolyzer 裂解炉,热解器
pyromechanism 火工机构
pyrometer 高温计
pyrometry 测高温学,高温测定法
Pyron 派朗
pyrophoric 自燃,自燃的
pyroshock environment 爆炸冲击环境
pyroshock simulator 爆炸冲击仿真器
pyroshock test 爆炸冲击试验
pyrotechnic 爆炸的,信号弹的,辉煌灿烂的
pyrotechnic composition 烟火剂
pyrotechnic device 烟火装置
pyrotechnic train 烟火药道
pyrotechnics 1.火工品,火工装置,烟火设备 2.烟火制造术,烟火施放法
pyrradiometer 全辐射计
PZ Pickup Zone 搭载区

Q

Q 1. Quantity 数量 2. Quality 质量 3. Question 问题
Q alloy 镍铬合金
Q meter 品质因数测量仪
Q stops 1. 动压表 2. 动压限动装置
Q&A Question and Answer 问答
QA 1. Queue Access 队列存取 2. Quality Assurance 质量保障,质量保证
QAB Quality Asssurance Board 质量保证委员会
QAD Quick Attach Detach 快速装卸
QAIR Quality Assurance Inspection Requirements 质量保证检查要求
QAK Quick-Attach Kit 快速连接装备,工具箱
QAL 1. Quality 质量 2. Quartz Aircraft Lamp 飞机用石英灯
QAM 1. Quadrature Amplitude Modulation 正交调幅 2. Quantized Amplitude Modulation 量化调幅 3. Queued Access Method 排队存取法 4. Quality Asssurance Manual 质量保证监控器
QAO Quality Assurance Office 质量保证办公室
QAP 1. Quality Assurance Procedure 质量保证程序 2. Quality Assurance Program 质量保证大纲
QAR 1. Quality Assurance Represntative 质量保证代表 2. Quick Access Recorder 快速存取记录器
QAS Quality Assurance Standard 质量保证标准
QASK Quadrature Amplitude Shift Keying 正交移幅键控
QASS Quality Assurance Standards for Suppliers 供应商质量保证标准
QAT Quadruple Arinc Transmitter 四倍速艾瑞克发射机
QAVC Quiet Automatic Volume Control 噪声自动音量控制
QB Quiet Background 静背景
QBA Quebec Air 魁北克航空公司
Q-ball 球状大气压力受感器
QC 1. Quality Check 质量检查 2. Quality Control 质量控制,质量管理 3. Quick Change 快速更改
QCI Quality Conformance Inspection 质量合格检查
QCGAT Quiet Clean General Aviation Turbofan 低噪声低污染通用航空涡轮风扇发动机
Q-character Q 特征位,标志位
QCL Quality Control Level 质量管理水平,品质控制水准
QCM 1. Quality Control Manual 质量控制手册 2. Quartz Crystal Monitor 石英晶体监控器
Q-code 缩语电码
QCS 1. Quality Control System 质量控制系统 2. Quality Control Specifications 质量控制规范
QCSMF Quadruple Cladding Single Mode Fiber 四包层单模光纤
QD 1. Quadrant Depression 俯角 2. Quick Donning 快速完成 3. Quality Documentation 质量文档
QDB Quality Directors Board 质量指导委员会
QDC Quick Dependable Communication 快速可靠通信
QDI Quality Data Integration (System) 质量数据综合(系统)
QDR 1. Quality Data Retrieval 质量数据检索 2. Quality Deficiency Report 质量缺陷报告
QE 1. Quality Engineering 质量工程 2. Quality Evaluation 质量评估
QEC Quadrantal Error Correction 象限误差修正
QEP 1. Quality Enhancement Program 质量提高计划 2. Quality Examination Program 质量检查计划
QET Quick Engine Test 发动机快速试车
QF Quick Firing 急促射,速射
QFA Quick Firing Ammunition 速射武器弹药
QFD Quality Function Deployment 质量功能部署
QFE Quiet,Fuel Efficient 低噪声并节油的
Q-feel 1. 动压载荷感觉器,动压感力器 2. 与动压成正比的操纵感力
QFI Qualified Flying Instructor 合格的飞行教员
QG Quality Gates 质量门
QI Quality Inspection 质量检查
QIC Quality Insurance Chain 质量保证系统
QIG Quality Improvement Group 质量改进小组
QIP Quality Inspection Point 质量检查点
QIR Quality Improvement Request 质量改进请求
QM 1. Quality Management 质量管理 2. Quadrature Modulation 正交调制 3. Qualification Matrix 鉴定表 4. Quality Manual 质量手册 5. Quality Memorandum 质量备忘录
QMA Qualification Maintainability Analysis 维修性定性分析
QMC Quality Monitor Console 质量监控台
QMF Quadrature Mirror Filter 正交镜像滤波器
QMI Qualification Maintainability Inspection 质量维修性检查
QML Qualified Manufacturers List 合格制造商目录

QMP   Quality Management Plan 质量管理大纲
QMS   Quality Management System 质量管理系统
QMSO   Quality Management System Oversight 质量管理系统监督
QNCATC   Queen Noor Civil Aviation Training Centre 努尔皇后民航培训中心
QO   Quality Organization 质量组织
QOP   Quality Operating Procedures 质量操作程序
QOPR   Qualitative Operational Requirement 定性操作要求
QOS   Quality Of Service 服务质量
QP   1. Qualified Product 合格产品 2. Quality Planing 质量策划 3. Quality Procedure 质量程序
QPD   Quality Procedural Document 质量程序性文件
QPL   Qualified Products List 合格产品目录,质量合格产品清单
QPM   Quality Project Management 质量项目管理
QPP   Qualitfication Program Plan 鉴定项目大纲
QPR   Quality Problem Report 质量问题报告
QPSK   Quadrature Phase Shift Keying 四相相移键控,正交相移键控
QR   1. Quality and Reliability 质量和可靠性 2. Quality Requirement 质量要求 3. Quantity Required 需要量 4. Quiet Radar 静默雷达,无信号雷达 5. Qualification Review 合格鉴定评审
QRA   Quality Reliability Assurance 质量可靠性保证
QRC   Quick-Reaction Capability 快速反应能力
QRDN   Quality Requirements Discrepancy Notice 质量不符合要求通知
QRE   Quality Responsible Engineer 质量责任工程师
QRGA   Quadrupole Residual Gas Analyzer 四极剩余的气体分析器
QRH   Quick Referencee Handbook 快速参考资料手册,快速检查手册
QRI   Quick Reaction Interceptor 快速反应截击机
QRMC   Quick Response Multicolor Copier 快速反应多色复印机
QRS   Quality Requirement System 质量要求制度
QRT   Quiet Radio Transmission 静噪无线电传输
QS   1. Quality System 质量体系 2. Quasi-Synchronous 准同步
QSH   Queit Short Haul 低噪声短程运输机
QSR   1. Quality Survey Report 质量调查报告 2. Quarterly Service Reports 季度服务报告
QSTOL   Quiet Short Take-Off and Landing 低噪声短距离起降飞机,短距起飞和着陆
QTL   Quick Turnaround Limits 快速周转极限
QTM   Quantization Table Modification 基于量化表修改
QTP   1. Qualification Test Plan 鉴定试验大纲 2. Qualification Test Procedure 鉴定试验程序
QTR   1. Quarter 四分之一 2. Quiet Tail Rotor 低噪声尾桨 3. Qualification Test Report 合格测试报告,鉴定试验报告
QTS   Quarts 夸脱(容量单位)
QTY   Quantity 数量
quad   1. 四心线组 2. 四角形,中庭
quaded cable   扭绞四芯电缆
quadrant electrometer   象限静电计
quadrantal deviation   象差
quadrantal points   象限基点
quadratic   平方的,二次方的
quadratic aircraft   二次飞机
quadratic element   二次元
quadratic equation   二次方程
quadratic error   二次误差
quadratic fit   二次型
quadratic loop   平方环
quadratic performance index   二次型性能指标
quadratic programming   二次规划
quadratic regulator   二次型调节器
quadratron   热阴极四极管
quadrature-axis output impedance   交轴输出阻抗
quadricycle landing gear   四轮的起落架
quadriplane   四翼飞机
quadrode   1. 四点式起落架 2. 四点式起落架的飞机 3. 四极管
quadrotor   四旋翼
quadruple camera   四镜头照相机
quadruple gun   四联机枪
quadruple launcher   四弹发射装置
quadruple register   四重寄存器
quadruple star   四合星
quadruple turret   四联装炮塔
quadruplex   四倍的,四重的,四路多工的
quadruplex system   四路多工制,四路传输制
quadrupole   四极透镜,四极的,四倍
quadrupole lens   四极透镜
quadrupole mass filter   四极滤质器
quadrupole mass spectrometer   四极质谱仪,四极质谱分析器
qualified   1. 有资格的,胜任的,适当的,合格的 2. 过检定的,得到许可的 3. 有限制的,有条件的
qualification   资格,条件,鉴定试验
qualification process   鉴定过程
qualification test   1. 鉴定试验 2. 定型试验
qualified product   合格品,限定
qualified supplier list   合格供应商目录
qualifier   1. 合格的物 2. 修饰词,限定词
qualifying bit   限定位

qualitative agreement	定性协议
qualitative behavior	定性行为
qualitative implication	定性的影响
qualitative information	定性信息
qualitative limitation	质量限制
qualitative perception	质量感
qualitative physical model	定性物理模型
qualitative test	定性测试
quality	1.质,质量 2.性质,品质,特性,性能,成色
quality analysis	质量分析
quality assessment	质量评价
quality assurance data system	质量保证数据系统
quality assurance mode	质量保证模式
quality assurance system	质量保证体系
quality assurance system analysis review	质量保证体系分析评审
quality audit	质量审核,质量监督,品质审核
quality auditor	质量审核员
quality base method	质底法
quality certification	品质认证
quality control test	质量控制试验
quality diagnostics	质量诊断
quality feedback	质量反馈,品质反馈
quality improvement	质量改进
quality index	质量指标,品质指标
quality loop	质量环,品质环
quality loss	质量损失,质量消耗
quality manual	质量手册
quality metric	质量度量
quality of space product	航天产品质量
quality plan	质量计划,质量大纲,品质计划
quality policy	质量方针,品质方针
quality record tracing card	质量记录跟踪卡
quality specification	质量标准,质量说明书,技术品质规格
quality surveillance	质量监督
quality test	质量检验,质量试验
quality value	质量价值
quality-related costs	质量成本
quantification	定量,量化
quantify	确定……的数量,定量
quantitative analysis	定量分析
quantitative command	数量指令
quantitative comparison	定量比较
quantitative criterion	定量标准
quantitative datum	定量数据
quantitative information	定量信息
quantitative limitation	数量限制
quantitative measure	数量测定
quantitative perception	数量感
quantitative solution	数量(定量)解法
quantity	量,数量,油量
quantity base method	量底法
quantization	1.分层,量化 2.量子化 3.数字化
quantization error	量化误差
quantization noise	量化噪声
quantization strategy	量化策略
quantization unit	量化单位
quantize	1.量子化 2.数字转换
quantized	量化的
quantized field theory	量子场论
quantized information	量化信息
quantum	定量,总数
quantum chemistry	量子化学
quantum clock	量子钟
quantum dot	量子点
quantum efficiency	量子效率
quantum electronics	量子电子学
quantum noise	量子噪声
quantum noise limited detector	量子噪声限制检波器,量子噪声限制接收机
quarry	采石场
quarter chord	四分之一弦
quarter chord point	四分之一弦长点,四分之一翼弦点
quarter orbit coupling	四分之一轨道耦合
quarter wave length plate	四分之一波片
quarter-phase	四分之一相位(90°相位差),正交相
quarter-round	四分之一
quarter-wave	四分波长线
quarter-wave mode	四分波长模式
quarter-wave transformer	四分之一波长变换器
quartet	四件一套
quartz	石英,水晶
quartz crystal filter	石英晶体滤波器
quartz glass	石英玻璃
quartz glass thermometer	石英玻璃温度计
quartz lamp	石英灯
quartz oscillator	石英晶体振荡器
quartz pendulous accelerometer	石英摆式加速度表
quartz plate	石英片
quartz reaction chamber	石英反应室
quartz resonator	石英谐振器
quartz thruster	石英推进器
quartz tube	石英管
quartz-ceramics	石英陶瓷
quartz-iodine lamp	石英碘灯
quasar	类星体,类星射电源
Quasat	【天】夸萨特,意大利微波天文卫星
quasi static	准静态的,似静态的
quasi static process	准静态过程

quasi stationary	准稳的,准静止的	query	疑问,质问,查询,询问
quasi stationary oscillation	准稳振荡	question and answer mode	问答式
quasi stationary state	准稳态,拟稳态	question vector	提问向量
quasi-cyclic code	准循环码	questionnaire	调查表,征求意见表
quasi-equilibrium	准平衡	queuing theory	排队论
quasigeoid	似大地水准面	queue	队,排队,队列
quasi-homogeneous alignment	准沿面排列	queue control block	队列控制块,排队控制程序
quasi-image homing head	准成像导引头	queue data set	队列数据集,队列数据组
quasi-linearization	拟线性化	queue register	排列寄存器
quasi-longitudinal propagation	准纵向传播	queue search limit	队列检索限制
quasi-Monte Carlo method	拟蒙特卡罗法	queue-back chain	队列反向链
quasi-neutral plasma	准中性等离子体	quick action rotary fastener	快卸转动锁扣
quasi-Newton method	拟牛顿法	quick action valve	速动阀,快动阀门
quasi-one-dimensional analysis	准一维分析	quick change real-time	快速更换的实时操作
quasi-one-dimensional model	准一维模型	quick display	快速显示
quasi-periodic	准周期的	quick draw graphics system	快速制图系统
quasi-periodic oscillation	准周期震荡	quick fuze	瞬发引信
quasi-periodic vibration	准周期振动	quick mission	快速任务
quasi-plane	准平面的	quick reference list	快速参照一览表
quasi-resonance	准共振	quick release coupling	快脱联轴节,快卸接头
quasi-shear stress	准剪应力	quick-burning powder	快速燃烧火药
quasi-sinusoid	准正弦波	quick-firing ammunition	速射武器弹药
quasi-stable adjustment	拟稳平差	quickness	急速,迅速
quasi-stationarity	准稳定(性)	quick-opening cluster adapter	快速开放集群适配器
quasi-stationary	准平稳,似稳的	quick-opening valve	急开阀
quasi-steady burning	准稳定燃烧	quick-operating	快速操作的,快动的
quasi-steady case	准稳定情况	quick-release box	脱离锁
quasi-steady control	准稳定控制	quick-response	快速响应的,灵敏的
quasi-steady flame	准稳定火焰	quick-seal coupling	速密耦合
quasi-steady manner	准稳定的方式	quicksilver	1.汞,水银 2.汞锡合金
quasi-steady-state wind	准平稳风	quick-stop	急停
quasi-stellar object	类星天体	quicktrans	快速空运
quasistellar radio source	类星射电源	quick-wear part	快损零件
quasi-transverse propagation	准横传播	quiescent current	静态电流,零值电流
quasi-two-dimensional	准二维的,拟二维的	quiescent flow	静态流量
quaternary code	四进制码,四元码	quiescent fluid	静止流体
quaternion	1.四(个),四元素 2.四元法	quiescent scenario	静止场景
quaternion estimation	四元数估计	quiescent state	静止状态
quaternion hypersphere	四元数超球面	quiet automatic gain control	无噪声自动增益控制
quaternion method	四元数法	quiet propeller	静止螺旋桨
quaternion normalization	四元数归一化	quiet sun	静稳太阳,宁静太阳
quench	熄灭,猝熄,淬火	quiet sun noise	静稳太阳噪声,宁静太阳噪声
quench correction	猝灭校正	quiet tunnel	静风洞
quench crack	淬火裂纹,烧裂,淬裂	quill shaft	套管轴,挠性短轴,中空轴
quench hardening	淬火硬化,急冷硬化	quotation	1.行市,定价,行情 2.引证,引语 3.行情表,估价单,报价表
quenching	猝熄,淬灭,急冷,淬火,阻塞		
quenching and tempering	淬火及回火调质处理	quote	1.报价 2.引述 3.举证,引用
quenching effect	猝灭效应	quotient	1.商 2.系数 3.份额

R

R 1. Radar 雷达 2. Radial 径向线 3. Rate 比率 4. Red 红色的 5. Registered 注册的 6. Repair 修理 7. Reserve 保留,预订

R(t) Reliability 可靠性,可靠度

R&D Research and Development 研究和开发

R&M Reliability and Maintainability 可靠性与维修性

R/A Radio Altimeter 无线电高度表

R/C Rate of Climb 爬升率

R/D Rate of Descent 下降速度,垂直下降率

R/RTA Removal/Replacement Task Analysis 拆装任务分析

R/T 1. Radio Transmitter 无线电发射 2. Receiver Transmitter 收发机,接收发送器 3. Rejection Tag 拒收标签

R/W Read/Write 读/写

RA 1. Radio Altimeter 无线电高度表 2. Radio Altitude (Height) 无线电高度 3. Ram Air 冲压空气 4. Release Authority 发放管理机构 5. Release Authorization 发放审定 6. Reliability Analysis 可靠性分析 7. Revision Authorization 版本授权表 8. Roll Axis 横滚轴

RAA Regional Airline Association 支线航空公司协会

RAAP Risk Analysis and Abatement Plan 风险分析与降低计划

racetrack procedure 直角航线程序

RACF Resource Access and Control Facility 资源存取和控制工具

RAD 1. Radar 雷达 2. Ram Air Duct 冲压空气进气道 3. Rapid Application Development 快速应用开发

RADANT Radome Antenna 天线罩

radar 雷达

radar absorbing structure 雷达波吸收结构

radar altimeter 雷达高度计,无线电高度表,雷达高度表

radar beam-riding guidance device 雷达波束导引制导装置

radar calibration 雷达校准与定标

radar cross-section 雷达有效截面积

radar fuze 雷达引信

radar instrumentation network 雷达测量网

radar look-direction 雷达视向

radar monitoring 雷达监控,雷达监视

radar ranger 雷达测距器

radar reflectivity 【电】雷达反射率

radar remote sensing 雷达遥感

radar remote sensing satellite 雷达遥感卫星

radar resolution 雷达分辨率,雷达解析度

radar return 雷达回波,雷达反射

radar scatterometer 雷达散射计

radar search 雷达搜索

radar separation 雷达间隔

radar stereo-viewing 雷达立体观测

radar tracking 雷达跟踪

radar tracking station 雷达跟踪站

radar transmitter 雷达发射机

radar trap 雷达陷阱

radar vectoring 雷达引导

radial distortion 径向畸变,辐射畸变差

radial forging 径向锻造

radial resolution 径向分辨率

radial-flow turbine 径流式涡轮

radiance 辐亮度,辐射率

radiant correction 辐射校正

radiant energy 辐射能

radiant existence 辐射出射度

radiant flux 辐射通量

radiant flux density 辐射通量密度,辐照度

radiated emission 【物】辐射,发射

radiated susceptibility 辐射敏感度

radiation 辐射,放射,发光,射线,辐射能

radiation belt 辐射带

radiation belt model 辐射带模式

radiation body 辐射体

radiation coefficient 辐射系数

radiation cooling 辐射冷却,放射冷却

radiation damage 辐射损伤,放射线损伤

radiation dose 辐射剂量

radiation dose rate 辐射剂量率

radiation emission measurement 辐射发射测量

radiation hardened component 抗辐射加固器件

radiation hardening 增强抗辐射性

radiation intensity 辐射强度

radiation measurement 辐射测量

radiation medicine 放射医学

radiation protection 辐射防护,放射线保护

radiation resistant coating 耐辐射涂层

radiation resolution	辐射分辨率
radiation sensitivity measurement	辐射敏感度测量
radiation temperature	辐射温度
radiative refrigerator	辐射致冷器
radiative thermal protection	辐射热保护
radiator	辐射体,辐射器,散热器
radio aggregation communication system	无线电集群通信系统
radio altimeter	无线电高度表,无线电测高计
radio beacon	无线电信标,雷达信标
radio compass error	无线电罗盘误差
radio direction-finding	无线电测向
radio distance-measuring	无线电测距
radio emission	无线电发射
radio frequency	无线电频率
radio frequency sensor	无线电频率传感器
radio fuze	无线电引信
radio guidance	无线电制导
radio interferometer	无线电干涉仪
radio measurement	无线电测量
radio position fixing method	无线电定位法
radio silence	无线电静寂
radio telecommand	无线电遥控
radio telemetry	无线电遥测
radio tracking system	1.无线电跟踪系统 2.无线电测量系统
radio transmission reconnaissance satellite	无线电传输型侦察卫星
radio wave refraction correction	电波折射修正
radio zero calibration	无线校零
radioactive cloud	放射性云
radioactive fallout	放射性沉降灰,放射性回降物,放射性沉降物
radioactive measurement	放射性测量
radio-cycle match	载频周期匹配
radiographic inspection	X射线检查,射线照相检查,射线透照检验
radiometer method	辐射计法
radiometric balance	辐射平衡
radiometric calibration device	辐射定标装置
radiometric resolution	辐射测量分辨率
radome	雷达天线罩,整流罩,雷达罩
RAG Reliability Analysis Group	可靠性分析组
rail carriage	轨道手推车
rail transfer	轨道转运
rail transporter	铁路运输车
railway platform truck	铁路平板车
railway transfer	铁路转运
rain echo attenuation compensation technique	雨回波衰减补偿技术
RAM Random Access Memory	随机存取存储器
ram-air turbopump	风动泵
ramjet engine	冲压喷气发动机,冲压式喷射发动机
ramp method of ablation test	斜坡烧蚀试验
random code fuze	随机码引信
random drift rate	随机漂移率
random error	随机误差,随机性错误,偶然误差
random fault	1.随机故障 2.偶然故障
random spectrum	随机谱
random vibration	随机振动,不规则振动
random vibration environment	随机振动环境
random vibration test	随机振动试验
random walk angle	随机游动角
range	1.航程,射程 2.范围,距离,区域,幅度 3.靶场 4.波段,量程 5.无线电指向标 6.测距 7.分级,归类
range accuracy	测距精度
range acquisition	距离测定
range acquisition time	距离捕获时间
range ambiguity	距离模糊
range curvature	距离弯曲
range cutoff	距离截止
range cutoff characteristic	距离截止特性
range direction	距离方向
range error	测距误差,距离误差,量程误差,射程误差
range factor	航程因子
range rate accuracy	测速精度
range rate error	距离变化率误差
range reference pole	距离标
range resolution	距离分辨率
range sighting	定距瞄准
range sum	距离和
range sum rate	距离和变化率
range tracking	距离跟踪
range walk	距离游动
range-while-search	边搜索边测距
ranging coverage	作用距离
ranging reticle	测距环,测距光环
ranging system	测距系统
rapid decompression	迅速减压
rapid decompression chamber	快速减压舱
rapid solidification	快速凝固
rapidly solidified material	快速凝固材料
RAPLOT Radar Plotting	雷达标图
RAPS Routin Area Picture Sheets	管线区图页
rarefied aerodynamics	稀薄空气动力学
rarefied gas flow	稀疏气体流
RAS	1. Radar-Absorbing Structure 雷达波吸收结构 2. Radar Advisory Service 雷达咨询服务 3. Rectified

Air Speed 修正空速 4. Remote Addressed Service 远程访问服务器 5. Repair Approval Sheet 修理批准单

RAT　1. Ram Air Temperature 冲压空气温度 2. Ram Air Turbine 冲压空气涡轮

rate　价格,比率,速率

rate damping　速率阻尼

rate gyroscope　速率陀螺,速率陀螺仪

rate integrating gyro　速率积分陀螺仪

rate of climb　上升率,上升垂直速度,爬升率

rate of closed hole　闭孔率

rate of discharge　放电速度,流率

rate of gas generation　发气率

rate of pitch　俯仰角速度

rate of roll　滚转角速率

rate of yaw　偏航速率

rate strap-down inertial guidance system　速率捷联式惯性制导系统

rate table　速率转台

rated capacity　额定容量

rated load　额定负载,额定负荷

rated operating conditions　额定运行条件

rated power of solar array　太阳电池阵额定功率

rated thrust　额定推力

rated voltage　额定电压

rate-of-climb indicator　爬升速率指示计,飞机爬升速率测定仪

rate-of-rise fixed temperature detector　差定温探测器

ratio of solar absorptance to emittance　吸收发射比

raw data　原始数据

RBL　Right Buttock Line 右剖面线

RC　1. Ratio Changer 比率转换器 2. Recommitment of Changes 再提交更改 3. Recurring Cost 重复成本 4. Reference Chord 参考弦 5. Rockwell Collins 罗克韦尔-柯林斯公司(航空电子系统)

RCB　Radio Control Bus 无线电控制总线

RCCP　Rough Cut Capacity Planning 粗切割能力计划

RCE　Radio Control Equipment 无线电控制设备

RCL　Radio Communications Link 无线电通信链

RCM　Reliability Centralized Maintenance 以可靠性为中心的维修

RCMA　Reliability Centralized Maintenance Analysis 以可靠性为中心的维修分析

RCO　Rotable Components Office 周转件办公室

RCU　Remote Control Unit 遥控器

RCV　Restrictor Check Valve 限流单向活门

RCVR　Receiver 接收器,接收机

RD　1. Raw Data 原始数据 2. Release Databases 发放数据库 3. Requirements Document 要求文件 4. Route Discontinuity 航路不连续

RDC　Request for Discrepancy Check 偏差检查需求

RDMI　Radio Distance Magnetic Indicator 无线电距离磁指示器

RDP　Radar Data Processing 雷达数据处理

RDSS　Radio Determination Satellite Service 无线电定位卫星服务

RDT　Reliability Development Testing 可靠性研发试验

RDU　Receiver/Decoder Unit 接收/解码装置

RE　Reverse Engineering 逆向工程

reachability　可达性,能达到性

REACT　Rain Echo Attenuation Compensation Technique 雨回波衰减补偿技术

reaction bond　反应结合

reaction turbine　反应式涡轮,反动式汽轮机

reaction wheel　反动式叶轮

reaction wheel control　反作用轮控制

reactive ion beam etching　反应离子束刻蚀

reactive ion etching　反应离子刻蚀

readaption　重新适应

readied segment escape　待发段逃逸

readiness　1. 待机 2. 战备状态,备用状态 3. 准备,准备就绪

readiness rate　准备完好率

readout clock　读出时钟

ready-to-eat food　即食食品

real antenna　真实天线

real aperture radar　真实孔径雷达

real aperture side-looking radar　真实孔径侧视雷达

real exposure time　实际曝光时间

real gas　真实气体,实际气体,非理想气体

real gas effect　真实气体效应,实在气体效应

real line subscriber　实线用户

real-time analysis　实时分析,快速分析

real-time application software　实时应用软件

real-time ballistic camera　实时弹道相机

real-time command　实时指令

real-time communication management program　实时通信管理程序

real-time data processing　实时数据处理

real-time display　实时显示

real-time measurement　实时测量

real-time network management program　实时网络管理程序

real-time operation software　实时操作软件

real-time output　实时输出

real-time processing computer　实时处理计算机

real-time recording　实时记录

real-time simulation　实时仿真,实时模拟

real-time system software　实时系统软件

real-time telemetry 实时遥测
reaming 铰孔,铰削
rear fin 尾翼
rear section 尾段
rebalance 重新平衡
RECAP Reliability Evaluation and Corrective Action Program 可靠性评定与纠正措施大纲
receiver channel combination 接收机通道合并
receiver protection 接收机保护
receptance 位移导纳
recession thickness 烧蚀厚度
RECHRG Recharger 充电器
recirculation valve 再循环阀,再循环门
recirculation zone 回流区
reconfigurable technology 可重组技术
reconnaissance airplane 侦察机
reconnaissance satellite 侦察卫星
reconnaissance subsystem 侦察分系统
reconnaissance system for radar 雷达侦察系统
reconstruction 重构,重建,重塑
recording measuring instrument 记录(测量)仪表
recording mode 记录方式
recoverable capsule 返回器
recoverable module 可恢复模块
recoverable photo reconnaissance satellite 回收型照相侦察卫星
recoverable satellite 返回式卫星
recoverable spacecraft 返回式航天器,可回收太空船
recovery area 回收区
recovery cassette 回收片盒
recovery coefficient 恢复系数,回收系数,采收率,复原系数
recovery mode 回收方式
recovery parachute 回收伞
recovery subsystem 回收分系统
recovery telemetry 回收遥测
recovery temperature 恢复温度
RECT Rectifier 整流器
rectangular cadmium-nickel battery 方型镉镍蓄电池
rectangular robot 直角坐标机器人
rectangular wind tunnel 矩形截面风洞
rectification error 整流误差
rectifier 校正仪,整流器,整流管
rectifying apparatus 纠正仪
recursion keeping 回归保持
recursive orbit 回归轨道
REDARS 1. Reference Engineering Data Automated Retrieval System 参考工程数据自动检索系统
2. Reference Engineering Drawing Automated Retrieval System 基准工程制图自动检索系统
reduced smoke propellant 少烟推进剂,微烟推进剂
reduced thrust take-off 减推力起飞
redundancy 冗余,(可靠性)余度,过多,多余
redundancy actuator 余度舵机
redundancy architecture 余度结构
redundancy management 余度管理
redundancy technology 余度技术
redundant information 冗余信息
reentry 再进入,重新入场
reentry altitude 再入高度
reentry angle 再入角
reentry blackout zone 再入黑障区
reentry control 再入控制
reentry environment 再入环境
reentry gas dynamics 再入气体动力学
reentry heating 再入加热
reentry load 再入载荷
reentry phase 再入段
reentry plasma sheath 再入等离子鞘
reentry point 再入点
reentry telemetry 再入遥测
reentry thermal protection 再入防热
reentry trajectory 再入轨道
reentry vehicle 再入飞行器
reentry velocity 再入速度
REF Reference 参考,基准
reference electrode 参比电极,参考电极
reference ellipsoid 参考椭球
reference frequency system 频率基准系统
reference material 参考资料
reference spectrum 参考谱
reference standard 参考标准,比照标准
REFL Reflection 反射系数
REFL PWR Reflection Power 反射功率
reflectance 反射率,反射比
reflectance characteristics 反射特性
reflectance coefficient 反射系数
reflectance factor 反射因子,反射率因数
reflectance spectral feature 反射波谱特征
reflected shock operation 反射型运行
reflected surface 反射表面
reflection 反射,反映
reflection of wave 波的反射
reflection plate method 反射平板法
reflection right-angle prism device 反射直角棱镜装置
reflective index 反射指数
reflective infrared 反射红外
reflective optical system 反射光学系统,光反射器

reflectivity	反射率,反射能力,反射性
reflectometer method	反射计法
refractive and reflective optical system	折反射式光学系统
refractive optical system	折射式光学系统
refractory alloy	高熔点合金
refrasil	石英玻璃纤维状材料
refueling platform	加油平台
refueling pod	加油吊舱
regeneration system	再生系统
regenerative cooling	再生冷却,回收冷却法
regenerative life support system	再生式生命保障系统
registration	注册,登记
registration certificate	注册证书,注册证明
regressive burning	减面燃烧,减推力燃烧
regular aerodrome	主降机场
regulation of metrological verification	计量检定规程
regulator	调节器,调定器,稳压器
rehearsal	练习,演习
rehearsal model	合练模型
rehearsal of all region	全区合练
rehearsal outline	合练大纲
rehydratable beverage	复水饮料
rehydratable food	复水食品
rehydratable packaging	复水包装
reinforced bulkhead	加强框,气密框
reinforced phase	强化相
reinforced plastic	增强塑料
rejection notice of verification	检定结果(不合格)通知书
REL　Relative	相关的
relative aperture	相对孔径,口径比
relative error	相对误差
relative flying height	相对飞行高度
relative measurement	相对测量,比较测量
relative radiometric calibration	相对辐射定标
relative spectral response	相对光谱响应
relative spectral sensitivity	相对光谱灵敏度
relative speed	相对速度,相对速率
relaxation equation	弛豫方程
relaxation phenomena	弛豫现象
relaxed static stability control	放宽静稳定性控制
relay optical system	中继光学系统
release circle	投弹圆
release slant range	投弹斜距,投弹斜距离
relevant fault	关联故障
reliability	可靠性
reliability allocation	可靠性分配
reliability and maintainability	可靠性和维修性
reliability assurance program	可靠性保证大纲
reliability block diagram	可靠性框图
reliability design	可靠性设计
reliability growth	可靠性增长
reliability growth management	可靠性增长管理
reliability growth test	可靠性增长试验
reliability index	可靠性指标,可靠性指数
reliability model	可靠性模型
reliability monitoring	可靠性监控
reliability of satellite in orbit test	卫星在轨测试交付可靠度
reliability of space product	航天产品的可靠性
reliability prediction	可靠性预计
reliability program plan	可靠性工作计划
reliability test	可靠性试验,可靠性测试
reliability verification	可靠性验证
reliability verification test	可靠性验证试验
reliable life	可靠寿命
relief valve	安全阀,减压阀
remote center compliance device	远心柔顺装置,遥轴顺应性装置
remote commanding	远程发令
remote exploding	远程爆炸
remote focusing	遥控调焦
remote monitor and control	远程监控
remote operating system	远程操作系统
remote sensing camera	遥感相机
remote sensing data base of geographical information	地理信息遥感数据库
remote sensing image	遥感影像,遥感图像
remote sensing picture	遥感图像
remote sensing satellite	遥感卫星
remote sensing subsystem	遥感分系统
remote sensor	遥测传感器
remote setting	遥调
remote-control rover	遥控自行装置
remotely piloted vehicle	遥控飞行器,遥控无人飞机
REOP　Reoperate	重新运行,恢复工作
reorientation	重新定向
repair	修理,检修,改正,修正
repair joint efficiency	修补接头效率
repairable item	可修复产品
REPD　Repaired	修理的,修理好的
REPDN　Reproduction	复制,再生产
repeatability	可重复性,反复性
repeatability of measurement	测量重复性
repetitive noise	重复噪声
replaceable unit	可更换单元
replay program	复演程序

replenishment 补充注油,补充,填满
reply pulse 应答脉冲
repressurization 复压
REPT Reference Engineering Photo Templates 基准工程照片样板
REQ 1. Request 需求 2. Required/Requirement 要求
RER Recommitment Evaluation Request 再提交评价要求
rescue action 救助活动
rescue spacecraft 救生飞船
rescue training 救生训练,抢险救援训练
research aircraft 研究机,试验机
research flight test 研究性试飞
reserve battery 备用电池
residual error 残留误差
residual gas analyzer 残余气体分析仪,剩余气分析器
residual hazard 残毒
residual life 剩余寿命
residual liquid level 剩余液位
residual magnetic moment 剩磁矩
residual nuclear radiation 剩余核辐射,残留核辐射
residual strain 残余应变
residual strength 剩余强度,残余强度
resilience-proof device 防回弹装置
resin content 树脂含量
resin matrix composite 树脂基复合材料
resistance 阻力,抗性,抵抗,阻抗,电阻
resistance brazing 电阻钎焊
resistance soldering 电阻钎焊
resistance spot welding 电阻点焊
resolution capability of fuze 引信分辨力
resolution of film 胶片解像力
resolution of range ambiguity 距离模糊率
resolution of real aperture 真实孔径分辨率
resolution of synthetic aperture 合成孔径分辨率
resonance 谐振,共振,共鸣
resonance frequency 共振频率,谐振频率
resonant element 共振元件
resources module 资源舱
respiratory decompensation 呼吸代偿障碍
response control 响应控制
responsive quantum efficiency 响应量子效率
RESTR Restriction 限制
restricted area 限航区,限制区
restricted space 限航空域
restricted three-body problem 限制三体问题
restrictor 限制器,节流口,阻尼孔
result data handling report 结果数据处理报告
result of a measurement 测量结果

resultant error of ground aiming 地面瞄准总误差
RET Reliability Evaluation Test 可靠性评定试验
reticle 标线,十字线,分度线
reticle stabilizing time 稳环时间
RETROF Retro Fire 制动发动机点火
retractable landing gear 收放式起落架
retreating blade 后行桨叶,后斜叶片
retro module 减速制动发动机舱
retro nozzle 反推喷管
retro rocket 反推火箭,制动火箭,减速火箭
retro-angle 制动角
retrofit 改装,改型,改进
retrogradation 倒退,后退,逆行,制动,减速
retrogressive orbit 逆行轨道
retroreflector 后向反射器,反光镜
return 返回,收回,返航,回程,(雷达)回波(信号),反射信号,应答(信号)
return altitude 返回高度
return angle 返回角
return course 返回过程
return flight 返航
return mode 返回方式
return phase rescue 返回段救生
return point 返回点
return subsystem 返回分系统
return system 返回系统
return technique 返回技术
return trajectory 返回轨道
return transfer oscillation 回输振荡
return velocity 回流速度
returning site control 返回落点控制
reusable rocket engine 重复使用火箭发动机
reusable thermal protection material 可重复使用防热材料
REV Revision 修订(本),修正
reverberation chamber 混响室
reverberation field 混响声场
reverberation test 混响试验
reversal 反向,逆转,反转,改变方向,反效
reversal nozzle 反向推力喷管
reversal procedure 反向程序,回转程序
reverse bias 反偏压,反向偏压
reverse charge 反充电
reverse flow combustor 回流燃烧室
reversed flow region 反流区
reversible boosted mechanical control 可逆助力机械操纵
rework 重做,改写,重写,返工
Reynolds equation 雷诺方程

Reynolds number correction　雷诺数校正
Reynolds stress　雷诺应力
RF　Radio Frequencey 射频,无线电频率
RFA　Request For Assistance 辅助设备需求
RFC　1. Request For Change 更改请求 2. Request For Comments 请求评注 3. Request For Conformity（制造）符合性检查请求单
RFD　Request For Deviation 偏离申请
RFI　1. Radio Frequency Interference 射频干扰 2. Request For Information 征求信息
RFP　Request For Proposals 报价单申请
RFR　Request For Reclassification 重新定级需求
RFS　Request For Service 服务需求
RFTP　Request For Techincal Proposal 技术招标书
RFU　Radio Frequency Unit 射频装置
RGLTR　Regulator 调节器,控制器
RI　1. Receiving Inspection 接收检验 2. Remove Item 拆卸项目 3. Replacement Item 更换项
ribbon parachute　带状降落伞,带式快速降落伞
RIC　Resistance Inductance Capacitance 电阻电感电容
ricochet　跳弹,跳飞
ride quality control　乘坐品质控制
RIDGE　Repository for Integrated Data Gathering and Exchange 综合数据会合和交换库
RIFI　Radio Interference Field Intensity 射频干扰场强
rigid multibody dynamics　多刚体动力学
rigid pavement　刚性道面
rigid plate　刚性板块
rigid polyurethane foams　硬质聚氨脂泡沫塑料
rigid rotor　刚性转子,刚性转翼,整体转筒
rigid spacecraft dynamics　刚性航天器动力学
rigid support　刚性支承
rigidity　刚度,刚性,稳定性,硬度
RII　Required Inspection Item 必检项目
RIIP　Required Inspection Item Program 必检项目方案
RIL　Red Indicating Light 红色指示灯
RIM　Receipt, Inspection and Maintenance 接收、检查和维护
ring rolling　环锻
ring slot parachute　环缝伞
RIP　Rapid Implementation Prototype 快速实现样机
RIS　Reliability Information Station 可靠性信息站
RISC　Reduced Instruction Set Computing 精简指令集计算
risk analysis　风险分析
RIT　Reliability Improvement Test 可靠性改进试验
RIU　Radio Interface Unit 无线电接口装置
rivet bonding　胶铆连接
riveted structure　铆接结构
riveting　铆接,铆,铆接法
riveting test　铆接试验
RJ　Regional Jet 支线喷气飞机
RKS　Record Keeping System 记录保存系统
RL　1. Reference Line 参考线,基准线 2. Revision Letter 修正字母 3. Route Legs 航路段
RLS　Requirements List for Serialization 系列化要求表
RLV　Relieve 释放
RLY　Relay 中继,重放,转播
RMDD　Raw Material Design Definition 毛料设计定义
RMI　1. Radio Magnetic Indicator 无线电磁指示器 2. Raw Material Item 原材料项
RMMS　Reliability and Maintainability Management System 可靠性维修性管理系统
RMP　1. Radio Management Panel 无线电管理板 2. Reliability Monitoring Program 可靠性监控大纲 3. Remote Maintenance Panel 远距维修板 4. Rigging Methodology Plan 装调技术大纲 5. Risk Management Plan 风险管理大纲
RMS　1. Raw Material Stock 原材料库 2. Reliability, Maintainability and Safety 可靠性、维修性和安全性
rms noise　均方根噪声值
RNL　Renewal 更新,翻新
RNLT　Running Light 航行灯
RNP　Required Navigation Performance 要求的导航性能
RNTP　Radio Navigation Tuning Panel 无线电导航调谐板
RO　1. Radio Operator 无线电操作员 2. Relation Object 关联对象 3. Roll Out 滑出
road transfer　公路转运
robot　机器人,自动机,遥控设备
robot system　自动控制系统
robotics　机扑学,机扑工程学
ROC　Rate Of Climb 爬升率
rocket body　箭体
rocket body diameter　箭体直径
rocket body structure　箭体结构
rocket engine　火箭发动机,喷气发动机
rocket engine high altitude test　火箭发动机高空试验
rocket engine plume　火箭发动机羽流
rocket exhaust noise　火箭排气噪声
rocket extract　火箭牵引
rocket launcher　火箭发射器,火箭发射机
rocket motor　火箭发动机
rocket plume test　火箭羽焰试验
rocket propellant　火箭推进剂
rocket propulsion system　火箭推进系统

rocket sled test 火箭滑车试验
rocket-ground port check 箭-地接口检查
Rockwell hardness test 洛氏硬度试验
ROD Rate Of Descent 下降率
rod-type structure 杆系结构
roll angle 滚转角,倾侧角
roll attitude 滚转姿态,坡度
roll attitude acquisition 滚动姿态捕获
roll deformation angle 横摇变形角
roll forging 滚锻
roll forming 滚弯成形,辊轧成形,成形轧制,轧膜
roll motion 横滚运动
roll spot welding 滚点焊
rolling 压延,轧制,滚动,横摇,滚的,滚转,滑行,卷绕
rolling moment 滚转力矩,滚动力矩
rolling rate oscillation 滚转速率振荡,滚转速振荡
rolling resonance 滚动共振
rolling subsidence mode 滚转收敛模态
roll-up type solar array 卷式太阳电池阵
roll-yaw coupling 滚动-偏航耦合
roll-yaw coupling control 滚动-偏航耦合控制
ROM Read Only Memory 只读存储器
root and tip loss factor 叶端损失系数
root chord 翼根弦
root-mean-square noise current 均方根噪声电流
root-mean-square noise voltage 均方根噪声电压,噪声的有效电压
rope-sled test for fuze 引信柔性滑轨试验
rotable nozzle 转动喷管
rotary dynamic derivative 旋转动导数
rotary forging 回旋锻造
rotary inverter 旋转变流机
rotary joint 旋转连接
rotating cardioid pattern 旋转心脏形方向图
rotating disk atomized powder 旋转盘雾化粉末
rotating parachute 旋转伞
rotating pylon 可转挂架,可转挂梁
rotating rocket testing 旋转弹试验
rotating stall 旋转失速
rotation impulse 旋转冲量
rotation speed 旋转速度
rotation velocity control 转速控制
rotational flow 旋流
rotative capacity 耐回转能力
rotor 转子,旋翼,转动体
rotor advance ratio 旋翼前进比
rotor angular momentum 相对角动量,相对动量矩
rotor balancing 转子平衡,转护平衡
rotor blade 旋翼桨叶,旋转机翼,转动叶片

rotor brake system 旋翼刹车系统
rotor coning 旋翼锥度
rotor disk area 旋翼桨盘面积,桨盘面积
rotor disk loading 旋翼桨盘载荷
rotor hub 旋翼桨毂
rotor inflow ratio 旋翼入流比
rotor solidity 旋翼实度
rotor thrust 旋翼拉力
rotor tower test 旋翼塔实验
rotor vortex ring 旋翼涡环
rotor wake 旋翼尾流
rotor windmill braking 旋翼风车制动
rotorcraft in transportation category 运输类旋翼机
rough strip 粗糙带
rough surface 粗糙表面
roughening 使粗糙,打毛
roughness 粗糙度
route 航路,路线,路程,通道,航线
route forecast 航路预报
route segment 航段
routine maintenance 日常维修,定期维修,例行维修
routine test 常规试验
router 路由器
rover 自行装置,越野车
RP 1. Reliability Program 可靠性大纲 2. Review Process 审查过程 3. Routing Protocal 路由协议
RPC Remote Procedure Call 远距程序调用
RPI Rapid Process Improvement 快速工艺改进
RPL Replaceable Parts List 可替换件清单
RPO Rejection Purchase Order 拒收订购单
RPP Reliability Program Plan 可靠性项目大纲
RPR Reliability Prediction Report 可靠性预测报告
RPU Receiver-Processor Unit 接收处理机组件
RR 1. Rapid Revision 快速修正 2. Regulatory Requirements 规章要求 3. Release Records 发放记录 4. Revision Record 修正记录 5. Roll Rate 横滚率
RRB Retrofit Review Board 改型评审委员会
RRI Router Reference Implementation 路由器基准执行程序
RRO Rejection Rework Order 拒收返工指令
RRP Rapid Revision Process 快速修正过程
RRTR Risk Reduction Test Report 减小风险试验报告
RS 1. Regional Standard 区域标准 2. Release Status 发放状态 3. Repair Stations 修理站,维修基地 4. Reset 恢复 5. Residual Strength 剩余强度 6. Resolver 旋转变压器 7. Ruled Surface 直纹面
RSCU Representative, Supplier, Customer, and User 代表、供应商、客户和用户
RSP 1. Required Surveillance Performance 要求的监视

性能 2. Reversion Select Panel 修正选择板
RSPL Recommended Spare Parts List 推荐备件清单
RSR Reliability and Safety Report 可靠性安全性报告
RSS Radar Signal Simulator 雷达信号模拟器
RSSI Received Signal Strength Indicator 接收信号强度指示器
RSVR Reservior 储存器
RT 1. Radio Telecommunication 无线电电信 2. Real Time 实时 3. Receiver Transmitter 收发机 4. Regression Testing 回归测试 5. Rejection Tag 拒收标签,拒收单
RTA Receiver Transmitter Antenna 收发机天线
RTCA Radio Technical Commission of Aeronautics 航空无线电技术委员会
RTFR Reliability, Trouble and Failure Report 可靠性、故障、失效报告
RTM Radio Transmission Module 无线电发送模块
RTO Reject Take Off 中断起飞
RTOS Real-Time Operating System 实时操作系统
RTP 1. Radio Tuning Pannel 无线电调谐板 2. Reliability Test Plan 可靠性试验大纲
RTR 1. Reliability Test Requirement 可靠性试验要求 2. Rotor 转子
RTRY Rotary 旋转
RTU Radio Tuning Unit 无线电调谐单元
RTV Return To Vendor 返回卖方
RU Rack Unit 导轨架安装装置
rudder 舵,操纵方向舵,方向舵

rudder actuator 方向舵传动器
rudder pedal 方向舵脚蹬
rudderon 偏转副翼
ruddervator 方向升降舵
run length encoding 行程长度编码,游程编码
running crack 速移裂纹
run-up time 启动时间
runway 跑道,起重机走道,吊车滑道,飞机跑道
runway capacity 跑道容量
runway center line light 跑道中线灯
runway edge light 跑道边灯,跑道边界灯
runway end identification light 跑道末端灯,跑道端识别灯,跑道尾识别灯
runway end safety area 跑道头安全区
runway limiting weight 跑道限制重量
runway marking 跑道标志,跑道号标志
runway pavement strength 道面强度
runway shoulder 跑道路肩
runway threshold light 跑道入口灯
rupture on orbit 在轨破裂
RV Relief Valve 释放活门,安全活门,减压活门
RVDT Rotary Variable Differential Transducer 旋转可变差动传感器
RVR Runway Visual Range 跑道目视距离,跑道灯目视距离
RWM Read-Write Memory 读写存储器
RWS Release Work Sheet 发放工作单
RZ Return to Zero 归零

S

S　1. Second 秒 2. Small 小的 3. South 南 4. Stowage 储物舱
S wave　地震横波
S/C　Step Climb 分阶段爬升
S/D　System Display 系统显示
S/N　Serial Number 系列号
S/O　Shutoff 切断，关车
S/R　Safety/Reliability 安全性/可靠性
S/SDD　System/Subsystem Design Documentation 系统/分系统设计文档
S/SED　System/Software Engineering Documentation 系统/软件工程文档
S/SS　System/Subsystem Specification 系统/分系规范
S/UT　Subroutine/Unit Testing 子程序/单元测试
S/W　Software 软件
SA　1. Safety Altitude 安全高度 2. Saturn-Apolio 土星火箭和阿波罗飞船（美国）3. Selective Availablity 选择可用性，用户精度选择机制（对非美国防部规定的用户，该机制起作用时，全球定位系统导航精度将降低）4. Shaft Angle 轴角 5. Single-Aisle 单过道的（指客舱内布局）6. Solar Array 太阳电池阵 7. Spacecraft Adapter 航天器对接连接器 8. Stand-Alone 独立操作的 9. Status Accounting 状态纪实 10. Steering Actuator 转弯作动筒 11. Structural Audit 结构（疲劳）审核 12. Submerged-Arc 潜入弧（焊接）13. Supportabilitiy Analysis 支援能力分析 14. Surface-to-Air 地对空 15. System Analysis 系统分析 16. System Architecture 系统构造
SAA　1. Service Access Area 服务接近区域 2. Society of Airline Analysts 航空公司分析员协会（美国）3. Special Assignment Airlift 专机空运 4. Swiss Aerobatic Association 瑞士特技飞行协会
SAAA　Sport Aircraft Association of Australia 澳大利亚体育（运动）飞机协会
SAAHS　Stability-Augmentation Attitude Hold System 增稳与姿态保持系统
SAAPA　South African Airways Pilots' Association 南非航空公司驾驶员协会
SAARU　Secondary Attitude Air Data Reference Unit 第二大气数据姿态基准装置
SABAR　satellites and balloons and rockets 卫星、气球与火箭
Sabin　赛（宾）（声吸收单位，1平方英尺面积上吸收的全部声能。无公制单位）
SABMIS　Surface-to-Air Ballistic-Missile Interception System 地对空弹道导弹截击系统
sabot　1. 次口径炮弹软壳，超高速穿甲弹弹心外壳 2. 火箭支撑环（沿发射筒滑动，并与火箭一起射出）3. （风洞内）模型支撑环 4. 镗杆，衬套
SABS　South America Broadcast Satellite 南美广播卫星
SAC　1. Soaring Association of Canada 加拿大滑翔协会 2. Space Activities Commission 空间活动委员会 3. Space Applications Centre 空间应用中心 5. Special Airworthiness Certificates 特殊适航证 4. Standard Atmospheric Condition 标准大气条件 5. Standing Advisory Committee 常设咨询委员会 6. Supplemental Air Carrier 辅助空运机 7. System Anti-Collision 防撞系统
SACCA　Scottish Advisory Council for Civil Aviation 苏格兰民航咨询委员会
SACCS　Strategic Air Command Control System 战略空军司令部管理系统
SACEUR　Supreme Allied Commander Europe 欧洲盟军最高司令（北大西洋公约组织）
SACLANT　Supreme Allied Commander, Atlantic 大西洋盟军最高司令（北大西洋公约组织）
SACLOS　Semi-Automatic Command to Line Of Sight 瞄准线半自动控制（导弹制导方法，导弹发射后被控制进入射手视力范围内）
SACMA　Suppliers of Advanced Composite Materials Association 先进复合材料供应商协会（美国）
SACNET　Secure Automatic Communication Network 保密自动通信网络，安全自动通信网络
SaCo　Samarium Cobalt 钐钴磁钢，钐钴
SACP　Surface Air Courte Portee (Missile) （地/水）面对空短程导弹
sacrificial corrosion　牺牲（性）腐蚀（用抗腐性较差的金属包覆在主体金属外部作为腐蚀防护层）
SACU　Stand Alone Compensator Unit 独立补偿单元
SAD　1. Safety and Arming Device （引信）保险和解除保险装置 2. Spares Advance Data 备件前期数据 3. Stress Analysis Data 应力分析数据
SADA　Software Airplane Distribution Application 软件机上分配程序
saddle　机架，机座

SADM　Standby Air Data Module 备用大气数据组件
SAE　1. Simple Arithmetic Expression 简单算术表达式 2. Society of Automotive Engineers 汽车工程师学会，机动车工程师学会 3. Stand Alone Equipment 独立设备 4. Supersonic Aircraft Engine 超声速飞机发动机
SAF　1. Specific Access Funtion 特定接入功能 2. System Administration Facility 系统管理设备
SAFCO　Sinotrans Airfreight Forwarding Company 中国航空货运代理公司
SAFE　1. Survival And Flight Equipment 救生与飞行设备 2. System for Automated Flight Efficiency 自动飞行数据系统
safe　1. 安全的，保险的，可靠的 2. 保险箱
safe and arming device　安全和解除保险装置
safe burst height　（原子弹的）安全爆炸高度
safe capture trajectory　安全捕获轨迹
safe condition　安全状况
safe consciousness　安全意识
safe ejection envelope　1. 救生性能包线 2. 安全弹射包线
safe ignition device　安全发火机构
safe landing　安全着陆
safe landing rate　安全着陆速度
safe landing site　安全的着陆地点
safe life　安全寿命，安全生活
safe life design　安全寿命设计
safe mode　安全模式
safe operation　安全操作
safe release altitude　安全投弹高度
safe separation　安全距离
safe state　安全状态
safe touchdown　安全着陆
safe trajectory　安全轨迹
safe working pressure　安全工作压力，设计压力
safe zone　安全区域
safeguard　防护设施，安全装置，防护器，保证措施
safeguard team　安全保卫组
safety　1. 安全，安全性 2. 安全措施
safety analysis　安全性分析
safety and reliability program　安全性可靠性大纲
safety barrier　安全势垒
safety belt　安全带
safety card　安全卡
safety check　安全检查
safety clearance　容许间隙，安全间隙
safety clothing　安全服，防护服
safety coefficient　安全系数
safety command　安全指令
safety control　安全控制
safety control circuit　安全控制电路
safety control information　安全控制信息
safety control system　安全控制系统
safety controller　安全控制器，安全控制机构
safety corridor　安全管道
safety criterion　安全准则
safety critical engine variable　安全关键引擎变量
safety critical part　安全关键件
safety decision　安全判决
safety decision rule　安全判决准则
safety device　安全装置，保安装置
safety disc　安全片
safety distance　安全距离
safety dog　安全轧头
safety engineering　安全工程，安全技术
safety exploder　安全信管
safety explosive　安全炸药
safety factor　安全系数，安全因数，保险系数
safety fuel　安全燃料
safety gap　安全放电器，安全隙
safety goggles　安全眼镜，护目镜
safety grounding　安全接地
safety height　安全高度
safety helmet　安全帽
safety issue　安全问题
safety lever　保险杆（自动武器保险装置上的杠杆）
safety limit　安全极限，安全限度
safety load factor　安全载荷因数
safety management　安全管理
safety margin　安全裕度
safety mode　安全模式
safety net　安全网，安全信息网
safety pin　安全别针，安全销机，保险丝，保险销
safety plug　安全塞，熔丝塞，熔线塞
safety pressure　安全压力
safety psychology　安全心理学
safety remote control　安全遥控
safety requirement　安全要求
safety speed　安全速度
safety system　1. 安全系统 2. 安全设备
safety thread　安全拉线
safety valve　安全阀
safety wire　保险丝（防止紧固件松动用的金属丝）
safety zone　1. 安全区 2. 保险道
safing　保险
SAFTP　System Aircraft Flight Test Plan 系统飞机试飞大纲
SAG　1. Standard Address Generator 标准地址发生器 2. Surface Action Group 水面行动大队 3. System

Analysis Group 系统分析组
sag 1. 倾斜 2. 下垂,下陷,下沉
sagger 匣钵
sagging 下垂,垂度
sagittal resolution 径向分辨率
Sagnac effect 萨奈克效应
SAGTP System Aircraft Ground Test Plan 系统飞机地面试验大纲
SAHRS Standby Attitude Heading Refernce System 备用姿态航向基准系统
SAI 1. Spherical Attitude Indicator 球形姿态显示器 2. Standby Airspeed Indicator 备用空速表 3. Standby Attitude Indicator 备用姿态指示仪 4. System Analysis Item 系统分析项目 5. System Architecture and Interface 系统特性和界面
SAICS Standard Avionics Integrated Control System 标准航空电子综合控制系统
sail 1. 航行,滑翔飞行 2. 水上飞机在水面航行(尤指在有风的情况下) 3. 帆,帆状物 4. 太阳帆 5. 指挥台围壳(潜艇艇壳上方的突出部分,亦称 fin, bridge fin, 舰桥整流罩,过去称为 conning tower 指挥台) 6. 翼尖小翼 7. 〈口语〉雷达天线
sail area 帆面积
sail characteristic acceleration 帆特征加速度
sail wing 帆翼
sailcraft 太阳帆航天器
sailing 航行,航行法
sailing flight 滑翔飞行
sailplane 翱翔机,高性能滑翔机
SAL System Address Label 系统地址标签
sal ammoniac 氯化铵(碳锌电池中电解质)
sale price 销售价格
sale representative 销售代表
sales outlet 销售渠道
salinity 含盐量
SALR Saturated Adiabatic Lapse Rate 饱和空气绝热(温度)垂直梯度,饱和空气绝热(温度)直减率
SALS 1. Seperate-Access Landing System 可选取(飞机)间隔的着陆系统 2. Shipborne Aircraft Landing System 舰载飞机着陆系统 3. Short Approach Light System 近距进场照明系统
salt spray chamber 盐水喷雾室
salt spray test 盐雾试验
salt water corrosion 盐水腐蚀
salt-fog test 盐雾试验
salvage 废物利用,废品处理,修补,抢救
salvo 1. 齐射 2. 齐投 3. 电子干扰措施同时并举
salvo area 齐射毁伤面积
salvo launch 连续发射

salvo-train bombing 连续齐投投弹
SAM 1. Safety and Arming Mechanism 保险及爆炸机构,爆炸及不爆炸机构 2. School of Aerospace Medicine 航空航天医学院(美国空军) 3. Sound Absorbing Material(s) 消声材料,吸声材料 4. Space Assembly and Maintenance 空间装配和维护 5. Special Air Missions 特种空中作战(中队),特殊飞行任务(中队) 6. Standard Avionics Modules 标准航空电子组件 7. Standard(ized) Assembly Module 标准(化)装配模式组件 8. Stratospheric Aerosol Measurement 平流层气溶胶测量仪 9. Surface to Air Missile 地空导弹
SAMA Small Aircraft Manufacturers Association 小飞机制造商协会
samarium 【化】钐(稀土元素,符号 Sm)
same source calibration 同源校准
SAMEX Shuttle Active Microwave Experiment 航天飞机有源微波实验
SAMI Service Access Multiplexer Interface 业务接入复用器接口
SAMIS Structural Analysis and Matrix Interpretive System 结构分析与矩阵判读系统
SAMOS 1. Satellite and Missile Observation Station 卫星和导弹观测站 2. Satellite and Missile Observation System 萨莫斯卫星,卫星导弹观测系统(美国早期照相侦察卫星) 3. Silicon and Aluminum Metal Oxide Semiconductor 硅和铝金属氧化物半导体
sample 1. 样品 2. 采样
sample and hold 采样保持
sample collector 取样器
sample design 样品设计
sample extraction 样本提取
sample gas 样气
sample holder 样品架
sample inspection 试件检验
sample interval 取样间隔
sample length 样本长度
sample line 样本线
sample point 采样点
sample population 样本总体
sample rate 抽样率
sample return 样本取回
sample return mission 样本取回任务
sample size 样本量,样本大小
sample stream 样本流
sample support 样品支撑
sample surface 表面抛光
sample temperature 样品温度
sample test 抽样试验
sample thickness 样品厚度

sample time   抽样时间
sample time interval   采样时间间隔
sample train   取样系列
sample tube   导管样件
sample volume   样品体积
sample zone   样品区
sampled data control system   取样控制系统
sampled data system   取样数据系统
sampled phase-locked loop   取样锁相环路
sampler   取样器,采样器
sampler and coder   1.采样编码器 2.采编器
sampling   取样
sampling circuit   采样电路
sampling condition   采样条件
sampling element   采样元件
sampling frequency   采样频率
sampling holder   采样保持器
sampling inspection   抽样检验
sampling instant   抽样瞬间
sampling interval   采样间隔
sampling lead   取样器
sampling life test   抽样寿命试验
sampling line   取样管路
sampling orifice   抽样口
sampling oscilloscope   采样示波器
sampling period   采样周期
sampling plan   取样方案
sampling point   采样点
sampling port   采样口
sampling probe   采样探针
sampling pulse   采样脉冲
sampling rate   采样频率
sampling scheme   抽样方案
sampling switch   采样开关,转换器
sampling system   采样系统
sampling technique   抽样技术
sampling test   采样试验
sampling theorem   采样定理
sampling time   取样时间
sampling with replacement   有放回取样
sampling without replacement   无放回取样
SAMS   1. Satellite Automatic Monitoring System 卫星自动监测系统 2. Shuttle Attached Manipulator System 航天飞机附属机械臂系统 3. Shuttle Automated Management System 航天飞机自动化管理系统 4. Space Assembly, Maintenance and Servicing 空间装配、维修和服务 5. Stratospheric and Mesospheric Sounder 平流层和中间层探测器
SAMSO   Space and Missile Systems Organization 航天与导弹系统组织(美国空军武器系统研究部)
SAMSON   Strategic Automatic Message Switching Operational Network 战略文电自动转接网
SAMTEC   Space and Missile Test Center 空间和导弹测试中心
San Marco Range   圣马科航天发射场(意大利)
sand and dust test   沙尘试验
sand blast test   吹砂试验
sand blasting   喷砂
sand casting   砂模烧铸
sand clock   沙漏
sand glass   沙漏
sand grain   砂粒
sand mold casting   砂型铸造
sand storm   沙暴
sandblast   喷砂器
sandblasting   吹砂(抛光)
Sandwich   夹芯结构,夹层结构,夹芯材料制造
sandwich construction   夹层结构
sandwich construction repair technique   夹层结构修补技术
sandwich material   夹层材料,夹芯材料
sandwich moulding   夹芯模塑
sandwich panel   夹层板
sandwich plate   夹层板
sandwich structure   夹层结构
SAO   Single Association Object 单相关控制对象
SAP   1. Service Access Points 服务接近点 2. Service Advertising Protocol 服务公告协议 3. Stabilized Approach 稳定进近
SAPAM   System Application and Processes for Aircraft Manufacturing 飞机制造的系统应用和工艺
SAR   1. Search and Rescure 搜救,搜索与救援 2. Selected Acquisition Report 选购项目报告(美国国防部按季度发布的武器选购状况报告) 3. Selected Air Reservist 选定的海军航空兵后备队员 4. Semi-Active Radar 半主动雷达(自导引),半主动雷达(寻的) 5. Stand Alone Radar 独立雷达 6. Stress Analysis Report 应力分析报告 7. Synthetic Aperture Radar 合成孔径雷达 8. System Analysis Report 系统分析报告
SARBE   Search and Rescue Beacon Equipment 搜索救援信标台(包括军用及民用)
SARCAP   Search and Rescue Combat Air Patrol 有救援任务的空中战斗巡逻
SAREX   Search and Rescue Exercise 搜寻与援救演习
SARH   Search & Rescue Homing 搜索与援救归航
SARI   Small Airport Runway Indicator 小型机场跑道指示器
SARIE   Semi-Automatic Radar Indentification Equip-

ment 半自动雷达识别设备
sarin 甲氟磷酸异丙酯
Saros 沙罗周期
SARP 1. Semi-automatic Radar Plotting 半自动雷达标图 2. Signal Auto Radar Processing System 信号雷达自动处理系统
SARPS Standard and Recommended Practices 标准及推荐措施，标准和推荐实践
SARS 1. Simulated Airborne Radar System 模拟机载雷达系统 2. Support and Restraint System 支承与限动系统
SARSAT 1. Search and Rescue Satellite 搜索与救援卫星 2. Search and Rescue Satellite Aided Tracking 搜索和救援卫星辅助跟踪计划（加拿大、美国、苏联、法国合作的全球空、海难卫星监视计划）3. Synthetic Aperture Radar Satellite 合成孔径雷达卫星
SARUS Search And Rescue Using Satellites 用卫星搜索和营救
SAS 1. Satellite Application Section 卫星应用小组（美国国家海洋与大气管理局）2. Satellite Automation System 卫星自动化系统，卫星自动控制系统 3. Single Audio System 单独音响系统 4. Small Angle Scattering 小角度散布 5. Small Astronomy Satellite 小天文卫星（美国天文卫星，属探险者卫星系列中的卫星）6. Software Accomppishment Summary 软件完成情况总结 7. Space Adaptation Syndrome 空间适应性综合症 8. Special Air Service 特种空军勤务（英国）9. Stability Augmentation System 增稳系统 10. Stall Avoidance Subsystem 防失速分系统 11. Station Address Set 站位设置 12. Surface Active Substance 表面活性材料 13. Survival Avionics System 救生电子设备
SASE 1. Semi Automatic Support Equipment 半自动保障设备 2. Specific Application Service Element 专门应用服务单元 3. Stand Alone Synchronization Equipment 独立同步设备
SASP 1. Science and Application Space Platform 科学和应用空间平台 2. Single Advanced Signal Processor 单独高级信号处理器
SASS 1. Saturn Automatic Software System "土星"火箭自动软件系统 2. Seasat-A Satellite Scatteromter 海洋卫星散射计 3. Strategic Airborne Surveillance System 机载战略监视系统
SAT Static Air Temperature 静温
SATAF Site Alteration Task Force 发射场改造专门工作组
SATC Semiautomatic Air Traffic Control 半自动空中交通管制
SATCOM Satellite Communication 卫星通信

SATCOMA Satellite Communiation Agency 卫星通信局（美国）
satellite 1. 卫星，人造卫星 2. 卫星机场
satellite aberration 卫星光行差
satellite access 卫星接入
satellite acquisition 卫星捕获
satellite almanac 卫星历书
satellite altimetry 卫星测高
satellite and launch vehicle matching 星箭匹配
satellite application 卫星应用
satellite area monitoring 卫星普查
satellite area-mass ratio 卫星面积质量比
satellite assembly 卫星总装
satellite attitude control 卫星姿态控制
satellite automonitor system 卫星自动监视系统
satellite autonomy 卫星自主性
satellite base data collection subsystem 星载数据收集分系统
satellite battery 人造卫星用电池
satellite beacon 卫星信标
satellite broadcasting 卫星广播
satellite bus line 卫星总线
satellite business 卫星商业系统
satellite channel 卫星通路
satellite channel mode 卫星信道方式
satellite climatology 卫星气候学
satellite clock bias 星钟偏差
satellite clock correction parameter 星钟修正参数
satellite clock time reference 星钟时间基准
satellite cloud picture 卫星云图
satellite cluster 卫星组
satellite communication 卫星通信
satellite communication center 卫星通信中心
satellite communication concentrator 卫星通信集中器
satellite communication control facility 卫星通信控制设施
satellite communication controller 卫星通信控制器
satellite communication earth station 卫星通信地面站
satellite communication terminal 卫星通信终端（设备）
satellite community reception 卫星公共接收站，卫星公共接收系统
satellite computer terminal 卫星计算机终端
satellite configuration 卫星构型
satellite constellation 卫星座
satellite control center 卫星控制中心
satellite control function 卫星控制功能
satellite counterweight 卫星配重
satellite coverage 卫星覆盖范围
satellite data 卫星数据，卫星资料

satellite data area　卫星数据区
satellite data communication　卫星数据通信
satellite data exchange　卫星数据交换
satellite data modulator　卫星数据调制器
satellite data processing system　卫星数据处理系统
satellite data recorder　卫星数据记录器
satellite data system　卫星数据系统
satellite data transmission system　卫星数据传输系统
satellite database　卫星数据库
satellite datum　卫星数据
satellite day　卫星白天
satellite defense　卫星防御
satellite delay compensation unit　卫星延迟补偿器
satellite design lifetime　卫星设计寿命
satellite discharging　卫星放电
satellite dish　碟形卫星天线
satellite Doppler navigation　卫星多普勒导航
satellite Doppler navigation system　卫星多普勒导航系统
satellite Doppler positioning　卫星多普勒定位
satellite Doppler shift measurement　卫星多普勒(频移)测量
satellite downlink　卫星下行链路
satellite early warning system　卫星(早期)预警系统
satellite earth observation　卫星对地观测
satellite earth resource exploration　卫星地球资源勘探
satellite earth station　卫星地球站
satellite earth terminal　卫星地面终端
satellite eclipse　卫星蚀(卫星进入行星或另一卫星影锥中,完全或部分地从观测者视野消失的现象)
satellite electric simulator　卫星电性等效器
satellite engineering　卫星工程
satellite environment　卫星环境
satellite ephemeris　卫星星历表
satellite equatorial orbit　赤道轨道卫星
satellite equipment connecting　卫星设备连接
satellite fairing　卫星整流罩
satellite for orientation navigation and geodesy　定向导航和测地卫星
satellite formation　卫星编队
satellite fragmentation　卫星碎片
satellite fundamental frequency　卫星基频
satellite galaxy　伴星系
satellite general layout　卫星总体布局
satellite geodesy　卫星大地测量
satellite gravimetry　卫星重力测量
satellite ground station　卫星地面站
satellite ground support　卫星地面保障
satellite ground track　卫星地面轨迹

satellite guidance　卫星制导
satellite identification　卫星识别
satellite image　卫星图像
satellite information　卫星信息
satellite information flow　卫星信息流
satellite information processor　卫星信息处理机
satellite infrared spectrometer　卫星红外分光计
satellite inspection network　卫星检测网络
satellite inspection technique　卫星检测技术
satellite inspector　卫星监视器
satellite interceptor　卫星截击器
satellite interceptor system　卫星截击系统
satellite interface message processor　卫星接口信息处理机
satellite killer　反卫星武器
satellite laboratory　卫星实验室
satellite laser ranger　卫星激光测距仪
satellite laser ranging　卫星激光测距
satellite launch　卫星发射
satellite launch site　卫星发射场
satellite launching center　卫星发射中心
satellite launching vehicle　人造卫星运载火箭
satellite library information network　卫星数据库信息网络
satellite lifetime　卫星寿命
satellite line　卫星航线
satellite link　卫星链路
satellite loading building　卫星加注厂房
satellite meteorological observation　卫星气象观测
satellite meteorology　卫星气象学
satellite minicomputer　卫星小型(计算)机
satellite mock up　卫星模装
satellite motion simulator　卫星运行模拟器
satellite multiple access　卫星多路存取
satellite navigation　卫星导航
satellite navigation computer　卫星导航计算机
satellite navigation earth station　卫星导航地面站
satellite network　卫星网(络)
satellite night　卫星黑夜
satellite observation　卫星观测
satellite observatories　卫星观测系统
satellite ocean surveillance　卫星海洋监视
satellite operating lifetime　卫星工作寿命
satellite operation program　卫星运行程序
satellite optical link　卫星光链路
satellite optical surveillance station　卫星光学监控站
satellite orbit　卫星轨道
satellite orbital altitude　卫星轨道高度
satellite orbital lifetime　卫星轨道寿命

satellite overall design	卫星总体设计
satellite packet transmission	卫星分组传输
satellite paper tape transfer	卫星纸带传送
satellite pattern	卫星配制图
satellite perturbance motion	卫星摄动运动
satellite perturbation	卫星摄动
satellite photo map	卫星照片图
satellite photogrammetry	卫星摄影测量
satellite photograph	卫星相片
satellite photographic study	卫星照片判读
satellite photography	卫星摄影
satellite piggyback environment	卫星搭载环境
satellite platform	卫星平台
satellite point	卫星点
satellite position prediction and display	卫星位置预报和显示
satellite potential	卫星电位
satellite power control device	星上电源控制设备
satellite power system	卫星动力系统
satellite problem	卫星问题
satellite processor	卫星处理机
satellite radio	卫星广播
satellite radio interferometry	卫星射电干涉测量
satellite ranging	卫星测距
satellite ranging navigation	卫星测距导航
satellite reconnaissance	卫星侦察
satellite recovery	卫星回收
satellite relay	卫星中继,卫星转发
satellite remote sensing	卫星遥感
satellite remote sensing system	卫星遥感系统
satellite rendezvous	卫星会合
satellite return program	卫星返回程序
satellite sealed container hoisting tool	卫星密封容器吊具
satellite search and rescue	卫星搜索与救援
satellite signal	卫星信号
satellite simulator	卫星模拟器
satellite simulation load	卫星仿真负载
satellite solar power station	卫星太阳能发电站
satellite sounding	卫星探测
satellite space station	卫星空间站
satellite stand	卫星停放平台
satellite station	卫星观测站
satellite subsystem	卫星分系统
satellite subsystem design	卫星分系统设计
satellite surveillance radar	卫星监视雷达
satellite surveying and mapping	卫星测绘
satellite switch	星上切换
satellite synchronous controller	卫星同步控制器
satellite system	1.卫星系统 2.辅助系统
satellite system design	卫星总体设计
satellite system engineering	卫星系统工程
satellite system monitor station	卫星系统监视站
satellite system software	卫星系统软件
satellite system test software	卫星系统测试软件
satellite technology	卫星技术
satellite telecommunication	卫星通信
satellite telemetry	卫星遥测
satellite temperature	卫星温度
satellite test center	卫星试验中心
satellite test language	卫星测试语言
satellite test sequence	卫星测试程序
satellite thermal design	卫星热设计
satellite thermal vacuum test	卫星热真空试验
satellite town	卫星城镇
satellite tracking	卫星跟踪
satellite tracking antenna	卫星跟踪天线
satellite tracking program	卫星跟踪程序
satellite tracking station	卫星跟踪站
satellite transfer	卫星转运
satellite transponder	卫星转发器
satellite triangulation	卫星三角测量
satellite vehicle	卫星运载火箭
satellite weapon	卫星(式)武器
satellite zenith angle	卫星天顶角
satellite-borne radar	(卫)星载雷达
satellite-pair	伴偶
satelloid	准卫星(飞行在行星大气层的轨道中,需作连续或间断机动的一种卫星)
satelloon	气球卫星
Satmex Satellite of Mexico	墨西哥(通信)卫星
SATNAV	1. Satellite Navigation 卫星导航 2. Satellite-assisted Navigation 卫星(辅助)导航
SATNET satellite network	卫星网
SATS	1. Satellite Automated Test System 卫星自动化测试系统 2. Shuttle Avionics Test System 航天飞机电子设备测试系统 3. Small Airfield for Tactical Support 战术支援小型机场(美国海军陆战队) 4. Small Applications Technology Satellite 小型应用技术卫星
saturable reactor	助磁式电抗器,饱和电抗器,饱和扼流圈
saturated adiabatic lapse rate	饱和绝热递减率
saturated air	饱和空气(含有最大量水汽的空气)
saturating capacity	饱和容量
saturation characteristics	饱和特性(曲线)
saturation constraint	饱和限制
saturation control	饱和控制
saturation current	饱和电流

saturation diving　饱和潜水
saturation exposure　饱和曝光量
saturation function　饱和函数
saturation level　饱和度
saturation magnetic induction　饱和磁感强度
saturation nonlinearity　饱和非线性
saturation parameter　饱和参量
saturation recording　饱和记录
saturation temperature　饱和温度
saturation vapour pressure　饱和蒸汽压
saturation voltage　饱和电压
Saturn　1.土星运载火箭（美国）2.【天】土星
Saturn atmosphere　土星大气
Saturn automatic software system　"土星"（火箭）自动软件系统
SAU　Storage Access Unit 存储存取部件
saucerman　外星人
saunter　以久航速度飞行，闲逛
sausage　风向袋，圆柱形系留气球
save　救，节约，储存
SAW　Surface Acoustic Wave 声表面波
SAW convolver　表面声波卷积器
SAW delay line　表面声波延迟线
SAW filter　表面声波滤波器
SAW interdigital transducer　表面声波叉指换能器
sawcut　（机翼前缘的）锯齿（槽）前缘切口
SAWF　Surface Acoustic Wave Filter 声表面波滤波器
SAWRS　Supplementary Aviation Weather Reporting Station 辅助航空气象报知站（美国家海洋与大气管理局）
sawtooth pattern　锯齿形曲线（卫星导航的一种误差曲线）
sawtooth waveform　锯齿波形
SAX　Simple API for XML 简易应用编程接口
SB　1.S-bands 波段 2.Service Bulletin 服务通报，服务通告，勤务通报，使用情况通告 3.Sideband 边带，边频带 4.Silver Brazen 银焊 5.Sonic Boom（超声速）音爆 6.Speed Brake 减速板 7.Spin Brake 停转制动器 8.Starboard 右舷，右侧 9.Status Board 飞机状况板
SBA　1.Small Businesses Association 小型企业协会（美国）2.Spot Beam Aerial，Spot Beam Antenna 点波束天线 3.Standard Beam Approach 标准波束引导进场
SBAC　1.Society of British Aerospace Companies 英国航空航天公司联合会 2.Society of British Aircraft Con-structors 英国飞机制造商联合会（现已被前者代替）
SBC　1.Single Board Computer 单板机 2.Slave Bus Controller 辅助总线控制器 3.Small Bayonet Cap 小装药管帽 4.Sonic Boom Committee 音爆委员会
SBD　1.Schematic Block Diagram 原理方块图 2.Schottky barrier diode 肖特基势垒二极管
SBES　Service Bulletin and Effecitivity System 服务通报和有效性系统
SBFET　Schottky barrier FET 肖特基势垒场效应晶体管
SBI　Service Bulletin Index 服务通告索引
SBN　Service Bulletin Number 服务通报编号
SBR　1.Signal-to-Background Ratio 信号背景比 2.Space-Based Radar 天基雷达
SBS　Stimulated Brillouin Scattering 受激布里渊散射
SC　1.Satellite Computer 卫星计算机，辅助计算机 2.Sequence Controller 顺序控制器 3.Service Ceiling 使用升限 4.Shaped Charge 锥形装药，聚能装药 5.Short Circuit 短路 6.Side Cowl 侧整流罩 7.Single Crystal 单晶 8.Software Control 软件控制 9.Space Command 外层空间司令部（美国）10.Speed Control（system）速度控制（系统）11.Special Committee 专门委员会 12.Special Condition 专门条件 13.Specification Change 规范更改 14.Stable Configuration 固定构型 15.Subgrade Code 地基强度标准（国际民航组织规定的）16.Swivel Coupling 转动接头
SCA　1.Shuttle Carrier Aircraft 航天飞机驮运机（美国波音747运输机改装而成）2.Simulation Control Area 模拟管制区 3.Single Crystal Alloy 单晶合金
SCADC　Standard Central Air Data Computer 标准中央大气数据计算机（美国）
scalability　可度量性，可缩比性
scalar　纯量，标量（区别于向量）
scalar dissipation rate　标量耗散率
scalar equation　无向量方程
scalar function　纯量函数，标量函数
scalar height　标高
scalar network analyzer　标量网络分析仪
scalar parameter　标量参数
scalar product　无向积
scalar quantity　标量
scalarization　标量化
scale　1.刻度盘，比例尺，标尺 2.刻度，标度，分划 3.按比例复制 4.等级，级别 5.规模，大小 6.（复数）天平，秤 7.鳞状物 8.铁屑，水锈，水垢
scale calculation　标定计算
scale change　1.标度改变 2.刻度变化
scale effect　尺度效应
scale error　标度误差
scale factor　刻度因子，尺寸比例系数，比例因子
scale factor stability　标度因素稳定性

scale height 太阳大气标高,标高
scale model 比例模型,缩比模型
scale model test 缩比模型试验
scale of length 长度比例尺
scale parameter 标度参量
scale weight 秤砝码
scaled 按比例(放大或缩小)的
scale-of-ten 1.十进位(的)计数元件系统 2.十进制
scale-of-two 1.二进制(记数法) 2.二进位(的)双稳计数元件系统
scaler 1.定标器,定标电路 2.(脉冲)计数器 3.换算器,换算电路 4.刮除器
scale-up 按比例增加,按比例放大
scaling 缩放比例
scaling factor 换算因数,比例因数
scaling function 标度函数
scaling law 比例法则
scaling of model 模型缩放
scaling parameter 换算因数,尺度参数
scaling relation 尺度关系,比例关系
scaling theory 标度理论
scaling-down 按比例缩小
scallop 成扇形
scalloped segment 瓜瓣
scan 扫描,扫掠
scan angle 扫描角
scan angle monitor 扫描角监控器
scan converter tube 扫描转换管
scan expansion lens 扫描扩展透镜
scan flywheel 扫描飞轮
scan line corrector 扫描线校正器
scan pattern 扫描模式
scan platform 扫描平台
scan rate 扫描频率
scan width 扫频宽度
scan-digitizing 扫描数字化
scandium 【化】钪(化学元素,符号 Sc)
scan-image 扫描影像
scanner 1.扫描器,扫描器,扫探天线 2.析像器,多点测量仪 3.雷达操作员
scanning angle 扫描角
scanning direction finding 搜索测向法
scanning earth sensor 扫描地球敏感器
scanning efficiency 扫描效率
scanning electron microscopy 扫描电子显微镜学
scanning field 扫描场
scanning field of view 扫描视场
scanning frequency measurement 搜索测频法
scanning horizon sensor 扫描地球敏感器

scanning laser fuze 扫描式激光引信
scanning microwave radiometer 扫描微波辐射计
scanning microwave spectrometer 扫描微波波谱仪
scanning multi-channel microwave radiometer 扫描多通道微波辐射计
scanning overlap coefficient 扫描重叠率
scanning radiometer 扫描辐射仪
scanning range 扫描范围
scanning rate 扫探速度,仪表扫视速率
scanning spectrometer 扫描分光仪,扫描光谱仪
scarer 驱鸟装置
scarf cloud 幞状云
scarfed 斜割的,有斜角切口的,斜面嵌接的
SCAS 1.Spacecraft Adapter Simulator 航天器对接器模拟器 2.Stability and Control Augmentation System 稳定性和操纵性增强系统
SCAT 1.Security Control of Air Traffic 空中交通安全管制 2.Speed Control, Approach/Takeoff 进近和起飞速度调节系统 3.Special Category 特殊类别 4.Supersonic Commercial Air Transport 超声速商用运输机
SCATANA Security Control of Air Traffic and Air Navigation Aids 空中交通安全控制与导航设备(美国联邦航空局,用于国防紧急需要时的规定和指令)
SCATHA Spacecraft Charging at High Altitudes 高空充电实验卫星,美国科学卫星
scatter 1.散布,分散,扩散,散射(辐射的) 2.(亚原子粒子因碰撞而)轨迹改变 3.(目标)弹着点分布 4.(数据)分散度
scatter bomb 散飞性燃烧弹
scatter communication 散射通信
scatter density 散射体密度
scatter point 散乱点
scatter propagation 散射传播
scattered cloud 散云
scattered field 散射场
scattering 散射
scattering body 散射体
scattering coefficient 散射系数
scattering cross section 散射截面
scattering effect in the atmosphere 大气散射效应
scattering image 散射图像
scattering loss 散射损耗
scattering matrix 散射矩阵
scattering power 散射功率
scatterometer 散射计
scavenge pipe 回油管
scavenge pump 换气泵,回油泵
SCB Selected Call Bandwidth 选定的呼叫带宽

SCC 1. Satellite Communication Concentrator 卫星通信集线器 2. Satellite Communications Controller 卫星通信控制器 3. Security Consultative Committee 安全咨询委员会 4. Single Conductor Cable 单导线电缆 5. Standing Consultative Commission 常设协商委员会 6. Supplier Collaboration Center 供应商合作中心 7. Supervisor Control Console 监督控制台 8. Supply Control Center 补给控制中心 9. Synchronous Communications Controller 同步通信控制器

SCCD Surface Charge Coupled Devices 表面电荷耦合器件

SCCF Satellite Comunication Control Facility 卫星通信控制设备

SCCS 1. Switching Control Center System 交换控制中心系统 2. Sub-Contract Control Sheet 子合同控制单

SCD 1. Selective Cell Discard 选择性信元丢弃 2. Single Channel Demultiplexer 单信道解复用器 3. Specification Control Document 规范控制文件 4. Specification Control Drawing 规范控制图 5. Subject to Captain's Discretion 由机长确定 6. System Category Diagram 系统分类图

SCDU Satellite Control Data Unit 卫星控制数据装置

SCE 1. Service Creation Environment 业务创建环境 2. Signal Combining Equipment 信号组合设备 3. Signal Conditioning Equipment 信号适配设备 4. Signal Conversion Equipment 信号变换装置

scenario 1.剧情说明 2.方案,作战想定 3.景像,情况

scene 物景,图幅,像幅,帧

scene generator 景像发生器

SCET Satellite Control Earth Terminal 卫星控制地球终端站

SCF 1. Satellite Control Facility 卫星控制设施 2. Solution Crystal Facility 溶液晶体生长设备

SCH 1. Simplified Combined Hharness 简化组合式背带系统 2. Sonobuoy Cable Hold 声呐浮标电缆舱(自动驾驶选择器)

schedule 表,图表,时间表,目录,一览表,计划,程序

schedule flight 航班飞行

scheduled departure 原定出发时间

scheduled flight 定период飞行,航班

scheduled maintenance 定期检修,预定维护

scheduled service 定期航班

scheduled speed 额定速率

scheduler 程序机,程序装置

scheduling 编目录,编制时间表,程序,工序,调度,编制进度表

scheduling algorithm 调度算法

scheduling model 调度模型

scheduling point 调度点

scheduling strategy 调度策略

scheduling technique 调度技术

scheduling variable 调度变量

schematic 原理图,图解视图

schematic diagram 示意图,原理图

schematic representation 图示

scheme 图解,计划,组合,体制

schlieren 纹影,纹影仪,纹影法

schlieren apparatus 纹影仪

schlieren photograph 数值纹影图

schlieren picture 纹影图片

schlieren system 纹影系统,纹影仪,施利伦系统

schlieren-interferometer 纹影干涉仪

SCHLT Searchlight 探照灯

Schmidt camera 施密特照相机

Schmidt trigger 施密特触发器

Schottky barrier 肖特基势垒

Schottky barrier diode 肖特基势垒二极管

Schottky defect 肖特基缺陷(指原子从晶格散逸)

Schottky diode 肖特基势垒二极管

Schottky effect 肖特基效应

Schottky integrated injection logic 肖特基集成注入逻辑

Schottky solar cell 肖特基太阳电池

Schuler pendulum 舒勒摆

Schuler principle 舒勒原理

SCIB Selective Channel Input Bus 选择通道输入总线

SCID Software Configuration Index Drawing 软件构型目录图

science satellite 科学卫星

scientific instrument 科学仪表

scientific measurement 科学计量

scientific method 科学方法

scientific notation 科学记数法

scientific observation 科学观察

scientific satellite 科学卫星

scintillation 1.(雷达屏幕上目标迅速移动引起的)回波起伏 2.闪光,闪烁

scintillation counter 闪烁计数器

scintillation counter telescope 闪烁计数器望远镜

scintillation detector 闪烁探测器

scintillation error 振幅起伏误差

scintillation interference 起伏干扰

scintillation proportional detector 闪烁正比探测器

scintillation spectrometer 闪烁分光计,闪烁谱仪

scintillator 闪烁体

scintillator solution 闪烁体溶液

scintillometer 闪烁计,闪烁计数器

scissor tail rotor 剪刀式尾桨

scissors 剪子,剪刀
SCL 1. Soliton Compression Laser 孤子压缩激光器 2. Software Configuration Library 软件构型库 3. Software Control Library 软件控制库 4. Speed Control Logic 速度控制逻辑 5. System Control Language 系统控制语言
sclerometer 加压刻痕硬度计
scleroscope 肖氏硬度计,验硬器
SCM 1. Selected Coding Method 选定的编码法 2. Selected Communication Mode 选定的通信模式 3. Service Circuit Module 服务电路模块 4. Signal Conditioner Module 信号调节器组件 5. Simplified Configuration Management 简化构型管理 6. Software Configuration Management 软件构型管理 7. Spoiler Control Module 扰流板控制组件 8. Stabilizer Control Module 安定面控制组件 9. Supply Chain Management 供应链管理 10. System Certification Matrix 系统合格审定表
SCML Supplier Custom Module List 供应商定制模块表
SCMP Software Configuration Management Plan 软件构型管理大纲
SCMR Surface Composition Mapping Radiometer 表面成分图像辐射仪
SCN 1. Satellite Control Network 卫星控制网络 2. Self Contained Navigation 自主式导航 3. Specification Control Navigation 规范控制导航 4. Specification Change Notice 规范更改通知 5. Subscriber Connection Network 用户连接网络
SCO 1. Section Controller 截面控制器 2. Senior Control Operator 控制室高级操纵员 3. Synchronous Connect-Oriented 面向连接同步
SCOB 1. Scattered Clouds Or Better 散云或更好天气 2. Selective Channel Output Bus 选择通道输出总线
scoop 1. 犀斗形进气口,进风斗,犀斗 2.(滑油)收油池
scoop-proof connector 防斜插连接器
scope 1. 显示器,显示器荧光屏 2. 阴极射线示波器,阴极射线管 3. 观测仪器 4.(活动)范围,视野 5.(导弹的)射程
scope background 显示器荧光屏的背景
scopodromic 航向对准目标的,自导引的
score 得分,划线,记分,刻痕,伤痕
Scorpion "蝎子"战斗机,天蝎座
scotoma 盲点(视觉的)
scotopia 暗适应
scotopic adaptation 暗适应
scotopic vision 微光视觉
scouring 1. 侵蚀,冲刷 2.(管道的)冲洗 3.(复数)冲洗下来的污垢

scout 侦察机,搜索救援机
scout-bombing 侦察轰炸
scouting front 空中搜索正面
SCP 1. Sphedcal Candle Power 球面烛光 2. Station Communications Processor 站通信处理机 3. Suppler Change Proposal 供应商更改建议 4. Suppler Custom Part 供应商定制件 5. System Concept Paper 系统设计原理论文
SCPC Single Channel Per Carrier 单路单载波,每路单载波
SCPL 1. Senior Commercial Pilots Licence 高级商用飞机驾驶员执照 2. Suppler Custom Parts List 供应商定制零件表
SCR 1. Selective Circuit Reservation 选择性电路预留 2. Semiconductor Controlled Rectifier 可控半导体整流器 3. Signal to Crosstalk Ratio 信号串话比 4. Silicon Controlled Rectifier 硅可控整流器 5. Software Change Request 软件更改要求 6. Specification Change Request 规范更改申请 7. Supplier Change Request 供应商更改要求 8. System Change Request 系统更改需求
scramble 紧急起飞
scrambler 扰码器
scrambling 1. 置乱 2. 紧急起飞
scramjet 超声速燃烧冲压喷气发动机
scramjet engine 超燃冲压发动机
scrap 切屑,碎片,废料,报废
scrap rate 废弃率,报废率
scrap view 局部视图
scrape 擦伤,刮痕,刮削声,刮,擦
scratch 条痕,擦痕
scratch file (计算机的)暂存文件,擦除文件
screaming 发出尖叫声
screeching 呼啸,振荡燃烧
screeding 抹平(用于黏胶操作)
screen 1. 屏,隔板 2. 银幕,炎光屏 3. 帘栅极 4. 滤网,筛 5. 遮护,屏蔽 6. 屏护队(为保护舰船免遭攻击而配置的舰艇、潜艇和飞机) 7. 起飞着陆假想障碍物
screen burn 荧光屏烧毁
screen filter 网式过滤器,网式滤油器,滤网
screen grid 屏栅极
screen printing 丝网印刷
screen speed 起落越障速度
screen voltage 帘栅极电压
screened 遮蔽的,屏蔽的
screened pair 屏蔽双股线(电缆),有接地屏蔽的双芯电缆
screening 1. 屏蔽,防波,遮护 2. 筛选
screening test 筛选试验,甄别试验

screw 螺旋，螺钉
screw compressor 螺旋式压缩机，螺式压气机
screw dislocation 螺形位错
screw filter 螺旋滤波器
screw pitch gauge 螺矩规，螺纹规
screw pump 螺旋泵
screwjack 螺旋千斤顶，螺旋动作筒
scribe mark 划线刻痕
scriber 刻图仪
scribing 划片，刻图
script 手写体
SCRJ Supersonic Combustion Ramjet 超声速燃烧冲压发动机
scroll 翻动，卷起
scrub 1.擦洗 2.（飞行计划的）撤销，取消 3.（飞行人员的）淘汰
scrubber 擦洗器，涤气器
SCS 1. Satellite Communication System 卫星通信系统 2. Sea Control Ship 海上控制舰 3. Selective Calling System 选择呼叫系统 4. Signal Corps School 通信员学校（美国）5. Single Crystal Sapphire 单晶蓝宝石 6. Society for Computer Simulation 计算机模拟学会（美国）7. Software Coding Standard 软件编码标准 8. Space Communication Satellite 空间通信卫星（日本）9. Spacecraft Control System 航天器控制系统 10. Speed Control System 速度控制系统 11. Stabilization Control System 稳定控制系统 12. Survivable Communication Satellite 可生存通信卫星 13. Survivable Control System 生存控制系统
SCSI Small Computer System Interface 小型计算机系统接口
SCT 1. Single Channel Transponder 单通道应答机 2. Staff Continuation Ttraining 业务人员在职训练（英国民用航空局）3. Surface Charge Transistor 表面电荷晶体管
SCTS System Component Technical Specification 系统部件技术规范
SCU 1. Seat Control Unit 座椅控制组件 2. Secondary Control Unit 辅助控制装置 3. Signal Conditioner Unit 信号调节器组件 4. Signal Conversion Unit 信号转化装置 5. Start Converter Unit 起动整流器组件 6. Supplemental Control Unit 辅助控制组件
scud 碎雨云
scupper 排水孔
SCV 1. Solar Constant Variations 太阳常数变化探测仪 2. Solenoid Controlled Valve 电磁活门 3. Surge Control Valve 喘振控制活门
SCW Super Critical Wing 超临界机翼
SD 1. Shared Data 共享数据 2. Shipping Document 货运单 3. Shut Down 关断，停车 4. Side Display 侧显示器 5. Smoke Detector 烟雾探测器 6. Specification Document 规范说明文档 7. Standardization Directory 标准化指南 8. Standardization Document 标准化文件 9. Storm Detection 暴风雨探测 10. Supplemental Data 增补数据 11. Supportability Discrepance 支援能力缺少 12. System Design 系统设计 13. System Display 系统显示
SDA 1. Shared Data Administration 共享数据管理部门 2. Source Data Automation 源数据自动化 3. System Design Alternative 备用系统设计
SDAC Systen Data Acquisition Concentrator 系统数据获取集中器
SDAI Standard Data Access Interface 标准数据访问接口
SDAS 1. Scientific Data Analysis System 科学数据分析系统 2. Source Data Automation System 源数据自动化系统
SDAT Symbolic Device Allocation Table 符号设备分配表
SDAU 1. Safety Data and Analysis Unit 安全数据与分析装置 2. Special Data Acquisition Unit 专用数据获取组件
SDBY Standby 待机状态，候补（旅客）
SDC 1. Safety Design Criteria 安全性设计准则 2. Signal Data Converter 信号数据转换器 3. Situation Display Console 情况显示器操作台（预警）4. Software Development Center 软件发展中心 5. Space Defense Center 空间防御中心（属美国空军航天防御司令部）6. Synchro-to-Digital Converter 同步器数字信号转换器
SDCR System Design Compliance Report 系统设计符合性报告
SDCU Smoke Detection Control Unit 烟雾探测控制组件
SDD 1. Standard Disk Drive 标准硬盘文档 2. Synthetic Dynamic Display 综合动态显示 3. System Description Document 系统说明书，系统描述文档 4. System for Distributed Database 分布式数据库系统
SDDD Software Design Description Document 软件设计说明书
SDE 1. Software Development Environment 软件开发环境 2. Storage Distribution Element 存储分配部件 3. Submission/Delivery Entity 交发/投送实体
SDET Special Detailed Test 特殊详细检查
SDF 1. Samarium Doped Fiber 掺钐光纤 2. Saudi Dedicated Air Freight（Saudi Arabia）专用航空货机（沙特）3. Saving and Describing Facility 保存和待役设施 4. Self Destruct Fuse 自毁引信 5. Service Data

Function 业务数据功能 6. Simplified Directional Facilities 简易定向设备 7. Software Development Folders 软件开发文件夹

SDFC Space Disturbance Forecast Center 空间扰动预报中心(美国)

SDG Speed Decreasing Gearbox 减速器,减速齿轮箱

SDHE Spacecraft Data Handling Equipment 航天器数据处理设备

SDI 1. Site Diversity Interface 位置分集接口 2. Source Data Identifier 源数据识别符 3. Source/Destination Identifier 源目标识别器,源/端识别符 4. Special Detailed Inspection 特殊详细检查 5. Strategic Defense Initiative 主动战略防御 6. Subscriber Distribution Interface 用户分配接口

SDIO Strategic Defense Initiative Organization 战略防御倡议组织(美国)

SDIP Specifically Designated Intelligence Position 特定情报阵地

SDL Software Development Library 软件开发库

SDLS Shuttle Derived Launch System 航天飞机派生运载系统

SDM 1. Service Data Management 业务数据管理 2. Shared Data Management 共享数据管理 3. Short Data Message 短数据消息 4. Space Division Multiplex 空分复用 5. Speaker Drive Module 扬声器驱动组件,话筒驱动模块 6. Subscriber Demodulator Module 用户调制模块

SDMA Space Division Multiple Access 空分多址

SDMP Shared Data Management Procedures 共享数据管理程序

SDMS Software Development Maintenance System 软件研制维修系统

SDN 1. Secondary Distribution Network 二级分配网 2. Software Defined Network 软件定义网 3. System Descriptive Note 系统说明书

SDO 1. Serial Data Output 串行数据输出 2. Serial Digital Output 串行数字输出

SDOM Sub Domain 子域

SDP 1. Shared Data Procedures 共享数据程序 2. Service Data Point 业务数据点 3. Service Difficulty Program 维修问题大纲 4. Service Discovery Protocol 服务发现协议 5. Service Domain Processor 业务域处理器 6. Shut Down Processor 关断处理器 7. Signal Data Processor 信号数据处理机 8. Software Development Plan 软件开发大纲 9. Software Design Plan 软件设计计划 10. Standard Data Processor 标准数据处理机 11. Surveillance Data Processing 监视数据处理 12. System Definition Phase 系统定义阶段 13. System Display Panel 系统显示板

SDQA Shared Data Quality Assurance 共享数据质量保证

SDR 1. Service Difficult Report 使用困难报告 2. Signal Data Recorder 信号数据记录器 3. Software Design Requirements 软件设计要求 4. Software Design Review 软件设计评审 5. Special Drawing Requirements 特殊图要求 6. Special Drawing Rights 特殊抽签权(在民航经营管理中,用于处理民航公司间的费用问题) 7. Splash Detection Radar 溅落点探测雷达 8. System Description Report 系统说明报告 9. System Design Report 系统设计报告 8. System Development Requirement 系统研制要求

SDRC Structural Dynamics Research Corporation 结构动力学研究会

SDRL 1. Subcontract Data Requirements List 转包合同数据需求清单 2. Supplier Data Requirements Lists 供应商数据要求表

SDS 1. Shared Data Standard 共享数据标准 2. Shop Distribution Standard 车间配发标准 3. Simulation Data Subsystem 模拟数据子系统 4. Space Division Switch 空分交换 5. Surveillance Distribution System 监视分配系统 6. System Decription Section 系统说明部分 7. Switched Data Service 交换数据业务

SDSF Short Duct Seperation Flow 短涵道分离流

SDSS Synergistic Defence Suppression System 配合防卫镇压系统

SDT 1. Short Data Transfer 短数据传递 2. Simplified Documentation Team 简化文件组

SDTM Structure Description Training Manual 系统描述训练手册

SDTP Software Development Tool Plan 软件开发方法大纲

SDU 1. Satellite Data Unit 卫星数据组件 2. Sensor Display Unit 卫星数据装置 4. Service Data Unit 服务数据装置 3. Selective Decode Unit 选择解码装置(二次监视雷达) 5. Signal Display Unit 信号显示器

SDW Specific Design Work 特定设计工作

SE 1. Service Engineering 服务工程 2. Shared Environment 共享环境 3. Single Engine 单发 4. Space Exploration 空间探索 5. Spares Engineering 备件工程 6. Support Equipment 保障设备 7. System Engineering 系统工程

sea anchor 海锚
sea boat 小艇
sea breeze 海风
sea cell battery 海水电池
sea clutter 海面杂波
sea coloring agent 海水染色剂
sea duty 海上任务

sea effect 海洋效应
sea floor spreading 海底扩张
sea fog 海雾,平流雾
sea gravimeter 海洋重力仪
sea level 海平面
sea level specific impulse 海平面比冲
sea level thrust 海平面推力
sea marker 海上救援标志
sea rescue 海上救援
sea shock 海底地震
sea sickness 晕船
sea skimming 掠海飞行
sea smoke 海面蒸汽雾
sea state 海面状况,海洋状况
sea survival 海上生存
SEAC Signalling Engineering and Administration Center 信令工程和管理中心
SEAD Suppression of Enemy Air Defence 对敌防空体系的压制
seadrome 水上机场,海面机场
Seafac Systems Engineering Avionics Facility 系统工程航空电子设备
seal 1.密封装置 2.密封件
seal cavity 密封腔
seal clearance 缝隙油膜
seal cover 密封盖,密封罩
seal design 密封设计
seal leakage 密封泄漏
seal material 密封材料
seal structure 密封结构
sealant 密封胶(剂),密封材料
sealed 封存的,铅封的
sealed cabin 气密舱,密封座舱
sealed cabin simulator 密封舱模拟器
sealed cadmium-nickel battery 密封镉镍蓄电池
sealed cell 密封电池
sealed headpiece 密闭头盔
sealed helmet 密闭头盔
sealed lead acid battery 密封铅酸蓄电池
sealed module 密封舱
sealed room elevating operation platform 封闭空闭升降平台
sea-level 海平面
sea-level condition 海平面条件
sea-level flight 海平面飞行
sea-level mode 海平面模式
sea-level pressure 海平面大气压(力)
sea-level static thrust 海平面静态推力
sea-level test 海平面测试

sea-level thrust 海平面推力
sealing 封口
sealing riveting 密封铆接
sealing test 石蜡密封度试验
seam 缝,接缝
seam weld 滚焊,缝焊,线焊
seam welding 缝焊
seaplane 水上飞机
seaplane basin 水上飞机起落区域
seaplane landing taxiing 水上飞机着水滑行
seaplane take-off taxiing 水上飞机起飞滑行
seaplane tank 水上飞机水箱,水上飞机试验水槽
seaplane tender 水上飞机勤务支援船
seaplane trim 水上飞机调整
sear notch （航炮)扣机齿,击发卡等
search 搜索,寻找
search algorithm 搜索算法
search area 搜寻范围
search barrier 搜索屏障
search capability 搜索能力
search coil 探测线圈
search domain 检囊
search field 检索字段
search function 搜索功能
search gas 探索气体,示漏气体
search method 搜索法
search mission 搜索任务(对指定地面目标搜索的空中侦察任务)
search pattern 搜索模式
search procedure 搜寻程序
search rate 搜索频率
search space 搜索空间
search strategy 搜索策略
search technique 搜索技术
searching area 搜索区
searchlight 探照灯
Seasat 海洋资源卫星
seasoning 干燥处理
seat 座椅
seat belt 座椅安全带
seat bucket 椅盆
seat configuration （机舱)座椅布局
seat cushion 座垫
seat kilometers 客座公里
seat pack 座席式降落伞,座式伞包
seat pack dinghy 座包式充气救生船
seat pitch 座椅排距
seating 座套,基座,阀座,座位数
seating arrangement 席位安排

seating comfort   座椅的舒适性
seating configuration   座位配置
seating density   客座密度
seat-mile   客运里程
sea-water activated battery   海水激活电池
seaworthiness   适海性
SEB   1. Seat Electronic Box 座椅电子盒 2. Service Evaluation Bulletin 服务评估通告 3. Source Evaluation Board 原始资料评价委员会(英国)
SEC   Secondary Electron Conduction 二次电子导电
secant   正割,割线
secant method   正割法
secant modulus   割线模量
secant ogive   割面尖拱(以小角度过渡到圆柱体的尖拱)
SECDEF   Secretary of State for Defense 国防部长(美国)
second anode   第二阳极
second bonding   二次胶接
second cosmic velocity   第二宇宙速度
second critical speed   第二临界转速
second cut   二道纹
second generation   第二代,改进型
second hand   秒针
second harmonic generation   二次谐波产生
second in command   副驾驶(员),(民航)副机长
second law of thermodynamics   热力学第二定律
second moment of area   断面惯性矩,截面二次矩,截面惯性矩
second of arc   弧秒
second officer   二副
second order   二级的,二阶的
second order perturbation   二级微扰
second order system   二次系统
second pilot   副驾驶(员),副机长
second source   第二电源,第二资源
second stage   第二级
second surface mirror   二次表面镜
secondary   1.次级的,二次的,二要的,副的 2.伴星
secondary air   二股流
secondary airsickness   继发性空晕病
secondary area   次要地区,次铰合面
secondary battery   1.蓄电池 2.二次电池
secondary bomb damage   间接轰炸效果
secondary breakup   二次破碎
secondary cell   二次电池
secondary channel   辅助信道
secondary coil   次级线圈
secondary collapse   继发性虚脱

secondary component   辅助成分,辅助部件
secondary control   辅助控制
secondary control unit   辅助控制器,(第)二级控制器
secondary cosmic radiation   次级宇宙辐射
secondary cosmic rays   次级宇宙线
secondary damage   继发损伤,间接破坏
secondary effect   副效应,继发效应,次生效应
secondary electric power supply   辅助电源,二次电源(补充主电源用的电源或指蓄电池的贮能装置)
secondary electrical power source   二次电源
secondary electron   次级电子
secondary electron conduction   次级电子导电
secondary electron emission   二次电子放射
secondary emission   二次发射
secondary end of a session   对话次端
secondary energy ratio   次要功比值
secondary explosion   后续爆炸,间接爆炸(由空中攻击引起的地面目标的爆炸)
secondary failure   1.从属失效 2.派生故障
secondary field   次生(电)场
secondary flat   次平面
secondary flight controls   辅助飞行控制器
secondary flow   二次流
secondary flow loss   二次流损失
secondary fluid   二次流体
secondary frequency   副频率
secondary front   副锋(在斜压冷气团内形成的锋)
secondary glider   二级滑翔机,二次滑翔机
secondary great circle   次大圆,副大圆(垂直于主大圆的大圆,如子午线)
secondary impact   二次撞击
secondary ion mass spectroscopy   二次离子质谱
secondary ionization   次级电离
secondary line   支线,应急制动管路
secondary loop   次级回路
secondary loss   二次损耗
secondary magnetization   次生磁化(强度)
secondary mass flow   二次流
secondary mirror   副镜
secondary nozzle   副喷嘴
secondary optimization   次级优化
secondary peak   次级峰值
secondary phase factor   次级相位因数
secondary power   副电源
secondary power supply   二次电源
secondary power system   辅助动力系统
secondary radar   二次雷达
secondary radiation   次级辐射,二次辐射
secondary radiator   二次辐射器

secondary remanent magnetization 次生剩磁
secondary separation 二次分离
secondary spatial disorientation 继发性空间定向障碍
secondary standard 副标准,工作标准
secondary standard solar cell 二级标准太阳电池
secondary star 次星
secondary stress 次应力(由加载后的挠度引起的,如一端加载的柱弯曲时产生的应力)
secondary structure 二次结构,次级结构
secondary surveillance radar 副监视雷达,二次监视雷达
secondary target 次要目标
secondary vortex 二次涡
secondary wave 次级波
secondary winding 次级绕组,次级线圈,副线圈
second-class failure of satellite 整星二级故障
second-order loop 二阶环路
second-order system 二次系统
second-order term 二阶项
second-order transition 二级相变
secrecy capacity 保密容量
secrecy system 保密体制,保密系统
secret command 保密指令
secret communication 保密通信
secret communication system 保密通信系统
secret grade 加密等级
SECS 1. Sensor Engagement Controller Subsystem 传感器衔接控制器分系统 2. Severe Environment Control System 恶劣环境控制系统 3. Solar Electric Communication Satellite 太阳能通信卫星 4. Space Environmental Control System 空间环境控制系统
section 1. 截面,剖面 2. 段,部分(尤指机体主要部分,如机头 nose section,但 wing section 又指机翼剖面) 3. 组,小分队(尤指两架作战飞机的小分队) 4. 型材(有标准截面的原材料,区别于 sheet, billets 或 strip) 5. 舱(导弹或火箭的主要部分,如制导舱) 6. 班(比排还小的组织,在空军则指双机组.美国把陆军的班称为 squad)
section drag 翼剖面曳力
section line 管界线,截面线,区划线,剖面线
section map 断面图
section modulus 剖面模量,截面模量
section shape 截面形状
sectional area 截面积
sector 1. 扇形,扇形物 2. 区,段,部分,部门 3. (防空)防御地段,扇区 4. (雷达扫描)扇(形)区 5. 航线段(民航机在两检点之间的段) 6. (无线电指向标的)扇形空间(即 A,N 区,又称 quadrant)
sector boundary 扇形边界

sector controller 防空兵器控制官
sector display 扇形显示
sector distance 航段距离
sector gear 扇形齿轮
sector mode 区段方式
sector operations center 区域作战中心
sector radio marker 扇区无线电指向标
sector scan 扇形扫探,区域扫描
sector SETAC 扇形塔康(导航系统)
sector structure 扇形结构
sectorial harmonics 扇形谐函数
secular aberration 长期光行差
secular acceleration 长期加速度
secular changes 长期变化,长缓变化
secular motion 长期运动
secular parallax 长期视差
secular perturbation 长期摄动
secular polar motion 长期极移
secular torque 长期力矩
secure 1. 安全的,保险的 2. 保密的(如保证密码不致被敌方破译,信号不受干扰等) 3. 关断发动机 4. 夺取并保持(空中优势)
secure communication 安全通信,保密通信
secure failure set 安全故障集
secure voice communication system 保密语音通信系统
secure voice cord board 保密话音转接
security 1. 保密,保险 2. 安全,保卫 3. 系留 4. (旅客登机前的)安全检查
security equipment (通信)保密设备
security restore 安全恢复(程序)
SED Supplier Engineering Definition 供应商工程定义
sedimentation 沉降
sedimentation analysis 沉降分析法
SEDIS Surface Emitter Detection Identification System 地面发射器侦察识别系统
SEE System and Electronics Engineering 系统和电子工程
Seeback Effect 塞贝克效应
seed 导火管组件,点火燃料组件
seed crystal 晶种
seeing 视宁度,星象宁静度
seeing disk 视宁圆面
seeing image 视宁像
seeker 自导导弹,自动导引弹头
seeker azimuth orientation 导引头方位测定
seeking 1. 寻找 2. 自导引,寻的 3. 定向
seepage 渗漏,渗出量
seesaw hub 跷板式桨毂
seesaw rotor 交替转翼

see-through plate 透明板
SEF 1. Severely Errored Frame 严重误码帧 2. Support Entity Function 支持实体功能
segment 1. 弓形体，扇形体 2. (分割的)部分，片，段，块 3. 起飞航线中的航线段 4. 航线上每两个航路点之间的航线段 5. 导弹飞行中的航线段(如加速段，巡航段，末段) 6. 圆缺，球缺 7. 整流子片
segment boundary 航段范围
segment display 节段显示
segment search argument 段搜索自变量
segmentation 分段，分割，整劣片
segmented circle 扇形标志(在无塔台的机场上用这种目视标志来提供起落航线的信息)
segmented encoding law 折线编码律
segmented mirror telescope 拼合镜面望远镜
segmented solid rocket motor 分段式固体火箭发动机
segregation 偏析，隔离
segregation coefficient 分凝系数
SEI 1. Software Engineering Institute 软件工程研究所 2. Standby Engine Indicator 备用发动机指示器
Seignette salt 酒石酸钾钠，塞格涅特盐
seismic mass 地震震动质量
seismic sea wave 地震海啸，地震津波
seismic seiche 地震假潮
seismograph 地震仪，地震计
seismometer 地震仪，地震计
seismometry 测震学
seismosociology 地震社会学
seismotectonics 地震构造学
selected interchangeability 选择互换性，有限互换性
selected methods for attacking right target (用精确制导炸弹)攻击确定目标的选定方法
selection 1. 选择，选拔 2. (工作状态的)转换，(系统的)接通 3. 选择的，精选物，选集
selection criterion 选择指标
selection effect 选择效应
selection integrated evaluation 选拔综合评定
selection method 选择法
selective arming fuze 瞬发延期两用引信
selective availability 用户精度选择能力(对非美国防部规定的用户，控制全球定位系统导航精度的能力)
selective availability error 用户选择误差(对不同用户规定的全球定位系统的导航误差)
selective cavity 选择谐振器
selective channel 选择通道
selective channel input bus 选择通道输入总线
selective channel output bus 选择通道输出总线
selective chopper radiometer 选择斩波辐射计
selective combiner 选择合并器

selective compliance assembly robot arm 平面关节型机器人
selective diffusion 选择扩散
selective doping 选择掺杂
selective epitaxy 选择外延
selective fading 选择性衰落(电离层密度变化引起信号的失真，故衰落随频率而变化)
selective feathering 选择性顺桨(指飞行员操纵选择开关使螺旋桨顺桨)
selective hardening 局部淬火
selective interference 窄带干扰，选择性干扰
selective interrogation 选择询问
selective inversion 选择性反转
selective inversion technique 选择性反演技术
selective ionization 选择性游离
selective jammer 选择性干扰机
selective level meter 选频电平表
selective loading (货物的)选择性装载(安排货物装载顺序时，考虑到卸货方便)
selective oxidation 选择氧化
selective pitch 选距
selective radiator 选择性辐射体
selectivity 选择性，选择能力
Select-vision 全息磁带录像的注册名
selenium 【化】硒(化学元素，符号 Se)
seleno- (词头)1. 月亮的 2. 硒
selenocentric coordinate 月心坐标
selenodesy 月面测量学
selenograph 月面图
selenographic coordinate 月面坐标
selenographic coordinate system 月面坐标系
selenographic latitude 月面纬度
selenographic longitude 月面经度
selenography 月面学，月理学
selenoid orbit 月球卫星轨道
selenology 月球学
self alignment 自对准
self breaking time 自制动时间
self correction 自校准
self discharge 自放电
self focusing 自聚焦
self induction 自感，自感应
self information 自信息
self insurance 自保
self loading 自举电路
self locking 自锁的，自闭的
self locking nut 自锁螺母
self mode-locking 自锁模
self monitoring 自监控

self navigation	自动导航
self noise	自噪声
self psychological regulation	自我心理调节
self repairing	自检修的
self scattering	自散射
self sufficiency	自足性,独立性(指系统的)
self-absorption	自吸收
self-accurate area simulation	自准区仿真
self-acting	自动的
self-adapting computer	自适应计算机
self-adaptive	自适应的
self-adaptive control system	自适应控制系统
self-adjusting	自调整的
self-aligned isolation process	自对准隔离工艺
self-aligned MOS integrated circuit	自对准 MOS 集成电路
self-aligning bearing	自位轴承,自调心轴承
self-alignment	自动调准
self-balancing	自平衡
self-bias	1.自动偏压 2.自动偏移,自偏移 3.自偏流
self-calibration	自检校
self-centering	自动定中心
self-clearing	自动退弹
self-cocking action	(引信)自动起爆
self-compensation static calibration	自补偿静校
self-contained navigation	自主导航,自备式导航
self-contamination	自污染
self-correcting	自校正的
self-correcting wind tunnel	自修正风洞
self-demagnetization	自动去磁
self-destruct	自毁
self-destruct device	(安全)自毁装置
self-destruction	自毁
self-destruction command	自毁指令
self-destruction permissible	允许自毁
self-destruction time	自毁时间
self-destructor	自毁机构
self-diffusion	自扩散
self-directing missile	自导引导弹
self-discharge	自放电
self-discharge rate	自放电率
self-equilibrating	自平衡
self-erecting	(陀螺)自竖立,自扶正
self-excitation	自激
self-excitation oscillation	自激振荡
self-excited generator	自励发电机
self-excited vibration	自激振动
self-focusing optical fiber	自聚焦光学纤维
self-gravitation	自吸引
self-guidance	自导引
self-guided ballistic phase	弹道自导段
self-hold	自保,自持
self-ignition	自燃
self-induced vibration	自感振动
self-inductance	自身电感
self-isolation	自隔离
self-learning	自主学习
self-learning system	自学习系统
self-locking valve	自锁阀
self-lubricating material	自润滑材料
self-magnetic	自磁的
self-maintained discharge	自持放电
self-optimizing control	自寻优控制
self-organization layer	自组织层
self-organizing	自组织
self-organizing system	自组织系统
self-orientating	自定向的
self-orthogonal convolutional code	自正交卷积码
self-oscillation	自激振荡
self-oscillator	自激振荡器
self-phasing array	自相阵
self-planning	自规划
self-protection arms	自卫武器
self-protection jamming	自保护干扰
self-pulsation	自脉动
self-pulse modulation	自脉冲调制
self-quenched counter	自猝灭计数管
self-quenching	自消灭
self-recorder	自动记录器
self-recording	自记录的
self-registering	自记录的
self-repairing	自修复
self-repairing system	自修复系统
self-reproducing system	自繁殖系统
self-rescue system	自救系统
self-restoring bootstrapped drive circuit	自恢复自举驱动电路
self-routing	自选路由
self-scanned array	自扫描阵列
self-scanned imaging system	自扫描成像系统
self-scanned linear array	自扫描线性阵列
self-scanned photosensing array	自扫描光敏阵列
self-screening jamming	自卫干扰
self-screening range	1.自卫距离 2.再屏蔽距离
self-sealed riveting	自封铆接
self-sealing coupling	自封接头
self-similarity	自相似性
self-start	自动起动

self-start system	自起动系统
self-sufficiency	自给自足
self-sustained burning	自持燃烧
self-sustained combustion	自持燃烧
self-sustained detonation	自持爆炸
self-tapping screw	自攻螺钉
self-test	自检测
self-tracking automatic lock-on circuit	自跟踪自动锁定电路,自跟踪自动截获电路
self-tuning	自动调谐
self-tuning controller	自校正控制器
self-tuning regulator	自校正调节器
selsyn	同步机,自动同步机,同步装置
selvedge	(织物的)织边,布边
SEM	1. Scanning Electron Microscope 扫描电子显微镜器 2. Standard Electronic Module 标准电子组件
semiactive	半主动的
semiactive attitude stabilization	半主动姿态稳定(被动和主动姿态稳定的组合)
semiactive control	半主动控制
semiactive fuze	半主动式引信
semiactive homing guidance	半主动自导引,半主动寻的制导
semiactive proximity fuze	半主动式近炸引信,半主动式引信
semiactive tracking system	半主动跟踪系统
semi-angle	半角
semi-armor-piercing shell	半穿甲弹
semi-articulated rotor	1.半铰接式旋翼 2.半刚接式旋翼 3.跷板式旋翼
semiautomatic	半自动的
semiautomatic control	半自动控制
semiautomatic machine tool	半自动机床(给予指令后开始运转,完成一个循环后,必须再给指令,重新开始运转)
semiautomatic operation	半自动操作
semiautomatic test	半自动测试
semiautomatic test equipment	半自动测试设备
semiautomatic tracking	半自动跟踪
semiautomatic weapon	半自动武器
semi-ballistic reentry	1.半弹道式再入 2.弹道升力再入
semicantilever	半悬臂
semicircle	半圆
semicircular canal	半规管(内耳的)
semicircular deviation	半圆罗差
semiconducting glass	半导体玻璃
semiconductive ceramics	半导体陶瓷
semiconductor	半导体
semiconductor balance	半导体天平
semiconductor detector	半导体探测器
semiconductor diode	半导体二极管
semiconductor laser	半导体激光器
semiconductor sensor	半导体敏感器
semiconductor thermoelectric cooling module	半导体温差制冷电堆
semi-custom IC	半定制集成电路
semidefinite	半定的
semi-detached binary	半接双星
semidiameter	半径
semidiameter correction	半径修正
semidiurnal	半日的
semi-empirical equation	半经验方程
semi-empirical formula	半经验公式
semi-empirical model	半经验模式
semi-focal chord	半焦弦,半通径
semi-lift reentry	半升力再入
semimajor axis	(椭圆)长半径,长半轴
semimetal	半金属
semiminor axis	(椭圆)短半径,短半轴
semi-monocoque	半硬壳式
semi-monocoque structure	半硬壳式结构
semi-orbit	半轨道
semi-permanent magnetic material	半永磁材料
semi-permeable membrane	半透膜
semi-regular variable	半规则变量
semi-revolution	半周,半转
semi-rigid model	半刚性模型
semispan	半翼展
semi-submerged carriage	半埋式悬挂(导弹紧贴机身半露于机身外面)
semitrailer	半挂车
semitransparent mirror	半透明镜
semitransparent photocathode	半透明光电阴极
semivariance	半方差
SEMP	System Engineering Management Plan 系统工程管理计划,系统工程管理大纲
send	发送
send console	发送控制台
send in error	误发
send only	只发送
send-back system	回送系统
sender	1.发送器,发射机,发码器 2.天线引向器
sending node	发送节点
senior management	高级管理
senior manager	高级管理人员
senior navigator	领航主任,领航长
senior pilot	1.高级飞行员(美国空军) 2.航空兵副中

队长(英国空军)
senior supervisor 主任监督员(即维修军士长)
sensation 感觉
sense 侦测,感应,感觉
sense amplifier 读出放大器
sense amplifier circuit 读放电路
sense bit 读出位
sense indicator 检测指示器,方向指示器
sense of polarization 极化旋向
sensible atmosphere (可量度出气动阻力的)可感知大气
sensible energy 显热
sensible enthalpy 显焓
sensible heat 显热,变温热(使某物质改变温度而不改变物理状态的热)
sensible heat transfer 显热传递
sensible horizon 合理地平线,测者真地平
sensible temperature 感受温度
sensing 1.感觉 2.检测,探测,信号感受 3.判定,确定 4.测向,偏航显示 5.炸点(或弹着点)距离和方向的判定,观测射弹 6.读出
sensing device 敏感装置
sensing element 敏感元件,传感元件
sensistor 硅电阻,正温度系数热敏电阻
sensitive altimeter 灵敏高度表
sensitive area 敏感地区,敏感区
sensitive equipment 敏感负荷
sensitive material 感光材料
sensitive relay 灵敏继电器,微电流继电器
sensitive switch 微动开关
sensitive threshold 灵敏阈
sensitivity 敏感度,灵敏度(输出被输入除,即系统对单位激励的反应程度)
sensitivity analysis 灵敏度分析,敏感性分析
sensitivity coefficient 灵敏度系数
sensitivity computation 灵敏度计算
sensitivity curve 灵敏度曲线
sensitivity equation 灵敏度方程
sensitivity factor 灵敏度系数
sensitivity factor of notch 缺口敏感系数
sensitivity function 灵敏度函数
sensitivity matrix 灵敏度矩阵
sensitivity of contact fuze 触发引信灵敏度
sensitivity-time control 灵敏度时间控制
sensitization 激活,活化,使产生放射性
sensitizing 感光
sensitometry 感光测定
sensor 1.传感器,敏感元件 2.探测装置,雷达 3.接收机,接收器

sensor array 1.感测器阵列 2.阵列传感器
sensor calibration 传感器标定
sensor coverage 感知覆盖
sensor element 1.传感元件 2.读出元件
sensor failure 传感器失效
sensor fault 传感器故障
sensor fusion 传感器融合
sensor information 遥感信息
sensor location 传感器定位
sensor model 传感器模型
sensor network 传感器网络
sensor node 传感器节点
sensor noise 传感器噪声
sensor output 传感器输出
sensor reading 传感器读数
sensor response 传感器响应
sensor signal 传感器信号
sensor system 敏感元件系统
sensorimotor performance 感觉运动操作
sensorineural deafness 感觉神经性耳聋
sensory conflict hypothesis 感觉冲突假说
sensory control 传感信息控制
sensory deprivation 感觉剥夺,感觉消失
sensory discrimination 感觉辨别
sensory feedback 感觉反馈
sensory information storage 感觉信息储存
sensory perception 感知觉
sensory stimulation 感觉刺激
sentence articulation 句子清晰度
sentence error probability 误句概率
SEO Senior Executive Officer 高级执行官
SEOS Synchronous Earth Observation Satellite 同步轨道对地观测卫星(美国)
SEP 1. Signaling End Point 信令终结点 2. Spherical Error Probability 球形误差概率 3. System Engineering Process 系统工程过程
separability 可分性
separable function 可分函数
separate 分开,隔开
separate system 可分系统
separated angle of difference beam 差波束分离角
separated boundary layer 分离边界层
separated flow 分离气流
separated region 分离区域
separating force 分离力
separation 1.(气流)分离 2.(空中交通管制的)间隔 3.(火箭级的)脱离,分离 4.脱离时间 5.退役 6.(截击机与目标之间的)距离
separation area 气流分离区

separation attitude	分离姿态
separation bolt	分离螺栓
separation bubble	分离泡
separation condition	分离条件
separation connector	分离脱落式接插件,脱落插座
separation control	分离控制
separation criterion	分离准则
separation device	分离机构,分隔装置
separation distance	分离距离
separation eddy	分离涡流
separation flow	分离流动
separation line	分隔线
separation loss	分离损耗
separation method	分离方法
separation minima	最小间隔、距离或高度差
separation nut	脱落螺母,分离螺母
separation of boundary layer	边界层分离
separation of variables	变量的分离,分离变量法
separation plane	分离面
separation point	(气流)分离点
separation principle	分离原理
separation process	分离过程
separation rate	分离率
separation region	隔离区
separation standards	间隔标准(飞机间的最小距离)
separation system	分离系统
separation technique	分离技术
separation test vehicle	分离试验用飞行器
separation theorem	分离定理
separation time	分离时间
separation value	分离值
separation velocity	分离速度
separation zone	分隔带
separator	分离器,隔板,分隔标志,区分符号
separatrix	分界线
SEPC Seccondary Electrical Power Contactor 第二电源接触器	
SEPP Stress Evaluation Prediction Program 应力评估预测程序	
SEPS Solar Electric Propulsion System 太阳能电推进系统	
SEPST Solar Electric Propulsion System Technology 太阳能电推进系统技术	
septate waveguide	隔膜波导
sequence	数列,序列
sequence call	顺序呼叫,连呼
sequence division modulation telemetry	序列分割调制遥测
sequence of measurement	测定次序
sequence suspension	顺序装挂法
sequence testing	顺序测试
sequence valve	顺序活门
sequenced ejection	1.依次弹射(机上各乘员按规定次序先后离机) 2.程序弹射(弹射过程中,自动控制各个机构的动作顺序,保持一定的时间差和先后次序)
sequencer	1.程序装置 2.定序器
sequency	序数
sequential algorithm	顺序算法
sequential analysis	序贯分析
sequential collation of range	不间断校对距离,距离顺序整理系统
sequential computer	时序计算机,顺序计算机
sequential control	程序控制,时序控制
sequential decoding	序列译码,序贯译码
sequential decomposition	顺序分解
sequential design	序贯设计
sequential estimation	序列估计
sequential hierarchy	时序层次,顺序层次
sequential least squares estimation	序贯最小二乘估计
sequential lobing	时序波瓣控制
sequential optimization	顺序优化
sequential probability ratio test	序贯概率比检测
sequential sampling	序贯取样
sequential sampling inspection	序贯抽样检验
sequential system	时序系统
SERC Science & Engineering Research Council 科学和工程研究委员会	
serial	系列,串联的,连接的
serial and parallel scan	串并联扫描
serial data	串行数据
serial high-density recording	串行高密度记录
serial mode	串行方式
serial number	序号,串联数,串联号,序数
serial port	串行端口
serial program	串行程序
serial scan	串联扫描
series	1.连续,系列,序列 2.级数 3.系,族 4.串联
series expansion	级数展开
series fuzing system	串联引爆系统
series loading	串联加载
series loss	串联损耗
series machine	串行(计算)机
series production	批(量)生产
series representation	系列代表性
series resistance	串联电阻
series solution	级数解
series to parallel converter	串行到并行转换器,串并转换器

series transmission　串联传输
series-parallel fuzing system　串并联引爆系统
serrate　锯齿形的
SERV INPH　Service Interphone 维护用内话
servation　保存
serve　供应,分发,服务,符合
server　服务器
service　1.(装备的)使用 2.军种 3.(设备的)保养,维修 4.服务,局,署 5.业务,勤务,服务
service area　无线电服务范围,供电区,有效工作区,辅助面积
service ceiling　使用升限,实用升限
service channel　勤务联络通路
service check　运行检查
service competition　服务竞争(改变服务种类及水平而保持收费水平的竞争)
service contract　服务合约
service door　勤务舱门
service engineer　维护工程师
service engineering　服务工程,服务性工程
service environment spectrum　使用环境谱
service evaluation　服务评估通告
service firing　1.战斗发射,部队试射 2.使用运转
service grade　服务质量,服务等级,服务百分度,服务程度
service inspection　使用检查
service instruction　运行规程,操作规程,使用规程,维护规程,使用说明书,业务规章
service kit　成套工具箱,成套维护工具
service life　使用寿命,使用期限,营运寿命
service module　服务舱,辅助设备舱
service practice firing　实弹演习射击
service provider　业务提供
service segment　服务段
service system　服务系统
service tank　供给箱
service telephone　勤务电话
service temperature　使用温度
service test　运行试验,操作试验,性能试验,工作试验,使用检查
service time　服务时间
service tower　使用塔,勤务塔
service zone　服务区
serviceability　适用性,速役性
servicer　导弹发射准备台
service-test model　性能试验模型
servicing　维护,维修,保养
servicing operation　整备作业
servo　1.伺服系统,伺服机构 2.助力器 3.航机

servo actuator　伺服作动器,伺服执行机构
servo aeroelasticity　伺服气动弹性
servo altimeter　伺服式高度表
servo amplifier　伺服放大器
servo control　伺服控制
servo link　伺服传动装置
servo load　舵机负载
servo loop　伺服回路,伺服系统,助力操纵系统
servo mechanism　伺服机构
servo motor　伺服电机
servo rudder　伺服舵(随动舵)
servo step size　伺服步距角(跟踪输入步距角的能力)
servo system　伺服系统
servo tab　伺服调整片
servo table　伺服转台
servo test　伺服试验
servo track　伺服(磁)道,伺服轨道
servo tracking　随动跟踪
servo valve　伺服阀
servo-cylinder　伺服油缸
servodrive　伺服传动,伺服传动装置
servodyne　伺服系统的动力传动装置
servoflap　伺服操纵片,随动操纵片
servomechanism　伺服机构
servomechanism tester　伺服机构试验器
servomotor　伺服电动机,伺服马达
SES　1. Satellite Earth Station 卫星地球站 2. Ship Earth Stations 舰船地球站 3. Shuttle Engineering System 航天飞机工程系统 4. Shuttle Engineering Simulator 航天飞机工程模拟器 5. Signal Encrypt System 信号加密系统 6. Software Executor System 软件执行系统 7. Solar Environment Simulator 太阳环境模拟器 8. Space Environment Simulation 空间环境模拟 9. System Electrical Schematic 系统电气原理图
SESP　System Engineering Standards and Procedures 系统工程标准和程序
sesquiplane　翼半双翼机
session layer　会晤层,对话层,会话层
set　1.装置,设备,接收机 2.台,套,组,副,集(合) 3.使处于某种状态(或位置),调整,调谐,置位 4.安置,安排,配置 5.固定,定型,凝结,凝固 6.(应变超过弹性极限造成的)永久变形 7.制定,提出
set control　1.控制支架 2.凝结控制
set of equation　联立方程
set point　设定值,给定值
set theory　集合论
setback　延迟,阻碍,将指针拨回,逆转,电刷后边,后退,减速,下降,下插
SETI　Search for Extraterrestrial Intelligence 地外文明

探测计划

SETOLS Surface Effect Take-Off and Landing System 地面效应起飞着陆系统

Set-on 安置，就位

SETP Society of Experimental Test Pilots 试飞员协会（美国）

setpoint 定位点

set-reset flip-flop 置位-复位触发器

SETS Synchronous Equipment Timing Source 同步设备定时源

setting 1.安装，设置 2.调整，调定，装定（在某一位置）3.调定值，调定位置，（机翼，襟翼，尾翼等的）安装角 4.底座，底脚 5.环境，背景 6.凝固

setting accuracy 定位精度，调整精度

setting angle 定位角，装置角

setting circle 定位度盘，度盘调整

setting hammer 压印锤

setting-out of junction 交叉点放样，旋放样

setting-out of main axis 主轴线测设

setting-out survey 放样测量

settling 起飞下沉（直升机起飞时，由于旋翼失去地面效应而造成的）

settling chamber 稳定段

settling time 调整时间，过渡过程时间，（瞄准具）稳环时间

set-top box 置顶盒

setup 1.装置，设备 2.安装，装配 3.调整，调定 4.计划，方案 5.机构，组织体系

SEU 1. Seat Electronics Unit 座椅电子组件 2. Sensor Electronics Unit 传感器电子装置

seven holes probe 七孔探头

seven pipe connector 七管连接器

severe damage 严重破坏

severe storm 暴风

severe turbulence 严重颠簸

severe type 重型

severe weather 恶劣天气

severity 严重程度，严肃，严格，厉害，猛烈

SEWT Simulator for Electronic Warfare Training 电子战训练模拟器

sexadecimal 十六进制的

sexagesimal 六十进制的

sextant 六分仪

sextet 六位字节

sextuple star 六合星

SF 1. Safety Factor 安全系数 2. Sequence Flasher 顺序闪光灯 3. Single Failure 单个故障 4. Spot Face 局部整面 5. Square Feet 平方英尺

SFAR Special Federal Aviation Regulations 特别联邦航空条例

SFC 1. Secondary Flow Control 二股气流控制 2. Shop Floor Control 车间现场控制，车间作业控制 3. Side Force Control 侧力控制 4. Specific Fuel Consumption 具体的燃油消耗量 5. Synchronous Forward Command 同步前向命令

SFCC 1. Side Facing Crew Cockpit 有侧座机组的驾驶舱（指有飞行工程师仪表板的）2. Slat/Flap Control Computer 缝翼/襟翼控制计算机

SFCG Space Frequency Coordination Group 空间频率协调组

SFCS 1. Safety Flight Control System 安全飞（行）控（制）系统 2. Secondary Fight Control System 辅助飞（行）控（制）制系统 3. Survivable Flight Control System 应急飞（行）控（制）系统

SFDF Subsystem Fault Detection Function 子系统故障检测功能

SFE Seller Furnished Equipment 卖方配置设备，供应商配置设备

SFEA Survival and Flight Equipment Association 生存和飞行设备协会

sferics 大气无线电干扰研究，天电学，天电测定法

SFHA System Functional Hazard Assessment 系统功能危险评估

SFI 1. Service Feature Instance 业务特征事例 2. Synthetic Flight Instructor 合成飞行训练仪

SFIT System Functionality Integration Team 系统功能集成组

SFL 1. Safe Fatigue Life 安全疲劳寿命 2. Sequenced Flashing Lights 顺序闪光灯 3. Substrate Fed Logic 衬底馈电逻辑

SFOF Space Flight Operations Facility 空间飞行操作工具

SFP 1. Selected Flight Path 预选航线 2. Simplified Flight Planning 简化飞行计划 3. Simulated Flight Plan 模拟飞行计划 4. Space Flight Project 航天计划，航天工程 5. Special Flight Permit 航空器特许飞行证

SFPCA Society For the Preservation of Commercial Aircraft 商用飞机保护协会（美国）

SFPM Surface Feet Per Minute 每分钟表面英尺

SFTP System Functional Test Procedure 系统功能试验程序

SFTR System Functional Test Report 系统功能试验报告

SFTS 1. Service Flying Training School 军种飞行训练学校 2. Spaceflight Talecommunications System 航天通信系统 3. Synthetic Flight Trainig System 综合飞行训练系统

SFU 1. Standby Filter Unit 备用过滤器单元 2. Store

and Forward Unit 存储转发单元
SFX Sound Effects 音响效果,声效应
SGC Signaling Grouping Channel 信令分组信道
SGD Structured General Document 结构化文档
SGEMP System Generated Electromagnetic Pulse 系统本身产生的电磁脉冲
SGJP Satellite Graphic Job Processor 卫星图形作业处理程序
SGL 1. Signal 信号 2. Single 单一的
S-glass fibre S(型)玻璃纤维
SGML 1. Standard Generalized Mark-up Language 标准通用标记语言 2. Standard Graphic Mark-up Language 特征语言
SGS Saint-Gobain-Sully (Windshields and Opening Windows) 赛特-古班-苏里公司(风档玻璃和通风窗)
SGU Symbol Generator Unit 字符产生器部件
SH 1. Sample and Hold 试用和保持 2. Sheet 张,图表 3. Shilling 先令(货币单位) 4. Showers 阵雨(国际民航组织)
shattering effect 破裂效应
shade 遮光板
shaded-pole motor 罩极电动机
shading 阴影
shadow 阴影,阴影照片
shadow band 影带
shadow cone 影锥
shadow definer 影符
shadow factor 荫蔽因数,阴影系数,阴影因子
shadow factory (经伪装的)军工厂
shadow mask 障板,阴罩
shadow price 影子价格
shadow zone 阴影区
shadowgraph 阴影照片,阴影图
shadowgraph system 阴影照相系统
shadowgraph technique 阴影法
shadowing analysis 阴影分析
shaft 轴,杆
shaft angle 轴交角
shaft horsepower 轴马力,轴功率
shaft power 轴功率
shaft turbine (涡轮轴发动机的)动力涡轮
shaft vibration 轴振动
shaker 振动器,振动试验器
shallow discharge 表面放电
shallow energy level 浅能级
shallow junction technology 浅结工艺
shallow shell 扁壳
shallow water 浅水区
Shannon entropy 香农熵
Shannon equation 香农公式
Shannon theory 香农理论
shape 形状,外形,模型,型材,形成,成型
shape change 形变
shape control 形状控制
shape design 外形设计
shape error 形状误差
shape factor 形状因数
shape function 形状函数
shape memory alloy 形状记忆合金
shape optimization 形状优化
shape oscillation 形状振荡
shape parameter 形状参数
shape transition 形状过渡
shaped charge 火箭蜂窝形药柱,破甲装药
shaped charge liner 锥形装药填料
shaped demolition charge 锥形爆破装药,空心爆破装药
shaped-charge armor-penetrating warhead 聚能装药穿甲弹头
shape-memory polymer 形状记忆高分子
shaper 1. 牛头刨床 2. 整形器 3. 脉冲形成电路
shaping 刨削
shaping circuit 整形电路
shaping network 整形网络
shaping time constant 成形时间常数
share 共享,均分,分摊
shared memory 共享存储器
sharing 分配,分摊
sharing of information 信息共享
sharing resource 共享资源
shark chaser 驱鲨剂
shark deterrent 防鲨物
sharp corner 锐角
sharp decrease 急剧下降
sharp edge 锐边
sharp image 清晰影像
sharp increase 急剧上升
sharp pulse 尖脉冲
sharp step 陡阶跃
sharp tuning 微调,细调
sharp-edged gust 突发阵风
sharpen 削尖,磨快
sharpening 锐化
sharpness 1. 锐度 2. 清晰度
shaving 刮削,剃齿
Shavit 彗星号(以色列运载火箭)
SHC Server Headend Control 服务器前端控制
SHEA Safety, Health and Environmental Affairs 安全、

健康和环境事务
shear 剪切,切变
shear angle 剪切角
shear band 剪切带
shear buckling 剪切失稳,剪切翘曲
shear center 剪切中心,剪心
shear coaxial injector 剪切同轴喷嘴
shear creep 剪切蠕变
shear deformation 剪切应变
shear dislocation 剪切位错
shear flow 剪切流
shear force 剪力,剪切力
shear lag 剪切迟滞
shear layer 剪切层
shear load 剪切载荷
shear melting 剪切熔融
shear modulus 剪切模数,剪切弹性系数
shear nut 拉剪螺母
shear plan (机身等的)切面放样
shear rate 剪切速率
shear slide 剪切滑坡
shear spinning 1.变薄旋压 2.强力旋压
shear stiffness 剪切刚度
shear strength 剪切强度,抗剪强度
shear stress 剪(切)应力
shear test 剪切试验
shear wave 等容波,切变波
shearing 剪断
shearing elastic modulus 剪切弹性模量
shear-thinning 剪切变稀
sheath 鞘,护套,(电子管)屏极,(电缆)铠装
shed 卸载,机库,工棚
shed vortex 脱体流
shedding 脱落,蜕落,电负荷卸载
sheet 膜,层,薄板,张,片,单子,表格,图纸
sheet dimension 图幅尺寸
sheet forming 板料成形
sheet resistance 薄层电阻
sheet thickness 薄板厚度
sheet-metal forming 钣金成形
shelf 货架,器材架,贮存架
shelf life 搁置寿命,贮藏期限,适用期
shelf support 搁板销,搁板座
shell 1.壳,壳体,套,罩 2.(常指口径大于29毫米的)炮弹
shell element 壳体单元
shell fragment 炮弹碎片
shell mold casting 壳型铸造
shell star 气壳星

shell structure 壳结构
shell theory 壳体理论
shell tracer 炮弹曳光剂
shellac 虫胶,紫胶,虫胶清漆,天然清漆
shell-destroying tracer 炮弹曳光引爆剂
shell-proof 防弹的
shelter 掩蔽所,掩蔽部,掩蔽,保护
Sherardizing 粉末镀锌,粉镀(一种防腐蚀处理技术)
SHF Super High Frequency 超高频
SHG Second Harmonic Generation 次谐波产生
shield 盾,防护物
shield cable 屏蔽电缆,铠装电缆,带护套的钢索
shield case 屏蔽罩
shielded bearing 屏蔽轴承,防污轴承
shielding 防护,屏蔽
shielding angle 遮蔽角
shielding efficacy measurement of shielding room 屏蔽室屏蔽效能测量
shielding factor 屏蔽系数
shielding ground 屏蔽地线
shielding structure 屏蔽结构
shift 位移,移动,转接,转换,移泊,值班,班
shift factor 移位因子
shift function 移位功能
shift reaction 变换反应
shift register 移位寄存器
shift-in character 移入字符(计算机代码扩充字符)
shim 填隙片,垫片
shimmy 摇晃,摆动,跳动,振动
shimmy damper 减摆器
ship 1.船,飞船 2.运送,船运
ship motion 船舶运动
ship position accuracy 船位精度
shipboard 船舷
shipbody deformation 船体变形
shipborne aircraft landing system 舰载飞机着陆系统
shipborne radar 船载雷达
shipment 船装货,发货
shipment element 装船要素
shipper 托运人,发货人
ship-to-air missile 舰空导弹
ship-to-ship missile 舰对舰导弹
shipyard 造船厂,修船厂
SHM 1. Shared Memory 共享存储器 2. Simple Harmonic Motion 简谐运动 3. Structural Health Monitoring 结构安全监控
shock 激波,冲波,冲击波,冲击
shock absorber 1.缓冲装置 2.减震器
shock absorption 减震,消振,缓冲

shock acceleration	冲击加速度
shock angle	冲击角
shock capturing method	激波捕捉法
shock closer	冲击闭合器
shock compression	激波内压缩
shock cord	减震橡筋绳,弹力索,缓冲绳
shock damage	冲击损伤
shock environment	冲击环境
shock excitation	冲击激励
shock expansion method	激波膨胀波法
shock fitting	激波拟合
shock fitting method	激波装配法
shock front	激波阵面,激波波前,激波锋
shock height	冲击高度
shock interaction	激波相交
shock isolator	震动隔离器,减震器,冲击隔离器,隔振器
shock layer	激波层
shock load	冲击负荷
shock location	冲击位置
shock loss	激波损失
shock machine	冲击机
shock motion	冲击运动
shock mount	隔震座架,防振安装座,减震器
shock noise	冲击噪声
shock oscillation	激波振荡
shock overpressure	激波超压,冲击波阵面超压
shock polar	激波极曲线
shock position	冲波位置
shock pressure	冲击压力
shock proof	防震的
shock pulse	冲击脉冲
shock reflection	激波反射
shock resistant	抗震的
shock response spectrum	冲击响应谱
shock smearing	激波涂抹
shock spectrum	激振频谱
shock stall	冲击波失速,激波失速
shock strength	激波强度
shock strength parameter	冲击强度参数
shock strut	缓冲支柱,减震支柱
shock test	冲击试验,振动试验,抗震试验
shock test machine	冲击试验机
shock tube	激波管,震动管
shock tunnel	激波风洞
shock velocity	激波速度
shock vibration	冲击振动
shock wave	1.激波 2.冲击波
shock wave angle	波角
shock wave front	激波波前,激波锋
shock wave structure	震波结构
shock wave-boundary layer interaction	激波-边界层干扰
shock-absorber	减震器,缓冲器,(起落架的)缓冲支柱
shock-expansion	震波膨胀
shock-expansion method	震波膨胀法
shock-layer	激波层
shock-tube	冲击波管
shockwave	冲(击)波,激波
shockwave reflection	冲击波反射
SHODOP	Short Range Doppler 近程多普勒导弹弹道测量系统
shoe	垫块,滑块,承座,包板,插座,桩靴,闸瓦,履带节
shoot	斜沟,送料糟
shooting method	射击法
shooting range	射击场,靶场
shooting star	流星
shop maintenance	车间维修
shop replaceable unit	车间可更换单元
shop visit rate	返修率
SHOPO	Supplier History and Open Purchase Orders 供应商历史和公开采购订单
Shoran	Short Range Navigation 短程导航,近程导航,近距离导航系统
Shore hardness	回跳硬度,肖氏硬度
shoreline	海岸线
shore-to-ship missile	岸舰导弹
short	短(用于表示所需的进近类型)
short axis	短轴
short base-line interferometer	短基线干涉仪
short burst	短速射,(高射炮)短点射
short channel effect	短沟效应
short circuit current	短路电流
short circuit current density	短路电流密度
short circuit protection	短路保护
short circuit termination	短路终端
short circuit turn pickoff	短距匝式传感器
short distance arming	短距离武装
short duration	短期间
short duration acceleration	短时间加速度
short haul	短程飞机
short period perturbation	短周期扰动
short period variable	短周期变量
short periodic perturbation	短周期摄动
short pulse	短脉冲
short range	近程
short range ballistic missile	近程弹道导弹
short round	短头弹

short takeoff 短距起飞
short takeoff and landing 短距起飞和着陆,短距起落
short takeoff and landing airplane 短距起落飞机
short term memory 短时记忆
short time horizon coordination 短时程协调
short trouble 短路故障
short wave fadeout 短波衰落
short-baseline system 短基线测轨系统
short-circuited 短路的
shortcoming 短处
shorten 弄短,变短
shortened 1.短路的 2.缩短的
shortest path 最短航迹
shortest path problem 最短路径问题
shortest route 最短路径
shortfall 不足之量,差额
short-haul 短距离的
short-period mode 短周期模态
short-period oscillation 短周期振动
short-range coverage acquisition 近距截获(目标)
short-range dogfight missile 近距格斗导弹
short-range guidance 近距制导
short-range navigation 近程导航
short-range transport 1.短程运输 2.(在正常巡航条件下飞行距离在1 200海里以下的)短程运输机
short-term frequency stability 短期频率稳定度
short-term prediction 短时预测
short-term stability 短期稳定度
shortwave 短波
short-wave communication 短波通信
short-wave infrared 短波红外
shot 1.发射,射击,投放 2.(火箭发动机)起动 3.(火箭、导弹的)飞行 4.弹丸,子弹(尤指猎枪子弹),炮弹 5.射程,范围 6.拍摄
shot blasting 喷丸清理,喷丸处理
shot cleaning 喷丸清理(工艺)
shot mark 弹痕
shot noise 散粒效应噪声
shot pattern 射弹散布
shot peen forming 喷丸成形
shot proof 防弹的
shot term repeatability 短期重复性
shot-peening 喷丸强化
shoulder 肩,侧安全道,道肩
shoulder belt 1.肩带 2.轴肩螺栓,杆肩螺栓(螺纹部分直径略小于杆部,用于固定塑料件及不可过紧的零件)
shoulder harness 肩背带,肩保险带
shoulder safety belt 肩部安全带

shoulder wing 高翼
show 显示,展出
show fight 表现斗志
shower counter 簇射计数器
shower particle 簇射粒子
showerhead 簇射头(液体燃料火箭喷嘴)
shrapnel 1.炮弹碎片,弹片 2.榴霰弹,子母弹
shrink fit 冷缩配合,热压配合
shrink rule 缩尺收缩定律
shrinkage 1.(铸件的)冷缩 2.收缩,收缩量,缩率
shrinkage allowance 收缩余量,允许收缩量
shrinkage crack 缩裂
shrinkage strain 收缩应变
shroud 1.罩,护罩,(火箭)整流罩 2.屏,屏蔽,幕 3.盖板,护板 4.(叶片的)冠,围带
shroud test 包罩试验
shrouded blade 带冠叶片
shrouded fins (炸弹)环形安定器
shrouded impeller 封闭式叶轮(离心压气机叶轮)
shuddering 抖动
shunt 分流,分路,并联,并联的,并励的,并联电阻,分流器
shunt connection 分流连接
shunt plug 分流器插头
shunt regulator 并联调整器
shunt resistance 并联电阻
shunt resistor 分流电阻(器)
shut off valve 截璃,闭锁阀
shutdown 1.关车 2.(故障造成)空中停车
shutdown point mass 关机点质量
shutoff 关闭装置
shutter 1.(照相机的)快门,光闸,光阀 2.百叶窗,节气门,(脉冲式喷气发动机的)片式活门,(可调喷口的)调节片 3.断路器
shutter efficiency 快门效率
shuttle 1.航天飞机 2.梭子,往复移送装置
shuttle bombing 穿梭轰炸
shuttle flight experiment 往返飞行实验
shuttle imaging radar 航天飞机成像雷达
shuttle imaging spectrometer 航天飞机成像光谱仪
shuttle remote manipulator system 往返飞行器遥控系统
shuttle service 短程运输
shuttle valve 选择滑阀
SI 1. Selective Interrogation 选择询问 2. Servicing Instruction 维护(使用)说明书 3. Simple Interest 单利 4. Software Identification 软件标识 5. Special Identification 特殊标识 6. Speed Intervention 速度干预 7. Standby Instruments 备用仪表 8. Straight In

(approach)直接进场(着陆) 9. Stereo Imaging 立体成像 10. Substitute Item 代用项 11. Supporting Interrogator 支援问答器 12. System Integration 系统综合

SIA 1. Singapore International Airlines 新加坡国际航空公司 2. Standard Instument Approach 标准仪表进近 3. System Interconnect Assembly 系统内连组件

SIAM Signal Information And Monitoring 信号信息和监控

SIB 1. Security Information Base 安全信息库 2. Super Information Base 超级信息库

SIC 1. Simple Interference Cancellation 简单干扰消除 2. System Isolation Contactor 系统隔离接触器

SIC&DH Scientific Instrument Computer and Data Handling 科学仪器计算机和数据处理

SID 1. Selected Item Drawing 选择项图表 2. Simulator Interface Device 模拟器接口装置 3. Software Interface Drawing 软件接口图 4. Spares Input Document 备件输入文档 5. Standard Instrument Departure 标准仪表离场 6. Structure Inspection Document 结构检查文件 7. Sudden Ionospheric Disturbance 意外的电离层干扰

side 边,侧
side burning 侧面燃烧
side car （飞艇的）侧吊舱
side clearway 侧净空
side control stick 1. 侧置驾驶杆 2. 侧杆
side crosstalk 半面串音
side elevation 侧视图
side float 侧浮筒,稳定浮筒
side force 侧力
side gun 机身外侧炮,机身外侧机枪
side hole 边孔
side intersection 侧方交会
side lap 侧边搭接
side load 边荷载
side lobe 旁瓣
side looking radar 侧视雷达
side number 舷号
side overlap 旁向重叠
side ramp 舷门跳板
side slip 侧滑
side tone 侧音
side tone ranging 侧音测距
side view 侧视图
side wall 侧壁
side wind 侧风
sideband 边带,边频率
side-by-side 并列式布置
sideline noise 边线噪声

sidelobe blanking 旁瓣对隐,旁瓣抑制
sidelobe cancellation 旁瓣对消,旁瓣抑制
side-looking 侧视
side-looking radar 1. 侧视雷达 2. 旁视雷达
side-looking sonar 侧视声呐
sidereal 恒星的
sidereal clock 恒星钟
sidereal day 恒星日
sidereal hour angle 恒星时角,春分点时角
sidereal month 恒星月
sidereal period 恒星周期,运行周期
sidereal system 恒星系
sidereal time 恒星时
sidereal year 恒星年
siderostat 定星镜
sideslip 侧滑(尤指内侧滑)
sideslip angle 侧滑角
sidestep 向旁侧避让,回避
sidestick 压边条
sidewall boundary layer 侧壁边界层
sidewall compression 侧压

SIDL System Identification List 系统识别表

SIDS 1. Sensor Integration Display System 传感综合显示系统 2. Standard Instrument Departure System 标准仪表离场飞行系统 3. Stellar Inertial Doppler System 天文惯性多普勒导航系统

SIE Single Instruction Execute 单指令执行

sieve 筛
sieve analysis 筛分析法

SIF 1. Selective Identification Facility(Feature)选择识别装置(特性) 2. Spares Information File 备件信息文件 3. Standard Integration Filter 标准组合滤波器 4. Standard Interchange Format 标准相互交换格式 5. System Integration Facility 系统综合设施

SIG 1. Signal 信号 2. Signature 特征 3. Significant 主要的,重要的 4. Simplified Inertial Guidance 简易惯性制导,简易惯性导航 5. Stellar Inertial Guidance 天文惯性导航,天文惯性制导

signal 信号
sight 1. 瞄准,观测 2. 瞄准具,表尺,观测器 3. 视力,视觉 4. 视界,视野
sight control 瞄准具操纵装置
sight fire control 瞄准射击控制
sight gauge 观测计(一种简单的观测液面高低的器具)
sight graduation 瞄准具分划
sight line coordinates 瞄准线坐标系
sight tracking line 瞄准跟踪线
sight tube 观察孔
sight-guided missile 光学瞄准制导导弹

sighting 1. 瞄准，观测 2. 发现（目标）
sighting centring 照准点归心
sighting diagram for bombing 轰炸瞄准图
sighting display unit 瞄准显示装置
sighting distance 视距
sighting line method 瞄直法
sighting pendant 观测吊架，瞄准锤
sighting point 照准点
sighting principle of absolute motion 绝对运动瞄准原理
sighting principle of relative motion 相对运动瞄准原理
sighting station 瞄准站
sighting target 视标
sighting-tube 望筒，窥管
sightline spin rate 瞄准线旋转角速度
sign 符号，标记，记号
sign bit 符号位
sign pattern 记号模型
signal 信号，符号，指令，预兆，发信号，发暗号
signal amplitude 信号幅度
signal analyzer 信号分析仪
signal area 信号区
signal attenuation 信号衰减
signal beam 信号束
signal board 信号板
signal cable for loading 加注信号电缆
signal code 信号代码
signal conditioner 信号调节器
signal conditioning 信号处理
signal conditioning equipment 信号适配设备
signal conversion 信号转换
signal converter 信号转换器
signal cover 信号覆盖范围
signal data 信号数据
signal data converter 信号数据转换器
signal delay 信号延迟
signal detection and estimation 信号检测估计
signal diode 信号二极管
signal distributor 信号分配器
signal electrode 信号电极（控制蓄电池充电程度的辅助电极）
signal element 信号码元，信号单元
signal enabling 启动信号，允许信号
signal encrypt system 信号加密系统
signal environment density 信号环境密度
signal fading 信号衰减
signal fading badly 信号严重衰减
signal film 信号胶片
signal flow graph 信号流图

signal format 信号格式
signal generator 信号发生器
signal ground 信号用接地
signal imitation 信号模仿，假信号响应
signal intelligence 通信情报
signal intensity 信号强度
signal interface 信号接口
signal interface unit 信号接口装置
signal leading edge 信号前沿
signal level 信号电平
signal/noise ratio 信号噪声比
signal node 信号结（点）
signal normalization 信号标准化，信号整形
signal overflow 信号溢出
signal panel 信号显示（面）板
signal pattern 信号模式
signal phase 信号相位
signal pistol 信号枪
signal plane 信号层
signal preservation 信号保存
signal processing 信号处理
signal processing element 信号处理单元
signal processor 信号处理机
signal pulsing 信号脉冲调制
signal quality detector 信号质量检测
signal race 信号追赶
signal rate 信号速率，信号发送速率
signal ratio 信号比
signal reading 信号读出
signal reception 信号接收
signal reconstruction 信号重构
signal regeneration 信号再生，信号整形
signal reshaping 信号再生，信号重新形成
signal resolver 信号分解器
signal return 信号返回
signal separation filter 信号分离滤波器
signal set 信号组
signal shaping 信号整形
signal simulator 模拟信号源
signal slip ring 信号模拟器
signal sorting 信号分类
signal source 信号源
signal space 信号空间
signal strength 信号强度
signal swing 信号摆幅
signal to clutter ratio 信号杂波比，信杂比
signal to crosstalk ratio 信号串扰比
signal to distortion ratio 信号失真比
signal to interference ratio 信号干扰比

signal to jamming ratio	信号干扰比
signal to noise ratio	随机杂波信杂比,信号噪声比
signal tracing	信号追踪
signal trailing edge	信号后沿
signal transducer	信号转换器
signal transition	信号变换,信号跃迁,信号过渡
signal unit	信号单位,信号元
signal unit format	信号单元格式
signal unit indicator	信号单元指示符
signalling	发信号,信号传输
signalling exchange	信令交换机
signalling rate	信号(发送)速率,信号发送速度
signalling sign	信号(发送)符号
signalling time slot	信令时隙
signalman	信号工
signal-noise ratio	信噪比
signal-plus-noise to noise ratio	信号加噪声对噪声之比
signatory	签署者,签字人
signature	1.性能,特性 2.签字
signature identification	特征标识
significance bit	有效位
significance level	显著性水平
significant change	显著变化
significant difference	显著性差异
significant impact	重大影响
significant improvement	显著改善
significant problem report	重大问题报告
significant result	显著性结果
significant weather	显著天气,重要天气

SIGS Stellar Inertial Guidance System 星座惯性制导系统

SIIP Structural Inspection Item Program 结构检查项目方案

SIL 1. Service Information Letter 服务信息函,服务情况信件 2. Speech Interference Level 语言干扰级别 3. System Integration Laboratory 系统综合实验室

silane	硅烷
silastic	硅橡胶
silencer	消声器,噪声抑制器
silent aircraft	静音飞机
silent practice	(高炮)哑射
silent target	无声目标
silent weapon	无声武器
silica	硅氧,二氧化硅
silica colloidal polishing	二氧化硅乳胶抛光
silica fibre	二氧化硅纤维,硅石纤维
silica gel	硅胶
silica glass	石英玻璃
silica layer	硅氧层
silica particle	二氧化硅粒子
silica phenolic	硅酚醛
silica powder	硅微粉
silica shell	二氧化硅壳层
silica window	石英玻璃观察孔
silicate	硅酸盐
silicate paint	硅酸盐漆
silicide	硅化物
silicomethane	甲硅烷
silicon	【化】硅(化学元素,符号 Si)
silicon anodization	硅阳极(氧)化处理,硅阳极电镀
silicon assembler	硅汇编程序
silicon avalanche photodiodes	硅雪崩光敏二极管
silicon carbide	碳化硅
silicon carbide fiber reinforced silicon carbide matrix composite	碳化硅纤维增强碳化硅复合材料
silicon chip	硅(基)片,硅芯片
silicon compiler	硅版编辑器
silicon controlled rectifier	可控硅整流器
silicon detector	硅二极管,硅检波器
silicon die	硅片
silicon diode	硅二极管
silicon dioxide	二氧化硅
silicon doping	硅掺杂,硅添加杂质
silicon foundry	硅铸造,硅制作
silicon gate	硅栅
silicon gate complementary MOS	硅栅互补金属氧化半导体,硅栅互补 MOS
silicon gate N-channel technique	硅栅 N 沟道技术
silicon gate self-aligned technology	硅栅自对准工艺
silicon hydride	硅烷
silicon monosulfide	一硫化硅
silicon monoxide	一氧化硅
silicon nitride	氮化硅
silicon on insulator	绝缘体上硅薄膜
silicon on sapphire	蓝宝石上硅薄膜
silicon oxynitride	氮氧化硅
silicon photo transistor	硅光电晶体管
silicon photodiode	硅光电二极管
silicon photoelectric diode	硅光电二极管
silicon precision alloy transistor	硅精密合金晶体管
silicon rectifier	硅整流器
silicon ribbon growth	硅带生长
silicon semiconductor triode	硅半导体三极管
silicon solar cell	硅太阳(能)电池
silicon target vidicon	硅靶视像管
silicon transistor	硅晶体管
silicon transistor logic circuit	硅晶体管逻辑电路
silicon wafer	硅片

silicone 1.硅酮 2.硅有机树脂硅酮,有机硅树脂,硅树脂
silicone adhesive 有机硅黏合剂,硅酮胶
silicone coating 硅酮涂层
silicone grease 硅脂
silicone oil 硅油
silicone polymer 硅酮聚合物,硅氧烷聚合物
silicone resin 硅氧树脂
silicone resin paint 硅树脂涂料
silicone rubber 硅橡胶
silicone varnish 硅酮树脂清漆
siliconizing 扩散渗硅,硅化处理
silk-screen printing 丝网印刷
silo 1.筒仓,仓库 2.导弹地下仓库,发射井
siloxane 硅氧烷
silumin 硅铝明,硅铝合金
silver 【化】银(化学元素,符号 Ag)
silver screen 银幕
silver solder 银焊料
silver-disk pyrheliometer 银盘日射强度计
silvering 银纹(有机玻璃上的微裂纹)
silylation 甲硅烷基化
SIM 1. Satellite Interpretation Message 卫星判读信息 2. Synchro Interface Module 同步接口模块 3. System Implementation Methodology (Baan technique) 系统实现方法(巴恩技术)
sim 1. simulation 模拟 2. simulator 模拟器
SIMA Slot Interleaved Multiple Access 时隙交错多址接入
SIMDS Single Instruction Multiple Data Stream 单指令多数据(流)
similar analysis 相似性分析
similar condition 相似条件
similar flow 相似流动
similarity 相似,相似性
similarity criterion 相似性准则
similarity law 相似法则,相似律
similarity parameter 相似参数
similarity solution 相似解
similarity theory 相似理论,相似原理
similitude 相似
SIMP 1. Satellite Information Message Protocol 卫星信息报文协议 2. Satellite Interface Message Processor 卫星接口报文处理机
simple 单纯的,简单的
simple harmonic quantity 正弦量
simple lens 单透镜
simple radial equilibrium 简单径向平衡
simple stress 简单应力,单纯应力(只有拉伸、压缩或剪切)
simple substance 单质
simple system 单一系统
simplex 单通道系统,单工(通道),无余度系统
simplex algorithm 单纯形算法
simplex burner 单路燃油喷嘴
simplex method 单纯形法
simplicity 操作简单
simplification 简化,化简
simplified inertial guidance system 简易惯性制导系统
simplified model of flutter 颤振简化模型
simplify 简化,精简
simply supported beam 简单支撑梁
SIMS Single Instruction Multiple Data Streams 单指令多数据流
SIMUL 1. Simulation 模拟 2. Simultaneously 同时地
simulant 模拟的,拟态的
simulated aircraft 仿真飞行器
simulated altitude test facility 模拟高空试验设备
simulated attack profile 模拟攻击概览图
simulated data reduction program 模拟数据压缩程序
simulated drill 仿真演练
simulated drop 模拟投放
simulated flight 1.仿真飞行 2.模拟飞行
simulated forced landing 模拟强迫着陆,模拟迫降
simulated human 模拟人体
simulated launch 仿真发射
simulated line 仿真线
simulated motion 模拟运动
simulated operation 模拟操作
simulated result 模拟结果
simulated satellite flight program 卫星仿真飞行程序
simulated weightlessness training 模拟失重训练
simulation 仿真,模拟
simulation algorithm library 仿真算法库
simulation architecture 仿真体系结构
simulation clock 仿真时钟
simulation code 模拟码
simulation environment 仿真环境
simulation experiment 模拟试验
simulation experiment mode library 仿真实验模式库
simulation facility 模拟装置
simulation graphic library 仿真图形库
simulation job 模拟作业
simulation knowledge base 仿真知识库
simulation laboratory 仿真实验室
simulation method 模拟法
simulation model library 仿真模型库
simulation of space debris 太空碎片仿真

simulation prediction	仿真预测
simulation process time	仿真过程时间
simulation program	仿真程序
simulation result	仿真结果
simulation run	仿真运行
simulation scenario	仿真想定
simulation software	仿真软件,模拟软件
simulation study	仿真研究,模拟研究
simulation support system	仿真支持系统
simulation system	仿真系统,模拟系统
simulation technique	仿真技术
simulation technology	仿真技术
simulation test	模拟试验,仿真试验
simulation time	仿真时间
simulation tool	模拟工具
simulator	1.模拟器(尤指飞机或导弹的飞行模拟装置),模拟装置 2.练习器
simulator induced syndrome	模拟器病
simulator model	模拟装置模型
simulator sickness	模拟器病
simultaneous differential equation	联立微分方程
simultaneous observation	同步观测同时观测
simultaneous optimization	同步优化
simultaneous track processor	并行轨道处理机
SIN Satellite Inspection Network	卫星检测网
SINCGARS Single Channel Ground and Airborne Radio Subsystem	单通道地空无线电分系统
sine	正弦
sine dwell test	正弦定频试验
sine sweep test	正弦扫描试验
sine tuning	正弦调谐
sine vibration test	正弦振动试验
sine wave	正弦波
singing margin	振鸣稳定度,(自)振鸣边际
single	单一的,单次的,唯一的
single attribute	单属性
single axle table	单轴转台
single bit	单码位
single bit error	单比特错误
single board processor	单板处理机
single chamber dual thrust rocket engine	单室双推力火箭发动机
single chamber liquid rocket engine	单推力室液体火箭发动机
single channel per burst	每分帧单路
single channel per carrier	每路单载波,单路单载波
single chip microcomputer	单片微型计算机
single chip processor	单片处理机,单片机
single command	单一指令
single component	单组分,单个部件
single component balance	单分量天平
single crystal	单晶
single crystal blade	单晶叶片
single crystal casting	单晶铸造
single crystal superalloy	单晶高温合金
single crystalline silicon solar cell	单晶硅太阳电池
single curvature	单曲率
single domain particle	单畴颗粒
single end	单端,单向
single event burnout	单粒子烧毁事件
single event effect	单粒子事件效应
single event functional interrupt	单粒子功能中断事件
single event latchup	单粒子锁定事件
single event multiple bit upset	单粒子多位翻转
single event upset	单粒子翻转事件
single expansion ramp nozzle	单膨胀斜面喷管
single fault	单一故障
single filament	单丝
single firing	一次烧成
single frequency	单频
single frequency laser	单频激光器
single hard error	单粒子硬错误
single heterojunction laser	单异质结激光器
single hop	单跳电路
single level process	单级过程
single loop control	单回路控制
single loop coordination strategy	单环协调策略
single measurement	单次测量
single mode operation	单模工作
single module	单模块
single pair	单配对
single particle	单粒子
single pass	单程,单次扫描,单通道
single phase	单相
single phase flow heat transfer	单相流换热
single pixel position	单像素位置
single point failure	单点失效
single pulse	单脉冲,单冲量
single pulse laser	单脉冲激光器
single rotor	单转子
single sampling	单次取样
single sensor	单传感器
single shot probability	单发命中概率
single side band	单边带
single sided board	单面印板
single solar cell	单体太阳电池
single spool	单路式
single stage rocket	单级火箭

single stage to orbit	单级入轨
single star and multistars navigation	单星及多星导航
single star navigation	单星导航
single stub tuner	单短线调谐器
single test	单项测试
single threading	单线处理
single tone command	单音指令
single tube manometer	单管压力计
single value nonlinearity	单值非线性
single vehicle	单辆汽车
single-antenna	单天线
single-aperture model	单孔模型
single-arm spectrometer	单臂谱仪
single-axis	单轴
single-axis accelerometer	单轴加速度表
single-axis attitude stabilization	单轴姿态稳定
single-band	单频带
single-base propellant	单基药
single-board computer	单板计算机,单板机
single-carriage clip-in assembly	单弹插夹式挂弹装置
single-carriage rack	单弹挂弹架
single-chamber multistage-thrust motor	单室多推力发动机
single-channel analyser	单道分析器
single-chip computer	单片计算机,单片机
single-crystal alloy	单晶合金
single-cycle	单循环
single-degree-of-freedom	单自由度
single-degree-of-freedom gyro	单自由度陀螺
single-degree-of-freedom system	单自由度系统
single-detector homing head	单元探测导引头
single-element detector	单元探测器
single-fail operative	单故障工作
single-frequency	单频
single-harmonic distortion	单谐波失真
single-hole	单孔
singleinput-multioutput	单输入多输出
single-layer	单层
single-line spectroscopic binary	单谱分光双星
single-link robot	单链路机器人
single-mode	单模
single-pass gain	单程增益,单通道增益
single-point constraint	单点约束
single-point grounding	单点接地,一点接地
single-pole double-throw	单刀双掷
single-pole single-throw	单刀单掷
single-pulse	单脉冲
single-rotor	单转动轴
single-seat	单座
single-segment transportation	分段运输
single-shaft	单杆的
single-shot	(枪、炮)单发的
single-shot kill probability	单发摧毁概率
single-shot probability	单发命中概率
single-side abrupt junction	单边突变结
single-sideband modulation	单边带调制
single-sideband reference level	单边带参考电平
single-sideband voice	单边带话音
single-simplex communication	单工通信
single-spar wing	单梁(机)翼
single-stage	单级
single-stage compressor	单级压气机
single-stage rocket	单级火箭
single-stage turbine	单级涡轮
single-stage vehicle	单级入轨运载器
single-stage-to-orbit	单级入轨
single-stage-to-orbit launch vehicle	单级入轨运载器
single-station system	单站制
single-station triggering	单站触发
singular arc	奇异弧线
singular attractor	奇异吸引子
singular control	奇异控制
singular differential equation	奇异微分方程
singular linear system	奇异线性系统
singular perturbation	奇异摄动法
singular state	奇(异)态
singular surface	奇异面
singular value	奇异值
singularity	1.奇点 2.奇异性
singularity of the universe	宇宙奇点
sink	1.汇,潭 2.下沉,降底 3.吸热器,换能器
sink capacity	库容量
sink current	反向电流
sink flow	汇流
sink information	汇集信息
sink nodes	宿节点
sink rate	下沉率,散热器速度
sinking	下沉,下降
sinking speed	下降速度,下沉速度
Sinope	【天】木卫九
SINS	1. Ships Inertial Navigation System 舰船惯性导航系统 2. Stellar Inertial Navigation System 天文惯性导航系统 3. Strapped-down Inertial Navigation System 捷联式惯性导航系统
sinter	渣,熔渣
sintered plate	烧结式极板
sintered type cadmium-nickel battery	烧结式镉镍蓄电池

sintering 烧结
sinus barotrauma 航空鼻窦炎
sinusoid 正弦曲线
sinusoidal 正弦曲线形,简谐运动的特征
sinusoidal distribution 正弦(式)分布
sinusoidal frequency shift keying 正弦频移键控
sinusoidal function 正弦函数
sinusoidal impulse 正弦脉冲
sinusoidal quantity 正弦量
SIO 1. Serial I/O 串行输入/输出 2. Service Information Octet 服务信息组
SIP 1. Satellite Information Processor 卫星信息处理机 2. Structure Inspection Plan 结构检查大纲 3. Structural Inspection Procedure 结构检查程序
SIR 1. Signal Interference Ratio 信扰比 2. Snow and Ice on Runway 跑道积雪和结冰 3. Special Inspection Request 特殊校验要求,特殊检查申请 4. Special Inspection Requirements 特殊检查要求 5. Structural Integrity Requirements 结构完整性要求 6. Structural Irregularity Report 结构缺陷报告 7. Submission Information Rate 信息送出率
siren 警报器
SIRS Satellite Infrared Radiation Spectrometer 卫星红外辐射光谱仪
SIS 1. Service Interphone System 勤务内话系统 2. Signaling Interworking Subsystem 信令互通子系统 3. Spares Investment Study 备件投资研究 4. Stall Identification System 失速识别系统 5. System Interrupt Supervisor 系统中断管理程序
SISDS Single Instruction, Single Data Stream 单指令单数据(流)
SIT 1. System Integration and Test 系统综合和试验 2. System Integration Team 系统综合团队
SITA Societe Internationale de Telecommunications Aeronautique 国际航空通信学会(法国)
site equipment change rate 发射场设备更换率
site fault rate 发射场故障率
site remanence 原地剩磁
site testing 选址
SITF 1. Shuttle Infrared Telescope Facility 航天飞机红外望远镜设备(美国) 2. Space Infrared Telescope Facility 空间红外望远镜设施(美国大型天文物理研究卫星)
SITP 1. Structural Integrity Test Plan 结构完整性试验大纲 2. System Integrity Test Procedure 系统综合试验程序
SITP/D System Integrity Test Plan/Description 系统综合试验大纲/说明
SITR 1. Structural Integrity Test Report 结构完整性试验报告 2. System Integration Test Report 系统综合试验报告
situation board 空情标图板
situation display 状况显示
SIU 1. Satellite Interface Unit 卫星接口单元,卫星界面装置 2. Server Interface Unit 服务器接口组件 3. Signaling Interface Unit 信令接口单元 4. Static Inverter Unit 静变流机组件 5. Subscriber Interface Unit 用户接口单元 6. System Interface Unit 系统接口部件
SIWL Single Isolated Wheel Load 单个分离机轮载荷
six 六
six bit code 六单位制(电)码
six-axis 六轴
six-component balance 六分力天平
six-degree-of-freedom 六自由度
six-degree-of-freedom model 六自由度模型
size 尺寸
size distribution 粒度分布,粒径分布
size range 粒度范围,块度范围,尺寸范围
size-distance illusion 大小距离错觉
size-distance perception 大小距离知觉
SJR Signal to Jamming Ratio 信号干扰比
SK Sketch 草图
SKEDMAINT Scheduled Maintenance 定期维修
SKEDOVHL Scheduled Overhaul 定期大修
skeg 浮筒护底板,滑橇尾鳍
sketch 草图,略图,画草图,示意图,简图,设计图,概略
sketch survey 草测
sketch-map 草图,略图,示意图
sketchmaster 转绘仪
skew 变形,时滞,倾斜
skew aileron 斜轴副翼
skew angle 倾斜角,斜交角
skew distortion 偏斜失真
skew ray 扭曲光线,折曲射线
skewed sensor geometry 斜置技术,惯性元件斜置布局,惯性元件斜置布局技术
skewed sensor technology 斜置技术
skewness 偏斜度
skew-symmetric matrix 斜对称矩阵
skew-symmetry 反对称性
ski landing gear 雪撬起落架,滑橇式起落架
skiatron 记录暗迹的阴极射线管
skid landing gear 滑橇式起落架
skidding 机轮打滑,拖胎
skiddometer 摩擦因数测定器
skiing glider 滑翔滑雪员(滑雪员背负升力面或翼伞)
ski-jump ramp 滑橇甲板

skimmer　水上飞机
skin　表皮，蒙皮
skin depth　趋肤深度
skin drag　摩擦阻力
skin effect　趋肤效应，集肤效应
skin paint　飞机蒙皮回波
skin stretch forming　蒙皮拉伸成形
skin thickness　表皮厚度
skin tracking　1.反射跟踪 2.表皮跟踪 3.雷达跟踪
skin-friction coefficient　表面摩擦因数
skin-friction drag　表面摩擦阻力
skin-friction force　表面摩擦力量
skip　1.跳跃式再入(大气层) 2.料斗，起重箱，料车
skip area　跳跃区域，盲区
skip bomb　用于跳弹轰炸的炸弹
skip distance　越程，跳跃距离
skip trajectory　跳跃轨迹
skip-fading　(电波的)跳跃衰落
skiplane　滑橇式起落架飞机，雪上飞机
skirt　裙，边缘
SKT　Socket 套筒
SKU　Site Key User 现场关键用户
sky background　天空背景
sky background power　天空背景功率
sky brightness　天空亮度
sky compass　天文罗盘，天罗盘
sky conditions　云量
sky diver　跳伞运动员
sky horn　天空刺叭
sky patrol　巡天观测
sky radiance　天空亮度
sky simulator　星空模拟器
sky survey　巡天观测
skycap　机场行李搬运员
Skydrol　斯凯德罗尔液压油(一种酯基的合成液压油，不可燃)
sky-hitching　搭乘免费飞机(旅行)
skyjack　空中劫持，劫机
Skylab　空间实验室，空中实验室
skylift　空运，空中运输
skyline　天涯，地平线，以天空为背景的轮廓
skyliner　旅客班机
Skymaster　巨型客机
Skynet　天网
skyphone　(民用飞机上供旅客使用的)空中电话
sky-wave interference　天波干扰
skywayman　劫持飞机者
SL　1. Sea Level 海平面 2. Sensitivity Level 响应级 3. Service Letter 服务信函 4. Solid 固体的 5. Station Line 站位线 6. Strobe Light 频闪灯
SLA　1. Service Level Agreement 服务水平协议 2. Site Level Aggregator 位置层集合器 3. Small Light Adapter 小灯适配器
slab　1.平板，厚片，混凝土路面，把……分成厚片，用石板铺 2.扁钢坯扁钢锭 3.音节，长字节
slab laser　条形激光器
slab-sided　侧面平坦的
slack　1.粉煤 2.松弛，松劲 3.减弱，减慢
slack variable　松弛变量
slacken　放松，减缓
slackness　疏忽，懒散，松弛，呆滞
SLAET　Society of Licensed Aircraft Engineers and Technologists 航天器工程师和技术专家协会
slag　渣，炉渣
SLAM　1. Side Load Arresting Mechanism 侧向载荷抑制机构 2. Stand-off Land Attack Missile 火力圈外发射的对地攻击导弹
slam acceleration　急剧加速，猛推油门
slamming　(小车式起落架的)粗猛接地
SLAMMR　Side Looking Airborne Multi Mission Radar 机载侧视多用雷达
slant　倾斜，斜线
slant range　斜距
slant range resolution　斜距分辨率
slant visibility　斜程能见度
SLAR　1. Side Looking Airborne Radar 机载侧视雷达 2. Slant Range 斜距
slash mark　松弛标记
slat　1.前缘缝翼 2.狭板，板条
slave　从动，随动，从属的，次要的，从动装置，随动的，从属
slave antenna　伺服天线，从动天线
slave node　从属节点
slave processor　从属处理机
slave station　副台，从属台，从动台
slaved system　从属系统
slaving　从属，从动
slaving principle　役使原理，从属原理
SLCC　1. Signaling Link Control Common 公共信令链路控制器 2. Software Life Cycle Cost 软件寿命期成本
SLCM　1. Sea Launched Cruise Missile 海上发射巡航导弹 2. Submarine(sea) Launched Cruise Missile 潜射巡航导弹
SLCSAT　Submarine Laser Communication Satellite 潜艇激光通信卫星(美国)
SLDVS　Scanning Laser Doppler Velocimeter System 扫描激光多普勒速度计系统
SLE　Service Life Evaluation 服务寿命评定，使用寿命

评定
sled 滑板,滑橇
sledplane 滑橇式起落架飞机,雪上飞机
sleep deprivation 剥夺睡眠,缺乏睡眠
sleep disturbance 睡眠障碍,睡眠失调
sleet 雨夹雪,冻雨
sleeve 1.套筒,衬筒,衬套 2.套管,套筒 3.导流管
sleeve valve 套阀,滑阀
sleeve-stub antenna 套筒短柱天线
slender body 细长物体,细长体
slender body theory 细长体理论
slender theory 细长体理论
slenderness 细长度
slenderness ratio 长细比,长径比
SLEP Service Life Extension Program 延(长使用)寿(命)计划
slew 上下选页
slew rate 转换速率
slew time 旋转时间
slewing 旋臂,回转
SLF Surface Lock Function 操纵面锁定功能
SLFCS Survivable Low Frequency Communications System 长期型低频通信系统
SLGR Small, Lightweight GPS Receiver 轻小型全球定位卫星接收机,袖珍 GPS 接收机
SLI System Level Interface 系统级界面
slice 片,薄片
slicer 1.切片器,分割器 2.限制器,脉冲限幅器
slicing 1.切片(一种工艺) 2.限幅
slide escape 滑道逃逸
slide navigation rule 领航计算尺
slide screw tuner 滑动螺钉调谐器
slide valve 滑阀
slider 滑动片,滑动触头,滑触头
sliding guide 导轨
sliding joint 滑动接合
sliding load 滑动负载
sliding pulser 滑移脉冲产生器
sliding vane rotary vacuum pump 旋片真空泵
sliding-mode 滑动模态
sling 吊索,吊链
sling load 起吊载荷
slinger ring 吊环
slinger ring atomizer 甩油盘
slip band 滑移带
slip casting 注浆成型
slip flow 滑移流
slip function 滑脱函数
slip gauge 滑规(一种厚度精密的超薄片)

slip joint 滑动接头,伸缩性连接
slip line 滑线,滑动线
slip plane 滑移面
slip ratio 滑移比率
slip ring 滑环
slip surface 滑面
slip tank 副油箱
slip vector 滑动向量
slippage 滑移
slippage mark 滑动记号
slipper 滑动件,游标
slipping time 滑行时间
slipping torque 滑动扭矩
slipping turn 侧滑转弯飞行
slip-ring rocket launcher 滑环式火箭发射器
slipstream 滑流,绕流
slit 缝隙,裂缝
slitless spectrograph 无缝摄谱仪
sliver 未烧尽的火药渣粒
SLM Standard Length Message 标准长度信息
SLMG Self Launching Motor Glider 自主起飞动力滑翔机
SLO Service Length Objectives 服务级目标
slope 倾斜,斜度,坡度
slope break 坡度转折
slope discriminator 斜率鉴频器
slope distance 斜距
slope efficiency 斜率效率
slope of fall 降落斜度
slope of lift curve 升力线斜率
slope theodolite 坡面经纬仪
slope transmission 提升传输
slosh 打,击,晃动,激荡,漏出
slosh barrier 晃动挡板
slosh cycle 晃动周期
slosh damping 晃动阻尼
slosh force 晃动力
slosh frequency 晃动频率
slosh mass 晃动质量
slosh suppression 晃动抑制
slosh torque 晃动力矩
sloshing (指液体推进剂或油液在容器或管道内的)晃荡,晃动
slot 1.裂口,切口 2.槽,缝隙
slot antenna 缝隙天线
slotted aileron 开缝副翼
slotted antenna array 裂缝阵天线,缝隙阵天线
slotted blade 开缝叶片(用以控制气流分离)
slotted coaxial antenna 同轴天槽天线

slotted flap 开缝襟翼
slotted line 开槽线,测量线
slotted rectangular waveguide antenna 矩形波导开槽天线
slotted wall 开槽壁
slotted waveguide antenna 波导缝隙阵天线
slotted-tube grain 管槽形药柱
slotting 插削
slow cook-off 慢速烤燃
slow decompression 缓慢减压
slow filling control 小流量加注控制
slow nova 慢新星
slow roll 慢滚
slow rotation room 慢旋转房
slow scan television 慢扫描电视
slow state 慢变状态
slow subsystem 慢变子系统
slow twist 缓斜膛线
slow varying parameter 缓变参数
slow-blow fuse 慢断熔丝,慢断保险丝(允许高额电流通过的时间超过规定值后才烧断的熔丝)
slower-than-real-time simulation 近实时仿真
slow-sweep noise-modulated jamming 慢扫频噪声调制干扰
slow-wave line 慢波线
SLP 1. Service Logic Processing 业务逻辑处理 2. Single Link Procedure 单链路规程 3. Space Limited Payload 有限空间商载 4. Speed Limit Point 速度限制点
SLR Service Level Requirement 服务等级要求
SLS 1. Safe Life Structure 安全寿命结构 2. Side Lobe Suppression 边波束抑制,旁瓣波束抑制 3. Signaling Link Selection 信令链路选择
SLST Sea-Level Static Thrust 海平面静推力,发动机地面静推力
SLT 1. Subscriber Line Terminal 用户线路终端 2. Supplier Lead Time 供应商提前期
sludge 软泥,淤泥,残渣,油泥,污水,沉淀物,金属碎屑
sludge chamber 污泥池淤渣池
slug 金属弹心,芯子
slug riveting 无头铆钉铆接
slugging 缓慢的
slung load 吊装(直升机下方用一根吊索所吊起的载荷)
slurry 淤浆,黏合剂,泥浆
slurry fuel 悬浮燃料
slurry infiltration 料浆浸渗法
slush 油灰,淤泥,煤泥,水泥沙浆,溅湿
SM 1. Small 小 2. Shared Mode 共享模式 3. Standard Manual 标准件手册 4. Station Manager 站长 5. Statute Miles 英里 6. Summer 夏季 7. Supplementary Modes 辅助方式
SM&P Supply Management and Procurement 供应管理和采购
SMA 1. Scheduled Maintenance Analysis 定期维修分析 2. Scalable Modular Architecture 可伸缩模块化结构 3. Segment Minimum Altitude 航段最低高度 4. Single Motor Actuator 单电机作动筒 5. Systems Management Architecture 系统管理体系结构
SMAE System Management Application Entity 系统管理应用实体
small aircraft 小型飞机,小型航空器
small arms 轻武器
small asymmetric nose testing 小不对称弹头试验
small circle 小圆,小圈
small end 连杆上端,连杆小端,小端,小头
small missile telecamera 小型导弹电视摄象机
small perturbation 小扰动方程
small perturbation equation 小扰动方程
small perturbation potential equation 小扰动位势方程
small satellite 小卫星
small-amplitude-slosh liquid dynamics 小幅液体晃动力学
small-bore 小口径的
small-scale model 小尺寸模型
small-scale test 小规模测试
small-signal analysis 小信号分析
small-signal gain 小信号增益
SMAP Service Management Access Point 业务管理接入点
SMART Standard Modular Avionics Repair and Test 标准组件化电子修理和测试
smart bomb 可控爆炸轰炸
smart fuze 灵巧引信
smart instrument 智能仪器
smart material 机敏材料
smart sensor 灵巧敏感器
smart structure 智能结构
smart terminal 灵巧终端(设备),弱智能终端
smart weapon 灵巧武器(指制导武器)
SMATV Satellite Master Antenna Television 卫星共用天线电视
SMAW Shielded Metalarc Welding 有保护的金属电弧焊
smaze 烟霾(烟和雾的组合物)
SMB 1. Server Message Block 服务器消息块 2. Small and Medium Businesses 中小型企业
SMC 1. Scheduled Maintenance Checks 定期维修检查

2. Seat Mile Cost 座英里成本 3. Standard Mean Chord 标准平均弦(长) 4. Surface Movement Control 地面活动管制 5. System Mnagement & Communication 系统管理和通信

SMD 1. Scheduling Management Display 调度管理显示 2. Surface Mounted Device 外装设备

SME 1. Small/Medium Enterprise 中小型企业 2. Small Message Entity 短消息实体 3. Signaling Message Encription 信令消息加密 4. Subject Matter Expert 主题专家

smear 涂,敷

smectic liquid crystal 近晶相液晶

smectic phase (液晶)层列相

SMES Small and Medium Enterprises 中小型企业

SMF 1. Service Management Function 业务管理功能 2. Single Mode Fiber 单模光纤 3. Software Maintenance Function 软件维护功能 4. Standard MIDI File 标准的数字化乐器接口文件 5. Sub-Multiframe 子复帧

SMG Statistical Multplexing Gain 统计复用增益

SMGCS Surface Movement Guidance and Control System 活动面运动导引和控制系统

SMI 1. Special Maintenance Instruction 特殊维护说明 2. Standard Message Identifiers 标准信息标识符

SMIP Scheduled Maintenance and Inspection Program 定期维修与检验方案

SMIR Software Migration and Installation Request 软件转移和安装需求

Smith chart 史密斯圆图,史密斯图(解电力传输线问题的标准算图)

SMM 1. Service Management Module 服务管理模块 2. System Management Model 系统管理模式

SMMR Surface Missile Medium Range 中程地表导弹

SMMW Submillimeter Wave 亚毫米波

smog 工业污染雾

SMOH Since Major Overhaul 自翻修后算起(在此词后加上 L.E 或 R.E 表示左发动机或右发动机)

smoke 烟雾

smoke bomb 烟雾炸弹

smoke box 烟箱,烟室

smoke detector 烟雾探测器

smoke flow method 烟流法

smoke hood 防烟面罩

smoke number 1.排烟数 2.发烟数

smoke particle 烟粒子

smoke pot 发烟罐

smoke tunnel 烟风洞

smoke wind tunnel 烟风洞

smokeless propellant 无烟推进剂

SMOM System Maintenance and Operating Manual 系统维修和操作手册

smooth 平稳,顺利

smooth function 平滑函数

smooth surface 平滑表面

smooth transition 顺利过渡,平稳过渡

smooth treatment of shock wave 激波光滑处理

smooth tube 光滑管

smoothen 使平滑

smoothness 光滑性

SMP 1. Service Management Point 业务管理点 2. Simple Management Protocol 简单管理协议 3. Structure Maintenance Programmer 结构维修大纲 4. Structural Modification Program 结构改装大纲 5. Symmetric Multiprocessor 对称多处理机

SMPL Supplier Module Parts List 供应商模块零件表

SMR 1. Signal to Multipath Ratio 信号多径比 2. Solid Moderated Reactor 固体慢化堆 3. Specialized Mobile Radio 专用移动无线电 4. Surface Movement Radar 地面活动雷达

SMRD Spin Motor Rotation Detector 旋转型发动机旋转检测器

SMRS Specialized Mobile Radio System 专用移动无线电系统

SMS 1. Safety Management System 安全管理系统 2. Security Management System 保安管理系统 3. Spectrum Monitoring System 频谱监控系统 4. Stores Management Set 外挂物管理设备 5. Synchronous Meteorological Satellite 同步气象卫星 6. System Management Server 系统管理服务器

SMT 1. Scheduled Maintenance Task 计划维护任务 2. Servo Mount 伺服器安装 3. Square Mesh Tracking 正方网格跟踪 4. Store Material Transfer 仓库器材调动 5. System Maintenance Test 系统维护测试

SMTI Selective Moving Target Indicator 可选择的运动目标指示器

SN 1. Serial Number 序号 2. Sequence Number 序列号 3. Service Network 服务网络 4. Standard Notes 标准注语 5. Stick Nudger 自动推杆器 6. Stock Number 存贮号

Sn 锡

SNA 1. Standard Network Architecture 标准化网络体系结构 2. System Network Analysis 系统网络分析 3. System Network Architecture 系统网络结构

snake drill 蛇形钻杆

snake mode 蛇形飞行

snaking 摆头,蛇行飞行

snap report 快速报告,任务执行情况初步报告

snap ring 卡圈,卡环,开口环

snap roll	急滚,快滚
snap shot	快速射击
snap-action	快动作的,速动的,瞬间的
snap-action switch	快速开关
snap-shoot sight	快速射击瞄准具
snapshot algorithm	快照算法
snap-up	快速拉起
snatch	抓住,攫取
SNC	1. Standard Navigation Computer 标准导航计算机 2. Sub Network Connection 子网连接 3. Synchronous Network Clock 同步网络时钟
SNDV	Strategic Nuclear Delivery Vehicles 战略核武器投射飞行器
sneak circuit analysis	潜在通路分析
sneak condition	潜在状态
sneeze gas	喷嚏性毒气
Snell law	斯涅耳定律(折射率定律)
SNET	Supplier Network 供应商网站
SNI	1. Service Node Interface 业务节点接口 2. Subscriber Network Interface 用户网络接口 3. Switching Network Interface 交换网络接口 4. Synchronous Network Interface 同步网接口
snips	平头剪,铁丝剪
SNM	1. Signaling Network Management 信令网管理 2. Simple Network Management 简单网络管理 3. Switching Network Module 交换网络模块
SNMP	Simple Network Management Protocol 简单网络管理议定书
snout	机头部分,锥形进口
snow accumulation	积雪
snow gauge	雪量计,雪湿度仪
snow pellets	雪丸(水和冰的细小不透明颗粒,比冰雹软,包括软雹和霰)
snowfall intensity	降雪强度
snowplane	摩托滑雪车,滑行雪车
SNP	1. Statistical Network Processor 统计网络处理机 2. Synchronous Network Processor 同步网络处理机
SNR	Schedule Negotiation Record 计划谈判记录
SNS	Stellar Navigation System 天文导航系统
snubber	减振器,缓冲器,消声器,锚链制止器
SO	1. Second Officer 领航员 2. Shipping Order 运货单,发货单 3. Supportability Objectives 支援能力目标
SOA	1. Semiconductor Optical Amplifier 半导体光放大器 2. Speed of Approach 进近速度,进近着陆速度 3. Start of Address 地址起始
soap bubble test	肥皂泡试验
soar	1. 滑翔,高飞 2. 特殊运行的航空编队
soaring	翱翔,滑翔
SOB	1. Side of Body 机身侧面 2. Souls on Board 机上人员
SOBs	Second Observer 第二观察员
SOC	1. Satellite Orbit Control 卫星轨道控制 2. Scoiety 协会,社会 3. Sector Operations Center 地区(分区)作战中心 4. Service and Overhaul Change 维修和大修更改 5. Start of Climb 开始爬升 6. Statement Of Compliance 型号资料审查表 7. Statement Of Conformity (制造)符合性声明 8. System Operation Center 航班系统运行中心 9. System Operations Control 系统运行控制
Society of Automotive Engineers	汽车工程师学会
socio-cybernetics	社会控制论
sock	开伞袋,伞衣套
socket	插座,插口,管座
socket connector	接插件
socket flange	插座式法兰
SOCS	Spacecraft Orientation Conrtol System 航天器定向控制系统
sodar	声雷达
sodium	1.【化】钠(化学元素,符号Na) 2.〈口语〉钠灯(同 sodium light)
sodium borohydride	硼氢化钠
sodium cloud	钠云
sodium iodide detector	碘化钠探测器
sodium light	钠光
sodium vapo(u)r lamp	钠蒸汽灯
sodium-sulfur battery	钠硫电池
SOF	1. Safety Officer of Flying 飞行安全员 2. Safety Of Flight 飞行安全
sofar	Sound Fixing And Ranging 声波定位与测距
SOFR	Safety Of Flight Report 飞行安全报告
soft clay	软质黏土,低烧结黏土,陶土
soft computing	软计算
soft constraint	软约束
soft decision	软判决
soft error rate	软误差率
soft failure	软失效
soft fire	缺氧火焰,文火
soft ignition	软点火
soft impact	软着陆
soft iron	软铁
soft key	软键
soft landing	软着陆
soft landing spacecraft	软着陆航天器
soft lunar landing	月球软着陆
soft magnetic ferrite	软磁铁氧体材料
soft magnetic materials	软磁材料
soft particle	软粒子(低能粒子)
soft radiation	软辐射

soft rubber	软质橡胶,软硫化橡胶
soft spring	软簧
soft stop	软停机
soft switch modulator	软管调制器
soft target	1.软目标,不加固的目标 2.非军事目标,软目标
soft valve	热离子管,热发射电子管
soft weapon	软武器(非直接杀伤目标的武器,如诱惑干扰弹)
soft x-rays	软 X 射线
soften	弄软,软化
soft-landing capsule	软着陆舱
softly fail	故障软化
software	软件,软设备,软件技术
software agent	软件代理
software architecture	软件架构
software behavior	软件行为
software component	软件构件
software design	软件设计
software design document	软件设计文件
software development	软件开发,软件研制
software development notebook	软件开发簿
software engineering	软件工程
software error	软件错误
software estimation	软件评估
software failure	软件故障
software fault	软件故障
software function	软件功能
software maintainability	软件可维修性
software maintenance	软件维护
software methodology	软件方法学
software package	软件包,程序包
software package of computer aided design	计算机辅助设计软件包
software pirate	软件非法翻印者,侵犯软件版权
software program	软件程序
software quality assurance	软件质量保证
software reliability	软件可靠性
software repository	软件存储库
software requirement	软件需求
software safety	软件安全性
software simulation	软件模拟
software system	软件系统
software testing	软件测试
software tool	(计算机)软件工具
software-compatible	软件兼容的
SOH	1. Scheduled Outage Hours 计划停用时数 2.Section Overhead 段开销 3. Start Of Heading 起始航向
SOI	1. Seats Occupied Information 座位占用情况 2.Standard Operating Instruction 标准运行指令 3. System Operator Instructions 系统操作员说明书
solar absorber	太阳辐射能吸收器
solar absorptance	太阳吸收率
solar absorption factor	太阳吸收系数
solar active region	太阳活动区
solar activity	太阳活动性
solar altitude	太阳高度
solar angle	太阳角
solar antapex	太阳背点
solar apex	太阳向点,太阳奔向点
solar apparent diameter	太阳视直径
solar array	太阳能电池阵
solar array deployment	太阳阵展开
solar array rate stability	太阳电池阵速率稳定度
solar array-battery system	太阳电池-蓄电池系统
solar battery	太阳能电池
solar calendar	太阳历,阴历
solar calibrator	太阳定标器
solar cell	太阳电池,单体太阳电池
solar cell array	太阳电池阵列
solar cell array wing	太阳电池阵翼
solar cell basic plate	太阳电池底板
solar cell module	太阳电池组合件
solar cell module area	太阳电池组件面积
solar cell panel	太阳电池组合板
solar cell temperature	太阳电池温度
solar collector area	太阳能集热器面积
solar concentrator	太阳能集中器
solar constant	太阳常数
solar converter	太阳能转换器
solar corona	日冕
solar corpuscular emission	太阳微粒发射
solar corpuscular rays	太阳微粒射线
solar cosmic ray	太阳宇宙线
solar cycle	太阳活动周
solar daily variation	太阳日变化
solar day	太阳日
solar disk	日轮日面日盘太阳圆面
solar distance	日地距
solar disturbance	太阳扰动
solar eclipse	日蚀,日食
solar eclipse limit	日食限
solar electromagnetic radiation	太阳电磁辐射
solar electron event	太阳电子事件
solar elevation angle	太阳高度角
solar energy	太阳能
solar escape velocity	第三宇宙速度

solar exposure	紫外线照射
solar flare	耀斑,日辉,太阳耀斑
solar flare proton	太阳耀斑质子
solar flux	日辐射通量
solar grade silicon solar cell	太阳级硅太阳电池
solar heat	太阳热
solar heating	太阳能供暖
solar illumination	太阳照度
solar infrared radiation	太阳红外辐射
solar irradiance	太阳辐射照度
solar latitude	太阳纬度
solar limb	日面边缘
solar longitude	太阳经度,日面经度
solar luminosity	太阳发光度
solar magnetograph	太阳磁像仪
solar module	太阳电池组件
solar neutrino unit	太阳中微子单位
solar observatory	太阳观测台
solar orbit	太阳轨道
solar oscillation	太阳振荡
solar paddles	太阳电池帆板
solar panel	太阳能电池板,太阳电池板
solar parallax	太阳视差
solar particle event	太阳粒子事件
solar particle radiation	太阳粒子辐射
solar patrol	太阳巡视
solar photovoltaic energy system	太阳光伏能源系统
solar physics	太阳物理学
solar plasma	太阳等离子体
solar power	太阳能电源
solar power generation	太阳能发电
solar power plant	太阳能电站
solar power satellite	太阳能发电卫星
solar power station	太阳能电站
solar powered aircraft	太阳能飞机
solar pressure	太阳光压,太阳辐射压力
solar probe	太阳探测器
solar probe spacecraft	太阳探测宇宙飞船
solar prominence	日珥
solar propulsion	太阳能推进,太阳动力
solar proton	太阳质子
solar proton event	太阳质子事件
solar proton monitor	太阳质子监视仪
solar pumping	日光泵浦
solar radiation	太阳辐射
solar radiation disturbance torque	太阳辐射干扰力矩
solar radiation factor	太阳辐射角系数
solar radiation perturbation	光压摄动
solar radiation pressure	太阳辐射压力,太阳风
solar radiation pressure perturbation	太阳光压摄动
solar radiation satellite	太阳辐射卫星
solar radiation simulator	太阳辐射模拟器
solar radiation stabilization	太阳辐射稳定
solar radio burst	太阳射电爆发
solar radio emission	太阳射电辐射
solar reflector	太阳光反射器
solar rocket	太阳能火箭
solar rotation	太阳自转
solar sail	太阳帆,太阳反射器
solar sail orientation control	太阳帆定向控制
solar sail spacecraft	太阳帆航天器
solar satellite	太阳卫星
solar sensor	太阳敏感器
solar service	太阳服务
solar simulation	太阳仿真试验
solar simulator	太阳模拟器
solar spectrograph	太阳摄谱仪
solar spectrum	太阳光谱
solar spectrum irradiancy	太阳光谱辐照度
solar storm	日暴
solar structure	太阳结构
solar system	太阳系
solar telescope	太阳望远镜
solar term	节气
solar tide	太阳潮
solar time	太阳时
solar tower	太阳塔(一种太阳望远镜)
solar ultraviolet radiation	太阳紫外辐射
solar wind	太阳风,海陆风
solar wind magnetic field	太阳风磁场
solar X-ray	太阳 X 射线
solar year	太阳年
solarimeter	日射强度计,太阳热量计
solar-terrestrial physics	日地物理学
solar-terrestrial relationship	日地关系
solar-terrestrial space	日地空间
solar-type star	太阳型星
solder	焊料,焊锡,焊钎接,钎焊
solder side	焊接面
solderability test	可焊性试验
soldered splice	焊接接头
soldering	焊接
solderless connection	非焊接接线,非焊接连接(导线连接的一种方法)
solderless splice	非焊接接头
solenoid	1.螺线管,筒形线圈 2.电磁铁
solenoid operated directional valve	电磁换向阀
solenoid valve	电磁阀

solfatara 硫质气孔
solid angle 立体角
solid body 固体,实体
solid carbon 实心炭棒
solid circle 实心圆
solid concentration 固体浓度
solid conductor 实心导线
solid core 实心铁心
solid earth geophysics 固体地球物理学
solid earth tide 固体潮
solid electron beam 实心电子束
solid electronics 固体电子学
solid failure 固定失效,固定故障
solid filling 固体装药,火药装药
solid fuel 固体燃料
solid hydrogen 固态氢
solid line 实线
solid loading 固体负荷
solid lubricant 固体润滑剂
solid mechanics 固体力学
solid model 实体模型
solid motor 固体发动机,固体燃料火箭发动机
solid panel 厚镶板
solid particle 固体微粒,实体颗粒
solid phase 固相
solid phase epitaxy 固相外延
solid propellant 固体推进剂
solid propellant booster 固体燃料起飞发动机
solid propellant combustion 固体推进剂燃烧
solid propellant engine 固体推进剂发动机
solid propellant gas generator 固体推进剂气体发生器
solid propellant motor assembly building 固体发动机总装厂房
solid propellant rocket 固体燃料火箭
solid propellant rocket engine 1.固体火箭发动机 2.固体推进剂火箭发动机
solid propellant rocket motor 固体(推进剂)火箭发动机
solid residue 固体残渣
solid rocket 固体火箭
solid rocket booster 固体火箭助推器
solid rocket fuel 固体火箭燃料
solid rocket motor 火箭发动机
solid rocket oxidizer 固体火箭氧化剂
solid rocket propellant 固体火箭推进剂
solid rocket propellant binder 固体火箭推进剂黏合剂
solid rocket propellant plasticizer 固体火箭推进剂增塑剂
solid solubility 固溶度

solid solution 固溶体
solid solution semiconductor 固溶体半导体
solid state 固态
solid state computer 固态计算机
solid state detector 固态探测器
solid state electronics 固体电子学
solid state laser 固体激光器
solid state magnetron 固态磁控管
solid state microwave source 固态微波源
solid state power amplifier 固态功率放大器
solid state relay 固体继电器,固态继电器
solid state software 固态软件
solid surface 固体表面
solid tantalum electrolytic capacitor 固体钽电解电容器
solid target 固体靶
solid track detector 固体径迹探测器
solid wall 实心墙
solid waste 固体废物,固态废物
solid wire 单线,实线
solid-beam efficiency 立体波束效率
solid-fuel 固体燃料
solidification 凝固
solidification front 凝固前沿
solidification process 凝固过程
solidification rate 凝固速率
solidification temperature 凝固温度
solidified jet testing 固化喷流试验
solidity 固态,固体,实度,稠度
solid-liquid interface 固液界面
solid-state circuit 固态电路
solid-state device 固态器件
solid-state memory 固态存储器
solid-state modulator 固态调制器
solid-state radar 固态雷达
solid-state refrigerator 固体致冷器
solid-state sensor 固态敏感器
solidus 固相线
solion 溶液离子电化学换能器
solitary wave 孤立波
soliton 孤子
soliton laser 孤子激光器
solo 单飞,第一次单飞
solo flight 单飞
SRMS Solar Radiation Monitoring Satellite 太阳辐射监测卫星(美国)
SOLS Schedule Order Listing System 进度订单列表系统
solstice 二至点(冬至和夏至)
solstitial colure 二至圈

solubility 溶(解)度,溶性,可溶解性
soluble organic fraction 可溶性有机分数
soluble resin content 可溶性树脂含量
solute 溶质
solute balance 溶质衡算
solute distribution 溶质分布
solute field 溶质场
solution algorithm 求解算法
solution approach 方案步骤
solution domain 解域
solution error 解题误差
solution hardening 固溶淬火,溶液硬化
solution heat treatment 固溶热处理
solution point 解点
solution procedure 求解步骤
solution space 解空间
solution surface 积分曲面
solution technique 溶液技术
solvate 溶剂化物
solvent 溶剂,解决方法
solvent cement 溶剂胶
solvent cleaning 溶剂清洗
solver 解算机,解算装置
SOM 1. Seats Occupied Message 座位占用电报 2. Software Operator Manual 软件操作手册 3. Start Of Message 消息起始 4. Standard On-line Module 标准联机组件 5. System Object Model 系统对象模型
somatic data 人体数据
somatogravic illusion 躯体重力错觉
somatogyral illusion 躯体转动错觉
somesthesis 躯体觉
SON Statement of Operational Need (采购军用品时由军方提出的)使用需求声明书,运行需求说明书
sonar Sound Operation Navigation And Range 声呐(声波导航及测距设备)
sonar array 声呐基阵
sonar background noise 声呐背景噪声
sonar cable 声呐电缆
sonar capsule 声呐盒反射高频声波器
sonar detection 声呐探测
sonar echo simulator 声呐回波模拟器
sonar navigation 声呐导航
sonar receiver 声呐接收机
sonar self noise 声呐自身噪声
sonar surveillance system 声呐监视系统
sonar transducer 声呐换能器
sonar transmission 声呐传输
sonar transmitter 声呐发射机
sonde 探测器,探棒,探空气球,控空火箭

SONIC Spares Ordering and Non-stop Inventory Control (System) 备件订货和无库存控制(系统)
sonic bang 声击
sonic barrier 声障,声垒,音障
sonic boom 声震,声爆,音爆
sonic drilling 超声钻孔
sonic line 声速线
sonic nozzle 声速喷嘴
sonic point 声速点
sonic soldering 声焊
sonic speed 声速,声速
sonic velocity 声速
sonic wave 声波
sonication 声波降解法
sonics 声能学
sonobuoy 声呐浮标
sonobuoy pattern 声呐浮标群
sonobuoy reference system 声呐浮标参考系统
soot 烟灰,煤烟
soot particle 灰粒
SOP 1. Simulated Operating Procedure 模拟操作过程 2. Simulated Output Program 模拟输出程序 3. Specification of Operation 运行规范 4. Standard Operation Procedure 标准使用程序
SOPA Standard Operating Procedure Amplified 标准操作扩展程序
sophisticated weapons 尖端武器
SOR 1. Sale Office Record 出票点记录 2. Serial Output Register 串行输出寄存器
sorbite 索氏体,索拜体
sorption 吸着(作用),吸附
sorption pump 吸附泵
sorption trap 吸附阱
SORT Specific Operation Requirement 指定的使用要求
sortie duration 战斗飞行持续时间
sortie number 出击数
sortie plot 航摄区域图
sortie rate 出动架次率
SOS 1. 国际遇险求救信号 2. Sidewall Overhead Stowage 旅客座椅上方行李柜 3. Service Operation System 业务运行系统
SOSUS Sound Surveillance System 水声监视系统(美国海军在某些海域,如格陵兰、冰岛、英国之间海底部署的水听器阵列)
SOTAS Stand-Off Target Acquisition System 远距离目标探测系统
SOTD Stabilized Optical Tracking Device 稳定光学跟踪装置

SOTS	Synchronous Orbiting Tracking Station 同步轨道跟踪站
sound advance	音响超前距
sound attenuation	声衰减
sound carrier	伴音载波,音频载波
sound channel	音频通道
sound communications	声通信
sound environment test chamber	声环境实验室
sound exposure level	声曝级
sound field	声场
sound image	声像(虚声源)
sound insulating chamber	隔声室
sound intensity	声(音)强度
sound level	声级(用分贝表示)
sound level meter	噪声计,噪声测试器
sound navigation and ranging	音响导航和测距,声音导航及搜索
sound power	声功率
sound pressure	声压
sound pressure level	声压级
sound probe	声探头,探声器
sound ray	声线
sound recognition system	声音识别系统
sound shadow region	声影区
sound source	声源
sound spectrum	声谱
sound speed	声速
sound strip	伴音剥离
sound synthesizer	声音综合器
sound track	声音合成器
sound transmission	声音传播
sound velocity	声速
sound wave	声波,纵向压力波
sounder	音响器,发声器
sounding	探测,探空
sounding balloon	探空气球
sounding meteorograph	探空气象仪
sounding rocket	探空火箭
sounding rocket test	探空火箭试验
Soundwich	消声蜂窝层结构(商用名)
source	源,电源,信源
source array	震源组合
source decoder	信源译码器
source encoding	信源编码
source flow	源流
source language	源语言(用来编写源程序的语言)
source nodes	源节点
source noise	源噪声
source of energy	能源
source packet	源包
source program	源程序
source-region electromagnetic pulse	源区电磁脉冲
south bound node	南向交点
south celestial pole	南天极
south magnetic pole	磁南极
south pole	南极,南极地带
southern hemisphere	南半球
SOV	Shut Off Valve 切断阀
SOW	Statement Of Work 工作说明
SP	1. Serialization Plan 系列化大纲 2. Software Procedure 软件程序 3. Space 空间,间隔 4. Spare Parts 备件 5. Special Performance 特殊性能 6. Static Port 静压口 7. Static Pressure 静压
SPA	1. Seaplane Pilots Association 水上飞机飞行员协会(美国) 2. Service Panel Adapter 维修面板适配器 3. Servo-Power Assembly 伺服动力装置 4. Servo Pre-Amplifier 前置伺服放大器 5. System Performance Analysis 系统性能分析
SPAAG	Self Propelled Anti Aircraft Gun 自行高射炮
SPACCS	Space Command and Control System 空间(飞行)指令和控制系统
space	航天技术,空间,宇宙空间,空间电视,空间遥测
space adaptation syndrome	空间适应综合征
space age	太空时代
space anemia	航天贫血症
space anthropometry	航天人体测量学
space astrometry	空间天体测量学
space astronomical observation	空间天文观测
space astronomy	空间天文学
space attack	航天攻击
space available	航班座位未满
space based laser	天基激光武器
space based radar	航天器雷达
space biology	航天生物学,空间生物学
space bone osteoporosis	航天骨疏松
space booster	航天助推器,航天运载火箭
space boots	航天靴
space borne refrigerator	星载致冷器
space capsule	航天舱,宇宙飞船座舱
space chamber	空间容腔
space character	空白字符
space charge current	空间电荷电流
space charge density	空间电荷密度
space charge layer	空间电荷层
space charge limitation	空间电荷限制
space charge wave	空间电荷波
space chemistry	空间化学
space colony	空间殖民地

space command center	航天指挥中心
space communication network	空间通信网
space cone	定瞬轴锥面,不动锥面,空间极迹锥面
space conflict	空间冲突
space control	空间控制
space control and navigation	航天控制与导航
space crew	航天乘员组
space crew enter spacecraft	航天员进舱
space crime	空间犯罪
space data acquisition network	空间数据采集网
space debris	空间垃圾,空间碎片
space debris environment	太空碎片环境
space defence	航天防御,防天
space diagram	空间图
space discrimination	空间鉴别
space diversity	空间分集
space division multiple access	空间分区多次进入
space division switching	空分切换
space division switching system	空间分配转换系统
space docking dynamics	空间对接动力学
space drugs	航天药物
space dynamic factors and survival	航天动力因素与救生
space electronics	空间电子学
space endocrinology	航天内分泌学
space engine	空间发动机
space engineering	航天工程
space engineering system	航天工程系统
space environment	空间环境
space environment adaptation training	航天环境适应性训练
space environment control and life support system	航天环境控制与生命保障系统
space environment effect	空间环境效应
space environment engineering	航天环境工程
space environment forecast	空间环境预报
space environment model	空间环境模式
space environment simulation	空间环境仿真
space environment test	空间环境试验
space environment warning	空间环境报警
space environmental biology	空间环境生物学
space environmental medicine	航天环境医学
space erectable antenna	空间可展开天线,空间可竖起天线
space ergonomics	航天功效学,空间功效学
space escape	航天救生
space experiment operation training	空间实验操作训练
space exploration	宇宙空间探索,星际探索
space exploration satellite	空间探测卫星
space extraterrestrial life exploration	航天地外生命探索
space factory	空间工厂
space filtering	空间滤波
space flight	航天,宇宙飞行
space flight doctor	航天医师
space flight environment	航天飞行环境
space flight environment simulation facilities	航天环境仿真设备
space flight feeding	航天飞行膳食
space flight principle	航天飞行原理
space flight stress	航天应激状态
space flight united headquarter	航天联合指挥部
space food	宇宙食物,宇宙食品
space food management	航天食品管理
space frame	空间构架,三维构架(一种由管和梁连接而成的结构)
space free flight unit manipulator	空间自由飞行器机械手
space freight pod	(可分离的)航天货舱
space fuel cell system	空间燃料电池系统
space geology	空间地质学
space gloves	航天手套
space greenhouse	太空温室
space guidance navigation and control	航天制导导航和控制
space harmonics	空间谐波
space heat flux	空间外热流
space heater	空间电热器,空间加热器,加热防潮设备
space helmet	航天头盔
space hematology	航天血液学
space ignition test	空间点火试验
space industry	空间产业,航天工业
space inertial reference equipment	空间惯性参考装置
space infrastructure	空间基础设施
space intersection	空间交会
space junk	空间垃圾
space laboratory	空间实验室
space landing site	航天着陆场
space lattice	空间点阵
space launch vehicle	航天运载器
space launch vehicle technology	航天运载器技术
space launching base	航天发射基地
space launching center	航天发射中心
space launching complex	航天发射综合设施
space launching site	航天发射场
space launching technology	航天发射技术
space law	空间法
space life science	空间生命科学

space loss 空间损耗
space lunar exploration 航天月球探测
space management 空间管理
space manipulator 空间机械臂
space mapping 空间制图
space marching method 空间推进法
space material 航天材料
space medical monitoring and support facilities 航天医监医保设备
space medicine 航天医学
space medicine database 航天医学数据库
space medicine engineering 航天医学工程
space medicine simulation test facilities 航天医学仿真试验设备
space medico-engineering 航天医学工程,空间医学工程
space medico-engineering facilities 航天医学工程设施
space microbiology 空间微生物学
space mirror 空间镜
space mission 航天任务
space motion sickness 航天运动病,空间运动病
space motion sickness susceptibility 航天运动病易感性
space myopia 空间近视
space neural science 航天神经科学
space neurobiology 空间神经生物学
space noise 空间噪声
space nuclear power 空间核电源
space object 空间物体
space observation 空间观测,太空观测
space operation center 航天操作中心
space operational medicine 航天实施医学
space optics 空间光学
space orientation 空间定向
space oxygen generation 空间制氧
space particle radiation 空间粒子辐射
space pathology 航天病理学
space perception 空间感受
space pharmaceutics 航天药剂学
space pharmacokinetics 航天药物动力学
space pharmacology 航天药理学,空间药理学
space photogrammetry 航天摄影测量,太空摄影测量
space photography 航天摄影
space physics 空间物理学
space physics exploration 空间物理探测
space physiological stress 航天生理应激
space physiology 航天生理学
space physiology and medicine 航天生理医学
space planetary exploration 航天行星探测
space plasma 空间等离子体

space platform 空间站
space pointing error 空间指向误差
space power 空间动力
space power source 航天能源
space power system 宇宙发电装置,航天动力系统
space probe 太空探测器
space program 太空计划
space propulsion 航天推进
space psychology 航天心理学
space race 太空竞赛
space radar 航天雷达
space radiation 空间辐射
space radiation environment 空间辐射环境
space ration 航天给养
space reactor 空间反应堆
space recipe 航天食谱
space reconnaissance 航天侦察
space remote sensing 航天遥感
space remote sensing for archaeology 航天遥感考古
space remote sensor 航天遥感器
space rendezvous 空间交会
space rescue 航天救援,航天救生
space research institute 航天科学研究机构
space research service 空间研究业务
space resection 空间后方交会
space return technology 航天返回技术
space robot 1.航天机器人 2.空间机器人
space rocket engine 空间火箭发动机
space science 宇宙科学
space segment 空间段
space ship 宇宙飞船,船天飞船
space shuttle 航天飞机
space shuttle imaging radar 航天飞机成像雷达
space shuttle main engine 航天飞机主发动机
space shuttle orbiter 空间往返轨道飞船
space sickness 空间飞行不适应症
space simulation 空间仿真
space simulator 空间模拟器,太空模拟器
space sleeping bag 航天睡袋
space station 宇宙空间站,太空电台
space station launch 空间站发射
space station remote manipulator system 空间站遥控机械手系统
space structure 空间结构
space suit 宇宙飞行服,宇宙服
space suit circulation system 航天服循环系统
space suit pressure schedule 航天服压力制度
space surveillance 航天监视
space surveillance and tracking system 空间监视与跟踪

系统
space surveillance system 空间监视系统
space system 空间系统
space system engineering 航天系统工程
space technology 1.航天技术 2.空间技术
space telecommand earth station 空间电信指挥地面站
space telemetry 航天遥测
space telemetry and control technology 航天测控技术
space telescope 太空望远镜
space test diet 航天飞行试验膳食
space toxicology 空间毒理学,航天毒理学
space tracking 航天跟踪
space tracking and data acquisition 航天测控与数据采集
space tracking and data acquisition network 航天测控与数据采集网
space tracking network 空间跟踪网
space tracking telemetry and control center 航天测控中心
space transportation 航天运输
space transportation system 航天运输系统
space travel 太空旅行,宇宙航行
space truss 空间桁架
space tug 空间拖船
space vector 空间矢量
space velocity 空速
space warfare 天战
space wave 空间波
space weapon 航天武器
space weather 空间天气
space-based space surveillance 空间基空间监视
space-based telescope 天基望远镜
spaceborne 航天器上的,航天器遥测
spaceborne computer 空间飞行器载计算机,宇宙机载计算机
spaceborne radar 航天(器载)雷达,星载雷达
spacebus 空间客车
space-charge grid 空间电荷栅极
space-charge region 空间电荷区
SPACECOM Space Communication 空间通信
spacecraft 航天器,航天器电视,航天器技术,航天器遥测设备
spacecraft assembly and test building 航天器装配测试厂房
spacecraft camera 航天器摄影机
spacecraft clock 航天器时钟
spacecraft component 航天器部件
spacecraft configuration 航天器(总体)布局,航天器构型

spacecraft contamination 航天器污染
spacecraft design 航天器设计
spacecraft detection 航天器探测
spacecraft docking 航天器对接
spacecraft dynamics 航天器动力学
spacecraft electronic equipment 航天器电子设备
spacecraft engineering 航天器工程
spacecraft environment 航天器环境
spacecraft environment contaminant 航天器环境污染
spacecraft environment engineering 航天器环境工程
spacecraft instrument 航天器仪表
spacecraft internal torque 航天器内力矩
spacecraft landing 航天器着陆
spacecraft launching 航天器发射
spacecraft launching base 航天器发射基地
spacecraft launching complex 航天器发射场
spacecraft launching site 航天器发射场
spacecraft long term operation management 长期运行管理
spacecraft manufacturing 航天器制造
spacecraft manufacturing engineering 航天器制造工程
spacecraft manufacturing technology 航天器制造工艺
spacecraft material 航天器材料
spacecraft model 航天器模型
spacecraft orbit 航天器轨道
spacecraft power supply 航天器电源
spacecraft power system 航天器电源系统
spacecraft prelaunch test 航天器发射前试验
spacecraft propulsion 航天器推进
spacecraft recovery 航天器回收
spacecraft recovery system 航天器回收系统
spacecraft reentry 航天器再入
spacecraft returning technique 航天器返回技术
spacecraft sterilization 航天器消毒
spacecraft structural strength 航天器结构强度
spacecraft structure 航天器结构
spacecraft structure system 航天器结构系统
spacecraft system 航天器系统
spacecraft technology 航天器技术
spacecraft test 航天器试验
spacecraft thermal control 航天器热控制
spacecraft thermal control system 航天器热控系统
spacecraft transfer 航天器转运
spacecraft-rocket unite check 航天器火箭联合检查
spaced armor 屏蔽装甲
space-fixed coordinate system 空间固连坐标系
space-ground communication 天地通信
space-ground communication system 天地通信系统
space-ground equipment 航天地面设备

space-ground integrating test	天地对接试验
spacelab	航天实验室
spaceport	航天港
space-qualified	空间适用的
spacer	垫片,隔离片,定位架
space-rated	适用于空间
spaceship	航天飞机,宇宙飞船
spaceship sterilization	飞船消毒
spacesuit	航天服
space-tapered array	变距阵
spacetime	空间和时间的
space-type diet	航天膳食,航天飞行膳食
spacewalk	太空行走,空间行走
spacing	间隔,间距,距离

SPADATS Space Detection And Tracking System 空间探测和跟踪系统

SPADS Satellite Position And Display System 卫星定位和显示系统

spaghetti	漆布绝缘管
spaghetti tubing	细麻布绝缘套管
spallation	剥落,散裂,分裂,蜕变

SPAM Satellite Processor Access Method 卫星处理机存取方法

span	1.量程 2.翼展 3.跨度,跨距,范围
span loading	翼展荷重,翼展载荷

SPAR 1. Slight Precision Approach Radar 辅助精密进近雷达 2. Special Passenger Arrangement 特殊旅客安排

spare kit	备用工具箱
spare level	备用级
spare parts	航材支援
spare track	备用磁道
spark	1.火花,电火花 2.发火花
spark chamber	火花室
spark channel	火花塞室通路
spark detector	火花探测器
spark discharge	火花放电
spark energy	火花能量
spark erosion	电火花侵蚀,电火花腐蚀
spark ignition	火花点火
spark leak detector	火花检漏器
spark machining	电火花加工
spark photography	闪光照相
spark plug	火花塞
spark plug resistor	火花塞电阻(器),电嘴电阻(器)
spark source mass spectrometry	火花源质谱
spark test	火花试验
spark-erosion machining	电火花加工
spark-erosion perforation	电火花穿孔
spark-erosion wire cutting	电火花线切割
sparking	发火花,点火
sparking plug	火花塞,点火电咀

SPAS 1. Safety and Performance Analysis System 安全性能分析系统 2. Safety Performance Analysis Subsystem 安全性能分析子系统

Spasur space-surveillance (system) 空间监视(系统)

SPATE Special Purpose Automatic Test Equipment 专用自动测试设备

spatial and temporal distribution	时空分布
spatial calibration	空间配准
spatial correlation	空间相关性
spatial data	空间数据
spatial development	空间发展
spatial dimension	空间维度
spatial discretization	空间离散化
spatial disorientation	空间迷航,空间定向障碍
spatial distribution	空间分布
spatial domain	空间域
spatial filtering	空间滤波
spatial frequency	空间频率
spatial location	空间位置
spatial mode	空间模式
spatial organization	空间组织,空间结构
spatial orientation	空间定向
spatial orientation trainer	空间定向训练器
spatial resolution	空间分辨率
spatial separation	空间隔离
spatial variation	空间变化,区域差异
spatiography	宇宙物理学
spationaut	航天员,宇航员
spats	轮罩,机轮整流罩

SPC 1. Semi Permanent Connection 半永久性连接 2. Signaling Point Code 信令点代码 3. Space 空间,间隔 4. Spare Parts Classification 备件分类 5. Statistical Processes Control 统计过程控制

SPCD Specification Document 规范文档

SPCM Specification-Make 规范编制

SPCO Specification-Outsid 超出规范

SPD Speed 速度

SPDL Supplier Part Data List 供应商零件数据表

SPDS Space-based Position Determination System 星基位置测定系统

SPE 1. Seller Purchased Equipment 卖方购买设备 2. Special Purpose Equipment 专用设备 3. Spherical Probable Error 球概率误差 4. Special Purpose Entity 特别目的机构 5. Synchronous Payload Envelope 同步净负荷区包迹

speaker	扬声器,话筒

SPEC　1. Special 专门的，特殊的 2. Scram Prevention Evaluation Checklist 防止紧急停止评定清单 3. Specification 规格，规范 4. Speech Predicrive Encoding Communication 语音预测编码通信
special altitude diet　特殊高空膳食
special ammunition　特种弹药
special attention　特别关注
special checkout equipment for satellite　卫星专用测试设备
special fares　特殊票价
special flight　特殊飞行，特殊任务飞行
special flight permit　特许飞行证
special material　特种材料
special nuclear material　特种核材料
special peripheral equipment　专用外部设备
special perturbation　特殊摄动
special physical examination　专项体检
special protective garment　特殊防护服
special purpose computer　专用计算机
special rule zone　特殊规则区
special shell　特种炮弹
special simulation technique　特殊仿真技术
special telemetry equipped missile　装有特种遥测设备的导弹
special test facility　专用测试设备
special tooling　1.专用工艺装备 2.专用工装
special VFR　特殊目视飞行规则（目视飞行的最低气象条件低于正常标准）
special weapon　特殊武器
special weapon emergency separation system　特种武器应急投放系统
special weapon overflight guide　特种武器飞越引导
special weapons disposal　特种武器处理
specialize　专门研究，专攻
species　物种，种类
species composition　物种组成，种类成分
specific capacitance　单位电容
specific capacity　比容量
specific charge　荷质比
specific commodity rate　单位商品销售率
specific conductance　电导率
specific consumption　单位消耗量
specific design　特定设计
specific efficiency　1.比冲效率 2.冲量效率
specific energy　比能量
specific enthalpy　比焓
specific excess power　单位剩余功率
specific force　比力
specific fuel consumption　燃料消耗率
specific heat　比热，比热容
specific heat calorimeter　比热容测定仪
specific heat capacity　比热容
specific heat ratio　比热比
specific humidity　比湿度
specific impulse　比推力，比冲量
specific impulse efficiency　比冲效率
specific impulse of blast wave　冲击波比冲量
specific mass　质量密度，比质量
specific optical density　比光密度
specific power　单位功率，比功率
specific propellant consumption　推进剂比消耗率
specific range　给定航程，特定飞行距离
specific search　全结构检索专用检索
specific speed　比速，比转数
specific stiffness　比刚度
specific strength　比强度，单位强度
specific thrust　比推力，单位推力
specific volume　比容积
specific weight　比重
specification　详细说明，说明书，指定，规格，规范，技术要求，技术条件
specification limit　规格界限，规范说明书，技术说明规范
specified　规定的，额定的，给定的
specified value　规定值
specify　具体说明，详细说明，规定，指定
specimen　样品，样本，标本
specimen geometry　标本几何学
speckle　散斑
speckle effect　斑点效应
speckle noise　斑点噪声
spectacles　眼镜特技
spectra　光谱，谱
spectral absorptance　光谱吸收比
spectral analysis　光谱分析
spectral bandwidth　光谱带宽
spectral channel　光谱波道
spectral classification　光谱分类
spectral content　光谱成分
spectral density　谱密度
spectral density matrix　谱密度矩阵
spectral detectivity　光谱探测率
spectral domain　谱域
spectral emissivity　光谱发射率，光谱比辐射率
spectral estimate　谱估计
spectral feature　光谱特征
spectral filter　光谱过滤器
spectral filtering　光谱滤波

spectral intensity 光谱强度,频谱辐射强度
spectral interference 光谱干扰
spectral irradiance 光谱辐照度
spectral line 光谱线,谱线
spectral matrix 频谱矩阵
spectral method 谱方法
spectral noise equivalent power 光谱噪声等效功率
spectral peak 光谱灵敏度峰值
spectral power 谱功率
spectral purity 频谱纯度
spectral radiance 光谱辐射率,光谱辐射强度
spectral radiant flux 光谱辐射能量
spectral range 频谱范围,光谱范围
spectral reflectance 光谱反射系数
spectral reflection characteristics 光谱反射特性
spectral reflection factor 光谱反射因数
spectral region 光谱区,光谱范围
spectral resolution 光谱分辨率,频谱分辨率
spectral response 光谱响应
spectral sensitivity 光谱灵敏度
spectral sensitivity curve 分光感度曲线
spectral set 谱集
spectral signature 光谱特征
spectral transmission factor 光谱透射因数
spectral transmittance 光谱透射
spectral type 光谱型
spectral width 谱宽
spectrogram 光谱图
spectrograph 摄像仪,分光摄像仪
spectrographic analysis 光谱分析
spectroheliograph 太阳单色光照相仪,太阳摄谱仪
spectroheliography 太阳单色光照相术
spectrohelioscope 太阳光谱观测镜,太阳单色光观测镜
spectrometer 分光计,分光仪,光谱仪,摄谱仪
spectrometric analysis 光谱测定分析
spectrometry 光谱测定
spectrophotometer 分光光度计
spectrophotometry 分光光度测定法,分光测光学
spectropyrheliometer 太阳辐射能光谱仪,太阳分光热量计
spectroradiometry 分光辐射度学
spectroscope 分光仪,分光镜
spectroscope amplifier 谱仪放大器
spectroscopic binary 分光双星
spectroscopic measurement 分光测定
spectroscopic method 光谱法
spectroscopic parallax 分光视差
spectroscopic temperature measurement 光谱法温度测量

spectroscopy 光谱学
spectrum 光谱,谱,频谱,波谱,交叉谱
spectrum analysis 频谱分析
spectrum analyzer 频谱分析仪
spectrum characteristic 波谱特性
spectrum cluster 波谱集群
spectrum density 谱密度
spectrum emissivity 发射率
spectrum estimate 谱估计
spectrum feature space 波谱特征空间
spectrum filtering 光谱滤波
spectrum hours 力谱小时
spectrum index 谱指数
spectrum intensity 谱辐射强度
spectrum inversion 频谱翻转
spectrum irradiance 光谱辐照度
spectrum management computer 光谱管理计算机
spectrum of electromagnetic wave 电磁波谱
spectrum of radiation 辐射光谱
spectrum of turbulence 扰动范围,湍流谱
spectrum pattern 谱图
spectrum purity 频谱纯度
spectrum response curve 波谱响应曲线
spectrum roll-off 频谱滚降
spectrum sensitivity 光谱灵敏度
spectrum slope 谱斜率
spectrum smoothing 谱平滑
spectrum stabilizer 稳谱器
spectrum stripping 剥谱
spectrum synthesis shock 冲击谱合成
spectrum utilization 频谱利用
spectrum variable 光谱变量
spectrum-luminosity diagram 光谱光度图
specular reflectance 镜面反射率
specular reflection 镜面反射
specular reflector 镜面反射器
SPED Spares Prices Estimatilng Database 备件价格评估数据库
speech communication 语言通信
speech encoding 语音编码
speech interference level 语言干扰级
speech inversion 1.语言频率倒演 2.通信频率倒换
speech network 语音网络
speech processing 语音处理
speech recognition 通话识别
speed 速率,速度
speed adjustment 速度调节
speed brake 减速刹车
speed change 变速

speed check tape	测速带
speed control	速度控制
speed control range	速度控制范围
speed course	速度测定基线
speed deceleration	速度阻滞
speed difference	1. 转差 2. 转速差
speed factor	速度系数,速度因数,速度因子
speed for best rate of climb	最佳爬升率速度
speed for maximum endurance	久航速度
speed for maximum range	最大航程速度
speed for steepest climb	陡升速度
speed governor	调速器,限速器
speed hang-up	转速悬挂
speed line	速率线,测速位置线,加速气管
speed memory	速度记忆
speed of light	光速
speed of response	响应速度
speed of sound	声速
speed profile	速度曲线
speed range	速度范围,转速范围
speed reduction	减速
speed reference system	速度参照系统
speed shutdown	速度关机
speed stability	速度安定性
speed synchronizer	转速同步器
speed up	加速
speed variation	速率变化
speed vector	速度向量
speedometer	速度表,转速表
speed-up capacitor	加速电容
spent bullet	乏弹(指枪、炮弹在飞行末端,冲力已尽)
spent fuel	废燃料,用过的燃料,乏燃料
SPET	Solid Propellant Electric Thruster 固体推进剂电推进器
SPF	1. Service Port Function 业务端口功能施 2. Specific Pathogen Free 无菌的 3. Storage Protect Feature 存储保护特性
SPGG	Solid Propellant Gas Generator 固体推进剂燃气发生器
SPH	Supplier Planning Horizon 供应商计划水平线
sphere	球,球面,球体,范围
sphere of influence	势力范围,影响范围
sphere surface	球面
spheric approximation	球面近似法
spherical aberration	球面像差
spherical albedo	球面反照率
spherical angle	球面角
spherical array	球面阵
spherical astronomy	球面天文学
spherical blast	球状爆炸波
spherical cap	球冠
spherical charge	球形装药
spherical component	球状子系
spherical coordinates	球面坐标
spherical divergence compensation	球面发散补偿
spherical error probable	圆概率误差,圆公算偏差
spherical galaxy	球状星系
spherical gas bottle	球形气瓶
spherical geometry	球面几何学
spherical harmonics	球谐函数,球面调和
spherical pyranometer	球形总日射表
spherical shape	球形
spherical shell	球形壳体
spherical subsystem	球状次系
spherical triangle	球面三角形
spherical trigonometry	球面三角(学,法)
spherical vessel	球形容器
spherical wave	球面波
sphericity	球形,球面
spherics	球面几何学,球面三角学
spheriflex hub	球形柔性桨毂
spherodizing	球状处理,球状化退火处理
spheroid	球体,回转椭圆体
spherometer	球面仪
sphygmography	脉搏描记法
sphygmomanometer	血压计
SPI	1. Short-Pulse Insertion 短脉冲输入 2. Special Position Identification 专用位置识别,特殊位置识别 3. Special Process Instruction 特殊位置标识
spicules	冰针,针状物
spider	1. 转子支架 2. 电枢辐臂
spigot	插口,套口接头,管端凹凸槽接合,栓塞,龙头
spike	尖峰信号,测试信号,钉,尖端,脉冲,点火燃料组件(反应堆技术)
spike burst	尖峰爆发
spike deconvolution	脉冲反褶积
spike inlet	带中心锥的进气口
spike leakage energy	波尖漏过能量
spike nozzle	尾锥喷管
spill burner	回油式喷燃器
spill door	溢流门
spill effect	溢出效应
spillage	溢出量,泄漏量,溢出
spilling	溢出,溅出
spillover	溢出,溢出量
SPILS	Stall Protection and Incidence Limiting System 失速防护及倾角限制系统
spin	1. 螺旋,尾旋 2. 旋转,自旋 3. 旋转成形

spin angle	自旋角
spin axis	1.(陀螺的)自旋轴,旋转轴,陀螺轴 2.螺旋轴
spin axis of gyro	陀螺自转轴
spin chute	反螺旋伞
spin decoupling	自旋去耦
spin down	旋转减慢
spin dynamics	旋转动力学
spin forming	旋压成形
spin motion	自转运动
spin motor	自旋发动机,旋转稳定用发动机,(导弹的)自旋火箭发动机
spin rate	滚转角速度,旋转角速度
spin rate control quantity	转速控制量
spin scan cloud camera	自旋扫描摄云相机
spin scanning earth sensor	自旋扫描地球敏感器
spin scanning horizon sensor	自旋扫描地平仪(靠星体自旋对地球扫描的红外地平仪)
spin shaping	旋压成形
spin stabilization	自旋稳定性,自旋稳定
spin stabilized satellite	自旋稳定卫星
spin table	旋转台
spin tuned magnetron	旋转调谐磁控管
spin tunnel	旋转风洞,螺旋风洞
spin up	起旋,加速旋转
spin vector	自旋向量
spin wave	自旋波
spindle	1.轴,心轴 2.杆,柱 3.(硬式飞艇的)锚杆构架(系留用)
spindle galaxy	纺锤状星系
spine	锥形骨架
spine robot	龙骨式机器人
spinel type ferrite	尖晶石型铁氧体
spinner	螺旋桨整流罩
spinner magnetometer	旋转磁强计
spinning	1.自旋,旋转 2.旋压(成形) 3.进入螺旋
spinning missile	自旋导弹
spinning nose dive	螺旋俯冲
spinning rate sensor	自旋速率敏感器
spinning with reduction	1.交薄旋压 2.旋薄
spinodal curve	旋节线
spin-stabilized rocket	旋转稳定火箭
spiral	螺旋
spiral angle	(螺旋齿轮的)螺旋角
spiral antenna	螺旋天线
spiral arm	旋臂
spiral bevel gear	螺旋伞齿轮,螺旋斜齿轮,锥形螺旋齿轮
spiral curve location	缓和曲线测设
spiral dive	急盘旋下降
spiral galaxy	螺旋形星云,旋涡星系
spiral glide	盘旋下降
spiral instability	螺旋不稳定性
spiral lens	螺旋透镜
spiral mode	螺旋模态
spiral orbit	螺旋轨道
spiral structure	旋涡结构
spiral target	螺旋目标
spiral target boundary	螺旋目标边界
spiral trajectory	螺旋轨迹
spiral transfer	螺旋转移
spiral winding	螺旋缠绕
spiraling rate	径向扩张率
spiral-phase antenna	螺旋相位天线
spiroid gear	锥涡轮
spirometer	肺量计
spirometry	肺量测定法
SPKR	Speaker 话筒,喇叭
SPL	Spare Parts List 备件清单
splash	泼溅,飞溅
splash lubrication	泼溅润滑
splash point	弹着点,溅点
splashdown	溅落
splashdown area	溅落区域
splashdown zone	溅落海域
splashing	水面溅落
splashing type pressure transducer	溅射式压力传感器
splatter	邻信道干扰,边带泼刺声
splice	1.(板形结构的)搭接,拼接 2.(钢索的)编接,绞接
spline	花键,花键接合用花键连接,开键槽,曲线尺
spline function	样条函数
split cameras	分束照相机
split charge	瞬时充电
split charter	并装包机(两家公司联合包租一架飞机)
split flap	裂开襟翼,分裂襟翼
split pair	劈分线对
split phase coding	分相编码
split vertical photography	分行竖直摄影
split-folded wave guide	分离折叠波导
splitter	分路器,分配器,分解器,分流隔板
splitter plate	导流板,分流板
splitting	分光
splitting parameter	分裂参数
SPM	1. Seats Protected Message 座位保留文电 2. Self Phase Modulation 自相位调制 3. Signal Processing Module 信号处理组件 4. Stabilizer Position Module 安定面位置模块(组件) 5. Surface Position Monitor 活动

面位置监控器

SPOH Since Part(ial) Overhaul 自局部翻修后(计算使用时间用)

spoiled material 废旧器材,废料

spoiler 扰流板,阻流板,气流防护板

spoke 辐条,轮辐

spoking (雷达荧光屏的)黑白扫描线混乱,交替干扰

spongy 海绵状的

sponson 机侧突出部,水上安定器,翼梢浮筒舷台,船旁保护装置,船侧凸出部

spontaneous detonation 自发爆炸,自炸

spontaneous emission 自发发射

spontaneous nystagmus 自发性眼球震颤

spoof 使迷失方向,使迷航

spoofer 采用电子或战术欺骗手段的飞机,诱骗装置

spoofing 电子欺骗,电子迷惑

spool 卷轴,线轴,线圈架,线圈,卷筒,卷,盘,直管段

spool turbojet 铜管式涡轮喷气发动机

sport parachute 运动伞

sport physiology 运动生理学

spot 1.点,光点 2.地点,部位 3.定位,确定位置,发现,探测目标 4.校测,观测弹着,确定射击偏差(以供校射) 5.机场上指定的着陆位置 6.(在航母甲板上)舰载机排成紧密队形(准备弹射起飞) 7.电子冲击荧光管表面的光亮区 8.地面防空监视哨(英国在第二次世界大战中用的民防用语) 9.瞄准式干扰

spot annealing 局部退火

spot beam 点波束

spot elevation 独立高程点

spot facing 锪平台,刮孔口平面

spot height 地面点高度,高程点

spot hover 现场盘旋,定点悬停

spot jammer 选择性干扰机

spot jamming 瞄准干扰

spot net 点网络

spot size 黑子大小,点尺寸

spot welding 点焊

spotlight 聚光灯

spotlight synthetic aperture radar 聚束

spotter 观空观察员,观察机,航空观察员

spotting 确定位置,瞄准目标,校射,弹着观测

spot-weld bonding structure 胶接点焊结构

SPP 1. Sensor Pointing Platform 遥感器定向平台 2. Sequence Packet Protocol 顺序包协议 3. Service Provision Point 服务提供点 4. Signal Processing Peripheral 信号处理外部设备 5. Special Purpose Processor 专用处理机 6. Speech Path Processing 话路处理 7. Standard Passenger Payload 标准旅客商载

SPPL Supplier Planning Parts List 供应商计划零件表

SPR 1. Sending Packet Rate 分组信息发送率 2. Service and Performance Report 勤务和性能报告 3. Spares Parts Release (Form) 备用零件发放(表格) 4. System Performance Report 系统性能报告

sprag clutch 斜撑离合器

spray 喷射,水沫,浪花,喷雾,雾化,喷溅

spray angle 喷雾角度

spray atomization and deposition 雾化喷射沉积

spray attack 喷撒攻击

spray bar 喷油管

spray characteristic 喷雾特性

spray coating 喷涂

spray coating foaming 喷涂发泡

spray combustion 喷雾燃烧

spray cone 喷射雾锥

spray cone angle 喷雾锥角

spray device 喷洒器

spray dome 喷水水墩

spray drying 喷射干燥

spray formed material 喷射成形材料

spray forming 喷射成形

spray jet 喷淋器,喷水口,喷雾嘴

spray particle 喷淋粒子

spray particle size 喷淋粒子大小

spray process 喷射成形

spray resistance 喷雾阻力

spray-cone angle 喷射锥角

spraying coating 喷涂包覆

spraying lacquer 喷漆

spraying painting 喷涂

spraying-ceramic seal ring 喷涂陶瓷密封环

spread 伸开,伸长,伸展,展开,涂胶量

spread fading channel 扩散衰落信道

spread spectrum 扩展频谱,频谱扩展

spread spectrum communication 扩频通信

spread spectrum multiple access 扩谱多址

spreader 扩展器,撒布器

spreading rate 扩张(速)率

spreading resistance 扩展电阻

spreadsheet 电子表格,数据表

spring 1.弹簧 2.跃出,裂开

spring analogy 弹簧近似

spring back 弹性回复

spring constant 弹簧常数

spring cover 弹簧口盖

spring equinox 春分

spring model 弹簧模型

spring rate 弹簧比率,弹簧刚度,弹簧刚性系数

spring recoil adapter 弹簧后座缓冲器,弹簧制退器

spring separation device	弹簧分离机构
spring sheet-holder	弹簧定位销
spring stiffness	弹簧刚度
spring tab	弹簧式补偿片
springback	弹性后效

SPRR Small Part Replacement Request 小零件替换需求

SPS 1. Sensor Processing Subsystem 传感器处理子系统 2. Service Process Solution 服务处理结果,服务进程方案 3. Service Provider System 服务供应商系统 4. Software Product Specification 软件产品规范 5. Speech Path Subsystem 话路子系统 6. Standard Position Service 标准定位服务 7. Surge Prevention System 防喘振系统

spur gear	直齿轮,圆柱齿轮
spurious correlation	伪相关
spurious effect	寄生效应
spurious output	乱真输出
spurious response	杂散响应/假信号/附加响应
sputter	飞溅,溅射
sputter ion pump	离子溅射原泵(一种真空泵)
sputtering	溅射,喷镀
sputtering deposition	溅射沉积
sputtering etching	沉射刻蚀

SQ 1. Service Quality 服务质量 2. Software Quality 软件质量 3. Squelch 静噪

SQA 1. Service Quality Agreement 服务质量协议 2. Software Quality Assurance 软件质量保证 3. System Queue Area 系统排队区

SQAP Software Quality Assurance Plan 软件质量保证大纲

SQBP Service Quality/Billing Processor 服务质量/填表处理器

SQC Statistical Quality Control 统计质量控制

SQD Service Quality Data 服务质量数据

SQE Software Quality Engineering 软件质量工程

SQL Structured Query Language 结构询问语言

SQM Square Meter 平方米

SQP Signal Quality Parameter 信号质量参数

SQPP Software Quality Program Plan 软件质量项目大纲

squad	班,组
squadron	空军中队,团体,中队,中队飞行队形
squall	1. 飑(间歇的强风)2. 持续几分钟的阵风,水平方向达1千米以上
square	正方形,街区,平方
square cavity	方腔
square cross-section	方形断面
square cylinder	方柱
square error	平方误差
square groove	方形轧槽
square matrix	方阵,矩形矩阵
square meter	平方米
square method	方格法
square parachute	方形(降落)伞
square plate	方形板
square resistance	方块电阻
square rod	方棒
square root	平方根,二次根
square sail	横帆
square search	方形搜索,展开方形搜索
square thread	直角螺纹,方螺纹
square wave	矩形波,方波
square-law compensation	平方律补偿
square-law detection	平方律检波
squarer	矩形波形成器,平方电路
square-wave generator	方波发生器
squaring shears	龙门剪床,平台剪切机
squeal	尖叫
squealer	声响器
squealer tip	凹槽状叶顶
squeeze	挤,压
squeeze bottle	塑料挤瓶
squeeze casting	模压铸造
squeeze riveting	压缩铆接法
squeezed state	压缩态
squib	小型烟火装置,爆炸帽
squidding	灯泡状态
squint	斜视角,斜视
squint angle	斜视角
squint mode	斜视模式
squirrel cage	鼠笼绕组,鼠笼式的
squirrel-cage induction motor	鼠笼式电动机
squirt	喷射,喷流

SR 1. Search and Rescue 搜寻与救援 2. Senior 年长的,高等的 3. Serialization Requirements 系列化要求 4. Service Request 服务需求 5. Sir 先生 6. Slant Range 斜距 7. Special Repair 专门修理 8. Sunrise 日出

SRA 1. Short Range Altitude 短程高度 2. Special Rules Area 特殊规则区 3. Specific Range Air 特定范围大气 4. Surveillance Radar Appoach 监视雷达进近

SRAAM Short Range Air-to-Air Missile 近距空空导弹(英国的试验性红外寻的导弹)

SRADD Software Requirements And Design Description 软件要求和设计说明

SRAM 1. Short-Range Attack Missile 近程攻击导弹 2. Static Random Access Memory 静态随机存取存储器

SRARM Short-Range Anti-Radiation Missile 近程反辐射导弹

SRB Solid Rocket Booster 固体火箭助推器/Source Route Bridge 源路由网桥

SRBM Short-Range Ballistic Missile 近程弹道导弹（射程约1 000千米以内）

SRC 1. Secondary Radar Code 二次雷达密码 2. Sensitivity Range Control 灵敏度范围控制 3. Service Response Center 服务响应中心 4. Science Research Council 科学研究委员会 5. Shop Resident Control 车间居留控制 6. Short Range Clamp 近程箝位 7. Spares Release Card 备件发放卡 8. Special Release Card 特殊发放卡

SRD 1. Software Requirement Document 软件要求文件 2. System Requirements Document 系统要求文件

SRDM Scanning Radiometer Data Manipulator 扫描辐射计数据处理器

SRE 1. System Responsible Engineer 系统责任工程师 2. Surveillance Radar Element 监视雷达元件

SRI Space Research Institute 1.空间研究院（保加利亚的国家空间活动中心，设在索菲亚）2.空间研究所（奥地利的主要空间科学研究机构）

SRL 1. Serial 顺序的 2. Significant Rework Log 重大返工记录 3. Stability Return Loss 稳定回损 4. System Reference Library 系统引用库

SRM 1. Send Routing Message 发送选路消息 2. Session and Resource Manager 会晤和资源管理器 3. Signaling Route Management 信令路由管理 4. Structure Repair Manual 结构修理手册

SRN Short-Range Navigation 近距导航系统

SRO Senior Reactor Operator 反应堆高级操纵员

SROA Schedule Requirements and Order Analysis (System) 计划要求和订货分析（系统）

SROC Short-Range Omnidirection Ceacon 近距离全向信标

SRP 1. Seat Reference Point 座位基准点 2. Selected Reference Point 选择基准点 3. Service Related Problem 维修相关问题 4. Snared Resources Processing 共享资源处理 5. Single Rotation Propeller 单旋转螺旋桨 6. Spatial Reuse Protocol 空间复用协议 7. Standard Repair Procedure 标准修理程序

SRR 1. Search and Rescue Region 搜寻与援救地区 2. Synchronous Round Robin 同步循环 3. System Requirements Review 系统要求评审

SRS 1. Seat Reservation System 订座系统 2. Secondory Radio Station 辅助无线电台 3. Series 系列 4. Software Requirements Specification 软件要求规范 5. Software Requirements Standard 软件要求标准 6. Speed Reference System 速度基准系统 7. Stimulated Raman Scattering 受激拉曼散射 8. Synchronous Relay Satellite 同步中继卫星

SRSR System Reliability and Safety Report 系统可靠性安全性报告

SRT 1. Satellite Receiver/Transmitter 卫星通信收发机 2. Standard Rate of Turn 标准转弯率 3. Subscriber Response Time 用户响应时间

SRTP System Rig Test Plan 系统台架试验大纲

SRTR System Rig Test Report 系统台架试验报告

SRU 1. Self Recording Unit 自动记录装置 2. Shop Replaceable Unit 车间可更换件，车间可替换装置

SRV 1. Safety Relief Valve 安全释放活门 2. Stopped-Rotor Vehicle 停转式旋翼飞行器

SRZ 1. Special Rules Zone 特殊规则地带 2. Survillance Radar Zone 监视雷达地带

SS 1. Service Standard 服务标准 2. Set Screw 定位螺丝钉 3. Single Shot 单冲 4. Sliding Scale 滑动标尺 5. Solid State 固体 6. Spares Support 备件支援 7. Speed Schedule 8. Static Strength 静强度 9. Stick Shaker 振杆器 10. Strength Specification 强度规范 11. Sunset 日落 12. Support Service 支援服务 13. Support Specifiation 支援规范 14. System Schematics 系统原理图，系统简图 15. System Specification 系统规范

SS&CR Stress Standards and Certification Requirements 应力标准和审定要求

SSA 1. Secured Systems Access 保护系统入口 2. Sector Safety Altitude 扇形区安全高度 3. System Safety Assessment 系统安全评估 4. Subsystem Allowed 允许的子系统

SSAL Simpilfied Short Approach Light 简式短进近灯光系统

SSB 1. Single Side Band 单边带 2. Split System Breaker 分离系统断电器 3. System Setup Box 系统起动盒 4. Symmeric Switched Broadband 对称交换宽带

SSBJ Supersonic Business Jet 超声速公务机

SSC 1. Self Sealing Coupling 自封接头 2. Service Support Center 服务保障中心 3. Single Stroke Chime 单击谐音 4. Spread Spectrum Communication 扩频通信 5. System Support Center 系统保障中心

SSCC Spin Scan Cloud Camera 自旋扫描摄云相机

SSCMP System Simulation Configuration Management Plan 系统模拟构型管理大纲

SSCNS Ships Self-Contained Navigation System 舰船自主式导航系统

SSCP 1. Service Specific Convergence Protocol 业务特定会聚协议 2. Service Switching and Control Point 业务交换控制点 3. System Service Control Point 系统服务控制点

SSD 1. Service Support Data 业务支撑数据 2. Shared Security Data 共享保密数据 3. Structure Safety Design 结构安全性设计 4. System Safety Document 系统安全性文件

SSE 1. Safe Shutdown Earthquake 地震安全停堆 2. South South East 南南东 3. System Supervisory Equipment 系统监控设备

SSEB Source Selection Evaluation Board 供方选择评价委员会

SSEC 1. Static Source Error Calibration 静压源误差校准 2. Static Source Error Correction 静压源误差修正

SSF 1. Service Switching Function 业务交换功能 2. Super-Saver Fares 超级省钱票价 3. System Support Facility 系统保障设施

SSFDR Solid-State Flight Data Recorder 固态飞行数据记录器

SSGS Standardized Space Guidance System 标准航天制导系统

SSI 1. Secondary Side Inspection 二次侧检查 2. Server Side Include 服务器端包含 3. Small-Scale Integration 小规模集成 4. Structural Significant Item 结构重要项目,重要结构项

SSIC Small Scale Integrated Circuit 小规模集成电路

SSID Supplementary Structural Inspection Document 结构检查补充文件,补充结构检查文件

SSKP Single Shot Kill Probability 单发杀伤概率

SSL Sound Suppression Liner 消声衬

SSM 1. Servo System Monitor 伺服系统监控器 2. Ship Short Memo 发运简要备忘录 3. Sign Status Matrix 信号状态基体 4. Solid State Memory 固态存储器 5. Surface-to-Surface Missile 面(地面或水面)对面导弹 6. System Schematics Manual 系统电路图手册,系统原理图手册

SSME Space Shuttle Main Engine 航天飞机主发动机

SSMO Safety Supervision Management Office 安全监督管理办公室

SSN 1. Seasonal 季节性的 2. Single Star Network 单星网络 3. Subscriber Switching Network 用户交换网络 4. Subsystem Number 子系统号码 5. Switched Service Network 交换业务网

SSO Single Sign On 单点登陆

SSOG Satellite Systems Operations Guide 卫星系统操作指南

SSOS Severe Storms Observational Satellite 强风暴观察卫星

SSP 1. Service Switching Point 业务交换点 2. Source Selection Plan 供货源选择计划 3. Supplier Specification Plan 供应商规范计划

SSPA 1. Solid State Phase Array 固态相控阵 2. Solid State Power Amplifier 固态功率放大器

SSPD Single Source of Product Data 单一产品数据源

SSPF Space Station Processing Facility 空间站加工设备

SSPI Single Source of Product Item 产品单一物料项源

SSPM 1. Software Standards and Procedures Manual 软件标准和程序手册 2. System Safety Program Manager 系统安全计划管理员

SSPO Support System Project Office 支援系统计划处

SSPP 1. System Safety Program Plan 系统安全性项目计划 2. System Security Program Plan 系统防护项目计划

SSR 1. Software Specification Review 软件规范评审 2. Special Service Requirement 特殊服务请求 3. Static Shift Register 静态移位寄存器 4. System Simulation Report 系统模拟报告

SSS 1. Solid State Switch 固态电门 2. Subscriber Service System 用户业务系统 3. Switched Star System 交换式星形系统 4. Switching Subsystem 交换子系统 5. System Simulation Specification 系统模拟规范 6. System/Subsystem Specification 系统/分系统规范

SSSM Subset-Specified Sequential Machine 子集规格时序机

SSSS Satellite Strategic Surveillance System 卫星战略监视系统

SST 1. Soliton-Supported Transmission 孤子支持的传输 2. Stainless Steel 不绣钢 3. Supersonic Transport 超声速运输机

SSTC Specialized System Test Contractor 专一系统测试承包人

SSTDMA 1. Satellite Switched Time-Domain Multiple Access 星上切换时分多址 2. Satellite System TDMA 卫星系统时分多址

SSTF Space Station Training Facility 空间站训练设施(美国)

SSTOL 1. Super Short Takeoff and Landing 超短距起落飞机 2. Supersonicand Short Takeoff and Landing 超声速短距起落飞机

SSTP System Simulation Test Plan 系统模拟试验大纲

SSU 1. Side Stick Unit 侧杆单元 2. Single Subscriber Unit 单用户单元 3. Space Switching Unit 空分交换单元 4. Subsequent Signal Unit 辅助顺序信号装置

SSUS Spinning Solid Upper Stage 自旋固体火箭上面级

SSVP System Simulation Validation Plan 系统模拟确认大纲

SSW 1. South South West 南南西 2. Standby Service Water 备用厂用水

ST 1. Seat 座位 2. Self Test 自测 3. Stabilizer Trim 安

定面配平 4. Start 开始 5. Statistics 统计 6. Steering Tiller 前轮转弯手柄 7. Street 街道 8. Stratus 层云 9. Summer Time 夏令时间 10. System Testing 系统测试

STA 1. Station 站,站位 2. Straight In Approach 直接进近

STAB 1. Stabilizer 安定面 2. Stabilize or Stabilization 稳定,安定性

stab detonator　刺发雷管
stabilator　全动式水平尾翼,全动平尾
stabilisation　稳定,稳压
stabilise　稳定
stabilitron　稳频管
stability　稳定性,稳定度
stability analysis　稳定性分析,稳度分析
stability analysis method　稳定性分析方法
stability augmentation　稳定性增强,稳定性的增长
stability augmentation system　稳定性增强装置,稳定性的增长系统
stability behavior　稳定性能
stability boundary　稳定性范围
stability characteristic　稳定性特征
stability condition　稳定条件,稳态
stability criterion　稳定度判据
stability curve　稳性曲线
stability derivatives　安定性导数,稳定性导数
stability domain　稳定域
stability factor　安定性系数,稳定因素
stability line　稳定性曲线
stability margin　稳定裕度
stability model　稳性试验模型
stability of gyroscope　陀螺稳定性
stability parameter　稳定性参数
stability problem　稳定性问题
stability property　稳定性
stability region　稳定域
stability robustness　稳定鲁棒性
stability test　稳定度试验
stability theorem　稳定性定理
stability theory　稳定理论
stability threshold　稳定阈度
stabilivolt　稳压管
stabilizability　可稳性
stabilization　稳定,安定
stabilization loop　稳定回路
stabilization margin of reentry vehicle　再入飞行器稳定裕度
stabilization process　稳定化进程
stabilization system　稳定系统

stabilization system test　稳定系统测试
stabilize　稳定,安定,保持平衡,使稳定,给……装安定器,消除内应力
stabilized master oscillator　稳定的主控振荡器
stabilized platform　稳定平台
stabilized power supply　稳定电源供应
stabilizer　1.安定面(尤指水平安定面),安定器 2.(化学反应的)稳定剂 3.(次要构件,如操纵面等内部的)低密度填心(如蜂窝结构) 4.(稳定用的)陀螺子系统(如雷达天线)
stabilizing　稳定化
stabilizing parachute　稳定伞(用于使空投伞稳定下降)
stabilizing skirt　稳定缘
stabilizing treatment　稳定化处理
stabilotron　稳频管
stable cavity　稳定谐振腔
stable combustion　稳燃
stable condition　稳定条件,稳定状态
stable crack　稳定裂缝
stable design　稳定设计
stable detonation　稳定爆炸
stable element　稳定元件
stable equilibrium　稳定平衡
stable fault　稳定性故障
stable isotope　稳定同位素
stable manifold　稳定流形
stable matrix　稳定矩阵
stable motion　稳定运动
stable operation　稳定运行,稳定操作
stable orbit　稳定轨道
stable platform　稳定平台
stable point　稳定点
stable region　稳定域
stable resonator　稳定谐振腔
stable stability　正安定性
stable system　稳定系统
stable tracking range　稳定跟踪距离
stable zone　稳定区

STAC 1. Single Terminal Access Controller (Software Utility) 单一终端入口控制器 2. Supersonic Transport Aircraft Committee 超声速运输机委员会(英国,1956—1962年)

stack　排气管,整齐地堆起,成堆
stacked dipole antenna　堆叠偶极天线
stacked loop antenna　多层环形天线
stacked profiles map　叠加剖面图
stacked solar cell　叠层太阳电池
stacked-beam radar　多层波束雷达
stacking density　堆积密度

stacking fault 堆垛层错
STACS Satellite Telemetry and Computer System 卫星遥测和计算机系统
STADAN Space Tracking and Data Acquisition Network 空间跟踪和数据采集网（美国）
stadia 视距,视距仪
stagnation 滞止
stage 阶段,时期
stage combustion 分级燃烧
stage cost 阶段成本
stage distance 级间距离
stage efficiency 分段效率,级效率
stage rocket 多级火箭
stage separation 多级分离
stage separation test 级间分离试验
staged combustion cycle 分级燃烧周期
staged combustion rocket engine 1.补燃火箭发动机 2.分级燃烧火箭发动机
staged crew 区间乘务员,接班空勤组
staged cutoff 分级关机
staged start 分级起动
stagger 斜罩,翼阶
stagger angle 安装角
stagger tuning 参差调谐,串联调谐
stagger wire 斜罩张线
staggered pulse train 参差脉冲列
staggered tuning 参差调谐
staging 级的分离,级的配置
staging area 集结待命地区
staging base 飞机中间停留基地
staging point 中途站
staging point for a rocket 火箭的分级点
stagnant hypoxia 郁滞性缺氧,循环障碍性缺氧
stagnant zone 滞留区
stagnation 停滞,滞止
stagnation condition 滞止状态
stagnation enthalpy 滞止焓
stagnation line 滞止线,驻点线
stagnation point 驻点,滞流点
stagnation pressure 滞止压力
stagnation pressure loss 滞止压力损失
stagnation pressure recovery 滞止压力恢复
stagnation temperature 驻点温度,滞止温度
stagnation zone 滞止区
stainless steel 不锈钢
stainless steel tube 不锈钢管
stake 钣金桩砧
stall 气流分离,失速,气体分离
stall accident 失速事故

stall condition 失速条件
stall control 失速控制
stall delay 失速延迟
stall flutter 失速颤动,失速颤振
stall inception 失速先兆
stall line 失速线,失速边界
stall margin 失速范围,失速界限
stall pattern 失速型态
stall point 失速点
stall propagation 失速传播
stall protection system 失速预防系统
stall speed 失速速度
stall warning 失速警报,失速预警,失速告警
stall warning system 失速警报装置,失速预警系统
stall zone 失速区
stalled pressure 极限压力
stalled torque 制动转矩,失速转矩,逆转转矩
stand 台,座
standard 标准规格,标准,标准器
standard air munitions package 标准空运军需品集装箱
standard antenna method 标准天线法
Standard ARM 制式武器
standard arrival route 标准进场航线
standard artillery atmosphere 标准弹道大气
standard artillery zone 标准弹道带
standard atmosphere 标准大气
standard cell 标准单元
standard cell method 标准单元法
standard class 标准等级
standard code 标准守则
standard conditions 标准条件,标准状态
standard coordinate 标准坐标
standard cosmological model 标准宇宙模型
standard cycle 标准循环
standard data format 标准数据格式
standard data set 标准数据集
standard deviation 均方根偏差,标准偏差
standard dispersion （弹道）标准散布
standard electronic module 标准电子组件
standard empty weight 标准空机重量
standard error 标准误差
standard field strength method 标准场强法
standard form 标准形式
standard gravity 标准重力（标准地球引力加速度时的重力）
standard grid 标准滤线栅
standard instrument departure 标准仪表（飞行起飞）离场
standard item 标准项目

standard load 正常负载
standard Loran 标准罗兰
standard mean chord 标准平均弦
standard method 标准方法,标准措施
standard model 标准样品,样机,标准形式
standard model test 标准模型试验
standard muzzle velocity 标准初速
standard operating procedures 标准工作程序
standard option 标准选件,标准选择
standard oxygen gear 标准供氧装备
standard parallel 标准纬圈,标准纬线
standard pitch 标准螺距
standard pressure altitude 标准气压高度
standard reentry trajectory 标准再入轨道
standard refraction 标准折射
standard signal generator 标准信号发生器
standard solar cell 标准太阳电池
standard specific impulse 标准比冲
standard star 标准星
standard stopping cross section 标准阻止截面
standard structure 标准组织
standard temperature 标准温度
standard temperature and pressure 标准温压
standard terminal arrival route 标准航站进场航路
standard testing motor 标准试验发动机
standard time 标准时间
standard trajectory 标准弹道
standard uncertainty 标准不确定度
standard white 标准白
standardisation 标准化
standardise 标准化
standardization program 标准化程序
standardized product 标准化产品
standby 1.备用的,应急的,待命的 2.备用品,应急设备 3.一等战备状态 4.(无线电呼叫信号)准备收报 5.(无线电代码)"我要暂停几秒钟" 6.如果有座位,可以登机
stand-by channel 备用通道
stand-by current 静态电流
stand-by emergency flight control system 备用飞行操纵系统
stand-by maintenance 辅助维护
standby mode 做好准备状态,备用状态
standby readiness 待机准备完好率
standby redundancy 置换式余度,备用冗余
standby regulator 备用调节器
standby reserve 待编后备队(美国)
standby station 备用站
standby time 待命时间

standing water 积水
standing wave 驻波,定波
standing wave antenna 驻波天线
standing wave detector 驻波检测器
standing wave meter 驻波表
standing wave ratio 驻波比
standoff 偏离预定航线
stand-off 1.基准距 2.离岸行驶
stand-off bomb 巡航炸弹巡航炸弹
stand-off distance 相隔距离
stand-off missile 空区外发射的空对地导弹
stand-off weapon 防区外发射武器
standpoint 立场,观点
Stanton number 斯坦顿数,史丹东数
staple fibre 切段纤维,人造短纤维
STAR Standard Terminal Arrival Route 标准进场路线,标准到场航线
star acquisition 星捕获
star atlas 星图集,星图
star catalog 星历表
star cloud 恒星云,星云
star cluster 星团,星群
star density 恒星密度
star field 星空背景
star grain 星形药柱,星形内孔燃烧的药柱
star identification 星识别
star magnitude 星等
star map 星地图,星图
star mapper 星测绘仪,星图制作者
star model 星体模型,恒星模型
star network 星状网,星形网,径向网
star recognition 星识别(对目标星与伪星进行判别)
star scanner 星体扫描装置
star search 星搜索
star sensor 星光传感器
star simulator (恒)星仿真器,(恒)星模拟器
star topology 星形布局
star tracker 星跟踪仪,星体跟踪仪
star tracking 星跟踪
star tracking guidance 星光制导
star vector 星光矢量
star wars program 星球大战计划(美国战略防御倡议计划的俗称)
star washer 星形垫圈
starboard 右侧,右舷,右
stardust 星尘
stare vision 凝视觉
starflex hub 星形柔性桨毂
staring homing head 凝视导引头

Starlab	恒星实验室(美国空间跟踪和定向实验设备)
starlight	星光
starlight scope	星光镜
starquake	星震
STARS	1. Silenttactical Attack/Reconnaissance System 无辐射战术攻击/侦察系统 2. Standard Terminal Arrival Route System 标准进场航路系统 3. Standard Terminal and Arrival Reporting System 标准航站及进场飞机报知系统 4. Surveillance Target Attack Radar System 监视目标攻击雷达系统
starspot	恒星黑子
starstreak	星光导弹(英国肖特公司生产的短程导弹)
start	起动
start ambient condition	起动环境
start level	起动级,起动电平
start of heading character	标题开始字符
start of text character	正文开始字符
start peak pressure	起动压力峰
start photographing	开拍
start point	起始点
start sensor	起动传感器
start signal	起始信号,启动信号
start switch	起动电门
start system	起动系统
start time	启动时间
start up cost	筹备费
start value	初始值
start vector	启动向量
starter	起动机
starter battery	起动用蓄电池组
starter gap	起动间隙
starter magneto	起动磁电机
starter system	初始系统初始值系统
starting	起动
starting chamber	起动点火室
starting coil	起动线圈(点火用)
starting current	起动电流
starting datum mark	起测基点
starting level	起始电平
starting load	起动载荷
starting phase	起始相
starting pressure	起动压力
starting symbol	起始符
starting torque	起动转矩,起动力矩
starting torque-frequency characteristic	起动矩频特性
starting up	开机,起动
start-stop teleprinter	起止式电传机
start-stop time	起停次数
start-stop type	启停式
startup	开车,起动发动机
startup characteristic	启动特性
startup time	启动时间
starute	星伞
state	振荡态
state component	状态分量
state constraint	状态受限
state diagram	状态图
state equation	状态方程
state estimation	状态估计
state event	状态事件
state feedback	状态反馈
state flight	专机飞行
state flip-flop	状态触发器
state information	状态信息,状态资讯
state minimization	状态极小化
state of charge	电荷状态
state of flag	旗帜国
state parameter	状态参数
state reduction	状态化简
state space description	状态空间描述
state trajectory	状态轨迹,状态轨道迹
state transition	状态转换
state transition matrix	状态转移矩阵
state uncertainty	状态不确定性
state variable	状态变量
state vector	状态向量
stated value	设定价值法定价格
state-estimator	状态判定器
state-feedback	状态反馈
statement	1. 语句 2. 陈述 3. 命题
state-space	状态空间
state-space method	状态空间法
state-space model	状态空间模型
static	1. 静止的,静力的,静压的,静电的 2. 静态,静力,静压,静电 3. 静电干扰 4. 天电干扰
static air-temperature indicator	大气静温表
static analysis	静态分析
static balance	静平衡
static balance test	静平衡试验
static calibration accuracy	静校准确度
static calibration of balance	天平静校
static calibration precision	静校精密度
static capacity	静电容
static ceiling	1. 静升限 2. 气球平衡高度(镇重全部投弃后的高度)
static charge	静电荷
static compliance	静态柔度
static condition	静止条件,静止状况

static control	静态控制,静电控制,静态控件
static conversion	静态转换
static cooling	静电冷却
static correction	静校正
static decoupling	静态解耦
static defect	静压损耗,静压亏损
static derivative	静导数
static discharger	静电放电器
static earth sensor	静态地平仪,静态地球敏感器(无活动部件,利用辐射平衡原理工作的地平仪)
static electricity	静电
static electricity resistance	防静电性
static equilibrium	静力平衡
static firing	静态射击,静态点火试验
static friction	静摩擦力
static friction torque	静摩擦力矩
static gearing ratio	静态传动比,舵面偏度与角位之比
static head	静压头,皮托管的静压受感器
static horizon sensor	静态地球敏感器
static input-output model	静态投入产出模型
static instability	静力学不稳定性,静态不稳定性
static interference	天电干扰,静电干扰
static inverter	静变流机
static lift	(气球的)静升力
static line	固定开伞索,静压操纵管路
static load	静载荷
static load test	静负载试验,静载荷试验
static map	静态图
static margin	静态冗余度,静安定度
static measurement	静态测量
static missile	静力试验导弹
static moment	静力矩
static photographic resolution	静态摄影分辨率
static pin	固定销
static position	静态定位
static pressure	静压强
static pressure gradient	静压梯度
static pressure gradient along tunnel axis	轴向静压梯度
static pressure head	静压头
static pressure lag	静压迟滞
static pressure measurement	静压测量
static pressure tap	静压测量点,静压接嘴
static pressure transducer	静压传感器
static rail	拉伞绳固定导轨
static random access memory	静态随机存取存储器
static regulator	静压调节器
static screen	静电屏蔽
static seal	静密封,衬垫,填料,衬料
static sensor	静态传感器
static soaring	静态滑翔,静力翱翔
static stability	静态稳定性,静安定性,静稳定
static stability compensation control system	静稳定性补偿控制系统
static stability marging	静稳定裕度
static stiffness	静态刚度
static strength	静强度
static stress	静应力,静胁强
static structure testing machine	结构静力试验机
static synchronizing torque characteristic	静态整步转矩特性
static system	静压系统
static temperature	静温,无扰动流温度(温度计随流体一起运动测出的温度)
static test	(结构的)静力试验,静态试验
static thrust	静推力
static tracking error	静态跟踪误差
static tube	静压管
static weapon	静止性化学武器
static wick	静电放电刷
statical mechanical magnification	静态机械放大倍数
static-base alignment	静基座对准
static-dynamic balance	静动组合天平
staticizer	串并行转换器,静化器
static-load test	静载荷试验
staticon	视像管,静象管
statics	静力学,天电干扰
station	站,台,位置,岗位,基地,测量点,坐标点
station acquisition	定点捕获
station error	控制站误差
station host	站主机
station keeping	1.定点保持 2.位置保持
station keeping system	位置保持系统
station keeping window	1.位置保持窗口 2.定点保持窗口
station location	无线电台位置
station location error	站址误差
station of departure	始发站
station set	话机,用户终端设备
station time	固定时间,停留时间
station visual zone	地球站可视范围
stationary	静止的,稳定的,驻留的,定常的
stationary battery	固定电池组
stationary blade	固定叶片
stationary channel	平稳信道
stationary echo	固定回波
stationary front	1.(气象的)静止锋 2.(激波的)滞止锋
stationary orbit	静止轨道,同步轨道
stationary plasma thruster	稳态等离子推进器

stationary point 1.驻点 2.平稳点
stationary process 平稳过程,恒定程序,定常过程
stationary random process 平稳随机过程
stationary satellite 静止卫星,定点卫星
stationary satellite navigation 静止卫星导航,定点卫星导航
stationary state 定态,静止状态
stationary target 固定目标
stationary test 平稳性检验
stationary value 平稳值
stationary wave 驻波
stationing accuracy 定点精度
stationkeeping 保持编队位置
statistic pattern recognition 统计模式识别
statistical accelerometer 加速度统计记录器
statistical analysis 统计分析
statistical astronomy 统计天文学
statistical communication theory 统计通信理论
statistical correlation 统计相关
statistical data 统计数据
statistical decision 统计判决
statistical decision theory 统计判定理论,统计决策论
statistical discrepancy 统计差异
statistical distribution 统计分布
statistical ecology 统计生态学
statistical error 统计误差
statistical estimation 统计估计
statistical mechanics 统计力学
statistical method 统计方法
statistical model 统计模型
statistical moment 统计动差
statistical parallax 统计视差
statistical property 统计特性
statistical quality control 统计质量控制
statistical sample 统计样本
statistical sampling 统计取样,统计抽样
statistical test 统计检验
statistical tolerance interval 统计容区(间)
statistical tolerance limits 统计容限
statistics 统计学
statokinetic stability 平衡运动稳定性
stator 静子(相对转子而言)
stator blade 静子叶片
stator casing 静子机匣
stator vane 整流叶片,定翼叶片
statoscope 微动气压计,灵敏气压计,高差仪
Statsionar 静止卫星(苏联静止轨道通信卫星)
status 1.状态,态势,情况 2.身份,地位 3.目标判定(尤指判定仅在雷达上显示的目标是敌、友或未知)

status board 飞机状况登记板
status word 状态字
statute mile 法定英里
stay 1.侧撑杆(尤指起落架的) 2.支撑 3.停留 4.抑制,阻止 5.(起落架的)撑杆 6.横向支撑件
stay time 停留时间
stay-up pressure suit 高空停留用加压服
STB System Test Bed 系统试验台
STBY Standby 备用的,待命的
STC 1. Self Test Complete 完成自测 2. Sensitivity Time Control 灵敏度时间控制 3. Supplemental Type Certificate(FAA) 补充(附加)型号合格证(FAA)
STCM Stabilizer Trim Control Module 安定面配平控制模块
STD 1. Scheduled Time of Departure 预计离开时间,预计离港时间 2. Standard 标准 3. Supplementary Takeoff Distance 辅助起飞距离 4. System Technical Description 系统技术说明
STDMA Self-organizing Time Division Multiple Access 自组织时分多址(数据(自动)传输(中继)器)
STE 1. Service Test Environment 业务测试环境 2. Service Testing Equipment 业务测试设备 3. Signaling Terminal 信令终端 4. Subscriber Test Equipment 用户测试设备 5. System Terminal Equipment 系统终端设备
steadiness 稳定
steady acceleration 等加速度
steady combustion 稳态燃烧
steady descent 稳定下降
steady detonation 稳态爆震波
steady dive 稳定俯冲
steady flight 定常飞行,稳定飞行
steady flow 定常流,稳定流动
steady jet 稳定射流
steady motion 定态运动
steady operation 稳定运行
steady pressure 稳态压力
steady pull-up method 稳定拉起法
steady solar simulator 稳态太阳仿真器
steady state 定常状态
steady state creep 稳态蠕变
steady state deviation 稳态偏差
steady state error 稳态误差
steady state life 稳态寿命(火箭发动机连续点火的累积时间)
steady state measurement 稳态测量
steady state performance 稳态性能
steady state power consumption 稳态功耗
steady straight flight method 定常直线飞行法

steady thermal state 稳定热状态
steady turn method 稳定转弯法
steady voltage 平稳电压
steady wind 稳定风
steady-flow 稳定流
steady-state condition 稳态条件
steady-state cosmology 稳恒态宇宙论
steady-state descent 稳定下降
steady-state error 稳态误差
steady-state flow 稳态流
steady-state operation 稳态运行
steady-state pressure 稳态压力
steady-state solution 稳态解
steady-state specific impulse 稳态比冲
steady-state temperature 稳态温度
steady-state value 稳态值
steady-state vibration 稳态振动(持续的周期振动)
stealth 1.隐身,隐形 2.隐形技术 3.隐形的
stealth aircraft 隐身飞机,隐形飞机
stealth material 隐形材料
stealth target 隐身目标
stealth technique 1.隐形技术 2.隐身技术
stealth technology 隐形技术
steam 水蒸气
steam catapult 蒸汽弹射器
steam ejection system 蒸汽弹射系统
steam oxidation 水汽氧化
steam turbine 汽轮机,蒸汽轮机,蒸汽透平
steel 钢铁
steel beam 钢梁
steel ring 钢环,钢圈
steel section 型钢
steel tube 钢管
steep gliding turn 大坡度下滑转弯
steep spectrum 陡谱
steep terrain 陡峭地形
steep-spectrum source 陆谱源
steer 1.转向 2.操舵装置
steerable antenna 可控天线
steering 操纵,控制
steering angle 转向角
steering command 操纵信号
steering computer 驾驶用计算机
steering equation 导引方程
steering instruction 导引指令
steering line 操纵绳
steering order 控制指令,导引指令
steering program 导引程序
steering unit 转向器,转向装置

STELLAR Star Tracker for Economical Long Life Attitude Reference 廉价长寿命姿态基准星跟踪器
stellar 星的,恒星的,星球的
stellar aberration 星体光行差,恒星光行差
stellar activity 恒星活动
stellar association 【天】星协
stellar astronomy 恒星天文学
stellar atmosphere 恒星大气
stellar camera 恒星摄影机
stellar complex 恒星复合体
stellar dynamics 恒星动力学
stellar envelope 恒星包层
stellar evolution 恒星演化
stellar guidance 天文制导,星光制导,天文导航
stellar inertial bombing system 天文惯性轰炸系统
stellar inertial guidance 恒星惯性制导,天文惯性导航
stellar kinematics 恒星运动学
stellar luminosity 恒星发光度
stellar magnetic field 恒星磁场
stellar magnitude 星等
stellar map matching 星图匹配导航
stellar motion 恒星运动
stellar occultation 恒星掩星
stellar orbit 恒星轨道
stellar parallax 恒星视差,日心视差
stellar physics 恒星物理学
stellar population 星族
stellar radar 星体雷达
stellar radiation 星体辐射
stellar radiometric detector 星体辐射探测器
stellar rotation 恒星自转
stellar sensitometer 恒星敏感器
stellar sensor 星光探测器
stellar spectrum 恒星光谱
stellar system 恒星系统
stellar tracker 星体跟踪仪
stellar ultraviolet radiation 恒星紫外辐射
stellar wind 星风
Stellite 钴铬钨钼合金,司太立特硬质合金
stem 1.晶体管管座 2.电子管心柱
stem region 干区
step 1.步,阶梯,台阶,踏板 2.步骤,阶段 3.(运载火箭的)级(与 stage 同义,但常用 stage) 4.(从某一高度层向另一高度层)上升的航段(需经空中管制员批准) 5.断阶(水上飞机浮筒或船身平底外形的突变部分,用来提高滑水性能,使其易于起飞)
step angle 步进角,步距角
step attenuator 步进衰减器
step change 阶跃变化

step frequency	步进频率
step function	阶跃函数
step index	1.步长指数 2.阶跃折射率
step index fiber	突变光纤
step motion electric drive	步进电驱动
step pitch	步距
step rate	步进率
step recovery diode	阶跃恢复二极管
step response	阶跃响应
step rocket	多级火箭
step stress test	步进应力试验
step switch	步进开关
step timer	步进定时器
step tracking	步进跟踪
step type multimode fiber	阶跃型多模光纤
step type optical fiber	阶跃型光导纤维
step voltage	阶跃电压
step waveform	阶跃波形
step width	步长
step out	失步,失调,时差
step-and-repeat system	分步重复系统
Stepanov method	斯蒂帕诺夫方法
step-by-step approach	逐步逼近法
step-by-step cut-off test	逐步截尾试验
step-by-step refinement	逐步求精法
step-by-step switch	步进式交换机
step-by-step tracking	步进跟踪
step-control	阶跃操纵
step-coverage	台阶覆盖
step-down	逐级下降
stepless control	无级控制,连续控制,无级调节,平滑调节,连续调节,均匀调节
stepped climb	分段爬高(每两段中有缓慢上升或水平飞行来分隔)
stepped seating	阶梯式座位配置
stepper	步进电动机,分档器,分节器
stepper motor	步进电机
stepping angle	步进角
stepping relay	步进继电器
stepping switch	步进开关
stepwise refinement	逐步细化法(一种系统开发方法)
stepwise regression	逐步回归
steradian	立体弧度,球面(角)度(立体角的单位)
stereo display	立体显示
stereo camera	立体摄影机
stereo comparator	1.立体坐标量测仪 2.体视比较仪
stereo image alternator	立体图像转换器
stereo pair	立体图像对
stereo photogrammetric survey	立体摄影测量
stereo triangulation	立体三角测量
stereogram	1.实体图,立体图 2.立体照片,体视照片 3.多边形
stereogrammic organization	多层结构
stereographic coverage	立体照片拍摄地区
stereographic projection	球面投影
stereomapping	立体测图
stereometer	立体量测仪
stereometry	立体几何(学),立体测绘学,体积测定
stereopair	立体像对
stereophone	立体声
stereophonic broadcasting	立体声广播
stereophonic record	立体声唱片
stereophonic TV	立体声电视
stereophotogrammetry	立体摄影测量
stereoplotter	立体测图仪
stereosat	立体卫星
stereoscope	立体镜
stereoscopic map	视觉立体地图
stereoscopic model	立体模型
stereoscopic pair	立体照相的一对影像
sterilize	消毒
stern attack	尾部攻击,尾后攻击
stern wave	尾波(水上飞机以低速滑行时在尾部前方形成的波)
stern-chase launching	尾追发射
sternpost	机身尾柱,船尾柱
stevenson screen	斯蒂文森百叶箱
stewardess	女乘务员,女管理人,女管家
STF	1. Solliton Transmission Fiber 孤子传输光纤 2. Staff 职员
STI	1. Special Technical Instruction 专用技术说明书 2. Statistics Time Interval 统计时间间隔
stibium	【化】锑(化学元素,符号 Sb)
stick	杆,棒,手把,手柄,粘,贴
stick force	黏附力,杆力
stick force per gram	每克驾驶杆力
stick pusher	自动推杆器
stick-free	松杆
sticking coefficient	黏附系数
sticking probability	黏附概率
stick-up lettering	粘贴注记
stiffened plate	加强板
stiffener	加强杆,加强板,加强件
stiffening shell	加筋壳结构
stiffness	1.刚度,刚性 2.安定性,稳定性 3.劲度
stiffness criterion	刚度准则
stiffness index	刚度因子,强度指数
stiffness matrix	刚度矩阵,劲度矩阵

stiffness ratio	刚度比
stiffness reduction	刚度折减
stilb	熙提（亮度单位）
still picture broadcasting	静止图像广播
still-air range	无风航程
stimulate	刺激，模拟
stimulated absorption	受激吸收
stimulated Brillouin scattering	受激布里渊散射
stimulated emission	受激发射
stimulated Raman scattering	受激拉曼散射
stimulus	1.激励,刺激 2.被测(的)量
sting	尾支臂,探针
stinger	刺,支臂,飞机尾炮
stink bomb	臭炸弹
Stirling cycle	斯特林循环
Stirling refrigerator	斯特林致冷器
STIU	Stabilizer Trim Interface Unit 安定面配平联接装置
STL	Scalability Test Laboratory 可量测性实验室
STLDD	Software Top Level Design Document 软件顶层设计文档
STLO	Science and Technology Liaison Office 科学技术联络处
STLT	Stern Light 尾灯
STM	1. Selective Traffic Management 选择性业务管理 2. Statute Miles 法定哩 3. Storm 风暴 4. Structural Test Model 结构强度试验模型 5. System Test Module 系统测试组件
STMU	Special Test and Maintenance Unit 专用试验和维护设备
STN	1. Station 站，站位，台 2. Switched Telecommunications Network 远程通信交换网络 3. Switched Telephone Network 电话交换网络
STO	1. Short Take Off 短程起飞,短距起飞 2. Solar Terrestrial Observatory 日地观测台
stochastic	随机的
stochastic acceleration	随机加速
stochastic approach	随机方法
stochastic control theory	随机控制理论
stochastic differential equation	随机微分方程
stochastic finite automation	随机有限自动机
stochastic grammar	随机文法
stochastic matrix	随机矩阵
stochastic parameter	随机参数
stochastic process	随机过程
stochastic property	随机特性
stochastic pushdown automaton	随机下推自动机
stochastic sampling	随机抽样
stochasticity	随机性
stockpile	库存(量),存货
stoichiometric	化学计算的,化学数量的
stoichiometric composition	化学计量成分,理想配比成分
stoichiometric mixture ratio	化学计算混合比
stoichiometric ratio	化学计量比
stoichiometry	化学计算法,化学计量学
STOL	Short Take-Off and Landing 1.短场起降 2.短距起落飞机
stony meteorite	石状陨石
stooge	无目的飞行,未遭遇的出击或巡逻
stop	停止,暂停,停机
stop band	抑止频带,阻带
stop drill	钻头
stop hole	止裂孔（在裂纹端部钻孔以止住裂纹扩展）
stop nut	(终端)止动螺帽
stop way	停止道
stopover	中途着陆,中途停留
stopping	停止,停车
stopping criterion	停止准则,停止判据
stopping cross section	阻止截面
stopping distance	阻止距离
stopping power	阻止本领
stopway	跑道头保险道
storability	可存储性
storable oxidizer	可贮存氧化剂
storable propellant	可贮存推进剂
storable propellant rocket engine	可贮存推进剂火箭发动机
storage	1.存贮 2.仓库 3.存储器,储存量
storage battery	1.蓄电池 2.二次电池
storage capacitance	储存电容
storage duration	存储期
storage function	存储函数
storage life	存放寿命,储存寿命
storage load	贮存载荷
storage modulus	储能模量
storage oscilloscope	储存示波器
storage period	存放期
storage reliability	贮存可靠度
storage requirement	存储要求
storage system	储存系统
storage tank	储气罐,贮罐
storage test	贮存试验
storage traveling wave tube	储频行波管
storage tube	储存管
storage zone	储存区
storage-type heater	蓄热式加热器
store management system	存储管理系统

store model　存储模式
store-carrying capacity　载弹量
stored-program control exchange　程控交换机
stores management system　储备管理系统
storm　暴风雨,风暴
storm observation satellite　风暴观测卫星(美国)
Stormsat　Storm Satellite 风暴卫星(美国)
storm-time variation　暴时变化
STOVL　Short Take-Off Vertical Landing 短距起飞、垂直着陆(飞机)
STOW　System for Take-Off Weight 起飞重量与飞机重心(显示)系统(传感器在各起落架上)
stow　贮藏,收藏
stow position　存放位置
stowage time　装载时间
stowage volume　装载量
STP　1. Satellite Tracking Program 卫星跟踪程序 2. Short-Term Planning 制订短期计划 3. Software Test Plan 软件测试大纲 4. Software Test Procedures 软件测试程序 5. Solar Terrestrial Programme 日地探测计划(西欧) 6. Space Test Program 空间试验计划,航天试验计划(美国国防部的) 7. Standard Temperature and Pressure 标准温度和标准压力(指气温和气压) 8. System-Test Procedure 系统试验程序
STPD　Standard Temperature, Pressure, Dry 标准温度(0℃)、标准气压和干燥气体状态
STR　1. Short Time Rating 短期使用额定值 2. Simulation Test Report 模拟试验报告 3. Software Test Report 软件试验报告 4. Software Trouble Report 软件故障报告 5. Standard Test Rack 标准试验架 6. Steer 操纵 7. Storage Rack 存放架 8. Strength Test Report 强度试验报告 9. Sustained Turn Rate 持续稳定转弯角速度 10. Systems Technology Radar 系统技术雷达
STRA　Simultaneous Turn Round Actions 模拟转盘操作
straddle　夹叉射击,夹叉弹
strafe　扫射,低空攻击
straight advancing klystron　直射速调管
straight channel　直槽
straight flow　直流式
straight line　直线
straight line path　直线路径
straight pipe　直管
straight pipeline　直管道
straight roller bearing　普通滚柱轴承
straight run　直馏产品
straight section　直线节
straight segment　直线段
straight tube　直管
straight wing　直线机翼,平直机翼
straight-forward dialog　直通对话
straight-forward network　直通网络
straight-in approach-IFR　仪表飞行规则的直线进近
straight-in approach-VFR　目视飞行规则的直线进近
straight-in landing　直线进近着陆
straight-line　直线的,线性的
straight-through operation　直通型运行
strain　应变,(应力下的)变形
strain accumulation　应变积累
strain energy　应变能
strain energy density　应变能密度
strain gage balance　应变式天平
strain gage balance with sting　支杆式应变天平
strain hardening　应变硬化
strain increment　应变增量
strain rate　变形速度
strain response　应变响应
strain sensor　应变传感
strain step　应变阶跃
strain viewer　应力测验仪
straingauge　应变仪
strainometer　应变仪
strain-rate　应变率
strake　导流片,整流片
strake wing　边条翼
strakelet　小箍带,小护板
strand　绳股,绳水股条
strand burner　条状燃烧室
stranded conductor　绞合导线
stranded wire　多股绞合线
strange attractor　奇异吸引体
strap　条,带,搭接片,连接条,用带子扎紧
strapdown　捷联式
strapdown gyroscope　捷联(式)陀螺
strapdown inertial guidance　捷联式惯性制导
strapdown inertial guidance system　捷联式惯性制导系统
strapdown inertial measurement unit　捷联式惯性测量装置
strapdown inertial navigation　捷联式惯性导航
strapdown inertial navigation system　捷联式惯性导航系统
strapdown system　捷联系统
strap-on booster　搭接助推器,捆绑式助推器
strap-on launch vehicle　捆绑火箭
strap-on rocket　捆绑式火箭发动机,外挂式火箭发动机
strap-on structure　捆绑结构

strapped-down 捷联式的
strapping 1. 搭接,捷接,捆带条,皮带材料 2. 多腔磁控管空腔间的导体耦合系统
strategic action 战略行动
strategic aeromedical evacuation 战略空运伤病员
strategic air transport 战略空运
strategic air warfare 战略空中作战
strategic airlift 战略空运
strategic area 战略区
strategic attack 战略攻击
strategic bomber 战略轰炸机
strategic communication satellite 战略通信卫星
Strategic Defense Triad 战略防御三位一体(美国)
strategic function 策略函数
strategic layer 策略层
strategic level 战略级
strategic map 战略地图
strategic missile 战略导弹
strategic nuclear weapon 战略核武器
strategic orbital system 战略轨道系统
strategic plan 战略计划
strategic reconnaissance 战略侦察
strategic satellite system 战略卫星系统
strategic system 战略系统
strategic transport 战略运输机
strategic warning 战略性警报
strategic warning post-decision time 战略警报决策后行动时间(收到警报作出决定后到敌行动开始之间的时间)
strategy 战略,战略学
stratification 分层,成层,分层结
stratified flow 成层流
stratiform cloud (成)层状云
stratify 分层
stratocumulus 层积云(符号 Sc)
stratopause 平流层顶
stratosphere 平流层,同温层,平流支
stratospheric airship 平流层飞艇
stratospheric sounding unit 平流层探测装置
stratostat 平流层气球
stratus 层云(符号 St)
stray 迷航飞机,掉队飞机,错运
stray bullet 流弹
stray delay 杂散延迟
stray element 杂散元件
stray inductance 杂散电感
stray light 杂光
stray loss 杂散损耗
stray radiation 杂散辐射

STRC 1. Science and Technology Research Center 科学技术研究中心 2. Strategic Training Route Complex 战略训练航线综合设施
streak 条纹,加条纹
streak camera 超快扫描照相机
streak image 拖尾图像
streak line 条纹线,流脉
streak photograph 条纹照片
streaking 图像拖尾
stream 流,气流
stream cipher 流密码
stream contamination of wind tunnel 风洞气流污染
stream direction 流向标志
stream function 流线函数,流量函数,流函数
stream line 流线,日冕射线
stream surface 流面
stream tube 流管(管壁是流线)
streamer 流管,流束
streamer chamber 流光室
streamer discharge 流注放电
streamline section 流线型截面
streamline wire 流线型张线
strength 力,力量,强度,浓度
strength criterion 强度较核规范
strength limit of short time in high temperature 短时高温强度极限
strength of source flow 源流强度
strength specification 强度规范
strength test 强度试验
strengthen 加强,巩固
stress 胁强,应力,应力状态,着重,强调
stress amplitude 应力幅
stress analysis 应力分析
stress bar 应力杆
stress concentration 应力集中
stress concentration factor 应力集中系数
stress condition 应力条件
stress constraint 应力约束
stress corrosion 应力腐蚀
stress cycle 应力循环(应力随时间变化的完整循环)
stress dislocation 应力位错
stress distribution 应力分布
stress field 应力场,胁强场
stress function 应力函数
stress glut 应力过量
stress intensity factor 应力强度因子
stress level 应力水平,应力级
stress model 应力模型
stress raiser 应力集中器

stress ratio	应力比
stress relaxation	应力缓和,应力松驰
stress release boot	人工脱黏
stress relief	应力解除
stress relief annealing	除内应力退火
stress relief flap	人工脱黏层
stress relieving	应力消除热处理,消除应力
stress response	应激反应,压力反应
stress resultant	应力综合
stress rupture	应力破坏
stress screening	应力筛选
stress severity factor method	应力严重系数法
stress state	应力状态,应激状态
stress tensor	应力张量
stress testing	压力测试,强度测试
stress trajectory	应力迹线
stress wave	应力波
stress wrinkle	应力皱损,应力翘曲
stressed skin	受力蒙皮
stress-induced defect	应力感生缺陷
stressometer	应力仪
stress-rupture plasticity	持久塑性
stress-strain curve	应力-应变曲线
stretch	伸展,伸张,伸张,拉伸
stretch bending	拉弯
stretch effect	拉伸作用
stretch forming	伸张成形
stretch sensitivity	牵张敏感度
stretchability	拉伸性,抽伸性
stretch-draw forming	拉伸压延成形
stretcher	担架,拉伸机,延伸器
stretch-wrap forming	拉弯成形
striation	辉纹,条纹
stricken area	杀伤区
strike	攻击,进攻,突击队,打击,突击
strike aircraft	强击机
strike force	突击力量,打击力量(由突击部队组成,不一定限于航空兵)
strike orientation	走向定向
strike photography	命中摄影术
striker	撞击器,触发器,撞针
striking velocity	接地速度,弹着速度
striking voltage	起弧电压
string	行,串
string grammar	串文法
string language	串语言
stringer	桁梁,翼梁
stringer structure	桁条式结构
strip aerial triangulation	航带法空中三角测量
strip camera	航线式相机,航带式相机,条幅(航带)摄影机
strip chart	记录仪纸带
strip fin	带状肋片
strip line	带状线
strip map	条幅式航线地图
strip mapping	条带状测绘
strip plot	航线图
strip theory	切片理论
stripcoat	剥落的外层,剥落的保护层
stripe	条纹,条子
stripe type laser	条形激光器
stripping of photoresist	去胶
strobe	选通脉冲,闸门
strobe light	频闪灯(高强度闪光灯标)
strobe marker	选通标志
strobe pulse	选通脉冲
stroboscope	频闪观测器,闪光仪
stroboscopic effect on distance measurement	测距频闪效应
stroke	行程,冲程,笔划,划线,斜号
strongback	导弹容器
strongpoint	受力接头,加强接头
strontium	【化】锶(化学元素,符号 Sr)
strontium barium niobate	铌酸锶钡
strop	引导绳,吊带
Strouhal number	斯特劳哈尔数,斯特劳黑尔数
STRS System Test Rig Specification	系统试验台架规范
structural adhesive	结构黏结剂,结构黏合剂
structural analysis	结构分析
structural bond	结构键
structural ceramics	结构陶瓷
structural coefficient	结构系数
structural component	结构零件,结构组件,结构部件
structural concept	结构概念
structural constraint	结构限制
structural control	结构控制,构造控制
structural damage	结构损伤
structural damping	构造衰减,结构衰减
structural deformation	结构变形
structural design	结构设计(指在设计结构中的全面工作)
structural dynamics	结构动力学
structural elasticity	结构弹性
structural element	结构元件
structural equation	结构方程式
structural factor	构造因素
structural failure	结构断裂,结构破损

structural fatigue 结构疲劳
structural flexibility 结构柔性
structural flow 结构流
structural frame 结构构架
structural framework 结构骨架
structural heat loss 结构热损失
structural heat transfer test 结构传热试验
structural integrity 结构完整性
structural load 结构荷重
structural mass 结构质量
structural material 结构材料
structural member 结构部件,结构元件
structural mode 结构模式
structural mode control 结构模态控制系统
structural model 结构模型
structural modeling 结构建模
structural optimization 构架最佳化
structural parameter 结构参数
structural passability 结构可通性
structural performance 结构特性
structural perturbation 结构摄动
structural problem 结构性问题
structural property 结构特性
structural rating 结构品级(号)
structural response 结构响应
structural section 构造断面
structural similarity 结构相似性
structural stability 结构稳定性
structural stiffness 结构刚度
structural strength 结构强度
structural support 结构支持
structural system 结构系统
structural thermal external pressure test 结构热外压试验
structural thermal low pressure test 结构热低压试验
structural thermal protection test 结构热防护试验
structural thermal stability test 结构热稳定性试验
structural thermal vibration transfer 结构热振动传递
structural titanium alloy 结构钛合金
structural type transducer 结构式传感器
structural vibration 结构振动
structural weight 结构
structure 结构,构造,结构件,结构物,组织,机构
structure defect 缺陷结构
structure design 结构设计
structure detection 结构检测
structure diagram 结构示意图
structure division 结构部分
structure model 结构模型

structure of composite material 复合材料结构
structure perturbation approach 结构摄动法
structure size 结构尺寸
structure static strength test 结构静强度试验
structure stiffness test 结构刚度试验
structure subsystem 结构分系统
structure-borne noise 结构噪声
structured program 结构化程序
structured programming 结构化程序设计
structured uncertainty 结构不确定性
strut 支柱,撑杆,(发动机吊舱的)吊架
strut attachment 支承装置
STS 1. Speed Trim System 速度配平系统 2. Status 状态 3. System Technical Specification 系统技术规范
STSI Space Telescope Science Institute 空间望远镜科学研究所(美国)
STT 1. Set-Top Terminal 机顶终端 2. Single Target Track 单目标追踪
STU 1. Set Top Unit 机顶装置 2. Short Ton Units 短吨单位 3. Signaling Terminal Unit 信令终端单元 4. Statistical Unit 统计单元 5. Storage Unit 存储单元 6. Student 学生
stub 短线,残端,票根,存根,柱脚,金属基础
stub plane 短翼机,根段翼,中翼
stub wing 短翼
stud 双头螺栓,柱螺栓,柱,销,双端螺栓
studio 播音室,摄影室
study 学习,研究
study phase 研究时段
stuff 云,密云,资料,原料
STV Structure Test Vehicle 结构强度试验飞行器
STW 1. Steward 男乘务员 2. Stewardess 女乘务员
stylus 尖笔,记录针,描画针
styrene 苯乙烯,苯代乙撑,苯次乙基
SU Signal Unit 信号装置
SUA Special Use Airspace 专用空域
sub Substitute 取代
subarray 子阵
subassembly 分组件,局部装配,组件,部件,部件装配,辅助装置
subauroral zone 亚极光带
sub-bit code 副比特码
subcarrier 副载波
subcarrier demodulator 副载波解调器
subcarrier discriminator 副载波鉴频器
subcarrier modulator 副载波调制器
subcarrier oscillator 副载波振荡器
subcenter 1. 分中心,子中心 2. (空袭警报)探知站
subchannel 分通道,子通道

subclass	次类,次分类
sub-cloud car	飞艇云下吊篮
subcommutator	副交换子,副时分开关
subcomponent	分组件,分装配件
subconsole	副控制台
subcontractor	转包合同商,分包合同商,转包商
subcooled	低温冷却的
subcritical	亚临界
subcritical pressure	亚临界压力
subcritical rotor	亚临界转子
subcrystalline	亚晶粒
subdivide	再分,细分
subdivisional organization	再分结构
subdomain	子区域
subduction	消减
subduction belt	俯冲带
subduction zone	俯冲带
subduction-type geothermal belt	消减型地热带
subdwarf	亚矮星
subelement	子元件,子元素
subface	底面
subframe	副帧,子框架,子帧
subframe synchronization	副帧同步
subgiant	次巨星,亚巨星
subgrade	路基,地基
subgravity	亚重力,次重力
subgravity illusion	亚重力错觉
subgroup signalling	子群信令
subharmonic	分谐波,次谐波
subharmonic frequency	次谐波频率
subharmonic response	次谐波响应
subject	题目,原因
subjective evaluation	主观评价
subjective information	主观信息
subjective measure	主观测量
subjective probability	主观概率
subjective tolerance level	主观耐受水平
sublattice	子格
sublayer	子层,次层
sublimation	升华,凝华,升华法
sublimation pump	升华泵
sublimator	升华器
subliminal stimulus	阈下刺激
sublunar point	月下点,月亮地理位置
submarine	潜水艇,水下的
submarine cable	海底电缆
submarine geomorphology	海底地貌学
submarine rocket	潜艇火箭
submarine striking force	核攻击潜艇舰队
submarining	潜水现象
submatrices	submatrix 的复数
submatrix	子阵
submerged condition	水下状态
submerged inlet	埋入式进气口
submerged nozzle	水中喷管,水下喷咀
submergence ratio of nozzle	喷管潜入比
submillimeter wave	亚毫米波
subminiature	超小型
subminiaturization	超小型化
submissile	子导弹,导弹的一级,弹头杀伤元件
submission	提出,提交
submit	提交,委托
submodel	子模型
submodulator	副调制器,辅助调制器
submunition	子弹药
subnet	子网络
suboptimal control	次优控制
suboptimal solution	次优解
suboptimal system	次优系统
suboptimality	次优性
suborbital flight	亚轨道飞行
suborbital trajectory	亚轨道弹道,亚轨道
suboxide	低值氧化物
subphase	子阶段
subpolar region	副极区
subpopulation	子总体
subproblem	子问题,部分问题,次要问题
subprogram	子程序,次程序
sub-projectile	子导弹,子射弹
subreflector	副反射器
subregion	子区域
SUBROC	Submarine Rocket 沙布洛克反潜火箭,反潜导弹
subroutine	子程序
subsample	子样本
subsatellite	子卫星,月球卫星
subsatellite point	卫星投影点
subscriber	用户
subscriber line interface circuit	用户专线接口电路
subscriber loop	用户回路
subscript	下标,标记,索引
subsection	分段,分部
subsector	分区,区
subset	用户电话机,子集
subsidence	沉降,下沉,沉陷,阻尼
subsidiary	子公司,辅助者
subsidy	津贴,补贴
subsolar point	太阳直射点日下点
subsonic aerodynamics	亚声速空气动力学
subsonic aircraft	亚声速飞机

subsonic diffuser	亚声速扩压器
subsonic flight	亚声速飞行
subsonic flow	亚声速流
subsonic leading edge	亚声速前缘
subsonic ramjet	亚声速冲压喷气发动机
subsonic reaction	亚声速反应
subsonic speed	亚声速
subsonic trailing edge	亚声速后缘
subsonic turbine	亚声速涡轮
subsonic velocity	亚声速
subsonic wind tunnel	亚声速风洞
subspace	子空间
subspace constraint	子空间约束
substage	子级
substance	1.实体,资产 2.材料,物质,实质
substandard	标准以下的,低于法定标准的
substantiation	证明,证实,核实
substar	星下点
substep	分步,子步
substitute	代替者,代替物,代用品,以……代替
substitute character	替换字符
substitute parts	替代件
substitution	置换,代替物
substitution error	替代误差
substitution method of measurement	替代测量方法
substitutional diffusion	替位扩散
substitutional impurity	替代式杂质
substorm	亚暴
substrate	1.基片,基体,衬底(太阳电池基体材料,如硅片) 2.感光胶层 3.真晶格
substrate bias	衬底偏置
substrate fed logic	衬底馈电逻辑
substrate temperature	衬底温度(真空蒸发工艺中被淀积基体的温度)
substratosphere	副平流层,亚同温层,亚平流层
substructure	底层结构,基体结构,子结构
subsynchronous layer	次同步层
subsynchronous whirl	准同步回旋,准同步振动
subsystem	分系统,子系统
subsystem test	分系统测试
subtask	子任务,程序子基元
subtense traverse	视差导线
subtrack	星下点轨迹,子轨道
subtract	减去,扣掉
subtrees	子树
subwoofer	低音炮,超低音
subzero treatment	低温处理,零度下处理
succeed	继承,接替,成功,完成
successive adjustment	序贯平差
successor	后继机,后继方案
succinic acid	丁二酸,琥珀酸
suckdown	吸入,吸下
suction	吸,吸入,虹吸,抽气,吸力,真空度,负压吸管
suction control	吸入控制器
suction effect	吸气作用,吸入效应,吸力效应
suction flap	吸气襟翼
suction gauge	吸力计,真空计
suction peak	吸力峰值
suction performance	抽气率
suction pressure	吸入压力
suction pump	抽吸泵,真空泵
suction rate	吸收率
suction side	吸力面,进口侧,吸入端
suction slot	(附面层)吸除缝,吸气缝
suction specific speed	汽蚀比转速
suction surface	吸力面,真空面
suction system	吸入系统
suction velocity	吸入速度
sudden change	突变
sudden commencement	急始
sudden frequency deviation	(短波)频率急偏
sudden inflight incapacitation	飞行中突然失能
sudden ionospheric disturbance	突发电离层骚扰
sudden phase anomaly	突发相位异常
sudden stoppage	突然停车(活塞式发动机在飞行中在螺旋桨旋转一圈内停车或涡轮喷气发动机在一秒钟内停车,说明发动机内部有严重损坏)
sufficiency	充足,满足
suffix code	后缀码
suggest	建议,提议
suit	服装,飞行服
suit bladder	代偿服扁平充气囊
suit oxygen supply	服装供氧
suit pressure regulator	服装压力调节器
suit ventilation	服装通风
suit ventilator	服装风机
suite	1.配套件 2.一般通用设备(指同类飞机所需的各种机载设备)
sulfate	硫酸盐
sulfating	硫酸化
sulfide	硫化物
sulfur	硫磺,硫磺色
sulfur content	含硫量,硫分
sulfur dioxide	二氧化硫
sulfur monoxide	一氧化硫
sulfuric acid	硫酸
sulphide	硫醚,硫化物
SUM	1. Software User's Manual 软件用户手册 2.Summary 总结,摘要 3.Summer 夏季
sum beam	和波束

summarise	摘要,概要
summary	1.摘要,提要 2.即时的,速决的
summation	总和
summer flying gear	夏季飞行服
summer solstice	夏至,夏至点
summer time	夏令时
summing amplifier	相加放大器
summing unit	求和装置
sum-of-squares	平方和
sump	池,坑,沉淀槽,油槽,油箱,贮槽
sun	太阳
sun acquisition	太阳捕获
sun angle	日照角
sun compass	太阳罗盘
sun gear	太阳齿轮
sun glint	太阳反辉区
sun line	太阳位置线
sun period	太阳周期
sun presence sensor	太阳出现传感器
sun sensor	日光传感器,太阳传感器
sun shield	遮阳板
sun synchronous satellite	太阳同步卫星
sun tracker	太阳跟踪仪
sun vector	太阳矢量
sunblazer	太阳探测器
sundial	日晷
sunk costs	沉没成本
sunlight	日光
sunlit	受日照的
sun-moon perturbation	日月摄动
sunphotometer	太阳光度计
sunrise	日出
sunseeker	寻日器,向日仪
sunset	日落
sunshine recorder	日照计,日照仪
sunspot	太阳黑子,日斑
sunspot group	日斑组
sun-synchronous	太阳同步
sun-synchronous orbit	太阳同步轨道
sun-synchronous orbit satellite	太阳同步轨道卫星
SUPARCO	Space and Upper Atmosphere Research Commission 空间和高层大气研究委员会
super cryogenic temperature preforming	超低温预成形
super group	超群
super low flight	超低空飞行
super master group	超主群
super real time simulation	超实时模拟
super spin	超自旋(细长体双自旋卫星在转移轨道上采用的一种姿态稳定方案)
superadiabatic lapse rate	超绝热递减率
superaerodynamics	稀薄气体空气动力学
supercarrier	超重型航空母舰,超重型运输机
supercavitation flow	超空涡流
super-cluster	超星系团
supercode	超码
supercommutation	超转接
supercomputer	巨型计算机,超大型计算机
superconducting coil	超导线圈
superconducting electronics	超导电子学
superconducting gyro	超导陀螺
superconducting magnet	超导磁体
superconducting metal	超导金属
superconductive gravimeter	超导重力仪
superconductive magnetometer	超导磁力仪
superconductivity	超导电性
superconductor	超导(电)体
superconductor detector	超导探测器
supercool	过度冷却
supercooled	过冷的
supercooled water droplet	过冷水滴
supercorona	超日冕
supercritical	超临界
supercritical aerofoil profile	超临界翼型
supercritical airfoil profile	超临界翼型
supercritical condition	超临界状态
supercritical cryogenic storage	超临界低温贮存
supercritical environment	超临界环境
supercritical flow	超临界流
supercritical gas	超临界气体
supercritical hydrogen	超临界氢
supercritical Mach number	超临界马赫数
supercritical nitrogen	超临界氮
supercritical pressure	超临界压力
supercritical wing	超临界机翼
supercruise	超声速巡航
supercruiser	超声速巡航飞机
superdirectivity	超方向性
superellipse	超椭圆
superfluidity	超流动性
superfluorescence	超辐射
supergiant	超巨星
supergranular cell	超米粒泡
supergranule	超米粒
superheat	过热
superheated vapour	过热蒸汽
superheater	过热器
superheterodyne receiver	超外差接收机
superhuge computer	巨型计算机
superhybrid composite	超混杂复合材料
superimpose	添加,重叠

superimposed circuit 叠加电路	supersonic missile 超声速导弹
superimposed ringing 叠加振荡	supersonic mode 超声速飞行状态
superimposed ripple 叠置波痕	supersonic model 超声速模型
superior conjunction 优良连接,上合	supersonic nozzle 超声速喷嘴
superior performance 性能优越	supersonic plate ablation test 超声速平板烧蚀试验
superior planets 地(轨)外行星	supersonic propeller 超声速螺旋桨
superior product 优势产品	supersonic range 超声波段
superior transit 天体上中天,上中天	supersonic region 超声速区域
superiority 优越,超越,优越性,优势	supersonic speed 超声速
superjumbo 超大型飞机(指运输机)	supersonic through-flow stage 超声速通流级
superlattice 超点阵,超结构	supersonic transport 超声速运输机
supermini 超小	supersonic turbine 超声涡轮
supernickel 超镍合金	supersonic vehicle 高速飞行器
supernova 超新星	supersonic wind tunnel 超声速风洞
superplastic forging 超塑性锻造	supersonics 超声速空气动力学,超声速飞行学
superplastic forming 超塑性成形,超塑性模锻	superstall 严重失速,严重气流分离
superplastic forming/diffusion bonding 超塑性成形/扩散连接	superstratosphere 超平流层(平流层以上的大气层)
superplasticity 超塑性,超塑性学	superstructure 上层结构,上部结构
superposition 1.叠加,重叠,叠置 2.叠加原理	super-synchronous orbit 超同步轨道
superposition principle 叠加原理	super-synchronous satellite 超同步卫星
superposition theorem 叠加定理	supertwisted birefringent effect 超扭曲双折射效应
superpressure 超压,余压	supervised training 监督训练
superquick and delay fuze 瞬发与延期两用引信	supervisor 控制器,管理程序,管理器,管理机,监控装置,管理者,监督人,管理人
superquick fuze 瞬发引信	supervisory computer control system 计算机监控系统
superradar 超远程警戒雷达	supervisory relay 监视继电器
superradiance 超辐射	supine position 仰卧姿势
superradiant laser 超辐射激光器	supplement 增补,补充
superrefraction 超折射	supplemental carrier 补充运载工具
superregeneration receiver 超再生接收机	supplemental type certificate 补充型号合格证
supersaturation 过饱和	supplementary channel broadcasting 附加信道广播
superscript 上标	supplementary contour 助曲线,辅助等高线
supersonic aerodynamics 超声速空气动力学	supplier 供给装置,气源,供应商,厂商
supersonic aircraft 超声速飞机	supply 供应,提供
supersonic bomb proximity fuze 超声速炸弹近炸引信	supply chain 供应链,供给链,供需链
supersonic business jet 超声速商务喷气飞机	supply pressure 供应压力,供给压力
supersonic combustion 超声速燃烧	supply spool 供片卷筒
supersonic combustion ramjet 超声速燃烧冲压发动机	supply system 供应系统,包干制
supersonic combustion ramjet engine 超声速燃烧冲压喷气发动机	support 支持,援助,供养
	support agreement 赡养协议
supersonic compressor 超声速压缩机,超声速压缩器	support bracket 支架,支撑托架
supersonic cruise 超声速巡航	support interference correction 支架干扰修正
supersonic diffuser 超声扩压器	support material 支撑材料,支架材料
supersonic ejector 超声速引射器	support measure 支护措施
supersonic engine 超声波发动机	support plate 支撑板,托板
supersonic firing 超声速射击	support stiffness 支承刚度
supersonic flight 超声速飞行	support sting 尾支杆
supersonic flow 超声速流	support structure 支撑结构
supersonic inlet 超声速进气口,超声速进气道	support strut 支柱
supersonic jet 超声速喷气机	support system 保障系统

support time	支护时间
supportability	保障性
supporting aircraft	支援飞机
suppose	推想,假设,以为,认为
suppress	镇压,抑制,隐瞒
suppressant	抑止剂
suppressed carrier system	抑制载波系统
suppression	抑制,镇压
suppression method	抑制方法
suppressive	压制性的,抑制性的
suppressor	1.抑制器,消除器,消声器,消声喷管,抑制栅极 2.抑制剂 3.抑制栅板(用于防止电气设备、电气设备或电路泄漏)
suppressor grid	抑制器,消除器
suppressor vacuum gauge	抑制型真空计
suprathermal ion detector	超热离子探测器
SUPT	Support 支持,支援
SUPV	Supervision or Supervisor 监督,检查,检查员
surcharge	附加费
surface	表面,平面
surface acoustic wave	表面声波
surface activated bonding	表面活性化结合
surface adsorption	表面吸附
surface analysis instrument	表面分析仪
surface antenna	面天线
surface area	表面面积
surface barrier detector	面垒探测器
surface boundary layer	表面边界层地表边界层
surface burning	表面燃烧
surface burst	地面爆炸
surface catalysis	表面催化
surface channel	表面沟道
surface charge	表面电荷
surface charging	表面充电
surface cleaning	表面清理
surface coating	表面涂覆
surface combustion	表面燃烧
surface concentration	表面浓度
surface contact diode	面接触二极管
surface contamination	表面污染
surface contamination meter	表面污染测量仪
surface cooler	表面式冷却器,表面冷却器
surface corrosion	表面腐蚀
surface deformation	曲面变形
surface density	表面密度
surface depletion layer	表面耗尽层
surface deposition	表面淀积
surface diffusion	表面扩散
surface discharge	表面放电
surface displacement	表面位移
surface distance	表面距离
surface echo	表面回波
surface effect	地(表)面效应
surface effect vehicle	气垫飞行器
surface element	表面元素
surface elevation	地面高程
surface energy	表面能
surface evolution	曲面演化
surface feature	地表特征
surface finish	表面光制,表面粗糙度
surface flow visualization	表面流动显示
surface force	表面力
surface hardening	表面淬火,表面硬化
surface hardness	表面硬度
surface heat flow	地表热流
surface heat flux	表面热流密度
surface heat transfer	表面热流
surface heat treatment	表面热处理
surface integrity	表面质量,表面完整性
surface inversion layer	表面反型层
surface irregularity	表面不均匀性
surface layer	表层
surface level	表面能级,地面高程
surface loading	表面负荷表面加载
surface mesh	表面网格
surface missile	地(水)面发射的导弹
surface model	曲面模型
surface modification	表面改性
surface morphology	表面形态
surface mounted component	表面安装部件
surface mounting inductor	表面安装电感器
surface mounting technology	表面安装技术
surface movement	表层移动
surface of position	位置面
surface of zero velocity	零速度面
surface perturbation	形貌扰动
surface plate	平台,平板,划线台
surface porosity	表面孔隙度
surface potential	表面势
surface pressure distribution	曲面压力分布
surface profile	表面轮廓
surface property	表面性质
surface quenching	表面淬火
surface reaction	表面反应
surface reaction control	表面反应控制
surface recession	表面退离,表面破坏
surface recombination	表面复合
surface resistivity	表面电阻系数
surface roughness	表面粗糙度,表面光洁度
surface shape	表面形貌

surface slope 水面坡度,地面坡度,表面坡度
surface state 表面态
surface structure 表面结构,表层结构
surface target 陆上或海上目标,(地球表)面上目标
surface target effect 表面目标效应
surface temperature 表面温度,地表温度
surface tension 表面张力
surface tension gradient 表面张力梯度
surface tension propellant tank 表面张力推进剂贮箱
surface thermocouple 表面热电偶
surface treating agent 表面处理剂
surface treatment 表面处理
surface velocity 表面速度
surface wave 表面波
surface wetness fraction 表面湿润系数
surface wind 地表风,地面风
surface zero 原子弹爆心投影点,地面爆炸中心(同 ground zero)
surface-flow 表面流
surface-friction drag 表面摩擦阻力
surface-to-air missile 地对空导弹
surface-to-mass ratio 表面质量比
surfacing 表面处理,表面加工,表面平整,镀面,覆面,堆焊
surfactant 表面活性剂,表面去垢剂
surge 波动,浪涌,大浪,高涨
surge box 恒压箱
surge control 喘振控制
surge current 浪涌电流
surge frequency 喘振频率
surge limit 喘振边界
surge line 喘振线,喘振边界
surge margin 喘振裕度,喘振边界
surge point 喘振点,喘点
surge voltage 浪涌电压
surge wave 涌波
surging 1.出现喘振 2.风洞气流速度、流量及压力不规则或低频脉动(一种故障) 3.燃油(在油箱内)晃动
surging shock 喘振冲击
surgistor 电涌限制电阻
surplus factor 多余因数
surplus value 剩余价值
surprise decompression 突然减压,意外减压
surrogate 替代者,代理人
surrogate model 代理模式,替代模型
surveillance 监视,监督
surveillance approach 监视进场,监视雷达引导进场
surveillance radar 监视雷达,搜索雷达
surveillance system 观测系统
survey 测量,测定,俯瞰,检查

survey adjustment 测量平差
survey grid 测网
survey line 测线
survey mark 测量标志
survey organization 调查机构
survey point 测点
survey station 测站
survey vessel 测量船
surveying 测定,测量,测量学
surveying and mapping 测绘学
surveying camera 测量相机
surveyor 测量员,勘测员,测量员,鉴定人
survival time 存活时间
survivability 耐受性,生存性
survivability analysis 生存性分析
survivable environment 生存环境
survivable flight control system 高生存力飞(行)控(制)系统
survival 幸存,救生
survival ability 生存能力
survival bag 救生包,生存袋
survival capsule 救生球
survival equipment 救生装备
survival facilities 救生设备
survival fatality 跳伞后等待救援时死亡
survival food 救生食品
survival gear 救生装备
survival kit 救生工具,救生设备
survival kit regulator 降落伞氧气调节器
survival pack 救生物品包
survival parachute 救生伞
survival radio 救生电台
survival radio beacon 救生电台信标
survival ration 救生口粮
survival training 救生训练,生存训练(救生训练的内容之一)
survival vest 救生背心
survivor 生存者,幸存者
SUS Sillicon Unilateral Switch 硅单向开关
susceptance 电纳
susceptibility 敏感性,磁化系数
suspected fault 可疑故障,推测故障
suspend 中止,暂时停止,吊,悬挂,悬置,悬浮
suspended sonde 悬挂式探空仪
suspended underwing unit 悬挂后翼装置
suspension 中止悬吊,悬挂,磁悬液
suspension and release equipment 悬挂投放装置
suspension device 悬挂装置
suspension line 悬挂线,架空线
suspension loop 吊环

suspension unit	悬挂装置
sustain	遭受,受,支撑,支持,维持,继续,证实,确认
sustainability	持续性,能维持性,永续性
sustained acceleration	持续性加速度
sustained arc	持续电弧
sustained combustion	持续燃烧
sustained flight	稳定持久飞行,(火箭)巡航飞行
sustained motor	持续运转发动机
sustained oscillation	持续振荡,自持振荡
sustainer	主发动机,巡航发动机
sustainer motor	1.主发动机 2.续航发动机
sustaining	维持
sustaining orientation	保持定向

SUT System Under Test 受试系统

SV 1. Safety Valve 安全活门 2. Selector Valve 选择活门 3. Service 保养(勤务),服务 4. Servo Valve 伺服活门 5. Shuttle Valve 两用活门 6. Solenoid Valve 电磁活门 7. Specific Volume 比容 8. Start Valve 启动活门

SVC 1. Service 服务 2. Signaling Virtual Channel 信令虚信道 3. Supervisory Channel 监控信道 4. Switched Virtual Calling 交换式虚呼叫 5. Switched Virtual Circuit 交换式虚电路 6. Switched Virtual Connection 交换式虚连接

SVD 1. Safe Vertical Distance 安全垂直距离 2. Seat Video Display 座椅视频显示器 3. Simultaneous Voice on Data 语音数据同时传输

SVFR Special Visual Flight Rules 特殊目视飞行规则

SVI Smoke Volatility Index 烟挥发指数

SVR Slant Visual Range 斜视距离

SW 1. Short Wave 短波 2. South West 西南 3. Stall Warning 失速警告 4. Switch 开关,电门

swallowing capacity	(涡轮机的)临界流量

SWAP Severe Weather Avoidance Plan 避开恶劣天气方案

swap	互换
swarm	群,密集
swashplate	旋转倾转盘,倾斜盘
swath	行,条
swath steering range	侧视范围
sway	摆动,摇动

SWBD switchboard 开关,配电盘,交换机

SWBS Summary Work Breakdown Structure 简要工作分解结构

SWC 1. Stall Warning Card 失速警告卡 2. Surge Withstand Capability 冲击电压承受能力 3. Switching Center 交换中心

SWE 1. Software Engineering 软件工程 2. Stress Wave Emission 应力波传播

sweat cooling	发汗冷却,发散冷却
sweating	发汗,通过多孔表面的蒸发
sweating manikin	出汗假人
sweat-out material	发汗材料
sweep	1.(机翼或尾翼)后掠(有时为前掠) 2.(变几何形翼的)变后掠周期 3.(螺旋桨桨叶的)角偏差 4.(雷达天线)扫探角 5.(阴极射线管)光点扫描 6.(对地面目标)扫射 7.覆盖连接的若干频段 8.(使用技术手段)侦破隐藏的监视装置
sweep amplitude	扫描幅度
sweep angle	掠角,扫描角
sweep back angle	后掠角
sweep forward angle	前掠角
sweep frequency oscillator	扫频振荡器
sweep jammer	扫频式干扰机
sweep method	扫描
sweep period	扫描周期
sweep time	扫描时间
sweep wave form	扫描波形
sweepforward	前掠形,前掠角
sweeping generator	扫描发生器
swept	后掠的,箭形的,扫掠的,摆动的,偏移的,倾斜的
swept back wing	后掠机翼,后掠翼
swept forward wing	前掠翼
swept frequency interferometer	扫频干涉仪
swept frequency reflectometer	扫频反射计
swept generator	扫频发生器
swept waist support	后掠式腰支杆
swept wing	后掠翼
swerve	1.转弯 2.偏向,偏差 3.屈折,折射

SWG Standard Wire Gauge 标准线规,标准线径规

swing	1.摇摆,左右晃动 2.(起动发动机时)扳螺旋桨 3.转动飞机(校罗盘)(指将飞机在地面原地做360°的回转) 4.无线电作用距离畸变(夜间效应) 5.(偏离飞机中心线的发动机失效时)飞机突然偏航
swing angle	旋角
swing arm	摇动臂
swing bearing	摆动轴承
swing nose	机头可折动的
swing scanning earth sensor	摆动扫描地球敏感器
swing-by trajectory	借力飞行轨道
swinging base	摆动基座

SWIP Superweight Improvement Program 超重改进计划

swirl	旋流,涡流,涡旋
swirl afterburner	旋流加力燃烧室
swirl atomizer	旋流式喷嘴
swirl burner	旋流燃烧器
swirl chamber	涡流室,涡流式燃烧室
swirl cup	旋流杯
swirl defect	旋涡缺陷

swirl injector 离心喷嘴	**symbolic pattern** 符号模式
swirl intensity 旋流强度	**symbolization** 符号化
swirl number 旋流数	**symbology** 象征学,象征法
swirl ratio 涡流比	**symmetric aerofoil** 对称翼型
swirl vanes 旋流器叶片	**symmetric channel** 对称信道
swirler 旋流器	**symmetric matrix** 对称矩阵
swirling flow 回旋流	**symmetric mode** 对称模式
swirlmeter 旋进流量计	**symmetric region** 对称区域
SWIS Stall-Warning and Identification System 失速告警和识别系统	**symmetric source** 对称信源
	symmetrical network 对称网络
switch 1.电门,开关,转换电门,转换器 2.转换,转接 3.转换注意力(指飞行员) 4.转机种	**symmetrical point** 对称点
	symmetrical profiling 对称剖面法
switch cam 超针闸刀	**symmetrize** 使对称,使匀称
switch command 开关指令	**symmetry** 对称(性),整齐,匀称
switch element 开关元件	**symmetry axis** 对称轴
switch time 开关时间	**symmetry check** 对称性检查
switch to internal power 转(内)电	**symmetry condition** 对称条件
switch tongue assembly 拨动舌组件	**symmetry plane** 对称平面,对称面
switchblade 道岔,开关闸刀	**symmetry property** 对称性
switchboard 配电盘,配电板,配电屏,开关板	**sympathetic actuation** 感应引爆
switched capacitor filter 开关电容滤波器	**sympathetic aerial detonation** 空中感应爆炸
switched network 开关网络,交换网络	**sympathetic flare** 相应耀斑
switchgear 转换装置,开关装置	**synchro** 同步机,同步器
switching 电门运用,变换,转换,交换	**synchro converter** 同步变流机
switching circuit 开关电路	**synchro receiver** 自整角接收机
switching diode 开关二极管	**synchro transformer** 自整角变压器
switching loss 开关损耗	**synchro transmitter** 自整角发送机
switching mode power supply transformer 开关电源变压器	**synchroniser** 同步器,同步装置,整步器
	synchronism 同步,同期
switching module 交换模块	**synchronization** 同步,同步化,同步现象
switching network 交换网络,接续网络	**synchronization technique** 同步技术
switching section 转接段	**synchronization technology** 同步技术
switching system 开关系统	**synchronization time** 同步时间
switching time 切换时间,开关时间	**synchronized broadcasting** 同步广播
switching transistor 开关晶体管	**synchronized discrete address beacon system** 同步离散地址信标系统
swivel nozzle 可转向喷管,转向喷管	
SWL Safe Working Load 安全工作载荷	**synchronizer** 同步器同步装置
SWR Standing Wave Ratio (Voltage) 驻波比(电压)	**synchronizing** 同步,同步化,同步现象
SWY Stopway (跑道头)应急停机地带,跑道保险道	**synchronizing gear** (射击)协调装置,同步机构
SX Sheet Explosive 薄片炸药	**synchronizing speed** 同步速度
syllable 字节,音节	**synchronizing torque** 整步转矩,同步转矩
sylphon 膜盒,波纹管	**synchronous altitude** 同步高度(卫星轨道)
symbiosis 共生现象	**synchronous baseline** 同步基线
symbiotic star 共生星	**synchronous bombing** 用协调式瞄准具的轰炸
symbol 符号,记号	**synchronous communication interface** 同步通信接口
symbol manipulation 符号处理	**synchronous controller** 同步控制器
symbol rate 符号率	**synchronous data network** 同步数据网络
symbol synchronization 码元同步	**synchronous data transfer** 同步数据传送
symbolic layout method 符号布图法	**synchronous data transmission** 同步数据传输
symbolic model 符号模型	**synchronous detection** 同步检波

synchronous detector	同步检波器
synchronous gyro motor	同步陀螺电机
synchronous high-speed camera	同步高速摄影机
synchronous idle character	同步空闲字符（计算机传输控制字符）
synchronous optical bombsight	协调式光学轰炸瞄准具
synchronous orbit	同步卫星轨道
synchronous plane	同步平面
synchronous satellite	同步卫星
synchronous transfer mode	同步传送方式
synchropter	同步交叉式双旋翼直升机
synchroscope	同步示波器同步指示仪
synchrotron radiation	同步加速器辐射
syncope	晕厥
syncrude	合成油
syndicate	企业联合组织，辛迪加
syndrome	1. 校正子 2. 伴随式 3. 综合症，并发症
synergic ascent	最佳上升
synergic curve	协合最佳曲线，最佳上升曲线
synergistic effect	协同效应
Synjet	合成喷气燃料
synodic month	朔望月
synodic period	会合周期，朔望周期
synoptic background	天气背景
synoptic meteorology	天气图学
syntactic analysis	句法分析
syntactic foam	复合泡沫塑料
syntactic pattern recognition	句法模式识别，结构模式识别
synthesis	综合，综合物，拼合，合成法
synthesis technique	综合技术
synthesized signal generator	合成信号发生器
synthesized sweep generator	合成扫频发生器
synthesizer	合成器，综合器
synthetic antenna	合成天线
synthetic aperture	综合孔径
synthetic aperture antenna	合成孔径天线
synthetic aperture radar	合成孔径雷达
synthetic aperture side-looking radar	合成孔径侧视雷达
synthetic array	合成天线阵
synthetic data processing	数据综合处理
synthetic display	综合显示
synthetic dynamic display	综合动态显示器
synthetic interferometer radar	合成孔径干涉雷达
synthetic oil	合成油
synthetic resins	合成树脂
synthetic rubber	合成橡胶
synthetic spectra	合成光谱
synthetic stereo image	合成立体图像
synthetic training	综合训练
synthetic video	综合视频影像
system	系统，系，体系，制（度），（方）式，装置
system aggregation	系统集结
system analysis	系统分析
system application	系统应用程序
system approach	系统方法
system architecture	系统体系结构（系统各单元之间的结构和关系）
system assembly	系统装配
system availability	系统可用性
system behavior	系统行为
system capacity	系统容量，系统运能
system certification	系统认证
system characteristic	系统特性
system chart	系统框图，系统流程图
system communication module	系统通信模块
system compatibility test	系统兼容性测试
system component	系统组件，系统单元
system concept	系统概念
system configuration	系统配置，系统结构
system constraint	系统约束
system convention	系统规定
system cost	系统成本
system data bus	系统数据总线
system delay	系统延迟
system description	系统描述
system design	系统设计
system design kit	系统设计成套工具
system design review	系统设计评审
system designer	系统设计人员
system development	系统开发
system development cycle	系统开发周期，系统研制周期
system diagnosis	系统诊断
system dynamics	系统动力学
system effectiveness	系统效能
system efficiency	系统效率
system element	系统元件
system energy	系统能量
system engineer	系统工程师
system engineering	系统工程
system equation	系统方程式
system evaluation	系统评价
system exploration	系统探索
system failure	系统故障
system frequency	系统频率
system gain	系统增益
system generated electromagnetic pulse	系统感生电磁脉冲

system homomorphism 系统同态	system program 系统程序,系统程式
system identification 系统识别	system reaction time 系统反应时间
system input 系统输入	system reliability 系统可靠性
system integration 系统集成	system requirement 系统需求
system integration test 系统联试	system response 系统响应
system interface 系统接口	system risk 系统风险
system isomorphism 系统同构	system safety 系统安全性
system level 系统级	system science 系统科学
system loss factor 系统损耗系数	system state 系统状态
system maintainability 系统可维护性	system test 系统试验,系统测试
system margin 系统余量	system theory 系统理论
system matrix 系统矩阵	system uncertainty 系统不确定性
system model 系统模型,系统模式	system validation 系统确认
system modelling 系统模造	system voltmeter 系统电压表
system of astronomical constants 天文常数系统	systematic code 系统码
system operation 系统操作	systematic drift rate 系统漂移率
system outage 系统中断	systematic error 系统错误,系统误差,固定误差
system output 系统输出	systematic fault 系统性故障
system parameter 系统参数,体系参数	systematic review 系统评价
system performance 系统性能,系统业绩	systematic study 系统研究
system planning 系统计划	systematic uncertainty 系统不确定度
system point 系统点	systematics 系统学,分类学
system power 系统功率	systematology 系统学,体系学
system pressure 系统压力	systemic factor 系统性因素
system principle 系统性原则	system-level test 系统级试验

T

T　1. Torget 目标,对象,靶,靶机 2. Temperature 温度 3. Tesla 特斯拉(磁通密度单位) 4. Thickness 厚度 5. Thrust 推力 6. Time 时间 7. Ton 吨 8. Top 顶部 9. Torque 力矩 10. Tower 指挥塔台 11. Tracking 跟踪 12. Trainer 教练员,教练机,练习器,教练设备 13. Transfer load 转港装载 14. Transmitter 传感器 15. Transmitting 发射 16. Transport 运输,运输机 17. Trim 配平 18. Turn 转弯,转向

T bar system　T 形标志线
T junction　T 形接头
T rail　T 形轨,T 形导轨
T section　1. T 形截面 2. T 形部分 3. T 字钢,T 形钢
T&A　Test and Adjust 测试和校准
T&C　Telemetry and Command 遥测和指令
T&DA　Tracking and Data Acquisition 跟踪和数据采集
T&DDM　Techniacl and Drawing Document Management 技术和图样文件管理
T&DRE　Tracking and Data Relay Experiment 跟踪和数据中继实验
T&E　Test and Evaluation 试验和评定
T&EL　Test and Evaluation Laboratory 测试鉴定实验室
T&G　Tracking and Guidance 跟踪和制导
T&PPCR　Tool and Production Planning Change Record 工装和生产大纲更改记录
T&V　Test and Validation 测试和确认
T/C　Top-of-Climb 爬升顶点
T/D　Top-of-Descent 下降起始点
T/L　Top-Level 最高级的,顶层
T/MGS　Transportable/Mobile Ground Station 地面机动站
T/O　Takeoff 起飞
T/R　1. Transformer/Rectifier 变压整流器 2. Transmitter/Receiver 发射接收机
T/R box　收/发开关
T/S　Target Seeker 1. 导引头 2. 自导引导弹
T/W　Thrust/Weight Ratio 推重比
TA　1. Target Alert 目标警报 2. Technical Assistance 技术支援 3. Telescoped Ammunition 弹壳套进式炮弹 4. Telegraphic Address 电报地址 5. Terrain Avoidance 地形回避 6. Territorial Army 本土陆军(英国) 7. Time of Arrival 到达时间 8. Traffic Advisory 交通咨询 9. Transition Altitude 过渡高度 10. Trend Analysis 趋势分析 11. Twin-Aisle 双过道 12. Terrain Avoidance 地形回避
TA&CE　Technical Analysis and Cost Estimate 技术分析和费用估计
TAAF　Test Analyze And Fix 试验分析和整理
TAAS　Three Axis Attitude Sensor 三轴姿态敏感器
TAB　Tape Automated Bonding 带式自动键合
tab　1. 调整片,补翼 2. 标牌,标签 3. 组合件,接头 4. (喷管中的)扰流片 5. 带式自动焊接 7. 平衡蝶片
tabbed flap　调整片式襟翼
table　1. 表,图表 2. x y 坐标台
table fine adjust　转台精调
table tracking microsyn　转台跟踪微动同步器
tabletop　桌面
tablock　(螺帽的)保险片
TABMS　Tactical Air Battle Management System 战术空中作战管理系统
Tabs　Telephone Automated Briefing System 电话自动询问飞行简令系统
tabular display　表格显示器
tabulate　把……制成表,列表显示
TABV　Theater Airbase Vulnerability 战区空军基地易遭攻击的弱点
tabwasher　调整垫片,保险片
TAC　1. Tactical Air Commands 战术空军 2. Technology Application Center 技术应用中心 3. Terminal Area Chart 航站区域图,终端区电子图表 4. Thermosetting Asbestos Composite 热固型石棉复合材料 5. Trim Augmentation Composite 配平增效计算机 6. Turbo-Alternator Compressor 涡轮交流发电机压缩机
TACAIR　Tactical Air 战术空军,战术航空兵
TACAMO　Take Charge and Move Out 塔卡莫空中转信飞机
TACAN　Tactical Air Navigation 战术空中导航系统,"塔康"系统
TACBE　Tactical Beacon 战术信标台
TACBL　Tactical Bomb Line 战术轰炸
TACCO　Tactical Coordinator 战术协调员
TACCAR　Time-Average Clutter Coherent Airborne Radar 时间平均杂波相干机载雷达
TACCS　1. Tactical Air Command and Control Spe-

cialist 战术空军指挥控制专家 2. Tactical Command and Control System 战术指挥和控制系统 3. Thomson-CSF Air Command and Control System 汤姆逊空军指挥控制系统（法国"米达斯"防空系统的一部分）

TACDACS Target Acquisition and Data Collection System 目标截获与数据汇集系统

TACEVAL Tactical Evaluation 战术使用评定

TACFAX Tactical Digital Facsimile 战术数字传真

TACG Tactical Air Ccontrol Group 战术空军控制大队

tachogenerator 测速发电机，转速表发电机，转速表传感器

tachometer 转速表

tachometric 测速的，转速的

tachometric aiming 测速瞄准

tachometric sights 测速瞄准器

tachyon 速子

Tacit Rainbow "沉默彩虹"反雷达导弹

TACIU Test Access Control Interface Unit 测试通道控制界面装置

Tacjam Tactical Jamming 战术干扰

tack 1. 定位搭焊 2. 抢风航行

tack coat 黏合层

tack rag 沾有稀释剂的不掉毛的软布

tack weld 暂时点焊

tack welding 定位焊

tackifier 增黏剂

TACLAN Tactical Landing System 战术着陆系统

TACMS Tactical Missile System 战术导弹系统

TACNAV Tactical Navigation 战术导航

TACOMSAT Tactical Communications Satellite 战术通信卫星

TACOS Tactical Airborne Countermeasure or Strike 空中战术干扰或突击

TACP Theater（Tactical）Air Control Party 战区（战术）空军控制组

TACR Tactical Reconnaissance 战术侦察

TACS 1. Tactical Air Control System 战术空军控制系统 2. Thruster Attitude Control System 推力器姿态控制系统 3. Tracking and Command Station 跟踪控制站

TACSATCOM Tactical Satellite Communication 战术卫星通信

TACSUPWING Tactical Support Wing 战术支援联队

TACT Transonic Aircraft Technology 跨声速飞机技术

TACTASS Tactical Towed Array Sonar System 战术拖曳声呐阵列系统

TACTEC Totally Advanced Communications Technology 全新通信技术

tactical 战术的，作战的，战斗的

tactical aeromedical evacuation 战术航空医疗后送

tactical air command center 战术航空兵指挥中心

tactical air coordinator 战术空军空中协调员

tactical air defence system 战术防空系统

tactical air operations center 战术空军作战中心

tactical air weapons control system 战术空军武器控制系统

tactical aircraft 战术飞机

tactical aircraft shelter 战术飞机掩体

tactical assault weapon 战术突击武器

tactical avionics system 战术航空电子系统

tactical ballistic missiles 战术弹道导弹

tactical bomb line 战术轰炸安全线

tactical communication satellite 战术通信卫星

tactical control 战术控制

tactical effectiveness missile test 导弹战术效果测定

tactical employment 战术使用

tactical environment 战术环境

tactical fighter 战术战斗机

tactical high-altitude penetrator 战术高空突防飞机

tactical information 战术信息

tactical information exchange sytem 战术信息交换系统

tactical intelligence 战术情报

tactical level 战术级

tactical maneuver 战术机动

tactical map 战术地图

tactical missile 战术导弹

tactical missile inertial reference plat-tactical plot forms 战术导弹惯性基准平台

tactical mission 战术任务

tactical plot 战术标图

tactical propulsion 战术推进

tactical reconnaissance 战术侦察

tactical resolution 战术决议

tactical situation display 战术态势显示器

tactical sonar range 声呐战术距离

tactical sortie 战术任务飞行

tactical support bomb 战术支援炸弹

tactical system 战术体系

tactical troop transport 战术部队的运输

tactifs tactical integrated flight system 战术综合飞行指示系统

tactile cue 触觉提示

tactile discrimination 触觉分辨

tactile display 触觉显示器，触摸显示器

tactile faceplate 触摸式荧光屏

tactile information 触觉信息

TACTS Tactical Aircrew Combat Training System 战术空勤人员战斗训练系统

TAD 1. Target Assembly Data 预定装配数据 2. Technology Availability Date 技术到手日期 3. Terrain Awareness Display 地形识别系统 4. Time Available for Delivery 有效投放（炸弹等）时间 5. Turboalternator Drive 涡轮交流发动机传动装置 6. Temporary Additional Duty 临时兼任职务

TADAR Target Acquisition, Designation and Aerial Reconnaissance 目标搜索和空中侦察

TADDS Target Alert Data Display Sets 目标警告参数显示装置

TADI Time Assigned Digital Interpolation 时间分配数字插空

TADIL Tactical Digital Information Link 战术数字信息链路

TADOC Transportable Air Defence Operations Center 移动式防空作战中心

tadpole profile 蝌蚪形翼型

tadpole target director post 目标指示站

TADS 1. Tactical Air Defence System 战术防空系统 2. Target Acquisition and Designation Sight 目标截获和指示瞄准具

TAERS Telecommunications And Earth Resources Stations 通信和地球资源站

TAF 1. Tactical Air Force 战术航空兵 2. Terminal Area Forecast 航站区天气预报，终端区域预报 3. Thermal Acoustic Fatigue 热声疲劳

TAFI Turn-Around Fault Isolation 再次出动准备中的故障隔离，来回飞行中故障隔离

TAFIIS Tactical Air Force Integrated Information System 战术空军综合信息系统

TAFS Terrain Aviodance and Following System 地形回避及跟踪系统

TAFSEG Tactical Air Force Systems Engineering Group 战术航空兵系统工程组

TAG 1. Tactical Airlift Group 战术空运大队 2. Telegraphist Air Gunner 报务员兼空中射击员 3. Telescoped Ammunition Gun 发射弹壳套进式炮弹的航炮 4. Thrust Alleviated Gyroscope 推力阻尼陀螺

TAGC Track Automatic Gain Control 跟踪自动增益控制

TAGS Text And Graphics System 电文和图像系统

Tags Technology for Automated Generation of Systems 新一代自动化系统的工程技术

TAH Transfer and Hold 转换并保持

TAI 1. Thermal Anti-Icing 热除冰（装置） 2. Total Active Inventory 在用（作战）飞机总数 3. True Airspeed Indicator 真空速指示器 4. True Airspeed Indicator 真空速表

tai feathers 浮动式喷口调节片

TAID Thrust Augmented Improved Delta 加大推力改进型黛儿塔（美国运载火箭）

tai-first-ayout 前尾式布局，鸭式布局

taikonaut 中国航天员

tail 1. 尾部，尾翼 2. 尾随，跟踪 3. 尾后截击 4. 飞机部 5. 尾后亮迹 6. 脉冲后的尖头信号 7. 彗尾，流星尾 8. 火箭发动机（弹道导弹的）

tail approach 尾后接敌，尾后进入攻击

tail bomb fuze 炸弹弹底引信

tail boom 1. （飞艇或飞机的）尾部悬臂，尾撑 2. 尾梁 3. 尾桁，尾桁架

tail buffet 尾抖振

tail bumper 护尾装置

tail control 尾翼控制

tail decalage 纵向翼角差

tail defensive system （轰炸机的）尾部射击装置

tail dock 尾部台架，尾部检修棚

tail down angle 1. 擦地角 2. 后坐角

tail drag 1. （飞艇的）尾部系留滑块 2. （飞机着陆时）尾部擦地

tail end 末端

tail fan 尾扇

tail fin 垂直安定面，垂直尾翼

tail first 鸭式（飞机），前尾式（飞机）

tail float 机尾浮筒（用于支承浮筒式水上飞机）

tail grab 尾翼舵夹

tail group 尾翼组

tail gun 机尾炮，机尾机枪

tail guy 飞艇尾部系留索

tail heaviness 尾重

tail incidence 机尾接地事故

tail lift 1. 举升后箱板 2. 尾翼升力 3. 车尾升降台

tail load 尾翼载荷

tail logo 尾部标识字

tail number 尾号

tail package sling 尾翼包装箱吊具

tail parachute 防螺旋伞

tail pipe 排气管（涡轮螺桨或涡轮发动机），尾喷管

tail position 尾部

tail radar 尾部雷达，护尾雷达

tail rotor 1. 尾桨 2. 直升机尾旋翼 3. 尾部螺旋桨

tail rotor gust lock 尾桨控制锁

tail section 尾段

tail section heater 尾端加温器

tail section heating gas distribution console 尾端加温控制台

tail section heating line system 尾端加温管路系统

tail setting angle （水平）尾翼安装角

tail skid 尾橇

tail sling	尾刺
tail spin	尾旋
tail structure	尾部结构
tail surface	尾翼面
tail tab	（炸弹）尾翼调整片
tail unit	尾翼组件
tail view	后视图
tail wagging	火箭发动机摆动
tail warning radar	护尾雷达，机尾警戒雷达
tail wheel	尾轮
tail wheel landing gear	后三点起落架
tail wind	顺风
tail wing	尾翼
tailboard	1.可取下的尾板 2.（车辆的）后挡板
tail- body combination	尾翼机身组合体，尾翼弹身组合体
tail-chase	尾后追击
tailcone	1.（飞机的）尾端整流锥 2.尾锥体
tail-dive	尾冲
tailed delta	带水平尾翼的三角翼飞机
tail-end charlie	1.尾部射击员 2.（单机跟进队形的）最后一架飞机
taileron	尾部升降副翼，全动平尾副翼，差动平尾
tailhooker	（航母）尾钩手
tailing finger	起动电极
tailless aircraft	无尾飞机
tailless airplane	无尾飞机
tailless configuration	无尾布局
tailless fighter	无尾战斗机
tail-off	发动机关闭
tail-on	脱离
tailor made	设计和制造的，特制的
tailored contact surface operation	缝合接触面运行
tailored fuel	特制燃油，专用燃油
tailoring	剪裁
tailpipe	排气管
tailpipe nozzle	长尾喷管
tailplane	1.水平安定面 2.全动水平尾翼
tailplane tank	平尾油箱
tailsitter	直立式飞机，立式垂直起降飞机
tailskid	尾橇
tailskid shoe	尾橇垫块
tailslide	尾冲
tailspin	尾旋，螺旋
tail-warning radar	护尾雷达
tail-warning search set	护尾器，护尾雷达
tail-warning set	护尾器
tailwheel	1.（后三点飞机的）尾轮 2.（前三点飞机的）辅助尾轮
tailwind	顺风
tain	锡箔
tained	保存的，保留的
TAINS	Tercom And Inertial Navigation System 地形匹配及惯性导航系统
TAIR	Terminal Area Instrumentation Radar 航站区仪表着陆雷达
TAIRCW	Tactical Air Control Wing 战术空军空中控制联队
TAIS	Tactical Air Intelligence System 战术空军情报系统
Taiyo	太阳卫星
take off	起飞
take off noise	起飞噪声
take off rating	1.（活塞式发动机的）起飞额定功率 2.（涡轮喷气发动机或涡扇发动机的）起飞额定推力
take off run	起飞滑跑距离
take-off	1.（飞机的）起飞 2.起跳
take-off absolute time	起飞绝对时
take-off and landing strip	升降带
take-off angle	1.出射角 2.离源角 3.起飞角度
take-off balance field length	起飞平衡场长
take-off boost	起飞进气压力
take-off clearance	起飞许可
take-off climb surface	起飞爬升面
take-off condition	起飞条件
take-off decision speed	起飞决断速度
take-off distance	起飞距离
take-off drift	起飞漂移
take-off drift measurement	起飞漂移量测量
take-off field	起飞跑道
takeoff gross weight	起飞总量，发射总重
takeoff mode	起飞方式
take-off performance	起飞性能
take-off phase	起飞阶段时间
take-off power	起飞功率
take-off procedure	起飞过程
take-off ramp	起飞弹射器，（火箭的）倾斜发射装置
take-off relative time	起飞相对时
take-off run	起飞滑跑
take-off safety speed	起飞安全速度
take-off speed	起飞离地速度
take-off thrust	起飞推力
take-off weight	起飞重量
take-off zero	起飞零点
take-off zone	起飞地带
TAKEOVER	自动驾驶仪已断开
takes	排管
Takesaki Launch Site	竹崎发射场（日本种子岛宇宙中

心的探空火箭发射场）

TAL Transatlantic Landing 跨大西洋着陆
talbot 塔尔博脱（MKS 制光能单位）
TALC Tactical Airlift Center 战术空运中心
TALCM Tactical Air-launched Cruise missile 空中发射的战术巡航导弹
Talcott level 太尔各特水准器
Talcott method 太尔各特法
talking robot 会说话的机器人
talk-through 通过主台通信
Tally Ho （空中截击代语）我已看见目标
Tally Ho Heads Up （空中截击代语）我已看到目标,但不能攻击
Tally Ho Pounce （空中截击代语）我已看见目标,准备攻击
TALONS Tactical Airborne Loran System 机载战术罗兰（近距导航）系统
TAM 1. Target Activated Munitions 目标触发的弹药,目标引爆的弹药 2. Task Analysis Methodology 任务分析方法 3. Technical Acknowledgement Message 技术通知文电
tamed frequency modulation 平滑调频
tamper 1. 干预 2. 损害
tandem 1. 前后直排,串列,串联 2. 直通联接
tandem actuator 串列式液压助力器,双腔液压助力器
tandem bicycle gear 自行车式起落架
tandem blade 串列叶片
tandem boost 串联式助推火箭
tandem booster 串联助推器
tandem charge 串联装药
tandem helicopter 双旋翼纵列式直升机
tandem launch 串联（多星）发射,一箭多星发射
tandem main gears 串列式主起落架
tandem network 汇接网络
tandem rocket 串联多级火箭
tandem rotors 纵列式旋翼
tandem seating 纵列座椅配置
tandem vehicle 串联式飞行器
tandem warhead 串列式弹头
tandem wheel gear 纵列机轮起落架
tandem-fan engine 串列风扇发动机
tandem-rotor helicoptor 纵列双旋翼直升机
tandem-wing aircraft 纵列双翼机
Tanegashima Space Centre 种子岛宇宙中心
tang （刀、锉等的）柄脚
tangency point 切点
tangent modulus 切线模量
tangent of camber 中弧切线
tangent off set method 切线支距法

tangent ogive 切面尖模
tangent plane 切面相切平面
tangent resolution 切向分辨率
tangent torque 正交力矩
tangent-cone method 切锥法
tangential acceleration 切向加速度
tangential component 切向分量
tangential distortion 切向畸变
tangential ellipse 切向转移椭圆
tangential hole 切线孔
tangential inlet 1. 切向入口 2. 切线进口
tangential landing 切向着陆
tangential lens distortion 切向畸变
tangential sensitivity 正切灵敏度
tangential target approach 相切直线接近目标法
tangential velocity 切向速度
tangent-tail empennage 与（导弹）弹体相切的水平尾翼
tangent-wedge method 切楔法
Tango T 字布,T 形标志
tank 箱,容器,罐,槽
tank attitude error 油箱姿态误差
tank bay 油箱舱
tank bottom 油罐底
tank capacity 1. 油罐容量 2. 舱柜容量 3. 振荡回路电容
tank farm 油罐群
tank head 油罐端板,压头槽
tank pressure 油罐压力
tank pressurization 油箱增压
tank radius 油罐半径
tank sealer 油箱密封剂
tank size 油箱大小
tank temperature 水箱温度
tank volume 储罐容积
tank wall 舱壁
tankage 1. 各油箱总容量 2. 燃油箱组,容器设备
tanker 空中加油机
tanker airplane 空中加油机
tanker refueling system 空中加油系统
TANS 1. Tactical Air Navigation System 战术空中导航系统 2. Terminal Area Navigation System 航站区导航系统
Tansei 淡青卫星（日本科学卫星和技术试验卫星）
TAOC Tactical Air Operations Center 战术空军作战中心
TAOS Technology for Autonomous Operational Survivability 自主生存技术卫星
TAP Termianl Area Productivity 终端区域生产率
tape 1. 胶带 2. 磁带

tape automated bond package　带式自动化键合封装
tape control　用磁带线穿孔纸带的自动控制
tape drive　1.磁带驱动器 2.纸带机
tape inclined position winding　带倾斜缠绕
tape instrument　带式仪表
tape plane winding　带重叠缠绕
tape recorder　磁带(记录)机,磁带录音机
tape register　磁带寄存器
tape resident　磁带驻留程序
tape teleprinter　纸带式电传机
tape winding　带缠绕成形
tapelayer　预浸料坯带布放设备(计算机控制的复合材料件的生产设计)
taper　1.(机翼或其他翼面的)尖根比,根梢比 2.(机翼的)梯形度,斜削度 3.电位器电阻分布特性
taper ratio　1.锥形比 2.锥度比
taper reamer　锥度铰刀
taper tap　斜削丝锥,锥度丝锥
tapered lens　圆锥透镜
tapered sheet　变厚度板
tapered waveguide　渐变截面波导,锥面波导
tape-wrapped carbon phenolic　带缠碳酚醛
tapped resistor　抽头电阻器
tapping cable　抽头电缆
tapping point　抽头点,分支点
TAPS　1. Telemetry Automatic Processing System 遥测自动处理系统 2. Tercom Aircraft Positioning System 地形匹配飞机定位系统 3. Turbulent Air Penetration Speed 颠簸穿过速度 4. Two-Axispointing System 双轴定向系统
taps　油门(杆)
TAR　1. Tactical Air Request 战术空中支援申请 2. Terrain Aviodance Radar 地形回避雷达,防撞雷达 3. Thermal Analysis Report 热分析报告 4. Threat Avoidance Receiver 威胁回避接收机 5. Type Approval Record 型号批准记录
Taran　1. Tactical Radar And Navigation 战术雷达与导航 2. Test And Repair As Necessary 按需测试和修理
Tarcap　Target Combat Aircraft Practice 作战飞机在目标上空的演习
TARCS　Test And Repair Control System 测试与修理控制系统
TARE　Telemetry Automatic Reduction Equipment 遥测数据自动压缩设备
tare　皮重,包装质量
tare correction　自重修正
tare weight　空重
TAREWS　Tactical Air Reconnaissance and Electronic Warfare Support 战术空中侦察及电子战支援(无人机)

target　1.目标,对象 2.靶,靶机 3.(雷达)反射脉冲
target acceleration　目标加速度
target acceleration command　目标加速度命令
target acceleration direction　目标加速度方向
target acceleration vector　目标加速度矢量
target acquisition　目标捕获,目标截获
target acquisition and sighting system　目标截获与瞄准系统
Target Agena　阿金纳目标星
target aircraft　目标飞行器
target allocation　(防空武器的)目标分配
target altitude　目标高度
target approach point　(飞机)进场检查点
target area　1.目标区 2.靶面积
target aspect system　目标方位系统
target attribute　目标属性
target bearing　目标方位
target blip　目标信号标记,(荧光屏上)目标脉冲信号,目标反射脉冲
target body　目标实体
target capacity　目标容量
target capture　目标捕获
target celestial body　目标天体
target complex　目标群
target computer　1.目标计算机 2.下位机
target crossover frequency　目标交叉频率
target cueing　目标指示
target date　预定(行动)日期
target datum　目标基准面
target designation　目标指示
target detecting device　目标探测装置
target detection　目标探测
target director post　目标引导哨
target discrimination　1.目标鉴别 2.目标分辨力
target dossier　目标资料卷宗,目标档案
target drone　目标靶机
target echo　目标回波
target electromagnetic signature　目标电磁特性
target engagement　弹目交会,瞄准交战
target error angle　目标偏差角,制导角误差
target fluctuation　目标起伏
target following　目标跟踪
target g load　目标g负载
target geolocation　目标定位
target glint　目标闪烁
target hardness　目标硬度,目标加固程度
target homing system　寻的系统
target identification　目标识别,目标标识

target identification head	目标识别头
target identification system	目标识别系统
target illumination radar	目标照射雷达
target illustration print	目标详图
target indication	目标指示
target information system	目标信息系统
target intercept computer	目标拦截计算机
target interception	目标拦截
target language	目标语言
target lock-on	目标锁定
target machine	目标机
target marker	目标标志
target material	靶物
target miss	脱靶,未命中目标
target missile	靶弹
target model	目标模型
target noise	目标噪声
target of opportunity	1.临时目标,意外目标 2.核打击临时目标
target orbit	目标轨道
target panel	靶板
target parameter	目标参数
target pattern	靶图,目标航线,靶机航线,攻击航线
target pipper	(雷达瞄准具的)目标光点
target planet	目标行星
target plate	靶板
target platform	目标平台
target plotter	目标标图员
target point	目标点
target position	目标位置
target positioning	目标定位
target power	靶功率负荷
target practice analysis	实弹演习射击效果分析
target practice cartridge	射击练习弹
target pressure	目标压力
target pressure distribution	目标压力分布
target price	目标价格
target radar	目标雷达
target range	1.靶场,射击场 2.目标距离
target range measurement	靶场测量
target recognition	目标识别
target reflectivity	目标反射率,目标反射性
target response	(核武器爆炸后的)目标反应
target return	目标回波
target reverser	双折流板式反推装置
target run	进入目标航路
target satellite	目标卫星
target scaled model	目标缩比模型
target scattering characteristics	目标散射特性
target scattering matrix	目标散射矩阵
target search head	目标搜索头,目标搜索器
target seeker error	导引头误差,目标位标器误差
target sensing technique	寻的技术,自导引技术,自导引设备
target sensor	目标传感器
target set	目标集
target signal	目标(回波)信号
target signature measurement	目标特性测量
target simulation	目标仿真
target simulator	目标模拟器,目标仿真器
target site	靶位点
target spacecraft	目标航天器(交会对接的追踪目标)
target spacing	目标间隔
target specimen	目标样本
target speed	目标速度
target spot	(雷达瞄准具的)目标标志,目标光点
target state estimator	目标状态估计器
target tape	目标带
target task	目标任务
target temperature	目标温度
target track	目标轨道,目标跟踪
target track correction	目标跟踪校正
target tracker	目标跟踪器
target tracker illuminator	目标跟踪照射雷达
target tracking	目标跟踪
target trajectory	1.目标轨道 2.目标弹道
target tug	靶机拖曳机
target type	目标类型
target value	目标值
target variable	目标变量,指标变量
target velocity	目标速度
target velocity vector	目标速度矢量
target window	目标窗
target-activated	目标(特性或行为)引爆的
target-activated area-denial weapon	目标引爆的区域限制武器
target-following radar	目标跟踪雷达
target-following task	目标跟踪任务
targeting	1.瞄准,把……定为目标 2.把目标资料输入制导软件 3.目标分配
targeting hand-off system	目标自动显示系统
targeting pod	瞄准吊舱
target-seeking bomb	自导引炸弹
target-seeking device	1.寻的装置,导引头 2.目标位标器
target-sighting angle	目标观测角
tariff	航运收费表
tarmac	1.硬铺面场地,硬地 2.机库前沥青停机坪

TARN Turn Around 再次飞行准备,回程飞行准备,调头回程飞行

TARP Test and Repair Processor 测试和修理处理机

TARPS Tactical Aircraft Reconnaissance Pod System 战术飞机空中侦察吊舱系统

TARS Turn-Around Ranging Stations 往返测距站

TARSP Tactical Air Radar Signal Processor 战术空军雷达信号处理器

TAS 1. Target-Acquisition System 目标截获系统 2. Technological Alarming System 技术预警系统 3. Telemetry Acquisition System 遥测捕获系统 4. Three-Axis Stabilization 三轴稳定 5. True Air Speed 真空速

TAS/SAT True Air Speed/Static Air Temperature 真空速/大气静温

TASC Touch-Activated Screen Control 触摸致动荧光屏控制

TASD Trajectory And Signature Data 弹道与特征数据

TASES Tactical Airborne Signal Exploitation System 战术机载信号利用系统

TASI Time Assignment Speech Interpolation 话音插空技术

task area 任务区

task cycle 作业周期

task force (海军)特混编队,(陆军及空军)特遣部队

task level language 作业级语言

task optimization 任务优化

task organized 按任务编成的

task performance 任务绩效,工作效能

task plan 任务计划

task planner 任务计划者

task program 作业程序

task programming 作业编程

task scheduling 任务调度

task sequence 任务序列

task variables 作业变量

tasked 为完成某一特定任务的,特遣的

tasking 派遣(任务)

task-to-task communication 任务-任务通信

TASM 1. Tactical Air-to-Surface Missile 战术空对地导弹 2. Tactical Anti-ship Missile 战术反舰导弹

TASS Tactical Air Support Squadron 战术空中支援中队

TAT 1. Tactical Armament Turret(直升机的)战术武器转塔 2. Technical Acceptance Team 技术验收组 3. Thrust-Augmented Thor 增大推力的雷神火箭 4. Total Air Temperature 大气总温 5. Turn-Around Time 再次出动准备时间,往返飞行时间

TATCA Terminal Air Traffic Control Automation 终端空中交通管制自动化

TATF Terminal Automation Test Facility 航站自动化试验装置

TATS Tactical Aircraft Training System 战术飞机训练系统

TAU Thousand Astronomical Unit 千天文单位探测器(美国一种预计将飞行一千个天文单位距离的行星际探测器)

TAV Transatmospheric Vehicle 跨大气层飞行器

TAWC Tactical Air Warfare Center 战术空中作战中心

TAWS Terrain Awareness and Warning System 地形识别和警告系统

TAWCS Tactical Air Weapons Control System 战术航空武器控制系统

taxi 滑行

taxi circuit 滑行路线

taxi pattern 滑行路线图

taxi radar 机场地面调度雷达

taxi test 滑行试验

taxi-clearance 滑行许可

taxi-holding position 滑行待命点

taxiing control 滑行控制

taxiing load 滑行载荷

taxiing-guidance sign 滑行引导标志

taxiing-guidance system 滑行引导系统

taxilane 滑行线

taxilight 滑行灯

taxiway 滑行道

taxiway center line light 滑行道中线灯

taxiway edge light 滑行道边灯

taxiway lights 滑行道灯

taxiway marking 滑行道标志

taxiway strip 滑行带

Taylor system 泰勒制

TBA Total Blade Area 桨叶总面积

TBB Transfer Bus Breaker 转换汇流条接触器

TBBF Top Baseband Frequency 最高基带频率

TBC 1. Tactical Bombing Competition 战术轰炸比赛 2. Tailored-Bloom Chaff 特制模糊干扰片 3. Thermal Barrier Coating 热防护涂层 4. To Be Confirmed 待确认

TBD 1. To Be Defined 待定义 2. To Be Determined 待定

TBE Timebase Error 时基误差

TBF 1. Tail Bomb Fuze 炸弹尾部引信 2. Time Between Failures (Faults) 故障间隔时间,失效间隔时间

TBG True Bearing 真方位

TBGR Tactical Bomb Group 战术轰炸机大队

TBH Truck-Bed Height 卡车车厢高度

TBI Target Bearing Indicator 全目标方位指示器

TBIG Terbium Iron Garnet 试铁石榴石

TBM 1. Tactical Ballistic Missile 战术弹道导弹 2. Theater Ballistic Missile 战区弹道导弹

TBO Time Between Overhaul 翻修间隔时间,翻修时限

TBPA Torso Back Protective Armour 护背甲

TBR 1. To Be Resolved 待解决 2. Torpedo-Bomber Reconnaissance 鱼雷轰炸机侦察

TBRP Timebase Recurrence Period 时基循环周期

TBS 1. Tailored Business Streams 精简作业流 2. To Be Specified 待规定 3. To Be Supplied 待提供

TC 1. Technical Control 技术控制 2. Terrain Clearance 离地高度 3. Terminal Computer 终端计算机 4. Test Cell 试车台 5. Test Complete 试验完成 6. Thermal Control 热控制 7. Thermocouple 热电偶 8. Time Constant 时间常数 9. Top Centricity 顶部中心线 10. Total Cost 总成本 11. Translating Cowl 平动整流罩 12. Type Certificate 型别合格证,定型证书 13. Type Certification 型号合格审定 14. Twin Coordinator 双坐标方位仪

TC&R Tracking, Command and Ranging 跟踪、指令和测距

TCA 1. Technical Collaboration Agreement 技术合作协定 2. Telecommande Automatique 自动遥控 3. Temperature Control Amplifier 温控放大器 4. Terminal Communication Adapter 终端通信适配器 5. Terminal Control Area 航站管制区,终端管制区 6. Time of Closest Approach 最近进场时间 7. Throttle Control Assembly 油门杆操纵组件 8. Track Crossing Angle 航迹交叉角 9. Transport Category Airplane 运输类飞机 10. Turbine Cooling Airflow 润轮冷却空气流

TCAS Traffic Alert and Collision Avoidance System 空中交通告警及防撞系统

TCB 1. Type Certification Basis 型号合格审定基础 2. Type Certification Board 型号合格审定委员会

TCC 1. TC Configuration 适航取证构型 2. Thermal-Control Coating 热控涂层 3. Thrust Control Computer 推力控制计算机 4. Titanium-Coated Carbon 涂钛石墨 5. Tracking and Control Center 跟踪和控制中心 6. Turbine Case Cooling 涡轮机匣冷却

TCCF Tactical Combat Control Facility 战术空战控制装置

TCCO Temperature Controlled Crystal Oscillator 温控晶振器

TCCS Thrust Control Computer System 推力控制计算机系统

TCD Type Certificate Data 型号合格证数据

TCDM Type Certification Document Management 型号合格审定文件管理

TCDS Type Certificate Data Sheet 型号合格证数据单

TCF Terrain Clearance Floor 地形净空最低标准

TCG Time Code Generator 时间码产生器

TCH Threshold Crossing Height 入口处穿越高度

TCI Tape-Controlled Inspection 磁带控制检查

TCLBS Ttropical Constant-level Balloon System 热带定高气球系统

TCM 1. Technical Coordination Meeting 技术协调会 2. Technical Telemetry Code Modulation 遥测编码调制 3. Time Compression Multiplex 时间压缩多路传输 4. Total Cost Management 总成本管理 5. Trajetory Correction Manoeuvre 弹道修正机动 6. Trim-Control Module 配平操纵组件

TCMA Time-Coordinated Multiple Access 时间协调多址,时间一致多址

TCML Target Coordinate Map Locator 地图上目标坐标定位器

TCMS Test Content Management System 试验内容管理系统

TCO Tape-Controlled Oscillator 磁带控制振荡器

TCP 1. Thrust Center Position 推力中心位置 2. Transfer-of-Control Point 管制交接点 3. Transmission Control Program 发送控制方案

TCQ Throttle Control Quadrant 油门操纵杆弧座

TCR 1. Technical Cost Review 技术成本评审 2. Telemetry, Command and Ranging 遥测、指令和测距 3. Telemetry Compression Routine 遥测压缩程序

TCS 1. Tactical Control Squadron 战术控制中队 2. Tanking Control System 燃料加注控制系统 3. Telemetry and Command System 遥测和指令系统 4. Thermal Control System 热控系统,温控系统 5. Tilt-Control Switch(机翼)倾转操纵电门 6. Touch Control Steering 触摸控制操纵 7. Tool Control Status 工装控制状态

TCT Type Certificate Team 型号合格审定组

TCTO Time-Compliance Technical Order 限时技术规程,时间符合性技术指令

TCU 1. TACAN Control Unit 塔康控制装置 2. Telephone Conversion Unit 电话转换装置 3. Temperature Control Unit 温控装置 4. Towering Cumulus 塔状积云

TCV 1. Temperature Control Valve 温度控制阀 2. Terminal Configured Vehicle 综合布局飞行器

TCXO 1. Temperature Compensated Crystal Oscillator 温度补偿晶体振荡器 2. Temperature Controlled

Crystal Oscillator 温控晶体振荡器

TD 1. Target Discrimination 目标识别 2. Technical Data 技术数据 3. Technical Directive 技术指示 4. Technical Docuementation 技术文档 5. Test Directive 试验提示 6. Thor-Delta 特德卫星 7. Time Delay 延时,延时电路 8. Tool Description 工装说明 9. Tool Design 工装设计 10. Touchdown 着地,接地 11. Transposition Docking 换位对接 12. Tunnel Diode 隧道二极管

TD&D Technical Data and Documents 技术数据和文档

TD&E 1. Tactics Development and Evaluation 战术研究和鉴定

TDA 1. Tunnal Diode Amplifier 隧道二极管放大器 2. Top-Down Approach 自上而下法 3. Type Design Approval 型号设计更改

TDAP Touchdown Aim Point 接地瞄准点

TDAS 1. Test-Data Acquisition System 试验数据采集系统 2. Tracking and Data Acquisition Satellite 跟踪和数据采集卫星

TDASS Tracking and Data Acquisition Satellite System 跟踪和数据采集卫星系统

TDB 1. Temperature Dry Bulb 干球温度 2. The Data Base 数据库

TDC 1. Target Designation Control 目标指示控制 2. Target Designator Control 目标指示器控制 3. Technical Development Center 技术发展中心 4. Telemetry Data Center 遥测数据中心 5. Through-Deck Cruiser 连续甲板(全甲板)巡洋舰 6. Time Delay Closing 延时结束 7. Type Design Change 型号设计更改

TDCS Traffic Data Collection System 空中交通数据收集系统

TDD 1. Target Detecting Device 目标探测装置 2. Tool Design Datasets 工装设计数据集

TDE Tool Design Engineering 工装设计工程

TDF Tactical Digital Facimile 战术数字传真

TDI 1. Telecommunications Data Interface 远程通信数据接口 2. Time Delay and Integration 延时积分 3. Triple Display Indicator 三重显示器

TDIL Tactical Digital Information Link 战术数字信息传输线

TDL Transistor Diode Logic 晶体管二极管逻辑

TDLS Tower Data Link System 塔台数据链系统

TDM 1. Technology Development Mission 技术开发任务 2. Time Division Modulation 时分调制 3. Time Division Multiplex 时分多路复用 4. Tracking Data Message 跟踪数据通报

TDMA Time Division Multiple Access 时分多路存取,时分多址

TDMC Time Division Multiplex Channel 时分多路复用通道

TDMS Thermal Desorption Mass Spectrometry 热解析质谱

TDO Tornado 1."狂风"(飞机名)2.龙卷风

TDoA Time Difference of Arrival 到达时差

TDOP Timing Dilution Of Precision 测时精度因子,测时精度冲淡系数,时间精度淡化

TDP 1. Technical Data Package 全套技术资料,技术资料包 2. Tracking Data Processor 跟踪数据处理机

TDR 1. Technical Despatch Reliability 技术放飞可靠率 2. Tool Design Request 工装设计需求 3. Tool Design Review 工装设计评审 4. Transponder 应答机

TDRS Tracking and Data Relay Satellite 跟踪与数据中继卫星

TDS 1. Tactical Data System 战术数据系统 2. Tactical Drone Squadron 战术无人机中队 3. Terminal Display System 终端显示系统 4. Tool Design Standard 工装设计标准 5. Training Depot Station 训练供应站 6. Translation and Docking Simulator 平移和对接仿真器

TDST Tower Data Services Terminal 塔台数据服务终端

TDU 1. Television Display Unit 电视显示装置 2. Terminal Display Unit 终端显示装置

TDWR Terminal Doppler Weather Radar 航空港多普勒气象雷达,终端多普勒气象雷达

TDY Temporary Duty 临时任务

TDZ Touchdown Zone 接地区

TDZE Touchdown Zone Elevation 接地区标高

TDZL Touchdown Zone Lights 接地区灯光设备

TE 1. Tactical Evaluation 战术研究 2. Tangent Elevation 瞄准角,仰角 3. Tool Engineering 工装工程 4. Trailing Edge 后缘

TE wave transverse electric wave 横电波,H波

TEA 1. Torque Equilibrium Attitude 力矩平衡姿态 2. Transferred-Electron Amplifier 转移电子放大器 3. Triethyl Alcohol 三乙基醇

teach pendant 示教盒

teaching programming 示教编程

Teak 麻栗树(高空核爆炸代号)

Teal Cameo "水鸭石雕"(美国研制无人机的计划名)

team objective 团队目标

team theory 队论

TEAMS Tactical Evaluation And Monitoring System 战术评定及监控系统

tear and wear 磨损,正常耗损

tear bomb 催泪(炸)弹

tear gas   催泪性毒气
tear ridge   （材料破坏断口的）撕裂棱
tearaway connector   快速断开接头
teardown   拆卸分解，拆散
teardown inspection   分解检查
teardrop   1.修正角转变着陆航线 2.水滴状的
teardrop canopy   水滴状座舱盖
tearing   图像撕裂
tearoff cap   易拉伞包
TEBOM   Tool Engineering Bill Of Material 工装工程物料表
TEC   1. Thermo-Electric Cooler 热电冷却器 2. TOS Evaluation Center 泰罗斯业务卫星计算中心
TECHED   Technical Editor 技术编辑
technetronic   电子技术化的
technical activity   技术活动
technical approach   1.技术因素分析法 2.技术途径
technical area   技术（准备）区
technical assistance   技术援助
technical authority   技术权威
technical capability   技术能力
technical challenge   技术挑战
technical community   技术社区
technical datum   技术参数
technical deficiency   技术缺陷
technical delay   技术性晚点起飞，技术性延迟
technical despatch reliability   技术放飞正常率，技术放飞可靠率
technical division   技术部
technical electrics   技术电气
technical error   技术误差
technical failure   技术故障
technical failure risk   技术故障风险
technical fellow   高级技术成员
technical fire control   技术性射击控制
technical grounding   工艺接地
technical incident   技术性事故征候
technical index   技术指标
technical issue   技术问题
technical know-why   技术知识，工作原理
technical manual   1.技术规范，技术规程 2.技术手册
technical orders   技术法规，技术规程
technical position   技术阵地
technical preparation building   技术厂房
technical program   技术程序
technical quality team   技术质量组
technical readiness   技术准备完好率
technical representative   技术代表
technical risk   技术风险

technical safety team   技术安全组
technical seminar   技术研讨会
technical sequence for satellite test   卫星测试技术流程
technical service crew   技术勤务组
technical service of aircraft   飞机的例行技术检查
technical servicing truck   修理车
technical site   技术阵地
technical specialist   专业技术人才
technical staff   技术人员
technical standard order   技术标准规定
technical stop   技术（性）停留
technical support   技术支持，技术援助
technical support mount   工艺支架
technical survey   技术检查
technical team   技术团队
technician   1.技师，技术员 2.技巧纯熟的人
technique   1.技巧，技术 2.手法
technique for cooling infrared detector   红外探测器的致冷技术
technique of close coupled cancard configuration   近耦合鸭式布局技术
technocrat   1.技术统治论者 2.家政治论者 3.技术统治
technological break   工艺分离面
technological compensation   工艺补偿
technological constraint   技术约束
technological ensemble   技术总体效果
technological experiment satellite   技术试验卫星
technological parameter   工艺参数
technological test satellite manipulation   技术实验卫星机械手
technologically useful life   技术可用寿命
technologist   1.技术专家 2.工艺学家
technology   1.技术 2.工艺 3.术语
technology advance   技术进步
technology alternative   1.替代技术 2.可选择的技术 3.可供选择的技术
technology assessment   技术评定
technology assessment method   技术评定方法
technology challenge   技术难关
technology demonstration   技术示范
technology demonstrator   技术示范者
technology development   技术开发
technology evaluation   技术评估
technology freeze   技术冻结
technology impact   技术影响
technology level   1.技术水平 2.技术层面
technology maturation   技术成熟
technology objective   技术目标

technology program	技术方案
technology readiness	技术准备
technology roadmap	技术路线图
technology transfer	技术转让
technology transmissible	可转让的技术
technology upgrade	技术升级
technomania	技术热
Techroll	泰克罗尔喷管
Techsat	技术卫星
TECR　Technical Reason	技术原因
tectonic map	大地构造图
tectonic stress	大地构造应力,地壳应力
tectonophysics	大地构造物理学

TED　1. Tactical Evaluation Display 战术判断显示器 2. Task Execution Documentation 任务执行文档 3. Tool and Equipment Drawing 工具和设备图纸 4. Total Energy Detector 总能量探测器 5. Trailing Edge Down 后缘下垂 6. Transferred-Electron Device 电子转移装置

TEDAR　Telemetered Data Reduction 遥测数据处理

TEE　1. T字接头 2. Tubular Extendible Element 可延伸管状部分

tee connector	三通接头
tee gearbox	T形传动机构
tee light	T字灯
teepee	1.圆锥形帐篷 2.超视距雷达
teeter	1.倾斜 2.轴向变形 3.变形
teetering rotor	跷跷板式旋翼,半刚接旋式翼
TEF　Total Environment Facility	总环境设备
Teflon	特氟隆,聚四氟乙烯
TEHP　Thrust Equivalent Horsepower	推力当量马力
TEI　Text Element Identifiers	主题单元标识符
tektites	玻璃陨石

TEL　1. Telebrief 地面对讲通信,直接通话 2. Telephonic 用电话传送的 3. Tetraethyl Lead 四乙基铝

Telamon	泰拉蒙无人机

Telar　Transport Erector Launcher Radar 带有升降架、发射架和雷达的运输车

TELATS　Tactical Electronic Locating and Targeting System 战术电子定位与目标信息输入系统

telebrief system	电话下达简令的制度
telechirics	遥控系统
Telecom	电信卫星
telecommand	遥控,遥控指令
telecommand channel service	遥控信道业务
telecommand equipment	遥控设备
telecommand master console	遥控主控台
telecommand packet	遥控信道业务
telecommand segment	遥控分机
telecommand segmentation layer	遥控分段层
telecommand session	遥控期
telecommand station	遥控站
telecommand sub-console	遥控分控台
telecommand subsystem	1.遥控分系统 2.指令分系统
telecommand terminal	遥控终端
telecommand transfer frame	遥控转移帧
telecommand transfer layer	遥控转移(传送)层
telecommunication(telecom)	1.电信 2.通信 3.电信公司
telecommunication access method	通信存取方法
telecommunication control unit	通信控制器
telecommunication network	远程通信网
telecommunication route	通信路由
telecommunication satellite	通信卫星
telecommunication satellite space station	通信卫星空间站
telecommunication service	通信服务,通信业务
telecommunication software architecture	通信软件系统结构
telecommunication system	通信系统,电信系统
telecompass	远距罗盘
telecon	电话会议
teleconference	电话会议
teleconferencing	卫星电话会议,通信会议
telecontrol	遥控,远程控制
telecontrol device self-check	遥控装置自检
telegraph	电报
telegraph automatic relay equipment	电报自动中继设备
telegraph network	电报网
telemail-telephone set	书写电话机
telemaintenance service	远程维护服务(业务)
telemanagement	远程管理
telematics	远程通信及信息处理
telemechanic system	遥控机械装置
telemechanics	遥控力学,远动学,遥控机械学
telemechanique	远动技术
telemedicine	远距离医学
telemeteorograph	遥测气象仪
telemeter	遥测仪,测距仪
telemetered measurement	遥测
telemetered parameter	遥测参数
telemetering	遥测
telemetering buoy	遥测浮标
telemetering communication system	遥测通信系统
telemetering signal blackout	遥测信号中断
telemetering system of frequency division type	频分(制)

遥测系统
telemetering system of time division type 时分(制)遥测系统
telemetering vehicle 遥测车
telemetry 1.遥测 2.遥测学 3.遥测技术
telemetry and command system 遥测和指令系统
telemetry and monitor network 遥测监视网
telemetry and remote control system 遥测遥控系统
telemetry and tracking data time zero alignment 遥测外测数据时间零点对齐
telemetry antenna on-board 箭上遥测天线
telemetry capacity 遥测容量
telemetry checkout system 遥测检测系统
telemetry compression routine 遥测压缩程序
telemetry computer systems 计算机遥测系统
telemetry computer word 遥测计算机字
telemetry data processing 遥测数据处理
telemetry data reduction 遥测数据处理
telemetry earth station 遥测地面站
telemetry encryption 遥测加密
telemetry errors 遥测误差
telemetry format 遥测格式
telemetry frame 遥测帧
telemetry front end 遥测前端
telemetry fuze 遥测引信
telemetry head 遥测弹头
telemetry implement plan 遥测实施方案
telemetry information 遥测信息
telemetry(information) simulation 遥测(信息)仿真
telemetry missile 1.遥测试验(导)弹 2.遥测(导)弹
telemetry parameter 遥测参数
telemetry program 遥测大纲
telemetry receive station 遥测(接收)站
telemetry standards 遥测标准
telemetry station 遥测站
telemetry subsystem 遥测分系统
telemetry symbol 遥测字符
telemetry system 遥测系统
telemetry, tracking and command system 遥测、跟踪与遥控系统,测控系统
telemicroscope 1.望远显微镜 2.遥测显微器 3.显微望远镜
telemonitoring 远程监控
TELENAV Television Navigation 电视导航
teleological system 目的系统
teleoperation 遥操作技术
teleoperator 遥控机械手
teleports 1.传送点 2.瞬间移动
telepresence 1.遥现 2.遥现技术

telepresence adaptive robotics 遥现自适应机器人技术
teleprinter 传真复印机,电传打字电报机
teleprocessing 远程信息处理,远距程序控制,遥控加工
teleprocessing system 远程(信息)处理系统
teleran television radar air navigation 电视雷达空中导航
telerobot 遥控机器人
telerobotics 1.遥控机扑技术 2.遥控机器人
Telesat 电信卫星(加拿大国内通信卫星,又名 Anik 卫星)
telescope 1.(空中加油的)伸缩套管 2.(旋翼的)折叠,缩进 3.(螺旋桨的)截尖 4.望远镜 5.远距离电磁辐射收集装置
telescope gauge 伸缩规
telescope structure 1.嵌入构造 2.叠扇构造
telescoped ammunition 弹壳套进式炮弹
telescopic landing gear 支柱式起落架
telescoping 1.用伸缩或折叠法 2.桨叶叶尖截短
telescoptic tube 伸缩套管
telescra mble 扰频干扰通信
telethermometer 遥测温度计
telethermoscope 遥测温度器
teletraffic 长途电话业务
television 电视
television automatic tracking 电视自动跟踪
television broadcasting 1.电视播送 2.广播电视学 3.帧电视传送
television camera 电视摄像机
television command 电视制导指令
television command guidance assembly 电视指令导引装置
television guidance 电视制导
television homing head 电视自动寻的导引头,电视寻的头
television measurement 电视测量
television pick-up house 电视摄影间
television signal 电视信号
television target identification 电视目标识别
television telescope 电视望远镜
television tracking 电视跟踪
television tracking measurement system 电视跟踪测量系统
telex 用户电报
telint telemetry intelligence 遥测情报(信息)
tell 1.分辨,辨别 2.(颜色、声音等)显示,识别
telling 轨迹报知
telluric(current) method 大地电流法
telluroid 似地球面,近似地形面
TELNET Telecommunication Network 远程通信网

telpak 宽频带通信通道
TELS Turbine Engine Loads Simulator 涡轮发动机载荷模拟器
TELTRAC Telemetry Tracking System 遥测跟踪系统
TEM 1. Transmission Electron Microscope 透射电子显微镜 2. Transverse Electromagnetic 横(向)电磁波 3. Technical Error Message 技术查错信息 4. Temperature 温度 5. Temporary 暂时的 6. Temporal 暂时的,当时的
TEM cell 横(向)电磁波室
TEM wave 横(向)电磁波
TEMI Terminating Equipment Mockup Input 终端设备样机输入
TEMISAT Telespazio Micro Satellite 泰米卫星
TEMP 1. Temperature 温度 2. Test and Evaluation Master Plan 试验和鉴定总计划
temper 1.(钢材的)回火 2. 硬度,韧度(影响钢硬度的)含碳量
temper brittleness 回火脆性
temperature 1. 温度 2. 气温 3. 使用温度范围
temperature accountability 温度决定因素
temperature aloft forecast 高空温度预报
temperature altitude 温度高度
temperature and humidity control system 温度与湿度控制系统
temperature capability 承温能力
temperature change 1. 温度变化 2. 气温变化 3. 温度的变换 4. 变温
temperature change rate 变温率
temperature coefficient 温度系数
temperature condition 1. 温度情况 2. 气温条件
temperature contour 温度等值(高)线
temperature control 温度控制,温控
temperature control coating 温控涂层
temperature control system 1. 温度控制系统 2. 温控系统
temperature control valve 温控阀
temperature controlled panel 温度控制板
temperature controlled shroud 温度控制屏
temperature correction 温度修正
temperature correlation 温度关联
temperature cycle 温度循环
temperature cycling test 温度循环试验
temperature dependence 1. 温度特性 2. 温度依赖性 3. 温度依赖性行为 4. 随温度变化的规律
temperature dependency 1. 温度依存性 2. 黏温关系 3. 温敏度 4. 温度关系
temperature difference 1. 温差 2. 温度差
temperature distortion 1. 温度畸变 2. 受热变形

temperature distribution 1. 温度分布 2. 温度场
temperature drop 1. 温降 2. 温度下降 3. 温度降差 4. 温度降
temperature effect 温度效应
temperature efficiency 1. 温度效率 2. 温效
temperature exponent 温度指数
temperature field 温度场
temperature fluctuation 1. 温度波动 2. 温度波动度 3. 温度涨落 4. 温度起伏
temperature gradient 温度梯度
temperature history 1. 温变史 2. 温度历程 3. 温度随时间的变化 4. 温度变化过程
temperature humidity infrared radiometer 温度湿度红外辐射计
temperature increase 温升
temperature jump 温度跃变
temperature level 温度水平
temperature limit 1. 应用温度范围 2. 温度极限
temperature limitation 温度限制
temperature measurement 1. 温度测量,测温 2. 水温测定
temperature of airflow 气流温度
temperature peak 温度峰值
temperature perturbation 温度波动
temperature plume 羽焰温度
temperature probe 1. 温度探针 2. 温度探测器
temperature profile 1. 温度曲线 2. 温度剖面图 3. 温度轮廓
temperature profile recorder 温度廊线记录器
temperature ramp 温度斜坡
temperature range 温度范围
temperature ratio 温度比
temperature recovery coefficient 温度恢复系数
temperature recovery factor 温度恢复因子,温度恢复因数
temperature reduction 温度下降
temperature region 温度区
temperature resolution 温度分辨率
temperature rise 温度上升
temperature rise during ascent 主动段温升
temperature rise ratio 加温比
temperature self-compensation 温度自补偿
temperature sensitive paint 示温涂料
temperature sensitivity of pressure 压强温度敏感系数
temperature sensitivity of burning rate 燃速温度敏感系数
temperature sensitivity 1. 温度灵敏度 2. 温度灵敏度系数 3. 温度感受性
temperature sensor 温敏元件,温度传感器

temperature shift 温度变化
temperature-compensated voltage limited charge 温度补偿限压充电
temperature shock test 温度冲击试验
temperature spread 温度差距(同一断面中最高和最低温度之差)
temperature status 温度状态
temperature stress 1.温差应力 2.热应力 3.温度力
temperature structure 温度结构
temperature test 温度试验
temperature transmitter 温度变送器
temperature traverse 温度横向分布,用位移机构测量温度分布
temperature value 温度值
temperature variation 温度变化
temperature-dependent viscosity 热变黏度
temperature-sensitive paint 热敏涂料
tempering 回火
template 样板
template base 模板库
template matching 模板匹配
templating 绘制情况图
TEMPO Technical Military Planning Operation 军事技术计划活动
temporal change 暂时变化
temporal evolution 时间演化
temporal integration (听觉)瞬时整合
temporal periodicity 时间周期
temporal resolution 1.时间分辨率 2.瞬时分辨力 3.瞬时清晰度
temporal variation 时间变化
temporary altitude 暂时高度
temporary capture 临时捕获
temporary failure 1.短暂失效 2.暂时故障
temporary flight restriction 临时性飞行限制
temporary grounding 暂时停飞,临时停飞
temporary point 临时停留点
temporary setting 临时装订,临时调定
temporary threshold shift 暂时性听阈偏移
Tempsat 坦波卫星(美国空军军事试验卫星)
tenacity 韧性
tenary 1.三个(一套)的,三重的,三元的 2.三种状态$(0,1,x)$的仪器
tend type landing forecast 趋势型着陆预报
tendency 趋势
tenet 1.原则 2.信条
Tenley 三军战术通信保密通话系统(美国国家保密局)
Tenma 天马卫星(日本天文卫星)

Tenrai "天籁"飞机
tensegrity 1.张拉整体 2.无一定尺寸限制的结构
tensegrity structure 张拉整体结构
tensile compressive stiffness matrix 拉压刚度矩阵
tensile dislocation 张力错位
tensile modulus 拉伸模量,抗张模量
tensile or compressive strength of matrix 基体拉压强度
tensile property 抗拉性能
tensile sample 拉伸试样
tensile strength 抗拉强度,拉伸强度,抗张强度
tensile strength of fiber 纤维拉伸强度
tensile stress 张应力
tensile test 拉伸试验
tensiometer 张力计
tension 1.张力,拉力 2.紧张,不安 3.电压
tension arm 张力臂
tension axis 张力轴
tension control 张力控制
tension field 张力场
tension force 张力
tension level 紧张度
tension loss 1.丢磅 2.张力损失
tension member 受拉杆件
tension regulator 张力调节器
tension shell 张力层
tension weight 1.张力重锤 2.张紧配重
tensioned membrane 张拉膜结构
tensioner 张力保持器
tensor 磁张线,张量
tensor of stress 应力张量
tensor permeability 张量磁导率
tensor product 张量积
tensor viscosity 张量黏性
tentative value 暂定值
tenuous plasma 稀薄等离子体
TEO Transferred-Electron Oscillator 转移电子振荡器
TEP Tactical Electronic Plot 战术电子标图
tephigram (大气的)温熵图
TEPIGEN Television Picture Generator 电视图像产生器
TER Triple Ejector Rack 弹射式三弹
tera 太(拉),兆
teracycle 太周
terahertz 太赫
Tercom 1. terrain comparison 地形比较 2. terrain contour-matching 地形匹配(导航技术)
terdenary 十三进制的
terebenthine 精制松节油(用作火箭发动机的燃料)
TEREC Tactical Electronics Reconnaissance 战术电子

侦察(设备)
TERLS Thumba Equatorial Rocket Launching Station 顿巴赤道附近火箭发射站(印度)
term　　1.术语 2.学期 3.期限 4.条款
terminal　　1.航站 2.终点站 3.终端 4.导弹飞行的末段 5.接线柱,管接头
terminal accuracy　　末段引导精度
terminal airport　　起止点航空港
terminal airspace　　1.终端空域 2.机场空域
terminal alternate　　备降航空站
terminal apron　　航站大楼停机坪
terminal area　　航站区,候机楼楼前区
terminal area chart　　机场区航图,航空港区域航图
terminal attenuator　　终端衰减器
terminal ballistics　　终点弹道学,末端弹道学
terminal boom　　根段梁
terminal capacity　　航站运载能力(吞吐量)
terminal clearance capacity　　航站容量
terminal communication adapter　　终端通信适配器
terminal communications　　航站通信设备
terminal condition　　1.边界条件 2.终端条件
terminal constraint　　终端约束
terminal control　　末段控制
terminal control area　　终端管制区,航站管制区
terminal defense　　末段防御
terminal descent　　末段下降
terminal (flight path) angle　　末端弹道倾角
terminal forecasts　　航站天气预报
terminal guidance　　末段制导
terminal imaging radar　　末段成像雷达
terminal link　　终端环节,终端链路
terminal manoeuvring area　　航站机动区空港周围的管制空域
terminal normal shock　　终端正激波
terminal phase　　末段,末期
terminal phase intercept　　末段截击,末段拦截
terminal point　　端点(航站)
terminal profile　　终结符轮廓
terminal radar service area　　航站雷达勤务区
terminal radial velocity　　终端径向速度
terminal shock　　结尾正激波
terminal state　　终止状态
terminal string　　终结符串
terminal strip　　接线条
terminal velocity　　1.极限速度 2.物体在空气中自接线条
terminal voltage　　端电压
terminally guided submissile　　末段制导子导弹
terminate　　1.使终止 2.使结束 3.解雇

terminating aircraft　　结束飞行的飞机
termination　　结束,终止
termination proof　　终止性证明
termination shock　　1.边界激波 2.终端激波
termination type power meter　　终端式功率计
terminator　　1.终端负载 2.推力终止器 3.明暗界线 4.晨昏线
terminology　　1.术语,术语学 2.用辞
TERN Terminal and Enroute Navigation 机场区及航线导航
ternary　　三进制
ternary code　　三进制码,三元码
ternary pulse circuit　　三元脉冲电路
terne plate　　镀锡(铅)低碳薄钢板
terotechnology　　设备综合工程(学),设备经济管理技术
TERPROM Terrain Profile Matching 地形剖面匹配系统
terps terminal instrument procedures　　终端仪表程序
terra　　1.地 2.地球 3.土地 4.月球高地
terra in a ltitude　　地面海拔高度
terra in echo　　地面回波
terra in-aided navigation　　地形辅助导航
terrain　　1.地形,地区,地带 2.领域,场所,范围
terrain altitude　　离地高度
terrain avoidance　　地形回避
terrain avoidance radar　　地形回避雷达
terrain camera　　地物相机
terrain clearance　　地形回避雷达
terrain clearance radar　　地形跟踪雷达
terrain clutter　　地形散射干扰,地形杂乱回波
terrain comparison　　地形对比
terrain correction　　地形(影响)校正
terrain database　　地形数据库
terrain feature　　地貌特征
terrain following　　地形跟踪
terrain following control　　地形跟踪控制
terrain following controller　　地形跟踪控制器
terrain following system　　地形跟踪系统
terrain generation　　地形生成
terrain map　　地形图
terrain matching　　地形匹配
terrain matching guidance　　地形匹配制导
terrain model　　地形模型
terrain orientation　　按地形对正
terrain profile　　1.地形剖面 2.地形轮廓
terrain referenced navigation　　地形基准导航
terrain scattering coefficient　　地形散射系数
terrain sensing unit　　地形敏感装置
terrain storage　　地形存储

terrain tracking radar	地形跟踪雷达
terrain-avoidance radar	地形回避雷达
terrain-following radar	地形跟踪雷达
terrain-profile recorder	地形剖面记录仪
terran	地球人，人类
terrestrial branch	地支
terrestrial camera	地面摄影机
terrestrial coordinate system	地球坐标系
terrestrial dynamical time	地球力学时
terrestrial gravitational perturbation	地球引力摄动
terrestrial guidance	陆上制导
terrestrial infrastructure	地面基础设施
terrestrial interferometry	地球干涉测度学
terrestrial network	地面网
terrestrial planet	类地行星
terrestrial radiation	地面辐射，地球辐射
terrestrial radio	地面无线电
terrestrial reference guidance	地球基准制导，地形识别导航
terrestrial refraction	地面折射
terrestrial scintillation	地面闪光
terrestrial service	地面业务
terrestrial space	地球空间，近地太空
terrestrial spectroscopy	地球谱学
terrestrial station	地面（微波）站
terrestrial surface radiation	地表面辐射
terrestrial triangle	大地三角形
territorial sovereignty	领土主权
territory	1.领土，领域 2.范围 3.地域 4.版图
TERS	1. Triple Ejector Racks 弹射式三弹挂弹架 2. Tropical Earth Resources Satellite 热带地球资源卫星（印度尼西亚的遥感卫星）
tertiary creep	第三阶段蠕变
TES	1. Test and Evaluation Squadron 试验和鉴定中队 2. Thermal Emission Spectrometer 热辐射摄谱仪 3. Tropospheric Emission Spectrometer 对流层发射光谱仪
TESAC	Training and Evaluation System for Active Countermeasures 积极（电子）对抗（干扰）的训练和鉴定系统
tesla	泰斯拉（磁通密度单位）
tesla meter	泰斯拉计
tesseral harmonics	田谐函数
tesseral harmonic coefficient	田谐系数
test	1.测试 2.试验 3.检验
test accessories kit	测试配件
test aircraft	试验飞机
test airspace	试验空域
test altitude	测试高度
test and analysis technology	测试分析技术
test and check program	测试检查程序
test and diagnostics	测试和诊断（程序）
test apparatus	试验装置，试验仪器
test area	测试范围
test article	试验件，试验样品
test assembly	试验组件
test at high attack angle	大攻角试验
test ban	禁止大气层核试验条约
test battery	试验蓄电瓶
test bed	试验台
test bench	试验台
test block	1.试样 2.试验台
test card	测试卡
test case	测试用例
test case generator	测试用例生成器
test cell	1.试验舱，试验台，试验间 2.测试储存单元
test cell pressure	试验间压力
test center	测试中心
test chamber	试验舱，测试舱
test channel	试验信道
test clip	试验夹
test club	台架试车桨
test condition	测试条件，试验状态
test coupon	1.附连试验板 2.试样
test data	测试数据
test datum	测试数据（单数）
test diamond	菱形测量区
test driver	1.测试驱动程序 2.试车驾驶员
test duration	1.测试时间 2.试验持续期 3.试验时间
test entry	测试开始
test environment	测试环境
test equipment	测试设备
test facility	1.试验设备，测试设备 2.试验装置
test facility transfer	测试设备转运
test fire	试验火，试发射
test firing	试射，试验点火，点火试验
test flight	试飞，测试飞行
test floor	试验台
test flow	测试流程
test fluid	测试液
test frame	1.测试框架，测验句式 2.试验格式
test furnace	1.试验炉 2.烤瓷炉测试
test gas	试验气体
test goal	测试目标
test group	发射工作队
test in assembly hall	总装厂测试
test in launch area	发射阵地测试
test in technical centre	技术阵地测试

test instrumentation subsystem　测试仪表分系统
test laboratory　测试实验室
test liquid　试验液体
test logic　测试逻辑
test maneuver　试验演习
test mass　试验质量
test matrix　1.测试矩阵 2.检验矩阵,试验矩阵
test medium　试验培养基
test method　测试方法
test mixture　试验混合物
test mode　测试方式
test model　测试模型
test motor　试验电动机
test nozzle　测试喷嘴
test object　1.被试品 2.测试对象
test objective　试验目的
test of assembly　组件试验
test of significance　显著性检验
test operation console for satellite　卫星测试操作控制台
test panel　1.试验盘 2.检验格框
test parameter　测试参数
test pattern　试验显示图像,测试图形
test period　试验周期,测试周期
test piece　试验样品,试件
test pilot　试飞员
test plan　1.试验计划 2.测试计划
test plate　检光板
test point　测试点
test port　测试孔
test pressure　试验压力
test pressure range　试验压力范围
test problem　1.测试题目 2.测试问题
test procedure　1.试验程序 2.检验法 3.检查法
test profile　试验剖面
test program　测试程序
test range　测试范围
test receiver　测试接收机
test record　测试唱片,测试记录
test report　测试报告
test residual　检测残差
test result　试验结果
test rig　1.试车台架,试验台 2.试验装置,试验设备
test rotor　试验转子
test run　1.调试试车,试验运行 2.试飞,试航,试进入
test sample　1.试样 2.试棒
test scenario　测试场景
test seal　检验盖章
test section　1.(风洞的)试验段 2.(飞行试验台的)试件工作段 3.测试区

test segment　测试片段
test sequence　测试顺序
test series　1.试验系列 2.检测系列
test set　1.试验装置,试验器具 2.测试机组
test setup　测试设置
test severity　试验严峻性
test signal　测试信号
test software　测试软件
test specification　测试规范
test specimen　试样,试件
test spectrum　试验频谱
test stand　火箭或飞弹的静止试验支架
test statistic　检验统计量
test strategy　测试策略
test subject　测试主题
test surface　探测面
test table　测试工作台
test tailoring　1.试验(规范和标准)取舍 2.试验剪裁
test tape　测试带
test target　测试目标
test team　试验组
test technique　测试技术
test technology　测试技术
test time　测试时间
test tolerance　试验允许偏差
test tone　测试音
test tube　试管
test unit　1.试验装置 2.试验设备
test value　测试值
test vane　试验十字板
test vector　测试向量
test vehicle　试验飞行器
test vessel　试验船
test with cable　有线测试
testability　可测试性
testability design　测试性设计
testbed　试验台
tester　测试仪
testing campaign　试验阶段周期
testing condition　测试条件
testing dummy　试验假人
testing facility　1.试验装置 2.测试装置
testing field　试验场,试飞机场
testing heat flux　试验外热流
testing machine　试验机
testing period　试验周期,测试周期
testing rig　试验钻机
testing time　1.测试时间 2.检查时间
test-launch　试射

test-launch operation rules	测试发射操作规程		
test-launch preplan	测试发射预案		
test-section	测试件		

test-launch operation rules 测试发射操作规程
test-launch preplan 测试发射预案
test-section 测试件
TET Turbine Entry Temperature 涡轮进口温度
tether 1. 范围 2. 系链 3. 拴绳
tether dynamics 绳索动力学
tether line 牵绳
tethered aerostat 大型系留浮空器
tethered balloon 系留气球
tethered balloon profiler 系留气球大气廓线仪
tethered satellite 系留卫星，系绳卫星
tethered satellite dynamics 绳系卫星动力学
tethered satellite system 系留卫星系统
tethered system 1. 空间绳系 2. 绳索系统
tethersonde 系留气球探空仪
tethersonde profiler 系留气球大气廓线仪
Tethys 1.【天】土卫三 2. 特提斯海，古地中海
TETNO Tetrahedron Not Operating 机场角锥形风标不工作
tetraethyl lead 四乙基铅
tetrahedron 角锥形风标
tetroon 等容气球
tetryl 1. 特屈儿 2. 三硝基苯甲硝胺（用作炸药或弹药）3. 四硝基炸药
TETWOG Turbine-Engine Testing Working Group 涡轮发动机测试工作组
TEU Trailing Edge Up 后缘向上
TEWA Threat Evaluation and Weapon Assignment（空袭）威胁判断与（防空）武器分派
Textolite 夹布胶木，织物酚醛塑料
texture 纹理，织构
texture analysis 纹理分析
texture enhancement 纹理增强
textured cell 绒面电池
textured visuals 逼真图像
TF 1. Terrain Following 地形跟踪 2. Tool Fabrication 工装制造 3. Trip Fuel 航班燃油
TFB Tower Fly-by 飞越塔台
TFC Total Final Consumption 最后总计消耗量
TFDM Tactical Fighter Dispensing Munition 战术歼击机（战斗机）配备弹药
TFDU Thin-Film Deposition Unit 薄膜淀析装置
TFEC Tactical Fighter Electronic Comba 战术战斗机电子作战
TFEL Thin-Film Electroluminescence 薄膜电致发光
TFELD Thin-Film Electroluminescent Display 薄膜电致发光显示器
TFG 1. Tactical Fighter Goup 战术战斗机大队 2. Thrust Floated Gyroscpe 推力悬浮陀螺仪

TFM Traffic Flow Management 交通流量管理
TFMS Tactical Frequency Management System 战术频率管理系统
TFO Time to First Overhaul 首次翻修期限
TFOV 1. Total Field of View 总视界 2. Tracking Field of View 跟踪视场
TFP Taskflow Process 任务流过程
TFR 1. Temporary Flight Restriction 临时飞行限制 2. Terrain-Following Radar 地形跟踪雷达 3. Total Fuel Remaining 总剩余燃油量 4. Transfer 变换器，发射机
TFS Tactical Fighter Squadron 战术战斗机中队
TFSF Time to First System Failure 系统首次发生故障时间
TFSUSS Task Force on Scientific Users of Space Station 空间站科学用户工作组
TFT 1. Thin Film Transistor 薄膜晶体管 2. Trim for Take-off 起飞配平
TFTP Trivial File Transfer Protocol 普通文档传输协议
TFTS 1. Tactical Fighter Training Squadron 战术战斗机训练中队 2. Terrestrial Flight Telephone System 陆上飞行电话系统
TFW Tactical Fighter Wing 战术战斗机联队
TFWC Tactical Fighter Weapons Center 战术战斗机武器中心
TG 1. Tachometer Generator 测速发电机 2. Terminal Guidance 末段制导 3. Transfer Gearbox 传动齿轮箱 4. Transport Geschwader 运输机联队
TGA Thermogravimetric Analysis 热解重量分析
TGB Transfer Gearbox 传动齿轮箱
TGCR Tactical Generic Cable Replacement 普通战术电缆的替代器
TGCS Travel Group Charters 旅游团体包机
TGG Third-Generation Gyro 第三代陀螺仪
TGL Touch-and-Go Landing 着陆后连续起飞
TGLV Terminal Guidance for Lunar Vehicle 月球飞行器末制导
TGM Training Guided Missile 训练导弹
TGP Twin-Gyro Platform 双陀螺平台
TGS 1. Taxiing Guidance System 滑行引导系统 2. Telemetry Ground Station 遥测地面站 3. Transportable Ground Station 机动地面站
TGSM 1. Terminally Guided Submissile 末段制导子导弹 2. Terminally Guided Submunition 末端制导子弹头
TGT 1. target 目标 2. Turbine Gas Temperature 涡轮燃气温度
TGW Terminally Guided Weapon 末段制导武器

TH　　1. Tooling Hole 工具孔　2. True Heading 真实航向　3. Total Height 总高　4. Tyrosine Hydroxylase 酪氨酸羟化酶

THAAD　　1. Theater High Altitude Area Defense 战场高空区域防御系统　2. Terminal High Altitude Area Defense 末端高空区域防御系统（简称萨德反导系统）

THAD　　Terminal Homing Accuracy Demonstrator 末段寻的精度演示（导）弹，末段自导引精度演示（导）弹

THAR　　Tyre Height Above Runway 轮胎距跑道高度

THAWS　　Ttactical Homing And Warning System 战术归航和报警系统

THDG　　True Heading 真航向

theater　　战区，战区的

theater range　　战区航程

theater-independent air operation　　战区独立空中战役

THEED　　Transmission High Energy Electron Diffraction 透射高能电子衍射

transmission line　　传输线

thematic cartography　　专题地图学

thematic map　　主题图

thematic mapper　　专题制图仪

thematic mapping　　专题制图

theme image　　专题图，主题图

theodolite　　经纬仪

theodolite for calibration　　标校经纬仪

theodolite traverse　　经纬仪导线

theodolite with laser ranging　　激光测距经纬仪

theorem　　定理，原理

theoretical aerodynamics　　理论空气动力学

theoretical all factor　　理论填充因数

theoretical allowable level　　理论容许水平

theoretical analysis　　理论分析

theoretical and experimental investigation　　理论和试验研究

theoretical approach　　理论研究法

theoretical calculation　　理论计算

theoretical capacity　　理论容量

theoretical characteristic velocity　　理论特征速度

theoretical correction method　　理论修正法

theoretical curve　　理论曲线

theoretical density　　理论密度

theoretical distance　　理想距离

theoretical error　　理论误差

theoretical framework　　理论框架

theoretical gravity　　理论重力

theoretical lethality index　　理论杀伤指数

theoretical method　　理论方法

theoretical model　　理论模型

theoretical origin　　理论零（原）点

theoretical point　　理论点

theoretical prediction　　理论预测

theoretical result　　理论结果

theoretical specific impulse　　理论比冲

theoretical study　　理论研究

theoretical thrust　　理论推力

theoretical thrust coefficient　　理论推力系数

theoretical trajectory　　理论弹道

theoretical value　　理论值

theoretician　　理论家，精通于理论的人

theory　　1. 理论　2. 原理　3. 学说　4. 推测

theory model　　理论模型

theory of errors　　误差理论

theory of large scale system　　大系统理论

theory of launch vehicle motion　　运载火箭运动理论

theory of launch vehicle trajectory　　运载火箭轨道理论

theory of optimal control　　最优控制理论

theory of small perturbations　　小扰动理论

theory of sphere-heavens　　浑天说

theory solution　　溶液理论

theory trajectory　　轨道理论

thermal　　1. 热的　2. 热气流，地方性上升气流　3. 利用热气流作翱翔飞行

thermal absorption　　热吸收

thermal acoustic fatigue　　热声疲劳

thermal actuation element　　热驱动元件

thermal ageing test　　热老化试验

thermal agitation noise　　热噪声

thermal analysis　　热分析

thermal anticing　　热除冰，加温防冰

thermal balance　　1. 热平衡　2. 热量平衡

thermal balance of satellite　　卫星热平衡

thermal balance test　　热平衡试验

thermal balancing test of satellite　　卫星热平衡试验

thermal barrier　　1. 热障　2. 热防护装置

thermal barrier coating　　热障涂层

thermal battery　　1. 热电池　2. 温差电池

thermal blooming　　1. 热加膜　2. 热散焦

thermal boundary　　热障

thermal boundary condition　　热边界条件

thermal boundary layer　　1. 热边界层　2. 温度边界层

thermal breakdown　　热击穿

thermal breeder　　热增殖反应堆

thermal capacitance　　1. 热电容　2. 热容

thermal capacitance ratio　　热容率

thermal choking　　热堵塞

thermal circuit breaker　　热断电器，过电流断路器

thermal cleaning　　热清洗

thermal coeffient of expansion　　热胀系数

thermal comfort	热舒适
thermal compensation alloy	热补偿合金
thermal condition	热力工况
thermal conductance of heat pipe	热管的热导
thermal conduction	热传导,导热
thermal conductivity	1.热传导系数 2.导热率 3.导热性
thermal conductivity gauge	热传导真空计
thermal conductivity vacuum gauge	热传导真空计
thermal connector	热敏接触器
thermal contact	热敏(开关)接点
thermal contact resistance	接触热阻
thermal control	1.热控制 2.温度控制
thermal control adhesive coating	热控带
thermal control coating	温控涂层
thermal control louver	热控百叶窗
thermal control paint	涂料型热控涂层
thermal control subsystem	热控分系统
thermal-control surface	热控表面
thermal control system	温控系统
thermal convection	热对流
thermal cueing unit	热感示装置,热标志装置
thermal cycle	热循环
thermal cycling test	热循环试验
thermal decomposition	热分解,熟解
thermal decomposition deposition	热分解淀积
thermal decomposition epitaxy	热解外延
thermal decomposition product	热分解产物
thermal degradation	热降解
thermal deicing	加温除冰
thermal design	热设计
thermal design requirement	热设计指标
thermal desorption mass spectrometry	热解吸质谱(学)
thermal destruction	热破坏
thermal diffusion	1.热扩散 2.热抗散
thermal diffusion coefficient	热扩散系数
thermal diffusivity	热扩散率(系数),导热系数,导热性
thermal diode	热二极管
thermal dissipation	热耗散
thermal dissociation	热离解
thermal drift	热漂移
thermal effect	热效应
thermal efficiency	热效率
thermal electron	热电子
thermal element	1.热元件 2.热敏元件
thermal emission	热发射,热辐射
thermal emittance	热发射
thermal energy	1.热能 2.温热热能法 3.热能专业
thermal environment	热环境
thermal equation	热力方程
thermal equation of state	热力状态方程式
thermal equilibrium	热平衡
thermal evolution	热演化
thermal exchange	热交换
thermal excitation	热激励
thermal expansion	热膨胀
thermal expansion coefficient	热膨胀系数
thermal expansion molding	热膨胀模成形
thermal exposure	热照射
thermal fatigue	热疲劳
thermal field	1.温度场 2.热场
thermal flux	1.热流 2.热通量
thermal gradien	温度梯度,热梯度
thermal gradiometer	热梯度仪
thermal grease	导热脂
thermal heating	动力加热
thermal history	热经历
thermal hydrogen	热氧脆
thermal image	热像图
thermal image guided version	热成像制导型
thermal image homing head	热成像导引头
thermal imager	热成像器,热成像装置
thermal imagery	热影像,热像
thermal imaging telescope	热成像望远镜
thermal impact	热冲击
thermal inertia	1.热惰性 2.热惯性 3.热惯量
thermal inertia mapper	热惯量成像卫星系统
thermal infrared	热红外
thermal input	热输入
thermal instability	热不稳定
thermal insulating layer	隔热层
thermal insulating material	绝热材料
thermal insulation	隔热,绝热,热绝缘
thermal insulation material	隔热材料
thermal insulator	绝热器,绝热体
thermal kill	热杀伤
thermal lag	热惯性
thermal lift	1.热升力 2.气团升力
thermal load	1.热负荷 2.热载荷
thermal loss	1.热损失 2.热耗
thermal management	热管理
thermal management subsystem	热管理子系统
thermal mapping scanner	热容量成像扫描仪
thermal margin	热余裕
thermal match	热匹配
thermal mathematical mode	热数学模型
thermal measurement	热测量
thermal model	热(试验)模型

thermal motion	热运动
thermal node	热结点
thermal noise	热(激)噪声,散粒效应噪声
thermal noise power	热噪声功率
thermal optical property	热光学性质
thermal oxidation	热氧化
thermal performance	热力性能
thermal picture synthesiser	热成像合成器
thermal plasma	热等离子体
thermal plasma clouds	热等离子体云
thermal point defect	热点缺陷
thermal pollution	热污染
thermal power	1.火力 2.火力发电 3.热功率 4.热电站
thermal process	热过程
thermal profile	1.温控技术 2.热剖面
thermal property	热特性
thermal protection	热防护
thermal protection coating	防热涂层
thermal protection graphite material	石墨防热材料
thermal protection layer	隔热防护层
thermal protection shield substructure	防热层背壁结构
thermal protection structure	防热结构
thermal protection system	防热系统,热防护系统
thermal protection system test	热防护系统试验
thermal protection tile	防热瓦
thermal protective material	1.防热材料 2.保温材料
thermal protective system	防热系统
thermal pulse	热脉冲
thermal pyrolysis	热裂解
thermal radiation	热辐射
thermal radiator	1.热辐射计 2.热辐射器 3.热辐射体
thermal reactor	热增殖反应堆
thermal relaxation	1.热驰豫 2.热松驰
thermal relay	热继电器
thermal relief valve	热安全活门
thermal resistance	热阻,加热损失
thermal resistance coefficient	热阻系数
thermal resistance layer	热阻层
thermal response	1.热反应 2.热响应
thermal runaway	1.热失稳 2.热失效 3.发热性过热 4.热致破坏,热致击穿 5.带电晶体管中的过热发散
thermal shock	热冲击,热震,热激波
thermal shock test	热冲击试验
thermal shroud	隔热罩
thermal simulation	热模拟
thermal soak	热浸
thermal spot	热斑
thermal spraying	热喷涂
thermal stability	1.热稳定性,耐热性 2.热稳性
thermal stability difference	热稳定性差异
thermal state	热状态
thermal state equation	热状态方程
thermal steady state	热稳定状态
thermal steady state condition	热稳定状态
thermal strain	1.热应变 2.热胁变
thermal strength	热强度
thermal strength test	热强度试验
thermal stress	热应力,温度应力
thermal structure	热结构
thermal structure analysis	热结构分析
thermal structure material	热结构材料
thermal suppression head	热力抑制压头
thermal switch	热开关
thermal tempering	热回火
thermal test	热试验
thermal thicket	近热障
thermal transpiration	热发散,热流逸
thermal vacuum	真空热
thermal vacuum chamber	热真空舱
thermal vacuum environment	真空热环境
thermal vacuum test	热真空试验
thermal velocity	热速度
thermal wave	热波
thermal weapon	热辐射武器
thermal wind	热成风
thermal wrinkling	热皱损,热折皱
thermal X-ray	热X射线
thermalization	热化
thermalize	使热化
thermal-lensing compensation	热透镜补偿
thermalloy	热合金
thermally activated battery	热激活电池
thermally expanded metal	热膨胀加工的金属
thermal-protect ablation material	烧蚀防热材料
thermal-shock crack	热冲击裂纹
thermal-structure test	热结构试验
thermal-type munitions	热型弹药
thermate	(燃烧弹中的)混合燃烧剂
thermel	热电温度计
thermie	兆卡
thermionic cathode	热阴极
thermionic converter	热离子转换器
thermionic emission	1.热离子发射 2.热电子发射
thermionic energy generator	热离子发电器
thermionic integrated micro modules	热离子集成微型组件
thermionic rectifier	热离子整流器

thermionic tube	热(阴极)电子管
thermionic valve	1.热离子管,热阴极电子管 2.(美)真空管
thermistor	热敏电阻
thermistor detector	热敏电阻探测器
thermistor self-heating	热敏电阻自热
thermit	铝热剂,高热剂
thermit bomb	铝热剂燃烧(炸)弹,高热燃烧(炸)弹
thermite incendiary(aerial) bomb	高热燃烧航空炸弹
THERMO	Thermal and Hydrodynamic Experiment Research Module in Orbit 热力学和流体动力学实验研究轨道舱
thermoacoustic environment	热声环境
thermoacoustic instability	热声不稳定
thermo-baro chamber	温压室,温压容器
thermo-breakdown	热击穿
thermo-calibration wind tunnel	热校测风洞
thermocapillary convection	热毛细对流
thermochemical ablation	热化学烧蚀
thermochemical equilibrium	热化学平衡
thermochemical model	热化学模型
thermochemical parameter	热化学参数
thermochemistry	热化学
thermochromic film	热变色胶片
thermochromic tube	热变色管
thermochromism	热致变色
thermocline	斜温层,温跃层
thermocompression bonding	热压焊
thermo-conditional suit	调温服
thermocouple	热电偶,温差电偶
thermocouple measurement	热电偶测量
thermocouple plug	热电偶塞
thermocycling	热循环
thermodynamic analysis	热力学分析
thermodynamic calculation	热力计算
thermodynamic cycle	热力循环
thermodynamic cycle analysis	热力循环分析
thermodynamic datum	热力学数据
thermodynamic efficiency	热力效率
thermodynamic engineering	热力学工程
thermodynamic environment	热动力环境
thermodynamic equilibrium	热动平衡
thermodynamic information	热力学信息
thermodynamic model	热力学模型
thermodynamic parameter	热力学参数
thermodynamic potential	1.热力学势 2.热力势
thermodynamic property	热力学性质
thermodynamic relation	热力学关系
thermodynamic state	热力状态
thermodynamic suppression head	热力抑制压头
thermodynamic system	1.热力学体系 2.热力学系统
thermodynamic variable	1.热力学变量 2.热力学参数
thermodynamics	热力学
thermoelectric actinograph	热电直接辐射计
thermoelectric converter	热电转换器
thermoelectric couple	热电偶,温差电偶
thermoelectric device	热电器件
thermoelectric effect	热电效应
thermoelectric generator	温差发电器
thermoelectric module	热电组件
thermoelectric module efficiency	热电器件效率
thermoelectric performance	热电性能
thermoelectric power	1.温差电势率 2.热电动力
thermoelectric power generation	热电发电
thermoelectric power generator	温差发电器
thermoelectric property	热电性质
thermoelectric refrigerator	温差电致冷器
thermoelectric system	热电系统
thermoform	热成型
thermogram	自记温度图,温谱图,温度过程线
thermographic camera	温度记录照相机,红外线照相机
thermography	温度记录法,热敏图成像,红外线照相术
thermogravimetric analysis	1.热解重量分析 2.热重分析
thermohydrometer	温差比重计
thermoionic diode	热离子二极管
thermoionic power generator	热离子发电器
thermoionic reactor	热离子反应堆
thermoluminescence detector	热致发光探测器
thermoluminescent dosemeter	热致发光剂量计
thermolysis	1.热解 2.散热作用
thermomagnetic curve	热磁曲线
thermomagnetic separation	热磁分离
thermo-mapping technique	热图技术
thermomechanical fatigue	热机械疲劳
thermomechanical treatment	形变热处理
thermometer screen	温度计百叶箱
thermometry	1.温度测量 2.温度测定法 3.计温学
thermo-molecular vacuum gauge	热分子真空计
thermonuclear action	热核作用
thermonuclear warhead	热核弹头
thermonuclear weapon	热核武器
thermo-optical property	热光性质
thermopause	热层顶
thermophoresis	1.热泳 2.热迁移
thermo-photovoltaic device	热光伏器件
thermophysical property	热物理性质
thermopile	1.热电堆 2.温差电堆

thermopile detector	热电堆探测器
thermoplastic	热塑(性)的
thermoplastic composite	热塑性复合材料
thermoplastic forming	热塑性成形
thermoplastic plastic	热塑性塑料
thermoplastic polymer resin	热塑料聚合树脂
thermoplastic recording	热塑性记录
thermoplastic resin	热塑性树脂
thermoplastic resin composite	热塑性树脂复合材料
thermoplasticity	热塑性
thermoplastics	热塑性塑料
thermoprene	环化橡胶
thermoprobe	测温探针,热探针
thermoprobe method	热探针法
thermoreceptors	温度感受器
thermorelay	热继电器,温差电偶继电器
thermoremanence	热顽磁,热剩磁
thermoremanent magnetization	热剩磁
thermosensitive fluid	热敏液体
thermoset	1.热固(性)的,热变定的 2.热固性,热变定性 3.热固性材料 4.热变定塑料
thermosetting	热固性,热固的
thermosetting composite	热固性复合材料
thermosetting plastics	热固性塑料
thermosetting resin composite	热固性树脂复合材料
thermosonic bonding	热超声焊
thermosphere	热层
thermospheric density	热层密度
thermospheric wind	热层风
thermostabilized food	热稳定食品
thermostat	温度自动调节器,恒温器
thermo-strength	热强度
thermosyphon	1.热虹吸 2.热对流系统 3.热虹吸器
thermotropic model	正温模型
Thevenins law	戴维宁定理,等效发生器定律
thick airfoil	厚翼形
thick boundary layer	厚边界层
thick film circuit	厚膜电路
thick film hybrid integrated circuit	厚膜混合集成电路
thick film ink	厚膜浆料
thick film integrated circuit	厚膜集成电路
thick film pressure transducer	厚膜压力传感器
thick plate	厚板
thicken	1.使变厚 2.使模糊 3.使……变复杂
thickened fuel	增稠燃油
thickener	增稠剂,胶化剂,稠化剂
thickening technology	增密工艺,增稠工艺
thickfilm	厚膜
thickness	1.U 机翼的 厚度 2.稠度,深度 3.浓度
thickness distance	最大厚度位置
thickness distribution	厚度分布
thickness gauge	厚度规,测厚仪
thickness lines	等厚度线
thickness of skin	皮肤厚度
thickness of the tube	炮管厚度
thickness parameter	厚度参数
thickness ratio	1.相对厚度 2.厚弦比
thickness variable	变厚度
thick-walled shell	厚壳弹
thimble	1.(钢索端头的)套环 2.(千分卡的)调整旋钮 3.疙瘩形雷达整流罩
thin	1.薄的 2.稀薄的 3.微弱的
thin airfoil	薄翼
thin airfoil theory	薄翼理论
thin disk	薄盘
thin electrode	薄焊条
thin film	薄膜
thin film circuit	薄膜电路
thin film deposition	薄膜淀积
thin down	变弱
thin film electroluminescence	薄膜电致发光
thin film hybrid integrated circuit	薄膜混合集成电路
thin film integrated circuit	薄膜集成电路
thin film lubrication	薄膜润滑,油膜润滑
thin film resistance thermometer	薄膜电阻温度计
thin film solar cell	薄膜太阳电池
thin film technology	薄膜技术
thin film thermoelectric module	薄膜热电器件
thin gate oxide	薄栅氧化层
thin ice	薄冰
thin layer	薄层,薄膜层
thin layer assumption	薄层假定
thin limit	薄极限
thin line	薄弱防线,防线薄弱处
thin liquid	稀液
thin polymer	薄层聚合材料
thin route	交通量不大的航线(通常指洲际航班)
thin route earth station	稀路由地面站
thin route master station	稀路由主站
thin route satellite communication	稀路由卫星通信
thin shell	薄壳
thin shell structure	薄壳结构
thin shock-layer theory	薄激波层理论
thin wall	薄壁
thin walled shell	1.薄壳弹 2.薄壁壳体
thin wire communication	稀线式通信
thin-case bomb	薄壳炸弹
thin-film transistor	薄膜晶体管

think tank　智囊团，思想库
thinned array antenna　疏化阵天线
thinner(s)　稀释剂，稀料
thin-walled beam　薄壁梁
thin-walled structure　薄壁结构
thioformaldehyde　硫代甲醛
THIR　Temperature Humidity Infrared Radiometer 温湿度红外辐射仪
third body motion　第三体运动
third contact　生光(日、月食)
third cosmic velocity　第三宇宙速度
third electrode　第三电极
third generation　第三代
third level　第三极
third line servicing　三级维修
third mode　第三模态
third order　1.三等 2.第三级的，三阶的
third order intercept point　三阶截距点
third party　第三者，第三方
third party risks　第三者风险
third phase　第三相
third quarter　下弦(月相)
third stage　1.第三阶段 2.第三级
third-angle projection　第三象限投影
third-body effect　第三体影响
third-body perturbation　第三体引力
third-class failure of satellite　整星三级故障
third-level carrier　第三类航空公司
thixotropic propellant　触变推进剂
thixotropy　摇溶现象，触变性
THL　Tailplane Hinge Line 尾翼铰接轴线
Thomson theorem　1.汤姆孙定理 2.开尔文定理
THOR　Tiered Hierarchy Overlayed Research 多层次重叠研究
Thor　雷神(美国运载火箭和中程弹道导弹)
Thor Able　艾布尔运载火箭(美国)
Thor-Able Star launch vehicle　艾布尔星运载火箭(美国)
Thor-Agena　阿金纳运载火箭(美国)
Thor-Altair　牵牛星运载火箭(美国)
Thor-Burner　博纳运载火箭(美国)
thorium cloth　钍布，钍屏蔽层
Thornaby bag　遇难机组应急物品空投袋
Thornel　松耐(碳和石墨纤维的商品名)
thou　英毫(一英寸的千分之一)
thought-controlled computer　思维控制计算机
THP　1.Thrust Horsepower 推力马力 2.Turbo-Hydraulic Pump 涡轮液压泵
THR　1.Threshold 阈值，跑道入口 2.Thrust 推力

thread　1.线 2.螺纹 3.思路 4.玻璃纤维
thread block　线程块
thread chaser　螺纹梳刀
thread gauge　螺纹规
thread insert　攻螺纹插入(指用钢制的螺钉拧入铝或其他软材料的无螺纹的孔内)
threading the needle　穿针引线(精确地穿越某一小区域)
thread-type oil seal　螺纹型滑油密封件
threat　1.敌防空措施，特指防空雷达、地空导弹、高炮和战斗机 2.威胁，恐吓 3.凶兆
threat boundaries　威胁边界线
threat cloud　1.威胁群 2.威胁云
threat density　威胁密度
threat evaluation　空袭威胁估计
threat exposure　1.曝光威胁 2.泄露威胁
threat hardware　威胁性装备
threat integration　对威胁的综合判断
threat level　威胁等级，威胁程度
threat library　威胁数据库
threat region　威胁地区
threat simulation　威胁模拟
threat situation　威胁态势
threat source　1.威胁源 2.潜在威胁
threat tube　威胁管道，威胁通道
three channel monopulse　三通道单脉冲
three control aeroplane　三轴操纵飞机
three dimensional integrated circuits　三维集成电路
three dimensional motion　三维运动，三度空间运动
three dimensional warning　三维告警
three dimensional wind tunnel　三维风洞
three dimensional wing　三维弹翼
three greens　〈口语〉起落架已放下并上锁(指三个绿灯都亮)
three phase flow　三相流
three poster　三喷管的
three primary colors　三基色
three shaft　三轴发动机
three shoting method　三次猜试法
three stars navigation　三星导航
three state controller　三位控制器
three states logic　三态逻辑
three step controller　三位控制器
three winch system　三绞车系统
three-axis　三轴的
three-axis attitude stabilization　三轴姿态稳定
three-axis attitude　三轴姿态
three-axis attitude indicator　三轴姿态指示器
three-axis automatic pilot　三轴(三通道)自动驾驶仪

three-axis control	三轴控制
three-axis satellite	三轴稳定卫星
three-axis stabilization	三轴稳定
three-axis stable platform	三轴稳定平台
three-axis turntable	三轴转台
three-axle table	三轴转台
three-body	三体的
three-body problem	三体问题
three-body reaction	三体反应
three-colour photometry	三色测光
three-degree-of-freedom	三自由度
three-dimension navigation	三维导航
three-dimension quartz	三向石英
three-dimensional	1.三维的 2.立体的 3.真实的
three-dimensional accuracy	三维位置精度
three-dimensional analysis method	三维分析法
three-dimensional analysis	三维分析
three-dimensional boundary	三维边界
three-dimensional boundary layer	三维附面层
three-dimensional display	三维显示
three-dimensional effect	1.空间效应 2.三维效应
three-dimensional error	三维位置误差
three-dimensional flow	1.三维流 2.三元流动
three-dimensional geometry	三维几何
three-dimensional method	三维方法
three-dimensional model	三维模型
three-dimensional modeling	1.三维模型 2.三维建模
three-dimensional nozzle	三元喷管
three-dimensional numerical model	三维数值模型
three-dimensional radar	三坐标雷达
three-dimensional roughness	1.三维粗度 2.三维粗糙度
three-dimensional simulation	三维模拟
three-dimensional space	三维空间
three-dimensional sphere	三维球面
three-dimensional structure	1.三维结构 2.立体构成 3.智力三维结构模型
three-dimensional surface	三维面
three-dimensionality	1.三维性 2.立体
three-direction quartz fiber reinforced quartz composite	三向石英复合材料
three-direction quartz fiber reinforced SiO₂ composite	三向石英增强二氧化硅复合材料
three-element	三元的,三元件的
three-hole	三孔的
three-impulse	三冲量
three-level structure	三能级
three-level system	三能级系统
three-loop	三环
three-parameter	三参量
three-phase current	三相交流电
three-phase equilibriumx	三相交流电
three-point landing	三点着陆
three-point method	三点法
three-point mooring	主点系留
three-pole switch	三极开关,三刀开关
three-pole, single-throw switch	三极单控开关,三刀单掷开关
three-probe method	三探针法
three-resistance coating	三防涂料
three-segment	三段式
three-stage	三级的
three-stage compressor	三段式压缩机
three-state	三态的
three-step	三步式
three-stream engine	三流发动机
three-terminal network	三端网络
threshold	1.阈,界限,门限,临界值 2.起始,开端 3.跑道入口,跑道头 4.失速点飞行,失速点,5.可听阈,闻阈,声的阈值 6.可取得最低舒适程度巡航的当量空速 7.始限值 8.初始反应点
threshold ablation	阈烧蚀
threshold breakdown	阈击穿
threshold contrast	最低对比度,阈对比,对比感阈
threshold current	阈值电流
threshold current density	阈电流密度
threshold curve	限界曲线
threshold decoding	阈译码
threshold detector	阈探测器
threshold displacement	跑道头限界距离
threshold effect	1.门槛效应 2.阈效应
threshold energy	1.能限 2.阈能
threshold extension	门限扩展
threshold extension technique	门限扩展技术
threshold illuminance	阈照度,亮度阈限,最低照明度
threshold level	阈电平
threshold lights	跑道入口灯
threshold limit value	限值,最大限值
threshold logic circuit	阈逻辑电路
threshold of audibility	1.可闻阈 2.听觉临界
threshold of discomfort	不适阈
threshold of feeling	感觉阈
threshold of sensitive	灵敏阈
threshold of stress intensity factor	应力强度因子阈值
threshold pressure	1.阈压力 2.临界压力
threshold rate	阈值频率
threshold speed	1.速度阈值,速度临界值 2.跑道入口处飞行速度
threshold temperature	临界温度

threshold test	门限检测
threshold value	1.阈值 2.门限值 3.界限值
threshold voltage	阈值电压
threshold wavelenth	阈值波长
thresholds	限额
THRH	Throttle Hold 油门保持
throat	1.窄路 2.口孔 3.(喷管的)临界截面,喷管喉道
throat area	颈截面
throat control	喉道控制
throat diameter	1.喉径 2.炉喉直径
throat flow	孔喉流动
throat heating	喉道加热
throat insert	喉衬
throat location	喉部位置
throat microphone	喉头送话器
throat radius	1.喉道半径 2.喉口半径
throat section	喉管段
throat temperature	炉顶温度
throat to port area ratio	喉通面积比
throat velocity	喉部速度
throatable	喉道可变的
throatless chamber	(火箭发动机的)无喉道推力室
throatless shear(s)	无弯喉剪床
throttle	1.推力控制手柄 2.油门,油门杆 3.节流阀,节流门,节气门 4.节流 5.收油门
throttle back	收油门
throttle characteristics	1.节流特性 2.油门特性
throttle controller	节气门控制器
throttle friction	油门摩擦锁
throttle icing	节气门结冰
throttle input	油门投入
throttle level	油门杆
throttle lever angle	油门杆角度
throttle lock	油门锁
throttle position	1.节气门位置 2.油门位置
throttle ratio	节流比
throttle setting	节流阀调整
throttle tension	油门锁紧度
throttling capability	节流能力
through	全程
through canopy ejection	穿盖弹射
through connection point	转接点
through connection station	转接站
through deck	直通甲板
through fare	全程票价
through flang	穿通式法兰
through hole	通孔
through line	直通线
through path	直通电路,正向通路
through rate	全程运费
through traffic	全程交通
throughcanopy ejection	穿盖弹射
throughflow	1.通流 2.直流 3.表层流
throughflow combustor	直流燃烧室
throughput	1.流量 2.吞吐量
throughput capacity	吞吐量
throughput of vacuum pump	真空泵抽气量
throughput rate	1.吞吐率 2.生产率
through-rate	通过速率
through-thickness	全厚度
throughway	车辆通道
throw	1.曲柄,曲拐 2.偏心距,行程 3.(航面的)偏度,摆幅 4.投掷,发射
throw weight	投掷重量
throw-over control	转换操纵
thruflight inspection	中途检查,短停检查
thruput	throughput吞吐量,通过量,流量
thrust	1.推力,侧向拉(压)力,轴向(压)力 2.刺 3.(猛)推,冲,插,碰撞
thrust angle	推力角
thrust augmentation	推力增大,加力
thrust augmentation ratio	推力增大比率
thrust available	可用推力
thrust axis	推力轴(线)
thrust balance	1.推力平衡 2.转子轴向力平衡
thrust bearing	推力轴承(承受轴向力),止推轴承
thrust buildup	推力增大,起飞增推过程
thrust build-up time	推力建立时间
thrust calibration	推力标定
thrust chamber	1.(火箭的)推力仓 2.推力室,燃烧室
thrust chamber area contraction ratio	推力室面积收缩比
thrust chamber mixture ratio	推力室混合比
thrust chamber specific impulse	推力室比冲
thrust chamber valve	推力室阀
thrust characteristic	推力特性
thrust coefficient	1.推力系数,(计算螺旋桨性能的)拉力系数 2.推力系数因子
thrust command	推力指令
thrust component	拉力分量
thrust control computer	推力控制计算机
thrust control system	推力控制系统
thrust curve	推力曲线
thrust cutback	推力收回
thrust cutback angle	推力收回角
thrust cutoff	发动机停车
thrust decay	推力衰减,推力减小
thrust decay impulse	后效冲量

thrust decay phase 后效段	**thrust rating** 推力定额
thrust direction 推力方向	**thrust rating computer** 推力自动管理计算机,推力状态计算机
thrust distribution 推力分布	
thrust drop-off time 推力消失时间	**thrust rating panel** 推力调定仪表板
thrust duration 1.推力持续时间 2.火箭发动机工作时间	**thrust regulator** 推力调节器
	thrust required 需用推力
thrust efficiency 推进效率	**thrust requirement** 推力需求
thrust equivalent horsepower 推进功率,推进马力	**thrust reserve** 推力储备,备用推力
thrust error 推力误差	**thrust response** 推力反应情况
thrust face (螺旋桨桨叶的)拉力面	**thrust reverser** 反推力装置,推力储备,备用推力
thrust force 推力	**thrust reversing rating** 反推力状态
thrust frontal area 迎面推力	**thrust section** 发动机舱,推进系统
thrust generation 推力产生	**thrust specific fuel consumption** 燃油消耗率,耗油率
thrust horsepower 推进马力,推进功率	**thrust spoiler** 推力扰流器,减推装置
thrust impulse 拉力冲量	**thrust stand** 1.推力试验台 2.发动机试验台架
thrust increment 推力增量	**thrust structure** 推力结构
thrust lapse 推力下降	**thrust surface** 推力面
thrust level 发动机额定推力	**thrust tail-off** 推力消失,推力减小
thrust lever 推力杆,油门杆	**thrust termination** 推力终止,(发动机)关机
thrust line 推力线	**thrust termination mechanism** 推力终止机构
thrust line adjustment 推力线调整	**thrust termination port** 止推孔,推力终止孔
thrust line deviation 推力线偏斜	**thrust termination pressure** 推力终止压强
thrust line eccentricity 推力线横移	**thrust termination time** 推力终止时间
thrust load 1.轴向载荷 2.推力负载	**thrust terminator** (固体火箭发动机的)推力终止器,消推器
thrust load cell 推力传感器	
thrust loading 推力载荷	**thrust to mass ratio** 推(力)质(量)比
thrust loss 推力损失	**thrust to weight ratio** 推重比
thrust loss factor 推力损失因素	**thrust transducer** 推力传感器
thrust management 推力管理	**thrust unit** 推力装置
thrust margin 推力余量	**thrust value** 推力值
thrust measurement 推力测量	**thrust vector** 推力矢量
thrust measurement rake 推力测量耙	**thrust vector control** 推力矢量控制
thrust meter 推力计	**thrust vector direction** 推力矢量方向
thrust misalignment 推力偏心	**thrust vectoring** 推力换向
thrust misalignment angle 推力偏心角	**thrust wall** 止推墙
thrust model 推力模型	**thrust weight ratio** 推(力)重(量)比
thrust moment 推力力矩	**thrust wire** 拉力张线
thrust nozzle 推力喷管	**thrust-assisted helicopter** 有助推装置的直升机
thrust nozzle expansion 推力喷管扩张	**thrust-augmented wing** 加力机翼,增推机翼
thrust orbit 推力轨道	**thrust-chamber** 推力室
thrust per frontal area 单位迎面推力	**thrust-chamber pressure** 推力燃烧室压力
thrust performance 推力性能	**thruster** 推力器,小推力发动机
thrust phase intercept (弹道)主动段截击	**thruster cartridge** 推力器药筒,推力器燃爆筒
thrust plate 1.配流盘 2.推力板 3.冲掩体 4.止推片	**thruster discharge** 推力器放电
thrust platform 推力架	**thruster technology** 推斥技术
thrust power 推力功率	**thruster type** 推进器型式
thrust profile 推力分布	**thruster valve** 推进器控制阀
thrust pulse 推力脉冲	**thrust-vector control** 推力矢量控制,推力向量控制
thrust range 推力范围	**thrust-vectoring** 推力矢量

thrust-vectoring engine　1.推力转向发动机 2.推力矢量发动机
THSA　Trimmable Horizontal Stabilizer Actuator 可配平水平安定面驱动器
THT　Transient Heat Transfer 不稳定传热，瞬时传热
thumbprint　拇指的指纹，个性特征
thumbscrew　1.手拧螺钉 2.蝶形螺钉
thumbstick　小操纵杆，姆指杆
thumbswitch　拨动式电门
thunderstorm　雷暴，大雷雨
thunderstorm effect　雷暴（雷雨）效应
thunderstreak discharge sensitivity measurement　雷电放电敏感度测量
thyratron　闸流管，充气三极管
thyrector diode　1.变阻齐纳二极管 2.可变电阻的硅二极管
THZ　terahertz 太赫，兆赫
TI　1. Target Identification 目标识别 2. Target Indicator 目标指示器 3. Task Interval 任务间隔 4. Terminal Interface 终端介面 5. Test Interpretation 试验解释 6. Thermal Infrared 热红外
Ti alloy　钛合金
TIA　1. Telecommunications Industry Association 电信工业协会 2. Type Inspection Approval 型号检查批准 3. Type Inspection Authorization 型号检查核准书，授权进行型号检验
TIALD　Thermal Imaging Airborne Laser Designator 热成像机载激光照射器
TIAS　True Indicated Airspeed 指示真空速
tiator　1.激光指示器 2.激光指示 3.激光照射器
TIB　Technical Intelligence Bureau 技术情报局
TIC　1. Tantalum Integrated Circuit 钽集成电路 2. Target Insertion Controller 目标信息输入控制器 3. Target Intercept Computer 目标拦截计算机 4. Technical Information Center 技术情报中心 5. Total Interference Criterion 总干扰标准
ticket agent　客票代理商
ticket counter　机场售票及办登记手续的柜台
ticket coupon　机票联单
ticket desk　旅客整理物品的桌子
ticket destination　票上目的地
ticket origin　票上出发地
ticket sale　售票
TICM　Thermal-Imaging Common Modules 热成像通用组件，热成像通用模件
TID　Tactical Information Display 战术信息显示器
tidal deformation　潮汐形变
tidal factor　潮汐因子
tidal friction　潮汐摩擦
tidal hypothesis　潮汐假说
tidal motion　潮汐运动
tidal oscillation　潮汐振荡
tidal wave　1.潮汐波，潮浪 2.海啸
tiddleywinks effect　小物体抛开作用，抛石效应
tide logic　潮汐学
tide perturbation　潮汐摄动
tide-generating force　引潮力
tide-generating potential　引潮位
TIDP　Telemetry and Image Data Processing 遥测与图像资料处理
TIE　Transcontinental Interconnection Experiment 跨州互连实验
tie　结构连接件
tie bar　旋翼拉杆
tie down care　地面勤务交接
tie line　直达中继线
tie point　连接点
tie rod　连接杆件
tied gyro　扣结式陀螺仪
tiedown　1.露天系留 2.系紧固定 3.系留装置
tiedown diagram　货物系紧方法示意图
tiedown test　系留试验
tiered hierarchy overlayed research　多层次重叠研究
TIES　Tactical Information Exchange System 战术情报交换系统
TIF　1. Take-off Inhibit Function 起飞时禁用功能 2. Telephone Interference Factor 电话干扰因素 3. Text Interchange Format 电文交换格式
TIG　1. Time of Ignition 点火时间
tig　1.轻碰 2.戏弄
Tiger　Terrifically Insensitive to Ground Effect Radar 地物效应极不敏感的雷达(亦称"虎"式雷达)
Tigercat　"豹猫"(英国地空导弹制导雷达)
tight beam radar　窄波束雷达
tightened inspection　从严检查
tightening　1.拧紧 2.使绷紧 3.扣紧
tightness　紧密，坚固
tightness test　气密性试验
tile　1.薄膜基片，厚膜基片 2.表面热保护系统的独立部件，防热瓦
TILS　Tactical Instrument Landing System 战术仪表着陆系统
tilt　倾斜角
tilt angle　倾转角
tilt angle of photograph　相片倾角
tilt correction　倾斜改正
tilt displacement　倾斜位移
tilt observation　倾斜观测

tilt rotor 倾转旋翼机
tilt rotor aircraft 1.转向旋翼航空器 2.倾转旋翼航空器
tilt table 倾斜台
tilt table response 倾斜台反应
tilt test table 倾斜试验台
tilt wing 偏转翼
tilting fuselage 机身转向的垂直起落飞机
tilting head 自动倾斜器,旋转倾转盘
tilting propeller 可偏转螺桨,可倾斜螺桨
tilting-engine/jet/propeller/wing 发动机、喷口、螺旋桨、机翼转向的垂直起落飞机
tiltmeter 倾斜仪
tilt-rotor aircraft 倾转旋翼机
TIM Transducer Interface Module 传感器接口模块
Timation Time Navigation Satellite 蒂马申卫星(美国海军导航卫星)
time 1.时间 2.次数 3.节拍 4.倍数
time section 时间剖面
time analyser 时间分析器
time and frequency measurement 时间频率测量
time and superquick fuze 定时与瞬发两用引信
time assignment speech interpolation 时分话音内插法
time average 时间平均值
time axis 1.时间轴 2.极轴
time base 时基
time base error 时基误差
time between failures 故障间隔时间,失效间隔时间
time between faults 故障间隔时间
time between overhauls 1.翻修时限 2.翻修寿命
time blanking 时间消隐
time bomb 定时炸弹
time break 定时开关,爆炸信号
time check 对时,对表
time circle 时间圆
time climb 爬升时间
time code format 时间码格式
time code generator 时间码发生器
time code translator 时间码译码器
time compression 时间压缩
time compression multiplex 时间压缩多路传输
time connection 时间连接
time constant 时间常数,时间常量
time constant of an electric motor 电动机时间常数
time constant of detector 探测时间常数
time constant of servo 航机时间常数
time constant term 时间常数项
time consuming 1.耗费时间的 2.旷日持久的
time control pulse 定时脉冲

time controlled gain 时间控制增益
time controlled overhaul 控时翻修
time correction 1.引信定时修正量 2.时间修正
time cross-section 时间剖面图
time delay 1.时延 2.延时,延迟
time delay circuit 延时电路
time delay correction 时延修正
time delay feature 延时装置
time delay integration device 时间延迟积分器件
time delay system 时滞系统
time dependence 1.时间相依 2.时间依赖性
time depth conversion 时深转换
time derivative 时间导数,时间微商
time determination 测时
time dilation 时间变慢效应
time division exchange 时分交换机
time division modulation 时分调制
time division multiple access 时分多址
time division multiple carrier 时分多重载波
time division multiplex 时分多路复用,时分多路,时分多工
time division multiplex channel 时分多路复用通道
time division multiplex communication 时分多路通信
time division multiplexed data bus integration techniques 时分多工传输数据总线的综合技术
time division multiplexer 时分多路复用器
time division multiplexing 时分多路复用
time division multiplexing scheme 时分多路复用方案,时分多路转换方案
time division multiplexing telemetry 时分多路遥测
time division multiplexing telemetry system 时分多路传输遥测系统
time division switching system 时分制交换机
time domain 时域,时间范畴
time domain 时域,时间域
time domain automatic network analyzer 时域自动网络分析仪
time domain matrix method 时域矩阵法
time domain measurement 时域测量
time domain model reduction method 时域模型降阶法
time enablement 允许时间
time evolution 时间演化
time expansion chamber 时间扩展室
time fire 定时射击,空炸射击
time for positioning 定位时间
time for pumping down 抽气时间
time for roughing 粗抽时间
time frame 时间范围,时间间隔
time fuze 1.时间引信 2.定时引信

time hack 预定时间
time history 1.时间历程 2.时间推移 3.随时间的变化 4.时间关系曲线图
time history duplication 时间历程复现
time hopping spread spectrum 跳时扩频
time horizon 1.时间范围 2.时间层
time in service 服役时间,使用时间,(飞机的)空中时间
time increment 时间增量
time integration 时间积分
time interleaving 时间交错
time interval 时间间隔
time interval of interest 目的层段
time invariant 时不变的,时不变
time jitter 时间抖动
time keeping 守时
time keeping system 时间记录系统
time lag 1.延时,时滞,滞后 2.(炸弹)落下时间滞差
time level 时间级
time line 时间线,等时线
time location 时间区位
time mean bleed 平均引气量
time measurement 1.测时 2.时间测定
time mechanical fuze 机械定时引信
time modulation 时间调制
time modulation sideband 时间调制边带
time moment 时间矩
time monitor 时间监视器
time multiplex 时分多路复用
time multiplexed system 时分多路复用系统
time multiplexer communication channel 时分多路复用通信信道
time of arrival 运到时间
time of consciousness 意识时间
time of day clock 日历钟
time of flight (炮弹或导弹的)飞行时间
time of incapacitation 失能时间
time of observation target 观测目标时间
time of operation 作用时间
time of perigee passage 过近地点时刻
time of perihelion passage 过近日点时间
time of recovery 恢复时间
time of satellite perigee passing 卫星过近地点时刻
time of unconsciousness 丧失意识时间
time of useful consciousness 有效意识时间
time on target 1.攻击目标时间 2.对目标照相时间 3.核武器爆炸时间 4.协同射击时间
time optimal control system 最快速控制系统
time out 超时

time out circuit 超时电路
time over 超时
time over target 飞越目标上空时间
time period 时期,期间
time phase 时相
time point 时间点
time profile 1.时间轮廓 2.时间表 3.时间剖面
time projectile (定时爆炸的)空炸炮弹
time projection chamber 时间投影室
time proof 长寿命的,耐久的
time pulse 时钟脉冲
time pulse distribution 时钟脉冲分配
time range 1.时间范围,时间段 2.时间间断
time rate 1.计时工资率 2.时间变化率
time reference scanning beam 时间基准扫描波束
time reference signal 标准时间信号
time reference system 时间基准系统
time relay 时间继电器
time removal 定时拆卸
time reserve 1.备用时间 2.缺氧维持时间,有效意识时间
time resolution 时间分辨率
time response 时间响应,时间特性
time reversal 时间逆转
time saving 节约时间
time scale 时标,时间刻度
time scale factor 时标因子,时间比例尺
time schedule controller 时序控制器
time selection 时间选择
time separation (空中交通管制的)时间间隔
time sequence 时序
time series 1.时间序列 2.时间数列
time series analysis 时间序列分析
time series method 时间序列法
time service 时间服务
time shared bus 分时总线
time shared multiplexor 分时多路转换器
time sharing 时间划分,分时
time sharing control 分时控制
time sharing system 分时系统
time shell 空炸炮弹
time shutdown 时间关机
time signal 1.时间信号 2.时间标记
time slice 时间切片
time span 时间跨度
time speed scale 地速时间标尺
time stagger removal 错时拆卸
time standard 时间标准
time synchronism 时间同步

time tick 时间记号,报时信号,报时滴答声	**time-lag action** 延时作用
time to first overhaul 第一次大修期	**time-lag system** 时滞系统
time to go 到达所需时间	**time-lapse photographyx** 定时摄影术
time to synchronization 同步时间	**time-limit failure** 时限故障,暂现故障
time tolerance 时间容差	**timeline** 时间轴,时间线
time trade-off 时间折中	**timely warning** 即时报警
time transfer 授时	**time-marching** 时间推进
time unit 时间单位	**time-of-flight** 渡越时间
time unitization 时统	**time-of-flight mass spectrometer** 飞行时间质谱仪
time variant scaling 时变比例	**time-of-flight neutron spectrometer** 飞行时间中子谱仪
time variation 1.时间变化 2.时间变化性	**time-of-flight spectrometer** 飞行时间能谱分析仪
time window 时间窗	**time-optimal** 时间最优的
time zero 1.计时起点 2.时间零点	**timeout** 1.超时 2.暂时休息 3.工间休息
time zone 时区	**time-program command** 时间程序指令
time-accurate 时间精确的	**timer** 1.计时器,定时器 2.计时员 3.时间标记,时间单元 4.时间统一系统,时统
time-amplitude converter 时间幅度变换器	**timer fuze** 时间引信,定时引信
time-averaged 按时间平均的	**time-response** 1.时间响应 2.时间特性
time-averaged pressure 时均压强	**timescale** 1.时间量程 2.时标
time-averaged velocity 时均速度	**time-sensitive target** 时限性目标
time-bandwidth product 时间带宽(乘)积	**time-sharing executive system** 分时执行系统
time-base circuit 时基电路	**timeswitch** 定时开关,计时电门
time-change item 限时更换件,定时更换件	**time-tag** 时间标签
time-coincidence command 时间符合指令	**time-term** 时间项
time-compliance technical order 限时技术规程	**time-to-go** 剩余时间
time-consuming process 费时的过程	**time-variable filtering** 时变滤波
time-coordinated multiple access 时间协调多址,时间一致多址	**time-varying** 时变的
timed delivery 计时投弹	**time-varying model** 时变模型
timed programme 定时程序	**time-varying parameter** 时变参数
time-delay action 延时作用	**time-varying spectral density** 时变谱密度
time-delay command 延时指令	**time-varying system** 时变系统
time-delay switch 延时开关	**time-varying trajectory** 时变轨迹
time-delay telemetry 延时遥测	**time-zone disease** 时区病
time-dependent method 时间相关法	**timing** 1.定时,计时 2.同步,协调 3.定时器
time-distance curve 时距曲线	**timing adjustment** 协调调整
time-division command 时分指令	**timing angular velocity** 调谐转速
time-division telemetry 时分遥测	**timing blasting fuze** 定时起爆引信
time-division telemetry system 时分遥测系统	**timing capacitor** 定时电容
time-domain 时域,时间域	**timing center** 时统中心
time-domain equalizer 时域均衡器	**timing circuit** 定时电路,时标电路
time-domain response 时域响应	**timing clock** 打卡钟,出勤记录钟
time-extension sample 延时样本	**timing consideration** 时间考虑
time-frequency code 时频码	**timing constraint** 定时限制
time-frequency modulation 时频调制	**timing diagram** 计时图,时序图
time-frequency-phase modulation 时频相调制	**timing discriminator** 定时鉴别器
time-fuzed bomb 定时炸弹	**timing equipment** 时统设备
time-integration 时间累积	**timing error** 定时误差
time-invariant system 定常系统,时不变系统	**timing filter amplifier** 定时滤波放大器
time-lag 时间间隔	**timing gauge** 协调量规

timing impulse 定时脉冲
timing information 定时信息
timing interval 定时间隔
timing jitter 定时信号的跳动
timing margin 时间余量
timing method 调速方式
timing model 时序模型
timing parallax 定时视差
timing pulse 定时脉冲,时钟脉冲
timing pulse distributor 同步脉冲分配器
timing pulse generator 定时脉冲发生器,时标脉冲发生器
timing pulse shaper 定时脉冲整形器,时钟脉冲整形器
timing recovery 定时恢复
timing sequence 1.时序 2.时标序列
timing sequence mechanism 时序机构
timing signal 定时信号
timing signal control panel 时统信号控制台
timing simulation 时序模拟
timing substation 时统分站
timing system 1.定时系统 2.时间统一系统
timing zero 时间统一系统零点,时统零点
TIMM Thermionc Integrated Micromodules 热离子集成微型组件
TIMPS Tactical Integrated Mission Planning Station 战术综合任务计划站
TIMS 1. Tailored Inspection Maintenance System 简单检查维修系统 2. Thermal Infrared Mapping Spectrometer 热红外成像光谱仪 3. Thermal Infrared Multispectral Scanner 热红外多光谱扫描仪 4. Tool Inventory Management System 工装库存管理
tin 1.【化】锡(化学元素,符号 Sn) 2.(低碳钢板的)镀锡 3.涂焊锡 4.马口铁 5.罐头,装罐
tin indium oxide 氧化铟锡
tin strip 预制金属板简易跑道
TINA Thermal-Imaging Navigation Aid 热成像导航设备
tind 点燃
ting 1.叮叮声 2.铃的响声
tinned wire 镀锡线
tinplague 锡瘟,锡疫
TINS 1. Thermal-Imaging Navigation System 热成像导航系统 2. Trains Inertial Navigation System 序列惯性导航系统
tint 1.色彩 2.浅色
TIP 1. Technical Information Panel 技术情报组 2. Test Integration Plan 综合测试计划 3. TIROS Information Processor 泰罗斯卫星信息处理机 4. Tool Index Plan 工艺装备定位计划 5. Tracking and Impact Prediction 跟踪与碰撞点预测
tip 1.翼尖 2.侦察照相机绕飞机横轴旋转的角度 3.翼尖油箱 4.轻拍
tip aerofoil 翼尖
tip aileron 翼尖副翼
tip back angle 防擦地角
tip blade 翼尖后掠旋翼
tip blade clearance 叶片间隙
tip cargo 翼尖货物
tip chord 1.翼尖翼弦 2.梢弦
tip clearance 1.翼尖距离 2.叶尖间隙
tip clearance height 叶尖间隙高度
tip clearance leakage 叶端泄漏损失
tip clearance vortex 车叶尖余隙涡流
tip cropping 截尖
tip displacement 末端位移
tip dragger 翼尖扰流片
tip drive 翼尖驱动
tip droop 翼尖下垂,翼尖下折
tip float 翼尖浮筒,稳定浮筒
tip gap 间隙
tip leakage 叶端漏泄
tip leakage flow 间隙泄漏流
tip leakage vortex 顶部间隙泄漏涡
tip loss 翼尖损失
tip loss facfor 旋翼桨损失系数
tip mass 末端质量
tip node 端节点
tip path plane 1.桨尖轨迹平面 2.挥舞不变平面
tip plate 顶板
tip pod 翼尖吊舱,翼尖外挂油箱
tip radius 齿顶圆角半径
tip rake 翼尖斜削
tip relief 1.修像 2.齿端修整 3.齿形修缘
tip seal 叶顶密封
tip shroud 围带
tip speed 桨尖速度
tip stall 翼尖失速
tip surface 齿顶曲面
tip tank 翼尖油箱
tip trailing vortex 翼尖尾涡
tip velocity 叶顶速度
tip vortex 1.梢涡 2.翼尖涡流
TIPIS Tactical Information Processing and Interpretation System 战术信息处理与破译系统
tip-off 1.在玻璃泡抽真空后真空管隐藏的最后部分 2.(导弹脱离发射装置时)翻倒,脱轨
tipover 翻倒倾覆
tip-path plane (旋翼机)桨尖面,桨叶尖旋转平面

tipping 包尖
TIPS Total Integrated Pneumatic System 总体气动系统,综合气动总系统
tipsail 翼尖小翼
tip-speed ratio 1.(螺旋桨的)叶尖相对速度 2.(旋翼的)工作状态特性系数,尖速比
tip-tilt mirror 自适应光学
TIR 1. Target-Illuminating Radar 目标照射雷达 2. Terminal Imaging Radar 末段成像雷达 3. Thermal Infrared 红外热成像 4. Total Indicator Reading 仪表总读数 5. Traffic Information Radar 空中交通信息雷达 6. Type Inspection Report 型号检查报告
tire 轮胎
tire model 轮胎模型
tire test 轮胎测试
TIRS Tether Initiated Recovery System 系绳回收系统
TIRSS Theater Intelligence, Reconnaissance and Surveillance Study 战区情报、侦察和监视研究
TIS 1. Tactical Intelligence Squadron 战术情报中队 2. Target Information System 目标信息系统 3. Thermal-Imaging System 热成像系统 4. Tracking Information Subsystem 跟踪信息分系统
TISEO Target-Identification System Electro-Optical 光电目标识别系统
tissue fluid vaporization 体液蒸发
tissue hypoxia 组织性缺氧
TIT Turbine Inlet Temperature 涡轮进口温度
tit 控制按钮
Titan 1. 大力神(美国运载火箭) 2.【天】土卫六
titanium 【化】钛(化学元素,符号 Ti)
titanium alloy 钛合金
titanium aluminide 钛铝化合物
titanium aluminum alloy 钛铝合金
titanium aluminum vanadium alloy 钛铝钒合金
titanium bar 泰坦锭
titanium dioxide 二氧化钛
titanium frame 钛架
titanium matrix composite 钛基复合材料
titanium vanadium alloy 钛钒合金
title block 标题栏
titles 飞机使用者名称,飞机业主名称
TJ Turbojet 涡轮喷气发动机
TJAG The Judge Advocate General 军法总监
TJS Tactical Jamming System 战术干扰系统
TK Track Angle 航迹角
TKE Track Angle Error 航迹角误差
TKR Tanker 加油机,加油车
TKSC Tsukuba Space Centre 筑波宇宙中心
TL 1. Target Length 目标长度 2. Terminal Location 终端定位 3. Testing Laboratory 检测实验室 4. Thermoluminescence 热致发光 5. Thrust Limit 推力极限 6. Total Life 总寿命 7. Towing Lug 牵引环
TLA 1. Throttle Lever Angle 油门杆角度 2. Towed Linear-Array 拖曳式线列阵
TLBR Tactical Laser Beam Recorder 战术激光束记录装置
TLC Telecommunication 长途通信
TLD 1. Technical Log Defect 记入技术日志的故障或缺陷 2. Thermoluminescent Device 热致发光装置
TLE 1. Thin-Layer Electrophoresis 薄层电泳 2. Trunk Line Equipment 转发器
TLG Tail Landing Gear 尾起落架
TLI Translunar Insertion 进入飞向月球的轨道
TLM Telemetry 遥测,远距离测量
TLS 1. Tactical Ladar Seeker 战术激光雷达导引头 2. Tactical Landing System 战术着陆系统 3. Target Level of Safety 安全性目标等级 4. Translunar Shuttle 跨越月球的航天飞机
TLSI Technical Log Special Inspection 履历本专门检查,记入技术日志的特种检查
TLSS Tactical Life Support System 战术飞机生命保障系统
TM 1. Tactical Missile 战术导弹 2. Telemetry 遥测 3. Test Method 试验方法 4. Thematic Mapper 专题测绘仪,专题制图仪,专题成像仪 5. Thermal Model 热试验模型,温控星 6. Torque Motor 力矩电机 7. Transcendental Meditation 超前构想 8. Transfer Mechanism 转换机构 9. Transverse Magnetic 横向磁场 10. Tropical Maritime 热带海上的
TM wave Transverse Magnetic wave 横磁波,E 波
TMA 1. Terminal Manoeuvring Area 航站空域,航站空中管制区,航站起落和滑行地带 2. Thrust Mode Annunciator 推力方式信号器 3. Trimethylamine 三甲胺
TMC 1. Telecommunication Maintenance Center 远程通信维护中心 2. Thrust Management Computer 推力管理计算机
TMCC Time-Multiplexed Communications Channel 时间多路复用通信信道
TMD 1. Tactical Munition Dispenser 战术武器分配器,武器集束弹箱 2. Theater Missile Defense 战区导弹防御战术弹药箱,战术子母弹箱
TMDE Test, Measurement and Diagnostic Equipment 测试、计量与诊断设备
TME 1. Total Mission Energy 任务总能量 2. Tranportation Mechanical Equipment 运输机械设备
TMEL Trimethyl-Ethyl Lead 三甲基乙基铅
TMF 1. Thrust Management Function 推力管理功能

2. True-Mass Flowmeter 真实质量流量计,实有质量流量计

TMG　1. Target Moving 移动目标,目标移动 2. Thermal Meteoroid Garment 加温防宇宙尘服 3. Track Made Good 航向保持正确,地面起降点连接直线

TMI　TRMM Microwave Imager 热带降雨测量卫星微波成像仪

TMIA　Trans-Mars Injection Assembly 火星飞行发射装置

TMIS　1. Technical and Management Information Systems 技术和管理信息系统 2. Technicians Maintenance Information System 技术人员维修信息系统

TML　Tetramethyl Lead 四甲基铅

TMM　1. Tantalum Manganese-oxide Metal 钽锰金属 2. Thermal Math Model 热数学模型 3. Tailored Materials Management 改进物料管理

TMN　True Mach Number 真马赫数

TMO　Traffic Management Office 空中交通管理处

TMP　1. Transverse-Magnetized Plasma 横向磁化等离子体 2. Twin Machine-gun Pod 双联机枪吊舱

TMPROC　Telemetry Processing 遥测数据处理

TMR　1. Topex Microwave Radiometer 托佩克斯微波辐射计 2. Triple Modular Redundant 三重模块式备分,三重模块冗余

TMS　1. Telemetry Modulation System 遥测调制系统 2. Test and Monitoring Station 测试与监控站 3. Thermal Management System 热管理系统 4. Thrust-Management System 推力管理系统 5. Traffic Management System 交通管理系统

TMSA　1. Technical Marketing Society of America 美国技术市场学会 2. Trainer-Mission Simulator Aircraft 教练机任务模拟机

TMSS　Tactical Munitions Safety Study 战术弹药保险研究

TMT　Thermomechanical Treatment 形变热处理,热机械处理

TMU　Traffic Management Unit 交通管理装置

TMV　True Mean Value 真平均值

TMXO　Tactical Miniature Crystal Oscillator 战术小型晶体振荡器

TN　1. Tactical Navigation 战术领航 2. True North 真北 3. Trunk Number 中继线编号

TNA　Time of Nearest Approach 最近进场时间

TNF　Theater Nuclear Forces 战区核部队

TNM　1. Tactical Nuclear Missile 战术核导弹 2. Transmission Network Management 传输网络管理

TNR　1. thinner 稀释剂 2. trainer 教练机,练习器

TNS　1. Technical News-Sheet 技术通信 2. Technical Numbering Standard 技术标号标准

TNT　1. trinitrotoluene 三硝基甲苯,梯恩梯炸药 2. Turn Not-Turned 协调不协调

TNTV　tentative 试验性的,暂时的,试行的

TNW　1. Tactical Nuclear Warfare 战术核作战 2. Theater Nuclear Weapon 战区核武器

TO　1. Tactical Observer 战术观察员 2. Take Off 起飞 3. Technical Officer 技术职员 4. Technical Operations 技术操作 5. Technical Order 技术规程,技术说明 6. time-off 起飞(发射)时间 7. Tool Order 工装指令 8. Transfer Orbit 转移轨道 9. Try Out 经过试验

TO EPR　Take-Off Engine Pressure Ratio 起飞发动机压力比

TO&E　Table of Organization and Equipment 编制及装备一览表

to/from　向台/背台

to/from indication　向/背台指示

to/from indicator　向/背台指示器

TOA　1. Time of Arrival 到达时间 2. Total Obligational Authority 拨款总授权 3. Transportation Operating Agencies 运输机构

toboggan　(空中加油的)下滑配合

TOC　1. Tactical Operation Center 战术作战中心 2. Target Optimization Control 目标最佳化控制 3. Technical Order Compliance 技术规程修改 4. Time of Crash 坠毁时间 5. Time Out Circuit 超时电路 6. Top Of Climb 爬升顶点 7. Total Operating Cost 运作总费用 8. Transfer Of Control 控制转移

TOCS　Terminal Operations Control System 航站营运控制系统

TOD　1. Take-Off Distance 起飞距离 2. Technical Objective Document 技术项目文件 3. Time Of Day 时刻 4. Time Of Delivery 交货时间 5. Time Of Departure 起程时间,起飞时间 6. Top Of Descent 下降最低点

TODA　Take-Off Distance Available 可用起飞距离

TODR　Take-Off Distance Required 需用起飞距离

TOE　Ton of Oil Equivalent 当量油料吨数

toe brakes　脚刹车,机轮刹车

toed in　发动机轴线内偏

toed out　发动机轴线外偏

toe-in angle　前束角

TOF　Take-Off Fuel 起飞燃油量

TOFL　Take-Off Field Length 起飞场长

TOFP　Take-Off Flightpath 起飞飞行路线(航迹线)

toggle　1. 手动投弹,人工投弹 2. 拨动电门,双态元件,触发器 3. 肘节,扭力臂 4. 套环,套索钉 5. 反复电路 6. 紧线钳

toggle circuit　触发电路

toggle flip-flop 双稳触发器,计数型触发器
toggle speed 计时速度
toggle switch 触发器,乒乓开关,钮子开关
togglier 进行人工投弹的轰炸员
TOGW Take-Off Gross Weight 起飞总重
toilet 厕所,盥洗室
token 标志,记号,令牌
token bus 1.总线网令牌传递方式 2.令牌总线(一种网络通信总线)
token cabling system 令牌传递环敷线系统
token ring 1.环状网令牌传递方式 2.令牌环
tol 1.Time On Line 公差标注 2.Tolerance 宽大,宽容,忍耐力,容差,公差 3.Toluene 甲苯
TOLA Take-Off and Landing Analysis 起飞和着陆分析
Told card 起飞着陆数据卡
tolerability 容忍度
tolerance 1.容差 2.公差 3.耐受剂量 4.容错 5.容限,耐力,耐受性 6.最大允许误差(仪表的) 7.机场上检查随身携带物品尺寸的样尺与物品尺寸的差数
tolerance allocation 公差分配
tolerance analysis 容差分析
tolerance dose 容许(辐射)剂量,耐受剂量
tolerance endpoint 耐受终点
tolerance fault 容错
tolerance frequency 容限频率
tolerance interval 容许区间(间隔)
tolerance level 允许(辐射)剂量级
tolerance limit 耐受极限,容许极限,公差极限
tolerance range 1.耐量范围 2.容差范围
tolerance time 耐受时间
toll call 长途通话
toll dispatcher desk 长途分发台
toll service observation board 长途业务检查台
toll telephone 长途电话
toll-connecting trunk 长途通信中继线
toluene 甲苯
tombac 1.顿巴黄铜 2.黄铜伸缩管
tomography X线断层摄影术
TOMS Total Ozone Mapping Spectrometer 总臭氧测绘分光计
TOMUIS Total Ozone Mapping with UV Imaging Spectrometer 总臭氧紫外成像光谱仪
ton 吨
ton force 吨力
ton of refrigeration 致冷吨数
Tonal "托纳尔"直升机
tonal balance 1.音调匀称 2.色调匀称
tone 1.音调,音响信号 2.单调的,单色的 3.专指发射空空导弹时的音响信号
tone level 音平
tone ranging 侧音测距
tone ranging system 侧音测距系统
toner 1.调色剂,增色剂 2.色粉
TOO Target Of Opportunity 临时目标,意外目标
tool 工具,刀具,机床,器械
tool bit 硬质刀具
tool center point 工具中心点
tool design 工艺装备设计
tool mark 刀痕,工具痕迹
tool steel 工具钢
toolbox 工具箱
tooling 1.机加工,刀具加工,切削 2.工艺装备,工具,仪器 3.机床安装
tooling probation 工艺装备试验
tooling verification 工艺装备验证
toolmaker 1.刀具制造工,制造、维修机床的技工 2.刀具制造厂
toolroom 1.工具室 2.高精密度超净车间
tooth method 锯齿法
tooth wheel 齿轮
TOP 1.Take-Off Power 起飞功率,起飞推力 2.Technical Operating Procedure 技术操作程序 3.Total Obscuring Power 总遮蔽能力
top beam 顶梁
top chord 上弦杆
top coating 外涂层,外保护层
top cover 上层掩护
top dead center 上死点
top dressing 喷洒化学品,施顶肥
top edge 1.上边缘 2.上沿
top hat 1.礼帽状的 2.礼帽形钢型材
top link chute guide 上除链导
top loading 顶上加载
top management 1.高管理层 2.董事会
top nozzle 上喷嘴
top of climb 上升最高点
top overhaul 气缸在位翻修
top panel 1.上面板 2.顶面板
top region material 顶区材料
top secret 绝密
top surface 上表面,顶面
top tube 上管
top view 顶视图
top wall 上磐
top-coat 1.面漆 2.表面涂层
top-down 自上而下,自顶向下
top-down design 自上而下设计

top-down development	自上而下开发
top-down method	自顶向下法
top-down testing	自上而下测试
TOPEX	Topography Experiment Satellite 海洋地貌实验卫星,托佩克斯卫星
TOPO	Topographical Position Satellite 地形测量卫星
topocentric	以观察者为中心的
topographic correction	地形(影响)校正
topographic map	地形图
topographical reconnaissance	地形侦察
topography	地形,地貌,地势,地形学,地形测量学
topological retrieval	拓扑检索
topological sorting	拓扑排序,拓扑分类
topological structure	拓扑结构
topology	1.拓扑学 2.地志学 3.局部解剖学
toponomastics	地名学
toponymy	1.地名学 2.部位命名法 3.地名之研究
topping	1.最高级的 2.顶层 3.拔顶,蒸去轻油 4.液体火箭发动机中液体推进剂的一种工作循环
topping cycle	闭式循环
topping off	加满(油箱),补充加注
topping up	充气
topple	陀螺漂移,陀螺倾倒
topple axis	倾覆轴
toppled	(陀螺)倾倒的
TOPS	Thermoelectric Outer-Planet Spacecraft 外行星温差航天器
top-side	1.正面 2.板面
top-side sounder	顶视探测仪
topside sounder satellite	顶视探测卫星
top-temperature control	限温装置
TOR	1. Tactical Operations Room 战术作战指挥室 2. Take-Off Run 起飞滑跑距离 3. Technical Operating Report 技术操作报告 4. Tentative Operational Requirement 暂定工作要求 5. Terms Of Reference 参考项,基准项 6. Time Of Release 投弹时间
torr	托(真空压强单位,1托相当于0.133千帕或1毫米水银柱的压强)
TORA	Take-Off Run Available 起飞滑跑可用距离
torch	1.火把,火炬 2.手电筒 3.启发之物
torch igniter	火炬点火器
torch injector	中心管
torching	1.排气冒火,火焰外伸 2.(超高真空技术中的)除气
toric	复曲面的
toric combiner	复曲面组合玻璃
toric lens	复曲面透镜
toric surface	复曲面
tornado	龙卷风
toroid	1.环形线圈 2.圆环面 3.螺旋管
toroidal	环形,环形的,喇叭口形的,螺旋管形的
toroidal coil	环形线圈
toroidal oscillation	环形振荡
toroidal permeability	环磁导率
toroidal structure	环状结构
toroidal vanes	环形导流片
torpedo	1.鱼雷 2.水雷 3.油井爆破筒 4.电鳐
torpedo director	鱼雷指挥仪
torpedo dropping point	鱼雷投放点
torque	1.转矩 2.扭矩 3.项圈 4.金属领圈
torque balance accelerometer	力矩平衡加速计
torque balance capture	力矩平衡归零
torque box	扭矩盒
torque brake	扭矩刹车装置
torque coefficient	扭转力矩系数,扭矩系数
torque coil	转矩线圈
torque component	扭矩分力
torque control	转矩控制,力矩调节
torque distribution	力矩分配
torque dynamometer	扭力测功器
torque effect	扭矩效应
torque feedback loop	力矩反馈回路
torque feedback test	力矩反馈试验
torque generator	力矩器
torque horsepower	轴马力
torque indicator	扭矩表
torque input	扭矩输入
torque link	扭力臂,防扭臂
torque magnetometer	磁转矩计
torque measuring gear	测扭齿轮,扭力测量齿轮
torque motor	1.力矩电动机 2.力矩马达 3.扭矩马达
torque output	扭力输出
torque ripple	转矩波动
torque ripple coefficient	力矩波动系数
torque stand	发动机试验台
torque synchro	力矩式自整角机
torque transducer	转矩传感器
torque tube	万向轴管,扭力管
torque vector	转矩矢量
torque wrench	扭力扳手,限力扳手
torque-angular displacement characteristic	矩角位移特性
torque-measuring mechanism	测扭装置
torquemeter	测扭机构,测扭器
torquer	转矩发生器,力矩马达,修正电动机,转矩计,扭矩传感器,力矩器
torquer temperature coefficient	力矩器温度系数
torque-set screw	调定扭矩螺钉

torque-torsional angle curve 扭矩-扭角曲线
torquing current 加矩电流
TORR Take-Off Run Required 起飞滑跑所需距离
torsion 1.扭转,扭力,扭矩 2.扭转变形
torsion balancex 扭转振动
torsion bar 扭力杆
torsion bar accelerometer 扭杆式加速度计
torsion box 扭矩盒,翼盒
torsion coupling 扭转耦联
torsion flutter 弯扭颤振速率
torsion frequency 扭振频率
torsion mode 扭转振型
torsion moment 1.扭矩 2.扭力矩
torsion pendulum 1.扭秤 2.扭转摆
torsion stiffness 扭转刚度
torsion stress 扭转应力
torsion strength 抗扭强度
torsion swing 扭摆旋转
torsion test 扭转试验
torsion type rate gyro 扭杆式速率陀螺仪
torsional 扭转的
torsional accelerometer 扭杆式加速度计
torsional deflection 扭转变形
torsional flutter 扭转颤振
torsional instability 扭转失稳
torsional load 扭转载荷
torsional mode 扭振模式
torsional moment 1.扭矩 2.扭转力矩
torsional oscillation 扭转振荡
torsional stiffness 扭转刚度
torsional vibration 扭转振动
torsional stress 扭转应力
torsion-bar tab 弹簧式补偿片,扭杆式补偿片
torso dummy 假人
torso harness 躯干背带
torus 1.花托 2.圆环面
torus reflector 环面反射器
TOS 1. Tiros Operational Satellite 泰罗斯业务卫星 2. Transfer Orbit Stage 转移轨道级
TOSS 1. Take-Off Safety Speed 起飞安全速度 2. TIROS Operational Satellite System 泰罗斯业务卫星系统
toss 抛,掷,拉起轰炸,突然抬起
toss bomb release 拉起投弹
toss bombing 拉起轰炸
toss bombing 1.上仰轰炸 2.拉起轰炸
toss delivery 拉起投弹
TOT 1. Time on Target 诸兵器协同射击目标的时间,飞机预定(或实际)攻击目标的时间,飞机预定(或实际)对目标照相的时间,核武器在预定爆心投影点上爆炸的时间 2. Total 总的,全部的
tot 1.合计 2.少量
total 总数,合计
total absorptance 总吸收率
total air 总空气量
total air temperature 总气温
total amount 1.总数 2.总价 3.总计
total amount of reentry aerodynamic heating 再入气动总加热量
total area 总面积
total azimuth lead angle 总修正角侧向分量,方向总修正角
total blade-width ratio 总桨叶宽度比
total cloud cover 总云量
total communication 全接入通信
total concentration 总浓度
total correcting angle 1.总修正角 2.总提前角
total cost 总成本
total cover-degree 总盖度
total current 全电流
total curvature 总曲率
total curvature correction 总弹道弯曲修正角
total delay 总延时
total digital aceess 全数字接入
total distance 总距离,全距离
total dose 总剂量
total drag 总阻力
total duration 总历时
total eclipse 全食
total efficiency 1.总效率 2.全效率
total electron 总电子
total electron content 电子总含量
total elevation lead angle 总高低修正角,总修正角垂直分量
total energy 总能量
total energy compensation 总能量补偿
total energy control 总能量控制
total energy detector 总能检测器
total enthalpy 总焓
total enthalpy probe 总焓探针
total exclusion zone 全面禁区
total excursion 行程
total failure 总体破坏
total fatigue life 疲劳总寿命
total field 总场
total field of view 总视场,覆盖区总宽
total flight service 飞行勤务总量
total flow 总流量

total force 总力
total fuel 总燃料
total fuze running time variation 定时引信作用时间总偏差
total gross weight 总毛重
total harmonic distortion 总谐波失真
total head 总压，全压头
total heat 焓
total heat flux 总热通量
total heat load 总热负荷
total holding time 全部占用时间
total hydrocarbon 总烃
total impulse 总冲量
total input 总输入
total input energy 总输入能量
total inversion 全反转
total ionizing dose 总电离剂量
total length 1.全长 2.总长度
total length closing error of traverse 导线全长闭合差
total lift 总升力
total linear momentum 总线性动量
total loss 1.全损 2.总损失
total mass 总质量
total mass after loading 加注后总质量
total noise rating 总噪声等级
total number 总数
total number of cell 细胞总数
total number of grid hole 栅格孔总数
total obligational authority 1.国防五年计划的总经费 2.某财年的经费
total operating time 总工作时间
total operations 全部运营业务
total output 总输出
total ozone mapping system 臭氧总量图示系统
total power 总功率
total power radiometer 全功率辐射计
total pressure 总压力，全压力，全压强
total pressure control 总压控制
total pressure distortion 总压畸变
total pressure distribution 总压分布
total pressure drop 总压力降
total pressure gauge 全压真空计
total pressure gradient 全压力增减率
total pressure impulse 压强总冲
total pressure inlet distortion 进气总压畸变
total pressure loss 总压力损失
total pressure loss parameter 总压损失参数
total pressure probe 总压管，皮托管
total pressure ratio 全压力比
total pressure recovery 总压力恢复
total probability 总概率
total propeller width ratio 螺旋桨总宽度比
total pulses 总脉冲次数
total quality control 全面质量控制
total quality management 全面质量管理
total radiator 全辐射体
total reaction 全部反应
total residue 总残留量
total solar energy monitor 总太阳能监测器
total solid 总固体量
total sulfur 全硫
total surface 总面积
total surface area 总表面积
total system 全套系统
total system energy 1.总能系统，总系统 2.全能量系统，整体系统 3.全能系统，综合系统
total temperature 1.总温（度）2.驻点温度
total temperature change 总温度变化
total temperature measurement 总温度测量
total temperature of airflow 总气流温度
total temperature probe 总温探头
total temperature transducer 总温传感器
total thrust 总推力
total time 总时间
total time of flight 总飞行时间
total traffic 总通信量
total transfer 1.全转移 2.总计转移 3.全额转让
total transfer time 总传输时间
total transition 总跃迁
total uncertainty 总体不确定
total utility 1.总效用 2.总效果
total variance 总方差，总变异数
total variation 全变差
total variation decreasing scheme 1.全变差下降格式 2.TVD 格式
total velocity variation 总速度变化
total volume 1.总体积 2.总容积 3.总成交量
total weight 1.总重量 2.设备总重
total workforce 劳动力总量
total-annular eclipse 全环食
totality 1.总和 2.全食
totalizing（measuring）instrument 累计式仪器
totalled 完全毁坏的，无法修理的
totally bound set 全有界集，预紧集
totally depleted semiconductor detector 全耗尽半导体探测器
total-package procurement 一揽子采购，通盘采购
total-pressure head 总压头

total-pressure tube 皮托管，总压管
TOTE Tracker Optical Thermally Enhanced 红外及光学跟踪装置
tote board 表格式的显示板
TOTLZ totalizer 加法器
touch screen 1.触敏控制板 2.触屏，触摸显示屏
touch screen technology 屏幕接触技术
touch sensitive panel 1.触敏控制板 2.触敏显示屏
touch up （装备出厂前的）修饰
touchdown 1.着陆，降落 2.触地
touchdown aim-point 接地目标点
touchdown condition 着陆条件
touchdown point 着陆点
touchdown position 着陆位置
touchdown rate of descent 接地时的下降率
touchdown speed 着陆速度
touchdown velocity 着陆速度
touchdown zone 1.着陆接地区 2.接地地带
touchdown zone elevation 1.接地区海拔高度
touchdown zone light 接地带灯
touchpad 触摸屏设备，触摸板
touchwire 接触丝
toughened glass 钢化玻璃
toughness 韧性，韧度
tour marketing 开拓旅游市场
tour operator 旅行社，航空旅游代理商
tour organizer 航空旅游组团人
tour package 全套旅游服务
tour wholesaler 旅游批发商，旅游代理商
tourist traffic 旅游客运量
Tournesol 向日葵卫星（法国空间物理探测卫星）
TOVS Tiros Operational Vertical Sounder 泰罗斯卫星实用垂直分布探测器
TOW 1.Take-Off Weight 起飞重量 2.Time Of Week 周时（用于 GPS 上） 3.Tube-Launched Optically-tracked Wire-guided 导管发射、光学跟踪、有线制导（导弹）
tow 1.拖 2.麻的粗纤维 3.拖曳所用之绳
tow dart 箭形拖靶
towbar 飞机牵引杆
towed array frigate 带拖曳式线列阵的护卫舰
towed array passive sonar 拖曳式线列阵被动声呐
towed bird system 吊舱系统
towed body 遥感装置
towed boom 拖架
towed glider 被牵引的滑翔机
tower crane 塔式起重机
tower shaft 塔式传动轴
tower telescope 塔式望远镜

towhook 牵引钩
towing brake 牵引刹车
towing equipment 牵引设备，拖航设备
towing eye 牵引环
towing sleeve 袋形拖曳
towing sonar 拖曳式声呐
towing tank 拖曳水槽
towing target 拖曳
Townend ring 汤恩环罩
Townsend avalanche 汤森离子雪崩
Townsend discharge 汤森放电
toxic attack 毒剂攻击
toxic cluster 集束毒气炸弹
toxic gas 毒气
toxic smoke compound 毒烟剂
toxic warfare 毒物（剂）战
toxication vertigo 药物中毒性眩晕
Toxic-B B 毒剂（苏联的一种空投或布撒的最普通的致死性战剂）
toxicity 毒性
toxicology 毒理学
toxicological weapon 毒性武器
TP 1.Tail Pipe 尾管 2.Target Parameter 目标参数 3.Target Practice 打靶练习 4.Technical Publication 技术出版物 5.Teleprocessing 遥处理 6.Test Pilot 试飞员 7.Test Point 测试点 8.Test Product 试验产品 9.Thermoplastics 热塑（性）塑料 10.Total Parts 全部零部件 11.Total Pressure 全压，总压 7.Transport Pilot 运输机驾驶员 8.Turbopump 涡轮泵 9.Turboprop 涡轮螺桨发动机
TP/TK Tip Tank 翼尖油箱
TPA 1.Taildragger Pilots Association 后三点飞机飞行员协会 2.Target Practice Ammunition 打靶练习弹药 3.Technical Publication Association 技术出版物协会 4.Traffic-Pattern Altitude 起落航线高度 5.Turbine Powered Aircraft (Airplane) 涡轮动力飞机
T-PAD Trunnion Pin Attachment Device 耳轴销连接装置
TPAR Tactical Penetration Aid Rocket 战术突防火箭
TPC 1.Total Program Cost 项目总费用，总计划成本 2.Traveler Part Card 运送车零件卡
TPD 1.Technical Publication Document 技术发布文档 2.Test Procedure Drawings 试验程序图
TPDU Transport Protocol Data Unit 传输议定数据装置
TPDR transponder 1.异频雷达收发机 2.转调器
TPE Tracking and Pointing Experiments 跟踪和定向实验
TPF Terminal Phase Final 1.弹道末段 2.最后阶段

TPG TV Picture Generator 电视图像产生器

TPI 1. Target Position Indicator 目标位置指示器 2. Terminal Phase Initiation 弹道末段开始，最后阶段开始

TPL 1. Terminal Permission List 终端许可清单 2. Tool Parts List 工装零件表 3. Trans-Pacific Landing 跨太平洋着陆

TPM 1. Technical Performance Managemnt 技术性能管理 2. Technical Performance Measures 技术性能测量 3. Terrain Profile Matching 地形剖面匹配

TPMU Tire Pressure Monitor Unit 轮胎压力监控装置

TPN 1. Technical Procedure Notice 技术程序通知 2. Tone-corrected Perceived Noise 音校正可感知噪声 3. Two-Position Nozzle 双位喷管

TPP 1. Technology Program Plan 技术项目计划 2. Testability Program Plan 可测试性项目大纲 3. Tip Path Plane 桨尖旋转面 4. Total-Package Procurement 整套采购

TPR 1. Temperature Profile Recorder 温度廓线记录器 2. Thermoplastic Rubber 热塑性橡胶

TPS 1. Technical Problem Solving 技术问题的解决 2. Telemetry Processing Station 遥测数据处理站 3. Thermal Picture Synthesizer 热成像合成装置 4. Thermal Protection System 防热系统，热防护系统 5. Tooling Procedure Specification 工装程序规范 6. Tons Per Squareinch 吨/平方英寸 7. Transitory Protection System 临时防护系统 8. troops 部队

TPU Terminal Position Update 末端位置修正

TPW Technical Publication Workstation 技术出版物制作处

TPWG Test Planning Working Group 制订试验计划工作组

TQA 1. Throttle Quadrant Assembly 油门杆弧座组件 2. Tool Quality Assurance 工装质量保证

TQC Total Quality Control 全面质量控制

TQCS Time，Quality，Cost，Service 周期、质量成本、服务

TQE Technical Quality Evaluation 技术质量鉴定

TQM Total Quality Management 全面质量管理

TQMS Total Quality Management System 全面质量管理系统

TR 1. Tactical Reconnaissance 战术侦察 2. Tail Rotor（直升机）尾桨 3. Target Recognition 目标识别 4. Technical Regulation 技术规程 5. Technical Review 技术评审 6. Temporary Reverser 临时修订 7. Test Report 试验报告 8. Thrust Reverser 反推力装置 9. Time-delay Relay 延时继电器 10. Time to Retrofire 制动发动机点火前 11. Transformer Rectifier 变压整流罩

TRA 1. Temporary Reserved Airspace 临时备用空域 2. Throttle Resolver Angle 油门杆分解角度 3. Thrust Reduction Altitude 减推力高度

TRAC 1. Telescoping Rotor Aircraft 旋翼伸缩式航空器，旋翼伸缩直升机 2. Terminal Radar Approach Control 终端雷达进近控制 3. Trials Recording and Analysis Console 试射记录与分析操作台

TRACA Total Radar Aperture Control Antenna 雷达总孔径控制天线

TRACALS Traffic-Control Approach and Landing System 空中交通管制进场着陆系统

trace 1. 扫描线 2. 图形显示上见到的数据线 3. 电子数据处理诊断技术 4. 踪迹，轨迹

trace equalization 道间均衡

traceability 溯源性，可追溯性

TRACER Tropospheric Radiometer for Atmospheric Chemistry and Environment Research 对流层大气化学和环境研究辐射计

tracer 1. 曳光剂，曳光管，曳光弹 2. 示踪物 3. 故障寻找器

tracer armor-piercing shell 曳光穿甲弹

tracer bullet 曳光穿甲弹

tracer control 曳光弹进行的射击控制

tracer display 示迹(线)显示，示踪(线)显示

tracer element 曳光剂

tracer fire 曳光弹射击

tracer flare 曳光照明弹

tracer fuze 曳光引信

tracer gas 1. 示踪气体 2. 探漏气体 3. 检测气体

tracer line gunsight 示迹线射击瞄准，示迹线瞄准具

tracer machining 仿形加工

tracer particle 示踪粒子

tracer projectile 曳光弹

tracer rocket 曳光火箭

tracer round 曳光弹

tracer sensing 曳光弹着观察

tracer sight 示迹线瞄准具

tracer-line snapshoot gunsight 示迹线快速射击瞄准，示迹线快射瞄准具，按弹迹快射瞄准具

tracing 1. 描绘，标图 2. 示踪，跟踪，寻找故障，辨别目标 3. 透写，透写图

track 1. 航迹，航线 2. 跟踪 3. 桨尖运动轨迹 4. 轮距，浮筒间距 5. 导轨，滑轨 6. 准，对准 7. 导电通路 8. 航迹角，画出航迹线，观察到航迹角

track angle 航迹角

track angle error 偏航角

track clearance 航路许可

track configuration 磁迹位形

track correlation 按航迹识别

track crossing angle 1.航迹角 2.航迹交叉角
track detecter 径迹探测器
track file 跟踪文件
track fusion 航迹融合
track handover 跟踪交接
track identity 目标属性
track infrared system 红外跟踪系统
track initiator 跟踪始发人
track intervals 航迹间隔
track length 轨道长度
track lock lever 座椅锁定手柄
track made good 航迹保持正确
track marker 跟踪标记
track No. (指挥引导系统中空中目标的)批号
track oriented 按航迹对正
track plotter 航迹记录器,航迹标图员
track production 建立跟踪信息档案
track radar set 跟踪雷达
track repetition 轨道重复时间
track separation 1.航路间隔 2.轨道间隔
track sled 滑轨车
track smoothing 航迹平滑
track symbology (雷达或其他电子显示器上的)跟踪目标符号,航迹符号
track telling 情报传递
track update 跟踪更新
trackability 可跟踪性
tracker 跟踪装置,跟踪雷达,跟踪操作员
tracking 1.跟踪 2.同锥度调整,(螺旋桨叶的)同轨迹调整,共面调整 3.精确跟踪目标
tracking accuracy 跟踪精度
tracking algorithm 跟踪算法
tracking and control center 跟踪和控制中心
tracking and data acquisition 自动跟踪与目标指示数据的获取
tracking and data relay satellite system 跟踪与数据中继卫星系统
tracking and data-relay satellite 跟踪与数据中继卫星
tracking and impact prediction 跟踪与弹着预测
tracking angle acceleration 跟踪角加速度
tracking angle acceleration without degradation 保精度跟踪角加速度
tracking angular rate 跟踪角速度
tracking angular rate without degradation 保精度跟踪角速度
tracking ball 跟踪球
tracking capacity 跟踪容量
tracking comparator 跟踪比较器
tracking condition prediction 观测预报

tracking controller 跟踪控制员
tracking data processing 外测数据处理
tracking distortion 循纹失真
tracking error 1.跟踪误差 2.循迹误差
tracking field of view 跟踪视场
tracking filter 跟踪滤波器
tracking filtering 跟踪滤波
tracking flag 旗标杆
tracking guidance 跟踪制导
tracking helmet 带瞄准具的飞行头盔
tracking index 瞄准光点,中心光点
tracking(information) simulation 外测(信息)仿真
tracking line 1.跟踪瞄准线 2.跟踪线
tracking mode 跟踪方式
tracking performance 跟踪性能
tracking performance requirement design 跟踪性能设计
tracking point inconsistency correction 跟踪点不一致修正
tracking radar 跟踪雷达
tracking range 跟踪范围
tracking rate 跟踪速率
tracking receiver 跟踪接收机
tracking ship 跟踪(测量)船
tracking signal blackout 跟踪信号中断
tracking station 1.跟踪站 2.测量站
tracking station of satellite 卫星跟踪站
tracking subsystem 跟踪分系统
tracking system 外测体制
tracking task 跟踪任务
tracking telemetering and control system 跟踪测控系统
tracking telemetry and command 跟踪遥测和指令
tracking telemetry and command station 跟踪遥测指令发送站
tracking telemetry and command subsystem 测控分系统
tracking telescope 跟踪望远镜
tracking vehicle 跟踪车
track-production area (雷达站的)跟踪信息区
trackway 预制道面
track-while-scan 1.边扫描边跟踪 2.边搜索边跟踪
track-while-scan radar 边扫描边跟踪雷达
Tracon Terminal Radar Approach Control 机场区雷达引导进场控制
TRACS Terminal Radar and Control System 航站雷达和控制系统
traction 1.附着力 2.牵引
traction battery 牵引用蓄电池
tractor 牵引机
tractor beam 牵引光束
tractor driver 牵引车驾驶员

tractor rocket escape system　牵引火箭救生系统
trade offs　权衡,补偿,抵销,折中
trade space　权衡空间
trade-off　1.权衡,折中 2.比较评定 3.协调 4.权衡利弊,选择
trade-off analysis　权衡分析
trade-off study　权衡研究
trading　1.交易 2.贸易 3.购物
trading company　贸易公司
trading decision　交易决策
traditional approach　传统方法
traditional architecture　传统建筑
Tradoc Training & Doctrine Command　训练与条令司令部
traffic　1.交通量 2.机场吞吐量 3.飞行中的一架飞机 4.运输量 5.通信量
traffic collision　撞车
traffic control　交通管理
traffic count　交通量计数
traffic density　1.业务量密度 2.占线密度 3.运输密度
traffic flow　交通流量
traffic information　资料流量资讯
traffic light signal　空中交通信号灯
traffic mix　流量混合比率
traffic pattern　1.交通模式 2.起落航线
traffic rate　通信量速率
traffic stream　车流
TRAG　Training Air Group 训练飞行大队
trail　1.(炸弹等落体的)退曳长,(射击)落后量 2.跟踪,追踪,尾随 3.顺风倾向 4.纵队队形 5.尾迹,踪迹,痕迹 6.转向距,稳定距
trail angle　1.支柱倾斜角 2.轮轴架角
trail correction　(炸弹)退曳修正量
trail formation　纵队编队
trail length　1.放伞长度,机伞距离 2.拖伞钢索长度
trail rope　1.气球拖绳 2.地面系留索
trailer　1.拖车 2.公路运输拖车
trailing aerial　拖曳天线
trailing edge　1.(翼型的)后缘 2.(脉冲)后沿,(脉冲)下降边
trailing sweep　1.后缘后掠角 2.向后缘偏离,后掠
trailing vortex　尾随涡
trailing-edge angle　后缘角
trailing-edge flap　后缘襟翼
train　1.训练 2.齿轮系 3.连续投下的炸弹 4.流星余迹 5.瞄准,对准
train bombing　连续投弹轰炸
trainability　瞄准能力
trainer　1.教练机 2.训练设备,练习器 3.培训程序 4.教练员
trainer helicopter　教练直升机
training aids　训练辅助手段,训练器材与参考资料
training carrier　训练航空母舰
training comprehensive evaluation　训练综合评定
training department　训练部门
training devices for space-flight simulation　空间飞行模拟训练器
training evaluation　训练评价
training flight　训练飞行
training handbook　训练手册
training manual　训练手册
training missile　训练导弹
training process　培训进程
training session　1.练习课 2.培训课程 3.训练项目
training set　1.教练组 2.训练集 3.训练区
training test in launch site　发射场合练测试
training vector　训练向量
train-singles release　单发连续投弹,单连投
trajectory　【物】轨道,弹道,轨迹
trajectory analysis　轨迹分析
trajectory arc　弹道弧,轨迹弧
trajectory bias　弹道偏差
trajectory calculation　轨道计算,轨迹计算
trajectory characteristic　弹道特性
trajectory control　轨迹控制
trajectory correction　轨道修正
trajectory data　弹道诸元
trajectory database　弹道库
trajectory deflection angle　弹远偏角
trajectory design　轨道设计
trajectory drop　1.弹道降低量 2.弹道降低修正量
trajectory equation　1.轨迹方程 2.弹道方程
trajectory generation　轨迹生成
trajectory history　弹道随时间的变化
trajectory information　弹道数据
trajectory length　轨迹长度
trajectory measurement　弹道测量
trajectory measurement system　弹道测量系统
trajectory model　轨迹模式
trajectory optimization　弹道优化,弹道最佳化
trajectory parameter　弹道参数
trajectory planner　轨道计划程序
trajectory planning　轨迹规划
trajectory prediction　弹道预测
trajectory prediction accuracy　弹道预测精确度
trajectory prediction algorithm　轨迹预测算法
trajectory reconstruction　弹道重构
trajectory scorer　弹道记录器

trajectory shift （导弹受推力装置影响的）弹道偏差量
trajectory simulation 弹道仿真
trajectory solution 轨迹解算
trajectory tilt angle 弹道倾角
trajectory tracking 轨迹跟踪
trajectory tracking system 1.外弹道测量系统 2.外测系统
tram 1.长臂规，梁规（即trammel bar 的简称）2.〈口语〉用长臂规测量
TRAM 1.Target-Recognition Attack Multisensor 目标识别攻击复式传感器 2.Test Reliability And Maintainability 试验可靠性和维修性
TRAMAR Tropical Rain Mapping Radar 热带降雨成像雷达
tramline pointer(s) 指针位置标线
tramlines 飞机引导线
trammel bar 长臂规，梁规
trammels 椭圆规，横木规，梁规
tramming 1.找出活塞上死点 2.手推车 3.用圆规测量
trance 1.恍惚 2.出神 3.着迷，入迷
tranquilizer 制导系统增稳装置，弹上控制系统敏感元件
trans 1.传动装置 2.变速箱
transaction 1.交易 2.事务 3.办理
Transall Transporter Allianz "协同"（法德合作生产的运输机）
Transat 子午仪改型卫星
TRANSATEL Transportable Satellite Telecommunications 移动式卫星通信
transatmospheric 跨大气层的
transatmospheric vehicle 空天飞机
transattack period 核突击持续时间
transborder 跨国境的
transceiver 无线电收发机
transceiving 无线电通信
transcoding 变换编码
transconductance 跨导
transcowl （风扇反推装置的）前后传动机构
transcribe 1.转录 2.抄写
transcription 1.抄写 2.抄本 3.誊写
transcription method 转录法
transcrystaline crack 穿晶裂纹
transcutaneous electrostimulation 经皮（肤）的电刺激
transducer 1.变换器，换能器，变流器 2.传感器，发射器 3.传送系统
transducer baseline 换能器基线
transducer gain 换能器增益
transducer with short-circuit turn 短路匝传感器
transductor 饱和电抗器，磁放大器

trans-earth injection 1.进入飞向地球的轨道 2.地球转移轨道射入
transfer 1.转移，迁移 2.转换，变换 3.传递，传输 4.中转，转机
transfer alignment 传递对准
transfer clock 传输时钟
transfer datum 基准面的换算
transfer ellipse 椭圆（轨道），椭圆转移轨道
transfer equation 转移方程
transfer flight 调机飞行
transfer frame 传送帧
transfer function 传递函数，转换函数
transfer function matrix 传递函数矩阵
transfer function method 传递函数法
transfer impedance 传递阻抗
transfer line 输送管路输电线，传输线
transfer matrix 1.转移距阵 2.转换矩阵
transfer matrix method 传递矩阵法
transfer mobility 传递导纳
transfer module 过渡舱，对接过渡舱
transfer of control 移交控制，指挥交接
transfer of control point 管制移交点
transfer officer （航母）调度军官
transfer orbit 转移轨道，过渡轨道
transfer orbit spin stabilization 转移轨道自旋稳定
transfer orbit stage 转移轨道级
transfer orbit three-axis stabilization 转移轨道三轴稳定
transfer passenger 旅客中转
transfer path 转移路径
transfer port 1.传送孔 2.输气孔
transfer punch 仿模冲床
transfer rate 1.传送率 2.转移速率
transfer ratio 转换率
transfer stage 转移阶段
transfer standard 传递标准
transfer time 1.传递时间 2.转移时间
transfer trajectory 转移轨道
transfer trajectory characteristic 转移轨道特征
transfer trajectory design 转移轨道设计
transfer velocity 传输速率
transferred-electron device 转移电子器件
transform 1.改变，使……变形 2.转换
transform fault 转换断层
transformation 1.转化 2.转换 3.改革 4.变形
transformation algorithm 变换算法
transformation equation 变换方程
transformation error 转换错误
transformation grammar 转换文法

transformation matrix 变换矩阵
transformation matrix of coordinates 坐标变换矩阵
transformation of coordinate 坐标变换
transformation of state variable 状态变量变换
transformation ratio 变压比
transformed value 变换值
transformer 1.变压器,变换器 2.纠正仪耦合放大器
transformer coupling amplifier 变压器耦合放大器
transgranular fracture 穿晶断裂
transhipment 空运中转
transient 瞬变的,过渡的,不稳定的态
transient acceleration 瞬时加速度,撞击加速度
transient agent 暂时性毒剂
transient analysis 暂态分析
transient behavior 1.瞬时行为 2.瞬态过程
transient control 瞬态控制
transient creep 暂态蠕变
transient datum 瞬态数据
transient deviation 瞬态偏差
transient distortion 瞬时失真
transient effect 瞬时效应
transient elements 瞬时根数
transient error 瞬态误差
transient fault 瞬时故障
transient force 瞬时力
transient heat flux method 瞬变热流法
transient heat transfer 不稳定传热,非稳态传热
transient maneuver 瞬态操作
transient maneuvering phase 瞬态操纵阶段
transient mode 1.暂态模态 2.过渡模式
transient operating condition 瞬变工况
transient orbital heat flux 瞬态外热流
transient path 瞬时航迹
transient peak 瞬时峰值
transient performance 瞬态性能
transient process 过渡过程
transient process characterisitic curve 暂态特性曲线
transient radiation effects on electronics 电子器件暂态辐射效应
transient response 瞬态响应
transient shock test 瞬态冲击试验
transient simulation 瞬态模拟
transient stage 1.暂态 2.瞬态级
transient startup 瞬态启动
transient state 瞬态
transient surface temperature probe 瞬态表面温度探头
transient test 暂态测试
transient testing 暂态测试
transient thermal behaviour 瞬时热状态(随时间而改变的传热过程)
transient thermal test 瞬态热测试
transient time 过渡过程时间
transient trimmer 瞬态配平机构
transient vertigo 一时性眩晕
transient vibration 瞬态振动
transient vibration environment 瞬态振动环境
transient X-ray source 暂现X射线源
transistor 晶体(三极)管,半导体(三极)管
transistor logic 晶体管逻辑
transistor radio 半导体收音机
transistor voltage regulator 晶体管(式)电压调节器
transistored radar 晶体管雷达
Transit 子午仪卫星
transit 1.飞越,经过 2.经纬仪,中星仪 3.中天,凌日 4.运输,运输线 5.地面运送,6.通过(管制空域)
transit circle 子午环
transit compass 经纬仪罗盘
transit hall 转载间
transit heading 过渡航向
transit height 过渡飞行高度
transit instrument 中星仪,子午仪
transit load 过路乘载
transit method 中天法
transit mode 过境模式
transit passenger 过境旅客
transit route 转接路由
transit speed 过渡飞行速度
transit stop 中途停留站,经停站
transit time 1.过渡时间,中继时间 2.传递时间 3.渡越时间
transit-fueling 转注
transition 1.过渡,转变 2.改装 3.变调
transition altitude 转换高度
transition architecture 过渡架构
transition area 过渡空域
transition band 过渡带
transition behavior 转变行为
transition boundary layer 过渡边界层
transition condition 过渡态
transition criterion 转移准则
transition curve 1.过渡曲线,介曲线 2.缓和曲线,转变曲线
transition curve location 缓和曲线测设
transition delay 1.过渡延迟 2.转换时间延迟
transition diagram 转移图
transition distance 过渡距离
transition down 向下过渡
transition duct 转移涵道

transition duration	过渡时间
transition envelope	过渡包线
transition flight	过渡状态飞行
transition flow	过渡流
transition front	相变前沿
transition function	转换函数
transition height	过渡高
transition layer	过渡(高度)层
transition length	缓和长度
transition level	过渡(高度)层最低高度
transition maneuver	转移机动飞行
transition manoeuvre	过渡状态机动
transition matrix	跃迁矩阵
transition mechanism	过渡机制
transition mode	1.过渡状态 2.过渡模式 3.临界电流模式
transition model	转换模型
transition module	过渡舱
transition option	转换期权
transition phase	过渡段
transition phenomenon	跃迁现象
transition piece	过渡连接件
transition point	转换点,转移点,转变点,过渡点,临界点
transition prediction	转换预测
transition probability	1.转移概率 2.跃迁概率
transition problem	转换问题
transition process	转变过程
transition radiation detector	穿越辐射探测器
transition radius	转位半径
transition rate	1.跃迁率 2.能量转变率
transition region	渡越区
transition section	渐变段
transition stability	转变稳度
transition state	过渡态
transition strip	联络道,联络地带
transition temperature	转变温度
transition up	向上过渡
transition velocity	过渡流速
transition zone	1.转变区 2.转换区
transitional flow	过渡流
transitional mode	过渡模式
transitional regime	过渡政权
transitional surface	过渡面
transitioning	过渡
translate	平移,移位,调动
translating	直线移动的
translating centerbody	可调进气锥
translating nozzle	转向喷口

translation	平移,调换,转换
translation bearing	滑转轴承
translation frequency	变换频率
translation motion	平动
translation parameters	平移参数
translation rocket	分离系统火箭发动机
translation stage	位移平台
translation table	转换表
translation vector	平移向量
translational control	转译控制
translational engine	平移发动机
translational flight	1.前飞 2.由一地飞至另一地
translational maneuver	平移机动
translational motion	平移运动
translational temperature	平动温度
translatory resistance derivatives	平移空气动力导数
translucent	半透明的
translucent rime	半透明冰
translunar	1.从地球到月球的(空间) 2.月(球轨道)外的
transmeridian flight	跨时区飞行
transmissibility	传递率,传输率,传递能力
transmissible pressure gauge	电远传压力表
transmission	1.传动装置,减速器,变速器 2.(直升机的)主减速器 3.(电磁辐射的)传播 4.传输,信号传输 5.透射
transmission anomaly	传播异常
transmission cable	传输电缆
transmission channel	1.传输通路 2.传输信道
transmission coefficient	1.透射系数 2.传输系数
transmission control character	传输控制字符
transmission cost	1.传输成本 2.输电成本
transmission deviations	传输偏差
transmission device	传动机构
transmission efficiency	传输效率
transmission factor	1.传输因数,传递系数 2.透射因数
transmission grating	透射光栅
transmission high energy electron diffraction	透射高能电子衍射
transmission loss	传输损耗
transmission matrix	透射矩阵
transmission modes	传输模,传输波型
transmission of detonation	传爆
transmission path	传输通道
transmission power	1.传动功率 2.发射功率
transmission probability	传输概率
transmission time	传输时间
transmission zero	传输零点
transmissive density	透射密度

transmissivity 透射率,透射性,透射系数
transmissometer 透射仪,视距测量计,能见度仪
transmit 传送,发送,发射
transmitive atomization 传递雾化
transmittance 1.透射比 2.(电磁辐射的)透射率 3.(光的)透明性,透光度
transmitted pulse 发射脉冲
transmitted wave 1.透射波 2.发射波
transmitter 1.发射机 2.送话器 3.变送器
transmitter chain 发射机链
transmitter power efficiency 发射机功率效率
transmitting antenna 发射天线
transmultiplexer 复用变换器
transom 船上板,(水上飞机浮筒的)尾横板,横档
transonic acceleration 跨声速加速度
transonic area law 跨声速面积律
transonic blading 跨声速叶片组
transonic compressor 跨声速压气机
transonic compressor rotor 跨声速压气机转子
transonic dip 跨声速凹坑
transonic flight 跨声速飞行
transonic flow 跨声速流
transonic flutter 跨声速颤振
transonic operation mode 透明工作方式
transonic rotor 跨声速转子
transonic speed 跨声速度
transonic transport 跨声速运输机
transonic tunnel 跨声速风洞
transonic turbine 跨声速涡轮
transonic wave 超声波
transonic wind tunnel 跨声速风洞
transonic wind-tunnel test 跨声速风洞试验
transonic wing 跨声速机翼
transosonde 平移探仪,平移探空气球
transparency 1.透明,透明度 2.幻灯片 3.有图案的玻璃
transparent 透明的
transparent ceramics 透明陶瓷
transparent code 透明码
transparent contention 透明竞争技术
transparent ferroelectric ceramics 透明铁电陶瓷
transparent foil 网纹片
transparent membrane 透明膜
transparent plasma 透明等离子体
transparent positive 透明正片
transparent radar reflection coating 透明雷达反射涂层
transparent structure 透波结构
transpiration 1.蒸发,汽化 2.排出,流出 3.流逸
transpiration cooling 发散冷却,发汗冷却,蒸发冷却
transpiration cooling material 发散冷却材料
transplanetary space 行星外空间
trans-Plutonian planet 冥外行星
transponder 应答机,应答器,脉冲转发器
transponder beacon 应答信标
transponder jamming 应答式干扰
transponder operating capacity 应答台工作容量
transponder technology 射频应答器技术
transport 1.运输机 2.运输,运输工具
transport aircraft/airplane 运输机
transport and hoisting load 运输和起吊载荷
transport category 运输种类
transport category aircraft 运输机类的飞机
transport condition 运输条件
transport equation 迁移方程,传递方程,输运方程
transport factor 转运因子
transport joint 1.运输枢纽 2.拆装主接头,大部件接头
transport lag 传输延迟
transport layer 传输层
transport mechanism 输片机构
transport process 1.输运过程 2.迁移过程 3.传送过程
transport property 输运性质
transport protocol 传送协议,点到点传输协议
transport rate 输送率
transport speed 输送速度
transport test 运输试验
transport theorem 流体传输定律
transport wander (陀螺的)视进动,视在漂移
transportability (可)运输性
transportability problem item 运输性问题产品
transportable air defence operations center 移动式防空作战中心
transportable earth station 可移动式地球
transportation 1.运输 2.运输系统 3.运输工具 4.流放
transportation environment 运输环境
transportation layer 传输层,传送层
transportation rate 运输费用
transportation system 1.运输系统 2.交通物流系统
transportation test 运输试验
transporter 运载工具,运输装置,运输器,一网络传输器
transporter tower (机动基地的)飞艇系留塔
transporter-erector-launcher 运输起竖发射车
transpose 1.调换 2.移项 3.颠倒顺序
transputers 高速运算集成电路发射机
transuranium elements 铀后元素,超铀元素
transversal wave 横波
transversality 1.横截性 2.横断

transversality condition 横截性条件
transverse 1.横向的 2.横梁,横轴
transverse acceleration 横向加速度
transverse antenna fuze 横向天线引信
transverse arrangement 横向布局
transverse axis 横轴
transverse bulkhead 横隔板,隔框
transverse direction 横向
transverse displacement 横向位移
transverse electric and magnetic mode 横电磁波模式
transverse electric mode 横电波模式
transverse filter 横向滤波器
transverse flux 横向磁场
transverse force 横向力
transverse injection 1.横向喷注 2.横向射流
transverse latitude-longitude inertial navigation system 横向经纬度惯导系统
transverse load 横向载荷
transverse magnetic mode 横磁波模式
transverse member 横向构件
transverse Mercator 横轴墨卡托投影
transverse mode 横模
transverse mode locking 横模锁定
transverse mode selection 横模选择
traverse network 导线网
transverse overlap 横向重叠率
transverse propagation 横向传播
transverse shear 横切力
transverse span 横向跨距
transverse strengthening 横向强化
transverse velocity 横向速度
transverse vibration 横向振动
transverse wave 横波
transverse-flow effect 横向流效应
TRAP 1. Terminal Radiation Airborne Measurements Program 机载终端辐射测量计划(此词不能缩写成 TRAMP) 2. Tracker Analysis Program 跟踪仪分析计划
trap 1.隔栅,陷井 2.收集器 3.药柱挡板 4.挡焰器 5.陷波器 6.阱 7.拦阻成功,拦阻着舰 8.陷波电路 9.油气收集器 10.过渡器 11.集油腔 12.防回火滤网
trap state 陷波态
trap weight 拦阻着舰的最大允许重量
TRAPATT Trapped Plasma Avalanche Triggered Transit 俘获等离子体雪崩触发渡越
trapeze bar 1.吊篮横杆 2.高架横杆
trapeze beam 悬吊梁
trapezium distortion 不规则四边形失真,不规则四边形畸变,梯形失真

trapezoid 1.梯形 2.不规则四边形
trapezoidal distortion 梯形畸变
trapezoidal modulation 梯形调制,不规则四边形调制
trapezoidal section 梯形截面
trapezoidal velocity 梯形速度
trapezoidal wave 梯形波
trapezoidal wing 梯形机翼
trapped electrons 捕获电子
trapped fuel 不能抽出的燃油
trapped orbit 捕获轨道
trapped particle 捕获粒子
trapped plasma avalanche triggered transittime diode 俘越二极管,俘获等离子体雪崩触发渡越时间二极管
trapped protons 捕获质子
trapped radiation 捕获辐射
trapped radiation model 捕获辐射模式
trapping 捕获
trapping region 捕获区
travel 1.旅行 2.游历 3.漫游
travel agent 旅行社,旅行代理商
travel distance 1.行进距离 2.流动距离 3.疏散距离
travel documents 旅行文件
travel group charter 旅行团包机
travel pod 旅行吊舱
travel time 传播时间,行程时间
travel time curve 走时曲线
traveler 1.旅行者 2.旅客 3.旅行推销员
traveling solvent zone method 移动溶液区溶法
traveling standard 搬运式标准
traveling wave 行波
travelling wave aerial 行波天线
travelling wave antenna 行波天线
traveling wave flutter 行波光颤振
traveling wave tube 行波管
travelling ionospheric disturbance 电离层干扰
travelling wave maser 行波激射器
travelling wave phototube 光电行波管
travelling wave tube 行波管
travelling wave tube amplifier 行波管放大器
traverse 1.转动 2.曲折航行 3.移测法,导线测量法 4.(测量用的)导线,横截线 5.横穿风 6.移位测量,沿横向(或径向)测气流参数分布
traverse angle 导线折角
traverse error 方向瞄准误差
traverse firing angle 发射旋转角,横向射角
traverse flying 沿曲折航线飞行
traverse leg 导线边
traverse point 导线点
traverse survey 导线测量

traverser 1.否认者 2.横越者 3.转盘
tray 1.托盘 2.文件盒 3.隔底匣 4.(无线电的)发射箱
TRB Technical Review Board 技术评审委员会
TRC 1.Thrust Rating Computer 推力额定值计算机 2.Thrust Reverser Cowl 反推整流罩
TRCS Test and Repair Control System 测试与修理控制系统
TRD 1.Torsional Resonance Damper 扭转谐振阻尼器 2.Transit Routing Domain 通场区域
TRE 1.Target-Rich Environment 多目标环境 2.Telecommunications Research Establishment 通信研究院 3.Transponder Equipment 应答机设备
tread 1.轮距 2.胎面,行驶面 3.轨迹 4.踏板,梯级,级宽
treadle (脚)踏板
treadlemill 1.跑台 2.跑步机
treat 1.治疗 2.对待 3.探讨 4.视为
treatment 1.治疗,疗法 2.处理,对待
treatment after plating 镀后处理
TREDS Tactical Reconnaissance Exploit Demonstration System 战术侦察研究验证系统
TREE 1.Test and Repair of Electronic Equipment 电子设备的试验和修理 2.Transient Radiation Effects on Electronics 瞬时辐射对电子部件的影响
tree code 树码
tree expansion 展开树形图
tree frontier 树缘
tree grammar 树文法
tree language 树语言
tree representation 树表示
tree structure 树结构
tree topology 树形拓扑结构
tree-top flight 掠地飞行
trefoil 三伞集束
TREL Transient Radiation Effects Laboratory 瞬时辐射效应实验室
trellis code 格码
trellis structure 篱状结构,格子结构
TREM Tropics Rain Exploring Mission 热带降雨探测卫星
trench 1.沟槽密封装置 2.槽形过道
trenched 槽形通道的
trend 1.趋势,倾向 2.走向
trend analysis 趋势分析
trend drift 趋向漂移
trend line 1.趋势线 2.走向线
trend method 趋势法
trending 趋势分析
TRF 1.Threat Radar Frequency 威胁雷达频率 2.Tuned Radio Frequency 调谐无线电频率
TRIAC Test Resources Improvement Advisory Council 试验场所改进咨询委员会
Triad 1.三体卫星 2.三位一体战略力量
triad 1.三元组,三素组,三脉冲组 2.三价原子,三价基 3.三次相似面晶体
triage 伤员分类,紧急调查伤亡情况
trial 1.试验,试用,试车 2.试验性的,试制的
trial function 试探函数,测试函数
trial maneuver 试航操纵
trial vector 试向量
triaminotrinitrobenzene 三氨基三硝基甲苯
triangle 1.三角(形) 2.三角关系
triangle constraint 三角形约束
triangle method 三角法
triangle of velocities 速度三角形
triangular lattice 三角晶格
triangular libration point 三角平动点
triangular matrix 三角矩阵
triangular membrane 三角形膜片
triangular parachute 三角形降落伞
triangular pattern 三角形航线
triangular pulse 三角脉冲
triangular system 三角形系统
triangular velocity 三角形速度
triangular window 三角形窗
triangulation 三角测量
triangulation balloon 三角测量标志气球
triangulation chain 三角链
triangulation location 三角定位法
triangulation station 三角测量点
triaxial angular accelerometer 三轴角加速度计
triaxial control 三轴控制
triaxial ellipsoid 三轴椭球体
triaxial fluxgate magnetometer 三轴饱和式磁强计
triaxial magnetometer 三轴磁强计
triaxials 1.三轴的 2.三维的 3.空间的
triazine 三嗪
triboelectrification 摩擦生电
tribology 摩擦学
tribometer 摩擦计
tribrach 三角基座
tributary station 分支局
tributyl phosphate 磷酸三丁酯
tri-camera photography 三相机照相(侦察照相时三架相机同时摄影,照相有部分重叠)
trichlorethylene 三氯乙烯
trichromatic 三色的
trickle charge 连续补充电,点滴充电法,涓流充电

trickle charger　连续补充电器
trickle loading　鱼贯登机
tricycle　前三点的
tricycle landing gear　前三点起落架
Trident　三叉戟（1.美国潜地远程弹道导弹 2.英国一种运输机的名称）
TRIG　Trigger 触发器，触发脉冲
trig point　对称点
trigger　1.扳机，射击按钮，弹射手柄 2.起动器，触发器，引爆器 3.触发脉冲 4.触发，启动
trigger bomb　着发炸弹，触发炸弹
trigger button　1.起动按钮 2.射击按钮，发射按钮
trigger delay　触发延迟
trigger electrode　触发极
trigger flip-flop　计数触发器
trigger for H-bomb　氢弹的起爆弹
trigger gap　起动隙缝
trigger gear　触发装置
trigger mechanism　1.触发机构，打火机构 2.引信
trigger switch　扳机电门，射击按钮
trigger tube　触发管
triggering　触发
triggering device　发射装置，扳机
triggering transformer　触发变压器
triglycine sulfide　硫酸三甘肽
trigonometric function　三角函数
trigonometric identity　三角恒等式
trigonometric leveling　三角高程测量
trigonometric leveling network　三角高程网
trigonometric parallax　三角视差
tri-inspection system　三检制
trijet　三发喷气式飞机
trike　1.三轮机架 2.用三轮机架的超轻型飞机
trike unit　三轮机架
trilateration　三边测量
trilateration network　三边网
trilateration radar　三基地雷达
trilateration range and range rate system　三边测距和测距变化率系统
trilateration survey　三边测量
trillion　1.万亿，兆，太（拉）
TRIM　Time-Related Instruction Management 及时指令管理（系统）
trim　1.调整，修整 2.调谐，微调，垫整 3.精密修整 4.配平，配平状态 5.配平度 6.俯仰角，纵倾，吃水差 7.配平机构，配平装载 8.座舱内部修饰物
trim air　空调热气
trim angle　1.纵倾角 2.配平角
trim angle of attack　配平攻角

trim coil　调整线圈
trim condition　俯仰船况
trim control　配平操纵
trim cord　调整片操纵索
trim die　修整模
trim drag　配平阻力
trim for take off　起飞配平
trim panel　装饰面板
trim pitch　俯仰配平
trim point　平衡点
trim position　配平位置
trim size　完工尺寸
trim tab method　调整片法
trim tank　配平油箱
trim template　修整用的型板
trim thrust　动力配平
trimetrogon camera　三镜头航空相机
trimetrogon photography　三镜头摄影术
trimmer　调整片，修正片，配平装置
trimmer capacitor　微调电容器
trimmer potentiometer　微调电位器
trimming effect mechanism　调整片效应机构
trimming inductor　微调电感器
trimming strip　修正片
trimming system　配平系统
trimming tanks　配平油箱
TRMP　Technical Risk Management Plan 技术风险管理大纲
TRMS　Technical Requirements Management System 技术要求管理系统
trinitrate　三硝酸酯
trioxide　三氧化物
trip　1.一次航行 2.分离机构，断开机构 3.启动电路保护系统 4.飞行，航班，航程
trip advisory　飞行简报
trip assignment　1.交通流分配 2.旅次分派
trip check　货运实施前检查
trip demand　出行需求
trip fuel　一次飞行所需燃料
trip insurance　航班保险
trip line　1.脱扣线 2.浮锚拉索，锚爪拉索
trip thread　引燃线
trip time　一次飞行时间
tripartite system　三重雷达系统
trip-free circuit breaker　自由释放断电器，自由脱扣断电器
triphibian　1.三栖飞机 2.三军飞机
triplane　三翼机
triple　1.三座椅组件 2.三倍的，三重的，三合一的

triple collision	三体碰撞
triple ejector rack	弹射式三弹挂弹架
triple galaxy	三重星系
triple junction	1.三联点 2.三重接头 3.三向连接构造
triple launcher	三联发射装置
triple modular redundancy	三重余度
triple modulation telemetering system	三重调制遥测系统
triple point	1.三相点 2.三重点
triple probe	三探针
triple release	三连投
triple seat	三联座椅
triple star	三合星
triple-base propellant	三基推进剂
triple-deck	三层
tripledecke	三层舱大型客机
triple-junction cell	多结伏电池
triple-output GPU	三输出地面电源设备
triplet	1.三个一组 2.三连音符 3.三元组中的一个
triple-tandem	三重串联
triple-tube	三管
triplex	三余度并联系统
triplex concept	三余度原则
tripod	1.三喷口推力转向涡扇发动机 2.三脚架,三角架,三脚支撑物
tripropellant	三(组)元推进剂
tripropellant rocket engine	三组元(推进剂)火箭发动机
tri-scanner antenna	三扫描器天线
trisonic wind tunnel	三声速风洞
tristar	三星牌旅行车
tristate buffer	三态缓冲器
tristate logic	三态逻辑
tritium	【化】氚(化学元素,符号 T),超重氢
tritium unit	氚单位
Triton	1."特里顿"(法国海军雷达) 2.海卫一
trivalent element	三价元素
trivial solution	1.明显解 2.平凡解
TRK	track 航迹
TRL	Transistor Resistor Logic 晶体管电阻逻辑
TRM	1. Technical Requirement Manual 技术规范,技术要求规程 2. Time Release Mechanism 定时释放(投放)机构 3. Transmission Resource Management 传输资源管理
TRMM	Tropical Rainfall Measuring Mission 热带降雨测量卫星
trochoid	旋轮线,摆线
trochoidal focusing mass spectrometer	余摆线聚焦质谱仪
trochoidal vacuum pump	余摆线真空泵
troland	特罗兰得(网膜照度单位)
troll(ing)	随机航线飞行
trolley	机场车辆,运输车,空中吊运车
trombone	可调 U 形同轴线,可调 U 形波导节
troop	1.军队 2.组 3.群 4.多数
troop parachute	伞兵伞
troop-carrying helicopter	部队运输直升机
troostite	屈氏体
tropical air mass	热带气团
tropical conditions	热带条件
tropical continental	热带大陆性
tropical earth resource satellite	热带地球资源卫星
tropical exposure	高温曝晒
tropical maritime	热带海洋性
tropical month	分至月
tropical orbit	回归轨道
tropical revolving storm	热带风暴
tropical trials	热带条件试验
tropical year	回归年
tropopause	对流层顶
troposcatter communication	对流层散射通信
troposphere	对流层
tropospheric delay correction	对流层延迟修正
tropospheric refraction correction	对流层折射修正
tropospheric scatter	对流层散射
tropospheric scatter communication	对流层散射通信
troubleshooting	故障判断,故障诊断
troubleshooting time	故障判断和排除时间
trough	1.低压槽 2.无引力点 3.水槽,油槽 4.曲线上的凹点 5.激波后的尾流
trouser	起落架支柱整流罩
TRP	1. Terminal Rendezvous Point 终端会合点 2. Threat-Recognition Processor 威胁识别处理器 3. Thrust-Rating Panel 推力状态仪表板 4. Triple Propeller 三螺旋桨 5. Tuition Refund Program 助学金计划
TRR	1. Test Readiness Review 试验准备状态评审 2. Test Rejection and Repair 试验驳回和补救 3. Total Removal Rate 总拆卸率 4. Tyre Rolling Radius 轮胎滚动半径
TRREQ	Track Required 应飞航迹(航迹角)
TRRR	Trilateration Range and Range-Rate 三边距离和目标距离变化率
TRS	1. Tactical Reconnaissance Sensor 战术侦察传感器 2. Tactical Reconnaissance Squadron 战术侦察中队 3. Tactical Reconnaissance System 战术侦察系统 4. Teleoperator Retrieval System 遥控机构回收系统 5. Tetrahedral Research Satellite 四面体研究卫星

6. Transverse Rupture Stress 横向断裂应力
7. Tropical Storm 热带风暴

TRSA Terminal Radar Service Area 航站雷达勤务区

TRT turret 炮塔

TRT 1. Takeoff Rated Thrust 起飞额定推力 2. Total Running Time 总运转时间

TRTG Tactical Radar Threat Generator 战术雷达威胁产生器

TRTT Tactical-Record Traffic Tele-typewriter 战术记录空中交通电传打字机

TRU Transformer Rectifier Unit 变压整流装置,变压整流器

truck 车架式起落架

truck-bed height 卡车车厢高度

TRUD Time Remaining Until Dive 俯冲前剩余时间

true airspeed 真空速

true altitude 真高,真高度

true anomaly 真近点角

true bearing 真方位

true centrifugal casting 离心铸造

true color image 真彩色影像

true cost 真实成本

true course 真航向

true cut 正确的切割

true equator 真赤道

true equinox 真春分点

true heading 真航向

true height 真高

true latitude 真实纬度

true meridian 真子午线,真子午圈

true model 真实模型

true north 真北

true place 真位置

true pole 真极

true position 真实位置

true power 实际功率

true prime vertical 真东西圈,真卯酉圈

true rotation 真自转周

true state 真状态

true system 真实系统

true temperature 真实温度

true track 真航迹角

true value 真值

true variance 真变异数

true vertical 真垂线

true-historic display 真实历程显示器

trumpet 1. 喇叭 2. 喇叭声

truncated nozzle 截短喷嘴

truncation 1. 截断 2. 切掉顶端

truncation error 截尾误差,截断误差

trunk 1. 通风道,主管 2. 本体,主干,主线 3. 干线飞机,干线航空公司

trunk air carrier 干线航空公司

trunk carriers 航空公司集团

trunk liner 干线客机

trunk offering 长途接入

trunk route 1. 干线 2. 战略空运路线 3. 行李舱,存物间

trunk service 地面运输业务

trunk subscriber 中继用户

trunked radio system 集群制无线通信系统

trunks 主要航空公司集群

trunnion 1. 耳轴 2. 炮耳

truss 桁架,构架

truss element 1. 桁架单元 2 桁架杆元

truss structure 桁架结构

TRUST Television Relay Using Small Terminals 使用小型终端的电视中继

trust region 信赖域

trust region algorithm 信赖域算法

truth table 真值表电子数据处理技术

TRV Thermal Relief Valve 热控释压活门

TRVR Touchdown Runway Visual Range 接地时跑道能见度

try 1. 尝试 2. 努力 3. 试验

try-again missile 能再次进入的导弹

tryptique 1. 空运入境免税证书 2. 飞机加油及其他勤务用的国际信用卡

TS 1. Technical Solution 技术解决方案 2. Technical Specification 技术规格,技术规范 3. Tensil Strength 抗拉伸强度 4. Terminology Standard 术语标准 5. Testing Standard 试验标准 6. Thermal Switch 热点门 7. Throttle Split 油门分叉 8. Thunderstorm 雷暴 9. Time Source 时间源 10. Tracking System 跟踪系统 11. Transattack Survivability 受攻击后的生存力 12. Transport Service 运输服务

TSA 1. Tactical Stealth Aircraft 战术隐形飞机 2. Tail Strike Assembly 尾端放电组件 3. Technical Service Agreement 技术服务协议

TSAP Transport Service Access Point 运输服务接近点

TSAT Technology Satellite 技术卫星

TSC 1. Tactical Satellite Communication 战术卫星通信 2. Thermal Stress Cracking 热应力裂纹 3. Thermal Surface Coating 防热表面涂层 4. Transporable Satellite Communication 移动卫星通信 5. Transportation Systems Center 运输系统中心(美国运输部)

TSCC Telemetry Standards Coordination Committee 遥测标准协调委员会

TSCLT Transportable Satellite Communications Link Terminal 移动式卫星通信链路终端

TSD Tactical Situation Display 战术空情显示

TSDF Target System Data File 目标系统数据文件存储器

TSDU Transport Service Data Unit 运输服务数据装置

TSE 1. Target State Estimator 目标状态估计器 2. Total System Error 系统总误差

TSEM Total Solar Energy Monitor 总太阳能监测器

TSF 1. Tactical Strike Fighter 战术攻击战斗机 2. Technical Supply Flight 技术供应飞行 3. Time to System Failure 出现故障前的时间 4. Two-Seater Fighter 双座战斗机

TSFC Thrust Specific Fuel Consumption 推力燃油消耗率(耗油率)

TSH Thermal Suppression Head 热力抑制压头

TSI 1. Time Since Installation 安装后使用时间 2. True (air) Speed Indicator 真空速表 3. Turn and Slip Indicator 转弯侧滑仪

Tsiolkovski formula 齐奥尔科夫斯基公式

TSIP Trimble Standard Interface Protocol 可调标准界面议定书

TSIR Total System Integration Responsibility 系统一体化总职责

TSLW Tyre Speed Limit Weight 轮胎速度限重

TSM 1. Time Scheduled Maintenance 定期维修 2. TIROS Stratospheric Mapper 泰罗斯卫星平流层探测资料映射模块

TSMO Time Since Major Overhaul 大翻修后(使用)时间

TSN Time Since New 从新算起的使用时间,制造出厂后(使用)时间

TSO 1. Technical Service Order 技术服务法规 2. Techincal Standard(s) Order 技术标准法规,技术标准规定(英国民航局) 3. Time Since Overhaul 翻修后工作时间,翻修后使用时间

TSOA Technical Standard Order Approval 技术标准规定项目批准书

TSOR Tentative Specific Operational Requirement 暂行具体使用要求,意向性具体使用要求

TSP Turret Stabilized Platform 炮塔稳定平台

TSPI Time, Space, Position Information 时间、空间、位置信息

TSPR Total System Performance Responsibility 系统性能可靠性

TSR 1. Tactical Strike Reconnaissance 战术、攻击与侦察 2. Telephone Service Representative 电话销售代表 3. Terminate and Stay Resident 终止并驻留 4. Torpedo Spotter Reconnaissance 鱼雷弹着观察员侦

察 5. Trade Study Report 贸易研究报告 6. Twin Side-by-side Rotor 横向并列双旋翼

TSS 1. Tactical Strike System 战术(对地)攻击系统 2. Technology Suport and Services 技术支持和服务 3. Tethered Satellite System 系留卫星系统,绳系卫星系统 4. Thrust Sensitive Signal 推力敏感信号 5. Time Sharing System 分时系统 6. Tromso Satellite Station 特罗姆瑟卫星站 7. Trouble Shooting System 故障搜索系统 8. Tunable Solid State 可调固态

TSSC Technical Supply Subcommittee 技术供应分组委员会

TSSCS Tactical Synchronous Satellite Communication System 战术同步卫星通信系统

TSSLS Titan Standardized Space Launch System 大力神标准化航天发射系统

TSSTS Tactical Sight System Training Simulator 战术瞄准系统训练模拟器

TST 1. Threshold Sampling Time 抽样起始时间 2. Transonic Transport 超声速运输机 3. Trouble Shooting Time 故障检修时间

TSTM Time Source Transition Module 时间源转换模块

TSTO Two Stage To Orbit 两级入轨

TSTR tester 检验器,测定器

Tsukuba Space Center 筑波航天中心(日本)

tsunami 海啸

TSUS Tariff Schedules of the US 美国(关)税率表

TSV Temperature Safety Valve 温度安全阀

TSW 1. Tactical Supply Wing 战术供应联队 2. Temperature Switch 温度开关 3. Test Switch 试验开关 4. Transfer Switch 转换开关

TT 1. Tactical Training 战术训练 2. Target Time 到达目标时间 3. Target Towing 拖靶的 4. Task Ticket 工作记录单 5. Tele-Typewriter 电传打字机 6. Test Tools 测试工具 7. Total Temperature 总温 8. Total Time 总使用时间 9. Tracking Telescope 跟踪望远镜 10. Trade Test 使用试验 11. Transit Time 飞越时间 12. Transportation Time 运输时间 13. Trim Tab 配平调整片 14. Trim Template 修整型板 15. Type Testing 型式试验 16. Turnround Time 再次出动准备时间

TT&C Tracking, Telemetry and Control 测控、跟踪、遥测与控制

TT&C application software 测控应用软件

TT&C coverage 测控覆盖率

TT&C event 测控事件

TT&C network simulation 测控网仿真

TT&C of boost phase 主动段测控

TT&C of initial phase 起始段测控

TT&C of injection phase 入轨测控

TT&C of in-orbit phase 运行段测控
TT&C of return phase 返回段测控
TT&C plan 测控计划
TT&C procedure 1.测控事件序列 2.测控程序
TT&C real-time software 测控实时软件
TT&C requirement 测控要求
TT&C simulation system 测控仿真系统
TT&C software 测控软件
TT&C station 测控站
TT&C station computer 测控站计算机
TT&C subsystem 测控分系统
TT&C system 1.跟踪测控系统 2.测控系统
TT&C system design 测控总体设计
TT&C system engineering 测控系统工程
TT&C task 测控任务
TT&C task analysis 测控任务分析
TTA Total Transducer Assembly 全压传感器组件
T-tail T形尾翼
TTC 1. Technical Training Center 技术训练中心 2. Technical Training Command 技术训练司令部 3. Top-Temperature Control 最高温度控制 4. Tracking, Telemetry and Command 跟踪、遥测和指挥
TTC&M Tracking, Telemetry, Command and Monitoring 跟踪、遥测、指挥和监控
TTCC Telemetering, Timing, Command and Control 遥测、定时、指挥和控制
TTCP The Technical Cooperation Program 技术合作计划
TTD Ttemporary Travel Document 临时传阅文件
TTF 1. Tanker Task Force 加油机特遣部队 2. Target Towing Flight 拖靶飞行 3. Test To Failure 破坏试验 4. Threat Training Facility 威胁训练装置 5. Time to Failure 发生故障时间,用到失效的时间
TTFF Time To First Fix 首次定位时间,首次调整期限
TTFO Time To First Overhaul 首次翻修期限
TTG Time To Go 发射时间,起飞时间,剩余时间
TTH Tactical Transport Helicopter 战术运输直升机
TTP Time To Protection 保护成功前时间
TTS 1. Test and Training Satellite 试验和训练卫星 2. Thrust Termination System 推力终止系统 3. Time To Station 到台时间,到达时间
TTSMOH Total Time Since Major Overhaul 大翻修后总(使用)时间
TTSN Total Time Since New 制造出厂后总(使用)时间
TTT 1. Tailplane Trimming Tank 平尾平衡油箱 2. Telmetry Telecommand Tracking 遥测遥控跟踪 3. Time Temperature Tolerance 容许温度时间 4. Total Temperature Transducer 总温传感器
TTU 1. Thrust Termination Unit 推力终止装置 2. Triplex Transducer Unit 三联传感器
TTV Teleoperator Transfer Vehicle 遥操作器转移飞行器
TTW 1. Time of Tension to War 战争前的紧张时间 2. Transition To War 转入战时体制
TTWS Terminal Threat Warning System 航站威胁警戒系统
TTY Teletypewriter 电传打字机
tub 1.浴盆形构件 2.副油箱,外挂油箱
Tuballoy 贫化铀,铀废料
tube 1.大型现代喷气客机 2.内胎 3.真空管,热离子管 4.电视机
tube area 管板
tube assembly 管路总成
tube bank 1.管束,管排 2.管族
tube coil 盘管
tube combustor 管形燃烧室
tube diameter 管口直径
tube erosion 炮膛磨蚀
tube exit 管道出口
tube flow forming 筒形变薄旋压
tube grain 管形药柱
tube height 管高度
tube length 管长
tube of diameter 管径
tube oil 内流油
tube orifice 炮管导气孔
tube rake 排管
tube resonator 管式谐振器
tube shear spinning 1.锥形变薄旋压 2.剪切旋压
tube size 管尺寸
tube spinning 筒形变薄旋压
tube tunnel 管风洞
tube type arc heater 管状电弧加热器
tube wall 管壁
tube yawmeter 管式偏航计
tube-filled discharger 导管式发射装置
tubeless tyre 无内胎轮胎
tubercle 1.结节 2.小瘤 3.小块茎
tuberculation 结瘤腐蚀
tubing 管系,管路,管道
tubo-annular 环管式燃烧室
Tubsat Technical University of Berlin Satellite 柏林技术大学卫星
tubular combustor 1.分管燃烧室 2.单管燃烧室
tubular rivet 管状柳钉,空心柳钉

tubular structure 管状结构
tubular thrust chamber 管束式推力室
TUCD Tool Unit Configuration Definition 工装单元构型定义
tuck 1.食物 2.船尾突出部 3.缝褶子 4.抱膝式跳水 5.活力 6.鼓声
tuft 1.一簇 2.丛生植物 3.一丛
tufting 粘贴丝线
tug 1.牵引,拖曳,拖航 2.牵引车 3.牵引机,拖靶机 4.空间拖船
tulip valve 漏斗形排气门
tumble 1.滚转,翻转,翻倒,滚翻 2.滚筒清理,滚筒抛光 3.前后滚翻
tumble limit 翻转限制器
tumble system 翻转系统
tumblehome 内倾
tumblehome line 内倾线
tumbler 1.拨动式开关 2.机心 3.关锁臂 4.回向机构 5.转筒,滚筒 6.转臂,拔杆开关 7.齿轮换向器
tumbler switch 转换开关
tumbling loop 翻滚回路
tumbling test 翻滚试验
tunable beam approach 可调谐波束控
tune 调谐,协调
tuned amplifier 调谐放大器
tuned oscillation gyro 调谐振动陀螺
tuned oscillator 调谐振荡器
tuned reflectometer 调配反射计
tuned rotor gyro 调谐转子陀螺
tuned speed 调谐速度
tungsten 【化】钨(化学元素,符号 W)
tungsten carbide 碳化钨
tungsten carbide core 碳化钨弹心
tungsten electrode 钨电极,钨极
tungsten emitter 钨质发射体
tungsten nozzle 钨喉衬
tungsten oxide 氧化钨,二氧化钨
tuning 调谐,调整
tunnel 1.隧道 2.风洞 3.轴向整流罩
tunnel diode 隧道二极管,江崎二极管,透纳二极管
tunnel effect 隧道效应
tunnel floor 隧道底板
tunnel pipe 隧道管
tunnel shock 风洞激波
tunnel speed 风洞气流速率
tunnel test 风洞试验
tunnel vision 管状视界,管状视野,管视
tunnel wall 风洞壁
tunneling hot electron transfer amplifier 隧穿热电子转移放大器
tunneling wave 隧道波
TUP Technology Utilization Program 技术转让计划
TURB turbulence 紊流,湍流
turbidity 混浊度,混浊性
turbine 涡轮(机),叶轮机
turbine aerodynamics 涡轮空气动力学
turbine bearing 涡轮轴轴承
turbine blade 涡轮叶片
turbine booster 涡轮增压器
turbine cascade 涡轮叶栅
turbine chamber 水轮机室,涡轮室
turbine cooling 涡轮冷却
turbine design 涡轮设计
turbine designer 水轮机设计师
turbine disc 涡轮盘
turbine discharge 涡轮机流量
turbine efficiency 涡轮效率
turbine end 涡轮端
turbine engine 汽轮机,涡轮发动机
turbine engine airborne integrated data system monitoring 燃气轮发动机机载综合数据系统监视
turbine engine monitoring system 燃气轮发动机监视系统
turbine entry 1.透平入口 2.涡轮入口
turbine entry temperature 涡轮进口温度
turbine exit 涡轮出口
turbine flow 涡轮流量
turbine inlet 涡轮机进口
turbine inlet temperature 涡轮进口温度
turbine map 涡轮特性图,涡轮特性曲线
turbine model 涡轮机模型
turbine noise 舰艇机舱噪声
turbine nozzle 1.导向器 2.涡轮静子 3.涡轮喷嘴环
turbine outlet 透平出口
turbine passage 透平通流部分
turbine performance 涡轮性能
turbine power 涡轮功率
turbine pressure 涡轮压力
turbine pressure ratio 涡轮压力比
turbine rotor 涡轮转子
turbine section 涡轮段
turbine shroud 涡轮叶冠
turbine speed 涡轮转速
turbine stage 涡轮级
turbine stator 涡轮静子
turbine vane 涡轮叶片
turbine wheel 涡轮转子
turbine-airscrew unit 涡轮螺旋桨组

turbo	涡轮增压机
turbo molecular pump	涡轮分子泵
turboblower	涡轮式鼓风机
turbocharger	涡轮增压器
turboexpander	膨胀涡轮,制冷涡轮
turbofan	涡轮风扇
turbofan engine	1.涡轮风扇发动机 2.内外涵发动机
turbofan jet	涡轮风扇飞机
turbofan prop	涡轮桨扇
turbogenerator	涡轮发电机
turbojet	涡轮喷气,涡轮喷气飞机
turbojet cycle	涡轮喷气发动机循环
turbojet engine	1.涡轮式喷射引擎 2.涡轮喷气发动机 3.喷气涡轮机
turbojet engine	涡轮喷气发动机
turbomachine	涡轮机
turbomachinery	叶轮机械,叶轮机
turbomachinery aerodynamics	涡轮机械空气动力学
turbomachinery design	涡轮机设计
turbomachinery designer	涡轮机械设计师
turbomachinery system	涡轮机械系统
turbomachinery test	涡轮机械测试
turbopause	湍流层顶
turboprop	涡轮螺桨,涡轮螺旋桨飞机
turboprop engine	涡轮螺旋桨发动机
turbopump	涡轮泵
turbopump feed system	涡轮泵输送系统,泵压式输送系统
turbopump power density	涡轮泵比功率
turbopump system cycle efficiency	涡轮泵系统循环效率
turbopump-feed liquid rocket engine	泵压式液体火箭发动机
turbopump-feed system	泵压式供应系统
turboramjet	涡轮冲压式喷气发动机
turborocket	涡轮火箭
turboshaft	涡轮轴
turboshaft engine	涡轮轴发动机
turbosphere	湍流层
turbostarter	涡轮起动机
turbosupercharger	涡轮增压器
turbulence	1.紊流,湍流 2.紊流率,湍流度
turbulence characteristic	湍流特征
turbulence closure	1.湍封闭 2.乱流闭合
turbulence closure model	湍封闭模式
turbulence dispersion	液滴的湍流扩散
turbulence energy	紊流动能
turbulence generation	紊流生成
turbulence intensity	1.湍流强度 2.扰动强度
turbulence kinetic energy	湍流动能
turbulence length	湍流长度
turbulence length scale	湍流长度尺度
turbulence level	1.湍流强度 2.湍流级
turbulence model	湍流模型
turbulence modeling	湍流模型
turbulence number	紊流度
turbulence plot	紊流(标)图
turbulence screen	湍流网
turbulence sphere	湍流球
turbulence statistic	湍流统计场
turbulence structure	湍流结构
turbulent boundary layer	1.湍流边界层 2.紊两面层
turbulent bursts	流突发
turbulent clouds fear	对紊流云惧怕
turbulent combustion	紊流燃烧
turbulent diffusion	湍流扩散
turbulent diffusivity	湍流扩散系数
turbulent dispersion	湍流扩散
turbulent dissipation	湍流耗散
turbulent exchange	湍流交换
turbulent flame	紊流火焰,紊流燃烧
turbulent flow	1.湍流 2.紊流
turbulent fluctuation	1.紊流脉动 2.紊动
turbulent heating	湍流加热
turbulent intensity	湍流强度
turbulent jet	紊动射流
turbulent kinetic energy	湍流动能
turbulent length	动荡长度
turbulent length scale	湍流尺度
turbulent mixing	紊流混合
turbulent model	紊流模型
turbulent motion	1.涡动 2.湍动
turbulent region	紊流区
turbulent screen	整流格栅,整流网
turbulent separation	湍流分离
turbulent shear	湍动剪切
turbulent spot	湍流斑
turbulent stress	湍流应力
turbulent structure	湍流结构
turbulent transition	层流紊流过渡段
turbulent transport	湍流输运
turbulent velocity	紊流速度
turbulent viscosity	紊流黏度
turbulent flows	湍流
turbulent wake	1.湍流尾流 2.紊流尾流
Turing machine	图灵机
Turing test	图灵实验
turkey	设计低劣、性能很差的飞机

Turkish air force　土耳其空军
Turksat　土耳其卫星
turn　1. 盘旋 2. 转弯 3. 变化
turn and bank indicator　转弯侧滑仪
turn and slip　转弯侧滑仪，针球仪
turn angle　1. 转角 2. 转角度 3. 回转测角
turn around ranging station　往返测距站，回转测距站
turn condition　转弯状态
turn coordination　转弯协调
turn coordinator　转弯协调器
turn error　回转误差
turn process　转体过程
turn radius　转弯半径
turn rate　1. 转弯角速度 2. 旋转率
turn ratio　1. 匝数比 2. 匝比
turn table　回转平台
turn the control wheel　压杆
turnaround　1. 再次飞行准备，回程准备 2. 再次出动准备时间，停放时间 3. 当日往返（任务）
turnaround cycle　往返周期
turnaround maintenance　再次出动前维修
turnaround time　1. 处理时间 2. 往返时间 3. 周转时间 4. 再次出动准备时间，停放时间
turnaround training　再次出动前训练
turnback　1. 中途返航 2. 转180°
turn-back　转身，折回
turnbuckle　松紧螺套
turn-in point　开始转弯点，攻击转弯点
turning　1. 转动降落伞以面向偏流 2. 手摇曲轴以旋转发动机
turn(ing) errors　转弯误差
turning in hover　悬停回转
turning point　转折点
turning quality problem to zero　质量问题归零
turning vane in corner　拐角导流片
turnkey　由一个承包商完全包办的工程
turnkey approach　交钥匙方式，启钥方式
turnkey launch service　交钥匙式发射服务
turnoff　1. 转弯滑出跑道 2. 脱离跑道点，跑道与滑行道的连接点 3. 脱离跑道用的滑道
turnoff lights　滑出口标志灯
turnover service　短停维护
turnover structure　翻倒防护结构
turnover time　1. 周转时间 2. 翻转时间
turnover voltage　翻转电压
turnstile antenna　绕杆式天线
turn-table　1. 转台，回转台 2. 电唱盘
turret　1. 炮塔 2. 活动射击装置
turret limits　炮塔转动范围

TUSLOG　The United States Logistics Group 美国援外使团后勤组
tutorial　个别指导
TV　1. Television 电视 2. Terminal Velocity 终点速度，末速度 3. Test Vehicle 试验用的飞行器 4. Throttle Valve 节油活门 5. Thrust Vectoring 推力换向 6. Total Volume 总容积
TV AT　Television Air Trainer 电视空中训练器
TV channel　电视频道
TV command　电视指令制导
TV control range　电视制导距离
TV guidance　电视制导
TV guided glide bomb　电视制导滑翔炸弹
TV homing　电视自导引，电视制导，电视引导
TV measurement　电视测量
TV SAT　电视卫星
TV seeker　电视导引头
TV terminal guidance　电视末段制导
TV tracking　电视跟踪
TV with dual sound programmes　双伴音电视
TVBC　Turbine Vane and Blade Cooling 涡轮叶片和叶刀冷却
TVBS　Television Broadcasting Satellite 电视广播卫星
TVC　Thrust Vector Control 推力向（矢）量控制
TVCS　Thrust Vector Control System 推力向量控制系统
TVD　1. Terminal Velocity Dive 极限速度俯冲 2. Toxic Vapor Disposal 毒气处理
TVdrop station　电视分出站
TVDU　Television Display Unit 电视显示装置
TVE　Total Vertical Error 总垂直误差
TVLA　Tuned Vertical Line Array 调谐声呐垂直线阵
TVM　Track-Via-Missile 导弹跟踪
TVRO　Television Receive Only 电视单收
TVS　1. Telemetry Video Spectrum 遥测视频频谱 2. Transaction Visibility System 交易可见性系统
TVSC　Television Sound Channel 电视伴音通道
TVSG　Television Signal Generator 电视信号发生器
TVSU　Television Sight Unit 电视瞄准装置
TW　1. Tail Wind 顺风 2. Team Work 工作团队 3. Tool Wagon 工具车 4. Twin Wire 双心导线 5. Typhoon Warning 台风警报
TW&AA　Tactical Warning and Attack Assessment 战术报警和攻击评估
TWC　Tungsten Wire Composite 钨丝复合材料
TWCP　Tape-Wrapped Carbon Phenolic 带缠碳酚醛
TWD　Touchwire Display 接触丝显示
TWDL　Terminal Weather Data Link 终端气象数据链
tweak　1. 扭 2. 拧 3. 焦急

TWEB Transcribed Weather Broadcast 预先录制的气象广播
tween 1.吐温 2.补间动画 3.非离子活性剂
twenty-minute rating 20分钟额定值
TWI Tail Warning Indicator 尾部警报指示器
twilight 曙暮光
twilight effect 曙暮光效应
twilight zone 半阴影区
twin 1.双发动机飞机 2.装两种发动机的飞机
twin barrel(ed) 双管(的),二联装(的)
twin boundary 孪晶间界
twin crystals 孪晶,双晶
twin engine 1.双联发动机 2.双发动机
twin failure 孪生故障
twin jet 双发喷气式飞机
twin paradox 时间相对变慢效应
twin quasar 双类星体
twin tail 双垂尾
twin turret 二联装炮塔
twin vertical fin 双垂尾
twin vertical tail 双垂尾
twin-aisle aircraft 双通(过)道飞机
twin-contact tyre 凹形胎面轮胎,双凸条轮胎
twin-delta wing 双(后掠)三角翼
twin-engine aircraft 双引擎飞机
twin-engine airliner 双引擎客机
twin(-finned) tail 双垂尾,双垂直尾翼
twin-float seaplane 双浮筒水上、机
twinkle roll 编队横滚
twin-spool 双转子(发动机)
twist 1.扭转 2.(绕纵轴的)滚转 3.搓合,绞合 4.弯曲,扭曲 5.使成螺旋形,自身旋转并作曲线运动
twist and steer 先滚转再转弯
twist angle 1.扭转角 2.捻转角
twist blade 扭叶片
twist coupling 扭转耦合
twist distribution 捻度分布
twist rotor 扭曲转子
twist test 1.捻度试验 2.扭转试验
twisted blade 1.弯扭叶片 2.复合倾斜叶片
twisted nematic mode 扭曲向列模式
twisted pair 双芯电缆,双扭线,扭绞二股电缆
twisted pair cord 双绞线
twisted rope 加捻绳
twisting mode 扭动模式
twisting moment 扭矩
twitch factor 抽搐因数
twizzle 旋转认动
two channel monopulse 双通道单脉冲

two degree-of-freedom gyro 两自由度陀螺仪
two dimensional display 二维显示
two dimensional flow 二维流
two dimensional nozzle 二维喷管
two dimensional wind tunnel 二维风洞
two dimensional wing 二维弹翼
two dimensional wing test 二维弹翼试验
two fixed-center problem 双不动中心问题
two freedom 双边自由航行协定
two frequency gas laser 双频气体激光器
two hand pursuit test 双手追踪试验
two impulse transfer 双冲量轨道转移
two inceptor control 双接收器控制
two phase combustion 两相燃烧
two phase flow 两相流
two phase flow heat transfer 两相流换热
two point boundary value problem 两点近值问题
two segment approach 两段下滑进场
two side limit cycle 双边极限环
two stars navigation 双星导航
two state controller 二位控制器
two stream hypothesis 二星流假说
two wire loudspeaking 二线场声
two-axis 两轴
two-axis automatic pilot 双轴自动驾驶仪
two-axis homing head 双轴导引头,双轴寻的头
two-axis stable platform 双轴稳定平台
two-band 双频段
two-bay biplane 两厢双翼机
two-body 二体
two-body motion 二体运动
two-body problem 二体问题
two-cable sling load 双索吊载
two-channel 双通道
two-color thermometry 比色测温法
two-colour diagram 两色图
two-colour photometry 两色测光
two-colour pyrometer 双(比)色高温计
two-component injector element 双组元喷嘴
two-control aircraft 两轴操纵航空器
two-cycle 两冲程的
two-degree of freedom gyro 二自由度陀螺仪
two-degree-of-freedom 二自由度
two-dimensional airfoil 二元翼型
two-dimensional array 1.二维阵列 2.二维数组
two-dimensional compression 二维压缩
two-dimensional coupling 二维耦合
two-dimensional flow 二维流动
two-dimensional geometry 二维几何

two-dimensional hyperbolic Bessel func-tion 二维双曲线贝塞尔函数
two-dimensional image 二维图像
two-dimensional inlet 二维进气道
two-dimensional model 二维模型
two-dimensional modeling 二维模拟
two-dimensional space 二维空间
two-dimensional wind tunnel 二维风洞
two-element 双元件
two-element radio interferometer 双天线射电干涉仪
two-equation 两方程
two-equation model 两方程模型
two-fluid model 双流体模型
two-gun turret 双联装炮塔
two-hand coordination tests 双手协调试验
two-image photogrammetry 双像摄影测量
two-impulse 双脉冲
two-layer 双层
two-layer model 双层模型
two-level bridge 双高度登机桥
two-link robot 双摆机器人
two-loop 1.双回路的 2.双闭环
two-man rule 双人规定
two-meal service 供应两餐的服务
two-medium photogrammetry 双介质摄影测量
two-mode 双模式
two-mode system 双模系统
two-parameter 双参数
two-phase 1.两相 2.两阶段 3.二相 4.双相
two-phase algorithm 分步算法
two-phase boundary 两相边界
two-phase flow 两相流
two-phase mixture 1.两相体 2.两相混合物
two-photon 双光子
two-photon absorption 双光子吸收
two-photon fluorescence method 双光子荧光法
two-place 双孔
two-player 1.双人 2.两方
two-point 二接点,双接点
two-point boundary 两点边界
two-point boundary value problem 两点边值问题
two-point boundary-value 两点边界值
two-port network 二口网络,双口网络
two-position propeller 两距位螺旋桨
two-probe method 双探针法
two-pulse rocket 两级火箭
two-rate oleo 两级油液减震支柱
two-reflector antenna 双反射天线
two-ribbon flare 双带耀斑

two-runway 双跑道
two-seat 双人位
two-segment 两段式
two-shaft 双轴
two-shaft engine 双轴涡轮发动机
two-side limit cycle 双边极限环
two-sided lapping 双面细磨
two-sided sequence 双边序列
two-spectra binary 双谱分光双星
two-speed 双速的
two-speed supercharger 双速增压器
two-spool 双转子的
two-stage 1.两级 2.双级
two-stage amber 两级琥珀色夜航训练法
two-stage igniter 两级点火器
two-stage light gun 二级轻气炮
two-stage shaped charge 两级聚能装药
two-stage supercharger 两级增压器
two-stage supercharging system 二级增压系统
two-stage-to-orbit 两级入轨
two-state 1.双稳 2.双态
two-step supercharger 两级操纵增压器
two-stream 双流束
two-stroke engine 二冲程发动机,二行程发动机
two-target game 双靶游戏
two-temperature 双温
two-terminal network 二端网络
two-time scale system 双时标系统
two-tone photographic reconnaissance satellite 双音型照相侦察卫星(苏联)
two-way adjustable gyro alignment plate 双向可调陀准直板
two-way carrier acquisition time 双向载波捕获时间
two-way communication 双向通信
two-way range rate measurement 双向测速
TWP Technical Work Program 技术工作项目
TWR 1. Tail-Warning Radar 尾部警戒雷达 2. Threat-Warning Reciever 尾部警报接收机 3. Thrust-Weight Ratio 推重比 4. Tower 指挥塔台 5. Turbulence Weather Radar 紊流气象雷达
TWS 1. Tail Warning Set 机尾警戒雷达(装置) 2. Technical Work Statement 技术工作说明 3. Threat-Warning System 尾部警戒系统 4. Track-While-Scan 边搜索边跟踪
TWSC Thin-Wall Steel Case 薄壁钢盒,薄壁钢壳体
TWSS TOW Weapon Subsystem "陶"式武器分系统
TWT 1. Transonic Wind Tunnel 跨声速风洞 2. Travelling Wave Tube 行波管
TWTA Travelling Wave Tube Amplifier 行波管放大器

TWU	Tactical Weapons Unit 战术武器部队
TWX	Theater War Exercise 战区战争演习
TWY	Taxiway 滑行道
TWYL	Taxiway Link 滑行道联络道
twystron	行波调管
TXPDR	Transponder 应答机
TY	Total Yield 总产量,总收率

tying launch rocket 捆绑式运载火箭
TYP 1. typical 典型的 2. type (of aircraft)飞机型别
typal map 类型地图
type 1.（飞机的）型别,型号 2. 机种
type A evaluation A 类评定
type A standard uncertainty A 类标准不确定度
type B evaluation B 类评定
type B standard uncertainty B 类标准不确定度
type certificate 型号合格证,型号生产许可证
type certificate data sheet 1. 型号许可证规范资料 2. 型号合格证数据单
type certification special condition 型号审定专用条件
type conversion 改装许可
type designation 型号命名字,型式认定,机型名称
type engine 标准类型的发动机
type inspection authority 定型审查权限,定型检查当局
type k 1. 定 K 式 2. K 型
type load 标准装载
type of aircraft 机型
type of camera 摄像机的类型
type of cell 电池类型
type of control 控制形式
type of decision 决策类型
type of discharge 放电模式
type of disturbance 搅动型式
type of engine 发动机型号
type of failure 故障类型
type of feed 装弹方式
type of film 电影类型
type of flight 1. 航班种类 2. 飞行类型
type of flow 流类型
type of formation 编队类型
type of interaction 相互作用形式,交互类型
type of measurement 测量类型
type of model 模型的类别
type of motion 运动方式
type of object 1. 物体类型 2. 对象类型 3. 天体类型
type of operation 1. 操作方式 2. 工作方式
type of part 部件类型
type of polymer 聚合物的类型
type of radar 雷达类型
type of sensor 传感器类型
type of service 服务类型
type of space 空间类型
type of spacecraft 宇宙飞船类型
type of task 任务类型
type of term 术语类型
type of transmitter 发射机类型
type record 定型记录
type test 定型试验
TYPH typhoon 台风
typical 典型的,标准的
typical example 典型例子
typical response 典型回应
typical result 典型值
typical section 1. 典型横切面 2. 标准断面
typical time 典型时间
typical value 1. 典型值 2. 范值 3. 代表值 4. 平均数
tyre 1. 轮胎 2. 轮箍
tyre inflation pressure 轮胎充气压力
tyre pressure 轮胎气压
tyre size 轮胎规格,轮胎尺寸

U

U 1. Unit 单位 2. Unspecified 未指明的,未详细说明的 3. Upper 较上面的,高空的

U shaped U 型

U.N. United Nations 联合国

UA 1. Unattended 无人值守的 2. Uncontrolled Airspace 非管制空域 3. United Airlines 联合航空公司 4. Unit Assembly 单元装配 5. Unit of Action 行动单元 6. Unmanned Aircraft 无人(驾驶)飞机 7. Unusual Attitude (Unusual Attitude Maneuver) 异常姿态机动 8. User Agent 用户代理

UAAA Ultralight Aircraft Association of Australia 澳大利亚超轻型飞机协会

UAB Until Advised By 直到接到……通知为止

UAC 1. United Aircraft Corporation 美国联合飞机公司 2. Upper-Airspace Control 高层空域管制 3. Urban Area Coverage 城市覆盖区

UAI 1. United Aircraft International 国际联合飞机公司 2. Universal Azimuth Indicator 通用方位指示器

UANC Upper-Airspace Navigation Chart 高层空域导航图

UAR 1. Unattended Radar 无人值守雷达 2. Upper Air Route 上层航线 3. Upper Atmosphere Research 高空大气研究

UARS 1. Unattended Radar Station 自动(无人操纵)雷达站 2. Unmanned Air Reconnaissance System 无人驾驶侦察机(航空侦察系统) 3. Upper Atmosphere Research Satellite 高层大气研究卫星

UART Universal Asynchronous Receiver/ Transmitter 通用异步收发机

UAS 1. Unavailable Second 不可用秒 2. Understanding, Agreement, Support 理解,协同,支持 3. University Air Squadron 大学飞行中队 4. United Aviation Services 联合航空服务公司 5. Unmanned Aerospace Surveillance 无人宇宙空间监视 6. Unmanned Aircraft System 无人(驾驶)飞机系统 7. Upper Air Space 高空空域

UATP Universal Air Travel Plan 环球(万国)空中旅行计划

UAV 1. Unmanned Aerial Vehicle 无人(驾驶)飞行器 2. Uniform Annual Values 年均值

UB 1. Usage Block 使用部分 2. Utility Bus 公用汇流条

UBE (Ubee, Ubie) Ultra Bypass Engine 1. 超高涵道比发动机 2. 无涵道风扇发动机

U-bomb 铀原子弹

UC 1. Unclassified 未分类的 2. Undercarriage 起落架 3. Undercharge 装药(充电)不足 4. Under Construction 正在施工 5. Unit Check 单元测验 6. Upper Center 上中 7. Upward Compatibility 向上兼容 8. User Class 用户等级

UCA Unmanned Combat Aircraft 无人驾驶作战飞机

UCD Use Case document 使用范例文档

Uchinoura Space Center 内之浦宇宙空间观测所

UCI 1. Unit Construction Index 单位结构指数 2. User Class Identifier 用户类别标识符 3. User Computer Interface 用户计算机界面

UCL 1. Up Command Link 上行指令链路 2. Upper Control Limit 控制上限,管制上限,上控制界限

UCNI Unified Communication, Navigation and Identification System 通信导航识别综合系统

U-code U 码

uctuated frequency 上下波动的频率

uctuated frequency condition 上下波动的频率条件下

uctuation amplitude 波动幅度

uctuation datum 浮动基准面

uctuation level 浮动水平

UD 1. Document Unit 文件单元 2. Universal Dipole 通用偶极子 3. Unplanned Derating 非计划降低出力 4. Upper Deck 上舱 5. User Data 用户数据

UDF 1. Unducted Fan 无涵道风扇,桨扇(发动机) 2. Unit Derating Factor 机组降低出力系数 3. User-Defined Function 用户定义功能

UDMH Unsymmetrical Dimethyl Hydrazine 偏二甲肼

UDMH railway tank transporter 偏二甲肼铁路油槽运输车

UDMU Universal Decoder Memory Unit 通用解码器存储单元

Udop Utra High Frequency Doppler Velocity and Position System 超高频多普勒速度和位置测量系统

UDRE User Differential Range Error(全球定位系统的)用户差分距离误差

UDS 1. Unidirectional Solidification 定向(单向)凝固 2. Universal Data Set (System) 通用数据装置(系统) 3. Un-supported (Unstocked) Dispersed (Dispersal) Site 无仓库的疏散地

UDT Update 更新,修改

UE 1. Unit Equipment (Establishment) 部队装备(设施),用户设备 2. Under-Excitation 鼓励不做

UEJ Unattended Expendable Jammer(无人照管)投弃式干扰机

UEO Unit Emplaning Officer 机载作战装备指挥官

UER 1. Unsatisfactory Equipment Report 不合格装备报告 2. Unscheduled Engine Removal 发动机提前拆卸,非计划发动机拆换率

UERE User Equivalent Range Error(全球定位系统的)用户等效距离误差

UF 1. Unavailable Factor 不可用系数 2. Under Frequency 欠频,频率过低 3. User-Friendly 用户友好

UFA Until Further Advised 在另行通知前

UFC Unified Fuel Control 燃油统一控制(系统),统一燃油控制(系统)

UFDR Universal Flight Data Recorder 通用型飞行参数记录仪,通用飞行数据记录器

UFF 1. Ultimate Fallback Facility 最终后备设施 2. Ulster Freedom Fighters 阿尔斯特自由战士 3. United Freedom Front 自由战线联盟 4. Universal Flip-Flop 通用触发器

UFO 1. UHF Follow-On UHF 后继星(美国军事通信卫星,用于取代"舰队通信卫星") 2. Unidentified Flying Object 不明飞行物,飞碟

UFOlogist 飞碟研究者,飞碟学家,不明飞行物研究家

UFS 1. Ultimate Factor of Safety 极限安全系数 2. Unit Fiber Structure 单元光纤结构

UFTAS (Uftas) Uniform Flight Test Analysis System 统一试飞(飞行试验)分析系统(美国空军飞行试验中心)

UGM Underwater Guided Missile 水下导弹

UGND Underground 地下

UGS Upgraded Silo(Program)抗毁力提高的地下发射井(计划)

UGT 1. Underground Test 地下试验 2. Urgent 紧急的

UH Upset Head 铆钉墩头

UHB 1. Ultra High Barrier 超高的障碍 2. Ultrahigh Brightness 超高亮度

UHC Unburned Hydrocarbons 未燃的碳氧化合物(烃)

UHD Ultra-High Density 超高密度

UHF Ultra High Frequency 超高频

UHPT Undergraduate Helicopter Pilot Training 直升机飞行学员训练

UHS Ultra-High Speed 超高速

UHT 1. Ultra-High Temperature 超高温 2. Unit Horizontal Tail 全动式平尾

Uhuru 自由卫星

UHV 1. Ultra-High Vacuum 超高真空 2. Ultra-High Voltage 超高(电)压

UI 1. Unit Interval 单位间隔 2. Unlock Instruction 开锁指令 3. Unrecognized Information 不可识别信息 4. Urgent Interrupt 紧急中断 5. User Identification 用户识别符 6. User Interface 用户接口,用户界面

UIC 1. Upper Information Center 高空信息中心(国际民航组织) 2. U-Interface Circuit U 接口电路 3. User Identification Code 用户识别码

UID 1. Unique Identifier 唯一标识符 2. Universal Identifier 通用标识符 3. User Identifier 用户标识符 4. User Identification 用户识别 5. User Interface Document 用户界面文档

UID atomization UID 雾化

UID dynamics simulation UID 动态仿真

UID elastic damper UID 弹性减震器

UIDS User Interface Development System 用户接口开发系统

UIF 1. Unfavourable Information File 相反情报档案 2. Unumbered Information Frame 非编号信息帧

UIL 1. Upper Integrity Limit 整合上限 2. User Interface Language 用户接口语言

UIM 1. Unique Interface Module 专用接口模块 2. Universal Identity Module 通用识别模块 3. User Identification Module 用户识别模块

UIP 1. Upgrading Instructor Pilot 升级训练的飞行教员 2. User Identification Program 用户标识程序 3. User Interface Program 用户接口程序

UIPE User Instantaneous Phase Error(全球定位系统的)用户瞬时相位误差

UIR Upper-flight Information Region 高空飞行情报区

UIS 1. Universal Information Services 通用信息业务 2. Upper Information Service 高空情报服务 3. Upper Information System 高空信息(情报)系统 4. Urban Information System 城市信息系统 5. User Interface System 用户接口系统 6. User-In-Service 用户在使用业务

UIT Ultraviolet Imaging Telescope 紫外成像望远镜

UJT Unijunction Transistor 单结晶体管

UK United Kingdom 英国

UKACC UK Air Cargo Club 英国空运(航空货运)俱乐部

UKADGE (UKAdge) UK Air Denfence Ground Environment "尤卡其"防空系统(英国地面防空系统)

UKADR United Kingdom Air Defence Region "尤卡德尔"防空区(英国防空区)

UKF-based estimator 基于 UKF 的估计

UKHS UK Hovercraft Society 英国(联合王国)气垫飞行器协会

UKIRT 1. UK Infrared Telescope 英国红外望远镜 2. UK Instrument Rating Test 英国仪表额定(性能)

测试

UKISC United Kingdom Industrial Space Committee 英国工业空间委员会

UKLF UK Land Forces 英国(联合王国)陆军部队

UKMF UK Mobile Force 英国(联合王国)机动部队

UKN Unknown 不明的,未知的

UKOOA UK Offshore Operators Association 英国(联合王国)沿海空运公司协会

UKRAOC 1. UK Regional Air Operations Center 英国(联合王国)地区空中作战中心 2. UK Royal Army Ordnance Corps (英国)皇家陆军军需部队

UKS United Kingdom Satellite 英国卫星

UKWMO UK Warning and Monitoring Organization 英国(联合王国)警戒和监控组织

UL 1. Underwriter's Laboratory 风险担保人试验研究室 2. Unexpected Loss 非预期损失 3. Unload 卸货 4. Up-Link 上行链路 5. Upper Limit 上限

ULA 1. Ultra-Light Aircraft 超轻型飞机 2. Uncommitted Logic Array 自由逻辑阵列 3. Uniform Linear Array 均匀线性阵列 4. Universal Logic Array 通用逻辑阵列

ULAA 1. Ultra-Light Aircraft Association 超轻型飞机协会(澳大利亚) 2. Ukrainian Library Association of America 美国乌克兰人图书馆协会

ULANA Unified Local Area Network Architecture 统一局部网结构

ULB Underwater Locator Beacon 水下定位信标机

ULC 1. Unit Load Container 单位载荷集装箱 2. Universal Logic Card 通用逻辑卡

ULD 1. Underwater Locating Device 水下定位装置 2. Unit Load Device 单位载荷运载器具 3. Universal Language Description 通用语言描述

ULEA Ultra-Long Endurance Aircraft 超长续航时间飞机,超久航飞机

ULI 1. Ultra-Low Interstitial 超低空隙 2. Universal Logic Implementation 通用逻辑执行过程 3. Unmanned Lunar Impact 无人探测器月球硬着陆

ULL 1. Ullage 漏损(量) 2. Unbundled Local Loop 非捆绑式本地环路 4. Unit Local Loading 单位局部加载 3. Uncomfortable Loudness Level 不舒适响度级

ULLA Ultra-Low-Level Airdrop (System) 超低空空投(系统)

ullage 1.(油箱的)剩余容积 2. 气枕,气隙(指贮箱中液体上面的气体) 3.(油箱的)耗去油量 4. 漏损量

ullage engine 1. 残留液收集发动机 2. 加压余量燃料引擎 3. 油箱油面上部的空间,漏损(量),损耗(量)

ullage gas 气隙气体

ullage height 油面高度

ullage manoeuvre 1. 残留液收集机动 2. 加压余量燃料操纵(火箭)

ullage pressure 气隙压力

ullage rocket 气垫增压火箭

ullage space 油箱内油液上方的剩余空间

ullage volume 损耗量

ULMS 1. Ultra Long-Range Missile Submarine 超远程导弹潜艇 2. Undersea/Underwater Launched Missile System 水下发射导弹系统 3. Undersea Long-Range Missile System 水下远程导弹系统

ULS Ultra Low Speed 超低速

ULS diesel 超低速内燃机

ULS jet 超低速喷气机

ULSA Ultra-Low Sidelobe Antenna 超低旁瓣天线

ULSIC Ultra Large Scale Integrated Circuit 超大规模集成电路

ultimate factor of safety (UFS)极限(最大允许)安全因数

ultimate failure 终极破坏

ultimate life 终极寿命(即最大允许寿命)

ultimate load 1. 极限载荷,最大载荷,临界载荷 2. 设计载荷

ultimate pressure 1. 极限压力 2. 极限压强

ultimate production aircraft 最新投入批生产的飞机

ultimate range ballistic missile 极限射程弹道导弹,最远程弹道导弹

ultimate strain 极限应变

ultimate strength 极限强度,强度极限,断裂强力

ultimate strength of the laminate 叠层的极限强度

ultimate stress 极限应力

ultimate vacuum 极限真空,极限真空度

ultra 超,过

ultra pure water 超纯水

ultra small aperture terminal 超小口径终端

ultra-bypass engine 1. 超高涵道比发动机 2. 超循环发动机

ultrafast opto-electronics 超快光电子学

ultrafiltration 超过滤,超过滤作用

ultrafine 1. 超精细的,超细的,特细的 2. 超细电解粉末

ultrahazardous activity 特别危险的活动(空间法用语),超危险作业

ultra-high strength steel 超高强度钢

ultra-high-density seating 超密(集)座椅布局

ultra-high-speed photography 超高速摄影

ultra-high-vacuum 超高真空

ultra-high-vacuum physical vapor deposition 超高真空物理气相沉积

ultralight 1. 超轻型的 2. 超轻型飞机

ultralight aircraft 超轻型飞机

ultralight airplane 超轻型飞机
Ultralights 超轻型飞机
Ultralightweight 超小型的,超微型的
ultra-long-range guided weapon 超远程制导武器
ultra-low airdrop 超低空空投
Ultramicroscope 超高倍显微镜
Ultraminiature 超小型的,超微型的
ultra-modern weapons 最新式武器,超现代武器
ultraprecision machining 超精(密)加工
ultrashort light pulse 超短光脉冲
ultra-short take-off 超短距起飞(有垂直起飞能力的飞机进行短距离滑跑起飞,也可称为短距起飞,飞行员可选择不同起飞方式)
ultrasonic 1.超声的,超声速的 2.超声波
ultrasonic bonding 1.超声波焊接,超声键合 2.超声波焊接法
ultrasonic cleaning 1.超声波清洗 2.超声波净化 3.超声波消磁
ultrasonic computerized tomography 超声计算机断层成像
ultrasonic crack detector 1.超声(波)裂纹探测器 2.超声波探伤仪
ultrasonic defectoscope 1.超声波探伤仪 2.超声波检验器
ultrasonic delay line 超声(波)延迟线
ultrasonic detection and measurement 超声(波)检测
ultrasonic detector 1.超声波探伤仪 2.超声波探测器
ultrasonic diffraction modulator 超声衍射调制器,声光调制器
ultrasonic Doppler blood flow imaging 超声多普勒血流成像
ultrasonic Doppler blood flowmeter 1.超声多普勒血流仪 2.多普勒超声血流量计
ultrasonic flaw detection 1.超声波探伤 2.超声波裂纹探测 3.超声波检测
ultrasonic frequency 1.超声频率 2.超音频,超声频
ultrasonic full waveform 超声波全波形
ultrasonic gas atomized powder 超声气体雾化粉末
ultrasonic guides for the blind 超声导盲器
ultrasonic inspection 超声(波)探伤,超声(波)检查
ultrasonic levelmeter 超声物位计
ultrasonic levitation 超声悬浮
ultrasonic light 超声光
ultrasonic machining 超声(波机械)加工
ultrasonic measurement 1.超声波测量 2.超声波测距
ultrasonic method 1.超声波法 2.超声致光衍射
ultrasonic nozzle 超声波喷嘴
ultrasonic ozone spectrometer 紫外臭氧光谱仪
ultrasonic perforating 超声波穿孔

ultrasonic remote sensing 超声波遥感
ultrasonic rolling 超声辊轧,超声波滚压
ultrasonic scanner 超声波扫描仪
ultrasonic signal 超声波信号
ultrasonic solar spectrometer 超声波太阳光谱仪
ultrasonic soldering 超声波软钎焊,超声焊接
ultrasonic testing 1.超声波检测,超声波试验 2.超声波探伤
ultrasonic transducer 1.超声换能器,超声波传感器 2.超声波探头
ultrasonic welding 超声波焊接,超声波熔接
ultrasonics 1.超声波学 2.超声技术
ultrasound 超声,超声波
ultrasound datum 超声数据
ultrasound echo 超声回波
ultrasound transducer 超声换能器
ultra-thin solar cell 超薄太阳能电池
ultraviolet 1.紫外线 2.紫外线辐射 3.紫外线的
ultraviolet absorbance monitor 1.紫外吸收监测器 2.紫外吸光度监察
ultraviolet astronomy 紫外天文学
ultraviolet detector 紫外线探测器
ultraviolet differential absorption radar 紫外差分吸收激光雷达
ultraviolet excess 紫外超
ultraviolet exposure 1.紫外线辐射 2.紫外线曝光
ultraviolet imagery 紫外线成像
ultraviolet imaging laser system 紫外激光成像系统
ultraviolet intensity 紫外线强度
ultraviolet laser 紫外激光器
ultraviolet radiation 紫外线,紫外线辐射,紫外线照射法
ultraviolet spectrometer 1.紫外光谱仪 2.紫外分光计
ultraviolet stratospheric imaging spectrometer 平流层紫外成像光谱仪
ultraviolet target radiometer 紫外目标辐射计
ultraviolet telescope 紫外望远镜
ultraviolet test 紫外试验
ultraviolet-excess object 紫外超天体
ULV 1.Upper Limit of Video 视频上限 2.Ultra-Low-Volume 超低容量
Ulysses 尤里塞斯探测器
UM 1.Undermentioned 下述的 2.Unified Messaging 统一传信 3.Unit Manufacture 单元制造 4.Unit of Measure 计量单位 5.Unmatched 不匹配的 6.Urgent Measure 加急电报
UMA 1.Unmanned Aircraft 无人(驾驶)飞机 2.Unscheduled Maintenance Actions 不定期维修活动 3.Unscheduled Maintenance Analysis 不定期维修

分析

umbilic(al) 1.地面缆线和管道,脐带式管缆 2.脱落插头 3.航空旅客桥,登机桥,登机走廊(俗称)

umbilical cable 1.脐带式电缆 2.连接电缆

umbilical connector 脐带连接器,自动脱落连接器

umbilical cord 1.(导弹发射前检测用的)操纵缆、控制缆 2.(航天器发射前用的)供应缆,脐带式管缆

umbilical cord tower 脐带塔

umbilical fitting (导弹)发射控制电缆接头

umbilical life support system 脐带式生命保障系统

umbilical mast 脐带塔

umbilical plug 脱落插头

umbilical socket 脱落插座,脐塞

umbilical stiffness 脐刚度

umbilical tower 1.脐带式管线塔架 2.缆塔,供应管缆塔,控制塔

umbra 本影,暗影,蚀时的地球

umbra eclipse 本影食

umbrella antenna 伞形天线

Umbriel 【天】天卫二

UME User Management Entity 用户管理实体

UMIN 1. Unaccompanied Infant 无随行婴儿 2. University Medical Information Network 大学医学信息网络

UML Unfied Modeling Language 标准建模语言

UMLS Universal Microwave Landing System 通用微波着陆系统

UMP Unscheduled Maintenance Program 不定期维修方案

UMS 1. Unified Messaging Service 统一传信服务 2. Urine Monitoring System 尿监测系统 3. Unmanned Multifunction Satellite 多功能无人卫星站 4. User Modifiable Software 用户可修改软件

UMT 1. Universal Military Training 普遍军(事)训(练) 2. Universal Mount 通用安装架

umtee(s) 参加普遍军训者

UMTS 1. Universal Military Training and Service 通用军训和服务 2. Universal Mobile Telecommunications Service 通用移动通信业务 3. Universal Mobile Telephone Service 通用移动电话业务

UN 1. Unable 不能够 2. Union 联合体 3. Upstream Node 上行节点 4. Urban Network 城市网

unable 不能执行指示,不能够

unaccelerated flight 无加速度飞行(包含三种情况:一是直线等速飞行;二是平直飞行,在垂直于飞行轨迹方向上除地球引力外无加速度,但可能有飞行方向的加速度;三是失重飞行,没有重力加速度)

unaccompanied baggage 非随机行李

unaccompanied children 无(成)人陪同儿童,全票儿童

unaligned coherence 未对齐的连贯性

un-ambiguity operating range 无模糊作用距离

Unamsat 自治大学卫星

unanimity rule 1.等价规则(国际航空运输协会的一个基本规则,即甲地到乙地与乙地到甲地的票价应相同,提供同等服务的各航空公司票价一致) 2.一致同意原则(空间法用语) 3.一致性法则

unanticipated event 意外事件

unanticipated fault 意外故障

unarmed helicopter 无武器直升机,非武装直升机

unarming 未解除保险

unassigned node 未指定的节点

unassisted escape 无助力离机

unassisted quick-donning capacity 迅速自行穿脱性能

unattended earth station 无人值守地球站,无人值守地面站

unattended operation 无人值守,无人值班的运行,自动操作,无监督的操作

unattended radar 无人值守雷达,自动雷达

unattended relay station 无人值守中继站

unattended repeater 无人增音站

unaugmented 1.(涡轮喷气发动机或涡轮风扇发动机)未装加力燃烧室的,未使用加力的 2.无增稳系统的

unavailable 手头没有的,不能用的,不是现成的,得不到的

Unavia 尤纳维亚(意大利国立研究和发展飞机制造工艺的机构)

unbalance 不平衡,失衡,失配

unbalanced cell 放电不均匀的电池

unbalanced field 非平衡机场,无界域

unbalanced turn 非协调转弯

unbiased estimation 无偏估计

unblocking field 解阻场

unblocking temperature 解阻温度

unblown 1.未吹气(装有上翼面吹气装置、吹气襟翼的短距起落飞机或其他动力吹气系统在飞行中不使用吹气的飞行状态) 2.(活塞式飞机)无增压器的

unbounded datum 无限的数据

unburned flow 未燃流

unburned fuel 未燃烧的燃料

unburned gas 未燃烧气体,未燃尽气体

unburned material 未燃烧的材料

unburned mixture 未燃烧混合物

unburnt 未燃的,不完全燃烧的

UNC 1. Uncertain 不确定的 2. United Nations Command 联合国军 3. Universal Navigation Computer 通用导航计算机

uncaging time 开锁时间

uncaging zero 开锁零位

uncalibrated angle bias　无标定的角偏差
uncalibrated hot-wire measurement　未校准的热线测量
uncambered blade　不成弧形的叶片
uncambered wing　无弯度机翼
uncatalyzed matrix　非催化矩阵
uncertain　情况不明（飞机分类中安全情况不明的一类，在超过预计到达时间或通报到达时间30分钟仍收不到无线电应答者即属情况不明阶段），渺茫的，不可预测的
uncertain case　情况不明的案例
uncertain control effectiveness matrix　不确定的控制效率矩阵
uncertain environment　不确定环境
uncertain knowledge　不确定信息
uncertain parameter　不确定参数
uncertain parameter effect　不确定参数的影响
uncertainty　不确定度，不确定性
uncertainty（of measurement）　（测量）不确定度
uncertainty of gyro drift　（陀螺）漂移不定性
uncertainty phase　不肯定阶段，不明阶段
uncharged drop　不带电荷的下降
unchecked baggage　未经检查的行李（手提物品和放在客舱内的行李）
UNCL　Unified Numerical Control Language 统一数控语言
unclassified　1.无密级（保密等级中的一级，即不需保全措施但仍可能因其他理由控制使用）2.不分类的
uncoated aluminum　未涂层铝
uncoated powder　未涂覆的粉末
uncocked position　1.（投弹机构的）非准备投弹状态 2.（枪炮击发装置的）非待发状态 3.（建制部队中的）分队 4.（特遣部队中的）支队 5.（供应物资的）标准单位，基数
uncommanded lateral motion　非指令横向运动
uncommanded lateral-directional motion　非指令侧向运动
uncommanded motion　非指令运动
uncompensated acidosis　缺代偿性酸中毒，非代偿性酸中毒
uncompensated alkalosis　缺代偿性碱中毒，非代偿性碱中毒
uncompensated range　未补偿的范围
uncompensated range bias　未补偿的偏差范围
uncompensatory zone　（指人体温度的）不能代偿范围
unconfined condition　无约束条件
unconfined detonation　无侧限爆轰
unconfined region　无侧限区
unconfined volume　无侧限量
unconfirmed removal　未证实拆卸

unconsciousness　意识丧失，失去知觉
unconsciousness threshold　意识丧失阈限
unconstrained case　无约束的情况下
uncontrolled airspace　无空中交通管制的空域
uncontrolled case　失去控制的案例
uncontrolled ditching　失去操纵的水上迫降
uncontrolled flow　1.无控制流动，非可流 2.敞喷
uncontrolled flutter　失去控制的颤振
uncontrolled jet　不受控制的飞机
uncontrolled mean location　不受控制的平均位置
uncontrolled mosaic　非控制点镶嵌图（未用地面控制点或其他方法定向的照片拼接图，不能用来准确地度量距离和方位）
uncontrolled reentry　不受控制的再入
uncontrolled state　非控制状态
uncontrolled vortex　不受控制的漩涡
unconventional control　非常规的控制
unconventional flight　非常规飞行
unconventional flight control　非传统的飞行控制
unconventional loss core　非常规核心损失
uncooled test　非制冷试验
uncooled tip　非制冷提示
uncooled turbine　非制冷机
uncooperative　无源的（不发射电磁波的），（飞机）未装应答机的
uncoordinated maneuver　不协调动作
uncorrected result　未修正结果
uncorrelated encounter　不相关的遭遇
uncorrelated model　不相关的模型
uncorrelated observation　非相关观察
uncouple　1.拆开，脱开，分开 2.去耦 3.解去联系
uncoupled　1.非耦合的，解耦合的 2.不受控制的
uncoupled modes　1.非耦合模态 2.非耦合振荡模态
UNCTAD　United Nations Conference on Trade And Development 联合国贸易和发展会议
UNCTLD airspace　uncontrolled airspace 无（空中交通管制的）空域
UND　1.Undated 无日期的 2.Under 在……下，欠
undamaged aircraft　无损坏的飞机
undamaged input rate　无损输入速度（速率陀螺能承受的最大输入速率）
undamaged panel　无损坏的控制板
undamped　1.非阻尼的（指振动），非减震的 2.不衰减的，不减幅的 3.未受潮的，不受潮湿影响的
undamped natural frequency　1.等幅振荡固有频率，等幅固有频率 2.无阻尼自然频率
undamped navigation mode　无阻尼导航模态（惯导系统的水平和方位回路均不加校正网络的工作模态）
undamped second-order system　无阻尼二阶系统

undamped system 无阻尼系统
undeflected control 偏转控制
under slung load （机身下）悬吊载重，吊载
underactuated control 欠驱动控制
underbelly 机身下部，机腹下部
underbody （导弹）尾部，底部
underbreathing 呼吸不足，肺换气不足
undercompounded generator 欠复励发电机（并联磁场效应大于串联磁场，输出电压随负载增大而降低的发电机）
under-convergence 欠会聚
undercool （使）过度冷却
undercurrent relay 低电流继电器
undercut 1.（焊接中的）咬边焊，下陷 2.截槽，掏槽 3.空刀，底切
undercut cavity 凹腔
underdamped response 欠阻尼响应
underdamping 欠阻尼，弱阻尼，不完全衰减
underdeck spray （运载火箭）发射台冲水冷却
underestimate 低估
underestimation 低估，估计不足
underexpanded jet 膨胀不足喷流
underexpanded nozzle 欠膨胀喷管
underexpanded sonic injection 欠膨胀高速注射
underexpansion 不完全膨胀
underexposure 曝光不足
underfloor gun 机身下部炮，机身下部机枪
underfrequency 频率过低，降低的频率，欠频
undergraduate 学员，尤指飞行学员
underground power room 地下电源间
underground test 地下试验
underheated emission 欠热发射
underlie 1.位于……之下 2.构成……基础
underpredict 偏低预测
underprediction 偏低预测
undershoot 1.射击未及（目标）2.投弹未及（目标）3.（着测）目测低 4.负脉冲信号，负尖峰 5.下冲，欠冲 军事单位
underside 1.下侧 2.下部表面 3.底面 4.阴暗面
understandability 可理解性
understeer 1.转向不足 2.对驾驶盘反应迟钝
understeer gradient 转向梯度
undersurface 下表面，底面，液面下的
undertesting 欠试验
undervoltage 电压不足，低电压，欠（电）压
underwater ballistic measurement 水下弹道测量
underwater camera 水下摄影机
underwater ejection 水下弹射
underwater vehicle 水下作业车

underwing 后翼，下翅翼下
undistorted sector 无失真的部门
undisturbed flow 无扰流，稳态流
undoped pigment 无掺杂的颜料
undulate 波状的，呈波浪型的
UNDV Undervoltage 电压不足，欠压
unenhanced case 应用案例
unequal precision measurement 非等精度测量
uneven expansion 不均匀的膨胀
unexposed specimen 未曝光的样品
UNF Union Nautique Francaise 法国航海联盟
unfavorable pilot 不适宜的飞行员
unfilled plate 未填充板
unfocused array 非聚焦阵列
unfocused laser 未聚焦激光器
unfocused SAR 非聚束合成孔径雷达
unfocused synthetic antenna 非聚焦合成天线
unforced case 自然情况下
UNI User Network Interface 用户网络接口
unit 1.单一 2.单元
uniform 1.制服 2.统一，均匀
unidirectional fibrous composite materials 单向纤维复合材料
unidirectional prepreg tape 单向预浸带
unidirectional tow 单向牵引
unified calibration of standard model 标模统校
unified formulation 统一制定
unified microwave system 微波统一系统
unified S-band 统一S波段
unified S-band system S波段统一系统
uniform density 均衡分布密度，均匀密度
uniform distribution 1.均匀分布，等分布 2.连续型均匀分布 3.离散型均匀分布
uniform distribution case 均匀分布的情况
uniform field 均强场
uniform flow 等流，等速流，均匀流
uniform fuel 均匀的燃料
uniform gas-pressure 均匀气压
uniform grid 均匀网格
uniform grid implementation 均匀网格实现
uniform inflow 均匀来流
uniform inlet 均匀的入口
uniform magnetic field 均匀磁场，均强磁场
uniform mesh 均等网格
uniform plasma 均匀等离子体
uniform pressure 均匀压力，等压力
uniform rate 1.统一运费率 2.相同的速度
uniform region 均质区域
uniform sampling 均匀取样，均稀抽样

uniform size 均一性
uniform stage 统一的阶段
uniform temperature 一定温度,均匀温度,恒温
uniform thermal field 均匀热场
uniform-area 均匀区域
uniform-area section 均匀区域
uniformity 1.均匀性,一致性,一致,一律 2.统一,无变化
unilateral decoupling 单向解耦
unimolecular decomposition-rate 单分子分解率
unimorph 单向变形
uninstalled thrust 卸载推力
unique fault 唯一的故障
unique opportunity 唯一的机会
unique solution 唯一解
uniqueness 唯一性,独特性
unit(of measurement) (测量)单位
unit aircraft 部队飞机,单位所属飞机
unit area 单位面积
unit area saving 单位面积的节省
unit cell 1.单元,晶胞,晶格单位 2.(单元)干电池 3.单位栅元
unit circle 单位圆
unit cost 单位成本
unit deformation 单位变形
unit direction 单元方向
unit effective scattering area 单位有效散射面积
unit feedback 单位反馈,全反馈
unit flying hours 装置飞行时间
unit frontal area 单位迎风面积
unit handling cost 标准件托运费用
unit impulse 单位脉冲
unit impulse function 单位冲激函数
unit length 单位长度
unit level test 单元测试
unit load device 单元装载设备
unit load method 单位载荷法,单位负载法
unit mass 单位质量
unit mass of air 空气质量单位
unit matrix 单位矩阵
unit of fire 弹药基数
unit of issue 1.基数 2.计量单位
unit of measurement 计量单位,容积单位,测量单位
unit of time 时间单位
unit price 单价
unit process 1.基本过程,单元过程 2.单元反应 3.单元作业
unit production cost 单位生产成本
unit quaternion 单位四元数

unit quaternion hypersphere 单位四元数的超球面
unit response 单位响应,部件响应
unit response function 单位响应函数
unit span 单元跨度
unit sphere 单位球面
unit step function 单位阶跃函数
unit step response 单位阶跃响应
unit tank volume 单罐容量
unit test 单元测试,单元试验
unit testing 1.单位测试,单元测试,部件测试 2.单元测试法
unit time 单位时间,单件计时件
unit type hangar 分段式机库,隔间式机库
unit value 单位价值,单值
unit vector 单位矢量,单位向量
unit volume 单位容积,单位体积
unit weight 单位重量
unit width 单位宽度,单宽
unitary datum 单一数据
unitary experimental datum 单一的实验数据
unitary matrix 幺正矩阵(一元矩阵),酉矩阵,单式矩阵
unitary transform 酉变换(算子),单一变换
unitaryx 弹头,不同于子母弹药或集束弹药
unit-ball 单位球
United Kingdom Air Defence Region 英国(联合王国)防空区
unitization 1.拼装(把许多小件物品装进一个标准货柜中) 2.单元化,成组化,规格化 3.联合,联合经营
unitized load 整体货物(指装在集装箱内的,固定在货盘上的,以及各种可以做为一个单件处理的货物),单元化货物
unit-sample sequence 单位取样序列
unit-vector normalization 单位矢量归一化
unity 1.同一,统一,一致 2.单元,单位
univariate (统)单变(量)的,一元的
univariate study 单因素研究
Universal Air Travel Plan 万国航空旅行计划卡(一种信用卡的名称,由航空公司发行,供经常乘飞机旅行的乘客使用),环球航空旅游计划
universal counter 通用计数器
universal decoder memory unit 通用译码器存储单元
universal gravitational constant 万有引力常数
universal horizon 宇宙视界
Universal Kriging 泛克立格法
universal low-thrust 通用的低推力
universal method of photogrammetric mapping 全能法测图
universal motor 通用电动机,交直流两用电动机

universal multiple bomb rack	通用复式炸弹架
universal plotting chart	通用天文领航作业图（美国）
universal polar stereographic projection	全能经纬仪，通用经纬仪，通用极球面投影
universal tester	通用测试仪，万能试验仪
universal theodolite	通用经纬仪，万能经纬仪
universal time	世界时，国际标准时间
universal time coordinated	协调世界时，国际标准时间
universal timing disc	万用定时盘（连接在发动机曲轴上，刻有360°角的刻度和悬垂指针的盘，定时用的）
universe	总体，宇宙，银河系，全域
unjammable radar	抗干扰雷达
UNK　Unknown	不明，不详
unknown disturbance	未知扰动
unknown number of target	未知目标数
unknown parameter	1.未知参数，未知参量　2.未知向量
unknown target	未知目标
unknown vector	未知向量
unlatch	1.打开卡爪　2.非锁存　3.解开
UNLGTD　Unlighted	无照明的
unlike-doublet	异质
unlike-doublet element	异质元素
unlike-impinging injector element	互击式喷嘴
unlimited ceiling	无限云高（云幂）
unlimited-crossover case	无限的交叉情况
unload	1.退弹　2.（从发射架上）取下导弹　3.卸载，放出，排出　4.减小过载　5.放电
unload motor of platform	1.平台卸载电机　2.平台稳定电机
unloaded bearing	卸下轴承
unloaded state	空载状态，卸载状态，卸载状态下
unloading cavity	1.卸荷腔　2.平衡腔
unloading valve	卸荷阀，卸载活门
unlock	（航炮）开膛，开锁，脱开，解除
unlocked	（自动武器）开锁的，开膛的，未锁定的
unlooped umbilical	开环运动脐
unmanned	非载人的，无人驾驶的
unmanned aerial vehicle	1.无人驾驶飞行器　2.无人驾驶飞机，无人机　3.无人控制的，无人操纵的，无人值班的
unmanned air	无人驾驶飞机
unmanned air vehicle	无人驾驶飞机，无人机
unmanned aircraft	无人驾驶飞机，无人机
unmanned aircraft system	无人飞机系统
unmanned capsule	无人舱
unmanned combat	无人作战
unmanned exploration	无人驾驶探索
unmanned flight	不载人飞行，无人飞行，无人驾驶飞行器
unmanned helicopter	无人驾驶直升机
unmanned mission	无人任务
unmanned multifunction satellite	无人多用途卫星
unmanned recovery vehicle	返回式无人飞行器，无人回收飞行器
unmanned scientific satellite	无人科学卫星
unmanned spacecraft	无人航天器，无人宇宙飞船，无人飞行器
unmanned system	无人系统
unmanned turret	自动控制炮塔，无人炮塔
unmanned vehicle	1.无人载具　2.无人驾驶车辆　3.无人驾驶飞行器　4.实现战车无人化
unmask	1.失去隐蔽点，被探测点　2.去屏蔽，无屏蔽
unmixed reactant	未混合的反应物
unmixed reactant surface	未混合的反应物表面
unmixed surface	未混合的表面
unmixedness	（燃料与空气在燃烧室气流中的）不混合度
UNMKD　Unmarked	1.未（加）标志的（如障碍物）　2.无标记
unmodeled dynamics	未建模动态
unmodeled external torque	未建模外转矩
unmodeled force	模型中未计及的作用力
unmodulated case	未调制的情况下
un-occupied focus	空置的焦点
UNOSC　United Nations Outer Space Committee	联合国外层空间委员会
unpaved runway	无道面跑道
unperturbed case	未扰动的情况下
unperturbed solution	无扰的解决方案
unperturbed two-body	无扰的两体
unpowered return	无动力返回
unpowered-flight phase	被动段
unprepared airfield	无(铺筑)道面的机场，无永久性跑道的机场
unpressurized bay	非增压舱
unpropelled submunition	无动力推进的子弹药，无推进装置的子弹药
unpublished route	非公开航线（仅供飞行员使用）
unpumpable fuel	不能泵出的燃油
unrefuelled range	不进行空中加油的航程
unregulated-voltage DC-bus system	无调压直流母线系统，不受监管的高压直流总线系统
unreinforced ablator	非增强烧蚀材料，未增强的热防护材料
unrestrictive launch	非限制性的发射
unrotated projectilex	中和地面发射的非制导火箭的统称

unsaturated control 不饱和控制	
unscheduled engine removal 非计划换发,非计划更换发动机,不定期发动机撤换	
unscheduled engine removal rate 不定期发动机撤换率,提前换发率	
unscheduled landing 非计划内着陆	
unscheduled maintenance 非计划维修,不定期维护,非预定维修(美国空军用语,指不在计划之内,不是预防性的维修,例如排除故障,发动机提前更换)	
unsealed strip 无防水铺筑道面的跑道	
unseparated flow 无分离流动	
unshrouded turbine 开式涡轮	
unspillable cell 液密式蓄电池,无泄漏蓄电池	
unstable aerofoil 1.不安定(机)翼 2.不安定翼型	
unstable aircraft 不稳定飞机	
unstable and stable manifold 不稳定和稳定流形	
unstable behavior 不稳定行为	
unstable cavity 不稳定谐振器,非稳腔	
unstable combustion 不稳定燃烧	
unstable combustor 不稳定燃烧室	
unstable direction 不稳定方向	
unstable discharge 不稳定的放电	
unstable discharge phenomenon 不稳定的放电现象	
unstable equilibria 不稳定平衡点	
unstable equilibrium 不稳定平衡	
unstable flame 不稳定框架	
unstable frequency 不稳定的频率	
unstable manifold 非稳定流形	
unstable missile 不稳定导弹	
unstable motion 不稳定运动	
unstable nutation 不稳定的章动	
unstable nutation mode 不稳定的章动模式	
unstable operation 不稳定工作,不稳定运转	
unstable orbit 不稳定的轨道	
unstable periodic orbit 不稳定周期轨道	
unstable pole-zero 不稳定的零极点	
unstable preferential-diffusion 不稳定的优先扩散	
unstable process 非平稳过程,不稳定过程	
unstable region 不稳定区域,不稳定锁模区域	
unstable resonator 不稳定谐振腔,不稳定谐振器	
unstable situation 不稳定的状态	
unstable source 不稳定的来源	
unstable structure 不稳定结构,不稳定组织,非稳定性结构	
unstable system 不稳定系统,不稳定的制度	
unstable wave 不稳定波	
unstable wave number 不稳定的波数	
unstart 不起动(超声进气道激波达不到设计位置),未启动	
unstart detection 不起动检测
unstart prevention 不起动预防
unstarted condition 未启动条件
unsteadiness 不稳定性,非定常性,易变性
unsteady activity 非定常的活动
unsteady aerodynamic response 非定常气动力响应
unsteady aerodynamics 非定常空气动力学
unsteady analysis 非定常的分析
unsteady approach 非定常的方法
unsteady behavior 非定常的行为
unsteady blade loading 叶片的非定常载荷
unsteady boundary 非定常边界
unsteady boundary condition 非定常边界条件
unsteady burning 不稳定燃烧
unsteady calculation 非定常计算
unsteady cavitation 不定常空化,非定常空泡
unsteady combustion 不稳定燃烧
unsteady compressor 不稳定的压缩机
unsteady datum 不稳定的基准
unsteady design 非定常设计
unsteady design airfoil 翼型非定常设计
unsteady detonation 非稳定爆轰
unsteady effect 不稳定的影响
unsteady ejector 不稳定的喷射器
unsteady expansion 非定常膨胀
unsteady flow 非定常流动,非恒稳流,非定常流
unsteady flow in pipe 非定常管流
unsteady flowfields 非定常流场
unsteady force 非稳定力
unsteady heat 非定常热
unsteady heat release 非定常的热释放
unsteady inlet 不稳定的入口
unsteady instrumentation 不稳定的仪器
unsteady lift 非定常升力
unsteady loading 非稳定载荷
unsteady measurement 非稳态测量
unsteady model 非稳态模型
unsteady moment 非定常的时刻
unsteady moment coefficient 非稳态力矩系数
unsteady motion 不稳定运动,非定常运动,非稳态运动
unsteady nozzle 不稳定的喷嘴
unsteady performance 非定常性能
unsteady phenomenon 不稳定现象
unsteady pressure 不稳定压力,非定常压力,脉动压力
unsteady pressure amplitude 非定常压力振幅
unsteady pressure distribution 非定常压力分布
unsteady pressure force 非定常压力
unsteady pressure measurement 非定常压力测量
unsteady pressure response 非定常压力响应

unsteady pump　非定常泵
unsteady pump cavitation　非定常水泵气蚀
unsteady reactor　不稳定的反应器
unsteady separation　非定常分离
unsteady simulation　非定常数值模拟
unsteady solution　非定常解
unsteady term　非稳态项
unsteady thrust　非定常推力
unsteady velocity　非定常速度
unsteady viscous effect　非定常黏性效应
unsteady vortex structure　非定常涡结构
unsteady vorticity　非定常涡量
unsteady-flow　不稳定流,不稳定流动,非定常流,非恒定流
unsteady-pressure　非定常压力
unsteady-pressure datum　非定常压力数据
unsteady-pressure data　非定常压力数据
unstick　(固定翼飞机)起飞离地点,松杆
unstick run　(地面或水面)起飞滑跑
unstick speed　起飞离地(离水)速度,离地速度
unstick speed ratio　飞行速度与离地(水)速度之比
unstrained member　不受力构件
unstretched laminar burning velocity　无拉伸层流燃烧速度
unstructured grid　非结构化网格,非结构网格
unstructured mesh　非结构化网格,非结构网格
unstructured methodology　非结构化方法
unstructured uncertainty　非结构不确定性
unsupervised learning　非监督学习
unsupported site　无后勤保障的起降场地
unswept　未扫清的,未波及的
unswept ramp　未波及的斜坡
unsymmetric thrust　不对称推力
unsymmetrical　(原指化学上分子结构的不对称,现亦用于航空)偏位的,不匀称的,不平衡的
unsymmetrical dimethylhydrazine　偏二甲肼
unsymmetrical flight　不对称飞行
UNT　Undergraduate Navigator Training 领航员本科(初级)训练(美国空军)
UNTSO　United Nations Truce Supervisory Organization 联合国停战监督组织
untwist　(机翼的)负扭转,反向扭曲,解开,拆开
unusable fuel　无用燃油,不可用燃油
unwarned exposed　事先未告知的暴露部队
unwrap　打开,解开
UOC　1. Ultimate Operational Capability 极限作战能力 2. Universal Output Computer 通用输出计算机 3. Useable On Code 在代码上可用的
UOL　1. Underwater Object Locator 水下物体定位器 2. United Olympic Life 奥林匹亚生命联盟
uorescence　荧光
uorescence image　荧光图像
uorescence intensity　荧光强度
uorescence line　荧光线
uorescence line shape　荧光线形状
uorescence signal　荧光信号
uorescence yield　荧光产量
UOS　1. Ultraviolet Ozone Spectrometer 紫外臭氧光谱计 2. Unless Otherwise Shown (Specified/Stated) 除非另有规定/说明 3. User Operations Support 用户操作保障 4. User-Out-of-Service 用户业务中断
UOSAT　University of Surrey Satellite 萨里大学卫星(英国萨里大学及该大学的萨里卫星技术公司研制的系列廉价小卫星)
UOW　Unit Of Work 工作单元
UP　1. Unguided Projectile 非制导射弹 2. Universal Platform 通用平台 3. Unplated 未镀的 4. Unrotated Projectile 无旋转稳定的射弹 5. Upper 较上面的,高空的 6. User Part 用户部分
up and away　站起来就走(对能垂直起飞的动力升力系统的描述),空中加速时
up front control panel　1. 前上方控制板 2. 正前方控制板
up range　首区
up time　1. 能工作时间,可用时间,正常运行时间 2. 作业时间 3. 可用状态
UPAS　Unit Production Assembly Schedule 单架生产装配进度计划
UPC　1. Unit Production Cost 单位生产成本 2. Universal Product Code 通用产品代码 3. Unmanned Payload Carrier 无人有效载荷运载工具 4. Usage Parameter Control 使用参数控制
up-conversation　上变频,上转换
up-command link　上行指令链路
UPD　1. Update 修正状态(此词多见于惯导系统控制显示器面板)更新 2. Unpaid 未付款的
update　1. 更新,革新,使现代化 2. 修改(电子设备输入新的记忆,雷达图像或新的信息等) 3. 修改航行数据(在航线上任何一点进行定位以修改航向等) 4. 改进型,改型(写在飞机型号之后,如 P-3 Update Ⅱ,即 P-3 改Ⅱ) 5. 最新资料
update cycle　更新周期
up-data link　上行数据链路
update point　更新点
update rate　(数据)更新速率,更新速度,校正速度,修正率
update rule　更新规则
update time　调准时间,更新时间

updraft carburetor 上吸式汽化器，上吸式化油器

updraught carburettor 前上方驾驶仪表板（位于战斗机座舱内前上方，平视显示器下方，用于控制通信导航识别等功能）

upgear 收好起落架（美国指挥用语，意指请注意起落架是否收好）

upgrade 提升，提高，升舱，升级，上坡，上限

upgrade mission 升级任务

upgrading 舱位升级（航空公司安排经济舱的客人到头等舱）

UPI 1. Undercarriage Position Indicator 起落架位置指示器（英国用法）2. User Personal Identification 用户个人识别

UPK Upkeep Period 维修期

upkeep 1. 保养，养护（泛指预防硬件恶化的措施，类似维修，通常只用于通用航空界）2. Upkeep 专炸水坝的掠水炸弹

upleg 上升段

uplift 1. 可减载重（在军用货机上指在飞行中可投掷的，民用飞机上指可卸除的）2. 外加燃油量（指飞机中由加油机加入的，但有时也指在本基地以外地点加入的燃油量）3. 装机货物，装机燃油 4. 提高，增加 5. 扬压

uplink 1. 向上传输 2. 上行链路 3. 上行线路

upload 1. 向上负载，逆载，载出 2. 上传

UPM 1. Undefined Protocol Machine 未定义协议机 2. Unit Production Manager 单位生产管理人员 3. Update Protocol Model 更新协议模型

upper airfoil 上机翼

upper airfoil surface 上机翼表面

upper airway flight 高空飞行

upper atmosphere 高层大气

upper bound 上界，最大值，上限

upper boundary 上界，上部边界

upper duct 上管

upper interface 上界面

upper limit 上限，上极限，上限尺寸，最大限度

upper pair of vane 叶片上对

upper panel 上翼片，上方控制面板，上门，上面板

upper part 上部，上半部分

upper portion 上部，上半部分

upper right 右上方，右上角

upper section 上部，上半部分

upper shaft 上端轴

upper side 上盘，上面，上侧，上方

upper stage 上面级，上一阶段

upper stage rocket engine 上面级发动机

upper stage rocket motor 上面级发动机

upper state 上态，上边频，高能态

upper surface 上翼面，上表面，气泡表面，上平面

upper tab 上片

upper triangular matrix 上三角矩阵

upper tube 上管

upper vane 上部叶片

upper wall 上盘，顶壁

upper wing 上翼，脱离翼

upper-surface transition 上表面的过渡

uprated cruise 大巡游

UPRM Univeral Platform Resource Management 通用平台资源管理器

UPS Uninterruptible Power Supply 不间断供电电源

upset 颠倒，推翻，倾覆，扰乱

upset rate 翻转率

upset recovery 破坏恢复

UPSMS UPS Management System 不间断供电电源管理系统

up-stream 上升流

upstream baffle-to-nozzle 上游挡板的喷嘴

upstream blade row 上游叶片排

upstream boundary 上游边界

upstream cavity 上游腔

upstream cavity trench 上游腔槽

upstream combustion 上游的燃烧

upstream combustion ramjet mode 上游燃烧冲压模式

upstream compressor 上游压缩机

upstream compressor spool 上游压缩机滑阀

upstream cylindrical port 上游圆柱口

upstream density 上游密度

upstream direction 上游方向

upstream domain 上游区域

upstream edge 上游边缘

upstream end 进入端，进汽端

upstream face 入水面，上游面

upstream flap 上游瓣

upstream flow 上游流

upstream influence 上游影响

upstream influence length 上游影响长度

upstream injection 上游注入

upstream interaction 上游的相互作用

upstream placement 上游位置

upstream plasma 上游等离子

upstream port 上游端口

upstream portion 上游部分

upstream position 上游位置

upstream pressure 进口压力，上游压力

upstream pressure perturbation 上游压力扰动

upstream propagation 上游传播

upstream region 上游区

upstream rotor 上游转子
upstream rotor blade 上游转子叶片
upstream sheath 上游鞘
upstream side 上游侧
upstream slot 上游槽
upstream stator 上游定子
upstream straight section 上游直管段
upstream turbomachinery 上游涡轮机
upstream vane 上游叶片
upstream vane wake 上游叶片尾迹
upstream wake 上游尾迹
upstream wall 上游坝墙
upstream work 上游工作
upstroke 活塞上行,上挑的笔迹,上升冲程,上行冲程
uptake 吸收,摄取,上风口,摄入
upward acceleration 向上的加速度
upward or downward acceleration 向上或向下的加速度
upwash 上洗流,上洗,升流,上游离子化,气流上洗
upwash correction 上升流的修正
upwash field 升流场
upwash flow 上升流
upwind method 迎风方法
upwind scheme 迎风格式
uranian satellite 天卫
uranium 【化】铀(化学元素,符号 U)
uranium density 铀密度
uranium distribution 铀的分布
uranium fluoride 氟化铀
uranium loading 铀萃取量
Uranus 【天】天王星
urban area 城市地区,城区
urban environment 城市环境
URE User Range Error 用户测距误差
urgent servicing 紧急维修
urine collection 小便收集,尿采集
urine collection bag 尿收集袋
URL Uniform Resource Locator 统一资源定位器
urn 缸,瓮,壶
US ton 短吨,美吨
US& RAeC United Service and Royal Aero Club 联合勤务队及皇家航空俱乐部(英国)
usability 可用性,适用性
usable part 使用部分
USAFE United States Air Force in Europe 美国驻欧洲空军
USAFSS United States Air Force Security Service 美国空军安全处
usage 1. 使用,运用 2. 用法 3. 用率 4. 已使用(飞行)的小时数

USASATCOMA United States Army Satellite Communications Agency 美国陆军卫星通信局
USAT 1. Ultra Small Aperture Terminal 超小口径终端 2. United States Army Transport 美国部队运输
USB 1. Upper Sideband 上边带 2. Upper-Surface Blowing 上翼面吹气
USC 1. Ultra Sonic Cleaning 超声波清洗 2. United States Code 美国代码 3. Universal Service Circuit 通用业务电路 4. User Service Center 用户服务中心
USCG United States Coast Guard 美国海岸警卫队
USCGS U. S. Coast and Geodetic Survey 美国海岸与大地测量局
USDAO United States Defense Attaché Office 美国国防武官处
USDOC (Usdoc) United States Department Of Commerce 美国商业部
USDOCO United States Documents Office(r)美国文献处(官)
USE 1. Ultrasonic Scanning Equipment 超声波扫描设备 2. Unconditionally Stable Extended 无条件稳定扩展 3. Unified Search Environment 统一的搜索的环境 4. Unified Stochastic Engine 统一的随机的引擎 5. User System Emulation 用户系统仿真
use case 1. 用例,用况 2. 使用案例,使用实例
use of aileron 使用副翼
use of cavity 利用腔
use of compliant wall 使用兼容的墙
use of digital technology 利用数字技术
use of fuel 燃料的用量
use of helium 使用氦气
use of nanoaluminum 使用纳米铝
use of outer space 外层空间的利用
use of wing 利用机翼
useful casualty levels 有效杀伤程度
useful consciousness 有效意识
useful consciousness duration 有效意识时间
useful consciousness limit 有效意识极限,储备时间极限
useful energy 有用能
useful life 有效期,有效寿命,使用寿命
useful lift 有用浮力(指轻于空气航空器的总重量与结构重量之差),有效升力
useful load 有用载重(包括业载、空勤组、设备、油料及压舱),有效载荷
useful power 有效功率
useful work 有效功,有用功
usefulness 有效性,有用性
USENET User Network 用户网络
user 用户

user application 用户应用
user charges 用户支付的费用,用户费
user datum 用户数据
user equipment 用户设备
user implementation 用户实现
user input 用户输入
user interface 用户界面
user need 用户需要
user terminal 用户终端
user-friendly interface 用户友好接口
users manual 用户手册
user-user protocol 用户-用户规程
USG 1. US Gallon 美加仑 2. United States Gauge 美国标准(线,量)规
USGPM US gal./min 每分钟(流量为)……美加仑
USGS US Geological Survey 美国地质调查局
USHGA United States Hang Gliding Association 美国悬挂式滑翔协会
USIS Ultraviolet Stratospheric Imaging Spectrometer 平流层紫外成像光谱仪
USL 1. Upper Specification Limit 规格上限 2. U.S. Laboratory 美国实验室(美国空间站上的)
USLANTCOM United States Atlantic Command 美国大西洋总部
USML United States Microgravity Laboratory 美国微重力实验室
USMS 1. United States Maritime Service 美国海运服务 2. United States Marshalls Service 美国调度服务 3. Utility Systems Management System 通用系统管理系统
USNO US Naval Observatory 美国海军天文台
USNTPS USN Test Pilot School 美国海军试飞员学校
USPA 1. United Security Professionals Association, Inc. 美国安全专业人员协会有限公司 2. United States Parachute Association 美国跳伞协会
USPS United States Postal Service 美国邮政部
USS 1. Ultraviolet Solar Spectrometer 紫外太阳光谱仪 2. United States Standard 美国标准 3. User Services Support 用户服务保障
USSA 1. United States Salvage Association 美国海难援救协会 2. United States Standard Atmosphere 美国标准大气
USSFIM United States Standard Flight Inspection Manual 美国标准飞行检查手册
USSG United States Standard Gauge 美国标准(线)规
USSP Universal Sensor Signal Processor 通用传感器信号处理机
USSR 1. Ultra Simple Skirmish Rules 超简单的冲突的规则 2. Underground Security Systems Research 地下安全系统的研究
USSS 1. United States Secret Service(US Treasury Departement) 美国特勤局 2. United States Social Security 美国社会保障 3. United States Steamship 美国轮船 4. Unmanned Sensing Satellite System 无人遥感卫星系统
USSST 1. United States Salt Spray Test 美国盐雾试验,美国盐喷洒测试 2. United States Standard Screw Thread 美国标准螺纹
USSTAF United States Strategic and Tactical Air Forces 美国战略与战术空军
USTO Ultra-Short Takeoff 超短距起飞
USV 1. United States Volunteers 美国志愿兵 2. Unmanned Surface Vehicle 无人艇,无人地面车辆
USW 1. Ultrasonic Welding 超声焊 2. Ultrashort Wave 超短波 3. Uranium Separative Work 铀分离功
UT 1. Ultra Thin 超薄型的 2. Ultrasonic Test 超声波检验 3. Unit Test 单元测试 4. Universal Time 世界时 5. Universal Trainer 通用练习器 6. Usability Testing 适用性测试 7. User Terminal 用户终端 8. Utility Tool 有用工装
UTA 1. Unit Training Assembly 部队集训 2. Unmanned Tactical Aircraft 无人驾驶战术飞机
Utail U形尾翼
UTC Universal Time Coordinated 世界协调时
UTC offset 世界协调时与当地时间的差值
UTDF Universal Tracking Data Format 通用跟踪数据格式
utilidor 1.通用输送管道 2.保温管道
utilisation 利用
utilise 利用,使用
utility 1.飞机的诚用型 2.在美国联邦航空局的飞机类型分类中,表示限制特技飞行的 3.功用,效用 4.有用的物体或器械 5.公用事业公司 6.公用事业
utility curve 效用曲线
utility finish 实用完成
utility function 效用融合
utility fusion 效用函数
utility fusion method 实用的融合方法
utility glider 辅助滑翔机,滑翔教练机,通用滑翔机
utility improvement 效用改进
utility management 效用管理
utility metric 效用度量
utility rate 效用率
utility space 效用空间
utility system 通用系统
utility theory 效用理论
utility value 效用值,利用价值
utilization 利用率,飞行时间利用率

utilization efficiency 利用效率

utilization rate 利用率

utilize 利用,使用

UTM 1. Universal Transverse Mercator 统一横轴墨卡托投影 2. Unsuccessful Transfer Message 信息传输失败 3. Utmost 极端的,极限,最大限度的,最大可能

UTM grid navigation 统一横轴墨卡托投影格网导航

UTP 1. Unit Test Pilot 部队试飞员 2. Universal Trunk Processor 通用中继处理机 3. Universal Trunk Protocol 通用中继协议 4. Unshielded Twisted Pair 非屏蔽双绞线 5. Utility Tape Processor 应用磁带处理机

U-tube manometer U 型压力计

U-turn maneuver 掉头机动

UUM 1. Underwater-to-Underwater Missile 水下对水下导弹,潜对潜导弹 2. Unification of Unit of Measurement Panel 量度单位统一小组

UUPI Ultrasonic Undercarriage Position Indicator 超声波起落架位置指示器

UUT 1. Unit Under Test 受试装置,被试件 2. Universal User Tracking 普遍的用户跟踪

UV 1. Ultraviolet 紫外线 2. Under-Voltage 电压不足,欠压 3. Unmanned Vehicle 无人驾驶飞行器 4. Underwater Vehicle 水下航行器 5. Upper-sideband Voice 上边带语音

UV photoelectron spectroscopy 紫外光电子谱(学)

UV radiometer 1. 紫外线辐射测量卫星(美国) 2. 紫外辐射仪

UV spectroirradiometer 紫外分光辐射计

UVAS 1. Ultraviolet Astronomical Satellite 紫外天文卫星 2. Unmanned Vehicle for Aerial Surveillance 空中监视无人机

UVDIAL Ultraviolet Differential Absorption Lidar 紫外差分吸收激光雷达

UVDS Ultraviolet Flame Detector System 紫外线火焰探测系统

UVEPROM Ultraviolet Erasable Programmable Read Only Memory 紫外线可擦除可编程只读存储器

UVEROM Ultraviolet Erasable Read-Only Memory 紫外线可擦只读存储器

UVL 1. Ultraviolet Laser 紫外线激光器 2. Ultraviolet Light 紫外线灯

UVPI Ultraviolet Plume Instrument 紫外羽焰探测仪

UVS 1. Ultraviolet Spectrometer 紫外线分光仪(光谱仪) 2. Ultraviolet Spectrometry 紫外线光谱测定法

UVV Ultimate Vertical Velocity 极限垂直速度

UW 1. Ultrasonic Wave 超声波 2. Unconventional Warfare 非常规战争

UWB 1. Ultra-Wide Band 超宽带 2. Underwater Burst 水下爆炸

UWR 1. Ultrapure Water Rinsing 超纯水漂洗 2. Underwater Range 水下射程,水下射击场

UXB Unexploded Bomb 未爆炸弹

UX-cored solder UX 芯焊料

UXO Unexploded Ordnance 未爆弹药,未爆炸武器

UXPLD Unexploded (Bomb) 未爆(炸弹)

UY Unit Years 机组使用年

V

V 1. Valve 活门 2. VIP Baggage 要客行李 3. Vertical 垂直的 4. Velocity 速度 5. Volt 伏特,伏 6. Voltage 电压 7. VOR Airway 航路
V antenna V形天线
V curve V形曲线
V family of launch vehicle V系列运载火箭
V imaging plane V成像平面
V slit type sun sensor 缝式太阳敏感器
V&V Verification & Validation 验证,认可
V/H Velocity and Horizontal 垂直和水平
VA 1. Variable Attenuator 可变阻尼器 2. Video Amplifier 视频放大器 3. Virginia 弗吉尼亚州 4. Virtual Address 虚拟地址 5. Visual Approach 目视进近 6. Volcanic Ash 火山灰 7. Volt Ampere 伏安
VAB Voice Answer Back 话音应答
VAC 1. Vacation 假期 2. Vacancy 空位 3. Vaccuum 真空 4. Value Added Carrier 增值运营商 5. Value-Added Chain 增值链 6. Video Amplifier Channel 视频放大器通道 7. Visual Approach Chart 目视进近图 8. Volts Alternating Current 交流电压
vacancy 空位
vacancy cluster 空位团
vacancy flow 空位流
VACBI 1. Video and Computer Based Instruction 录像和计算机辅助教学 2. Visual and Computer Based Instruction 目视和计算机教学
vaccm relay 真空继电器
vacuo 真空
vacuum 真空,真空系统
vacuum accumulator 真空贮气筒
vacuum arc 真空电弧
vacuum bag moulding 真空袋成型
vacuum bakeout 真空烘烤
vacuum balance 真空天平
vacuum bell jar 真空钟罩
vacuum booster pump 真空助力泵
vacuum brake 真空制动器
vacuum brazing 真空钎焊,真空焊接
vacuum breakdown 真空击穿
vacuum capacitor 真空电容器
vacuum casting 真空铸造
vacuum chamber 真空室,真空箱
vacuum chamber pressure 真空室压力
vacuum charging and discharging test 真空充电与放电试验
vacuum coating 真空镀膜,真空涂层
vacuum cold welding 真空冷焊
vacuum cold welding test 真空冷焊试验
vacuum condition 真空条件,真空环境
vacuum cup stopcock 旋塞阀的真空杯
vacuum degassing 真空除气,真空脱气
vacuum degree 真空度
vacuum deposited coating 真空沉积涂层
vacuum deposition 1. 真空镀膜 2. 真空淀积
vacuum discharge 真空放电
vacuum dissipator 真空喷雾器
vacuum dry friction 真空干摩擦
vacuum electron beam welding 真空电子束焊
vacuum electron device 真空电子器件
vacuum evaporation 真空蒸发,真空涂膜
vacuum evaporation technology 真空蒸发工艺
vacuum facility background pressure 真空设施的背景压力
vacuum fluorescent display 真空荧光显示器
vacuum forming 真空成形
vacuum gas filling 真空充气
vacuum gauge 真空计
vacuum glue pouring 真空灌胶
vacuum hardening 真空淬火
vacuum heat treatment 真空热处理
vacuum impregnation 真空浸渍
vacuum impregnation degassing 真空脱气
vacuum jet 真空喷射泵
vacuum jet exhauster 真空喷射器
vacuum leak detecting system 真空检漏系统
vacuum leak detection 真空检漏
vacuum level 真空度,真空能级
vacuum lock chamber 真空锁定室
vacuum manometer 真空压力计,真空压力表
vacuum metal flexible pipe 真空金属软管
vacuum metal hard pipe 真空金属硬管
vacuum multi-layer insulation 真空多层绝热
vacuum oil filling 真空充油
vacuum optical test bench 真空光学试验台
vacuum pan 真空锅
vacuum pipe socket 真空管插座

vacuum port 真空口
vacuum potting 真空浸渍
vacuum pressure valve 真空压力阀,呼吸阀
vacuum pump 真空泵
vacuum pumping system 真空排气系统
vacuum putty 真空封泥
vacuum quenching 真空淬火
vacuum relief valve 真空安全阀,真空减压阀
vacuum retort furnace 真空干馏炉
vacuum seal 真空密封
vacuum specific impulse 真空比冲
vacuum sublimation 真空升华
vacuum system 真空系统,真空装置
vacuum tap 真空接口
vacuum test 真空试验,真空检验
vacuum thermogravimetry 真空热重分析法
vacuum thrust 真空推力
vacuum thrust-to-weight ratio 真空推力重量比
vacuum tube casting 真空插管浇注
vacuum tunnel 真空式风洞
vacuum ultraviolet 真空紫外线
vacuum ultraviolet radiation 真空紫外辐射
vacuum valve 1.真空管,电子管 2.真空阀
vacuum vapor plating 真空蒸发镀膜
vacuumometer 真空计,低压计
vacuum-pumping 真空排气,抽真空
VAD 1. Value-Added Distributor 增值分销商 2. Velocity Azimuth Display 速度方位显示器
VADS Value-Added Data Service 增值数据业务
V-aerial V形天线
VAFB Vandenberg Air Force Base 范登堡空军基地
VAI 1. Variable Area Intake 高截面进气口 2. Video Assisted Instruction 视频辅助指令
valence (化合)价,原子价
valence band 价(电子)带
valence bond (化合)价键,价键耦合
valence crystal 价键晶体
valence electron 价电子
validate 使生效,批准,使合法化,证实
validation 生效,在空白机票上盖章生效,批准,确认
validation case 验证案例
validation mission 验证任务
validation of type certificate 型号认可证书
validation phase 生效阶段
validation range 验证范围
validation test 确认试验
validity 有效性,真实性
valley-span antenna 山谷天线
valuation 估价,评价

valuation charges 声明价值费
valuation model 评价模式
value 值
value chain 价值链
value decomposition 值分解,值算法,值分解法
value engineering 价值工程,工程经济学
value function 价值函数
value of apoapsis radius 远拱点半径值
value of coef 价值系数
value of heat 热量值
value of objective 价值目标
value of on-orbit servicing 在轨服务价值
value of pressure 压力值
value of state 值状态
value of tension 张力值
value of thrust 推力值
value of time 时间价值
value of upper limit 上限值
value of variable 变量值
value problem 边值问题
value proposition 价值主张,价值定位
values 数值,用途,用处,价值观
valve 活门,阀,电子管,真空管
valve duration 气门开启持续时间
valve gear 阀动装置
valve hood 阀盖,阀帽
valve lag 阀滞后,气门迟关
valve lead 气门提前打开导气程阀门
valve operation 阀操动
valve petticoat 阀衬
valve response 阀门响应
valve rigging 阀装置
valve seat 阀座
valve stem 气门杆,阀杆
valve timing gear 阀定时装置
valve timing mechanism 气门定时机构
valve with electrically motorized operation 电动阀
valving element 阀元件
VAM 1. Video Access Module 视频接入模块 2. Video Administration Module 视频管理模块 3. Virtual Access Method 虚拟存取法 4. Visual Approach Monitor 目视进近监控仪
VAN Value-Added Network 增值网
Van Allen belt 范艾伦带,范艾位带
Van Atta reflector 范爱塔反射器
Van de Graaff 范德格拉夫起电机
Van der Waals equation 范德瓦尔斯方程
Van der Waals forces 范德瓦尔斯力
vanadium 【化】钒(化学元素,符号 V)

vanadizing　渗钒
vane　1. 叶片,轮叶 2. 风向标 3. 静子叶片
vane actuator　燃气舵致动器
vane angle　桨角,叶轮角
vane carrier　叶片
vane cascade　叶栅
vane count　叶片数
vane frequency　叶片频率
vane passage　叶片通道
vane pitch　叶片节距,叶栅栅距
vane potential　叶片潜力
vane pressure　叶片压力
vane pump　旋板泵,叶轮泵,叶片泵
vane response　叶片响应
vane row　叶片排
vane servomotor　轮叶接力器
vane shaft　叶片轴
vane span　叶片跨度
vane stagger　叶片斜罩
vane surface　叶片表面
vane surface pressure　叶片表面压力
vane trailing-edge shock　叶片后缘激波
vane wake　叶尾迹
vaned diffuser　叶片式扩散器
vaneless diffuser　无叶扩压器,无叶片式扩散器,无导连的扩压器
vaneless space　无叶片空间
vanish　消失
vanishing line　消失线,没影线
vanishing point　消失点,隐没点
vanishing point control　合点控制,灭点控制
vantage　有利情况,优势
VAP　1. Verification & Acceptance Plan 识别和接受计划 2. Video Access Point 视频接入点 3. Videotex Access Point 可视图文接入点 4. Visual Aids Panel 目视设备信号板 5. Visual Approach Procedure 目视进近程序
VAPI　Visual Approach Path Indicator 目视进近航路指示器
vapor　蒸汽,水汽
vapor barrier clothing　防湿服
vapor bubble　蒸汽泡
vapor cavity　蒸汽空泡
vapor cleaning　蒸汽净洗
vapor compression　蒸汽压缩
vapor content　含汽量
vapor crystal facility　气相晶体设施
vapor cycle air conditioning system　蒸汽循环空调系统
vapor cycle cooling system　蒸汽循环冷却系统

vapor density　蒸汽密度
vapor density coefficient　蒸气密度系数
vapor deposition　汽相沉积
vapor equilibrium　汽相平衡
vapor expansion　蒸汽膨胀
vapor lock　蒸汽汽塞,汽封
vapor mass　蒸气质量
vapor phase　蒸汽相
vapor phase epitaxial growth　汽相外延生长
vapor phase epitaxy　汽相外延
vapor phase reflow soldering　气相再流焊
vapor pressure of liquid aluminum　液态铝蒸气压力
vapor refrigeration compressor　气相冷冻压缩机
vapor region　蒸汽区
vapor screen technique　蒸汽屏法
vapor volume　蒸汽容积
vaporization　蒸发,汽化
vaporization model　汽化模型
vaporization rate　汽化速度,蒸发速度
vaporize　使蒸发,汽化
vaporizer　蒸发器,汽化器,喷雾器
vaporizer exit　蒸发器出口
vaporizer housing　汽化器壳体
vaporizer wall　管壁
vaporizing ammonia electrothermal system　电热系统蒸氨
vaporizing combustor　蒸发式燃烧室
vapor-screen method of flow visualization　蒸汽屏显示
vapothorax　蒸汽胸
vapour cycle　蒸汽循环
vapour degreasing　蒸汽除油,蒸汽脱脂
vapour deposition coating　气相沉积涂层
vapour lock　汽塞,汽封,汽嘴
vapour pressure　蒸汽压力,水汽压
vapour tension　蒸汽压,汽压
vapour trail　水汽尾迹,雾化尾迹
vapour-phase inhibitor　汽相阻化剂,气相缓蚀剂
VAPS　1. Virtual Applications Programming Software 虚拟应用程序编程软件 2. Virtual Avionics Prototyping System 航空电子虚拟样机系统
VAQ　VHF Avionics Qualification 甚高频航空电子设备资格
VAR　1. Value Added Reseller 增值转卖 2. Volt-Ampere Reactive 无功伏安
varactor　1. 变容二极管 2. 变抗器,可变电抗器
varactor diode　变容二极管
varactor tuner　变容二极管调谐器
variability　变率,可变性,变化性
variable　1. 可变的,可调的 2. 变量,变数,变项

variable beam antenna of fuze 可变波束引信天线	**variable resonance** 可变谐振
variable camber 1.可变的曲度 2.可变弯度 3.变曲面	**variable specific** 可变比
variable camber flap 可变弯度瓣	**variable speed** 变速,变转速
variable camber wing 可变弧高机翼	**variable stability airplane** 可变稳定性飞机
variable cant 变倾斜	**variable stability flight control** 变稳飞行控制
variable capacitance diode 可变电容二极管	**variable stability helicopter** 变稳定直升机
variable capacitor 可变电容器	**variable star** 变星,变光星
variable coefficient 可变系数	**variable stator** 可变定子
variable control 可变控制	**variable stator vane** 可变静子叶片,可调静子叶片
variable control effector 变量的控制效应	**variable structure system** 可变结构系统
variable cross 变截面	**variable sweep** 变翼,可变翼
variable cross section chamber 变截面室	**variable swept back wing** 变后掠翼
variable cycle engine 可变周期发动机,变循环发动机	**variable swept wing** 变后掠翼
variable datum boost control 可变数据升压控制	**variable thermal conductance heat pipe** 变热导热管
variable delivery pump 变量输送泵	**variable threshold logic** 可变阈逻辑
variable density wind tunnel 变密度风洞	**variable thrust** 可调推力,可变推力
variable discharge turbine 变流量涡轮	**variable thrust liquid rocket engine** 变推力液体火箭发动机
variable displacement pump 变量泵,变排量泵	**variable time bomb** 可变时间炸弹
variable drive ratio mechanism 变传动比机构	**variable time fuze** 变时引信
variable energy 可变能量	**variable trajectory range control** 可变轨迹范围控制
variable entropy 变熵	**variable wing** 可变机翼,变几何形状机翼
variable expansion 变膨胀	**variable wing camber control** 机翼变弯度控制
variable flow restriction 可变流量限制	**variable X-ray source** 变量X射线源
variable frequency AC power system 变频交流电源系统	**variable-area** 可变面积,可变区域,可变截面的
variable frequency vibration test 变频振动试验	**variable-area exhaust nozzle** 可变面积抽气喷管,可调面积排气装置
variable geometry 可变几何形状	**variable-area nozzle** 可变截面喷管
variable geometry aircraft 变几何形状飞机	**variable-area wing** 可变机翼面积
variable geometry combustor 变几何燃烧室	**variable-position cowl lip** 可变位整流罩前缘
variable geometry design 变几何设计	**variables** 变量
variable geometry inlet 可调进气道	**variance** 方差,偏离值,分歧,差异,变化
variable group 变量组	**variance analysis** 方差分析
variable horizon 可变地平线	**variance curve** 方差曲线
variable inductor 可变电感器,可调电感器,可变电感线圈	**variance method** 方差法
variable inlet guide vane 可变进气导向叶片	**variance of unit weight** 单位权方差,方差因子
variable interaction 变量的相互作用	**variance-covariance matrix** 方差-协方差矩阵
variable length code 变长码	**variance-covariance propagation law** 方差-协方差传播律
variable model 参数模型	**variant** 变体,变异的,不同的
variable nebula 可变星云	**variation** 变化,变量,偏差
variable nozzle 变截面喷嘴,可调喷嘴	**variation coefficient** 变差系数,变分系数,变化系数
variable object 可变对象	**variation flap** 变异瓣
variable overhead 可变间接费	**variation method** 变分法
variable parameter simulation 变参数仿真	**variation of acoustic velocity** 声波速度变化
variable pitch propeller 变距螺旋桨,可变节距推进器,可变螺距螺旋桨	**variation of flow parameter** 流动参数的变化
variable porosity wind tunnel 变量的孔隙率风洞	**variation requirement** 变化的要求
variable radio source 射电变源,变射电源	**variation trend** 变化趋势
variable ratio 可变比值的	**variational equation** 变分方程
variable reluctance resolver 变磁阻旋转变压器	

variational method 变分法
variational operation 变分运算
variational principle 变分原理
varicose mode 对称模式
variety 多样化,变化,种类
variety of fuel 燃料种类
variety of sensor 各种传感器
variomat 变线仪
variometer 变感器,可变电感器,升降速度仪
varioscale projection 变比例投影
various angle of attack 不同角度的攻击
various aspect 各方面
various binder 各种黏合剂
varnish 清漆,上漆
varsol 烃类溶剂
vary 1. 改变,变更,多样化 2. 变化,不同
VAS 1. VISSR Atmospheric Sounder 可见光红外自旋扫描辐射计大气探测器 2. Value-Added Service 增值服务 3. Vortex Advisory System 涡咨询系统
VASI Visual Approach Slope Indicator 目视进近坡度指示器
VASIS Visual Approach Slope Indicator System 目视进近坡度指示系统
vasodepressor syncope 血管减压神经性晕厥
vasovagal attack 血管迷走神经性发作
vasovagal syncope 血管迷走神经性晕厥
VAST 1. Versatile Avionics System Tester 航空电子系统多样测试器
vast distance 辽阔的距离
VAT 1. Value-Added Tax 增值税 2. Visual Audio Tool 视觉音频工具
VATLS Visual Airborne Target Locator System 机载可视目标定位器系统
VAW Vertical Assistance Window 垂直援助窗口
VAWT Vertical Axis Wind Turbine 立轴风力机
V-beam radar V形波束雷达
V-belt V形波带,V形皮带
VBS Vertical Beam Sensor 垂直波束传感器
VBV 1. Variable Bleed Valves 可变引气活门 2. Variable Bypass Valve 可变旁通活门
VC 1. Valuable Cargo 贵重货物 2. Variable Camber 可变的曲度 3. Ventilation Controller 通风控制器 4. Vertical Circle 地平径圈 5. Video Compressor 视频压缩器 6. Video Conference 电视会议 7. Video Controller 视频控制器 8. Virtual Call 虚呼叫 9. Virtual Channel 虚信道 10. Virtual Circuit 虚电路 11. Virtual Connection 虚连接 12. Virtual Container 虚容器 13. Volume of Compartment 机舱容积
VCC 1. Video Cable Communication 视频电缆通信 2. Video Capture Card 图像捕获卡 3. Video Control Center 视频控制中心 4. Virtual Call Capability 虚呼叫能力 5. Virtual Call Control 虚呼叫控制 6. Virtual Centralized Controller 虚拟集中控制器 7. Virtual Channel Connection 虚信道连接
VCD 1. Voltage Controlled Device 电压控制装置 2. Vortex Control Device 涡流控制器
VCE 1. Virtual Call Evolution 虚呼叫演变 2. Virtual Channel Entity 虚信道实体
VCHP Variable Conductance Heat Pipe 可变热导热管
VCID Voice Controlled Interactive Device 声控交互式设备
VCL Visual Check List 目视检查清单
VCM 1. Value Chain Mangement 价值链管理 2. Version Compatibility Matrix 版本兼容性表 3. Video Camera Module 摄像机模块
VCO Voltage Controlled Oscillator 电压控制振荡器
VCOS Visible Caching Operating System 可视高速缓存操作系统
VCR 1. Video Cassette Recorder 盒式磁带录像机 2. Visual Control Room 目视操纵室
VCS 1. Ventilation Control System 通风控制系统 2. Video Conference Service 视频会议业务 3. Virtual Circuit Switch 虚电路交换 4. Visual Call Sign 可视电话呼号 5. Voice Communication System 话音通信系统 6. Volume Control System 容积控制系统 7. Voter Comparator Switch 表决比较器开关
VCSS Voice Communication Switching System 无线电通话转接系统
VCU 1. Valve Control Unit 活门控制组件 2. Video Control Unit 视频控制组件 3. Voice Channel Unit 话音信道单元 4. VDL Control Unit 甚高频数据链控制装置
VD 1. Vertical Display 垂直显示 2. Vertical Distance 垂直距离 3. Video Decoder 视频译码器 4. Virtual Data 虚拟数据 5. Virtual Device 虚拟设备
VDA 1. Validation of Design Approval 设计认可批准书 2. Video Distributing Amplifier 视频分配放大器 3. Virtual Data Access 虚拟数据存取 4. Voltage Drop Analysis 电压降分析
VDC 1. Video Display Controller 视频显示控制器 2. Voltage Direct Current 直流电压
VDD 1. Version Description Document 改版说明文件 2. Virtual Device Driver 虚拟设备驱动程序 3. Visual Display Data 可视化显示数据 4. Voice Data Display 音频数据显示器
VDDL Virtual Data Description Language 虚拟数据描述语言
VDE Voice Data Entry 话音数据输入

VDI 1. Vertical Deviation Indicator 垂直偏差指示器 2. Video Data Interrogator 显示数据询问器 3. Video Device Interface 视频设备接口 4. Video Display Interface 视频显示器接口

VDL VHF(Very High Frequency) Data Link 甚高频数据链

VDM 1. Video Display Metafile 视频显示元文件 2. Vienna Development Method 维也纳开发方法

VDMA Variable Destination Multiple Access 接收站可变多址访问

VDOP Vertical Dilution of Precision 垂直精度扩散因子

VDP 1. Vehicle Deadlined for Parts 航空器因零件损坏而停用 2. Vertical Data Processing 垂向数据处理 3. Video Display Processor 视频显示处理机 4. Visual Descent Point 目视下降点

VDS 1. Variable Depth Sonar 变深声呐 2. Video Display System 视频显示系统 3. Voice Data Service 话音数据通信业务

VDT 1. Variable Density Tunnel 可变密度风洞 2. Video Data Terminal 可视数据终端 3. Video Dial Tone 视频拨号音 4. Video Display Terminals 视频显示终端 5. Voice Data Trunking 话音数据中继

VDU 1. Video Display Unit 视频显示组件 2. Video Distribution Unit 视频分配组件 3. Visual Display Unit 目视显示装置

VE 1. Value Engineering 价值工程 2. Velocity East 向东速度 3. Video Expander 视频扩展器 4. Virtual Enterprise 虚拟企业 5. Visual Exempted 免除目视

VECP Value Engineering Change Proposal 价值工程更改建议

vectodyne 推力换向式垂直起飞飞机
vector 1. 矢量,向量 2. 引导,引向目标
vector aiming method 向量瞄准法
vector change 向量变化
vector component 矢量分量
vector computer 矢量计算机,向量计算机
vector cross product 矢量叉积
vector data 矢量数据
vector diagram 矢量图
vector far field monitor 向量的远场监测器
vector field 矢量场,向量场
vector flight control 矢量控制
vector function 向量函数,矢量函数
vector generator 矢量发生器
vector gunsight 矢量瞄准器
vector immittance meter 矢量导抗测量仪
vector Lyapunov function 向量李雅普诺夫函数
vector magnetometer 矢量磁力仪
vector magnitude 矢量幅度
vector maneuver 矢量机动
vector measurement 矢量测量
vector network analyzer 矢量网络分析仪
vector plotting 矢量绘图
vector quantity 矢量,向量
vector result 结果向量
vector separation 矢量分离
vector sight 矢量瞄准器
vector solution 矢量解法
vector steering 航线操纵,推力矢量控制
vector sum 矢量和
vector theory 向量理论
vector warhead emission velocity 弹头发射速度矢量
vectored attack 导向攻击
vectored thrust 矢量推力
vectored thrust engine 定向推力发动机,推力转向式发动机
vectoring 引导,定向
vectoring in forward flight 向前飞行的航向
vectoring nozzle 导向喷管
vectoring post 引导哨
vectoring radar 引导雷达
vee cone V形锥体
vee gutter V形排水槽,三角槽
vee tail V型尾翼
vee tube V形管
veeder counter 速力计,测程器,测程仪
veer 1. 调向,变向 2. 放出,顺转
veer off 偏离航向,偏离规定方向
veering 风向按时针方向变化
Vega 【天】织女星
vegetation 植被,生长,草木
vegetation canopy 植被冠层
vegetation map 植被图
vehicle 1. 车辆,交通工具 2. 传播媒介,工具
vehicle aerodynamics 飞行器空气动力学
vehicle agent 飞行器代理
vehicle attitude 飞行器姿态
vehicle axes 飞行器轴
vehicle center 飞行器中心
vehicle concept 概念飞行器
vehicle configuration 飞行器配置
vehicle control 飞行器控制
vehicle coordinate system 运载体坐标系
vehicle definition 飞行器定义
vehicle design 飞行器设计
vehicle drag 飞行器阻力
vehicle dynamics 飞行器动力学

vehicle forebody 飞行器前部	**velocity coefficient** 速度系数
vehicle frame 飞行器架	**velocity command** 速度指令
vehicle gross weight 飞行器总重量	**velocity component** 分速度,速度分量
vehicle height 飞行器高度	**velocity contour** 速度分布图,等流速线
vehicle inertia 飞行器惯性	**velocity control** 速度控制,航速控制
vehicle interaction 飞行器相互作用	**velocity coordinate system** 速度坐标系
vehicle management 飞行器管理	**velocity correction** 速度修正
vehicle management system 飞行器管理系统	**velocity coupling** 速度耦合
vehicle maneuver 机动飞行器	**velocity cutoff** 速度关机
vehicle mass 飞行体质量,飞行器质量	**velocity datum** 速度数据
vehicle mass ratio 飞行体质量比,飞行器质量比	**velocity deception jamming** 速度欺骗干扰
vehicle model 飞行器模型,飞行器型号	**velocity defect** 速度缺陷
vehicle motion 飞行器运动	**velocity deficit** 速度亏损
vehicle name 飞行器名称	**velocity derived by differential** 微分求速
vehicle performance 飞行器性能	**velocity diagram** 速度图,速度分解图,速度曲线
vehicle position 飞行器位置	**velocity direction** 速度指令方向取反
vehicle probe 探测机	**velocity discontinuity** 速度不连续性
vehicle property 飞行器性能	**velocity dispersion** 速度弥散度,速度分散
vehicle response 飞行器响应	**velocity distribution** 速度分布
vehicle rock 飞行器滚摇	**velocity disturbance** 速度干扰
vehicle solution 飞行器解决方案	**velocity divergence** 速度散度
vehicle speed 飞行器速度	**velocity ellipsoid** 速度椭球
vehicle state 飞行器状态	**velocity error** 速度误差
vehicle station 机载电台	**velocity error angle** 速度误差角
vehicle surface 飞行器表面	**velocity estimate** 速度估计
vehicle system 飞行器系统	**velocity estimation** 速度估计,速度估算
vehicle technology 飞行器技术	**velocity feedback** 速度反馈
vehicle trajectory 飞行器轨迹	**velocity field** 速度场
vehicle type 飞行器类型	**velocity field measurement** 速度场测量
vehicle velocity 飞行器速度	**velocity filtering** 速度滤波
vehicle velocity vector 飞行器速度矢量	**velocity fluctuation** 速度脉动,速度起伏,速度波动
vehicle volume 飞行器流量	**velocity gain** 速度增量
vehicle weight 飞行器重量	**velocity gate** 速度门
vehicle-derived navigation data 飞行器导航的数据来源	**velocity gradient** 速度梯度
veil cloud 幔云	**velocity head** 速度头
Vela 维拉卫星	**velocity increase** 速度增加
vela sensor 贝拉传感器	**velocity increment** 速度增量
velcro 1.尼龙搭扣 2.黏合扣,魔术贴	**velocity indication coherent integrator** 速度指示相干积分器
velcro strength 尼龙的强度	**velocity information** 速度信息
velcro strip 尼龙搭扣带	**velocity integration** 速度积分
velocimeter 速度计,水中声速测定仪	**velocity jump** 速度跃变,速度跳变
velocimetry 速度测量学	**velocity jump gun** 速度跳变电子枪
velocity 速率,速度	**velocity limit** 速度极限
velocity ambiguity 速度模糊	**velocity loss** 速度损失
velocity amplitude 速度振幅	**velocity magnitude** 速度值
velocity blast contour 速度冲击轮廓	**velocity measurement** 速度测量,测速
velocity boundary layer 速度边界层	**velocity microphone** 速率式话筒
velocity change 速度变化,速率变化	**velocity model** 速度模型
velocity characteristics 速度特性	

velocity node 速度节点	**velocity-vector** 速度矢量
velocity of advance 推进速度	**velocity-vector roll** 速度向量卷
velocity of air 空气流速	**velocity-vector roll rate** 速度矢量滚转率
velocity of collision 碰撞速度	**velours** 棉绒,丝绒,天鹅绒
velocity of free recoil 自由后坐速度	**VEN** 1. Ventilation 通风 2. Virtual Enterprise Network 虚拟企业网 3. Virtual Equipment Number 虚拟设备号码
velocity of impact 冲击速度	
velocity of recession 退行速度	
velocity off carrier 离机速度	**vena** 静脉
velocity orbit 速度的轨道	**vena contracta** 射流紧缩
velocity peak 峰值速度	**vendor** 供应商
velocity perturbation 速度波动	**vendor audit** 供应商审计
velocity potential 速度势,流速势	**vendor list** 供应商清单
velocity profile 流速分布图,速度剖面	**vendor profile** 供货商档案
velocity quotient 速度系数	**veneer** 单板,饰面,薄木片,面层
velocity ratio 速度比,相对速度	**V-engine** V型发动机
velocity reduction 速度降低	**vent** 通气孔,通风孔
velocity region 速度区	**vent cap** 通气口盖
velocity relaxation 速度放松	**vent centerline** 喷口中心线
velocity response 速度反应	**vent hem** 排气孔边
velocity response spectrum 速度反应谱	**vent patch** 排气孔补片
velocity result 速度结果	**vent plug** 液孔塞,通风孔塞
velocity signal 速度信号	**vent thrust** 排气推力
velocity signature 速度特征	**vent torque** 排气扭矩
velocity space 速度空间	**vent valve** 排气阀,通气阀
velocity spray 速度喷雾	**vented cadmium-nickel battery** 镉镍电池排出
velocity steady-state 速度稳定状态	**vented cell** 排气式蓄电池
velocity surface 速度面	**ventilated clothing** 通风服
velocity tapering 速度渐变	**ventilated flight clothing** 通风飞行服
velocity term 速度项	**ventilated shock** 通风的冲击
velocity to height meter 速度、高度表	**ventilated suit** 通风服
velocity to height ratio 速高比	**ventilating garment** 通风服
velocity tracking 速度跟踪	**ventilating model** 通气模型
velocity transducer 速度传感器	**ventilating pressure cabin** 通风式增压座舱
velocity triangle 速度三角形	**ventilating wall** 通气壁
velocity turn 速度转	**ventilation** 1.空气流通 2.通风,换气
velocity uniformity 速度均匀性	**ventilation factor** 通风因子
velocity value 速度值	**ventilation helmet** 通风头盔
velocity variation 变速	**venting cycle** 排气循环
velocity vector 速度矢量	**ventral** 机身下部的,机腹的
velocity/height ratio 速度/高度比	**ventral container** 机腹集装箱
velocity-based analysis 基于速度的分析	**ventral fin** 腹鳍,腹侧翼
velocity-based gain-scheduling 基于速度的增益调度	**ventral pallet** 腹侧托盘
velocity-based gain-scheduling approach 基于速度的增益调度方法	**ventral pod** 腹仓
	ventral radar 机腹雷达
velocity-based linearization 速率线性化	**ventral tank** 机身下油箱
velocity-based linearization family 基于线性化族速度	**venture** 冒险,风险,风险企业
velocity-coupled response 速度耦合响应	**venturi** 文氏管
velocity-coupling 速度耦合	**venturi meter** 文氏管流量计
velocity-coupling model 速度耦合模型	**Venturi tube** 文氏管

Venus 【天】金星
Venus seismology 金星震学
venusian atmosphere 金星的大气层
Venusian orbit image radar 金星轨道成像雷达
venusian surface 金星的表面
VEP Video Entertainment Player 视频娱乐放像机
verbal brief 语言简介
Verdan versatile digital analyser 计数式计算机通用数字分析器
vergence 1.透镜焦度 2.构造转向 3.倾向,朝向
verification 验证,检验
verification certificate 验单
verification scheme 检查简图
verifier 1.检验器,计量器 2.核对员
verify 证实,检验
vernal equinox 春分点
verneuil 焰溶法
vernier 游标,游尺
vernier adjustment 微调,游标调整,游尺调节
vernier engine 微调发动机,游标发动机
vernier rocket engine 微调火箭发动机
vernier rocket motor 游动火箭发动机
versine 正矢
versine function 正矢函数
version 改型,变体,版本,方案
verta plane 垂直起落飞机
vertex 台风转向点,极点,顶
vertical 垂直的
vertical acceleration 垂直加速度
vertical air photograph 垂直航空照片
vertical angle 垂直角
vertical ascent 垂直上升
vertical assembly and test building 垂直装配和测试厂房
vertical axis 垂直轴,纵轴
vertical bank 垂直盘旋
vertical board 立式配电板
vertical bomb 垂直炸弹
vertical camera 直立式摄影机
vertical checkout 垂直测试
vertical circle 垂直圈,垂直圆
vertical clearance 垂直间隙,纵间距
vertical climb 垂直爬升
vertical component 垂直分量
vertical control network 高程控制网
vertical control point 高程控制点
vertical control survey 高程控制测量
vertical cylinder 立式汽缸
vertical decelerator 垂直减速器,冲击塔

vertical decomposition 纵向分解
vertical density profile 剖面密度
vertical development 纵向发展
vertical direction 垂直方向
vertical displacement 竖直位移
vertical engine 竖型发动机,立式发动机
vertical envelopment 垂直包围
vertical epipolar line 垂核线
vertical epipolar plane 垂核面
vertical exaggeration 垂直扩大,垂直夸张,纵比例尺放大
vertical fin 垂直安定面,垂直尾翼
vertical force 垂直分力,垂直力
vertical gradient of gravity 重力垂直梯度
vertical ground 垂直地面
vertical guidance 竖直制导
vertical gust 垂直突风
vertical gyro 垂直陀螺仪
vertical gyroscope 垂直陀螺仪
vertical hinge 垂直铰
vertical hub 垂直中心
vertical impact 垂向冲击
vertical incidence sounding 垂直入射的探测
vertical index level 垂直指数水平
vertical injection logic 垂直注入逻辑
vertical interval 垂直间隔,竖向间隔
vertical interval test signal 帧扫描期插入测试信号
vertical junction solar cell 垂直结太阳电池
vertical landing 垂直登陆
vertical launch 垂直发射
vertical launch booster 垂直发射助推器
vertical launch vehicle 垂直发射车
vertical layout 竖向定位
vertical lift-to-drag 垂直升力和阻力
vertical line 垂直线,铅垂线
vertical linear polarization 垂直线极化
vertical load 垂直载荷,垂直负载
vertical missile packaging building 导弹垂直总装厂房
vertical mixer 立式搅拌机
vertical motion 垂直运动,竖向运动
vertical navigation 垂直导航
vertical nystagmus 垂直性眼球震颤
vertical orientation 垂直定向
vertical parallax 上下视差,垂直视差,纵视差
vertical perception 垂直知觉
vertical photograph 垂直航摄照片
vertical photography 垂直航空摄影,垂直摄影
vertical pincer 垂直夹击
vertical plane 垂直面,垂直投影面,竖直平面

vertical plane rigid-body model	垂直平面刚体模型
vertical polarization	垂直极化,垂直偏振
vertical position	竖直位置,垂直位置
vertical positioning error	垂直定位误差
vertical pressure gradient	垂直压力梯度
vertical probable error	垂直误差
vertical profile	垂直断面
vertical rate	垂直速率
vertical redundancy check	垂直冗余校验
vertical reference	垂直基准,垂直基准线
vertical refraction coefficient	垂直折光系数
vertical refraction error	垂直折光差
vertical region	垂直区域
vertical replenishment	垂直补给,立式补给
vertical reverse	垂直急跃升转弯
vertical riser	垂直立管
vertical rivet	垂直铆钉
vertical rivet number	垂直的铆钉数
vertical rolling scissors	垂直滚动的剪刀
vertical seeking ejection seat	垂直寻求弹射座椅
vertical separation	1.垂直间隔 2.竖向分离
vertical separation ratio	垂直分离比
vertical shock	竖震
vertical situation	垂直状况
vertical sounder	垂直探测器
vertical speed	垂直速度,升降速度
vertical spin tunnel	垂直螺旋风洞
vertical stabilizer	垂直安定面,垂直尾翼
vertical stacking	垂直叠加
vertical state transportation of launch vehicle	运载器垂直运输
vertical stiffeners	垂直加固件
vertical stress	垂直应力
vertical strip	垂直带
vertical strut	垂直杆
vertical symmetry plane	纵向对称平面
vertical tail	垂直尾翼
vertical tail area	垂直机尾面积,垂直失速区
vertical takeoff	垂直起飞
vertical takeoff and landing airplane	垂直起落飞机
vertical tape instrument	垂直带式仪表
vertical temperature profile radiometer	温度垂直廓线辐射仪
vertical thrust	垂直推力
verticle test	垂直测试
vertical translation	垂直位移
vertical trunk circumference	躯干垂直围
vertical tube	垂直管
vertical turn	垂直转弯,垂直旋转
vertical velocity	竖向速度,垂直速度
vertical vibration	垂直振动,垂向振动
vertical visibility	垂直能见度
vertical wall	垂直壁
vertical wind shear	垂直风剪切,风的垂直切变
verticle wind tunnel	立式风洞
vertical wing	垂直翼
vertical-coverage pattern	垂直作用范围
verticality adjustment	垂直度调整
vertically pointing Doppler radar	垂直指向多普勒雷达
vertical-position indicator	垂直位置指示器
vertical-sync signal	竖直同步信号,帧同步信号
vertices	至高点
vertiginous epilepsy	眩晕性癫痫
vertigo	眩晕
vertipad	垂直起降点
vertiplane	直升飞机
vertiport	垂直升降机场,垂直起降机场
very early universe	极早期宇宙
very high	很高
very high frequency	甚高频
very high frequency direction finder	甚高频测向器
very high frequency omni-range	甚高频全向信标
very high performance	甚高性能
very high resolution radiometer	甚高分辨率辐射仪
very high speed integrated circuit	超高速集成电路
very high speed photography	甚高速摄影
very large array	甚大阵,超大型基阵
very long baseline interferometer	甚长基线干涉仪
very rarefied plasma	甚稀薄等离子体
very recognition signal	非常识别信号
very small aperture terminal	甚小孔径地球站
very small satellite	甚小型卫星
very-large-scale integration	超大规模集成
VES	1. Video Encoding Standard 视频编码标准 2. Video Entertainment System 录像娱乐系统
vesicant（agent）	发泡剂
vessel	船只,器皿
vessel approach and berthing system	舰船停靠系统
Vesta	【天】灶神星,四号小行星
vestage axial compressor	轴流压气机遗迹
vestibular autorotation test	前庭自动旋转试验
vestibular function	前庭机能
vestibular function asymmetry hypothesis	前庭功能不对称假说
vestibular function caloric test	前庭机能变温试验
vestibular function disturbance	前庭功能紊乱
vestibular function examination	前庭功能检查
vestibular function test	前庭机能实验

vestibular function training 前庭功能训练
vestibular nystagmus 前庭性眼震
vestibular origin illusion 前庭性错觉
vestibular physiology 前庭生理学
vestibular spinal reflex 前庭脊髓反射
vestibulo-cerebellar response 前庭小脑反应
vestibulo-ocular reflex 前庭眼反射
vestibulo-ocular response 前庭动眼反应
vestibulo-perceptional response 前庭知觉反应
vestibulo-proprioceptive illusion 前庭本体性错觉
vestibulo-spinal response 前庭脊髓反应
vestibulo-vegetative reflex 前庭植物神经反射
vestigial lobe 退化的叶
vestigial sideband 残留边带
VF 1. Visual Field 视野 2. Visual Flight 目视飞行 3. Video Frequency 视频频率 4. Voice Frequency 音频
VFA Voltage Feedback Amplifier 电压反馈放大器
VFD 1. Vacuum Fluorescent Display 真空荧光显示 2. Voice Frequency Dialing 话频拨号
VFO Variable Frequency Oscillator 变频振荡器
VFR Visual Flight Rules 目视飞行规则
VFR on top 云上目视飞行
VFR terminal area charts 目视飞行终端区图
VFR tower 目视飞行指挥塔台
VFT Verification of Flight Test 试飞验证
VG 1. Vacuum Generator 真空发生器 2. Validity Generalisation 有效推广 3. Variable Geometry 可变几何形状 4. Velocity Generator 测速发电机 5. Velocity Gravity 超重力 6. Vertical Gyro 垂直陀螺 7. Video Graphy 图文电视 8. Vortex Generator 涡流产生器
VGA Video Graphics Adapter 视频图形适配器
VGI Vertical Gyroscope Indicator 垂直陀螺仪指示器
V-groove isolation V形槽隔离
V-groove MOS field effect transistor V型槽的MOS场效应晶体管
VGS 1. Video Graphics System 视频图形显示系统 2. Visual Guidance System 视觉导航系统
VGSI Visual Glide Slope Indicator 目视下滑道指示器
VH Very High 很高
VHF Very High Frequency 甚高频
VHF band 甚高频带
VHF communication 甚高频通信,特高频通信
VHF NAV VHF Navigation 甚高频导航
VHF omnidirectional radio range 甚高频全向信标
VHFDF Very High Frequency Direction Finder 甚高频测向仪
VHFOR Very High Frequency Omnirange 甚高频全向信标
VHPIC Very High Performance Integrated Circuit 超高性能集成电路
VHRR Very High Resolution Radiometer 甚高分辨率辐射计
VHRSR Very High Resoltion Scanning Radiometer 甚高分辨率扫描辐射剂
VHSIC Very High Speed Integrated Circuit 超高速集成电路
VI 1. Virtual Image 虚拟图像 2. Visual Inspection 目视检查
V-I characteristic curve of cell 电池伏安特性曲线
V-I characteristic curve of solar cell 太阳能电池伏安特性曲线
via hole 通孔
viability 1. 生活力,耐久性 2. 健全性
vial 管形瓶,水准器,玻璃管
VIB Velocity of Initial Buffet 起始抖动速度
vibramat 弹性玻璃丝垫
vibrate 1. 振动 2. 摆动 3. 犹豫 4. 激动
vibrating acoustical test 振动声学测试
vibrating beam accelerometer 振梁加速度计
vibrating quartz blade sensor 振动石英片传感器
vibrating rectification error 振动校正误差
vibrating reed gyro 振簧式陀螺
vibrating rotor gyro 振动陀螺
vibrating string accelerometer 振弦式加速计
vibrating voltage regulator 振动式电压调节器
vibrating wire converter 振动弦式变换器
vibrating wire force transducer 振弦式力传感器
vibrating-wire rate sensor 振动线速度传感器
vibration 振动,颤动
vibration absorber 减震器,振动阻尼器
vibration absorption material 防振材料
vibration acceleration 振动加速度
vibration amplitude 振动强度,振幅
vibration analysis 振动分析
vibration beam accelerometer 振梁式加速度计
vibration component 振动元件
vibration control 振动控制
vibration damage 受震破坏
vibration damping 减振,振动阻尼
vibration drift rate 振动漂移率
vibration energy 振动能量
vibration environment 振动环境
vibration frequency 振动频率,波动频率
vibration indicator 振动指示器,振动仪
vibration isolation 隔振,振动绝缘
vibration isolator 隔振体,隔振器,减振器
vibration level 振动级
vibration machine 振动器,振动机

vibration meter 测振仪,振动计
vibration milling 振动磨
vibration minimization 振动最小化
vibration mitigation 减振
vibration modal frequency 振动模态频率
vibration mode 振动方式,振动模式
vibration noise 振动噪声
vibration perception 振动知觉
vibration pick-up 拾振仪,振动传感器
vibration power 振动功率
vibration problem 振动问题
vibration protection 振动防护
vibration reduction 减振
vibration reduction problem 减振问题
vibration response 振动响应
vibration severity 振动烈度
vibration shape 振型
vibration signal 振动信号
vibration suppression 减震
vibration test 振动试验,耐震性试验
vibration test rocket 火箭振动试验
vibration testing 振动试验,振动测试
vibrational distribution 振动分布
vibrational equilibrium 振动平衡
vibrational excitation 振动激励
vibrational ground state 振动基态
vibrational mode 振动模式
vibrational relaxation 振动弛豫
vibrational temperature 振动温度
vibrator 振动器,振动子
vibrator-type voltage regulator 振动式电压调节器
vibratory hub 振动轮毂
vibratory hub load 振动轮毂负载
vibratory level 振动水平
vibratory load 振动荷载
vibratory motion 振动
vibratory response 振动响应
vibratory torque control 振动扭矩控制
vibroacoustic environment 声振环境
vibrograph 振动记录仪,示振器
vibrometer 振动计,测振器
vibropendulous error 振摆误差
VIC Very Important Cargo 极重要货物
vice chairman 副会长,副委员长,副议长,副主席
vice president 副总裁,副总统,副总经理
vicinity 附近,邻近
vicious cycling 恶性循环
VID 1. Video Image Display 视频图像显示 2. Virtual Image Display 虚拟图像显示器

vide 请见,参阅
video 视频,电视电话,影像
video amplifier 视频放大器
video camera 摄影机
video cassette recorder 盒式磁带录像机
video cipher 视频密码
video compression 视频压缩
video detector 视频检波器
video disk 视盘
video display 视频显示
video extractor 视频提取
video head 视频磁头
video infrared spin scan radiometer 视频红外自旋扫描辐射仪
video integrator and processor 视频积分处理器
video link 视频波道
video map 视频地图
video mapping 视频扫描指示,频谱扫描指示
video processing circuit 视频信号处理电路
video processor 视频处理器
video record 录像片
video recording head 录像头
video reproducing head 放像头
video signal 视频信号,图像信号
video solar magnetograph 太阳视频磁像仪
video tape 录像带
video tape recorder 录像机,磁带录像机
video track 视频磁迹,图像跟踪
video transmission 图像传输,视频传输
videogrammetry 视频测量
videogrammetry result 视频测量结果
videogrammetry test 视频测量试验
videotex 单向传视,视频图文,显示图像
vidicon 光导摄像管,视像管
vidicon camera 光导摄像管相机
VIEW 1. Video Information Exchange Window 视频信息交换窗口 2. Virsual Informative Environment Workstation 虚拟交互环境工作站
view angle 视角
view finder tube 寻像管
viewfinder 瞄准器,取景器,拾影器
viewgraph 视图
viewing 检视,观察
viewing angle 观察角
viewing area 观察区,观察面积
viewing range of the seeker 自导引系统截获目标的距离
viewing station 观察站
viewpoint 1. 观点,意见,角度 2. 视角 3. 视点

VIEWS Visual Imaging Electromagnetic Window System 视觉成像电磁窗口系统
viff 1. 突然转向能力 2. 突然转向
vigilance 警惕性,警觉,警戒
vigilance performance 警戒能力
Vigilant 警醒的
vignette 装饰图案,小插图
VIGV Variable Inlet Guide Vanes 可变进口导向叶片
vik 湾
viking 北欧海盗,船边计程仪
viking orbiter 维金轨道飞船
viking program 海盗计划
viking project 海盗项目
VIM 1. Vacuum-Induction Melting 真空感应熔炼 2. Vendor Information Manual 供应商信息手册 3. Vendor-Independent Messaging 与销售商无关的消息
VIMS Video Management System 视频管理系统
vinyl cyanide 丙稀腈,乙烯基氰
violate 1. 违反 2. 妨碍 3. 侵犯
violation 1. 违反,妨碍,侵犯 2. 违犯,违背
violation of sovereignty 侵犯主权
violent galaxy 激变星系
violet 1. 紫罗兰 2. 蓝紫色,紫罗蓝色
violet band 紫外波段
violet cell 紫电池
violet solar cell 紫光太阳电池
VIP 1. Very Important Person/Passenger 重要人物/旅客 2. Video Information Provider 视频信息提供者 3. Video Interface Processor 视频接口处理机 4. Virtual Interface Processor 虚拟接口处理机 5. Visual Identification Point 目视轰炸标志点 6. Vision Information Processing 视觉信息处理 7. Visual Image Processor 视频图像处理机 8. Visual Integrated Presentation 目视综合显示 9. Visual, Intelligent and Personal 可视化,智能化和个人化 10. Voice Information Processor 话音信息处理机 11. Voice Interactive Phone 交互式音频电话
virga 幡状云,雨幡
virgin aluminum 原铝
virgin aluminum size 原铝的大小
virgin material 纯净原材料
virgin medium 空白媒体,未用媒体
Virgo 1. 维尔格铬镍钨系合金钢 2. 室女星座,室女宫
Virial theorem 维里定理
VIRR Visible Infrared Radiometer 可见光红外辐射计
VIRS Visble Infrared Scanner 可见光红外扫描仪
VIRSR Visible Infrared Scanning Radiometer 可见光红外扫描辐射计

VIRSS Visual and Infrared Sensor Systems 目视与红外探测系统
virtual ablation method 虚拟消融方法
virtual axis 虚轴
virtual bomb 虚弹
virtual cathode 虚阴极,假想阴极
virtual change 虚拟的变化
virtual channel 虚拟信道
virtual channel access service 虚拟信道接入业务
virtual channel access sublayer 虚拟信道接入层
virtual channel data unit 虚拟信道数据单
virtual circuit 虚拟电路
virtual common memory 虚拟公共存储器
virtual containment 虚拟的容器
virtual control 虚拟控制
virtual displacement 虚位移,假位移
virtual environment 虚拟环境
virtual fill 虚拟填充
virtual geomagnetic pole 虚地磁极
virtual gravity 虚重力,假重力
virtual height 虚高,有效高度
virtual image display 虚像显示
virtual inertia 虚拟惯性,表现惯性
virtual leader 虚拟领袖
virtual leak 虚漏
virtual level 虚能级,假想电平
virtual machine 虚拟机,虚拟计算机
virtual memory 虚拟存储器
virtual pressure 虚压
virtual reality 虚拟现实
virtual reflection height 虚拟反射高度
virtual source 虚声源
virtual storage 虚拟存储器
virtual stress 原始应力,感应力
virtual structure 虚结构
virtual structure approach 虚拟结构的方法
virtual structure rotation 虚拟结构的旋转
virtual target 视在目标
virtual temperature 虚温,等效温度
virtual terminal 虚拟终端
virtual wire 虚拟线
virtual work 虚功
virtual world 虚拟世界
VIS 1. Video Interface System 视频接口系统 2. Visual Instruction Set 视算指令集 3. Visual Instrumentation Subsystem 可见测试仪器子系统 4. Visual Interactive Simulation 可视交互式模拟 5. Voice Information Service 语音信息服务
viscoelastic 黏弹性

viscoelastic damper	黏弹性阻尼器
viscoelastic damping	黏弹性阻尼
viscoelastic element	黏弹性元件
viscoelastic material	黏弹性材料,黏弹性物质
viscoelasticity	黏弹性
viscometer	黏度计
viscosimeter	黏度计,黏滞计
viscosity	黏滞性,黏度
viscosity coefficient	黏性系数
viscosity index	黏度指数
viscosity manometer	黏滞压力计
viscosity measurement	黏度测量
viscosity of fluid	流体黏性
viscosity solution	黏性溶液
viscosity vacuum gauge	黏滞真空计
viscosity valve	黏度阀门
viscous aquaplaning	黏滞水飘
viscous damper	黏性阻尼器
viscous damping	黏滞阻尼
viscous diffusion	黏性扩散
viscous dissipation	黏性耗散,黏滞耗散
viscous drag	黏性阻力
viscous effect	黏滞效应,黏滞作用
viscous energy dissipation	黏滞的能量耗散
viscous factor	黏滞因素
viscous flow	黏滞流,黏性流
viscous flow aerodynamics	黏性空气动力学
viscous fluid	黏性流体,粘力
viscous flux	黏性通量
viscous force	黏性力
viscous forebody drag	黏性前体阻力
viscous heating	黏性发热
viscous interaction	黏性干扰效应
viscous interaction parameter	黏性干扰参数
viscous interference	黏性干扰
viscous layer	黏滞层
viscous loss	黏滞损失
viscous pressure drag	黏性压差阻力
viscous remanence	黏滞剩磁
viscous remanent magnetization	黏性剩余磁化
viscous shock	黏性激波
viscous shock-layer equation	黏性激波层方程
viscous stability	黏性稳定性
viscous stress	黏性应力
viscous sublayer	层流底层,黏性底部
viscous term	黏性项
viscous wake	黏性尾迹
viscous wall	黏壁
viscous-inviscid interaction	黏性无黏性干扰
viscous-layer	黏性层
visibility	能见度,视距,可见度
visibility factor	可见度系数
visibility limit	能见度限度,可见限度
visibility meter	能见度计,视度计
visibility polygon	可见多边形
visibility region	可见区域
visible and infrared spin scan radiometer	可见光和红外自旋扫描辐射仪
visible arc	明弧,弧段可见扇形区
visible emission	可见光发射
visible flame	可见火焰
visible horizon	视地平线,可见地平线
visible image	可见像,视像
visible light	可见光,可见光全息照相术
visible light spectrum	可见光光谱
visible line	可见谱线,轮廓线,外形线
visible plume tracer technique	可见烟流追踪法
visible radiation	可见光辐射,可见光
visible range	可视区,能见距离
visible resolution	目视分辨率
visible spectrum	可见光谱
visible spectral domain	可见光谱域
visible spectral remote sensing	可见光遥感
visible wavelength	可见波长
visible-band	可见光波段
visible-IR radiometer	可见红外辐射仪
vision	视力,视觉,洞察力
vision acceleration	视觉加速度
vision computer	视觉计算机
vision dim	视力模糊
vision disorder	视觉障碍
vision flight	飞行视觉
vision metrology	视觉测量
vision model	视觉模型
vision sensor	视觉传感器
vision standard	视力标准
vision system	视觉系统,目视系统
visionary	幻影的,空幻
vision-based control	基于视觉的控制
vision-based measurement	基于视觉的测量
vision-based sensor	基于视觉传感器
vision-based state	基于视觉的状态
visionics	视觉学
visit and search	临检与搜索
Visol	乙烯基异丁醚
visor	1.遮阳板 2.护目镜 3.头盔滤光镜
VISSR	Visible Infrared Spin Scan Radiometer 可见光红外自旋扫描辐射计

VISTA Virtual Integrated Software Testbed for Avionics 航空电子虚拟综合软件试验台
visual 1.目视的,视觉的,可见的 2.直观的
visual accommodation 视力调节,视力适应
visual acuity 1.视觉灵敏度,视敏度 2.视力
visual adjustable 目视可调
visual aids 助视器
visual alerting device 目视告警装置
visual analytics 可视化分析
visual and optical countermeasures 视觉与光学的对策
visual and physical motion 视觉和身体的运动
visual angle 视角
visual approach 目视进场着陆
visual approach path indicator 目视进近航道指示器
visual approach slope indicator 目视进场下滑道指示器
visual approach slope indicator light system 目视进近坡度灯光系统
visual attention 融合关注度,视觉注意力
visual autokinesis 视自体动作
visual axis 视轴
visual background 视觉背景
visual balance 视觉平衡
visual binary 目视双星
visual blurring 视力模糊
visual camouflage 可见光伪装
visual capacity 视觉能力,目视能力
visual carrier frequency 图像载频
visual check 目视检查
visual clean 目视清洁,视觉清洁
visual coding 视觉编码
visual command link 目视跟踪指令线路,目视跟踪指令制导系统
visual condition 视觉条件
visual contact 有视觉接触
visual contrast 视觉对比
visual control 目测检查,外观检查
visual cue 目视信号,视觉信号,目视感示
visual detection range 视觉检测范围
visual discrimination 视力分辨
visual discrimination test 目视鉴别试验,视觉鉴别试验
visual discriminatory acuity 视觉辨别敏度
visual display 直观显示设备,可视显示
visual distress 视力障碍,视觉困扰
visual Doppler indicator 多普勒雷达目视显示器
visual envelope 目视范围
visual environment 视觉环境
visual extinction meter 可见光消光计
visual fatigue 视觉疲劳
visual field 视场,视野
visual flight 目视飞行
visual flight rules 目视飞行规则
visual gain 视觉增益
visual head-up image 目视抬头影像
visual holding 视觉保持
visual identification 目视识别
visual illusion 视觉性错觉,目视错觉
visual information 视觉信息,可见信息,目视观察情报
visual inspection 1.目视检查,肉眼检查 2.外观检查
visual interpretation 目视判读
visual lag 视觉延迟
visual laydown delivery 可视瞄准投放
visual lead 视觉引导
visual lead time 视觉引导的时间
visual magnitude 视觉等级,目视星等
visual manoeuvring area 视觉操纵区域
visual meteorological condition 目视气象条件
visual monitoring 目视监测
visual observables 视觉可见
visual observation 目视观测
visual observator 目视观测者
visual/optical countermeasures 目视/光学对策
visual orientation 目测方位,目测定向,目视定向
visual perception 视觉
visual performance 视觉功能
visual photometry 1.目视测光 2.目视光度学
visual presentation 视觉显示
visual protective device 视觉防护装备
visual range 目视范围,视程,能见距离
visual reference 目视基准,目视参考物
visual report 可视报表
visual reporting post 目视报知站
visual resolution 目视鉴别率
visual response 视反应
visual scene 视觉景色,图像景色
visual separation 目视隔离
visual sighting 目视瞄准,用光学瞄准具瞄准
visual signal 视觉信号
visual signalling apparatus 视觉信号装置
visual simulation 视觉模拟
visual stimulus 视觉刺激
visual system 视景系统
visual target acquisition system 目视目标截获系统
visual task 视觉作业
visual time 视听时代
visual time delay 视觉延迟
visual variable 视觉变量
visual warning 视觉告警
visual zenith telescope 目视天顶仪

visualization 1.可视化,显影 2.目测
visualization of cavitation 空间可视化
visualization picture 可视化图像
visualize 形象,预见,目测检验直观化
visually coupled 视觉结合
visual-perception 视觉感知
visuals 形象化景象
visual-scene 视觉场景
visual-scene background 视觉场景的背景
visuometer 视野计
vis-viva formula 活力方程
VIT Video Image Terminal 视频图像终端
vital action 生活作用
vital area 1.防空要区,军事要区 2.目标要害部位
Viterbi algorithm 维特比算法
Viterbi decoding 维特比译码
vitiate 削弱,破坏,损害
vitiation 破坏,损害
vitiation air 污染空气
vitiator 破坏者
vitreous 1.透明,琉态 2.玻璃质,玻璃质的 3.玻璃状,玻璃状的
vitrifying 玻璃化
vitroceramics 玻璃陶瓷
VIU Video Interface Unit 视频界面装置
vivacity 活泼,快活,有生气
VKIFD Von Karman Institute for Fluid Dynamics 沃·卡尔曼流体动力学学会
VL 1.Variable Loss 可变损耗 2.Vertical Lower 垂直下方 3.Very Low 极低 4.Virtual Laboratory 虚拟实验室 5.Virtual Link 虚拟连接
VLA 1.Very Large Array 极大望远镜阵 2.Very Light Aeroplane 甚轻型飞机 3.Very Low Altitude 超低空
VLADD Visual Low Angle Drogue Delivery 视角较小的风向标传递
VLAN Virtual Local Aera Network 虚拟局域网
Vlasov equation 符拉索夫方程
VLBI Very Long Baseline Interferometer 甚长基线干涉量度法
VLCHV Very Low Cost Harrassment Vehicle 极廉价空中骚扰飞行器
VLD Variable Length Data 可变长数据
VLED Visible Light Emitting Diode 可见光二极管
VLF 1.Variable Length Field 变长字段 2.Very Low Frequency 甚低频
VLP 1.Vertical Linear Polarization 垂直线极化 2.Video Long Play 长时间放像 3.Video Long Player 长时间图像放映器
VLR 1.Very Long Range 甚远程,超远程 2.Visited Location Registor 访问位置寄存器
VLS Vertical Launch System 垂直发射系统
VLSI Very Large Scale Integration 超大规模集成
VLSIC Very Large Scale Integrated Circuit 超大规模集成电路
VM 1.Velocity Meter 速度表 2.Velocity Modulation 速度调制 3.Vendors Manuals 供应商手册 4.Video Mail 视频邮件 5.Virtual Machine 虚拟机 6.Virtual Manufacturing 虚拟制造 7.Virtual Memory 虚拟存储器 8.Volatile Memory 易失存储器 9.Voltmeter 电压表
VMA 1.Valid Memory Address 有效存储地址 2.Virtual Memory Address 虚拟存储地址 3.Visual Manoeuvring Area 视觉操纵区域
VMC 1.Visual Manoeuvring Circles 视觉操纵圈 2.Visual Meteorological Conditions 目视气象条件 3.Visual Minimum Condition 可视最小条件 4.Voice Messaging Coder 语音消息编码器
VMDI Vector Miss-Distance Indicator 矢量脱靶距离指示器
VMF 1.Variable Message Format 可变信息格式 2.Visibility Management File 视度管理文件
VMM 1.Virtual Machine Manager 虚拟机管理程序 2.Virtual Machine Monitor 虚拟机监控程序
VMO Variable Metering Orifice 可调计量孔
VMS 1.Variable Message Sign 可变消息符号 2.Vehicle Management System 飞行器管理系统 3.Vehicle Monitoring System 飞行器监视系统 4.Vehicle Motion Sensor 飞行器运动传感器 5.Velocity Measuring System 速度测量系统 6.Vertical Motion Simulator 垂直运动模拟器 7.Video Modulation System 视频调制系统 8.Virtual Memory System 虚拟存储器系统 9.Voice Mail Server 语音信箱服务器 10.Voice Mail Service 语音信箱业务 11.Voice Mail System 语音邮件系统 12.Voyage Management System 航程管理系统
VMSC 1.Vendor Make Subcontract 卖方作出的转包合同 2.Visited Mobile-services Switching Center 受访移动业务交换中心
VNAV Vertical Navigation 垂直面导航,垂直导航
VNIR Visible and Near Infrared 可见光和近红外
VNR VHF Navigation Receiver 甚高频导航接收机
VNRT Very Near Real Time 甚近实时
VNS Visual Network Station 可视网络站
VO 1.Verbal Order 口头命令 2.Version Object 版本对象 3.Virtual Organization 虚拟组织
VOA 1.Velocity of Arrival 到达速度 2.Variable Optical Attenuator 可变光衰减器
VOC 1.Volatile Organic Compounds 挥发性有机化合

物 2. Variable Output Circuit 可变输出电路
vocoder 1. 声码器,自动语言合成器 2. 语音信号分析合成系统声码器
VOCOM Voice Communication 话音通信
VOD 1. Vertical Obstruction Data 垂直障碍物数据 2. Video On Demand 视频需求 3. Voice Over DDN 用DDN传话音
VODACOM Voice Data Communication 话音数据通信
VOGAD Voice Operated Gain Adjusting Device 音控增益调节设备
voice 1. 声音 2. 话语 3. 用送话器递情报
voice activation 话音激活
voice clearance 语音间隙
voice coil 音圈,声圈
voice command system 话音指令控制系统
voice communication 话音通信,电话通信
voice data processing 语音数据处理
voice encryption equipment 语音加密设备
voice frequency 音频
voice grade channel 语音波段电路
voice interface 音频接口
voice mail 语音邮件
voice message unit 语音信息装置
voice operated switch 话音控制开关
voice recorder 话音记录器
voice-band transmission 话带传输
voice-frequency facility terminal 音频设备终端
voiceless homing 非语音归航
voice-operated device anti-singing 音控防鸣器
voice-operated relay 音控继电器
void 空隙,气孔,作废,无效的
void area index 空隙面积指数
void content 空隙容积
void fraction 1. 空白部分 2. 空隙率
VOL Vertical on-board Landing 舰上垂直降落
vol plane 惯性飞行
volatile 1. 易失的 2. 挥发的,易挥发掉的
volatile condensable material 挥发性凝材料
volatile content 挥发物含量
volatile memory 易失存储器
volcanic ash cloud 火山灰云
volitional (wilful) act 意志(故意)动作
Volmet 日常地对空气象信息广播
VOLMET broadcast 航空气象广播
volt 伏特
voltage 电压
voltage accuracy 电压精度
voltage amplifier 电压放大器
voltage controlled avalanche oscillator 电压控制雪崩振荡器
voltage controlled oscillator 压控振荡器
voltage drift 电压漂移
voltage efficiency 电压效率
voltage gradient 电压梯度,比电压
voltage regulator 调压器,倒管稳压器,电压第器
voltage responsivity 红外探测器的电压响应度
voltage sensor 电压敏感器
voltage source inverter 电压源型逆变器
voltage spike 电压尖峰
voltage stability 电压稳定度
voltage stabilizing diode 稳压二极管
voltage stabilizing transformer 稳压变压器
voltage standing wave ratio 电压驻波比
voltage surge 电压浪涌,电压冲击
voltage temperature coefficient 电压温度系数
voltage temperature coefficient of a solar cell 太阳电池的电压温度系数
voltage temperature cut-off 截止电压温度
voltage window 电压窗口
voltage-fed aerial 电压馈电天线
voltage-tuned magnetron 电压调谐磁控管
voltmeter 伏特计,电压表
volume collision rate 体积碰撞率
volume compensator 音量补偿器,体积补偿器
volume electrical resistance 体积电阻
volume flight charge 飞行的电荷量
volume flow rate 体积流率,体积量
volume flow rate of a vacuum pump 真空泵抽气速率
volume of fire 火力
volume scattering 体散射
volume unit 响度单位
volume velocity 体积速度
volumetric burning rate 体积的燃烧速率
volumetric efficiency 容积效率
volumetric heat intensity 容热强度
volumetric loading 容量负荷,定容加料
volumetric loading fraction 体积装填分数
volumetric performance 体积动态
volumetric specific energy 体积比能量
volumetric specific power 体积比功率
voluntary accommodation 随意调节
volute 螺旋管
VOM 1. Volt-Ohm Milliammeter 万用表 2. Volt-Ohm Meter 伏特-欧姆计
vomiting agent 呕吐性毒剂
VOP 1. Valued as in Original Policy 估价如原保险单所载 2. Video Object Plane 视频对象平面 3. Voice Over Packet 语音封包

VOR 1. Visual Omnirange 目视飞行全向指向标
2. VHF Omnidirectional Range 甚高频全向信标
VOR beacon 甚高频全向信标
VOR scanned array 电扫式伏尔天线阵
VOR test facility 甚高频全向无线电信标测试设施
VOR test signal 甚高频全向无线电信标测试信号
VOR/DME VHF Ommidirectional Range/Distance Measuring Equitment 甚高频全向信标测距器
VOR/ILS VHF Omnidirectional Range/Instrument Landing System 甚高频全向无线电信标仪表着陆系统
VOR/MB VHF Omnidirectional Range/ Marker Beacon 甚高频全向信标/信标机
vortex 涡流,旋涡
vortex breakdown 涡旋破碎
vortex breakdown location 涡破裂位置
vortex breakdown point 涡破裂点
vortex burst 涡旋崩解
vortex burst trajectory 涡破裂的轨迹
vortex bursting 涡破裂
vortex center 涡旋中心
vortex chamber 涡流室
vortex combustor 旋涡燃烧室
vortex configuration 涡结构,涡配置
vortex containment 涡壳
vortex core 涡核,旋涡中心
vortex effect 涡流效应
vortex element 涡旋元件
vortex filament 涡丝
vortex flap 涡流襟翼
vortex flow 涡流
vortex generator 涡流发生器
vortex hazard 涡旋危害
vortex interaction 涡相互作用
vortex lattice 涡流栅
vortex lattice method 涡串法,涡列法
vortex lift 涡升力
vortex lifting 涡升力
vortex line 涡线,旋线
vortex location 涡位置
vortex method 涡方法
vortex of uniform size 大小均匀的涡
vortex pair 涡对,涡动副,旋涡偶
vortex precession flowmeter 旋进式漩涡流量计
vortex region 涡柳,旋涡区
vortex ring 涡环,涡环性能
vortex segment 涡段
vortex separation 涡流分离
vortex shedding 涡流发散

vortex shedding flowmeter 涡街流量计
vortex sheet 涡面,旋涡层
vortex sound 涡声
vortex span 涡跨度
vortex street 涡道
vortex strength 涡量度
vortex structure 旋卷构造
vortex suppression device 涡抑制装置
vortex surface sheet 涡面
vortex system 涡系
vortex theory 涡理论
vortex trail 涡旋尾迹
vortex transport 涡运输
vortex tube 涡管,涡流管
vortex wake 涡粒
vortex-acoustic coupling 涡声耦合
vortex-blade 涡流叶片
vortex-filament 涡丝
vortex-filament model 涡模型
vortex-generator 涡流发生器
vortex-induced drag 涡阻
vortex-induced wind 风涡激
vortex-lattice method 涡格法
vortex-pair 对涡
vortex-ring state 满环状态
vortex-shedding 旋涡脱落
vortex-shedding frequency 旋涡脱落频率
vortical disturbance 旋涡的干扰
vortical flow 旋涡流
vortical motion 涡动,涡旋运动
vortical structure 旋卷构造,涡结构
vortical wake 旋涡尾流
vortical wave 旋涡的波
vorticity 旋涡状态
vorticity datum 涡度基准
vorticity disturbance 涡度扰动
vorticity effect 涡度效应
vorticity effect extent 涡度的影响程度
vorticity field 旋涡场
vorticity flux 涡通量
vorticity sheet 旋涡片
vorticity transport model 涡量输运模型
vorticity transport 旋涡转移
vorticity wave 涡旋波
vortil(l)on 涡流器,旋涡发生器
VOS 1. Virtual Operating System 虚拟操作系统
2. Voice Operated Switch 声控开关
Voskhod 上升号计划
Vostok 上升号飞船

VOT 1. VHF Omnirange Test 甚高频全向导航测试 2. Voice Operated Transmission 声控传输

voter 选举人，表决器

voter threshold 表决器阈

voting system 表决系统

voyage 航行，航空，航海，航程

voyage repair 航程检修

Voyager "旅行者"空间探测器

VP 1. Valve Point 阀点 2. Vapor Pressure 蒸汽压 3. Variable Pitch 可变桨距 4. Vector Processor 矢量处理机 5. Vent Pipe 通气管 6. Vertical Profile 垂直断面 7. Video Phone 可视电话 8. Video Processor 视频信号处理机 9. Virtual Path 虚通路 10. Virtual Processor 虚拟处理机 11. Vision Processing 视频处理

VPC 1. Valve Position Control 阀门开度控制 2. Vertical Parity Check 垂直奇偶校验 3. Video Processing Cabinet 视频处理柜 4. Virtual Path Connection 虚通路连接

VPDM Virtual Product Development Management 虚拟产品开发管理

VPH Virtual Path Handler 虚通道处理

VPI 1. Vapour Phase Inhibitor 汽相防锈剂 2. Vacuum Pressure Impregnation 真空压力浸渍 3. Vertical Path Integrator 垂直轨迹积分器 4. Virtual Path Identifier 虚通路标识符

VPK Vehicles with Protection Kits 装有保护工具箱的飞行器

VPM 1. Vehicles Per Mile 每英里车辆数 2. Virtual Product Management 虚拟产品管理 3. Voice Path Management 话路管理 4. Volts Per Meter 伏特/米

VPN 1. Vendor Part Number 供应商零件号 2. Virtual Path Network 虚通道网络 3. Virtual Personal Network 虚拟个人网 4. Virtual Private Network 虚拟专用网络

VPS Vital Production System 核心生产系统

VQ 1. Vector Quantization 矢量量化 2. Virtual Queuing 虚队列 3. Voice Quality 音量

VR 1. Variable Resistor 可变电阻器 2. Vendor Requirements 协作厂要求 3. Virtual Reality 虚拟现实 4. Virtual Route 虚路由 5. Voltage Regulator 调压器 6. Voice Recorder 录音器

VR simulation Virtual Reality Simulation 虚拟现实仿真

VRAM Video Random Access Memory 视频随机存取储存器

VRB Voice Rotating Beacon 音频旋转信标

VRC 1. Vertical Redundancy Check 垂直冗余校验 2. Version Release Control 改版发放控制 3. Visible Record Computer 可见记录计算机

VRF Visual Recording Facility 目视记录设备

VRM 1. Variable Range Marker 可变距离标志 2. Voice Recognition Module 语音识别模块

VRP 1. Vehicle Routing Problem 车载路由问题 2. Video RISC Processor 视频精简指令集计算机处理器 3. Visual Reporting Post 目视报知站

VRU 1. Velocity Reference Unit 速度参考装置 2. Video Reproducer Unit 视频复制组件 3. Voice Response Unit 音频响应装置

VS 1. Variable Store 可变存储器 2. Velocity Search 速度搜索 3. Vent Scoop 通气进气口 4. Vertical Speed 垂直速度 5. Video Server 视频服务器 6. Video Signal 视频信号 7. Virtual Storage 虚拟存储 8. Virtual System 虚拟系统

VSA Visual Software Agent 可视软件代理

VSAM 1. Vestigial Sideband Amplitude Modulation 残留边带调幅 2. Virtual Storage Access Method 虚存取法

VSAT Very Small Aperture Terminal 甚小口径终端

VSB 1. Vendor Service Bulletin 供应商服务通报 2. Vestigal Side Band 残留边带 3. VHF Survival Beacon 甚高频救生信标

VSC 1. Vacuum System Controller 真空系统控制器 2. Video Scan Converter 视频扫描变换器 3. Voice Signal Converter 声音信号转换器

VSCF Variable Speed Constant Frequency 变速恒频

VSCS 1. Vertical Stabilizer Control System 垂直安定面控制系统 2. Voice Switching and Control System 语音开关和控制系统

VSD 1. Variable Speed Drive 变速传动 2. Vertical Situation Display 垂直状态显示 3. Vision & Strategy Document 远景与战略文件

VSE 1. Vehicle System Engineering 飞行器系统工程 2. Virtual Server Environment 虚拟服务器环境

VSI 1. Velocity and Steering Indicator 速率和转弯指示器 2. Vertical Speed Indicator 垂直速度指示器 3. Vertical System Incorporation 垂直系统公司 4. Video Sweep Integrator 视频扫描积累器

VSL Vertical Speed Limit 垂直速度限制

VSM 1. Variable Speed Modem 变速调制解调器 2. Vertical Separation Minimum 最小垂直间隔 3. Vestigial Sideband Modulation 残留边带调制 4. Video Service Module 视频服务模块 5. Video Sweep Modulation 视频扫描调制 6. Virtual Service Management 虚拟服务管理 7. Virtual Shared Memory 虚拟共享存储器 8. Voice Server Module 语音服务器模块

VSR 1. Very Short Range 极短距,甚近程 2. Vertical Speed Required 垂直速度的要求 3. Vibratory Stress

Relief 振动时效
VSS 1. Vapor Suppression System 蒸汽弛压系统 2. Video Storage System 视频存储系统 3. Video Sub System 视频子系统 4. Virtual Support Subsystem 虚拟支援子系统
VSSC Vikram Sarabhai Space Centre 维克拉姆（印度）航天中心
VST 1. Video System Test 视频系统测试 2. Virtual Storage 虚拟存储器
VSTC Verification of Supplementary Type Certification 补充型号认可证
VSTOL Vertical or Short Take Off and Landing 垂直短距起飞落
VSTT Variable Speed Training Target 变速训练靶
VSV Variable Stator Vane 可调静叶
VSW 1. Vertical Speed and Windshear 垂直速度与风切变 2. Very Short Wave 超短波
VSWR Voltage Standing Wave Ratio 电压驻波比
VT 1. Variable Thrust 可变推力 2. Vertical Tail 垂直尾翼 3. Vertical Tabulation 纵向制表 4. Video Terminal 视频终端 5. Verification Test（合格性的）验证试验 6. Visual Telephony 可视电话 7. Virtual Terminal 虚拟终端 8. Virtual Tributary 虚拟支路
VT fuze Variable Time Fuze 无线电引信,变时引信
VTA 1. Variable Transfer Address 可变转移地址 2. Vertical Track Alert 垂直航迹预警 3. Vertex Time of Arrival 到达顶点的时候 4. Video Terminal Adapter 视频终端适配器
VTAS Visual Target Acquisition System 目视目标探测系统
VTC 1. Validation of Type Certificate 型号认可证 2. Videotex Terminal Control 可视图文终端控制 3. Visual Terminal Chart 目视航站图
VTIP Visual Target Identification Point 目视目标识别点
VTIR Visible and Thermal Infrared Radiometer 可见光与热红外辐射计
VTK Vertical Track 垂直航迹
VTL 1. Variable Threshold Logic 可变阈值逻辑 2. Virtual Tape Library 虚拟磁带库
VTM 1. Vapour Tension Meter 蒸汽压计 2. Virtual Teller Machine 虚拟出纳机
VTOHL Vertical Take Off and Horizontal Landing 垂直起飞与水平着陆
VTOL Vertical Take Off and Landing 垂直起落
VTOVL Vertical Take Off and Vertical Landing 垂直起飞和垂直着陆
VTPR Vertical Temperature Profile Radiometer 垂直温度分布辐射计
VTR 1. Variable Takeoff Rating 可变起飞速率 2. Video Tape Recorder 磁带录象机 3. Video Tape Reproducer 录像带复制器
VTS 1. Vacuum Toilet System 真空洗手间系统 2. Verification Test System 鉴定试验系统 3. Vertical Test Stand 垂直试验台 4. Vibration Test System 振动试验系统 5. Video Teleconference System 电视电话会议系统 6. Virtual Terminal Services 虚拟终端服务 7. Virtual Terminal System 虚拟终端系统
vulcanite 硬质橡胶
Vulkollan 氨基甲酸乙酯橡胶,聚胺酯材料
vulnerability 易损性,脆弱性
vulnerability assessment 易损性评估
vulnerability factor 脆弱性因素
vulnerable quarter 致命方向,易被攻破方向
VV 1. Velocity Vector 速度向量 2. Vertical Velocity 垂直速度 3. Vertical Visibility 垂直能见度 4. Vice Versa 反过来,否则
VWS 1. Vertical Wind Sheer 垂直风切变 2. Virtual Working System 虚拟工作系统

W

W　　1. Wall 侧壁 2. Wardrobe 衣帽间，储物间 3. Warhead 核武器弹头 4. Watts 瓦,瓦特 5. Waves 波 6. Weapon 武器 7. West 西 8. Wire 导线,金属线 9. We 我们 10. White 白色的 11. Weight 重量

W engine　　W 型发动机

W comp　　Wind component 风的分量

W/L　　Weapon Loading 武器悬挂,装弹,挂弹,武器安装

W/V　　Wind Voltage 风速向量,风矢量

WA　　1. Work Authorization 工作审定 2. Wind Angle 风角 3. Wing Area 机翼面积 4. Waveform Analyzer 波形分析器 5. Wireless Access 无线接入

WAAAM　　Wide-Area Anti-Armour Munitions 大面积反坦克弹

WAAF　　Women's Auxiliary Air Force 空军妇女辅助队

WAACS　　Western Airway and Air Communication Service 西方航空与空中通信业务

WAAS　　1. Wide Area Active Surveillance 广域有源监视 2. Wide Area Augmentation System 广域增强系统 (Method of Differential GPS 差分全球定位系统方法)

WAC　　1. Wake Analysis and Control 尾流分析与控制 2. Weapon Aiming Computer 武器瞄准计算机 3. Wide Area Centrex 广域集中用户交换机 4. World Aerobatic Championships 世界特技飞行冠军赛 5. World Aeronautical Chart 世界航空图

WACA　　World Airlines Club Association 世界航空公司俱乐部协会

WACCC　　Worldwide Air Cargo Commodity Classification 世界空运货物分类法

WACO　　World Air Cargo Organization 世界航空货运组织

WACRA　　World Airline Customer Relations Association 世界航空公司客户关系协会

WAD　　Workload Assessment Device 工作量估算仪

WADGNSS　　Wide Area Differential Global Navigation Satellite System 广域差分全球导航卫星系统

WADS　　Wide Area Differential System 广域差分系统

wafer　　1. 晶片,芯片 2. 半导体用的单晶片 3. 薄片

WAFS　　World Area Forecast System 世界范围天气预报系统

wagner beam　　瓦格纳梁

WAI　　Wing Anti-Ice 机翼防冰

WAIPS　　Wing Anti-Ice Pressure Sensor 机翼防冰压力传感器

WAIV　　Wing Anti-Ice Valve 机翼防冰阀

wake　　1. 尾流,尾迹 2. 伴流

wake boundary　　尾流边界,伸流边界,迹流边界

wake vortex　　1. 尾涡,尾涡流 2. 尾列

wake-vortex avoidance　　尾涡规避

walking beam　　1. 摇臂 2. 摆动梁

walkway　　1. 大飞机或飞艇的狭窄通道 2. 有遮盖的人行道 3. 轻便梯

walkway girder　　过道梁

wall boundary　　壁面边界

wall boundary layer　　壁面边界层

wall cavity　　墙身空腔

wall cell　　壁光电池

wall condition　　壁面条件

wall constraint　　风洞的边壁约束

wall cooling　　器壁冷却管壁冷却

wall density　　壁密度

wall displacement　　壁面移位

wall heat　　壁热

wall heat flux　　壁面热通量

wall heat loss　　壁面热损失

wall heat transfer　　壁传热

wall hole　　壁孔

wall hole region　　壁孔区域

wall jet　　面射流

wall loss　　器壁热损失

wall material　　墙体材料

wall microphone　　墙体扩音器

wall pressure　　壁面压力

wall pressure distribution　　壁面压力分布

wall pressure increase　　壁面压力增加

wall pressure information method　　壁压信息法

wall pressure profile　　壁压力分布图

wall roughness　　壁面粗糙度

wall sensor　　壁式感测器

wall shear　　壁面剪切

wall shear stress　　壁面切应力,壁面剪切应力

wall static pressure　　壁面静压

wall temperature　　壁温

wall temperature distribution　　壁温分布

wall thermocouple　　管壁热电偶

wall-bounded flow 壁面边界流
wall-density 壁密度
walled plain 环形低地
wall-interference 管壁干涉
wallowing 倾斜摆动,摇摆
WAMS Wide Area Message System 广域消息系统
WAN Wide Area Network 广域网
wand 条形码读入器
wander 漂移,偏移
WAR Weight Analysis Report 重量分析报告
war materiel 战争物资
war of annihilation 歼灭战
war of elimination 绝灭式战争
war of possession 占领式战争
War Office 陆军部(英国)
war readiness materiel 战略物资
war reserve 战备,战争物资储备
war reserve missile 战备导弹
WARC 1. World Administrative Radio Conference 世界无线电行政大会 2. World Administrative Radio Consortium 世界无线电协议管理组织
warehouse space 仓库容积
warehouse-in inspection 入库检验
warehouse-out inspection 出库检验
warhead assembly 整装弹头,整装战斗部
warhead body tube 弹头壳体,战斗部壳体
warhead booster 弹头助爆药,战斗部助爆药
warhead carrier 弹头运载工具
warhead damage agent 弹头杀伤体
warhead damage volume 战斗部杀伤范围
warhead destructiveness 弹头破坏效果
warhead emission pattern 弹头爆炸能量分布图
warhead lethality 战斗部杀伤力,弹头杀伤力
warhead match 弹头匹配
warhead section 战斗部
warhead telemetry 弹头遥测
warhead yield 弹头威力
warm gas thruster 热气推力器,热气体小推力发动机
warm-up drift 1. 预热式频率漂移,输电功率恢复时频率漂移 2. 加热漂移,加热偏移
warm-up time 预热时间
warning area 1. 危险区 2. 警告区
warning indicator 警告信号器
warning light 警告灯
warning order 预先号令,准备命令
warning panel 警告板
warning receiver 警告接收机
warning red 紧急警报
warning system 报警系统

warning time 报警时间
warning white 解除警报
warning yellow 空袭警报
warranty 保证书,保单
war-readiness spares kit 武器系统战略成套零件箱
warship 军舰,战舰
wartime flight rate 战时出勤率
wartime rate 战时飞行强度,战时飞机利用率
WAS Weapon-Aiming System 武器瞄准系统
WASA Work Area Self Assessment 工作区域自我评价
wash-out 1. 冲洗,因故障降落 2. 外洗角
washout filter 冲洗过滤器
waspaloy 沃斯帕洛依镍基耐高温耐蚀合金
wastage rate 淘汰率
waste collection and management system 废物收集与管理系统
waste collector 废物收集器
waste gas exhaust tower 废气排放塔
waste gate 废气门
waste heat 废热,余热
waste liquid treatment station 废液处理厂
waste management system 废物处理系统
waste material 废料
waste particle 废颗粒
waste stream 废气流
waste treatment 废物处理
waste water collector 废水收集器
WAT Weight, Altitude, Temperature 重量、高度、温度
WAT curve 重高温曲线
watch dog timer 监视定时器
Watchdog Coastal Defence System "警犬"海岸防御系统
water ballast 水载压
water barrier 1. 水拦阻带 2. 水障
water bomber helicopter 消防直升机
water chiller 水冷却器
water column instrumentation radar 水柱测量雷达
water condition 海水状况,水文条件
water content 湿度,含水量,含水率
water dispenser 水分配器
water drinker 饮水器
water droplet trajectory 水滴轨迹
water equivalent depth 当量水深
water evaporator 水蒸发器
water filter 水过滤器
water gauge 1. 水柱压力计 2. 水位计,水位标尺 3. 水表
water hammer 水击,水锤

water heater	热水器
water immersion	水浸水浸装置
water immersion test	浸水试验
water injection	注水,喷水
water jet machining	高压水射流加工
water line	吃水线
water load	水上负载,水负荷
water management system	水管理系统
water mass fraction	水的质量分数
water mixture	水混合物
water mole	水摩尔
water mole fraction	水摩尔分数
water separator	脱水器水分离器
water steam	水蒸气
water sublimator	水净化器
water supply pressure regulator	供水压力调节器
water system	1.水系 2.供水系统
water tank of simulated weightlessness	仿真失重水槽
water test	水试验,液压试验
water tunnel	水洞
water turbine	水轮机
water twister	水扭转器
water vapor dilution	水蒸气稀释
water-blending ratio	水混合比
water-blending weight	水混合重
water-cooled combustor	水冷式燃烧器
water-cooled facility	水冷设施
water-cooled garment	水冷服,液冷服
water-cooled suit	水冷服
water-displacing fluid	驱水剂
waterhammer phenomenon	水锤现象
waterhammer pressure	水锤调压计
water-immersion simulation test	水浸仿真试验
water-supply pressure	供水压力
water-tunnel model	水槽模型
WATOG	World Airlines Technical Operations Glossary (Airline Industry Standard) 世界航空技术营运术语(航空公司行业标准),全球航空公司技术应用词汇
wattage rating	额定功率,额定瓦特数
watt-hour efficiency	瓦时效率
wattmeter	瓦特计,功率表
wave activity	1.波活动性 2.波活动
wave amplitude	波幅,波振幅
wave angle	波角
wave coefficient	波系数
wave controller	波控制器
wave crest	波峰
wave dynamics	波动动力学
wave equation	波动方程
wave equation migration	波动方程偏移
wave form synthesizer	波形合成器
wave front reconstruction	波前再现
wave front sensor	波前传感器
wave growth	波增长
wave guide tee	T型波导
wave height	波浪高度
wave impedance	波阻抗
wave instability	波不稳定性
wave interaction	波相互作用
wave lift	波升力
wave making resistance	兴波阻力
wave mode	波模式
wave motion	波动
wave parameter	波浪参数
wave pattern	1.波动图形 2.波型 3.波动图式 4.波模型
wave pressure	1.波压 2.波浪压力
wave propagation	波的传播,电波传播
wave radiation	电波辐射
wave rotor	波转子
wave segment	波段
wave soldering	波峰焊接法
wave speed	波速,波纹速度,波速率
wave structure	波结构
wave tube	波管
wave unstart	波急停
wave velocity	波速
wave-absorbing controller	吸波控制器
waveform	1.波形 2.波形图
wavefront	波前,波阵面
waveguide	波导,波导管
waveguide communications system	波导通信系统
waveguide gas laser	波导式气体激光器
waveguide iris	波导膜片
waveguide mode suppressor	波导模型抑制器
waveguide modes	波导振荡模
waveguide switch	波导开关
waveguide window	波导输出窗
wavelength axis	波长轴
wavelength constant	波长常数
wavelength division multiplexer	波分多路复用器
wavelength division multiplexing	波分复用,波长区分多路复用
wavelength range	波长范围
wave-off light	复飞指示灯
waverider design	乘波体设计
waverider length	乘波体长度
wave-rotor performance	波转子性能

wave-speed variation 波速度变化
waviness 1. 波纹度，波度 2. 波浪 3. 卷曲度
wavy behavior 波浪状表现
wavy wall 波形墙
wavy wall roughness 波浪壁粗糙程度
way 1. 路，航路 2. 航行，航程 3. 方法 4. 船台 5. 滑道
way station 中间站，小站
waybill tax 运单税
waypoint navigation 航路点导航
waypoint passage 航点通道
waypoint path 航点的路径
WB 1. Weather Bomber 气象探测飞机 2. Wide Band 宽带 3. Wide Beam 宽波束 4. Wing Body 机翼，机身 5. Wire Bundle 线束
WBAG Wire Bundle Assembly Group 线束装配组
WBC Weight and Balance Computer 重量与平衡计算机
WBH Weight and Balance Handbook 重量与平衡手册
WBL Wing Buttock Line 机翼纵剖面线
WBM Weight and Balance Manual 重量与平衡手册
WBRR Wire Bundle Revision Record 线束修正手册
WBS 1. Weight and Balance System 重量与平衡系统 2. Work Breakdown Structure 工作分解结构
WBSD Work Breakdown Structure Document 工作分解结构文件
WBSE Work Breakdown Structure Element 工作分解结构单元
WC 1. Wind Component 风向量 2. Work Center 工作中心
WCP 1. Weather Communication Processor 气象通信处理器 2. Weight Control Plan 重量控制大纲 3. Wing Chord Plane 机翼弦平面，翼弦平面
WCS 1. Wavelength Channel Selector 波长信号选择器 2. Wireless Communication System 无线通信系统 3. World Coordinate System 世界坐标系统
WCT Worst Cases Testing 最坏情况测试
WD 1. Wind Direction 风向 2. Wiring Diagram 线路图
WDC World Data Center 世界数据中心
WDEL Weapons Development and Engineering Laboratories 武器发展与工程实验室
WDM Wiring Diagram Manual 线路图册
WDNS Weapon-Delivery and Navigation System 武器投放与导航系统
WDS 1. Weapon Direction System 武器导引系统 2. Western Defense System 西部防御系统
WEA Weather 气象
WEAA Western European Airports Association 西欧航空港协会，西欧机场协会
WEAAC Western European Airports Authorities Conference 西欧机场管理局会议
WEAAP Western European Association for Aviation Psychology 西欧航空心理学协会
weak detection spot 探测薄弱地段
weak interference 弱干扰
weapon 1. 武器，军械，凶器，武装 2. 工具 3. 侧翼（空中编队的最左侧或右侧的战斗正面）4.（带翼状标志的）飞行胸章
weapon accuracy 武器精准度
weapon and electrical maintainer 武器与电气维修人员
weapon and electrical rating 军械电气员
weapon bay 1. 武器舱 2. 弹舱
weapon bus 武器车
weapon carrier 1. 武器载机 2. 武器挂架
weapon console 武器控制台
weapon control switchboard 武器控制板
weapon control system simulator 武器控制系统模拟器
weapon debris 武器碎片
weapon direction system 武器制导系统
weapon director 武器操纵台长
weapon fit 武器配备
weapon handling 武器的维护
weapon laser system 激光武器系统
weapon line 武器投射安全线
weapon mode 武器投射工作状态
weapon pack 武器发射器
weapon pattern effectiveness 武器投射散布效果
weapon pod 武器吊舱
weapon readiness states 兵器战斗准备状态
weapon release 武器投放
weapon release line 轰炸线，投弹线
weapon release profile 武器投放剖面图
weapon release system 武器投放系统
weapon station 武器站
weapon system 武器系统
weapon system coordination drawing 武器系统协调图
weapon system directive 武器系统指令
weapon system readiness test 武器系统准备状态试验
weapon target line 武器目标线
weapon-aiming 武器瞄准
weapon-aiming reference 武器瞄准基准线
weapon-aiming system 武器瞄准系统
weaponeering 确定使用武器数量
weaponry 1. 武器，武器装备，武器系统 2. 武器设计制造学
weapons assignment 武器分配
weapons carrier 1. 武器运载工具，武器运载机 2. 兵器车，军械车
weapons department 武器部门

weapons effectiveness testing 武器效能测试	**weight manifest** 重量记载单
weapons hold 禁止射击	**weight missile** 重型导弹
weapons of mass destruction 大规模杀伤武器,大规模毁灭性武器	**weight model** 重量模型
weapons of precision 精确武器	**weight of installation** 装置重量
weapons selector 1.武器选择器 2.武器选择电门	**weight penalty** 重量带来的损失
wear analysis 磨损分析	**weight per horsepower** 单位马力重量
wear rate 磨损率	**weight per unit thrust** 单位推力重量
wear test 磨损试验	**weight predictor** 重预测
weathercock stability 风标安稳定性	**weight ratio** 体重比,重量比
weathering 风化,自然时效	**weight reciprocal of figure** 图形权倒数
weathervane 风向标	**weight reduction** 减轻重量
web 梁腹板	**weight saving** 重量减轻
WEC World Energy Conference 世界能源会议	**weight set** 砝码组
WECPNL Weighted Equivalent Continuous Perceived Noise Level 加权当量连续感觉噪声级	**weight to power ratio of solar array** 太阳电池阵的重量功率比
wedge 楔,楔形物	**weight zone** 弹重级数,弹体重量分类
wedge angle 楔角,尖角,研磨角	**weight(ing) coefficient** 加权系数
wedge bonding 楔形接合,楔焊	**weighted arithmetic average** 加权算术平均数
wedge corner 楔形角	**weighted average surrogate model** 加权平均代理模型
weigh in 称重量	**weighted battery of test** 检验标准加权组合
weighing cell 称重传感器	**weighted combination** 加权组合
weighing mass 称量	**weighted curve** 加权曲线
weight and balance 重量与平衡	**weighted mean** 加权平均值
weight and balance computation 重量和平衡计算	**weighted norm** 加权范数
weight and balance sheet 载重与平衡表	**weighted stack** 加权叠加
weight asymmetry 重量不对称	**weighted sum** 加权和
weight break 重量计价暴落点	**weighted total delay cost** 加权延迟总成本
weight breakdown 重量分组,重量划分	**weightedbar** 稳定杆
weight charge 按重量计的运费	**weighting coef** 权重系数
weight class 重量级	**weighting factor** 加权因子
weight coefficient 重量系数	**weightlessness adaptation** 失重适应
weight component 重量分量	**weightlessness countermeasures** 失重对抗措施
weight control 体重控制	**weightlessness physiological effect** 失重生理效应
weight enumerator 重量枚举器	**weightlessness simulation** 失重仿真
weight estimation 重量估算	**weightlessness simulation test** 失重仿真试验
weight factor 1.加权因子,加权系数 2.重量系数	**weight-limited** 有重量限制的
weight flow 重量流量	**weight-out** 重量已达限
weight fraction 重量百分率,重量分数	**weld** 焊接
weight function 加权函数	**weld bead** 焊缝,焊接熔化部位
weight gain 重量增加	**weld bonding** 胶接点焊
weight gradient 重量梯度	**weldability** 可焊性,焊接工艺性
weight group 重量组	**weldableness** 焊接性
weight growth 增重	**weldbonding** 实焊粘接
weight growth factor 增重因素	**welded patch** 焊修补片
weight in running order 发动机运转状态重量	**welded steel blade** 钢板焊接桨叶,空心钢质桨叶
weight increase 增重	**welded structure** 焊接结构
weight load factor 重量运载系数	**welded wing** 间隔不变的双机横队
weight loss 1.重量损失 2.失重	**welding crack sensibility** 焊接裂纹敏感性
	welding deformation 焊接变形

welding jig　焊接夹具
welding machine　焊接机
welding rod　焊条
welding stress　焊接应力
well-balanced system　平衡的系统
well-deployed　已经部署多时的，有作战使用经验的
well-develop　良好发展
well-document　充分证据
well-documented experiment　证据充分的实验
well-resolved simulation　高清晰模拟
WEM　1. Warning Electronic Module 报警电子微型组件 2. Welfare of (Air Force) Enlisted Men (空军)士兵福利待遇
WEMA　Western Electronics Manufacturers Association 西方电子设备制造商协会
WEP　Weapon Effect Planning 武器效能计划
WES　Warning Electronic System 电子告警系统
West Ford　西福特卫星（美国通信卫星）
Westar　西联星（美国国内通信卫星）
WESTE　Weapons Effectiveness and System Test Environment 武器效能与系统试验环境
WET　1. Weapons Effectiveness Testing 武器效能测试 2. Weight Effectiveness Testing 有效重量试验
wet　1. 有外挂油箱的，油箱在外的，机翼内有油箱的 2. 有喷水(加力)的 3. 物体表面与之接触的 4.〈口语〉实弹
wet anti-exposure suit　湿型抗暴露服
wet area　潮湿区域
wet emplacement　湿式发射阵地，湿定位
wet filter　湿式过滤器
wet firing　战斗射击，实弹射击
wet fuel rocket　液体燃料火箭
wet start　湿起动
wet suit　湿式潜水服
WETF　Weightless Environment Training Facility 失重环境训练设施
wet-oxygen oxidation　湿氧氧化
wet-run anti icing　保温防冰
wetted area　沾润面积，浸润面积
wetted fibre　湿纤维
wetted surface　湿润面，流体浸润(接触)表面
wetting angle　润湿角
wetting liquid　浸润性液体
WEU　Warning Electronics Unit 电子警告组件，电子警告装置
WF　Weather-reconnaissance Fighter 气象侦察战斗机
WFA　Wide Frequency Antenna 宽频天线
WFD　Widespread Fatigue Damage 广布疲劳损伤
WFM　Work Flow Management 工作流管理

WFOV　Wide Field Of View 宽视界
WFPI　Wing Flap Position Indicator 襟翼位置指示器
WG　1. Water Gauge 水表 2. Wave Guide 波导 3. Weight Guarantees 重量担保 4. Working Group 工作组
WGN　White Gaussian Noise 白高斯噪声
WGS　1. World Geodetic Survey 世界大地测量勘查 2. World Geodetic System 世界大地测量系统，世界测地系统
WH　Watt-Hour 瓦特小时
wheel scanning horizon sensor　飞轮扫描地球敏感器
wheel speed　轮速
whiffletree　横杠
whip　轴跳动
whip antenna　鞭状天线
whip stall　尾冲失速
whirling arm　旋转臂，旋转杆
whirling arm testing　旋臂机试验
whirling mode　旋涡模式
whirltower　旋翼试验塔
whirlwind　旋风
whirlybird　直升机
whisker　晶须，金属须，触须线
White Cloud　白云无反射面
white hot　白热的
White Sands　白沙导弹靶场
white tailed　1. 空白机尾的 2. 待租的，待售的（机尾上没有标志，喷涂何种标志由承租人或买主决定）
white-light　白光
white-noise　白噪声
white-noise excitation　白噪声激发
white-noise sequence　白噪声序列
Whitney punch　惠特尼冲孔机
Whittaker-Shannon sampling theorem　惠特克-香农采样定理
Whitworth　惠特华斯螺纹
whizzer　电视变焦镜头
whizzkid　文职分析员或顾问
whizzo　武器系统操作员或军官
WHO　World Health Organization 世界卫生组织
whole body counter　体内计数器
whole engine　整个引擎
whole injection jet　整个注射喷口
whole ion　整个离子
whole mission　整个任务
whole period　整个阶段
whole process　整个过程
whole structure　整个结构
whole thing　整个事情

whole transfer　整体传输
whole transfer trajectory　整体传输轨道
whole-range distance　全程距离
whole-range point　全程点
whole-surface coverage　整个表面覆盖
WI　1. Wallops Island 沃洛普斯岛 2. Welding Institute 焊接研究所
wi　无线
WIA　Wounded In Action 作战负伤
WIC　1. Warning Information Correlation 告警信息对比分析 2. Work In Cycle 进入循环
wick　1. 静电放电刷 2. 油门 3. 管芯
wicking　1. 灯芯 2. 芯吸作用 3. 灯芯材料 4. 导火线
WIDE　Wide-angle Infinity Display Equipment 广角无限显示设备
wide angle radiometer　广角辐射仪
wide approach　大航线进场着陆
wide area differential GPS　1. 广域差分 GPS 2. 宽域差分 GPS
wide area net　广域网, 远程网
wide area network　广域网
wide deck　宽底座汽缸
wide range　宽波段的, 宽量程的
wide variety　1. 电池 2. 品种繁多
wide-angle　广角
wide-angle radar　宽波束雷达
wide-angle reflection　大角度反射
wide-area　广域
wideband　宽带, 宽频
wideband active repeater satellite　宽带有源转发卫星
wideband amplifier　宽频带放大器
wideband gain　宽带增益
wideband networking waveform　宽带网络波形
wideband source　宽带信号
wideband tape recorder　宽频带磁记录器
wideband video tape recorder　宽带录像机
wide-body　宽体
wide-body aircraft　宽体飞机
wide-distribution　广泛分布
wide-field integration　广角集成
wide-speed-range　宽转速范围
widespread use　广泛使用
width　宽度, 广度
width ratio　宽比
Wien bridge　维杜式振荡器
WIG　Wireless Internet Gateway 无线因特网网关
wiggler　波动器
wiggle strip　波纹板
will not fire　（代码, 发给对空观察员或申请单位以确认地面炮火）不参与打击该目标
willbe　将会发生
Williot　威辽特图
Wilson formula　威尔逊公式
Wimet　外梅特合金
WIN　1. WWMCCS Intercomputer Network 全球军事指挥与控制系统计算机联网 2. Wireless Intelligent Network 无线智能网 3. Worldwide Intelligent Network 全球智能网
winch　绞车, 卷扬机
winch controller　绞车操作员
winch suspension　绞车吊架
winching　绞车作业
winchman　绞车员
wind　风卫星
wind across　风的正侧分量
wind angle　风角
wind axes　风轴
wind blast　气流吹袭
wind comfort criterion　风舒适度标准
wind condition　风况
wind cone　风向袋
wind correction angle　风修正角（航向与航迹角之差）
wind deflection correction　风偏修正量
wind diagram　航行（速度）三角形
wind effect　1. 风效应 2. 风力作用
wind ellipse　风椭圆
wind energy　风能
wind error　风误差
wind gradient　风梯度
wind gust　阵风
wind magnitude　风量值
wind map　风图
wind marker　风标
wind measurement　测风
wind measurement from satellite　卫星测量风
wind model　风模式
wind plane　风平面
wind power　风力
wind profile　风剖面
wind rate　风量
wind rate of cabin　座舱大气风速
wind realization　风实现
wind recording gear　风向风速记录仪
wind rotor　冷却风扇, 吹风机
wind shield　风挡, 挡风板
wind shield anti-fogging system　风挡防雾系统
wind shield rain removal system　风挡除雨系统
wind shift　风向突变

wind speed	风速
wind star	风向图
wind state	风的状况
wind strength	风力
wind tee	丁字风向标
wind tetrahedron	角锥风向标
wind tunnel	风洞
wind tunnel axis system	风洞坐标系
wind tunnel balance	1.风洞天平 2.气动力天平
wind tunnel body	风洞本体
wind tunnel energy ratio	风洞能量比
wind tunnel flow rate	风洞流量
wind tunnel free-flight test	风洞自由飞实验
wind tunnel operation	风洞运行
wind tunnel program control	风洞程序控制
wind tunnel simulation capability	风洞仿真能力
wind tunnel test	风洞试验
wind tunnel test data base	风洞试验数据库
wind tunnel-computer integration	风洞计算机一体化
wind turbine	1.风力发动机 2.风力涡轮机
wind uncertainty	风不确定性
wind vane	风向标
wind velocity	风速
wind wave channel	风浪槽
WINDMG	Wind Magnitude 风值
WINDR	Wind Direction 风向
wing aerodynamics	机翼空气动力学
wing axis	机翼轴,机翼轴线
wing box	翼盒
wing cell	翼组
wing chord	机翼弦
wing configuration	翼配置
wing control	翼控制
wing cross section	机翼的横截面
wing deformation	机翼变形
wing design	机翼设计
wing drag	翼阻力,机翼阻力
wing flap	襟翼
wing flow	翼流
wing flow testing	机翼流试验
wing flutter	翼颤
wing gap	翼隙
wing gun	机翼炮,机翼机枪
wing gun mounting	机翼枪架
wing half	半翼
wing heavy	机翼自动倾斜,翼重
wing inboard multiple weapon pylon	机翼内侧多种武器外挂梁
wing incidence	机翼倾角
wing leading-edge	机翼前沿
wing level	机翼水平
wing leveller	机翼水平仪
wing lift	机翼升力
wing load	翼载荷
wing loading	翼载
wing model	机翼模型
wing motion	翼运动
wing panel	翼片
wing pivot	主翼控制器,转掠翼
wing plan	机翼平面形
wing planform	机翼平面形状
wing position	机翼位置
wing profile	机翼剖面,翼型
wing radiator	翼载散热器
wing reaction	机翼反作用力
wing reference area	机翼参考面积
wing rib	翼肋
wing root	翼根
wing rotation	旋转翼
wing section	1.翼型 2.翼段 3.翼剖面,翼切面
wing segment	机翼段
wing semi-span	机翼半翼展
wing separation	机翼分离
wing setting	机翼安装角
wing shape	翼状
wing size	翼大小
wing skin	机翼蒙皮
wing slot	翼缝
wing span	翼展
wing spar	翼梁
wing structural model	机翼结构模型
wing structure	机翼结构
wing strut	机翼斜支柱
wing surface	翼面
wing sweep	机翼后掠角
wing system	翼系统
wing tank	舷顶边舱
wing tip	机翼翼端
wing tip carriage	翼尖外挂
wing tip missile	翼稍携带导弹
wing tip suspension	翼尖悬挂
wing tip vortex	翼尖旋涡
wing weight	机翼重量
wing-body connection	翼身连接
wing-box front spar	翼盒前梁
wing-box rear spar	翼盒后梁
wing-control missile	翼控式导弹,转动翼式导弹
wing-drop	翼下降

wing-drop event	翼下降事件
wing-drop motion	翼下降运动
wing-drop problem	翼下降问题
winged bomb	有翼炸弹
winged helicopter	带翼直升机,有翼直升机
winged missile	有翼导弹,飞航式导弹
wing-fuselage	机翼机身组合体
wing-fuselage optimization	机翼机身优化
wing-in-ground effect	地面效应
wingless configuration	无翼式布局
wingless missile	无翼导弹
winglet	小翼,小翅
winglet chord	小翼弦
winglet design	小翼设计
wing-located fuel bladder	翼内软油箱,翼内燃油囊
wing-low	翼低
wingman	僚机驾驶员
wing-mounted engine	翼置发动机
wing-rock	机翼摇晃
wing-rock motion	机翼摇晃运动
wing-rock tendency	机翼摇晃倾向
wing-root	翼根
wing-rotor system	翼转子系统
wing-section	机翼截面
wings-level condition	机翼水平状态
wingtip missile	翼稍携带的导弹
wingtip rake	翼尖斜削,翼尖斜削度
wingtip response	翼尖响应
wingtip system	翼梢系统
wingtip tank	翼尖油箱
wingtip vortex	翼端涡
WIP	1. Weapon Insert Panel 武器信息输入板 2. Weight Improvement Program 重量改进计划 3. Work In Process 有序工作,正在进行的工作
wipe technique	白化技术
wire	电线,金属丝
wire antenna	天线
wire bead	钢丝撑轮圈
wire braid	钢丝编织
wire bundle	导线束
wire cloth	金属丝网
wire command guidance	有线指令制导
wire command guidance system	有线指令制导系统
wire communication	有线通信
wire explosion	电线爆炸
wire group	导线束,导线组
wire guidance	引线制导,引线引导
work mode	工作模式
wire link	有线联系,有线通信
wire lockingring	线锁环
wire loop	线环,钢丝圈
wired program	插线程序
wired remote control	有线遥控
wired storage	磁线储存
wired telemetry	有线遥测
wired-guided torpedo	有线制导鱼雷,线导鱼雷
wire-drawing bench	拉丝台
wireframe model	线架模型
wire-guided	引线引导的
wireless	1. 无线的 2. 无线电
wireless communication	无线通信
wireless machine	无线电通信飞机
wireless telegraphy	无线电报
wiresonde	有线探空仪
wire-tunnels	线隧道
wireway	电缆管道,电缆束管道
wire-wound	绕线式
wiring	布线,配线,接线
wiring board	接线板
wiring tunnel	电缆管道
WIRS	Wire Information and Release System 线束信息和发放系统
WISI	World Index of Space Imagery 全球空间图像索引
WISR	Weight and Inertial Status Report 重量与惯量状态报告
WITS	Weather Information Telemetry System 气象信息遥测系统
WJAC	Women's Junior Air Corps 妇女初级航空队
WL	1. Water Level 水线 2. Wave Length 波长 3. Wideband Limiter 宽带限幅器
WLDP	Warning Light Display Panel 报警灯显示板
WLI	Water Landing Impact 水上降落撞击
WM	1. Wattmeter 瓦特表 2. Wiring Manual 线路图手册,布线手册
WMC	1. World Meteorological Center 世界气象中心 2. World Meteorological Congress 世界气象大会
WMD	1. Weapons of Mass Destruction 大规模杀伤兵器 2. World Meteorological Day 世界气象日
WMO	World Meteorological Organization 世界气象组织
WMT	WXR Mount 气象雷达安装架
WN	Week Number 周号
WNDSHR	Windshear 风切变
WO	1. Warning Order 报警指令 2. Work Operate 工作操作 3. Work Order 工作指令 4. Written Off 注销,报废 5. Wash-Out 过渡阶段
WO ATO	Without Assisted Takeoff 不用起飞助推器

WOA	Weapons Orientation Advanced 武器发展动向
wobble	1. 摆动 2. 震颤 3. 不稳定运动 4. 变量
wobble plate	旋转倾转盘
wobble pump	手摇泵
wobbulator	摇频信号发生器,频率摆动,射线偏斜
WOC	1. Wing Operations Center 联队作战中心 2. Wireless Optical Communication 无线光通信
WOFF	Weight Of Fuel Flow 燃油流量重量
womanaut	女航天员,女宇航员
WOPL	Wear-Out Parts List 消耗件清单
word of memory	内存信息
word rate	字频率,字速率,字速
work addition	工作增加
work capacity evaluation	工作能力评定
work coefficient	工作系数
work distribution	工作分配
work effectiveness	工作有效度
work environment	工作环境
work exchange	工作交流
work extraction	工作提取
work force	工作力
work frequency	工作频率
work hardening	加工硬化,加工强化
work interaction	工作互动
work lights	工作灯
work load	1. 工作负荷 2. 工作载
work package	1. 工序组 2. 单位工作量
work potential	工作潜力
work process	1. 加工过程 2. 工作进程 3. 工作工程
work production	工作生产
work ratio	工作效率
work relation	工作关系
work space	工作空间
work station	工作地,工作站
work study	工作研究
work transfer	工作交接
worker agent	工人代理
working field of view	工作视场
working fluid	液压油,工缀
working load	工作负荷,工作载荷
working medium	工组质
working medium of wind tunnel	风洞工作介质
working point	工作点
working pressure/temperature	工作压力/温度
working section	工作段
working space	作业空间
working standard	工作标准
working standard solar cell	工作标准太阳电池
work-rest cycle	作息制度
world aeronautical charts	世界航图
World Geographic Reference System	世界地理参考系统
world model	全球模型
World Weather Watch	世界天气监视网
worldwide gravimetric basic point	世界重力基点
worldwide reference system	全球参考系统
worldwide satellite communication	全球卫星通信
worst cold case	低温工况
worst condition analysis	最坏情况分析
worst hot case	高温工况
WOTAN	Wavelength-agile Optical Transport and Access Network 波长灵活的光传送和接入网
wounding power	杀伤力
WOW	Weight on Wheels 轮档重量
WP	1. Waterproof 防水 2. Waypoint 航路点 3. Word Processing 字处理
WPA	WiFi Protected Access 无线保护访问
WPFC	World Precision-Flying Championships 世界精确飞行锦标赛
WPMS	Work Plan Management System 工作计划管理系统
WPN	weapon 武器
WPR	Waypoint Position Report 航路点位置报告
WPT	waypoint 航路点
WPU	Weapon Programming Unit 武器程序装置
WR	1. Weather Radar 气象雷达 2. Wing Root 翼根
WRA	Weapon Replaceable Assembly 武器可更换组件
WRAF	Woman's Royal Air Force 皇家空军妇女队
wrap angle	包角
wrap contact	绕接接触件
wrap-around boost motors	环列式助推器
wrap-around fin	卷弧翼
wrap-around type solar cell	卷包式太阳电池
wrapped around folding fin	环列折叠尾翼
wrapped connection	绕线连接
wrapper	包装机
wrap-round engine	环列式满轮冲压发动机
wrap-round windscreen	半球形风挡,整体模压的前驾驶舱
WRCS	1. Weapons Release Computer Set 武器投掷计算机装置 2. Weapons Release Control System 武器发射控制系统
WRE	Weapons Research Establishment 武器研究院
wreckage trajectory analysis	残骸轨迹分析
wrecker	机场抢救车
WRG	Wiring 接线,布线
write	存入,记录输入的信息
wrong manifold path	错误歧管路径

wrought aluminum alloy 变形铝合金
wrought magnesium alloy 变形镁合金
wrought superalloy 变形高温合金
wrought titanium alloy 变形钛合金
WRP　1. Wing Reference Plan 机翼基准平面 2. Wireless Routing Protocol 无线路由选择协议
WRR　Wire Rework Record (Form) 导线重制记录（表格）
WRS　Wide-area Reference Station 广域基准站
WRSK　War-Readiness Spares Kit 备战备用工具箱
WRT　WXR Receiver/Transmitter 气象雷达接收/发射机
WS　1. Warning System 报警系统 2. Water Supply 供水 3. Wheel Speed 轮速 4. Windscreen 风挡 5. Wind Shear 风切变 6. Wing Span 翼展 7. Wind Speed 风速 8. Wing Station 机翼站位 9. Work Station 工作站
WSAP　Weapon System Aquisition Process 武器系统筹措过程
WSAT　Weapon System Acceptance Test 武器系统验收试验
WSC　Weapon System Controller 武器系统控制器
WSCC　Weapon System Configuration Control 武器系统外形调整
WSCP　Warning and System Control Panel 警告和系统控制面板
WSCS　1. Weapon System Communication System 武器系统通信系统 2. Wheel Steering Control System 机轮转弯控制系统
WSD　1. Warning and System Display 报警和系统显示器 2. Work Statement Definition 工作说明定义
WSDC　Weapon System Design Criteria 武器系统设计准则
WSDCU　Wideband Satellite Delay Compensation Unit 宽带卫星延迟补偿器
WSDL　Weapon-System Data-Link 武器系统数据传输设备，武器系统数据传输线
WSDP　Weapon System Development Plan 武器系统研制计划
WSDRN　Western Satellite Data Relay Network 西部卫星数据中继网
WSECL　Weapon System Equipment Component List 武器系统部件一览
WSEG　Weapon-Systems Evaluation Group 武器系统鉴定组
WSEM　Weapon System Evaluation Missile 武器系统鉴定用导弹
WSEP　Waste Solidification Engineering Prototype 废液固化工程原型
WSF　Work Station Function 工作站功能

WSFC　1. Wallops Space Flight Center 沃洛普斯航天中心 2. White Sands Field Center 白沙战地中心
WSFO　Weather Service Forecast Office 气象局预报处
WSI　1. Water-Separometer Index 水分器指标 2. Weather Severity Index 天气严重性指数 3. Wind Shear Indicator 风切变指示器
WSMC　1. Weapons System Management Code 武器系统管理代号 2. Western Space and Missile Center 西部航天与导弹中心
WSMIS　Weapon System Management Information System 武器系统管理信息系统
WSMR　White Sands Missile Range 白沙导弹靶场
WSMS　Wind Shear Monitor System 风切变监视系统
WSO　1. Weapon Systems Officer 武器系统军官 2. Weapon System Operator 武器系统操作 3. Weather Service Office 气象服务室 4. World Space Organization 世界空间组织
WSON　World-wide Satellite Observing Network 全球卫星观测网
WSPG　White Sands Proving Ground 白沙试验场
WSPO　Weapon System Project Office 武器系统计划处
WSPS　Worldwide Standard Port System 全球标准接口系统
WSR　1. Weapon System Reliability 武器系统可靠性 2. Weather Surveillance Radar 气象监视雷达
WSRN　Western Satellite Research Network 西部卫星研究网
WSRT　Weapons System Reliability Test 武器系统可靠性试验
WSSC　Weapon-System Support Center 武器系统支援中心
WSSG　Warning System Symbol Generator 警告系统符号生成器
WSSP　Weapon-System Support Program(me) 武器系统器材技术保障计划
WSSS　Weapon Storage and Security System 武器储存与安全系统
WST　1. Weapons Systems Test 武器系统试验 2. Weapon System Trainer 武器系统训练装置，驾驶训练和飞行战术训练装置 3. Wireless Subscriber Terminal 无线用户终端 4. World Satellite Terminal 世界卫星终端
WSTA　Wing Station 机翼站位
WSU　Wheel Spin Up 机轮加速
WT　1. Water Tank 水箱 2. weight 加权，重量，砝码 3. Wind Tunnel 风洞 4. Wireless Telegraphy 无线电报
WTGC　Work Type Group Code 工作类型组别代码
WTI　Weapon and Tactics Instructor 武器及战术教员
WV　1. Wind Velocity 风速 2. Working Voltage 工

作电压

WVAC Working Voltage Alternate Current 交流工作电压

WVDC Working Voltage Direct Current 工作直流电压

WW 1. War Weary 作战磨损的 2. Wheel Well 轮舱 3. Wing Web 机翼辅助梁 4. With Weapons 带有武器的

WWABNCP Worldwide Airborne National Command Post 全球国家空中指挥所

WWACPS Worldwide Airborne Command Post System 世界空中指挥所系统

WWMCCS Worldwide Military Command and Control System 全球军事指挥与控制系统

WWW 1. World Weather Watch 世界天气监视网 2. World Wide Web 全球信息网,世界广域网,万维网

X

X 1. extra 额外的,格外的 2. cross 交叉 3. extension 电话分机,延长 4. exclusive 不包括,除外 5. trans 转换
X axis 横坐标轴,X轴,横坐标
X bracing 1.十字(撑)架,X形拉条 2.交叉(X形)联接,交叉(X形)支撑,剪刀撑
X direction X轴向,横轴向
X engine X型发动机
X plates 水平偏转板,X板(阴极射线管内的)
X position X位置
X ray analysis X射线分析
X ray burster X射线爆发源
X ray flaw detection, X ray inspection X光探伤
X ray pulsar X射线脉冲星
X ray shield X射线屏蔽层
X ray star X射线星
XA Auxiliary Amplifier 辅助放大器
xact exact 准确的
XARM Cross Arm 横臂(撑,担),线担
Xaser X射线激射器
XBAG Excess Baggage 超重行李
Xbar 1.指引杆,横臂,横杆 2.横排灯
Xbsw 纵横制接线器
XC Cross Connect 交叉连接
XCH 1.Cross Channel 跨频道 2.Exchange 交换兑换
X-Channel X波道
XCTB Cross-Connect Test Bed 交叉连接测试台
XCVR Transceiver Unit 无线电收发机
X-D X Direction X轴向
XDCR Transducer 传感器,变化器
XE 1.Experimental Engine 实验型发动机 2.External Electrical 外接电源
Xe Xenon 氙(惰性气体)
Xenobiology 外空生物学
Xenon 【化】氙(化学元素,符号 Xe)
xenon propellant 氙推进剂
xerography 静电复印(术),干印(术),(磁记录)干印图,硒板摄影,硒鼓复印
XES X-Ray Energy Spectrometer(医)X射线能量分光计
Xfer Transfer 转移
Xfmr Transformer 1.变压器 2.变换器 3.变量器 4.互(传)感器

XGA Extended Graphics Array 扩展图像阵列
XGAM Experimental Guided Air Missile 实验型航空导弹
Xgr Exchanger 1.X交换器(机,剂) 2.换(放,散)热器
XH 1.Experimental Helicopter 实验型直升机 2.Extra Heavy 超重 3.Extra High 超高
XIO Execute Input/Output 执行输入/输出
XL 1.Crystal 晶体 2.Extra Large 特大 3.Extra Long 极长
X-line X坐标线,横轴线
XM 1.External Master 外部主机 2.Extra Marker 外加指点标
Xmit Transmit 传输,发射,发报,发信
X-Mitter Transmitter 无线电发射机
XML Extensible Markup Language 可扩展的标志语言
XMM Extended Memory Manager 延伸储存器管理程序
Xmtr Transmitter 发报机,发射机,送话器
XNS Xerox Network System 静电印刷网络系统
XO 1.Exchange Order 换票证 2.Executive Officer(舰上)副中队长,副舰长,(海军航空站)副站长
X-parallax X视差
XPD X Transponder (also XPDR, XPNDR, TPR)应答机
XPDR Transponder 应答机
Xpl Explosion 爆炸
Xpld Exploded 已爆炸
Xpln Explain 解释
XPNDR Transponder 发送-应答机,询问机
XPP Transmit Personality Prom 发送终端专用可编程只读存储器
XPS X-ray Photoelectron Spectros-Copy X射线光电子谱
XR Experimental Rotor 实验型旋翼机
X-ray 1.X光,X射线 2.射线望远镜 3.用X光检查
X-ray astronomy X射线天文学
X-ray background radiation X射线本底辐射
X-ray computerized tomography X射线计算机层析成像法
X-ray detector X射线探测器
X-ray diffraction X射线衍射
X-ray flux X射线通量
X-ray inspection X射线检验,X射线探伤

X-ray irradiation　X射线照射
X-ray lithography　X射线光刻
X-ray microanalysis　X射线微分析
X-ray photoelectron　X射线光电子能谱
X-ray photoelectron spectroscopy　X射线光电子光谱学
X-ray photogrammetry　X射线摄影测量
X-ray photon　【医】X(射)线光子,X(射)线辐射量子
X-ray proportional counter　X射线正比计数器
X-ray radiation　X射线辐射
X-ray radiography　【医】X射线摄影术
X-ray source　【医】X(射)线源
X-ray target　X射线靶
X-ray telescope　X射线望远镜
X-ray topography　X射线局部厚层断层摄影法
XRD　X-Ray Diffractometer X射线衍射仪
XRT　X Ray Topography　X射线局部厚层断层摄影法
XS　1. Atmospherics 天电 2. Excess 超过 3. Expenses 费用,开支
XSCC　China Xian Satellite Control Center 中国西安卫星测控中心
XSECT　Cross Section 横截面,横断面
XSLC　China Xichang Satellite Launch Center 中国西昌卫星发射中心
XSM　Experimental Strategic Missile 实验性战略导弹
Xsmn　Transmission 传输,发射
XSSDU　Expedited Session Service Data Unit 加速会晤业务数据单元
Xstr　Transistor 晶体管
XT　Reception (Transponder to TCAS)接收(应答机到防撞系统)
Xtal　Crystal 晶体
XTD　Cross Track Distance 偏航距
X-wave　异常波
X-Y mount　X-Y轴结构基座
XY plane　XY平面
X－Y recorder　X-Y坐标记录器
Xylene　二甲苯
Xylonite　硝酸纤维素塑料
XYP　X－Y Plotter X－Y坐标绘图仪
XYR　X－Y Recorder X－Y坐标记录器
XZ plane crossing　【医】XZ平面交叉

Y

Y 1. yellow 黄色的 2. yaw 偏航 3. why 为什么 4. you 你,你们 5. economy class 普通舱 6. Y-axis Y 轴,纵轴 7. year 年
Y joint Y(叉)形接头,Y 形结合
Y junction Y 接头
YA 1. Yard 码 2. 政府民航当局 3. Yaw Axis 偏航轴
yag laser 激光器
yagi antenna 八木天线
yankee escape system 美式撤离系统
yankee paper machine 杨琪抄纸机,杨琪造纸机
yard 1. 码 2. 编组站,档场
yarn 纱,纱线
yarn abrader 纱线耐磨试验仪
yaw 偏航,左右摇摆
yaw angle 偏航角,偏流角,偏转角,偏斜角
yaw axis 偏航轴,偏荡轴
yaw damper 偏航阻尼器,偏航调节板,减摆器
yaw damper computer 偏航阻尼计算机
yaw damper control panel 偏航阻尼控制面板
yaw rudder 方向舵
yawhead 偏航传感器
yaw(ing) moment 偏航力矩
yawing moment coefficient 偏航力矩系数
yawing moment due to ailerons 偏转副翼产生的偏航力矩
yawing moment due to rolling velocity 由滚转角速度引起的偏航力矩
yawmeter 偏航计,测向计,偏流计,偏转仪
Y-axis 纵坐标轴,Y 轴
YB Year Book 年鉴
Yb 镱
YBD Y-Branching Device Y 型分路器
Y-channec Y 信道
Y-connection Y 形连接头,星形连接头
Y-cut crystal Y 切割晶体
YCZ Yellow Caution Zone 黄色注意区,黄灯注意区
YD 1. Yards 码(长度单位) 2. Yaw Damper 偏航阻尼器
YDA 1. Yesterday 昨天 2. Yaw Damper Actuator 偏航阻尼器作动器

YDC Yaw-Damper Computer 偏航阻尼器计算机,偏摆阻尼器耦合器
YDS Yaw Damper System 偏航阻尼系统
yellow alert 预备警报
yellow caution zone 黄灯警戒区
YES 1. Yield Estimation Subsystem 产量预报子系统 2. Yankee Escape System 美式撤离系统
yield 1. 产量 2. 产额回收率 3. 成品率 4. 屈服极限 5. 二次放射系数 6. 产出 7. 委弃,让予
yield factor of safety 屈服安全系数
yield failure 屈服破坏
yield function 屈服函数
yield limit 屈服极限,屈服点,屈服极限流动性范围
yield load 屈服载荷
yield point 屈服点,流动点,屈服温度,软化点
yield stress 屈服应力,屈服强度
yielding 永久变形,产生,形成;流动性的,可压缩性的
YIG Yttrium Iron Garnet 钇铁石榴石材料
yig resonator 钇铁石榴石谐振器
YLT Yellow Light 黄色灯
YM 1. Your Message 你方电报 2. Yawing Moment 偏航力矩
YMS Yield Management System 收益式电脑作业管理系统
Yoke 1. 轭,轭状物 2. 套 3. 束缚,支配 4. 磁轭 5. 人孔压板 6. 偏传线圈 7. 卡钳
YOS Year Of Service 飞机服役年数
Yo-Yo 溜溜球,运输机装车溜槽
Yo-Yo technique 电缆收放技术
YP Yield Point 屈服点,流动点
YR Yaw Rate 偏航率
YS 1. Yield Strength 屈服强度 2. Yielding Stress 屈服应力
YSAS Yaw Stability Augmentation System 偏航稳定增强系统
YSE Yaw Steering Error 偏航驾驶误差
yttrium aluminate laser 铝酸钇激光器
yttrium aluminium garnet crystal 钇铝石榴石晶体
yttrium aluminium garnet laser 钇铝石榴石激光器
Yuri-3a 尤利-3a

Z

Z 1. zone 地带,地区,范围,层 2. 晶带,色区 3.【计】存储区,区段 4. zero 零 5. Z-axis Z 轴
Z component Z 轴向分量
Z direction Z 方向
Z section Z 形钢,Z 型截面,之字钢
Z technology Z 技术(芯片上沿 Z 轴的红外传感器凝视焦面阵)
ZA 1. Zero-Adjusted 调零 2. Zone Aerial 分区天线
ZAB 1. Zaragoza Air Base (Spain) 萨拉戈萨的空军基地(西班牙) 2. Zinc-Air Battery 锌空气电池
ZAF Zero Alignment Feature 零校准装置
ZAM Zoning and Accesss Manual 区域和口盖手册
Zamak 查马克合金(一种锌基压铸或锻铸合金)
Zap 1.〈口语〉砸坏,或使不能使用 2. 一次进入就使目标(通常指空中目标)失去战斗力 3. Zapp Flap 易操纵襟翼,小操纵力襟翼,后退式襟翼
ZAR 1. Zaire 扎伊尔 2. Zeus Acquisition Radar 宙斯探测雷达
Z-beacon Z 信标
Z-correction Z 修正量
Z-count 1. Z 计数 2. 子帧计算
ZD 1. Zero Defects 无故障,无缺陷 2. Zero Dispersion 零色散 3. Zone Distance 航段距离
Z-domain Z 域
ZE 1. Zero Effect 无效应,零效应 2. Zero Evaluation 区域评价
ZEC Zinc Electrochemica Cell 锌电池
zeeman effect 塞曼效应
zelling 零长导轨发射
zellon 四氯乙烯
zellweger tester 泽尔韦格尔测试仪
Zener 1. 齐纳 2.(Zener) Voltage 齐纳电压
Zener breakdown 齐纳击穿
Zener current 齐纳电流
Zener diode 齐纳二极管
zenit launch vehicle 天顶号运载火箭
zenith 顶点,天顶
zenith camera 天顶相机
zenith distance 天顶距
zenith star 天顶星
zenith nadir axis 天顶天底轴
zenith telescope 天顶仪
zenithal rains 天顶雨

zenographic 木星表面的
zenographic coordinates 木星地理坐标系,木面坐标
zeolite 沸石
zeolite absorption pump 沸石吸附泵
zero 零,零点,零位,零度
zero adjust 零位调整
zero attitude 零姿态
zero carrier 零振幅载波
zero channel threshold 零道阈
zero defects concepts 无缺陷概念,完美思想
zero detection probability 零探测概率
zero doppler line 零多普勒线
zero drift 零位偏移,零位漂移
zero drift current 零漂移电流
zero error 零点误差,零位误差,零误差
zero fuel weight 零燃油重量
zero gap semiconductor 零带隙半导体
zero gravity 零重心,无重力,失重
zero ground speed maneuver (直)原地机动(地速为零)
zero hour 零时条件,零时状态
zero in 1. 归零校正,瞄准具校正 2. 同步调零
zero incidence wing 零攻角机翼
zero initial bias 零始偏压
zero launch 零长轨发射
zero launcher 无导轨发射装置
zero length 起始长度,零长度
zero length spring gravimeter 零长弹簧重力仪
zero level of homing head 导引头零位电平
zero lift 零升力
zero lift angle 零升力角
zero lift moment 零升力力矩
zero lift drag 零升阻力
zero liquid level 零液位
zero meridian 零子午线
zero momentum mode 零动量方式
zero momentum system 零动量系统
zero offset 零点误差,零位误差,零误差,零点偏移(速度陀螺仪在无输入信号下的输出)
zero point 零点
zero range calibration 距离校零
zero reader 零读数器,零位指示器
zero relative level point 零相对电平点
zero setting 调到零点,定零位

zero steady-state error system	零稳态误差系统
zero suppression	【计】消零（将无用的零位消去）
zero temperature coefficient point	零温度系数点
zero terminal	零线端子
zero testing	零位测试
zero time reference	零时基准,计时起点
zero zone	零区,零时区
zero-age main sequence	零龄主序
zero-based budget	零基预算
zero-crossing detector	零交点检测器
zero-crossing discriminator	过零鉴别器
zeroed time	零时化
zero-equilibrium receiver	零平衡接收机
zero-error system	零误差系统
zero-feet attack	超低空攻击
Zero-G	零过载零重力失重状态
Zero-G maneuverability	失重条件下的机动性
zeroing	调零,零位调整
zero-initial-length spring	零长弹簧
zero-input response	零输入响应
zero(-length) launcher	零长发射器,零长发射架,无导轨发射装置
zero-lift angle of attack	零升力迎角
zero-lift chord	零升力翼弦,零升力弦
zero-lift moment	零升力矩
zeroload	空载,无载
zero-mean	零平均值
zero-memory channel	零记忆信道
zero-memory source	零记忆信源
zero-offset processing	零偏置处理
zero-order	零序,零次,零价,零级,零阶的
zero-order elimination kinetics	零级消除动力学
zero-order holder	零阶保持器
zero-phase effect	零相位效应
zero-point vibration	零点振动
zero-pressure	零压力
zero-pressure balloon	零压气球
zero-shifting technique	零点移位法
zero-spin	零自旋
zero-sum	零和
zero-thrust pitch	零推力螺距
zero-velocity	零速度
zero-velocity surface	零速度面
zero-vibration	零点振动
zero-zero	1.能见度为零 2.零零条件（着陆时云幕和能见度都为零的条件）3.零高度零速度（指弹射跳伞）
zero-zero ejection test	双零弹射试验
ZF	1. Zero Frequency 零频率 2. Zone Fuel 航段耗油
ZFC	Zero Failure Criteria 零故障准则,零失误准则
ZFCG	Zero Fuel Center of Gravity 无燃油重心
ZFT	Zero Flight Time 零飞行时间,无飞行时间
ZFW	Zero Fuel Weight 零燃油重量
ZGI	Zonal Guidance Interval 区域间隔
ZHR	Zenithal Hourly Rate 天顶每时出现率
ZI	1. Zonal Index 带指数,带指标 2. Zero Input 零输入,无输入 3. Zero Inventory 零库存 4. Zone Interception 分区截击
ZIA	Zone Integration Area 客舱综合区
zigzag	1.锯齿形(的),Z(之)字形(的),曲折(的,物),(作)锯齿状 2.交错 3.变压器中的一种绕组连接方式 4.盘旋弯曲
zigzag filter	曲折滤波器
zigzag slow wave line	曲折线慢波线
zinc	【化】锌（化学元素,符号 Zn）
zinc blende lattice structure	闪锌矿晶格结构
zinc chloride type dry cell	氯化锌型干电池
zinc chromate	铬酸锌
zinc chromate primer	铬酸锌底漆
zinc sulfide	硫化锌
zinc-carbon dry cell	锌-碳干电池
zinc-chlorine battery	锌-氯电池
zincify	镀锌,在……上包以锌,在……中加锌
zinc-oxygen battery	锌-氧蓄电池
zinc-silver battery	锌-银蓄电池
ZIP	Zone Inspection Plan 区域检查大纲
zip fastener	拉链
zipper	拉锁,拉链
zirconium	锆
ZM	1. Z Marker Z 指点标 2. Zambia 赞比亚 3. Zone Marker 区域指点标 4. Zone Melting 区熔 5. Zone Metering 分区计量
ZMA	Zone Multicast Address 区(域)多播地址
ZMDW	Zero-Material-Dispersion Wavelength 零材料色散波长
Z-meter	阻抗表,Z 表
ZN	Zone 地带,地区
ZNC	Zone Network Controller 局域网控制器
ZNM	Zone Nautical Miles 航段里程（海里）
ZNR	Zinc Oxide Nonlinear Resistor 氧化锌非线性电阻
ZO	Zero Output 零输出
zodiac	1.黄道带（在黄道南北各宽 9°的带,分为十二分,称十二宫,主要行星和月球均在此带内运动）,十二宫图 2.苏迪亚牌手表 3. Zodiac 佐迪阿克电阻合金 4.左迪阿克铜镍锌合金
zodiacal counterglow	黄道对日照
zodiacal light	黄道光
zodiacal signs	黄道十二宫

ZOE　Zero Energy 零能量
zonal circulation　纬向气流
zonal harmonics　带调和，带谐函数，球带调和数，带函数
zonal wind　纬向风带状气流
zond　苏联向金星的星际空间发射的探测火箭
zond spacecraft　"探测器"航天器，向金星星际空间发射的太空飞机
zone　地带，地区，区域，范围，划分区域，稳定区域，层，圈，环带
zone melting　区域熔融，区域精炼法，区域熔化，区域熔融
zone melting and refining　区熔
zone of action　作用区域，接触区
zone of avoidance　隐带
zone of intersection　交叉地带，交叉区
zone of protection　保护区域，保护范围
zone of totality　全食带
zone plate　波带片，环板，波域片
zone time　1.区时，区域时间，地方时间 2.航段飞行时间 3.地带时 4.标准时
zoning　1.分区制 2.区域划分，区域化，分区规划 3.地带划分 4.透镜天线相位波前修整 5.环带生长机场区域，注记带分局，布局分区，分区规划，分区取样，分区，制区，划区域，规定
zoom　1.急速上升 2.图像电子放大，图像变比 3.摄像机移动 4.直线爬升，垂直爬升 5.跃升 6.变焦距
zoom lens　1.可变焦距的镜头 2.可变焦距透镜，变焦距透镜
zoom stereoscope　可变焦距立体镜
zoom system　变焦系统
zooming　1.快速上升的 2.缩放 3.图像放大

zoot suit　阻特装
ZP　1.Zero Potential 零电位 2.Zero Power 零功率
ZROC　Zero Rate Of Climb 爬升率为零，等速爬升
ZS　Zero Shift 零偏移
ZSA　1.Zero Set Amplifier 调零放大器 2.Zonal Safety Analysis 区域安全性分析
ZSD　Zebra Stripe Display 条纹显示器
ZT　1.Zero Time 开始时间，航段飞行时间 2.Zone Time 区时
ZTC　1.Zero Temperature Coefficient 零点温度系数 2.Zone Temperature Controller 区域温度控制器
ZTD　Zenith Total Delay 总延迟
ZTDL　Zero-Thrust Descent and Landing 零推力下降和着陆
ZTM　Zero-Type Mechanism 回零型机构
Z/TME　ZULU Time 世界标准时间，格林威治平时
Z-transform　Z变换
ZULU（Zulu）Time　Greenwich Mean Time（GMT）格林尼治平时，世界时，国际标准时间
ZW　Zero Wear 无磨损
zwick tensile tester　兹维克强力试验仪
ZWG　Zonal Working Group 地区工作组
ZXCFAR　Zero Crossing Constant False Alarm Rate 零交叉恒定虚警率
zyglo　荧光探伤，荧光透视法，荧光探伤器
zyglo inspection　油浸探伤，荧光探伤法
zyglo penetrant method　杰格罗油浸探伤法
ZZ　Zig-Zag 锯齿形
ZZA　Zamak Zinc Alloy 锌基压铸合金
ZZD　Zig-Zag Diagram 折线图
ZZR　Zigzag Riveting 交错铆接
ZZV　Zero-Zero Visibility 能见度极差

附 录

附录1 公制和英美制度量衡对照表

公制	英美制
长度	**长度**
1 毫米 millimeter (mm) =0.001 米	1 英寸 inch (pl. inches) =2.540 0 厘米
1 厘米 centimeter (cm) =0.01 米=0.393 7 英寸	1 英尺 foot (pl. feet) =12 英寸=0.304 8 米
1 米 meter (m) =3.280 8 英尺=1.093 6 码	1 码 yard =3 英尺=0.914 4 米
1 千米 kilometer (km) =1 000 米=0.621 4 英里	1 英里 mile =1 760 码=1.609 3 千米
地积	**水程长度**
1 公亩 are (a) =100 平方米=0.024 7 英亩	1 链 cable length, cable's length =185.2 米
1 公顷 hectare (ha) =100 公亩=2.471 1 英亩	1 海里 sea mile =10 链=1.852 千米
容积	**地积**
1 毫升 milliliter (ml) =0.001 升	1 英亩 acre=40.468 6 公亩
1 升 liter (l) =0.260 0 加仑(美)=0.260 0 加仑(英)	**质量(常衡)**
质量	1 盎司 ounce =28.349 6 克
1 毫克 milligram (mg) =0.001 克	1 磅 pound=16 盎司=0.453 9 千克
1 克 gram (g) =1 000 毫克	1 长吨(英吨)long ton=2 240 磅=1016.047 0 千克
1 千克 kilogram (kg) =1 000 克	1 短吨(美吨)short ton=2 000 磅=907.184 9 千克
1 吨 ton (t) =1 000 千克	**液量**
1 克=0.035 2 盎司(常衡)	1 加仑 gallon =4.546 升(英)=3.785 升(美)
1 千克=2.204 6 磅(常衡)	
1 吨=0.984 2 长吨(英吨)=1.102 3 短吨(美吨)	

附录2 希腊字母应用对照表

大写	小写	英语名称	中文注音	数学、物理意义
A	α	Alfa	阿尔法	角度；系数
B	β	Bita	贝塔	磁通系数；角度；系数
Γ	γ	Gama	伽马	电导系数（小写）
Δ	δ	Delta	德尔塔	变动；密度；屈光度
E	ε	Epsilon	伊普西龙	对数之基数
Z	ζ	Zita	泽塔	系数；方位角；阻抗；相对黏度；原子序数
H	η	Yita	艾塔	磁滞系数；效率（小写）
Θ	θ	Sita	西塔	温度；相位角
I	ι	Yota	约塔	微小；一点儿
K	κ	Kapa	卡帕	介质常数
Λ	λ	Lamda	兰布达	波长（小写）；体积
M	μ	Miu	缪	微（千分之一）；放大因数（小写）
N	ν	Niu	纽	磁阻系数
Ξ	ξ	Ksai	克西	
O	ο	Omikron	奥密克戎	
Π	π	Pai	派	圆周率＝圆周÷直径≈3.141 6
P	ρ	Rou	柔	电阻系数（小写）；密度
Σ	σ	Sigma	西格马	总和（大写），表面密度；跨导（小写）
T	τ	Tao	套	时间常数
Υ	υ	Yupsilon	宇普西龙	位移
Φ	φ	Fai	佛爱	磁通；角
X	χ	Hai	海	
Ψ	ψ	Psai	普西	角速；介质电通量（静电力线）
Ω	ω	Omiga	欧米伽	欧姆（大写）；角速（小写）；角

附录3 美国飞机命名体系

美国飞机一般以三位字母+编号命名,其中三位字母分别代表:
(1)状态前缀(Status Prefix);
(2)基本任务(Basic Mission);
(3)飞机机型(Vehicle Type)。
注:附加任务(Modified Mission)是可选项,位于基本任务的左侧,只有当需要对基本任务做出修改时才会使用。

1. 首字母代表状态前缀(Status Prefix)

当下列字母用作航空器名称前缀时,表示该航空器的状态。
G:永久停飞、地上训练使用(Permanently Grounded);
J:暂时性特殊测试(Special Test, Temporary);
N:永久性特殊测试(Special Test, Permanent);
X:实验机(Experimental);
Y:原型机(Prototype);
Z:规划设计(Planning)。

2. 次字母代表基本任务(Basic Mission)

当下列字母用作航空器命名时,表示该航空器的主要或基本任务。
A:攻击(Attack);
B:轰炸(Bomber);
C:运输(Transport);
E:特殊电子战(Special Electronic Installation);
F:空中战斗(Fighter);
L:激光测试(Laser);
O:前线观测(Observation);
P:海上巡逻(Patrol);
R:侦察(Reconnaissance);
S:反潜(Antisubmarine);
T:教练(Trainer);
U:通用(Utility);
X:实验(Research)。

3. 末字母代表飞机机型(Vehicle Type)

当下列字母用作航空器命名时,表示该航空器的类型。
G:滑翔机(Glider);
H:直升机(Helicopter);
Q:无人驾驶飞行器(Unmanned Aerial Vehicle);
S:航天飞机(Spaceplane);
V:垂直/短距起降飞机(VTOL/STOL);
Z:浮空器(Lighter-Than-Air Vehicle)。

4. 附加任务(Modified Mission)

当下列字母用作航空器名称前缀时,表示执行的任务是附加在基本任务以外的不同任务。

A:攻击(Attack);

C:运输(Transport);

D:指引、引向(Director);

E:特殊电子战(Special Electronic Mission);

F:空中战斗(Fighter);

H:搜索与救援(Search and Rescue);

K:空中加油(Tanker);

L:极端天气任务(Cold Weather);

M:多任务(Multi-mission);

O:侦测(Observation);

P:巡逻(Patrol);

Q:无人操作(Drone);

R:侦察(Reconnaissance);

S:反潜(Antisubmarine);

T:训练(Trainer);

U:通用(Utility);

V:人员输送(Staff);

W:天气观测(Weather)。

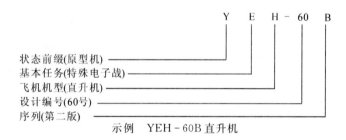

示例　YEH-60B直升机

附录4 美国导弹和火箭命名体系

美国导弹和火箭一般以三位字母＋编号命名，其中三位字母分别代表：
(1)发射载台/环境(Launching Environment)；
(2)主要任务(Mission)；
(3)武器的类型(Type)。

1. 首字母代表发射载台/环境

字母	代表	详释
A	Air	空中载台发射
B	Multiple	多平台发射
C	Coffin or Container	路基平台、倾斜发射
F	Individual	单兵携带、单兵发射
G	Runway	跑道发射
H	Silo-Stored	井下保存
L	Silo-Launched	地下井发射
M	Mobile	陆基移动发射、车辆发射
P	Soft Pad	陆基临时固定发射平台发射
R	Surface ship	水面载台/舰艇发射
S	Space	太空发射
U	Underwater	水下潜射

2. 次字母代表武器的主要任务

字母	代表	详释
C	Transport	运输
D	Decoy	诱饵/迷惑武器，情报战武器
E	Electronic/Communication	特殊电子战武器，通信
G	Surface Attack	对地攻击武器、打击车辆/舰船/地面目标
I	Aerial/Space Intercept	防空拦截、打击空中目标
L	Launch Detection/Surveillance	探测/侦查
M	Scientific/Calibration	校准
N	Navigation	导航
Q	Drone	遥测
S	Space	太空战、空间目标摧毁
T	Training	训练
U	Underwater attack	水下攻击、反潜
W	Weather	天气侦测

3. 末字母代表武器的类型

字母	代表	详释
B	Booster	助推器、起飞发动机
L	Launch Vehicle	运载火箭
M	Guided Missile	导弹
N	Probe	探测器
R	Rocket	火箭、不可控式火箭、自由落体火箭
S	Satellite	卫星、人造卫星